HANDBUCH DER PHYSIK

HERAUSGEGEBEN VON

S. FLÜGGE

BAND XXIX

OPTISCHE INSTRUMENTE

MIT 678 FIGUREN

SPRINGER-VERLAG
BERLIN · HEIDELBERG · NEW YORK
1967

ISBN-13: 978-3-642-46078-4 e-ISBN-13: 978-3-642-46077-7
DOI: 10.1007/978-3-642-46077-7

Softcover reprint of the hardcover 1st edition 1967

Library of Congress Catalog Card Number A 56-2942

Titel-Nr. 5768

Inhaltsverzeichnis.

Optical Calculations and Optical Instruments, an Introduction. By WALTER T. WELFORD, Reader in Applied Optics, Ph. D., D. Sc. (London), Physics Department, Imperial College, London (Great Britain). (With 29 Figures) 1

Méthode de calcul des systèmes optiques. Par ANDRÉ MARÉCHAL, Professeur à la Faculté des Sciences, Délégué Général à la Recherche scientifique et technique, Paris (France), EDGAR D. HUGUES, Directeur technique de la Societé C.E.R.C.O., Neuilly/Seine (France) et PIERRE GIVAUDON, Ancien élève de l'école polytechnique, Noisy-le-Roi (France). (Avec 29 Figures) 43

A. Introduction 43

B. Une méthode de calcul 44

I. Les phases successives 44

II. L'étude du système dans le domaine paraxial 45

III. Etude du système dans le troisième ordre 50

IV. La réduction des aberrations 52

a) Généralités 52

b) Variation locale de chemin optique en fonction des divers paramètres . . 54

c) Changement de sphère de référence et expression générale de la variation de chemin optique aberrant 57

d) Elaboration des dérivées partielles du chemin optique aberrant 59

e) Réduction des aberrations chromatiques 62

V. Les schémas de calcul 63

C. Conclusion 66

Theorie der geometrisch-optischen Bildfehler. Von Professor Dr. HELMUT MARX, Institut für Theoretische Physik der Universität Mainz und Optische Werke Ernst Leitz GmbH., Wetzlar (Deutschland) 68

Einleitung 68

I. Konstruktionsdaten eines rotationssymmetrischen optischen Systems 70

II. Strahldaten; Übergangs- und Brechungs-Formeln 79

III. Gaußsche Näherung; Bildlot-Aberration 92

IV. Iterations-Formelsatz für die Bildhöhen-Aberration windschiefer Strahlen . . 101

V. Iterations-Formelsatz für die meridionale, sagittale und tangentiale Bildhöhenaberration 123

VI. Eikonale; Bildfehler-Koeffizienten 139

VII. Durchrechnungs- und Rekursionsformeln für die Bildfehler-Koeffizienten beliebiger Ordnung 152

VIII. Strenge Darstellung der meridionalen, sagittalen und tangentialen Bildhöhen-Aberration als Funktion der Strahlanfangsdaten und der Langeschen Konstruktionsdaten; Näherungsdarstellungen 166

Das photographische Objektiv. Von Professor Dr. phil. GG. FRANKE, Physikalisches Institut der Universität Gießen und Optische Werke Ernst Leitz GmbH., Wetzlar (Deutschland). (Mit 20 Figuren) 192

Spiegelnde Prismen. Von Dr.-Ing. HELMUT NAUMANN, Leiter der Patentabteilung und des literarischen Büros der Optischen Werke G. Rodenstock, München (Deutschland). (Mit 55 Figuren) 211

Teleskope. Von Dr. rer. nat. KLAUS BAHNER, Hauptobservator an der Landessternwarte Heidelberg-Königstuhl, Heidelberg (Deutschland). (Mit 80 Figuren) 227

A. Optische Abbildung in Fernrohren 228

I. Einige Gesetze der optischen Abbildung 228

a) Geometrisch-optische Beziehungen 228

b) Wellenoptische Ergänzungen 238

II. Abbildungsfehler 243

a) Monochromatische Fehler 245

b) Farbabhängige Fehler 253

c) Zusammenwirken der Einzelfehler 256

B. Optische Systeme für Fernrohre 259

I. Anordnungen mit Reflexionsoptik 259

a) Klassische Reflektoren 259

b) Weiterentwicklung der Reflektoren 264

c) Schmidt- und Meniskus-Systeme 272

II. Systeme mit Refraktionsoptik 282

C. Realisierung der optischen Elemente 291

D. Konstruktionselemente der Fernrohre 313

I. Zelle und Rohr 313

II. Achsenmontierung 322

III. Bewegungseinrichtungen 337

Literatur 341

Mikroskope. Von Dr. phil. HEIN C. CLAUSSEN, Wissenschaftlicher Mitarbeiter der Ernst Leitz GmbH Wetzlar und Lehrbeauftragter an der Universität Marburg (Deutschland). (Mit 97 Figuren) 343

I. Grundlagen 343

II. Normale Mikroskope 350

III. Besondere Geräte 393

IV. Stereoskopische Mikroskope 414

Literatur 417

Diffraction Gratings. By Professor Dr. GEORGE W. STROKE, Head of the Electro-Optical Sciences Laboratory, Electrical Engineering Dept., The University of Michigan, Ann Arbor, Michigan (United States of America). (With 261 Figures) 426

A. General survey 426

I. Historical review 426

II. Diffraction by periodic structures and gratings 433

III. Types of gratings 434

B. Spectroscopic gratings 444

I. The two aspects and the three theoretical descriptions of the diffraction of light by gratings 444

II. Geometrical optics theory of grating image — formation and luminosity of gratings in spectrometers and spectrographs 454

III. Electromagnetic theory of diffraction and of image formation by gratings . 504

IV. Physical optics theory of spectroscopic image formation by gratings (Fourier-transform formulation) 514

V. Light efficiency and "blaze" of gratings 570

VI. Testing of gratings 609

VII. Ruling of optical gratings and production of grating-replicas 664

C. Modulated interference-gratings in wave-front reconstruction imaging and optical image filtering 735

Optische Meßmethoden. Von Professor Dr. phil. REINHART SCHULZE, Direktor des Institutes für Technische Optik der Technischen Hochschule Stuttgart (Deutschland). (Mit 107 Figuren) 755

I. Einleitung 755
II. Allgemeine Grundsätze 758
a) Über die Anordnungen 758
b) Über den Beobachter 762
III. Urnormal Lichtwellenlänge 766
a) Anschluß von Endmaßen an die Lichtwellenlänge 767
b) Anschluß von Strichmaßen an die Lichtwellenlänge 782
c) Ausführung von Strichmaßstäben 785
IV. Mikroskopische Geräte und Verfahren 786
V. Fernrohrgeräte und Verfahren zur Fluchtungsprüfung 809
VI. Interferenzmeßmethoden für die mikroskopische Oberflächenprüfung 827

Literatur 832

Sachverzeichnis (Deutsch-Englisch) 835
Subject Index (English-German) 847
Index (Français) 860

Optical Calculations and Optical Instruments, an Introduction.

By

W. T. Welford.

With 29 Figures.

1. Introduction. The physical principles underlying the design of optical instruments are very few in number and yet the process of design has developed to a high level of sophistication and precision. The reason for this may be seen by comparing optical design with other disciplines of applied physics, e.g. electronic circuit design or the design of pressure vessels. In all cases there is a first order or linear theory which may be used to obtain a more or less crude approximation to the final design; in circuit design the next stage would usually be the experimental adjustment or trimming of the actual circuit elements, partly to allow for the approximations in the design and partly to allow for manufacturing tolerances on the components; a pressure vessel would usually have such large factors of safety that no further refinement would be needed; but in the optical case the results from the linear theory, the so-called paraxial or Gaussian optics, must then be further refined by taking account of aberrations and finally by exact raytracing before the design is constructed. The Gaussian theory is represented by the linear terms in a power series in several variables which represents the behaviour of the optical system and the aberrations are higher terms in this series; the raytracing process yields the sum of the series for chosen values of the variables. In some physical disciplines there is no equivalent of exact raytracing and in others such a process would be pointless because of manufacturing tolerances on components, but it is possible to make optical systems with great precision and so an elaborate process of optical design can be justified.

The process of optical design is essentially one of cut-and-try approximation. It is possible to determine the performance of a specified system by the abovementioned process of computing linear properties and aberrations, and the aberrations can be obtained either as terms in the series expansion or as differences between raytracing results and the linear approximation; however, it is not yet possible to design a system to have given properties by a *systematic* process of successive approximation, still less by use of closed formulae. One reason for this is that the aberrations of a system depend on its constructional parameters in a very complex way, but a deeper reason is that not enough is at present known about how well corrected an optical system can possibly be. To clarify this second point, it is known that a triplet photographic objective of the type invented by H. Dennis Taylor some seventy years ago can be designed to have all the primary or first order aberrations precisely zero, since there are enough degrees of freedom in the main constructional parameters, namely the powers and shapes of the three components and the dispersions of the glasses. However, if the subsidiary parameters are varied, i.e. thicknesses and refractive indices, designs with different higher order aberration residuals can be obtained and it is not yet known how to decide whether a design represents the best obtainable or whether, by some further slight change, a further improvement would result. The same holds, a fortiori, for more complex designs.

It is fairly certain that it is not possible to reduce all aberrations exactly to zero in a system containing only a finite number of spherical surfaces and thus a further question arises as to the amounts of aberrations which are tolerable and how these should be balanced among the different aberration types. Some progress has been made with such problems in recent years but we are still far from knowing how to choose an optimum in any given case. Thus there are two major uncertainties in optical design, a) to what extent can aberrations be reduced in a given kind of optical system and b) how should the remaining residuals be distributed among different aberration types in order to obtain the best result? To these can be added a third, the question of how well such a system can be manufactured, but we are not here concerned with this.

In this article we shall outline the theory of geometrical optics as it applies to optical instruments and show how the properties of optical systems are computed. Detailed expositions of various aspects of the subject are given in other parts of this Volume and also in other Volumes of this Handbuch; here we are concerned with the broad outlines of the general topic of optical calculations and optical systems.

2. The domain of geometrical optics. In designing optical instruments we use chiefly geometrical optics; the scalar wave theory is used in assigning aberration tolerances and, of course, in interference devices such as thin-film systems and diffraction gratings the wave theory is necessarily used in more detail. However, by optical instruments we mean essentially image forming systems, since it is these which require elaborate optical design in the sense discussed in Sect. 1 above. We therefore begin by discussing the concepts, ray and geometrical wavefront, which are used and showing how they are related to the concepts of physical optics.

The intuitive concept of a "ray of light" is easily grasped if we restrict ourselves to isotropic, homogeneous media with sharp boundaries; if we add SNELL's law (which we take for convenience to include reflection as a special case of refraction) and such postulates as reversibility of ray directions and non-interference of intersecting rays we have immediately the whole basis of the geometrical optics of optical instruments. That is to say, we can build up the linear theory and the theory of aberrations without any further appeal to laws of physics. When we come to the question of aberration tolerances it is convenient to introduce another concept, that of optical path length, and another physical law, FERMAT's principle of stationary optical path along a ray. By means of these it is possible to express aberrations in terms of varying optical path lengths, and then varying phases of disturbances arriving in the neighbourhood of an image point. This latter is essentially a concept deriving from physical optics but it is found to be applicable within the domain of geometrical optics. Also it can be shown that SNELL's law and FERMAT's principle are substantially equivalent, i.e. the whole of geometrical optics can be deduced from either. These inter-relations are indicated diagrammatically in Fig. 1. Considering a pencil of rays from a single point source, it is possible to mark on each ray a point at a given optical path length W from the source; these points then lie on a geometrical wavefront since they represent a surface which the light would reach in the time W/c after leaving the source, where c is the velocity of light in vacuo. The concept of a geometrical wavefront then leads to the idea of wavefront aberration, which is expressed by the departure from the ideal spherical shape of a wavefront produced by an optical system. This is essentially not very different from the concept of optical path difference, but it is easier to visualize and to explain.

The approach to geometrical optics sketched above is simple and intuitively satisfactory to the engineer but it lacks rigour; the physicist would prefer a theory more directly derived from physical optics. Such a derivation has been given; starting with a differential equation for either scalar or vector waves it is possible to carry out a limiting process in which the wavelength tends to zero; the result is another differential equation in which the dependent variable is the optical path length W:

$$\left(\frac{\partial W}{\partial x}\right)^2+\left(\frac{\partial W}{\partial y}\right)^2+\left(\frac{\partial W}{\partial z}\right)^2=n^2(x, y, z), \tag{2.1}$$

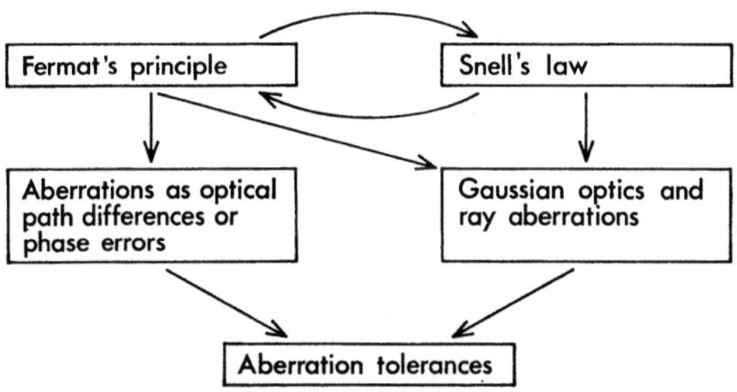

Fig. 1. Relations between the main concepts in the theory of optical systems.

where n is the refractive index, assumed to be variable either stepwise or continuously.

The surfaces

$$W(x, y, z)=\text{constant} \tag{2.2}$$

are the geometrical wavefronts, since in the above mentioned derivation these surfaces are surfaces of constant phase before the limiting process is carried out.

Furthermore the unit vector

$$\mathbf{s}=\frac{1}{n}\operatorname{grad} W \tag{2.3}$$

represents the direction of the Poynting vector, if the process began with electromagnetic waves, so that $\mathbf{s}$ is the ray direction, or direction of energy transport.

This limiting procedure has been described by several authors[1]; it is possible to carry the process further and obtain the notion of cones of rays transporting a constant flux of energy, from which the inverse square law follows. The rays and wavefronts can be shown to be orthogonal and SNELL's law and FERMAT's principle can be obtained. Finally it is shown by BORN and WOLF[2] that if the higher order terms in the limiting process are considered carefully the approximation of geometrical optics is invalid near the edges of shadows and near foci. It is, of course, at just these places that marked diffraction effects are observed. Fig. 2 illustrates diagrammatically these steps.

[1] For example, M. BORN and E. WOLF, Principles of Optics, 2nd edit. London and New York 1964; A. SOMMERFELD, Optics (Vol. IV of Lectures on Theoretical Physics, translated by O. LAPORTE and P. A. MOLDAUER). New York: Academic Press 1954.

[2] Loc. cit.

A different approach to this process of deriving geometrical optics from electromagnetic theory was originated by LUNEBERG[1] and it has been systematized and further developed by KLINE and KAY[2]. In this approach the propagation of certain kinds of discontinuities in the electromagnetic field is considered and it is shown that these obey the postulates of geometrical optics and they can therefore be identified with geometrical wavefronts. This approach is shown by KLINE and KAY to lead to the formulation of geometrical optics in anisotropic media and, with suitable extensions, to produce a sequence of approximate diffraction theories of considerable utility in scattering and diffraction problems.

Thus for the purposes of instrumental optics we may regard geometrical optics as proven and valid within limits which can be understood from such work as that described above. To recapitulate, we use in geometrical optics the concepts "ray" and "wavefront"; the rays are the loci along which the light energy is transported; the wavefronts are surfaces orthogonal to all the rays which originated at a single point source. The wavefronts are obtained as surface at constant optical path length $W=\int n\,ds$ along the rays from the source point[3]. The rays are refracted and reflected at discontinuities of refractive index in directions given by SNELL's law, but the geometrical optics approximation tells us nothing of the relative intensities in the reflected and refracted beams. Also the ray paths obey FERMAT's principle, a sufficiently rigorous form of which for most purposes is that, for a physically possible light path between two points A and B, the optical path length from A to B is stationary; i.e., neighbouring paths which are not possible ray paths have the same optical path length, to the first order of small quantities.

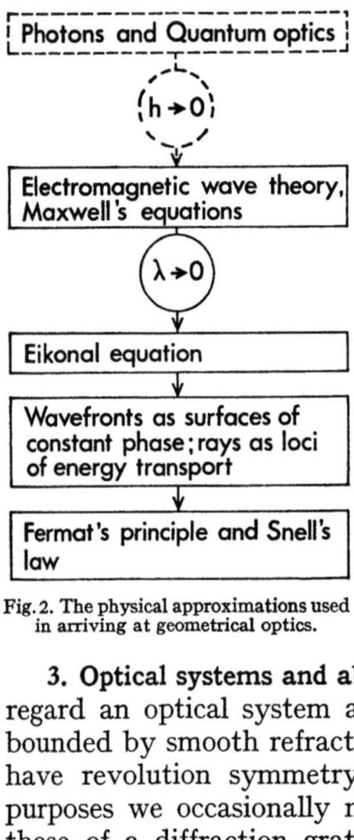

Fig. 2. The physical approximations used in arriving at geometrical optics.

3. Optical systems and aberrations. For most purposes in this chapter we can regard an optical system as comprising a sequence of uniform isotropic media bounded by smooth refracting or reflecting surfaces; furthermore, it usually will have revolution symmetry about an axis. However, for instrumental optics purposes we occasionally need to consider anisotropic media, surfaces such as those of a diffraction grating and, infrequently at present, inhomogeneous or non-uniform media. Excluding these special cases for the moment, the optical system can be defined as a distribution $n(x, y, z)$ of refractive index throughout a certain region of space which is bound rather arbitrarily by edges of lenses, metal mounts etc. Thus the first problem in optical calculation, that of finding out how a given optical system performs, is in principle already solved; we merely have to obtain the appropriate solutions of the partial differential equation of the geometrical wavefronts, i.e., Eq. (2.1). These solutions, written symbolically as in Eq. (2.2), then yield the ray components, for if $\mathbf{s}$ is a unit vector along a ray through (x, y, z) then

$$n\mathbf{s} = \operatorname{grad} W(x, y, z), \tag{3.1}$$

as in Eq. (2.3).

[1] R. K. LUNEBURG: Mathematical theory of optics. Berkeley, U.S.A.: University of California Press 1965.

[2] M. KLINE and W. KAY: Electromagnetic theory and geometrical optics. London: John Wiley & Sons 1965.

[3] Here as elsewhere we consider only isotropic media.

It is, of course, found to be impracticable to obtain such solutions of Eq. (2.1), which is generally known as the eikonal equation, except in trivially simple cases; the approach adopted in numerical work is to calculate the ray paths directly and then to determine the wavefront shapes from these if they are required to be known. We discuss this process in more detail in Sect. 6. Returning to the eikonal equation, the particular utility of this equation and certain others related to it is in obtaining general information about the kinds of aberrations which may be present in optical systems. This descriptive knowledge of aberration types is essential to optical computations of all kinds.

Before going into details of the theory of aberration types we discuss the ways in which aberrations may be defined and the relationships between the different definitions. From the instrumental point of view the fundamental concept is that of a pencil from a point object O passing through an optical system and

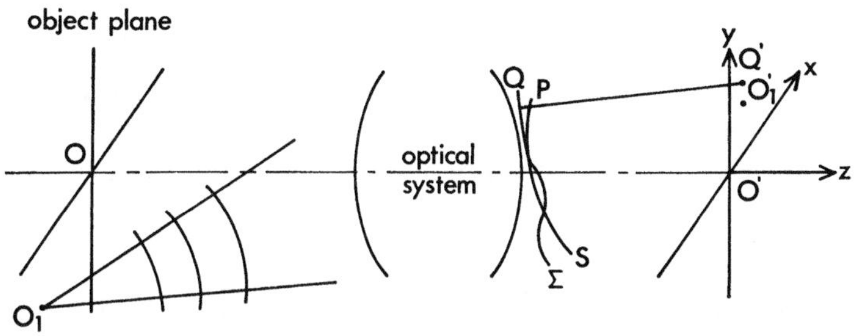

Fig. 3. The definitions of aberration as either ray deviation or wavefront deformation.

forming a more or less imperfect image of the point object; by the term "pencil" we can now understand the ensemble of rays and wavefronts from O, limited in extent by a diaphragm or the rim of a lens within the system. Referring to Fig. 3, let O_1 be the object point and let O_1' be the "ideal" image point, i.e. that point through which all the rays would pass if the system had no aberrations. Then we can describe a portion S of a sphere, centred on O_1' and of any convenient radius, which obviously represents a wavefront of the pencil as it would be if, again, there were no aberrations; this surface is called a *reference sphere*. We introduce for convenience rectangular axes $O'\,xyz$ of which the $x-y$ plane may be thought of as defining an ideal image plane, although as yet we make no assumptions about image flatness.

Let an actual, aberrated, wavefront from O_1 be Σ, chosen arbitrarily to meet S at the z-axis. Let a ray of the pencil pass through Q on Σ, P on S and Q' on the $x-y$ plane. Then the two most important ways of defining aberrations are (a) in terms of the vector $O_1'Q'$, of which the x and y components are the transverse ray aberrations, and (b) in terms of the optical path length from P to Q, i.e. $n\,PQ$, which is the wavefront aberration. Using (x, y), the coordinates of P, to specify the particular ray of the pencil, it is clear that the wavefront aberration is a function of x and y and so it may be written $W(x, y)$; it is simply related to the solutions of the eikonal equation. Let the coordinates of O_1' and Q' be (X_0, Y_0) and (X, Y). Then the components of transverse ray aberration are easily shown to be related to the wavefront aberration as follows:

$$\left.\begin{aligned} X-X_0 &= -\frac{R'}{n}\frac{\partial W}{\partial x} \\ Y-Y_0 &= -\frac{R'}{n}\frac{\partial W}{\partial y} \end{aligned}\right\} \qquad (3.2)$$

where $R' = PQ'$. These equations may, for most practical purposes be simplified by taking R' as constant and equal to R, the radius of the reference sphere.

Eq. (3.2) shows that the ray aberration components are, to a sufficient approximation for most purposes, proportional to the components of the gradient of the wavefront aberration.

Each of these two definitions of aberration has particular advantages in certain circumstances and it is convenient to be able to refer to either in discussing optical design.

The theory of aberration types was proposed by W. R. HAMILTON. In terms our present notation it imposes certain restrictions on the possible forms of the aberration function. To simplify the argument we consider wavefront aberrations only, but the extension to ray aberrations is immediate. The wavefront aberration is a function of the object position and of position of the ray in the pupil, but because of rotational symmetry we need consider only object points along one radial direction in the field: then we write the aberration as $W(x, y, \eta)$, where η is a coordinate which is a measure, angular or linear, of position in the field[1]. Then W is written as a power series in the three variables, but on account of the axial symmetry of the optical system not terms all in the series are possible; it is found that only the terms $x^2 + y^2$, $y\eta$ and η^2 and powers and products of these may occur. The power series then takes the form

$$\left.\begin{aligned} W(x, y, \eta) = {} & a_1(x^2+y^2) + a_2 y\eta \\ & + b_1(x^2+y^2)^2 + b_2 y\eta(x^2+y^2) + b_3 y^2\eta^2 + b_4\eta^2(x^2+y^2) + b_5\eta^3 y \\ & + c_1(x^2+y^2)^3 + \cdots . \end{aligned}\right\} \qquad (3.3)$$

Here terms on the right which do not contain the pupil variables x, y have been omitted, since they correspond only to a uniform increment in phase over the wavefront. The sets of terms corresponding to different powers of the new variables x^2+y^2, $y\eta$ and η^2 are the sets of aberrations of different orders. There is a large number of different sets of names and symbols for these and we shall not attempt to list them all; for our purposes the terms of the fourth degree (those with coefficients b_1 to b_5)are the *primary aberration terms*, the sixth degree terms (coefficients $c_1 \ldots$) are secondary terms and so on. The corresponding terms for ray aberrations would be of degrees three and five, respectively.

The terms with coefficients a_1 and a_2 have a special significance; it is shown in detailed treatments that these terms vanish when the reference sphere to which the aberrations are referred has its centre at a suitably chosen point. In terms of ray aberrations, a_1 and a_2 vanish if a correct choice is made of the ideal image point from which the aberrations are measured. In fact if only these terms are taken into account we obtain a first approximation to the geometrical optics of the symmetrical optical system, the approximation known as paraxial optics or Gaussian optics.

As usually explained, *Gaussian optics*, so called after C. F. GAUSS, who in 1841 gave the first systematic exposition of it, is the study of the properties of the system when only rays very near to the axis of symmetry are considered. It is, however, necessary to express mathematically the approximation involved and this can be done by saying that only linear and quadratic terms in the aperture and field coordinates are to be allowed in the expressions for wavefront shapes and

[1] The aberrations, are of course, functions of the positions of object and image along the optical axis, but in the application of the Hamilton theory to practical optical design this dependence appears in a rather different way: the aberration coefficients which occur in the power series development of $W(x, y, \eta)$ are seen to be functions of these positions.

correspondingly, shapes of refracting surfaces. The reasoning in the previous paragraph shows why it is just these terms which are involved in the paraxial approximation.

The next set of terms (coefficients b_1 to b_5), are those corresponding to the five primary aberrations, spherical aberration, coma, astigmatism, field curvature and distortion, first systematically discussed by L. SEIDEL in 1856 (see Sect. 5 below).

Detailed descriptions of the primary aberrations in terms of ray patterns and of wavefront shapes can be found in many standard texts[1]; the actual light flux distribution in a point image formed with some aberration or combination of aberrations differs considerably from what would be predicted by the simple geometrical optics considerations on account of diffraction effects. Some excellent photographs and diagrams showing this comparison in several cases are available[2]. The important point to be made here is that the differences between geometrical optics and physical fact are greatest when the aberrations are small, i.e. when the deviations from sphericity of the wavefront do not exceed a few wavelengths. In the extreme case of no aberrations the geometrical image reduces to a single point, whereas the physical image is the Airy diffraction pattern.

4. Gaussian optics. The optics of the paraxial region of the symmetrical optical system forms the basis of the first approach to the design of optical systems. The concepts of the focal length and the power of a system and the location of its principal points belong essentially to Gaussian optics. The notions of relative aperture, entrance and exit pupils, vignetting and field coverage strictly involve regions more remote from the axis, at any rate in the vast majority of optical systems, but nevertheless they may very usefully be discussed in terms of Gaussian optics at some stages in optical design. Thus Gaussian optics describes an ideal or aberration-free system and the initial specification of a system to be designed would naturally be in terms of its paraxial properties.

One approach to Gaussian optics has been to consider it as the description of the operation of an ideal system in making one-to-one transformations between rays in object space and rays in image space. If the requirements of perfect imaging of points, straight lines and planes are satisfied then such an ideal system behaves like a symmetrical optical system within its Gaussian region. This approach is not particularly useful to the optical designer since it makes no contact with the physical aspects of the optical system, i.e. its reflecting and refracting surfaces.

A more informative approach is to assume axial symmetry and to study the behaviour of a system under the restrictive approximations mentioned in Sect. 3, i.e. that terms describing surface or wavefront shapes of higher than quadratic power in the aperture and field are to be neglected. Thus if a surface (or a wavefront from an axial object point) has the form

$$z = c_2 r^2 + c_4 r^4 + \cdots, \tag{4.1}$$

where r is the radial distance from the axis and z is a coordinate along the axis, it follows from the Gaussian restriction that only the term $c_2 r^2$ is to be considered. Then in the Gaussian approximation all refracting surfaces are described by one parameter, their axial curvature, so that distinctions between spheres, paraboloids, etc. vanish and all surfaces may equally be said to be spherical. The same applies

[1] See e.g. L. C. MARTIN and W. T. WELFORD, Technical Optics, Vol. I, second ed. London: Pitman 1966 or M. BORN and E. WOLF, Principles of Optics, second edit. London: Pergamon Press 1964.

[2] M. CAGNET, M. FRANÇON and J. C. THRIER: Atlas of optical phenomena. Berlin-Göttingen-Heidelberg: Springer 1962.

to wavefronts, and it is then easily shown that all wavefronts are spherical and so form perfect point images; extended images are also found to be free from field curvature and distortion.

The definitions of the principal foci, as the images of axial object points at infinity, and of the principal points, as the axial points on planes of unit magnification, follow naturally and the two focal lengths, f and f' are defined as the distances from the principal planes to the respective foci, as in Fig. 4.

The concepts of object space and image space are needed to permit a complete algebraical description of paraxial image formation. As is customary, the light is supposed to travel from left to right, so that objects would be thought of as on the left of the optical system, but the object space extends through the system to the right, so as to allow for virtual objects, i.e. images formed by a preceding system. Similarly the image space extends to the left, and both spaces overlap completely.

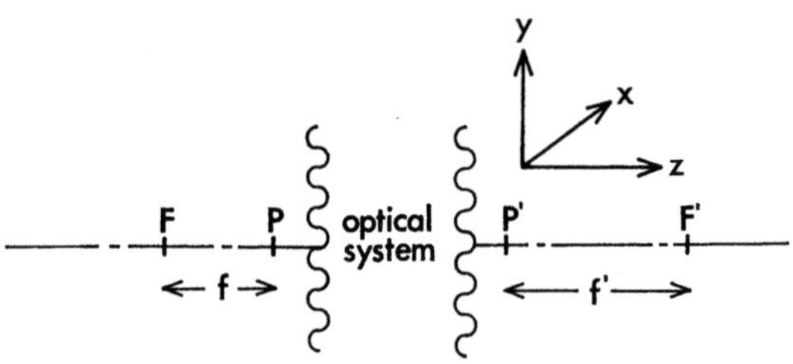

Fig. 4. The foci and principal points of a symmetrical optical system.

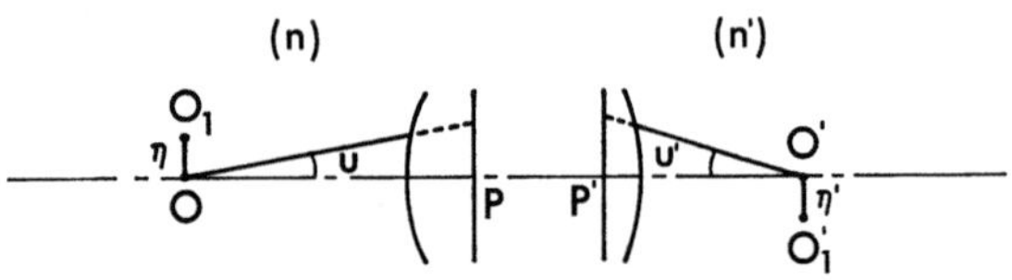

Fig. 5. The Smith-Helmholtz-Lagrange magnification law.

It is then possible to introduce *conjugate distance equations*, which relate the positions of axial object and image points. For example, if z is the distance from the object side focus to an object point O (i.e. $z = FO$) and z' is the corresponding distance in image space, we have

$$z z' = f f', \tag{4.2}$$

usually known as NEWTON's conjugate equation. Also, if l and l' are the conjugate distances measured from the principal planes (l: PO, l': $P'O'$) then

$$\frac{f'}{l'} + \frac{f}{l} = 1. \tag{4.3}$$

The transverse magnification m_T can also be found, in terms of these quantities:

$$m_T = \frac{l'}{l} = -\frac{f}{z} = -\frac{z'}{f'}. \tag{4.4}$$

It is of some interest that these results and others derivable from them can be obtained from the fact of rotational symmetry of the optical system together with the reasonable assumption that a ray in object space corresponds to one and only one ray in image space. No knowledge of the structure of the system, or of physical principles, e.g. FERMAT's principle, needs to be involved.

Further results follow from FERMAT's principle (or by the use of the law of refraction together with some assumptions about the structure of the system); these are, the relationship between the two focal lengths f and f' and the power K,

$$-\frac{n}{f} = \frac{n'}{f'} = K, \tag{4.5}$$

(where n and n' are the indices of object and image space) and the Smith-Helmholtz-Lagrange magnification law; the latter gives the transverse magnification in terms of the paraxial convergence angles of the axial pencil (see Fig. 5):

$$m_T = \eta'/\eta = \frac{n\,u}{n'\,u'}. \tag{4.6}$$

When Eq. (4.6) is re-written in the form

$$n u \eta = n' u' \eta' \tag{4.7}$$

it reveals the nature of the quantity $n u \eta$ as an invariant through the optical system. It is denoted by the symbol H.

The preceding results contain the essentials of the Gaussian optics of the general symmetrical optical system, as far as is needed for optical design purposes. The next stage involves the specific structure of the optical system, i.e. the curvatures, refractive indices, thicknesses and separations of the components. The procedures in determining Gaussian properties are then of two kinds. We can either obtain closed formulae or we can use numerical iterative procedures. The first approach is nowadays used only for certain very important special cases, namely the single reflecting or refracting surface, the thick lens and the combination of two general systems. The second approach, which is, of course, paraxial ray tracing in one form or another, is used to obtain the Gaussian properties of any given system by iterative calculations through the successive elements.

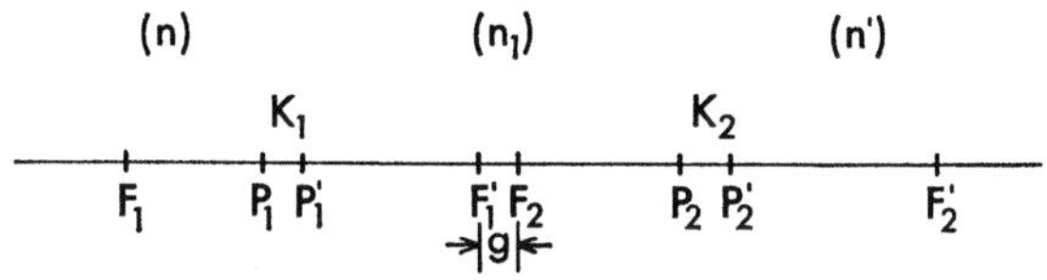

Fig. 6. Combining two optical systems of powers K_1 and K_2.

The power of *a single refracting surface* is $(n'-n)\,c$, where c is the curvature of the surface and n and n' are the refractive indices of the media on either side. (It is convenient to note here that the case of a reflecting surface in a medium n is nowadays always treated by putting $n' \equiv -n$; this convention, together with the conventions of coordinate geometry for distances, as in Fig. 4, gives consistent formulae for reflection and refraction.) The principal points coincide at the refracting surface, by inspection. The nodal points, which are two axial conjugate points N and N' such that a ray in object space through N passes through N' in image space at the same convergence angle, coincide at the centre of curvature.

If *two arbitrary systems* of powers K_1 and K_2 are to be combined, as in Fig. 6, let g be the distance between their neighbouring focal points, F_1' and F_2; then the power K of the combined system is

$$K = -g \frac{K_1 K_2}{n_1}, \tag{4.8}$$

where n_1 is the refractive index of the intermediate medium. The position of F', the second principal focus of the combined system, is given by

$$F_2' F' = \frac{n_1 n'}{g K_2^2} = -\frac{n' K_1}{K_2 K}. \tag{4.9}$$

Other expressions are available for these and related quantities in terms of the distance $P_1' P_2$ between, adjacent principal planes. From such expressions the very useful formulae for the power of a thin lens, a thick lens and a pair of separated thin lenses are found. However, as mentioned above, it is not modern practice to produce explicit formulae on these lines for a larger number of elements.

Paraxial raytracing is useful not merely for calculating Gaussian properties of given systems; it forms the first step in calculating the primary aberration terms, and also the higher order aberration terms when these are used, which is rather rarely nowadays; paraxial raytracing, possibly carried out graphically, is also an essential aid in certain auxiliary aspects of optical instrument design such as the placing of folding mirrors and prisms, condenser design, light flux calculations and so forth. There would be little point in describing all the very large number of different systems of paraxial raytracing which have evolved, since the general availability of digital computers has caused many of the reasons for the evolution to lose their force. The choice now would depend on the particular system of primary aberration calculation to be used and again no strong reasons appear to operate against the use of what is substantially L. SEIDEL's original formulation.

Before describing one method of paraxial raytracing we note that the raytracing process also has a function in connection with the stops or apertures which always must be present in an optical system. Except in trivial cases, the aperture stop (and the field stop if such exists) operates outside the Gaussian region but it is usual to treat the theory of stops paraxially, at any rate as a first approxima-

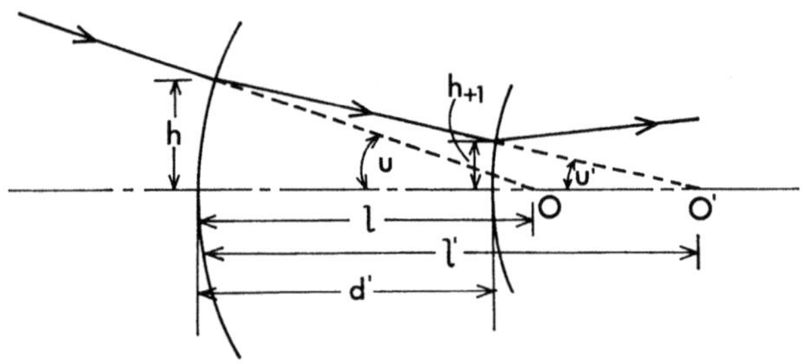

Fig. 7. Paraxial raytracing.

tion, since valuable design information is obtained in this way. It would be difficult and clumsy to deal with stops by the closed formula approach but the raytracing method is very convenient.

Fig. 7 illustrates the most commonly used raytracing method; it applies to meridian rays only, i.e. rays which lie in a meridian plane (through the axis of symmetry), since all the behaviour of the system in the paraxial region is easily deduced from such rays. At a surface of curvature c the incoming ray is incident at height h and meets the axis at O, the intersection length being l; the ray is specified by the two coordinates h and u, where

$$u = h/l; \tag{4.10}$$

the quantity u is generally called the convergence angle, but bearing in mind the paraxial restriction it is better to think of it as a ratio; its sign is then given by those of h and l, both positive in Fig. 7. The transition to the refracted ray is given by

$$n'u' - nu = hK, \tag{4.11}$$

where K is the power of the surface, so that $K = (n' - n)c$. At the next surface we have

$$h_{+1} = h - d'u'. \tag{4.12}$$

These two formulae are used alternatively in succession through the system to trace the ray. Eq. (4.11) is generally called the *refraction equation* and Eq. (4.12) the *transfer equation*, the names having an obvious significance. We do not go into the specialized questions of notation and numerical subscripts to indicate the successive surfaces and spaces; these are very much a matter of individual taste and convenience. Of much more importance are the questions, how the ray trace starts and finishes, in fact how one decides which rays are to be traced, and here we have to return to the question of aperture and field stops.

According to the usual Abbe theory, the *aperture stop* is defined as that aperture which is effective in limiting the angular size of the entrance pencil from the axial object point. Thus, looking into the system from the object point O, all the axial apertures such as lens rims and metal rings could in principle be seen by the portions of the system to their left, i.e. they would appear as images in object space; the image which subtends the smallest angle at O (see Fig. 8) say $A_1 B_1$, is evidently the image of the aperture stop $A B$. This image is called the *entrance pupil* and the image in the other side of the system, $A_1 B_2$, is the *exit pupil*; it evidently has the corresponding relationship with O'. For ray tracing purposes it is useful to have the concept of a pupil in every intermediate image space, just as there is a series of intermediate images of O.

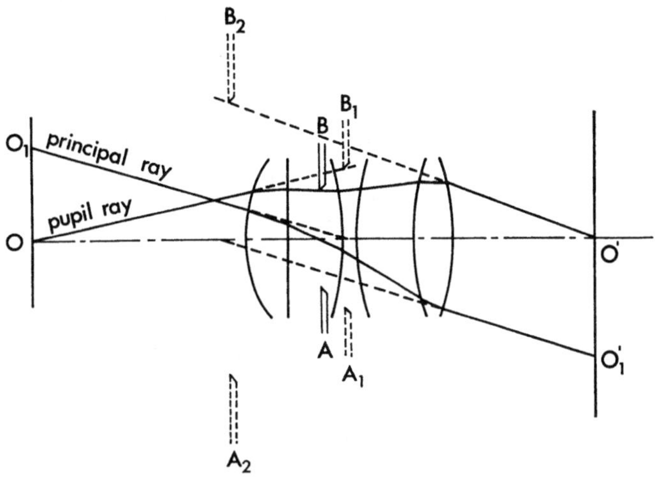

Fig. 8. The aperture stop and the pupils.

The entrance and exit pupils bound the pencils of light from off-axis object points which can pass through the optical system; from the off-axis object point O_1 (Fig. 8) the principal ray is that ray which enters the system through the centre of the entrance pupil and which consequently passes through the centres of the aperture stop and the exit pupil. The rest of the off-axis pencil surrounds the principal ray.

There must be in any image forming system an effective aperture stop which limits the aperture of the system, but the corresponding entity for limiting the field of view, the field stop, does not always exist physically and its role in optical computations is generally rather secondary. The extreme field of view may be limited at the computing stage by a variety of considerations which have little to do with optical performance, whereas the exact size and position of the aperture stop are often vital factors in image formation.

We thus arrive at the definition in Gaussian optics of the aperture, field of view and position of pupils and it is these three things which are needed for the paraxial raytracing which is to be done as a preliminary to assessing aberrations.

It is usual to trace a paraxial ray from the axial object point to the rim of the entrance pupil; this is often called the *pupil ray*. Although, as pointed out above, the aperture stop diameter is usually well beyond the Gaussian region, yet it is permissible to trace this ray by Eqs. (4.11) and (4.12) or some equivalent equations, since the variables h and u which describe the ray position appear only linearly in these equations and the process may thus be considered as a scaling-up in the direction perpendicular to the axis of the system of a truly paraxial ray. Alternatively, we may suppose that the raytracing process is really only the determination of the successive intermediate axial image points and the variables h and u are mere computing conveniences. However, the values of h and u at each surface for a ray at (nominally) full aperture are needed for most systems of calculating the primary aberrations, and this is the real reason for tracing the pupil ray.

A further ray, the *principal ray* from the object point at full field angle, is also needed for the primary aberration calculation. Here the various systems differ considerably; in some cases this second ray is not actually traced but the data needed are obtained by some auxiliary calculations on the pupil ray data. This was in fact SEIDEL's approach and the relevant formulae are known as the Seidel difference formulae. It is easy to see that such a calculation must in principle be possible, since the pupil ray and the principal ray may both be regarded as

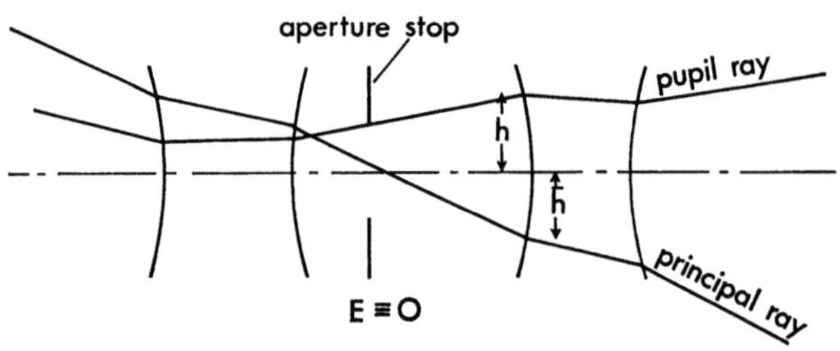

Fig. 9. Computing the eccentricity parameter E.

rays from axial object points, O for the pupil ray and the centre of the pupil for the principal ray; now a knowledge of the pupil ray is in effect a statement of a conjugate distance relation at each refracting surface, so that the image formation for the centre of the pupil is also known at each surface.

The paraxial angle of incidence of the ray in Fig. 7 is $hc-u$, so that the law of refraction at the surface gives

$$n'(hc-u')=n(hc-u). \tag{4.13}$$

This quantity is known as the *refraction invariant* and it is denoted by A; the value of A at each surface for the pupil ray enters into the primary aberration formulae and also the value of the refraction invariant for the principal ray, which we denote by $\bar{A}$, is required. *Seidel's difference formulae* for the values of $\bar{A}$ may now be summarized as follows.

An auxiliary quantity E is defined at each surface by

$$E=\frac{1}{H}\cdot\frac{\bar{h}}{h}, \tag{4.14}$$

where $\bar{h}$ is the incidence height of the principal ray (Fig. 9) and H is the Lagrange invariant for the whole system[1] [Eq. (13)]. Then the value of the refraction invariant for the principal ray at each surface is

$$\bar{A}=\frac{H}{h}(1+AhE). \tag{4.15}$$

[1] In the present context the quantity H clearly links the pupil ray and the principal ray, since if the convergence angle in any space is known, H gives the size of the intermediate image in that space.

The values of E are found from the fact that the difference between the values at successive surfaces is

$$E_{+1}-E=\frac{d}{n h h_{+1}}, \tag{4.16}$$

where d is the axial separation of the surfaces and n is the refractive index of the intervening medium. But from Eq. (4.14) the value of E at the aperture stop (considered as a ghost surface with no refractive index change) is identically zero, so that by starting from the stop and working outwards the values of E at all surfaces can be found.

The above summary applies to the particular need to compute $\bar{A}$ for primary aberrations but in fact other applications can be made. The other quantities describing the principal ray are given by

$$\bar{h}=HhE, \tag{4.17}$$

$$n\bar{u}=\frac{H}{h}(nhuE-1), \tag{4.18}$$

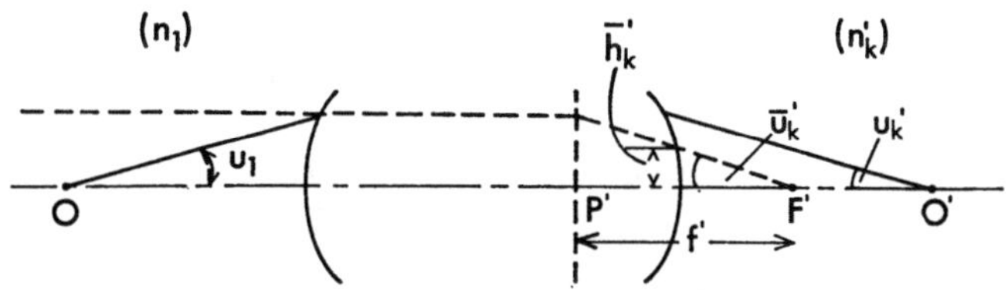

Fig. 10. Use of the Seidel difference formulae for computing the focal length.

where we have adopted the uniform notation of a bar to denote quantities belonging to the principal ray. These results can now be used to compute the focal length of a system for which a pupil ray has been traced. If the pupil ray came from an object point at infinity there is no problem, since the second focal length is given by

$$f'=h_1/u'_k, \tag{4.19}$$

where h_1 is the incidence height at the first surface and u'_k is the convergence angle of the ray in the final image space. If, on the other hand, the pupil ray comes from an object point at a finite distance we can use the Seidel difference formulae to find the data of the ray from an infinitely distant object, as in Fig. 10, where this ray is shown in broken line. For, if this second ray is regarded as the principal ray then the pupil in object space is at infinity and the value of E at the first surface is

$$E=\frac{1}{n_1 u_1 h_1}. \tag{4.20}$$

Then we can find E_k at the final surface by the difference formula [Eq. (4.16)], and so $\bar{u}'_k$ is obtained from Eq. (4.18); thus the focal length f' is found from $f'=h_1/\bar{u}'_k$. In a similar way the back focal length, i.e. the distance from the last surface to F', is obtained as the quantity $\bar{h}'_k/\bar{u}'_k$.

In connection with these and later applications it is worth noting that a change in the value of E at a given surface, due to a change in position of the stop in a space adjacent to that surface, is reflected as the same additive (or subtractive) change to the values of E at all other surfaces. Thus if the E's have once been calculated the effect of a change in stop position appears as the same increment ΔE at all surfaces. This elegant property permits the use of the parameter E, sometimes called the *eccentricity*, to obtain simple formulae showing the effect of changes of stop position on the primary aberrations. It is implied in such a change that the diameter of the stop changes as it moves along the axis in such a way as to keep the size of the initially incident axial pencil constant.

We shall not discuss here all the different formulations of Gaussian optics which have been proposed, as they are all descriptions of what is essentially a simple linear process when considered in terms of the appropriate ray coordinates. One interesting formulation which ought to be mentioned was by T. SMITH in a series of papers starting in 1930; SMITH used matrices to represent the refraction and transfer operations and so was able to represent the Gaussian properties of a system of k surfaces as the product of $2k-1$ two-by-two matrices. The method has recently been revived by BROUWER[1], who gave extensive applications to plane mirrors and to the representation of aberration equations. The paraxial raytracing equations (4.11) and (4.12) can be put in matrix form as

$$\begin{pmatrix} h \\ n'u' \end{pmatrix} = \begin{pmatrix} 1 & 0 \\ K & 1 \end{pmatrix} \begin{pmatrix} h \\ nu \end{pmatrix} \tag{4.21}$$

and

$$\begin{pmatrix} h_{+1} \\ n'u' \end{pmatrix} = \begin{pmatrix} 1 & -d'/n' \\ 0 & 1 \end{pmatrix} \begin{pmatrix} h \\ n'u' \end{pmatrix}; \tag{4.22}$$

it can be seen that a continued product of matrices representing alternately refraction and transfer will yield eventually the equation

$$\begin{pmatrix} h_k \\ n_k' u_k' \end{pmatrix} = \begin{pmatrix} a_{11} & a_{12} \\ a_{21} & a_{22} \end{pmatrix} \begin{pmatrix} h_1 \\ n_1 u_1 \end{pmatrix}. \tag{4.23}$$

The matrix (a_{ij}) clearly has unit determinant, so that it contains three independent quantities; of these the power of the system is easily found to be a_{21} and the other two variables are simply related to the positions of the principal planes with respect to the refracting surfaces. There does not appear to be any great advantage to this representation of Gaussian properties other than compactness. However, BROUWER's matrix treatment of mirrors offers a neat way of solving problems in which an image has to be produced in a prescribed position and orientation.

5. The primary aberrations. Many systems of formulae have been published for the primary aberrations and it would not be possible or desirable to review them all here; we shall, however, state some simplifying generalizations about these formulae and we shall quote certain systems to facilitate the discussion.

The first generalization to be made is that all systems derive the aberrations by a summation of terms over the separate surfaces[2] of an optical system; consequently these terms are fundamentally of the nature of wavefront aberration contributions (or contributions to a characteristic function), since wavefront aberrations are, in primary aberration approximation, additive surface by surface. This is true even of systems which purport to compute, say, transverse ray aberrations[3]; in such cases the individual surface contributions have to be multiplied by magnification factors to make them additive and the factors are such as to make the products essentially of the form of wavefront aberration contributions.

---

[1] W. BROUWER: Matrix methods in optical instrument design. New York: V. A. Benjamin 1964.

[2] Sometimes the system is thought of as a set of separated thin lenses, rather than individual refracting surfaces, but this does not affect the discussion.

[3] See, e.g., A. E. CONRADY, Applied Optics, and Optical Design, Vol. 1. Oxford 1929, reprinted 1960 by Dover Publications, Inc., New York.

The second generalization is that the surface contribution terms for all systems are mathematically identical, apart from simple numerical factors and perhaps a factor which is some integral power of the Lagrange invariant H. This is because the primary aberration terms are of the fourth degree jointly in the aperture and field angle, *as measured by paraxial quantities*, and the several possible paraxial variables which may be used are all linearly interrelated. Thus no matter which combination of paraxial variables may be used in the primary aberration formulae they are all equivalent[1].

These generalizations are relevant to the question, which of the many developments of aberration theory shall the would-be optical designer study? The answer to this question is, in primary aberration approximation it does not matter, since they are all the same, but for higher order aberrations there are several possibilities which will be mentioned in the next section.

The theoretically most satisfying approach to aberration theory is undoubtedly through one of the forms of characteristic function, a potential function of which the derivatives are, in effect, aberration coefficients. However, this development is somewhat remote from the practicalities of optical design and it may appear more straightforward to obtain aberration formulae directly from raytrace data. We discuss the latter approach in this section and in Sect. 8 we give a brief summary of the characteristic function method.

The formulae we shall give for the Seidel aberrations are simple and they enable the user to see how the structure of an optical system is related to its aberrations, but on the other hand the extension to higher order aberrations is impracticably complicated. By contrast the eikonal development (see Sect. 8) offers the possibility of a consistent and systematic development of higher order aberration terms.

We now give briefly one system of *Seidel aberration calculations*, to provide a basis for more detailed discussions. We take fractional coordinates (x, y) in the exit pupil, i.e. the scale is chosen so that $x^2+y^2=1$ gives the rim of the pupil; similarly we use a fractional field coordinate η, so that $\eta=1$ corresponds to the extreme field. It is implied that the field and aperture are related by an assigned value of the Lagrange invariant H. Then, recalling Eq. (3.3), we can write the Seidel terms for a complete system as follows

$$\left.\begin{aligned} W(x, y, \eta) = {} & \tfrac{1}{8} S_{\mathrm{I}} (x^2+y^2)^2 + \tfrac{1}{2} S_{\mathrm{II}} (x^2+y^2)\, y\eta + \\ & + \tfrac{1}{4} S_{\mathrm{III}} (x^2+3y^2)\,\eta^2 + \tfrac{1}{4} S_{\mathrm{IV}} (x^2+y^2)\,\eta^2 + \tfrac{1}{2} S_{\mathrm{V}}\, y\eta^3, \end{aligned}\right\} \tag{5.1}$$

where S_{I} to S_{V} are the aberration terms to be expressed as sums over the refracting surfaces. The numerical coefficients are inserted to simplify the formulae. The quantities S_{I} to S_{V} are known as the *Seidel sums*; they clearly represent values of the primary aberrations at maximum aperture and field. The surface terms can be expressed in terms of the ray coordinates illustrated in Fig. 7; re-defining these for convenience we have,

n, n',	refractive indices in object space and image space of the current surface;
c,	curvature of surface;
h,	incidence height of paraxial pupil ray at full aperture;
u, u',	convergence angles of pupil ray in object space and image space;
$A, \bar{A}$,	refraction invariants of pupil ray and principal ray.

[1] The same is not true for higher order aberration formulae, and this is one of the many difficulties in deriving and applying such formulae.

It is understood, of course, that these quantities should have a subscript or some other indication that they apply in turn to successive surfaces, but as mentioned in Sect. 4, we are not here concerned to develop notation in great detail. The expressions for the Seidel sums for a system of spherical surfaces are then

$$S_{\mathrm{I}}=\Sigma A^2 h\Delta\left(\frac{u}{n}\right), \tag{5.2}$$

$$S_{\mathrm{II}}=\Sigma A\bar{A} h\Delta\left(\frac{u}{n}\right), \tag{5.3}$$

$$S_{\mathrm{III}}=\Sigma \bar{A}^2 h\Delta\left(\frac{u}{n}\right), \tag{5.4}$$

$$S_{\mathrm{IV}}=-H^2\Sigma c\Delta\left(\frac{1}{n}\right), \tag{5.5}$$

$$S_{\mathrm{V}}=\Sigma\frac{\bar{A}^3}{A} h\Delta\left(\frac{u}{n}\right)-H^2\Sigma\frac{\bar{A}}{A}c\Delta\left(\frac{1}{n}\right). \tag{5.6}$$

Here the summations extend over all the surfaces; the symbol Δ indicates the increment on refraction, e.g. $\Delta\left(\frac{1}{n}\right)=\frac{1}{n'}-\frac{1}{n}$.

These expressions are, apart from differences in notation, very similar to those given by SEIDEL. The refraction invariant of the principal ray, $\bar{A}$, can be obtained either by tracing a separate principal ray or by the Seidel difference formulae, Eqs. (4.14) to (4.16); in some formulations these are substituted explicitly in the expressions for the Seidel sums, but this detracts from the simplicity of the formulae as given above.

The terms S_{I} to S_{V} correspond in order to *spherical aberration, coma, astigmatism, field curvature and distortion.* Certain generalizations can immediately be drawn from these expressions as they stand. For example, in the absence of astigmatism ($S_{\mathrm{III}}=0$) the quantity S_{IV}/H^2 is the curvature of the final image surface and it may be seen that this quantity, known as the *Petzval curvature,* does not involve the object and image positions but depends only on the structure of the system. The other aberrations, however, depend strongly on the conjugates. Spherical aberration and the Petzval curvature are independent of the position of the aperture stop, since the quantity $\bar{A}$ does not appear in S_{I} and S_{IV}. The other aberrations can, however, be significantly changed in magnitude by an axial shift of the aperture stop and these changes are neatly expressed by means of the eccentricity E, Eq. (4.14), in terms of $HE=\bar{h}/h$. A displacement in which the size of the aperture stop is adjusted so as to ensure that the size of the axial pencil is unchanged corresponds to the same increment δE to the values of E at all surfaces. The resulting increments in the Seidel sums are

$$\delta S_{\mathrm{II}}=H\delta E\cdot S_{\mathrm{I}}; \tag{5.7}$$

$$\delta S_{\mathrm{III}}=(H\delta E)^2\cdot S_{\mathrm{I}}+2H\delta E\cdot S_{\mathrm{II}}; \tag{5.8}$$

$$\delta S_{\mathrm{V}}=(H\delta E)^3\cdot S_{\mathrm{I}}+3(H\delta E)^2\cdot S_{\mathrm{II}}+H\delta E\cdot(3S_{\mathrm{III}}+S_{\mathrm{IV}}). \tag{5.9}$$

From these equations it follows that in the absence of spherical aberration the coma does not depend on the stop position, but conversely, if S_{I} is non-zero, then a stop position can be found for which the coma vanishes. Similar rules apply to astigmatism and distortion. Of course, these simple results apply only to the primary aberrations and they are very limited in application because usually there is only a rather restricted range of possible positions of the aperture stop, for practical reasons. However, in systems of small aperture and extended field the manipulation of the position of the aperture stop often provides a useful extra design parameter.

Further insight into the dependence of the primary aberrations on the form of an optical system is provided by sets of equations which give the Seidel sums of thin lenses. The *thin lens approximation* is of limited utility in the detailed design of many optical systems because the process of putting in the necessary working thicknesses disturbs either the aberration balance or the Gaussian properties, but it can be very useful in deriving general principles and there are certain classes of systems, in which the field and aperture are moderate, where thin lens methods give adequate results. Many sets of thin lens formulae have been published, differing mainly in the choice of the variables used to specify the Gaussian conjugates and the "bending", i.e. ratio of curvatures, for a lens of given power; most simply these variables might be the reciprocal of the object side conjugate and the curvature of the first surface, and the rest of the Gaussian properties are fixed by the given power; the Seidel spherical aberration and the coma with the aperture stop at the thin lens are then obtained as rather complex polynomials in these quantities, the coefficients being functions of the refractive index of the lens. Many alternative sets of variables have been proposed, these usually being linear functions of those mentioned above and of the power; the resulting equations for S_{I} and S_{II} then take simpler forms and it is easier to see certain general modes of dependence of the aberrations on conjugates and bending.

The utility of the thin-lens approach is more obvious in considering the remaining aberrations, since it is found that S_{III} is equal to $H^2 K$, where as usual H is the Lagrange invariant and K the power, again for the aperture stop at the lens; furthermore S_{IV} is found to be $H^2 K/n$ and S_{V} is identically zero, again for the stop at the lens. The values of S_{II}, S_{III} and S_{V} when the stop is not at the lens are found by means of stop-shift equations such as Eqs. (5.7) to (5.9). C. G. WYNNE has collected together[1] and systematized a number of results of thin lens theory which are useful in lens design. Many authors have given the standard solution for the Taylor triplet type of photographic objective, based on thin lens theory[2], and numerous applications have been made to telescope objective design.

So far we have restricted the discussion to spherical surfaces, but the extension to *non-spherical surfaces* is very simple. Of the different approaches available, the most natural seems to be to represent an aspheric by the equation of a sphere together with an added term for the aspherizing; thus we have

$$z = \tfrac{1}{2} c\,(x^2 + y^2) + (\tfrac{1}{8} c^3 + G)\,(x^2 + y^2)^2 + \cdots \tag{5.10}$$

where the term $G\,(x^2+y^2)^2$ represents the aspherizing. The degree of approximation in Eq. (5.10) corresponds to the Seidel approximation, since terms in the fourth power of the aperture are included, so that the aspherizing gives just one additional parameter. The increment to S_{I} due to the aspherizing, or *figuring*, is found to be

$$S_{\mathrm{I}}^{*} = 8 \Sigma G h^4 \Delta(n), \tag{5.11}$$

where the summation extends over all aspherized surfaces.

An aspherizing has no effect on the other Seidel sums if the stop is at the surface in question, since these all depend on lower powers than 4 of the aperture

[1] WYNNE, C. G.: Optica Acta **8**, 255 (1961).

[2] See e.g. T. SMITH, Trans. Opt. Soc., Lond. **24**, 168 (1922/23); M. BEREK, Grundlagen der praktischen Optik. Berlin: W. de Gruyter & Co. 1930; R. E. STEPHENS, J. Opt. Soc. Amer. **38**, 1032 (1948).

coordinate, but if the stop is remote from the surface then the primary spherical aberration will induce increments in coma, astigmatism and distortion, since the oblique pencils pass eccentrically through the surface. By analogy with Eqs. (5.7) to (5.9) these increments are

$$S_{\mathrm{II}}^{*}=HE\cdot S_{\mathrm{I}}^{*}=(\bar{h}/h)\,S_{\mathrm{I}}^{*},\tag{5.12}$$

$$S_{\mathrm{III}}^{*}=(HE)^{2}\cdot S_{\mathrm{I}}^{*},\tag{5.13}$$

$$S_{\mathrm{V}}^{*}=(HE)^{3}\cdot S_{\mathrm{I}}^{*}.\tag{5.14}$$

Such a figuring can sometimes with advantage be thought of as separated from a lens or mirror and applied to a thin plate of zero Gaussian power. The elementary theory of the Schmidt camera is most easily put in such terms.

The role of the primary aberrations in the optical design of image-forming systems will become clearer in the later sections of this chapter; here we can summarize it by saying that in certain systems of small aperture and field angle it may be sufficient to reduce the primary aberrations to suitably chosen tolerance values, but in most systems it is necessary also to take account of higher order aberrations and to achieve a balance; however, in all cases it is necessary to have the primary aberrations under control, in the sense that parameters must be available to the designer which enable him to *choose* non-zero but small primary aberration residuals to partly cancel effects of higher order terms.

6. Higher order aberrations and ray tracing. In order to evaluate the geometrical optics performance of an optical system completely it is clearly necessary to go beyond the simple but limited processes of paraxial raytracing and primary aberration calculation. The most direct approach is to trace exact or finite rays through the system and to reduce the residual ray aberration to chosen tolerance levels. The alternative approach is to calculate higher order aberrations than the Seidel terms of Sect. 5 and to reduce these to tolerance levels. Both systems have been extensively advocated and used but it seems fair to state at present the raytracing method is mostly used.

There are several reasons for the popularity of raytracing as the chief design tool. In the first place, although higher order aberration theory seems particularly attractive to the mathematician it is often found that the formulae obtained are so complicated as to be unusable by any but the small band of enthusiasts who originally derived them. Several examples of well-developed systems of higher order aberration formulae are now to be found in the literature[1] and in some cases a single expression occupies a whole page. Secondly, it is always a hope of the optical designer to use an aberration formula not merely for numerical calculation but to gain insight into the way in which an aberration contribution depends on the parameters of the optical system; this can be done with the primary aberration terms but there does not seem to be much hope of using higher order terms in this way, except perhaps the spherical aberration terms, since the formulae are too complicated; any higher order term will involve not only the constructional parameters and the Gaussian ray data but also some or all of the corresponding aberrations of all lower orders.

A final drawback to the use of higher order aberrations is that the terms seem to converge rather slowly to the exact aberrations, as found by raytracing, so that one must always resort to raytracing for a final check. In fact no general

[1] See, for example, the contribution by H. MARX to the present volume, or H. A. BUCHDAHL, Optical Aberration Coefficients, Oxford 1954. A recent survey is given by J. FOCKE, Progress in Optics, Vol. IV, p. 3. Amsterdam: North-Holland 1965.

theorems are known on the convergence of aberration expansions, although it seems unlikely that they do not converge.

Raytracing is essentially a simple process. Let P and P_{+1} in Fig. 11 be points of incidence of a ray on two successive surfaces and let $\boldsymbol{r}$ and $\boldsymbol{r}'$ be unit vectors along the ray before and after refraction at P; let $\boldsymbol{a}$ be the unit normal at P, let $\boldsymbol{b}$ be the position vector of P from the origin O and let $\boldsymbol{b}_{+1}$ be the corresponding vector for P_{+1}. We suppose $\boldsymbol{r}$ and P known, so that $\boldsymbol{a}$ is known from the form of the refracting surface; then $\boldsymbol{r}'$ can be found from the law of refraction

$$n'\boldsymbol{r}'\times\boldsymbol{a}=n\boldsymbol{r}\times\boldsymbol{a}. \tag{6.1}$$

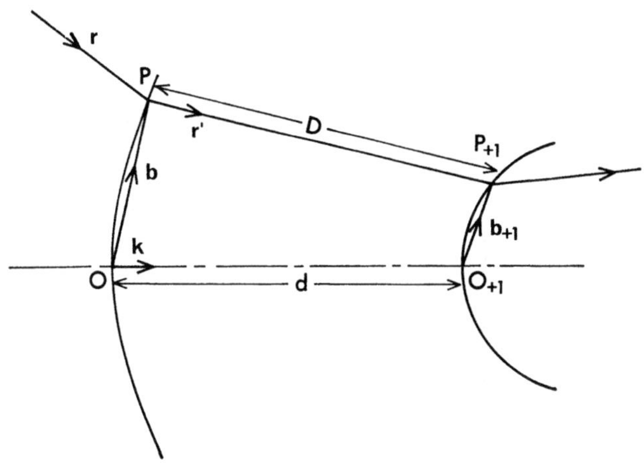

Fig. 11. Finite ray tracing; the ray shown can be a skew ray.

The distance D along the refracted ray to the next surface can be found from the equation

$$\boldsymbol{b}+D\boldsymbol{r}'-\boldsymbol{b}_{+1}=d\boldsymbol{k}, \tag{6.2}$$

where $\boldsymbol{k}$ is a unit vector along the axis and d is the axial separation, together with an equation representing the known form of the refracting surface.

This outline may be filled in by a variety of systems of formulae for numerical work. Those found in the earlier literature were developed to facilitate logarithmic computation and, later on, the use of desk calculating machines. Considerable ingenuity was devoted to the development of methods which retained numerical precision under different conditions and a universal feature was that special methods were used for tracing meridian rays, i.e. rays which intersect the optical axis, since skew rays involved longer formulae. The almost universal availability of fast electronic computers has changed the situation to the extent that the difference in time and cost between tracing meridian and skew rays is of no significance and so a single programme is used for all rays. There would be no point in giving full details of the formulae here since they are such as would naturally be developed from the discussion of Fig. 11[1]. It may be noted that the transfer process, i.e. the determination of D from Eq. (6.2), involves the solution of a quadratic equation if the refracting surface is spherical or is a quadric of revolution; if it is an aspheric of higher degree an iterative process is involved.

[1] See, for example, L. C. MARTIN and W. T. WELFORD, Technical Optics, Vol. I, 2nd edit. London: Pitman 1966.

The question of the mode of representing an *aspheric surface* is of some interest. A quadric of revolution is most naturally described by an equation of the form

$$z = \tfrac{1}{2} c (x^2 + y^2 + \varepsilon z^2), \tag{6.3}$$

where the parameter ε is negative for a hyperboloid, zero for a paraboloid, positive for an ellipsoid and, of course, unity for a sphere. Other aspherics might be represented by the equation of a sphere with an added power series beginning with a fourth degree term, to represent the additional figuring; this would be the obvious extension of the method for primary aberrations, as in Eq. (5.10). However, aspherics with little Gaussian power, such as the surfaces of Schmidt plates and similar components, are not very well represented by such an equation; it is found that for these "nearly flat" surfaces it is better to use a simple power

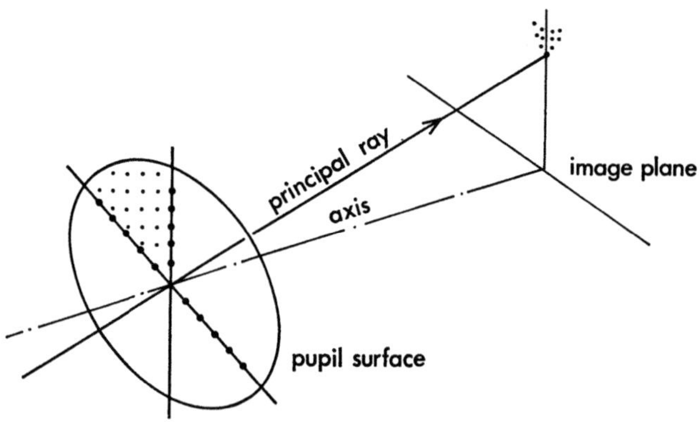

Fig. 12. Generation of a spot diagram.

series for the whole equation and then it is often the case that quite heavy figurings can be represented by only two terms, quadratic and fourth degree, with adequate accuracy. The use of fast computers for raytracing has caused a major change in optical design techniques on account of the ease with which large numbers of skew rays can be traced. Previously it had been the custom to rely on meridian raytracing almost entirely, supplemented by certain formulae (see Sect. 10) which are used for calculating astigmatism of narrow pencils round finite principal rays, but now it is possible to trace several skew rays of each oblique pencil at negligible cost. The assessment of performance can thus be based on entirely new considerations, as we shall see. To be more specific, an experienced calculator can compute the path of a skew ray through one surface in about 10 minutes with an electromechanical desk calculator[1] but a modern electronic computer needs only a small fraction of a second for the same process.

There are several ways in which the information from such raytracing can be used in designing; the choice is to some extent a matter of individual preference and it also depends on whether the computer raytracing is being used as part of a process of automatic optical design (see Sect. 12), or as an aid to conventional designing, in which the optical designer uses a mixture of intuition and logic to evolve changes in the system. We now describe some of these ways of using raytracing results.

The immediately obvious way of presenting raytracing results is in the form of the so-called *spot diagram;* points (circles or crosses) are marked to show the

[1] The author has no experience of the electronic desk calculators which are becoming available at the time of writing.

intersection with the image plane of a set of rays belonging to one oblique pencil, so that it may be regarded as a simultaneous plot of the two components of transverse ray aberration. If a large number of ray intersections is plotted (about 100 or more) the resulting diagram evidently represents in some sense the point image in the geometrical optics approximation, the density of points being interpreted as light intensity (Fig. 12). This is usually done by tracing actually only a small number of rays, (say 10) and finding the other points by interpolation. In effect this means that polynomials are fitted to represent the components of ray aberrations as functions of pupil position.

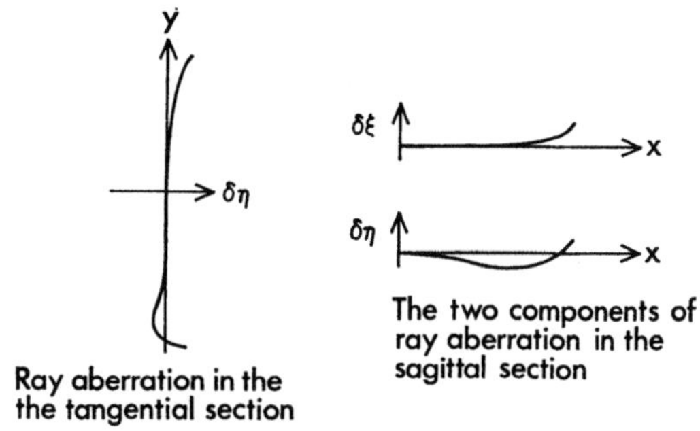

Fig. 13. Plots of transverse ray aberration; $\delta\xi$ and $\delta\eta$ are the components of aberration, x and y are the exit pupil coordinates.

We are thus led to a second representation in which the components of transverse ray aberration are plotted as functions of pupil position for, usually, different azimuths in the pupil (Fig. 13).

Fast computers make it entirely practicable to calculate optical path lengths along the rays by direct summation of the ray segments between successive points of incidence; if these segments are summed to the intersection with a reference sphere (Fig. 3) we obtain the wavefront aberration as a function of pupil position and this may be plotted as level lines of constant aberration (Fig. 14) or as sections of the wavefront in various azimuths (Fig. 15).

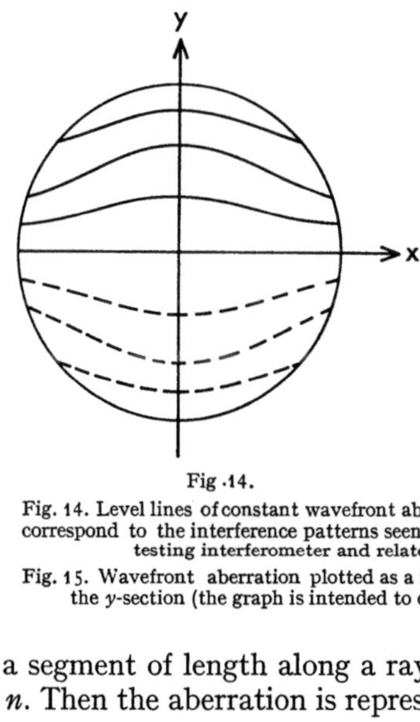

Fig .14.

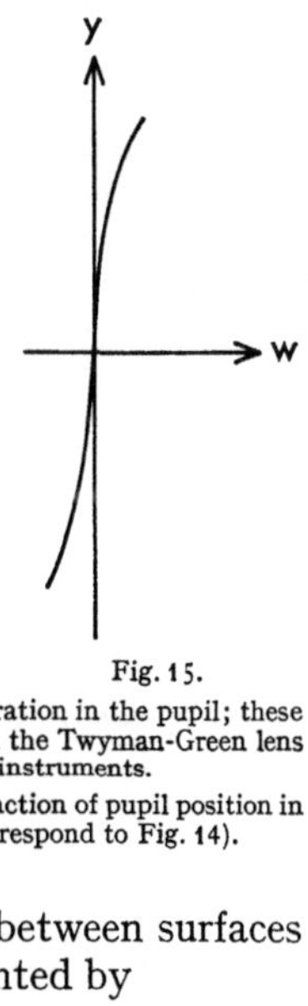

Fig. 15.

Fig. 14. Level lines of constant wavefront aberration in the pupil; these correspond to the interference patterns seen in the Twyman-Green lens testing interferometer and related instruments.

Fig. 15. Wavefront aberration plotted as a function of pupil position in the y-section (the graph is intended to correspond to Fig. 14).

In symbols, let D represent a segment of length along a ray between surfaces in a medium of refractive index n. Then the aberration is represented by

$$W(x, y) = \Sigma n (D_{pr} - D_a), \tag{6.4}$$

where the subscript pr denotes distances measured along the principal ray and subscript a denotes another ray, this latter meeting the exit pupil plane at (x, y); the summation is taken from the object point to the reference sphere in the image space.

Finally, the raytracing results may be used as the starting point for a computation of some kind of "figure of merit", perhaps to be used in the process of automatic design. For example, the mean square distance of the points in the spot diagram from the principal ray may be used or the Strehl intensity ratio or the optical transfer function may be computed. We refer to these again in Sect. 11.

The results of raytracing give the total aberration, whether in ray or optical path form, whereas the primary aberration calculations give only the first terms of the aberration expansion. Thus the differences between raytracing results and primary aberrations represent the sums of all higher order aberration residuals. Any chosen skew ray of an oblique pencil will in general have a mixture of different kinds of aberration but by suitable choice of the rays to be traced it is fairly easy to make a rough separation. For example, rays of an axial pencil show only spherical aberration of various orders, a principal ray shows distortion only, oblique rays in the meridian section show field curvature and coma, and oblique rays in the sagittal section show astigmatism, coma and certain higher order aberrations.

The fitting of polynomials to the aberrations, mentioned above, can be carried out on either the ray or wavefront aberrations; if the terms in the polynomials are suitably chosen they may be regarded as approximately representing the corresponding terms in the aberration expansion. While the coefficients will not be exactly the same as the true aberration coefficients such a representation has the advantage that it will almost certainly fit the raytracing results better than the true aberration expansion would, if taken to the same number of terms.

7. Chromatic aberrations and optical materials. So far we have not introduced the complicating fact that the refractive index is a function of the frequency of the light, so that all Gaussian and aberrational properties of an optical system will in general also be functions of frequency. The difficulties which arise in correcting chromatic aberrations are due in considerable part to the particular form of the functional dependence on frequency, so we find it convenient to discuss at the same time this dependence in the actual optical media which have to be used.

The point is conveniently illustrated by the elementary problem of an achromatic doublet consisting of two thin lenses in contact. Let the refractive indices of the two lenses be $n_1(\lambda)$ and $n_2(\lambda)$[1] and let Δ_{ab} denote the change (in refractive index, say) on going from λ_a to λ_b. Then it is easily shown that the condition that the power of the doublet be constant for all wavelengths in a given range is

$$\frac{\Delta_{ab} n_1}{\Delta_{ab} n_2} = C, \tag{7.1}$$

where C is a constant and where λ_a and λ_b can take any values over the assigned range. This requires that a relationship such as

$$n_1(\lambda) = C n_2(\lambda) + \text{const.} \tag{7.2}$$

shall hold, i.e. that the two glasses shall have dispersion curves which are linearly related. It is well known that no two glasses exist which satisfy such a relation to suitable tolerances, a fact sometimes vividly called the "irrationality of dispersion", so that it is impossible to make a perfect doublet achromat. The departure from this condition gives rise to residual chromatic effects known as "*secondary spectrum*".

The above discussion illustrates two points, first that chromatic aberration can arise in the Gaussian or paraxial region, as opposed to the monochromatic aberrations, and secondly that in correcting chromatic aberration we often have to consider effects dependent quadratically as well as linearly on wavelength changes, the former being secondary spectrum effects.

Linear chromatic effects are usually rather easily controlled but in general little can be done about secondary spectrum beyond choice of glasses to approach

[1] We use wavelength rather than frequency as the variable in conformity with tradition; the vacuum wavelength is understood.

as near to fulfillment of a condition like Eq. (7.1) as possible. Thus the optical glass manufacturers list the properties of their optical glasses in such a way as to facilitate this choice when necessary, and much work has gone into the manufacture of glasses with more nearly matching dispersions.

This basic problem of a thin system achromatized for power, i.e. the telescope objective, can also be tackled by using more than two glasses to decrease the secondary spectrum. The most thorough investigation on these lines is that by HERZBERGER and MCCLURE[1], who investigated the so-called *superachromats*; these are three-component objectives with back focal length which is constant throughout the visible spectrum to within about 0.02% of the equivalent focal length, about one-tenth of the corresponding figure for ordinary doublet achromats. Unfortunately the powers of the individual components are so large that other aberrations, including chromatic variation of spherical aberration, begin to get out of control, so that it is unlikely that the superachromat offers much advantage in any cases of practical importance. As for the monochromatic aberrations, it is possible to deal either with a ray or wavefront representation of chromatic aberration. The most obvious way is simply to re-trace the same rays, using the refractive indices appropriate to a chosen range of wavelengths; from the results either ray or wavefront aberrations in the different wavelengths can be obtained; this is entirely practicable nowadays with the availability of fast computers. However, there is still a need for manageable formulae to give approximate values for use in preliminary design work, to complement the primary aberration formulae. For such approximate formulae secondary spectrum effects are ignored and the chromatic aberration is represented by the difference between the properties of the system at two suitably chosen wavelengths.

The most general result of this kind is obtained by differentiating Eq. (6.4):

$$\Delta_{ab} W(x, y) = \Sigma(\Delta_{ab} n) \cdot (D_{\mathrm{pr}} - D_{\mathrm{a}}), \tag{7.3}$$

where Δ_{ab} denotes the change in refractive index or in wavefront aberration corresponding to a change in wavelength as above. In this equation it can be seen that terms like $n \Delta_{ab} D_{\mathrm{pr}}$ have been omitted, an approximation which is justified by FERMAT's principle, since the change in ray paths with wavelength is small[2].

From Eq. (7.3) or otherwise it is possible to obtain expressions for *the first order or Gaussian changes with colour*; these are changes in focussing distance and in magnification which are called primary longitudinal and transverse chromatic aberration. As for the primary aberrations, they may be expressed either as ray aberrations or as wavefront shifts; sums of terms over the surfaces of the system are used, analogously with Eqs. (5.2) to (5.6), and the actual expressions may be found in standard texts. These two sums are traditionally associated with the primary aberrations as the main group of "aberrations" which must be held under control in any system which is to have reasonably good aberration correction. It is worth stressing, however, that the longitudinal and transverse chromatic "aberrations" are strictly in the Gaussian region, a fact of which the importance is made manifest in two ways: first, by the comparatively large magnitude of uncorrected chromatic aberrations, e.g. the change in focal length of a singlet lens through the visible spectrum is about 2% of its focal length, and secondly by the fact that the colour correction depends mainly on the choice of the powers of the individual components and is to a large extent independent of the shapes of the components. It is furthermore found that very little can be

---

[1] M. HERZBERGER and N. R. MCCLURE: Appl. Optics **2**, 553 (1963).

[2] A. E. CONRADY: Month. Notices Roy. Astron. Soc. **64**, 182, 458 (1904); see also R. A. HERMANN, Geometrical Optics. Cambridge 1900.

done to improve secondary chromatic effects (secondary spectrum and its off-axis analogues) in a given type of lens design by subtle choice of glasses. In this respect the chromatic effects are similar to field curvature, whereas the other aberrations can be controlled by choice of lens shapes.

The most interesting developments in optical glasses have taken place in the visible region of the spectrum; some optical glass catalogues now list upwards of 200 different glass types, and the great majority of these appear mainly for facility in achieving chromatic correction in the visible and near infra-red region (as far as 2 nm wavelength approximately). By contrast, in the ultraviolet only four refracting materials are of any practical importance, lithium fluoride, fluorite (CaF_2), fused silica and crystal quartz, the first two being low dispersion materials and the others high dispersion; very little sophisticated optical design is possible with such a restricted range, particularly since the dispersions vary rapidly. In the infra-red there are more materials but there is very little variety in dispersion. Thus both in the infra-red and ultraviolet most optical systems use reflection for image formation, which solves both the problem of chromatic aberration and that of adjustment, since the system can be aligned in the visible region. It is fair to point out, however, that there are considerable limitations to what can be done with reflecting systems, mainly owing to obstruction of the light beams by mirrors on return paths; thus we do not find any analogue in reflecting optics of the elaborate wide field, wide aperture photographic objectives; the nearest approach, the Schmidt system (with its derivatives), has a relatively restricted angular field and even then the image surface is very inconveniently situated. It thus turns out that the chief applications of reflecting systems are in astronomy, where great physical size is needed to obtain light-gathering power; in this case the fact that mirrors can be made in larger sizes than lenses for reasons of physical stability, homogeneity of glass, etc., is not the only consideration; the maximum tolerable aberrations in a well-corrected system are of the order of magnitude of a wavelength of wavefront deformation whatever the physical size of the system, so that a given design may be substantially perfectly corrected with a focal length of, say, 100 mm, but if it is linearly scaled up in size by a factor 10 it could have very large aberrations. This difficulty may be overcome to some extent by using non-spherical surfaces to improve the monochromatic aberration correction, but nothing can be done about chromatic aberrations, so that the chromatic correction of refracting systems becomes increasingly difficult as the size increases. This is therefore a further reason why reflecting optics are used for very large optical systems.

It can be seen from the above that the dispersive properties of optical glasses and other optical materials constitute a serious obstacle to the further development of optical design. It would be of the utmost significance for the optical instrument industry if optical materials were developed with dispersions which matched better than those at present available, but it must be admitted that the prospects for such a development seem very slim.

8. The characteristic function and the eikonal. In this section we shall explain briefly the use of the characteristic function mentioned in Sect. 5 (p. 15) and show how the results connect with other approaches to aberration theory. The development has been given by several authors[1] but the expositions do not connect well with optical design interests.

[1] See, e.g. G. C. STEWARD, The symmetrical optical system. Cambridge 1928; J. L. SYNGE, Geometrical optics, an introduction to HAMILTON's method. Cambridge 1937; R. J. PEGIS, The modern development of Hamiltonian optics. Progress in Optics, Vol. I, p. 3—29. Amsterdam: North-Holland 1961.

Let $P(x, y, z)$ and $P'(x', y', z')$ be two points in object and image space; in general one and only one ray can pass through them, since they are not conjugates. HAMILTON defined the *characteristic function* $W(x, y, z, x', y', z')$ as the optical path length along the ray from P to P' and he showed that its partial derivatives with respect to the coordinates gave the corresponding direction cosines of the two segments of the ray in object and image space[1]. Thus if W is known all the properties of the system are known.

A slightly different function is more convenient for most purposes; this is the *eikonal*, so called by H. BRUNS although it was also used by HAMILTON. The eikonal is defined as follows: in Fig. 16, O and O' are fixed points in object and image space, usually but not necessarily on the axis, and a ray has direction cosines (L, M, N), (L', M', N') in the two spaces. Perpendiculars through O and O' are constructed to the two parts of the ray, meeting it at Q and Q' and the eikonal E is defined as the optical path length from Q to Q' considered as a function of (L, M, L', M'). The partial derivatives of E with respect to these direction cosines then give, again apart from trivial factors, the coordinates of P and P', the points of intersection of the ray with fixed planes through O and O'.

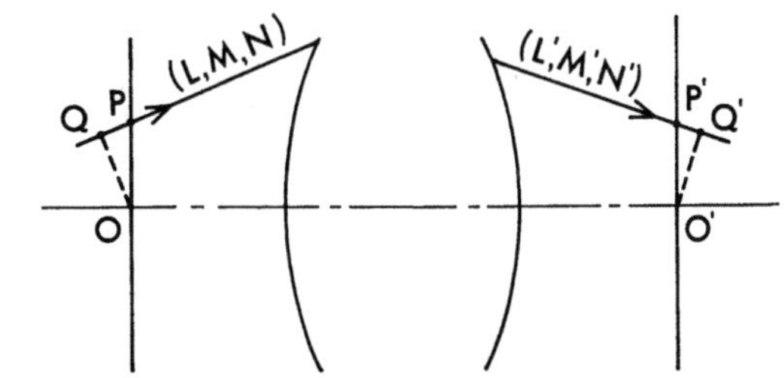

Fig. 16. Definition of the eikonal.

In the systematic development of aberration theory on these lines O and O' are axial conjugates with magnification G and we further define the pupil magnification to be S; then symmetrical variables α, β, γ are defined as follows:

$$\left.\begin{aligned} (S-G)^2\alpha &= L^2+M^2-2S(LL'+MM')+S^2(L'^2+M'^2), \\ (S-G)^2\beta &= L^2+M^2-(S+G)(LL'+MM')+SG(L'^2+M'^2), \\ (S-G)^2\gamma &= L^2+M^2-2G(LL'+MM')+G^2(L'^2+M'^2). \end{aligned}\right\} \quad (8.1)$$

It is assumed that E can be expanded as a power series in α, β, γ and it can be shown that if E is a function of γ alone then there are no aberrations, in fact this is partly the reason for the choice of these variables. The significance of this can be made clear by putting in the paraxial approximation, when it can be seen that, again apart from irrelevant scale factors, α, β, and γ correspond respectively to the expressions x^2+y^2, ηy and η^2 which appeared in Eq. (3.3) as the variables for the aberration expansion. Of course, the equivalence is not exact beyond the Gaussian region but it can now be seen that γ is simply a field variable, so that it would naturally be expected that a system has no aberrations if E is a function of γ alone.

It is then possible to determine the coefficients of the powers of γ in the expression for E if the system has no aberrations. Any increments to these will correspond to a kind of aberration, even if only powers of γ appear, namely spherical aberration of the pupil points; this is not generally counted as an aberration since it is possible to have perfect image formation even though the pupil is aberrated, but it is necessary to take it into account in the general theory because pupil aberrations transform into image aberrations if the object and image conjugates are changed.

[1] Strictly they are the direction cosines multiplied by the refractive indices; for this general explanation we shall ignore these factors in what follows.

In the further development of the eikonal theory a power series for E in terms of α, β and γ is assumed and it is then possible to obtain a very general transformation showing how the terms in this power series of all orders change as the object and stop positions are changed, i.e. we can find how all orders of aberration change with G and S. To see how this can be done we note that (L, M, L', M') represent uniquely one ray of the quadruple infinity of rays passing through the system. Fig. 17 shows the ray in image space, cutting the pupil and image planes at Q' and P'. The eikonal power series gives the aberration terms as different combinations of the variables (α, β, γ), which are themselves functions of the direction cosines. To obtain the transformation to new pupil and image positions we have to express this expansion in terms of new variables $(\alpha', \beta', \gamma')$, corresponding to the new G', S', these being given in terms of the *same direction cosine variables* by equations like (8.1). Thus it is in principle simple to express (α, β, γ) in terms of $(\alpha', \beta', \gamma')$, the results being

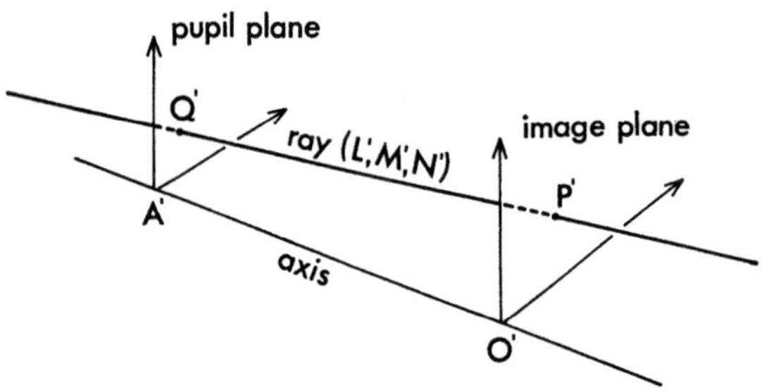

Fig. 17. Derivation of the changes in aberrations due to stop and conjugate shifts.

$$\left.\begin{aligned}
(S-G)^2\alpha &= (G'-S)^2\alpha' - 2(S'-S)(G'-G)\beta' + (S'-S)^2\gamma';\\
(S-G)^2\beta &= (G'-G)(S'-S)\alpha' - \{(S'-G)(G'-S) + (G'-G)(S'-S)\}\beta' +\\
&\quad + (S'-G)(S'-S)\gamma';\\
(S-G)^2\gamma &= (G'-G)^2\alpha' - 2(S'-G)(G'-G)\beta' + (S'-G)^2\gamma'.
\end{aligned}\right\} \tag{8.2}$$

These values can be substituted to obtain a new series for E in terms of $(\alpha', \beta', \gamma')$; the rearranged coefficients contain functions of the old and new pupil and image magnifications and so explicit expressions for the new aberration terms as functions of the old ones are obtained. This analysis was first carried out by T. SMITH and it is reproduced in the previously mentioned references.

This still does not, of course, give any information about the actual magnitudes of the individual aberrations; these can be obtained, as before, as surface-by-surface contributions for primary aberrations. The eikonal theory offers no short-cuts to obtaining the higher order terms.

The very general transformation obtained by SMITH for all orders of aberration coefficients does not seem to be of much practical interest for optical design purposes, although it has significance for the deeper question, what kind of aberration correction is in principle possible for an optical system. However, the form of the transformation which is obtained when only primary aberrations are considered is of interest and we reproduce it here. In order to do this it is necessary to re-write the general expression for the aberrations of a symmetrical optical system [see Eq. (3.3)] in a more general form:

$$W(\alpha, \beta, \gamma) = b_1\alpha^2 + b_2\alpha\beta + b_3\beta^2 + b_4\alpha\gamma + b_5\beta\gamma + b_6\gamma^2, \tag{8.3}$$

where

$$\alpha = (x^2 + y^2), \quad \beta = \xi x + \eta y, \quad \gamma = \xi^2 + \eta^2, \tag{8.4}$$

and (x, y) are exit pupil coordinates and (ξ, η) are image plane coordinates; these are now essentially paraxial quantities and b_1 to b_6 are aberration coefficients having the dimensions of reciprocal length cubed, since W represents optical path aberration. We have now put in the term in b_6, which represents spherical aberration of the pupil, since it will transform into other aberrations when the image plane is shifted. Also we allow the Gaussian image point to have a non-zero ξ coordinate, for similar reasons.

Now let the image and stop magnifications be respectively G and S and suppose both are changed to G' and S'; the aberrations will then be represented by

$$W(\alpha,\beta,\gamma)=b_1'\alpha^2+b_2'\alpha\beta+b_3'\beta^2+b_4'\alpha\gamma+b_5'\gamma\beta+b_6'\gamma^2, \tag{8.5}$$

where b_1' to b_6' are new aberration coefficients but α,β,γ represent the variables given by Eq. (8.4) on the new exit pupil and image planes. The transformation then gives the new aberration coeffieients in terms of the old as follows:

$$\left.\begin{aligned}(S'-G')^4 b_1' = {} & (S-G')^4 b_1+(S-G')^3(G-G')\,b_2+(S-G')^2(G-G')^2(b_3+b_4)+\\ & +(S-G')(G-G')^3 b_5+(G-G')^4 b_6;\end{aligned}\right\} \tag{8.6}$$

$$\left.\begin{aligned}(S'-G')^4 b_2' = {} & -4(S-G')^3(S-S')\,b_1-(S-G')^2\{3(S-S')(G-G')+\\ & +(S-G')(G-S')\}\,b_2-2(S-G')(G-G')\{(S-G')(G-S')+\\ & +(S-S')(G-G')\}(b_3+b_4)-(G-G')^2\{3(S-G')(G-S')+\\ & +(S-S')(G-G')\}\,b_5-4(G-G')^3(G-S')\,b_6;\end{aligned}\right\} \tag{8.7}$$

$$\left.\begin{aligned}(S'-G')^4 b_3' = {} & 4(S-G')^2(S-S')^2 b_1+2(S-G')(S-S')\{(S-G')(G-S')+\\ & +(S-S')(G-G')\}\,b_2+\\ & +\{(S-G')(G-S')+(S-S')(G-G')\}^2 b_3+\\ & +4(S-G')(G-G')(S-S')(G-S')\,b_4+2(G-G')(G-S')\times\\ & \times\{(S-G')(G-S')+(S-S')(G-G')\}\,b_5+\\ & +4(G-G')^2(G-S)^2 b_6;\end{aligned}\right\} \tag{8.8}$$

$$\left.\begin{aligned}(S'-G')^4 b_4' = {} & 2(S-G')^2(S-S')^2 b_1+(S-G')(S-S')\{(S-G')(G-S')+\\ & +(G-G')(S-S')\}\,b_2+2(S-G')(S-S')(G-G')(G-S')\,b_3+\\ & +\{(G-G')^2(S-S')^2+(S-G')^2(G-S')^2\}\,b_4+\\ & +(G-G')(S-S')\{(G-G')(S-S')+(G-S')(S-G')\}\,b_5+\\ & +2(G-G')^2(G-S')^2 b_6;\end{aligned}\right\} \tag{8.9}$$

$$\left.\begin{aligned}(S'-G')^4 b_5' = {} & -4(S-G')(S-S')^3 b_1-(S-S')^2\{3(G-S')(S-G')+\\ & +(S-S')(G-G')\}\,b_2-2(S-S')(G-S')\{(S-G')(G-S')+\\ & +(S-S')(G-G')\}(b_3+b_4)+\\ & -(G-S')^2\{3(G-G')(S-S')+(S-G')(G-S')\}\,b_5-\\ & -4(G-G')(G-S')^3 b_6;\end{aligned}\right\} \tag{8.10}$$

$$\left.\begin{aligned}(S'-G')^4 b_6' = {} & (S-S')^4 b_1+(S-S')^3(G-S')\,b_2+(S-S')^2(G-S')^2(b_3+b_4)+\\ & +(S-S')(G-S')^3 b_5+(G-S')^4 b_6.\end{aligned}\right\} \tag{8.11}$$

It will be observed that the above transformation is quite different in character from that more usually used to give the effect of changes in stop position; in the latter case, Eqs. (5.7) to (5.9) or some equivalent apply and they give the changes as *increments* to the *actual aberrations* at full aperture and field. In the present, more general, transformation we have a set of linear relations between *aberration coefficients*. The aberration of the stop (b_6) must appear because it takes part in changes involving the image plane (change in G) but as can easily be seen it does not come into the calculation if only the stop plane is changed. Eqs. (5.7) to (5.9) do not give the effect of a change of object position but this is, of course, implied by the dependence of the Seidel sums themselves on the Gaussian conjugates.

It is also possible to obtain expressions in terms of the eccentricity parameter E for changes in aberrations produced by object shifts[1]; these contain the pupil aberrations and they are, of course, equivalent to Eqs. (8.6) to (8.11) but with $S' \equiv S$.

9. Aplanatism and isoplanatism. In considering the formation of optical images according to scalar wave theory, i.e. the ordinary physical optics approximation, the concept of *isoplanatism* is almost always implicit although it is not often discussed. The concept applies to images of extended objects; it comprises broadly the notion that the aberrations are constant over a certain finite region of the field, i.e. the pencils from each point in the object have, to within a certain tolerance, the same aberrations. The assumption of isoplanatism is necessary to all calculations of images of periodic or non-periodic structures so far published.

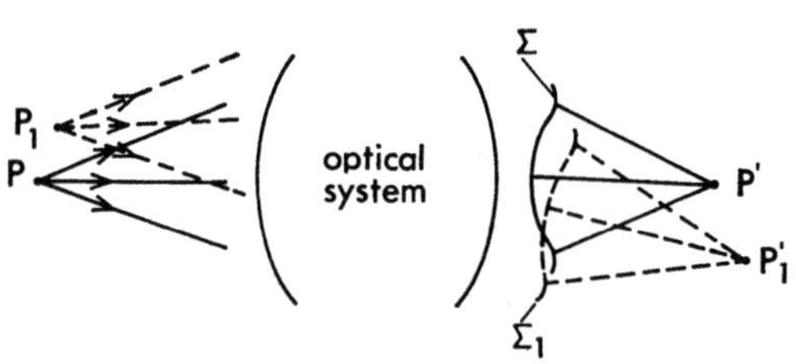

Fig. 18. Definition of isoplanatism.

The same concept appears in a different way in connection with the aberrations affecting object points near the axis of a symmetrical optical system. Leaving aside the various orders of spherical aberration, which are of course present in the axial pencil, the most important aberrations near the axis are those depending *linearly* on the field angle, i.e. primary coma and a series of higher order coma terms which, in the notation used in Eq. (3.3), would be written

$$(x^2+y^2)^2 y\eta, \ (x^2+y^2)^3 y\eta, \ldots. \tag{9.1}$$

Such aberrations clearly have a bad effect on images very near the axis, particularly since they produce an asymmetric point image, and an optical system which is corrected for these terms is said to be *isoplanatic*. If there is in addition no spherical aberration it is said to be *aplanatic*. In an isoplanatic system the first aberrational changes in moving away from the axis are those depending on the square of the field variable, such as astigmatism and field curvature, so that a region near the axis of reasonably stable image formation is to be expected.

Many different forms of isoplanatism condition have been published; they are expressed in terms of the properties of a pencil of rays of large aperture from a point object and they give the condition that the pencil shall remain unchanged in shape for a small movement of the object. Thus in Fig. 18 the object point P is imaged as the (aberrated) image P' formed by the wavefront Σ; a small movement of P to P_1 produces a shift of P' and Σ to P_1' and Σ_1 and the system will be isoplanatic in this region if Σ_1 is congruent with Σ, neglecting quantities of the order of the square of PP_1.

The most general such condition is probably T. SMITH's *optical cosine law*[2], which is a condition for congruence of the two pencils of rays without any assumptions of symmetry or freedom from aberration. In its general form it is of little practical interest but it can be specialized to the axial region, when it becomes the Abbe sine condition if it is assumed that no spherical aberration is

[1] C. G. WYNNE: Proc. Phys. Soc. Lond. **65** B, 429 (1952).
[2] T. SMITH: Trans. Opt. Soc., Lond. **24**, 31 (1922/23).

present. In the case when spherical aberration is present several equivalent conditions were given by BEREK, all derived from the Staeble-Lihotzky condition[1], and also by DI JORIO[2]. In all these cases only conditions for isoplanatism were given, but it is of much greater interest to the optical designer to have an expression which gives the magnitude of any departure from isoplanatism, since this can then be used as a tolerance condition.

Confining our attention now to the axial region it is clear that such an expression would give the coma arising in the system at small distances from the axis, i.e. it would give, from Eq. (9.1) the amount of aberration of the form $f(r)\cos\varphi\cdot\eta$, where η is the field variable, $\cos\varphi$ is the pupil azimuth and $f(r)$ is an (odd) function of pupil radius which is some linear combination of r^3, r^5 and so on. Expressions of this kind were given by CONRADY[3] and we give now substatially his formulation. Let O and O' be axial conjugates (Fig. 19) and let a finite ray from O have convergence angles U and U' in object and image space; as indicated, there may be some spherical aberration. Let the Gaussian off-axis conjugates be O_1 and O_1', with heights η and η' as indicated. We suppose that the distance of O' from the exit pupil is l' but that the finite ray meets the axis at a distance L' from the pupil. Let u and u' be convergence angles of a paraxial ray from O to O'; as is usual it is supposed that the paraxial ray and the finite ray coincide in object space, so that $u \equiv \sin U$ but usually $u' \neq \sin U'$. Then the coma as an optical path difference or wavefront deformation which appears in the pencil from O, is given by

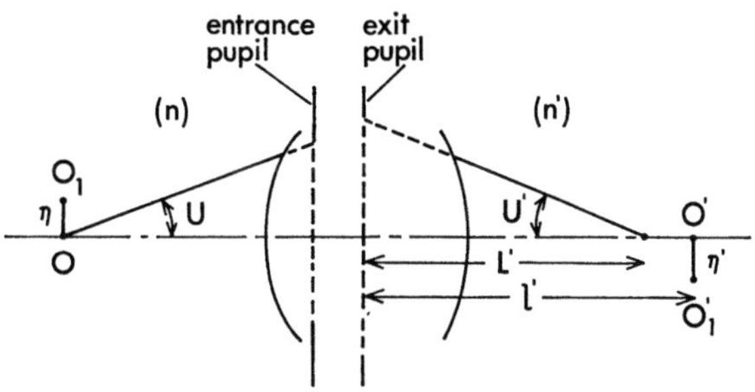

Fig. 19. The isoplanatism condition.

$$W_{\text{coma}} = H\cos\varphi\left\{\frac{L'}{l'}\,\frac{\sin U'}{u'} - \frac{\sin U}{u}\right\}, \tag{9.2}$$

where H is the Lagrange invariant, given by $nu\eta$, for the values of u and η in the case in question.

This result is of great importance in the design of optical systems of small field and large aperture, such as microscope objectives, and Conrady made numerous ingenious applications of it. For example, it can be seen that the position of the stop is a relevant variable, since L' and l' are measured from the exit pupil, so that the stop position for zero coma in the presence of spherical aberration can be found. Of course, this equation applies only to the particular finite ray which has been traced, so that it would be conceivable that a system could be isoplanatic for one radius in the pupil but not elsewhere. In systems like microscope objectives of large numerical aperture several rays might be tested.

To some extent the use of the isoplanatism test might now be replaced by simply a large amount of raytracing, since this is so easy to do with an electronic computer, but there still seems to be a good field of application for it in the earlier stages of an optical design. Also, in its simplest form as the Abbe sine condition,

[1] M. BEREK: Grundlagen der praktischen Optik. Berlin: W. de Gruyter & Co. 1930.
[2] M. DI JORIO: J. Opt. Soc. Amer. **39**, 305 (1949).
[3] A. E. CONRADY: Applied Optics and Optical Design, Part I. Oxford 1929, reprinted 1960 by Dover Publ. Inc. New York. See also W. T. WELFORD, A Guide to Instrument Design, p. 141 (ed. J. THOMSON). London: Taylor and Francis 1963.

it has important bearings on the possible degree of correction of aberrations and on the photometry of optical instruments.

10. Optical systems without axial symmetry. Most attention has been devoted to the symmetrical optical system in the literature since most optical systems are intended to be axially symmetric, but there are several interesting fields of application of non-axially symmetric image-forming systems; for example, spectroscopic instruments involving dispersing prisms and gratings, and anamorphotic lens systems, which are used for producing images with varying magnifications in different directions in the image plane; also the image formation in the immediate neighbourhood of a principal ray in an axially symmetric optical system may be included under this heading. We therefore discuss in this section some general aspects of such systems.

Very little of interest can be said about systems with no symmetry at all. The two important classes of restricted symmetry are those systems having respectively two planes of symmetry (intersecting at right angles in an axis) and one plane of symmetry. The first class is exemplified by systems of cylindrical lenses as used in anamorphotic projectors, the second by spectroscopic systems and by the imagery round a principal ray; in the latter case the plane of symmetry is that containing the principal ray and the axis of the main optical system, in spectroscopic systems it is the plane containing the direction of dispersion.

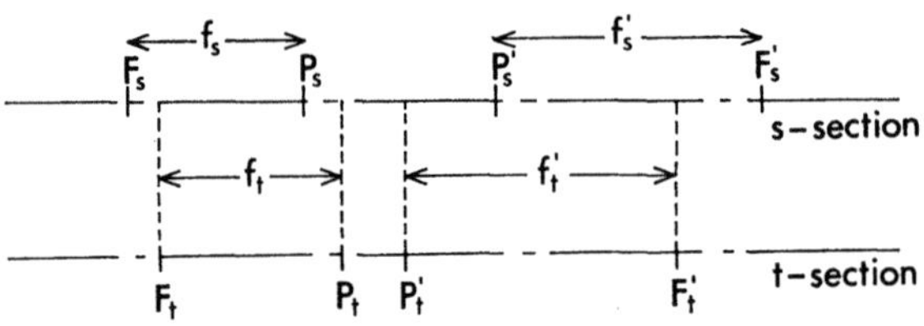

Fig. 20. Gaussian properties of a system with two planes of symmetry.

For convenience in discussion we shall refer to a chosen principal ray in systems with one plane of symmetry as the "axis"; we can then discuss the nature of the image formation in the neighbourhood of this axis and draw appropriate parallels with systems with two planes of symmetry and with completely symmetrical systems. It then turns out that if we consider only object points actually on (not near) the axis there is no difference between the systems with two planes of symmetry and those with one in the Gaussian approximation. Both behave as systems with one set of Gaussian properties in the plane of symmetry and another set in the orthogonal plane. For convenience we denote these sections by subscripts s and t respectively, these letters standing for "sagittal" and "tangential" in agreement with the usual terminology for image formation round a principal ray. Then each section has its own set of principal points and foci, as in Fig. 20, where the longitudinal locations are intended to correspond and the only relationship between the two sections is that implied by the equation connecting the two focal lengths [Eq. (4.5)], i.e. we have

$$\frac{n}{f_s} = -\frac{n'}{f_s'}; \quad \frac{n}{f_t} = -\frac{n'}{f_t'}. \tag{10.1}$$

In general such a system has axial astigmatism; there are two sets of axial conjugates which give stigmatic images, but since they are found as roots of a pair of simultaneous quadratic equations the two sets may coincide or they may have complex axial coordinates, so that for practical optics, the number of pairs of conjugates which agree in both sections can be two, one or zero. When the axial conjugates coincide the axial image has no astigmatism but the transverse magnification is in general different in the s and t sections; the image formation is then said to be *anamorphotic*. Further special conditions have to be satisfied if there are to be equal magnifications in both sections. The axial image cannot be stigmatic for all conjugate positions unless the principal planes and foci coincide in both sections. The principal applications for cylindrical optical systems exemplifying these ideas are in spectacle lenses and in certain cine-projection systems. Some recently proposed spectacle lenses with continuously varying power in the vertical plane are, in effect, systems with one plane of symmetry.

In ordinary optical systems the image formation in the neighbourhood of a principal ray is generally astigmatic. Two very simple conjugate distance relations are used for tracing the successive positions of the conjugates in the s and t sections along the ray, to determine the final image as a pair of astigmatic focal lines. Let I and I' be the angles of incidence and refraction of the principal ray at a surface, let c_s and c_t be the curvatures in the sagittal and tangential sections and let s, s', t and t' be the conjugate distances from the point of incidence. The equations, due to THOMAS YOUNG and J. C. STURM are

$$\left.\begin{aligned} \frac{n'}{s'}-\frac{n}{s}&=(n'\cos I'-n\cos I)\,c_s;\\ \frac{n'}{t'}\cos^2 I'-\frac{n}{t}\cos^2 I&=(n'\cos I'-n\cos I)\,c_t. \end{aligned}\right\} \qquad (10.2)$$

By means of these equations the shape of the surfaces in the final image region on which the sagittal and tangential focal lines are formed can be found. They are so simple and the final result has such immediate graphic application for the lens designer that some form of this calculation is almost always included in computer programmes for ray tracing.

To pursue the subject of image formation in cylindrical systems beyond the Gaussian region immediately brings in considerably more complication than in the case of axially symmetrical systems, since there are sixteen primary aberrations instead of five; we refer to recent publications for more details[1].

11. Aberration tolerances. It seems to be clearly established by experience that it is impossible to reduce the aberrations precisely to zero in any but trivial optical systems, although this has not yet been rigorously proved. It is therefore essential to establish tolerances for aberrations, so that the optical designer knows when to stop. The subject of aberration tolerances is chiefly governed by results from scalar diffraction theory, so that it is in detail beyond the scope of this chapter, but we give here a brief indication of the general principles and of the kinds of results obtained.

The first point is that the image of a single bright point object is the Airy pattern (Fig. 21), if the system has no aberrations and has a circular aperture stop; this is a diffuse circular patch surrounded by diffraction rings and it bears little relation to the infinitely small, infinitely bright point of light, which would be predicted by a naive application of geometrical optics. The simplest system of optical tolerances is based on a study of the effects of aberrations on the Airy pattern. It is found that different aberration types all affect the Airy pattern in more or less the same way provided the amounts of aberration are rather small. Here "small" means that the deviation of the wavefront at the exit pupil from a reference sphere centred on the ideal image point shall be of the order of a quarter of a wavelength or less. In fact a general tolerance of $\pm\lambda/4$ deformation from a true

[1] For example, C. G. WYNNE, Proc. Phys. Soc. Lond. B **67**, 529 (1954), gives the primary aberration theory; finite ray tracing methods for toric surfaces are given by W. WEINSTEIN, J. Res. Nat. Bur. Stds., **54**, 251 (1955) and for gratings by W. T. WELFORD, Optica Acta, **9**, 389 (1962). The general primary aberration theory of gratings is given by W. T. WELFORD, Progress in Optics, **4**, 241 (1965).

sphere is the celebrated *Rayleigh limit*, proposed by Lord RAYLEIGH on the basis of a study of the special case of change of focus as an aberration; there is a plausible line of argument that within the Rayleigh limit all the secondary Huygens wavelets will arrive in the region of focus with contributions to the complex amplitude which all have the same sign. In fact it is found that the kind of aberration has some influence on the magnitude of the tolerance, i.e. whether it is spherical aberration, coma, etc. or a mixture, but that the general effect is at first similar in all cases, namely a drop in the intensity of the central maximum

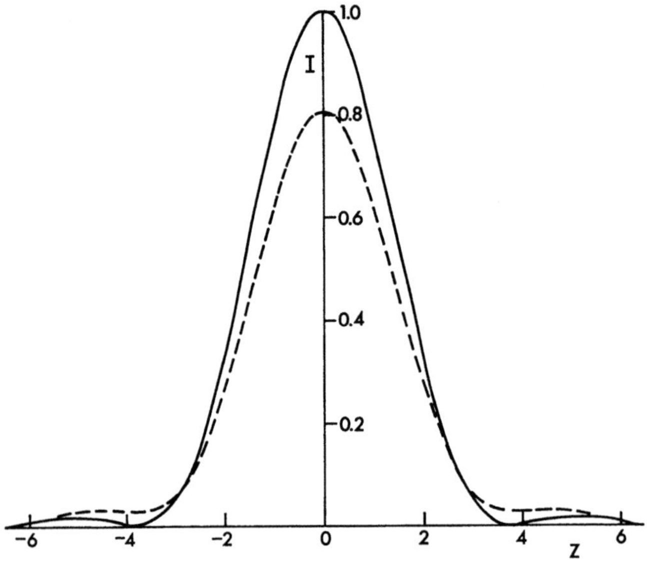

Fig. 21. The Airy pattern (full line) and the effect of a change of focus of $\lambda/4$ at the edge of the pupil (broken line). The intensity in the in-focus pattern is

$$I = \left(\frac{2J_1(z)}{z}\right)^2$$

where the abscissa z is a dimensionless coordinate given by $z = (2\pi/\lambda)\,n u \eta$; here u is the covergence angle in the image forming pencil and η is the distance from the centre of the image. The broken line corresponds to a Strehl intensity ratio 0.805.

of the Airy pattern together with a spreading of the light to the other regions of the pattern, so that the minima are no longer zeroes.

The lowered central intensity, expressed as a fraction of the original, is known as the *Definitionshelligkeit*, a term coined by K. STREHL in 1894; it is usually translated as the "*Strehl intensity ratio*". A Strehl intensity of 0.8 is generally accepted as the tolerance limit for systems of high quality. It corresponds very nearly to the Rayleigh limit for the simplest case, change of focus. This figure has the merit that the aberrations producing it are invariably "small" in the sense explained above and it turns out that the relation between aberrations and Strehl intensity ratio can then be expressed very simply in terms of the mean-square aberration over the pupil. A survey of tolerances on these lines is given elsewhere in the Handbuch[1].

Tolerances based on the Strehl intensity ratio are suitable for application to optical systems in which it is possible to achieve almost perfect aberration correction over the chosen field of view. Such systems are microscope objectives and

[1] A. MARÉCHAL: Handbuch der Physik, Bd. XXIV, S. 147. Berlin-Göttingen-Heidelberg: Springer 1956.

telescope objectives, for example. The notion can be extended to cover applications to spectroscopic systems by considering effects of small aberrations on line images[1]. In each case the result is a set of tolerances for different aberration types. The formulae when properly expressed enable the designer to take advantage of partial cancelling of one aberration by another of different kind; to mention the simplest example, if a change of focus is allowed in the presence of primary spherical aberration, as is almost always the case, then the tolerance is one wavelength, four times what would have been expected from the $\lambda/4$ rule. This principle of aberration balancing is, of course, in considerable application in e.g. wide-field photographic objectives, where higher order aberrations are partly cancelled by primary, although the final result may not be within the Strehl intensity tolerance of 0.8.

It is customary to say that an optical system is "Rayleigh limit corrected" or that it is "diffraction limited" if its Strehl intensity exceeds 0.8 or, at any rate, is near that figure. For the very important class of systems which are not diffraction limited the Strehl intensity ratio has little significance since the degradation of the Airy pattern by larger aberrations is much too complex to be described by a single parameter. The lower state of correction of such systems is not necessarily a very bad thing; often they may be intended to be used in conjunction with a detector or receptor of limited spatial resolution. For example, an F/1.8 photographic objective may be used with a photographic emulsion capable of resolving structure of about 20 μ in size, whereas its Airy disc, if it were well enough corrected to produce a recognizable one, would be about 2 μ in diameter; therefore it would be pointless to attempt to correct this objective to the diffraction limit.

A system of tolerances is thus needed for dealing with such cases. The procedure had been until fairly recently to assume that geometrical optics gives a reasonable indication of the image quality for "large" aberrations; the meaning of "large" here is even vaguer than that of "small" in the case of the Strehl intensity ratio; it is usually said to be "several times" the corresponding Strehl tolerance, but even so it must be said that in detail the point image still deviates appreciably from the geometrical optics prediction for aberrations so large as to be unacceptable for most conceivable systems. Nonetheless most optical designers use criteria based on ray aberrations in the main stages of automatic optical design (i.e. by means of electronic computers) because they are easy to formulate and apply.

More recently the use of modulation transfer to assess the performance of non-diffraction-limited systems has developed[2]. The optical system is supposed to form an image of an extended object which may be imagined as a transparency of sinusoidally varying transmission, varying between transparent and completely opaque. Then whatever the aberrations of the optical system the image will similarly have a sinusoidal intensity distribution but with reduced contrast, i.e. the minima will not be completely black, and, if there are coma-type aberrations, with phase shift, i.e. the maxima will not be at the correct geometrical image points[3]. The reduction in contrast and the phase shift can be written as a single complex number which is a function of the spatial frequency ("lines per millimetre") in the object; the number, called the modulation transfer function

---

[1] W. T. Welford: Optica Acta **10**, 121 (1963).

[2] The nomenclature has proliferated and the author makes no attempt to adjudicate between contrast transfer, response, optical transfer and many others.

[3] Strictly the image is only a similar sinusoidal image if the object is incoherently illuminated. Harmonics can appear if the illumination is partially coherent.

or a similar term, is also a function of the aberrations and thus for a system such as a photographic objective it depends on position in the field and orientation of the lines in the test object. Clearly in any discussion of modulation transfer it is necessary to assume isoplanatism, as pointed out in Sect. 9.

The theory and method of calculation of modulation transfer, originated by P. M. DUFFIEUX and O. SCHADE and further developed by H. H. HOPKINS, is given by several authors[1]. The computation involves an integration over the pupil which is similar to that involved in computing the Airy pattern, although the powers of the pupil variables involved are decreased by unity. However it has been shown[2] that a geometrical optics approximation is possible for large aberrations which gives quite useful results. This bears out the intuitive use of ray aberrations mentioned above. The reason is evidently that the fine detail in the individual Airy disc is lost when an extended image is formed.

The modulation transfer can be shown to vanish identically for frequencies above a certain critical value, namely $2n \sin U/\lambda$ cycles per unit length, where U is the convergence angle of the image forming beam and n is the refractive index of the medium in which the image is formed. For lower frequencies the behaviour depends on the aberrations but for any given frequency the modulation cannot be higher than for an aberrationless system. Thus some interesting comparisons with electrical filter circuits are possible. The applications for the optical designer are to those medium quality systems which have to work in conjunction with electrical systems which are frequency limited, e.g. lenses for television cameras or for photographing cathode ray tubes; in these cases the design ought to maximize the modulation transfer up to the cut-off frequency, ignoring completely what happens beyond that point. In practice the calculation is very lengthy, even using the geometrical optics approximation, since it is necessary to allow for an irregular, vignetted pupil shape, and it seems to be used as a final check rather than as a working tolerance in the course of a design. Indeed it is probable that almost all designers rely on a simple function of transverse ray aberrations to provide a routine working tolerance, e.g. the mean square transverse ray aberration for a given object point, comparing this with a multiple of the Airy disc radius.

12. Optical design. In this section we discuss the general problem of the design of optical systems and show how the material of the previous sections is applied. It must be emphasied that, as stated earlier, we are not attempting to give a detailed description of design techniques but merely to indicate general trends of thought and current practice. Thus this whole chapter is in one way an introduction to the more specialized articles in this volume of the Handbuch and in another way it covers an even broader field and relates a variety of different approaches to optical design.

Within the scope of this chapter there are two distinct parts to the design of optical systems, which we may call respectively the Gaussian design and the aberrational design. *Gaussian design* involves at the most elementary level merely the choice of focal length and aperture, but almost invariably more complex considerations are involved, such as the overall length of the system, design of prisms and mirrors, appropriate positioning of aperture and field stops and so forth. *Aberrational design* involves the choice of a suitable number of components and glass types and the determination of curvatures etc. to give

[1] For example, M. BORN and E. WOLF, Principles of Optics 3rd edit. Oxford: Pergamon Press 1964; K. MIYAMOTO, Progress in Optics, Vol. I, p. 31. Amsterdam: North-Holland 1961.

[2] See K. MIYAMOTO, loc. cit.

adequate aberration correction. These design stages are inter-connected by various considerations, some non-optical in character, such as cost of manufacture, some of an engineering nature such as ease of manufacture of spherical or aspherical surfaces, and some concerned with the sheer possibility or otherwise of attaining good correction with a given Gaussian design.

The Gaussian design has to be based on the proposed use of the optical system. A useful example of the kind of consideration which arises is provided by the distinction between two kinds of light detectors: there are those in which the response depends on the flux density (illumination) in the image, such as the photographic emulsion or the retina, and those which respond according to the total incident flux, such as photomultipliers. In the former case high luminosity (i.e. speed in the photographer's sense) is obtained by having a short focal ratio in the image space, but in the latter case this is immaterial. Thus the kind of detector must be taken into consideration in the Gaussian design. At the other end of the system the nature of the light source may frequently have to be taken account of by the design of a suitable condenser system; the condenser has to provide adequate light flux and also uniformity of illumination over a certain area or angular aperture. In some systems, e.g. telescopes, photographic cameras etc., there is no condenser since natural illumination is used; here there will still be a question of light gathering power or "luminosity" which will determine the aperture of the system and, of course, the field of view has also to be specified. These and related topics cannot, in most cases, be simply stated as Gaussian requirements; what can be achieved is determined by the possibilities of aberration correction and this in turn depends on what materials will transmit the required spectral range, whether the materials are available in adequate size and quality and whether adequate mechanical tolerances can be maintained. Thus we make the point that the Gaussian design of an optical instrument may involve the optical designer proper, as well as the instrument designer and the prospective user.

An interesting illustration is provided by the present-day situation in the field of large astronomical telescopes[1]. At the time of writing several projects for telescopes of apertures in the 3 to 4 metre range are in train throughout the world. The astronomers (i.e. the users) want the usual kind of convertible instrument which can be used at prime, cassegrain or coudé focus. The instrument designers wish to use as short a focal ratio as possible for the primary mirror in order to shorten the main telescope "tube"; this facilitates the mechanical mounting and helps with the problems of thermal disturbances in the air in the tube. The optical designer has to take account of the effect of a shorter focus primary mirror on the usable field at the cassegrain focus and he also has to decide whether the system should be a conventional cassegrain (paraboloid-hyperboloid) or a Ritchie-Chrétien (both mirrors aspherized differently from the standard quadric shape, to give aplanatism at the cassegrain focus); in the former case the prime focus can be used direct, in the latter case a prime focus corrector system must be added, but this when made will give more field at the prime focus than the paraboloid on its own. Then further questions naturally arise, e.g. as to whether one should keep to the former system and add a prime focus corrector to the paraboloid; this leads on to similar systems for improving the field correction at the cassegrain focus. All these possibilities affect the mechanical design and the possible uses of the telescope, so that the optical designer must take account of these factors.

A rather different aspect of Gaussian design applies to what may be called periscopic systems, by which we mean systems such as periscopes in which an

[1] Cf. the contribution of K. Bahner in this volume.

object has to be re-imaged in such a way as to produce a real image at a given place with a given magnification. Generally there will also be some limitations on the positions of the pupils, on the diameters of the components and on the positions in which components can actually be placed. Thin lens theory can usually be applied as a first approximation in such cases. The simplest problem of this kind would be, given an object and an image position separated by a certain distance, to find the position and power of a lens to produce an image of given magnification; this trivial problem can easily become more complex if it is required that both object and image be real, for then it is found that a single lens will not serve for certain ranges of the magnification. If two lenses are allowed we can then fix either the powers or the positions and obtain a solution, or the extra freedom can be used to satisfy conditions of pupil imagery. Proceeding in this way we can rapidly reach considerable algebraic complexity and frequently such problems are solved by graphical approximations[1].

The aberrational optical design stage is, of course, much more difficult than the Gaussian design and it is here that the optical designer finds the greatest challenge. The process can again be divided into two sections; the first part involves the choice of the design in outline, i.e. the number of components and in some cases, glass types. It is chiefly at this stage that the skill of the designer is needed, since it is impossible to predict without considerable experience which general design type to start with. The best examples for this part of the problem come from the domain of photographic objectives, where aperture and field angle are both important. Here major advances such as the Cooke triplet, the Tessar and the Roussinov wide-angle objective have been achieved by the process well know to scientists of intuition combined with hard thinking, although it is easy enough to produce sound reasoning leading to the design once it has been invented. When a reasonable corpus of designs of a given type has been built up in patent specifications and elsewhere, the designer can estimate the possibilities of the type. Some collections of designs with specifications and aberration curves have been published[2] and these are useful as guidance material within the domain of known design types.

In this first stage of aberrational design some help is also to be obtained from certain well known optical devices and principles; we recapitulate some of these, not with the aim of systematic development of the subject but rather to show how use is made of apparently disconnected notions in starting a design; in this respect there is a clear parallel with electronic circuit design methods.

A cemented doublet may be corrected for longitudinal chromatic aberration[3] by suitable choice of the powers and dispersions of the components; if it is a "thin doublet" this correction holds good at all conjugates and it is likewise automatically corrected for transverse colour for all stop positions. It may then be corrected for (primary) spherical aberration, provided the index difference at the contact surface is large enough and of the right sign, by bending the whole lens, i.e. adding the same increment of curvature to all surfaces; this correction holds for isolated conjugates only[4]. By suitable choice of the refractive indices the doublet

---

[1] An interesting paper in which 2-, 3-, 4-lens systems are treated algebraically, taking account of the Petzval sum, was given by H. H. HOPKINS in: Proceedings of the Conference on Optical Instruments and Techniques (ed. K. J. HABELL), p. 133. London: Chapman & Hall 1962.

[2] The most recent is contained in A. COX, A system of lens design. London: Focal Press 1964.

[3] Apart from secondary spectrum.

[4] From Eq. (8.6) it follows that up to four different sets of conjugates can be found for which S_{I} is zero, since b' has a quartic dependence on G', the object magnification.

may then be made coma-free also, i.e. it is aplanatic, or alternatively the cement contact may be broken and the two components individually bent to give aplanatism. The aplanatism holds for all stop positions but only one pair of conjugates, in Seidel approximation. Such a doublet has astigmatism depending only on the total power for the stop position at the lens; the astigmatism does not change with stop position if the doublet is aplanatic. Similarly the field curvature depends substantially on the total power, since it is equal to $\Sigma K/n$ for a set of thin lenses of powers K and refractive indices n. These ideas are illustrated in Fig. 22.

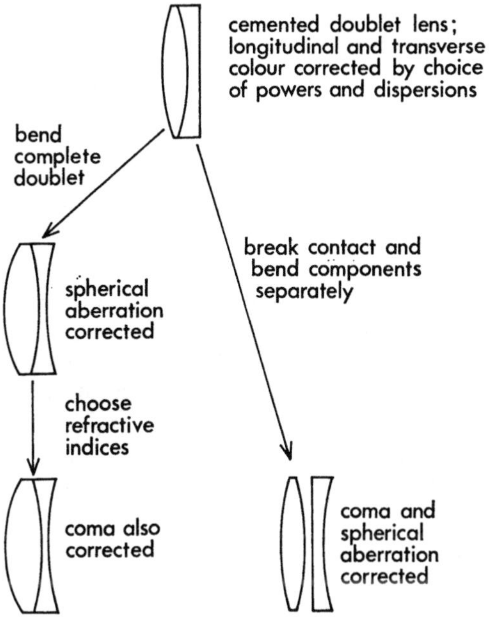

Fig. 22.

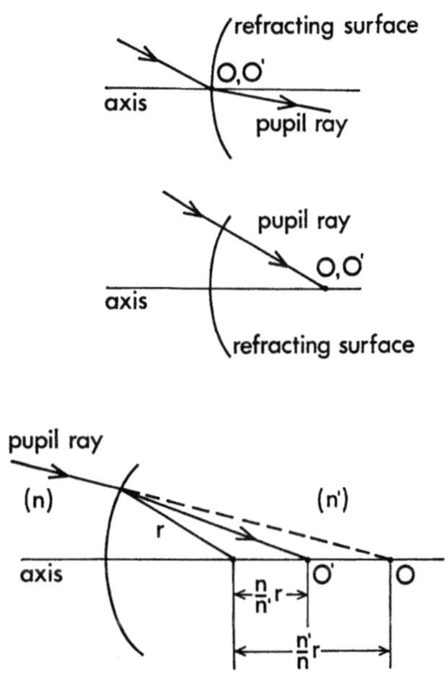

Fig. 23.

Fig. 22. Stages in designing an aplanatic achromatic doublet.

Fig. 23a–c. The three cases of aplanatic refraction at a single surface: (a) the object at the surface, (b) the object at the centre of curvature, (c) what is usually called "the aplanatic case".

There are three sets of conjugates for which a spherical refracting surface has aplanatic imagery, as in Fig. 23. In the first, object and image coincide at the refracting surface (a), in the second they coincide at the centre of curvature (b) and in the third they have the relationships indicated in the Figure (c). These all are perfectly aplanatic for indefinitely large aperture; in case (c) there is also zero primary astigmatism, but not in the other two cases. Cases (a) and (b) have obvious analogues for reflection. All three cases find uses in optical design; in particular, case (c) is valuable as a means of achieving increase or decrease of convergence in beams of large angular aperture without introducing spherical aberration, coma or primary astigmatism in the monochromatic image.

The contribution to S_{IV}, the primary field curvature, of thin lenses depends only on their power so that there is a natural tendency for a system of positive power to have positive field curvature. The thick meniscus (Fig. 24a) having equal curvatures, gives zero contribution to S_{IV} but yet has positive power, so such elements frequently appear in photographic objectives of large aperture (Fig. 24b). Mirrors of positive power have the opposite sign of field curvature from that of lenses, so that lenses and mirrors can be used in combination to produce a flat field. Another element used to decrease field curvature is the field-flattening lens; a field lens is a component placed nominally at (in practice near) a real image, where its main effects are to change the field curvature and to affect the pupil imagery; Fig. 25 shows a field lens as a component of a photographic objective, its main function being to contribute negative field curvature so as to flatten the field.

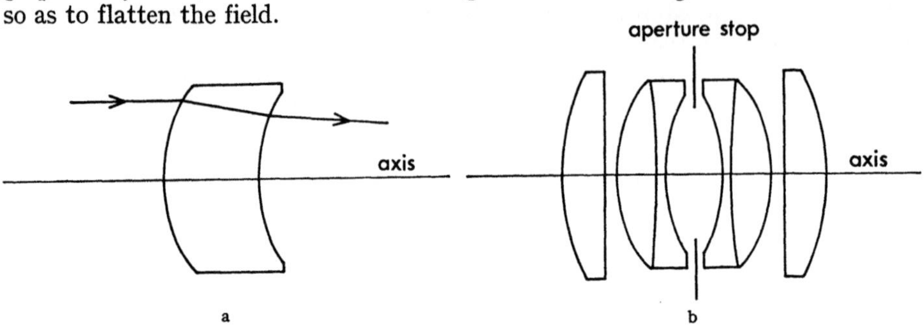

Fig. 24a and b. The thick meniscus with positive power and zero Petzval sum; (a) the meniscus (b) its use in the double Gauss type of objective.

The chromatic aberrations are peculiarly irritating to the optical designer since it is very simple to achieve first order correction but extremely difficult to do much more than this, i.e. to correct secondary spectrum, because of the form of the dispersions of optical glasses; we have already mentioned in Sect. 7 that the theoretical superachromats are not of much practical use because of the large monochromatic aberrations introduced. A device for obtaining first order correction depends on the principle of obtaining some approach to

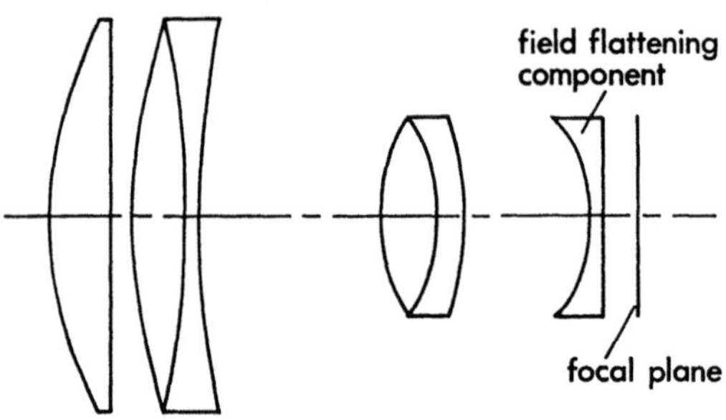

Fig. 25. A field flattening component in a photographic objective.

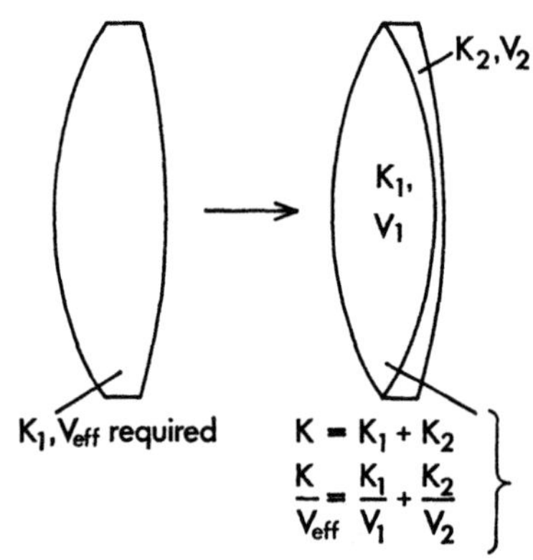

Fig. 26. A ghost surface (zero change of mean refractive index) inserted to give a chosen effective V-value.

monochromatic correction first and then correcting the colour without disturbing the monochromatic correction. The first order colour correction can be specified in terms of the dispersions to be given to the component thin lenses of the system, their refractive indices having already been specified by the monochromatic aberration correction. If no actual glasses are available with the required dispersions then a thin lens may be changed to a cemented doublet, as in Fig. 26; with suitably chosen powers of the components any required dispersion, positive or negative, may be imitated, using components of given dispersion. The mean refractive indices of the components may be chosen to be both equal to that of the original thin lens, so that there is no change in the monochromatic aberrations; the contact surface is then said to be a ghost surface.

As was indicated in Sect. 7 there is at present no possibility of eliminating secondary spectrum in purely refracting systems, and it is a major advantage of reflecting systems that there is no chromatic aberration. There is an interesting kind of mixed reflecting and refracting system, i.e. catadioptric system, in which quite powerful refractive elements occur but yet the colour correction is substantially complete. A good example is the telescope originated by SCHUPMANN some 60 years ago and recently re-introduced as a coronagraph[1]. A singlet objective lens forms a real image which is then reimaged at magnification −1 by a mirror

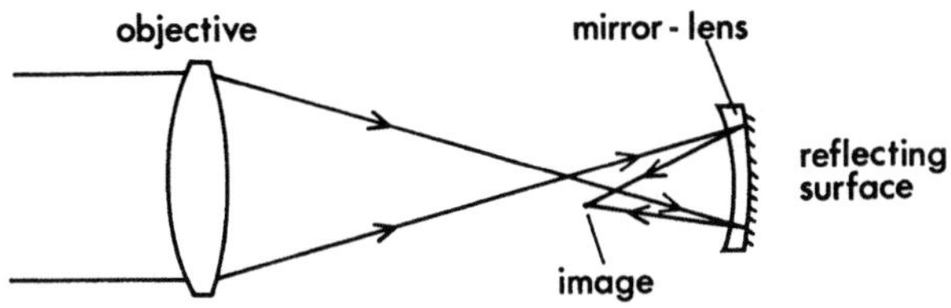

Fig. 27. The Schupmann telescope. The objective and the mirror-lens are made of the same glass and the secondary spectrum is negligible.

lens as in Fig. 27; the glass of the mirror lens is the same as that of the objective and if the powers are suitably chosen the longitudinal chromatic aberration is almost perfectly corrected.

In general, *reflecting systems* have to be used to produce colour-free images and we note here some points additional to those made in Sect. 7. In complex reflecting systems the returning light beams may strike the reflecting surfaces in the wrong order, so that in practice it is rare to find more than two components in a purely reflecting system. This offers rather poor possibilities for aberration correction, so one or both surfaces may be aspherised, as in the classical cassegrain telescope and its derivatives. For purely geometrical reasons such systems have a small field angle and they also tend to be troubled with stray light from scatter off mirror supports etc., in fact the purely optical design is considerably simpler than these associated design problems.

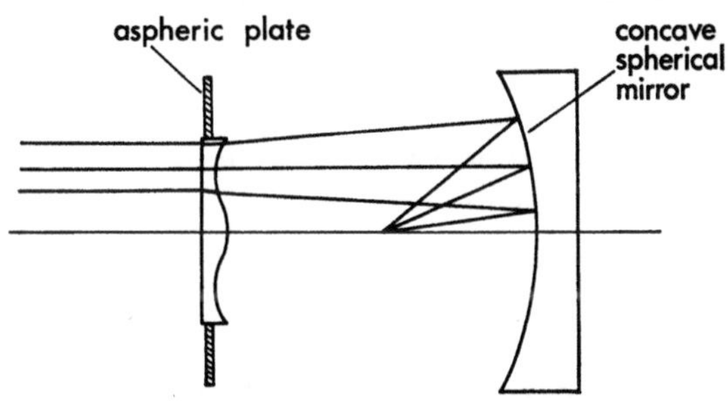

Fig. 28. Use of an aspheric in the Schmidt camera. The plate is at the aperture stop, so that it corrects the spherical aberration of the mirror without introducing oblique aberrations (in the primary aberration approximation).

The use of aspherizing gives the optical designer extra freedom, although it does not immediately yield strikingly improved results when added to an existing design. The most fertile concept in this connection is the *aspheric plate*, a thin glass plate of zero or small Gaussian power but with higher order terms in its profile. For example, in the *Schmidt camera* (Fig. 28) the aspheric plate is placed at the centre of curvature of a concave spherical mirror; the mirror alone, with the aperture stop at the centre of curvature, has no astigmatism or coma and the shape of the aspheric plate is chosen to cancel the spherical aberration. This application, in which the aspheric plate is at a pupil position so that it affects only spherical aberration[2] is probably the commonest, but subtler applications, in which the aspheric plate has several functions, are possible; for example in the Ritchey-Chrétien modification of the cassegrain telescope both mirrors have

[1] J. H. RUSH and G. K. SCHNABLE: Appl. Optics 3, 1347 (1964).

[2] There are, of course, certain higher order oblique and chromatic aberrations introduced.

aspherizings different from the classical quadrics of revolution so as to obtain aplanatism.

The second stage of aberrational optical design is that in which, having settled on a general outline of the system, the aberrations have to be reduced to chosen tolerance levels. This process depended until about 10 years ago on a laborious sequence of raytracing, using desk calculating machines, and the estimation of suitable changes in the system to produce improvements; this choice of changes was partly a matter of calculation based on finite differences, or in some cases differential formulae, and partly a matter of guesswork based on experience. It has now been taken over to a very large extent by very elaborate *computer programmes*; broadly speaking these programmes will start from a crudely designed system, trace a number of rays through it, estimate its performance in terms of some specified "merit function", trace more rays to find the effect of changes in parameters on the merit function and then calculate an optimum combination of changes; the cycle is then repeated until a chosen value of the merit function is attained.

From the above brief description it would seem rather easy to make an automatic design programme to calculate the best possible optical system within given limits of number of components, overall size etc. In fact it turns out that the development of automatic design programmes has been a major research activity in several centres for many years and it is certain that no programme at present available does all that could be done. There are several reasons for this and we shall close this chapter with a discussion of them and of the prospects for future developments.

The first difficulty is that we do not know how good a system is possible within given limits and, worse than that, there is no agreement as to how the quality of a system shall be specified except in the case of diffraction limited systems. It also turns out that various criteria of quality proposed by the specialists in image assessment (modulation transfer and related functions) are too difficult to use as working merit functions in an automatic design programme; in fact to compute modulation transfer reliably in an actual design, as opposed to a theoretical system with stated amounts of primary spherical aberration, etc., and with no vignetting, seems at present to be very difficult and the existing programmes do not always agree very well with each other[1]. Thus the working merit function in an automatic design programme is generally some kind of weighted mean square ray aberration or wavefront aberration; such functions must have approximate relationships with modulation transfer and the Strehl intensity ratio, but these relationships are almost certainly not monotonic in the interesting regions and the approximations are not well understood.

A further difficulty is that the merit function, whatever it may be chosen to be, probably has many minima separated by regions of higher values; thus, if the merit function is represented by a surface in $n+1$-dimensional space, where n is the number of parameters in the design available as variables, this surface has many valleys separated by hills; this is suggested by Fig. 29, which shows a contour map of such a surface, the contours being in units of the merit function. Now it may happen that a rough design starts at a point such as P and it is easily conceivable that a step-by-step process could eventually reach the point Q, but it is difficult to see how R could be attained. Thus the problem is to know how the

[1] The author is indebted to Dr. C. G. WYNNE and others who took part in the 1966 Rochester meeting on automatic optical design for this information and for much useful discussion on programming problems.

lowest minimum is to be reached and since, as pointed out above, we do not know how low this is, it is a difficult problem. Clearly it is necessary to adopt a policy of rather deliberately larger increments of change of parameter in some cases if it is desired to explore the whole domain, or, at any rate, more of it than has previously been studied by non-automatic methods.

One more difficulty in automatic design programmes is rather surprising; this is the problem of holding what are called the boundary conditions. This term does not signify quite the same thing as in e.g. the subject of differential equations; what is meant includes such things as holding the power and relative aperture constant, preventing lens edge thicknesses from going negative, limiting the overall length of the system and controlling many other things which the human designer would take care of without thinking about it. It is, of course, a familiar situation in large computer programmes but apparently the difficulties which arise in this way in automatic optical design are particularly acute.

The details of the aberration correction technique vary greatly as between programmes. In some cases an approximate primary aberration correction is obtained and the rest is done by raytracing. In other cases some or all of the secondary aberrations are computed by one of the schemes referred to in Sect. 6, in order to get a better approximation to correction before using raytracing. A third approach is to trace rays and fit polynomials to the raytracing results so as to give approximately the breakdown into different aberration types; if the rays are suitably chosen the fitting process involves only well-conditioned equations and the effect of parameter changes on these different aberration types can be found in the usual way. It will undoubtedly be a considerable time before one or other of these methods is clearly seen to be superior, and the same remark applies to the various methods used to choose the appropriate combination of parameter changes at each re-cycling of the design process. It is rather attractive to use higher order aberration contributions as well as the primary aberrations in the earlier stages, but the interpretation is by no means simple; this is because the contribution to a given higher order aberration which arises at a surface may be divided into two parts, usually called the *intrinsic* and the *induced* contributions. The intrinsic contribution is a function only of the curvature of the surface, the refractive indices on either side and the Gaussian properties of the incident pencil, but the induced contribution depends on the amount of aberration of lower orders which is present in the incident pencil. Thus there may be no intrinsic aberration at a certain surface but a change in aberrations at an earlier surface in the system can cause some induced aberration to appear. This situation, well known to optical designers, can cause difficulty in interpreting higher order aberrations and in deciding at which surface a change must be made in order to obtain a desired improvement.

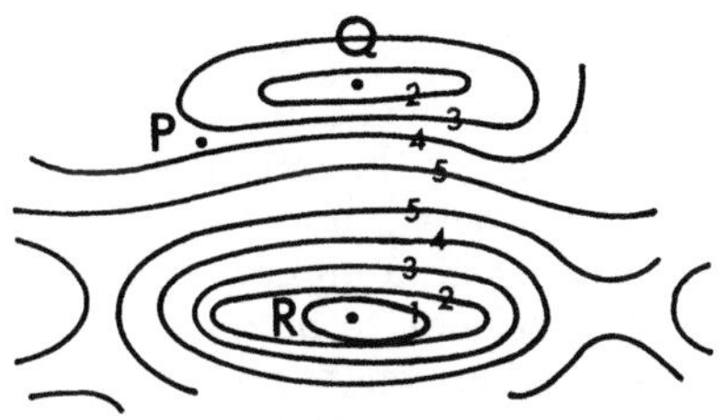

Fig. 29. Contour map of a merit function, showing two neighbouring minima.

The possibilities of aspheric surfaces in automatic design are only just beginning to be explored, probably because the practical difficulties of manufacture make the designer wary of introducing an aspheric except as a last resort. In that case the aspheric would have a rather clearly defined function, such as that of the corrector plate in a Schmidt camera, but the really interesting approach would be to permit unlimited aspherizing, to see just how much improvement is possible.

Programmes for automatic design can be divided into two classes. In the first class the number of parameters to be varied (n) i.e. refractive indices, curvatures, etc., is equal to or greater than the number of conditions to be satisfied (m), i.e. aberrations to be reduced to specified values. In the second class n may be less than m. The first kind is intrinsically simpler since it involves solving m equations with n unknown, $n \geqq m$, and this is always possible if the parameters are genuinely independent and have suitable ranges. To take a trivial example, an exact solution for the primary aberrations to be zero in a Taylor triplet is known to be possible and this would be a suitable problem for the first kind of programme. The second kind involves a real optimising procedure, since the best compromise solution, as indicated by the merit function, has to be found within the available conditions. It is to such programmes that the discussion relating to Fig. 29 belongs. An example would be to design a Taylor triplet for the best balance of total aberrations, since there are only just enough parameters to control the primary aberrations.

We conclude by re-emphasising that this chapter constitutes an attempt at a survey of the general field of optical calculations. We have made no attempt at completeness; the aim has been to indicate the general character of the process of optical design and to show the present-day trends, particularly as influenced by electronic computer techniques. Optical designing, in common with other applied science techniques, is being deeply changed by computer methods; the most obvious change is that of speed but a more subtle kind of change is in method; previously it has been possible to analyze the steps in approaching a design but now the steps are carried out by a computer under the control of a programme with many alternative paths, and the result is that we may obtain a good design but without knowing how it was done!

Méthode de calcul des systèmes optiques.

Par

A. Maréchal, E. Hugues, P. Givaudon.

Avec 29 Figures.

A. Introduction.

1. Le problème du calcul des systèmes optiques. Tout instrument d'optique est un groupement judicieux d'objectifs élémentaires, de prismes et de miroirs destiné à donner d'un espace objet d'étendue limitée une image; l'analyse de cette image est confiée à un récepteur qui peut être l'œil, une émulsion photographique, une cellule photo-électrique, un iconoscope etc. En général, l'utilisateur fixe un programme technique dans lequel se trouvent précisés:

— les champs objet et image
— les ouvertures
— les conditions d'éclairage
— les qualités de finesse du récepteur
— la sensibilité spectrale de la source et du récepteur
— l'encombrement (maximum souhaitable) du système
— parfois le prix de revient maximum.

Pour atteindre ce but avec plus ou moins de bonheur l'ingénieur opticien est alors chargé de fixer dans le détail le nombre de lentilles, les matériaux (indices et dispersions), les caractéristiques géométriques (diamètres, épaisseurs, rayons de courbure), et les tolérances de construction.

Dans la pratique, certains systèmes très simples peuvent se composer d'une seule lentille mais les systèmes de grande qualité comportent souvent de dix à vingt lentilles c'est le cas des microscopes, du périscope de marine, des objectifs à focale variable, etc.

On peut juger de l'embarras du calculateur devant le grand nombre des paramètres: pour chaque lentille on doit choisir en effet 2 rayons de courbure, 1 indice, 1 dispersion, 1 épaisseur et 1 intervalle d'air, soit 6 paramètres. Ceci entraîne bien entendu une grande diversité de solutions possibles pour résoudre le même problème.

Le calcul des combinaisons, destiné à la recherche des valeurs optimales à donner aux paramètres pour obtenir des résultats acceptables s'effectue maintenant en profitant de la puissance et de la rapidité des calculatrices électroniques à grande capacité.

2. L'évolution des techniques. Rappelons tout d'abord très brièvement les principales étapes de l'acquisition des connaissances de base indispensables à l'exécution des calculs:

— Expression des lois fondamentales auxquelles sont attachés les noms de Descartes (1637) et de Fermat.

— Correction de l'aberration sphèrique dans un doublet collé (Huygens 1678 et 1703).

— Découverte du chromatisme (Newton 1728).
— Résolution de l'achromatisme prévu par Euler, réalisé par Dollond.
— Etude des conditions de stationnarité du stigmatisme (Abbe-Herschel).
— Calcul par Seidel et plus tard par Schwarzschild des coefficients d'aberration du 3ième ordre.
— Calcul des aberrations d'ordre supérieur et en particulier du 5ième ordre par Kohlschütter (1908) et leur mise en application par Cruickshank.
— Enfin de nombreuses études modernes concernant le calcul automatique des retouches finales (Herzberger, Feder, Wynne, Girard, Baker, R. E. Hopkins).

L'introduction de machines à programme automatique, il y a une quinzaine d'années environ, marque une étape décisive dans l'évolution des techniques de calcul; alors qu'il fallait environ 10 minutes pour effectuer une réfraction dans un plan méridien aux calculateurs très entraînés armés seulement de table de logarithmes, les machines modernes effectuent une marche extra-méridienne beaucoup plus complexe en moins d'une seconde avec une sécurité excellente.

On comprend aisément que des méthodes très variées soient pratiquement utilisées par les calculateurs; par ailleurs la nature industrielle de ces travaux amène à une discrétion souvent préjudiciable à la rapidité des progrès. Nous proposons de décrire dans le Chap. B IV, une méthode dont les principaux mérites sont d'économiser au mieux le temps de calcul de la machine et de pouvoir être transposée ultérieurement à l'automatisme.

B. Une méthode de calcul.

I. Les phases successives.

3. On peut distinguer dans le calcul d'un système optique les étapes suivantes:

α) L'ébauche du système, où le calculateur décide de la nature des composantes. Dans cette première phase il est évident que l'expérience personnelle et la documentation jouent un rôle prépondérant. Néanmoins, quelques tentatives intéressantes ont été faites (H. H. Hopkins 1961) pour mettre au point une méthode rationnelle d'exploration rapide des systèmes possibles schématisés par des combinaisons de lentilles minces.

β) Les lentilles minces étant mises en place avec leurs puissances élémentaires la réduction des aberrations du 3ième ordre va nous permettre assez rapidement de déterminer les cambrures approximatives à donner aux diverses lentilles, et d'introduire ensuite les épaisseurs.

γ) Le schéma initial ainsi obtenu possède le plus souvent des aberrations encore inacceptables; on doit alors faire varier légèrement les nombreux paramètres du système de façon à réduire les aberrations à des valeurs tolérables; cette partie du calcul était extrêmement longue et fastidieuse avant l'introduction des machines à calculer automatiques; l'utilisation de ces machines a amené l'introduction de «fonctions de mérite» (Feder) qui permettent de chiffrer la qualité globale de l'image, par exemple à l'aide de la somme pondérée des carrés des aberrations transversales complétée éventuellement par des expressions qui tiennent compte de caractéristiques telles que l'encombrement, le prix de revient etc.; la détermination de la meilleure correction équivaut à la recherche du minimum d'une fonction qu'on peut évaluer numériquement pour chacun des ensembles des valeurs des paramètres de construction; on travaille en somme, en première approximation, à rendre minimum une forme quadratique dont les nombreux coefficients sont a priori inconnus. Ce procédé présente d'ailleurs une large part d'arbitraire puisque chaque calculateur peut à volonté choisir ses coefficients de pondération.

La méthode qui sera décrite permet essentiellement:

— de supprimer pratiquement les tâtonnements aveugles en tirant du calcul à la machine non seulement les valeurs des aberrations (exprimées sous forme de chemins optiques) mais aussi leurs dérivées par rapport aux paramètres de construction. Il est alors assez facile de choisir les paramètres utiles et de déterminer leurs variations pour obtenir une bonne correction des aberrations puisqu'on connaît l'influence de tous les paramètres sur la forme de la surface d'onde en divers points du champ. Le domaine d'application de la méthode a pour limite celui de la linéarité des aberrations par rapport aux paramètres de construction; il semble que cette linéarité soit meilleure pour les chemins optiques que pour les aberrations transversales.

— de juger instantanément de la qualité de l'image au vu des déformations de la surface d'onde; l'observation des courbes d'écart par rapport à la forme sphérique apporte un ensemble d'informations plus complet que le calcul d'une fonction de mérite.

— de connaître du même coup toutes les répliques à de légères variations des paramètres de construction (en particulier variations des indices) et les tolérances de montage de l'instrument (y compris les tolérances de centrage).

II. L'étude du système dans le domaine paraxial.

4. Définitions et notations. *α) Quantités liées à un dioptre* (Fig. 1). On appelle S le sommet et C le centre du dioptre. Dans un système formé de plusieurs dioptres, chaque dioptre est caractérisé par:

— son rang r

— son rayon de courbure R compté algèbriquement de S vers C ($R = SC$) ou l'inverse de son rayon c ($c = 1/R$)

— l'indice qui le précède n ou l'indice qui le suit n'

— la distance qui le sépare du dioptre précédent e, ou la distance qui le sépare du dioptre suivant e'.

En général nous déterminerons un dioptre par les quantités c_r, n'_r, e'_r.

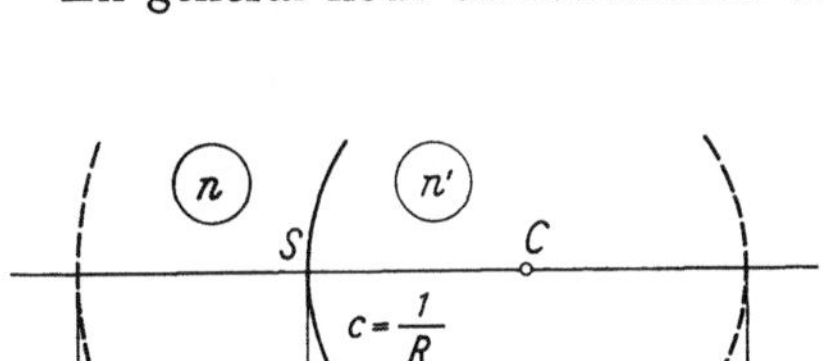

Fig. 1.

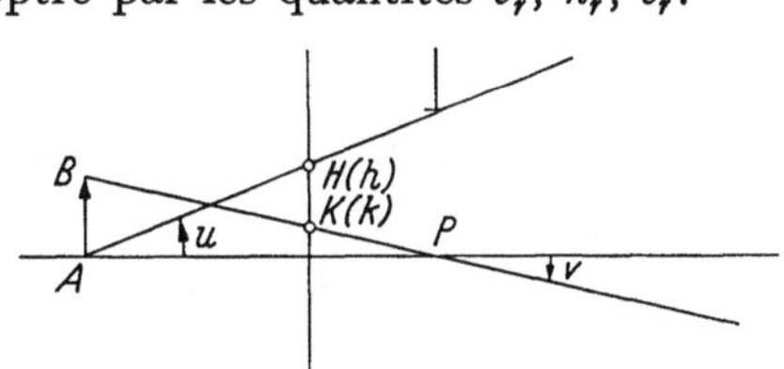

Fig. 2.

β) Quantités liées à un rayon (Fig. 2). Il est commode de définir un rayon paraxial par son angle polaire u et par un point H d'abcisse connue et d'ordonnée h; il est souvent avantageux de considérer un deuxième rayon défini par son angle polaire v et par un point K de même abcisse que le point H et d'ordonnée k.

On réserve en général les notations u et h à un «rayon d'ouverture» issu d'un point A de l'axe et les notations v et k à un rayon principal issu d'un point B en dehors de l'axe et passant par le centre de la pupille.

Au point de vue du calcul paraxial, les quantités u, v, h, k sont des infiniment petits du premier ordre.

On peut montrer que la quantité $l=n(hv-ku)$ ne dépend pas de l'abcisse des points H et K et qu'elle est invariante à travers un système optique. C'est l'invariant de Lagrange. Considérons en effet (Fig. 3), deux éléments AB et $A'B'$ conjugués, normaux à l'axe et situés dans deux milieux d'indice n et n'. Soient P et P' les positions de la pupille, Q et Q' les bords homologues de la pupille, π et π' deux plans quelconques perpendiculaires à l'axe. Les rayons

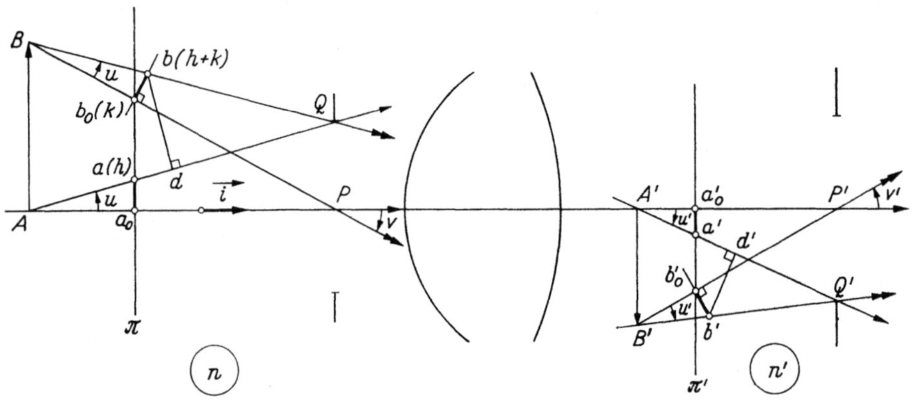

Fig. 3.

principaux issus de A et de B sont $AP\,A'P'$ et $BP\,B'P'$; les rayons marginaux issus de A de et B sont $AQ\,A'Q'$ et $BQ\,B'Q'$. On appelle a_0, a_0' et b_0, b_0' les intersections des rayons principaux et des plans π et π'; les surfaces d'onde des deux pinceaux issus de A et de B en a_0, a_0', et b_0, b_0' sont des éléments $a_0\,a$, $a_0'\,a'$ et $b_0\,b$, $b_0'\,b'$ normaux aux rayons principaux. Le principe de Fermat exprime que

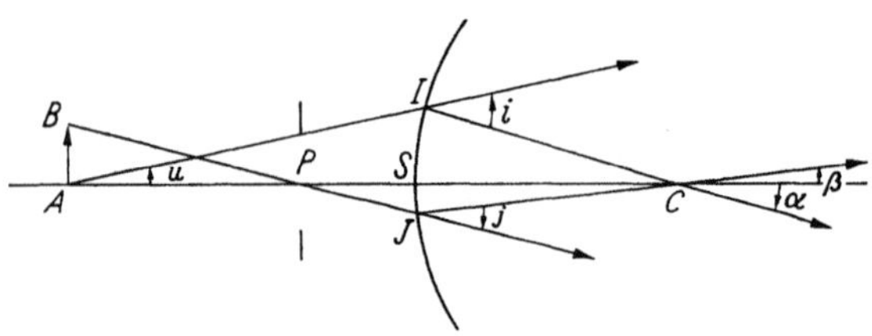

Fig. 4.

$$\left.\begin{aligned}(b\,b') &= (b_0\,b_0'),\\ (a\,a') &= (a_0\,a_0').\end{aligned}\right\} \qquad (4.1)$$

En appelant $\vec{i}$ le vecteur unitaire de l'axe on peut calculer la différence $(b_0\,b_0')-(a_0\,a_0')$:

$$(b_0\,b_0')-(a_0\,a_0')=n'\,\overrightarrow{a_0'\,b_0'}\cdot\vec{i}-n\,\overrightarrow{a_0\,b_0}\cdot\vec{i}=0. \qquad (4.2)$$

La comparaison de cette égalité avec les deux précédentes montre que $(a\,a')=(b\,b')$.

Si d et d' sont les projections de b et b' sur le rayon $AQ\,A'Q'$, l'égalité précédente se traduit par:

$$n\,\overline{ad}=n'\,\overline{a'd'}.$$

Il ne reste plus qu'à exprimer $\overline{ad}$ et $\overline{a'd'}$ en fonction de h, k, u, v, et h', k', u', v', pour démontrer que

$$l=n(h\,v-k\,u)=n'(h'v'-k'u'). \qquad (4.3)$$

γ) *Quantités liées à un dioptre et à un rayon* (Fig. 4). On appelle I et J les points d'intersection des rayons d'angle polaire u et v avec le dioptre.

Les ordonnées des points I et J sont appelées h et k. Les angles d'incidence et de réfraction en I et J sont appelés i et i', j et j'. Ils sont comptés à partir des normales IC et JC orientées dans le sens de propagation de la lumière incidente.

Les angles polaires des normales IC et JC sont appelés α et β. Ils sont comptés à partir de l'axe du système.

D'une manière générale au niveau du dioptre de rang r tous les éléments de l'espace objet sont notés avec indice et sans accent (u_r, v_r, i_r, j_r, A_r, B_r etc.) tandis que les éléments de l'espace image sont notés avec indice et accent (u'_r, v'_r, i'_r, j'_r, A'_r, B'_r etc.). Néanmoins les éléments propres à l'espace objet initial sont notés sans indice et sans accent (u, v, AB, etc.) et les éléments propres à l'espace image final sont notés sans indice et avec accent (u', v', $A'B'$ etc.).

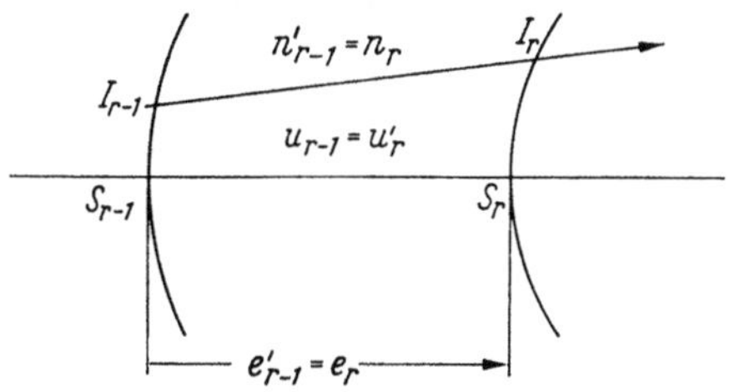

Fig. 5.

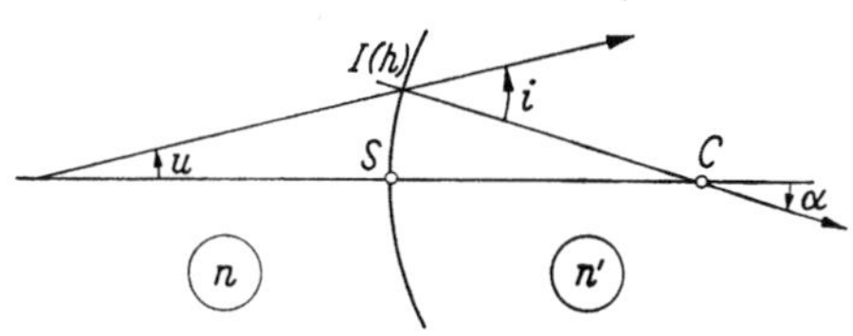

Fig. 6.

5. Marche d'un rayon paraxial. Les quantités h, u, n relatives au dioptre de rang r sont calculées au moyen des équations suivantes (Fig. 5):

$$\left.\begin{aligned} h_r &= h_{r-1} + e'_{r-1} u'_{r-1}, \\ u_r &= u'_{r-1}, \\ n_r &= n'_{r-1}. \end{aligned}\right\} \tag{5.1}$$

La réfraction d'un rayon paraxial s'exprime par les équations suivantes (Fig. 6):

$$\left.\begin{aligned} \alpha &= -h\,c, \\ i &= -\alpha + u, \\ i' &= \frac{n\,i}{n'}, \\ u' &= i' + \alpha. \end{aligned}\right\} \tag{5.2}$$

L'élimination de i, i' et α entre ces 4 équations permet de calculer directement u' en fonction de u:

$$u' = \frac{n}{n'}\,u - \frac{n'-n}{n'}\,h\,c. \tag{5.3}$$

Il faut signaler qu'un schéma de calcul très voisin de celui qui vient d'être décrit peut être utilisé pour étudier la marche d'un rayon voisin d'un rayon principal quelconque et atteindre la position des focales.

6. Calcul des éléments paraxiaux du système. On étudie la marche à travers le système de deux rayons particuliers: le premier est un «rayon d'ouverture» issu d'un point A de l'axe, le deuxième est un rayon principal issu d'un point B en dehors de l'axe. On se fixe des éléments initiaux h, k, u, v, dans un plan π du premier milieu (indice n) (Fig. 7).

La marche des rayons décrits au paragraphe précédent permet de calculer à chaque dioptre les quantités h_r, k_r, u'_r, v'_r, et les distances $x'_r = \overline{S_r A'_r}$ et $p'_r = \overline{S_r P'_r}$

qui fixent la position de l'image et de la pupille par rapport au dioptre (Fig. 8):

$$\left.\begin{aligned} x'_r &= -\frac{h_r}{u'_r}, \\ p'_r &= -\frac{k_r}{v'_r}. \end{aligned}\right\} \qquad (6.1)$$

Pratiquement pour fixer les conditions initiales et calculer la dimension de l'image nous allons distinguer deux cas suivant que l'objet est à l'infini ou à distance finie.

α) L'objet est à l'infini (Fig. 9). On choisit comme plan de départ le plan de la pupille d'entrée, considérée comme étant le dioptre de rang 0. Dans ces

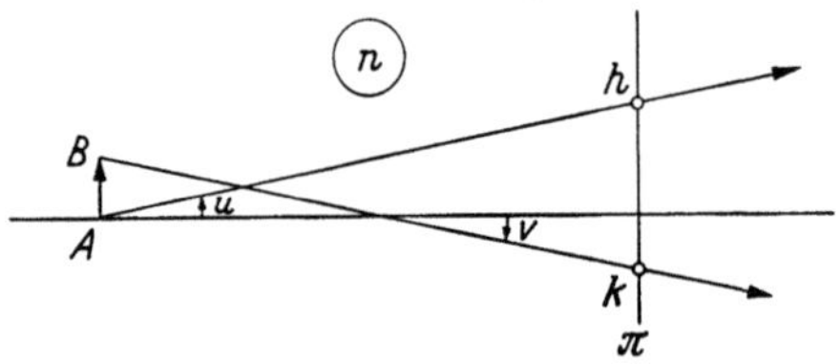

Fig. 7.

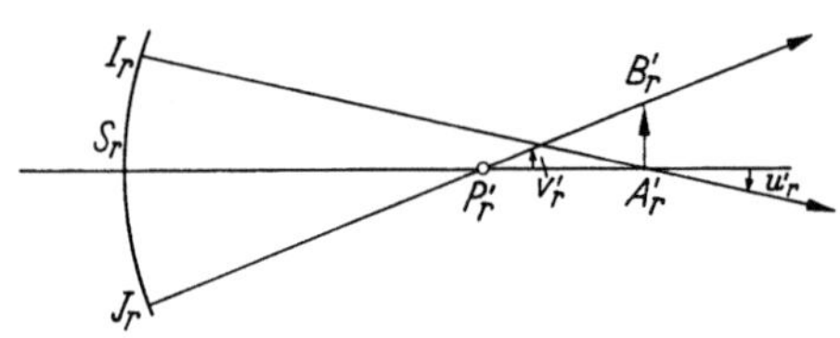

Fig. 8.

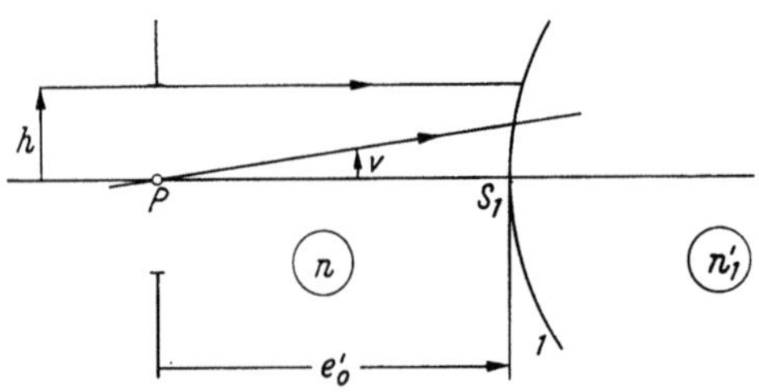

Fig. 9.

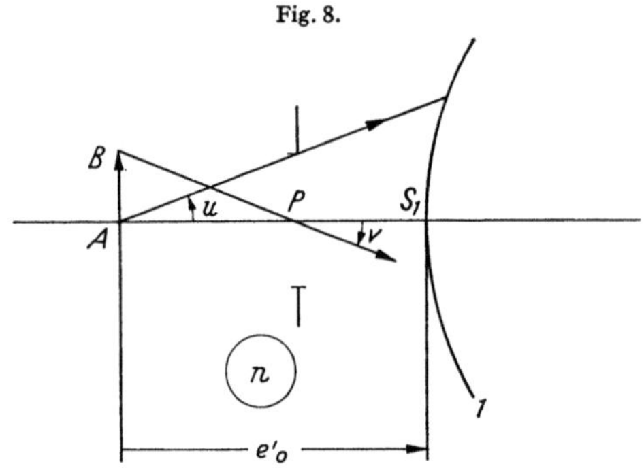

Fig. 10.

conditions h représente le rayon de la pupille d'entrée, les éléments initiaux sont alors:

$$\left.\begin{matrix} h_0 = h, & k_0 = 0, & n'_0 = n, \\ u'_0 = 0, & v'_0 = v, & e'_0 = \overline{P S_1}. \end{matrix}\right\} \qquad (6.2)$$

On calcule à chaque dioptre x'_r et p'_r définis par les Eq. (6.1). La distance focale image f'_r fixe la dimension de l'image. Elle se calcule au moyen de l'équation (6.3)

$$f'_r = -\frac{h}{u'_r}. \qquad (6.3)$$

Dans le cas d'un système afocal $u' = 0$ et l'on définit alors le grossissement

$$G = \frac{h}{h'} = \frac{n' v'}{n v}. \qquad (6.4)$$

β) L'objet est à distance finie (Fig. 10). On choisit comme plan π de départ le plan de l'objet, considéré comme étant le dioptre de rang 0. On appelle y et y' les dimensions de l'objet et de l'image, les éléments initiaux sont alors:

$$\left.\begin{matrix} h_0 = 0, & k_0 = y, & n'_0 = n, \\ u'_0 = u, & v'_0 = v, & e'_0 = \overline{A S_1}. \end{matrix}\right\} \qquad (6.5)$$

On calcule à chaque dioptre x'_r et p'_r définis par les Eq. (6.1).

Le grandissement $\gamma_r = y'_r/y$ fixe la dimension de l'image. Il se calcule au moyen de l'équation (6.6):

$$\gamma_r = \frac{n\,u}{n'_r\,u'_r}\,. \tag{6.6}$$

Il semble utile de signaler que pour traiter un système dans lequel l'objet et l'image sont à distance finie, il est judicieux de fixer les conditions initiales dans un espace tel que l'objet soit plus grand que l'image.

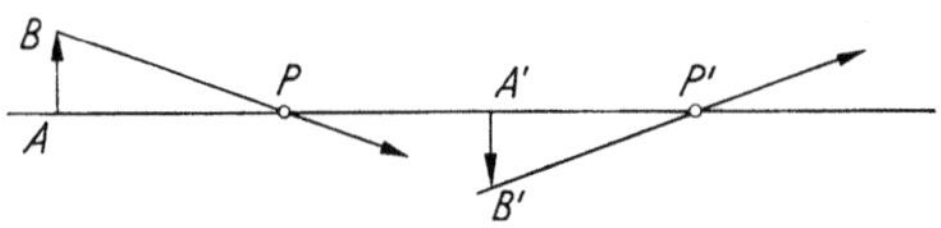

Fig. 11.

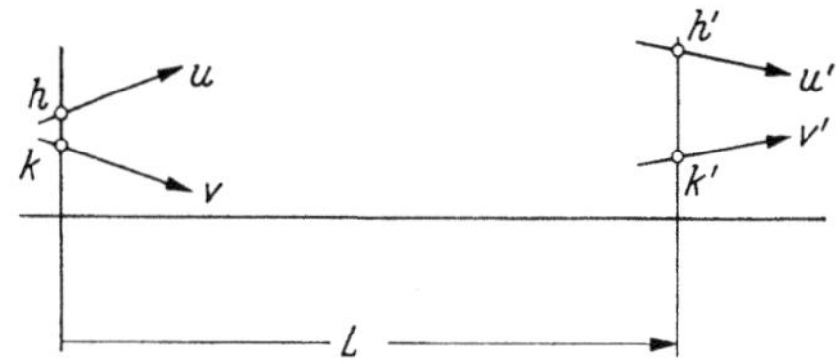

Fig. 12.

7. Eléments paraxiaux d'un système de lentilles minces. Le problème de la mise en place d'un système de lentilles minces répondant à certaines conditions trouve une solution élégante et féconde en considérant deux rayons paraxiaux.

H. H. Hopkins a donné en 1961 une solution du problème suivant: on fixe deux points A et A' que l'on veut conjuguer avec un certain grandissement et deux points P et P' qui représentent les positions des pupilles d'entrée et de sortie (Fig. 11); on cherche tous les couples de lentilles minces qui remplissent ces conditions. Nous allons donner la solution d'un problème un peu plus général (Fig. 12).

On considère deux plans distants d'une longueur L dans un milieu d'indice 1 (ces deux plans ne sont pas nécessairement conjugués); on connaît l'état de deux rayons paraxiaux quelconques au niveau de ces deux plans: h, u, h', u', pour le premier rayon; k, v, k', v', pour le deuxième rayon; on cherche tous les couples de lentilles minces qui remplissent ces conditions.

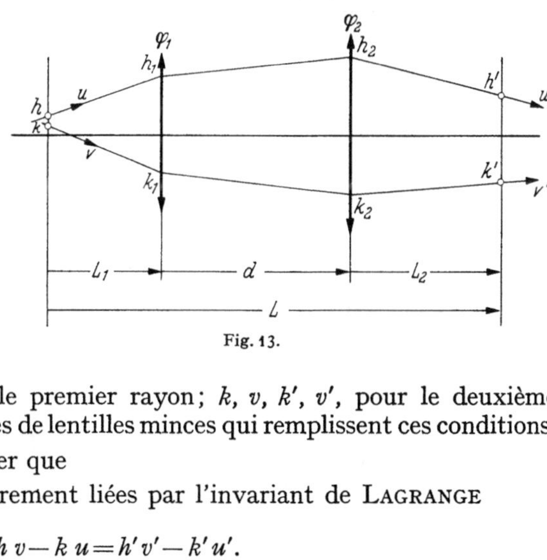

Fig. 13.

Remarquons pour commencer que

1. les données sont nécessairement liées par l'invariant de Lagrange

$$l = h\,v - k\,u = h'\,v' - k'\,u'.$$

2. le problème admet une infinité de solutions si

$$(h-h')\,(v-v') - (k-k')\,(u-u') \neq L(u\,v' - u'\,v)\,. \tag{7.1}$$

Lorsque la relation (7.1) n'est pas satisfaite, le système est réductible et de manière unique à une seule lentille.

Supposons remplie la relation (7.1) et appelons (Fig. 13) φ_1 et φ_2 les puissances des deux lentilles, Φ la puissance de l'ensemble, L_1 la distance du premier plan à la première lentille, d l'intervalle des deux lentilles, L_2 la distance de la deuxième lentille au deuxième plan de référence. En prenant L_1 comme paramètre, les différents éléments du système sont calulés en utilisant le schéma suivant:

$$\left.\begin{aligned}
&\Phi=\frac{u\,v'-v\,u'}{l},\\
&\begin{cases} h_1=h+L_1 u,\\ k_1=k+L_1 v,\end{cases}\\
&\begin{cases} h_2=l\,\dfrac{h'-h_1-u'(L-L_1)}{l-(h_1 v'-k_1 u')},\\ k_2=l\,\dfrac{k'-k_1-v'(L-L_1)}{l-(h_1 v'-k_1 u')},\end{cases}\\
&d=\frac{h_1 k_2-k_1 h_2}{l},\\
&L_2=\frac{h_2 k'-k_2 h'}{l}.\\
&\begin{cases} \varphi_1=\dfrac{l-(h_2 v-k_2 u)}{l d},\\ \varphi_2=\dfrac{l-(h_1 v'-k_1 u')}{l d}.\end{cases}
\end{aligned}\right\}\qquad (7.2)$$

Signalons pour terminer qu'un grand nombre de problèmes relatifs à la mise en place des lentilles minces et épaisses peut se résoudre par des méthodes de ce genre particulièrement bien adaptées à l'emploi de calculateurs électroniques.

III. Etude du système dans le troisième ordre.

8. Relations générales. Considérons un dioptre, centré sur l'axe, défini par son développement en série limité au terme du 4ième ordre

$$X=\frac{c}{2}(Y^2+Z^2)+\frac{\lambda+1}{8}c^3(Y^2+Z^2)^2,$$

dans lequel λ est un coefficient définissant la nature de la méridienne du deuxième degré (pour la sphère $\lambda=0$ et pour la parabole $\lambda=-1$).

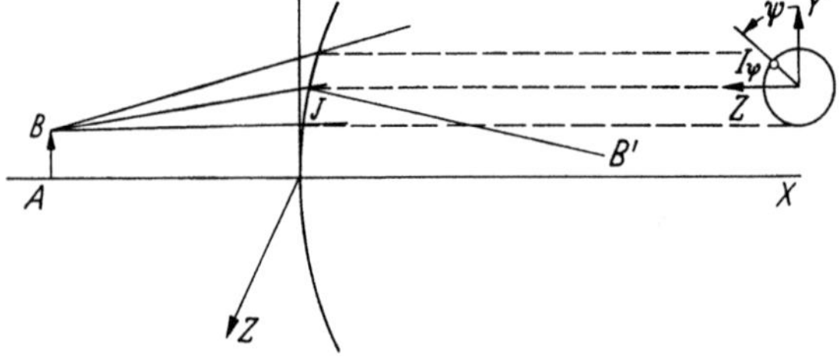

Fig. 14.

D'autre part, considérons un pinceau peu ouvert de rayons issus d'un point B voisin de l'axe et situé dans le plan méridien (Fig. 14).

Un rayon quelconque BI de ce pinceau peut se définir au niveau du dioptre par la hauteur k du rayon principal BJ, la hauteur h du rayon d'ouverture et l'angle ψ fixant son orientation dans la pupille; ainsi ses coordonnées sur le dioptre prendront les valeurs suivantes:

$$\begin{aligned}
&X=\frac{c}{2}(Y^2+Z^2)+\frac{\lambda+1}{8}c^3(Y^2+Z^2)^2,\\
&Y=k+h\cos\psi,\\
&Z=h\sin\psi.
\end{aligned}$$

L'étude du développement en série du chemin optique aberrant[1]:

$$n'\Delta' = (BJB') - (BIB')$$

limité aux termes du 4ième ordre conduit à l'expression suivante

$$n'\Delta' = S_{\mathrm{I}} + S_{\mathrm{II}} \cos\psi + S_{\mathrm{III}} \cos 2\psi + S_{\mathrm{IV}} + S_{\mathrm{V}} \cos\psi. \tag{8.1}$$

On exprime les sommes S en fonction de l, i, j, u, u', h, k, déjà définis et elles prennent en posant,

$$\Delta \frac{u}{n} = \frac{u'}{n'} - \frac{u}{n},$$

$$\mathcal{P} = -c\left(\frac{1}{n'} - \frac{1}{n}\right)$$

les formes suivantes:

$$\left.\begin{aligned}
S_{\mathrm{I}} &= -\frac{1}{8}(ni)^2 h \Delta \frac{u}{n} + \frac{n'-n}{8}\lambda c^3 h^4,\\
S_{\mathrm{II}} &= -\frac{1}{2}(ni)(nj)\, h \Delta \frac{u}{n} + \frac{n'-n}{2}\lambda c^3 h^3 k,\\
S_{\mathrm{III}} &= -\frac{1}{4}(nj)^2 h \Delta \frac{u}{n} + \frac{n'-n}{4}\lambda c^3 h^2 k^2,\\
S_{\mathrm{IV}} &= -\frac{1}{2}(nj)^2 h \Delta \frac{u}{n} + \frac{l^2 \mathcal{P}}{4} + \frac{n'-n}{2}\lambda c^3 h^2 k^2,\\
S_{\mathrm{V}} &= -\frac{1}{2}\left(\frac{nj}{ni}\right)^3 h \Delta \frac{u}{n} + \frac{1}{2}\frac{nj}{ni} l^2 \mathcal{P} + \frac{n'-n}{2}\lambda c^3 h k^3.
\end{aligned}\right\} \tag{8.2}$$

Il faut aussi noter une expression de S_{V} dont le calcul ne présente pas d'indétermination:

$$S_{\mathrm{V}} = \frac{1}{2}\frac{n}{n'} j(n'-n)\left[\frac{n}{n'} h(v+kc)^2 + c u k^2 + h v^2\right] + \frac{n'-n}{2}\lambda c^3 h k^3.$$

9. Application aux lentilles minces. Les expressions précédentes appliquées aux lentilles minces dont les puissances et les intervalles ont été déterminés en utilisant par exemple les formules (7.2) vont donner rapidement une idée de la qualité du système en fonction des cambrures Γ de chaque lentille

$$\Gamma = \frac{c_1 + c_2}{2}.$$

Le choix des cambrures est fait de façon à annuler ou à rendre minimum les sommes S_{I}, S_{II}, S_{III}, S_{IV}, S_{V}.

On introduit ensuite les épaisseurs de lentille nécessaires à une réalisation; les modifications apportées par les épaisseurs aux sommes S peuvent en général être compensées par de légères retouches sur les cambrures, suivies d'une homothétie de façon à conserver les puissances initiales.

Le système ainsi dégrossi présente certainement des aberrations importantes qu'il s'agit de connaître et de réduire par l'étude exacte du trajet de certains rayons judicieusement choisis.

[1] A MARÉCHAL: Optique Géométrique Générale. Handbuch der Physik, tome XXIV. p. 135. Berlin-Göttingen-Heidelberg: Springer 1956. — H. H. HOPKINS: Wave Theory of Aberrations. Oxford 1950.

IV. La réduction des aberrations.

a) Généralités.

10. Rappel de quelques notions connues. *α) Chemin optique entre un point O et un point O'.* On appelle chemin optique la quantité $L_{OO'}$ telle que:

$$L_{OO'}=(O\,O')=\sum n\,\varrho \tag{10.1}$$

en désignant par ϱ la longueur du parcours du rayon dans le milieu d'indice n. Si O et O' ne sont pas sur le rayon (Fig. 15), on convient de définir le chemin optique par:

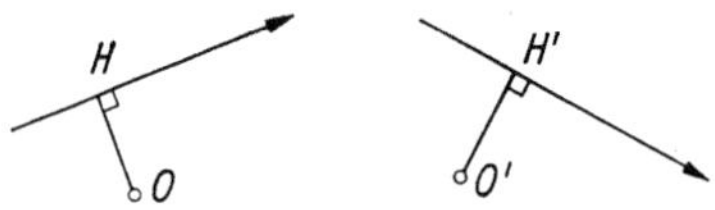

Fig. 15.

$$L_{OO'}=L_{HH'}. \tag{10.2}$$

H et H' étant les pieds des perpendiculaires abaissées de O et de O' sur le rayon. La raison de ce choix apparaîtra à la Sect. 10ε.

β) Principe de Fermat. Le trajet suivi par la lumière de O à O' est celui pour lequel le chemin optique est stationnaire par rapport à toutes les déformations possibles de la trajectoire. Autrement dit: les rayons lumineux sont les extrêmales du chemin optique.

Rappelons que si l'on considère deux chemins voisins (Fig. 16) suivis par la lumière, la différentielle dL s'exprime par:

$$dL=\overrightarrow{n'u'}\cdot\overrightarrow{dO'}-\overrightarrow{n\,u}\cdot\overrightarrow{dO}=n'\overline{O'H'}-n\,\overline{OH}. \tag{10.3}$$

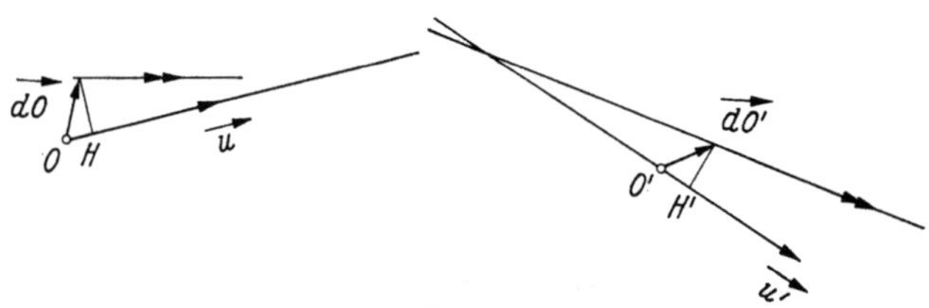

Fig. 16.

$\vec{u}$ et $\vec{u'}$ sont les vecteurs unitaires portés par le rayon dans l'espace objet et l'espace image, $\overrightarrow{dO}$ et $\overrightarrow{dO'}$ sont les déplacements élémentaires de O et O', $\overline{OH}$ et $\overline{O'H'}$ leurs projections sur les rayons.

γ) Théorème de Gouy. On peut transférer de l'espace objet à l'espace image une petite déformation de la surface d'onde; si Δ est la déformation dans l'espace objet, et Δ' la déformation dans l'espace image, on a:

$$n'\Delta'=n\Delta. \tag{10.4}$$

δ) Chemin optique aberrant. Le chemin optique aberrant $n'\Delta'$ est égal aux variations de chemin optique entre la source et le centre de la sphère de référence:

$$n'\Delta'=L_0-L. \tag{10.5}$$

L_0 est le chemin optique le long d'un rayon de référence; L est le chemin optique le long du rayon étudié.

ε) Relation entre les aberrations transversales et Δ'. Appelons O' le centre de la sphère de référence (Fig. 17); O' est le sommet du trièdre $O'x'y'z'$. Les cosinus directeurs du rayon émergent sont L', M', N'. Le point $I'(y', z')$ est le point d'intersection du rayon et du plan $O'y'z'$. H' est la projection de O' sur le rayon. Nous nous proposons de relier les aberrations transversales y' et z', même si elles sont grandes, aux dérivées partielles de L (ou Δ') par rapport aux paramètres M' et N' qui fixent l'orientation du rayon dans l'espace image.

Evaluons le chemin optique $(O\,I')$:

$$(O\,I') = L_{OH'} + n'\overline{H'I'} = L_{OH'} + n'(M'y' + N'z').$$

Considérons maintenant un rayon voisin perçant le plan $O'y'z'$ en $I' + dI'$ et calculons de deux façons la différentielle $d(O\,I')$:

$$d(O\,I') = dL_{OH'} + n'[M'dy' + N'dz' + y'dM' + z'dN'].$$

D'après la relation (10.3) on peut aussi écrire

$$d(O\,I') = n'\vec{u'}\cdot\overrightarrow{dI'} = n'(M'dy' + N'dz').$$

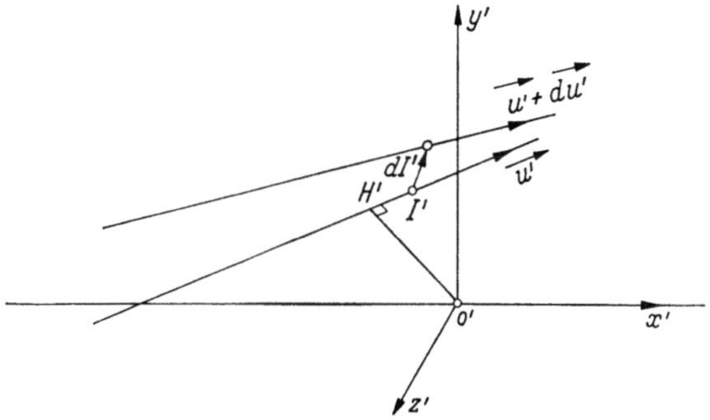

Fig. 17.

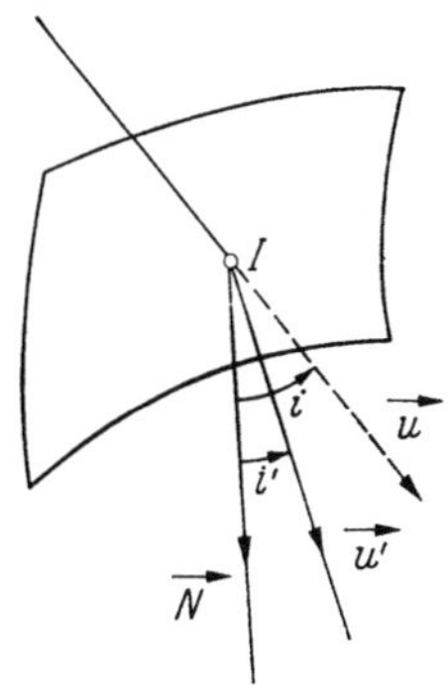

Fig. 18.

Il résulte de la comparaison de des égalités que

$$dL_{OH'} = -n'(y'dM' + z'dN').$$

Puisque $L_{OH'}$ n'est fonction que de M' et de N', on aura nécessairement

$$\frac{\partial L_{OH'}}{\partial M'} = -n'y', \qquad \frac{\partial L_{OH'}}{\partial N'} = -n'z'. \tag{10.6}$$

Finalement le résultat peut s'exprimer sous la forme suivante:

$$\left.\begin{aligned} y' &= -\frac{1}{n'}\frac{\partial L}{\partial M'} = \frac{\partial \Delta'}{\partial M'}, \\ z' &= -\frac{1}{n'}\frac{\partial L}{\partial N'} = \frac{\partial \Delta'}{\partial N'}. \end{aligned}\right\} \tag{10.7}$$

Cette relation, qui reste valable quand les quantités y' et z' sont grandes, nous incite à adopter la définition de L donnée à la Sect. 10α et à conserver l'usage de Δ' défini par (10.5) même si les aberrations sont importantes.

ζ) Expression vectorielle de la loi de la réfraction. Si $\vec{u}$ et $\vec{u'}$ sont les vecteurs unitaires des rayons incident et réfracté, $\vec{N}$ celui de la normale au point d'incident I (Fig. 18), on a

$$n'\vec{u'} = n\,\vec{u} + \mathcal{I}\vec{N} \tag{10.8}$$

en posant

$$\mathcal{I} = n'\cos i' - n\cos i. \tag{10.9}$$

η) Convention relative aux miroirs. Dans le cas des miroirs nous utiliserons la convention suivante qui permet de traiter les miroirs avec les mêmes équations que les dioptres et d'utiliser les mêmes programmes de calcul.

Les indices n'_r et les épaisseurs e'_r qui suivent une réflexion d'ordre impair sont négatifs, les autres sont positifs; les courbures sont toujours comptées suivant le même axe. La relation (10.8) justifie cette convention.

11. Principe de la méthode. Un système étant déterminé par ses paramètres optiques et géométriques (c_r, e'_r, n'_r pour le dioptre de rang r), on fait passer un certain nombre de rayons méridiens et extraméridiens pour chacun des points-objet B choisis dans le champ, de façon à connaître les surfaces d'onde dans l'espace image ainsi que les dérivées partielles de Δ' pour chacun des rayons et chacun des paramètres.

Soit O' le centre de la sphère de référence actuelle pour le faisceau de rayons issus de B (Fig. 19): O' est en général le point d'intersection du rayon moyen, passant par le centre de la pupille, avec le plan image F'.

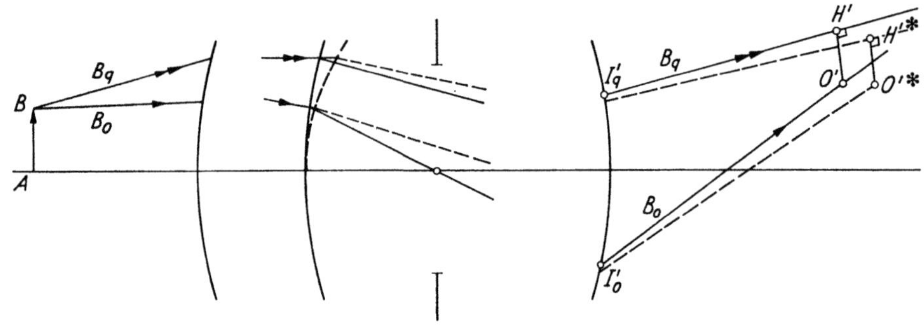

Fig. 19.

Appelons B_0 le rayon moyen et B_q un rayon quelconque du faisceau issu de B, I' l'intersection d'un rayon et du dernier dioptre, H' la projection de O' sur ce rayon. Nous désignerons les éléments futurs au moyen du signe *.

L'écart Δ'_{B_q} entre la surface d'onde réelle et la sphère de référence a pour valeur

$$n'\Delta'_{B_q} = [\sum n\,\varrho + n'\overline{I'H'}]_{B_0} - [\sum n\,\varrho + n'\overline{I'H'}]_{B_q}. \tag{11.1}$$

Si l'on fait subir à un paramètre p_r quelconque du système (la courbure sur la Fig. 19) une variation dp_r, le rayon futur B_q^* va se séparer du rayon actuel B_q au niveau de la variation et passera finalement en H'^* projection du centre de la sphère de référence future O'^*; O'^* est l'intersection du rayon moyen futur B_0^* et du plan image futur F'^*.

Les divers termes de l'expression de $n'\Delta'_{B_q}$ vont dans ces conditions subir des variations qu'il faut évaluer afin de connaître la nouvelle forme de la surface d'onde rapportée à la sphère centrée en O'^*.

A cet effet on évalue la variation locale de chemin optique produite par la variation dp_r du paramètre p_r; on transfère ensuite cette variation dans l'espace image en utilisant le théorème de GOUY; enfin on tient compte du déplacement de O' en O'^* du centre de la sphère de référence.

Autrement dit il nous faut évaluer les différences de chemins optiques

$$(BO'^*) - (BO') \quad \text{et} \quad (BH'^*) - (BH').$$

b) Variation locale de chemin optique en fonction des divers paramètres.

12. Méthode de calcul. On décompose le chemin optique total en trois fractions en plaçant de part et d'autre du paramètre variable p_r deux points: K en amont, K' en aval (Fig. 20).

Les chemins optiques actuel et futur s'écrivent respectivement

$$\left.\begin{aligned} L &= (BK) + (KK') + (K'H') \\ \text{et} \qquad L^* &= (BK) + (KK'^*) + (K'^*H'^*). \end{aligned}\right\} \tag{12.1}$$

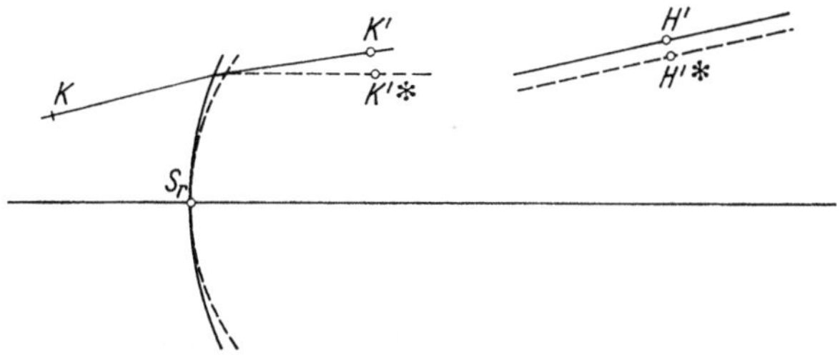

Fig. 20.

Pour étudier l'effet de la variation locale, on peut admettre que $K'K'^*$ et $H'H'^*$ sont perpendiculaires au rayon; dans ces conditions l'expression (10.3) montre que

$$(K'H') = (K'^*H'^*).$$

On en déduit que

$$dL = L^* - L = (KK'^*) - (KK') = (KK')^* - (KK');$$

c'est-à-dire

$$dL = d(KK')$$

ou

$$\frac{\partial L}{\partial p_r} = \frac{\partial (KK')}{\partial p_r}. \tag{12.2}$$

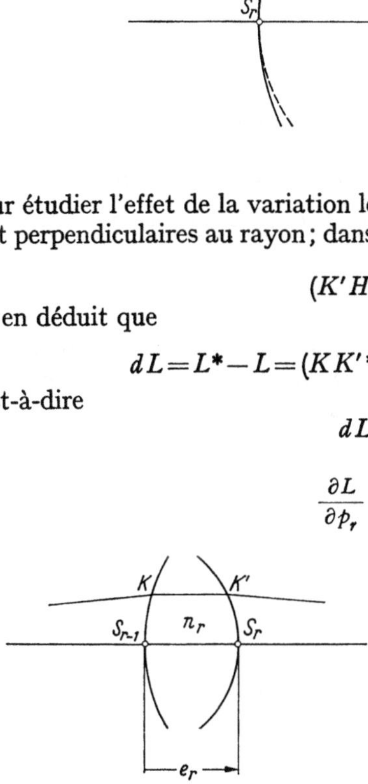

Fig. 21.

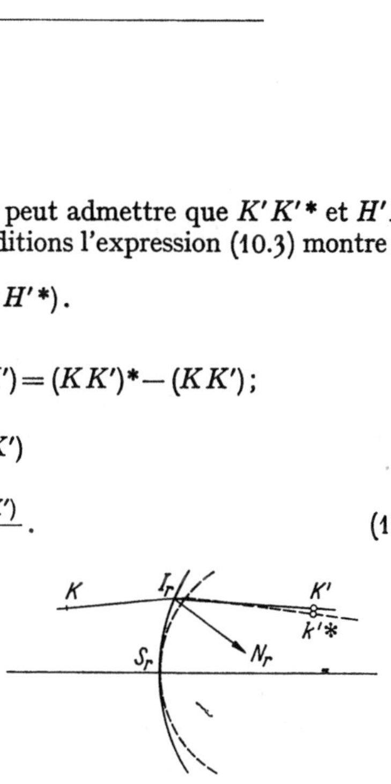

Fig. 22.

13. Calcul des dérivées partielles du chemin optique. *α) Cas des indices.* Choisissons pour points K et K' les points d'impact I_{r-1} et I_r du rayon sur les surfaces qui limitent le milieu d'indice n_r (Fig. 21) et rappelons que

$$\varrho_r = \overline{I_{r-1} I_r} = KK'.$$

Il est manifeste que

$$dL = \varrho_r \, dn_r.$$

On en déduit que

$$\frac{\partial L}{\partial n_r} = \varrho_r. \tag{13.1}$$

β) Cas des courbures (ou plus généralement des paramètres qui fixent la forme du dioptre). Choisissons pour points K et K' deux points quelconques, situés de part et d'autre du dioptre (Fig. 22) et évaluons dL selon la méthode décrite à la Sect. 12 en utilisant la relation (10.3)

$$d L = d\left[(K I_r) + (I_r K')\right]$$
$$= n_r \overrightarrow{u_r} \cdot \overrightarrow{d I_r} - n'_r \overrightarrow{u'_r} \cdot \overrightarrow{d I_r}.$$

Tenant compte de la relation (10.8), dL s'écrit

$$dL = -\mathcal{I}_r \overrightarrow{N_r} \cdot \overrightarrow{dI_r}.$$

On en déduit que

$$\frac{\partial L}{\partial c_r} = -\mathcal{I}_r \overrightarrow{N_r} \cdot \frac{\partial \overrightarrow{I_r}}{\partial c_r}. \tag{13.2}$$

Remarquons que nous pouvons choisir arbitrairement le déplacement $\overrightarrow{dI_r}$ puisque l'expression (13.2) ne fait intervenir que sa projection sur la normale $\vec{N}$; on peut en particulier prendre l'équation du dioptre sous la forme $f(X, Y, Z, c) = 0$, l'axe OX étant confondu avec celui de l'instrument et faire varier c en laissant Y et Z fixes; le déplacement $\overrightarrow{dI_r}$ est ainsi parallèle à l'axe et a pour valeur

$$|\overrightarrow{dI}| = \frac{\partial X}{\partial c} dc = -\frac{f'_c}{f'_x} dc.$$

Comme la normale $\vec{N}$ est orientée de façon que $\cos i$ et $\cos i'$ soient positifs, on écrit l'équation $f(X, Y, Z, c) = 0$ de telle sorte que la dérivée f'_x soit positive.

Dans ces conditions le produit scalaire $\vec{N} \cdot \overrightarrow{dI}$ devient

$$\vec{N} \cdot \overrightarrow{dI} = \frac{f'_x}{\sqrt{f'^2_x + f'^2_y + f'^2_z}} \left(-\frac{f'_c}{f'_x} dc\right) = -\frac{f'_c\, d_c}{\sqrt{f'^2_x + f'^2_y + f'^2_z}}.$$

La dérivée partielle s'écrit maintenant

$$\frac{\partial L}{\partial c_r} = \mathcal{I}_r \frac{f'_c}{\sqrt{f'^2_x + f'^2_y + f'^2_z}}. \tag{13.3}$$

Donnons quelques exemples choisis parmi les surfaces les plus courantes.

1. Sphère de courbure c.

$$\left.\begin{aligned} f(X, Y, Z, c) &\equiv 2X - c(X^2 + Y^2 + Z^2) = 0, \\ \frac{\partial L}{\partial c_r} &= -\mathcal{I}_r \frac{X_r}{c_r}. \end{aligned}\right\} \tag{13.4}$$

2. Sphère de faible courbure (en particulier le plan pour $c = 0$). La relation précédente est toujours valable, mais la précision des calculs sur X peut être insuffisante et l'on aura avantage à écrire:

$$\frac{\partial L}{\partial c_r} = -\mathcal{I}_r \frac{Y_r^2 + Z_r^2}{2}. \tag{13.5}$$

3. Cylindre circulaire de génératrices parallèles à OZ.

$$\left.\begin{aligned} f(X, Y, Z, c) &\equiv 2X - c(X^2 + Y^2) = 0, \\ \frac{\partial L}{\partial c_r} &= -\mathcal{I}_r \frac{X_r}{c_r}. \end{aligned}\right\} \tag{13.6}$$

La même relation vaut pour le cylindre de génératrices parallèles à OY.

4. Paraboloïde de courbure au sommet c.

$$\left.\begin{aligned} f(X,Y,Z,c) &\equiv 2X - c(Y^2+Z^2) = 0, \\ \frac{\partial L}{\partial c_r} &= -\mathscr{I}_r \frac{X_r}{c_r\sqrt{1+2c_r X_r}}. \end{aligned}\right\} \tag{13.7}$$

γ) Cas des épaisseurs. Choisissons pour points K et K' deux points arbitraires (Fig. 23); et convenons que pour faire varier l'intervalle e_r de la quantité de_r, nous imprimons à la fraction aval du système une translation de_r. L'évaluation du chemin optique aval peut se faire alors le long de l'ancien trajet à partir du point K'^*. On obtient

$$dL = d(KK') = n_r L_r de_r$$

en appelant L_r le cosinus directeur du rayon sur l'axe du système.

Finalement

$$\frac{\partial L}{\partial e_r} = n_r L_r. \tag{13.8}$$

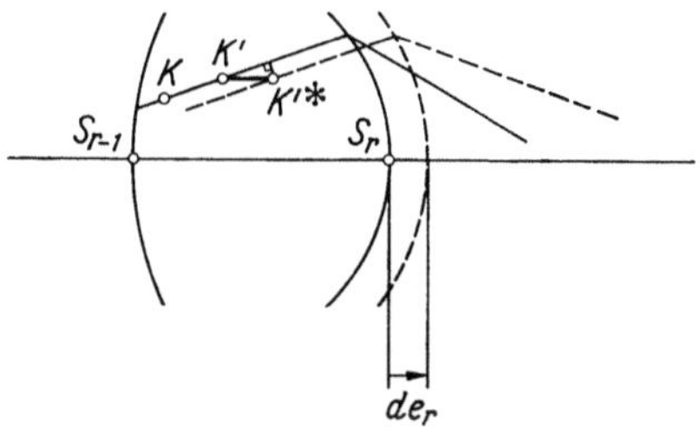

Fig. 23.

δ) Récapitulation des résultats. Rassemblons les expressions des dérivées partielles $\partial L/\partial p_r$:

$$\left.\begin{aligned} \frac{\partial L}{\partial n_r} &= \varrho_r, \\ \frac{\partial L}{\partial c_r} &= -\mathscr{I}_r \overrightarrow{N_r} \cdot \frac{\partial \overrightarrow{I_r}}{\partial c_r}, \\ \frac{\partial L}{\partial e_r} &= n_r L_r. \end{aligned}\right\} \tag{13.9}$$

Il faut bien noter que ces expressions ne concernent que les variations de chemin optique. Elles supposent que le chemin optique est compté dans l'espace image jusqu'à un point O' fixe et voisin du rayon émergent (Sect. 12). Nous utiliserons ces expressions dans la Sect. 14 pour calculer les dérivées partielles de l'écart aberrant $n'\Delta'$ en leur adjoignant des termes tenant compte du changement de sphère de référence et des aberrations transversales.

c) Changement de sphère de référence et expression générale de la variation de chemin optique aberrant.

14. Lorsqu'on fait varier un paramètre p_r le plan de mise au point F' et le point d'intersection O' du rayon moyen et du plan F' varient.

Sur la Fig. 24 F' est le plan de mise au point, O' l'intersection du rayon moyen B_0 et du plan F', J' l'intersection du rayon B_q et du plan F', H' la projection de O' sur le rayon B_q, $\overrightarrow{u'}$ le vecteur unitaire du rayon B_q et $\overrightarrow{u'_0}$ celui du rayon B_0; $\overrightarrow{O'J'}$ est donc le vecteur aberration transversale. Les quantités futures sont désignées par le signe *.

Appelons B la source et calculons de deux façons la différentielle $dL_{j'}$ du chemin optique $L_{j'} = (BJ')$.

En combinant les propriétés des expressions (10.3) et (13.9) on obtient

$$dL_{j'} = L_{j'}^* - L_{j'} = \frac{\partial L}{\partial p_r} dp_r + n' \overrightarrow{u'} \cdot \overrightarrow{dJ'}. \tag{14.1}$$

En exprimant le chemin optique $L_{J'}$ en fonction de $L_{H'}$, et en différentiant le résultat on obtient

$$L_{J'} = L_{H'} + n' \vec{u'} \cdot \overrightarrow{O'J'}$$

et

$$dL_{J'} = dL_{H'} + n' \overrightarrow{O'J'} \cdot \overrightarrow{du'} + n' \vec{u'} (\overrightarrow{dJ'} - \overrightarrow{dO'}). \tag{14.2}$$

La comparaison de (14.1) et de (14.2) donne la différentielle $dL_{H'}$ du chemin optique $(B\,O')$ le long du rayon B_q

$$dL_{H'} = \frac{\partial L}{\partial p_r} dp_r + n' \vec{u'} \cdot \overrightarrow{dO'} - n' \overrightarrow{O'J'} \cdot \overrightarrow{du'}. \tag{14.3}$$

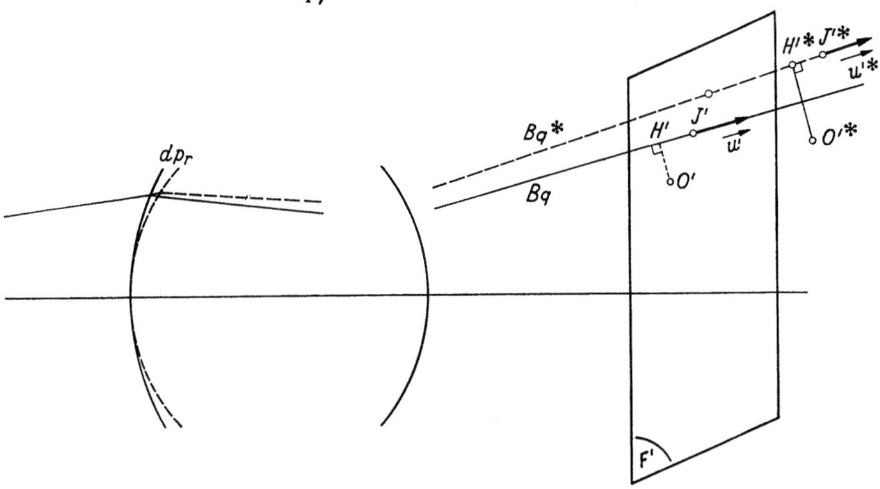

Fig. 24.

On calcule de même la différentielle $dL_{0'}$ du chemin optique (BO') le long du rayon B_0

$$dL_{0'} = \frac{\partial L_0}{\partial p_r} dp_r + n' \vec{u'_0} \cdot \overrightarrow{dO'}. \tag{14.4}$$

Après avoir rappelé que le chemin optique aberrant $n'\Delta'$ est tel que

$$n'\Delta' = L_0 - L,$$

on obtient l'expression suivante:

$$\boxed{\frac{\partial(n'\Delta')}{\partial p_r} = \left(\frac{\partial L_0}{\partial p_r} - \frac{\partial L}{\partial p_r}\right) + n' \frac{\partial \overrightarrow{O'}}{\partial p_r} \cdot (\vec{u'_0} - \vec{u'}) + n' \overrightarrow{O'J'} \cdot \frac{\partial \vec{u'}}{\partial p_r}.} \tag{14.5}$$

Cette expression appelle les commentaires suivants:

α) Elle tient compte

— de la variation locale de chemin optique: $\left(\frac{\partial L_0}{\partial p_r} - \frac{\partial L}{\partial p_r}\right)$,

— du changement de sphère de référence: $n' \frac{\partial \overrightarrow{O'}}{\partial p_r} \cdot (\vec{u'_0} - \vec{u'})$;

— de l'aberration transversale: $n' \overrightarrow{O'J'} \cdot \frac{\partial \vec{u'}}{\partial p_r}$.

Ce dernier terme est un terme correctif dont l'influence ne doit pas être considérée comme negligeable.

β) On verra plus loin (aux Sect. 15 et 16) comment élaborer les composantes des vecteurs

$$\frac{\partial \overrightarrow{O'}}{\partial p_r} \quad \text{et} \quad \frac{\partial \overrightarrow{u'}}{\partial p_r}.$$

γ) La traduction analytique de l'expression (14.5) ne présente pas de difficultés. Appelons

L'_0, M'_0, N'_0 les cosinus directeurs du rayon B_0,

L', M', N' les cosinus directeurs du rayon B_q,

$\frac{\partial x'}{\partial p_r}$ et $\frac{\partial y'}{\partial p_r}$ les composantes du vecteur $\frac{\partial \overrightarrow{O'}}{\partial p_r}$

$\delta y'$ et $\delta z'$ les composantes du vecteur $\overrightarrow{O'J'}$.

L'expression (14.5) prend alors la forme suivante:

$$\left.\begin{aligned} \frac{\partial(n'\Delta')}{\partial p_r} = \left(\frac{\partial L_0}{\partial p_r} - \frac{\partial L}{\partial p_r}\right) + n'\left[(L'_0 - L')\frac{\partial x'}{\partial p_r} + (M'_0 - M')\frac{\partial y'}{\partial p_r}\right] + \\ + n'\left[\delta y' \frac{\partial M'}{\partial p_r} + \delta z' \frac{\partial N'}{\partial p_r}\right]. \end{aligned}\right\} \tag{14.6}$$

d) Elaboration des dérivées partielles du chemin optique aberrant.

15. Calcul des composantes $\partial x'/\partial p_r$ et $\partial y'/\partial p_r$ du vecteur $\partial\overrightarrow{O'}/\partial p_r$. Nous utiliserons les dérivées partielles $\partial L/\partial p_r$ pour certains rayons judicieusement choisis.

Considérons sur la Fig. 25 le plan de l'objet et celui de la pupille:

— A représente un point au centre du champ; B est un point tel que l'objet AB puisse être considéré comme paraxial; C et D sont des points situés de plus en plus loin de l'axe.

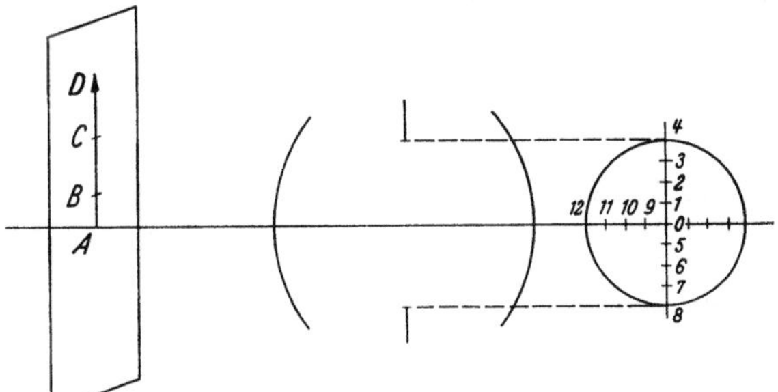

Fig. 25.

— Les indices q caractérisant la position du rayon dans la pupille sont tels que le rayon 0 correspond au centre, les rayons 1 et 5 à des rayons méridiens de faible ouverture, les rayons 2 et 6, 3 et 7, 4 et 8 à des rayons méridiens d'ouverture croissante, les rayons 9, 10, 11, 12 à des rayons extra-méridiens d'ouverture croissante.

— On notera que dans ces conditions les rayons A_1, B_0, B_1, B_5 sont pratiquement des rayons paraxiaux.

La composante de défaut de mise au point $\partial x'/\partial p_r$ est déduite des variations de chemin optique des rayons A_0 et A_1 en utilisant l'expression (14.6) dans laquelle on recherche la valeur $\partial x'/\partial p_r$ telle que $\frac{\partial(n'\Delta')}{\partial p_r}$, $\delta y'$ et $\delta z'$ soient nuls.

$$\frac{\partial x'}{\partial p_r} = -\frac{1}{n'} \frac{\frac{\partial L_{A_0}}{\partial p_r} - \frac{\partial L_{A_1}}{\partial p_r}}{1 - L'_{A_1}}. \tag{15.1}$$

La composante de variation de grandeur de l'image en un point quelconque du champ $\partial y'/\partial p_r$ est calculée de manière analogue:

$$\frac{\partial y'}{\partial p_r} = -\frac{1}{n'}\,\frac{\dfrac{\partial L_0}{\partial p_r}-\dfrac{\partial L_1}{\partial p_r}}{M_0'-M_1'} - \frac{L_0'-L_1'}{M_0'-M_1'}\,\frac{\partial x'}{\partial p_r},$$

ou encore en utilisant les rayons d'indices 1 et 5:

$$\frac{\partial y'}{\partial p_r} = -\frac{1}{n'}\,\frac{\dfrac{\partial L_1}{\partial p_r}-\dfrac{\partial L_5}{\partial p_r}}{M_1'-M_5'} - \frac{L_1'-L_5'}{M_1'-M_5'}\,\frac{\partial x'}{\partial p_r}. \tag{15.2}$$

La quantité

$$-\frac{1}{n'}\,\frac{\dfrac{\partial L_1}{\partial p_r}-\dfrac{\partial L_5}{\partial p_r}}{M_1'-M_5'}$$

représente la variation de grandeur de l'image sans défaut de mise au point; elle est très utile dans l'étude du chromatisme de grandeur.

Remarque. La composante de défaut de mise au point de la pupille $\partial p'/\partial p_r$ se calcule comme $\partial x'/\partial p_r$ en utilisant les rayons A_0 et B_0

$$\frac{\partial p'}{\partial p_r} = -\frac{1}{n'}\,\frac{\dfrac{\partial L_{A_0}}{\partial p_r}-\dfrac{\partial L_{B_0}}{\partial p_r}}{1-L_{B_0}'}. \tag{15.3}$$

16. Calcul des composantes $\partial M'/\partial p_r$ et $\partial N'/\partial p_r$ du vecteur $\partial \vec{u}'/\partial p_r$. Le vecteur $\partial \vec{u}'/\partial p_r$ intervient dans l'expression (14.5) par le terme correctif $\overrightarrow{B'J'}\cdot\partial\vec{u}'/\partial p_r$. Tenant compte de cette remarque, nous nous limiterons dans son calcul à une valeur approchée, dont la précision est en pratique suffisante.

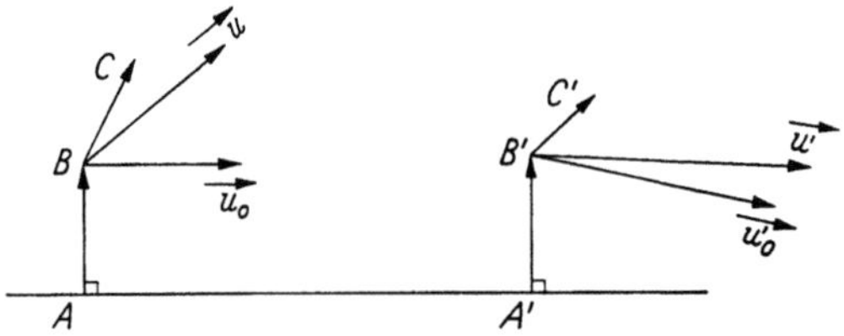

Fig. 26.

Considérons (Fig. 26) un élément $\overrightarrow{BC}$ et son image $\overrightarrow{B'C'}$; le vecteur unitaire image du rayon principal issu de B est $\vec{u}_0'$, celui d'un rayon quelconque $\vec{u}'$. Si l'élément $\overrightarrow{BC}$ est suffisamment petit pour que sa longueur puisse être assimilée à un infiniment petit du premier ordre, la différence $(BB')-(CC')$ est constante. On a alors

$$n'\,\overrightarrow{B'C'}\cdot(\vec{u}_0'-\vec{u}') = \text{constante}$$

et

$$d(\overrightarrow{B'C'})\cdot(\vec{u}_0'-\vec{u}') + \overrightarrow{B'C'}(d\vec{u}_0'-d\vec{u}') = 0.$$

Rappelant que les cosinus directeurs de $\vec{u}_0'$ sont L_0', M_0', 0, ceux de $\vec{u}'$, L', M', N', envisageons deux éléments $\overrightarrow{B'C'}$: le premier de composantes 0, $\Delta y'$, 0, et le deuxième 0, 0, $\Delta z'$. Nous en déduisons que

$$d(\Delta y')\,[M_0'-M'] + \Delta y'(dM_0'-dM') = 0$$

et

$$d(\Delta z')\,[-N'] + \Delta z'[-dN'] = 0.$$

Supposant finalement que le système est corrigé de la distorsion, nous écrivons que

$$\frac{d(\Delta y')}{\Delta y'}=\frac{d(\Delta z')}{\Delta z'}=\frac{d y_0'}{y_0'},$$

y_0' étant la dimension de l'image $A'B'$. Dans ces conditions

$$\left.\begin{aligned} dM' &= dM_0' - \frac{d y_0'}{y_0'}(M'-M_0'),\\ dN' &= -\frac{d y_0'}{y_0'}N'. \end{aligned}\right\}\tag{16.1}$$

Il reste à calculer dM_0'. Evaluons (Fig. 27) de deux manières différentes $\tan\vartheta_0'$, ϑ_0' étant l'angle sous lequel l'image est vue de la pupille de sortie; en restant dans le domaine paraxial:

$$\tan\vartheta_0'=\frac{M_0'}{L_0'}=\frac{y_0'}{x'-p'}.$$

On trouve en différentiant et en introduisant la variation paraxiale dp' de la pupille de sortie

$$dM_0'=\frac{L_0'M_0'^2}{y_0'}\left[dp'-dx'+\frac{L_0'}{M_0'}dy_0'\right].\tag{16.2}$$

Signalons au passage qu'on peut appliquer ce procédé de calcul au cas où la pupille est à l'infini en remplaçant p' par $\pi'=1/p'$ et dp' par

$$d\pi'=-y_0'^2L_0'^2/M_0'^2\,dp'.$$

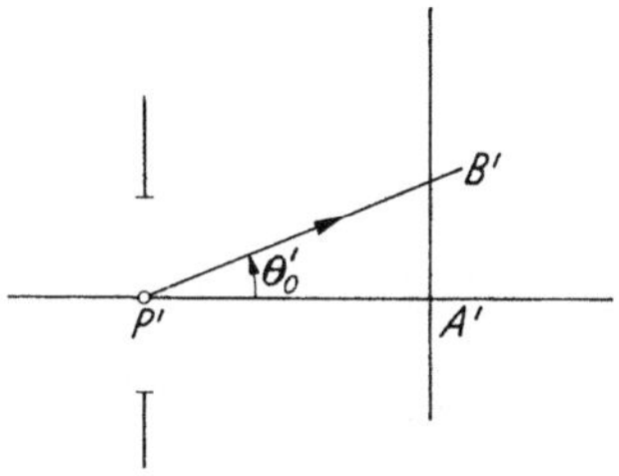

Fig. 27.

En regroupant (16.1) et (16.2) les dérivées partielles $\partial M'/\partial p_r$ et $\partial N'/\partial p_r$ prennent la forme suivante:

$$\left.\begin{aligned} \frac{\partial M'}{\partial p_r} &= \frac{L_0'M_0'^2}{y_0'}\frac{\partial p'}{\partial p_r}-\frac{L_0'M_0'^2}{y_0'}\frac{\partial x'}{\partial p_r}+\frac{L_0'^2M_0'-(M'-M_0')}{y_0'}\frac{\partial y_0'}{\partial p_r},\\ \frac{\partial N'}{\partial p_r} &= -\frac{N'}{y_0'}\frac{\partial y_0'}{\partial p_r}. \end{aligned}\right\}\tag{16.3}$$

17. Calcul des dérivées partielles de la distance focale, du grandissement et de la distorsion. *α) Cas de la distance focale f'.* On calcule pour le champ B la dérivée $(\partial y'/\partial p_r)_B$ au moyen de la relation (15.2). Comme on peut considérer que la distorsion est nulle pour ce champ, on obtient immédiatement

$$\frac{1}{f'}\frac{\partial f'}{\partial p_r}=\frac{1}{y'_{B_0}}\left(\frac{\partial y'}{\partial p_r}\right)_B.\tag{17.1}$$

β) Cas du grandissement. Par un raisonnement analogue on obtient:

$$\frac{\partial\gamma}{\partial p_r}=\frac{1}{y'_{B_0}}\left(\frac{\partial y'}{\partial p_r}\right)_B.\tag{17.2}$$

γ) Cas de la distorsion D. Pour un point objet à l'infini dans une direction ϑ,

$$D=\frac{y_0'}{f'\tan\vartheta}-1,$$

d'où

$$\frac{\partial D}{\partial p_r}=\frac{y_0'}{f'\tan\vartheta}\left[\frac{\partial y_0'}{\partial p_r}-\frac{y_0'}{f'}\frac{\partial f'}{\partial p_r}\right].\tag{17.3}$$

Pour un point objet à distance finie,

$$\vartheta=\frac{y_0'}{\gamma y_0}-1$$

d'où

$$\frac{\partial D}{\partial p_r}=\frac{1}{\gamma y_0}\left[\frac{\partial y_0'}{\partial p_r}-\frac{y_0'}{\gamma}\frac{\partial\gamma}{\partial p_r}\right].\tag{17.4}$$

18. Equation du dépouillement. On appellera équation du dépouillement la forme particulière que prend l'équation (14.6) quand on donne aux dérivées partielles $\partial M'/\partial p$ et $\partial N'/\partial p$ les valeurs que nous venons de calculer.

Dans le cas général d'un point quelconque du champ $y_0' \neq 0$ et l'équation (14.6) prend la forme suivante:

$$\left.\begin{aligned}
&\frac{\partial(n'\Delta)}{\partial p_r} = \frac{\partial L_0}{\partial p_r} - \frac{\partial L}{\partial p_r} + n'\left[K_x \frac{\partial x'}{\partial p_r} + K_y \frac{\partial y_0'}{\partial p_r} + K_p \frac{\partial p'}{\partial p_r}\right]\\
&\text{avec}\\
&K_x = (L_0' - L') - \frac{L_0' M_0'^2}{y_0'}\,\delta y',\\
&K_y = (M_0' - M') + [L_0'^2 M_0' + (M_0' - M')]\,\frac{\delta y'}{y_0'} - N'\frac{\delta z'}{y_0'},\\
&K_p = \frac{L_0' M_0'^2}{y_0'}\,\delta z'.
\end{aligned}\right\} \quad (18.1)$$

Dans le cas du point A situé sur l'axe $y_0' = 0$, l'équation (14.6) prend alors la forme suivante

$$\left.\begin{aligned}
&\frac{\partial(n'\Delta)}{\partial p_r} = \frac{\partial L_0}{\partial p_r} - \frac{\partial L}{\partial p_r} + n'\left[K_{A_x} \frac{\partial x'}{\partial p_r} + K_{A_y} \cdot \frac{1}{y_0'}\,\frac{\partial y_0'}{\partial p_r}\right]\\
&\text{avec}\\
&K_{A_x} = 1 - L', \qquad K_{A_y} = -M'\,\delta y'.
\end{aligned}\right\} \quad (18.2)$$

Les relations (18.1) et (18.2) appellent les remarques suivantes.

1. Le terme $1/y_0'\ \partial y_0'/\partial p_r$ qui intervient dans (18.2) est un terme paraxial, qui est égal à $1/f'\ \partial f'/\partial p_r$ ou $1/y\ \partial y/\partial p_r$ selon que l'objet est à l'infini ou à distance finie.

2. Les quantités K_x, K_y, K_p, K_{A_x}, et K_{A_y} sont liées au rayon q. Il y en a autant que l'on étudie de rayons en dehors des rayons moyens.

3. Les quantités $\partial L/\partial p_r$ sont liées au rayon q et au dioptre r. Leur nombre est égal au nombre de rayons étudiés multiplié par le nombre de dioptres.

e) Réduction des aberrations chromatiques.

19. Les aberrations chromatiques peuvent être traitées en utilisant les résultats que nous venons d'obtenir; elles apparaissent en effet comme un cas particulier de la variation des aberrations en fonction des indices.

Introduisons pour chaque matière du système l'indice n et le paramètre Δn qui caractérise la dispersion de cette matière entre les deux raies pour lesquelles on veut réaliser l'achromatisme. Les différents effets du chromatisme sont évalués de la manière suivante.

α) Cas du chromatisme axial. Le chromatisme axial $\Delta x'$ est tel que

$$\Delta x' = \sum \frac{\partial x'}{\partial n_r}\,\Delta n_r. \quad (19.1)$$

β) Cas du chromatisme de grandeur. Appelons $(\partial y'/\partial n_r)_\lambda$ la quantité

$$-\frac{1}{n'}\,\frac{\dfrac{\partial L_1}{\partial n_r} - \dfrac{\partial L_5}{\partial n_r}}{M_1' - M_5'}$$

qui intervient dans le calcul de $\partial y'/\partial n_r$, selon l'équation (15.2). Le chromatisme de grandeur $\Delta y'$ est tel que

$$\Delta y' = \sum \left(\frac{\partial y'}{\partial n_r}\right)_\lambda \Delta n_r. \quad (19.2)$$

γ) Cas de la variation chromatique des aberrations. Pour étudier ce cas, on utilise l'expression générale (14.6) mise sous la forme de l'équation de dépouillement (18.1). En introduisant les résidus de chromatisme axial $\Delta x'$ et de grandeur $\Delta y'$; l'écart chromatique de chemin optique aberrant $(n'\Delta')_\lambda$, prend la forme suivante,

$$(n'\Delta')_\lambda = n'[K_x \Delta x' + K_y \Delta y'] + \sum \left[\left(\frac{\partial L_0}{\partial n_r} - \frac{\partial L}{\partial n_r}\right) + n' K_p \frac{\partial p'}{\partial p_r}\right] \Delta n_r. \quad (19.3)$$

20. Les considérations précédentes permettent maintenant de préciser le procédé de calcul brièvement décrit à la Sect. 3. Après avoir fait passer dans le système un certain nombre de rayons, on obtient un nombre important d'informations non seulement sur l'état actuel du système, mais surtout sur son évolution en fonction des paramètres de construction. La réduction des aberrations devient alors une opération que l'on peut mener avec rigueur en profitant de la puissance des calculateurs électroniques, sans avoir recours à des essais empiriques.

V. Les schémas de calcul.

21. Marche d'un rayon. La marche d'un rayon paraxial a été décrite aux Sects. 5 et 6 et fournit tous les éléments pour calculer les aberrations du 3ième ordre dont les valeurs sont indiquées à la Sect. 8.

Nous allons indiquer le schéma de calcul permettant de calculer les éléments d'un rayon quelconque, subissant une réfraction sur une quadrique de révolution. Le schéma est aussi valable pour les miroirs, à condition d'utiliser la convention indiquée à la Sect. 10η.

La quadrique est représentée dans un trièdre $SXYZ$, où S est le sommet de la quadrique et SX l'axe du système (Fig. 28), par l'équation

$$\left.\begin{aligned} &f(X, Y, Z, c) \\ &\equiv X - \frac{c}{2}\left[(1+\lambda) X^2 + Y^2 + Z^2\right] = 0 \end{aligned}\right\} \quad (21.1)$$

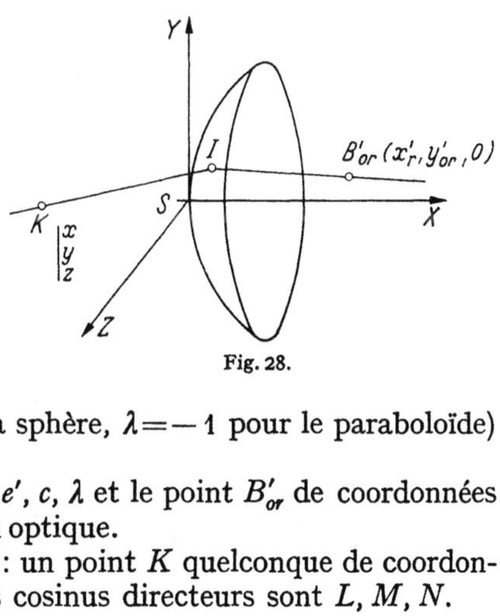

Fig. 28.

où λ caractérise la méridienne ($\lambda = 0$ pour la sphère, $\lambda = -1$ pour le paraboloïde) et c est la courbure au sommet.

Les données liées au dioptre sont: n, n', e', c, λ et le point B'_{or} de coordonnées x'_r, y'_{or}, o, jusqu'où doit être calculé le chemin optique.

Les données liées au rayon incident sont: un point K quelconque de coordonnées x, y, z et son vecteur unitaire dont les cosinus directeurs sont L, M, N.

On veut obtenir,

— les coordonnées X, Y, Z du point d'incidence I,

— les cosinus directeurs L', M', N' du rayon réfracté,

— les dérivées partielles locales $\partial\mathcal{L}/\partial e$, $\partial\mathcal{L}/\partial c$, $\partial\mathcal{L}/\partial n$,

— le chemin optique $\mathcal{L}$, jusqu'au point B'_{or}.

α) Formule de passage du dioptre de rang $r-1$ au dioptre de rang r. On choisit comme point K_r le point I_{r-1} d'incidence sur le dioptre précédent. Le passage d'un dioptre au suivant s'écrit alors de la manière suivante:

$$x_r = X_{r-1} - e'_{r-1}, \quad y_r = Y_{r-1}, \quad z_r = Z_{r-1}$$

et

$$L_r = L'_{r-1}, \quad M_r = M'_{r-1}, \quad N_r = N'_{r-1}.$$

β) Réfraction du rayon. Supprimons l'indice r pour alléger l'écriture. Le schéma de calcul va exprimer les opérations suivantes:

— On cherche la longueur ϱ telle que $\overrightarrow{KI}=\varrho\,\vec{u}$ et que I soit situé sur la quadrique; ceci nous amène à résoudre une équation du type

$$A\varrho^2-2B\varrho+C=0$$

et à ne conserver que la racine utile.

— On exprime la réfraction au moyen de la relation,

$$n'\overrightarrow{u'}=\overrightarrow{n\,u}+\mathcal{I}\vec{N}$$

dans laquelle

$$\mathcal{I}=n'\cos i'-n\cos i.$$

— Les cosinus directeurs de la normale $\vec{N}$ sont

$$\alpha=\frac{f'_x}{\sqrt{f'^2_x+f'^2_y+f'^2_z}},$$

$$\beta=\frac{f'_y}{\sqrt{f'^2_x+f'^2_y+f'^2_z}},$$

$$\gamma=\frac{f'_z}{\sqrt{f'^2_x+f'^2_y+f'^2_z}}.$$

Tenant compte de ces diverses considérations le schéma de calcul prend la forme suivante (le premier membre des équations représente l'élément calculé):

$$\left.\begin{aligned}A&=c(1+\lambda L^2),\\ B&=L-c[(1+\lambda)\,L\,x+M\,y+N\,z],\\ C&=c[(1+\lambda)\,x^2+y^2+z^2]-2x.\end{aligned}\right\}\tag{21.2}$$

$$\varrho=\frac{C}{B+\sqrt{B^2-A\,C}}.\tag{21.3}$$

$$\left.\begin{aligned}X&=x+L\,\varrho,\\ Y&=y+M\,\varrho,\\ Z&=z+N\,\varrho.\end{aligned}\right\}\tag{21.4}$$

$$\left.\begin{aligned}\alpha&=\frac{1-(1+\lambda)\,c\,X}{\sqrt{1-2\lambda\,c\,X+\lambda(\lambda+1)\,c^2X^2}},\\ \beta&=\frac{-c\,Y}{\sqrt{1-2\lambda\,c\,X+\lambda(\lambda+1)\,c^2X^2}},\\ \gamma&=\frac{-c\,Z}{\sqrt{1-2\lambda\,c\,X+\lambda(\lambda+1)\,c^2X^2}},\end{aligned}\right\}\tag{21.5}$$

$$\left.\begin{aligned}\cos i&=\alpha\,L+\beta\,M+\gamma\,N,\\ \sin^2 i&=1-\cos^2 i,\\ \sin^2 i'&=\frac{n^2}{n'^2}\sin^2 i,\\ \cos i'&=\sqrt{1-\sin^2 i'},\\ \mathcal{I}&=n'\cos i'-n\cos i.\end{aligned}\right\}\tag{21.6}$$

$$\left.\begin{aligned} n'L' &= nL + \mathscr{I}\alpha, \\ n'M' &= nM + \mathscr{I}\beta, \\ n'N' &= nN + \mathscr{I}\gamma. \end{aligned}\right\} \tag{21.7}$$

$$\left.\begin{aligned} &\frac{\partial \mathscr{L}}{\partial n} = \varrho, \\ &\frac{\partial \mathscr{L}}{\partial c} = -\mathscr{I}\frac{X}{c\sqrt{1-2\lambda c X+\lambda(\lambda+1)c^2X^2}} \quad \text{ou si } c=0, \quad \frac{\partial \mathscr{L}}{\partial c} = -\mathscr{I}\frac{Y^2+Z^2}{2}, \\ &\frac{\partial \mathscr{L}}{\partial e} = nL. \end{aligned}\right\} \tag{21.8}$$

$$\mathscr{L} = \sum n\,\varrho + n'[L'(x'_r - X) + M'(y'_{or} - Y) - N'Z]. \tag{21.9}$$

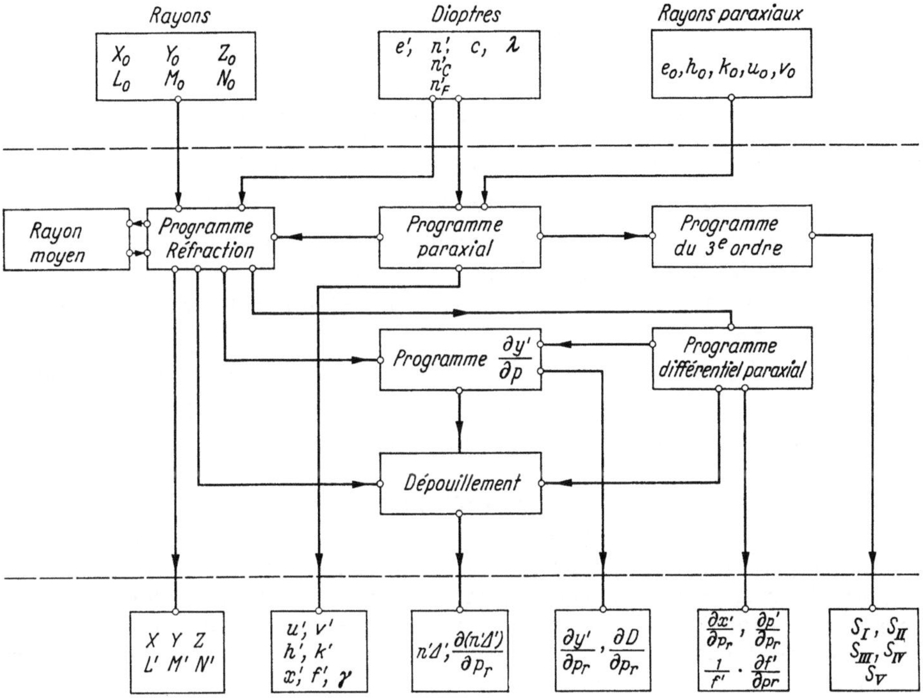

Fig. 29.

22. Diagramme d'exploitation. Le diagramme de la Fig. 29 résume les principales opérations, qui permettent d'exploiter le procédé de calcul qui vient d'être décrit.

α) Les données. Nous trouvons trois groupes de données, à savoir

— les données liées aux dioptres: e', n', c et λ et éventuellement les indices extrêmes du spectre à achromatiser (n'_c et n'_F par exemple),

— les données liées aux rayons paraxiaux e_0, h_0, k_0, u_0, v_0,

— les données liées aux rayons: le point $K_0(X_0, Y_0, Z_0)$ et les cosinus directeurs initiaux L_0, M_0, N_0.

β) Les programmes. Le diagramme comprend six programmes,

— le programme paraxial qui permet d'élaborer les éléments paraxiaux et en particulier x' qui fixe la position de l'image,

— le programme du 3ième ordre qui utilise les résultats du programme paraxial et fournit S_{I}, S_{II}, S_{III}, S_{IV}, S_{V},

— le programme «réfraction», qui fournit pour chaque rayon les coordonées X, Y, Z du point d'impact sur le dioptre, les cosinus directeurs du rayon réfracté L', M', N'; les dérivées partielles locales $\partial\mathcal{L}/\partial e$, $\partial\mathcal{L}/\partial c$, $\partial\mathcal{L}/\partial n$ et le chemin optique $\mathcal{L}$. Le passage des rayons d'un champ donné doit être précédé du passage du rayon moyen afin de connaître la dimension y'_0 de l'image et les cosinus directeurs finaux L'_0, M'_0, N'_0 du rayon moyen qui permettront d'élaborer les K de la Sect. 18, équations (18.1) et (18.2),

— le programme différentiel paraxial qui élabore $\partial x'/\partial p$, $\partial p'/\partial p$ et selon les cas $1/f'\ \partial f'/\partial p$ ou $1/\gamma\ \partial\gamma/\partial p$ à partir d'éléments fournis par les rayons A_0, A_1, et B_0, B_1 et B_5.

— le programme $\partial y'/\partial p$ qui élabore $\partial y'/\partial p$ et $\partial D/\partial p$ (dérivées partielles de la distorsion, voir Sect. 17) à partir d'éléments fournis par les rayons 1 et 5 de chaque champ,

— le programme «dépouillement» qui élabore d'une part l'écart aberrant $n'\Delta'$ et d'autre part les quantités $\frac{\partial(n'\Delta')}{\partial e}$, $\frac{\partial(n'\Delta')}{\partial c}$, $\frac{\partial(n'\Delta')}{\partial n}$.

γ) *Les résultats.* Regroupons les résultats obtenus. Nous y trouvons,

— tous les éléments paraxiaux classiques et toutes leurs dérivées partielles,

— les aberrations du 3ième ordre,

— la marche des rayons et la cible des points d'impact des rayons en chaque point du champ,

— toutes les dérivées partielles de dimension de l'image et de distorsion,

— la forme de la surface d'onde en chaque point du champ,

— toutes les dérivées partielles de l'écart aberrant.

C. Conclusion.

23. Nous avons donné ci-dessus les éléments essentiels d'une méthode moderne de calcul, particulièrement adaptée aux possibilités des machines automatiques; on peut bient entendu imaginer et mettre au point des variantes ou des compléments qui en améliorent l'efficacité. D'autre part, le calcul des fonctions de mérite (Feder[1] 1957) peut être combiné avec l'utilisation des dérivées partielles des aberrations.

Les progrès importants des méthodes de calcul ont conduit récemment à des résultats remarquables quant à la qualité des systèmes réalisés; l'obtention de ces résultats est d'ailleurs due en partie à une mise au point de nouveaux matériaux: on peut utiliser maintenant des verres très réfringents (dont l'indice est presque égal à 2 par exemple), qui permettent une correction plus facile des aberrations; il en résulte que dans de nombreux cas on s'approche des limites imposées par la diffraction et il est probable que des progrès dans les performances des récepteurs permettront bientôt de profiter pleinement des possibilités des optiques excellentes dont on pourra prochainement disposer.

Signalons à titre d'exemples quelques réalisations du domaine des objectifs de prises de vues.

α) Les champs angulaires attaignent maintenant des valeurs très importantes:

— 120° de champ total sans distorsion pour des objectifs de prise de vues aériennes destinées à la topographie,

[1] D. Feder: J. Opt. Soc. Amer. **47**, 902, 913 (1957).

— 110° de champ pour des objectifs de télévision et de cinéma (format 16 mm) de longueur focale $f=5{,}6$ mm et d'ouverture $f/1{,}8$.

β) Les ouvertures attaignent parfois $f/0{,}85$ pour des objectifs de prise de vues d'images radioscopiques sur vidicon (la longueur focale étant $f=50$ mm).

γ) La mise au point d'objectifs à focale variable (zooms) a conduit à des réalisations remarquables; le rapport des longueurs focales extrêmes atteint maintenant la valeur 10 dans un objectif ouvert à $f/2{,}2$ pour le cinéma de 16 mm. (Les longueurs focales extrêmes sont 12,5 et 125 mm.)

Ce ne sont là que quelques échantillons des réalisations actuelles, mais le rythme de création de systèmes nouveaux s'accélère d'année en année, en particulier pour faire face aux nouveaux problèmes posés à l'occasion des recherches spatiales ou de développements analogues.

Nota:

Cet article ayant été rédigé en novembre et décembre 1961, il faut noter que des progrès appréciables ont été obtenus depuis lors; les dérivées partielles paraxiales sont calculées directement par des formules finies; les dérivées partielles de la grandeur de l'image, des cosinus directeurs du vecteur $\partial\vec{u}/\partial p$, du chromatisme sont élaborées rigoureusement. La généralité du programme permet d'étudier les cas particuliers, tels que les systèmes afocaux et télécentriques. Un programme d'optimisation par la méthode des moindres carrés amortis a été enchaîné au programme principal pour lui donner toute son efficacité.

Theorie der geometrisch-optischen Bildfehler[1].

Von

Helmut Marx.

Einleitung.

Der Zusammenhang zwischen den Konstruktionsdaten eines optischen Systems und seinen Bildfehlern höherer Ordnung ist sehr verwickelt. Wir betrachten hier nur starre rotationssymmetrische Systeme brechender (oder spiegelnder) Flächen. Und wir betrachten nur den geometrisch-optischen Strahlenverlauf, sehen ab von der Beugung des Lichts. Auch von der Polarisation eines Lichtstrahls, von der Interferenz sich treffender Lichtstrahlen und von der Schwächung der Lichtströmung durch Absorption und Streuung in den durchströmten Medien und durch Reflexion an den brechenden Flächen (und unvollkommene Reflexion an den spiegelnden Flächen) sehen wir ab. Dann ist die Strömung der Lichtenergie im Bild allein bestimmt durch die aus dem Objekt kommende Lichtströmung und den geometrisch-optischen Verlauf der „Strahlen", längs denen die Lichtenergie strömt.

Bei der Berechnung optischer Systeme bemüht sich der optische Rechner, Linsen- (oder Spiegel-) Anordnungen zu finden, die eine gewünschte Abbildung vermitteln. In der Regel soll ein *endlich* ausgedehnter Objektbereich[2] durch *endlich* geöffnete Strahlenbündel in einer bestimmten, gewünschten Weise abgebildet werden; und meist soll mit Licht aus einem *endlichen* Wellenlängenbereich abgebildet werden. Der Objektbereich, den das System[3] soll abbilden können, ist gegebenenfalls (wie in der Photographie) räumlich ausgedehnt. Bekanntlich ist es nicht möglich, durch ein und dasselbe System *jeden* Punkt eines endlich ausgedehnten Raumgebiets durch ein endlich geöffnetes Strahlenbündel wieder in genau einen Punkt abzubilden (mit Ausnahme der Abbildung dieses Raumgebiets im Maßstab $n:n'$; n und n' sind die Brechungsindizes im Objektraum und im Bildraum); die Aufgabe, ein System zu berechnen, das mit endlicher Öffnung einen endlichen Objekt-*Raum streng punktförmig abbildet* (in einem von $n:n'$ verschiedenen Maßstab), ist prinzipiell unlösbar. Aber auch bei Beschränkung auf die Punkte einer endlichen Objekt-*Fläche* ist das Ideal, jeden Punkt dieser Fläche durch ein endlich geöffnetes Strahlenbündel in genau einen Punkt abzubilden, unerreichbar bei praktisch allen Aufgaben, denen der optische Rechner begegnet.

[1] Die Gleichungen dieses Beitrages sind nicht ziffernweise, sondern abschnittweise numeriert, so daß z.B. Gl. (3.47) die Gleichung 47 von Abschnitt III bedeutet; diese Gleichung steht in Ziff. 19. — Besondere Numerierungen tragen die Gleichungsgruppen 4A (Ziff. 22), 4B (Ziff. 24), IV (Ziff. 27), V (Ziff. 33), VII (Ziff. 46—47) und VIII (Ziff. 50—52). — Die Ziffern 12—14 in Abschnitt II, die Abschnitte III, IV und V, die Ziffern 37—39 und 46—47 in Abschnitt VI und VII, und der Abschnitt VIII stellen Originalarbeiten des Verfassers dar, die bei ihrem Eingang als Beitrag zu diesem Handbuch (Abschnitt II—VII im März 1963, Abschnitt VIII im Februar 1967) noch nicht veröffentlicht waren.

[2] Im Grenzfall einer unendlich entfernten Objekt-Fläche ist ein endlicher Bereich von g w gemeint; w ist der „Bildwinkel", den die von einem Objektpunkt kommenden Strahlen mit der Rotationsachse des optischen Systems bilden.

[3] Für „starres rotationssymmetrisches System brechender oder spiegelnder Flächen" sagen wir kürzer „(optisches) System"; für „brechende oder spiegelnde Flächen" sagen wir kürzer „Flächen".

Eine typische Aufgabe ist die: Ein bestimmter, kreisförmiger Objekt-Ebenen-Ausschnitt (*Feld* genannt) soll mit weißem Licht durch ein System aus brechenden Kugel- oder Plan-Flächen abgebildet werden auf eine Bild-Ebene, in einem bestimmten Maßstab und (meistens) unter Beachtung weiterer Bedingungen[1]. Dabei sollen die Linsenfassungen und die Blende des Systems solche Durchmesser haben, daß mindestens die Punkte der Feld-Mitte durch Strahlenbündel bestimmter Größe (*Öffnung* genannt) abgebildet werden. Schließlich sollen die unvermeidlichen Abweichungen von der idealen punktförmigen Abbildung (*Aberrationen* genannt) mehr oder weniger gut vernachlässigbar sein bei der Anwendung, für die das System vorgesehen ist. — Je nach den zur Verfügung stehenden Mitteln hat das System mehr oder weniger Freiheitsgrade; sollen nur Linsen mit Kugel- oder Plan-Flächen vorkommen, nicht mit asphärischen Flächen (die einen hohen Fertigungsaufwand erfordern), dann hat das System endlich viele Freiheitsgrade (*Konstruktionsdaten* genannt): die Radien r der brechenden Flächen (bei einer Planfläche ist $r=\infty$; bei einer Kugelfläche ist r positiv oder negativ, je nachdem ob das Licht von außen her oder von innen her auf die Kugelfläche trifft), die Abstände a der Flächen voneinander (längs der Rotationsachse des Systems gemessen), die Brechungsindizes $n=n(\lambda)$ der Medien zwischen den Flächen (λ ist die Wellenlänge des Lichts). In der Regel hat man als Medium Luft ($n=1$ normiert) oder Glas, bei dem die Funktion $n(\lambda)$ im Wellenlängenbereich des weißen Lichts je nach dem Anwendungsfall genügend genau dargestellt werden kann durch einen universellen Interpolationsansatz mit zwei oder mit drei oder mit vier von Glas zu Glas verschiedenen Koeffizienten. Ist k die Anzahl der brechenden Flächen, dann hat das System also rund $4k$ Freiheitsgrade bei der Abbildung mit weißem Licht und falls man $n(\lambda)$ genügend genau mit zwei Koeffizienten interpolieren kann. Zur vollständigen Beschreibung des Systems gehören außerdem noch Angaben über die Blende und andere Strahlenbegrenzungen (Linsenfassungen, Luken).

Für den optischen Rechner besteht das Problem: diese Konstruktionsdaten des Systems so zu wählen, daß die gewünschte Abbildung mit vernachlässigbaren Aberrationen erreicht wird. Der Zusammenhang zwischen den Konstruktionsdaten einerseits und den Aberrationen andererseits ist sehr verwickelt; und der optische Rechner hat es mit einer *einfachen*, wenn auch viel Rechenaufwand erfordernden Aufgabe *oder* einer sehr *schwierigen* Aufgabe zu tun, je nachdem ob er nur die Aberrationen verschiedener Strahlen bei *vorgegebenen* Konstruktionsdaten ausrechnen will, *oder* ob er die noch *unbekannten* Konstruktionsdaten des Systems berechnen will bei vorgegebenen Toleranzen für die Aberrationen. Für die letztere Aufgabe ist die Theorie heute noch nicht so weit entwickelt, daß der optische Rechner die gesuchten Konstruktionsdaten im wesentlichen durch Anwendung einiger Rechenschemata finden könnte. Zwar hat die Theorie dem optischen Rechner schon nützliche Rechenverfahren, strenge und Näherungs-Verfahren, zur Verfügung gestellt; diese allein reichen aber allzu oft nicht aus, und der optische Rechner ist auf seine Erfahrung, auf Probieren (trial and error) und auf sein Fingerspitzengefühl angewiesen. Allzu oft kommt beim optischen Rechnen der Fall vor, daß im Laufe der Berechnung eines Systems bestimmte Aberrationen noch zu groß sind und daß man *theoretisch nicht überblickt*, ob unter den gegebenen Umständen diese Aberrationen unvermeidlich sind, oder ob sie durch mäßige Änderungen der Konstruktionsdaten verringert werden könnten, oder ob ein völlig neuer Ansatz des Systems zum Ziel führt.

[1] Zum Beispiel kann die Entfernung zwischen Objekt und Bild vorgeschrieben sein, oder eine Mindestentfernung des Bilds von der letzten brechenden Fläche, oder der für das System zur Verfügung stehende Platz; oder der Aufwand zur Herstellung ist vorgeschrieben.

Man benötigt *überschaubare* Darstellungen des Zusammenhangs zwischen den Konstruktionsdaten und den Aberrationen einschließlich der Aberrationen höherer Ordnung. Auch beim Korrigieren mit Hilfe elektronischer Rechenmaschinen besteht das Bedürfnis nach solchen Darstellungen.

Nach vorbereitenden Rechnungen und Entwicklungen in den Abschnitten II bis V werden wir in Abschnitt V und Abschnitt VIII Darstellungen der Aberrationen höherer Ordnung finden, die deren verwickelten Zusammenhang mit den Konstruktionsdaten leichter diskutieren und besser überschauen lassen als etwa Darstellungen, die von Potenzreihenentwicklungen ausgehen.

Will man nur zu vorgegebenen Konstruktionsdaten die (bei den Potenzreihenentwicklungen auftretenden) Bildfehler-Koeffizienten höherer Ordnung berechnen, dann kann dies nach einem einfachen, für alle Bildfehler-Ordnungen gleichen Rekursions-Schema (Abschnitt VII) geschehen. Nach diesem Schema lassen sich auch die irreduziblen, aus einem Eikonal (Abschnitt VI) hergeleiteten Bildfehler-Koeffizienten beliebiger Ordnung berechnen, und die bei deren Entwicklung nach einer geeigneten Farbvariablen (Abschnitt I) auftretenden Koeffizienten.

I. Konstruktionsdaten eines rotationssymmetrischen optischen Systems.

1. Abstände. Die Rotationsachse des optischen Systems (kurz „optische Achse" oder „Achse" genannt) machen wir zur z-Achse eines rechtshändigen kartesischen Koordinatensystems x, y, z. In Zeichnungen und in der Vorstellung lassen wir die z-Achse von links nach rechts, die y-Achse von unten nach oben, und die x-Achse von vorn nach hinten laufen. Jede brechende (oder spiegelnde) Fläche hat *einen* Schnittpunkt mit der optischen Achse, den wir den *Scheitel* der Fläche nennen; oder die Fläche hat *keinen* Schnittpunkt mit der Achse. Im letzteren Fall denken wir uns für die Rechnung die Fläche ergänzt zu einer Fläche mit Scheitel; einen solchen nur gedachten Scheitel nennen wir „virtuell", im Gegensatz zu einem „reell" vorhandenen. Eine Fläche, die zwei Schnittpunkte mit der Achse hat, zerlegen wir in zwei Flächen mit je einem reellen Scheitel. So hat, jedenfalls für die Rechnung, jede brechende (oder spiegelnde) Fläche genau einen Scheitel.

Die Flächen numerieren wir in der Reihenfolge, in der sie von den Lichtstrahlen getroffen werden. Bei Systemen mit Spiegeln erhält dann eine Fläche, die von einem Lichtstrahl mehrmals getroffen wird, mehrere Nummern. Vorerst nehmen wir ein einziges, für das ganze System gemeinsames kartesisches Koordinatensystem an (später geben wir jeder Fläche ein eigenes Koordinatensystem, mit von Fläche zu Fläche verschiedenem Koordinatenanfangspunkt auf der optischen Achse). Die z-Koordinate des Scheitels S_v der Fläche Nr. v bezeichnen wir mit $z(S_v)$. Der ersten brechenden (oder spiegelnden) Fläche geben wir die Nr. 1 (später wird davor auch noch eine Fläche Nr. 0 eingeführt werden), mit k bezeichnen wir die Nummer der letzten Fläche. Wir setzen fest, daß die vom Objekt auf die erste Fläche zulaufenden Lichtstrahlen in der Zeichnung von links nach rechts laufen, daß ihre Richtung also eine positive z-Komponente hat. Dann ist bei den üblichen Linsen-Systemen (keine Spiegel, nur reelle Scheitel)

$$z(S_1) \leqq z(S_2) \leqq z(S_3) \leqq \cdots \leqq z(S_k), \tag{1.1}$$

und der reelle Bildraum befindet sich rechts von der Fläche Nr. k, der virtuelle links davon.

Den *Abstand* der Scheitel zweier aufeinanderfolgender Flächen Nr. v und Nr. $(v+1)$ bezeichnen wir mit a'_v oder mit a_{v+1}; und wir zählen ihn positiv oder negativ, je nachdem ob S_{v+1} rechts oder links von S_v liegt:

$$a'_v \equiv a_{v+1} \equiv z(S_{v+1}) - z(S_v). \tag{1.2}$$

Bei den üblichen Linsen-Systemen ist nach (1.1):

$$a'_v \geqq 0 \quad (v = 1 \ldots k-1). \tag{1.3}$$

2. Radien. Eine brechende (oder spiegelnde) *Kugelfläche* ist, genauer gesagt, ein zur optischen Achse rotationssymmetrischer Ausschnitt einer Kugel-Oberfläche: bei reell vorhandenem Scheitel eine Kugelkappe; bei nicht reell vorhandenem Scheitel eine Kugelzone, die wir uns für die Rechnung ergänzt denken zu einer Kugelkappe. Wegen der Rotationssymmetrie liegt der Kugel-Mittelpunkt, wie der Scheitel, auf der optischen Achse. Liegt nun von dem reellen oder virtuellen Scheitel S der Kugelfläche (Kugelkappe) aus gesehen der Kugel-Mittelpunkt C rechts, dann zählen wir den Kugel-*Radius* r positiv; liegt C links von S, dann zählen wir den Kugel-Radius r negativ:

$$r_v \equiv S_v C_v. \tag{1.4}$$

Hierbei und im folgenden bezeichnen wir, wenn A und B zwei Punkte sind, die auf der optischen Achse liegen, mit AB die zwischen A und B liegende Strecke *mit* Vorzeichen: AB ist positiv oder negativ, je nachdem ob B rechts oder links von A liegt:

$$AB \equiv \overline{AB} \cdot \operatorname{sign}[z(B) - z(A)]. \tag{1.5}$$

Mit $\overline{AB}$ bezeichnen wir die *ohne* Vorzeichen gemeinte Strecke zwischen A und B ($\overline{AB} \geqq 0$). Den von A nach B führenden Strecken-Vektor werden wir mit $\overrightarrow{AB}$ bezeichnen.

Eine *Planfläche* ist, genauer gesagt, ein zur optischen Achse rotationssymmetrischer Ausschnitt einer Ebene. Eine Planfläche (Nr. w) können wir auffassen als eine Kugelfläche mit dem Scheitel im Endlichen und dem Mittelpunkt im Unendlichen, also mit *unendlichem Radius*:

$$r_w = \infty. \tag{1.6}$$

Wenn wir im folgenden von „Kugelflächen" sprechen, dann meinen wir in der Regel Kugelflächen einschließlich Planflächen.

Rotationssymmetrische Flächen, die keine Kugelflächen sind, heißen *asphärische Flächen.* Eine asphärische Fläche des Systems (mit der Nr. v) hat entweder einen reellen Scheitel, oder sie ist ergänzt zu denken, so daß sie einen virtuellen Scheitel hat. Das Flächenelement beim Scheitel S_v hat *einen* Krümmungsmittelpunkt C_v (kurz *Mittelpunkt* genannt), der wie S_v auf der optischen Achse liegt. Den zugehörigen Krümmungsradius (mit Vorzeichen)

$$r_v \equiv S_v C_v \tag{1.4a}$$

nennen wir auch kurz den *Radius* der asphärischen Fläche.

3. Asphärische Koeffizienten. Mit den in Ziff. 1 eingeführten kartesischen Koordinaten kann die Flächengleichung einer asphärischen Fläche, ohne große Beschränkung der Allgemeinheit, in der Form

$$z - z(S) = \sum_{i=0}^{\infty} \bar{\alpha}^{(i)} \cdot (x^2 + y^2)^{i+1} \tag{1.7}$$

geschrieben werden. Dieser Ansatz schließt auch die Kugelflächen mit ein. Abgesehen von dem Grenzfall einer Planfläche (für die alle $\bar{\alpha}^{(i)}=0$ sind), stellt (1.7) aber erst mit unendlich vielen Gliedern eine Kugelfläche streng dar (und auch diese nur im Bereich der Kugelkappe einer Halbkugel). Beim Paraboloid dagegen hat (1.7) nur ein Glied:

$$z-z(S)=\bar{\alpha}^{(0)}\cdot(x^2+y^2). \tag{1.8}$$

Ein anderer Ansatz, der die Kugelfläche (und zwar im ganzen Bereich der Oberfläche der Vollkugel) mit einem einzigen Glied streng darstellt, verwendet anstelle von $\sqrt{x^2+y^2}$, der „Höhe" des Flächen-Punkts $P(x, y, z)$ über der optischen Achse, als unabhängige Variable die *Sehne* zwischen dem Scheitel S und dem Flächen-Punkt P. Mit der Abkürzung

$$p\equiv z-z(S) \tag{1.9}$$

für die „Pfeilhöhe" des Flächen-Punkts P ist diese Sehne

$$s=\sqrt{x^2+y^2+p^2}, \tag{1.10}$$

und unser Ansatz lautet[1]:

$$p=\sum_{i=0}^{\infty}\alpha^{(i)}\cdot(\tfrac{1}{2}s^2)^{i+1}. \tag{1.11}$$

$\alpha^{(i)}$ nennen wir den i-ten *asphärischen Koeffizienten*. Der Ansatz (1.11) mit nur dem allerersten Glied, dem Glied mit dem *nullten* asphärischen Koeffizienten $\alpha^{(0)}$:

$$p=\alpha^{(0)}\cdot\tfrac{1}{2}s^2 \tag{1.12}$$

stellt die *Kugel*fläche mit dem Radius

$$r=\frac{1}{\alpha^{(0)}} \tag{1.13}$$

dar, denn aus (1.12) und (1.10) folgt $x^2+y^2+(p-1/\alpha^{(0)})^2=(1/\alpha^{(0)})^2$. Dann und nur dann, wenn außer dem nullten *mindestens ein weiterer* asphärischer Koeffizient von null verschieden ist, hat man eine *asphärische* Fläche.

Die Beziehung (1.13) gilt, wie man leicht einsehen kann, auch für den Radius $r\equiv SC$ einer *asphärischen* Fläche:

$$r\equiv SC=\frac{1}{\alpha^{(0)}}, \tag{1.13a}$$

der nullte asphärische Koeffizient $\alpha^{(0)}$ ist gleich der Scheitel-Krümmung.

4. Brechungsindizes. Wir betrachten zunächst *eine* Wellenlänge des Lichts, etwa die der grünen Hg-Linie e (546,1 mμ). Für Licht dieser Wellenlänge hat das Medium zwischen der Fläche Nr. v und der Fläche Nr. $v+1$ einen Brechungsindex, den wir mit n'_v oder mit n_{v+1} bezeichnen. Dieser Brechungsindex hat bei Systemen ohne Spiegel positives Vorzeichen. Bei Systemen mit Spiegeln läßt sich, wie wir gleich sehen werden, das Reflexionsgesetz formal wie ein Brechungsgesetz schreiben, wenn man einem Teil der Brechungsindizes das negative Vorzeichen gibt (aber den Betrag läßt, so daß allemal $|n'_v|$ der normale, positiv gezählte, Brechungsindex des Mediums zwischen der Fläche Nr. v und der Fläche Nr. $v+1$ ist): Wir zählen n'_v positiv oder negativ, je nachdem ob die Strahlen auf dem Weg von der v-ten zur $v+1$-ten Fläche eine gerade Anzahl (0; 2; ...) oder eine ungerade Anzahl (1; 3; ...) von Spiegelungen hinter sich haben. So erhalten bei einer *brechenden* Fläche Nr. v n_v und n'_v gleiches Vorzeichen (so daß $n'_v/n_v=$

[1] Diesen Ansatz brachte der Verfasser in einem Vortrag auf der Tagung der Deutschen Gesellschaft für angewandte Optik, Wetzlar 1950.

$|n'_v|/|n_v|$ ist), während bei einer *spiegelnden* Fläche Nr. w n_w und n'_w *verschiedenes* Vorzeichen erhalten (bei gleichem Betrag, so daß $n'_w/n_w = -1$ ist). Das *Brechungsgesetz* mit den normalen, vorzeichenlosen Brechungsindizes $|n_v|$ und $|n'_v|$:

$$|n'_v| \cdot \sin \iota'_v = |n_v| \cdot \sin \iota_v \tag{1.14}$$

gilt also (wegen $n'_v/n_v = |n'_v|/|n_v|$) auch mit den vorzeichenbehafteten Brechungsindizes n_v und n'_v:

$$n'_v \cdot \sin \iota'_v = n_v \cdot \sin \iota_v; \tag{1.15}$$

ι ist der Einfallswinkel, ι' ist der Ausfallswinkel. Und das mit vorzeichenbehafteten ι und ι' geschriebene *Reflexionsgesetz*:

$$\iota'_w = -\iota_w \tag{1.16}$$

läßt sich (wegen $n'_w/n_w = -1$) formal wie das Brechungsgesetz (1.15) schreiben:

$$n'_w \cdot \sin \iota'_w = n_w \cdot \sin \iota_w. \tag{1.17}$$

5. Die Dispersion der optischen Gläser. Wir interessieren uns nun für die Änderung des Brechungsindex n mit der Wellenlänge λ (*Dispersion* genannt). Im Wellenlängenbereich des sichtbaren Lichts sind die Funktionen $n(\lambda)$ der verschiedenen *optischen Gläser* einander sehr ähnlich.

Sehr genau läßt sich im sichtbaren Bereich $n(\lambda)$ interpolieren durch einen Ausdruck, der *vier* von Glas zu Glas verschiedene Koeffizienten $k^{(0)}$, $k^{(1)}$, $k^{(2)}$, $k^{(3)}$ *linear* enthält (f_1, f_2, f_3 sind universelle Funktionen der Wellenlänge):

$$n = k^{(0)} + k^{(1)} f_1(\lambda) + k^{(2)} f_2(\lambda) + k^{(3)} f_3(\lambda); \tag{1.18}$$

das haben unabhängig voneinander M. HERZBERGER[1] und W. GEFFCKEN[2] mit recht unterschiedlichen Tripeln (f_1, f_2, f_3), die wir hier nicht diskutieren wollen, gezeigt. — Andere Medien als Glas und Luft ($n=1$) wollen wir nicht betrachten.

Nach (1.18) ist bei einem optischen Glas im ganzen sichtbaren Bereich $n(\lambda)$ bestimmt durch die Werte von n bei *vier* über den sichtbaren Bereich gleichmäßig verteilten Wellenlängen. Auch beim optischen Rechnen ist es üblich und meistens ausreichend, zur Beurteilung der Abbildung mit weißem Licht *vier* Wellenlängen herauszugreifen und nur Lichtstrahlen dieser vier Farben durchzurechnen. An Stelle der früher benutzten vier Wellenlängen der Linien C(H), D(Na) [oder d(He)], F(H), G'(H) bürgern sich bei der Berechnung photographischer Objektive in der letzten Zeit immer mehr die vier Wellenlängen der Linien c(Cd), e(Hg), f(Cd), g(Hg) ein[3]:

Linie (Element)	Wellenlänge	
c (Cd)	643,85 mμ	(1.19)
e (Hg)	546,07 mμ	
f (Cd)	479,99 mμ	
g (Hg)	435,83 mμ	

Die Brechungsindizes bei diesen vier Wellenlängen bezeichnen wir mit n_c, n_e, n_f, n_g. Nach (1.18) ist (mit drei anderen universellen Funktionen der Wellenlänge)

$$n - n_e = (n_c - n_e)\,\bar{f}_1(\lambda) + (n_f - n_e)\,\bar{f}_2(\lambda) + (n_g - n_e)\,\bar{f}_3(\lambda)$$

[1] M. HERZBERGER: Modern Geometrical Optics. New York 1958.
[2] W. GEFFCKEN: Vortrag auf der Tagung der Deutschen Gesellschaft für angewandte Optik, Mainz 1957 (abgedruckt in „Beiträge zur Angewandten Glasforschung", S. 255—268, Stuttgart 1959).
[3] Nach einem Vorschlag von F. WEIDERT, Z. techn. Phys. **7**, 304—309 (1926).

oder mit einer ersten, einer zweiten und einer dritten n-Differenz als Koeffizienten:

$$\left.\begin{aligned} n - n_e = \tfrac{1}{2}(n_f - n_c)\cdot g_1(\lambda) + \tfrac{1}{2}[(n_g - n_e) - (n_f - n_c)]\cdot g_2(\lambda) + \\ + \{[(n_g - n_f) - (n_f - n_e)] - [(n_f - n_e) - (n_e - n_c)]\}\cdot g_3(\lambda). \end{aligned}\right\} \quad (1.20)$$

Die dritte n-Differenz

$$\{[(n_g - n_f) - (n_f - n_e)] - [(n_f - n_e) - (n_e - n_c)]\} \equiv \Delta^3 n \quad (1.21)$$

und die zweite n-Differenz

$$\left.\begin{aligned} \tfrac{1}{2}\{[(n_g - n_f) - (n_f - n_e)] + [(n_f - n_e) - (n_e - n_c)]\} \\ = \tfrac{1}{2}[(n_g - n_e) - (n_f - n_c)] \equiv \Delta^2 n \end{aligned}\right\} \quad (1.22)$$

sind, wie wir sehen werden, klein gegen die erste n-Differenz

$$\tfrac{1}{2}[(n_f - n_e) + (n_e - n_c)] = \tfrac{1}{2}(n_f - n_c) \equiv \Delta^1 n. \quad (1.23)$$

Die bekannte Hartmannsche Interpolationsformel

$$n \approx l^{(0)} + \frac{l^{(1)}}{\lambda - l} \quad (1.24)$$

enthält *drei* von Glas zu Glas verschiedene Koeffizienten $l^{(0)}$, $l^{(1)}$, l. Der Ausdruck (1.24) für $n(\lambda)$ ist eine gebrochen lineare Funktion von λ (und damit auch der Frequenz $\nu = c/\lambda$ des Lichts) und insofern sehr einfach; er ist aber nicht linear in allen drei Koeffizienten.

Nach (1.24) ist[1]

$$n - 1 \approx (n_e - 1)\cdot\left[1 + k\cdot\frac{\lambda_e - \lambda}{\lambda - l}\right] \quad (1.25)$$

(mit n_e, k, l als Koeffizienten). Der Koeffizient l liegt bei den optischen Gläsern zwischen rund 140 und 240 mμ. Wie gleichmäßig (hinsichtlich der Dispersion) die vier Linien c, e, f, g über das Spektrum verteilt sind, sieht man daran, daß bei einem der vorkommenden l-Werte, nämlich bei

$$l_* = 180{,}3 \text{ m}\mu, \quad (1.26)$$

die Brechungsindizes für die vier Linien c, e, f, g praktisch äquidistant liegen: nach (1.25) ist, wenn man $l = l_*$ setzt,

$$n_g - n_f \approx n_f - n_e \approx n_e - n_c; \quad (1.27)$$

denn nach (1.19) und (1.26) ist

$$\left.\begin{aligned} (\lambda_e - \lambda_c)/(\lambda_c - l_*) &= -0{,}2109 = -1\cdot 0{,}2157 + 0{,}0048; \\ (\lambda_e - \lambda_e)/(\lambda_e - l_*) &= \quad 0 \quad = \quad 0; \\ (\lambda_e - \lambda_f)/(\lambda_f - l_*) &= \quad 0{,}2205 = \quad 1\cdot 0{,}2157 + 0{,}0048; \\ (\lambda_e - \lambda_g)/(\lambda_g - l_*) &= \quad 0{,}4314 = \quad 2\cdot 0{,}2175. \end{aligned}\right\} \quad (1.28)$$

Ein Glas mit der Eigenschaft (1.27) ist z.B. das Schott-Glas BaK4, bei dem $n_e = 1{,}571\,25$ und $n_e - n_c = 0{,}005\,01$; $n_f - n_e = 0{,}005\,23$; $n_g - n_f = 0{,}005\,01$ ist.

Hierdurch wird es nahegelegt, zur Beschreibung der Dispersion die Größe

$$w \equiv \frac{\lambda_e - \lambda}{\lambda - l_*} \quad \text{mit} \quad l_* = 180{,}3 \text{ m}\mu \quad (1.29)$$

[1] In dieser Ziff. 5 nehmen wir den Brechungsindex n *ohne* Vorzeichen.

anstelle der Wellenlänge λ als Maß für die (Spektral-) Farbe zu nehmen[1]; dieses Maß ist dem Dispersionsverlauf besser angepaßt, und es macht die Farben c, e, f, g praktisch äquidistant, wie wir an (1.28) gesehen haben:

$$\left.\begin{array}{ll} \text{Farbe} & w \\ c & -0{,}2157+0{,}0048 \\ e & 0 \\ f & 0{,}2157+0{,}0048 \\ g & 0{,}4314 \end{array}\right\} \tag{1.30}$$

Schreiben wir die Hartmannsche Interpolationsformel (1.25) auf w anstelle von λ um, dann erhalten wir einen auch in w gebrochen linearen Ausdruck für $n-1$:

$$n-1 \approx (n_e-1)\cdot\left[1+\frac{\alpha w}{1+\beta w}\right] \tag{1.31}$$

(mit n_e, α, β als Koeffizienten). Bei optischen Gläsern liegt der Koeffizient β zwischen rund $-0{,}2$ und $+0{,}1$ [nach (1.31), (1.29) und (1.25) ist $\beta=(l-l_*)/(l-\lambda_e)$]. Das im Nenner des Ausdrucks (1.31) neben der 1 stehende βw ist also klein gegen 1:

$$\left.\begin{array}{l} \beta w_c = +0{,}042\cdots-0{,}021 \\ \beta w_e = \quad 0 \\ \beta w_f = -0{,}044\cdots+0{,}022 \\ \beta w_g = -0{,}086\cdots+0{,}043 \end{array}\right\} \quad \text{für}\quad \beta=-0{,}20\cdots+0{,}10. \tag{1.32}$$

Zur Unterscheidung von dem Brechungsindex n, den das Glas wirklich hat, bezeichnen wir den interpolierten Wert mit $\bar{n}$. Nach (1.31) ist

$$n-1 \approx \bar{n}-1; \tag{1.33a}$$

$$\bar{n}-1 = (n_e-1)\cdot\left[1+\frac{\alpha w}{1+\beta w}\right]. \tag{1.33b}$$

α und β hängen in einfacher Weise zusammen mit zwei beim optischen Rechnen gebräuchlichen Kombination aus n_c, n_e, n_f, n_g, nämlich mit der „Abbeschen Zahl"

$$\nu \equiv \frac{n_e-1}{n_f-n_c} \tag{1.34}$$

und mit der „relativen Teildispersion"

$$\Theta \equiv \frac{n_g-n_e}{n_f-n_c}. \tag{1.35}$$

Und zwar folgt, wie wir gleich nachrechnen, aus (1.33b) und aus

$$-0{,}2<\beta<+0{,}1\,, \tag{1.36}$$

daß

$$\text{auf } 4^0/_{00} \text{ genau:}\qquad \frac{1}{0{,}4314\alpha} = \bar{\nu}\left[\equiv\frac{n_e-1}{\bar{n}_f-\bar{n}_c}\right] \tag{1.37}$$

und

$$\text{auf } 4^0/_{00} \text{ genau:}\qquad \frac{1}{1+0{,}4314\beta} = \bar{\Theta}\left[\equiv\frac{\bar{n}_g-n_e}{\bar{n}_f-\bar{n}_c}\right] \tag{1.38}$$

ist: Nach (1.33b) und (1.30) ist, wenn wir δ für 0,2157 und ε für 0,0048 schreiben, $\frac{\bar{n}_f-\bar{n}_c}{(n_e-1)\,\alpha} = \frac{\delta+\varepsilon}{1+(\delta+\varepsilon)\beta} + \frac{\delta-\varepsilon}{1-(\delta-\varepsilon)\beta} = \frac{2\delta}{1-(\delta\beta)^2} - \frac{4(\delta\beta)\,\varepsilon}{[1-(\delta\beta)^2]^2} + O(\varepsilon^2) = 2\delta$ mit

[1] Ebenso geht H. A. BUCHDAHL[2] vor.

[2] H. A. BUCHDAHL: Optical Aberration Coefficients, London 1954.

einem maximalen Fehler von $4^0/_{00}$, den man im β-Bereich (1.36) bei $\beta = -0{,}2$ hat; also ist $\frac{1}{2\delta\cdot\alpha} = \frac{n_e - 1}{\bar{n}_f - \bar{n}_c}$ auf $4^0/_{00}$ genau. Nach (1.33b) und (1.30) ist ferner $\frac{\bar{n}_g - n_e}{(n_e - 1)\alpha} = \frac{2\delta}{1 + 2\delta\beta}$, folglich $\frac{1}{1 + 2\delta\beta} = \frac{\bar{n}_g - n_e}{\bar{n}_f - \bar{n}_c}$ auf $4^0/_{00}$ genau.

Wir nehmen an, daß die Koeffizienten α und β so gewählt sind, daß die Interpolationsformel die Werte der Abbeschen Zahl ν und der relativen Teildispersion Θ gut wiedergibt. Dann ist also $\nu \approx \frac{1}{0{,}4314\,\alpha}$ und $(\Theta - 1) \approx \frac{1}{1 + 0{,}4314\beta} - 1$, und aus letzterem folgt wegen (1.36), daß

$$|\Theta - 1| < 0{,}1 \tag{1.39}$$

ist. Nach (1.35) und (1.22) ist $(\Theta - 1) = \Delta^2 n / \Delta^1 n$, also ist nach (1.39): $|\Delta^2 n| \ll \Delta^1 n$, wie bei Gl. (1.22) behauptet wurde.

Die bei Gl. (1.21) stehende Behauptung, daß $|\Delta^3 n| \ll \Delta^1 n$ ist, bedeutet, weil bei der Interpolation (1.33) $|n - \bar{n}| \ll \Delta^1 n$ ist, daß auch $|\Delta^3 \bar{n}| \ll \Delta^1 \bar{n}$ ist, was wir an (1.33b) nachprüfen können. Danach ist

$$\frac{\Delta^3 \bar{n}}{(n_e - 1)\alpha} = \delta\left(\frac{2}{1 + 2\delta\beta} + \frac{-3}{1 + \delta\beta} + \frac{1}{1 - \delta\beta}\right) + \varepsilon\left(\frac{-3}{(1 + \delta\beta)^2} - \frac{1}{(1 - \delta\beta)^2}\right) + O(\varepsilon^2),$$

also

$$\frac{\Delta^3 \bar{n}}{\Delta^1 \bar{n}} \approx \frac{\Delta^3 \bar{n}}{(n_e - 1)\,\delta\alpha} \approx -4\frac{\varepsilon}{\delta} = -0{,}09.$$

Hiermit stimmen die Messungen überein: bei den optischen Gläsern des Schott-Katalogs ist

$$\left[\frac{\Delta^3 n}{\Delta^1 n} \equiv\right] \frac{[(n_g - n_f) - (n_f - n_e)] - [(n_f - n_e) - (n_e - n_c)]}{\frac{1}{2}(n_f - n_c)} = -0{,}09 \pm 0{,}02. \tag{1.40}$$

In (1.40) soll „$= -0{,}09 \pm 0{,}02$" heißen: „hat bei den verschiedenen Gläsern Werte zwischen $-0{,}11$ und $-0{,}07$".

Mit „den optischen Gläsern des Schott-Katalogs" meinen wir die folgenden 226 Gläser: die Gläser des 1956 erschienenen Katalogs Nr. 350 der Jenaer Glaswerke Schott & Gen., Mainz; *einschließlich* der Gläser der ersten Nachtragsliste (Nr. 366) und der im September 1960 erschienenen zweiten Nachtragsliste (Nr. 371) zum Katalog Nr. 350; jedoch *ausschließlich* der folgenden drei Gläser: FKS1 ($\nu = 81{,}6$), SFS1 ($\nu = 21{,}2$), SFS6 ($\nu = 20{,}2$). Bei allen übrigen 226 Gläsern des Katalogs und der beiden Nachtragslisten liegt ν im Bereich

$$70 > \nu > 25. \tag{1.41}$$

Erfahrungsgemäß besteht ein enger Zusammenhang zwischen den verschiedenen relativen Teildispersionen $\left(\text{z.B. zwischen } \frac{n_g - n_f}{n_f - n_c} \text{ und } \frac{n_e - n_c}{n_f - n_c}\right)$, und ein recht enger Zusammenhang zwischen der relativen Teildispersion $\Theta \equiv \frac{n_g - n_e}{n_f - n_c}$ (z.B.) und der Abbeschen Zahl $\nu \equiv \frac{n_e - 1}{n_f - n_c}$: Bei den Gläsern des Schott-Katalogs ist, wie zwei Auftragungen auf Grund der Katalogwerte zeigen,

$$\frac{n_g - n_f}{n_f - n_c} = -2{,}5000\,\frac{n_e - n_c}{n_f - n_c} + 1{,}711 + \varepsilon_1; \qquad -0{,}007 \leqq \varepsilon_1 \leqq +0{,}007 \tag{1.42}$$

und

$$\frac{n_c - n_e}{n_e - 1} = -0{,}8416\,\frac{n_f - n_e}{n_e - 1} - 0{,}00104 + \varepsilon_2; \quad -0{,}00014 \leqq \varepsilon_2 \leqq +0{,}00014. \tag{1.43}$$

Aus (1.42) folgt durch einfache Umrechnung:

$$\left.\begin{aligned}\left[\frac{\Delta^3 n}{\Delta^1 n}\equiv\right] &\frac{[(n_g-n_f)-(n_f-n_e)]-[(n_f-n_e)-(n_e-n_c)]}{\frac{1}{2}(n_f-n_c)}\\ &=-0{,}286\,(\Theta-1)-0{,}089+2{,}3\,\varepsilon_1.\end{aligned}\right\}\qquad(1.44)$$

Aus (1.43) und (1.42) folgt

$$\left[\frac{\Delta^2 n}{\Delta^1 n}\equiv\right](\Theta-1)=(0{,}111_5+\varepsilon_1)-(0{,}0019_8-1{,}9\,\varepsilon_2)\,\nu. \qquad (1.45)$$

Ferner kommen, wie ein Blick auf den Übersichtsplan des Schott-Katalogs zeigt, bei den verschiedenen n_e-Werten recht unterschiedliche Wertebereiche von ν, und damit auch von $\Delta^1 n \equiv (n_f - n_c)/2$, vor. Die Gläser des Schott-Katalogs liegen in dem Streifen

$$0{,}030\,(n_e-1{,}29)<(n_f-n_c)<0{,}071\,(n_e-1{,}35), \qquad (1.46)$$

ohne überall die beiden Ränder dieses Streifens zu erreichen; bei diesen Gläsern liegen die Werte von $\nu\equiv(n_e-1)/(n_f-n_c)$, wie schon erwähnt, in dem Bereich (1.41); die n_e-Werte dieser Gläser liegen im Bereich

$$1{,}44<n_e<1{,}91. \qquad (1.47)$$

6. Dispersions-Koeffizienten. Nach der Interpolationsformel (1.31),

$$n-1\approx(n_e-1)\cdot[1+\alpha w/(1+\beta w)],$$

ist auch

$$n-1\approx(n_e-1)\cdot[1+\alpha w\cdot(1-\beta w+\beta^2 w^2)], \qquad (1.48)$$

weil $|(n_e-1)\cdot\alpha w\cdot\beta^3 w^3|<0{,}3\cdot 10^{-4}$ ist; denn nach (1.37), (1.41), (1.47) und (1.30) ist $|(n_e-1)\alpha w|<0{,}04$, und nach (1.32) ist $|\beta^3 w^3|<0{,}0007$. Der Interpolationsausdruck (1.48), ein Polynom dritten Grades in w, enthält nur drei Koeffizienten (n_e,α,β). Die beste Interpolation durch ein Polynom dritten Grades erreicht man mit *vier* Koeffizienten[1]:

$$n-1\approx\hat{n}-1; \qquad (1.49\text{a})$$

$$\hat{n}-1=(n_e-1)\cdot\left[1+\delta^{(1)}\cdot\left(\frac{w}{w_g}\right)+\delta^{(2)}\cdot\left(\frac{w}{w_g}\right)^2+\delta^{(3)}\cdot\left(\frac{w}{w_g}\right)^3\right]. \qquad (1.49\text{b})$$

Dieser Ansatz ist nahezu von der oben erwähnten Form (1.20), denn wir werden sehen, daß im wesentlichen $\delta^{(1)}$ durch $\Delta^1 n$, $\delta^{(2)}$ durch $\Delta^2 n$, und $\delta^{(3)}$ durch $\Delta^3 n$ gegeben ist. Wir lassen die Frage offen, wie genau das Polynom (1.49b) den Brechungsindex n im ganzen sichtbaren Gebiet interpoliert; uns werden nur die vier Farben c, e, f, g interessieren, für die (1.49b) exakt angepaßt werden kann.

Der Brechungsindex n könnte als Funktion der Farbvariablen w in Strenge durch die Potenzreihe[1]

$$n-1=(n_e-1)\cdot\left[1+\sum_{i=1}^{\infty}\delta^{(i)}\cdot\left(\frac{w}{w_g}\right)^i\right] \qquad (1.50)$$

dargestellt werden. Den in der Potenzreihe (1.50) auftretenden Koeffizienten $\delta^{(i)}$ nennen wir den i-ten *Dispersions-Koeffizienten.*

Wir interpolieren den Brechungsindex n durch das Polynom (1.49b) mit n_e, $\delta^{(1)}$, $\delta^{(2)}$, $\delta^{(3)}$ als Koeffizienten. Wir setzen $\delta^{(1)}$, $\delta^{(2)}$, $\delta^{(3)}$ aber nicht den gleichlautenden Koeffizienten der Potenzreihe (1.50) exakt gleich; sondern $\delta^{(1)}$, $\delta^{(2)}$, $\delta^{(3)}$

---

[1] Ebenso setzt H. A. BUCHDAHL[2] $n(\lambda)$ an.

[2] Siehe S. 75, Anm.[1] und [2].

sollen so gewählt sein, daß die Interpolationsformel (1.49b) die Brechungsindizes bei den vier uns interessierenden Farben c, e, f, g exakt wiedergibt:

$$\hat{n}_c = n_c; \quad \hat{n}_e = n_e; \quad \hat{n}_f = n_f; \quad \hat{n}_g = n_g. \tag{1.51}$$

Auch die so bestimmten Koeffizienten $\delta^{(1)}$, $\delta^{(2)}$, $\delta^{(3)}$ des Polynoms (1.49b) nennen wir *Dispersions-Koeffizienten.*

Die gemäß (1.23), (1.22), (1.21) aus n_c, n_e, n_f, n_g gebildeten Differenzen $\Delta^1 n$, $\Delta^2 n$, $\Delta^3 n$, berechnen sich [nach (1.49b), (1.30) und (1.51)] folgendermaßen aus $(n_e - 1)$, $\delta^{(1)}$, $\delta^{(2)}$, $\delta^{(3)}$:

$$\left.\begin{aligned} \frac{\Delta^1 n}{n_e - 1} &= 0{,}5000\,\delta^{(1)} + 0{,}0111\,\delta^{(2)} + 0{,}1252\,\delta^{(3)};\\ \frac{\Delta^2 n}{n_e - 1} &= 0{,}4889\,\delta^{(2)} + 0{,}3748\,\delta^{(3)};\\ \frac{\Delta^3 n}{n_e - 1} &= -0{,}0444\,\delta^{(1)} - 0{,}0226\,\delta^{(2)} + 0{,}7164\,\delta^{(3)}. \end{aligned}\right\} \tag{1.52}$$

Damit können die Zusammenhänge (1.44) und (1.45) auf die Koeffizienten $\delta^{(1)}$, $\delta^{(2)}$, $\delta^{(3)}$ umgeschrieben werden. Aus (1.44) wird:

$$\delta^{(3)} = -(0{,}141_6 + 0{,}018_3\,\varepsilon_1)\,\delta^{(2)} + 1{,}3_8\,\varepsilon_1 \cdot \delta^{(1)},$$

und aus (1.45):

$$0{,}975_3\,\delta^{(2)} + 0{,}72_2\,\delta^{(3)} = (0{,}111_5 + \varepsilon_1)\,\delta^{(1)} - (0{,}0019_8 - 1{,}9\,\varepsilon_2).$$

Aus diesen beiden Gleichungen folgt

$$\delta^{(3)} = -0{,}142\,\delta^{(2)} + 1{,}4\,\varepsilon_1 \cdot \delta^{(1)}; \tag{1.53}$$

$$\delta^{(2)} = 0{,}128\,\delta^{(1)} - (0{,}0022_7 - 2{,}2\,\varepsilon_2), \tag{1.54}$$

mit den Werten (1.42), (1.43) von ε_1 und ε_2 also

$$\delta^{(3)} = -0{,}142\,\delta^{(2)} \pm 0{,}010\,\delta^{(1)}; \tag{1.53a}$$

$$\delta^{(2)} = 0{,}128\,\delta^{(1)} - (0{,}0022_7 \pm 0{,}0003_1). \tag{1.54a}$$

Nach (1.41) liegt bei den Gläsern des Schott-Katalogs $\delta^{(1)}$ im Bereich

$$\tfrac{1}{70} \leqq \delta^{(1)} \leqq \tfrac{1}{25}; \tag{1.55}$$

denn $\delta^{(1)}$ ist näherungsweise gleich $1/\nu$, wie wir anschließend in Gl. (1.58) feststellen werden. — Das Glied mit $\delta^{(1)}$ in der Beziehung (1.54a) für $\delta^{(2)}$ ändert sich nach (1.55) um $(0{,}128\,\delta^{(1)})_{\max} - (0{,}128\,\delta^{(1)})_{\min} = 0{,}0033$; demgegenüber sind die Abweichungen $(\pm 0{,}0003_1)$ recht klein. (1.54a) bedeutet also einen recht engen Zusammenhang zwischen $\delta^{(2)}$ und $\delta^{(1)}$. — Dagegen kommen in dem Ausdruck (1.53a) für $\delta^{(3)}$ die Abweichungen $(\pm 0{,}010\,\delta^{(1)})$ an die Änderung des Glieds mit $\delta^{(2)}$ heran, so daß wir nicht ohne weiteres (1.53a) als einen Zusammenhang zwischen $\delta^{(3)}$ und $\delta^{(2)}$ ansehen können. Nach (1.53) und (1.54a) ist

$$\delta^{(3)} = -0{,}142\,[(0{,}128 - 10\,\varepsilon_1)\,\delta^{(1)} - (0{,}0022_7 \pm 0{,}0003_1)]. \tag{1.56}$$

Die Angabe $-0{,}007 \leqq \varepsilon_1 \leqq +0{,}007$ in (1.42) gilt für den *ganzen* ν-Bereich (1.41); bei den *einzelnen*, verschiedenen ν-Werten kommen unterschiedliche ε_1-Bereiche vor (wie eine Auftragung zeigt):

$\delta^{(1)} \approx 1/\nu =$	1/70	1/55	1/40	1/32,5	1/25
$\varepsilon_1 \geqq$	$-0{,}002$	$-0{,}007$	$-0{,}005$	$-0{,}002$	$+0{,}004$
$\varepsilon_1 \leqq$	$+0{,}006$	$+0{,}005$	$+0.004$	$+0{,}004$	$+0{,}007$

(1.57)

Mit diesen ε_1-Werten zeigt (1.56) einen lockeren Zusammenhang zwischen $\delta^{(3)}$ und $\delta^{(1)}$.

Aus der ersten Gl. (1.52) und (1.54a), (1.53a) folgt, daß im ν-Bereich (1.41)

$$\text{auf } 3\tfrac{1}{2}\text{‰ genau:} \qquad \delta^{(1)} = 2\,\frac{\Delta^1 n}{n_e - 1}\left[\equiv \frac{1}{\nu}\right] \tag{1.58}$$

ist. Aus der zweiten Gl. (1.52) und (1.53a) folgt, daß im ν-Bereich (1.41)

$$\text{auf } \pm 0{,}0075 \text{ genau:} \quad 0{,}871\,\nu\cdot\delta^{(2)} = \frac{\Delta^2 n}{\Delta^1 n}\,[\equiv(\Theta - 1)] \tag{1.59}$$

ist. Und aus der dritten Gl. (1.52) und (1.54a) folgt, daß im ν-Bereich (1.41)

$$\text{auf } \pm 0{,}0035 \text{ genau:} \quad 1{,}433\,\nu\cdot\delta^{(3)} = \frac{\Delta^3 n}{\Delta^1 n} + 0{,}0895 \tag{1.60}$$

ist.

Nach (1.56) und (1.57) liegt $\delta^{(3)}/\delta^{(1)}$ im Bereich

$$-0{,}014 \leqq \delta^{(3)}/\delta^{(1)} \leqq +0{,}016, \tag{1.61}$$

und nach (1.54a) und (1.55) liegt $\delta^{(2)}/\delta^{(1)}$ im Bereich

$$|\delta^{(2)}/\delta^{(1)}| \leqq 0{,}08. \tag{1.62}$$

Also ist

$$|\delta^{(3)}/\delta^{(1)}| \ll |\delta^{(2)}/\delta^{(1)}|_{\max} \ll 1. \tag{1.63}$$

II. Strahldaten; Übergangs- und Brechungs-Formeln.

7. Radiusvektor, optische Richtung. Die Lage eines bestimmten Strahls, der auf eine (brechende oder spiegelnde) Fläche Nr. v zuläuft, können wir beschreiben durch Angabe zweier Vektoren: erstens des *Ortsvektors*

$$\boldsymbol{a}_v \equiv \overrightarrow{O_v P_v}, \tag{2.1}$$

der von einem auf der optischen Achse liegenden Anfangspunkt O_v zu einem Punkt P_v des betrachteten Strahls führt; zweitens der *optischen Richtung*

$$\boldsymbol{n}_v \equiv |n_v|\cdot\vec{\varsigma}_v \qquad (\equiv n_v\cdot\vec{\sigma}_v), \tag{2.2}$$

das ist die mit dem absoluten Brechungsindex multiplizierte Richtung $\vec{\varsigma}_v$ des betrachteten Strahls[1]. Größen, die wie $\boldsymbol{a}_v$ und $\boldsymbol{n}_v$ die Lage eines Strahls beschreiben, nennen wir *Strahldaten.*

Als Anfangspunkt O_v können wir den Mittelpunkt C_v der Fläche Nr. v wählen (vorausgesetzt daß dieser im Endlichen liegt), und als Punkt P_v auf dem Strahl können wir dessen Auftreffpunkt D_v auf die Fläche Nr. v wählen. Den von $O_v = C_v$ nach $P_v = D_v$ führenden Ortsvektor $\boldsymbol{a}_v$ nennen wir den *Radiusvektor* $\boldsymbol{r}_v$:

$$\boldsymbol{r}_v \equiv \overrightarrow{C_v D_v}. \tag{2.3}$$

In dieser Ziff. 7 nehmen wir $\boldsymbol{r}_v$; $\boldsymbol{n}_v$ als Strahldaten *vor* der Fläche Nr. v, und $\boldsymbol{r}_v$; $\boldsymbol{n}'_v$ als Strahldaten *hinter* der Fläche Nr. v.

$$\boldsymbol{n}'_v \equiv |n'_v|\cdot\vec{\varsigma}\,'_v \qquad (\equiv n'_v\cdot\vec{\sigma}\,'_v) \tag{2.4}$$

---

[1] $\vec{\varsigma}$ ist die Strahlrichtung, d.h. die Richtung, in der die Energie längs dem Strahl strömt. $\vec{\sigma}$ ist damit identisch, außer bei negativen n; dort ist $\vec{\sigma} = -\vec{\varsigma}$.

ist die optische Richtung des von der Fläche Nr. v weglaufenden Strahls, $\vec{\zeta}_v'$ seine Richtung.

Wir führen, wie in Ziff. 1 beschrieben, ein kartesisches Koordinatensystem ein, dessen z-Achse mit der optischen Achse zusammenfällt und „von links nach rechts" läuft; die aus dem Objekt kommenden Strahlen haben eine Richtung mit positiver z-Komponente:

$$(\boldsymbol{n}_1)_z > 0. \tag{2.5}$$

Die x-, y-, z-Komponenten des Radiusvektors $\boldsymbol{r}_v$ bezeichnen wir mit x_v, y_v, z_v:

$$\boldsymbol{r}_v \equiv \{x_v; y_v; z_v\}, \tag{2.6}$$

und die der optischen Richtung $\boldsymbol{n}_v$ mit ξ_v, η_v, ζ_v:

$$\left.\begin{aligned} \boldsymbol{n}_v &\equiv \{\xi_v; \eta_v; \zeta_v\}; \\ \boldsymbol{n}_v' &\equiv \{\xi_v'; \eta_v'; \zeta_v'\}. \end{aligned}\right\} \tag{2.7}$$

8. Normaler Aufbau des optischen Systems. Als *linke* Umgebung der Fläche Nr. v bezeichnen wir diejenige ihrer beiden Umgebungen (auf der einen und auf der anderen Seite der Fläche), die beim Scheitel S_v links von der Fläche liegt; die *rechte* Umgebung liegt beim Scheitel S_v rechts von der Fläche. Wir sagen, daß ein Strahl „von links her" oder „von rechts her" die Fläche Nr. v erreicht, je nachdem ob er sie von ihrer linken Umgebung her oder von ihrer rechten Umgebung her erreicht. Und wir sagen, daß ein Strahl „nach rechts hin" oder „nach links hin" die Fläche Nr. v verläßt, je nachdem ob er sie nach ihrer rechten Umgebung hin oder nach ihrer linken Umgebung hin verläßt.

Ein Strahl, der von links her die Fläche Nr. v erreicht, läuft in der Regel auch „von links nach rechts" in dem Sinne, daß seine Richtung eine positive z-Komponente hat. Das muß aber nicht so sein; in einer meniskenförmigen dünnen Luftlinse z. B. kann der Fall vorkommen, daß ein Strahl die Austrittsfläche dieser Luftlinse zwar von ihrer linken Umgebung her erreicht, daß er dabei aber von rechts nach links läuft. Wir wollen (zunächst) zulassen, daß ein Strahl von links nach rechts *oder* von rechts nach links läuft *unabhängig davon*, ob er von links her oder von rechts her die nächste Fläche erreicht.

Linsen-Systeme sind normalerweise so aufgebaut, daß alle zur Abbildung vorgesehenen Strahlen jede Fläche des Systems von links her erreichen und nach rechts hin verlassen. Und Linsen-Spiegel-Systeme sind normalerweise so aufgebaut: daß die Strahlen nach einer geraden Anzahl (0; 2; ...) von Spiegelungen die Flächen nach rechts hin verlassen und von links her erreichen; und daß die Strahlen nach einer ungeraden Anzahl (1; 3; ...) von Spiegelungen die Flächen nach links hin verlassen und von rechts her erreichen. Den normalerweise nicht vorkommenden, aber denkbaren Fall, daß ein Strahl nach einer geraden Anzahl von Spiegelungen von rechts her, oder nach einer ungeraden Anzahl von Spiegelungen von links her auf brechende oder spiegelnde Flächen des Systems trifft, wollen wir ausschließen. Nach Ziff. 4 läßt sich diese Einschränkung so formulieren: Ein Strahl erreicht die Fläche Nr. v von links her *oder* von rechts her, *je nachdem ob* $n_v > 0$ *oder* $n_v < 0$ ist. Und ein Strahl verläßt die Fläche Nr. v nach rechts hin *oder* nach links hin, *je nachdem ob* $n_v' > 0$ *oder* $n_v' < 0$ ist. Setzen wir noch fest, daß im Auftreffpunkt D_v des Strahls auf die Fläche Nr. v die Flächennormale eine Richtung $\vec{\varphi}_v$ hat, die von der linken Umgebung der Fläche zu ihrer rechten Umgebung weist, dann läßt sich die genannte Einschränkung so formulieren: Es ist immer

$$\vec{\varphi}_v \, \boldsymbol{n}_v / n_v > 0; \quad \vec{\varphi}_v \, \boldsymbol{n}_v' / n_v' > 0. \tag{2.8}$$

Bei einer Kugelfläche Nr. v ist wegen (1.4) bei positivem r_v die Normalenrichtung $\vec{\varphi}_v \uparrow\downarrow \boldsymbol{r}_v$; und bei negativem r_v ist $\vec{\varphi}_v \uparrow\uparrow \boldsymbol{r}_v$. Wegen $\vec{\varphi}_v^2 = 1$; $\boldsymbol{r}_v^2 = r_v^2$ ist also

$$\vec{\varphi}_v = -\boldsymbol{r}_v / r_v, \tag{2.9}$$

so daß nach (2.8) bei Kugelflächen immer

$$\frac{\boldsymbol{r}_v \boldsymbol{n}_v}{r_v n_v} < 0; \quad \frac{\boldsymbol{r}_v \boldsymbol{n}'_v}{r_v n'_v} < 0 \tag{2.10}$$

ist.

9. Übergangsformeln. Beim Übergang des Strahls von der Fläche Nr. v zur Fläche Nr. $v+1$ läuft der Strahl geradlinig mit der optischen Richtung

$$\boldsymbol{n}'_v \equiv \boldsymbol{n}_{v+1}. \tag{2.11}$$

Diese Richtung hat also der vom Auftreffpunkt D_v auf der Fläche Nr. v zum Auftreffpunkt D_{v+1} auf der Fläche Nr. $v+1$ führende Streckenvektor $\overrightarrow{D_v D_{v+1}}$, der gleich $(-\boldsymbol{r}_v + \overrightarrow{C_v C_{v+1}} + \boldsymbol{r}_{v+1})$ ist:

$$[(\boldsymbol{r}_{v+1} - \boldsymbol{r}_v + \boldsymbol{e}_{v+1})\, \boldsymbol{n}_{v+1}] = 0, \tag{2.12}$$

wobei

$$\boldsymbol{e}'_v \equiv \boldsymbol{e}_{v+1} \equiv \overrightarrow{C_v C_{v+1}} \tag{2.13}$$

ist. Nach Ziff. 1 und 2 ist, wenn wir

$$\boldsymbol{e}_{v+1} \equiv e_{v+1} \cdot \boldsymbol{k} \tag{2.14}$$

setzen ($\boldsymbol{k} \equiv$ Richtung der z-Achse $\equiv$ Richtung der optischen Achse),

$$e_{v+1} = a_{v+1} + r_{v+1} - r_v. \tag{2.15}$$

Durch (2.12) ist der Radiusvektor $\boldsymbol{r}_{v+1}$ des Auftreffpunkts D_{v+1} nur insoweit bestimmt, als (2.12) besagt, daß D_{v+1} auf dem Strahl mit den Daten $(\boldsymbol{r}_v; \boldsymbol{n}'_v)$ liegt. Daß D_{v+1} der Auftreffpunkt auf der Fläche Nr. $v+1$ ist, bedeutet, daß seine Koordinaten der Gl. (1.11) für die Fläche Nr. $v+1$ genügen. Diese Flächengleichung können wir wegen (1.9), (1.10) und (1.4a) so schreiben:

$$z_{v+1} + r_{v+1} = \sum_{i=0}^{\infty} \alpha_{v+1}^{(i)} \left(\frac{x_{v+1}^2 + y_{v+1}^2 + (z_{v+1} + r_{v+1})^2}{2} \right)^{i+1}, \tag{2.16}$$

oder wegen (1.13a) und (2.6) so:

$$\boldsymbol{r}_{v+1}^2 = r_{v+1}^2 - 2 r_{v+1} \cdot \sum_{i=1}^{\infty} \alpha_{v+1}^{(i)} \left[\tfrac{1}{2} (\boldsymbol{r}_{v+1} + \boldsymbol{k} r_{v+1})^2 \right]^{i+1}. \tag{2.17}$$

Im Fall einer Kugelfläche Nr. $v+1$ ist $\alpha_{v+1}^{(i)} = 0$ für $i \geqq 1$, also:

$$\boldsymbol{r}_{v+1}^2 = r_{v+1}^2, \tag{2.18}$$

was wir in diesem Fall auch unmittelbar aus (2.3) und (1.4) hätten schließen können.

Bei einer Kugelfläche Nr. $v+1$ ist $\boldsymbol{r}_{v+1}$ durch (2.12) und (2.18) noch nicht eindeutig bestimmt: (2.18) gilt sowohl für den Radiusvektor des wirklichen Auftreffpunkts D_{v+1}, als auch für den Radiusvektor des zweiten Schnittpunkts E_{v+1}, den die Strahl-Gerade mit der Kugel-Oberfläche hat, von der die Fläche Nr. $v+1$ ein Ausschnitt ist. Durchläuft man die Strahl-Gerade in Strahlrichtung $\vec{\zeta}_{v+1}$, dann erreicht man aber D_{v+1} von der Umgebung der *einen* Seite der Fläche her, und E_{v+1} von der Umgebung der *anderen* Seite der Fläche her. Die für D_{v+1} gültige Beziehung (2.10):

$$\frac{\boldsymbol{r}_{v+1}\boldsymbol{n}_{v+1}}{r_{v+1}n_{v+1}} < 0 \tag{2.19}$$

gilt daher für E_{v+1} nicht. Also bestimmen die Formeln (2.12), (2.18), (2.19) die Strahldaten $\boldsymbol{r}_{v+1}$ und $\boldsymbol{n}_{v+1} \equiv \boldsymbol{n}'_v$ vor der Kugelfläche Nr. $v+1$ eindeutig aus den Strahldaten $\boldsymbol{r}_v$ und $\boldsymbol{n}'_v$ hinter der Fläche Nr. v. Diese Formeln nennen wir die *Übergangs-Formeln*, gültig für den Übergang von einer beliebigen Fläche zu einer Kugelfläche.

Bei einer asphärischen Fläche Nr. $v+1$ ist, je nach deren Gestalt und je nach der Lage des ankommenden Strahls, $\boldsymbol{r}_{v+1}$ durch (2.12) und (2.17) eindeutig oder nicht eindeutig bestimmt. Die erste Beziehung (2.8), angewandt auf die Fläche Nr. $v+1$, schränkt die Mehrdeutigkeit zwar ein, garantiert aber nicht immer die eindeutige Bestimmung von $\boldsymbol{r}_{v+1}$. Löst man (2.12) und (2.17) nach $\boldsymbol{r}_{v+1}$ in der Weise auf, daß man eine zu der Lösung führende Iteration angibt, dann ist durch den Beginn dieser Iteration die Lösung eindeutig bestimmt, und man kann durch eine geeignete Vorschrift über den Beginn der Iteration erreichen, daß sie zu der richtigen Lösung führt. Von dieser Vorschrift abgesehen, sind die Formeln (2.12), (2.17) die *Übergangs-Formeln*, gültig für den Übergang von einer beliebigen Fläche zu einer asphärischen Fläche.

10. Brechungsformeln. Trifft ein Strahl eine brechende Fläche (die Nr. der Fläche lassen wir jetzt weg) in einem Punkt mit einer Flächennormalen-Richtung $\vec{\varphi}$, dann wird er nach dem Snelliusschen Brechungsgesetz[1] in der Weise gebrochen, daß

$$\boldsymbol{n}' \cdot \sin \sphericalangle (\boldsymbol{n}', \vec{\varphi}) = \boldsymbol{n} \cdot \sin \sphericalangle (\boldsymbol{n}, \vec{\varphi}) \tag{2.20}$$

ist. Und zwar bleibt der Strahl [im Fall $\sphericalangle (\boldsymbol{n}, \vec{\varphi}) \neq 0$ und $\neq 180°$] in der durch den einfallenden Strahl und die Flächennormale gebildeten „Einfallsebene":

$$[\boldsymbol{n}' \vec{\varphi}] \,||\, [\boldsymbol{n} \vec{\varphi}],$$

und in der Einfallsebene überquert der Strahl bei der Brechung die Flächennormale:

$$[\boldsymbol{n}' \vec{\varphi}] \uparrow\uparrow [\boldsymbol{n} \vec{\varphi}]. \tag{2.21}$$

(2.20) bedeutet, daß $|[\boldsymbol{n}' \vec{\varphi}]| = |[\boldsymbol{n} \vec{\varphi}]|$ ist. Hieraus und aus (2.21) folgt, daß $[\boldsymbol{n}' \vec{\varphi}] = [\boldsymbol{n} \vec{\varphi}]$, also

$$[(\boldsymbol{n}' - \boldsymbol{n}) \vec{\varphi}] = 0 \tag{2.22}$$

ist. Auch im Fall $\sphericalangle (\boldsymbol{n}, \vec{\varphi}) = 0$ oder $= 180°$ gilt (2.22), wegen $\boldsymbol{n} \,||\, \vec{\varphi} \,||\, \boldsymbol{n}'$.

Bei einer Kugelfläche folgt wegen (2.9) weiter:

$$[(\boldsymbol{n}' - \boldsymbol{n})\, \boldsymbol{r}] = 0. \tag{2.23}$$

Man kann leicht nachrechnen: daß durch (2.23), durch die aus (2.4) folgende Gleichung

$$\boldsymbol{n}'^2 = n'^2 \tag{2.24}$$

und durch die zweite Beziehung (2.10):

$$\frac{\boldsymbol{r}\,\boldsymbol{n}'}{r\,n'} < 0 \tag{2.25}$$

[1] Vgl. auch (1.15); $|\iota|$ ist entweder $= \sphericalangle (\boldsymbol{n}, \vec{\varphi})$ oder $= 180° - \sphericalangle (\boldsymbol{n}, \vec{\varphi})$.

$\boldsymbol{n}'$ eindeutig bestimmt ist. Die Formeln (2.23), (2.24), (2.25) bestimmen also die Strahldaten $\boldsymbol{r}$, $\boldsymbol{n}'$ hinter einer Kugelfläche eindeutig aus den Strahldaten $\boldsymbol{r}$, $\boldsymbol{n}$ vor dieser Kugelfläche. Diese Formeln nennen wir die *Brechungs-Formeln*, gültig für die Brechung an einer Kugelfläche.

Bei einer asphärischen Fläche betrachten wir zunächst den Schnittpunkt $\tilde{C}$ der Flächennormalen (im Auftreffpunkt D) mit der optischen Achse, und wir stellen fest, daß

$$\tilde{\boldsymbol{r}} \equiv \overrightarrow{\tilde{C}D} \tag{2.26}$$

die Richtung $\pm\vec{\varphi}$ hat:

$$\tilde{\boldsymbol{r}} \parallel \vec{\varphi}, \tag{2.27}$$

so daß nach (2.22)

$$[(\boldsymbol{n}'-\boldsymbol{n})\,\tilde{\boldsymbol{r}}]=0 \tag{2.28}$$

ist. Durch (2.28), durch die aus (2.4) folgende Gleichung

$$\boldsymbol{n}'^2 = n'^2 \tag{2.24a}$$

und durch die aus (2.8) und (2.27) folgende Beziehung

$$\operatorname{sign}\frac{\tilde{\boldsymbol{r}}\boldsymbol{n}'}{r\,n'} = \operatorname{sign}\frac{\tilde{\boldsymbol{r}}\boldsymbol{n}}{r\,n} \tag{2.29}$$

ist $\boldsymbol{n}'$ eindeutig bestimmt aus $\boldsymbol{n}$ und $\tilde{\boldsymbol{r}}$. Wir müssen noch feststellen, wie man $\tilde{\boldsymbol{r}}$ bestimmt. Für eine gedachte Verschiebung $d\boldsymbol{r}$ des Auftreffpunkts D, die in der brechenden Fläche verläuft, gilt $\tilde{\boldsymbol{r}}\,d\boldsymbol{r}=0$, weil $d\boldsymbol{r}\perp\vec{\varphi}$ und $\vec{\varphi}\parallel\tilde{\boldsymbol{r}}$ ist. Setzen wir

$$\tilde{\boldsymbol{r}} \equiv \overrightarrow{\tilde{C}D} \equiv \{x;\,y;\,\tilde{z}\}, \tag{2.30}$$

und $\boldsymbol{r}\equiv\overrightarrow{CD}\equiv\{x;\,y;\,z\}$ [nach (2.6)], dann ist also $\tilde{z}\,dz=-x\,dx-y\,dy$ oder wegen (1.10), (1.9), (1.4a): $\tilde{z}\,dz=-d(\tfrac{1}{2}s^2)+(z+r)\,d(z+r)$, folglich

$$\tilde{z}-z-r=-d(\tfrac{1}{2}s^2)/dz. \tag{2.31}$$

Schreiben wir

$$z+r=\mathrm{F}(\tfrac{1}{2}s^2) \tag{2.32}$$

für (2.16), dann ist nach (2.31) also

$$\tilde{z}=z+r-1/\mathrm{F}'(\tfrac{1}{2}s^2), \tag{2.33}$$

wobei $\mathrm{F}'(\;)\equiv d\mathrm{F}/d(\;)$ ist. Nach (2.16), (2.32) ist

$$\mathrm{F}'(\tfrac{1}{2}s^2)=\sum_{i=0}^{\infty}(i+1)\,\alpha^{(i)}\,(\tfrac{1}{2}s^2)^i,$$

so daß aus (2.30), (2.33) und $s^2=(\boldsymbol{r}+\boldsymbol{k}r)^2$ folgt:

$$\tilde{\boldsymbol{r}}=\boldsymbol{r}+\boldsymbol{k}\left\{r-1\Big/\sum_{i=0}^{\infty}(i+1)\,\alpha^{(i)}\,[\tfrac{1}{2}(\boldsymbol{r}+\boldsymbol{k}\,r)^2]^i\right\}. \tag{2.34}$$

Die Formeln (2.34), (2.28), (2.24a), (2.29) bestimmen die Strahldaten $\boldsymbol{r}$, $\boldsymbol{n}'$ hinter einer asphärischen Fläche eindeutig aus den Strahldaten $\boldsymbol{r}$, $\boldsymbol{n}$ vor dieser Fläche. Diese Formeln nennen wir die *Brechungs-Formeln*, gültig für die Brechung an einer asphärischen Fläche.

Wir hatten Brechungsindizes *mit* Vorzeichen eingeführt, damit wir eine Reflexion formal wie eine Brechung behandeln können. Bei einer spiegelnden

Fläche ist, auf Grund unserer Festsetzung über das Brechungsindex-Vorzeichen,

$$n' = -n. \tag{2.35}$$

Wir zeigen, daß die für die *Brechung* an Kugelflächen nachgewiesenen „Brechungsformeln" (2.23), (2.24), (2.25) auch für die *Reflexion* an Kugelflächen gelten. Die Brechungsformel (2.25) gilt nach Ziff. 8 wegen des normalen Aufbaus des optischen Systems. Die Brechungsformel (2.24) gilt wegen (2.4). Und die Brechungsformel (2.23) bedeutet wegen (2.9), (2.2), (2.4), (2.35), daß $[(\vec{\varsigma}\,' - \vec{\varsigma})\vec{\varphi}] = 0$ ist: der Strahl bleibt in der Einfallsebene, überquert darin die Flächennormale, und es ist $\sin \sphericalangle (\vec{\varsigma}\,', \vec{\varphi}) = \sin \sphericalangle (\vec{\varsigma}, \vec{\varphi})$; die Brechungsformel (2.23) beschreibt also auch die Reflexion richtig. Wie in Ziff. 10 schon gesagt, ist durch (2.23), (2.24), (2.25) $\boldsymbol{n}'$ eindeutig bestimmt. Wir sehen, daß die Brechungsformeln (2.23), (2.24), (2.25) auch bei der Reflexion die Strahldaten $\boldsymbol{r}$, $\boldsymbol{n}'$ hinter einer Kugelfläche eindeutig aus den Strahldaten $\boldsymbol{r}$, $\boldsymbol{n}$ vor dieser Kugelfläche bestimmen. — Ebenso bestimmen die Brechungsformeln (2.34), (2.28), (2.24a), (2.29) auch bei einer spiegelnden asphärischen Fläche $\boldsymbol{r}$, $\boldsymbol{n}'$ eindeutig aus $\boldsymbol{r}$, $\boldsymbol{n}$.

11. Übersetzungsschlüssel bei Kugelflächen. Im Folgenden werden wir uns auf rotationssymmetrische Systeme von *Kugelflächen* beschränken. Dann haben wir nach Ziff. 9 und 10 die *Übergangsformeln*

$$[(\boldsymbol{r}_{v+1} - \boldsymbol{r}_v + \boldsymbol{e}_{v+1})\,\boldsymbol{n}_{v+1}] = 0; \qquad \boldsymbol{r}_{v+1}^2 = r_{v+1}^2; \qquad \frac{\boldsymbol{n}_{v+1}\,\boldsymbol{r}_{v+1}}{n_{v+1}\,r_{v+1}} < 0 \tag{2.36a}$$

und die *Brechungsformeln*

$$[(\boldsymbol{n}' - \boldsymbol{n})\,\boldsymbol{r}] = 0; \qquad \boldsymbol{n}'^2 = n'^2; \qquad \frac{\boldsymbol{r}\,\boldsymbol{n}'}{r\,n'} < 0. \tag{2.37a}$$

Wir sehen, daß die Brechungsformeln ähnlich gebaut sind wie die Übergangsformeln. Wir können die Übergangsformeln („Üb.") umwandeln in die Brechungsformeln („Br."), indem wir

$$\left.\begin{array}{llllllll} \text{ersetzen:} & \boldsymbol{r}_v & \boldsymbol{n}_{v+1} & \boldsymbol{r}_{v+1} & \boldsymbol{e}_{v+1} & -r_v & +n_{v+1} & -r_{v+1} \quad (\text{Üb.}) \\ & & & & & & & \qquad\quad \downarrow \\ \text{durch:} & \boldsymbol{n} & \boldsymbol{r} & \boldsymbol{n}' & 0 & +n & -r & +n' \quad (\text{Br.}). \end{array}\right\} \tag{2.38}$$

In diesem *Übersetzungsschlüssel* (2.38) könnten wir die Übersetzung $-r_v \to +n$ weglassen, wenn es uns nur auf die Umwandlung der Übergangsformeln (2.36a) in die Brechungsformeln (2.37a) ankäme. Wir wollen aber zu diesen Formeln noch zusätzliche Aussagen hinzunehmen, und zwar

zu den Übergangsformeln:

$$\boldsymbol{n}_{v+1}^2 = n_{v+1}^2; \qquad \frac{\boldsymbol{r}_v\,\boldsymbol{n}_{v+1}}{r_v\,n_{v+1}} < 0; \qquad \boldsymbol{r}_v^2 = r_v^2; \tag{2.36b}$$

und *zu den Brechungsformeln*:

$$\boldsymbol{r}^2 = r^2; \qquad \frac{\boldsymbol{n}\,\boldsymbol{r}}{n\,r} < 0; \qquad \boldsymbol{n}^2 = n^2. \tag{2.37b}$$

Wir könnten in dem Übersetzungsschlüssel (2.38) auch $+r_v$ statt $-r_v$, $+r_{v+1}$ statt $-r_{v+1}$, und $+r$ statt $-r$ schreiben, wenn es uns nur auf die Umwandlung von (2.36a, b) in (2.37a, b) ankäme. Wir wollen aber auch noch Aussagen über *achsennahe Strahlen* hinzunehmen, und zwar

zu den Übergangsformeln:

$$\left.\begin{array}{l} \text{zählt man} \quad (x_v;\, y_v),\ (\xi_{v+1};\, \eta_{v+1}),\ (x_{v+1};\, y_{v+1}) \\ \qquad\qquad \text{als kleine Größen erster Ordnung,} \\ \text{dann sind} \quad (z_v + r_v),\ (\zeta_{v+1} - n_{v+1}),\ (z_{v+1} + r_{v+1}) \\ \qquad\qquad \text{kleine Größen zweiter Ordnung;} \end{array}\right\} \tag{2.36c}$$

und *zu den Brechungsformeln*:

$$\left.\begin{array}{l}\text{zählt man}\quad (\xi;\eta),\quad (x;y),\quad (\xi';\eta') \\ \qquad\qquad\qquad\qquad\qquad \text{als kleine Größen erster Ordnung,} \\ \text{dann sind}\quad (\zeta-n),\ (z+r),\ (\zeta'-n') \\ \qquad\qquad\qquad\qquad\qquad \text{kleine Größen zweiter Ordnung.}\end{array}\right\}\quad (2.37\,\mathrm{c})$$

Der Übersetzungsschlüssel (2.38) wandelt (2.36a, b, c) um in (2.37a, b, c); nach (2.7) und (2.6) ist

$$\boldsymbol{n} \equiv \{\xi;\eta;\zeta\};\qquad \boldsymbol{r} \equiv \{x;y;z\}. \qquad (2.39)$$

Zur Begründung der Aussagen (2.36c); (2.37c): „Achsennahe" Strahlen sind Strahlen, die in der Nähe der optischen Achse verlaufen (genauer gesagt: deren Auftreffpunkte D_w auf die Fläche Nr. w in der Nähe des Flächenscheitels S_w liegen[1], gemessen an dem Radius dieser Fläche), und die nahezu parallel zur optischen Achse verlaufen. — Bei einem achsennahen Strahl sind also erstens x und y klein; wir zählen sie als kleine Größen erster Ordnung (und den Radius r, dem gegenüber x und y klein sind, als Größe nullter Ordnung). Hieraus folgt: $z = \pm\sqrt{r^2-(x^2+y^2)} = \pm r + 0^{(2)}$ [mit $0^{(N)}$ bezeichnen wir eine kleine Größe N-ter Ordnung]; und zwar gilt

$$z = -r + 0^{(2)},$$

weil beim achsennahen Strahl D nahe S liegt, also $\boldsymbol{r} \equiv \overrightarrow{CD} \approx \overrightarrow{CS} = -\boldsymbol{k}r$ und damit $z \approx -r$ ist. — Bei einem achsennahen Strahl ist zweitens $\boldsymbol{n} \equiv \{\xi,\eta;\zeta\}$ nahezu parallel $\pm\boldsymbol{k}$, also sind ξ und η klein; wir zählen auch sie als kleine Größen erster Ordnung (und den Brechungsindex n, dem gegenüber ξ und η klein sind, als Größe nullter Ordnung.) Hieraus folgt: $\zeta = \pm\sqrt{n^2-(\xi^2+\eta^2)} = \pm n + 0^{(2)}$; und zwar gilt

$$\zeta = +n + 0^{(2)},$$

weil beim achsennahen Strahl D nahe S liegt, also [vgl. Ziff. 8] $\vec{\varphi} \approx \boldsymbol{k}$ und damit nach (2.8) $\zeta/n \approx \vec{\varphi}\,\boldsymbol{n}/n > 0$ ist. — Nicht nur die Konstruktionsdaten r und n, sondern auch e (und damit a) zählen wir als Größen nullter Ordnung.

Leitet man aus den Übergangsformeln (2.36a, b, c) eine Aussage A ab, und übersetzt man anschließend in dem Ergebnis alle Größen gemäß dem Übersetzungsschlüssel (2.38), dann erhält man eine neue Aussage A^+, die man aus den Brechungsformeln (2.37a, b, c) ebenso hätte ableiten können wie A aus (2.36a, b, c). So kann man sich die Ableitung von Aussagen A^+ aus den Brechungsformeln ersparen, wenn man A^+ durch bloße Anwendung des Übersetzungsschlüssels gewinnen kann aus Aussagen A, die man schon aus den Übergangsformeln abgeleitet hat. Zum Beispiel hat man nach Auflösung von (2.36a) nach $\boldsymbol{r}_{k+1}$, indem man den Übersetzungsschlüssel (2.38) anwendet, die Auflösung von (2.37a) nach $\boldsymbol{n}'$ dastehen und braucht sie nicht mehr durchzuführen.

Mit den Bezeichnungen[2]

$$\boldsymbol{r}_{v+\frac{1}{2}} \equiv \boldsymbol{n}'_v (\equiv \boldsymbol{n}_{v+1});\qquad (\mp) r_{v+\frac{1}{2}} \equiv n'_v (\equiv n_{v+1});\qquad \boldsymbol{o}_{v+\frac{1}{2}} \equiv \boldsymbol{e}'_v (\equiv \boldsymbol{e}_{v+1}) \qquad (2.40)$$

[1] Auch im Fall *virtueller* Scheitel gibt es achsennahe Strahlen, entweder reell vorhandene oder virtuell konstruierte.

[2] In (2.41) gilt das obere oder das untere (eingeklammerte) Vorzeichen, je nachdem ob man in (2.40) das obere oder das untere Vorzeichen nimmt. Das obere Vorzeichen entspricht dem Übersetzungsschlüssel (2.38).

können wir die Übergangsformeln (2.36a) und die Brechungsformeln (2.37a) *gleichlautend* schreiben: Die Formeln[1]

$$[(\boldsymbol{r}_{v+1} - \boldsymbol{r}_v + \boldsymbol{o}_{v+\frac{1}{2}})\, \boldsymbol{r}_{v+\frac{1}{2}}] = 0; \qquad \boldsymbol{r}_{v+1}^2 = r_{v+1}^2; \qquad {}_{(}\pm{}_{)} \frac{\boldsymbol{r}_{v+\frac{1}{2}} \boldsymbol{r}_{v+1}}{r_{v+\frac{1}{2}} r_{v+1}} > 0 \tag{2.41}$$

sind wegen (2.40) bei *ganzzahligem* v die Übergangsformeln (2.36a), und bei halbzahligem v ($\equiv w - \frac{1}{2}$; w ganzzahlig) verallgemeinerte Brechungsformeln:

$$[(\boldsymbol{n}_w' - \boldsymbol{n}_w + \boldsymbol{o}_w)\, \boldsymbol{r}_w] = 0; \qquad \boldsymbol{n}_w'^2 = n_w'^2; \qquad \frac{\boldsymbol{r}_w \boldsymbol{n}_w'}{r_w n_w'} < 0, \tag{2.37A}$$

nach denen mit $\boldsymbol{o}_w \neq 0$ die brechende oder spiegelnde Fläche eine ähnliche Wirkung hat wie die gezackte Fläche einer Fresnelschen Stufenlinse. Die Formeln (2.37A) mit

$$\boldsymbol{o}_w = 0 \quad (w \text{ ganzzahlig}) \tag{2.42}$$

sind identisch mit den Brechungsformeln (2.37a). — Der Lauf eines Strahls durch ein optisches System ist bestimmt durch den Formelsatz (2.36a); (2.37a) [Übergangs- und Brechungsformeln]. Mit den Bezeichnungen (2.40) ist der Lauf des Strahls schon durch den *halben Formelsatz* (2.41) bestimmt[2].

12. Komplexe Strahldaten; zugehörige Übergangs- und Brechungsformeln. C. CARATHÉODORY hat, in einem etwas anderen Zusammenhang[6], die zur Rotationsachse des optischen Systems senkrechten Komponenten V_y und V_x eines Vektors $\boldsymbol{V}$ zu einer komplexen Zahl $(V_y + i V_x)$ zusammengefaßt und damit gerechnet; viele Formeln schreiben sich mit diesen komplexen Zahlen einfacher. Bevor wir diese komplexe Schreibweise einführen, um *unsere* Formeln und Umrechungen zu vereinfachen, führen wir noch andere Strahldaten als $\boldsymbol{n}$ und $\boldsymbol{r}$ ein.

Wir setzen

$$\boldsymbol{n} \equiv n \cdot \vec{\sigma} \equiv n \cdot \{-G_x;\ -G_y;\ \overline{G}\} \tag{2.43}$$

und

$$[\boldsymbol{r}\,\boldsymbol{n}] \equiv r\,n \cdot \{J_y;\ -J_x;\ \overline{J}_x\}; \qquad (\boldsymbol{r}\,\boldsymbol{n}) \equiv -r\,n \cdot \overline{J}_y. \tag{2.44}$$

$\overline{J}_y$ und $\overline{J}_x$ sind nicht die y- und die x-Komponente eines Vektors, sondern werden nur deshalb so geschrieben, weil wir später auch $\overline{J}_y$ mit $\overline{J}_x$ zu der komplexen Zahl $(\overline{J}_y + i\overline{J}_x)$ zusammenfassen wollen, ebenso wie wir die Komponenten J_y und J_x zu der komplexen Zahl $(J_y + i J_x)$ zusammenfassen werden. J_y und J_x sind die y- und die x-Komponente des Vektors $[\boldsymbol{k}[\boldsymbol{r}\,\boldsymbol{n}]]/(n r)$.

Wegen $\boldsymbol{n}^2 = n^2$ und (2.43) ist

$$G_y^2 + G_y^2 + \overline{G}^2 = 1, \tag{2.45}$$

und wegen $[\boldsymbol{r}\,\boldsymbol{n}]^2 + (\boldsymbol{r}\,\boldsymbol{n})^2 = \boldsymbol{r}^2 \boldsymbol{n}^2 = r^2 n^2$ und (2.44) ist

$$J_y^2 + J_x^2 + \overline{J}_y^2 + \overline{J}_x^2 = 1. \tag{2.46}$$

---

[1] Siehe Fußnote 2, S. 85.

[2] Mit dem *unteren* (eingeklammerten) Vorzeichen wurde dieser halbe Formelsatz in [3] eingeführt. Schon früher hatte M. LANGE[4] einander sehr ähnliche Brechungs- und Übergangs-Formeln (mit den Komponenten von $\boldsymbol{n}$ und $\boldsymbol{r}$ als Strahldaten, und nach diesen aufgelöst) veröffentlicht; er hatte aber diese Ähnlichkeit nicht so weit getrieben, daß der ganze Formelsatz in zwei gleiche Hälften zerfällt. — Der dem halben Formelsatz mit dem *oberen* Vorzeichen entsprechende Übersetzungsschlüssel (2.38) wurde in [5] angegeben und angewendet.

[3] H. MARX: Z. Physik **131**, 411 (1952).

[4] M. LANGE: Diss. Rostock 1909 (Leipzig: B. G. Teubner 1909).

[5] H. MARX: Optica Acta **1**, 127—140 (1954).

[6] C. CARATHÉODORY: Sitzungsber. Bayer. Akad. Wiss. (Math.-naturw. Abt.), 1943, 199—216.

Aus $\boldsymbol{n}[\boldsymbol{r}\,\boldsymbol{n}]=0$ und (2.43), (2.44) folgt:

$$\overline{G}\,\overline{J}_x = G_x J_y - G_y J_x. \tag{2.47}$$

Danach sind $\overline{G}$, $\overline{J}_y$, $\overline{J}_x$ bis auf die Vorzeichen von $\overline{G}$ und $\overline{J}_y$ gegeben durch G_y, G_x, J_y, J_x. Wegen $\frac{\boldsymbol{r}\,\boldsymbol{n}}{r\,n}<0$ und (2.44) ist immer

$$\overline{J}_y > 0. \tag{2.48}$$

In der Regel ist auch $\overline{G}>0$. Denn andernfalls würde nach (2.43) und Ziff. 4 der Strahl nach einer geraden Anzahl (0; 2; ...) von Spiegelungen von rechts nach links, oder nach einer ungeraden Anzahl (1; 3; ...) von Spiegelungen von links nach rechts laufen. Das ist, auch bei dem in Ziff. 8 definierten „normalen" System-Aufbau, nicht ausgeschlossen, kommt aber in der Regel nicht vor. Von nun an wollen wir den möglichen Fall $\overline{G}<0$ ausschließen, also immer

$$\overline{G} > 0 \tag{2.49}$$

annehmen. Bei achsennahen Strahlen ist sicher $\overline{G}>0$, denn bei achsennahen Strahlen ist nach (2.36c) und (2.43), (2.44):

$$\left.\begin{aligned} G_y &= 0^{(1)}; & G_x &= 0^{(1)}; & \overline{G} &= 1 + 0^{(2)};\\ J_y &= 0^{(1)}; & J_x &= 0^{(1)}; & \overline{J}_y &= 1 + 0^{(2)} \end{aligned}\right\} \tag{2.50}$$

und nach (2.47)

$$\overline{J}_x = 0^{(2)}. \tag{2.51}$$

Die erste Übergangsformel (2.36a) und die erste Brechungsformel (2.37a) schreiben wir gemäß (2.43), (2.44) auf G_y, G_x, J_y, J_x um:

$$\left.\begin{aligned} r_{v+1}\cdot(J_y)_{v+1} &= r_v\cdot(J_y)'_v - e'_v\cdot(G_y)'_v;\\ r_{v+1}\cdot(J_x)_{v+1} &= r_v\cdot(J_x)'_v - e'_v\cdot(G_x)'_v;\\ r_{v+1}\cdot(\overline{J}_x)_{v+1} &= r_v\cdot(\overline{J}_x)'_v; \end{aligned}\right\} \tag{2.52}$$

$$\left.\begin{aligned} n'\cdot J'_y &= n\cdot J_y;\\ n'\cdot J'_x &= n\cdot J_x;\\ n'\cdot \overline{J}'_x &= n\cdot \overline{J}_x. \end{aligned}\right\} \tag{2.53}$$

Wegen $\boldsymbol{n}_{v+1}\equiv\boldsymbol{n}'_v$ und (2.43) ist

$$\left.\begin{aligned} (G_y)_{v+1} &\equiv (G_y)'_v;\\ (G_x)_{v+1} &\equiv (G_x)'_v;\\ \overline{G}_{v+1} &\equiv \overline{G}'_v. \end{aligned}\right\} \tag{2.54}$$

$(G_y)_v$, $(G_x)_v$, $(J_y)_v$, $(J_x)_v$ können wir (im Fall $r_v \neq \infty$) als Strahldaten vor der Fläche Nr. v nehmen; durch diese vier Größen ist (im Fall $r_v\neq\infty$) die Lage des Strahls vor der Fläche Nr. v vollständig beschrieben [$\boldsymbol{n}$ ist nach (2.43) gegeben durch G_y, G_x und $\overline{G}$; $n^2\cdot\boldsymbol{r}=[\boldsymbol{n}[\boldsymbol{r}\,\boldsymbol{n}]]+\boldsymbol{n}(\boldsymbol{r}\,\boldsymbol{n})$ nach (2.43), (2.44) durch G_y, G_x, J_y, J_x und $\overline{G}$, $\overline{J}_y$, $\overline{J}_x$; $\overline{G}$, $\overline{J}_y$, $\overline{J}_x$ nach (2.45) bis (2.49) durch G_y, G_x, J_y, J_x]. Und $(G_y)'_v$, $(G_x)'_v$, $(J_y)'_v$, $(J_x)'_v$ können wir (im Fall $r_v\neq\infty$) als Strahldaten hinter der Fläche Nr. v nehmen.

Der Weg eines Strahls durch das optische System, beschrieben durch die Strahldaten G_y; G_x; J_y; J_x, ist bestimmt durch: die ersten beiden Formeln (2.52), die $(J_y; J_x)_{v+1}$ aus $(J_y; J_x)'_v$ und $(G_y; G_x)'_v$ bestimmen; die ersten beiden Formeln (2.54), die $(G_y; G_x)_{v+1}$ mit $(G_y; G_x)'_v$ identifizieren; die ersten beiden Formeln

(2.53), die $(J_y; J_x)'$ aus $(J_y; J_x)$ bestimmen; und noch abzuleitende Formeln, die anschließend $(G_y; G_x)'$ bestimmen.

Mit den komplexen Zahlen

$$G \equiv G_y + i G_x; \qquad G' \equiv G'_y + i G'_x \tag{2.55}$$

und

$$\left.\begin{aligned} J &\equiv J_y + i J_x; & J' &\equiv J'_y + i J'_x \\ \bar{J} &\equiv \bar{J}_y + i \bar{J}_x; & \bar{J}' &\equiv \bar{J}'_y + i \bar{J}'_x \end{aligned}\right\} \tag{2.56}$$

lassen sich die letztgenannten, jeweils ersten beiden Gleichungen (2.52), (2.54), (2.53) so schreiben:

$$r_{v+1} J_{v+1} = r_v J'_v - e'_v G'_v; \tag{2.57a}$$

$$G_{v+1} \equiv G'_v; \tag{2.57b}$$

$$n' J' = n J. \tag{2.57c}$$

Die noch fehlende Formel für G' lautet

$$\frac{G' + J'}{\bar{G}' + \bar{J}'} = \frac{G + J}{\bar{G} + \bar{J}} \tag{2.57d}$$

und läßt sich folgendermaßen ableiten: Nach (2.44) und (2.56) ist $\boldsymbol{r} n J = (y + i x)\zeta - z(\eta + i\xi)$ und $-\boldsymbol{r} n \bar{J} = z\zeta + (y - i x)(\eta + i\xi)$, also nach (2.55) und (2.43):

$$\left.\begin{aligned} \boldsymbol{r}(G + J) &= (y + i x)\,\bar{G} + (\boldsymbol{r} + z)\,G; \\ \boldsymbol{r}(\bar{G} + \bar{J}) &= (\boldsymbol{r} - z)\,\bar{G} + (y - i x)\,G. \end{aligned}\right\} \tag{2.58}$$

Hieraus folgt, weil [wegen $(x^2 + y^2 + z^2) = \boldsymbol{r}^2 = r^2$]

$$\frac{y + i x}{r - z} = \frac{r + z}{y - i x} \tag{2.59}$$

ist,

$$\frac{G + J}{\bar{G} + \bar{J}} = \frac{y + i x}{r - z}. \tag{2.60}$$

Entsprechend ist

$$\frac{G' + J'}{\bar{G}' + \bar{J}'} = \frac{y + i x}{r - z}. \tag{2.61}$$

Aus (2.60) und (2.61) folgt die obige Formel (2.57d). — Die im Nenner der rechten Seite von (2.57d) stehenden Größen $\bar{G}$ und $\bar{J}$ lassen sich auf G und J zurückführen: Nach (2.45), (2.46), (2.47) und (2.55), (2.56) ist

$$\left.\begin{aligned} &\bar{G}\bar{G}^* + G G^* = 1; \\ &\bar{J}\bar{J}^* + J J^* = 1; \\ &\bar{G}\bar{J}^* + G J^* \text{ reell.} \end{aligned}\right\} \tag{2.62}$$

Da $\bar{G}$ reell ist, hätten wir statt $\bar{G}\bar{G}^*$ auch $\bar{G}^2$ schreiben können. Es ist also $\bar{G}^2 = 1 - G G^*$ und wegen (2.49):

$$\left.\begin{aligned} \bar{G} &= \underset{+}{\sqrt{1 - G G^*}}; \\ \bar{G}' &= \underset{+}{\sqrt{1 - G' G'^*}}. \end{aligned}\right\} \tag{2.57e}$$

Nach der dritten Aussage (2.62) ist $\bar{G}(\bar{J}^* - \bar{J}) + GJ^* - G^*J = 0$. Hieraus, aus der zweiten Gleichung (2.62) und aus (2.48) folgt:

$$\left.\begin{aligned} \bar{J} &= \sqrt[+]{1 - J\,J^* - \bar{J}_x^2} + i\bar{J}_x; & \bar{J}_x &= \frac{GJ^* - G^*J}{2i\bar{G}}; \\ \bar{J}' &= \sqrt[+]{1 - J'\,J'^* - \bar{J}_x'^2} + i\bar{J}_x'; & \bar{J}_x' &= \frac{G'J'^* - G'^*J'}{2i\bar{G}'}. \end{aligned}\right\} \tag{2.57f}$$

Die Formeln (2.57a, b) nennen wir die *komplexen Übergangsformeln*, (2.57c, d) die *komplexen Brechungsformeln*.

13. Windschiefe Invariante; Meridianstrahlen. Einfacher als nach den beiden rechten Formeln (2.57f), sind $\bar{J}_x$ und $\bar{J}_x'$ gemäß den Formeln

$$\left.\begin{aligned} (\bar{J}_x)_v &= \frac{1}{r_v n_v} \cdot I; \\ (\bar{J}_x)_v' &= \frac{1}{r_v n_v'} \cdot I \end{aligned}\right\} \tag{2.57g}$$

gegeben durch I, die sogenannte *windschiefe Invariante*:

Wegen (2.52) und (2.53) ist

$$r_1 n_1 (\bar{J}_x)_1 = r_1 n_1' (\bar{J}_x)_1' = r_2 n_2 (\bar{J}_x)_2 = r_2 n_2' (\bar{J}_x)_2' = \cdots (\equiv I); \tag{2.63}$$

die Größe $r n \bar{J}_x$ hat bei einem bestimmten Strahl überall im System (vor und hinter allen Flächen) den gleichen Wert; diesen Wert bezeichnen wir mit I. Anders ausgedrückt, ist die Größe

$$I \equiv r n \bar{J}_x \tag{2.64}$$

invariant gegenüber dem Übergang:

$$I_{v+1} = I_v' \tag{2.65}$$

und der Brechung:

$$I_w' = I_w. \tag{2.66}$$

Aus (2.63) und (2.57f) folgt (2.57g) mit

$$I = r_1 n_1 (\bar{J}_x)_1 = \left(r n \frac{GJ^* - G^*J}{2i\bar{G}}\right)_1. \tag{2.57h}$$

So sind $(\bar{J}_x)_v$ und $(\bar{J}_x)_v'$ direkt gegeben durch Strahlendaten im Objektraum (vor der ersten Fläche).

Bei einem Strahl mit $I = 0$ ist vor und hinter jeder Fläche $[\boldsymbol{r}\,\boldsymbol{n}]\,\boldsymbol{k} = 0$, wie man an (2.64), (2.44) sieht. Bei jeder Fläche liegen also die Richtung $\boldsymbol{k}$ der optischen Achse, der von der optischen Achse zum Strahl führende Radiusvektor $\boldsymbol{r}$, und die optische Richtung $\boldsymbol{n}$ des Strahls in *einer* Ebene. Der Strahl mit $I = 0$ liegt also mit jedem seiner von Fläche zu Fläche geraden Teilstücke in einer „Meridianebene", d.h. in einer die optische Achse enthaltenden Ebene. Und zwar liegt er mit allen seinen Teilstücken in ein und derselben Meridianebene (abgesehen von dem längs der optischen Achse laufenden Strahl, der überall jeder beliebigen Meridianebene angehört), weil die Meridianebene, in der eines seiner Teilstücke liegt, zugleich die Einfallsebene bei der nächsten Fläche ist, sodaß auch das anschließende Teilstück in derselben Meridianebene bleibt.

Bei einem Strahl mit $I \neq 0$ ist vor und hinter jeder Fläche $[\boldsymbol{r}\,\boldsymbol{n}]\,\boldsymbol{k} \neq 0$, die Vektoren $\boldsymbol{k}, \boldsymbol{r}, \boldsymbol{n}$ liegen also nicht in einer gemeinsamen Ebene. Der Strahl mit $I \neq 0$ liegt also mit keinem seiner Teilstücke in einer Meridianebene. Ein solcher

Strahl heißt „windschiefer Strahl". Die Invariante I heißt „windschiefe Invariante", weil sie dann und nur dann einen von 0 verschiedenen Wert hat, wenn der Strahl windschief ist.

Untersuchen wir die Abbildung eines Objektpunktes, dann können wir wegen der Rotationssymmetrie des abbildenden Systems ohne Beschränkung der Allgemeinheit annehmen, daß der Objektpunkt in der y, z-Ebene liegt. Diese Meridianebene, die alle zu untersuchenden Objektpunkte enthält, nennen wir die „Hauptmeridianebene". — Liegt der Objektpunkt nicht auf der optischen Achse, dann verlaufen alle von ihm ausgehenden nicht-windschiefen Strahlen in der Hauptmeridianebene. Liegt der Objektpunkt auf der optischen Achse, dann sind alle von ihm ausgehenden Strahlen nicht-windschief, und wir können wegen der Rotationssymmetrie die Untersuchung beschränken auf die in der Hauptmeridianebene verlaufenden Strahlen.

Diese in der Hauptmeridianebene, also in der y, z-Ebene, verlaufenden Strahlen nennen wir „Meridianstrahlen". Bei einem solchen Strahl verschwinden die x-Komponenten der Strahldaten $\boldsymbol{r}$ und $\boldsymbol{n}$. Nach (2.43), (2.44) ist dann auch

$$\left.\begin{aligned} G_x &= 0; & G'_x &= 0;\\ J_x &= 0; & J'_x &= 0;\\ \bar{J}_x &= 0; & \bar{J}'_x &= 0. \end{aligned}\right\} \tag{2.67}$$

Die in (2.55), (2.56) definierten G, J und $\bar{J}$ sind bei einem Meridianstrahl also reell, nicht nur $\bar{G}$. Die „komplexen" Übergangs- und Brechungs-Formeln (2.57a, b, c, d) enthalten dann nur reelle Größen. Mit den darin vorkommenden Strahldaten $G=G_y$; $J=J_y$ und $G'=G'_y$; $J'=J'_y$ hängen die in (2.57d) im Nenner auftretenden Größen $\bar{G}$; $\bar{J}=\bar{J}_y$ und $\bar{G}'$; $\bar{J}'=\bar{J}'_y$ nach (2.57e), (2.57f) jetzt einfacher zusammen:

$$\left.\begin{aligned} \bar{G} &= \underset{+}{\sqrt{1-G^2}}; & \bar{G}' &= \underset{+}{\sqrt{1-G'^2}};\\ \bar{J} &= \underset{+}{\sqrt{1-J^2}}; & \bar{J}' &= \underset{+}{\sqrt{1-J'^2}}. \end{aligned}\right\} \tag{2.68}$$

Die Größen G und J nennen wir „Sinus-Größen", und die Größen $\bar{G}$ und $\bar{J}$ nennen wir „Cosinus-Größen", weil nach (2.68) bei einem Meridianstrahl G der Sinus eines Winkels σ ist, von dem $\bar{G}$ der Cosinus ist:

$$G = \sin\sigma; \quad \bar{G} = \cos\sigma, \tag{2.69}$$

und J der Sinus eines Winkels ι, von dem $\bar{J}$ der Cosinus ist:

$$J = \sin\iota; \quad \bar{J} = \cos\iota. \tag{2.70}$$

Nach (2.43), (2.44) ist σ der (mit Vorzeichen genommene) Winkel zwischen dem Meridianstrahl und der optischen Achse, und ι der (mit Vorzeichen genommene) Brechungswinkel.

14. Höhen und Lote. Wir nennen den Streckenvektor, der von einem bestimmten Punkt A der optischen Achse zum Strahl führt: eine *Höhe* $\boldsymbol{H}_A$, wenn dieser Vektor auf der Achse senkrecht steht; und ein *Lot* $\boldsymbol{L}_A$, wenn dieser Vektor auf dem Strahl senkrecht steht. Den Vektor

$$\boldsymbol{Q}_A \equiv \bar{G}\cdot\boldsymbol{H}_A \tag{2.71}$$

nennen wir ein *Zweierlot*; er hat die Richtung der Höhe $\boldsymbol{H}_A$ und ist wie diese *zwei*dimensional (nämlich ohne z-Komponente), und er hat bei Meridianstrahlen die Länge des *Lotes* $\boldsymbol{L}_A$:

Ist P irgendein Punkt auf dem Strahl, und ist

$$\boldsymbol{a} \equiv \overrightarrow{AP}, \tag{2.72}$$

dann gilt: $(\boldsymbol{H}_A - \boldsymbol{a}) \parallel \vec{\sigma}$ und $\boldsymbol{H}_A \perp \boldsymbol{k}$, folglich:

$$\boldsymbol{H}_A = [\boldsymbol{k}[\boldsymbol{a}\vec{\sigma}]]/(\boldsymbol{k}\vec{\sigma}), \tag{2.73}$$

also nach (2.71):

$$\boldsymbol{Q}_A = [\boldsymbol{k}[\boldsymbol{a}\vec{\sigma}]]. \tag{2.74}$$

Nehmen wir als Punkt P den Fußpunkt des Lotes von A auf den Strahl, dann ist $\boldsymbol{a} = \boldsymbol{L}_A$. Nach (2.74) gilt also:

$$\boldsymbol{Q}_A = [\boldsymbol{k}[\boldsymbol{L}_A\vec{\sigma}]]. \tag{2.75}$$

Bei Meridianstrahlen ist $[\boldsymbol{L}_A\vec{\sigma}] \perp \boldsymbol{k}$, also $\boldsymbol{Q}_A^2 = [\boldsymbol{L}_A\vec{\sigma}]^2 = \boldsymbol{L}_A^2$.

Nehmen wir als Punkt P den Auftreffpunkt D des Strahls auf die brechende oder spiegelnde Fläche, und nehmen wir als Punkt A den Mittelpunkt C der Kugelfläche, dann ist $\boldsymbol{a} = \boldsymbol{r}$, und nach (2.74) gilt:

$$\boldsymbol{Q}_C = [\boldsymbol{k}[\boldsymbol{r}\vec{\sigma}]], \tag{2.76}$$

nach (2.44) also:

$$\boldsymbol{Q}_C = r \cdot \{J_x; J_y; 0\}. \tag{2.77}$$

Nach (2.74) und (2.72) ist $\boldsymbol{Q}_A - \boldsymbol{Q}_B = [\boldsymbol{k}[(\overrightarrow{AP} - \overrightarrow{BP})\vec{\sigma}]] = [\boldsymbol{k}[(\overrightarrow{AB})\vec{\sigma}]]$, wegen $(\overrightarrow{AB}) = (AB)\,\boldsymbol{k}$ und (2.43) also:

$$\boldsymbol{Q}_A - \boldsymbol{Q}_B = (AB) \cdot \{G_x; G_y; 0\}. \tag{2.78}$$

Wir fassen die y- und x-Komponente des Zweierlots $\boldsymbol{Q}_A$ zu einer komplexen Größe zusammen:

$$Q_A \equiv (\boldsymbol{Q}_A)_y + i(\boldsymbol{Q}_A)_x, \tag{2.79}$$

die wir ein *komplexes Lot* nennen. A kann z.B. der Scheitel S oder der Mittelpunkt C sein. Das komplexe Lot Q_S nennen wir auch das *Scheitellot*, und das komplexe Lot Q_C das *Mittelpunktslot*. Nach (2.77) und (2.56) ist

$$Q_C = r \cdot J, \tag{2.80}$$

die Sinus-Größe J also gleich dem Mittelpunktslot, dividiert durch den Radius.

Mit den komplexen Größen (2.79) und (2.55) lautet (2.78):

$$Q_A - Q_B = (AB) \cdot G. \tag{2.81}$$

Nehmen wir als Punkt A den Scheitel S und als Punkt B den Mittelpunkt C, dann ist $(AB) = r$, und nach (2.81) und (2.80) gilt:

$$Q_S = r \cdot (G + J). \tag{2.82}$$

$(G+J)$ tritt im Zähler der Formel (2.57d) auf, die als einzige der vier komplexen Übergangs- und Brechungsformeln (2.57) nicht-linear in G und J ist. Nach (2.57d) und (2.82) ist

$$\frac{Q'_S}{\overline{G}' + \overline{J}'} = \frac{Q_S}{\overline{G} + \overline{J}}. \tag{2.57d'}$$

Bei einem Meridianstrahl sind die Scheitellote Q_S und Q'_S reell und dem Betrag nach gleich den Loten, die vom Scheitel S auf den einfallenden Strahl und den aus-fallenden Strahl gefällt sind. — Auch in der Übergangsformel (2.57a) tritt, wenn man e'_v nach (2.15) durch a'_v ausdrückt, $r \cdot (G+J) = Q_S$ auf:

$$(Q_S)_{v+1} = (Q_S)'_v - a'_v G'_v. \tag{2.57a'}$$

III. Gaußsche Näherung; Bildlot-Aberration[1].

15. Paraxiale und Gaußsche Näherung. *Drei* der vier komplexen Übergangs- und Brechungs-Formeln, (2.57a, b, c), sind *linear* in den Strahldaten G und J; die *vierte* Formel, (2.57d), ist *nicht-linear*. Nach (2.50), (2.51) und (2.55), (2.56) ist bei achsennahen Strahlen:

$$\left.\begin{aligned} G &= 0^{(1)}; & \overline{G} &= 1 + 0^{(2)}; \\ J &= 0^{(1)}; & \overline{J} &= 1 + 0^{(2)}, \end{aligned}\right\} \tag{3.1}$$

nach (2.57a, b, c, d) also:

$$r_{v+1} J_{v+1} = r_v J'_v - e'_v G'_v; \tag{3.2a}$$

$$G_{v+1} \equiv G'_v; \tag{3.2b}$$

$$n' J' = n J; \tag{3.2c}$$

$$G' + J' = G + J + 0^{(3)}. \tag{3.2d}$$

Die nicht-lineare vierte Formel wird also linear in den Strahldaten G; J und G'; J' (die kleine Größen erster Ordnung sind), wenn wir die mit $0^{(3)}$ bezeichnete kleine Größe dritter Ordnung vernachlässigen. So werden alle vier Übergangs- und Brechungs-Formeln homogen linear in den Strahldaten, bestimmen diese aber nur näherungsweise. Die durch den linearisierten Formelsatz bestimmten Näherungswerte für G; J bezeichnen wir mit $\hat{g}$; $\hat{j}$:

$$r_{v+1} \hat{j}_{v+1} = r_v \hat{j}'_v - e'_v \hat{g}'_v; \tag{3.3a}$$

$$\hat{g}_{v+1} \equiv \hat{g}'_v; \tag{3.3b}$$

$$n' \hat{j}' = n \hat{j}; \tag{3.3c}$$

$$\hat{g}' + \hat{j}' = \hat{g} + \hat{j}. \tag{3.3d}$$

Ist, was wir annehmen wollen, $G_1 = \hat{g}_1 + 0^{(3)}$ und $J_1 = \hat{j}_1 + 0^{(3)}$, dann gilt nach (3.2) und (3.3) bei allen Flächen-Nummern v:

$$\left.\begin{aligned} G_v &= \hat{g}_v + 0^{(3)}; & G'_v &= \hat{g}'_v + 0^{(3)}; \\ J_v &= \hat{j}_v + 0^{(3)}; & J'_v &= \hat{j}'_v + 0^{(3)}. \end{aligned}\right\} \tag{3.4}$$

Die durch (3.3) bestimmte Näherung nennen wir die *paraxiale* Näherung, weil sie nach (3.4) um so besser den wirklichen Strahlenverlauf wiedergibt, je näher er *bei der Achse* liegt. Ein „Paraxialstrahl" ist ein Strahl, der unendlich nahe bei der Achse liegt[3].

Wir nähern noch weiter, indem wir die Dispersion der Gläser vernachlässigen. In den paraxialen Übergangs- und Brechungs-Formeln (3.3) kommen als Konstruktionsdaten vor: die Radien r der Kugelflächen, deren Mittelpunkts-Ent-

[1] Ohne komplexe Schreibweise wurden die reduzierten Bildlote und deren Aberrationen gegenüber der kollinearen Abbildung der Lote in [2] eingeführt.

[2] H. MARX: Optik **4**, 97—148 (1948).

[3] Bei virtuellen Scheiteln gibt es nur virtuell konstruierte Paraxialstrahlen.

fernungen e, und die Brechungsindizes $n=n(\lambda)$. Im sichtbaren Wellenlängen-Bereich ist nach Ziff. 5 bei optischen Gläsern

$$n=n_e+0(10^{-2}), \tag{3.5}$$

wobei n_e der Brechungsindex bei der Wellenlänge $\lambda_e=546{,}1$ mμ der grünen Hg-Linie e ist. Verstehen wir unter $0(\Gamma)$ eine Größe, die $\sim\Gamma$ *oder* kleiner ist, dann gilt (3.5) auch für Luft. Andere Medien als Luft oder Glas sollen nicht vorkommen. Mit der Bezeichnung

$$\left.\begin{array}{ll} m & \text{für } n_e; \\ 0^{[2]} & \text{für die Größenordnung von } (n-n_e) \end{array}\right\} \tag{3.6}$$

lautet (3.5):

$$n=m+0^{[2]}; \tag{3.5a}$$

in den Abschnitten IV, V und VIII werden wir $0^{[2]}$ ebenso wie $0^{(2)}$ als „kleine Größe zweiter Ordnung" zählen[1]. — Nach (3.3c), (3.5a) und (3.1), (3.4) ist

$$m'\hat{j}'=m\hat{j}+0^{[2]+(1)}, \tag{3.3c'}$$

mit der Bezeichnung

$$0^{[A]+(B)} \quad \text{für} \quad 0^{[A]}\cdot 0^{(B)}. \tag{3.7}$$

Wir vernachlässigen nun die Dispersion der Gläser: Wir streichen bei (3.3c') das Glied $0^{[2]+(1)}$, machen also die paraxialen Übergangs- und Brechungs-Formeln Wellenlängen-unabhängig. Die hierdurch bestimmten Näherungswerte für $\hat{g};\hat{j}$ bezeichnen wir mit $g;j$:

$$r_{v+1}j_{v+1}=r_v j'_v-e'_v g'_v; \tag{3.8a}$$

$$g_{v+1}\equiv g'_v; \tag{3.8b}$$

$$m'j'=mj; \tag{3.8c}$$

$$g'+j'=g+j. \tag{3.8d}$$

Ist, was wir annehmen wollen, $\hat{g}_1=g_1+0^{[2]+(1)}$ und $\hat{j}_1=j_1+0^{[2]+(1)}$, dann gilt nach (3.3a, b, c', d) und (3.8) bei allen Flächen-Nummern v:

$$\left.\begin{array}{ll} \hat{g}_v=g_v+0^{[2]+(1)}; & \hat{g}'_v=g'_v+0^{[2]+(1)}; \\ \hat{j}_v=j_v+0^{[2]+(1)}; & \hat{j}'_v=j'_v+0^{[2]+(1)}, \end{array}\right\} \tag{3.9}$$

also wegen (3.4):

$$\left.\begin{array}{ll} G_v=g_v+0^{(3)}+0^{[2]+(1)}; & G'_v=g'_v+0^{(3)}+0^{[2]+(1)}; \\ J_v=j_v+0^{(3)}+0^{[2]+(1)}; & J'_v=j'_v+0^{(3)}+0^{[2]+(1)}. \end{array}\right\} \tag{3.10}$$

Die durch (3.8) bestimmte Näherung nennen wir die *Gaußsche* Näherung. g und j nennen wir die Gaußschen Näherungswerte von G und J, oder kurz deren „Gaußsche Werte". Den monochromatischen Brechungsindex $m=n(\lambda_e)$ fassen wir auf als den Gaußschen Wert des Wellenlängen-abhängigen Brechungsindex $n=n(\lambda)$.

Die Übergangsformeln (3.8a, b) für die Gaußschen Werte $g;j$ der Strahldaten $G;J$ stimmen mit den Übergangsformeln (3.2a, b) für die wirklichen Werte $G;J$ überein. In der Gaußschen Näherung beschreiben daher die Daten $g'_v;j'_v$ hinter der Fläche Nr. v *ein gleiches Strahlstück* wie die Daten $g_{v+1};j_{v+1}$ vor der Fläche

[1] Das entspricht der in der englichen Sprache üblichen Zählung der fünf Seidelschen monochromatischen Bildfehler *und* der zwei Seidelschen Farbfehler als „primary aberrations" (Bildfehler dritter Ordnung). — Bei den Potenzreihenentwicklungen in Abschnitt VI und VII dagegen werden wir Größen $\sim 0^{[2]}$ wie die Konstruktionsdaten n, r und a als Größen nullter Ordnung zählen.

Nr. $v+1$. Dieses *Gaußsche* Strahlstück zwischen den Flächen Nr. v und Nr. $v+1$ weicht aber im allgemeinen ab von dem zugehörigen *wirklichen* Strahlstück zwischen den Flächen Nr. v und Nr. $v+1$, das durch die Strahldaten G'_v; J'_v oder auch durch die Strahldaten G_{v+1}; J_{v+1} gegeben ist. In der Gaußschen Näherung setzt sich ein Strahl zwar aus lauter einzelnen Strahlstücken zusammen, die jeweils von einer Fläche zur nächsten geradlinig verlaufen; diese Stücke setzen sich aber bei den brechenden (oder reflektierenden) Flächen nicht gemäß den strengen Brechungsformeln aneinander. — Gaußsche Näherungswerte werden wir noch für weitere Größen, nicht nur für G und J, einführen. Das ist dann jeweils eine Größe, die sich auf ein Strahlstück zwischen zwei Flächen (Nr. v und Nr. $v+1$) bezieht, und die als Funktion von G'_v; J'_v; n'_v und Konstanten[1] oder auch als Funktion von G_{v+1}; J_{v+1}; n_{v+1} und Konstanten[1] ausdrückbar ist. Die gleiche Funktion von g'_v; j'_v; m'_v und den Konstanten[2] oder von g_{v+1}; j_{v+1}; m_{v+1} und den Konstanten[2] stellt dann den Gaußschen Näherungswert der betrachteten Größe dar.

16. Kollineare Abbildung der komplexen Lote. Für das Scheitellot Q_S schreiben wir fortan kürzer Q:

$$Q \equiv Q_S. \tag{3.11}$$

Nach (2.82) ist

$$Q = r \cdot (G + J). \tag{3.12}$$

Den Gaußschen Näherungswert hiervon bezeichnen wir mit q:

$$q \equiv r \cdot (g + j). \tag{3.13}$$

Wir schreiben die Formeln (3.8) der Gaußschen Näherung auf die Variablen g und q anstelle von g und j um. Wegen (2.15) und mit

$$\varrho \equiv \frac{1}{r} \tag{3.14}$$

wird aus (3.8a, b, c, d):

$$q_{v+1} = q'_v - a'_v g'_v; \tag{3.15a}$$

$$g_{v+1} \equiv g'_v; \tag{3.15b}$$

$$m'(g' - \varrho q') = m(g - \varrho q); \tag{3.15c}$$

$$q' = q. \tag{3.15d}$$

Dieser Formelsatz (3.15) versagt nicht bei einer Planfläche, im Gegensatz zu dem Formelsatz (3.8).

Ein Meridianstrahl trifft die optische Achse in einem Punkt E, dessen Entfernung von einem anderen Achsenpunkt A

$$(AE) = Q_A / G \tag{3.16}$$

ist. Denn nach (2.81) ist $(AE) \cdot G = Q_A - Q_E$; und weil der Achsen-Punkt E auch auf dem Strahl liegt, ist $Q_E = 0$. Insbesondere gilt für die vom Scheitel S aus gemessene „Schnittweite" (SE):

$$(SE) = Q/G \tag{3.17}$$

mit $Q \equiv Q_S$. Die Gaußsche Näherung liefert für die Strahldaten Q und G des Strahlstücks vor einer bestimmten Fläche die Näherungswerte q und g, für die Schnittweite (SE) dieses Strahlstücks also den Näherungswert

$$s = q/g. \tag{3.18}$$

---

[1] Größen, die — im Gegensatz zu G; J und n — von den Strahldaten und der Wellenlänge des Lichts unabhängig sind (wie z. B. r oder $\mathring{q}$ oder m).

[2] die nicht ausgewechselt werden

Aus den Übergangsformeln (3.15a, b) für q, g [die mit den Übergangsformeln (2.57a'), (3.2b) für Q, G übereinstimmen] folgt eine Übergangsformel für die Gaußsche Schnittweite s *allein*:

$$s_{v+1} = s'_v - a'_v \tag{3.19a}$$

[die mit der Übergangsformel $(SE)_{v+1} = (SE)'_v - a'_v$ für die wirkliche Schnittweite (SE) übereinstimmt]. Aus den Brechungsformeln (3.15c, d) für q, g folgt eine Brechungsformel für die Gaußsche Schnittweite s *allein*:

$$m'\left(\frac{1}{s'} - \varrho\right) = m\left(\frac{1}{s} - \varrho\right) \tag{3.19b}$$

[während die wirkliche Schnittweite $(SE)'$ nach einer Brechung sich im allgemeinen *nicht allein* aus der Schnittweite (SE) vor der Brechung ergibt]. In der Gaußschen Näherung werden nach (3.19a, b) alle von *einem* Achsenpunkt ausgehenden Strahlen nach Brechungen oder Spiegelungen wieder in *einen* Achsenpunkt zusammenlaufen (reell oder virtuell).

Die in Ziff. 15 durch (3.8) definierte Gaußsche Näherung läßt aber im allgemeinen *nicht* die von einem Punkt *außerhalb* der Achse ausgehenden Strahlen wieder in einem Punkt zusammenlaufen. Die in Ziff. 15 eingeführte „Gaußsche Näherung" liefert also *nicht* die unter dem Namen „Gaußsche Abbildung" bekannte kollineare Abbildung der *Höhen*, bei der die Höhe $(\boldsymbol{H}_M)'_w$ (welche die Strahlen hinter der Fläche Nr. w über einem Punkt M haben) eine lineare Funktion der Höhen $(\boldsymbol{H}_A)_v$ und $(\boldsymbol{H}_B)_v$ ist (die die Strahlen vor der Fläche Nr. v über zwei verschiedenen Punkten A und B haben). Sondern nach der „Gaußschen Näherung" (3.8) hat man eine kollineare Abbildung der *Zweierlote*: das Zweierlot $(\boldsymbol{Q}_M)'_w$ ist eine lineare Funktion der Zweierlote $(\boldsymbol{Q}_A)_v$ und $(\boldsymbol{Q}_B)_v$. Das bedeutet nach (2.79), daß das komplexe Lot $(Q_M)'_w$ eine lineare Funktion der komplexen Lote $(Q_A)_v$ und $(Q_B)_v$ ist. Und dies folgt daraus, daß nach (2.80), (2.81) die verschiedenen komplexen Lote Linearkombinationen der komplexen Strahldaten G und J sind, und daß deren Gaußsche Näherungswerte nach (3.8) durch lineare Formeln miteinander verknüpft sind.

17. Objektlot, Pupillenlot; Gaußsche Koeffizienten. Wir nehmen an, hinter der Fläche Nr. b befinde sich in einer achsensenkrechten Ebene die Blende des optischen Systems. Den Achsenpunkt dieser Blenden-Ebene bezeichnen wir mit P'_b. Im (reellen oder virtuellen) Objektraum gibt es nun einen Achsenpunkt P_1, den die Flächen Nr. 1 bis b nach der Gaußschen Näherung (3.19a, b) scharf in den Blenden-Achsenpunkt P'_b abbilden; die Formeln (3.19a, b) bestimmen $s_1 = (S_1 P_1)$ aus $s'_b = (S_b P'_b)$ allein. Diesen Punkt P_1 nennen wir den Achsenpunkt der Eintrittspupille. Unter der „Eintrittspupille" versteht man das bei der Gaußschen Abbildung des Blendenraums in den Objektraum entstehende Bild der Kreisfläche der Blendenöffnung; hiermit wollen wir uns aber nicht befassen. Wir führen die *Eintrittspupille* ein als eine Planfläche mit dem Punkt P_1 als Scheitel. Diese Planfläche zählen wir als Fläche Nr. 0; an ihr werden die Lichtstrahlen weder gebrochen noch gespiegelt:

$$\left.\begin{aligned} r_0 &= \infty; \\ n_0 &= n'_0 \;(\equiv n_1). \end{aligned}\right\} \tag{3.20}$$

Die Lage der Eintrittspupille beschreiben wir durch den Abstand

$$a'_0 \equiv a_1 \equiv S_0 S_1 \tag{3.21}$$

zwischen dem Achsenpunkt $S_0 = P_1$ der Eintrittspupille und dem Scheitel S_1 der Fläche Nr. 1. Je nachdem ob der Achsenpunkt der Eintrittspupille im reellen oder im virtuellen Objektraum liegt, ist a_0' positiv oder negativ. Die Formeln (3.15a, b) gelten auch für $v=0$, die Formeln (3.15c, d) gelten auch bei der Fläche Nr. 0.

Wir wollen die Abbildung von Objektpunkten betrachten, die in *einer* achsensenkrechten Ebene liegen, die wir die *Objektebene* nennen und kurz mit *OE* bezeichnen. Damit wir den Grenzfall einer ins Unendliche hinausgerückten Objektebene nicht als Sonderfall zu behandeln brauchen, beschreiben wir die Lage eines Objektpunkts nicht durch seine kartesischen Koordinaten x_{OE} und y_{OE} in der Objektebene (diese werden unendlich, wenn die *OE* ins Unendliche hinausrückt), sondern durch

$$X_0 \equiv \frac{m_0}{a_0} \cdot x_{OE}; \qquad Y_0 \equiv \frac{m_0}{a_0} \cdot y_{OE}, \tag{3.22}$$

wobei

$$a_0 \equiv S_{OE} S_0$$

die Entfernung zwischen dem Achsenpunkt S_{OE} der Objektebene und dem Achsenpunkt S_0 der Eintrittspupille ist. (Wie in Ziff. 13 schon gesagt, können wir ohne Beschränkung der Allgemeinheit die Untersuchung beschränken auf Objektpunkte in der y, z-Ebene. Wir werden aber erst später dementsprechend $x_{OE}=0$ und damit $X_0=0$ setzen.)

Die Lage eines Strahls im Objektraum können wir beschreiben durch

$$X_0, Y_0 \quad \text{und} \quad x_0, y_0; \tag{3.23}$$

x_0, y_0 sind die x-, y-Koordinaten des Punkts, in dem der Strahl die Eintrittspupille (Fläche Nr. 0) trifft.

Nach (2.79), (2.71) und (3.11) ist das Eintrittspupillen-Scheitellot

$$Q_0 = \overline{G}_0 \cdot (y_0 + i x_0) \tag{3.24}$$

und das Objektebenen-Scheitellot

$$u_0 \equiv Q_{(S_{OE})} = \overline{G}_0 \cdot (y_{OE} + i x_{OE}). \tag{3.25}$$

Q_0 nennen wir kurz das *Eintrittslot*, und $u_0 \equiv Q_{(S_{OE})}$ kurz das *Objektlot*. Das mit m_0/a_0 multiplizierte Objektlot,

$$U_0 \equiv \frac{m_0}{a_0} \cdot u_0, \tag{3.26}$$

nennen wir das *reduzierte Objektlot*. Statt durch (3.23) können wir den Strahl im Objektraum auch beschreiben durch das reduzierte Objektlot

$$U_0 = \overline{G}_0 \cdot (Y_0 + i X_0) \tag{3.27a}$$

und das Eintrittslot

$$Q_0 = \overline{G}_0 \cdot (y_0 + i x_0). \tag{3.27b}$$

Nach (2.81) ist $(u_0 - Q_0) = a_0 \cdot G_0$, also wegen (3.26):

$$G_0 = \frac{1}{m_0} U_0 - \frac{1}{a_0} Q_0. \tag{3.28}$$

So sind die Strahldaten G; Q vor der Fläche Nr. 0 Linearkombinationen des reduzierten Objektlots U_0 und des Eintrittslots Q_0. Da die Gaußsche Näherung (3.15) die Näherungswerte g; q aller Strahldaten G; Q linear miteinander ver-

verknüpft, sind also nicht nur[1]

$$g_0 = G_0; \quad q_0 = Q_0, \tag{3.29}$$

sondern alle g_v; q_v und g'_v; q'_v Linearkombinationen des reduzierten Objektlots U_0 und des Eintrittslots Q_0:

$$\left.\begin{aligned} g_v &= \overset{\times}{g}_v U_0 + \overset{\circ}{g}_v Q_0; & g'_v &= \overset{\times}{g}{}'_v U_0 + \overset{\circ}{g}{}'_v Q_0; \\ q_v &= \overset{\times}{q}_v U_0 + \overset{\circ}{q}_v Q_0; & q'_v &= \overset{\times}{q}{}'_v U_0 + \overset{\circ}{q}{}'_v Q_0. \end{aligned}\right\} \tag{3.30}$$

Die Koeffizienten $\overset{\times}{g}, \overset{\circ}{g}, \overset{\times}{q}, \overset{\circ}{q}$ dieser Linearkombinationen nennen wir *Gaußsche Koeffizienten*. Sie sind Funktionen der Konstruktionsdaten des optischen Systems:

Aus (3.15) und (3.30) mit $Q_0 = 0$ folgt, daß die für die g; q gültigen Formeln auch für die $\overset{\times}{g}$; $\overset{\circ}{q}$ gelten:

$$\overset{\times}{q}_{v+1} = \overset{\times}{q}{}'_v - a'_v \overset{\times}{g}{}'_v; \tag{3.31a}$$

$$\overset{\times}{g}_{v+1} \equiv \overset{\times}{g}{}'_v; \tag{3.31b}$$

$$m'(\overset{\times}{g}{}' - \varrho \overset{\times}{q}{}') = m(\overset{\times}{g} - \varrho \overset{\times}{q}); \tag{3.31c}$$

$$\overset{\times}{q}{}' = \overset{\times}{q}. \tag{3.31d}$$

Und mit $U_0 = 0$ folgt aus (3.15) und (3.30) das Entsprechende für die $\overset{\circ}{g}$; $\overset{\circ}{q}$:

$$\overset{\circ}{q}_{v+1} = \overset{\circ}{q}{}'_v - a'_v \overset{\circ}{g}{}'_v; \tag{3.32a}$$

$$\overset{\circ}{g}_{v+1} \equiv \overset{\circ}{g}{}'_v; \tag{3.32b}$$

$$m'(\overset{\circ}{g}{}' - \varrho \overset{\circ}{q}{}') = m(\overset{\circ}{g} - \varrho \overset{\circ}{q}); \tag{3.32c}$$

$$\overset{\circ}{q}{}' = \overset{\circ}{q}. \tag{3.32d}$$

Nach (3.30), (3.29) und (3.28) ist

$$\left.\begin{aligned} \overset{\times}{g}_0 &= \frac{1}{m_0}; & \overset{\circ}{g}_0 &= -\frac{1}{a_0}; \\ \overset{\times}{q}_0 &= 0; & \overset{\circ}{q}_0 &= 1. \end{aligned}\right\} \tag{3.33}$$

Die Formeln (3.31), (3.32) und (3.33) bestimmen die Gaußschen Koeffizienten $\overset{\times}{g}, \overset{\circ}{g}, \overset{\times}{q}, \overset{\circ}{q}$ aus den Konstruktionsdaten r, m, a des optischen Systems. Man sieht, daß die Gaußschen Koeffizienten (wie die Konstruktionsdaten) reelle Größen nullter Ordnung sind.

Aus (3.31a, b) und (3.32a, b) folgt

$$(\overset{\circ}{q}\overset{\times}{g} - \overset{\times}{q}\overset{\circ}{g})_{v+1} = (\overset{\circ}{q}\overset{\times}{g} - \overset{\times}{q}\overset{\circ}{g})'_v,$$

und aus (3.31c, d) und (3.32c, d) folgt

$$(m\overset{\circ}{q}\overset{\times}{g} - m\overset{\times}{q}\overset{\circ}{g})'_v = (m\overset{\circ}{q}\overset{\times}{g} - m\overset{\times}{q}\overset{\circ}{g})_v.$$

Also ist die Größe

$$l \equiv m(\overset{\circ}{q}\overset{\times}{g} - \overset{\times}{q}\overset{\circ}{g})$$

invariant gegenüber Brechung und Übergang:

$$l'_v = l_v;$$

$$l_{v+1} = l'_v.$$

[1] Im Objektraum lassen wir die Gaußschen Werte g_0 und q_0 der Strahldaten G_0 und Q_0 mit ihren wirklichen Werten übereinstimmen. Später schließen wir die Gaußsche Näherung auch in anderer Weise an den wirklichen Strahlenverlauf an.

Nun ist nach (3.33) $l_0 = m_0(\overset{\circ}{q}_0 \overset{\times}{g}_0 - \overset{\times}{q}_0 \overset{\circ}{g}_0) = 1$. Die Größe l hat daher vor und hinter jeder Fläche den Wert 1:

$$m(\overset{\circ}{q}\,\overset{\times}{g} - \overset{\times}{q}\,\overset{\circ}{g}) = 1. \tag{3.34}$$

18. Gaußsche Bildebene. Meist soll das optische System dazu dienen, Punkte einer Objekt*ebene* (OE) scharf und verzeichnungsfrei abzubilden in eine Bild*ebene* (BE). Diese *ideale Abbildung* ist aber, von Ausnahmefällen abgesehen, in Strenge nicht realisierbar. So läßt man *Aberrationen* gegenüber der idealen Abbildung zu in einem geringen Ausmaß, das bei der vorgesehenen Anwendung des optischen Systems noch nicht stört:

$$\left.\begin{aligned} y_{BE} &= \mathrm{V}\cdot y_{OE} + \text{zulässige Aberrationen};\\ x_{BE} &= \mathrm{V}\cdot x_{OE} + \text{zulässige Aberrationen}; \end{aligned}\right\} \tag{3.35}$$

$\mathrm{V} = \mathrm{const.}$

In der Gaußschen Näherung (3.19a, b) wird der Achsenpunkt S_{OE} der Objektebene nach den Brechungen oder Spiegelungen an allen Flächen des Systems scharf abgebildet in einen Achsenpunkt, den wir mit S_{GBE} bezeichnen; die Formeln (3.19a, b) bestimmen $s'_k = (S_k S_{GBE})$ aus $s_1 = (S_1 S_{OE})$ allein. Diesen Punkt S_{GBE} nennen wir den Achsenpunkt der Gaußschen Bildebene; die *Gaußsche Bildebene* (GBE) ist die im Punkt S_{GBE} auf der Achse senkrecht stehende Ebene.

Als Bildebene BE, z.B. als Filmebene bei der photographischen Abbildung, wird eine Ebene gewählt, in der die Aberrationen gegenüber der idealen Abbildung am geringsten sind. Diese optimale Bildebene BE fällt gewöhnlich nicht mit der Gaußschen Bildebene GBE zusammen, liegt aber in deren Nähe (weil in der Gaußschen Näherung S_{OE} scharf in S_{GBE} abgebildet wird). Da man aus den Aberrationen in der GBE in einfacher Weise auf die Unschärfe und Verzeichnung in der benachbarten BE und auch auf die Lage dieser optimalen BE schließen kann, können wir die Abbildung der OE in die *optimale* Bildebene BE beschreiben durch die Aberrationen in der *Gaußschen* Bildebene GBE.

Bei der Gaußschen Näherung (3.8) weicht, von der Dispersion ($n \neq m$) abgesehen, nur *eine* dieser vier Formeln (3.8) ab von den strengen Formeln (2.57a, b, c, d). Deshalb sind die Abweichungen des wirklichen Strahlenverlaufs gegenüber dem Strahlenverlauf nach der *Gaußschen Näherung* einfacher darzustellen, als wenn wir von einer Näherung mit *zwei* abweichenden Formeln ausgehen würden. Dies wäre der Fall, wenn wir von der in Ziff. 16 erwähnten kollinearen Abbildung der *Höhen* statt von der durch (3.8) gegebenen kollinearen Abbildung der komplexen Lote ausgehen würden. Die kollineare Abbildung der Höhen (auch „Gaußsche Abbildung" genannt) ergibt sich aus Übergangs- und Brechungs-Formeln, die nicht wie (3.8) die Sinus-Größen G und J linear miteinander verknüpfen, sondern die „Tangens-Größen" $G/\overline{G}$ und $J/\overline{G}$. Zwei der vier strengen Übergangs- und Brechungs-Formeln (2.57a, b, c, d) lassen sich linear in $G/\overline{G}$ und $J/\overline{G}$ schreiben:

$$r_{v+1}\left(\frac{J}{\overline{G}}\right)_{v+1} = r_v\left(\frac{J}{\overline{G}}\right)'_v - e'_v\left(\frac{G}{\overline{G}}\right)'_v; \tag{2.57a*}$$

$$\left(\frac{G}{\overline{G}}\right)_{v+1} = \left(\frac{G}{\overline{G}}\right)'_v; \tag{2.57b*}$$

zwei Formeln können aber nicht linear in $G/\overline{G}$ und $J/\overline{G}$ geschrieben werden:

$$n'\,\overline{G}'\cdot\left(\frac{J}{\overline{G}}\right)' = n\,\overline{G}\cdot\left(\frac{J}{\overline{G}}\right); \tag{2.57c*}$$

$$\frac{\left(\frac{G}{\overline{G}}\right)' + \left(\frac{J}{\overline{G}}\right)'}{1 + \overline{J}'/\overline{G}'} = \frac{\left(\frac{G}{\overline{G}}\right) + \left(\frac{J}{\overline{G}}\right)}{1 + \overline{J}/\overline{G}}. \tag{2.57d*}$$

Bei diesen *zwei* Formeln müßten also Vernachlässigungen vorgenommen werden, um einen in den Tangens-Größen linearen Formelsatz und damit die kollineare Abbildung der Höhen zu erhalten.

Letzten Endes interessieren wir uns für die Aberrationen gegenüber der idealen Abbildung der Objektebene in die Gaußsche Bildebene, also für eine Abweichung gegenüber der kollinearen Abbildung der Höhen und nicht der komplexen Lote. Wir wählen aber den kleinen Umweg über eine Untersuchung der Abweichungen gegenüber der kollinearen Abbildung der komplexen Lote, weil diese Abweichungen einfacher darzustellen sind.

19. Bildlot-Aberration, Flächenanteile. Ein aus der Mitte S_{OE} der Objektebene kommender Strahl (bei dem also $U_0=0$ ist) trifft nach (3.30), (3.18) vor der v-ten Fläche die optische Achse bei einer Schnittweite (SE), die in Gaußscher Näherung gleich

$$\mathring{s}_v = \frac{\mathring{q}_v}{\mathring{g}_v} \tag{3.36a}$$

ist; entsprechend ist

$$\mathring{s}'_v = \frac{\mathring{q}'_v}{\mathring{g}'_v}. \tag{3.36b}$$

Die im Punkt $\mathring{E}_v$ (mit der Entfernung $\mathring{s}_v$ von S_v) auf der optischen Achse senkrechte Ebene nennen wir die Gaußsche Zwischenbildebene vor der v-ten Fläche, und die im Punkt $\mathring{E}'_v$ achsensenkrechte Ebene (mit dem Abstand $\mathring{s}'_v$ von S_v) die Gaußsche Zwischenbildebene hinter der v-ten Fläche. Hinter der letzten (k-ten) Fläche sprechen wir, wie in Ziff. 18, einfach von der Gaußschen Bildebene; $\mathring{E}'_k$ ist identisch mit dem in Ziff. 18 genannten Achsenpunkt S_{GBE} der Gaußschen Bildebene. Vor der nullten Fläche haben wir die Objektebene; $\mathring{E}_0$ ist identisch mit dem Achsenpunkt S_{OE} der Objektebene.

Das $\mathring{E}_v$-Lot des wirklichen Strahls vor der v-ten Fläche nennen wir sein dortiges *Bildlot* und bezeichnen es mit u_v:

$$u_v \equiv (Q_{(\mathring{E}_v)})_v; \tag{3.37a}$$

das $\mathring{E}'_v$-Lot des Strahls hinter der v-ten Fläche nennen wir das Bildlot des Strahls hinter der v-ten Fläche und bezeichnen es mit u'_v:

$$u'_v \equiv (Q_{(\mathring{E}'_v)})'_v. \tag{3.37b}$$

Wegen $\mathring{E}'_v \equiv \mathring{E}_{v+1}$ ist

$$u'_v \equiv u_{v+1}. \tag{3.38}$$

Nach (3.37a), (3.36a) und (2.81) ist

$$u_v = \frac{1}{\mathring{g}_v}(\mathring{g}\,Q - \mathring{q}\,G)_v; \tag{3.39a}$$

entsprechend ist

$$u'_v = \frac{1}{\mathring{g}'_v}(\mathring{g}\,Q - \mathring{q}\,G)'_v. \tag{3.39b}$$

In Gaußscher Näherung haben nach (3.39) und (3.30) u_v und u'_v die Werte

$$(u_v)_{\mathfrak{G}} = \frac{1}{\mathring{g}_v}(\mathring{g}\,\check{q} - \mathring{q}\,\check{g})_v \cdot U_0;$$

$$(u'_v)_{\mathfrak{G}} = \frac{1}{\mathring{g}'_v}(\mathring{g}\,\check{q} - \mathring{q}\,\check{g})'_v \cdot U_0,$$

nach (3.34) also die Werte

$$\left.\begin{aligned}(u_v)_{\mathfrak{G}} &= \frac{-1}{m_v \overset{\circ}{g}_v} \cdot U_0; \\ (u'_v)_{\mathfrak{G}} &= \frac{-1}{m'_v \overset{\circ}{g}'_v} \cdot U_0.\end{aligned}\right\} \tag{3.40}$$

Das $(-m_v \overset{\circ}{g}_v)$-fache des *Gaußschen* Bildlots $(u_v)_{\mathfrak{G}}$ ist also gleich dem reduzierten Objektlot, das nach (3.26), (3.33) gleich dem $(-m_0 \overset{\circ}{g}_0)$-fachen des Objektlots u_0 ist:

$$U_0 = -m_0 \overset{\circ}{g}_0 \cdot u_0. \tag{3.41a}$$

Das $(-m_v \overset{\circ}{g}_v)$-fache des *wirklichen* Bildlots u_v bezeichnen wir mit U_v:

$$U_v \equiv -m_v \overset{\circ}{g}_v \cdot u_v; \tag{3.42a}$$

entsprechend:

$$U'_v \equiv -m'_v \overset{\circ}{g}'_v \cdot u'_v \; [= U_{v+1}]. \tag{3.42b}$$

U_v und U'_v nennen wir die ***reduzierten Bildlote*** vor und hinter der v-ten Fläche. Ihre Gaußschen Näherungswerte

$$\left.\begin{aligned}(U_v)_{\mathfrak{G}} &= -m_v \overset{\circ}{g}_v \cdot (u_v)_{\mathfrak{G}}; \\ (U'_v)_{\mathfrak{G}} &= -m'_v \overset{\circ}{g}'_v \cdot (u'_v)_{\mathfrak{G}} \; [= (U_{v+1})_{\mathfrak{G}}]\end{aligned}\right\} \tag{3.43}$$

sind, wie gesagt, gleich dem reduzierten Objektlot U_0:

$$(U'_v)_{\mathfrak{G}} = (U_v)_{\mathfrak{G}} = U_0. \tag{3.44}$$

Die reduzierten Bildlote U_v und U'_k des wirklichen Strahls weichen im allgemeinen von dem Gaußschen Wert (3.44) ab; diese Abweichungen

$$\left.\begin{aligned}\Delta U_v &\equiv U_v - (U_v)_{\mathfrak{G}} = U_v - U_0; \\ \Delta U'_v &\equiv U'_v - (U'_v)_{\mathfrak{G}} = U'_v - U_0\end{aligned}\right\} \tag{3.45}$$

nennen wir die *Bildlot-Aberrationen* vor und hinter der v-ten Fläche.

Nach (3.42), (3.39) ist

$$\left.\begin{aligned}U_v &= m_v (\overset{\circ}{q} G - \overset{\circ}{g} Q)_v; \\ U'_v &= m'_v (\overset{\circ}{q} G - \overset{\circ}{g} Q)'_v.\end{aligned}\right\} \tag{3.46}$$

Die Bildlot-Aberration $\Delta U'_k \equiv U'_k - U_0$ hinter der letzten (k-ten) Fläche des Systems ist wegen $U'_v \equiv U_{v+1}$ und $U_0 = U'_0$:

$$\Delta U'_k = \sum_{v=1}^{k} (U'_v - U_v); \tag{3.47}$$

zu $\Delta U'_k$ trägt die v-te Fläche den Summanden $(U'_v - U_v)$ bei, den wir den *Anteil* dieser Fläche an der Bildlot-Aberration nennen. Nach (3.46) und (3.32c, d) ist dieser Anteil, ohne die Flächennummer v geschrieben,

$$(U' - U) = m(\varrho \overset{\circ}{q} - \overset{\circ}{g}) \cdot (Q' - Q) - \overset{\circ}{q} \cdot [m'(\varrho Q' - G') - m(\varrho Q - G)]. \tag{3.48}$$

Setzen wir, entsprechend (3.30),

$$j_v = \overset{\times}{j}_v U_0 + \overset{\circ}{j}_v Q_0; \qquad j'_v = \overset{\times}{j}{}'_v U_0 + \overset{\circ}{j}{}'_v Q_0, \tag{3.49a}$$

dann wird nach (3.13) und (3.30)

$$\overset{\circ}{q} = r \cdot (\overset{\circ}{g} + \overset{\circ}{j}); \qquad \overset{\times}{q} = r \cdot (\overset{\times}{g} + \overset{\times}{j}), \tag{3.49b}$$

und dann können wir $m\mathring{j}$ für $m(\varrho\mathring{q}-\mathring{g})$ schreiben. Wegen (3.12) können wir mJ für $m(\varrho Q-G)$ schreiben. So wird aus (3.48):

$$(U'-U)=m\mathring{j}\cdot(Q'-Q)-\mathring{q}\cdot(m'J'-mJ). \tag{3.50}$$

Damit ist der Anteil einer Fläche an der Bildlot-Aberration $\Delta U_k'$ zurückgeführt auf:

1. den bei Paraxialstrahlen verschwindenden, bei wirklichen Strahlen aber vorhandenen Unterschied zwischen den Scheitelloten Q und Q' vor und hinter der Fläche $\left[\text{nach (2.57d') ist } \frac{Q'}{G'+J'}=\frac{Q}{G+J}\right]$;

2. den bei Strahlen der Hauptfarbe e verschwindenden, bei Strahlen anderer Farbe aber vorhandenen Unterschied zwischen mJ und $m'J'$ [nach (2.57c) ist $n'J'=nJ$].

Die Summe der Flächen-Anteile (3.50) ergibt nach (3.47) die Bildlot-Aberration $\Delta U_k'\equiv U_k'-U_0$, also die Aberration von

$$U_k'=-m_k'\mathring{g}_k'\cdot\bar{G}_k'(y_{GBE}+i\,x_{GBE}) \tag{3.51}$$

gegenüber

$$U_0=-m_0\,\mathring{g}_0\cdot\bar{G}_0(y_{OE}+i\,x_{OE}). \tag{3.52}$$

(3.51) folgt aus (3.42b), (3.37b), (2.79), (2.71) und $\mathring{E}_k'\equiv S_{GBE}$; (3.52) folgt aus (3.41a) und (3.25).

IV. Iterations-Formelsatz für die Bildhöhen-Aberration windschiefer Strahlen.

20. Der Faktor α oder β. Im Ausdruck (3.50) für den Anteil $(U'-U)$ einer Fläche an der Bildlot-Aberration des Systems ist der erste Summand

$$\mathrm{A}\equiv m\,\mathring{j}\,(Q'-Q) \tag{4.1}$$

nach (3.12) und (3.2d) bei achsennahen Strahlen eine kleine Größe dritter Ordnung:

$$\mathrm{A}=0^{(3)}. \tag{4.2}$$

Beim zweiten Summanden

$$\mathrm{B}\equiv-\mathring{q}\,(m'J'-mJ) \tag{4.3}$$

kann man wegen (3.2c) die Größe

$$\beta\equiv\left(\frac{m'}{n'}-\frac{m}{n}\right) \tag{4.4}$$

als Faktor auftreten lassen:

$$\mathrm{B}=-\beta\cdot\mathring{q}\cdot nJ. \tag{4.5}$$

Nach (3.5a), (1.47) ist

$$\beta=0^{[2]}, \tag{4.6}$$

also nach (3.1) und (3.7):

$$\mathrm{B}=0^{[2]+(1)}. \tag{4.7}$$

Der Ausdruck (4.5) stellt $\mathrm{B}=0^{[2]+(1)}$ dar als ein Produkt mit Faktoren niederer, jedoch nicht negativer Ordnung: $0^{[2]}$; $0^{(0)}$; $0^{(1)}$.

Auch $\mathrm{A}=0^{(3)}$ werden wir auf Größen niederer Ordnung zurückführen. Wir werden die Rechnung so führen, daß in dem Ausdruck für A die Größe

$$\alpha\equiv\left(\frac{1}{n'}-\frac{1}{n}\right) \tag{4.8}$$

als gemeinsamer Faktor auftritt. Wir wollen nämlich α später als eine kleine Größe behandeln.

Bei einer Glas-Glas-Fläche (Kittfläche) ist nach (1.47)

$$|\alpha| < 0{,}2 \tag{4.9a}$$

und bei einer Glas-Luft-Fläche

$$0{,}3 < |\alpha| < 0{,}5. \tag{4.9b}$$

Bei einer Spiegelfläche an Glas ist nach (1.47)

$$1{,}0 < |\alpha| < 1{,}4 \tag{4.10a}$$

und bei einer Spiegelfläche an Luft

$$|\alpha| = 2. \tag{4.10b}$$

Nun schließen wir Spiegelflächen aus. Dann ist $|\alpha| < 0{,}5$; α hat um so kleinere Werte, je näher n' bei n liegt.

Im Hinblick darauf, daß wir α später als eine kleine Größe behandeln, nennen wir α oder eine Größe $\sim 0(\alpha)$ eine (kleine) Größe „ersten Grades", im Gegensatz zu einer Größe erster Ordnung wie z.B. Q, deren Größenordnung unabhängig ist von der von α, und die in dem Maße klein wird, wie der Strahl sich der Achse nähert (vgl. Ziff. 11). Eine Größe, die wie z.B. $J' - J = \alpha \cdot n J$ sowohl $\sim 0(\alpha)$ als auch $\sim 0^{(1)}$ ist, nennen wir eine Größe ersten Grades und erster Ordnung. Eine Größe $\sim 0(\alpha^\gamma) \cdot 0^{(\omega)}$ nennen wir eine Größe γ-ten Grades und ω-ter Ordnung[1].

In dem abzuleitenden Ausdruck für A werden wir neben dem gemeinsamen Faktor α nur Faktoren $\sim 0(\alpha^0)$ auftreten lassen, keine Faktoren negativen Grades, die im Fall kleiner α groß werden wie z.B. $\frac{1}{m' - m}$. Als Größen $\sim 0(\alpha^0)$, also nullten Grades, zählen wir: die Konstruktionsdaten n, $1/n$; r, $1/r$; e, a und die Gaußschen Koeffizienten $\mathring{g}, \check{g}$; $\mathring{\jmath}, \check{\jmath}$ und $\mathring{q}, \check{q}$ (Größen nullter Ordnung), die Sinusgrößen G; J und das Scheitellot Q (Größen erster Ordnung), ferner durch Addition, Subtraktion und Multiplikation gebildete Kombinationen dieser Größen — solange nicht irgendwelche Beziehungen einen Grad > 0 nachweisen[2].

Aus der abzuleitenden Formel für A, mit der wir eine Formel für den Flächenanteil

$$(U' - U)_v = \mathrm{A}_v + \mathrm{B}_v \tag{4.11}$$

der Bildlot-Aberration erhalten, und zusätzlichen Formeln beabsichtigen wir zunächst einen Formelsatz zu bilden, der die Bildhöhen-Aberration windschiefer Strahlen mit den Strahl-Anfangsdaten und geeigneten Konstruktionsdaten in Zusammenhang bringt. Es wird ein Formelsatz sein, der bei iterativer Anwendung die Bildhöhen-Aberration beliebig genau aus den Strahl-Anfangsdaten und den Konstruktionsdaten ergeben würde. Bei Zugrundelegung der Gaußschen Näherung wird schon der erste Iterationsschritt nicht nur die *Seidelschen Bildfehler*, sondern

---

[1] Schreiben wir $\Gamma \sim 0(\varepsilon)$ oder $\Gamma = 0(\varepsilon)$, dann meinen wir, daß die Größe Γ von der Größenordnung von ε *oder* kleiner ist. Entsprechend ist mit $0^{(\omega)}$ eine Größe gemeint, die von ω-ter Ordnung klein *oder* kleiner ist. Für eine Summe aus Größen $\sim 0(\alpha^\gamma) \cdot 0^{[\psi] + (\omega)}$ schreiben wir ebenfalls $0(\alpha^\gamma) \cdot 0^{[\psi] + (\omega)}$.

[2] Zum Beispiel folgt aus $J' - J' = \alpha \cdot n J$, daß $J' - J = 0(\alpha) \cdot 0^{(1)}$ ist. Man kann auch nachrechnen, daß $G' - G = 0(\alpha) \cdot 0^{(1)}$ ist. — Die Zählung der Konstruktionsdaten n; ϱ; a als Größen nullten Grades ist nicht bei allen optischen Systemen gut brauchbar. Unsere Entwicklungen sind an dem in Ziff. 21 im Anschluß an Gl. (4.31) beschriebenen Fall orientiert, der bei den meisten photographischen Objektiven vorliegt.

auch noch *in erster Näherung für kleine* $\alpha \equiv \left(\frac{1}{n'} - \frac{1}{n}\right)$ *und kleine* $\beta \equiv \left(\frac{m'}{n'} - \frac{m}{n}\right)$ *alle Bildfehler höherer Ordnung* liefern. [Zu diesem Zweck werden wir den Faktor α in dem Ausdruck für A auftreten lassen, so wie im Ausdruck (4.5) für B der Faktor β auftritt.] — Diesen Iterations-Formelsatz werden wir (in Abschnitt V) zu einem Iterations-Formelsatz für die meridionale, sagittale und tangentiale Bildhöhen-Aberration umformen[1].

Die Formeln des genannten *ersten* Iterationsschritts stellen, ähnlich wie bei der bekannten Seidelschen Näherung, den Flächenanteil $(U' - U)_v$ der Bildlot-Aberration dar durch einen Ausdruck, der nur die Konstruktionsdaten bei der Fläche Nr. v und die Gaußschen Näherungswerte von Strahldaten bei der Fläche Nr. v enthält[2]. Insofern sind diese Formeln ähnlich übersichtlich wie die der Seidelschen Näherung. Sie liefern aber nicht nur wie diese die Seidelschen Bildfehler, sondern auch noch einen wesentlichen Teil der Bildfehler höherer Ordnung, in erster Näherung für kleine α und β alle Bildfehler höherer Ordnung.

Die Seidelsche Näherung berücksichtigt bei der Darstellung von $(U' - U)_v$, nach einer Potenzreihenentwicklung, alle Glieder der Ordnungen $0^{(3)}$ und $0^{[2]+(1)}$, und vernachlässigt alle Glieder höherer Ordnung. Die Seidelsche Näherung für $(U' - U)_v$ ist, von der Entwicklung nach $0^{[2]}$ abgesehen, eine erste Näherung für *kleine* Strahldaten G; J. Die in dieser Näherung auftretenden „Seidelschen Bildfehler" spielen aber bekanntlich noch bei Strahlen eine wesentliche Rolle, bei denen (z. B). $|J|$ Werte bis etwa 0,7 annimmt[3]. Auch dann noch ist der Flächenanteil $(U' - U)_v$ *qualitativ* durch die Seidelsche Näherung gegeben. Zwar können in der Summe aller Flächenanteile, weil Anteile verschiedenen Vorzeichens einander aufheben, die Beiträge höherer Ordnung überwiegen. Jedoch hat die Wirkung der Änderung eines einzelnen Konstruktionsdatums auf diese Summe in grober Näherung den Seidelschen Wert, so daß der optische Rechner beim Korrigieren eines Systems die Seidelsche Näherung nicht nur bei achsennahen Strahlen, sondern bei praktisch allen Strahlen mit Nutzen anwenden kann. Freilich reicht diese grobe Näherung allein nicht mehr aus, sobald Aberrationen höherer Ordnung dieser Strahlen korrigiert werden sollen. — In ähnlicher Weise, so ist zu erwarten, treten die in erster Näherung für kleine α (und β)

---

[1] *Zusatz bei der Korrektur:* Hieraus werden wir in Abschnitt VIII eine strenge Darstellung der meridionalen, sagittalen und tangentialen Bildhöhen-Aberration als Funktion der Strahl(anfangs)daten und der (Langeschen) Konstruktionsdaten herleiten. Die Formeln des genannten ersten Iterationsschritts (die wir aus dem Iterations-Formelsatz am Ende von Abschnitt V unmittelbar erhalten) und genauere Darstellungen der meridionalen, sagittalen und tangentialen Aberration (die wir in Abschnitt VIII aus der strengen Darstellung ableiten werden) und auch die strenge Darstellung selbst (zu der wir im Abschnitt VIII kommen) sind recht übersichtlich aufgebaut und erlauben, auch in höherer Ordnung das Zustandekommen und Zusammenwirken der Aberrationen analytisch zu verfolgen.

[2] H. SLEVOGT hat wohl als erster gezeigt, daß die Aberrationen auch in höherer Ordnung unter Umständen in dieser Form näherungsweise dargestellt werden können (H. SLEVOGT, Optik **15**, 597—598, 1958). Er hat eine Näherung dieser Art für monochromatische Meridianstrahl-Aberrationen höherer Ordnung angegeben, die sich bei Fernrohrobjektiven bewährt hat. Die SLEVOGTsche Näherung liefert aber *nicht* in erster Näherung für kleine α alle Bildfehler höherer Ordnung, weil (in unserer Schreibweise) für $(U' - U) = \mathrm{A}$ der Ausdruck

$$\mathrm{A} = m \mathring{\jmath} \cdot r \cdot (\operatorname{asd} G + \operatorname{asd} J - \operatorname{asd} G' - \operatorname{asd} J')$$

mit der „Arcus-Sinus-Differenz"

$$\operatorname{asd} S \equiv \arcsin S - S$$

benutzt wird, bei dem *nicht* α oder eine Größe $\sim 0(\alpha)$ als gemeinsamer Faktor auftritt.

[3] Nach (2.70) bedeutet $|J| = 0{,}7$ bei Meridianstrahlen, daß $|\iota| = 45°$ ist.

auftretenden Bildfehler höherer Ordnung auch bei nur mäßig kleinen α-Werten deutlich und ausgeprägt in Erscheinung[1].

In der nächsten Ziff. 21 führen wir die vorgesehene Umrechnung von $\mathrm{A} \equiv m\,\mathring{j} \cdot (Q' - Q)$ streng durch. Das liefert uns eine Formel für $(U' - U) = \mathrm{A} + \mathrm{B}$, die wir dann zu einem (in Ziff. 22 zusammengestellten) Iterations-Formelsatz für die Bildlot-Aberration windschiefer Strahlen ergänzen. Diesen Formelsatz werden wir in Ziff. 23 und 24 auf *mittlere Strahldaten*, in Ziff. 25 auf *Langesche Konstruktionsdaten* und in Ziff. 26 und 27 zu einem Formelsatz für die *Bildhöhen-Aberration* umschreiben.

21. Ein strenger Ausdruck für den Flächenanteil der Bildlot-Aberration; ein Iterations-Formelsatz für alle vorkommenden Variablen. Für das Verhältnis der Scheitellote vor und hinter einer Fläche gilt nach (2.57d'):

$$\frac{Q'}{Q} = \frac{\bar{G}' + \bar{J}'}{\bar{G} + \bar{J}}, \tag{4.12}$$

also ist

$$\frac{Q' - Q}{Q + Q'} = \frac{(\bar{G}' - \bar{G}) + (\bar{J}' - \bar{J})}{(\bar{G} + \bar{G}') + (\bar{J} + \bar{J}')}. \tag{4.13}$$

Die Differenzen $(\bar{G}' - \bar{G})$ und $(\bar{J}' - \bar{J})$ im Zähler formen wir um. Da

$$\begin{aligned}(\bar{G}' - \bar{G}) \cdot (\bar{J}_y + \bar{J}_y') &= \boldsymbol{k}(\vec{\sigma}' - \vec{\sigma}) \cdot \vec{\varphi}(\vec{\sigma} + \vec{\sigma}') = -[\boldsymbol{k}(\vec{\sigma} + \vec{\sigma}')][\vec{\varphi}(\vec{\sigma}' - \vec{\sigma})] \\ &= (G_y + G_y')(J_y' - J_y) + (G_x + G_x')(J_x' - J_x),\end{aligned}$$

$$\begin{aligned}(G + G')^*(J' - J) &= (G_y + G_y')(J_y' - J_y) + (G_x + G_x')(J_x' - J_x) + \\ &\quad + i[(G_y + G_y')(J_x' - J_x) - (G_x + G_x')(J_y' - J_y)]\end{aligned}$$

und wegen (2.47), (2.53)

$$(\bar{G} + \bar{G}') \cdot (\bar{J}_x' - \bar{J}_x) = -[(G_y + G_y')(J_x' - J_x) - (G_x + G_x')(J_y' - J_y)]$$

ist, können wir

$$(\bar{G}' - \bar{G}) = (G + G')^* \cdot \frac{J' - J}{\bar{J}_y + \bar{J}_y'} + i(\bar{G} + \bar{G}') \cdot \frac{\bar{J}_x' - \bar{J}_x}{\bar{J}_y + \bar{J}_y'} \tag{4.14}$$

setzen. Da ferner

$$\begin{aligned}(\bar{J}_y' - \bar{J}_y) \cdot (\bar{J}_y + \bar{J}_y') &= \bar{J}_y'^2 - \bar{J}_y^2 \\ &= -(J_y'^2 - J_y^2) - (J_x'^2 - J_x^2) - (\bar{J}_x'^2 - \bar{J}_x^2),\end{aligned}$$

$$\begin{aligned}(J + J')^*(J' - J) &= +(J_y'^2 - J_y^2) + (J_x'^2 - J_x^2) + \\ &\quad + i\langle (J_y + J_y')(J_x' - J_x) - (J_x + J_x')(J_y' - J_y)\rangle\end{aligned}$$

[1] *Zusatz bei der Korrektur:* Diese *erste* oder „primäre" Näherung (1), eine „sekundäre" Näherung (2) und eine „differentielle" Näherung (3) werden sich in Abschnitt VIII als drei Zwischenstufen zwischen der Seidelschen Näherung (0) und der dort hergeleiteten strengen Darstellung (4) der Bildhöhenaberration ergeben. Aus (4) erhält man (3), hieraus (2), hieraus (1), hieraus (0) durch vier Näherungsschritte *unterschiedlicher Art*, so daß sich jeweils auch der Bereich der gültigen Aussagen einengt, dieser Bereich bei (1) also größer ist als bei (0). — Daß wir α (und damit gemäß Ziff. 21 auch β) als eine kleine Größe behandeln werden, bedeutet nicht, daß wir unabhängig von der Entwicklung nach Bildfehler-Ordnungen auch nach α entwickeln werden. Sondern wir werden zur Herleitung der Formelsätze für die primäre und die sekundäre Näherung nur nach denjenigen Größen erster oder höherer *Ordnung* entwickeln, die *zugleich auch* Größen ersten oder höheren *Grades* sind. Wir teilen damit z. B. die Größen zweiter Ordnung je nach ihrem Grad in größere (vom Grad 0) und kleinere (vom Grad $\geqq 1$) ein und *verhindern, daß nach den größeren entwickelt wird.* So bringt die Behandlung von α als kleiner Größe nicht eine *zusätzliche* Näherung herein, sondern sie verhindert umgekehrt einen Teil der Vernachlässigungen, die bei den herkömmlichen Entwicklungen nach Bildfehler-Ordnungen gemacht werden.

und wegen (2.53):

$$\langle\ \rangle=0$$

ist, können wir

$$(\bar{J}'_y-\bar{J}_y)=-(J+J')^*\cdot\frac{J'-J}{\bar{J}_y+\bar{J}'_y}-(\bar{J}_x+\bar{J}'_x)\cdot\frac{\bar{J}'_x-\bar{J}_x}{\bar{J}_y+\bar{J}'_y}$$

und

$$(\bar{J}'-\bar{J})=-(J+J')^*\cdot\frac{J'-J}{\bar{J}_y+\bar{J}'_y}+i(\bar{J}+\bar{J}')\cdot\frac{\bar{J}'_x-\bar{J}_x}{\bar{J}_y+\bar{J}'_y} \tag{4.15}$$

setzen.

Aus (4.13), (4.14), (4.15) folgt:

$$\frac{Q'-Q}{Q+Q'}=\frac{(G+G')^*-(J+J')^*}{(\bar{G}+\bar{G}')+(\bar{J}+\bar{J}')}\cdot\frac{J'-J}{\bar{J}_y+\bar{J}'_y}+i\,\frac{\bar{J}'_x-\bar{J}_x}{\bar{J}_y+\bar{J}'_y}, \tag{4.16}$$

daher nach (4.11), (4.1), (4.3) und (2.53), (4.8), (4.4):

$$\left.\begin{aligned}(U'-U)&=m\mathring{\jmath}\cdot(Q'-Q)-\mathring{q}\cdot(m'J'-mJ);\\(Q'-Q)&=\alpha\cdot\frac{Q+Q'}{\bar{J}_y+\bar{J}'_y}\left(\frac{(G+G')^*-(J+J')^*}{(\bar{G}+\bar{G}')+(\bar{J}+\bar{J}')}\,nJ+i\,n\bar{J}_x\right);\\(m'J'-mJ)&=\beta\cdot nJ.\end{aligned}\right\} \tag{4.17}$$

Dieser Ausdruck für den Flächenanteil $(U'-U)$ der Bildlot-Aberration gilt in Strenge. In ihm kommen vor:

1. Größen nullter Ordnung und nullten Grades (vgl. Ziff. 20):

$$\tfrac{1}{2}(\bar{J}_y+\bar{J}'_y)=1+0^{(2)};\quad \tfrac{1}{4}(\bar{G}+\bar{G}'+\bar{J}+\bar{J}')=1+0^{(2)}; \tag{4.18a}$$

$$m\,\mathring{\jmath}=0^{(0)};\qquad \mathring{q}=0^{(0)}. \tag{4.18b}$$

2. α, eine Größe nullter Ordnung und ersten Grades[1]:

$$\alpha_v=0(\alpha)\cdot 0^{(0)}. \tag{4.18c}$$

3. Größen erster Ordnung und nullten Grades:

$$nJ=0^{(1)};\quad (G+G'-J-J')^*=0^{(1)};\quad (Q+Q')=0^{(1)}. \tag{4.18d}$$

4. Eine Größe zweiter Ordnung und nullten Grades:

$$n\bar{J}_x=0^{(2)}. \tag{4.18e}$$

5. β, nach Ziff. 15 als Größe zweiter Ordnung gezählt; wir zählen außerdem β als Größe ersten Grades:

$$\beta_v=0(\alpha)\cdot 0^{[2]}. \tag{4.18f}$$

Die in Ziff. 20 erwähnte erste Näherung für kleine α und kleine β ist nach (4.18f) einfach eine erste Näherung für kleine α.

Bei einer Glas-Luft-Fläche Nr. v ($n'_v=1$) ist nach (4.4) und (4.8) die Größe $\beta_v[=(1-n_e/n)_v]$ das Produkt aus $\alpha_v[=(1-1/n)_v]$ und der reziproken Abbeschen Zahl

$$N_v\equiv 1\Big/\left(\frac{n-1}{n-n_e}\right)_v. \tag{4.19a}$$

---

[1] Unter $0(\alpha)$ verstehen wir summarisch die Größenordnung *aller* vorkommenden α-Werte [genauer gesagt, eine Größe, die von dieser Größenordnung oder kleiner ist (vgl. S. 102, Anm. 1)].

Daß wir bei einer Glas-Luft-Fläche β als Größe ersten Grades zählen, ließe sich daher auch ableiten aus der Festsetzung, die Abbesche Zahl als Größe nullten Grades zu zählen. Und daß wir bei einer Glas-Luft-Fläche β als Größe zweiter Ordnung zählen, ließe sich ableiten aus der Festsetzung, die reziproke Abbesche Zahl [die nach Ziff. 5 Werte hat, die klein gegen 1 sind] als Größe zweiter Ordnung zu zählen.

Aus der Festsetzung, die reziproken Abbeschen Zahlen N als Größen nullten Grades und zweiter Ordnung zu zählen:

$$N = 0^{[2]}, \tag{4.19b}$$

leitet sich auch bei einer Kittfläche (Nr. w) ab, daß β_w eine Größe ersten Grades und zweiter Ordnung ist: Nach (4.4) und (4.19a) ist $-\beta_w = (1-1/n)'_w \times N'_w - (1-1/n)_w N_w$, wegen $N = 0^{[2]}$ und $(1-1/n) = 0(\alpha)$ also $\beta_w = 0(\alpha)\cdot 0^{[2]}$.

Wir kehren zu Gl. (4.17) zurück und sehen, daß

$$\left.\begin{aligned} (U'-U) &= 0(\alpha)\cdot 0^{(3)} + 0(\alpha)\cdot 0^{[2]+(1)}; \\ (Q'-Q) &= 0(\alpha)\cdot 0^{(3)}; \\ (m'J'-mJ) &= 0(\alpha)\cdot 0^{[2]+(1)} \end{aligned}\right\} \tag{4.20}$$

ist. Hieraus und aus (2.57a, b), (3.12); (3.15), (3.13); (3.29) folgt, daß

$$\left.\begin{aligned} \Delta G &\equiv G - g = 0(\alpha)\cdot 0^{(3)} + 0(\alpha)\cdot 0^{[2]+(1)}; \\ \Delta J &\equiv J - j = 0(\alpha)\cdot 0^{(3)} + 0(\alpha)\cdot 0^{[2]+(1)}; \\ \Delta Q &\equiv Q - q = 0(\alpha)\cdot 0^{(3)} + 0(\alpha)\cdot 0^{[2]+(1)} \end{aligned}\right\} \tag{4.21}$$

ist.

Im Ausdruck (4.17) kommt die nach (2.57g) zur windschiefen Invarianten I proportionale Größe $n\bar{J}_x$ vor:

$$n\bar{J}_x = \varrho\cdot I. \tag{4.22}$$

Nach (2.57h) und (3.12) ist

$$I = \left(n\,\frac{G\,Q^* - G^*\,Q}{2i\bar{G}}\right)_1,$$

wegen $G_1 = G'_0 = G_0$ und $Q_1 = Q'_0 - a'_0 G'_0$; $Q'_0 = Q_0$ also auch

$$I = \left(n\,\frac{G\,Q^* - G^*\,Q}{2i\bar{G}}\right)_0.$$

Hieraus folgt wegen (3.28):

$$I = -n_0\cdot\frac{(U_0)_y(Q_0)_x}{m_0\,\bar{G}_0}, \tag{4.23}$$

wenn wir, wie in Ziff. 13 begründet, nur Objektpunkte in der y,z-Ebene betrachten; bei diesen ist nach (3.22)

$$X_0 = 0, \tag{4.24a}$$

nach (3.27a) also

$$(U_0)_x = 0. \tag{4.24b}$$

Um die in (4.17) vorkommende Größe $n\bar{J}_x$ in einfacher Weise auf die „reduzierte Objekthöhe“ Y_0 und das Eintrittslot Q_0 zurückzuführen, bedarf es nach (4.22), (4.23) und (3.27a) keiner Iteration oder Näherung. Danach ist in Strenge

$$(n\bar{J}_x)_v = \varrho_v I = -\varrho_v\cdot\frac{n_0}{m_0}\cdot Y_0\cdot(Q_0)_x. \tag{4.25}$$

$\bar{J}_y$ und $\bar{J}$ sind nach (2.57f) gegeben durch J und $\bar{J}_x$:

$$\left.\begin{aligned}\bar{J}_y &= \underset{+}{\sqrt{1 - J J^* - \bar{J}_x^2}}\,; \\ \bar{J} &= \bar{J}_y + i \bar{J}_x\,;\end{aligned}\right\} \tag{4.26}$$

$\bar{G}$ ist nach (2.57e) gegeben durch G:

$$\bar{G} = \underset{+}{\sqrt{1 - G G^*}}\,. \tag{4.27}$$

Entsprechende Formeln gelten für $\bar{J}'_y, \bar{J}'$ und $\bar{G}'$.

So sind die mit einem Querstrich versehenen Größen im Ausdruck (4.17) in einfacher Weise gegeben durch G; J, G'; J' und objektseitige Strahldaten. Um den angekündigten Iterations-Formelsatz zu bekommen, werden wir zu (4.17), (4.25), (4.26), (4.27) noch die folgenden Formeln für G; J, G'; J' und Q hinzunehmen:

$$\left(\frac{Q}{\mathring{q}}\right)_{v+1} - \left(\frac{Q}{\mathring{q}}\right)'_v = -\,(t_{v+1} - t_v)\, U'_v\,, \tag{4.28a}$$

wobei

$$t \equiv -\frac{\overset{\times}{q}}{\mathring{q}} \tag{4.28b}$$

ist;

$$\mathring{q} \cdot G = +\frac{1}{m}\, U + \mathring{g}\, Q\,; \qquad \mathring{q} \cdot G' = +\frac{1}{m'}\, U' + \mathring{g}'\, Q'\,; \tag{4.29}$$

$$\mathring{q} \cdot J = -\frac{1}{m}\, U + \mathring{j}\, Q\,; \qquad \mathring{q} \cdot J' = -\frac{1}{m'}\, U' + \mathring{j}'\, Q'\,. \tag{4.30}$$

(4.29) und (4.30) folgen aus (3.46) und (3.12), (3.49). (4.28) folgt aus (3.46), (3.32a), (3.31a) und (3.34).

Die in (4.28a) auftretende Differenz

$$t_{v+1} - t_v = \frac{a'_v / m'_v}{\mathring{q}_v\, \mathring{q}_{v+1}} \tag{4.31}$$

ist der sogenannte Seidelsche Abstand zwischen der v-ten und der $(v+1)$-ten Fläche. Er unterscheidet sich von dem reduzierten Abstand a'_v/m'_v zwischen der v-ten und der $(v+1)$-ten Fläche um den Faktor $(\mathring{q}_v\, \mathring{q}_{v+1})$, der im Fall $a_0 = \infty$ (Objektebene im Unendlichen) bei vielen optischen Systemen, z.B. bei den meisten photographischen Objektiven, nicht sehr stark von 1 abweicht. An solche optischen Systeme ist bei unseren Entwicklungen in erster Linie gedacht. Nur bei solchen Systemen ist die in Ziff. 20 eingeführte Zählung der Konstruktionsdaten n, ϱ, a als Größen nullten Grades gut brauchbar; aus dieser Zählung folgt nämlich im Fall $a_0 = \infty$ nach (3.33) und (3.32), daß $\mathring{g} = 0(\alpha)$ und $\mathring{q} = 1 + 0(\alpha)$ ist, dem Grad nach also $\mathring{q}$ nicht stark von 1 abweicht. Wir richten uns nur nach solchen Systemen (schließen andere aber nicht ausdrücklich aus). Der Seidelsche Abstand $(t_{v+1} - t_v)_{a_0 = \infty}$ kann dann gut anstelle des Abstandes $a'_v = S_v S_{v+1}$ als Maß dafür genommen werden, wie weit die Flächen Nr. v und Nr. $v+1$ auseinander liegen.

Wir stellen noch fest: Ist $\bar{S}_v$ das durch die Flächen Nr. $(v-1)$ bis Nr. 0 im *Objektraum* (in der Regel virtuell) entworfene Gaußsche Bild des Scheitels S_v der Fläche Nr. v, dann ist im Fall $a_0 = \infty$; $n_0 = 1$:

$$t_{v+1} - t_v = (\bar{S}_v\, \bar{S}_{v+1})\,;$$

der Seidelsche Abstand der Flächen Nr. v und Nr. $v+1$ ist in Gaußscher Näherung der Abstand, den die Scheitel dieser Flächen für einen Beobachter im

Objektraum zu haben scheinen. Nach (3.30) und (4.28b) ist nämlich

$$q_v' = q_v = \mathring{q}_v (Q_0 - t_v U_0), \tag{4.32}$$

nach (3.17), (3.28) im Fall $a_0 = \infty$; $n_0 = 1$ also

$$(S_0 \overline{S}_v) = [Q_0 / U_0]_{q_v = 0} = t_v; \tag{4.33}$$

t_v ist in Gaußscher Näherung der Abstand, den für einen Beobachter im Objektraum der v-te Flächenscheitel vom Achsenpunkt der Eintrittspupille zu haben scheint. — Hieraus oder auch direkt aus (4.28b), (3.33) folgt

$$t_0 = 0. \tag{4.33a}$$

Wie die Formeln (4.17) und (4.25) bis (4.30) das reduzierte Bildlot U_k' und damit die Bildlot-Aberration $\Delta U_k' = U_k' - U_0$ eines Strahls iterativ aus den Anfangsdaten des Strahls bestimmen, zeigen wir im nächsten Abschnitt. Um der Übersicht willen beschränken wir uns von nun an[1] auf den Fall

$$a_0 = \infty; \quad n_0 = 1, \tag{4.34}$$

den man z.B. bei der photographischen Abbildung unendlich oder sehr weit entfernter Objekte in Luft hat.

22. Zusammenfassung. 1. Vorgegeben sind das reduzierte Objektlot

$$U_0 = (U_0)_y \tag{4A.1a}$$

und das Eintrittslot

$$Q_0 = (Q_0)_y + i (Q_0)_x \tag{4A.1b}$$

eines Strahls, und damit auch seine reduzierte Objekthöhe

$$Y_0 = U_0 / \underset{+}{\sqrt{1 - U_0^2}} \tag{4A.1c}$$

und die windschiefe Invariante

$$I = -Y_0 \cdot (Q_0)_x. \tag{4A.1d}$$

2. Ein Iterationsschritt geht aus von Näherungswerten der reduzierten Bildlote $U_v' \equiv U_{v+1}$ und der Scheitellote Q_v, Q_v':

$$\left.\begin{matrix} U_1'; & U_2'; & \dots & U_k' \\ Q_1, Q_1'; & Q_2, Q_2'; & \dots & Q_k, Q_k' \end{matrix}\right\} \tag{4A.2a}$$

und er bestimmt verbesserte Näherungswerte dieser Größen. Beim ersten Iterationsschritt kann man z.B. ausgehen von der Gaußschen Näherung U_0 für U_1'; U_2'; ... U_k' und der Gaußschen Näherung $q_1 = q_1' = \mathring{q}_1 (Q_0 - t_1 U_0)$; $q_2 = q_2' = \mathring{q}_2 (Q_0 - t_2 U_0)$; ... $q_k = q_k' = \mathring{q}_k (Q_0 - t_k U_0)$ für Q_1, Q_1'; Q_2, Q_2'; ... Q_k, Q_k'. — Die verbesserten Näherungswerte

$$\left.\begin{matrix} \mathfrak{U}_1'; & \mathfrak{U}_2'; & & \mathfrak{U}_k' \\ \mathfrak{Q}_1, \mathfrak{Q}_1'; & \mathfrak{Q}_2, \mathfrak{Q}_2'; & \dots & \mathfrak{Q}_k, \mathfrak{Q}_k' \end{matrix}\right\} \tag{4A.2b}$$

liefert der Iterationsschritt 3, 4, 5.:

3. Durch U und Q sind G und J gegeben:

$$\left.\begin{aligned} \mathring{q} \cdot G &= + \frac{1}{m} U + \mathring{g}\, Q; \\ \mathring{q} \cdot J &= - \frac{1}{m} U + \mathring{j}\, Q, \end{aligned}\right\} \tag{4A.3a}$$

[1] Mit Ausnahme von Ziff. 35—37, wo wir auch $n_0 \neq 1$ zulassen werden.

und damit auch $\overline{G}$ und $\overline{J}$:

$$\left.\begin{aligned} &\overline{G}=\overset{+}{\sqrt{1-GG^*}}\,; \\ &\overline{J}_x=\frac{\varrho}{n}\cdot I\,; \\ &\overline{J}_y=\overset{+}{\sqrt{1-(JJ^*+\overline{J}_x^2)}}\,; \\ &\overline{J}=\overline{J}_y+i\,\overline{J}_x. \end{aligned}\right\} \qquad (4\text{A}.3\,\text{b})$$

Entsprechende Formeln gelten für G'; J' und $\overline{G}'$; $\overline{J}'$.

4. Der verbesserte Näherungswert der Scheitellot-Aberration an einer Fläche ist

$$(\mathfrak{Q}'-\mathfrak{Q})=\alpha\cdot\frac{Q+Q'}{\overline{J}_y+\overline{J}_y'}\left(\frac{(G+G'-J-J')^*}{\overline{G}+\overline{G}'+\overline{J}+\overline{J}'}\,nJ+i\,n\overline{J}_x\right), \qquad (4\text{A}.4\,\text{a})$$

und der verbesserte Näherungswert des Anteils dieser Fläche an der Bildlot-Aberration:

$$(\mathrm{U}'-\mathrm{U})=m\,\mathring{\jmath}\cdot(\mathfrak{Q}'-\mathfrak{Q})-\mathring{q}\cdot\beta\cdot nJ. \qquad (4\text{A}.4\,\text{b})$$

5. Damit erhält man die verbesserten Näherungswerte U_1'; U_2'; ... U_k' der reduzierten Bildlote:

$$\mathrm{U}_v'=U_0+\sum_{\tau=1}^{v}(\mathrm{U}'-\mathrm{U})_\tau \qquad (4\text{A}.5\,\text{a})$$

und, falls für einen weiteren Iterationsschritt benötigt, die verbesserten Näherungswerte $\mathfrak{Q}_1$, $\mathfrak{Q}_1'$; $\mathfrak{Q}_2$, $\mathfrak{Q}_2'$; ... $\mathfrak{Q}_k$, $\mathfrak{Q}_k'$ der Scheitellote:

$$\left.\begin{aligned} &(\mathfrak{Q}/\mathring{q})_0=Q_0\,; \\ &(\mathfrak{Q}/\mathring{q})_0'=Q_0\,; \\ &(\mathfrak{Q}/\mathring{q})_w=(\mathfrak{Q}/\mathring{q})_{w-1}'-(t_w-t_{w-1})\,\mathrm{U}_{w-1}'\,; \\ &(\mathfrak{Q}/\mathring{q})_w'=(\mathfrak{Q}/\mathring{q})_w+(\mathfrak{Q}'-\mathfrak{Q})_w/\mathring{q}_w. \end{aligned}\right\} \qquad (4\text{A}.5\,\text{b})$$

23. Mittlere Strahldaten. Die Anzahl der in dem Iterations-Formelsatz (4A) vorkommenden Strahlvariablen werden wir verringern können.

Zuvor stellen wir fest, von welcher Größenordnung 0_{N+1} die Fehler $\mathfrak{d}\mathrm{U}\,(\equiv\mathrm{U}-U_{\text{streng}})$ und $\mathfrak{d}\,\mathfrak{Q}\,(\equiv\mathfrak{Q}-Q_{\text{streng}})$ *nach* dem N-ten Iterationsschritt sind, wenn 0_N die Größenordnung der Fehler $\mathfrak{d}U(\equiv U-U_{\text{streng}})$ und $\mathfrak{d}\,Q\,(\equiv Q-Q_{\text{streng}})$ *vor* dem N-ten Iterationsschritt war:

$$\mathfrak{d}U=0_N; \qquad \mathfrak{d}Q=0_N. \qquad (4.35)$$

Aus (4A.3a) und (4A.3b) folgt

$$\left.\begin{aligned} &\mathfrak{d}G=0_N; && \mathfrak{d}J=0_N; \\ &\mathfrak{d}\overline{G}=0^{(1)}\cdot 0_N; && \mathfrak{d}\overline{J}_y=0^{(1)}\cdot 0_N; \qquad \mathfrak{d}\overline{J}_x=0. \end{aligned}\right\} \qquad (4.36)$$

Nach (4A.4a), (4A.4b) ist dann

$$\left.\begin{aligned} &\mathfrak{d}(\mathfrak{Q}'-\mathfrak{Q})=0(\alpha)\cdot 0^{(2)}\cdot 0_N; \\ &\mathfrak{d}(\mathrm{U}'-\mathrm{U})=0(\alpha)\cdot 0^{(2)}\cdot 0_N+0(\alpha)\cdot 0^{[2]}\cdot 0_N, \end{aligned}\right\} \qquad (4.37)$$

also haben nach (4A.5a), (4A.5b) die Fehler $\mathfrak{d}\mathrm{U}$ und $\mathfrak{d}\,\mathfrak{Q}$ *nach* dem N-ten Iterationsschritt die Größenordnung

$$0_{N+1}=0(\alpha)\,\{0^{(2)}+0^{[2]}\}\cdot 0_N. \qquad (4.38)$$

Jeder einzelne Iterationsschritt verbessert also die Näherungswerte der reduzierten Bildlote und der Scheitellote in der Weise, daß er bei den Fehlern dieser Näherungswerte die Ordnung um 2 und den Grad um 1 erhöht.

Die Gaußschen Näherungswerte weichen nach (4.21) von den wirklichen Werten ab um Fehler dritter Ordnung und ersten Grades. Legt man bei der Iteration (4A) dem ersten Iterationsschritt die Gaußsche Näherung zugrunde:

$$0_1 = 0(\alpha) \cdot \{0^{(2)} + 0^{[2]}\}\, 0^{(1)}, \tag{4.39a}$$

dann liefert der erste Iterationsschritt eine Näherung, bei der Fehler fünfter Ordnung und zweiten Grades vernachlässigt sind:

$$0_2 = 0(\alpha^2) \cdot \{0^{(2)} + 0^{[2]}\}^2\, 0^{(1)}, \tag{4.39b}$$

und der zweite Iterationsschritt eine Näherung, bei der Fehler siebter Ordnung und dritten Grades vernachlässigt sind:

$$0_3 = 0(\alpha^3) \cdot \{0^{(2)} + 0^{[2]}\}^3\, 0^{(1)}. \tag{4.39c}$$

Die Ausdrücke (4A.3b), (4A.4a), (4A.4b) für $(Q' - Q)$ und $(U' - U)$ enthalten neben Konstruktionsdaten und Gaußschen Koeffizienten, welche bei einem bestimmten optischen System konstante Werte haben, die folgenden von den Strahl-Anfangsdaten abhängigen sieben Variablen:

$$Q;\, G;\, J, \quad Q';\, G';\, J' \quad \text{und} \quad I. \tag{4.40}$$

Wir rechnen jetzt nach, daß man zur Darstellung von $(Q' - Q)$ und $(U' - U)$ mit den vier Variablen

$$Q^{\cdot} \equiv \tfrac{1}{2}(Q + Q'); \quad G^{\cdot} \equiv \tfrac{1}{2}(G + G'); \quad J^{\cdot} \equiv \tfrac{1}{2}(J + J') \quad \text{und} \quad I \tag{4.41}$$

auskommt[1] und somit $(Q' - Q)$ und $(U' - U)$ einfacher darstellen kann. Auf die neuen Variablen (4.41) werden wir anschließend den ganzen Iterations-Formelsatz umschreiben.

$Q^{\cdot}$ nennen wir das *mittlere Scheitellot*; $G^{\cdot}$ und $J^{\cdot}$ nennen wir *mittlere Sinus-Größen*. Mit den mittleren Sinus-Größen hängen die in (4A.4a) auftretenden *mittleren Cosinus-Größen*

$$\overline{G}^{\cdot} \equiv \tfrac{1}{2}(\overline{G} + \overline{G}'); \quad \overline{J}^{\cdot} \equiv \tfrac{1}{2}(\overline{J} + \overline{J}') \tag{4.42}$$

in ähnlicher Weise zusammen, wie $\overline{G}$ mit G und wie $\overline{J}$ mit J:

$$\overline{G}^{\cdot} = \underset{+}{\sqrt{L - G^{\cdot} G^{\cdot *}}} \tag{4.43a}$$

und

$$\overline{J}_x^{\cdot} = \frac{1}{2}\left(\frac{1}{n} + \frac{1}{n'}\right) \varrho \cdot I; \tag{4.43b}$$

$$\overline{J}_y^{\cdot} = \underset{+}{\sqrt{L - J^{\cdot} J^{\cdot *} - \overline{J}_x^{\cdot 2}}}; \tag{4.43c}$$

$$\overline{J}^{\cdot} = \overline{J}_y^{\cdot} + i\, \overline{J}_x^{\cdot}, \tag{4.43d}$$

wobei

$$L \equiv \left(\frac{\overleftrightarrow{\sigma} + \overleftrightarrow{\sigma}'}{2}\right)^2 \tag{4.44}$$

gemäß

$$L = 1 - \left(\frac{n' - n}{n + n'}\right)^2 \Big/ \left(\frac{1}{J^{\cdot} J^{\cdot *} + \overline{J}_x^{\cdot 2}} - \frac{1}{L}\right) \tag{4.45}$$

---

[1] Man könnte $(Q' - Q)$ und $(U' - U)$ sogar durch die *zwei* Variablen $G^{\cdot}$ und $J^{\cdot}$ allein ausdrücken [oder durch zwei andere Variablen wie z. B. $Q^{\cdot}$ und $U^{\cdot} \equiv \tfrac{1}{2}(U + U')$].

durch $J^{\cdot}$, $\bar{J}_x^{\cdot}$ und $\frac{n'-n}{n+n'}$ gegeben ist. Man sieht, daß

$$L = 1 - \left(\frac{n'-n}{n+n'}\right)^2 (J^{\cdot} J^{\cdot *} + \bar{J}_x^{\cdot 2}) + 0(\alpha^2)\cdot 0^{(4)} = 1 + 0(\alpha^2)\cdot 0^{(2)} \tag{4.46}$$

ist.

(4.43 b) folgt aus der zweiten Gl. (4A.3 b). Übrigens ist nach (2.47) auch

$$\bar{G}^{\cdot}\bar{J}_x^{\cdot} = G_x^{\cdot} J_y^{\cdot} - G_y^{\cdot} J_x^{\cdot}. \tag{4.47}$$

(4.43 a) und (4.43 c) in Verbindung mit (4.44) folgen aus

$$\begin{aligned}(\bar{G}+\bar{G}')^2 &= (\boldsymbol{k}\,\vec{\sigma} + \boldsymbol{k}\,\vec{\sigma}')^2 = (\vec{\sigma}+\vec{\sigma}')^2 - [\boldsymbol{k}(\vec{\sigma}+\vec{\sigma}')]^2 \\ &= (\vec{\sigma}+\vec{\sigma}')^2 - (G_y + G_y')^2 - (G_x + G_x')^2\end{aligned}$$

und

$$\begin{aligned}(\bar{J}_y+\bar{J}_y')^2 &= (\vec{\varphi}\,\vec{\sigma} + \vec{\varphi}\,\vec{\sigma}')^2 = (\vec{\sigma}+\vec{\sigma}')^2 - [\vec{\varphi}(\vec{\sigma}+\vec{\sigma}')]^2 \\ &= (\vec{\sigma}+\vec{\sigma}')^2 - (J_y + J_y')^2 - (J_x + J_x')^2 - (\bar{J}_x + \bar{J}_x')^2;\end{aligned}$$

(4.45) folgt aus

$$(\bar{J}_y \pm \bar{J}_y')^2 = (\vec{\sigma} \pm \vec{\sigma}')^2 - \left(\frac{n \pm n'}{n+n'}\right)^2 \cdot 4[J^{\cdot} J^{\cdot *} + \bar{J}_x^{\cdot 2}]$$

und

$$\bar{J}_y^2 - \bar{J}_y'^2 = J'J'^* + \bar{J}_x'^2 - JJ^* - \bar{J}_x^2 = \frac{n-n'}{n+n'}\cdot 4[J^{\cdot} J^{\cdot *} + \bar{J}_x^{\cdot 2}],$$

wonach

$$\{(\vec{\sigma}+\vec{\sigma}')^2 - 4[\;\;]\}\left\{(\vec{\sigma}'-\vec{\sigma})^2 - \left(\frac{n'-n}{n+n'}\right)^2\cdot 4[\;\;]\right\} = \left(\frac{n'-n}{n+n'}\right)^2 \cdot (4[\;\;])^2,$$

also wegen $L \equiv (\vec{\sigma}+\vec{\sigma}')^2/4 = 1 - (\vec{\sigma}'-\vec{\sigma})^2/4$:

$$\left(\frac{n'-n}{n+n'}\right)^2 \frac{1}{1-L} + \frac{1}{L} = \frac{1}{[J^{\cdot} J^{\cdot *} + \bar{J}_x^{\cdot 2}]}$$

ist.

Auf die mittleren Sinus-Größen $G^{\cdot}$; $J^{\cdot}$, die mittleren Cosinus-Größen $\bar{G}^{\cdot}$; $\bar{J}^{\cdot}$ und das mittlere Scheitellot

$$Q^{\cdot} = r(G^{\cdot} + J^{\cdot}) \tag{4.48}$$

umgeschrieben, lauten die Ausdrücke (4A.4 a), (4A.4 b) für $(Q'-Q)$ und $(U'-U)$:

$$\left.\begin{aligned}(Q'-Q) &= -2\,\frac{n'-n}{n+n'}\cdot\frac{Q^{\cdot}}{\bar{J}_y^{\cdot}}\left(\frac{G^{\cdot *} - J^{\cdot *}}{\bar{G}^{\cdot} + \bar{J}^{\cdot}}\,J^{\cdot} + i\bar{J}_x^{\cdot}\right); \\ (U'-U) &= m\mathring{\jmath}\cdot(Q'-Q) - \mathring{q}\cdot 2\,\frac{\dfrac{m'}{n'} - \dfrac{m}{n}}{\dfrac{1}{n} + \dfrac{1}{n'}}\cdot J^{\cdot}.\end{aligned}\right\} \tag{4.49}$$

Der für $(Q'-Q)$ und $(U'-U)$ gewonnene Ausdruck (4.49), (4.43) mit den *mittleren* Größen $Q^{\cdot}$; $G^{\cdot}$; $\bar{G}^{\cdot}$; $J^{\cdot}$; $\bar{J}^{\cdot}$ ist einfacher als der Ausdruck (4A.4 a), (4A.4 b), (4A.3 b) mit den ungestrichenen *und* den gestrichenen Größen Q, Q'; G, G'; J, J'; $\bar{G}, \bar{G}'$; $\bar{J}, \bar{J}'$. Die gemäß (4.45) implizit gegebene Größe L ist wegen (4.46) leicht zu diskutieren. — Die Umformung auf die mittleren Größen setzt $(n+n') \neq 0$ voraus; auch deshalb mußten wir *Spiegelflächen ausschließen.*

Aus der Formel (4A.3 a) für J und der entsprechenden Formel für J' folgt, daß

$$\mathring{q}\cdot\tfrac{1}{2}(mJ + m'J') = -U^{\cdot} + m\mathring{\jmath}\cdot Q^{\cdot} \tag{4.50}$$

ist mit dem *mittleren Bildlot*

$$U^{\bullet} \equiv \tfrac{1}{2}(U+U'). \tag{4.51}$$

Also ist

$$\frac{\frac{m}{n}+\frac{m'}{n'}}{\frac{1}{n}+\frac{1}{n'}}\mathring{q}\cdot J^{\bullet} = -U^{\bullet}+m\mathring{\jmath}\cdot Q^{\bullet}, \tag{4.52a}$$

und nach (4.48):

$$G^{\bullet} = \varrho\, Q^{\bullet} - J^{\bullet}; \tag{4.52b}$$

so lassen sich $G^{\bullet}$ und $J^{\bullet}$ zurückführen auf $U^{\bullet}$ und $Q^{\bullet}$.

Um einen Iterations-Formelsatz zu erhalten, der bei einem Iterationsschritt aus Näherungswerten $U^{\bullet}$; $Q^{\bullet}$ des mittleren Bildlots und des mittleren Scheitellots verbesserte Näherungswerte $\mathrm{U}^{\bullet}$; $\mathrm{Q}^{\bullet}$ und auch U'_k liefert, brauchen wir anstelle der Formeln (4A.5a) und (4A.5b) eine Formel für $\mathrm{U}^{\bullet}_v$ und U'_v, und eine Formel für $\mathrm{Q}^{\bullet}_w$. Mit den Bezeichnungen

$$\sum_{\tau=1}^{v-\frac{1}{2}}(\,)_\tau \equiv (\,)_1+(\,)_2+\cdots+(\,)_{v-1}+\tfrac{1}{2}(\,)_v \tag{4.53a}$$

und

$$\mathrm{U}'_{v-\frac{1}{2}} \equiv \mathrm{U}^{\bullet}_v \tag{4.53b}$$

ist nach (4A.5a) und (4.51):

$$\left.\begin{aligned} &\mathrm{U}'_v = U_0 + \sum_{\tau=1}^{v}(\mathrm{U}'-\mathrm{U})_\tau \\ &(v=\tfrac{1}{2},\ 1,\ 1\tfrac{1}{2},\ 2,\ \ldots\ k-\tfrac{1}{2},\ k) \end{aligned}\right\} \tag{4.54}$$

und nach (4A.5b) und (4.41):

$$\left(\frac{\mathrm{Q}^{\bullet}}{\mathring{q}}\right)_w = Q_0 + \sum_{\tau=1}^{w-\frac{1}{2}}\left(\frac{\mathrm{Q}'-\mathrm{Q}}{\mathring{q}}\right)_\tau - \sum_{\tau=1}^{w}(t_\tau - t_{\tau-1})\,\mathrm{U}'_{\tau-1}. \tag{4.55}$$

In Ziff. 24 stellen wir den auf die *mittleren* Größen umgeschriebenen Iterations-Formelsatz zusammen. Dabei verwenden wir: anstelle von n und $m \equiv n_e$ die reziproken Brechungsindizes

$$\nu \equiv \frac{1}{n}; \quad \mu \equiv \frac{1}{m}, \tag{4.56}$$

für ν/μ die Abkürzung

$$\bar{\nu} \equiv \frac{\nu}{\mu} \equiv \frac{m}{n}, \tag{4.57}$$

anstelle des Gaußschen Koeffizienten $\mathring{q}$ seinen Reziprokwert

$$c \equiv 1/\mathring{q}, \tag{4.58}$$

und für $m\mathring{\jmath}$ die Abkürzung $\mathring{\varkappa}$:

$$\mathring{\varkappa} \equiv m\mathring{\jmath}. \tag{4.59}$$

Mit ν bezeichnen wir also fortan nicht wie in Abschnitt I die Abbesche Zahl, sondern den reziproken Brechungsindex. Von der Abbeschen Zahl werden wir im Folgenden keinen Gebrauch machen.

c nennen wir die *reziproke Höhe*; nach (4.58), (3.30) ist c_v die Höhe über dem nullten Flächenscheitel, dividiert durch die Höhe über dem v-ten Flächenscheitel, und zwar bei einem aus der Mitte der Objektebene ($U_0 = 0$) kommenden Paraxialstrahl der Farbe e.

24. Zusammenfassung. 1. Vorgegeben sind das reduzierte Objektlot

$$U_0 = (U_0)_y \tag{4B.1a}$$

und das Eintrittslot

$$Q_0 = (Q_0)_y + i(Q_0)_x \tag{4B.1b}$$

eines Strahls und damit auch seine reduzierte Objekthöhe

$$Y_0 = U_0 / \underset{+}{\sqrt{1 - U_0^2}} \tag{4B.1c}$$

und die windschiefe Invariante

$$I = -Y_0 \cdot (Q_0)_x. \tag{4B.1d}$$

2. Ein Iterationsschritt geht aus von Näherungswerten der mittleren Bildlote $U_v^{\bullet}$ und der mittleren Scheitellote $Q_v^{\bullet}$:

$$\left.\begin{matrix} U_1^{\bullet};\; U_2^{\bullet};\; \ldots\; U_k^{\bullet} \\ Q_1^{\bullet};\; Q_2^{\bullet};\; \ldots\; Q_k^{\bullet} \end{matrix}\right\} \tag{4B.2a}$$

und er bestimmt verbesserte Näherungswerte dieser Größen und der reduzierten Bildlote. Beim ersten Iterationsschritt kann man z.B. ausgehen von der Gaußschen Näherung U_0 für $U_1^{\bullet}; U_2^{\bullet}; \ldots U_k^{\bullet}$ und der Gaußschen Näherung $(Q_0 - t_1 U_0)/c_1$; $(Q_0 - t_2 U_0)/c_2; \ldots (Q_0 - t_k U_0)/c_k$ für $Q_1^{\bullet}; Q_2^{\bullet}; \ldots Q_k^{\bullet}$. — Die verbesserten Näherungswerte

$$\left.\begin{matrix} \mathrm{U}_1^{\bullet};\; \mathrm{U}_1';\; \mathrm{U}_2^{\bullet};\; \mathrm{U}_2';\; \ldots\; \mathrm{U}_k^{\bullet};\; \mathrm{U}_k' \\ \mathrm{Q}_1^{\bullet};\; \quad \mathrm{Q}_2^{\bullet};\; \quad \ldots\; \mathrm{Q}_k^{\bullet} \end{matrix}\right\} \tag{4B.2b}$$

liefert der Iterationsschritt 3., 4., 5.:

3. Durch U und $Q^{\bullet}$ sind $J^{\bullet}$ und $G^{\bullet}$ gegeben:

$$\left.\begin{aligned} \frac{\overline{\nu} + \overline{\nu}'}{\nu + \nu'} \cdot J^{\bullet} &= -c(U^{\bullet} - \mathring{\varkappa}\, Q^{\bullet}); \\ G^{\bullet} &= \varrho\, Q^{\bullet} - J^{\bullet} \end{aligned}\right\} \tag{4B.3a}$$

und damit auch $\overline{J}^{\bullet}$ und $\overline{G}^{\bullet}$:

$$\overline{J}_x^{\bullet} = \tfrac{1}{2}(\nu + \nu')\, \varrho \cdot I; \tag{4B.3b}$$

$$\left(\frac{\nu' - \nu}{\nu + \nu'}\right)^2 \cdot \frac{1}{1 - L} + \frac{1}{L} = \frac{1}{[J^{\bullet} J^{\bullet *} + \overline{J}_x^{\bullet 2}]}; \tag{4B.3c}$$

$$\left.\begin{aligned} \overline{J}_y^{\bullet} &= \underset{+}{\sqrt{L - [J^{\bullet} J^{\bullet *} + \overline{J}_x^{\bullet 2}]}}; \\ \overline{J}^{\bullet} &= \overline{J}_y^{\bullet} + i\, \overline{J}_x^{\bullet}; \\ \overline{G}^{\bullet} &= \underset{+}{\sqrt{L - G^{\bullet} G^{\bullet *}}}. \end{aligned}\right\} \tag{4B.3d}$$

4. Der verbesserte Näherungswert der Scheitellot-Aberration an einer Fläche ist

$$(\mathrm{Q}' - \mathrm{Q}) = 2\frac{\nu' - \nu}{\nu + \nu'} \cdot \frac{Q^{\bullet}}{\overline{J}_y^{\bullet}} \left(\frac{G^{\bullet *} - J^{\bullet *}}{\overline{G}^{\bullet} + \overline{J}^{\bullet}}\, J^{\bullet} + i\, \overline{J}_x^{\bullet} \right), \tag{4B.4a}$$

und der verbesserte Näherungswert des Anteils dieser Fläche an der Bildlot-Aberration:

$$(\mathrm{U}' - \mathrm{U}) = (\mathrm{Q}' - \mathrm{Q}) \cdot \mathring{\varkappa} - 2\frac{\overline{\nu}' - \overline{\nu}}{\nu + \nu'} \cdot J^{\bullet}/c. \tag{4B.4b}$$

5. Damit erhält man die verbesserten Näherungswerte $U_1^{\bullet}$; U_1'; $U_2^{\bullet}$; U_2'; ... $U_k^{\bullet}$; U_k' der mittleren und der reduzierten Bildlote:

$$\left.\begin{aligned}
U_1^{\bullet} &= \overbrace{U_0'}^{U_0} + \tfrac{1}{2}(U'-U)_1; & U_1' &= U_1^{\bullet} + \tfrac{1}{2}(U'-U)_1;\\
U_2^{\bullet} &= U_1' + \tfrac{1}{2}(U'-U)_2; & U_2' &= U_2^{\bullet} + \tfrac{1}{2}(U'-U)_2;\\
&\vdots & &\vdots\\
U_k^{\bullet} &= U_{k-1}' + \tfrac{1}{2}(U'-U)_k; & U_k' &= U_k^{\bullet} + \tfrac{1}{2}(U'-U)_k
\end{aligned}\right\} \tag{4B.5a}$$

und, falls für einen weiteren Iterationsschritt benötigt, die verbesserten Näherungswerte $Q_1^{\bullet}$; $Q_2^{\bullet}$; ... $Q_k^{\bullet}$ der mittleren Scheitellote:

$$\left.\begin{aligned}
[c_1 Q_1^{\bullet}] &= [\overbrace{c_0}^{1}\overbrace{Q_0^{\bullet}}^{Q_0}] + \tfrac{1}{2}c_0 \overbrace{(Q'-Q)_0}^{0} - (t_1 - \overbrace{t_0}^{0})\overbrace{U_0'}^{U_0} + \tfrac{1}{2}c_1(Q'-Q)_1;\\
[c_2 Q_2^{\bullet}] &= [c_1 Q_1^{\bullet}] + \tfrac{1}{2}c_1 (Q'-Q)_1 - (t_2 - t_1) U_1' + \tfrac{1}{2}c_2(Q'-Q)_2;\\
&\vdots\\
[c_k Q_k^{\bullet}] &= [c_{k-1} Q_{k-1}^{\bullet}] + \tfrac{1}{2}c_{k-1}(Q'-Q)_{k-1} - (t_k - t_{k-1}) U_{k-1}' + \tfrac{1}{2}c_k(Q'-Q)_k.
\end{aligned}\right\} \tag{4B.5b}$$

25. Langesche Konstruktionsdaten. Den Iterations-Formelsatz (4B) werden wir noch weiter umformen: in dieser Ziff. 25 durch Einführung Langescher Konstruktionsdaten, damit nicht in dem Formelsatz *mehr* von System zu System verschiedene Konstanten stehen als das System Freiheitsgrade hat; anschließend in Ziff. 26 durch Umstellung auf bildseitige Strahldaten als vorgegebene Strahldaten, damit die Darstellung der Bild*höhen*-Aberration nicht komplizierter ist als die der Bildlot-Aberration.

In (4B) sind alle von den Strahl-Anfangsdaten abhängigen „Variablen" mit großen Buchstaben bezeichnet. Mit kleinen Buchstaben sind alle „Konstanten" bezeichnet, die bei einem bestimmten optischen System konstante Werte haben (oder nur Wellenlängen-abhängig sind):

$$\nu,\ t,\ \varrho; \tag{4.60a}$$

$$\bar{\nu},\ c,\ \mathring{\varkappa}. \tag{4.60b}$$

Das sind mehr Konstanten als das System Freiheitsgrade hat. Von den „Konstruktionsdaten" $\nu \equiv \frac{1}{n(\lambda)}$, a, $\varrho \equiv \frac{1}{r}$ liegen nach (4.34) von vornherein fest:

$$\nu_0' = 1; \qquad a_0 = \infty, \tag{4.61}$$

die übrigen $3k$ Konstruktionsdaten

$$\nu_1' \ldots \nu_k', \qquad a_1 \ldots a_k, \qquad \varrho_1 \ldots \varrho_k \tag{4.62}$$

können unabhängig voneinander variiert werden und stellen so die *verschiedenen* optischen Systeme dar. Zählen wir die Funktion $\nu_v' = 1/n_v'(\lambda)$ als 1 Freiheitsgrad, dann hat also ein optisches System $3k$ Freiheitsgrade. Bei einem *bestimmten* optischen System liegen diese Freiheitsgrade fest, die Konstruktionsdaten (4.62) sind Konstanten.

Die Abstände $a_1 \ldots a_k$ kommen in dem Formelsatz (4B) nicht vor, aber die Seidelschen Abstände t der Flächen von der Eintrittspupille. Nach (4.33a) liegt von vornherein fest:

$$t_0 = 0, \tag{4.63}$$

und nach Ziff. 21 beschreiben die Seidelschen Abstände $t_1 \ldots t_k$ ebenso gut wie die Abstände $a_1 \ldots a_k$ die Lage der Flächenscheitel Nr. 1 bis Nr. k. Wir können also auch die $3k$ Größen

$$\nu'_1 \ldots \nu'_k, \quad t_1 \ldots t_k, \quad \varrho_1 \ldots \varrho_k \tag{4.64}$$

als Konstruktionsdaten nehmen; sie beschreiben das optische System vollständig, und sie können unabhängig voneinander variiert werden.

Von den im Formelsatz (4B) vorkommenden Konstanten (4.60a, b) können also ν, t, ϱ weitgehend beliebige, voneinander unabhängige Werte haben (abgesehen von ν'_0 und t_0). Die anderen Konstanten $\bar{\nu}$, c, $\mathring{\varkappa}$ sind aber mit den ν, t, ϱ fest verbunden:

$$\bar{\nu} = \frac{\nu(\lambda)}{\mu}; \quad \mu \equiv \nu(\lambda_e); \tag{4.65}$$

$$\left.\begin{aligned}
(m\mathring{g})_1 &\equiv (m\mathring{g})'_0 = (m\mathring{g})_0 = 0,\\
c_1 &= c_0 = 1,\\
(m\mathring{g})_2 &\equiv (m\mathring{g})'_1 = (m\mathring{g})_1 + \left(\frac{1}{\mu'_1} - \frac{1}{\mu_1}\right)\varrho_1/c_1,\\
c_2 &= c_1 + (t_2 - t_1)(m\mathring{g})_2,\\
(m\mathring{g})_3 &\equiv (m\mathring{g})'_2 = (m\mathring{g})_2 + \left(\frac{1}{\mu'_2} - \frac{1}{\mu_1}\right)\varrho_2/c_2,\\
c_3 &= c_2 + (t_3 - t_2)(m\mathring{g})_3,\\
&\vdots\\
c_k &= \cdots;
\end{aligned}\right\} \tag{4.66}$$

$$\left.\begin{aligned}
\mathring{\varkappa} &= \frac{\varrho}{\mu c} - (m\mathring{g})\\
\text{oder}\quad &= \frac{\varrho}{\mu' c} - (m\mathring{g})'.
\end{aligned}\right\} \tag{4.67}$$

Die Formeln (4.66) folgen aus (4.61), (3.33), (4.58), (3.32c, d) und (3.32a), (4.31). Die Formel (4.67) folgt aus (4.59), (3.49).

Danach bedarf es einer *Durchrechnung von der Fläche Nr. 1 bis zur Fläche Nr. v*, um c_v und $\mathring{\varkappa}_v$ aus den Konstruktionsdaten ν, t, ϱ zu berechnen.

Nach M. Lange[1] kann man aber ν, $\check{g}$, $\mathring{g}$ als Konstruktionsdaten nehmen (die also das System vollständig beschreiben und unabhängig voneinander variiert werden können) und erhält *ohne Durchrechnung* aus diesen Konstruktionsdaten die ν, a, ϱ und alle Gaußschen Koeffizienten:

$$t = -\frac{m'\check{g}' - m\check{g}}{m'\mathring{g}' - m\mathring{g}}; \quad m \equiv 1/\nu(\lambda_e); \tag{4.68a}$$

$$c = m(\check{g} + t\mathring{g}) \quad \text{oder} \quad = m'(\check{g}' + t\mathring{g}'); \tag{4.68b}$$

$$\check{q} = -t/c; \tag{4.68c}$$

$$\mathring{q} = 1/c; \tag{4.68d}$$

$$a'_v = m'_v(t_{v+1} - t_v)/(c_v c_{v+1}); \tag{4.68e}$$

$$\varrho = c(m'\mathring{g}' - m\mathring{g})/(m' - m). \tag{4.68f}$$

Schwierigkeiten bereitet hierbei der Grenzfall einer Planfläche, bei dem $(m'\mathring{g}' - m\mathring{g}) = 0$ und $(m\check{g}' - m\check{g}) = 0$ ist, der Ausdruck (4.68a) für t also unbestimmt

[1] M. Lange: Z. Instrumentenkd. **31**, 307—313 (1911).

wird; das erschwert die Kontrolle darüber, daß die Seidelschen Flächenabstände $(t_{v+1}-t_v)$ und damit die Abstände a'_v in einer ausführbaren Größenordnung liegen[1]. — Die Formeln (4.68a) und (4.68b) folgen aus (3.34); t und c hatten wir gemäß (4.68c, d) definiert; (4.68e) und (4.68f) folgen aus (4.31) und (3.32c, d). — Solche Konstruktionsdaten wie die von LANGE eingeführten, aus denen sich nämlich die ν, a, ϱ und alle Gaußschen Koeffizienten *ohne Durchrechnung* ergeben, nennen wir *Langesche Daten*.

Die in dem Iterations-Formelsatz (4B) auftretenden Konstanten (4.60a, b) sind nach (4.28b), (4.57), (4.58), (4.59) einfache Kombinationen aus ν, (a), ϱ und Gaußschen Koeffizienten. Wir könnten also im Formelsatz (4B) alle Konstanten ν, t, ϱ; $\bar{\nu}$, c, $\mathring{\varkappa}$ bei der Fläche Nr. v mit Hilfe von (4.68) ausdrücken durch Langesche Daten ν, $\overset{\times}{g}$, $\mathring{g}$ bei der Fläche Nr. v, nämlich durch ν_v, $\overset{\times}{g}_v$, $\mathring{g}_v$ und ν'_v, $\overset{\times}{g}'_v$, $\mathring{g}'_v$ allein. Wir hätten dann in dem Formelsatz nur ebensoviele Konstanten stehen, wie das optische System Freiheitsgrade hat (abgesehen von den von vornherein festgelegten Konstanten: $\nu'_0=1$; $\overset{\times}{g}'_0=1$; $\mathring{g}'_0=0$); alle vorkommenden Konstanten könnten unabhängig voneinander variiert werden. Wir werden aber die Langeschen Daten ν, $\overset{\times}{g}$, $\mathring{g}$ nicht einführen, weil der Ausdruck (4.68a) für t in dem durchaus normalen Grenzfall einer Planfläche unbestimmt würde, geringfügige Variationen der Konstruktionsdaten ν, $\overset{\times}{g}$, $\mathring{g}$ also leicht zu unausführbaren Werten der Seidelschen Flächenabstände $(t_{v+1}-t_v)$ führen würden.

Man kann diese Schwierigkeit beseitigen (oder vielmehr, wie wir sehen werden, gegen eine geringere Schwierigkeit eintauschen) dadurch, daß man den nur in beschränktem Ausmaß variierbaren Seidelschen Abstand t als Konstruktionsdatum nimmt, wie wir das bei (4.64) getan hatten; man sieht dann von vornherein nur ausführbare Variationen des Konstruktionsdatums t vor. Die Daten (4.64) sind aber keine Langeschen Daten: zur Berechnung der ν, a, ϱ und der Gaußschen Koeffizienten aus den ν, t, ϱ bedarf es einer Durchrechnung. Welche anderen Daten als ν, ϱ bilden zusammen mit t Langesche Daten? Den reziproken Brechungsindex ν müssen wir beibehalten, weil auch er nur in sehr beschränktem Ausmaß variierbar ist und wie t als Konstruktionsdatum unmittelbar unter Kontrolle bleiben muß. Es ist also die Frage, mit welchem Datum zusammen ν und t Langesche Daten sind. Im Zusammenhang mit dem Formelsatz (4B) empfiehlt es sich, eine darin schon auftretende Konstante zu ν und t hinzuzunehmen. Nach (4.60a, b) kommt c oder $\mathring{\varkappa}$ in Frage (die Konstante $\bar{\nu}$ ist nur eine Abkürzung für ν/ν_e). Ebenso wie mit ϱ zusammen, bilden ν und t auch mit $\mathring{\varkappa}$ zusammen *keine* Langeschen Daten, wohl aber mit der reziproken Höhe c zusammen:

$$\mathring{q}=1/c; \tag{4.69a}$$

$$\overset{\times}{q}=-t/c; \tag{4.69b}$$

$$(m\,\mathring{g})_{v+1}\equiv(m\,\mathring{g})'_v=\frac{c_{v+1}-c_v}{t_{v+1}-t_v}; \tag{4.69c}$$

$$(m\,\overset{\times}{g})_{v+1}\equiv(m\,\overset{\times}{g})'_v=c_v-t_v(m\,\mathring{g})'_v \quad \text{oder} \quad =c_{v+1}-t_{v+1}(m\,\mathring{g})_{v+1}; \tag{4.69d}$$

$$a'_v=\frac{t_{v+1}-t_v}{\mu'_v c_v c_{v+1}}; \qquad \mu\equiv\nu(\lambda_e); \tag{4.69e}$$

$$\varrho=c\,\frac{(m\,\mathring{g})'-(m\,\mathring{g})}{m'-m}; \qquad m\equiv 1/\mu; \tag{4.69f}$$

$$\left.\begin{aligned}\mathring{\varkappa}&=m\,\varrho/c-(m\,\mathring{g})=m'\varrho/c-(m\,\mathring{g})'\\&=-\frac{\mathring{g}'-\mathring{g}}{\mu'-\mu}.\end{aligned}\right\} \tag{4.69g}$$

---

[1] H. MARX: Optik 7, 91—95 (1950).

Hierbei bereitet der Grenzfall eines verschwindenden Abstands zweier benachbarter Flächen Schwierigkeiten; mit $(t_{v+1}-t_v)=0$ wird $(c_{v+1}-c_v)=0$, der Ausdruck (4.69c) für $(m\mathring{g})'_v$ also unbestimmt. Hiervon sehen wir zunächst ab. — Damit auch die Krümmung ϱ_k der letzten Fläche durch (4.69f) und (4.69c) dargestellt werden kann, müssen t_{k+1} und c_{k+1} definiert sein. Dazu führen wir im (reellen oder virtuellen) Bildraum eine nichtbrechende Planfläche Nr. $(k+1)$ ein, deren Seidelscher Abstand t_{k+1} von der Eintrittspupille gleich 0 ist:

$$\left.\begin{aligned} n'_{k+1} &= n_{k+1} \equiv n'_k; \\ \varrho_{k+1} &= 0; \\ t_{k+1} &= 0. \end{aligned}\right\} \tag{4.70}$$

Diese Fläche heißt die „Austrittspupille" des optischen Systems; ihr Achsenpunkt ist — das bedeutet $t_{k+1}=0$ — in Gaußscher Näherung das Bild des Achsenpunkts der Eintrittspupille.

Einige der Größen v, t, c haben von vornherein feste Werte:

$$\left.\begin{aligned} &v'_0 = 1; \\ &t_0 = 0; \quad t_{k+1} = 0; \\ &c_0 = 1; \quad c_1 = 1. \end{aligned}\right\} \tag{4.71}$$

Die übrigen $3k$ Größen

$$v'_1 \ldots v'_k; \quad t_1 \ldots t_k; \quad c_2 \ldots c_{k+1} \tag{4.72}$$

sind die Konstruktionsdaten des Systems; sie beschreiben das System vollständig und können unabhängig voneinander variiert werden, von dem Grenzfall $(t_{v+1}-t_v)=0$ abgesehen. Nach (4.69) sind die v, t, c Langesche Daten. Die unter (4.69) genannten Schwierigkeiten, die der Grenzfall $(t_{v+1}-t_v)=0$ bereitet, sind nicht so groß [wie die oben genannten Schwierigkeiten, die bei den Langeschen Daten $v, \check{g}, \mathring{g}$ der Grenzfall einer Planfläche bereitet]. Zwar wird mit $(t_{v+1}-t_v)\approx 0$ auch $(c_{v+1}-c_v)\approx 0$, so daß [vgl. den Ausdruck (4.69c) für $(m\mathring{g})'_v$] bei Änderungen die Kontrolle darüber erschwert ist, daß die Größen $(m\mathring{g})$ und damit die Flächenkrümmungen $\varrho = c\,\frac{(m\mathring{g})'-(m\mathring{g})}{m'-m}$ in einer brauchbaren Größenordnung liegen. Aber erstens ist der Spielraum der brauchbaren $(m\mathring{g})$- und ϱ-Werte größer [als der Spielraum der brauchbaren t- und a-Werte, deren Kontrolle bei den Langeschen Daten $v, \check{g}, \mathring{g}$ erschwert ist]. Und zweitens variiert man beim Korrigieren eines optischen Systems die Abstände a oder die Seidelschen Abstände t nicht so oft wie die Brechkräfte der Flächen, so daß meistens der Nenner in (4.69c) fest bleibt und man nur die Variation der im Zähler stehenden Daten c mit Vorsicht zu handhaben hat [— bei den Langeschen Daten $v, \check{g}, \mathring{g}$ dagegen hätte man die in (4.68a) im Zähler stehenden Daten $\check{g}$ *und zugleich* die im Nenner stehenden Daten $\mathring{g}$ *im wesentlichen voneinander abhängig* zu variieren, damit die Brechkräfte sich ändern, aber die Seidelschen Abstände t sich nur unwesentlich ändern; nur *in einem kleinen Spielraum* könnten die Konstruktionsdaten $\check{g}$ und $\mathring{g}$ wirklich *unabhängig voneinander* variiert werden].

Wenn wir nun durch die Langeschen Daten v, t, c alle anderen Konstanten aus dem Formelsatz (4B) eliminieren, verwenden wir die Abkürzung

$$\mu \equiv v_e, \tag{4.73}$$

und die Abkürzung c^- für den *Differenzenquotienten* von c nach t:

$$(c^-)_v \equiv \frac{c_v - c_{v-1}}{t_v - t_{v-1}}; \tag{4.74a}$$

$$(c^-)'_v \equiv \frac{c_{v+1} - c_v}{t_{v+1} - t_v}. \tag{4.74b}$$

Nach (4.69c) ist

$$c^- = m\,\mathring{g} \tag{4.75}$$

und

$$(c^-)_{v+1} = (c^-)'_v. \tag{4.76}$$

Schreiben wir einfacher

$$c_v^- \quad \text{für} \quad (c^-)_v; \tag{4.77a}$$

$$c_v^+ \quad \text{für} \quad (c^-)'_v, \tag{4.77b}$$

dann ist also

$$c_{v+1}^- \equiv \frac{c_{v+1} - c_v}{t_{v+1} - t_v} \equiv c_v^+. \tag{4.78}$$

Nur in den Absätzen 3 und 4 des in Ziff. 24 zusammengestellten Formelsatzes (4B) kommen andere Konstanten als ν, t, c vor: $\bar{\nu}, \mathring{\varkappa}, \varrho$ in (4B.3a); ϱ in (4B.3b); $\bar{\nu}, \mathring{\varkappa}$ in (4B.4b). Mit Hilfe von (4.69) auf ν, t, c umgeschrieben, lauten diese Gleichungen:

$$\left.\begin{aligned} &\mu \equiv \nu_e; \qquad c_{v+1}^- \equiv c_v^+ \equiv \frac{c_{v+1} - c_v}{t_{v+1} - t_v}, \\ &\frac{\frac{\nu}{\mu} + \frac{\nu'}{\mu'}}{\nu + \nu'} \cdot J^{\cdot} = -c \cdot \left(U^{\cdot} + \frac{\mu' c^+ - \mu c^-}{\mu' - \mu} Q^{\cdot}\right), \\ &\frac{\frac{\nu}{\mu} + \frac{\nu'}{\mu'}}{\nu + \nu'} \cdot G^{\cdot} = +c \cdot \left(U^{\cdot} + \frac{\nu c^- + \nu' c^+}{\nu + \nu'} Q^{\cdot}\right), \end{aligned}\right\} \tag{4B.3a*}$$

$$\frac{2}{\nu + \nu'} \cdot \bar{J}_{\varkappa} = +c \cdot \frac{c^+ - c^-}{\frac{1}{\mu'} - \frac{1}{\mu}} \cdot I; \tag{4B.3b*}$$

$$(\mathrm{U}' - \mathrm{U}) = -\frac{\mu' c^+ - \mu c^-}{\mu' - \mu} \cdot (\mathrm{Q}' - \mathrm{Q}) - 2 \cdot \frac{\frac{\nu'}{\mu'} - \frac{\nu}{\mu}}{\nu + \nu'} \cdot \frac{1}{c} \cdot J^{\cdot}. \tag{4B.4b*}$$

Der Differenzenquotient $c_{v+1}^- \equiv c_v^+ \equiv (c_{v+1} - c_v)/(t_{v+1} - t_v)$ ist die sogenannte *reduzierte Brechkraft* des Teilsystems der Flächen Nr. 1 bis Nr. v; der aus der Mitte der unendlich fernen Objektebene kommende paraxiale Meridianstrahl der Farbe e hat nach (4.69c), (3.30), (3.10), (2.69) zwischen der v-ten und der $(v+1)$-ten Fläche die „Neigung"

$$-m'_v \sin \sigma'_v = -c_v^+ \cdot Q_0.$$

$F'_v \equiv 1/c_v^+$ nennt man die *reduzierte Brennweite* des Teilsystems der Flächen Nr. 1 bis Nr. v, $f'_v \equiv m'_v F'_v \equiv 1/(\mu'_v c_v^+)$ seine *Brennweite*, $(\mu'_v c_v^+)$ seine *Brechkraft*.

Die in (4B.3b*) auftretende Differenz $(c^+ - c^-)_v$ ist die reduzierte Brechkraft der Fläche Nr. v; ihre Summe über alle Flächen ist die reduzierte Brechkraft des Systems:

$$c_k^+ = \sum_{v=1}^{k} (c^+ - c^-)_v. \tag{4.79}$$

Die in (4B.3a*) und (4B.4b*) auftretende Differenz $(\mu' c^+ - \mu c^-)_v$ ist die Brechkraft der Fläche Nr. v; ihre Summe über alle Flächen ist die Brechkraft des

Systems:

$$(\mu_k' c_k^+) = \sum_{v=1}^{k} (\mu' c^+ - \mu\, c^-)_v. \tag{4.80}$$

Wenn im Bildraum Luft ($\mu_k' = 1$) ist, sind beide Summen gleich.

Nach (4.75), (4.69f) und (4.8), (4.58) ist die reduzierte Brechkraft einer Fläche

$$(c^+ - c^-) = (m' - m)\, \varrho/c = -m\, m' \alpha_e \cdot \mathring{q}\, \varrho; \tag{4.81}$$

sie ist nach Ziff. 20 also eine Größe nullter Ordnung und ersten Grades. Und nach (4.75), (4.69g) ist die Brechkraft einer Fläche

$$(\mu' c^+ - \mu\, c^-) = -(\mu' - \mu)\, \mathring{\varkappa} = -\alpha_e \cdot m(\mathring{q}\, \varrho - \mathring{g}); \tag{4.82}$$

auch sie ist nach Ziff. 20 eine Größe nullter Ordnung und ersten Grades.

Der in der zweiten Zeile (4B.3a*) und in (4B.4b*) auftretende Quotient

$$\frac{\mu' c^+ - \mu\, c^-}{\mu' - \mu} = -\mathring{\varkappa} \tag{4.82a}$$

und der in (4B.3b*) auftretende Quotient

$$\frac{c^+ - c^-}{\frac{1}{\mu'} - \frac{1}{\mu}} = \varrho/c \tag{4.81a}$$

sind Größen nullter Ordnung und nullten Grades.

26. Bildseitige Strahldaten als vorgegebene Strahldaten. Die Bildlot-Aberration $\varDelta U_k'$ ist (vgl. Ziff. 17 und 19) die Differenz zwischen dem reduzierten Objektlot

$$U_0 = \overline{G}_0 \cdot Y_0 \tag{4.83}$$

und dem reduzierten Bildlot

$$U_k' = -c_k^+ \cdot \overline{G}_k' (y_{GBE} + i\, x_{GBE}). \tag{4.84}$$

Scharfe und verzeichnungsfreie Abbildung der unendlich fernen Objektebene in die Gaußsche Bildebene bedeutet, daß das Verhältnis zwischen der reduzierten Objekthöhe Y_0 und der Bildhöhe $(y_{GBE} + i\, x_{GBE})$ bei allen Strahlen das gleiche ist. Bei Paraxialstrahlen der Farbe e (von denen wir annehmen, daß sie unter den Strahlen vorkommen) ist $U_k'/U_0 = 1$ und $\overline{G}_0 = \overline{G}_k' = 1$, nach (4.83), (4.84) also $-c_k^+ \cdot (y_{GBE} + i\, x_{GBE})/Y_0 = 1$. Bei scharfer und verzeichnungsfreier Abbildung ist also die Differenz zwischen der *reduzierten Objekthöhe*

$$T_0 \equiv Y_0 + i\, X_0 = Y_0 = U_0/\overline{G}_0 \tag{4.85}$$

und der *reduzierten Bildhöhe*

$$T_k' \equiv -c_k^+ (y_{GBE} + i\, x_{GBE}) = U_k'/\overline{G}_k' \tag{4.86}$$

gleich 0 für alle Strahlen. Diese Differenz

$$\varDelta T_k' \equiv T_k' - T_0 \tag{4.87}$$

nennen wir die *Bildhöhen-Aberration* eines Strahls. Als Funktion der reduzierten Objekthöhe Y_0, des Eintrittslots Q_0 und der Wellenlänge des Lichts beschreibt die Bildhöhen-Aberration $(T_k' - T_0)$ die *Abweichung von der idealen*, nämlich scharfen und verzeichnungsfreien, *Abbildung* der Objektebene *OE* in die Gaußsche Bildebene *GBE*.

Wir können uns aber auch die Lichtrichtung umgekehrt denken und die Abbildung der *GBE* in die *OE* untersuchen. Das heißt, wir können die reduzierte Bildhöhe T_k' reell vorgeben und uns interessieren für die Bildhöhen-Aberration

$(T_k'-T_0)$ als Funktion der reduzierten Bildhöhe T_k', eines weiteren (komplexen) Strahldatums im Bildraum und der Wellenlänge des Lichts. Aus der Abweichung von der idealen Abbildung der *GBE* in die *OE* kann man in einfacher Weise auf die Unschärfe und Verzeichnung schließen, mit der die *OE* im Bildraum abgebildet wird.

Neben der reduzierten Bildhöhe kann man als Strahldatum im Bildraum das *Austrittslot*

$$P_{k+1} \equiv c_{k+1} Q_{k+1} \tag{4.88}$$

nehmen, das nach (4.32), (4.58), (4.70) in Gaußscher Näherung gleich dem Eintrittslot Q_0 ist. — Die mit dem Langeschen Datum c_v multiplizierten Scheitellote Q_v, Q_v' und $Q_v^{\bullet}$ nennen wir die *reduzierten Scheitellote* P_v, P_v' und $P_v^{\bullet}$:

$$\left.\begin{aligned} P_v &\equiv c_v Q_v;\\ P_v^{\bullet} &\equiv c_v Q_v^{\bullet} = \tfrac{1}{2}(P_v + P_v');\\ P_v' &\equiv c_v Q_v', \end{aligned}\right\} \tag{4.89}$$

das reduzierte Austrittspupillen-Scheitellot $P_{k+1} = P_{k+1}^{\bullet} = P_{k+1}'$ kurz das „Austrittslot".

Daß wir jetzt die reduzierte Bildhöhe $T_k' \equiv T_{k+1}$ und das Austrittslot P_{k+1} als unabhängige Strahlvariablen nehmen (und nicht die reduzierte Objekthöhe T_0 und das Eintrittslot $Q_0 = P_0$), hat folgenden Grund: Wenn man, wie bei dem Iterations-Formelsatz (4B), U_0 (oder $T_0 = U_0/\underset{+}{\sqrt{1-U_0^2}}$) und Q_0 vorgibt und die Bildlot-Aberration $(U_k'-U_0)$ bestimmt, dann kann man hieraus *allein* nicht in Strenge die Bildhöhen-Aberration $(T_k'-T_0)$ bestimmen. Durch U_k' allein ist nämlich G_k' und damit die Cosinus-Größe

$$\bar{G}_k' = \underset{+}{\sqrt{1 - G_k' G_k'^{*}}} \tag{4.90}$$

nicht gegeben, die man zur Bestimmung von $(T_k'-T_0)$ noch benötigt:

$$\left.\begin{aligned} (T_k'-T_0) &= \frac{(U_k'-U_0)}{\bar{G}_k'} + \left(\frac{1}{\bar{G}_k'} - \frac{1}{\bar{G}_0}\right) U_0;\\ \bar{G}_0 &= \underset{+}{\sqrt{1-U_0^2}}. \end{aligned}\right\} \tag{4.91}$$

Wenn man dagegen T_k' und P_{k+1} vorgibt und die Bildlot-Aberration $(U_k'-U_0)$ bestimmt, dann kann man hieraus allein in Strenge die Bildhöhen-Aberration $(T_k'-T_0)$ bestimmen. Da nämlich die Objektebene im Unendlichen liegt, ist $\bar{G}_0 = \underset{+}{\sqrt{1-G_0 G_0^{*}}}$ durch U_0 allein gegeben und damit auch $(T_k'-T_0)$ durch T_k', P_{k+1} und $(U_k'-U_0)$ allein:

$$(T_k'-T_0) = \frac{(U_k'-U_0)}{\bar{G}_0} + \left(\frac{1}{\bar{G}_k'} - \frac{1}{\bar{G}_0}\right) U_k'; \tag{4.92a}$$

$$\bar{G}_0 = \underset{+}{\sqrt{1-U_0 U_0^{*}}}; \qquad U_0 = U_k' - (U_k'-U_0); \tag{4.92b}$$

$$\left.\begin{aligned} \bar{G}_k' = \underset{+}{\sqrt{1-G_k' G_k'^{*}}}; \qquad G_k' &= c_{k+1} U_k' + c_{k+1}^{-} P_{k+1},\\ n_k' &= 1 \text{ vorausgesetzt}; \end{aligned}\right\} \tag{4.92c}$$

$$\left.\begin{aligned} &c_{k+1} \cdot U_k' = S \cdot \left[\underset{+}{\sqrt{1-(c^- P_x)_{k+1}^2 - (c^- P_y)_{k+1}^2 (1-S^2)}} - (c^- P_y)_{k+1} \cdot S\right],\\ &\text{wobei } S \equiv (c_{k+1} T_k') / \underset{+}{\sqrt{1+(c_{k+1} T_k')^2}} \text{ und } T_k' \text{ reell vorgegeben ist.} \end{aligned}\right\} \tag{4.92d}$$

Die Formel für G'_k folgt unter der Voraussetzung, daß im Bildraum wie im Objektraum Luft ist:

$$n_{k+1} \equiv n'_k = 1, \tag{4.93}$$

aus der Formel (4B.3a*) für $G_v^{\cdot}$, die auch auf die Fläche Nr. $v=k+1$ anwendbar ist. Auf den Fall (4.93) werden wir uns bis zum Ende von Abschnitt V und in Abschnitt VIII beschränken. — (4.92d) folgt aus $U'_k = \bar{G}'_k \cdot T'_k$ und (4.92c). Die Formel für U'_k wird einfacher, wenn man neben der reduzierten Bildhöhe T'_k die *Austrittshöhe*

$$H_{k+1} \equiv (P/\bar{G})_{k+1} = c_{k+1} \cdot (y_{k+1} + i\, x_{k+1}) \tag{4.94}$$

anstelle des Austrittslots P_{k+1} vorgibt:

$$\frac{1}{\bar{G}'_k} = \underset{+}{\sqrt{1 + \underline{G}'_k \underline{G}'^*_k}}; \qquad \underline{G}'_k = c_{k+1} T'_k + c^-_{k+1} H_{k+1}; \tag{4.92c*}$$

$$U'_k = \bar{G}'_k \cdot T'_k = T'_k \Big/ \underset{+}{\sqrt{1 + \underline{G}'_k \underline{G}'^*_k}}. \tag{4.92d*}$$

Nach (2.47), (2.57g) und (3.12) ist die windschiefe Invariante

$$\left.\begin{aligned} I &= n_v (G_x Q_y - G_y Q_x)_v / \bar{G}_v \\ &= n'_v (G_x Q_y - G_y Q_x)'_v / \bar{G}'_v, \end{aligned}\right\} \tag{4.95}$$

nach (3.46), (4.58), (4.89) also

$$\left.\begin{aligned} I &= \frac{n_v}{m_v} (U_x P_y - U_y P_x)_v / \bar{G}_v \\ &= \frac{n'_v}{m'_v} (U_x P_y - U_y P_x)'_v / \bar{G}'_v, \end{aligned}\right\} \tag{4.96}$$

und wegen (4.93), (4.86) und $(T'_k)_x = 0$:

$$I = -T'_k (P_{k+1})_x. \tag{4.97}$$

In der nächsten Ziffer stellen wir den Iterations-Formelsatz für die Bildhöhen-Aberration zusammen. Außer in der Umstellung auf die reduzierte Bildhöhe und die Austrittshöhe als Strahl-Anfangsdaten, und außer in der Verwendung der Langeschen Daten ν; t; c, unterscheidet sich dieser Formelsatz von dem obigen Formelsatz (4B) darin: daß anstelle des Scheitellots Q das reduzierte Scheitellot $P \equiv c\, Q$ verwendet ist (mit dem sich der Formelsatz etwas einfacher schreiben läßt), und daß die am Ende von (4B) stehenden Formeln für $P^{\cdot} \equiv c\, Q^{\cdot}$ an den Anfang des Iterationsschritts gestellt sind.

27. Zusammenstellung des Iterations-Formelsatzes für die Bildhöhen-Aberration. 1. Vorgegeben sind:

die reduzierte Bildhöhe $T_{k+1} = (T_{k+1})_y$
und die Austrittshöhe $H_{k+1} = (H_{k+1})_y + i\,(H_{k+1})_x$ (IV.1a)

eines Strahls, und damit auch:

die Tangens-Größe $\underline{G}_{k+1} = (c\,T + c^- H)_{k+1}$,
die Cosinus-Größe $\bar{G}_{k+1} = 1 \Big/ \underset{+}{\sqrt{1 + \underline{G}_{k+1} \underline{G}^*_{k+1}}}$, (IV.1b)

das reduzierte Bildlot $U_{k+1} = \bar{G}_{k+1} T_{k+1}$,
das Austrittslot $P_{k+1} = \bar{G}_{k+1} H_{k+1}$ (IV.1c)

und

$$\left.\begin{aligned}&\text{die windschiefe Invariante} \qquad I = -T_{k+1}(P_{k+1})_x\\ &\qquad\qquad [\text{oder} = -U_{k+1}(H_{k+1})_x].\end{aligned}\right\} \qquad \text{(IV.1d)}$$

Die Objektebene liegt im Unendlichen. Im Objektraum und im Bildraum ist Luft.

2. Ein Iterationsschritt geht aus von $\Delta U_{k+1}=0$ und von Näherungswerten für die Bildlot-Aberrationen $\Delta U_v^{\cdot} \equiv U_v^{\cdot} - U_{k+1}$; $\Delta U_v \equiv U_v - U_{k+1}$ und für die reduzierten Scheitellot-Aberrationen $(P'-P)_v$ bei den Flächen Nr. $v=k\ldots 1$:

$$\left.\begin{matrix}\Delta U_k^{\cdot}; & \Delta U_k; & \Delta U_{k-1}^{\cdot}; & \Delta U_{k-1}; & \ldots & \Delta U_1^{\cdot}; & \Delta U_1,\\ (P'-P)_k; & & (P'-P)_{k-1}; & & \ldots & (P'-P)_1. & \end{matrix}\right\} \qquad \text{(IV.2a)}$$

Beim ersten Iterationsschritt kann man z.B. von der Gaußschen Näherung ausgehen, also 0 für $\Delta U_v^{\cdot}$, ΔU_v und $(P'-P)_v$ nehmen. Verbesserte Näherungswerte

$$\left.\begin{matrix}\Delta \mathrm{U}_k^{\cdot}; & \Delta \mathrm{U}_k; & \Delta \mathrm{U}_{k-1}^{\cdot}; & \Delta \mathrm{U}_{k-1}; & \ldots & \Delta \mathrm{U}_1^{\cdot}; & \Delta \mathrm{U}_1,\\ (\mathrm{P}'-\mathrm{P})_k; & & (\mathrm{P}'-\mathrm{P})_{k-1}; & & \ldots & (\mathrm{P}'-\mathrm{P})_1 & \end{matrix}\right\} \qquad \text{(IV.2b)}$$

liefert der Iterationsschritt 3., 4.:

3. Durch

$$\left.\begin{aligned}&U_v^{\cdot} = U_{k+1} + \Delta U_v^{\cdot}\\ \text{und}\qquad &P_v^{\cdot} = P_{k+1} - \sum_{\tau=v+\frac{1}{2}}^{k} (P'-P)_\tau - t_v U_{k+1} + \sum_{\tau=v}^{k} (t_{\tau+1} - t_\tau)\,\Delta U_{\tau+1}\end{aligned}\right\} \qquad \text{(IV.3a)}$$

(wonach beim ersten Iterationsschritt, wenn er von der Gaußschen Näherung ausgeht, $U_v^{\cdot} = U_{k+1}$ und $P_v^{\cdot} = P_{k+1} - t_v U_{k+1}$ ist) sind $J_v^{\cdot}$ und $G_v^{\cdot}$ gegeben:

$$\left.\begin{aligned}&\mu \equiv \nu_e; \qquad c_{v+1}^{-} \equiv c_v^{+} \equiv \frac{c_{v+1} - c_v}{t_{v+1} - t_v};\\ &\frac{\dfrac{\nu}{\mu} + \dfrac{\nu'}{\mu'}}{\nu + \nu'} \cdot J^{\cdot} = -c \cdot U^{\cdot} - \frac{\mu' c^{+} - \mu c^{-}}{\mu' - \mu} P^{\cdot};\\ &\frac{\dfrac{\nu}{\mu} + \dfrac{\nu'}{\mu'}}{\nu + \nu'} \cdot G^{\cdot} = +c \cdot U^{\cdot} + \frac{\nu c^{-} + \nu' c^{+}}{\nu + \nu'} P^{\cdot},\end{aligned}\right\} \qquad \text{(IV.3b)}$$

und damit auch $\bar{J}^{\cdot}$ und $\bar{G}^{\cdot}$:

$$\frac{2}{\nu + \nu'} \cdot \bar{J}_x^{\cdot} = c \cdot \frac{c^{+} - c^{-}}{\dfrac{1}{\mu'} - \dfrac{1}{\mu}} \cdot I; \qquad \text{(IV.3c)}$$

$$\left(\frac{\nu' - \nu}{\nu + \nu'}\right)^2 \cdot \frac{1}{1-L} + \frac{1}{L} = \frac{1}{[J^{\cdot} J^{\cdot *} + \bar{J}_x^{\cdot 2}]}; \qquad \text{(IV.3d)}$$

$$\left.\begin{aligned}&\bar{J}_y^{\cdot} = \sqrt[+]{L - [J^{\cdot} J^{\cdot *} + \bar{J}_x^{\cdot 2}]};\\ &\bar{J}^{\cdot} = \bar{J}_y^{\cdot} + i\,\bar{J}_x^{\cdot};\\ &\bar{G}^{\cdot} = \sqrt[+]{L - G^{\cdot} G^{\cdot *}}.\end{aligned}\right\} \qquad \text{(IV.3e)}$$

4. Der verbesserte Näherungswert der reduzierten Scheitellot-Aberration an einer Fläche ist

$$(\mathrm{P}'-\mathrm{P}) = 2\frac{\nu'-\nu}{\nu+\nu'}\cdot\frac{P^\cdot}{\bar{J}_y^\cdot}\left(\frac{G^{\cdot *}-J^{\cdot *}}{\bar{G}^\cdot+\bar{J}^\cdot}\,J^\cdot + i\bar{J}_x^\cdot\right), \tag{IV.4a}$$

und der verbesserte Näherungswert des Anteils dieser Fläche an der Bildlot-Aberration:

$$(\mathrm{U}'-\mathrm{U}) = -\frac{1}{c}\left[\frac{\mu' c^+ - \mu c^-}{\mu'-\mu}\,(\mathrm{P}'-\mathrm{P}) + 2\frac{\frac{\nu'}{\mu'}-\frac{\nu}{\mu}}{\nu+\nu'}\,J^\cdot\right]. \tag{IV.4b}$$

Damit erhält man die verbesserten Näherungswerte der Bildlot-Aberrationen $\Delta\mathrm{U}_{k+\frac{1}{2}} \equiv \Delta\mathrm{U}_k^\cdot$; $\Delta\mathrm{U}_k$; ... $\Delta\mathrm{U}_{1\frac{1}{2}} \equiv \Delta\mathrm{U}_1^\cdot$; $\Delta\mathrm{U}_1$:

$$\left.\begin{array}{l} \Delta\mathrm{U}_w = -\sum\limits_{\tau=w}^{k} (\mathrm{U}'-\mathrm{U})_\tau \\ (w = k+\frac{1}{2};\ k;\ \ldots\ 1\frac{1}{2};\ 1). \end{array}\right\} \tag{IV.4c}$$

In (IV.3a) und (IV.4c) ist bei halbzahligem $w = v+\frac{1}{2}$:

$$\sum_{\tau=v+\frac{1}{2}}^{k} (\ )_\tau \equiv \tfrac{1}{2}(\ )_v + (\ )_{v+1} + (\ )_{v+2} + \cdots + (\ )_k. \tag{IV.4d}$$

Beim letzten Iterationsschritt braucht man die Formel (IV.4c) nur zur Bestimmung von $\Delta\mathrm{U}_1$ anzuwenden.

5. Nach dem letzten Iterationsschritt 3., 4. erhält man aus der Bildlot-Aberration

$$(U_{k+1} - U_0) = -\Delta\mathrm{U}_1 \tag{IV.5a}$$

die Bildhöhen-Aberration $(T_{k+1} - T_0)$:

$$\left.\begin{array}{l} U_0 = U_{k+1} - (U_{k+1} - U_0); \\ \bar{G}_0 = \underset{+}{\sqrt{1 - U_0 U_0^*}}; \\ (T_{k+1} - T_0) = \dfrac{(U_{k+1} - U_0)}{\bar{G}_0} - \left(\dfrac{1}{\bar{G}_0} - \dfrac{1}{\bar{G}_{k+1}}\right) U_{k+1}. \end{array}\right\} \tag{IV.5b}$$

War der erste Iterationsschritt von der Gaußschen Näherung ausgegangen, dann liefern N Iterationsschritte die Bildhöhen-Aberration $(T_{k+1} - T_0)$ bis auf einen Fehler $(3+2N)$-ter Ordnung und $(1+N)$-ten Grades.

V. Iterations-Formelsatz für die meridionale, sagittale und tangentiale Bildhöhen-Aberration.

28. Brechungs-Sinus und Bildhöhen-Aberration eines windschiefen Strahls an einer Fläche. Ebenso wie wir die Bildlot-Aberration $(U_{k+1} - U_0)$ des Systems als Summe ihrer Flächenanteile $(U' - U)_v$ dargestellt haben:

$$(U_{k+1} - U_0) = \sum_{v=1}^{k} (U' - U)_v, \tag{5.1}$$

läßt sich — anders als wir es in Abschnitt IV getan haben — auch die Bildhöhen-Aberration $(T_{k+1} - T_0)$ als Summe ihrer Flächenanteile $(T' - T)_v$ darstellen:

$$(T_{k+1} - T_0) = \sum_{v=1}^{k} (T' - T)_v; \tag{5.2}$$

dabei sind

$$T_v \equiv U_v/\overline{G}_v, \quad T'_v \equiv U'_v/\overline{G}'_v \tag{5.3a}$$

die *reduzierten Bildhöhen* vor und hinter der Fläche Nr. v. Nach Ziff. 19 ist T_v das $(-m_v \mathring{q}_v)$-fache der komplexen Höhe des Strahls in der Gaußschen Zwischenbildebene (GZ) vor der v-ten Fläche:

$$T_v = -c_v^- \cdot [(y_{GZ})_v + i(x_{GZ})_v] = T'_{v-1}. \tag{5.3b}$$

Zunächst schreiben wir den Ausdruck (4.95) für die windschiefe Invariante I um auf die reduzierte Bildhöhe T_v und den *reduzierten Brechungs-Sinus*[1]

$$\mathsf{K}_v \equiv n_v J_v, \tag{5.4a}$$

der *vor* und *hinter* der v-ten Fläche den gleichen Wert hat:

$$\mathsf{K}'_v \equiv n'_v J'_v = n_v J_v \equiv \mathsf{K}_v, \tag{5.4b}$$

so daß auch der *mittlere* reduzierte Brechungssinus

$$\mathsf{K}^{\bullet}_v \equiv \tfrac{1}{2}(\mathsf{K}_v + \mathsf{K}') \tag{5.4c}$$

denselben Wert hat:

$$\mathsf{K}_v = \mathsf{K}^{\bullet}_v = \mathsf{K}'_v. \tag{5.4d}$$

Aus (4.95), (4A.3a) und (5.3a), (5.4b) folgt

$$\frac{1}{\mathring{\varkappa}_v}(T_x \mathsf{K}_y - T_y \mathsf{K}_x)_v = I = \frac{1}{\mathring{\varkappa}_v}(T_x \mathsf{K}_y - T_y \mathsf{K}_x)'_v. \tag{5.5}$$

Die in (4.59) eingeführte Abkürzung $\mathring{\varkappa}$ für $m\mathring{j}$ können wir auffassen als einen zum reduzierten Brechungs-Sinus K gehörigen Gaußschen Koeffizienten: Nach Ziff. 15 ist

$$\varkappa \equiv m\, j \tag{5.6a}$$

der Gaußsche Näherungswert von $\mathsf{K} \equiv n\, J$, und

$$\varkappa' \equiv m'\, j' \tag{5.6b}$$

der Gaußsche Näherungswert von $\mathsf{K}' \equiv n'\, J'$. Nach (3.8c) sind diese beiden Gaußschen Näherungswerte der reduzierten Brechungs-Sinus vor und hinter einer Fläche einander gleich:

$$\varkappa' = m' j' = m\, j = \varkappa, \tag{5.6c}$$

ebenso wie es nach (5.4b) auch die beiden reduzierten Brechungs-Sinus vor und hinter der Fläche sind. Nach (5.6a), (3.13), (3.30), 3.49) ist

$$\varkappa_v = (m_v \overset{\times}{j}_v)\, U_0 + (m_v \mathring{j}_v)\, Q_0 \tag{5.7a}$$

oder [vgl. (3.30)]

$$\varkappa_v = \overset{\times}{\varkappa}_v\, U_0 + \mathring{\varkappa}_v\, Q_0 \tag{5.7b}$$

mit den Gaußschen Koeffizienten

$$\overset{\times}{\varkappa} = m\overset{\times}{j} \quad \text{und} \quad \mathring{\varkappa} = m\mathring{j}. \tag{5.7c}$$

[1] Die Sinus-Größe J_v nennen wir den „Brechungs-Sinus" vor der v-ten Fläche; bei Meridianstrahlen ist nach (2.70), (2.44) J_v der Sinus des Brechungswinkels ι_v vor der v-ten Fläche, nämlich des mit Vorzeichen genommenen Winkels zwischen dem einfallenden Strahl und der Flächennormale. J'_v ist entsprechend der „Brechungs-Sinus" hinter der v-ten Fläche. $J^{\bullet}_v$ nennen wir den „mittleren Brechungs-Sinus" an der v-ten Fläche.

Wir führen noch

$$\left.\begin{array}{l}\overline{\mathsf{K}}\equiv n\overline{J}; \quad \overline{\mathsf{K}}'\equiv n'\overline{J}';\\ \overline{\mathsf{K}}^{\cdot}\equiv\frac{1}{2}(\overline{\mathsf{K}}+\overline{\mathsf{K}}')\end{array}\right\} \tag{5.8a}$$

ein. Wegen $n'\overline{J}'_x=n\overline{J}_x$ ist

$$\overline{\mathsf{K}}_x=\overline{\mathsf{K}}^{\cdot}_x=\overline{\mathsf{K}}'_x, \tag{5.8b}$$

und nach (2.57g):

$$(\overline{\mathsf{K}}_x)_v=\varrho_v\cdot I. \tag{5.9}$$

Aus den beiden Ausdrücken (5.5) für die windschiefe Invariante I folgt

$$\frac{T'_x-T_x}{\mathsf{K}_x}=\frac{T'_y-T_y}{\mathsf{K}_y}, \tag{5.10}$$

also ist der Quotient aus dem Flächenanteil $(T'-T)_v$ der Bildhöhen-Aberration und dem reduzierten Brechungs-Sinus K_v reell:

$$(T'-T)_v=(\text{reelle Größe})\cdot\mathsf{K}_v. \tag{5.11}$$

Den Ausdruck (IV.4a, b) für die Bild*lot*-Aberration $(U'-U)_v$ an der Fläche Nr. v rechnen wir nun auf die Bild*höhen*-Aberration $(T'-T)_v$ um.

Im folgenden treten häufig *bisher verwendete Variablen V dividiert durch $\overline{G}$* auf. Dafür führen wir die Abkürzung $\underline{V}$ ein:

$$\underline{V}\equiv\left(\frac{V}{\overline{G}}\right). \tag{5.12a}$$

Später werden auch Variationen δV oder dV dividiert durch $\overline{G}$ auftreten. Dafür schreiben wir in entsprechender Weise $\underline{\delta V}$ bzw. $\underline{dV}$:

$$\underline{\delta V}\equiv\left(\frac{\delta V}{\overline{G}}\right); \quad \underline{dV}\equiv\left(\frac{dV}{\overline{G}}\right), \tag{5.12b}$$

nicht zu verwechseln mit $\delta\underline{V}\equiv\delta(V/\overline{G})$ und $d\underline{V}\equiv d(V/\overline{G})$. Ein an $\underline{\delta V}$, $\underline{dV}$ oder $\underline{V}$ unten gesetzter Flächenindex τ und ein oben gesetzter Strich (′) gelten für V und für $\overline{G}$:

$$\left.\begin{array}{ll}\underline{V}_\tau\equiv\left(\frac{V}{\overline{G}}\right)_\tau\equiv\frac{V_\tau}{\overline{G}_\tau}; & \underline{V}'_\tau\equiv\left(\frac{V}{\overline{G}}\right)'_\tau\equiv\frac{V'_\tau}{\overline{G}'_\tau};\\ \underline{\delta V}_\tau\equiv\left(\frac{\delta V}{\overline{G}}\right)_\tau\equiv\frac{\delta V_\tau}{\overline{G}_\tau}; & \underline{\delta V}'_\tau\equiv\left(\frac{\delta V}{\overline{G}}\right)'_\tau=\frac{\delta V'_\tau}{\overline{G}'_\tau};\\ \underline{dV}_\tau\equiv\left(\frac{dV}{\overline{G}}\right)_\tau\equiv\frac{dV_\tau}{\overline{G}_\tau}; & \underline{dV}'_\tau\equiv\left(\frac{dV}{\overline{G}}\right)'_\tau\equiv\frac{dV'_\tau}{\overline{G}'_\tau}.\end{array}\right\} \tag{5.13}$$

Ein an $\underline{\delta V}$, $\underline{dV}$ oder $\underline{V}$ oben gesetzter Punkt (˙) dagegen bedeutet die Mittelbildung für die unterstrichene Größe als Ganzes:

$$\underline{V}^{\cdot}_\tau\equiv\frac{1}{2}(\underline{V}_\tau+\underline{V}'_\tau) \quad \equiv\frac{1}{2}\left(\frac{V_\tau}{\overline{G}_\tau}+\frac{V'_\tau}{\overline{G}'_\tau}\right); \tag{5.14a}$$

$$\left.\begin{array}{l}\underline{\delta V}^{\cdot}_\tau\equiv\frac{1}{2}(\underline{\delta V}_\tau+\underline{\delta V}'_\tau)\equiv\frac{1}{2}\left(\frac{\delta V_\tau}{\overline{G}_\tau}+\frac{\delta V'_\tau}{\overline{G}'_\tau}\right);\\ \underline{dV}^{\cdot}_\tau\equiv\frac{1}{2}(\underline{dV}_\tau+\underline{dV}'_\tau)\equiv\frac{1}{2}\left(\frac{dV_\tau}{\overline{G}_\tau}+\frac{dV'_\tau}{\overline{G}'_\tau}\right);\end{array}\right\} \tag{5.14b}$$

$$d\underline{V}^{\cdot}_\tau\equiv d(\underline{V}^{\cdot}_\tau) \quad \equiv\frac{1}{2}\left(d\frac{V_\tau}{\overline{G}_\tau}+d\frac{V'_\tau}{\overline{G}'_\tau}\right). \tag{5.14c}$$

Mit dieser Bezeichnung ist die reduzierte Bildhöhe

$$\left.\begin{aligned} T_v &\equiv (U/\overline{G})_v \equiv \underline{U}_v; \\ T'_v &\equiv (U/\overline{G})'_v \equiv \underline{U}'_v \end{aligned}\right\} \tag{5.15}$$

und das im folgenden auftretende Produkt

$$\frac{1}{2}\left(\frac{1}{\overline{G}_v} + \frac{1}{\overline{G}'_v}\right)\cdot \mathsf{K}_v = \underline{\mathsf{K}}^{\boldsymbol{\cdot}}_v. \tag{5.16}$$

Aus (5.15) folgt

$$\left.\begin{aligned} (\underline{U}'-\underline{U}) &= \frac{1}{2}\left(\frac{1}{\overline{G}}+\frac{1}{\overline{G}'}\right)\cdot(U'-U) + \frac{1}{2}\left(\frac{1}{\overline{G}'}-\frac{1}{\overline{G}}\right)\cdot(U+U') \\ &= \frac{1}{2}\left(\frac{1}{\overline{G}}+\frac{1}{\overline{G}'}\right)\cdot\left[(U'-U) - \frac{U^{\boldsymbol{\cdot}}}{\overline{G}^{\boldsymbol{\cdot}}}(\overline{G}'-\overline{G})\right]. \end{aligned}\right\} \tag{5.17}$$

Für $(\overline{G}'-\overline{G})$ setzen wir den Ausdruck (4.14) ein, umgeschrieben auf K und $\overline{\mathsf{K}}_x$ anstelle von J und $\overline{J}_x$, und erhalten:

$$(\underline{U}'-\underline{U}) = \frac{1}{2}\left(\frac{1}{\overline{G}}+\frac{1}{\overline{G}'}\right)\cdot\left[(U'-U) - (\nu'-\nu)\frac{U^{\boldsymbol{\cdot}}}{\overline{J}^{\boldsymbol{\cdot}}_y}\left(\frac{G^{\boldsymbol{\cdot}*}}{\overline{G}^{\boldsymbol{\cdot}}}\mathsf{K} + i\overline{\mathsf{K}}_x\right)\right]. \tag{5.18a}$$

Entsprechend ist

$$(\underline{P}'-\underline{P}) = \frac{1}{2}\left(\frac{1}{\overline{G}}+\frac{1}{\overline{G}'}\right)\cdot\left[(P'-P) - (\nu'-\nu)\frac{P^{\boldsymbol{\cdot}}}{\overline{J}^{\boldsymbol{\cdot}}_y}\left(\frac{G^{\boldsymbol{\cdot}*}}{\overline{G}^{\boldsymbol{\cdot}}}\mathsf{K} + i\overline{\mathsf{K}}_x\right)\right]. \tag{5.18b}$$

Mit den Abkürzungen $\mathring{\varkappa}$ für $-(\mu' c^+ - \mu c^-)/(\mu'-\mu)$ und $\overline{\nu}$ für ν/μ [vgl. (4.82a), (4.57)] und wegen

$$J^{\boldsymbol{\cdot}} = \frac{\nu+\nu'}{2}\mathsf{K}; \quad \overline{J}^{\boldsymbol{\cdot}}_x = \frac{\nu+\nu'}{2}\overline{\mathsf{K}}_x \tag{5.19}$$

ist nach (IV.4b):

$$c\cdot(U'-U) = \mathring{\varkappa}\cdot(P'-P) - (\overline{\nu}'-\overline{\nu})\cdot\mathsf{K}, \tag{5.20}$$

nach (IV.4a):

$$(P'-P) = (\nu'-\nu)\cdot\frac{P^{\boldsymbol{\cdot}}}{\overline{J}^{\boldsymbol{\cdot}}_y}\left(\frac{G^{\boldsymbol{\cdot}*}-J^{\boldsymbol{\cdot}*}}{\overline{G}^{\boldsymbol{\cdot}}+\overline{J}^{\boldsymbol{\cdot}}}\mathsf{K} + i\overline{\mathsf{K}}_x\right), \tag{5.21}$$

und nach der zweiten Zeile (IV.3b):

$$c\cdot U^{\boldsymbol{\cdot}} = \mathring{\varkappa}\cdot P^{\boldsymbol{\cdot}} - \overline{\nu}^{\boldsymbol{\cdot}}\cdot\mathsf{K}\,. \tag{5.22}$$

Damit erhalten wir für die Bildhöhen-Aberration $(\underline{U}'-\underline{U})_v$ und die Scheitelhöhen-Aberration $(\underline{P}'-\underline{P})_v$ an der Fläche Nr. v die Ausdrücke:

$$\left.\begin{aligned} c\cdot(\underline{U}'-\underline{U}) = \mathring{\varkappa}\cdot(\underline{P}'-\underline{P}) + (\nu'-\nu)\cdot\frac{\overline{\nu}^{\boldsymbol{\cdot}}}{\overline{J}^{\boldsymbol{\cdot}}_y}\left(\frac{G^{\boldsymbol{\cdot}*}}{\overline{G}^{\boldsymbol{\cdot}}}\mathsf{K} + i\overline{\mathsf{K}}_x\right)\underline{\mathsf{K}}^{\boldsymbol{\cdot}} - \\ -(\overline{\nu}'-\overline{\nu})\cdot\underline{\mathsf{K}}^{\boldsymbol{\cdot}}; \end{aligned}\right\} \tag{5.23a}$$

$$(\underline{P}'-\underline{P}) = (\nu'-\nu)\cdot\frac{P^{\boldsymbol{\cdot}}}{\overline{J}^{\boldsymbol{\cdot}}_y}\left(\frac{G^{\boldsymbol{\cdot}*}-J^{\boldsymbol{\cdot}*}}{\overline{G}^{\boldsymbol{\cdot}}+\overline{J}^{\boldsymbol{\cdot}}} - \frac{G^{\boldsymbol{\cdot}*}}{\overline{G}^{\boldsymbol{\cdot}}}\right)\underline{\mathsf{K}}^{\boldsymbol{\cdot}}. \tag{5.23b}$$

Mit Hilfe von (2.60), (2.61) und (2.58) kann man nachrechnen, daß

$$\frac{P^{\boldsymbol{\cdot}}}{c}\left(\frac{G^{\boldsymbol{\cdot}*}-J^{\boldsymbol{\cdot}*}}{\overline{G}^{\boldsymbol{\cdot}}+\overline{J}^{\boldsymbol{\cdot}}} - \frac{G^{\boldsymbol{\cdot}*}}{\overline{G}^{\boldsymbol{\cdot}}}\right) = -(r+z)\cdot\frac{\overline{G}^{\boldsymbol{\cdot}2}+G^{\boldsymbol{\cdot}}G^{\boldsymbol{\cdot}*}}{\overline{G}^{\boldsymbol{\cdot}}},$$

folglich

$$\frac{\underline{P'-P}}{\underline{\mathsf{K}}^{\boldsymbol{\cdot}}} \text{ reell} \tag{5.24}$$

ist. Nach (5.23a) ist, weil $\left(\frac{G^{\cdot *}}{\overline{G}^{\cdot}} \mathsf{K} + i\,\overline{\mathsf{K}}_x\right)$ reell ist [vgl. (4.14)], mit $\frac{P'-P}{\underline{\mathsf{K}}^{\cdot}}$ auch

$$\frac{U'-U}{\underline{\mathsf{K}}^{\cdot}} \quad \text{reell}, \tag{5.25}$$

in Übereinstimmung mit (5.11).

29. Sagittale und tangentiale Variationen. Am Ende des Abschnitts IV haben wir die reduzierte Bildhöhe $T_{k+1} \equiv \underline{U}_{k+1}$ und die Austrittshöhe $H_{k+1} \equiv \underline{P}_{k+1}$ als unabhängige Strahldaten eingeführt. Ohne Beschränkung der Allgemeinheit konnten wir die Bildhöhe als reell annehmen (der Bildpunkt liegt dann in der y, z-Ebene):

$$\underline{U}_{k+1} = (\underline{U}_{k+1})_y. \tag{5.26}$$

Die Austrittshöhe

$$\underline{P}_{k+1} = (\underline{P}_{k+1})_y + i\,(\underline{P}_{k+1})_x \tag{5.27}$$

ist reell oder komplex, je nachdem ob der Strahl meridional ist oder nicht.

Wegen der Spiegelsymmetrie des optischen Systems sind, in Abhängigkeit von $(\underline{U}_{k+1})_y$, $(\underline{P}_{k+1})_y$ und $(\underline{P}_{k+1})_x$ betrachtet,

$$\text{ungerade Funktionen von } (\underline{P}_{k+1})_x\text{:} \quad \text{alle } \xi_v \text{ und } x_v; \tag{5.28}$$

und

$$\left.\begin{aligned} \text{gerade Funktionen von } (\underline{P}_{k+1})_x\text{:} \quad & \text{alle } \eta_v \text{ und } y_v, \\ & \text{und alle } \zeta_v \text{ und } z_v. \end{aligned}\right\} \tag{5.29}$$

Hieraus folgt [vgl. (2.43), (2.44) und (3.12), (3.46)], daß bei den komplexen Variablen G, J; $\overline{G}, \overline{J}$; Q, U und damit[1] bei allen oben eingeführten komplexen Variablen W die *Realteile gerade* Funktionen und die *Imaginärteile ungerade* Funktionen von $(\underline{P}_{k+1})_x$ sind:

$$W_y \text{ gerade;} \tag{5.30a}$$

$$W_x \text{ ungerade.} \tag{5.30b}$$

Die windschiefe Invariante I können wir nach (5.9) auffassen als den Imaginärteil von $(r\,\overline{\mathsf{K}})$; sie ist ungerade.

Häufig genügt es beim optischen Rechnen, außer den Meridianstrahlen nur die den Meridianstrahlen *unendlich benachbarten* windschiefen Strahlen zu untersuchen. Aus deren Aberrationen kann man *weitgehend* auf die Aberrationen *aller* windschiefen Strahlen schließen, wie wir in Ziff. 38 sehen werden.

Dem Meridianstrahl

$$(\underline{U}_{k+1})_y; \quad (\underline{P}_{k+1})_y; \quad (\underline{P}_{k+1})_x = 0 \tag{5.31a}$$

ist der windschiefe Strahl

$$(\underline{U}_{k+1})_y; \quad (\underline{P}_{k+1})_y; \quad \delta(\underline{P}_{k+1})_x \tag{5.31b}$$

„*sagittal* benachbart" und der Meridianstrahl

$$(\underline{U}_{k+1})_y; \quad (\underline{P}_{k+1})_y + d(\underline{P}_{k+1})_y; \quad 0 \tag{5.31c}$$

---

[1] Vgl. die Definitions-Gleichungen (4.89), (4.94), (5.3a), (5.4), (5.8) und (5.12a) für P, H, T, K, $\overline{\mathsf{K}}$ und die unterstrichenen Variablen. — Mit den „komplexen Variablen" sind im folgenden gemeint: die in Abschnitt II und III eingeführten Variablen G, J; $\overline{G}, \overline{J}$; Q, U; die in Abschnitt IV und Ziff. 28 eingeführten Variablen P, K, $\overline{\mathsf{K}}$; und zugehörige unterstrichene Variablen wie z.B. $\underline{U} \equiv T$, $\underline{P}_{k+1} \equiv H_{k+1}$, $\underline{\mathsf{K}}$. Und zwar sind gemeint: alle diese Variablen ohne oberen Index, alle mit einem Strich ('), und alle mit einem Punkt (˙).

„*tangential* benachbart"; die Variationen

$$\delta(\,)\ \text{und}\ d(\,)\ \text{sind unendlich klein.} \tag{5.32}$$

Mit den Strahlen (5.31 b) erfassen wir alle den Merdianstrahlen unendlich benachbarten windschiefen Strahlen. Einen solchen Strahl (5.31 b) nennen wir kurz einen „δ-Strahl". Den Meridianstrahl (5.31 a) nennen wir den zu dem δ-Strahl (5.31 b) *zugehörigen* Meridianstrahl.

Aus (5.30a) und (5.32) folgt, daß ein δ-Strahl und sein zugehöriger Meridianstrahl bei allen komplexen Variablen W *gleiche Realteile* W_y haben:

$$(W_y)_{(\delta)} = W_y; \tag{5.33a}$$

die auf einen δ-Strahl sich beziehende Größe bezeichnen wir hier mit dem zusätzlichen Index (δ), während die auf den zugehörigen Meridianstrahl sich beziehenden Größe ohne zusätzlichen Index geschrieben ist. Die Variation einer Variablen W bei Übergang von einem Meridianstrahl zu einem sagittal benachbarten Strahl nennen wir die „sagittale Variation" von W und bezeichnen sie mit δW:

$$\delta W \equiv W_{(\delta)} - W. \tag{5.34}$$

Nach (5.33a) ist bei allen komplexen Variablen W

$$\delta W_y = 0, \tag{5.35}$$

folglich

$$\delta W = i \cdot \delta W_x. \tag{5.36}$$

Aus (5.30b) und (5.32) folgt, daß bei einem Meridianstrahl

$$W_x = 0, \tag{5.37}$$

folglich

$$W = W_y \tag{5.38}$$

ist, und daß

$$\delta W_x \quad \text{proportional} \quad \delta(\underline{P}_{k+1})_x \tag{5.39}$$

ist; der Proportionalitätsfaktor ist eine Funktion von $(\underline{U}_{k+1})_y$ und $(\underline{P}_{k+1})_y$. Nach (5.37) ist

$$(W_x)_{(\delta)} = \delta W_x. \tag{5.33b}$$

Alle ohne zuästzlichen Index geschriebenen (d.h. auf einen Meridianstrahl sich beziehenden) Variablen W sind also reell, alle Variationen δW rein-imaginär (wenn nicht null). Oder, was dasselbe bedeutet, der Realteil von $W_{(\delta)}$ ist gleich W, der Imgainärteil von $W_{(\delta)}$ gleich δW (durch i).

Wir werden in diesem Abschnitt V und in Abschnitt VIII nur noch Meridianstrahlen und δ-Strahlen betrachten, brauchten also nur noch mit Meridianstrahl-Variablen W und sagittalen Variationen δW zu rechnen. Wir werden aber auch mit tangentialen Variationen dW rechnen, weil die Formeln für die sagittale *und* die tangentiale Variation der Bildhöhen-Aberration, die wir ableiten werden, die Diskussion des Zusammenhangs zwischen den Konstruktionsdaten und dem Astigmatismus höherer Ordnung ermöglichen.

30. Flächenanteil der sagittalen Bildhöhen-Aberration. Nach (5.25) und (5.24) ist

$$\frac{\delta\underline{U}' - \delta\underline{U}}{\delta\underline{K}^{\boldsymbol{\cdot}}} = \frac{\underline{U}' - \underline{U}}{\underline{K}^{\boldsymbol{\cdot}}} \tag{5.40a}$$

und

$$\frac{\delta\underline{P}' - \delta\underline{P}}{\delta\underline{K}^{\boldsymbol{\cdot}}} = \frac{\underline{P}' - \underline{P}}{\underline{K}^{\boldsymbol{\cdot}}}. \tag{5.40b}$$

Der Flächenanteil $(\delta \underline{U}' - \delta \underline{U})$ der *sagittalen Bildhöhen-Aberration*[1] ist also das $(\delta \underline{K}^{\bullet}/\underline{K}^{\bullet})$-fache des Flächenanteils $(\underline{U}' - \underline{U})$ der meridionalen Bildhöhen-Aberration. Für diesen gilt nach (5.23a, b):

$$c \cdot (\underline{U}' - \underline{U}) = \mathring{\varkappa} \cdot (\underline{P}' - \underline{P}) + \left[+ (\nu' - \nu) \cdot \frac{\bar{\nu}^{\bullet} \mathrm{K}}{\bar{J}^{\bullet}} \frac{G^{\bullet}}{\bar{G}^{\bullet}} - (\bar{\nu}' - \bar{\nu}) \right] \cdot \underline{\mathrm{K}}^{\bullet}; \tag{5.41a}$$

$$\mathring{\varkappa} \cdot (\underline{P}' - \underline{P}) = \qquad - (\nu' - \nu) \cdot \frac{\mathring{\varkappa} P^{\bullet}}{\bar{J}^{\bullet}} \left(\frac{G^{\bullet}}{\bar{G}^{\bullet}} - \frac{G^{\bullet} - J^{\bullet}}{\bar{G}^{\bullet} + \bar{J}^{\bullet}} \right) \cdot \underline{\mathrm{K}}^{\bullet}, \tag{5.41b}$$

also ist

$$c \cdot (\delta \underline{U}' - \delta \underline{U}) = \mathring{\varkappa} \cdot (\delta \underline{P}' - \delta \underline{P}) + \left[+ (\nu' - \nu) \cdot \frac{\bar{\nu}^{\bullet} \mathrm{K}}{\bar{J}^{\bullet}} \frac{G^{\bullet}}{\bar{G}^{\bullet}} - (\bar{\nu}' - \bar{\nu}) \right] \cdot \delta \underline{\mathrm{K}}^{\bullet}; \tag{5.42a}$$

$$\mathring{\varkappa} \cdot (\delta \underline{P}' - \delta \underline{P}) = \qquad - (\nu' - \nu) \cdot \frac{\mathring{\varkappa} P^{\bullet}}{\bar{J}^{\bullet}} \left(\frac{G^{\bullet}}{\bar{G}^{\bullet}} - \frac{G^{\bullet} - J^{\bullet}}{\bar{G}^{\bullet} + \bar{J}^{\bullet}} \right) \cdot \delta \underline{\mathrm{K}}^{\bullet}. \tag{5.42b}$$

Von den Formeln (5.41a, b) für den Flächenanteil $(\underline{U}' - \underline{U})$ der meridionalen Bildhöhen-Aberration werden wir keinen weiteren Gebrauch machen. Die meridionale Bildhöhen-Aberration $(\underline{U}_{k+1} - \underline{U}_0)$ ist nämlich als Summe der Flächenanteile $(\underline{U}' - \underline{U})$ durch den Ausdruck (5.41a, b) und zusätzliche Formeln nicht so einfach dargestellt, wie durch den Formelsatz (IV) in Ziff. 27, wenn wir diesen auf Meridianstrahlen spezialisieren.

Die Formeln (5.41a, b) haben wir gebraucht, um in einfacher Weise den Ausdruck (5.42a, b) für den Flächenanteil $(\delta \underline{U}' - \delta \underline{U})$ der sagittalen Bildhöhen-Aberration $(\delta \underline{U}_{k+1} - \delta \underline{U}_0)$ zu erhalten. Dieser Ausdruck dient auch zur Darstellung der tangentialen Bildhöhen-Aberration (Ziff. 31 und 32). Mit dem sagittalen Flächenanteil $(\delta \underline{U}' - \delta \underline{U})$ steht nämlich der tangentiale Flächenanteil $(d\underline{U}' - d\underline{U})$ in einer einfachen Beziehung (Ziff. 31).

Wir könnten zwar auch den Formelsatz (IV) auf sagittale Variationen spezialisieren, die sagittale Bildhöhen-Aberration also auf die Flächenanteile $(\delta U' - \delta U)$ der sagittalen Bildlot-Aberration zurückführen; daran würden wir aber die Darstellung der tangentialen Bildhöhen-Aberration nicht so einfach anschließen können.

Wir führen also die *meridionale* Bildhöhen-Aberration $(\underline{U}_{k+1} - \underline{U}_0)$ auf die Flächenanteile $(U' - U)$ der meridionalen Bild*lot*-Aberration zurück, und die *sagittale* Bildhöhen-Aberration $(\delta \underline{U}_{k+1} - \delta \underline{U}_0)$ auf die Flächenanteile $(\delta \underline{U}' - \delta \underline{U})$ der sagittalen Bild*höhen*-Aberration[2]. Dagegen werden wir im Anschluß hieran die *tangentiale* Bildhöhen-Aberration $(d\underline{U}_{k+1} - d\underline{U}_0)$ weder auf die Flächenanteile $(d\underline{U}' - d\underline{U})$ der tangentialen Bildhöhen-Aberration, noch auf die Flächenanteile $(d\bar{U}' - d\bar{U})$ der tangentialen Bildlot-Aberration zurückführen, sondern[2] auf die Flächenanteile

$$(d\underline{U}' - d\underline{U}) \equiv \left(\frac{dU'}{\bar{G}'} - \frac{dU}{\bar{G}} \right)$$

der „tangentialen Bild*winkel*-Aberration"

$$(d\underline{U}_{k+1} - d\underline{U}_0) \equiv \left(\frac{d\,U_{k+1}}{\bar{G}_{k+1}} - \frac{d\,U_0}{\bar{G}_0} \right) = \left(\frac{d\,U_{k+1}}{\bar{G}_{k+1}} - d\sigma_0 \right).$$

[1] Für die sagittale Variation der Bildhöhen-Aberration sagen wir kurz „sagittale Bildhöhen-Aberration". Für die Bildhöhen-Aberration der Meridianstrahlen (von denen die sagittalen Variationen ausgehen) sagen wir kurz „meridionale Bildhöhen-Aberration". Für die später noch vorkommende tangentiale Variation der Bildhöhen-Aberration werden wir kurz „tangentiale Bildhöhen-Aberration" sagen.

[2] Abgesehen von einer Umformung in Ziff. 32, die sich aber bei Strahlen der Hauptfarbe e nicht auswirkt.

σ_0 ist der sogenannte „Bildwinkel", das ist in dem hier vorliegenden Fall der unendlich entfernten Objektebene der Winkel, den der Meridianstrahl im Objektraum mit der optischen Achse bildet. Durch den Bildwinkel σ_0 könnte die Lage des Objektpunkts beschrieben werden; wir haben sie nach (3.22) oder (3.27a) durch die reduzierte Objekthöhe $\underline{U}_0 = \operatorname{tg}\sigma_0$ oder das reduzierte Objektlot $U_0 = \sin\sigma_0$ beschrieben.

Während $d\underline{W}$ und $\underline{dW}$ sich im allgemeinen unterscheiden, sind $\delta\underline{W}$ und $\underline{\delta W}$ immer gleich:

$$\delta\underline{W} = \delta\frac{W}{\overline{G}} = \frac{\delta W}{\overline{G}} = \underline{\delta W}, \tag{5.43}$$

weil $\overline{G}$ auch bei windschiefen Strahlen immer reell, also

$$\delta\overline{G} = 0 \tag{5.44}$$

ist.

31. Sagittale Bildhöhen-Aberration und tangentiale Bildwinkel-Aberration an einer Fläche. Die mit dem Übergang von dem Meridianstrahl (5.31a) zu dem tangential benachbarten Strahl (5.31c) verbundene Änderung der Variablen W bezeichnen wir mit

$$dW \tag{5.45}$$

und nennen sie die „tangentiale Variation" von W.

Führen wir bei Meridianstrahlen gemäß (2.69) und (2.70) den *Strahlwinkel* σ und den *Brechungswinkel* ι ein:

$$G = \sin\sigma; \qquad \overline{G} = \cos\sigma, \tag{2.69}$$

$$J = \sin\iota; \qquad \overline{J} = \cos\iota, \tag{2.70}$$

dann können wir für die einzige nicht-lineare der vier Übergangs- und Brechungs-Formeln (2.57a, b, c, d), nämlich für

$$\frac{G' + J'}{\overline{G}' + \overline{J}'} = \frac{G + J}{\overline{G} + \overline{J}}, \tag{2.57d}$$

einfach schreiben:

$$\sigma' + \iota' = \sigma + \iota, \tag{5.46}$$

eine bekannte (bei der „trigonometrischen" Durchrechnung von Meridianstrahlen benutzte) Formel.

Aus (5.46) folgt

$$d\sigma' + d\iota' = d\sigma + d\iota \tag{5.47}$$

und hieraus wegen (2.69), (2.70):

$$\frac{dG'}{\overline{G}'} + \frac{dJ'}{\overline{J}'} = \frac{dG}{\overline{G}} + \frac{dJ}{\overline{J}}. \tag{5.48}$$

Nun folgt aus (4A.3a) in einfacher Weise:

$$\left.\begin{aligned} \mathring{\varkappa} G &= c^- J + \varrho\, U; \\ \mathring{\varkappa} G' &= c^+ J' + \varrho\, U'. \end{aligned}\right\} \tag{5.49}$$

Damit können wir (5.48) auf dU, dU' und $d\mathsf{K} \equiv n\, dJ$ umschreiben:

$$\frac{dU'}{\overline{G}'} - \frac{dU}{\overline{G}} = -r\left[\left(\frac{c^+}{n'\overline{G}'} - \frac{c^-}{n\overline{G}}\right) - \mathring{\varkappa}\,\frac{n'\overline{J}' - n\overline{J}}{n\overline{J}\cdot n'\overline{J}'}\right]\cdot d\mathsf{K}. \tag{5.50}$$

Andererseits folgt aus der ersten Gleichung (2.44):

$$\frac{r\,n J_x + z\xi}{\zeta} = x = \frac{r\,n' J_x' + z\xi'}{\zeta'},$$

also

$$\frac{\xi'}{\zeta'} - \frac{\xi}{\zeta} = \frac{r}{z}\cdot\frac{\zeta' - \zeta}{\zeta\zeta'}\cdot n J_x, \tag{5.51}$$

und aus der zweiten Gleichung (2.44) in Verbindung mit der ersten Gleichung (2.37a):

$$\frac{r}{z}\,(\zeta' - \zeta) = -\,(n'\overline{J}_y' - n\overline{J}_y), \tag{5.52}$$

und damit

$$\frac{\xi'}{\zeta'} - \frac{\xi}{\zeta} = -\,\frac{n'\overline{J}_y' - n\overline{J}_y}{\zeta\zeta'}\cdot n J_x$$

oder wegen $\xi \equiv -nG_x$; $\zeta \equiv n\overline{G}$:

$$\frac{G_x'}{\overline{G}'} - \frac{G_x}{\overline{G}} = \frac{n'\overline{J}_y' - n\overline{J}_y}{n\overline{G}\cdot n'\overline{G}'}\cdot n J_x. \tag{5.53}$$

Mit Hilfe der aus (4A.3a) folgenden Gleichungen

$$\left.\begin{aligned} \mathring{\varkappa}\, G_x &= c^- J_x + \varrho\, U_x; \\ \mathring{\varkappa}\, G_x' &= c^+ J_x' + \varrho\, U_x' \end{aligned}\right\} \tag{5.54}$$

können wir (5.53) auf U_x, U_x' und $\mathsf{K}_x \equiv n J_x$ umschreiben:

$$\frac{U_x'}{\overline{G}'} - \frac{U_x}{\overline{G}} = -r\left[\left(\frac{c^+}{n'\overline{G}'} - \frac{c^-}{n\overline{G}}\right) - \mathring{\varkappa}\,\frac{n'\overline{J}_y' - n\overline{J}_y}{n\overline{G}\cdot n'\overline{G}'}\right]\cdot \mathsf{K}_x, \tag{5.55}$$

woraus wegen (5.25) übrigens auch

$$\frac{U'}{\overline{G}'} - \frac{U}{\overline{G}} = -r\left[\left(\frac{c^+}{n'\overline{G}'} - \frac{c^-}{n\overline{G}}\right) - \mathring{\varkappa}\,\frac{n'\overline{J}_y' - n\overline{J}_y}{n\overline{G}\cdot n'\overline{G}'}\right]\cdot \mathsf{K} \tag{5.56}$$

folgt. (5.55) und (5.56) gelten für beliebige windschiefe Strahlen. Wir wenden (5.55) auf δ-Strahlen an und erhalten wegen (5.33a, b) und (5.38), (5.36):

$$\frac{\delta U'}{\overline{G}'} - \frac{\delta U}{\overline{G}} = -r\left[\left(\frac{c^+}{n'\overline{G}'} - \frac{c^-}{n\overline{G}}\right) - \mathring{\varkappa}\,\frac{n'\overline{J}' - n\overline{J}}{n\overline{G}\cdot n'\overline{G}'}\right]\cdot \delta\mathsf{K}. \tag{5.57}$$

Nach (5.50) und (5.57) ist

$$\frac{\underline{dU'} - \underline{dU}}{\underline{d\mathsf{K}}^{\bullet}} - \frac{\underline{\delta U'} - \underline{\delta U}}{\underline{\delta\mathsf{K}}^{\bullet}} = \mathring{\varkappa}\cdot\frac{r}{n\,n'}\;\frac{n'\overline{J}' - n\overline{J}}{\overline{G}^{\bullet}}\left(\frac{\overline{G}\,\overline{G}'}{\overline{J}\,\overline{J}'} - 1\right). \tag{5.58}$$

Man kann nachrechnen, daß dies

$$= (\nu' - \nu)\,\frac{\mathring{\varkappa}}{c}\;\frac{P^{\bullet}}{\overline{J}^{\bullet}}\;\frac{G^{\bullet} - J^{\bullet}}{G^{\bullet}(\overline{J}\,\overline{J}')} \tag{5.59}$$

ist:

$$\frac{n'\bar{J}'-n\bar{J}}{n'-n}=\frac{n\,n'}{n'-n}\,\frac{\cos\iota'\sin\iota-\cos\iota\sin\iota'}{n\sin\iota}$$

$$=\frac{n\,n'}{n'-n}\,\frac{(\sin\iota-\sin\iota')\cdot\cos\frac{\iota'-\iota}{2}}{n\sin\iota\quad\cdot\cos\frac{\iota+\iota'}{2}}$$

$$=\cos\frac{\iota'-\iota}{2}\Big/\cos\frac{\iota+\iota'}{2};$$

$$\bar{G}\bar{G}'-\bar{J}\bar{J}'=\cos\sigma\cos\sigma'-\cos\iota\cos\iota'$$

$$=\tfrac{1}{2}\left[\cos(\sigma+\sigma')-\cos(\iota+\iota')\right]$$

$$=-\sin^2\frac{\sigma+\sigma'}{2}+\sin^2\frac{\iota+\iota'}{2}$$

$$=-\left[\left(\frac{\sin\sigma+\sin\sigma'}{2}\right)^2-\left(\frac{\sin\iota+\sin\iota'}{2}\right)^2\right]\Big/\cos^2\frac{\iota'-\iota}{2},$$

also

$$\frac{n'\bar{J}'-n\bar{J}}{n'-n}\,(\bar{G}\bar{G}'-\bar{J}\bar{J}')=-\left[G^{\cdot 2}-J^{\cdot 2}\right]/\bar{J}^{\cdot},$$

woraus, wegen $r(G^{\cdot}+J^{\cdot})=P^{\cdot}/c$, (5.59) folgt.

Nicht nur den *Mittelwert* $\bar{J}^{\cdot}$ aus $\bar{J}$ und $\bar{J}'$ können wir, nach (IV.3e), auf $J^{\cdot}$ und L zurückführen:

$$\bar{J}^{\cdot}=\underset{+}{\sqrt{L-J^{\cdot 2}}},\qquad(5.60)$$

sondern auch das in (5.59) im Nenner auftretende *Produkt* aus $\bar{J}$ und $\bar{J}'$:

$$(\bar{J}\bar{J}')=\cos\iota\cdot\cos\iota'$$

$$=1-\sin^2\frac{\iota'-\iota}{2}-\sin^2\frac{\iota+\iota'}{2}$$

$$=\cos^2\frac{\iota'-\iota}{2}-\left(\frac{\sin\iota+\sin\iota'}{2}\right)^2\Big/\cos^2\frac{\iota'-\iota}{2},$$

also

$$(\bar{J}\bar{J}')=L-J^{\cdot 2}/L,\qquad(5.61)$$

denn nach (4.44) ist

$$L=\frac{1}{2}\left[1+\cos(\iota'-\iota)\right]=\cos^2\frac{\iota'-\iota}{2}.$$

Mit (5.60), (5.61) und dem aus (4.45) für Meridianstrahlen folgenden Ausdruck für L,

$$L=1-\left(\frac{\nu'-\nu}{\nu+\nu'}\right)^2\Big/\left(\frac{1}{J^{\cdot 2}}-\frac{1}{L}\right),\qquad(5.62)$$

sind $\bar{J}^{\cdot}$ und $(\bar{J}\bar{J}')$ und $J^{\cdot}$ zurückgeführt.

Nach (5.58) und (5.59) besteht zwischen $(\underline{\delta U'}-\underline{\delta U})_v$, der sagittalen Bildhöhen-Aberration an der Fläche Nr. v, und $(\underline{dU'}-\underline{dU})_v$, der „tangentialen Bildwinkel-Aberration an der Fläche Nr. v“, die Beziehung:

$$c\cdot\left(\frac{\underline{dU'}-\underline{dU}}{\underline{d\mathsf{K}^{\cdot}}}-\frac{\underline{\delta U'}-\underline{\delta U}}{\underline{\delta\mathsf{K}^{\cdot}}}\right)=(\nu'-\nu)\,\frac{\mathring{\varkappa}P^{\cdot}}{\bar{J}^{\cdot}}\,\frac{G^{\cdot}-J^{\cdot}}{\bar{G}^{\cdot}(\bar{J}\bar{J}')}.\qquad(5.63\,\mathrm{a})$$

Damit ist auch

$$\mathring{\varkappa}\cdot\left(\frac{\underline{dP'}-\underline{dP}}{\underline{d\mathsf{K}^{\cdot}}}-\frac{\underline{\delta P'}-\underline{\delta P}}{\underline{\delta\mathsf{K}^{\cdot}}}\right)=(\nu'-\nu)\,\frac{\mathring{\varkappa}P^{\cdot}}{\bar{J}^{\cdot}}\,\frac{G^{\cdot}-J^{\cdot}}{\bar{G}^{\cdot}(\bar{J}\bar{J}')}.\qquad(5.63\,\mathrm{b})$$

Denn nach der zweiten Gl. (4A.3a) ist

$$c\,dU = \mathring{\varkappa}\,dP - m\,dJ$$

und

$$c\,\delta U = \mathring{\varkappa}\,\delta P - m\,\delta J,$$

folglich

$$\frac{\mathring{\varkappa}\,dP}{dJ} - \frac{c\,dU}{dJ} = \frac{\mathring{\varkappa}\,\delta P}{\delta J} - \frac{c\,\delta U}{\delta J}$$

und entsprechend

$$\frac{\mathring{\varkappa}\,dP'}{dJ'} - \frac{c\,dU'}{dJ'} = \frac{\mathring{\varkappa}\,\delta P'}{\delta J'} - \frac{c\,\delta U'}{\delta J'},$$

daher

$$\mathring{\varkappa}\cdot\left(\frac{\underline{dP'-dP}}{\underline{d\mathsf{K}^{\boldsymbol{\cdot}}}} - \frac{\underline{\delta P'-\delta P}}{\underline{\delta\mathsf{K}^{\boldsymbol{\cdot}}}}\right) = c\cdot\left(\frac{\underline{dU'-dU}}{\underline{d\mathsf{K}^{\boldsymbol{\cdot}}}} - \frac{\underline{\delta U'-\delta U}}{\underline{\delta\mathsf{K}^{\boldsymbol{\cdot}}}}\right). \tag{5.64}$$

Die Beziehungen (5.63a), (5.63b) geben die Möglichkeit, die Darstellung der tangentialen Bildhöhen-Aberration in einfacher Weise an die Darstellung der sagittalen Bildhöhen-Aberration anzuschließen.

32. Modifizierte Flächenanteile der sagittalen und tangentialen Aberrationen. Die in dem Ausdruck (5.42a, b) für den sagittalen Flächenanteil $(\delta\underline{U}' - \delta\underline{U})$ auftretende Größe $\delta\underline{\mathsf{K}}^{\boldsymbol{\cdot}}$ steht mit $\delta\underline{U}^{\boldsymbol{\cdot}}$ und $\delta\underline{P}^{\boldsymbol{\cdot}}$ nicht in einer einfachen Beziehung (wenn nicht $\bar{\nu} = \bar{\nu}'$ ist[1]), wohl aber mit $\underline{\delta(U/\bar{\nu})^{\boldsymbol{\cdot}}}$ und $\underline{\delta(P/\bar{\nu})^{\boldsymbol{\cdot}}}$: Nach der zweiten Gl. (4A.3a) und (5.4) ist

$$\left.\begin{aligned}
\bar{\nu}\,\mathsf{K} &= -c\,U + \mathring{\varkappa}\,P; & \bar{\nu}\,\delta\mathsf{K} &= -c\,\delta U + \mathring{\varkappa}\,\delta P; & \bar{\nu}\,d\mathsf{K} &= \cdots;\\
\bar{\nu}'\,\mathsf{K} &= -c\,U' + \mathring{\varkappa}\,P'; & \bar{\nu}'\,\delta\mathsf{K} &= -c\,\delta U' + \mathring{\varkappa}\,\delta P'; & \bar{\nu}'\,d\mathsf{K} &= \cdots,
\end{aligned}\right\} \tag{5.65}$$

folglich

$$\left.\begin{aligned}
(\bar{\nu}\,\mathsf{K})^{\boldsymbol{\cdot}} &= -c\cdot U^{\boldsymbol{\cdot}} + \mathring{\varkappa}\cdot P^{\boldsymbol{\cdot}};\\
\underline{\delta\mathsf{K}}^{\boldsymbol{\cdot}} &= -c\,\underline{\delta(U/\bar{\nu})}^{\boldsymbol{\cdot}} + \mathring{\varkappa}\,\underline{\delta(P/\bar{\nu})}^{\boldsymbol{\cdot}};\\
\underline{d\mathsf{K}}^{\boldsymbol{\cdot}} &= -c\,\underline{d(U/\bar{\nu})}^{\boldsymbol{\cdot}} + \mathring{\varkappa}\,\underline{d(P/\bar{\nu})}^{\boldsymbol{\cdot}}.
\end{aligned}\right\} \tag{5.66}$$

Die nach (5.65) auch richtige Gleichung

$$\mathsf{K}^{\boldsymbol{\cdot}} = -c\cdot(U/\bar{\nu})^{\boldsymbol{\cdot}} + \mathring{\varkappa}\cdot(P/\bar{\nu})^{\boldsymbol{\cdot}}$$

wollen wir nicht in Betracht ziehen.

Mit den Variablen

und

$$\left.\begin{aligned}
K &\equiv \bar{\nu}\,\mathsf{K} = m\,J\\
K' &\equiv \bar{\nu}'\,\mathsf{K} = m'\,J'
\end{aligned}\right\} \tag{5.67}$$

anstelle von K wird aus (5.66):

$$\left.\begin{aligned}
K^{\boldsymbol{\cdot}} &= -c\cdot U^{\boldsymbol{\cdot}} + \mathring{\varkappa}\cdot P^{\boldsymbol{\cdot}};\\
\underline{\delta(K/\bar{\nu})}^{\boldsymbol{\cdot}} &= -c\,\underline{\delta(U/\bar{\nu})}^{\boldsymbol{\cdot}} + \mathring{\varkappa}\,\underline{\delta(P/\bar{\nu})}^{\boldsymbol{\cdot}};\\
\underline{d(K/\bar{\nu})}^{\boldsymbol{\cdot}} &= -c\,\underline{d(U/\bar{\nu})}^{\boldsymbol{\cdot}} + \mathring{\varkappa}\,\underline{d(P/\bar{\nu})}^{\boldsymbol{\cdot}}.
\end{aligned}\right\} \tag{5.68}$$

Gleichungen wie (5.68), die für die durch $(\bar{\nu}\,\overline{G})$ dividierten Variationen δW oder dW ebenso lauten wie für die Ausgangsgrößen W, schreiben wir im folgen-

[1] Im Fall der Hauptfarbe e ist $\bar{\nu} = \bar{\nu}' (= 1)$ und die im folgenden eingeführte Größe $K = \mathsf{K}$.

den nicht dreifach, sondern wir fassen sie jeweils zusammen, indem wir

$$\left.\begin{aligned} &d^0W \quad \text{für} \quad W; && d^0W' \quad \text{für} \quad W' \\ &d^iW \quad \text{für} \frac{\delta W}{\bar{\nu}\bar{G}}; && d^iW' \quad \text{für} \quad \frac{\delta W'}{\bar{\nu}'\bar{G}'} \\ &d^1W \quad \text{für} \frac{d W}{\bar{\nu}\bar{G}}; && d^1W' \quad \text{für} \quad \frac{d W'}{\bar{\nu}'\bar{G}'} \end{aligned}\right\} \tag{5.69}$$

schreiben und vereinbaren:

$$\text{Gleichungen mit } d^\varepsilon \text{ gelten für } \varepsilon=0;\ i;\ 1. \tag{5.70}$$

Ein an $d^\varepsilon W$ oben rechts gesetzter Punkt ($^\cdot$) soll die Mittelbildung für $(d^\varepsilon W)$ als Ganzes bedeuten:

$$d^\varepsilon W^\cdot \equiv (d^\varepsilon W)^\cdot \equiv \tfrac{1}{2}(d^\varepsilon W + d^\varepsilon W'). \tag{5.71}$$

So läßt sich (5.68) zusammenfassen zu:

$$d^\varepsilon K^\cdot = -c\, d^\varepsilon U^\cdot + \mathring{\varkappa}\, d^\varepsilon P^\cdot. \tag{5.72}$$

Die Formeln (5.42a, b) für den sagittalen Flächenanteil schreiben wir nun auf d^iU, d^iP, d^iK anstelle von $\delta\underline{U}$, $\delta\underline{P}$, $\delta\underline{K}$ um:

$$c\cdot(d^iU'-d^iU) = \mathring{\varkappa}\cdot(d^iP'-d^iP) + \frac{\nu'-\nu}{\bar{\nu}^\cdot}\cdot\frac{K^\cdot}{\bar{J}^\cdot}\cdot\frac{G^\cdot}{\bar{G}^\cdot}\cdot d^iK^\cdot; \tag{5.73a}$$

$$\mathring{\varkappa}\cdot(d^iP'-d^iP) = -\frac{\bar{\nu}'-\bar{\nu}}{\bar{\nu}^\cdot}\cdot\mathring{\varkappa}\cdot d^iP^\cdot - \frac{\nu'-\nu}{\bar{\nu}^\cdot}\cdot\frac{\mathring{\varkappa}P^\cdot}{\bar{J}^\cdot}\left[\frac{G^\cdot}{\bar{G}^\cdot} - \frac{G^\cdot-J^\cdot}{\bar{G}+\bar{J}^\cdot}\right] d^iK^\cdot. \tag{5.73b}$$

Die entsprechenden Formeln für den tangentialen Flächenanteil erhalten wir ebenfalls aus (5.42a, b), durch Anwendung der Beziehungen (5.63a, b):

$$c\cdot(d^1U'-d^1U) = \mathring{\varkappa}\cdot(d^1P'-d^1P) + \frac{\nu'-\nu}{\bar{\nu}^\cdot}\cdot\frac{K^\cdot}{\bar{J}^\cdot}\cdot\frac{G^\cdot}{\bar{G}^\cdot}\cdot d^1K^\cdot; \tag{5.74a}$$

$$\left.\begin{aligned} \mathring{\varkappa}\cdot(d^1P'-d^1P) = -\frac{\bar{\nu}'-\bar{\nu}}{\bar{\nu}^\cdot}\cdot\mathring{\varkappa}\, d^1P^\cdot - \frac{\nu'-\nu}{\bar{\nu}^\cdot}\cdot\frac{\mathring{\varkappa}P^\cdot}{\bar{J}^\cdot}\left[\frac{G^\cdot}{\bar{G}^\cdot} - \frac{G^\cdot-J^\cdot}{\bar{G}^\cdot+\bar{J}^\cdot} - \right.\\ \left. - \frac{G^\cdot-J^\cdot}{\bar{G}^\cdot(\bar{J}\,\bar{J}')}\right] d^1K^\cdot \end{aligned}\right\} \tag{5.74b}$$

Für $d^\varepsilon U^\cdot$ und $d^\varepsilon P^\cdot$ gelten, wie man leicht nachrechnen kann, die Übergangsformeln:

$$d^\varepsilon U^\cdot_{v+1} - d^\varepsilon U^\cdot_v = \tfrac{1}{2}(d^\varepsilon U' - d^\varepsilon U)_v + \tfrac{1}{2}(d^\varepsilon U' - d^\varepsilon U)_{v+1}; \tag{5.75a}$$

$$d^\varepsilon P^\cdot_{v+1} - d^\varepsilon P^\cdot_v = \tfrac{1}{2}(d^\varepsilon P' - d^\varepsilon P)_v - (t_{v+1}-t_v)\, d^\varepsilon U_{v+1} + \tfrac{1}{2}(d^\varepsilon P' - d^\varepsilon P)_{v+1}. \tag{5.75b}$$

Die Gleichungen, welche die meridionale, sagittale und tangentiale Bildhöhen-Aberration mit den vorgegebenen Strahldaten und den Langeschen Konstruktionsdaten in Zusammenhang bringen, haben wir jetzt alle bis auf zwei Formeln, die, bei vorgegebener tangentialer Variation, für $d^1 U_{k+1}$ und für $d^1 P_{k+1}$ gelten, und bis auf die Beziehung zwischen der tangentialen Bildwinkel-Aberration $(d^1 U_{k+1} - d^1 U_0)$ und der tangentialen Bildhöhen-Aberration $(d\underline{U}_{k+1} - d\underline{U}_0)$. Mit $(d^i U_{k+1} - d^i U_0)$ erhalten wir unmittelbar die sagittale Bildhöhen-Aberration $(\delta\underline{U}_{k+1} - \delta\underline{U}_0)$. Und die Beziehung zwischen der meridionalen Bildlot-Aberration $(U_{k+1} - U_0)$ und der meridionalen Bildhöhen-Aberration $(\underline{U}_{k+1} - \underline{U}_0)$ ist durch (IV.5b) gegeben. — Bei der tangentialen Variation d ist nach (5.31a, c):

$$d\,\underline{U}_{k+1} = 0. \tag{5.76a}$$

Hieraus folgt, wie man leicht nachrechnen kann:

$$d^1 U_{k+1} = -(c^- UG\, d\underline{P})_{k+1}; \tag{5.76b}$$

$$d^1 P_{k+1} = -(c^- PG\, d\underline{P})_{k+1} + d\underline{P}_{k+1}. \tag{5.76c}$$

Das sind die zwei noch benötigten Formeln für $d^1 U_{k+1}$ und $d^1 P_{k+1}$; neben $d\underline{U}_{k+1}=0$ geben wir $d\underline{P}_{k+1}$ vor. Die noch fehlende Gleichung für die tangentiale Bildhöhen-Aberration lautet:

$$\bar{G}_0^2 \cdot (d\underline{U}_{k+1} - d\underline{U}_0) = (d^1 U_{k+1} - d^1 U_0) + (c^- UG\, d\underline{P})_{k+1}; \tag{5.76d}$$

sie folgt aus (5.76a), (5.76b) und daraus, daß $U_0 = G_0$, folglich $d\underline{U}_0 = d^1 U_0/\bar{G}_0^2$ ist.

33. Iterations-Formelsatz für die meridionale, sagittale und tangentiale Bildhöhen-Aberration. 1. Vorgegeben sind:

$$\left.\begin{array}{ll} \text{die reduzierte Bildhöhe} & \underline{U}_{k+1}, \\ \text{die meridionale Austrittshöhe} & \underline{P}_{k+1}, \\ \text{eine sagittale Variation mit } \delta\underline{U}_{k+1}=0 \text{ und imaginärem} & \delta\underline{P}_{k+1}, \\ \text{eine tangentiale Variation mit } d\underline{U}_{k+1}=0 \text{ und reellem} & d\underline{P}_{k+1} \end{array}\right\} \tag{V.1a}$$

und damit auch:

$$\left.\begin{array}{ll} \text{die Tangensgröße} & \underline{G}_{k+1} = (c\underline{U} + c^-\underline{P})_{k+1}, \\ \text{die Cosinusgröße} & \bar{G}_{k+1} = 1/\overset{+}{\sqrt{1+(\underline{G}_{k+1})^2}}, \\ \text{die Sinusgröße} & G_{k+1} = \bar{G}_{k+1}\underline{G}_{k+1}, \\ \text{das reduzierte Bildlot} & U_{k+1} = \bar{G}_{k+1}\underline{U}_{k+1}, \\ \text{das Austrittslot} & P_{k+1} = \bar{G}_{k+1}\underline{P}_{k+1}. \end{array}\right\} \tag{V.1b}$$

Die Objektebene liegt im Unendlichen. Im Objektraum und im Bildraum ist Luft.

2. Ein Iterationsschritt geht aus von $\Delta^\varepsilon U_{k+1} = 0$ und von Näherungswerten für

$$\Delta^\varepsilon U_v^\cdot \equiv d^\varepsilon U_v^\cdot - d^\varepsilon U_{k+1}; \quad \Delta^\varepsilon U_v \equiv d^\varepsilon U_v - d^\varepsilon U_{k+1} \quad \text{und} \quad (d^\varepsilon P' - d^\varepsilon P)_v \tag{V.2a}$$

[$\varepsilon = 0,\ i,\ 1$; $d^0 W \equiv W$] bei den Flächen Nr. $v = k, k-1, \ldots 1$; diese Näherungswerte bezeichnen wir ebenfalls mit $\Delta^\varepsilon U_v^\cdot$; $\Delta^\varepsilon U_v$ und $(d^\varepsilon P' - d^\varepsilon P)_v$. Beim ersten Iterationsschritt kann man z.B. von der Näherung 0 für alle Größen (V.2a) ausgehen.

Verbesserte Näherungswerte

$$\Delta^\varepsilon \mathrm{U}_v^\cdot; \quad \Delta^\varepsilon \mathrm{U}_v \quad \text{und} \quad (d^\varepsilon \mathrm{P}' - d^\varepsilon \mathrm{P})_v \tag{V.2b}$$

liefert der Iterationsschritt 3., 4.:

3. Durch

$$\left.\begin{array}{l} d^\varepsilon U_v^\cdot = d^\varepsilon U_{k+1} + \Delta^\varepsilon U_v^\cdot \\ \text{und} \\ d^\varepsilon P_v^\cdot = d^\varepsilon P_{k+1} - \sum\limits_{\tau=v+\frac{1}{2}}^{k} (d^\varepsilon P' - d^\varepsilon P)_\tau - \\ \qquad - t_v d^\varepsilon U_{k+1} + \sum\limits_{\tau=v}^{k} (t_{\tau+1} - t_\tau)\, \Delta^\varepsilon U_{\tau+1}, \end{array}\right\} \tag{V.3a}$$

wobei

$$\left.\begin{aligned} d^0U_{k+1}&\equiv U_{k+1}; & d^iU_{k+1}&=0; & d^1U_{k+1}&= \qquad -(c^- UG\, d\underline{P})_{k+1} \\ d^0P_{k+1}&\equiv P_{k+1}; & d^iP_{k+1}&=\delta\underline{P}_{k+1}; & d^1P_{k+1}&=d\underline{P}_{k+1}-(c^- PG\, d\underline{P})_{k+1} \end{aligned}\right\} \tag{V.3b}$$

ist, sind $d^\varepsilon K_v^{\cdot}$, $J^{\cdot}$ und $G^{\cdot}$ gegeben:

$$\left.\begin{aligned} c_{v+1}^- &\equiv c_v^+ \equiv \frac{c_{v+1}-c_v}{t_{v+1}-t_v}; \\ \bar\nu &\equiv \frac{\nu}{\nu_e}; \qquad \mathring{\varkappa} \equiv -\frac{\nu_e' c^+ - \nu_e c^-}{\nu_e' - \nu_e}; \end{aligned}\right\} \tag{V.3c}$$

$$\left.\begin{aligned} d^\varepsilon K^{\cdot} &= -c\, d^\varepsilon U^{\cdot} + \mathring{\varkappa}\, d^\varepsilon P^{\cdot}; \\ J^{\cdot} &= \frac{\nu^{\cdot}}{\bar\nu^{\cdot}} K^{\cdot}; \\ G^{\cdot} &= \frac{\nu^{\cdot}}{\bar\nu^{\cdot}} c\, U^{\cdot} + \frac{(\nu c^-)^{\cdot}}{\bar\nu^{\cdot}} P^{\cdot}; \end{aligned}\right\} \tag{V.3d}$$

und damit auch $(\bar J \bar J')$, $\bar J^{\cdot}$ und $\bar G^{\cdot}$:

$$\left.\begin{aligned} L &= 1-\left(\frac{\nu'-\nu}{\nu+\nu'}\right)^2 \Big/ \left(\frac{1}{J^{\cdot 2}}-\frac{1}{L}\right); \\ (\bar J \bar J') &= L - J^{\cdot 2}/L; \\ \bar J^{\cdot} &= \underset{+}{\sqrt{L-J^{\cdot 2}}}; \\ \bar G^{\cdot} &= \underset{+}{\sqrt{L-G^{\cdot 2}}}. \end{aligned}\right\} \tag{V.3e}$$

4. Für die verbesserten Näherungswerte $\Delta^\varepsilon \mathrm{U}_{v+\frac{1}{2}} \equiv \Delta^\varepsilon \mathrm{U}_v^{\cdot}$; $\Delta^\varepsilon \mathrm{U}_v$ und $(d^\varepsilon \mathrm{P}' - d^\varepsilon \mathrm{P})_v$ gelten die Formeln:

$$\left.\begin{aligned} c(\mathrm{U}'-\mathrm{U}) &= \frac{\nu'-\nu}{\bar\nu^{\cdot}}\,\frac{\mathring{\varkappa} P^{\cdot}}{\bar J^{\cdot}}\,\frac{G^{\cdot}-J^{\cdot}}{\bar G^{\cdot}+\bar J^{\cdot}}\cdot K^{\cdot} - \frac{\bar\nu'-\bar\nu}{\bar\nu^{\cdot}}\cdot K^{\cdot} \\ \mathring{\varkappa}(\mathrm{P}'-\mathrm{P}) &= \frac{\nu'-\nu}{\bar\nu^{\cdot}}\,\frac{\mathring{\varkappa} P^{\cdot}}{\bar J^{\cdot}}\,\frac{G^{\cdot}-J^{\cdot}}{\bar G^{\cdot}+\bar J^{\cdot}}\cdot K^{\cdot} \\ c(d^i\mathrm{U}'-d^i\mathrm{U}) &= -\frac{\nu'-\nu}{\bar\nu^{\cdot}}\,\frac{cU^{\cdot}}{\bar J^{\cdot}}\,\frac{G^{\cdot}}{\bar G^{\cdot}} d^iK^{\cdot} + \frac{\nu'-\nu}{\bar\nu^{\cdot}}\,\frac{\mathring{\varkappa} P^{\cdot}}{\bar J^{\cdot}}\,\frac{G^{\cdot}-J^{\cdot}}{\bar G^{\cdot}+\bar J^{\cdot}}\, d^iK^{\cdot} - \frac{\bar\nu'-\bar\nu}{\bar\nu^{\cdot}}\mathring{\varkappa}\, d^iP^{\cdot} \\ \mathring{\varkappa}(d^i\mathrm{P}'-d^i\mathrm{P}) &= -\frac{\nu'-\nu}{\bar\nu^{\cdot}}\,\frac{\mathring{\varkappa} P^{\cdot}}{\bar J^{\cdot}}\,\frac{G^{\cdot}}{\bar G^{\cdot}} d^iK^{\cdot} + \frac{\nu'-\nu}{\bar\nu^{\cdot}}\,\frac{\mathring{\varkappa} P^{\cdot}}{\bar J^{\cdot}}\,\frac{G^{\cdot}-J^{\cdot}}{\bar G^{\cdot}+\bar J^{\cdot}}\, d^iK^{\cdot} - \frac{\bar\nu'-\bar\nu}{\bar\nu^{\cdot}}\mathring{\varkappa}\, d^iP^{\cdot} \\ c(d^1\mathrm{U}'-d^1\mathrm{U}) &= -\frac{\nu'-\nu}{\bar\nu^{\cdot}}\,\frac{cU^{\cdot}}{\bar J^{\cdot}}\,\frac{G^{\cdot}}{\bar G^{\cdot}} d^1K^{\cdot} + \frac{\nu'-\nu}{\bar\nu^{\cdot}}\,\frac{\mathring{\varkappa} P^{\cdot}}{\bar J^{\cdot}}\left(\frac{G^{\cdot}-J^{\cdot}}{\bar G^{\cdot}+\bar J^{\cdot}} + \frac{G^{\cdot}-J^{\cdot}}{\bar G^{\cdot}(\bar J \bar J')}\right) d^1K^{\cdot} - \frac{\bar\nu'-\bar\nu}{\bar\nu^{\cdot}}\mathring{\varkappa}\, d^1P^{\cdot} \\ \mathring{\varkappa}(d^1\mathrm{P}'-d^1\mathrm{P}) &= -\frac{\nu'-\nu}{\bar\nu^{\cdot}}\,\frac{\mathring{\varkappa} P^{\cdot}}{\bar J^{\cdot}}\,\frac{G^{\cdot}}{\bar G^{\cdot}} d^1K^{\cdot} + \frac{\nu'-\nu}{\bar\nu^{\cdot}}\,\frac{\mathring{\varkappa} P^{\cdot}}{\bar J^{\cdot}}\left(\frac{G^{\cdot}-J^{\cdot}}{\bar G^{\cdot}+\bar J^{\cdot}} + \frac{G^{\cdot}-J^{\cdot}}{\bar G^{\cdot}(\bar J \bar J')}\right) d^1K^{\cdot} - \frac{\bar\nu'-\bar\nu}{\bar\nu^{\cdot}}\mathring{\varkappa}\, d^1P^{\cdot}, \end{aligned}\right\} \tag{V.4a}$$

$$\left.\begin{aligned} \Delta^\varepsilon \mathrm{U}_w &= -\sum_{\tau=w}^{k} (d^\varepsilon \mathrm{U}' - d^\varepsilon \mathrm{U})_\tau \\ &(w = k+\tfrac{1}{2};\ k;\ \ldots\ 1\tfrac{1}{2};\ 1). \end{aligned}\right\} \tag{V.4b}$$

Beim letzten Iterationsschritt braucht man (V.4a) nur zur Bestimmung der $(d^\varepsilon \mathrm{U}' - d^\varepsilon \mathrm{U})$ und (V.4b) nur zur Bestimmung von $\Delta^\varepsilon \mathrm{U}_1$ anzuwenden.

5. Nach dem letzten Iterationsschritt 3., 4. erhält man mit den drei Aberrationen

$$(d^\varepsilon U_{k+1} - d^\varepsilon U_0) = -\Delta^\varepsilon \mathrm{U}_1 \qquad [\varepsilon = 0, i, 1;\ d^0 U \equiv U] \tag{V.5a}$$

die meridionale Bildhöhen-Aberration $(\underline{U}_{k+1}-\underline{U}_0)$, die sagittale Bildhöhen-Aberration $(\delta\underline{U}_{k+1}-\delta\underline{U}_0)$ und die tangentiale Bildhöhen-Aberration $(d\underline{U}_{k+1}-d\underline{U}_0)$:

$$\left.\begin{aligned} U_0 &= U_{k+1}-(U_{k+1}-U_0);\\ \overline{G}_0 &= \underset{+}{\sqrt{1-U_0^2}}; \end{aligned}\right\} \tag{V.5b}$$

$$\left.\begin{aligned} (\underline{U}_{k+1}-\underline{U}_0) &= \frac{(U_{k+1}-U_0)}{\overline{G}_0} - \left(\frac{1}{\overline{G}_0}-\frac{1}{\overline{G}_{k+1}}\right) U_{k+1}\\ (\delta\underline{U}_{k+1}-\delta\underline{U}_0) &= (d^i U_{k+1}-d^i U_0)\\ (d\,\underline{U}_{k+1}-d\,\underline{U}_0) &= \frac{(d^1 U_{k+1}-d^1 U_0)}{\overline{G}_0^2} + \frac{1}{\overline{G}_0^2}(c^- U\, G\, d\,\underline{P})_{k+1}. \end{aligned}\right\} \tag{V.5c}$$

Bei diesem Iterations-Formelsatz ist nicht daran gedacht, ihn iterativ zur numerischen Strahldurchrechnung anzuwenden (als Ersatz etwa für die „trigonometrische" Strahldurchrechnung einschließlich der Astigmatismus-Rechnung). Damit, daß wir von einem Iterations-Formelsatz sprechen, wollen wir zum Ausdruck bringen: 1. Dieser Formelsatz bestimmt die Aberrationen genauer, als man die in die Formeln eingehenden Größen zu kennen braucht. 2. Der Formelsatz würde die Aberrationen beliebig genau bestimmen, wenn die in die Formeln eingehenden Größen beliebig genau bekannt wären. — Der zweite Punkt bedeutet, daß der Formelsatz alle Zusammenhänge streng darstellt[1]. Der erste Punkt gibt uns die Möglichkeit, ausgehend von einer sehr einfachen Näherung, durch einmalige (nicht numerische, sondern analytische) Anwendung der Iterationsformeln eine einfache Näherungsdarstellung der meridionalen, sagittalen und tangentialen Bildhöhen-Aberration zu erhalten, deren Gültigkeitsbereich über den der Seidelschen Näherung hinausgeht:

34. Die primäre Näherung. Die Näherung 0 für die Größen (V.2a) ist nicht für alle diese Größen die Gaußsche Näherung. Für $\Delta^0 U_v^{\cdot}$, $\Delta^0 U_v$ und $(d^0 P'-d^0 P)$ ist die Näherung 0 die in Abschnitt III eingeführte „Gaußsche Näherung", in der man eine kollineare Abbildung der Lote hat. Die Näherung 0 für $\Delta^i U_v^{\cdot}$, $\Delta^i U_v$ und $(d^i P'-d^i P)$ entspricht der in Abschnitt III erwähnten kollinearen Abbildung der Höhen, die auch „Gaußsche Abbildung" genannt wird. Die Näherung 0 für $\Delta^1 U_v^{\cdot}$, $\Delta^1 U_v$ und $(d^1 P'-d^1 P)$ entspricht weder der kollinearen Abbildung der Lote, noch der kollinearen Abbildung der Höhen.

Die wahren Werte der Aberrationen (V.2a) sind von der Größenordnung

$$0_1\cdot 0(d^e),$$

wobei

$$0_1 = 0(\alpha)\cdot\{0^{(2)}+0^{[2]}\}\,0^{(1)} \tag{5.77a}$$

ist und wir

$$\left.\begin{aligned} &0(d^0) && \text{für} && 0(1),\\ &0(d^i) && \text{für} && 0(\delta)\equiv 0\left(\frac{\delta W}{W}\right),\\ &0(d^1) && \text{für} && 0(d)\equiv 0\left(\frac{dW}{W}\right) \end{aligned}\right\} \tag{5.78}$$

[1] *Zusatz bei der Korrektur:* Hiervon werden wir in Abschnitt VIII ausgehen, um eine *strenge*, mit Hilfe linearer Operatoren explizit hingeschriebene Darstellung der meridionalen, sagittalen und tangentialen Bildhöhen-Aberration als Funktion der Strahlanfangsdaten und der Langeschen Konstruktionsdaten herzuleiten.

schreiben. Gegenüber den wahren Werten der Aberrationen (V.2a) haben also die Näherungswerte 0 einen Fehler der Größenordnung $0_1 \cdot 0(d^e)$. Legt man bei einer Anwendung des Formelsatzes (V) die Näherung 0 für die Größen (V.2a) zugrunde und führt nur *einen* Iterationsschritt durch, dann erhält man die verbesserten Näherungswerte (V.2b) mit einem Fehler der Größenordnung

$$0_2 \cdot 0(d^e),$$

wobei

$$0_2 = 0(\alpha^2) \cdot \{0^{(2)} + 0^{[2]}\}^2 \, 0^{(1)} \tag{5.77b}$$

ist[1]. Man erhält dann auch die *meridionale, sagittale und tangentiale Bildhöhen-Aberration* (V.5c) bis auf Fehler der Größenordnung (5.77b), also *bis auf Fehler zweiten Grades und fünfter Ordnung* [von $0(d^e)$ abgesehen]. In dieser durch *einen* Iterationsschritt erhaltenen Näherung sind also alle Bildfehler dritter Ordnung *und* in erster Näherung für kleine α alle Bildfehler höherer Ordnung berücksichtigt. Wir nennen diese Näherung die *primäre Näherung*.

Da nach (4.81) $c^- = 0(\alpha)$; nach (4.4), (4.18f)

$$\frac{\nu}{\mu} - 1 = 0(\alpha) \cdot 0^{[2]}$$

und nach (4.46)

$$L - 1 = 0(\alpha^2) \cdot 0^{(2)}$$

ist, können wir in (V.3a, d, e) 0 für $d^1 U_{k+1}$; μ für ν, 1 für $\bar{\nu}^{\boldsymbol{\cdot}}$ und 1 für L setzen bei dem einen Iterationsschritt, der zu der primären Näherung führt, ohne an der Größenordnung (5.77b) des Fehlers dieser Näherung etwas zu ändern [wie man an (V.4a) in Verbindung mit (V.3) sieht]. Dann wird mit den Abkürzungen

$$c_{v+1}^- \equiv c_v^+ \equiv \frac{c_{v+1} - c_v}{t_{v+1} - t_v}; \qquad \mathring{\varkappa} \equiv -\frac{\mu' c^+ - \mu c^-}{\mu' - \mu}; \qquad \mathring{\chi} \equiv \frac{\mu c^- + \mu' c^+}{\mu + \mu'} \tag{5.79a}$$

nach (V.3):

$$d^i K^{\boldsymbol{\cdot}} = \mathring{\varkappa}\, \underline{\delta P}_{k+1}; \qquad d^1 K^{\boldsymbol{\cdot}} = \mathring{\varkappa}\, \underline{dP}_{k+1}; \tag{5.79b}$$

$$P^{\boldsymbol{\cdot}} = P_{k+1} - t\, U_{k+1}; \tag{5.79c}$$

$$\left.\begin{aligned} K^{\boldsymbol{\cdot}} = \frac{1}{\mu^{\boldsymbol{\cdot}}} J^{\boldsymbol{\cdot}} &= \mathring{\varkappa}\, P_{k+1} - (c + t\mathring{\varkappa})\, U_{k+1}; \\ \frac{1}{\mu^{\boldsymbol{\cdot}}} G^{\boldsymbol{\cdot}} &= \mathring{\chi}\, P_{k+1} + (c - t\mathring{\chi})\, U_{k+1}; \end{aligned}\right\} \tag{5.79d}$$

$$\left.\begin{aligned} \bar{J}^{\boldsymbol{\cdot}} &= \underset{+}{\sqrt{1 - J^{\boldsymbol{\cdot}2}}}; \\ \bar{G}^{\boldsymbol{\cdot}} &= \underset{+}{\sqrt{1 - G^{\boldsymbol{\cdot}2}}}, \end{aligned}\right\} \tag{5.79e}$$

nach (V.4)[2]:

$$\left.\begin{aligned} (U_{k+1} - U_0) &= \sum_1^k \left\{(\nu' - \nu) \cdot \left[\frac{\mathring{\varkappa}}{c} \frac{P^{\boldsymbol{\cdot}}}{\bar{J}^{\boldsymbol{\cdot}}} \frac{G^{\boldsymbol{\cdot}} - J^{\boldsymbol{\cdot}}}{\bar{G}^{\boldsymbol{\cdot}} + \bar{J}^{\boldsymbol{\cdot}}} \right] - (\bar{\nu}' - \bar{\nu}) \cdot \frac{1}{c}\right\} \cdot K^{\boldsymbol{\cdot}} \\ (d^i U_{k+1} - d^i U_0) &= \sum_1^k \left\{(\nu' - \nu) \cdot \left[-\frac{U_{k+1}}{\bar{J}^{\boldsymbol{\cdot}}} \frac{G^{\boldsymbol{\cdot}}}{\bar{G}^{\boldsymbol{\cdot}}} + \frac{\mathring{\varkappa}}{c} \frac{P^{\boldsymbol{\cdot}}}{\bar{J}^{\boldsymbol{\cdot}}} \frac{G^{\boldsymbol{\cdot}} - J^{\boldsymbol{\cdot}}}{\bar{G}^{\boldsymbol{\cdot}} + \bar{J}^{\boldsymbol{\cdot}}} \right] - (\bar{\nu}' - \bar{\nu}) \cdot \frac{1}{c}\right\} d^i K^{\boldsymbol{\cdot}} \\ (d^1 U_{k+1} - d^1 U_0) &= \sum_1^k \left\{(\nu' - \nu) \cdot \left[-\frac{U_{k+1}}{\bar{J}^{\boldsymbol{\cdot}}} \frac{G^{\boldsymbol{\cdot}}}{\bar{G}^{\boldsymbol{\cdot}}} + \frac{\mathring{\varkappa}}{c} \frac{P^{\boldsymbol{\cdot}}}{\bar{J}^{\boldsymbol{\cdot}}} \left(\frac{G^{\boldsymbol{\cdot}} - J^{\boldsymbol{\cdot}}}{\bar{G}^{\boldsymbol{\cdot}} + \bar{J}^{\boldsymbol{\cdot}}} + \frac{G^{\boldsymbol{\cdot}} - J^{\boldsymbol{\cdot}}}{\bar{G}^{\boldsymbol{\cdot}} \bar{J}^{\boldsymbol{\cdot}2}} \right) \right] - (\bar{\nu}' - \bar{\nu}) \cdot \frac{1}{c}\right\} d^1 K^{\boldsymbol{\cdot}} \end{aligned}\right\} \quad (5.7$$

[1] Vgl. den Anfang der Ziff. 23.

[2] Wie bei (5.79d) setzen wir 1 für $\bar{\nu}^{\boldsymbol{\cdot}}$.

und nach (V.5)[1] mit

$$\left.\begin{aligned} \Gamma &\equiv \underset{+}{\sqrt{1-(U_{k+1})^2}}; \quad G_{k+1} = (c^{-} P + c\,U)_{k+1}: \\ \Gamma^3 \cdot (\underline{U}_{k+1} - \underline{U}_0) &= (U_{k+1} - U_0) + \tfrac{1}{2}(G_{k+1}^2 - U_{k+1}^2)\,U_{k+1} \\ (\delta \underline{U}_{k+1} - \delta \underline{U}_0) &= (d^i U_{k+1} - d^i U_0), \\ \Gamma^2 \cdot (d\underline{U}_{k+1} - d\underline{U}_0) &= (d^1 U_{k+1} - d^1 U_0) + c_{k+1}^{-}\, U_{k+1}\, G_{k+1}\, d\underline{P}_{k+1}. \end{aligned}\right\} \quad (5.79\mathrm{g})$$

Ohne an der Größenordnung (5.77b) des Fehlers dieser primären Näherung (5.79) etwas zu ändern, können darin die Strahldaten U_{k+1} und P_{k+1} gegen U_0 und P_0 ausgewechselt werden, und auch $\delta\underline{P}_{k+1}$ und $d\underline{P}_{k+1}$ gegen $\delta\underline{P}_0$ und $d\underline{P}_0$. Dann sind mit (5.79) die meridionale Bildhöhen-Aberration $(\underline{U}_{k+1} - \underline{U}_0)$, die sagittale Bildhöhen-Aberration $(\delta\underline{U}_{k+1} - \delta\underline{U}_0)$ und die tangentiale Bildhöhen-Aberration $(d\underline{U}_{k+1} - d\underline{U}_0)$ in *primärer Näherung* als Funktionen der objektseitigen Strahldaten U_0 (Bildwinkel-Sinus), P_0 (Eintrittslot), $\delta\underline{P}_0$, $d\underline{P}_0$ dargestellt[2].

Die Formeln (5.79f) zeigen deutlich, wie sehr für das Zustandekommen von Bildfehlern höherer Ordnung Flächen verantwortlich sind, bei denen große Brechungswinkel auftreten: Bei diesen Flächen werden die in den Nennern stehenden mittleren Brechungs-Cosinus $\bar{J}^{\cdot}$ deutlich kleiner als 1 und verstärken damit die Wirkung der Seidelschen Flächenanteile, die gegeben sind durch die Summanden von (5.79f) mit $\bar{J}^{\cdot} = 1$ und $\bar{G}^{\cdot} = 1$ in den Nennern. Da allgemein $\bar{J}^{\cdot}$ von Fläche zu Fläche durchschnittlich stärker schwankt als $\bar{G}^{\cdot}$ [weil $G' - G \approx -(\nu' - \nu) \cdot n J = 0(\alpha) \cdot n J$ ist[3]], und da in den Nennern $\bar{G}^{\cdot}$ weniger wirksam vorkommt als $\bar{J}^{\cdot}$ (abgesehen von den Gliedern, bei denen $U_{k+1} G^{\cdot}$ im Zähler steht), kommt es vor allem auf die Wirkung der Abweichungen an, die $\bar{J}^{\cdot}$ von 1 hat. — So finden wir die den optischen Rechnern lange bekannte Faustregel, daß große *Brechungswinkel* starke Bildfehler höherer Ordnung hervorrufen, als Teilaussage unserer Theorie wieder. Darüber hinaus beschreiben die Formeln (5.79f) diese Wirkung quantitativ (in primärer Näherung). — Diese Faustregel spricht von großen Brechungswinkeln und nicht von großen Flächenkrümmungen. In vielen Fällen treten große Brechungswinkel auf an Flächen, deren Krümmungen groß sind; das erweckt den Eindruck, als würden stark gekrümmte Flächen starke Bildfehler höherer Ordnung hervorrufen. Nach der Faustregel und nach unserer Theorie sind es aber die Brechungswinkel und nicht die Flächenkrümmungen, auf die es primär ankommt.

VI. Eikonale; Bildfehler-Koeffizienten.

35. Das Eikonal $\mathscr{E}$. Wir betrachten nun wieder beliebige, meridionale und windschiefe, Strahlen. Um der Übersicht willen bleiben wir aber dabei, die Objektebene im Unendlichen anzunehmen. Wir geben die reduzierte Bildhöhe $\underline{U}_{k+1}$ zusammen mit einer geeigneten Pupillen-Variablen vor und interessieren uns, nach wie vor, für die Bildhöhen-Aberration als Funktion der vorgegebenen Strahldaten und der Konstruktionsdaten des Systems. Im Objektraum und im Bildraum lassen wir jetzt auch andere Medien als Luft zu.

---

[1] Bis auf Vernachlässigungen der Größenordnung (5.77b); es ist $G_{k+1} = U_{k+1} + 0(\alpha) \cdot 0^{(1)}$.

[2] In Ziff. 38 werden wir sehen, daß man aus den *meridionalen* und *sagittalen* Bildhöhen-Aberationen weitgehend auf die Bildhöhen-Aberrationen *aller* windschiefen Strahlen schließen kann. — Die Darstellung der *tangentialen* neben der sagittalen Bildhöhen-Aberration dient zur Diskussion des Astigmatismus höherer Ordnung.

[3] Bei einer in Luft stehenden Linse mit den Flächen Nr. 1 und 2 ist in Gaußscher Näherung $-(G_2^{\cdot} - G_1^{\cdot})/(J_2^{\cdot} - J_1^{\cdot}) = \alpha_2/(2-\alpha_2) = \frac{1}{5} \ldots \frac{1}{3}$ für $\alpha_2 = \frac{1}{3} \ldots \frac{1}{2}$ $[n_{\mathrm{Glas}} = 1{,}50 \ldots 2{,}00]$.

Nur für die Rechnung im nächsten Absatz nehmen wir: als Fläche Nr. 0 nicht die Eintrittspupille, sondern eine nicht-brechende Kugelfläche mit endlichem Radius, deren Scheitel im reellen oder virtuellen Objektraum liegt; und als Fläche Nr. $(k+1)$ nicht die Austrittspupille, sondern eine nicht-brechende Kugelfläche mit endlichem Radius, deren Scheitel im reellen oder virtuellen Bildraum liegt. Im übernächsten Absatz lassen wir wieder $r_0 \to \infty$ und $r_{k+1} \to \infty$ gehen und die Fläche Nr. 0 mit der Eintrittspupille, die Fläche Nr. $k+1$ jedoch mit der Gaußschen Bildebene (der wir dann aber nicht mehr die Nr. $k+1$ geben) zusammenfallen.

Zwischen der Fläche Nr. 0 und der Fläche Nr. $k+1$ legt ein Strahl den sogenannten *Lichtweg*

$$\mathscr{E} \equiv \sum_{v=0}^{k} n'_v \cdot \overrightarrow{D_v D_{v+1}} \tag{6.1}$$

zurück (D_v ist der Auftreffpunkt des Strahls auf die Fläche Nr. v). Gehen wir zu einem Nachbarstrahl über und bezeichnen alle damit verbundenen Änderungen mit d, dann ist wegen $\overrightarrow{D_v D_{v+1}} = \boldsymbol{r}_{v+1} - \boldsymbol{r}_v + \boldsymbol{e}'_v = l'_v\,\boldsymbol{n}'_v$ (vgl. Ziff. 9) und wegen $\boldsymbol{n}^2 = n^2 : (l\boldsymbol{n})\,d\,(l\boldsymbol{n}) = n^2\,l\,d\,l;\, d\,(\boldsymbol{n}'_v \cdot \overrightarrow{D_v D_{v+1}}) = n'^2_v\,d\,l'_v = \boldsymbol{n}'_v\,d\,(l'_v \boldsymbol{n}'_v) = \boldsymbol{n}_{v+1}\,d\boldsymbol{r}_{v+1} - \boldsymbol{n}'_v\,d\boldsymbol{r}_v$, also

$$d\mathscr{E} = -\boldsymbol{n}'_0\,d\boldsymbol{r}_0 - \sum_{v=1}^{k} (\boldsymbol{n}'_v - \boldsymbol{n}_v)\,d\boldsymbol{r}_v + \boldsymbol{n}_{k+1}\,d\boldsymbol{r}_{k+1}.$$

Wegen des Brechungsgesetzes (2.23), wonach $\boldsymbol{n}'_v - \boldsymbol{n}_v = \lambda_v\,\boldsymbol{r}_v$ ist, und wegen $\boldsymbol{r}_v^2 = r_v^2$ ist $(\boldsymbol{n}'_v - \boldsymbol{n}_v)\,d\boldsymbol{r}_v = \lambda_v\,\boldsymbol{r}_v\,d\boldsymbol{r}_v = 0$, folglich

$$d\mathscr{E} = -\boldsymbol{n}_0\,d\boldsymbol{r}_0 + \boldsymbol{n}_{k+1}\,d\boldsymbol{r}_{k+1}. \tag{6.2}$$

Lassen wir nun $r_0 \to \infty$ und $r_{k+1} \to \infty$ gehen, dann wird $dz_0 = 0$ und $dz_{k+1} = 0$, und aus (6.2) wird:

$$d\mathscr{E} = -\eta_0\,dy_0 - \xi_0\,dx_0 + \eta_{k+1}\,dy_{k+1} + \xi_{k+1}\,dx_{k+1}.$$

Die Fläche Nr. 0 lassen wir jetzt wieder mit der Eintrittpupille zusammenfallen. Die Fläche Nr. $k+1$ lassen wir mit der Gaußschen Bildebene zusammenfallen; wir schreiben dann aber y' und x' für y_{k+1} und x_{k+1} (und η'; ξ' für η_{k+1}; ξ_{k+1}), und behalten die Nummer $k+1$ wieder der Austrittspupille vor. Dann ändert sich also der zwischen der Eintrittspupille und der Gaußschen Bildebene zurückgelegte Lichtweg um

$$d\mathscr{E} = -\eta_0\,dy_0 - \xi_0\,dx_0 + \eta'\,dy' + \xi'\,dx', \tag{6.3}$$

wenn man von einem Strahl zu einem Nachbarstrahl übergeht, und wenn sich dabei die Eintrittspupillen-Höhe um $\{dx_0;\ dy_0;\ 0\}$ und die Bildhöhe um $\{dx';\ dy';\ 0\}$ ändert.

Nun können wir durch die Bildhöhe $\{x';\ y';\ 0\}$ und die Eintrittspupillen-Höhe $\{x_0;\ y_0;\ 0\}$ einen Strahl eindeutig kennzeichnen, die vier Variablen y_0, x_0, y', x' also als *vorgegebene Strahldaten* nehmen. Der Lichtweg $\mathscr{E}$ ist im allgemeinen von Strahl zu Strahl verschieden, er ist eine Funktion der Strahldaten (y_0, x_0, y', x'):

$$\mathscr{E} = \mathscr{E}(y_0, x_0, y', x'). \tag{6.4}$$

Als Funktion dieser Variablen betrachtet, nennt man $\mathscr{E}$ auch die Eikonalfunktion oder kurz das *Eikonal*. Man sieht an (6.3), daß die *vier* Funktionen

$$\left.\begin{aligned} \eta_0 &= \eta_0(y_0, x_0, y', x'); & \xi_0 &= \xi_0(y_0, x_0, y', x');\\ \eta' &= \eta'(y_0, x_0, y', x'); & \xi' &= \xi'(y_0, x_0, y', x') \end{aligned}\right\} \tag{6.5}$$

alle aus der *einen* Eikonalfunktion (6.4) ableitbar sind:

$$\left.\begin{aligned} \eta_0 &= -\,\partial\mathcal{E}/\partial y_0; & \xi_0 &= -\,\partial\mathcal{E}/\partial x_0; \\ \eta' &= +\,\partial\mathcal{E}/\partial y'; & \xi' &= +\,\partial\mathcal{E}/\partial x'. \end{aligned}\right\} \tag{6.6}$$

Zwischen den vier Funktionen (6.5) bestehen also recht enge Kopplungen.

36. Eikonal-Koeffizienten. Führen wir Polarkoordinaten R_0, Ψ_0 und R', Ψ' anstelle von y_0, x_0 und y', x' ein:

$$\begin{aligned} y_0 &\equiv R_0 \cos \Psi_0; & x_0 &\equiv R_0 \sin \Psi_0; \\ y' &\equiv R' \cos \Psi'; & x' &\equiv R' \sin \Psi', \end{aligned}$$

und betrachten $\mathcal{E}$ als Funktion dieser Polarkoordinaten:

$$\mathcal{E} = \mathcal{E}[R_0, \Psi_0, R', \Psi'],$$

dann können wir feststellen: Wegen der Rotationssymmetrie des optischen Systems ändert der Lichtweg $\mathcal{E}$ sich nicht, wenn man, bei konstantem R_0 und R', Ψ_0 um ψ und zugleich Ψ' um ψ ändert. Und wegen der Spiegelsymmetrie des optischen Systems ändert $\mathcal{E}$ sich nicht, wenn man das Vorzeichen von Ψ_0 und zugleich das Vorzeichen von Ψ' umkehrt. — Daraus folgt, daß bei einem bestimmten optischen System $\mathcal{E}$ eindeutig durch R_0, R' und $|\Psi_0 - \Psi'|$ gegeben ist, daß also $\mathcal{E}$ eine eindeutige Funktion von

$$\left.\begin{aligned} A &\equiv R'^2 & &\equiv y'^2 + x'^2; \\ B &= R_0 R' \cos(\Psi_0 - \Psi') & &\equiv y_0 y' + x_0 x'; \\ C &\equiv R_0^2 & &\equiv y_0^2 + x_0^2, \end{aligned}\right\} \tag{6.7}$$

drei Kombinationen der vier Strahldaten (y_0, x_0, y', x'), ist:

$$\mathcal{E} = \mathcal{E}(A; B; C). \tag{6.8}$$

Bei Systemen aus Kugelflächen (einschließlich Planflächen), die wir hier betrachten, können die Komponenten y_v, x_v der Radiusvektoren und η_v, ξ_v der optischen Richtungen als Potenzreihen nach den Eintrittspupillen- und Bildebenen-Koordinaten y_0, x_0, y', x' angesetzt werden; es lassen sich (vgl. Abschnitt VII) Durchrechnungs- und Rekursionsformeln für die Koeffizienten dieser Potenzreihen angeben. Da nach (6.1) $\mathcal{E} = \sum n_v \cdot (D_v D_{v+1}) = \sum n_v^2 [(z+r)_{v+1} - (z+r)_v + a_v']/\zeta_v'$ ist, und da $(z+r) = \varrho(y^2 + x^2)/(1 + \underset{+}{\sqrt{1 - \varrho^2(y^2+x^2)}})$ und $1/\zeta = \nu/\underset{+}{\sqrt{1 - \nu^2(\eta^2 + \xi^2)}}$ ist, kann mit den y_v, x_v und η_v, ξ_v auch die Eikonalfunktion $\mathcal{E} = \overset{+}{\mathcal{E}}(y_0, x_0, y', x')$ in eine Potenzreihe nach y_0, x_0, y', x' entwickelt werden.

Nicht selbstverständlich ist die Feststellung, daß dann $\mathcal{E} = \mathcal{E}(A; B; C)$ auch in eine Potenzreihe nach $A; B; C$ entwickelt werden kann. Allein daraus, daß eine Größe nach (y_0, x_0, y', x') entwickelbar ist, und daß diese Größe eine Funktion der drei quadratischen Kombinationen $(A; B; C)$ von (y_0, x_0, y', x') ist, folgt noch nicht die Entwickelbarkeit dieser Größe nach $(A; B; C)$, sondern nach einem Beweis von C. CARATHÉODORY[1] deren Entwickelbarkeit nach

$$A, B, C \quad \text{und} \quad D \equiv x_0 y' - y_0 x' = \sqrt{AC - B^2}. \tag{6.9}$$

Danach ist also, weil man wegen $D^2 + B^2 = AC$ entweder mit der ersten Potenz von D oder mit der ersten Potenz von B auskommt:

$$\mathcal{E} = \mathfrak{P}_0(A; B; C) + D \cdot \mathfrak{P}_1(A; B; C) \tag{6.10}$$

[1] C. CARATHÉODORY: Geometrische Optik. Berlin 1937 [§ 57].

oder

$$\mathcal{E}=\mathfrak{Q}_0(A\,;D\,;C)+B\cdot\mathfrak{Q}_1(A\,;D\,;C)\,, \tag{6.11}$$

wobei $\mathfrak{P}_0$, $\mathfrak{P}_1$ und $\mathfrak{Q}_0$, $\mathfrak{Q}_1$ Potenzreihen nach ihren Argumenten sind. Nun hatten wir aber festgestellt, daß wegen der Rotations- und Spiegel-Symmetrie der Lichtweg $\mathcal{E}$ eine *eindeutige* Funktion von $(A\,;B\,;C)$ ist. Hieraus folgt, daß $D\cdot\mathfrak{P}_1(A\,;B\,;C)=0$ ist[1] in dem Ausdruck (6.10) für $\mathcal{E}$, daß $\mathcal{E}$ also eine Potenzreihe nach $(A\,;B\,;C)$ ist. — Für den Ausdruck (6.11) folgt, daß in den Potenzreihen $\mathfrak{Q}_0$ und $\mathfrak{Q}_1$ nur gerade Potenzen von D vorkommen, daß also

$$\mathcal{E}=\mathcal{E}_0(A\,;C\,;D^2)+B\cdot\mathcal{E}_1(A\,;C\,;D^2) \tag{6.12}$$

ist, wobei $\mathcal{E}_0$ und $\mathcal{E}_1$ Potenzreihen nach ihren Argumenten sind.

Den Koeffizienten des Glieds mit $A^\alpha B^\beta C^\gamma$ in der Potenzreihe $\mathcal{E}(A\,;B\,;C)$ bezeichnen wir mit $\mathcal{E}^{\alpha\beta\gamma}$:

$$\mathcal{E}=\sum_{\alpha=0}^{\infty}\sum_{\beta=0}^{\infty}\sum_{\gamma=0}^{\infty}\mathcal{E}^{\alpha\beta\gamma}\cdot A^\alpha B^\beta C^\gamma\,. \tag{6.13}$$

A, B, C und D sind Größen zweiter Ordnung. $A^\alpha B^\beta C^\gamma$ ist also eine Größe der Ordnung $\omega=2(\alpha+\beta+\gamma)$. Den zugehörigen Koeffizienten $\mathcal{E}^{\alpha\beta\gamma}$ nennen wir einen „Eikonal-Koeffizienten ω-ter Ordnung"; er selbst ist eine Größe nullter Ordnung[2], er ist nur multipliziert mit einer Größe ω-ter Ordnung. Die Eikonal-Koeffizienten sind Funktionen der Wellenlänge des Lichts, wenn wir Strahlen verschiedener Wellenlänge betrachten. In der Potenzreihe (6.13) kommen $(\omega+2)(\omega+4)/8$ verschiedene Eikonal-Koeffizienten ω-ter Ordnung vor:

$$\left.\begin{array}{l} 1;\ 3;\ 6;\ 10;\ 15;\ 21;\ldots\ \text{Eikonal-Koeffizienten}\\ 0;\ 2;\ 4;\ \ 6;\ \ 8;\ 10;\ldots\ \text{-ter Ordnung.}\end{array}\right\} \tag{6.14}$$

Auch die Koeffizienten $\mathcal{E}_0^{\alpha\gamma\varepsilon}$ und $\mathcal{E}_1^{\alpha\gamma\varepsilon}$ der in (6.12) stehenden Potenzreihen $\mathcal{E}_0$ und $\mathcal{E}_1$ nennen wir „Eikonal-Koeffizienten"; nach (6.12) ist

$$\mathcal{E}=\sum_{\alpha=0}^{\infty}\sum_{\gamma=0}^{\infty}\sum_{\varepsilon=0}^{\infty}(\mathcal{E}_0^{\alpha\gamma\varepsilon}+\mathcal{E}_1^{\alpha\gamma\varepsilon}\cdot B)\,A^\alpha C^\gamma D^{2\varepsilon}\,. \tag{6.15}$$

$\mathcal{E}_0^{\alpha\gamma\varepsilon}$ ist ein Eikonal-Koeffizient der Ordnung $2(0+\alpha+\gamma+2\varepsilon)$, $\mathcal{E}_1^{\alpha\gamma\varepsilon}$ ist ein Eikonal-Koeffizient der Ordnung $2(1+\alpha+\gamma+2\varepsilon)$. In (6.15) kommen, ebenso wie in (6.13), $(\omega+2)(\omega+4)/8$ verschiedene Eikonal-Koeffizienten ω-ter Ordnung vor.

Differenziert man die Reihe (6.13) [oder die Reihe (6.15)] gemäß (6.6) nach y_0 und nach x_0, um die „Neigung" $(\eta_0+i\xi_0)$ des Strahls im Objektraum und damit das reduzierte Objektlot U_0 zu erhalten:

$$-\frac{n_0}{m_0}\cdot U_0=(\eta_0+i\xi_0)=-\left(\frac{\partial\mathcal{E}}{\partial y_0}\right)_{x_0y'x'}-i\left(\frac{\partial\mathcal{E}}{\partial x_0}\right)_{y_0y'x'}, \tag{6.16}$$

dann verschwinden aus der Reihe (6.13) die Glieder mit den Koeffizienten $\mathcal{E}^{\alpha00}$ [oder aus der Reihe (6.15) die Glieder mit den Koeffizienten $\mathcal{E}_0^{\alpha00}$], weil diese Glieder von y_0 und x_0 nicht abhängen. Bei den Gliedern $(\omega-1)$-ter Ordnung der

---

[1] In dem Ausdruck (6.10) für $\mathcal{E}$ kommt, neben den eindeutigen Potenzreihen $\mathfrak{P}_0(A\,;B\,;C)$ und $\mathfrak{P}_1(A\,;B\,;C)$, eine zweideutige Funktion von $(A\,;B\,;C)$ vor: $D\equiv x_0y'-y_0x'=\sqrt{AC-B^2}$. Ein Strahl, bei dem $D\neq0$ ist (das ist ein windschiefer Strahl) und sein Spiegelbild z. B. an der y, z-Ebene unterscheiden sich zwar nicht in den Werten von $A\,;B\,;C$ und nicht im Betrag von D, wohl aber im Vorzeichen von D. Bei diesen beiden Strahlen wäre, trotz der gleichen Werte von $(A\,;B\,;C)$, $\mathcal{E}$ unterschiedlich, wenn nicht $\mathfrak{P}_1=0$ wäre.

[2] $\mathcal{E}^{001}$ ist eine Größe $\sim0^{[2]}$, die wir aber in Abschnitt VI und VII als Größe nullter Ordnung zählen, nicht wie in Abschnitt III bis V als Größe zweiter Ordnung.

gemäß (6.16) erhaltenen Reihe für $-\frac{n_0}{m_0}\cdot U_0=(\eta_0+i\xi_0)$ tritt also 1 Eikonal-Koeffizient weniger auf als bei den Gliedern ω-ter Ordnung der Reihe (6.13) oder (6.15). So treten

$$\text{bei den Gliedern}\quad \left.\begin{array}{l} 2;\ 5;\ 9;\ 14;\ 20;\ldots \text{ Koeffizienten} \\ 1;\ 3;\ 5;\ \ 7;\ \ 9;\ldots \text{-ter Ordnung} \end{array}\right\} \qquad (6.17)$$

der Reihe für $-\frac{n_0}{m_0}U_0=(\eta_0+i\xi_0)$ auf.

Nehmen wir für $\mathscr{E}$ die Reihe (6.13), dann ist nach (6.16)

$$\left.\begin{aligned} \frac{n_0}{m_0}\cdot U_0 = -(\eta_0+i\xi_0) = \sum_{\alpha=0}^{\infty}\sum_{\beta=0}^{\infty}\sum_{\gamma=0}^{\infty}\big[(\beta+1)\,\mathscr{E}^{\alpha(\beta+1)\gamma}\cdot(y'+ix')+{}\\ +2(\gamma+1)\,\mathscr{E}^{\alpha\beta(\gamma+1)}\cdot(y_0+ix_0)\big]A^{\alpha}B^{\beta}C^{\gamma};\\ A\equiv(y'^2+x'^2);\qquad B\equiv(y_0y'+x_0x');\qquad C\equiv(y_0^2+x_0^2), \end{aligned}\right\} \qquad (6.18)$$

womit wir für die reduzierte Objekthöhe

$$\underline{U}_0=U_0/\sqrt[+]{1-U_0U_0^*/m_0^2} \qquad (6.19)$$

und (was uns letzten Endes interessiert) für die Bildhöhen-Aberration $(\underline{U}_{k+1}-\underline{U}_0)$ als Funktion der reduzierten Bildhöhe

$$\underline{U}_{k+1}=-c_{\overline{k+1}}(y'+ix') \qquad (6.20)$$

und der Eintrittshöhe

$$\underline{P}_0=(y_0+ix_0) \qquad (6.21)$$

einen Ausdruck erhalten, der nach (6.17) in einer Näherung erster Ordnung nur 2 Koeffizienten; in einer Näherung dritter Ordnung nur (2+5) Koeffizienten; in einer Näherung fünfter Ordnung nur (2+5+9) Koeffizienten; in einer Näherung siebter Ordnung nur (2+5+9+14) Koeffizienten (und so fort) enthält. Diese Koeffizienten sind Funktionen der Wellenlänge des Lichts und der Konstruktionsdaten des optischen Systems, bei einem bestimmten System also Funktionen der Wellenlänge. Es sind in den verschiedenen Näherungen weit weniger Koeffizienten, als allein aufgrund der Entwickelbarkeit von U_0 nach den Real- und Imaginärteilen von $\underline{U}_{k+1}$ und $\underline{P}_0$ anzunehmen ist. So haben sich die am Ende von Ziff. 35 genannten, durch die Ableitung aus einer Eikonal-Funktion gegebenen Kopplungen *und* die Rotations- und Spiegelsymmetrie des optischen Systems ausgewirkt.

In Ziff. 38 werden wir sehen: Diese verhältnismäßig geringe Anzahl von Koeffizienten in den verschiedenen Näherungen bringt es mit sich, daß die Bildhöhen-Aberrationen aller windschiefen Strahlen weitgehend ersichtlich sind aus den meridionalen und sagittalen Bildhöhen-Aberrationen.

Vorher wollen wir noch eine andere Eikonal-Darstellung der Bildhöhen-Aberration $(\underline{U}_{k+1}-\underline{U}_0)$ kennenlernen, bei der *nicht* die mit der obigen Eikonal-Darstellung verbundene Komplikation auftritt, daß man die Potenzreihe (6.18) in den Wurzelausdruck (6.19) einzusetzen hat, um die Objekthöhe $\underline{U}_0$ zu erhalten; sondern bei der man direkt die Objekthöhe $\underline{U}_0$ und nicht das Objektlot U_0 als Potenzreihe nach der Bildhöhe und einer Pupillen-Variablen erhält.

37. Das Eikonal $\bar{\mathscr{E}}$. Das Eikonal $\mathscr{E}$ hatten wir eingeführt als den Lichtweg zwischen dem Durchstoßpunkt D_0 des Strahls durch die Eintrittspupille und dem Durchstoßpunkt D_{GBE} des Strahls durch die Gaußsche Bildebene. Da seine

Änderung bei Übergang zu einem Nachbarstrahl nach (6.3)

$$d\mathscr{E} = -\eta_0\, dy_0 - \xi_0\, dx_0 + \eta'\, dy' + \xi'\, dx' \tag{6.22}$$

ist, erhalten wir auch für die Änderung des Lichtwegs $(\mathscr{E} + \eta_0 y_0 + \xi_0 x_0)$ einen einfachen Ausdruck:

$$d(\mathscr{E} + \eta_0 y_0 + \xi_0 x_0) = y_0\, d\eta_0 + x_0\, d\xi_0 + \eta'\, dy' + \xi'\, dx'. \tag{6.23}$$

$(\mathscr{E} + \eta_0 y_0 + \xi_0 x_0)$ ist der Lichtweg zwischen dem Fußpunkt D_* des Lots $\boldsymbol{L}_0$, das von der Mitte S_0 der Eintrittspupille auf den Strahl gefällt ist, und dem Durchstoßpunkt D_{GBE} durch die Gaußsche Bildebene. Denn $(\eta_0 y_0 + \xi_0 x_0) = \boldsymbol{n}_0(\overrightarrow{S_0 D_0}) = \boldsymbol{n}_0(\overrightarrow{D_* D_0})$ ist der Lichtweg zwischen D_* und D_0.

Der Höhenvektor $\boldsymbol{H}_0 \equiv \overrightarrow{S_0 D_0}$ hat x_0; y_0; 0 als Komponenten:

$$\boldsymbol{H}_0 = \{x_0;\, y_0;\, 0\}. \tag{6.24}$$

Die Komponenten des Lotvektors $\boldsymbol{L}_0 \equiv \overrightarrow{S_0 D_*}$ bezeichnen wir mit x_*; y_*; z_*:

$$\boldsymbol{L}_0 = \{x_*;\, y_*;\, z_*\}. \tag{6.25}$$

Wegen $(\boldsymbol{n}_0 \boldsymbol{H}_0) = \boldsymbol{n}_0(\overrightarrow{D_* D_0})$ und $\overrightarrow{D_* D_0} \parallel \boldsymbol{n}$ ist $\overrightarrow{D_* D_0} = (\boldsymbol{n}_0 \boldsymbol{H}_0)\, \boldsymbol{n}_0 / n_0^2$, daher

$$\left.\begin{aligned} \boldsymbol{L}_0 &= \boldsymbol{H}_0 - (\boldsymbol{n}_0 \boldsymbol{H}_0)\, \boldsymbol{n}_0 / n_0^2 \\ &= [[\boldsymbol{n}_0 \boldsymbol{H}_0]\, \boldsymbol{n}_0] / n_0^2, \end{aligned}\right\} \tag{6.26}$$

also

$$\left.\begin{aligned} y_* &= [y_0 \zeta_0^2 - (x_0 \eta_0 - y_0 \xi_0)\, \xi_0] / n_0^2; \\ x_* &= [x_0 \zeta_0^2 + (x_0 \eta_0 - y_0 \xi_0)\, \eta_0] / n_0^2. \end{aligned}\right\} \tag{6.27}$$

Diese Komponenten des Lotvektors $\boldsymbol{L}_0$ treten auf, wenn wir die in (6.23) stehenden Differentiale des Real- und Imaginärteils des (mit n_0/m_0 multiplizierten) reduzierten Objekt*lots*

$$\frac{n_0}{m_0}\, U_0 = -(\eta_0 + i\, \xi_0) \tag{6.28}$$

umrechnen auf die Differentiale des Real- und Imaginärteils der (mit n_0/m_0 multiplizierten) reduzierten Objekt*höhe*

$$\frac{n_0}{m_0}\, \underline{U}_0 = -(\eta_0 + i\, \xi_0)\, \frac{n_0}{\zeta_0} \equiv -(\bar{\eta} + i\, \bar{\xi}). \tag{6.29}$$

Aus (6.29) und aus $\zeta_0^2 = n_0^2 - \eta_0^2 - \xi_0^2$ folgt, daß

$$\left.\begin{aligned} d\eta_0 &= [d\bar{\eta} - \bar{\xi}(\bar{\eta}\, d\bar{\xi} - \bar{\xi}\, d\bar{\eta}) / n_0^2]\, \zeta_0^3 / n_0^3; \\ d\xi_0 &= [d\bar{\xi} + \bar{\eta}(\bar{\eta}\, d\bar{\xi} - \bar{\xi}\, d\bar{\eta}) / n_0^2]\, \zeta_0^3 / n_0^3 \end{aligned}\right\} \tag{6.30}$$

ist. Damit wird aus (6.23):

$$\begin{aligned} d(\mathscr{E} + \eta_0 y_0 + \xi_0 x_0) = {} & [y_0 - (x_0 \bar{\eta} - y_0 \bar{\xi})\, \bar{\xi} / n_0^2] \left(\frac{\zeta_0}{n_0}\right)^3 d\bar{\eta} + \\ & + [x_0 + (x_0 \bar{\eta} - y_0 \bar{\xi})\, \bar{\eta} / n_0^2] \left(\frac{\zeta_0}{n_0}\right)^3 d\bar{\xi} + \\ & + \eta'\, dy' + \xi'\, dx', \end{aligned}$$

nach (6.27) und (6.29) also

$$\left.\begin{aligned} d(\mathscr{E} + \eta_0 y_0 + \xi_0 x_0) &= \frac{\zeta_0}{n_0}\, y_*\, d\bar{\eta} + \frac{\zeta_0}{n_0}\, x_*\, d\bar{\xi} + \eta'\, dy' + \xi'\, dx' \\ &= \bar{y}\, d\bar{\eta} + \bar{x}\, d\bar{\xi} + \eta'\, dy' + \xi'\, dx', \end{aligned}\right\} \tag{6.31}$$

wobei

$$\bar{y} \equiv \frac{\zeta_0}{n_0} y_*; \quad \bar{x} \equiv \frac{\zeta_0}{n_0} x_* \tag{6.32}$$

die mit ζ_0/n_0 multiplizierten x- und y-Komponenten des Lotvektors $\boldsymbol{L}_0$ sind, der von der Mitte der Eintrittspupille zum Strahl führt. In einer Näherung erster Ordnung stimmen $(\bar{y}+i\bar{x})$ und $(y_0+i x_0)=\underline{P}_0$ überein; $(\bar{y}+i\bar{x})$ kann also ebenso gut wie $\underline{P}_0$ als Pupillenvariable genommen werden[1].

Bei einem Meridianstrahl ist $\bar{x}=0$, und $\bar{y}$ hat die anschauliche Bedeutung des „Lot-Höhen-Lotes", das man folgendermaßen erhält: Das Lot $\boldsymbol{L}_0$ führt von der Eintrittspupillen-Mitte S_0 Strahl-senkrecht zum Punkt D_* des Strahls[2]. Der Punkt D_* liegt in einer achsensenkrechten Ebene, deren Achsenpunkt wir mit S_* bezeichnen. Die „Lot-Höhe" führt von S_* Achsen-senkrecht zum Punkt D_* des Strahls. Das „Lot-Höhen-Lot" führt von S_* Strahl-senkrecht zum Strahl.

Aus (6.31) folgt unmittelbar: Führen wir

$$\left.\begin{aligned} \bar{\mathscr{E}} &\equiv (\mathscr{E}+\eta_0 y_0+\xi_0 x_0)-(\eta_0 y_*+\xi_0 x_*) \\ &= (\mathscr{E}+\eta_0 y_0+\xi_0 x_0)-(\bar{\eta}\bar{y}+\bar{\xi}\bar{x}) \end{aligned}\right\} \tag{6.33}$$

als neues Eikonal ein, dann ist dessen Änderung

$$d\bar{\mathscr{E}}=-\bar{\eta}\,d\bar{y}-\bar{\xi}\,d\bar{x}+\eta' d y'+\xi' d x'. \tag{6.34}$$

$\bar{\mathscr{E}}$ ist eine Funktion der vorgegebenen Strahldaten, als die wir die Bildvariablen y', x' und die Eintrittspupillen-Variablen $\bar{y}$, $\bar{x}$ nehmen können:

$$\bar{\mathscr{E}}=\bar{\mathscr{E}}(\bar{y},\bar{x},y',x'). \tag{6.35}$$

Nach (6.34) ist

$$\left.\begin{aligned} \bar{\eta}&=-\partial\bar{\mathscr{E}}/\partial\bar{y}; & \bar{\xi}&=-\partial\bar{\mathscr{E}}/\partial\bar{x}; \\ \eta'&=+\partial\bar{\mathscr{E}}/\partial y'; & \xi'&=+\partial\bar{\mathscr{E}}/\partial x'. \end{aligned}\right\} \tag{6.36}$$

Die Betrachtungen und Rechnungen, die wir in Ziff. 36 im Anschluß an die in Ziff. 35 eingeführte Eikonalfunktion $\mathscr{E}=\mathscr{E}(y_0, x_0, y', x')$ durchgeführt hatten, lassen sich auf die neue Eikonalfunktion $\bar{\mathscr{E}}=\bar{\mathscr{E}}(\bar{y},\bar{x},y',x')$ übertragen: Man braucht in Ziff. 36 nur y_0, x_0 und $\mathscr{E}$ durch $\bar{y}$, $\bar{x}$ und $\bar{\mathscr{E}}$ zu ersetzen; ferner η_0, ξ_0, U_0 und $\underline{P}_0$ durch $\bar{\eta}$, $\bar{\xi}$, $\underline{U}_0$ und

$$\bar{P}\equiv(\bar{x}+i\,\bar{y}). \tag{6.37}$$

Nicht übertragbar ist die Gl. (6.19), die in Ziff. 36 die Eikonaldarstellung der Bildhöhen-Aberration $(\underline{U}_{k+1}-\underline{U}_0)$ kompliziert machte, und die wir nun durch Einführung des neuen Eikonals überflüssig gemacht haben.

Das neue Eikonal $\bar{\mathscr{E}}$ ist der Lichtweg, den der Strahl zwischen dem Fußpunkt D_{LPL} des „Lot-Pfeilhöhen-Lots" und dem Durchstoßpunkt D_{GBE} durch die Gaußsche Bildebene zurücklegt. Das Lot-Pfeilhöhen-Lot erhält man folgendermaßen: Das Lot $\boldsymbol{L}_0$ führt von S_0 Strahl-senkrecht zum Fußpunkt D_*. Von diesem Fußpunkt aus kommt man Achsen-parallel zu einem Punkt $\bar{S}_*$ der Eintrittspupille. Die „Lot-Pfeilhöhe" führt von $\bar{S}_*$ Achsen-parallel zum Punkt D_* des Strahls. Das „Lot-Pfeilhöhen-Lot" führt von $\bar{S}_*$ Strahl-senkrecht zum Strahl.—

[1] Da die vom Objektraum in den Bildraum gelangenden Strahlenbündel in sehr vielen Fällen nicht durch die Blende des optischen Systems, sondern durch die Vignettierung begrenzt werden, ist bei der Wahl einer Variablen für die Öffnung (Apertur) weder die Blendenhöhe gegenüber der Höhe in irgendeiner Pupille, noch eine Pupillenhöhe gegenüber einer anderen (mit ihr in einer Näherung erster Ordnung übereinstimmenden) Variablen vorzuziehen.

[2] Im Gegensatz zu der Höhe $\boldsymbol{H}_0$, die von S_0 Achsen-senkrecht zum Strahl führt.

Zwischen dem Fußpunkt D_{LPL} des Lot-Pfeilhöhen-Lots und dem Fußpunkt D_* des Lots $\boldsymbol{L}_0$ liegt der Lichtweg $\boldsymbol{n}_0 \overrightarrow{D_{LPL} D_*} = \boldsymbol{n}_0 \cdot (D_{LPL} D_*) = n_0 \cdot z_* \cos(\boldsymbol{n}, \boldsymbol{k}) = \zeta_0 z_*$, also wegen $\boldsymbol{n}_0\{x_*; y_*; z_*\} = 0$ der Lichtweg $-(\eta_0 y_* + \xi_0 x_*)$. Folglich ist $(\mathscr{E} + \eta_0 y_0 + \xi_0 x_0) - (\eta_0 y_* + \xi_0 x_*) \equiv \bar{\mathscr{E}}$, wie behauptet, der Lichtweg zwischen D_{LPL} und D_{GBE}.

38. Die Bedeutung der meridionalen und sagittalen Bildhöhen-Aberrationen für alle windschiefen Strahlen. Entwickeln wir das Eikonal $\bar{\mathscr{E}}$ entsprechend (6.15) nach Potenzen von $(y'\bar{x} - x'\bar{y})^2$, dann können wir leicht feststellen, daß die Bildhöhen-Aberrationen aller windschiefen Strahlen weitgehend ersichtlich sind aus den meridionalen und sagittalen Aberrationen $(\underline{U}_{k+1} - \underline{U}_0)_{\text{mer}}$ und $(\delta \underline{U}_{k+1} - \delta \underline{U}_0)/\delta \bar{x}$. — Von nun an nehmen wir wieder $n_0 = 1$ an.

Ohne Beschränkung der Allgemeinheit können wir $x' = 0$ annehmen. Dann ist nach (6.15), (6.9) und (6.7), umgeschrieben auf $\bar{y}$, $\bar{x}$ und $\bar{\mathscr{E}}$:

$$\bar{\mathscr{E}} = \sum_{\alpha=0}^{\infty} \sum_{\gamma=0}^{\infty} \sum_{\varepsilon=0}^{\infty} (\bar{\mathscr{E}}_0^{\alpha\gamma\varepsilon} + \bar{\mathscr{E}}_1^{\alpha\gamma\varepsilon} \cdot y'\bar{y})\, y'^{2\alpha} (\bar{y}^2 + \bar{x}^2)^{\gamma} (y'\bar{x})^{2\varepsilon}. \tag{6.38}$$

Hieraus folgt nach (6.36) für die Bildhöhen-Aberration

$$(\underline{U}_{k+1} - \underline{U}_0) = (\bar{\eta} + i\bar{\xi} - c_{\overline{k+1}} \cdot y')$$

der Ausdruck:

$$\left.\begin{aligned}
(\underline{U}_{k+1} - \underline{U}_0) = &- c_{\overline{k+1}} \cdot y' - \\
&- \sum_{\alpha=0}^{\infty} \sum_{\gamma=0}^{\infty} \sum_{\varepsilon=0}^{\infty} \Big\{ \bar{\mathscr{E}}_1^{\alpha\gamma\varepsilon} \cdot y' + \\
&+ 2(\bar{\mathscr{E}}_0^{\alpha\gamma\varepsilon} + \bar{\mathscr{E}}_1^{\alpha\gamma\varepsilon} \cdot y'\bar{y}) \cdot \frac{\gamma(\bar{y} + i\bar{x})}{(\bar{y}^2 + \bar{x}^2)} + \\
&+ 2(\bar{\mathscr{E}}_0^{\alpha\gamma(\varepsilon+1)} + \bar{\mathscr{E}}_1^{\alpha\gamma(\varepsilon+1)} \cdot y'\bar{y}) \cdot (\varepsilon+1)\, y'^2 (i\bar{x}) \Big\} \times \\
&\times y'^{2\alpha} (\bar{y}^2 + \bar{x}^2)^{\gamma} (y'\bar{x})^{2\varepsilon}.
\end{aligned}\right\} \tag{6.39}$$

Die in diesem Ausdruck stehenden Koeffizienten $\bar{\mathscr{E}}_0^{\alpha 00}$ fallen heraus, weil sie mit $\gamma = 0$ multipliziert sind. Wir nennen die verbleibenden Koeffizienten $\bar{\mathscr{E}}_0^{\alpha\gamma\varepsilon}$ und $\bar{\mathscr{E}}_1^{\alpha\gamma\varepsilon}$ mit Ausnahme von $\bar{\mathscr{E}}_1^{000}$ „Bildfehler-Koeffizienten"; [anstelle von $\bar{\mathscr{E}}_1^{000}$ nennen wir $(c_{\overline{k+1}} + \bar{\mathscr{E}}_1^{000})$ einen „Bildfehler-Koeffizienten". Dann und nur dann, wenn alle diese Bildfehler-Koeffizienten (die Funktionen der Wellenlänge λ des Lichts sind) verschwinden, ist $(\underline{U}_{k+1} - \underline{U}_0) = 0$ für alle Strahldaten y', $\bar{y}$, $\bar{x}$ (und λ). In dem Maße, wie die Bildfehler-Koeffizienten klein sind, haben die meridionalen und windschiefen Strahlen kleine Bildhöhen-Aberrationen $(\underline{U}_{k+1} - \underline{U}_0)$.

Entsprechend wie in Ziff. 36 können wir feststellen, daß es

$$\left.\begin{matrix}
2;\ 5;\ 9;\ 14;\ 20;\ \ldots & \text{Bildfehler-Koeffizienten} \\
1;\ 3;\ 5;\ \ \ 7;\ \ \ 9;\ \ldots & \text{-ter Ordnung}
\end{matrix}\right\} \tag{6.40}$$

gibt. Den Bildfehler-Koeffizienten $\bar{\mathscr{E}}_{\o}^{\alpha\gamma\varepsilon}$ nennen wir einen Bildfehler-Koeffizienten der Ordnung $\psi = [2(\o + \alpha + \gamma + 2\varepsilon) - 1]$; er steht in den Gliedern ψ-ter Ordnung der Reihe (6.39). Den Bildfehler-Koeffizienten $(c_{\overline{k+1}} + \bar{\mathscr{E}}_1^{000})$ nennen wir einen Bildfehler-Koeffizienten erster Ordnung.

Ist die meridionale Bildhöhen-Aberration $(\underline{U}_{k+1} - \underline{U}_0)_{x'=0;\,\bar{x}=0} = 0$ *und* die sagittale Bildhöhen-Aberration $(\delta \underline{U}_{k+1} - \delta \underline{U}_0)/\delta \bar{x} \equiv \left[\frac{\partial}{\partial \bar{x}} (\underline{U}_{k+1} - \underline{U}_0)\right]_{x'=0;\,\bar{x}=0} = 0$, dann ist nach (6.39)

$$\left.\begin{aligned} c_{k+1}^{-}+\bar{\mathscr{E}}_1^{000}&=0,\\ \bar{\mathscr{E}}_1^{\alpha 00}&=0 \quad \text{für} \quad \alpha\geqq 1;\\ \bar{\mathscr{E}}_0^{\alpha\gamma 0}=0 \quad \text{und} \quad \bar{\mathscr{E}}_1^{\alpha\gamma 0}&=0 \quad \text{für} \quad \gamma\geqq 1;\\ \bar{\mathscr{E}}_0^{\alpha\gamma 1}=0 \quad \text{und} \quad \bar{\mathscr{E}}_1^{\alpha\gamma 1}&=0, \end{aligned}\right\} \tag{6.41}$$

und bei den (nicht in sagittaler Nachbarschaft der Meridianstrahlen verlaufenden) windschiefen Strahlen bleiben die Bildhöhen-Aberrationen übrig:

$$\left.\begin{aligned} (\underline{U}_{k+1}-\underline{U}_0)=-\sum_{\alpha=0}^{\infty}\sum_{\gamma=0}^{\infty}\sum_{\varepsilon=2}^{\infty}\Big\{\bar{\mathscr{E}}_1^{\alpha\gamma\varepsilon}\cdot y'+\\ +2(\bar{\mathscr{E}}_0^{\alpha\gamma\varepsilon}+\bar{\mathscr{E}}_1^{\alpha\gamma\varepsilon}\cdot y'\bar{y})\cdot\left(\frac{\gamma(\bar{y}+i\bar{x})}{(\bar{y}^2+\bar{x}^2)}-\frac{\varepsilon}{i\bar{x}}\right)\Big\}\cdot y'^{2\alpha}(\bar{y}^2+\bar{x}^2)^{\gamma}(y'\bar{x})^{2\varepsilon}. \end{aligned}\right\} \tag{6.42}$$

Diese Aberrationen sind nur noch von siebter Ordnung klein; alle Bildfehler-Koeffizienten erster, dritter, fünfter Ordnung sind aus dem Ausdruck verschwunden. Von den 14 Bildfehler-Koeffizienten siebter Ordnung ist nur ein einziger, $\bar{\mathscr{E}}_0^{002}$, übriggeblieben:

$$(\underline{U}_{k+1}-\underline{U}_0)=\frac{4}{i}\,\bar{\mathscr{E}}_0^{002}\,y'^4\bar{x}^3+0^{(9)}. \tag{6.43}$$

Von den 20 Bildfehler-Koeffizienten neunter Ordnung sind drei übriggeblieben: $\bar{\mathscr{E}}_0^{102}$, $\bar{\mathscr{E}}_0^{012}$, $\bar{\mathscr{E}}_1^{002}$. — Wir können also feststellen: Mit den meridionalen und sagittalen Bildhöhen-Aberrationen verschwinden weitgehend auch die Bildhöhen-Aberrationen der übrigen windschiefen Strahlen.

Verschwinden die meridionalen und sagittalen Aberrationen $(\underline{U}_{k+1}-\underline{U}_0)_{\text{mer}}$ und $(\delta\underline{U}_{k+1}-\delta\underline{U}_0)/\delta\bar{x}$ nicht, und sind sie als Funktionen der Meridianstrahl-Daten y' und $\bar{y}$ (und der Wellenlänge des Lichts) bekannt, dann sind damit auch die in (6.41) aufgezählten (nun aber nicht verschwindenden) Bildfehler-Koeffizienten bekannt. Von dem Ausdruck (6.39) für die Bildhöhen-Aberrationen aller windschiefen Strahlen bleibt dann unbekannt nur der in (6.43) rechts stehende Beitrag $\frac{4}{i}\,\bar{\mathscr{E}}_0^{002}y'^4\bar{x}^3+0^{(9)}$. Aus den meridionalen und sagittalen Aberrationen $(\underline{U}_{k+1}-\underline{U}_0)_{\text{mer}}$ und $(\delta\underline{U}_{k+1}-\delta\underline{U}_0)/\delta\bar{x}$ als Funktionen von y' und $\bar{y}$ sind also die Aberrationen $(\underline{U}_{k+1}-\underline{U}_0)$ aller windschiefen Strahlen weitgehend ersichtlich.

Die meridionalen und sagittalen Aberrationen $(\underline{U}_{k+1}-\underline{U}_0)_{\text{mer}}$ und $(\delta\underline{U}_{k+1}-\delta\underline{U}_0)$ können, auch wenn sie als Funktionen *anderer* Strahldaten als y' und $(\bar{y};\delta\bar{x})$ bekannt sind, als Maß für die windschiefen Aberrationen $\left[(\underline{U}_{k+1}-\underline{U}_0)-\left(\frac{4}{i}\bar{\mathscr{E}}_0^{002}y'^4\bar{x}^3+0^{(9)}\right)\right]$ genommen werden, da die letzteren mit den ersteren verschwinden.

39. Die Abweichung der Neigung von ihrem idealen Wert. Bei dem zuerst eingeführten Eikonal $\mathscr{E}$ war die Komplikation aufgetreten, daß die Bildhöhen-Aberration

$$(\underline{U}_{k+1}-\underline{U}_0)$$

sich nicht linear in den Eikonal-Koeffizienten $\mathscr{E}^{\alpha\beta\gamma}$ [oder $\mathscr{E}_0^{\alpha\gamma\varepsilon}$, $\mathscr{E}_1^{\alpha\gamma\varepsilon}$] ausdrücken läßt, bedingt durch den Wurzelausdruck (6.19) für $\underline{U}_0$. Statt das Eikonal $\bar{\mathscr{E}}$ einzuführen, können wir diese Komplikation auch dadurch vermeiden, daß wir als Maß für den Unterschied zwischen der reduzierten Objekthöhe $\underline{U}_0$ und der reduzierten Bildhöhe $\underline{U}_{k+1}$ nicht deren Differenz nehmen, sondern die Differenz zwischen einer geeigneten Funktion von $\underline{U}_0$ und der gleichen Funktion von $\underline{U}_{k+1}$, nämlich

$$\left(\frac{\underline{U}_{k+1}}{\sqrt{1+\underline{U}_{k+1}\underline{U}_{k+1}^*}}-\frac{\underline{U}_0}{\sqrt{1+\underline{U}_0\underline{U}_0^*}}\right). \tag{6.44}$$

Der Subtrahend ist nichts anderes als das reduzierte Objektlot $\underline{U}_0=-(\eta_0+i\,\xi_0)$. Für den Minuenden können wir, weil wir die Bildhöhe reell vorgeben, $\underline{U}_{k+1}/\sqrt{1+\underline{U}_{k+1}^2}=-c_{\overline{k}+1}\,y'/\sqrt{1+(c_{\overline{k}+1}\,y')^2}$ schreiben. Die Differenz (6.44) ist also gleich

$$\left[(\eta_0+i\,\xi_0)-(c_{\overline{k}+1}\,y')/\sqrt{1+(c_{\overline{k}+1}\,y')^2}\right].$$

Wir geben also $y_0+i\,x_0$ und $y'+i\,x'$ vor, lassen x' verschwinden und nehmen die Abweichung

$$\Delta N_0\equiv[N_0-(N_0)_{id}] \tag{6.45}$$

der „Neigung"

$$N_0\equiv(\eta_0+i\,\xi_0) \tag{6.46}$$

von ihrem idealen Wert

$$(N_0)_{id}=(c_{\overline{k}+1}\,y')/\sqrt{1+(c_{\overline{k}+1}\,y')^2} \tag{6.47}$$

als Maß für die Bildhöhen-Aberration.

Nach (6.18), (6.47) und (6.45) ist

$$\left.\begin{aligned}\Delta N_0=&\sum_{\alpha=0}^{\infty} c_N^{(\alpha)}\cdot y'^{\,2\alpha+1}-\\ &-\sum_{\alpha=0}^{\infty}\sum_{\beta=0}^{\infty}\sum_{\gamma=0}^{\infty}\left[\bar\beta\,\mathscr{E}^{\alpha\bar\beta\gamma}\cdot y'+2\bar\gamma\,\mathscr{E}^{\alpha\beta\bar\gamma}(y_0+i\,x_0)\right]\cdot\\ &\cdot y'^{\,2\alpha}(y'\,y_0)^{\beta}(y_0^2+x_0^2)^{\gamma}\end{aligned}\right\} \tag{6.48}$$

mit den Abkürzungen

$$\left.\begin{aligned}&\bar\beta\equiv\beta+1;\quad \bar\gamma\equiv\gamma+1;\\ &c_N^{(\alpha)}\equiv-\binom{-1/2}{\alpha}(c_{\overline{k}+1})^{2\alpha+1}.\end{aligned}\right\} \tag{6.49}$$

(6.48) stellt die Aberration ΔN_0 als Funktion der vorgegebenen Strahldaten y', y_0, x_0 durch einen Ausdruck dar, der *linear* in den Eikonal-Koeffizienten $\mathscr{E}^{\alpha\beta\gamma}$ ist. Die Eikonal-Koeffizienten $\mathscr{E}^{\alpha 00}$ kommen in (6.48) nicht vor; sie waren bei der Differentiation (6.16) herausgefallen. Die verbliebenen Koeffizienten $\mathscr{E}^{\alpha\beta\gamma}$, mit Ausnahme der Koeffizienten $\mathscr{E}^{\alpha 10}$, nennen wir „Bildfehler-Koeffizienten" und bezeichnen sie auch mit $\Delta\mathscr{E}^{\alpha\beta\gamma}$. Anstelle der $\mathscr{E}^{\alpha 10}$ nennen wir die Größen

$$[\mathscr{E}^{\alpha 10}-c_N^{(\alpha)}]\equiv\Delta\mathscr{E}^{\alpha 10} \tag{6.50}$$

„Bildfehler-Koeffizienten". Dann und nur dann, wenn alle Bildfehler-Koeffizienten $\Delta\mathscr{E}^{\alpha\beta\gamma}$ verschwinden, ist die Aberration $\Delta N_0=0$ für alle Strahldaten y', y_0, x_0. Wir nennen $\Delta\mathscr{E}^{\alpha\beta\gamma}$ einen Bildfehler-Koeffizienten der Ordnung $\psi=[2(\alpha+\beta+\gamma)-1]$; er steht in Gliedern ψ-ter Ordnung der Reihe für ΔN_0:

$$\left.\begin{aligned}\Delta N_0=-\sum_{\alpha=0}^{\infty}\sum_{\beta=0}^{\infty}\sum_{\gamma=0}^{\infty}\left[\bar\beta\,\Delta\mathscr{E}^{\alpha\bar\beta\gamma}\cdot y'+2\bar\gamma\,\Delta\mathscr{E}^{\alpha\beta\bar\gamma}(y_0+i\,x_0)\right]\cdot\\ \cdot y'^{\,2\alpha}(y'\,y_0)^{\beta}(y_0^2+x_0^2)^{\gamma},\end{aligned}\right\} \tag{6.51a}$$

$$=-\left(\frac{\partial}{\partial y_0}+i\,\frac{\partial}{\partial x_0}\right)\Delta\mathscr{E} \tag{6.51b}$$

mit

$$\Delta\mathcal{E} \equiv \sum_{\alpha=0}^{\infty}\sum_{\beta=0}^{\infty}\sum_{\gamma=0}^{\infty} \Delta\mathcal{E}^{\alpha\beta\gamma}\cdot y'^{2\alpha}(y'y_0)^{\beta}(y_0^2+x_0^2)^{\gamma}. \tag{6.51 c}$$

Die in (6.51 b, c) stehenden, durch die Differentiation $\left(\frac{\partial}{\partial y_0}+i\frac{\partial}{\partial x_0}\right)$ aber herausfallenden Koeffizienten $\Delta\mathcal{E}^{\alpha 00}$ können wir mit den Eikonal-Koeffizienten $\mathcal{E}^{\alpha 00}$ identifizieren.

Nicht nur die Koeffizienten $\mathcal{E}^{\alpha\beta\gamma}$, sondern auch die Koeffizienten $\Delta\mathcal{E}^{\alpha\beta\gamma}$ der Reihe (6.51 c) für $\Delta\mathcal{E}$ wollen wir „Eikonal-Koeffizienten" der Ordnung $2(\alpha+\beta+\gamma)$ nennen. Je nachdem ob wir daran denken, daß $\Delta\mathcal{E}^{\alpha\beta\gamma}$ ein Koeffizient der Reihe (6.51 c) für $\Delta\mathcal{E}$ ist, oder daß $\Delta\mathcal{E}^{\alpha\beta\gamma}$ ein Koeffizient des Ausdrucks (6.51 a) für ΔN_0 ist, nennen wir $\Delta\mathcal{E}^{\alpha\beta\gamma}$ einen Eikonal-Koeffizienten der Ordnung $2(\alpha+\beta+\gamma)$ oder einen Bildfehler-Koeffizienten der Ordnung $[2(\alpha+\beta+\gamma)-1]$. Diese Bildfehler- oder Eikonal-Koeffizienten $\Delta\mathcal{E}^{\alpha\beta\gamma}$ nennen wir genauer „irreduzible" Bildfehler- oder Eikonal-Koeffizienten — im Gegensatz zu den im folgenden vorkommenden „reduziblen" Koeffizienten, zwischen denen lineare Beziehungen bestehen, so daß sie auf eine geringere Anzahl von Koeffizienten zurückführbar sind.

In Abschnitt VII werden wir ein Rechenschema zusammenstellen, das bei vorgegebenen Konstruktionsdaten (n, r, a) des optischen Systems die bei der einfachen Potenzreihenentwicklung der Aberration ΔN_0 nach y', y_0, x_0:

$$\left.\begin{aligned} \Delta N_0 = & \sum_{\substack{\sigma=0\\ \left(\sigma+\tau \text{ ungerade}\atop v \text{ gerade}\right)}}^{\infty}\sum_{\tau=0}^{\infty}\sum_{v=0}^{\infty} \Delta N_0^{\sigma\tau,v}\cdot(y')^{\sigma}(y_0)^{\tau}(x_0)^{v} + \\ & + i\sum_{\substack{\sigma=0\\ \left(\sigma+\tau \text{ gerade}\atop v \text{ ungerade}\right)}}^{\infty}\sum_{\tau=0}^{\infty}\sum_{v=0}^{\infty} \Delta N_0^{\sigma\tau,v}\cdot(y')^{\sigma}(y_0)^{\tau}(x_0)^{v} \end{aligned}\right\} \tag{6.52}$$

auftretenden „reduziblen Bildfehler-Koeffizienten" $\Delta N_0^{\sigma\tau,v}$ auszurechnen erlaubt. Damit erhält man auch die bei der einfachen Potenzreihenentwicklung von $\Delta\mathcal{E}$ nach y', y_0, x_0:

$$\Delta\mathcal{E} = \sum_{\substack{\sigma=0\\ \left(\sigma+\tau \text{ gerade}\atop v \text{ gerade}\right)}}^{\infty}\sum_{\tau=0}^{\infty}\sum_{v=0}^{\infty} \Delta\mathcal{E}^{\sigma\tau,v}\cdot(y')^{\sigma}(y_0)^{\tau}(x_0)^{v} \tag{6.53}$$

auftretenden „reduziblen Eikonal-Koeffizienten" $\Delta\mathcal{E}^{\sigma\tau,v}$ mit Ausnahme der $\Delta\mathcal{E}^{\sigma 0,0}$. Denn nach (6.53), (6.52) und (6.51 b) ist

$$\left.\begin{aligned} &\text{sowohl} && \Delta\mathcal{E}^{\sigma\tau,v} = -\frac{1}{\tau}\Delta N_0^{\sigma(\tau-1),v}, && \text{wenn } \tau\neq 0 \text{ ist;} \\ &\text{als auch} && \Delta\mathcal{E}^{\sigma\tau,v} = -\frac{1}{v}\Delta N_0^{\sigma\tau,(v-1)}, && \text{wenn } v\neq 0 \text{ ist.} \end{aligned}\right\} \tag{6.54}$$

Und aus den somit berechenbaren *reduziblen* Eikonal-Koeffizienten $\Delta\mathcal{E}^{\sigma\tau,v}$ (ohne die $\Delta\mathcal{E}^{\sigma 0,0}$) lassen sich die *irreduziblen* Eikonal-Koeffizienten $\Delta\mathcal{E}^{\alpha\beta\gamma}$ (ohne die $\Delta\mathcal{E}^{\alpha 00}$) ermitteln, wie wir in der nächsten Ziffer sehen werden.

40. Umrechung der reduziblen in die irreduziblen Koeffizienten. Drücken wir die Entwicklungsvariablen (y'), (y_0), (x_0) von (6.53) durch die Entwicklungsvariablen (y'^2), $(y'y_0)$, $(y_0^2+x_0^2)$ von (6.51 c) aus:

$$[(y')^{\sigma}(y_0)^{\tau}]_{\sigma+\tau \text{ gerade}} = (y'^2)^{\frac{\sigma-\tau}{2}}(y'y_0)^{\tau}; \tag{6.55}$$

$$[(x_0)^{v}]_{v \text{ gerade}} = \left[(y_0^2+x_0^2) - \frac{(y'y_0)^2}{(y'^2)}\right]^{v/2}, \tag{6.56}$$

und setzen wir dies in (6.53) ein, dann erhalten wir:

$$\Delta\mathcal{E}=\sum_{\substack{\sigma=0\\ \left(\substack{\sigma+\tau\text{ gerade}\\ v\text{ gerade}\\ \varkappa\leqq v/2}\right)}}^{\infty}\sum_{\tau=0}^{\infty}\sum_{v=0}^{\infty}\sum_{\varkappa=0}^{\infty}(-1)^{\varkappa}\binom{v/2}{\varkappa}\Delta\mathcal{E}^{\sigma\tau,v}\cdot(y'^2)^{\frac{\sigma-\tau}{2}-\varkappa}\cdot(y'y_0)^{\tau+2\varkappa}\cdot(y_0^2+x_0^2)^{\frac{v}{2}-\varkappa}$$

$$=\sum_{\substack{\sigma=0\\ \left(\substack{\sigma-\beta\text{ gerade}\\ \varkappa\leqq\beta/2}\right)}}^{\infty}\sum_{\beta=0}^{\infty}\sum_{\gamma=0}^{\infty}\sum_{\varkappa=0}^{\infty}(-1)^{\varkappa}\binom{\gamma+\varkappa}{\varkappa}\Delta\mathcal{E}^{\sigma(\beta-2\varkappa),(2\varkappa+2\gamma)}\cdot(y'^2)^{\frac{\sigma-\beta}{2}}\cdot(y'y_0)^{\beta}\cdot(y_0^2+x_0^2)^{\gamma}$$

$$=\sum_{\substack{\alpha=-\infty\\ \left(\substack{\alpha\geqq-\beta/2\\ \varkappa\leqq+\beta/2}\right)}}^{+\infty}\sum_{\beta=0}^{\infty}\sum_{\gamma=0}^{\infty}\sum_{\varkappa=0}^{\infty}(-1)^{\varkappa}\binom{\gamma+\varkappa}{\varkappa}\Delta\mathcal{E}^{(2\alpha+\beta)(\beta-2\varkappa),(2\varkappa+2\gamma)}\cdot(y'^2)^{\alpha}(y'y_0)^{\beta}(y_0^2+x_0^2)^{\gamma}.$$

Durch Koeffizientenvergleich mit (6.51c) stellen wir fest[1]:

$$\left.\begin{aligned}\Delta\mathcal{E}^{\alpha\beta\gamma}=\sum_{\varkappa=0}^{\leqq\beta/2}(-1)^{\varkappa}\binom{\gamma+\varkappa}{\varkappa}\Delta\mathcal{E}^{(2\alpha+\beta)(\beta-2\varkappa),(2\varkappa+2\gamma)}\\ \text{für }\alpha\geqq 0\ (\text{und }\beta\geqq 0;\ \gamma\geqq 0)\end{aligned}\right\}\qquad(6.57\text{a})$$

und

$$\left.\begin{aligned}\sum_{\varkappa=0}^{\leqq\beta/2}(-1)^{\varkappa}\binom{\gamma+\varkappa}{\varkappa}\Delta\mathcal{E}^{(2\alpha+\beta)(\beta-2\varkappa),(2\varkappa+2\gamma)}=0\\ \text{für }-\frac{\beta}{2}\leqq\alpha<0\ (\text{und }\beta\geqq 2;\ \gamma\geqq 0),\end{aligned}\right\}\qquad(6.57\text{b})$$

wobei in der oberen Grenze der Summe „$\leqq\beta/2$" bedeutet: $\beta/2$, wenn β gerade ist; $(\beta-1)/2$, wenn β ungerade ist.

(6.57a) ist die Formel, mit der man aus den reduziblen Eikonal-Koeffizienten $\Delta\mathcal{E}^{\sigma\tau,v}$ (einschließlich oder ausschließlich der $\Delta\mathcal{E}^{\sigma 0,0}$) die irreduziblen Eikonal-Koeffizienten $\Delta\mathcal{E}^{\alpha\beta\gamma}$ (einschließlich bzw. ausschließlich der $\Delta\mathcal{E}^{\alpha 00}$) berechnen kann.

(6.57b) gibt die Zusammenhänge an, die zwischen den bei der einfachen Potenzreihenentwicklung auftretenden Koeffizienten $\Delta\mathcal{E}^{\sigma\tau,v}$ bestehen. Die durch *eine* Gl. (6.57b) miteinander verknüpften Koeffizienten $\Delta\mathcal{E}^{\sigma\tau,v}$ haben die gleiche Ordnung $\omega=2(\alpha+\beta+\gamma)$. In ω-ter Ordnung hat man $\omega(\omega+2)/8$ Beziehungen (6.57b). Diese Anzahl ist die Differenz zwischen der Anzahl $(\omega+2)(\omega+4)/8$ der irreduziblen Eikonal-Koeffizienten $\Delta\mathcal{E}^{\alpha\beta\gamma}$ der Ordnung ω, und der Anzahl $(\omega+2)^2/4$ der reduziblen Eikonal-Koeffizienten $\Delta\mathcal{E}^{\sigma\tau,v}$ der Ordnung ω:

Für	$\omega=$	2	4	6	8	10		
ist	$(\omega+2)(\omega+4)/8=$	3	6	10	15	21	$[\Delta\mathcal{E}^{\alpha\beta\gamma}]$;	(6.58)
	$\omega(\omega+2)/8=$	1	3	6	10	15	[(6.57b)];	
	$(\omega+2)^2/4=$	4	9	16	25	36	$[\Delta\mathcal{E}^{\sigma\tau,v}]$.	

In ähnlicher Weise kann man auch eine Formel erhalten, mit der man die in dem Ausdruck

$$\Delta N_0=-\left(\frac{\partial}{\partial y_0}+i\frac{\partial}{\partial x_0}\right)\Delta\mathcal{E}\qquad(6.51\text{b})$$

mit

$$\Delta\mathcal{E}=\sum_{\alpha=0}^{\infty}\sum_{\gamma=0}^{\infty}\sum_{\varepsilon=0}^{\infty}(\Delta\mathcal{E}_0^{\alpha\gamma\varepsilon}+\Delta\mathcal{E}_1^{\alpha\gamma\varepsilon}\cdot y'y_0)\cdot y'^{2\alpha}(y_0^2+x_0^2)^{\gamma}(y'x_0)^{2\varepsilon}\qquad(6.51\text{c*})$$

[1] H. MARX: Optik 4, 141–142 (1948).

auftretenden irreduziblen Eikonal-Koeffizienten $\Delta\mathscr{E}_0^{\alpha\gamma\varepsilon}$ und $\Delta\mathscr{E}_1^{\alpha\gamma\varepsilon}$ berechnen kann aus den reduziblen Eikonal-Koeffizienten $\Delta\mathscr{E}^{\sigma\tau,v}$:

$$\Delta\mathscr{E}_\varnothing^{\alpha\gamma\varepsilon} = \sum_{\varkappa=0}^{\varepsilon} (-1)^\varkappa \binom{\gamma+\varkappa}{\varkappa} \Delta\mathscr{E}^{(\varnothing+2\alpha+2\varepsilon)\,(\varnothing+2\gamma+2\varkappa),\,(2\varepsilon-2\varkappa)}; \tag{6.57a*}$$

dies gilt für $\alpha \geqq 0$ (und $\gamma \geqq 0$; $\varepsilon \geqq 0$; $\varnothing = 0$ oder 1). In diesem Zusammenhang erhält man die Beziehungen (6.57b) in der Form:

$$\left.\begin{aligned} &\sum_{\varkappa=0}^{\varepsilon} (-1)^\varkappa \binom{\gamma+\varkappa}{\varkappa} \Delta\mathscr{E}^{\,\varnothing+2\alpha+2\varepsilon)\,(\varnothing+2\gamma+2\varkappa),\,(2\varepsilon-2\varkappa)} = 0 \\ &\text{für } -\varepsilon \leqq \alpha < 0 \text{ (und } \gamma \geqq 0;\ \varepsilon \geqq 1;\ \varnothing = 0 \text{ oder } 1). \end{aligned}\right\} \tag{6.57b*}$$

41. Entwicklung nur nach objektseitigen Strahldaten. Die Koeffizienten $\Delta N_0^{\sigma\tau,v} = -(\tau+1)\,\Delta\mathscr{E}^{\sigma(\tau+1),v} = -(v+1)\,\Delta\mathscr{E}^{\sigma\tau,(v+1)}$ [aus denen man gemäß (6.57a) die irreduziblen Bildfehler-Koeffizienten $\Delta\mathscr{E}^{\alpha\beta\gamma}$ oder gemäß (6.57a*) die irreduziblen Bildfehler-Koeffizienten $\Delta\mathscr{E}_\varnothing^{\alpha\gamma\varepsilon}$ berechnen kann] treten in (6.52) bei der Potenzreihenentwicklung von ΔN_0 nach den Strahldaten y', y_0, x_0 auf. y' ist ein Strahldatum im Bildraum; y_0 und x_0 sind Strahldaten im Objektraum. Der angekündigte Formelsatz zur Berechnung der reduziblen Bildfehler-Koeffizienten $\Delta N^{\sigma\tau,v}$ ist, weil die Entwicklungsvariablen y', y_0, x_0 teils dem Bildraum und teils dem Objektraum angehören, nicht so einfach, wie wenn die Entwicklungsvariablen ein und demselben Raum angehören würden.

Nur dem Objektraum gehören zum Beispiel die Strahldaten $\eta_0 = -U_0$ (reell vorgegeben) und y_0; x_0 an, nach denen wir die Bildhöhe $y' + i\,x'$ und damit die Bildhöhen-Aberration

$$(\underline{U}_{k+1} - \underline{U}_0) = \left[-c_{k+1}^{-}(y' + i\,x') + \eta_0/\sqrt{1-\eta_0^2}\right] \equiv \Delta M_{k+1} \tag{6.59}$$

entwickeln können:

$$\left.\begin{aligned} \Delta M_{k+1} = &\sum_{\substack{\sigma=0\\ \binom{\sigma+\tau \text{ ungerade}}{v \text{ gerade}}}}^{\infty} \sum_{\tau=0}^{\infty} \sum_{v=0}^{\infty} \Delta M_{k+1}^{\sigma\tau,v} (\eta_0)^\sigma (y_0)^\tau (x_0)^v + \\ &+ i \sum_{\substack{\sigma=0\\ \binom{\sigma+\tau \text{ gerade}}{v \text{ ungerade}}}}^{\infty} \sum_{\tau=0}^{\infty} \sum_{v=0}^{\infty} \Delta M_{k+1}^{\sigma\tau,v} (\eta_0)^\sigma (y_0)^\tau (x_0)^v. \end{aligned}\right\} \tag{6.60}$$

ΔM_{k+1} ist, nach (6.59), die Abweichung der „reduzierten“ (mit $-c_{k+1}^{-}$ multiplizierten) Bildhöhe

$$M_{k+1} \equiv -c_{k+1}^{-}(y' + i x') \quad [\equiv \underline{U}_{k+1}] \tag{6.61}$$

von ihrem idealen Wert

$$\left.\begin{aligned} (M_{k+1})_{id} &= -\eta_0/\sqrt{1-\eta_0^2} \quad [\equiv \underline{U}_0] \\ &= \sum_{\alpha=0}^{\infty} c_M^{(\alpha)} \cdot \eta_0^{2\alpha+1} \quad \text{mit} \quad c_M^{(\alpha)} \equiv (-1)^{1+\alpha} \binom{-1/2}{\alpha}. \end{aligned}\right\} \tag{6.62}$$

Zunächst, in Ziff. 42 bis 46, wollen wir einen Formelsatz zur Berechnung der in (6.60) auftretenden Koeffizienten $\Delta M_{k+1}^{\sigma\tau,v}$ aufstellen. Danach, in Ziff. 47, werden wir diesen Formelsatz so abwandeln, daß man die reduziblen Bildfehler-Koeffizienten $\Delta N_0^{\sigma\tau,v}$ und damit die irreduziblen Bildfehler-Koeffizienten $\Delta\mathscr{E}^{\alpha\beta\gamma}$ oder $\Delta\mathscr{E}_\varnothing^{\alpha\gamma\varepsilon}$ berechnen kann.

Auch die Koeffizienten $\Delta M_{k+1}^{\sigma\tau,v}$ wollen wir im folgenden „Bildfehler-Koeffizienten" nennen.

In dem folgenden Abschnitt VII werden wir, nach Einführung der „Vektorskalare" in Ziff. 43, den Querstrich ($\bar{\ }$) und die Unterstreichung ($\underline{\ }$) zur Bezeichnung des Skalarteils und des Vektorteils eines Vektorskalars verwenden. Die bisher mit Querstrich oder Unterstreichung versehenen Buchstaben kommen in Abschnitt VII nicht vor.

Außerdem werden wir in Abschnitt VII die Gaußsche Bildebene als Fläche Nr. $(k+1)$ zählen, nicht die Austrittspupille. Die Eintrittspupille zählen wir nach wie vor als Fläche Nr. 0. Die Flächen Nr. 0 und Nr. $(k+1)$ sind nicht-brechende Planflächen ($r=\infty$; $n'=n$).

VII. Durchrechnungs- und Rekursionsformeln für die Bildfehler-Koeffizienten beliebiger Ordnung.

42. Übergangs- und Brechungsformeln. Die Aufgabe, die Bildfehler-Koeffizienten beliebiger Ordnung zu bestimmen, führen wir zurück auf die Aufgabe, bei Potenzreihenentwicklungen die Koeffizienten eines Produkts aus den Koeffizienten seiner beiden Faktoren zu bestimmen. Die Zurückführung der einen Aufgabe auf die andere ist in einfacher Weise möglich, weil man alle Gleichungen, die den Verlauf der Strahlen in Systemen aus Kugel- und Planflächen beschreiben, entweder linear oder bilinear oder quadratisch in geeigneten Variablen schreiben kann.

Zur Berechnung der Bildfehler-Koeffizienten $\Delta M_{k+1}^{\sigma\tau,v}$ hat man eine Durchrechnung (von Fläche zu Fläche) und eine Rekursion (von Ordnung zu Ordnung) durchzuführen, und zwar läßt sich — wie wir sehen werden — die ganze Durchrechnung und Rekursion durch wiederholte Anwendung von nur vier einfachen Formeln durchführen[1]:

wenn $\sigma+\tau+v$ ungerade ist $(\sigma+\tau+v\geqq 1)$

$b_v^{\sigma\tau,v}$	zu bestimmen aus:	$\langle h_v b_v\rangle^{\sigma\tau,v}=0$	$(v\geqq\frac{1}{2})$
$h_{v+\frac{1}{2}}^{\sigma\tau,v}$	zu bestimmen aus:	$h_{v+\frac{1}{2}}^{\sigma\tau,v}-h_{v-\frac{1}{2}}^{\sigma\tau,v}=b_v^{\sigma\tau,v}$	$(v\geqq\frac{1}{2})$,

wenn $s+t$ und u gerade sind $(s+t+u\geqq 2)$

$h_{v+\frac{1}{2}}^{st,u}$	zu bestimmen aus:	$\langle h_{v+\frac{1}{2}} h_{v+\frac{1}{2}}\rangle^{st,u}=0$	$(v\geqq 0)$
$b_v^{st,u}$	zu bestimmen aus:	$h_{v+\frac{1}{2}}^{st,u}-h_{v-\frac{1}{2}}^{st,u}=b_v^{st,u}$	$(v\geqq\frac{1}{2})$.

Die Klammer $\langle\ \rangle^{\sigma\tau,v}$ mit den oberen Indizes bezeichnet eine nach Cauchy benannte Produktensumme über drei Indizes, die folgendermaßen erklärt ist:

$$\langle AB\rangle^{\sigma\tau,v}=\sum_{s=0}^{\sigma}\sum_{t=0}^{\tau}\sum_{u=0}^{v}A^{st,u}B^{(\sigma-s)(\tau-t),(v-u)}.$$

Wir gehen aus von Übergangs- und Brechungs-Formeln für einen windschiefen Strahl, die nicht nach den Unbekannten aufgelöst sind, sondern als quadratische, bilineare und lineare Beziehungen zwischen den Strahldaten stehenbleiben. Und zwar gehen wir aus von dem *halben Formelsatz* (2.41) mit dem unteren

[1] H. Marx: Z. Physik **131**, 408—419 (1952); Optik **4**, 97—148 (1948). — Im folgenden tritt als oberer Index der griechische Buchstabe v auf, der nicht mit dem stets als unterer Index erscheinenden lateinischen Buchstaben v zu verwechseln ist.

(eingeklammerten) Vorzeichen:

$$[(\boldsymbol{r}_{v+\frac{1}{2}}-\boldsymbol{r}_{v-\frac{1}{2}}+\boldsymbol{o}_v)\,\boldsymbol{r}_v]=0; \tag{7.1a}$$

$$\boldsymbol{r}^2_{v+\frac{1}{2}}=r^2_{v+\frac{1}{2}} \quad (\mathrm{sign}\,\boldsymbol{r}_v\boldsymbol{r}_{v+\frac{1}{2}}=-\,\mathrm{sign}\,r_v r_{v+\frac{1}{2}}), \tag{7.1b}$$

durch den $\boldsymbol{r}_{v+\frac{1}{2}}$ bestimmt ist aus $\boldsymbol{r}_{v-\frac{1}{2}}$ und $\boldsymbol{r}_v$. Als Strahldaten haben wir in dem halben Formelsatz (7.1 a, b) die Radiusvektoren $\boldsymbol{r}$ und die optischen Richtungen $\boldsymbol{n}$ genommen:

$$\boldsymbol{r}_v\equiv\begin{cases}\boldsymbol{r}_v\equiv\{x;\,y;\,z\}_v, & \text{wenn } v \text{ ganzzahlig ist;}\\ \boldsymbol{n}_{v+\frac{1}{2}}\equiv\{\xi;\,\eta;\,\zeta\}_{v+\frac{1}{2}}, & \text{wenn } v \text{ halbzahlig ist;}\end{cases} \tag{7.2}$$

als Konstruktionsdaten haben wir [vgl. (2.40) mit dem unteren Vorzeichen] die Radien r, die Brechungsindizes n und die Mittelpunkts-Entfernungen e genommen:

$$r_v\equiv\begin{cases}r_v, & \text{wenn } v \text{ ganzzahlig ist;}\\ n_{v+\frac{1}{2}}, & \text{wenn } v \text{ halbzahlig ist;}\end{cases} \tag{7.3}$$

$$\boldsymbol{o}_v\equiv\begin{cases}0, & \text{wenn } v \text{ ganzzahlig ist;}\\ \boldsymbol{e}_{v+\frac{1}{2}}\equiv\{0;\,0;\,e\}_{v+\frac{1}{2}}, & \text{wenn } v \text{ halbzahlig ist.}\end{cases} \tag{7.4}$$

Später werden wir anstelle der Mittelpunkts-Entfernung e den Scheitel-Abstand a als Konstruktionsdatum nehmen[1].

Die Achsenrichtungen des (rechtshändigen) kartesischen Koordinatensystems bezeichnen wir mit $\boldsymbol{i}, \boldsymbol{j}, \boldsymbol{k}$:

$$\left.\begin{aligned}&\boldsymbol{i}=\text{Richtung der } x\text{-Achse } (\perp \text{ zur Hauptmeridianebene});\\ &\boldsymbol{j}=\text{Richtung der } y\text{-Achse („meridionale Richtung“)};\\ &\boldsymbol{k}=\text{Richtung der } z\text{-Achse } (\uparrow\uparrow \text{ zur optischen Achse}).\end{aligned}\right\} \tag{7.5}$$

43. Der halbe Formelsatz in Vektorskalar-Schreibweise. Wir zerlegen den Dreiervektor $\boldsymbol{r}$ in einen Summanden $\underline{h}$ (Zweiervektor) senkrecht zur optischen Achse und einen Summanden $\pm\boldsymbol{k}\cdot\bar{h}$ ($\bar{h}$ skalar) parallel zur optischen Achse:

$$\boldsymbol{r}_v\equiv\underline{h}_v+\boldsymbol{k}\bar{h}_v, \quad \text{wenn } v \text{ ganzzahlig ist, also } \boldsymbol{r}_v \text{ ein Radiusvektor ist;} \tag{7.6}$$

$$\boldsymbol{r}_v\equiv\underline{h}_v-\boldsymbol{k}\bar{h}_v, \quad \text{wenn } v \text{ halbzahlig ist, also } \boldsymbol{r}_v \text{ eine optische Richtung ist.} \tag{7.7}$$

Ferner setzen wir

$$\boldsymbol{o}_v\equiv-\boldsymbol{k}\,\bar{\varepsilon}_v, \quad \text{wenn } v \text{ ganzzahlig ist;} \tag{7.8}$$

$$\boldsymbol{o}_v\equiv+\boldsymbol{k}\,\bar{\varepsilon}_v, \quad \text{wenn } v \text{ halbzahlig ist.} \tag{7.9}$$

Vektorielle Multiplikation der Gl. (7.1 a) mit $\boldsymbol{k}$ ergibt dann:

$$(\underline{h}_{v+\frac{1}{2}}-\underline{h}_{v-\frac{1}{2}})\,\bar{h}_v+(\bar{h}_{v+\frac{1}{2}}-\bar{h}_{v-\frac{1}{2}}+\bar{\varepsilon}_v)\,\underline{h}_v=0, \tag{7.10}$$

und für Gl. (7.1 b) kann man schreiben:

$$\underline{h}^2_{v+\frac{1}{2}}+\bar{h}^2_{v+\frac{1}{2}}=r^2_{v+\frac{1}{2}} \quad (\mathrm{sign}\{\bar{h}_v\bar{h}_{v+\frac{1}{2}}-\underline{h}_v\underline{h}_{v+\frac{1}{2}}\}=\mathrm{sign}\,r_v r_{v+\frac{1}{2}}). \tag{7.11}$$

Nach (7.6), (7.7) und (7.2) ist

$$\bar{h}_v=+z_v; \quad \underline{h}_v=\{x;\,y;\,0\}_v, \quad \text{wenn } v \text{ ganzzahlig ist} \tag{7.6a}$$

und

$$\bar{h}_v=-\zeta_{v+\frac{1}{2}}; \quad \underline{h}_v=\{\xi;\,\eta;\,0\}_{v+\frac{1}{2}}, \quad \text{wenn } v \text{ halbzahlig ist.} \tag{7.7b}$$

---

[1] In Ziff. 45; dort werden wir d_v für $a_{v+\frac{1}{2}}$ schreiben (v halbzahlig).

Nach (7.8), (7.9) und (7.4) ist

$$\bar{\varepsilon}_v = 0, \qquad \text{wenn } v \text{ ganzzahlig ist} \tag{7.8a}$$

und

$$\bar{\varepsilon}_v = e_{v+\frac{1}{2}}, \qquad \text{wenn } v \text{ halbzahlig ist.} \tag{7.9a}$$

Wir fassen nun den (senkrecht zur optischen Achse liegenden) Zweiervektor $\underline{h}$ und den Skalar $\bar{h}$ durch ein dazwischen gesetztes Additionszeichen zu einer gemischten Größe zusammen, die wir einen „Vektorskalar" nennen und mit h bezeichnen:

$$h \equiv \underline{h} + \bar{h}. \tag{7.12}$$

$\underline{h}$ nennen wir den „Vektorteil" von h; $\bar{h}$ nennen wir den „Skalarteil" von h.

Zwei solche Vektorskalare[1] $f \equiv \underline{f} + \bar{f}$ und $g \equiv \underline{g} + \bar{g}$ nennen wir dann und nur dann einander gleich:

$$f = g, \tag{7.13a}$$

wenn sie im Vektorteil und im Skalarteil übereinstimmen:

$$\underline{f} = \underline{g}; \quad \bar{f} = \bar{g}. \tag{7.13b}$$

Einen Vektorskalar $S \equiv \underline{S} + \bar{S}$ nennen wir dann und nur dann die Summe aus zwei Vektorskalaren $A \equiv \underline{A} + \bar{A}$ und $B \equiv \underline{B} + \bar{B}$:

$$S = A + B, \tag{7.14a}$$

wenn

$$\underline{S} = \underline{A} + \underline{B}; \quad \bar{S} = \bar{A} + \bar{B} \tag{7.14b}$$

ist. Einen Vektorskalar $P \equiv \underline{P} + \bar{P}$ nennen wir dann und nur dann das Produkt aus zwei Vektorskalaren $A \equiv \underline{A} + \bar{A}$ und $B \equiv \underline{B} + \bar{B}$:

$$P = A\,B, \tag{7.15a}$$

wenn

$$\underline{P} = \bar{A}\underline{B} + \underline{A}\bar{B}; \quad \bar{P} = \underline{A}\underline{B} + \bar{A}\bar{B} \tag{7.15b}$$

ist. (Mit $\underline{A}\underline{B}$ ist das skalare Produkt aus $\underline{A}$ und $\underline{B}$ gemeint.) Aus (7.14) folgt für die Addition das kommutative Gesetz

$$A + B = B + A \tag{7.14c}$$

und das assoziative Gesetz

$$(A + B) + C = A + (B + C). \tag{7.14d}$$

Aus (7.15) folgt für die Multiplikation das kommutative Gesetz

$$A\,B = B\,A, \tag{7.15c}$$

nicht aber das assoziative Gesetz. Aus (7.14) und (7.15) folgt das distributive Gesetz

$$A(B + C) = A\,B + A\,C. \tag{7.15d}$$

Ist W ein Vektorskalar, dann soll $\underline{W}$ immer den Vektorteil von W und $\bar{W}$ immer den Skalarteil von W bedeuten. Einen Vektorskalar f, dessen Skalarteil $\bar{f}$ gleich 0 ist, identifizieren wir mit dem Zweiervektor $\underline{f}$; und einen Vektorskalar g, dessen Vektorteil $\underline{g}$ gleich 0 ist, identifizieren wir mit dem Skalar $\bar{g}$. Wir sehen also einen Zweiervektor oder einen Skalar als Sonderfall eines Vektorskalars an.

[1] Die Buchstaben f, g und A, B bezeichnen im folgenden *beliebige* vektorskalare Größen; in der alten Bedeutung kommen A, B und g nicht mehr vor.

Jetzt können wir für die Übergangs- oder Brechungs-Formeln (7.10), (7.11) schreiben:

$$\underline{(h_v b_v)} = 0, \tag{7.10a}$$

wobei

$$b_v \equiv h_{v+\frac{1}{2}} - h_{v-\frac{1}{2}} + \bar{\varepsilon}_v \tag{7.10b}$$

ist; und

$$\overline{(h_w h_w)} = r_w^2. \tag{7.11a}$$

Die Vorzeichenbeziehung bei (7.11) lautet jetzt (mit w für $v+\frac{1}{2}$):

$$\operatorname{sign}(\bar{h}_w \bar{h}_{w-\frac{1}{2}} - \underline{h}_w \underline{h}_{w-\frac{1}{2}}) = \operatorname{sign} r_w r_{w-\frac{1}{2}};$$

wir benützen davon aber für später nur eine Teilaussage, die sich auf den mit der optischen Achse zusammenfallenden Strahl ($\underline{h}_0 = 0$; $\underline{h}_{\frac{1}{2}} = 0$) bezieht: $\operatorname{sign} \bar{h}_w \bar{h}_{w-\frac{1}{2}} = \operatorname{sign} r_w r_{w-\frac{1}{2}}$, daher $\operatorname{sign}(\bar{h}_w / r_w) = \operatorname{sign}(\bar{h}_{\frac{1}{2}} / r_{\frac{1}{2}}) = \operatorname{sign}(-\boldsymbol{k}\, \boldsymbol{r}_{\frac{1}{2}} / r_{\frac{1}{2}}) = -1$. Hieraus und aus (7.11a) folgt:

$$\bar{h}_w = -r_w, \quad \text{wenn} \quad \underline{h}_0 = 0; \quad \underline{h}_{\frac{1}{2}} = 0 \quad \text{ist.} \tag{7.11b}$$

Wir sind von Vektor-Formeln zu Vektorskalar-Formeln übergegangen, weil sich daran die Potenzreihenentwicklungen übersichtlicher vornehmen lassen.

44. Potenzreihendarstellung eines Vektorskalars. Als unabhängige, vorgegebene Strahldaten wählen wir $\underline{h}_0 \equiv \{x_0;\, y_0;\, 0\}$, den Höhenvektor in der Eintrittspupille, und $\underline{h}_{\frac{1}{2}} \equiv \{x_+;\, y_+;\, 0\} \equiv \{\xi_0;\, \eta_0;\, 0\}$, den Neigungsvektor im Objektraum. Nach wir vor nehmen wir die Objektebene im Unendlichen an, so daß die verschiedenen Neigungsvektoren $\underline{h}_{\frac{1}{2}}$ den verschiedenen Objektpunkten eineindeutig zugeordnet sind ($n_0 = 1$ vorausgesetzt). Ohne Beschränkung der Allgemeinheit können wir $\xi_0 = 0$ setzen; wir haben dann

$$y_0, \quad x_0 \quad \text{und} \quad y_+ \quad (\equiv \eta_0) \tag{7.16}$$

als unabhängige Variablen. Später, in Ziff. 47, werden wir nicht y_+, sondern die Bildhöhe y' vorgeben, um einen Formelsatz zur Berechnung der reduziblen Bildfehler-Koeffizienten $\Delta N_0^{\sigma\tau,v}$ zu erhalten. Jetzt beabsichtigen wir, einen Formelsatz zur Berechnung der in Ziff. 41 eingeführten Bildfehler-Koeffizienten $\Delta M_{k+1}^{\sigma\tau,v}$ aufzustellen.

Die Wellenlänge λ des Lichts ist eine vierte unabhängige Variable. Sie kommt über die Brechungsindizes (r_v mit halbzahligem v) herein, deren Wellenlängen-Abhängigkeit fast alle im folgenden vorkommenden Koeffizienten zu Funktionen von λ macht. Später (in Ziff. 46) werden wir auch nach Potenzen der in Abschnitt I eingeführten Variablen w entwickeln, die ein geeignetes Maß für die Farbe des Lichtstrahls ist (anstelle von λ). Im sichtbaren Bereich können die Brechungsindizes in Strenge als Potenzreihen nach w angesetzt werden. Polynome dritten Grades in w geben die Brechungsindizes bei den vier Farben c, e, f, g exakt wieder.

Alle Zweiervektoren $\underline{h}_v$ denken wir uns in Potenzreihen nach den drei Größen (7.16) entwickelt:

$$\underline{h}_v = \sum_{\substack{\sigma=0 \\ (\sigma+\tau+v \text{ ungerade})}}^{\infty} \sum_{\tau=0}^{\infty} \sum_{v=0}^{\infty} \boldsymbol{h}_v^{\sigma\tau,v} \cdot (\boldsymbol{y}_+^{\sigma}\, \boldsymbol{y}_0^{\tau}\, \boldsymbol{x}_0^{v}), \tag{7.17a}$$

ebenso alle Skalare $\bar{h}_v$:

$$\bar{h}_v = \sum_{\substack{\sigma=0 \\ (\sigma+\tau \text{ gerade}; \\ v \text{ gerade})}}^{\infty} \sum_{\tau=0}^{\infty} \sum_{v=0}^{\infty} h_v^{\sigma\tau,v} (\boldsymbol{y}_+^{\sigma}\, \boldsymbol{y}_0^{\tau}\, \boldsymbol{x}_0^{v}); \tag{7.17b}$$

dabei ist

$$\left.\begin{array}{lll}(\boldsymbol{y}_+^\sigma\,\boldsymbol{y}_0^\tau\,\boldsymbol{x}_0^v)\equiv & & \\ \boldsymbol{j}\cdot y_+^\sigma\,y_0^\tau\,x_0^v, & \text{wenn } \sigma+\tau \text{ ungerade,} & v \text{ gerade ist} \\ \boldsymbol{i}\cdot y_+^\sigma\,y_0^\tau\,x_0^v, & \text{wenn } \sigma+\tau \text{ gerade,} & v \text{ ungerade ist} \\ y_+^\sigma\,y_0^\tau\,x_0^v, & \text{wenn } \sigma+\tau \text{ gerade,} & v \text{ gerade ist} \\ [\boldsymbol{i}\boldsymbol{j}]\cdot y_+^\sigma\,y_0^\tau\,x_0^v, & \text{wenn } \sigma+\tau \text{ ungerade,} & v \text{ ungerade ist.}\end{array}\right\} \tag{7.18}$$

Die Darstellung (7.17a, b), (7.18) begründen wir so: Bei den hier betrachteten Systemen aus Kugelflächen können wir voraussetzen, daß alle x_v, y_v (das sind die x- und y-Komponenten aller $\underline{h}_v$) in einem bestimmten Intervall der Variablen (7.16) als Potenzreihen nach diesen Variablen darstellbar sind. Drehung eines Strahls (7.16) um 180° um die optische Achse ändert die Vorzeichen der drei Variablen (7.16) und aller anderen x_v, y_v; also kommen in (7.17a) nur Glieder mit ungerader Ordnung $(\sigma+\tau+v)$ vor. Spiegelung des Strahls (7.16) an der Hauptmeridianebene (y, z-Ebene) ändert nur die Vorzeichen von x_0 und allen anderen x_v; also kommen in der Potenzreihenentwicklung für x_v nur Glieder mit ungerader Potenz v vor, und in der Potenzreihenentwicklung für y_v nur Glieder mit gerader Potenz v. Folglich treten bei x_v nur Glieder mit geradem $\sigma+\tau$ und ungeradem v, bei y_v nur Glieder mit ungeradem $\sigma+\tau$ und geradem v auf. Das leistet gerade die Darstellung (7.17a), (7.18) von $\underline{h}_v\equiv\boldsymbol{i}\,x_v+\boldsymbol{j}\,y_v$. Die Darstellung (7.17b), (7.18) von $\bar{h}_v$ folgt aus (7.17a), (7.18) wegen (7.11) und wegen $\boldsymbol{i}\boldsymbol{j}=0$. Die Definition (7.18) für ungerades $\sigma+\tau$ und ungerades v wird ohne Einfluß sein; wir setzen nämlich fest[1]:

$$h_v^{\sigma\tau,v}=0, \quad \text{wenn } \sigma+\tau \text{ ungerade, } v \text{ ungerade ist.} \tag{7.17c}$$

Durch (7.17a, b), (7.18) und (7.17c) sind die Koeffizienten $h_v^{\sigma\tau,v}$ für alle Indizes[2] $\sigma\tau$, v mit ganzen, nicht negativen Zahlen σ; τ; v definiert. Wir können nun, als Zusammenfassung von (7.17a) und (7.17b), für den Vektorskalar h_v die Potenzreihe

$$h_v=\sum_{\sigma=0}^{\infty}\sum_{\tau=0}^{\infty}\sum_{v=0}^{\infty} h_v^{\sigma\tau,v}(\boldsymbol{y}_+^\sigma\,\boldsymbol{y}_0^\tau\,\boldsymbol{x}_0^v) \tag{7.17d}$$

mit eindeutig bestimmten Koeffizienten $h_v^{\sigma\tau,v}$ schreiben[3].

Im folgenden kommen außer den h_v nur solche Vektorskalare vor, die durch Addition und Multiplikation aus den h_v und aus Konstanten hervorgegangen sind. Daher können alle vorkommenden Vektorskalare eindeutig in der Form (7.17d), (7.18) dargestellt werden, und für alle gilt eine Gl. (7.17c). — Aus zwei *solchen* Vektorskalaren A und B bilden wir nun die Summe $(A+B)$ und das Produkt $(A\cdot B)$, und wir führen durch Umordnen, Ausmultiplizieren und Koeffizientenvergleich die Koeffizienten $(A+B)^{\sigma\tau,v}$ der Summe und die Koeffizienten $(A\cdot B)^{\sigma\tau,v}$ des Produkts zurück auf die Koeffizienten $A^{st,u}$ und $B^{st,u}$ der beiden Summanden

[1] Physikalisch bedeutet diese Festsetzung, daß wir uns hier mit Lichtoptik, nicht mit Elektronenoptik befassen.

[2] Ein *Index-Tripel* $\sigma\tau, v$ nennen wir kürzer einen *Index* $\sigma\tau, v$.

[3] Betrachten wir nur *eine* Wellenlänge des Lichts, dann wird der Wert eines bestimmten Koeffizienten $h_v^{\sigma\tau,v}$ durch eine einzige Zahl beschrieben, während der Wert eines bestimmten Vektorskalars h_v durch drei Zahlen beschrieben wird.

bzw. Faktoren: Es ist

$$\sum_{\sigma=0}^{\infty}\sum_{\tau=0}^{\infty}\sum_{v=0}^{\infty}(A+B)^{\sigma\tau,v}(\boldsymbol{y}_+^\sigma\,\boldsymbol{y}_0^\tau\,\boldsymbol{x}_0^v)$$

$$=\sum_{\sigma=0}^{\infty}\sum_{\tau=0}^{\infty}\sum_{v=0}^{\infty}A^{\sigma\tau,v}(\boldsymbol{y}_+^\sigma\,\boldsymbol{y}_0^\tau\,\boldsymbol{x}_0^v)+\sum_{\sigma=0}^{\infty}\sum_{\tau=0}^{\infty}\sum_{v=0}^{\infty}B^{\sigma\tau,v}(\boldsymbol{y}_+^\sigma\,\boldsymbol{y}_0^\tau\,\boldsymbol{x}_0^v)$$

$$=\sum_{\sigma=0}^{\infty}\sum_{\tau=0}^{\infty}\sum_{v=0}^{\infty}(A^{\sigma\tau,v}+B^{\sigma\tau,v})(\boldsymbol{y}_+^\sigma\,\boldsymbol{y}_0^\tau\,\boldsymbol{x}_0^v),$$

daher

$$(A+B)^{\sigma\tau,v}=A^{\sigma\tau,v}+B^{\sigma\tau,v}. \tag{7.19}$$

Und es ist

$$\sum_{\sigma=0}^{\infty}\sum_{\tau=0}^{\infty}\sum_{v=0}^{\infty}(A\cdot B)^{\sigma\tau,v}(\boldsymbol{y}_+^\sigma\,\boldsymbol{y}_0^\tau\,\boldsymbol{x}_0^v)$$

$$=\sum_{s=0}^{\infty}\sum_{t=0}^{\infty}\sum_{u=0}^{\infty}A^{st,u}(\boldsymbol{y}_+^s\,\boldsymbol{y}_0^t\,\boldsymbol{x}_0^u)\cdot\sum_{S=0}^{\infty}\sum_{T=0}^{\infty}\sum_{U=0}^{\infty}B^{ST,U}(\boldsymbol{y}_+^S\,\boldsymbol{y}_0^T\,\boldsymbol{x}_0^U)$$

$$=\sum_{S=0}^{\infty}\sum_{T=0}^{\infty}\sum_{U=0}^{\infty}\sum_{s=0}^{\infty}\sum_{t=0}^{\infty}\sum_{u=0}^{\infty}A^{st,u}B^{ST,U}(\boldsymbol{y}_+^S\,\boldsymbol{y}_0^T\,\boldsymbol{x}_0^U)(\boldsymbol{y}_+^s\,\boldsymbol{y}_0^t\,\boldsymbol{x}_0^u).$$

Da nach Voraussetzung die Koeffizienten $A^{st,u}$ und $B^{ST,U}$ die Eigenschaft (7.17c) haben; und da nach (7.18) das skalare Produkt $(\boldsymbol{y}_+^S\,\boldsymbol{y}_0^T\,\boldsymbol{x}_0^U)\cdot(\boldsymbol{y}_+^s\,\boldsymbol{y}_0^t\,\boldsymbol{x}_0^u)=0$ ist (wegen $\boldsymbol{ij}=0$), wenn $S+T$ und u ungerade, U und $s+t$ gerade sind, oder wenn $S+T$ und u gerade, U und $s+t$ ungerade sind; verschwinden alle Glieder mit ungeradem $\sigma+\tau$ und zugleich ungeradem v ($\sigma\equiv S+s$; $\tau\equiv T+t$; $v\equiv U+u$). Nach (7.18) ist bei den übrigen Gliedern $(\boldsymbol{y}_+^S\,\boldsymbol{y}_0^T\,\boldsymbol{x}_0^U)(\boldsymbol{y}_+^s\,\boldsymbol{y}_0^t\,\boldsymbol{x}_0^u)=(\boldsymbol{y}_+^\sigma\,\boldsymbol{y}_0^\tau\,\boldsymbol{x}_0^v)$, wenn nicht $A^{st,u}\,B^{ST,U}=0$ ist aufgrund der Eigenschaft (7.17c) der Koeffizienten $A^{st,u}$ und $B^{ST,U}$. So wird

$$\sum_{\sigma=0}^{\infty}\sum_{\tau=0}^{\infty}\sum_{v=0}^{\infty}(A\cdot B)^{\sigma\tau,v}(\boldsymbol{y}_+^\sigma\,\boldsymbol{y}_0^\tau\,\boldsymbol{x}_0^v)$$

$$=\underset{\substack{\text{ohne Glieder}\\ \text{mit zugleich}\\ \text{ungeradem }\sigma+\tau\\ \text{und ungeradem }v}}{\sum_{\sigma=0}^{\infty}\sum_{v=0}^{\infty}\sum_{\tau=0}^{\infty}}\left(\sum_{s=0}^{\sigma}\sum_{t=0}^{\tau}\sum_{u=0}^{v}A^{st,u}B^{(\sigma-s)(\tau-t),(v-u)}\right)(\boldsymbol{y}_+^\sigma\,\boldsymbol{y}_0^\tau\,\boldsymbol{x}_0^v),$$

daher

$$(A\cdot B)^{\sigma\tau,v}=0,\qquad\text{wenn }\sigma+\tau\text{ ungerade, }v\text{ ungerade ist} \tag{7.20a}$$

und

$$(A\cdot B)^{\sigma\tau,v}=\langle AB\rangle^{\sigma\tau,v},\quad\text{wenn nicht zugleich }\sigma+\tau\text{ ungerade, }v\text{ ungerade ist,} \tag{7.20b}$$

wobei

$$\langle AB\rangle^{\sigma\tau,v}\equiv\sum_{s=0}^{\sigma}\sum_{t=0}^{\tau}\sum_{u=0}^{v}A^{st,u}B^{(\sigma-s)(\tau-t),(v-u)} \tag{7.20c}$$

eine Cauchysche Produktensumme über drei Indizes ist.

45. Durchrechnungs- und Rekursionsformeln für die Koeffizienten. Aus den Übergangs- oder Brechungs-Formeln (7.10a, b), (7.11a) folgt nach Ziff. 44 unmittelbar:

wenn $\sigma+\tau+v$ ungerade ist $(\sigma+\tau+v\geqq 1)$

$$\langle h_v\, b_v\rangle^{\sigma\tau,v}=0 \qquad (v\geqq\tfrac{1}{2}) \tag{7.21a}$$

$$h_{v+\frac{1}{2}}^{\sigma\tau,v}-h_{v-\frac{1}{2}}^{\sigma\tau,v}=b_v^{\sigma\tau,v} \qquad (v\geqq\tfrac{1}{2}), \tag{7.21b}$$

wenn $s+t$ und u gerade sind $(s+t+u\geqq 2)$

$$\langle h_{v+\frac{1}{2}}\, h_{v+\frac{1}{2}}\rangle^{st,u}=0 \qquad (v\geqq 0) \tag{7.22a}$$

$$h_{v+\frac{1}{2}}^{st,u}-h_{v-\frac{1}{2}}^{st,u}=b_v^{st,u} \qquad (v\geqq\tfrac{1}{2}) \tag{7.22b}$$

und aus (7.11b), (7.10b):

$$h_{v+\frac{1}{2}}^{00,0}=-r_{v+\frac{1}{2}} \qquad (v\geqq 0) \tag{7.23a}$$

$$b_v^{00,0}=d_v \equiv \bar{\varepsilon}_v-r_{v+\frac{1}{2}}+r_{v-\frac{1}{2}} \qquad (v\geqq\tfrac{1}{2}). \tag{7.23b}$$

Die r_v sind die Brechungsindizes (v halbzahlig) und die Radien (v ganzzahlig), die d_v sind die Flächenabstände (v halbzahlig) und die „Brechungsdifferenzen" (v ganzzahlig). [Der Flächenabstand $d_{w+\frac{1}{2}}$, w ganzzahlig, ist der Abstand zwischen dem w-ten und dem $(w+1)$-ten Flächenscheitel. Die Brechungsdifferenz d_v, v ganzzahlig, ist gleich dem Brechungsindex vor der v-ten Fläche minus dem Brechungsindex hinter der v-ten Fläche.] Die $h_v^{00,0}$, $b_v^{00,0}$ stellen also die Konstruktionsdaten des optischen Systems (einschließlich der Bildebene) dar.

Wegen (7.17c) ist ferner

wenn $s+t$ und u ungerade sind: $(s+t+u\geqq 2)$

$$h_{v+\frac{1}{2}}^{st,u}=0 \qquad (v\geqq 0) \tag{7.24a}$$

$$b_v^{st,u}=0 \qquad (v\geqq\tfrac{1}{2}). \tag{7.24b}$$

Bekannt sind, wegen der Potenzreihenentwicklung nach $y_+\equiv(\underline{h}_{\frac{1}{2}})_y$; $y_0\equiv(\underline{h}_0)_y$; $x_0\equiv(\underline{h}_0)_x$ und wegen $x_+\equiv(h_{\frac{1}{2}})_x=0$; $r_0=\infty$, die folgenden zum Objektraum gehörigen Koeffizienten:

$$h_0^{10,0}=0; \quad h_0^{01,0}=1; \quad h_0^{00,1}=1. \tag{7.25a}$$

$$h_0^{\sigma\tau,v}=0, \quad \text{wenn } \sigma+\tau+v>1 \text{ ist.} \tag{7.25b}$$

$$h_{\frac{1}{2}}^{10,0}=1; \quad h_{\frac{1}{2}}^{01,0}=0; \quad h_{\frac{1}{2}}^{00,1}=0. \tag{7.26a}$$

$$h_{\frac{1}{2}}^{\sigma\tau,v}=0, \quad \text{wenn } \sigma+\tau+v \text{ ungerade} >1 \text{ ist.} \tag{7.26b}$$

Gesucht sind die zur Gaußschen Bildebene gehörigen Koeffizienten $h_{k+1}^{\sigma\tau,v}$, und zwar beschränken wir uns auf diejenigen, deren Indizes $\sigma\tau, v$ in dem endlichen Intervall

$$\sigma\leqq\mathfrak{S}; \quad \tau\leqq\mathfrak{T}; \quad v\leqq\mathfrak{U} \tag{7.27}$$

liegen.

Der Gang der Rechnung ist, wie wir gleich einsehen werden, folgender: Einerseits fängt man mit der Anwendung der Formeln (7.21a, b), (7.22a, b) bei den kleinsten Nummern v an und gelangt in $\frac{1}{2}$-Schritten bis zur Nummer $v=k+\frac{1}{2}$; insofern sind die Formeln (7.21a, b), (7.22a, b) „Durchrechnungs"-Formeln. Andererseits fängt man bei den kleinsten Ordnungen $\omega(\equiv\sigma+\tau+v)$ an und gelangt in 1-Schritten bis zur größten Ordnung $\omega=\Omega(\equiv\mathfrak{S}+\mathfrak{T}+\mathfrak{U})$; insofern sind die Formeln (7.21a, b), (7.22a, b) „Rekursions"-Formeln. Bei einer bestimmten Ordnung

$$\omega=\sigma+\tau+v \tag{7.28}$$

wendet man die Formeln (7.21a, b), (7.22a, b) für alle zum Bereich (7.27), (7.28) gehörigen Indizes $\sigma\tau, v$ an, wobei innerhalb dieses Bereiches die Reihenfolge der verschiedenen $\sigma\tau, v$ gleichgültig ist.

Es ist ratsam, bei konstanter Nummer v die Rechnung (7.21a, b) (7.22a, b) für alle zum Bereich (7.27) gehörigen Indizes durchzuführen und erst dann zur nächsten Nummer überzugehen[1]. — Wir stellen nun fest: Hält man beim Rechnen die Durchrechnungs-Reihenfolge (von kleineren zu größeren v-Werten) und die Rekursions-Reihenfolge (von kleineren zu größeren ω-Werten) ein, und berechnet man, wenn nicht $v=0$ ist,

$$\text{erst } b_v^{\sigma\tau,v}, \text{ dann } h_{v+\frac{1}{2}}^{\sigma\tau,v}, \qquad \text{wenn } \sigma+\tau+v \text{ ungerade ist;} \tag{7.29}$$

$$\text{erst } h_{v+\frac{1}{2}}^{st,u}, \text{ dann } b_v^{st,u}, \qquad \text{wenn } s+t \text{ und } u \text{ gerade sind,} \tag{7.30}$$

dann kann man mit Hilfe der

Formel:	(7.21a)	(7.21b)	(7.22a)	(7.22b)	
berechnen:	$b_v^{\sigma\tau,v}$	$h_{v+\frac{1}{2}}^{\sigma\tau,v}$	$h_{v+\frac{1}{2}}^{st,u}$	$b_v^{st,u}$	; (7.31)
	$\sigma+\tau+v$ ungerade		$s+t$ und u gerade		

dann ist nämlich die Bestimmung der übrigen in diesen Formeln stehenden Koeffizienten jeweils schon vorausgegangen.

So gelangt man — ausgehend von $h_v^{00,0}$, $b_v^{00,0}$ gemäß (7.23a, b) und von $h_0^{\sigma\tau,v}$, $h_{\frac{1}{2}}^{\sigma\tau,v}$ gemäß (7.25a, b), (7.26a, b) und nach Nullsetzen der in (7.24a, b) genannten Koeffizienten — durch wiederholte Anwendung der Formeln (7.21a, b), (7.22a, b) in der obengenannten Reihenfolge schließlich zu den gesuchten, zum $\sigma\tau, v$-Intervall (7.27) gehörigen Koeffizienten $h_{k+1}^{\sigma\tau,v}$. Hieraus lassen sich[2] die Bildfehler-Koeffizienten $\Delta M_{k+1}^{\sigma\tau,v}$ ($\sigma+\tau+v$ ungerade) bestimmen:

$$\left.\begin{aligned}\Delta M_{k+1}^{\sigma 0,0} &= C\cdot h_{k+1}^{\sigma 0,0} - c_M^{(\alpha)};\\ \Delta M_{k+1}^{\sigma\tau,v} &= C\cdot h_{k+1}^{\sigma\tau,v}, \quad \text{wenn} \quad (\tau, v)\neq(0,0) \text{ ist,}\end{aligned}\right\} \tag{7.32}$$

wobei

$$C=-1/h_{k+1}^{10,0}(\lambda_e) \tag{7.32a}$$

und

$$c_M^{(\alpha)}=(-1)^{1+\alpha}\binom{-1/2}{\alpha}; \quad \alpha=\frac{\sigma-1}{2} \tag{7.32b}$$

ist.

Im Objektraum ist Luft:

$$n_0=1. \tag{7.33}$$

---

[1] So hat man bei der Berechnung der Bildfehler-Koeffizienten $\Delta M_{k+1}^{\sigma\tau,v}$ weniger Zwischengrößen zu speichern, als man schließlich zu speichern hätte, wenn man bei konstantem Index $\sigma\tau, v$ die Rechnung für alle Nummern $v\,(\leq k+\frac{1}{2})$ durchführen und erst dann zum nächsten Index übergehen würde. Die letztere Reihenfolge wird aber bei der Berechnung der Bildfehler-Koeffizienten $\Delta N_0^{\sigma\tau,v}$ unumgänglich sein.

[2] Nach Ziff. 41; es ist $y'\equiv(\underline{h}_{k+1})_y$, $x'\equiv(\underline{h}_{k+1})_x$ und $c_{k+1}^{-}=1/h_{k+1}^{10,0}(\lambda=\lambda_e)$.

Ist das optische System „auf Brennweite 1 reduziert", was keine Beschränkung der Allgemeinheit bedeutet:

$$h_{k+1}^{10,0}(\lambda_e) = 1, \tag{7.34}$$

dann ist

$$\left.\begin{aligned} \Delta M_{k+1}^{\sigma 0,0} &= -h_{k+1}^{\sigma 0,0} + (-1)^{\alpha}\binom{-1/2}{\alpha}; \quad \alpha = \frac{\sigma-1}{2}; \\ \Delta M_{k+1}^{\sigma\tau,v} &= -h_{k+1}^{\sigma\tau,v}, \quad \text{wenn } (\tau, v) \neq (0,0) \text{ ist.} \end{aligned}\right\} \tag{7.32*}$$

Auch bei Planflächen sind die Formeln (7.21a, b), (7.22a, b) anwendbar; daß nach (7.23a) bei einer Planfläche $h^{00,0} = \infty$ ist, behindert nicht die im folgenden beschriebene Berechnung von $b^{\sigma\tau,v}$ aus $\langle hb\rangle^{\sigma\tau,v} = 0$ und von $h^{st,u}$ aus $\langle hh\rangle^{st,u} = 0$.

Die Berechnung von $b^{\sigma\tau,v}$ aus $\langle hb\rangle^{\sigma\tau,v} = 0$:

Hat man die (für eine bestimmte Wellenlänge des Lichts berechneten) Zahlenwerte der Größen $h^{st,u}$ und $b^{st,u}$ jeweils nach ihrer Berechnung tabelliert:

Tabelle 1

				⋮
				...
			...	$h^{00,3}$

				...
			...	$h^{10,2}$
		...	$h^{01,2}$	$h^{00,2}$

				...
			...	$h^{20,1}$
		...	$h^{11,1}$	$h^{10,1}$
	...	$h^{02,1}$	$h^{01,1}$	$h^{00,1}$

				...
			...	$h^{30,0}$
		...	$h^{21,0}$	$h^{20,0}$
	...	$h^{12,0}$	$h^{11,0}$	$h^{10,0}$
...	$h^{03,0}$	$h^{02,0}$	$h^{01,0}$	$h^{00,0}$

Tabelle 2

$b^{00,0}$	$b^{01,0}$	$b^{02,0}$	$b^{03,0}$	...
$b^{10,0}$	$b^{11,0}$	$b^{12,0}$	...	
$b^{20,0}$	$b^{21,0}$	...		
$b^{30,0}$	...			
...				

$b^{00,1}$	$b^{01,1}$	$b^{02,1}$	...
$b^{10,1}$	$b^{11,1}$	...	
$b^{20,1}$	...		
...			

$b^{00,2}$	$b^{01,2}$	...
$b^{10,2}$	...	
...		

$b^{00,3}$	...
⋮	

so berechnet man $b^{\sigma\tau,v}$ aus $\langle hb\rangle^{\sigma\tau,v} = 0$ in folgender Weise:

Man denkt sich Tabelle 1 so über Tabelle 2 geschoben, daß paarweise diejenigen Koeffizienten $h^{st,u}$ und $b^{ST,U}$ übereinanderliegen, deren Indizessumme $st, u + ST, U = \sigma\tau, v$ ist.

Dann bildet man das Produkt aus dem ersten Zahlenpaar $(h^{\sigma\tau,v}\, b^{00,0})$, addiert dazu das Produkt aus dem zweiten Zahlenpaar, addiert dazu das Produkt aus dem dritten Zahlenpaar und so fort, bis man jedes übereinanderliegende Zahlenpaar multipliziert und hinzuaddiert hat, außer dem letzten $(h^{00,0}\, b^{\sigma\tau,v})$.

Diesen letzten (Produkt-)Summanden kann man nicht berechnen, weil $b^{\sigma\tau,v}$ ja die gesuchte Unbekannte ist. Nun soll aber die ganze (Cauchysche Produkt-) Summe $\langle hb\rangle^{\sigma\tau,v} = 0$ sein, deshalb dividiert man die Summe ohne den letzten Summanden durch $-h^{00,0}$ und erhält so $b^{\sigma\tau,v}$.

Die Berechnung von $h^{st,u}$ aus $\langle hh\rangle^{st,u} = 0$

erfolgt ganz ähnlich; nur kommt jetzt die Unbekannte $h^{st,u}$ im ersten Summanden $(h^{st,u}\, h^{00,0})$ und im letzten Summanden $(h^{00,0}\, h^{st,u})$ der Summe $\langle hh\rangle^{st,u}$ vor. Deshalb dividiert man die Summe ohne den ersten und letzten Summanden durch $-2h^{00,0}$ und erhält so $h^{st,u}$.

46. Entwicklung nach der Farbvariablen W. Entwickeln wir außer nach den drei Strahldaten y_+, y_0, x_0 auch nach der in Abschnitt I eingeführten Farbvariablen w oder nach

$$W \equiv \frac{w}{w_g} \quad \left[\equiv \frac{\lambda_e-\lambda}{\lambda-l_*}\Big/\frac{\lambda_e-\lambda_g}{\lambda_g-l_*}\right], \tag{7.35}$$

dann erhalten wir für die in der Reihe

$$\underline{h}_{k+1} = \sum_{\substack{\pi=0\\(\sigma+\tau+v\\ \text{ungerade})}}^{\infty} \sum_{\sigma=0}^{\infty} \sum_{\tau=0}^{\infty} \sum_{v=0}^{\infty} h_{k+1}^{[\pi]\sigma\tau,v} \cdot W^{\pi} (\boldsymbol{y}_+^{\sigma}\, \boldsymbol{y}_0^{\tau}\, \boldsymbol{x}_0^{v}) \tag{7.36}$$

auftretenden Koeffizienten $h_{k+1}^{[\pi]\sigma\tau,v}$, in ganz entsprechender Weise wie für die $h_{k+1}^{\sigma\tau,v}$, den im folgenden zusammengestellten Formelsatz (VII).

Jeden Vektorskalar A setzen wir als Potenzreihe nach y_+, y_0, x_0 und W an:

$$\left.\begin{aligned} A &= \sum_{\pi=0}^{\infty}\sum_{\sigma=0}^{\infty}\sum_{\tau=0}^{\infty}\sum_{v=0}^{\infty} A^{[\pi]\sigma\tau,v}\cdot W^{\pi}(\boldsymbol{y}_+^{\sigma}\,\boldsymbol{y}_0^{\tau}\,\boldsymbol{x}_0^{v});\\ &\qquad (\boldsymbol{y}_+^{\sigma}\,\boldsymbol{y}_0^{\tau}\,\boldsymbol{x}_0^{v}) \quad \text{wie in (7.18) definiert};\\ A^{[\pi]\sigma\tau,v} &= 0, \quad \text{wenn } \sigma+\tau \text{ ungerade, } v \text{ ungerade ist.} \end{aligned}\right\} \tag{7.37}$$

Dann ist

$$(A+B)^{[\pi]\sigma\tau,v} = A^{[\pi]\sigma\tau,v} + B^{[\pi]\sigma\tau,v} \tag{7.38}$$

und

$$(A\cdot B)^{[\pi]\sigma\tau,v} = 0, \quad \text{wenn } \sigma+\tau \text{ ungerade, } v \text{ ungerade ist;} \tag{7.39a}$$

$$\left.\begin{aligned}(A\cdot B)^{[\pi]\sigma\tau,v} = \langle AB\rangle^{[\pi]\sigma\tau,v}, \quad &\text{wenn nicht zugleich}\\ &\sigma+\tau \text{ ungerade, } v \text{ ungerade ist;}\end{aligned}\right\} \tag{7.39b}$$

wobei

$$\langle AB\rangle^{[\pi]\sigma\tau,v} \equiv \sum_{p=0}^{\infty}\sum_{s=0}^{\infty}\sum_{t=0}^{\infty}\sum_{u=0}^{\infty} A^{[p]st,u} B^{[\pi-p](\sigma-s)(\tau-t),(v-u)} \tag{7.39c}$$

eine Cauchysche Produktensumme über vier Indizes ist.

Wir schreiben kürzer

$$A^{[\pi]} \quad \text{für} \quad A^{[\pi]00,0}. \tag{7.40}$$

Die in der Reihe

$$n = n(W) = \sum_{\pi=0}^{\infty} n^{[\pi]}\cdot W^{\pi} \tag{7.41a}$$

auftretenden Koeffizienten $n^{[\pi]}$ stehen mit den in (1.50) auftretenden Dispersions-Koeffizienten $\delta^{(i)}$ in dem Zusammenhang:

$$\left.\begin{aligned} n^{[0]} &= n_e;\\ n^{[\pi]} &= (n_e-1)\cdot\delta^{(\pi)}, \quad \text{wenn } \pi \geqq 1 \text{ ist.}\end{aligned}\right\} \tag{7.41b}$$

Mit $n_v \equiv r_{v-\frac{1}{2}}$ (v ganzzahlig) sind die in (7.23a, b) vorgegebenen Konstruktionsdaten r und d zum Teil Funktionen von W:

$$\left.\begin{aligned} r_v &= \sum_{\pi=0}^{\infty} r_v^{[\pi]}\cdot W^{\pi}, &&\text{wenn } v \text{ halbzahlig ist;}\\ r_v &= r_v^{[0]}, &&\text{wenn } v \text{ ganzzahlig ist;}\end{aligned}\right\} \tag{7.42a}$$

$$\left.\begin{aligned} d_v &= d_v^{[0]}, &&\text{wenn } v \text{ halbzahlig ist;}\\ d_v &= \sum_{\pi=0}^{\infty} d_v^{[\pi]}\, W^{\pi}, &&\text{wenn } v \text{ ganzzahlig ist;}\end{aligned}\right\} \tag{7.42b}$$

$$d_v^{[\pi]} = r_{v-\frac{1}{2}}^{[\pi]} - r_{v+\frac{1}{2}}^{[\pi]}. \tag{7.42c}$$

Wir nehmen an, daß im Objektraum Luft ist:

$$r_{\frac{1}{2}} = 1, \tag{7.43a}$$

und daß das optische System auf Brennweite 1 reduziert ist:

$$h_{k+1}^{[0]1 0,0} = 1. \tag{7.43b}$$

Die Rekursion beginnt mit:

$$\boxed{\begin{aligned} h_{v+\frac{1}{2}}^{[\pi]0 0,0} &= -r_{v+\frac{1}{2}}^{[\pi]} \\ b_v^{[\pi]0 0,0} &= d_v^{[\pi]} \end{aligned}} \quad \begin{aligned} &(v \geqq 0) \\ &(v \geqq \tfrac{1}{2}). \end{aligned} \tag{VII.1}$$

Die Durchrechnung beginnt mit:

$$\boxed{\begin{aligned} &h_0^{[0]1 0,0} = 0;\; h_0^{[0]0 1,0} = 1;\; h_0^{[0]0 0,1} = 1; \\ &h_0^{[\pi]\sigma\tau,v} = 0, \quad \text{wenn } \sigma+\tau+v > 1 \text{ oder } \pi > 0 \text{ ist;} \end{aligned}} \tag{VII.2}$$

$$\boxed{\begin{aligned} &h_{\frac{1}{2}}^{[0]1 0,0} = 1;\; h_{\frac{1}{2}}^{[0]0 1,0} = 0;\; h_{\frac{1}{2}}^{[0]0 0,1} = 0; \\ &h_{\frac{1}{2}}^{[\pi]\sigma\tau,v} = 0, \quad \text{wenn } \sigma+\tau+v \text{ ungerade und} \\ &\qquad\qquad\quad \text{dabei } \sigma+\tau+v > 1 \text{ oder } \pi > 0 \text{ ist.} \end{aligned}} \tag{VII.3}$$

Von vornherein werden die „elektronenoptischen“ Koeffizienten gleich 0 gesetzt:

wenn $s+t$ und u ungerade sind:

$$\boxed{\begin{aligned} h_{v+\frac{1}{2}}^{[\pi]s t,u} &= 0 \\ b_v^{[\pi]s t,u} &= 0 \end{aligned}} \quad \begin{aligned} &(v \geqq 0) \\ &(v \geqq \tfrac{1}{2}). \end{aligned} \tag{VII.4}$$

Die Durchrechnungs- und Rekursionsformeln lauten:

wenn $\sigma+\tau+v$ ungerade ist:

$$\begin{aligned} &b_v^{[\pi]\sigma\tau,v} \text{ zu bestimmen aus:} \\ &h_{v+\frac{1}{2}}^{[\pi]\sigma\tau,v} \text{ zu bestimmen aus:} \end{aligned} \quad \boxed{\begin{aligned} &\langle h_v b_v\rangle^{[\pi]\sigma\tau,v} = 0 \\ &h_{v+\frac{1}{2}}^{[\pi]\sigma\tau,v} - h_{v-\frac{1}{2}}^{[\pi]\sigma\tau,v} = b_v^{[\pi]\sigma\tau,v} \end{aligned}} \quad \begin{aligned} &(v \geqq \tfrac{1}{2}) \\ &(v \geqq \tfrac{1}{2}) \end{aligned} \tag{VII.5}$$

wenn $s+t$ und u gerade sind:

$$\begin{aligned} &h_{v+\frac{1}{2}}^{[\pi]s t,u} \text{ zu bestimmen aus:} \\ &b_v^{[\pi]s t,u} \text{ zu bestimmen aus:} \end{aligned} \quad \boxed{\begin{aligned} &\langle h_{v+\frac{1}{2}} h_{v+\frac{1}{2}}\rangle^{[\pi]s t,u} = 0 \\ &h_{v+\frac{1}{2}}^{[\pi]s t,u} - h_{v-\frac{1}{2}}^{[\pi]s t,u} = b_v^{[\pi]s t,u} \end{aligned}} \quad \begin{aligned} &(v \geqq 0) \\ &(v \geqq \tfrac{1}{2}). \end{aligned} \tag{VII.6}$$

Durch wiederholte Anwendung dieser vier Durchrechnungs- und Rekursions-Formeln erhält man schließlich die auf die Gaußsche Bildebene (Fläche Nr. $k+1$) bezogenen Koeffizienten $h_{k+1}^{[\pi]\sigma\tau,v}$ ($\sigma+\tau+v$ ungerade), und damit[1] die Bildfehler-Koeffizienten $\Delta M_{k+1}^{[\pi]\sigma\tau,v}$:

$$\boxed{\begin{aligned} &\Delta M_{k+1}^{[0]\sigma 0,0} = -h_{k+1}^{[0]\sigma 0,0} + (-1)^{\frac{\sigma-1}{2}} \binom{-1/2}{\frac{\sigma-1}{2}}; \\ &\Delta M_{k+1}^{[\pi]\sigma\tau,v} = -h_{k+1}^{[\pi]\sigma\tau,v}, \quad \text{wenn } (\pi,\tau,v) \neq (0,0,0) \text{ ist;} \end{aligned}} \tag{VII.7}$$

[1] [Nach (6.61), (6.62), (6.59); es ist $c_{\overline{k+1}} = 1$, da das optische System auf Brennweite 1 reduziert ist.]

das sind die Koeffizienten bei der Potenzreihenentwicklung der Bildhöhen-Aberration[1] ΔM_{k+1}:

$$\left.\begin{aligned} \Delta M_{k+1} = & \sum_{\substack{\pi=0 \\ \left(\sigma+\tau \text{ ungerade} \atop v \text{ gerade}\right)}}^{\infty} \sum_{\sigma=0}^{\infty} \sum_{\tau=0}^{\infty} \sum_{v=0}^{\infty} \Delta M_{k+1}^{[\pi]\sigma\tau,v} \cdot W^{\pi} y_{+}^{\sigma} y_0^{\tau} x_0^{v} + \\ & + i \sum_{\substack{\pi=0 \\ \left(\sigma+\tau \text{ gerade} \atop v \text{ ungerade}\right)}}^{\infty} \sum_{\sigma=0}^{\infty} \sum_{\tau=0}^{\infty} \sum_{v=0}^{\infty} \Delta M_{k+1}^{[\pi]\sigma\tau,v} \cdot W^{\pi} y_{+}^{\sigma} y_0^{\tau} x_0^{v}. \end{aligned}\right\} \quad (7.44)$$

Die in Ziff. 41 eingeführten Bildfehler-Koeffizienten $\Delta M_{k+1}^{\sigma\tau,v}$ sind die Reihen

$$\boxed{\Delta M_{k+1}^{\sigma\tau,v} = \sum_{\pi=0}^{\infty} \Delta M_{k+1}^{[\pi]\sigma\tau,v} \cdot W^{\pi}.} \quad \text{(VII.8)}$$

Bei der Hauptfarbe (e) ist $\Delta M_{k+1}^{\sigma\tau,v} = \Delta M_{k+1}^{[0]\sigma\tau,v}$.

47. Die Berechnung der reduziblen und der irreduziblen Bildfehler-Koeffizienten. Wir entwickeln nun nach

$$y' \equiv y_{k+1} \quad (7.45)$$

anstelle von $y_{+} \equiv \eta_0$ (und nehmen $x' \equiv x_{k+1} = 0$ an, anstelle von $x_{+} \equiv \xi_0 = 0$), und wir fragen nach einem Formelsatz zur Berechnung der in der Reihe

$$\{\xi_0; \eta_0; 0\} \equiv \underline{h}_{\frac{1}{2}} = \sum_{\substack{\pi=0 \\ \left(\sigma+\tau+v \atop \text{ungerade}\right)}}^{\infty} \sum_{\sigma=0}^{\infty} \sum_{\tau=0}^{\infty} \sum_{v=0}^{\infty} h_{\frac{1}{2}}^{[\pi]\sigma\tau,v} \cdot W^{\pi} (\boldsymbol{y}'^{\sigma} \boldsymbol{y}_0^{\tau} \boldsymbol{x}_0^{v}) \quad (7.36^*)$$

auftretenden Koeffizienten $h_{\frac{1}{2}}^{[\pi]\sigma\tau,v}$, weil wir daran interessiert sind, Formeln zur Berechnung der reduziblen Bildfehler-Koeffizienten $\Delta N_0^{\sigma\tau,v}$ zu erhalten (um damit nämlich die irreduziblen Bildfehler-Koeffizienten $\Delta \mathscr{E}^{\alpha\beta\gamma}$ und $\Delta \mathscr{E}_{\varnothing}^{\alpha\gamma\varepsilon}$ berechnen zu können). Daß im Objektraum Luft ist, und daß das optische System auf Brennweite 1 reduziert ist, behalten wir bei.

Die Indizes $[\pi]\sigma\tau,v$ beziehen wir jetzt bei allen Vektorskalaren A auf die Entwicklung nach W; y'; y_0; x_0:

$$\left.\begin{aligned} & A = \sum_{\pi=0}^{\infty} \sum_{\sigma=0}^{\infty} \sum_{\tau=0}^{\infty} \sum_{v=0}^{\infty} A^{[\pi]\sigma\tau,v} \cdot W^{\pi} (\boldsymbol{y}'^{\sigma} \boldsymbol{y}_0^{\tau} \boldsymbol{x}_0^{v}); \\ & (\boldsymbol{y}'^{\sigma} \boldsymbol{y}_0^{\tau} \boldsymbol{x}_0^{v}) \text{ entsprechend wie in (7.18) definiert;} \\ & A^{[\pi]\sigma\tau,v} = 0, \text{ wenn } \sigma+\tau \text{ ungerade, } v \text{ ungerade ist.} \end{aligned}\right\} \quad (7.37^*)$$

Die Aussage in der zweiten Zeile von (VII.3) ist die einzige Aussage des Formelsatzes (VII), die *bei der neuen Bedeutung der Indizes* unzutreffend ist; sie bezieht sich auf die Koeffizienten $h_{\frac{1}{2}}^{[\pi]\sigma\tau,v}$, die in Ziff. 46 die bekannten Koeffizienten der Reihe (7.37) für $A = \underline{h}_{\frac{1}{2}}$ waren, die aber jetzt die gesuchten Koeffizienten der Reihe (7.36*) sind.

Die erste Zeile von (VII.3) trifft, wie wir gleich sehen werden, noch zu:

$$\boxed{h_{\frac{1}{2}}^{[0]10,0} = 1; \qquad h_{\frac{1}{2}}^{[0]01,0} = 0; \qquad h_{\frac{1}{2}}^{[0]00,1} = 0.} \quad \text{(VII.3*)}$$

[1] Vgl. (6.59).

Die anderen $h_{\frac{1}{2}}^{[\pi]\sigma\tau,v}$ mit ungeradem $\sigma+\tau+v$ sind jetzt unbekannt. Statt dessen sind, wegen der Entwicklung nach $y' \equiv (\underline{h}_{k+1})_y$ und wegen $x' \equiv (\underline{h}_{k+1})_x = 0$, bekannt:

$$h_{k+1}^{[0]10,0} = 1; \quad h_{k+1}^{[0]01,0} = 0; \quad h_{k+1}^{[0]00,1} = 0; \tag{7.46a}$$

$$h_{k+1}^{[\pi]\sigma\tau,v} = 0, \quad \text{wenn } \pi > 0 \text{ oder } \sigma+\tau+v > 1 \tag{7.46b}$$

ist. Aus (7.46a) folgt (III.3*), weil $\{0^{(2)}+0^{[2]}\}0^{(1)} = \Delta N_0 = -y' + \eta_0 + \{0^{(2)}+0^{[2]}\}0^{(1)}$ ist [vgl. (6.45), (6.46) und (6.47) mit $c_{k+1}^{-} = 1$]. — Die Koeffizienten $b_v^{[0]\sigma\tau,v}$; $h_{v+\frac{1}{2}}^{[0]\sigma\tau,v}$ mit $\sigma+\tau+v=1$ können also gemäß dem Formelsatz (VII) durchgerechnet werden.

Wir nehmen nun an, daß bei konstantem Index

$$[\pi]\sigma\tau,v \tag{7.47}$$

die Rechnung für alle Nummern v durchgeführt wird, bevor zum nächsten Index übergegangen wird. Dann ist also die Rechnung für alle von (7.47) verschiedenen Indizes

$$[p]\,st,u \quad \text{mit} \quad p \leqq \pi; \quad s \leqq \sigma; \quad t \leqq \tau; \quad u \leqq v$$

schon abgeschlossen, wenn die Rechnung für den Index (7.47) durchgeführt wird. Beginnt man nun, wenn $\pi > 0$ oder $\sigma+\tau+v > 1$ ist ($\sigma+\tau+v$ ungerade), die Durchrechnung mit dem bekannten Anfangswert $h_0^{[\pi]\sigma\tau,v} = 0$, zusammen mit dem *falschen* Wert

$$*h_{\frac{1}{2}}^{[\pi]\sigma\tau,v} = 0 \tag{7.48a}$$

an Stelle des unbekannten Anfangswertes $h_{\frac{1}{2}}^{[\pi]\sigma\tau,v}$; und berechnet man daraus mit Hilfe von (VII.5) nacheinander *vorläufige* (falsche) Werte für $b_{\frac{1}{2}}^{\cdots}, h_1^{\cdots}, b_1^{\cdots}, h_{1\frac{1}{2}}^{\cdots}, \ldots$ $b_{k+\frac{1}{2}}^{\cdots}, h_{k+1}^{\cdots}$, die wir mit

$$*b_v^{[\pi]\sigma\tau,v}, \quad *h_{v+\frac{1}{2}}^{[\pi]\sigma\tau,v} \tag{7.48b}$$

bezeichnen; dann erhält man *im allgemeinen nicht* den Wert 0 für $h_{k+1}^{[\pi]\sigma\tau,v}$, wie es nach (7.46b) sein sollte, sondern einen von 0 verschiedenen Wert

$$*h_{k+1}^{[\pi]\sigma\tau,v}. \tag{7.48c}$$

Damit haben wir aber das Negative des *richtigen* Anfangswertes $h_{\frac{1}{2}}^{[\pi]\sigma\tau,v}$ erhalten:

$$h_{\frac{1}{2}}^{[\pi]\sigma\tau,v} = -\,*h_{k+1}^{[\pi]\sigma\tau,v}, \tag{7.49a}$$

und hiermit die *richtigen* Zwischenwerte $b_{\frac{1}{2}}^{\cdots}, h_1^{\cdots}, b_1^{\cdots}, h_{1\frac{1}{2}}^{\cdots}, \ldots b_{k+\frac{1}{2}}^{\cdots}$:

$$\left.\begin{aligned} b_v^{\cdots} &= *b_v^{\cdots} + b_v^{[0]10,0} \cdot h_{\frac{1}{2}}^{\cdots}; \\ h_{v+\frac{1}{2}}^{\cdots} &= *h_{v+\frac{1}{2}}^{\cdots} + h_{v+\frac{1}{2}}^{[0]10,0} \cdot h_{\frac{1}{2}}^{\cdots}. \end{aligned}\right\} \tag{7.49b}$$

Denn nach (VII.5) ist, wenn wir Δh_v für $(h_v^{\cdots} - *h_v^{\cdots})$ und Δb_v für $(b_v^{\cdots} - *b_v^{\cdots})$ schreiben:

$$\Delta h_v \cdot b_v^{[0]00,0} + h_v^{[0]00,0} \cdot \Delta b_v = 0;$$

$$\Delta h_{v+\frac{1}{2}} \quad - \quad \Delta h_{v-\frac{1}{2}} \quad = \Delta b_v,$$

und nach (VII.5) ist

$$h_v^{[0]10,0} \cdot b_v^{[0]00,0} + h_v^{[0]00,0} \cdot b_v^{[0]10,0} = 0;$$

$$h_{v+\frac{1}{2}}^{[0]10,0} \quad - \quad h_{v-\frac{1}{2}}^{[0]10,0} \quad = b_v^{[0]10,0},$$

also gelten für die Δb_v; $\Delta h_{v+\frac{1}{2}}$ die gleichen homogen linearen Durchrechnungsformeln wie für die $b_v^{[0]10,0}$; $h_{v+\frac{1}{2}}^{[0]10,0}$. Sie unterscheiden sich nur in den Anfangswerten:

$$\Delta h_0 = 0; \qquad \Delta h_{\frac{1}{2}} = h_{\frac{1}{2}}^{[\pi]\sigma\tau,v};$$

$$h_0^{[0]10,0} = 0; \qquad h_{\frac{1}{2}}^{[0]10,0} = 1,$$

also ist

$$\frac{\Delta b_v}{b_v^{[0]10,0}} = \frac{\Delta h_{v+\frac{1}{2}}}{h_{v+\frac{1}{2}}^{[0]10,0}} = \frac{h_{\frac{1}{2}}^{[\pi]\sigma\tau,v}}{1},$$

wie in (7.49b) behauptet. Nach der zweiten Gl. (7.49b), angewandt auf $v = k + \frac{1}{2}$, ist wegen (7.46a), (7.46b): $0 = {}^*h_{k+1}^{\cdots} + h_{\frac{1}{2}}^{\cdots}$, wie in (7.49a) behauptet.

Wir haben also die Vorschrift:

1. Wenn $\sigma + \tau + v = 1$ und $\pi = 0$ ist, kann gemäß dem Formelsatz (VII) durchgerechnet werden.

2. Wenn $\sigma + \tau + v$ ungerade und dabei $\sigma + \tau + v > 1$ oder $\pi > 0$ ist, beginnt man die Durchrechnung mit den Anfangswerten (VII.2), (VII.3) und führt sie gemäß (VII.5) bis zum Bildraum durch. Man erhält so für $h_{\frac{1}{2}}^{[\pi]\sigma\tau,v}$; $b_{\frac{1}{2}}^{[\pi]\sigma\tau,v}$; $h_1^{[\pi]\sigma\tau,v}$; ... $b_{k+\frac{1}{2}}^{[\pi]\sigma\tau,v}$; $h_{k+1}^{[\pi]\sigma\tau,v}$ im allgemeinen falsche Werte:

$${}^*h_{\frac{1}{2}}^{[\pi]\sigma\tau,v}; \quad {}^*b_{\frac{1}{2}}^{[\pi]\sigma\tau,v}; \quad {}^*h_1^{[\pi]\sigma\tau,v}; \quad \ldots \quad {}^*b_{k+\frac{1}{2}}^{[\pi]\sigma\tau,v}; \quad {}^*h_{k+1}^{[\pi]\sigma\tau,v},$$

aus denen man aber die richtigen Werte berechnen kann:

$$\boxed{\begin{aligned} h_v^{[\pi]\sigma\tau,v} &= {}^*h_v^{[\pi]\sigma\tau,v} - h_v^{[0]10,0} \cdot {}^*h_{k+1}^{[\pi]\sigma\tau,v} \\ b_v^{[\pi]\sigma\tau,v} &= {}^*b_v^{[\pi]\sigma\tau,v} - b_v^{[0]10,0} \cdot {}^*h_{k+1}^{[\pi]\sigma\tau,v} \end{aligned}} \qquad \begin{aligned} &(v \geqq \tfrac{1}{2}); \\ &(v \geqq \tfrac{1}{2}). \end{aligned} \qquad \text{(VII.5*)}$$

3. Wenn $s + t$ und u gerade sind, kann gemäß dem Formelsatz (VII) durchgerechnet werden.

4. In jedem der Fälle 1, 2, 3 führt man bei konstantem Index $[\pi]\sigma\tau, v$ die Durchrechnung vom Objektraum bis zum Bildraum durch, und geht erst dann zum nächsten Index über.

Mit den nach (VII.3*) und (VII.4) bekannten und den nach (VII.3), (VII.5), (VII.5*) berechneten Koeffizienten $h_{\frac{1}{2}}^{[\pi]\sigma\tau,v}$ hat man die Koeffizienten der Reihe (7.36*) für $\underline{h}_{\frac{1}{2}}$, den Neigungsvektor im Objektraum. Daraus erhält man[1] die Bildfehler-Koeffizienten $\Delta N_0^{[\pi]\sigma\tau,v}$:

$$\boxed{\begin{aligned} &\Delta N_0^{[0]\sigma 0,0} = h_{\frac{1}{2}}^{[0]\sigma 0,0} - \binom{-1/2}{\frac{\sigma-1}{2}}; \\ &\Delta N_0^{[\pi]\sigma\tau,v} = h_{\frac{1}{2}}^{[\pi]\sigma\tau,v}, \quad \text{wenn } (\pi, \tau, v) \neq (0, 0, 0) \text{ ist;} \end{aligned}} \qquad \text{(VII.7*)}$$

das sind die Koeffizienten bei der Potenzreihenentwicklung der Aberration ΔN_0:

$$\left.\begin{aligned} \Delta N_0 = &\sum_{\substack{\pi=0 \\ (\sigma+\tau \text{ ungerade} \\ v \text{ gerade})}}^{\infty} \sum_{\sigma=0}^{\infty} \sum_{\tau=0}^{\infty} \sum_{v=0}^{\infty} \Delta N_0^{[\pi]\sigma\tau,v} \cdot W^{\pi} y'^{\sigma} y_0^{\tau} x_0^{v} + \\ &+ i \sum_{\substack{\pi=0 \\ (\sigma+\tau \text{ gerade} \\ v \text{ ungerade})}}^{\infty} \sum_{\sigma=0}^{\infty} \sum_{\tau=0}^{\infty} \sum_{v=0}^{\infty} \Delta N_0^{[\pi]\sigma\tau,v} \cdot W^{\pi} y'^{\sigma} y_0^{\tau} x_0^{v}. \end{aligned}\right\} \qquad (7.50)$$

[1] [Nach (6.45), (6.46) und (6.47) mit $c_{k+1}^{-} = 1$ (das optische System ist auf Brennweite 1 reduziert).]

Die in Abschnitt VI eingeführten reduziblen Bildfehler-Koeffizienten $\Delta N_0^{\sigma\tau,v}$ sind die Reihen

$$\boxed{\Delta N_0^{\sigma\tau,v} = \sum_{\pi=0}^{\infty} \Delta N_0^{[\pi]\sigma\tau,v} \cdot W^{\pi}.} \tag{VII.8*}$$

Bei der Hauptfarbe (e) ist $\Delta N_0^{\sigma\tau,v} = \Delta N_0^{[0]\sigma\tau,v}$.

Aus den $(\psi+1)(\psi+5)/4$ reduziblen Bildfehler-Koeffizienten $\Delta N_0^{\sigma\tau,v}$ der Ordnung ψ kann man die $(\psi+1)(\psi+7)/8$ irreduziblen Bildfehler-Koeffizienten $\Delta\mathscr{E}^{\alpha\beta\gamma}$ der gleichen Ordnung berechnen, oder auch die $(\psi+1)(\psi+7)/8$ irreduziblen Bildfehler-Koeffizienten $\Delta\mathscr{E}_{\varrho}^{\alpha\gamma\varepsilon}$ derselben Ordnung:

$$\begin{aligned}
&\Delta\mathscr{E}^{\alpha\beta\gamma} = \sum_{\varkappa=0}^{\leq\beta/2} (-1)^{\varkappa} \binom{\gamma+\varkappa}{\varkappa} \Delta\mathscr{E}^{(2\alpha+\beta)(\beta-2\varkappa),(2\varkappa+2\gamma)};\\
&\Delta\mathscr{E}_{\varrho}^{\alpha\gamma\varepsilon} = \sum_{\varkappa=0}^{\varepsilon} (-1)^{\varkappa} \binom{\gamma+\varkappa}{\varkappa} \Delta\mathscr{E}^{(\varrho+2\alpha+2\varepsilon)(\varrho+2\gamma+2\varkappa),(2\varepsilon-2\varkappa)}\\
&\text{mit}\quad \Delta\mathscr{E}^{\sigma\tau,v} = \begin{cases} -\frac{1}{\tau}\,\Delta N_0^{\sigma(\tau-1),v} & (\tau \neq 0),\\ -\frac{1}{v}\,\Delta N_0^{\sigma\tau,(v-1)} & (v \neq 0).\end{cases}
\end{aligned} \tag{VII.9}$$

VIII. Strenge Darstellung der meridionalen, sagittalen und tangentialen Bildhöhenaberration als Funktion der Strahlanfangsdaten und der Langeschen Konstruktionsdaten; Näherungsdarstellungen[1].

Dieser Abschnitt schließt inhaltlich an Abschnitt V an[2]. Ausgehend von dem Iterations-Formelsatz (V), werden wir die meridionale, die sagittale und die tangentiale Bildhöhenaberration explizit als Funktion der Strahlanfangsdaten und der Langeschen Konstruktionsdaten darstellen. Mit Hilfe linearer Operatoren läßt sich in Strenge die meridionale Aberration als Kettenprodukt und die sagittale und tangentiale Aberration als Kettenbruch darstellen, wobei die einzelnen Elemente des Kettenprodukts oder Kettenbruchs die Aberrationsbeiträge der einzelnen Flächen des optischen Systems sind.

48. Reduzierte Pupillenlote. Zunächst führen wir, in jedem der Medien des optischen Systems, neben dem dortigen reduzierten (Zwischen-)Bildlot U auch ein reduziertes (Zwischen-)Pupillenlot V ein, und zwar für windschiefe Strahlen (U und V komplex). In Ziff. 49 werden wir dann für Meridianstrahlen die Flächenanteile $(U'_\tau - U_\tau)$ und $(V'_\tau - V_\tau)$ der Bildlot-Aberration und der Pupillenlot-Aberration als Funktionen $2\varphi_\tau$ und $2\psi_\tau$ der mittleren Lote $U_\tau^{\bullet}$ und $V_\tau^{\bullet}$ bestimmen; wir werden die Bildlotaberration vor der Fläche Nr. τ als Funktion von U_τ und V_τ, und die Bildlotaberration hinter der Fläche Nr. τ als Funktion von U'_τ und V'_τ betrachten und nach dem Zusammenhang zwischen diesen beiden Funktionen fragen; es wird sich zeigen: zwischen diesen beiden Aberrationsfunktionen besteht, mit linearen Operatoren geschrieben, ein inhomogen linearer Zusammenhang, vermittelt durch die am Anfang der Ziff. 49 bestimmten Funktionen φ_τ und ψ_τ.

[1] Abschnitt VIII wurde im Februar 1967 abgeschlossen und angefügt.

[2] Auch die Voraussetzungen und Benennungen sind die gleichen wie dort, und die Bezeichnungen außer den Buchstaben $u, v, \varepsilon, \iota, \sigma, \varsigma, \varphi$ und T, die wir hier mit anderer Bedeutung als dort einführen werden.

In Ziff. 25 hatten wir die im reellen oder virtuellen Bildraum gelegene „Austrittspupille" eingeführt, deren Achsenpunkt in Gaußscher Näherung das Bild des Achsenpunkts der Eintrittspupille ist. Und in Ziff. 26 hatten wir das reduzierte Austrittspupillen-Scheitellot $P_{k+1} \equiv c_{k+1} Q_{k+1}$ eingeführt, das in Gaußscher Näherung gleich dem Eintrittspupillen-Scheitellot Q_0 ($=P_0$) ist. Ganz entsprechend führen wir jetzt auch in den Medien zwischen Objektraum und Bildraum „Zwischenpupillen" ein, deren Achsenpunkte in Gaußscher Näherung Bilder des Achsenpunkts der Eintrittspupille sind. Im Medium zwischen der Fläche Nr. τ und der Fläche Nr. $\tau+1$ (oder in der virtuellen Fortsetzung dieses Mediums) sei $Z'_\tau \equiv Z_{\tau+1}$ der Achsenpunkt der dortigen Zwischenpupille. Dann ist nach Ziff. 14 das komplexe Lot $(Q_{(Z'_\tau)})'_\tau = Q'_\tau - (S_\tau Z'_\tau) G'_\tau$ oder $(Q_{(Z_{\tau+1})})_{\tau+1} = Q_{\tau+1} - (S_{\tau+1} Z_{\tau+1}) G_{\tau+1}$. Da (vgl. Ziff. 25) die Zwischenpupille wie die Austrittspupille den Seidelschen Abstand 0 von der Eintrittspupille hat, ist [vgl. (4.31) und (3.32a)] $-t_\tau = \frac{(S_\tau Z'_\tau)/m'_\tau}{\mathring{q}_\tau \mathring{q}_{(Z'_\tau)}}$; $\mathring{q}_{(Z'_\tau)} = \mathring{q}_\tau - (S_\tau Z'_\tau)\, \mathring{g}'_\tau$, also $(S_\tau Z'_\tau) = -t_\tau\, \mathring{q}_\tau m'_\tau \cdot \mathring{q}_{(Z'_\tau)}$; $\mathring{q}_{(Z'_\tau)} = \mathring{q}_\tau / (1 - t_\tau\, \mathring{q}_\tau \times m'_\tau\, \mathring{g}'_\tau)$. Da nun $Q'/\mathring{q} \equiv P'$ und [s. (3.46)] $\mathring{q} m' G' - m' \mathring{g}' Q' = U'$ ist, erhalten wir für das „reduzierte Pupillenlot"

$$V'_\tau \equiv (Q_{(Z'_\tau)})'_\tau / \mathring{q}_{(Z'_\tau)} \tag{8.1}$$

den Ausdruck $V'_\tau = (1 - t_\tau\, \mathring{q}_\tau\, m'_\tau\, \mathring{g}'_\tau)\, P'_\tau + t_\tau (U'_\tau + m'_\tau\, \mathring{g}'_\tau\, Q'_\tau)$, also

$$V'_\tau = P'_\tau + t_\tau\, U'_\tau. \tag{8.2}$$

Dieses reduzierte Zwischen-Pupillenlot ist in Gaußscher Näherung gleich dem Eintritts-Pupillenlot $P_0 = Q_0$:

$$(V'_\tau)_{\mathfrak{G}} = P_0, \tag{8.3}$$

denn nach (4.32) ist $(P'_\tau)_{\mathfrak{G}} \equiv q'_\tau / \mathring{q}_\tau = Q_0 - t_\tau U_0$, und nach (3.44) ist $(U'_\tau)_{\mathfrak{G}} = U_0$. — Mit $V'_\tau \equiv (Q_{(Z'_\tau)})'_\tau / \mathring{q}_{(Z'_\tau)}$ ist

$$V_{\tau+1} \equiv (Q_{(Z_{\tau+1})})_{\tau+1} / \mathring{q}_{(Z_{\tau+1})} \tag{8.4}$$

identisch:

$$V'_\tau = V_{\tau+1}. \tag{8.5}$$

Entsprechend wie wir für V'_τ den Ausdruck (8.2) abgeleitet haben, könnten wir für $V_{\tau+1}$ auch den Ausdruck

$$V_{\tau+1} = P_{\tau+1} + t_{\tau+1}\, U_{\tau+1} \tag{8.6}$$

ableiten; dieser Ausdruck folgt aber auch aus (8.5), (8.2) und der Beziehung (4.28a), nach der

$$P_{\tau+1} - P'_\tau = -(t_{\tau+1} - t_\tau)\, U'_\tau \tag{8.7}$$

ist.

Die im vorigen Absatz für jedes der Medien zwischen Objektraum und Bildraum eingeführten Definitionen und Rechnungen gelten freilich auch für den Objektraum und für den Bildraum. Die Gln. (8.1) bis (8.7) gelten dann auch für $\tau=0$ und für $\tau=k$. Aus den Definitionen oder formal aus (8.2) und

$$t_0 = 0 \tag{8.8}$$

folgt, daß V'_0 (wofür wir auch V_0 schreiben) und P_0 identisch sind:

$$V'_0 \equiv V_0 = P_0. \tag{8.9}$$

Entsprechend folgt aus den Definitionen, oder formal aus (8.6) und

$$t_{k+1} = 0, \tag{8.10}$$

daß V_{k+1} und P_{k+1} identisch sind:

$$V_{k+1} = P_{k+1}. \tag{8.11}$$

Aus (8.2) und (8.6) folgt, daß bei jeder Fläche

$$(V' - V) = (P' - P) + t(U' - U) \tag{8.12}$$

und

$$V^{\cdot} = P^{\cdot} + t U^{\cdot} \tag{8.13}$$

ist.

49. Differenz-Operatoren; inhomogen linearer Zusammenhang zwischen den meridionalen Aberrationsfunktionen vor und hinter einer Fläche. Um die Flächenanteile $(U' - U)$ und $(V' - V)$ der meridionalen Bildlot- und Pupillenlot-Aberration durch $U^{\cdot}$ und $V^{\cdot}$ auszudrücken, müssen wir erstens die in den Ausdrücken (V.4a) für $(U' - U)$ und $(P' - P)$ stehenden Variablen $P^{\cdot}$, $K^{\cdot}$; $J^{\cdot}$, $G^{\cdot}$; $\bar{J}^{\cdot}$, $\bar{G}^{\cdot}$ durch $U^{\cdot}$ und $V^{\cdot}$ ausdrücken. Zweitens müssen wir den Ausdruck für $(P' - P)$ ersetzen durch einen für $(V' - V)$.

Zunächst ist nach (8.13) unmittelbar

$$P^{\cdot} = V^{\cdot} - t U^{\cdot}. \tag{8.14}$$

Hiermit folgt aus (V.3d):

$$K^{\cdot} = \frac{\bar{\nu}^{\cdot}}{\nu^{\cdot}} J^{\cdot} = \mathring{\varkappa} V^{\cdot} + \check{\varkappa} U^{\cdot} \quad \text{mit} \quad \check{\varkappa} = -c - t\mathring{\varkappa}; \quad \mathring{\varkappa} = -\frac{\nu_e' c^{+} - \nu_e c^{-}}{\nu_e' - \nu_e}; \tag{8.15}$$

$$\frac{\bar{\nu}^{\cdot}}{\nu^{\cdot}} G^{\cdot} = \breve{\chi} V^{\cdot} + \hat{\chi} U^{\cdot} \quad \text{mit} \quad \hat{\chi} = +c - t\breve{\chi}; \quad \breve{\chi} = +\frac{\nu' c^{+} + \nu c^{-}}{\nu + \nu'}. \tag{8.16}$$

Und nach (V.3e) ist

$$\bar{J}^{\cdot} = \sqrt{L - J^{\cdot 2}} \quad \text{mit } L \text{ nach (8.19) und } J^{\cdot} \text{ nach (8.15)}; \tag{8.17}$$

$$\bar{G}^{\cdot} = \sqrt{L - G^{\cdot 2}} \quad \text{mit } L \text{ nach (8.19) und } G^{\cdot} \text{ nach (8.16)}; \tag{8.18}$$

$$\left.\begin{aligned} &L = 1 - \left(\frac{\nu' - \nu}{\nu + \nu'}\right)^2 J^{\cdot 2}/(1 - J^{\cdot 2} - \epsilon), \quad \text{wobei} \\ &\epsilon \approx \left(\frac{\nu' - \nu}{\nu + \nu'}\right)^2 J^{\cdot 4}/(1 - J^{\cdot 2}) \quad \text{klein gegen } J^{\cdot 2} \text{ ist;} \\ &\text{in Strenge ist } \epsilon = \frac{1}{2}\left[E - \sqrt{E^2 - 4\left(\frac{\nu' - \nu}{\nu + \nu'}\right)^2 J^{\cdot 4}}\,\right] \\ &\text{mit } E = 1 - J^{\cdot 2} - \left(\frac{\nu' - \nu}{\nu + \nu'}\right)^2 J^{\cdot 2}. \end{aligned}\right\} \tag{8.19}$$

In dem Ausdruck für $(U' - U)$, der schon in (V.4a) steht, und in dem Ausdruck für $(V' - V)$, den wir nach (8.12) sofort aus den beiden Ausdrücken (V.4a) für $(U' - U)$ und $(P' - P)$ erhalten:

$$(U' - U) = +\frac{\mathring{\varkappa}}{c} \frac{\nu' - \nu}{\bar{\nu}^{\cdot}} \frac{P^{\cdot}}{\bar{J}^{\cdot}} \frac{G^{\cdot} - J^{\cdot}}{\bar{G}^{\cdot} + \bar{J}^{\cdot}} K^{\cdot} - \frac{1}{c} \frac{\bar{\nu}' - \bar{\nu}}{\bar{\nu}^{\cdot}} K^{\cdot}; \tag{8.20a}$$

$$(V' - V) = -\frac{\check{\varkappa}}{c} \frac{\nu' - \nu}{\bar{\nu}^{\cdot}} \frac{P^{\cdot}}{\bar{J}^{\cdot}} \frac{G^{\cdot} - J^{\cdot}}{\bar{G}^{\cdot} + \bar{J}^{\cdot}} K^{\cdot} - \frac{t}{c} \frac{\bar{\nu}' - \bar{\nu}}{\bar{\nu}^{\cdot}} K^{\cdot} \tag{8.20b}$$

sind für $P^{\cdot}$, $K^{\cdot}$; $J^{\cdot}$, $G^{\cdot}$; $\bar{J}^{\cdot}$, $\bar{G}^{\cdot}$ die Ausdrücke (8.14) bis (8.19) eingesetzt zu denken. So sind $(U' - U)$ und $(V' - V)$ Funktionen von $U^{\cdot}$ und $V^{\cdot}$:

$$\tfrac{1}{2}(U_\tau' - U_\tau) = \varphi_\tau(U_\tau^{\cdot}; V_\tau^{\cdot}); \tag{8.21a}$$

$$\tfrac{1}{2}(V_\tau' - V_\tau) = \psi_\tau(U_\tau^{\cdot}; V_\tau^{\cdot}). \tag{8.21b}$$

Wir betrachten nun die *zwei*parametrige Schar der aus den verschiedenen Objektpunkten kommenden und mit verschiedenem Eintrittspupillenlot ins optische System einfallenden *Meridianstrahlen.* Als Parameter der Schar können wir das (reelle) Objektlot U_0 und das (reelle) Eintrittspupillenlot V_0 wählen oder auch — wo das zweckmäßiger ist — das Zwischenbildlot U_τ und das Zwischenpupillenlot V_τ vor irgendeiner Fläche Nr. τ.

Die Aberration

$$\Delta U_\tau \equiv U_\tau - U_{(0)} \tag{8.22a}$$

des Bildlots U_τ eines Strahls vor der Fläche Nr. τ gegenüber dem Gaußschen Näherungswert U_0:

$$U_{(0)} = U_0, \tag{8.22b*}$$

oder gegenüber einem *anderen* Näherungswert $U_{(0)}$, ist (i. allg.) von Strahl zu Strahl verschieden, d.h. ΔU_τ ist eine Funktion der beiden Strahlparameter. Wir interessieren uns für die Aberration ΔU_τ im Medium Nr. τ als *Funktion* der im gleichen Medium sichtbaren Parameter U_τ und V_τ:

$$\Delta U_\tau = F_\tau(U_\tau, V_\tau). \tag{8.23}$$

Und wir fragen zunächst danach, wie man aus der Aberrations*funktion* $F_\tau(u, v)$ vor der Fläche Nr. τ die Aberrations*funktion* $F_{\tau+1}(u, v)$ hinter der Fläche Nr. τ bestimmen kann[1]. — Für $F_{\tau+1}$ können wir auch F'_τ schreiben:

$$F_{\tau+1}(u, v) \equiv F'_\tau(u, v), \tag{8.24}$$

da $U_{\tau+1}$ mit U'_τ, $V_{\tau+1}$ mit V'_τ, und $\Delta U_{\tau+1} \equiv U_{\tau+1} - U_{(0)}$ mit der Aberration

$$\Delta U'_\tau \equiv U'_\tau - U_{(0)} \tag{8.25}$$

identisch ist, die wir als eine Funktion F'_τ von U'_τ und V'_τ ansetzen:

$$\Delta U'_\tau = F'_\tau(U'_\tau, V'_\tau). \tag{8.26}$$

Nach (8.26), (8.23) und (8.21a) ist wegen $(\Delta U'_\tau - \Delta U_\tau) = (U'_\tau - U_\tau)$:

$$F'_\tau(U'_\tau, V'_\tau) - F_\tau(U_\tau, V_\tau) = 2\varphi_\tau(U^\bullet_\tau; V^\bullet_\tau), \tag{8.27}$$

woraus wegen (8.21a) und (8.21b) weiter folgt:

$$\left.\begin{aligned} &F'_\tau\left(U^\bullet_\tau + \varphi_\tau(U^\bullet_\tau; V^\bullet_\tau),\ V^\bullet_\tau + \psi_\tau(U^\bullet_\tau; V^\bullet_\tau)\right) - \\ &- F_\tau\left(U^\bullet_\tau - \varphi_\tau(U^\bullet_\tau; V^\bullet_\tau),\ V^\bullet_\tau - \psi_\tau(U^\bullet_\tau; V^\bullet_\tau)\right) = 2\varphi_\tau(U^\bullet_\tau; V^\bullet_\tau). \end{aligned}\right\} \tag{8.28}$$

Zwischen den Funktionen F_τ und F'_τ besteht also die Beziehung:

$$F'_\tau\left(u + \varphi_\tau(u; v), v + \psi_\tau(u; v)\right) - F_\tau\left(u - \varphi_\tau(u; v), v - \psi_\tau(u; v)\right) = 2\varphi_\tau(u; v). \tag{8.29}$$

Wenn wir im folgenden für $\varphi_\tau(u; v)$ und $\psi_\tau(u; v)$ einfacher φ und ψ schreiben, lassen wir aber nicht außer acht, daß φ und ψ Funktionen von $(u; v)$ und keine Konstanten sind. Da φ und ψ nicht konstant sind, kann man für $F'_\tau(u + \varphi, v + \psi)$ *nicht* $e^{\varphi\frac{\partial}{\partial u} + \psi\frac{\partial}{\partial v}} F'_\tau(u, v)$ schreiben; bekanntlich führt bei *konstanten* Δu und

[1] Für die Argumente der von nun an betrachteten *Funktionen* schreiben wir u und v, und wir identifizieren u und v je nach den Umständen mit den einen oder anderen Strahldaten. In der alten Bedeutung werden die Buchstaben u und v nicht mehr vorkommen (in Abschnitt III war u das nicht-reduzierte Bildlot; bis Abschnitt VII war v die Flächennummer).

Δv der Verschiebungs-Operator

$$\mathrm{e}^{\left(\Delta u \frac{\partial}{\partial u} + \Delta v \frac{\partial}{\partial v}\right)} \equiv \sum_{N=0}^{\infty} \frac{1}{N!} \left(\Delta u \frac{\partial}{\partial u} + \Delta v \frac{\partial}{\partial v}\right)^N \tag{8.30}$$

eine Funktion $f(u, v)$ in die Funktion $f(u+\Delta u, v+\Delta v)$ über:

$$\mathrm{e}^{\left(\Delta u \frac{\partial}{\partial u} + \Delta v \frac{\partial}{\partial v}\right)} f(u, v) = f(u+\Delta u, v+\Delta v). \tag{8.31}$$

Wir bezeichnen den Operator, der eine Funktion $F(u, v)$ überführt in die Funktion $F\big(u+\varphi(u; v),\ v+\psi(u; v)\big)$, wobei φ und ψ vorgegebene Funktionen von u und v sind, zunächst mit $[\varphi, \psi]$:

$$[\varphi, \psi] F(u, v) \equiv F\big(u+\varphi(u; v), v+\psi(u; v)\big). \tag{8.32}$$

Dieser Operator $[\varphi, \psi]$ ist linear: führt er $F_{(1)}(u, v)$ in $\Phi_{(1)}(u, v)$ und $F_{(2)}(u, v)$ in $\Phi_{(2)}(u, v)$ über, dann führt er $(\text{const} \cdot F_{(1)} + \text{konst} \cdot F_{(2)})$ in $(\text{const} \cdot \Phi_{(1)} + \text{konst} \cdot \Phi_{(2)})$ über.

Für (8.29) können wir jetzt schreiben:

$$[+\varphi, +\psi] F'(u, v) - [-\varphi, -\psi] F(u, v) = 2\varphi. \tag{8.29a}$$

Zwischen den Aberrationsfunktionen $F_\tau(u, v)$ und $F'_\tau(u, v)$ besteht also, mit linearen Operatoren geschrieben, ein *inhomogen linearer* Zusammenhang. φ und ψ sind kleine Größen dritter Ordnung und ersten Grades, und $[\varphi, \psi] F'(u, v) \equiv F'(u+\varphi, v+\psi)$ ist näherungsweise gleich $F'(u, v) + \varphi \frac{\partial}{\partial u} F'(u, v) + \psi \frac{\partial}{\partial v} F'(u, v)$:

$$[\varphi, \psi] \approx 1 + \varphi \frac{\partial}{\partial u} + \psi \frac{\partial}{\partial v}. \tag{8.33}$$

$[\varphi, \psi]$ ist *exakt* gleich 1 plus einem Operator, den wir vorübergehend mit ω bezeichnen wollen:

$$\omega \equiv [\varphi, \psi] - 1; \qquad \omega F'(u, v) = F'(u+\varphi, v+\psi) - F'(u, v). \tag{8.34}$$

Wir wollen für diesen Operator ω eine einprägsame, seiner Wirkung entsprechende Schreibweise einführen. In Anlehnung an die gelegentlich vorkommende Vektor-Schreibweise

$$\frac{\partial}{\partial \mathfrak{r}} f(\mathfrak{r}) \quad \text{für} \quad \operatorname{grad} f(\mathfrak{r}),$$

also

$$\mathfrak{a} \frac{\partial}{\partial \mathfrak{r}} \quad \text{für} \quad a_x \frac{\partial}{\partial x} + a_y \frac{\partial}{\partial y} + a_z \frac{\partial}{\partial z},$$

wobei

$$\mathfrak{a} \equiv \{a_x, a_y, a_z\} \quad \text{und} \quad \mathfrak{r} \equiv \{x, y, z\}$$

ist, können wir

$$\frac{(\varphi, \psi)\,\partial}{\partial(u, v)} \text{ für } \quad \varphi \frac{\partial}{\partial u} + \psi \frac{\partial}{\partial v}$$

schreiben, und damit für (8.33) auch

$$[\varphi, \psi] \approx 1 + \frac{(\varphi, \psi)\,\partial}{\partial(u, v)} \tag{8.33'}$$

schreiben. Und in Anlehnung hieran setzen wir

$$[\varphi, \psi] \equiv 1 + \frac{(\varphi, \psi)\,\partial^+}{\partial^+(u, v)}, \tag{8.35}$$

schreiben also $(\varphi,\psi)\,\partial^+/\partial^+(u,v)$ für den Operator ω, der die Differenz $F'(u+\varphi, v+\psi)-F'(u,v)$ bildet, wenn man ihn auf $F'(u,v)$ wirken läßt:

$$\frac{(\varphi,\psi)\,\partial^+}{\partial^+(u,v)}F'(u,v)=F'\big(u+\varphi(u;v),\,v+\psi(u;v)\big)-F'(u,v). \tag{8.36}$$

Nach (8.35) und (8.33′) ist

$$\frac{(\varphi,\psi)\,\partial^+}{\partial^+(u,v)}\approx\frac{(\varphi,\psi)\,\partial}{\partial(u,v)}. \tag{8.33a}$$

Wir können (8.33a) vervollständigen und *exakt* den linearen Operator $(\varphi,\psi)\,\partial^+/\partial^+(u,v)$ durch die Multiplikatoren $\varphi(u;v)$ und $\psi(u;v)$ und die Differentialoperatoren $\partial/\partial u$ und $\partial/\partial v$ ausdrücken:

$$\frac{(\varphi,\psi)\,\partial^+}{\partial^+(u,v)}=\sum_{N=1}^{\infty}\frac{1}{N!}\sum_{s=0}^{N}\binom{N}{s}\varphi^{N-s}\psi^s\left(\frac{\partial}{\partial u}\right)^{N-s}\left(\frac{\partial}{\partial v}\right)^s. \tag{8.37}$$

Das erste Glied der Reihe $\sum_{N=1}^{\infty}$ auf der rechten Seite dieser Gleichung ist identisch mit der rechten Seite der Näherung (8.33a). Die Gl. (8.37) folgt aus der Taylor-Entwicklung von $F'(u+\varphi, v+\psi)$ nach φ und ψ:

$$\begin{aligned}\left[1+\frac{(\varphi,\psi)\,\partial^+}{\partial^+(u,v)}\right]F'(u,v)=F'(u+\varphi,v+\psi)&=\sum_{r=0}^{\infty}\sum_{s=0}^{\infty}\frac{1}{r!}\frac{1}{s!}\frac{\partial^{r+s}F'(u,v)}{\partial u^r\,\partial v^s}\varphi^r\psi^s\\&=\sum_{N=0}^{\infty}\frac{1}{N!}\sum_{s=0}^{N}\binom{N}{s}\frac{\partial^N F'(u,v)}{\partial u^{N-s}\,\partial v^s}\varphi^{N-s}\psi^s.\end{aligned}$$

Wir werden sehen, daß man in sehr guter Näherung bloß das erste Glied ($N=1$) der Reihendarstellung (8.37) für $(\varphi,\psi)\,\partial^+/\partial^+(u,v)$ zu nehmen braucht, daß man also, vereinfacht gesagt, in sehr guter Näherung ∂/∂ für ∂^+/∂^+ nehmen kann.

Will man trotzdem in aller Strenge Ausdrücke, in denen der lineare Operator $(\varphi,\psi)\,\partial^+/\partial^+(u,v)$ vorkommt, analytisch diskutieren, dann braucht man ebenfalls nicht die komplizierte Reihe (8.37) heranzuziehen. Sondern dann kann man sich diesen linearen Operator durch (8.36) definiert denken, wonach seine Wirkung auf eine Funktion $F'(u,v)$ einer variablen Differenz-Bildung gleichkommt.

Neben dem „oberen Differenz-Operator" $(\varphi,\psi)\,\partial^+/\partial^+(u,v)$ führen wir auch einen „unteren Differenz-Operator" $(\varphi,\psi)\,\partial^-/\partial^-(u,v)$ ein:

$$\left.\begin{aligned}\frac{(\varphi,\psi)\,\partial^+}{\partial^+(u,v)}F'(u,v)&\equiv F'\big(u+\varphi(u;v),\,v+\psi(u;v)\big)-F'(u,v);\\\frac{(\varphi,\psi)\,\partial^-}{\partial^-(u,v)}F(u,v)&\equiv F(u,v)-F\big(u-\varphi(u;v),\,v-\psi(u;v)\big).\end{aligned}\right\} \tag{8.36a}$$

Man sieht, daß

$$+\frac{(+\varphi,+\psi)\,\partial^+}{\partial^+(u,v)}\equiv-\frac{(-\varphi,-\psi)\,\partial^-}{\partial^-(u,v)} \tag{8.38}$$

ist, und daß mit

$$\frac{(\varphi,\psi)\,\partial^+}{\partial^+(u,v)}\approx\frac{(\varphi,\psi)\,\partial}{\partial(u,v)} \tag{8.33a}$$

auch

$$\frac{(\varphi,\psi)\,\partial^-}{\partial^-(u,v)}\approx\frac{(\varphi,\psi)\,\partial}{\partial(u,v)} \tag{8.33b}$$

ist.

Für die Gl. (8.29) oder (8.29a), die $F'(u,v)$ mit $F(u,v)$ verknüpft, können wir nunmehr schreiben:

$$\left[1+\frac{(\varphi,\psi)\,\partial^+}{\partial^+(u,v)}\right]F'(u,v)-\left[1-\frac{(\varphi,\psi)\,\partial^-}{\partial^-(u,v)}\right]F(u,v)=2\varphi. \tag{8.39}$$

Diese Gleichung können wir nach $F'(u,v)$ auflösen:

$$F'_\tau(u,v)=\left[1+\frac{(\varphi_\tau,\psi_\tau)\,\partial^+}{\partial^+(u,v)}\right]^{-1}\left\langle 2\varphi_\tau+\left[1-\frac{(\varphi_\tau,\psi_\tau)\,\partial^-}{\partial^-(u,v)}\right]F_\tau(u,v)\right\rangle. \tag{8.39a}$$

Das Inverse $\left[1+\frac{(\varphi_\tau,\psi_\tau)\,\partial^+}{\partial^+(u,v)}\right]^{-1}$ lassen wir in den Formeln so stehen. Das erleichtert analytische Diskussionen und Umformungen eher, als daß es sie erschwert. Wir wollen aber feststellen, daß

$$\left[1+\frac{(\varphi_\tau,\psi_\tau)\,\partial^+}{\partial^+(u,v)}\right]^{-1}=\left[1-\frac{(\varphi'_\tau,\psi'_\tau)\,\partial^-}{\partial^-(u,v)}\right] \tag{8.39a'}$$

ist, wobei die Funktionen $\varphi'_\tau=\varphi'_\tau(u,v)$ und $\psi'_\tau=\psi'_\tau(u,v)$ durch eine Umkehrung auf die Funktionen $\varphi_\tau(u,v)$ und $\psi_\tau(u,v)$ zurückgeführt werden können, nämlich durch Auflösung der folgenden oberen zwei Gleichungen (8.39a″) nach u und v, und durch Subtraktion dieser Umkehrung $u=u[u';v']$ bzw. $v=v[u';v']$ von u' bzw. v':

$$\left.\begin{aligned} u+\varphi_\tau(u;v)&=u'; & v+\psi_\tau(u;v)&=v'\\ u'-\varphi'_\tau(u';v')&=u; & v'-\psi'_\tau(u';v')&=v.\end{aligned}\right\} \tag{8.39a''}$$

Bei der durch (8.14) bis (8.21) gegebenen Struktur der Funktionen φ_τ und ψ_τ ist die Auflösung der oberen zwei Gleichungen (8.39a″) nach u und v leicht durchzuführen, und wir könnten die Funktionen φ'_τ und ψ'_τ explizit hinschreiben; wir tun das aber nicht, weil wir diese Darstellung von φ'_τ und ψ'_τ nicht benötigen. — Die Gl. (8.39a′), wonach aus

$$\left[1+\frac{(\varphi,\psi)\,\partial^+}{\partial^+(u,v)}\right]F'(u,v)=\Phi(u,v) \tag{8.39A}$$

folgt, daß

$$F'(u,v)=\left[1-\frac{(\varphi',\psi')\,\partial^-}{\partial^-(u,v)}\right]\Phi(u,v) \tag{8.39B}$$

ist mit den in (8.39a″) definierten Funktionen φ' und ψ', ist noch zu beweisen: Nach (8.39A) ist

$$F'\big(u+\varphi(u;v),\,v+\psi(u;v)\big)=\Phi(u,v).$$

Hieraus wird, mit den in (8.39a″) eingeführten Variablen u' und v' anstelle von u und v,

$$F'(u',v')=\Phi\big(u'-\varphi'(u';v'),\,v'-\psi'(u';v')\big)$$

oder, indem wir für u' und v' jetzt u und v schreiben,

$$F'(u,v)=\Phi\big(u-\varphi'(u;v),\,v-\psi'(u;v)\big).$$

Dies besagt auch die Gl. (8.39B), die also aus (8.39A) folgt, w.z.b.w.

Denken wir uns nun die Gl. (8.39a) für $\tau=k;\ k-1;\ \ldots;\ 2;\ 1$ hingeschrieben und $F_k\equiv F'_{k-1};\ F_{k-1}\equiv F'_{k-2};\ \ldots;\ F_3\equiv F'_2;\ F_2\equiv F'_1$ daraus eliminiert, dann haben wir den Zusammenhang zwischen F'_k und F_1 da stehen:

$$F'_k(u,v)={}^{\tau=k}\left\|\left[1+\frac{(\varphi_\tau,\psi_\tau)\,\partial^+}{\partial^+(u,v)}\right]^{-1}\left\langle 2\varphi_\tau+\left[1-\frac{(\varphi_\tau,\psi_\tau)\,\partial^-}{\partial^-(u,v)}\right]\right\|_{\tau=1}F_1(u,v). \tag{8.40}$$

Hierbei ist mit ${}^{\tau=k}|\ldots|_{\tau=1}$ gemeint, daß das zwischen den vertikalen Strichen Stehende erst mit k für τ hinzuschreiben ist, dahinter mit $k-1$ für τ, dahinter mit $k-2$ für τ, und so fort, schließlich mit 1 für τ. Dahinter steht dann $F_1(u, v)$. Hiernach müßten noch alle vorher aufgemachten und noch *nicht* geschlossenen Klammern geschlossen werden, aber es führt zu keinen Mißverständnissen und ist übersichtlicher, wenn man sich das erspart. Beim Ausschreiben der Gl. (8.40) z.B. sind hinter $F_1(u, v)$ noch k spitze Klammern $\rangle$ zu denken. Im Zweifelsfall kann dort, wo die noch nicht geschlossenen Klammern alle geschlossen zu denken sind, ein Schlußstrich $\|$ geschrieben werden, wie wir es z.B. am Ende des folgenden Absatzes tun werden.

Mit ${}_{\tau=1}|\ldots|^{\tau=k}$ ist gemeint, daß das zwischen den vertikalen Strichen Stehende erst mit 1 für τ, dahinter mit 2 für τ, dahinter mit 3 für τ hinzuzuschreiben ist, und so fort, schließlich mit k für τ. Summe $\sum(\ldots)$ und Produkt $\prod(\ldots)$ z.B. stellen sich in dieser Wiederholungsschreibweise ${}_{\tau=1}|\ldots|^{\tau=k}$ so dar:

$$b+\sum_{\tau=1}^{k} a_\tau = b\ {}_{\tau=1}| + a_\tau|^{\tau=k} = b+a_1+a_2+a_3+\cdots+a_k$$

oder kürzer geschrieben:

$$b+\sum_1^k a = b\ {}_1| + a|^k = b+a_1+a_2+a_3+\cdots+a_k,$$

und

$$B\cdot\prod_{\tau=1}^{k} A_\tau = B\ {}_{\tau=1}|\cdot A_\tau|^{\tau=k} = B\cdot A_1\cdot A_2\cdot A_3\cdot\ldots\cdot A_k$$

oder kürzer geschrieben:

$$B\cdot\prod_1^k A = B\ {}_1|\cdot A|^k = B\cdot A_1\cdot A_2\cdot A_3\cdot\ldots\cdot A_k.$$

Die Bezeichnungen ${}_{\tau=k}|\ldots|^{\tau=1}$ und ${}^{\tau=1}|\ldots|_{\tau=k}$ sind reserviert für Umkehrungen. Während $b\ {}_1| + a|^k$ und

$$b\ {}^k| + a|_1 = b+a_k+a_{k-1}+a_{k-2}+\cdots+a_1$$

einander gleich sind, und auch $B\ {}_1|\cdot A|^k$ und

$$B\ {}^k|\cdot A|_1 = B\cdot A_k\cdot A_{k-1}\cdot A_{k-2}\cdot\ldots\cdot A_1,$$

bedeuten ${}_1|\ldots|^k$ und ${}^k|\ldots|_1$ bei abwechselndem Multiplizieren und Addieren wie z.B. bei (8.40) etwas Verschiedenes:

$${}^k|A'\langle a+A|_1 F\| + b = A'_k\langle a_k + A_k A'_{k-1}\langle a_{k-1}+A_{k-1}\ldots A'_1\langle a_1+A_1F\rangle\ldots\rangle\rangle + b;$$
$${}_1|A'\langle a+A|^k F\| + b = A'_1\langle a_1 + A_1 A'_2\langle a_2+A_2\ldots A'_k\langle a_k+A_kF\rangle\ldots\rangle\rangle + b.$$

Bei der Wahl (8.22b*) für $U_{(0)}$ ist nun $F_1(u, v)$ bekannt:

$$F_1(U_1, V_1) = \varDelta U_1 \equiv U_1 - U_{(0)} = U_1 - U_0 = 0, \tag{8.41*}$$

und dann wird aus (8.40):

$$F'_k(u, v) = {}^{\tau=k}\left|\left[1+\frac{(\varphi_\tau, \psi_\tau)\,\partial^+}{\partial^+(u, v)}\right]^{-1}\left\langle 2\varphi_\tau + \left[1-\frac{(\varphi_\tau, \psi_\tau)\,\partial^-}{\partial^-(u, v)}\right]\right|\right|_{\tau=1} 0, \tag{8.42*}$$

worin $\varphi_\tau = \varphi_\tau(u; v)$ und $\psi_\tau = \psi_\tau(u; v)$ bekannt sind [s. oben die Gln. (8.14) bis (8.21b), durch die $\varphi_\tau(u; v)$ und $\psi_\tau(u; v)$ bestimmt sind, wenn man dort u und v

für $U^{\cdot}$ und $V^{\cdot}$ schreibt, und alle übrigen Größen mit dem Flächenindex τ versieht]. Außer u und v gehen in $\varphi_\tau(u;v)$ und $\psi_\tau(u;v)$ nur Langesche Konstruktionsdaten ν, t, c aus der nächsten Umgebung der Fläche Nr. τ ein [s. (8.14) bis (8.21 b); hierin ist $\bar{\nu}\equiv\nu/\nu_e$ und $c_\tau^- \equiv \frac{c_\tau - c_{\tau-1}}{t_\tau - t_{\tau-1}}$, $c_\tau^+ \equiv \frac{c_{\tau+1} - c_\tau}{t_{\tau+1} - t_\tau}$]. Ähnlich wie bei der Darstellung der Seidelschen Bildfehler in der Langeschen Form hat man bei der Darstellung (8.42*) nacheinander einen Flächenanteil Nr. 1, einen Flächenanteil Nr. 2, einen Flächenanteil Nr. 3, und so fort bis zum Flächenanteil Nr. k, wobei in jeden dieser Flächenanteile jeweils nur Langesche Konstruktionsdaten aus der nächsten Umgebung der zugehörigen Fläche eingehen (neben u und v). Bei SEIDEL-LANGE sind diese Flächenanteile Summanden, hier Wiederholungs-Stücke |...| mit abwechselnder Addition und Operatormultiplikation. — So ist mit (8.42*), worin die bekannten Funktionen φ_τ und ψ_τ eingesetzt zu denken sind, die Bild*lot*-Aberration $(U_{k+1}-U_0)=\Delta U_k'=F_k'$ als Funktion von $u=U_k'$ und $v=V_k'$ und der Langeschen Konstruktionsdaten des optischen Systems dargestellt, und zwar in Strenge und mit Hilfe von linearen Operatoren explizit in der Form eines Kettenprodukts (bei dem man abwechselnd zu addieren und zu multiplizieren hat).

50. Meridionale Bildhöhenaberration. Bei der Darstellung (8.42*) fehlt noch der Übergang von der Bild*lot*-Aberration $(U_{k+1}-U_0)$ zu einer Größe, die die Abweichung von der idealen, Gaußschen Abbildung beschreibt. Auch ist das Bildlot $U_k'\equiv U_{k+1}$ als unabhängige Variable nicht geeignet, weil es nicht wie die Bild*höhe* $\underline{U}_{k+1}$ eine Koordinate für die verschiedenen Punkte der Gaußschen Bildebene (GBE) ist. Wir nehmen aber nicht $\underline{U}_{k+1}$ als unabhängige Variable, sondern

$$U_+ \equiv \frac{\underline{U}_{k+1}}{\sqrt{1+\underline{U}_{k+1}^2}}; \tag{8.44}$$

und wir nehmen nicht $(\underline{U}_{k+1}-\underline{U}_0)$ als Maß für die Abweichung von der idealen Gaußschen Abbildung, sondern $(U_+ - U_0)$, das führt zu einfacheren Formeln.

Während $(\underline{U}_{k+1}-\underline{U}_0)$ angibt, wie im Raum vor der ersten Fläche der *Tangens* des Bildwinkels[1] abweicht von dem Wert, den er bei idealer, Gaußscher Abbildung haben müßte in Anbetracht der Strahlhöhe in der GBE, gibt $(U_+ - U_0)$ an, wie der *Sinus* des Bildwinkels abweicht von dem Wert, den er bei Gaußscher Abbildung haben müßte in Anbetracht der Strahlhöhe in der GBE. Die Sinus-Abweichung $(U_+ - U_0)$ beschreibt die Abweichung von der Gaußschen Abbildung ebenso gut wie die Tangens-Abweichung $(\underline{U}_{k+1}-\underline{U}_0)$. Wir nennen deshalb nicht nur $(\underline{U}_{k+1}-\underline{U}_0)$, sondern auch $(U_+ - U_0)$ eine „meridionale Bildhöhenaberration".

Die Differenz zwischen U_{k+1} und U_+ ist, nach (8.44) und wegen $\underline{U}\equiv U/\overline{G}$,

$$\left.\begin{aligned}(U_+ - U_{k+1}) &= \left(\frac{1}{\sqrt{1-G_{k+1}^2+U_{k+1}^2}}-1\right)U_{k+1}\\ &= \frac{(G_{k+1}^2-U_{k+1}^2)\,U_{k+1}}{\sqrt{1-G_{k+1}^2+U_{k+1}^2}\,\left(1+\sqrt{1-G_{k+1}^2+U_{k+1}^2}\right)};\end{aligned}\right\} \tag{8.45 a}$$

hierin ist

$$G_{k+1} = c_{k+1}^- V_{k+1} + c_{k+1} U_{k+1}. \tag{8.45 b}$$

(8.45 a, b) stellt $(U_+ - U_{k+1})$ als Funktion von U_{k+1} und V_{k+1} dar:

$$(U_+ - U_{k+1}) = \varphi_+(U_{k+1}; V_{k+1}), \tag{8.46}$$

[1] Der „Bildwinkel" ist [in dem nach (4.34) vorliegenden Fall $a_0=\infty$] der Winkel, den vor der ersten Fläche der Strahl mit der optischen Achse bildet.

oder anders geschrieben:

$$(U_+ - U_k') = \varphi_+ (U_k'; V_k'). \tag{8.46'}$$

Betrachten wir, wie oben, $\Delta U_k'$ als Funktion F_k' von U_k' und V_k', und betrachten wir entsprechend ΔU_+ als Funktion F_+ von U_+ und V_+:

$$U_+ - U_{(0)} \equiv \Delta U_+ = F_+ (U_+, V_+), \tag{8.47}$$

wobei wir

$$V_+ \equiv V_{k+1} \tag{8.48}$$

setzen, dann können wir, wegen $(U_+ - U_k') = (\Delta U_+ - \Delta U_k')$, für (8.46') schreiben:

$$F_+ (U_+, V_+) - F_k'(U_k', V_k') = \varphi_+ (U_k'; V_k'), \tag{8.49}$$

woraus wegen (8.46') und (8.48) weiter folgt:

$$F_+ \big(U_k' + \varphi_+ (U_k'; V_k'), V_k'\big) - F_k'(U_k', V_k') = \varphi_+ (U_k'; V_k')$$

oder

$$F_+ \big(u + \varphi_+ (u; v), v\big) - F_k'(u, v) = \varphi_+ (u; v). \tag{8.50}$$

Es ist also, mit der Abkürzung φ_+ für $\varphi_+ (u; v)$,

$$\left[1 + \frac{(\varphi_+, 0)\,\partial^+}{\partial^+ (u, v)}\right] F_+ (u, v) - F_k'(u, v) = \varphi_+, \tag{8.50a}$$

oder nach $F_+ (u, v)$ aufgelöst[1]:

$$F_+ (u, v) = \left[1 + \frac{(\varphi_+, 0)\,\partial^+}{\partial^+ (u, v)}\right]^{-1} \langle \varphi_+ + F_k'(u, v) \rangle, \tag{8.50b}$$

und wir können jetzt durch Einsetzen des Ausdrucks (8.40) für $F_k'(u, v)$ den Zusammenhang zwischen $F_+ (u, v)$ und $F_0(u, v)$ $[\equiv F_1(u, v)]$ hinschreiben:

$$\left.\begin{aligned} F_+ (u, v) = \left[1 + \frac{(\varphi_+, 0)\,\partial^+}{\partial^+ (u, v)}\right]^{-1} \Big\langle \varphi_+ + \\ + \left\| \left[1 + \frac{(\varphi_\tau, \psi_\tau)\,\partial^+}{\partial^+ (u, v)}\right]^{-1} \Big\langle 2\varphi_\tau + \left[1 - \frac{(\varphi_\tau, \psi_\tau)\,\partial^-}{\partial^- (u, v)}\right] \right\|_{\tau=1}^{\tau=k} F_0(u, v) \Big\| . \end{aligned}\right\} \tag{8.51}$$

Bei der Wahl (8.22b*) für $U_{(0)}$ ist [s. (8.41*)] $F_0(u, v) = 0$, und wir erhalten

$$\left.\begin{aligned} F_+ (u, v) = \left[1 + \frac{(\varphi_+, 0)\,\partial^+}{\partial^+ (u, v)}\right]^{-1} \Big\langle \varphi_+ + \\ + \left\| \left[1 + \frac{(\varphi_\tau, \psi_\tau)\,\partial^+}{\partial^+ (u, v)}\right]^{-1} \Big\langle 2\varphi_\tau + \left[1 - \frac{(\varphi_\tau, \psi_\tau)\,\partial^-}{\partial^- (u, v)}\right] \right\|_{\tau=1}^{\tau=k} 0 \Big\| , \end{aligned}\right\} \tag{8.51*}$$

eine Darstellung der Bildhöhenaberration $(U_+ - U_0) = F_+$ als Funktion der Strahldaten $u = U_+$ und $v = V_{k+1}$, und der Langeschen Konstruktionsdaten; u, v und die Langeschen Konstruktionsdaten treten in den bekannten, eingesetzt zu denkenden Funktionen φ, ψ auf.

Mit einem Stern (*) sind ab Gl. (8.22b*) alle die Gleichungen versehen, bei denen $U_{(0)} = U_0$ angenommen ist. Bei den anderen Gleichungen ist die Wahl von $U_{(0)}$ noch frei. So können wir (8.51) nach $F_0(u, v)$ auflösen:

$$\left.\begin{aligned} F_0(u, v) = \left\|_{\tau=1} \left[1 - \frac{(\varphi_\tau, \psi_\tau)\,\partial^-}{\partial^- (u, v)}\right]^{-1} \Big\langle -2\varphi_\tau + \left[1 + \frac{(\varphi_\tau, \psi_\tau)\,\partial^+}{\partial^+ (u, v)}\right] \right\|^{\tau=k} \times \\ \times \Big\langle -\varphi_+ + \left[1 + \frac{(\varphi_+, 0)\,\partial^+}{\partial^+ (u, v)}\right] F_+ (u, v), \end{aligned}\right\} \tag{8.52}$$

[1] Aus (8.39a'), (8.39a'') folgt, daß $[1 + (\varphi_+, 0)\,\partial^+/\partial^+ (u, v)]^{-1}$ auf die Form $[1 - (\varphi_+', 0)\,\partial^-/\partial^- (u, v)]$ gebracht werden kann; die Funktion φ_+' ließe sich leicht aus der durch (8.45), (8.46) gegebenen Funktion φ_+ bestimmen.

können

$$U_{(0)} = U_+ \tag{8.53}$$

wählen, so daß nach (8.47)

$$F_+ (U_+ ; V_+) = \Delta U_+ \equiv U_+ - U_{(0)} = U_+ - U_+ = 0 \tag{8.54}$$

wird, und erhalten

eine strenge Darstellung der Bildhöhenaberration $(U_+ - U_0) = -F_0$ als Funktion der objektseitigen Strahldaten $u = U_0$; $v = V_0$ (U_0 = Bildwinkelsinus; V_0 = Eintrittspupillenlot) und der Langeschen Konstruktionsdaten des optischen Systems:

$$-F_0(u, v) = {}_{\tau=1}\left|\left[1 - \frac{(\varphi_\tau, \psi_\tau)\, \partial^-}{\partial^-(u, v)}\right]^{-1} \left\langle 2\varphi_\tau + \left[1 + \frac{(\varphi_\tau, \psi_\tau)\, \partial^+}{\partial^+(u, v)}\right]\right|^{\tau = k} \varphi_+ , \tag{VIII.1}$$

wobei

$$\left.\begin{aligned} \varphi_+ = \frac{G+u}{\bar{u}}\,\frac{G-u}{1+\bar{u}}\,u \quad \text{ist mit} \quad & G = c_{k+1}^- v + c_{k+1} u; \\ & \bar{u} = \sqrt{1 + u^2 - G^2}, \end{aligned}\right\} \tag{VIII.2}$$

und

$$\left.\begin{aligned} \varphi_\tau &= \left(+\frac{\mathring{\varkappa}}{c}\,\frac{\nu'-\nu}{\nu+\nu'}\,\frac{P^\cdot}{\bar{J}^\cdot}\,\frac{G^\cdot - J^\cdot}{\bar{G}^\cdot + \bar{J}^\cdot}\,J^\cdot - \frac{1}{c}\,\frac{\bar{\nu}'-\bar{\nu}}{\nu+\nu'}\,J^\cdot\right)_\tau; \\ \psi_\tau &= \left(-\frac{\overset{\times}{\varkappa}}{c}\,\frac{\nu'-\nu}{\nu+\nu'}\,\frac{P^\cdot}{\bar{J}^\cdot}\,\frac{G^\cdot - J^\cdot}{\bar{G}^\cdot + \bar{J}^\cdot}\,J^\cdot - \frac{t}{c}\,\frac{\bar{\nu}'-\bar{\nu}}{\nu+\nu'}\,J^\cdot\right)_\tau \end{aligned}\right\} \tag{VIII.3}$$

mit:

$$\left.\begin{aligned} &\frac{\bar{\nu}+\bar{\nu}'}{\nu+\nu'}\,J^\cdot = \mathring{\varkappa}\,v + \overset{\times}{\varkappa}\,u; \quad \overset{\times}{\varkappa} = -c - t\mathring{\varkappa}; \quad \mathring{\varkappa} = -\frac{\nu_e' c^+ - \nu_e c^-}{\nu_e' - \nu_e}; \\ &\frac{\bar{\nu}+\bar{\nu}'}{\nu+\nu'}\,G^\cdot = \breve{\chi}\,v + \hat{\chi}\,u; \quad \hat{\chi} = +c - t\breve{\chi}; \quad \breve{\chi} = +\frac{\nu c^- + \nu' c^+}{\nu+\nu'}; \\ &P^\cdot = v - tu; \qquad c_\tau^+ = \frac{c_{\tau+1} - c_\tau}{t_{\tau+1} - t_\tau}; \\ &\bar{\nu} = \nu/\nu_e; \qquad c_\tau^- = \frac{c_\tau - c_{\tau-1}}{t_\tau - t_{\tau-1}}; \\ &\bar{J}^\cdot = \sqrt{1 - J^{\cdot 2} - \iota}; \\ &\bar{G}^\cdot = \sqrt{1 - G^{\cdot 2} - \iota}; \\ &\iota \text{ ist klein gegen } J^{\cdot 2}: \\ &\iota = \left(\frac{\nu'-\nu}{\nu+\nu'}\right)^2 J^{\cdot 2}/[1 - J^{\cdot 2}/(1-\iota)] \quad \text{(implizit)} \\ & = \frac{l}{2} - \sqrt{\left(\frac{l}{2}\right)^2 - \left(\frac{\nu'-\nu}{\nu+\nu'}\right)^2 J^{\cdot 2}} \quad \text{(explizit)} \\ &\text{mit} \quad l = 1 - J^{\cdot 2} + \left(\frac{\nu'-\nu}{\nu+\nu'}\right)^2 J^{\cdot 2}. \end{aligned}\right\} \tag{VIII.4}$$

Außer u und v, die ohne Index bleiben, sind in (VIII.4) wie in (VIII.3) alle Größen mit dem Index τ versehen zu denken (abgesehen von den beiden, schon mit den Indizes $\tau-1$; τ; $\tau+1$ beschrifteten Definitionsgleichungen für die

Differenzenquotienten c_τ^- und c_τ^+). Die Langeschen Konstruktionsdaten sind:

$$\left.\begin{array}{llllllll} \nu_1=1\,; & \nu_1'\equiv\nu_2\,; & \nu_2'\equiv\nu_3\,; & \ldots\,; & \nu_{k-2}'\equiv\nu_{k-1}\,; & \nu_{k-1}'\equiv\nu_k\,; & \nu_k'=1\,, & \\ t_0=0\,; & t_1\,; & t_2\,; & \ldots\,; & t_{k-1}\,; & t_k\,; & t_{k+1}=0\,, & \\ c_0=1\,; & c_1=1\,; & c_2\,; & \ldots\,; & c_{k-1}\,; & c_k\,; & c_{k+1}\,. & \end{array}\right\} \quad \text{(VIII.5)}$$

In dem Wiederholungs-Stück oder „Flächenanteil“ $|\ldots|$ des Ausdrucks (VIII.1) treten neben dem „additiven Flächenanteil“ $2\varphi_\tau$ die linearen Operatoren

$$\left[1+\frac{(\varphi_\tau,\psi_\tau)\,\partial^+}{\partial^+(u,v)}\right] \quad \text{und} \quad \left[1-\frac{(\varphi_\tau,\psi_\tau)\,\partial^-}{\partial^-(u,v)}\right]^{-1} \qquad \text{(VIII.6)}$$

auf. Man kann in (VIII.1)

$$\left.\begin{array}{lll} \left[1+\dfrac{(\varphi_\tau,\psi_\tau)\,\partial^+}{\partial^+(u,v)}\right] & \text{durch} & \left[1+\left(\varphi_\tau\dfrac{\partial}{\partial u}+\psi_\tau\dfrac{\partial}{\partial v}\right)\right] \quad \text{und} \\ \left[1-\dfrac{(\varphi_\tau,\psi_\tau)\,\partial^-}{\partial^-(u,v)}\right]^{-1} & \text{durch} & \left[1-\left(\varphi_\tau\dfrac{\partial}{\partial u}+\psi_\tau\dfrac{\partial}{\partial v}\right)\right]^{-1} \quad \text{ersetzen} \end{array}\right\} \quad \text{(VIII.6a)}$$

in einer sehr guten Näherung: im monochromatischen Fall kommt von der siebten Bildfehler-Ordnung an eine Ungenauigkeit herein, und zwar beträgt die in der siebten Bildfehler-Ordnung bei der Fläche Nr. τ entstehende Ungenauigkeit weniger als das $0{,}4\cdot\left(\frac{n'-n}{n'+n}\right)_\tau^2$-fache des Glieds siebter Ordnung des additiven Flächenanteils $2\varphi_\tau$, das sind

bei Glas-Luft-Flächen	mit $n_{\text{Glas}}=1{,}91$:	$<3{,}9\%$;
	mit $n_{\text{Glas}}=1{,}667$:	$<2{,}5\%$;
	mit $n_{\text{Glas}}=1{,}44$:	$<1{,}3\%$;
und bei Kittflächen:		$<0{,}8\%$

des Glieds siebter Ordnung von $2\varphi_\tau$.

Diese geringfügige Ungenauigkeit, die beim Ersetzen der Operatoren $\left[1\pm\frac{(\varphi_\tau,\psi_\tau)\,\partial^\pm}{\partial^\pm(u,v)}\right]^{\pm1}$ durch $\left[1\pm\left(\varphi_\tau\frac{\partial}{\partial u}+\psi_\tau\frac{\partial}{\partial v}\right)\right]^{\pm1}$ hereinkommt, ließe sich sogar zu ihrem größten Teil kompensieren dadurch, daß man in der Darstellung (VIII.1) zugleich den additiven Flächen-Anteil

$$2\varphi_\tau \quad \text{durch} \quad \left[1-\frac{1}{2}\left(\varphi_\tau^2\frac{\partial^2}{\partial u^2}+2\varphi_\tau\psi_\tau\frac{\partial^2}{\partial u\,\partial v}+\psi_\tau^2\frac{\partial^2}{\partial v^2}\right)\right]2\varphi_\tau \quad \text{ersetzt.} \quad \text{(VIII.6b).}$$

Denn nach (8.39) und (8.37), (8.38) ist

$$\left.\begin{array}{l} \left[1+\left(\varphi\dfrac{\partial}{\partial u}+\varphi\dfrac{\partial}{\partial v}\right)+\dfrac{1}{2}\left(\varphi^2\dfrac{\partial^2}{\partial u^2}+2\varphi\psi\dfrac{\partial^2}{\partial u\,\partial v}+\psi^2\dfrac{\partial^2}{\partial v^2}\right)+\cdots\right]F'- \\ -\left[1-\left(\varphi\dfrac{\partial}{\partial u}+\psi\dfrac{\partial}{\partial v}\right)+\dfrac{1}{2}\left(\varphi^2\dfrac{\partial^2}{\partial u^2}+2\varphi\psi\dfrac{\partial^2}{\partial u\,\partial v}+\psi^2\dfrac{\partial^2}{\partial v^2}\right)-\cdots\right]F-2\varphi=0, \end{array}\right\} \text{(8.55)}$$

die durch (VIII.6a) entstehende Ungenauigkeit in erster Näherung also ein Summand

$$+\frac{1}{2}\left(\varphi^2\frac{\partial^2}{\partial u^2}+2\varphi\psi\frac{\partial^2}{\partial u\,\partial v}+\psi^2\frac{\partial^2}{\partial v^2}\right)(F'-F)$$

neben -2φ, für den wir wegen (8.55) näherungsweise auch

$$+\left(\varphi^2\frac{\partial^2}{\partial u^2}+2\varphi\psi\frac{\partial^2}{\partial u\,\partial v}+\psi^2\frac{\partial^2}{\partial v^2}\right)\varphi$$

schreiben können:

$$\left.\begin{aligned}&\left[1+\left(\varphi\frac{\partial}{\partial u}+\psi\frac{\partial}{\partial v}\right)\right]F'(u,v)-\left[1-\left(\varphi\frac{\partial}{\partial u}+\psi\frac{\partial}{\partial v}\right)\right]F(u,v)\\&=\left[1-\frac{1}{2}\left(\varphi^2\frac{\partial^2}{\partial u^2}+2\varphi\psi\frac{\partial^2}{\partial u\,\partial v}+\psi^2\frac{\partial^2}{\partial v^2}\right)\right]2\varphi+\cdots.\end{aligned}\right\}\tag{8.56}$$

Nun wollen wir die unter (VIII.6a) genannten Prozentzahlen nachweisen. Im monochromatischen Fall ($\nu=\nu_e\equiv\mu$; $\bar{\nu}=1$) ist nach (VIII.3)

$$\left.\begin{aligned}\varphi&=+\overset{\circ}{\varkappa}\cdot\frac{\mu'-\mu}{\mu+\mu'}\frac{1}{c}\frac{P^\cdot}{J^\cdot}\frac{G^\cdot-J^\cdot}{G^\cdot+J^\cdot}J^\cdot;\\\psi&=-\overset{\times}{\varkappa}\cdot\frac{\mu'-\mu}{\mu+\mu'}\frac{1}{c}\frac{P^\cdot}{J^\cdot}\frac{G^\cdot-J^\cdot}{G^\cdot+J^\cdot}J^\cdot,\end{aligned}\right\}\tag{8.57}$$

also wird mit

$$\partial_\tau\equiv\overset{\circ}{\varkappa}_\tau\frac{\partial}{\partial u}-\overset{\times}{\varkappa}_\tau\frac{\partial}{\partial v}\tag{8.58}$$

und

$$\gamma_\tau\equiv\left(\frac{\mu'-\mu}{\mu+\mu'}\frac{1}{c}\frac{P^\cdot}{J^\cdot}\frac{G^\cdot-J^\cdot}{G^\cdot+J^\cdot}J^\cdot\right)_\tau\tag{8.59}$$

aus (8.56):

$$[1+\gamma\partial]F'(u,v)-[1-\gamma\partial]F(u,v)=[1-\tfrac{1}{2}\gamma^2\partial^2]2\overset{\circ}{\varkappa}\gamma+\cdots.\tag{8.56a}$$

Der hierbei durch „$+\cdots$“ angedeutete Rest ist von neunter Ordnung und viertem Grad klein. Rechts steht der additive Flächenanteil ($2\overset{\circ}{\varkappa}\gamma$) minus einer Korrektur $\frac{1}{2}\gamma^2\partial^2(2\overset{\circ}{\varkappa}\gamma)=(2\overset{\circ}{\varkappa}\gamma)\cdot\frac{1}{2}\gamma\,\partial^2\gamma$, die nach (8.59) von siebter Ordnung und drittem Grad klein ist, und die mit (VIII.6a) *vernachlässigt* wird. Wir wollen feststellen, welche Ungenauigkeit das in der siebten Bildfehler-Ordnung hereinbringt. Dann müssen wird das Glied siebter Ordnung dieser Vernachlässigung vergleichen mit dem Glied siebter Ordnung des additiven Flächenanteils ($2\overset{\circ}{\varkappa}\gamma$), der bei dieser Vernachlässigung rechts stehen bleibt. Wir denken uns $\gamma=\gamma(u,v)$ nach u und v, Größen erster Ordnung und nullten Grades, entwickelt:

$$2\overset{\circ}{\varkappa}\gamma(u,v)=2\overset{\circ}{\varkappa}(\gamma^{(3)}+\gamma^{(5)}+\gamma^{(7)}+\gamma^{(9)}+\cdots),$$

wobei die $\gamma^{(\Omega)}$ homogene Polynome Ω-ter Ordnung in u und v sind. $\gamma^{(\Omega)}$ nennen wir das Glied Ω-ter Ordnung von $\gamma=\gamma(u,v)$. Nach (8.59) ist γ eine Größe dritter Ordnung und ersten Grades, $\gamma^{(\Omega)}$ also eine Größe Ω-ter Ordnung und ersten Grades. Mit dem Glied $2\overset{\circ}{\varkappa}\gamma^{(7)}$, einer Größe siebter Ordnung und *ersten Grades,* ist — wie gesagt — das Glied siebter Ordnung der bei der Näherung (VIII.6a) vernachlässigten Korrektur $(2\overset{\circ}{\varkappa}\gamma)\cdot\frac{1}{2}\gamma\,\partial^2\gamma$ zu vergleichen, das von siebter Ordnung und *drittem Grad* klein ist. Dieses Glied — wir bezeichnen es mit $2\overset{\circ}{\varkappa}\pi^{(7)}$ — ist

$$2\overset{\circ}{\varkappa}\pi^{(7)}=(2\overset{\circ}{\varkappa}\tilde{\gamma})\cdot\frac{1}{2}\tilde{\gamma}\,\partial^2\tilde{\gamma}\quad\text{mit}\quad\tilde{\gamma}=\frac{\mu'-\mu}{\mu+\mu'}\frac{1}{c}\frac{P^\cdot}{1}\frac{G^\cdot-J^\cdot}{2}J^\cdot;$$

$$P^\cdot, G^\cdot, J^\cdot \text{ wie in (VIII.4).}$$

Zu vergleichen ist $2\overset{\circ}{\varkappa}\pi^{(7)}$ mit $2\overset{\circ}{\varkappa}\gamma^{(7)}$, d.h. wir haben das Verhältnis $(2\overset{\circ}{\varkappa}\pi^{(7)})/(2\overset{\circ}{\varkappa}\gamma^{(7)})$ zu bilden: Nach (8.59) und (VIII.4) ist

$$\begin{aligned}\gamma=\frac{\mu'-\mu}{\mu+\mu'}\frac{1}{c}\frac{P^\cdot}{1}\frac{G^\cdot-J^\cdot}{2}J^\cdot\cdot\frac{1}{L}\bigg[1&+\frac{3}{4}\frac{J^{\cdot2}}{L}+\frac{1}{4}\frac{G^{\cdot2}}{L}+\frac{5}{8}\left(\frac{J^{\cdot2}}{L}\right)^2+\\&+\frac{2}{8}\frac{J^{\cdot2}}{L}\frac{G^{\cdot2}}{L}+\frac{1}{8}\left(\frac{G^{\cdot2}}{L}\right)^2+\cdots\bigg],\end{aligned}$$

wegen

$$\frac{1}{L} = 1 + \left(\frac{m'-m}{m+m'}\right)^2 J^{\cdot 2} + \left[1 + \left(\frac{m'-m}{m+m'}\right)^2\right]\left(\frac{m'-m}{m+m'}\right)^2 J^{\cdot 4} + \cdots$$

also

$$\gamma^{(7)} = \frac{\mu'-\mu}{\mu+\mu'}\,\frac{1}{c}\,\frac{P^{\cdot}}{1}\,\frac{G^{\cdot}-J^{\cdot}}{2}\,J^{\cdot}\cdot\left(\frac{5}{8}J^{\cdot 4} + \frac{2}{8}J^{\cdot 2}G^{\cdot 2} + \frac{1}{8}G^{\cdot 4} + \bar{\pi}\right)$$

mit dem vernachlässigbar kleinen positiven

$$\bar{\pi} \equiv \left(\frac{m'-m}{m+m'}\right)^2 J^{\cdot 2}\left\langle\left[\frac{7}{4} + \left(\frac{m'-m}{m+m'}\right)^2\right] J^{\cdot 2} + \frac{1}{2}G^{\cdot 2}\right\rangle.$$

So wird

$$\frac{2\,\mathring{\varkappa}\,\pi^{(7)}}{2\,\mathring{\varkappa}\,\gamma^{(7)}} = \frac{\frac{1}{2}\tilde{\gamma}\,\partial^2\tilde{\gamma}}{\frac{5}{8}J^{\cdot 4} + \frac{2}{8}J^{\cdot 2}G^{\cdot 2} + \frac{1}{8}G^{\cdot 4} + \bar{\pi}}\,. \tag{8.60}$$

Nun ist $P = cQ = \frac{c}{\varrho}\,(G^{\cdot} + J^{\cdot})$, also

$$\tilde{\gamma} = \frac{\mu'-\mu}{\mu+\mu'}\,\frac{1}{\varrho}\,\frac{G^{\cdot 2}-J^{\cdot 2}}{2}\,J^{\cdot}. \tag{8.61}$$

Da

$$\frac{\bar{\nu}+\bar{\nu}'}{\nu+\nu'}\,\partial J^{\cdot} = \left(\mathring{\varkappa}\frac{\partial}{\partial u} - \overset{\times}{\varkappa}\frac{\partial}{\partial v}\right)(\mathring{\varkappa}v + \overset{\times}{\varkappa}u) = 0 \tag{8.62a}$$

und wegen $G^{\cdot} = \frac{\varrho}{c}\,P^{\cdot} - J^{\cdot}$; $\partial J^{\cdot} = 0$; $\overset{\times}{\varkappa} = -c - t\mathring{\varkappa}$:

$$\partial G^{\cdot} = \left(\mathring{\varkappa}\frac{\partial}{\partial u} - \overset{\times}{\varkappa}\frac{\partial}{\partial v}\right)\frac{\varrho}{c}\,(v - tu) = \varrho \tag{8.62b}$$

ist, folgt aus (8.61): $\frac{1}{2}\tilde{\gamma}\,\partial^2\tilde{\gamma} = \left(\frac{\mu'-\mu}{\mu+\mu'}\right)^2 \frac{G^{\cdot 2}-J^{\cdot 2}}{4}\,J^{\cdot 2}$, und aus (8.60) wird:

$$\frac{2\mathring{\varkappa}\pi^{(7)}}{2\mathring{\varkappa}\gamma^{(7)}} = \left(\frac{m'-m}{m+m'}\right)^2 \frac{2(G^{\cdot 2}-J^{\cdot 2})J^{\cdot 2}}{5J^{\cdot 4} + 2J^{\cdot 2}G^{\cdot 2} + G^{\cdot 4} + 8\bar{\pi}}\,. \tag{8.63}$$

Rechts im Nenner ist $8\bar{\pi}$ vernachlässigbar, $8\bar{\pi}$ ist überdies positiv, also ist $\left|\frac{2\mathring{\varkappa}\pi^{(7)}}{2\mathring{\varkappa}\gamma^{(7)}}\right| \Big/ \left(\frac{m'-m}{m+m'}\right)^2$ kleiner als der größte Wert von $\left|\frac{2(G^{\cdot 2}-J^{\cdot 2})J^{\cdot 2}}{5J^{\cdot 4}+2J^{\cdot 2}G^{\cdot 2}+G^{\cdot 4}}\right|$, der gleich $\frac{2}{5}$ ist:

$$\left|\frac{2\mathring{\varkappa}_\tau\pi_\tau^{(7)}}{2\mathring{\varkappa}_\tau\gamma_\tau^{(7)}}\right| < 0{,}4\cdot\left(\frac{n_e'-n_e}{n_e+n_e'}\right)_\tau^2. \tag{8.64}$$

Das Glied siebter Ordnung der durch (VIII.6a) bei der Fläche Nr. τ entstehenden Ungenauigkeit, einer Größe siebter Ordnung und dritten Grades, verhält sich also zu dem Glied siebter Ordnung des additiven Flächenanteils $2\varphi_\tau$ wie weniger als $0{,}4\cdot\left(\frac{n_e'-n_e}{n_e+n_e'}\right)_\tau^2$ zu 1. Damit sind die unter (VIII.6a) genannten Prozentzahlen, die Zahlenbeispiele für (8.64) sind, nachgewiesen. $\left(\frac{n_e'-n_e}{n_e+n_e'}\right)_\tau^2$, eine Größe nullter Ordnung und zweiten Grades, ist das Reflexionsvermögen der Fläche Nr. τ (bei senkrechtem Einfall), die Näherung (VIII.6a) ist monochromatisch also in der *siebten* Bildfehler-Ordnung — vereinfacht gesagt — so gut, wie das Reflexionsvermögen klein ist.

Der nach (VIII.6a) auftretende inverse Operator

$$\left[1 - \left(\varphi_\tau\frac{\partial}{\partial u} + \psi_\tau\frac{\partial}{\partial v}\right)\right]^{-1} = 1 + \sum_{N=1}^{\infty}\left(\varphi_\tau\frac{\partial}{\partial u} + \psi_\tau\frac{\partial}{\partial v}\right)^N \tag{8.65}$$

kann in sehr guter Näherung gleich

$$\left[1+\left(\varphi_\tau \frac{\partial}{\partial u}+\psi_\tau \frac{\partial}{\partial v}\right)+\left(\varphi_\tau \frac{\partial}{\partial u}+\psi_\tau \frac{\partial}{\partial v}\right)^2\right] \tag{8.65'}$$

gesetzt werden; dann entsteht nämlich in der siebten Bildfehler-Ordnung keine zusätzliche Ungenauigkeit. Eine solche entsteht erst, wenn man (8.65) durch

$$\left[1+\left(\varphi_\tau \frac{\partial}{\partial u}+\psi_\tau \frac{\partial}{\partial v}\right)\right] \tag{8.65''}$$

annähert.

Die Gl. (8.65) bedeutet eine Anwendung der bekannten Operatorgleichung

$$[1-\omega]^{-1}=1+\sum_{N=1}^{\infty}\omega^N, \tag{8.65a}$$

wonach bei einem kleinen Operator ω aus

$$[1-\omega]F'(u,v)=\Phi(u,v) \tag{8.65A}$$

folgt, daß

$$F'(u,v)=\left[1+\sum_{N=1}^{\infty}\omega^N\right]\Phi(u,v) \tag{8.65B}$$

ist: Bei Gl. (8.65 A) sprechen wir von einem „kleinen" Operator ω, wenn $F'\approx\Phi$ ist und man aus einer Näherung $F'_{(N)}$ für F' eine *verbesserte* Näherung $F'_{(N+1)}$ mit Hilfe der Iterationsformel

$$F'_{(N+1)}=\Phi+\omega F'_{(N)}$$

erhält. Hieraus folgt mit $F'_{(0)}=\Phi$: $F'_{(1)}=\Phi+\omega\Phi$; $F'_{(2)}=\Phi+\omega\Phi+\omega^2\Phi$; und so fort, also $F'=F'_{(\infty)}=\left[1+\sum_{N=1}^{\infty}\omega^N\right]\Phi$.

Die Summe $[1+\sum\omega^N]$ kann man auch als Produkt schreiben:

$$\left[1+\sum_{N=1}^{2^s-1}\omega^N\right]=\prod_{r=0}^{s-1}\left[1+\omega^{(2^r)}\right], \tag{8.65b}$$

wobei es auf die Reihenfolge der Faktoren nicht ankommt. Das Produkt kommt mit s Faktoren aus, wenn es die Summe von 2^s Summanden darstellen soll.

51. Sagittale und tangentiale Bildhöhenaberration. Eine der strengen Darstellung (VIII) der meridionalen Bildhöhenaberration entsprechende, strenge Darstellung der sagittalen und der tangentialen Bildhöhenaberration wollen wir in dieser Ziffer herleiten. Obwohl wir, wie bei der Herleitung von (VIII), uns nicht auf den monochromatischen Fall zu beschränken brauchten, wollen wir uns doch von nun an auf diesen Fall ($n=n_e\equiv m$, also $\nu=\nu_e\equiv\mu$, $\bar\nu=1$) beschränken; erstens um der besseren Übersicht willen; zweitens, weil es ratsam ist, für die Farbfehler eine gesonderte Entwicklung durchzuführen[1].

[1] Eine, bei der die Farbabhängigkeit der Brechungsindizes an einer Fläche allein von der auch bei (VIII) auftretenden Größe $\frac{\bar\nu'-\bar\nu}{\nu+\nu'}$ erfaßt wird $\Big[$indem $\frac{\nu'-\nu}{\nu+\nu'}=\frac{\mu'-\mu}{\mu+\mu'}+\frac{2\mu\mu'}{\mu+\mu'}\times\frac{\bar\nu'-\bar\nu}{\nu+\nu'}$; $\frac{\bar\nu+\bar\nu'}{\nu+\nu'}=\frac{2}{\mu+\mu'}-\frac{\mu'-\mu}{\mu+\mu'}\,\frac{\bar\nu'-\bar\nu}{\nu+\nu'}$ und $\frac{\nu c^-+\nu' c^+}{\nu+\nu'}=\frac{\mu c^-+\mu' c^+}{\mu+\mu'}+(c^+-c^-)\frac{\mu\mu'}{\mu+\mu'}\times\frac{\bar\nu'-\bar\nu}{\nu+\nu'}$ gesetzt wird$\Big]$, und bei der dann nach $\frac{\bar\nu'-\bar\nu}{\nu+\nu'}$ entwickelt wird. Diese Entwicklung führen wir in diesem Handbuch-Beitrag nicht durch.

Die in Ziff. 49 für beliebige *F*arben eingeführte *F*unktion F bezeichnen wir im *m*onochromatischen Fall mit M und nennen sie die *m*eridionale Aberrations-Funktion. Daneben führen wir noch eine *s*agittale Aberrations-Funktion S und eine *t*angentiale Aberrations-Funktion T ein.

Wir betrachten, wie in Ziff. 49 und 50, die zweiparametrige Schar aller Meridianstrahlen, ordnen jedem dieser Meridianstrahlen einen sagittal benachbarten Strahl und einen tangential benachbarten Strahl zu, und interessieren uns für $d^i U_\tau / d^i V_\tau$ als Funktion von (U_τ, V_τ), und für $d^1 U_\tau / d^1 V_\tau$ als Funktion von (U_τ, V_τ):

$$\left.\begin{aligned} d^i U_\tau / d^i V_\tau &= S_\tau (U_\tau, V_\tau); \\ d^1 U_\tau / d^1 V_\tau &= T_\tau (U_\tau, V_\tau). \end{aligned}\right\} \tag{8.66}$$

Für $S_{\tau+1}$ und $T_{\tau+1}$ schreiben wir auch S'_τ und T'_τ:

$$\left.\begin{aligned} d^i U'_\tau / d^i V'_\tau &= S'_\tau (U'_\tau, V'_\tau); \\ d^1 U'_\tau / d^1 V'_\tau &= T'_\tau (U'_\tau, V'_\tau). \end{aligned}\right\} \tag{8.66'}$$

Wir werden feststellen, wie die beiden Funktionen $S_\tau(u, v)$ und $S'_\tau(u, v) \equiv S_{\tau+1}(u, v)$ miteinander zusammenhängen, und wie die beiden Funktionen $T_\tau(u, v)$ und $T'_\tau(u, v) \equiv T_{\tau+1}(u, v)$ miteinander zusammenhängen. Das wird uns zu einer strengen Darstellung der sagittalen und der tangentialen Bildhöhenaberration als Funktion der Strahlanfangsdaten U_0, V_0, $\underline{\delta V_0}$, $\underline{d V_0}$ und der Langeschen Konstruktionsdaten führen.

Die Formeln (V.4a) mit[1]

$$\left.\begin{aligned} U^\cdot &= u; \qquad P^\cdot = v - t u; \\ K^\cdot &= \mathring{\varkappa} \cdot v + \overset{\times}{\varkappa} \cdot u; \\ d^i K^\cdot &= \mathring{\varkappa}\, d^i V^\cdot + \overset{\times}{\varkappa}\, d^i U^\cdot; \\ d^1 K^\cdot &= \mathring{\varkappa}\, d^1 V^\cdot + \overset{\times}{\varkappa}\, d^1 U^\cdot; \\ (\bar{J}\bar{J}^\cdot) &= L - J^{\cdot 2}/L; \qquad L = 1 - \iota \end{aligned}\right\} \tag{8.67}$$

und mit $G^\cdot$, $J^\cdot$; $\bar{G}^\cdot$, $\bar{J}^\cdot$; ι gemäß (VIII.4) stellen die meridionale, sagittale und tangentiale Bildlot- und Scheitellot-Aberration *an einer Fläche* dar als Funktion von $u = U^\cdot$, $v = V^\cdot$; $d^i U^\cdot$, $d^i V^\cdot$; $d^1 U^\cdot$, $d^1 V^\cdot$. Diese Formeln schreiben wir zunächst um auf Formeln für die Bildlot- und Pupillenlot-Aberration an der Fläche, in denen wir zur Vereinfachung der Ausdrücke den in Ziff. 50 eingeführten Differentialoperator

$$\partial_\tau \equiv \mathring{\varkappa}_\tau \frac{\partial}{\partial u} - \overset{\times}{\varkappa}_\tau \frac{\partial}{\partial v} \tag{8.58}$$

und den Differentialoperator

$$\theta \equiv u \frac{\partial}{\partial u} + v \frac{\partial}{\partial v} \tag{8.68}$$

verwenden [der Differentialoperator θ wirkt auf eine Funktion von u und v in der Weise, daß bei einer Entwicklung dieser Funktion nach Potenzen von u und v die Glieder erster Ordnung mit 1, die Glieder zweiter Ordnung mit 2, die Glieder dritter Ordnung mit 3, die Glieder vierter Ordnung mit 4, und so fort, multipliziert werden; von dem Differentialoperator ∂ hatten wir in Ziff. 50 festgestellt, daß

$$\partial G^\cdot = \varrho \tag{8.62b}$$

[1] Die Glieder mit $d^i P^\cdot$ und $d^1 P^\cdot$ in (V.4a) verschwinden im monochromatischen Fall.

und

$$\partial J^{\cdot} = 0, \tag{8.62a}$$

also auch

$$\partial \iota = 0 \tag{8.62a'}$$

und

$$\partial \bar{J}^{\cdot} = 0 \tag{8.62a''}$$

ist]. Lassen wir nämlich die Differentialoperatoren ∂_τ und θ auf die bei allen Formeln (V.4a) vorkommende[1] Funktion

$$\left.\begin{aligned} &\varepsilon_\tau(u;v) \equiv \left(\frac{\mu'-\mu}{\mu+\mu'}\,\frac{1}{c}\,\frac{P^{\cdot}}{\bar{J}^{\cdot}}\,\frac{G^{\cdot}-J^{\cdot}}{\bar{G}^{\cdot}+\bar{J}^{\cdot}}\right)_\tau \\ \text{mit}\quad &J^{\cdot}, G^{\cdot};\ P^{\cdot};\ \bar{J}^{\cdot}, \bar{G}^{\cdot} \quad \text{gemäß (VIII.4)} \end{aligned}\right\} \tag{8.69}$$

wirken, dann erhalten wir:

$$\partial_\tau\,\varepsilon_\tau = \left(\frac{\mu'-\mu}{\mu+\mu'}\,\frac{1}{\bar{J}^{\cdot}}\,\frac{G^{\cdot}}{\bar{G}^{\cdot}}\right)_\tau \tag{8.69a}$$

und

$$\theta\,\varepsilon_\tau = \left(\frac{\mu'-\mu}{\mu+\mu'}\,\frac{1}{c}\,\frac{P^{\cdot}}{\bar{J}^{\cdot}}\,\frac{G^{\cdot}-J^{\cdot}}{\bar{G}^{\cdot}(\bar{J}\bar{J}')}\right)_\tau \tag{8.69b}$$

[mit $J^{\cdot}$, $G^{\cdot}$; $P^{\cdot}$; $\bar{J}^{\cdot}$, $\bar{G}^{\cdot}$ gemäß (VIII.4) und mit $(\bar{J}\bar{J}')$ gemäß (8.67)], zwei Ausdrücke, die in den sagittalen und tangentialen Formeln (V.4a) vorkommen[1]. So wird aus (V.4a):

$$\left.\begin{aligned} \tfrac{1}{2}(U'-U) &= \qquad\qquad\qquad + \mathring{\varkappa}\,\varepsilon\cdot J^{\cdot} \\ \tfrac{1}{2}(V'-V) &= \qquad\qquad\qquad - \overset{\times}{\varkappa}\,\varepsilon\cdot J^{\cdot} \\ \tfrac{1}{2}(d^i U'-d^i U) &= -u(\partial\varepsilon)\,d^i J^{\cdot} + \mathring{\varkappa}\,\varepsilon\,d^i J^{\cdot} \\ \tfrac{1}{2}(d^i V'-d^i V) &= -v(\partial\varepsilon)\,d^i J^{\cdot} - \overset{\times}{\varkappa}\,\varepsilon\,d^i J^{\cdot} \\ \tfrac{1}{2}(d^1 U'-d^1 U) &= -u(\partial\varepsilon)\,d^1 J^{\cdot} + \mathring{\varkappa}(\varepsilon+\theta\,\varepsilon)\,d^1 J^{\cdot} \\ \tfrac{1}{2}(d^1 V'-d^1 V) &= -v(\partial\varepsilon)\,d^1 J^{\cdot} - \overset{\times}{\varkappa}(\varepsilon+\theta\,\varepsilon)\,d^1 J^{\cdot} \end{aligned}\right\} \tag{8.70a}$$

mit

$$\left.\begin{aligned} &\varepsilon = \frac{\mu'-\mu}{\mu+\mu'}\,\frac{1}{c}\,\frac{P^{\cdot}}{\bar{J}^{\cdot}}\,\frac{G^{\cdot}-J^{\cdot}}{\bar{G}^{\cdot}+\bar{J}^{\cdot}}; \\ &P^{\cdot} = v - t u; \\ &\frac{2}{\mu+\mu'}\,G^{\cdot} = \mathring{\chi}\,v + \overset{\times}{\chi}\,u; \quad \overset{\times}{\chi} = +c - t\mathring{\chi}; \quad \mathring{\chi} = +\frac{\mu c^- + \mu' c^+}{\mu+\mu'}; \\ &\frac{2}{\mu+\mu'}\,J^{\cdot} = \mathring{\varkappa}\,v + \overset{\times}{\varkappa}\,u; \quad \overset{\times}{\varkappa} = -c - t\mathring{\varkappa}; \quad \mathring{\varkappa} = -\frac{\mu' c^+ - \mu c^-}{\mu'-\mu}; \\ &\frac{2}{\mu+\mu'}\,d^i J^{\cdot} = \mathring{\varkappa}\,d^i V^{\cdot} + \overset{\times}{\varkappa}\,d^i U^{\cdot}; \qquad c_\tau^+ = \frac{c_{\tau+1}-c_\tau}{t_{\tau+1}-t_\tau}; \\ &\frac{2}{\mu+\mu'}\,d^1 J^{\cdot} = \mathring{\varkappa}\,d^1 V^{\cdot} + \overset{\times}{\varkappa}\,d^1 U^{\cdot}; \qquad c_\tau^- = \frac{c_\tau - c_{\tau-1}}{t_\tau - t_{\tau-1}}; \\ &\bar{G}^{\cdot} = \sqrt{1 - G^{\cdot 2} - \iota}; \\ &\bar{J}^{\cdot} = \sqrt{1 - J^{\cdot 2} - \iota}; \quad \iota \text{ wie in (VIII.4)}; \\ &\partial_\tau \equiv \mathring{\varkappa}_\tau \frac{\partial}{\partial u} - \overset{\times}{\varkappa}_\tau \frac{\partial}{\partial v}; \\ &\theta \equiv u\frac{\partial}{\partial u} + v\frac{\partial}{\partial v}, \end{aligned}\right\} \tag{8.70b}$$

[1] (V.4a) ist auf den monochromatischen Fall spezialisiert zu denken.

und zwar gelten die Formeln (8.70a, b) für einen Meridianstrahl, der vor der betrachteten Fläche die reduzierten Bild- und Pupillenlote U und V, und hinter der Fläche die reduzierten Bild- und Pupillenlote U' und V' hat, wenn man in (8.70a, b)

$$u = U^{\cdot}\,[\equiv \tfrac{1}{2}(U+U')]; \qquad v = V^{\cdot}\,[\equiv \tfrac{1}{2}(V+V')] \tag{8.70c}$$

setzt.

Die noch nicht bewiesene Gl. (8.69a) folgt daraus, daß

$$\varepsilon \Big/ \frac{\mu'-\mu}{\mu+\mu'} = \frac{1}{c}\,\frac{P^{\cdot}}{\overline{J}^{\cdot}}\,\frac{G^{\cdot}-J^{\cdot}}{\overline{G}^{\cdot}+\overline{J}^{\cdot}} = \frac{1}{\varrho}\,\frac{G^{\cdot}+J^{\cdot}}{\overline{J}^{\cdot}}\,\frac{G^{\cdot}-J^{\cdot}}{\overline{G}^{\cdot}+\overline{J}^{\cdot}}$$

$$= \frac{1}{\varrho}\left(1-\frac{\overline{G}^{\cdot}}{\overline{J}^{\cdot}}\right) = \frac{1}{\varrho}\left(1-\frac{\sqrt{1-G^{\cdot 2}-\iota}}{\sqrt{1-J^{\cdot 2}-\iota}}\right)$$

ist, und daß $\partial J^{\cdot} = 0$, $\partial\iota = 0$ und $\partial G^{\cdot} = \varrho$ ist. — Die ebenfalls noch nicht bewiesene Gl. (8.69b) folgt aus dem soeben errechneten Ausdruck

$$\varepsilon \Big/ \frac{\mu'-\mu}{\mu+\mu'} = \frac{1}{\varrho}\left(1-\frac{\overline{G}^{\cdot}}{\overline{J}^{\cdot}}\right) \tag{8.69'}$$

für ε, aus (V.3e) und daraus, daß

$$\theta \equiv u\,\frac{\partial}{\partial u} + v\,\frac{\partial}{\partial v} = J^{\cdot}\left(\frac{\partial}{\partial J^{\cdot}}\right)_{G^{\cdot}} + G^{\cdot}\left(\frac{\partial}{\partial G^{\cdot}}\right)_{J^{\cdot}} \tag{8.68'}$$

ist: Zunächst folgt aus der ersten Gl. (V.3e) $\Big[$durch Wegdifferenzieren der Konstanten $\left(\frac{\nu'-\nu}{\nu+\nu'}\right)^2\Big]$, daß

$$\frac{1}{2}\,J^{\cdot}\,\frac{\partial L}{\partial J^{\cdot}} = \frac{(L-1)\,L}{L-J^{\cdot 2}/L}$$

ist. Aus der dritten und vierten Gl. (V.3e) und aus (8.68') folgt, daß

$$\frac{\overline{J}^{\cdot}}{\overline{G}^{\cdot}}\,\theta\,\frac{\overline{G}^{\cdot}}{\overline{J}^{\cdot}} = \frac{1}{2}\,\frac{1}{\overline{G}^{\cdot 2}}\,\theta\,\overline{G}^{\cdot 2} - \frac{1}{2}\,\frac{1}{\overline{J}^{\cdot 2}}\,\theta\,\overline{J}^{\cdot 2}$$

$$= \left(\frac{1}{2}\,\frac{1}{\overline{G}^{\cdot 2}}\,\theta L - \frac{G^{\cdot 2}}{\overline{G}^{\cdot 2}}\right) - \left(\frac{1}{2}\,\frac{1}{\overline{J}^{\cdot 2}}\,\theta L - \frac{J^{\cdot 2}}{\overline{J}^{\cdot 2}}\right)$$

ist. Setzen wir hier für $\frac{1}{2}\,\theta L = \frac{1}{2}\,J^{\cdot}\,\frac{\partial L}{\partial J^{\cdot}}$ den Ausdruck $(L-1)\,L/(L-J^{\cdot 2}/L)$ ein, dann wird

$$\frac{\overline{J}^{\cdot}}{\overline{G}^{\cdot}}\,\theta\,\frac{\overline{G}^{\cdot}}{\overline{J}^{\cdot}} = -\,\frac{G^{\cdot 2}-J^{\cdot 2}}{\overline{G}^{\cdot 2}\,\overline{J}^{\cdot 2}}\,\frac{L-J^{\cdot 2}}{L-J^{\cdot 2}/L},$$

nach der zweiten und dritten Gl. (V.3e) also

$$\theta\,\frac{\overline{G}^{\cdot}}{\overline{J}^{\cdot}} = -\,\frac{G^{\cdot 2}-J^{\cdot 2}}{\overline{J}^{\cdot}\,\overline{G}^{\cdot}\,(\overline{J}\,\overline{J}')}$$

$$= -\,\frac{\varrho}{c}\,\frac{P^{\cdot}}{\overline{J}^{\cdot}}\,\frac{G^{\cdot}-J^{\cdot}}{\overline{G}^{\cdot}(\overline{J}\,\overline{J}')}.$$

Nach (8.69') schließlich ist $\theta\varepsilon = -\,\frac{\mu'-\mu}{\mu+\mu'}\,\frac{1}{\varrho}\,\theta\,\frac{\overline{G}^{\cdot}}{\overline{J}^{\cdot}}$, woraus mit dem zuletzt erhaltenen Ausdruck für $\theta\,\frac{\overline{G}^{\cdot}}{\overline{J}^{\cdot}}$ die zu beweisende Gl. (8.69b) folgt.

Denken wir uns nun in die dritte und vierte Gl. (8.70a) den Ausdruck (8.70b) für $d^i J^{\cdot}$,

$$d^i J^{\cdot} = \mu^{\cdot}\,\overset{\circ}{\varkappa}\,\frac{d^i V + d^i V'}{2} + \mu^{\cdot}\,\overset{\times}{\varkappa}\,\frac{d^i U + d^i U'}{2},$$

eingesetzt, dann haben wir zwei in $d^i U$, $d^i V$, $d^i U'$, $d^i V'$ homogen lineare Gleichungen — oder nach deren Division durch $d^i V$ zwei in

$$\frac{d^i U}{d^i V}\,; \qquad \frac{d^i U'}{d^i V'}\cdot\frac{d^i V'}{d^i V}\,; \qquad \frac{d^i V'}{d^i V}$$

inhomogen lineare Gleichungen, aus denen wir $d^i V'/d^i V$ eliminieren können, so daß wir einen Zusammenhang zwischen $d^i U/d^i V$ und $d^i U'/d^i V'$ erhalten. Und zwar kommt, wenn man noch das von der vierten Gl. (8.70a) hereinkommende v durch $\frac{2}{\mu+\mu'}\,J^{\cdot} = \mathring{\varkappa} v + \overset{\times}{\varkappa} u$ eliminiert, heraus:

$$\left.\begin{aligned} &\frac{(1-J^{\cdot}\,\partial\varepsilon)\,d^i U'/d^i V'}{1+\frac{\overset{\times}{\varkappa}}{\mathring{\varkappa}}\cdot d^i U'/d^i V'} - \frac{(1+J^{\cdot}\,\partial\varepsilon)\,d^i U/d^i V}{1+\frac{\overset{\times}{\varkappa}}{\mathring{\varkappa}}\cdot d^i U/d^i V} \\ &\qquad\qquad = (\mu+\mu')\,\mathring{\varkappa}\left[-u\,(\partial\varepsilon)+\mathring{\varkappa}\,\varepsilon\right]. \end{aligned}\right\} \tag{8.71 S}$$

Ganz entsprechend erhält man aus der fünften und sechsten Gl. (8.70a) und dem Ausdruck (8.70b) für $d^1 J^{\cdot}$:

$$\left.\begin{aligned} &\frac{(1-J^{\cdot}\,\partial\varepsilon)\,d^1 U'/d^1 V'}{1+\frac{\overset{\times}{\varkappa}}{\mathring{\varkappa}}\cdot d^1 U'/d^1 V'} - \frac{(1+J^{\cdot}\,\partial\varepsilon)\,d^1 U/d^1 V}{1+\frac{\overset{\times}{\varkappa}}{\mathring{\varkappa}}\,d^1 U/d^1 V} \\ &\qquad\qquad = (\mu+\mu')\,\mathring{\varkappa}\left[-u\,(\partial\varepsilon)+\mathring{\varkappa}\,(\varepsilon+\theta\,\varepsilon)\right]. \end{aligned}\right\} \tag{8.71 T}$$

(8.71 S) und (8.71 T) gelten für einen Meridianstrahl samt dem ihm zugeordneten sagittalen Nachbarstrahl und dem ihm zugeordneten tangentialen Nachbarstrahl — wobei dieser Meridianstrahl vor der betrachteten Fläche die Daten U und V, und hinter der Fläche die Daten U' und V' hat —, wenn man gemäß (8.70c) $u = U^{\cdot}$ und $v = V^{\cdot}$ nimmt.

Wir wollten $d^i U_\tau/d^i V_\tau$ als Funktion S_τ von (U_τ, V_τ) und $d^1 U_\tau/d^1 V_\tau$ als Funktion T_τ von (U_τ, V_τ) betrachten, und wollten feststellen, wie die beiden *Funktionen* S_τ und $S'_\tau \equiv S_{\tau+1}$ miteinander zusammenhängen, und wie die beiden *Funktionen* T_τ und $T'_\tau \equiv T_{\tau+1}$ miteinander zusammenhängen.

In (8.71 S) mit $u = U^{\cdot}$ und $v = V^{\cdot}$ treten *diejenigen* Verhältnisse $d^i U/d^i V$ und $d^i U'/d^i V'$ sagittaler Variationen auf, die bei dem dort betrachteten Meridianstrahl mit den Daten U, V vor der Fläche und den Daten U', V' hinter der Fläche (samt dessen sagittalem Nachbarstrahl) vorliegen. In (8.71 S) ist also

$$\begin{aligned} d^i U/d^i V &= S\,(U, V) \\ &= S\left(U^{\cdot}-\tfrac{1}{2}(U'-U),\ V^{\cdot}-\tfrac{1}{2}(V'-V)\right) \\ &= S\left(u-\mathring{\varkappa} J^{\cdot}\,\varepsilon(u;v),\ v+\overset{\times}{\varkappa} J^{\cdot}\,\varepsilon(u;v)\right) \\ &= \left[1-\frac{(\mathring{\varkappa} J^{\cdot}\,\varepsilon,\ -\overset{\times}{\varkappa} J^{\cdot}\,\varepsilon)\,\partial^-}{\partial^-(u,v)}\right] S\,(u,v) \end{aligned}$$

oder, mit den Abkürzungen

$$\left.\begin{aligned} (\gamma\,\partial^+) &\equiv \frac{(+\mathring{\varkappa}\gamma,\ -\overset{\times}{\varkappa}\gamma)\,\partial^+}{\partial^+(u,v)}\,; \qquad (J^{\cdot}\varepsilon\,\partial^+) \equiv \left((J^{\cdot}\varepsilon)\,\partial^+\right); \\ (\gamma\,\partial^-) &\equiv \frac{(+\mathring{\varkappa}\gamma,\ -\overset{\times}{\varkappa}\gamma)\,\partial^-}{\partial^-(u,v)}\,; \qquad (J^{\cdot}\varepsilon\,\partial^-) \equiv \left((J^{\cdot}\varepsilon)\,\partial^-\right) \end{aligned}\right\} \tag{8.72}$$

[entsprechend der oben eingeführten Akbürzung ∂ für $\mathring{\varkappa}\,\frac{\partial}{\partial u} - \overset{\times}{\varkappa}\,\frac{\partial}{\partial v}$, also $(\gamma\,\partial)$ für $(+\mathring{\varkappa}\gamma,\ -\overset{\times}{\varkappa}\gamma)\,\partial/\partial(u,v)$]:

$$d^i U/d^i V = \left[1-(J^{\cdot}\,\varepsilon\,\partial^-)\right] S\,(u,v),$$

und entsprechend

$$d^i U'/d^i V' = \left[1+(J^{\cdot}\,\varepsilon\,\partial^+)\right] S'(u,v).$$

Damit wird aus (8.71 S):

$$\left.\begin{aligned}&(1-J^{\cdot}\,\partial\varepsilon)\,[1+(J^{\cdot}\,\varepsilon\,\partial^{+})]\,\frac{S'(u,v)}{1+\frac{\overset{\times}{\varkappa}}{\mathring{\varkappa}}\,S'(u,v)}-\\&-(1+J^{\cdot}\,\partial\varepsilon)\,[1-(J^{\cdot}\,\varepsilon\,\partial^{-})]\,\frac{S(u,v)}{1+\frac{\overset{\times}{\varkappa}}{\mathring{\varkappa}}\,S(u,v)}\\&=(\mu+\mu')\,\mathring{\varkappa}\,[\mathring{\varkappa}\,\varepsilon-u\,(\partial\varepsilon)]\,.\end{aligned}\right\}\qquad(8.73\,S)$$

Ganz entsprechend wird aus (8.71 T):

$$\left.\begin{aligned}&(1-J^{\cdot}\,\partial\varepsilon)\,[1+(J^{\cdot}\,\varepsilon\,\partial^{+})]\,\frac{T'(u,v)}{1+\frac{\overset{\times}{\varkappa}}{\mathring{\varkappa}}\,T'(u,v)}-\\&-(1+J^{\cdot}\,\partial\varepsilon)\,[1-(J^{\cdot}\,\varepsilon\,\partial^{-})]\,\frac{T(u,v)}{1+\frac{\overset{\times}{\varkappa}}{\mathring{\varkappa}}\,T(u,v)}\\&=(\mu+\mu')\,\mathring{\varkappa}\,[\mathring{\varkappa}\,(\varepsilon+\theta\,\varepsilon)-u\,(\partial\varepsilon)]\,.\end{aligned}\right\}\qquad(8.73\,T)$$

Zwischen den sagittalen Aberrationsfunktionen $S(u,v)$ vor einer Fläche und $S'(u,v)$ hinter der Fläche besteht also, mit linearen Operatoren geschrieben, ein *gebrochen linearer* Zusammenhang; ebenso zwischen den tangentialen Aberrationsfunktionen $T(u,v)$ vor und $T'(u,v)$ hinter der Fläche; und wir können unmittelbar in der Form eines *Kettenbruchs* den Zusammenhang zwischen $S_1(u,v)$ und $S'_k(u,v)$, und ebenso den zwischen $T_1(u,v)$ und $T'_k(u,v)$ hinschreiben[1]:

$$\left.\begin{aligned}\frac{-1}{S_1(u,v)}={}_1\Big|\,\frac{\mathring{\varkappa}}{\overset{\times}{\varkappa}}+1\big/\!\!\big\langle\,[1-(J^{\cdot}\,\varepsilon\,\partial^{-})]^{-1}\,(1+J^{\cdot}\,\partial\varepsilon)^{-1}\,\big\langle(\mu+\mu')\,\mathring{\varkappa}\,[\mathring{\varkappa}\,\varepsilon-u\,(\partial\varepsilon)]+\\+(1-J^{\cdot}\,\partial\varepsilon)\,[1+(J^{\cdot}\,\varepsilon\,\partial^{+})]\big/\!\!\big\langle-\frac{\overset{\times}{\varkappa}}{\mathring{\varkappa}}+\Big|^{k}\,\frac{-1}{S'_k(u,v)}\,;\end{aligned}\right\}\qquad(8.74\,S)$$

$$\left.\begin{aligned}\frac{-1}{T_1(u,v)}={}_1\Big|\,\frac{\overset{\times}{\varkappa}}{\mathring{\varkappa}}+1\big/\!\!\big\langle\,[1-(J^{\cdot}\,\varepsilon\,\partial^{-})]^{-1}\,(1+J^{\cdot}\,\partial\varepsilon)^{-1}\,\big\langle(\mu+\mu')\,\mathring{\varkappa}\,[\mathring{\varkappa}\,(\varepsilon+\theta\,\varepsilon)-u\,(\partial\varepsilon)]+\\+(1-J^{\cdot}\,\partial\varepsilon)\,[1+(J^{\cdot}\,\varepsilon\,\partial^{+})]\big/\!\!\big\langle-\frac{\overset{\times}{\varkappa}}{\mathring{\varkappa}}+\Big|^{k}\,\frac{-1}{T'_k(u,v)}\,.\end{aligned}\right\}\qquad(8.74\,T)$$

Die Ausdrücke (8.73 S) bis (8.74 T) versagen bei verschwindend kleinem $\mathring{\varkappa}$, nicht aber bei verschwindend kleinem $\overset{\times}{\varkappa}$. Sie lassen sich leicht so umformen, daß sie bei verschwindend kleinem $\mathring{\varkappa}$ brauchbar werden, dann aber bei verschwindend kleinem $\overset{\times}{\varkappa}$ versagen[2]:

$$\left.\begin{aligned}&(1-J^{\cdot}\,\partial\varepsilon)/[1+(J^{\cdot}\,\varepsilon\,\partial^{+})]\left[\frac{\mathring{\varkappa}}{\overset{\times}{\varkappa}}+S'(u,v)\right]-\\&-(1+J^{\cdot}\,\partial\varepsilon)/[1-(J^{\cdot}\,\varepsilon\,\partial^{-})]\left[\frac{\mathring{\varkappa}}{\overset{\times}{\varkappa}}+S(u,v)\right]\\&=-(\mu+\mu')\,\overset{\times}{\varkappa}\,[\overset{\times}{\varkappa}\,\varepsilon+v\,(\partial\varepsilon)]\,;\end{aligned}\right\}\qquad(8.73\,s)$$

$$\left.\begin{aligned}&(1-J^{\cdot}\,\partial\varepsilon)/[1+(J^{\cdot}\,\varepsilon\,\partial^{+})]\left[\frac{\mathring{\varkappa}}{\overset{\times}{\varkappa}}+T'(u,v)\right]-\\&-(1+J^{\cdot}\,\partial\varepsilon)/[1-(J^{\cdot}\,\varepsilon\,\partial^{-})]\left[\frac{\mathring{\varkappa}}{\overset{\times}{\varkappa}}+T(u,v)\right]\\&=-(\mu+\mu')\,\overset{\times}{\varkappa}\,[\overset{\times}{\varkappa}\,(\varepsilon+\theta\,\varepsilon)+v\,(\partial\varepsilon)]\,,\end{aligned}\right\}\qquad(8.73\,t)$$

---

[1] Auch bei einem Operator A schreiben wir A/B für $A\,B^{-1}$.

[2] Mit ()/[][] ist ()/⟨[][]⟩ gemeint. — Es lassen sich auch Ausdrücke angeben, die weder im Fall $\mathring{\varkappa}=0$ noch im Fall $\overset{\times}{\varkappa}=0$ versagen. Den Fall, daß $\mathring{\varkappa}_\tau=0$ und zugleich $\overset{\times}{\varkappa}_\tau=0$ ist, gibt es nicht.

folglich

$$\left.\begin{aligned}-S_1(u,v)={}_1\Big|[1-(J^{\cdot}\varepsilon\,\partial^-)]^{-1}\Big\langle+\frac{\mathring{\varkappa}}{\check{\varkappa}}+1\Big/\!\Big\langle(1+J^{\cdot}\partial\varepsilon)^{-1}\Big\langle-(\mu+\mu')\check{\varkappa}[\check{\varkappa}\varepsilon+v(\partial\varepsilon)]+\\+(1-J^{\cdot}\partial\varepsilon)\Big/\!\Big\langle-\frac{\mathring{\varkappa}}{\check{\varkappa}}+[1+(J^{\cdot}\varepsilon\,\partial^+)]\Big|^k[-S_k'(u,v)];\end{aligned}\right\}\quad(8.74$$

$$\left.\begin{aligned}-T_1(u,v)={}_1\Big|[1-(J^{\cdot}\varepsilon\,\partial^-)]^{-1}\Big\langle+\frac{\mathring{\varkappa}}{\check{\varkappa}}+1\Big/\!\Big\langle(1+J^{\cdot}\partial\varepsilon)^{-1}\Big\langle-(\mu+\mu')\check{\varkappa}[\check{\varkappa}(\varepsilon+\theta\,\varepsilon)+\\+v(\partial\varepsilon)]+(1-J^{\cdot}\partial\varepsilon)\Big/\!\Big\langle-\frac{\mathring{\varkappa}}{\check{\varkappa}}+[1+(J^{\cdot}\varepsilon\,\partial^+)]\Big|^k[-T_k'(u,v)].\end{aligned}\right\}\quad(8.74$$

Wir hatten jedem der Meridianstrahlen einen sagittal benachbarten Strahl und einen tangential benachbarten Strahl zugeordnet, hatten aber noch nicht festgelegt, welchen Punkt der sagittale, und welchen Punkt der tangentiale Nachbarstrahl mit dem Meridianstrahl gemeinsam haben soll. Wir könnten beide Nachbarstrahlen aus dem gleichen Objektpunkt kommen lassen, aus dem auch der Meridianstrahl kommt: $\delta\underline{U}_0=0$, $d\underline{U}_0=0$. Die Bildhöhen-Aberrationen $(\delta\underline{U}_{k+1}-\delta\underline{U}_0)$ und $(d\underline{U}_{k+1}-d\underline{U}_0)$ treten dann in der Gaußschen Bildebene auf. Oder wir können den sagittalen und den tangentialen Nachbarstrahl durch den gleichen Punkt der Gaußschen Bildebene gehen lassen, durch den auch der Meridianstrahl geht:

$$\delta\underline{U}_{k+1}=0,\qquad d\underline{U}_{k+1}=0;\tag{8.75}$$

die Bildhöhen-Aberrationen $(\delta\underline{U}_{k+1}-\delta\underline{U}_0)$ und $(d\underline{U}_{k+1}-d\underline{U}_0)$ treten dann im Objektraum auf. Die sagittalen und tangentialen Bildfehler eines optischen Systems werden auf die eine Weise ebenso gut beschrieben wie auf die andere. Wir wählen die Festlegung (8.75), weil das zu einfacheren Formeln führt.

Wegen $\delta\underline{U}_{k+1}=\underline{\underline{\delta U}}_{k+1}=d^i U_{k+1}$ folgt aus (8.75): $d^i U_{k+1}=0$, also $d^i U_k'/d^i V_k'=0$ oder

$$S_k'(u,v)=0.\tag{8.75 S}$$

Und wegen $\underline{U}_{k+1}=U_{k+1}/\sqrt{1-G_{k+1}^2}$; $G_{k+1}=(c^- V+c\,U)_{k+1}$ folgt aus (8.75): $(1-c^- VG)_{k+1}\,d\,U_{k+1}+(c^- UG)_{k+1}\,d\,V_{k+1}=0$, also $d^1 U_k'/d^1 V_k'=d\,U_{k+1}/d\,V_{k+1}=(-c^- UG)_{k+1}/(1-c^- VG)_{k+1}$ oder

$$\left.\begin{aligned}T_k'(u,v)&=\left(\frac{-c^- uG}{1-c^- vG}\right)_{k+1}\\ \text{mit}\quad G_{k+1}&=c_{k+1}^-\,v+c_{k+1}\,u.\end{aligned}\right\}\tag{8.75 T}$$

Die beiden Ausdrücke (8.75 S, T) für $S_k'(u,v)$ und $T_k'(u,v)$ sind in (8.74 S, T) rechts einzusetzen, dann sind (8.74 S) und (8.74 T) mit $u=U_0$; $v=V_0$ Gleichungen, die

$$S_1(U_0,V_0)=d^i U_0/d^i V_0$$

und

$$T_1(U_0,V_0)=d^1 U_0/d^1 V_0$$

darstellen für den Fall (8.75), den wir gewählt haben. In diesem Fall ist die sagittale Bildhöhenaberration $(\delta\underline{U}_{k+1}-\delta\underline{U}_0)=-\delta\underline{U}_0=-d^i U_0=-S_1(U_0,V_0)\cdot d^i V_0$, und die tangentiale Bildhöhenaberration $(d\underline{U}_{k+1}-d\underline{U}_0)=-d\underline{U}_0=-d\,(U_0/\sqrt{1-U_0^2})=-(dU_0)/\sqrt{1-U_0^2}^{\,3}=-d^1U_0/(1-U_0^2)=-T_1(U_0,V_0)\cdot d^1V_0/(1-U_0^2)$. — So erhalten wir[1]:

[1] Die Darstellung (VIII) der meridionalen Bildhöhenaberration (U_+-U_0) wird, auf den monochromatischen Fall spezialisiert, in den nun folgenden Formelsatz (VIII.1M) bis (VIII.4′) mit aufgenommen. Diesen Formelsatz als Ganzes bezeichnen wir dann im Text mit (VIII′).

Eine strenge Darstellung der meridionalen Bildhöhenaberration $(U_+ - U_0) = -M_0(u, v)$, der sagittalen Bildhöhenaberration $(\delta \underline{U}_{k+1} - \delta \underline{U}_0) = -S_0(u, v) \cdot \underline{\delta V_0}$ und der tangentialen Bildhöhenaberration $(d\underline{U}_{k+1} - d\underline{U}_0) = -T_0(u, v) \cdot \underline{dV_0}/\sqrt{1-u^2}$ [für den Fall $\delta \underline{U}_{k+1} = 0$; $d\underline{U}_{k+1} = 0$] als Funktionen der objektseitigen Strahldaten $u = U_0$; $v = V_0$; $\underline{\delta V_0}$; $\underline{dV}_0$ und der Langeschen Konstruktionsdaten des Systems:

$$-M_0(u, v) = {}_1\big|\, [1 - (\gamma\partial^-)]^{-1} \langle 2\{\mathring{\varkappa}\gamma\} + [1 + (\gamma\partial^+)] \,\big|^k \varphi_+ ; \tag{VIII.1 M}$$

$$\left.\begin{aligned} -1/S_0(u, v) = {}_1\Big|\, \frac{\overset{\times}{\varkappa}}{\mathring{\varkappa}} + 1 \Big/\!\!\Big\langle\, [1 - (\gamma\partial^-)]^{-1} (1 + \partial\gamma)^{-1} \big\langle 2\{\mathring{\varkappa}\breve{\gamma} - u(\partial\breve{\gamma})\} + \\ + (1 - \partial\gamma)\,[1 + (\gamma\partial^+)] \Big/\!\!\Big\langle - \frac{\overset{\times}{\varkappa}}{\mathring{\varkappa}} + \Big|^k \infty ; \end{aligned}\right\} \tag{VIII.1 S}$$

$$\left.\begin{aligned} -1/T_0(u, v) = {}_1\Big|\, \frac{\overset{\times}{\varkappa}}{\mathring{\varkappa}} + 1 \Big/\!\!\Big\langle\, [1 - (\gamma\partial^-)]^{-1} (1 + \partial\gamma)^{-1} \big\langle 2\{\mathring{\varkappa}(\breve{\gamma} + \theta\breve{\gamma}) - u(\partial\breve{\gamma})\} + \\ + (1 - \partial\gamma)\,[1 + (\gamma\partial^+)] \Big/\!\!\Big\langle - \frac{\overset{\times}{\varkappa}}{\mathring{\varkappa}} + \Big|^k \frac{1}{\varphi_T} , \end{aligned}\right\} \tag{VIII.1 T}$$

wobei

$$\left.\begin{aligned} &\varphi_+ = \frac{G+u}{\bar{u}} \, \frac{G-u}{1+\bar{u}} \, u ; \qquad \varphi_T = \frac{c_{\overline{k+1}}\, uG}{\bar{v}} ; \\ &\bar{u} = \sqrt{1 + u^2 - G^2} ; \qquad \bar{v} = 1 - c_{\overline{k+1}}\, vG ; \qquad G = c_{\overline{k+1}}\, v + c_{k+1} u \end{aligned}\right\} \tag{VIII.2'}$$

und

$$\left.\begin{aligned} &\gamma = J^{\cdot} \cdot \varepsilon ; \qquad \breve{\gamma} = \frac{\mu + \mu'}{2} \mathring{\varkappa} \cdot \varepsilon ; \\ &\varepsilon = \frac{\mu' - \mu}{\mu + \mu'} \, \frac{1}{c} \, \frac{P^{\cdot}}{\bar{J}^{\cdot}} \, \frac{G^{\cdot} - J^{\cdot}}{\bar{G}^{\cdot} + \bar{J}^{\cdot}} ; \end{aligned}\right\} \tag{VIII.3'}$$

$$\left.\begin{aligned} &P^{\cdot} = v - tu ; \\ &\frac{2}{\mu + \mu'} J^{\cdot} = \mathring{\varkappa} v + \overset{\times}{\varkappa} u ; \quad \overset{\times}{\varkappa} = -c - t\mathring{\varkappa} ; \quad \mathring{\varkappa} = -\frac{\mu' c^+ - \mu c^-}{\mu' - \mu} ; \\ &\frac{2}{\mu + \mu'} G^{\cdot} = \mathring{\chi} v + \overset{\times}{\chi} u ; \quad \overset{\times}{\chi} = +c - t\mathring{\chi} ; \quad \mathring{\chi} = +\frac{\mu' c^+ + \mu c^-}{\mu' + \mu} ; \\ &\bar{J}^{\cdot} = \sqrt{L - J^{\cdot 2}} ; \qquad c_\tau^+ = \frac{c_{\tau+1} - c_\tau}{t_{\tau+1} - t_\tau} ; \\ &\bar{G}^{\cdot} = \sqrt{L - G^{\cdot 2}} ; \qquad c_\tau^- = \frac{c_\tau - c_{\tau-1}}{t_\tau - t_{\tau-1}} ; \\ &L = 1 - \left(\frac{\mu' - \mu}{\mu + \mu'}\right)^2 J^{\cdot 2} / [1 - J^{\cdot 2}/L] \quad \text{(implizit)} \\ & = 1 - \frac{l}{2} + \sqrt{\left(\frac{l}{2}\right)^2 - \left(\frac{\mu' - \mu}{\mu + \mu'}\right)^2 J^{\cdot 2}} \quad \text{(explizit)} \\ &\qquad \text{mit } l = (1 - J^{\cdot 2}) + \left(\frac{\mu' - \mu}{\mu + \mu'}\right)^2 J^{\cdot 2} ; \\ &\theta \equiv u \frac{\partial}{\partial u} + v \frac{\partial}{\partial v} ; \\ &\partial_\tau \equiv \mathring{\varkappa}_\tau \frac{\partial}{\partial u} - \overset{\times}{\varkappa}_\tau \frac{\partial}{\partial v} ; \\ &(\gamma_\tau \partial_\tau^+) \equiv (+\mathring{\varkappa}_\tau \gamma_\tau, -\overset{\times}{\varkappa}_\tau \gamma_\tau)\, \partial^+ / \partial^+ (u, v) ; \\ &(\gamma_\tau \partial_\tau^-) \equiv (+\mathring{\varkappa}_\tau \gamma_\tau, -\overset{\times}{\varkappa}_\tau \gamma_\tau)\, \partial^- / \partial^- (u, v) \end{aligned}\right\} \tag{VIII.4'}$$

ist. Der Übersicht halber ist in den meisten Formeln der Index τ (die Flächen-Nr.) weggelassen. Die Langeschen Konstruktionsdaten sind in (VIII.5) aufgezählt.

Der hier zusammengestellte Formelsatz gilt für den monochromatischen Fall: $\nu=\mu$.

Für den in (VIII.1 S) stehenden Flächenanteil $|\ldots|$ kann auch

$$\left.\begin{aligned}&\Big|1\big/\!\!\big\langle[1-(\gamma\partial^-)]^{-1}\big\langle+\frac{\mathring{\varkappa}}{\overset{\times}{\varkappa}}+1\big/\!\!\big\langle(1+\partial\gamma)^{-1}\big\langle-2[\overset{\times}{\varkappa}\hat{\gamma}+v(\partial\hat{\gamma})]+\\&\qquad+(1-\partial\gamma)\big/\!\!\big\langle-\frac{\mathring{\varkappa}}{\overset{\times}{\varkappa}}+[1+(\gamma\partial^+)]\big/\!\!\big\langle\Big|\end{aligned}\right|\right\}\quad\text{(VIII.1 }s\text{)}$$

gesetzt werden, und für den in (VIII.1 T) stehenden Flächenanteil $|\ldots|$ kann

$$\left.\begin{aligned}&\Big|1\big/\!\!\big\langle[1-(\gamma\partial^-)]^{-1}\big\langle+\frac{\mathring{\varkappa}}{\overset{\times}{\varkappa}}+1\big/\!\!\big\langle(1+\partial\gamma)^{-1}\big\langle-2[\overset{\times}{\varkappa}(\hat{\gamma}+\theta\hat{\gamma})+v(\partial\hat{\gamma})]+\\&\qquad+(1-\partial\gamma)\big/\!\!\big\langle-\frac{\mathring{\varkappa}}{\overset{\times}{\varkappa}}+[1+(\gamma\partial^+)]\big/\!\!\big\langle\Big|\end{aligned}\right\}\quad\text{(VIII.1 }t\text{)}$$

gesetzt werden, wobei

$$\hat{\gamma}=\frac{\mu+\mu'}{2}\overset{\times}{\varkappa}\cdot\varepsilon \qquad \text{(VIII.3'')}$$

ist. Die Ausdrücke (VIII.1 s) und (VIII.1 t) versagen nicht bei verschwindend kleinem $\mathring{\varkappa}$. Sie lassen aber nicht wie die in (VIII.1 S) und (VIII.1 T) stehenden Ausdrücke die „additiven Flächenanteile“ $2\{\mathring{\varkappa}\breve{\gamma}-u(\partial\breve{\gamma})\}\equiv 2\varphi^{(i)}$ und $2\{\mathring{\varkappa}(\breve{\gamma}+\theta\breve{\gamma})-u(\partial\breve{\gamma})\}\equiv 2\varphi^{(1)}$ hervortreten, deren Summen $\sum_1^k 2\varphi^{(i)}$ und $\sum_1^k 2\varphi^{(1)}$ in primärer Näherung (s. Ziff. 52) die sagittale Aberration $(-S_0)$ und die tangentiale Aberration $(-T_0-\varphi_T)$ darstellen.

52. Näherungsdarstellungen. Zwischen der *strengen* Darstellung der (meridionalen, sagittalen und tangentialen) Bildhöhenaberration einerseits und der *Seidelschen* Näherungsdarstellung andererseits liegen verschiedene *Zwischenstufen* der Näherung, auf die wir schrittweise von der strengen Darstellung aus kommen wollen.

α) *Differentielle Näherung.* Ersetzt man in den strengen Ausdrücken (VIII.1 M, S, T) überall ∂^- und ∂^+ durch ∂, also die vorkommenden Differenz-Operatoren durch Differential-Operatoren, dann kommt man zu einer sehr guten Näherung; in Ziff. 50 hatten wir für die meridionale Bildhöhenaberration die hereinkommende Ungenauigkeit nachgerechnet. Wir wollen im folgenden den Fehler der Näherung mehr summarisch betrachten, indem wir wie in Abschnitt IV und V seine Größenordnung nach (Bildfehler-)Ordnung und Grad zählen. Wir wollen dabei aber eine als Faktor auftretende Potenz p von $\left(\frac{\mu'-\mu}{\mu+\mu'}\right)_\tau=\left(\frac{n_e-n_e'}{n_e+n_e'}\right)_\tau\equiv\bar{\alpha}_\tau\left[=\frac{\alpha_\tau}{\mu_\tau+\mu_\tau'}\right]$ nicht einfach als $0(\alpha^p)$ zählen, sondern wollen den Faktor $\bar{\alpha}_\tau^p$ ausdrücklich nennen, weil ein solcher Faktor sicherer als $0(\alpha^p)$ zu kleinen Werten führt: bei Glas-Luft-Flächen sind die Werte von $\bar{\alpha}$ nur mäßig klein ($|\bar{\alpha}|=\frac{1}{5}\ldots\frac{1}{3}$ für $n_{\text{Glas}}=1{,}50\ldots 2{,}00$) und nicht sehr klein, so daß durch Summationen eine Größe $0(\alpha^p)$ einiges von der Kleinheit eingebüßt haben kann[1], die das einzelne Glied mit dem Faktor $\bar{\alpha}_\tau^p$ noch hat. — An Hand der Rechnung auf S. 177—179 können wir feststellen, daß beim Ersetzen der Differenz-Operatoren $\gamma\partial^+$ und $\gamma\partial^-$

[1] Ebenso wie bei der üblichen summarischen Zählung nach Bildfehler-Ordnungen damit zu rechnen ist, daß eine durch Summationen zustandegekommene Größe $0^{(\Omega)}$ einiges von der Kleinheit eingebüßt haben kann, die ein einzelnes Glied mit einer Potenz G^Ω oder J^Ω (z.B.) noch hat: dann nämlich, wenn auch bei Werten $|J|\sim 0{,}5$ (z.B.) die Entwicklungen und die Zählung nach Bildfehler-Ordnungen vorgenommen werden; gerade dann interessiert man sich aber für die Bildfehler höherer Ordnung, entwickelt z.B. eine Theorie der Bildfehler 5. Ordnung und vernachlässigt dabei summarisch Größen $\sim 0^{(7)}$. Solche Entwicklungen sind bekanntlich brauchbar und nützlich, obwohl die Zählung bedenklich ist.

durch den Differential-Operator $\gamma\partial$ im meridionalen Fall ein Fehler der Größenordnung $\bar{\alpha}_\tau^3\cdot 0^{(7)}$ hereinkommt, wenn wir die bei einem Flächenanteil (Nr. τ) gemachten Vernachlässigungen, *additiv neben dem additiven Flächenanteil* $2\{\mathring{\varkappa}\gamma\}_\tau$ *betrachtet*, als Fehler der Näherung zählen. Im sagittalen und tangentialen Fall zählen wir den Fehler der Näherung in entsprechender Weise, nur mit dem Unterschied, daß wir die vor (VIII.1 M, S, T) angegebene Multiplikation mit $\delta\underline{V}_0$ bzw. $d\underline{V}_0$, mit der S_0 und T_0 auf die sagittale und tangentiale Bildhöhenaberration umgerechnet werden:

$$(\delta\underline{U}_{k+1}-\delta\underline{U}_0)=-S_0(u,v)\cdot\delta\underline{V}_0;$$

$$(d\underline{U}_{k+1}-d\underline{U}_0)=-T_0(u,v)\cdot d\underline{V}_0/\sqrt{1-u^2},$$

bei der Zählung mitberücksichtigen. Das heißt, wir multiplizieren die bei einem Flächenanteil der Darstellung (VIII.1 S) bzw. (VIII.1 T) gemachten Vernachlässigungen, additiv neben dem additiven Flächenanteil $2\{\mathring{\varkappa}\breve{\gamma}-u(\partial\breve{\gamma})\}$ bzw. $2\{\mathring{\varkappa}(\breve{\gamma}+\theta\breve{\gamma})-u(\partial\breve{\gamma})\}$ betrachtet, noch mit $0(\delta\underline{V}_0)=0^{(1)}$ bzw. $0(d\underline{V}_0)=0^{(1)}$ $[0(d^i)$ und $0(d^1)$ ignorieren wir; vgl. Ziff. 34], um diese Vernachlässigungen so zu bewerten, wie sie sich bei der gesuchten Bildhöhenaberration auswirken. Dann ist der Fehler der Näherung, bei der ∂^+ und ∂^- durch ∂ ersetzt sind, im sagittalen und tangentialen Fall wie im meridionalen Fall von der Größenordnung $\bar{\alpha}_\tau^3\cdot 0^{(7)}$. Auch wenn man in der Darstellung (VIII.1 s); (VIII.1 t), die bei einer Fläche mit verschwindend kleinem $\mathring{\varkappa}$ gebraucht wird, ∂^+ und ∂^- durch ∂ ersetzt, ist die bei dieser Fläche (Nr. τ) hereinkommende Ungenauigkeit — so bewertet, wie sie sich bei der Bildhöhenaberration des Systems auswirkt — von der Größenordnung $\bar{\alpha}_\tau^3\cdot 0^{(7)}$.

β) *Sekundäre Näherung.* Ein nächster Näherungsschritt, der zu einer „sekundären" Näherung führt, ist der, in (VIII.1 M, S, T) nach allen neben einer 1 stehenden Größen (die von zweiter Ordnung und erstem Grad klein sind) zu entwickeln, und auch nach $\check{\varkappa}/\mathring{\varkappa}$ (was zwar eine Größe nullter Ordnung und nullten Grades ist[1], die aber neben dem Reziproken einer Größe zweiter Ordnung und ersten Grades steht), und dann Glieder zu vernachlässigen, die — additiv neben dem additiven Flächenanteil $2\{\}_\tau$ betrachtet — von der Größenordnung $\bar{\alpha}_\tau\cdot 0(\alpha^2)\cdot 0^{(7)}$ sind [nach der Multiplikation mit $0^{(1)}$ im sagittalen und im tangentialen Fall]. Diese Entwicklung wird übersichtlicher, wenn wir von den Formeln (8.39); (8.73 S); (8.73 T) ausgehen, die zu der Darstellung (VIII.1 M); (VIII.1 S); (VIII.1 T) geführt hatten, und wenn wir zuvor die

γ) *Primäre Näherung*

betrachten. Bei der „primären" Näherung sind die Vernachlässigungen bei einem Flächenanteil Nr. τ [wieder wie unter α) und β) betrachtet] von der Größenordnung $\bar{\alpha}_\tau\cdot 0(\alpha)\cdot 0^{(5)}$. Wir hatten die primäre Näherung schon im Anschluß an die Iterationsformeln in Abschnitt V kennengelernt, wollen sie jetzt aber den Formeln (8.39); (8.73 S); (8.73 T) entnehmen. Nach diesen Formeln ist

$$M'-M=2\mathring{\varkappa}\gamma-2\gamma\,\partial M^{\boldsymbol{\cdot}}+\bar{\alpha}^3\cdot 0^{(7)} \qquad (8.76M)$$

$$=2\mathring{\varkappa}\gamma \qquad +\bar{\alpha}\cdot 0(\alpha)\cdot 0^{(5)} \qquad (8.77M)$$

mit $M=M(u,v)$; $M'=M'(u,v)$; $M^{\boldsymbol{\cdot}}\equiv\frac{1}{2}(M+M')$ und mit $\mathring{\varkappa}$, γ, ∂ wie in (VIII');

$$S'-S=2\{\mathring{\varkappa}\breve{\gamma}-u(\partial\breve{\gamma})\}\cdot\left[1+2\frac{\check{\varkappa}}{\mathring{\varkappa}}S^{\boldsymbol{\cdot}}\right]+2(\partial\gamma)S^{\boldsymbol{\cdot}}-2\gamma\partial S^{\boldsymbol{\cdot}}+\bar{\alpha}\cdot 0(\alpha^2)\cdot 0^{(6)} \qquad (8.76S)$$

$$=2\{\mathring{\varkappa}\breve{\gamma}-u(\partial\breve{\gamma})\}+\bar{\alpha}\cdot 0(\alpha)\cdot 0^{(4)} \qquad (8.77S)$$

[1] Den Fall, daß $\mathring{\varkappa}$ verschwindend klein ist, dürfen wir, wie eine genauere Betrachtung zeigt, ignorieren bei dieser Entwicklung; die Endformeln (VIII.1 a, b, c) gelten auch für diesen Fall.

mit $S=S(u,v)$; $S'=S'(u,v)$; $S^{\cdot}\equiv\frac{1}{2}(S+S')$ und mit $\overset{\times}{\varkappa}$, $\breve{\gamma}$ wie in (VIII');

$$\left.\begin{aligned}T'-T=2\{\mathring{\varkappa}(\breve{\gamma}+\theta\breve{\gamma})-u(\partial\breve{\gamma})\}\cdot\left[1+2\frac{\overset{\times}{\varkappa}}{\mathring{\varkappa}}T^{\cdot}\right]+2(\partial\gamma)T^{\cdot}-2\gamma\partial T^{\cdot}+\\+\bar{\alpha}\cdot 0(\alpha^2)\cdot 0^{(6)}\end{aligned}\right\}\quad(8.76\,T)$$

$$=2\{\mathring{\varkappa}(\breve{\gamma}+\theta\breve{\gamma})-u(\partial\breve{\gamma})\}+\bar{\alpha}\cdot 0(\alpha)\cdot 0^{(4)}\qquad(8.77\,T)$$

mit $T=T(u,v)$; $T'=T'(u,v)$; $T^{\cdot}\equiv\frac{1}{2}(T+T')$ und mit θ wie in (VIII').

Die Gln. (8.77 M, S, T) sind die Formeln für $(M'-M)_\tau$, $(S'-S)_\tau$, $(T'-T)_\tau$ in *primärer* Näherung. Man sieht sofort, daß wegen $M'_k=-\varphi_+$; $S'_k=0$; $T'_k=-\varphi_T$

$$\left.\begin{aligned}-M_\sigma&=\sum_{\tau=\sigma}^{k}2\{\mathring{\varkappa}\gamma\}_\tau+\varphi_++0(\alpha^2)\cdot 0^{(5)};\\-S_\sigma&=\sum_{\tau=\sigma}^{k}2\{\mathring{\varkappa}\breve{\gamma}-u(\partial\breve{\gamma})\}_\tau+0(\alpha^2)\cdot 0^{(4)};\\-T_\sigma&=\sum_{\tau=\sigma}^{k}2\{\mathring{\varkappa}(\breve{\gamma}+\theta\breve{\gamma})-u(\partial\breve{\gamma})\}_\tau+\varphi_T+0(\alpha^2)\cdot 0^{(4)}\end{aligned}\right\}\quad(8.76')$$

ist ($\sigma=1\ldots k+1$). Dies sind die Formeln für M_σ, S_σ, T_σ in *primärer* Näherung. Die Formeln für M_ς, S_ς, T_ς in *sekundärer* Näherung erhält man, indem man in (8.76 M, S, T) rechts für $M^{\cdot}$, $S^{\cdot}$, $T^{\cdot}$ die Ausdrücke der primären Näherung (8.76') einsetzt, dann über die Flächen Nr. ς bis k summiert und $M'_k=-\varphi_+$, $S'_k=0$, $T'_k=-\varphi_T$ setzt. Die

δ) *Seidelsche Näherung*

entsteht aus der primären Näherung dadurch, daß im Ausdruck (VIII.3') für ε die in den Nennern stehenden Cosinus-Größen $\bar{J}^{\cdot}$ und $\bar{G}^{\cdot}$ gleich 1 gesetzt werden, und daß in den Ausdrücken (VIII.2') für φ_+ und φ_T die in den Nennern stehenden Größen $\bar{u}$ und $\bar{v}$ gleich 1 gesetzt werden; insgesamt wird bei der meridionalen Bildhöhenaberration, und ebenso bei der sagittalen und bei der tangentialen Bildhöhenaberration, eine Größe $\sim 0(\alpha)\cdot 0^{(5)}$ vernachlässigt.

So kommen wir zu den im folgenden zusammengestellten Formelsätzen für die *primäre* Näherung und für die *sekundäre* Näherung. Der Formelsatz für die *differentielle* Näherung geht aus der strengen Darstellung (VIII') unmittelbar dadurch hervor, daß dort ∂^- und ∂^+ durch ∂ ersetzt werden. Auch den Formelsatz für die *Seidelsche* Näherung brauchen wir nicht gesondert hinzuschreiben; er geht aus dem Formelsatz für die primäre Näherung dadurch hervor, daß $\bar{J}^{\cdot}$, $\bar{G}^{\cdot}$, $\bar{u}$, $\bar{v}$ durch 1 ersetzt werden.

Für die *primäre* Näherung hat man anstelle von (VIII.1 M, S, T) die Formeln:

$$\left.\begin{aligned}-M_0(u,v)&=\sum_{\tau=1}^{k}2\varphi_\tau+\varphi_++0(\alpha^2)\cdot 0^{(5)}; &\varphi&=\mathring{\varkappa}\cdot\gamma;\\-S_0(u,v)&=\sum_{\tau=1}^{k}2\varphi_\tau^{(i)}\qquad+0(\alpha^2)\cdot 0^{(4)}; &\varphi^{(i)}&=\mathring{\varkappa}\cdot\breve{\gamma}-u(\partial\breve{\gamma});\\-T_0(u,v)&=\sum_{\tau=1}^{k}2\varphi_\tau^{(1)}+\varphi_T+0(\alpha^2)\cdot 0^{(4)}; &\varphi^{(1)}&=\mathring{\varkappa}(\breve{\gamma}+\theta\breve{\gamma})-u(\partial\breve{\gamma})\\&\text{mit }\mathring{\varkappa},\gamma,\breve{\gamma},\partial,\theta\text{ und }\varphi_+,\varphi_T\\&\text{gemäß (VIII.2'), (VIII.3'), (VIII.4').}\end{aligned}\right\}\quad(\text{VIII.1}\,p)$$

Für die *sekundäre* Näherung hat man anstelle von (VIII.1 M, S, T) die Formeln

$$\left.\begin{aligned}
-M_0(u,v) &= \sum_{\tau=1}^{k} (M'-M)_\tau + \varphi_+ ;\\
-S_0(u,v) &= \sum_{\tau=1}^{k} (S'-S)_\tau ;\\
-T_0(u,v) &= \sum_{\tau=1}^{k} (T'-T)_\tau + \varphi_T ;
\end{aligned}\right\} \qquad \text{(VIII.1 a)}$$

$$\left.\begin{aligned}
\tfrac{1}{2}(M'-M)_\tau &= \varphi_\tau + \gamma_\tau\,\partial_\tau \sum_{\tau+\frac{1}{2}}^{k+\frac{1}{2}} 2\varphi + \bar{\alpha}_\tau \cdot 0(\alpha^2)\cdot 0^{(7)};\\
\tfrac{1}{2}(S'-S)_\tau &= \varphi_\tau^{(i)} + [2\psi_\tau^{(i)} + (\partial_\tau\gamma_\tau) + \gamma_\tau\,\partial_\tau] \sum_{\tau+\frac{1}{2}}^{k+\frac{1}{2}} 2\varphi^{(i)} + \bar{\alpha}_\tau \cdot 0(\alpha^2)\cdot 0^{(6)};\\
\tfrac{1}{2}(T'-T)_\tau &= \varphi_\tau^{(1)} + [2\psi_\tau^{(1)} + (\partial_\tau\gamma_\tau) + \gamma_\tau\,\partial_\tau] \sum_{\tau+\frac{1}{2}}^{k+\frac{1}{2}} 2\varphi^{(1)} + \bar{\alpha}_\tau \cdot 0(\alpha^2)\cdot 0^{(6)}
\end{aligned}\right\} \qquad \text{(VIII.1 b)}$$

mit

$$\left.\begin{aligned}
&\varphi = +\mathring{\varkappa}\cdot\gamma;\\
&\varphi^{(i)} = +\mathring{\varkappa}\cdot\breve{\gamma} \qquad -u(\partial\breve{\gamma}); \qquad \psi^{(i)} = -\overset{\times}{\varkappa}\cdot\breve{\gamma} \qquad -v(\partial\breve{\gamma});\\
&\varphi^{(1)} = +\mathring{\varkappa}(\breve{\gamma}+\theta\breve{\gamma}) - u(\partial\breve{\gamma}); \qquad \psi^{(1)} = -\overset{\times}{\varkappa}(\breve{\gamma}+\theta\breve{\gamma}) - v(\partial\breve{\gamma});\\
&\sum_{\tau+\frac{1}{2}}^{k+\frac{1}{2}} 2\varphi^{()} \equiv \varphi_\tau^{()} + 2\varphi_{\tau+1}^{()} + 2\varphi_{\tau+2}^{()} + \cdots + 2\varphi_{k-1}^{()} + 2\varphi_k^{()} + \varphi_{k+1}^{()};\\
&\qquad \varphi_{k+1} = \varphi_+; \qquad \varphi_{k+1}^{(i)} = 0; \qquad \varphi_{k+1}^{(1)} = \varphi_T;\\
&\mathring{\varkappa}, \overset{\times}{\varkappa}, \gamma, \breve{\gamma}, \partial, \theta \text{ und } \varphi_+, \varphi_T \text{ gemäß (VIII.2'), (VIII.3'), (VIII.4').}
\end{aligned}\right\} \qquad \text{(VIII.1 c)}$$

Die Ausdrücke (VIII.1 b) für die Flächenanteile $(M'-M)_\tau$, $(S'-S)_\tau$, $(T'-T)_\tau$ enthalten in dieser sekundären Näherung zwar Summen über die Flächenanteile $2\varphi^{()}$ der nachfolgenden Flächen und damit nicht nur Konstruktionsdaten aus der Umgebung der Fläche Nr. τ. Aber bei dem Flächenanteil $(M'-M)_\tau$ der *meridionalen* Aberration ist es eine Summe über die nachfolgenden Flächenanteile der *gleichen meridionalen* Aberration (nur in geringerer Näherung); und das Entsprechende gilt für den sagittalen und den tangentialen Flächenanteil (VIII.1 b). — Daher können wir zur Darstellung von $-M_0(u,v)$ jetzt wieder ein Kettenprodukt einführen, und wir können die Darstellung von $-S_0(u,v)$ und $-T_0(u,v)$ auf die gleiche Form bringen in dieser sekundären Näherung; der Flächenanteil Nr. τ enthält dann nur Konstruktionsdaten aus der Umgebung der Fläche Nr. τ:

$$\left.\begin{aligned}
-M_0(u,v) &= {}_1| + \varphi \;\; + [1 \qquad\qquad\qquad\qquad + 2\gamma\partial] \langle + \varphi \;\; |^k + \varphi_+ + 0(\alpha^3)\cdot 0^{(7)};\\
-S_0(u,v) &= {}_1| + \varphi^{(i)} + [1 + 4\psi^{(i)} + 2(\partial\gamma) + 2\gamma\partial] \langle + \varphi^{(i)} |^k \qquad + 0(\alpha^3)\cdot 0^{(6)};\\
-T_0(u,v) &= {}_1| + \varphi^{(1)} + [1 + 4\psi^{(1)} + 2(\partial\gamma) + 2\gamma\partial] \langle + \varphi^{(1)} |^k + \varphi_T + 0(\alpha^3)\cdot 0^{(6)};\\
&\qquad \varphi,\ \varphi^{(i)}, \ldots \text{ wie in (VIII.1 c).}
\end{aligned}\right\} \qquad \text{(VIII.1 a b)}$$

Weitere Umformungen und Diskussionen der hier gegebenen Darstellungen der Bildhöhenaberration werden in der Zeitschrift Optica Acta veröffentlicht[1].

[1] H. Marx: Optica Acta (1967 in Vorbereitung).

Das photographische Objektiv.

Von

Gg. Franke.

Mit 20 Figuren.

1. Einleitung. Das photographische Objektiv hat die Aufgabe, von dem Objektraum ein ebenes perspektivisches Bild zu entwerfen. Nur in besonderen Fällen ist die Bildfläche nicht eben, sondern zylindrisch oder sphärisch gewölbt. Dagegen wird häufig auch das Objekt eben sein, wenn nämlich eine Reproduktion von einem flächenhaften Gegenstand gemacht werden soll. Nach der geometrischen optischen Abbildung wird der Objektraum von einem optischen System wieder in ein räumliches Gebilde transformiert. Die Projektion dieses Bildraumes auf eine Ebene ist indessen kein optisches, sondern ein geometrisches Problem, für welches die Regeln der Perspektive maßgeblich sind. Das optische System ist daran durch den Ort seiner Pupillen beteiligt. Die Größe der Pupillen bestimmt die Definition der auf dem Auffangschirm erhaltenen Bildpunkte, also ihre Schärfe. Die besonderen Anforderungen, die an ein photographisches Objektiv gestellt werden, beruhen darauf, daß die reale optische Abbildung der ideellen Abbildung, die durch eine gebrochene lineare Funktion vermittelt wird, nicht in Strenge genügt. Diese Abweichungen, die Bildfehler, müssen bei einem photographischen Objektiv oft bis zu hohen Aperturen (Öffnungsverhältnissen) und gleichzeitig bis zu großen Bildwinkeln auf unmerkliche oder wenigstens nicht störende Beträge herabgedrückt werden.

Die Theorie des photographischen Objektivs ist weitgehend identisch mit der Theorie der Bildfehler dritter Ordnung, also der Seidelschen Theorie. Für die Berechnung eines brauchbaren Objektives größeren Öffnungsverhältnisses reicht die Seidelsche Theorie freilich bei weitem nicht aus. Aber sie liefert doch eine gute Systematik für die Klassifizierung und für das Verständnis der Wirkungsweise optischer Systeme. Hinzu kommt ihre Bedeutung für die Berechnung eines Objektives, weil ein optisches System nur dann befriedigende Leistungen zeigt, wenn die fünf Seidelschen Bildfehlerkoeffizienten annähernd auf Null gebracht worden sind. Dabei müssen gewisse kleine Restbeträge angestrebt werden, um auch die Einflüsse der Bildfehlerkoeffizienten höherer Ordnung zu berücksichtigen. Wenn also die Seidelsche Theorie auch nicht unmittelbar ein befriedigend korrigiertes Objektiv liefert, so liegen dessen Konstruktionsdaten doch in der Nähe der mit Hilfe der Bildfehlertheorie dritter Ordnung gefundenen Werte. Gleichzeitig erhält man aber aus der Berechnung der Bildfehlerkoeffizienten dritter Ordnung wertvolle Hinweise auf den Einfluß jeder einzelnen Fläche auf den jeweiligen Fehler. Die anschließende Durchrechnung des Verlaufes einzelner Strahlen vom Objekt bis zum Bild ist jedoch unerläßlich, weil nur auf diese Weise ein Urteil über die Güte der Strahlenvereinigung in der Bildebene gewonnen werden kann. Diese Durchrechnung enthält natürlich implizit den Einfluß der Bildfehlerkoeffizienten aller Ordnungen. Die getrennte Betrachtung der Koeffizienten dritter Ordnung aber enthüllt dem Rechner die in dem System ruhenden Möglichkeiten einer weiteren Verbesserung der Strahlenvereinigung.

2. Lochkamera. In den Anfangszeiten der Photographie verwendete man als Objektive einfache Systeme, wie sie von der Camera lucida her gebräuchlich waren, also Fernrohrobjektive. Solche Objektive wurden mit verhältnismäßig langen Brennweiten und kleinen Gesichtsfeldern benutzt. Von einer Entwicklung photographischer Objektive kann man erst sprechen, als man begann, sich von dieser Überlieferung zu lösen. Hier soll aber die Betrachtung der photographischen Objektive nicht von historischen, sondern von systematischen Gesichtspunkten aus geführt werden.

Eine Systematik hat mit der Lochkamera zu beginnen, bei der auf einem Auffangschirm ein reelles Bild ohne Linsen als Folge der geradlinigen Ausbreitung des Lichtes entsteht. Dieses Bild erscheint aber aus zwei Gründen unscharf. Der erste ist die endliche Ausdehnung des Loches. Das Bild eines einzelnen leuchtenden Objektpunktes ist die direkte Projektion des Loches von dem Objektpunkt aus auf die Auffangfläche. Daraus folgt, daß diese rein geometrische Unschärfe um so geringer ist, je kleiner der Durchmesser des Loches gewählt wird. Der zweite Grund für die Bildverbreiterung ist die Beugung an der Begrenzung des Loches. Sie wirkt sich um so stärker aus, je kleiner der Durchmesser der Öffnung wird. Aus diesem gegensätzlichen Einfluß einer Verkleinerung der Öffnung darf man erwarten, daß es für den Durchmesser des Bildchens einen günstigsten Lochdurchmesser gibt. Er hängt von der Entfernung des Objektes und von dem Abstand l der Auffangebene von der Öffnung ab, also von der „Bildweite". Für unendlich weit entfernte Gegenstände ist dieser günstigste Lochdurchmesser bei Licht von der Wellenlänge $\lambda = 0{,}00055$ mm

$$d = 0{,}038\sqrt{l}$$

(l und d in mm gemessen). Damit wird der Durchmesser des Zerstreuungsscheibchens

$$2\varrho = 0{,}36\sqrt{l}\,10^{-3}\,\text{mm}.$$

Das mit Hilfe der Lochkamera erzeugte Bild ist frei von den Bildfehlern im üblichen Sinne.

3. Einfache Objektive. Die unverkittete Einzellinse als photographisches Objektiv findet sich nur bei sehr billigen Kameras. Wegen der Farbabweichung und des sphärischen Fehlers ist ihre Anwendung auf ganz geringe Öffnungsverhältnisse beschränkt. Wollte man der Linse eine Form geben, bei der der sphäri-

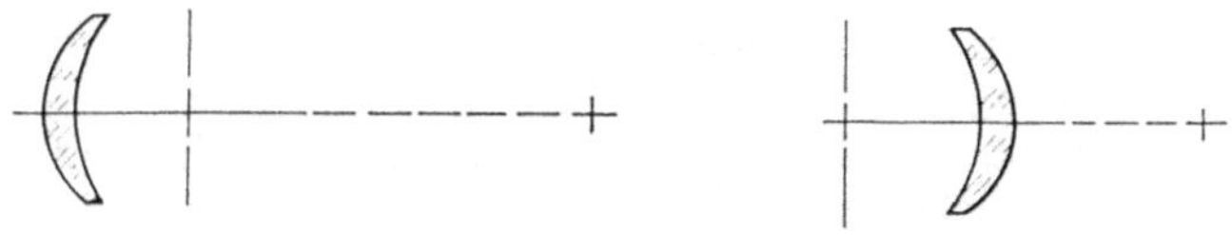

Fig. 1. Einzellinse mit Hinterblende.

sche Fehler ein Minimum ist, so stiegen die Fehler des Bildfeldes mit wachsendem Abstand von der Mitte sehr schnell an. Man bevorzugt deswegen eine Form, bei welcher die Bildqualität im Bildfeld nicht so schnell abfällt und nimmt dafür einen etwas größeren sphärischen Fehler in der Bildmitte in Kauf. Erreicht wird das, indem man einen Verstoß gegen die Sinusbedingung durch entsprechende Durchbiegung der Linse einführt und dann in einem Abstand vor oder hinter der Linse eine Blende anordnet (Fig. 1). Die Linse bekommt so die Form eines Meniskus, der seine hohle Seite der Blende zukehrt. Für solche Menisken ist charakteristisch, daß bei ihnen der Astigmatismus besonders kleine Werte annimmt.

Eine wesentliche Verbesserung bringt die Einführung einer zweiten Linse zur Korrektion des Farbfehlers. Dazu muß diese zweite Linse als Zerstreuungslinse ausgeführt werden und kann entweder mit der ersten Linse verkittet (Fig. 2) oder auch alleinstehend benutzt werden. Verzichtet man auf die Verkittung, so gewinnt man damit die Möglichkeit, auch einen zweiten Fehler zu beeinflussen. Diese Bauart erlaubt Öffnungsverhältnisse bis zu 1:8. Ordnet man zwei gleiche Menisken symmetrisch zur Blende an, so daß sie ihre hohlen Flächen der Blende zukehren (Fig. 3), so bleiben hinsichtlich des Astigmatismus die Vorteile erhalten,

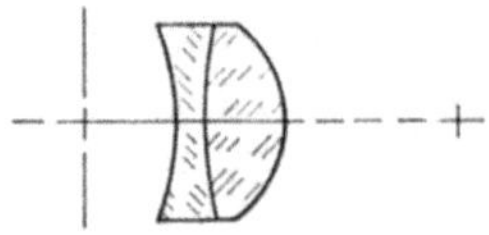

Fig. 2. Achromatische Einzellinse mit Vorderblende.

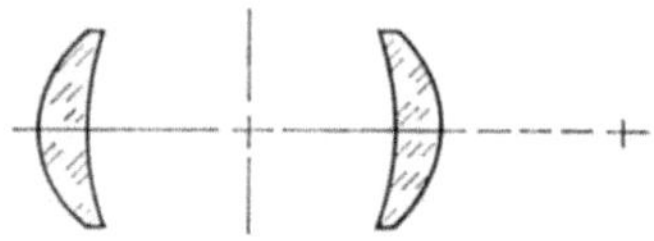

Fig. 3. Symmetrisches Objektiv.

die auch den einzelnen Meniskus schon auszeichneten. Wegen der symmetrischen Anordnung heben sich aber weiterhin alle diejenigen Bildfehler auf, die auf eine Unsymmetrie im Aufbau zurückzuführen sind. Für den Hauptstrahl bedeutet das, daß die Verzeichnung und ihre Farbvariation beseitigt sind (Farbquerfehler). Hält man auch für Ding- und Bildebene Symmetrie ein, d.h. wählt man den Abbildungsmaßstab 1:1, so gilt das gleiche auch für die Sinusbedingung. Dieser Vorteil bleibt sogar erhalten, jedenfalls für die Praxis hinreichend genau, wenn man von der Symmetrie abweicht. Damit verbleiben als Hauptmängel eines solchen Systems nur noch der Farblängsfehler, der sphärische Fehler und die Petzval-Krümmung. Davon läßt sich der Farblängsfehler dadurch beseitigen, daß man beide Menisken einzeln achromatisiert (Fig. 4). Dabei läßt sich auch der sphärische Fehler vermindern und man hat mit diesen symmetrischen Aplanaten einen Objektivtypus, der sich bis auf die Petzval-Krümmung für mäßige Öffnungsverhältnisse recht gut auskorrigieren läßt.

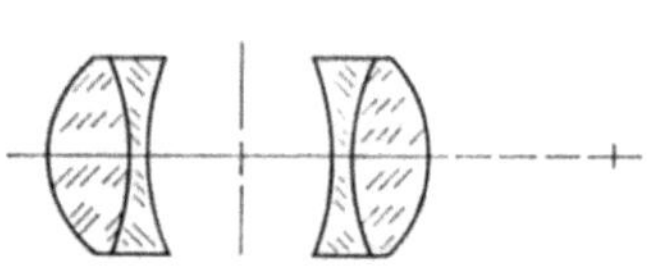

Fig. 4. Doppelanastigmat.

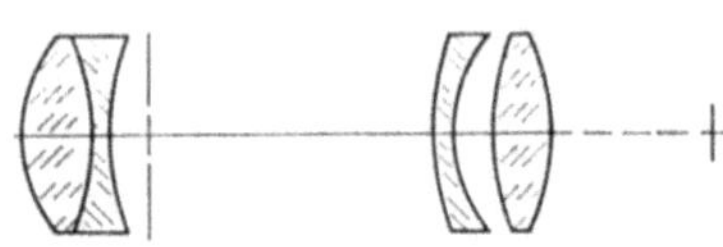

Fig. 5. Petzval-Objektiv.

4. Petzval-Objektiv. Durch einen Verzicht auf den symmetrischen Aufbau kann man das Öffnungsverhältnis erheblich steigern. Die Blende rückt dicht an das vordere Teilsystem; dadurch verliert man zwar die Möglichkeit, den Astigmatismus mit dem Vorderglied zu beeinflussen, dafür aber steht nun das Hinterglied ausschließlich zur Verfügung. Nachträglich wird dann mit dem Vorderglied die Beseitigung des sphärischen Fehlers und die Erfüllung der Sinusbedingung für das ganze System bewerkstelligt. So ist das berühmte Petzvalsche Porträtobjektiv (Fig. 5) aufgebaut, bei dem das Öffnungsverhältnis bis auf 1:3,2 gebracht werden konnte. Das Vorderglied läßt sich verkittet oder unverkittet ausführen, wobei sich bei einem unverkitteten Vorderglied die Möglichkeit bietet, den sphärischen Restfehler in den Zonen zu beseitigen. Auch das Hinterglied, das nun also stark astigmatisch und wenig sphärisch wirkt, läßt verschiedene Bauarten zu. Der einzige Mangel des Petzvalschen Porträtobjektivs ist die starke Wölbung des Bildfeldes, wodurch das brauchbare Feld eingeengt wird. Als Projektions-Kinoobjektiv ist es auch heute noch sehr gut, weil sich mit den modernen hochbrechenden Gläsern das Öffnungsverhältnis leicht noch höher treiben läßt und für diesen

Zweck die Brennweite gegenüber dem Felddurchmesser im allgemeinen groß wird. Als Aufnahmeobjektiv ist bei Porträtaufnahmen eine lange Brennweite wegen der angenehmeren Perspektive erwünscht. Angesichts der geringen Empfindlichkeit der damals verwendeten photographischen Schichten kam das hohe Öffnungsverhältnis einem dringenden Wunsche der Benutzer entgegen.

5. Hebung der Petzval-Krümmung. Die bisher angestellte Betrachtung läßt die Petzval-Krümmung als den hartnäckigsten unter den Bildfehlern erscheinen. Auf zwei Wegen hat man versucht, ihm entgegenzutreten: Durch eine geeignete Glaswahl und durch besondere Maßnahmen im gesamten Aufbau des Objektivs.

α) Glaswahl. Zu der Forderung nach bestimmten Glaseigenschaften wird man durch folgende Überlegung geführt.

Die Petzval-Krümmung ist gegeben durch

$$P = \sum \frac{\varphi_i}{n_i},$$

wobei die φ_i die Brechkräfte der einzelnen Linsen, die n_i die Brechwerte ihrer Gläser sind. Bei einem aus dünnen Linsen zusammengesetzten Linsenpaar gilt für die Farbabweichung $\Delta\Phi$ der Gesamtbrechkraft Φ:

$$\Delta\Phi = \frac{\varphi_1}{\nu_1} + \frac{\varphi_2}{\nu_2},$$

wobei die ν_i die Abbeschen Zahlen der Gläser sind. Verlangt man, daß der Farbfehler und gleichzeitig auch die Petzval-Krümmung beseitigt werden sollen, so muß sowohl

$$\frac{\varphi_1}{\nu_1} + \frac{\varphi_2}{\nu_2} = 0$$

als auch

$$\frac{\varphi_1}{n_1} + \frac{\varphi_2}{n_2} = 0$$

sein. Daraus folgt $\nu_1/\nu_2 = n_1/n_2$ oder auch $\nu_1/n_1 = \nu_2/n_2$, d.h. daß das Glas mit dem höheren ν-Wert (also der geringeren Farbzerstreuung) auch den höheren Brechwert besitzen muß. Nun zeigt aber ein Blick auf die Glaskataloge, daß die allgemeine Tendenz genau umgekehrt verläuft. In der Mitte des vorigen Jahrhunderts beschränkten sich die verfügbaren Gläser ausschließlich auf solche, bei denen der höhere ν-Wert mit einem niedrigeren Brechwert verbunden war. Erst aus dieser hier dargelegten Erkenntnis heraus wurde an die Glashütten die Forderung herangetragen, „neue Gläser" zu entwickeln, die das entgegengesetzte Verhalten zeigten. Zwar führten diese Bemühungen zu einem Erfolg, trotzdem aber ergibt sich für das Verhältnis n_1/n_2 ein Wert, der nicht sehr stark von 1 abweicht. Wenn der gleiche Wert auch in dem Verhältnis von ν_1 und ν_2 eingehalten werden muß, so sind also die Gläser hinsichtlich der Farbzerstreuung nicht allzusehr voneinander verschieden. Man ist also zu einem Kompromiß gezwungen und kann entweder den Farbfehler gut auskorrigieren, wobei man sich mit einer mäßigen Verbesserung des Feldes durch ein Herabsetzen der Petzval-Summe begnügt; oder man legt einen größeren Nachdruck auf das Feld und gibt hinsichtlich der Farbfehler nach. Nun haben diese „Neuachromate" einen schwerwiegenden Nachteil. Sie lassen sich nicht sphärisch auskorrigieren. Das liegt daran, daß bei einem verkitteten Achromaten alter Bauart die Kittfläche zerstreuend wirkt, und diese zerstreuende Wirkung ist für die Beseitigung der sphärischen Abweichung notwendig. Bei einem Neuachromaten aber ist die Kittfläche selbst sammelnd und kann deswegen die sphärischen Fehler der sammelnden

Außenflächen nicht verringern, sondern sie erhöht sie sogar. Man muß also einen Neuachromaten mit einem Altachromaten kombinieren, wobei die Kittfläche in dem Altachromaten die sphärische Abweichung des ganzen Systems kompensieren muß, während die Kittfläche in dem Neuachromaten für die Verbesserung des Astigmatismus herangezogen werden kann. Dazu müssen die beiden Glieder noch einen gewissen Abstand bekommen. Man gelangt so zu einer Bauart, die als *Doppelanastigmat* bezeichnet wird (ähnlich Fig. 5).

β) Linsenstellung. Der andere Weg, die Petzval-Krümmung zu beseitigen, und zwar in einer vollkommeneren Weise, als es durch Einführung der neuen Glasarten möglich ist, benutzt den Umstand, daß der Einfluß einer Linse oder Fläche auf die Petzval-Summe von ihrer Stellung im Strahlengang ganz unabhängig ist. Das geht sofort aus der Formel für die Petzval-Krümmung hervor, in der die Stellung der Linse im System gar nicht vorkommt. Für die Gesamtbrennweite eines Linsensystems aber ist die Stellung der Linse sogar entscheidend; denn dafür gilt

Fig. 6. Smyth-Linse.

$$\Phi = \varphi_1 + h_{21}\varphi_2 + h_{31}\varphi_3 + \cdots.$$

So hat eine Linse im Brennpunkt des Systems den Faktor $h_{i1} = 0$, ändert also die Brennweite des Systems überhaupt nicht. Auf die Petzval-Summe jedoch wirkt sie mit φ_i/n_i. Es liegt also nahe, in der Brennebene eine hinreichend starke Zerstreuungslinse aufzustellen und damit die Petzval-Wölbung vollständig zu beseitigen. Man bezeichnet eine solche Linse dicht vor der Bildebene als *Smyth-Linse* (Fig. 6). Sie hat z.B. beim Petzval-Objektiv Anwendung gefunden, um es für ein größeres Bildfeld brauchbar zu machen.

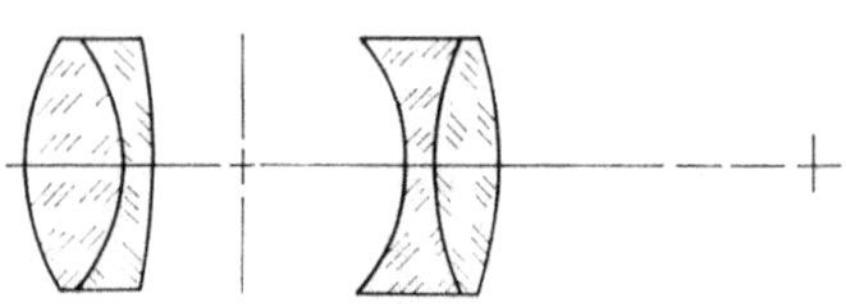

Fig. 7. Teleobjektiv.

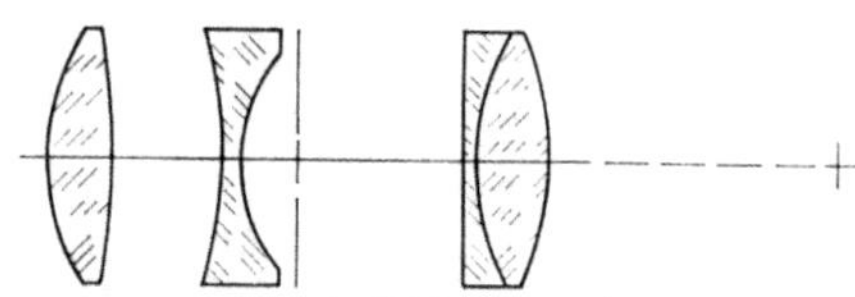

Fig. 8. Triplett (Tessar).

6. Teleobjektiv und Triplett. Schiebt man die Smyth-Linse aus der Brennebene heraus auf das Objektiv zu, so gewinnt sie einen immer stärkeren Einfluß auf die Brennweite und zwar im Sinne einer Verlängerung derselben, ohne im selben Maße auch die Schnittweite zu vergrößern. Man bekommt dabei ein Objektiv, bei dem die Gesamtlänge vom ersten Linsenscheitel bis zur Brennebene kleiner als die Brennweite wird (*Teleobjektiv*, Fig. 7), gleichzeitig aber auch die Petzval-Schale geebnet ist. Schließlich kann diese starke Zerstreuungslinse auch als Mittellinse in das System aufgenommen werden. Das ist zuerst in der sogenannten Cooke-lens geschehen, einem System, das wir heute als *Triplett* bezeichnen, weil es drei einzeln stehende Glieder enthält, die voneinander durch einen Abstand getrennt sind. Tatsächlich gelingt es auf diese Weise, unter Verwendung von nur drei einzelnen Linsen sowohl die Farbfehler als auch alle fünf Seidelschen Bildfehler in befriedigendem Maße für ein Öffnungsverhältnis bis zu etwa 1:4 zu beseitigen. Das erreichbare Bildfeld liegt bei 30°. Geht man mit dem Öffnungsverhältnis oder dem Bildfeld weiter, so machen sich die Zonenfehler stark bemerkbar und setzen die Bildqualität herab.

Zu einem ganz ähnlichen Aufbau kam Rudolph bei der Entwicklung des *Tessars* (Fig. 8). Man kann ihn als ein Triplett mit verkittetem Hinterglied ansprechen. Tatsächlich hat er sich aus dem Doppelanastigmaten entwickelt, bei dem die Verkittung des Vordergliedes, eines Altachromaten, aufgegeben und zwischen den beiden Linsen ein Abstand eingeführt wurde. Dabei blieb im Hinterglied der Neuachromat mit seiner sammelnden Kittfläche erhalten. Diese sammelnde Kittfläche erweist sich außerdem noch als sehr günstig für die Verbesserung der sphärischen Zonenfehler. Von dem Triplett gibt es eine große Zahl von Abarten, darunter auch solche, bei denen die Vorderlinse oder sogar die mittlere Zerstreuungslinse verkittet sind.

7. Gauß-Typ. In der Systematik steht dem Triplett der *Gauß-Typ* nahe, wenn sie auch beide auf den ersten Blick völlig verschieden zu sein scheinen. Die Entwicklung hat ebenfalls nicht vom Triplett zum Gauß-Typ geführt, sondern ist einen ganz anderen Weg gegangen. Der innere Zusammenhang ergibt sich nur unter dem Gesichtspunkte einer systematischen Einordnung der verschiedenen Typen. Beim Gauß-Typ wird nämlich die zerstreuende Wirkung der Mittellinse

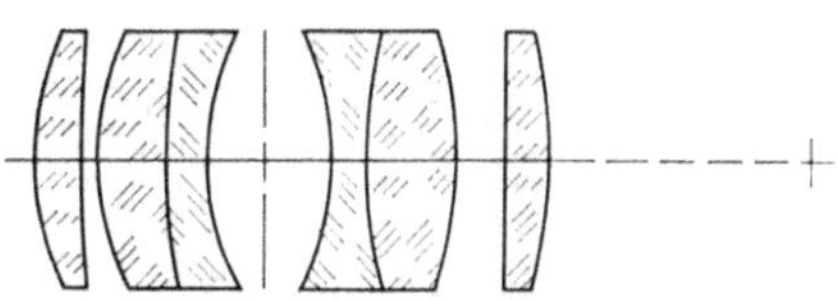

Fig. 9. Gauß-Objektiv.

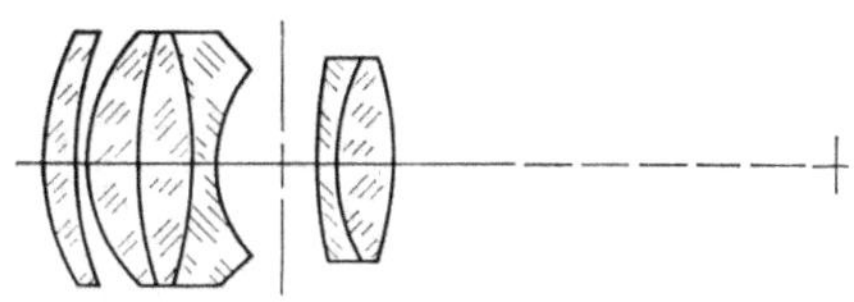

Fig. 10. Sonnar.

nicht durch eine bikonkave Linse bewerkstelligt, sondern durch zwei einander zugekehrte, konkave Glasluftflächen; sie gehören zu dicken Menisken, deren sammelnde Flächen jeweils dem Vorder- und Hinterglied zugerechnet werden können (Fig. 9). Man kann also sagen, daß die Luftabstände des Tripletts mit Glas ausgefüllt sind. Wir haben damit genau wie beim Triplett an der Stelle kleiner h-Werte, wo ihr Einfluß auf die Brennweite gering ist, zerstreuende Flächen, welche die Petzval-Summe herabsetzen. Die Hohlflächen schließen die Blende ein und die Hauptstrahlen treffen diese Flächen unter relativ kleinen Winkeln. Dadurch ist diese Bauart von vornherein hinsichtlich einer astigmatischen Korrektion günstiger als alle anderen Typen. Denn sie enthält fast durchweg Flächen, welche geringe astigmatische Fehler verursachen und leicht durch eine oder zwei astigmatisch überkorrigierende Flächen kompensiert werden können. Aber auch in bezug auf die Hebung der sphärischen Abweichung ist der Gauß-Typ vorteilhaft, weil die Korrektion dieses Bildfehlers ohne zonale Abweichungen erfolgt. Das bedeutet, daß man das Öffnungsverhältnis erhöhen kann, und man erreicht damit 1:1,4. Lediglich die Korrektion der Koma bereitet einige Schwierigkeiten. Das liegt daran, daß an den der Blende benachbarten Hohlflächen die Strahlen des zu einem außeraxialen Objektpunkt gehörenden Bündels mit sehr unterschiedlichem Einfallswinkel auftreffen. Man muß also darauf ein besonderes Augenmerk richten.

Die Familie des *Sonnars* steht gewissermaßen zwischen dem Gauß-Typ und dem Triplett. Während nämlich im Vorderglied bereits der Übergang zur Gauß-Hälfte vollzogen ist, gehört der hinter der Blende stehende Teil der Triplettgruppe an. Mit dem Gauß-Typ hat der Sonnartyp die Eigenschaft gemeinsam, sich für hohe Öffnungsverhältnisse auskorrigieren zu lassen, ja, als die älteren Verwandten des Gauß-Objektivs haben die Sonnare den Zug zu den extremen Öffnungsverhältnissen erst recht begründet.

8. Spezialobjektive. Man kann die bisher besprochenen Objektive als Universalobjektive bezeichnen, weil sie den meisten Bedürfnissen der Photographie gerecht werden. Das brauchbare Bildfeld entspricht ungefähr der Brennweite, wobei natürlich dieses brauchbare Bildfeld mit unterschiedlicher Qualität ausgezeichnet wird. In der Gebrauchsphotographie, mit den meist räumlich ausgedehnten Objekten, ist aber das Abbildungsverhältnis verschieden weit entfernter Gegenstände — d.h. also die Perspektive — von maßgeblicher Bedeutung. Dadurch entsteht das Bedürfnis nach Objektiven verschiedener Brennweite, teils länger, teils kürzer als die Norm für das vorliegende Bildformat. Das bedeutet, daß bei den langen Brennweiten der Bildwinkel kleiner, bei kurzen Brennweiten größer wird, wenn man mit einem gegebenen Format arbeitet. Nun wachsen aber die Bildfehler des Astigmatismus und der Wölbung mit $\tan^2 w$ an, die Verzeichnung sogar mit $\tan^3 w$. Andererseits nehmen alle Fehler proportional mit der Brennweite zu. Daraus folgt, daß die astigmatischen Fehler in den Bildecken mit wachsender Brennweite linear abnehmen und umgekehrt bei kurzen Brennweiten entsprechend linear anwachsen. Nun eignen sich aber die verschiedenen Bauarten verschieden gut für die Korrektion der Bildfehler im Bildfeld, und

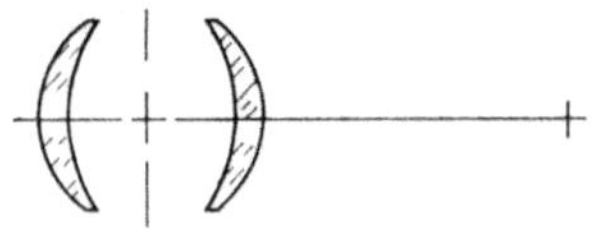

Fig. 11. Hypergon.

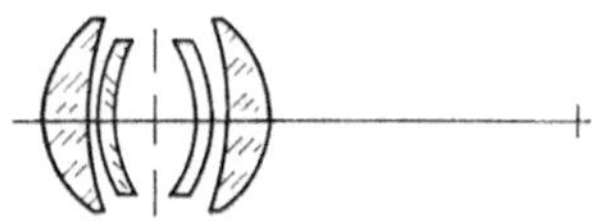

Fig. 12. Topogon.

deshalb pflegt man auch die Objektive nicht durch eine proportionale Vergrößerung oder Verkleinerung auseinander hervorgehen zu lassen, sondern man benutzt den jeweils günstigsten Typ, bei dem das Format sich gerade gut korrigieren läßt. In extremen Fällen kann man bei solchen Sonderobjektiven vielfach auch auf ein hohes Öffnungsverhältnis verzichten.

α) *Weitwinkelobjektive.* Bei den Weitwinkelobjektiven geht man von einer Anordnung aus, bei der zwei sammelnde Menisken im Minimum des Astigmatismus beiderseits symmetrisch zur Blende stehen. Die Symmetrie sorgt für Freiheit von Verzeichnung, die aber ohnehin für jedes der einzelnen Glieder unbedenklich ist. Mit diesem als Hypergon bezeichneten Objektiv wurde ein Bildfeld von 135° erreicht, allerdings ist es weder auf den Öffnungsfehler hin noch auf die Farben hin korrigiert. Indem RICHTER die sammelnden Menisken durch zwei zerstreuende ergänzte, gelang es ihm, das Öffnungsverhältnis auf 1:6,3 und höher zu bringen. Die Verzeichnung ist dabei noch so unbedeutend, daß das Objektiv für photogrammetrische Zwecke brauchbar ist. Allerdings war dabei das Bildfeld auf 100° herabgesetzt worden. Durch die Verwendung weiterer Linsen, deren Flächen alle hohl zur Blende hin sind und durch das Einführen von Kittflächen gelangte man zu Objektiven hohen Öffnungsverhältnisses für die Photogrammetrie. Der Verzicht auf Symmetrie führte für die Gebrauchsphotographie zu Weitwinkelobjektiven, die sich in ihrem Aufbau dem Gauß-Typus nähern. Man begibt sich allerdings des Vorteils der „natürlichen Verzeichnungsfreiheit", die den symmetrischen Aufbau auszeichnet. Für extreme Weitwinkelobjektive ist diese Forderung aber ohnehin nicht aufrecht zu erhalten. Denn die Beleuchtungsstärken, die ein als Lambert-Strahler leuchtendes Objekt in den Ecken des Bildfeldes im Verhältnis zur Mitte erzeugt, ist bei einer aberrationsfreien Abbildung gegeben durch $I/I_0 = \cos^4 w$, nimmt also mit dem Bildwinkel sehr stark ab. Bei einem Bildfeld von $2w = 90°$, also in den Ecken, ist die verbleibende Beleuchtungsstärke nur noch ein Viertel von der in der Mitte. Dieser Abfall, die „natürliche Vignettierung", läßt sich durch eine entsprechende künstliche Vignettierung der Bildmitte kompensieren, solange der Bildwinkel nicht zu groß wird. Für ein Weitwinkelobjektiv von 180° versagt dieses Mittel. Dazu kommt noch, daß eine Projektion der Halbkugel auf eine Ebene nur möglich wird, wenn eine starke Verzeichnung in Kauf genommen werden kann. Die Einführung einer solchen Verzeichnung macht aber auch den angegebenen Wert des Helligkeitsabfalles zum Bildrande hin hinfällig und man behält dadurch in den Ecken eine ausreichende Beleuchtungsstärke. Als Beispiel dafür sei das Objektiv in Fig. 14 gegeben. Charakteristisch ist die vorgezogene Zerstreuungslinse, welche als Meniskus ausgebildet wird und die Winkel des Hauptstrahles gegen die Achse gleich beim Eintritt in das System herabsetzt. Gleichzeitig wird von dieser Zerstreuungslinse ein Bild des Gegenstandes in einer gewissen Entfernung vor dem Objektiv entworfen und das Hauptglied des Objektives entwirft von diesem virtuellen Bild ein reelles Bild auf der Auffangebene. Das bedeutet aber, daß das Hauptobjektiv etwas höher in der Apertur beansprucht wird, als das ohne die Zerstreuungslinse der Fall wäre. Es werden also bei einem solchen Weitwinkelobjektiv durch die

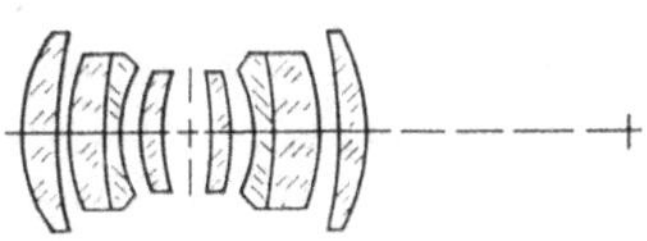

Fig. 13. Weitwinkel-Gauß-Objektiv.

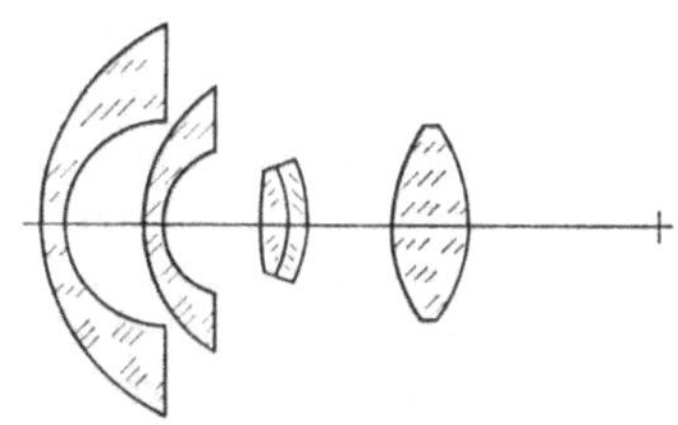

Fig. 14. Extremes Weitwinkelobjektiv.

Zerstreuungslinse die höheren Anforderungen hinsichtlich des Feldes durch etwas höhere Anforderungen hinsichtlich der Apertur ersetzt. Schließlich bewirkt die Zerstreuungslinse auch noch eine Verlängerung der Schnittweite, weil der hintere Hauptpunkt zurückverlegt wird. Man bezeichnet solche Objektive als Retrofokus-Objektive; sie spielen überall da eine besondere Rolle, wo ein großer Abstand zwischen dem letzten Linsenscheitel und der Auffangebene eingehalten werden muß.

β) Lange Brennweiten, Teleobjektive. Das Gegenstück zu den Weitwinkelobjektiven sind die Systeme langer Brennweite, bei denen ein größerer Abbildungsmaßstab auf dem Negativ erreicht wird. Mit der Verringerung des Bildwinkels bei gegebenem Format gehen die Anforderungen hinsichtlich der Korrektion des Feldes zurück, andererseits nehmen aber alle linearen Abweichungen proportional zu der Verlängerung der Brennweite zu. Es hätte nun keinen rechten Sinn, die Durchmesser der Unschärfekreise ebenfalls mit der Brennweite wachsen zu lassen. Denn dann könnte man ebensogut von der mit kurzer Brennweite gemachten Aufnahme eine entsprechend starke Vergrößerung herstellen. Die Benutzung einer langen Brennweite für die Aufnahme ist, abgesehen von der anderen Perspektive nur dann gerechtfertigt, wenn die Zerstreuungskreise in dem gleichen Verhältnis zum Korn der Emulsion stehen, wie bei der Aufnahme mit kurzer Brennweite. Das ist also bei der Wahl der langen Brennweite zu bedenken. Damit wachsen also die Anforderungen an die Korrektion der angulären Abweichungen. Glücklicherweise wird im allgemeinen bei längeren Brennweiten auf ein extremes Öffnungsverhältnis verzichtet, schon um das Gewicht nicht zu stark anwachsen zu lassen. Die mitunter unbequeme Länge des ganzen Systems versucht man dadurch zu verringern, daß man sie als Teleobjektive ausbildet.

Das Teleobjektiv besteht aus einem sammelnden Vorderglied und einem zerstreuenden Hinterglied; zwischen beiden liegt ein im Verhältnis zur Brennweite merklicher Abstand. Bei einem solchen Aufbau ist die gesamte Baulänge vom

Scheitel der ersten Fläche bis zur Bildebene kleiner als die Brennweite. Der wesentliche Nachteil dieser Bauart liegt darin, daß das Öffnungsverhältnis des Vordergliedes für sich allein betrachtet, schon recht hoch sein muß, auch wenn das Öffnungsverhältnis des gesamten Systems nur mäßig bleibt. Denn dieses Öffnungsverhältnis des Vordergliedes wird in demselben Maßstab von der Zerstreuungslinse heruntergesetzt, wie die Brennweite verlängert wird. Außerdem werden die Restfehler des Vordergliedes ebenfalls infolge der vergrößernden Wirkung des zweiten Gliedes erhöht, so daß also zu ihrer Kompensation das Hinterglied entsprechend höher beansprucht werden muß. Allerdings bringt die Kombination eine wesentliche Verbesserung der Petzval-Summe mit sich.

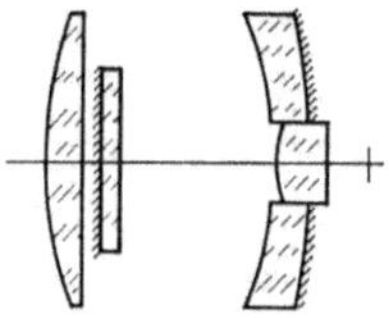

Fig. 15. Spiegelobjektiv.

Bei sehr langen Brennweiten macht sich das sogenannte sekundäre Spektrum bemerkbar, dessen Korrektion besondere Schwierigkeit bereitet. Das ist mit einer der Gründe, die die Entwicklung von Spiegelsystemen als brennweitige Aufnahmeobjektive gefördert haben. In der Kombination von Spiegel und Linsen erlauben solche Systeme eine völlige Beseitigung der Farbfehler und der Petzval-Summe; die Baulänge kann sehr stark reduziert werden. Dafür ist die Unterdrückung von falschem Licht schwierig, die ringförmige Pupille erschwert das Anbringen einer Blende und die Tiefenschärfe nimmt beim Abblenden nicht zu. Schließlich ist das Zentrieren der Spiegel schwierig und kostspielig. So haben sich die Spiegelobjektive trotz ihrer erheblichen Vorzüge in der Praxis nicht recht durchsetzen können.

9. Variofokale Systeme. *α) Mechanischer Ausgleich.* Der Wechsel der Brennweiten muß insbesondere für Kinoaufnahmen schnell möglich sein. Deswegen wurde hier zuerst ein System gefordert, bei dem ein Wechsel der Brennweite

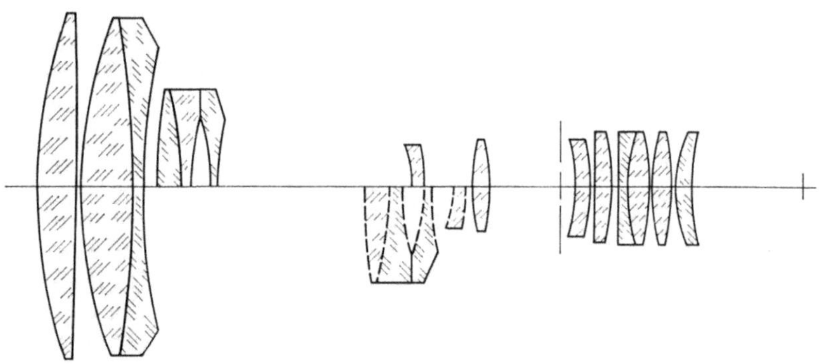

Fig. 16. Vario-Objektiv mit mechanischem Ausgleich.

durch die einfache Betätigung eines Knopfes möglich ist, wobei also das einmal eingestellte Bild trotz des Brennweitenwechsels scharf bleiben soll. Objektive dieser Art heißen variofokale Systeme (Zoom-lens). Sie werden heute auch für das Kleinbildformat der Stehbildkameras angeboten. Für ihren optischen Aufbau lassen sich zwei Wege beschreiten. Wüllner stellte sich 1888 die Aufgabe, eine achsensenkrechte Ebene in eine andere im Abstande L abzubilden, wobei der Abbildungsmaßstab stetig veränderlich sein und dabei die Abbildung unverändert scharf bleiben soll. Die von ihm hergeleiteten Formeln enthalten keine Voraussetzungen über die gegenseitige Lage von Ding- und Bildebene, sie gelten also auch für virtuelle Abbildungen. Man benötigt für eine solche Abbildung mindestens zwei Linsen, die gegeneinander verschoben werden müssen. Die beiden Verschiebungen stehen in keinem linearen Verhältnis zueinander. Es kann also höchstens die eine Verschiebung durch eine Schraubenbewegung erfolgen, während die Verschiebung der anderen Linse nach einem quadratischen Gesetz geht. Bei der Anwendung dieses Grundgedankens auf die Anlage eines photographischen Objektives pflegt man eine Frontlinse vor den verschiebbaren Linsen anzuordnen. Sie dient der Scharfstellung bei verschiedenen Objektentfernungen und bildet zunächst einmal die Dingebene in einer geringen Entfernung hinter dem Objektiv ab. Man wählt dann zweckmäßigerweise für die Maßstabsänderungen zerstreuende Linsen. Für eine Schmalfilm-Kinokamera (8 mm) läßt sich unter Verwendung von 14 Linsen ein Öffnungsverhältnis von 1:1,4 und ein Brennweitenbereich von 8 bis 48 mm erreichen. Die Schwierigkeit bei der Korrektion liegt in der Beseitigung der Verzeichnung. Indessen bleibt das Bildfeld gegenüber den Brennweiten in jedem Falle klein, so daß diese Verzeichnung noch in erträglichen Grenzen bleibt. Grundsätzlich sind die variofokalen Systeme deswegen auf Brennweitenbereiche beschränkt, die gegenüber der Bilddiagonalen als groß anzusehen sind. Die nicht lineare Verschiebung der einen von den beiden Linsen bereitet mechanische Schwierigkeiten, für deren Überwindung besondere konstruktive Maßnahmen getroffen werden müssen. Da die Konstanz der Bildlage durch die Stellung der beiden Linsen erreicht wird, bezeichnet man diese Bauart als den mechanischen Ausgleich.

β) Optischer Ausgleich. Einen ganz anderen Weg, der hauptsächlich von Bergstein verfolgt worden ist, benutzt der sogenannte optische Ausgleich. Zwischen zwei verschieblichen Linsen, die auch sammelnd oder zerstreuend gewählt werden können, befindet sich eine fest angeordnete Linse. Dann lassen sich die Brennweiten der Linsen so bestimmen, daß bei einer gemeinsamen Verschiebung der beiden Linsen, die beiderseits der feststehenden Linse angeordnet sind, die Bildebene erhalten bleibt, wenigstens in einem für die praktischen Bedürfnisse ausreichendem Maße. Diese Lösung erfüllt also die Aufgabe nicht streng, und es hängt von der Geschicklichkeit des Rechners ab, eine solche Wahl der Brennweiten zu treffen, daß die Abweichungen von der Schärfe klein bleiben. Man bezeichnet diese Bauart als den optischen Ausgleich.

10. Perspektive. Es ist nicht ganz korrekt, von der Perspektive eines photographischen Bildes oder gar eines Objektives zu sprechen. Denn die in dem Bilde festgehaltene Darstellung des Dingraumes ist hinsichtlich der Perspektive ausschließlich durch den Aufnahmestandpunkt, genauer durch die Lage des Mittelpunktes der Eintrittspupille gegenüber den Gegenständen im Dingraum bestimmt. Bei photographischen Objektiven liegt die Eintrittspupille immer im Objektiv oder mindestens in seiner unmittelbaren Nähe. Die von den einzelnen Punkten des Dingraumes nach der Mitte der Eintrittspupille hinzielenden Strahlen, die Hauptstrahlen, liefern mit ihren Durchstoßpunkten durch die Einstellebene eine perspektivische Darstellung des Dingraumes. Dabei ist als Einstellebene diejenige achsensenkrechte Ebene anzusehen, die zu der Auffangebene hinter dem Objektiv konjugiert ist. Das in der Einstellebene liegende fiktive Bild wird von dem Objektiv auf der Auffangebene entworfen und gegebenenfalls von der photographischen Schicht festgehalten, wobei die Größe des aufgefangenen Bildes durch den zwischen Einstell- und Auffangebene geltenden Abbildungsmaßstab β' gegeben ist.

Nun werden aber wegen der endlichen Größe der Pupille alle diejenigen Punkte des Dingraumes, die nicht in der Einstellebene liegen, nur undeutlich, unscharf, wiedergegeben. Denn die Eintrittspupille nimmt einen von dem Dingpunkt ausgehenden Strahlenkegel auf, dessen Spitze eben in dem Dingpunkt liegt und dessen Basis von der Eintrittspupille selbst gebildet wird. Dieser Strahlenkegel besitzt in der Einstellebene einen Querschnitt, der sich als die Projektion der Eintrittspupille auf die Einstellebene von dem Dingpunkt aus ergibt. Das ist der objektseitige Zerstreuungskreis der für den betrachteten Objektpunkt unscharfen Abbildung. Sowohl für die Untersuchung der Perspektive als auch der Schärfentiefe ist die Benutzung des Begriffes der Einstellebene besonders bequem.

Der Bildfeldwinkel w, der sich aus der Größe y des Objektes in der Einstellebene und aus ihrer Entfernung E von der Eintrittspupille ergibt, ist für einen Beobachter mit dem Auge am Ort der Eintrittspupille zugleich der Gesichtsfeldwinkel, unter welchem ihm das Objekt erscheint. Es ist $\tan w = y/E$. Bezieht man die

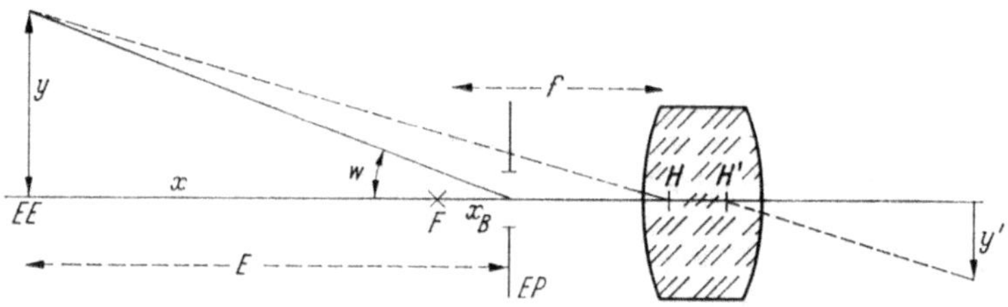

Fig. 17. Zur Perspektive photographischer Bilder. *EE* Einstellebene; *EP* Eintrittspupille; *F* dingseitiger Brennpunkt; *H*, *H'* Hauptpunkte.

Entfernungen auf den vorderen Brennpunkt, so ist $E = x - x_B$, wobei x_B der Abstand der Eintrittspupille von dem vorderen Brennpunkt ist und x die Entfernung der Einstellebene von diesem Brennpunkt. Damit ist

$$\tan w = \frac{y}{x - x_B}.$$

Bei der Abbildung der Einstellebene auf die Auffangebene wird $y' = y\,\beta'$. Nach den Newtonschen Abbildungsgleichungen ist $x = f/\beta'$ und $x_B = f/\beta'_B$, wobei β'_B die Pupillenvergrößerung bezeichnen soll. Damit läßt sich $\tan w$ auch schreiben

$$\tan w = \frac{y'}{\left(1 - \frac{\beta'}{\beta'_B}\right) \cdot f}.$$

Bei der Betrachtung des fertigen Bildes aus dem Abstand E' erscheint die Bildstrecke y' unter dem Winkel $\tan w' = y'/E'$. Soll das Bild einen „natürlichen" Eindruck machen, so müssen die beiden Winkel w und w' gleich sein. Daraus folgt für den perspektivisch „richtigen" Abstand

$$E' = \left(1 - \frac{\beta'}{\beta'_B}\right) f.$$

Bei der Aufnahme von Objekten in größerer Entfernung ist der Abbildungsmaßstab β' sehr klein und kann vernachlässigt werden. Solche Aufnahmen sollten also aus einem Abstand betrachtet werden, der gleich der Brennweite des Aufnahmeobjektives ist. Das wird allerdings vielfach nicht möglich sein, weil die Augenakkommodation so kurze Betrachtungsabstände nicht zuläßt. Dann kann man entweder eine Lupe gleicher Brennweite benutzen oder man stellt ein im Maßstab v vergrößertes Bild her und betrachtet es aus dem Abstand $v \cdot f$, wobei man nun v so wählen kann, daß ein bequemes Beobachten möglich ist. Bei Projektionsbildern kann man die Vergrößerung durch den Projektionsabstand und die Brennweite des zur Projektion dienenden Objektives ausdrücken:

$v = E_p/f_p$, wobei E_p der Abstand des Projektionsschirmes von dem vorderen Brennpunkt des Projektionsobjektives und f_p seine Brennweite bedeutet. Dann ist der richtige Betrachtungsabstand gegeben durch

$$E' = \frac{f_a}{f_p} E_p \qquad (f_a \text{ Brennweite des Aufnahmeobjektives}).$$

Bei Aufnahmen in annähernd natürlicher Größe, also im Maßstab um $\beta' = -1$ herum, spielt das Glied β'/β'_B eine wesentliche Rolle. Allerdings ist β'_B für die photographischen Objektive jeweils verschieden und liegt etwa zwischen $0{,}6 < \beta'_B < 1{,}4$. Man kann also den günstigsten Betrachtungsabstand einer photographischen Aufnahme nur angeben, wenn man diesen Wert der Pupillenvergrößerung kennt und außerdem noch die Brennweite des Aufnahmeobjektives und den Abbildungsmaßstab. Zum Glück ist aber das Auge gegen geringe Abweichungen von dem günstigsten Betrachtungsabstand nicht empfindlich.

Hiernach ist die Perspektive selbst von den Eigenschaften der abbildenden Optik ganz unabhängig, wie ja schon aus dem Umstande hervorgeht, daß sie bereits in der Einstellebene vorgefunden werden kann. Der Unterschied des mit langer und kurzer Aufnahmebrennweite wiedergegebenen Raumes ergibt sich erst dadurch, daß üblicherweise mit einem Wechsel des Aufnahmeobjektives auch der Aufnahmestandpunkt geändert wird. Denn man pflegt den Vergleich so vorzunehmen, daß man das Hauptmotiv, z.B. ein schmiedeeisernes Tor, mit beiden Objektiven durch einen passend gewählten Abstand im Bild auf die gleiche Größe bringt. Dann tritt bei einer Aufnahme mit kurzer Brennweite der Hintergrund, in unserem Beispiel etwa ein Gebäude, gegenüber dem Tor zurück, während bei einer Aufnahme mit einem Objektiv langer Brennweite das Gebäude sehr viel größer wiedergegeben wird. So wirkt also die Verwendung von Objektiven kurzer Brennweite raumdehnend, die von Objektiven langer Brennweite aber raumraffend, das Bild verflachend. Betrachtet man beide Vergleichsaufnahmen aus der jeweilig „richtigen" Betrachtungsentfernung, so würde der eigene Aufnahmestandpunkt leicht zu erkennen sein, und danach könnte man die wahren Größenverhältnisse besser abschätzen. Gerade das pflegt aber nicht der Fall zu sein.

11. Schärfentiefe. Das Urteil über die mehr oder weniger ausreichende Schärfe eines Bildes hängt sehr wesentlich von der Vergrößerung ab, mit welcher das Bild betrachtet wird. Selbst ein sauber ausgeführter Druck mit scheinbar scharf konturierten Buchstaben zeigt unter der Lupe oder einem schwachen Mikroskop unscharfe Konturen. Man muß also über die bei der Betrachtung des Bildes anzuwendende Vergrößerung eine Vereinbarung treffen, wenn überhaupt ein allgemein gültiges Urteil über die Schärfe gefunden werden soll. Für diese Zwecke ist die Festlegung sinnvoll, daß die Beurteilung der Schärfe bei der Vergrößerung oder dem Betrachtungsabstand zu erfolgen hat, bei welcher der perspektivisch richtige Bildeindruck entsteht. Dann ist der Durchmesser des Zerstreuungskreises durch die Erkennbarkeit der Unschärfe gegeben. Für diesen Wert liegen allerdings keine zuverlässigen Zahlen vor; als Erfahrungswert wird ein Betrag von 2 Bogenminuten angesehen, also etwa 1/1500. Dann dürfen also die Durchmesser der von einem Objektpunkt ausgehenden und von der Eintrittspupille begrenzten Strahlenkegel in der Einstellebene den Betrag von $\zeta = E/1500$ nicht übersteigen. Bezeichnet man die größte zulässige Entfernung mit E_F, die kürzeste mit E_N, so ergibt sich aus der Abbildung

$$\frac{\xi}{D} = \frac{E_F - E}{E} \quad \text{und} \quad \frac{\zeta}{D} = \frac{E - E_N}{E}.$$

Führt man die Blendenzahl $z=f/D$ ein, so ist

$$\frac{E_F}{E}=1+\frac{\zeta}{D}=1+\frac{\zeta z}{f}, \qquad \frac{E_N}{E}=1-\frac{\zeta z}{f}.$$

Für die Fern- und Nahebene, welche die Grenzen des Tiefenschärfebereiches bilden, erhält man damit

$$E_{F,N}=E\left(1\pm\frac{\zeta z}{f}\right)$$

und kann darin für ζ den oben angegebenen Wert einsetzen. Für die Wahl eines geeigneten Wertes von ζ kann man aber auch noch anders argumentieren. Bedenkt man, daß photographische Bilder doch nur in wenigen Fällen aus dem perspektivisch richtigen Betrachtungsabstand angesehen werden, sondern mei-

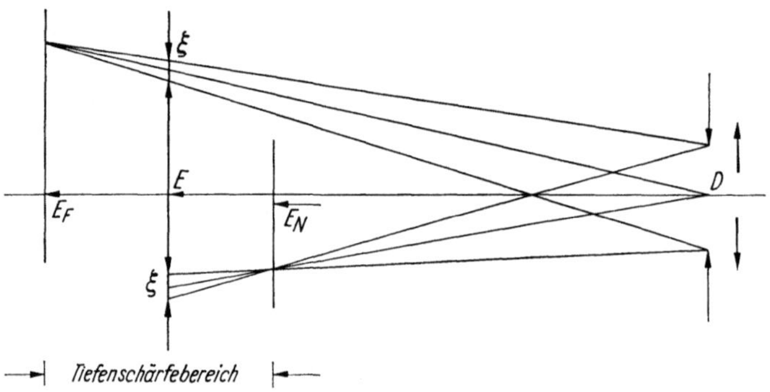

Fig. 18. Die Tiefenschärfe. E Entfernung der Einstellebene; E_F Entfernung der Fernebene; E_N Entfernung der Nahebene ζ Durchmesser des dingseitigen Zerstreuungsscheibchens.

stens aus dem Abstand der deutlichen Sehweite oder dem konstanten Abstand von einem Projektionsschirm, so ist die Wahl eines festen, von der Aufnahmebrennweite unabhängigen Wertes des bildseitigen Zerstreuungskreises gerechtfertigt. Denn die mit verschiedenen Aufnahmebrennweiten erzeugten Bilder sollen gleich scharf erscheinen. Nun ist aber $\zeta=\zeta'/\beta'$, wofür man wegen $\beta'=f/(E+f)$ auch

$$\zeta=\zeta'\frac{E+f}{f}$$

schreiben kann. Nun hat man zwei Fälle zu unterscheiden. Entweder wird eine Fernaufnahme gemacht, bei welcher der Aufnahmestandpunkt nicht geändert wird, oder man wählt bei einer Nahaufnahme den Aufnahmeabstand so, daß das Hauptmotiv bei der Verwendung der langen Brennweite in der gleichen Größe erscheint wie von dem näheren Aufnahmestandpunkt aus unter Verwendung einer kurzen Aufnahmebrennweite. Im ersten Fall ist E konstant und groß gegen f. Damit errechnet sich für den Schärfentiefebereich

$$E_F-E_N=2\frac{z\zeta'}{f^2}E$$

mit einer quadratischen Zunahme bei abnehmender Brennweite.

Im zweiten Fall (Nahaufnahme mit jeweils gleicher Größe des Hauptobjektes) ergibt sich

$$E_F-E_N=2\frac{z\zeta'}{f\beta'}E,$$

d.h. der Schärfentiefebereich nimmt nur linear mit der Brennweitenverkleinerung zu.

Die zulässigen Durchmesser der Zerstreuungskreise sind von dem zuständigen Normenausschuß in Abhängigkeit vom Bildformat festgelegt worden. Darüber gibt die nebenstehende Tabelle Auskunft.

Tabelle.

Negativformat	Diagonale	Zulässiger Durchmesser des Unschärfekreises
3,6 × 4,8	6	0,010
6,2 × 8,3	10,4	0,015
7,5 × 10,36	12,8	0,015
16 × 22	27,2	0,025
18 × 24	30,0	
24 × 24	33,9	0,033
24 × 36	43,3	
45 × 60	75,0	0,050
60 × 60	84,9	0,075
60 × 90	108,2	0,100
90 × 120	150,0	0,100

Abgesehen von den kleinsten Formaten entsprechen die empfohlenen Werte etwa der Regel $\zeta' = \text{Bilddiagonale}/1500$.

Teilweise wird aber auch ein schärferer Maßstab angelegt und dann ein entsprechend kleinerer Wert für den zulässigen Durchmesser des Unschärfekreises angenommen. Dadurch ergeben sich in den veröffentlichten Schärfentiefetabellen Unterschiede. Meist ist der der Berechnung der Tabelle zugrunde gelegte Wert angegeben. Ferner wird zuweilen auch die Entfernung nicht von der Eintrittspupille aus, sondern von der Filmebene aus oder von der Kamerarückwand aus gerechnet.

12. Entzerrung. Wir haben bisher stillschweigend vorausgesetzt, daß die Bildebene senkrecht zur optischen Achse angeordnet ist. Im allgemeinen aber — und es gibt Apparate, die dafür eingerichtet sind, kann die Auffangebene auch gegen die achsensenkrechte Ebene verkippt werden. Dann tritt ein besonderer Fall der perspektivischen Verzerrung ein, der im folgenden behandelt werden soll. Man beobachtet diese perspektivische Verzerrung oft, wenn bei der Aufnahme von Gebäuden, von Bildern und dergleichen die Kamera nicht horizontal gehalten worden ist. Dann konvergieren alle vertikalen Linien, eine Erscheinung, die bei Architekturaufnahmen als stürzende Linien besonders gefürchtet ist, weil bei der unmittelbaren Betrachtung unter gleichen Verhältnissen eine solche Konvergenz der vertikalen Geraden im allgemeinen gar nicht beobachtet wird. In unserer Wahrnehmung wird der optische Effekt durch die Erfahrung sofort korrigiert, und erst, wenn dabei ganz ungewöhnliche Verhältnisse auftreten, denen gegenüber die Erfahrung versagt, fällt die Konvergenz der vertikalen Linien bei der direkten Beobachtung auf. Auf dem photographischen Bild hingegen wirken die stürzenden Linien schon bei kleinen Neigungen unnatürlich.

Zur genaueren Untersuchung der Verhältnisse betrachten wir die Fig. 19. Die Objektebene trage ein quadratisches Netz und sei gegen die Einstellebene um den Winkel α gekippt. Dann liegt der Konvergenzpunkt aller parallel zu der Papierebene liegenden Geraden der Objektebene OE auf der Parallelen durch die Mitte des Hauptpunktes. Mithin ist $l = E \tan\alpha = (x - f) \cot\alpha$. Dafür kann man schreiben

$$l = f(1/\beta' - 1) \cot\alpha .$$

Wir wollen nun voraussetzen, daß der im Bildfeld erscheinende Bereich der Objektebene ganz innerhalb der Schärfentiefe liegt. Bildseitig ist dann

$$l' = \beta' l = f(1 - \beta') \cot\alpha .$$

Bei einer Aufnahme mit starker Verkleinerung, also bei einer Architekturaufnahme ist $l' = f \cot\alpha$.

Man strebt natürlich einen möglichst großen Wert von l' an, damit stürzende Linien nicht mehr auffallen. Die Formel zeigt, daß es dafür besser ist, die Aufnahme mit einem Objektiv langer Brennweite zu machen als mit einem kürzerer Brennweite. Um nun das Objekt in gleicher Größe auf das Bild zu bekommen, muß man den Aufnahmeabstand rund proportional zu den Brennweiten vergrößern. Damit wird aber auch der erforderliche Kippwinkel kleiner, den ich einführen muß, um das Objekt gegebener Höhe in das Bildfeld zu bekommen. Der Konvergenzpunkt rückt rund proportional zu dem Quadrat der Brennweite weg. Oder umgekehrt ausgedrückt: Bei der Benutzung einer kurzen Aufnahmebrennweite ist die Gefahr, stürzende Linien zu erhalten, unverhältnismäßig viel stärker als bei der Verwendung einer längeren Brennweite. Auf die Frage, wie die affine Verzerrung eines in der Dingebene quadratischen Netzes rückgängig gemacht werden kann, soll hier nicht eingegangen werden. Es ist die Grundaufgabe der Photogrammetrie und dort in allgemeinster Weise behandelt worden.

Im Nahbereich überschreitet bei hinreichendem Öffnungsverhältnis der abgebildete Teil der schräg liegenden Objektebene den Tiefenschärfebereich des Objektivs. Dann

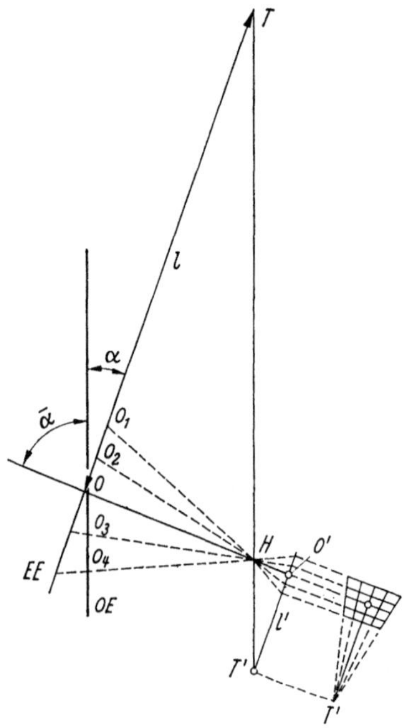

Fig. 19.

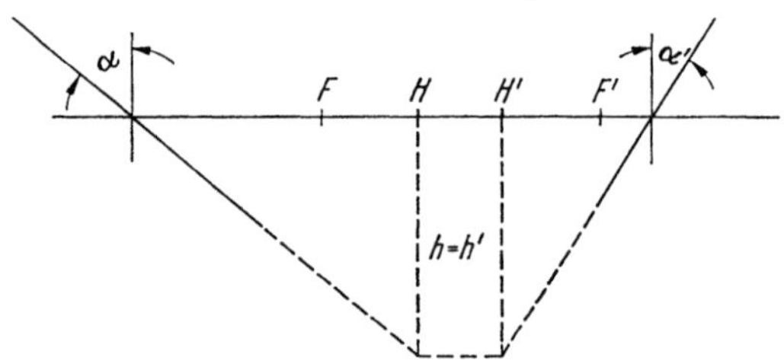

Fig. 20.

Fig. 19. *EE* Einstellebene; *OE* Objektebene; *T* objektseitiger Fluchtpunkt; *T'* bildseitiger Fluchtpunkt.

Fig. 20. Die Scheimpflug-Bestimmung.

wird es notwendig, auch die Auffangebene gegen die optische Achse zu kippen. Die Gaußsche Optik verlangt für den Winkel zwischen den Ebenen und der Achse das Einhalten der Beziehung $\cot\alpha' = \beta' \cot\alpha$, wobei β' die laterale Vergrößerung in den Durchstoßpunkten der Achse durch die beiden konjugierten Ebenen bedeutet. Scheimpflug hat dafür die Formulierung gegeben: Die beiden Ebenen, Objektebene und Bildebene, müssen die zugehörigen Hauptebenen des Objektivs in zwei parallelen Geraden im gleichen Abstand von der Achse schneiden. Diese Beziehung sichert also eine gleichmäßig scharfe Abbildung der Objektebene auf die Auffangebene.

13. Die Strahlungsvermittlung. Die von den einzelnen Objektpunkten ausgehende Strahlung unterschiedlicher Intensität und spektraler Verteilung bringt auf der lichtempfindlichen Schicht photoelektrische oder photochemische Wirkung hervor, die später als Bild sichtbar gemacht werden. Wegen der im allgemeinen abweichenden spektralen Empfindlichkeit der lichtabsorbierenden Schicht gegenüber dem normalen Auge sind die Begriffe der Lichttechnik für die Behandlung der „Helligkeit" photographischer Bilder nicht korrekt. Zum mindesten ist, wenn die bewährten Begriffe der Lichttechnik dabei gebraucht werden, zu beachten, daß der spektrale Verlauf der Empfindlichkeit für die photographische Schicht schon mit der Emulsion schwankt, fast immer bis ins tiefe Ultraviolett reicht und manche Schichten auch für das infrarote Gebiet sensibilisiert sind. Der geometrische Strahlenfluß wird aber natürlich nicht betroffen; es handelt sich immer nur um das Verhältnis der wirksamen Strahlung zu der Gesamtstrahlung, welches für Auge und photographische Schicht verschieden ist.

Ein Flächenelement $d\sigma$ des diffus strahlenden Objektes sendet in die Eintrittspupille den Strahlungsstrom (Lichtstrom)

$$\Phi = d\sigma \cdot B\,\pi \sin^2 u\,.$$

Das Objektiv übermittelt diesen Strahlungsstrom mit einem gewissen Verlust infolge von Absorptionen und Reflexionen dann der Austrittspupille und diese strahlt auf das Bildelement $d\sigma'$ den Strahlungsstrom:

$$\Phi' = d\sigma' B' \pi \sin^2 u'.$$

Beide Strahlungsströme müssen bis auf den Verlustfaktor gleich sein: $\Phi' = \tau\,\Phi$. Nun ist aber bei einer aberrationsfreien optischen Abbildung $d\sigma' = \beta'^2 d\sigma$ und wegen der Gültigkeit der Sinusbedingung

$$\sin u' = \frac{1}{\beta'} \sin u\,.$$

Mithin ist $B' = \tau B$. Die Bestrahlungsstärke ist

$$E' = \frac{\Phi'}{d\sigma'} = \tau\,B\,\pi \sin^2 u'\,.$$

Für unendlich weit entfernte Objekte in der Mitte des Gesichtsfeldes ist wegen

$$\sin u' = \frac{D}{2f} = \frac{1}{2z}$$

die Bestrahlungsstärke

$$E' = \tau\,B\,\frac{\pi}{4}\,\frac{1}{z^2}\,.$$

Bei nahe gelegenen Objekten ergibt sich dafür

$$E' = \tau\,B\,\pi\,\frac{1}{4z^2\left(1 - \frac{\beta'}{\beta'_B}\right)^2 + \beta'^2}\,.$$

wofür man in praxi immer mit genügender Genauigkeit setzen darf

$$E' = \tau\,B\,\pi\,\frac{1}{4z^2(1-\beta')^2}\,.$$

Bisher war stillschweigend vorausgesetzt worden, daß es sich um ein auf der Achse gelegenes Flächenelement handelt, und es erhebt sich nun noch die Frage, wie sich der Einfluß des Bildwinkels auf die Beleuchtungsstärke in der Bildebene auswirkt. Nimmt man die Aufnahmeentfernung als groß an gegenüber dem Durchmesser der Eintrittspupille, dann ist der Raumwinkel, unter dem die Eintrittspupille von $d\sigma$ aus erscheint,

$$dw = \frac{D^2\pi}{4} \cos w \, \frac{1}{l^2} = \frac{D^2\pi}{4}\,\frac{\cos^3 w}{E^2}\,.$$

Außerdem ist die Strahlungsstärke unter dem Winkel w gegen die Normale bei einem diffusen Strahler $B\cos w$. Damit wird also der Lichtstrom

$$\Phi = \frac{B D^2 \pi}{4 f^2} \frac{\cos^4 w}{\left(\frac{1}{\beta'} - \frac{1}{\beta'_B}\right)^2} d\sigma .$$

Bildseitig ist die Beleuchtungsstärke damit

$$E'_w = \frac{\tau B \pi}{4 z^2} \frac{\cos^4 w}{\left(\frac{1}{\beta'} - \frac{1}{\beta'_B}\right)^2} \frac{d\sigma}{d\sigma'} .$$

Die Abhängigkeit vom Bildwinkel w ergibt sich so zu

$$K = \frac{E'_w}{E'_0} = \frac{\cos^4 w}{\left(1 - \frac{\beta'}{\beta'_B}\right)^2} \frac{d\sigma}{d\sigma'} \cdot \beta'^2 .$$

Ist das System verzeichnungsfrei, so gilt

$$d\sigma' = \beta'^2 d\sigma$$

und man bekommt

$$K = \cos^4 w \frac{1}{\left(1 - \frac{\beta'_0}{\beta'_B}\right)^2} .$$

Andererseits kann eine Verzeichnung

$$\beta'^2_w < \frac{d\sigma'}{d\sigma} < \beta'^2$$

die relative Beleuchtungsstärke am Rande des Feldes verbessern, ja bei extremen Winkelobjektiven mit einem Bildwinkel w gegen 90° ist eine tonnenförmige Verzeichnung die Voraussetzung für das Zustandekommen eines auffangbaren Bildes. Setzen wir ein verzeichnungsfreies Objektiv voraus, so bekommt man schon bei einem Objektiv mit einem halbseitigen Bildfeld von $w = 45°$ einen Abfall der Beleuchtungsstärke von dem Faktor $\cos^4 w = 0{,}25$. Zu dieser sogenannten „natürlichen Vignettierung" tritt nun noch als weiterer Faktor, der die Beleuchtungsstärke außerhalb der Bildmitte herabsetzt, der Umstand, daß bei allen photographischen Objektiven am Rande des Bildfeldes der Lichtkegel durch Fassungsränder beschnitten, vignettiert wird. Theoretisch ist diese „künstliche Vignettierung" durch eine entsprechende Dimensionierung der Linsen leicht zu beheben, wenn auch praktisch kaum durchzuführen. Die künstliche Vignettierung ist natürlich stärker bei den Weitwinkelobjektiven und bei solchen Systemen, die gegenüber der Brennweite verhältnismäßig lang gebaut sind. Um die unterschiedlichen Beleuchtungsstärken sowohl der künstlichen als auch der natürlichen Vignettierung für die Verwendung des Objektives in der Kamera unschädlich zu machen, führt man gelegentlich eine Lichtschwächung der Bildmitte ein, z.B. durch ein am Rande abgeschwächtes Filter, oder auch durch eine zentrale Sternblende. Solche Maßnahmen können indessen auf die Form des Beugungsscheibchens und des Zerstreuungsscheibchens bei unscharfer Einstellung einen Einfluß haben.

Die Bestrahlungsstärke in der Auffangebene ist maßgeblich für die photochemische oder photoelektrische Wirkung. Der Unterschied liegt nur darin, daß in dem einen Fall die photochemische Wirkung durch eine bestimmte Lichtmenge (Lichtstrom · Zeit) ausgelöst wird, in dem anderen Fall der Lichtstrom laufend Elektronen freisetzt, so daß für die Dauer der Beleuchtung ein elektrischer Strom fließt, dessen Stärke von der Bestrahlungsstärke und der Flächengröße abhängt. In der Photographie hat man, um eine bestimmte Schwärzung zu erhalten, zu bilden:

$$E\,t = \text{const} \cdot \frac{t}{z^2}\,.$$

Mit wachsender Blendenzahl muß man die Belichtungszeit quadratisch anwachsen lassen. Die Erfahrung bestätigt diese Regel bis auf die kleinsten Werte der Blendenzahl. Es wird immer wieder von Praktikern versichert, daß beim Übergang von der Blendenzahl 2 auf die Blendenzahl 1,4 die Belichtungszeit nicht halbiert werden darf, wenn man ausreichend belichtete Bilder erhalten will. Die Ursachen dieser Erscheinung sind noch nicht geklärt.

Es war schon erwähnt worden, daß die spektrale Empfindlichkeitsverteilung der photographischen Schicht von der des Auges erheblich abweicht. Vor allem ist es bisher nicht möglich gewesen, nach der kurzwelligen Seite hin eine bessere Übereinstimmung der Empfindlichkeitskurven für das Auge und die photographischen Emulsionen zu erzielen. Das hat seine Auswirkungen vor allem bei der Farbenphotographie, wo der ultraviolette Anteil des Lichtes dem blauen Farbton zugeschlagen wird. Wechselnde Stärke des ultravioletten Gehalts im Licht könnte dann zu unterschiedlichen Wiedergaben des Blau führen, das dadurch oft unnatürlich wirkt. Man ist deswegen bestrebt, die spektrale Durchlässigkeit der photographischen Objektive durch eine geeignete Glaswahl und durch zusätzliche Filter soweit mit dem Film abzustimmen, daß merkliche Farbverfälschungen nicht auftreten können. Im roten Gebiet ist es verhältnismäßig einfach, die Filmempfindlichkeit der des Auges anzugleichen.

Der Verlust an Leuchtdichte im Objektiv, den wir durch den Faktor τ in den Formeln berücksichtigt haben, ist durch Absorption und Reflexion bedingt. Die Absorption ist aber gering und äußert sich meist nur in einer mehr oder weniger merklichen leichten Gelbfärbung des Objektives, weil die in modernen Objektiven verwendeten Glasarten hohen Brechwertes sich nicht immer ganz farbfrei herstellen lassen. Stärker ist schon der Verlust durch Reflexionen an den einzelnen Flächen, vor allem den Grenzflächen gegen Luft. Sie mußten früher je nach dem Brechwert des Glases gemäß den Fresnelschen Formeln mit 4 bis 7% an jeder Fläche berücksichtigt werden. Rechnet man im Mittel mit 5%, so ist der Verlustfaktor bei n Flächen $\tau = 0{,}95^n$. Das ist bei 10 Flächen ein Wert von $\tau = 0{,}6$. Die modernen Verfahren der Reflexminderung durch das Aufdampfen von dünnen Interferenzschichten auf die Glasoberflächen setzen die Intensität des reflektierenden Lichtes erheblich herab. Bei den meisten Flächen bleibt ein Rest von 0,1 bis 0,5%, im Mittel etwa 0,3%. An den Außenflächen des Systems, die einer mechanischen Beanspruchung durch Wischen und Reiben besonders ausgesetzt sind, muß man einen harten Belag wählen, dessen Wirkung etwas geringer ist und noch etwa 1% des auffallenden Lichtes reflektiert. Ein belegtes System mit 10 Flächen hat danach einen Verlustfaktor von $\tau = 0{,}99^2 \cdot 0{,}997^8 \approx 0{,}96$, der tatsächlich auch praktisch zu erreichen ist. Der Vorteil des Belegens beruht nicht so sehr auf der Steigerung der Beleuchtungsstärke in der Filmebene, als vielmehr auf der Minderung des Streulichtes, welches durch doppelte Reflexionen auf den Film geworfen wird. Bei dem angenommenen System mit 10 Flächen gegen Luft ohne Vergütung, sind es etwa 9% des auffallenden Lichtes, die sich kontrastmindernd dem geometrisch-optischen Bilde überlagern.

14. Bildgüte. Die Beurteilung der Objektive hinsichtlich ihrer Leistung kann von verschiedenen Gesichtspunkten aus erfolgen. Man kann den Korrektionszustand vermessen und wird dabei ein Urteil über die Leistung des Rechners gewinnen, der dieses System entworfen hat. Man wird danach aber nur wenig über die praktische Bewährung des Objektives aussagen können. Denn die praktische Bewährung wird ganz wesentlich von dem Zweck der damit gemachten photographischen Aufnahme beeinflußt. An eine künstlerische Porträtaufnahme werden ganz andere Erwartungen geknüpft als an eine Luftbildaufnahme. In der Beurteilung einer Aufnahme, bei der es nicht auf die exakte Wiedergabe der feinsten Einzelheiten ankommt, sind nach den Untersuchungen von KRÖBEL Kontrast und Auflösung in dem Bilde weitgehend austauschbar. Dagegen verlangt eine technisch-wissenschaftliche Aufnahme ein Maximum an Informationen. Eine physikalische Überprüfung eines Objektives kann natürlich die jenseits der technischen Grenze liegenden Komponenten nicht erfassen. Die muß sich auf die Messung der physikalischen Größen beschränken. Das nun leistet die Kontrastübertragungsfunktion, welche den Zusammenhang zwischen der Strichbreite eines gleichmäßig geteilten Gitters und dem Kontrastfaktor der Abbildung angibt. Dabei wird von einem solchen Gitter mit bekanntem Kontrast durch das Prüfobjektiv ein Bild entworfen, in welchem der verbleibende Kontrast gemessen wird. Das Verhältnis der Kontrastwerte des Bildes gegenüber denen des Objektes ist der Übertragungsfaktor für die jeweils benutzte Linienfrequenz. Die Kontrastübertragungsfunktion ist abhängig von der Einstellgenauigkeit, der Apertur, dem Ort des Bildfeldes und der Richtung der Rasterstriche gegenüber dem Meridianschnitt sowie schließlich von der Farbe. Man hat es also tatsächlich mit einer großen Zahl von Kontrastübertragungsfunktionen zu tun, die sich nur schwer herabdrücken läßt. Der wesentliche Vorteil bei der Benutzung der Kontrastübertragungsfunktion als Aussage über die Leistung des Objektivs liegt darin, daß sie für die verschiedenen Glieder einer Übertragungskette multiplikativ zusammentreten, so daß man die Kontrastübertragungsfunktion der gesamten Kette aus denen der einzelnen Glieder leicht berechnen kann.

Spiegelnde Prismen.

Von

H. Naumann.

Mit 55 Figuren.

1. Allgemeines. Spiegelnde Prismen erteilen einem Lichtbündel eine andere Richtung, eine andere Lage (z.B. durch Parallelversetzung) und der dabei vermittelten Abbildung eine Drehung oder Wendung. Sie wirken gelegentlich durch die Brechung, meist durch innere Reflexion, wobei Totalreflexion bevorzugt wird, und können grundsätzlich durch eine Spiegelanordnung ersetzt werden, haben aber den Vorteil, daß alle spiegelnden Flächen unverrückbar an einem einzigen Werkstück vereinigt sind, von Ausnahmen abgesehen. Hinsichtlich der optischen Abbildung wirken fast alle Prismen als dicke Planplatten.

2. Einteilung [*1, 2, 3, 4*]. Jedem Prisma kommt die Kombination je einer der folgenden Eigenschaften zu:

I. Weiterleitung des Lichtes in gleicher Richtung.
II. Knickung um 30°.
III. Knickung um 45°.
IV. Knickung um 60°.
V. Knickung um 90°.
VI. Knickung um 120°.
VII. Knickung um 180°.

Prismen, die den Strahlengang um einen dazwischen liegenden Winkel knicken, lassen sich z.B. durch zeichnerische Interpolation leicht bestimmen.

VIII. Knickung um einen veränderlichen Winkel.

A. Keine Änderung der Bildlage.
B. Seitenvertauschung in der Versetzungs- oder Knickebene.
C. Seitenvertauschung in der dazu senkrechten Ebene.
D. Bilddrehung um 180°, d.h. Bildaufrichtung.

1. Keine Parallelversetzung, Knickung des Strahls in einer Ebene.
2. Konstante Parallelversetzung.
3. Veränderliche Parallelversetzung.

Sehr oft wird eine Kombination solcher räumlicher Änderungen durch mehrere verschieden gestaltete Prismen geboten, die nach ihrem Erfinder oder ersten Benutzer (oft unheitlich) benannt werden. Die meisten Prismen bestehen aus einem Glasstück, einige müssen aus zwei Teilen verkittet, andere aus zwei Teilen in einer Fassung zusammengehalten werden.

3. Abmessungen. Die Lichtweglänge im Glas e wird im folgenden als Vielfaches vom Durchmesser d eines zylindrisch begrenzten durchgeleiteten Lichtbündels angegeben. Ist n die Brechzahl des Prismenwerkstoffes, dann gilt wie bei einer Planplatte für den äquivalenten Luftweg e_0

$$e_0 = e/n.$$

Für konvergente oder divergente Bündel sind Einsparungen an der Größe der Prismen möglich, die zweckmäßig graphisch ermittelt werden. Die Winkel der Prismenflächen gegeneinander sind den Abbildungen zu entnehmen, ebenso Länge l, Breite b, Höhe h und Bemerkungen hinsichtlich Verspiegelung, Glasart oder Befestigung.

4. Grundformen. Gewisse Grundformen kehren bei vielen Prismen als Bestandteile immer wieder. Diese sind:

das rechtwinklig-gleichschenklige Prisma oder der „Halbwürfel";

das Dachkantprisma;

der Tripelspiegel;

das Pentagonprisma.

α) Das rechtwinklig-gleichschenklige Prisma. Es dient zum Knicken des Strahlengangs um 90°; für seine Abmessungen (vgl. Fig. 1) gilt:

$$l = b = h = e = d.$$

Ein Bündel, das in Luft die Apertur $\sin u$ hat, wird total reflektiert, wenn

$$\sin u = n \cdot \sin(45° - \varepsilon'),$$

wobei ε' der Grenzwinkel der Totalreflexion ist.

Man braucht also Prismen hoher Brechzahl, wenn man weit geöffnete Bündel total reflektieren will. Ist kein Glas mit hinreichend hohem n vorhanden, muß man die Hypotenuse verspiegeln.

β) Die Dachkante. Zweimalige Spiegelung gibt seitenrichtige Bilder: Dies wird im Dachkantspiegel benutzt (Fig. 2). Zwei unter 90° zueinander stehende Planspiegelflächen werfen in der senkrecht zur Dachkante stehenden Ebene

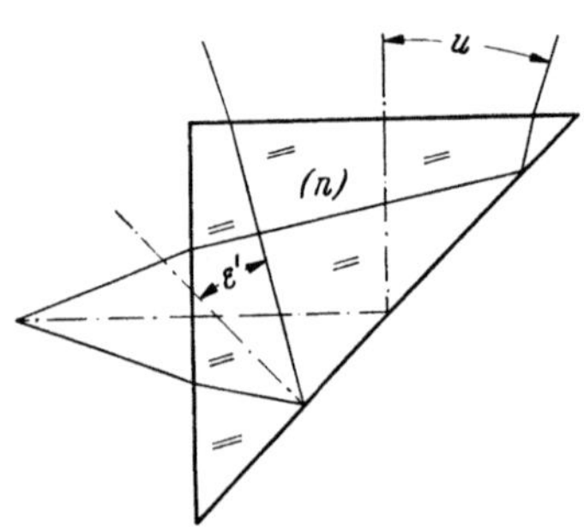

Fig. 1. Rechtwinklig-gleichschenkliges Prisma („Halbwürfel").

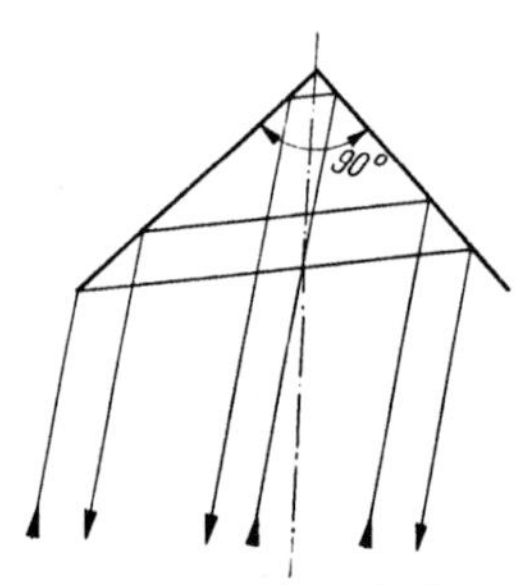

Fig. 2. Dachkantspiegel.

alle Strahlen in die Herkunftsrichtung zurück. In der die Dachkante enthaltenden Ebene wirkt die Anordnung nur als einfacher Spiegel, es gilt das Reflexionsgesetz. Ist d der Bündeldurchmesser, so ist die Breite b des Einzelspiegels $b = d/\sqrt{2}$; für schief unter dem Winkel α gegen die Winkelhalbierende einfallende Strahlen muß sein:

$$b = d \cdot \cos(45° - \alpha)/\cos 2\alpha.$$

Sehr wichtig ist genaue Innehaltung des 90°-Winkels, weil sonst ein Bildsprung stattfindet oder Doppelbilder auftreten. Eine Dachkante wird nur dann aus zwei getrennten Spiegeln aufgebaut, wenn sie sehr groß ist, im allgemeinen wird sie einem Prisma so angeschliffen, daß an der Dachkante innere Reflexion eintritt. Meist werden rechtwinklig-gleichschenklige Prismen benutzt, deren Hypotenuse die Dachkante trägt; in der Reflexionsebene findet Knickung um 90° statt, der Reflexionswinkel an den beiden die Dachkante bildenden Flächen beträgt 60°, so daß auch bei praktisch weit geöffneten Bündeln Totalreflexion eintritt und weder Verspiegelung noch hochbrechendes Glas erforderlich ist.

Allerdings kann die Dachkante infolge der Halbierung des Bündels die Auflösung, d.h. die Abbildungsgüte, gefährden und darüber hinaus bei etwas unsymmetrischer Lage durch Interferenzwirkung das Bild weiter verschlechtern[1].

Für ein rechtwinkliges Dachkantprisma (Fig. 3) gilt $e = d\sqrt{3}$. Die freien Spitzen der Dachkante können parallel zur gegenüberliegenden Kathete abgeschnitten werden, ebenso ein Teil an der beiden Katheten gemeinsamen Kante.

Eine Drehung der beiden spiegelnden Flächen um die Dachkante

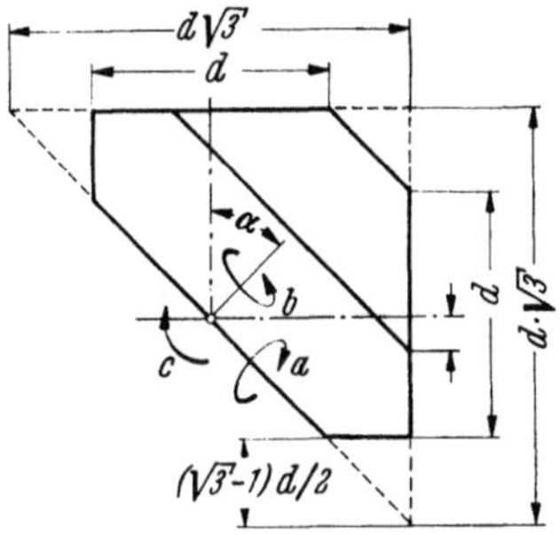

Fig. 3. Rechtwinklig-gleichschenkliges Dachkantprisma.

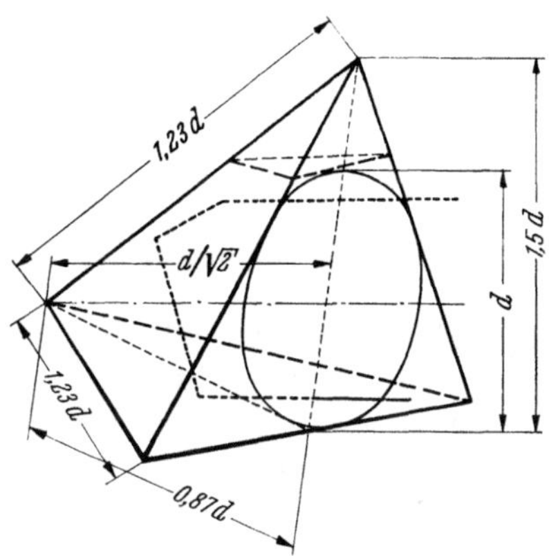

Fig. 4. Tripelspiegel.

(Achse *a*) hat keinerlei geometrische Einwirkung auf den Strahlengang; die Dachkante braucht nur in sich sehr genau zu sein. Drehung um Achse *b*, d.h. das Reflexionslot, bewirkt Bilddrehung um den doppelten Winkel; die Kante muß also parallel zur Ebene der Bündelachsen bleiben. Eine Drehung der Dachkante oder des Prismas um Achse *c* (senkrecht zur Zeichenebene) hat denselben Einfluß wie bei einer einfachen spiegelnden Fläche; der Knickwinkel ändert sich um den doppelten Betrag, die Bildlage bleibt erhalten.

γ) Der Tripelspiegel. Der Tripelspiegel (Fig. 4), die räumliche Erweiterung der Dachkante, besteht aus drei unter 90° aufeinander stehenden Spiegelflächen, die wegen der besseren Innehaltung der Winkel einem Prisma in Form einer Würfelecke angeschliffen werden. Alle Strahlen, aus welcher Richtung sie auch in den Tripelspiegel fallen, werden in ihre Herkunftsrichtung zurückgeworfen. Daraus folgt, daß der Tripelspiegel selbst nie gegenüber dem Strahlengang ausgerichtet zu werden braucht. Blickt man in ihn hinein, so erscheint das Auge stets im Schnittpunkt der drei Dachkanten.

Die überstehenden Ecken des als Würfelecke gedachten Tripelspiegels können so abgeschnitten werden, daß er sechseckig erscheint. Der Lichtweg ist $e = \sqrt{2}\,d$.

Der Spiegel wird bei Rückstrahlern an Fahrzeugen, Signalen usw. angewandt. Gelegentlich wird nur ein Ausschnitt aus dem Tripelspiegel benutzt, der „Tripelstreifen", der als Zusammenbau eines Halbwürfelprismas mit einem Dachkant-

[1] G. Joos: Zeiss-Nachr. **4**, H. 9, 221 (1943). — H. Steps: Optik **3**, 483 (1948).

prisma betrachtet werden kann und zur Einrichtung von Scheinwerfern dient (Fig. 5).

δ) Das Pentagon-Prisma. Zu den grundlegenden Prismenformen gehört das Fünfeck- oder Pentagon-Prisma, dessen Abmessungen aus Fig. 6 hervorgehen. Der Knickwinkel *k* des Lichtbündels von 90° zuzüglich der durch die Fertigung bedingten Ungenauigkeiten ist von der Drehlage des Prismas innerhalb der Zeichenebene unabhängig. Nachteilig ist, daß die beiden spiegelnden Flächen versilbert werden müssen; $e = 3{,}41\,d$. Eine Bilddrehung oder Seitenvertauschung findet nicht statt.

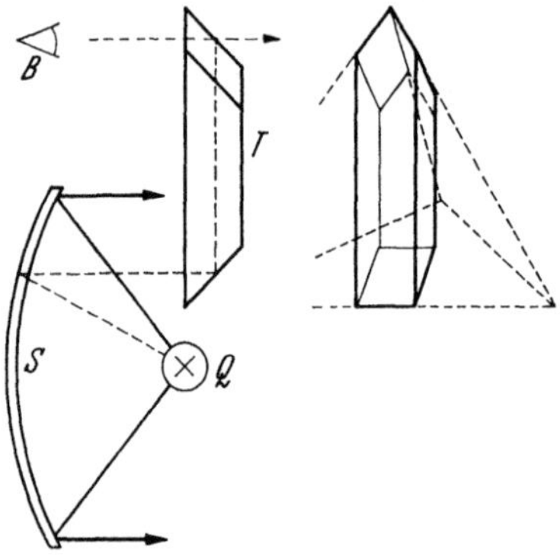

Fig. 5. Tripelstreifen. *Q* Lichtquelle; *S* Scheinwerferspiegel; *T* Tripelstreifen; *B* Beobachter.

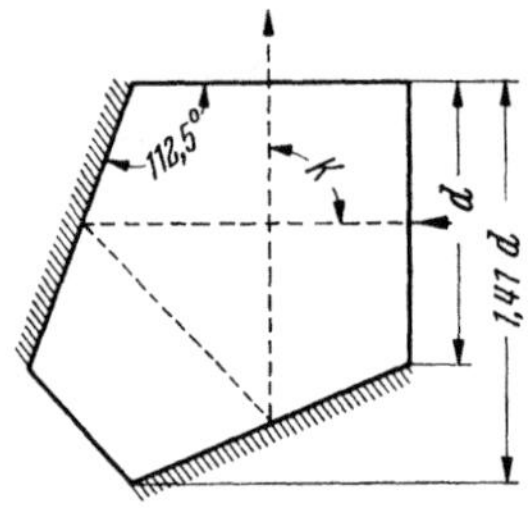

Fig. 6. Pentagonprisma.

Eine verspiegelte Fläche kann gegebenenfalls durch eine Dachkante ersetzt werden. Dann wird $e = 4{,}16\,d$. Das Prisma vertauscht die Seiten in der senkrecht zur Knickebene liegenden Richtung (Fig. 7, Prisma nach GOULIER).

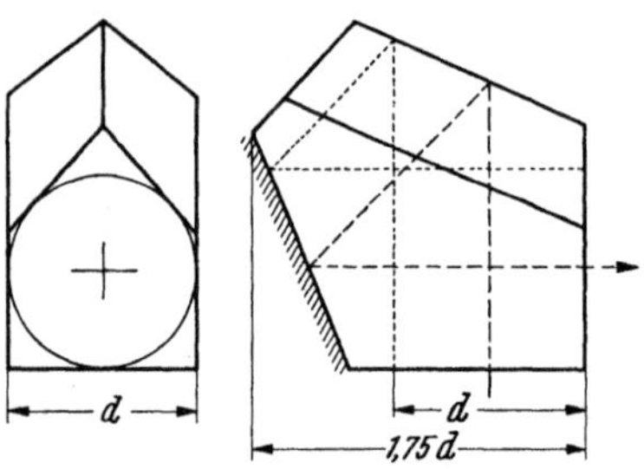

Fig. 7. Pentagonprisma mit Dachkante nach GOULIER.

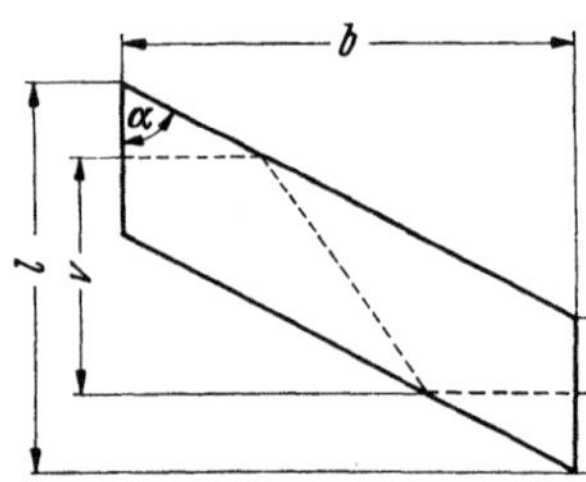

Fig. 8. Rhomboidprisma mit von 45° abweichendem Winkel α.

5. Spezielle bzw. zusammengesetzte Prismen. Für die hier folgende Besprechung der wichtigsten Prismenformen siehe die Vorbemerkungen unter 2. Nach den dort gegebenen Symbolen I bis VIII, A bis D, 1 bis 3 sind im folgenden die behandelten Prismen geordnet.

I. A. 1. Die Planplatte bewirkt Verlängerung des mechanischen Weges gegenüber dem optischen Weg um $\Delta = l(n-1)/n$, wenn l die Plattendicke ist. Sie wird eingeschaltet, wenn in einem von zwei zusammengehörigen Lichtwegen ein Abstandsunterschied herbeigeführt oder auch ausgeglichen werden soll.

I. A. 2. Das Rhomboidprisma gestattet Drehbarkeit des austretenden Strahls um den eintretenden als Achse oder umgekehrt, etwa in binokularen Geräten zur Anpassung an verschiedene Augenabstände. Bei wenig geöffneten Bündeln kommt man mit Totalreflexion aus, sonst müssen die 45°-Flächen verspiegelt werden. Wird das Prisma schräg gelegt (Fig. 8), dann können auch größere

Aperturen mit Totalreflexion beherrscht werden. Die Versetzung v ist

$$v = 2d\cos^2\alpha;$$

die Lichtweglänge

$$e = d\,(\cot\alpha + \tan\alpha);$$

ferner gilt

$$l = v + d; \qquad b = d\,(\tan\alpha + \cot\alpha\cos 2\alpha).$$

I. A. 3. Veränderliche Versetzung erreicht man durch zwei Halbwürfel, die durch mechanische Mittel gegeneinander verschiebbar sind (Fig. 9).

Dasselbe, insbesondere einschließlich des Wertes $v=0$ und negativer Werte, leistet eine zur optischen Achse neigbare Planplatte, sofern der mit der Neigung veränderliche Astigmatismus nicht stört.

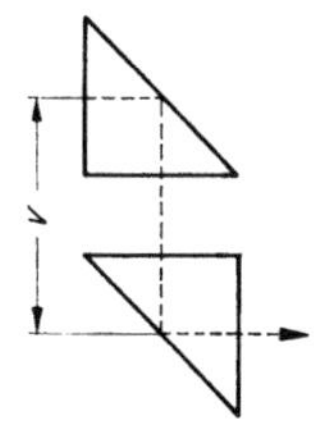

Fig. 9. Zwei Halbwürfel gestatten veränderliche Versetzung.

I. B. 1. a) Eine Spiegelung: Prisma nach Dove oder Amici (Fig. 10), auch nach Delaborne. Es gilt

$$l = d\,[\tan\alpha + 1/\tan(\alpha - \alpha')],$$

wobei

$$\sin\alpha = n\cdot\sin\alpha'.$$

Unter der kürzeren der parallelen Trapezseiten bleibt ein dreieckiger Raum vom Strahlengang frei von der Höhe

$$t = \frac{d}{2}\,[1 - \tan\alpha\tan(\alpha - \alpha')].$$

Der Lichtweg ist

$$e = d/\sin(\alpha - \alpha').$$

Das Prisma darf nur im telezentrischen Strahlengang benutzt werden, weil durch die Brechung an den schiefen Seitenflächen Astigmatismus auftritt. Üblich ist $\alpha=45°$.

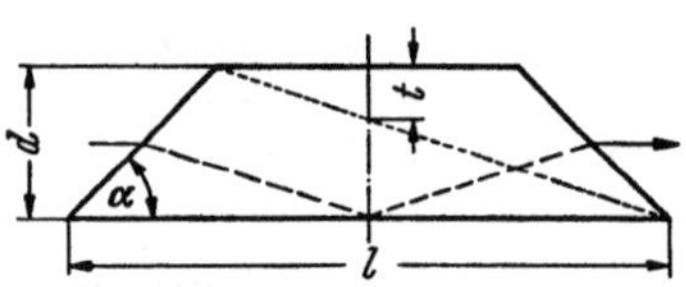

Fig. 10. Prisma nach Dove bzw. Amici.

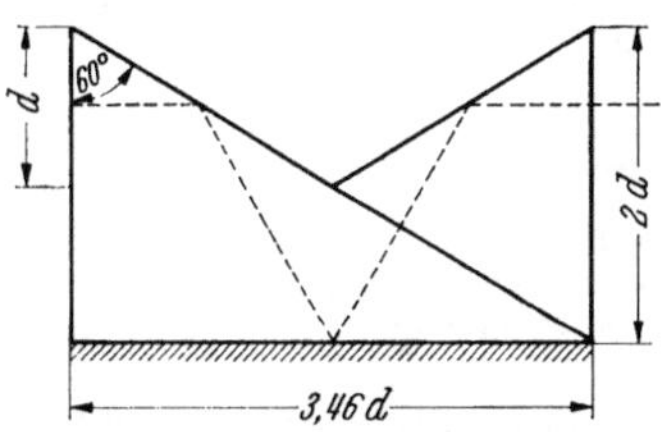

Fig. 11. Prisma nach Abbe bzw. König.

b) Drei Spiegelungen: Prisma von Abbe bzw. König. Es ist aus zwei Teilen zu verkitten; außerdem muß die Grundfläche verspiegelt werden (Fig. 11). Der Lichtweg ist

$$e = 5{,}20\,d.$$

Das Prisma kann auch im divergenten oder konvergenten Strahlengang benutzt werden; der Totalreflexionswinkel des benutzten Glases zuzüglich des halben Öffnungswinkels des Bündels darf 60° nicht überschreiten.

c) Fünf Spiegelungen: Das Prisma nach Schmidt oder Pechan[1] besteht aus zwei Teilen, die lose bzw. mit kleinem Luftraum aufeinander liegen (Fig. 12). Der Lichtweg ist $e=4{,}62d$. Beide Prismenteile sind in einer geeigneten Fassung sorgfältig zu lagern und zu befestigen. Ebenfalls fünf Spiegelungen, aber keine praktische Bedeutung hat das sehr umfangreiche Prisma nach Huet.

[1] D.R.P. 108685 vom 27. Juni 1899.

Alle Prismen unter I.B.1 drehen das Bild um den doppelten Winkel, wenn sie um die optische Achse gedreht werden. Sie finden Verwendung für die Bildaufrichtung in Rundblickfernrohren und ähnliche Zwecke.

I. B. 2. Seitenvertauschung in der Versatzebene. a) Drei Spiegelungen: Zusammenbau eines Halbwürfels mit einem Pentagon-Prisma. Beide Teile können getrennt oder verkittet sein. Ist $v > d$ die Versetzung, so ist $e = 3{,}41 + v$.

b) Prisma nach SPRENGER oder LEMAN; es hat den Vorteil, aus einem Stück zu bestehen. In dieser Form wird es selten benutzt (Fig. 13); $e = 3{,}46d$.

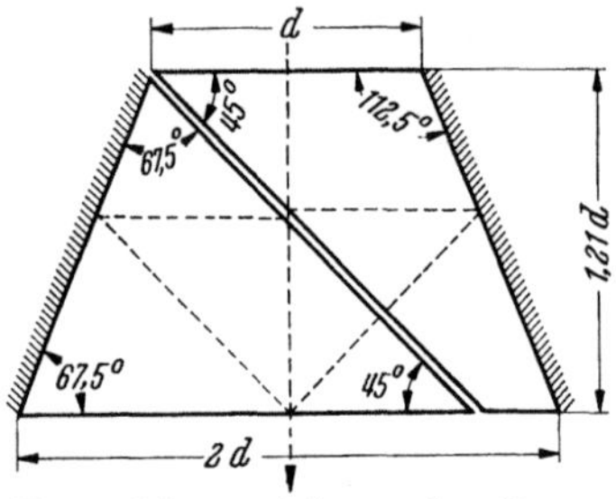

Fig. 12. Prisma nach SCHMIDT bzw. PECHAN.

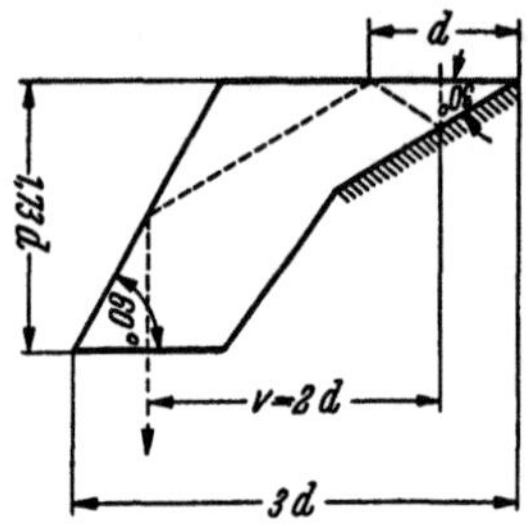

Fig. 13. Prisma nach SPRENGER bzw. LEMAN.

c) Prismen mit fünf Reflexionen sind von HUET angegeben worden, haben aber unwirtschaftlich lange Glaswege.

I. C. 1. Die hier zu nennenden Lösungen s. unter I.B.1; beide Fälle sind identisch.

I. C. 2. Seitenvertauschung senkrecht zur Versatzebene. Hier gelten die Lösungen wie unter I.A.2, jedoch wird eine reflektierende Fläche durch eine Dachkante

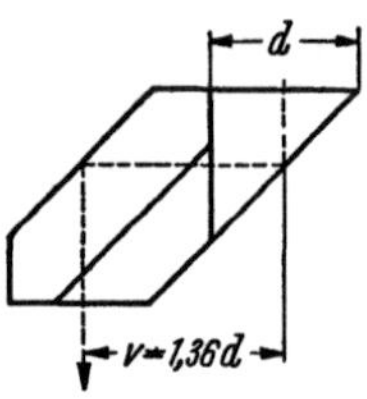

Fig. 14. Halbwürfel und Dachkantprisma.

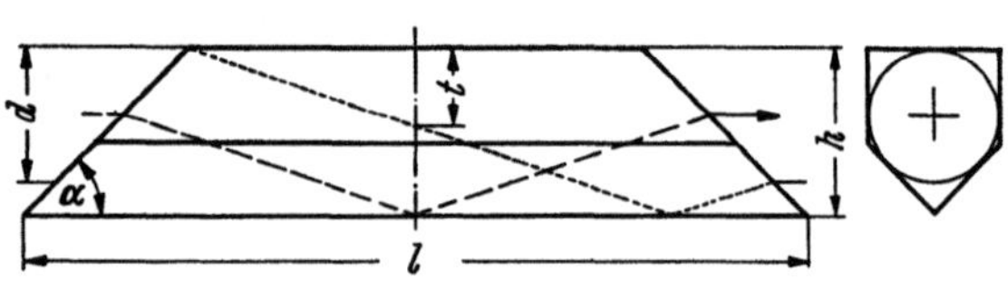

Fig. 15. Prisma nach AMICI mit Dachkante.

ersetzt (Fig. 14). Man beachte, daß der Glasweg im Halbwürfelprisma d, im Dachkantprisma aber $d\sqrt{3}$ beträgt. Wenn beide Prismen verkittet werden, ist die Versetzung $1{,}36d$.

I. D. 1. Bildaufrichtung in nicht geknicktem Strahlengang bieten alle Lösungen aus I.B.1, wenn eine spiegelnde Fläche durch eine Dachkante ersetzt wird:

a) Prisma nach AMICI, auch nach WIRTH (Fig. 15): Es ist

$$l = \sqrt{2}\, d\, [\tan\alpha + \cot(\alpha - \alpha')] .$$

Eine freie Stelle im Rücken des Prismas hat die Höhe

$$t = 0{,}83\, d \cdot [1 - \tan\alpha \tan(\alpha - \alpha')] ;$$

$h = 1{,}21\,\mathrm{d}$; $e = \sqrt{2} \cdot d/\sin(\alpha - \alpha')$. Auch hier tritt Astigmatismus auf, wenn das Prisma nicht im telezentrischen Strahlengang benutzt wird.

b) Das Prisma nach ABBE bzw. KÖNIG kann die Dachkante an jeder der drei spiegelnden Flächen tragen, zweckmäßig an der Grundfläche, weil dann deren

Verspiegelung erspart wird. Das Prisma wird zur Bildaufrichtung in Fernrohren benutzt, wenn keine Achsenversetzung und keine allzu große Zunahme des mechanischen Durchmessers gegenüber dem Bündeldurchmesser erwünscht ist (Fig. 16). Die Totalreflexion an den beiden oberen Planflächen kann durch Spiegelung an zwei vom Prisma getrennten Metallspiegeln ersetzt werden, was gelegentlich Justiervorteile bringt (LEITZ[1]).

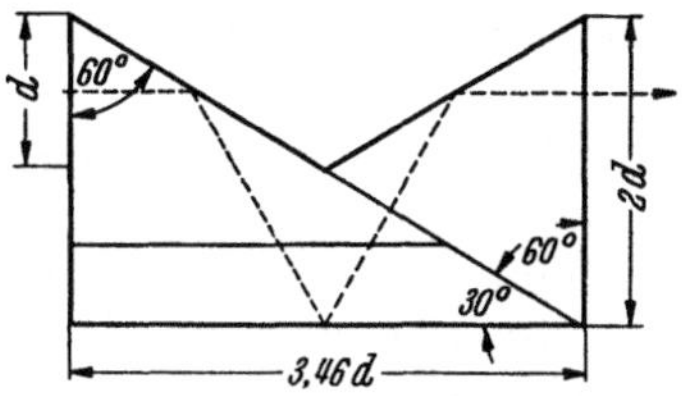

Fig. 16. Prisma nach ABBE bzw. KÖNIG mit Dachkante.

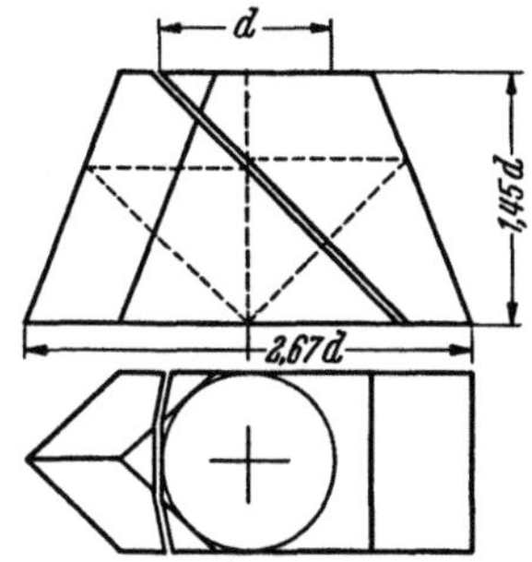

Fig. 17. Prisma nach PECHAN mit Dachkante.

c) Fünf Spiegelflächen: Das Prisma von PECHAN kann ebenfalls eine Dachkante erhalten. Es kommt in Frage, wenn eine kurze mechanische Baulänge wertvoll ist, ein langer Glasweg aber nicht schadet (Fig. 17). Zu dieser Klasse gehört auch ein Prisma von ZEISS[2].

I. D. 2. Bildumkehr mit seitlicher Versetzung. a) In nahezu allen Fernrohren wird diese Aufgabe durch die „Porro-Prismen erster Art" gelöst (Fig. 18). Die

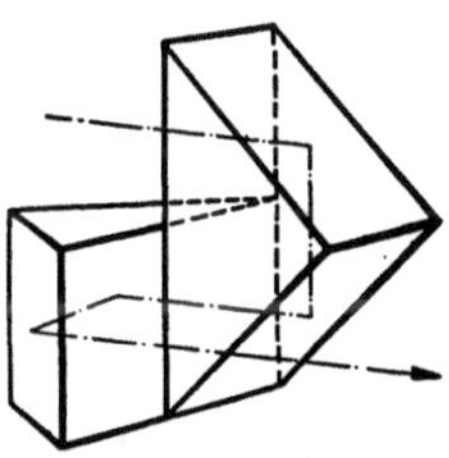

Fig. 18. Porro-Prismen erster Art.

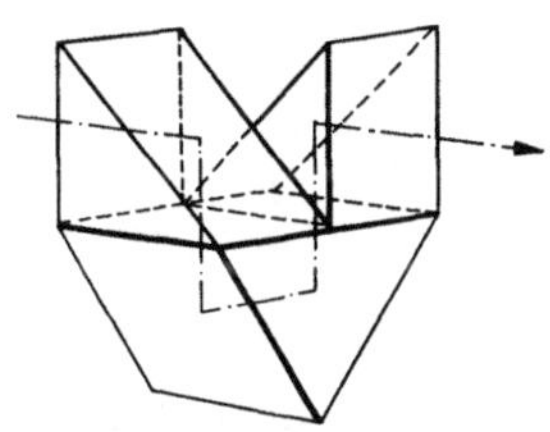

Fig. 19. Porro-Prismen zweiter Art.

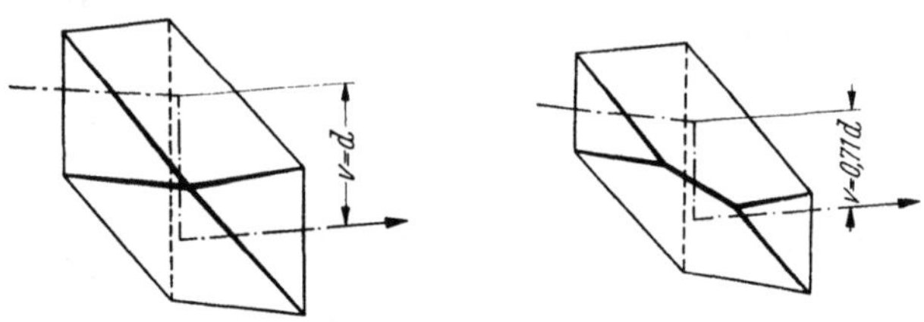

Fig. 20. Verkürzung des Porro-Prismas zweiter Art.

Prismen brauchen nicht mit den halben Hypotenusen aufeinander zu liegen, sondern können beliebig weit auseinandergezogen sein. Ist der Strahlengang nicht zylindrisch begrenzt, so können die Prismen verschieden groß ausgeführt, ein langer Strahlengang auf fast ein Drittel seiner Länge „zusammengefaltet" werden. Es ist $e=4d$, die Versetzung $v=\sqrt{2}d$.

b) Die „Porro-Prismen zweiter Art" zeigt Fig. 19. Hier ist $v=d$, $e=4d$. Häufig werden die Prismen als zwei spiegelbildlich gleiche Teile nach Fig. 20 ausgeführt, die eine geringere Bauhöhe gestatten, nämlich $1{,}71d$ statt $2d$; der

[1] Deutsches Patent 801542 vom 2. August 1949.
[2] D.R.P. 518143 vom 4. August 1929.

Lichtweg ist dann $e = 3{,}42d$. Durch Abstandsänderung kann der Betrag der Versetzung seitlich vergrößert werden. Eine Weiterentwicklung sind die Tetraeder-Prismen nach HUET[1].

c) Das Prisma von LEMAN oder SPRENGER mit Dachkante hat den Vorteil, aus einem einzigen Glasstück zu bestehen. Die Justierung liegt ein für allemal im Glase fest und hängt von keinerlei Befestigungsmittel ab. Es wird deshalb z.B. in Zielfernrohren benutzt, wo die erhebliche Versetzung auch von Vorteil ist; $e = 4{,}34\,d$ (Fig. 21).

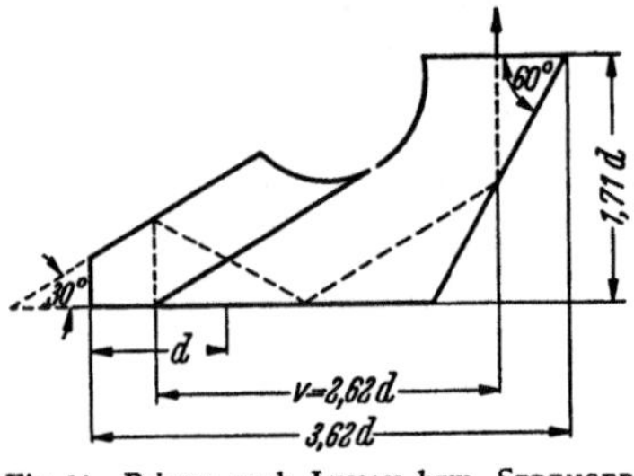

Fig. 21. Prisma nach LEMAN bzw. SPRENGER mit Dachkante.

d) Ein weiteres Prisma von HUET entspricht grundsätzlich dem von LEMAN, hat zwei Spiegelungen mehr und somit noch höhere Versetzung, $e = 8{,}27d$. Dies Prisma wird selten benutzt (Fig. 22). Hierzu sind weitere ähnliche Prismen nach ZEISS zu erwähnen[2].

Schließlich kann ein Pentagon-Prisma mit einem Halbwürfelprisma vereinigt werden, wobei einer dieser Teile eine Dachkante erhält (RODENSTOCK[3], MOELLER):

e) Dachkante am Pentagon-Prisma (Fig. 23). $v = d$; $e = 5{,}21\,d$.

f) Dachkante am Halbwürfel (Fig. 24). $v = 1{,}37d$; $e = 5{,}15\,d$.

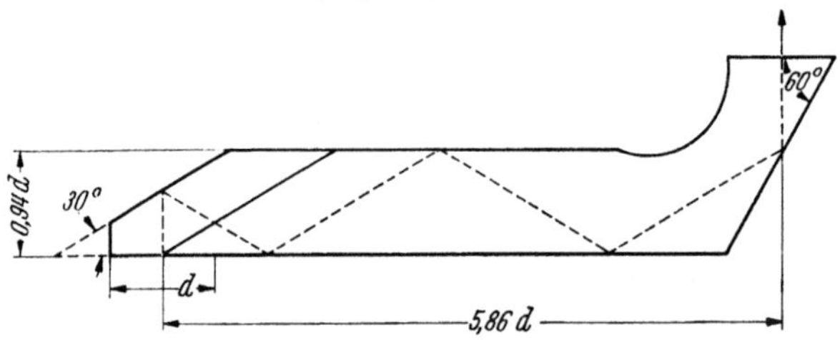

Fig. 22. Prisma nach HUET mit Dachkante.

Die erste Form ist der zweiten vorzuziehen, da meist nur eine Fläche zu verspiegeln ist. Natürlich können beide Elemente getrennt und in größerem Abstand voneinander angeordnet werden.

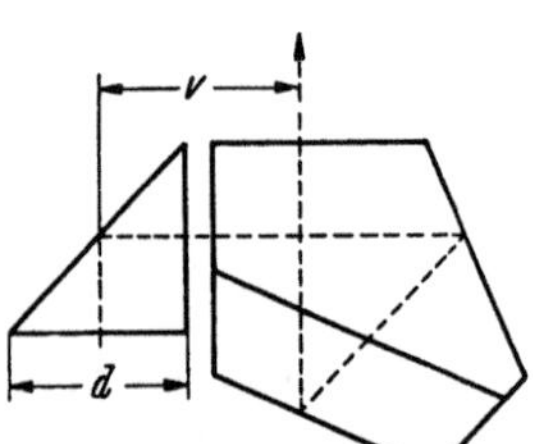

Fig. 23. Halbwürfel und Pentagonprisma mit Dachkante.

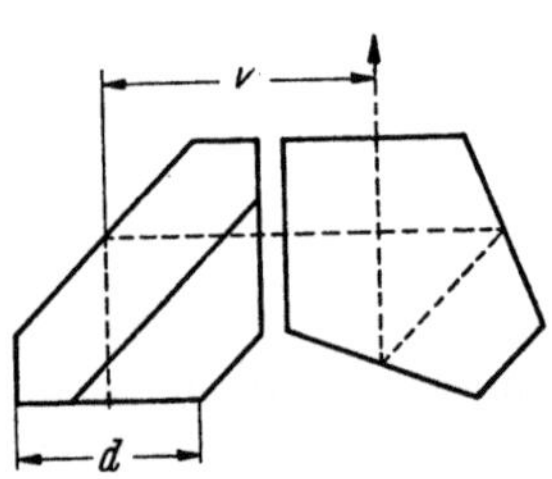

Fig. 24. Dachkantprisma und Pentagonprisma.

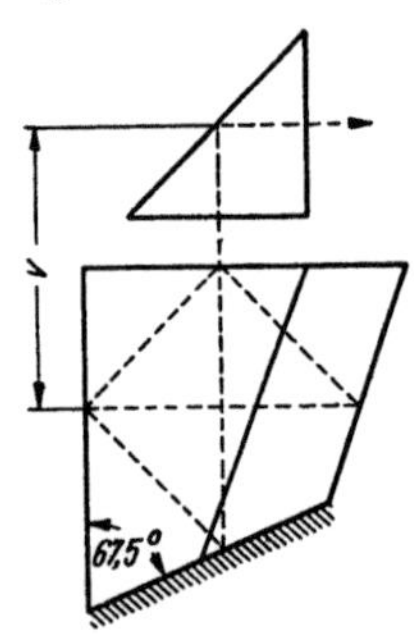

Fig. 25. Prisma nach SCHMIDT.

g) Prisma von SCHMIDT: Halbwürfel mit Vierseitprisma (Fig. 25); $v = 1{,}21\,d$, $e = 7{,}1\,d$. Die Prismen dürfen nicht verkittet werden, weil sonst keine Totalreflexion stattfindet.

[1] D.R.P. 104149 vom 26. April 1898.

[2] Gebrauchsmuster 980017 und Patentanmeldung vom 24. Juni 1925.

[3] D.R.P. 444823 vom 7. Juni 1925.

I. E. 1. Bilddrehung um 90° ohne Seitenvertauschung erfordert zwei oder vier Reflexionen und ist mit einem Prisma ohne Achsenversetzung nicht durchzuführen. Man kann z.B. zwei DOVEsche Prismen ohne Dachkante hintereinander anordnen und das zweite um 45° drehen, so daß das Bild um 90° stürzt; das erfordert eine große Baulänge.

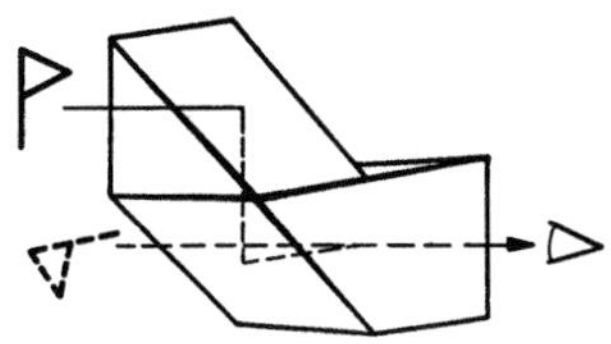

Fig. 26. Dreifachprisma zur Bilddrehung.

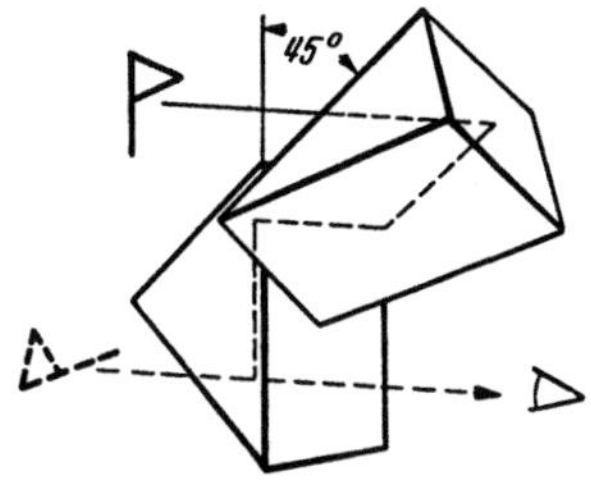

Fig. 27. Bilddrehung mit Porro-Prismen.

a) Eine einfache Anordnung mit Versetzung zeigt Fig. 26. Das vom Strahlengang vermittelte Bild wird hierbei im Gegenzeigersinne um 90° gedreht; für eine Drehung im Zeigersinne ist das Prisma spiegelbildlich aufzubauen. Die Versetzung $v=\sqrt{2}d$ wird kleiner, wenn die Prismen nach Fig. 20 gestaltet werden.

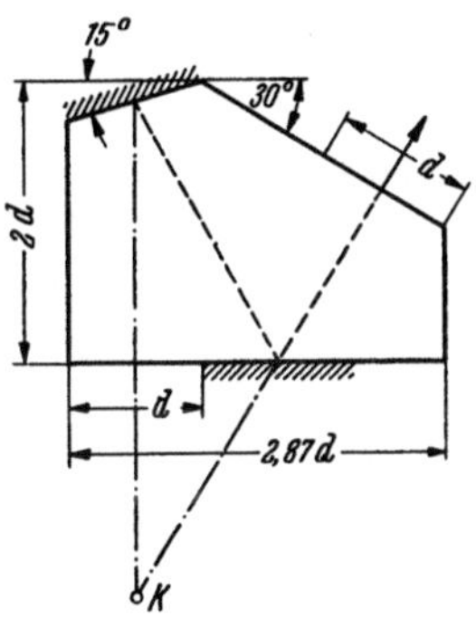

Fig. 28.

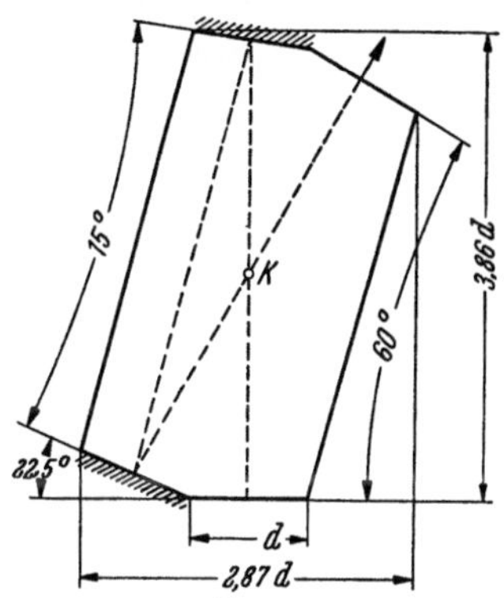

Fig. 29.

Fig. 28 und 29. Um 30° knickende Prismen.

b) Ohne Seitenvertauschung arbeitet ein aus dem Porro-Satz erster Art abzuleitender Aufbau nach Fig. 27. Durch Ändern des 45°-Winkels kann die Bilddrehung variiert werden.

II.A.1. Ablenkung des Bündels um 30° ohne Bilddrehung erfordert zwei Spiegelungen in einer Ebene, wobei der Knickpunkt K der Bündelachse außerhalb oder innerhalb des Prismas liegen kann (Fig. 28 und 29). $e=5{,}46d$ bzw. $=11{,}33d$. Beide Prismen sind sehr groß, die Aufgabe löst man besser mit zwei getrennten Spiegeln.

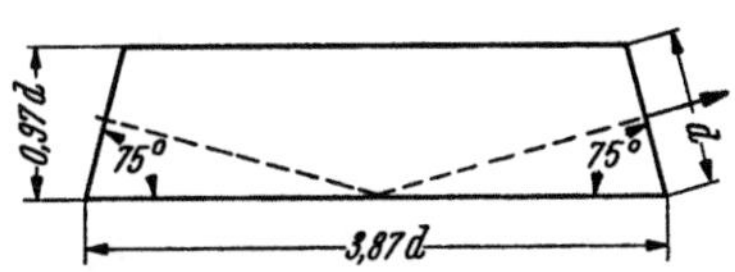

Fig. 30. Um 30° knickendes, seitentauschendes Prisma.

II.A.2. Knickung um 30° und Versetzung aus der Knickebene ist durch zwei getrennte Prismen aus II.A.1 und I.A.2 ohne weiteres möglich. Natürlich können solche Kombinationen auch verkittet werden.

II.B.1. Knickung um 30° mit Seitentausch kann durch einen Spiegel der Länge $3{,}9d$ bewirkt werden, oder durch ein Prisma nach Fig. 30 mit einer Reflexion; es ist $e=3{,}74d$. Lösungen mit drei Spiegelungen sind leicht konstruierbar.

II.C.1. Seitentausch senkrecht zur Knickebene fordert eine Dachkante in den Prismen nach Fig. 28 oder 29.

II.D.1. Bildaufrichtung mit Knickung um 30° ist durch ein Prisma der Form II.B.1, aber mit Dachkante, leicht durchführbar. Es wird dann $e=4{,}56d$ (Fig. 31).

III.A.1. Knickung eines Strahlenganges um 45° ohne Bildlagenbeeinflussung kommt vor, um den senkrechten oder waagerechten Strahlengang eines Gerätes

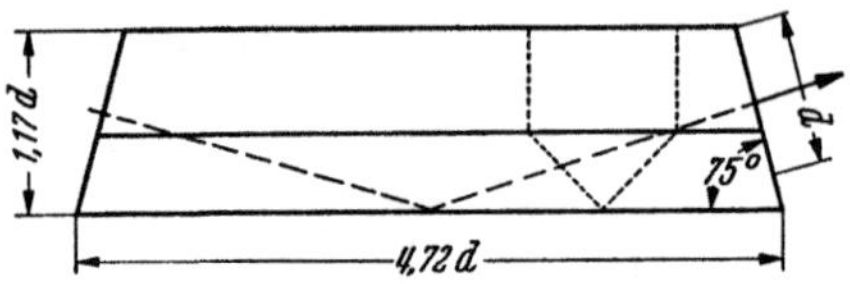

Fig. 31. Um 30° knickendes bildaufrichtendes Prisma. Die gestrichelten Linien deuten seinen Querschnitt an.

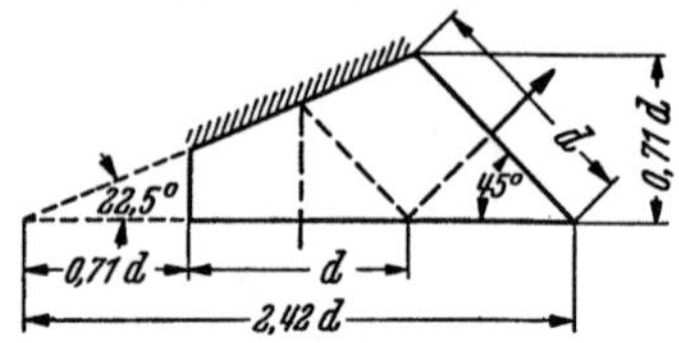

Fig. 32. Prisma nach Bauernfeind.

um 45° nach oben in das bequem gehaltene Auge richten zu können. Hier kann der Knickpunkt in oder neben dem Prisma liegen.

a) Prisma nach Bauernfeind (Fig. 32); $l=2{,}42d$; davon kann $0{,}71d$ abgeschnitten werden. $e=1{,}71d$.

b) Eine andere Form zeigt Fig. 33; $e=3{,}41d$. Das Prisma kommt ohne verspiegelte Flächen aus, sofern die Einschränkungen aus Ziff. 4α beachtet werden.

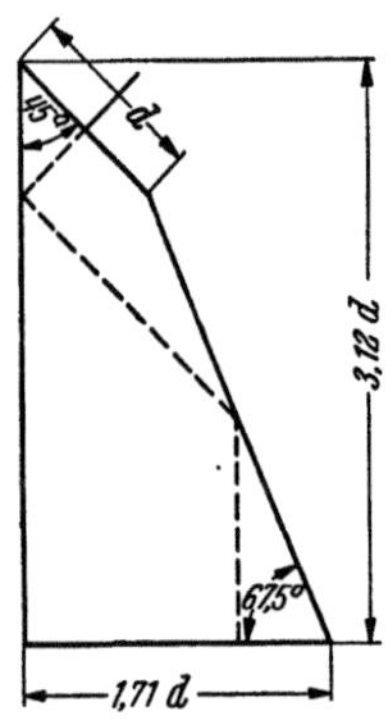

Fig. 33. Um 45° knickendes Prisma mit außen liegendem Knickpunkt.

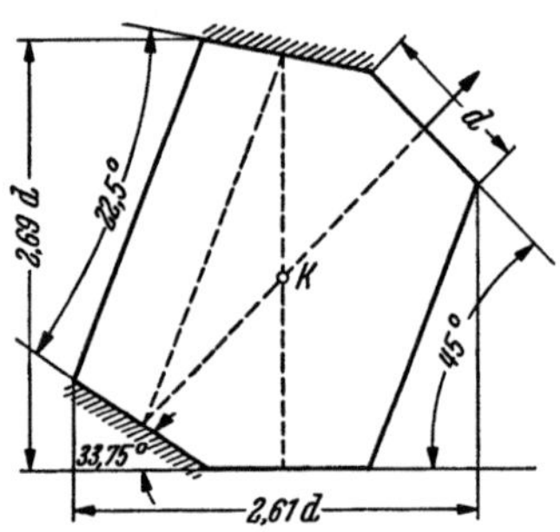

Fig. 34. Um 45° knickendes Prisma mit innerem Knickpunkt.

c) Für ein Prisma mit im Innern liegendem Knickpunkt (Fig. 34), das zwei verspiegelte Flächen haben muß, gilt $e=7{,}46d$.

d) Ein Prisma mit viermaliger Spiegelung verlegt den Knickpunkt besonders weit vom Prisma weg (Fig. 35); $e=2{,}71d$.

III.B.1. Knickung um 45° bei Seitenvertauschung in der Knickebene kann

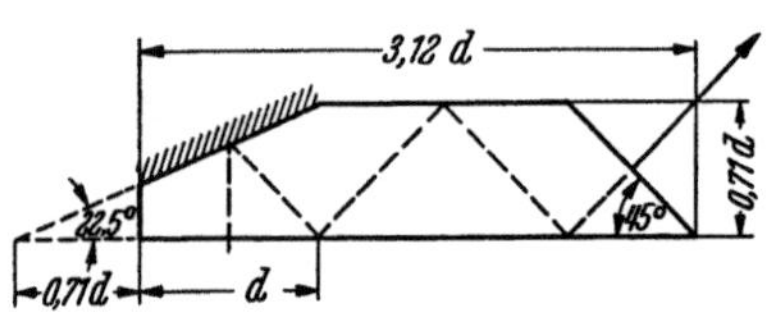

Fig. 35. Um 45° knickendes Prisma mit weit außen liegendem Knickpunkt.

a) durch einen Spiegel unter 67,5° der Länge $2{,}62d$ geleistet werden. Dasselbe leistet ein Prisma wie das in Fig. 30 gezeichnete, jedoch mit den geänderten Maßen $l=2{,}62d$, $e=2{,}41d$, $h=0{,}92d$ und einem Winkel von 67,5° statt 75°.

b) Drei Reflexionen hat das Prisma nach Schmidt (Fig. 36), $e=2{,}41d$.

III.C.1. Seitenvertauschung senkrecht zur Knickebene erfordert zwei Reflexionen in der letzteren, davon eine an einer Dachkante. Es ist also den Prismen unter III.A.1 eine Dachkante anzufügen.

III.D.1. Vollständige Bildaufrichtung bei Knickung um 45° ist bei Mikroskopen und Fernrohren häufig. Dazu ist eine ungerade Anzahl von Reflexionen, davon eine an einer Dachkante, erforderlich.

a) Dem Doveschen Prisma ähnlich ist das in Fig. 37 abgebildete mit $e = 3{,}55\,d$.

b) Drei Spiegelungen hat das Prisma nach SCHMIDT (Fig. 38), $e = 2{,}54\,d$. Das Prisma ist in Fernrohren wegen seiner kleinen Baulänge und seines kurzen Lichtweges beliebt.

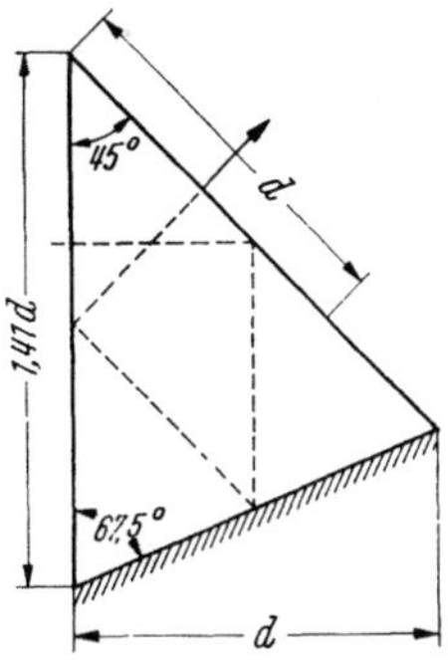

Fig. 36. Prisma nach SCHMIDT.

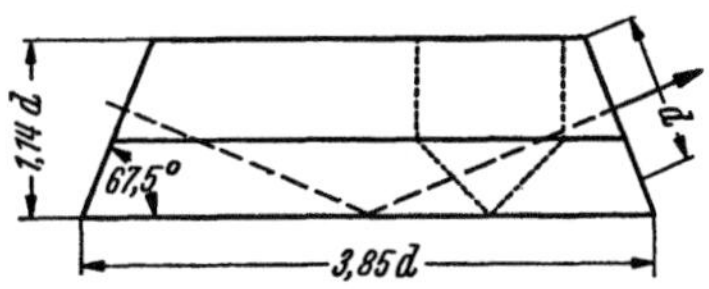

Fig. 37. Um 45° knickendes Prisma mit Bildaufrichtung. Querschnitt durch gestrichelte Linien angedeutet.

III.E.1. Bilddrehung um 90° bei Knickung des Strahlenganges um 45° löst man durch Kombination bekannter Prismen.

IV.A.1. Für eine Knickung um 60° gelten grundsätzlich die Lösungen wie für 30 und 45°, nur mit anderen Winkeln, die leicht zu ermitteln sind. Dasselbe gilt für die Abmessungen und Lichtwege.

IV.B.1. Knickung um 60° mit Seitenvertauschung ist durch einen Spiegel der Länge $2d$ oder durch ein gleichseitiges Prisma möglich; die Spitze kann abgeschnitten werden, $e = 1{,}73\,d$.

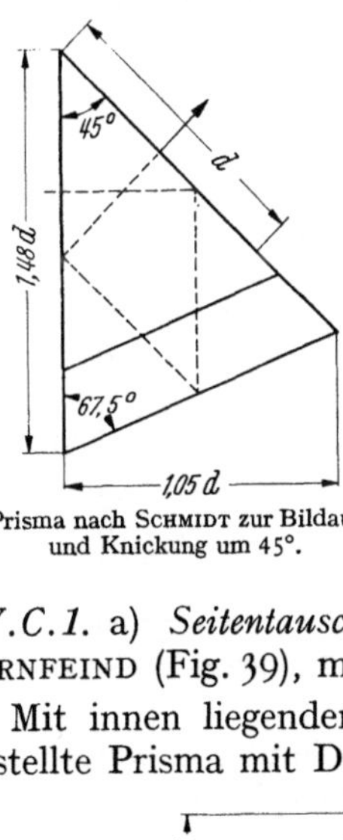

Fig. 38. Prisma nach SCHMIDT zur Bildaufrichtung und Knickung um 45°.

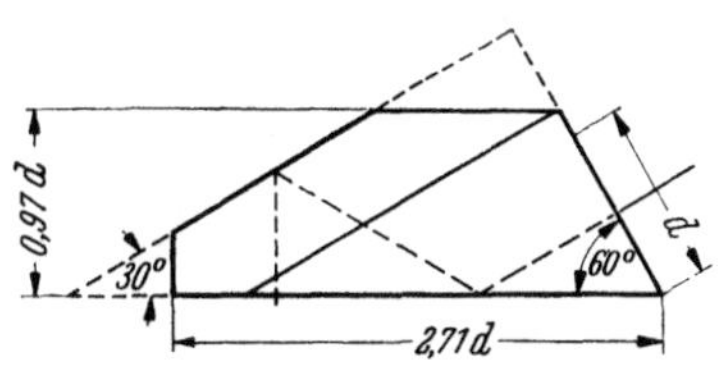

Fig. 39. Prisma nach BAUERNFEIND.

IV.C.1. a) *Seitentausch senkrecht zur Knickebene* liefert das Prisma nach BAUERNFEIND (Fig. 39), mit Dachkante versehen, $e = 1{,}81\,d$.

b) Mit innen liegendem Knickpunkt der Bündelachsen ist das in Fig. 34 dargestellte Prisma mit Dachkante möglich.

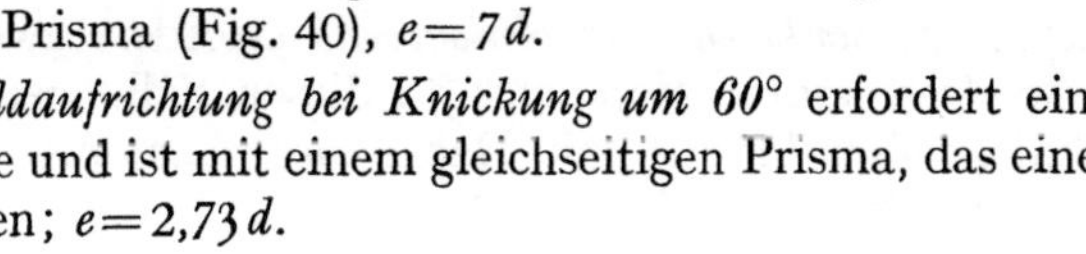

Fig. 40. Um 60° knickendes seitenvertauschendes Prisma.

c) Mit vier Reflexionen ergibt sich ein sehr langes, den Knickpunkt weit hinauslegendes Prisma (Fig. 40), $e = 7\,d$.

IV.D.1. Bildaufrichtung bei Knickung um 60° erfordert eine Spiegelung an einer Dachkante und ist mit einem gleichseitigen Prisma, das eine solche aufweist, leicht zu erzielen; $e = 2{,}73\,d$.

V.A.1. Rechtwinklige Abknickung eines Strahlenganges ist die häufigste Aufgabe eines Prismas. Ohne Seitenvertauschung löst sie das Pentagon-Prisma (Fig. 6). Andere Lösungen bietet das Viereckprisma nach WOLLASTON (Fig. 41), $e=4{,}82d$.

Man kann dasselbe Prisma auch mit vier Spiegelungen benutzen, dann müssen freilich die beiden stumpfwinklig zueinander stehenden Flächen verspiegelt und außerdem die Totalreflexionsbedingungen eingehalten werden (Fig. 42); $e=6{,}85d$.

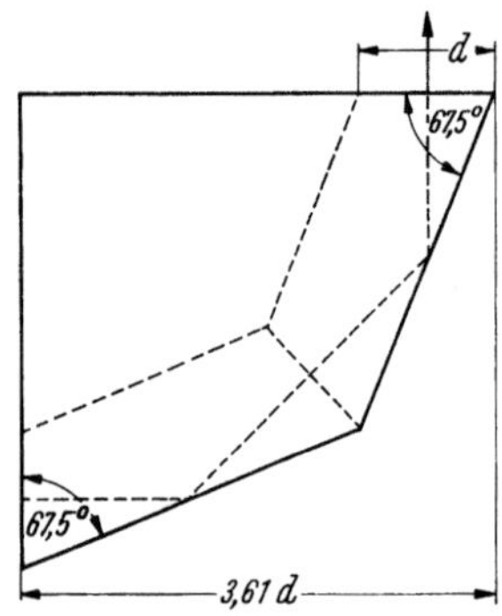

Fig. 41. Viereckprisma nach WOLLASTON.

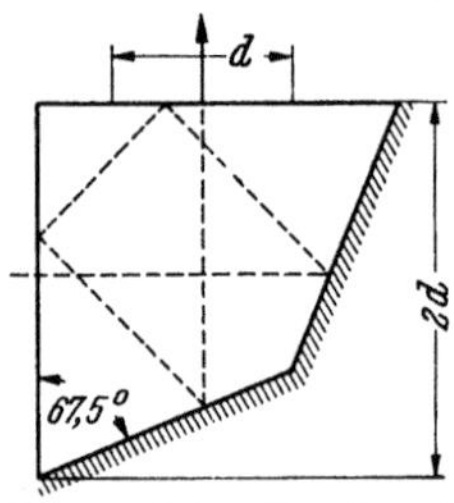

Fig. 42. Wollaston-Prisma in anderer Anwendung.

V.B.1. Abknickung um 90° und Seitenvertauschung leistet ein Planspiegel der Länge $1{,}41\,d$ oder ein Halbwürfelprisma.

V.C.1. Seitentausch senkrecht zur Knickung bedingt eine Dachkante an dem Pentagon-Prisma oder dem Prisma nach WOLLASTON.

V.D.1. Vollständige Bildaufrichtung bei 90° Knickung liefert ein Dachkantprisma; s. Ziff. 4β.

V.E.1. Bilddrehung um 90° bei Knickung um 90° liefern zwei Halbwürfel, die zu einem Ganzen vereinigt sein können wie die Hälfte eines Porro-Prismas zweiter Art (Fig. 20).

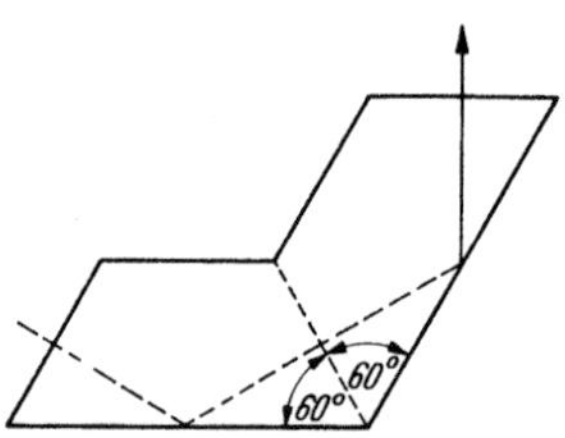

Fig. 43. Knickung um 120° durch zwei gleichseitige 60°-Prismen.

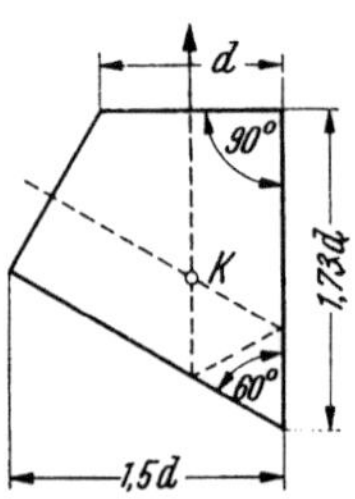

Fig. 44. Um 120° knickendes Vierseitprisma.

VI.A.1. Knickung eines Strahlenganges um 120° kommt vor, wenn ein langer Strahlengang auf engem Raum zusammengedrängt, beispielsweise ein von oben nach unten verlaufender Strahlengang auf einer pultartig angeordneten Fläche aufgefangen werden soll. Man knickt zweimal um je 60° durch Prismen mit einer Reflexion oder Planspiegel; ein Beispiel gibt Fig. 43. Eine Lösung mit innen liegendem Knickpunkt zeigt Fig. 44, $e=3{,}46d$.

VI.B.1. Seitenvertauschung in der Reflexionsebene erreicht man am einfachsten mit einem Planspiegel der Länge $1{,}15d$.

VI.C.1. Seitentausch senkrecht zur Knickebene erfordert zwei Spiegelungen, davon eine an einer Dachkante. Man setzt für diesen Fall einen Halbwürfel mit einem um 30° knickenden Prisma mit Dachkante zusammen oder ein Dachkantprisma mit einem Prisma nach Fig. 30.

VI.D.1. Für volle Bildaufrichtung bei Knickung um 120° gibt es eine Reihe von Lösungen. Grundsätzlich setzt man zwei Prismen zusammen, deren Knicksumme 120° beträgt; davon richtet das eine das Bild auf, während das andere die Bildlage nicht beeinflußt; oder man läßt das eine in der Knickebene, das andere senkrecht dazu die Bildseiten tauschen. Am einfachsten und kleinsten ist ein Dachkantprisma nach Fig. 45.

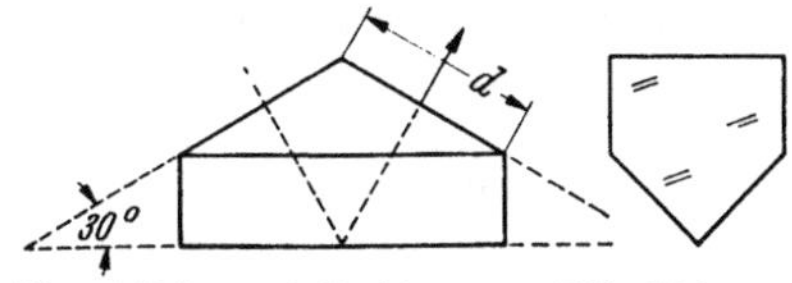

Fig. 45. Prisma mit Dachkante zur Bildaufrichtung.

Bei Knickungen des Strahlenganges um mehr als 90°, insbesondere bei 180°, ist für die Frage der Seiten- und Höhenvertauschung wichtig, ob die Knickungsebene senkrecht steht oder waagerecht liegt. Im letzteren Fall wird der Betrachter sich umdrehen und seinen Blick ebenfalls in der waagerechten Ebene wenden, und es gelten folgende Spiegelungsregeln (s. Fig. 46): Ein Bild werde auf einer Durchlichtwand A entworfen und vom Standpunkt B_1 aus betrachtet. Dann werden

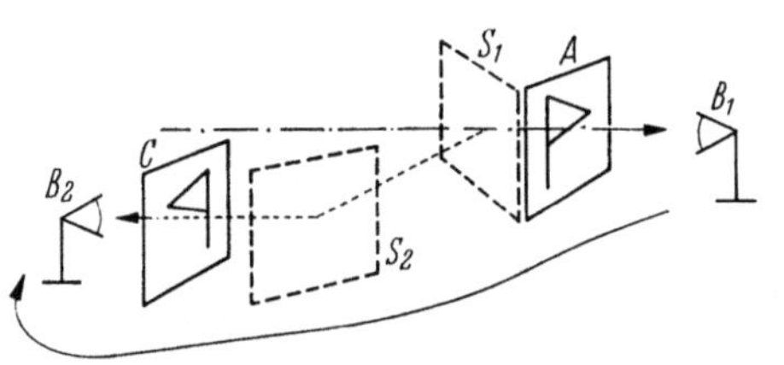

Fig. 46.

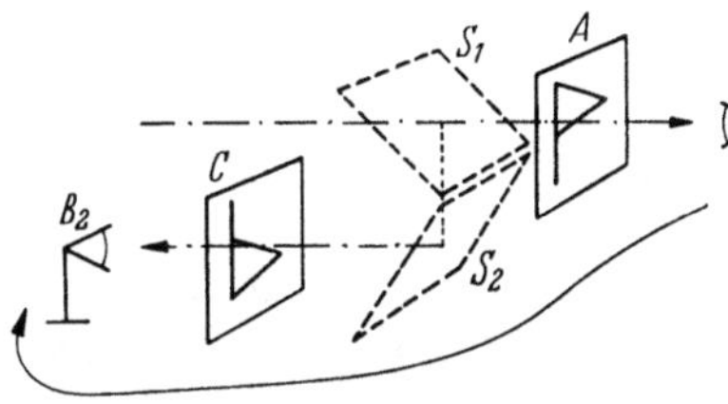

Fig. 47.

Fig. 46 und 47. Zum Problem der Bildumkehr bei Knickung um mehr als 90°.

die Spiegel S_1 und S_2 eingefügt, die das Bild auf die Wand C werfen; der Beobachter begibt sich nach B_2 und sieht das Bild in derselben Lage wie auf A, weil Knickung des Strahlenganges und Bewegung des Beobachters in der gleichen Ebene stattfanden. Vollzieht sich die Knickung in der senkrechten Ebene (Fig. 47), so wird das Bild umgekehrt. Bilden Knickebene und Wendeebene des Beobachters einen Winkel α miteinander, so dreht sich das Bild um 2α. Es sind also je nach der Lage der Ebenen bilddrehende oder -aufrichtende Mittel einzufügen bzw. wegzulassen.

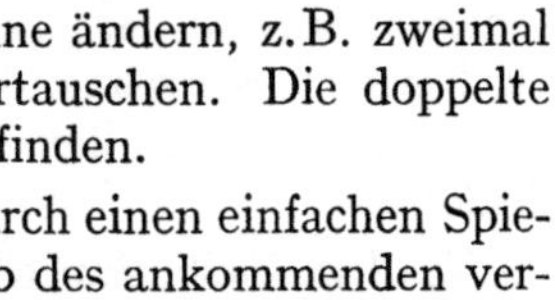

Fig. 48. Knickung um 180° mit Versetzung.

VII.A.1. Rückwurf, also Knickung um 180°, unter Beibehaltung der Bildlage leisten zwei Spiegel, Halbwürfelprismen oder sonst zwei Prismen, die zusammen um 180° knicken und die Bildlage im gleichen Sinne ändern, z.B. zweimal das Bild aufrichten oder zweimal dieselben Seiten vertauschen. Die doppelte Knickung muß aber in der waagerechten Ebene stattfinden.

VII.B.1. Seitentausch bei Knickung um 180° wird durch einen einfachen Spiegel bewirkt; soll der rückgeführte Strahl nicht innerhalb des ankommenden verlaufen, dann sind drei Spiegelungen in der Knickebene nötig, die in vielen Möglichkeiten angeordnet werden können, beispielsweise in einem gleichseitigen Prisma mit 120° Spitzenwinkel (Fig. 48), $e=1{,}24d$. Die Knickung kann in der waagerechten wie in der senkrechten Ebene stattfinden.

VII.C.1. Höhentausch bei Knickung um 180° geschieht durch zwei Spiegelungen, von denen eine an einer Dachkante stattfinden muß; das Ganze ist ein Tripelspiegel (vgl. Ziff. 4γ). Wie die Knickwinkel verteilt sind und welche Elemente im einzelnen benutzt werden, ist gleichgültig.

VII.D.1. Volle Bildaufrichtung leisten zwei Spiegelungen in der senkrechten Ebene, also z.B. ein Teilprisma eines Porro-Prismensatzes erster Art nach Fig. 18.

VII.E.1. Bilddrehung um 90° leisten die eben erwähnten Elemente, wenn sie unter 45° aufgestellt sind.

VIII.A. Knickung um einen veränderlichen Winkel ohne Beeinflussung der Bildlage erfordert, daß von zwei spiegelnden Flächen eine schwenkbar ist. Man benutzt entweder zwei Planspiegel oder eine Prismenanordnung nach Fig. 9, sofern eine seitliche Versetzung statthaft ist.

Für kleine Knickwinkel insbesondere beiderseits des einbezogenen Nullwertes kommen brechende flache Prismen in Frage, die, zu zweien angeordnet,

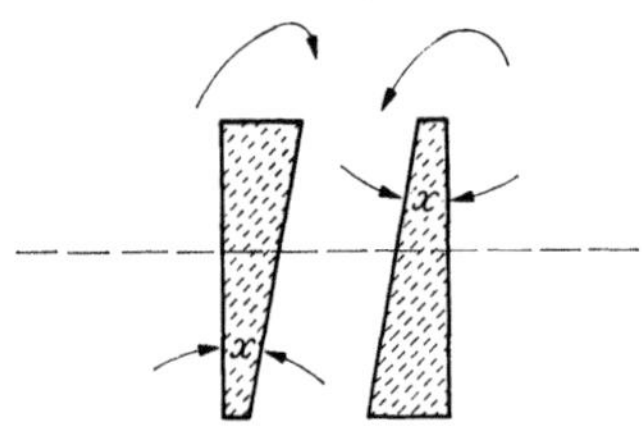

Fig. 49. Abatscher Doppelkeil.

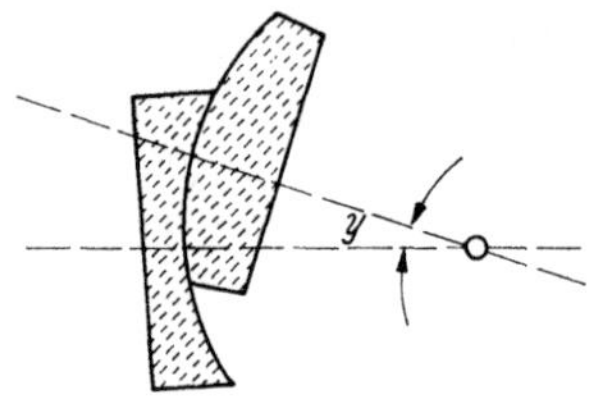

Fig. 50. Veränderbares Prisma aus Konkav- und Konvexlinse.

im Gegensinne um die optische Achse gedreht werden (Abatscher Doppelkeil, Fig. 49). Ist x der Prismenwinkel, n die Brechzahl des Prismenwerkstoffes und 2α die gegenseitige Verdrehung der Prismen aus der gezeichneten Nullage heraus, dann ist die Strahlablenkung β

$$\beta = 2(n-1)\cdot x\cdot \sin\alpha .$$

Ebenso kann man eine plankonkave und eine plankonvexe Linse — sphärisch oder zylindrisch — von entgegengesetzt gleicher Brechkraft gegeneinander ver-

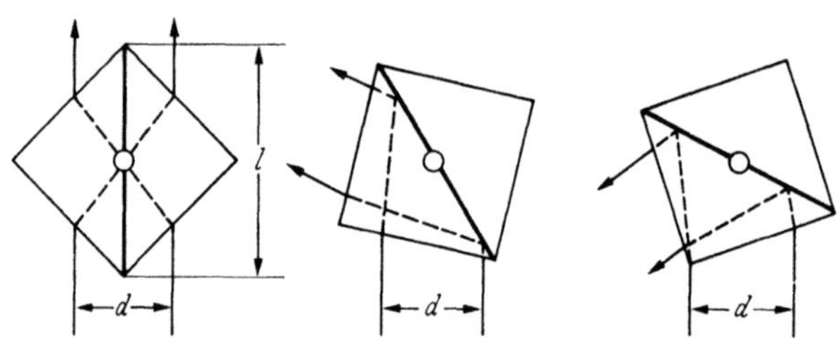

Fig. 51. Drehprisma zur veränderbaren Knickung.

schieben oder in der gekrümmten Fläche verdrehen (Fig. 50). Es entsteht dann ein brechendes Prisma, dessen Kantenwinkel gleich dem Verdrehwinkel y ist.

In all diesen Fällen ist der mit der Brechung verbundene Farbfehler zu bedenken.

VIII.B. Seitentausch in der Knickebene ist durch einen einzigen Spiegel leicht herbeizuführen. Soll der Nullwert einbezogen und gegebenenfalls symmetrisch überschritten werden, dann legt man zwei Prismen nach AMICI (Fig. 10) oder zwei Halbwürfel mit den Hypotenusen zusammen (Fig. 51). Wenn sie verspiegelt werden, sind ±120°-Knickung erreichbar.

VIII.C.; *VIII.D.* Bei Seitentausch in der Knickebene oder bei voller Bildaufrichtung sind die Prismenanordnungen unter I—VII, C. D. so aufzulösen, daß eine einzelne Spiegelfläche schwenkbar angeordnet werden kann, falls man nicht die dort angegebenen Prismen mit einer der Anordnungen aus *VIII.A.* verbindet.

6. Fertigungsfehler bei Prismen. Jedes noch so sauber geschliffene und polierte Prisma zeigt Herstellungsfehler, die sich auf den Strahlengang auswirken. Wenn die Prismenflächen zwar eben sind, die Winkel aber Fehlerbeträge gegenüber dem beabsichtigten Wert enthalten, ergeben sich Abweichungen in der Richtung des austretenden Strahls; hierfür seien einige wichtige Beispiele im folgenden zusammengestellt. Dabei seien α, β, γ usw. die Fehlerbeträge.

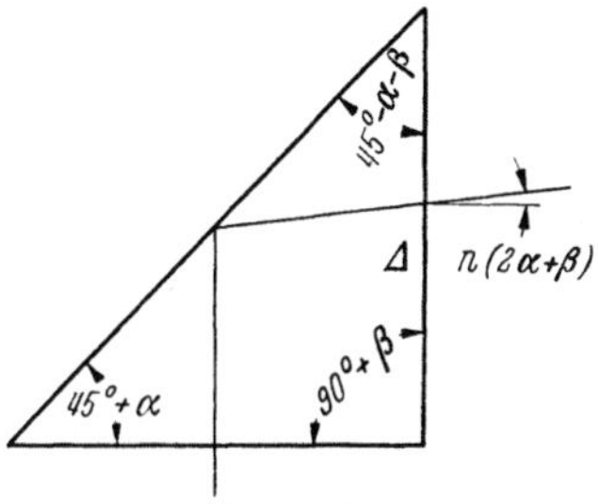

Fig. 52. Winkelfehler am Halbwürfelprisma.

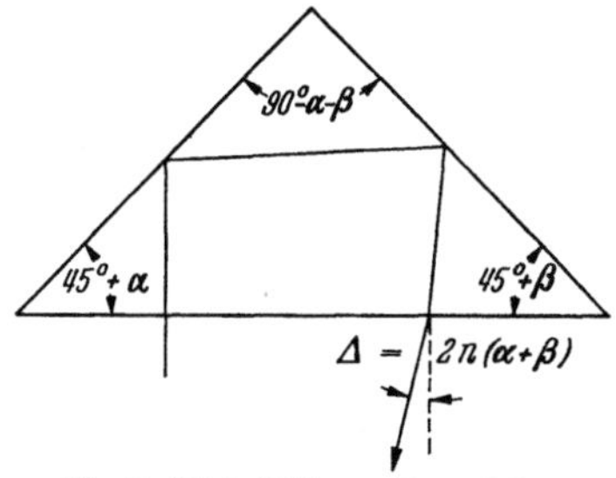

Fig. 53. Winkelfehler am Porro-Prisma.

I. Gleichschenklig-rechtwinkliges Prisma (Halbwürfel). Beide Katheten werden rechtwinklig vom Strahl durchsetzt, wenn $\alpha = -\beta/2$, also die Gleichschenkligkeit gewahrt ist (Fig. 52).

II. Porro-Prisma. Hier verschwindet ein Fehler in der Parallelrichtung des ankommenden und rückwirkenden Strahls, wenn $\beta = -\alpha$, d.h. wenn die Rechtwinkligkeit gewahrt ist; die Schenkel brauchen nicht gleich lang zu sein (Fig. 53).

III. Pentagon-Prisma. Drehen des Prismas in der Zeichenebene ändert am Betrag der Strahlenabknickung nichts; s. den Winkel x in Fig. 54. Von Einfluß auf den Knickwinkel sind alle Prismenwinkel in ungefähr gleichem Maße.

IV. Beim Dove-Prisma verschwindet ein Richtungsfehler, wenn die Seitenwinkel gleich sind; der Mittelstrahl wird freilich unter Umständen versetzt.

Diese Erscheinungen beziehen sich auf Fehler der Winkel zwischen benachbarten brechenden oder spiegelnden Flächen innerhalb der Zeichenebene.

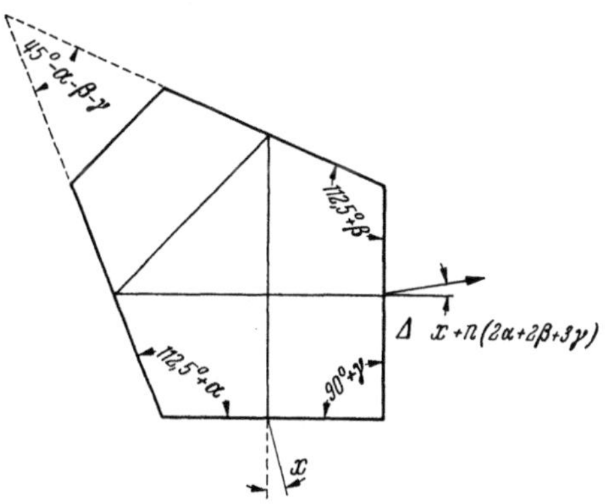

Fig. 54. Winkelfehler am Pentagon-Prisma.

Die Prismen wurden dabei also als Säulen mit genau parallelen Kanten betrachtet. Eine oder mehrere Flächen eines Prismas können aber geneigt zum Lot auf der Zeichenebene stehen. Dieser sog. „Pyramidalfehler“ bewirkt außer Ablenkungen in der Knickebene auch solche aus dieser Ebene heraus sowie Bilddrehungen.

Das Prisma werde so justiert gedacht, daß Ein- und Austrittsfläche senkrecht zur Knickebene orientiert bleiben; dann braucht nur die Neigung spiegelnder Flächen betrachtet zu werden. Findet an einer solchen Fläche die Reflexion unter dem Winkel α an Stelle von α_0 statt und ist sie um den kleinen Winkel g gekippt, so sind die Ausweichungen h des Strahls aus der Zeichenebene (Fig. 55) $h = 2g\cos\alpha_0$ oder für

$\alpha_0 =$	15°	22,5°	30°	45°	60°	67,5°
$h/g =$	1,932	1,848	1,732	1,414	1,000	0,765.

Die Seitenablenkung s im Sinne einer erhöhten Knickung ist stets $s = g \cdot h$, also quadratisch von g abhängig. Besitzt ein Prisma mehrere spiegelnde Flächen, so addieren sich die Strahlausweichungen s; die h können sich unter günstigen Umständen aufheben.

Als dritter Fehler kommt hinzu, daß die Flächen nicht eben sein müssen, sondern eine sehr flache Krümmung, also eine Brechkraft aufweisen, die meist positiv und oft zylindrisch ist. Das Prisma stellt eine Vereinigung von einer Linse mit Hohlspiegeln dar und bildet ein optisches System der Brechkraft Φ, die sich ergibt aus

$$\Phi = (n-1)\left(\frac{1}{r_1} + \frac{1}{r_2}\right) + \sum_s \frac{2}{r_s \cos\alpha},$$

wobei r_1, r_2 die (sehr großen) Radien der Ein- und Austrittsfläche, r_s die der Spiegelflächen darstellen. Diese Formel gilt für die Brechkraft in der Knickebene; senkrecht dazu ist die Brechkraft

$$\Phi_s = (n-1)\frac{1}{r_1 + r_2} + \sum_s \frac{2\cos\alpha}{r_s}.$$

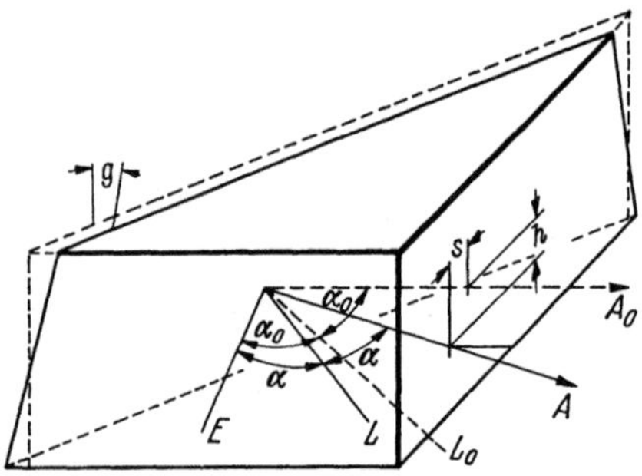

Fig. 55. Pyramidalfehler. E einfallender, A_0 fehlerfrei reflektierter Strahl, L_0 Lot, α_0 Reflexionswinkel. — A, L, α dieselben bei Pyramidalfehler durch Kippwinkel g. — s, h Seiten- und Höhenablenkung.

Sind die Flächen nicht Kugelflächen, sondern torisch oder zylindrisch, so sind die Krümmungswerte in den betrachteten Ebenen zugrunde zu legen. Die Glasdicken kann man vernachlässigen; praktisch wird man die Gesamtbrechkraft durch Messen der Brennweite des ganzen Prismas bestimmen, wozu es (für vorkommende Brennweiten von etwa 10 m bis 10 km) besondere Methoden gibt[1].

Literatur.

[1] NAUMANN, H.: Optik für Konstrukteure, 2. Aufl. Knapp: Düsseldorf 1959.
[2] FIORENTINI, A.: Vocabolario dei prismi ottici. Pubbl. Ist. Naz. Ottica, Firenze **14**, No. 3 (1959).
[3] GARDNER, A.: Optical design. Bureau of Standards, Washington, No. 550. 1927.
[4] FLÜGGE, H.: Praxis der geometrischen Optik. Göttingen 1962.

[1] R. PELZ: Z. Instrumentenkde. **58**, 495 (1938). — J. FLÜGGE: Messung optischer Grundgrößen. Karlsruhe 1953.

Teleskope.

Von

K. Bahner.

Mit 80 Figuren.

1. Einleitung. Die Meßaufgabe der praktischen Astronomie heißt ganz allgemein: Bestimmung der Strahldichte an der scheinbaren Himmelskugel und u.U. ihrer zeitlichen Änderung in Abhängigkeit von den beiden Ortskoordinaten, der Wellenlänge und dem Polarisationszustand. Meßverfahren und Interpretation werden dabei sehr vereinfacht, wenn man eine einschränkende Auswahl der Unbekannten trifft und über die anderen vernünftige Annahmen macht oder Vorkenntnisse benutzt. Stets verlangt die Messung eine Zuordnung von Himmelsausschnitten und Strahlungsempfängern, die — im optischen Wellenlängenbereich zwischen 300 und 10000 nm — durch Abbildung mit einem *Fernrohr* geliefert wird[1].

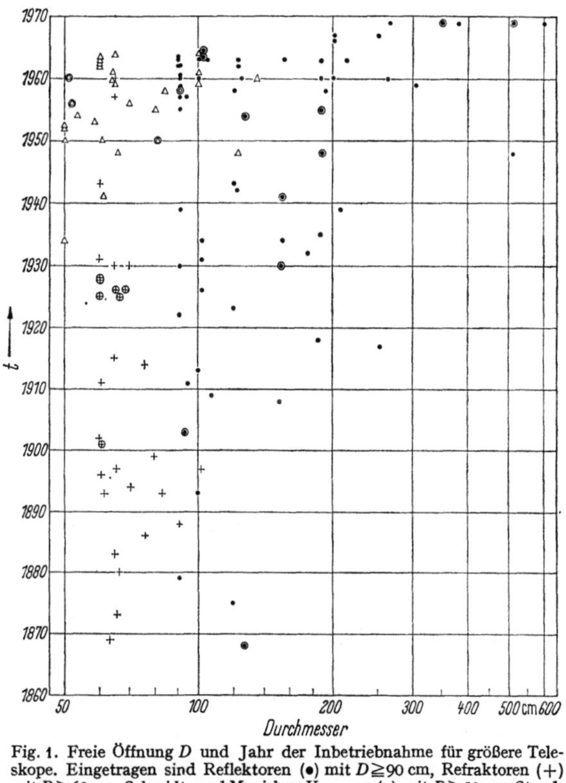

Fig. 1. Freie Öffnung D und Jahr der Inbetriebnahme für größere Teleskope. Eingetragen sind Reflektoren (•) mit $D \geqq 90$ cm, Refraktoren (+) mit $D \geqq 60$ cm, Schmidt- und Meniskus-Kameras (▵) mit $D \geqq 50$ cm. Standort auf der Südhalbkugel ist durch Umkreisung (○) gekennzeichnet. Einige *Projekte* von Großinstrumenten sind schematisch unter 1969 eingetragen.

Für die „geometrischen" Unbekannten (Koordinaten an der Sphäre) genügen bei nicht zu schwachen Objekten verhältnismäßig kleine Instrumente, weil hier die optisch inhomogene Erdatmosphäre die Grenze für die Winkelauflösung setzt. Bei der Messung der „physikalischen" Größen (Strahlungsflußdichte aus einem gewählten Raumwinkelelement) bringt dagegen die Vergrößerung der Auffangfläche ein günstigeres Signal/Rausch-Verhältnis. Da hohe zeitliche Diskrimination selten verlangt wird, sind allerdings bei reinem Signalrauschen Fernrohrdurchmesser und Integrationszeit in gewissen Grenzen austauschbar; in der Praxis ergibt sich jedoch, daß für viele Probleme

[1] 1 nm $= 10^{-9}$ m $= 10$ Å; 1 μm $= 10^{-6}$ m.

nur große Instrumente sinnvoll eingesetzt werden können. Die zeitliche Entwicklung des Fernrohrbaus in den letzten 100 Jahren (Fig. 1) zeigt dementsprechend neben einer allgemein zunehmenden Punktdichte auch ein Anwachsen der Abmessungen für die Größt-Instrumente. Ob allerdings bis etwa 1990 ein Teleskop von etwa 10 m Öffnung zustandekommt, wie die Extrapolation nahelegt, ist noch nicht abzusehen.

Seit den letzten Handbuch-Zusammenfassungen über astronomische Fernrohre [*1*], [*2*] vor etwa 30 Jahren hat sich das Gewicht der astronomischen Beobachtungstätigkeit weiter nach der „physikalischen" Seite verschoben. Das drückt sich auch in der instrumentellen Entwicklung aus: große Refraktoren werden kaum noch gebaut, dafür sind die mittelgroße Schmidt-Kamera, das „photometrische" 1-m-Teleskop und auch schon der 2-m-Reflektor die „Arbeitspferde" der praktischen Astronomie geworden. Daneben bestehen konkrete Pläne für den Bau mehrerer Spiegelteleskope der 3- bis 5-m-Klasse [*3*][1]; an einem 6-m-Instrument[2] wird gearbeitet. Die folgende Darstellung rückt dementsprechend die Spiegelinstrumente stärker in den Vordergrund. Bewußt ausgeklammert wurden spezielle astrometrische Instrumente sowie die Ballon- und Satellitenteleskope. Historische Darlegungen finden sich in [*4*], [*5*]. Zusammenfassende Darstellungen zum Gegenstand dieses Artikels sind in [*6*] bis [*10*] gegeben, z.T. mit Tabellen für die Daten[3] der größeren Forschungsinstrumente.

A. Optische Abbildung in Fernrohren.

I. Einige Gesetze der optischen Abbildung.

a) Geometrisch-optische Beziehungen.

2. Abbildung in der Gaußschen Näherung. In der geometrischen Optik können wir eine repräsentative Auswahl der von einem Objektpunkt ausgehenden Strahlen mit wiederholter Anwendung des Reflexions- und des Brechungsgesetzes durch ein abbildendes System hindurch verfolgen. Die Stelle der (passend definierten) engsten Einschnürung des austretenden Bündels ist der „Bildpunkt". Dieses Verfahren der Strahldurchrechnung ist bei der Konstruktion optischer Systeme üblich und seit der Einführung von programmgesteuerten Rechenautomaten auch für größere Strahl-Anzahlen praktikabel. Für Übersichtszwecke, und auch um den Einfluß von Änderungen der Systemparameter zu erfassen, möchte man jedoch *direkte* Verknüpfungen der Systemdaten mit Ort und Größe von Objekt und Bild haben, die möglichst auch noch die Güte der Strahlenvereinigung angeben sollen. Das ist nur mit Näherungsverfahren möglich.

Die *Gauß*sche Näherung beschränkt sich auf eine so enge Umgebung der Achse eines optischen Systems („paraxiales Gebiet"), daß für die Winkel, die die Strahlen mit der Achse bilden, Sinus = Tangens = Arcus gesetzt werden kann. In der Gauß-Näherung werden homozentrische Bündel des Dingraums in homozentrische Bündel des Bildraums übergeführt, und es existieren keine Abbildungsfehler[4]. Wir betrachten von vornherein nur den Fall, daß der Brechungsindex vor und hinter dem System gleich eins ist; die zusammengehörenden Haupt- und

[1] Siehe auch Ground-Based Astronomy. A Ten-Year Program. National Academy of Sciences-National Research Council. Washington 1964.

[2] Trans. IAU XII B, 130 (1966).

[3] Siehe auch K. BAHNER, Optische Instrumente. In: LANDOLT-BÖRNSTEIN, N. S. Gr. VI, Bd. 1 (H. H. VOIGT, Hrsg.). Berlin-Heidelberg-New York 1965.

[4] Allenfalls kann man Defokussierung und falschen Bildmaßstab als „Abbildungsfehler" im Gauß-Gebiet ansehen.

Knotenpunkte fallen also zusammen. Es seien (Fig. 2) x die Abstände von den Brennpunkten F, s die von den Flächenscheiteln (Schnittweiten), g die von den Hauptpunkten H (Gegenstands- bzw. Bildweite) in Achsrichtung gemessen, y die Querabstände senkrecht zur Achse; Größen im *Bild*raum werden durch x', s' usw. gekennzeichnet. Eine dem Nichtfachmann in allen denkbaren Fällen einleuchtende Vorzeichenfestsetzung scheint es in der geometrischen Optik nicht zu geben. Wir vereinbaren, daß sammelnde Elemente positive (bildseitige) Brennweite f haben und umgekehrt; die allgemeinere, auch für mehrgliedrige Systeme gültige Vorzeichendefinition der Brennweite benutzt allerdings die *Bildorientierung* (s. unten). Strecken in Lichtrichtung werden positiv, gegen die Lichtrichtung

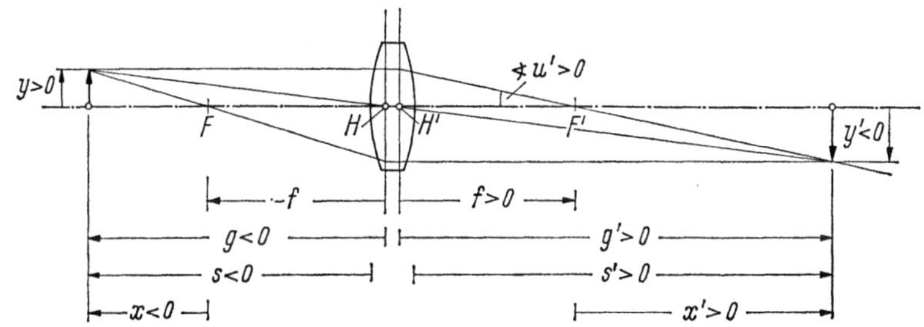

Fig. 2. Bildkonstruktion im paraxialen Gebiet. Bezeichnung der Größen. Vorzeichenkonvention.

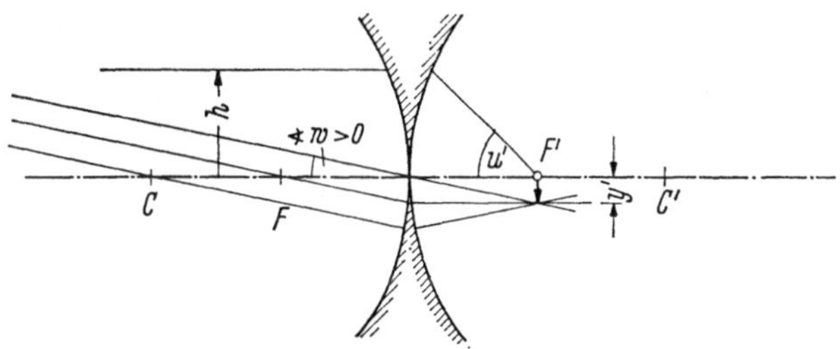

Fig. 3. Abbildung eines fernen ausgedehnten Objekts mit einem Hohlspiegel. Definition der Brennweite.

negativ gezählt. Strecken senkrecht zur Achse werden nach oben positiv genommen. Winkel mit der Achse sind positiv, wenn der Strahl von links oben nach rechts unten läuft. Krümmungsradien R sind von der Fläche zum Zentrum gerichtet: bei einer Bikonvexlinse ist $R_1 > 0$, $R_2 < 0$, bei einem Hohlspiegel $R < 0$ und infolgedessen $f = -R/2$. Bei Spiegeln ist es übersichtlich, den Strahlengang nach der Reflexion umzuklappen, wobei die Spiegelfläche zweimal dargestellt wird; damit bleibt die konventionelle Lichtfortpflanzung von links nach rechts auch im Bildraum erhalten (Fig. 3).

Den *Bildort* gibt die Newtonsche Abbildungsgleichung

$$x \cdot x' = -f^2.$$

Sie ist handlicher als die vertraute „Linsenformel"

$$-\frac{1}{g} + \frac{1}{g'} = \frac{1}{f}.$$

Für $x = -c \cdot f$ folgt $x' = f/c$, einem Dingabstand von c Brennweiten vor dem dingseitigen Brennpunkt entspricht ein Bildabstand von $1/c$ Brennweiten hinter dem bildseitigen Brennpunkt (der auch bei $f < 0$ den Abstand f vom zugehörigen Hauptpunkt hat!).

Die *Bildgröße* ist gegeben durch die Lateralvergrößerung

$$\frac{y'}{y}=\beta=-\frac{x'}{f}=\frac{f}{x}=\frac{g'}{g};$$

dabei bedeutet negatives β umgekehrtes Bild.

Die völlig legitime Bildkonstruktion der Fig. 2 darf nicht dazu verführen, die im Gauß-Gebiet unbestimmten Winkelfunktionen für einen weiteren Bereich nach geometrischer Anschauung festzulegen. Es ist zwar paraxial $f=h/u'$ (h ist die Einfallshöhe eines dingseitig achsenparallelen Strahls, u' sein bildseitiger Schnittwinkel mit der Achse); außerhalb der Gauß-Optik gibt es aber keine Haupt*ebenen* mehr, und $f=h/\tan u'$ ist keine brauchbare Beziehung zu einer allgemeinen Definition der Brennweite, sie widerspräche auch der Sinusbedingung. Die Brennweite läßt sich mit dem *Abbildungsmaßstab* für ein fernes Objekt definieren, der die Winkelgröße w im Dingraum mit der linearen Bildgröße y' verknüpft (Fig. 3).

Im Gauß-Bereich ist $f=-y'/w$, was wir (in Anlehnung an die Zentralprojektion etwa in einer Lochkamera) erweitern zu

$$f=-\frac{y'}{\tan w}.$$

Zwar sind auch dabei scharfe Abbildung und ebenes Bildfeld vorausgesetzt, doch läßt sich diese Definition wenigstens unmittelbar in die astronomische Praxis übertragen.

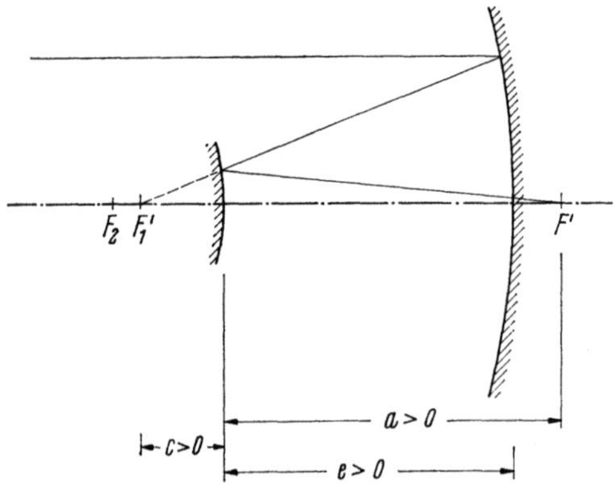

Fig. 4. Gauß-Abbildung im Zwei-Spiegel-System.

Für ein System aus zwei Elementen ergeben sich die Systemdaten aus den Daten der Elemente im paraxialen Bereich nach Fig. 4 (mit Rücksicht auf die wichtigsten Anwendungen als Zwei-Spiegel-System gezeichnet). Es ist

$$f=\frac{a}{c}f_1=m f_1.$$

m ist die sog. Sekundärvergrößerung des Systems. Das zweite Element (hier mit $f_2<0$) bildet F_1' nach F' ab:

$$\frac{1}{a}=\frac{1}{c}+\frac{1}{f_2};\qquad a=\frac{c f_2}{f_2+c}=\frac{c f}{f_1}$$

und damit

$$f=\frac{f_1 f_2}{f_1+f_2-e};\qquad a=\frac{(f_1-e)f_2}{f_1+f_2-e}. \tag{2.1}$$

$e=f_1-c$ ist der Scheitelabstand der Elemente.

3. Begrenzung von Strahlenbündeln. Die Strahlenbündel in optischen Instrumenten haben endliche Querschnitte. Die seitliche Begrenzung wird durch die Fassungsränder der optischen Elemente bewirkt oder auch durch eigens eingesetzte Blenden, die wir ebenfalls als zentriert annehmen wollen. Die Bündelbegrenzungen bestimmen nicht nur den Energietransport durch das Instrument und die Größe des Bildfeldes, sie können auch die Abbildungsfehler wesentlich beeinflussen.

Vom Schnittpunkt der Achse mit der Objektebene oder der Bildfläche aus gesehen erscheint eine der Blenden oder Fassungen unter dem kleinsten Winkel; sie wird als *Aperturblende* bezeichnet. Welche von mehreren vorhandenen Begrenzungen als Aperturblende wirkt, kann vom benutzten Strahlengang abhängen. Für Bildpunkte außerhalb der Achse schneiden u.U. die Ränder von weiteren Blenden in den Strahlengang hinein (Vignettierung), was zu einem feldabhängigen Lichtverlust und zu einer Änderung der Bildstruktur führt.

Alle *vor* der Aperturblende liegenden optischen Glieder entwerfen ein Bild von ihr im Objektraum; dieses ist die *Eintrittspupille* des Systems (die also auch

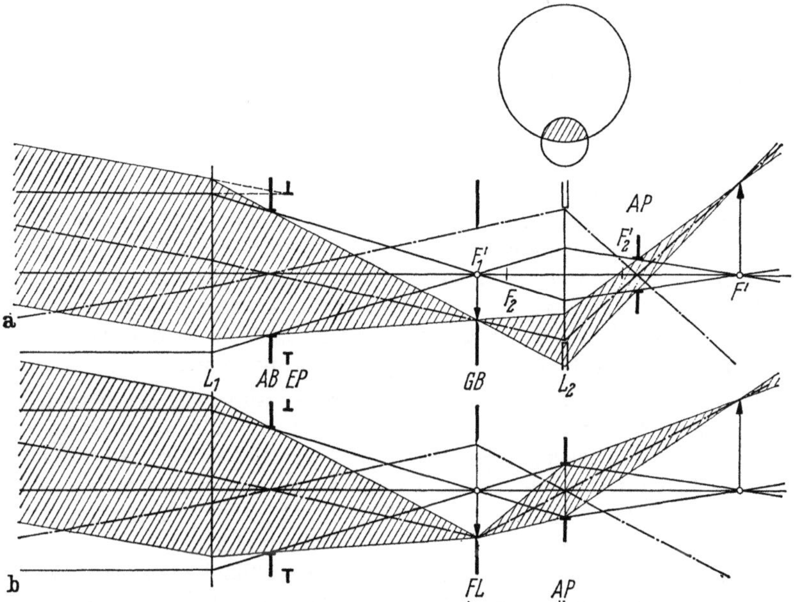

Fig. 5a u. b. System aus zwei dünnen Elementen (Objektiv L_1 mit Hinterblende, Nachvergrößerungslinse L_2). Abbildungsstrahlengang ———, Hauptstrahlengang — · — und geneigtes Bündel (schraffiert). *AB* Aperturblende, *EP* Eintrittspupille (hier virtuell), *AP* Austrittspupille, *GB* Gesichtsfeldblende. a Trotz großen Durchmessers vignettiert der Rand von L_2 das geneigte Bündel (der Bündelquerschnitt in der Ebene von L_2 ist für die Pfeilspitze eingezeichnet). b Die Feldlinse *FL* in der Zwischenbildebene bildet die Aperturblende in die Vergrößerungslinse L_2 ab. Nur der Pupillenstrahlengang wird geändert, der Abbildungsstrahlengang bleibt unbeeinflußt. Bei kleinerem Durchmesser von L_2 ist Vignettierungsfreiheit erreicht.

virtuell sein kann). Die *hinter* der Aperturblende liegenden Glieder erzeugen auf gleiche Weise die *Austrittspupille* im Bildraum. Ein- und Austrittspupille sind so auch Bilder voneinander. Im allereinfachsten Fall — als dünn angenommenes Fernrohrobjektiv ohne weitere Blende — ist die Linsenfassung zugleich Aperturblende, Eintritts- und Austrittspupille.

Eine Blende am Ort des Bildes wirkt als *Gesichtsfeldblende*; sie wird als *Eintrittsluke* in die Objektebene und als *Austrittsluke* in weitere Bildflächen abgebildet (z.B. bei Umkehr- oder Verlängerungssystemen).

Jeder Strahl durch die auf der optischen Achse liegenden Mittelpunkte der Pupillen wird als *Hauptstrahl* bezeichnet. Hauptstrahlen spielen für geneigt einfallende Bündel eine ähnliche Rolle wie die Achse für Axialbündel. Die Pupillenmitten sind als Kreuzungspunkte der Hauptstrahlen die Projektionszentren des Systems. Der Winkel w eines Hauptstrahls gegen die optische Achse heißt *Hauptstrahlneigung*, bildseitig — w' — auch Bild(feld)winkel oder Gesichtsfeldwinkel. Der Winkel u', den ein Strahl vom Rand der Austrittspupille zum Achsenpunkt des Bildes mit der Achse bildet, ist der Öffnungs- oder *Aperturwinkel*. Häufig charakterisiert man die Bündelkonvergenz auch durch die Öffnungszahl $N=f/D$ oder das Öffnungsverhältnis $1:N$; der Durchmesser der Eintrittspupille D wird bei Fernrohren freie Öffnung genannt.

Bei komplizierten Anordnungen ist es erforderlich, neben dem *Abbildungsstrahlengang* auch den Pupillenstrahlengang oder *Hauptstrahlengang* zu betrachten und sich über die Vignettierung Rechenschaft abzulegen (Fig. 5). Die Vignettierungsverhältnisse überblickt man am besten, wenn man alle Ränder, Blenden und Mittenabschattungen mit den davor liegenden Elementen in den Dingraum abbildet; der effektive Bündelquerschnitt ist (für fernes Objekt) durch die Parallelprojektion dieser Bilder unter der jeweiligen Hauptstrahlneigung bestimmt.

4. Die Seidel-Näherung. Beim Übergang zu den von der Praxis geforderten nichtverschwindenden Apertur- und Bildwinkeln wird der Geltungsbereich der Gauß-Optik verlassen. Strahlenbündel, die im Dingraum homozentrisch sind, sind es im Bildraum allgemein nicht mehr. Zur Ermittlung der Abbildungseigenschaften von Flächen und Flächenfolgen ist man auf die „trigonometrische Durchrechnung" einzelner oder vieler Strahlen angewiesen.

Mittels Reihenentwicklung der Winkelfunktionen und Abbrechen nach wenigen Gliedern erhält man jedoch analytische Näherungen, die einen Überblick über die *Abbildungsfehler* — die Abweichungen von der Strahlenvereinigung im Gaußschen Bildpunkt — in Abhängigkeit von den Pupillenkoordinaten und der Hauptstrahlneigung gestatten. Die größte Bedeutung hat die Seidelsche Theorie erlangt, die die Glieder dritter Ordnung in y/f noch mitnimmt.

α) *Deformation von Flächen.* Im Bereich dieser Näherung kann man die Abbildung dadurch beeinflussen, daß man den optischen Flächen rotationssymmetrische systematische Abweichungen von der Kugelfläche gibt, die Flächen *deformiert.* Die Kugelfläche (als einzige Fläche an starren Körpern, die sich durch gegenseitiges Schleifen genau herstellen läßt) spielt in der praktischen Optik eine ausgezeichnete Rolle. Die Gleichungen für nichtsphärische Rotationsflächen werden deshalb zweckmäßig auf eine passende Kugelfläche bezogen. Wählt man dafür die Scheitel-Schmiegungskugel, so wird die Abbildung im paraxialen Bereich durch die Deformation nicht beeinflußt.

Für einen Kreis (als Meridiankurve der Kugelfläche mit Krümmungsradius R) mit Scheitel im Nullpunkt und Mittelpunkt bei $\xi_M = R$, $\eta_M = 0$ ist

$$\xi = \frac{\eta^2}{2R} + \frac{\eta^4}{8R^3} + \frac{\eta^6}{16R^5} + \cdots.$$

Für wenig von der Kugel abweichende Flächen kann man setzen

$$\xi = \frac{\eta^2}{2R} + (1+b)\,\frac{\eta^4}{8R^3};$$

in dieser Näherung wird die Deformation durch *eine* Konstante b ausreichend beschrieben. Die verschiedenen Werte von b ergeben folgende Meridiankurven:

$b=0$	Kreis
$0>b>-1$	Ellipse (große Achse = ξ-Achse)
$b=-1$	Parabel
$b<-1$	Hyperbel
$b>0$	Ellipse (kleine Achse = ξ-Achse).

In der betrachteten Näherung ist $b=-\varepsilon^2$, wenn ε die numerische Exzentrizität des Kegelschnitts darstellt.

Die Abweichung der deformierten Fläche von der Scheitel-Schmiegungskugel ist

$$\Delta\xi = b\cdot\frac{\eta^4}{8R^3}$$

und im Falle einer Spiegelfläche

$$\Delta\xi = b\cdot\frac{h^4}{64f^3};\qquad \Delta\xi_{\max} = b\cdot\frac{f}{1024N^4}.$$

Die geringsten Abweichungen[1] von der deformierten Fläche hat diejenige Kugel, die sie im Scheitel berührt und am Rand schneidet; für sie gilt

$$\Delta\xi = b\cdot\frac{\eta^4 - H^2\eta^2}{8R^3};\qquad H = \eta_{\max} = \frac{D}{2}.$$

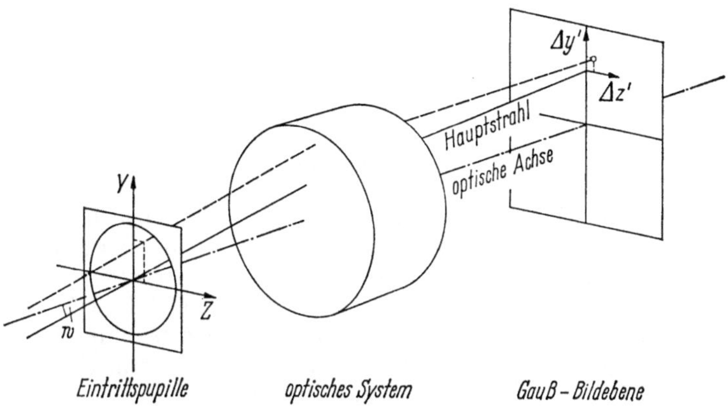

Fig. 6. Bezeichnungen für die Seidel-Näherung.

Für gekrümmte Flächen ist die Konstante b handlich, weil brennweitenunabhängig; sie ist aber bei deformierten *Plan*flächen nicht definiert. In jedem Falle kann man die Auf- oder Abtragung mit $\Delta\xi = \varkappa\cdot\eta^4$ beschreiben. Für Flächen mit Krümmungsradius R ist

$$\varkappa = \frac{b}{8R^3}.$$

β) *Abweichung von der Strahlenvereinigung in einem Punkt.* Die Strahlenvereinigung läßt sich im Seidel-Bereich so beschreiben (Fig. 6): Der Hauptstrahl von einem fernen Punkt habe die Neigung w gegen die Achse des Systems; ein zu ihm paralleler Strahl habe in der Eintrittspupille die Koordinaten Y im Meridional- und Z im Sagittalschnitt (die Meridionalebene enthält den Hauptstrahl und die optische Achse, der Sagittalschnitt steht senkrecht darauf und geht durch den Durchstoßpunkt des Hauptstrahls). Alle Längen seien in der Einheit f = Systembrennweite angegeben. Der Durchstoßpunkt des betrachteten Strahls durch die paraxiale Bildebene hat dann vom Gaußschen Bildpunkt die Abstände

$$\left.\begin{aligned}
\Delta y' &= -\tfrac{1}{2}Y(Y^2+Z^2)\textstyle\sum \mathbf{I}\\
&\quad+\tfrac{1}{2}(3Y^2+Z^2)\tan w \textstyle\sum \mathbf{II}\\
&\quad-\tfrac{1}{2}Y\tan^2 w \textstyle\sum \mathbf{III}\\
&\quad+\tfrac{1}{2}\tan^3 w \textstyle\sum \mathbf{V}\\
&\text{im Meridionalschnitt und}\\
\Delta z' &= -\tfrac{1}{2}Z(Y^2+Z^2)\textstyle\sum \mathbf{I}\\
&\quad+YZ\tan w \textstyle\sum \mathbf{II}\\
&\quad-\tfrac{1}{2}Z\tan^2 w \textstyle\sum \mathbf{IV}
\end{aligned}\right\}\qquad (4.1)$$

[1] Zum allgemeinen Fall des minimalen Differenz*volumens* bei beliebiger Deformation s. TH. W. J. UNTI, Appl. Optics 5, 319 (1966).

im Sagittalschnitt. In diesen Formeln für die *Querabweichungen* fallen die Glieder erster Ordnung (die Gauß-Glieder) heraus; Glieder zweiter und überhaupt gerader Ordnung entstehen aus Symmetriegründen nicht (beim Übergang von w in $-w$, Y in $-Y$ und Z in $-Z$ muß auch $\Delta y'$ in $-\Delta y'$ und $\Delta z'$ in $-\Delta z'$ übergeführt werden). Die „Seidel-Koeffizienten" **I** bis **V** lassen sich für jede Fläche aus Krümmungsradius, Scheitelabständen, Brechungsindizes und Blendenlage errechnen. Für eine Folge von Flächen 1 bis ν ist $\mathbf{I}_{\text{Syst}} = \sum_\nu \mathbf{I}_\nu$, usw.; die Fehler sollen also hinreichend klein sein. Wegen der Wahl der Längeneinheit charakterisieren die Koeffizienten unmittelbar die *Winkel*aberrationen.

Man bezeichnet den Abbildungsfehler, der gegeben ist durch

I	als Sphärische Aberration,
II	Koma,
III, IV	Astigmatismus und Bildwölbung,
V	Verzeichnung.

In dem Verzeichnungs-Term der Gl. (4.1) treten die Pupillen-Koordinaten nicht auf; die Schärfe der Abbildung ist also nicht gestört, sondern nur ihre Ähnlichkeit. Auch die Koeffizienten **III** und **IV** müssen nicht einzeln verschwinden, damit ein scharfes Bild zustande kommt. Wenn $\mathbf{I} = \mathbf{II} = 0$ und $\mathbf{III} = \mathbf{IV}$, so erhält man $\Delta y' = \text{const} \cdot Y$, $\Delta z' = \text{const} \cdot Z$ für jedes w und damit einen Bildpunkt unabhängig von Y, Z. Er liegt aber allgemein auf einer *gekrümmten* Bildschale und nur für $w = 0$ oder $\mathbf{III} = \mathbf{IV} = 0$ auf der Gaußschen Bildebene.

Die Ausdehnung der statt eines Bildpunktes entstehenden Zerstreuungsfigur für einen fernen Objektpunkt ist hauptsächlich bestimmt durch den Pupillenrand $Y^2 + Z^2 = D^2/4$. Für die *isolierten* Abbildungsfehler sind die folgenden Grössen zweckmäßig [mit $N = f/D$ = Öffnungszahl = reziprokes Öffnungsverhältnis; $\mathbf{IIIa} = \frac{1}{2}(\mathbf{III} - \mathbf{IV})$; $\mathbf{IVa} = \frac{1}{2}(\mathbf{III} + \mathbf{IV})$]:

$$\left.\begin{array}{ll}
\boldsymbol{S} = \dfrac{1}{16} \cdot \dfrac{f}{N^3} \cdot \sum \mathbf{I} & \text{Sphärische Querabweichung (Radius des Zerstreuungskreises in der Gaußschen Bildebene)} \\[2ex]
\Delta x'_S = \dfrac{1}{8} \dfrac{f}{N^2} \cdot \sum \mathbf{I} & \text{Sphärische Längsabweichung} \\[2ex]
\boldsymbol{K} = \dfrac{3}{8} \dfrac{f}{N^2} \tan w \cdot \sum \mathbf{II} & \text{Querabweichung (Durchmesser) der meridionalen Koma} \\[2ex]
\boldsymbol{A} = \dfrac{1}{4} \dfrac{f}{N} \tan^2 w \cdot \sum \mathbf{IIIa} & \text{Astigmatische Querabweichung (Radius auf der Schale der mittleren Bildkrümmung)} \\[2ex]
\Delta x'_A = \dfrac{1}{2} f \tan^2 w \sum \mathbf{IIIa} & \text{(Halbe) astigmatische Differenz} \\[2ex]
R_m = -\dfrac{f}{\sum \mathbf{IVa}} & \text{Krümmungsradius der Schale mittlerer Bildkrümmung.}
\end{array}\right\} \quad (4.2)$$

Die Vorzeichen der Seidel-Summen bestimmen dabei die Richtung der Aberration. $\sum \mathbf{I} > 0$ bedeutet sphärische Unterkorrektion, Schnittweite der Randstrahlen kleiner als die Schnittweite der Axialstrahlen; $\sum \mathbf{II} > 0$: Komafigur radial nach innen gerichtet; $\sum \mathbf{IIIa} > 0$: Tangentialschale liegt innerhalb der Sagittalschale; $R_m > 0$: Bildschale gegen die Lichtrichtung erhaben. Für astronomische Anwendungen ist oft die Angabe von Queraberrationen im *Winkelmaß* $\mathcal{S} = \boldsymbol{S}/f$, $\mathcal{K} = \boldsymbol{K}/f$, $\mathcal{A} = \boldsymbol{A}/f$ vorteilhaft.

Für einen Kugelspiegel mit Eintrittspupille im Spiegelscheitel (**I** $= +0{,}25$, **II** $= -0{,}5$, **IIIa** $= +1{,}0$, **IVa** $= +1{,}0$) sind die Seidel-Aberrationen in Fig. 7 dargestellt, und zwar die Queraberration im Winkelmaß als Funktion der Haupt-

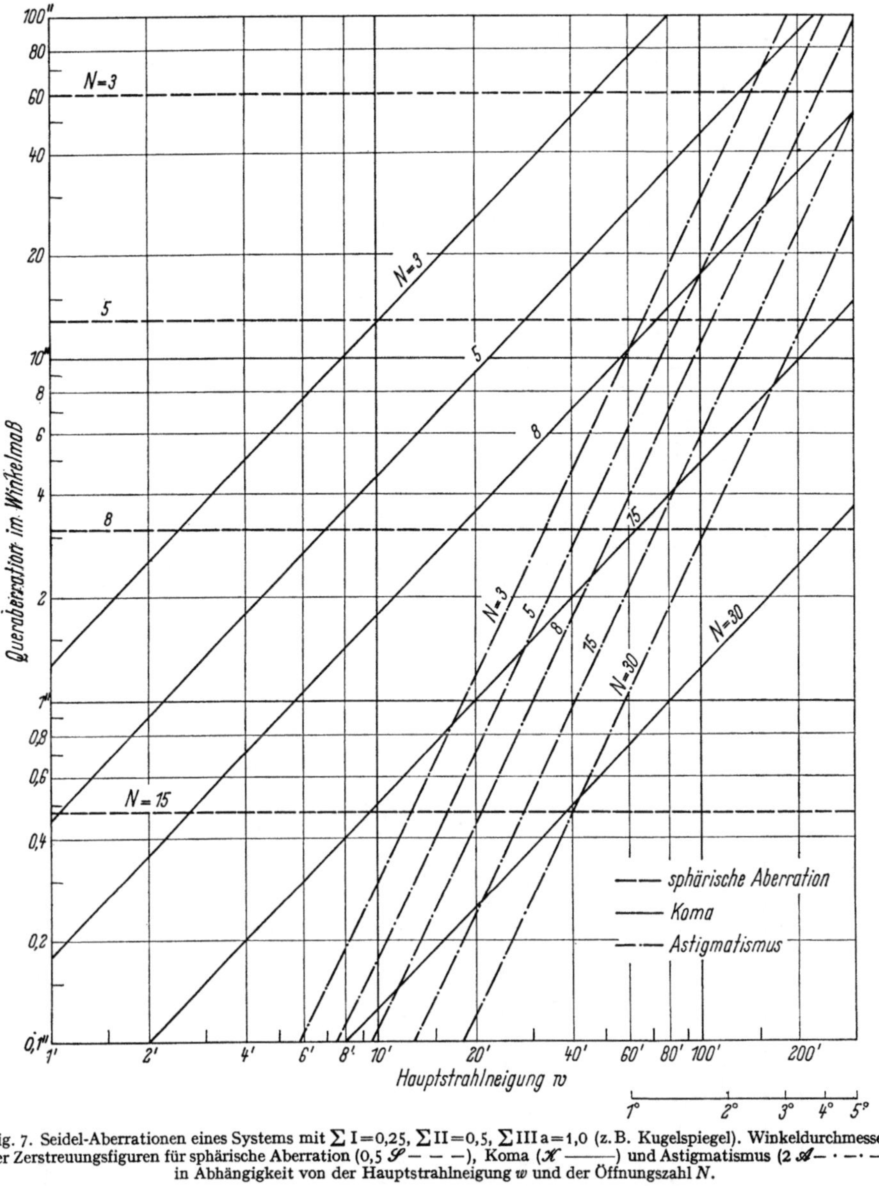

Fig. 7. Seidel-Aberrationen eines Systems mit $\sum$ I $=0{,}25$, $\sum$ II $=0{,}5$, $\sum$ IIIa $=1{,}0$ (z. B. Kugelspiegel). Winkeldurchmesser der Zerstreuungsfiguren für sphärische Aberration ($0{,}5\,\mathcal{S}$ — — —), Koma ($\mathcal{K}$ ———) und Astigmatismus ($2\,\mathcal{A}$ — · — · —) in Abhängigkeit von der Hauptstrahlneigung w und der Öffnungszahl N.

strahlneigung mit der Öffnungszahl als Parameter. Dabei ist für die sphärische Aberration $0{,}5\,\mathcal{S}$ (als Durchmesser der Zerstreuungsscheibe in der optimalen Bildfläche, s. Ziff. 9), für den Astigmatismus $2\mathcal{A}$ (als Durchmesser des Zerstreuungskreises auf der Schale mittlerer Bildkrümmung) angegeben. Die Werte beziehen sich also nicht auf die gleiche Lage der Empfängerfläche. Für die außeraxialen Fehler $\mathcal{K}$ und $\mathcal{A}$ gelten sie auch bei einem deformierten Spiegel, z. B. einem Rotationsparaboloid.

Tabelle 1.

	ν	R_ν	$e_{\nu-1,\nu}$	h_ν/h_1	$p_\nu;(\vartheta_\nu)$	I	II	III a	P	IV a
1. Kugelspiegel (Blende im Scheitel)	1	−2,000	—	1,000	−2,000	+0,250	−0,500	+1,000	−1,000	+1,000
2. Parabolspiegel (Blende im Scheitel)	1	−2,000	—	1,000	−2,000	+0,250	−0,500	+1,000	−1,000	+1,000
$b = -1{,}000$	1*	—	—	—	(0,000)	−0,250	0,000	0,000	0,000	0,000
						$\sum$ 0,000	−0,500	+1,000	−1,000	+1,000
3. Cassegrain-System	1	−0,500	—	—	−0,500	+16,000	−8,000	+4,000	−4,000	+4,000
Blende im Hauptspiegel $b_1 = -1{,}000$	1*	—	—	—	(0,000)	−16,000	0,000	0,000	0,000	0,000
Sekundärvergrößerung $m=4$	2	+0,150	0,194	0,225	−0,917	−4,219	+3,867	−3,545	+13,333	+6,224
$b_2 = -2{,}778$	2*	—	—	—	(+0,861)	+4,219	+3,633	+3,128	0,000	+6,256
						$\sum$ 0,000	−0,500	+3,583	+9,333	+16,480
4. Ritchey-Chrétien-System	1	−0,667	—	—	−0,667	+6,750	−4,500	+3,000	−3,000	+3,000
Blende im Hauptspiegel $b_1 = -1{,}081$	1*	—	—	—	(0,000)	−7,296	0,000	0,000	0,000	0,000
Sekundärvergrößerung $m=3$	2	+0,267	0,244	0,267	−0,958	−2,133	+2,044	−1,959	+7,500	+3,582
$b_2 = -5{,}023$	2*	—	—	—	(+0,917)	+2,679	+2,456	+2,251	0,000	+4,502
						$\sum$ 0,000	0,000	+3,292	+4,500	+11,084
5. Schmidt-Kamera	0*	—	—	—	(0,000)	−0,250	0,000	0,000	0,000	0,000
	1	−2,000	2,000	1,000	0,000	+0,250	0,000	0,000	−1,000	−1,000
						$\sum$ 0,000	0,000	0,000	−1,000	−1,000
6. Schmidt-Cassegrain-System nach BAKER	0*	—	—	—	(0,000)	−0,622	0,000	0,000	0,000	0,000
Sekundärvergrößerung $m = 1{,}538$	1	−1,300	1,394	1,000	0,094	+0,910	+0,085	+0,008	−1,538	−1,522
$b_1 = +0{,}016_5$	1*	—	—	—	(1,394)	+0,015	+0,021	+0,029	0,000	+0,058
$b_2 = 0$	2	+1,300	$0{,}422_5$	0,350	0,350	−0,304	−0,106	−0,037	+1,538	+1,464
						$\sum$ 0,000	0,000	0,000	0,000	0,000

Bei Zwei-Spiegel-Systemen ist $h_2/h_1 = a$ wegen $f = 1$.

γ) Größe der Seidel-Koeffizienten. Die Seidel-Summen $\sum \mathbf{I}$ bis $\sum \mathbf{IVa}$ für einige wichtige Systeme sind in Tabelle 1, für viele weitere in [*11*], [*12*][1] gegeben.

Die Berechnung der Koeffizienten für jede Fläche aus den Systemdaten ist eine der Hauptaufgaben der geometrischen Optik[2]. Die für Optikrechnungen üblichen Formelsätze sind wegen des schrittweisen Aufbaus verschiedener Hilfsgrößen nicht besonders übersichtlich; für den einfachen Fall von einer oder zwei Spiegelflächen sind unten leidlich überschaubare Ausdrücke angegeben. Der Koeffizient $\mathbf{I}_\nu$ hängt ab vom Krümmungsradius der ν-ten Fläche, von der Änderung des Brechungsindex an ihr, und von der Einfallshöhe, die ein ursprünglich achsenparalleler Strahl mit $h_1=1$ auf ihr hat. Die weiteren Koeffizienten gehen aus $\mathbf{I}_\nu$ mit einem Multiplikator p_ν hervor, der zusätzlich vom Scheitelabstand zur vorhergehenden Fläche und von der Lage der Eintrittspupille beeinflußt wird. Schließlich tritt noch — **III** und **IV** verknüpfend — der „Petzval-Term" $\boldsymbol{P}_\nu$ auf, in den nur Krümmungsradius und Brechzahländerung eingehen, unabhängig von der Blendenlage und von der Stellung der Fläche im System. Es ist

$$\left.\begin{array}{ll} \mathbf{II}_\nu = p_\nu\, \mathbf{I}_\nu & \\ \mathbf{III\,a}_\nu = p_\nu\, \mathbf{II}_\nu; & \mathbf{III}_\nu = 3\,\mathbf{III\,a}_\nu + \boldsymbol{P}_\nu \\ \mathbf{IV}_\nu = \mathbf{III\,a}_\nu + \boldsymbol{P}_\nu; & \mathbf{IV\,a}_\nu = 2\,\mathbf{III\,a}_\nu + \boldsymbol{P}_\nu \\ \mathbf{V}_\nu = p_\nu\, \mathbf{IV}_\nu. & \end{array}\right\} \tag{4.3}$$

Die Deformationen der Flächen liefern die durch einen Stern bezeichneten Anteile:

$$\left.\begin{array}{ll} \mathbf{II}^*_\nu = \vartheta_\nu \cdot \mathbf{I}^*_\nu & \\ \mathbf{III\,a}^*_\nu = \vartheta_\nu \cdot \mathbf{II}^*_\nu; & \mathbf{III}^*_\nu = 3\vartheta_\nu \cdot \mathbf{II}^*_\nu \\ \mathbf{IV}^*_\nu = \vartheta_\nu \cdot \mathbf{II}^*_\nu; & \mathbf{IV\,a}^*_\nu = 2\vartheta_\nu \cdot \mathbf{II}^*_\nu \\ \mathbf{V}^*_\nu = \vartheta_\nu \cdot \mathbf{IV}^*_\nu. & \end{array}\right\} \tag{4.4}$$

Hier ist der Multiplikator ϑ_ν anzuwenden.

Im folgenden ist wieder die Systembrennweite als Längeneinheit angenommen, dazu fernes Objekt und Brechungsindex $n=1$ im Ding- und Bildraum. Der Abstand der Eintrittspupille vom ersten Flächenscheitel sei s_p (<0, wenn die Blende *vor* dem Scheitel liegt). Dann gilt für eine sammelnde Spiegelfläche:

$$\left.\begin{array}{ll} \mathbf{I} = \frac{1}{4}; & \mathbf{I}^* = \frac{b}{4} \\ \vartheta = -s_p; & p = -2 - s_p \\ \boldsymbol{P} = -1. & \end{array}\right\} \tag{4.5}$$

Für das Zwei-Spiegel-System ist mit den Bezeichnungen[3] von Fig. 4 (S. 230)

$$\left.\begin{array}{lll} \mathbf{I}_1 = \frac{m^3}{4}; & \mathbf{I}^*_1 = \frac{b_1 m^3}{4} & \\ \vartheta_1 = -s_p; & p_1 = \vartheta_1 - \frac{2}{m}; & \boldsymbol{P}_1 = -\frac{1}{f_1} \\ \mathbf{I}_2 = -\frac{a}{4}(m-1)(m+1)^2; & \mathbf{I}^*_2 = -b_2 \frac{a}{4}(m-1)^3 & \\ \vartheta_2 = -\frac{e}{a} - s_p; & p_2 = \vartheta_2 - \frac{2}{a(m+1)} & \\ \boldsymbol{P}_2 = -\frac{1}{f_2}. & & \end{array}\right\} \tag{4.6}$$

[1] Siehe auch H. Siedentopf, Leistung der Fernrohre. In: Landolt-Börnstein, N. S. Gr. VI, Bd. 1 (H. H. Voigt, Hrsg.). Berlin-Heidelberg-New York 1965.

[2] Dazu s. etwa [*13*] bis [*18*] und die Artikel von Welford und Marx in diesem Band.

[3] Von den Größen m, e, a sind nur zwei unabhängig wegen $a + me = 1$.

Mit diesen Gleichungen beherrscht man auch noch Systeme aus einem oder zwei Spiegeln und einer dünnen brechkraftlosen, aber mit einem Profil vierter Ordnung deformierten *Korrektionsplatte* als „nulltem" Element in der Eintrittspupille. Wie bei der Behandlung des Schmidt-Teleskops (Ziff. 20) gezeigt wird, kompensiert eine Platte mit dem Profil $\Delta\xi = \varkappa\eta^4$ die sphärische Aberration eines Kugelspiegels, wenn

$$\varkappa = \frac{1}{32(n-1)f^3}$$

ist (Platte nach dem Rand zu dicker werdend; n ist ihr Brechungsindex). Für $f=1$ ergibt sich so

$$\mathbf{I}_0^* = -8\varkappa(n-1).$$

Da die Platte wegen ihrer Lage die außeraxialen Fehler nicht beeinflußt, tritt für sie in den Seidel-Summen nur dieser Term $\mathbf{I}_0^*$ auf. Ist im resultierenden System neben der sphärischen Aberration auch die Koma gehoben, kann man die Pupille stets formal verschieben, weil der Astigmatismus dann blendenunabhängig ist. Ohne diese manchmal etwas künstlich wirkende Pupillenverschiebung beschreibt die „Plattendiagramm-Analyse" nach BURCH[1] Anordnungen aus Spiegeln und Korrektionsplatten.

Beide Darstellungsweisen führen zu den Bedingungsgleichungen

$$\left.\begin{aligned} P - aQ + \mathbf{I}_0^* &= 0 \\ eQ - s_{\text{pl}}\,\mathbf{I}_0^* + \tfrac{1}{2} &= 0 \\ -\frac{e^2}{a}Q + s_{\text{pl}}^2\,\mathbf{I}_0^* + \frac{1-e}{a} &= 0 \end{aligned}\right\} \tag{4.7}$$

für das sukzessive Verschwinden von sphärischer Aberration, Koma, und Astigmatismus im Zwei-Spiegel-System mit Korrektionsplatte. Dabei ist zur Abkürzung

$$P = \frac{m^3}{4}(b_1+1)$$
$$Q = \frac{(m-1)^3}{4}\left[b_2 + \left(\frac{m+1}{m-1}\right)^2\right]$$

gesetzt. s_{pl} ist der Abstand der Platte vom Primärspiegelscheitel. Die Platte kann auch aus einer virtuellen Position ($s_{\text{pl}} > 0$) in das System hinein abgebildet werden, unter Änderung der Profilkonstante $\varkappa$ entsprechend $1:\beta^4$ (β ist die Lateralvergrößerung bei der Abbildung der Platte).

Der Geltungsbereich der Seidel-Theorie ist für die in der Astronomie praktisch vorkommenden Spiegelsysteme recht groß. Bei Linsensystemen (und auch bei Spiegelsystemen mit großen Apertur- und Bildwinkeln) werden die vernachlässigten Glieder höherer Ordnung merklich; höhere Näherungen sind aber in der Fernrohr-Praxis wenig verwendet worden[2]. Gewißheit über den erreichten Korrektionszustand eines Systems bietet die Durchrechnung einer ausreichenden Anzahl von Strahlen.

b) Wellenoptische Ergänzungen.

5. Beugungsfigur der Kreispupille. Nach der Strahlenoptik müßte ein ideal abbildendes System von einem Dingpunkt ein punktförmiges Bild liefern. Die Wellenoptik korrigiert diese Vorstellung und beschreibt die Lichtverteilung in der Nähe des Bildpunktes als Ergebnis der *Beugung* an der Apertur (s. dazu z.B. [*19*] bis [*23*]).

[1] C. R. BURCH: Monthly Notices Roy. Astron. Soc. **102**, 159 (1942); s. auch LINFOOT [*19*].

[2] Vgl. hierzu den Artikel von MARX in diesem Bande.

Für den einfachsten Fall, daß die Eintrittspupille ein hoher Spalt der Breite D und das Ding ein ferner Punkt ist, gilt für die Bestrahlungsstärke I in der Brennebene bei nicht zu kleinem N und in der Nähe des geometrischen Bildpunkts

$$I(y')=\left(\frac{\sin l}{l}\right)^2; \quad \Delta x'=0,$$

mit der normalisierten Koordinate

$$l=\frac{\pi}{N}\cdot\frac{y'}{\lambda}$$

(dabei bedeuten $N=f/D$ die Öffnungszahl, λ die Wellenlänge, y' den Abstand vom Bildpunkt senkrecht zu den Spaltkanten). Es wird also $I=1$ für $l=0$ und $I=0$ für $l=\pi, 2\pi\ldots$, d.h. erstmals für $y'/f=\lambda/D$; ist die Blende m Wellenlängen breit, so ist die erste Nullstelle $1/m$ Radian vom Maximum entfernt.

Im bei astronomischen Instrumenten wichtigsten Fall, der *kreisförmigen* Pupille, ergibt sich

$$I(\varrho')=\left(\frac{2J_1(q)}{q}\right)^2; \quad \Delta x'=0 \qquad (5.1)$$

mit $q=\frac{\pi}{N}\cdot\frac{\varrho'}{\lambda}$ und $\varrho'^2=y'^2+z'^2$. Mißt man die seitlichen Abstände durch den zugehörigen Winkel im Dingraum α, so muß man $q=\pi\cdot D\cdot\frac{\alpha}{\lambda}$ setzen, d.h. $q=152{,}3\,\frac{D}{\lambda}\cdot\alpha$ (D in cm, λ in nm, α in Bogensekunden). J_1 ist die Bessel-Funktion erster Ordnung. Ihre Nullstellen liegen bei

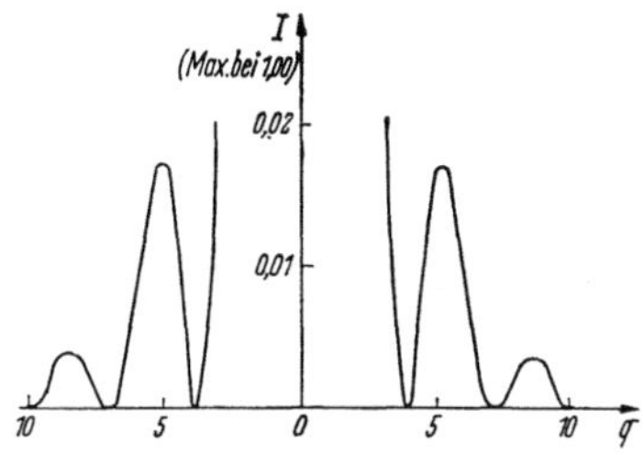

Fig. 8. Verteilung der Bestrahlungsstärke ($I_{max}=1$) in der Bildebene für den inneren Teil der Beugungsfigur eines fernen Punktobjekts bei ungestörter Kreispupille. $q=\frac{\pi}{N\cdot\lambda}\cdot\varrho'$.

$$q_0=3{,}83;\quad 7{,}02;\quad 10{,}17;\quad 13{,}32;\quad \ldots$$

$$\frac{q_0}{\pi}=1{,}22;\quad 2{,}23;\quad 3{,}24;\quad 4{,}24;\quad \ldots.$$

Man erhält also ein zentrales „Beugungsscheibchen" mit einer Folge von konzentrischen, nach außen rasch schwächer werdenden hellen Ringen, die durch nahezu äquidistante Kreise mit Bestrahlungsstärke Null getrennt sind (Fig. 8). Der Radius des ersten Nullkreises ist im folgenden mit ϱ_0' bezeichnet; es gilt

$$\varrho_0'=1{,}22\,N\lambda.$$

Eine Tabelle für (5.1) gibt FRANÇON [*20*], S. 274.

Längs der Achse ergibt sich für die Bestrahlungsstärke

$$I(\Delta x')=\left(\frac{\sin p/4}{p/4}\right)^2; \quad \varrho'=0, \qquad (5.2)$$

wobei $p=\frac{\pi}{2\lambda N^2}$ der normalisierte Abstand von der Brennebene ist. Die Nullstellen liegen bei $p=\pm 4\pi,\ 8\pi\ldots.$ Die vollständige Verteilung im bildpunktnahen Raum ist in Fig. 9 dargestellt, die einen Quadranten des die Achse enthaltenden Schnitts wiedergibt.

Von der gesamten einfallenden Strahlung L_∞ tritt durch einen zentrierten Kreis mit Radius ϱ' in der Bildebene der Bruchteil

$$\frac{L\varrho'}{L_\infty}=1-J_0^2(q)-J_1^2(q); \quad \Delta x'=0.$$

Daraus folgt, daß das zentrale Scheibchen 83,8% und der erste, zweite, dritte Ring 7,2%, 2,8%, 1,5% der Gesamtstrahlung enthalten. Für andere Einstellebenen ist der Bruchteil der Fig. 10 zu entnehmen.

Die durch die Beugung gesetzte *Auflösungsgrenze* für die Abbildung zweier gleichheller „selbstleuchtender" Punkte (inkohärente Schwingung) gilt nach

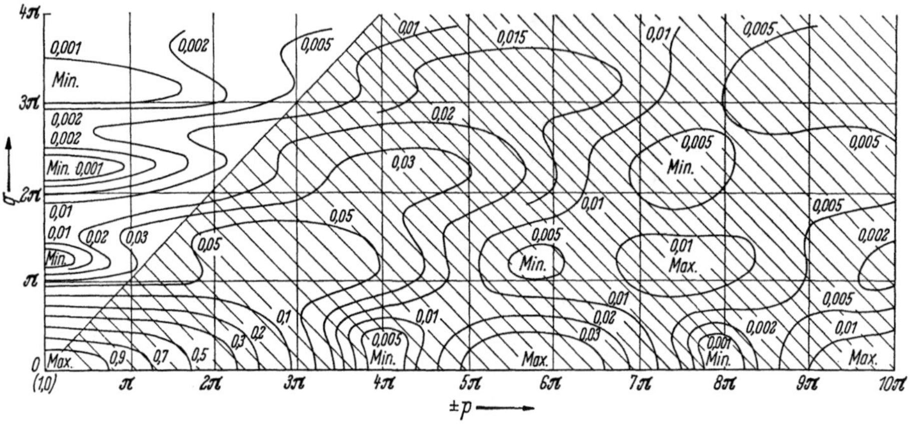

Fig. 9. Isophoten der Beugungsverteilung in einer die Achse enthaltenden Ebene (Bedingungen wie in Fig. 8). Die vollständige räumliche Verteilung entsteht durch Spiegelung an der q- und Rotation um die p-Achse. Der nach geometrischer Anschauung lichterfüllte Kegel ist schraffiert. $p = \frac{\pi}{2N^2\lambda}\,\Delta x'$. Nach LINFOOT und WOLF [1].

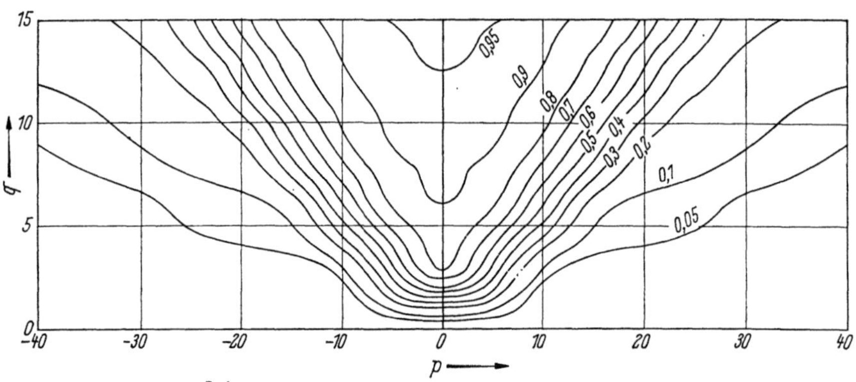

Fig. 10. Strahlungsfluß-Anteil $\frac{L\varrho'}{L_\infty}$ durch zentrierte Kreisflächen in Abhängigkeit von ϱ' und $\Delta x'$ (Bedingungen wie in Fig. 8 und 9). Nach WOLF [2].

RAYLEIGH dann als erreicht, wenn das Maximum der einen Beugungsfigur in das erste Minimum der anderen fällt, also für

$$\Delta\varrho' = \varrho_0' = 1{,}22\,N\lambda. \tag{5.3}$$

Dem entspricht ein Winkelabstand im Dingraum von

$$\Delta\alpha_R = 1{,}22\,\frac{\lambda}{D}\ \text{Radian}; \qquad \Delta\alpha_R = 0'',0252\,\frac{\lambda}{D}\ (\lambda \text{ in nm, } D \text{ in cm}).$$

Das *lineare* Auflösungsvermögen in der Bildebene hängt also ab vom *Öffnungsverhältnis*, die *Winkel*auflösung vom *Durchmesser* der Eintrittspupille. Für

[1] E. H. LINFOOT and E. WOLF: Proc. Phys. Soc. Lond. B **66**, 145 (1953).
[2] E. WOLF: Proc. Roy. Soc. A **204**, 533 (1951).

visuelle Beobachtung ($\bar{\lambda}=555$ nm) ist

$$\Delta\alpha_R = \frac{14''}{D} \quad (D \text{ in cm}).$$

Bei dem so bestimmten Abstand der Punkte hat das „Lichtgebirge" — Bestrahlungsstärke senkrecht zur Bildfläche aufgetragen — eine Einsattelung zwischen den Höchstwerten auf etwa 74% der Gipfelhöhe. Wenn der Nachweis geringerer Kontraste möglich ist, kann die Rayleigh-Grenze durchaus unterschritten werden; Ähnlichkeit zwischen Ding und Bild besteht ohnehin nicht, der Abstand der Maxima ist kleiner als der der geometrischen Bildpunkte. Der Sattel verschwindet bei $\Delta\varrho'=0{,}95\,N\lambda$ (theoretisches Auflösungsvermögen nach STREHL).

Für „beleuchtete" Punkte ist das Auflösungsvermögen schlechter; bei völliger Kohärenz verschwindet der Sattel bei $\Delta\varrho'=1{,}46\,N\lambda$.

Man sollte sich darüber klar sein, daß alle wie immer definierten „Auflösungsgrenzen" an spezielle Voraussetzungen gebunden sind; sie sind brauchbar vor allem im Vergleich verschiedener optischer Anordnungen untereinander. Prinzipiell läßt sich aus einer hinreichend genauen Photometrie des Lichtgebirges auf Detail auch unterhalb der Beugungsgrenze schließen, wie es in der Radioastronomie häufig geschieht. Eine physikalische Grenze ist gegeben durch das Rauschen von Signal und Empfänger, das die Entzerrung für Detail unter $\Delta\varrho'\approx N\lambda$ immer unsicherer macht.

6. Andere Pupillenformen, Apodisation. Der Fall einer ungestörten kreisförmigen Eintrittspupille kann zwar für Refraktoren angenommen werden; bei Spiegelfernrohren ist die Pupille jedoch meist durch Sekundärspiegel oder Kassette und durch die zugehörigen Haltestreben teilweise verdeckt. Dies führt zu einer Änderung des Beugungsbildes.

Wird von der Pupille mit Durchmesser D_1 ein innerer kreisförmiger Teil mit Durchmesser D_2 abgedeckt, so verkleinern sich der Radius ϱ_0' für das erste Minimum und der Lichtanteil im zentralen Beugungsscheibchen zugunsten der hellen Ringe. In der Fokalfläche gilt

Tabelle 2.

	Volle Apertur		Schmale Randzone	
	$\varrho'\cdot\frac{1}{N\lambda}$	I	$\varrho'\cdot\frac{1}{N\lambda}$	I
Max. 0. Ordnung	0	1,0	0	1,0
1. Nullstelle	1,22	0	0,77	0
Max. 1. Ordnung	1,62	0,017	1,22	0,16
2. Nullstelle	2,23	0	1,76	0
Max. 2. Ordnung	2,67	0,0041	2,23	0,090
3. Nullstelle	3,24	0	2,76	0
Max. 3. Ordnung	3,69	0,0016	3,24	0,057
4. Nullstelle	4,24	0	3,76	0

$$I=\frac{1}{(1-\varepsilon^2)^2}\left[\frac{2J_1(q)}{q}-\varepsilon^2\frac{2J_1(\varepsilon q)}{\varepsilon q}\right]^2; \qquad \Delta x'=0$$

mit $\varepsilon=D_2/D_1$. Die Minima sind gegeben durch $J_1(q_{\min})=\varepsilon\,J_1(\varepsilon q_{\min})$. Für den extremen Fall der schmalen Randzone sind die Werte in Tabelle 2 denen für ungestörte Pupille gegenübergestellt.

Die Bruchteile der Gesamtenergie im Beugungskern und im ersten hellen Ring in Abhängigkeit von ε zeigt Fig. 11.

Die Verteilung längs der Achse ist gegeben durch

$$I = \left[\frac{\sin\left(\frac{1}{4}(1-\varepsilon^2)\,p\right)}{\frac{1}{4}p}\right]^2; \qquad \varrho' = 0.$$

Die vollständige Lichtverteilung im bildpunktnahen Raum für zwei Werte von ε ist der Fig. 32, S. 370, des Beitrags von CLAUSSEN in diesem Band zu entnehmen, wo die Koordinaten mit u und v statt p und q bezeichnet sind. Bei $\varepsilon = 0{,}25$ sind die Abweichungen gegenüber $\varepsilon = 0$ noch nicht sehr groß; bei stärkerer Mittenabschattung werden die Unterschiede zwischen der wahren Lichtverteilung und der geometrischen Auffassung (hohler Kegel) auffällig: hier wächst die Tiefenerstreckung $\Delta x'$ des brauchbaren Fokus merklich, und die Dunkelstellen auf der Achse liegen um den Faktor $\frac{1}{1-\varepsilon^2}$ weiter von der Brennebene entfernt als bei freier Apertur.

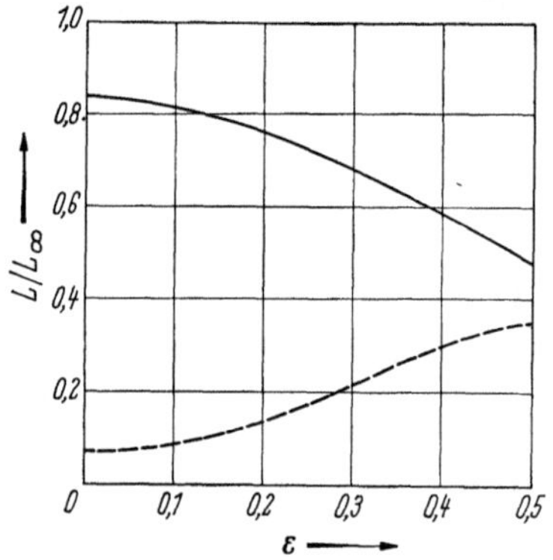

Fig. 11. Bruchteil des Strahlungsflusses im zentralen Beugungsscheibchen $\left(\frac{L\varrho_0'}{L_\infty}\text{ ———}\right)$ und im ersten hellen Beugungsring $\left(\frac{L\varrho_1' - L\varrho_0'}{L_\infty}\text{ – – –}\right)$ für eine Kreispupille mit Mittenabschattung $\varepsilon = D_2/D_1$. Fernes Punktobjekt, Bildebene.

Die Abschattung durch zentral über der Öffnung liegende Balken erzeugt die bekannten Beugungsstrahlen senkrecht zur Balkenrichtung. Ein vierarmiges Haltekreuz ruft also eine vierstrahlige, ein dreiarmiges aber eine sechsstrahlige Beugungsfigur hervor. Qualitative Darstellungen für den inneren Teil sind in [1] wiedergegeben. Die Strahlen im äußeren Teil der Beugungsfigur kann man erforderlichenfalls beseitigen durch Masken über den Haltestreben, die von Kreisbögen begrenzt sind. Eine Form mit minimalem Lichtverlust hat COUDER angegeben[2]. Der innere Teil der Beugungsfigur wird wenig geändert, das zuvor in den Strahlen konzentrierte Licht jedoch kreissymmetrisch verteilt.

Ein partielles Unterdrücken der hellen Beugungs*ringe* ist mit einer geeigneten Form der Pupille möglich. So ist die Bestrahlungsstärke in den Maxima längs der *Diagonalen* einer quadratischen Apertur wesentlich kleiner als in den entsprechenden Ringen der Kreisöffnung; das hat man benutzt, um schwache Begleiter heller Objekte sichtbar zu machen[3]. Solche Möglichkeiten einer Beeinflussung der Beugungsfigur durch ortsabhängige Änderung der Schwingungsamplitude[4] oder -phase in der Pupille sind in jüngerer Zeit systematisch untersucht worden [*24*]. Anwendungen in der Astronomie hat SINTON[5] beschrieben; für die leichter realisierbare Beeinflussung durch Blenden gibt TUVIKENE[6] Beispiele. Bei allen diesen Verfahren zeigt sich, daß mit einer Verkleinerung des Beugungskerns eine Vergrößerung des Lichtanteils in den Ringen einhergeht und umgekehrt. Da für große astronomische Fernrohre bei Beobachtung vom Erdboden aus die Beugungsgrenze ohnehin nicht erreicht wird, haben die Methoden in der Praxis nur eine begrenzte Bedeutung.

[1] J. SCHEINER u. S. HIRAYAMA: Abhandl. Akad. Berlin 1894, Anhang; s. auch [*7*]. — E. EVERHART and J. W. KANTORSKI: Astron. J. **64**, 455 (1959).

[2] A. COUDER: Astronomie **48**, 25 (1934), abgedruckt in [*31*] II, S. 620; s. auch R. H. HARDIE, Publ. Astron. Soc. Pacific **76**, 173 (1964).

[3] A. COUDER et P. JAQUINOT: C. R. Acad. Sci., Paris **208**, 1639 (1939).

[4] R. STRAUBEL: Zeeman Verhandlingen 1935, S. 302.

[5] W. M. SINTON: J. Opt. Soc. Amer. **42**, 284 (1952).

[6] L. M. TUVIKENE: Opt. Spectr. **10**, 144 (1961).

II. Abbildungsfehler.

Ein reales Teleskop entwirft von einem Stern allgemein nicht die in Ziff. 5 und 6 beschriebene Beugungsfigur; es zeigt Abbildungsfehler. Diese haben ihre Ursachen a) in der Geometrie der optischen Flächen (solche Abbildungsfehler werden hier stets nur in der in Ziff. 4 eingeführten Seidel-Näherung betrachtet) und im nicht idealen Justierzustand eines mehrgliedrigen Systems; b) in der Wellenlängenabhängigkeit des Brechungsindex von optischen Medien (bei heterochromatisch strahlenden Quellen merklich); c) in der Inhomogenität dieser Medien und der unvollkommenen Ausführung der Flächen; schließlich d) in der optischen Inhomogenität der Atmosphäre, durch die hindurch beobachtet werden muß, von der Stratosphäre bis zur nächsten Umgebung der Kuppel und der Luft im Fernrohr selbst. Bei den Fehlerursachen a) und b) ist eine analytische Behandlung möglich, während c) zum Teil und d) ausschließlich statistisch zu beschreiben sind.

7. Maße für die Fehlergröße. Für die Größe der Abbildungsfehler haben sich verschiedene Maßzahlen eingebürgert. Im Geltungsbereich der geometrischen Optik bevorzugt der Benutzer meist die *Queraberration* $\Delta y'$ oder $\Delta y'/f$, die schon in den Seidel-Ausdrücken (4.1) auftrat. Die rechnende Optik gibt oft *Längsaberrationen* $\Delta x' = \Delta s'$ an; für einen fernen Objektpunkt ist $\Delta s' = \frac{f}{h} \Delta y' = 2N \Delta y'$ für einen Strahl vom Blendenrand.

Einen Übergang zur physikalischen Optik vermittelt die *Wellenaberration*[1]. Eine *Wellenfläche* ist eine Fläche gleicher Phase der Lichtschwingung oder eine Fläche gleicher optischer Weglänge von der Quelle aus; sie ist zugleich eine Normalfläche zu den geometrischen Lichtstrahlen. Außerhalb der Erdatmosphäre sind die Wellenflächen im Licht eines Fixsterns mit großer Annäherung Ebenen senkrecht auf der Richtung zum Stern. Bei der idealen Abbildung in einen Punkt sind die Wellenflächen Kugelausschnitte mit dem Bildpunkt als Zentrum. Die Differenz zwischen einer solchen, durch die Lage des Empfängers ausgewählten „Referenzkugel" und der wirklichen Wellenfläche ist die Wellenaberration.

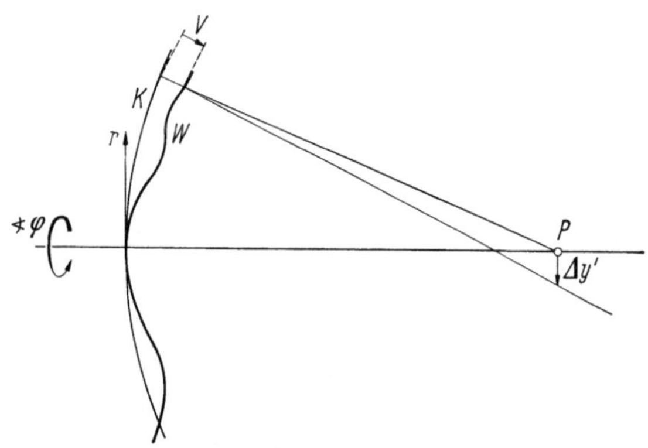

Fig. 12. Wellenaberration und geometrische Queraberration.

In Fig. 12 sei W die Wellenfläche, K die Referenzkugel mit Zentrum in P und Radius R_W. Die Wellenaberration V ist der Abstand zwischen K und W in Richtung $K \rightarrow W$ längs der Normalen oder längs des Radius von K (dieser Unterschied ist bedeutungslos; oft wird V sogar parallel zur Achse gemessen, da selbst bei $N=3$ noch $\cos \arcsin \frac{r}{R} \geqq 0{,}985$ ist). Dann gilt im Meridian

$$\Delta y' = -R_W \frac{dV}{dr}; \quad V(r) = -\frac{1}{R_W} \int_0^r \Delta y' \, dr \quad \text{mit} \quad V(0) = 0. \tag{7.1}$$

Für die geometrischen Abbildungsfehler ist V als Funktion des Abstands von der Bündelachse r und einem Azimutwinkel φ angebbar, meist mit der Normierung

[1] Dazu s. besonders [25].

$0 \leqq r \leqq 1$ und der Betrachtung von V in der Austrittspupille; in der verwendeten Näherung ändert sich allerdings V nicht beim Fortschreiten längs der Normalen. Für das ferne Punktobjekt gilt also auch (mit Einfallshöhe h)

$$V(h) = -\frac{1}{f^2} \int_0^h \Delta s' \, h \, dh. \tag{7.2}$$

Als Einheit für V wird gewöhnlich die Lichtwellenlänge λ gewählt.

Bei Defokussierung und sphärischer Aberration entfernt sich die deformierte Wellenfläche einseitig von der Referenzkugel und ist durch die Angabe des Maximalwertes am Bündelrand charakterisiert. Bei Koma und Astigmatismus ist die Wellenfläche antisymmetrisch zur Referenzkugel und kann durch

$$V_{\max} = \pm \text{const.} \cdot \lambda$$

beschrieben werden. Bei unregelmäßigem Verlauf gibt der Abstand zweier die Wellenfläche einschließender konzentrischer Kugeln „die Wellenaberration", wobei der Mittelpunkt optimal zu wählen ist. Im Falle einer Spiegelfläche erzeugt ein Flächenfehler der Höhe Δ eine Aberration von $V \approx 2\Delta$ auf der Wellenfläche; für eine brechende Fläche ist $V \approx (n-1) \cdot \Delta$ (Fläche gegen Luft).

Für jeden der geometrischen Abbildungsfehler ist mit $V_{\max}$ das Verhältnis der Queraberration $\Delta y'$ zur Größe des idealen Beugungsbildes (charakterisiert durch ϱ_0') festgelegt. Bei einer gegebenen optischen Anordnung (mit fixierter *Winkel*-aberration $\Delta y'/f$, charakterisiert durch Öffnungszahl N und Seidel-Koeffizienten $\sum$ **I** bis $\sum$ **IV**) *wächst* also die Wellenaberration mit den Abmessungen des Systems.

Wenn die Wellenaberration einen glatten Verlauf (ähnlich der von sphärischer Aberration dritter Ordnung) hat, so ist bei kleiner Aberration schon mit der Angabe des *Maximal*wertes $V_{\max}$ die Abbildungsqualität ausreichend gekennzeichnet. Ist das Rayleigh-Kriterium

$$V_{\max} \leqq \frac{\lambda}{4}$$

erfüllt, braucht man bei visueller Beobachtung noch keine merkliche Bildverschlechterung zu befürchten; das Beugungsbild bleibt im wesentlichen erhalten. Allerdings ist das Kriterium *nicht* ausreichend für den Fall von kurzwelligen Zonenfehlern oder gesetzmäßigen Schwankungen der Wellenfläche mit kleiner örtlicher Periode: hier sind auch bei $V_{\max} = \lambda/4$ erhebliche Auswirkungen auf die Beugungsfigur mit entsprechender Verschlechterung der Abbildung möglich[1]. Für kleine Aberrationen hat STREHL im Anschluß an RAYLEIGH die *Definitionshelligkeit* Z als beugungstheoretisches Fehlermaß eingeführt. Sie ist gegeben durch

$$Z = \frac{I}{I_0},$$

wobei I die Bestrahlungsstärke im Maximum der aberrationsbehafteten Beugungsfigur ist, I_0 die im Zentrum der idealen Beugungsfigur für die gleiche Apertur. Da sich Form und Durchmesser des zentralen Beugungsscheibchens bei schwacher Aberration wenig ändern, ist Z ein Maß für die Verteilung der Gesamtenergie zwischen diesem Kern (der die Abbildung vermittelt und das Auflösungsvermögen bestimmt) und den außerhalb liegenden Teilen der Beugungsfigur (die den Kontrast verschlechtern). Nach Übereinkunft betrachtet man $Z = 80\%$ als Grenzwert für im Sinne der Wellenoptik gute optische Systeme (Rayleigh-

[1] Ein besonders „bösartiges" Beispiel gibt R. BARAKAT, J. Opt. Soc. Amer. **55**, 572 (1965).

Strehl-Kriterium[1]). Bei visueller Beobachtung tritt dabei noch keine wesentliche Bildverschlechterung ein. Im Fall der sphärischen Aberration dritter Ordnung entspricht $Z=0{,}8$ einer Wellenaberration von $V_{\max}=\lambda/4$.

a) Monochromatische Fehler.

In den Ziffern 8 bis 11 werden die Abbildungsfehler der Gauß- und Seidel-Näherung als axiale und außeraxiale Fehler einfacher Elemente mit sphärischen Flächen eingeführt. Die Bezeichnungen und ihre Toleranzen gelten aber für bestimmte Gesetzmäßigkeiten der Wellenaberration unabhängig von ihrer Ursache; variiert z.B. der Krümmungsradius mit dem Azimutwinkel auf einem Spiegel, so entsteht schon in der Achse ein astigmatisches Bild. Toleranzen sind für die isolierten Fehler angegeben.

8. Defokussierung. Ist der Abstand der Empfängerfläche vom Bildpunkt Δ, so wird der Radius des geometrischen Zerstreuungsscheibchens

$$\Delta y' = -\frac{\Delta}{2N}.$$

Die Wellenaberration ergibt sich als Abstand zweier sich berührender Kugeln, deren Zentren um Δ auseinanderliegen:

$$V = \frac{r^2}{2R_W^2}\cdot\Delta; \qquad V_{\max} = \frac{1}{8N^2}\cdot\Delta$$

am Rand des Bündels. Die Auswirkung auf das Beugungsbild kann man den Fig. 9 und 10 entnehmen. Für $p/\pi=1$ sinkt die relative Intensität auf der Achse auf 0,8; das entspricht einer Strehl-Toleranz von

$$|\Delta| \leqq 2N^2\lambda \tag{8.1}$$

für Defokussierung bei Abwesenheit aller anderen Fehler. Hier ist also $V_{\max}=\lambda/4$ für $Z=80\%$, wie bei der sphärischen Aberration. Für ein Objektiv mit $N=3{,}2$ darf demnach die Fokuseinstellung höchstens $20\,\lambda$ ($\approx 0{,}01$ mm) falsch sein, wenn das Rayleigh-Strehl-Kriterium nicht verletzt sein soll.

Beim Durchbiegen der photographischen Platte im Schmidt-Teleskop treten Abweichungen zwischen Ist- und Soll-Empfängerfläche von mindestens dieser Größenordnung auf. Normale Photoplatten haben merklich größere Abweichungen von einer Planfläche; Spiegelglasplatten halten die Toleranzbedingung gerade ein.

9. Sphärische Aberration. Für ein Bündel mit nicht verschwindendem Durchmesser hängt die Schnittweite s' von der Einfallshöhe h ab, und zwar liegt bei sammelnden Linsen oder Spiegeln mit sphärischen Flächen der Bildpunkt für die Randzone *vor* dem paraxialen Bildpunkt: Sphärische *Unter*korrektion. Aus dem Seidel-Ausdruck (4.1) können wir entnehmen, daß die Querabweichung der Aberration mit der dritten Potenz des Öffnungswinkels wächst, aber von der Hauptstrahlneigung unabhängig ist. Es genügt demnach, sie in der Achse zu betrachten, obwohl sie im ganzen Feld existiert. Von der Blendenlage ist der Öffnungsfehler unabhängig.

Die Einhüllende der Strahlen in der Nähe des Bildpunktes ist die wohlbekannte Kaustik der sphärischen Aberration, deren „Kelch" bei Unterkorrektion gegen den Lichteinfall geöffnet ist. Das Bündel hat seine engste Einschnürung in einer Auffangfläche, die vom paraxialen Bildpunkt nach dem Schnittpunkt der Randstrahlen zu verschoben ist, und zwar um $^3/_4$ dieser Strecke; der Radius des Zerstreuungskreises ist hier nur noch $^1/_4$ von dem in der Gauß-Bildebene. Allerdings bleibt dieses geometrisch-optische Resultat für kleine

[1] Vgl. auch Fig. 2 des Artikels von CLAUSSEN auf S. 344 in diesem Band.

Aberrationen, bei denen die Lichtverteilung maßgeblich von der Beugung beeinflußt wird, nicht erhalten.

Die rechnende Optik macht das Ergebnis von trigonometrischen Strahldurchrechnungen häufig als Diagramm Einfallshöhe h gegen sphärische Längsabweichung $\Delta s'$ anschaulich (Fig. 38, S. 284). Bei Linsensystemen mit größerem Öffnungsverhältnis erweist sich dabei die Seidel-Näherung als nicht ausreichend; zu berücksichtigen ist auch die „sekundäre" Aberration fünfter Ordnung (der Ausdruck für $\Delta y'$ enthält ein von w unabhängiges Glied mit $1/N^5$). Sind die Anteile der dritten und der fünften Ordnung am Pupillenrand entgegengesetzt

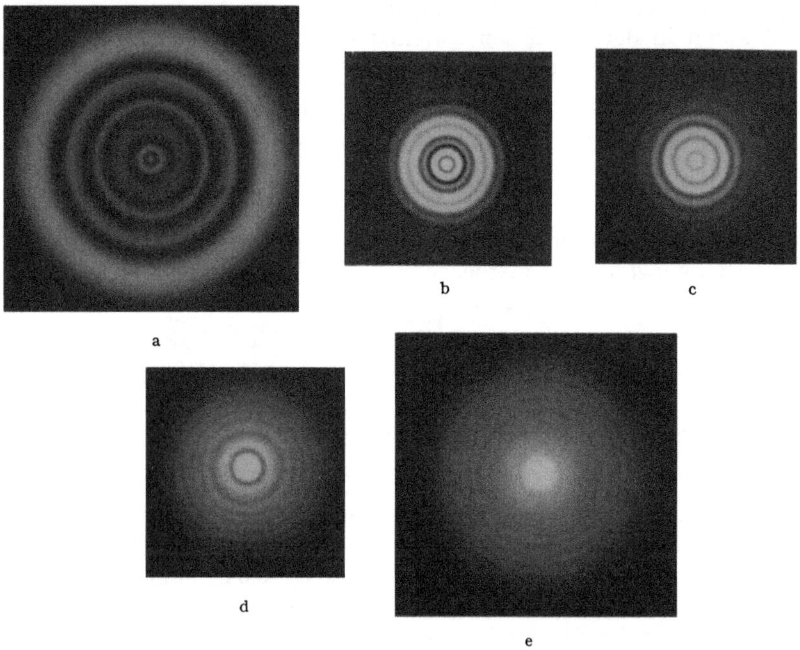

Fig. 13a—e. Beugungsfiguren einer Kreispupille für ein Strahlenbündel mit sphärischer Aberration dritter Ordnung, beim Verschieben des Empfängers durch die Fokalregion. Bild a entspricht dem Marginalfokus, Bild d dem Paraxialfokus. Nach Cagnet et al. [*45*].

gleich, so spricht man von „Hebung der sphärischen Aberration", obwohl dabei ein „Zonenfehler" mit Maximum bei $1/\sqrt{2}$ des Pupillenradius übrigbleibt.

Auch die Verringerung der sphärischen Aberration durch Fokusänderung können wir als eine solche Kompensation beschreiben; die Defokussierung wird dabei als „sphärische Aberration erster Ordnung" betrachtet ($\Delta y'$ proportional $1/N$, feldunabhängig).

Die *Wellenaberration* der primären sphärischen Aberration stellt sich als Rotationsparaboloid vierter Ordnung dar. Aus (7.1) und (5.1) folgt der Zusammenhang

$$V_{\max} = \frac{1}{8} \cdot \frac{\boldsymbol{S}}{N}, \qquad \frac{V_{\max}}{\lambda} = \frac{\boldsymbol{S}}{6{,}6\,\varrho_0'};$$

d.h. wenn die maximale Seidel-Querabweichung das 6,6fache vom Radius des ersten dunklen Beugungsringes beträgt, ist die Wellenaberration am Rand gerade $1\,\lambda$. Vom optimalen Fokus aus wird

$$\frac{V_{\max}}{\lambda} = \frac{1}{26{,}2} \cdot \frac{\boldsymbol{S}}{\varrho_0'}.$$

Hinsichtlich des *Beugungsbildes* stellen wir fest, daß sich seine Form bei schwacher Aberration wenig ändert, wohl aber die Intensitätsverteilung (Fig. 13): in der Mitte des zentralen Scheibchens sinkt die relative Intensität auf 80% für $V_{max}=\lambda/4$ (Rayleigh-Grenze). Im Paraxial-Fokus und außerhalb erscheinen die inneren Ringe zu hell gegenüber der idealen Abbildung; im Marginal-Fokus und überhaupt innerhalb der besten Einstellung ist die Mitte der Beugungsfigur zu schwach, und die Beugungsringe nehmen mit zunehmender Ordnung an Helligkeit zunächst zu (Merkregel: Bei sphärischer *Unter*korrektion *intrafokal* heller *Rand*, extrafokal helle Mitte).

Die Verhältnisse seien für den *Kugelspiegel* näher betrachtet: hier gilt bei fernem Dingpunkt (Fig. 14)

$$\Delta x'(\vartheta)=\overline{F'A}=\overline{F'C}-\overline{AC}=f-\frac{f}{\cos\vartheta}$$

(f ist die paraxiale Brennweite). Mit $\sin\vartheta=\frac{h}{2f}$ wird

$$\frac{1}{\cos\vartheta}=\left(1-\frac{h^2}{4f^2}\right)^{-\frac{1}{2}}$$

$$=1+\frac{h^2}{8f^2}+\frac{3}{128}\frac{h^4}{f^4}\cdots,$$

also

$$\Delta x'(h)=-\frac{1}{8}\frac{h^2}{f}$$

und

$$\Delta y'=-\frac{1}{8}\frac{h^3}{f^2}$$

in der Seidel-Näherung. Durch Vergleich mit dem Seidel-Ausdruck (4.1) finden wir so $\mathbf{I}=+0{,}25$.

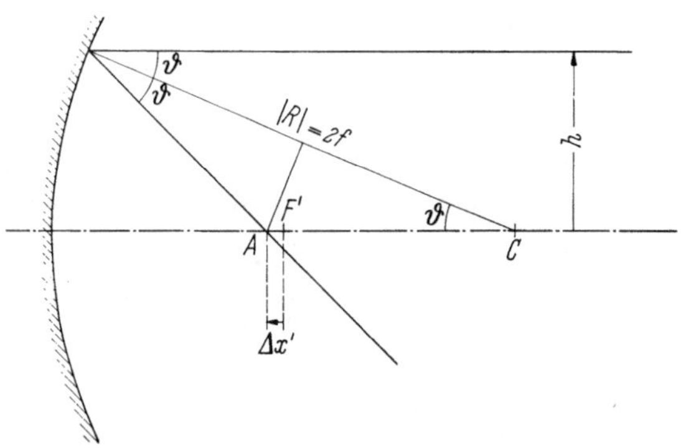

Fig. 14. Sphärische Aberration beim Kugelspiegel.

Betrachtet man die Wellenfläche in der Pupille, so kann man Gl. (7.1) schreiben

$$V(h)=-\frac{1}{f}\int_0^h \Delta y'\,dh.$$

Die Wellenaberration wird also

$$V(h)=\frac{1}{32}\frac{h^4}{f^3}, \tag{9.1}$$

wenn $V(0)=0$, mit dem Maximalwert

$$V_{max}=\frac{1}{32}\frac{H^4}{f^3}=\frac{1}{512}\frac{f}{N^4}, \quad \text{wobei} \quad H=h_{max}=\frac{D}{2}.$$

Wählen wir eine andere Referenzkugel durch Verschieben des Empfängers längs der Achse um $\Delta<0$, so läßt sich die Wellenaberration verkleinern:

$$-\Delta y'=\frac{h}{f}\left(\frac{h^2}{8f}+\Delta\right); \qquad V=\frac{h^4}{32f^3}+\Delta\cdot\frac{h^2}{2f^2}. \tag{9.2}$$

Die Wellenaberration verschwindet für den Rand, wenn $\Delta=-\frac{H^2}{16f}$, wenn der gewählte Fokus also in der *Mitte* zwischen Paraxial- und Randstrahlenfokus liegt. Dann ist

$$V(h)=\frac{1}{32f^3}(h^4-h^2H^2).$$

Die größte Aberration liegt nun bei $h = H/\sqrt{2}$:

$$|V_{\max}| = \frac{1}{128} \cdot \frac{H^4}{f^3},$$

sie beträgt also nur $1/4$ des für den Paraxialfokus gefundenen Wertes. In der Seidel-Approximation bleibt die sphärische Aberration des Kugelspiegels demnach unschädlich im Sinne des Rayleigh-Kriteriums, solange (bei Einstellung auf den optimalen Fokus)

$$\frac{H^4}{f^3} \leqq 32\lambda \quad \text{oder} \quad \frac{D^4}{f^3} = \frac{D}{N^3} \leqq 512\lambda.$$

Setzen wir $\lambda = 0{,}5\ \mu\text{m}$, so ist die Parabolisierung eines Fernrohrspiegels überflüssig für

$$N \geqq 3{,}4\sqrt{D} \quad (D \text{ in cm}).$$

Eine störende sphärische Aberration kann man beseitigen, indem man die *Hälfte* der Wellenaberration vom Spiegel abträgt, also

$$\Delta\xi = \frac{h^4}{64 f^3}$$

und somit Deformationskonstante $b = -1$, womit die wohlbekannte Fokaleigenschaft des Rotationsparaboloids wiedergefunden ist.

10. Koma. Für weitgeöffnete Bündel mit nichtverschwindender Hauptstrahlneigung kommt als weiterer Abbildungsfehler die *Koma* hinzu. Die Strahlenverteilung im schiefen Bündel ist nach der Reflexion oder Brechung nicht mehr rotationssymmetrisch zu ihrem Hauptstrahl, sondern hat nur noch die Meridionalebene zur Symmetrieebene. Für eine *Einzelfläche* (aber nur für diese!) kann man die Strahlverteilung im komatischen Bündel als einen exzentrischen Ausschnitt aus dem Strahlenverlauf der sphärischen Aberration beschreiben. Koma kann rein beobachtet werden an einem Parabolspiegel, bei dem die Ebene der Eintrittspupille durch den Brennpunkt geht; auch bei einem Parabolspiegel mit Blende im Spiegel tritt bei großem Öffnungsverhältnis und mäßigem Bildwinkel nahezu reine Koma auf (s. Fig. 22). Das Bild eines außeraxialen Punktes erscheint als radial gerichtetes Büschel mit starker Lichtkonzentration an der Spitze. Den Seidel-Ausdrücken kann man entnehmen, daß die Ausdehnung der Zerstreuungsfigur proportional zum Bildwinkel und zum Quadrat des Öffnungsverhältnisses wächst. Solange die sphärische Aberration nicht korrigiert ist, hängt der Koma-Koeffizient **II** von der Lage der Eintrittspupille ab.

Aus den Formeln (4.1) für $\Delta y'$ und $\Delta z'$ folgt, daß beim Durchlaufen einer Zone in der Eintrittspupille mit $Y^2 + Z^2 = \text{const}$ die Strahlen von einem fernen außeraxialen Dingpunkt die Gauß-Ebene in einem Kreis durchstoßen, der *zweimal* überdeckt wird; die Strahlen schneiden sich nur noch paarweise, ein Bildpunkt kommt auch für eine schmale Zone nicht mehr zustande. Die Mittelpunkte der Durchstoßkreise liegen im Meridionalschnitt, und zwar um den doppelten Kreisradius ($\varrho = \frac{1}{3}\boldsymbol{K}$) vom Gauß-Bildpunkt entfernt (Fig. 15). Die Gesamtheit der Kreise hat zwei gemeinsame Tangenten, die im Gauß-Bildpunkt einen Winkel von 60° einschließen.

Bei korrigierter sphärischer Aberration läßt schon das *axiale* Strahlenbündel erkennen, ob die Abbildung komafrei ist: es muß dazu die *Sinus-Bedingung* erfüllen. Im uns interessierenden Fall des fernen Objekts lautet sie

$$\frac{h}{\sin u'} = \text{const} = f.$$

Verlängert man die im Ding- und im Bildraum liegenden Abschnitte des gleichen Lichtstrahls bis zu ihrem Schnitt, so müssen die Schnittpunkte demnach auf einer Kugelfläche mit dem Brennpunkt als Zentrum und der Brennweite als Radius liegen. Hier wird also nicht nur verlangt, wie bei der Hebung der sphärischen Aberration, daß die *Schnittweiten* s' von der Einfallshöhe h unabhängig sind; die verschiedenen Zonen sollen vielmehr auch noch den gleichen *Abbildungsmaßstab* liefern, d.h. die gleiche Brennweite f haben. Mit Erfüllung der Sinusbedingung wird übrigens auch Koma höherer Ordnung gehoben.

Ein optisches System, das frei von sphärischer Aberration ist und der Sinusbedingung genügt, nennt man seit ABBE *aplanatisch*. Aus der anschaulichen Bedeutung der Sinusbedingung ist sofort zu entnehmen, daß eine einzelne Spiegelfläche nicht aplanatisch sein kann: Brennpunkt und Krümmungsmittelpunkt müßten zusammenfallen.

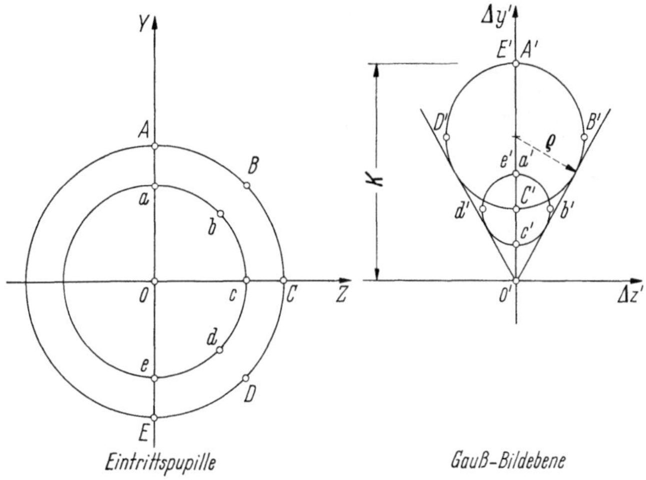

Fig. 15. Zerstreuungsfigur bei Seidel-Koma.

Die *Wellenaberration* bei reiner Koma ist von der Form $V(r) = \text{const} \cdot r^3 \cos\varphi$, wobei φ der Azimutwinkel ist (von der Meridionalebene aus gezählt). Die deformierte Wellenfläche wölbt sich also im Meridian nach entgegengesetzten Seiten von der Referenzkugel weg. Der Zusammenhang mit der geometrischen Aberration ist gegeben durch

$$V_{\max} = \pm\frac{1}{6}\cdot\frac{\boldsymbol{K}}{N}; \qquad \frac{V_{\max}}{\lambda} = \pm\frac{1}{4{,}9}\cdot\frac{\boldsymbol{K}}{\varrho_0'}.$$

Im *Beugungsbild* macht sich schwache Koma durch einseitige Verstärkung der hellen Ringe bemerkbar; stärkere führt zu einer Verformung, die es schließlich der geometrischen Zerstreuungsfigur ähnlich macht (Fig. 16 und 17). Die Strehl-Toleranz für eine komatische Wellenfläche ist $V_{\max} = 0{,}2\,\lambda$. Nun ist die oben gegebene Form der Wellenaberration vom Gaußschen Bildpunkt aus betrachtet; das Maximum der Beugungsfigur ist jedoch radial verschoben. Mit dem neuen Zentrum der Referenzkugel wird die Aberration

$$V(r) = \text{const}\cdot\left(r^3 - \tfrac{2}{3}r\right)\cos\varphi.$$

Man kann dies als partielle Kompensation der Koma durch einen Verzeichnungs-Term erster Ordnung, $V \sim r\cdot\cos\varphi$ (geneigte Ebene) ansehen. Bezogen auf den

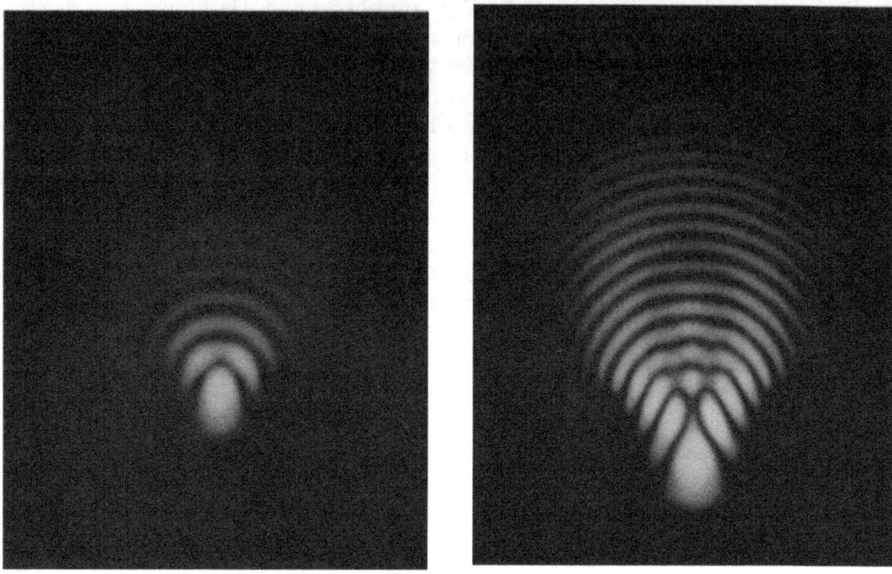

a b

Fig. 16a u. b. Beugungsbilder einer Kreispupille für Bündel mit Seidel-Koma (a) vom Betrag $V_{max} = \pm 1\,\lambda$, (b) $V_{max} = \pm 2{,}5\,\lambda$ (bezogen auf die im Gauß-Bildpunkt zentrierte Referenzkugel); Empfänger in der Paraxial-Bildebene. Nach CAGNET et al. [45].

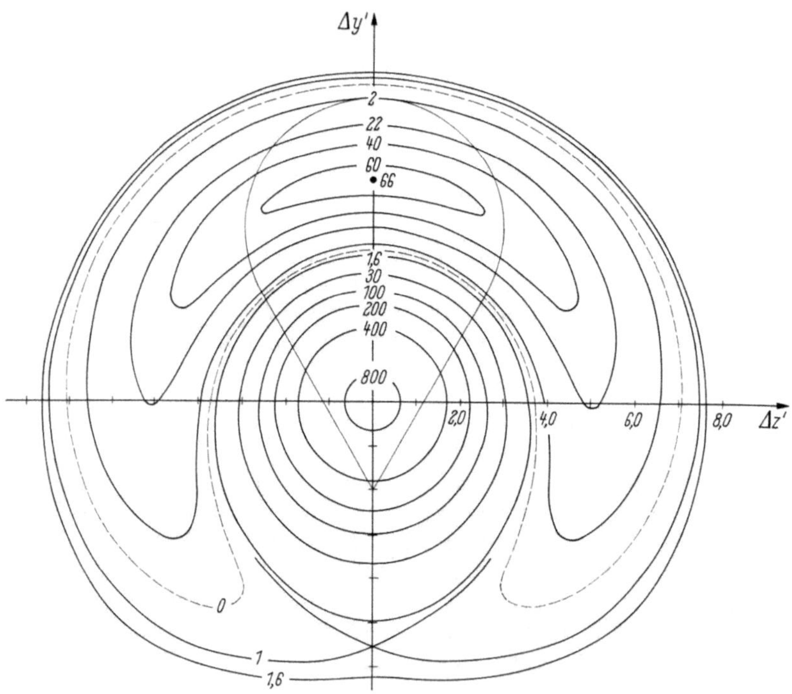

Fig. 17. Isophoten im Beugungsbild für Koma ($V = 0{,}48\,\lambda\,(r^3 - \frac{2}{3}r)\cos\varphi$). Bestrahlungsstärke im Zentrum der *ungestörten* Beugungsfigur $I_0 = 1000$; Definitionshelligkeit $Z = 0{,}88$. $\Delta y'$ und $\Delta z'$ sind in der Einheit $q = \frac{\pi}{N} \cdot \frac{\varrho'}{\lambda}$ angegeben. Der Umriß der Seidel-Zerstreuungsfigur ($\boldsymbol{K} = 2{,}35\,\varrho_0'$) ist eingetragen. Nach NIJBOER[1].

[1] B. R. A. NIJBOER: Diss. Groningen 1942.

paraxialen Bildpunkt kann man die Toleranz verdreifachen auf

$$|V_{\max}| \leqq 0{,}6\,\lambda .$$

Für einen Parabolspiegel oder Kugelspiegel mit Eintrittspupille im Scheitel ($\sum \mathbf{II} = -0{,}5$) bedeutet diese Grenze, daß

$$\frac{D}{N^2} \cdot \tan w \leqq 19{,}2\,\lambda \qquad (10.1)$$

sein muß, d.h. mit $\lambda = 0{,}5\,\mu\mathrm{m}$:

$$\tan w \lesssim 1 \cdot 10^{-3}\frac{N^2}{D}; \qquad w \lesssim 200'' \cdot \frac{N^2}{D} \quad (D \text{ in cm}).$$

11. Astigmatismus und Bildwölbung. Auch wenn ein schiefes Bündel ganz in einem Gebiet nahe dem Hauptstrahl verläuft — so, daß bei axialem Hauptstrahl die Gauß-Optik gälte —, tritt ein Abbildungsfehler auf. Die wirksamen Krümmungsradien in den beiden Hauptschnitten sind nämlich verschieden, und das Bündel hat nur zwei Symmetrieebenen. Ein meridionales (bei Behandlung des Astigmatismus sprechen die Optiker meist von *tangential*) und ein sagittales *Büschel* mit gleichem Hauptstrahl liefern getrennte Bildpunkte, deren Gesamtheit für ein Dingfeld auf je einer Rotationsfläche (Bildschale) liegt. Als Büschel bezeichnet man eine *ebene* Mannigfaltigkeit von zusammengehörenden Strahlen. Die Bildschalen berühren einander und die Gauß-Bildebene in der Achse. Das volle (immer noch enge) Bündel gibt an keiner Stelle mehr einen scharfen Bildpunkt, sondern auf den beiden Bildschalen je einen kurzen Strich, eine *Brennlinie.* Die Brennlinie auf der Bildschale der tangentialen Büschel hat sagittale Richtung, was die Terminologie nicht vereinfacht. Der Bündelquerschnitt im Bildraum ändert sich beim Fortschreiten in Lichtrichtung stetig. Auf einer „mittleren" Bildschale ist er ein Kreis, davor und dahinter eine Ellipse, deren große Achse sich beim Durchgang durch die mittlere Schale um 90° dreht, und deren Exzentrizität in den Brennlinien die Maxima, im Zerstreuungskreis das Minimum hat (vgl. Fig. 25, S. 267).

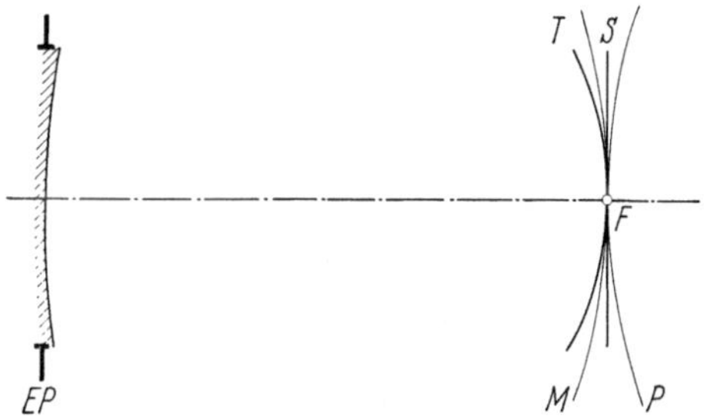

Fig. 18. Lage der tangentialen (T) und der sagittalen (S) Schale, der Schale mittlerer Bildkrümmung (M) und der Petzval-Schale (P) bei einem Hohlspiegel mit Eintrittspupille im Scheitel.

Solange in einem System die Koma nicht korrigiert ist, hängt der Astigmatismus von der Blendenlage ab. Reinen Astigmatismus können wir an den Zwei-Spiegel-Aplanaten beobachten, oder mit guter Näherung an zweilinsigen Fernrohrobjektiven im monochromatischen Licht. Den Seidel-Formeln entnehmen wir, daß der Radius des Zerstreuungskreises proportional zur Öffnung unabhängig von der Brennweite und zum Quadrat des Bildwinkels wächst. Der halbe Abstand der beiden Bildschalen, die *astigmatische Differenz,* ist dem Quadrat des Bildwinkels proportional. In der Seidel-Näherung sind die Bildschalen Kugelflächen; ihre Krümmungen sind durch $\sum \mathbf{III}$ (tangentiale Schale) und $\sum \mathbf{IV}$ (sagittale Schale) gegeben. Die oben erwähnte mittlere Schale ist die Schale der mittleren Krümmung (Fig. 18). Die Petzval-Schale hat die Krümmung $\sum \boldsymbol{P}$, die in einfacher Weise von den Systemdaten abhängt:

$$\sum \boldsymbol{P} = \sum_i \frac{1}{n_i f_i}. \qquad (11.1)$$

In dieser Schreibweise bezieht sich jeder Summand i auf ein optisches *Element*, also eine Linse oder spiegelnde Fläche; deren Abstände, die Pupillenlage, die Durchbiegung von Linsen, die Deformation von Flächen gehen in $\sum \boldsymbol{P}$ nicht ein. Die Brennweite f von Linsen ist ohne Berücksichtigung der Dicke zu rechnen, bei Spiegelflächen ist $n=-1$ zu setzen. Nach (4.3) fallen die tangentiale und die sagittale Schale auf die Petzval-Schale zusammen, wenn die astigmatische Differenz beseitigt werden kann (z.B. durch Deformation von Spiegelflächen, Verschieben der Pupille); die Petzval-Summe gibt also die verbleibende Bildkrümmung bei gehobenem Astigmatismus. Ist die *mittlere* Bildschale plan („Bildfeldebnung im übertragenen Sinne"), so liefert die Petzval-Summe den Koef-

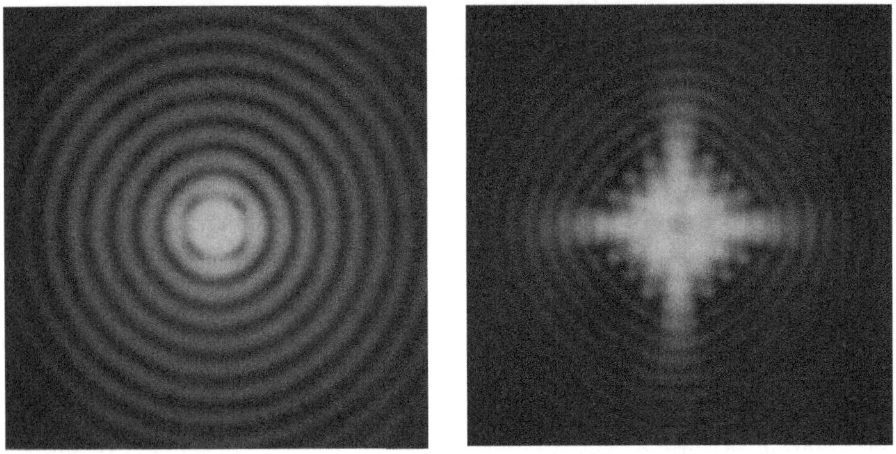

a b

Fig. 19a u. b Beugungsfiguren einer Kreispupille für astigmatische Bündel; Empfänger auf der mittleren Bildschale. Wellenaberration links $V_{max}=0{,}23\,\lambda$, rechts $V_{max}=1{,}6\,\lambda$. Nach CAGNET et al. [45].

fizienten der astigmatischen Differenz. Nur für einen Anastigmaten ($\sum$ **IIIa** $=0$) führt die Forderung $\sum \boldsymbol{P}=0$ zur Bildfeldebnung; dazu müssen $\sum$ **III** und $\sum$ **IV** einzeln verschwinden.

An trigonometrischen Rechnungen für Linsensysteme werden wieder die Grenzen der Seidel-Näherung deutlich. Im klassischen Diagramm — Hauptstrahlneigung w gegen sagittale bzw. tangentiale Längsabweichung Δs, Δt (Fig. 43, S. 290) — bleibt bei komplizierteren Anordnungen kaum noch etwas von den gleichmäßig gekrümmten Schalen übrig.

Die Wellenaberration des Astigmatismus ist gegeben durch

$$V(r)=\text{const}\cdot r^2\cos 2\varphi;$$

die Wellenfläche krümmt sich also im Meridionalschnitt nach der einen, im Sagittalschnitt nach der entgegengesetzten Seite von der Referenzkugel (mit Zentrum auf der mittleren Schale) weg. Zur geometrischen Aberration bestehen die Beziehungen

$$V_{max}=\pm\frac{1}{4}\cdot\frac{\boldsymbol{A}}{N}, \qquad \frac{V_{max}}{\lambda}=\pm\frac{1}{3{,}3}\cdot\frac{\boldsymbol{A}}{\varrho_0'}.$$

Im *Beugungsbild* führt schwacher Astigmatismus zu einer Aufhellung der Ringe in den Hauptschnitt-Azimuten (bei Beobachtung im mittleren Fokus), stärkerer erzeugt kreuzförmige Isophoten oder ein auf der Spitze stehendes Quadrat (Fig. 19 und 20).

Damit $Z \geqq 80\%$, muß $|V_{\max}| \leqq 0{,}17\,\lambda$ sein; Kompensation durch eine niedere Ordnung gibt es beim Astigmatismus nicht mehr. Für den Kugelspiegel führt die Toleranzgrenze zu der Bedingung

$$\frac{D}{N}\tan^2 w \leqq 2{,}7\,\lambda$$

oder für $\lambda = 0{,}5\ \mu\text{m}$:

$$\tan w \lesssim 0{,}01\sqrt{\frac{N}{D}}; \quad w \lesssim 2400''\sqrt{\frac{N}{D}} \quad (D \text{ in cm}).$$

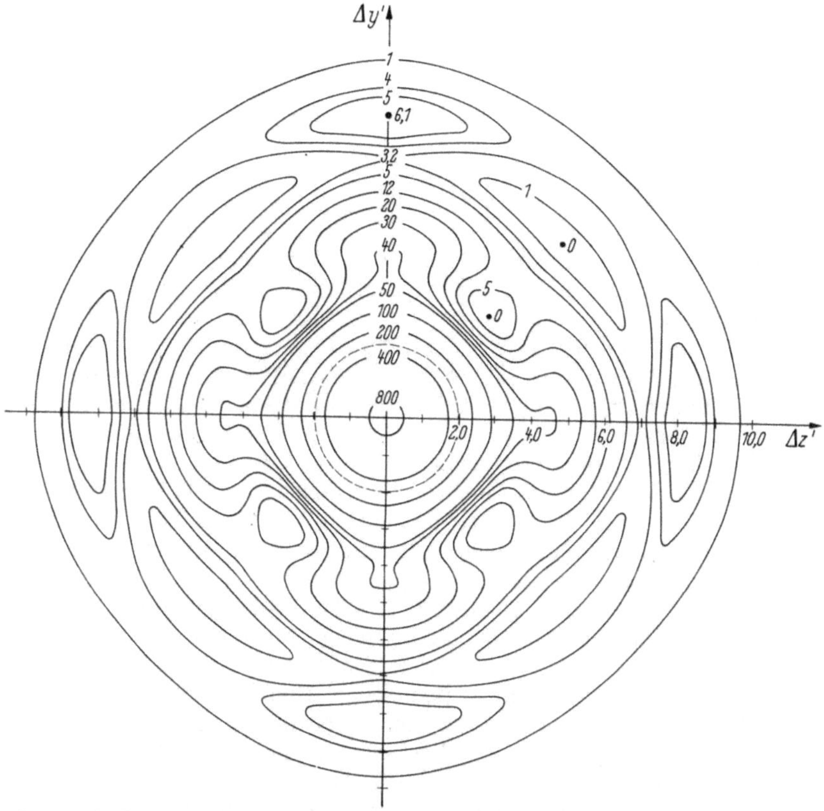

Fig. 20. Isophoten im Beugungsbild für Astigmatismus (Bezeichnungen wie Fig. 17) auf der Schale mittlerer Krümmung; $V = 0{,}16\,\lambda r^2 \cos 2\varphi$. Definitionshelligkeit $Z = 0{,}84$. Gestrichelt: Seidel-Zerstreuungskreis ($\boldsymbol{A} = 0{,}52\,\varrho_0'$).

b) Farbabhängige Fehler.

12. Primäre chromatische Aberrationen. Der Brechungsindex n der für Refraktionsoptik verwendeten Stoffe variiert mit der Wellenlänge (bei „normaler Dispersion“ wächst n mit abnehmendem λ). Die „Stärke“ (Brechkraft, reziproke Brennweite) einer *dünnen* Einzellinse ist

$$\frac{1}{f} = \varphi = (n-1)\left(\frac{1}{R_1} - \frac{1}{R_2}\right)$$

(R_1, R_2 sind die — vorzeichenbehafteten — Krümmungsradien der Flächen, die groß gegen die Linsendicke vorausgesetzt wurden); die Brennweite hängt damit über den Brechungsindex von der Wellenlänge ab. Der Bildort ist gegeben durch

$$\frac{1}{s'} = \frac{1}{s} + \varphi.$$

Führen wir für die wellenlängenabhängige Variation das Differenzenzeichen δ ein, so können wir schreiben

$$\delta\left(\frac{1}{s'}\right)=\delta\varphi=\frac{\varphi}{\nu} \quad \text{mit} \quad \frac{1}{\nu}=\frac{\delta\varphi}{\varphi}=\frac{\delta n}{n-1}, \tag{12.1}$$

wobei die δ-Größen als klein angenommen sind. $\delta(1/s')$ ist unabhängig von der Dingweite s. Die Materialkonstante ν (gelegentlich auch V geschrieben) wird als Abbesche Zahl oder reziproke relative Dispersion bezeichnet.

Bei fernem Dingpunkt und Einstellung des Empfängers in den Fokus der Wellenlänge λ_1 ergibt sich für λ_2 eine Längsabweichung der chromatischen Aberration von

$$\Delta x'=\delta f=\frac{1}{\nu}\cdot f$$

und ein Durchmesser des Zerstreuungskreises

$$2\varrho_{\mathrm{Chr}}=N\cdot\Delta x'=\frac{1}{\nu}\cdot D \tag{12.2}$$

unabhängig von der Brennweite; ν ist für λ_1, λ_2 zu bilden. Bei Einstellung auf einen mittleren Fokus fallen die Zerstreuungskreise für λ_1 und λ_2 zusammen und sind halb so groß.

Diese chromatische Aberration des Bildorts steht in einer gewissen Analogie zur sphärischen Aberration: sie ist im ganzen Feld vorhanden und von der Pupillenlage unabhängig; statt von der Einfallshöhe hängt hier die Schnittweite von der Wellenlänge ab.

Bei einer dünnen Linse, die mit der Pupille zusammenfällt, sind die verschiedenfarbigen Bildraum-Hauptstrahlen eines schiefen Bündels identisch. Für jede Einstellfläche liegen also wenigstens die Zentren der chromatischen Zerstreuungskreise auch für außeraxiale Dingpunkte aufeinander; die — wenn auch unscharfen — Bilder sind für alle Farben gleich groß. Das gilt nicht mehr für eine dicke Linse oder eine andere Blendenlage: neben dem Farblängsfehler des Bildorts gibt es einen *Farbquerfehler* der Bildgröße, auch chromatische Vergrößerungsdifferenz, Farbvergrößerungsfehler, laterale oder transversale chromatische Aberration genannt. Wegen der weitgehenden Unabhängigkeit vom Farbortsfehler stellt sie eine zweite *primäre* chromatische Aberration dar (und darf nicht mit der Querabweichung des Farbortsfehlers verwechselt werden!).

In der technischen Optik ist es üblich, die Abbesche Zahl ν für bestimmte Wellenlängen zu bilden. Als Standardwellenlängen werden benutzt

*	1014,0 nm
A'	768,2
C	656,3
d	587,6
e	546,1
F	486,1
g	435,8
h	404,7
**	365,0

Für Instrumente zum visuellen Gebrauch wählt man $\nu=\frac{n_d-1}{n_F-n_C}$. Dieses ν ist für normale Gläser von der Größenordnung 50; ohne Beseitigung der chromatischen Schnittweitendifferenz ist also nach (12.2) überhaupt kein Instrument mit Linsenoptik für einen größeren Wellenlängenbereich zu bauen.

Die „Achromatisierung“ gelingt bei einem dünnen Zwei-Linsen-System mit zwei verschiedenen Glassorten von differierendem ν. Allgemein gilt für das dünne System $\Phi=\varphi_1+\varphi_2$ (Φ Brechkraft des Systems; Indizes bezeichnen die Linsen). Das System ist achromatisch mit

$$\varphi_1=\frac{\nu_1}{\nu_1-\nu_2}\cdot\Phi,\quad \varphi_2=-\frac{\nu_2}{\nu_1-\nu_2}\cdot\Phi; \tag{12.3}$$

dann ist nämlich die Achromasiebedingung

$$\frac{\varphi_1}{\nu_1}=-\frac{\varphi_2}{\nu_2}\quad\left(\text{allgemein}\sum_m\frac{\varphi_m}{\nu_m}=0\right) \tag{12.4}$$

erfüllt. Die Schnittweiten — für beliebigen Dingabstand — sind für C- und F-Licht gleich groß. Da die Abbeschen Zahlen für beide Glieder positiv sind, muß notwendig eine der beiden Linsen sammelnd, die andere zerstreuend wirken. Durch die Achromasiebedingung ist nur die Brechkraft der Linsen festgelegt; über ihre *Form* ist nichts ausgesagt, sie bleibt für die Korrektion weiterer Abbildungsfehler frei.

Transversale Farbfehler existieren hier nicht, auch für nicht mit den Linsen zusammenfallende Blende (auf die Voraussetzung des *dünnen* Systems sei nochmals hingewiesen). Für ein dickes System kann dagegen laterale chromatische Aberration auch bei gehobener Schnittweitendifferenz auftreten, was man sich als Analogie zum Verstoß gegen die Sinusbedingung vorstellen kann (Abhängigkeit des Abbildungsmaßstabs von der Wellenlänge, durch Pupillenverschiebung beeinflußbar).

13. Chromatische Restfehler. Das Zusammenlegen der Bildorte für zwei Wellenlängen im Achromaten ergäbe völlige Freiheit vom Farbortsfehler, wenn der *Verlauf* der Dispersion — bei unterschiedlicher Größe — bei beiden Glasarten der gleiche wäre. Das ist jedoch nicht der Fall; der zweilinsige Achromat zeigt infolgedessen ein *„sekundäres Spektrum“*. Als Maßzahl für den Dispersionsverlauf dient die „relative Teildispersion“

$$\vartheta=\frac{\nu_{1,2}}{\nu_{2,3}}=\frac{n_3-n_2}{n_2-n_1}$$

einer Glasart (Indizes bezeichnen die Wellenlänge); für den visuellen Bereich wählt man

$$\vartheta=\frac{n_g-n_F}{n_F-n_C}\,.$$

Für die dritte Wellenlänge ist damit

$$\Delta\Phi=\sum_m\vartheta_m\frac{\varphi_m}{\nu_m}$$

und

$$-\Delta s_k'=s_k'^2\sum_{m=1}^{k}\vartheta_m\frac{\varphi_m}{\nu_m}\,,$$

wobei über die Linsen zu summieren ist. Für fernes Objekt folgt beim Zweilinser

$$\Delta f=\frac{\vartheta_1-\vartheta_2}{\nu_2-\nu_1}\cdot f.$$

Erwünscht wären also zwei Glasarten mit stark differierendem ν [damit nach (12.3) die Einzelstärken klein bleiben] bei gleichem ϑ. Für die meisten Gläser gilt aber angenähert die empirische Relation

$$\vartheta=A+B\nu$$

mit $A \approx 0{,}65$, $B \approx -0{,}0018$ für die Wellenlängen C, F, g. Gewöhnliche Achromate haben demnach eine Längsabweichung des sekundären Spektrums von etwa $2\,^0/_{00}$ der Brennweite bei 436 nm, wenn für 486 und 656 nm korrigiert ist, und zwar nahezu unabhängig von der Wahl der Glasarten.

Vom Standpunkt der *Wellenoptik* aus ist die longitudinale chromatische Aberration als eine Defokussierung nach Ziff. 8 anzusehen. Bei Abbildung mit heterochromatischem Licht geht die Wellenlängenabhängigkeit der Empfänger-Empfindlichkeit entscheidend in die Auswirkung auf das effektive Beugungsbild ein und führt zu einer merklichen Unbestimmtheit der Toleranzen.

Setzt man die halbe Längsaberration zwischen F und C für eine *einfache* Kron-Linse ($\nu = 60$) mit der Rayleigh-Toleranz gleich, so ergibt sich als Bedingung $N \lesssim 80\,D$; da die Augenempfindlichkeit bei diesen Wellenlängen gegenüber dem Maximum schon merklich abgefallen ist, kann man die Toleranz wohl verdoppeln und

$$N \gtrsim 40\,D \qquad (D \text{ in cm}) \tag{13.1}$$

zulassen.

Beim klassischen *Achromaten* liegt der Scheitel der Farbkurve (s. Fig. 37 auf S. 283) etwa $0{,}0005\,f$ innerhalb des Fokus für C und F, und die Bedingung für Einhalten des Beugungskriteriums wird

$$N \gtrsim D \qquad (D \text{ in cm}) \tag{13.2}$$

(der Zahlenfaktor wird meist als 1,1 angegeben).

Außer dem direkten Einfluß auf Bildort und Bildgröße hat die Wellenlängenabhängigkeit des Brechungsindex noch eine weitere Folge: die Maßnahmen, die zur Hebung der Seidel-Aberrationen ergriffen werden, sind nur für eine bestimmte Wellenlänge genau richtig; für andere ergeben sich chromatische Reste der geometrischen Abbildungsfehler. Bei den Objektiven astronomischer Refraktoren ist die *chromatische Differenz der sphärischen Aberration* (Gauß-Fehler) viel diskutiert (und zuweilen in ihrer Wirkung wohl überschätzt) worden (Fig. 38 auf S. 284). Die chromatischen Variationen werden — wegen der Kleinheit von $\delta n/n$ — erst merklich, wenn die Seidel-Approximation ohnehin nicht mehr ausreicht, und liegen außerhalb des Bereichs der hier angestrebten Übersicht.

c) Zusammenwirken der Einzelfehler.

14. Gesamt-Qualitätsbewertung. Die Abbildungsqualität im Fernrohr wird im allgemeinen nicht von einem einzelnen Fehler bestimmt, sondern durch das Zusammenwirken von Beugung, Abbildungsfehlern, Unvollkommenheiten der Ausführung sowie Momentanzustand des optischen Systems und der Erdatmosphäre. In das Beobachtungsresultat gehen auch die Eigenschaften des Strahlungsempfängers ein. Je nach der vorliegenden Meßaufgabe können recht verschiedene Anforderungen an die Abbildung gestellt werden. Es ist deshalb nicht zu erwarten, daß sich ein Teleskop hinsichtlich seiner optischen Qualität durch eine universelle Gütezahl[1] kennzeichnen läßt.

Man kann die Abbildung vollständig beschreiben durch das „Lichtgebirge", die zweidimensionale Verteilung der Bestrahlungsstärke im Bild eines fernen

[1] Zur Einführung in Probleme der Bildgüteparameter und als Übersicht der neueren Entwicklung können dienen: Proc. NBS Semicentennial Symposium on Optical Image Evaluation (Nat. Bur. Stand., Circ. 526). Washington 1954. — Beiträge verschiedener Autoren in J. Opt. Soc. Amer. **46**, No. 9 (1956); Appl. Optics **2**, No. 4 (1963) und **3**, No. 1 (1964). — E. H. LINFOOT: Qualitätsbewertung optischer Bilder. Braunschweig 1960. — N. S. KAPANY and J. J. BURKE: J. Opt. Soc. Amer. **52**, 1351 (1962). Vgl. hierzu auch den einleitenden Beitrag zu diesem Bande von W. T. WELFORD.

Punktes, der ja astronomisch ein höchst reales Objekt ist. Die Verteilung ändert sich im allgemeinen mit der Hauptstrahlneigung w, dem Öffnungsverhältnis $1:N$ und der Fokuseinstellung $\Delta x'$. Es ist gerade der Vorteil der Seidel-Idealisierung, daß sie für die geometrisch-optische Approximation diese Abhängigkeiten leicht überschaubar erfaßt. Ist das Objekt nicht punktförmig, so muß (im hier stets nur betrachteten inkohärenten Fall) seine Strahlungsdichte-Verteilung mit der Punktbild-Verteilung gefaltet werden, um die Bildverteilung zu ergeben.

Eine zweite Beschreibungsmöglichkeit liefert die *Pupillenfunktion*, die Phase und Amplitude der Lichtschwingung in der Austrittspupille angibt. Sie enthält außer der Wellenaberration $V(Y,Z)$ noch die Amplitudenverteilung, die bei der Betrachtung von Fernrohroptik gewöhnlich vernachlässigt wird, aber z.B. für Apodisationsprobleme entscheidend wichtig ist. Aus der Beugungstheorie der Abbildung ist bekannt, daß die Amplitudenverteilung in der Bildfläche die Fouriertransformierte der Pupillenfunktion ist; ihr quadrierter Betrag ergibt die Punktbild-Verteilung.

Die Abbildung ist schließlich drittens zu charakterisieren durch die *optische Übertragungsfunktion*[1], die die Kontrastverringerung und Phasenverschiebung im Bild von Cosinusgittern beschreibt, abhängig von deren Ortsfrequenz $\frac{1}{y}$, $\frac{1}{z}$. Als Fouriertransformierte der Punktbild-Verteilung hat sie den Vorteil, daß einer *Faltung* mehrerer unabhängiger Verteilungen zur Erzeugung der resultierenden Gesamt-Verteilung eine *Multiplikation* der entsprechenden Übertragungsfunktionen entspricht. Die optische Übertragungsfunktion ist zugleich die autokorrelierte Pupillenfunktion. Die drei genannten Funktionen hängen in mathematisch einfacher Weise miteinander zusammen und liefern die gleiche Information über den Abbildungsprozeß. Obwohl die Diskussion der Übertragungsfunktion — für die Praxis meist beschränkt auf ihren Betrag, die Modulations-Übertragungsfunktion — in der gegenwärtigen optischen Literatur einen breiten Raum einnimmt, sind Anwendungen auf die Fernrohr-Abbildungen noch selten[2]. Man beschreibt diese aber häufig durch die Punktbild-Verteilung oder die Wellenaberration.

Für das Punktbild kann das unter Voraussetzung der Kreissymmetrie ganz grob durch die Angabe eines geeignet definierten mittleren Radius geschehen, z.B. der Technischen Konstante nach LEHMANN (s. Ziff. 31). Brauchbarer sind Werte für die *Lichtkonzentration*, den Bruchteil vom Gesamtstrahlungsfluß, der durch zum Bild zentrierte Lochblenden von verschiedenem Durchmesser tritt. Eine detailliertere Beschreibung, die auch unsymmetrische Verteilungen erfaßt, ist im Bereich der geometrischen Optik mit dem *Durchstoßdiagramm* möglich (Fig. 21). Es gibt graphisch die Durchstoßpunkte einer großen Zahl von Strahlen durch eine ausgewählte Empfängerfläche wieder. Damit die Flächendichte der Punkte ein Maß für die Bestrahlungsstärke im Bild darstellt, muß jeder Strahl den gleichen Flächenanteil der Eintrittspupille repräsentieren. Das Diagramm kann aus den Systemdaten berechnet werden (HERZBERGERs spot diagram[3]). Experimentell läßt es sich mit den verallgemeinerten Hartmann-Methoden (s. Ziff. 31) gewinnen; die Schwierigkeiten der direkten Photometrie des Lichtgebirges werden so umgangen.

---

[1] Einen Überblick gibt H. H. HOPKINS in K. J. HABELL (Hrsg.), Proc. Conf. Opt. Instr. Techn. London 1961, p. 480. London 1962.

[2] R. M. SCOTT, and R. V. SHACK: Predictions of Performance by Modulation Transfer Technique, Preprints 5. Conf. ICO Stockholm 1959.

[3] M. HERZBERGER: J. Opt. Soc. Amer. **37**, 485 (1947). — D. G. HAWKINS and E. H. LINFOOT: Monthly Notices Roy. Astron. Soc. **105**, 334 (1945).

Von der Pupillenfunktion interessiert hauptsächlich der Phasenteil, die Wellenaberration. Die Angabe der Maximalaberration ist nur sinnvoll bei kleiner Aberration oder regelmäßigem Verlauf. Zu einer Beschreibung etwa in der Form „Genauigkeit $\lambda/10$" muß stets gesagt werden, ob damit der Maximal- oder der rms-Wert der Wellenaberration V oder des Flächenfehlers Δ gemeint ist; ein Spiegel mit $\Delta_{rms}=\lambda/10$ kann u. U. Maximalaberrationen bis $V_{max}=1\,\lambda$ haben! Die mittlere Neigung von Wellenflächen-Elementen entspricht geometrisch der mittleren Querabweichung im Punktbild. Neigungen als Funktionen der Pupillenkoordinaten können mit dem Foucaultschen Schneidenverfahren (s. Ziff. 32) qualitativ sichtbar gemacht oder aus Hartmann-Daten berechnet werden. Interferometrische Prüfverfahren ergeben unmittelbar eine Höhenlinienkarte der Wellenfläche, wie sie aus den Neigungen durch Integration ebenfalls zu erhalten ist.

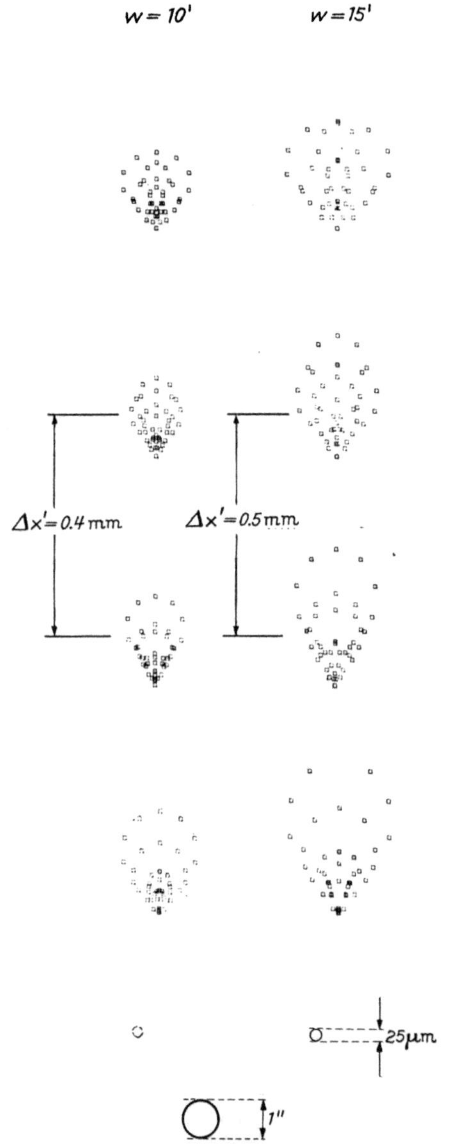

Fig. 21. Durchstoßdiagramm für ein Cassegrain-Teleskop mit $f=1620$ cm, $N=7{,}5$, $N_{prim}=2{,}6$. Für jede der beiden Hauptstrahlneigungen ($w=10'$, $w=15'$) ist die Durchstoßfigur für vier äquidistante Empfängerflächen angegeben. Zusätzlich eingetragen sind Kreise mit 25 μm und 1'' Durchmesser. Nach SCHULTE[1].

Nur wenn die kombinierten Abbildungsfehler so klein sind, daß die Form des Beugungskerns im wesentlichen erhalten bleibt, ist die Qualitätsbewertung mit der einen Maßzahl der Strehlschen Definitionshelligkeit Z sinnvoll. Nach VÄISÄLÄ [*26*][2] gilt in ausreichender Näherung

$$Z = 1 - \frac{4\pi^2}{\lambda^2}\,\overline{V^2},$$

wobei das mittlere Abweichungsquadrat $\overline{V^2}$ optimal auf die Referenzkugel zu beziehen ist, für die $\overline{V}=0$. Im Falle hinreichend schwacher Aberration (dazu muß V_{rms} merklich kleiner sein als $\lambda/2\pi$) ist demnach die Bildqualität von der Form der Wellenfläche unabhängig und nur durch das mittlere Schwankungsquadrat der Wellenaberration bestimmt. Für $Z \geqq 80\%$ folgt $V_{rms} \leqq \lambda/14$; für „vernünftige" Verläufe von V setzt die Rayleigh-Bedingung $V_{max} \leqq \lambda/4$ (jetzt bei einseitig berührender Referenzkugel) eine ähnliche Grenze.

Für größere, statistisch unregelmäßige Störungen der Wellenfläche läßt sich die Abbildung mit zwei Parametern charakterisieren, wenn man die Form des

[1] D. H. SCHULTE: Appl. Optics **2**, 141 (1963).
[2] Siehe auch A. MARÉCHAL, Rev. d'Opt. **26**, 257 (1947).

Störungs-*Spektrums*, also eine Beziehung zwischen Amplitude und „Wellenlänge" der Störungskomponenten *vorgibt*. Neben das Schwankungsquadrat tritt nun eine charakteristische Länge, die etwa beschreibt, unterhalb welcher Grenze die Komponenten der Welligkeit keine merkliche Amplitude mehr haben. Je größer diese Amplituden insgesamt werden, desto mehr tritt im Punktbild die von der Gesamtapertur herrührende Beugungsfigur zurück gegen einen Halo, dessen Abmessungen von der vorherrschenden „Wellenlänge" der Störung bestimmt sind[1].

B. Optische Systeme für Fernrohre.

I. Anordnungen mit Reflexionsoptik.

a) Klassische Reflektoren.

Bei den klassischen Reflektoren werden die aus der Geometrie bekannten Fokaleigenschaften der Kegelschnitte ausgenutzt, um die sphärische Aberration zu beseitigen und damit wenigstens für die Achse stigmatische Abbildung ferner Objekte zu erreichen. Üblich sind die Anordnungen Newton- bzw. Primärfokus, Cassegrain- und Coudé-Fokus, die oft untereinander austauschbar in einem Instrument vereinigt sind, wobei der gleiche Hauptspiegel verwendet wird.

15. Primär- und Newton-Fokus. Die einfachste überhaupt mögliche Anordnung des Strahlungsempfängers ist die im Primärfokus eines Rotationsparaboloids. Bei kleineren und mittelgroßen Instrumenten (bis $D \approx 200$ cm) wird statt dessen meist die Newton-Anordnung verwendet; ihr 45°-Planspiegel, der die Fokalfläche etwa an die besser zugängliche Rohrwand verlegt, hat dann einen kleineren Durchmesser als der Strahlungsempfänger selbst, und die Mittenabschattung bleibt geringer.

Das Öffnungsverhältnis liegt bei größeren Instrumenten allgemein zwischen 1:3 und 1:5, wobei neuerdings die kleineren Öffnungszahlen bevorzugt werden, weil das kurze Rohr stabiler ist und einen kleineren Kuppeldurchmesser erfordert. Für $N \lesssim 3$ wird allerdings die Herstellung der optischen Fläche schwierig. Die Weiterleitung des weitgeöffneten Strahlenbündels in Spektrographen und Photometern ist nicht immer bequem; wenn zur Anpassung des Öffnungsverhältnisses zusätzliche optische Glieder eingeführt werden müssen, geht der Vorteil des direkten Fokus wieder verloren. Für Photographie ist die Fokustoleranz nach (8.1) ziemlich eng.

Der alles beherrschende Bildfehler ist die außeraxiale *Koma* (Fig. 22). Es gilt

$$\boldsymbol{K} = \frac{3}{16} \cdot \frac{1}{N^2} y'; \qquad \mathcal{K} = \frac{3}{16} \cdot \frac{1}{N^2} w, \tag{15.1}$$

wobei $\boldsymbol{K}$ und y' bzw. $\mathcal{K}$ und w jeweils in den gleichen Einheiten auszudrücken sind. Läßt man eine Queraberration von 1″ zu, so beträgt der brauchbare Bildfelddurchmesser

$$2w(\text{Koma } 1'') \approx 11'' \cdot N^2,$$

bei $N = 3{,}3$ also 2 Bogenminuten. Fordert man $\boldsymbol{K} \leqq 40$ µm (angepaßt an die Auflösung höchstempfindlicher Platten), so wird der Radius des brauchbaren Bildfeldes wenig über 2 mm beim gleichen Öffnungsverhältnis, unabhängig von der Brennweite. Das Beugungskriterium (10.1) ergibt

$$\tan w \lesssim 10^{-3} \frac{N^2}{D} \qquad (D \text{ in cm})$$

[1] H. Scheffler: Z. Astrophys. **55**, 1 (1962); **58**, 170 (1964).

Fig. 22. Ausschnitt aus einer Himmelsaufnahme im Primärfokus eines Parabolspiegels (Öffnungsverhältnis 1:4). Am oberen Bildrand ist $w = 1°,7$.

oder

$$y' \lesssim 10^{-3}\,\text{cm} \cdot N^3.$$

Wird bei größerem Achsabstand auch die Bildfeldwölbung merklich, so verzerrt sich der doppelt umlaufene Koma-Kreis einer Zone (s. Fig. 15 auf S. 249) zu einer zweiblätterigen Figur[1] (Fig. 23); beim Zusammenwirken aller Zonen treten die vier Tangenten deutlich hervor (Fig. 22 am oberen Ende).

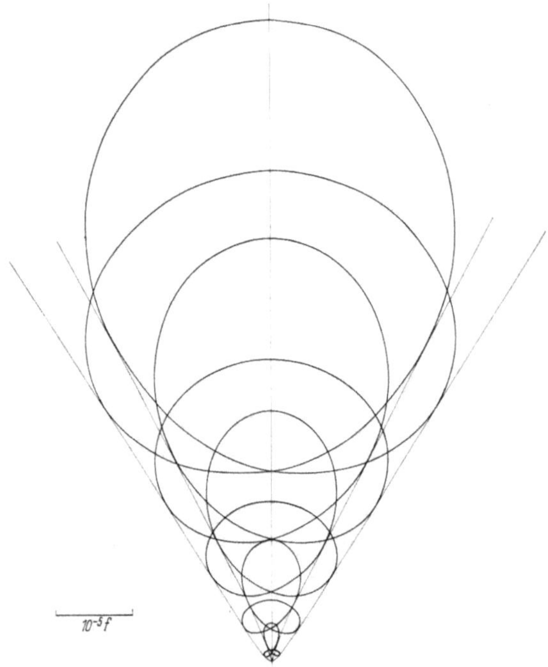

Fig. 23. Zerstreuungsfigur für einen fernen Punkt bei Abbildung durch ein Rotationsparaboloid; $\tan w = 0{,}01$ ($w = 34',5$). Zonen mit $h/f = 0{,}02$, 0,04, 0,06, 0,08 und 0,10 (entspr. $N = 5$). Nach Thiersch.

Der Planspiegel eines Newton-Systems darf keine merkliche Krümmung haben, weil sonst (wegen der Schrägstellung des Spiegels) Astigmatismus auf der Achse eingeführt wird vom Betrag

$$A = \frac{0{,}7\,L^2}{2\,N\,R},$$

wobei R der Krümmungsradius des „Plan"-Spiegels, L der Abstand Planspiegel—Fokus ist.

16. Cassegrain- und Coudé-Fokus. Werden größerer Abbildungsmaßstab, schlankeres Lichtbündel und besser zugängliche Fokalfläche gewünscht, so läßt sich der Parabolspiegel zum Cassegrain-System erweitern (Fig. 24). Durch

[1] F. Thiersch: Die Reflexion eines Parallelstrahlenbündels am Paraboloid. Diss. T. H. München, Halle 1914.

den konvexen Sekundärspiegel vor dem Primärfokus wird die Systembrennweite vergrößert, ohne daß die Baulänge wächst (Prinzip des Teleobjektivs). Der Hauptspiegel muß in der ohnehin abgeschatteten Mitte durchbohrt werden, wenn man nicht die Nasmyth-Anordnung mit einem zusätzlichen 45°-Planspiegel verwenden will. Die praktisch angewandten Öffnungsverhältnisse liegen etwa zwischen 1:10 und 1:20.

Im Bereich der Gauß-Optik gelten [vgl. Gl. (2.1)] folgende Beziehungen (mit Brennweite des Primärspiegels f_1, des Sekundärspiegels $f_2 < 0$, Systembrennweite f, $\overline{F_1' S_2} = c$, $\overline{S_2 F'} = a$, $\overline{S_1 S_2} = e$, $\overline{S_1 F'} = a - e = g > 0$ für Bild hinter dem Haupt-

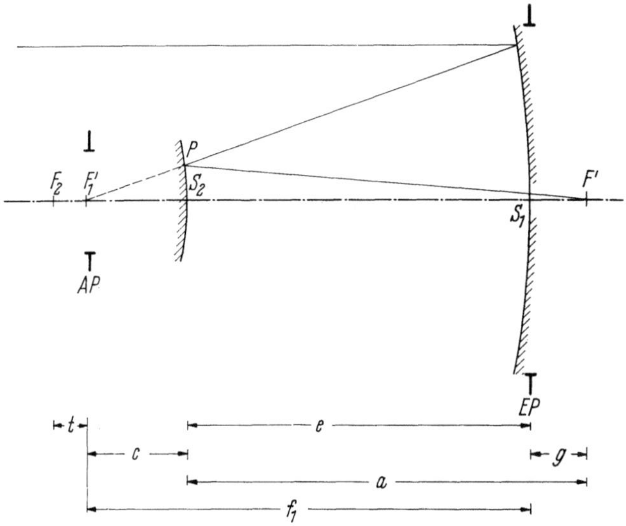

Fig. 24. Geometrie des Cassegrain-Systems.

spiegel, Sekundärvergrößerung $m = a/c$): Bei bekanntem f_1, f_2, e ist

$$f = m f_1 = \frac{f_1 f_2}{f_1 + f_2 - e}; \qquad a = \frac{(f_1 - e) f_2}{f_1 + f_2 - e}.$$

Zuweilen sind die Beziehungen

$$f = a + m e; \qquad f_1 = \frac{a}{m} + e; \qquad f_2 = -\frac{a}{m - 1}$$

nützlich. Soll ein vorhandener Primärspiegel zum Cassegrain-System ergänzt werden (f_1, m und g vorgegeben), so folgt

$$e = \frac{m f_1 - g}{m + 1}; \qquad f_2 = -\frac{m}{m^2 - 1} (f_1 + g).$$

Ist der Hauptspiegel die Eintrittspupille, so liegt die Austrittspupille um

$$\Delta_p = c \cdot \left(1 - \frac{g}{f - e}\right)$$

hinter dem Sekundärspiegel-Scheitel; sie hat den Durchmesser $\frac{a}{f - e} \cdot D_1$.

Variationen der Systemdaten führen zu einer Bildverschiebung

$$d a = m^2 \left[d f_1 + \frac{(m - 1)^2}{m^2} d f_2 - d e\right].$$

Wird zum Fokussieren der Scheitelabstand e geändert, so gilt (für festes f_1, f_2)

$$dg = -(m^2+1)\,de; \quad df = \frac{f}{f_1+f_2-e}\,de; \quad dm = -\frac{m^2(m-1)}{a}\,de, \tag{16.1}$$

d.h. Sekundärspiegel und Bild wandern gleichsinnig, und beim Verkleinern des Scheitelabstandes wächst die Systembrennweite. Setzt man $f_1+f_2-e=-t$, wobei $t=\frac{a}{m\cdot(m-1)}>0$ der Abstand der Brennpunkte von Primär- und Sekundärspiegel ist, so kann man die Änderung der Systembrennweite ausdrücken durch

$$\frac{df}{f} = -\frac{dt}{t}.$$

Der minimale Durchmesser des Sekundärspiegels (nur die Achse frei von Vignettierung) ist

$$D_2 = \frac{a}{f}\,D_1.$$

Liegt das Bild im Scheitel des Primärspiegels ($g=0$), muß der Sekundärspiegel für einen vignettierungsfreien Bildfeldwinkel w den Durchmesser

$$D_2 = \frac{1}{m+1}\,(D_1+2wf) \tag{16.2}$$

haben; allgemein ist

$$D_2 = \frac{a}{f}\cdot D_1 + 2we.$$

Der Lichtverlust durch diese Mittenabschattung ist nur von der Größenordnung 10%; bei größeren Teleskopen dürfte auch die Veränderung der Beugungsfigur (s. Ziff. 6) kaum zu einer Bildverschlechterung führen.

Eine Eigenheit des Cassegrain-Systems ist die sog. *Tagblindheit*: Licht von Strahlungsquellen außerhalb des Bildfeldes kann, am Sekundärspiegel vorbei einfallend, den Strahlungsempfänger direkt erreichen. Besonders störend ist diese Feldaufhellung bei Beobachtung ausgedehnter Objekte (Mond) oder gegen hellen Hintergrund; der Bildkontrast wird dabei u.U. merklich verringert. Bei weiterführenden Strahlengängen (z.B. auch Okularbeobachtung) kann man sich mit einer Blende nahe der Austrittspupille helfen, die das Bild des Sekundärspiegels genau umschließt. Andernfalls ist die Fremdlichtbeseitigung mit besonderen Blenden im Fernrohr (am zweckmäßigsten schwach konischen Rohren parallel zur optischen Achse) möglich, wobei aber für größere Felder entweder Vignettierung oder künstliche Vergrößerung der Mittenabschattung zugelassen werden müssen.

Die bisherigen Betrachtungen gelten für alle nach dem Cassegrain-Grundtyp aufgebauten Zwei-Spiegel-Anordnungen, also auch für Ritchey-Chrétien- oder Baker-Schmidt-Systeme. Das Cassegrain-System im strengen Sinne hat parabolischen Primärspiegel; will man die Abbildung eines axialen Bündels mit nichtverschwindendem Aperturwinkel stigmatisch machen, so muß auch der Sekundärspiegel deformiert werden. Für die Abbildung nach F_1' (Fig. 24) ist die sphärische Aberration gehoben; das gilt auch für die Abbildung nach F', wenn $PF'-PF_1'=$ const unabhängig von der Einfallshöhe ist (Prinzip des konstanten Lichtwegs). Die gesuchte Meridiankurve ist also eine Hyperbel mit den (geometrischen!) Brennpunkten F' und F_1'. Die numerische Exzentrizität ist $\frac{a+c}{a-c}=\varepsilon=\frac{m+1}{m-1}$, die Deformationskonstante

$$b_{2,c} = -\left(\frac{m+1}{m-1}\right)^2.$$

Die *Fokustoleranz* in der Bildfläche ist entsprechend dem kleineren Öffnungsverhältnis gegenüber dem Primärfokus um m^2 vergrößert, doch gilt dessen strengere Toleranz wegen (16.1) unverändert für den Abstand des Sekundärspiegelscheitels e. Beim Einstellen einer anderen Bildlage und Systembrennweite durch Variation von e nach Gl. (16.1) tritt sphärische Aberration auf vom Betrag

$$\sum \mathbf{I} = -\frac{1}{f} m^2(m^2-1)\,de;$$

die Strehl-Toleranz beträgt somit

$$|de| \leqq \frac{128 N^4}{m^2(m^2-1)}\lambda; \qquad |dg| \lesssim 0{,}008\ \mathrm{cm}\cdot N^2 N^2_{\mathrm{prim}}.$$

Die außeraxiale *Koma* des Systems ist ebenso groß wie die eines entsprechenden Kugel- oder Parabolspiegels ($\sum \mathbf{II} = -0{,}5$); die Gln. (15.1) bleiben also gültig (wobei $N = f/D$ jetzt die Öffnungszahl des Systems ist). Der *Astigmatismus* wird dagegen vergrößert: liegt die Bildfläche im Scheitel des Primärspiegels (was in grober Näherung meist erfüllt ist), so ist der Astigmatismus das m-fache von dem des äquivalenten Parabolspiegels. Allgemein gilt

$$\sum \mathbf{III\,a} = \frac{f-e}{a} = \frac{mf+g}{f+mg}.$$

Die bei einem Zwei-Spiegel-System mögliche *Dezentrierung*[1] bewirkt weitere Abbildungsfehler. Im für Fernrohre wichtigsten Fall der Parallelversetzung des Sekundärspiegels senkrecht zur optischen Achse um den kleinen Betrag δ erfolgt zunächst eine Verschiebung des Bildes um

$$\Delta y' = -(m-1)\,\delta,$$

d.h. nach der entgegengesetzten Seite. Weiter tritt im ganzen Bildfeld gleichförmige *Koma* auf vom Betrag (für das Cassegrain-System)

$$\boldsymbol{K}_\delta = \frac{3}{16}\,\frac{m(m^2-1)}{N^2}\,\delta,$$

zusätzlich zur außeraxialen Koma. Das Strehl-Kriterium (Wellenaberration $V \leqq 0{,}6\,\lambda$) führt für $\lambda = 500$ nm auf eine Toleranzgrenze von

$$\delta \lesssim 10^{-3}\ \mathrm{cm}\cdot N^3_{\mathrm{prim}},$$

also von einigen Zehntelmillimetern für reale Instrumente. Die Toleranzen für die Sekundärspiegel-Position parallel und senkrecht zur Achse sind damit etwa ebenso groß wie die Toleranzen für Fokus und seitlichen Achsabstand (wegen außeraxialer Koma) bei Abbildung durch den Hauptspiegel allein.

Die Dezentrierungs-Koma eines Zwei-Spiegel-Systems mit beliebiger Deformation der Flächen (bei stigmatischer Abbildung in der Achse) kann man ausdrücken durch

$$\boldsymbol{K}_\delta = \frac{3}{32}\,\frac{(m-1)^2}{N^2}\,[2+(1-b_2)(m-1)]\,\delta \tag{16.3}$$

für das auf den unverrückten Primärspiegel axial einfallende Bündel.

Die kaum verwendete *Gregory*-Bauweise erzielt die Brennweiten-Verlängerung durch einen konkaven Sekundärspiegel jenseits des Primärfokus. Die Meridian-

[1] A. E. Conrady: Monthly Notices Roy. Astron. Soc. **79**, 384 (1919). — A. Maréchal: Rev. d'Opt. **29**, 1 (1950). — H. Slevogt: Optik **20**, 488 (1963); s. auch A. Baranne, J. Observateurs **49**, 75 (1966).

kurve für Hebung der sphärischen Aberration ist hier eine Ellipse in der Umgebung des Hauptscheitels. Die Baulänge ist gegenüber dem Cassegrain-System vergrößert, das Bild aufgerichtet.

Die vielbenutzte *Coudé-Anordnung* ist optisch ein Cassegrain-System, mit N meist im Bereich zwischen 30 und 40. Durch Planspiegel wird das Strahlenbündel nach Reflexion am Sekundärspiegel so abgelenkt, daß seine Achse mit der Stundenachse des Fernrohrs zusammenfällt. Damit ist eine raumfeste, vom bewegten Instrument unabhängige Lage der Bildfläche gegeben, innerhalb deren das Bild mit dem Fernrohr um die Achse rotiert. Für punktförmige Objekte in der Achse ist so die Verwendung fest aufgestellter Strahlungsempfänger ohne weiteres möglich; für ausgedehnte Objekte kann man die Bildrotation aufheben (z.B. durch ein mit der halben Geschwindigkeit der Stundenachse angetriebenes Dove-Prisma). Die möglichen Anordnungen der Planspiegel hängen vom Montierungstyp ab und werden im Zusammenhang mit diesem in Abschnitt D II (S. 322) aufgeführt.

b) Weiterentwicklung der Reflektoren.

17. Quasi-Cassegrain-Systeme. Die sphärische Aberration eines Zwei-Spiegel-Systems kann auch gehoben werden, ohne daß die Spiegel *einzeln* aberrationsfrei sind; Korrektion für die Achse ist bei beliebig vorgegebenem b_1 *oder* b_2 stets möglich. Ausgehend vom Cassegrain-System mit $b_{1,c} = -1$, $b_{2,c} = -\left(\frac{m+1}{m-1}\right)^2$ führt die Bedingung gleicher Änderung des optischen Weges auf

$$\frac{\Delta b_1}{\Delta b_2} = \frac{a}{f}\left(\frac{m-1}{m}\right)^3, \qquad (17.1)$$

wobei

$$b_1 = b_{1,c} + \Delta b_1; \qquad b_2 = b_{2,c} + \Delta b_2.$$

Der gleiche Ausdruck folgt aus der ersten der Gln. (4.7) mit $\mathbf{I}_0^* = 0$.

Anordnungen, die sich dieser Freiheit der Deformation bedienen, werden als Quasi-Cassegrain-Systeme bezeichnet. Während bei den klassischen Spiegelsystemen die Meridiankurven exakte Kegelschnitte sind, deren Fokaleigenschaften unmittelbar ausgenutzt werden, treffen hier die Bezeichnungen Ellipsoid oder Hyperboloid gemäß Ziff. 4α nur in der Näherung dritter Ordnung zu; über die Größe der Glieder höherer Ordnung ist damit nichts ausgesagt.

Von einiger praktischer Bedeutung ist der Fall

$$b_1 = 0; \qquad b_2 = \frac{f}{a}\left(\frac{m}{m-1}\right)^3 - \left(\frac{m+1}{m-1}\right)^2;$$

damit kann der undeformierte Hauptspiegel einer Schmidt-Kamera als Lichtsammler für Spektrographen oder Photometer nutzbar gemacht werden. Der Sekundärspiegel ist ein abgeplattetes Rotationsellipsoid.

Die Prüfung der Spiegel soll erleichtert werden mit den Deformationen

$$b_2 = 0; \qquad b_1 = -1 + \frac{a}{f}\,\frac{(m^2-1)(m+1)}{m^3}$$

(Dall-Kirkham-System)[1]. Der Hauptspiegel ist ein Ellipsoid, das bei gängiger Auslegung zu etwa $^3/_4$ parabolisiert ist. Die aus (4.6) folgende Koma beider Anordnungen ist deutlich größer als die des echten Cassegrain[2].

[1] A. R. KIRKHAM: Sci. Amer. **185**, 118 (1951).

[2] R. T. JONES: J. Opt. Soc. Amer. **44**, 630 (1954). — R. R. WILLEY: Sky and Telescope **23**, 191 (1962).

18. Aplanatische Zwei-Spiegel-Systeme. In einer wichtigen Arbeit hat SCHWARZSCHILD [*13*] Ausdrücke für die Seidel-Aberrationen des Zwei-Spiegel-Systems mit beliebiger Deformation der Flächen und daraus die Bedingungen für ihr Verschwinden abgeleitet. Auf diesen Schwarzschild-Formeln beruhen die Gln. (4.6). Die einzige Anordnung, bei der alle vier wesentlichen Fehler dritter Ordnung gehoben sind, hat einen zerstreuenden Primärspiegel mit $D_1 < D_2$ und ist als Fernrohr-Objektiv unbrauchbar. Das sog. „Schwarzschild-System" läßt Restastigmatismus zu, hat aber Feldebnung im übertragenen Sinne. Es ist ein Cassegrain-Typ mit *sammelndem* Sekundärspiegel, also $0 < m < 1$, und Bildfläche zwischen den Spiegeln. Nachteile sind die große Baulänge ($\sim 1{,}7f$), die starke Deformation des Hauptspiegels und die kräftige Abschattung durch den Sekundärspiegel. Auf einer *gekrümmten* Bildfläche läßt sich der Astigmatismus beseitigen, wenn $e = 2f$ (Schwarzschild-Anastigmat [1]). Die ungünstigen Eigenschaften werden dabei allerdings noch verstärkt. Beide Typen haben keine praktische Bedeutung erlangt; das Schmidt-System bietet eine elegantere (wenn auch prinzipiell verwandte) Lösung für einen Anastigmaten mit großem Öffnungsverhältnis und großem Feld.

Verlangt man nur Aplanasie, also $\sum \mathbf{I} = \sum \mathbf{II} = 0$, so ergibt sich aus SCHWARZSCHILDs Formeln mit $m > 1$ der Ritchey-Chrétien-Typ [*27*] [2], der neuerdings auf großes Interesse stößt; im Bereich der Gauß-Optik ist er identisch mit dem Cassegrain-System. Weitere Darlegungen zu den aplanatischen Spiegelsystemen [3] finden sich in [*11*], [*12*], [*28*]. Unter den unendlich vielen nach (17.1) möglichen Paaren b_1, b_2 ist gerade eins, bei dem neben der sphärischen Aberration auch die Koma beseitigt ist. Aus (4.7) kann man entnehmen, daß der Sekundärspiegel dazu die Deformation

$$b_2 = b_{2,c} - \frac{2f}{e(m-1)^3} \tag{18.1}$$

bekommen muß ($b_{2,c}$ ist wieder die Deformation des echten Cassegrain nach Ziff. 16). b_2 wächst also für m nahe 1 stark an. Die Deformation des Hauptspiegels b_1 hat, weil in der Aperturblende liegend, auf die Koma des Systems keinen Einfluß. Die sphärische Aberration wird gehoben, wenn für den Primärspiegel nach (17.1)

$$b_1 = -1 - \frac{2a}{e m^3} = -1 - 2 \cdot \frac{f - em}{e m^3}. \tag{18.2}$$

Der Hauptspiegel ist also mäßig, der Hilfsspiegel stark hyperbolisch deformiert. Der Primärfokus des Ritchey-Chrétien-Hauptspiegels allein hat sphärische Aberration (Überkorrektion) von

$$\mathcal{S} = \frac{1}{64} \cdot \frac{b_1 + 1}{N^3}$$

im Paraxialfokus und ein Viertel davon im Optimalfokus. Für die Zwei-Spiegel-Aplanate hat SCHWARZSCHILD auch die Ansätze für die Meridiankurven bis zu beliebig hohen Ordnungen mitgeteilt; sie folgen aus der Forderung nach stigmatischer Abbildung eines fernen Achsenpunktes, d.h. nach gleicher optischer Weglänge für das axiale Bündel, und der Sinusbedingung.

Wie man der Tabelle 1 auf S. 236 entnehmen kann, sind Astigmatismus und Bildwölbung des Ritchey-Chrétien-Systems merklich größer als beim äquivalenten Parabolspiegel. Da aber die Koma beseitigt ist, werden die Bilder von außeraxialen

---

[1] A. COUDER: C. R. Acad. Sci., Paris **183**, 1276 (1926 II).

[2] G. W. RITCHEY: Trans. Opt. Soc., Lond. **29**, 197 (1927); s. auch A. BARANNE, J. Observateurs **49**, 75 (1966).

[3] H. THEISSING u. O. ZINKE: Optik **3**, 451 (1948).

Punkten *symmetrisch* vergrößert, wenn als Bildfläche die Schale der mittleren Bildkrümmung genommen wird; sie hat den Radius

$$R_m = -\frac{af}{(m^2-1)e+f}$$

(hohl gegen die Lichtrichtung). Die astigmatische Querabweichung ist

$$\mathcal{A} = \frac{1}{8} \cdot \frac{2f-e}{a} \cdot \frac{1}{N} \tan^2 w;$$

ihr Anwachsen begrenzt das nutzbare Feld. Den Seidel-Koeffizienten des Astigmatismus kann man in der wegen allgemein $g \ll f$ handlichen Form

$$\sum \mathbf{IIIa} = \frac{2f-e}{2a} = \frac{(m+\frac{1}{2})f+\frac{1}{2}g}{f+mg} \tag{18.3}$$

schreiben.

Infolge der stärkeren Deformation des Sekundärspiegels ist die Zentrierempfindlichkeit des Ritchey-Chrétien-Typs gegenüber dem äquivalenten Cassegrain-System etwas vergrößert [nach Gl. (16.3)]; Dezentrierung wirkt sich jedoch wegen des allgemein kleineren N stärker aus als bei einem üblichen Cassegrain-Teleskop.

Bei der Auslegung eines Ritchey-Chrétien-Systems[1] sind Kompromisse nötig. Der Forderung nach kurzer Baulänge und damit kleinem N_{prim} steht entgegen, daß für den Astigmatismus genähert $\sum \mathbf{IIIa} \sim m$ und damit $\mathcal{A} \sim \frac{1}{N_{\text{prim}}} \cdot w^2$ gilt. Die Dezentrierungs-Toleranz ist wesentlich von N_{prim} bestimmt; die Herstellung großer Primärspiegel in erstklassiger optischer Qualität ist derzeit auf den Bereich

Tabelle 3. *Größere Ritchey-Chrétien-Systeme.*

	D_1 cm	f cm	N	N_{prim}	m	Feld	D_2/D_1	Jahr
U.S. Naval	102	691	6,8	4,0	1,7	bis 1°,5 ⌀	0,4	1934
Siding Spring (zu Canberra) . .	102	820	8,0	3,5	2,3	1° ⌀		1964
Kitt Peak 84″ . .	213	1620	7,6	2,6	2,9	42′×52′		1964
Herstmonceux . .	249	3490	14	3,0	4,6	20′ ⌀	0,21	1967
Kitt Peak 150″ . .	380	3420	9	2,8	3,2	36′ ⌀	(0,33)	geplant
ESO 3,5 m	350	2890	8,3	3,0	2,8	30′ ⌀	(0,33)	geplant
Cerro Tololo . . .	152	1140	7,5	3,5	2,1	1°,5 ⌀*	(0,41)	1967

* Mit Korrektor.

$N_{\text{prim}} \gtrsim 2{,}5 \ldots 3$ beschränkt. Für die Abschattung durch den Sekundärspiegel und vor allem dessen Deformation ist großes m günstig. Die System-Öffnungszahl sollte nicht über $N \approx 9$ wachsen, damit optimale Belichtung noch in vernünftigen Zeiten erreichbar ist. Bei zu langer Brennweite begrenzt schließlich das handhabbare Plattenformat das Feld. Eine Zusammenstellung von Systemdaten ausgeführter oder geplanter Instrumente gibt Tabelle 3. Den von SCHULTE[2] mitgeteilten Durchstoßdiagrammen kann man entnehmen, daß die Seidel-Näherung für ein System mit $N_{\text{prim}} = 2{,}6$, $N = 7{,}5$ die Bildfehler noch vernünftig darstellt; im besten Fokus erreicht der Durchmesser des Zerstreuungsscheibchens 0″,7 am Rand eines Feldes von 30′ Durchmesser (Fig. 25).

[1] Siehe auch J. LANDI-DESSY, Mem. Soc. Astronom. Ital. **34**, 411, 415 (1963).

[2] D. H. SCHULTE: Appl. Optics **2**, 141 (1963).

Mit zusätzlichen optischen Gliedern läßt sich das Ritchey-Chrétien-System noch verbessern. Der Astigmatismus ist mit einer dünnen asphärischen *Korrektionsplatte* (vgl. Ziff. 20) kurz vor dem Fokus zu beseitigen. Dadurch wird allerdings die Aplanasie geringfügig gestört; stimmt man die drei Deformationen nach (4.7) optimal aufeinander ab, kann man volle anastigmatische Korrektion erzielen[1]. Das System entspricht nun einem Baker-Schmidt-Teleskop (vgl. Ziff. 21β) mit virtueller Eintrittspupille, deren reelles Bild kurz vor dem Fokus liegt. Fügt man noch eine Ebnungslinse (vgl. Ziff. 20δ) hinzu, so ergibt sich eine von den wesentlichen Fehlern dritter Ordnung freie Anordnung, bei der Bildfehler höherer Ordnung und die chromatischen Variationen der Seidel-Aberrationen (besonders des Astigmatismus) das nutzbare Feld begrenzen[2]. Die von SCHULTE mitgeteilten Durchstoßdiagramme zeigen, daß für das letzte Instrument der Tabelle 3 die Durchmesser der Zerstreuungsfiguren in einem Feld von 1°,5 ∅ im Wellenlängenintervall von 340 bis 660 nm unter 0″,5 bleiben. Korrektionsplatte und Ebnungslinse werden aus Quarz hergestellt. Im achsnahen Gebiet kann man das Teleskop auch *ohne* dioptrische Glieder als reines Zwei-Spiegel-System benutzen (etwa für Photometrie aller Wellenlängen), weil die dann auftretende sphärische Aberration sehr gering bleibt.

Auch mit fokusnahen dioptrischen Systemen mit nur sphärischen Flächen, also abgewandelten Ebnungslinsen, läßt sich bei passender Auslegung der Astigmatismus verringern[3] und so das nutzbare Bildfeld erweitern, mit zweilinsigen Anordnungen[4,5] ebenfalls bis mindestens $w=45'$.

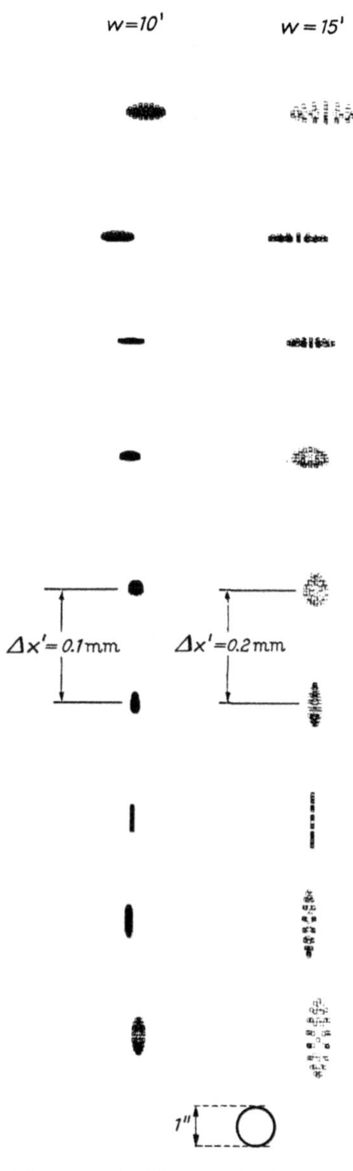

Fig. 25. Durchstoßdiagramm für ein Ritchey-Chrétien-System mit $f=1620$ cm, $N=7{,}5$, $m=2{,}9$. Angegeben sind die Zerstreuungsfiguren für mehrere äquidistante Einstellflächen bei $w=10'$ und $w=15'$ und ein Kreis von 1″ Durchmesser. Nach SCHULTE.

19. Feldkorrektoren. Parabolspiegel sind für große Öffnungen realisierbar, haben aber bei großen Öffnungsverhältnissen ein sehr kleines Bildfeld (vgl. Ziff. 15, S. 259). Die in Ziff. 20 behandelten Schmidt-Kameras und ihre Abwandlungen geben zwar sehr gute Abbildung über große Felder, erfordern aber dioptrische Elemente von der Größe der Eintrittspupille und sind daher in ihrer freien Öffnung beschränkt. Um große Reflektoren photographisch auszunutzen, hat man versucht, wenigstens ein kleines Feld mit dioptrischen Zusatzsystemen kleinen Durchmessers zu korrigieren.

[1] S. C. B. GASCOIGNE: Observatory **85**, 79 (1965).
[2] D. H. SCHULTE: Appl. Optics **5**, 309 (1966).
[3] H. KÖHLER: Studien für das ESO-Projekt 1963/64 (unpubliziert); s. auch [3].
[4] C. G. WYNNE: Appl. Optics **4**, 1185 (1965).
[5] S. ROSIN: Appl. Optics **5**, 675 (1966). — R. WILSON: Optik (1967) (im Druck).

Zwei- oder dreilinsige Korrektoren dieser Art zu parabolisierten Hauptspiegeln sind von Ross [*29*] eingeführt worden; sie können afokal oder mit geringer negativer Brechkraft gebaut werden. Die Seidel-Näherung für dünne Elemente gibt nur einen ersten Anhalt für den Aufbau eines solchen Systems, weil Fehler höherer Ordnung, transversale Farbfehler und chromatische Variation der Seidel-Aberrationen ebenfalls eine Rolle spielen. Aus der Analyse von Ross geht hervor, daß die gleichzeitige Hebung von Koma und Astigmatismus einerseits und sphärischer Aberration andererseits mit einem dünnen Dublett nicht möglich ist, ohne unzulässige Krümmungen der Flächen einzuführen. Das System behält deshalb einen kleinen Betrag von sphärischer Unterkorrektion. Später hat Ross einen zum Hauptspiegel hohlen Meniskus in einigem Abstand vor dem Dublett hinzugefügt, der die sphärische Korrektion wesentlich verbessert. Das korrigierbare Feld wächst mit der Öffnungszahl des *Hauptspiegels* und wird hauptsächlich durch das Anwachsen der Fehler höherer Ordnung begrenzt; für das Arbeiten in einem breiten Wellenlängenbereich spielt auch die transversale Farbabweichung eine Rolle.

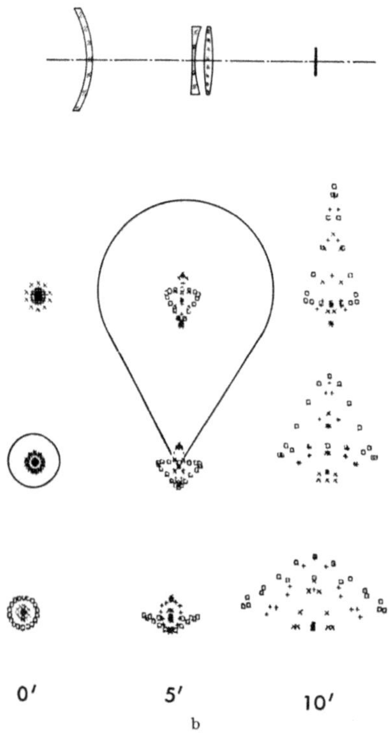

Fig. 26. a Aufbau eines dreilinsigen Korrektionssystems nach Ross. b Durchstoßdiagramm für einen der Ross-Korrektoren ($N_{\text{Syst}}=4{,}7$) zum Palomar-Spiegel. Bildfeldwinkel $w=0'$, $5'$, $10'$; drei verschiedene Einstellebenen. Der Kreis um die Durchstoßfigur für den besten Fokus auf der Achse hat $1''$ Durchmesser; bei $w=5'$ ist die Seidel-Koma des Hauptspiegels allein eingezeichnet.

Ross-Korrektoren gehören seit ihrer Erfindung zur Standard-Ausrüstung der großen Reflektoren. Ein dreilinsiger Korrektor für den Palomar-Spiegel z.B. bringt das System von $N=3{,}3$ auf $N=4{,}7$ und wird für ein Feld von $15'$ Durchmesser benutzt, wobei die geometrischen Zerstreuungsscheibchen im größeren Teil des Feldes unter $1''$ bleiben (Fig. 26).

In recht allgemeiner Weise hat Paul [*28*] die Möglichkeiten der Bildfehlerkorrektur in Systemen aus zwei oder drei dünnen Elementen untersucht. Dabei wird wieder von den Schwarzschildschen Formeln ausgegangen, und es werden unter anderem die Zwei-Spiegel-Systeme, Kombinationen von einem Spiegel und einem afokalen Dublett, Systeme aus drei Spiegeln oder einem Spiegel und zwei deformierten Korrektionsplatten behandelt. Eine Analyse und Erweiterung der Ross-Anordnung hat auch Wynne [*30*] gegeben; er hält am parabolischen Hauptspiegel fest und führt neben dem (nicht notwendig afokalen) Dublett ein drittes Element ein, das ein Sekundärspiegel oder eine asphärische Korrektionsplatte sein kann. Schließlich hat Wynne[1] auch einen verbesserten Ross-Korrektor berechnet, bei dem — unter Verwendung eines hochbrechenden Lanthan-Kron-Glases für das Hinterglied — ein größeres nutzbares Feld erzielt wird.

Das von einem Ross-Dublett ausreichend korrigierte Feld kann beträchtlich erweitert werden, wenn das Zusatzsystem nur Koma und Astigmatismus beseitigt

[1] Siehe Fußnote 6, S. 269.

und die Hebung eines Restes von sphärischer Aberration einem anderen Element überlassen wird. Bei dem „Reflektor-Korrektor" von BAKER[1] wird das achromatische Dublett (optimalisiert für Koma und Astigmatismus und mit positiver Brechkraft wegen $\sum \boldsymbol{P}=0$) mit einer ringförmigen Korrektionsplatte im eintretenden Strahlengang kombiniert, die die restliche sphärische Aberration beseitigt (Fig. 27). Bei einem resultierenden Öffnungsverhältnis von etwa 1:4 ergibt sich gute Korrektion über ein ziemlich großes ebenes Feld (bis $5° \times 5°$). Der parabolische Primärspiegel ohne Korrektor kann weiter in den klassischen Kombinationen benutzt werden. Der bedenklichste Bildfehler ist laterale chromatische Aberration, die für weit auseinanderliegende Farbbereiche verschiedene Korrektoren nötig macht. Die Anordnung erscheint geeignet für mittelgroße Instrumente, da ja die Korrektionsplatte die volle Apertur haben muß. Das System ist mit 60 cm (Nashville[2]) und 90 cm (Kap-Sternwarte[3]) freier Öffnung ausgeführt worden.

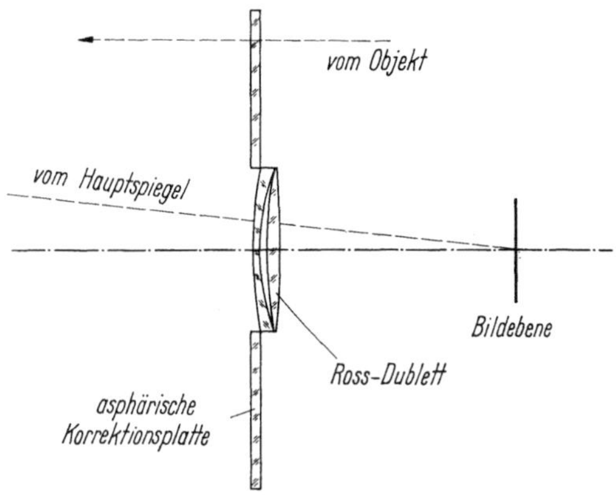

Fig. 27. Reflektor-Korrektor nach BAKER.

Es ist auch möglich, die für die sphärische Überkorrektion nötige Deformation auf den Hauptspiegel zu legen[4]. Sie hat keinen Einfluß auf die außeraxialen Fehler, wenn die Eintrittspupille im Spiegel liegt; der nun hyperbolische Hauptspiegel allein bildet aber selbst in der Achse nicht mehr stigmatisch ab. Die Kombination Hyperbolspiegel — Dublett ist später noch mehrfach vorgeschlagen worden; z.B. hat SONNEFELD ab 1935 derartige Systeme als Großkameras entwickelt (nach KÖHLER [*12*]). Das Durchstoß-Diagramm eines von ROSIN[5] gerechneten Korrektors zeigt ausgezeichnete Abbildung über ein Feld von 1° Durchmesser bei $N=10$. Ein allgemeineres Interesse für astronomische Zwecke haben Korrektoren zu hyperbolischen Primärspiegeln erst im Zusammenhang mit dem Ritchey-Chrétien-System gefunden, dessen Hauptspiegel ja im richtigen Sinne überdeformiert ist (s. Ziff. 18); auf diese Kombinationsmöglichkeit hatte schon PAUL [*28*] hingewiesen. Die mit Korrektoren vom Ross-Typ oder abgewandeltem Aufbau gegebenen Möglichkeiten hat WYNNE[6] diskutiert. BARANNE[7] berechnete einen Korrektor aus drei Linsen mit einer asphärischen Fläche, der bei $N=3$ für ein Feld von 1° Durchmesser Bilder mit weniger als 1″ Zerstreuungsscheibchen liefert (Fig. 28). Die Systeme sind kompakt und leicht realisierbar, haben aber kleine Farbquerfehler und sind nur für einen beschränkten Wellenlängenbereich brauchbar. Eine allerdings aufwendige Korrektionsanordnung von BELOROSSOVA et al.[8] gibt sogar ausreichende Bildqualität über ein Feld von fast $2° \times 2°$ bei $N=3{,}7$.

---

[1] J. G. BAKER: Paper presented to the A.A.S. meeting at Columbus (Dez. 1947) (unpubl.) und in [*31*] III, 1. — Siehe auch WYNNE [*30*]; W. KÜHN: Jenaer Jb. 1954, Teil 2, S. 352

[2] C. K. SEYFERT: Mitt. Astron. Ges. 1955, 77 (1956).

[3] D. S. EVANS: Observatory **84**, 222 (1964).

[4] Schon R. A. SAMPSON, Monthly Notices Roy. Astron. Soc. **73**, 524 (1913).

[5] S. ROSIN: J. Opt. Soc. Amer. **51**, 331 (1961).

[6] C. G. WYNNE: Appl. Optics **4**, 1185 (1965).

[7] A. BARANNE: Publ. Haute-Provence **7**, No. 9 (1964); J. Observateurs **49**, 75 (1966).

[8] T. S. BELOROSSOVA, D. D. MAKSUTOV, N. V. MERMAN u. M. A. SOSNINA: Mitt. Pulkovo **23**, H. 5 (Nr. 175), 162 (1964).

Eine vom Ross-Korrektor verschiedene Möglichkeit zur Vergrößerung des nutzbaren Feldes bietet eine *Folge* von asphärischen *Korrektionsplatten* mit gegenseitigem Abstand im konvergenten Strahlengang nahe dem Fokus. Die Plattendurchmesser bleiben hier auch für große Instrumente in einem realisierbaren Bereich. Als Vorteil dieses Korrektor-Typs ist die Freiheit von primären chromatischen Aberrationen anzusehen; die Wahl des Materials ist also von der optischen Rechnung her nicht eingeschränkt. Die Kombination eines Hauptspiegels mit zwei Platten war schon von PAUL [28] und WYNNE [30] behandelt worden. Ein von MEINEL[1] vorgeschlagener Korrektor benutzt drei Platten und erreicht damit ausreichende Bildqualität in einem Feld von etwa 1° ⌀ bei $N=3$. Zur Bildfeldebnung wäre eine weitere Linse vorzusehen. Die Anordnung ist nicht an eine bestimmte Deformation des Hauptspiegels gebunden und wird neuerdings auch für den (in der Achse nicht stigmatischen) Primärfokus eines Ritchey-Chrétien-Hauptspiegels diskutiert. So haben für die beiden größten Instrumente der Tabelle 3 SCHULTE[2] bzw. KÖHLER[3] Korrektoren vom Meinel-Typ berechnet (Fig. 29). SCHULTE gibt auch Daten für zwei andere Korrektor-Typen: nach ROSIN mit stärker hyperbolischem Hauptspiegel und achromatischem Dublett sowie nach WYNNE mit Ritchey-Chrétien-Hauptspiegel, einer asphärischen Platte und achromatischem Dublett.

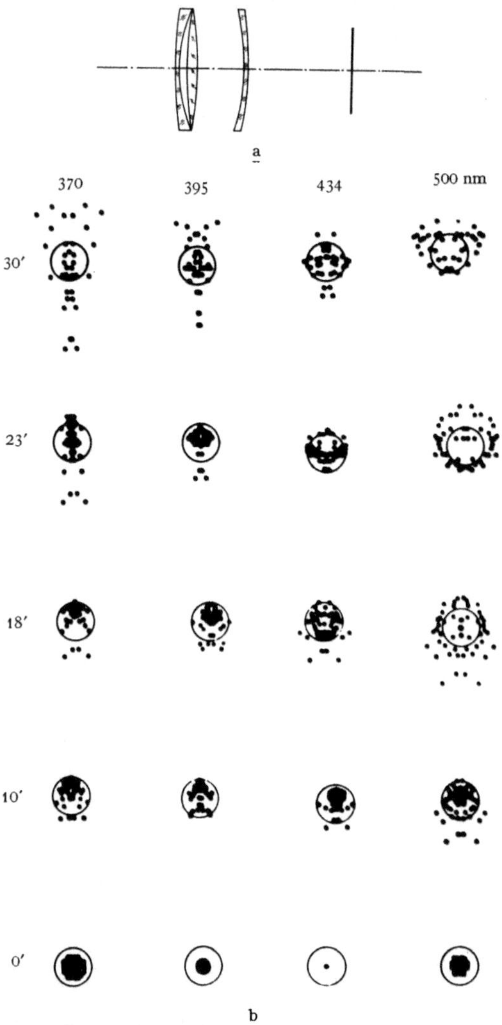

Fig. 28a u. b. Afokaler Korrektor mit einer asphärischen Fläche (Typ V von BARANNE) für den Primärfokus eines Ritchey-Chrétien-Hauptspiegels mit $N=3$. a Aufbau des Zusatzsystems. b Durchstoßdiagramme auf einer gekrümmten Bildfläche für verschiedene Bildwinkel und Wellenlängen. Der Kreis hat 0,″5 Durchmesser. Für die äußeren Wellenlängen ist die Bildqualität durch Nachfokussieren etwas zu verbessern. Der Korrektor ist vignettierungsfrei bis $w=23'$.

Auf dem Gebiet der Feldkorrektoren läuft die Entwicklung im Augenblick so rasch, daß noch keine zusammenfassende Übersicht[4] möglich ist; es gibt auch keine einhellige Meinung, ob generell Systeme mit sphärischen Flächen oder mit dünnen Korrektionsplatten vorzuziehen sind. Bei vergleichbarer monochromatischer Abbildungsqualität erscheint der Meinel-Typ aufwendiger als der modifizierte Ross-Typ, gestattet dafür aber das Arbeiten im gesamten photographisch

[1] A. B. MEINEL: Astrophys. J. **118**, 335 (1953).
[2] D. H. SCHULTE: Appl. Optics **5**, 313 (1966).
[3] H. KÖHLER: Studien zum ESO-Projekt 1964/65 (unpubl.).
[4] Siehe z. B. die Diskussionen in [3].

interessierenden Wellenlängenbereich ohne Wechsel der optischen Elemente. Praktische Erfahrungen mit den neuen Korrektoren liegen noch nicht vor.

Mit nur einer asphärischen Platte kann man beim Ritchey-Chrétien-Hauptspiegel neben der sphärischen Aberration auch die Koma beseitigen durch passende Wahl der „Stärke" und des Abstands vom Fokus. Das entspricht einer Lösung $\sum \mathbf{I} = \sum \mathbf{II} = 0$ der Gln. (4.5) mit $b < -1$ und $s_p > 0$, d.h. virtueller Eintritts-

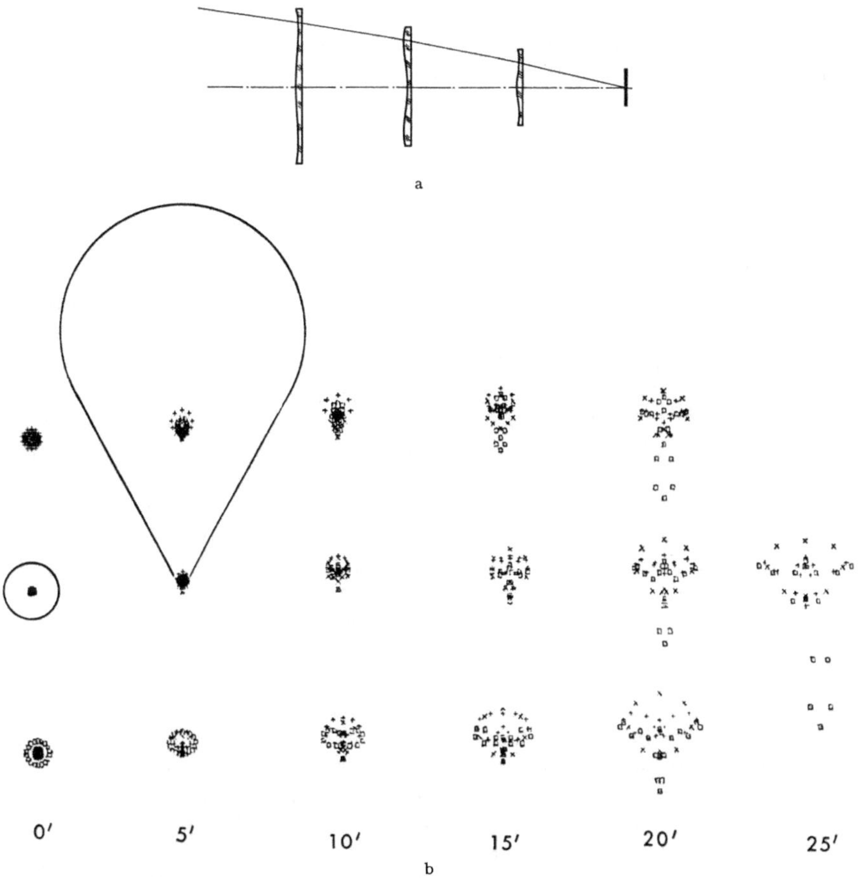

Fig. 29a u. b. Korrektionssystem aus drei asphärischen Platten für den Primärfokus eines Ritchey-Chrétien-Hauptspiegels mit $N=2{,}8$ (nach SCHULTE). a Aufbau des Systems (die Profile der Korrektionsplatten sind stark überhöht); b Durchstoßdiagramme für verschiedene Bildwinkel und Fokuseinstellungen bei $\lambda = 420$ nm. Der Kreis hat 1″ Durchmesser; bei $w = 5'$ ist die Seidel-Koma des Primärspiegels allein eingezeichnet.

pupille hinter dem Spiegel, die reell in die Nähe des Fokus abgebildet wird. Das brauchbare Feld wird vom Astigmatismus begrenzt und ist mindestens so groß wie bei einem Parabolspiegel mit Ross-Korrektor; die Platte ist jedoch hinsichtlich der chromatischen Fehler günstiger[1]. Auch zu Quasi-Cassegrain-Anordnungen sind Feldkorrektoren angegeben worden[2]. Das Wynne-Rosin-System[3] benutzt zu einem parabolischen Hauptspiegel einen sphärischen Sekundärspiegel in Cassegrain-Anordnung mit verhältnismäßig großem Öffnungsverhältnis. Damit ist die sphärische Überkorrektion gegeben, die für das einwandfreie Arbeiten des dioptri-

[1] S. C. B. GASCOIGNE: Observatory **85**, 79 (1965).
[2] WYNNE [*30*]; H. KÖHLER: Optik **17**, 485 (1960).
[3] S. ROSIN: Appl. Optics **3**, 151 (1964).

schen Korrektors nötig ist. Bei einem solchen Instrument wären neben einem korrigierten Feld die Foci des klassischen Reflektors nach wie vor mit reiner Spiegeloptik verfügbar, was z.B. für Infrarot-Anwendungen erwünscht ist. Rechnungen für Großinstrumente scheinen noch nicht vorzuliegen.

Eine gewisse Verwandtschaft zu Feldkorrektoren zeigen Systeme, die das Öffnungsverhältnis eines Spiegelteleskops vergrößern. Die resultierende Brennweite ist dabei durch $f = N_{\text{Syst.}} \cdot D_1$ gegeben. Lichtstarke Anordnungen dieser Art (mit $N_{\text{Syst.}} = 1 \ldots 2$) sind für Flächenphotometrie mit sehr schmalbandigen Filtern oder Perot-Fabry-Interferometern vorteilhaft und haben deshalb Kollimator und Kameraobjektiv mit parallelem Strahlengang dazwischen. Um den Durchmesser der Zusatzoptik klein zu halten, führt man eine Feldlinse nahe der ursprünglichen Bildfläche ein. Gebaut wurden Systeme für den Newton-Fokus (reducteur focal, COURTÈS[1]) und den Cassegrain-Fokus (MEINEL[2]).

c) Schmidt- und Meniskus-Systeme.

20. Schmidt-Teleskop.

α) *Aufbau und Wirkungsweise.* Die Anforderungen der Himmelsphotographie — großes Bildfeld bei großem Öffnungsverhältnis — hatten durch Weiterentwicklung photographischer Objektive zu Systemen geführt, die trotz erheblichen Aufwands nur in einem begrenzten Wellenlängengebiet brauchbar sind und deren freie Öffnung kaum über etwa 50 cm hinaus getrieben werden kann. Eine viel elegantere Lösung gab B. SCHMIDT[3] mit seinem „lichtstarken komafreien Spiegelsystem", das oft als die bedeutendste optische Erfindung der ersten Hälfte des 20. Jahrhunderts bezeichnet wird. Eine typische Schmidt-Kamera liefert bei $N=3$ und dem Felddurchmesser $2w = 6°$ Bilder, deren Qualität überwiegend von Luftunruhe oder Plattenkorn, nicht aber von Abbildungsfehlern bestimmt ist. Hatte SCHWARZSCHILD die Deformation der Flächen unabhängig von ihren Fokaleigenschaften gemacht, so ging SCHMIDT einen Schritt weiter und löste die Deformation völlig von der abbildenden Fläche, indem er sie auf eine dünne, im Gauß-Bereich brechkraftlose „Korrektionsplatte" übertrug. Gleichzeitig mit der Angabe des Prinzips demonstrierte SCHMIDT, daß die erforderlichen asphärischen Flächen (mit nach herkömmlichen Begriffen extrem starker Deformation) auch wirklich mit der notwendigen Genauigkeit hergestellt werden können[4].

Durch die Entkoppelung von Spiegelfläche und Deformation gewinnt der Abstand der Eintrittspupille vom Spiegelscheitel Einfluß auf die Koma des Systems, wie man den Gln. (4.5) entnehmen kann: Aus $\sum \mathbf{II} = 0$ folgt $2 + s_p(b_1 + 1) = 0$, was mit $b_1 = -1$ (Hebung der sphärischen Aberration durch Parabolisierung) unerfüllbar ist. Dagegen lassen sich im Schema für das Ein-Spiegel-System mit zusätzlicher Korrektionsplatte in der Eintrittspupille (Beitrag $\mathbf{I}_0^*$) s_p und b_1 so bestimmen, daß $\sum \mathbf{II} = 0$ und $\sum \mathbf{IIIa} = 0$; die Lösung ist $s_p = -2$ und $b_1 = 0$. Damit auch $\sum \mathbf{I} = 0$, muß $\mathbf{I}_0^* = -\frac{1}{4}$ sein. Die sphärische Aberration wird sozusagen „nachträglich" beseitigt; die Deformation der Korrektionsplatte hat keinen Einfluß auf die außeraxialen Fehler, weil sie in der Pupille liegt.

Ganz anschaulich kann man sich die Wirkungsweise der Schmidt-Kamera so vorstellen (Fig. 30): Die Eintrittspupille eines sphärischen Hohlspiegels liegt in der achsensenkrechten Ebene durch den Krümmungsmittelpunkt C. Dadurch verliert die Achse ihre ausgezeichnete Stellung, und auch geneigte Bündel bleiben

[1] G. COURTÈS: Ann. Astrophys. **23**, 133 (1960).
[2] A. B. MEINEL: Astrophys. J. **124**, 652 (1956).
[3] B. SCHMIDT: Mitt. Hamburg **7**, 15 (1932).
[4] R. SCHORR: Mitt. Hamburg **7**, 175 (1936).

symmetrisch zu ihrem Hauptstrahl; Koma und Astigmatismus können somit nicht auftreten. Die sphärische Aberration beseitigt man, indem man dem Bündel die entsprechende Wellenaberration mit entgegengesetztem Vorzeichen in der Eintrittspupille aufprägt. Das geschieht mit einer Korrektionsplatte, deren Profil aus Gl. (9.1) abzuleiten ist oder aus Gl. (9.2), wenn man nicht am Gauß-Fokus festhält und eine geringe Paraxial-Brechkraft der Platte zuläßt. Das Bild liegt auf einer zur Spiegelfläche konzentrischen Kugel mit deren halbem Krümmungsradius; der Strahlungsempfänger ist dieser Krümmung anzupassen. Um auch außeraxiale Bündel voll auszunutzen, muß man dem Spiegel eine größere Öffnung geben als der Korrektionsplatte.

Man kann schließlich die Schmidt-Kamera auch ansehen als Schwarzschild-Anastigmaten (s. Ziff. 18) mit *ebenem*, aber deformiertem Primärspiegel; die

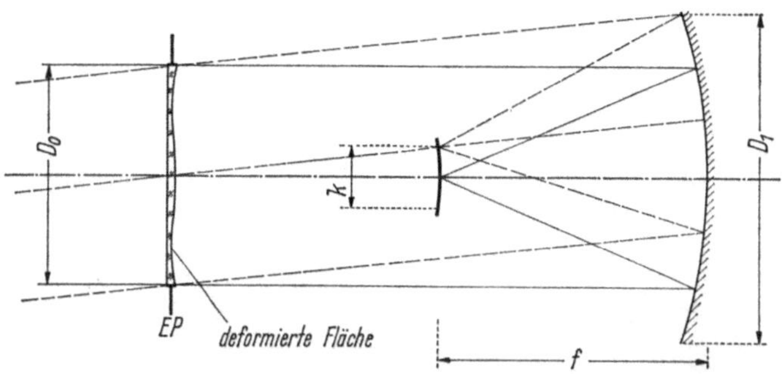

Fig. 30. Aufbau der Schmidt-Kamera. Gegenüber einem realen System sind Bild- und Öffnungswinkel etwa um einen Faktor zwei vergrößert; das Profil der Korrektionsplatte ist stark überhöht.

Deformation des Sekundärspiegels (der dem Hauptspiegel der Schmidt-Kamera entspricht) verschwindet in diesem Fall. Um überhaupt Licht einfallen zu lassen, was bei diesem Zwei-Spiegel-System nicht möglich wäre, ersetzt man den deformierten Planspiegel durch die Korrektionsplatte. Angaben zum Prinzip und zur optischen Weiterentwicklung der Schmidt-Kamera, z. T. mit weiteren Literaturhinweisen, sind in [*11*], [*12*], [*19*][1] zu finden.

Ein typischer Vertreter der mittelgroßen Schmidt-Astrokamera ist das Instrument des Warner and Swasey Observatory in Cleveland[2] mit einer Korrektionsplatte von 61 cm und einem Spiegel von 91 cm ⌀; die Brennweite beträgt 214 cm, das benutzte Feld hat etwas über 5° ⌀. Die größten Schmidt-Teleskope sind der „Big Schmidt" des Palomar Observatory mit 122/183/307 cm und 6,°5 × 6,°5 Feld[3] sowie das Universal-Teleskop des Karl-Schwarzschild-Observatoriums Tautenburg mit 134/200/400 cm und 3,°4 × 3,°4 Feld[4], das auch als Quasi-Cassegrain-System benutzt werden kann.

β) *Profil der Schmidt-Platte.* Eine ebene Wellenfläche erleidet durch Einsetzen einer Platte mit der Dicke g und dem Brechungsindex n die Verzögerung $V = (n-1)\,g$. Für den Kugelspiegel ist die Wellenaberration des Öffnungsfehlers nach Gl. (9.2)

$$V = \frac{h^4}{32 f^3} + \varDelta \frac{h^2}{2 f^2}$$

[1] B. Strömgren: Vjschr. Astron. Ges. **70**, 65 (1935). — C. Carathéodory: Elementare Theorie des Spiegelteleskops von B. Schmidt. Leipzig u. Berlin 1940. — F. E. Ross: Astrophys. J. **92**, 400 (1940). — L. Oterma: Turku Inform. No. 9 (1955). — I. S. Bowen: Kap. 4 in [*9*].

[2] J. J. Nassau: Astrophys. J. **101**, 275 (1945).

[3] R. G. Harrington: Publ. Astron. Soc. Pacific **64**, 275 (1952).

[4] N. Richter: Sterne **37**, 89 (1961).

(Δ ist der Abstand des Referenzkugel-Zentrums vom Gauß-Bildpunkt). Zur Kompensation ist eine variable Dicke der Korrektionsplatte

$$g(h) = \frac{1}{n-1}\left[\frac{h^4}{32f^3} + \Delta\frac{h^2}{2f^2}\right] + g(0)$$

nötig; dabei ist vorausgesetzt, daß ein Strahl auch nach Durchgang durch die Platte achsenparallel bleibt und damit die Einfallshöhe h für Platte und Spiegel gilt.

Statt Δ führen wir eine Formkonstante des Plattenprofils a ein:

$$a = -\frac{16f}{H^2}\cdot\Delta \quad \text{mit} \quad H = h_{\max} = \frac{D_0}{2},$$

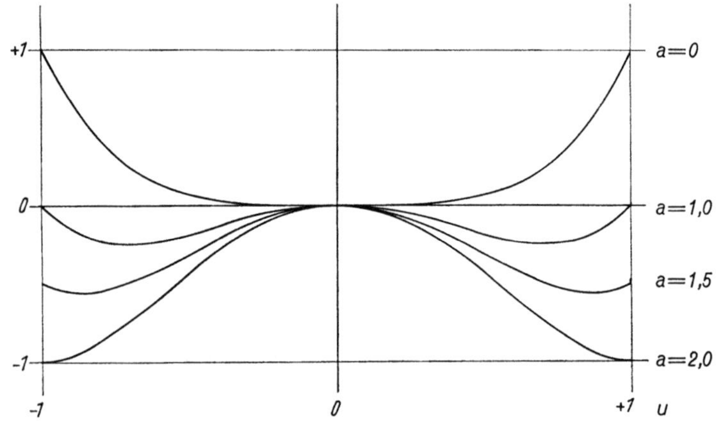

Fig. 31. Profilfunktion $(u^4 - a u^2)$ von Schmidt-Korrektionsplatten für verschiedene Werte des Formparameters a.

außerdem den normierten Achsabstand $u = h/H$; damit wird

$$g(h) = \frac{1}{512}\cdot\frac{1}{n-1}\cdot\frac{D_0}{N^3}\,(u^4 - a u^2) + g(0)\,. \tag{20.1}$$

Für einige Werte von a sind die Klammerausdrücke in Fig. 31 wiedergegeben. Um die von der Korrektionsplatte verursachten Farbfehler klein zu halten (siehe Abschnitt ζ dieser Ziffer), wählte SCHMIDT $a = 1{,}5$. Obwohl die Platte als „stark deformiert" bezeichnet wird, ist ihre Abweichung von einer Planparallel-Platte gering: für $a = 1{,}5$, $N = 3$, $n = 1{,}5$, $D_0 = 100$ cm ist das Maximum von $g(h) - g(0)$ erst 0,08 mm!

γ) *Bildfeld und Vignettierung.* Die Anordnung der Kassette im eintretenden Strahlenbündel führt zu einer Kopplung zwischen der Öffnungszahl N, der relativen Abschattung k/D_0 (k ist der Kassettendurchmesser) und dem Winkeldurchmesser des Bildfelds $2w$:

$$\frac{k}{D_0} \approx \frac{2w}{57}\,N \qquad (w \text{ in Grad}) \tag{20.2}$$

(ohne Berücksichtigung eines Kassettenrandes). Systeme mit großem Feld sind also nur bei kleiner Öffnungszahl möglich. Um das Feld vom Durchmesser k vignettefrei zu beleuchten (Fig. 30), muß der Spiegel eine Öffnung von

$$D_1 = D_0\left(1 + 2\frac{k}{D_0}\right)$$

haben; der Wert der Klammer ist meist nahe 1,5.

Bei diesem Aufbau des Systems wird manchmal als störend empfunden, daß der Kassettenschatten in außeraxialen Bündeln exzentrisch liegt und bei Defokussierung der Bildaufbau unsymmetrisch wird. Da die Lage der Eintrittspupille keinen Einfluß auf die Fehler dritter Ordnung hat, nachdem die sphärische Aberration beseitigt ist, kann man die Blende bei festgehaltener Korrektionsplatte auch in die Höhe der Kassette legen. Dann müssen für vignettefreies Feld Spiegel und Korrektionsplatte den gleichen Durchmesser

$$D_{0,1}=D+k$$

(D der Blendendurchmesser) bekommen. Allerdings werden bei dieser Anordnung, weil Eintrittspupille und Korrektionsplatte nicht mehr zusammenfallen, die Restfehler größer (insbesondere die chromatische Aberration vom Bildwinkel abhängig).

δ) *Ebnungslinse.* Die kräftige Bildfeldwölbung des Schmidt-Systems (Krümmungsradius der Fokalfläche gleich Brennweite) wäre ohne die außerordentlich hohe Bildqualität kaum akzeptiert worden. Bei längeren Brennweiten lassen sich dünne Photoplatten über passend geformte Stempel durchbiegen; unterhalb von $f\approx 50$ cm ist man auf Film angewiesen. Eine Alternative bietet die *Ebnungslinse* nach P. SMITH, die schon SCHMIDT in seiner ersten Notiz erwähnte.

Bildebnung wird erreicht durch Einsetzen einer Plankonvexlinse, mit dem Krümmungsradius

$$R_E=\frac{n-1}{n}f \qquad (20.3)$$

(n Brechungsindex des Linsenmaterials) und der Randdicke Null, unmittelbar vor der Emulsion. Das ist eine Anwendung der Formel für den auf Luft reduzierten Glasweg bei hinreichend kleinen Sehnen und Pfeilhöhen. Schreibt man in (20.3) statt der Kamerabrennweite f den Krümmungsradius der Bildfläche R, so gilt die Gleichung allgemein, mit Mittendicke Null auch für hohle Flächen. Im anastigmatischen Schmidt-System kann man die Ebnungslinse auch als Glied betrachten, das die Petzval-Summe (11.1) zum Verschwinden bringt; der Petzval-Term ist dem des Hauptspiegels entgegengesetzt gleich. Damit der Abbildungsstrahlengang nicht gestört wird, muß die Linse mit dem Bild zusammenfallen.

Durch Einsetzen einer realen Ebnungslinse entstehen Abbildungsfehler[1], und zwar Koma, die durch eine kleine Verschiebung der Korrektionsplatte auf den Spiegel zu verringert werden kann (s. Abschnitt η dieser Ziffer), und sphärische Aberration, die aber bei Öffnungsverhältnissen um 1:3 noch nicht ins Gewicht fällt. Die Farbfehler des Systems (s. Abschnitt ζ dieser Ziffer) werden vergrößert.

Eine eingehende Untersuchung [*19*] ergibt, daß das nachträgliche Einsetzen einer Ebnungslinse in ein für gekrümmte Platten optimalisiertes System zu einer merklichen Verschlechterung der Abbildung führt. Wird die Kamera jedoch für ausschließliche Verwendung mit der Ebnungslinse zusammen als Einheit entworfen, so steht ihre Bildqualität dem ursprünglichen System kaum nach; lediglich die Farbfehler sind geringfügig vergrößert.

ε) *Toleranzen.* Die *Fokustoleranz* beträgt bei dem großen Öffnungsverhältnis nach dem Beugungskriterium etwa 0,01 mm; auch bei geometrisch-optischer Betrachtungsweise sind nur wenige 0,01 mm zugelassen. Eine *Querversetzung* der

[1] F. E. ROSS: Astrophys. J. **92**, 400 (1940).

Kassette um δ wirkt sich am Rand des Bildfeldes aus wie eine Defokussierung

$$\Delta x' = w \cdot \delta \approx 0{,}05\,\delta;$$

die Toleranz beträgt also wenige 0,1 mm[1].

Gegen kleine *Neigungen* der Korrektionsplatte ist das System weitgehend unempfindlich wegen der Kleinheit der außeraxialen Fehler. Auch *Durchbiegung* der Platte ist ungefährlich, da die Verzögerung der Wellenfront durch die Dicke des Glaswegs bestimmt wird.

Eine *Translation* der Schmidt-Platte in ihrer Ebene um δ führt zu feldunabhängiger Koma: aus der Differenz zwischen den Wellenaberrationen der zentrierten und der verschobenen Platte folgt

$$\Delta V_{\max} = \frac{1}{64 N^3}\,\delta$$

oder

$$\boldsymbol{K}_\delta = \frac{3}{32 N^2}\,\delta; \qquad \mathcal{K}_\delta = \frac{3}{32 N^2}\,\frac{\delta}{f}.$$

Bei einem Teleskop mit $D_0 = 1$ m, $N = 3$ ergibt sich eine Ausdehnung der Koma-Figur von 0,″7 = 0,01 mm für eine Querversetzung von 1 mm. Eine Verschiebung

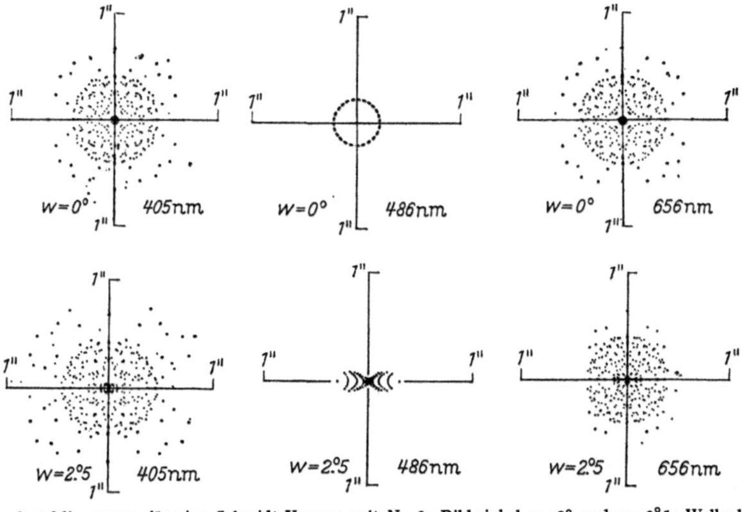

Fig. 32. Durchstoßdiagramme für eine Schmidt-Kamera mit $N=3$; Bildwinkel $w=0°$ und $w=2{,}^\circ 5$; Wellenlängen 405, 486, 656 nm (nach Linfoot [19]). Statt des (stigmatischen) axialen Bildes bei der Korrektionswellenlänge 486 nm ist der Umfang des zentralen Beugungsscheibchens für eine freie Öffnung von $D_0 = 60$ cm eingetragen. Eine Mittenabschattung ist nicht berücksichtigt.

der Korrektionsplatte *längs der Achse* um Δ wirkt für die Hauptstrahlneigung w wie eine Querversetzung $\delta = \Delta \cdot \tan w$ und führt damit zu *außeraxialer* Koma; es ist

$$\boldsymbol{K}_\Delta = \frac{\Delta}{2f}\,\boldsymbol{K}_s,$$

wobei $\boldsymbol{K}_s$ die Koma des sphärischen Spiegels mit Blende im Scheitel (Fig. 7 auf S. 235) bedeutet.

---

[1] Ein Verfahren zur Justierung der Kassettenneigung beschreiben D. W. Dewhirst and G. G. Yates, Observatory 74, 71 (1954); s. auch H. Haffner, Observatory 74, 210 (1954); W. Blatt: Askania-Warte 18, H. 57, 10 (1961).

Wenn man die Korrektionsplatte als dünn betrachtet, hat die Schmidt-Kamera im Seidel-Bereich keine Abbildungsfehler außer der Bildfeldwölbung (s. Tabelle 1 auf S. 236). Die geometrischen Fehler höherer Ordnung und die chromatischen Fehler (Fig. 32) sind so klein, daß sie nur bei großen Dimensionen oder bei extremem Feld bzw. Öffnungsverhältnis merklich werden:

ζ) *Farbfehler der Korrektionsplatte.* Die Korrektionsplatte mit dem Profil (20.1) prägt dem eintretenden Bündel die geforderte Wellenaberration

$$V = \frac{1}{512} \cdot \frac{f}{N^4} (u^4 - a u^2)$$

auf bei der Wellenlänge λ, für die der Brechungsindex n ist. Bei einer anderen Wellenlänge λ' mit Brechungsindex n' liefert die gleiche Platte

$$V' = \frac{n'-1}{n-1} V.$$

Es bleibt also ein wellenlängenabhängiger unkompensierter Rest der sphärischen Aberration (Gauß-Fehler) vom Betrag

$$\delta V = V' - V = \frac{n'-n}{n-1} V = \frac{1}{\nu} V,$$

wobei die langwellige Seite unterkorrigiert ist und umgekehrt. Der Faktor von V ist der Kehrwert der Abbeschen Zahl ν, gebildet für λ und λ'. Die zugehörige Queraberration beträgt

$$\delta y' = \frac{1}{\nu} f \frac{dV}{dh} = \frac{f}{\nu} \cdot \frac{(4u^3 - 2au)}{256 N^3}.$$

Der kleinste Zerstreuungskreis ergibt sich für denjenigen Wert von a, der zu dem kleinsten Betrag des Extremwerts für die durch $(4u^3 - 2au)$ gegebene normierte Steigung des Profils führt. Im Bereich $0 \leqq u \leqq 1$ ist $a_{\text{opt}} = 1{,}5$ mit den Extremwerten ∓ 1 bei $u = 0{,}5$ und $u = 1$. Das Minimum der Plattendicke (neutrale Zone) liegt bei $u = 0{,}866$. Die so bestimmte Platte gibt für die sphärochromatische Aberration einen Radius des Zerstreuungsscheibchens von

$$|\delta y'(\lambda, \lambda')| = \frac{f}{\nu} \cdot \frac{1}{256 N^3};$$

der Winkeldurchmesser ist

$$\frac{2\delta y'}{f} = \frac{1}{\nu} \cdot \frac{1610''}{N^3}.$$

Bei einer Korrektionsplatte aus UBK 7 ist $\nu = 100$ zwischen 386 und 430 bzw. 430 und 490 nm und führt bei einem Instrument mit $N = 3$ zu einem Zerstreuungskreis von 0,″6 Durchmesser.

Eine achsensenkrecht in den konvergenten Strahlengang gestellte planparallele Platte (z.B. Farbfilter) von der Dicke d und der Brechzahl n_F verursacht neben der paraxialen Fokusversetzung $\Delta s' = +d \cdot \left(1 - \frac{1}{n_F}\right)$ auch eine sphärische Überkorrektion

$$\Delta s'_{\text{Pl}} = \frac{n_F^2 - 1}{n_F^3} \cdot d \cdot \frac{h^2}{2f^2}; \qquad \Delta s'_{\max} \approx 0{,}047 \frac{d}{N^2} \quad \text{für} \quad n_F = 1{,}5 \ldots 1{,}6$$

und

$$V_{\text{Pl}} = \frac{n_F^2 - 1}{n_F^3} \cdot d \cdot \frac{h^4}{8 f^4}$$

bzw. nach passender Defokussierung

$$V_{\mathrm{Pl}} = \frac{n_F^2 - 1}{n_F^3} \cdot d \cdot \frac{1}{128 N^4} (u^4 - a u^2).$$

Damit ist eine Kompensation des Gauß-Fehlers möglich: Gleichsetzung von δV und V_{Pl} ergibt

$$d = \frac{n^3}{n_F^2 - 1} \cdot \frac{f}{4\nu}; \quad \nu \approx 0{,}66 \frac{f}{d}.$$

Um die diesem ν entsprechende Wellenlängendifferenz verschiebt die Planplatte die Korrektion nach der langwelligen Seite. Bei „photographisch" korrigierten Schmidt-Platten führen dicke Filter so zu einer weiteren Verschlechterung der ultravioletten Bilder.

η) *Abbildungsfehler außerhalb der Achse.* Die Korrektionsplatte, die ja im Gegensatz zum Spiegel eine ausgezeichnete Achse besitzt, führt zu (allerdings kleinen) *außeraxialen Aberrationen.* Einmal bewirkt der schräge Durchgang durch die Platte ein Anwachsen der Wellenverzögerung, das mit $\sin^2 w$ geht; zum anderen durchstößt ein Strahl, dem auf dem Spiegel die Einfallshöhe h entspricht (bezogen auf den schiefen Hauptstrahl), die Platte bei $h/\cos w$ (Koordinate in der Platte gemessen). Beide Effekte zusammen führen zu einer Zerstreuungsfigur mit der maximalen Querabweichung

$$\frac{\Delta y'(w)}{f} = \frac{\sin^2 w}{32 n N^3};$$

für das betrachtete Teleskop mit 1:3 ergibt sich bei $w = 3°$, $n = 1{,}5$: $\frac{2\Delta y}{f} = 0{,}''9$.

Die hier berücksichtigten Glieder von Reihenentwicklungen reichen bei extremen Öffnungsverhältnissen nicht aus. Vollständige Ausdrücke fünfter Ordnung sind von BAKER[1] angegeben worden.

21. Nebenformen der Schmidt-Kamera. Die Schmidtsche Erfindung eröffnete neue Möglichkeiten für den Bau von Astrokameras mit großem Feld und großem Öffnungsverhältnis. Es setzten daher schon bald Versuche ein, die beiden Nachteile des Schmidt-Teleskops — starke Bildwölbung und große Baulänge gleich der doppelten Brennweite — zu beseitigen oder die als schwierig angesehene Herstellung der Korrektionsplatte zu umgehen. Viele der vorgeschlagenen Typen sind in [*11*] und [*12*] aufgeführt. Im ganzen bevorzugt man jedoch heute den „klassischen" Schmidt-Spiegel, jedenfalls für die Verwendung als große Astrokamera.

α) *Wright-Väisälä-System.* WRIGHT[2] und VÄISÄLÄ[3] zeigten, daß man die Eintrittspupille mit der Korrektionsplatte auch in die Ebene durch den Brennpunkt legen kann, wenn man den Spiegel ebenfalls deformiert. Dann bleibt das System komafrei, bekommt jedoch merklichen Astigmatismus; es wird Bildfeldebnung im übertragenen Sinne erreicht. Die Baulänge ist gleich der Brennweite. Aus den Seidel-Formeln (4.7) folgt für $s_p = -1$ und $\sum \mathbf{I} = 0$, $\sum \mathbf{II} = 0$: $b_1 = +1$; $\mathbf{I}_0^* = -\frac{1}{2}$. Der Hauptspiegel bekommt demnach die entgegengesetzte Deformation wie bei einem Paraboloid (die Krümmung *wächst* nach dem Rand zu); die „Stärke" der Korrektionsplatte ist doppelt so groß wie beim Schmidt-System. Mit diesen Werten wird $\sum \mathbf{IIIa} = +\frac{1}{2}$, also halb so groß wie beim Kugelspiegel mit Blende im Scheitel, und $\sum \mathbf{IVa} = 0$, d.h. die Schale der mittleren Bildkrümmung ist

[1] J. G. BAKER: Proc. Amer. Phil. Soc. **82**, 323 (1940) = Harvard Reprint No. 198; s. auch LINFOOT [*19*].
[2] F. B. WRIGHT: Publ. Astron. Soc. Pacific **47**, 300 (1935).
[3] Y. VÄISÄLÄ: Astron. Nachr. **259**, 197 (1936).

eben. Das brauchbare Bildfeld ist damit mehr als doppelt so groß wie bei einem Ritchey-Chrétien-Teleskop, kann aber mit dem der Schmidt-Kamera nicht konkurrieren. Ein Instrument dieses Typs steht in Turku[1].

β) Schmidt-Cassegrain-Anordnungen. Einen Typ mit verkürzter Baulänge ohne Kompromisse hinsichtlich der Abbildungsfehler hat BAKER[2] angegeben (Fig. 33). Die Verwendung von zwei Spiegelflächen in Cassegrain-artiger Anordnung und einer Korrektionsplatte gestattet die Beseitigung der vier wesentlichen Seidel-Fehler. Die Fokalfläche liegt etwas innerhalb des Hauptspiegelscheitels und ist durch einen Durchbruch verhältnismäßig gut zugänglich. Die Scheitelkrümmungen von Primär- und Sekundärspiegel müssen entgegengesetzt gleich sein, damit anastigmatische Bildebnung möglich wird (Petzval-Bedingung). Für $g \approx 0$ (vgl. Fig. 24 auf S. 261) ist damit die Sekundärvergrößerung auf $m \approx 1{,}6$ festgelegt, mit entsprechend starker Mittenabschattung nach (16.2).

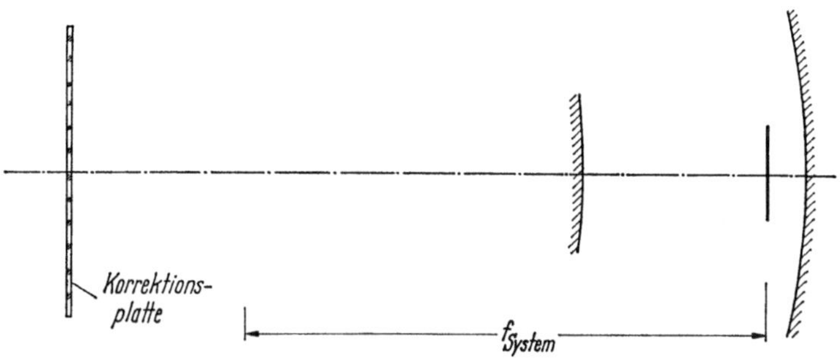

Fig. 33. Schmidt-Cassegrain-Kamera nach BAKER. Gegenüber einem realen System sind Bild- und Öffnungswinkel etwa verdoppelt.

In der Seidel-Näherung wird das Zwei-Spiegel-System zusammen mit der Pupillenposition so bestimmt, daß Koma und Astigmatismus verschwinden. Die Korrektionsplatte in der Eintrittspupille beseitigt die sphärische Aberration; die Aperturblende kann im resultierenden anastigmatischen System so verschoben werden, daß sich günstige Vignettierungsverhältnisse ergeben. Mit der Forderung $\sum \mathbf{II} = \sum \mathbf{IIIa} = 0$ ist das Zwei-Spiegel-System noch nicht festgelegt, man kann z.B. die Lage der Korrektionsplatte oder die Deformation eines Spiegels in gewissen Grenzen vorschreiben. BAKER gibt Daten für Systeme mit besonders kurzer Baulänge bzw. Verzeichnungsfreiheit, oder läßt eine der Spiegelflächen *sphärisch.* Die Deformation des anderen Spiegels ist dabei so gering, daß sogar *zwei* sphärische Flächen plus Korrektionsplatte ein brauchbares aplanatisches System mit geringem Restastigmatismus ergeben (SLEVOGT [*11*]). Die „Stärke" der Korrektionsplatte ist in allen Fällen ein mehrfaches von der des äquivalenten Schmidt-Teleskops (vgl. die Beiträge $\mathbf{I}_0^*$ in Tabelle 1, S. 236), mit entsprechendem Anwachsen der Restfehler.

Ein Baker-Schmidt-System ist das „ADH-Teleskop" des Boyden Observatory[3]; das Feld hat 4°,8 ⌀ bei $N = 3{,}7$ (nominell), der Abstand der Korrektionsplatte vom Hauptspiegelscheitel beträgt $1{,}4f$.

22. Meniskus-Systeme. Die Herstellung einer asphärischen Korrektionsplatte ist schwierig, wenn man sie von den herkömmlichen optischen Fertigungsmethoden her beurteilt. Dieser Gesichtspunkt sollte bei Einzelstücken der Groß-

[1] Y. VÄISÄLÄ: Turku Inform. No. 6 (1950).
[2] J. G. BAKER: Proc. Amer. Phil. Soc. **82**, 339 (1940) = Harvard Reprint No. 199; s. auch LINFOOT [*19*].
[3] E. M. LINDSAY: Irish Astron. J. **2**, 140 (1953).

Astrooptik nicht zu schwer wiegen; für eine Serienanfertigung kleinerer Instrumente hat er eine gewisse Berechtigung. Nach Bekanntwerden des Schmidt-Systems versuchte man dementsprechend, die Korrektionsplatte durch dioptrische Glieder mit *sphärischen* Flächen zu ersetzen. Eine praktikable Lösung wurde fast gleichzeitig an mehreren Stellen gefunden; in der Literatur ist sie mit den Namen MAKSUTOV[1] und BOUWERS [*32*] verknüpft. Über weitere Systeme mit sphärischen Korrektions-Elementen haben SLEVOGT [*11*] und KÖHLER [*12*] berichtet. Bei den Meniskus-Teleskopen wird die sphärische Unterkorrektion des Kugelspiegels kompensiert durch die Überkorrektion eines verhältnismäßig dicken, stark durchgebogenen *Meniskus* mit schwacher negativer Brechkraft (Fig. 34). Das ist möglich, weil der Öffnungsfehler schon bei einer Linse günstigster Form ein Vielfaches von dem eines Spiegels gleicher Brennweite beträgt und mit der Durchbiegung noch anwächst. Allerdings sind die Glieder fünfter und höherer Ordnung für Spiegel und Mensikus verschieden; es entsteht also ein

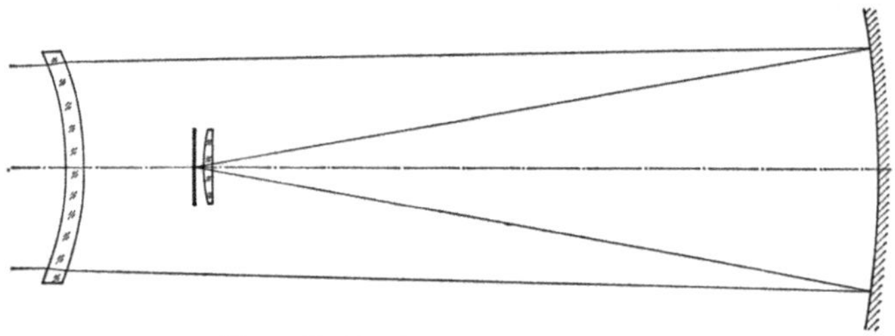

Fig. 34. Meniskus-Kamera mit Ebnungslinse.

Zonenfehler der sphärischen Aberration, der bei großem Öffnungsverhältnis merklich wird. Die Angleichung gelingt um so besser, je dicker der Meniskus ist und je kleiner infolgedessen seine Durchbiegung bleibt. Bei den von MAKSUTOV angegebenen Systemen (mit Meniskendicke gleich $^1/_{10}$ Durchmesser) erreicht die sphärische Aberration 0,″9 bei $N=3$, nach einer empirischen Beziehung

$$\mathcal{S}_{\text{zon}} = 123'' \cdot \frac{1}{N^{4,5}}$$

im Optimalfokus; sie ist durch schwache Deformation einer Meniskenfläche weitgehend zu beseitigen. Der Meniskus ist *achromatisch*, d.h. paraxial frei von Farbortsfehler, wenn

$$\Delta R = d \cdot \frac{n^2 - 1}{n^2}$$

(ΔR Differenz der Krümmungsradien für die Flächen, d Dicke des Meniskus, n Brechungsindex). Er hat noch chromatische Differenz der sphärischen Aberration, die aber bei 1:3 kaum ins Gewicht fällt. Durch passende Wahl des Abstands Meniskus—Spiegel ist eine weitgehende Koma-Kompensation möglich; nach MAKSUTOVs Formeln liegt diese Distanz in der Größenordnung 1,3 ... 1,4f.

Für eine Kamera mit großem Feld ist es erforderlich, die außeraxialen Bildfehler (vor allem die transversale chromatische Aberration) herabzusetzen. Das gelingt in geometrisch einleuchtender Weise dadurch, daß man das System streng *konzentrisch* auslegt: die Krümmungsmittelpunkte der Meniskusflächen und des Spiegels fallen mit dem Pupillenzentrum zusammen (Fig. 35). Die Anordnung ist damit völlig ohne ausgezeichnete Achse, der Bildaufbau ist im gesamten Feld gleichmäßig, und der mögliche Bildfeldwinkel wird nur durch die Vignettierung

[1] D. D. MAKSUTOV: J. Opt. Soc. Amer. **34**, 270 (1944).

begrenzt. Allerdings hat der konzentrische Meniskus primären Farbortsfehler neben der zonalen und der sphärochromatischen Aberration. In der Praxis ist eine Kompromißlösung zwischen dem achromatischen und dem konzentrischen Fall möglich, die gegenüber der Schmidt-Kamera den Vorteil der kürzeren Baulänge hat. Zu einer weiteren Steigerung der Bildqualität (vor allem hinsichtlich des Farbvergrößerungsfehlers) ist vorgeschlagen worden, den Meniskus durch ein zweilinsiges dioptrisches Korrektionssystem zu ersetzen, wobei gleichzeitig der Hauptspiegel elliptisch deformiert wird[1].

Meniskus-Astrokameras sind mit 50 cm Öffnung und 70 cm Spiegeldurchmesser in der Sowjetunion mehrfach gebaut worden; die größte in Abastumani[2] hat einen 70 cm-Meniskus bei 98 cm Spiegeldurchmesser und 210 cm Brennweite, der Bildfelddurchmesser beträgt nahezu 5°.

Als Großkamera hat das Meniskus-System nicht die gleiche Bedeutung gewonnen wie das Schmidt-Teleskop. Für freie Öffnungen über etwa 50 cm ist die Herstellung der verhältnismäßig tiefen Menisken kaum einfacher als die von Schmidt-Platten, und eine Durchmesser-Grenze wird eher erreicht. Unentbehrlich ist der Meniskus als Korrektionsglied in Super-Schmidt-Systemen (s. Ziff. 23) und Spektrographen-Kameras.

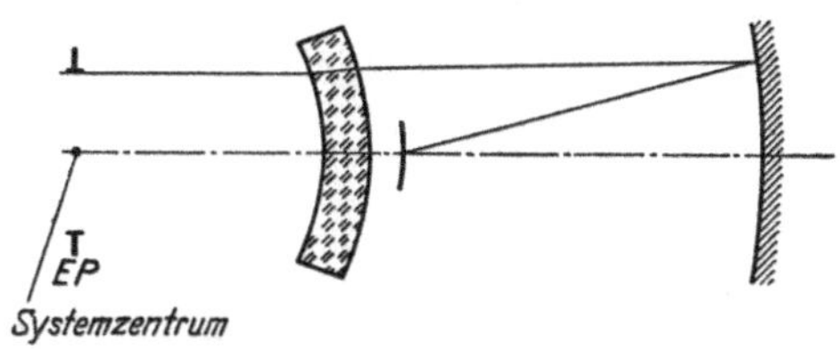

Fig. 35. Konzentrisches Meniskus-System.

In Analogie zum Baker-Schmidt-System sind auch *Meniskus-Cassegrain*-Anordnungen möglich. Astrokameras mit $N=4{,}5$ und einem Feld von etwa 4° ⌀ hat WALAND[3] berechnet. Sie können aber auch mit mäßigen Apertur- und Bildwinkeln als visuelle Fernrohre[4] benutzt werden, wobei die verspiegelte Mitte der Meniskus-Hinterfläche als Sekundärspiegel dient. Gegenüber dem zweilinsigen Objektiv (s. Ziff. 24) hat dieser Typ den Vorteil verschwindender Farbfehler (die Rayleigh-Bedingung ist über einen großen Wellenlängenbereich einzuhalten) und kurzer Baulänge, gegenüber dem klassischen Cassegrain-System den des geschlossenen Rohres ohne Haltestreben für den Sekundärspiegel.

23. Super-Schmidt-Kameras. Die Schmidt-Korrektionsplatte gestattet die völlige Beseitigung auch der höheren Ordnungen von sphärischer Aberration selbst für sehr großes Öffnungsverhältnis, doch werden dabei die außeraxialen Fehler merklich. Beim konzentrischen Meniskus als Korrektor ist die Abbildung von der Hauptstrahlneigung unabhängig, doch läßt sich die sphärische Aberration nicht bis zu höheren Ordnungen heben. Für extreme Anforderungen an Feld und Öffnungsverhältnis ist die Kombination der beiden Korrektionsmittel naheliegend und auch schon früh angegeben worden[5]. Die Schmidt-Platte in der Eintrittspupille korrigiert nur die sphärischen Reste, ihre Deformation ist viel kleiner als bei reinen Schmidt-Systemen.

Zur Verringerung der farbabhängigen Fehler des Meniskus kann die Korrektionsplatte aus zwei Elementen (Gläser mit verschiedener Dispersion) zusammengesetzt werden, deren Trennfläche jedoch nicht konzentrisch liegt; damit sie die

[1] T. S. BELOROSSOVA, D. D. MAKSUTOV, N. V. MERMAN u. M. A. SOSNINA: Trudy 15. Astrometr. Konf. S.S.S.R. (Pulkovo 1960), S. 215 (1963).
[2] R. I. KILADZE: Bull. Abastumani Astrophys. Obs. **24**, 35 (1959).
[3] R. L. WALAND: J. Opt. Soc. Amer. **51**, 359 (1961).
[4] N. W. MERMAN: Mitt. Pulkovo **20**, H. 6 (No. 159), 144 (1958).
[5] BOUWERS [*32*]. — D. G. HAWKINS and E. H. LINFOOT: Monthly Notices Roy. Astron. Soc. **105**, 334 (1945).

Symmetrie möglichst wenig stört, sollen die beiden Glasarten den gleichen Brechungsindex bei einer mittleren Wellenlänge haben.

Ein Beispiel für ein solches ultralichtstarkes Weitwinkel-Objektiv ist das der Harvard-Meteorkamera von BAKER[1] mit zwei symmetrisch zum Systemzentrum liegenden Menisken und $D=30$ cm, $f=20$ cm, $N_{eff}=0{,}82$, $2w_{max}=52°$ (Fig. 36a). Im photographischen Wellenlängenbereich ist die Lichtkonzentration selbst am Bildrand bemerkenswert (80% in einem Kreis mit 50 μm Durchmesser).

Auch ohne Meniskus lassen sich Schmidt-Systeme zu großen Bild- und Aperturwinkeln treiben: die ebenfalls von BAKER entwickelte Optik einer Kamera zur Satellitenphotographie (Fig. 36b) hat ein dreigliedriges symmetrisches Korrektionssystem mit zwei sphärischen und vier asphärischen Flächen[2]. Durch Verwendung von Spezialgläsern ist das System für drei weit auseinanderliegende

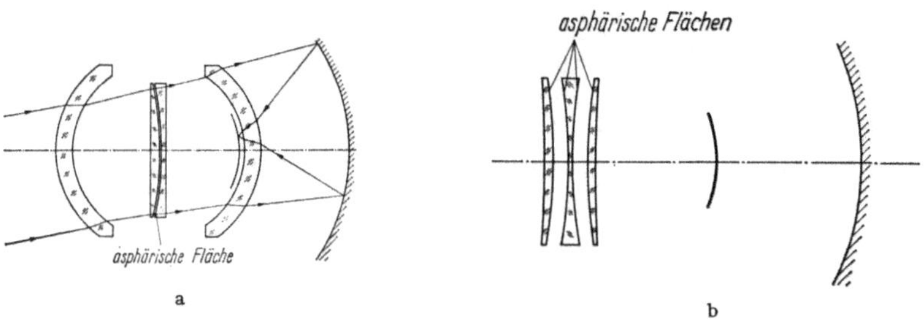

Fig. 36a u. b. Ultralichtstarke Weitwinkel-Spiegelobjektive. a Optik der Harvard-Meteor-Kamera (Super-Schmidt-System nach BAKER). b Optik der Baker-Nunn-Kamera mit dreigliedrigem Korrektionssystem in der Eintrittspupille.

Wellenlängen korrigiert und damit vom Ultraviolett bis zum Rot brauchbar. Bei $f=51$ cm und $N\approx 1$ gehen selbst am Rand des Feldes von $5°\times 30°$ noch 80% des Lichts in einen Kreis von 20 μm Durchmesser.

II. Systeme mit Refraktionsoptik.

24. Achromatische Objektive.

α) *Chromatische Korrektion.* Die Toleranzbedingung für die dünne Einzellinse (13.1) führt zu extrem kleinen Öffnungsverhältnissen (wie sie vor Erfindung des Achromaten ja auch angewendet wurden); schon bei mäßiger Öffnung muß ein einigermaßen lichtstarkes Objektiv *achromatisiert* werden. Eine vollständige Beschreibung des sekundären Spektrums auf der Achse gibt die *Farbkurve* (Fig. 37). Die normalen zweilinsigen Achromate (Fraunhofer-Typ) sind durch die mit E bezeichnete Kurve charakterisiert. Bei den für astronomische Fernrohre üblichen Öffnungsverhältnissen um 1:15 ist die Toleranzgrenze (13.2) für größere Instrumente deutlich überschritten, und von etwa $D=30$ cm an macht das sekundäre Spektrum das Bild unbefriedigend. Die Energieverteilung im *Beugungsbild* für heterochromatisches Licht haben LINFOOT und WOLF[3] untersucht.

Eine Reduktion des sekundären Spektrums ist dringend erwünscht; dagegen sind die üblichen Fernrohrobjektive hinreichend dünn, und die Eintrittspupille fällt mit den Linsen nahezu zusammen: *Farbquerfehler* können deshalb weitgehend unberücksichtigt bleiben. Durch Auswahl von im ν, ϑ-Diagramm günstig liegenden Gläsern läßt sich das sekundäre Spektrum um etwa $^1/_3$ vermindern (Halbapo-

[1] F. L. WHIPPLE: Harvard Meteor Program. Cambridge 1947 = Harvard Reprint Ser. II No. 19; s. auch W. R. BRADFORD, Observatory **76**, 172 (1956); J. DAVIS, Quart. J. Roy. Astron. Soc. **4**, 74 (1963).

[2] K. G. HENIZE: Sky and Telescope **16**, 108 (1957).

[3] E. H. LINFOOT and E. WOLF: Monthly Notices Roy. Astron. Soc. **112**, 452 (1952).

chromate, Kurve AS) oder bei Beschränkung auf $N \gtrsim 20$ auch halbieren (PAULY[1]). Allerdings bleibt dabei die ν-Differenz — für möglichst kleine ϑ-Differenz — ebenfalls klein; die Einzellinsen müssen also stärkere Brechkraft bekommen, die Innenflächen werden stark gekrümmt, die Korrektion der geometrischen Bildfehler wird schwieriger, die Zentrierempfindlichkeit größer. Die Verwendung von drei Linsen aus verschiedenen Gläsern gestattet es, die Schnittweiten für drei Wellenlängen zusammenzulegen (die Farbkurve ähnelt dann einer Parabel dritten Grades). Das übrigbleibende „tertiäre" Spektrum ist aber nicht notwendig geringer als das sekundäre eines Zweilinsers. Ein Vorteil des Dreilinsers ist, daß die Krümmungen der Flächen kleiner bleiben (oder bei gleicher Krümmung Gläser

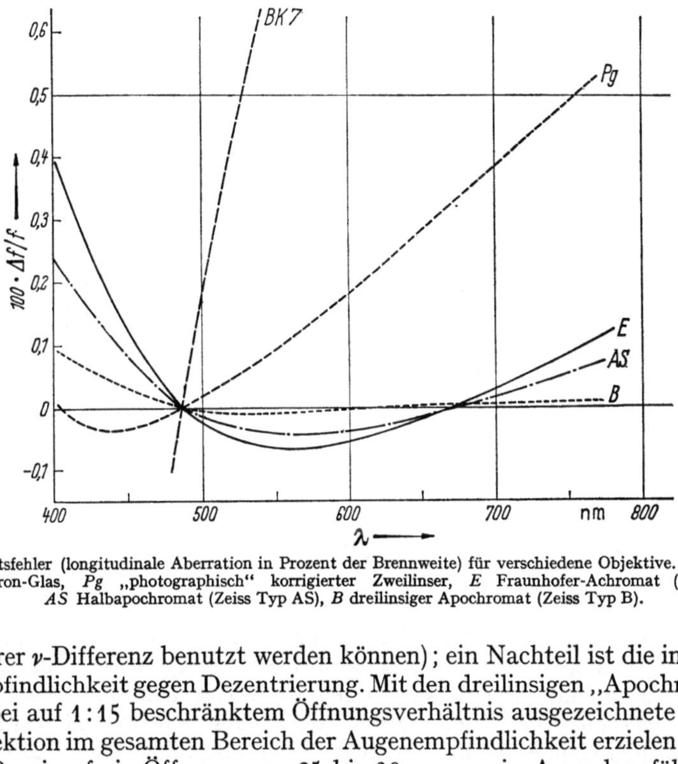

Fig. 37. Farbortsfehler (longitudinale Aberration in Prozent der Brennweite) für verschiedene Objektive. *BK* 7 Einfache Linse aus Kron-Glas, *Pg* „photographisch" korrigierter Zweilinser, *E* Fraunhofer-Achromat (Zeiss Typ E), *AS* Halbapochromat (Zeiss Typ AS), *B* dreilinsiger Apochromat (Zeiss Typ B).

mit kleinerer ν-Differenz benutzt werden können); ein Nachteil ist die immer noch große Empfindlichkeit gegen Dezentrierung. Mit den dreilinsigen „Apochromaten"[2] läßt sich bei auf 1:15 beschränktem Öffnungsverhältnis ausgezeichnete chromatische Korrektion im gesamten Bereich der Augenempfindlichkeit erzielen (Kurve B, Fig. 37). Da eine freie Öffnung von 25 bis 30 cm nur in Ausnahmefällen[3] überschritten wird, reichen sie jedoch nicht in die Größenordnung, in der sie wirklich interessant wären. Die durch die Fortschritte der Glasherstellung gegebenen Möglichkeiten für große visuelle Refraktoren sind von BAKER[4] diskutiert worden. Daß selbst auf diesem klassischen Gebiet die Entwicklung noch nicht abgeschlossen ist, zeigen die „Superachromate" HERZBERGERS[5], die sich als Folge der Darstellbarkeit der Dispersionskurven für alle optischen Werkstoffe mit vier Parametern ergeben.

Der Scheitel der Farbkurve wird für visuelle astronomische Zweilinser meist vom Maximum der Augenempfindlichkeit (555 nm) etwas nach größeren Wellenlängen hin verschoben (~575 nm); der F- und der C-Fokus fallen dann nicht

[1] M. WOLF: Z. Instrumentenkde 19, 1 (1899).
[2] H. D. TAYLOR: Monthly Notices Roy. Astron. Soc. 54, 328 (1894).
[3] Astron. J. 69, 685 (1964).
[4] J. G. BAKER: Appl. Optics 2, 111 (1963).
[5] M. HERZBERGER and N. R. MCCLURE: Appl. Optics 2, 553 (1963).

mehr genau zusammen. Für Photographie mit nicht farbsensibilisierten Platten vereinigt man F und h, was zu einem Scheitel der Farbkurve etwa bei G' (434 nm) führt. Zur Änderung der Farbkorrektur (von visuell nach photographisch oder umgekehrt) hat man gelegentlich afokale Zusatzsysteme relativ kleinen Durchmessers verwendet[1], die allerdings zu einer erheblichen chromatischen Vergrößerungsdifferenz im Systembrennpunkt führen.

Eine ganz andere Art der Achromatisierung hat Schupmann mit dem *Medial-Fernrohr*[2] vorgeschlagen. Hier werden die sammelnde und die zerstreuende Komponente mit einer Feldlinse aufeinander abgebildet und so in optischen Kontakt gebracht. Verwendet man die gleiche Glassorte für beide Komponenten,

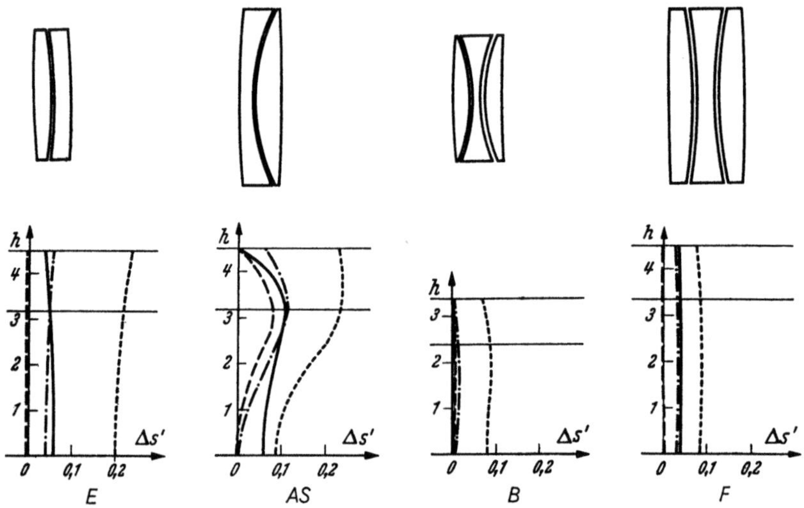

Fig. 38. Aufbau und axiale Korrektion für visuelle Zeiss-Objektive (außer den Typen der Fig. 37 ist unter F der „Schwerflint-Apochromat" aufgeführt). Die Systemschnitte sind für eine Brennweite von 25 cm gezeichnet; in den Korrektionsdarstellungen sind $\Delta s'$ und h auf $f = 100$ bezogen. Wellenlängen: ------ 436 nm, — · — · — 486 nm, — — — 546 nm, ——— 656 nm. Nach König/Köhler [8].

so läßt sich bei passender Wahl der Stärke vollkommene Achromasie erzielen; dabei geht allerdings die zur Abbildung nötige Konvergenz des Bündels verloren und muß durch eine *Spiegel*fläche (die von Natur achromatisch ist) wiederhergestellt werden. Die Kombination Negativkomponente—Hohlspiegel kann dabei wesentlich kleiner als das Positivglied sein und bildet einen Korrektor nach Art eines Mangin-Spiegels. Die Möglichkeiten des Medials hat neuerdings Baker[3] diskutiert und auch darauf hingewiesen, daß mit ähnlichen Korrektoren die Farbkurven existierender großer Refraktoren verbessert werden könnten.

β) Sonstige Abbildungsfehler. Durch die Achromasiebedingung (12.3) ist nur die *Stärke* der Komponenten eines zweilinsigen Objektivs festgelegt; ihre *Form* kann — zur Beseitigung anderer Abbildungsfehler — noch variiert werden. Als eine der zu erfüllenden Bedingungen wählt man stets die Hebung der *sphärischen Aberration*. Sie wird erreicht, wenn die Kaustik der positiven Komponente bei Abbildung eines fernen Punktes gleich der (virtuellen) Kaustik der zerstreuenden Linse ist, die ein aus dem *System*brennpunkt kommendes Bündel hat. Die sphärische Aberration einer Linse ändert sich mit ihrer *Durchbiegung*; zu jeder Durchbiegung der einen Komponente gehören allgemein zwei Durchbiegungswerte der

[1] Siehe z. B. F. E. Ross, Astrophys. J. **76**, 184 (1932).
[2] L. Schupmann: Die Medial-Fernrohre. Leipzig 1899.
[3] J. G. Baker: Astron. J. **59**, 74 (1954).

anderen, die zur Kompensation führen. In einem gewissen Bereich ist die Krümmung der ersten Fläche — unter Einhaltung der vorgesehenen Brennweite, der Achromasie und der axialen Korrektur — noch frei wählbar. Dieser Freiheitsgrad wird genutzt, um eine weitere Bedingung zu erfüllen. Meist wird dafür Korrektion der Koma gewählt; man spricht dann (historisch wohl zu Unrecht) von Einhaltung der Fraunhofer-Bedingung. Andere Möglichkeiten sind Hebung der sphärischen Aberration für eine zweite, verhältnismäßig kleine Dingweite (Herschel-Bedingung, ermöglicht Prüfung mit Testobjekt in endlicher Entfernung) oder Verkittbarkeit (die inneren Krümmungsradien sind gleich; die Bedingung ist für größere Astro-Objektive bedeutungslos, weil diese wegen Verspannungsfreiheit stets unverkittet sein müssen). Fordert man Hebung der sphärischen Aberration für eine zweite Wellenlänge (Gauß-Bedingung), so wird man auf einen Typ mit vier gegen den Lichteinfall erhabenen Flächen geführt. Bei den klassischen Zweilinsern wird allerdings der Gauß-Fehler vom sekundären Spektrum so stark übertroffen, daß seine Berücksichtigung nicht gerechtfertigt ist.

Die axiale Korrektion der Fernrohrobjektive wird traditionsgemäß in einem Diagramm Schnittweitendifferenz $\Delta s'$/Einfallshöhe h mit Kurven für verschiedene Wellenlängen graphisch dargestellt (Fig. 38). Man erkennt, wie sehr das sekundäre Spektrum überwiegt; nur beim Halbapochromaten sind der Zonenfehler und die chromatische Differenz der sphärischen Aberration merklich — eine Folge der starken Krümmung der Innenfläche. Die *außeraxialen Abbildungsfehler* sind bei Okularbeobachtung kaum störend, wegen des relativ kleinen Gesichtsfelds. Der Seidel-Koeffizient für Koma liegt für Objektive vom Fraunhofer-Typ etwa zwischen 0 und $+0{,}2$, ist also deutlich kleiner als der eines Parabolspiegels ($-0{,}5$). Der Astigmatismus dünner Objektive ($\sum \mathbf{IIIa} = +1$) ist ebenso groß wie der entsprechender Spiegel, die Krümmung der mittleren Bildschale jedoch merklich größer. Zu beachten ist, daß die Grenzen der Seidel-Näherung bei Linsensystemen viel früher erreicht werden als bei Spiegeln.

25. Okulare.

α) *Visueller Gebrauch astronomischer Fernrohre.* Das von einem Linsen- oder Spiegelobjektiv entworfene Bild kann man mit freiem Auge betrachten (Fig. 39). Hat das Auge den Abstand B vom Bild (Weite des deutlichen Sehens), so ist die Fernrohrvergrößerung Γ (definiert als $\Gamma = w_2/w_1$) gegeben durch $|\Gamma| = f_{\text{obj}}/B$. Das vom Objektiv aufgesammelte Licht wird dabei jedoch schlecht ausgenutzt: die wirksame Öffnungszahl ist bei dieser Anordnung $N_{\text{eff}} = B/d$ (d ist der Durchmesser der Augenpupille), und mit $B = 25$ cm, $d \leqq 0{,}7$ cm folgt $N_{\text{eff}} \geqq 35$. Das Betrachten eines größeren Feldes macht zudem Translationen des Auges erforderlich. Für bessere Anpassung oder stärkere Vergrößerung muß man das Auge näher an das Bild heranbringen und dafür eine Betrachtungsoptik, ein *Okular*, verwenden.

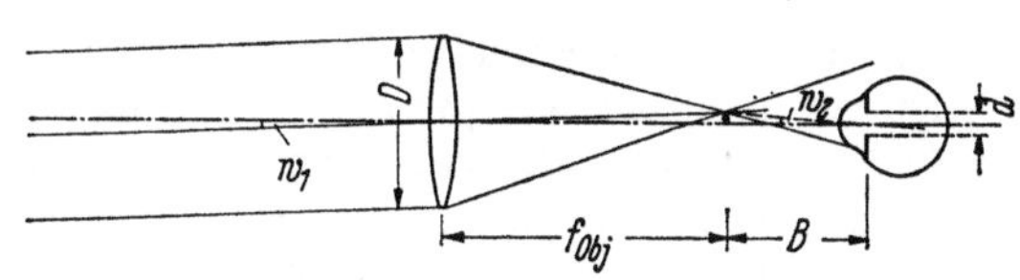

Fig. 39. Visuelle Fernrohrbeobachtung ohne Okular.

Liegen Objekt und Bild (bei fernakkommodiertem Auge des Beobachters) in großer Entfernung, so bildet das „astronomische“ oder Keplersche Fernrohr (Fig. 40) ein afokales oder teleskopisches System. Man entnimmt der Abbildung, daß

$$\frac{f_{\text{obj}}}{f_{\text{okl}}} = \frac{D}{P} = -\frac{w_2'}{w_1}$$

ist; P ist der Durchmesser der Austrittspupille, die hier auch Okularkreis oder Ramsden-Kreis genannt wird. Wegen der benutzten Gauß-Näherung ist die

Winkelfunktion nicht spezifiziert. Ein axialer und ein außeraxialer Stern mit dem Winkelabstand w_1 erscheinen dem Beobachter unter dem Winkelabstand w_2'; das Verhältnis $\Gamma = w_2'/w_1$ wird deshalb als Fernrohrvergrößerung bezeichnet. $\Gamma < 0$ beim astronomischen Fernrohr charakterisiert das umgekehrte Bild.

Für größere Winkel w muß erfüllt sein

$$\frac{\tan w_2'}{\tan w_1} = \text{const},$$

damit die Abbildung verzeichnungsfrei erscheint. Zwischen Ding und Bild findet dann eine Ähnlichkeitstransformation in einer fernen, achsensenkrechten Ebene statt. Diese Definition der Verzeichnungsfreiheit ist aber nur für den Fall des fest auf ein Objekt gerichteten Fernrohrs sinnvoll. Wird beim freihändigen Gebrauch

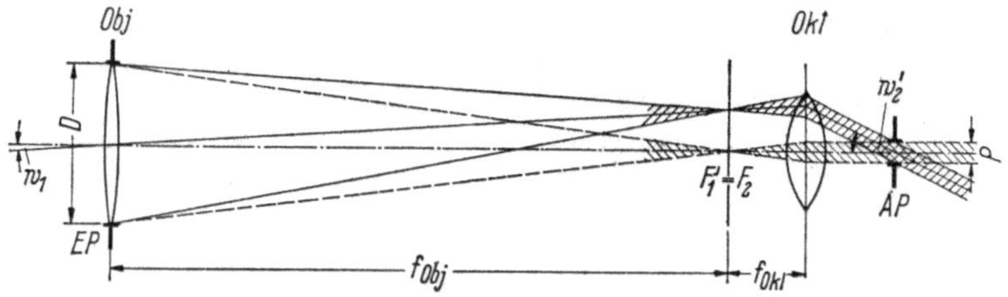

Fig. 40. Abbildung im Keplerschen Fernrohr.

das Instrument zum Abtasten eines größeren Objekts geschwenkt, so treten beim verzeichnungsfreien Fernrohr „Bildverbiegungen" auf; hier wäre $w_2'/w_1 = \text{const}$ vorteilhafter.

Damit das Auge den gesamten Lichtstrom zugeführt bekommt, muß $d \geqq P$ sein, wenn Austritts- und Augenpupille zusammenfallen, also

$$|\Gamma| \geqq \Gamma_p = \frac{D}{d}.$$

Γ_p ist die sog. Normalvergrößerung (Danjon und Couder [6] benutzten den treffenden Ausdruck äquipupillare Vergrößerung). Der Wert von d hängt von der Sehfeldleuchtdichte ab und liegt bei astronomischen Anwendungen zwischen etwa 0,2 cm (Mond) und 0,7 cm (schwächste Objekte). Ist $|\Gamma| < \Gamma_p$ so wird das Objektiv durch die — zentriert vorausgesetzte — Augenpupille abgeblendet auf $D_{\text{eff}} = d \cdot |\Gamma|$.

Das Gesichtsfeld wird, um einen vignettierten Rand zu vermeiden, gewöhnlich durch eine Blende vom Durchmesser G in der gemeinsamen Fokalebene von Objektiv und Okular begrenzt. Den scheinbaren (Winkel-)Radius des Gesichtsfeldes gibt $\tan w_2' = \frac{G}{2f_{\text{okl}}}$, den wahren im Objektraum $\tan w_1 = -\frac{G}{2f_{\text{obj}}} = \frac{\tan w_2'}{\Gamma}$. Liegt die Augenpupille nicht am Ort der Austrittspupille, so kann Vignettierung des Feldes eintreten (Schlüssellochbeobachtung).

Damit die durch Beugung am Objektiv gegebene Auflösungsgrenze nach (5.3) sicher erreicht wird, sollte dem Radius des ersten dunklen Rings (mit $w_1 = 14''/D$, D in cm) ein scheinbarer Gesichtswinkel von mindestens $w_2' \approx 2$ Bogenminuten entsprechen; dies führt auf $|\Gamma_B| = 8{,}6\,D \approx 10\,D$ und $P_B = 0{,}1$ cm. Bei großen Instrumenten ist diese sog. „förderliche Vergrößerung" (nach Danjon und Couder besser „auflösende Vergrößerung") allerdings unrealistisch; bei kleineren mag sie, z. B. für Meßzwecke, um ein mehrfaches überschritten werden. Die für Γ_B erforderliche Okularbrennweite ist $f_{\text{okl}, B} = 0{,}1\, N$ (cm), stärkere Okulare als $f_{\text{okl}} = 15$ mm sind also für visuelle Standardrefraktoren mit $N = 15$ kaum erforderlich. Die handelsüblichen Okularsätze reichen meist bis $f = 4$ mm herab.

Bei vielen visuellen Anwendungen von Fernrohren werden in der Bildebene Fadenkreuze oder Mikrometerfäden angebracht. Bei dunklem Feld und beleuchteten Fäden wirken sie wie ein selbstleuchtendes Objekt, das nach allen Seiten strahlt; von den Elementen des Bildes dagegen gehen nur begrenzte Strahlenbündel aus. Für Meßzwecke ist es erforderlich, daß von koinzidierenden Punkten von Bild und Faden auch die Bündel zusammenfallen und so gleiche Teile des Okulars und des Auges durchsetzen. Dies kann durch eine körperliche Blende erreicht werden, die sich mit der Austrittspupille deckt. Wegen der Farbfehler von Okular und Auge sollte das zur Fadenbeleuchtung verwendete Licht eine ähnliche spektrale Zusammensetzung haben wie das des Objekts; rote Fadenbeleuchtung ist also ausgesprochen ungünstig. Bei dunklen Fäden im aufgehellten Feld ist das Zusammenfallen der Bündel nicht ohne weiteres zu erreichen, jedoch ergibt sich bei Bisektion des Sternscheibchens mit dem dunklen Faden automatisch korrekte Lichtführung, weil hierbei das benutzte Stück des Fadens vom Stern selbst beleuchtet wird.

β) *Aufbau und Eigenschaften der Okulare.* Ein Okular wird mit großen Hauptstrahlneigungen w_2' benutzt. Das Öffnungsverhältnis für das zu jedem Bildpunkt gehörende Bündel ist gleich dem des Objektivs, das Verhältnis von Durchmesser zu Brennweite des Okulars jedoch — wegen des erforderlichen Bildfelds — viel größer. Die Pupillen des Okulars fallen mit dem Objektivrand und der Austrittspupille des Fernrohrs zusammen. Ein Okular ist stets ein *dickes* System. Dadurch wird die *laterale* chromatische Aberration merklich: die Hauptstrahlneigung w_2' und damit die Vergrößerung Γ sind wellenlängenabhängig. Die Bezeichnungen chromatische Vergrößerungsdifferenz oder Farbvergrößerungsfehler entsprechen gerade diesem Anwendungsfall. Bei einer einfachen Okularlinse unterscheiden sich die Bildgrößen für die Grenzen des visuellen Bereichs um Beträge von der Größenordnung Prozent; ein außeraxialer Stern erscheint also als radial gerichtetes Spektrum (Rot innen). Die *longitudinale* chromatische Aberration der einfachen Linse könnte in vielen Fällen vernachlässigt werden, da die Bedingung (13.1) $f \leqq N^2/40$ cm für $N = 15$ ohne weiteres erfüllt ist (ein Okular muß um so besser korrigiert sein, je *länger* die Brennweite ist!). Auch sphärische Aberration tritt für $N = 15$ kaum in Erscheinung; die merklichsten monochromatischen Abbildungsfehler sind *Bildwölbung* (die durch die Akkommodationsmöglichkeit des Auges in gewissen Grenzen ausgeglichen werden kann) und kissenförmige *Verzeichnung* (die beim Betrachten unbemerkt bleibt, aber für Mikrometermessungen ebenso wie die Wölbung unerwünscht ist). Für die kleinen Öffnungszahlen von Spiegelteleskopen im Primärfokus sind allerdings die üblichen Okulare nicht ausreichend korrigiert.

Zur Hebung der chromatischen Vergrößerungsdifferenz ist es nur erforderlich, die *Brennweite* des Okulars zu achromatisieren; Achromasie der *Schnittweiten* ist nicht erforderlich. Die verschiedenfarbigen austretenden Hauptstrahlen, die zu einem heterochromatischen eintretenden Hauptstrahl gehören, sind dann zwar u. U. gegeneinander versetzt, aber jedenfalls parallel; das Auge nimmt die Versetzung der Bündelachsen nicht wahr. Die Achromasie der Brennweite ist schon mit zwei Linsen aus dem *gleichen* Material zu erreichen. Für ein Zweilinsensystem gilt (2.1)

$$f = \frac{f_1 f_2}{f_1 + f_2 - e},$$

wobei e der Linsenabstand ist. Infolge der Dispersion sei für eine zweite Wellenlänge $f_1^* = f_1(1+\varepsilon)$, $f_2^* = f_2(1+\varepsilon)$; damit $f^* = f$, muß sein

$$\frac{f_1 f_2}{f_1 + f_2 - e} = \frac{(1+2\varepsilon) f_1 f_2}{(1+\varepsilon)(f_1+f_2) - e},$$

also

$$e = \frac{f_1 + f_2}{2}. \tag{25.1}$$

Dieses Verfahren zur Beseitigung der chromatischen Vergrößerungsdifferenz hat bereits HUYGENS bei dem nach ihm genannten Okular (Fig. 41b) angewendet. Es besteht aus zwei Plankonvexlinsen (Planseiten zum Auge), die meist nach dem

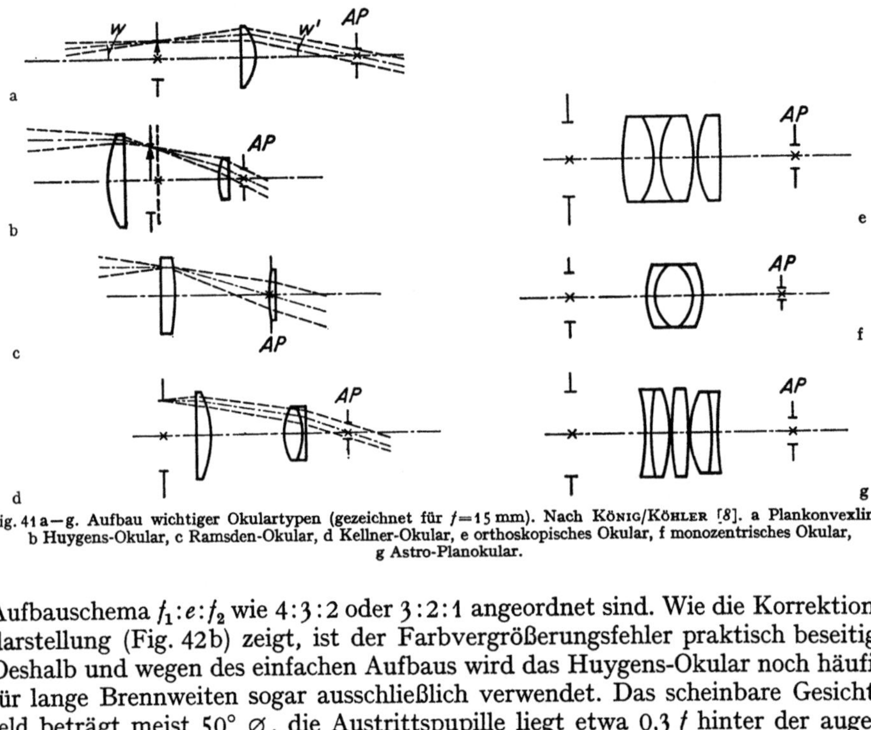

Fig. 41a—g. Aufbau wichtiger Okulartypen (gezeichnet für $f = 15$ mm). Nach KÖNIG/KÖHLER [8]. a Plankonvexlinse, b Huygens-Okular, c Ramsden-Okular, d Kellner-Okular, e orthoskopisches Okular, f monozentrisches Okular, g Astro-Planokular.

Aufbauschema $f_1 : e : f_2$ wie 4:3:2 oder 3:2:1 angeordnet sind. Wie die Korrektionsdarstellung (Fig. 42b) zeigt, ist der Farbvergrößerungsfehler praktisch beseitigt. Deshalb und wegen des einfachen Aufbaus wird das Huygens-Okular noch häufig, für lange Brennweiten sogar ausschließlich verwendet. Das scheinbare Gesichtsfeld beträgt meist 50° ⌀, die Austrittspupille liegt etwa 0,3 f hinter der augenseitigen Linse. Die Reflexe sind auch ohne Vergütung der Flächen so schwach, daß das Okular gewöhnlich als reflexfrei bezeichnet wird. Zur Verwendung mit Mikrometern ist es nicht geeignet, weil die Bildfläche zwischen den beiden Linsen schlecht zugänglich ist; für etwa dort angebrachte Fäden ist das Okular auch gar nicht chromatisch korrigiert, ebensowenig wie für die Gesichtsfeldblende (ein helles Feld erscheint blaugesäumt).

Für Mikrometermessungen ist es erforderlich, daß die Bildebene außerhalb des Okulars liegt. In der älteren Literatur wird eine solche Bauart als „positiv" bezeichnet; das Huygens-Okular ist in dieser Terminologie „negativ", trotz positiver Brennweite. Das erste korrigierte Mikrometerokular ist das von RAMSDEN (Fig. 41c) mit $f_1 = f_2 = e = f$, womit die Huygens-Bedingung ebenfalls erfüllt ist. Die beiden Plankonvexlinien kehren einander die gewölbten Seiten zu. Das Bild liegt in der objektivseitigen, die Austrittspupille in der augenseitigen Linse. Innerhalb der Gauß-Optik ändert die vordere Linse als reine Feldlinse den Abbildungsstrahlengang nicht. Sie dient lediglich dazu, das Objektiv in die hintere, vergrößernde Linse abzubilden, deren Durchmesser infolgedessen nicht mehr den Felddurchmesser beeinflußt.

In der reinen Form ist das Ramsden-Okular für die Praxis unbrauchbar; einmal weil die Augenpupille nicht an den Ort der Austrittspupille gebracht werden kann, zum anderen weil Staub und Kratzer auf der ersten Fläche zusammen mit dem Bild scharf gesehen werden. Als Abhilfe hat man die Linsen etwas näher zusammengerückt, wodurch Bildfläche und Pupille günstiger liegen. Dabei geht allerdings die strenge Korrektion der chromatischen Vergrößerungsdifferenz verloren, und der Okulartyp spielt deshalb keine Rolle mehr. Eine wesent-

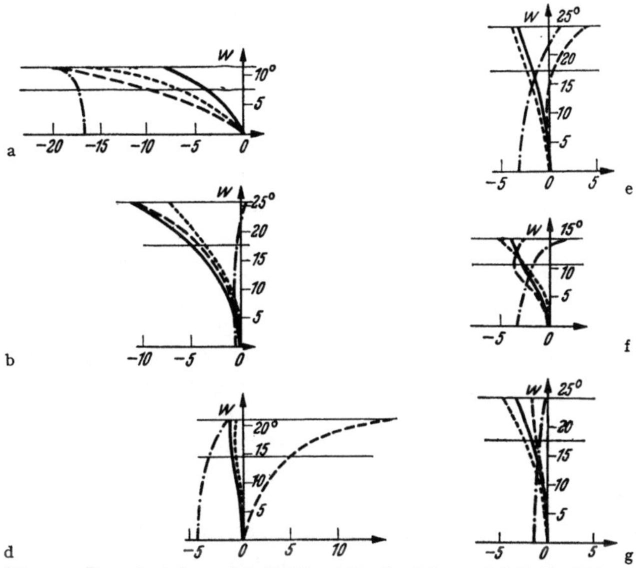

Fig. 42a–g. Korrektion von Fernrohrokularen. Für die Korrektionsdarstellung wird die Strahlrichtung umgekehrt und die Abbildung eines augenseitigen fernen Objekts in die Fernrohr-Bildfläche betrachtet. Angegeben sind als Funktion der Hauptstrahlneigung w (entspricht dem subjektiven Bildradius w_2' in Fig. 40): sagittale ——— und meridionale – – – Längsabweichung (bezogen auf $f = 100$); Verzeichnung ------ in Prozent; chromatische Vergrößerungsdifferenz zwischen 486 und 656 nm — · — · — in Promille. Nach König/Köhler [8]. a Plankonvexlinse, b Huygens-Okular, d Kellner-Okular, e orthoskopisches Okular, f monozentrisches Okular, g Astro-Planokular.

liche Verbesserung brachte das vom modifizierten Ramsden-Typ abgeleitete Kellnersche Okular (Fig. 41d), bei dem das augenseitige Element aus zwei Linsen mit verschiedenen Gläsern verkittet ist. Dadurch läßt sich der Farbvergrößerungsfehler ohne Einhalten der Huygens-Bedingung weitgehend korrigieren (Fig. 42d). Das Okular hat bemerkenswert geringe Verzeichnung (Kellner bezeichnete es deshalb als „orthoskopisch"), aber ungünstige Bildwölbung. Es wird meist für ein Bildfeld von 40° ∅ ausgelegt. Eine Variante hat das Kollektiv *vor* der Bildfläche wie das Huygens-Okular; dabei rückt aber das Bild wieder unangenehm nahe an die Feldlinse heran.

Für stärkere Vergrößerungen ist zu fordern, daß Bildfläche und Austrittspupille möglichst weit — d.h. von der Größenordnung der Brennweite — von der ersten bzw. letzten Fläche des Okulars abliegen. In dieser Hinsicht günstig ist das „orthoskopische Okular" nach Abbe mit dreifacher verkitteter Feld- und einfacher augenseitiger Linse. Es wird mit einem Bildfeld von 40° ∅ gebaut und als reflex-, farb- und verzeichnungsfrei bezeichnet; für Mikrometermessungen ist es besonders geeignet. Großen Bild- und Augenabstand haben auch die dreilinsigen verkitteten Okulare, die man sich als von der einfachen Linse abgeleitet vorstellen kann. Verbreitet ist eine als „monozentrisch" bezeichnete Form (Fig. 41f) von STEINHEIL, die für Felder von 25 bis 30° ∅ recht gute Korrektion liefert und reflexfrei ist; daneben gibt es ein echtes monozentrisches System (mit gemeinsamem Krümmungsmittelpunkt für alle Flächen) mit ähnlichen Eigenschaften, ebenfalls von STEINHEIL berechnet. Daß sich bei erhöhtem Aufwand auch die Korrektion noch verbessern läßt, zeigt das bei Zeiss entwickelte Astro-Plan-Okular aus fünf Linsen, das für 40 bis 50° Felddurchmesser gebaut wird. Die Erfle-Okulare und ähnliche Typen, deren Gesichtsfeld 70° und mehr beträgt, sind bei Astroinstrumenten höchstens für Sucher von Interesse.

Nähere Angaben über Okulartypen[1] sind u. a. in [8], § 17 zu finden.

26. Linsenobjektive für Astrokameras. An die Objektive der astronomischen Kameras stellt man viel höhere Anforderungen hinsichtlich Öffnungsverhältnis und Bildfeld als an die in Ziff. 24 behandelten Fernrohrobjektive, deren Grenze mit dem „Himmelskartenrefraktor" ($f=343$ cm, $N=10$, Feld $2°\times 2°$) etwa erreicht war. Für große Astrokameras fordert man etwa $N=5$ und $w_{\max}=4°$, was bei Brennweiten von der Größenordnung 2 m einen beträchtlichen optischen Aufwand erfordert. Kleinere Kameras werden mit noch größeren Feldern gebaut; ihre Objektive schließen an die normaler Photoapparate an (vgl. etwa FLÜGGE [16]). Die Seidel-Koeffizienten stellen bei den großen Werten von $1:N$ und w für Linsenoptik die Strahlenvereinigung nicht ausreichend dar. Die Konstruktionsdaten der Objektive müssen deshalb, von einem Grundtyp ausgehend, mit Strahldurchrechnungen verbessert werden. Die erzielte Korrektion wird traditionell in Diagrammen anschaulich gemacht, die die Zonenfehler (axiale Schnittweite als Funktion der Einfallshöhe, möglichst für mehrere Wellenlängen) und die Lage der Bildflächen (sagittale und tangentiale Längsabweichung als Funktion der Hauptstrahlneigung, Fig. 43) angeben; eine vergleichbar anschauliche Darstellung für Koma gibt es nicht. Das Durchstoßdiagramm beginnt sich auch hier durchzusetzen. Wegen der begrenzten Auflösung der Photoplatten gibt man die Querabweichungen meist im linearen Maß und fordert z. B. eine Lichtkonzentration von 100% in einem Kreis von 25 μm.

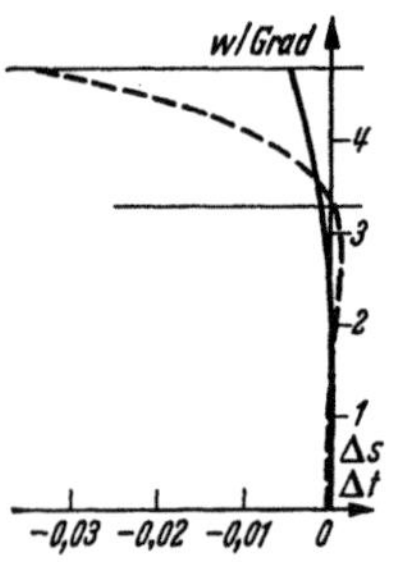

Fig. 43. Feldkorrektion des Sonnefeld-Vierlinsers. Dargestellt sind die sagittale ——— und die tangentiale — — — Längsabweichung für die Brennweite $f=100$ in Abhängigkeit vom Bildwinkel w.

Trotz der Konkurrenz des Schmidt-Teleskops werden Linsenobjektive noch in gewissem Umfang für Zwecke der Positionsbestimmung und der photographischen Photometrie benutzt. Als Vorteile sind das große, ebene und leicht zugängliche Bildfeld und die geringe Baulänge der Kamera anzusehen. Das älteste Objektiv, das den Forderungen nach Feld und Lichtstärke nahekam, war das „Porträt-Objektiv" von PETZVAL, wegen seiner Herkunft aus zwei Achromaten mit großem Luftabstand auch „doublet lens" genannt. Die moderne Entwicklung setzt mit dem Triplett (Cooke lens) von TAYLOR[2] ein, der Urform vieler Kamera-Anastigmate. Gegenwärtig werden Großobjektive nur noch vierlinsig hergestellt. Zu erwähnen sind die Bauart von ROSS[3], der Sonnefeld- und der Askania[4]-Vierlinser (Fig. 44). Sie sind außer für den „photographischen" Bereich mit Farbscheitel bei 430 nm auch für 550 nm bzw. 650 nm korrigiert worden. Bei 40 bzw. 50 cm freier Öffnung und $N=4\ldots 7$ zeichnen sie Felder von $6°\times 6°$ bis $10°\times 10°$ aus.

[1] Siehe auch H. H. SELBY, Telescope Eyepieces. In [31] III.

[2] H. D. TAYLOR: Monthly Notices Roy. Astron. Soc. **64**, 613 (1904); s. auch SCHWARZSCHILD [13].

[3] F. E. ROSS: J. Opt. Soc. Amer. **5**, 123 (1921); s. auch S. VASILEVSKIS, Publ. Astron. Soc. Pacific **76**, 14 (1964).

[4] H. SCHMIDT: Askania-Warte **21**, H. 63, 3 (1964).

Während das Ross-Objektiv symmetrisch ist, sind die beiden anderen Bauweisen aus dem Triplett durch Aufspalten der Vorder- bzw. Hinterlinse hervorgegangen. Mittels Deformation einer Fläche werden die Zonenfehler der sphärischen Aberration klein gehalten. Die Verzeichnung ist sehr klein.

Wenn bei diesen Vierlinsern auch die monochromatische Korrektion sehr hoch getrieben werden kann, so bleibt doch der schwerwiegende Nachteil des sekundären Spektrums bestehen. Dieses ist von der gleichen Größenordnung wie beim zweilinsigen Achromaten; ein Objektiv mit einer Brennweite von wenigen Metern hat also über den photographischen Wellenlängenbereich eine Fokusvariation von Millimetern, während bei $N=5$ die Rayleigh-Fokustoleranz nur $\pm 25\ \mu m$ beträgt.

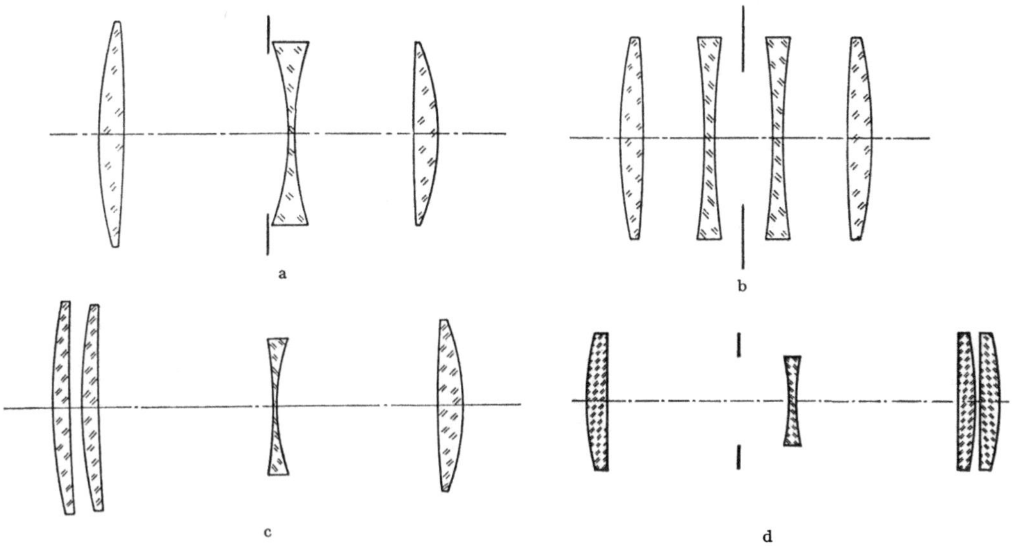

Fig. 44a—d. Aufbau von Astrographen-Objektiven: a Taylor-Triplett, b Ross-Vierlinser, c Sonnefeld-Vierlinser, d Askania-Vierlinser.

Die Bilder der schwächsten Sterne haben zwar nur Durchmesser von der Größenordnung 30 µm, und die Aufnahmen machen den Eindruck großer Schärfe, doch beruht das z.T. nur auf einem „Filtereffekt": bereits für Wellenlängen wenige 100 Å außerhalb der fokussierten wird das Licht auf einen so großen Zerstreuungskreis ausgebreitet, daß die Bestrahlungsstärke auf der Platte unter die Schwärzungsschwelle sinkt. Die helleren Sterne haben dementsprechend kräftig wachsende Bilddurchmesser, was für manche photometrischen Verfahren nicht unerwünscht ist. Die Energiebilanz ist aber viel ungünstiger als bei der Schmidt-Kamera oder beim Reflektor, und die Grenzhelligkeiten sind bis zu zwei Größenklassen schlechter, weil die Defokussierung die Hintergrundschwärzung nicht beeinflußt.

C. Realisierung der optischen Elemente.

27. Werkstoffe für Refraktionsoptik. Das Material für brechende Elemente wird in erster Linie nach *optischen* Gesichtspunkten ausgewählt. Benutzt werden hauptsächlich *Gläser,* für das Ultraviolett auch Quarz. In den neuerdings für die Astronomie erschlossenen Wellenlängenbereichen (ferneres Ultraviolett durch Raketen und Satelliten, ferneres Infrarot durch neuentwickelte Empfänger) gewinnen andere Werkstoffe Bedeutung, die aber hier nicht behandelt werden. Auch von den klassischen Materialien können nur einige Eigenschaften erwähnt werden, die besonders für die Herstellung großer optischer Elemente wichtig sind. Zusammenfassende Übersichten[1] finden sich etwa in [33], [34], ausführliche Daten[2] in den Glaskatalogen der Hersteller.

In die optischen Rechnungen gehen zunächst der Brechungsindex n und die Abbesche Zahl ν ein, die gewöhnlich als n_d und ν_d (vgl. Ziff. 12) angegeben werden.

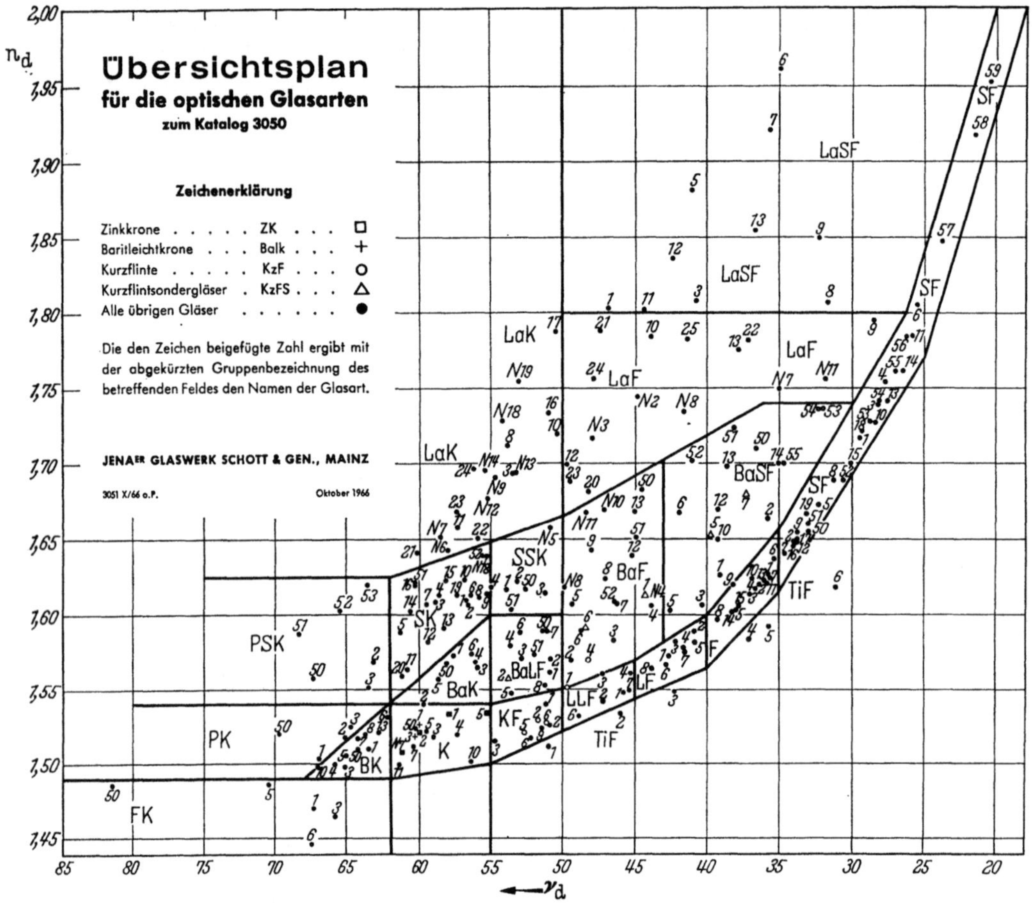

Fig. 45. n, ν-Diagramm für optische Gläser.

Traditionell nennt man Gläser mit großer Dispersion ($\nu_d \leqq 50 \ldots 55$) Flintgläser, solche mit kleinerer Dispersion Krongläser. Durch Zusätze zu diesen Bezeichnungen wird auf Besonderheiten der chemischen Zusammensetzung oder der physikalischen Eigenschaften hingewiesen. Infolge systematischer Entwicklung über viele Jahrzehnte bedecken die optischen Gläser ein ziemlich großes Feld im n, ν-Diagramm (Fig. 45).

Um über das sekundäre Spektrum von Objektiven etwas aussagen zu können, ist die Kenntnis der relativen Partialdispersion

$$\vartheta_\lambda = \frac{n_\lambda - n_F}{n_F - n_C}$$

[1] Siehe auch J. G. BAKER, Appl. Optics 2, 111 (1963).

[2] O. LINDIG: Gläser und Kunststoffe. In: Landolt-Börnstein, 6. Aufl., Bd. II, Teil 8. Berlin, Göttingen, Heidelberg: Springer 1962.

(vgl. Ziff. 13) erforderlich. In der von HERZBERGER[1] vorgeschlagenen Klassifikation für optische Gläser treten zwei entsprechend gebildete Werte auf, nämlich

$$P^* = \frac{n_F - n^*}{n_F - n_C} \quad \text{und} \quad P^{**} = \frac{n_F - n^{**}}{n_F - n_C},$$

daneben $\mu_F = n_F - 1$ und $\delta = n_F - n_C$. Mit diesen vier Parametern ist der Verlauf des Brechungsindex über einen großen Wellenlängenbereich weitgehend festgelegt.

Neben diesen die Strahlenbrechung charakterisierenden Daten ist vor allem die *Absorption* zu berücksichtigen. Im Infrarot sind optische Gläser meist bis mindestens 2,5 μm gut durchlässig. Die nach dem Ultraviolett zunehmende Absorption besonders der Flintgläser macht sich beim visuellen Gebrauch höchstens durch eine schwache Gelbfärbung bemerkbar, schließt aber manche Anwendungen völlig aus. Durch hohe Reinheit der Ausgangsstoffe und „saubere" Herstellungsverfahren ist es gelungen, die Brauchbarkeitsgrenze immer weiter ins Ultraviolette zu verschieben (Fig. 46); hier erscheint eine Revision überkommener Ansichten angezeigt. Für ältere Großobjektive ist allerdings die Gesamt-Transmission[2] besonders im „photographischen" Wellenlängenbereich recht gering. Bei großen Glaswegen können sich auch schwache Absorptionsbänder[3] bemerkbar machen, die mit Darstellungen wie in Fig. 46 gar nicht erfaßt werden. Große Prismen werden zwar kaum noch verwendet, doch können auch in Großobjektiven oder fiberoptischen Lichtleitern entsprechende Glaswege auftreten.

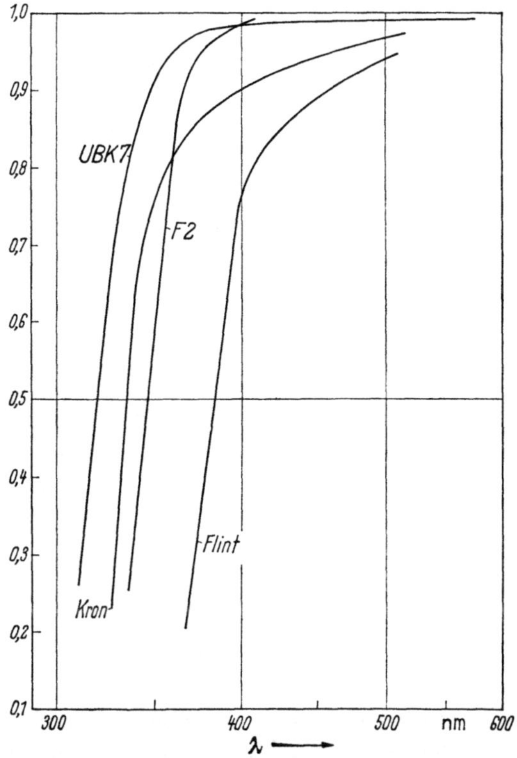

Fig. 46. Reintransmissionsgrad optischer Gläser am kurzwelligen Ende des Durchlaßbereichs für 2,5 cm Schichtdicke. Außer einem „klassischen" Kron- bzw. Flintglas sind die Typen UBK 7 (UV-Kron) und F2 (Flint) des Jenaer Glaswerks Schott & Gen., Mainz, eingetragen.

Zum Lichtverlust in optischen Systemen trägt auch die *Reflexion* an den Grenzflächen zwischen Luft und Glas bei, die bei senkrechter Inzidenz $\left(\frac{n-1}{n+1}\right)^2$ pro unvergütete Fläche beträgt (Fig. 47). Bei einem Vierlinser z.B. werden dadurch immerhin etwa 30% der einfallenden Energie dem Nutzstrahlengang entzogen; ein Teil davon erreicht als Streulicht die Bildfläche.

An Glas für hochwertige optische Systeme ist eine Reihe weiterer Forderungen zu stellen. Man verlangt *Homogenität*, d.h. Konstanz des Brechungsindex innerhalb des benutzten Glasvolumens. Ein Gebiet mit um Δn abweichender Brechzahl

[1] M. HERZBERGER: J. Opt. Soc. Amer. **53**, 661 (1963).
[2] Siehe z.B. F. L. WHIPPLE, Publ. Astron. Soc. Pacific **47**, 305 (1935).
[3] Siehe z.B. C. T. ELVEY and R. S. ZUG, Astrophys. J. **70**, 243 (1929).

und der Erstreckung d in Lichtrichtung bewirkt eine lokale Wellenaberration $V=d\cdot\Delta n$. Die Fortschritte in der Glasherstellung machen es heute möglich, auch für große Scheiben $\Delta n \leq 5\cdot 10^{-6}$ zu garantieren (Objektivscheiben-Qualität). Damit wird die inhomogenitätsbedingte Wellenaberration für d bis zur Größenordnung Zentimeter weitgehend unschädlich. In der Blütezeit der großen Refraktoren mußte dagegen der Optiker versuchen, durch „lokale Retusche" an einer oder mehreren Flächen die damals viel größeren Brechungsindexschwankungen zu kompensieren: wenn die Inhomogenität Δn die ganze Glasdicke d ausfüllt, so gilt für ein dünnes System

$$V=\Delta n\cdot d+(n-1)\cdot\Delta d,$$

wobei Δd die Tiefe der Retusche ist. Die Beziehung zeigt auch, mit welcher Genauigkeit bei der Prüfung von Platten auf Homogenität die Flächenfehler Δd bestimmt oder beseitigt werden müssen.

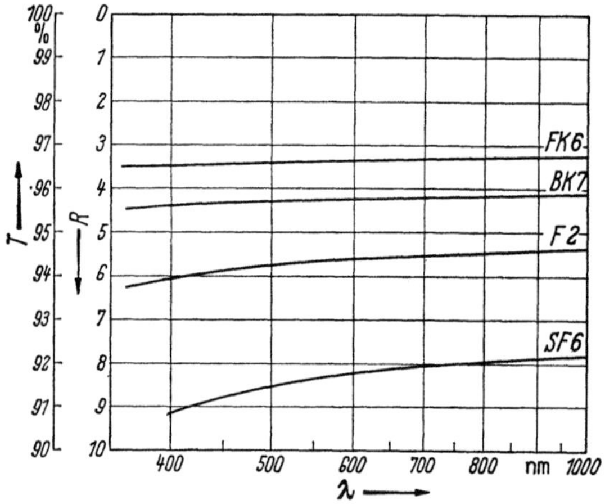

Fig. 47. Strahlenteilung an einer Luft-Glas-Fläche bei senkrechter Inzidenz für verschiedene Glastypen. T durchgelassener, R reflektierter Anteil.

Auch bei völliger materieller Homogenität des Glases kann eine Inhomogenität des Brechungsindex durch Variation der *Temperatur* innerhalb des Glasvolumens entstehen. Der Temperaturkoeffizient der Brechzahl hängt von der Glaszusammensetzung ab und ist selbst wieder temperatur- und auch wellenlängenabhängig. Für Flintgläser wird bei 20 °C und die d-Linie $\Delta n/\Delta t=15\cdot 10^{-6}$ grad^{-1} erreicht. Bei hohen Anforderungen ist also u. U. Temperaturhomogenität innerhalb weniger 0,1° notwendig.

Optisches Glas soll frei von *inneren Spannungen* sein. Elastische Spannungen entstehen bei zu schnellem Abkühlen der Glasschmelze, das die Einstellung des Temperaturgleichgewichts in jedem Augenblick verhindert. Durch gesteuerte langsame Abkühlung („Feinkühlen") sucht man die Spannungen unter die Toleranzgrenze zu drücken und gleichmäßig zu verteilen.

Der Spannungszustand bewirkt optische Anisotropie und läßt sich mit den Methoden der Spannungsoptik analysieren; bei der Prüfung zwischen gekreuzten Polarisatoren werden „Isoklinen" beobachtet. In einer gut gekühlten flachen Kreisscheibe ergeben sich die Extreme der Fortpflanzungsgeschwindigkeit für Lichtschwingung in Richtung des Scheibenradius (langsame Komponente) und senkrecht dazu. Der Gangunterschied ε von entsprechenden austretenden Wellenflächen ist wieder $\varepsilon=\Delta n\cdot d$; neben Δn kann auch ε/d (gewöhnlich in nm/cm angegeben) die Doppelbrechung in einem Material beschreiben. Nach COUDER [*35*] wächst bei der Kreisscheibe ε etwa mit dem Quadrat des Mittenabstands. Solange am Rand $\varepsilon \lesssim 100$ nm, ändert sich die Beugungsfigur eines Objektivs nur unwesentlich; da ε/d bis etwa 5 ... 10 nm/cm herabgedrückt werden kann, ist auch bei ziemlich dicken Scheiben kaum mit einer Bildverschlechterung durch Doppelbrechung zu rechnen. Neben dieser Differenz der Brechungsindizes für ordentlichen und außerordentlichen Strahl verursachen Spannungen auch eine Variation des mittleren Brechungsindex, die ein Vielfaches der Differenz betragen kann. Schließlich neigt nicht genügend spannungsfreies Glas zu *Formänderungen*, wenn das Zusammenwirken der Spannungskräfte geändert wird, z.B. beim Zerschneiden oder Abtrennen von Teilen oder bei Temperaturänderung. Mit der Angabe der Doppelbrechung sind die *Spannungen* selbst nicht ausreichend erfaßt, weil der Brewster-Koeffizient (nm/cm · bar) mit der Glaszusammensetzung variiert und für schwersten Flint sogar das Vorzeichen wechselt.

Die übrigen Glasfehler — Schlieren, Blasen, Fremdkörpereinschlüsse — lassen sich für viele Glasarten so weit vermeiden, daß sie nicht stören. Bei Fernrohroptik ist mehr als für viele andere Anwendungszwecke auf *Verwitterungsbeständigkeit* zu achten.

Die Möglichkeiten der *Reflexverminderung* an Glas-Luft-Flächen durch Aufbringen dünner Schichten mit geeignetem Brechungsindex werden für die astronomische Großoptik nur zögernd genutzt. Hartvergütung erfordert Erhitzen auf einige 100 °C, und der Belag ist nur durch Neupolieren wieder zu entfernen; solcher Behandlung möchte der Optiker seine mit viel Mühe figurierten Flächen nicht aussetzen. Die Korrektionsplatten-Reflexe bei Schmidt-Teleskopen hat man durch Vergütung zu verringern gesucht[1]. Anwendungsmöglichkeiten dünner Schichten zur Reflexionsminderung, für „dielektrische Spiegel", als Filter und Strahlenteiler hat BAUMEISTER[2] erörtert.

28. Werkstoffe für Spiegel.

α) *Spiegelträger*. Spiegel in astronomischen Fernrohren sind Oberflächenspiegel, die fast stets mit einer aufgedampften Aluminiumschicht versehen sind. Da es bei ihnen wesentlich auf die Erhaltung der Flächenform ankommt, spielen die mechanischen und thermischen Stoffeigenschaften die Hauptrolle bei der Materialauswahl für den Träger. Polierfähigkeit, Mikrostruktur und Porosität braucht man kaum noch zu berücksichtigen, weil die Technik des Aufbringens von Oberflächenschichten aus geeigneten Stoffen ständig Fortschritte macht. Die Brauchbarkeit verschiedener Materialien als Spiegelwerkstoff[3] ist in [*36*], [*44*] diskutiert.

Die wichtigste mechanische Eigenschaft betrifft die Deformation unter dem Eigengewicht; in sie geht neben einem Formfaktor die Materialkonstante E/γ ein (E Elastizitätsmodul, γ spezifisches Gewicht), deren Streubereich allerdings — selbst für so verschiedene Stoffe wie Glas, Stahl und Aluminium — nicht sehr groß ist (Tabelle 4). Erwünscht ist weiter niedrige Dichte, um das Gewicht des Spiegels und damit der tragenden Elemente und des ganzen Instruments klein halten zu können. Aus dem gleichen Grund ist man übrigens bei sehr großen Spiegeln zu Rippenkonstruktionen gezwungen, selbst auf Kosten der Steifigkeit.

Während der Beobachtungszeit sollen sich Spiegel möglichst nahe am Temperaturgleichgewicht mit der Umgebung befinden. Bisher versucht man stets, sie wenigstens den langfristigen Schwankungen der Außentemperatur folgen zu lassen; für große Spiegel wird so die Geschwindigkeit des Temperaturausgleichs entscheidend wichtig. Die als „Temperaturleitfähigkeit" bezeichnete Größe $\frac{\lambda}{c\cdot\varrho}$ (λ Wärmeleitfähigkeit, c spezifische Wärme, ϱ Dichte) bestimmt — neben dem Oberflächensprung — die Abklingzeit nach einem thermischen Schock. Eine Gütezahl für das thermische Übergangsverhalten muß dazu noch den Ausdehnungskoeffizienten α berücksichtigen, von dem die Größe der resultierenden

---

[1] 2 m-Universalteleskop Tautenburg.
[2] P. W. BAUMEISTER: Astron. J. **69**, 338 (1964).
[3] Siehe auch A. B. MEINEL in [*9*].

Tabelle 4.

	$\varrho\,(\text{g}\cdot\text{cm}^{-3})$ $\gamma\,(\text{p}\cdot\text{cm}^{-3})$	$E\,(\text{p}\cdot\text{cm}^{-2})$	$\frac{E}{\gamma}$	$\lambda\,\left(\frac{\text{cal}}{\text{cm}\cdot\text{sec}\cdot\text{grad}}\right)$	$c\,\left(\frac{\text{cal}}{\text{g}\cdot\text{grad}}\right)$	$\alpha\,(\text{grad}^{-1})$	$\frac{\lambda}{c\cdot\varrho}$	$\frac{\lambda}{c\cdot\alpha}$	$\frac{\lambda}{\varrho\cdot c\cdot\alpha}$	$\frac{E}{\alpha}\cdot\frac{\lambda}{c\cdot\varrho}$
Kronglas	2,5	$7{,}5\cdot10^{8}$	$3{,}0\cdot10^{8}$	0,0018	0,17	$75\cdot10^{-7}$	0,0042	$1{,}4\cdot10^{3}$	560	$0{,}42\cdot10^{12}$
Duran 50 (Pyrex)*	2,23	6,3	2,83	0,0028	0,18	32	0,0070	4,9	2200	1,4
Quarz	2,20	7,4	3,36	0,0032	0,17	5,6	0,0086	34	15000	11
CER-VIT** (C 101)	2,50	9,4	3,27	0,0040	0,217	$0\pm1{,}5$	0,008	—	—	—
Spiegelbronze	8,6	8,0	0,93	0,20	0,08	186	0,29	130	16000	12
Stahl	7,7	21	2,73	0,11	0,12	110	0,12	83	11000	23
Aluminium	2,70	7,1	2,61	0,50	0,215	240	0,86	97	36000	25
Beryllium	1,82	29	15,7	0,38	0,52	124	0,40	59	33000	93

Da es sich um technische Substanzen handelt, muß man den Daten eine gewisse Unsicherheit zubilligen.

* Für „Pyrex" der Corning Glass Works werden je nach Zusammensetzung etwas verschiedene Werte angegeben; eine für Spiegel empfohlene Sorte hat Eigenschaften, die mit den angeführten von „Duran 50" des Glaswerks Schott & Gen. Mainz praktisch übereinstimmen.

** Owens-Illinois, Inc.

Deformation abhängt. So haben SCOTT [36] $\frac{\lambda}{c\cdot\alpha}$, COUDER[1] $\frac{1}{K}=\frac{\lambda}{\varrho\cdot c\cdot\alpha}$, MAKSUTOV [44] $\psi=\frac{E}{\alpha}\cdot\frac{\lambda}{c\cdot\varrho}$ zur Beurteilung von Spiegelwerkstoffen benutzt. Außer solchen Materialkonstanten und Masse bzw. Volumen ist auch hier die Form wichtig. Rippen- oder Zellenstrukturen haben geringere Wärmekapazität und größere Oberfläche als Massivscheiben, doch verringern geschlossene Zellen die Wärmeleitung, Durchbrüche für passive Luftzirkulation aber die mechanische Steifigkeit.

Glas ist lange Zeit der meistverwendete Spiegelwerkstoff gewesen, doch ist es für große Spiegel wenig geeignet wegen seiner ungünstigen thermischen Eigenschaften. Der (damals chemisch versilberte) Glasspiegel hatte nach seiner Einführung in die Fernrohrtechnik 1856 verhältnismäßig rasch den bis dahin üblichen Spiegelmetall-Spiegel verdrängt. Als Hauptvorteile galten das geringe Gewicht und die leichte Bearbeitbarkeit nach bewährten Verfahren; beim Erneuern der Verspiegelung wurde die optische Figur nicht geändert. Das Reflexionsvermögen (Fig. 48) ist viel höher als beim Bronzespiegel. Vom Glas als Spiegelträger verlangt man weder optische Homogenität noch hohe Lichtdurchlässigkeit, es soll aber möglichst spannungsfrei sein. Als Toleranz wird oft ein Gangunterschied der induzierten Doppelbrechung von weniger als 30 nm/cm angegeben; für verschiedene Glasarten oder etwa Quarz bedeutet das jedoch verschiedene Spannungen. Blasen und stärkere Schlieren stören nur, wenn die optische Fläche sie anschneidet.

Die Temperaturübergangseffekte machten sich bei den nach 1900 entstehenden größeren Glasspiegeln sehr unangenehm bemerkbar. Es war deshalb ein erheblicher Fortschritt, als es gelang, große Spiegelscheiben aus Glas mit niedrigem Ausdehnungskoeffizienten herzustellen. Solche *Borosilikat-Gläser* (Handelsnamen

[1] A. COUDER: Vistas in Astronomy 1, 372 (1955).

Pyrex, Duran usw.) wurden für Haushaltwaren und chemische Apparaturen verwendet; die ersten großen Fernrohrspiegel aus Pyrex wurden für die Instrumente des Palomar Observatory gegossen. Seit etwa 1935 sind fast alle größeren Teleskopspiegel aus Borosilikat-Glas hergestellt worden. Die Gütezahlen liegen um einen Faktor 3 bis 4 günstiger als bei Spiegelglas.

Eine weitere Verringerung der thermischen Deformation, die vor allem für große Instrumente erwünscht ist, läßt sich mit *Quarz* als Spiegelwerkstoff erreichen. Geschmolzener natürlicher Quarz und aus chemischen Reaktionen hervorgegangenes Siliziumdioxid unterscheiden sich kaum; die Gütezahlen sind um einen Faktor 7 bis 8 besser als bei Borosilikat-Glas. Quarzspiegel lassen sich mit üblichen Schleif- und Poliermethoden auf sehr hohe Flächenqualität bringen.

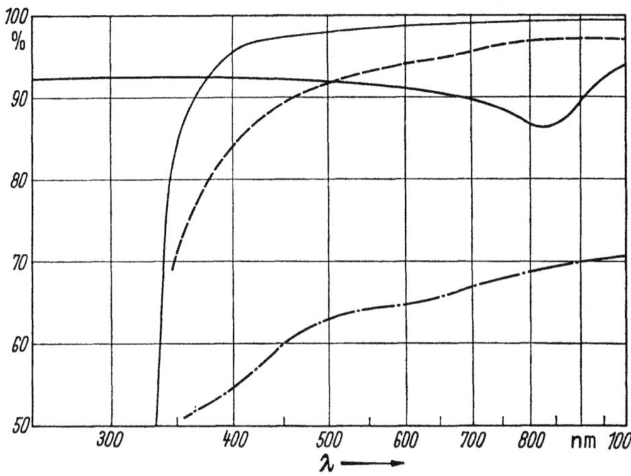

Fig. 48. Reflexionsvermögen von Oberflächenspiegeln. ——— Aluminium, frisch aufgedampft, ——— Silber, frisch aufgedampft, — — — Silber, chemisch niedergeschlagen, — · — · — Spiegelbronze, poliert.

Geschmolzener Quarz kann allerdings nicht in hinreichend großen Volumen gegossen werden; die Herstellung großer Scheiben ist erst seit wenigen Jahren dadurch möglich, daß man kleinere Stücke miteinander verschmilzt. Der größte fertigbearbeitete Quarzspiegel (Ende 1964) mit 157 cm Durchmesser und 27 cm Gesamt-Dicke[1] besteht aus vier parallelen Schichten, während ein anderes Verfahren die Scheiben aus sechseckigen Prismen zusammensetzt. Arbeiten zur Produktion von Quarzscheiben bis zu 4 m Durchmesser sind im Gang. Für größere Spiegel wird Quarz z.Z. so eindeutig bevorzugt, daß alle Großprojekte Quarzspiegel vorsehen.

Neuerdings ist es auch gelungen, Werkstoffe für Spiegelscheiben mit extrem kleinem Ausdehnungskoeffizienten auf *Glas*-Basis zu produzieren. Das als Glas erschmolzene Material wird dabei durch einen kontrollierten Entglasungsprozeß in einen teilkristallinen Zustand übergeführt. Je nach Größe der entstehenden Kristallite sind die Scheiben mehr oder weniger durchsichtig. „CER-VIT" (Owens-Illinois, Inc.) wird mit einem Ausdehnungskoeffizienten von $0 \pm 1{,}5 \cdot 10^{-7}\ \text{grad}^{-1}$ im interessierenden Temperaturbereich angeboten. Hinsichtlich anderer Eigenschaften (auch Bearbeitbarkeit[2]) ist es Quarz nicht unterlegen; die gute Durchsichtigkeit gestattet Inspektion und Spannungsprüfung. Bisher (Ende 1966) ist die Anfertigung von Spiegelscheiben mit etwa 1 m Durchmesser bekannt geworden.

[1] K. Aa. Strand: Appl. Optics **2**, 1 (1963).
[2] R. E. Cox: Sky and Telescope **32**, 159 (1966).

Nach einigen Jahrzehnten absoluter Vorherrschaft von Glasspiegeln wurde die Herstellung großer Spiegel aus *Metall* wegen der günstigen thermischen Eigenschaften wieder erwogen. Ein Nachteil der Metallspiegel ist die Schwierigkeit, den Spannungszustand zu kontrollieren. Über die Langzeit-Formbeständigkeit liegen noch keine ausreichenden Erfahrungen vor. Der Verwendung von *Aluminiumlegierungen* mit ihren guten mechanischen und thermischen Eigenschaften stand zunächst die schlechte Oberflächenbeschaffenheit und Polierbarkeit entgegen. Einen Ausweg bietet die Beschichtung mit Kanigen (Handelsname für amorphes Nickel-Phosphid), das chemisch auf viele Grundmetalle niedergeschlagen werden kann und sich gut polieren läßt. Ein verrippter Aluminiumspiegel von 160 cm Durchmesser und 25 cm Dicke dient als abbildendes System im Sonnenteleskop des Kitt Peak National Observatory[1]. Die hohe Wärmeleitfähigkeit macht es hier möglich, die absorbierte Energie durch Kühlung wegzuschaffen. Der vorgeformte Metallkörper trägt einen 0,13 mm starken Kanigenüberzug, der dann nach optischen Methoden weiterbearbeitet wurde. Die Flächenqualität eines erstklassigen Quarz- oder Glasspiegels scheint sich noch nicht erreichen zu lassen. Die günstigen Eigenschaften von *Stahl* hat vor allem Maksutov hervorgehoben, der auch praktische Erprobungen vorgenommen hat. Ein Reflektor mit 70 cm-Stahlspiegel ist im Astronomischen Hauptobservatorium Pulkovo aufgestellt[2].

Bemerkenswert ist das große E/γ von Beryllium, das aus diesem Grunde als Spiegelmaterial diskutiert wird[3]. Seiner allgemeinen Anwendung stehen aber die Giftigkeit und der hohe Preis entgegen, selbst wenn alle technologischen Probleme gelöst werden.

β) *Verspiegelungen.* Die *Aluminisierung* von Oberflächenspiegeln durch Aufdampfen im Vakuum[4,5] war ein erheblicher Fortschritt in der Fernrohr-Technologie. Zwar hat im sichtbaren Wellenlängenbereich eine frisch niedergeschlagene Silberschicht ein geringfügig höheres Reflexionsvermögen als eine aufgedampfte Aluminiumschicht (Fig. 48), doch wird dies mehr als kompensiert durch die Unempfindlichkeit und Beständigkeit der Aluminium-Verspiegelung und ihre gleichmäßig hohe Reflexion im astronomischen UV. Für diese Vorteile nimmt man den apparativen Aufwand einer Vakuumanlage (für große Teleskope fast immer unmittelbar beim Instrument) ohne weiteres in Kauf. Zusammenfassende Übersichten der Apparaturen und Arbeitsverfahren und der Film-Eigenschaften sind in [*37*], [*38*] gegeben. In einem Vakuum von mindestens 10^{-4} Torr wird Reinstaluminium an Wolframwendeln zunächst geschmolzen und dann bis zu einem ausreichenden Dampfdruck erhitzt; die Atome fliegen praktisch ohne Zusammenstöße mit dem Restgas zu der einige Dezimeter entfernten Spiegelfläche. Die Schicht soll nicht viel stärker sein, als zur Unterdrückung der Lichtdurchlässigkeit nötig ist (etwa 200 nm). Der zunächst weiche Aluminiumfilm „härtet" an der Luft durch Bildung einer Oxidschicht, die in wenigen Wochen bis zu einer Dicke von 5 bis 10 nm wächst; dabei kann die Reflexion im Ultraviolett etwas nachlassen. Die Eigenschaften der Aluminiumschicht sind ziemlich stark von den Aufdampfbedingungen abhängig[6,7]. Das beste UV-Reflexionsvermögen und

---

[1] A. K. Pierce: Appl. Optics **3**, 1337 (1964).

[2] N. N. Mikhelson: Mitt. Pulkovo **24**, H. 1 (No. 177), 12 (1964).

[3] N. N. Mikhelson: Mitt. Pulkovo **21**, H. 3 (No. 162), 153 (1958). — A. B. Meinel in [*9*]. — W. P. Barnes, Jr.: Appl. Optics **5**, 1883 (1966).

[4] J. Strong: Astrophys. J. **83**, 401 (1936).

[5] R. C. Williams and G. B. Sabine: Astrophys. J. **77**, 316 (1933).

[6] G. Hass: J. Opt. Soc. Amer. **45**, 945 (1955). — G. Hass and J. E. Waylonis: J. Opt. Soc. Amer. **51**, 719 (1961).

[7] W. M. Sinton, H. L. Johnson and K. L. Hallam: Publ. Astron. Soc. Pacific **71**, 239 (1959).

zugleich die größte Alterungsbeständigkeit bekommt man beim raschen Aufdampfen (Größenordnung 10 sec) normal zur Fläche in einem Vakuum von wenigen 10^{-5} Torr, wodurch kompakte, gut haftende Filme entstehen. Solche Verspiegelungen vertragen das Waschen mit Wasser und Detergentien.

Spiegel in Fernrohren, besonders wenn die Fläche nach oben zeigt, sind einem verstärkten mechanischen und u. U. auch chemischen Angriff ausgesetzt. Man hat deshalb versucht, den spiegelnden Film durch eine weitere, besonders widerstandsfähige Schicht zu schützen. Siliziummonoxid hat den Nachteil, unterhalb von 500 nm zu absorbieren; bei langsamem Aufdampfen unter verhältnismäßig hohem Druck ist es bis 300 nm brauchbar[1]. Im oxidierenden Restgas[2] werden höhere Siliziumoxide niedergeschlagen, die hinsichtlich der Absorption günstiger sind; die Prozesse sind z. T. Firmeneigentum. In der Beobachtungspraxis unter ungünstigen klimatischen Bedingungen haben sich Schutzschichten nicht immer bewährt; die Technologie des Gebietes macht aber immer noch Fortschritte. Sehr harte Oberflächenschichten erhält man auch durch kontrollierte elektrochemische Oxidation des Aluminiums ähnlich dem Eloxalverfahren. Beim Entfernen von Schutz- und Spiegelschichten muß man berücksichtigen, daß Gläser gegen chemischen Angriff keineswegs unempfindlich sind[3].

29. Zur Herstellung optischer Flächen. Die Bearbeitungsverfahren für optische Flächen sind auch in der dem Astronomen zugänglichen Literatur verschiedentlich dargestellt worden (s. z. B. [*31*], [*44*]), sie sollen hier nicht behandelt werden. Erwähnt sei, daß das Ergebnis des Grob- und Fein*schleifens* (mit starren Schalen) i. allg. eine sphärische Fläche ist, mit Gruben von wenigen μm Tiefe bedeckt. Diese Schleifgruben werden durch das anschließende *Polieren* beseitigt. Über die dabei wirksamen Prozesse sind recht verschiedene Ansichten geäußert worden; eine zusammenfassende Darstellung gibt HOLLAND [*39*]. Beim Polieren können Flächen auch mäßig (d. h. bis zu Abtragungen von einigen μm) *deformiert* werden, da die Polierschalen nicht starr sind. Größere Abweichungen von der Kugelgestalt kann man erzeugen durch abwechselndes Schleifen mit kleinen Schalen und Polieren, durch Schleifen mit flexiblen Werkzeugen[4], oder durch Ausnutzen der elastischen Durchbiegung bei dünnen Korrektionsplatten[5].

Die erreichbare Flächengenauigkeit hängt von den Abmessungen des Werkstücks und der Größe der Deformation ab. Nach TEXEREAU[6] ist ein 2 m-Paraboloid mit $N = 4 \ldots 5$ etwa die Grenze, bis zu der man noch eine einigermaßen normale Beugungsfigur (Einhaltung der Rayleigh-Strehl-Bedingung) erwarten kann. Die Flächendefekte lassen sich zwanglos einteilen in Fehler der Figur und solche der Oberfläche. Bei den Oberflächen-Unvollkommenheiten (mangelhafte Politur, Kratzer, Haarrisse) sind die Störungen der Wellenfläche größenordnungsmäßig ebenso breit wie tief, bedecken aber nur ein kleines Areal. Sie führen zu Streulicht, das bis in große Winkelabstände vom Bild reicht, das aber bei Astro-Teleskopen allgemein nicht in dem Maße stört wie etwa beim Koronographen. Die Figur-Fehler haben dagegen Steigungen der Wellenaberration von der Größenordnung 10^{-5} bis 10^{-6} über verhältnismäßig große Gebiete und beeinflussen deshalb vor allem den inneren Teil der Beugungsfigur. Neben Abweichungen der Gesamt-Flächengestalt — bei großen optischen Elementen auch beeinflußt von

[1] G. HASS and N. W. SCOTT: J. Opt. Soc. Amer. **39**, 179 (1949).
[2] G. HASS: J. Opt. Soc. Amer. **39**, 532 (1949).
[3] G. E. MANVILLE: Observatory **80**, 237 (1960).
[4] H. W. COX: J. Sci. Instrum. **16**, 257 (1939).
[5] R. SCHORR: Z. Instrumentenkde. **56**, 336 (1936) beschreibt das Verfahren von B. SCHMIDT; s. auch E. EVERHART, Appl. Optics **5**, 713 (1966).
[6] J. TEXEREAU: Appl. Optics **2**, 23 (1963).

der Lagerung und dem thermischen Zustand — treten als typische Bearbeitungsfehler die „Zonen“ und eine Welligkeit oder Buckligkeit auf (vgl. Fig. 53 auf S. 306). Die rotationssymmetrischen Zonenfehler können entstehen, wenn die gegenseitige Translation von Werkzeug und Werkstück nur klein ist oder wenn mit Schalen kleinen Durchmessers gearbeitet wird. Die Welligkeit hat offenbar eine charakteristische Steigung, die von der Poliertechnik abhängt und vom Spiegeldurchmesser kaum beeinflußt wird[1]. Die erreichten Bestwerte liegen bei einigen 10^{-7}, was zusammen mit charakteristischen Abmessungen von einigen cm zu Wellenaberrationen von wenigen 10 nm führt. Auch durch verlängertes Polieren sind diese Buckel nicht zu beseitigen; bei der genannten Größe haben sie allerdings kaum noch Einfluß auf die Beugungsfigur.

30. Objektivprüfung. Die geforderte Flächengenauigkeit ist nur zu erreichen, wenn in den späteren Bearbeitungsstadien eine ständige Kontrolle mit *optischen Prüfmethoden* stattfindet. Mit der Abweichung der austretenden Wellenfläche von der Referenzkugel für ein fernes Punktobjekt sind unmittelbar Ort und Tiefe der noch erforderlichen Abtragung gegeben. Für die Qualitätsbewertung des fertigen Produkts kommen auch andere Gütekriterien (vgl. Ziff. 14) in Betracht.

Während der Bearbeitung sollte so weit wie möglich im Laboratorium geprüft werden. Nur bei sehr großen Spiegeln ist es wohl nötig, die letzten Retuschen nach Tests an der im Instrument montierten Optik vorzunehmen[2]. An die thermische Homogenität der Luft im Prüfraum werden hohe Anforderungen gestellt, weil bei den langen Lichtwegen die Brechungsindexänderung $\Delta n/\Delta t \approx -1 \cdot 10^{-6}/°\mathrm{C}$ merklich wird. Zweckmäßig erscheint die Anordnung mit vertikalem Strahlengang in temperaturgesteuerter Umgebung[3]; in einem Extremfall[4] hat man die Prüfung in einem Vakuumtank vorgenommen. Läßt sich die geforderte Homogenität nicht erreichen, so ist man — ebenso wie bei der Prüfung im fertigen Instrument mit natürlichem Stern — auf Verfahren mit *zeitlicher Integration* angewiesen, z. B. photographische Aufnahmen mit Belichtungszeiten von der Größenordnung Minuten.

Gegenüber den auf die Bestimmung der Wellenaberration abzielenden Methoden ist die direkte Betrachtung eines fokalen Sternbilds mit kurzbrennweitigem Okular zurückgetreten, obwohl auch sie Aussagen über die Abbildungsqualität gestattet. Die Änderung des Bündelquerschnitts vor und hinter dem Brennpunkt läßt sphärische Aberration erkennen; erscheint eine Mittenabblendung (Sekundärspiegelschatten) *inner*halb des Fokus zu groß und umgekehrt, so liegt sphärische *Unter*korrektion vor. Aus dem Aussehen der intra- und extrafokalen *Beugungsfigur* kann ein geübter Beobachter weitgehende Schlüsse über den Korrektions- und Justierzustand eines Objektivs ziehen[5]. Vor allem die Hersteller großer Refraktorobjektive haben sich dieses Verfahrens bedient; bei großen Spiegeln ist es wegen der kleinen Winkelausdehnung des Beugungsbildes nicht anwendbar.

Die Zahl der Prüfmethoden im engeren Sinne und ihrer Varianten ist so groß[6], daß in den folgenden Kapiteln nur einige wichtige dargelegt werden können. Sie lassen sich zwanglos in drei Gruppen einordnen:

---

[1] J. TEXEREAU: Ciel et Terre **66**, 57 (1950).

[2] I. S. BOWEN: Publ. Astron. Soc. Pacific **62**, 91 (1950); s. auch Fußnote 3.

[3] D. S. BROWN: Sky and Telescope **29**, 350 (1965).

[4] Sky and Telescope **27**, 345 (1964).

[5] H. D. TAYLOR: The Adjustment and Testing of Telescope Objectives. Newcastle upon Tyne 1946. Eine frühere Ausgabe ist übersetzt in Z. Instrumentenkde. **14**, 113, 153, 189 (1894).

[6] Fundgruben sind die Bände von [*31*], die Zeitschrift Sky and Telescope und die weitere ernsthafte Amateurliteratur. Zusammenstellungen haben gegeben: G. YVON: Contrôle des Surfaces Optiques — Qualité des Systèmes Optiques. Paris 1925. — V. RONCHI: La prova dei sistemi ottici. Bologna 1925; siehe auch die einschlägigen Abschnitte in [*6*], [*44*].

1. das Hartmann-Verfahren mit seinen modernen Verallgemeinerungen, das die Lichtstrahlen der geometrischen Optik mit engen Bündeln annähert;

2. die Schneidenprüfung nach FOUCAULT, bei der die Steigung der Wellenaberration in einer gewählten Richtung in Helligkeitswerte umgesetzt wird;

3. die interferometrischen Methoden, mit denen man im anschaulichsten Fall unmittelbar eine Schichtliniendarstellung der Wellenaberration erhält.

Dabei gibt die Hartmann-Methode den Gesamtverlauf der Wellenaberration quantitativ, aber nur für eine beschränkte Anzahl von ausgewählten Stellen. Die Elimination von statistischen Schwankungen durch zeitliche Integration ist ohne weiteres möglich. Das Foucaultsche Verfahren erfaßt die ganze Fläche, aber zunächst nur qualitativ. Zum Nachweis von lokalen Störungen der Wellenfläche ist es höchst empfindlich; Foucault- und Hartmann-Methode können sich so vorteilhaft ergänzen. Ihre volle Leistungsfähigkeit gewinnt die Schneidenprüfung, wenn sie als Nullmethode angewendet wird. Interferometrische Testverfahren vereinigen — wenigstens potentiell — die Vorzüge der beiden genannten Methoden; bei passend ausgewähltem Aufbau hält sich auch der apparative Aufwand in Grenzen. Alle Prüfmethoden arbeiten am besten für *kleine* systematische und zufällige Abweichungen der Wellenfläche von der Kugelform, bei Fernrohrobjektiven also mit Abbildung eines fernen axialen Punktes. Ein solcher ist im Labor nur für kleinere Objektiv-Durchmesser mit einem Kollimator zu realisieren. Für mittelgroße Objektive ist noch die Prüfung in Autokollimation möglich, die allerdings einen einwandfreien Planspiegel vom Durchmesser der Optik erfordert. Bei sehr großen Spiegeln ist man darauf angewiesen, mit einem *nahen* Punktobjekt — gewöhnlich unter Abbildung 1:1 — zu prüfen. Die dabei auftretende sphärische Aberration kann man für Hartmann- und Interferometer-Methoden rechnerisch berücksichtigen; für die Schneidenmethode ist die Kompensation mit zusätzlichen optischen Gliedern möglich. Beide Verfahren gestatten auch die Prüfung von Teilsystemen, die für keine Objektlage ein stigmatisches Bild liefern (z.B. Ritchey-Chrétien-Hauptspiegel).

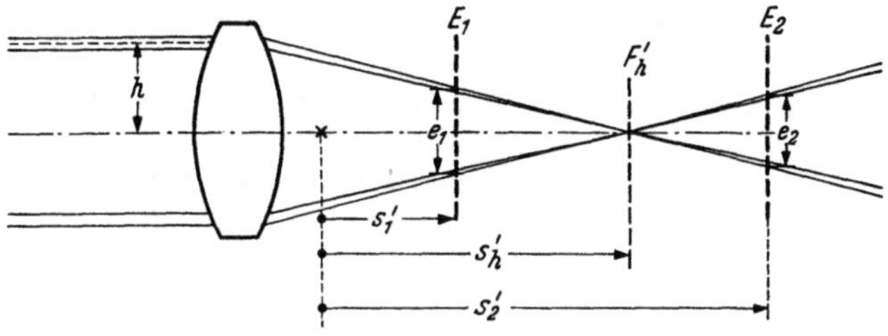

Fig. 49. Geometrie der Hartmann-Methode.

31. Objektivprüfung nach Hartmann. Die Hartmann-Methode [*40*] ist in der Astronomie ein Standard-Verfahren vor allem zur Beurteilung fertiger Systeme. Unmittelbares Meßergebnis sind *Schnittweiten* in Abhängigkeit von Azimut und Einfallshöhe. Vor das Objektiv wird eine „Hartmann-Blende" gestellt, die auf einer Anzahl von Durchmessern symmetrisch angeordnete Lochpaare für verschiedene Einfallshöhen trägt (Fig. 49). Die bei Beleuchtung mit einem natürlichen oder künstlichen Stern entstehenden Teilbündel sieht man als Strahlen im Sinne der geometrischen Optik an. Ihre Durchstoßpunkte durch die intra- und extrafokalen Einstellebenen E_1, E_2 werden photographisch aufgezeichnet und die Abstände e_1, e_2 dieser Durchstoßpunkte für jedes Lochpaar mit hoher Genauigkeit gemessen. Unter der *Voraussetzung*, daß sich die Strahlen jedes

Paares auf der Achse schneiden, gilt

$$\frac{s'_h - s'_1}{s'_2 - s'_h} = \frac{e_1}{e_2}$$

und damit

$$s'_h = s'_1 + \frac{e_1}{e_1 + e_2}(s'_2 - s'_1).$$

Die Schnittweiten s' sind von einem willkürlichen Bezugspunkt (× in Fig. 49) aus gemessen.

Als Lochdurchmesser werden 2 bis 5‰ der Brennweite empfohlen. Bei Linsenobjektiven muß mit monochromatischem Licht oder wenigstens einem eingeengten Wellenlängenbereich um den Scheitel der Farbkurve herum gearbeitet werden, damit die chromatischen Restfehler nicht die Meßgenauigkeit zu sehr herabsetzen. Die Untersuchung mit einem natürlichen Stern erfordert zwar guten Luftzustand und saubere Nachführung, doch funktioniert die zeitliche Integration immerhin besser als bei anderen Methoden. Im Labor kann man bei kurzen Brennweiten mit künstlichem Stern in großem endlichen Abstand prüfen. Arbeitet man in Autokollimation, so kommt die Hartmann-Blende zweckmäßig dicht an den Planspiegel. Die Fehler des Objektivs erscheinen hier verdoppelt.

Bei der Prüfung im Krümmungsmittelpunkt eines Hohlspiegels ist für deformierte Flächen die Anordnung der Durchstoßpunkte der Anordnung der Löcher in der Hartmann-Blende wegen der sphärischen Aberration nicht mehr ähnlich. Für einen Parabolspiegel sind die gemessenen Schnittweiten zu verringern um

$$\Delta s'_h = \frac{h^2}{2f} + \frac{h^4}{16f^3} + \cdots,$$

wenn der künstliche Stern im paraxialen Krümmungsmittelpunkt steht. Für beliebig deformierte Flächen gilt in der Seidel-Näherung

$$\Delta s'_h = -b \cdot \frac{h^2}{2f}.$$

Zur Reduktion auf den Fokus müssen die resultierenden Längsaberrationen durch 4 dividiert werden.

Der Verlauf der s'_h mit h läßt sphärische Aberration und Zonen erkennen. Eine Abhängigkeit der s'_h vom Azimut auf dem Objektiv ist auf „Astigmatismus auf der Achse" (mangelnde Rotationssymmetrie) zurückzuführen. Aus den Längsabweichungen kann man die Wellenaberration herleiten mit

$$V(h) = \frac{1}{f^2}\int_0^h h(s'_h - s'_w)\,dh.$$

Dabei definiert s'_w das Zentrum der Referenzkugel. Das Objektiv wird nach einer meist nicht sehr großen Anzahl von Flächenausschnitten beurteilt; die Integration ist nur sinnvoll, wenn die Wellenfläche durch die Tangenten in diesen Ausschnitten vernünftig repräsentiert ist.

Bei großen Objektiven sind die Beugungsscheibchen so klein, daß man die Bildqualität geometrisch-optisch beschreiben kann; es ist dann sinnvoll, eine *mittlere* Querabweichung in der besten Bildfläche anzugeben. Eine solche ist die „Technische Konstante" T nach Lehmann[1]:

$$T = \frac{2 \cdot 10^5}{f^2} \cdot \frac{\sum h^2 |s'_h - s'_m|}{\sum h}.$$

[1] H. Lehmann: Z. Instrumentenkde. **22**, 103, 325 (1902); **23**, 289 (1903).

Die Queraberrationen jeder Zone $(s'_h - s'_m) h/f$ haben dabei h (proportional zur Zonenfläche) als Gewicht bekommen; durch Multiplikation mit $200000/f$ wird der mittlere Radius etwa in Bogensekunden ausgedrückt. Zur Bestimmung der durch s'_m gegebenen optimalen Einstellebene sind verschiedentlich Rechenvorschriften angegeben worden, doch erscheinen bei dem ohnehin begrenzten Wert der technischen Konstanten und den einschränkenden Voraussetzungen umständliche Rechnungen kaum sinnvoll. Für erstklassige Großoptik verlangt und erreicht man für T Werte von höchstens wenigen 0,1. Ausführlich beschriebene Beispiele[1] sind in [*6*], [*26*] gegeben.

Die klassische Hartmann-Methode kann nach ihren Voraussetzungen nur meridionale Komponenten der Querabweichung liefern. Lokale Fehler — besonders bei größeren Systemen — führen aber dazu, daß sich die „Strahlenpaare" i. allg. überhaupt nicht schneiden. Es erscheint daher zweckmäßig, die Strahlen zwischen den Einstellebenen *einzeln* zu verfolgen[2]. Bei allerdings stark erhöhtem Rechenaufwand läßt sich so für die beste Bildfläche ein Durchstoßdiagramm angeben, aus dem die Lichtkonzentration unmittelbar zu entnehmen ist. BECK und FEHLKAMM[3] verwenden eine Blende, deren Löcher auf den Gitterpunkten eines *quadratischen* Netzes liegen; dadurch repräsentiert jeder Strahl den gleichen Anteil der Objektivfläche und geht mit dem gleichen Gewicht in die Reduktion ein. Die beiden Aufnahmen werden zweckmäßig auf der gleichen Platte nach verdrehungsfreier Verschiebung gemacht. Gemessen werden die rechtwinkligen Koordinaten y, z der Durchstoßpunkte relativ zum Schwerpunkt aller Punkte. In der durch ε gegebenen Bildfläche ist die Querabweichung vom Schwerpunktsstrahl in y-Richtung gegeben durch

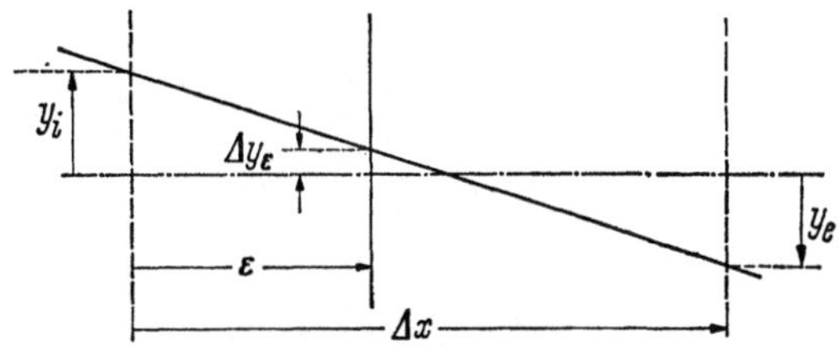

Fig. 50. Strahlverlauf beim „absoluten" Hartmann-Test.

$$\Delta y_\varepsilon = \frac{\varepsilon}{\Delta x}(y_e - y_i) + y_i$$

(Fig. 50); eine entsprechende Gleichung gilt in z-Richtung. Durch Variation von ε, die auch graphisch erfolgen kann, läßt sich die Bildlage mit der besten Lichtkonzentration bestimmen.

Eine verallgemeinerte Hartmann-Methode haben auch MAYALL und VASILEVSKIS[4] bei der Untersuchung des Lick-Spiegels (D = 300 cm) während der letzten Bearbeitungsstadien verwendet. Beobachtet wurde mit einem natürlichen Stern. Die auf Durchmessern liegenden Löcher der Blende sind auf wenige 0,1 mm genau positioniert. Die Durchstoßpunkte werden für nur eine Plattenlage in radialer und tangentialer Richtung ausgemessen. Für ein Rotationsparaboloid haben die Durchstoßpunkte den Zentrumsabstand

$$\varrho = \Delta x \cdot \frac{h}{f}\left(1 + \frac{h^2}{4f^2} \cdots\right),$$

wenn der Stern in der optischen Achse steht. Deren genaue Lage ist aber unbekannt, und es muß ein Koma-Term nach (4.1) eingeführt werden. Dagegen ist

[1] Siehe auch J. S. PLASKETT, Publ. Victoria **1**, 39ff. (1922).
[2] R. KINGSLAKE: Trans. Opt. Soc., Lond. **29**, 133 (1928).
[3] H. G. BECK: Jenaer Jb. 1962, 75. — G. FEHLKAMM: Jenaer Jb. 1962, 87; Jenaer Rdsch. **7**, 116 (1962).
[4] N. U. MAYALL and S. VASILEVSKIS: Astron. J. **65**, 304 (1960).

die kleine Dezentrierung der auf die Eintrittspupille projizierten Blende vernachlässigbar. Durch Ausgleichung wird das optimale Paraboloid ermittelt, das die Quadratsumme der Querabweichungen (Gewicht h) zum Minimum macht. Zu bestimmen sind sechs Unbekannte (drei für die Verbesserung von Nullpunkt und Orientierung, zwei für den Durchstoßpunkt der optischen Achse auf der Platte, dazu ein Skalenfaktor für die Verbesserung von Δx und f). Die radialen und tangentialen Reste ergeben die Neigungen der Flächenelemente; Summation längs Radien und längs Kreisen $h = \text{const}$ führt zu Aberrationsprofilen und — nach

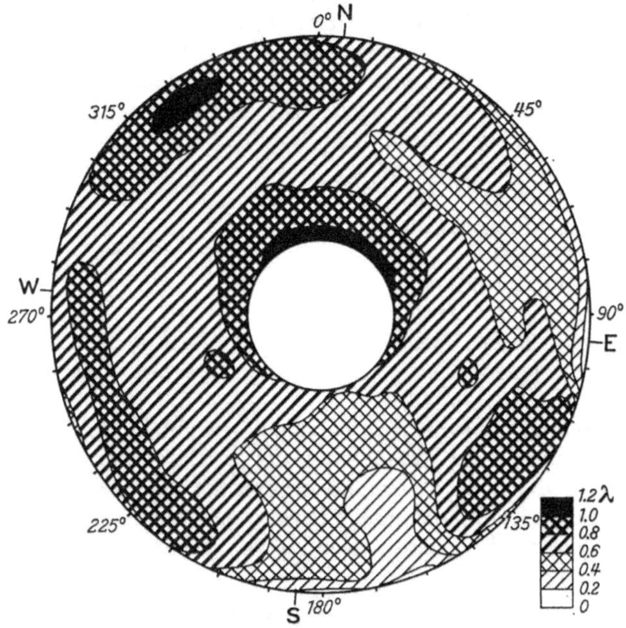

Fig. 51. Schichtlinien-Karte des Lick 120-inch-Spiegels, abgeleitet aus einem verallgemeinerten Hartmann-Test. Bezogen auf eine mittlere Fläche sind die Abweichungen fast überall kleiner als $\pm 0{,}4\ \lambda$. Vom Licht eines axialen Sterns werden 70 bzw. 95% in einem Kreis von 25 bzw. 50 µm ($\approx 0\overset{\prime\prime}{,}7$) Durchmesser konzentriert. Nach MAYALL und VASILEVSKIS.

Verschiebung auf ein gemeinsames Bezugsniveau — zu einer Höhenkarte der Flächenfehler (Fig. 51) als Unterlage für die weitere Bearbeitung. Für den Fokus resultiert schließlich ein Durchstoßdiagramm und damit die Punktbild-Verteilung weitgehend unabhängig von Luftunruhe und Diffusionslichthof. Die vollständigen Querabweichungen sind im Mittel um einen Faktor 1,4 größer als ihre radialen Komponenten; man muß also schließen, daß bei dieser großen und entsprechend flexiblen Scheibe die beiden Komponenten gleich groß sind.

32. Schneidenmethode nach Foucault. Die Foucault-Methode [*41*] zeichnet sich durch Anschaulichkeit und hohe Empfindlichkeit bei einfachsten Hilfsmitteln aus[1]. Üblich ist die strahlenoptische Beschreibung, bei der Beugungsphänomene als „Störungen“ unterdrückt werden. In Fig. 52 sei W die Wellenfläche eines homozentrischen Bündels. Das beobachtende Auge (Pupille P, u. U. mit Fernrohr; auch Kamera) nimmt das ganze Bündel auf und sieht eine helle Scheibe von der Form des Bündelquerschnitts. Schiebt man in einer achsensenkrechten intrafokalen Ebene I einen scharfkantigen Schirm („Schneide“) in das Bündel hinein, so werden die Strahlen auf der Schneidenseite nach und nach

[1] Siehe auch [*44*], [*31*].

abgefangen, und der „Schatten" wandert in der gleichen Richtung wie die Schneide; bei Schneidenbewegung in der extrafokalen Ebene E ist die Schattenbewegung umgekehrt. Beim Durchgang durch das Krümmungszentrum C von W erfolgt der Übergang von Hell zu Dunkel plötzlich (oder jedenfalls, weil das Bild in C nichtverschwindenden Durchmesser hat, gleichmäßig über die ganze Fläche). Damit ist zunächst eine sehr scharfe Methode zur Bestimmung des Krümmungszentrums für den schneidensenkrechten Schnitt von W gewonnen, die viel zur Festlegung des Fokus benutzt wird.

Für eine Störung der Wellenfläche U wird beim Durchgang der Schneide durch C der „linke Abhang" zuerst verdunkelt, dann die ungestörte Fläche W und schließlich der „rechte Abhang". Sind die Querabweichungen von der gleichen Größenordnung wie der Bilddurchmesser in C, so treten kontinuierliche Übergänge in der Leuchtdichte-Verteilung auf, und die Wellen*aberration* erscheint als Relief, beleuchtet von der der Schneide gegenüberliegenden Seite. Der Schneidenbewegung entspricht eine Bewegung der fiktiven Lichtquelle L in Pfeilrichtung (Fig. 52). Der Schnittpunkt der Schneide mit der Bündelachse ist das Zentrum der Referenzkugel.

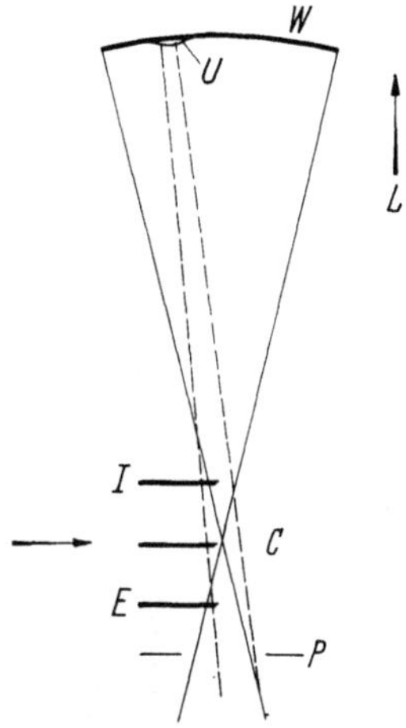

Fig. 52. Geometrische Interpretation der Schneidenprüfung nach FOUCAULT.

Für eine Beugungsbeschreibung betrachtet ZERNIKE[1] als Störung der Wellenfläche ein Cosinus-„Phasengitter" (z.B. hervorgerufen durch die Welligkeit eines Spiegels), das zu zwei seitlichen Beugungsbildern mit 90° Phasenvoreilung gegen das Zentralbild führt. Deckt die Schneide eins davon ab, so läßt die Interferenz der Wellensysteme von Zentral- und anderem Seitenbild das Gitter als Amplitudengitter erscheinen, dessen Extrema auf den Nullstellen des Phasengitters liegen. Das anschließend an diese Vorstellung entwickelte Phasenkontrast-Verfahren[2] mit einer Phasenplatte für die nullte Ordnung des Zentralbildes zeigt zwar kleine Wellenaberrationen quantitativ und mit hoher Empfindlichkeit, ist aber für größere Aberrationen wenig geeignet und stößt bei Großoptik auf die gleichen Stabilitätsprobleme wie die direkten Interferometermethoden. Eine vollständigere wellentheoretische Behandlung der Schneidenmethode, deren Ergebnisse auch für den Praktiker von Nutzen sind, hat LINFOOT [*19*] gegeben. Danach ist zwar die strahlenoptisch ermittelte Leuchtdichte-Verteilung auf dem Objektiv für schwache Aberrationen verschiedentlich zu korrigieren, doch bleibt die „gefühlsmäßige" Interpretation von lokalen Fehlern im wesentlichen richtig.

Für einen stigmatisch abbildenden Strahlengang ist die Methode so empfindlich, daß lokale Störungen der Wellenfläche von wenigen 0,01 λ sicher erfaßt werden können. Auch sphärische Aberration und „Astigmatismus auf der Achse" (mangelnde Rotationssymmetrie der Wellenfläche) lassen sich bis zu Beträgen nachweisen, die die Abbildungsqualität praktisch nicht mehr beeinflussen. Sphärische Aberration führt zu Lichtverteilungen, die auf Profile wie in Fig. 31 (S. 274) schließen lassen. Axialer Astigmatismus ist nicht zu erkennen, wenn die Schneide einer der Hauptachsen parallel ist. Beträgt der Winkel zwischen Schneide und Hauptachsen 45°, so liegt im mittleren Fokus die Trennlinie zwischen Hell und Dunkel *senkrecht* zur Schneide; sie dreht sich, wenn diese in x-Richtung durch

[1] F. ZERNIKE: Monthly Notices Roy. Astron. Soc. **94**, 377 (1934).
[2] Siehe auch C. R. BURCH, Monthly Notices Roy. Astron. Soc. **94**, 384 (1934); **95**, 548 (1935); **96**, 438, 711 (1936).

den Fokalbereich geht. Auch schwacher isolierter Astigmatismus ist daran zu erkennen, daß bei gewahrter Symmetrie längs der Schneidensenkrechten die schneidenparallele Lichtverteilung unsymmetrisch ist.

Bei der Prüfung eines Fernrohrobjektivs im Fokus mit natürlichem Stern als Lichtquelle überlagert sich dem vom Objektiv herrührenden Aberrations-Relief die Phasenstörung der Luft-Inhomogenität und verhindert das visuelle Erkennen feinerer Einzelheiten. Durch photographische Integration (HARTMANN [*40*]) mit Belichtungszeiten von der Größenordnung Minuten kann man das wechselnde

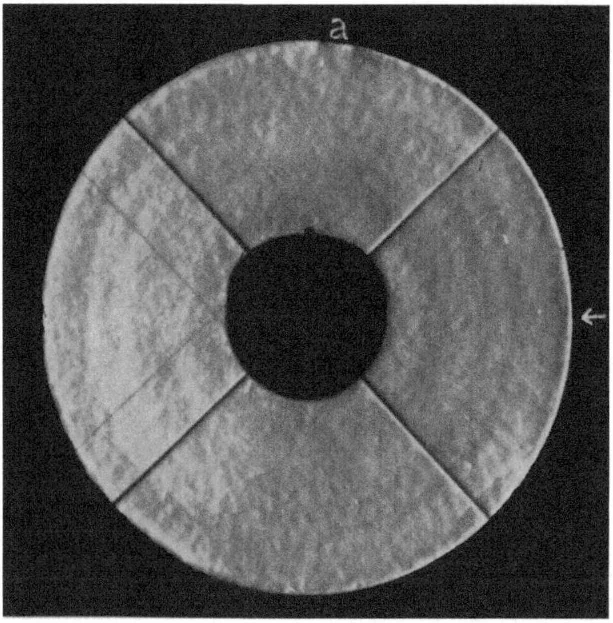

Fig. 53. Fokogramm eines Parabolspiegels (Observatoire de Haute-Provence) mit $D=193$ cm, $N=5$; natürlicher Stern (Zenitdistanz 28°), Belichtungszeit 2 min. Die Zonen haben Maximal-Wellenaberrationen von $\frac{1}{7}\lambda$, die kleinen Buckel charakteristische Abweichungen von $\frac{1}{30}\lambda$. Die lokale Deformation bei a wird von der seitlichen Abstützung des Spiegels verursacht. Der Pfeil bezeichnet die Eintauchrichtung der Schneide. Nach TEXEREAU.

Schlierenbild ausmitteln; das Aberrations-Relief bleibt — allenfalls mit vermindertem Kontrast — erhalten (Fig. 53). Bei der Aufnahme muß der Stern sorgfältig auf der Schneide gehalten werden. Eine geeignete Kamera beschreibt TEXEREAU[1].

Im Labor lassen sich unmittelbar nur die Kugelfläche mit Schneide und künstlichem Stern nebeneinander im Krümmungsmittelpunkt bzw. das Rotationsellipsoid mit Stern im einen, Schneide im zweiten (geometrischen!) Brennpunkt so prüfen, daß die Sollfläche ein ebenes Relief zeigt (Nullmethode) und damit die hohe Empfindlichkeit des Verfahrens ausnutzbar wird. Bei großem Öffnungsverhältnis und kurzer Brennweite eines sphärischen Spiegels muß mit Strahlenteiler gearbeitet werden, um Außeraxialfehler zu vermeiden. Die Erkennbarkeit feinen Details wächst mit abnehmendem Durchmesser der Sternblende bis zu größeren Bruchteilen des Beugungsdurchmessers; ebenso wächst aber auch die relative Intensität der an Kanten und stark gegen die Wellenfront geneigten Flächen entstehenden Beugungssäume und -streifen. Der Stern muß — besonders wenn während des Polierens mit unbelegter Spiegelfläche gearbeitet

[1] J. TEXEREAU: Sky and Telescope **28**, 345 (1964).

wird — eine hohe Leuchtdichte haben. Die Lichtstärke des Tests läßt sich erhöhen, wenn statt eines Sterns ein enger Spalt parallel zur Schneide verwendet wird; über die Spalthöhe dürfen sich die Abbildungsfehler nicht ändern. Beim Spalt ist auch die zuweilen störende Projektion von Augeninhomogenitäten gemildert. Der Realität von Erscheinungen mit geringem Kontrast sollte man sich stets durch Bewegen der Schneide und des Auges vergewissern.

Zu dem klassischen Schneidenverfahren gibt es Nebenformen, z.B. mit Stern und Loch, Stern und kleiner Kreisblende, Spalt und Faden. Diese symmetrischen Methoden lassen keinen Reliefeindruck entstehen; bei ihnen erscheinen alle Objektivstellen hell, die zur Abbildung in den gewählten Punkt beitragen bzw. nicht beitragen. Auch die Schattenbild-Verbiegungen eines außerhalb von C (Fig. 52) durch das Bündel geführten Fadens lassen Abbildungsfehler erkennen. Den bewegten Faden kann man hier durch ein feststehendes grobes Parallel- oder Kreuzgitter ersetzen[1].

Bei der wichtigen Laboratoriums-Prüfung von *deformierten* Flächen läßt sich für kleine Abweichungen von der Kugel zwar im Krümmungsmittelpunkt der allgemein glatte Verlauf der Fläche feststellen, doch gestatten die Profile der sphärischen Aberration kein unmittelbares Urteil über die Größe der Deformation. Für stark deformierte Flächen — wie Parabolspiegel mit $N=3$ bis 5 — sind die Querabweichungen im Krümmungszentrum so viel größer als der ideale Bilddurchmesser, daß Feinheiten in den harten Kontrasten völlig untergehen.

Den Verlauf der Deformation kann man mit RITCHEYS[2] Zonenblenden-Verfahren prüfen, bei dem die Längsabweichung der sphärischen Überkorrektion ausgemessen wird. Dazu wird der Spiegel bis auf zwei Ausschnitte einer schmalen Zone auf einem schneidensenkrechten Durchmesser abgedeckt. Der Schnittpunkt der Strahlen von der Zonenhöhe h mit der Achse ist (mit den üblichen Symmetrievoraussetzungen) dadurch gegeben, daß beide Zonenausschnitte beim Eintauchen der Schneide *gleichzeitig* dunkel werden. Die Beurteilung erfordert den photometrischen Vergleich zweier getrennt liegender Flächen, der durch die Beugungssäume an den Maskenrändern zusätzlich erschwert wird. Die Längsabweichung eines Parabolspiegels mit Lichtquelle im paraxialen Krümmungsmittelpunkt ist

$$\Delta x = \frac{h^2}{2f} + \frac{h^4}{16 f^3} + \cdots;$$

für beliebig deformierte Flächen gilt $\Delta x = -b \cdot \frac{h^2}{2f}$. Das Verfahren ist weniger genau als die Nullmethoden.

Beim Zonenblenden-Verfahren wird das Zentrum derjenigen Referenzkugel bestimmt, die die (im Spiegel liegende) Wellenfläche in der Zone h berührt. Schärfer läßt sich der Krümmungsmittelpunkt eines Meridiankurven*elements* erfassen, doch ist dafür die genaue Messung in zwei Koordinaten nötig[3]. Bezogen auf den paraxialen Krümmungsmittelpunkt gibt ein einseitiger Spiegelausschnitt mit dem (vorzeichenbehafteten) Achsabstand h ein Bild des Krümmungsmittelpunkts auf der Kaustik

$$\Delta x = -\frac{3}{2} b \frac{h^2}{f}; \quad \Delta y = \frac{1}{2} b \frac{h^3}{f^2}$$

für die fehlerfreie Fläche. Man arbeitet mit Spalt im paraxialen Krümmungsmittelpunkt und Zonenblende und benutzt zum Ausmessen einen in x- und y-

[1] FOUCAULT [*41*]. — F. JENTZSCH: Phys. Z. **29**, 66 (1928). — G. SCHULZ: Ann. Physik **85**, 189 (1928). Das Verfahren von M. L. LENOUVEL, Rev. Opt. **2**, 211, 315 (1924) stellt den Übergang zum Scherungsinterferometer nach RONCHI (vgl. Ziff. 33) dar.

[2] G. W. RITCHEY: Astrophys. J. **19**, 53 (1904).

[3] R. PLATZECK and E. GAVIOLA: J. Opt. Soc. Amer. **29**, 484 (1939).

Richtung verschiebbaren Faden, oder photographische Aufzeichnung ohne weitere Optik in einer größeren Zahl von Ebenen $x = \text{const}$ (so bei der Prüfung des Palomar 5 m-Spiegels).

Will man Parabolspiegel oder andere einen fernen Objektpunkt stigmatisch abbildende Systeme im Labor mit der Nullmethode prüfen, so kann man die *Autokollimationsanordnung* (Fig. 54) verwenden. Der Planspiegel muß hohe Genauigkeit haben. Von RITCHEY stammt das klassische Prüfschema: Der Objektivspiegel wird zunächst als genauer Kugelspiegel hergestellt, mit Nullprüfung im Zentrum. Er dient nun zur Kontrolle des Planspiegels nach dem Verfahren von Fig. 55, wobei neben Zonen und lokalen Fehlern auch eine allgemeine schwache Krümmung festgestellt wird. Ein großer Krümmungsradius R_P verursacht in C die astigmatische Differenz

$$2\Delta x_A = \frac{4L^2 \sin\alpha \tan\alpha}{R_P}.$$

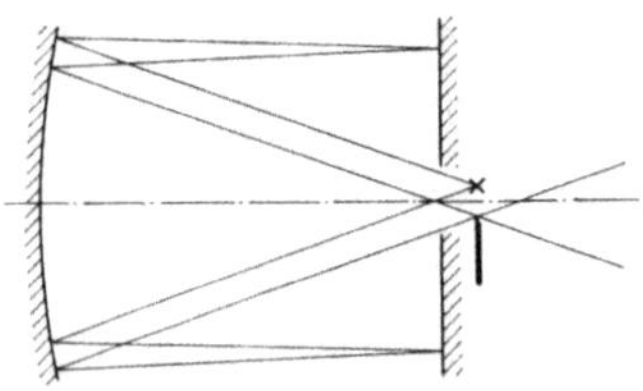

Fig. 54. Schneiden-Prüfung eines Parabolspiegels in Autokollimation.

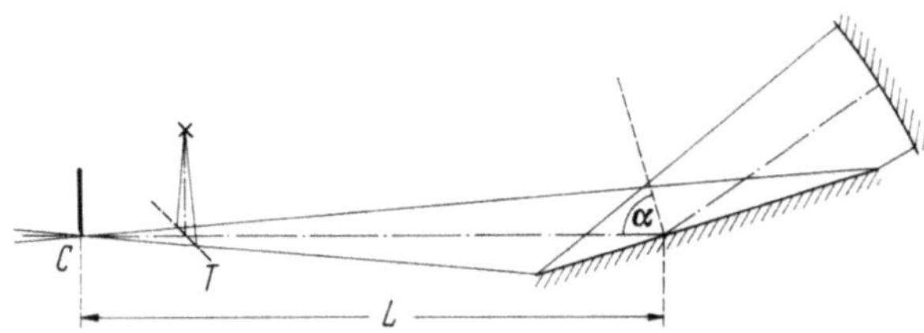

Fig. 55. Prüfung eines Planspiegels nach COMMON-RITCHEY.

Anschließend wird der bisher sphärische Spiegel unter Prüfung in Autokollimation parabolisiert.

Bei deformierten Einzelflächen mit $b \neq -1$ ist dieses Verfahren nicht anwendbar, und auch für große Parabolspiegel erscheint der Aufwand recht groß. Stellt man an die Flächengenauigkeit sehr hohe Ansprüche, so wird die Genauigkeitsforderung für den Planspiegel unerfüllbar. Diese Schwierigkeiten lassen sich mit den *Kompensationsmethoden* umgehen, die Null-Tests für beliebig deformierte Hohlflächen liefern. Bei Prüfung in der Nähe des Krümmungsmittelpunkts wird hierbei die sphärische Überkorrektion mit der Unterkorrektion eines *kleinen* Zusatzsystems aufgehoben. Die Strahlen eines homozentrischen Bündels müssen — bei doppeltem Durchgang — das Zusatzsystem in solcher Höhe und mit solcher Neigung verlassen, daß sie den zu prüfenden Spiegel senkrecht treffen. Die Flächen des Zusatzsystems sollen Herstellung und Prüfung mit großer Genauigkeit gestatten.

Die Kompensation mit einer dünnen asphärischen Korrektionsplatte beschreibt GASCOIGNE[1]. Hohlspiegel als Kompensationselement haben BURCH[2] und MAKSUTOV [*44*][3] verwendet; auch ROSS[4] diskutiert diese Anordnung. Bei der Kompensation mit Linsen (COUDER[5], ROSS[4], DALL[6]) muß man mit monochromatischem Licht arbeiten. Wegen der Verschiedenheit der höheren Ordnungen bleiben bei der Kompensation Reste, die für größere Spiegel zusätzlich berücksichtigt werden müßten. Sie lassen sich bis auf unschädliche Beträge beseitigen, wenn Korrektionssystem und Spiegel in Kontakt gebracht werden. Bei einer von

[1] S. C. B. GASCOIGNE: Monthly Notices Roy. Astron. Soc. **100**, 462 (1940).

[2] C. R. BURCH: Monthly Notices Roy. Astron. Soc. **96**, 438 (1936).

[3] D. D. MAKSUTOV: Mitt. Pulkovo **21**, H. 1 (No. 160), 5 (1957). — M. A. SOSNINA: Mitt. Pulkovo **21**, H. 3 (No. 162), 137 (1958).

[4] F. E. ROSS: Astrophys. J. **98**, 341 (1943).

[5] A. COUDER: Rev. Opt. **6**, 49 (1927).

[6] H. E. DALL: J. Brit. Astron. Assoc. **57**, 201 (1947) u. [*31*] III.

OFFNER[1] gegebenen Anordnung geschieht das durch eine Feldlinse nahe dem Krümmungsmittelpunkt, die die aberrationserzeugende Korrektionslinse auf den Spiegel abbildet. Man erreicht so sehr genaue Kompensation mit verhältnismäßig kleinen Zusatzsystemen, die z.B. interferometrisch geprüft werden können. Das Verfahren läßt sich an beliebige Deformationen anpassen und liefert auch eine handliche Prüfmethode für den hyperbolischen Hauptspiegel des Ritchey-Chrétien-Systems[2].

33. Interferometrische Prüfung. Die Interferometer-Verfahren haben erst neuerdings breitere Anwendung gefunden, obwohl die Flächenkontrolle mit dem Prüfglas und auch die Twyman-Greensche Form des Michelson-Interferometers

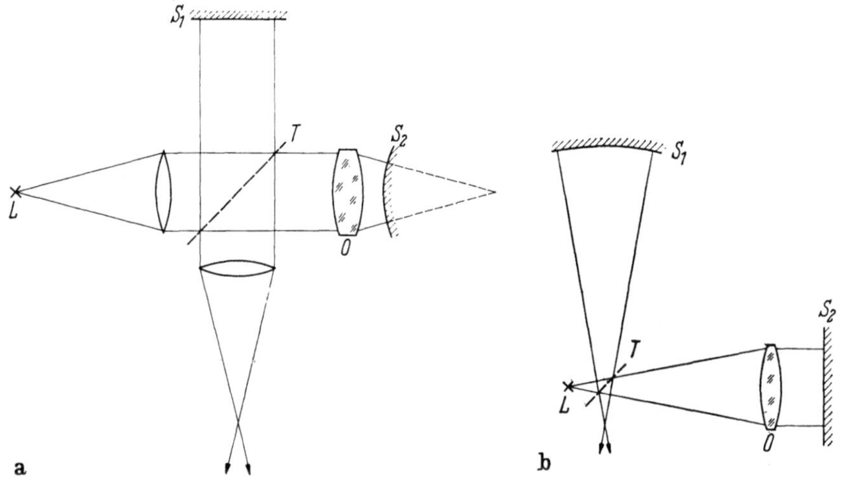

Fig. 56. a Twyman-Green-Interferometer; b Variante zur Prüfung größerer Objektive.

dem Optiker schon länger vertraut sind. Bei beiden findet ein Vergleich zweier Flächen mit Fizeau-Streifen statt. Frühe Beispiele einer interferometrischen bzw. durch Interferenz verschärften Prüfung von größeren Objektiven sind die Methoden von MICHELSON[3] und von VÄISÄLÄ [*26*], beide mit punktweiser Abtastung der Wellenfläche. Besonders übersichtlich ist die Objektivprüfung nach TWYMAN und GREEN[4] mit dem in Fig. 56a gezeigten Aufbau. Dicht hinter dem zu prüfenden Objektiv O steht ein fehlerfreier Kugelspiegel S_2, dessen Krümmungsmittelpunkt in den Brennpunkt von O fällt. Ist auch O fehlerfrei, so kehrt eine einfallende Planwelle nach Durchlaufen von $O-S_2-O$ als ebene Wellenfront zurück und erzeugt durch Überlagerung mit der von der fehlerfreien Vergleichsfläche S_1 reflektierten Planwelle ein gleichmäßig ausgeleuchtetes Gesichtsfeld, dessen Helligkeit bei monochromatischer Quelle L von der relativen Phasenlage der beiden Wellensysteme abhängt. Ist O fehlerhaft, so weicht die zurückkehrende Wellenfront zwischen O und dem Teiler T um die doppelte Wellenaberration $2V$ von der Ebene ab (O wird zweimal durchsetzt). Für die resultierenden Interferenzstreifen gilt (vereinfacht) die Parameterdarstellung $V(Y,Z)=k\cdot\frac{\lambda}{2}$. Dabei sind Y, Z rechtwinklige Koordinaten in der Austrittspupille; k ändert sich nur ganzzahlig, kann aber einen Bruch für die Phasendifferenz enthalten. Bei dieser idealen

[1] A. OFFNER: Appl. Optics **2**, 153 (1963).
[2] D. H. SCHULTE: Appl. Optics **2**, 141 (1963).
[3] A. A. MICHELSON: Astrophys. J. **47**, 283 (1918).
[4] C. CANDLER: Modern Interferometers. London 1951; s. auch TWYMAN [*33*].

Justierung des Interferometers entstehen so sehr anschauliche quantitative Isohypsenkarten der Wellenaberration (Fig. 57). Die Streifenformen für die Seidel-Aberrationen hat Kingslake[1] behandelt.

Um bei kleinen Fehlern die Pupille besser zu füllen, benutzt man statt dieser Einstellung „auf Ringe" oft die Justierung „auf Streifen" durch Kippung von S_1 oder Translation von S_2; die interferierenden Wellenfronten bilden dann einen Keil mit dem Keilwinkel $\varkappa$, und bei fehlerfreiem Prüfling O ist das Gesichtsfeld von zur Keilkante parallelen und äquidistanten geraden Streifen durchsetzt ($V_\varkappa = \varkappa \cdot Y = k \cdot \frac{\lambda}{2}$ bei Kippung um die Z-Achse). Bei fehlerhaftem Objektiv gilt $V(Y, Z) - \varkappa \cdot Y = k \cdot \frac{\lambda}{2}$. Kleine Abweichungen von der Geradlinigkeit der Streifen können unmittelbar als *Profile* der Aberrationsfläche gedeutet werden, denn es ist z.B. für den mittleren Streifen $Y \approx \frac{1}{\varkappa} \cdot V(0, Z)$ mit $k = 0$, $Y \approx 0$.

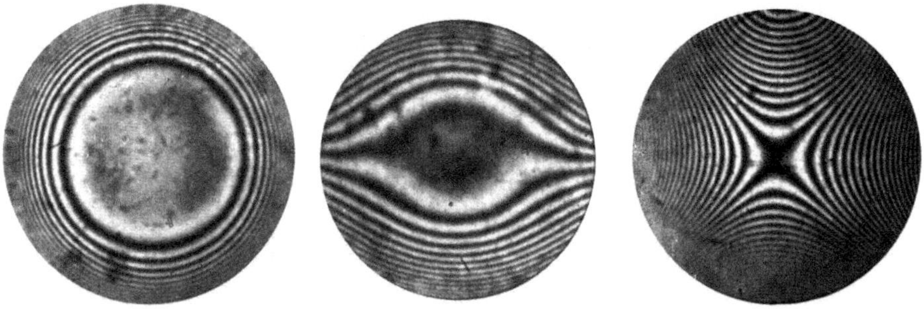

a b c

Fig. 57a—c. Interferogramme im Twyman-Green-Interferometer. Die Wellenflächen haben a sphärische Aberration; b Koma; c Astigmatismus. Nach Maréchal.

Bei hohem apparativem Aufwand kann man so die Wellenaberration für die ganze Apertur quantitativ erfassen. Ebenso wie die zwei einwandfreien Vergleichsflächen müssen auch die übrigen Elemente des Interferometers mindestens die gleiche Öffnung haben wie das zu prüfende Objektiv; für Fernrohroptik kommt das Twyman-Green-Interferometer also kaum in Frage. Die Abmessungen werden günstiger, wenn man die Teilerplatte T im nichtparallelen Strahlengang benutzt und zwei *sphärische* Wellenfronten miteinander vergleicht (Fig. 56b; die Anordnung ist schon von Twyman angegeben, jedoch mehrfach nacherfunden worden). Selbst von der Forderung gleicher optischer Länge in den beiden Armen des Interferometers kann man abgehen, wenn man als Lichtquelle einen Laser mit hinreichender Kohärenzlänge benutzt. Einen Hohlspiegel an der Stelle von $O + S_2$ könnte man so mit einer wesentlich kleineren Vergleichsfläche prüfen. Wegen der sehr hohen Stabilitätsforderung bleibt jedoch die Prüfung astronomischer Großoptik mit einem getrenntarmigen Interferometer auch dann noch schwierig.

Brauchbare Interferometer erhält man dagegen, wenn man *beide* sich überlagernden Wellenfronten vom Prüfling erzeugen läßt und so das Streifensystem stabilisiert. Verlaufen dazu noch die zusammengehörenden Strahlen der beiden Komponenten nahe beieinander, so werden auch die Anforderungen an die Homogenität der Luft geringer. Bei geeignetem Aufbau gibt es statt einer Grenze für den Durchmesser des Prüflings allenfalls noch eine solche für sein Öffnungs-

[1] R. Kingslake: Trans. Opt. Soc. **27**, 94 (1925/26); s. auch Maréchal [*42*] u. Françon [*20*].

verhältnis. In den *Scherungs*-Interferometern[1] werden derartige Anordnungen realisiert durch Vergleich der vom Prüfling kommenden Wellenfläche mit sich selbst nach Aufteilung und anschließender Translation, Umklappung oder Drehung der einen Komponente gegenüber der anderen. Eine weitere Möglichkeit ist die „radiale Scherung" (relative Maßstabsänderung), die schließlich auf die Erzeugung der zweiten Wellenfläche aus einem engen und daher praktisch aberrationsfreien Bündel führt.

Werden bei Interferometern mit *lateraler* Scherung die Komponenten einer ebenen Wellenfläche in ihrer Ebene gegeneinander verschoben oder die einer sphärischen Wellenfläche um ihren Krümmungsmittelpunkt auseinander geschwenkt, so ergibt die Überlagerung ein gleichmäßiges Gesichtsfeld. Sind Aberrationen V vorhanden, so entstehen Streifen für

$$V(Y,Z)-V(Y+a,Z)=k\lambda\approx a\frac{\partial V}{\partial Y}$$

für hinreichend kleine a, wobei die Scherung vom Betrag a in Y-Richtung liegt. Längs eines Streifens ist also die *Neigung* der Wellenfläche gegen die Referenzfläche konstant; von einem Streifen zum nächsten ändert sie sich um λ/a. Die Streifen sind demgemäß Isophoten des entsprechenden Foucault-Schneidenbildes. Sichtbar gemacht werden nur Gradienten in Scherungsrichtung. Periodische Störungen der Wellenfläche mit der Periode a bleiben ohne Einfluß.

Nun gilt für eine Wellenfläche mit sphärischer Aberration dritter Ordnung

$$V\sim(Y^2+Z^2)^2 \quad\text{und}\quad \frac{\partial V}{\partial Y}\sim Y(Y^2+Z^2).$$

Im Scherungsinterferogramm haben die Streifen also die gleiche Form wie im Twyman-Interferogramm die einer Wellenfläche mit Koma, für die ja $V\sim Y\cdot(Y^2+Z^2)$ ist. Analog verhalten sich Koma und Astigmatismus.

Durch eine zusätzliche *Kippung* der einen Komponente gegen die andere kann die Pupille wieder mit Streifen parallel zur Kippachse gefüllt werden. Für symmetrische Aberrationen wie Defokussierung und Öffnungsfehler entsteht ein symmetrisches Streifenbild, wenn die Kippachse die Z-Achse ist, wegen

$$a\frac{\partial V}{\partial Y}-\varkappa Y=k\lambda;$$

bei Kippung um die Y-Achse = Scherungsrichtung ist die Interferenzfigur $\left(a\frac{\partial V}{\partial Y}-\varkappa Z=k\lambda\right)$ antisymmetrisch. Auch die reine Translation einer sphärischen Wellenfront wirkt wie Scherung plus Kippung um die Z-Richtung. Eine eingehende Diskussion des Scherungsinterferometers gibt BATES[2]; ein Reduktionsverfahren, das aus dem Interferogramm die Form der Wellenfläche ohne Symmetrievoraussetzungen abzuleiten gestattet, hat SAUNDERS[3] mitgeteilt.

Das früheste Interferometer mit lateraler Scherung war das von WAETZMANN[4]. Eine dicke Planparallelplatte bewirkt Bündeltrennung und Scherung, wobei zur Kompensation eine Autokollimationsanordnung nötig ist. Neuerdings hat MURTY[5] ein ähnliches Interferometer beschrieben, das durch Verwendung einer Laser-Quelle auch im einfachen Durchgang Interferenzen liefert, und dazu

[1] Eine Übersicht gibt O. BRYNGDAHL, Applications of Shearing Interferometry. In: Progress in Optics, Vol. IV (Hrsg. E. WOLF). Amsterdam 1965.

[2] W. J. BATES: Proc. Phys. Soc., Lond. **59**, 940 (1947).

[3] J. B. SAUNDERS: J. Res. Nat. Bur. Stand. B **65**, 239 (1961).

[4] W. BARTHOLOMEYCZYK: Z. Instrumentenkde. **70**, 212 (1962).

[5] M. V. R. K. MURTY: Appl. Optics **3**, 531 (1964).

eine Modifikation für Großoptik vorgeschlagen. BATES[1] gab ein auf der Mach-Zehnder-Anordnung beruhendes Scherungsinterferometer an, das DREW[2] und BROWN[3] unter Verzicht auf Universalität vereinfacht haben. Das Brownsche Instrument ist als verkittete Einheit mit fixierter Scherung und Kippung für Routineprüfungen geeignet. Aus dem Köstersschen Interferenz-Doppelbildprisma hat SAUNDERS neben umklappenden[4] auch scherende[5] Interferometer entwickelt. Sie stellen ebenfalls kleine, festverkittete Einheiten dar, die zur Prüfung mit natürlichem Stern in den Okularstutzen des Teleskops eingesetzt werden können.

Sehr gering ist der apparative Aufwand bei der Schattenstreifenmethode von RONCHI[6], die — jedenfalls für höhere Gitterfrequenzen — als Vielfach-Scherungsinterferometrie anzusehen ist. Ein kurz vor dem Bildpunkt in den Strahlengang gebrachtes Gitter mäßiger Periode (Größenordnung 10 bis 100 Linien/mm) erzeugt

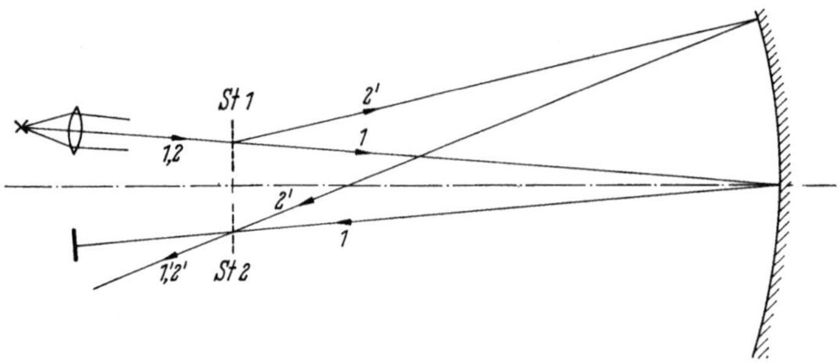

Fig. 58. Streuplatten-Interferometer nach BURCH.

seitliche Beugungsbilder, von denen die versetzten Bündel ausgehen. Bei der Prüfung von Objektiven[7] bleibt man allerdings meist im geometrisch-optischen Bereich.

Um die Wirkung des *radial* scherenden Interferometers deutlich zu machen, sei die Wellenaberration in der Austrittspupille nach Polarkoordinaten (r, Φ) entwickelt mit Gliedern

$$V(r, \Phi) = A\, r^i \cos^j \Phi .$$

Für eine um den Faktor s radial vergrößerte Wellenfront ist die Aberration

$$V_e = A \left(\frac{r}{s}\right)^i \cos^j \Phi$$

und damit die Differenz der beiden Wellenflächen

$$\Delta V = A\, r^i \cos^j \Phi \cdot \left(1 - \frac{1}{s^i}\right);$$

ist die radiale Maßstabsänderung hinreichend groß, so liefert das Streifensystem $(\Delta V = k \cdot \lambda)$ im gemeinsamen Feld eine Schichtliniendarstellung der Wellen-

[1] W. J. BATES: Proc. Phys. Soc., Lond. **59**, 940 (1947).
[2] R. L. DREW: Proc. Phys. Soc., Lond. B **64**, 1005 (1951).
[3] D. BROWN: Proc. Phys. Soc., Lond. B **67**, 232 (1954).
[4] J. B. SAUNDERS: In: STRONG [*43*], Appendix C.
[5] J. B. SAUNDERS: Astron. J. **69**, 449 (1964); J. Res. Nat. Bur. Stand. C **68**, 155 (1964).
[6] V. RONCHI: Atti Fond. Ronchi **17**, 93 (1962) mit umfangreichem Literaturverzeichnis; Kurzfassung: Appl. Optics **3**, 437 (1964).
[7] J. A. ANDERSON and R. W. PORTER: Astrophys. J. **70**, 175 (1929). — A. R. KIRKHAM: In [*31*] I. — A. A. SHERWOOD: J. Brit. Astron. Ass. **68**, 180 (1958); Atti Fond. Ronchi **15**, 340 (1960).

aberration wie beim Twyman-Interferometer. Eine Übersicht der Realisierungsmöglichkeiten gibt MURTY[1].

Autokollimationsverfahren in Analogie zu dem von Fig. 56b benutzen statt der Teilerplatte ein Element, das kohärente Bündel mit verschiedenem Öffnungswinkel, aber zusammenfallenden Achsen erzeugt. Das enge Referenzbündel bildet die Lichtquelle auf den Prüfling ab und wird aberrationsfrei reflektiert; das weite Testbündel füllt die Spiegelapertur. Im zurückkehrenden Strahlengang wird auch das Referenzbündel auf großen Öffnungswinkel gebracht. Die Interferenzfigur entspricht der eines radial scherenden Interferometers mit sehr großem s. Bemerkenswert einfach ist das von BURCH[2] angegebene Interferometer, bei dem das weite Bündel durch *Streuung* an geeigneten (z.B. teilmattierten) Flächen *St 1*, *St 2* erzeugt wird (Fig. 58). Die Verwendung eines Streuplatteninterferometers zur Prüfung größerer Spiegel erwähnt SCOTT [*36*]. DYSON[3] benutzt zur Bündeltrennung eine Linse aus doppelbrechendem Material mit verschiedenen Brennweiten für ordentlichen und außerordentlichen Strahl oder polarisierende Strahlenteiler in Kombination mit $\lambda/4$-Platten, MURTY[4] die Beugung an Zonenplatten. Kann man einen kleinen seitlichen Abstand der Bündelachsen zulassen, so sind auch rein geometrisch-optische Anordnungen mit einem Kösters-Prisma möglich[5].

D. Konstruktionselemente der Fernrohre[6].

In diesem Kapitel wird auf existierende oder im Bau befindliche Teleskope mit Kurzbezeichnungen hingewiesen. Literaturangaben zu vielen Einzelheiten bei älteren Instrumenten finden sich in KING [*4*]; weitere Quellen sind:

Bergedorf Schmidt	O. HECKMANN: Mitt. Astron. Ges. 1954, 57 (1955).
Cambridge 36-inch	R. O. REDMAN: Quart. J. Roy. Astron. Soc. **1**, 10 (1960).
Caracas Schmidt	H. SCHMIDT: Askania-Warte **18**, H. 57, 1 (1961).
Caracas 1-m	H. KÖHLER: Optik **17**, 485 (1960).
Cleveland Schmidt	J. J. NASSAU: Astrophys. J. **101**, 275 (1945).
Haute-Provence 1,93-m	A. COUDER: Ann. d.'Astrophys. **23**, 311 (1960).
Herstmonceux 98-inch	R. v. d. R. WOOLLEY: Quart. J. Roy. Astron. Soc. **3**, 249 (1962). — G. M. SISSON: Observatory **84**, 183 (1964).
Kitt Peak 36-inch	A. B. MEINEL: In [*9*].
Kitt Peak 84-inch	A. B. MEINEL: In [*9*].
Kitt Peak 150-inch	D. L. CRAWFORD: Sky and Telescope **29**, 268 (1965).
Lick 120-inch	W. W. BAUSTIAN: In [*9*].
McDonald 82-inch	Contrib. McDonald No. 1.
Mt. Wilson 60-inch	G. W. RITCHEY: Astrophys. J. **29**, 198 (1909).
Palomar 200-inch	I. S. BOWEN: In [*9*].
Richmond Hill	R. K. YOUNG: J. Roy. Astron. Soc. Canad. **28**, 97 (1934).
Tautenburg	N. RICHTER: Sterne **37**, 89 (1961).
U.S. Naval 61-inch	K. AA. STRAND: Appl. Optics **2**, 1 (1963); Vistas in Astronomy **8**, 9 (1966).
Victoria 72-inch	J. S. PLASKETT: Publ. Victoria **1**, No. 1, 7 (1919).

I. Zelle und Rohr.

34. Fassungen für Refraktionsoptik. Die Fassung kleinerer optischer Teile macht kaum Schwierigkeiten. Dagegen sind bei den größeren dioptrischen Elementen Kräfte wechselnder Richtung bis zu 100 kp bei engen Toleranzen und

[1] M. V. R. K. MURTY: Appl. Optics **3**, 853 (1964).

[2] J. M. BURCH: Nature, Lond. **171**, 889 (1953), und in Interferometry (Nat. Phys. Lab. Symposium No. 11), S. 277. London 1960. — Siehe auch J. DYSON in STRONG [*43*], App. B; A. H. SHOEMAKER and M. V. R. K. MURTY, Appl. Optics **5**, 603 (1966).

[3] J. DYSON: J. Opt. Soc. Amer. **47**, 386 (1957); s. auch W. H. STEEL, Two-Beam Interferometry. In: E. WOLF (ed.), Progress in Optics, Vol. 5 (1966).

[4] M. V. R. K. MURTY: J. Opt. Soc. Amer. **53**, 568 (1963).

[5] D. S. BROWN: J. Sci. Instrum. **39**, 71 (1962).

[6] Hierzu s. auch A. B. MEINEL, Design of Reflecting Telescopes. In [*9*].

ausschließlich am Rand der Scheiben aufzunehmen. Dem Konstrukteur kommt dabei zugute, daß brechende Glieder gegen kleine Biegungen unempfindlich sind; die optische Weglänge wird dadurch in erster Näherung ja nicht geändert. Es ist deshalb möglich, noch ziemlich große Objektive mit starren Dreipunktlagerungen (je axial und radial) zu fassen. Die Fassung muß zwangsfrei sein und darf auch über die differierende thermische Ausdehnung von Glas und Fassungsmaterial keine Verspannung hervorrufen. Für Kronglas mit einem Ausdehnungskoeffizienten $\alpha_G = 8 \cdot 10^{-6}$ in einer Stahlfassung mit $\alpha_F = 11 \cdot 10^{-6}/°\mathrm{C}$ tritt zwischen $-30°$ und $+20°$ eine Ausdehnungsdifferenz von 0,15 mm auf 1 m Länge auf. Man läßt entweder Temperaturspiel $\Delta_t = D(\alpha_F - \alpha_G)(t - t_{\min})$ (Δ_t Spiel bei der Temperatur t, D Objektivdurchmesser, $t_{\min}$ niedrigste Temperatur, bei der die Fassung noch nicht klemmt) oder hält die Scheiben radial mit zwei festen und einem federnden Punkt. Höheren Ansprüchen werden Kompensationsfassungen gerecht, mit gegeneinanderarbeitenden Gliedern aus Metallen von verschiedenem Ausdehnungskoeffizienten[1] oder auch mit Wellrohrkörpern, die mit einer Flüssigkeit hoher Ausdehnung gefüllt sind[2]. Die Voraussetzung für Kompensationsverfahren — thermische Homogenität — ist bei den verhältnismäßig dünnen Objektivscheiben einigermaßen erfüllt. Zur Vermeidung der lokalen Spannungen in der Nähe der Auflagepunkte hat man bei großen Kameraobjektiven auch Zwischenkörper aus gummiartigem Kunststoff benutzt[3]. Einige der Entlastungsverfahren, wie sie für Spiegelscheiben üblich sind, ließen sich auf die Randlagerung großer Objektive ohne weiteres übertragen, doch sollte das Fassungsgewicht stets möglichst niedrig bleiben.

Die Fassungen mehrlinsiger Objektive müssen das Zentrieren der Einzellinsen gestatten und den Justierzustand im Betrieb erhalten. Bei photographischen Großobjektiven ist es zweckmäßig, wenn man Translationen und Kippungen der Einzelglieder unabhängig von der (zufälligen) Linsenberandung vornehmen kann[4]; die einzuhaltenden Toleranzen gehen herab bis zu 10 µm und 10″.

Die Justierung selbst erfolgt entweder optisch mittels der Reflexbilder an den Flächen (Fluchtfernrohr) oder mechanisch mit Drehtisch und Meßuhr[3].

35. Lagerung großer Spiegel.

α) Anforderungen. Von einer Spiegelfassung verlangt man, daß sie die Figur der optischen Fläche und ihre Position relativ zum Rest des Systems in allen Lagen des Instruments erhält. Auch bei Temperaturänderung darf sie keinen Zwang auf den Spiegel ausüben. In aller Schärfe stellt sich das Fassungsproblem bei den Hauptspiegeln für große und sehr große Reflektoren, mit Gewichten bis zur Größenordnung 10 t.

Von den Freiheitsgraden des starren Körpers macht einer — Rotation um die optische Achse, die zugleich Symmetrieachse ist — keine besonderen Schwierigkeiten; der Spiegel muß demnach eine definierte Lage bekommen hinsichtlich Translation längs der optischen Achse, zwei Translationen in seiner Ebene, und zwei Rotationen um in dieser Ebene liegende Achsen. Dabei müssen die definierenden Elemente kleine Bewegungen in den anderen Freiheitsgraden zulassen, für differentielle thermische Expansion oder auch Justierzwecke. Die Kraftrichtung muß innerhalb einer Halbkugel variieren können; auch bei Verbiegungen der tragenden Struktur darf kein Zwang auftreten.

Die Spiegellagerung wird weiter dadurch kompliziert, daß nur kleine Spiegel als starr angesehen werden dürfen. Größere verbiegen sich optisch merklich unter

---

[1] R. STEINHEIL: Z. Instrumentenkde. **14**, 170 (1894); s. auch [*44*].
[2] M. JOHN: Askania-Warte **18**, H. 57, 6 (1961).
[3] L. G. DETERDING: Appl. Optics **1**, 403 (1962).
[4] W. BLATT u. E. MALLWITZ: Askania-Warte **20**, H. 62, 8 (1963).

ihrem Eigengewicht, wenn sie nicht auf mehr Punkte gelagert werden als zur Definition der Position erforderlich sind. Die überschüssigen Punkte dürfen dabei nicht starr sein. Biegende Kräfte im Spiegel treten dann nicht auf, wenn jedes Volumenelement mit der seinem Gewicht entsprechenden Kraft unterstützt wird; die optimale Verteilung der Stützpunkte und der zugehörigen Kräfte läßt sich aus dieser Bedingung für mäßige Punktzahlen aber nicht herleiten. Bei Neigungsänderung muß die Verteilung auf axiale und radiale Komponenten richtig erfolgen. Der korrekte Entlastungszustand soll möglichst auch bei kleinen Verrückungen des Spiegels erhalten bleiben („Astasie"). Auch die entlastenden Kräfte müssen so eingeleitet werden, daß keine Behinderung von Bewegungen in anderen Freiheitsgraden auftritt[1].

Ein echtes Schwimmen wäre für einen Hohlspiegel mit vertikaler Achse keine ideale Entlastung: der konstante Druck auf die Rückfläche berücksichtigt nicht die ausgeschliffene Kalotte, und das System ist nur hinsichtlich Translationen senkrecht zur Achse astatisch.

β) Achsenparallele Kräfte. Den bei weitem überwiegenden Einfluß auf die Spiegelfigur haben Kräfte parallel zur optischen Achse. Die maximale Durchbiegung b einer dünnen auf der Rückseite unterstützten zylindrischen Platte unter dem Einfluß des Eigengewichts ist gegeben durch

$$b = \text{const} \cdot \frac{\gamma}{E} \cdot \frac{D^4}{d^2} \cos z$$

(mit Wichte γ, Elastizitätsmodul E, Durchmesser D, Dicke d, Neigung gegen die Horizontale z). Dabei sind Glieder ab d^2/D^2 vernachlässigt. Die Konstante hängt von der Unterstützungs-Geometrie ab. Daneben gehen wesentlich ein die Materialkonstante γ/E (s. Tabelle 4) und ein Formfaktor $(D^2/d)^2$. Es gibt danach kein Verhältnis von Durchmesser zu Dicke, das genügend Biegungssteifheit unabhängig von den Abmessungen liefert; eine Scheibe mit $D=500$ und $d=50$ zeigt die gleiche Durchbiegung wie eine mit $D=10$ und $d=2\cdot 10^{-2}$. In seiner grundlegenden Arbeit macht COUDER [*35*] zunächst Angaben über das Profil einer dünnen durchgebogenen Platte, die in einem zum Rand konzentrischen Kreis unterstützt wird. Ein Vergleich mit strengeren Rechnungen[2] für eine „dicke" Platte ergibt befriedigende Übereinstimmung. Durch Superposition wird weiter die Durchbiegung für mehrere unterstützende Kreisringe ermittelt; die Kraftverteilung hängt dabei davon ab, ob Minimalisierung der entstehenden Krümmung (z.B. für Planspiegel) oder der sphärischen Aberration dritter Ordnung (für Hohlspiegel, wenn Nachfokussierung zugelassen ist) angestrebt wird[3]. Die erforderliche Zahl der Stützpunkte pro Kreis folgt aus dem maximal zulässigen Durchhang. Mit einer Toleranz von $V_{\max}=7\cdot 10^{-6}$ cm und $E/\gamma = 3\cdot 10^{8}$ cm (Glas, Quarz, ...) ergeben sich schließlich halbempirische Maximalwerte des Formfaktors D^2/d für verschiedene Anordnungen der Stützpunkte, die von $D^2/d \leqq 125$ cm bei drei Randpunkten über $D^2/d \leqq 460$ cm bei sechs Rand- und drei inneren Punkten mit gleicher Belastung bis $D^2/d \leqq Z > 2400$ cm bei 33 Punkten auf drei Kreisen wachsen. Aus diesen Angaben[4] kann man entnehmen, daß für eine Stützpunkt-Anzahl $n \geqq 12$ die Größe $\frac{D^2}{d\cdot n} \approx 70$ cm nahezu konstant ist; demnach ist für einen charakteristischen Punktabstand

$$\frac{D}{\sqrt{n}} \lesssim 8{,}3\sqrt{d} \qquad (D,\ d \text{ in cm})$$

[1] F. G. PEASE: Publ. Astron. Soc. Pacific **44**, 308 (1932).

[2] J. P. O. SILBERSTEIN and B. I. GREEN: Observatory **80**, 234 (1960).

[3] Zur Wahl der Kreisradien s. auch T. V. BUDNIKOVA u. V. M. FRIDMAN, Mitt. Pulkovo **24**, H. 1 (No. 177), 119 (1964).

[4] Siehe auch MAKSUTOV [*44*].

zu setzen. Die maximale Durchbiegung einer *ausgedehnten* Platte (ohne Berücksichtigung von Randeffekten) unter dem Eigengewicht ist $b = \frac{\gamma}{E} \cdot \frac{a^4}{d^2}$, wenn in den Gitterpunkten eines Quadratnetzes mit Abständen $2a$ unterstützt wird. Mit $E/\gamma = 3 \cdot 10^8$ cm und der Forderung $b \leqq 3 \cdot 10^{-6}$ cm folgt $2a \lesssim 10\sqrt{d}$. Bei den größten existierenden Spiegeln werden die Couderschen Grenzwerte überschritten.

γ) Kräfte in der Scheibenebene. Die Kräfte in der Scheibenebene spielen gegenüber den axialen nur eine kleine Rolle. COUDER betrachtet die Profiländerung, die die Pressung des Eigengewichts an einer auf der Mantelfläche stehenden Scheibe konstanter Dicke verursacht, und findet, daß sie bis $D \approx 3$ m vernachlässigbar ist. SCHWESINGER[1] berücksichtigt auch die durch die ausgeschliffene Kalotte entstehenden Biegmomente und die vollständige Verteilung der Normalkraft K am Scheibenumfang. Als Formparameter wählt er $\varkappa = \frac{D^2}{16 d \cdot f}$. Damit die Scheibe getragen wird, muß $\int_0^{2\pi} K(\varphi) \cos\varphi \, d\varphi$ gleich dem Scheibengewicht sein (φ ist der Azimutwinkel auf dem Spiegel, von der Lotrichtung aus gezählt). Bei der idealen Entlastung mit $K \sim \cos\varphi$ sollte selbst eine Scheibe von $D = 5$ m noch keine schädlichen Aberrationen zeigen. Allerdings sind dazu in der „oberen" Hälfte *Zug*kräfte nötig, die man erst ganz neuerdings anwendet[2]. Beschränkt man sich auf radiale Druckkräfte, so ist die günstigste Verteilung $K \sim 1 + \cos\varphi$; sie ist anzunähern durch den hydrostatischen Druck einer Flüssigkeit, in die die Scheibe gerade ganz eintaucht. Wenig schlechter ist die Verteilung $K = \text{const}$ in der unteren und $K = 0$ in der oberen Hälfte, die sich bei der reibungsfreien Aufhängung in einem völlig biegsamen Band einstellt. Diese bei Optikern zur Prüfung beliebte Unterstützungsart zeichnet sich durch das Fehlen von Astigmatismus aus. Dagegen führt die Entlastung $K \sim \cos\varphi$ auf der unteren, $K = 0$ auf der oberen Hälfte zu einer astigmatischen Deformation, weil weder der Verkürzung des vertikalen noch der Verlängerung des horizontalen Krümmungsradius Kräfte entgegenwirken. SCHWESINGER gibt in einem Diagramm die bei resultierenden Wellenaberrationen $V_{\text{rms}} = \lambda/15$ und $\lambda/30$ maximal zulässigen Spiegeldurchmesser in Abhängigkeit vom Formfaktor $\varkappa$ für die verschiedenen Unterstützungsarten.

Bei sehr großen Spiegeln verteilt man auch die achsensenkrechten Kräfte auf viele Stützpunkte, die die ganze Fläche erfüllen. Dazu wird die Schwerebene[3] durch Bohrungen in der Spiegelrückseite zugänglich gemacht. Zur Gewichtsersparnis können die Entlastungen für die verschiedenen Komponenten vereinigt werden (Palomar 200-inch), was jedoch zu einer komplizierten Justierung führt.

Eine astatische Entlastung für Kräfte in der Spiegelebene ermöglicht die Definition der radialen *Position* in der (für den Cassegrain-Strahlengang meist ohnehin vorhandenen) zentralen Bohrung, womit die fragwürdige Kompensation der thermischen Ausdehnungsdifferenz zwischen Spiegel und Fassung über den ganzen Spiegeldurchmesser hinweg umgangen wird. Die Randunterstützung darf axiale Bewegungen des Spiegels nicht hemmen. Das ist besonders wichtig für Schmidt-Teleskope, bei denen die Hauptspiegelposition über Temperatur-Kompensationsglieder nach *vorn* definiert ist.

δ) Ausführung von Entlastungssystemen[4]. Die Verteilung des Gewichts auf viele Stützpunkte kann bei kleineren Spiegeln noch mit einer nachgiebig-elastischen

---

[1] G. SCHWESINGER: J. Opt. Soc. Amer. **44**, 417 (1954).

[2] R. M. SCOTT: Appl. Optics **1**, 387 (1962); auch Herstmonceux 98-inch.

[3] Zur optimalen Angriffstiefe s. V. S. CHERNINA, Mitt. Pulkovo **24**, H. 1 (No. 177), 125 (1964).

[4] Dazu s. auch G. M. SISSON, Vistas in Astronomy **3**, 92 (1960) und A. B. MEINEL in [*9*], 38ff.

Unterlage (Filzplatte o.ä.) geschehen, doch hat der Spiegel dann keine definierte Position mehr. Das Entlastungssystem nach GRUBB erzwingt die gleichmäßige Verteilung des Gewichts auf mehr als drei diskrete Punkte dadurch, daß die Auflageflächen an den Ecken von Dreiecken angeordnet sind; die Dreiecksplatten werden in ihren Schwerpunkten unterstützt und können um die Stützpunkte Kippbewegungen ausführen, ebenso wie die Auflageplatten um die Eckpunkte der Dreiecke. Das Grubb-System ist nicht astatisch; die Kraftverteilung ist durch die Geometrie festgelegt, so daß eine Justierung der entlastenden Kräfte nicht ohne weiteres möglich ist. Die Anordnung ist mit neun Auflagepunkten für Spiegel bis $D=190$ cm gebaut worden (Richmond Hill). Einige Verbreitung hat auch ein System mit 18 Punkten[1] gefunden, bei dem je zwei Dreiecksplatten auf einem gleicharmigen Hebel mit radialer Achse sitzen (Fig. 59). Mit diesem zweietagigen Aufbau dürfte allerdings die Leistungsgrenze des Systems erreicht sein, jenseits derer Reibung und Biegung in der Entlastungseinrichtung selbst die Wirkungsweise beeinträchtigen. Bewegungsfreiheit in axialer Richtung soll das 18 Punkte-System des Caracas-Schmidt liefern[2].

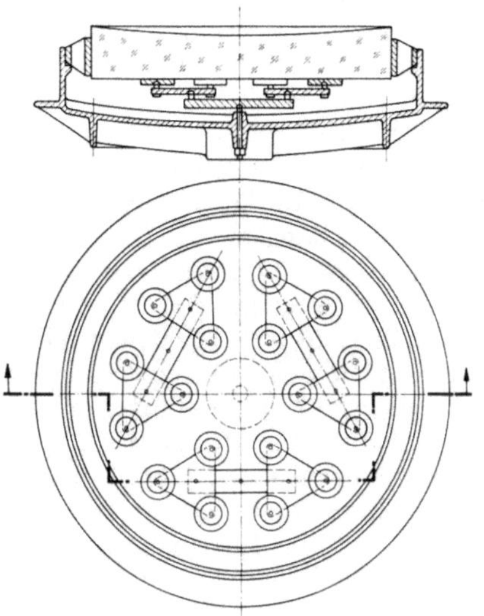

Fig. 59. Grubbsches Entlastungssystem mit 18 Auflagepunkten. Nach MEYER.

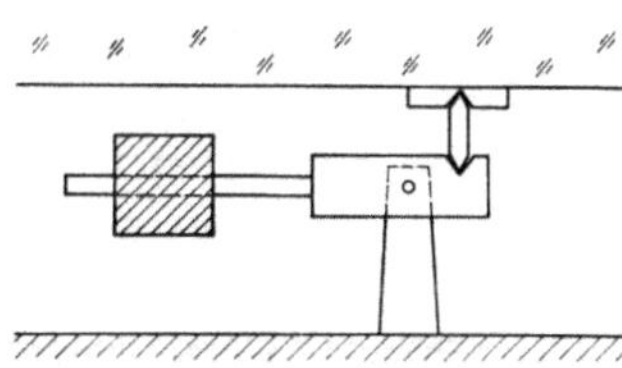

Fig. 60. Prinzip der Gewichtshebel-Entlastung von LASSELL.

Bei großen Spiegeln wurde bis vor kurzem ausschließlich das Entlastungssystem nach LASSELL[3] angewendet. Hier wird die entlastende Kraft an jedem Stützpunkt unabhängig von allen anderen durch einen zweiarmigen Gewichtshebel eingeführt (Fig. 60); sie kann empirisch justiert werden. Das System gestattet die Einstellung der Spiegelneigung (und in geringem Umfang der axialen Position), indem drei der Entlastungshebel mikrometrisch einstellbar fixiert werden. Die Auflageplatten müssen dem Spiegel Freiheit zu radialen Bewegungen lassen, sollen aber gegen Abgleiten nahe $z=90^\circ$ gesichert und möglichst selbstzentrierend sein. Auch zur radialen Entlastung des Spiegelrandes ist das Gewichtshebelsystem vielfach verwendet worden, obwohl es zu einer ungünstigen Kraftverteilung am Umfang führt.

Sehr große und damit zwangsläufig flexible Spiegel sind empfindlich gegen Abweichungen von der korrekten Kraftverteilung[4], wie sie selbst bei richtiger Justierung der Entlastung nach einer Neigungsänderung durch Haftreibung entstehen können. So arbeitete die axiale Entlastung für den Palomar 200-inch erst

[1] F. MEYER: Z. Instrumentenkde. **50**, 58 (1930). — J. H. HINDLE in [*31*] I.
[2] M. JOHN: Askania-Warte **18**, H. 57, 6 (1961).
[3] G. W. RITCHEY: Astrophys. J. **5**, 143 (1897).
[4] Dazu s. auch V. A. PALMOV u. V. A. PUPYREV, Mitt. Pulkovo **24**, H. 1 (No. 177), 145 (1964).

zufriedenstellend, nachdem man die Reibung auf 0,1% der Kraftkomponente verringert hatte. Ein so großer Spiegel hat eine definierte Figur nur zusammen mit seinem Entlastungssystem und wird auch auf diesem bearbeitet. Da nur kleine Drehwinkel um die Hebelachsen verlangt werden, liegt die Verwendung von Federgelenken nahe.

Reibungsschwierigkeiten werden weitgehend vermieden mit *pneumatischen* und *hydraulischen* Entlastungssystemen, die darüber hinaus auch zu einer Gewichtsersparnis führen können. Bei der Luftdrucklagerung (U.S.-Naval 61-inch; Herstmonceux 98-inch) wird die axiale Unterstützung durch auf die Spiegelrückseite wirkenden Überdruck geliefert. Die Kraft verteilt sich dabei zwangsläufig gleichmäßig über die Fläche. Ist der Spiegel so groß, daß die variable Dicke berücksichtigt werden muß, wird eine Unterteilung in konzentrische Bereiche notwendig. Die Druckvariation mit $\cos z$ muß durch eine Steuerung (Analogrechner) eingegeben werden. Der Ausgangsdruck wird so eingestellt, daß auf den drei definierenden Auflageflächen eine passende Restkraft verbleibt. Wenn der Luftdruck durch einen elastischen Schlauch oder Beutel ausgeübt wird, muß man dafür sorgen, daß dieser die Spiegelrückseite nicht thermisch isoliert. Mit *Unter*druck läßt sich mit dem pneumatischen System auch eine korrekte Entlastung für Sekundärspiegel erzielen, die mit der Spiegelfläche nach unten hängen.

Die Spiegelunterstützung ist ferner mit *hydraulischen Druckkörpern* ausgeführt worden. Der Öldruck wird am einfachsten durch ein Analogsystem mit Gewichtshebel gesteuert. Mit dem hydraulischen System erreicht man leicht und reibungsfrei günstige Übersetzungsverhältnisse. Ein Vorteil der radialen Öldruck-Entlastung ist, daß sie dem Spiegel infolge der Biegsamkeit der Balgen genügend axiale Bewegungsfreiheit läßt; ein Klemmen ist auch bei kleinen Verkippungen des Spiegels ausgeschlossen. Pneumatische und hydraulische Systeme erscheinen besonders geeignet für eine Steuerung oder Regelung der Entlastungseinrichtung zur Optimalisierung der Spiegelfigur, wie sie für große Instrumente diskutiert wird. Schließlich hat man zur radialen Entlastung auch den hydrostatischen Auftrieb verwendet (U.S. Naval 61-inch). Dabei lassen quecksilbergefüllte Schläuche zwischen der Mantelfläche des Spiegels und der Zelleninnenwand den Spiegel schwimmen; bei richtiger Bemessung der Schlauchbreite ist die Kraftverteilung $K \sim 1 + \cos \varphi$ realisierbar.

36. Rohrmontierung. Zur gegenseitigen Fixierung der optischen Komponenten und der Strahlungsempfänger und zu ihrer Verbindung mit dem Achsensystem dient die *Rohrmontierung*. Von ihr wird verlangt, daß sie die Position großer Massen (bis zu vielen Tonnen bei den größten Instrumenten) unabhängig von der Angriffsrichtung der Schwerkraft relativ zueinander auf größenordnungsmäßig 0,1 mm in radialer und bis 0,01 mm in axialer Richtung definiert. Bei kleinen und mittelgroßen Instrumenten kann man diese Forderung noch durch Materialfestigkeit und passende Konstruktion zu erfüllen suchen; bei großen Teleskopen ist man auf *Kompensation* der unvermeidlichen Biegungen angewiesen. Auch *Entlastung* für einzelne Elemente ist angewendet worden. Wenn diese passiven Systeme zu schwer werden oder ihre Genauigkeit nicht ausreicht, bleibt noch der Ausweg einer gesteuerten oder geregelten Anordnung (u. U. auch intermittierende Regelung durch Eingriff des Beobachters).

Große Spiegelteleskope benutzen heute allgemein das von SERRURIER[1] für den Palomar 200-inch entwickelte System, mit dem die relative Biegung für zwei Elemente praktisch beseitigt werden kann (Fig. 61). An einem kräftigen Mittelstück mit quadratischem Querschnitt sind über Strebensysteme aus je

[1] M. SERRURIER: Civ. Engng. (Easton, Pa.) **8**, 524 (1938).

vier V-förmigen Gliedern („A-beams") zwei Halterungsringe angebracht. Am Mittelstück greift die Deklinationsachse an; die Halterungsringe tragen Spiegelzelle bzw. Primärfokusempfänger oder Sekundärspiegel. Durch die Konstruktion wird zunächst erreicht, daß auch bei Biegung die Ringebenen stets parallel zueinander und zur Zentralebene des Mittelstücks bleiben (Parallelogramm-Führung). Mit passender Dimensionierung der Streben kann man schließlich — unter der Voraussetzung des gleichen Biegungsgesetzes — die Biegungsbeträge auf beiden Seiten gleich groß machen, und zwar auch bei recht verschiedenen Längen der V-Träger (Lick 120-inch; Kitt Peak 84-inch). Die optisch geforderten Biegungstoleranzen lassen sich mit dem Serrurier-System auch für große Instrumente erfüllen: bei Palomar 200-inch entfernt sich der Durchstoßpunkt der Hauptspiegelachse durch das Ross-Korrektionssystem nicht mehr als 0,25 mm von seiner mittleren Lage, bei etwa 17 m Rohrlänge.

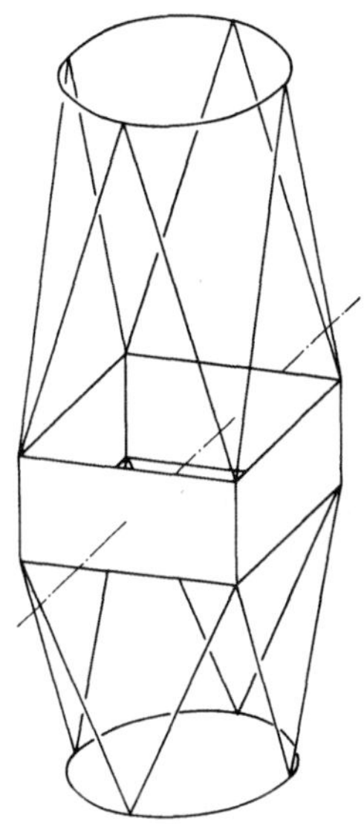

Fig. 61. Biegungskompensierte Rohrkonstruktion nach SERRURIER.

Bis zu Spiegeldurchmessern von etwa 2 m lassen sich photographische Einrichtungen oder Spektrographen im Primärfokus noch von der Rohrwand her bedienen (McDonald 82-inch). Bei größeren Instrumenten baut man eine Beobachtungskabine in das Rohr ein (erstmals beim Palomar 200-inch, Fig. 78). Bei etwas kleineren Dimensionen (Lick 120-inch, Herstmonceux 98-inch) hat die Kabine ovalen Querschnitt und ist exzentrisch drehbar angebracht. Damit läßt sich die Abschattung kleiner halten und trotzdem in allen Lagen eine einigermaßen bequeme Beobachterposition erreichen. Für das letztgenannte Instrument ist auch für den *Cassegrain-Fokus* ein mit dem Rohr verbundener Beobachterplatz hinter der Spiegelzelle vorgesehen; die Anordnung scheint sich für Großinstrumente allgemein einzubürgern.

Für die optimale Ausnutzung der Beobachtungszeit ist es wichtig, zwischen den verschiedenen in einem größeren Teleskop inkorporierten optischen Anordnungen schnell und ohne Neujustierung wechseln zu können. Beim Palomar 200-inch ist die Primärfokus-Kabine fest eingebaut; die verschiedenen Sekundärspiegel befinden sich nahe an ihren Arbeitspositionen, in die sie bei Bedarf eingeschwenkt werden können. Der Lick 120-inch gestattet den Austausch des vordersten Rohrabschnitts mit der Kabine gegen einen Sekundärspiegel-Träger. Schnelles und bequemes Wechseln zwischen den Sekundärspiegeln für den Ritchey-Chrétien- und den Coudé-Fokus erlaubt die Wende-Ring-Anordnung (Kitt Peak 84-inch, Fig. 70). Die beiden Spiegel sind Rücken an Rücken montiert; das Haltekreuz greift an einen Ringträger an, der innerhalb des vorderen Rings der Rohrkonstruktion liegt und um zwei Zapfen geschwenkt werden kann.

Bei Schmidt-Kameras sind *drei* Elemente (Korrektionsplatte, Hauptspiegel und Kassette) relativ zueinander zu fixieren. Da sie wegen des Streulichtschutzes ohnehin ein geschlossenes Rohr haben müssen, hat man stets auch den Rohrkörper tragend oder mittragend ausgebildet. Die nötige Biegungssteifigkeit wird erzielt durch eingezogene Ringe und Stege, doppelwandige Ausführung mit Kastenversteifung, oder eine Kombination dieser mit Entlastungs- und Kompensationsmaßnahmen. Um die Fokussierung innerhalb der optisch geforderten Toleranz (0,01 mm) zu halten, hat man den Spiegel (oder auch Spiegel und Kassettenträger) astatisch entlastet und ihren gegenseitigen Abstand und die Neigung durch drei temperaturkompensierte Stäbe definiert, die auch die thermisch bedingte Brennweitenänderung des Spiegels (u.U. verzögert) mit ausgleichen können.

37. Maßnahmen gegen Temperatureffekte. Man hat bis jetzt nicht ernsthaft versucht, Fernrohre bei konstanter Temperatur zu betreiben. Damit bei größeren Instrumenten die thermischen Effekte klein bleiben, ist man jedoch bestrebt, eine Aufheizung während des Tages zu verhindern und möglichst den ganzen Kuppel-Innenraum, mindestens aber die thermisch trägen optischen Elemente etwa auf Nacht-Temperatur zu halten. Bisher wurden dazu überwiegend passive Anordnungen benutzt[1]. Ein Betrieb nahe dem Temperaturgleichgewicht mit der Umgebung wird erleichert durch Auswahl eines Aufstellungsortes mit nur geringer Abnahme der Lufttemperatur während der Nacht.

Eine homogene Temperaturänderung beeinflußt bei Linsen- und Spiegelobjektiven die *Brennweite* und damit Bildort und Abbildungsmaßstab. Für Spiegel gilt $f_t = f_0(1 + \alpha t)$, wobei α der lineare Ausdehnungskoeffizient des Spiegelmaterials ist (vgl. Tabelle 4, S. 296). Bei Linsensystemen ergibt sich eine kompliziertere Abhängigkeit von Material- und Formgrößen[2]. Der „effektive Ausdehnungskoeffizient" für die Brennweite beträgt meist wenige 10^{-5}; er läßt sich durch passende Wahl der Glasarten verringern. In die Änderung der *Fokuseinstellung* geht nur die Differenz zur Ausdehnung des Rohrs ein. Hier ist Kompensation möglich und z B. bei Schmidt-Kameras üblich.

Die Annahme einer homogenen Temperaturverteilung ist für größere Teleskope kaum erfüllt; oft sind sie schon bei Beginn der Beobachtung wärmer als die Außenluft, deren Temperatur noch absinkt. Es müssen daher ständig Ausgleichsvorgänge ablaufen, die zu Temperaturgradienten und damit zu Änderungen der Flächenform und Inhomogenität des Brechungsindex führen[3]. Selbst wenn man die Tageserwärmung völlig vom Instrument fernhalten könnte, führte ein Luftmassenwechsel mit Kaltfrontdurchgang zu einem thermischen Schock, dessen freies Abklingen bei großen Teleskopen Tage dauert. Hier liegt gegenwärtig eine der einschneidendsten Einschränkungen für den Betrieb sehr großer Fernrohre[4]. Mit Spiegeln aus Material mit extrem kleinem Ausdehnungskoeffizienten (vgl. Ziff. 28α), wie es für jetzt projektierte Großinstrumente verfügbar ist, sollten die Effekte wesentlich verkleinert werden können.

An großen Linsenobjektiven beobachtet man allgemein sphärische Unterkorrektion[5], wenn das Glas wärmer ist als die umgebende Luft. Abhilfe bringt Ventilation der Objektivzelle[6]; auch Anpassung der sphärischen Korrektion an einen charakteristischen Temperaturverlauf ist versucht worden[7].

Bei Spiegeln[8] ändert ein Temperaturgradient parallel zur optischen Achse in erster Näherung nur den Krümmungsradius der Spiegelfläche. Temperaturdifferenzen zwischen Vorder- und Rückfläche können sich bei Nicht-Gleichgewicht von Spiegel und Umgebung dadurch einstellen, daß die Wärmeabgabe an der Vorderfläche weniger behindert ist als an der Rückseite, die dem selbst wieder thermisch trägen Hinterbau der Spiegelzelle gegenübersteht. Die Vorderfläche ist dann im Betrieb die kältere, mit Differenzen Δt von der Größenordnung einige 0,1 Grad. Eine planparallele Platte der Dicke d nimmt unter dem Einfluß eines konstanten

[1] Dazu s. CH. FEHRENBACH in [3].
[2] F. L. O. WADSWORTH: Monthly Notices Roy. Astron. Soc. **63**, 573 (1903).
[3] A. COUDER [35]; s. auch G. G. SLIUSAREV, Opt. Spectr. **6**, 134 (1959). — W. P. BARNES jr.: Appl. Optics **5**, 701 (1966). — J. N. HANSON: J. Opt. Soc. Amer. **55**, 916 (1965).
[4] Dazu s. z. B. die Beiträge und Diskussionen in [3].
[5] Siehe z. B. P. A. IANNA, Vistas in Astronomy **6**, 93 (1965).
[6] F. SCHLESINGER: Proc. Nat. Acad. Washington **1**, 13 (1915).
[7] R. SCHORR: Vjschr. astron. Ges. **68**, 189 (1933).
[8] Hierzu s. besonders COUDER [35].

axialen Temperaturgradienten den Krümmungsradius $R = \frac{d}{\alpha \Delta t}$ (hohl gegen die kalte Seite) an; die (kleine) Krümmung überlagert sich der der optischen Fläche so, daß eine Brennweitenänderung

$$|\Delta f| = \frac{2f^2 \alpha}{d} \Delta t \tag{37.1}$$

resultiert.

Der „Randeffekt", ein Zonenfehler vor allem der äußeren Spiegelpartien, ist die Folge eines radialsymmetrischen Temperaturgradienten *senkrecht* zur optischen Achse. Der Gradient beruht auf dem zusätzlichen Wärmeverlust durch die Mantelfläche der Spiegelscheibe. Der Randeffekt hängt in schwer zu übersehender Weise von Form und Fassung des Spiegels und seinem Spannungszustand ab[1]. Bei einem frei in Luft sich abkühlenden niedrigen Kreiszylinder[2] sollte der Rand absinken, sich also sphärische Überkorrektion des Randes ergeben; dabei nimmt der Randeffekt mit der Spiegeldicke zu. Reale Teleskope zeigen teils Über-, teils Unterkorrektion bei fallender Temperatur. Bei der verrippten Scheibe des Palomar-Teleskopes wird der Randabfall des gegenüber der Umgebung zu warmen Spiegels auf die rasche Abkühlung und Kontraktion des äußeren soliden Glasrings zurückgeführt.

Um die thermischen Effekte klein zu halten, sind im wesentlichen zwei Wege eingeschlagen worden. Einmal kann man den Spiegel in möglichst guten thermischen Kontakt mit der umgebenden Luft bringen; man versucht dann, die Wärmeabgabe von Vorder- und Rückseite gleich zu machen und die des Randes — durch Isolierung — zu verhindern. Bei der Kontrolle des Wärmeflusses spielt der für beide Seiten verschiedene Transportmechanismus eine Rolle: an der verspiegelten Vorderseite wirkt hauptsächlich Konvektion, während an der Rückfläche Strahlung und Leitung zur Zelle überwiegen. Das Verspiegeln der Rückseite (als Maßnahme zur Symmetrierung gedacht) unterdrückt nur die Strahlung und ist ziemlich nutzlos. Allgemein ist es für das thermische Verhalten hier vorteilhaft, den Spiegel so leicht wie mechanisch vertretbar zu machen. Das günstige Verhältnis Oberfläche/Volumen eines Rippenspiegels fördert den schnellen Temperaturausgleich. Beim Palomar-Spiegel hat es sich bewährt, nach einem Kaltfrontdurchgang die Stahlkonstruktion hinter dem Spiegel mit Ventilatoren rasch um die Hälfte des Lufttemperatur-Abfalls abzukühlen[3]; eine ähnlich günstige Wirkung solcher Ventilatoren war beim Lick-Reflektor nicht festzustellen[4].

Wenn bei sehr großen Spiegeln die letzte Retusche der Flächen in situ mit Prüfung an natürlichen Sternen vorgenommen wird, ergibt sich eine gewisse Anpassung der Figur an einen charakteristischen Temperaturgang (Palomar, Lick[5]).

Die zweite Möglichkeit, vor allem für Instrumente mit geschlossenem Rohr, ist eine weitgehende thermische *Isolierung* der hinteren Fernrohrteile mit dem Spiegel. Bei stark behinderter Wärmeabgabe (oder sogar Wärmezufuhr[6]) ändert sich die Spiegeltemperatur so langsam, daß eine brauchbare Figur erhalten bleibt. Wird der Wärmeinhalt von Spiegel und Zelle im wesentlichen über die Spiegel-Vorderseite abgegeben, so ist der axiale Gradient offenbar eine vernünftige Annäherung an die realen Verhältnisse; die für das Tautenburger Teleskop[7] mitge-

[1] Siehe etwa J. TEXEREAU, Sky and Telescope **28**, 345 (1964).
[2] Quantitativ behandelt von P. WELLMANN, Z. Astrophys. **35**, 104 (1954).
[3] I. S. BOWEN in [*9*].
[4] S. VASILEVSKIS in [*3*].
[5] Dazu s. auch J. TEXEREAU, Fußnote 1; D. S. BROWN: Sky and Telescope **29**, 350 (1965).
[6] A. COUDER: Vistas in Astronomy **1**, 372 (1955).
[7] N. RICHTER: Mitt. Tautenburg Nr. 12 (1964).

teilte Abhängigkeit der Brennweite von der Temperaturdifferenz am Spiegel folgt jedenfalls weitgehend der Beziehung (37.1). Die Bildqualität ist hier bis zu einer Differenz $\Delta t \lesssim 1.4$ °C nicht merklich gestört.

Besonders bei diesem zweiten Verfahren muß man damit rechnen, daß die verschieden temperierte und deshalb optisch inhomogene Luft im Rohr die Abbildungsqualität beeinflußt. Wirksame Abhilfe dagegen (und gegen die Störung durch aus dem Kuppelspalt abziehende Warmluft) bringt eine Installation, wie sie COUDER[1] beim großen Haute-Provence-Instrument verwirklicht hat. Mit kräftigen Ventilatoren wird hier in der Kuppelöffnung und im Rohr eine Strömung von außen nach innen erzwungen. In das Rohr tritt Luft nur von außerhalb der Kuppel am Oberende ein und wird durch einen porösen inneren Tubus gesaugt; so entsteht eine laminare Strömung mit längs der Achse bis zum Spiegel auf Null abnehmender Geschwindigkeit. Schließlich kann man die über dem warmen Spiegel entstehende zellulare Konvektion unmittelbar an seinem Außen- und Innenrand absaugen.

II. Achsenmontierung.

38. Anforderungen und Elemente. Die Montierung gestattet dem Beobachter, das Fernrohr auf das gewählte Objekt zu richten und diesem bei seiner scheinbaren Bewegung — für astronomische Objekte wesentlich parallel zum Äquator — zu folgen. In einfachster Weise wird dies durch Drehmöglichkeit um zwei zueinander senkrechte Achsen erreicht, deren eine der Erdachse parallel ist (äquatoreale Aufstellung, *parallaktische* Montierung). Mit dieser Achsenlage ist eine enge Beziehung zum Koordinatensystem an der Sphäre gegeben: die Drehwinkel der beiden Achsen entsprechen unmittelbar Stundenwinkel und Deklination; die Achsen werden entsprechend als Stundenachse oder Polachse und als Deklinationsachse bezeichnet. Zur Kompensation der Erdrotation genügt eine *gleichförmige* Drehung um die Stundenachse allein, wobei ein mit dem Fernrohr fest verbundener Empfänger seine Orientierung relativ zur Sphäre behält.

Diese Vorteile sind so schwerwiegend, daß bisher alle größeren Forschungsinstrumente parallaktisch montiert sind. Bei großen Teleskopen wird der Nachteil spürbar, daß die Stundenachse im allgemeinen schief zur Gravitationsrichtung liegt und damit Größe und Richtung der biegenden Kräfte sowie die Lastverteilung in den Deklinationslagern sich ändern. Moderne technische Hilfsmittel für Koordinatenübertragung und -wandlung und automatische Nachführung lassen auch Instrumente mit zur Erdschwere orientierten Achsen als möglich erscheinen, wie sie ja bei vielen großen Radioteleskopen benutzt werden[2]. Diese „azimutale“ Aufstellung hat den Vorteil, daß das Lager der Vertikalachse nur konstante axiale und das der Horizontalachse nur konstante radiale Kräfte aufzunehmen hat; die biegenden Kräfte sind für die Achsen konstant und können für Rohr und Spiegel nur in einer Halbebene variieren. Allerdings liegt der Instrumentenpol, in dem die Transformation vom äquatorealen in das azimutale Koordinatensystem eine Unstetigkeitsstelle hat, sehr ungünstig im Zenit. Für die Rotationen des Bildfelds relativ zum Rohr und des Teleskops um die Achsen ergeben sich also gerade im bevorzugten Beobachtungsgebiet große Geschwindigkeiten und Beschleunigungen. In dieser Hinsicht wäre es günstiger, den Instrumentenpol in den Horizont zu legen (Horizontal-Rahmenmontierung), doch tritt dabei wieder eine Achse mit variabler Neigung und Lastverschiebung im Lager auf.

[1] A. COUDER: Ann. Astrophys. **23**, 311 (1960); s. auch R. O. REDMAN, Quart. J. Roy. Astron Soc. **1**, 13 (1960).

[2] P. B. FELLGETT: Occasional Notes Roy. Astron. Soc. **3**, 143 (1956). — E. H. RICHARDSON in [*3*].

Für die Anforderungen zur Nachführung bei verschiedenen Orientierungen der Achsen hat VASILEVSKIS[1] anschauliche Diagramme gegeben; auch im günstigsten Fall bleiben allgemein ungleichförmige Antriebe um beide Achsen und ungleichförmige Rotation des Bildfelds. Während also azimutale Montierungen bei Lichtsammlern für Photometrie und Spektrographie anwendbar scheinen, passen sie wenig zu der gegenwärtigen Tendenz, auch mit Großinstrumenten größere Felder zu photographieren.

Solange die Aufgaben von Tubus und Montierung von passiven Systemen ohne gesteuerte oder geregelte Elemente übernommen werden, steht der Montierungsbau für größere Instrumente unter dem Zwang einer Gewichts-Eskalation: beim Lick 120-inch wiegt der Spiegel 3,5 t, die Spiegelzelle 5,5 t, das Rohr 33 t, Gabel und Stundenachse 83 t; beim Palomar 200-inch sind ein Rohr von 110 t und eine Stundenachsen-Konstruktion von 350 t nötig, um einen Spiegel von 13 t zu tragen. Dementsprechend ist die Entwicklung der Montierungen entscheidend vom Fortschritt der Lagerungstechnik beeinflußt worden. Lange Zeit war die erforderliche Führungsgenauigkeit (angestrebt Bruchteile von 1 Bogensekunde) nur mit Gleitlagern zu erreichen, die bei größeren Instrumenten Entlastungseinrichtungen nötig machten. Neben Rollen mit Gewichtshebel- oder Feder-Andruck hat man auch den hydrostatischen Auftrieb eines Schwimmkörpers in Quecksilber benutzt (Mt. Wilson 60-inch, 100-inch). Elegantere Lösungen sind mit *Wälzlagern* möglich (Victoria 72-inch), die für große Instrumente bei der Herstellung „vorgespannt" werden, um schlotterfreien Gang unter Last zu erzielen. Mit Wälzlagern läßt sich das Anlaufmoment um die Polachse auch bei größeren Instrumenten auf ∼10 kpm herabdrücken. Beachtlich sind die Leistungen der einseitigen Deklinationslager bei Englischen Achsen- oder Polachsen-Montierungen, wo sie neben axialer und radialer Last noch ein kippendes Moment (sämtlich variabel) aufnehmen müssen (McDonald 82-inch).

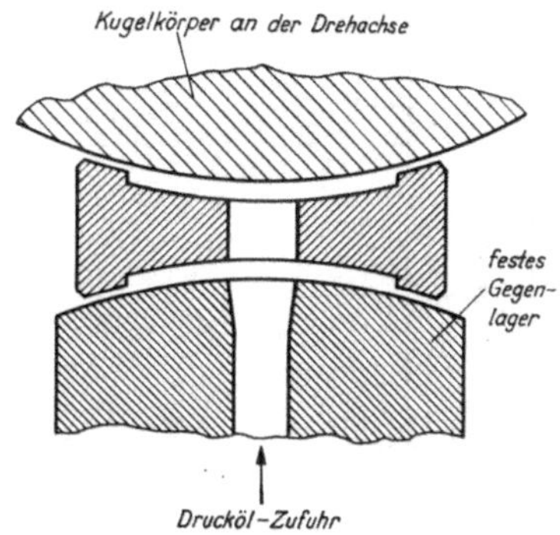

Fig. 62. Schema eines selbstzentrierenden Öldruck-Lagers.

Für sehr große Instrumente, bei denen das Gewicht der bewegten Teile in die Größenordnung von 100 t und mehr kommt, wäre das erforderliche Antriebsmoment um die Stundenachse auch mit Wälzlagern unzulässig hoch. Für den Palomar 200-inch wurde deshalb eine *hydraulische* Lagerung entwickelt, die seitdem bei fast allen größeren Instrumenten in ähnlicher Form verwendet worden ist. Öldrucklager eignen sich für die fest gelagerte Stundenachse; sie sind keine passiven Systeme mehr, das Instrument muß gegen Bewegung bei nicht ausreichendem Öldruck verriegelt werden. Die Wirkungsweise zeigt schematisch Fig. 62: die Lagerflächen der Achse (Zylinder oder Kugeln) werden zu hochpräzisen Rotationsfiguren bearbeitet; sie ruhen auf Tragkörpern (pads) mit genau passenden und zentrierten Gegenflächen. Zwischen die Lagerflächen wird viskoses Öl unter einem Druck von einigen 10 kp/cm² gepreßt, das an den Rändern mit einigen l/min herausquillt. Dabei hebt sich die Last um etwa 0,1 mm. Die Tragkraft des Lagers ist (Innenfläche × voller Öldruck + Fläche der Randzone × effektiver Öldruck dort). Die tragende Fläche ist verhältnismäßig groß, und lokale Deformationen (die bei Wälzlagern mit Punkt- oder Linienberührung unvermeid-

[1] S. VASILEVSKIS in [3].

Tabelle 5.

Montierung	Bewegung beschränkt	Unerreichbare Gebiete	Gegengewicht zum Momentenausgleich um α-Achse	Last zwischen den Lagern für	Feste Achsschnittpunkte	Planspiegel für Coudé-Anordnungen *
1. Deutsche Montierung	+	—	+	—	$\alpha\delta$, δT	$2f$
2. Polachsen-Montierung	—	—	+	—	$\alpha\delta$, δT	$2f$
3. Gabel-Montierung	—	—	—	δ	$\alpha\delta = \delta T$	$\langle 1d\rangle$, $\langle 1d\,2f\rangle$, $3\ldots 4f$
4. Englische Achsen-Montierung . . .	+	ZuK	+	α	$\alpha\delta$, δT	$2f$
5. Englische Rahmen-Montierung . . .	+	P, ZuK	—	α, δ	$\alpha\delta = \delta T$	$\langle 1d\rangle$, $\langle 1d\,2f\rangle$, $4f$
6. Hufeisen-Montierung	+	$\lvert t\rvert \gtrsim 7^h$	—	α, δ	$\alpha\delta = \delta T$	$\langle 1d\rangle$, $\langle 1d\,2f\rangle$
7. Knickrahmen-Montierung	+	ZuK	+	α, δ	δT	$\langle 1d\,2f\rangle$, $4f$

Abkürzungen: P Himmelspol; ZuK Zirkumpolarstern in unterer Kulmination; α Stundenache; δ Deklinationsachse; T Fernrohrachse; f fester Planspiegel; d Planspiegel mit Drehantrieb; $\langle\ \rangle$ nur für beschränkten Deklinationsbereich.

* Für den Fall, daß die Achse des austretenden Bündels mit der Stundenachse zusammenfällt (vgl. S. 336).

lich sind) treten nicht auf. Es ergeben sich so kleine Lagerreibungswerte, daß u. U. besondere Dämpfungsmaßnahmen notwendig sind gegen Schwingungen, die z. B. durch Windstoß angeregt werden (Palomar 200-inch). Das genannte Teleskop braucht für die Stunden-Nachführung ein Drehmoment von 7 kpm.

Hydraulische Systeme gestatten es auch, eine *Kugel* als Lagerungselement zu verwenden. Fällt der Kugelmittelpunkt mit dem Schwerpunkt der bewegten Teile zusammen, so genügt eine Unterstützung symmetrisch zum Schwerpunktslot, und die Aufspaltung in axiale und radiale Kräfte ist unnötig. Körperliche Achsen haben nur noch Führungskräfte aufzunehmen und sind damit entlastet. Für die Stundenachse ist die Kugel-Öldruck-Lagerung beim Bergedorf-Schmidt und seitdem bei den in Jena gebauten Großinstrumenten (2 m-Teleskope in Tautenburg und für Pirkuli/Kaukasus und Ondrejov) sowie dem Caracas 1 m-Reflektor angewendet worden. Eine einzige Kugel als tragendes Element für beide Achsen ist bei der „pol-universalen Montierung“ von Zeiss-Oberkochen vorgesehen.

Für die parallaktischen Montierungen hat sich eine Reihe von Formen herausgebildet; einige ihrer Eigenschaften sind in Tabelle 5 aufgeführt. Die „deutschen“ Montierungstypen 1 bis 3 haben einseitige, die „englischen“ 4 bis 7 beidseitige Lagerung der Stundenachse. Symmetrische Ausführungen (3, 5, 6) brauchen kein Gegengewicht, was sie für Großinstrumente besonders geeignet macht. Beidseitigen Angriff der Deklinationsachse am Rohrkörper mit resultierender günstiger Spannungsverteilung hat außer ihnen auch die Knickrahmenmontierung. Die Realisierungsmöglichkeiten für den Coudé-Strahlengang hängen davon ab, wie sich Fernrohr-, Deklinations- und Stundenachse schneiden. Bei der historischen Entwicklung der Montierungstypen ist angestrebt worden, die Bewegung des Fernrohrs von der Montierung her nicht zu beschränken und möglichst den ganzen Himmel zugänglich zu machen. Nur bei den größten Instrumenten ab etwa 2,5 m Öffnung kann der Beobachter von fest mit dem Rohr verbundenen Plätzen aus arbeiten; bei kleineren müssen die Fokuspositionen einigermaßen bequem von außen erreichbar sein. Zum Cassegrain-Fokus gelangt man vorzugsweise mit einer Hebebühne, der aber unter Umständen Teile der Montierung (Pfleiler, Gabel, Gegengewicht) im Wege sind. Beschreibt man die Lage des Empfängers in rechtwinkligen Koordinaten (vom Achsenschnittpunkt aus X nach Süden, Y nach Westen, Z nach oben positiv), so ist seine Position gegeben durch

$$X = A \sin\varphi \sin t - B \cos h \cos Az$$
$$= A \sin\varphi \sin t + B(\cos\varphi \sin\delta - \sin\varphi \sin\delta \cos t)$$
$$Y = -A \cos t - B \cos h \sin Az$$
$$= -A \cos t - B \cos\delta \sin t$$
$$Z = A \cos\varphi \sin t - B \sin h$$
$$= A \cos\varphi \sin t - B(\sin\varphi \sin\delta + \cos\varphi \cos\delta \cos t).$$

Dabei sind A die Länge der Deklinationsachse zwischen Stundenachse und Rohrachse, B die Länge des Rohrs zwischen Deklinationsachse und Empfänger; t, δ, Az, h, φ bezeichnen Stundenwinkel, Deklination, Azimut, Höhe und (positiv angenommene) geographische Breite.

39. Deutsche Montierungstypen. Bei den deutschen Typen ist die Stundenachse nur auf einer Seite gelagert, Deklinationsachse und Rohr hängen infolgedessen über. Bei der hauptsächlich für Refraktoren verwendeten klassischen deutschen Montierung (Fig. 63) liegt die verhältnismäßig kurze Stundenachse auf einer senkrechten Säule oder einem Pfeiler auf. An ihrem oberen Ende ist die Deklinationsachse gelagert; das Gewicht des Fernrohrs an deren einem Ende wird durch das Gegengewicht am anderen balanciert. Auf ein weiteres Gegengewicht, das den Schwerpunkt der bewegten Teile zwischen die Stundenlager und in die Säulenmitte brächte, verzichtet man oft bei größeren Polhöhen. Das Okularende des Rohrs führt zwar große Bewegungen aus, bleibt aber leicht erreichbar. Für die bei großen Instrumenten zweckmäßige Hebebühne ist ausreichend Platz (Fig. 64).

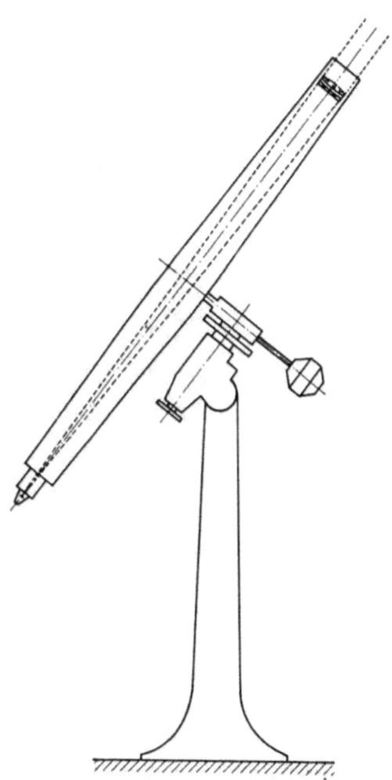

Fig. 63. Deutsche Montierung für einen Refraktor.

Da das Unterende des Rohrs an die Säule anstoßen kann, muß für Objekte über einer Grenz-Deklination ($\delta \gtrsim \varphi$) beim Meridiandurchgang das Fernrohr von der West- auf die Ostseite der Säule gebracht werden. Danach erscheinen das Bildfeld relativ zum Rohr und das Rohrende relativ zum Beobachter um 180° verdreht; photographische Aufnahmen über den Meridian hinweg sind hier nicht möglich. Für kleinere Refraktoren wird eine Coudé-Form mit zwei fixierten Plan-Umlenkspiegeln benutzt, bei der das untere Ende des Fernrohrs in die Deklinations- und Stundenachse verlegt ist (Fig. 65). Neben den Vorteil der feststehenden Bildfläche (in der das Bild mit der Stundenachse rotiert) tritt hier der weitere, daß das Fernrohr beim Meridiandurchgang nicht mehr durchgeschlagen werden muß.

Der Zwang zum Durchschlagen entfällt auch, wenn man den Achsenschnittpunkt über die Säule so weit überhängen läßt, daß das Rohr auch noch bei

$\delta = 90°$ eine volle Umdrehung um die Polachse ausführen kann. Dazu wird entweder die Säule abgewinkelt und bekommt für eine ausreichende Länge die Richtung der Polachse (Kniesäulenmontierung), oder die Achse wird über das obere Lager hinaus verlängert. In beiden Fällen hat ein Bauelement in Polachsenrichtung dabei Freiträgercharakter (Fig. 66) und muß ein konstantes Biegemoment aufnehmen. Die Sammelbezeichnung *Polachsenmontierung* erscheint für diese Typen angemessen.

Fig. 64. Refraktor ($D = 65$ cm) der Sternwarte Babelsberg (1915); Hebebühne in der untersten Position.

Ein frühes Beispiel ist die Repsoldsche Astrographen-Montierung. Andere Formen, bei denen der untere Teil der Säule weggefallen ist, sind in neuerer Zeit für Spiegelteleskope bis $D = 2$ m mehrfach gebaut worden (Kitt Peak 36-inch, Victoria 48-inch, Siding Spring 40-inch, Baku/Pirkuli und Ondrejov 2 m, Fig. 67). Das Rohr ist frei beweglich, und im Cassegrain-Fokus ist genügend Platz für Zusatzapparate. Das Gegengewicht kann man von der Deklinationsachse wegverlegen (Fig. 68) und, von einem die schiefe Säule außen umschließenden Rohr getragen, in der Nähe des Fernrohrfußes anbringen. Ein Coudé-Fokus ist mit zwei im Rohr bzw. in der Stundenachse fixierten 45°-Planspiegeln leicht zu realisieren; die körperliche Deklinationsachse schrumpft dann auf ein einseitiges Ring- oder Plattenlager zusammen.

Auch die *Zeiss-Montierung* (Entlastungsmontierung nach MEYER)[1] ist in ihrer verbreitetsten Form der Polachsen-Montierung verwandt. Bei ihr ist die Ableitung der Kräfte von den führenden Lagern getrennt. Das Rohr sitzt auf einem Bock außerhalb der Deklinationsachse und ist um sie durch zwei Gegengewichte balanciert. Ein weiteres großes Gewicht am unteren Ende der Stundenachse entlastet die Deklinationsachse und beseitigt zugleich das den Pfeiler kippende Moment. Seit der Entwicklung hochpräziser Wälz- und hydraulischer Lager ist diese Konstruktion überholt, die mit hohem Aufwand zu leicht und genau laufenden Instrumenten geführt hatte.

Als symmetrisches Gegenstück zur Polachsenmontierung kann man die *Gabelmontierung* ansehen (Fig. 69). Die überhängende Stundenachse ist in zwei Arme aufgeteilt, zwischen denen das Rohr an den Deklinations-Achszapfen hängt. Die Gabelmontierung wird gern für Spiegelteleskope verwendet, bei denen der Rohrschwerpunkt ohnehin nahe dem Hinterende liegt. Für das Arbeiten im Cassegrain-Fokus kann man so genügend Raum vorsehen, ohne daß die Gabelarme ungebührlich lang werden (Fig. 70). Die Armlänge wird auch davon bestimmt, ob in niedrigen geographischen Breiten der meridianseitige Horizont erreichbar und in hohen der Spiegel bei senkrechter Stellung des Rohres herausnehmbar sein soll. Für Objekte in der Nähe des Pols ist bei kleineren Instrumenten das Beobachten besonders in großen Stundenwinkeln unbequem; hier ist die Polachsen-Montierung eindeutig überlegen. Abhilfe bringt der Nasmyth-Strahlengang mit Lichtaustritt durch die Deklinationsachse, allerdings um den Preis einer

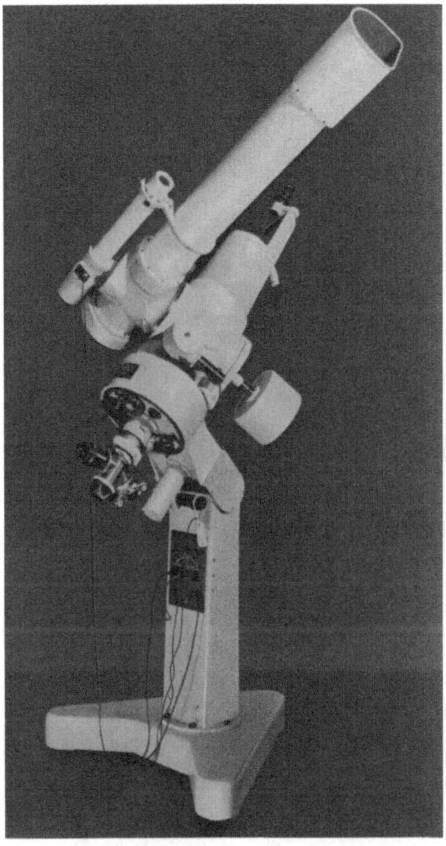

Fig. 65.

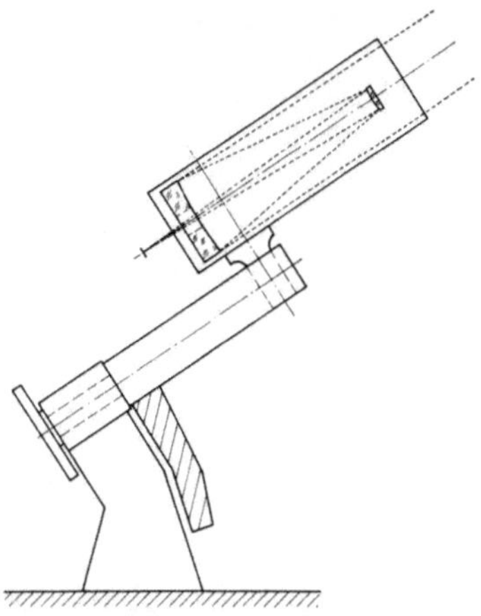

Fig. 66.

Fig. 65. Moderner Coudé-Refraktor. Mit dem Planspiegel in der Stundenachse kann der Strahlengang wahlweise nach der Nord- oder Südseite gelenkt werden. Werkfoto Carl Zeiss Oberkochen.

Fig. 66. Schema einer Polachsen-Montierung (schraffiert: Gegengewicht).

[1] F. MEYER: Z. Instrumentenkde. **50**, 58 (1930).

zusätzlichen Spiegelfläche. Die Empfängerwege sind bei der Gabelmontierung klein; bei der Nasmyth-Anordnung beschreibt der Fokus den gleichen Kreis um die Stundenachse für alle Deklinationen.

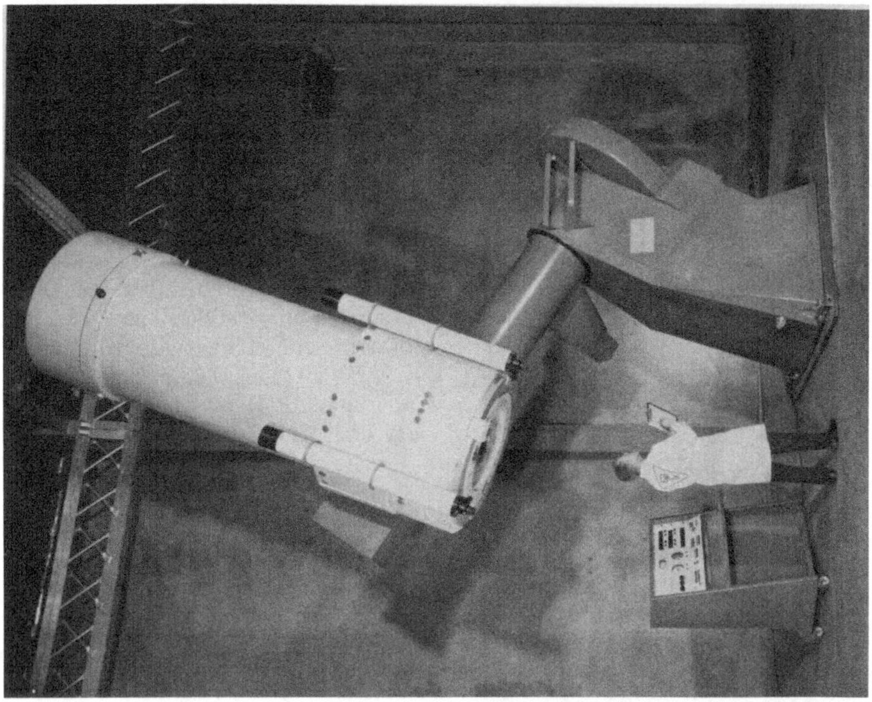

Fig. 68. 40-inch-Reflektor (Siding Spring Station des Mount Stromlo Observatory, 1964) mit Ritchey-Chrétien- und Coudé-Strahlengang. Die tragende Säule ist hohl zur Aufnahme der Stundenachse; das Rohr zum Gegengewicht umschließt sie. Werkfoto: Boller & Chivens, a Division of the Perkin Elmer Corporation.

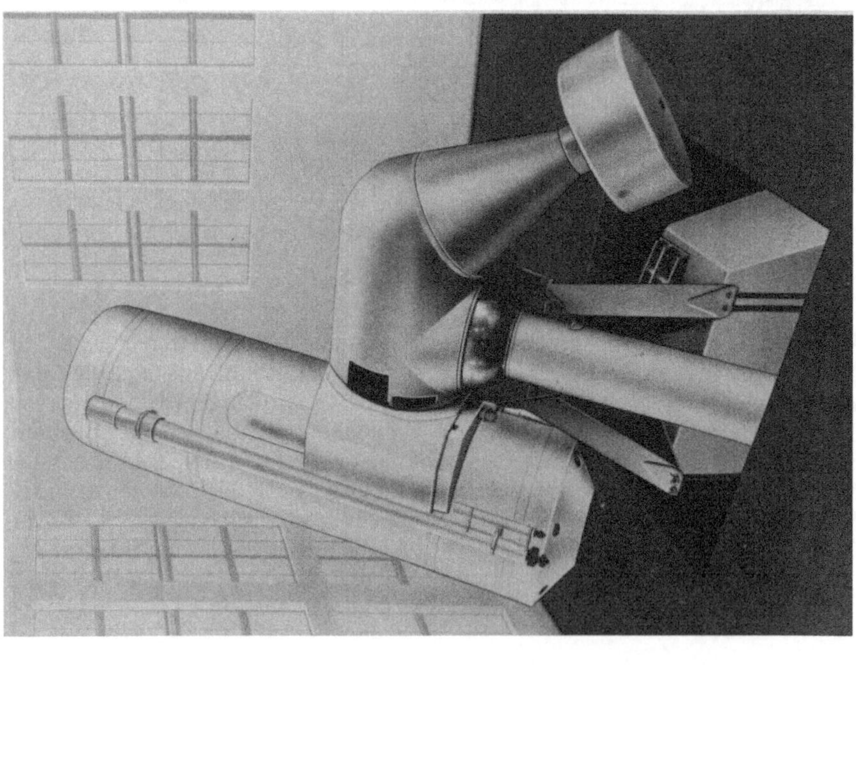

Fig. 67. Reflektor für die Sternwarte Pirkuli/Kaukasus (1966) mit $D=200$ cm, $N_{\text{prim}}=4{,}5$. Polachsen-Montierung mit Kugel-Öldruck-Lagerung der Stundenachse. Das von der Deklinationsachse wegverlegte Gegengewicht verschiebt den Schwerpunkt der bewegten Teile zum Kugelmittelpunkt. Werkfoto Jenoptik Jena GmbH.

Die Gabelmontierung wird auch für große Instrumente (Fig. 71, 72) benutzt (Palomar Schmidt, Tautenburg 2 m, Kitt Peak 84-inch, Krim 2,6 m, Lick 120-inch). Dabei wird die Durchbiegung der Gabelarme[1] merklich. In der Gabelebene bleibt die Wirkung gering; senkrecht dazu verschiebt die Änderung der Tangentenrichtung an der Spitze des Freiträgers den momentanen instrumentellen Pol, der damit stundenwinkelabhängig wird. Die unangenehmste Auswirkung ist Rotation des Bildfeldes bei Aufnahmen in hohen Deklinationen. Aktive Kompensation dieser Biegung durch eine zusätzliche Steuerung in Deklination ist bei geeigneter Justierung der Polachse möglich.

Die Gabelbiegung kann stark reduziert werden durch den Aufbau auf einer großen äquatorparallelen Scheibe (polar disk), mit großer Armdicke senkrecht zur Gabelebene (Herstmonceux 98-inch, Fig. 73). Die Scheibe — mit 1 m Dicke und 6,7 m ∅ bei dem erwähnten Teleskop — dient gleichzeitig als Stundenlagerung; Umfang und Unterseite sind entsprechend bearbeitet und werden mit definierenden und entlastenden Öldruck-Kissen unterstützt. Bei dieser Scheiben-Montierung gibt es keine körperliche Polachse mehr, die ein kippendes Moment aufnehmen könnte. Der Schwerpunkt aller bewegten Teile muß also über der Unterstützungsfläche liegen, was mit abnehmender geographischer Breite ($|\varphi| \lesssim 30°$) Schwierigkeiten macht.

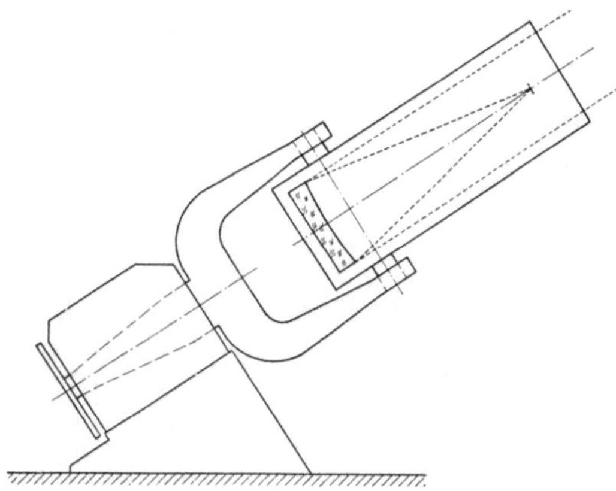

Fig. 69. Gabel-Montierung für einen Reflektor; Primärfokus-Strahlengang.

In gewissem Umfang gilt das auch von einer Ausführung der Gabelmontierung, bei der von einer Kugel um den Schwerpunkt der bewegten Teile diejenige Zone realisiert wird, die das Schwerpunktslot enthält. An der genau gearbeiteten Kugelzone greifen die hydraulischen Tragkörper an (Bergedorf Schmidt, Tautenburg). Azimut und Höhe der Polachse sind durch Verschieben des nur führenden Lagers an ihrem unteren Ende kräftefrei einstellbar, womit die betriebsmäßige Justierung der Instrumenten-Polhöhe zur optimalen Anpassung[2] für photographische Aufnahmen möglich wird.

Da sich bei der Gabelmontierung Fernrohr- und Stundenachse in einem festen Punkt schneiden, ist ein Coudé-Strahlengang mit nur einem zusätzlichen Planspiegel möglich (Drei-Spiegel-System, s. Fig. 77). Der Spiegel muß um $45° - \frac{\delta}{2}$ gegen die Rohrachse geneigt sein und dementsprechend jeweils um die Hälfte einer Deklinationsbewegung verstellt werden. Im Rohr ist an der Äquatorseite ein Schlitz für den Lichtdurchtritt nötig, der den Mittelkörper von Großinstrumenten mechanisch schwächt. Für $|\delta| \gtrsim 50°$ wird die Reflexion unangenehm flach. Die endgültige Beschränkung für δ ergibt sich durch das Eintreten des Hauptspiegels selbst in den Strahlengang bei $|\delta| = 90° - \beta$; β ist der Winkel, unter dem der Spiegel- oder Zellenradius von der Deklinationsachse aus erscheint, mit Zuschlag für den Bündeldurchmesser. Großes Öffnungsverhältnis des Haupt-

[1] S. Vasilevskis: Astron. J. **67**, 464 (1962).
[2] Siehe Fußnote 3 von S. 337.

spiegels und Deklinationslagerung nahe dem Hinterende des Rohrs wirken hier ungünstig.

Fig. 70. 84-inch-Teleskop ($N_{\text{prim}}=2{,}6$; Ritchey-Chrétien- oder Coudé-System) des Kitt Peak National Observatory (1963). Schräg durch die Gabel führt ein Rohr als Streulichtschutz für den Coudé-Strahlengang mit drei festen Planspiegeln. Die Sekundärspiegel sind mittels Wendering leicht auswechselbar. Foto KPNO.

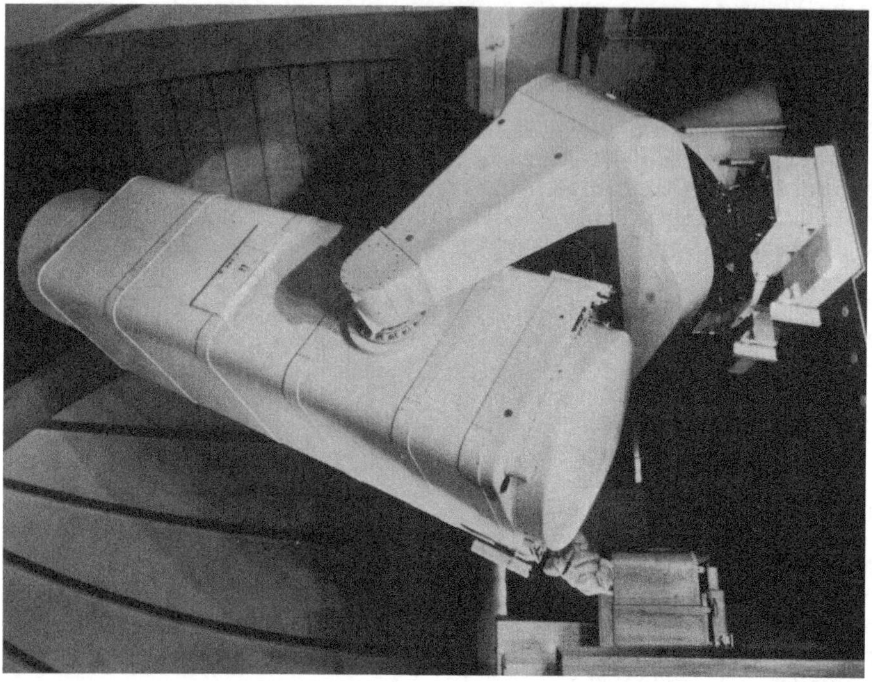

Fig. 71. 2-m-Universalteleskop (1960) der Sternwarte Tautenburg (Thüringen). Der sphärische Hauptspiegel ($D=200$ cm, $f=400$ cm) kann zusammen mit einer Korrektionsplatte ($D_0=134$ cm) eine Schmidt-Kamera bilden, oder mit Sekundärspiegeln Quasi-Cassegrain- bzw. Coudé-Systeme mit 21 m bzw. 92 m Brennweite. Werkfoto Jenoptik Jena GmbH.

In dem mit *einem* Hilfsspiegel nicht zugänglichen Gebiet kann man den Strahlengang mit drei Planspiegeln um den Hauptspiegel herumlenken (Fünf-

Fig. 72. 120-inch-Reflektor des Lick Observatory (1959). Um den Rohrschwerpunkt und damit die Deklinationsachse näher an den Hauptspiegel zu bringen, sind an der Spiegelzelle zusätzliche Gewichte angebracht. Lick Observatory Photograph.

Fig. 73. 98-inch-Reflektor für das Royal Greenwich Observatory Herstmonceux (1967). Werkfoto Sir Howard Grubb Parsons & Co. Ltd.

Spiegel-Coudé, s. Fig. 77), die in der Ebene von Rohr und Stundenachse liegen. Der erste Planspiegel muß wieder einen Antrieb auf const $-\frac{\delta}{2}$ bekommen, der zweite sitzt auf einem Ausleger auf der Gabel und der dritte in der Stundenachse. Statt dessen bevorzugt man oft eine Anordnung mit drei oder vier *festen* Planspiegeln, die das Lichtbündel in der Gabelebene um den Hauptspiegel führen: der erste lenkt das Bündel in die δ-Achse, der zweite wird von einem nicht mitrotierenden Ausleger auf dieser getragen, der dritte liegt in der Stundenachse (Kitt Peak 84-inch, Fig. 70). Will man den ziemlich flachen Einfall auf dem dritten Planspiegel vermeiden, kann man einen vierten Spiegel einführen und so alle unter 45° benutzen (Tautenburg).

Das Drei-Spiegel-System bekommt einen größeren Arbeitsbereich in Deklination, wenn das Strahlenbündel auf der *Pol*seite aus dem Rohr geführt wird. Der Fokus liegt dabei entweder über dem Instrument oder muß durch einen festen Planspiegel verlegt werden (Cambridge 36-inch). Für den vorwiegend untersuchten Teil des Himmels ist der Einfallswinkel auf dem Planspiegel günstiger, und der Hauptspiegel kann nicht mehr in den Strahlengang geraten.

40. Englische Montierungstypen. Die englischen Montierungen sind durch zweifache Lagerung der Polachse beiderseits der Last charakterisiert. Dadurch bleiben ihre Biegungen viel kleiner als bei den überhängenden Polachsen der deutschen Typen. Bringt man die Deklinationsachse an der Stelle der Stundenachse an, wo deren elastische Linie parallel zur Lagerverbindung ist, so wirkt sich die Biegung auf Einstellung und Nachführung nicht aus. Die unsymmetrische Form ist die *Englische Achsenmontierung*, bei der das Rohr einseitig am Ende der Deklinationsachse sitzt (Fig. 74). Das Teleskop muß beim Meridiandurchgang nicht durchgeschlagen werden; demgegenüber fällt der Nachteil kaum ins Gewicht, daß je nach Länge der Deklinationsachse und Breite des Polpfeilers Zirkumpolarsterne nahe der unteren Kulmination nicht beobachtbar sind. Viele Astrographen und größere Schmidt-Teleskope (Cleveland, Ann Arbor, Bloemfontein, Tonantzintla) haben eine unsymmetrische englische Montierung, aber auch Reflektoren mit $D \approx 2$ m (Victoria, Perkins, McDonald, die Grubb-Parsons 75-inch-Serie von Richmond Hill bis Kottamia).

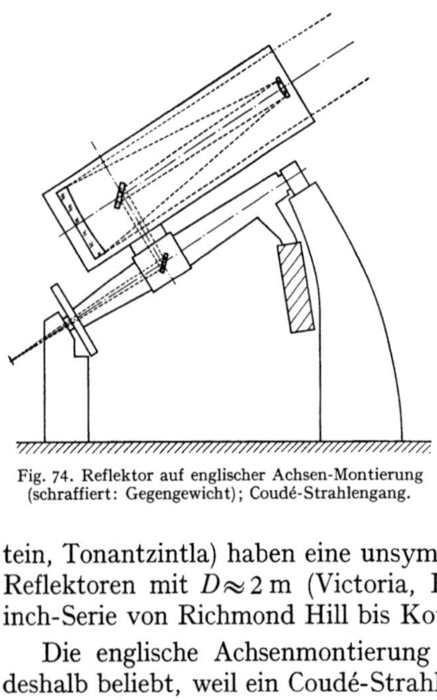
Fig. 74. Reflektor auf englischer Achsen-Montierung (schraffiert: Gegengewicht); Coudé-Strahlengang.

Die englische Achsenmontierung ist für solche Spiegelteleskope vor allem deshalb beliebt, weil ein Coudé-Strahlengang mit zwei festen 45°-Planspiegeln zu verwirklichen ist (McDonald 82-inch): der erste lenkt das Bündel in die Deklinationsachse, der zweite ist in der Stundenachse angebracht. Die körperliche Deklinationsachse muß dazu bereits vor der Mitte der Stundenachse enden, doch sind die Probleme der kurzen Lagerung mit überhängender Last lösbar. Das Gegengewicht gleicht nur noch die Drehmomente um die Stundenachse aus, beseitigt aber nicht mehr das Kippmoment in der Deklinationslagerung. Man kann es deshalb an den polseitigen Pfeiler verlegen, wo es die Bewegung einer Hebebühne weniger beeinträchtigt (McDonald 82-inch, Fig. 75).

Fig. 75. 82-inch-Reflektor (1939) des McDonald Observatory. Foto: McDonald Obs.

Fig. 76. 100-inch-Reflektor des Mt. Wilson Observatory, 1917.

Die symmetrische Form des englischen Typs ist die *Englische Rahmen-Montierung*. Bei ihr ist die Polachse zwischen den Lagern als Rahmen ausgebildet, in dem das Rohr an Deklinations-Achszapfen hängt. Vom mechanischen Standpunkt aus ist sie ideal, da hinsichtlich beider Achsen symmetrisch und ohne freitragende Elemente. Ihr entscheidender Nachteil ist, daß ein mehr oder weniger großes Gebiet um den Himmelspol unzugänglich bleibt. Das spielt für äquator-

Fig. 77. 200-inch-Teleskop (1948) des Palomar Observatory. Außer dem Primär- und dem Cassegrain-Strahlengang sind die Lichtwege für den Drei- und den Fünf-Spiegel-Coudé markiert. Zeichnung R. W. Porter.

nahe Standorte keine Rolle, hat aber sonst die Anwendung verhindert. Die bemerkenswerte Ausnahme ist der Mt. Wilson 100-inch (Fig. 76); für dieses damals größte Instrument waren die prinzipiellen Vorteile der Montierung ausschlaggebend. Für den Coudé-Strahlengang bieten sich bei der Rahmenmontierung die gleichen Möglichkeiten wie bei der Gabelmontierung.

Eine Modifikation der englischen Rahmenmontierung ist beim Palomar 200-inch verwendet worden, die *Hufeisen-Montierung* (Fig. 77, 78). Sie geht aus der Rahmenmontierung hervor, wenn man das polseitige Lager stark vergrößert und die so entstehende Scheibe senkrecht zur Deklinationsachse bis über die Mitte ausschneidet. Das Rohr kann bis $\delta = 90°$ in die Scheibe hineinschwenken; ein größerer Stundenwinkelbereich beiderseits $t=12^h$ bleibt unzugänglich, weil hier das Lager nicht mehr trägt. Der Durchmesser des „Hufeisens" beträgt 14 m. Für den geplanten Kitt Peak 150-inch ist eine ähnliche Montierung vorgesehen (Fig. 79),

Fig. 78. Primärfokus-Kabine des Palomar 200-inch.

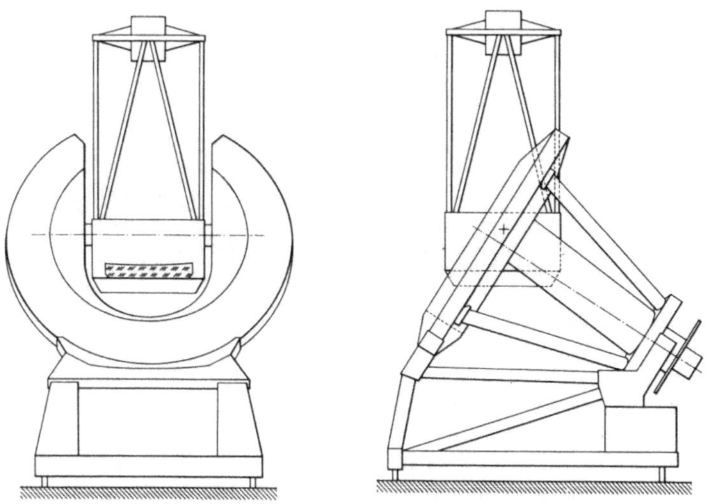

Fig. 79. „Hufeisen"-Montierung für den 150-inch-Reflektor (im Bau) des Kitt Peak National Observatory; Ansicht von Norden und von Westen. Stunden- und Deklinationsachse schneiden sich nicht, das Coudé-Schutzrohr muß diese Versetzung ausgleichen.

bei der die Deklinationsachse in der Ebene des Hufeisens selbst liegt. Die Scheibe muß so tief ausgeschnitten sein, daß das Hinterende des Rohres bei $\delta \approx 0°$ durch ihre Ebene schwenken kann. In Deklination ist die Bewegung über den Pol hinaus nicht behindert und damit auch das Gebiet der unteren Kulmination mit der Einstellung $t'=t\pm 12^h$, $\delta'=180°-\delta$ zugänglich.

Bei dem zuletzt genannten Instrument wird auch eine neuartige, für symmetrische Montierungen geeignete Coudé-Anordnung realisiert. Der erste Planspiegel wirft das Bündel in die δ-Achse, der zweite zur Stundenachse, wie beim Kitt Peak 84-inch. Der dritte, in der Verlängerung der Stundenachse liegende Planspiegel lenkt das Bündel in eine neue Ausgangsrichtung, die horizontal und im spitzen Winkel zum Meridian verläuft (etwa nach NO oder NW für einen Standort auf der Nordhalbkugel). Bei Rotation des Teleskops um die Stundenachse muß

Fig. 80. Doppel-Astrograph ($D=50$ cm, $f=375$ cm), Bauart Askania (1963). Werkfoto Carl Zeiss Oberkochen.

die Normale des dritten Planspiegels auf dem Mantel eines schiefen Kreiskegels laufen, dessen Achse die Winkelhalbierende zwischen Polachse und Ausgangsrichtung ist und dessen Basiskreis in einer Ebene senkrecht zur Polachse liegt. Mit diesem Antrieb (Rotation plus Kippung) eines verhältnismäßig kleinen Spiegels gewinnt man günstige Einfallswinkel auf allen Flächen und eine vorteilhafte Achslage des Coudé-Spektrographen. Das System kann ohne Umbau für alle Deklinationen benutzt werden; der Fernrohrtubus ist nicht durch einen Schlitz geschwächt.

Eine weitere Nebenform ist die *Knickrahmen-Montierung*[1] (Caracas Schmidt und Astrograph). Hier macht man den Pol dadurch beobachtbar, daß man den Rahmen aus seiner Ebene herauswölbt; die Deklinationsachse schneidet die Stundenachse nicht, sondern steht so weit von ihr ab, daß das Rohr bei $\delta=90°$

[1] C. KÜHNE: Astron. J. **62**, 267 (1957).

über das polseitige Lager hinwegsehen kann (Fig. 80). Damit wird Momentenausgleich um die Stundenachse notwendig, der bei den Caracas-Instrumenten durch Gegengewichte nahe den Pfeilern erzielt ist. Das Hinterende des Rohrs führt nur kleine Bewegungen aus. Coudé-Systeme sind mit Knickrahmen-Montierungen noch nicht ausgeführt worden, erscheinen aber ohne weiteres möglich.

III. Bewegungseinrichtungen.

41. Nachführung. Der Stundenantrieb[1] soll mit einer Umdrehung pro Sterntag die Wirkung der Erdrotation aufheben. Das geschieht heute fast ausschließlich mit hochkonstantem Frequenzgenerator, Leistungsverstärker, Synchronmotor und Schneckenradgetriebe. Mit dieser Anordnung ist eine konstante, lastunabhängige Drehzahl ohne Regulatoren auf der Leistungsseite gegeben. Für Beobachtungsprogramme, die das längere Verfolgen eines Objekts erforderlich machen, möchte man dem Antrieb abweichende Geschwindigkeiten[2] geben (ohne Abgehen von der Konstanzforderung), die vorwiegend langsamer sind als die Sternzeitgeschwindigkeit. Mit einer solchen — empirisch oder durch Vorausberechnung[3] bestimmten — Anpassung der Antriebsgeschwindigkeit kann man die $d\alpha/dt$-Komponente infolge von Bewegung des Objekts (Mond, Planeten), Refraktion, Biegung des Rohrs und der Achsen und Aufstellungsfehlern kompensieren, aber auch Spektren verbreitern und ausgedehnte Objekte abtasten. Die Kompensation in *zwei* Koordinaten erfordert einen Antrieb mit einstellbarer Geschwindigkeit auch um die Deklinations-Achse. Bei Instrumenten mit großem Abbildungsmaßstab und großem Feld können die genannten Störungsursachen noch zu einer merklichen Rotation des Bildfeldes führen.

Als Quelle der Steuerfrequenz wurden mechanische Oszillatoren (Stimmgabel, Saite) mit elektrischer Erregung verwendet; heute bevorzugt man rein elektrische, u.U. quarzstabilisierte Generatoren. Die Genauigkeitsforderung von 10^{-4} bis $10^{-5} \approx 1$ sec/Tag ist verhältnismäßig leicht zu erfüllen. Die Oszillatorfrequenz wird gewöhnlich bis zur lokalen Netzfrequenz heruntergeteilt, die übrigens in vielen Fällen für kleinere Instrumente auch direkt zum hinreichend frequenzkonstanten Antrieb dienen kann. Eine Variation der Antriebsgeschwindigkeit erzielt man meist durch Änderung der Steuerfrequenz. Es wird entweder der Oszillator verstimmt (z.B. elektromagnetisch gesteuerte Änderung der Saitenspannung, Palomar 200-inch) oder die konstantgehaltene Grundfrequenz durch *kontinuierliche* Phasenänderung variiert[4]. Die mechanischen Lösungsmöglichkeiten mit stufenlos veränderlichem Getriebe oder Differential mit einstellbarer Zusatzdrehzahl zwischen Antriebsmotor und Schnecke sind demgegenüber zurückgetreten. Der Fortschritt der elektronischen Technik gestattet auch digitalisierte Lösungen[5].

Die Oszillatorfrequenz steuert einen Leistungsverstärker, dessen Ausgangsspannung dem Antriebsmotor zugeführt wird. Die für den Stundenantrieb benötigte mechanische Leistung ist bei modernen Instrumenten klein (der Palomar 200-inch z.B. braucht nur $^1/_{12}$ PS, bei 500 t bewegter Last); die Anforderung an elektrische Leistung geht also kaum über 100 W hinaus. Die Drehzahl des Motors

---

[1] Dazu s. R. R. McMath and O. C. Mohler: Telescope Driving Mechanisms. In Telescopes [9].
[2] R. R. McMath: Publ. Michigan **7**, No. 1 (1937), besonders S. 42ff.
[3] E. S. King: Ann. Harvard **41**, No. 6 (1902). — S. Arend: Obs. Roy. Belg. Monographies No. 2 (1951).
[4] J. H. DeWitt Jr., C. K. Seyfert and R. H. Hardie: Publ. Astron. Soc. Pacific **70**, 293 (1958). — V. M. Konshin u. N. N. Mikhelson: Mitt. Pulkovo **22**, H. 4 (No. 169), 158 (1961).
[5] D. E. Trumbo: Astron. J. **69**, 560 (1964) (Abstr.).

wird durch Getriebe bis in die Größenordnung 1 U/min untersetzt. Das Getriebe schließt meist auch die Umwandlung von mittlerer Zeit in Sternzeit ein, wozu eine Übersetzung von $1:\frac{366{,}2422}{365{,}2422}=1:1{,}0027379$ notwendig ist. Die Approximation $\frac{366}{365}$, mit $\frac{6\times 61}{5\times 73}$ zu realisieren, benutzt nur ein Zahnradpaar mit ungewöhnlichen Zähnezahlen; ihr Verhältnis 1:1,0027397 verursacht einen Fehler von $2\cdot 10^{-6}$ ($\approx$0,2 sec/Tag). Die noch bessere Annäherung $\frac{4029}{4018}=\frac{51\times 79}{49\times 82}=$ 1,0027377 mit $2\cdot 10^{-7}$ Abweichung ist für Fernrohrantriebe kaum erforderlich[1]. Mit dieser Anordnung kann man für den Sternzeitantrieb des Instruments Normalzeitoszillatoren verwenden, was wohl den Vorzug gegenüber nach Sternzeit schwingenden Oszillatoren verdient.

Während die Genauigkeitsforderungen für das Getriebe bis zum letzten Zahnradpaar verhältnismäßig leicht zu erfüllen sind, lassen sie sich mit dem Schneckenradantrieb für die Stundenachse selbst gerade noch einhalten. Für hohe Ansprüche wird man verlangen, daß an jeder Stelle die Summe aller Fehler relativ zu einer größeren Umgebung unter 1″ bleibt und ihre maximale Änderungsgeschwindigkeit merklich unter der Refraktionsänderung im Zenit von 1 sec/Std; die periodischen Fehler während einer Schneckenumdrehung sollen wenige 0″,1 nicht übersteigen und die lokalen Abweichungen weniger Auslenkung als die langsame Komponente der Richtungsszintillation verursachen. Wenn man die Möglichkeiten einer ständigen Regelung der Antriebsgeschwindigkeit ausnutzen kann, bleibt als Hauptforderung die Freiheit von allen unregelmäßigen oder kurzperiodischen Fehlern mit einer Zeitskala von der Größenordnung Minuten und weniger. Da einem Winkel von 0″,5 am Rand eines Schneckenrades von 40 bis 200 cm Radius nur eine Bogenlänge von 1 bis 5 μm entspricht, ist die angestrebte Genauigkeit gegen ein hohes oder variables Reibungsmoment um die Stundenachse nicht zu erreichen. Die Zähnezahl des Schneckenrades liegt meist zwischen 288 und 1440 (Vorzugswert 720), so daß die Schnecke 5 bis 1 min für eine Umdrehung braucht. Um die engen Toleranzen einzuhalten, muß auch der *Eingriff* der Schnecke wohldefiniert sein. Toter Gang zwischen Schnecke und Schneckenrad kann durch ein konstantes Drehmoment um die Stundenachse (am einfachsten mit Gewichtszug) beseitigt werden, das zugleich ein axiales Spiel der Schnecke verhindert. Andere Möglichkeiten sind Teilen und Auseinanderfedern der Schnecke oder Simultaneingriff von zwei Schnecken. Die Eingriffsgeometrie bleibt auch bei Plan- oder Radialschlag des Schneckenrades erhalten, wenn man die Schneckenlagerung durch das Schneckenrad selbst führt und die Schnecke nur in ihrer Achsrichtung raumfest definiert. Zum Schutz vor Beschädigungen des Antriebs wird bei großen Instrumenten das axiale Drucklager der Schnecke oder ihr Lagerrahmen durch Reibung gehalten; bei Überschreiten einer bestimmten axialen Kraft kann die Schnecke ausweichen (Palomar 200-inch, Lick 120-inch).

Der häufigste *Fehler* eines sonst richtig konstruierten Stundenantriebs ist eine periodische Abweichung mit der Schneckendrehzahl. Die Prüfung geschieht am einfachsten durch Photographie der Spur eines Sternes mit nicht zu großer Deklination nahe dem Meridian bei langsamem Antrieb des Instrumentes um die δ-Achse. Hat der periodische Fehler Sinusverlauf, so kann er durch eine periodische axiale Bewegung der Schnecke (Führung eines exzentrischen Punktes auf einer schräg angestellten Ebene)[2] oder durch eine periodisch ungleichförmige Drehgeschwindigkeit (exzentrische Anordnung eines mit Schneckendrehzahl um-

[1] Weitere Annäherungen gibt E. WIEDEMANN, Z. Instrumentenkde. **74**, 322 (1966).
[2] J. HARTMANN: Astron. Nachr. **158**, 1 (1902).

laufenden Zahnrads)[1] kompensiert werden. Stimmen die Steigung der Schnecke und die Teilung des Schneckenrades nicht überein, so ergibt sich ein sägezahnförmiger periodischer Fehler[2]. Eine Prüfanordnung für den fortschreitenden Fehler eines Schneckenantriebs hat REICHE[3] beschrieben.

Während der periodische Fehler des Antriebs stets klein gehalten werden sollte, können glatt verlaufende Abweichungen über größere Teile des Schneckenradumfangs eher toleriert werden, besonders wenn eine *automatische Nachführung* durch lichtelektrische Abtastung der Objekt-Position benutzt wird. Seit Einführung der Photomultiplier macht die Konstruktion solcher Nachführsysteme keine Schwierigkeiten mehr. Sie können vier Empfänger verwenden, denen Sternlicht von einer vierseitigen reflektierenden Pyramide zugeführt wird[4]. In der Soll-Position erhält jeder Empfänger den gleichen Anteil des Strahlungsflusses; fällt das Sternbild nicht auf die Pyramidenspitze, so ergeben sich Signaldifferenzen für gegenüberliegende Empfänger, die die Feinbewegung des Fernrohrs im richtigen Sinne aktivieren. Nur einen Empfänger braucht die Anordnung mit einer in der Bildebene rotierenden Halbblende[5,6]. Solange das Sternbild auf dem Rotationszentrum liegt, ist das Signal nicht mit der Rotationsfrequenz moduliert; bei Abweichung davon geben Phase und Amplitude der Modulation Regelgrößen für Richtung und Betrag der Positionskorrektur. Statt des ganzen Fernrohrs kann man auch — wie beim visuellen Führen — die Kassette auf einer Schlittenführung steuern[7] oder ziemlich schnelle Komponenten durch Neigungsänderung einer planparallelen Platte im Strahlengang kompensieren[6]. Eine Abschätzung der Photonenstatistik ergibt, daß lichtelektrische Nachführungen etwa bis zur gleichen Grenzhelligkeit arbeiten sollten, bei der man mit dem Auge noch sicher pointieren kann[8]. Naheliegend ist der Gedanke, die über eine hinreichende Zeit integrierten Korrekturen als Steuergröße für eine inkrementelle Änderung der Antriebs*frequenz* zu benutzen.

42. Koordinaten-Einstellung und Anzeige. Zum *Einstellen* eines Objekts benötigt man schnelle Bewegungen um beide Achsen. Bei kleineren Instrumenten kann das Richten noch mit der Hand geschehen, bei größeren benutzt man Motorantriebe. Üblich sind drei diskrete Geschwindigkeitsstufen für Grob- bzw. Feineinstellung und Pointieren (slew, set, guide). Bei modernen Instrumenten strebt man eine möglichst hohe Grobgeschwindigkeit an, die aber dem Trägheitsmoment der bewegten Teile angepaßt sein muß. So erlaubt das gedrungene Rohr des Kitt Peak 84-inch eine Grobgeschwindigkeit von 120°/min; bei sehr großen Instrumenten ist man nicht über 45°/min hinausgegangen (Palomar 200-inch, Lick 120-inch). Die Pointierbewegung liegt meist bei einigen Bogensekunden pro Sekunde für kleinere bis zu unter 1″/sec für große Instrumente. Damit ist — für drei Geschwindigkeitsstufen — ein Stufenverhältnis von etwa 60 gegeben. Zur besseren Anpassung an Abbildungsmaßstab und Feldgröße in den verschiedenen Foci kann man die Geschwindigkeiten in einem gewissen Bereich wählbar machen (z.B. Lick 120-inch). Auch Bewegungen mit variabler Geschwindigkeit — etwa

[1] R. H. HARDIE and C. M. BALLARD: Publ. Astron. Soc. Pacific **74**, 242 (1962).
[2] G. E. KRON: Publ. Astron. Soc. Pacific **72**, 505 (1960).
[3] S. REICHE: Z. Instrumentenkde **72**, 237 (1964).
[4] G. ALTER: Z. Instrumentenkde. **49**, 553 (1929).
[5] H. W. BABCOCK: Astrophys. J. **107**, 73 (1948). — P. E. ARGYLE: Publ. Victoria **11**, 201 (1959). — Y. A. SABININ u. P. V. NIKOLAYEV: Mitt. Krim **24**, 203 (1960).
[6] H. W. BABCOCK, B. H. RULE and J. S. FASSERO: Publ. Astron. Soc. Pacific **68**, 256 (1956).
[7] A. E. WHITFORD and G. E. KRON: Rev. Sci. Instr. **8**, 78 (1937). — R. H. WEITBRECHT: Rev. Sci. Instr. **28**, 122 (1957).
[8] P. B. FELLGETT: Occasional Notes Roy. Astron Soc. **3**, 143 (1956).

analoggesteuert durch die Auslenkung eines Steuerknüppels — sind gebaut worden. Für die langsamen Zusatzbewegungen (Refraktionsausgleich, bewegte Objekte, Spektrenverbreiterung) wählt man in Rektaszension eine Variation der Stundenantriebs-Geschwindigkeit bis zu einigen Prozent, in Deklination einen Antrieb mit bis zu 1″/sec.

Die Motoren können meist von einem festen Schaltpult und für die niederen Geschwindigkeitsstufen zusätzlich von den Beobachtungsstationen aus geschaltet werden. Die Betätigungsorgane (Drucktasten, „Paddel", Steuerknüppel) müssen selbstausschaltend sein. Für Primärfokus-Kabinen sind zusätzliche „Totmann-Sicherungen" und Notstop-Ausschalter sinnvoll. Gegenseitige Verriegelung widersprechender Bewegungen und Endlagen-Ausschalter (Horizont) sind selbstverständlich. Eine Beschleunigungsbegrenzung läßt sich durch Schwungmassen auf der hochtourigen Seite der Motoren verhältnismäßig leicht erreichen. Die Grob-Bewegung in Stundenwinkel kann bei Instrumenten bis etwa zur 2 m-Klasse mit dem gleichen Schneckenradgetriebe wie der Sternzeit-Stundenantrieb vorgenommen werden. Die größten Teleskope verwenden zwei parallele Getriebe für die schnellste bzw. alle übrigen Geschwindigkeiten, die zwar beide dauernd im Eingriff sind, von denen aber nur jeweils eins treibt. Eine konservativere Bauweise sieht zwischen Schneckenrad und Stundenachse eine Klemme vor, die für die schnelle Bewegung des Instruments gelöst wird. Ein Vorteil dieser Konstruktion ist, daß das Schneckenrad gleichmäßig über seinen Umfang und nicht, wie bei der starren Verbindung, bevorzugt in der Meridiangegend abgenutzt wird. Zur Einstellung und zum Antrieb in Deklination werden Schnecken- oder Zahnradgetriebe, für die langsamen Bewegungen auch die Tangentenschraube verwendet. Eine konstante „Vorspannung" zur Erzielung eines definierten Eingriffs ist hier ebenfalls üblich. Ist das Instrument klemmenlos und damit dauernd fest mit den Getrieben verbunden, muß eine besondere Möglichkeit zum Ausbalancieren vorgesehen werden.

Die Anzeige der Drehwinkel um die beiden Achsen, Stundenwinkel t und Deklination δ, erfolgt klassisch mit Teilkreisen. Dabei rotiert der t-Kreis üblicherweise mit dem Instrument und hat einen raumfesten Index; beim δ-Kreis ist auch die entgegengesetzte Anordnung gebaut worden (für Ablesung vom Okularende aus). Die beim Äquatoreal des 19. Jahrhunderts üblichen feingeteilten Kreise mit Nonius oder Mikrometer sind bei größeren Instrumenten bald verhältnismäßig groben, auch aus der Entfernung leicht ablesbaren gewichen. Entlastung für den Beobachter bringt ein zweiter Kreis auf der Stundenachse, der unmittelbar die Rektaszension $\alpha = \Theta - t$ (Θ ist die Sternzeit) anzeigt. Für Instrumente mit Klemme kann dieser Kreis vom kontinuierlich laufenden Schneckenrad mitgenommen werden; er muß so eingestellt sein, daß der Index des t-Kreises die Sternzeit gibt. Abgelesen wird an einem vom Instrument mitgeführten Index am Nullpunkt des t-Kreises.

Moderne Teleskope nutzen die Möglichkeiten der Winkel-*Fernübertragung*, die (mehrstufige) Drehfeldsysteme oder auch optische bzw. magnetische Kodierscheiben (besonders für Digitalausgabe) bieten. Eine Genauigkeit von der Größenordnung 1 Bogenminute ist verhältnismäßig leicht zu erreichen und für die meisten Zwecke ausreichend. Eine Anzeige auf $0\rlap{.}'1$ ist technisch möglich, aber nur sinnvoll, wenn die Korrekturen für Refraktion, Biegung usw. mit einem Rechner angebracht werden. Die Umwandlung des Stundenwinkels in Rektaszension ist im Übertragungsweg — durch analoge oder digitale Differenzbildung mit der Sternzeit — zu bewerkstelligen. Bei der Einstellung des Instruments mit Handtaster bevorzugen die meisten Beobachter die Analog-Anzeige auf einem Zifferblatt, bei der auch ein rasch sich ändernder Winkel noch erfaßbar bleibt. Die

Umwandlung der Drehwinkel in elektrische Signale öffnet ein weites Feld für Koordinaten-Vorwahl, Wiederholungsschaltungen, komplette Programmsteuerungen, Koordinatentransformation zur automatischen Nachführung der (azimutal montierten) Kuppel etc.

Literatur.

[1] Handbuch der Astrophysik, Bd. I/1. Berlin 1933.

[2] Handbuch der Experimentalphysik, Bd. 26. Leipzig 1937.

[3] CRAWFORD, D. L. (Hrsg.): The Construction of Large Telescopes (Int. Astr. Union Symposium No. 27). 1967.

[4] KING, H. C.: The History of the Telescope. London 1955.

[5] RIEKHER, R.: Fernrohre und ihre Meister. Berlin 1957.

[6] DANJON, A., et A. COUDER: Lunettes et Télescopes. Paris 1935.

[7] DIMITROFF, G. Z., and J. G. BAKER: Telescopes and Accessories. Philadelphia and Toronto 1945.

[8] KÖNIG, A., u. H. KÖHLER: Die Fernrohre und Entfernungsmesser, 3. Aufl. Berlin-Göttingen-Heidelberg 1959.

[9] KUIPER, G. P., and B. M. MIDDLEHURST (Hrsg.): Telescopes (Stars and Stellar Systems Vol. I). Chicago 1960.

[10] MICZAIKA, G. R., and W. M. SINTON: Tools of the Astronomer. Cambridge (Mass.) 1961.

[11] SLEVOGT, H.: Über eine Gruppe von aplanatischen Spiegelsystemen. Z. Instrumentenkde. **62**, 312 (1942).

[12] KÖHLER, H.: Die Entwicklung der aplanatischen Spiegelsysteme. Astron. Nachr. **278**, 1 (1949).

[13] SCHWARZSCHILD, K.: Untersuchungen zur geometrischen Optik I—III. Abhandl. Ges. Wiss. Göttingen, Math.-Phys. Kl., N. F. **4**, No. 1—3 (1905) = Mitt. Göttingen 9—11 (1905).

[14] BEREK, M.: Grundlagen der praktischen Optik. Berlin u. Leipzig 1930.

[15] MARÉCHAL, A.: Imagerie géometrique, Aberrations. Paris 1952.

[16] FLÜGGE, J.: Das photographische Objektiv. Wien 1955.

[17] PICHT, J.: Grundlagen der geometrisch-optischen Abbildung. Berlin 1955.

[18] CONRADY, A. E.: Applied Optics and Optical Design. New York 1957—1960.

[19] LINFOOT, E. H.: Recent Advances in Optics. Oxford 1955.

[20] FRANÇON, M.: Interférences, diffraction et polarisation. In: Handbuch der Physik, Bd. 24. 1956.

[21] MARÉCHAL, A., et M. FRANÇON: Diffraction. Structure des Images. Influence de la Cohérence de la Lumière. Paris 1960.

[22] BARAKAT, R.: The Intensity Distribution and Total Illumination of Aberration-Free Diffraction Images. In: Progress in Optics, Vol. I (Editor E. WOLF). Amsterdam 1961.

[23] BORN, M., and E. WOLF: Principles of Optics, 2. ed. Oxford and London 1964.

[24] JAQUINOT, P., and B. ROIZEN-DOSSIER: Apodisation. Progress in Optics, Vol. III, 29 (1964).

[25] HOPKINS, H. H.: Wave Theory of Aberrations. Oxford 1950.

[26] VÄISÄLÄ, Y.: Neue Methoden zur Untersuchung der Objektive. Diss. Helsinki 1922 = Ann. Univ. Turkuensis Ser. A **1** No. 2 (1922).

[27] CHRÉTIEN, H.: Le télescope de Newton et le télescope aplanétique. Rev. Opt. **1**, 13, 49 (1922).

[28] PAUL, M.: Systèmes correcteurs pour réflecteurs astronomiques. Rev. Opt. **14**, 169 (1935).

[29] ROSS, F. E.: Astrometry with Mirrors and Lenses. Astrophys. J. **77**, 243 (1933). — Lens Systems for Correcting Coma of Mirrors. Astrophys. J. **81**, 156 (1935).

[30] WYNNE, C. G.: Field Correctors for Parabolic Mirrors. Proc. Phys. Soc., Lond. B **62**, 772 (1949).

[31] INGALLS, A. G. (Hrsg.): Amateur Telescope Making. 1935 (I). Amateur Telescope Making Advanced. 1937 (II). Amateur Telescope Making (Book Three). 1953 (III).

[32] BOUWERS, A.: Achievements in Optics. New York and Amsterdam 1946.

[33] TWYMAN, F.: Prism and Lens Making, 2. ed. 2. Abdr. London 1957.

[34] KREIDL, N. J., and J. L. ROOD: Optical Materials. In: R. KINGSLAKE (Hrsg.), Applied Optics and Optical Engineering, Vol. 1. New York and London 1965.

[35] COUDER, A.: Recherches sur les déformations des grands miroirs. Bull. Astron. 2. Ser. (Mém.) **7**, 201, 353 (1931).

[36] SCOTT, R. M.: Optical Engineering. Appl. Optics **1**, 387 (1962).

[37] HOLLAND, L.: Vacuum Deposition of Thin Films. London and New York 1956.

[38] HASS, G.: Mirror Coatings. In: R. KINGSLAKE (ed.), Applied Optics and Optical Engineering, Vol. 3. New York and London 1965.

[39] HOLLAND, L.: The Properties of Glass Surfaces. New York 1964.

[40] HARTMANN, J.: Objektivuntersuchungen. Z. Instrumentenkde. **24**, 1, 33, 97 (1904). — Untersuchungen über das 80-cm-Objektiv des Potsdamer Refraktors. Publ. Potsdam **15**, Nr. 46 (1908). — Die Korrektur des Potsdamer 80-cm-Objektivs. Z. Instrumentenkde. **29**, 217 (1909).

[41] FOUCAULT, L.: Mémoire sur la construction des télescopes en verre argenté. Ann. Paris (Mém.) **5**, 197 (1859).

[42] MARÉCHAL, A.: Optique géometrique générale. In: Handbuch der Physik, Bd. 24. 1956.

[43] STRONG, J.: Concepts of Classical Optics. San Francisco and London 1958.

[44] MAKSUTOW, D. D.: Technologie der astronomischen Optik. Berlin 1954. (Übersetzung von Izgotovlenie i Issledovanie Astronomicheskoi Optiki. Leningrad-Moskva 1948).

[45] CAGNET, M., M. FRANÇON u. J. C. THRIERR: Atlas optischer Erscheinungen. Berlin-Göttingen-Heidelberg: Springer 1962.

Mikroskope.

Von

H. C. CLAUSSEN.

Mit 97 Figuren.

1. Einleitung. Mikroskope sind, wie das Wort sagt, solche optischen Instrumente, die in Verbindung mit dem Auge dazu dienen, kleine Dinge zu betrachten[1]. Dabei wird die Bezeichnung Mikroskop auf jene optischen Instrumente beschränkt, bei denen die Beobachtungsobjekte dem Auge oder das Auge dem Objekt beliebig angenähert werden können. So ist eine Abgrenzung gegen die Teleskope gegeben, die zur Beobachtung ferner Objekte bestimmt sind. Natürlich gibt es auch Objekte im Zwischenbereich und entsprechende Übergänge zwischen Mikroskopen und Fernrohren.

Wenn man kleine Gegenstände betrachtet, nähert man sie dem Auge möglichst weit. Man vergrößert den Sehwinkel, bis das auf der Netzhaut entstehende Bild des Objektes aufgelöst wird. Eine Grenze für die Annäherung an das Auge ist durch die Akkomodationsfähigkeit des Auges gegeben (*Nahpunkt*). Bei noch größerer Annäherung werden die Objekte nicht mehr scharf abgebildet. Ein kurzsichtiger Beobachter ist hierbei einem Normalsichtigen gegenüber im Vorteil; er kann seine Brille abnehmen und das Objekt dann wesentlich weiter nähern. Der maximale Sehwinkel, unter dem ein bestimmtes Objekt gesehen werden kann, ist also für jeden Beobachter verschieden. Um diese Schwierigkeiten zu umgehen, hat man sich darauf geeinigt, einen Beobachter als „normal“ anzusehen, dessen Nahpunkt 250 mm vom Auge entfernt liegt. Man nennt diese Entfernung in der deutschen Literatur zumeist konventionelle oder deutliche Sehweite. Besser sollte man sie als Bezugssehweite bezeichnen, weil alle Vergrößerungsangaben in der Mikroskopie auf ein Objekt bezogen werden, das bei Beobachtung ohne Instrument in dieser Entfernung gesehen wird.

Für das Auflösungsvermögen des Auges gilt, daß zwei Objekte eine Bogenminute voneinander entfernt erscheinen müssen, um noch getrennt gesehen zu werden. Diese Bogenminute entspricht bei der Bezugssehweite von 250 mm einer Länge von 73 μm. Zur Beobachtung von Objekten in und unter dieser Größenordnung sind also jedenfalls „Mikroskope“ anzuwenden.

I. Grundlagen.

2. Grenze der Vergrößerungsfähigkeit von Mikroskopen. Vor einer Besprechung des Aufbaus solcher Mikroskope ist zu ermitteln, ob ihre Vergrößerung beliebig weit zu treiben ist oder ob bestimmte Grenzen aus physikalischen Gründen nicht überschritten werden können.

Beugungsbild und laterale Auflösung. Jedes punktförmige selbstleuchtende Objekt wird durch ein Objektiv bekanntlich als Beugungsscheibchen abgebildet (HERSCHEL, 1830). Der Durchmesser d des ersten dunklen Ringes eines solchen

[1] mikrós (griechisch) = klein, skopein (griechisch) = betrachten.

Beugungsscheibchens hängt von der Wellenlänge λ des Lichtes und von der Apertur

$$A = n \cdot \sin u \tag{2.1}$$

des Objektives ab, und zwar ist bei fehlerfreier Abbildung (AIRY, 1835)

$$d_l = 1{,}22 \frac{\lambda}{A}. \tag{2.2}$$

Wenn das selbstleuchtende Objekt auf der Achse liegt, ist u der Winkel des Aperturstrahles und n die Brechzahl des Mediums, in dem u gemessen wird (Fig. 1).

Die Intensitätsverteilung des Beugungsscheibchens in der Bildebene hat die Form (Fig. 2a)

$$I = \left(\frac{2\,J_1(z)}{z}\right)^2. \tag{2.3}$$

Dabei ist J_1 die Besselsche Funktion 1. Ordnung.

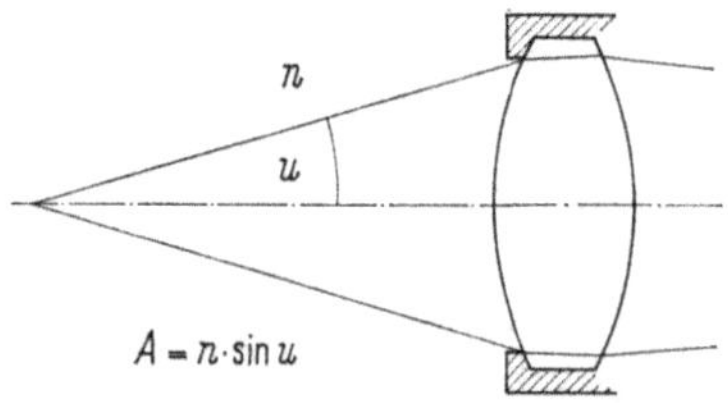

Fig. 1. Apertur (numerische Apertur) A eines Objektivs.

Betrachtet man zwei nahe beieinander liegende, mit gleicher Intensität strahlende selbstleuchtende Punkte, deren Strahlung man als inkohärent annehmen kann, so entstehen zwei gleiche Beugungsbilder, deren Intensitäten sich dort addieren, wo die Beugungsbilder aufeinander fallen.

Nach einem Vorschlag von RAYLEIGH definiert man zwei Bilder so lange als getrennt erkennbar, wie die Intensität zwischen den beiden um wenigstens 20% absinkt. Das ist bei zwei Beugungsscheibchen nach Gl. (2.3) der Fall, wenn das Maximum der Intensität des einen Beugungsscheibchens auf dem ersten Minimum der Intensität des zweiten liegt. Der Intensitäts-

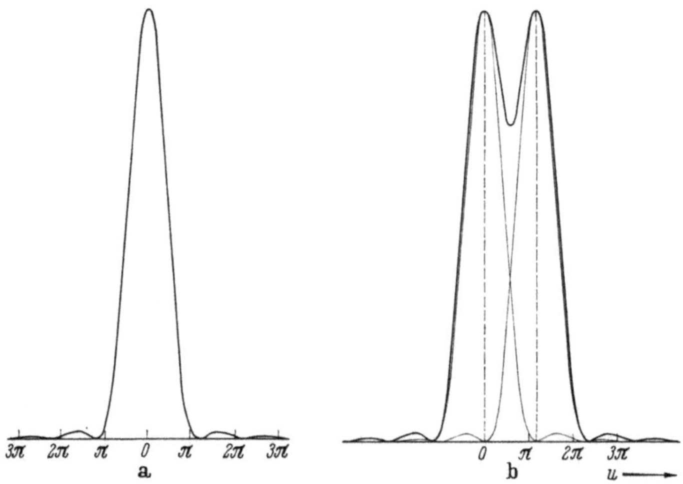

Fig. 2a u. b. Intensitätsverteilung im Beugungsbild a eines Selbstleuchters, b von 2 Selbstleuchtern im Abstande von 0,61 λ/A.

sattel zwischen beiden Maxima ist dann 26.4% niedriger als die Intensität der Maxima der resultierenden Beugungsfigur, deren Abstand gerade halb so groß ($\varepsilon = 0{,}5$) ist wie der Durchmesser d des ersten dunklen Ringes (Fig. 2b). Man muß aber bedenken, daß dieser Wert $\varepsilon = 0{,}5$ nur ein Richtwert ist, der je nach Art des Empfängers über- und unterschritten werden kann.

Das Auflösungsvermögen eines Objektives beschreibt man durch den kleinsten Abstand, bei dem dieses Objektiv zwei senkrecht zur optischen Achse liegende Punkte noch getrennt abbildet. Es ist also im Falle von zwei inkohärent strahlenden Punkten (Selbstleuchter) und unter der Voraussetzung, daß der Intensitätssattel zwischen den beiden Punktbildern 26,4% tief ist

$$\delta_i = 0{,}61\,\frac{\lambda}{A}. \tag{2.4}$$

Mikroskopische Objekte sind im allgemeinen keine inkohärent strahlenden Selbstleuchter, sondern beleuchtete Objekte. So ist die Frage zu klären, ob zwei kleine, von einer Lichtquelle beleuchtete Löcher als inkohärente Strahler angesehen werden können. Wenn sie nämlich als kohärente oder teilkohärente Strahler angesehen werden müssen, wird die Intensität im Bilde nicht gleich der Summe der Amplitudenquadrate $A_1^2 + A_2^2$ der beiden Teilbilder sein, sondern man wird ein weiteres Glied berücksichtigen müssen, das den Phasen Rechnung trägt. BEREK (1926, 1929), VAN CITTERT (1934), ZERNIKE (1938), HOPKINS und BARHAM (1950) und HOPKINS (1951, 1953, 1957) haben den Einfluß von Teilkohärenz geklärt und kamen über die Bestimmung eines Phasen-Kohärenz-Faktors zu dem Ergebnis, daß die „wirksame Lichtquelle" und damit das Verhältnis von Beleuchtungsapertur zu Beobachtungsapertur

$$\sigma = \frac{A_{\text{Bel}}}{A_{\text{Obj}}} = \frac{n_{\text{Bel}} \cdot \sin u_{\text{Bel}}}{n_{\text{Obj}} \cdot \sin u_{\text{Obj}}} \tag{2.5}$$

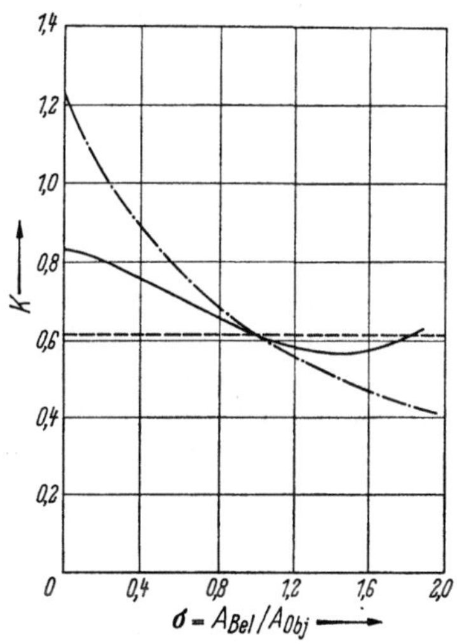

Fig. 3. Abhängigkeit des Auflösungsvermögens vom Kohärenzgrad.

für den Kohärenzgrad maßgebend ist. Bei Erfüllung der Forderung, daß die Intensität zwischen den beiden Maxima um 26,4% absinkt, geben HOPKINS und BARHAM (1950) die Zahl K an, die bei Teilkohärenz anstelle von 0,61 in (2.4) zu setzen ist

$$\delta_t = K \cdot \frac{\lambda}{A}. \tag{2.6}$$

Fig. 3 zeigt K in Abhängigkeit vom Verhältnis σ von Beleuchtungs- zu Beobachtungsapertur (ausgezogene Kurve) und ohne Berücksichtigung der Beleuchtungsapertur, $K = 0{,}61$ (gestrichelte Kurve). Die strichpunktierte Kurve gibt die Werte nach der Abbeschen Theorie, wenn die Auflösung durch $A = \frac{1}{2}(A_{\text{Obj}} + A_{\text{Bel}})$ bestimmt wird.

Im Bereich σ von 0 bis 2 ändert sich K zwischen 0,82 und 0,57. Die Auflösung ist also im Bereich relativ kleiner Beleuchtungsaperturen bis etwa 30% schlechter und bei Beleuchtungsaperturen, die größer als die Beobachtungsapertur sind, ein wenig besser als bei rein inkohärenter Strahlung. Die Größenordnung der Auflösung ändert sich nicht.

Die Angabe gilt für jede beliebige Art der Beleuchtung, wenn nur die Beleuchtungsapertur — vom Objekt aus gesehen — erfüllt ist. Sie ist unabhängig vom Korrektionsgrad der Beleuchtung.

Als Apertur ist, je nachdem ob man sich für das Auflösungsvermögen im Bild oder im Objekt interessiert, die bild- oder objektseitige Apertur (Fig. 4) einzusetzen. Bei hinreichender Einhaltung der Sinusbedingung

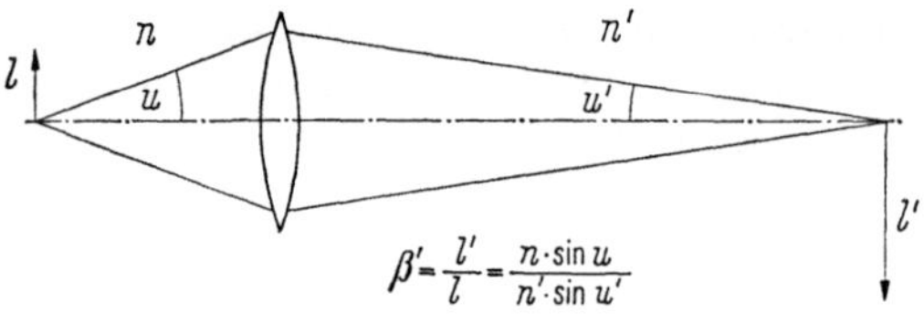

Fig. 4. Zusammenhang zwischen objekt- und bildseitiger Apertur.

$$\frac{l'}{l} = \beta' = \frac{n \sin u}{n' \sin u'} = \frac{A}{A'} \tag{2.7}$$

gilt also bei Mikroskopobjektiven für die objektseitige Apertur $A = \beta' A'$. Üblich ist es, das Auflösungsvermögen von Mikroskopobjektiven als Länge im Objekt anzugeben.

Das Auflösungsvermögen wird gesteigert durch Vergrößerung der Apertur und durch Verkleinerung der Wellenlänge. Bei subjektiver Beobachtung im „weißen" Licht sieht man als Schwerpunkt im allgemeinen das Maximum der Augenempfindlichkeit bei 550 nm an, er liegt aber bei etwas kürzerer Wellenlänge. (Für Messung und Rechnung wird man die diesem Maximum sehr benachbarte grüne Hg-Linie e, $\lambda = 546{,}1$ nm, verwenden.) Damit ist aber für visuelle Instrumente eine Verbesserung der Auflösung nur durch Erhöhung der Apertur möglich. Der Winkel u wird nie ganz 90° erreichen, weil das Objektiv stets einen gewissen Abstand vom Objekt hat. In der Praxis gilt $\sin u \leqq 0{,}95$. Dieser Wert ist der Maximalwert, wenn sich an irgendeiner Stelle zwischen Objektiv und Objekt eine Luftschicht ($n = 1$) befindet. Strahlen, die mit höherer Apertur aus dem Objekt austreten, würden bei Eintritt in die Luftschicht so gebrochen oder totalreflektiert, daß sie nicht in das Objektiv eintreten könnten (Fig. 5a). Höhere Aperturen kann

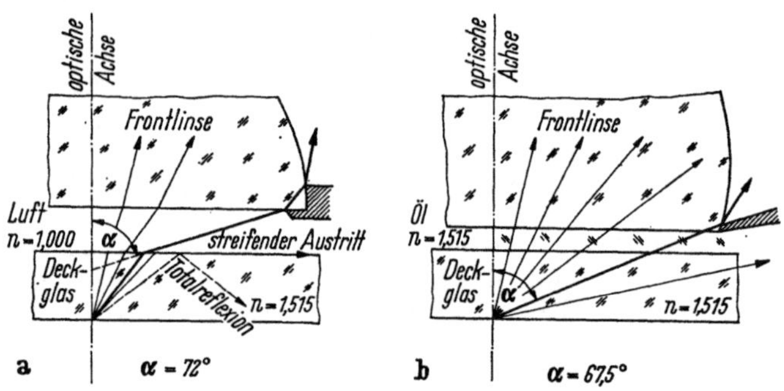

Fig. 5a u. b. Objektiv-Aperturen. Schematische Darstellung des Strahlenverlaufs zwischen Objekt und Frontlinse bei a einem Trockenobjektiv: $A = 1{,}00 \sin 72° = 0{,}95$, b einer Ölimmersion: $A = 1{,}515 \sin 67{,}5° = 1{,}40$.

man erreichen, wenn man den Zwischenraum zwischen Objekt und Objektiv mit Medien mit höheren Brechzahlen ausfüllt. Bei biologischen Objekten, die zumeist in ein Kittmittel eingebettet und mit einer dünnen Glasscheibe (Deckglas) abgedeckt sind, ist es nicht notwendig, mit der Brechzahl des Zwischenmediums über die von Kittmittel und Glas hinauszugehen. So hat sich für diesen Zweck seit langem ein „Immersionsöl" mit einer Brechzahl von $n = 1{,}515$ entsprechend der eines mittleren Kronglases eingebürgert. Ursprünglich benutzte man eine Auflösung von Cedernholzöl in Xylol, heute verwendet man andere Medien mit niedrigerem Dampfdruck, die nicht eindicken und ihre Brechzahl nicht durch Abdampfen des Lösungsmittels verändern. Mit solchem „Immersionsöl" müßten Aperturen von $1{,}515 \times 0{,}95 = 1{,}44$ erreicht werden können; in der Praxis kommt man nicht über $A = 1{,}40$ hinaus (Fig. 5b).

Bei Lebendbeobachtungen in Wasser wird das Auflösungsvermögen durch die Brechzahl des Wassers $n=1{,}333$ und $A=1{,}20$ begrenzt.

Immersionsmittel mit höheren Brechzahlen als 1,515 werden zur Mikroskopie nur in Sonderfällen benutzt, weil sie zumeist Eigenschaften haben, die sie zu einer Dauerverwendung ungeeignet erscheinen lassen, wie z. B. chemische Agressivität oder unangenehmen Geruch.

Fig. 6 gibt den Zusammenhang zwischen in der Mikroskopie erreichbaren Aperturen und zugehöriger Auflösung. Man sieht, daß das laterale Auflösungsvermögen, also das Auflösungsvermögen in der Objektebene, im besten Falle etwa eine halbe Wellenlänge ist. Zwei Objektdetails, die näher aneinander liegen, können nicht mehr getrennt gesehen werden. Alle Objekte, die kleiner sind, werden auf jeden Fall und unabhängig von ihrer wahren Form und Größe als Beugungsscheibchen so abgebildet, wie es der Objektivapertur entspricht.

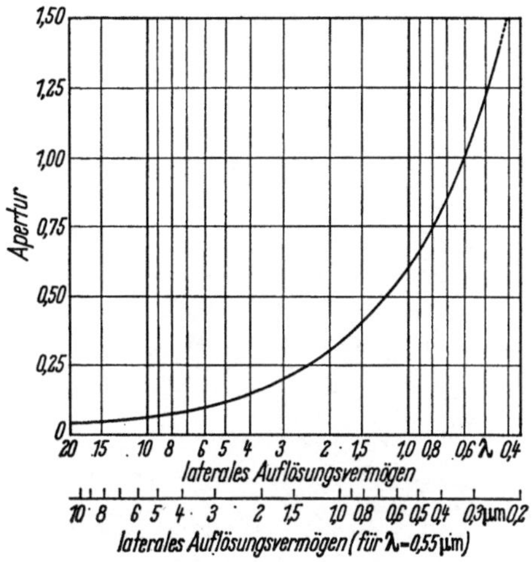

Fig. 6. Auflösungsvermögen von Mikroskopobjektiven in Wellenlängen bzw. in µm für grünes Licht ($\lambda=550$ nm) in Abhängigkeit von der Apertur.

3. Förderliche Vergrößerung. Der kleinste mit dem Mikroskop auflösbare Abstand im Objekt ist nach Gl. (2.4) $\delta=0{,}61\ \lambda/A$. Dieser Abstand muß vergrößert werden, damit er auch vom Auge aufgelöst werden kann. Bei einem Pupillendurchmesser von 2 mm und hohem Kontrast im Objekt ist das Auflösungsvermögen des Auges eine Bogenminute. Sie entspricht, wie bereits gesagt, einer Länge von 73 µm im Abstande von 250 mm vom Beobachter. Auf diesen Abstand beziehen wir auch das mikroskopische Bild und erhalten für $\lambda=0{,}550$ µm

$$V_{\min}=\frac{73\,A}{0{,}61\times 0{,}55}=220\,A\,. \tag{3.1}$$

Unter den angegebenen Bedingungen ist also die notwendige Mindestvergrößerung das 220fache der Objektivapertur. Bei geringerer Vergrößerung wird das Auge mit Sicherheit nicht alles Detail, das im Bilde vorhanden ist, wahrnehmen können. Im allgemeinen ist die Austrittspupille des Mikroskopes wesentlich kleiner als 2 mm, der Kontrast im Bild liegt weit unter 1. So kommt man in der Praxis zu der Feststellung, daß die Gesamtvergrößerung des Mikroskopes dann richtig gewählt ist, wenn sie etwas mehr als das Doppelte bis Vierfache des errechneten Minimalwertes erreicht. Diesen Vergrößerungsbereich nennt man den Bereich der förderlichen Vergrößerung:

$$V_{\text{förderlich}}=500\ A_{\text{Obj}}\ \text{bis}\ 1000\ A_{\text{Obj}} \tag{3.2}$$

(vgl. ABBE, 1882/1883).

Man versteht nach dem oben Gesagten, daß die Festlegung eines solchen Bereiches nicht mehr als ein grober Anhalt sein kann. Bei geringem Bildkontrast wird man z. B. die untere Grenze etwas unterschreiten, bei hohem Kontrast über die obere Grenze hinausgehen. Man wird diese Grenze ebenfalls absichtlich überschreiten, wenn z. B. beim Druck eines mikroskopischen Bildes ein Raster überlagert wird oder wenn man für Meßzwecke den Sehwinkel zwischen zwei Meßpunkten vergrößern will.

Die angegebenen Werte gelten in der Mikrophotographie oder Mikroprojektion natürlich stets für das mit dem Auge betrachtete Bild und unter Beachtung des Beobachtungsabstandes. In der Mikrophotographie oder bei Übertragung des Bildes durch Fernsehen ist das Auflösungsvermögen der Schicht oder des Fernsehschirmes zusätzlich zu beachten. Es wird im allgemeinen in Linien/mm angegeben.

4. Axiales Auflösungsvermögen. Die nächste Frage gilt der Ausdehnung der Beugungsfigur in Richtung der optischen Achse und der grundsätzlichen Möglichkeit, zwei hintereinander liegende Objektpunkte getrennt zu sehen. LOMMEL (1885, 1886) schuf die Grundlagen zur Berechnung der räumlichen Struktur eines Beu-

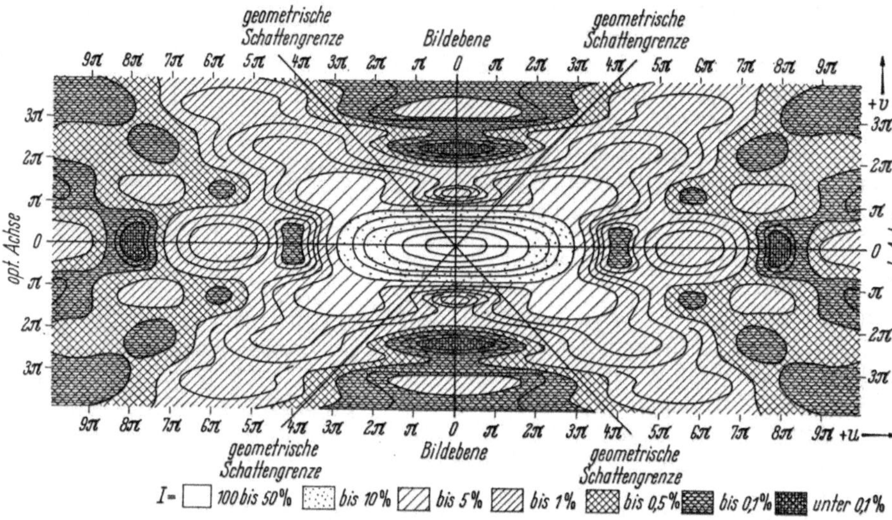

Fig. 7. Verteilung der Lichtintensitäten in der Nähe eines Bildpunktes bei aberrationsfreier Abbildung. Isophotendarstellung (nach BEREK 1926 mit Korrekturen nach LINFOOT und WOLF, 1953). Die Variablen u und v sind definiert durch die Gleichungen $u = A^2 z \cdot 2\pi/\lambda$, $v = A\, r\, 2\pi/\lambda$, wobei $r = \sqrt{x^2+y^2}$ ist. Dabei ist $u/4\pi$ die Ordnungszahl der Minima beim Defokussieren und v/π die Ordnungszahl der Minima in der Bildebene, seitlich des Zentrums. $|v/u| \lesseqgtr 1$ bedeutet Lage eines Punktes im geometrischen Licht- bzw. Schattenbereich, $|v/u| = 1$ ist die geometrische Schattengrenze. Vgl. Fig. 32.

gungsbildes, die BEREK (1926, 1927) auf das Mikroskop anwendete. ZERNIKE und NIJBOER (1949), LINFOOT und WOLF (1953, 1956) und TSCHUNKO (1965) veröffentlichten genauer berechnete Bilder, die die Intensitätsverteilung eines Beugungsbildes in einem Schnitt entlang der optischen Achse zeigen. Fig. 7 basiert auf der Angabe von LINFOOT und WOLF[1].

Aus dem Abstand der beiden ersten Minima,

$$d_a = \frac{4n\lambda}{A_{\mathrm{Obj}}^2}, \tag{4.1}$$

wobei n der Brechungsindex des Einbettungsmediums für das Objekt ist, folgerte BEREK (1927) theoretisch

$$\delta_a = \frac{4n\lambda}{A_{\mathrm{Obj}}^2}\,\tau + \frac{n \cdot \lambda}{A_{\mathrm{Obj}} \cdot V}\, l. \tag{4.2}$$

Dabei sind τ und l physiologische Konstanten und zwar $\tau = \frac{1}{8}$ und $l = 0{,}34$; l ist eine Bogenlänge = Sehweite × Sehwinkel, V die Gesamtvergrößerung des Mikroskops.

[1] Die Bereksche Darstellung ist häufig falsch übernommen worden, und zwar sind die geometrischen Schattengrenzen, die unter 45° zu den Achsen der Darstellung laufen müssen, unter einem falschen Winkel eingezeichnet.

Berek bestätigte die Beziehung durch eingehende Messungen am Mikroskop und wählte zur Beobachtung jeweils die Beleuchtungsapertur ein wenig größer als die Beobachtungsapertur.

Fig. 8 zeigt den Zusammenhang zwischen axialem Auflösungsvermögen und Gesamtvergrößerung eines Mikroskops in Abhängigkeit von verschiedenen Objektivaperturen.

Bei Anwendung schwacher Gesamtvergrößerungen kann der Tiefenbereich δ_a durch Betätigung der Augenakkomodation zwischen den Grenzen S_1 und S_2 für die Sehweite wesentlich vergrößert erscheinen und zwar um einen Betrag Δ, der zu δ_a zu addieren ist:

$$\Delta = n \frac{S^2}{V^2} \left(\frac{1}{S_1} - \frac{1}{S_2} \right). \qquad (4.3)$$

Hierbei ist S die auf 250 mm normierte Sehweite (vgl. Ziff. 1).

Für den Akkomodationsbereich von ∞ bis 250 mm ergibt sich in Abhängigkeit von V:

V	10	40	100	400	1000	2500	4000
Δ in µm	3600	230	36	2,3	0,36	0,06	0,02

Nach den Ergebnissen von Ziff. 2—4 sind Lichtmikroskope so zu planen, daß sie laterale Vergrößerungen von etwa 1—1500 × zur Abbildung und Messung von Objekten in der Bildebene und bis etwa 2500 × zur Bestimmung von Längen in Richtung der optischen Achse ermöglichen.

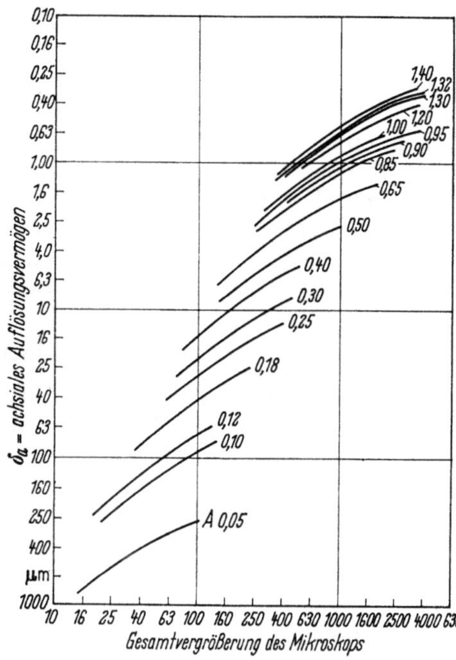

Fig. 8. Axiales Auflösungsvermögen eines Mikroskops.

5. Historische Bemerkung. Im Jahre 1873 veröffentlichte Ernst Abbe seine „Beiträge zur Theorie des Mikroskops und der mikroskopischen Wahrnehmung" und in der Folge weitere Arbeiten zu diesem Thema, die in Band 1 seiner gesammelten Abhandlungen zu finden sind. Nach seinem Tode veröffentlichten Lummer und Reiche (1910) das, was man heute die Abbesche Theorie der Bildentstehung im Mikroskop nennt, nach Niederschriften in den Vorlesungen von Abbe. Dieser wies erstmalig auf die Verschiedenheit der physikalischen Grundlagen bei der Abbildung von Selbstleuchtern und Nichtselbstleuchtern hin, er unterschied zwischen der „primären" Interferenzerscheinung am Objekt und der durch die Objektivapertur hervorgerufenen „sekundären" hinter dem Objektiv und zeigte prinzipiell den Weg zur Berechnung der Lichtverteilung im Bild von Nichtselbstleuchtern mit Hilfe von drei Integrationen und zwar je einer über die Objektstrukturen und die freie Öffnung der Objektive, die Amplituden und Phasen berücksichtigt, und einer dritten über die Lichtquelle, die die Intensitäten berücksichtigt. Die Lichtverteilung im Bilde von Selbstleuchtern wird mit Hilfe von zwei Integrationen und zwar über den Teil der von der Objektivöffnung durchgelassenen Wellenfläche nach Amplituden und über die Ausdehnung des Selbstleuchters nach Intensitäten bestimmt. Abbe wählte als nichtselbstleuchtende Objekte Gitter, die er mit Büscheln der Apertur 0 beleuchtete, und kam zu seinen bekannten Angaben für die Auflösung bei gerader Beleuchtung $d = \lambda / A_{\mathrm{Obj}}$ und bei schräger Beleuchtung mit einer Neigung gleich der Apertur des Objektivs $d=\lambda/2A_{\mathrm{Obj}}$. Auf Grund seiner Überlegungen entwickelte er den nach ihm benannten Beleuchtungsapparat mit einer in dessen unterer Brennebene exzentrisch verschiebbaren Irisblende und empfahl ihre Verwendung grundsätzlich auch für nicht gitterförmige Objekte. Schon im Jahre 1878 nahm ROBERT KOCH und 1880—1882 der Histologe ALTMANN gegen diese Vorschrift Stellung und KOCH empfahl, den Abbeschen Kondensor mit hoher Beleuchtungsapertur zu verwenden.

Fast gleichzeitig mit ABBEs erster Veröffentlichung geht HELMHOLTZ (1874) von der Abbildung eines Lichtpunktes aus und gibt das Auflösungsvermögen für Luft, also $n=1$, und für eine quadratische Apertur zu $\delta=0{,}5\ \lambda/A$ an. Nur in einem Nachwort geht er auf die gerade erschienene Abbesche Arbeit ein. Ihm folgt LORD RAYLEIGH (1896). Erst nach ABBE hat WOLFKE (1912, 1913) den Einfluß hoher Beleuchtungsaperturen auf die Lichtverteilung im Bilde berechnet. BEREK (1924—1950) stieß bei einer Beschäftigung mit der Frage der Bildentstehung im Mikroskop sehr bald auf die grundlegenden Unterschiede zwischen der Abbildung von Selbstleuchtern und Nichtselbstleuchtern und formulierte auf Grund der Arbeiten von MANDELSTAM (1911) und VON LAUE (1914) eine Äquivalenzbedingung für die Abbildung von Nichtselbstleuchtern, die mit hoher Apertur beleuchtet werden, im Vergleich zu der von Selbstleuchtern (1929). Die schon zitierten Arbeiten von VAN CITTERT, ZERNIKE, HOPKINS und BARHAM sind unabhängig davon.

II. Normale Mikroskope.

6. Einstufige Mikroskope: Lupen und Lesegläser. Will man ein Objekt unter größerem Sehwinkel betrachten als es im Nahpunkt erscheint, kann man im einfachsten Falle eine Sammellinse zwischen Auge und Objekt halten. Akkomodiert

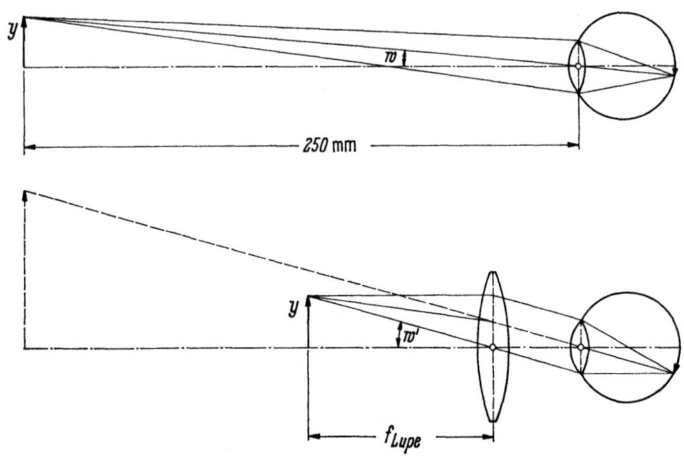

Fig. 9. Lupenvergrößerung. Skizze zu Gl. (6.1).

man dann auf Unendlich, muß das Objekt in den vom Auge entfernten Brennpunkt dieser Sammellinse gebracht werden, und es ergeben sich aus Fig. 9 folgende Beziehungen für die Vergrößerung V des Sehwinkels:

$$\left.\begin{aligned} \tan w &= \frac{y}{250\ \mathrm{mm}}; \quad \tan w' = \frac{y}{f_{\mathrm{Lupe}}}; \\ V &= \frac{\tan w'}{\tan w} = \frac{250\ \mathrm{mm}}{f_{\mathrm{Lupe}}}. \end{aligned}\right\} \qquad (6.1)$$

Eine in der geschilderten Weise benutzte Linse nennt man *Lupe* und V entsprechend die *Lupenvergrößerung*.

Man sieht ohne weiteres, daß die so definierte Lupenvergrößerung unabhängig ist vom Ort des Auges und unabhängig davon, ob mit unbewegtem oder mit blickendem, also um den Augendrehpunkt bewegtem Auge, beobachtet wird.

Anders wird es, wenn man annimmt, daß mit Nahakkomodation oder mit fehlsichtigem Auge beobachtet wird, was sicher sehr häufig der Fall ist. Bei Nahakkomodation oder nicht korrigierter Kurzsichtigkeit des Beobachters rückt das Objekt entsprechend näher an die Lupe heran, bei Weit- oder starker Alterssichtigkeit über den Brennpunkt hinaus von der Lupe weg. Will man sehr genaue Angaben über die subjektive Sehwinkelvergrößerung machen, so ist zu bedenken, daß man den Scheitel des Sehwinkels bei ruhendem Auge in die Eintrittspupille des Auges und bei blickendem Auge in den Augendrehpunkt legen muß. Von diesen Punkten aus wäre die Sehweite s'_a des Auges zu messen (Fig. 10).

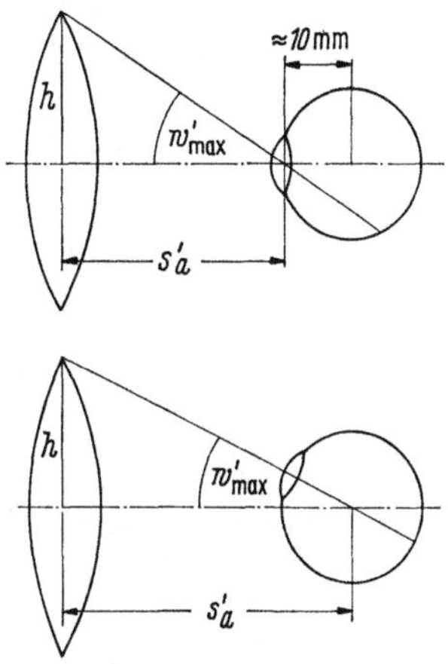

Fig. 10. Lupe. Maximaler Sehwinkel bei unbewegtem Auge (oben) und blickendem Auge (unten).

Im allgemeinen genügt eine wesentlich gröbere Annäherung. Man betrachte die Lupe als eine dünne Linse und gebe ihre Brechkraft in Dioptrien an. Dann wird aus (6.1) wegen

$$D_{\text{Lupe}} = 1000 \text{ mm} : f_{\text{Lupe}}$$

$$V = \frac{D_{\text{Lupe}}}{4}, \qquad (6.2)$$

wobei D_{Lupe} in dptr gemessen wird.

Ein nicht korrigiertes fehlsichtiges oder nahakkomodiertes Auge kann man ersetzt denken durch ein normales auf Unendlich akkomodiertes Auge und eine davorgeschaltete Linse, die der Fehlsichtigkeit oder Akkomodation entspricht. Die Brechkraft dieser Linse mißt man gleichfalls in dptr und addiert sie zur Brechkraft der Lupe. So erhält man eine Ersatzlupe, die die individuelle Vergrößerung des Sehwinkels bezogen auf ein ohne Instrument gesehenes Objekt in 250 mm Entfernung angibt. Diese Art der Betrachtung ist zulässig, wenn die Linse mit der zusätzlichen Brechkraft den Abstand Null von der Lupe hat, das Auge also sehr nahe an der Lupe steht. Akkomodation auf 250 mm entspricht einer Brechkraft von 4 dptr. In diesem Falle erhält man $V_s = V+1$, und man sieht, daß die Wirkung der Akkomodation bzw. Fehlsichtigkeit bei schwachen Lupen wesentlich stärker auffällt als bei starken. Aus diesem Grunde wird bei Lupen mit einer Vergrößerung $V \geqq 4$ im allgemeinen die Lupenvergrößerung angegeben, während man bei schwächeren Lupen die Angabe der Brennweite bevorzugt.

Wo liegen Apertur- und Sehfeldbegrenzung bei einer Lupe? Üblicherweise ist der Durchmesser der Lupe größer als die Eintrittspupille des Auges. So begrenzt letztere die Apertur, und die Iris des Auges ist Aperturblende. Das Sehfeld wird durch den Rand der Lupe begrenzt, der aber nicht scharf abgebildet werden kann, weil man das Auge bei der Beobachtung möglichst nahe an die Lupe heranbringt, um den Sehwinkel zu vergrößern. Der Sehwinkel ergibt sich aus

$$\tan w'_{\max} = \frac{h}{s'_a}. \qquad (6.3)$$

Dabei ist $w'_{\max}$ der halbe Sehwinkel, h die Eintrittshöhe des Grenzstrahles in die Lupe und s'_a bei unbewegtem Auge der Abstand der Eintrittspupille des Auges und bei blickendem Auge der Abstand des Augendrehpunktes von der Lupe (Fig. 10).

Bei ganz starken Lupen wird ihr Durchmesser kleiner als der der Augenpupille. Wählt man, um einfach rechnen zu können, ein Glas mit $n=1,5$, dann ist bei einer plankonvexen Linse der Radius der konvexen Fläche $r=f/2$, bei einer bikonvexen Linse mit gleichen Flächenradien $r=f$.

Eine 250fache Lupe ($f=1$ mm) würde also als plankonvexe Linse mit einem Radius $r=\frac{1}{2}$ mm und als bikonvexe Linse mit $r=1$ mm ausgeführt werden müssen. Die größten Durchmesser ergeben sich, wenn die plankonvexe Linse eine Halbkugel und die bikonvexe Linse eine Vollkugel ist. Selbst wenn die Korrektion in diesen Fällen befriedigend wäre, würden die Linsendurchmesser kleiner als die der normalen Augenpupille, die man mit 4 mm annehmen darf. Hier wäre also die Öffnung der Lupe die Aperturblende.

Der Holländer A. VAN LEEUWENHOOK (1632—1723) hat mit derartigen „Mikroskopen", die er selbst herstellte, gearbeitet. P. H. VAN CITTERT (nach BOEGEHOLD, 1958) hat eines dieser Mikroskope nachgemessen und gibt folgende Daten an: Linse bikonvex mit beiderseitig $r=0,75$ mm, Mittendicke $d=1,1$ mm. Die Linse ist zwischen zwei Stahlplättchen gefaßt, von denen die objektseitige eine Öffnung von 0,5 mm und die augenseitige eine solche von 0,8 mm Durchmesser hat. Nimmt man $n=1,52$ an, so ist $s'_{\infty}=0,48$ mm der Arbeitsabstand, $f=0,96$ mm und $V=295\times$. Die größtmögliche Öffnung wäre in diesem Falle 1,5 mm. Sie ist offensichtlich der Abbildungsfehler wegen auf 0,5 mm begrenzt worden. Der Aperturwinkel ergibt sich damit aus $\tan u=0,25:0,48$ zu $u=27,5°$ und die Apertur $A=\sin u=0,46$ (vgl. Ziff. 2). Sehr unbequem sind der kleine Arbeitsabstand und vor allem der kleine Abstand zwischen dem augenseitigen Linsenscheitel und dem Objekt, der nur $0,48+1,1=1,58$ mm beträgt. Um ein möglichst großes Feld zu übersehen, muß man das Auge sehr nahe an die Linse heranbringen und dann durch die kleine Öffnung hindurch wie durch ein Schlüsselloch beobachten, indem man das Auge auf einen Kreisbogen um die Lupe herum bewegt.

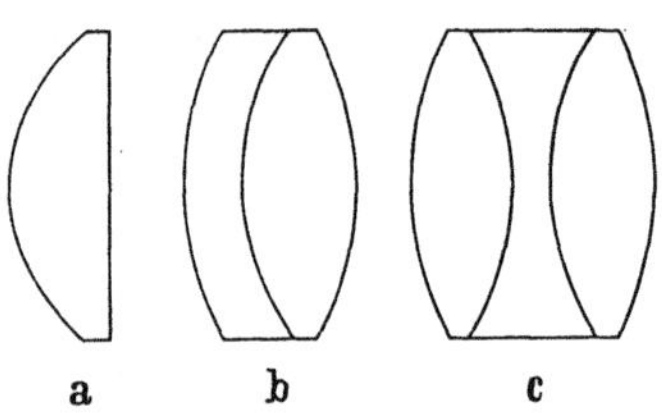

Fig. 11 a—c. Verschiedene typische Lupenformen.

Aus Korrektionsgründen wird man Lupen mit Vergrößerungen über $V=30\times$ als bikonvexe oder als plankonvexe Linse mit der Planseite zum Objekt hin ausführen, weil der Öffnungsfehler hier die größte Rolle spielt. Bei schwachen Lupen mit großem Sehwinkel ist besonders auf die Korrektion von Astigmatismus und Verzeichnung zu achten. Öffnungsfehler, Asymmetriefehler und chromatische Vergrößerungsdifferenz treten nicht störend in Erscheinung. Man verwendet Plankonvexlinsen mit der konvexen Seite zum Objekt hin und bei ganz schwachen Lupen, die einen Übergang zur Brille darstellen, zum Objekt hin durchgebogene Menisken. Aus der Brillenoptik ist bekannt, daß es optimale Durchbiegungen zur Korrektion des Astigmatismus gibt. Diese Überlegungen lassen sich auf Lupen übertragen.

Die Theorie fordert zur Beseitigung der Abbildungsfehler unsymmetrische einfache Linsen mit vorgeschriebener Durchblickrichtung. Bei falscher Blickrichtung werden die Abbildungsfehler dann aber sehr merklich, ohne daß der Benutzer sich über die falsche Benutzung seiner Lupe klar ist. Aus diesem Grunde verwendet man lieber symmetrische Linsen mit kleinerem Durchmesser und damit kleinerem Sehwinkel oder man spaltet die Linse auf und führt einen oder mehrere Kittradien ein, mit denen man den Astigmatismus so beeinflußt, daß die Bildqualität in beiden Durchblickrichtungen hinreichend gleich ist. Zwei solche Typen sind in Fig. 11 neben einer plankonvexen Linse dargestellt. Die dreilinsige Lupe ist völlig symmetrisch und unter dem Namen Steinheil-Lupe bekannt.

Bei Plankonvexlinsen liegt die eine Hauptebene im Scheitel der konvexen Fläche, das Objekt im zugehörigen Brennpunkt. Damit wird der Objektabstand also gleich der Brennweite. Er verkürzt sich mit der Akkomodation entsprechend der Newtonschen Abbildungsformel

$$z \cdot z' = -f^2 \tag{6.4}$$

um $z = -f^2/z'$, wenn z' die vom bildseitigen Brennpunkt aus gemessene Sehweite ist, auf die akkomodiert wird. Sie ist im Falle der Fig. 12 negativ.

Setzt man z' gleich der Nahpunktsentfernung des Auges, so ist der Wert von z gleich der Tiefe des Bereiches im Objektraum, die man durch Betätigung der Akkomodation von Unendlich bis zum Nahpunkt scharf einstellen kann.

Bei einer bikonvexen Linse liegen die Hauptpunkte im Inneren der Linse, und die Objektabstände werden kleiner.

Entsprechende Werte für plankonvexe und Steinheil-Lupen verschiedener Vergrößerungen sind in Tabelle 1 zusammengestellt. Sie werden ergänzt durch die Angabe der Entfernung B vom augenseitigen Linsenscheitel bis zum Objekt, wenn das Auge auf Unendlich eingestellt ist. Wenn man berücksichtigt, daß das Auge der Lupe möglichst genähert werden muß, um das Sehfeld zu vergrößern, sieht man leicht, daß die Verwendung von Lupen mit höherer als etwa 25facher Vergrößerung sehr unbequem sein muß.

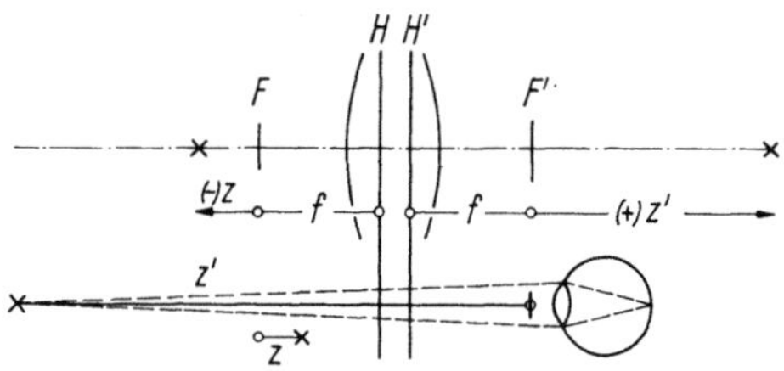

Fig. 12. Lupe. Objekt- und Bildlage zu Gl. (6.4).

Die Nachteile starker Lupen sind also: der freie Arbeitsabstand ist zumeist noch kleiner als die Entfernung des objektseitigen Brennpunktes vom vorderen Linsenscheitel; der Abstand zwischen Auge und Objekt ist sehr klein; wenn der Durchmesser der Lupe gleich dem Durchmesser der Augenpupille oder kleiner ist, ergibt sich Schlüssellochbeobachtung.

Tabelle 1. *Lupen: Schärfentiefe und Arbeitsabstände.*

	V / f(mm)	4× / 62,5	5× / 50	10× / 25	25× / 10	50× / 5	100× / 2,5
Schärfentiefe z_{250} bei Akkomodation von ∞ bis 250 mm	(mm)	15,6	10	2,5	0,40	0,10	0,025
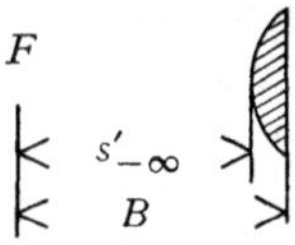	$s'_{-\infty} = f$ (mm)	62,5	50	25	10	5	2,5
	B (mm)	70	56	29	13	7	3,8
	$s'_{-\infty}$ (mm)	57	45	22	8,5	4	1,5
	B (mm)	80	65	32	12,5	6,5	4

Lesegläser. Ganz schwache Lupen mit geringerer Vergrößerung als $V = 4\times$ und Brennweite $f \geqq 62{,}5$ mm lassen sich mit großem Durchmesser herstellen.

Die Forderung, das Auge zur Vergrößerung des Sehfeldes nahe an die Linse heranzubringen, wird deshalb weniger wichtig. Man nennt derartige schwache Lupen, die für einen großen Augenabstand korrigiert sind, Lesegläser und benutzt sie z. B. zum Ablesen von Teilungen und zum Lesen kleiner Schrift. Im allgemeinen arbeitet man mit ihnen mit nahakkomodiertem Auge und entsprechend höherer Vergrößerung V_s als Lupenvergrößerung. Aus diesem Grunde sind auch Lesegläser mit z. B. 250 mm Brennweite und einer Lupenvergrößerung $V=1\times$ nicht sinnlos.

7. Zweistufiges Mikroskop: Grundsätzliches. Wegen der geschilderten Schwierigkeiten mit Lupen geht man für stärkere Vergrößerungen zu einer zweistufigen Abbildung über. Mit einem objektseitigen Linsensystem, dem Objektiv, entwirft man ein Bild des Objektes, das man durch ein augenseitiges zweites Linsensystem das Okular betrachtet.

Das Objektiv entwirft vom Objekt O ein vergrößertes reelles Bild O', das Zwischenbild genannt wird. Das Okular hat zwei Aufgaben: dieses Bild wie mit einer Lupe weiter zu vergrößern und mit dem Objektiv zusammen ein möglichst fehlerfreies Bild des Objektes im Unendlichen zu entwerfen. Gleichzeitig muß es die Objektivöffnung A in die Eintrittspupille des Auges, nach A', abbilden. Dann entsteht auf der Retina eines nach Unendlich blickenden Beobachters ein vergrößertes Bild O'' des Objektes. Wie Fig. 13 zeigt, ist dieses Bild seitenverkehrt und kopfstehend. Je nach der Entfernung des Zwischenbildes muß das Objekt im Brennpunkt des Objektives oder weiter von ihm entfernt liegen.

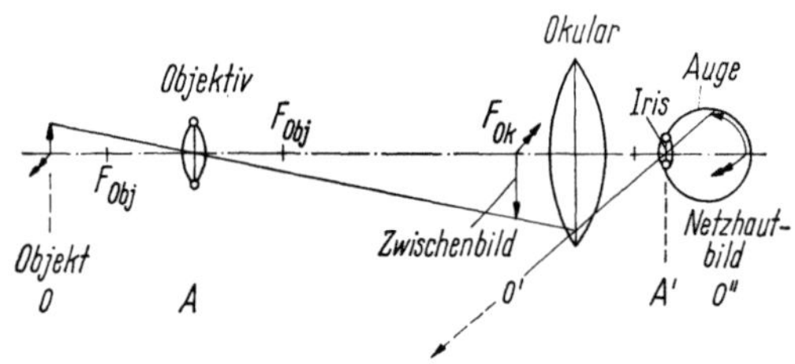

Fig. 13. Abbildung mit zweistufigem Mikroskop.

Die Gesamtvergrößerung eines Mikroskopes ist gleich dem Produkt aus dem Vergrößerungsmaßstab des vom Objektiv erzeugten Zwischenbildes und der Lupenvergrößerung des Okulars:

$$V_{\text{Mikroskop}} = M_{\text{Objektiv}} \times V_{\text{Okular}}\,. \tag{7.1}$$

Man beobachtet in der Mikroskopie das Objekt im allgemeinen von oben her. Die einfachste Möglichkeit, Objektive und Okulare gegeneinander zu fixieren, ist deshalb, die Objektive am unteren Ende eines Tubusrohres anzuschrauben und die Okulare am oberen Ende in das Rohr des Tubus einzustecken (Fig. 14). Die Entfernung vom Außenende des Anschraubgewindes zum anderen Ende des Tubusrohres wird „mechanische Tubuslänge" genannt. Sie wird von den Herstellern im allgemeinen angegeben, ihre Kenntnis ist aber von völlig untergeordneter Bedeutung, wie gleich gezeigt wird. Wesentlich ist nur, daß sie bei einem bestimmten Mikroskop nicht verändert wird und daß die Verlagerung des Bildes in die Zwischenbildebene durch Veränderung des Abstandes zwischen Objekt und Objektiv erfolgt.

Üblicherweise bringt man eine Anzahl von Objektiven verschiedener Vergrößerung auf einem Wechsler unter, der es erlaubt, durch eine Drehbewegung ein Objektiv gegen ein anderes auszutauschen. Durch Einführung eines solchen Wechslers, der meist aus einer Kugelkappe besteht und als Revolver bezeichnet wird, ergeben sich Einschränkungen für den Bau der Objektive, die an den Revolver geschraubt werden. Die Okulare in einem Revolver zu wechseln, ist nicht gebräuchlich.

Hat man für alle Objektive an einem Revolver die Entfernung vom Objekt zum Zwischenbild festgelegt, dann kann man ein Objektiv gegen das andere austauschen, ohne nachfokussieren zu müssen. Hält man zusätzlich die Entfernung des Zwischenbildes vom oberen Tubusrand konstant und baut man alle Okulare dementsprechend, dann ist auch ein Okularaustausch ohne Nachfokussierung möglich. Von wirklichem Interesse ist also die Kenntnis der Entfernung vom Objekt zum unteren Tubusrand, für die eine Anzahl von Objektiven „abgeglichen" ist, die Entfernung vom unteren Tubusrand zum Zwischenbild und vom Zwischenbild zum oberen Tubusrand, für die bestimmte Okulare untereinander „abgeglichen" sind.

Diese Längen sind bis jetzt nicht allgemein gültig festgelegt. Die in Fig. 14 in Klammern angegebenen Werte mögen einen Anhalt für das Erstrebenswerte und Erreichbare geben.

Betrachtet man ein Objektiv als dünne sammelnde Linse, dann läßt sich leicht angeben, welche Abbildungsmaßstäbe M bei den vorgeschlagenen Entfernungen möglich sind. Die geringste Vergrößerung mit einem als einfache dünne Linse gedachten Objektiv wird dann erreicht, wenn diese Linse in Höhe des unteren Tubusrandes sitzt. Dann ist das Verhältnis von Bildweite zu Objektweite und damit der Abbildungsmaßstab mit den Werten von Fig. 14 $M = 150\,\text{mm} : 45\,\text{mm} = 3{,}3 : 1$.

Objektive mit geringerer Vergrößerung können deshalb nur abgeglichen werden, wenn man sie als „Telesysteme" aufbaut. Objektive mit stärkerer Vergrößerung sind stets abgleichbar.

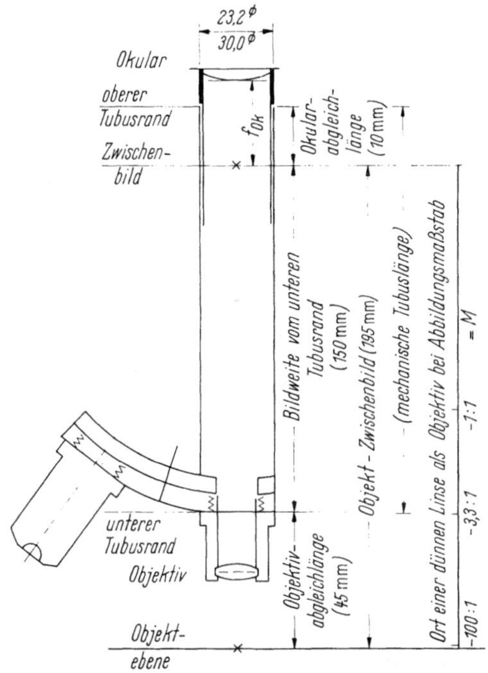

Fig. 14. Zweistufiges Mikroskop mit „abgeglichenen" Objektiven und Okularen.

Bei Objektiven, die nicht für endliche, sondern für unendliche Bildweite berechnet sind, bei denen das Objekt also im Brennpunkt liegt, wird das Zwischenbild mit einer Tubuslinse (Ziff. 14) in deren bildseitigem Brennpunkt erzeugt. Für beide Objektivarten sind die grundsätzlichen Überlegungen gleich; die Besonderheiten werden in Ziff. 14 besprochen.

8. Zweistufige Abbildung durch Objektive und Okulare. Wie in Ziff. 2 und 3 diskutiert, ist die Gesamtvergrößerung eines Mikroskops nach oben hin begrenzt. Man sollte die obere Grenze der förderlichen Vergrößerung auch für Meßzwecke nicht um mehr als 100% überschreiten. Bei einer maximalen Apertur von etwa 1,30 kommt man so im Extremfall in der Mikroskopie zu einer Gesamtvergrößerung von etwa 2500 ×. Die Untergrenze der Gesamtvergrößerung soll nach Möglichkeit im Bereich der Lupenvergrößerung liegen und den Wert 20 × noch unterschreiten. In besonderen Fällen werden Vergrößerungen bis 5 × hinunter angestrebt. Im allgemeinen wird von einem Mikroskop ein Vergrößerungsbereich zwischen etwa 20 × und 1000 × verlangt.

Diese Forderungen bedingen Serien von Objektiven und Okularen, die einzeln beliebig miteinander kombiniert werden können. Die Aperturen der Objektive müssen deren Vergrößerungen so angepaßt sein, daß bei Verwendung mittelstarker Okulare Gesamtvergrößerungen im Bereich der förderlichen Vergrößerung erreicht werden. Außerdem ist die Forderung nach möglichst weitgehender Einhaltung bestimmter Abgleichlängen zu beachten. Die Notwendigkeit, innerhalb großer Serien von Objektiven bzw. von Okularen die Einzelglieder austauschen zu können, ist ein ganz wesentliches Merkmal der mikroskopischen Optik.

Nach Ziff. 3 steckt in den auf die Apertur (und damit auf ein bestimmtes Objektiv) bezogenen Gesamtvergrößerungen von der Untergrenze mit 220 A bis zur oberen Grenze der förderlichen Vergrößerung mit 1000 A, die aber noch etwas überschritten werden kann, der Faktor 4,5. Es wird also sinnvoll sein, Okulare mit Vergrößerungen, die sich wie 1:4 bis 1:5 verhalten, zu verwenden. Aus den verschiedensten Gründen haben sich für Okulare in der Mikroskopie Vergrößerungen zwischen 5 × oder 6,3 × und 25 × eingebürgert. Wählt man von Okular zu Okular einen Vergrößerungsfaktor 1,25 × (z.B. DIN 58.886), so kommt man zu einer Reihe von acht Okularen mit den Vergrößerungen 5 ×, 6,3 ×, 8 ×, 10 ×, 12,5 ×, 16 ×, 20 ×, 25 ×. In den meisten Fällen wird eine Reihe mit dem Faktor 1,6 × und den Vergrößerungen 6,3 ×, 10 ×, 16 ×, 25 × ausreichen.

Tabelle 2.

M_{Obj}	A_{Obj}		dazu für $V = 1000\,A$ Okular	
	von	bis	von	bis
2,5 : 1 ×	0,05	0,08	20	32
4 : 1	0,10	0,16	25	40
6,3 : 1 ×	0,16	0,25	25	40
10 : 1	0,20	0,32	20	32
16 : 1 ×	0,28	0,50	20	32
25 : 1	0,45	0,65	20	25
40 : 1 ×	0,45	0,95	16	25
63 : 1	0,85	0,95	12,5	16
100 : 1 ×	0,95	1,36	10	12,5

× : Reihe mit Stufen 2,5:1.

Die Okulare mit der Aufgabe, die Vergrößerung in sinnvoller Abhängigkeit von der Apertur der Objektive zu variieren, überstreichen einen Vergrößerungsbereich von 1:5. Da die Gesamtvergrößerung der Mikroskope wie 1:500 variieren soll, sind Objektive notwendig, deren Vergrößerungen sich wie 1:100 unterscheiden und die oben gestellten Anforderungen an die für die Gesamtvergrößerungen notwendigen Aperturen erfüllen. Die maximale Objektivvergrößerung muß nach Festlegung der Okularvergrößerungen etwa $M = 100:1$ betragen, und es ergeben sich Reihen, die mit $M = 1:1$ usw. beginnen.

Auch bei den Objektiven ist eine geometrische Reihe (z.B. DIN 58.886) für die Vergrößerung nützlich. Man kann aber größere Stufensprünge wie etwa 2,5:1 (notfalls 1,6:1) einführen, um die Zahl der Objektive zu begrenzen. Die in der obenstehenden Tabelle genannten Objektive gehören solchen geometrischen Reihen an. Es darf aber nicht vergessen werden, daß die hier angestellten Überlegungen die Herstellung von Mikroskopen nicht von Anfang an gesteuert haben, und daß in der Mikroskopie manches nur aus der historischen Entwicklung heraus und manche andere Abweichung von der aufgestellten Regel nur vom Untersuchungsobjekt her zu verstehen ist.

Durch Benutzung nur von Vergrößerungswerten aus der genannten geometrischen Reihe (Normungszahlen nach ISO R3-1953 (E) oder DIN 323, Grundreihe R10) sind die Gesamtvergrößerungen wieder Zahlen dieser Reihe und man kann eine bestimmte Gesamtvergrößerung jeweils durch Kombination stärkerer Objektive und schwächerer Okulare wie schwächerer Objektive und stärkerer Okulare erreichen und damit verschieden nahe an die beiden Grenzen der förderlichen Vergrößerungen herangehen. Die in jeder Zehnerpotenz gültigen Werte der Reihe sind: 100, 125, 160, 200, 250, 320, 400, 500, 630, 800.

Die in Tabelle 2 angegebenen Objektive sind den Katalogen verschiedenster Firmen entnommen und sollen einen Eindruck vermitteln, in wie weit mit den üblichen Okularen der Bereich von minimaler bis oberer Grenze der förderlichen

Vergrößerung erreicht werden kann. Angegeben ist jeweils die Okularvergrößerung der geometrischen Auswahlreihe, mit der die Obergrenze der förderlichen Vergrößerung erreicht wird.

Kombiniert man ein beliebiges Objektiv mit einem beliebigen Okular (aus zusammenpassenden Reihen), dann muß die Abbildung so gut sein, daß gegenüber der durch die Beugung verursachten Unschärfe keine Abbildungsfehler zu erkennen sind. Dies bedeutet Einhaltung der Rayleigh-Grenze bzw. der Staebleschen Definitionshelligkeit[1]. Diese Forderung gilt mindestens für die Bildmitte, denn es ist ja möglich, ein Objekt so unter dem Objektiv zu verschieben, daß die interessierenden Teile des Objektes im Bereich der fehlerfreien Abbildung liegen.

9. Korrektionsmöglichkeiten und Aufbau von Objektiven. Die Felder der von den Objektiven entworfenen Bilder werden durch Blenden im Zwischenbild begrenzt.

In Abhängigkeit vom Okularaußendurchmesser hat man mit Sehfeldern von höchstens 18 bis 20 mm Durchmesser und solchen von höchstens 28 mm Durchmesser zu rechnen. Nimmt man die Austrittspupille der Objektive im Mittel 30 mm unter dem unteren Tubusrand an, dann bedeuten 18 bzw. 28 mm Sehfelddurchmesser im Okular Feldwinkel von $2\times 2{,}8°$ bzw. $2\times 4{,}5°$ für die Objektive. Die hier angenommene Lage der Austrittspupille ist ein Mittelwert, denn man legt die Aperturblende in die bildseitige Brennebene der Objektive und weicht hiervon nur bei ganz schwachen Objektiven ab.

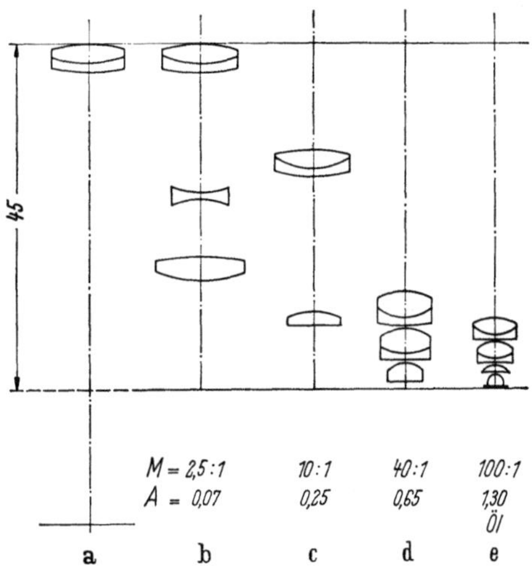

Fig 15. a—e. Achromatische Mikroskopobjektive.

Den kleinen Feldwinkeln der Objektive stehen Aperturwinkel bis zu $2\times 72°$ ($A=0{,}95$) gegenüber. Bei der Korrektion werden deshalb Öffnungsfehler und Koma nicht mehr mit Hilfe der Koeffizienten dritter Ordnung zu übersehen sein, wohl aber Bildfeldwölbung, Astigmatismus und Verzeichnung.

Zur Korrektion des sphärischen Fehlers ist es bei $A=0{,}12$ bis $0{,}15$ möglich, mit einem sammelnden Gliede in Luft auszukommen (Fig. 15a). Zu diesen niedrigen Aperturen gehören allerdings Objektive mit Vergrößerungen unter $M=4:1$, die in dieser einfachen Form nicht am Revolver abgeglichen werden können. Zwei sammelnde Glieder in Luft (Fig. 15c) ermöglichen eine Korrektion des Öffnungsfehlers bis etwa $A=0{,}25$. Eine weitere Erhöhung der Apertur ist nur dann möglich, wenn man objektseitig eine, im allgemeinen plankonvexe, Linse vorschaltet (Fig. 15d), deren bildseitige Fläche aplanatisch ist. Von der Aplanasie wird man zweckmäßigerweise so weit abweichen, wie es zum Ausgleich der restlichen Fehler der anderen Flächen notwendig ist. Eine solche Frontlinse erlaubt eine auskorrigierbare Erhöhung der Apertur um etwa den Faktor n^2, wobei n die Brechzahl des Frontlinsenglases ist. Bei Aperturen über ungefähr $A=0{,}45$ ist zu berücksichtigen, daß jedes optische Glied zwischen Objekt und Frontlinse, also z. B. das Deckglas, einen Einfluß auf den Öffnungsfehler hat (Überkorrektion) und bei der Korrektion zu berücksichtigen ist. Bei Aperturen über 0,75 ändern schon Abweichungen von einigen $^1/_{100}$ mm in der Deckglasdicke den Öffnungsfehler in unzu-

[1] Vgl. Ziff. 2 und den Beitrag von K. Bahner, S. 244 in diesem Band.

lässiger Weise. Um dem Mikroskopiker eine Möglichkeit zu geben, diese Fehler selbst zu korrigieren, baut man Objektive, bei denen der Abstand zwischen Linsengruppen verstellt werden kann (Fig. 16). Eine Verkleinerung des Linsenabstandes wirkt der durch ein zu dickes Deckglas eingeführten Überkorrektion entgegen, weil die Strahlen die stark unterkorrigierenden objektseitigen Flächen der Frontlinse dann in größerem Abstand von der Achse durchsetzen.

Erhöht man die Apertur der Objektive dadurch (Ziff. 2), daß man Objektiv und Präparat mit einer Flüssigkeit immergiert (Fig. 5b und 15e), so wird die Brechkraft der objektseitigen Fläche der Frontlinse entsprechend geringer. Unter einer „homogenen Immersion" versteht man ein Immersionsobjektiv, dessen Frontlinse möglichst genau denselben Brechungsindex besitzt wie die Immersionsflüssigkeit. Es ist aber nicht einzusehen, weshalb man auf die immergierten Flächen als Korrektionsmittel verzichten soll. Man wird deshalb im allgemeinen ebenso von der „Homogenität" der Immersion abweichen wie von der Aplanasie der bildseitigen Fläche.

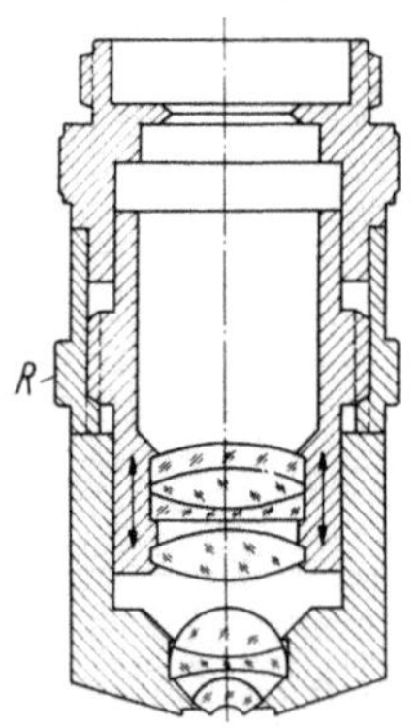

Fig. 16.

Fig. 17.

Fig. 16. Objektiv mit Korrektionsfassung. Durch Drehen am Rändelring *R* wird der Abstand der oberen Linsenglieder gegenüber den unteren geändert und damit die sphärische Korrektion beeinflußt.

Fig. 17. Monochromat (aus Quarzglas).

Immersionsmittel über $n = 1{,}515$ sind im allgemeinen nicht sinnvoll, weil die zu beobachtenden Objekte mit Medien abgedeckt werden, deren Brechzahlen nicht über 1,515 liegen und damit die Grenzapertur bestimmen. In Sonderfällen brauchen Objektive nur für eine Farbe korrigiert zu sein. Solche Fälle wären Fluoreszenzstrahlung einer Wellenlänge oder die Strahlung bestimmter Laser. Diese Objektive werden als *Monochromate* bezeichnet (Fig. 17). Im allgemeinen wird subjektiv und mit weißem Licht gearbeitet. Dann ist für das Maximum der Augenempfindlichkeit $\lambda = 550$ nm (bzw. für die Quecksilberlinie e ($\lambda = 546{,}1$ nm) optimal zu korrigieren und der Farbunterschied des Öffnungsfehlers, auch Gaußfehler genannt, möglichst weitgehend zu beseitigen[1]. Der Grad der Korrektion hängt davon ab, ob nur Materialien zur Verfügung stehen, deren relative Teildispersionen dem allgemeinen Gang der Gläser folgen oder ob davon abweichende Materialien wie verschiedene Fluoride und Alaune, die kubisch kristallisiert und optisch isotrop sind, und in den letzten Jahren auch einige Gläser zur Korrektion benutzt werden können. Man erhält im ersten Fall, also mit Normalgläsern, *Achromate* (Fig. 15), bei denen die Schnittweiten für Rot und Violett größer sind als für eine mittlere Wellenlänge (Fig. 18b). Den Restfehler nannte E. ABBE das sekundäre Spektrum und verminderte ihn durch Einführung von Linsen aus Fluorit (CaF_2). Derartige Objektive (Fig. 19) nannte er *Apochromate* (E. ABBE, 1886, 1890; M. VON ROHR, 1936). Ihr Charakteristikum ist ein Wendepunkt in der Kurve, die die Schnittweitenfehler in Abhängigkeit von der Wellenlänge zeigt (Fig. 18c). Eine Stellung zwischen Apochromaten und Achromaten nehmen die sogenannten *Fluoritobjek-*

[1] Bei Neumessung von optischen Daten sollten die leider noch häufig verwendete Mitte der Natrium-Doppellinie D ($\lambda = 589{,}3$ nm) und die Heliumlinie d ($\lambda = 587{,}6$ nm) zu Gunsten der Quecksilberlinie e ($\lambda = 546{,}1$ nm) verlassen werden (WEIDERT, 1926; FRANKE, 1951).

tive ein, bei denen das sekundäre Spektrum nicht ganz so gut korrigiert ist wie bei Apochromaten. Objektive, deren entsprechende Kurven zwei Wendepunkte zeigen, werden als *Superachromate* bezeichnet.

Eine chromatische Variation des paraxialen Abbildungsmaßstabes tritt dann auf, wenn die ν-Werte der Gläser im System nicht den Verhältnissen der Durch-

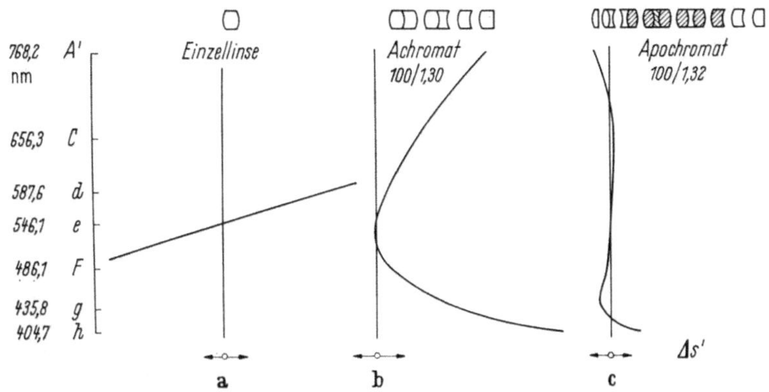

Fig. 18a–c. Farbkorrektion a einer Einzellinse, b eines Achromaten, c eines Apochromaten.

stoßhöhen der Strahlen angepaßt sind. Sie ist bei einfachen Systemen mit kleiner Frontlinse, die man nicht verkitten kann, unmöglich. So zeigen einfach aufgebaute starke Objektive im allgemeinen eine chromatische Vergrößerungsdifferenz, die für eine Ölimmersion von 2 mm Brennweite bis auf 1,3% steigen kann, wenn man die Vergrößerungen, wie jetzt üblich, für die Cadmiumlinien C' ($\lambda = 643{,}8$ nm und F' ($\lambda = 480{,}0$ nm) bestimmt.

Auch beim Auskorrigieren von Apochromaten treten diese Schwierigkeiten auf (z.B. H. Boegehold, 1958, S. 87). Ganz allgemein ist die Vergrößerung für kurzwelliges Licht stärker als für langwelliges. Man sieht also bei weißem Licht auf einem exzentrisch im Bildfeld liegenden undurchsichtigen Detail zur Bildmitte hin einen blauen und auf der entgegengesetzten Seite einen roten Saum.

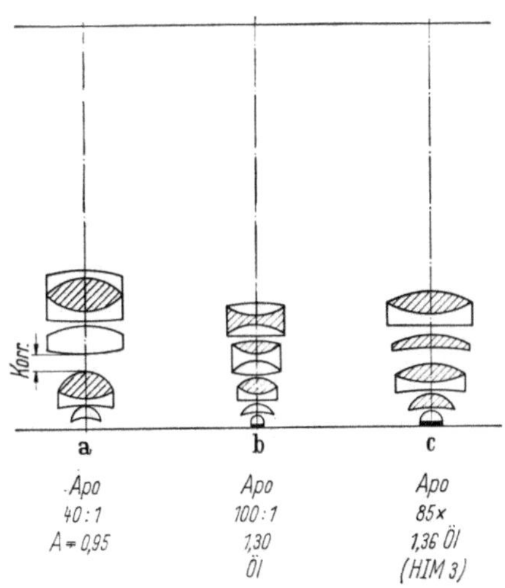

Fig. 19a–c. Apochromate. Die schraffiert gezeichneten Linsen bestehen aus Materialien mit Teildispersionen, die von denen der üblichen Gläser abweichen.

Objektive, die nur aus verkitteten Linsen bestehen und solche, bei denen die Linsen auch möglichst weit bildseitig angebracht sind, wie z. B. Objektive mit geebnetem Bildfeld, lassen sich auf eine verschwindend kleine chromatische Vergrößerungsdifferenz korrigieren. Dieser Gruppe steht dann aber eine andere Gruppe von viel benutzten Objektiven mit chromatischen Vergrößerungsdifferenzen bis zu etwa 1,3% gegenüber. Will man für alle Objektive einen einheitlichen Fehler erreichen, um ihn an anderer Stelle einheitlich zu korrigieren, wird man einen Wert zwischen 1,0 und 1,3% festlegen und die ν-Werte der Gläser in den Objektiven entsprechend verteilen.

Wegen der Größe der Aperturen von Mikroskopobjektiven ist es notwendig, erhebliche Sorgfalt auf die Korrektion der Asymmetriefehler zu verwenden. Die Beachtung der Isoplananasiebedingung (LIHOTZKY, 1919; STAEBLE, 1919) bei Entwurf eines Objektives läßt relativ leicht übersehen, ob für eine bestimmte Pupillenlage und für die geringen Feldwinkel bei Mikroskopen die Asymmetriefehler bei beliebig hohen Aperturen beseitigt sind. Der Öffnungsfehler ist in diesen Bündeln bis auf kleine Größen höherer Ordnung dem Öffnungsfehler auf der optischen Achse gleich.

Astigmatismus und Bildfeldwölbung sind wegen der Kleinheit der Feldwinkel schon bei Anwendung der Seidelschen Koeffizienten dritter Ordnung gut zu beherrschen.

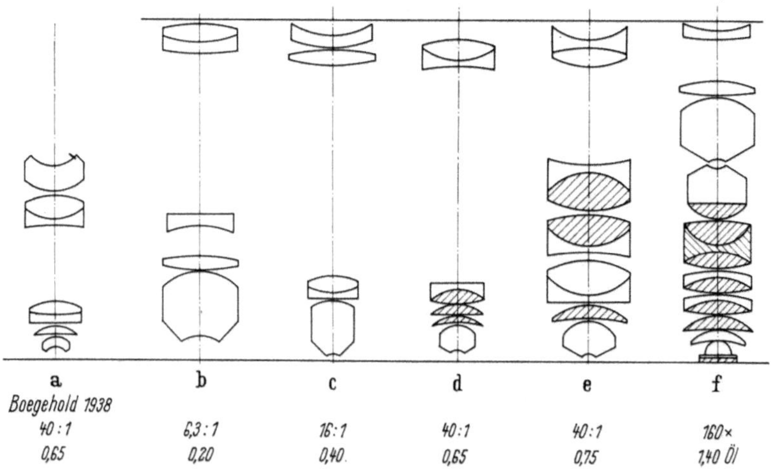

Fig. 20a—f. Planobjektive verschiedener Aperturen. Schraffiert gezeichnete Linsen wie bei Fig. 19.

Die Petzvalsumme für eine dünne Linse in Luft mit der Brechzahl n und den Radien r_1 und r_2 ergibt sich mit

$$\frac{1}{f} = (n-1)\left(\frac{1}{r_1} - \frac{1}{r_2}\right)$$

aus[1]

$$P_\nu = -\frac{1}{r_\nu}\left(\frac{1}{n'_\nu} - \frac{1}{n_\nu}\right) \quad \text{für eine Einzelfläche} \tag{9.1}$$

zu

$$\left.\begin{aligned} \Sigma P &= -\frac{1}{r_1}\left(\frac{1}{n} - 1\right) - \frac{1}{r_2}\left(1 - \frac{1}{n}\right) \\ &= \frac{1}{n \cdot f}. \end{aligned}\right\} \tag{9.2}$$

Dabei ist $1/\Sigma P$ der Scheitelradius der parabelähnlichen Bildschale, wenn das System frei von Astigmatismus ist.

Mikroskopobjektive bestehen aus mehreren dicken sammelnden Linsen, und man bekommt für solche Objektive im allgemeinen $\Sigma P = 1/f$ und eine Bildschale, deren Scheitelradius gleich der Brennweite des Objektivs ist. Ein Objektiv mit $M = 100:1$ hat etwa 2 mm Brennweite. So ist es klar, daß die von ihm erzeugten Bilder sehr stark gewölbt sind und man nur auf die Bildmitte oder eine Ringzone fokussieren kann (Fig. 24).

[1] M. BEREK. 1930, S. 62.

Nach einigen Versuchen zur Anwendung negativer Flächen und Linsen, die aber nur geringe Wirkung hatten, machte BOEGEHOLD (1938) den Vorschlag, dicke Menisken mit hoher negativer Petzvalsumme, aber positiver Brechkraft zur Ebnung der Bildfelder einzuführen. Er nannte die ersten derartigen Objektive für 18 mm Okularsehfeld *Planachromate*. Die Verwendung von „dicken Menisken" ist sehr ergiebig. Daher gab BOEGEHOLD mit seiner Veröffentlichung den Anstoß zu vielen weiteren Arbeiten, die noch im Fluß sind und bis zur Bildebnung von ganz kurzbrennweitigen Objektiven bis zu 28 mm okularem Sehfeld geführt haben (BLASCHKE, 1954; BOEGEHOLD, 1950, 1952; BOEGEHOLD und BERNHARDT, 1955; CLAUSSEN, 1956, 1964; OETTEL, 1951; WYNNE, 1961). Fig. 20 gibt eine Zusammenstellung einiger weniger typischer Objektivformen.

10. Korrektionsmöglichkeiten und Aufbau von Okularen. Mit den Okularen soll man das Zwischenbild wie mit einer Lupe betrachten können. Dabei besteht im Gegensatz zur Lupe hier die Möglichkeit, das Sehfeld scharf zu begrenzen, dadurch daß man ein bestimmtes Stück aus dem Zwischenbild durch eine Blende herausblendet. Mit der Vergrößerung des Okulars ist — wie bei der Lupe — seine Brennweite festgelegt. Weiter muß das Okular die Austrittspupille des Objektivs, die für das Okular Eintrittspupille ist, an einem Ort abbilden, an den die Eintrittspupille des Auges mühelos gebracht werden kann. Und zum dritten müssen die Linsen des Okulars in vernünftiger Entfernung vom Zwischenbild und damit vom Tubusrand liegen. Diese Forderungen können erfüllt werden, wenn man das Okular mit der Brennweite f aus mindestens zwei Linsen mit den Brennweiten f_1 und f_2 aufbaut und als dritte Variable den Hauptpunktsabstand e dieser beiden Linsen wählt (Fig. 21). Wenn man eine der Linsen ins Zwischenbild setzt und die andere im Abstand ihrer Brennweite darüber (Fig. 22a), bewirkt letztere, die man Augen-

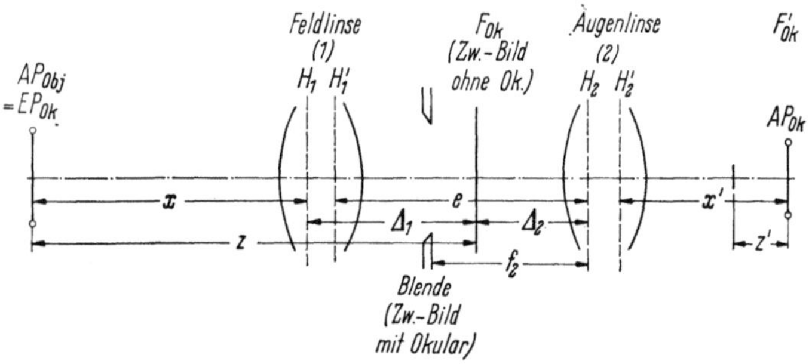

Fig. 21. Zweilinsiges Okular. Skizze zu Gl. (10.1).

linse nennt, die Vergrößerung, während die Linse im Zwischenbild zur Pupillenabbildung dient. Sie wird, weil sie alle im Feld ankommenden Strahlenbündel zum Auge hin sammelt, *Feldlinse* oder *Collectivlinse* genannt. Im allgemeinen vermeidet man es, eine Linse genau ins Zwischenbild zu setzen, weil dort die Feldblende angebracht werden muß. Eine Feldlinse unterhalb des Zwischenbildes ändert dessen Lage und Durchmesser. Untereinander abgeglichene Okulare (Fig. 22, s. auch Ziff. 7) müssen also keinesfalls Blenden im gleichen Abstand vom Tubusrand besitzen. Eine Feldlinse mit positiver Brechkraft verlagert das Zwischenbild nach unten und verkleinert es (Fig. 22b), während eine negative Feldlinse das Zwischenbild zum Tubusrand hin verlagert und vergrößert (Fig. 22d). Für jedes Okular läßt sich ein Feldlinsen-Faktor c angeben, der aussagt, wie stark die Wirkung der Feldlinse auf die Vergrößerung im Zwischenbild ist. Das ist für Meßzwecke interessant, denn man kann im Zwischenbild Maßstäbe, sogenannte Mikrometerplättchen, anbringen und damit Längen im Zwischenbild messen (vgl. Ziff. 12).

Der Zusammenhang zwischen der Brennweite f_1, f_2 der Einzellinse und dem Abstand e bei einem zweilinsigen Okular der Brennweite $f=250/V$ ergibt sich mit den Bezeichnungen der Fig. 21:

$$\left.\begin{aligned} e&=f+\Delta_1\left(\frac{f}{z}+\frac{x'}{f}\right);\\ f_2&=\frac{ef}{f+\Delta_1};\\ f_1&=f\cdot\frac{f_2-e}{f_2-f}.\end{aligned}\right\}\qquad(10.1)$$

Dabei wird für z (und x) eine mittlere Pupillenlage des Objektivs angenommen.

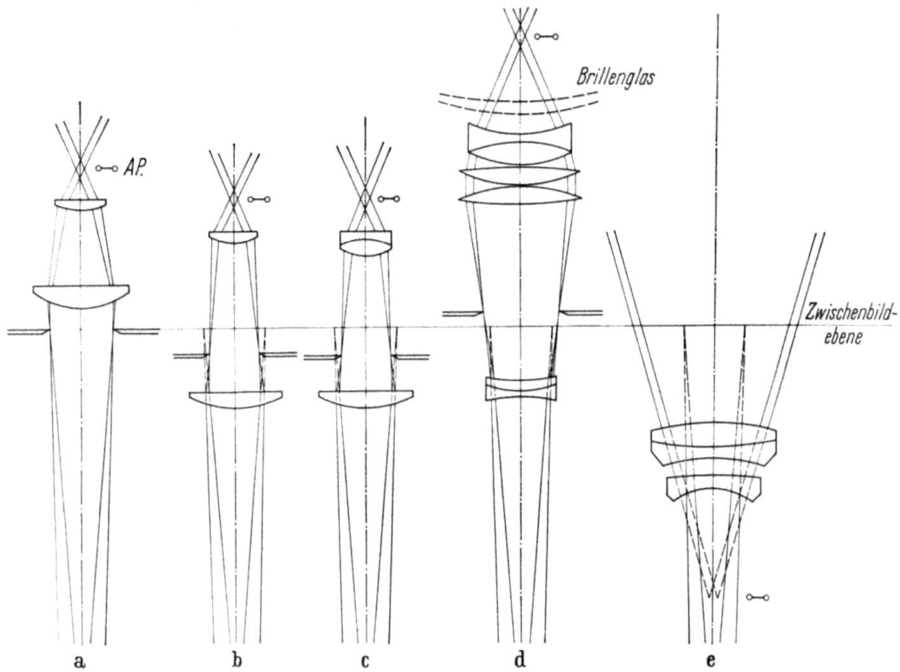

Fig. 22a—e. Untereinander abgeglichene Okulare. Die vom Objektiv erzeugten Bilder liegen in derselben Ebene, der Zwischenbildebene. a—c Okulare mit normalem Pupillenabstand, d Brillenträgerokular, e Negativokular.

Es ist selbstverständlich auch möglich, statt eines positiven Linsensystems ein negatives (Fig. 22e) als Okular zu benutzen. So, wie das als dünne Linse gedachte positive Okular um die Entfernung f augenseitig des Zwischenbildes sitzt, muß das negative objektseitig sitzen. Unangenehm ist, daß sich bei Negativ-Okularen die Austrittspupille unterhalb der Okulare befindet und damit für das Auge nicht zugänglich ist. Derartige Okulare scheiden deshalb für subjektive Beobachtung aus, sie sind „Projektive", und erfordern wegen ihrer Lage unter dem Zwischenbild die Verwendung besonderer Tuben.

Die objektseitigen Aperturen der Okulare sind gleich den bildseitigen Aperturen der Objektive. Der Zusammenhang zwischen objekt- und bildseitiger Apertur der Objektive ist gegeben durch (2.7).

Damit sieht man leicht aus Tabelle 2 (Ziff. 8), daß die Aperturen, für welche die Okulare korrigiert werden müssen, hauptsächlich zwischen $A=0{,}01$ und $0{,}03$ liegen. Der Öffnungsfehler und die chromatische Variation des Bildortes sind daher beim Korrigieren der Okulare von geringer Bedeutung. Diese Aussage gilt aber natürlich nur für Mikroskop-Okulare und macht es verständlich, daß diese Okulare nur dort verwendet werden können, wo die objektseitigen Aperturen entsprechend klein sind.

Die Durchmesser der Austrittspupille sind mit den Bezeichnungen der Fig. 23

$$\varnothing \; \mathrm{AP}_{\mathrm{Ok}} = 2h = 2f_{\mathrm{Ok}} \cdot \sin u'_{\mathrm{Obj}} = 2f_{\mathrm{Ok}} \cdot A'_{\mathrm{Obj}}. \tag{10.2}$$

Nimmt man die oben angegebenen Grenzwerte für die Apertur, dann liegen die Durchmesser der Austrittspupille für ein 5faches Okular zwischen 1 und 3 mm und für ein 25faches Okular zwischen 0,2 und 0,6 mm. Bei Beleuchtungsaperturen, die kleiner sind als die Objektivaperturen, erscheinen die Pupillen noch entsprechend kleiner. Die Kleinheit der Bündelquerschnitte im Okular und in der Austrittspupille stört bei der Beobachtung, weil schon kleinste Partikel auf den Linsenflächen oder im Auge deutlich sichtbare Schatten im Netzhautbild erzeugen. Im Kammerwasser des Auges schwimmende Teilchen können beim Mikroskopieren ganz erheblich stören.

Die üblichen Okulare für subjektive Beobachtung haben einen Außendurchmesser von 23,2 mm. Mit ihnen kann man, wenn man eine sammelnde Feldlinse sehr tief unter das Zwischenbild setzt, eine Fläche von höchstens 22 mm ⌀ aus dem Zwischenbild ausblenden und beobachten. Bei normaler Baulänge der Okulare werden Flächen mit 18 bis 20 mm ⌀ ausgeblendet. Diese Fläche nennt man das *okulare Sehfeld* und den Zahlenwert des in mm gemessenen Durchmessers die *Sehfeldzahl.* Okulare von 30 mm Außendurchmesser haben Sehfelder bis 28 mm Durchmesser.

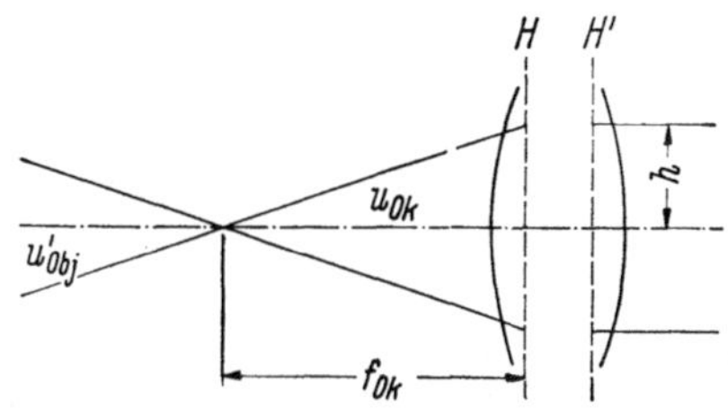

Fig. 23. Zu Gl. (10.2): Berechnung des Durchmessers einer Okular-Austrittspupille.

Mit blickendem Auge zu beobachten ist wegen der Kleinheit der Austrittspupillen beim Mikroskop nicht sinnvoll. Man muß den Kopf dauernd bewegen, um durch die Mikroskoppupille in die verschiedenen Richtungen zu blicken und ermüdet dadurch rasch. Die großen Bildwinkel der Fernrohrokulare sind aus diesem Grunde in der Mikroskopie nicht erstrebenswert.

Ohne die Augen zu drehen, kann man mühelos ein Feld von $2w_{\mathrm{Ok}} = 30°$ und als Grenzwert, der stark vom Beobachter abhängt, etwa 45° bis 55° übersehen. Es hat also wenig Sinn, Okulare zu bauen, mit denen diese Grenzen nach unten oder oben überschritten werden.

Mit

$$\tan w'_{\mathrm{Ok}} = \frac{\text{Sehfelddurchmesser}}{2\,f_{\mathrm{Ok}}} \tag{10.3}$$

ergeben sich für die maximalen Sehfelder von 18 bzw. 28 mm ⌀ schon bei 6,3fachen Okularen bildseitig Sehwinkel von 25° bis 39°. Schwächere Okulare haben unter diesem Gesichtspunkt also keinen Sinn. Mit einem 14,5fachen Okular würde bei 18 mm Sehfeld bereits ein Sehwinkel von 55° erreicht. Stärkere Okulare mit 30 mm Außendurchmesser sind deshalb optisch nicht notwendig.

Felder mit Sehwinkeln von etwa 12° bis 27,5° gegen die Achse sind für Okulare nicht sehr groß und in Verbindung mit den kleinen Aperturen der Okulare nicht schwer zu korrigieren. Am wichtigsten sind der Astigmatismus in Verbindung mit Bildfeldwölbung, die chromatische Vergrößerungsdifferenz und die Verzeichnung.

Bei den bekanntesten einfachen Okularen nach dem Huygens- oder dem Ramsden-Typ lassen sich Astigmatismus und chromatische Vergrößerungsdifferenz auskorrigieren, auch wenn man beide Linsen aus demselben Glas herstellt. Die chromatische Vergrößerungsdifferenz für $F'-C'$ läßt sich durch Verwendung von Flintglas in der Feldlinse auf 0,5 bis 0,6% bringen, und zwar ist, wenn man von einem fehlerfreien Bild auf der Augenseite ausgeht, im Zwischenbild die Vergrößerung für Blau größer als für Rot. Ein derartiges Okular kompensiert also die chromatische Vergrößerungsdifferenz der Objektive.

Durch Verkittung (Fig. 22c) und eventuelle Aufspaltung der Augenlinse (Fig. 22d) kann man Astigmatismus und chromatische Vergrößerungsdifferenz in weiten Bereichen beeinflussen. Am schwierigsten ist die Verzeichnung zu beherrschen. Da sie aber fast nur von dem Augenlinsenglied verursacht wird und Objektiv und Feldlinse des Okulars praktisch keine Verzeichnung bringen, werden Messungen mit Hilfe einer Teilung im Zwischenbild nicht gestört. Die durch die Petzvalsumme bestimmte Krümmung des Bildfeldes sollte nach Formel (9.2) proportional mit der Vergrößerung der Okulare zunehmen. Das ist nicht der Fall, und man weicht von dem zu erwartenden Werte ab, weil man die Brechkräfte zwischen Feldlinse und Augenlinse bei den Okularen verschiedener Vergrößerung unterschiedlich aufteilt.

11. Beispiel für die korrektionsmäßige Anpassung von Objektivserien und Okularserien aneinander. Jetzt ist die Frage zu klären, ob es möglich ist, für alle Objektive und Okulare einen bestimmten Korrektionszustand im Zwischenbild festzulegen, der von beiden Seiten in möglichst vielen Fällen zu erreichen ist und der dann eine fehlerfreie Abbildung in möglichst großen Feldbereichen garantiert.

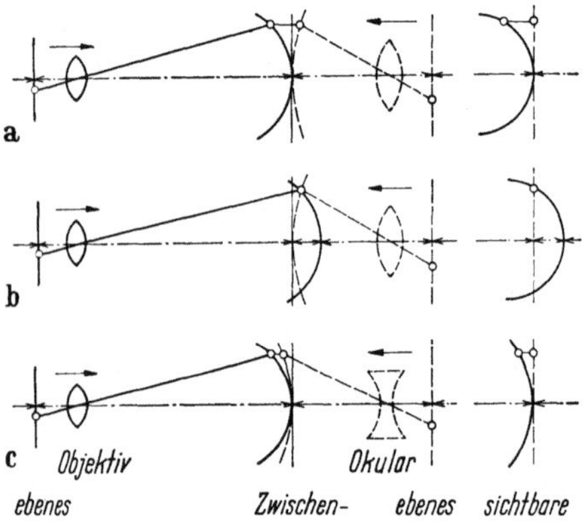

Fig. 24a—c. Die zwei Bildschalen im Zwischenbild eines Mikroskops liegen, bei Abwesenheit von Astigmatismus, dort, wo ein ebenes Objekt vom Objektiv und ein ebenes Bild durch das Okular abgebildet wird. Der Abstand korrespondierender Punkte auf dem vom Okular zu dem vom Objektiv entworfenen Zwischenbild gibt nach Größe und Richtung die Bildfeld-Wölbung, wie sie vom Beobachter gesehen wird. a und b Positives Okular: a Bild im Zentrum fokussiert. b In einer Randzone fokussiert. c Negativ-Okular im Zentrum fokussiert. Die Bilder erscheinen in den Gebieten scharf, in denen die Tiefenschärfe größer ist als der Abstand der beiden Bildschalen.

Der erste Schritt wäre die Einführung von Objektiven und Okularen, die für die gleiche chromatische Vergrößerungsdifferenz korrigiert sind. Diese muß, wie in Ziff. 9 besprochen wurde, zwischen 1% und 1,3% liegen. Wählt man den hohen Wert, dann ist der Unterschied gegenüber einfachen zweilinsigen Okularen sehr hoch und vermindert sich bei Verwendung von Flintglas in der Feldlinse auf etwa die Hälfte. Wählt man mit 1% den kleineren Wert, dann wird der Restfehler, besonders bei Verwendung von Flintglas in der Feldlinse, durchaus erträglich.

Zur Aufhebung der Bildfeldwölbung der Okulare, die für subjektive Beobachtung in Frage kommen, gibt es kein technisch vernünftiges Mittel. Betrachtet man die Verhältnisse zunächst in der Annahme, daß Objektive und Okulare frei von Astigmatismus seien, und rechnet bei den Objektiven von einem ebenen Objekt und bei den Okularen von einem ebenen Bild aus zum Zwischenbild hin, dann ergeben sich die in Fig. 24a und b dargestellten Verhältnisse. Man kann, da die Bildschalen von Objektiv und Okular der positiven Petzvalsumme sammelnder Linsensysteme wegen zu den Systemen hin gekrümmt sind, die Bildschalen nur mit ihren Scheiteln (Fig. 24a) oder, wenn man das Objekt näher an das Objektiv heranbringt, in einer Ringzone (Fig. 24b) aufeinander legen. Der

Beobachter wird beim Fokussieren entsprechend dem Abstand der Bildschale des Objektivs von der des Okulars nachstellen müssen, er bemerkt die rechts im Bilde dargestellte Bildwölbung.

Der durch die Tiefenschärfe bestimmte Bereich um die Berührungslinie der beiden Bildschalen herum wird dann scharf erscheinen.

Fig. 24c zeigt, daß bei Verwendung eines Negativokulars, dessen Petzvalschale entsprechend der negativen ΣP vom Okular weg gekrümmt ist, sich die sichtbare Wölbung des Bildes dadurch verkleinert. Eine volle Ebnung ist auf diesem Wege nur zu erreichen, wenn die Brennweiten von Objektiv und Okular etwa gleich sind, und das ist nur sehr selten der Fall. Trotz des Vergrößerungs- und damit auch Brennweitenverhältnisses der Okulare wie 1:4 oder 1:5 kann man durch die schon erwähnte unterschiedliche Verteilung der Brechkräfte zwischen Feld- und Augenlinse bei den Okularen erreichen, daß für fast alle Okulare die Werte für ΣP zwischen $+0{,}02$ und $+0{,}04$ liegen und man mit dem Mittelwert $\Sigma P = +0{,}03 \pm 0{,}01$ rechnen kann. Die astigmatismusfreie Bildschale ist damit zum Auge hin geöffnet. Um volle Ebnung zu erreichen, muß die Petzvalschale der Objektive entsprechend durchgebogen werden und der entgegengesetzten Richtung entsprechend den Wert $\Sigma P = -0{,}03$ bekommen.

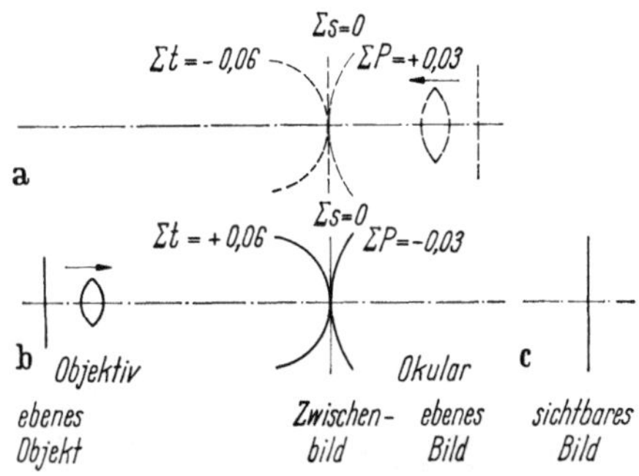

Fig. 25a—c. Bildebnung unter Berücksichtigung von ΣP und $\Sigma \Gamma$. Lage der tangentialen und sagittalen Bildschale sowie der Petzvalschale a des Okulars, b des Objektivs. Der Abstand entsprechender Schalen ist Null, so resultiert c ein ebenes Bild.

Dann hat man sich für den Astigmatismus einen geeigneten Wert auszusuchen, der von der Objektiv- und Okularseite her gut zu erreichen ist.

Führt man für die Okulare einen Astigmatismus $\Sigma \Gamma = -0{,}03$ ein, dann erhält man zwei Bildschalen, von denen die sagittale wegen $\Sigma s = \Sigma P + \Sigma \Gamma$ geebnet und die tangentiale wegen $\Sigma t = \Sigma P + 3\,\Sigma \Gamma$ zum Objektiv hin geöffnet ist (Fig. 25a).

Um die Bildschalen der Objektive genau auf die entsprechenden Bildschalen der Okulare legen zu können, müssen die Objektive unter Berücksichtigung der entgegengesetzten Richtung auf $\Sigma P = -0{,}03$ und $\Sigma \Gamma = +0{,}03$ korrigiert werden (Fig. 25b). Die einander entsprechenden Bildschalen fallen dann aufeinander und das vom Objektiv und Okular gemeinsam entworfene Bild ist eben und frei von Astigmatismus (Fig. 25c).

Man hat bei dem hier vorgeschlagenen Fehler des Zwischenbildes etwas mehr Mittel zur Beeinflussung der ΣP des Objektives aufzuwenden als zur Ebnung der Petzvalschale notwendig wäre. Bedenkt man aber, daß bei einem Objektiv von 2 mm Brennweite eine Petzvalsumme von $\Sigma P = +0{,}5$ zu erwarten ist, dann ist der durch diese Festlegung notwendige Mehraufwand von 6% zur Anpassung an die Okulare sehr gering.

Die Einführung einer negativen $\Sigma \Gamma$ auf der Okularseite bewährt sich, weil dadurch der Abstand einander entsprechender Bildschalen auch bei Benutzung kurzbrennweitiger Objektive mit nicht geebnetem Bildfeld deutlich verkleinert wird und damit eine Bildfeldebnung eintritt. Fig. 26 zeigt in derselben Art der Darstellung, wie in den vorhergehenden Abbildungen die Wirkung einer geeigneten Objektiv-Okular-Paarung am Beispiel starker Ölimmersionen. Der Maßstab für die Darstellung im Zwischenbild ist in der Achse und im Feld der gleiche, der Felddurchmesser ist 28 mm. Wenn man den Bereich der Schärfentiefe im Zwischenbild aus dem entsprechenden Bereich im Objekt berechnet, darf man nicht vergessen, daß die Tiefenvergrößerung das Quadrat der lateralen Vergrößerung ist. Der Kombination einer normalen Ölimmersion mit einem positiven Okular (Fig. 26a) ist wieder die mit einem Negativokular (Fig. 26b) gegenüber-

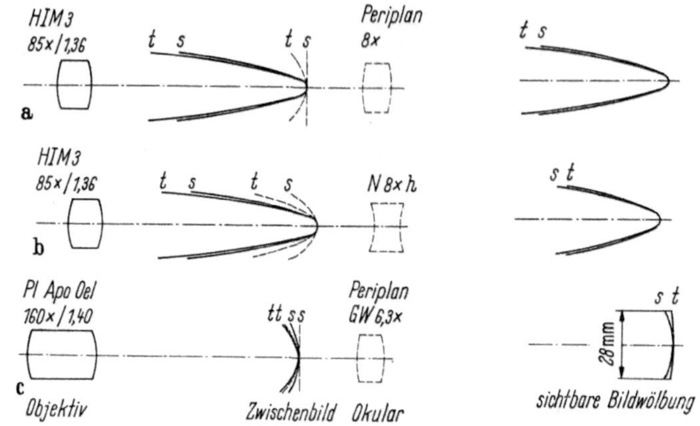

Fig. 26a—c. Korrektion der Bildfeldwölbung und des Astigmatismus nach Fig. 25 mit den Objektiven und Okularen, die für die Aufnahmen der Fig. 27 verwendet wurden. HIM3 (Fig. 19c), Pl Apo Oel 160×/1,40 (Fig. 20f), Periplan 8× (Fig. 22c), N 8× *h* (Fig. 22e), Periplan GW 6,3× (ähnlich Fig. 22c).

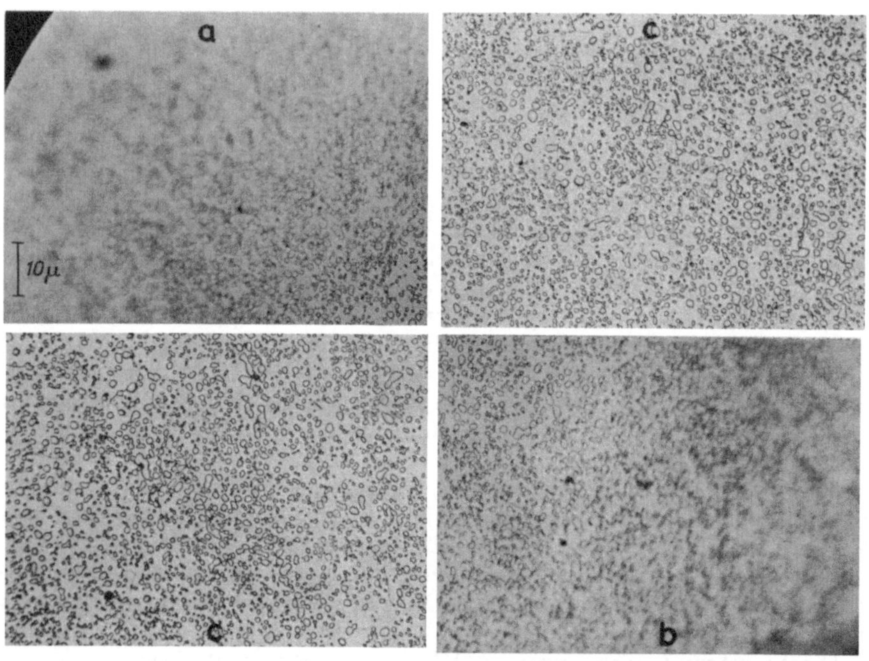

Fig. 27a—c. Photomontage von drei Mikrophotographien mit den Objektiv-Okular-Kombinationen nach Fig. 26a—c (Originalvergrößerung $M = 1000:1$, hier $M = 700:1$). Das Bild ist aus Vierteln der einzelnen Aufnahmen zusammengesetzt und zeigt die geringe bildebnende Wirkung des Negativokulars (b) und die große Wirkung des Planobjektivs (c) gegenüber der Kombination von normalem Objektiv und Positivokular (a).

gestellt, und man sieht, daß die Korrektion des Astigmatismus dort wesentlich mehr zur Ebnung des Bildes beiträgt als die negative Petzvalsumme des Okulars. Daß die Kombination des positiven Okulars mit einem entsprechenden Planobjektiv ein bei 28 mm Sehfeld-Durchmesser voll geebnetes Bild liefert, zeigen Fig. 26c und Fig. 27, in der die Teile a, b, c denen der Fig. 26 entsprechen.

Mit dem vorliegenden Beispiel wurde gezeigt, daß ein spezieller Korrektionszustand im Zwischenbild und nicht der Aufbau der Okulare für die Benutzung interessant ist. Aus diesem Grunde werden auch Okulare mit den verschiedensten Linsenanordnungen, die den gleichen Korrektionszustand besitzen und deshalb mit entsprechend korrigierten Objektiven zusammen benutzt werden sollen, mit einem bestimmten Namen belegt. Es bürgert sich mehr und mehr ein, alle Okulare, die Astigmatismus und chromatische Vergrößerungsdifferenz nicht beeinflussen, als *Huygens-Okulare* zu bezeichnen, weil sie korrektionsmäßig dieselben Eigenschaften wie das ursprünglich von HUYGENS angegebene Okular (Fig. 22b) besitzen.

Okulare, mit denen bestimmte Fehler im Zwischenbild, insbesondere die chromatische Vergrößerungsdifferenz, kompensiert werden, haben, firmenabhängig, besondere Bezeichnungen wie Kompensations-Okulare, Periplan-Okulare, Plan-Okulare, Komplanatische Okulare und andere mehr. Man darf annehmen, daß die Okulare eines Namens etwa gleiche Korrektionseigenschaften besitzen.

12. Mikrometerokulare und Anpassung der Okulare an Augenfehler. Der Blendenrand als senkrecht zum tangentialen Büschel liegendes Bildelement wird auf der tangentialen Schale eines Okulars scharf abgebildet. Bildmitte und Bildrand werden deshalb in verschiedenen Ebenen liegen. Bei stärkerer negativer $\Sigma\Gamma$ des Okulars liegt der Blendenrand objektseitig von der Brennebene (Fig. 28), in der man die schon erwähnte Strichplatte für Meßzwecke anbringt. Um Fehlsichtigen die Möglichkeit zur Fokussierung auf die Strichplatte zu geben, muß das Augenlinsenglied (2) verstellbar sein. Der Verstellbereich Δ hängt von dem Grad der Fehlsichtigkeit D (in dptr.) und der Brennweite f_2 des Augenlinsengliedes ab. Es ist hinreichend genau

$$\Delta = \frac{f_2^2 \cdot D}{1000}. \tag{12.1}$$

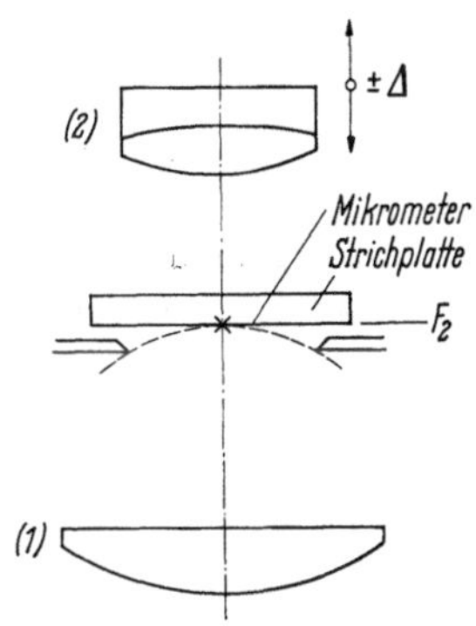

Fig. 28. Mikrometer-Okular.

Okulare mit verstellbarer Augenlinse können auch ohne Strichplatte zur Kompensation von Kurz- und Weitsichtigkeit verwendet werden. Astigmatismus der Augen ist durch Verstellen der Augenlinse selbstverständlich nicht zu korrigieren. Der Mikroskopiker muß in diesem Falle mit Brille arbeiten und benötigt dafür Okulare mit besonders großem Pupillenabstand (Fig. 22d), sog. Brillenträger-Okulare (z. B. CLAUSSEN 1958).

13. Spiegelobjektive. Die zur Farbkorrektion der Mikroskopobjektive zur Verfügung stehenden Gläser ermöglichen, wie bereits festgestellt wurde, keine vollkommen apochromatische Korrektion. Es liegt daher nahe, die farbfehlerfreie Reflexion zur Abbildung zu verwenden und Spiegelobjektive herzustellen. Dies um so mehr, als ja Spiegelobjektive für Fernrohre bestens bekannt sind und Spiegelsysteme als Dunkelfeldkondensoren verwendet werden. Nach frühen Anfängen (R. SMITH 1738) mit nicht aplanatischen Objektiven niedriger Aperturen und nach der Schwarzschildschen Veröffentlichung über die Theorie der Spiegelteleskope (K. SCHWARZSCHILD 1905) war D. D. MAKSUTOV (1932) der erste, der ein Spiegelobjektiv für Mikroskope vorschlug. In etwa den letzten 40 Jahren sind eine ungeheure Zahl von Vorschlägen zur Herstellung von Spiegelobjektiven der verschiedensten Art gemacht worden. Eine gute Übersicht über alle Typen gibt NORRIS (1955). Hier sollen nur die nach SCHWARZSCHILD benannten besprochen werden.

Diese Objektive bestehen aus zwei zum Objekt hin gekrümmten sphärischen Spiegelflächen (Fig. 29). Die Abbildung ist im Seidelschen Bereich frei von Öffnungsfehler, Koma und Astigmatismus, wenn beide Flächen konzentrisch sind und das Verhältnis der Radien r_1/r_2 Gl. (13.1) entspricht (GREY 1950). Mit den Vorzeichenregeln von FLÜGGE (1937) und den Bezeichnungen der Fig. 30,

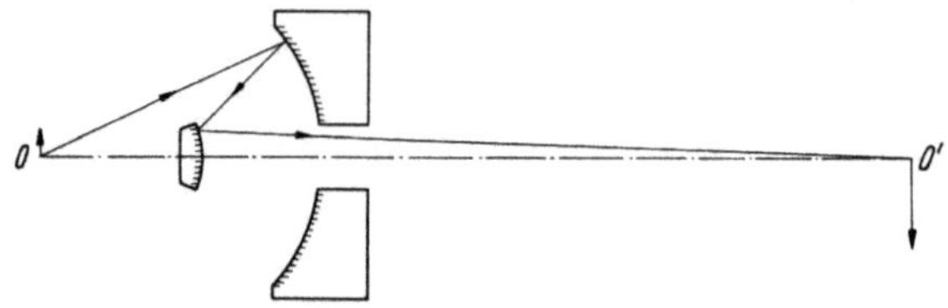

Fig. 29. Spiegelobjektiv nach SCHWARZSCHILD mit den beiden Spiegeln in Luft.

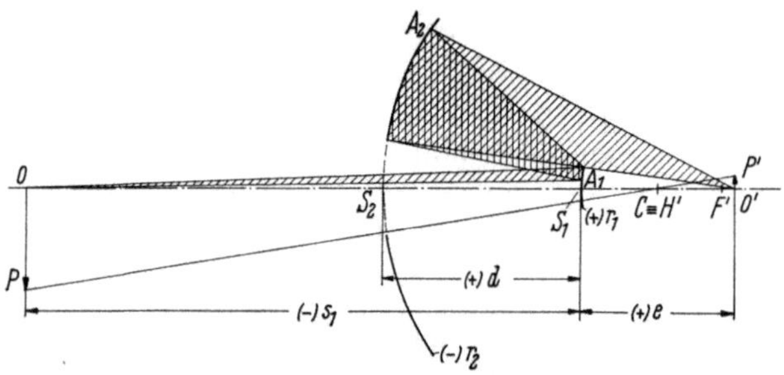

Fig. 30. Spiegelobjektiv nach SCHWARZSCHILD. Schemaskizze (RIESENBERG, 1956) mit den im Text benutzten Bezeichnungen. Objekt und Bild sind, wie beim Korrigieren üblich, vertauscht.

die wie üblich für Rechnung vom Bild zum Objekt gedacht ist, ergibt sich (RIESENBERG 1956) für das konzentrische Spiegelpaar

$$\frac{r_1}{r_2} = -\frac{3}{2} + \frac{r_1}{r_1 - s_1} + \sqrt{\frac{5}{4} - \frac{r_1}{r_1 - s_1}}\,; \tag{13.1}$$

$$f = \frac{1}{2}\,\frac{r_1 r_2}{r_1 + r_2}\,. \tag{13.2}$$

Wegen

$$\Sigma P = 2\,\frac{r_1 + r_2}{r_1 r_2} \tag{13.3}$$

ist der Scheitelradius für die zusammenfallenden Bildschalen also gleich der Brennweite f des Objektivs.

Der innere Teil der Apertur wird durch den objektseitigen Spiegel abgeschattet. Die Zentralabschattung t wird meistens definiert als das Verhältnis von minimaler zu maximaler Apertur der zur Abbildung beitragenden Strahlen; oft wird aber auch das Flächenverhältnis, also t^2, angegeben. RIESENBERG gibt als Näherung mit den Bezeichnungen der Fig. 30 an

$$t = \frac{S_1 A_1}{S_1 O'} : \frac{S_2 A_2}{S_2 O'} = \frac{s_1}{e \cdot M}\,. \tag{13.4}$$

Hier bedeutet e den „freien Objektabstand“ und M den Abbildungsmaßstab des Objektivs.

Die Zentralabschattung wird damit für das durch die Gln. (13.1) und (13.2) beschriebene konzentrische aplanatische Objektiv

$$t = \left(1 - \frac{r_1}{r_1 - s_1}\right) \cdot \left(\frac{r_1}{r_1 - s_1} + 2\sqrt{\frac{5}{4} - \frac{r_1}{r_1 - s_1}}\right)^{-1}. \tag{13.5}$$

Für den Fall $s_1 = \infty$ wird die Zentralabschattung $t = 1/\sqrt{5} = 0{,}477$ und $t^2 = 0{,}20$. Mit diesen Werten sind später ausgeführte Objektive zu vergleichen.

In Abhängigkeit von der durch die Abschattung entstehenden ringförmigen Apertur ändert sich die Intensitätsverteilung im Beugungsbild (Ziff. 2, 4). Der Kern wird schmäler, dafür treten um den Kern Ringe mit erheblicher Beleuchtungsstärke (Fig. 31, 32) auf, die für den Kern des Beugungsscheibchens einen Intensitätsverlust und für das gesamte Bildfeld einen Streulichtuntergrund bedeuten. Man wird deshalb, ganz besonders für die Photometrie, kleine Abschattungen anstreben müssen.

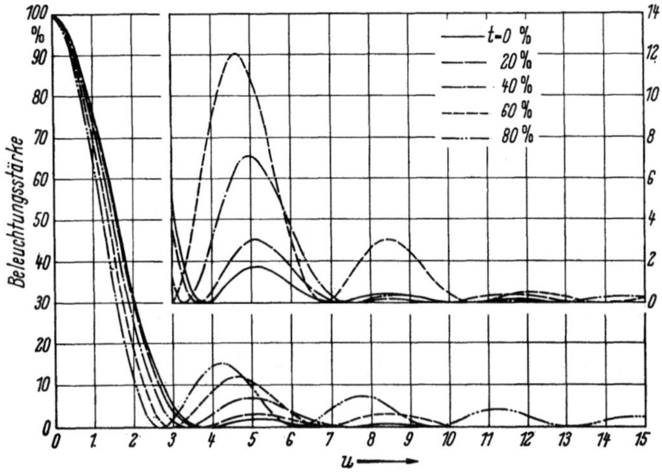

Fig. 31. Intensitätsverteilung in der Bildebene in Abhängigkeit vom Grad der zentralen Abschattung t (nach RIESENBERG, 1956).

Korrektion der Spiegelobjektive. Wir waren davon ausgegangen, daß Spiegelobjektive vollkommen apochromatisch seien. Damit sind sie nicht nur für den sichtbaren Bereich, sondern ebenso im UV und IR (Ziff. 21) interessant. Die Objektive müssen also für diese Bereiche „durchlässig" sein. Das Medium zwischen den Spiegeln darf nicht merkbar absorbieren und die Spiegeloberflächen müssen hinreichend reflektieren. Damit ist es erlaubt, z.B. für UV-Mikroskopie den Raum zwischen den Spiegeln mit Quarzglas, das in verschiedenen Reinheitsgraden bis etwa 230 nm gut durchlässig ist, zu füllen. Man hat dann nur dafür zu sorgen, daß die Ein- und Austrittsfläche des Objektivs zu Objekt und Bild konzentrisch sind und kommt zu einer Anordnung nach Fig. 33. Eine solche Anordnung ist für die Achspunkte zulässig, weil alle Strahlen dann ungebrochen in das Objektiv ein- und daraus austreten, die Dispersion des optischen Materials also nicht zur Wirkung kommt. Für schiefe Büschel treten dagegen Fehler auf. Bei biologischen Objekten wird das Präparat zumeist mit einem Deckglas abgedeckt und der Raum zwischen Objekt und Deckglas mit einem Einbettungsmittel gefüllt. Auch dadurch entstehen gegenüber dem reinen Spiegelsystem Farbfehler. Bei hohen Aperturen bringt die Einführung eines Deckglases einen merklichen sphärischen Fehler. Ein solcher, und zwar eine kräftige Zone des sphärischen Fehlers tritt bei den Spiegelsystemen selbst auf, wenn man den Bereich der Fehler dritter Ordnung verläßt und zu hohen Aperturen übergeht. Bei der Erweiterung des Wellenlängenbereiches, in dem Spiegelobjektive angewendet werden sollen, ist zu bedenken, daß ein bestimmter, z.B. in Schnittweitendifferenzen gemessener Fehler mit abnehmender Wellenlänge eine immer stärkere Verbiegung der Wellenfläche bedeutet. So kommt man dazu, auch bei Spiegelobjektiven nach weiteren Korrektionsmöglichkeiten zu suchen. Ein grundsätzliches Mittel ist Abweichung von der Konzentrizität der Spiegelflächen. Weiter wendet man asphärische Spiegelflächen (BURCH 1947) oder zusätzliche Linsen vor, zwischen oder hinter den

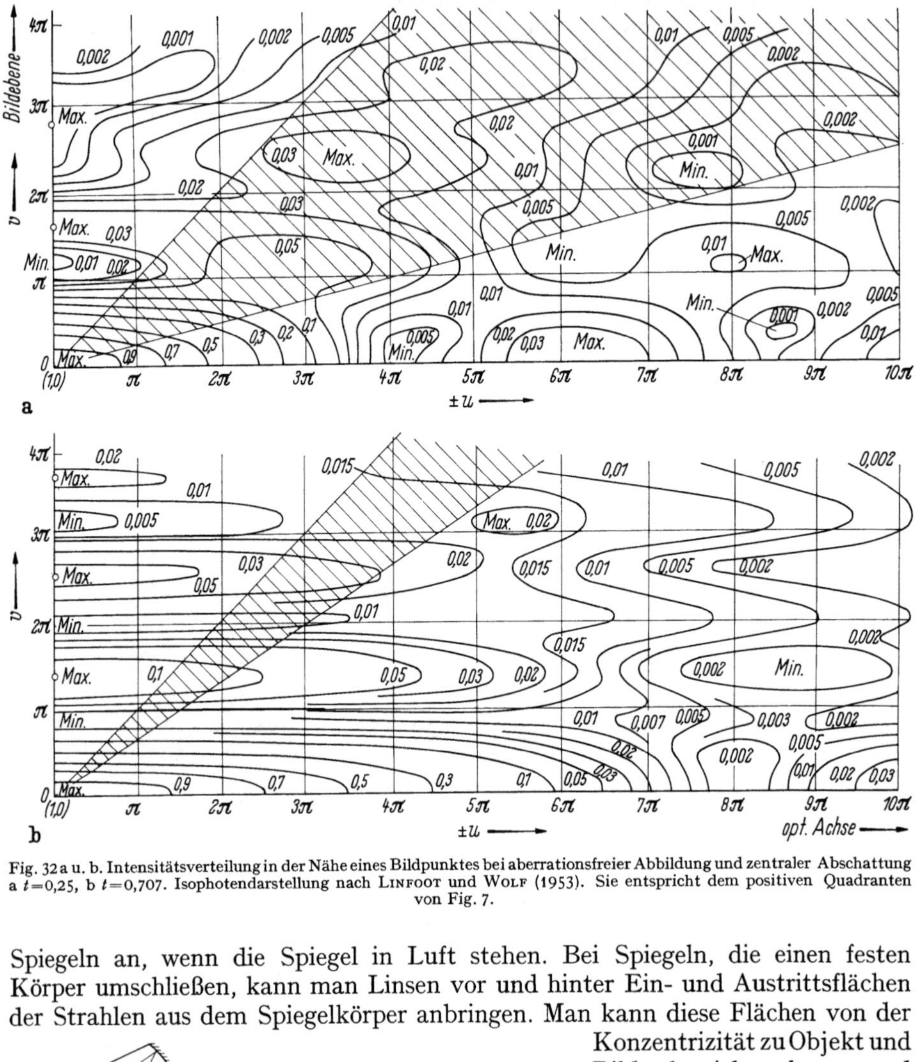

Fig. 32a u. b. Intensitätsverteilung in der Nähe eines Bildpunktes bei aberrationsfreier Abbildung und zentraler Abschattung a $t = 0{,}25$, b $t = 0{,}707$. Isophotendarstellung nach LINFOOT und WOLF (1953). Sie entspricht dem positiven Quadranten von Fig. 7.

Spiegeln an, wenn die Spiegel in Luft stehen. Bei Spiegeln, die einen festen Körper umschließen, kann man Linsen vor und hinter Ein- und Austrittsflächen der Strahlen aus dem Spiegelkörper anbringen. Man kann diese Flächen von der Konzentrizität zu Objekt und Bild abweichen lassen und man kann — jedenfalls im sichtbaren Bereich, wo genügend Materialien zur Verfügung stehen — die Spiegelkörper verkitten. Häufig wird man sich bei der Korrektion auf einen anwendungstechnisch sinnvollen Wellenlängenbereich beschränken, um nicht zu viele Mittel zur Korrektion verwenden zu müssen. Durch Vorschalten einer Plankonvexlinse kann man Immersionsobjektive bauen und dadurch die Apertur erhöhen.

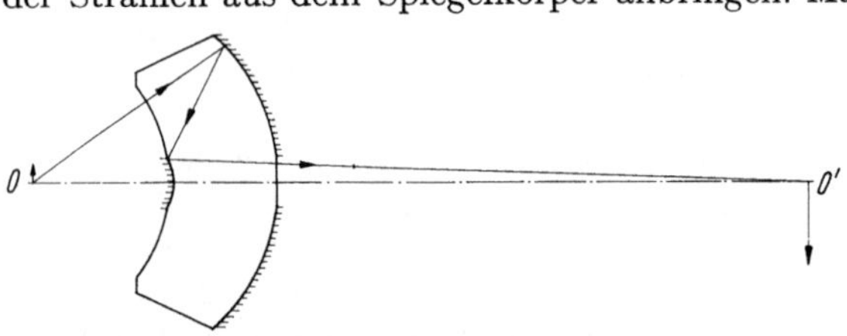

Fig. 33. Spiegelobjektiv nach SCHWARZSCHILD. Der Spiegelkörper besitzt 4 optisch wirksame Flächen: 2 spiegelnde und je eine zu Objekt und Bild konzentrische Ein- und Austrittsfläche.

Die förderliche Vergrößerung war in Ziff. 3 für weißes Licht mit dem Schwerpunkt bei $\lambda = 550$ nm bestimmt worden. Da der Durchmesser der Beugungsscheibchen proportional λ abnimmt, steigen die Werte für die förderliche Vergrößerung umgekehrt proportional zu λ an.

Sowohl das Auflösungsvermögen der Rezeptoren wie die Endvergrößerungen der Bilder müssen diesen Bedingungen sinnvoll angepaßt werden. Als Okulare in Verbindung mit dem Auge werden auch Linsenokulare verwendet, die nicht frei von Farbfehlern sind. Im UV stehen Leuchtschirme und Bildwandler und im IR Bildwandler zur Verfügung. Wenn die Bildwandler bis zum sichtbaren Bereich empfindlich sind, kann man den Bildwandler ins Zwischenbild setzen und seinen Leuchtschirm bei geeigneter Vergrößerung beobachten. Arbeitet man mit Photoschichten, kann man ohne Verwendung eines Okulars photographieren. Zu diesem Zweck muß man Objektive mit so hoher Eigenvergrößerung und so großem bildseitigem Sehfeld herstellen, daß bei Vergrößerung der Negative auf die gewünschte Endvergrößerung das Silberkorn der photographischen Schicht noch nicht stört. Wenn bei Beobachtung mit Bildwandlern oder bei Photographie ohne Okulare gearbeitet wird, soll das Zwischenbild selbstverständlich fehlerfrei sein, und man muß entsprechend korrigierte Okulare zum Einstellen im sichtbaren Bereich verwenden. Beim Korrigieren der Objektive darf man unterscheiden, ob sie nur in der Bildmitte, z.B. zur Photometrie, benutzt werden sollen oder zur Beobachtung größerer Felder. Im ersten Falle kann auf die Anwendung feldebnender Mittel verzichtet werden.

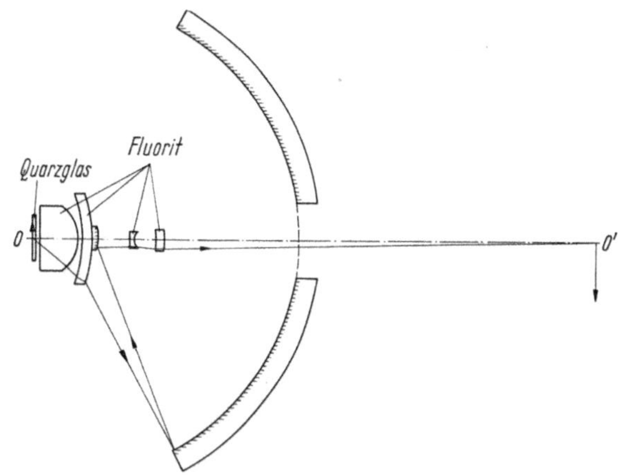

Fig. 34. Immersions-Spiegelobjektiv nach GREY (1949). $A = 1{,}10$; $t^2 = 0{,}07$.

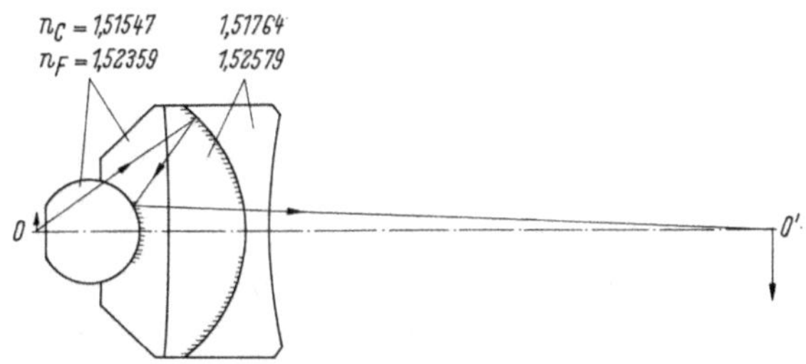

Fig. 35. Immersions-Spiegelobjektiv nach BOUWERS, BLAISSE und BULTHUIS (1951). $A = 1{,}15$; $t^2 = 0.03$.

Fig. 34 und 35 zeigen je einen Vertreter der besprochenen Objektive vom Schwarzschild-Typ mit zusätzlichen Korrektionsmitteln. Bei dem Objektiv nach GREY (1949) ist der Körper des konvexen Spiegels mit der in Lichtrichtung zweiten Linse verkittet und kann mit ihr zusammen wie eine gewöhnliche Linse gefaßt werden. Bei den einfachen Spiegelsystemen wird der konvexe Spiegel üblicherweise mit drei Metallstegen gehaltert, die die Öffnung verkleinern und deren Ränder Abbildungsstörungen hervorrufen. Das Objektiv (Fig. 35) nach BOUWERS, BLAISSE und BULTHUIS (1951) benutzt zur Korrektion des Öffnungsfehlers höherer Ordnung eine Kittfläche, an welche Materialien mit ganz geringem Unterschied der Brechzahlen grenzen. Sie wird von den abbildenden Strahlen dreimal durchsetzt. Die Verspiegelung auf der konvexen Spiegelfläche ist in der

Mitte vollständig reflektierend. Die Ringzone darum herum ist soweit halbdurchlässig verspiegelt, wie es für die Reflexion des Aperturstrahles notwendig ist. Die äußeren Aperturstrahlen werden also voll durchgelassen und teilreflektiert, die inneren teildurchgelassen und voll reflektiert. Mit dieser Maßnahme kann die Zentralabschattung merklich verkleinert werden.

14. Tuben und Tubuslinsen.

α) *Tuben.* In Ziff. 7 wurde horizontale Tischlage und vertikal stehender Tubus als üblich bezeichnet. Diese Aussage ist jetzt etwas zu modifizieren. Die Mikroskoptische sind häufig so hoch über den Arbeitstischen, daß es mühsam ist, senkrecht von oben in die Tuben zu blicken. Wo es der Präparate wegen möglich

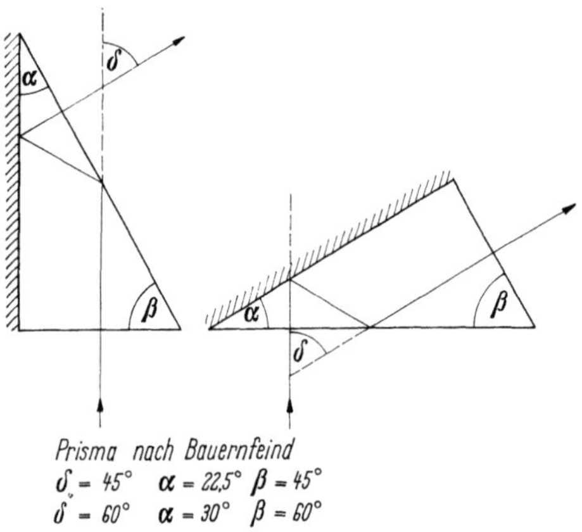

Fig. 36. Ablenkprisma für Mikroskoptuben.

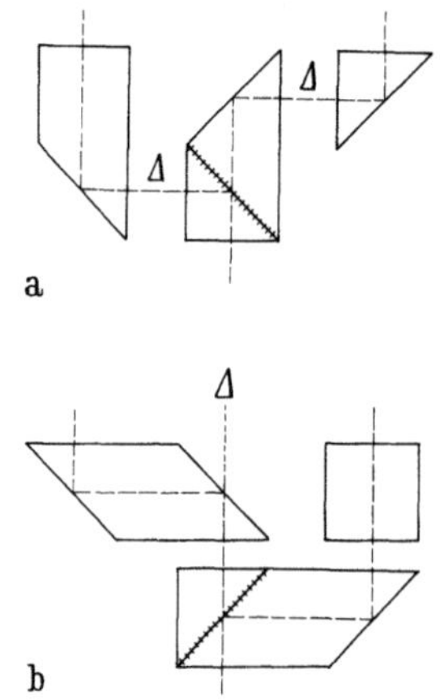

Fig. 37 a u. b. Prismensätze mit physikalischer Strahlenteilung für Binokulartuben.

ist, kann man Tisch und Tubus schräg stellen. Besser ist es, den Tubus abzuknicken. Weil die Bildlage hierbei nicht in einer Richtung umgekehrt werden soll, muß man statt eines einzigen Spiegels zur Ablenkung der Strahlen deren zwei verwenden. Im allgemeinen benutzt man dazu ein Prisma nach Fig. 36, das liegend oder stehend verwendet werden kann.

Derartige Prismen wirken auf die Korrektion wie eine senkrecht zum Strahlengang stehende Glasplatte. Sie verlängern die Bildweite für den Paraxialstrahl, wenn ihre Dicke d ist, um

$$\Delta s' = \frac{d(n-1)}{n} . \tag{14.1}$$

Da die Prismen im Tubus verwendet werden, wo Apertur und Sehwinkel klein sind, kann ihre Wirkung auf die Bildfehler in weiten Grenzen vernachlässigt werden.

So ist die Möglichkeit gegeben, auch Tuben für binokulare Beobachtung zu schaffen und sie gegen andere Tuben auszutauschen. Man teilt die gesamten Abbildungsbüschel an einer halbdurchlässigen Fläche in zwei identische auf (physikalische Strahlenteilung) und lenkt die beiden Büschel, und zwar meistens untereinander parallel, zu den beiden Augen des Beobachters. Üblich sind die zwei Anordnungen nach Fig. 37. Im Falle a wird durch Abstandsänderung an den mit Δ bezeichneten Stellen auf den Augenabstand des Beobachters (65 ± 10 mm) eingestellt, im Falle b durch Schwenken der Prismen um die mit Δ markierte Achse.

Im Falle a ändert sich bei Änderung des Augenabstandes die Tubuslänge um den halben Betrag der Abstandsänderung und muß in kritischen Fällen kompensiert werden. Im Falle b drehen sich im Okular angebrachte Meßmarken gegenüber dem Bild des Objektes und müssen, wo dies unerwünscht ist wie in der Polarisationsmikroskopie, automatisch zurückgedreht werden. Die notwendigen Prismendurchmesser ergeben sich aus der Form des lichtgefüllten Kegelstumpfes mit der Austrittspupille der Objektive als Deckfläche und dem Sehfeld als Basis. Die mechanische Länge der Binokularteile der Tuben ist, wenn man Reflexe an den Seitenflächen der Prismen vermeiden will, etwa das Dreifache des Sehfelddurchmessers. Damit ist der Raumbedarf in Richtung der optischen Achse groß und die Verlagerung des Bildes durch die Prismen um $\Delta s'$ nach (14.1) sehr erfreulich.

β) Tubuslinsen. Nicht immer haben die Tuben des Mikroskopes die Länge vom unteren Tubusrand bis zum Zwischenbild, die dem Rechnungswert der Objektive

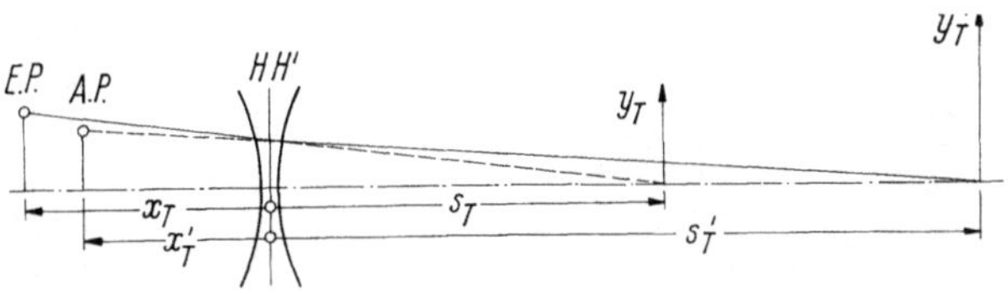

Fig. 38. Zusammenhang zwischen der Änderung von Bildort und Bildgröße und Pupillenort und Pupillengröße bei Abbildung durch eine Tubuslinse.

entspricht. Dies ist z.B. der Fall, wenn bei einem Polarisationsmikroskop ein Kalkspat-Analysator in Verbindung mit binokularer Beobachtung verwendet wird oder wenn die Objektive für unendliche Bildweite berechnet sind.

Im zweiten Falle erzeugen die Objektive nur mit einem weiteren sammelnden Glied zusammen ein reelles Zwischenbild. Dieses Glied sitzt möglichst dicht oberhalb des Objektivs im Tubus und wird Tubuslinse genannt. Der große Vorteil einer solchen Anordnung ist, daß man den Abstand zwischen Objektiv und Tubuslinse zufolge der Abbildung nach Unendlich in relativ weiten Grenzen variieren kann. Der Maßstab im Zwischenbild ist gleich dem Verhältnis der Brennweiten von Tubuslinse und Objektiv. Normiert man die Brennweite der Tubuslinse auch auf 250 mm, dann kann man die Vergrößerung im Zwischenbild wie folgt angeben:

$$M = \frac{f_{\text{Tubl}}}{f_{\text{Obj}}} = \frac{250}{f_{\text{Obj}}} \cdot \frac{f_{\text{Tubl}}}{250} = V_{\text{Obj}} \cdot q_\infty . \tag{14.2}$$

Man ordnet also dem Objektiv eine Lupenvergrößerung V_{Obj} und der Tubuslinse einen Vergrößerungsfaktor q_∞ zu, durch deren Multiplikation man den Maßstab im Zwischenbild erhält.

Bei diesen Objektiven wird deshalb die Lupenvergrößerung angegeben und man schreibt in Deutschland zur Unterscheidung z.B. $V = 40\times$ für ein solches Objektiv, während bei den für endliche Tubenlänge berechneten Objektiven der Maßstab des vom Objektiv in der Sollentfernung erzeugten Zwischenbildes in der Form $M = 40:1$ angegeben wird.

Bei Tubuslinsen, mit denen die Bildlage bei Objektiven verändert wird, die für endliche Bildweite berechnet sind, ist der Vergrößerungsfaktor q_e mit den Bezeichnungen der Fig. 38

$$q_e = \frac{y_T'}{y_T} = \frac{s_T'}{s_T} \tag{14.3}$$

und die zugehörige Brennweite der Tubuslinse, wenn ihre Lage zum ursprünglichen Bild durch s_T gegeben ist

$$f_{\text{Tubl}} = -\frac{s_T\, s'_T}{s'_T - s_T}, \tag{14.4}$$

bzw., wenn eine bestimmte Vergrößerung vorgegeben ist,

$$f_{\text{Tubl}} = -\frac{s'_T - s_T}{(q_e - 1)^2}\, q_e. \tag{14.5}$$

Weil in der Mikroskopie zunehmend mehr Wert auf Vergrößerungsreihen mit bestimmten Stufensprüngen gelegt wird, muß man die Längenänderung $(s' - s)_T$

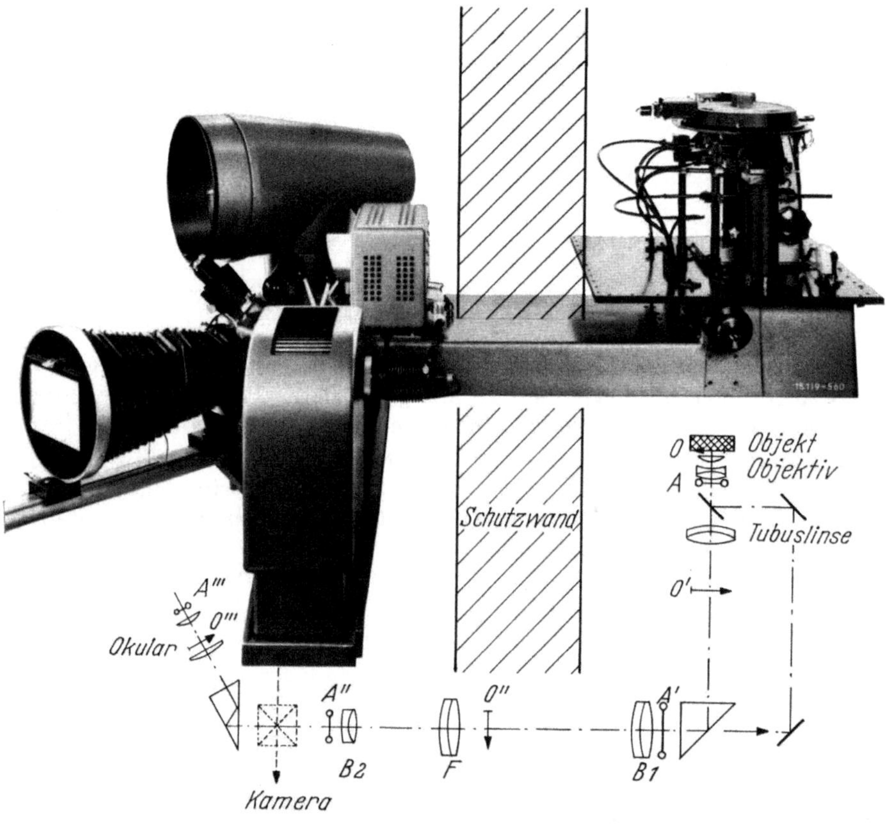

Fig. 39. Mikroskop für heiße Zellen. Oben: Rechts der Schutzwand das Stativ, links von vorn Lampenhaus, Kamera, Binotubus, Projektionsmattscheibe. Unten: Schemaskizze der Abbildungsoptik. Das Objekt O wird mit Objektiv und Tubuslinse nach O', mit der Tubuslinse $B1$ nach O'' und mit $B2$ ins Okular-Zwischenbild abgebildet. Die Abbildungen durch $B1$ und $B2$ sind bildumkehrend. Zur Pupillenabbildung A, A', A'' dient zusätzlich die Feldlinse F, die nahe bei O'' liegt und die Abbildung nur wenig beeinflußt. A'' liegt für das Okular nach Ort und Größe wie üblicherweise A des Objektivs. — Der Beleuchtungsstrahlengang ist nur mit seinem Endteil angedeutet. Er liegt im Instrument neben den gezeigten Abbildungslinsen und den Bedienungselementen in der Grundplatte.

der Tuben geeignet wählen. Wo dies nicht möglich ist, kann man die Tubuslinse als Telesystem aufbauen und damit H_T und H'_T in eine solche Entfernung vom ursprünglichen Zwischenbild legen, daß s'/s den gewünschten Wert annimmt. Mit solchen Tubuslinsen kann man auch bei einem vorgegebenen Tubus verschiedene Vergrößerungsfaktoren q einführen. Eine Verkleinerung ist nur dann sinnvoll, wenn die Objektive für einen hinreichend großen Bildwinkel korrigiert sind.

Zu beachten ist, daß die Tubuslinsen Zwischenglieder zu den gemeinsam auskorrigierten Gruppen der Objektive und Okulare sind und deshalb keine merklichen Fehler bringen dürfen. Die Tubuslinsen sitzen zwischen Objektiv und Okular und damit in dem Teil des Strahlenganges, der durch kleine Apertur- und Sehwinkel ausgezeichnet ist. Schwierigkeiten treten im allgemeinen nur dann auf, wenn die Tubuslinsen sehr große Telewirkung bekommen und ihre Einzelglieder deshalb starke Brechkräfte haben müssen.

Wichtig ist eine Beachtung der Pupillenabbildung, weil ja die Okulare für eine ganz bestimmte Pupillenlage korrigiert sind. Zur Regelung der Pupillenlage kann ein zusätzliches Glied im Tubuslinsensystem notwendig werden.

Die Austrittspupille der Objektive wird durch die Tubuslinse an dem anderen Ort in entsprechend geänderter Größe abgebildet.

Versteht man wie in Fig. 38 unter $(x-s)_T$ den Abstand des Zwischenbildes von der Austrittspupille des Objektives und unter $(x'-s')_T$ den entsprechenden Abstand des durch die Tubuslinse verlagerten Bildes der Pupille, dann ist die Pupillenvergrößerung bei einer Tubuslinse mit dem Vergrößerungsfaktor q_e

$$M_{P,e} = \frac{1}{q_e} \cdot \frac{(s'-x')_T}{(s-x)_T} \tag{14.6}$$

und entsprechend für eine Unendlich-Tubuslinse

$$M_{P,\infty} = \frac{f_{\text{Tubl}} - x'_T}{f_{\text{Tubl}}}. \tag{14.7}$$

Die Pupillenvergrößerung kann erheblichen Einfluß auf die notwendigen Querschnitte für Prismen in Binokulartuben und ähnliches haben.

Wenn eine starke Verlängerung des Tubus oder wenn eine vollkommene Bildaufrichtung notwendig sind, kann man, wie beim alten terrestrischen Fernrohr, eine Bildumkehrung mit Hilfe einer Zwischenabbildung einführen. Als wichtigste Fehler sind in diesem Falle die ΣP der beiden notwendigen Glieder, bildumkehrender Teil B und Feldlinse F, und der Gaußfehler zu nennen

Derartige Zwischenabbildungen verwendet man z.B. bei Mikroskopen, die in Räumen mit γ- oder α-Strahlern stehen (Fig. 39), um außerhalb des Strahlungsschutzes beobachten und photographieren zu können (GABLER 1959, LEITZ 1965). In Räumen mit γ-Strahlern verfärben sich normale Gläser braun. Es gibt nur wenige strahlenresistente Typen (OTLEY 1952, JAHN 1962, FISCHER 1962) und entsprechend keine sehr weitgehenden Möglichkeiten zur optischen Korrektion (FERGUSON 1956).

15. Pankratische Systeme. Unter pankratischen Systemen, auch Vario- oder Zoom-Optik, genannt, versteht man optische Systeme, in denen man einzelne Glieder oder Gruppen von Gliedern so gegeneinander verschiebt, daß sich bei fester Lage von Objekt und Bild die Brennweite und damit der Abbildungsmaßstab ändert. Man unterscheidet zwischen Pankratiken mit *optischem Ausgleich* und Pankratiken mit *mechanischem Ausgleich*, je nachdem, ob die Geschwindigkeitsverhältnisse der Linsenbewegungen im Verstellbereich konstant oder in bestimmter Weise veränderlich sind (Fig. 40). Beim optischen Ausgleich wird durch vermehrten Linsenaufwand erreicht, daß die Bildebene nur wenig oder hinreichend wenig wandert. Beim

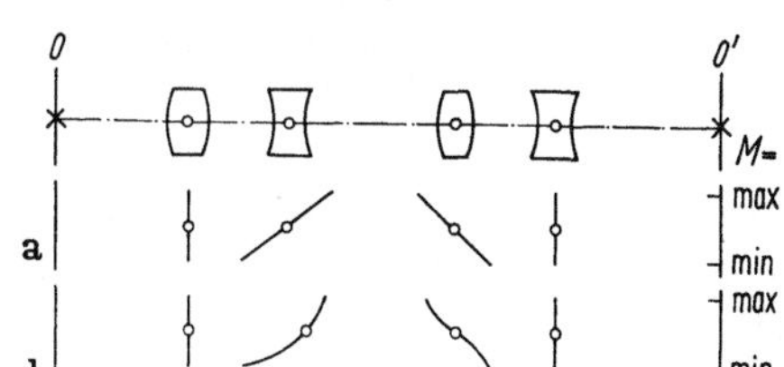

Fig. 40a u. b. Viergliedriges pankratisches System, bei dem die inneren Glieder verschoben werden. a Optischer Ausgleich, b mechanischer Ausgleich.

mechanischen Ausgleich bleibt die Bildebene exakt stehen, wenn das mechanische Getriebe für die Linsenbewegungen entsprechend aufwendig ist und genau arbeitet.

Die Größe des Verstellbereiches wird durch den Pankratikfaktor angegeben. Er ist der Quotient aus der stärksten und schwächsten Vergrößerung.

Beim Verstellen einer Pankratik ändert sich mit der Vergrößerung die Apertur sowie die Lage und Größe einer abgebildeten Pupille. Kann man die Pupillenlage hinreichend konstant halten, so ist es möglich, Pankratiken an jeder Stelle des Mikroskops zu verwenden. Man muß aber bedenken, daß der Aufwand höher ist als für ein nicht-pankratisches System und daß eine Linse, die zwei sehr nahe Nachbarn hat, nicht verschoben werden kann.

Die erste Anwendung in der Mikroskopie erfolgte wohl als Kondensor. Zur Zeit werden Pankratiken bevorzugt als Tubuslinsen und als Okulare sowie als Objektive mit kleinen Aperturen ausgeführt. Die dabei üblichen Pankratikfaktoren sind für einfache Okulare zweimal, für Projektionsokulare und Tubuslinsen zwischen dreimal und fünfmal und für Objektive zwischen dreimal und achtmal.

Die Verwendung pankratischer Systeme ist sicher dann sehr sinnvoll, wenn man ein Objekt z.B. in der Kinematographie oder beim Fernsehen formatfüllend abbilden will. Auch in der Technik zum Erreichen günstiger Arbeitsbedingungen sind solche Systeme bequem. Zur Beurteilung von Größenkonstanz bei Untersuchungsobjekten sind sie ungeeignet.

16. Beleuchtung. Die meisten mikroskopischen Objekte sind keine Selbstleuchter. Sie müssen also zur Beobachtung beleuchtet werden und zwar entsprechend den in Ziff. 2 aufgestellten Bedingungen. Die Objekte können durchsichtig, teildurchsichtig oder völlig undurchsichtig sein. Letztere müssen aus der Hälfte einer Vollkugel heraus beleuchtet werden, die das Objektiv symmetrisch umschließt, während die durchsichtigen und teildurchsichtigen Objekte aus der anderen Hälfte der Vollkugel beleuchtet werden (Fig. 41). So weit sie absorbieren, sieht man sie als Schattenbilder, wenn die Apertur der Beleuchtung wenigstens teilweise kleiner als die Beobachtungsapertur ist. Sie erscheinen dann dunkel im *hellen Feld*. Wenn die niedrigste Beleuchtungsapertur dagegen größer als die Beobachtungsapertur ist, kann nur gebeugtes Licht in das Objektiv eintreten. Orte, von denen Beugungslicht ins Objektiv kommt, werden im Bild deshalb hell auf *dunklem Feld* erscheinen. Entsprechend diesen Verhältnissen von Beleuchtungs- und Beobachtungsapertur spricht man von Hellfeldbeleuchtung und Hellfeldbeobachtung und Dunkelfeldbeleuchtung und Dunkelfeldbeobachtung.

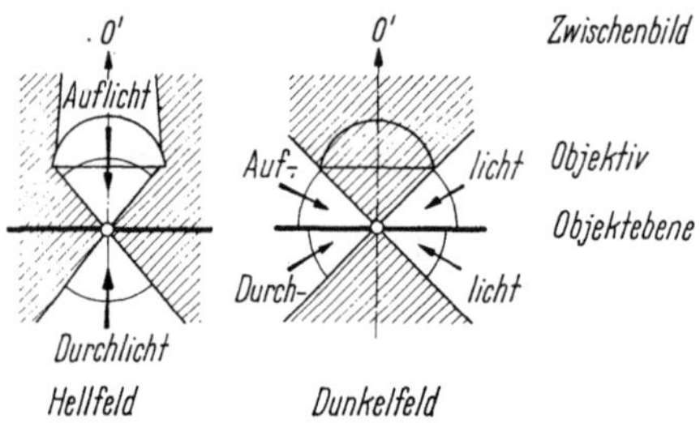

Fig. 41. Beleuchtungsbereiche für Hell- und Dunkelfeld im Durch- und Auflicht.

Je nachdem, ob man die Objekte *durch*leuchtet oder die Beleuchtung von der Beobachtungsseite her *auf* die Objekte richtet, spricht man von Durchlichtbeleuchtung und Durchlichtbeobachtung oder Auflichtbeleuchtung und Auflichtbeobachtung.

Die Art der medizinisch-biologischen Präpariertechnik, nämlich die Herstellung von Dünnschnitten mit Dicken zwischen im allgemeinen 5 und 20 μm und selektive Anfärbung der chemisch unterschiedlichen Bestandteile der ehemals lebenden Substanzen bringt es mit sich, daß die Durchlichtbeobachtung im Hellfeld die wichtigste Rolle spielt. Bei der Beobachtung von Lebensvorgängen ist es nur selten möglich, so dünne Schichten lebenden Gewebes abzupräparieren oder wachsen zu lassen, daß eine Durchlichtbeobachtung möglich ist (Intravital-Mikroskopie). Hier wird man meistens mit Auflicht beleuchten müssen. Das gleiche gilt für Untersuchungen von Metallen und Erzen, die man nur in den seltensten Fällen so dünn schleifen kann, daß sie einer Durchlichtbeobachtung zugänglich werden.

Bei der Durchlichtbeleuchtung sind die optischen Mittel für die Beleuchtung streng von denen für die Beobachtung getrennt. So ist es am einfachsten, mit der Besprechung der *Durchlichtbeleuchtung* zu beginnen.

Die Forderung lautet, mit Hilfe einer Lichtquelle das vom Objektiv und Okular abgebildete Feld mit Beleuchtungsaperturen zu beleuchten, die bei Hellfeld im wesentlichen kleiner und gleich der Beobachtungsapertur und bei Dunkelfeld nur größer als die Beobachtungsapertur sind.

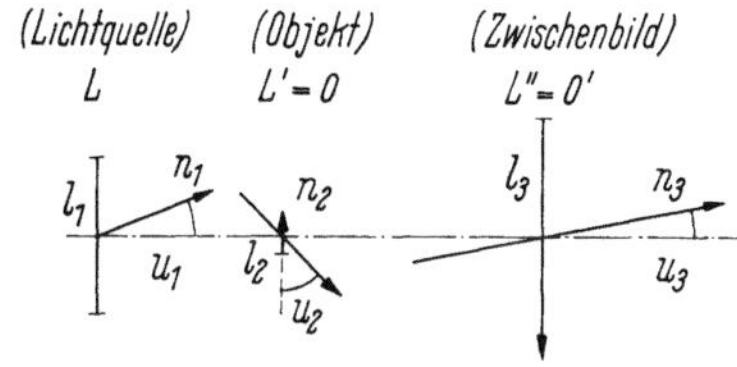

Fig. 42. n, l und u in Lichtquelle, Objekt und Zwischenbild.

Wenn Feld im Objekt und Beobachtungsapertur bekannt sind, läßt sich sofort angeben, wie groß die Lichtquelle sein muß, denn es besteht natürlich auch zwischen Lichtquellenebene, Objektebene und Bildebene (Fig. 42) der Zusammenhang

$$l_1\, n_1 \sin u_1 = l_2\, n_2 \sin u_2 = l_3\, n_3 \sin u_3 .$$

Dabei sind l_1 und l_2 und l_3 die Durchmesser von Lichtquelle, Objektfeld und Zwischenbild und $n \sin u = A$ die Aperturen eines Strahles in Lichtquelle und Objekt oder an einer anderen Stelle der Abbildung. Das Produkt von größtem Feld und zugehöriger Apertur auf der Abbildungsseite darf auf der Beleuchtungsseite niemals unterschritten werden und bestimmt die Größe der Lichtquelle in Abhängigkeit von der Apertur des optischen Systems, mit dem die Lichtquelle abgebildet wird. Hier kann man als Faustregel annehmen, daß ein größerer Aperturwinkel als $2 \times 45°$ jedenfalls bei Linsenoptik nur mit ungewöhnlichen Mitteln zu erreichen ist und daß sich alle Lichtquellen in Medien mit einer Brechzahl $n \approx 1$ befinden. Damit ist die Apertur auf der Seite der Lichtquellen $A = 0{,}70$. Unter diesen Voraussetzungen sind die Mindestgrößen der Lichtquellen festgelegt. Im Zwischenbild ist nach Ziff. 10 mit Aperturen bis $A = 0{,}03$ und Feldern bis zu 20 bzw. 28 mm Durchmesser zu rechnen. Dementsprechend ergeben sich für $l \cdot A$ Werte von etwa 0,60 bzw. 0,90. Das bedeutet unter der Voraussetzung, daß die Lichtquelle mit $A = 0{,}70$ abgebildet wird, notwendigerweise Lichtquellendurchmesser von 0,85 bzw. 1,3 mm. Die viel zu lesende Forderung nach „punktförmigen" Lichtquellen darf nur in diesem Sinn verstanden werden, daß Lichtquellen für die Mikroskopie gegenüber den Lichtquellen z. B. für Projektion klein sind. Die hier bestimmten Werte sind aber, wie wir sofort sehen werden, Minimalwerte, die nur in Sonderfällen erreicht werden können, weil für den Mikroskopiker sein Instrument ein Werkzeug ist, das möglichst wenig Bedienung erfordern soll. Nach den bisherigen Überlegungen müßte das Produkt $l \cdot A$ in der Objektebene bei jedem Objektivwechsel den Objektivaperturen bzw. den sich ändernden Feldgrößen entsprechend anders aufgespalten werden. Mit anderen Worten, die Lichtquelle müßte in jedem Falle mit einem anderen optischen System abgebildet werden und diese Systeme müßten gleichzeitig mit den Objektiven gewechselt werden. Diese Schwierigkeit umgeht man dadurch, daß man sich auf eine zweistufige (manchmal auch dreistufige) Abbildung der Lichtquelle beschränkt. Bei jeder Abbildungsstufe muß dann gleichzeitig das größte in Frage kommende Feld, also das vom schwächsten Objektiv abgebildete, und die höchste Objektivapertur in Rechnung gesetzt werden. Verlangt man z.B., daß für die Objektive $M=2{,}5:1$, $A=0{,}08$ bis zum Objektiv $M=10:1$, $A=0{,}25$ ein Beleuchtungssystem und für die Objektive $M=16:1$, $A=0{,}35$ bis $M=100$ (oder mehr)$:1$, $A=1{,}40$ das zweite Beleuchtungssystem dienen soll, dann ist das Produkt $l \cdot A$ diesmal auf die Objektebene bezogen, zu bilden aus:

$$(\text{Feld im Okular}: M_{\text{schwächstes Obj}}) \times A_{\text{stärkstes Obj}} = (l \cdot A)_{\text{Objektebene}}$$

Stufe I: $(20:2{,}5) \times 0{,}25 = 2{,}00$

Stufe II: $(20:16) \times 1{,}40 = 1{,}75$ für Sehfelddurchmesser 20 mm.

Diese Werte sind für Sehfelder von 28 mm mit 1,4 zu multiplizieren und man erhält Produkte $l \cdot A = 2{,}8$ bzw. 2,45.

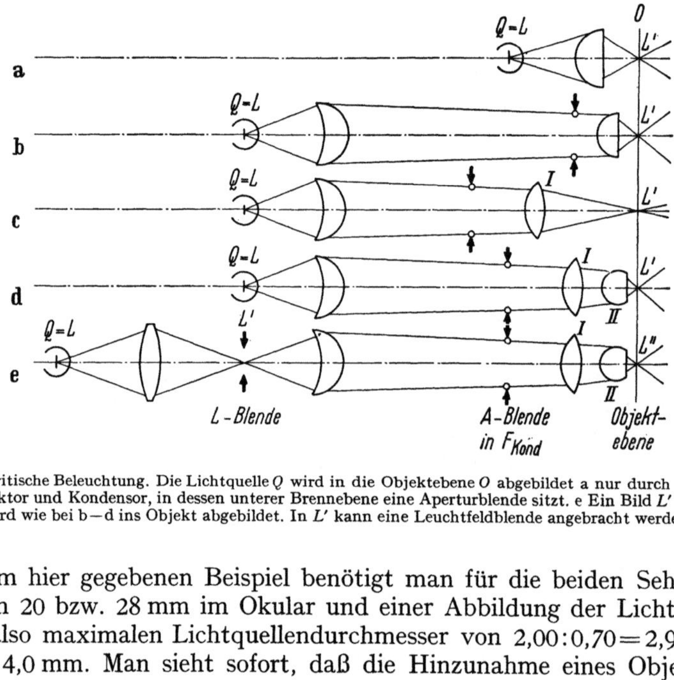

Fig. 43a—e. Kritische Beleuchtung. Die Lichtquelle Q wird in die Objektebene O abgebildet a nur durch den Kondensor b—d Mit Kollektor und Kondensor, in dessen unterer Brennebene eine Aperturblende sitzt. e Ein Bild L' der Lichtquelle wird wie bei b—d ins Objekt abgebildet. In L' kann eine Leuchtfeldblende angebracht werden.

Bei dem hier gegebenen Beispiel benötigt man für die beiden Sehfelddurchmesser von 20 bzw. 28 mm im Okular und einer Abbildung der Lichtquelle mit $A=0{,}70$ also maximalen Lichtquellendurchmesser von $2{,}00:0{,}70=2{,}9$ mm bzw. $2{,}8:0{,}70=4{,}0$ mm. Man sieht sofort, daß die Hinzunahme eines Objektivs mit $M=1:1$ einen 2,5fachen Lichtquellendurchmesser erfordern würde, wenn man die Aufteilung der beiden Beleuchtungsstufen nicht anders vornimmt. Diese ist so, wie sie hier vorgeführt ist, aus geometrischen und aus Korrektionsgründen sinnvoll. Deshalb wird im allgemeinen zu einem Objektiv $M=1:1$ ein besonderes Beleuchtungssystem verwendet.

Bei Dunkelfeldbeleuchtung wird innerhalb eines Hohlkegels beleuchtet. In diesem Falle bestimmt die Apertur der Außenwand des Kegels, $A_{\max}$, die Größe der Lichtquelle. Da man bei Dunkelfeld im allgemeinen zu jedem Objektiv ein eigenes Beleuchtungssystem benutzt, genügen die oben angegebenen Maße der Lichtquellen.

Die einfachste Beleuchtungsanordnung ist die sog. *kritische Beleuchtung*, bei der die Lichtquelle exakt ins Objekt abgebildet wird (Fig. 43a). Dabei wird vorausgesetzt, daß man die Lichtquelle näherungsweise als Ebene ansehen kann. Bildet man die Lichtquelle mit einer nahe der Lichtquelle angebrachten Linse in großer Entfernung, z.B. im Unendlichen ab, dann kann man in Objektnähe eine zweite Linse anordnen, die die Lichtquelle ins Objekt abbildet (Fig. 43b). Die der Lichtquelle nahe Linse nennt man den *Kollektor*, die objektnahe *Kondensor*. Für die im Objekt beleuchtete Fläche gilt dann für den Fall der Abbildung nach Unendlich:

$$\varnothing \text{ Leuchtfeld im Objekt} = \frac{f_{\text{Kond}}}{f_{\text{Koll}}} \cdot \varnothing \text{ Lichtquelle} \tag{16.1}$$

und die Beleuchtungsapertur im Objekt:

$$A_{\text{Kond}} = \frac{f_{\text{Koll}}}{f_{\text{Kond}}} \cdot A_{\text{Koll}}. \tag{16.2}$$

Im allgemeinen wird man die Lichtquelle mit dem Kollektor nicht streng nach Unendlich, sondern nur auf große Entfernung abbilden. Die Beziehungen (16.1) und (16.2) zeigen aber, daß grundsätzlich Lichtquelle, Kollektor und Kondensor eine Einheit sind, deren Glieder man nicht beliebig gegen andere umtauschen kann.

Spaltet man den Kondensor in zwei Glieder auf, dann kann man durch geeignete Verteilung der Brechkräfte erreichen, daß das objektfernere Glied der oben genannten Stufe I und das ganze System der Stufe II entspricht (Fig. 43c, d). Ein Kondensor, bei dem der hochaperturige Teil nicht abgeschraubt werden muß, sondern zur Seite geklappt werden kann, wird als Klappkondensor bezeichnet. Er ist, weil sehr bequem anwendbar, wohl der am meisten benutzte Kondensortyp.

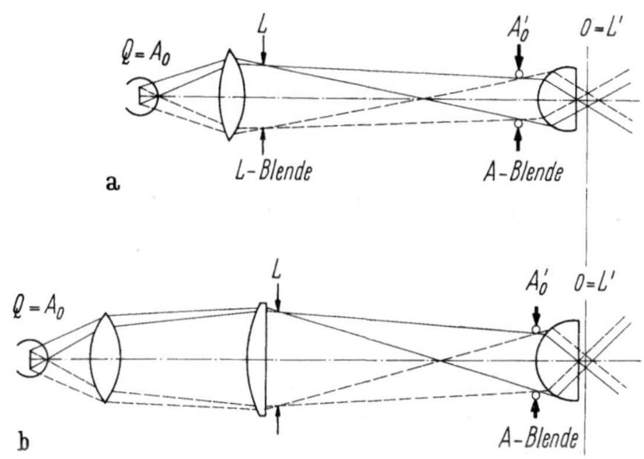

Fig. 44a u. b. Köhlersche Beleuchtung. Abbildung der Lichtquelle $Q = A_0$ in die Aperturblende A'_0 in der unteren Brennebene des Kondensors und Abbildung L' einer Leuchtfeldblende L durch den Kondensor ins Objekt O. a Abbildung der Lichtquelle mit dem Kollektor, b modifiziert durch Abbildung mit Kollektor und Hilfslinse.

Mit Hilfe einer im Kondensor eingebauten Blende ist es möglich, die Beleuchtungsapertur herabzusetzen. Man setzt diese Blende im allgemeinen in die untere Brennebene des Kondensors. Dann entspricht jedem Punkt der Blende ein in sich paralleles Büschel im Objektraum. Das Bild der Blende liegt in der bildseitigen Brennebene des Objektivs. Diese wird bei der Korrektion der Objektive als Aperturebene angenommen. Eine Begrenzung des beleuchteten Feldes ist bei dieser Art „kritischer Beleuchtung" nicht möglich. Nur dann, wenn man die Lichtquelle zweimal nacheinander abbildet, kann man eine im ersten Bild angebrachte Blende in die Objektebene abbilden und mit ihr das beleuchtete Feld regeln (Fig. 43e). Eine derartige Blende nennt man *Leuchtfeldblende* im Gegensatz zur Sehfeldblende, die das übersehbare Feld begrenzt. Eine Leuchtfeldblende ist nützlich, weil sie verhindert, daß ein großer nicht beobachteter, aber beleuchteter Teil des Objektes Licht ins Objektiv streut und durch die zugeführte Strahlung unnötig erwärmt wird. Abgesehen davon, daß ein solcher Aufbau mit Leuchtfeldblende recht lang wird, hat kritische Beleuchtung den Nachteil, daß jede Struktur der Lichtquelle (Ziff. 18) ins Objekt abgebildet wird.

A. Köhler (1893) schlug eine andere Beleuchtungsanordnung vor, die diese Nachteile vermeidet. Er bildet die Lichtquelle nicht in das Objekt, sondern in die Aperturblende des Kondensors und stattdessen eine Ebene nahe der Öffnung des Kollektors in das Objekt ab. Dann ist es möglich, in dieser Ebene eine Leuchtfeldblende anzubringen (Fig. 44a). Eine solche Anordnung wird als *Köhlersche Beleuchtung* bezeichnet. Die grundsätzlichen Überlegungen bezüglich der Lichtquellengröße und Kollektorapertur werden von der Änderung der Anordnung nicht berührt. Wenn aus räumlichen Gründen die Lichtquelle in sehr großer Entfernung vom Objekt angebracht werden muß, kann man die Anordnung etwas variieren, wie in Fig. 44b dargestellt ist. Der Kollektor muß wegen seiner notwendigerweise hohen Apertur nahe bei der Lichtquelle angeordnet werden, man kann ihn aber in eine Kollektorlinse und eine in der Praxis oft als *Hilfslinse* bezeichnete Linse aufspalten und diese Hilfslinse mit einer dort angebrachten Leuchtfeldblende durch den Kondensor ins Objekt abbilden.

Schließlich kann man Hilfslinse und Leuchtfeldblende mit dem Kondensor und seiner Aperturblende zu einer Einheit vereinigen (BEREK 1913), die *Zweiblenden-Kondensor* genannt wird. Allen diesen Anordnungen ist gemein, daß nicht die Lichtquelle, sondern die Öffnung einer die Lichtquelle abbildenden Linse ins Objekt abgebildet wird. Man kann deshalb diese Anordnungen mit Hilfslinse durchaus auch als *modifizierte Köhlersche Beleuchtung* bezeichnen. Zu bemerken ist, daß man es im allgemeinen vermeidet, die Leuchtfeldblende gar zu nahe an den Kollektor oder die Hilfslinse zu stellen, weil sonst jede Unsauberkeit auf diesen Linsen ins Objekt abgebildet wird.

Wegen Inhomogenität der Lichtquellen (Ziff. 18) führt man Streumittel so ein, daß sie bei kritischer Beleuchtung besonders die Leuchtdichteverteilung, bei Köhlerscher Beleuchtung die Abstrahlcharakteristik homogenisieren. Diese Streumittel bewirken eine effektive Vergrößerung der Lichtquelle.

Fig. 45a, b zeigt ein großes Mikroskop mit durch Hilfslinse modifizierter Köhlerscher Beleuchtung, Klappkondensor, Objektivrevolver, Tubuslinse, mit einem kombinierten Binokular- und Phototubus und mit Okular im Binokularteil. Der Strahlengang ist stark schematisiert dargestellt.

Bei *Auflichtbeleuchtung* im Hellfeld verwendet man im allgemeinen die Objektive gleichzeitig zur Abbildung und als Kondensoren für die Beleuchtung. Der kleinen Abstände zwischen Objekt und Objektiv wegen ist nämlich die Einspiegelung der Beleuchtung mit einem teildurchlässigen Spiegel (physikalische Strahlenteilung) zumeist nur hinter dem Objektiv möglich (Fig. 46a). Das hat den Nachteil, daß Reflexe erster Ordnung von den Linsenflächen ins Auge gelangen. Die in der Mikroskopie allgemein übliche Reflexminderung an den Glas-Luft-Flächen (z.B. WOLTER 1956) mit Hilfe von $\lambda/4$-Schichten hat hier also eine ganz besondere Bedeutung. Ehe man diese Möglichkeiten hatte, half man sich, wenn die Reflexe an den Linsenflächen störten, in der Weise, daß man die eine Hälfte der Objektivpupille zur Beleuchtung und die andere Hälfte zur Beobachtung benutzte (*geometrische Strahlenteilung*). Reflexe entstehen dabei nur in der für den Beobachter vom Strahlenteiler verdeckten Hälfte des Objektivs. Man ist dann aber darauf angewiesen, den Strahlenteiler genau in die Apertur des Objektivs zu legen (Fig. 46b), weil die Feldbüschel sonst (Fig. 46c) vignettiert werden. Ein anderer großer Nachteil solcher Anordnung ist, daß die Beobachtungsaperturen und damit die Auflösung in einer Richtung auf die Hälfte reduziert werden und der Schwerpunkt des Beleuchtungsbüschels schräg zum Schwerpunkt des Beobachtungsbüschels steht. Eine derartige schiefe Beleuchtung bewirkt, daß das Bild sich beim Fokussieren zu verschieben scheint.

Wenn man *physikalische Strahlenteilung* anwenden kann, ist man nicht darauf angewiesen, den Strahlenteiler in die Objektivpupille und diese möglichst oberhalb der letzten bildseitigen Linsenfläche zu legen. Es ist dann möglich, die bildseitige Brennebene der Objektive als Aperturblende zu benutzen und dadurch objektseitig telezentrischen Strahlengang der Hauptstrahlen zu erreichen. Das bedeutet, daß alle Mittelstrahlen der schiefen Büschel senkrecht auf die Objekte treffen und in sich reflektiert werden, während die gegen diese Hauptstrahlen geneigten Strahlen symmetrisch dazu reflektiert und vom Objektiv wieder aufgenommen werden. Fig. 47 zeigt die Abhängigkeit der Hauptstrahlneigung im Objekt in Abhängigkeit vom Blendenort. Infolge der sehr unterschiedlichen Brennweiten der Objektive ist es nicht ganz einfach, alle bildseitigen Brennebenen an die gleiche Stelle zu legen. Man wird sich deshalb häufig mit einer Annäherung an dieses Ideal begnügen müssen. Grundsätzlich läßt sich sagen, daß die Gleichlage der Brennebenen am leichtesten bei sehr kompliziert aufgebauten Objektiven, wie Planobjektiven für 28 mm Sehfeld, zu erreichen ist (CLAUSSEN 1964). Man hat dort ge-

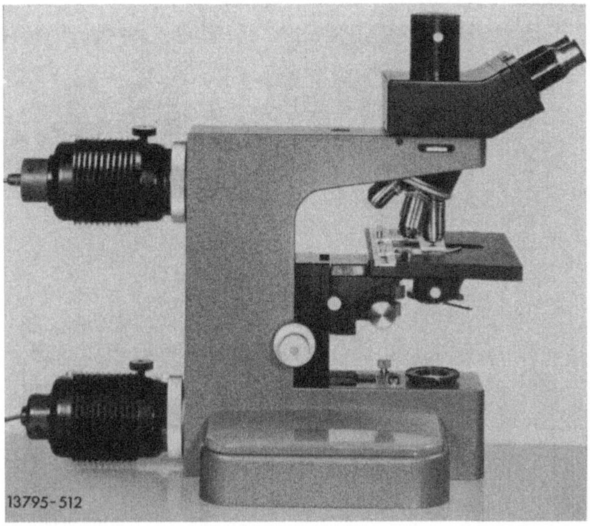

a

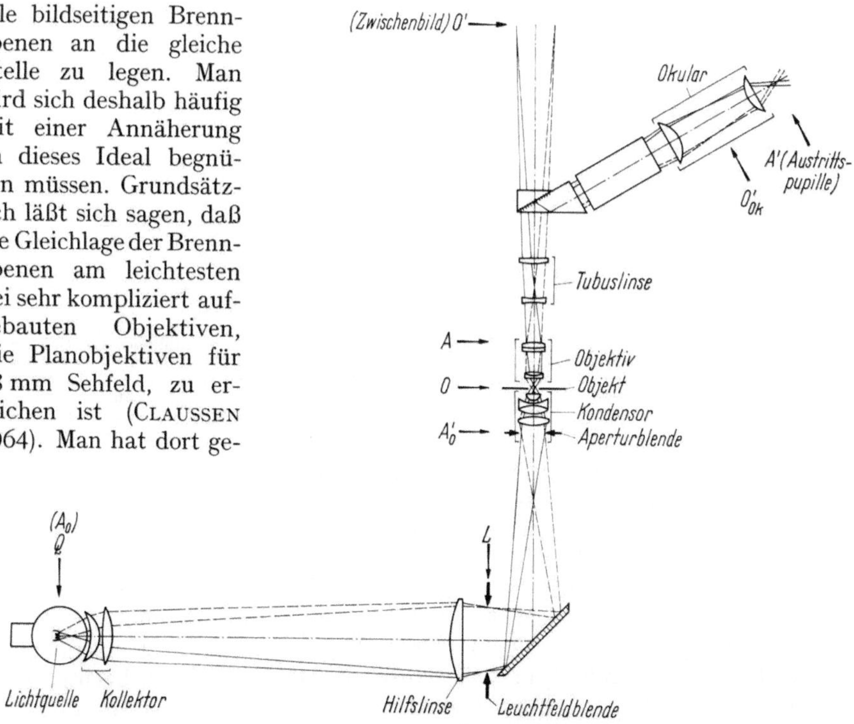

b

Fig. 45. a Mikroskop mit eingebauter Beleuchtung, b Schemazeichnung der Beleuchtungs- und Abbildungsoptik.

nügend Freiheiten, diese zusätzliche Forderung zu erfüllen. Fast hoffnungslos ist es bei den einfachsten Typen, die wieder den Vorteil haben, nur wenig reflektierende Glas-Luft-Flächen zu besitzen.

Bei Auflicht-Hellfeld-Beleuchtung sind ebenso wie in Durchlicht kritische und Köhlersche Beleuchtung möglich. Es ist natürlich verboten, das Objektiv

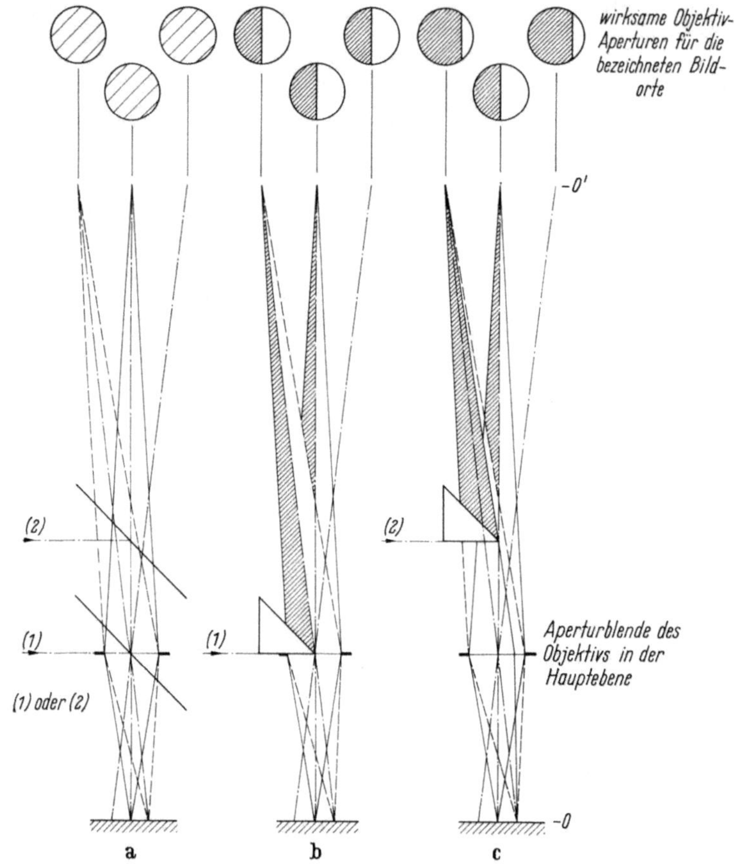

Fig. 46a—c. Physikalische und geometrische Strahlenteilung in und oberhalb der Aperturblende eines Objektivs bei Auflichtbeleuchtung. a Bei physikalischer Strahlenteilung werden unabhängig von Objektort und der Lage des Strahlenteilers die vollen Objektivaperturen zur Abbildung benutzt. b Bei geometrischer Strahlenteilung in der Aperturblende des Objektivs werden alle Objektpunkte mit der halben Pupille abgebildet. c Bei Lage des Strahlenteilers außerhalb der Aperturebene tritt zusätzlich und zwar senkrecht zur Kante des Strahlenteilers eine Vignettierung der Feldbüschel auf.

in seiner Funktion als Kondensor mit einer Aperturblende zu versehen und dort abzublenden. Deshalb ist es notwendig, die Aperturblende in den reinen Beleuchtungsteil zu legen. Fig. 48 gibt ein Beispiel für eine modifizierte Köhlersche Beleuchtung, wie sie sehr häufig benutzt wird. Die Lichtquelle A_0 wird mit dem Kollektor (K) und Hilfslinse (1) in die Aperturblende (A_0') und von dort mit Hilfslinse (2) in die bildseitige Brennebene des Objektivs abgebildet (A_0''). Das Objektiv bildet A_0'' über das Objekt, das die abbildenden Strahlen reflektiert, beim Rückweg der Strahlen in sich selbst, in das für den Beobachter sichtbare Bild A ($=A_0'''$) der Lichtquelle und damit — wie gefordert — in die bildseitige Brennebene des Objektivs als Aperturebene ab. Die Kollektoröffnung L wird durch die Hilfslinse (1) in die Leuchtfeldblende L' und von (2) und dem

Objektiv im Objekt $L''=O$ abgebildet. Das Objektbild O' entsteht oberhalb des halbdurchlässigen zu denkenden Spiegels.

Für *Dunkelfeldbeleuchtung* im Durch- und Auflicht sind kritische wie Köhlersche Beleuchtung möglich. Man wird Kondensoranordnungen den Vorzug geben, die am wenigsten Streulicht erzeugen. Die an den meisten Objekten entstehenden Beugungswellen sind nämlich so schwach, daß kleinste Mengen von Streulicht schon außerordentlich stören.

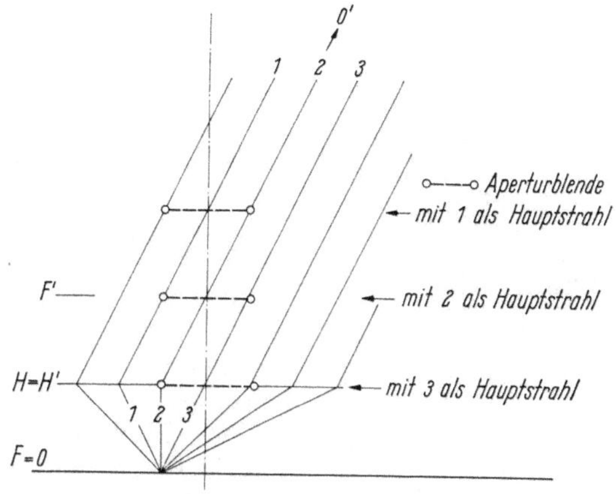

Fig. 47. Beeinflussung der Hauptstrahlneigung im Objektraum durch geeignete Wahl des Ortes der Aperturblende. Nur wenn die Aperturblende in der bildseitigen Brennebene F' des Objektivs liegt, sind die Hauptstrahlen (hier Strahl 2) objektseitig telezentrisch und werden am Objekt in sich selbst reflektiert. Strahl 1 kehrt auf dem Wege 3, und Strahl 3 auf dem Wege 1 ins Objektiv zurück. — Wählt man eine andere Ebene, wie beispielsweise die mit Strahl 1 oder 3 als Hauptstrahl, fallen diese schräg auf das Objekt und das von der Aperturblende durchgelassene Büschel wird nach Reflexion am Objekt nur noch teilweise vom Objektiv aufgenommen. (In Fig. 46 ist die Aperturblende nur zur Vereinfachung der Darstellung in die Hauptebene gelegt worden.

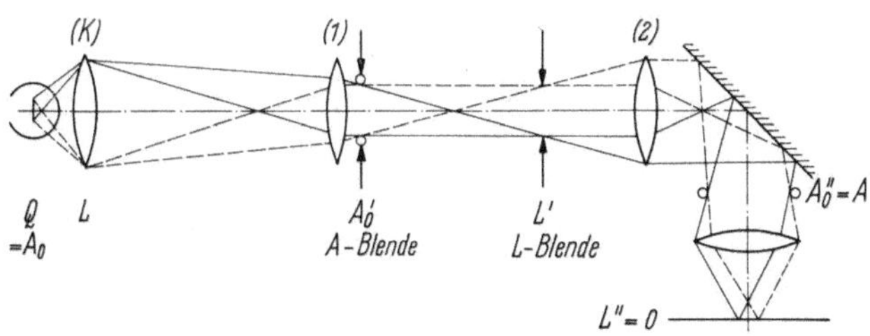

Fig. 48. Modifizierte Köhlersche Autlicht- Hellfeld-Beleuchtung (Erläuterung im Text).

Völlig unabhängig von der Anordnung auf einer geraden optischen Achse sind sogenannte *Lichtleiter* (Barta 1936), die das Licht in zylindrischen oder konischen Stäben oder Fasern aus Glas oder Kunststoff durch Totalreflexion an ihren Wänden weiterleiten und die zu mehreren gebündelt als *Faseroptik* bezeichnet werden (Kapany 1966, Reichel und Tiedeken 1964). Diese Bündel sind flexibel und geeignet für Beleuchtung und — je nach Anordnung, Querschnitt und Zahl der Einzelfasern im Bündel — auch zur Abbildung.

17. Korrektion von Kondensoren. Die Kondensoren sollen dieselben Felder mit den Aperturen beleuchten können, mit denen diese Felder durch die Objektive beobachtet werden. So gelten grundsätzlich dieselben Forderungen wie für das Korrigieren der Objektive, nur mit wesentlich größeren Toleranzen. Man sieht dies schon daraus, daß man ohne Störungen für die Objektträger etwa die zehnfache Dickenabweichung ($d=1{,}0\pm 0{,}2$ mm) erlaubt wie bei Deckgläsern ($d=$

0,17 ± 0,02 mm). Bei einfachsten Kondensoren wird der Öffnungsfehler nur so weit korrigiert, daß die Aperturen der Objektive gleichmäßig mit Licht gefüllt erscheinen. Bei höher korrigierten Kondensoren wird näherungsweise aplanatisch korrigiert und wegen der Abbildung der Sehfeldblende auf hinreichende Kleinheit von Gaußfehler und chromatischer Vergrößerungsdifferenz geachtet. Da im allgemeinen mit den Kondensoren nur der Blendenrand der Sehfeldblende abgebildet wird, der nur aus Teilen besteht, die bei gleichem Abstand von der Bildmitte tangential liegen, ist das Korrigieren von Verzeichnung und Astigmatismus überflüssig. Fig. 49, 50 zeigen einen Hellfeld-Klappkondensor in verschiedenen Ausführungsformen. Wenn in Sonderfällen noch höhere Forderungen an die Bildqualität gestellt werden, benutzt man auch im Durchlicht Mikroskop-Objektive und als Objekt-„Träger" Deckgläser oder besonders dünne Objektträger.

Fig. 49. Durchlicht-Hellfeldkondensor mit austauschbaren und ausklappbaren Kondensorköpfen.

Zur Dunkelfeldbeleuchtung kann man die Randpartie von normalen Kondensoren verwenden. Im Durchlicht müssen die zentralen Teile nur abgeblendet sein, während bei Auflichtbeleuchtung besondere Kondensoren benötigt werden, deren zentrale Teile fehlen müssen, weil die Objektive diesen Platz benötigen (Fig. 51a, 52a). Besser als derartige Ringlinsen sind Spiegelsysteme mit einer

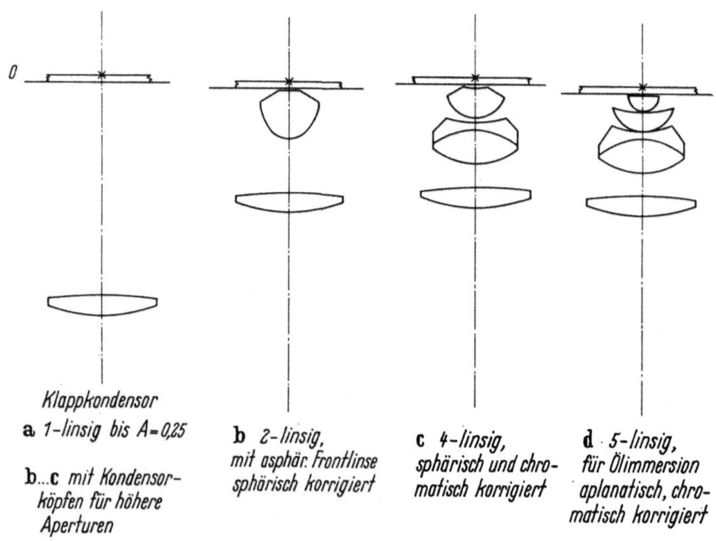

Fig. 50. Schema des optischen Aufbaus des Kondensors nach Fig. 49. Die Linsen sind in der Stellung gezeigt, in der die Leuchtfeldblende in das Objekt abgebildet wird.

paraboloiden oder mit zwei sphärischen Flächen (Fig. 51b, c; 52b). Die Abbildung mit solchen Spiegelsystemen ist in Ziff. 13 so ausführlich behandelt, daß hier darauf verzichtet werden kann.

a
b
c

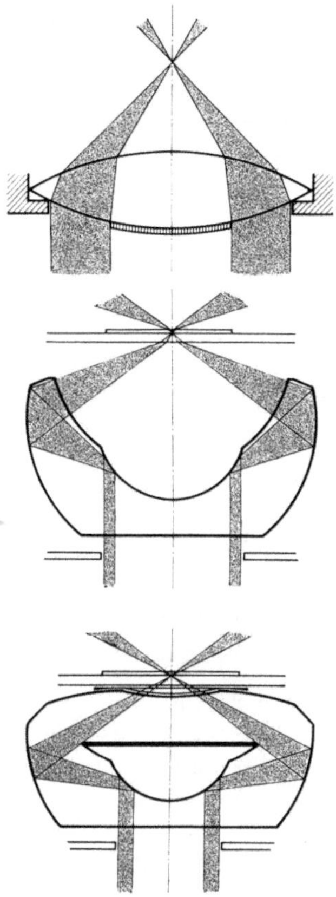

Fig. 51.

a
b

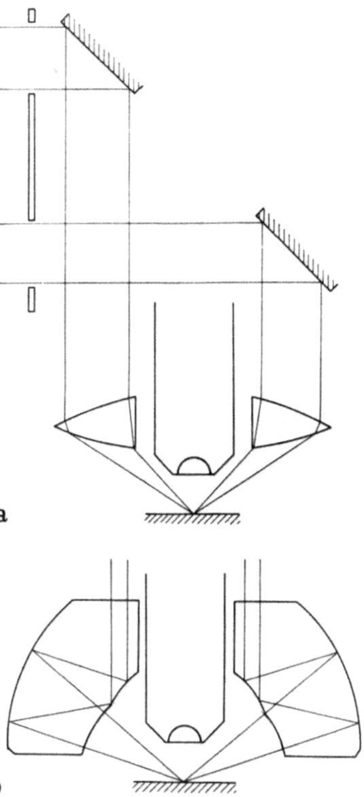

Fig. 52.

Fig. 51a—c. Durchlicht-Dunkelfeld-Kondensoren. a Hellfeldkondensor mit abgeblendetem Zentralbereich, b Spiegelkondensor für Aperturen $<1{,}0$, c Spiegelkondensor für Ölimmersion.

Fig. 52a u. b. Auflicht-Dunkelfeldkondensoren umschließen das abbildende Objektiv. a Ringlinse, b Spiegelkondensor.

18. Lichtquellen für die Mikroskopie. Wenn man bei einer bestimmten Vergrößerung mikroskopiert, ist nach Gl. (10.2) durch Objektivapertur und Gesamtvergrößerung die maximale Größe der Austrittspupille festgelegt. Sie kann noch durch Betätigung der Aperturblende auf der Beleuchtungsseite herabgesetzt werden und ist dann als ein gegebener und nicht veränderbarer Wert anzusehen. Die Fläche der Austrittspupille des Mikroskopes ist stets kleiner als die der Eintrittspupille des Auges. So kann man für das Mikroskop — bei subjektiver Beobachtung — feststellen, daß die Bestrahlungsstärke E_n auf der Netzhaut wenigstens im Verhältnis der Flächen der Pupillen

$$\frac{(\varnothing\,\mathrm{AP_{Mikroskop}})^2}{(\varnothing\,\mathrm{EP_{Auge}})^2} \tag{18.1}$$

abgesunken ist. Dazu kommen Verluste durch Absorption und Reflexion im Instrument und am Objekt. Die Bestrahlungsstärke E ist der Leuchtdichte L der Objekte (bei Selbstleuchtern) bzw. der Lichtquellen (bei Nichtselbstleuchtern) direkt proportional. So liegt die einzige Freiheit zur Erhöhung der Bestrahlungsstärke im Bild also in einer Erhöhung der Leuchtdichte der Lichtquellen, die zur Mikroskopie verwendet werden.

Ta-

Strahler		Lichtquellen für Mikroskopie: Beispiele (nach K. WEBER 1966)				Leuchtdichte regelbar		Mittlere Leuchtdichte [sb]
Der Strahler ist ein fester Körper	Glühwendel	Netzlampe	Kolben innen mattiert	220 V	15 W	nein		~ 20
		Niedervolt-lampe	Flachkernwendel	6 V	15 W	ja		1000
			Rundwendel	6 V	30 W	ja	bei 5,7 A	1700
							bei 6,1 A	2500
			Flachkernwendel	12 V	100 W	ja		3500
		Jodglühlampe		12 V	100 W	ja		4500
		Wolframband		6 V	100 W	ja		800
		Pointolite	abgeschlossene Bogenlampen ~		100 W	nein		1500
		Zirkonoxyd	abgeschlossene Bogenlampen =		100 W	nein		4000
		Sunlamp		10 V	300 W	nein		4500
		Kohlebogen		10 A =	550 W	nein		15000
Der Strahler ist ein Gas		Hg-Hochdrucklampe			125 W	nein		800
		Hg-Höchstdrucklampe		HBO 200 W	200 W	nein		25000
				HBO 100 W	100 W	nein		150000
		Hg-Höchstdruck-Kapillar-Lampe		SP 500	500 W	nein		25000
		Wasserstofflampen		WHS 200	200 W	(ja)		—
		Xenon-Höchstdruck-Lampe		XBO 150 W	150 W	nein		10000
				XBO 450 W	450 W	(ja)		35000
		Kondensierte Funkenentladung			(1 Ws)	—		10^7 über 1 µs

Die notwendige Größe der Lichtquellen wurde in Ziff. 16 abgeschätzt. Lichtquellen werden in Objekt- oder Aperturebene abgebildet und sollen diese Ebene möglichst gleichmäßig ausleuchten.

Um Lichtquellen für die Mikroskopie miteinander vergleichen zu können, ist nach K. WEBER (1965, 1966) die Kenntnis folgender Eigenschaften notwendig:

1. Die *mittlere Leuchtdichte* in dem optisch benutzten Bereich des Strahlers, bei Köhlerscher Beleuchtung also der von der Beleuchtungsaperturblende begrenzte Teil des Lichtquellenbildes.

2. Als *leuchtende Fläche* wird die Projektion der emittierenden Fläche senkrecht zur optischen Achse bezeichnet. Bei Glühlampen ist es der von der Wendel gefüllte Bereich, bei Gasentladungslampen die Ausdehnung des Plasmas.

3. Die *räumliche Verteilung der Leuchtdichte* wird im Bild der Lichtquelle punktweise ausgemessen. Ungleichmäßige Verteilung der Leuchtdichte wird bei kritischer Beleuchtung im Bild selbst sichtbar, bei Köhlerscher Beleuchtung ist die Aperturebene inhomogen ausgeleuchtet, was zu Störungen der Abbildung führt.

4. Die *Abstrahlcharakteristik* gibt die Lichtstärke der Lichtquelle als Funktion des Abstrahlwinkels an. Ungleichmäßigkeiten in der Abstrahlcharakteristik werden in der Kollektorebene sichtbar und bei Köhlerscher Beleuchtung ins Objekt abgebildet. Man mindert die Wirkung der Inhomogenitäten durch Einführung von genarbten, d.h. polierten Mattflächen oder feinen Mattflächen vor der Leuchtfeldblende.

belle 3

Leuchtende Fläche [mm²]	Räumliche Verteilung der Leuchtdichte	Abstrahlcharakteristik	Spektrale Strahldichteverteilung	Stabilität z: zeitlich r: räumlich	Lebensdauer h
~10 × 10			$T_f = 2500$ °K		1000
1,8 × 1,8 2,2 × 2,2 — 2,5 × 4,5 2,2 × 4,4	ungleichmäßig	ungleich-mäßig	$T_f = 2850$ °K $T_f = 3200$ °K $T_f = 3400$ °K $T_f = 3400$ °K —	mit stabilis. Netzgerät: stabil	100 100 25 25 50
8 × 2 2,5 ⌀ 1,6 ⌀ 3 ⌀	sehr gleichmäßig gleichmäßig gleichmäßig sehr gleichmäßig	gleichmäßig gleichmäßig bis A 0,4	$T_f = 2850$ °K $T_f = 3200$ °K $T_f = 3250$ °K $T_f = 3600$ °K		100 500 250 —
4 ⌀	gleichmäßig	gleichmäßig	$T_f = 4200$ °K		(1)
20 × 1 2,5 × 1,3 0,3 × 0,3	in der Säule gleichmäßig ungleichmäßig ungleichmäßig		Linien Linien und Kontinuum	z und r Instabilität möglich	4000 200 200
12 × 1 1 ⌀	in der Säule gleichmäßig gleichmäßig		 Kontinuumion UV	z: instab. r: stabil z und r stabil	50 500
2,5 × 1 2,4 × 1,2	ungleichmäßig ungleichmäßig		$T_f = 6000$ °K $T_f = 6250$ °K	z und r Instabilität möglich	1200 2000
2 × 2	ungleichmäßig		energiegleiches Spektrum 220...700 nm	von Funke zu Funke Φ_t und R unterschiedlich	—

5. Die *spektrale Verteilung der Strahldichte* wird für den jeweils interessierenden Spektralbereich festgelegt. Bei Kontinuumstrahlern, die etwa wie ein schwarzer Körper emittieren, genügt zur Kennzeichnung der spektralen Verteilung die Angabe der Farbtemperatur. Bei Strahlern, die einzelne Spektrallinien oder solche Linien über einem Kontinuum zeigen, ist Kenntnis der Verteilungskurve notwendig.

6. Die *Stabilität des Strahlers* ist wichtig in allen Fällen, in denen im mikroskopischen Bild photometriert werden soll (Ziff. 26). Unter zeitlicher Instabilität versteht WEBER Schwankungen der Leuchtdichte und unter räumlicher Instabilität z.B. das Wandern des Plasmas von Gasentladungslampen entlang der Elektrodenoberfläche und quer zur optischen Achse.

7. Die Kenntnis der *Lebensdauer* einer Lichtquelle wird häufig für ihre Anwendungsmöglichkeit entscheidend sein.

Beispiele für Lichtquellen sind in Tabelle 3 zusammengestellt. Man hat zwei Strahlertypen zu unterscheiden: solche, bei denen ein fester, glühender Körper die Strahlung emittiert, und solche, bei denen ein Gasplasma die Strahlungsquelle bildet. Der am meisten verwendete Festkörper-Strahler ist Wolfram als gewendelter Draht oder Band, dessen Schmelzpunkt bei 3653 °K liegt. Mit steigender Heiztemperatur T_w erhöht sich die Leuchtdichte L (Fig. 53), und die Lebensdauer sinkt durch Verdampfung und Kondensation des Dampfes an der Wand des Kolbens, in dem sich der Strahler befindet. Die spektrale Strahldichteverteilung läßt sich durch Angabe einer Farbtemperatur T_f beschreiben.

Durch mehr oder weniger starkes Aufheizen kann man somit verschiedene Leuchtdichten und, was z.B. für die Farbphotographie wesentlich ist, verschiedene Farbtemperaturen einstellen. Bei Wolframdrahtlampen ist zu beachten, daß

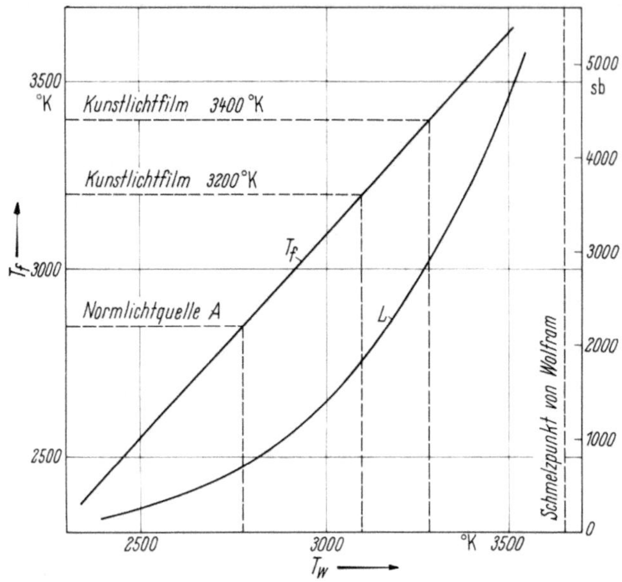

Fig. 53. Abhängigkeit der Leuchtdichte L und Farbtemperatur T_f eines Wolframdrahtes von seiner Heiztemperatur T_w.

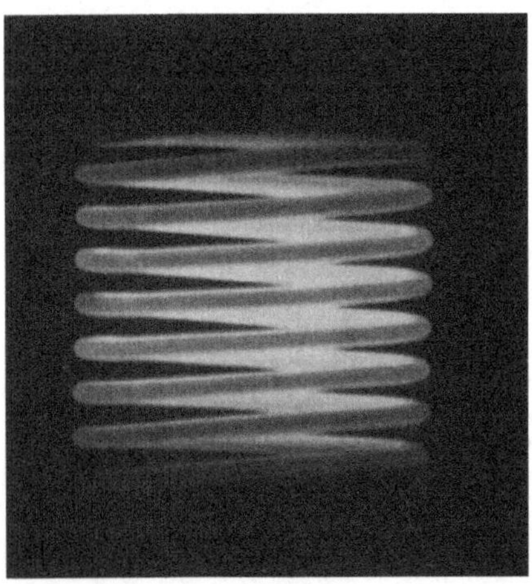

Fig. 54.

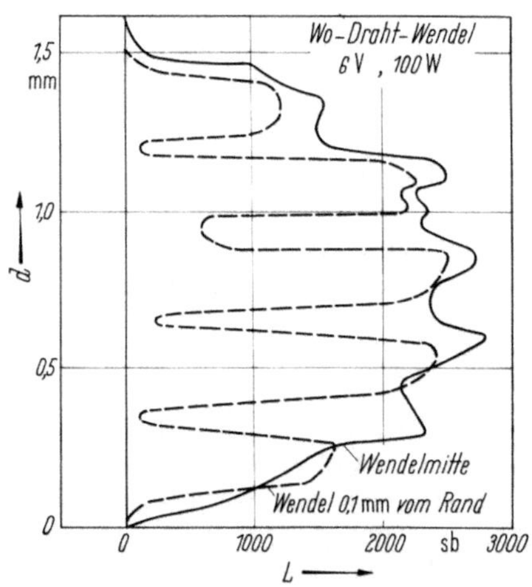

Fig. 55.

Fig. 54. Glühende Einfachwendel einer Lampe 6 V 30 W. Die Aufnahme zeigt die Inhomogenität der Leuchtdichte. Man erkennt, daß die Wendel im Innern am hellsten glüht. Durch die Geometrie der Wendel ist in verschiedenen Beobachtungsrichtungen der Flächenanteil der leuchtenden Drähte gegenüber den Lücken verschieden groß. So ergibt sich eine periodische Ungleichmäßigkeit der Abstrahlcharakteristik (K. WEBER).

Fig. 55. Leuchtdichteverteilung einer senkrecht stehenden Einfachwendel von 1,5 mm Durchmesser in der Wendelmitte und 0,1 mm vom Rand.

die einzelnen Windungen eines Drahtes in Abhängigkeit von der Spannungsdifferenz benachbarter Drahtstücke genügend Abstand haben müssen, um Kurzschlüsse innerhalb der Lampen zu vermeiden. Man verwendet deshalb im allgemeinen Niedervoltlampen mit dicken Drähten, die hinreichend stabil sind.

Fig. 54 zeigt eine solche glühende Wendel von 2,2 mm Durchmesser und etwa gleicher Länge. Als leuchtende Fläche wird man diese Maße angeben und diese Fläche z. B. bei Köhlerscher Beleuchtung in die Aperturblende abbilden. Die räumliche Verteilung der Leuchtdichte ist für eine 6 V, 100 W Einfachwendellampe in Fig. 55 dargestellt und zwar für einen Schnitt in Wendelmitte und 0,1 mm von ihrem Rande entfernt. Die Abstrahlcharakteristik für eine Lampe 12 V, 60 W zeigt Fig. 56. Man sieht die starke Inhomogenität von Abstrahlcharakteristik und Leuchtdichteverteilung in diesen Abbildungen sehr gut. Trotzdem werden

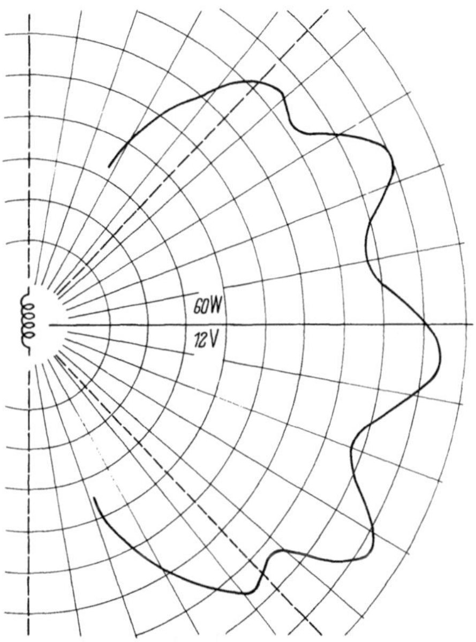

Fig. 56. Abstrahlcharakteristik einer Einfachwendel in einer Ebene durch die Wendelachse.

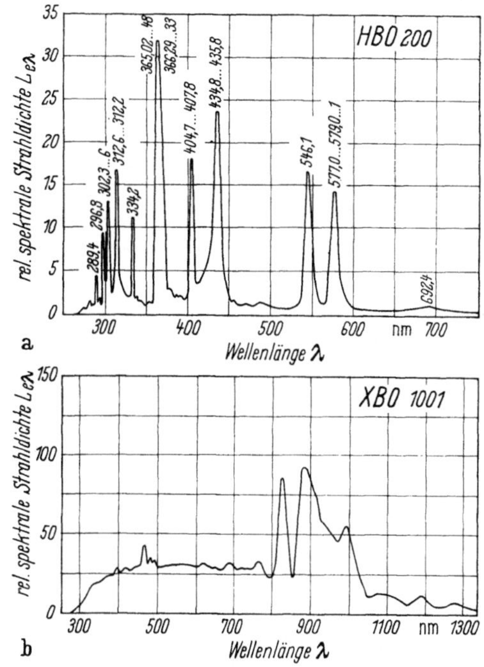

Fig. 57 a u. b. Relative spektrale Strahldichteverteilung einer a Quecksilber-, b Xenon-Höchstdrucklampe.

solche Lampentypen ebenso wie ein Typ mit Flachkernwendel und dadurch größerer Leuchtfläche sehr viel benutzt, denn die Leuchtdichte ist in breiten Grenzen regelbar. Die Rundwendellampe 6 V 30 W läßt sich z. B. auf die Farbtemperatur 3200 °K mit einer mittleren Leuchtdichte von 1700 sb und 100 h Lebensdauer und auf 3400 °K mit entsprechend 2500 sb und 25 h Lebensdauer einstellen. Zur Erhöhung der Lebensdauer verwendet man neuerdings eine Beigabe von Halogenen in die Gasfüllung der Lampe, mit deren Hilfe man ein Abscheiden des Wolframs am Glaskolben vermeidet und Wolfram zur Wendel zurückführt.

Eine sehr gleichmäßige Leuchtdichte und Abstrahlung zeigt die Gruppe Wolframbandlampe, Wolframpunktlicht (Pointolite), Zirkonoxyd und General-Electric-Sun-Lamp. Sie werden bevorzugt zur Eichung von Mikroskop-Photometern benutzt. Nur die Wolframbandlampe ist voll regelbar, ihre Leuchtdichte kann, wie ein Vergleich von Leuchtdichte und Lebensdauer mit Wolframdrahtlampen zeigt, aber nicht sehr hoch getrieben werden.

Der bekannte Kohlebogen mit seiner erheblich höheren Leuchtdichte ist zu Vergleichszwecken eingefügt; er ist zeitlich und räumlich inkonstant. Bei Gasentladungslampen sind Quecksilber- und Xenonlampen von größerem Interesse.

Quecksilberdampf strahlt in Hochdrucklampen ein Linienspektrum ab, das diese für Zwecke der Photometrie interessant macht, besonders weil die räumliche Verteilung der Leuchtdichte und die Abstrahlcharakteristik günstig sind. Man erreicht aber nur etwa 800 sb. Bei Höchstdrucklampen (30 bis 60 Atmosphären Gasdruck) wird das Linienspektrum von einem starken Untergrund überlagert (Fig. 57a, b), und man erreicht hier eine mittlere Leuchtdichte von 25000 und Leuchtdichten in der Nähe des Brennfleckes von 100000 sb. Diese Lichtquellen sind sehr klein, und man bildet sie zur Vergrößerung der leuchtenden Fläche gern durch einen Kugelspiegel, gegen dessen Zentrum sie etwas seitlich verscho-

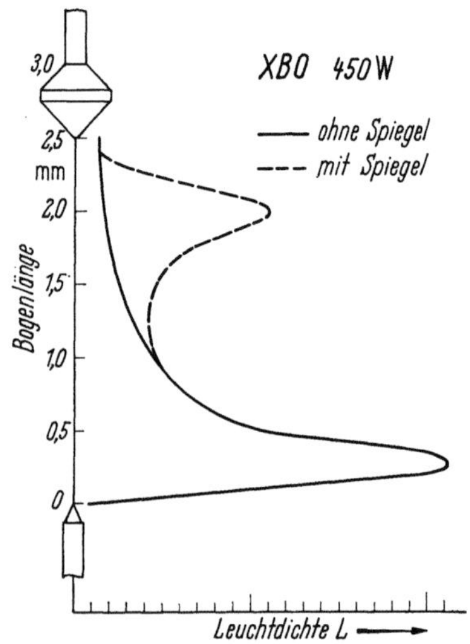

Fig. 58. Leuchtdichteverteilung einer Xenonhöchstdrucklampe.

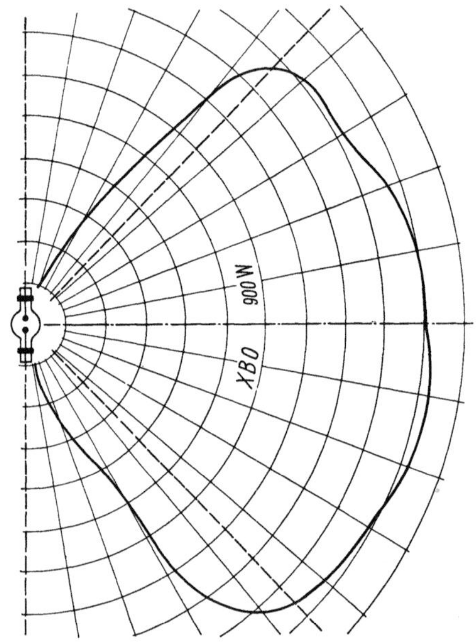

Fig. 59. Die Abstrahlcharakteristik einer Höchstdrucklampe ist gleichmäßig bis zu den Richtungen, in denen die Elektroden Schatten werfen.

ben sind, genau neben sich und kopfstehend ab. Einen Eindruck von Leuchtdichtenverteilung und Abstrahlcharakteristik eines solchen Strahlers vermitteln Fig. 58 und 59 für derartige Quecksilberlampen wie für Xenonlampen, die im sichtbaren und im UV ein Kontinuum zeigen mit Farbtemperaturen, die dem Tageslicht und damit auch dem Tageslichtfilm entsprechen.

Elektronenblitze bedienen sich einer Gasentladung in Xenon von etwa $^1/_{1000}$ sec Dauer und mit einer Leuchtdichte etwa wie die einer Xenon-Höchstdrucklampe. Ihre Abstrahlung ist sehr gleichmäßig und ihre Farbtemperatur entspricht der des Tageslichts. So werden Elektronenblitze gern zur Mikro-Farbphotographie verwendet. Zum Einstellen benötigt man eine besondere Leuchte, die entweder in den Blitz abgebildet und durch ihn hindurch gesehen wird, oder eine Leuchte, die über einen Strahlenteiler eingespiegelt wird und, vom Objekt aus gesehen, am gleichen Ort wie der Blitz zu liegen scheint.

19. Mikrophotographie, Mikroprojektion, Fernsehen. Zur Mikrophotographie wie zur Mikroprojektion wird das Zwischenbild mit dem Okular auf die photographische Schicht (bzw. die Schicht eines anderen Empfängers) oder auf eine Projektionswand projiziert. Die gewünschte Bildgröße kann dabei zwischen etwa 3 m Durchmesser für Projektion auf eine Wand und einer Diagonale von 5,5 mm bei 8 mm Kinofilm schwanken. Die Projektionsflächen stehen grundsätzlich nicht in unendlicher Entfernung vom Okular, für die diese korrigiert sind (Fig. 60a). Würde man mit unverändertem Mikroskop auf diese nahe Ebene fokussieren, müßte das Zwischenbild objektseitig vor dem Brennpunkt des Okulars liegen und das Objektiv würde mit falscher Bildweite benutzt werden (Fig. 60b). Zur Korrektur auf die vorliegende Projektionsentfernung benötigt man entweder ein Okular mit verstellbarer Augenlinse (Ziff. 12), die man so weit herausschraubt, daß das Bild der Okularblende gemeinsam mit dem Objektbild auf der Projek-

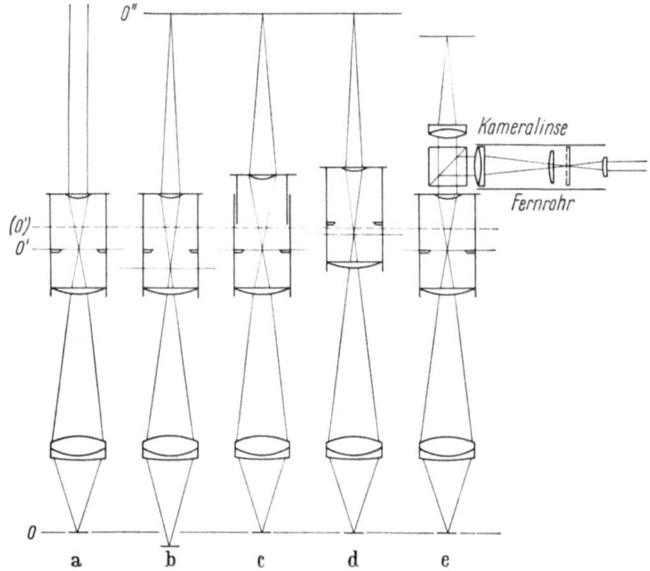

Fig. 60a—e. Erhaltung der richtigen Bildweite des Zwischenbildes (*O'*) bei Mikroprojektion und Mikrophotographie. Die aus dem Objektiv austretenden Strahlen müssen das Zwischenbild in (*O'*) entwerfen, es wird von der Feldlinse nach *O'* verlagert. a Normale Einstellung des Okulars bei subjektiver Beobachtung. b—e Scharfstellung des Bildes auf eine nicht in unendlicher Entfernung liegende Projektionsfläche *O''*, b durch Objektverschiebung, bei falscher Benutzung des Objektivs, c durch Verstellung der Okular-Augenlinse, d durch Verschieben des ganzen Okulars, e durch Benutzung einer Kameralinse.

tionswand scharf erscheint (Fig. 60c), oder man zieht das ganze Okular, nachdem man subjektiv fokussiert hat, so weit aus dem Tubus, bis das zuvor fokussierte Objektbild, nicht aber die Blende, auf der Projektionswand scharf erscheint (Fig. 60d).

Die Vergrößerung im Projektionsbild ist in diesen Fällen

$$M = V_{\text{Mikroskop}} \times \frac{\text{Projektionsentfernung}}{250\ \text{mm}}. \qquad (19.1)$$

Dabei ist die Projektionsentfernung von der Austrittspupille des Mikroskops aus zu messen. Die Projektionsentfernung bestimmt die Größe des Projektionsbildes, dessen Durchmesser bei der Bezugssehweite von 25 cm Projektionsentfernung gleich Sehfelddurchmesser des Okulars $\times V_{\text{Ok}}$ ist. Man sieht leicht, daß man bei den in Ziff. 8 u. 10 für Okulare angegebenen Daten und einem Bilddurchmesser von 3 m keine großen Projektionsentfernungen erreichen kann. Aus diesem Grunde werden speziell für die Mikroprojektion Okulare mit Vergrößerungen bis $V=1\times$ und einheitlichem Sehfelddurchmesser angeboten, bei denen die Austrittspupille zum Teil im Okular liegt.

Bei den notwendigen Projektionsentfernungen ergeben sich für verschiedene Okulare zum Teil ganz erhebliche Verschiebungen z der Bilder aus der Brennebene, der eigentlichen Bildebene. Sie sind für einige Okulare in Tabelle 4 nach $z \cdot z' = -f^2$ in mm berechnet, wobei z' die Projektionsentfernung ist.

Tabelle 4.

Okular f	1× 250	2,5× 100	5× 50	10× 25	25× 10
$z' = 5$ m	12,5	2	0,5	0,125	0,02
$z' = 250$ mm	250	40	10	2,5	0,4
$z' = 100$ mm	625	100	25	6,25	1

Im allgemeinen haben die Augenlinsen der Okulare etwas kürzere Brennweiten als das jeweilige Gesamtokular. Dadurch ist die notwendige Verstellung der Augenlinsen etwas geringer.

Wenn man mit unverändertem Mikroskop auf kurze Entfernung photographieren will, verwendet man Kameralinsen, in deren Brennebene das Bild entsteht (Fig. 60e). Die notwendige Brennweite der Kameralinsen ergibt sich aus

$$f_{\text{Kam}} = \frac{\varnothing \text{ Projektionsbild}}{\varnothing \text{ Okularsehfeld}} \cdot f_{\text{Okular}}, \tag{19.2}$$

ihre Vergrößerung wird als Kamerafaktor $p = f/250$ angegeben. Ein derartiger Aufbau gibt die Möglichkeit, zwischen Okular und Kameralinse einen Strahlenteiler zu schalten und das Bild mit Fernrohr statt auf Mattscheibe einzustellen.

Tabelle 5.

Nach FRENK (1960)	Lichtstrom in Lumen durch 18 mm Okular-Sehfeld	Schirmleuchtdichte in asb bei Schirmfläche 3 m²	Aufgelöste Punkte
Mikroprojektor mit XBO 501 mit 18 mm Sehfeld Objektiv Pl 10/0,25 Pl 80×/0,95	 60 lm 12 lm	 20 4	 1,8 Millionen 0,4 Millionen
Vidikon 260 Punkte/Zeile S-Orthikon 260 Punkte/Zeile	3,6 mlm 0,6 mlm	Fernseh-Projektor MW 6-2 13,3 MW 13-38 48 Eidophor (670)	 0,05 Millionen
		bei Bildröhre 1250 [maximale Fläche (Diagonale 65 cm) 0,2 m²] zum Vergleich: Vollmond auf weißes Papier 0,2	

Die Möglichkeiten des Ersatzes von Mikroprojektion mit sehr hoher Strahlenbelastung des Objektes durch Fernseheinrichtungen hat FRENK (1960) nach Messungen von K. WEBER zahlenmäßig abgeschätzt.

Mit dem Mikroskop werden in Abhängigkeit vom Verhältnis Apertur: Vergrößerung des Objektivs in 18 mm Sehfeld zwischen 0,4 und 21 Mill. Bildpunkte aufgelöst, bei 28 mm Sehfeld zwischen 1,2 und 4,3 Mill. Bildpunkte und beim Fernsehen mit 625 Zeilen Norm 0,05 Mill. Bildpunkte. Die Objektbelastung ist angegeben als Lichtstrom in Lumen durch 18 mm okulares Sehfeld und den Leuchtdichten der Projektionswand gegenübergestellt.

Die Zusammenstellung (Tabelle 5) zeigt, daß die Lichtströme durch 18 mm Sehfeld bei Mikroprojektion bis 10^4mal so groß sein müssen wie bei Schwarz-Weiß-Fernsehprojektion, um etwa dieselbe Bildschirm-Leuchtdichte zu erreichen.

Dafür ist die Zahl der Bildpunkte auf $^1/_{20}$ zurückgegangen und eine Farbinformation fehlt. Bei Farbfernsehen sind neuerdings ähnliche Lichtströme wie in der Tabelle angegeben nötig.

Die Objektbelastung ist für 1 mm² Fläche im Objekt zu bestimmen. Sie ist bei subjektiver Mikroskopie 50 mlm/mm² (WEBER 1966). Die oben angegebenen 60 lm bei Pl 10/0,25 entsprechen etwa 24 lm/mm² im Objekt, die 12 lm bei Pl 80×/0,95 300 lm/mm².

Während z.B. eine Stahlprobe im Auflicht gegenüber den bisher möglichen Lichtströmen völlig unempfindlich ist, kann man bei etwa 50 lm/mm² bei gefärbten Präparaten im Durchlicht schon Schwierigkeiten bekommen durch Erwärmung der Präparate, Entgasung der Kittmittel und Entfärbung der zum großen Teile nicht lichtechten Farbstoffe.

III. Besondere Geräte.

In den Ziff. 20—27 werden mikroskopische Geräte und Verfahren besprochen, die den Anwendungsbereich des Mikroskopes ganz wesentlich erweitert haben und zeigen, daß das Mikroskop auf dem Wege ist, vom reinen Beobachtungsinstrument immer mehr zu einem Meßgerät zu werden.

Der Kern aller dieser Instrumente ist das bisher besprochene Mikroskop, ob sein Arbeitsbereich aus dem Sichtbaren ins Ultraviolette oder Infrarot verlegt wird, ob die ultraviolette Strahlung zur Abtötung von Zellen oder zur Erregung von Fluoreszenz benutzt wird, ob man mit Hilfe von polarisiertem Licht, mit Phasenkontrast oder durch Verbindung mit einem Interferometer Phasendifferenzen im Objekt sichtbar macht und mißt oder ob man die Absorptionseigenschaften bestimmt.

Die im folgenden besprochenen Ausrüstungen dienen in den verschiedensten Kombinationen der Bestimmung physikalischer und über diese chemischer Eigenschaften der mikroskopischen Objekte. Sie werden in einer Reihenfolge besprochen, bei der nach Möglichkeit das einzelne Gerät oder die einzelne Methode beschrieben ist, ehe sie mit einer anderen kombiniert wird. Auf reine Beobachtungsmethoden, wie Intravitalmikroskopie oder die Beobachtung von Gefügeumwandlungen in Heiztischen, die z. B. gegenüber normalen Objektiven und Kondensoren solche mit großen Arbeitsabständen erfordern, kann, um den Rahmen des Artikels nicht zu sprengen, nicht eingegangen werden.

Zu den folgenden Abschnitten existiert eine große Menge Literatur, die sich mit der Verwendung der verschiedenen Geräte und Gerätekombinationen für bestimmte Aufgaben in der Physik, der Biologie oder der Technik beschäftigt. Bevorzugt werden solche Arbeiten zitiert, die zur Illustration des Gesagten dienen und durch Literaturangaben weiterhelfen können.

20. Phasenkontrasteinrichtungen. Der Phasenkontrast, ein fast ausschließlich in der Mikroskopie benutztes Verfahren, ist von H. WOLTER (1956) in Band 24 dieses Handbuches so ausführlich besprochen worden, daß es hier nur kurz zu erwähnen ist.

Nicht absorbierende Objekte, die nur die Phase des durch sie hindurchgehenden Lichtes beeinflussen, konnten üblicherweise nur im Dunkelfeld sichtbar gemacht werden, in dem jedes Beugungszentrum, also jede Inhomogenität auch in bezug auf die Lichtphase sichtbar wird, wenn das direkte Licht vollkommen von der Beobachtung ausgeschlossen ist.

ZERNIKE (1932, 1934) zeigte mit Hilfe der Abbeschen Theorie, also bei Beleuchtung mit unendlich ferner Lichtquelle, der Beleuchtungsapertur $A=0$ und für ein ideales Phasengitter mit kleinen Phasendifferenzen zwischen Balken und Lücken im Vergleich zu einem idealen Amplitudengitter, daß bei gleicher Phase des nullten Maximums, also für das nichtgebeugte Licht, sich die einzelnen anderen Maxima $m=\pm 1, \pm 2\ldots$ jeweils um $\varphi=90°$ unterscheiden (Fig. 61). Wenn man also in der Apertur des Objektivs das nullte Maximum gegen die Nebenmaxima gemeinsam in seiner Phase um 90° verschiebt, muß aus dem Bild eines Amplitudengitters ebenso das unsichtbare Bild eines Phasengitters werden, wie das unsichtbare Phasengitter einem Amplitudengitter gleich, also mit hellen und dunklen Balken abgebildet werden muß. Verzögert man das nullte Maximum gegenüber den Nebenmaxima um $\lambda/4$, spricht man von „positivem“ Phasenkontrast, in dem dickere Objekte und Objekte mit höherer Brechzahl dunkler als ihre Umgebung erscheinen. Im umgekehrten Fall, also bei Vorausschieben des nullten Maximums, hat man „negativen“ Phasenkontrast, in dem dickere Objekte heller als ihre Umgebung aussehen.

Die Phasenschiebung wird in der bildseitigen Brennebene der Objektive durch eine dünne Schicht bewirkt, die das Licht gegenüber dem Umfeld um 90° oder 3 × 90° verzögert, in letzterem Fall also um 90° vorschiebt. Die phasenschiebende Schicht auf dem Phasenplättchen muß dem Bild der Lichtquelle größengleich sein.

In der Praxis wird eine saubere Trennung von direktem und gebeugtem Licht nicht möglich sein, denn, wenn man statt eines Gitters ein normales Objekt benutzt und dadurch keine klar begrenzten Beugungsmaxima hat, sondern das gebeugte Licht die ganze Pupille füllt, wird man an der Stelle des Aperturblendenbildes mit der phasenschiebenden Schicht auch einen Teil des gebeugten Lichtes beeinflussen. Weiterhin benötigt man eine von Null deutlich unterschiedene Beleuchtungsapertur. Man verwendet, um die durch diese Änderungen gegenüber dem idealen Phasenkontrast auftretenden Störungen möglichst klein zu halten, als Blenden Lichtringe und entsprechend in der Objektivapertur Phasen*ringe*, die aber eine endliche Breite haben müssen. Dieser nicht ganz saubere Eingriff in die Pupille ist z.B. Ursache der als Halo bezeichneten Erscheinung, daß dunkle Objektdetails bei positivem Kontrast von einem hellen, und helle Objektdetails bei negativem Kontrast von einem dunklen Saum umgeben sind. Er wird bei abnehmender Ringbreite schmaler, weshalb man sehr darauf achtet, die Phasenringe nur möglichst wenig breiter zu machen als die Bilder der Lichtringe. An der Breite der Phasenringe liegt es ebenso, daß die Kontraste im großen Objektdetail nicht stimmen, weil von solchem Detail viel gebeugtes Licht auf den Phasenring fällt.

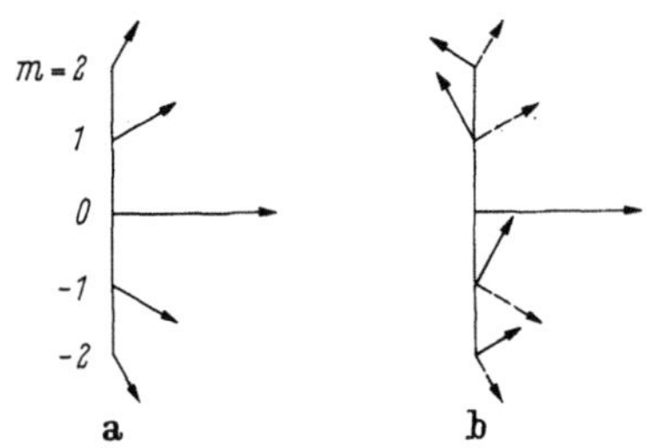

Fig. 61 a u. b. Schwingungszustand in den verschiedenen Spektren eines Amplitudengitters (a) und eines Phasengitters (b), ausgezogene Vektoren. Die Phasen sind als Winkel der Vektoren gegen die Horizontale dargestellt (nach ZERNIKE, 1942).

ZERNIKE zeigt schon, daß man zur Steigerung des Kontrastes die Amplitude des direkten Lichts im Phasenring mindern soll. Im allgemeinen wird man etwa 75% des direkten Lichtes absorbieren und nur da zur höheren Absorption übergehen, wo die Phasenschiebung im Objekt sehr klein ist.

Das normale „Phasenobjekt“ des Mikroskopikers weicht ganz erheblich vom idealen ab und die untersuchten Objekte sind häufig absorbierend. BRICE und KECK (1947) und KECK und BRICE (1949) haben für nahezu punktförmiges Objektdetail die zur Erzeugung von optimalem Kontrast im Bild notwendige Phasenschiebung und Absorption im Phasenring berechnet. Sie kommen entsprechend der Variation der Objekte zu großen Unterschieden für die wünschenswerten Eigenschaften der Phasenringe. Das würde Einrichtungen mit variabler Absorption und Phasenschiebung wie z.B. bei Polanret (OSTERBERG 1947) als günstig erscheinen lassen.

Die übliche Ausführung ist ein Kondensor mit einigen auf Revolver austauschbaren Ringblenden in seiner unteren Brennebene und Objektive mit Phasenringen, die diesen Lichtringen angepaßt sind (Fig. 62a). Da im allgemeinen die Dicke der beobachteten Objekte mit steigender Objektivvergrößerung abnimmt, kann man stärkere Objektive durch eine höher absorbierende Phasenschicht den Objekten anpassen. Im übrigen bevorzugen es die Mikroskopiker, ihre Objekte durch Verwendung geeigneter Einbettungsmittel an die Objektive anzupassen, statt eine große Zahl von Objektiven auszutauschen.

Zu schnellem Vergleich von Objektstellen im Phasenkontrast mit den bekannten Bildern in Hellfeld und Dunkelfeld war es, besonders in den Anfangszeiten der Phasenkontrastmikroskopie, wichtig, diesen Übergang von einer zur anderen Beobachtungsart schnell und ohne das Objekt aus dem Auge zu verlieren zu machen.

HEINE (1949) schlug hierfür die Verwendung eines modifizierten Dunkelfeldkondensors (Fig. 62b) vor, welcher zusätzlich zur Dunkelfeldbeleuchtung einen Lichtring erzeugte, der durch Heben und Senken des Kondensors in den Objektiven verschieden groß abgebildet werden konnte. Er erzeugte bei Abbildung auf einen Phasenring Phasenkontrast und im übrigen ein Hellfeld mit ringförmigen kleinen Beleuchtungsaperturen, wie man sie zur Beobachtung von Objekten mit geringem Kontrast benötigt. Da der Lichtring bei diesem Kondensor dicht unter dem Objekt entsteht, können die zugeordneten Phasenringe nicht in der bildseitigen Brennebene der Objektive liegen.

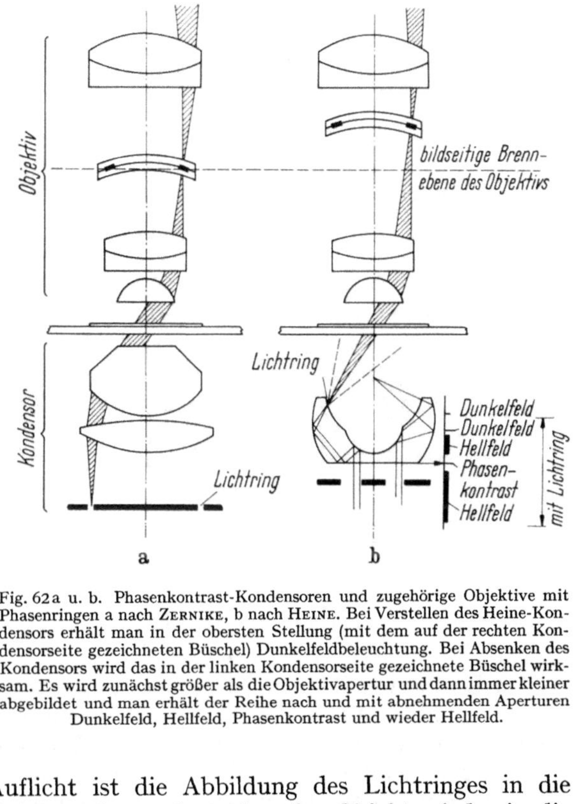

Fig. 62a u. b. Phasenkontrast-Kondensoren und zugehörige Objektive mit Phasenringen a nach ZERNIKE, b nach HEINE. Bei Verstellen des Heine-Kondensors erhält man in der obersten Stellung (mit dem auf der rechten Kondensorseite gezeichneten Büschel) Dunkelfeldbeleuchtung. Bei Absenken des Kondensors wird das in der linken Kondensorseite gezeichnete Büschel wirksam. Es wird zunächst größer als die Objektivapertur und dann immer kleiner abgebildet und man erhält der Reihe nach und mit abnehmenden Aperturen Dunkelfeld, Hellfeld, Phasenkontrast und wieder Hellfeld.

Bei Phasenkontrast im Auflicht ist die Abbildung des Lichtringes in die hintere Brennebene des Objektivs und von dort über das Objekt wieder in die hintere Brennebene notwendig. Man verliert durch die Absorption im Phasenring also viel Licht. Im allgemeinen liegen die hinteren Brennebenen der Objektive nicht in einer Ebene (Ziff. 9 u. 16), so daß ein Fokussieren auf der Beleuchtungsseite oder die Verwendung verschiedener Systeme zur Abbildung der Lichtringe in die Brennebenen der Objektive notwendig ist.

Dasselbe Problem ergibt sich, wenn man die Phasenringe nicht im Objektiv, sondern in einer Zwischenabbildung der Pupille anbringen will.

21. UV-Mikroskope, IR-Mikroskope, Mikro-Strahlenstich. UV- und IR-Mikroskope dienen zur Beobachtung von Objekten, die in diesen Bereichen absorbieren oder reflektieren. Die UV-Mikroskopie wird zusätzlich zur Erhöhung der Auflösung benutzt (KÖHLER 1904). Die nutzbaren Spektralbereiche hängen von der Durchlässigkeit der optischen Materialien für Beleuchtungs- und Abbildungsoptik und den Empfindlichkeitsbereichen der Rezeptoren ab. Eine klare Grenze zum kurzwelligen hin ist etwa 200 nm, wo ein Übergang von Luft ins Vakuum notwendig wäre. UV-Bildwandler sind etwa von 200 bis 700 nm brauchbar und IR-Bildwandler im Bereich des sichtbaren Lichtes und bis 1 µm. Im nahen UV und IR sind die normalen Objektive und Okulare brauchbar. Für die fernen Bereiche verwendet man Spiegelobjektive oder bei Linsenobjektiven Monochromate aus Quarzglas oder Ultra-Achromate aus UV-durchlässigen Materialien. Mit den Objektiven verwendete Okulare müssen entweder entsprechend farbkorrigiert sein oder man verzichtet auf eine zweistufige Abbildung und erzielt durch Verwendung größerer Bildweiten und unter Verzicht auf Okulare in einem Schritt Vergrößerungen, die den Aperturen der Objektive angepaßt sind.

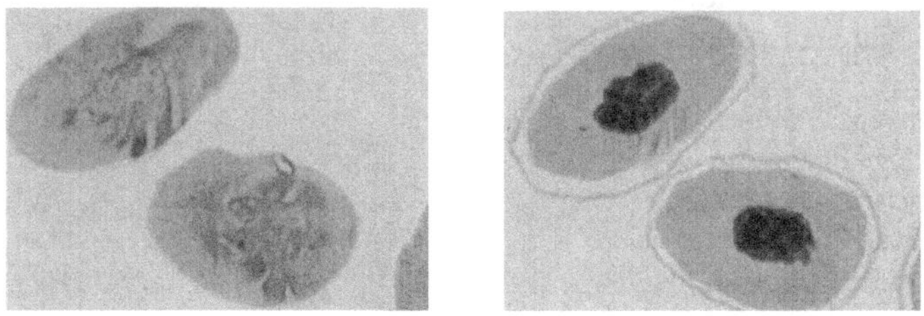

a b

Fig. 63a u. b. Froschbluterythrocyten im UV a bei 300 nm, b bei 250 nm (H. G. KRÜGER, 1961).

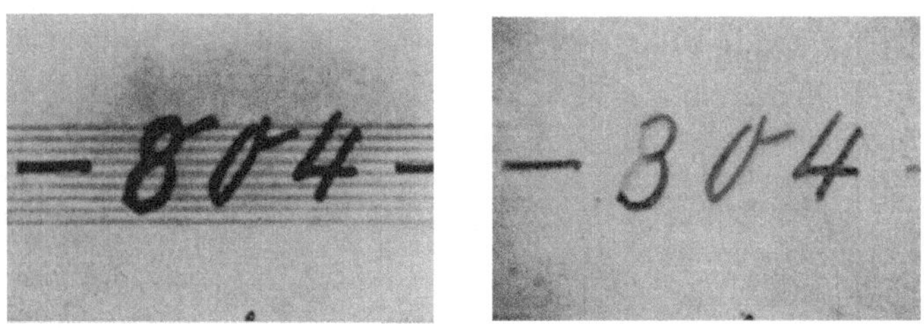

a b

Fig. 64a u. b. Schriftprobe a im sichtbaren Licht, b im IR zeigt die Änderung einer 3 in eine 8 (H. BECKER).

Im UV verwendet man als Immersionsmittel das UV-durchlässige Glycerin mit etwa 10% Wasser ($n_e = 1{,}47$). Die subjektive Beobachtung erfolgt über geeignete Bildwandler. Es ist sinnvoll, im UV direkt, also unter Umgehung des Bildwandlers und der mit ihm verbundenen Bildverschlechterung zu photographieren, während man im IR zu entscheiden hat, ob man die Schwierigkeiten mit Lagerung der IR-Photomaterialien oder die Bildverschlechterung im Bildwandler höher einschätzt. Im UV wurden früher als Rezeptoren Fluoreszenzschirme mit Zinksulfid benutzt, deren Wirkungsgrad etwa $^1/_{50}$ der eines Bildwandlers ist.

Als Beispiel für die Verschiedenheit der Absorption bei verschiedenen Wellenlängen im UV: Froschblut-Erythrocyten bei 300 nm, Fig. 63a, und bei 250 nm, Fig. 63b, und für Anwendung des IR eine Schriftprobe im sichtbaren Licht, Fig. 64a, und im IR, Fig. 64b. Der Vergleich zeigt hier die Änderung einer 3 in eine 8.

Mikro-Strahlenstich. Ultraviolette Strahlung schädigt lebendes Gewebe. Im allgemeinen bemüht man sich deshalb, dort mit möglichst kleinen Strahlungsdosen auszukommen. Anders ist es, wenn man einzelne Zellen gezielt schädigen will, z.B. um bei einem Ei festzustellen, welche Körperteile sich aus einer bestimmten Zelle entwickeln. Schädigt man diese Zelle durch Bestrahlung, werden entsprechende Ausfallserscheinungen auftreten. Die Aufgabe ist also, das Objekt zu beobachten und ein kleines Gebiet des Objektes mit UV zu bestrahlen. Lebende Objekte sind im allgemeinen Phasenobjekte. So ist Beobachtung im Phasenkontrast oder Dunkelfeld mit der Projektion von UV-Strahlung durch eine in ihrer Form variable Blende auf eine bestimmte Objektstelle zu kombinieren. Vgl. z.B. Tschachotin (1912), Seidel (1952), Gottschewski (1954), Uretz and Perry (1957), C. L. Smith (1964). Selbstverständlich ist hier nicht nur ultraviolettes Licht als schädigende Strahlung geeignet.

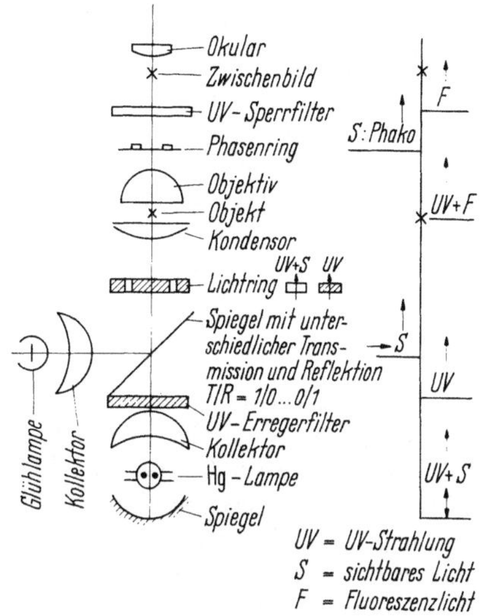

Fig. 65. Phasenkontrast-Fluoreszenz (Schemaskizze). Fluoreszenz: Aus der Strahlung der Hg-Lampe wird durch ein UV-Erregerfilter das zur Erzeugung der Fluoreszenz notwendige UV ausgesondert. Das Material um den Lichtring herum ist das des UV-Erregerfilters, also UV-durchlässig. Die UV-Strahlung erregt im Objekt sichtbares Fluoreszenzlicht und wird selbst durch das UV-Sperrfilter darüber nicht durchgelassen. Man sieht die Fluoreszenzstrahlung. Phasenkontrast: Das weiße Licht einer Glühlampe wird von der Seite eingespiegelt. Nur die Fläche des Lichtringes ist für weißes Licht durchlässig und wird auf den Phasenring abgebildet. Man beobachtet mit weißem Licht in Phasenkontrast. Beide Beobachtungen sind gleichzeitig möglich. Die Lichtanteile sind durch Ein- und Ausschaltung der Lichtquellen und Regulierung der Leuchtdichte der Glühlampe zu variieren.

22. Fluoreszenzmikroskopie. Bei Bestrahlung mit kurzwelligem Licht fluoreszieren gewisse Substanzen. Üblicherweise verwendet man in der Mikroskopie zur Bestrahlung kurzwelliges Blau oder Ultraviolett und erzeugt eine Fluoreszenzstrahlung im sichtbaren. Man hat aber durchaus auch an andere Wellenlängenpaarungen zu denken. Fluoreszieren die zu untersuchenden Substanzen selbst, spricht man von Primär- oder Eigenfluoreszenz, fluoreszieren sie nach Einfärbung mit geeigneten Farbstoffen, spricht man von sekundärer Fluoreszenz und man unterscheidet durch Beobachtung der Fluoreszenzerscheinung, ob die Objekte mit einem bestimmten Fluoreszenzfarbstoff anfärbbar sind oder nicht und in welcher Farbe sie fluoreszieren. Die benötigten Farbmengen sind so gering, daß sie lebende Substanzen im allgemeinen nicht schädigen.

Man beleuchtet durch ein geeignetes Filter mit kurzwelligem Licht und mit entsprechend lichtdurchlässiger Beleuchtungsoptik und filtert dieses Beleuchtungslicht vor dem Auge aus, während man das Fluoreszenzlicht möglichst weitgehend durchläßt. Dadurch sieht man die fluoreszierenden Objekte hell auf dunklem Grunde. Man versteht leicht, daß eine geeignete Wahl dieser Filter wesentlich für die Dunkelheit des Untergrundes ist, wobei das die Erregerstrahlung absorbierende Filter, das Sperrfilter, erheblichen Einfluß auf den durchgelassenen Anteil und damit die „Farbe“ des Fluoreszenzlichtes hat. UV-Strahlung kann mit farblosen Filtern gesperrt werden, Blaulicht nur mit Gelbfiltern. Dafür hat schon eine hochbelastete Glühlampe einen hinreichenden Anteil von Blaulicht, während für die UV-Strahlung aufwendigere Lampentypen benötigt werden.

Verwendet man Hellfeld-Beleuchtung mit hoher Apertur, wird die Anregungsintensität höher sein als bei gleichaperturiger Dunkelfeldbeleuchtung, bei der der zentrale Aperturbereich fehlt. Dafür ist hier die Gefahr ausgeschaltet, im Objektiv an Linsen oder Kittschichten Fluoreszenzlicht zu erzeugen. Das Ideal ist also ein Hellfeld-Kondensor hoher Apertur, bei dem man nach Belieben verschieden große zentrale Aperturbereiche ausblenden kann. Ein solcher Kondensor bietet zusätzlich die Möglichkeit für Phasenkontrastbeobachtung, wenn man z.B. die

Lichtringe nicht mit einem voll absorbierenden, sondern mit einem UV-durchlässigen, aber sichtbares Licht sperrenden Filter umgibt. Dann ist die gesamte Kondensorapertur UV-durchlässig, aber nur der Lichtring für sichtbares Licht (Fig. 65). Die Anteile von UV-Strahlung und Weißlicht ergeben die Mischung von Fluoreszenzanteil und Phasenkontrastanteil im Bild.

23. Polarisationsmikroskope. Zur Untersuchung doppelbrechender Substanzen wird das gewöhnliche Mikroskop durch je einen Polarisator vor und hinter der Objektebene ergänzt. Als Polarisatoren wurden früher fast ausschließlich Prismenanordnungen aus Kalkspat benutzt, die das auf sie fallende natürliche Licht in zwei zueinander senkrechte polarisierte Büschel aufspalten, das ordentliche Büschel ($n_\omega = 1{,}6584$) an einer Kittfläche reflektieren und das außerordentliche Büschel ($n_\varepsilon = 1{,}4864$) als linear polarisiertes Licht durchlassen. Sie sind heute weitgehend ersetzt durch Polarisationsfilter aus dichroitischem Material, bei denen die eine Welle hinreichend absorbiert wird. Der „Polarisationsgrad" von Polarisatoren ist definiert (z.B. HASE 1961) als

$$P = 100\left(1 - \frac{D_+}{D_\parallel}\right)\% \,. \tag{23.1}$$

Dabei ist $D_\parallel$ die Durchlässigkeit zweier gleicher Polarisatoren oder eines Polarisators gegen ein Polarisationsprisma in Parallelstellung der durchgelassenen Schwingungen und D_+ die Durchlässigkeit beider in Kreuzungsstellung für einen bestimmten interessierenden Spektralbereich. Der Polarisationsgrad von Polarisationsprismen ist etwa 99,99% im Bereich ihrer Durchlässigkeit von etwa 300 bis 2000 nm und gleicht damit dem von brauchbaren Filtern im Bereich von 380 nm bis 750 nm. Es gibt sowohl Prismen wie auch Polarisationsfilter, die speziell für weitere Durchlässigkeitsbereiche entwickelt sind. Ein Nachteil der Polarisationsfilter ist ihre durch die Absorption des Lichtes auftretende Erwärmung, die sie in Verbindung mit Lichtquellen hoher Intensität nur mit Vorpolarisatoren verwendbar machen, wenn man in einem solchen Fall nicht lieber zu Polarisationsprismen greift, bei denen es Formen gibt, die den reflektierten Teil des Lichtes austreten lassen.

Orthoskopische Beobachtung. Kreuzt man die Schwingungsrichtungen der beiden im Mikroskop eingebauten Polarisatoren, soll die dazwischen eingebaute Optik so geringe Spannungsdoppelbrechung haben, daß das Sehfeld auch bei heller Beleuchtung absolut dunkel ist. Eine Störung tritt dann nur noch durch Drehung der Polarisationsebene an stark gekrümmten Linsenflächen der Objektive auf. Auf die Beseitigung dieses Fehlers wird später eingegangen.

Bringt man eine doppelbrechende Platte auf einem drehbaren Tisch zwischen die beiden gekreuzten Polarisatoren, ist die zu beobachtende Intensität I in Abhängigkeit von der des einfallenden Lichtes I_0, vom Winkel ψ, den die im Kristall vorhandenen und aufeinander senkrechten Schwingungsrichtungen mit den Schwingungsrichtungen von Polarisator und Analysator bilden und von der Phasendifferenz φ der beiden Schwingungen im Kristall nach FRESNEL

$$I = I_0 \sin^2 2\psi \sin^2 \frac{\varphi}{2} \,. \tag{23.2}$$

Dabei ergibt sich φ aus dem Gangunterschied Γ der beiden Schwingungen als Produkt der Differenz der Brechungsindices in Beobachtungsrichtung, $n'_\gamma - n'_\alpha$, der Präparatdicke d und der Wellenlänge λ_0, für die die Brechungsindices bestimmt sind, zu

$$\varphi = 2\pi \frac{\Gamma}{\lambda_0} = 2\pi \frac{d}{\lambda_0}(n'_\gamma - n'_\alpha) \,. \tag{23.3}$$

Ist $\sin^2\frac{\varphi}{2}\neq 0$, ergibt sich für $\psi=0$ oder $\psi=\frac{\pi}{2}$ die Intensität $I=0$. Das Präparat wird in den Stellungen, in denen die Schwingungsrichtungen n'_γ und n'_α mit den Schwingungsrichtungen von Polarisator und Analysator zusammenfallen, der sogenannten Normal- oder Auslöschungsstellung, dunkel erscheinen. Für $\psi=\frac{\pi}{4}$ wird $I=I_0\sin^2\frac{\varphi}{2}$; das Präparat erscheint maximal hell.

Bei einer Umdrehung der Kristallplatte um die Mikroskopachse folgen also vier Dunkelstellungen nach jeweils 90°, unterbrochen von vier Bezirken mit Aufhellung und maximaler Helligkeit in den vier Diagonalstellungen, jeweils unter 45° zu den Normalstellungen.

Der Faktor $\sin^2\frac{\varphi}{2}$ wird ein Maximum, wenn

$$\Gamma=d(n'_\gamma-n'_\alpha)=(2m+1)\frac{\lambda}{2}, \tag{23.4}$$

und ein Minimum, wenn

$$\Gamma=d(n'_\gamma-n'_\alpha)=2m\frac{\lambda}{2}=m\cdot\lambda \tag{23.5}$$

ist. Eine Platte erscheint also dunkel, wenn die beiden Wellenzüge um eine ganze Wellenlänge, und hell, wenn sie um eine halbe Wellenlänge gegeneinander verschoben sind. Die im Kristall kohärenten und senkrecht zueinander schwingenden Wellen sind nicht interferenzfähig. Ihre Anteile nach der Schwingungsrichtung des Analysators sind dagegen interferenzfähig, wodurch die berechneten Helligkeiten und Dunkelheiten zu erklären sind.

Für den Fall der Diagonalstellung, also für $\psi=\pi/4$, ergibt sich, daß Platten mit einem Gangunterschied von $\Gamma=m\lambda$ eine lineare resultierende Schwingung parallel zur Schwingung des Polarisators erzeugen. Man nennt solche Platten λ-Platten. Für $\Gamma=(2m+1)\,\pi/2$, eine $\lambda/2$-Platte, ist die resultierende Schwingung ebenfalls linear, aber senkrecht zur Schwingung des Polarisators und für eine Platte mit $\Gamma=(2m+1)\,\lambda/4$, eine $\lambda/4$-Platte, ist die resultierende Schwingung zirkular und zwar für gerade m links-zirkular und für ungerade m rechts-zirkular.

Ist das Präparat keine Platte, sondern ein Keil und zwar in Diagonalstellung ($\psi=45°$), folgen einander bei monochromatischen Licht entsprechend den Formeln (23.4), (23.5) beginnend mit Dunkelheit an der Keilschneide helle und dunkle Streifen, bei denen der Abstand δ zwischen gleich hellen Streifen für einen Keilwinkel α

$$\delta=\frac{\cot\alpha}{n'_\gamma-n'_\alpha} \tag{23.6}$$

ist. Der Abstand der Streifen wächst mit der Wellenlänge, da die Änderung der Doppelbrechung mit der Wellenlänge im allgemeinen recht klein ist. Im weißen Licht fehlen die Wellenlängen, für die $\Gamma=m\lambda$ ist, völlig. So wird ein Keil von der Schneide her gesehen an der Stelle gelb erscheinen, wo sein Gangunterschied gerade der Wellenlänge des komplementären Blau gleich ist und wo deshalb diese Wellenlänge ausgelöscht ist. Dann folgt orange und rot dort, wo grün zum ersten Male ausgelöscht wird. Es folgt die ganze Reihe der Interferenzfarben, die so typisch sind, daß man bis zu der Stelle, wo Dunkelheit für 4λ-grün herrscht, den zugehörigen Gangunterschied angeben kann.

Die Farbreihe wird in „Ordnungen" geteilt, die jeweils einem Gangunterschied von 551 nm entsprechen. $\lambda=551$ nm ist etwa das Maximum der Augenempfindlichkeit und wird hier zum erstenmal ausgelöscht. Die zugehörige Interferenzfarbe ist ein Purpurrot, das sehr empfindlich bei Verkleinerung des Gangunterschiedes zu Rot und bei Vergrößerung des Gangunterschiedes zu Violett um-schlägt. Man nennt diese Farbe am Ende der ersten Ordnung daher „teinte sensible“. Auch die anderen Ordnungen, also z. B. bei 1102 nm, enden wegen des Fehlens von Grün mit Rottönen, die blasser sind und sich bei weitem nicht so empfindlich mit dem Gangunterschied ändern.

Die Interferenzfarben werden nur dann eindeutig sein, wenn Doppelbrechung $(n'_\gamma - n'_\alpha)$ und Dicke d eindeutig vorgegeben sind. Deshalb ist darauf zu achten, daß man bei kleinem Beleuchtungsaperturwinkel u beobachtet. Wegen der schiefen Dicke $d_u = d/\cos u$ erhält man schon bei Aperturen von 0,40 für die Randstrahlen etwa 10% höhere Gangunterschiede als für die Achse und damit eine erhebliche Mischung und Verweißlichung der Dispersionsfarben. Die Änderung der Doppelbrechung mit der Beobachtungsrichtung ist nicht allgemein abzuschätzen.

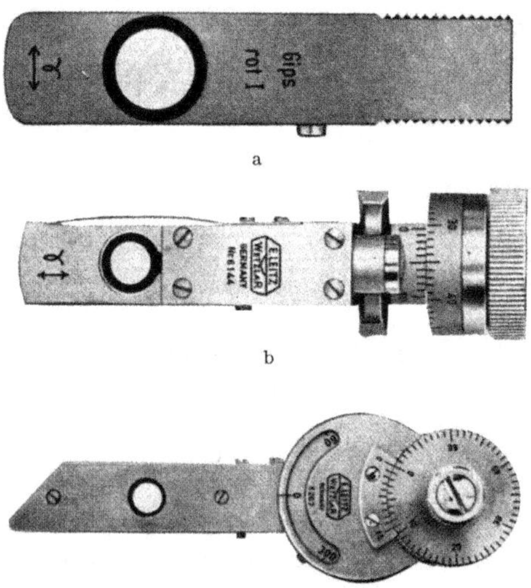

Fig. 66a—c. Kompensatoren für Polarisationsmikroskope a mit festem Gangunterschied ($\Gamma = 551$ nm), b Kippkondensator (nach Berek), c Kompensator mit azimutaler Drehung (nach Sénarmont).

Wenn man zwei doppelbrechende Platten aufeinander legt, kann man sie so gegeneinander drehen, daß die Schwingungsrichtungen der „schnelleren Wellen“ (n'_α) parallel liegen. Die Gangunterschiede werden sich addieren. Man kann die beiden Platten auch so legen, daß die Schwingungsrichtung der schnelleren Welle in der einen der langsameren (n'_γ) in der anderen parallel liegt. Dann werden sich die Gangunterschiede subtrahieren. Der resultierende Gangunterschied wird Null, wenn beide Platten gleichen Gangunterschied haben. Die Stellung der Präparate zueinander nennt man im ersten Fall Additions-, im zweiten Fall Subtraktionsstellung. Nach diesem Prinzip stellt man mit bekannten Kristallplatten fest, in welcher Richtung im Präparat die schnellere oder die langsamere Welle schwingt, und mißt man Gangunterschiede. Derartige Vorrichtungen mit bekannten Kristallplatten nennt man *Kompensatoren* und unterscheidet die im folgenden zusammengestellten Gruppen (Fig. 66). Sie werden mit $\psi = 45°$ zu den Polarisatoren und zwar zumeist in einen Schlitz im Tubus eingeschoben und auf ihnen, mit Ausnahme der Gruppe 2c, ist die Schwingungsrichtung für n'_γ angegeben. Dabei entsprechen sich $\gamma \| n'_\gamma \| \mathfrak{c} \| H_1$ bzw. $\alpha \| n'_\alpha \| \mathfrak{a} \| H_2$.

1. Kompensatoren mit festem Gangunterschied: Gipsplättchen Rot erster Ordnung ($\Gamma = 551$ nm) und Glimmerplättchen $\lambda/4$ ($\Gamma \approx 140$ nm).

Beide werden benutzt, um den relativen Charakter einer Schwingungsrichtung im Objekt festzustellen, d. h. festzustellen, welche Schwingungsrichtung im Objekt n'_α bzw. n'_γ entspricht. Beide Typen ergeben im Bereich verschiedener Gangunterschiede besonders gut erkennbare Interferenzfarben und werden deshalb nebeneinander gebraucht.

2. Kompensatoren mit variablem Gangunterschied:

a) Kompensationskeile (aus Quarz oder Gips). Man bringt sie am besten in einer Objektebene an (direkt auf dem Objekt oder im Zwischenbild, wenn das Mikroskop einen auf das Okular aufsetzbaren Analysator besitzt) und beobachtet gleichzeitig Objekt und Keil, der mit einer Teilung zum Ablesen des Gangunterschiedes an der Kompensationsstelle versehen sein muß.

b) Kipp-Kompensatoren (Kristallplatten oder Kombinationen von Kristallplatten) z.B. nach NIKITIN (1910), BEREK (1913), EHRINGHAUS (1931). Durch Drehen an einem Triebknopf, weshalb diese Kompensatoren im allgemeinen als Drehkompensatoren (Fig. 66b) bezeichnet werden, kippt man die Kristallplatten um eine zur Mikroskopachse senkrechte Achse, so daß ihr Gangunterschied in Richtung der Mikroskopachse variiert. Die Größe des Gangunterschiedes wird aus dem Kippwinkel errechnet.

Meßbereiche: a) bis etwa 5λ und b) bis etwa 100λ. Meßgenauigkeit für a): $\Delta\Gamma = \pm 3$ bis 5 nm. (In der dritten Ordnung bewirkt $0{,}1^\circ$ Kippung eine Gangunterschiedsänderung von etwa 10 nm).

c) Kompensatoren mit azimutaler Drehung (Fig. 66c), nämlich:

c1) nach SÉNARMONT, vgl. z.B. M. BEREK in RINNE-BEREK (1953, S. 317—323) durch Analyse der im Objekt entstandenen Schwingungsellipse mit einer $\lambda/4$-Platte. Die elliptische Schwingung wird durch den Kompensator in eine lineare verwandelt und das Azimut dieser Schwingung mit einem drehbaren Analysator bestimmt. Meßbereich: Eine Ordnung: $m\lambda$ bis $(m+1)\lambda$, $(m = 0, 1, 2 \ldots)$, Meßgenauigkeit: subjektiv etwa $\lambda/200$, mit zusätzlichen Hilfsmitteln wie Halbschattenplatten bis $\lambda/2000$.

c2) nach BRACE (1904) und KÖHLER (1921) mit Platten bekannter, sehr kleiner Doppelbrechung, $\Gamma x = \lambda/30$, $\lambda/20$, $\lambda/10$, $\lambda/4$. Messung durch azimutales Drehen des Plättchens unter Beibehaltung der Lage des Analysators. Meßbereich: zwischen 0 und Γx. Meßgenauigkeit: subjektiv $0{,}1\,\Gamma x$ und mit Hilfsmitteln bis etwa $0{,}01\,\Gamma x$.

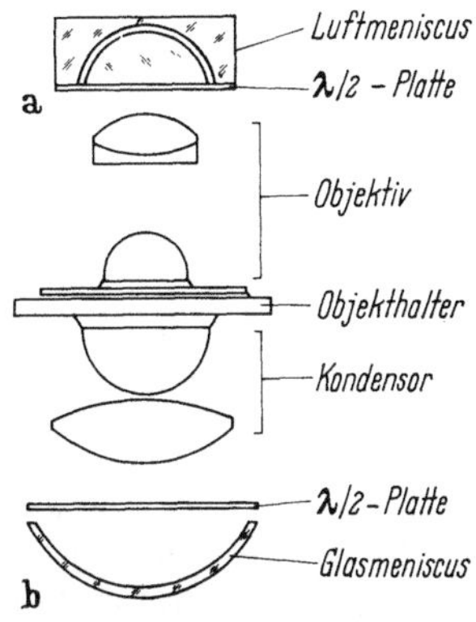

Fig. 67a u. b. Korrektion der Drehung der Polarisationsebene an Linsenflächen in Objektiv, Objekt und Kondensor nach INOUÉ (1957). Anordnung der Korrektionsglieder (a) und (b) oberhalb des Objektivs und unterhalb des Kondensors. Der Glas- oder Luftmeniskus führt für jedes Lichtbüschel denselben Betrag der Lichtdrehung ein wie die Linsen und der Objektträger. Durch die Halbwellenlängeplatten wird der Drehungssinn aber umgekehrt. So wird Korrektion über die volle Apertur erreicht. Alle Glieder müssen spannungsfrei sein.

Störungen durch Depolarisation sollen im Mikroskop — zwischen den Polarisatoren — kleiner sein als die zu beobachtenden bzw. zu messenden Gangunterschiede. Die Empfindlichkeitsgrenze des Auges liegt erfahrungsgemäß bei ca. 1 nm. Eine erhebliche Störung ist, wie schon gesagt, die Drehung der Polarisationsebene an stark gewölbten Linsenflächen, insbesondere an den halbkugelförmigen Frontlinsenflächen der Objektive (z.B. RINNE-BEREK 1953, S. 139). INOUÉ (1952), INOUÉ und HYDE (1957), KUBOTA und INOUÉ (1959) untersuchten diese Störung beugungsoptisch und empfahlen ein Korrektionssystem (Fig. 67), das allerdings sehr schwer spannungsfrei zu fertigen ist. WEBER und KORNDER (1962) schlugen stattdessen eine reflexerhöhende, dielektrische Schicht in Verbindung mit einer oder zwei $\lambda/2$-Platten vor.

Konoskopische Beobachtung. Zuvor war bereits festgestellt worden, daß zur Bestimmung des Gangunterschiedes in Richtung der Mikroskopachse mit kleinen Aperturen gearbeitet werden muß, weil der Gangunterschied sich mit der Richtung im Präparat ändert. Jeder Richtung im Präparat ist ein bestimmter Ort in der Aperturebene zugeordnet. So hat man nur die bildseitige Brennebene der Objektive zu beobachten, um die den einzelnen Richtungen im Präparat zugeordneten Interferenzerscheinungen zu sehen, wie sie bei orthoskopischer Beobachtung das ganze Bild füllen.

Als Beispiel wählen wir einen einachsigen Kristall, geschnitten genau senkrecht zu einer optischen Achse. Aus der Kristallphysik ist bekannt, daß er zwei

Hauptbrechungsindices hat, von denen der eine, der ordentliche, n_ω in jeder Richtung wirksam ist, während der außerordentliche n_ε nur in einer Ebene wirksam ist, in der man die Hauptdoppelbrechung $n_\varepsilon - n_\omega$ bestimmen kann. In der auf dieser Ebene senkrechten Richtung ist $n'_\varepsilon = n_\omega$, d.h. in dieser Richtung ist keine Doppelbrechung festzustellen, es ist die sogenannte optische Achse. Im Falle $n_\varepsilon > n_\omega$ spricht von optisch positiven, im Falle $n_\varepsilon < n_\omega$ von optisch negativen Kristallen. Die in den verschiedenen Richtungen wirksamen Brechungsindices kann man als Kugel und einbeschriebenes (negativer Kristall) oder als Kugel und umbeschriebenes Rotationsellipsoid (positiver Kristall) darstellen (Fig. 68). Die Schwingungsrichtungen der außerordentlichen und der ordentlichen Wellen liegen im Hauptschnitt (Ebenen, in denen die optische Achse liegt) und senkrecht dazu. Damit ist die Deutung der Erscheinungen, die ein solcher Kristall in der hinteren Brennebene eines Objektives bei gekreuzten Polarisatoren hervorruft, einfach.

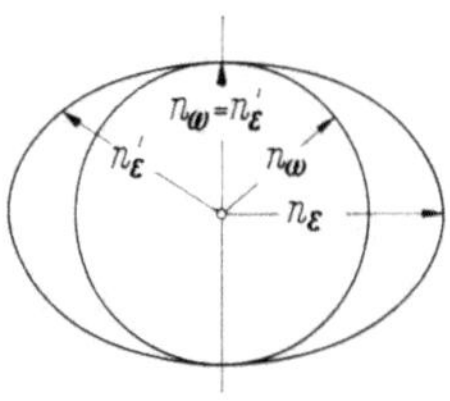

Fig. 68. Größe der Brechzahlen in Abhängigkeit von der Lichtrichtung in einem einachsigen, optisch positiven Kristall. n_ε ist der Extremwert des außerordentlichen Brechungsindex, wirksam in jeder Richtung senkrecht zur optischen Achse die durch $n_\omega = n'_\varepsilon$ gekennzeichnet ist.

a) Alle Stellen, in denen der Gangunterschied Null ist, müssen bei jeder Wellenlänge dunkel sein: Die Richtung der optischen Achse im Kristall, die in unserem Fall der Mikroskopachse entspricht, erscheint im Zentrum der Aperturebene.

b) Weiter müssen alle Stellen dunkel sein, deren Schwingungsrichtungen parallel zu einem der Polarisatoren liegen ($\psi = 0°$), während Stellen in Diagonalstellung ($\psi = 45°$) maximal hell sein müssen. Aus den Bemerkungen zu Fig. 68 versteht man, daß ein schwarzes Balkenkreuz mit Kreuzungspunkt am Ort der optischen Achse erscheinen muß (Fig. 69). Es wird sich beim Drehen des Präparates nicht bewegen, weil die zunächst in Diagonalstellung befindlichen Präparatstellen in Normalstellung wandern und umgekehrt.

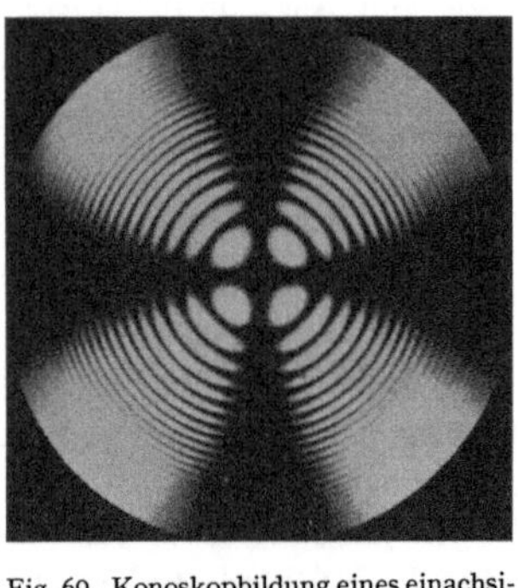

Fig. 69. Konoskopbildung eines einachsigen Kristalls, geschnitten senkrecht zu seiner optischen Achse. Der Durchstoßpunkt der optischen Achse im Bildzentrum und das Balkenkreuz sind auch in weißem Licht farblos. Die Ringe zeigen die Folge der Interferenzfarben, sie sind im monochromatischen Licht für $\Gamma = (1, 2, 3 \ldots)\,\lambda$ dunkel (nach H. HAUSWALDT).

c) In monochromatischem Licht müssen weiterhin alle Stellen dunkel sein, für die der Gangunterschied $\Gamma = (1, 2, 3, \ldots)\,\lambda$ ist. Diese Richtungen müssen auf konzentrischen Kreisen um die optische Achse liegen. Der erste Ring entspricht $\Gamma = \lambda$, der zweite $\Gamma = 2\lambda$ usf. Zu beachten ist, daß Γ nicht nur mit der Doppelbrechung im Kristall, sondern auch mit der Neigung gegen die Achse und zwar wie schon gesagt mit dem Cosinus des Aperturwinkels wächst. In weißem Licht sieht man konzentrische Ringe mit der beim Keil besprochenen Folge der Interferenzfarben.

Nach dem Gesagten ist es einfach, die Bilder von Fig. 70 zu deuten, die das Interferenzbild eines einachsigen Kristalls in Schnitte schief zur optischen Achse zeigen. Die Neigung der optischen Achse ist in der oberen Zeile kleiner als Beleuchtungs- und Beobachtungsapertur, in der unteren Zeile größer. Um sie wieder im Aperturbereich zu sehen, müßte man Objektiv und Kondensor mit höherer Apertur verwenden. Man beobachtet die bildseitige Aperturebene des Objektivs entweder nach Herausnahme des Okulars aus dem Tubus direkt oder mit einem Hilfsmikroskop, dessen Objektiv, die sogenannte Bertrand-Linse, man in den Tubus einschiebt, wobei das Zwischenbild der Objektivapertur in der normalen Zwischenbildebene des Mikroskops erzeugt wird, so daß die Okulare unverändert zur Beobachtung benutzt werden können. Weil man sozusagen im

Konus der Aperturbüschel durch das Objekt blickt, nennt man diese Art der Untersuchung die konoskopische Beobachtung im Gegensatz zur orthoskopischen bei Beobachtung der Objektebene.

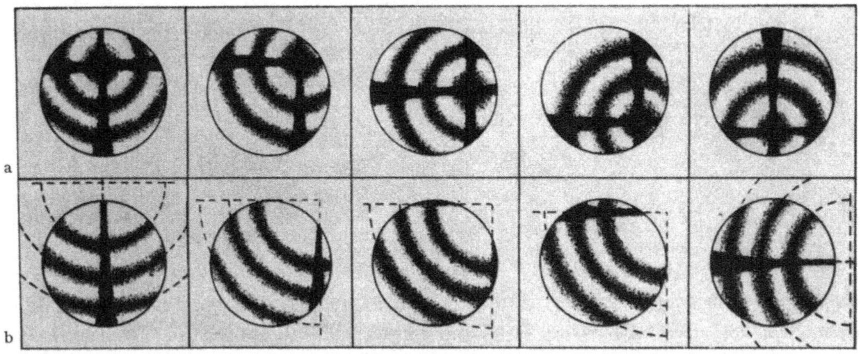

Fig. 70a u. b. Konoskop-Bild eines einachsigen Kristalls, zu beobachten in der bildseitigen Brennebene des Objektivs. Die Neigung der optischen Achse ist in a kleiner, in b größer als die Apertur des Beobachtungsobjektivs (nach K. CHUDOBA).

24. Geometrische Messungen mit dem Mikroskop[1]. Zur Messung bieten sich die Objektebene, das Zwischenbild und jede beliebige Ebene an, in die das Objektbild projiziert wird. Da die Aperturblende der Mikroskopobjektive in deren bildseitige Brennebene gelegt wird, müssen die Mittelstrahlen (Hauptstrahlen) der die einzelnen Objektpunkte abbildenden Büschel senkrecht auf der Objektebene stehen. Sie sind „objektseitig telezentrisch" (vgl. Fig. 47 auf S. 383). Beim Defokussieren wird ein Objektpunkt deshalb unscharf abgebildet, aber sein Mittelpunkt bleibt ortsfest und wandert nicht radial aus. Die Objektive haben im allgemeinen nur unbedeutende Verzeichnung, bis etwa 3% Unterschied in der Vergrößerung zwischen Bildmitte und Bildrand und nur die Augenlinsen der Okulare bringen häufig große Verzeichnung, worauf man bei Messungen im Projektionsbild achten muß.

α) *Längenmessung senkrecht zur Mikroskopachse.* Mikroskoptische haben, soweit sie als sogenannte Kreuztische eine Objektverschiebung in zwei aufeinander senkrechten Richtungen zulassen, zumeist Teilungen, mit einer Ablesemöglichkeit auf 0,1 mm über Längen von 10 bis 50 mm. Man legt die Bilder von Anfangs- und Endpunkt der zu messenden Strecke nacheinander auf eine Markierung in der Zwischenbildebene und liest die Verschiebung des Objektes am Tisch ab. Diese Methode ist nur bequem bei Strecken, die größer als ein Sehfeld im Objekt sind und die parallel zu einer Tischbewegung liegen. Kleinere Strecken mißt man im Zwischenbild mit einer dort angebrachten Teilung (Ziff. 12), die man mit einer Teilung in der Objektebene (Objektmikrometer) eicht, wenn es nicht genügt, die bekannte Intervallänge im Zwischenbild durch Division durch das Produkt Objektivvergrößerung × Tubusfaktor × Feldlinsenfaktor auf die entsprechende Länge im Objekt umzurechnen. Diese Messungen sind genau, wenn man Strecken mißt, die deutlich größer als das Auflösungsvermögen des gerade benutzten Objektivs sind. Objektlängen in der Nähe des Auflösungsvermögens mißt man infolge des Intensitätsverlaufs im Beugungsbild (vgl. Fig. 2 auf S. 344) zu groß. Man vergleiche hierzu die Arbeiten von CHARMAN (1963, 1964), DAVIES (1964), HELML u. SCHEDLING (1951), VIÉNOT (1956), WATRASIEWICZ (1965), WEINSTEIN (1955).

[1] Die Bemerkungen unter dieser Ziffer beziehen sich auf Mikroskope, die nicht speziell für Meßaufgaben konstruiert sind. Letztere werden im Artikel von R. SCHULZE in diesem Bande besprochen.

Eine besondere Art der Längenmessung ist in der *Kernspurplattenphysik*[1] üblich, es ist die Messung des Abstandes einzelner Silberkörner der Bahn des schnellen Teilchens von einer idealen Geraden, die nicht im Objekt vorhanden ist. Die Länge der Geraden ist wesentlich größer als ein Sehfeld. Da der Ablauf von Mikroskoptischen nicht hinreichend gerade ist, ersetzt man die üblichen

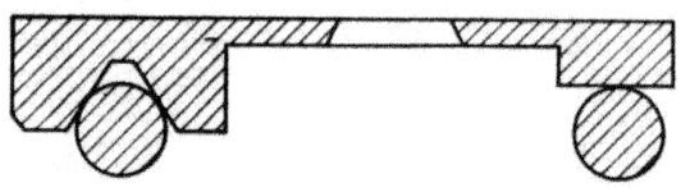

Fig. 71. Tischführung mit geradem Tischablauf durch Lagerung des Tisches auf zwei zylindrisch geschliffenen Stangen.

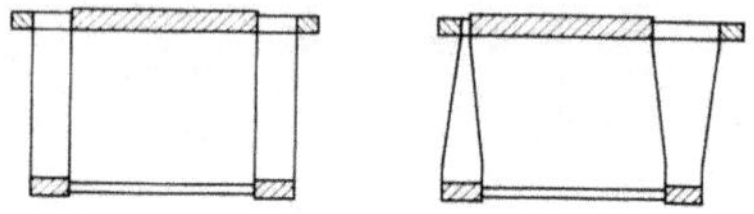

Fig. 72. Tischführung mit geradem Tischablauf durch Aufhängung des Tisches in Paaren von Bandfedern (PLAINEVAUX, 1953).

Führungen durch eine geschliffene, zylindrische Stange, auf der der Tisch in einem V-Lager bewegt wird (Fig. 71). Nach einem anderen Vorschlag (z.B. PLAINEVAUX 1953) hängt man den Tisch in Paaren von Bandfedern auf, die nur eine Bewegung in einer Richtung erlauben (Fig. 72) oder man verwendet normale Mikroskoptische und stützt (Fig. 73) die federnd aufgehängten Objektive an einer mit dem Tisch fest verbundenen planen Glasfläche ($r > 1$ km) ab (STODIEK 1955). Das Objektiv folgt damit den Tischbewegungen und entwirft das Objektbild im Zwischenbild um die Tischbewegung verschoben, während diese Verschiebung bei fester Anbringung des Objektivs mit der Objektivvergrößerung vergrößert wäre. Man setzt den Fehler also auf $1/V_{\text{Objektiv}}$ herab. Da man üblicherweise hier 100fache Objektive verwendet, ist die Wirkung beachtlich.

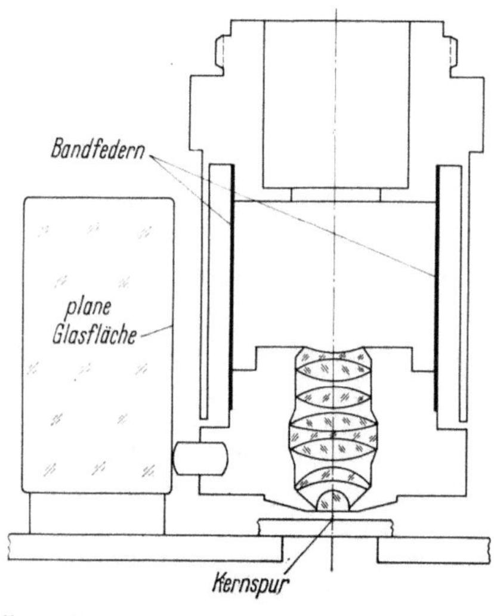

Fig. 73. Kompensation von Tischbewegungen durch Mitbewegen des Kopfes eines Mikroobjektivs, das mit Bandfedern an eine plane, auf dem Mikroskoptisch befestigte Glasplatte gedrückt wird (STODIEK, 1955).

Außer einer einfachen Teilung im Zwischenbild verwendet man Mikrometerokulare, bei denen man einen Faden oder Doppelfaden im Sehfeld mit einer Mikrometerschraube meßbar bewegt, eine dicke Glasplatte, die man kippt, um das Objektbild gegenüber einer festen Marke im Okular meßbar zu verlagern (T. CLAUSEN 1841) oder Okulare, in denen man Doppelbilder des Objektes erzeugt, die man durch Verstellen der Strahlenteilung in Kontakt bringt (z.B. DYSON 1960).

β) Längenmessung in Richtung der optischen Achse. Bei Längenmessungen in Richtung der optischen Achse ist das in Ziff. 4 bezüglich des axialen Auflösungsvermögens Gesagte zu beachten und zu bedenken, daß die feinen Bewegungen der Mikroskope Fokussier-, aber keine eigentlichen Meßeinrichtungen sind. Wo es möglich ist, wird man das Objekt so stellen, daß eine Lateralmessung möglich ist. Mißt man axial, ist die Verschiebung Δ_z zwischen Objekt und Objektiv mittels der Teilung am Feintrieb oder besser mit einer angesetzten Meßuhr zu bestimmen.

[1] Vgl. auch den Artikel von M. B. SHAPIRO: Nuclear Emulsions, in Band 45 dieses Handbuchs.

Die wahre Länge l_z im Objekt ergibt sich zu

$$l_z = \Delta_z \frac{n_{\text{Objekt}}}{n_{\text{Immersion}}},$$

dabei ist $n_{\text{Immersion}}$ die Brechzahl des Mediums, in dem das Objektiv verschoben wurde. In vielen Fällen wird man Interferenzmikroskope (Ziff. 25) zur Bestimmung axialer Längen verwenden.

γ) *Längenmessung schief zur Mikroskopachse.* Zur Bestimmung der wahren Länge von Objekten, die beliebig schief im Raum liegen, wie z.B. Kernspuren in Photoplatten, deren Korndichte bestimmt werden soll, muß man die Projektion ihrer Länge auf die Bildebene und die Höhendifferenz zwischen Anfangs- und Endpunkt bestimmen. Wegen der Bildfehler, wie Bildwölbung und Astigmatismus im Zwischenbild, kann die z-Koordinate nur in Bildmitte gemessen werden. Bei Verschieben des Objektes ändert sich die Höhenlage des Mikroskoptisches im µm-Bereich. Man nimmt deshalb die Unter- und die Oberseite der Emulsion als eben an und bestimmt die Endpunkte der Spur gegen diese beiden Ebenen, als deren Repräsentanten man tiefstes und höchstes Korn des Schleiers ansieht. Die aus diesen Werten bestimmte Höhendifferenz ist vor Berechnung der Spurlänge auf die ursprüngliche Schichtdicke der Emulsion vor dem Fixieren umzurechnen.

δ) *Winkelmessungen.* Winkelmessungen in der zur Mikroskopachse senkrechten Ebene führt man so durch, daß man den zu messenden Winkel mit seinem Scheitel in die Mikroskopachse legt und entweder das Objekt auf einem Drehtisch so dreht, daß zunächst der eine, dann der andere Schenkel des zu messenden Winkels mit einer im Zwischenbild markierten Richtung übereinstimmt oder daß man eine geeignete, eventuell drehbare Teilung in einem Okular benutzt. Bei Polarisationsmikroskopen (Ziff. 23) sind im Zwischenbild Fäden gespannt, die genau mit den Schwingungsrichtungen von Polarisator und Analysator übereinstimmen, so daß man auch Winkel zwischen Kristallflächen und den im Kristall möglichen Schwingungsrichtungen (Auslöschungsstellung) messen kann. Parallelstellung der Schnittkante einer Kristallfläche mit einer Schwingungsrichtung nennt man *gerade Auslöschung* im Gegensatz zur *schiefen Auslöschung* in allen anderen Fällen. Drehtische pflegen eine Teilung in 1° oder $1/_{10}{}^\circ$ zu haben. Mikrometerokulare, bei denen die Winkelteilung im Zwischenbild angebracht ist, erlauben Messungen bis auf einige Bogenminuten.

ε) *Flächenbestimmungen.* Flächenbestimmungen sind einfach, wenn die zu messenden Flächen so regelmäßig sind, daß man sie als Produkt zweier Längen in der Objektebene bestimmen kann, oder sie hinreichend gut mit der Fläche von Kreisbildern oder Rechteckbildern, die man im Zwischenbild anbringt oder erzeugt, vergleichen kann. Über unregelmäßige Flächen legt man im Zwischenbild Netzmikrometer und bestimmt die Fläche aus der Zahl der überdeckten Quadrate, oder man photographiert solche Objekte und planimetriert sie in der Photographie.

Die Bestimmung von Flächenanteilen hat sich vom Ausschneiden von Flächenanteilen aus einer Photographie und der Gewichtsbestimmung der zusammengehörigen Ausschnitte eines Gemengteiles über eine Längenmessung der Gemengteile entlang einer Netzteilung und zwar in zwei aufeinander senkrecht stehenden Richtungen (ROSIWAL 1898) zu einer Zählung von Trefferpunkten für die verschiedenen Gemengteile in einem über das Objekt gelegten Netz von Zielpunkten entwickelt. Fig. 74 kann die beiden Verfahren erläutern. Das schwarze Gemengteil bedeckt 19,4% der horizontalen Geraden und 21,3% der vertikalen Geraden, im Mittel 20,35% und es bedeckt 79 der 400 Kreuzungspunkte, also 19,85%, während der wirkliche Inhalt der schwarzen Flächen 20,0% ist.

Zur Messung der Längensummen benutzt man eine Verschiebung des Objektes auf einem sogenannten Integrationstisch (Fig. 75), bei dem die Drehung mehrerer Spindeln eine gleichmäßige Objektbewegung erzeugt, so daß man für jedes Gemengteil an einer bestimmten Spindel zu drehen hat, deren Drehweg der Weglänge in diesem Gemengteil proportional ist. Zur Erhöhung der Meßgenauigkeit kann man dieselbe Integrationsvorrichtung in eine Okularzwischenbildebene legen (z.B. SCHUCHARDT 1957). Die Punktzählung kann durch schrittweise Objektverschiebung mit einem Objekttisch (Point counter) und Zählung der Gemengteile, die in Bildmitte auf einem Strichkreuz liegen, erfolgen (z.B. CHAYES 1956), oder durch Zählung der Treffer für bestimmte Gemengteile auf einem geeigneten Zählnetz im Zwischenbild (z.B. HENNIG 1958).

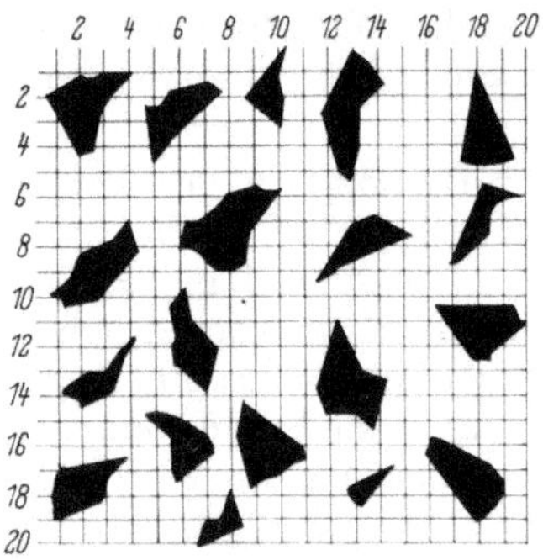

Fig. 74. Zur Bestimmung von Flächenanteilen eines Gemenges wird dieses mit einer Netzteilung überlagert und entweder die Länge der überdeckten Netzlinien gemessen oder die Zahl der überdeckten Kreuzungspunkte der Netzlinien ausgezählt und zur Länge aller Netzlinien oder zur Zahl aller Kreuzungspunkte ins Verhältnis gesetzt.

Der Übergang auf Volumina ist nur dann einfach, wenn man Objektschnitte oder Dünnschliffe hat, deren Dicke klein gegenüber der Objektgröße ist, so daß man für alle Objektteile dieselbe Dicke annehmen kann. Dann ergibt sich das Volumenverhältnis gleich dem Verhältnis der Längenanteile bei Messung mit Integrationstisch bzw. gleich dem Trefferverhältnis. Die Genauigkeit der Methoden ist, bei größerer Geschwindigkeit der Punktzählung, gleich, so daß sich diese Methode durchgesetzt hat.

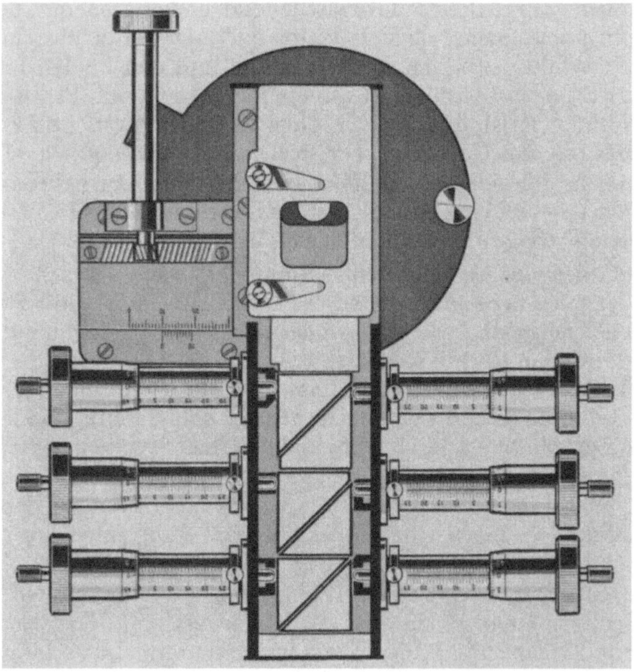

Fig. 75. Integrationstisch mit 6 Spindeln zur Bestimmung von 6 Gemengeteilen. — Drehung an einer beliebigen Spindel schiebt das der Spindel vorgelagerte Dreieck und bewirkt eine Bewegung des hell gezeichneten Tischteiles gegenüber dem dunklen.

Die eben geschilderten Verfahren erlauben es nur, das Verhältnis von Gemengteilen zu bestimmen. Sie geben keine Auskunft über die Teilchengrößen in einem Gemenge. Für eine solche Teilchengrößenbestimmung ist eine Einzelmessung der Teilchen und ihre Summation nach Größenklassen notwendig. Selbstverständlich verwendet man zur Vereinfachung der Messung Methoden wie bei der Flächenbestimmung beschrieben oder man photometriert die Querschnitte der Teilchen als Repräsentanten für ihr Volumen. Bei Benutzung von im Fernsehen bekannten Methoden kann man die mikroskopischen Bilder abtasten und die auszuwertenden Teilchen zählen und klassifizieren.

ζ) *Verwendung kippbarer Tische für Meßzwecke.* Da die Neigungswinkel auszumessender Kernspuren im allgemeinen klein gegen die Ebene des Emulsionsträgers sind und die speziellen Kernspurobjektive einen großen Arbeitsabstand haben, wurden zur Vereinfachung der Messung Kipptische vorgeschlagen, mit denen die Spuren senkrecht zur Mikroskopachse gestellt werden können. Ein solches Kippverfahren ist in der Polarisationsmikroskopie (FEDOROFF 1893) jetzt allgemein üblich, um optische und geometrische Richtungen von Mineralen im Gestein einzumessen. Man ergänzt den Dünnschliff auf seinem Objektträger durch zwei Kugelabschnitte aus Glas zu einer Vollkugel, die kardanisch aufgehängt wird und durch die man ohne wesentliche Ablenkungen der Beobachtungsrichtung und damit auch ohne Totalreflexion in einem sehr großen Winkelbereich beobachten und messen kann. Man nennt derartige Vorrichtungen Universal-Drehtische, vgl. BEREK (1924), REINHARD (1931), EMMONS (1943) die üblicherweise vier Drehachsen besitzen (Fig. 76), die sich in einem Punkte schneiden. Es gibt auch 2- und 5achsige Tische. Nimmt man die Drehachse des Mikroskoptisches hinzu, stehen 3, 5 und 6 Achsen zur Verfügung. Zur Berechnung der Winkel trägt man die gemessenen Achsen und Flächennormalen in ein stereographisches Netz ein, aus dem die resultierenden Winkel mit der bei Einmessung von Mineralen möglichen Genauigkeit von 1° bis 2° abgelesen werden können.

Fig. 76. Universal-Drehtisch (U-Tisch) mit 4 Achsen zum Aufsetzen auf einen Mikroskoptisch.

25. Interferenzmikroskope[1]. Die Interferenzmikroskopie wird vornehmlich als Meßverfahren, aber auch zur Beobachtung in der Mikroskopie zunehmend mehr verwendet. Man spricht im allgemeinen von *Interferenzmikroskopie*, wenn man ein System von Interferenzstreifen erzeugt, und von *Interferenzkontrast*, wenn ein Streifen das Sehfeld ausfüllt und man Hell-Dunkel-Effekte oder Interferenzfarben beobachtet, je nachdem, ob man im einfarbigen oder im weißen Licht arbeitet. Der Interferenzkontrast ist eine Beobachtungsmethode, die es gestattet, nicht absorbierende Objekte kontrastreich abzubilden. Bei Phasenkontrast treten Störungen dadurch auf, daß die räumliche Trennung von direktem und gebeugtem

[1] Diese Ziffer kann an Band 24 des Handbuches anschließen und zwar an den Artikel von FRANÇON (1956): Interférences, refraction et polarisation.

Licht nicht sauber ist. Bei Interferenzmikroskopie gibt es Schwierigkeiten bezüglich der Definiertheit der Referenzflächen.

Völlig ohne diese Störung arbeiten Mikroskope mit Michelson-Interferometer (Fig. 77), die im wesentlichen im Auflicht benutzt werden und solche nach MACH und ZEHNDER (Fig. 78), deren Domäne die Durchlichtmikroskopie ist. Durch völlige Aufspaltung der beiden miteinander interferierenden Wellenzüge ist es

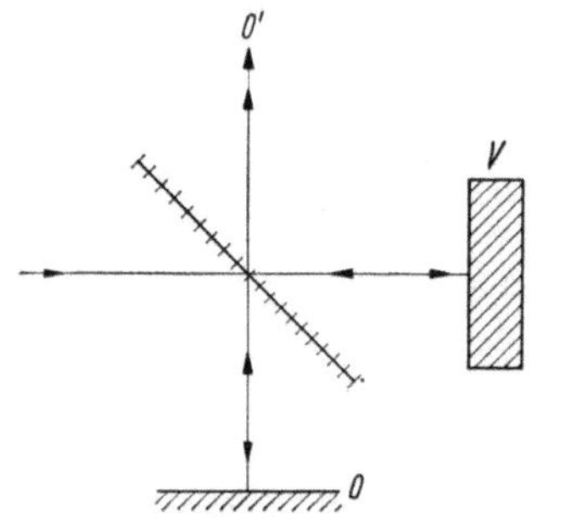

Fig. 77. Zweistrahl-Interferometer nach MICHELSON. O Objektfläche, V Vergleichsfläche.

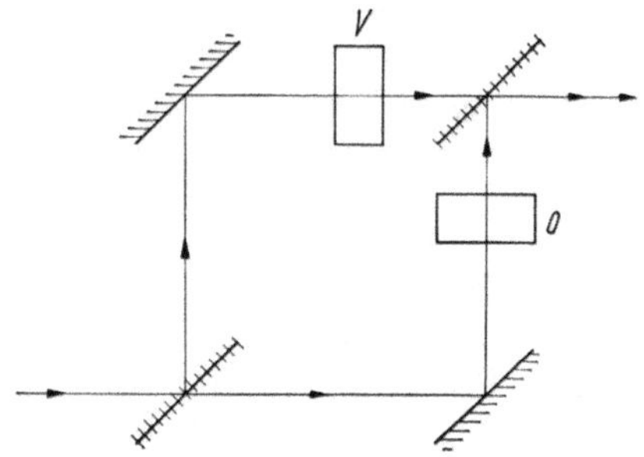

Fig. 78. Zweistrahl-Interferometer nach MACH-ZEHNDER. O Objekt, V Vergleichskörper.

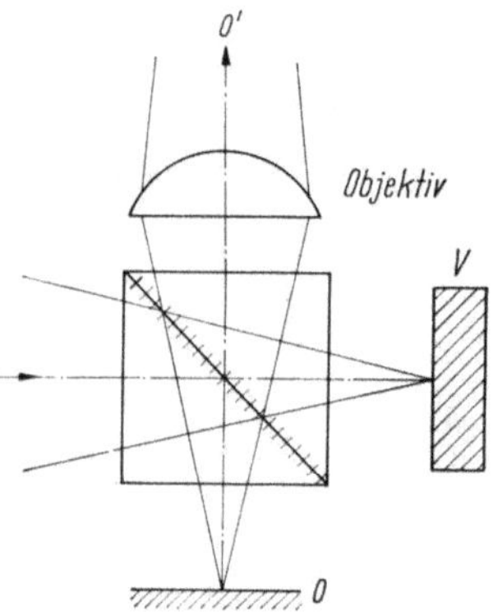

Fig. 79. Michelson-Interferometer. Schema einer Auflicht-Anordnung vor einem Mikroskop-Objektiv (LINNIK, 1938).

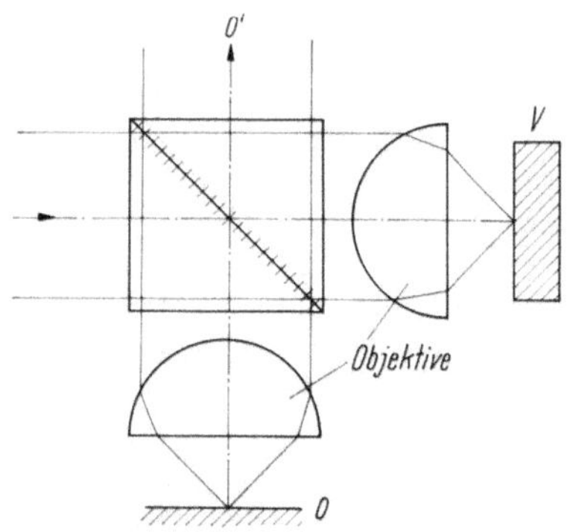

Fig. 80. Michelson-Interferometer. Schema einer Auflicht-Anordnung mit je einem Objektiv zur Beobachtung von Objektoberfläche O und Vergleichsfläche V (LINNIK, 1933).

möglich, im Referenzteil eine bekannte Beeinflussung der Wellenfront vorzunehmen, so daß die Interferenzerscheinungen dem Objekt eindeutig zugeordnet werden können.

Bei einer von SAGNAC, von LINNIK (1938) und von KRUG und LAU (1951) vorgeschlagenen Variante eines Michelson-Interferometers wird dieses mit Strahlenteilung und Referenzfläche vor das Objektiv gesetzt (Fig. 79), und man beobachtet gleichzeitig die Bilder von Objekt und Referenzfläche. Durch Kippen der Referenzfläche erzeugt man zur Beurteilung des Objektes geeignete Interferenzstreifen. Die Grenze für ein vor das Objektiv geschaltetes Michelson-Interferometer ist durch den Arbeitsabstand der Objektive gegeben und liegt etwa bei 20facher Objektivvergrößerung. Wünscht man höhere Objektivvergrößerungen, muß man die Strahlenteilung hinter das Beobachtungsobjektiv legen und die Referenzfläche mit einem hinreichend gleichen Objektiv beobachten (Fig. 80).

Solche Objektivpaare lassen sich in der Fertigung bei nicht allzu komplizierten Objektivformen leicht finden. Es ist empfehlenswert, Objektive zu verwenden, die für unendliche Bildweite korrigiert sind, und den Strahlenteiler zwischen Objektiv und Tubuslinse zu setzen.

Man verwendet diesen Mikroskoptyp im Auflicht und bevorzugt ihn zur Beurteilung von bearbeiteten Metalloberflächen, an denen man z.B. die Tiefe von Kratzern mißt.

Nach dem Mach-Zehnder-Typ lassen sich Interferenzmikroskope bauen, die für Durchlicht-Beobachtung prädestiniert sind, weil man einen ungestörten Referenzteil zur Verfügung hat und damit auch bei Präparaten, die das ganze Sehfeld füllen, eindeutige Aussagen über den Gangunterschied an jeder Stelle des Präparates erhalten kann. Derartige Meßmöglichkeiten sind in der Biologie zur Massebestimmung von Gewebeteilen von großer Wichtigkeit

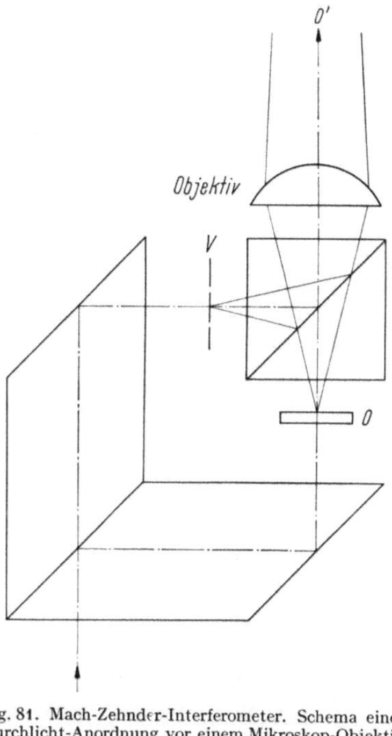

Fig. 81. Mach-Zehnder-Interferometer. Schema einer Durchlicht-Anordnung vor einem Mikroskop-Objektiv (KRUG und LAU, 1951).

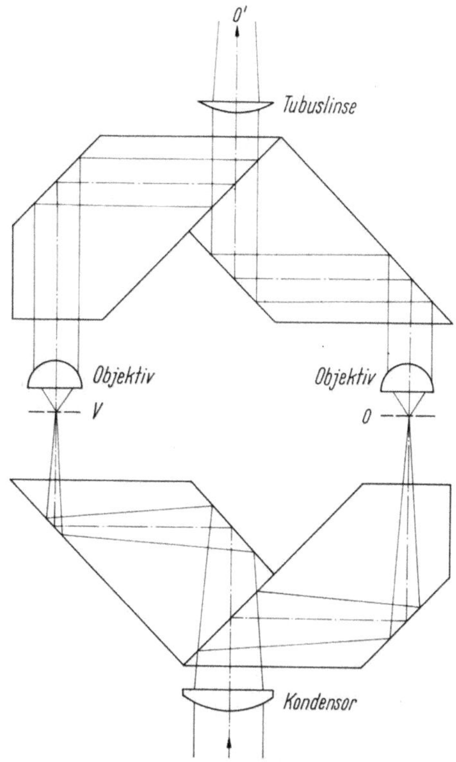

Fig. 82. Mach-Zehnder-Interferometer. Schema einer Durchlichtanordnung (HORN, 1955).

(vgl. z.B. BARER 1953 und DAVIES 1958). Fig. 81 und Fig. 82 geben Beispiele für ein vor ein Objektiv geschaltetes Mach-Zehnder-Interferometer nach KRUG und LAU (1951) und ein solches mit zwei Objektiven zwischen den Strahlenteilern nach HORN (1955). Der Einfluß der Beleuchtungsapertur auf die Meßgenauigkeit (INGELSTAM-JOHANSSON 1958) ist groß, weshalb man unter Verzicht auf höchste Auflösung mit kleinen Beleuchtungsaperturen, A etwa 0,3, arbeitet.

Prototyp einer weiteren Klasse von Mikro-Interferometern ist das *Polarisationsmikroskop* (Ziff. 23), mit dem man Interferenzen der beiden Wellenzüge von doppelbrechenden Substanzen dadurch beobachtet, daß man die zwei nicht interferenzfähigen zueinander senkrechten Schwingungen nach Rekombination am Analysator miteinander interferieren läßt.

Bildet man das Objekt durch geeigneten Aufbau des Interferometers zweimal dicht nebeneinander ab, kann man Interferenzen zwischen der Objektstelle und ihrer Umgebung erzeugen. BATES (1947) brachte beispielsweise einen Spiegel als ein mit einer spaltförmigen Lichtquelle beleuchtetes Objekt vor einem Mach-Zehnder-Interferometer an und verkippte den einen Ablenkspiegel so, daß über die zwei Zweige des Interferometers zwei Bilder des Objektes entstanden, deren beide Wellenfronten miteinander interferierten. Er bezeichnete diese Form eines Interferometers als *Shearing-Interferometer*.

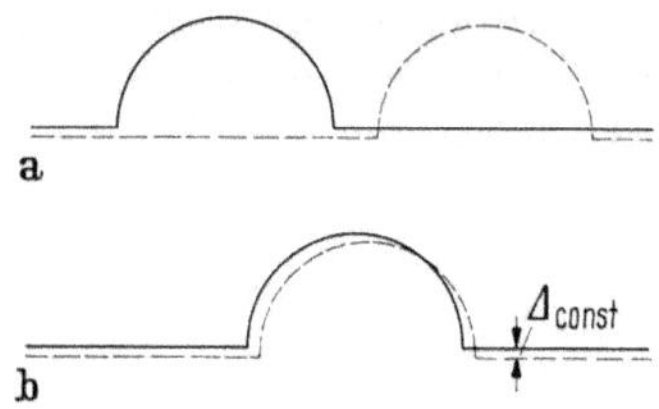

Fig. 83 a u. b. Lage der Wellenfronten bei a totalem, b differentiellem Shearing.

Wenn die Verschiebung der beiden Wellenfronten gegeneinander in der Bildebene größer als das gerade interessierende Objekt ist (Fig. 83 a), spricht man von *totalem* Shearing und beobachtet die Phasenunterschiede zwischen Objekt und Vergleichsstelle. Nur wenn diese Stelle objektfrei ist, kann man die Gangunterschiede des Objektes gegen einen Vergleichsraum mit bekannter Form und Brechzahl bestimmen wie bei den zuvor genannten Interferometern. Wählt man die Verschiebung der Wellenflächen gegeneinander etwa so groß, daß sie der Auflösungsgrenze des Mikroskops im Bild gleich ist (Fig. 83 b) spricht man von

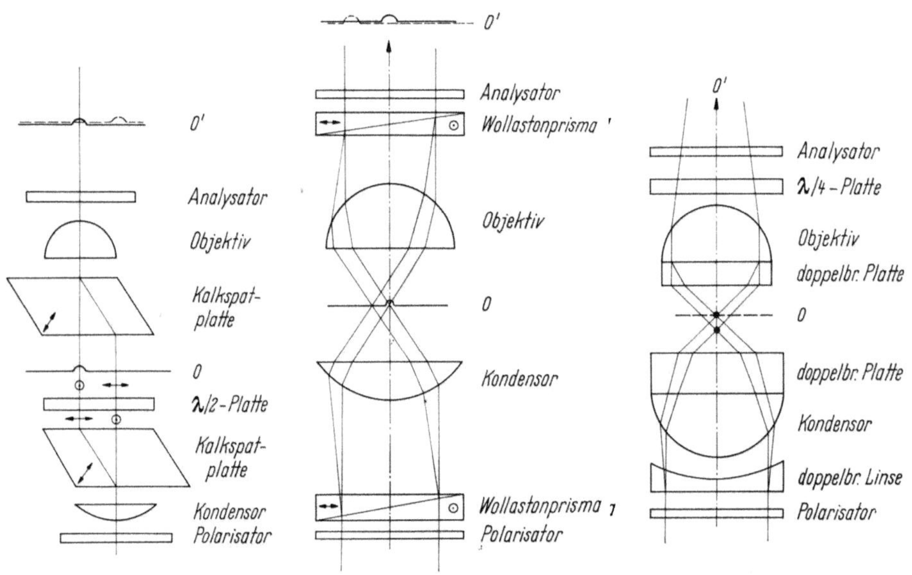

Fig. 84. Fig. 85. Fig. 86.

Fig. 84. Shearing-Polarisations-Interferometer (JAMIN, 1868 — LEBEDEFF, 1930). Aufspaltung und Strahlenvereinigung erfolgen mit Kalkspatplatten. Die Schwingungsrichtungen der beiden Strahlen werden zwischen den Platten mit einer $\lambda/2$-Platte um 90° gedreht, um gleiche Weglängen auf beiden Wegen zu erreichen.

Fig. 85. Shearing-Polarisations-Interferometer (SMITH, 1947, 1955; NOMARSKI, 1952). Aufspaltung und Strahlenvereinigung mit Wollastonprismen.

Fig. 86. Shearing-Polarisations-Interferometer (SMITH, 1947). Schema der Anordnung, bei der im Gegensatz zu den Anordnungen der Fig. 84 und 85 die miteinander interferierenden Büschel nicht lateral, sondern in der optischen Achse gegeneinander verschoben sind.

differentiellem Shearing (FRANÇON 1952) und beobachtet den Gradienten der Brechzahl im Objekt. Auch differentielles Shearing wird häufig als Interferenzkontrast bezeichnet und dadurch leider nicht von normaler Interferenz, bei der ein Streifen das Bild füllt, unterschieden.

In der Mikroskopie wählt man im allgemeinen Polarisationsmikroskope und in ihnen Kristalle und Kristallkombinationen zur Aufspaltung des Lichtes statt mechanischer Mittel wie die Verkippung eines Spiegels bei BATES. Als Beispiel seien in Fig. 84 der Aufbau nach JAMIN und LEBEDEFF (1930) und in Fig. 85 ein Aufbau nach SMITH (1947, 1955) und NOMARSKI (1952) mit lateraler Bildverdoppelung und in Fig. 86 ein Aufbau nach SMITH (1955) mit axialer Bildverdoppelung gebracht. Letzterer Typ wird als Doppelfokus-System bezeichnet. Alle Teilungswinkel und Objekte sind hier so gezeichnet, daß sich totale Verdoppelung ergibt.

Bei Shearing-Interferometern kann man statt des Objektes auch das Zwischenbild im Okular aufspalten, wie FRANÇON gezeigt hat.

Die Beobachtung reeller Keile und ihrer Interferenzen wird in der Mikroskopie wegen der Gefahr, die Objekte durch Auflegen einer Referenzfläche zu beschädigen, fast nur in der Metallmikroskopie gebraucht. Dort verwendet man den Vorschlag von TOLANSKY (1948), statt der üblichen Zweistrahl-Interferenzen mit Hilfe einer hochreflektierenden Referenzfläche mit geringer Absorption Vielstrahl-Interferenzen zu benutzen, bei denen die Interferenzstreifen wesentlich schmaler sind und sich besser einmessen lassen als bei Zweistrahl-Interferenz.

26. Mikroskop-Photometer und Mikroskop-Spektralphotometer. In der Mikrophotometrie und Mikrospektralphotometrie gibt es ebenso wie in der Makrophotometrie eine große Vielfalt von Instrumenten. Prototyp eines Mikroskopphotometers dürfte ein Zweistrahl-Photometer (Fig. 87) nach BEREK (1931) sein, bei dem der Abgleich der Photometerfelder visuell und zwar durch Kreuzen von Polarisationsprismen vorgenommen wurde. Der Übergang auf Einstrahl-Photo-

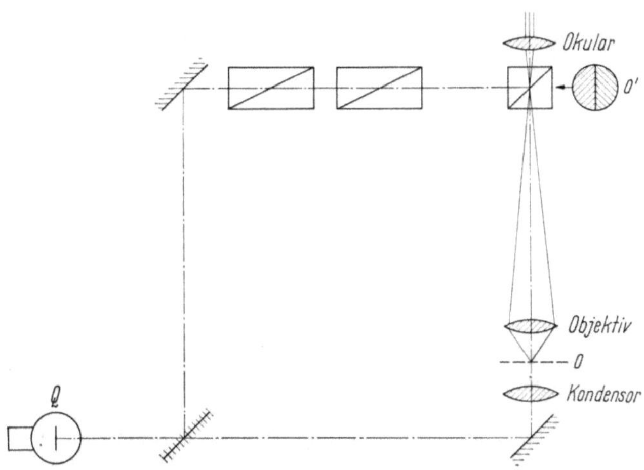

Fig. 87. Mikroskopphotometer. Schema der Zweistrahlenanordnung nach BEREK (1931). Vom Beleuchtungsbüschel wird ein Vergleichsbüschel abgetrennt. Der Helligkeitsabgleich erfolgt mit Polarisationsprismen. In der Bildebene O' werden in Halbfeldern das Objekt und das objektleere Vergleichsfeld auf gleiche Helligkeit gebracht.

meter (Fig. 88) mit elektrischer Messung durch Photozellen oder Photoelemente erfolgte sehr langsam. Bei diesem Typ blendete AGROSKIN (1959) (Fig. 89) die zu messende Fläche durch ein Loch in einem Spiegel aus. Über den Spiegel beobachtete er das Umfeld der Meßfläche, das Objekt im Meßfeld blieb unsichtbar. WEBER (1965) (Fig. 90) führte eine variable Meßblende ein, die er von hinten beleuchtet und in das Sehfeld einspiegelt, so daß Meßfeld und Umfeld beobachtet werden und die Wirkung der Ausblendung beurteilt werden kann.

Um die zeitliche und räumliche Inkonstanz von Gasentladungslampen (Ziff. 18) zu eliminieren, zweigten CASPERSSON, LOMAKKA und CARLSON (1960) durch physikalische Strahlenteilung einen Teil des Beleuchtungslichtes auf einen

Fig. 88. Mikroskopphotometer. Schema der Einstrahlanordnung. In $O'' = B$ wird ein Meßfeld ausgeblendet, darüber der elektrische Meßempfänger. Die Beobachtung des Objektes erfolgt wie in der Mikrophotographie (Fig. 60e) über ein Fernrohr. Das Meßfeld selbst ist nicht sichtbar, doch kann im Fernrohr in O'' eine der Blende entsprechende Marke angebracht werden.

Fig. 89. Mikroskopphotometer. Schema der Einstrahlanordnung nach AGROSKIN. Die Beobachtung des gesamten Feldes in O' erfolgt über einen Hohlspiegel und ein Fernrohr. Der gemessene Bereich B wird vom Hohlspiegel nicht reflektiert und erscheint deshalb als dunkler Fleck in O''.

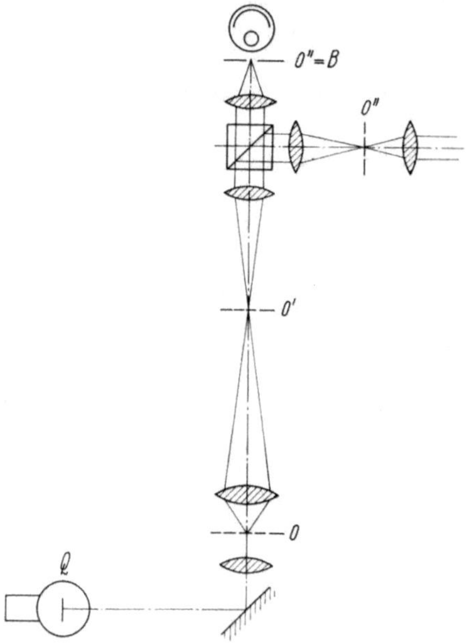

Fig. 88.

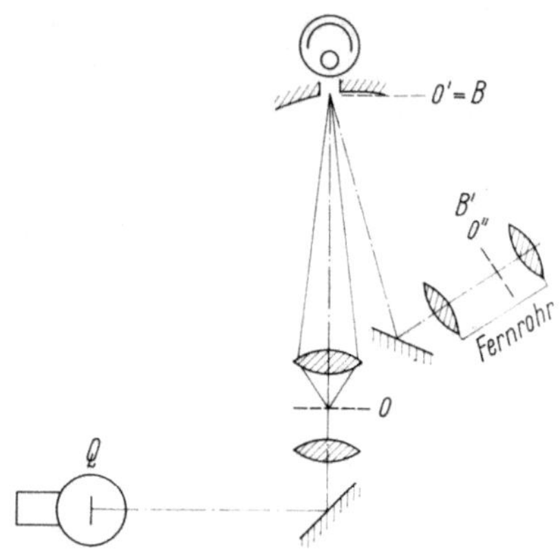

Fig. 89.

zweiten Empfänger ab und steuerten über ihn die Empfindlichkeit des Meßempfängers (Fig. 91). Eine volle Zweistrahlanordnung mit rotierenden Sektorenspiegeln zur Erzeugung von Wechsellicht erlaubt bei spektraler Lichtzerlegung vor dem Mikroskop zusätzlich eine Photometrie über das Spektrum (HANSEN 1961) (Fig. 92), während nach Vorschlägen z. B. von SEEDS und WILKINS (1950) ein Spektrograph mit seinem Eintrittsspalt in das Zwischenbild des Mikroskops gestellt wird und es damit möglich ist, bei Beleuchtung mit weißem Licht den auf den Eintrittsspalt fallenden Bildteil spektral zu zerlegen und so vom Objekt eine Serie von Spektrogrammen zu erzeugen (Fig. 93). Die subjektive Beobachtung des Bildes auf dem Spalt erfolgt im Phasenkontrast oder mit Bildwandler.

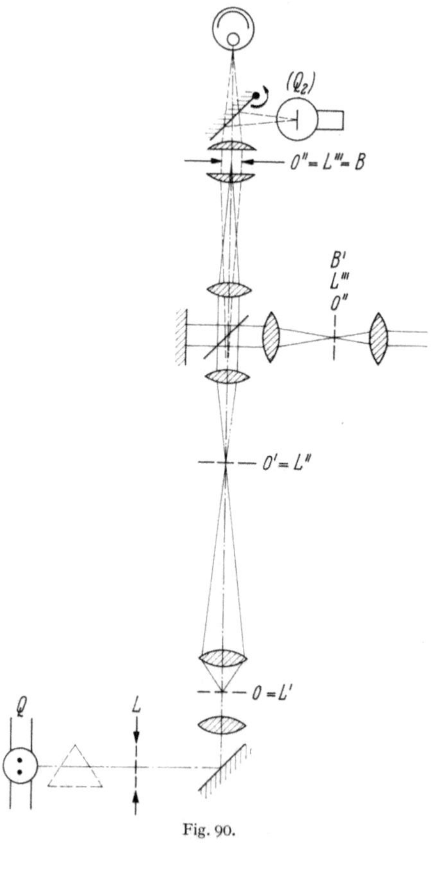

Fig. 90.

Fig. 90. Mikroskopphotometer. Schema der Einstrahlanordnung (nach K. WEBER, 1965) mit verstellbarer Meßblende B und Rückspiegelung dieser Blende in das Objektbild O''. Die Blende wird von der Lichtquelle Q_2 beleuchtet und erscheint als heller Fleck, der Ort, Größe und Form der Meßblende im Objektbild wiedergibt. Zur Vermeidung von Streulicht kann durch Blende L, die ebenfalls in O'' sichtbar wird, der beleuchtete Bereich im Objekt O begrenzt werden. Spektrale Zerlegung des Lichtes vor dem Mikroskop ist möglich.

Mikroskop-Photometer werden außer zu Absorptions- bzw. Reflexionsmessung auch zur Mikrofluorometrie — Messung der Fluoreszenz (vgl. Ziff. 22) —, bei der Auswertung von Mikro-Autoradiographien, wo sie die Auszählung der Silberkörner durch eine Messung des bei Dunkelfeldbeleuchtung abgebeugten Lichtes ersetzen, und als Mikro-Densitometer (Photometrie in einem Photo, z. B. von Spektrallinien) verwendet, um nur einige Beispiele zu nennen.

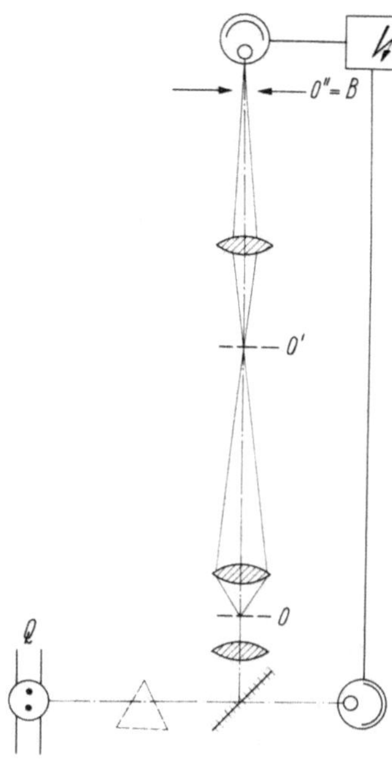

Fig. 91.

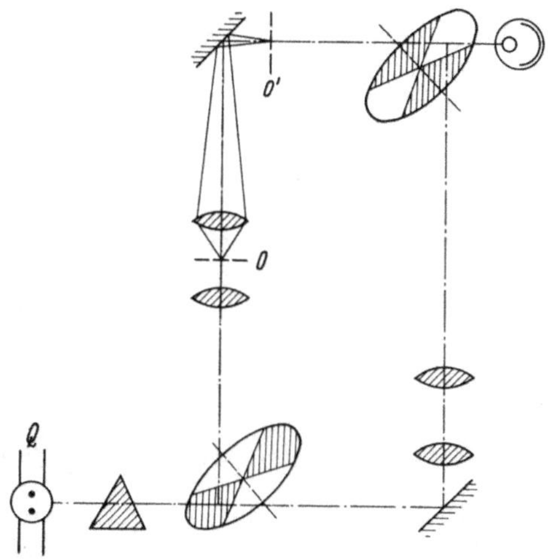

Fig. 92.

Fig. 91. Mikroskopphotometer. Schema der Einstrahlanordnung für elektrisches Messen mit Kompensation der Lichtstromschwankungen und spektraler Zerlegung vor dem Mikroskop (nach Caspersson, Lomakka und Carlson, 1960). Die Kompensationsschaltung ermöglicht die Verwendung von zeitlich und räumlich nicht gut konstanten Hochdruck-Gasentladungslampen.

Fig. 92. Mikroskopphotometer. Schema einer Zweistrahlanordnung mit Wechsellicht (Hansen, 1961). Bei Photometrie durch das Spektrum wird eine Wechsellichtschaltung benutzt, bei der in schnellem Wechsel durch rotierende Sektorenspiegel der Meßstrahl und der Vergleichsstrahl freigegeben werden. Der Quotient wird elektrisch gemessen.

Fig. 93. Mikroskopphotometer. Schema einer Einstrahlanordnung mit Spektrographenspalt in der Bildebene des Mikroskops (Seeds und Wilkins, 1950) und spektraler Zerlegung des auf den Spalt fallenden Bildteils. Die Beobachtung des ganzen Bildfeldes erfolgt in O', die des Spaltes und von ihm ausgeblendeten Bildteiles in O'' und zwar je nach Absorptionsbereich des Objektes direkt, mit Phasenkontrast oder mit Bildwandler. Die Blende L ermöglicht wie in Fig. 90 eine Reduktion von Streulicht.

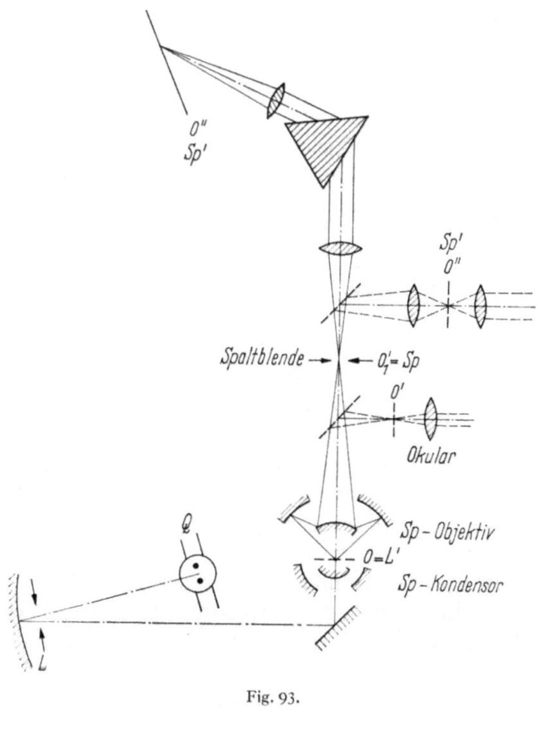

Fig. 93.

27. Brechzahlbestimmung mit dem Mikroskop. Weil die Kenntnis von Brechzahlen der Objekte auch für die Dickenbestimmung notwendig ist, hat man für ihre Messung eine ganze Reihe spezieller Methoden in der Mikroskopie ausgearbeitet. In der Mineralogie gilt die Merkregel für die sogenannte Beckesche Linie, daß bei zwei Mineralen, die direkten Kontakt miteinander haben, beim *H*eben des Tubus (= Er*h*öhung des Abstandes Objekt — Objektiv), ein *h*eller Lichtsaum in das *h*öher brechende Medium wandert. Diese Erscheinung ist gut sichtbar bis zu $\varphi \approx 60°$ und wird auch bei Brechzahlbestimmung durch Einbettung in Lösungen benutzt. Man mischt entweder zwei Lösungen bis zum Verschwinden der Beckeschen Linie und bestimmt dann die Brechzahl der Mischung mit einem Mikrorefraktometer oder man wählt eine Lösung, deren Variation der Brechzahl mit Änderung von Wellenlänge und Temperatur bekannt ist und variiert eine dieser beiden oder beide so lange, bis die Beckesche Linie verschwindet. Eine weit empfindlichere Methode zur Feststellung der Brechzahl-Gleichheit ist Beobachtung im Phasenkontrast (Ziff. 20) oder im Interferenzkontrast (Ziff. 25).

In doppelbrechenden Materialien bestimmt man beide Brechzahlen in Achsrichtung des Mikroskops nacheinander dadurch, daß man auf dem Polarisationsmikroskop (Ziff. 23) zuerst die eine und dann die andere Schwingung durch Parallelstellung zum Polarisator erzeugt und die Brechzahl wie geschildert bestimmt. Der Analysator ist hierbei ausgeschaltet.

IV. Stereoskopische Mikroskope.

28. Bei normalem stereoskopischem Sehen betrachtet man Objekte, die zwischen Nah- und Fernpunkt liegen, gleichzeitig mit beiden Augen, deren Abstand zwischen 55 und 75 mm zu liegen pflegt. Bei einem Augenabstand von 65 mm und einer Objektentfernung von ∞ bis 250 mm sehen die beiden Augen ein Objekt unter Winkeln von 0° bis 7,5° bzw. Aperturen von 0 bis 0,13 gegen die Symmetrieebene. Ein solcher Winkel ist klein gegenüber dem Aperturwinkel der meisten Mikroskopobjektive, und man kann leicht Einrichtungen schaffen, mit deren Hilfe man mit beiden Augen durch verschiedene Stellen der Objektivpupille blickt. Vorschläge zu einer geometrischen Strahlenteilung der Pupille sind alt. Wenn der Bildeindruck orthoskopisch sein soll, hat man dafür zu sorgen, daß das rechte Auge von rechts und das linke Auge von links auf das Bild des Objektes blickt, so wie die zugehörigen Aperturwinkel im Objektiv von rechts und links auf das Objekt gerichtet sind.

Vergegenwärtigt man sich, daß bei einem normalen Mikroskop der rechte Teil der Austrittspupille dem entsprechenden linken Teil der Objektivöffnung und umgekehrt der linke Teil der Austrittspupille dem rechten Teil der Objektivöffnung entspricht (Fig. 94), dann sieht man, daß man bei Beobachtung eines nicht seitenrichtigen Bildes durch einen Binokulartubus die nasalen Seiten der Pupillen abdecken muß, um mit einem normalen binokularen Mikroskop orthoskopisch sehen zu können. ABBE (1880) machte einen solchen Vorschlag (Fig. 94a) mit mechanischen Blenden, die genau in der Austrittspupille angebracht waren. Eine in der Austrittspupille angebrachte Blende stört, weil nun die Iris des Auges nicht an dieselbe Stelle gebracht werden kann. Einfacher ist es daher, den Augenabstand der Okulare so falsch einzustellen, daß die Augeniris die Pupillen beschneidet und man nur durch die äußeren Teile der Austrittspupille blickt. Die beiden geschilderten Methoden haben den Nachteil, die Strahlenteilung in der sehr kleinen Austrittspupille vorzunehmen. LIHOTZKY (1924) machte deshalb den Vorschlag einer Zwischenabbildung der Objektivpupille hinter die Strahlenteilung mit je einer großen Pupille in jedem Okulartubus. Er zeigte, daß es zur Erzielung von Stereo-Effekten genügte, nur auf einer Seite eine Pupille schief zu beschneiden, um gute stereoskopische Bilder zu erzeugen, und hatte dabei den Vorteil, eines der Teilbilder mit voller Apertur, also auch voller Auflösung, zu erzeugen.

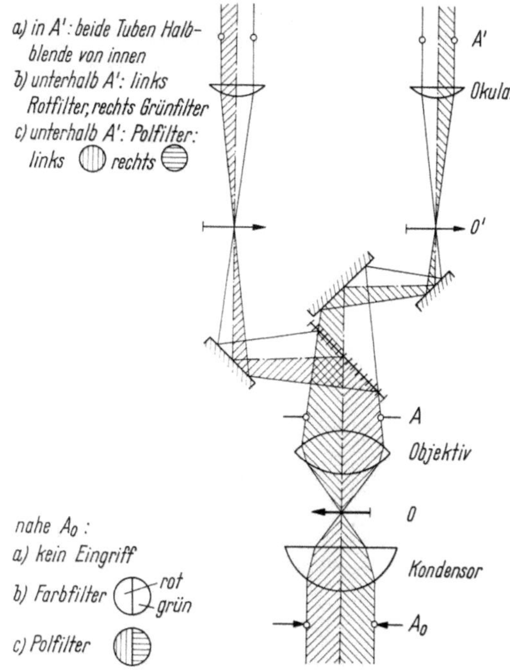

Fig. 94a—c. Verwendung eines Binokularmikroskops zu stereomikroskopischer Beobachtung a nach ABBE durch Abdecken der halben Austrittspupillen A' b nach HEIMSTÄDT und c nach LOOS durch Aufspaltung der Pupille A_0 des Kondensors.

Diese Verfahren konnten sich nicht durchsetzen. Eine Pupillenaufteilung in der Aperturebene des Kondensors scheint einfacher und wurde von HEIMSTÄDT (1931) auf der Basis der Anaglyphenmethode, also mit Rot-Grün-Aufteilung (Fig. 94b), und von LOOS (1939) für Pupillenteilung in zwei Teile mit Polarisationsfiltern vorgeschlagen, die in den beiden Hälften der Filter senkrecht aufeinander stehende Schwingungen erzeugen (Fig. 94c). Die Okulare tragen einheitlich grüne oder rote bzw. Polarisationsfilter mit einer einheitlichen Schwingungsrichtung, mit denen man die eine Hälfte der Pupille auslöscht. Die Teilungslinie der Filter im Kondensor muß dabei so liegen, daß die Hälften rechts-links-symmetrisch sind. Fig. 95 zeigt die Aufnahme eines Ereignisses in einer Kernspuremulsion, das mit einer derartigen Polarisationseinrichtung aufgenommen wurde. Man sieht, daß die Schärfentiefe, wie nach Ziff. 4 zu erwarten, sehr klein ist.

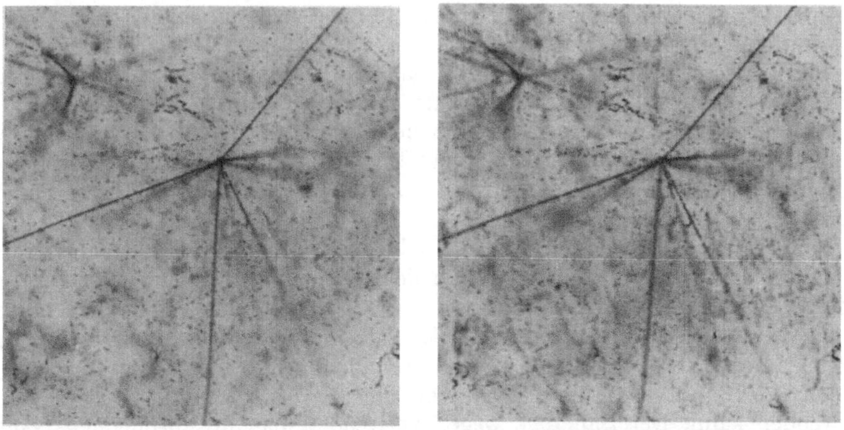

Fig. 95. Mikrostereoaufnahme eines kernphysikalischen Ereignisses bei etwa 1000facher Vergrößerung. Man betrachte das rechte Bild mit dem rechten, das linke Bild mit dem linken Auge. Man sieht dann drei Bilder nebeneinander, von denen das mittlere räumlich erscheint und die Verteilung der Spuren im Raum zeigt. Eigene Aufnahme mit Anordnung nach Fig. 94c.

Stereoaufnahmen sind auch möglich, wenn man mit einem gewöhnlichen monokularen Mikroskop arbeitet und das Objekt gegenüber der optischen Achse geeignet verkippt.

Üblicherweise bezeichnet man als Stereomikroskope eine Kombination von zwei Mikroskopen, die so zusammengebaut sind, daß man mit je einem Auge durch ein Mikroskop beobachtet, das unter einem gewissen Winkel und symmetrisch zum zweiten auf die Objektebene ausgerichtet ist. Die Winkel gegen die Symmetrieebene liegen zwischen etwa 4° und 8° und bestimmen die maximale

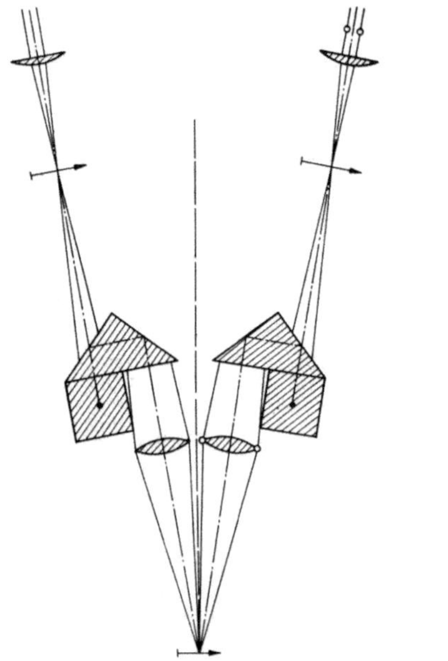

Fig. 96. Stereomikroskop (nach GREENOUGH) mit Objektivpaaren und Bildaufrichtung durch Prismen.

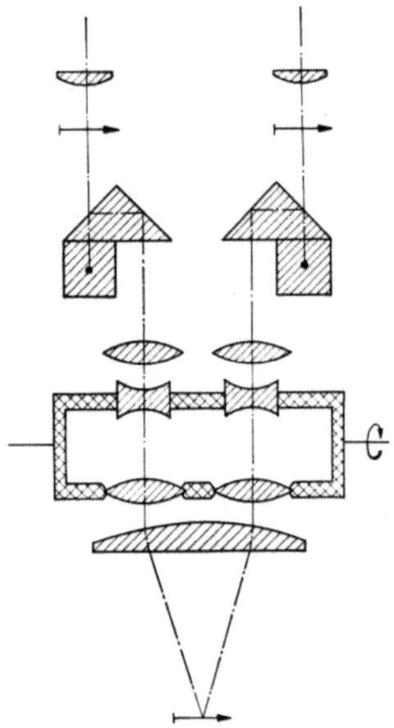

Fig. 97. Stereomikroskop mit gemeinsamem Hauptobjektiv und Einzelobjektiven in jedem Tubus. Dazwischen Galilei-Fernrohre zur Vergrößerungswechslung.

Apertur solcher Mikroskope (Fig. 96). Da diese Mikroskope für präparative Arbeiten benutzt werden, ist es notwendig, für eine vollkommene Bildaufrichtung, etwa durch Porroprismen, zu sorgen.

Den geringsten optischen Aufwand hat man, wenn man eine Serie von Objektivpaaren vorsieht, die gegeneinander gewechselt werden können. Man wird diese Objektive im allgemeinen als einfache Achromate ausführen, deren Sitz, wie in Ziff. 7 ausgeführt, durch die Objekt-Bildentfernung und die Vergrößerung bestimmt ist. Die freien Arbeitsabstände bei hohen Vergrößerungen werden bei diesem Typ recht klein.

Eine andere Möglichkeit ist es, das Objekt in den Brennpunkt eines großen gemeinsamen Objektivs zu legen, hinter dem man, für jeden Tubus getrennt, ein weiteres Objektiv anbringt, das ein Bild in der Zwischenbildebene des zugehörigen Okulars erzeugt. Man hat dann die Möglichkeit, zwischen dem gemeinsamen und den einzelnen Objektiven Fernrohre anzubringen (Fig. 97), die man in beiden Stellungen, vergrößernd und verkleinernd und zusätzlich ausgeschaltet, verwenden kann. Der Baulänge wegen verwendet man im allgemeinen Galilei-Fernrohre (REINERT 1938).

Die Vergrößerung der Grundanordnung ergibt sich aus dem Brennweitenverhältnis von Einzelobjektiv zu gemeinsamem Objektiv. Sie ist bei Zuschaltung des Fernrohres mit dessen Vergrößerung, die das Verhältnis der Brennweiten von sammelndem zu zerstreuendem Glied ist, zu multiplizieren oder zu dividieren, je nachdem, in welcher Position das Fernrohr zugeschaltet ist.

Die Vergrößerung des Grundstativs kann man durch Veränderung der Brennweite des gemeinsamen Objektivs verändern, z.B. durch Vorschalten von Linsen. Zugeschaltete Sammellinsen erhöhen die Vergrößerung und verkleinern den Arbeitsabstand, Negativlinsen bewirken das Gegenteil.

Der zur Korrektion notwendige Aufwand ist bei diesen Stativen höher. Besondere Probleme treten dadurch auf, daß die Pupillenlage des Objektivteils in den drei Schaltstellungen nur schwer gleich zu halten ist.

Stereoskopisch abgebildete Objekte werden meistens in schrägem Auflicht-Dunkelfeld beobachtet. Dabei dient eine getrennt aufgestellte Leuchte mit Kollektor als Lichtquelle zur Erzeugung von möglichst plastischer Beleuchtung. Im Durchlicht ist getrennte Beleuchtung für jedes einzelne Mikroskop notwendig oder eine Beleuchtung so großer Apertur, daß beide Objektive gemeinsam ausgeleuchtet werden. Häufig verwendet man als Lichtquelle eine mattweiße, von der Seite angestrahlte Fläche.

Literatur.

Lehrbücher und Handbücher.

Allen, R. M.: The microscope. 1940 New York, 286 p.

— Photomicrography, 2nd ed. 1958 Princeton, N.Y., London, 441 p.

Boegehold, H.: Das optische System des Mikroskops, 2. Aufl. 1959 Berlin, 192 S.

Clark, G. L.: The encyclopedia of microscopy. 1961 New York ,693 p.

Ehringhaus/Trapp: Das Mikroskop. (Seine wissenschaftlichen Grundlagen und seine Anwendung), 6. Aufl. 1966 Stuttgart, 132 S.

Françon, M.: Progress in microscopy, 1961 Oxford-London-New York-Paris, p. 295, 92 refs.

Freund, H.: Handbuch der Mikroskopie in der Technik, 8 Bd. in 15 Teilbänden. 1951—1966 erschienen 10 Teilbände, I.1 Instrumente der Durchlichtmikroskopie 1957, 736 S.; I.2 Instrumente der Auflichtmikroskopie, 1960, 744 S.; II.1—3 Mikroskopie der Bodenschätze, 1952, 1954, 1958, 2029 S.; IV.1,3,4 Mikroskopie der Silikate, 1955, 1964, 1965, 1821 S.; V.1—2 Mikroskopie des Holzes und des Papiers, 1951, 983 S.

Hauser, F., bearb. v. W. Oettel u. H. Gause: Das Arbeiten mit auffallendem Licht in der Mikroskopie, 2. Aufl. 1960, 173 S., 73 Ref.

Heunert, H. H., u. K. Philipp: Grundlagen der Schmalfilmtechnik. 1957, 200 S.

Langeron, M.: Précis de microscopie, 7. ed. 1949 Paris, 1430 p.

Malies, H. M.: Applied microscopy and photomicrography, 1959 London, 143 p.

Martin, L. C., and B. K. Johnson: Practical microscopy, 2 ed. 1949 London and Glasgow, 124 p.

Michel, K.: Die Mikrophotographie, 2. Aufl. Wiss. und angew. Photographie, Bd. X. 1962 Wien, 734 S.

— Die Grundzüge der Theorie des Mikroskops in elementarer Darstellung, 2. Aufl. 1964 Stuttgart, 355 S., 124 Lit.

Nedham, G. H.: The practical use of the microscope (including photomicrography). 1958 Springfield (Ill.), 493 p.

Oettel, W. O.: Grundlagen der Metallmikroskopie. 1959 Leipzig, 216 S.

Otto, L.: Durchlichtmikroskopie; Geräte und Verfahren. 1959 Berlin, 472 S.

Patzelt, V., u. F. Raaz: Das Mikroskop und seine Nebenapparate im Dienst der Naturwissenschaften, Medizin und Technik. 1950, Wien, 162 S.

Payne, B. O.: Microscope design and construction. 1954 York, 201 p.

Policard, A., M. Bessis et M. Locquin: Traité de microscopie. Instruments et techniques. 1957 Paris, 608 p.

Vickers, A. E. J. (Editor): Modern methods of microscopy. 1956 London, 114 p.

Literatur zu Ziff. 1 bis 5.

Abbe, E.: Über die Bestimmung der Lichtstärke optischer Instrumente. Jena. Z. Med. Naturwiss. **6**, 263—291 (1871) = Ges. Abh., Bd. 1, 14—44.

— Beiträge zur Theorie des Mikroskops und der mikroskopischen Wahrnehmung. Schultze Arch. mikr. Anat. **9**, 413—468 (1873) = Ges. Abh., Bd. 1, 45—100.

Abbe, E.: Über die Grenzen der geometrischen Optik. Sitzber. Jena. Ges. Med. Naturw. 71—109 (1880) = Ges. Abh., Bd. 1, 273—312.
— On the estimation of aperture in the microscope. J. Roy. Microscop. Soc. (2), **1**, 388—423 (1881), deutsch: Ges. Abh., Bd. 1, 325—374.
— The relation of aperture and power in the microscope. J. Roy. Microscop. Soc., (2), **2**, 300—309, 460—473 (1882); **3**, 790—812 (1883) deutsch: Ges. Abh., Bd. 1, 375—435.
— On the effect of illumination by means of wide-angled cones of light. J. Roy. Microscop. Soc. (2), **9**, 721—724 (1889) deutsch: Ges. Abh., Bd. 1, 473—477.
— Gesammelte Abhandlungen, Bd. 1: Abhandlungen über die Theorie des Mikroskops. Jena 1904, 486 S.
Altmann, R.: Über die Vorbemerkungen des Herrn Prof. Abbe zu seinen „Grenzen der geometrischen Optik". Arch. Anat. u. Physiol., Anat. Abt. 354—363 (1880).
— Über die Vorbemerkungen des Herrn Prof. Abbe zu seinen Grenzen der geometrischen Optik. (Fortsetzung das Mikroskop betreffend.) Arch. Anat. u. Physiol., Anat. Abt. 52—59 (1882).
Arnulf, A., O. Dupuy et F. Flamant: Phénomène de diffraction dans l'image de deux points lumineux en lumière cohérente. C. R. Acad. Sci., Paris **234**, 1273—1275 (1952).
Airy, G. B.: On the diffraction of an object-glass with circular aperture. Trans. Cambridge Phil. Soc. **5**, 283—291 (1835).
Berek, M.: Betrachtungen zur Darstellung des Abbildungsvorganges im Mikroskop und zur Frage des Auflösungsvermögens im Hellfeld und Dunkelfeld. Z. wiss. Mikrosk. **41**, 1—15 (1924).
— Ist die Unterscheidung von selbstleuchtenden und nichtselbstleuchtenden Objekten für die Auswirkungen im Abbildungsvorgang wesentlich? Ztr.-Ztg. Opt. u. Mech. **45**, 143—144 (1924).
— Über Kohärenz und Konsonanz des Lichtes. Z. Physik **36** 675—688, 824—838 (1926); **37**, 387—394, 420—450 (1926).
— Entwickelung und gegenwärtiger Stand der Lehre von der Abbildung im Mikroskop. Sitzber. Ges. Förder. ges. Naturw. Marburg **61**, 251—282 (1926).
— Grundlagen der Tiefenwahrnehmung im Mikroskop mit einem Anhang über die Bestimmung der obersten Grenze des unvermeidlichen Fehlers einer Messung aus der Häufigkeitsverteilung der zufälligen Maximalfehler. Sitzber. Ges. Förder. ges. Naturw. Marburg **62**, 189—223 (1927).
— On the extent to which real image formation can be obtained in the microscope. J. Roy. Microscop. Soc. **49**, 240—244 (1929).
— Über die wirkliche Abbildung von Nichtselbstleuchtern und ihre Grenzen. Z. Physik **53**, 483—493 (1929).
— Abbildung eines ultramikroskopischen Elements mit rotationssymmetrischem Störungsfeld durch ein Mikroskop beliebiger Apertur im Dunkelfeld. Ann. Physik **4**, 285—315 (1930).
— Zur Theorie der Abbildung im Mikroskop. Optik **1**, 475—486 (1946); **3**, 289—297 (1948); **4**, 377—390 (1949); **5**, 1—30, 144—159, 329—353 (1949); **6**, 1—7, 219—237 (1950).
Canals-Frau, D., et M. Rousseau: Influence de l'éclairage partillement cohérent sur la formation des images de quelques objets étendus opaques. Optica Acta **5**, 15—27 (1958), 9 réfs.
Cittert van, P. H.: Die wahrscheinliche Schwingungsverteilung in einer von einer Lichtquelle direkt oder mittels einer Linse beleuchteten Ebene. Physica, Haag **1**, 201—210 (1934).
— Kohaerenz-Probleme. Physica, Haag **6**, 1129—1138 (1939). 2 Lit.
— Degree of coherence. Physica, Haag **24**, 505—507 (1958), 2 refs.
De, M., and S. C. Som: Diffraction images of circular openings with partially coherent illumination. Optica Acta **9**, 17—31 (1962), 14 refs.
— Diffraction images of circular phase objects in partially coherent light. J. Opt. Soc. Amer. **53**, 779—787 (1963).
Françon, M.: Contribution à l'amélioration du pouvoir séparateur. Nuovo Cim. **9**, Suppl. No 3, 283—287 (1952).
Gandy, R. O.: Out-of-focus diffraction patterns for microscope objectives. Proc. Phys. Soc. Lond., B **67**, 825—831 (1954).
Goodbody, A. M.: The influence of condensor aperture on the resolution of a pair of parallel lines. Proc. Phys. Soc. Lond. B **70**, 361—368 (1957), 5 refs.
Helmholtz, H.: Die theoretische Grenze für die Leistungsfähigkeit der Mikroskope. Poggendorfs Ann., Jubelband 1874, 557—584.
Herschel, J. F. W.: Light. Encyclopaedia Metropolitana: or, the Universal dictionary of knowledge, II, 4, Second Division, p. 341—586, (1830).
Hopkins, H. H.: The concept of partial coherence in optics. Proc. Roy. Soc. Lond. A **208**, 263—277 (1951). 4 refs.

Hopkins, H. H.: On the diffraction theory of optical images. Proc. Roy. Soc. Lond. A **217**, 408—432 (1953).
— Application of coherence theory in microscopy and interferometry. J. Opt. Soc. Amer. **47**, 508—526 (1957).
—, and P. M. Barham: The influence of the condensor on microscopic resolution. Proc. Phys. Soc. Lond. B **63**, 737—744 (1950).
Koch, R.: Untersuchungen über die Ätiologie der Wundinfektionskrankheiten (1878). Ges. Werke 1912, Bd. 1, S. 61—108.
Laue, M. v.: Zur Theorie der optischen Abbildung. Ann. Physik (4) **43**, 165—168 (1914), 2 Lit.
Linfoot, E. H., and E. Wolf: Diffraction images in systems with an annular aperture. Proc. Phys. Soc. Lond. B **66**, 145—149 (1953).
— Phase distribution near focus in an aberration-free diffraction image. Proc. Phys. Soc. Lond. B **69**, 823—832 (1956).
Lummer, O., u. F. Reiche: Die Lehre von der Bildentstehung im Mikroskop von Ernst Abbe. Braunschweig 1910, 108 S., 17 Lit.
Mandelstam, L.: Zur Abbeschen Theorie der mikroskopischen Bilderzeugung. Ann. Physik (4) **35**, 881—897 (1911).
Maréchal, A.: Influence de l'ouverture de condenseur sur le contraste des images au microscope. Bull. microscopie appl. (2) **3**, 90 (1953).
—, et P. Lacomme: Une méthode pour améliorer le pouvoir séparateur des instruments d'optique basée sur l'amplification de phase. Nuovo Cim. **9**, Suppl. No. 3, 354—357 (1952).
Nomarski, G., et M. Rousseau-Gaudeix: Influence de l'éclairage semi-cohérent sur la limite de séparation. Cas d'une source annulaire. J. Phys. Radium **16**, 13 S—16 S (1955), 10 réfs.
Osterberg, H., and L. W. Smith: Diffraction theory for images of disc-shaped particles with Köhler illumination. J. Opt. Soc. Amer. **50**, 362—369 (1960), 3 refs.
Rayleigh Lord (Strutt, J. W.): On the theory of optical images, with special reference to the microscope. Phil. Mag. **42**, 167—195 (1896). Sci. papers IV, 235—260.
Richards, B., and E. Wolf: The Airy pattern in systems of high angular aperture. Proc. Phys. Soc. Lond. B **69**, 854—856 (1956), 6 refs.
Slansky, S.: Images d'un disque clair ou sombre en éclairage partiellement cohérent. Rev. d'Opt. **39**, 555—577 (1960), 17 réfs.
Smith, L. W.: Diffraction images of disc-shaped particles computed for full Köhler illumination. J. Opt. Soc. Amer. **50**, 369—374 (1960), 2 refs.
—, and H. Osterberg: Diffraction images of circular self-radiant discs. J. Opt. Soc. Amer. **51**, 412—414 (1961), 3 refs.
Som, S. C.: On partial coherence and the concept of "effective source" as used in instrumental optics. Optica Acta **10**, 179—186 (1963).
Tschunko, H. F. A.: Irradiance distribution in the errorfree diffraction image. J. Opt. Soc. Amer. **55**, 1—4 (1965), 4 refs.
Wolfke, M.: Allgemeine Abbildungstheorie selbstleuchtender und nichtselbstleuchtender Objekte. Ann. Physik (4) **39**, 569—610 (1912).
— Über die Abbildung eines Gitters außerhalb der Einstellebene. Ann. Physik (4) **40**, 194—200 (1913).
Zernike, F.: The concept of degree of coherence and its application to optical problems. Physica Haag **5**, 785—795 (1938).
—, et B. R. A. Nijboer: Théorie de la diffraction des aberrations. Théorie des images optiques, Paris 1949, p. 227—235.

Literatur zu Ziff. 6 bis 18.

Abbe, E.: Über einen neuen Beleuchtungsapparat am Mikroskop. Schultze Archiv mikr. Anat. **9**, 469—480 (1873) = Ges. Abh., Bd. 1, 101—112.
— Über neue Mikroskope. (Über Verbesserungen des Mikroskopes mit Hilfe neuer Arten optischen Glases.) Sitzber. Jena. Ges. Med. Naturw., **1886**, 107—128 = Ges. Abh., Bd. 1, 450—472, engl. Übers. H. A. Miers: J. Roy. Microscop. Soc. (2) **6**, 20—34 (1886).
— Über die Verwendung des Fluorits für optische Zwecke. Z. Instrumentenkde. 1—6 (1890) = Ges. Abh., Bd. 1, 478—486.
Barta, E.: Der Mikroilluminator. Z. wiss. Mikrosk. **52**, 276—292 (1935), 20 Lit.
Berek, M.: Über einen Zweiblendenkondensor für Polarisationsmikroskope. Verh. Ges. dtsch. Naturforsch. **2**, 1, 600—601 (1913).
— Grundlagen der praktischen Optik. 1930, Berlin u. Leipzig, 152 S.
Blaschke, W. S. S.: A simple form of flat-field microscope objective. J. Sci. Instrum. **31**, 173—174 (1954).

BOEGEHOLD, H.: Die Verbesserung des Bildfeldes der Mikroskopobjektive (Planachromate). Z. wiss. Mikrosk. **55**, 17—25 (1938).
— Das Bildfeld des Mikroskops. Jenaer Zeiss-Jahrb. **1950**, 19—33, 17 Lit.
— Neue Mikroskopobjektive und -okulare. Feingerätetechnik **1**, 114—117 (1952), 5 Lit.
— Das optische System des Mikroskops. 1958, Berlin, 192 S., 135 Lit.
—, u. H. BERNHARDT: Die Weiterarbeit an Mikroskopobjektiven und -okularen im Zeiss-Werk seit Abbes Tode. Jenaer Jahrb. **1955**, 2. Teil, 185—206, 8 Lit.
—, u. A. KÖHLER: Das Homal, ein System, welches das mikrophotographische Bild ebnet. Z. wiss. Mikrosk. **39**, 249—262 (1922).
BOUWERS, A., B. S. BLAISSE u. H. W. BULTHUIS: Katadioptisches System mit Hohlspiegel und Kittfläche. Niederländ. Priorität v. 22. 3. 1951; Oest. Pat. 177.937; US Pat. 2704417.
BURCH, C. R.: Reflecting microscopes. Proc. Phys. Soc. Lond. **59**, 41—46 (1947).
CLAUSSEN, C. H. (ab 1960: H. C.): Von der Optik im Mikroskop. Objektiv (Werkz. opt. Werke E. Leitz Wetzlar), Nr. 15, 21—28 (1956).
— Okulare für Brillenträger. Leitz Mitt. Wiss. Techn. **1**, 18—20 (1958).
— Microscope objectives with plano correction. Appl. Optics **3**, 993—1003 (1964).
CONRADY, A. E.: Applied optics and optical design. 1957, 1960 New York, 2 parts, 841 p.
DETERMANN, H.: Anwendung der Beugungstheorie auf rechnerisch korrigierte Mikroskopobjektive zur Beurteilung der Restfehler. Optica Acta **13**, 149—194 (1966), 6 Lit.
Deutsche Normen: Mikroskope. Vergrößerungen: Lupenvergrößerung-Maßstab. DIN 58886 (Mai 1960).
DRUCKS, H.: Bemerkung zur Theorie der Superachromate. Optik **23**, 523—534 (1965/66), 7 Lit.
FERGUSON, K. R.: The performance of radiation protected microscope objectives. US Atomic Energy Commission, Techn. Inf. Serv., Oak Ridge, Tenn., TID-5280 (Suppl. **1**) 95—105 (1956).
FISCHER, R.: Optisches Glas in Kernphysik und Kerntechnik. Isotopentechnik **2**, 360—363 (1962), 8 Lit.
FLÜGGE, J.: Zur Behandlung der geometrischen Theorie des Kugelspiegels. Z. Instrumentenkde. **57**, 495—500 (1937).
FRANKE, G.: Vorschläge für die Festsetzung von Meßlinien zur Charakterisierung optischer Daten. Optik **8**, 57—60 (1951).
FRENK, H.: Möglichkeiten und Grenzen der Fernsehmikroskopie. Radio Mentor 1960, 887—893.
GABLER, F.: Ein ferngesteuertes Mikroskop für die „heiße" Metallographie. Radex Rundschau 1958/59, 557—562.
GREY, D. S.: A new series of microscope objectives: II. Preliminary investigation of catadioptric Schwarzschild systems. J .Opt. Soc. Amer. **39**, 723—728 (1949).
— A new series of microscope objectives. III. Ultraviolet objectives of intermediate numerical aperture. J. Opt. Soc. Amer. **40**, 283—290 (1950).
JAHN, W.: Strahlenresistente optische Gläser. Glastechn. Ber. **35**, 479—483 (1962), 7 Lit.
KAPANY, N. S.: Fiber optics. Principles and applications. London and New York 1967, about 400 p.
KÖHLER, A.: Ein neues Beleuchtungsverfahren für mikrophotographische Zwecke. Z. wiss. Mikrosk. **10**, 433—440 (1893).
Leitz, E. GmbH: Großes Metallmikroskop MM 5 RT mit Fernbedienung. Kerntechnik **7**, 551 (1965).
LIHOTZKY, E.: Verallgemeinerung der Abbeschen Sinusbedingung für Systeme mit nicht gehobener Längenaberration. Sitzber. Akad. Wiss. Wien, math.-nat. Kl., Abt. IIa, **128**, 85—90 (1919).
MAKSUTOV, D. D.: Spiegelobjektiv. UdSSR Pat. 40.859, 1932.
MARÉCHAL, A.: Optique géométrique générale. In: Handbuch der Physik, Bd. 24, S. 44—170. 1956. 29 Lit.
NORRIS, K. P.: Development of reflecting microscopes. Research **8**, 94—101 (1955), 61 refs.
—, W. E. SEEDS and M. H. F. WILKINS: Reflecting microscopes with spherical mirrors. J. Opt. Soc. Amer. **41**, 111—119 (1951).
OETTEL, W.: Ein neuer Schritt in der Entwicklung der Auflichtmikroskopie. Jenaer Jahrb. **1951**, 37—49, 3 Lit.
OTLEY, K. O.: Effects of high-energy radiation upon glasses. Glass Industry **33**, 24—27 (1952), 42 refs.
PAYNE, B. O.: Observations on microscope condensors, objectives and eyepieces. J. Roy. Microscop. Soc. (3) **69**, 200—214 (1949); **70**, 317—341 (1950); **72**, 31—45 (1952), 32 refs.
REICHEL, W., u. R. TIEDEKEN: Faseroptik — ein neues Bauelement im optischen Gerätebau. Feingerätetechnik **13**, 39—46 (1964). 29 Lit.

REINERT, G. G.: Dunkelfeld- und Ultramikroskopie. Handbuch der mikroskopischen Technik, Bd. 13. Stuttgart 1942, 73 S., 30 Lit.
RICHTER, R.: Vorrichtung zur Beleuchtung mikroskopischer Objekte. DRP 620537 (1933).
RIESENBERG, H.: Das Spiegelmikroskop und seine Anwendungen. Jenaer Jahrb. **1956**, 30—78, 89 Lit.
— Ein nicht-rotationssymmetrisches Zweispiegelsystem mit korrigierter Bildfeldneigung. Jenaer Jahrb. **1965**, 15—20, 5 Lit.
ROHR, M. v.: Ernst Abbes Apochromate. Carl Zeiss Jena 1936, 64 S.
ROE, E. M. F.: Applications of the reflecting microscope. Research (London) **8**, 147—157 (1955), 55 refs.
SCHWARZSCHILD, K.: Untersuchungen zur geometrischen Optik. II. Theorie der Spiegelteleskope. Abh. Kgl. Ges. Wiss. Göttingen, math.-physik. Kl., N. F. **4**, 3—28 (1905).
SMITH, R.: Complete system of optics. London 1738.
STAEBLE, F.: Isoplanatische Korrektion und Proportionalitätsbedingung. (Zur Bedeutung der Abbeschen Sinusbedingung bei sphärisch nicht korrigierten Systemen endlicher Öffnung.) Sitzber. bay. Akad. Wiss., math.-physik. Kl. **1919**, 163—196.
WEBER, K.: Strahlungsquellen für die Mikrophotometrie. Acta Histochem., Suppl. **6**, 157—163 (1965), 4 Lit.
— Private Mitteilung 1966.
WEIDERT, F.: Die zweckmäßige Wahl der Wellenlängen und Lichtquellen für die Messung von Brechungsverhältnissen. Z. techn. Phys. **7**, 304—310 (1926).
WOLTER, H.: Optik dünner Schichten. In: Handbuch der Physik, Bd. 24, S. 461—554. 1956.
WYNNE, C. G.: Flat-field microscope objective. J. Sci. Instrum. **38**, 92—94 (1961).

Literatur zu Ziff. 20 bis 27.

AGROSKIN, L. S., and N. V. KOROLEV: Microscope-spectrophotometers. Optics Spectroscopy **6**, 544—545 (1959).
BARER, R.: Aspects of ultra-violet and infra-red microspectrography with the Burch reflecting microscope. Disc. Faraday Soc. **9**, 369—378 (1950).
— Learning about the invisible. (Ultra-violet and infra-red photomicrography). Phot. J. B **90**, 83—91 (1950).
— Determination of dry mass, thickness, solid and water concentration in living cells. Nature **172**, 1079 (1953), 3 refs.
—, and V. E. COSSLETT: Advances in optical and electron microscopy 1. London and New York 1966, 282 p.
BATES, W. J.: A wavefront shearing interferometer. Proc. Phys. Soc. Lond. **59**, 940—950 (1947).
BENARIE, M.: Détermination de la limite inférieure de comptage dans une granulometrie microscopique. Rev. opt. **45**, 55—80 (1966), 10 réfs.
BENNETT, A. H., H. OSTERBERG, H. JUPNIK, and O. W. RICHARDS: Phase microscopy, principles and applications. New York and London 1951, 320 p., 14 refs.
BEREK, M.: Zur Messung der Doppelbrechung hauptsächlich mit Hilfe des Polarisationsmikroskops. Zbl. Mineral. **1913**, 388—396, 427—435, 464—470.
— Mikroskopische Mineralbestimmung mit Hilfe der Universaldrehtischmethoden. Berlin 1924, 168 S.
— Das Problem der quantitativen Mikroskopie der Erzmineralien und seine Lösung. Z. Kristallogr. **76**, 1—22 (1931).
BESSIS, M., et G. NOMARSKI: Conditions de l'irradiation ultraviolette d'organites cellulaires. C. R. Acad. Sci., Paris **245**, 766—768 (1959), 4 réfs.
BEYER, H.: Theorie und Praxis des Phasenkontrastverfahrens. Techn. physik. Monogr. Leipzig **17**, 312 S. (1965).
BOGUTH, W., u. L. GOERTZ: Diffraktometrische Bestimmung des mittleren Erythrozytendurchmessers mit Hilfe des Mikroskopes. Z. wiss. Mikrosk. **66**, 24—31 (1964).
BRACE, D. B.: A half-shade elliptical polarizer and compensator. Phys. Rev. **18**, 70—88 (1904).
BRICE, A. T., and P. H. KECK: The most favorable absorption of the phase plate in phase contrast microscopy. J. Opt. Soc. Amer. **37**, 647—651 (1947).
BRYNGDAHL, O.: Applications of shearing interferometry. Progress in Optics **4**, 37—83 (1965), 74 refs.
BURRI, C.: Das Polarisationsmikroskop. 1950 Basel, 308 S., Lit.
CARLSON, L., and B. L. GLEDHILL: Studies on the dry mass of bull spermatozoal heads. Exptl. Cell Research **41**, 376—384 (1966), 27 refs.

CASPERSSON, T.: Cell growth and cell function. 1950 New York. 185 p. Bibliogr. 7 p.
—, G. LOMAKKA u. L. CARLSON: Eine Instrumentenausrüstung für quantitative Cytochemie. Acta Histochem. **9**, 139—156 (1960), 10 Lit.
CAUSLEY, D., and J. Z. YOUNG: Flying spot microscope use in particle analysis. Research, London **8**, 430—434 (1955).
CHARMAN, W. N.: Some experimental measurements of diffraction images in low-resolution microscopy. J. Opt. Soc. Amer. **53**, 410—414 (1963), 13 refs.
— Diffraction images of circular objects in high-resolution microscopy. J. Opt. Soc. Amer. **53**, 415—419 (1963), 12 refs. Vgl. J. Roy. Microscop. Soc. **82** (1963/64), 81—94 und 163—171.
CHAYES, F.: Petrographic modal analysis. 1956 New York, 113 p.
CLAUSEN, T.: Beschreibung eines neuen Micrometers. Astronom. Nachr. **18**, Spalte 95—96 (1841).
COONS, A. H.: I. Antibodies and antigens labelled with fluorescin. II. The diagnostic application of fluorescent antibodies. Z. allg. Patholog. **22**, 693—723 (1959).
DAVIES, C. N.: Shape of small particles. Nature **201**, 905—907 (1964), 6 refs.
DAVIES, H. G.: The determination of mass and concentration by microscope interferometry. General cytochemical methods (Ed. J. F. DANIELLI), vol. 1, 1958 New York.
DEELEY, E. M.: An integrating microdensitometer for biological cells. J. Sci. Instrum. **32**, 263—267 (1955).
DYSON, J.: Precise measuring by image-splitting. J. Opt. Soc. Amer. **50**, 754—757 (1960).
EHRINGHAUS, A.: Drehbare Kompensatoren aus Kombinationsplatten doppelbrechender Kristalle. Z. Kristallogr. **76**, 315—321 (1931).
EMMONS, R. C.: The universal stage (with five axes of rotation). Geol. Soc. Amer., Memoir **8**, 1943, 205 S.
ERÄNKÖ, O.: Quantitative methods in histology and microscopic histochemistry. Basel and New York: 1955. 160 p.
FEDOROW, E. v.: Universal-(Theodolith)-Methode in der Mineralogie und Petrographie. II. Krystallographische Untersuchungen. Z. Kristallogr. **22**, 229—268 (1894).
FISCHMEISTER, H.: Verfahren und Apparate der quantitativen Metallographie. Prakt. Metallogr. **2**, 251—263 (1965)., 31 Lit.
FRANÇON, M.: Contraste de phase et contraste par interférences. 1952 Paris, 264 p.
— Le microscope à contraste de phase et le microscope interférentiel. 1954 Paris, 150 p. Bibliogr. 13 p.
— Interférences, diffraction et polarisation. In: Handbuch der Physik, Bd. 24, S. 171—460. 1956, 90 réfs.
GAHM, J.: Quantitative polarisationsoptische Messungen mit Kompensatoren. Zeiss-Mitt. **3**, 153—192 (1964), 18 Lit.
GLAGOLEFF, A. A.: On the geometrical methods of quantitative mineralogic analysis of rocks. Trans Instn Econ. Mineral., Mosk. **59** (1933).
GOTTSCHEWSKI, G. H. M.: Die Methoden der Fluoreszenz- und Ultraviolett-Mikroskopie und Spektroskopie in ihrer Bedeutung für die Zellforschung. Mikroskopie **9**, 147—167 (1954).
HAASE, M.: Optische Eigenschaften neuerer Polarisationsfilter. Zeiss-Mitt. **2**, 173—181 (1960/62), 16 Lit.
HAITINGER, M., J. EISENBRAND u. G. WERTH: Fluoreszenzmikroskopie, 2. Aufl. 1959 Leipzig, 168 S., 413 Lit.
HALLIMOND, A. F.: Manual of the polarizing microscope 2. ed. 1953 York, 204 p., 128 refs.
HANSEN, G.: Energetische Grenzen der Mikro-Spektralphotometrie. Zeiss-Mitt. **2**, 117—124 (1961).
HARTSHORNE, M. H.: Crystals and polarizing microscope, third ed. 1960 New York, 557 p.
HAWKSLEY, P. G. W.: Theory of particle sizing and counting by track scanning. Brit. J. Appl. Phys., Suppl. **3**, 125—132 (1954), 16 refs.
HEINE, H.: Optisches Beobachtungsverfahren mit Phasenkontrast. DBP 821128 (Ernst Leitz GmbH), 1949.
HELML, F., u. J. A. SCHEDLING: Untersuchungen über die Radienabhähgigkeit der Durchmesserverfälschung, die bei der mikroskopischen Abbildung kugelförmiger Selenteilchen auftritt. Acta phys. Austriaca, **5**, 237—239 (1951), 4 Lit.
HENNIG, A.: Kritische Betrachtungen zur Volumen- und Oberflächenmessung in der Mikroskopie. Zeiss-Werkz. **6**, 78—86 (1958), 13 Lit.
HOLMES, D. A.: Exact theory of retardation plates. J. Opt. Soc. Amer. **54**, 1115—1120 (1964), 13 refs.
— Wave optics theory of rotary compensators. J. Opt. Soc. Amer. **54**, 1340—1347 (1964), 22 refs.
— Fluid immersion as a means for reducing resonance effects in rotary compensators. J. Opt. Soc. Amer. **55**, 209—210 (1965), 3 refs.

HORN, W.: Interferometer nach MACH-ZEHNDER für Durchlichtbeobachtung. DBP 1021595 (Ernst Leitz GmbH), 1955.
INGELSTAM, E., and L. P. JOHANSSON: Correction due to aperture in transmission interference microscopes. J. Sci. Instrum. **35**, 15—17 (1958), 18 refs.
— —, and C. F. BRUCE: Obliquity corrections in transmission interference microscopes. J. Sci. Instrum. **36**, 246 (1959), 5 refs.
INOUÉ, S.: Studies on depolarization of light at microscope lens surfaces. I. The origin of stray light by rotation at the lens surfaces. Exp. Cell Research **3**, 199—208 (1952), 8 refs.
—, and W. L. HYDE: Studies on depolarization of light at microscope lens surfaces. II. The simultaneous realization of high resolution and high sensitivity with the polarizing microscope. J. Biophys. Biochem. Cytol. **3**, 831—838 (1957), 11 refs.
JAMIN, M. J.: Sur un réfracteur différentiel pour la lumière polarisée construit par Lutz. C. R. Acad. Sci., Paris **67**, 814—816 (1868).
JOHNSON, B. K.: Microscopy in the region of wave-length $\lambda = 2{,}000$ Å to 1,000 Å. J. Roy. Microscop. Soc. **73**, 24—29 (1953).
KÄSEMANN, E.: Fortschritte in der Herstellung dichroitischer Flächenpolarisatoren. Optik **19**, 161—168 (1962), 7 Lit.
KECK, P. H., u. A. T. BRICE: Die günstigste Dimensionierung der Phasenplatte beim Phasenkontrast-Verfahren. Optik **5**, 31—47 (1949).
KING, J.: Fluorescence microscopy. Research (London) **8**, 191—199 (1955).
KÖHLER, A.: Mikrophotographische Untersuchungen mit ultraviolettem Licht. Z. wiss. Mikroskop. **21**, 129—165; 273—304 (1904).
— Ein Glimmerplättchen Grau I. Ordnung zur Untersuchung sehr schwach doppelbrechender Präparate. Z. wiss. Mikroskop. **38**, 29—42 (1921).
—, u. W. LOOS: Das Phasenkontrastverfahren und seine Anwendungen in der Mikroskopie. Naturwissenschaften **29**, 49—61 (1941).
—, u. M. v. ROHR: Eine mikrophotographische Einrichtung für ultraviolettes Licht. Z. Instrumentenkde **24**, 341—349 (1904).
KRUG, W., u. E. LAU: Neue Verfahren der Interferenzmikroskopie. Technik **6**, 122—128 (1951).
— — Ein Interferenzmikroskop für Durch- und Auflichtbeobachtungen. Ann. Phys. **8**, 329—340 (1951).
—, J. RIENITZ u. G. SCHULZ: Beiträge zur Interferenzmikroskopie. 1961 Berlin, 352 S., Lit. 20 S.
— — —, translated by J. DICKSON: Contributions to interference microscopy. 1964 London, 394 p., 26 p. bibliography.
KUBOTA, H., and S. INOUÉ: Diffraction images in the polarizing microscope. J. Opt. Soc. Amer. **49**, 191—198 (1959), 9 refs.
—, and H. SAITO: Further study on the diffraction images in the polarizing microscope. J. Opt. Soc. Amer. **50**, 1020—1024 (1960), 6 refs.
LAND, E. H.: Some aspects of the development of sheet polarizers. J. Opt. Soc. Amer. **41**, 957—963 (1951), 9 refs.
LEBEDEFF, A. A.: Interféromètre à polarisation et ses applications. Rev. Opt. **9**, 385—413 (1930).
LINNIK, W. P.: Ein Apparat für mikroskopisch-interferometrische Untersuchung reflektierender Objekte (Mikrointerferometer). C. R. Acad. Sci. U.R.S.S. Nouv. Sér. (1933).
— Mikrointerferometer. Optiko-mechanitscheskaja promyschlennestj (opt.-mech. Industrie) **8**, 9 (1938).
MELLORS, R. C.: Analytical cytology. Methods for studying cellular form and function. 1955 New York-Toronto-London. 516 p. Bibliogr.
MENZEL, E., u. K. SCHMIDT: Dickenmessung an Objekten geringer lateraler Ausdehnung mit Dreistrahlinterferenz. Z. angew. Phys. **6**, 409—414 (1954).
NAIRN, R. C. (Editor): Fluorescent protein tracing. 1962, Edinburgh and London, 280 p., 35 p. Bibliography.
NAORA, H.: Microspectrophotometry in visible light range. Handbuch der Histochemie, Bd. I,1 (Allgemeine Methodik), S. 192—219, 123 refs. 1958.
NEVES E CASTRO, M., M. FRANÇON et M. ROUSSEAU: Installation pour l'étude spectrale de la fluorescence en microscopie. Rev. Opt. **37**, 385—390 (1958).
NIKITIN, W.: (Kipp-Kompensator.) Z. Kristallogr. **47**, 378—379 (1910).
NOMARSKI, G.: Dispositif interférentiel à polarisation. Franz. Pat. 1059124 (Anm. 1952).
OSTER, G., and A. W. POLLISTER (Editors): Physical techniques in biological research. 1955, 1956, New York, Vol. I Optical techniques, 564 p. Vol. III Cells and tissues 728 p.
OSTERBERG, H.: The polanret microscope. J. Opt. Soc. Amer. **37**, 726—730 (1947).
—, and A. J. CARLAN: Determination of refractive indices by immersion methods and phase microscopy. Trans. Amer. Microscop. Soc. **77**, 340—353 (1958), 4 refs.

Piller, H.: Die Phasenkontrastmikroskopie als Hilfsmittel zur Bestimmung feinkörniger, speziell dünner, transparenter Minerale. Heidelberger Beitr. Mineral. u. Petrogr. **3**, 307—334 (1952).
— Bemerkungen über den Einfluß der Segmentgröße auf die Genauigkeit bei der Messung von Neigungswinkeln mit dem Universaldrehtisch. Mikroskopie **12**, 166—174 (1957), 11 Ref.
—, and K. v. Gehlen: On errors of reflectivity measurements and of calculations of refractive index *n* and absorption coefficient *k*. Amer. Mineral. **49**, 867—882 (1964), 21 refs.
Plainevaux, J. E. C.: Guidage par suspension sur lames élastiques. Nuovo Cim. **10**, 1451—1460 (1953).
Pollister, A. W.: Physical techniques in biological Research. IIIa Cells and tissues, 2nd ed. 1966 New York, London, 408 p.
Price, G. R., and S. Schwartz: Fluorescence microscopy. Aus G. Oster and A. W. Pollister, Physical techniques in biological research, vol. 3, chapt. 3. 1956, New York p. 91—148.
Reinhard, M.: Universal-Drehtischmethoden. 1931 Basel, 119 S.
Richards, O. W.: Fluorescence microscopy. In: Mellors, Anal. Cytol. **1955**, 5, 1—37.
— Measurement with phase and interference microscope. Appendix: Measurement of vertical depth. ASTM Spec. techn. publ. No. 257, 6—18 (1959), 38 refs.
Rinne, F., u. M. Berek: Anleitung zu optischen Untersuchungen mit dem Polarisationsmikroskop, 2. Aufl. 1954 Stuttgart, 366 S.
Rosenbusch, H., u. E. A. Wülfing: Mikroskopische Physiographie der petrographisch wichtigen Mineralien, 5. Aufl., Bd. I, 1 Untersuchungsmethoden. 1921/1924 Stuttgart, 847 S., Lit.
Rosiwal, A.: Über geometrische Gesteinsanalysen. Verh. K. K. geol. Reichsanstalt **1898**, 143—175.
Sandritter, W.: Ultraviolettmikrospektrophotometrie. In: Handbuch der Histochemie, Bd. I/1 (Allg. Methodik), S. 220—318. 1958. 21 S. Lit.
Schuchardt, E.: Das Integrationsverfahren in der mikroskopischen Technik. In: Handbuch Mikroskopie in der Technik, Bd. I/1, S. 563—588. 1957, 43 Lit.
Seeds, W. E., and M. H. F. Wilkins: Ultra-violet Microspectrographic studies of nucleoproteins and crystals of biological interest. Disc. Faraday Soc. **9**, 417—423 (1950).
Seidel, F.: Regulationsbefähigung der embryonalen Säugetierkeimscheibe nach Ausschaltung von Blastemteilen mit einem UV-Strahlenstichapparat. Naturwissenschaften **39**, 553—554 (1952).
Slater, P. N.: The production of circular apertures of known diameter for calibration purposes in high resolution microscopy. Optics Metrology **1960**, 269.
Smith, C. L.: Microbeam and partial cell irradiation. Intern. Rev. Cytol. **16**, 133—153 (1964), 89 refs.
Smith, F. H.: Improvements in or relating to microscopy. Brit. Pat. 639014 (appl. 1947).
— Microscopic interferometry. Research (London) **8**, 385—395 (1955), 19 refs.
Stodiek, W.: Arrangement for precise scattering measurements in nuclear emulsions. Nuovo Cim. **2**, 467—470 (1955).
Tchan, Y. T.: Phase-fluorescence microscopy. Nature **179**, 1243—1244 (1957).
Teucher, M.: Technik und Anwendung der Kernemulsionen. Ergeb. exakt. Naturw. **28**, 407—471 (1955), 4 S. Lit.
Tolansky, S.: Multiple-beam interferometry of surfaces and films. 1948 Oxford, 196 p.
Tschachotin, S.: Die mikroskopische Strahlenstichmethode, eine Zelloperationsmethode. Biol. Zentr. **32**, 623—630 (1912).
Uretz, R. B., and R. P. Perry: Improved ultraviolet microbeam apparatus. Rev. Sci. Instrum. **28**, 861—866 (1957).
Viénot, J. Ch.: Sur une mesure de profil en éclairage cohérent. Rev. Opt. **35**, 517—525 (1956), 6 réfs.
Walton, W. H.: Survey of the automatic counting and sizing of particles. Brit. J. Appl. Phys., Suppl. **3**, 121—125 (1954).
Watrasiewicz, B. M.: Image formation in microscopy at high numerical aperture. Optica Acta **12**, 167—176 (1965), 16 refs.
— Theoretical calculations of images of straight edges in partially coherent illumination. Optica Acta **12**, 391—400 (1965), 6 refs.
Weber, K.: Leitz-Mikroskop-Photometer MPV mit variabler Meßblende. Leitz-Mitt., Wiss. u. Techn. **3**, 103—107 (1965), 10 Lit.
—, u. F. Kornder: Polarisationseinrichtung. DBP 1.157.808, Anm. 29. 6. 62 (Ernst Leitz GmbH).
Weinberger, H., and J. Harris: Interference effect in quarterwave plates. J. Opt. Soc. Amer. **54**, 552—553 (1964), 1 ref.

WEINSTEIN, W.: Images of incoherenty illuminated bright and opaque discs. J. Opt. Soc. Amer. **45**, 1006—1008 (1955), 2 refs.

WEST, S. S., C. N. LOESER, and M. D. SCHOENBERG: Television spectroscopy of biological fluorescence. IRE Transact. Med. Electronics ME-**7**, 138—142 (1960).

WOLTER, H.: Schlieren-, Phasenkontrast- und Lichtschnittverfahren. In: Handbuch der Physik, Bd. 24, S. 555—645. 1956. 3 S. Lit.

ZERNIKE, F.: Einrichtung zur Verdeutlichung optischer Abbildungen. DRP 636168, Anm. 1932, (Carl Zeiss).

— Beugungstheorie des Schneidenverfahrens und seiner verbesserten Form, der Phasenkontrastmethode. Physica Haag **1**, 689—704 (1934), 9 Lit.

— Diffraction theory of the knife-edge test and its improved form, the phase-contrast method. Monthly Notices Roy. Astron. Soc. **94**, 377—384 (1934).

— Phase contrast, a new method for the microscopic observations of transparent objects. Part I and II. Physica Haag **9**, 686—698, 974—986 (1942).

— A precision method for measuring small phase differences. (3-Spalt- und 5-Spalt-Interferenzen.) J. Opt. Soc. Amer. **40**, 326—328 (1950).

ZWORYKIN, V. K., L. E. FLORY and R. E. SHRADER: Ultraviolet television microscopy. Electronics N.Y. **25**, 150 (1952).

Infrared microscope. Rev. Sci. Instrum. **37**, 245—246 (1966).

Literatur zu Ziff. 28.

ABBE, E.: Beschreibung eines neuen stereoskopischen Oculars. Z. Mikroskopie **2**, 207—234 (1880) = Carls Repertorium Exp.-Physik, **17**, 197—224 (1881) = Ges. Abh. **1**, 244—272 (1904).

GALBRAITH, W.: A method for high-power stereoscopic microscopy. Quart J. Mikroscop. Sci. **98**, 9—13 (1957).

GREGORY, R. L.: The "solid image" microscope. Research and Development **1961**, 101—102, 1 ref.

HEIMSTÄDT, O.: Die Anaglyphenmethode in der Stereomikroskopie. Zentr.-Ztg. Opt. Mech. **52**, 132—135 (1931).

LIHOTZKY, E.: Über Mikroskope und Doppelokulare für binokulare und stereoskopische Beobachtung. Z. wiss. Mikroskopie **41**, 305—324 (1924) = Mitt. Leitzwerke 35 (1924).

LOOS, W.: Stereoprojektion und Polarisationsfilter. Zeiss-Nachr. **3**, 39—53 (1939).

REINERT, G. G.: Mikroskop. DGM 1.447.446 (Carl Zeiss, Jena), 1938, DWP 6105 (G. G. Reinert).

WISE, F. C., F. E. J. OCKENDEN, and P. K. SARTORY: The binocular microscope: Its development, illumination and manipulation. 1950 London, 30 p.

Figurennachweis. Fig. 54 stellte Dr. KLAUS WEBER, Figuren 39, 45a, 49, 64, 66, 75 und 76 die Firma Ernst Leitz GmbH, Wetzlar, freundlicherweise zur Verfügung. — Fig. 63 wurde aus KRÜGER, H. G.: „Anwendung des Geradsicht-Spiegelmonochromators zum Aufbau eines registrierenden Spektralphotometers und eines UV-Mikroskops" entnommen. Fig. 69 stammt aus HAUSWALDT, H.: Interferenzerscheinungen an doppelbrechenden Kristallplatten im konvergenten polarisierten Licht, Magdeburg 1902, Fig. 70 aus CHUDOBA, K.: Mikroskopische Charakteristik der gesteinsbildenden Mineralien, Freiburg i. Br. 1932. Die Figuren 27 und 95 sind eigene Aufnahmen des Verfassers.

Diffraction Gratings.

By

GEORGE W. STROKE.

With 261 Figures.

A. General survey.

1. Scope of article. This article is more particularly devoted to fundamental properties of optical gratings, some of which have not yet been fully discussed in the literature. More generally, it is devoted to a rigorous treatment of basic principles which have been found to be applicable when dealing with practical or theoretical situations in the properties, ruling, testing and use of optical gratings. A list of references is given in the bibliography, to guide work towards additional applications. A brief historical review of the most relevant highlights in the development of the theory and ruling methods of gratings is given, and particular emphasis is laid on the most advanced applicable mathematical and experimental approaches.

I. Historical review.

2. Grating ruling and theory. The first gratings appear to have been produced by the American astronomer DAVID RITTENHOUSE in Philadelphia around 1786, and were described by him in the Transactions of the American Philosophical Society, 2, 201 (1786). Quite independently, JOSEPH VON FRAUNHOFER, some 37 years later, around 1821, invented and produced what appear to be the first ruled optical gratings, quite similar to those now in use in modern spectroscopy.

In his 1786 paper, DAVID RITTENHOUSE reports that he became interested in studying the (diffraction) effects that had been observed by looking through a silk handkerchief at a distant lamp: "In order to make my experiments with more accuracy, I made a square of parallel hairs about half an inch each way. And to have them nearly parallel and equidistant, I got a watchmaker to cut a very fine screw on two small pieces of brass wire. In the threads of these screws, 106 of which made one inch, the hairs were laid 50 or 60 in number. Looking through these hairs at a small opening in the window shutter of a dark room, 1/30 inch wide and three inches long,holding the hairs parallel to the slit, and looking toward the sky, I saw three parallel lines, almost equal in brightness, and on each side four or five other much fainter and growing more faint, coloured and indistinct, the farther they were from the middle line ...". Further, after describing angular measurements, RITTENHOUSE says: "I was surprised to find that the red rays are more bent out of their first direction, and the blue rays less ..." and notes that "By pursuing these experiments it is probable that new and interesting discoveries may be made, respecting the properties of this wonderful substance, light, ...".

Much of the early and frequently colorful history of grating ruling is known from descriptions by H. KAYSER (1900), C. F. MEYER (1934), C. CANDLER (1950), G. R. HARRISON (1949), H. D. and H. W. BABCOCK (1951), L. A. SAYCE (1953), J. STRONG (1960), V. RONCHI (1962) and others[1].

[1] DAVID RITTENHOUSE: Trans. Amer. Phil. Soc. **2**, 201 (1786). — JOSEPH VON FRAUNHOFER: Denkschr. kgl. Akad. Wiss. München **8**, 1—76 (1821/22). — Gilbert's Ann. Physik **74**, 337—378 (1823). — Edinburgh J. Sci. **7** (1827); **8** (1828). — H. KAYSER: Handbuch der

It is generally much less well known that JOSEPH VON FRAUNHOFER did not only invent the diffraction grating in 1821, and this deliberately by arranging a multiple array of slits, and not accidentally as has been sometimes described, but that he actually ruled optical gratings on metallic films (gold) deposited on glass as well as on glass itself, rather than merely producing his famous wire gratings by winding threads and wires around screws. Moreover, FRAUNHOFER introduced the use of ruling diamonds, studied the use of gratings in reflexion and in immersion, as well as in transmission, noted the similarity in the diffraction images produced by glass fibers and those produced by metallic wires of various nature, and observed and discussed polarization effects in the diffracted spectra.

In fact, FRAUNHOFER'S understanding of gratings, as described in his own papers[1], shows an amazing depth of perception and insight into the various grating diffraction problems, both theoretical and experimental. As pointed out by J. S. AMES[2] of the Johns Hopkins University, who translated some of FRAUNHOFER'S papers into English in 1898, some of FRAUNHOFER'S most important contributions to gratings involve: 1. the discovery of the principle of plane gratings, 2. the derivation of the grating equation, 3. the discussion of the effect upon spectra of a) periodic errors in the ruling of gratings, b) the shape of the grooves, c) the relative width of opaque and open spaces. An excellent history of grating ruling development and extensive references to the work of FRAUNHOFER and other pioneers is given in the outstanding article on diffraction gratings prepared by H. KAYSER for his Handbuch der Spektroskopie in 1900.

The principal innovations in the ruling of diffraction gratings are usually attributed to FRAUNHOFER, 1823, (loc. cit.), ROWLAND, 1882, MICHELSON, 1912, ANDERSON, 1914, R. W. WOOD, 1911–1944, H. D. and H. W. BABCOCK, 1951, J. STRONG, 1951, and G. R. HARRISON and G. W. STROKE, 1955, 1961, 1963[3].

Frequently, although with some notable exceptions, the principal innovators in the ruling of gratings appear also to have made important contributions to new developments

Spektroskopie, Vol. I. Leipzig: S. Hirzel 1900. — C. F. MEYER: Diffraction of Light, X-rays, and Material Particles. Ann. Arbor, Michigan: J. W. Edwards 1949. — C. CANDLER: Modern Interferometers. London: Hilger & Watts Ltd. 1951. — G. R. HARRISON: J. Opt. Soc. Amer. **39**, 413 (1949). This paper gives an excellent review of the early development of grating ruling, as well as an extensive list of additional references. — H. D. BABCOCK and H. W. BABCOCK: J. Opt. Soc. Amer. **41**, 776 (1951). — L. A. SAYCE: In: Endeavour, p. 210 (1953). — J. STRONG: J. Opt. Soc. Amer. **50**, 1148 (1960).

[1] JOSEPH VON FRAUNHOFER, Gesammelte Schriften, herausgeg. von E. LOMMEL. München: Verlag der K. Akademie 1888. — Denkschr. kgl. Akad. Wiss. München (loc. cit.) **8**, 1–76 (1821/22). — Gilbert's Ann. Physik **74**, 337–378 (loc. cit.).

[2] J. S. AMES: Prismatic and Diffraction Spectra. Memoirs by JOSEPH VON FRAUNHOFER. New York: Harper & Brothers 1898. Also see: H. KAYSER: Handbuch der Spectroscopie, Vol. I, Chap. IV, p. 395. Leipzig: S. Hirzel 1900.

[3] H. A. ROWLAND: Phil. Mag. (4), **13**, 469 (1882). — Physical Papers, p. 487 and 689. Baltimore: The Johns Hopkins Univ. Press 1902. — A. A. MICHELSON: Nature, Lond. **362** (January 11, 1912). — Univ. of Chicago Mag. **7**, 141 (1915). — Proc. Amer. Phys. Soc. **54**, 137 (1915). — J. Franklin Inst. **181**, 785 (1916). — Studies in Optics, p. 99. Chicago, Ill.: University Chicago Press 1927; reprinted in 1962. — J. A. ANDERSON: Glazebrook's Dictionary of Applied Physics. The manufacture and testing of diffraction gratings, Vol. IV, p. 30. 1923. — R. W. WOOD: Phil. Mag. **48**, 497 (1924). — H. D. BABCOCK and H. W. BABCOCK: J. Opt. Soc. Amer. **41**, 776 (1951). — H. W. BABCOCK: Applied Optics **1**, 415 (1962). — J. STRONG: J. Opt. Soc. Amer. **41**, 3 (1951); **50**, 1148 (1960). — G. R. HARRISON and G. W. STROKE: J. Opt. Soc. Amer. **45**, 112 (1955); **50**, 1153 (1960). — G. W. STROKE: J. Opt. Soc. Amer. **51**, 1321 (1961). This paper gives extensive references to the development of grating ruling and testing. — Traçage de grands réseaux de diffraction à l'aide d'un chariot mobile asservi par interférométrie photoélectrique. In: Optics in Metrology, edit. by P. MOLLET. Oxford: Pergamon Press 1960. — Ruling, Testing and Use of Optical Gratings for High-Resolution Spectroscopy. In: Progress in Optics, Vol. II, edit. by E. WOLF. Amsterdam: North-Holland Publishing Company 1963. — G. R. HARRISON: Proc. Amer. Phil. Soc. **102**, 438 (1958).

in the theory of gratings (FRAUNHOFER 1823, ROWLAND 1882, MICHELSON 1912, ANDERSON 1914, STROKE 1961, and so on).

In addition, well known contributions to the theory of optical gratings were made by Lord RAYLEIGH (1907), VOIGT (1911), IGNATOWSKY (1914), U. FANO (1938) and ARTMANN (1942) in particular, and the problem of gratings and their anomalies have also attracted the interest of several dozen other investigators, among them E. MASCART (1889), H. S. ALLEN (1902), R. W. WOOD (1912, 1935), A. SOMMERFELD, F. REICHE (1911, 1953), J. STRONG (1936, 1958), C. F. MEYER (1934), E. INGELSTAM and E. DJURLE (1953), A. MARÉCHAL (1958), LOHMANN (1959), V. TWERSKY (1956, 1961), BABCOCK and BABCOCK (1951), J. STRONG (1951, 1961), HARRISON and STROKE (1960, 1961), BABCOCK (1962), and others[1].

The recent availability of optical gratings in great numbers and of high quality has resulted in a new interest in the solution of the two theoretical problems of the diffraction of light by gratings: 1. the distribution of the diffracted light in the spectral diffraction patterns as formed by both perfect as well as by imperfect gratings, 2. the distribution of the diffracted energy among the waves of different orders. The spectral image formation problem can now be considered as largely solved (STROKE, 1960, 1961) and a particular simplicity has been introduced into the theoretical treatment of gratings (INGELSTAM and DJURLE, 1953, STROKE, 1960, 1961) by the use of Fourier transform description of diffraction and image formation as developed by MICHELSON (1927), DUFFIEUX (1946), MARÉCHAL (1947, 1960), H. H. HOPKINS (1950), E. L. O'NEILL (1956), ARSAC (1959, 1961) and others[2]. But the boundary value problem of the energy distribution among orders, as well as the problem of scattering, which are both determined by the detailed structure of the groove profile, still remain to be solved. A significant step towards a complete solution of the problem of blaze in optical gratings has been reported by STROKE (1963) when he pointed out theoretically, and verified experimentally, that high efficiencies in blazed optical gratings could only be attained to the extent that one could avoid polarization in the diffracted light by a suitable choice of groove profile, the choice being determined by consideration of the electromagnetic nature of diffraction and of the boundary conditions to be satisfied on the surface of the grating by the total field (incident + diffracted)[3,4].

Contemporary gratings ruled for high resolution work in the visible domain have recently attained practical perfection (HARRISON and STROKE, 1955, 1960, STROKE 1961) and the effect of grating imperfections on the spectral images is now well understood (STROKE 1960, 1961).

For a grating to be perfect every groove should have the same shape and depth and should be positioned at an exact multiple of the spacing constant *from the origin*. The grooves should also be parallel to each other. Early gratings were ruled on mechanical dividing engines in which one groove is ruled each

---

[1] Lord RAYLEIGH: Proc. Roy. Soc. (Lond. A) **79**, 399 (1907). — U. FANO: Ann. Physik **32**, 393 (1938). — K. ARTMANN: Z. Physik **119**, 529 (1942). — E. MASCART: Traité d'Optique, Vol. I, p. 373. Paris: Gauthier-Villars 1889. — H. S. ALLEN: Phil. Mag. **3**, 92 (1902); **6**, 559 (1903). — R. W. WOOD: loc. cit.; also Phil. Mag. **48**, 497 (1924). — A. SOMMERFELD: Optics. New York: Academic Press Inc. 1954. — F. REICHE: Rept. EM-61, Inst. Math. Sci. New York Univ. 1953. — J. STRONG: loc. cit.; also Phys. Rev. **49**, 291 (1936). — C. F. MEYER: loc. cit. — E. INGELSTAM and E. DJURLE: loc. cit. — A. MARÉCHAL: Optica Acta **5**, 70 (1958). — A. LOHMANN: Optica Acta **6**, 175 (1959). — V. TWERSKY: Inst. Radio Engnrs., Trans., PGAP, AP-4, p. 330 (1956). — On multiple scattering of Waves. Report of the U.S. of America National Committee to the XIII General Assembly of the International Scientific Radio Union, London, England, September 5 to 15 (1960): both these papers contain a very extensive list of references, including titles, in particular connection with the multiple scattering by gratings of cylinders.

[2] See references listed in Sect. 35.

[3] G. W. STROKE: Phys. Letters **5**, 45 (1963).

[4] See also chapter B V. and references in Sects. 60 to 71.

time the grating carriage has been advanced through one step by a screw. The screw is actuated through a gear, ratched-and-pawl system, once per diamond tool cycle. In this "stop and go" motion, the location of the groove on the grating depends on built-in mechanical perfection of the engine, friction, temperature and distorsions in particular. Practical perfection in spacing are considered to be attained when the grooves are properly placed to better than 1/10 of the shortest wavelength of interest. In the visible, this corresponds to some 0.02 microns or 10^{-6} inches. Local errors, especially when covering a large portion of the grating, contribute to scattered light. Local error, whether regular or periodic, should be kept under a few thousandths of a wavelength.

Most of the contemporary gratings have been ruled in aluminum films evaporated on perfectly flat or spherical glass blanks, ever since JOHN STRONG made the basic contribution to the production of gratings by developing the methods of vacuum deposition of metals in 1936. The shaping of ruling diamonds has also been perfected since the early work of H. D. and H. W. BABCOCK, and has greatly helped in the achievement of preferential energy distribution of the diffracted energy among orders (blazing of gratings). Gratings for the soft X-ray region are usually ruled directly on glass according to the method developed by M. SIEGBAHN in Sweden. Various thin-film overcoatings of the ruled gratings (for example $Al + MgF_2$) have been recently developed to increase reflectivities in the far ultraviolet (G. HASS, P. H. BERNING, R. P. MADDEN, P. G. WILKINSON, and others)[1].

Grating replication and its history will be more particularly discussed in Sect. 89.

3. Ruling engines. For over half a century, up to the 1940's, the Johns-Hopkins University was one the world's principal suppliers of very good gratings, and the contributions of ROWLAND, ANDERSON and WOOD are almost legendary. Comparable fame has been achieved by MICHELSON and his University of Chicago engines, even though his aim was to be satisfied with nothing less than superb gratings; as a result, only a very small number of gratings were produced by him and his successor H. G. GALE.

Two schools of grating ruling have been developing in the United States of America.

One of the schools is that of the *built-in mechanical perfection* originated by H. A. ROWLAND at the Johns-Hopkins University in Baltimore in 1882 and pursued there since then by J. A. ANDERSON (1911—1916), R. W. WOOD (1923 to 1950) and JOHN STRONG (1947) in particular. From this school has also grown the grating ruling activity at the Mount Wilson Observatory in Pasadena, where it was started by JACOMINI and J. A. ANDERSON (1912—1929) and then pursued by H. D. BABCOCK (with E. D. PRALL and O. SWANSON) (1929—1948) and more recently by H. W. BABCOCK (1948), as well as the engines at Jarrell-Ash Co. (D. K. RICE and WM. MACARTHUR) and at Baird-Atomic Inc. (W. BAIRD, W. LANGTON with F. GILLIS). Bausch & Lomb, Inc. (D. RICHARDSON, R. S. WILEY) has been successfully ruling gratings by mechanical means since about 1950 on one of the two Michelson engines, suitably improved, after it was obtained from the University of Chicago.

The other school is that of the *interferometrically controlled ruling of gratings*, which was first considered by A. A. MICHELSON in the early 1900's, but was successfully achieved only in 1955 by G. R. HARRISON and G. W. STROKE on the M.I.T. ruling engine. That engine had itself been started at the University

---

[1] See references in Sect. 69.

of Chicago by A. A. MICHELSON and was transferred to M.I.T. in 1948 after some work on it had been carried out by H. G. GALE (1930—1940). An improvement of almost one order of magnitude in the quality of gratings, some exceeding 10 inches in width, ruled under interferometric servo control on the M.I.T. engine, was reported by STROKE in 1961, who also showed these gratings to be some 100 times more perfect than the already very famous 6-inch wide gratings ruled by J. A. ANDERSON on ROWLAND'S engine in 1914. The high quality of the M.I.T. gratings is indeed required for the use of gratings at high angles (and high resolutions) rather than at low angles of 5 to 10° (and at low resolutions) for which ROWLAND'S gratings were in fact equally well adapted.

The principle of the M.I.T. interferometric control system have now been adapted to a 4-inch engine at the Bausch & Lomb Company in Rochester, N.Y., (R. F. JARRELL and G. W. STROKE, 1963), and to a 10-inch engine at the Jarrell-Ash Company in Newtonville, Mass. Similar control systems are being considered for at least one engine in Japan, and H. D. BABCOCK has adapted the principle of interferometric control to the "stop and go" motion of the Mount Wilson engine in 1961. A somewhat similar system has been completed in 1963 by P. CONNES, G. PIEUCHARD and R. GOULLEY for a small engine at JOBIN et YVON, ARCUEIL, France. Basically, the interferometric control system for grating ruling uses for spacing a "natural" ruler set up by an optical interference fringe field which is produced in monochromatic light between a reference mirror and a moving mirror mounted on the grating carriage. The reference mirror is maintained in fixed position with respect to the ruling diamond plane. Under interferometric control, the grating carriage is made to move approximately in phase with the diamond, one cycle of the diamond reciprocating motion corresponding to an advance of one groove on the grating. As the diamond travels across the grating blank in ruling a groove, it is made to produce an artificial periodic signal which measures its position in its own cycle. The interference fringe signals are produced at the rate of one fringe per $\lambda/2$ of grating advance and measure the *actual* grating advance. The fringe signal is continuously compared in phase to the diamond reference signal by electronic means and is made to stay in phase with it by appropriately accelerating or decelerating the moving carriage by a servo-motor through a differential.

Since 1950 there has been an ever increasing interest in high-resolution grating spectroscopy. The M.I.T. and the Mount Wilson engines are capable of ruling high-resolution gratings of widths exceeding 8-inches, and the Strong engine at the Johns-Hopkins University has been ruling very good gratings up to 6-inches in width. Moreover, the two commercial firms previously referred to, Bausch & Lomb and Jarrell-Ash, have been successful in ruling gratings of very good quality, sufficient for general spectroscopic work in the visible and ultraviolet domains. Since the early 1950's or so these firms have not only been meeting demands for high-quality gratings by their own production, but have also been successful in replicating several of the M.I.T. gratings. The engine built by Baird-Atomic in Cambridge, Mass. has been ruling 4-inch wide concave infrared gratings of good commercial quality since 1942, and is believed to be the first commercial engine to have operated successfully. A large engine at the University of Michigan has long been the world's prime supplier of infrared gratings of widths up to 18-inches, usable up to wavelengths of 200 microns or more. (H. M. RANDALL, E. F. BARKER with DE KOTINSKY and WEYRICH, see: H. M. RANDALL, 1938, 1956; E. F. BARKER and H. M. RANDALL, 1961, private communication.) Some existing mechanical engines might be adapted to the interferometrically controlled ruling of gratings, and controlled engines of completely novel designs are being planned for the ruling of very large astronomical gratings, and for further improvements of gratings, in particular for the reduction of scattered light and use in the ultraviolet domains.

Before relative perfection for their particular gratings was reported for the M.I.T. and Johns Hopkins gratings ruled on ROWLAND's engines, MICHELSON also obtained at least two very good 10-inch gratings (1905), the best of which got accidentally broken. Only photographs of spectra obtained with these gratings remain available, and they tend to indicate that MICHELSON's gratings were not of the same quality as the best of the contemporary gratings. This is also known with certainty, for the best of ROWLAND's gratings, of which interferograms were recently obtained (STROKE, 1961). In fact the argument concerning relative perfection in gratings is somewhat academic, since the requirements on quality have been consistently increasing as improvements on the production of gratings were proven to be achievable.

It has been becoming increasingly apparent that progress in the improvement of gratings is less the result of empiricism and colorful advances, than a result of scientific method and skillful use of the state of technology.

Besides the *M.I.T.*, *Mount Wilson*, *Jarrell-Ash*, *Bausch & Lomb*, *Johns-Hopkins*, *Baird-Atomic* and *University of Michigan* engines, which all produce very good gratings as described, several, other engines have been producing very good spectroscopic gratings since their completion. The most well known are the following: a) M. SIEGBAHN's engines at Nobel Institute in Stockholm (since 1930, 4- and 6-inch gratings and more particularly gratings ruled on glass for X-ray spectroscopy); b) Lord BLYTHSWOOD's engine (built by Otto Hilger) at the National Physical Laboratory in Teddington (since 1908, small gratings up to $3\frac{1}{2}$ inches in width); c) A Russian engine of the Rowland type, at the State Optical Institute in Leningrad, operating since sometime after 1945 (F. GERASIMOV and associates, plane and concave gratings up to 6 inches wide, comparable in characteristics to the good commercial gratings ruled by the Jarrell-Ash and Bausch & Lomb companies in the United States, coarse infrared echelettes up to 12.5 inches in width comparable to the echelettes ruled at the University of Michigan); d) an engine completed by E. LEIBHARDT (in the middle of the 1950's); and e) one by the Perkin-Elmer Corporation in 1961 (infrared gratings). Work on the attainment of new high-quality ruling engines is proceeding at *Carl Zeiss* in Oberkochen West Germany (G. HANSEN, H. VOLKMANN), in *Japan* (HITACHI) and in *France* (G. PIEUCHARD, R. GOULLEY and P. CONNES at JOBIN et YVON).

4. Recent trends. The attainment of high-resolution gratings has coincided, and indeed met, with the search for optical tools capable of analyzing hyperfine structures and isotope shifts in the optical domain. Similar demands are made on the resolving power of gratings in line-shape studies in astrophysics, as well as in the studies of plasmas, and of the light emitted from optical masers, at frequencies which range from the X-ray and far ultraviolet deep into the infrared domains. As a result of this work, a new interest has been developing in the theoretical understanding of the spectral resolving power and energy concentration capabilities of gratings. These had indeed been generally considered as fairly well understood, at least since Lord RAYLEIGH's and ROWLAND's times in the early 1900's. However, experiments with the newly available gratings, produced in great numbers in various academic and industrial laboratories, have not only demonstrated insufficiencies in the classical diffraction and image formation descriptions, but have clearly called for a review of the basis and applicability of the theoretical foundations used in such work. The possibility of improving on classical diffraction patterns, and indeed of considerably increasing the ultimate resolving power by appropriate shaping and coating of the diffracting apertures has been discussed by STRAUBEL, JACQUINOT, ROIZEN-DOSSIER and others[1]. In starting from a somewhat different point of view, BOUWKAMP and DE BRUINS, P. AIGRAIN, J. C. SIMON, BROUSSAUD and SPITZ, L. J. CHU, TORALDO DI FRANCIA and others[2] predicted and demonstrated the existence of super-resolving apertures. In fact, as a result of the development of image-formation theories in the Fourier-transform formulation by DUFFIEUX, MARÉCHAL and others, in particular, it has become conceivable to foresee ways of attaining unlimitted resolving powers with finite gratings under suitable noise-free conditions. Alternately, with the help of the theory of distributions developed by L. SCHWARTZ, J. ARSAC (1961) has shown that it is possible to calculate the steepest possible slope in the center of the image-forming diffraction patterns, that is the narrowest possible diffraction-pattern center which could be obtained with a finite wavefront of any imaginable shape. The limitations in spectral resolving power and the diffraction pattern imperfections arising from grating imperfections can now be fully predicted from interferometric measurements of the diffracted wave fronts as shown by STROKE. New grating improvements have thus become possible by singling out the grating errors that are particularly harmful in high-resolution work. In fact, as appears clearly with the help of the new image-formation theory, there are cases where it becomes possible to use even moderately imperfect gratings as if indeed they were absolutely perfect.

Further improvements and extensions of the resolution and luminosity attainable with gratings appear from these theoretical studies. Moreover, new improvements in the capability of blazing the gratings, that is of obtaining preferred directions of concentration of the diffracted energy should result when a complete formulation of the effect of groove-shapes on the energy distribution will have been fully developed (STROKE, 1963). The boundary-value problem, which manifests itself in the diffraction of light by optical gratings, calls for an energy-distribution theory based on a rigorous or well approximated solution of MAXWELL's electromagnetic equations. One can look forward to the possibility of extending the use of high-resolution gratings to spectral regions at lower wavelengths in the far ultraviolet and X-ray domains, which are not yet accessible to contempary gratings.

Completely new ways of extracting the spectroscopic information directly from the diffracted wave fronts may now be expected. Novel spectroscopic arrangements using grating mosaics, multiple slits (A. GIRARD, 1958, 1962, 1963) and several diffractions by the same or by successsive gratings, with or without wavelength-selective modulation have already shown considerable gains in resolution and luminosity over classical arrangements in spectrographs and spectrometers. Coarse echelons, echelette and echelle gratings, when crossed with order-sorting devices, have helped in significantly the dimensions of spectrographs and of the photographic plates by displacing the spectra in two-dimensional arrays (R. W. WOOD, 1947; G. R. HARRISON, 1949). Grating spectroscopy has been considered as having desirable advantages of large free spectral ranges (wavelength intervals obtained without overlap) at high resolutions and luminosities.

---

[1] See, for example, P. JACQUINOT and B. ROIZEN-DOSSIER: Apodization. In: Progress in Optics, Vol. III., edit. by E. WOLF. Amsterdam: North-Holland Publishing Company 1963; also references in Sect. 50.

[2] See, for example, J. C. SIMON, G. BROUSSAUD and E. SPITZ: C. R. Acad. Sci. Paris **248**, 2309 (1959). — A. I. KARTASHEV: Optika i Spektroskopia **9**, 204 (1960) (English Transl.).

II. Diffraction by periodic structures and gratings.

5. General. Throughout the entire frequency spectrum of electromagnetic waves, the inteιaction of a periodic structure of matter with an incident wave is found to produce a set of diffracted waves distributed in discrete angular diιections. The similarity in the angular spectra of diffracted waves in various wavelengths is well known. The distribution of the diffracted waves with angle is found to depend basically only on the period of the diffracting grating when that period is measured in wavelengths. Elementary particles, atoms and molecules, are also known to exhibit a wave character and to show the properties of reflection, refraction, interference and diffraction. The term *scattering* is frequently used in parallel with the term *diffraction*, in particular in electromagnetic theory. Waves like particles are subject to scattering. *Scattering* is generally understood to refer primarily to the interaction of particles and photons with other particles of atomic or molecular dimensions, and to the resulting deflection of the photons and particles from their original path, with oι without change in energy and momentum. The term *diffraction* is most generally used to refer to the scattering of waves by macroscopic aggregates of particles, that is by bodies or apertures of various shapes, but the same term is also used to refer to the diffraction by periodic gratings at both atomic and macroscopic dimensions. Scattering, on the other hand, is also used as the term describing the generally diffuse diffraction of light by microscopic particles or random surface deformations on mirrors, lenses and other optical elements. Here we shall concern ourselves principally with diffraction by periodic structures of particles and by gratings of aggregate matter.

Of course, there exist significant differences between the diffraction of various waves and particles by gratings of matter. A most notable difference might be expected to exist between the diffraction phenomena at wavelengths where the size of the atoms, molecules and crystal-lattice spacings encountered in matter is comparable to the wavelengths on one hand, and, on the other hand, the diffraction phenomena with optical gratings used at wavelengths which are *large* compared to atomic distances.

The electric and magnetic natuιe of the particles, as well as their energy and ability to penetrate into matter should also be expected to determine whether or not diffraction will actually be observed in a given case.

Not only the relative size of the wavelength, but also the nature of the particles, their charge, magnetic moment and energy, as well as their ability to penetrate the medium determine the appearance of diffraction of matter waves by penetration of particles into matter.

Similarly, there exists a smallest period of the periodic structure which will allow diffraction of real homogeneous waves. The smallest grating period is found to be equal to one half the shortest wavelength of the waves incident on the grating. In other words, the spacing constant of the diffraction grating must be greater than the wavelength (or more accurately than the half-wavelength) to give rise to diffraction.

6. Optical gratings. Ruled diffraction gratings are being used either in transmission or in reflection, principally for spectroscopic purposes. For *optical gratings* only the dielectric constant, permittivity and conductivity of the medium manifest themselves predominantly in the interaction of the electromagnetic field with the surface of the grating. The electromagnetic constants of the grating determine the boundary conditions which must be satisfied at the surface of the grating by the total electromagnetic field, that is by the incident, transmitted and diffracted waves. The diffraction of electromagnetic waves by gratings is thus found to represent an electromagnetic boundary value problem.

Waves ranging from short X-rays to long infrared waves can be diffracted by optical gratings. It is interesting to note that electrons too have been made to manifest their wave character and to produce a characteristic electromagnetic radiation by passing near the surface of optical grating (Smith-Purcell experiment).

Even though the grating spacing constant must exceed the half-wavelength for a grating to give rise to diffraction, diffuse *scattering* of light does result when electromagnetic waves interact with surfaces that present statistically irregular deformations that are small compared to the wavelength. Moreover, even when their surface deformations vary over distances that are small compared to the wavelength, the actual shape or profile of the individual grating grooves is found to play an essential role in determining the distribution of the diffracted energy among the various waves at the different discrete grating angles. In spectroscopic applications of gratings, the energy distribution of the light-source manifests itself in the angular distribution of wave of different frequencies as they leave the grating. It is of obvious interest to attempt to concentrate a large fraction of the energy at any one frequency in just one single direction, rather than letting it be split up among diffracted waves of different orders, which leave the grating different angles.

III. Types of gratings.

Gratings are not only used in spectroscopy, but for many other important applications. For example, gratings are being used as antennae in radio-astronomy, and in distance and angle measuring devices.

When interacting with electromagnetic radiations, the following type of gratings can be conveniently distinguished:

7. Spectroscopic gratings. In spectroscopy, gratings like prisms, are used to study the composition of light emitted from excited atoms, molecules, and other sources. Spectra are obtained by suitably separating the radiation emitted from the light-sources into its component frequencies. Optical gratings consist of equally spaced grooves ruled on plane or concave surfaces, and are used either in transmission or more generally in reflection, from the soft X-ray to the far infrared domains. Most usually gratings are used in spectrographic arrangements where their function is to form monochromatic spectral images of an illuminated entrance slit. The spectral lines formed in spectrographs are images of the light-intensity distribution corresponding to single transitions in the source. The wavelength coordinate in the spectral images is in a direction at right angles to the slit height in the object plane. The spectral images of an infinitely narrow slit illuminated by a purely single-frequency monochromatic radiation are described as spectral diffraction patterns. Ideally, the spectral diffraction patterns formed by an infinitely wide grating should be infinitely narrow. With a finite width even a perfect grating can at best produce diffraction patterns of a finite and easily calculable width. Gratings which can produce perfect diffraction patterns of finite width are called stigmatic. Resolving powers in excess of 10^6 have been obtained with single 10-inch wide gratings in the visible. Considerably higher resolving powers are achieved by successive diffraction from a single grating or from several gratings used in series or in parallel. With one limited exception, only *plane* gratings can be made to produce perfectly stigmatic spectra. *Concave* gratings, which are generally ruled to have a perfect spacing on the chord of a spherical or otherwise curved surface, generally produce spectral lines with considerable aberrations. Since they do not produce stigmatic spectra, concave gratings are not used when the attainment of very high resolution is required. However, unlike plane gratings, concave gratings have focussing properties and require neither collimatorsn or camera optics to produce the spectra. When additional reflection losses introduced by the collimating and focussing mirrors, that need to be associated to plane gratings, are not tolerable, concave gratings are given preference. This is particularly the case in certain for ultraviolet and X-ray applications.

Reflection gratings for use in the visible, near ultraviolet and near infrared domains are ruled with boat-shaped and chisel-shaped diamonds by pressing well-shaped grooves in metallic films, generally aluminum, evaporated on perfectly flat or spherical glass blanks. Spacing constants of 300 and 600 lines per millimeter are most frequently used, but 1200 lines per millimeter spacings are not uncommon. Occasionally finer spacings, such as 2400 lines/mm are desirable for special purposes. Ruled echelle gratings of high quality have been successfully produced with spacing-constants as large as 30 lines/mm, and a few 200 grooves/inch echelles were obtained by a lapping process.

Among the best large high-resolution gratings that have been produced so far are the gratings ruled under interferometric control on the M.I.T. ruling engines; they are flat, generally have ruled widths of 10 inches, groove lengths of 5 inches and spacing constants of 300 lines/mm. Interferometric tests show that the grooves on the best of these gratings are correctly placed within better than 1/10 wavelength in the visible, that is within about 0.03 microns or so, and parallel to each other to better than 0.04 second of arc over the entire grating surface. Perfect stigmatism can be achieved with such gratings.

To produce their highest resolving powers, gratings need to be used close to *autocollimation* at the highest possible angles of incidence and diffraction, that is at angles of 64 to 75° with respect to normal to the grating surface. In auto-collimation the angles of incidence and of diffraction are on the same side of the normal to the grating surface. At these high angles of 64 to 75° the beams are approaching grazing angles of incidence and of diffraction with respect to the grazing surface on the same side of the normal to the grazing surface. Only plane gratings are sufficiently stigmatic for high resolution purposes.

In the far ultraviolet and X-ray domains, the reflectivity of the gratings, whether ruled in metal or glass, increases considerably when the gratings are used almost as if they were mirrors, with the incident and diffracted beams also very close to grazing incidence on the grating surface, but now with the angle of diffraction on opposite side of normal. In this case only comparatively very small dispersions and resolving powers are being obtained, but the use of the grating "at grazing incidence" is the only known means for obtaining spectra at the far vacuum ultraviolet and X-ray wavelengths. Generally only concave gratings are primarily used in this way. The measurement of wavelengths in the X-ray domain, required in crystallography, was first obtained with concave gratings used at grazing incidence (COMPTON and DOAN, 1925). Soft X-ray spectra and wavelengths greater than 10 or 20 Å are studied with concave gratings generally ruled directly on glass surfaces. The diffraction of X-rays by optical gratings makes use of the fact that the refractive index of materials is generally slightly smaller than unity for X-rays and that total reflection of X-rays occurs on the vicinity of grazing incidence. Considerable advances in increasing the reflectivities of metallic film down to 1000 Å or less have been recently achieved (by G. HASS) by overcoating freshly deposited aluminum films with thin layers of magnesium fluoride or other dielectric films. The use of concave gratings at higher angles of incidence and the achievement of higher resolutions at very short wavelengths is becoming possible. As the reflectivities of surfaces are being further increased the stigmatism and very high resolutions that can be obtained only with those gratings is being extended toward the far vacuum ultraviolet X-ray domains.

Another important factor involved in the increasing of light efficiency obtainable with gratings is the possibility of preferentially concentrating the diffracted energy into single directions of diffraction for any given wavelength. Unlike prisms, with which the incident energy in a given wavelength appears in only one direction, gratings tend to split up the incident monochromatic energy into several waves, leaving the grating at different angles. By proper shaping of the groove profile and by a proper choice of the ratio of ruled width to grating spacing, it becomes possible to concentrate sometimes as much as 75 to 90 percent of the incident energy into a single order, in the case of a reflection grating. Gratings with such a directivity are called "blazed gratings".

Gratings are also used in transmission for special applications. With modern grating replication techniques, transmission phase-gratings of fairly good quality are not unusual. Transmission gratings can also be "blazed" for maximum directivity. A complete theory of blaze remains yet to be developed and needs to be based on electromagnetic theory. Experience shows that when the grating constants become small as compared to the wavelength (say $a \simeq 2$ to 3λ), the electromagnetic character of light starts to manifest itself predominantly. When the blaze depends markedly on the polarization of the incident wave with respect to the groove length, and it is necessary to use electromagnetic equations as a basis for any accurate prediction of directivity. There are also indications that one *cannot* consider a grating to be describable only as a phase grating for calculations of blaze.

Reflection gratings used in the *infrared* and far *infrared* are generally used in low orders, that is with spacing constants comparable to the wavelength. For such gratings, the attainment of good blaze is usually a greater problem than the achievement of adequate spacing regularity, which appear as very good at the long wavelengths in the one to several hundred micron domains. Coarse infrared gratings are ruled in silver or in solder. The gratings ruled for the infrared at the University of Michigan have spacing constants as coarse as 25 lines per inch (1 line per millimeter) and have been used for wavelengths out to 200 microns. Infrared gratings of 600 to 2400 lines per inch are in frequent demand. The largest infrared gratings have ruled surfaces of 18×10 inches.

8. Microwave gratings (phased arrays, interferometers, artificial dielectrics). Metallic cylinders or parabolic aerials are frequently disposed into regular grating-like arrays for use as receivers in the radio and microwave domains.

The outputs of the individual, spacially isolated receivers, are electrically combined so as to form what is effectively a radio-telescope of dimensions which exceeds considerably the maximum dimensions achievable with a single radio-telescope antenna, and therefore to attain very narrow beams and high angular resolutions. Moreover, even with grating elements in a spatially fixed position, it is possible to tilt the optical axis of the antenna by as much as 45° on either side of the zenith by appropriate phase switching between the grating elements. With the narrow beams and the possibility of tilting their direction, phased arrays are particularly useful in the survey of discrete radio sources and in the study of the details in their special energy distribution.

MILLS' cross[1] is a single-beam radio telescope consisting of 500 half-wave dipoles arranged in a right-angle cross, and extending over 1500 Å, all the dipoles are polarized in the same direction. The instrument was designed to operate at a wavelength of 3.5 meters and the beam has an angular diameter of about 50 min of arc. As a comparison, optical high-resolution gratings have diffraction patterns with angular diameters on the order of the second of arc. Another type of radio-astronomical grating, also introduced in Australia, has been designed by CHRISTIANSEN and WARBURTON. It is a multiple-aerial interferometer consisting of 32 parabolic aerials of $2m$ diameter, each fed in phase and spaced over a 700-ft. baseline running east and west. This aerial was designed to operate at the 21-cm hydrogen line. The polar diagram of grating-like aerials correspond to the diffracted waves of different orders. For this interferometer, the different beams or lobes are separated by about 1.7 degrees from each other and their width is only 3 min of arc in the EW plane. The width of the lobes in the NS plane is governed simply by the polar diagrams of the individual aerials. When used to study the radio emission from the sun, of which the angular diameter is about half a degree of arc, the interferometer is simply left to scan across the sun by the rotation of the earth.

The theory of phased arrays is basically very similar to that of gratings in the optical domain. Of course, the possibility of separately processing the electrical signals from the individual grating elements in the radio-domain introduces an element of flexibility which is unparalleled in ruled gratings used in the optical domain. The tilting of the beams in the radio gratings corresponds to a displacement of the grooves along the grating surface.

The reason why large radio-telescopes are used more generally than phased arrays in radio-astronomy, even though their resolving power is smaller, is that they are not as frequency selective as the arrays and also because they can respond to smaller energies in a given bandwidth. On the other hand, the advantages of signal processing from the individual elements in phased arrays has not been duplicated in single antennae.

Another interesting application of grating-like structures in the microwave domain is found in the possibility of forming artificial dielectrics with particular properties of refraction. For example, it is well known that stigmatism in the focussing of plane wave fronts could be achieved with media bounded by simple surfaces, if the refractive index of the medium could be continuously varied at will. A step in that direction in the microwave domain is provided by a stack of properly spaced, flat, perforated plates, in which the holes may or may not be regularly spaced. The dimensions of the holes are generally small compared to the wavelength. Gratings of perforated plates have been applied to the construction of microwave antennae for which their ability to focus with only flat plates forms an essential advantage[2].

9. Image-forming (Fresnel-zone) gratings. It is well known that a system of alternately transparent and opaque rings of increasing diameters, appropriately spaced on a flat plate, has the property of focussing incident plane or spherical waves, and of producing images comparable to those produced by a lens or a mirror of similar dimensions. The diameters of the rings are chosen so as to have the mean phase from one transparent zone to the next increase by 2π radian or one wavelength as seen from a focus, and thus to obtain constructive

[1] See, for example, Handbuch der Physik, Vol. 53 and 54, edit. by S. FLÜGGE. Berlin-Göttingen-Heidelberg: Springer 1959, 1962.

[2] See, for example, G. BROUSSAUD: Ann. Radioélectr. **10**, No. 39 (January 1955); **11**, No. 43 (January 1956).

interference from the elementary waves "diffracted" by the successive zones. For maximum efficiency, the total area of every transparent zone is made to be a constant by appropriately decreasing the radial width of the zones as one proceeds from the center toward the outer zones.

It is clear that more than one focus along the axis can be found for which the condition of constructive interference from the various zones will hold. A Fresnel-zone plate is thus a multiple-focus device. Moreover, it is also clear that the intensity at any given focus will be wavelength dependent. Zones plates might thus be expected to be suitable for spectroscopic purposes. This has indeed been found to be the case. Particularly important applications have been made in the microwave domain at millimeter waves, in the filtering of harmonics from harmonic generators. For example, with a generator driven at 50 kMc/sec (wavelength about 6 mm) from which the third harmonic at 150 kMc/sec is desired, a Fresnel-zone filter could reduce the power at the 50 kMc/sec fundamental and at the 100 kMc/sec first harmonic by 48 decibels respectively below the power at 150 kMc/sec.

In the basic form of the Fresnel-zone plate roughly one half of the transmitting area must be opaque in order to make the focusing phenomenon possible. This is because one needs an increase of 2π of the mean phase from one contributing zone to the next. However, if instead of suppressing the contributions from the opaque zones completely, one were to make these zones also fully transparent, but simultaneously introduces a phase difference of π (that is $\lambda/2$) in these zones with respect to the neighboring zones, the total amplitude at a focus would be doubled and the intensity quadrupled. Such a plate is called a phase-reversing half-zone plate.

Moreover, since the phase of each contributing zone varies slowly by π, it would seem that this arrangement would be less efficient than one in which the phase could be made to vary progressively over each zone and then made to jump back by π at the beginning of the following zone (in the case where all the zones are now fully transparent).

In the limit an ideal zone plate would thus resemble the well known Fresnel lenses used in lighthouses for example. However, one can show that only a 20 percent gain in intensity would be obtained at a focus of an ideal Fresnel lens when compared to a quarter period zone plate, in which the plate aperture is divided up into $\pi/2$ ($\lambda/4$) zones and phase jumps are introduced where they are appropriate.

Half-wave and quarter-wave plates for use in microwave domains have been made out of plastic materials by introducing one and respectively three different groove depths between successive 2π zones. At wavelength of 4 mm, for example, the groove depths for a quarter wave plate of refractive index $\lambda = 1.5$ may be of the order of 2, 4 and 6 mm respectively for the quarter wave zones, which would have a width of 2 mm or so in the middle of a 200 mm diameter lens, with a 200 mm focal angle. It is clear that such grooves are fairly straightforward to achieve.

An essential advantage of zone plates is that they can be used at very wide apertures compared to the "focal lengths". Aperture ratios of diameter/focal length equal to unity are not unusual.

Even though the "Fresnel" zone plate is usually attributed to Soret, R. J. Wood[1] pointed out that Lord Rayleigh constructed one four years earlier (1871) and also suggested the principle of phase reversing half-wave plates.

[1] R. W. Wood: Physical Optics, 3rd. edit. New York: MacMillan & Co. 1934.

10. Metrological gratings[1]. Light waves themselves, when used in interferometers, provide the most accurate ruler available in nature. The counting of fringes between fixed mirrors to measure their separation is well known. Ever since MICHELSON's time, and even before, there have been many suggestions on the possibility of using interferometers for measurement of *motion* and for the *control* of moving carriages in comparators and machines. It was only in 1955, however, that HARRISON and STROKE first demonstrated in practice that an interferometrically controlled moving carriage could in fact be positioned in accordance with the photoelectric fringe signal produced by an interferometer (HARRISON and STROKE, 1955, STROKE 1957). By measuring the positions of the grooves ruled on gratings produced on interferometrically servo-controlled M.I.T. ruling engine, it was possible to verify that, under favorable circumstances, the accuracy inherently available from the interference field had actually been achieved in the motion and position of the carriage. The measurement of groove positions on the ruled gratings is also carried with the help of an optical two-beam interferometer according to a method described by STROKE by comparing a wave front diffracted by the grating to a wave front reflected from a perfect reference mirror. A positioning accuracy of better than 10^{-6} inches (0.03 microns) has been demonstrated with the M.I.T. ruling engine.

When monochromatic light is made to pass through a two-beam interferometer with parallel mirrors, the intensity of the light beam is found to be modulated sinusoidally with respect to the mirror separation in the interferometer. With one mirror fixed and the other mirror moving, the instantaneous light flux intensity received, on a photoelectric receiver is a measure of the position of the moving mirror. One fringe cycle results from the motion through half a wavelength (except for a small correction which may result from the fact that light beams also traverse the interferometer with a finite range of angles with respect to the normal). In the visible domain, one fringe may correspond to about $^1/_4$ micron or 10^{-5} inches. With isotopic and optical maser light sources, the fringes are visible over great path differences, of several feet for electrodless Hg 198 discharge tubes and considerably more for optical masers. Of course, great care needs to be taken in the adjustments of interferometers to avoid systematic errors in the interpretation of the carriage motion by the fringe signals.

A fringe signal quite comparable to that obtained from optical or microwave interferometers can now be obtained when light is made to pass through a pair of periodically transparent amplitude or phase gratings, one of which is fixed and the other in motion (BURCH 1958, 1959, 1960, 1962). Moiré gratings for use in metrology are generally obtained by replication from the same ruled master, but they can also be produced or reproduced photographically. Typically the two gratings have their identical grooves arranged parallel to each other, and the light beam modulation is recorded on a photoelectric receiver, just like in real interferometers. One "moiré-fringe" cycle is obtained with a pair of metrological gratings for a displacement of one period of the grooves on the grating. (Groove spacings of 100 and more usually 1000 lines per inch are now generally used in metrological gratings.) When used in the range where 10^{-4} inch errors are tolerable, metrological gratings are clearly less sensitive to mechanical distorsions and deformations than they would be in the micro-inch range where optical interferometers are generally used. Moreover, since the fringe distance

[1] See, for example, J. M. BURCH: The Metrological Application of Diffraction Gratings. In: Progress in Optics, Vol. II., edit. by E. WOLF. Amsterdam: North-Holland Publishing Company 1963.

obtained with metrological gratings does not depend on the length of the light wave in the same wave as it depends on it in real interferometers, it is considerably less influenced by atmospheric variations of the wavelength. In real interferometers, compensation for atmospheric variations of the wavelength are generally required when the interferometers are used in air. However the pressure-effect computing and correction methods for use with interferometric servo-control and measurement are straight forward, and do not present any greater complexity than the other electronic circuits which need to be used both with moiré or real interference fringe systems to exact the metrological information. The principle advantage of moiré-gratings for metrology is their rather long length of fringe cycle, on the order of 10^{-2} to 10^{-3} inches or so, which are particularly suitable for obtaining the moderate precision on the order of 10^{-5} to 10^{-4} inches required in general mechanical production. Of course the 10^{-5} inches fringe cycles obtained in the middle of the optical domain with real interferometers, are much more suitable for the achievement of the precision of 10^{-7} to 10^{-6} inches, particularly when absolute accuracy is required, as in the production of optical diffraction gratings. Moiré-gratings of up to 14 ft. length have been achieved, and are believed to have an overall accuracy of 10^{-4} inches. Optical interferometers with photoelectric fringe signals have been used over distance of 1 foot or so, with precisions of 10^{-7} inches.

Radial gratings, when used to measure angles over the entire 2π radians, have no parallel in real interferometers. It is therefore difficult to assess the details of their accuracy, but it is believed to be on the order of ± 2 seconds with a typical $10^1/_2$ inch diameter grating and 32400 divisions. Optical interferometers using two mirror systems are known to be able to provide accuracies on the order of $\pm {}^1/_{500}$ second of arc over a range on the order of a few seconds of arc.

11. Radiation-generating gratings (Smith-Purcell-Salisbury effect). When a sheet electron beam is accelerated in vacuo and made to pass at grazing incidence over a metallic diffraction grating, normally to the grooves, it is found that electromagnetic radiation is generated under suitable conditions. With a 20000 groove per inch grating and an accelerating potential of 320 kV the output wavelength is about 20×10^{-6} inches (about 5000 Å) and emerges at an angle of about 20° with the grating surface (Fig. 1). Beam sheets of about 0.25 mm are used with gratings of such spacings, and "coherence" of the emitted radiation has been demonstrated for a region of about 200 grooves with a 15000 groove per inch grating. A 6-inch wide 4 groove per inch grating with grooves $^1/_4$ cm deep has been used to produce radiation of 1 cm wavelength with the help of 150 kV electrons.

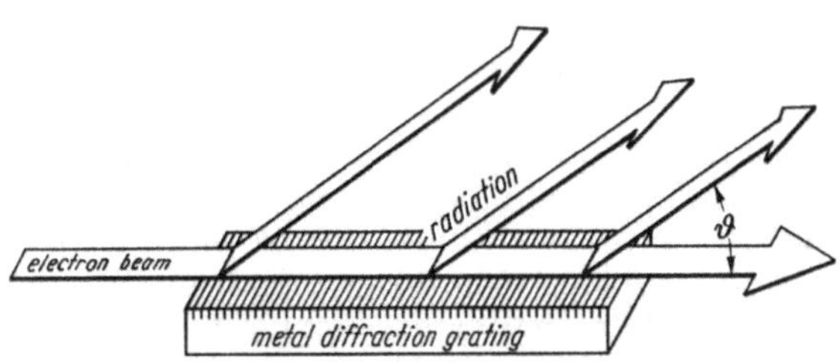

Fig. 1. Smith-Purcell-Salisbury effect. Radiation wavelength

$$\lambda = a\left(\frac{c}{V_e} - \cos\vartheta\right)$$

where: a = grating spacing constant, V_e = electron velocity, c = velocity of light. For a particular situation, where $1/a = 20000$ lines per inch., $\vartheta = 20°$, $V_e \approx 320000$ volts, the output wavelength is about 20 millionths of an inch.

The effect appears to have been invented independently by W. W. SALISBURY on one hand, and by E. M. PURCELL and S. J. SMITH on the other. The present applications of the effect are very much along the lines first described by SMITH and PURCELL in a letter to the Physical Review [92, 1069 (1953)] received September 25, 1953 and published on November 15, 1953. In that letter SMITH and PURCELL briefly describe the qualitative basis of the theory and of the effect and give experimental proof obtained with the help of a 48 mm wide grating: light in the 5000 Å region was photographed with electron energies ranging from 309 to 340 kV. In a patent (U.S. patent No. 2634372) entitled "Super High-Frequency Electromagnetic Wave Generator" granted on April 7, 1953 (and applied for on October 26, 1949) W. W. SALISBURY had suggested a somewhat comparable scheme according to which super high frequency electromagnetic waves (light) of the order of a millimeter or less in wavelength and in particular monochromatic light could be generated with the help of an electron beam interacting with an optical diffraction grating, provided that a "wave reflector" placed at a small angle above the grating were associated to the apparatus. The wave reflector appears to have been abandoned by SALISBURY in the device, described in 1961 and 1963, which uses and considerably extends the applications of the effect along the lines first described by SMITH and PURCELL.

According to W. W. SALISBURY, the "Smith-Purcell" effect may be understood qualitatively with the help of the electrostatic image theory already proposed by JAMES CLERK MAXWELL. In terms of this theory, the electron beam together with its image charges, reflected in the grating as in Fig. 2 forms a series of moving dipoles. The image charge appears to undulate, following the grating surface shape. The two charges thus appear to oscillate and therefore will tend to emit radiation at a frequency determined by the frequency of the oscillation and by the angle formed by the direction of the moving charges with the direction of observation of the emitted radiation. The wavelength λ of the radiation observed at the angle ϑ is related to the grating spacing constant a and to the electron velocity V_e by the equation

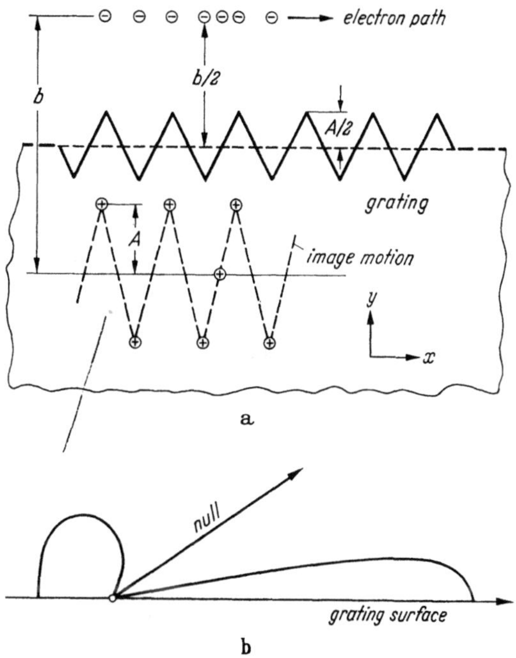

Fig. 2a and b. a) Dipole motion of electron and its image in Smith-Purcell-Salisbury effect. b) Relativistic radiation-power density distribution in Smith-Purcell-Salisbury effect.

$$\lambda = a\left[\frac{c}{V_e} - \cos\vartheta\right] \tag{11.1}$$

or equivalently by the equation

$$\lambda = a\left[\frac{1}{\beta} - \cos\vartheta\right] \tag{11.2}$$

which holds even for velocities where relativistic properties of the electron need to be considered (c = velocity of light, and $\beta = V_e/c$). Experiment appears to confirm these equations (Fig. 3).

In the case of gratings with non-sinusoidal profiles, the radiation produced will correspond to the various Fourier series terms in the expansion of the grating profile. Frequencies up to the seventh term have been observed by SALISBURY and W. A. EVANS with a grating of quasi-rectangular profile. The wavelengths observable under suitable conditions with gratings of non-sinusoidal profiles are given by

$$\lambda_n = \frac{a}{n}\left(\frac{1}{\beta} - \cos\vartheta\right) \tag{11.3}$$

where n = integer = number of the Fourier term in the expansion of the grating profile.

The radiated power has also been computed with the help of the electric dipole radiation theory, by taking into account the relativistic change of mass of the electron. Two cases are of interest.

a) The case when the N electrons in a beam add their energy independently (incoherently), for which the average radiated power is given by

$$\overline{P}_a = \sum_0^N \overline{P}_N = \frac{N e^2 A^2 f^4 \pi^3}{3 \varepsilon c^3} \tag{11.4}$$

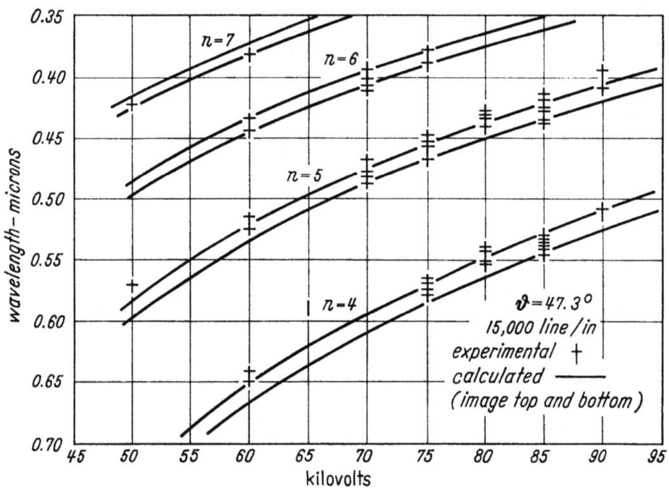

Fig. 3. Experimental wavelengths observed in Smith-Purcell-Salisbury effect.

where A comes from the equation

$$y = B + A \cos \omega t \tag{11.5}$$

describing the separation of the positive and negative charges in the model used, and e = charge of the electron, $\omega = 2\pi f$ = angular frequency of the motion, ε = permittivity of space.

b) The case when the electrons are "bunched", for which the average radiated power is given by

$$\overline{P}_b = \frac{(N e)^2 A^2 f^4 \pi^3}{3 \varepsilon c^3} \tag{11.6}$$

showing the great gain in intensity that should result from suitable bunching.

One also finds that the angular distribution of intensity should correspond to an equation of the form

$$P\,d\Omega = \frac{e^2 \pi^2 (N e)^2 A^2 f^4 [(1-\beta\cos\vartheta)^2 - (1-\beta)^2 \sin^2\vartheta \cos^2\varphi]}{2\varepsilon c^3 (1-\beta\cos\vartheta)^5}\, d\vartheta\, d\varphi, \tag{11.7}$$

where $d\Omega = d\vartheta\, d\varphi$ is the differential of the solid angle of observation. This equation may be interpreted as giving a measure of the fraction of the total radiated intensity that can be expected at a given angle, and it is somewhat comparable the equation giving the intensity distribution (blaze) in the spectra obtained from a grating when it is used for spectroscopic purposes. A maximum of intensity in the generated radiation should be found according to the above equation not only near 0° in the forward direction (in practice about 10 or 20° above the grating surface as a result of "shading" effects resulting from the finite groove depth), but also near 90° that is in the vicinity of the normal to the grating.

In the form produced by VARO Inc. under the direction of W. W. SALISBURY the radiation generating device using what may be called the Purcell-Salisbury-Smith effect can produce radiation not only in the visible, infrared and ultraviolet domains, but has also been used to produce centimeter wavelength radiation with a suitable grating. There is still some question as to the degree and type of coherence of the radiation emitted, when compared to radiations produced by optical masers for example. Interesting properties of this device seem to be the continuous tuning over a wide range of frequencies, as well as straightforward amplitude and frequency modulation possibilities.

12. Gratings as polarizers. While FRAUNHOFER already studied polarization effects in his gratings in 1821, it was HEINRICH HERTZ who appears to have first used a wire grid as a polarizer in 1888, when he was testing the properties of the newly discovered radio waves[1]. He found that a wire grid with 1 mm copper wires spaced 3 cm appart transmitted 66 cm radio waves when the electric vector in the incident beam was perpendicular to the wires. Alternately, the waves with the $\boldsymbol{E}$ vector parallel to the wires were reflected. (HERTZ's wire grating is preserved in the Deutsches Museum in München.) HERTZ's results were extended to the far infrared by DU BOIS and RUBENS in 1911. Noble metal wires of 25 micron diameter and larger were used for polarization at wavelength bands from some 24 to 314 microns. C. W. PETERS and W. K. PURSLEY (1954) reported a great complexity of the polarization effects depending, in particular, upon the ratio of grating space to wavelength. However, two years later, W. K. PURSLEY (1956) found that the general trend of the phenomena was preserved, as one would expect, when the wavelength, wire period and diameter were scaled up and down in the same ratio: experiments in both the 3 cm microwave and 80 to 500 micron infrared domains were performed, and compared with theoretical results given by IGNATOWSKY between 1905 and 1914. BIRD and PARRISH (1960) produced wire grids for use as transmission polarizers in the 2 to 15 micron infrared domain by metallic shadowing of the tops of the grooves in plastic replica gratings (spacing of 2160 grooves/mm). They also give a list of references and an analysis of past work. Significant theoretical contributions to the field of gratings of conducting cylinders and the theory of multiple scattering have been made by V. TWERSKY (1956, 1960), who also gives extensive references.

Much attention has been devoted to the study of polarization as a phenomenon in *optical* (ruled) gratings (FRAUNHOFER, 1821, R. W. WOOD, 1902 on, Lord RAYLEIGH, 1907, C. H. PALMER, 1952, 1956, 1961, E. A. YAKOVLEV and F. M. GERASIMOV, 1961 among others), and to the investigation of various discrepancies (such as Wood "anomalies") between experiments and theoretical predictions based on approximate, and in general scalar, descriptions of the dif-

[1] See, for example, G. R. BIRD and M. P. PARRISH jr.: The Wire Grid as a Near-Infrared Polarizer. J. Opt. Soc. Amer. **50**, 886 (1960). — C. W. PETERS and W. K. PURSLEY: J. Opt. Soc. Amer. **42**, 887 (1954).

fraction of light by optical gratings. These studies are examined again in Section B. VII.

Much less attention appears to have been devoted to taking into account the electromagnetic nature of the diffraction of light by gratings and the nature of the boundary conditions on the surface on the grating in an attempt of suitably exploiting them for the purpose of increasing the fraction of the incident light that can be concentrated in any given order, angle and wavelength (STROKE, 1961, 1963): see Section B.VII.

A remarkable case of polarization by a blazed reflection grating was predicted and verified in the 3.42 cm microwave domain by MARÉCHAL and STROKE in 1959[1]. It is found (MEECHAM[2], 1956, MARÉCHAL and STROKE, 1959, STROKE, 1960, 1963) that MAXWELL's electromagnetic equations are satisfied by the incident wave and only one diffracted wave returning exactly back in the direction of incidence, for a grating formed by rectangular groove steps $m\,\lambda/2$ deep, in autocollimation on the groove facet normal to the $m\,\lambda/2$ facet, when the electric $\boldsymbol{E}$ vector in the incident field is normal to the $m\,\lambda/2$ facet. The application of electromagnetic boundary value considerations to the improvement of the efficiency in blazed gratings, in particular by avoiding the 90° polarizing groove profile, were first reported by STROKE[3] in 1963. See also Section B.VII.

B. Spectroscopic gratings.

I. The two aspects and the three theoretical descriptions of the diffraction of light by gratings.

13. Introduction. It will become clear from the discussion in this chapter that the diffraction of light by gratings takes on two aspects: 1. *an electromagnetic aspect*, having to do with the very origin of diffraction of electromagnetic waves by a periodic boundary and with the directions and amplitudes of the various diffracted waves, 2. *a scalar aspect* having to do with the image forming properties of the various diffracted waves when they are limited in size by gratings of finite dimensions and limited in quality by grating imperfections.

Moreover, the spectroscopic application of gratings for the resolution of wave fronts of different wavelengths, and in general, for the analysis of light, involves certain *geometrical* characteristics of gratings.

Three rather different mathematical formulations are found to be appropriate for the solution of problems arising in the diffraction of light by gratings.

1. For the *determination of the amplitudes of the diffracted waves*, and, in general, for the improvements in the energy distribution among orders (blaze) of the gratings, boundary value methods and consideration of the electromagnetic boundary conditions are basically most appropriate. Under certain conditions, in particular for echelle gratings and for amplitude gratings, physical optics approximations may be sufficient.

2. For the *determination of the grating and wave front quality* and for the evaluation of the relation between grating and diffraction pattern imperfections, physical optics theory is particularly suited, when expressed in the Fourier-transform approximation of the rigourous HUYGENS' principle.

[1] A. MARÉCHAL et G. W. STROKE: C. R. Acad. Sci. Paris **249**, 2042 (1959).

[2] W. C. MEECHAM: J. Appl. Phys. **27**, 361 (1956). — W. C. MEECHAM and C. W. PETERS: J. Appl. Phys. **28**, 216 (1957).

[3] G. W. STROKE: Attainment of High Efficiencies in Blazed Optical Gratings by Avoiding Polarization in the Diffrated Light. Phys. Letters **5**, 45 (1963).

3. For the determination of the *spectroscopic characteristics of gratings*, such as resolving power, luminosity and overlapping of spectra, the approximations of geometrical optics are found to be particularly appropriate.

In this chapter, we shall put a particular emphasis on showing that the origin of the diffraction of light by gratings resides in the electromagnetic character of light in the boundary conditions that are to be satisfied by the total field, incident and diffracted, on the surface of the periodic structure which form the grating.

14. Distribution of light among diffracted waves of various orders. The grating equation. (Electromagnetic boundary-value problems; electromagnetic and phy-

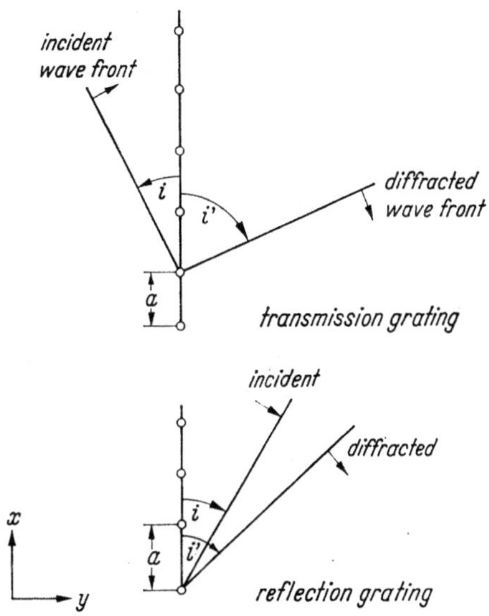

Fig. 4. Positive incident and diffracted wavefronts for transmission and reflection gratings.

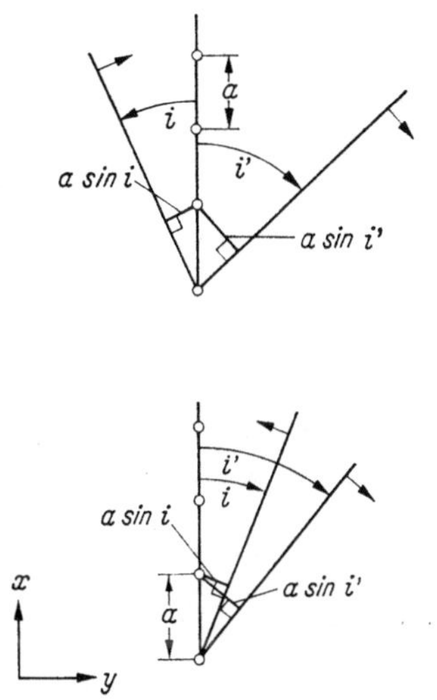

Fig. 5. Positive incident and diffracted wavefronts for transmission and reflection gratings.

sical optics description.) Perhaps the most important equation describing the diffraction of light by gratings is the so-called "grating equation"

$$\sin i + \sin i' = \frac{m\lambda}{a} \tag{14.1}$$

where i and i' are the angles formed by the incident and diffracted plane wave fronts with the mean surface of a grating (Fig. 4). a = grating spacing constant, λ = wavelength and m = any integer. The plane wavefronts are surfaces of constant phase. Eq. (14.1) is applicable to the principal plane normal to the groove length (Fig. 4).

The usual derivation of the grating equation with the help of HUYGENS' principle is straightforward, but *does not* prove that the diffracted waves exist in only a discrete set. In fact, when a plane wave is incident on a plane grating one generally assumes, correctly but without proof, that the diffracted waves are also plane and exist in only a discrete set. The proof of the existence of a discrete set of diffracted waves is given below. But first we recall the demonstration of the grating Eq. (14.1) according to HUYGENS' principle. According to HUYGENS' principle, if the incident wave front and the diffracted wave fronts are surfaces of constant phase, as they should be, then the geometry of Fig. 5

immediately gives

$$a \sin i + a \sin i' = m\lambda \tag{14.2}$$

from which the grating equation is obtained in its usual form of Eq. (14.1).

We now proceed with the proof of the existence of a discrete set of plane diffracted waves produced by the incidence of a plane wave on a grating. Ruled gratings are essentially two-dimensional structures. As such their surface S can be described by a function $S=f(x, y)$ which is independent of one of the coordinates, say z, but is periodic as a function of x,

$$S_{x+pa} = S_x \quad (p = \text{integer}). \tag{14.3}$$

The groove length is along z (Fig. 6). Next we recall that there exists a class of electromagnetic problems which are completely independent of one coordinate, say also z. This is the class of *two-dimensional* problems. It is known that two-dimensional electromagnetic problems are essentially scalar in nature. This does not mean that they can be identified to non-electromagnetic problems (for example to acoustical problems) without any further restrictions, but rather that two-dimensional problems can be expressed in terms of only a *single* dependent electromagnetic field variable (for example E_z or H_z). It can be further shown that the single dependent electromagnetic field variable in two-dimensional problems satisfies the Helmholtz wave equation

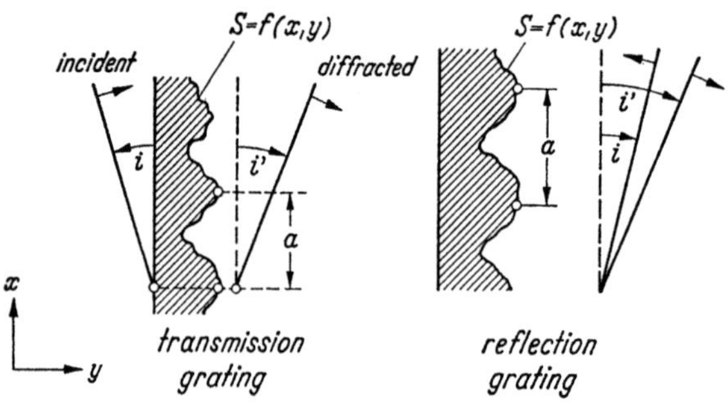

Fig. 6. Parameters defining diffraction by gratings.

$$\frac{\partial^2 V_z}{\partial x^2} + \frac{\partial^2 V_z}{\partial y^2} + \frac{\partial^2 V_z}{\partial z^2} + k^2 V_z = 0 \tag{14.4}$$

where V_z is identified to E_z or H_z respectively (V_z is the space dependent part of the field vector component. The factor $\exp(-i\omega t)$ has been cancelled out as usual).

A fundamental elementary solution of (14.4) can be written in the form

$$\exp[ik(x\sin\vartheta + y\cos\vartheta)], \tag{14.5}$$

which is the equation of a plane wave. [When ϑ is real, Eq. (14.5) represents a homogeneous plane wave; when ϑ is imaginary or complex, equation (14.5) represents an inhomogeneous plane wave or *evanescent* wave.]

A complete solution of equation (14.4) is formed by an angular spectrum of plane waves and can be represented by a Fourier integral:

$$\int_{-\infty}^{+\infty} A(\vartheta) \exp[ik(x\sin\vartheta + y\cos\vartheta)]\, d\vartheta, \tag{14.6}$$

where $A(\vartheta)$ are the amplitudes of the various waves. The coefficients $A(\vartheta)$ are generally complex, and are to be determined in order to solve any particular problem described by Eq. (14.6).

Now let us consider the diffraction of a plane incident wave by the grating surface S. Since we are dealing with electromagnetic waves, it is clearly necessary to investigate the diffraction of polarized waves. Moreover, in order to define a two-dimensional problem, we need to orient the field vectors which describes the incident wave so as to be parallel to the grooves of the grating. Therefore either $\boldsymbol{E}$ or $\boldsymbol{H}$ in the incident wave is to be parallel to z, for electric or magnetic waves respectively. The corresponding field components for an electric or magnetic incident wave are thus E_z^i or H_z^i respectively. (In case of unpolarized incident light, the solution of the diffraction problem is obtained by a linear superposition of the solutions for the $\boldsymbol{E}$ and $\boldsymbol{H}$ wave respectively. Alternately, if the $\boldsymbol{E}$ and $\boldsymbol{H}$ wave solutions are different, as indeed they generally are, a grating will polarize unpolarized incident light.) The only assumption about the grating that we are making is that the surface is periodic in x according to Eq. (14.3). Let the electromagnetic field vector component parallel to z in the incident wave be described by V_z^i. (V_z^i can be identified to either E_z^i or H_z^i for an electric or magnetic wave respectively.) We have for the incident wave:

$$V_z^i = \exp[-ik(x\sin i + y\cos i)]. \tag{14.7}$$

Here i is the angle of incidence (and $ik = \sqrt{-1}\,k$ of course). By symmetry, the diffracted field will also have z-components only. Let V_z^d be the diffracted field. Since the total field, incident plus diffracted, namely $V_z = V_z^i + V_z^d$ satisfies the Helmholtz wave Eq. (14.4), and since the incident field V_z^i satisfied the same equation, the diffracted field V_z^d must also satisfy the wave equation.

The diffracted field V_z^d can then be represented in its most general form by an *angular spectrum of plane waves*

$$V_z^d = \int_{-\infty}^{+\infty} V_z^d(i')\exp[ik(x\sin i' + y\cos i')]\,di' \tag{14.8}$$

where $V_z^d(i')$ is the amplitude of the plane "diffracted" wave corresponding to the angle of diffraction i'. According to Eq. (14.8) there exists an infinity of diffracted waves distributed in continuous angular directions. We shall now show that the periodic nature of the grating boundary and the boundary condition which must be satisfied by the *total* field on the boundary are restricting the continuous infinity of diffracted waves to an infinite set of discrete waves, distributed in only discrete angular directions according to the grating Eq. (14.1). This conclusion and the grating equation are independent of the particular nature of the boundary, and of the electromagnetic material characteristic of the grating, and depend only on the periodicity of the grating surface. The particular shape of the grating grooves also does not enter this conclusion.

According to electromagnetic theory, the total electromagnetic field

$$V_z = V_z^i + V_z^d \tag{14.9}$$

must satisfy certain boundary conditions on the surface of the grating. (For instance if $V_z = E_z$, where E_z represents the total electromagnetic field tangential to the surface of the grating, then $E_z = 0$ on the surface of a perfect conductor. If $V_z = H_z$, then $\partial H_z/\partial n$ must be zero on the surface of a perfect conductor, $\boldsymbol{n}$ being the normal to the surface.) Moreover, and this is the crucial step in the argument, the periodic nature of the grating boundary in the x-direction (with a period a) requires that the total electromagnetic field on the surface of the grating be the same for a displacement of an integral number of periods a in the x-direction. One must have

$$(V_z)_{x+pa} = (V_z)_x \tag{14.10}$$

on the grating surface, p being an integer. We have

$$(V_z^i + V_z^d)_{x+pa} = (V_z^i + V_z^d)_x \quad (p = \text{integer}). \tag{14.11}$$

In particular, Eq. (14.11) must be true for a displacement of one period.

Eqs. (14.7) and (14.8), when introduced into (14.11) give

$$\left.\begin{aligned} &\exp[-ik(x\sin i + y\cos i)]\exp[-ikpa\sin i] + \\ &\qquad + \int V_z^d(i')\exp[ik(x\sin i' + y\cos i')]\exp[ikpa\sin i']\,di' \\ &= \exp[-ik(x\sin i + y\cos i)] + \int V_z^d(i')\exp[ik(x\sin i' + y\cos i')]\,di'. \end{aligned}\right\} \tag{14.12}$$

After division by $\exp[-ik(x\sin i + y\cos i)]$ and factoring of $\exp[-ikpa\sin i]$, Eq. (14.12) gives

$$\left.\begin{aligned} &\exp[-ikpa\sin i]\{1 + \int V_z^d(i')\exp[ik(x+pa)(\sin i + \sin i')]\times \\ &\qquad\qquad \times\exp[iky(\cos i + \cos i')]\,di'\} \\ &= 1 + \int V_z^d(i')\exp[ikx(\sin i + \sin i')]\exp[iky(\cos i + \cos i')]\,di'. \end{aligned}\right\} \tag{14.13}$$

For any given k, p, a and any given angle of incidence i, the factor $\exp[-ikpa\sin i]$ is a constant, and for Eq. (14.13) to be satisfied, we must have

$$kpa(\sin i + \sin i') = n\,2\pi \quad (n = \text{integer}) \tag{14.14}$$

that is

$$\boxed{k(\sin i + \sin i') = m\frac{2\pi}{a}} \quad (m = \text{integer}). \tag{14.15}$$

Recalling that $k = 2\pi/\lambda$, Eq. (14.15) gives the result which we were seeking, namely

$$\boxed{\sin i + \sin i' = m\frac{\lambda}{a}} \quad (m = \text{integer}), \tag{14.1}$$

which is indeed the grating equation. It is important to note that all integer values of m are admitted, and therefore the angles i and i' can be not only real, but also imaginary.

We have just shown[1] that when a plane wave

$$V_z^i = \exp[-ik(x\sin i + y\cos i)] \tag{14.16}$$

is incident on a periodic surface (that is, on a grating of period a) the boundary conditions which are to be satisfied by the total field V_z on the surface of the grating give rise to a set of diffracted waves V_z^d such that

$$V_z^d = V_z - V_z^i \tag{14.17}$$

where

$$V_z^d = \sum_m (V_z^d)_m \exp[ik(x\sin i'_m + y\cos i'_m)] \tag{14.18}$$

and

$$\sin i + \sin i'_m = \frac{m\lambda}{a} \quad (m = \text{integer}). \tag{14.19}$$

Truly, *it is the boundary conditions and the periodicity of the grating which are at the origin of the diffraction of light by gratings.* We note that at least two grooves are required to form a grating.

[1] For an alternate proof, see Appendix II in R. PETIT, Rev. Opt. **42**, 263 (1963).

More generally, we can establish a grating equation for an electromagnetic wave incident from any direction (not necessarily with its propagation vector in the xy plane as before) and for the corresponding diffracted waves.

Let α, β, γ, be the direction cosines for the propagation vector $\boldsymbol{k}_i$ of the incident wave (Fig. 7), such that

$$\boldsymbol{k}_i = -k(\boldsymbol{i}_r)_i = -\frac{2\pi}{\lambda}(\boldsymbol{i}_r)_i = -k(\alpha \boldsymbol{i}_x + \beta \boldsymbol{i}_y + \gamma \boldsymbol{i}_z),$$

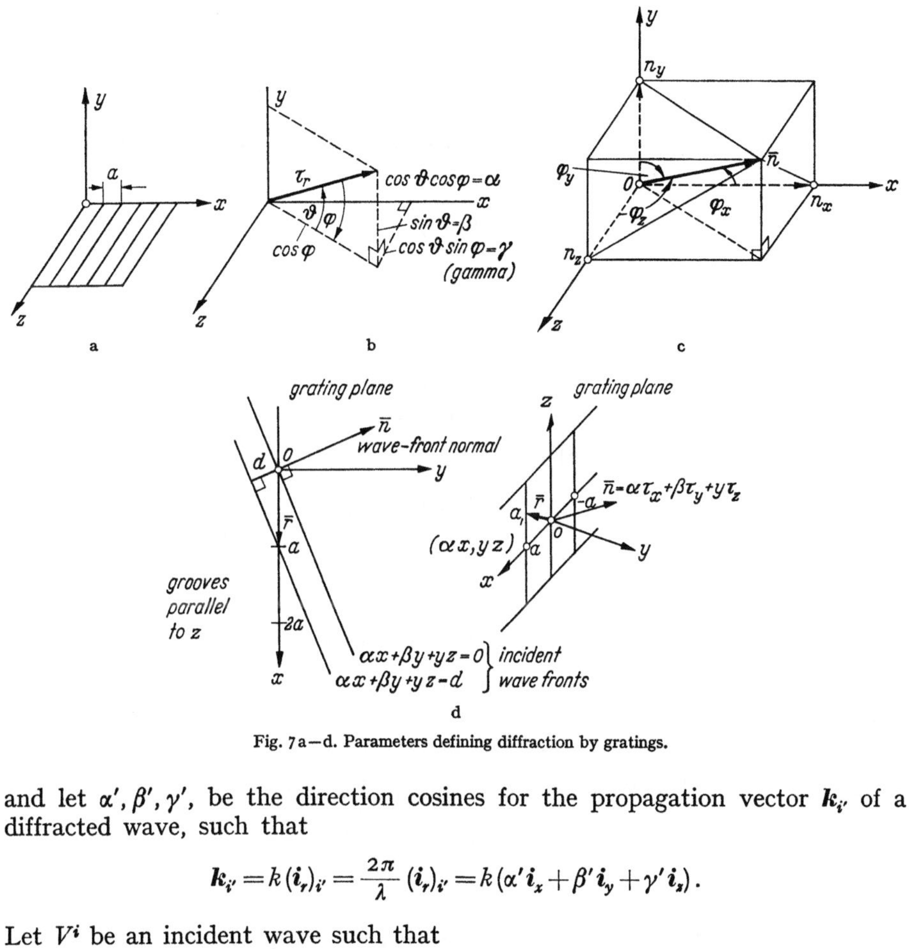

Fig. 7a—d. Parameters defining diffraction by gratings.

and let α', β', γ', be the direction cosines for the propagation vector $\boldsymbol{k}_{i'}$ of a diffracted wave, such that

$$\boldsymbol{k}_{i'} = k(\boldsymbol{i}_r)_{i'} = \frac{2\pi}{\lambda}(\boldsymbol{i}_r)_{i'} = k(\alpha' \boldsymbol{i}_x + \beta' \boldsymbol{i}_y + \gamma' \boldsymbol{i}_z).$$

Let V^i be an incident wave such that

$$V^i = \exp\left[-ik(\alpha x + \beta y + \gamma z)\right]$$

and let V^d be the diffracted field, such that

$$V^d = \iint V^d(\alpha', \beta') \exp\left[ik(\alpha' x + \beta' y + \gamma' z)\right] d\alpha' d\beta'$$

where

$$\begin{cases} \alpha = \cos\vartheta \cos\varphi \\ \beta = \sin\vartheta \\ \gamma = \cos\vartheta \sin\varphi \\ \alpha^2 + \beta^2 + \gamma^2 = 1 \end{cases} \quad \text{and} \quad \begin{cases} \alpha' = \cos\vartheta' \cos\varphi' \\ \beta' = \sin\vartheta' \\ \gamma' = \cos\vartheta' \sin\varphi' \\ \alpha'^2 + \beta'^2 + \gamma'^2 = 1. \end{cases}$$

V^i and V^d satisfy a Helmholtz wave equation of the form

$$\nabla^2 V + k^2 V = 0.$$

We now require not only that $(V^i + V^d)$ be periodic in x on the grating surace, but also that the periodicity in x with a period a be true for any z. We immediately obtain the requirement that

$$\gamma' = -\gamma, \tag{14.20a}$$

$$\alpha' + \alpha = \frac{m\lambda}{a}, \tag{14.20b}$$

with

$$\alpha^2 + \beta^2 + \gamma^2 = 1 \quad \text{and} \quad \alpha'^2 + \beta'^2 + \gamma'^2 = 1. \tag{14.20c}$$

These equation can be very easily derived from first principles, with the help of Fig. 7c and Fig. 7d. We have (Fig. 7c)

$$\boldsymbol{n} = \boldsymbol{i}_r$$

and

$$\boldsymbol{n} = n_x \boldsymbol{i}_x + n_y \boldsymbol{i}_y + n_z \boldsymbol{i}_z$$

with

$$n_x = \cos\varphi_x = \alpha$$

$$n_y = \cos\varphi_y = \beta$$

$$n_z = \cos\varphi_z = \gamma.$$

Let the incident wave normal be

$$\boldsymbol{n} = \alpha \boldsymbol{i}_x + \beta \boldsymbol{i}_y + \gamma \boldsymbol{i}_z$$

and the diffracted wave normal be

$$\boldsymbol{n} = \alpha' \boldsymbol{i}_x + \beta' \boldsymbol{i}_y + \gamma' \boldsymbol{i}_z.$$

In Fig. 7d, let the incident waves at 0 and at a be

$$\alpha x + \beta y + \gamma z = 0$$

and

$$\alpha x + \beta y + \gamma z = d,$$

respectively. Let the diffracted waves at 0 and at a be

$$\alpha' x + \beta' y + \gamma' z = 0$$

and

$$\alpha' x + \beta' y + \gamma' z = d,$$

respectively. On the first groove, at any point a_1 such that

$$\boldsymbol{r} = a \boldsymbol{i}_x + z \boldsymbol{i}_z$$

we have for the incident wave

$$d = \boldsymbol{n} \cdot \boldsymbol{r} = \alpha a + \gamma z$$

and for the diffracted wave

$$d' = \boldsymbol{n}' \cdot \boldsymbol{r} = \alpha' a + \gamma' z.$$

For constructive interference, we must have

$$d' + d = k\lambda,$$

that is

$$(\alpha' + \alpha) a + (\gamma' + \gamma) z = k\lambda.$$

This equation must be independent of the coordinate z, and we must therefore have the equations

$$\gamma' = -\gamma$$

and

$$\alpha' + \alpha = m\,\lambda/a$$

with

$$\alpha^2 + \beta^2 + \gamma^2 = \alpha'^2 + \beta'^2 + \gamma'^2 = 1$$

as stated in (14.20a), (14.20b) and (14.20c) above.

Eqs. (14.20a) and (14.20b) can be considered as expressing the "grating equation" in its most general form. Clearly when the incident wave has its propagation vector $\boldsymbol{k}_i$ in the xy plane, then $\gamma = 0$, $x = \sin i$, $x' = \sin i'$, and Eq. (14.20b) reduces to the grating Eq. (14.19) obtained before.

We shall later use Eq. (14.20) to calculate the *curvature of spectral lines* which appears when a long slit, parallel to z, is used in a spectrograph.

In vector form, with $\boldsymbol{a} = a\boldsymbol{i}_x$, and with $|\boldsymbol{k}_{i'}| = |\boldsymbol{k}_i| = 2\pi/\lambda$, Eqs. (14.20) give

$$\boldsymbol{a} \cdot (\boldsymbol{k}_{i'} - \boldsymbol{k}_i) = m 2\pi. \tag{14.20d}$$

We conclude by making two important observations:

1. It is clear that a complete solution of the problem of the diffraction of light by gratings requires the determination of the amplitudes $(V_z^d)_m$ of the various diffracted waves for a particular shape and nature of the surface and grooves, under any condition of incidence and for the various wavelengths. The amplitudes $(V_z^d)_m$ are generally complex. A general rigorous solution of the problem of diffraction of light by gratings for any groove shape and boundary condition is now being obtained in explicit form, and the mathematical problem is very formidable indeed[1]. Most basically, the amplitudes $(V_z^d)_m$ of the waves of various orders are determined by the boundary conditions on the surface of the grating, and more particularly by the detailed shape of the grooves. It is very important to observe that at least two neighbouring grooves and their shape need to be considered when investigating the energy distribution with order m in a grating. It is clearly not correct to consider the grooves of a grating as single slits acting independently, since the periodic nature of the grating is not accounted for in such an analogy. We have shown above how basic the periodic nature of the grating is to the phenomenon of diffraction and the determination of the amplitudes of the wave of different orders. The *distribution of the incident energy among the diffracted waves of various orders* form *one* of the important problems of the diffraction of light by grating. The mathematical solution of the energy distribution problem in gratings generally takes the form of a boundary value problem. It will be further described in Chaps. B III and B VII. Only in special cases are physical-optics approximations sufficient to determine the distribution of the diffracted light among the waves of various diffracted orders. Some of the special cases may, however, be of great particular interest in practice.

2. The grating Eqs. (14.1) and (14.20) have been derived for the case of a plane wave front incident on an infinitely extended grating. We shall show in Chap. B IV that the same equation is also applicable for a grating of finite width in the x-direction. Moreover, real gratings are not perfectly periodic; in fact they generally present both periodic and non-periodic groove position irregularities. As a result of the groove-positioning irregularities, there appear small deviations from flatness in the plane diffracted wave fronts. In turn,

[1] See Sects. 60–71.

the wave front imperfections result in imperfections in the diffraction pattern of finite size which correspond to the finite size of the diffracted wave fronts.

Indeed, when a finite grating is used with a collimating mirror or lens, the finite size of the diffracted wave fronts result in the formation of diffraction patterns of finite width identical to the pattern which would be obtained by reflection on a plane mirror of the same aperture. Irregularities in the grating spacing, which result in wave front deformations, create imperfections in the diffraction pattern quite identical to diffraction-pattern imperfections which would appear with imperfect wave fronts originating from any other optical element. In fact, there is no need for distinguishing a diffracted wave front from a reflected or transmitted wave front, as far as image formation is concerned. It is the general case for image formation by optical instruments, that only macroscopic aberration of the wave fronts (that is, wave front deviations extending over appreciable fractions of the aperture) result in imperfections in the vicinity of the centers of the diffraction patterns. This is also true for gratings. *The formation of diffraction patterns and images* by both perfect and imperfect gratings is the other important problem of the diffraction of light by optical gratings. It is described in the next section and will be more thoroughly described in Chap. IV.

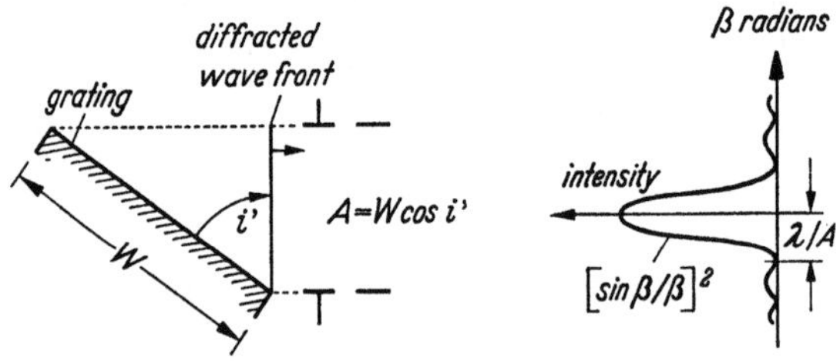

Fig. 8. Diffraction pattern formed by grating of width W, in autocollimation at angle $i=i'$.

Here we only wish to recall that the diffraction pattern corresponding to a wave front having a rectangular aperture of width A has the form of a $(\sin\beta/\beta)^2$ function with a first zero at $\beta=\lambda/A$, where $A=W\cos i'$ is the width of the aperture of the diffracted wave front as determined by a grating of ruled width W, used at an angle of diffraction i' (Fig. 8). One has:

$$\beta=\pi\frac{x}{x_0}=\pi\frac{x}{(\lambda/A)f}$$

where f is the focal-length of camera-mirror.

15. Spectroscopic image formation. (Scalar problem: physical optics and geometrical optics description.) When an electromagnetic wave is incident on a periodically deformed boundary, the scattered field consists of a set of waves distributed in only discrete angular directions. The angles of diffraction are determined by the angle of incidence, the wavelength and the period of the boundary deformation. The amplitude of the diffracted waves are determined by the particular shape of the groove profile, the boundary conditions to be satisfied at the surface of the grating, the groove spacing period, as well as by the angles of incidence and diffraction. In general, there is an infinite set of diffracted waves, corresponding to both real and imaginary angles (of both incidence and diffraction), and the amplitudes of the diffracted waves are generally complex.

When gratings are used for spectroscopic purposes, their function is to separate the diffracted wave fronts of different frequencies (or wavelengths) by angles which are sufficiently large to permit easy resolution by an appropriate optical system and receptor. Generally the diffracted wave fronts are focussed by a "camera" mirror or lens onto a photoelectric or photographic detector in the form of "spectral lines" which are images of an illuminated entrance slit placed in the focal plane of a "collimator" illuminating the grating. As a result of the finite size of real gratings, the diffracted wave fronts produce diffraction patterns of *finite* width in the focal plane of the camera mirror, rather than delta functions of intensity, as would be obtained with infinitely wide wave fronts and gratings. In addition, the finite size of the entrance slit requires a very different summation for the case of coherent illumination as compared to the more general case of incoherent illumination. Moreover, the groove positioning and depth irregularities, and the flatness defects, always encountered in real gratings result in further broadenings and irregularities in the spectral lines, and in fact in the appearance of spurious secondary maxima in the diffraction patterns. The spurious diffraction pattern maxima, resulting from grating imperfections, take on the aspect of spurious spectral lines variously called "satellites", "ghosts", and "grass".

When the spectroscopic characteristics of gratings are to be described, one is at first interested in determining the maximum resolution and luminosity achievable with a grating of a given size. Unlike on interferometers, which are also used for spectroscopic purposes, the resolution and luminosity of gratings are found to be independent of the overlapping of waves of different orders. A large free spectral range (wavelength range obtained without overlapping) forms a significant practical superiority of gratings over interferometers in otherwise comparable spectroscopic applications. The *resolution* of a grating has to do with the minimum angular separation of two wave fronts of neighboring wavelengths with respect to the angular size of the diffraction pattern. The *luminosity* has to do with the maximum number of photons per second per unit wavelength bandwidth that are detectable at a given resolution. Resolution and luminosity are clearly not independent quantities and both also depend on the width of the grating. However, at a given angle of diffraction, the resolution and luminosity are found to be independent on the groove spacing or grating constant. This is also true for the angular dispersion, that is, the angular separation between wave fronts of two given neighboring wavelengths. The grating spacing constant does have an effect on the spectroscopic characteristics of gratings; it governs the amount of overlap between spectra of different orders. Unlike prisms, but like interferometers, gratings are indeed known to produce more than one (diffracted) wave of a given wavelength. As a result, there will be varying amounts of overlap between spectra which may, or may not, present a problem in practice. (In general the amount of overlap in gratings is the same as that which appears in the use of a Fabry Pérot interferometer having a plate separation corresponding to the apparent grating groove depth in the direction of diffraction. However, while groove spacing on the order of a few microns and never more than a few tenths of a millimeter are generally used in optical gratings in the visible, at resolving powers in the 10^6 range, Fabry Pérot plate separation on the order of the centimeter are required for comparable resolutions. The overlap of spectra is thus some 10^3 times less critical in gratings than in interferometers).

16. Summary. In summary, it firstly has become clear that the *spectroscopic problems* appearing in the use of perfect gratings (resolution, luminosity, dispersion and spectral overlapping) are essentially *geometrical* in character and can be dealt with by the help of *geometrical optics theory*. This will be further developed in Chap. B II.

Secondly, the effects of grating imperfections on the spectral diffraction patterns, and indeed the problem of *spectroscopic image formation* by perfect or imperfect diffracted wave fronts depend only on the macroscopic character of the wave fronts (phase and amplitude distribution across rather extended portions of diffracted wavefronts). Spectroscopic image formation can be most elegantly dealt with in the Fourier-transform expression of *physical optics* and of HUYGENS' principle. This theory will be developed in Chap. B IV. More particularly, the rigorous form of HUYGENS' principle will be derived there, in order to clearly establish the foundations and regions of applicability of the Fourier-transform approximation. A theory of resolving power will also be found in the same Chap. B IV. The local character of wave fronts does have an effect on image formation, but local wave front aberrations and irregularities in the positioning of individual grooves primarily affect regions far removed from the diffraction pattern centers and are also at the origin of *scattered light*. When local wave front irregularities are comparatively small in number and when they present no periodic regularity, then their effect on the spectroscopic image quality and resolution is generally considerably less harmful than the effect of small-amplitude, but extended wave front deviations. Of course, simple relations between grating imperfections and wave front aberrations can be established (Ch. B IV). The reason why such great importance is attached to the theory of imperfect wave fronts and of their effects on the spectral image, is that the phase distribution in the diffracted wave fronts is immediately accessible to measurement, by placing the grating into one of the arms of a Michelson Twyman-Green interferometer and by comparing a diffracted wave front to a perfect wave front reflected from a reference mirror placed in the other arm of the interferometer (STROKE, 1955, 1961)[1] (Chap. B VI).

Thirdly and finally it should be clear that the *electromagnetic character* of the diffraction of light is most fully reaveled in the distribution of the diffracted energy among the waves of various orders. Most generally, the energy distribution and the (complex) amplitudes of the various diffracted waves depend on the boundary conditions that are to be satisfied on the surface of the grating by the total field, incident plus diffracted. The particular shape of the individual grooves plays a fundamental role in determining preferred directions of energy concentration. But it is essential to observe that the interaction of at least two grooves needs in general to be considered for any theoretical calculations of blaze, since at least two grooves are required to form a grating. The most general theory of energy distribution takes the form of an electromagnetic boundary value problem. It will be further developed in Chap. B V.

In many practical cases, in particular for blazed gratings having rectangular or quasi-rectangular groove profiles, approximate theories of a physical optics character may provide valuable approximations. This will be shown in Chap. B V.

II. Geometrical optics theory of grating image-formation and luminosity of gratings in spectrometers and spectrographs[2].

17. Reflection and transmission gratings. Optical gratings are used either in transmission (Fig. 9a) or in reflection (Fig. 9b). The incident plane wave fronts are generally obtained with the help of a collimator (lens or mirror) and the

[1] G. W. STROKE: J. Opt. Soc. Amer. **45**, 30 (1955); **51**, 1321 (1961); see also G. W. STROKE in: Progress in Optics, Vol. II (1963) (loc. cit.).

[2] For general reference to the spectroscopic concepts, see for example, R. A. SAWYER: Experimental Spectroscopy. New York: Prentice-Hall 1946, 1951. — G. R. HARRISON, R. C. LORD and J. R. LOOFBOUROW: Practical Spectroscopy. New York: Prentice-Hall 1948. — J. STRONG: Concepts in Classical Optics. San Francisco: W. H. FREEMAN & Co. 1958; also R. W. DITCHBURN: Light. New York: Interscience Inc. 1953, 1963.

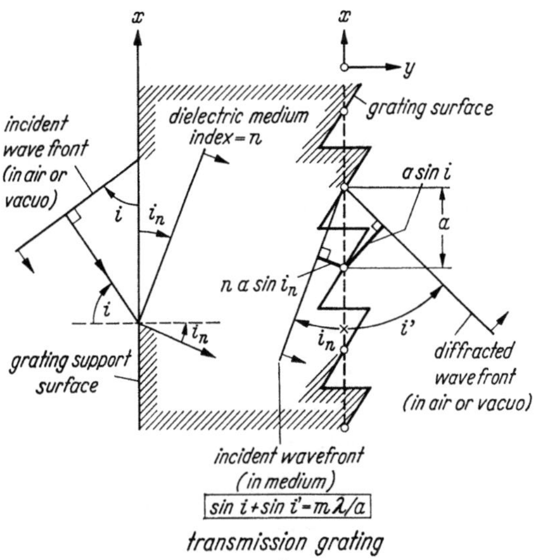

a

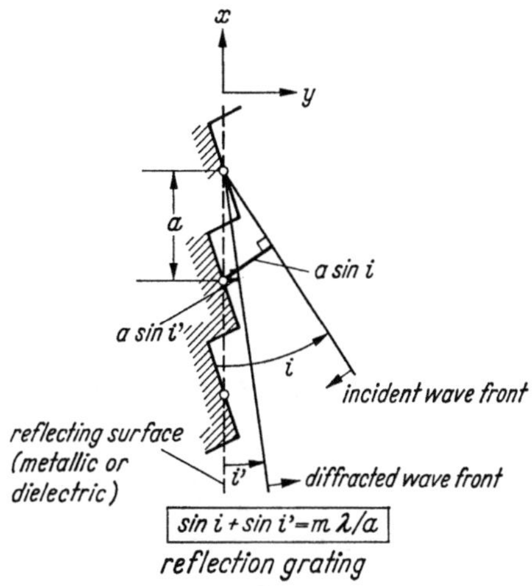

b

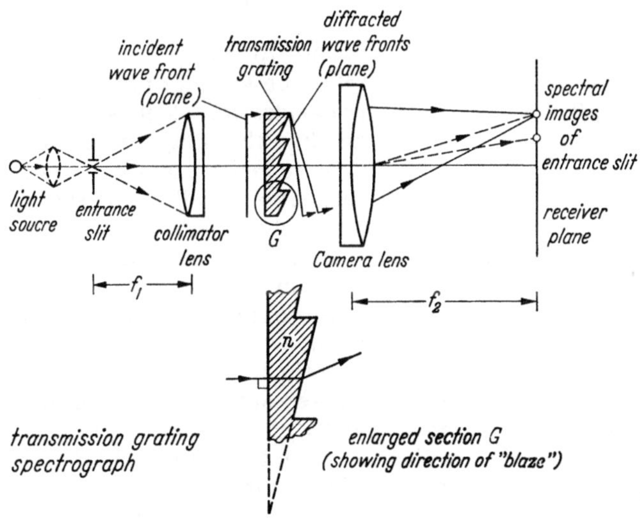

c

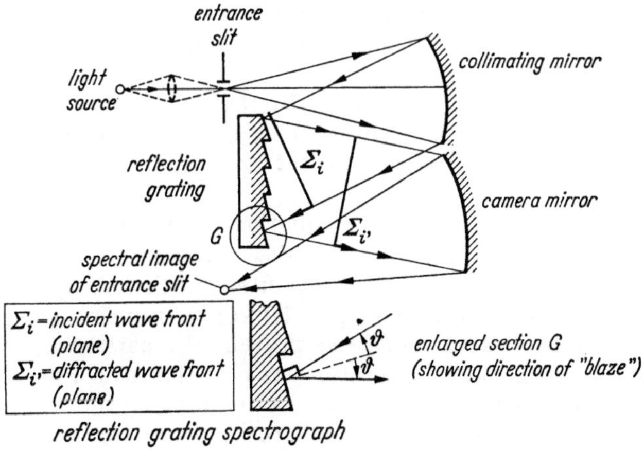

d

Fig. 9 a—d. a) Transmission grating parameters. b) Reflection grating parameters. c) Transmission grating spectrograph. d) Reflection grating spectrograph.

diffracted plane wave fronts are focussed onto a receiver (photoelectric tube or photographic plate) with the help of a camera lens or mirror (Fig. 9c, d). Perfectly stigmatic spectral images are obtained with plane gratings if both gratings and collimating and focussing optics are perfect.

We define perfection in a focussing system the property of transforming perfectly plane wave fronts into perfectly spherical wave fronts. When the collimator is also perfect and produces plane incident wave fronts, perfect stigmatism can be achieved in the spectroscopic arrangements shown in Fig. 9c and d.

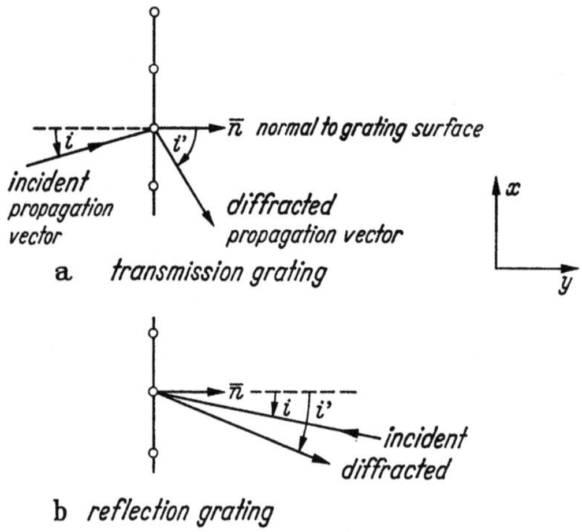

Fig. 10a and b. Transmission and reflection grating diffraction-parameters.

18. Grating equations for plane gratings. The grating equation

$$\boxed{\sin i + \sin i' = m\,\lambda/a} \tag{18.1}$$

or more generally

$$\boxed{\alpha + \alpha' = m\,\lambda/a}\,, \tag{18.2a}$$

$$\boxed{\gamma + \gamma' = 0} \tag{18.2b}$$

applies to both transmission and reflection gratings provided that the angles are taken as in Fig. 9a and Fig. 9b. The grating equation is quite general and basic to the calculation of all of the spectroscopic characteristics of gratings. For both transmission and reflection gratings, i and i' are the angles formed by the incident and diffracted waves with the mean grating surface that is with the plane parallel to xz. It will become clear from what follows that i and i' are to be taken positive for orientations of the waves with respect to the surface shown in Fig. 9a and Fig. 9b or equivalently in Fig. 4. The medium outside of the grating can have any refractive index, and it is in this medium that the angles of incidence and diffraction, i and i', are measured. It is also in this medium that the wavelength λ of Eqs. (14.1) and (18.2) is to be defined. We recall that $c = f\lambda$, where c = velocity of light, f = frequency of radiation, λ = free space wavelength. For a medium of refractive index n_0, in which the velocity of light is $v_0 = c/n_0$, the wavelength is $\lambda_0 = \lambda/n_0$, and we have $v_0 = f\lambda_0$. Of course it is the frequency f which is characteristic of an electromagnetic radiation.

Sometimes i and i' are taken as the angles formed by the propagation vectors (or "rays") with the normal $\boldsymbol{n}$ to the mean grating surface, as shown in Fig. 10. In this case, i and i' are positive when they are on the *same* side of the normal in a transmission grating.

In Sect. 14 we have established the grating equation rigorously with the help of the boundary conditions that are to be satisfied by the total electromagnetic field on the surface of the grating. We also pointed out that the grating equation could be simply established with the help of HUYGENS' principle—once the existence of plane diffracted waves had been proven. In particular, we derived the grating equation for a reflection grating, as also shown in Fig. 9b. Here we use HUYGENS' principle to demonstrate its power in deriving the grating equation for two additional uses, for *transmission* gratings and for gratings *immersed* in a medium of refractive index different from unity. Rather than making a general statement at this time about the applicability of the grating Eq. (18.1) or (18.2), we shall demonstrate its generality with the help of these two examples.

Let us first consider the *transmission grating* of Fig. 9a. Let the refractive index of the medium forming the grating with respect to the outside be n, and let the grating support surface be plane and parallel to xz, that is, parallel to the mean grating surface. Let the incident wave form an angle i with the support surface as usual (Fig. 9a). HUYGENS' principle, when applied to the grating surface immediately gives

$$n a \sin i_n + a \sin i = m\lambda. \quad (18.3)$$

The law of refraction, when applied to the grating support surface, gives

$$\sin i = n \sin i_n. \quad (18.4)$$

We thus immediately get from (18.3) and (18.4) the grating equation

$$\sin i + \sin i' = m\,\lambda/a \quad (18.5)$$

that is to be applied to the transmission grating as a whole, without requiring a detailed consideration of the propagation of the waves inside the medium. In fact, since Eq. (18.5) is independent of the refractive index, and as long as $n \geqq 1$ we shall indeed obtain diffracted waves emerging in the positive y direction, as shown in Fig. 9a.

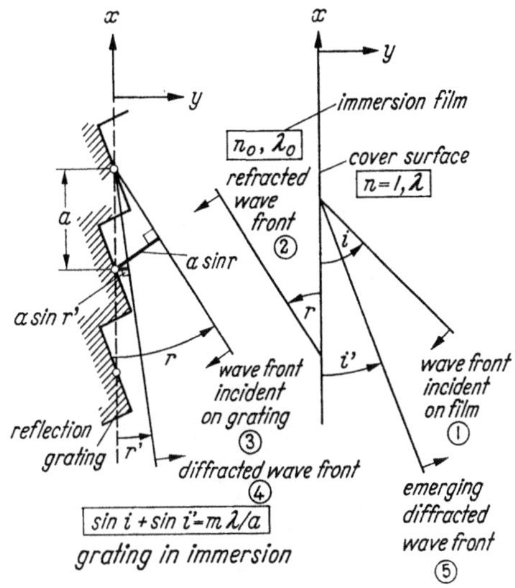

Fig. 11. Grating in immersion: parameters.

Next, we consider the diffraction by a *grating immersed into a medium* of refractive index n_0 (Fig. 11). For practical reasons, we examine the case of a reflection grating covered with a medium of refractive index n_0, such that the top surface of the medium n_0 is parallel to xz, the mean grating surface. (For instance, this can be achieved by covering the grating with a plane parallel glass plate and an immersion liquid of index n_0). Let the wavelength in the medium of index 1 be λ as usual, and let λ_0 be the wavelength in the medium n_0, such that

$$\lambda = n_0 \lambda_0. \quad (18.6)$$

By the law of refraction we have

$$\sin i = n_0 \sin r; \qquad \sin i' = n_0 \sin r'. \quad (18.7)$$

In the medium, the grating equation is:

$$a \sin r + a \sin r' = m\,\lambda_0/a. \quad (18.8)$$

By multiplying both sides of the equation by n_0, we immediately obtain, with the help of Eqs. (18.6) and (18.7):

$$\sin i + \sin i' = m\,\lambda/a \quad (18.9)$$

which is identical to Eq. (18.1).

We have thus obtained the remarkable conclusion that the grating equation applies in the form (18.1), or equivalently (18.2), not only to a *reflection* grating, but also to a *transmission* grating and to a grating covered by an *immersion film*, provided, of course, that the support surface of the transmission grating and the cover surface of the immersion film are parallel to the mean grating surface. We need to recall that i, i' and λ are to be measured in the same medium.

In Figs. 9, 10, and 11 the gratings are plane, and both the incident and the diffracted wave fronts are plane. Reflection gratings are also ruled on spherical surfaces and are used as "concave" gratings. These are discussed in Sect. 25.

We conclude by observing that special spectroscopic advantages of increased dispersion may be obtained by making the dielectric medium in the transmission grating (Fig. 9a) and the "immersion" grating (Fig. 11) into a *prism*, that is by introducing an appropriate angle between the support and mean grating surfaces (transmission grating) and between the cover and mean grating surface (immersion grating) respectively[1]. It is clear that the increase in dispersion (angular separation between diffracted beams of neighboring wavelengths) is due to the *prism*, and not to any new property of the grating or of the grating equation, of which the generality has been proven and demonstrated.

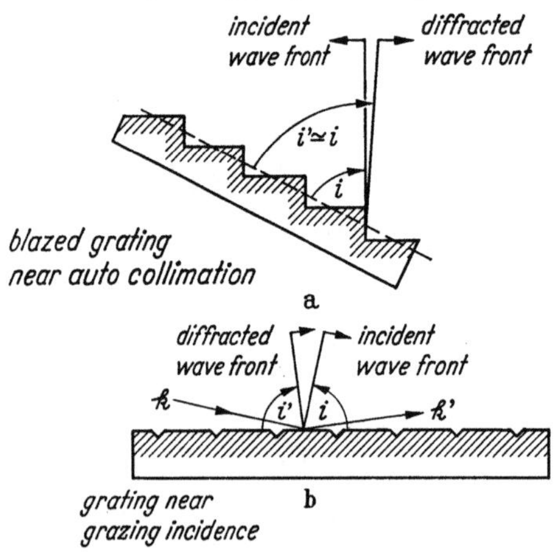

Fig. 12a and b. a) Grating in autocollimation. b) Grating near grazing incidence.

Let us also point out the important case when the angles of incidence and diffraction are equal, and are on the same side of the normal in a reflection grating. This is the case of *autocollimation* (Fig. 12a) when

$$i' \simeq i \tag{18.10}$$

and for which the grating equation takes the form

$$2 \sin i' = m\lambda/a. \tag{18.11}$$

Most blazed gratings with shaped, quasi-rectangular groove profiles, are used in the vicinity of autocollimation in such a way that the propagation vector $\boldsymbol{k}$ is almost normal to one or the other of the groove facets (Fig. 12a).

The use of gratings in autocollimation ($i \simeq i'$) for which the angles i' and i are on the *same* side of the normal (Fig. 12a) needs to be distinguished from the use of gratings in the vicinity of *grazing incidence* (Fig. 12b), where the magnitudes of i' and i are also equal, but now $i' \simeq -i$, that is i' and i are now on *opposite sides* of the normal. In grazing incidence it is of course necessary to use the grating equations in the form of Eq. (18.1).

19. Dispersion. For spectroscopic purposes, one is interested in separating the diffracted wave fronts of different wavelengths. In general, angular separation of wave fronts is desired. In Sect. 25 we shall deal with the angular directions of wave fronts with widely separated wavelengths and with the problem of overlapping spectra. Here we deal with the angular separation between wave fronts of closely neighboring wavelengths, which is also defined as *angular dispersion*.

The angular position i' of waves of different wavelengths is given in Eq. (18.1). We define as *angular dispersion* the quantity $di'/d\lambda$. From Eq. (18.1):

$$\sin i + \sin i' = m\lambda/a \tag{19.1}$$

[1] E. HULTHÉN and H. NEUHAUS: Ark. Fysik **8**, 343 (1954).

we have by differentiation

$$\cos i\, di + \cos i'\, di' = \frac{m}{a}\, d\lambda . \tag{19.2}$$

For a constant angle of incidence, $di = 0$, and we have for the dispersion

$$\frac{di'}{d\lambda} = \frac{1}{\cos i'}\,\frac{m}{a} = \delta . \tag{19.3}$$

With Eq. (19.1) we have

$$\delta = \frac{di'}{d\lambda} = \frac{\left(\frac{\sin i}{\cos i'} + \tan i'\right)}{\lambda} = \frac{\frac{1}{\cos i'}\,(\sin i + \sin i')}{\lambda} \tag{19.4}$$

and in autocollimation ($i' \simeq i$)

$$\delta = \frac{di'}{d\lambda} = \frac{2\tan i'}{\lambda} . \tag{19.5}$$

Thence follows the important conclusion that the *angular dispersion δ is independent of the grating constant*, and is the same for gratings of any spacing constant at given angles of incidence and diffraction. We note also that the dispersion increases with the tangent of the angle in autocollimation.

In the focal plane of a focussing system of focal length, the *linear dispersion* is

$$D = \frac{dl'}{d\lambda} = f\,\frac{di'}{d\lambda} \tag{19.6}$$

with

$$D = f\,\frac{1}{\cos i'}\,(\sin i + \sin i')/\lambda \tag{19.6a}$$

in general and

$$D = 2f\tan i'/\lambda \tag{19.6b}$$

in autocollimation.

It is important to observe that the wave fronts of neighboring wavelengths $\Delta\lambda$, dispersed by $\Delta i'$ according to (19.4) and (19.5) are *incoherent* with respect to each other, since they correspond to different frequencies in the source. We have

$$\Delta i' = \frac{di'}{d\lambda}\,\Delta\lambda \tag{19.7}$$

and

$$\Delta l' = \frac{dl'}{d\lambda}\,\Delta\lambda = f\,\Delta i' . \tag{19.8}$$

20. Resolving power. A complete treatment of resolving power requires the theory of physical optics (see Chap. B IV)[1]. In particular one needs to differentiate between coherent and incoherent illumination when using finite slit-widths. In what follows, incoherent illumination and infinitely narrow entrance slits are assumed. According to Lord RAYLEIGH (1888)[2] two neighboring spectral lines are called "resolved" if the diffraction pattern of one of the lines coincides with the first diffraction minimum of the other (Fig. 13). Now the diffraction patterns of two closely neighboring lines have practically the same shape. We know that the first diffraction minimum for a grating of rectangular width W is found at

$$u = \Delta i_0' = \frac{\lambda}{A} = \frac{\lambda}{W\cos i'} \tag{20.1}$$

[1] J. ARSAC: Transformation de Fourier et théorie des distributions, ch. 9, section 9.4. Paris: Dunod 1960.

[2] Lord RAYLEIGH: Scientific Papers, Vol. 1, p. 415. — R. W. DITCHBURN: Proc. Roy. Irish Acad. 39, A, 58 (1930).

where

$$A = W \cos i' \tag{20.2}$$

is the aperture width of the diffracted wave front. The minimum wavelength difference $\Delta\lambda$ which will be resolvable in terms of Rayleigh's "criterion" is found from Eqs. (19.4), (19.7), and (20.1) by rewriting Eq. (19.7) as

$$\Delta i' = \frac{\frac{1}{\cos i'}(\sin i + \sin i)}{\lambda}\,\Delta\lambda \tag{20.3}$$

and by requiring that

$$\Delta i' \geqq \Delta i'_0. \tag{20.4}$$

We get

$$\Delta\lambda = \frac{\lambda^2}{W(\sin i + \sin i')} \tag{20.5}$$

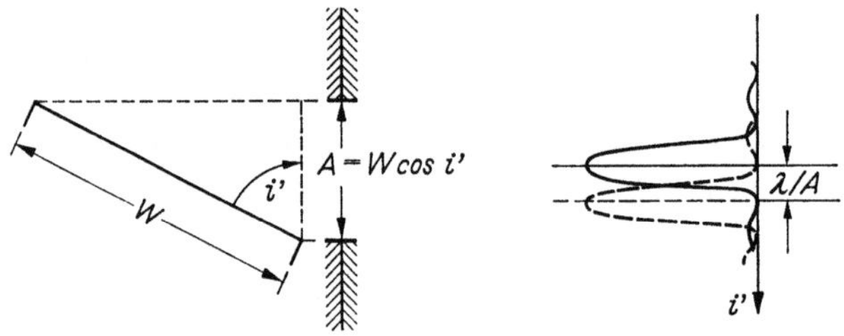

Fig. 13. Resolving "power" of gratings at Rayleigh limit.

and in the more usual form

$$\boxed{\text{R.P.} = \frac{\lambda}{\Delta\lambda} = \frac{W(\sin i + \sin i')}{\lambda}.} \tag{20.6}$$

When the grating is used in autocollimation, we have similarly

$$\boxed{\text{R.P.} = \frac{\lambda}{\Delta\lambda} = \frac{2W\sin i'}{\lambda}.} \tag{20.7}$$

The expression $\text{R.P.} = \lambda/\Delta\lambda$ is usually called "resolving power" in spectroscopy. Here again we note that the "resolving power" of a grating depends only on the grating width W and angle of use, as well as on the wavelength, but that it is independent of the number of grooves or grating constant. In fact, this result is not surprising.

The dispersion of a grating does not depend on its grating constant, and the diffraction pattern width (between its first minima) is also independent of the grating constant. Most fundamentally, the resolving power is, of course, simply determined by the width A of the diffracted wave front and the dispersion $\delta = di'/d\lambda$. In fact, combining Eqs. (20.2), (19.4), and (20.6), we see immediately that

$$\boxed{\text{Resolving power} = (\text{Aperture}) \times (\text{Angular Dispersion})}$$

or

$$\text{R.P.} = A \cdot \delta. \tag{20.8}$$

Eq. (20.8) conveniently summarizes the physical origin of the resolving power, and Eqs. (20.6) and (20.7) relate the resolving power to the relevant parameters.

It is important to note that A and δ are in general not independent variables since they both depend on i' and i.

It is remembered that the resolving power is sometimes given in textbook as R.P. $=NM$ where N is the total number of grooves on the gratings and m the order of diffraction. This relation might tend to be misleading by singling out the total number of grooves and the order as contributing to high resolving powers. The equation R.P. $=Nm$ is, of course, correct, and in agreement with Eqs. (20.6) and (20.7). Since $W=Na$, and $\sin i+\sin i'=m\lambda/a$, we can instead write Eq. (20.6) as

$$\text{R.P.}=\frac{N a m \lambda}{a \lambda}=Nm. \tag{20.9}$$

But while N, like W, is an independent variable, m depends on a and is equal to $a(\sin i \sin i')/\lambda$. Since we know that R.P. is independent of a at a given angle, we may write $Na=W$ and thus retrieve Eq. (20.6) which was derived from first principles above.

We conclude by note that *high resolving powers are obtained by using large gratings at high angles of incidence and diffraction, in the neighborhood of autocollimation.*

High angular dispersion is also obtained by using the gratings at high angles.

It is clear, though, that the aperture A of the grating will decrease with increasing angle, and therefore it might at first appear that a less efficient use of the light will be made when using gratings at high angles. Just the contrary is true, however, as we shall show in the next section, and also in Sect. 26, when dealing simultaneously with the problem of *luminosity and resolution.*

21. Dispersion, resolving power, aperture and luminosity of gratings. Let us merely recall here that what matters in the use of gratings in spectrographs and spectrometers is not the angular dispersion δ, but rather the linear dispersion $D=f\delta$ which gives the linear distance of two neighboring lines in the focal plane of the instrument. Limitations in resolution are ultimately set by granularity and noise in detectors, as well as by threshold sensitivities to observable light flux.

Noise limitations are always less serious when the spectral lines are widely spread out in comparison to the grain of the photographic plates, or when one can use rather large exit slits, (provided, of course here, that the slits have a small width when compared to the diffraction pattern width). Moreover, the linear dispersion is to be sufficiently great to provide "resolving power". Now, it is clear that a given linear dispersion D can be maintained within a wide range of variation of the angular dispersion δ and of the focal length f, provided only that their product $f\delta=D$ remains constant. How this variation has a bearing on our problem appears when the criterion of "luminosity" is introduced, as in Sect. 26. Luminosity has to do with the light flux reaching the detector, and we know that the flux F is proportional to HA/f^2, where $A=W\cos i'$ as before and H is the length of the grooves (on height of grating). We can write $F=k_f A/f^2$ where the constant k_f is given by $k_f=s h H B_\lambda \Delta\lambda T_G$. Here $s=$ entrance slit width, $h=$ entrance slit height parallel to H, $B_\lambda=$ brightness of spectrum as a function of λ, $\Delta\lambda=$ wavelength bandwidth admitted through slit and $T_G=$ transmittivity of grating. It is clear that if D is given, s and $\Delta\lambda$ do not vary with f. (h can in practice be maintained constant with f, or even increased with shorter f). We therefore find that we can make $k_f=k$, a constant, in the important practical case of high-resolution grating spectrometers and spectrographs. A very useful set of equations which permits the evaluation of the performance of gratings in spectrometers and spectrographs can now be given for the general case, as follows:

$$\left.\begin{array}{lll} \textit{Aperture width:} & A = W\cos i' & \text{(a)} \\ \textit{Linear dispersion:} & D = f\dfrac{1}{\cos i'}(\sin i + \sin i')/\lambda & \text{(b)} \\ \textit{Resolving power:} & \text{R.P.} = \dfrac{W(\sin i + \sin i')}{\lambda} = \dfrac{\lambda}{\Delta\lambda} & \text{(c)} \\ \textit{Flux:} & F = k_f A/f^2 = k_f \dfrac{1}{f}\dfrac{\text{R.P.}}{D} = k_f \dfrac{1}{f}\dfrac{1}{D}\dfrac{W(\sin i + \sin i')}{\lambda} & \text{(d)} \end{array}\right\} \quad (21.1)$$

(Grating ruled width $=W$, angle of incidence $=i$, angle of diffraction $=i'$, focal length $=f$).

In autocollimation we have

$$\left.\begin{array}{lll} \textit{Aperture witdh:} & A = W\cos i' & \text{(a)} \\ \textit{Linear dispersion:} & D = (2f\tan i')/\lambda & \text{(b)} \\ \textit{Resolving power:} & \text{R.P.} = \dfrac{2W\sin i'}{\lambda} = \dfrac{\lambda}{\Delta\lambda} & \text{(c)} \\ \textit{Flux:} & F = k_f A/f^2 = k_f \dfrac{1}{f}\dfrac{\text{R.P.}}{D} = k_f \dfrac{1}{f}\dfrac{1}{D}\dfrac{2W\sin i'}{\lambda} & \text{(d)} \end{array}\right\} \quad (21.2)$$

(grating ruled width $=W$, angle of incidence and diffraction $=i'$, focal length $=f$).

It is immediately apparent from these equations that while the flux F decreases with i' as $\cos i'$, this decrease may be more than compensated by the increase in flux with decreasing f, as $1/f^2$. In fact, when the flux is written as

$$F = k_f \frac{1}{f}\frac{1}{D}\frac{W(\sin i + \sin i')}{\lambda} \quad \text{(a)}$$

or as

$$F = k_f \frac{1}{f}\frac{1}{D}\frac{2W\sin i'}{\lambda} \quad \text{(autocollimation)} \quad \text{(b)} \qquad (21.3)$$

and if $D = f\frac{1}{\cos i'}(\sin i + \sin i') = \text{constant}$, respectively $D = (2f\tan i/\lambda) = \text{constant}$ are taken as a criterion, it becomes clear that an increase in i and i' will not only increase $\sin i$ and $\sin i'$, but also permit a reduction in f, both of which result in an increase of flux. The flux for resolving power at constant dispersion also increases with *angle* for the same reason, as can be seen from

$$\frac{F}{\text{R.P.}} = k_f \frac{1}{f}\frac{1}{D}. \qquad (21.4)$$

Numerical examples for high-resolution gratings are given in Sect. 26. It has, of course, been assumed that the grating is suitably blazed for the desired angles of use.

22. Overlapping orders. Unlike prisms, which disperse any wavelength of the spectrum into just one direction (Fig. 14a), a grating generally produces several wave fronts of a given wavelength and diffracts them into different directions, according to the grating equation

$$\sin i' = \frac{m\lambda}{a} - \sin i. \qquad (22.1)$$

A certain overlap exists between spectra of different orders, but is generally found to be only moderately disturbing, as will now be shown. In fact one of the principal advantages of high-resolution gratings over FABRY-PÉROT and other interferometric elements, when used in spectrometers, is that the free spectral range of gratings is not only independent of their dispersion and resolving power, but is also considerably greater than the free spectral range of interferometers for identical resolutions.

One defines as *free spectral range* in wavelenght units the quantity F_y, where

$$F_\lambda = \lambda_2 - \lambda_1 \qquad (\text{Å}) \tag{22.2}$$

or in wave number units (cm^{-1}) the quantity

$$F_\sigma = \sigma_1 - \sigma_2 \quad (\text{cm}^{-1}) \tag{22.3}$$

in which λ_1 and λ_2, respectively σ_2 (cm^{-1}) $=1/\lambda_2$ (cm) and σ_1 (cm^{-1}) $=$ $1/\lambda$ (cm) are the extreme wavelengths (respectively "wave numbers") in the spectral bandwidth of interest, which can be obtained in any given order *without* overlap. We assume $\lambda_2 > \lambda_1$.

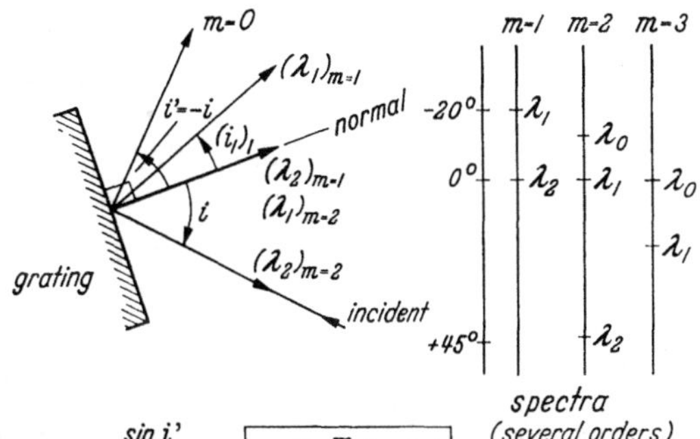

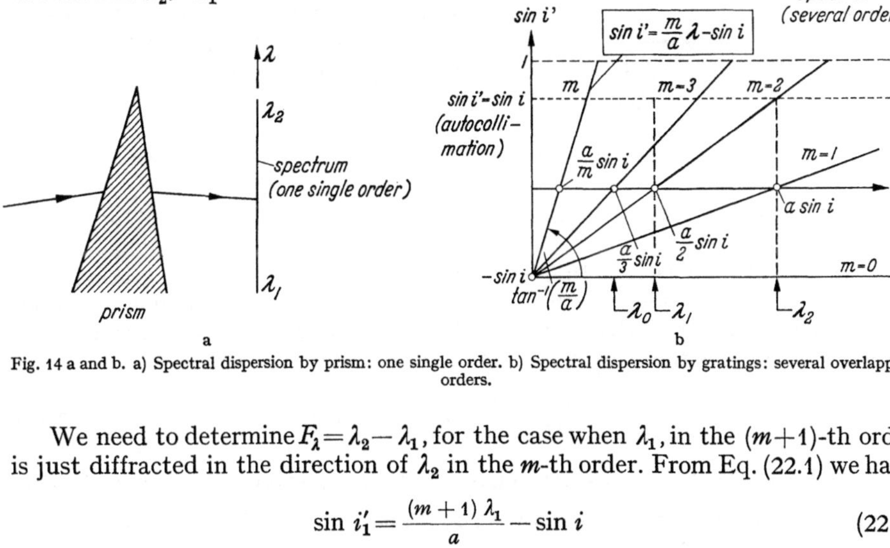

a b

Fig. 14 a and b. a) Spectral dispersion by prism: one single order. b) Spectral dispersion by gratings: several overlapping orders.

We need to determine $F_\lambda = \lambda_2 - \lambda_1$, for the case when λ_1, in the $(m+1)$-th order is just diffracted in the direction of λ_2 in the m-th order. From Eq. (22.1) we have

$$\sin i'_1 = \frac{(m+1)\,\lambda_1}{a} - \sin i \tag{22.4}$$

and

$$\sin i'_2 = \frac{m\,\lambda_2}{a} - \sin i\,. \tag{22.5}$$

We require that $\sin i'_2 = \sin i'_1$ and we must have

$$m\,(\lambda_2 - \lambda_1) = \lambda_1\,. \tag{22.6}$$

Therefore

$$F_\lambda = \lambda_2 - \lambda_1 = \frac{\lambda_1}{m} = \frac{\lambda_1\,\lambda_2}{a\,(\sin i + \sin i'_2)}\,, \tag{22.7}$$

with F_λ in angströms if λ_1 and λ_2 are also in angströms. Also,

$$F_\sigma = \sigma_1 - \sigma_2 = \frac{1}{a\,(\sin i + \sin i'_2)}\,. \tag{22.8}$$

Eq. (22.8) shows that the free spectral range F_σ depends only on the grating spacing constant a and on the angles of incidence and diffraction. In fact, at given angles of incidence and diffraction, the free spectral range F_σ can be chosen at will by choosing an appropriate grating constant a.

For example, in *autocollimation*, (Fig. 15), Eqs. (22.7) and (22.8) become

$$F_\lambda = \frac{\lambda_1 \lambda_2}{2a \sin i_2'} \tag{22.9}$$

and

$$F_\sigma = \frac{1}{2a \sin i_2'}, \tag{22.10}$$

where

$$t = a \sin i_2' \tag{22.11}$$

is the "depth" of a fully ruled groove in the blazed grating of Fig. 15 used in autocollimation at the angle i_2'. We have

$$F_\sigma = \frac{1}{2a \sin i_2'} = \frac{1}{2t} \tag{22.12}$$

in autocollimation [Eq. (22.11) defining t]. A 300 grooves/mm grating used in autocollimation at an angle of 64° has a free spectral range at $\lambda_2 = 5000$ Å ($\sigma_2 = 20\,000$ cm^{-1}) equal to

$$F_\sigma = \sigma_1 - \sigma_2 = \frac{1}{2 \times \frac{1}{300} \times 0{,}9} = \frac{300}{1.8} = 166.67 \text{ cm}^{-1}.$$

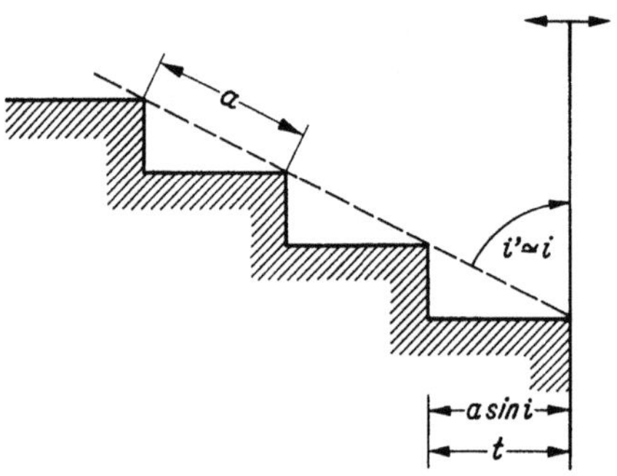

Fig. 15. Echelon, echelette and echelle grating parameters.

Thus, $\sigma_1 = 166.7 + 20\,000 \simeq 20\,167$, $\lambda_1 \simeq 4960$ Å and $F_\lambda \simeq 40$ Å. If the grating were 250 mm (10 inches) wide, its resolving power at that angle would be 900000. For comparison we recall that a Fabry-Pérot étalon also has a free spectral range $\Delta\sigma = 1/(2t)$ were t is the plate separation and a resolving power

$$[\text{R.P.}]_{\text{FP}} = \frac{\sigma}{\delta\sigma} = \frac{N\sigma}{\Delta\sigma} = N\sigma 2t, \tag{22.13}$$

where N is the number of effective beams. The same resolving power of 900000 is obtained at $\sigma = 20{,}000$ cm^{-1} with an etalon when $Nt = \frac{900\,000}{2 \times 20\,000} = 22.5$, where N number of "effective" beams. For $N = 22.5$, $t = 1$ cm. But the free spectral range is now only 1/2 cm^{-1} rather than 167 cm^{-1} as obtained with the grating at the same resolving power. Clearly, unlike in a grating (for which the free spectral range is independent of the resolving power), the free spectral range of a Fabry-Pérot étalon decreases with increasing resolving power. Moreover, the free spectral range of a grating is several orders of magnitude greater than the free spectral range of a Fabry-Pérot étalon at identical resolutions.

If the spectral bandwidth of the incident beam exceeds the free spectral range, a "pre-monochromator" or crossed dispersion are used to either limit the incident bandwidth to the free spectral range, or to separate the spectra of different orders.

When the comparatively large free spectral range obtained with gratings is adequate for a given study, or when pre-filtering is straightforward, the choice of grating constant is determined by practical considerations having to do with the quality of the grooves that can be ruled and with the blaze.

Properties of three groups of rather coarse gratings are considered in the next section. The deliberately great overlap of spectra in coarse well-blazed gratings can be used with advantage in association with crossed dispersion to reduce photographic plate dimensions.

23. Echelons, Echelettes and Echelle gratings[1]. "Escalier" is the French word for staircase, and a look at Fig. 15 points to some similarity between the profile of a blazed grating and a staircase.

Gratings with well blazed grooves and coarse spacings on the order of 10 mm, 1 mm and 0.1 mm have been named echelons, echellettes and echelles by MICHELSON[2], R. W. WOOD[3] and G. R. HARRISON[4], respectively. Their basic properties of resolving power, dispersion and free spectral range differ in no way from the general properties of all gratings. In fact, there is no theoretical reason for singling out these gratings from the general case of well blazed gratings, when dealing with their image-forming properties.

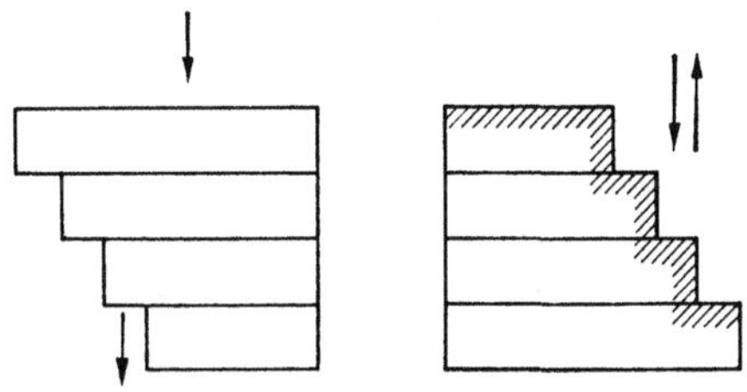

Fig. 16. Transmission and reflection echelons.

It is however of considerable practical importance to recognize that gratings with coarsely spaced grooves have the same resolving power and dispersion, for a given width and angle, as gratings of finer spacings. The spacing of echelons and of echelle gratings is made deliberately large with *respect to the wavelength*, so as to obtain such a great amount of overlap between the spectra that all wavelengths of interest leave the gratings within a very small angular range, on the order of 2° or so in a representative case. The comparatively wide groove face of echelons and echelles help in concentrating a large part of the diffracted light within the small angular range within which all of the spectral information is distributed.

Echelon gratings are singled out in practice because they are not ruled like gratings, but are formed by suitably assembling a stack of flat glass or quartz plates, as shown in Fig. 16. Only a very small number of echelons have been successfully completed in practice (MICHELSON, W. E. WILLIAMS). Plates from 1 to 20 mm in thickness have been used, principally for high resolution work in the visible (MICHELSON, WILLIAMS) before gratings could be ruled with sufficient regularity to permit the attainment of high resolutions (HARRISON and STROKE).

Echelette gratings are ruled blazed gratings with well shaped grooves of rectangular profile, and spacings up to 1 mm or so destined for use in the infrared. Their spacing is generally chosen to be comparable to the wavelength so as to permit work in low orders of diffraction for the purpose of avoiding overlap.

Echelle gratings are ruled blazed gratings with well shaped grooves of rectangular profile and spacings up to 0.1 mm or so, destined principally for use in the visible and ultraviolet domains. Like that of echelons, the spacing constant

[1] For general reference to the history, development and classical treatment, see for example, W. E. WILLIAMS: Applications of Interferometry. London: Methuen & Co. 1930, 1958, or C. CANDLER: Modern Interferometers. London: Hilger & Watts 1951.

[2] A. A. MICHELSON: Astrophys. J. **6**, 36 (1898).

[3] R. W. WOOD: J. Opt. Soc. Amer. **37**, 733 (1947).

[4] G. R. HARRISON: J. Opt. Soc. Amer. **39**, 522 (1949).

of echelle gratings is chosen to be large compared to the wavelength. (Some early echelle gratings have been produced by a lapping process by Bausch & Lomb, Inc.).

α) *Transmission echelon.* According to HUYGENS' principle, if the incident and diffracted wave fronts are to be plane, we must have (Fig. 17).

$$n \sin i - \sin i' = m\,\lambda/a \tag{23.1}$$

where n is the refractive index of the plates. By differentiation we have

$$\left.\begin{aligned} dn\,\sin i + n\cos i\,di \\ = -\cos i'\,di' = \frac{m}{a}\,d\lambda \end{aligned}\right\} \tag{23.2}$$

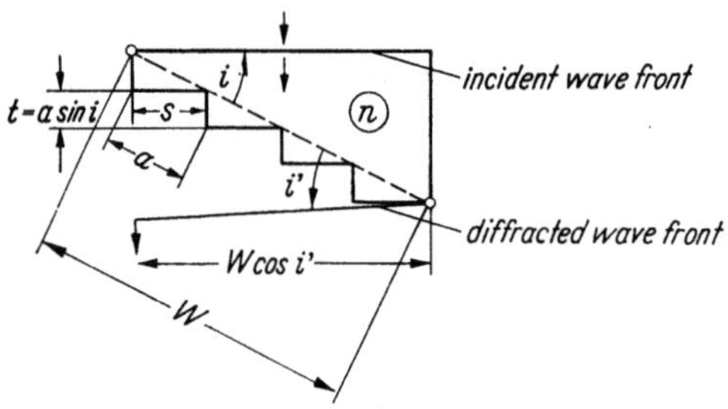

Fig. 17. Transmission echelon parameters.

and the dispersion is

$$\frac{di'}{d\lambda} = -\frac{1}{\cos i'}\left[\frac{m}{a} - \sin i\,\frac{dn}{d\lambda}\right]. \tag{23.3}$$

We also have

$$\Delta i' = \frac{di'}{d\lambda}\,\Delta\lambda. \tag{23.4}$$

To calculate the resolving power R.P. $=\lambda/\Delta\lambda$ we must have

$$|\Delta i'| = |\Delta i_0'| = \frac{\lambda}{W\cos i'} \tag{23.5}$$

where $\Delta i_0'$ is the angular position of the first minimum in the diffraction pattern corresponding to the width $W\cos i'$ of the diffracted wave front. We get

$$\left.\begin{aligned} \text{R.P.} &= \frac{\lambda}{\Delta\lambda} \\ &= W\left[\frac{m}{a} - \sin i\,\frac{dn}{d\lambda}\right] \end{aligned}\right\} \tag{23.6}$$

or

$$\left.\begin{aligned} \text{R.P.} = \frac{\lambda}{\Delta\lambda} = W\Big[\frac{n\sin i}{\lambda} - \\ - \frac{\sin i'}{\lambda} - \sin i\,\frac{dn}{d\lambda}\Big]. \end{aligned}\right\} \tag{23.7}$$

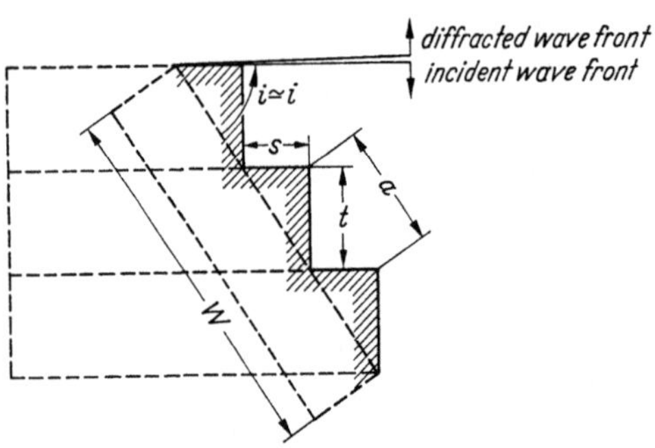

Fig. 18. Reflection echelon parameters.

In general, transmission echelons are used so that $i' \simeq i$ in Fig. 17. Then

$$\left.\begin{aligned} \text{R.P.} \cong \frac{W}{\lambda}\Big[(n-1)\sin i - \\ - \sin i\,\frac{dn}{d\lambda}\Big]. \end{aligned}\right\} \tag{23.8}$$

The free spectral range can be calculated from Eq. (23.1) as in the general case. One gets

$$F_\lambda = \lambda_2 - \lambda_1 = \frac{\lambda_1}{m} - \frac{(n_1 - n_2)\,t}{m} \tag{23.9}$$

where $t = a\sin i$ and n_1 and n_2 are the indexes for λ_1 and λ_2.

β) *Reflection echelons and echelles.* The properties of dispersion, resolving power and free spectral range of reflection echelons and echelles are strictly identical to those of reflection gratings in general. A reflection echelon and echelle gratings are generally used as a blazed reflection grating, very close to autocollimation, when $i' \simeq i$ (Fig. 18). We therefore have

$$\frac{di'}{d\lambda} = \frac{2\tan i'}{\lambda} \tag{23.10}$$

which can also be written as

$$\frac{d i'}{d\lambda} = \frac{2t}{\lambda s} \tag{23.11}$$

where

$$t = a \sin i' \tag{23.12}$$

and

$$s = a \cos i'. \tag{23.13}$$

The resolving power is

$$\text{R.P.} = \frac{\lambda}{\Delta\lambda} = \frac{2W \sin i'}{\lambda} \tag{23.14}$$

with $W = N a$, which can also be written as

$$\text{R.P.} = \frac{2Nt}{\lambda} \tag{23.15}$$

where N is the total number of plates in the echelon, or grooves of the echelle. The free spectral range is given in Eqs. (23.9) and (23.10), and can be written as

$$F_\lambda = \frac{\lambda_1 \lambda_2}{2t} \tag{23.16}$$

and

$$F_\sigma = \frac{1}{2t} \tag{23.17}$$

Particular advantages of deliberate overlap within a narrow angular domain with, however, still sufficient free spectral range can be obtained by a judicious choice of angle spacing constant. These advantages can be obtained with ruled "echelle" gratings and were first pointed out by HARRISON in 1949.

24. Curvature of spectral lines[1]. α) *Straight entrance slit.* The grating equation for plane waves incident from a direction defined by its direction cosines α, β, γ was shown to be

$$\gamma' = -\gamma \tag{24.1}$$

$$\alpha + \alpha' = m\lambda/a \tag{24.2}$$

$$\left.\begin{aligned}&\alpha^2+\beta^2+\gamma^2\\ &=\alpha'^2+\beta'^2+\gamma'^2=1\end{aligned}\right\} \tag{24.3}$$

where α', β', γ' are the direction cosines of the diffracted waves, a = grating spacing constant, λ = wavelength, m = an integer. The grooves of the plane grating are parallel to z (Fig. 19).

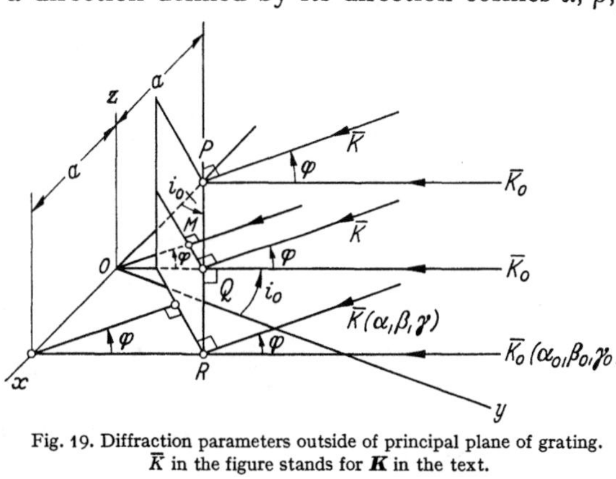

Fig. 19. Diffraction parameters outside of principal plane of grating. $\overline{K}$ in the figure stands for $\boldsymbol{K}$ in the text.

To every direction α, β, γ, (and α', β', γ') corresponds one point A (and A') in the focal plane of the collimator (and focussing mirror). Clearly [cf. Eq. (24.1)] the points A and A' are symmetrically placed with respect to the principal ($x\,y$) plane.

In general, the incident waves are made to originate at a straight slit parallel to the grooves, and perpendicular to the $x\,y$ plane, in the focal plane of the col-

[1] For general reference, see for example C. S. RUPERT: Slit curvature in grating monochromators employing single or multiple diffraction. J. Opt. Soc. Amer. **42**, 779–781 (1952); also R. MINKOWSKI: Astrophys. J. **96**, 306 (1942) ("Curvature of the lines in plane grating spectra".)

limator. The corresponding spectral images of the slit, formed by the diffracted waves in the focal plane of the focussing mirror, are found to be curved.

The curvature of the spectral lines can be most conveniently expressed by calculating the angle $\Delta i_0'$ formed in the principal plane between the *projection of the direction* $\boldsymbol{K}'$ (corresponding to A') and the *direction* $\boldsymbol{K}_0'$ (corresponding to A_0'). It is clear that for a straight entrance slit, $\boldsymbol{K}$ projects onto $\boldsymbol{K}_0$. However, $\boldsymbol{K}'$ does not project onto $\boldsymbol{K}_0'$. We recall that $\boldsymbol{K}_0$ and $\boldsymbol{K}_0'$ form the angles of incidence i_0 and of diffraction i_0' in the principal plane ($x\,y$ plane) of the grating, such that the equation

$$\sin i_0 + \sin i_0' = m\lambda/a \tag{24.4}$$

is satisfied.

We now show that $\boldsymbol{K}'$ does not project onto $\boldsymbol{K}_0'$, but rather onto a direction $\boldsymbol{K}_\varphi'$ forming an angle i_φ with the y axis in the $x\,y$ plane.

In Fig. 19, the point A of the straight entrance slit corresponds to the direction $\boldsymbol{K}$ obtained by inclining the direction $\boldsymbol{K}_0$ by φ about the line $\overline{PQR}$, where $\overline{PQR} \perp \boldsymbol{K}_0$. The line $\overline{PQR}$ forms an angle i_0 with the x-axis, and we recall that

$$\sin i_0 = \alpha_0. \tag{24.5}$$

The wave front incident from the direction $\boldsymbol{K}$ intersects the $x\,y$ plane in the line $\overline{PQR}$ and we have

$$\sin\varphi = \gamma. \tag{24.6}$$

Similarly, for the diffracted waves

$$\sin\varphi' = \gamma'. \tag{24.7}$$

But $\gamma' = -\gamma$, and therefore

$$\varphi' = -\varphi. \tag{24.8}$$

It is important to note that the angle φ' (corresponding to the direction $\boldsymbol{K}'$) is obtained by inclining $\boldsymbol{K}_0'$ by $-\varphi$ about a line $\overline{PQ'R'}$ which forms an angle i_φ' (not i_0') with the x-axis in the principal plane. The direction $\boldsymbol{K}_\varphi'$ in the $x\,y$ plane is perpendicular to $\overline{PQ'R'}$ and forms the angle i_φ' with the y-axis in that plane.

The angle i_φ' is obtained (Fig. 19) by observing that the diffracted wave front will be plane if the equation

$$\frac{a\sin i_\varphi'}{\cos\varphi'} = \frac{a\sin i_\varphi'}{\cos\varphi} = m\lambda - \frac{a\sin i_0}{\cos\varphi} \tag{24.9}$$

is satisfied. Using (24.4), Eq. (24.9) can also be written as

$$\frac{\sin i_\varphi' + \sin i_0}{\cos\varphi} = \frac{m\lambda}{a} = \sin i_0 + \sin i_0'. \tag{24.10}$$

since

$$m\lambda/a = \sin i_0 + \sin i_0'$$

according to Eq. (24.4).

We have

$$\boxed{\sin i_\varphi' - \cos\varphi \sin i_0' = \sin i_0 (\cos\varphi - 1)} \tag{24.11}$$

which defines the angle i_φ' in terms of i_0, i_0'. Eq. (24.11) and φ determine the curvature of spectral lines without approximations.

For the case where $i_\varphi' \simeq i_0'$ we can write

$$i_\varphi' = i_0' + \Delta i_0' \tag{24.12}$$

and

$$\sin i_\varphi' = \sin(i_0' + \Delta i_0') \simeq \sin i_0' + \Delta i_0' \cos i_0'. \tag{24.13}$$

For small φ

$$\cos\varphi \simeq 1-\varphi^2/2 \tag{24.14}$$

and Eqs. (24.11) to (24.14) give

$$\boxed{\Delta i_0' \simeq -\left(\frac{\sin i_0+\sin i_0'}{\cos i_0'}\right)\frac{\varphi^2}{2}\,.} \tag{24.15}$$

The angle φ can be expressed in terms of the height h of a point of the entrance slit above the center. We have

$$h \simeq f\varphi \tag{24.16}$$

where f is the focal length of the collimator (the camera mirror is assumed to have the same focal length). We have from (24.15)

$$\Delta x_0' = f\Delta i_0' \tag{24.17}$$

and with (24.16) we get

$$\boxed{\Delta x_0' \simeq -\left(\frac{\sin i_0+\sin i_0'}{\cos i_0'}\right)\frac{h^2}{2f}\,.} \tag{24.18}$$

Since $\Delta i_0'<0$, $\sin i_\varphi'<\sin i_0'$, and therefore the convexity of the spectral lines is towards the increasing wavelengths (Fig. 20). The curvature of the spectral lines is parabolic, but the parabola can be closely matched by a radius

$$\boxed{R \simeq f\frac{\cos i_0'}{\sin i_0'+\sin i_0}\,.} \tag{24.19}$$

In autocollimation ($i_0' \simeq i_0$)

$$R \simeq \frac{f}{2\tan i_0'} \tag{24.20}$$

and

$$\Delta x_0' \simeq -\tan i_0'\,\frac{h^2}{f}\,. \tag{24.21}$$

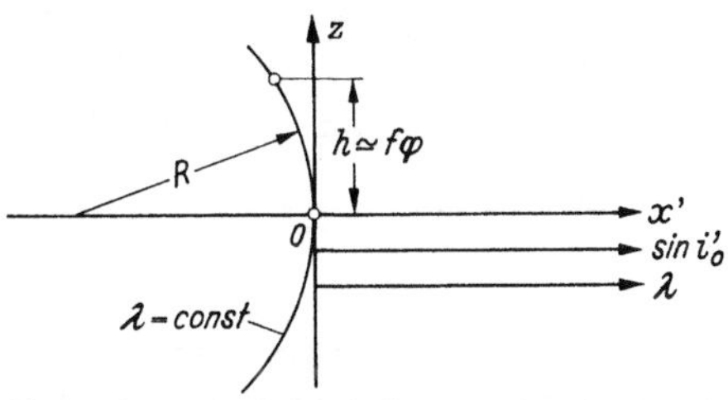

Fig. 20. Curvature of spectral lines in grating spectrographs and spectrometers.

Clearly, the slit curvature is wavelength dependent. In particular, for entrance and exit slits with fixed position (as is the case in monochromators), an exit slit of a curvature variable with wavelength must be used. (This undesirable feature can be overcome by the use of *circular* slits as described in the next section.)

If a straight exit slit were used in place of a curved exit slit (for example, in monochromator applications) the wavelength difference along the exit slit height is obtained by differentiating the grating equation

$$\sin i_0+\sin i_0' = m\lambda/a\,. \tag{24.4}$$

We have

$$\Delta\lambda = \frac{\cos i_0'}{\sin i_0+\sin i_0'}\,\Delta i_0' \tag{24.22}$$

and, with Eqs. (24.15) and (24.16),

$$\Delta\lambda \simeq -\lambda\frac{h^2}{2f^2}\,. \tag{24.23}$$

Curved slits are also required in spectroheliographs. For example, in a 2 meter high-resolution instrument using a grating around 64° in autocollimation, $R\simeq 500$ mm. For 100 mm long slits, $\Delta x_0' \simeq 2.5$ mm.

In a high resolution spectrograph, using the same grating, at the same angle, but now with a focal length of 10 meters, the radius becomes $R\simeq 2500$ mm and with 200 mm long slits, $\Delta x'_0 \simeq 0.2$ mm. (With 10 mm long slits, we still have $\Delta x'_0 \simeq 0.05$ mm and $\Delta\lambda \simeq -0.0625$ Å at 5000 Å.)

We finally note that for certain applications it is preferable to have both the entrance and the exit slit curved. In Fig. 21 it is clear that we can replace the straight entrance slit and curved exit slit of radius R by a curved entrance slit (of same radius R) and a straight exit slit.

When both the entrance and exit slits are curved as shown in Fig. 21 b, they will clearly have the same radius of curvature $2R$ when the angles i_0 and i'_0 corresponding to the slit centers are centered symmetrically about $(i_0+i'_0)/2$.

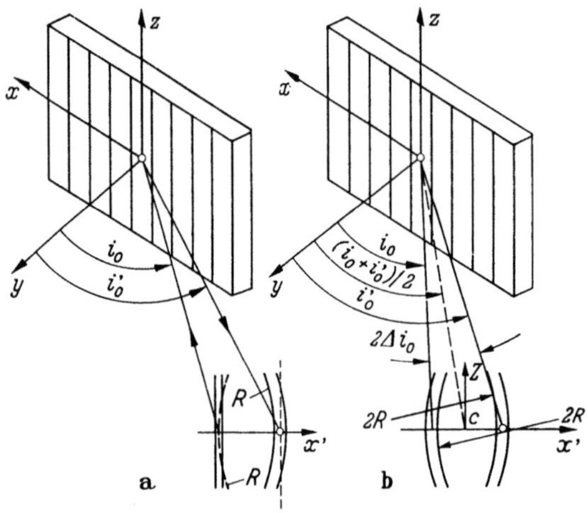

Fig. 21a and b. Curvature of spectral lines in grating spectrographs and spectrometers.

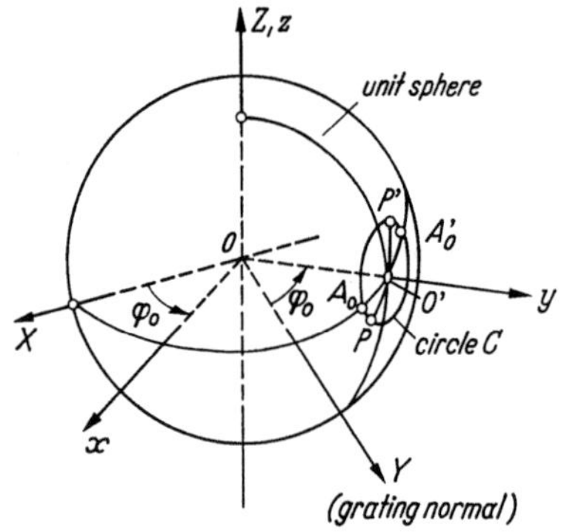

Fig. 22. Stigmatism of spectral lines on circle C, with a slit coincident with circle C (see also Fig. 25).

However, the curvature of the slit need not be $2R$. Moreover, the slit curvature can be made wavelength independent under certain conditions. This is shown in the next section.

β) *Circulary curved entrance and exit slits.* W. Fastie (1952)[1] found experimentally that good stigmatism could be achieved with long curved slits, if the entrance and exit slits are formed by portions of a *circle* passing through a source point A_0 and a corresponding spectral image point A'_0 both located in the principal plane of a plane grating. In fact the stigmatism can be shown to be perfect. A formal proof of this fact follows: Consider the unit sphere (Fig. 22) such that the grating is in the XZ plane with the grooves parallel to Z. The XY plane is the principal plane. The source point A_0 and spectral image point A'_0 are in the XY plane and satisfy the grating equation

$$\sin i_0 + \sin i'_0 = m\lambda/a \tag{24.4}$$

where $i_0 = \widehat{YOA_0}$, $if = \widehat{YOA'_0}$. Let an xyZ system be obtained by rotation of the XYZ system about Z by an angle φ_0 such that

$$\varphi_0 = \frac{i_0 + i'_0}{2}. \tag{24.24}$$

The Y-axis intersects the great circle passing through $A_0 A'_0$ at O'. Consider the little circle C centered at O' and passing through $A_0 A'_0$. We shall show that the

[1] W. G. Fastie: J. Opt. Soc. Amer. **42**, 647 (1952).

grating equation

$$\gamma' = -\gamma, \tag{24.1}$$

$$\alpha' + \alpha = \frac{m\,\lambda}{a}. \tag{24.2}$$

is satisfied on the little circle C (with α, γ, α', γ) defined with respect to the XYZ system as usual), provided that we apply (24.1), (24.2) to two points P and P' located on the intersection of a great circle, passing through O' ,with the little circle C. We recall that α, β, γ respectively α', β', γ', are projections of the unit vec-

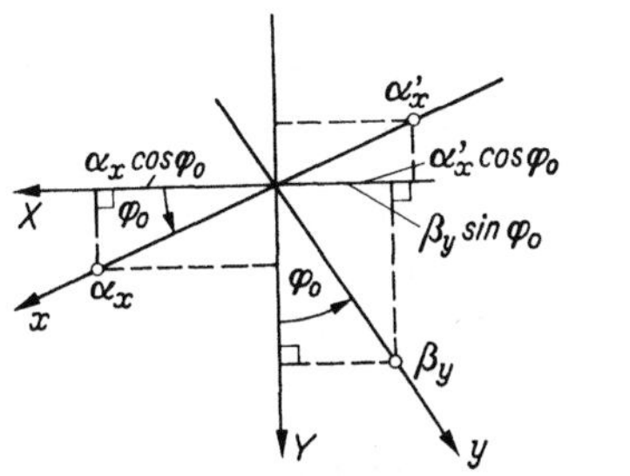

Fig. 23. Curvature of spectral lines: parameters for computation.

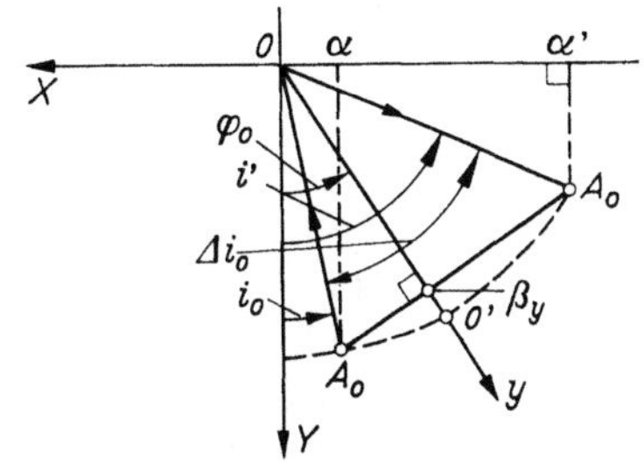

Fig. 24. Curvature of spectral lines: parameters for computation.

tor OP, respectively OP', on the $-X$, Y, Z axes. Similarly, $\alpha_x, \beta_y, \gamma_z$ respectively $\alpha_x, \beta_y, \gamma_z$, are the projections of these same vectors on the xyZ system. Clearly,

$$\gamma'_z = \gamma' \tag{24.25}$$

$$\gamma_z = \gamma \tag{24.26}$$

Also (Fig. 23)

$$\left.\begin{aligned} \alpha &= \alpha_x \cos\varphi_0 - \beta_y \sin\varphi_0, \\ \beta &= \alpha_x \sin\varphi_0 + \beta_y \cos\varphi_0, \end{aligned}\right\} \tag{24.27}$$

$$\left.\begin{aligned} \alpha' &= -\alpha_x \cos\varphi_0 - \beta_y \sin\varphi_0, \\ \beta' &= -\alpha_x \sin\varphi_0 + \beta_y \cos\varphi_0. \end{aligned}\right\} \tag{24.28}$$

For any pair of points P and P' on the intersection of the great circle through O' with the small circle C, we have, (Fig. 22),

$$\alpha'_x = -\alpha_x, \tag{24.29}$$

$$\beta'_y = \beta_y = \text{constant}, \tag{24.30}$$

$$\gamma'_z = -\gamma_z. \tag{24.31}$$

Particularly noteworthy is Eq. (24.30). With Eqs. (24.27) and (24.28) we have

that is

$$\left.\begin{aligned} \alpha' + \alpha &= (\alpha_x - \alpha_x)\cos\varphi_0 - (\beta_y + \beta_y)\sin\varphi_0, \\ -(\alpha' + \alpha) &= 2\beta_y \sin\varphi_0. \end{aligned}\right\} \tag{24.32}$$

(The minus sign results from the choice of direction of X in Fig. 22.) It remains to be shown that $2\beta_y \sin\varphi_0 = m\lambda/a$.

In Fig. 24, we have

$$\varphi_0 = \frac{i_0 + i'_0}{2} \tag{24.24}$$

$$\Delta i_0 = i'_0 - i_0 \tag{24.33}$$

$$\beta_y = \cos\frac{\Delta i_0}{2} = \cos\frac{i'_0 - i_0}{2}. \tag{24.34}$$

Therefore

$$\left.\begin{aligned} 2\beta_y \sin\varphi_0 &= 2\cos\left(\frac{i'_0 - i_0}{2}\right)\sin\left(\frac{i'_0 + i_0}{2}\right), \\ \text{i.e.}\quad 2\beta_y \sin\varphi_0 &= \sin i_0 + \sin i'_0 = \frac{m\lambda}{a}, \end{aligned}\right\} \quad (24.35)$$

q.e.d. The grating Eq. (24.1), (24.2) is therefore satisfied for any two points P and P' on the little circle C, passing through A_0 and A'_0, centered on O', and having a diameter $A_0 A'_0$. Appropriate portions of a circular slit of diameter $A A'_0$ can therefore be used as the entrance and exit slits: they are related by *perfect stigmatism.*

Clearly one circle (circular slit) centered on O' in fixed position can be used for any wavelength in a spectrometer, the change of wavelength being obtained by rotation of the grating about Z. The remarkable advantage of the circular slit C (or of portions of the circle C used for the entrance and exit slits) is that the *same slit curvature* is usable for any wavelength (Fig. 25). And, of course, the basic advantage of curved slits is that they permit the use of very long slits, which is of particular interest for increasing the luminosity in spectrometers, as was first pointed out by H. M. RANDALL and J. STRONG in 1931[1].

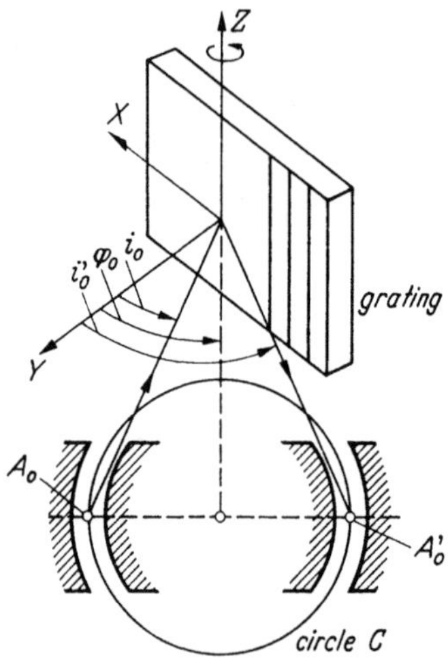

Fig. 25. Stigmatism of spectral lines on circle C, in case of a slit coincident with circle C.

25. Concave gratings[2,3]. Their aberrations and astigmatism. Gratings ruled on concave mirrors or on lenses have the property of acting as their own collimating and focussing systems. Most generally the spacing of the grooves is made to be perfectly periodic *on the chord* of a spherical mirror surface, and concave gratings can in principle be ruled on any ruling engine capable of ruling plane gratings. Non-periodic spacings have special advantages and can be obtained by programing the advance of an interferometrically controlled engine, for example. Unlike in plane gratings, where the finite groove length contributes to the luminosity of the gratings and spectra, without impairing the stigmatism, it is found that the finite groove length in concave gratings results in appreciable and generally irreducible astigmatism. In fact, except in one special case (when the grating is used with a collimator and forms spectra on its normal), concave gratings produce wave fronts with such considerable aberrations from sphericity (astigmatism) that the quality of the spectral images is not limited by diffraction or by the small aberrations in the principal section of the grating but rather by geometrical

---

[1] H. M. RANDALL and J. STRONG: Rev. Sci. Instrum. **2**, 585 (1931).

[2] Concave gratings were first proposed by H. A. ROWLAND in November 1882 and described by him in a paper entitled "On Concave gratings for optical purposeses. Amer. J. Sci. (3) **26**, 87–98 (1883); Phil. Mag. (5) **16**, 197–210 (1883). — Physical Papers, p. 492. Baltimore: The Johns Hopkins Un. Press. 1902.

[3] For general reference, see also H. KAYSER: Handbuch der Spektroskopie, Vol. I. Leipzig: Hirzel 1900; and T. NAMIOKA, in: Space Astrophysics, pp. 228–268, edit. by W. LILLER (The Un. of Michigan Inst. of Science and Techn. Series, New York: McGraw-Hill Book Co. 1961); and C. F. MEYER: Diffraction of Light, X-rays and material particles. Ann. Arbor Michigan: J. W. Edwards 1949; and C. CANDLER: Modern Interferometers. London: Hilger & Watts 1951; and A. MARÉCHAL: Imagerie Géométrique Aberrations. Paris: Revue d'Optique 1952.

aberrations. It will indeed become apparent that the limitations on the grating width, which can be uniformly ruled, in practice generally maintain the aberrations to tolerably low values in practical cases. The ruling limitations result from the appreciable as well as varying concavity (or convexity) of the grating surface and the fact that blaze of grooves change across the grating: indeed, the ruling diamond needs to rule in a fixed plane with a single setting to achieve the most uniform spacing of the grooves. Even if wider concave gratings were theoretically attainable, by ruling grooves with non-uniform spacings, for example, the widths of concave gratings would still be limited by heretofore unsolved ruling difficulties. As it is, achievable widths generally closely correspond to the maximum usable theoretical widths. When a choice needs to be made, gratings with short groove lengths and wider widths are preferable from the ruling standpoint; they also have a greater resolving power as long as they are well ruled and within the theoretical limits of aperture.

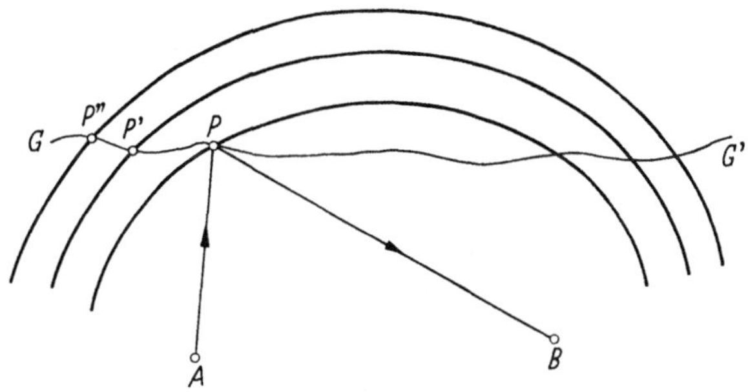

Fig. 26. Focussing grating.

The considerable practical importance of concave gratings, results, of course, from the absence of additional collimating and focussing optics, and from the absence of reflection losses associated to their use. This is of particular importance in the far vacuum ultraviolet (below 1200 Å) and X-ray regions where the loss of luminosity and generally poor image quality, due to astigmatism, is generally far outweighed by the gain in luminosity resulting from the absence of focussing (and collimating) optics. In the near ultraviolet, visible and ultraviolet regions, and generally when good stigmatism and high resolution are required, plane gratings have a greater luminosity, are simpler to use, and are nowadays much preferred to concave gratings. Plane gratings with perfectly uniform blaze and of theoretically unlimited dimensions can today be ruled to be perfectly stigmatic in practice. A 1-meter focal length spectrometer using a 10 inch wide, 5 inch high, stigmatic plane grating at 64° in the visible has been achieved, while the largest ruled area of an available 2 meter radius concave grating of similar focal length is 95 × 40 mm, and then only with a 10° blaze.

The basic image forming qualities of concave gratings were first predicted and verified by Rowland. They are simple to understand in principle, but the mathematical formalism involved in the treatment of concave gratings is generally complicated by the calculations of astigmatism and aberrations. As a rule, in view of the irreducible imperfections of concave gratings, reasonable approximations and simplicity of the argument are preferred to mathematical rigour in the calculation of the abberations. It is a remarkable fact that the uniformly periodic spacings along the *chord* of a spherical mirror is the most suitable positioning of the grooves on the resulting grating.

1. Focussing properties of gratings. Rowland already pointed out that a theoretically perfect grating for one position of the entrance and exit slits could be ruled on *any* surface, flat or otherwise. This surprising result holds only for a given wavelength. It can casily bc shown as follows. Consider two points A and B (Fig. 26) where B is the image of A. Consider a family of ellipsoids with their foci at A and B such that

$$\left.\begin{aligned} AP+PB&=m\lambda,\\ AP'+P'B&=(m+1)\lambda,\\ AP''+P''B&=(m+2)\lambda,\end{aligned}\right\}\qquad(25.1)$$

Eqs. (25.1) hold for any point of the particular ellipsoid. Let GG' be the intersection of a grating surface of any shape with the plane of the figure. Clearly, if P, P', P'' are grooves on the surface GG', then the light originating from A, and diffracted by P, P', P'', interferes constructively at B. The "Fresnel zone" created by the intersection of GG' with the ellipsoids forms a focussing zone grating with foci at A and B in the wavelength λ. A perfectly stigmatic focus B, for light on grating at A is obtained only on the wavelength λ, but neighboring wavelengths come to approximate foci in the neighborhood of B. The stigmatism between A and B holds only in the plane of the figure. However, we shall later show a particular case where true stigmatism can be achieved with a spherical grating of any groove length.

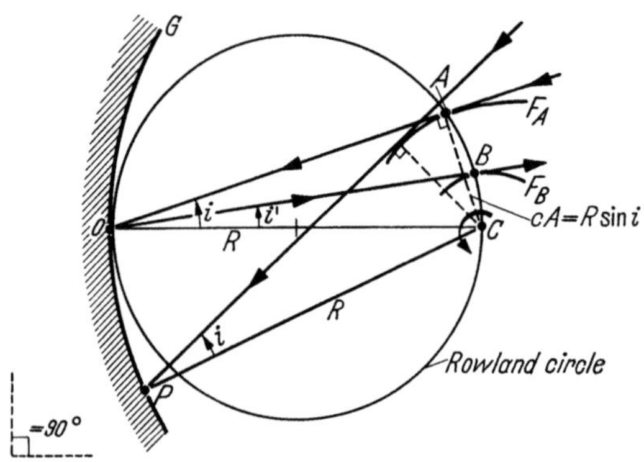

Fig. 27. Concave-grating Rowland circle computation (according to MARÉCHAL).

2. *Focussing property and grating equation for concave grating*[1]. We shall show that a concave grating has a focussing property comparable to that shown in the preceeding section for a grating of any shape. However, the focussing property is now only approximate and no perfect stigmatism is achieved in any focus. On the other hand, the foci obtained are of comparable quality for all permissible wavelengths. We shall further show that the general grating equation

$$\sin i+\sin i'=m\lambda/a \qquad (25.2)$$

also holds for the foci of a concave grating, provided that the spacing is periodic on a *chord* (parallel to a tangent through the apex O of the grating) and provided that the angle i and i' are measured at the apex (or pole).

Finally we shall find that the foci (source and spectra) of a concave grating are distributed on a circle having a diameter equal to the radius of the grating and osculating at the apex of the grating. This is the *Rowland circle*, first predicted by H. A. ROWLAND in 1883[2].

Before proceeding with the analytical search for the foci of a concave grating, let us show that a geometrical solution satisfying all these conditions is immediately obvious.

In Fig. 27, let C be the center of curvature of the concave grating G, let A be the source and let B be an image focus on ROWLAND's circle. Let $AOC=i$ and $\widehat{BOC}=i'$. If we now swing the entire right -angle triangles AOC and BOC about C, then the vertex O will describe the grating surface G, while A and B describe two curves F_A and F_B which are the envelopes of the directions of dif-

[1] See, e. g., general references, KAYSER, ROWLAND, MARÉCHAL, MEYER, loc. cit. on p. 472.
[2] loc. cit. footnote 2 on p. 472.

fraction. The intersections of F_A and F_B with ROWLAND's circle can be considered as foci. We note that F_A and F_B are portions of circles with radii $R \sin i$ and $R \sin i'$ respectively. In terms of geometrical optics, one is led to the conclusion that, within terms of second order, B is a focus for light originating from A and satisfying the grating equation in the plane of the figure. It remains to be shown that the spacing constant a of a concave grating needs to be uniform on a chord GG' (Fig. 28). Consider the geometry of Fig. 28. The tangential projection of a at O is equal to a, and at P it is rigorously equal to

$$a_i' = \frac{a}{\cos \varphi_i} \tag{25.3}$$

where

$$\varphi_i = \widehat{OCP}. \tag{25.4}$$

In order that Eq. (25.2)

$$a \sin i + a \sin i' = m\lambda \tag{25.2}$$

be equally applicable at O and P, we should have

$$a_i \sin i = a \sin i_0 \tag{25.5}$$

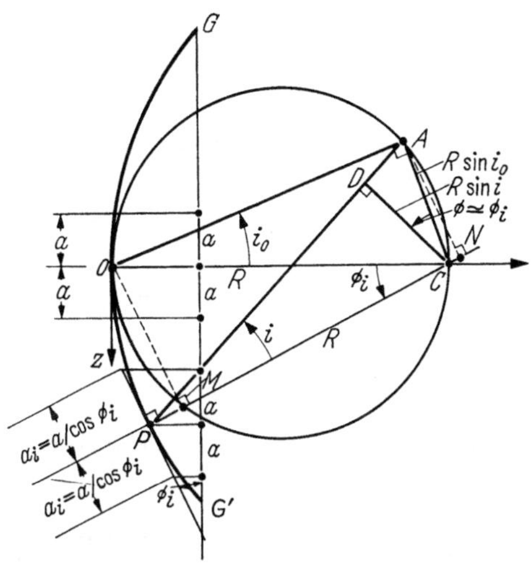

Fig. 28. Concave grating computation parameters.

where a_i is the ideal grating constant measured along the grating surface. Let us examine whether the actual grating constant a_i' permits to satisfy Eq. (25.5). We should have

$$a_i' \sin i \overset{?}{=} a \sin i_0,$$

or in consequence of (25.3):

$$\sin i \overset{?}{=} \sin i_0 \cos \varphi_i. \tag{25.6}$$

We must examine whether (25.6) is true. In the triangle CAD, the angle φ is very nearly equal to the angle φ_i, $\varphi \simeq \varphi_i$ and we have, to this approximation,

$$CD = R \sin i = CA \cos \varphi = R \sin i_0 \cos \varphi \simeq R \sin i_0 \cos \varphi_i,$$

and therefore we do indeed have

$$\sin i \simeq \sin i_0 \cos \varphi_i. \tag{25.7}$$

The degree of approximation in $\varphi \simeq \varphi_i$ will later permit to determine the aberration of the concave grating.

We now proceed with a more formal proof of the image formation by concave gratings. It should be clear from the preceeding discussion that no perfect stigmatism is achievable in concave gratings.

Let us consider the geometry of Fig. 29. To start with, we consider the principal section of the grating passing through A, B, O and the center of curvature C.

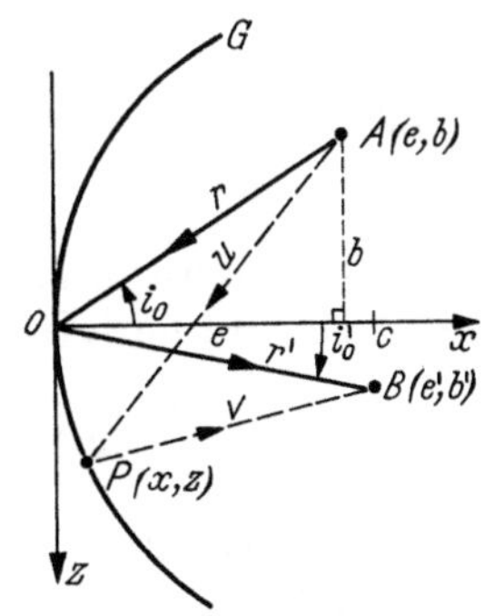

Fig. 29. Concave grating computation parameters.

Let A be the source and B the image as before, and let the coordinates of A be e, b, those of B: e', b'. Let $AO = r$, $AP = u$, $OB = r'$, $PB = v$. If P is on the n-th Fresnel zone and if B is the focus of the m-th order, we must have

$$u + v = r + r' \pm nm\lambda. \tag{25.8}$$

Let us assume that the spacing in the direction Z is periodic with a period a. Then the ordinate z of P is (Fig. 30)

$$z = n a \tag{25.9}$$

and (25.8) becomes

$$u + v = r + r' \pm m \frac{\lambda z}{a}. \tag{25.10}$$

If (25.10) could be satisfied for all grooves on the grating then B would be a perfectly stigmatic image of A.

We now examine the geometrical conditions implied by (25.10). We shall find that (25.10) indeed implies the grating Eq. (25.2), provided that the grating width in the z direction does not exceed a certain value which will be determined.

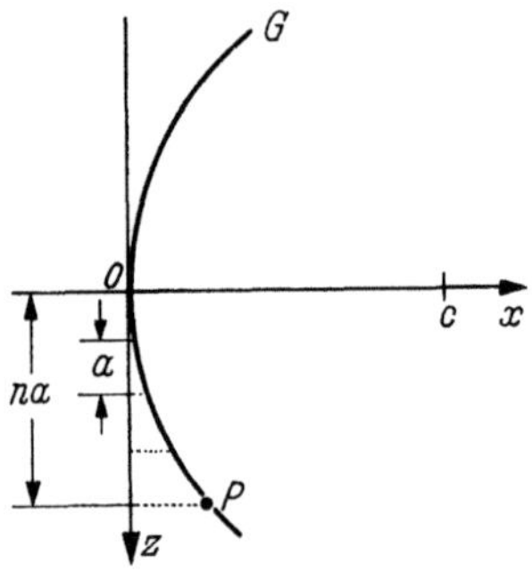

Fig. 30. Concave grating computation parameters.

In the xz plane, we have

$$u^2 = (z - b)^2 + (x - e)^2. \tag{25.11}$$

Expanding, and observing that $e^2 + b^2 = r^2$, we have

$$\left.\begin{aligned} u^2 &= z^2 + b^2 - 2zb + x^2 + e^2 - 2xe \\ &= r^2 + x^2 + z^2 - 2bz - 2ex. \end{aligned}\right\} \tag{25.12}$$

The equation of the grating in the xz plane is

$$2Rx = x^2 + z^2 \tag{25.13}$$

where $R = OC$ is the radius of grating. Eqs. (25.12) and (25.13) give

$$u^2 = \left(r - \frac{b z}{r}\right)^2 + \left(\frac{e}{r^2} - \frac{1}{R}\right) e z^2 + \left(1 - \frac{e}{R}\right) x^2. \tag{25.14}$$

For a large radius of curvature, x^2 is small and we shall approximate u^2 by neglecting the third term in (25.14). By expanding u^2 in the form

$$u^2 = r^2 \left(1 + \frac{b^2 z^2}{r^4} - \frac{2bz}{r^2} + \frac{e^2 z^2}{r^4} - \frac{e z^2}{R r^2}\right) \tag{25.15}$$

and by expanding the square root up to and including terms of the second order, we obtain as a further approximation

$$u \simeq \left(r - \frac{b z}{r}\right) + \frac{1}{2r}\left(\frac{e}{r^2} - \frac{1}{R}\right) e z^2. \tag{25.16}$$

Similarly

$$v \simeq \left(r' - \frac{b_1 z}{r_1}\right) + \frac{1}{2r}\left(\frac{e'}{r'^2} - \frac{1}{R}\right) e' z^2. \tag{25.17}$$

Eq. (25.10) must be satisfied by (25.16) and (25.17) to the order of approximation involved. We must have:

$$\left(-\frac{b}{r} - \frac{b'}{r'} \pm m \frac{\lambda}{a}\right) z + \left[\frac{e}{2r}\left(\frac{e}{r^2} - \frac{1}{R}\right) + \frac{e'}{2r'}\left(\frac{e'}{r'^2} - \frac{1}{R}\right)\right] z^2 = 0. \tag{25.18}$$

In order that (25.18) be satisfied for all z, the coefficients of z and z^2 must be independently equal to 0. Therefore, we obtain

$$\frac{b}{r} + \frac{b_1}{r_1} = \pm m \frac{\lambda}{a}, \tag{25.19}$$

$$\frac{e}{r}\left(\frac{e}{r^2} - \frac{1}{R}\right) + \frac{e'}{r'}\left(\frac{e'}{r'^2} - \frac{1}{R}\right) = 0 \tag{25.20}$$

as the conditions of approximate stigmatism between A and B. Eq. (25.19) determines the angles i_0 and i_0', namely the angles of incidence and of diffraction at O. We have

$$b = r \sin i_0, \quad b' = r' \sin i_0' \tag{25.21}$$

and (25.19) becomes the grating equation

$$\sin i_0 + \sin i_0' = m\lambda/a \tag{25.22}$$

which determines the directions of incidence and diffraction for the source A and the foci B as seen from O.

We finally want to determine explicit values for the positions r and r' of the source and foci respectively. This can be achieved with the help of (25.20).

In Fig. 29 we have

$$e = r \cos i_0, \quad e' = r' \cos i_0' \tag{25.23}$$

and (25.20) immediately gives

$$\frac{\cos^2 i_0}{r} + \frac{\cos^2 i_0'}{r'} = (\cos i_0 + \cos i_0') \frac{1}{R} \tag{25.24}$$

as the general equation relating the distances r and r' of the foci to O.

More generally, by requiring the invariance of the light path function

$$F = AP + PB + \frac{m\lambda}{a} z = u + v + \frac{m\lambda}{a} z \tag{25.25}$$

with respect to z, that is by writing

$$\frac{\partial F}{\partial z} = 0, \tag{25.26}$$

one finds (ROWLAND, BEUTLER, NAMIOKA) that we must have

$$r = R \cos i_0 \quad \text{and} \quad r' = R \cos i_0' \tag{25.27}$$

or

$$r = \infty \quad \text{and} \quad r' = \frac{R \cos^2 i_0'}{\cos i_0 + \cos i_0'}. \tag{25.28}$$

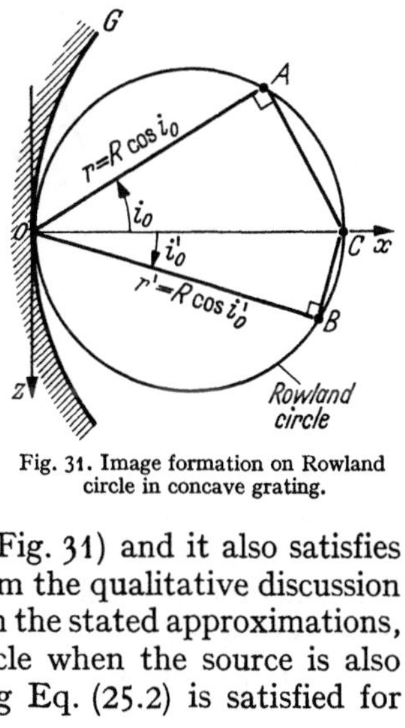

Fig. 31. Image formation on Rowland circle in concave grating.

Eq. (25.27) is clearly satisfied on ROWLAND's circle (Fig. 31) and it also satisfies the focussing Eq. (25.24). This was to be expected from the qualitative discussion preceeding the more formal proof. We thus find, within the stated approximations, that the best foci are obtained on the Rowland circle when the source is also placed on that circle. Moreover, the general grating Eq. (25.2) is satisfied for concave gratings.

Eq. (25.28) ,which also minimizes Eq. (25.25) shows that a good focus, and good spectra, will also be obtained when the source is removed to infinity, that is, when the concave grating is illuminated in collimated light (obtained from an auxiliary collimator).

In the case of collimated illumination, the foci B are not any more on the Rowland circle. A particularly important arrangement of this type is the *Wadsworth* mount in which the spectra B are formed on the normal OC to the grating. Clearly then $i_0' = 0$ and (25.28) gives

$$r' = \frac{R}{1 + \cos i_0}. \tag{25.29}$$

where $r' = OB$ (Fig. 32). As it happens, the Wadsworth mount of a concave grating has the singular property of presenting no astigmatism as it would be caused by the finite groove length in all other concave grating mounts.

We shall discuss the *astigmatism* of concave gratings in Subsect. 4 (p. 480). But first we need to show that the *aberration* in the principal plane of concave gratings is generally tolerably small with gratings of rulable dimensions.

3. Wavefront aberration in principal plane of concave gratings. When the grooves are regularly spaced on the *chord* of a concave grating, the aberrations in the principal plane are found to be negligibly small as compared to the astigmatism resulting from finite groove length. In fact, for gratings of rulable dimensions, as determined by diamond mechanisms and blaze, the aberrations is generally found to be on the order of only a small fraction of a wavelength in the visible, and frequently within the Rayleigh $^1/_4\,\lambda$ limit. This is the case of gratings used at angles about the normal, even when they attain 45° or more. Considerably larger aberration exist when gratings are used near grazing incidence in the vacuum UV.

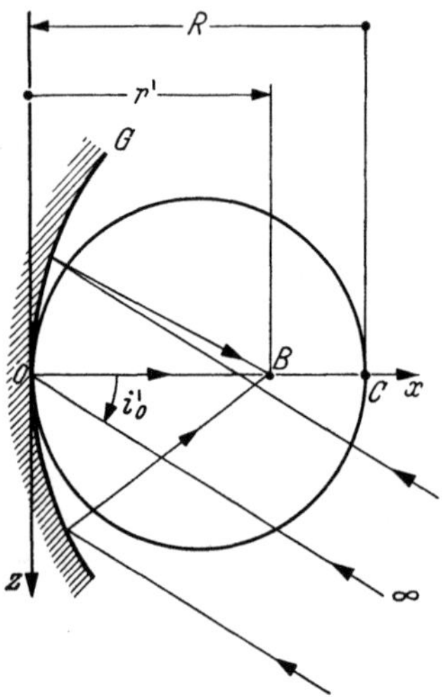

Fig. 32. "Wadsworth" image-formation with concave grating.

Consider Fig. 28. The ideal spacing constant a_i, measured along the grating surface, as a function of i should be

$$a_i \sin i = a \sin i_0. \tag{25.30}$$

But we found the actual grating constant is a_i' where

$$a_i' = \frac{a}{\cos\varphi_i}. \tag{25.31}$$

We need to determine

$$\Delta a = a_i - a_i' \tag{25.32}$$

as a function of φ_i and find the total spacing error by integration.

We shall want to determine the grating width for which the aberration Δa exceeds some specified tolerance, for example the Rayleigh limit. We shall find that the permissible width is very large indeed and of the order of the maximum width than can be ruled under usual conditions.

In Fig. 28, consider the auxiliary construction obtained by extending PC, dropping the perpendiculars AN and OM onto PC and by joining MA (not joined in Fig. 28).

We have (following MEYER[1], CANDLER[2])

$$\frac{\sin i_0}{\sin i} = \frac{AN/AM}{AN/AP} = \frac{AP}{AM}. \tag{25.33}$$

On ROWLAND's circle, of radius $R/2$, AM is a chord sustending an angle $(\pi/2 - i_0 + \varphi_i)$ at C. Therefore

$$AM = R\cos(i_0 - \varphi_i). \tag{25.34}$$

Since the angle PAM is very small, we have

$$AP - AM \simeq PM\cos i_0 \tag{25.35}$$

and since $OMC = \pi/2$,

$$PM = R(1 - \cos\varphi_i) \tag{25.36}$$

and

$$AP = PM\cos i_0 + AM. \tag{25.37}$$

[1] loc. cit. footnote 3 on p. 472.
[2] loc. cit. footnote 3 on p. 472.

Finally

$$\left.\begin{aligned}\frac{a_i}{a} &= \frac{\sin i_0}{\sin i} = \frac{AP}{AM} = \frac{\cos(i_0-\varphi_i)+(1-\cos\varphi_i)\cos i_0}{\cos(i_0-\varphi_i)}\\ &=1+(1-\cos\varphi_i)\frac{\cos i_0}{\cos(i_0-\varphi_i)}\,.\end{aligned}\right\} \tag{25.38}$$

For small values of φ_i

$$\cos\varphi_i \cong 1-\varphi_i^2/2\,, \qquad \sin\varphi_i \cong \varphi_i \tag{25.39}$$

and (25.38) becomes

$$\left.\begin{aligned}\frac{a_i}{a} &\cong 1+\frac{1}{2}\varphi_i^2\frac{\cos i_0}{\cos i_0-\varphi_i\sin i_0}\,,\\ \frac{a_i}{a} &\cong 1+\frac{1}{2}\varphi_i^2(1+\varphi_i\tan i_0)\,.\end{aligned}\right\} \tag{25.40}$$

that is

Therefore, since

$$a_i' = a/\cos\varphi_i \cong a(1+\varphi_i^2)\,, \tag{25.41}$$

we have

$$\Delta a = a_i - a_i' \cong \tfrac{1}{2} a\varphi_i^3\tan i_0 \tag{25.42}$$

which is the expression first obtained by MEYER (1934).

In order to find the total groove displacement at the edge of the ruled width to which corresponds an angle $(\varphi_i)_{\max}$ we need to integrate (25.42).

We observe that one groove a has a width $R\,d\varphi_i$ and therefore

$$\overline{\Delta a} = \tfrac{1}{2}\tan i_0 \int_0^{(\varphi_i)_{\max}} \varphi_i^3 R\,d\varphi_i \tag{25.43}$$

giving

$$\boxed{\overline{\Delta a} = \tfrac{1}{8} R\tan i_0\,(\varphi_i)_{\max}^4} \tag{25.44}$$

or, since

$$\sin\varphi_i = \frac{z}{R} \cong \varphi_i\,, \tag{25.45}$$

we can also write (25.44) as

$$\boxed{\overline{\Delta a} = \frac{1}{8}\tan i_0\frac{z_{\max}^4}{R^3}\,,} \tag{25.46}$$

In autocollimation

$$2\sin i_0 = m\lambda/a \tag{25.47}$$

and a displacement of one spacing a along the grating surface produces a wave-front aberration of $2a\sin i_0 = m\lambda$. For the aberration to be smaller than λ/p, the groove displacement must be smaller than $a/pm = \lambda/2p\sin i_0$. We must therefore have

$$\frac{1}{8}\tan i_0\frac{z_{\max}^4}{R^3} \leqq \frac{a}{pm} = \frac{\lambda}{2p\sin i_0} \tag{25.48}$$

that is

$$\boxed{z_{\max}^4 \leqq \frac{4R^3}{\sin^2 i_0/\cos i_0}\,\frac{\lambda}{p}\,.} \tag{25.49}$$

For example, let $p=4$, so that $\lambda/p=\lambda/4$ (Rayleigh tolerance) at $\lambda=5000$ Å $=5\times10^{-4}$ mm and consider a $R=6$ meter grating used at $i_0=36°$. We find $2z_{\max}=250$ mm.

More generally, MACH, STERN and EDLÉN (1932) find that the aberration can be written as

$$\overline{\Delta a} = \frac{z^4}{8R^3}(\sin i_0\tan i_0 + \sin i_0'\tan i)\,. \tag{25.50}$$

It is found that as long as the width of a concave grating is less than a value W_{opt} expressed as

$$W_{opt} = 3.55 \left[\frac{2.38\,\lambda R^3}{\sin i_0 \tan i_0 + \sin i_0' \tan i_0'}\right] \tag{25.51}$$

the resolving power (for an intensity peak saddle of $8/\pi^2$) is

$$R = \frac{W_{eff}\,(\sin i + \sin i')}{\lambda} \tag{25.52}$$

where

$$W_{eff} = 0.92\, W_{opt}\,. \tag{25.53}$$

For $W < W_{opt}$, W_{eff} at first decreases and then oscillates about a value $W_{eff} = 0.75 \times W_{opt}$ (Fig. 33). Unfortunately these results apply only to the principal plane of the grating. When light is incident on a grating with finite groove length, the

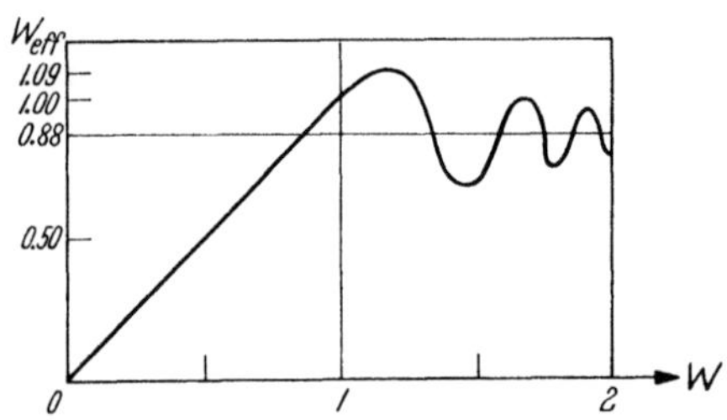

Fig. 33. Astigmatism of concave gratings: dependence of resolving-power parameter "W_{eff}" on grating width W (according to Mach, Stern and Edlén, 1932).

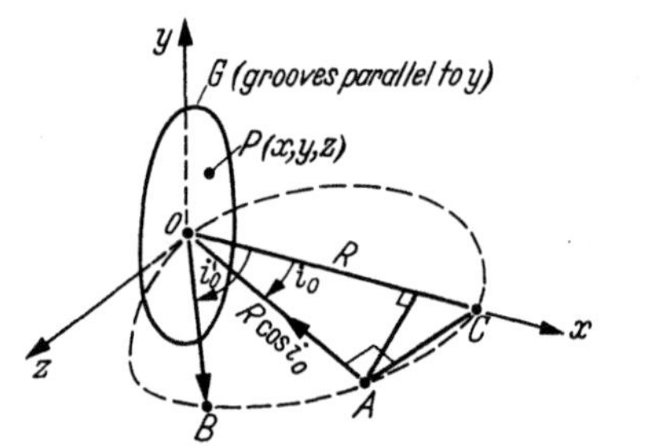

Fig. 34. Astigmatism of concave gratings: computation of effect of finite groove length.

image quality is limited by astigmatism rather than by the small aberration. The loss of resolving power due to astigmatism is not easy to express simply, but can generally be interpreted in terms of geometrical optics (Maréchal, Beutler, Namioka).

4. *Astigmatism in concave gratings, resulting from finite groove lengths.* We need to evaluate the variation of the light path function $AP + PB$ as a function of z and y (Fig. 34). The grating Eq. (25.2) requires that

$$\sin i_0 + \sin i_0' = m\lambda/a\,, \tag{25.54}$$

i.e. that

$$z(\sin i_0 + \sin i_0') = m\lambda \quad \text{if} \quad z = a\,. \tag{25.55}$$

The coordinates of A are

$$x_A = R\cos i_0, \qquad y_A = 0, \qquad z_A = R\cos i_0 \sin i_0, \tag{25.56}$$

and similar coordinates exist for B. The equation of the spherical grating is

$$x^2 + y^2 + z^2 - 2Rx = 0 \tag{25.57}$$

where x, y, z are the coordinates of P on the grating surface. We have

$$\left.\begin{aligned} \overline{AP}^2 &= (R\cos^2 i_0 - x)^2 + y^2 + (R\cos i_0 \sin i_0 - z)^2 \\ &= R^2\cos^2 i_0 \left[1 + \frac{2x}{R\cos^2 i_0} - \frac{2x}{R} - \frac{2z\sin i_0}{R\cos i_0}\right]. \end{aligned}\right\} \tag{25.58}$$

By expanding the square root, including terms of second order, we have

$$AP \simeq R\cos i_0 - z\sin i_0 + x\,\frac{\sin^2 i_0}{\cos i}\,. \tag{25.59}$$

With

$$x \simeq \frac{y^2+z^2}{2R} \tag{25.60}$$

we get

$$A\,P \simeq R\cos i_0 - z\sin i_0 + \left(\frac{y^2+z^2}{2R}\right)\frac{\sin^2 i_0}{\cos i_0} \tag{25.61}$$

and finally

$$\boxed{\begin{aligned} L = A\,P + P\,B \simeq R(\cos i_0 + \cos i_0') - z(\sin i_0 + \sin i_0') + \\ + \left(\frac{y^2+z^2}{2R}\right)\left(\frac{\sin^2 i_0}{\cos i_0} + \frac{\sin^2 i_0'}{\cos i_0'}\right). \end{aligned}} \tag{25.62}$$

Eq. (25.62) is particularly suitable for evaluating the astigmatism. We first note that in the vicinity of $z=0$, $\Delta L = m\lambda$ when $\Delta z = a$, except for the term

$$\Delta_{\text{Astigmatism}} = \frac{y^2}{2R}\left(\frac{\sin^2 i_0}{\cos i_0} + \frac{\sin^2 i_0'}{\cos i_0'}\right) \tag{25.63}$$

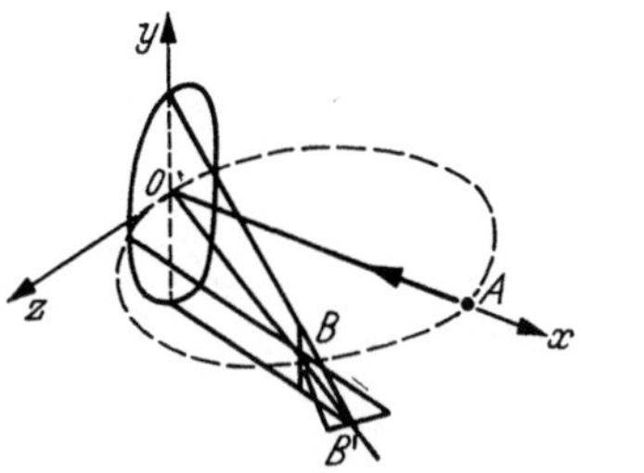

Fig. 35. Astigmatism of concave gratings: effect of finite groove length.

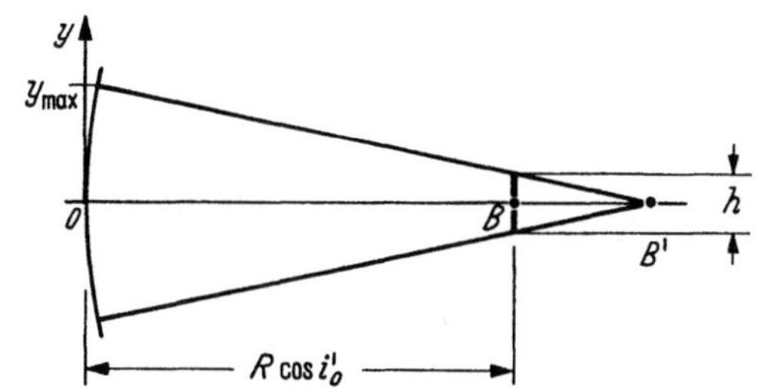

Fig. 36. Astigmatism of concave gratings: computation of effect of groove length.

which is a measure of the astigmatism of the wave front. We immediately note that the concave grating (used on ROWLAND's circle where the aberration in the principal plane was shown to be minimum) always presents an irreducible astigmatism.

We next use (25.63) to evaluate the distance between the two foci of the astigmatic wave front, and to calculate the length of the image element which results from the spreading out of the light from a point source (Figs. 35 and 36).

The reference sphere is centered at B, where

$$O\,B = R\cos i_0'. \tag{25.64}$$

This is the center of the *vertical* focus. There will be a *horizontal* focus at some distance $O\,B'$, such that generally $O\,B' > O\,B$ as appears from (25.62). From geometrical optics theory (MARÉCHAL) we have

$$\Delta_{\text{Astigmatism}} = \frac{y^2}{2}\left(\frac{1}{O\,B'} - \frac{1}{O\,B}\right) \tag{25.65}$$

where $O\,B' = s$ is the sagittal focal distance and $O\,B = t$ is the tangential focal distance. One finds

$$O\,B' = R\cos i_0'\left(\frac{1}{\cos^2 i_0' - \sin i_0 \tan i_0 \cos i_0'}\right) \tag{25.66}$$

and

$$O\,B' - O\,B = R\,\frac{\sin^2 i_0' + \sin^2 i_0 \cos i_0'}{\cos i_0 \cos i_0' - \sin^2 i_0}. \tag{25.67}$$

Clearly (Fig. 36) the height h of the tangential focal element is

$$\frac{h}{y_{\max}} = \frac{OB' - OB}{OB'} = 1 - \frac{OB}{OB'} \tag{25.68}$$

or

$$\frac{h}{y_{\max}} = \sin^2 i_0' + \frac{\sin^2 i_0 \cos i_0'}{\cos i_0}. \tag{25.69}$$

It is remarkable that the astigmatic spreading of a point into a line element is independent of the radius of the grating. For example, for the 6 meter radius grating above, for which we found a permissible width $2z = 250$ mm, let us consider the astigmatism for a groove length $2\,y_{\max} = 100$ mm. We get in autocollimation

$$h = 50 \times 2 \sin^2 i_0 = 100 \sin^2 i_0 = 32\ \text{mm}.$$

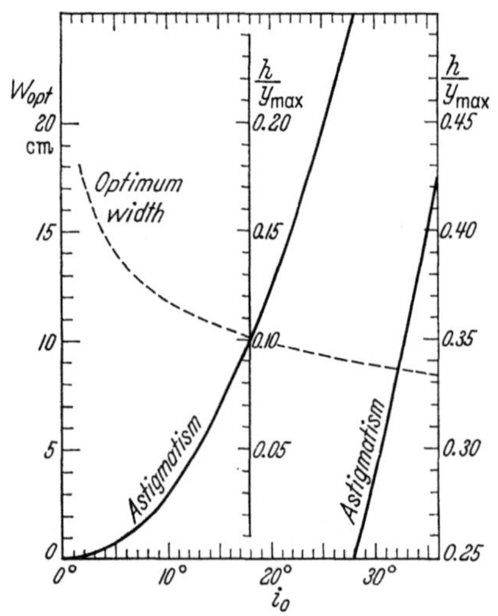

Fig. 37.

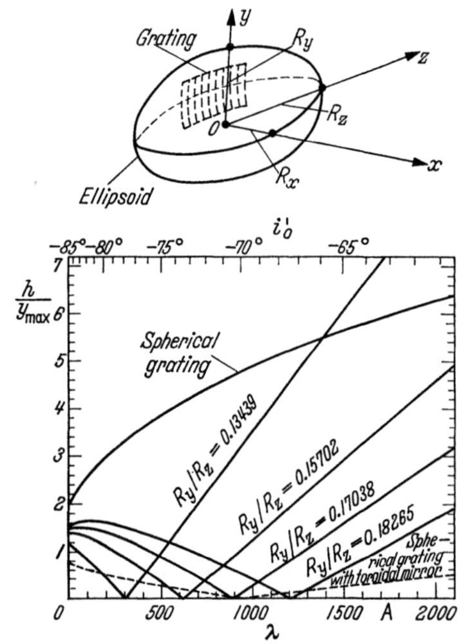

Fig. 38.

Fig. 37. Optimum grating width, W_{opt}, and length of astigmatic image for a point light source per unit length of grating ruling, $h/y_{\max}$, in the Rowland mounting of a concave grating (according to T. NAMIOKA, in *Space Astrophysics*, edited by WM. LILLER, McGraw Hill Book Co. 1961). A 1 m grating with 1200 grooves/mm and the first order are assumed. i_0 is the angle of incidence.

Fig. 38. Astigmatism per unit length of the grating rulings in a concave grating for a point light source, $h/y_{\max}$, in the grazing incidence mount ($i_0 = 85°$). A grating constant of 1/600 mm is assumed. In the figure the amount of astigmatism is plotted for a spherical concave grating, ellipsoidal gratings with various R_y/R_z values, as noted, and a combination of a spherical concave grating and a toroidal mirror (after T. NAMIOKA, in *Space Astrophysics*, edited by WM. LILLER, McGraw Hill Book Co., 1961).

The astigmatism is very considerable indeed. In *photographic* systems an appreciable light intensity loss clearly results from the astigmatism. The light loss exceeds the gain resulting from the absence of collimating and focussing optics. In systems with *photoelectric detection,* all the flux may still be collected and the astigmatism is tolerable from this standpoint.

Fig. 37 shows (after NAMIOKA[1]) the optimum grating width and the length of the astigmatic image for a point light source for unit length of grating ruling, $2h/2y_{\max}$ in the Rowland mounting. A 1-meter radius grating with 1200 grooves/mm and the first order are assumed. i_0 is the angle of incidence. Fig. 38 and 39 show comparable curves for gratings in grazing incidence.

[1] loc. cit. in footnote 3 on p. 472.

Finally, we prove that *no astigmatism* exists in the Wadsworth mounting of the grating, in which the spectra are formed on the normal and the incident beams are plane collimated wave fronts, with direction cosines α and β.

Referring to Fig. 34, the equation of the incident wave front is

$$\alpha x+\gamma z=0 \tag{25.70}$$

and the coordinates of the image B are $\varrho\alpha$, 0, $\varrho\gamma'$. We need to determine the path length variations from a point (M) in the incident wave front to the point B. Following MARÉCHAL[1], we have

$$\left.\begin{aligned} MPB = &-(\alpha x+\gamma z)+\\ &+[(\alpha'\varrho-x)^2+\gamma^2+\\ &+(\gamma'\varrho-z)^2]^{\frac{1}{2}}.\end{aligned}\right\} \tag{25.71}$$

We obtain, by expansion

$$\left.\begin{aligned} MPB\simeq &-\varrho+\\ &+(\gamma+\gamma')z-(\alpha+\alpha')x+\\ &+\frac{y^2+z^2(1-\gamma'^2)}{2\varrho}.\end{aligned}\right\} \tag{25.72}$$

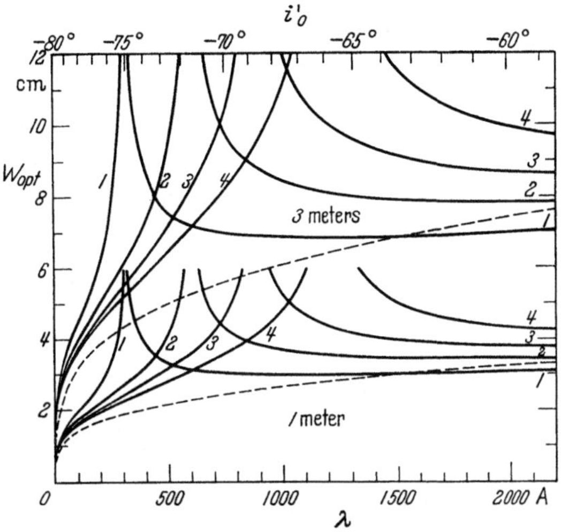

Fig. 39. Optimum grating width, $W_{\rm opt}$, for a concave grating, in the grazing incidence mounting ($i_0=80°$). The upper solid curves show $W_{\rm opt}$ of ellipsoidal concave gratings with $R_z^2/R_y=1$ m. The lower solid curves are for ellipsoidal concave gratings of $R_y/R_z=0.21254$, 0.23435, 0.25089, and 0.26643 respectively. The upper and lower dashed curves are for spherical concave gratings with $R=3$ m and $R=1$ m, respectively, where R is the radius of curvature of the grating. For ellipsoidal concave gratings, $R_x=R_y\neq R_z$, where R_x, R_y and R_z are the lengths of the semi-axes of the ellipsoid (according to T. NAMIOKA, in *Space Astrophysics*, edited by WM. LILLER, McGraw Hill Book Co., 1961).

The grating Eq. (25.2) can be written here in the form

$$(\gamma+\gamma')a=m\lambda. \tag{25.73}$$

If (25.73) is satisfied, the diffracted wave will be spherical if

$$(\alpha+\alpha')\frac{y^2+z^2}{2R}=\frac{y^2+z^2(1-\gamma'^2)}{2\varrho} \tag{25.74}$$

which gives the condition

$$\left.\begin{aligned} &\left.\begin{aligned}\gamma'&=0\\ \alpha'&=0\end{aligned}\right\}\text{: diffraction on normal,}\\ &\frac{1}{\varrho}=\frac{1+\alpha}{R}.\end{aligned}\right\} \tag{25.75}$$

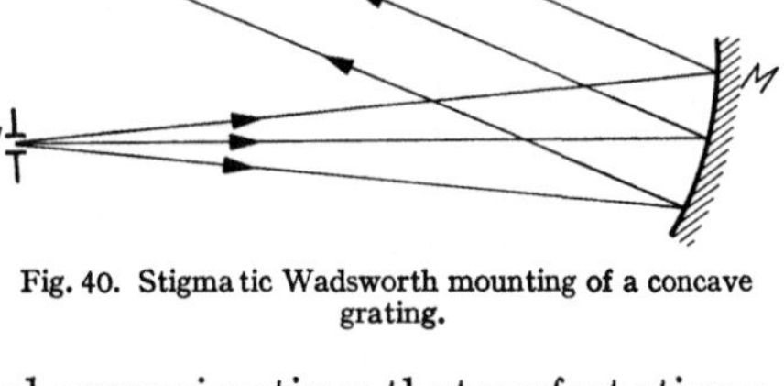

Fig. 40. Stigmatic Wadsworth mounting of a concave grating.

One concludes from (25.75) within the stated approximations that perfect stigmatism is achieved on the normal of a concave grating illuminated in collimated light. The focal distance $\varrho=r'$ obtained in (25.75) is identical to that obtained in (25.29), and we have

$$\varrho=r'=\frac{R}{1+\cos i_0}. \tag{25.76}$$

Fig. 40 shows the stigmatic Wadsworth arrangement of a concave grating G. For photoelectric work, G is rotated about O. The focal curve F_B is a parabola of Eq. (25.76) and approximate stigmatism is achieved on F_B in the vicinity of B.

[1] A. MARÉCHAL, loc. cit. in footnote 3 on p. 472.

Table 1. *Summary of some characteristics of the grating mountings.*

A 1 m, 1200 grooves/mm grating is assumed for a mounting except for the grazing incidence (2 m, 600 grooves/mm), Wadsworth (2 m, 1200 grooves/mm) and Ebert-Fastie mountings (plane). In the table the following abbreviations are used. *G* grating; *S* entrance slit; *S'* exit slit; *O* center of the Rowland circle; *P* vertical axis through the grating center; and *C*, vertical axis displaced by some distance from the grating center.

	Rowland	Abney	Wadsworth	Normal incidence	In plane Eagle	Off plane Eagle
Wavelength coverage with 1 A resolution	500∼8000 A	500∼8000 A	1000∼8000 A	500∼8000 A	500∼8000 A	Not suitable to gratings whose radius of curvature is shorter than 3 meters
Scanning mechanism	*G* & *S'* move along rails	*S* rotates about *O*	*G* rotates about *P*; *S'* moves along parabola	*S'* moves along Rowland circle	*G* moves along rail & rotates. *S'* rotates about *S*	
Fixed optical elements	Entrance slit	Exit slit, grating	Entrance slit, mirror	Entrance slit, grating	Entrance slit	
Complexity to achieve monochromator action	Medium	Simple	Complex	Simple	Medium	
Angle of incidence, α	Varies	Varies	Varies	Constant, near 0°	$\alpha \cong \beta_0$, varies	
Angle of diffraction, β_0	0°	0°	0°	Varies	$\beta_0 \cong \alpha$, varies	
Direction of diffracted rays	Varies	Constant	Varies	Varies	Varies	
Direction of line connecting entrance slit to grating center	Constant	Varies	Constant	Constant	Constant	
Astigmatism	Small	Small	Zero	Very small for $m=1$, small or medium for $m=-1$	Very small	
1 number with 1 A resolution in the vacuum ultraviolet region	7∼9	7∼9	13∼16	7∼10	6∼8	
Numbers of reflecting surfaces	1	1	2	1	1	

Table 1 (continued).

	Grazing incidence	Radius	Seya-Namioka	Johnson-Onaka	Ebert-Fastie
Wavelength coverage with 1 A resolution	10 (depends on α) 2000 A	500 ~ 8000 A	500 ~ 6000 A	500 ~ 1700 A	1200 ~ 8000 A
Scanning mechanism	S' moves along Rowland circle	G moves along Rowland circle	G rotates about P	G rotates about C	G rotates about P
Fixed optical elements	Entrance slit, grating	Entrance slit, exit slit	Entrance slit, exit slit	Entrance slit, exit slit	Entrance slit, exit slit, mirror
Complexity to achieve monochromator action	Medium	Simple	Very simple	Simple	Very simple
Angle of incidence, α	Constant, $\geqq 80°$	Varies	$\alpha - \beta_0 \cong 70°$, varies	Varies, near normal	Varies
Angle of diffraction, β_0	Large angles, varies	Varies	$\alpha - \beta_0 \cong 70°$, varies	Small angles, varies	Varies
Direction of diffracted rays	Varies	Varies	Constant	Varies	Constant
Direction of line connecting entrance slit to grating center	Constant	Varies	Constant	Varies	Constant
Astigmatism	Very large (for spherical grating); large (for ellipsoidal grating)	Very small	Large (for spherical grating); very small (for ellipsoidal grating)	Small	Very small
1 number with 1 A resolution in the vacuum ultraviolet region	30 ~ 60	7 ~ 8	10 ~ 14	11 ~ 16	4 ~ 6
Numbers of reflecting surfaces	1	1	1	1	1

5. *Concave grating mounts used in practice.* The various concave grating mounts used most frequently are sketched in Fig. 41, and some of the characteristics (after NAMIOKA) are described in Table 1. For a more general treatment of concave gratings, see in particular BEUTLER[1], NAMIOKA[2], as well as references in SAWYER[3], HARRISON, LORD and LOOFBOUROW[4], etc.

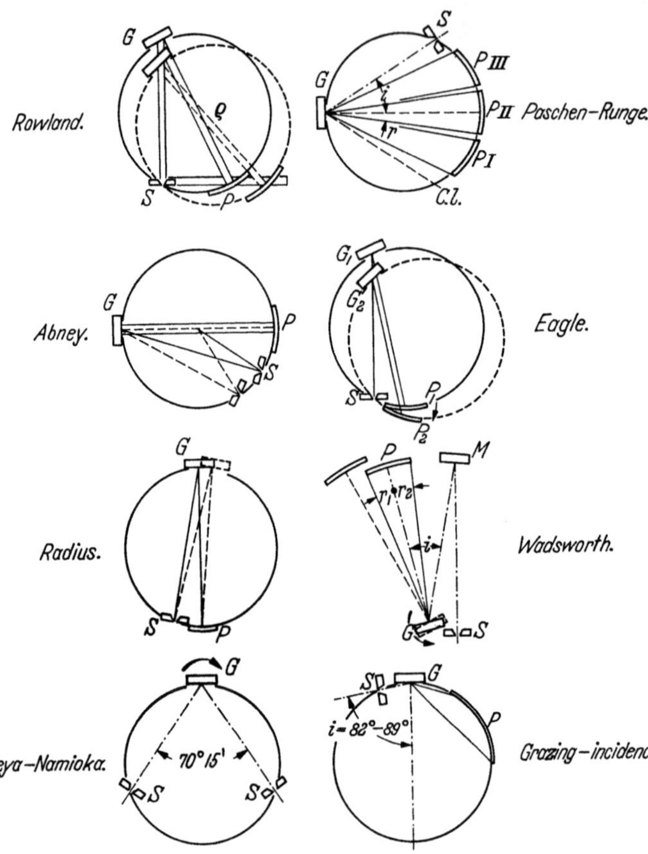

Fig. 41. Concave grating mountings (according to R. F. JARELL, in *Encyclopedia of Spectroscopy*, edited by G. L. CLARK, Reinhold Publishing Co., 1960).

26. Luminosity of gratings in spectrometers and spectrographs[5]. *1. Introduction.* A complete treatment of the luminosity of spectrometers and spectrographs generally aims at computing the flux reaching the detector in a given spectral bandwidth. Alternately one seeks to express the minimum wavelength bandpass $\Delta\lambda$, or the corresponding wavenumber difference $\Delta\nu$ which can be resolved by the spectrometer or spectrograph, in the presence of noise, granularity of the photo-

[1] H. G. BEUTLER: J. Opt. Soc. Amer. **35**, 311 (1945).

[2] loc. cit. in footnote 3 on p. 472.

[3] loc. cit. in footnote 2 on p. 454.

[4] loc. cit. in footnote 2 on p. 454.

[5] For general reference, see for example, J. STRONG: Concepts in Classical Optics. San Francisco: W. H. Freeman & Co. 1958, also P. JACQUINOT and CH. DUFOUR: J. Recherches CNRS, Lab. Bellevue (Paris) **6**, 91 (1948). — J. STRONG: J. Opt. Soc. Amer. **47**, 354 (1957). — P. JACQUINOT: J. Opt. Soc. Amer. **44**, 761 (1954). — G. W. STROKE: J. Opt. Soc. Amer. **51**, 1321 (1961). — G. W. STROKE and H. H. STROKE: J. Opt. Soc. Amer. **53**, 333 (1963); also P. H. VAN CITTERT: Z. Physik **65**, 547 (1930); **69**, 298 (1931).

graphic plates, finite slit widths (as limited also by diffraction) and threshold sensitivities of the detectors. Along with the entrance slit width, grating dimensions and focal length, the spectral dispersion $\Delta i'/\Delta\lambda$ is one of the fundamental factors which determines the luminosity and resolution.

To a large extent the theory relating to resolving power and luminosity can be given in terms of physical and geometrical optics.

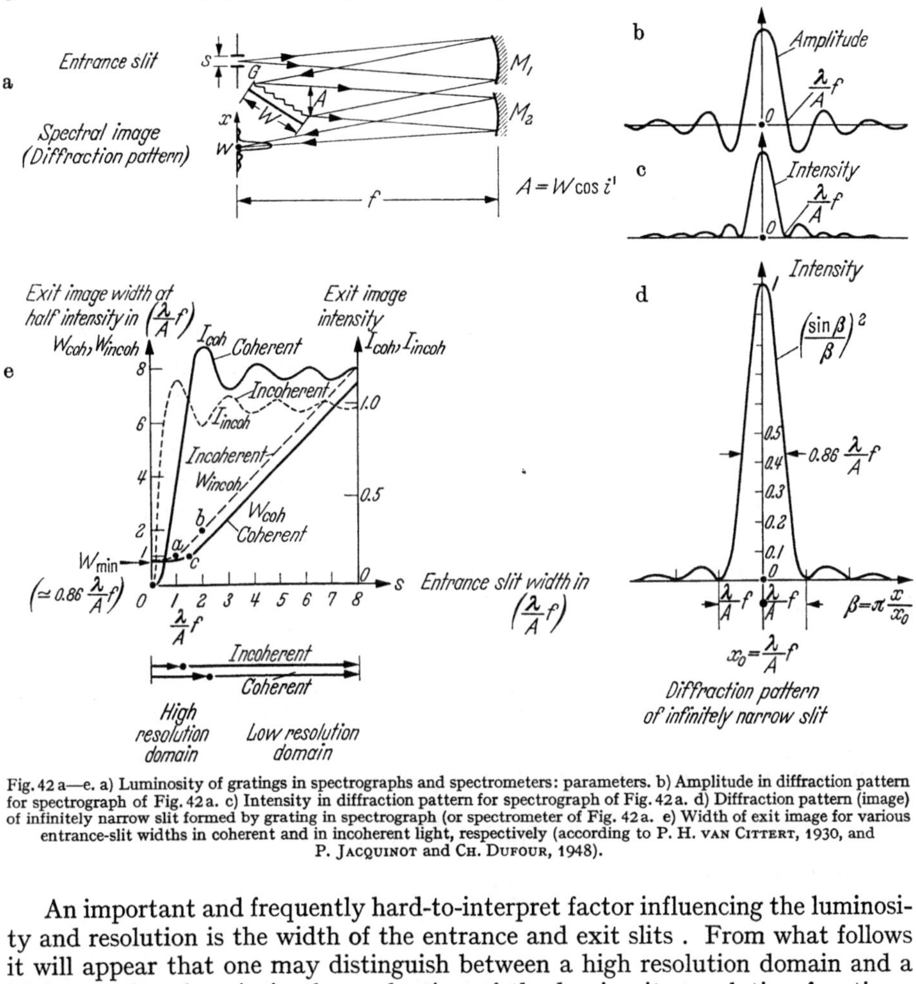

Fig. 42 a—e. a) Luminosity of gratings in spectrographs and spectrometers: parameters. b) Amplitude in diffraction pattern for spectrograph of Fig. 42 a. c) Intensity in diffraction pattern for spectrograph of Fig. 42 a. d) Diffraction pattern (image) of infinitely narrow slit formed by grating in spectrograph (or spectrometer of Fig. 42 a. e) Width of exit image for various entrance-slit widths in coherent and in incoherent light, respectively (according to P. H. van Cittert, 1930, and P. Jacquinot and Ch. Dufour, 1948).

An important and frequently hard-to-interpret factor influencing the luminosity and resolution is the width of the entrance and exit slits . From what follows it will appear that one may distinguish between a high resolution domain and a low resolution domain in the evaluation of the luminosity-resolution functions. One also needs to take into account the degree of coherence in the slit illumination, especially in the high-resolution domain.

It will become apparent that the problem of luminosity and resolution needs to separately consider the following parameters:

1. Spectrographs and spectrometers
2. Intensity (spectrographs) and flux (spectrometers)
3. Diffraction-limited spectral images (high resolution domains) and slit-broadened images (low resolution domains)
4. Limitations by imaging and limitations by detectors (granularity, slit widths)
5. Spectral energy limited sensitivity and detector limited sensitivity.

2. Effects of finite entrance slit width in high resolution domain. The finite aperture of the grating results in diffraction patterns of finite widths ($0.86\, f\lambda/A$ at half intensity) even with infinitely narrow slits (Fig. 42). It is a remarkable fact that the theoretical diffraction-pattern width is only slightly broadened when the entrance slit is opened to a finite width. Even when the entrance slit width becomes equal to the diffraction pattern width, the spectral image in incoherent light has been broadened by only 30% beyond its minimum width corresponding to an infinitely narrow slit (Fig. 42e). In fact, frequently there is not even a need for opening the slit that much in practice, in incoherent light. On the other hand, with coherent light the slit may be opened twice as much without a loss of resolution. For instance, in incoherent light the diffraction pattern is broadened by less than 10% for a slit width of about 70% of the diffraction pattern width (that is with a slit some $0.6\, f\lambda/\text{Å}$ wide), and this while attaining more than 90% of the maximum possible spectral line intensity (Fig. 43a).

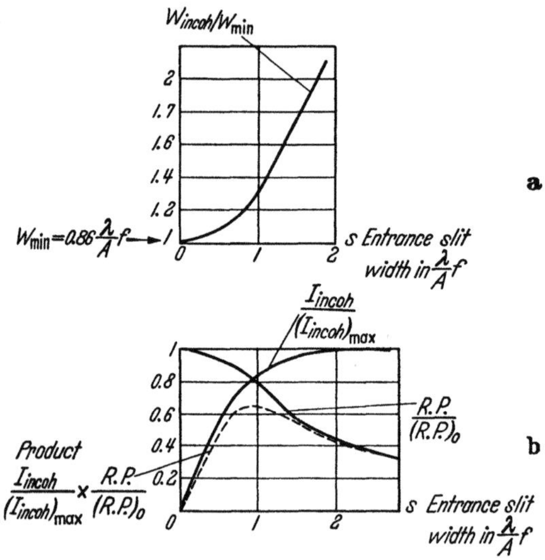

Fig. 43a and b. Effect of finite entrance slit-width in high-resolution domain (according to P. JACQUINOT and CH. DUFOUR, 1948), for the case of incoherent illumination. a) Width W_{incoh} of exit image as a function of entrance slit-width. b) Intensity and resolving-power ratios of exit image for various values of entrance slit-width.

The effect of the finite slit width can be immediately evaluated by recalling that a monochromatic (single frequency) spectral image in incoherent light is simply the convolution of the diffraction pattern with the distribution of intensity across the slit width.

We have for the diffraction pattern (Fig. 42d)

$$I(x)=\left(\frac{\sin x}{x}\right)^2 \tag{26.1}$$

and for the spectral image

$$I(x)=\int_{-s/2}^{+s/2} s(x')\, I(x-x')\, dx' \tag{26.2}$$

where

$$x=\beta\,\frac{x_0}{\pi} \tag{26.3}$$

and

$$x_0=\frac{\lambda}{A}\, f. \tag{26.3a}$$

However $s(x)=1$ for a uniformly illuminated slit. One obtains

$$I_{\text{incoh}}(x)=\int_{x'=-s/2}^{x'=+s/2} I(x'-x)\, dx' \tag{26.4}$$

and, with an appropriate change of variables

$$I_{\text{incoh}}(x)=\int_{-x-s/2}^{-x+s/2} I(y)\, dy. \tag{26.5}$$

The integration of Eq. (26.5) was first carried out for this case, as well as for the case of coherent illumination by P. H. VAN CITTERT[1] in 1930. The results are shown in Fig. 42. Similar calculations and further extensions, in particular for spectrometers using photoelectric receptors were given in 1948 by P. JACQUINOT and CH. DUFOUR[2]. Some of their results are shown in Fig. 43.

In the case of a coherently illuminated slit, the convolution needs to be carried out with the amplitude diffraction pattern

$$A(x) = \frac{\sin x}{x} \tag{26.6}$$

and the spectral image is the intensity image

$$I_{\text{coh}}(x) = A_{\text{incoh}}(x)\, A^*_{\text{incoh}}(x)$$

obtained by squaring $A(x)$. One has

$$I_{\text{coh}}(x) = \left[\int_{-x-s/2}^{-x+s/2} A(y)\, dy\right]^2. \tag{26.7}$$

One should note that the broadening of the diffraction patterns W_{coh} and W_{incoh} beyond the minimum width $W_{\min}$ (of the infinitely narrow slit) is a non-linear function in the domain of slit widths corresponding to the highest attainable resolving powers. One also notes that a comparatively much smaller broadening is obtained in this domain with coherent illumination than with incoherent illumination.

Practically the same minimum half-width of the diffraction pattern [$W_{\min} \simeq 0.86\, f\lambda/A = 0.86$ R.W. (resolution width) at half intensity] is obtained with both incoherent and coherent illumination. However, with coherent slit illumination, the half width of the spectral image grows at first only insignificantly with increasing slit opening, while the spectral image widens much more rapidly with incoherent slit illumination, as already noted by P. H. VAN CITTERT. For instance, an entrance slit of 1 R.W. will still yield a spectral image of about $1.07\, W_{\min}$ in coherent light, but $1.3\, W_{\min}$ in incoherent light. Slit widths between 0.5 to 2 R.W. are generally chosen for attaining maximum resolution in spectrometers and spectrographs. Theoretical results obtained by VAN CITTERT show that maximum intensity in incoherent illumination is attained with a slit width of 1 R.W. $= f\lambda/A$. The maximum intensity in coherent light is attained for an entrance slit of 2 R.W. $= 2f\lambda/A$, for which the line half-width has become $1.47\, W_{\min}$. One notes that almost twice as wide an entrance slit can be used in case of coherent illumination as in the case of incoherent illumination, for comparable resolution at maximum intensity. In practice therefore similar resolutions are obtained for both types of illumination, even though the appropriate slit widths are quite different. In photoelectric use of gratings, one should thus obtain twice as much flux with coherent as with incoherent illumination, at the same resolution, in this domain of slit widths between one and two Rayleigh widths.

According to JACQUINOT and DUFOUR, it is of interest to examine the product of the intensity and resolution as a function of entrance slit width, as shown in Fig. 40. For incoherent light the product of intensity and resolving power has a maximum for an entrance slit of one Rayleigh width ($s = f\lambda/A$), and both the resolving power and intensity are seen to have about 80% of their maximum value for this value of s.

---

[1] P. H. VAN CITTERT: Z. Physik **65**, 547 (1930); **69**, 298 (1931).

[2] loc. cit. in footnote 5 on p. 486.

In general, the conditions of slit illumination are not sufficiently well defined to permit a priori decision as to the optimum slit width. Experiments with spectrometers show that it is indicated to progressively reduce the slit width until the best resolution is obtained without excessive reduction of intensity. As seen from VAN CITTERT'S and JACQUINOT and DUFOUR'S curves, which are generally confirmed by experiment, practically theoretical resolution can be achieved with intensities of the order of about 70% in both incoherent and coherent light. For example, one finds that in a 1.8 meter spectrometer using a 190 mm wide grating at 64° in autocollimation, the best resolution, with still very high intensity at 4358 Å, is achieved with a 6 micron wide slit width ($f\lambda/A \simeq 10.5$ microns). One can infer that the illumination is incoherent in this case, which would be expected with the 2-inch long curved slits used in that instrument. On the other hand, in a 3.4 meter Ebert spectrograph using a 6 inch wide grating at low angle, for which $f\lambda/A \simeq 7.3$ micron at about 3200 Å the best slit width appears to be about 15 microns according to R. K. BREHM, which would tend to indicate "coherent" illumination. In the case of spectrographs, G. R. HARRISON, LORD and LOOFBOUROW[1] recommend that it is best to calculate the approximate slit width and start by setting the slit at some value less than this. Then by gradually opening the slit width and by observing the illumination behind the slit on a white card, it is found that a considerable increase in brightness is observed at the width of the slit which is optimum with the particular illumination used.

3. *Effects of finite entrance slit width in low-resolution domain.* It is seen (Fig. 42e) that a widening of the entrance slit in the low resolution domain does not increase the intensity in the spectral lines beyond its maximum, but it merely increases their width, and therefore will tend to lower the resolution linearly with the width of s. It should be clear that most of the preceding conclusions will be additionally modified when gratings are used in spectrometers.

4. *Effects of exit slit in photoelectric use of spectrometers.* In the use of a scanning exit slit of width s' a further convolution is involved between the spectral line and this slit. In practice, one frequently has $s'=s$, where s = width of the entrance slit. The line shapes as obtained with incoherent illumination in such arrangements are shown in Fig. 44a, according to JACQUINOT and DUFOUR, for different values of $s'=s$. The flux recorded in a spectrometer can be written as

$$\Phi_{\text{incoh}}(x) = \int_{x-s/2}^{x+s/2} I_{\text{incoh}}(y)\, dy \tag{26.8}$$

in terms of Eq. (26.5). Fig. 44b, c, d, shows the width at half intensity, W'_{inc}, as well as the ratio of the flux $\Phi_{\text{incoh}}/\Phi_0$ of the flux Φ_{incoh} passed through the exit to the flux Φ_0 that would be obtained according to geometrical optics, as a function to the entrance slit width. (The exit slit is assumed to be equal to the entrance slit). According to these curves, obtained by JACQUINOT and DUFOUR, as well as from the Φ_{coh}/Φ_0 curve which we have derived with the help of VAN CITTERT'S curves of Fig. 42, one clearly sees that photoelectric work will indeed give rather high flux densities with the slits that permit attainment of the vicinity of the theoretical resolving powers. One should not attach too much importance to the percentage of flux compared to the geometrical optics flux, since the latter represents an arbitrary value which has to be evaluated with respect to the source intensities. The criterion for evaluating either a spectrometer or a spectrograph should be *adequate* flux or intensity in the use of a particular detector, source and instrument, rather than comparative efficiencies.

[1] loc. cit. in footnote 2 on p. 454.

5. Illumination of spectrographs and spectrometers. Appropriate illumination of the spectroscopic instruments is required in order to fully attain their maximum resolving power and luminosity. Basically, it is necessary that beams originating from any point of the slit have a sufficient angular aperture to fill the collimating optics and grating.

The light flux that can be admitted into the instrument is determined by the ratio of grating aperture A to the focal length f of the collimator, by the width

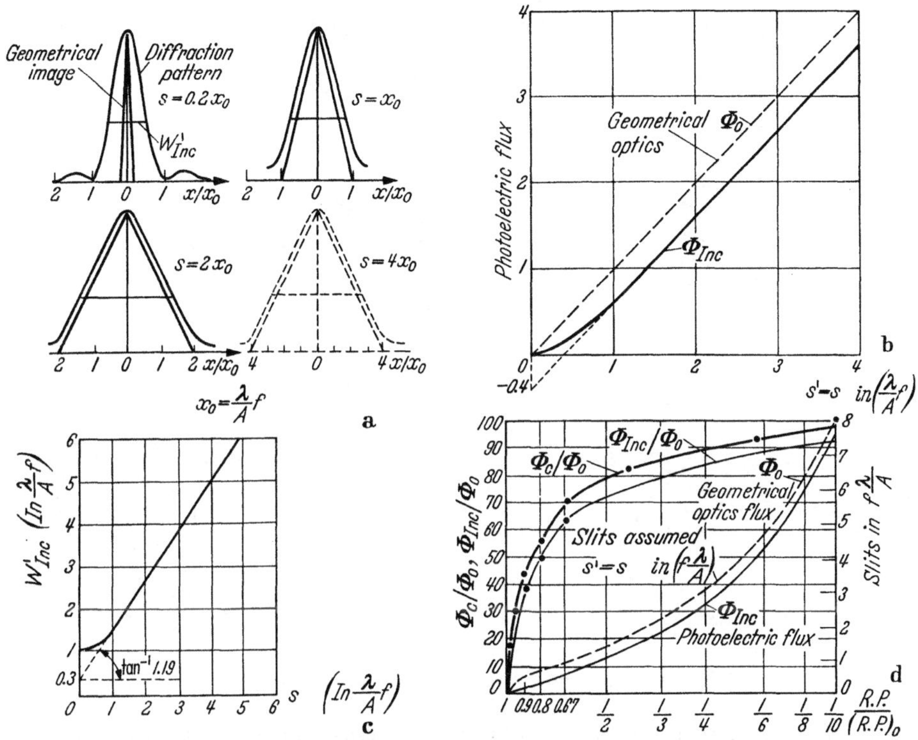

Fig. 44 a—d. a) Line shapes in photoelectric spectrometer in incoherent illumination (after JACQUINOT and DUFOUR). b) to d) Photoelectric spectrometer flux parameters (after JACQUINOT and DUFOUR).

of the entrance slit s and by the spectral brightness or luminance B_λ of the image of the source on the entrance slit. Here (see Fig. 45)

$$B_\lambda = \frac{dI}{dS\cos\alpha} = \frac{d\Phi/d\Omega}{dS\cos\alpha} = \frac{\text{watts/steradian}}{\text{cm}^2}.$$

It is well known (see A. MARÉCHAL, this Encyclopedia, Vol. 24, p. 58) that the brightness B_λ of an object or source remains conserved in optical image formation. As a result, provided only that the light from the source image on the slit fills the collimator, there is no preferred position for the source and illuminating lens, in front of the entrance slit, which will permit to increase the flux Φ entering the instrument.

6. Photometric concepts and definitions (energetic units). The following definitions and concepts are required and are established according to P. JACQUINOT[1]:

[1] P. JACQUINOT: Optique Ondulatoire, Editions Claude Hermann, Paris 1959.

Flux Φ (watts). This is the total flux in watts=joules/sec=energy/time=photons/second which is radiated from a source of area into a solid angle Ω at the wavelength λ (or in the spectral bandwidth $\Delta\lambda$).

Intensity I (watts/solid angle). Intensity describes the spatial distribution of flux. One has

$$I=\frac{d\Phi}{d\Omega}. \tag{26.9}$$

Solid angle (Steradians). It is defined (Fig. 45a) as the ratio of the area ds seen normally from a point O to the square of the distance r of the area to the point O:

$$d\Omega=\frac{ds_A}{r^2}=\frac{dS_A\cos\alpha}{r^2}. \tag{26.10}$$

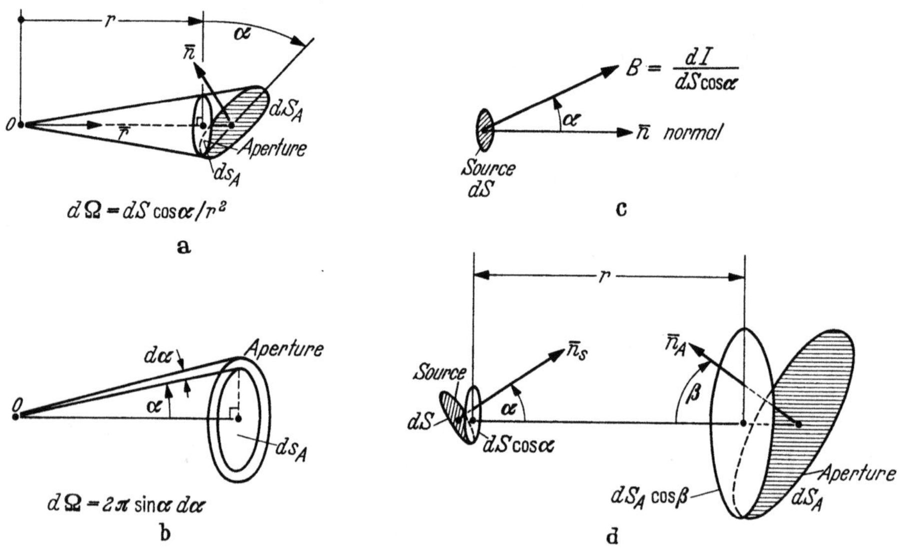

Fig. 45a—d. Spectrograph and spectrometer illumination parameters.

A useful angular relation for computing $d\Omega$ is shown in Fig. 45b. As seen from a point, the total solid angle about it is simply the area of the sphere divided by r^2, that is 4π.

Mean intensity I_m (watts/solid angle): One has

$$I_m=\frac{1}{4\pi}\int I\,d\Omega. \tag{26.11}$$

Brightness or luminance, B, (watts/solid angle/area): One has (Fig. 45c) with reference to the source area dS,

$$B_\lambda=\frac{dI}{dS\cos\alpha}. \tag{26.12}$$

Flux in an elementary beam, $d^2\Phi$: From above, one has

$$d\Phi=I\,d\Omega \tag{26.13}$$

and

$$d^2\Phi=dI\,d\Omega=B_\lambda\,dS\cos\alpha\,d\Omega. \tag{26.14}$$

Relation between source, aperture, brightness and flux: The following important relation is established between the source area dS, aperture area dS_A and flux

(Fig. 45d). One has

$$d^2\Phi = B_\lambda\, dS \cos\alpha\, d\Omega \tag{26.14}$$

and

$$d\Omega = \frac{dS_A \cos\beta}{r^2}. \tag{26.15}$$

Therefore

$$d^2\Phi = \frac{B_\lambda\, dS \cos\alpha\, dS_A \cos\beta}{r^2}, \tag{26.16}$$

and it is seen that the source area and aperture area have perfectly symmetrical roles.

"Etendue" of elementary beam, d^2U: The quantity

$$d^2U = \frac{dS \cos\alpha\, dS_A \cos\beta}{r^2} \tag{26.17}$$

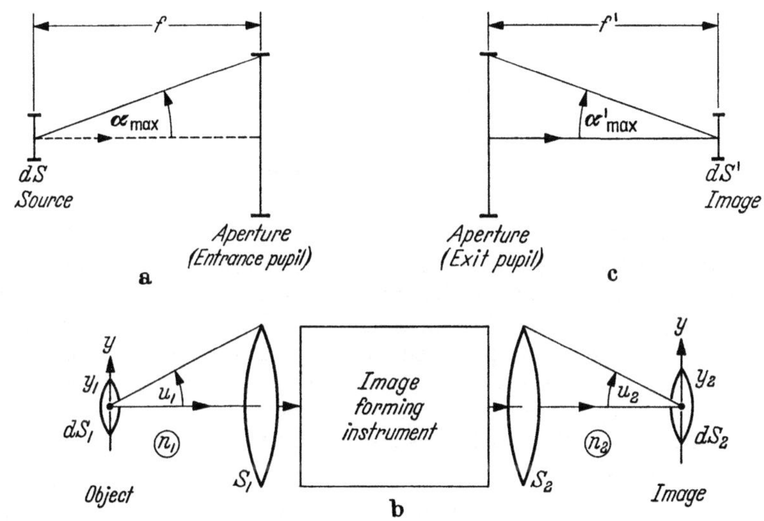

Fig. 46a—c. Spectrometer illumination and imaging parameters.

is defined as "étendue" of the elementary beam. It characterises the geometrical factor determining the flux entering an aperture from a source. The elementary flux is given by

$$d^2\Phi = B_\lambda d^2 U. \tag{26.18}$$

It is simply the product of the spectral luminance by the geometrical factor d^2U. One also has

$$d^2U = dS \cos\alpha\, d\Omega \tag{26.19}$$

and

$$d^2U = ds\, d\Omega. \tag{26.20}$$

Etendue of beam, dU: For a beam of finite solid angle Ω determined by an angle $\alpha_{\max}$ as seen from a source ds (Fig. 46a), one has from (26.20)

$$\left.\begin{aligned} dU &= \int_0^{\alpha_{\max}} dS \cos\alpha\, d\Omega = dS \int_0^{\alpha_{\max}} \cos\alpha\, d\Omega \\ &= dS \int_0^{\alpha_{\max}} \cos\alpha\, 2\pi \sin\alpha\, d\alpha; \end{aligned}\right\} \tag{26.21}$$

that is

$$\boxed{dU = \pi\, dS \sin^2 \alpha_{\max}.}$$

The quantity dU characterizes the source area dS and the angle α_{max} within which the flux may enter the aperture. The quantity dU is defined as the "étendue" of the beam. The total étendue of the space in front of the source is obtained by setting $\alpha_{max}=\pi/2$; one has $dU=\pi\, dS$.

Flux entering aperture from source, $d\Phi=B_\lambda\, dU$: The total flux $d\Phi$ at the wavelength λ entering an aperture seen under the angle α_{max} from a source dS is given by

$$d\Phi=B_\lambda\, dU. \tag{26.22}$$

One has

$$\text{Flux}=\text{luminance}\times\text{étendue}$$

or

$$\text{Flux}=\text{brightness}\times\text{étendue}.$$

Clearly, one can write for a source of finite dimensions S

$$\boxed{\Phi=B_\lambda U=\text{flux}} \tag{26.23}$$

with

$$\boxed{U=\pi S\sin^2\alpha_{max}=\text{étendue}} \tag{26.24}$$

as long as α_{max} can be taken as being the same from all points of the source.

Conservation of flux, brightness and luminance: It can be shown quite generally for any image-forming instrument that the three quantities of flux $d\Phi$, brightness or luminance B_λ and étendue dU remain conserved between the object and image space. This statement is true provided that there is no absorption or scattering loss. It should be quite clear from its definition that the *flux is conserved between an object and its image,* except for losses. The conservation of B_λ and dU apply to the case when the object and image space are in the same medium. In the general case, we have (Fig. 46b)

$$n_1 y_1 \sin u_1=n_2 y_2 \sin u_2 \tag{26.25}$$

according to ABBE's relation beetwen an object y_1, in the medium of index n_1, and its image y_2 in the medium of index n_2 (cf. MARÉCHAL's article in Vol. 24 of this Encyclopedia). By squaring, we have

$$n_1^2 y_1^2 \sin^2 u_1=n_2^2 y_2^2 \sin^2 u_2$$

i.e., with Eq. (26.21)

$$n_1^2\, dU_1=n_2^2\, dU_2. \tag{26.26}$$

One therefore finds that *the product $n^2\, dU$ is conserved between an object and its image.* Further, since $d\Phi=B\, dU$, one also finds that *the quantity B/n^2 is conserved between an object and its image.*

Illumination of image $I'=d\Phi'/dS'$ = flux/area: The illumination of the image, defined by I' can be immediately obtained by comparing Fig. 46c with Fig. 46a. One has, from Eqs. (26.21) and (26.22)

$$I'=\frac{d\Phi'}{dS'}=B_\lambda \pi \sin^2\alpha'_{max} \tag{26.27}$$

where $d\Phi'=d\Phi$ as shown in the previous paragraph. Eq. (26.27) shows the remarkable fact that the illumination in the image depends on the solid angle determined by the exit aperture, as it is seen from the image point, and not on the solid angle determined by the entrance aperture as seen from the source (or slit in a spectrograph). Of course, since (with $n_2 = n_1$) we have from ABBE'S relation

$$y_2^2 \sin^2 \alpha'_{\max} = y_1^2 \sin^2 \alpha_{\max},$$

we obtain

$$dS' \sin^2 \alpha'_{\max} = dS \sin^2 \alpha_{\max}$$

and therefore

$$I' = \frac{d\Phi'}{dS'} = B_\lambda \pi \frac{dS}{dS'} \sin^2 \alpha_{\max}. \tag{26.28}$$

When the focal length f' of the camera system is different from the focal length f of the collimator, one has

$$\frac{\text{Image}}{\text{Object}} = \frac{y'}{y} = \frac{f'}{f} = \sqrt{\frac{dS'}{dS}} \tag{26.29}$$

and

$$\frac{dS}{dS'} = \frac{f^2}{f'^2}.$$

Finally

$$\boxed{I' = \frac{d\Phi'}{dS'} = B_\lambda \pi \frac{f^2}{f'^2} \sin^2 \alpha_{\max}} \tag{26.30}$$

where $\alpha_{\max}$ refers to the entrance collimator aperture. In as much as one can afford to get a reduction in the image dS' with respect to dS, the illumination I′ is seen to increase with $(f/f')^2$. This is of particular importance in photographic instruments (spectrographs). In spectrometers, were the criterion of luminosity is flux rather than illumination of the image, the principle of conservation of flux does not favor having a shorter focal length f' in the camera system than the focal length f in the collimator.

Flux reaching image $d\Phi'$: In as much as there is no loss and $d\Phi' = d\Phi$, one has

$$\boxed{d\Phi' = B_\lambda \pi \, dS \sin^2 \alpha_{\max}} \tag{26.31}$$

showing the parameters relevant in spectrometers.

7. *Luminosity of spectrographs.* What matters in spectrographs is the intensity I' defined by Eq. (26.30) and equal to the flux per unit area in the image.

One needs to consider the efficiency $(T_G)_\lambda$ of the grating as defined by

$$(T_G)_{\lambda,m} = \frac{\text{Flux diffracted into wavelength } \lambda \text{ in order } m}{\text{Flux incident on grating}} \tag{26.32}$$

and also the image intensities I_{coh} or I_{incoh} as determined from VAN CITTERT'S graphs of Fig. 42. Let I_{coh} or I_{incoh} be written as I_{VC}. Eq. (26.30) gives

$$I'_\odot = \frac{d\Phi'}{dS'} = I_{\text{VC}} B_\lambda (T_G)_{\lambda,m} \pi \frac{f^2}{f'^2} \sin^2 \alpha_{\max} \tag{26.33}$$

where the symbol $\odot$ refers to the aperture of radius $\frac{1}{2}A$ inscribed into the aperture AH determined by the grating at the angle i (Fig. 47).

We first note that

$$\sin^2 \alpha_M = \frac{A^2}{4f^2} \cos^2 \alpha_M \tag{26.34}$$

and we have

$$I'_\odot = I_{\text{VC}} B_\lambda (T_G)_{\lambda,m} \left(\frac{\pi A^2}{4}\right) \frac{\cos^2 \alpha_M}{f'^2}. \tag{26.35}$$

In reality, the aperture determined by a grating is generally rectangular and has an area AH. One needs to determine the integral $\int_{\square} d\Omega$ that corresponds to this aperture, rather than the integral determined in (26.21). As a good approximation, one can obtain $I'_{\square}$ corresponding to the rectangular aperture simply by defining

$$I'_{\square} = \frac{AH}{\frac{\pi A^2}{4}} I'_{\odot}. \tag{26.36}$$

With this definition, we have

$$\boxed{I'_{\square} = I_{\mathrm{VC}} B_\lambda (T_G)_{\lambda,m} \frac{AH}{f'^2} \cos^2 \alpha_M.} \tag{26.37}$$

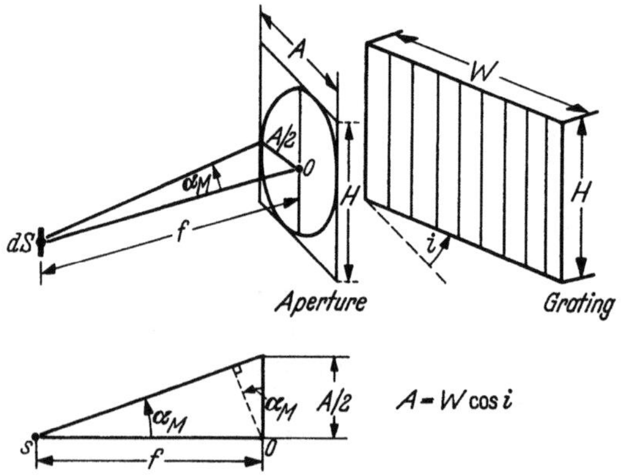

Fig. 47. Luminosity parameters in grating spectrographs.

In general, $\cos^2 \alpha_M$ may be taken as unity. Indeed, with an $f/2$ aperture ($A = f/2$), $\tan \alpha_M = \frac{1}{4}$, $\cos \alpha_M = 0.967$ and $\cos^2 \alpha_M \simeq 0.94$. With an $f/1$ aperture ($A = f$), $\cos^2 \alpha_M \simeq 0.72$. One has finally

$$\boxed{I'_{\square} = I_{\mathrm{VC}} B_\lambda (T_G)_{\lambda,m} \frac{AH}{f'^2}} \tag{26.38}$$

for the intensity in the spectral image in a spectrograph. I_{VC} is the factor which takes care of the entrance slit width in spectrographs.

8. *Luminosity of spectrometers.* For spectrometers, it is the flux rather than the intensity which matters. Instead of VAN CITTERT's coefficients, one now needs to use the $\Phi_{\mathrm{incoh}}/\Phi_0$ and $\Phi_{\mathrm{coh}}/\Phi_0$ coefficients from Fig. 44. Let $\Phi_J = \Phi_{\mathrm{incoh}}/\Phi_0$ in incoherent illumination, and $\Phi_J = \Phi_{\mathrm{coh}}/\Phi_0$ for the case of coherent illumination. From Eq. (26.31) and with similar considerations as those applicable to the calculation of Eq. (26.38), we have

$$\boxed{\Delta \Phi'_{\square} = \Phi_J B_\lambda \Delta\lambda \, (T_G)_{\lambda,m} \, sh \frac{AH}{f^2}} \tag{26.39}$$

for the flux reaching the detector in the spectral bandwidth $\Delta\lambda$. Here s is the slit width, in the direction of dispersion, and h the slit height, in a direction normal to the direction of dispersion. One notes that the flux involves the collimator focal length f rather than the camera focal length as in the intensity Eq. (26.39).

9. *Minimum usable slit width and maximum f-number in spectrometers and spectrographs.* Aside from the apparent flux limitations by a decreasing slit width according to VAN CITTERT[1] and JACQUINOT[2], one clearly cannot reduce the slit to a value which is small when compared to the wavelength. It is reasonable to investigate the consequence of this limitation. Let it be assumed that $s = n\lambda$ where n may be of the order of 2 to 4 in the limit. In the domain of scalar diffraction, let $s = \frac{\lambda}{A} f$ and let us first determine how the diffraction pattern cor-

[1] loc. cit. in footnote 1 on p. 489.
[2] loc. cit. in footnote 1 on p. 491.

responding to the slit fills the aperture A. We have, for the first minimum,

$$\frac{\lambda}{s}f=\frac{\lambda}{\frac{\lambda}{A}f}f=A,$$

and we note (Fig. 48a) that the aperture has a width of half the distance between the minima. Similarly, the slit has a width equal to half the width between the minima of the diffraction pattern corresponding to the slit.

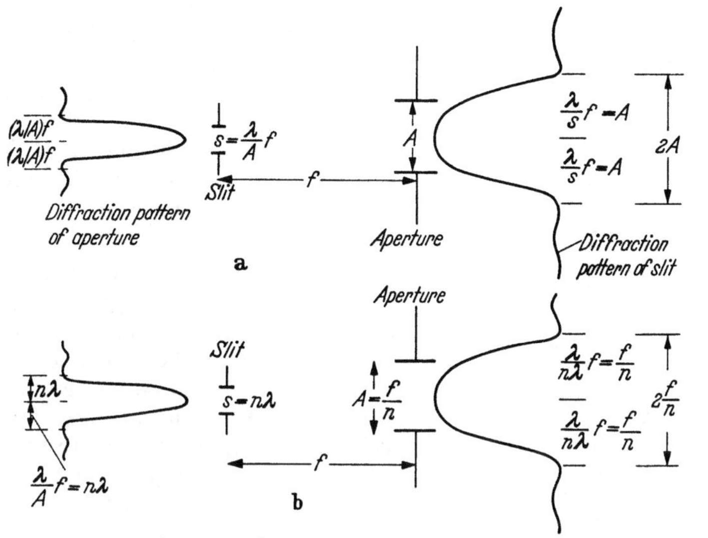

Fig. 48 a and b. Slit width parameters in grating spectrographs and spectrometers.

Next (Fig. 48b), one easily finds that the same ratio of aperture to slit diffraction pattern as before will be obtained with

$$A=\frac{f}{n}$$

when the slit width is $s=n\lambda$. In other words, the maximum f-number, corresponding to a slit n wavelengths wide, is equal to n. In practice f-numbers larger than $f{:}2$ may therefore not be attainable with instruments using slits according to these considerations.

10. Energy-limited resolving power. There are cases when the minimum detectable energy will be the limiting parameter in the operation of a spectrograph. It is of interest to relate the resolved bandpass $\Delta\lambda_R$ to the minimum detectable flux $\Phi_{\min}$. We first note that the reduction in resolving power R.P./(R.P.)$_0$, as a function of the increase in flux Φ_J/Φ_0 can be read off the graphs in Fig. 44. A formal relation can be established as follows, according to JOHN STRONG (1958)[1]. With the dispersion $di'/d\lambda$, one can write

$$\Delta\lambda=\frac{d\lambda}{di'}\Delta i'.$$

The slit width is then $s=f\Delta i'=f\frac{di'}{d\lambda}\Delta\lambda$. Eq. (26.39) becomes

$$\Delta\Phi'_{\square}=\Phi_J B_\lambda(T_G)_{\lambda,m}\Delta\lambda^2\frac{di'}{d\lambda}h\frac{AH}{f^2}$$

[1] loc. cit. in footnote 5 on p. 486.

and one has

$$\Delta\lambda = \sqrt{f\,\frac{(\Delta\Phi'_{\square})_{\min}/\Phi_J}{(T_G)_{\lambda,m}\,(di'/d\lambda)\,A\,H\,h}}\,. \tag{26.40}$$

$\Delta\lambda_{\min}$ represents the smallest "resolvable" wavelength bandpass for a given flux $(\Delta\Phi'_{\square})_{\min}$, dispersion $(di'/d\lambda)$ and other parameters.

The resolving power corresponding to the minimum detectable flux can be written as

$$(\text{R.P.})_{\min} = \frac{\lambda}{\Delta\lambda_{\min}} = \lambda\sqrt{\frac{(T_G)_{\lambda,m}\,(di'/d\lambda)\,A\,H\,h}{f\,(\Delta\Phi'_{\square})/_{\min}\Phi_J}}\,, \tag{26.41}$$

showing that the resolving power will increase with increasing dispersion $(di'/d\lambda)$, with increasing grating-width A and height H, with increasing slit height h, as well as with decreasing focal length f and threshold flux $(\Delta\Phi'_{\square})_{\min}$. Of course, Eqs. (26.40) and (26.41) are only applicable for $\text{R.P.} \leqq (\text{R.P.})_0$ where $(\text{R.P.})_0$ is the resolving power determined by diffraction, $[\text{R.P.} = 2W\sin i'/\lambda$ and $A = W\cos i']$.

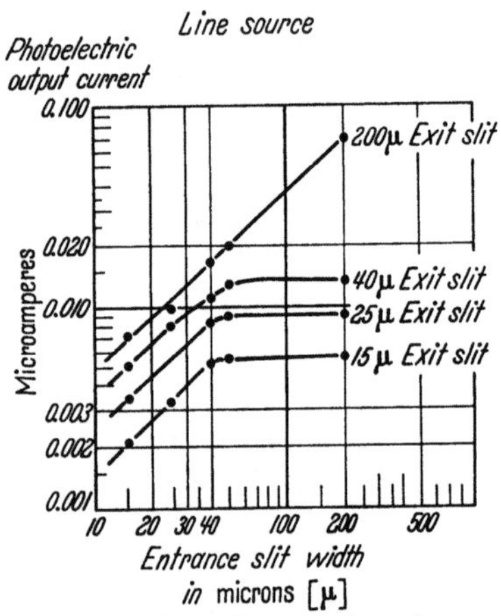

Fig. 49. Experimental flux values in grating spectrometer having an instrumental line-breadth of 25 μ (after R. F. JARRELL).

11. Spectrometers and spectrographs with instrumental line widths greater than those due to diffraction alone. When the aberrations of the optics cause the instrumental intensity distribution $I(x)$ to be considerable broader than the diffraction patterns, $I(x)$ the convolution in Eq. (26.2) needs to be carried out with $I_i(x)$ rather than with $I(x)$.

12. Spectrometers with unequal exit and entrance slits. Two cases need to be distinguished: (a) spectral line-sources, and (b) continuous sources.

a) Generally speaking, it is found that the flux recorded in a spectrometer, with a spectral line and a fixed exit slit, increases as the entrance slit is opened, only so long as the "breadth" of the spectral line is less than the nominal entrance slit width, plus any aberration and diffraction broadenings (R. F. JARRELL, 1960)[1]. Once this slit width is exceeded, the flux does not seem to increase. Fig. 49, according to JARRELL, shows the result of experimental flux measurements for an $f = 500$ mm spectrometer with a 2×2 inch 15000 groove/inch grating, and an instrumental line breadth of about 25 microns.

b) In case of continuous sources, the flux simply increases linearly with the entrance slit opening. When both slits are being opened, the flux will increase as the square of the slit openings.

Clearly, in both cases, the resolving power decreases as the slits are opened, according to geometrical optics, in this domain of large slit apertures.

13. Additional effects of slit widths and illumination are given in standard textbooks on Practical Spectroscopy, along with numerous references, for example, R. A. SAWYER (Experimental Spectroscopy), G. R. HARRISON, R. C. LORD and J. R. LOOFBOUROW (Practical Spectroscopy), G. L. CLARK (The Encyclopedia of Spectroscopy).

27. Luminosity of gratings in spectrometers using multiple slits (GOLLAY, GIRARD) or interferometric modulation (P. CONNES). *1. Basic concepts.* A considerable increase in luminosity, without the corresponding reduction of resolution that is characteristic of single-slit spectrometers, can be obtained with the help

[1] R. F. JARRELL, in: Encyclopedia of Spectroscopy, edit. by G. L. CLARK. New York: Reinhold Publishing Comp. 1960.

of multiple slit arrangements, such as those described by M. J. E. GOLLAY[1] in 1949 and 1951, and by A. GIRARD[2] in 1960, 1961 and 1963. Some basic similarity exists between the multi-slit modulated spectrometers and the SISAM Spectrometer, first described by P. CONNES in 1957[3], in which two gratings are placed in the two arms of a Michelson Twymann-Green interferometor, and where modulation is achieved by periodic oscillation of the beam splitter. GOLLAY's arrangements were the first ones to demonstrate the gains in luminosity with the help of a rotating two-dimensionally patterned multi-slit mask in 1949; an external computer was required to extract the spectral information, somewhat like in interferometric spectroscopy using Fourier-transform methods. In this sense, it falls somewhat outside the scope of this section. A multi-slit double monochromator using no moving parts was suggested by W. A. SHURCLIFF[4] also in 1949, but the possibility of attaining the significant gains in luminosity without any basic changes in the arrangement and operation of a conventional scanning monochromator appears to have been first suggested and demonstrated by A. GIRARD in 1960.

Recent descriptions of the Girard "grille" multi-slit spectrometer are given by the author in Applied Optics, **2**, 79 (1962) and in the Journal de Physique et le Radium, **24**, 139 (1963). Gains of luminosity in excess of 100 with no loss in resolution have been obtained in the infrared with a grating spectrometer using the arrangement shown in Fig. 52. The significant advantage of GIRARD's multi-slit spectrometer is not only that it permits to achieve the gain in luminosity without any basic change in the conventional scanning grating spectrometer, but also that it requires no computers to obtain the spectrum. The spectrum is obtained directly by scanning exactly as in a single-slit spectrometer. It is to be noted that this is also true for CONNE's SISAM spectrometer, although the basic principle there requires an interferometer, rather than masks, in addition to the grating.

The basic idea materialized in GIRARD's version of a multi-slit scanning spectrometer is that the triangular *slit scanning function*, characteristic of a single-entrance, single-exit slit spectrometer (Fig. 50) can be synthetized by an equivalent set of pairs of sinusoidally transparent entrance and exit slit masks, with the sub-masks of various spacial periods being arranged one above the other, as it were, along the height of the single slit which they replace (Fig. 50). The set of equivalent masks is generated from the single slit scanning function with the help of Fourier analysis. Inasmuch as the set of equivalent masks should have a periodic transparency, with positive and negative transmissions, and since negative transmissions do not exist in the ordinary sense, it is necessary to generate the synthetized set with the help of a suitable set of exit masks and a pair of complementary entrance masks having complementary transmissions along the direction of dispersion x (Fig. 51). Analysis verifies that the resolution obtained with the synthetized slit scanning function corresponds to the smallest period in the set, where e is the width of the single slit. However the luminosity is proportional to the product EH (Fig. 51b), which is considerably larger than the luminosity eH corresponding to the single slit of width e and height H, thus resulting in a significant gain in luminosity at no loss in resolution.

---

[1] M. J. E. GOLLAY: J. Opt. Soc. Amer. **39**, 437 (1949); **41**, 468 (1951).

[2] A. GIRARD: Optica Acta **7**, 8 (1960). — Applied Optics **2**, 79 (1963). — J. Physique Radium **24**, 139 (1963).

[3] P. CONNES: Optica Acta **4**, 136 (1957). — J. Phys. Radium **19**, 262 (1958). — These, Rev. Opt. **38**, 157, 416 (1959); **39**, 402 (1960).

[4] W. A. SHURCLIFF: J. Opt. Soc. Amer. **39**, 1048 (1949).

In practice the pair of complementary entrance slit masks is materialized in a single mask in which the opaque sections of the basic mask are made to be fully reflecting, and therefore to have the maximum "transmission" for the complementary mask. Similarly, the transmitting sections of the basic mask have minimum "transmission" for the complementary mask[1].

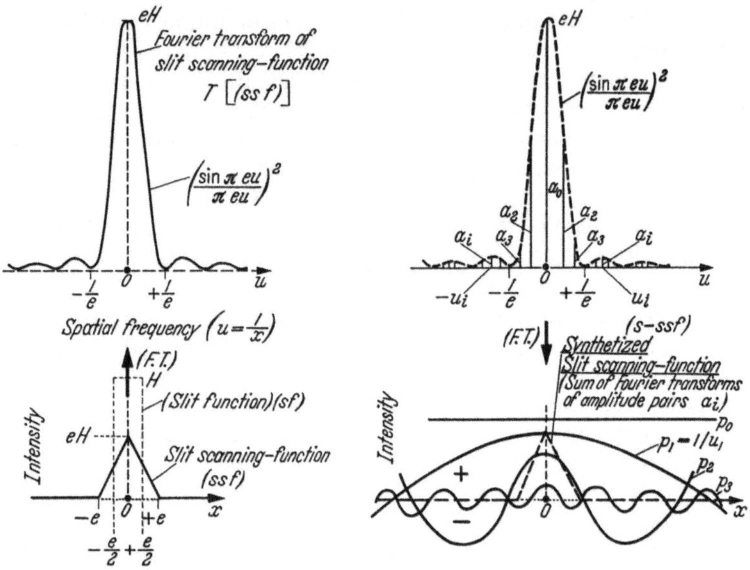

a b

Fig. 50 a and b. Fourier-transform interpretation of Girard multiple-slit spectrometer.

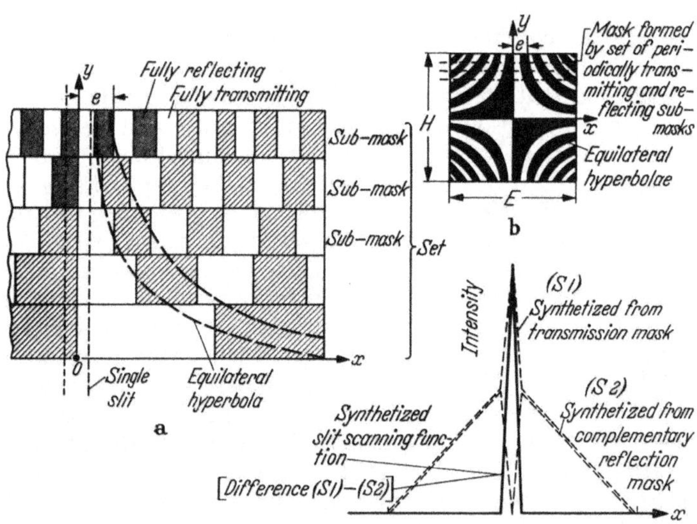

a b c

Fig. 51 a—c. Multiple-slit and slit-scanning function representation for Girard spectrometer.

[1] A pair of Fresnel-zone plates can be used in a transmission arrangement (STROKE, FALCONER, FUNKHOUSER, 1965). In this case, the entrance slit consists of two "positive" halves of a Fresnel zone, while the exit slit consists of a "positive" half and a "negative" half of a Fresnel-zone plate. Other types of "matched" filters may combine the "Girard" type hyperbolic filters with Fresnel-zone filters.

The principle of the Girard multi-slit spectrometer can be immediately understood with the help of the Fourier transform theory of image formation.

2. Fourier transform theory of multi-slit spectrometers. Let the entrance and exit slits of a conventional spectrometer have the same width e and height H, to which corresponds a slit function (sf) (Fig. 50a).

The slit scanning function (ssf), is obtained by convolution of the slit function with itself and we have

$$(\text{ssf}) = (\text{sf}) \otimes (\text{sf}). \tag{27.1}$$

In the spatial frequency domain (spatial frequency $u=1/X$) we have by Fourier transformation

$$T[(\text{ssf})] = T[(\text{sf})] \cdot T[(\text{sf})] \tag{27.2}$$

where $T[\;]$ indicates the Fourier transformation of the bracketed function. Now

$$T[(\text{sf})] = \frac{\sin \pi e u}{\pi e u} \tag{27.3}$$

in the case shown, so that

$$T[(\text{ssf})] = \left(\frac{\sin \pi e u}{\pi e u}\right)^2 \tag{27.4}$$

as represented in Fig. 50a.

It only remains to be argued that the $T[(\text{ssf})]$ function of (Fig. 50a) can be generated by a set of amplitude-pairs a_i at the frequencies $+u_i$, and $-u_i$, and the amplitude a_0 at the spatial frequency $u_0=0$.

By again reverting to the spatial domain, with the help of inverse Fourier transformation, one concludes the argument by noting that to each of these frequency pairs corresponds in the spatial domain a sinusoidally transparent mask having a period $p_i=1/u_i$.

The various sub-masks of frequency p_i are simply arranged one above each other, along the height of the single entrance and exit slits which they replace. Within the usual limitations of having only a finite number of sub-masks, the slit scanning function thus synthetized is fully equivalent to the single slit scanning function, with, however, a considerable gain in flux which is illustrated in Fig. 51, the gain being $G=H/2e$.

In practice, the set of sub-masks is formed by a set of periodic rectangular functions rather than sinusoidal functions. Moreover, the fact that negative transparencies are not available in the usual sense is taken care of by making the entrance slit fully reflecting (in the parts that are to be opaque in transmission). By a suitable chopping arrangement (Fig. 52) the entrance slit mask is used alternately in transmission and in reflection in order to permit generating the desired slit scanning function (Fig. 51c).

It is easily seen from Fig. 51 that the discrete set of sub-masks will have equal transmission lines along equilateral hyperbolae

$$y = \frac{2\pi N}{k x} \; (N = \text{integer}), \tag{27.5}$$

the transparency being

$$\tau(x, y) = \cos k x y. \tag{27.6}$$

The practical realization of the mask is shown in Fig. 51b. For the first term of the series development, one has by transmission through the entrance slit mask

$$\tau_{\text{in}} = \tfrac{1}{2}(1 + \cos k x y) \tag{27.7}$$

and by reflection from the "complementary" entrance slit mask

$$\tau_{\text{in, complem.}} = \tfrac{1}{2}(1 - \cos k\,x\,y) \tag{27.8}$$

as the slit scanning functions, which permit to obtain the desired synthetized slit scanning function shown in Fig. 51c.

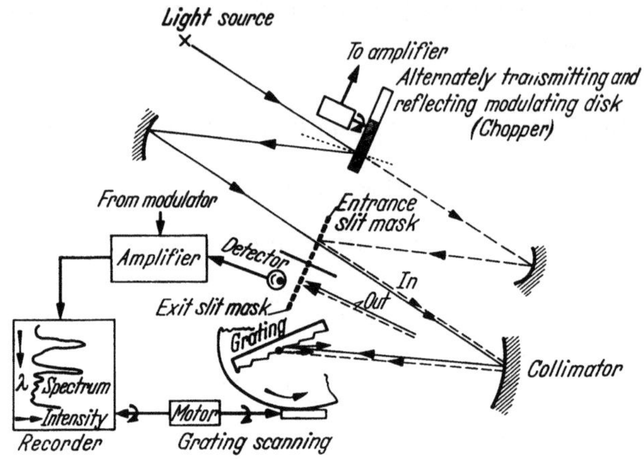

Fig. 52. Girard multiple-slit high-luminosity spectrometer.

It is clear that a perfectly stigmatic mounting is required in order to image the entrance slit mask on the exit slit mask via the grating. In particular, gratings with capable of generating perfectly flat wave fronts are required. A Littrow arrangement appears to give suitably small aberrations. A reproduction of a spectrum recorded is shown in Fig. 53, and a comparison of recordings with a single slit (0.05 × 30 mm) and 30 × 30 mm masks in Fig. 54. Masks have been obtained by photo-engraving on various supports such as CaF_2, NaCl, KRS5 for particular applications in the infrared. Apodization (Jacquinot, Roizen-Dossier)[1] of the grating aperture is used to achieve the most favorable instrument function.

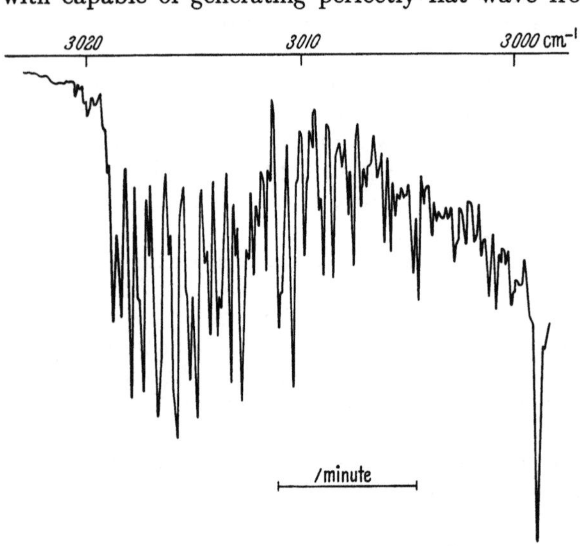

Fig. 53. Spectrum of methane (in emission) near 300 °C obtained in a Girard multiple-slit spectrometer (Gas cell dimensions: length = 15 cm, pressure = 50 mm mercury at 20 °C) [after Girard].

3. Interferometric modulation spectrometers. Fig. 55a shows the basic arrangement of the SISAM spectrometer as described by P. Connes (1958, 1959, 1961), and Fig. 55b and Fig. 55c shows two of the alternate schemes which he suggested. (SISAM = spectromètre interférentiel à sélection par amplitude de modulation). As described by Connes, it is seen that two identical gratings are

[1] In Progr. Optics, III, loc. cit. in footnote 1 on p. 432.

used in place of the two mirrors in a Michelson Tywman-Green interferometer. For a certain wavelength, the two gratings can be adjusted to be in autocollimation and the arrangement is then identical to the interferometer with the mirrors set parallel, to each other, so as to obtain a flat field. For this particular wavelength, a path difference variation (obtained by rotation of the compensating plate) will simply result in a flux modulation, just like in any such photoelectric interferometer (STROKE 1957). By oscillating the compensator plate in a periodic see-saw (saw-tooth) manner, on achieves the the desired modulation at any particular wavelength. The modulation is sinusoidal for a linear path difference variation. The other neighboring or more distant wavelengths are modulated considerably less or not at all (with the exception, of course, of overlapping wavelengths). By simultaneously rotating the two gratings $G1$ and $G2$, one obtains modulation of successive wavelengths. By using a demodulating amplifier similar to that of Fig. 54, one obtains a direct recording of the spectrum, just like in the multi-slit spectrometer. The flux gain at a given resolution, over that achievable in single slit spectrometers, can be very considerable (on the order of several thousand), but it depends on signal to noise considerations in the receptor as much as on

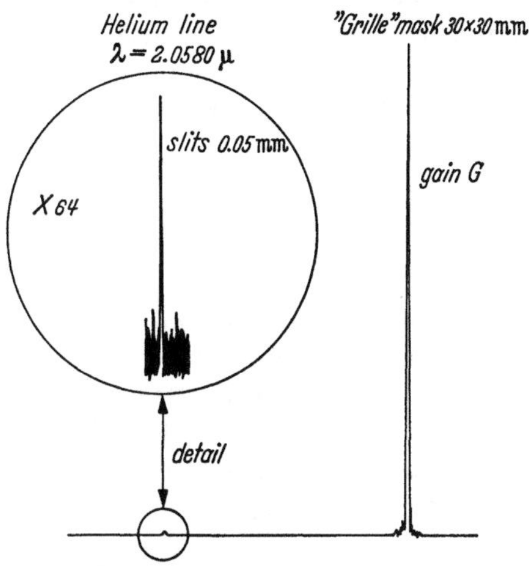

Fig. 54. Comparison of luminosity in a single-slit grating spectrometer (left) and a 30 × 30 mm "grille" multiple-slit Girard spectrometer (ratio of amplitudes = 30) [after GIRARD].

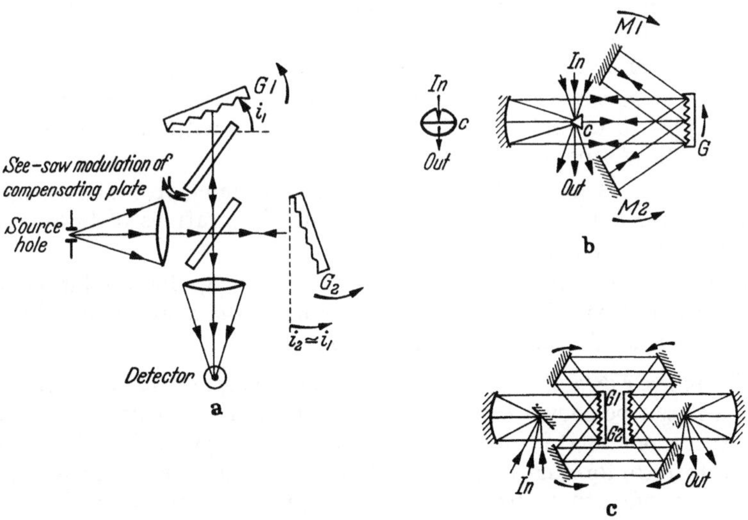

Fig. 55 a—c. Several forms of SISAM spectrometer (after P. CONNES).

geometry. In practice a gain of 80 has been achieved at 6000 cm^{-1} by CONNES in an experimental arrangement using small 76×65 mm gratings near 30° in autocollimation, with a source hole of 7.5×10^{-3} radians aperture. Like in the multi-slit spectrometer, apodization of the grating and interferometer apertures is used to achieve the most favorable instrument function.

28. Other methods of increasing the luminosity of gratings in spectrometers and spectrographs. The methods of the preceding Sect. 27 can be compared to several other methods which all aim at best exploiting the flux diffracted by a grating. These are examined in specialized literature dealing with the design principles of spectrometers and spectrographs[1]. Multi-pass use of gratings (RANK, HULTHÉN and others), the use of gratings in series (STROKE and STROKE)[1] as well as grating mosaics, are among the methods which have been used for increasing not only the luminosity but also the resolving powers attainable with gratings of given dimensions.

III. Electromagnetic theory of diffraction and of image formation by gratings.

29. Formation of images and energy distribution in optical gratings. α) *The two aspects of the diffraction of light by gratings.* Two aspects of the diffraction of light by gratings have been recognized ever since JOSEPH VON FRAUNHOFER[2] invented and ruled the first gratings for optical purposes in 1821:

1. the distribution of light of a given wavelength into several waves of different orders, leaving the grating at angles determined by the grating equation

$$\sin i + \sin i' = m\lambda/a \tag{29.1}$$

and

2. the formation, by each of the diffracted waves, of "spectral lines", which are monochromatic images of the entrance slit of the spectrometer or spectrograph.

FRAUNHOFER also recognized the effects of periodic ruling errors on the spectral images and noted remarkable polarization effects in the diffracted waves.

More recently, in particular as a result of work by A. MARÉCHAL and G. W. STROKE (1959, 1960, 1963)[3], the two aspects of diffraction of light that need to be distinguished can be singled out as follows:

1. the *electromagnetic aspect*, having to do with the very origin of the diffraction of electromagnetic waves by a periodic boundary, and with the direction, polarization and amplitudes of the various waves, and

2. a *scalar aspect*, having to do with the image-forming properties of the various diffracted waves, when they are limited in size by gratings of finite width and limited in quality by grating imperfections.

Moreover, the spectroscopic applications of gratings for resolutions of wave fronts of different wavelengths, and in general for the analysis of light, involves certain *geometrical characteristics* of gratings.

β) *Theoretical calculation of energy distribution and of spectral diffraction patterns.* Basically, it can be shown (Sects. 31 and 32) that the various "HUYGENS' principle" expressions used in optics (for example its "rigorous" vector form, or its Fourier-transform formulation) can be obtained from MAXWELL's electromagnetic equations, provided that certain important approximations are made. One of the fundamental approximations is that the HUYGENS' principle formulation applies only to the vicinity of the center of quasi-spherical image-forming wave front

---

[1] See, for example, G. W. STROKE and H. H. STROKE: J. Opt. Soc. Amer. **53**, 333 (1963).

[2] JOSEPH VON FRAUNHOFER: Denkschr. kgl. Akad. Wiss. München **8**, 1 (1821/22). — Gilbert's Ann. Physik **74**, 337 (1823). — Edinburgh J. Sci. **7**, (1827); **8** (1828).

[3] A. MARÉCHAL and G. W. STROKE: C R. Acad. Sci. Paris **249**, 2042 (1959). — G. W. STROKE: Rev. Opt. **39**, 291—398 (1960). — G. W. STROKE: Phys. Letters **5**, 45 (1963).

(SILVER, 1949[1]; MARÉCHAL[2], 1956; STROKE 1962[3]). Since the waves of various orders diffracted by a grating generally form large angles with respect to each other, and with respect to the "slit"-like apertures of the individual grooves which one would like to consider as a source, it is clear that HUYGENS' principle cannot in general be applied to the diffraction by the individual grooves. No good agreement between such predictions and experiment is found to exist in practice (MADDEN and STRONG[4], 1958; STROKE[5], 1960, 1961, 1963), and no such agreement should be expected (STROKE, 1960, 1963).

An exception is formed by certain of the echelle-type gratings (WOOD[6] 1947; HARRISON 1949[7]), with quasi-rectangular groove profiles, and coarse spacings (in terms of wavelength), used so that the various diffracted waves are almost superposed in autocollimation with the direction of the incidence, along the preferred direction of energy distribution (blaze). Here the angles between the various diffracted waves of interest are very small and the conditions of applicability of HUYGENS' principle appear to be more closely satisfied than when the gratings are used with the spectra distributed over a wide range of angles. However, even for the echelle-type gratings, MARÉCHAL and STROKE (1959) have shown the existence of a special case of polarization and the complete absence of any but the single autocollimated order, for a rectangular groove depth of an integral number of half-wavelengths used in normal incidence on one groove face, when the electric vector in the incident wave is normal to the other groove face. In this case the grating behaves like a perfect mirror, and no such behaviour can be predicted on the basis of a HUYGENS' principle approximation. However, electromagnetic theory and the boundary values to be satisfied by the total field (incident + diffracted) on the surface of the grating immediately give the exact solution for this case.

In general, the electromagnetic character of diffraction and polarization effects start to manifest themselves predominantly when the groove facets and spacings become comparable to the wavelengths. As shown by STROKE (1963), high efficiencies in blazed gratings can only be attained to the extent that one can avoid polarization in the diffracted light by an appropriate choice of groove profile, the choice being based on electromagnetic boundary-value considerations. Indeed, if the grating were a perfect polarizer for a given groove shape, angle and wavelength, it could at most throw 50 percent of the incident energy into the corresponding order, rather than close to 100 percent as appears possible with grooves chosen on the basis of electromagnetic theory. In particular, the 90° quasi-rectangular groove profile referred to above (MARÉCHAL and STROKE 1959) (STROKE 1960), had been tacitly assumed to be the ideal groove shape for gratings, since the pioneering contributions by RAYLEIGH (1888, 1907)[8] and TROWBRIDGE

[1] S. SILVER: In: Microwave Antenna Theory and Design, M.I.T. Radiation Lab. Series Volume No. 12, Chapt. 4 and 5. New York: McGraw Hill Book Co. 1949.

[2] A. MARÉCHAL: Optique Géométrique Générale. In: Handbuch der Physik, Vol. XXIV, edit. by S. FLÜGGE, Sect. A II. Berlin-Göttingen-Heidelberg: Springer 1956.

[3] G. W. STROKE: Abstract Proceedings, 6th Internat. Conference for Optics, München, August 20—25, 1962.

[4] R. P. MADDEN and J. STRONG: Appendix P in: J. STRONG's Concepts in Classical Optics. San Francisco: W. H. Freeman & Co. 1958.

[5] G. W. STROKE: Rev. Opt. **39**, 291—398 (1960). — J. Opt. Soc. Amer. **51**, 1321 (1961). — Ruling, Testing and Use of Optical Gratings for High-Resolution Spectroscopy: In: Progress in Optics, Vol. II, edit. by E. WOLF. Amsterdam: North-Holland Publishing Company 1963.

[6] R. W. WOOD: J. Opt. Soc. Amer. **37**, 733 (1947).

[7] G. R. HARRISON: J. Opt. Soc. Amer. **39**, 522 (1949).

[8] Lord RAYLEIGH: In: Encyclopedia Britannica, Vol. XXIV (1888). — Phil. Mag. **14**, 60 (1907) (Scientific Papers V, p. 405). — Proc. Roy. Soc. Lond. A **79**, 399 (1907) (Scientific Papers V, p. 388).

and WOOD (1910)[1]. It was shown by STROKE (1963) that the 90° groove profile is in fact the least desirable groove profile, because of its tendency to polarize the diffracted light. STROKE (1963) also showed that significant gains in the efficiency of blaced gratings are obtained by ruling 100° and 120° included-angle grooves, rather than the electromagnetically undesirable 90° grooves. In summary, therefore, it is now clear that calculations based on HUYGENS' principle should not be expected to give reliable agreement with energy distribution (blaze) of gratings, and in fact they do not agree with experiment, except in special cases (RAYLEIGH 1907; U. FANO[2] 1938; STROKE 1960).

On the other hand, HUYGENS' principle in its Fourier-transform formulation is a powerful tool for the prediction of the distribution of energy in the diffraction patterns formed by the diffracted wave fronts, in particular when they show small but disturbing imperfections, as a result of residual ruling or grating blank imperfections. HUYGENS' principle has been successfully used by STROKE (1959, 1960, 1961) to calculate the complex amplitude (and intensity) distribution in the spectral diffraction patterns by Fourier transformation of the phase distribution as it appears, in readily usable graphical form, in the diffracted wave-front interferograms (Sect. 59). The computed diffraction patterns are used to obtain spectral information and resolutions considerable in excess of those conceivable on the basis of Rayleigh-type criteria, since the knowledge of the intensity distribution in the diffraction patterns permits to solve the convolution integral of image formation: distribution of intensity in the recorded spectrum = convolution of intensity distribution in spectral diffraction pattern with the intensity distribution in the spectrum (which is sought).

The boundary value problem of diffraction and the distribution of energy among the diffracted orders is dealt with in Chapter B VII. Here we deal with the electromagnetic basis of the Fourier-transform formulation of HUYGENS' principle and of image formation.

30. Rigorous form of HUYGENS' principle for electromagnetic waves (Maxwell-Lorentz theory). In order to show the domain and limits of applicability of the so-called KIRCHHOFF-HUYGENS' principle in grating image formation, one needs to recall that the HUYGENS' principle can be derived from MAXWELL's electromagnetic equations, provided that certain approximations are made at appropriate steps in the derivation. The approximations made then clearly describe the domain and applicability of the Fourier-transform formulation of diffraction.

Before proceeding with the derivation of the approximate formulations of the diffraction theory of image formation, one needs to show that it is possible to derive without approximation, a rigorous form of HUYGENS' principle [Eq. (30.9)] from MAXWELL's equations. After that, an approximate form of the rigorous HUYGENS' principle is derived [Eq. (31.25)] and finally the desired Fourier-transform formulation is given in Eq. (32.1). The approximations involved will clearly appear in the course of the derivations.

Many authors have dealt with aspects of this problem[3]. Among them are H. A. LORENTZ (1896), LARMOR (1903), KOTTLER (1923), BAKER and COPSON (1939, 1950), BOUWKAMP (1940, 1954), BORN and WOLF (1959), E. WOLF (1951), CROZE and DARMOIS (1949), MARÉCHAL (1956), A. RUBINOWICZ (1957), MARÉCHAL and FRANÇON (1960)[4], ROBIEUX (1960). Extensive biblio-

[1] A. TROWBRIDGE and R. W. WOOD: Phil. Mag. **20**, 770 (1910).
[2] U. FANO: Ann. Physik **32**, 393 (1938).
[3] For early references, see for ex. C. J. BOUWKAMP: Rep. Progr. Phys. **17**, 35—100 (1954).
[4] A. MARÉCHAL et M. FRANCON: Diffraction, edit. Revue d'Optique, Paris 1960.

graphies are given by BAKER and COPSON[1], BOUWKAMP (1954) and RUBINOWICZ[2] among others. Here we follow in part the developments of LORENTZ, RUBINOWICZ and MARÉCHAL.

MAXWELL's equations, in augmented Gaussian form are

$$\operatorname{curl} \boldsymbol{H} = \frac{4\pi}{c} \dot{\boldsymbol{j}}_e + \frac{1}{c} \varepsilon_0 \frac{\partial \boldsymbol{E}}{\partial t}, \tag{30.1}$$

$$\operatorname{curl} \boldsymbol{E} = -\frac{4\pi}{c} \dot{\boldsymbol{j}}_m - \frac{1}{c} \mu_0 \frac{\partial \boldsymbol{H}}{\partial t} \tag{30.2}$$

where $\boldsymbol{j}_e$ = electric current density and $\boldsymbol{j}_m$ = magnetic current density. Let

$$\left.\begin{aligned} \boldsymbol{H} &= \boldsymbol{H} \exp(-i\omega t), \\ \boldsymbol{E} &= \boldsymbol{E} \exp(-i\omega t). \end{aligned}\right\} \tag{30.3}$$

In vacuo, where $\varepsilon_0 = 1$ and $\mu_0 = 1$, MAXWELL's equations are

$$\left.\begin{aligned} \operatorname{curl} \boldsymbol{H} &= \frac{4\pi}{c} \dot{\boldsymbol{j}}_e - \frac{i\omega}{c} \boldsymbol{E}, \\ \operatorname{curl} \boldsymbol{E} &= -\frac{4\pi}{c} \dot{\boldsymbol{j}}_m + \frac{i\omega}{c} \boldsymbol{H}. \end{aligned}\right\} \tag{30.4}$$

Let $\boldsymbol{E}'$, $\boldsymbol{H}'$ be an electromagnetic field that also satisfies MAXWELL's equations, and let $\boldsymbol{j}_e'$, $\boldsymbol{j}_m'$ be the corresponding electric, respectively magnetic current densities. We have

$$\operatorname{curl} \boldsymbol{H}' = \frac{4\pi}{c} \dot{\boldsymbol{j}}_e' - \frac{i\omega}{c} \boldsymbol{E}', \tag{30.5}$$

$$\operatorname{curl} \boldsymbol{E}' = -\frac{4\pi}{c} \dot{\boldsymbol{j}}_m' + \frac{i\omega}{c} \boldsymbol{H}'. \tag{30.6}$$

With the help of the well known vector identity $\operatorname{div}(\boldsymbol{E} \times \boldsymbol{H}') - \operatorname{div}(\boldsymbol{E}' \times \boldsymbol{H}) = \boldsymbol{H} \operatorname{curl} \boldsymbol{E} - \boldsymbol{E} \operatorname{curl} \boldsymbol{H}' - \boldsymbol{H} \operatorname{curl} \boldsymbol{E}' + \boldsymbol{E}' \operatorname{curl} \boldsymbol{H}$ and GAUSS' theorem

$$\int_R \operatorname{div} \boldsymbol{D}\, dv = \int_S \boldsymbol{D} \cdot \boldsymbol{n}\, dS$$

Eqs. (30.4) and (30.5) and (30.6) give the famous Lorentz integral equation

$$\int_S [(\boldsymbol{E} \times \boldsymbol{H}') \cdot \boldsymbol{n} - (\boldsymbol{E}' \times \boldsymbol{H}) \cdot \boldsymbol{n}]\, dS = \frac{4\pi}{c} \int_R [\boldsymbol{H} \boldsymbol{j}_m' - \boldsymbol{E} \boldsymbol{j}_e' - \boldsymbol{H}' \boldsymbol{j}_m + \boldsymbol{E}' \boldsymbol{j}_e]\, dv \tag{30.7}$$

where the volume R is enclosed by the surface S. The following equation (the so-called HELMHOLTZ-HUYGENS' equation) is obtained, still without any approximations:

$$\boldsymbol{E}_{P'} = \frac{c}{4\pi} \int_S \left[\frac{\boldsymbol{E}_m}{\boldsymbol{i}_m} \cdot (\boldsymbol{n} \times \boldsymbol{E}) + \frac{\boldsymbol{E}_e}{\boldsymbol{i}_e} \cdot (\boldsymbol{H} \times \boldsymbol{n}) \right] dS + \int_R [\boldsymbol{E}_m(\boldsymbol{j}_m) + \boldsymbol{E}_e(\boldsymbol{j}_e)]\, dv \tag{30.8}$$

and one obtains finally

$$\left.\begin{aligned} \boldsymbol{E}_{P'} = \int_S \Big[&-\frac{ik}{4\pi} (\boldsymbol{n} \times \boldsymbol{H}) \frac{e^{-ikr}}{r} - \frac{i}{k} \operatorname{grad} \operatorname{div} \frac{1}{4\pi} (\boldsymbol{n} \times \boldsymbol{H}) \frac{e^{-ikr}}{r} + \\ &+ \operatorname{curl} \frac{1}{4\pi} (\boldsymbol{n} \times \boldsymbol{E}) \frac{e^{-ikr}}{r} \Big] dS \end{aligned}\right\} \tag{30.9}$$

---

[1] B. B. BAKER and E. T. COPSON: The Mathematical Theory of HUYGENS' Principle. Oxford: Clarendon Press 1950.

[2] A. RUBINOWICZ: Die Beugungswelle in der Kirchhoffschen Theorie der Beugung. Polska Akademia Nauk, Warszawa, 1957.

where $k=2\pi/\lambda$, $i=\sqrt{-1}$ and $\boldsymbol{n}$=normal to S. Eq. (30.9) is a rigorous form of Huygens' principle for electromagnetic waves. It involves no approximations. It states that the electric field $\boldsymbol{E}_{P'}$ at a point P' can be obtained in terms of appropiate integration of the contributions at P' of the $\boldsymbol{E}$ and $\boldsymbol{H}$ field distribution on the surface S enclosing the point P'. It is clear that no other sources are distributed within the region R enclosed by S.

31. Approximate form of rigorous Huygens' principle. Eq. (30.9) is a rigorous expression of Huygens' principle obtained from Maxwell's equations without approximations. We next derive an approximate form of Huygens' principle which gives $\boldsymbol{E}_0$, the complex amplitude of the electric field, in the vicinity O' of the center of a quasi-spherical wave front Σ.

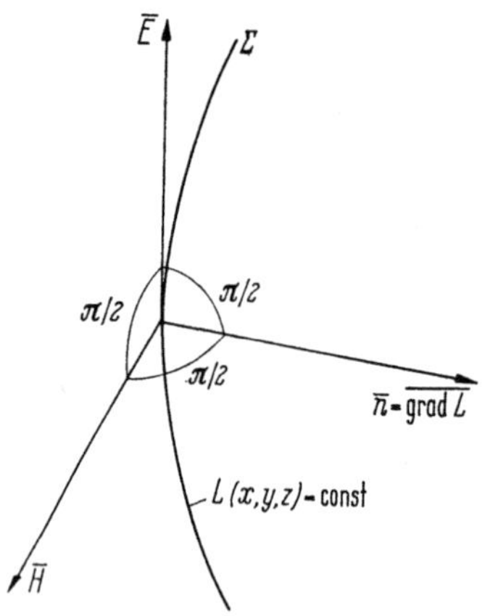

Fig. 56. Wavefront coordinates. Vectors are denoted by bars, in the text by bold type.

The equation to be derived is

$$\boldsymbol{E}_{O'} \simeq \frac{i}{\lambda} \frac{\mathrm{e}^{-ikR}}{R} \iint_{\Sigma} \boldsymbol{E}_0 \, \mathrm{e}^{ik\Delta} \, dS \tag{31.1}$$

where Δ is the wave-front aberration of the quasi-spherical wave front Σ from the spherical wave front S, of radius R and centered at O'.

The approximations involved in deriving Eq. (31.1) will be stated in the process of derivation.

First, one assumes that

$$\lambda \ll r, \tag{31.2}$$

that is that kr is large. Let the surface S be a wave front Σ described by

$$L(x, y, z) = \text{constant}. \tag{31.3}$$

One also may admit that S is very close but not necessarily coincident with Σ. The relation between $\boldsymbol{E}$, $\boldsymbol{H}$, Σ and the wave-front normal $\boldsymbol{n}$ is shown in Fig. 56. We next show, according to Maréchal and Françon (1960) that $\boldsymbol{E}$ and $\boldsymbol{H}$ are tangent to the wave front Σ [i.e. to the phase front $L(x, y, z)$=constant] and that they form an orthogonal triad with grad L, which is in the direction of $\boldsymbol{n}$ (in the vacuo grad $L=\boldsymbol{n}$). In the Heaviside-Lorentz system of rational units, Maxwell's equations in vacuo ($\mu=1$) can be written as

$$\operatorname{curl} \boldsymbol{H} = \frac{\varepsilon}{c} \frac{\partial \boldsymbol{E}}{\partial t}, \tag{31.4}$$

$$\operatorname{curl} \boldsymbol{E} = -\frac{1}{c} \frac{\partial \boldsymbol{H}}{\partial t}. \tag{31.5}$$

With the time dependence stated explicitly, one has in complex notation

$$\left.\begin{aligned} \boldsymbol{E} &= \boldsymbol{E}_0(x, y, z) \exp\left[+i\omega\left(t - \frac{L_E(x, y, z)}{c}\right)\right], \\ \boldsymbol{H} &= \boldsymbol{H}_0(x, y, z) \exp\left[+i\omega\left(t - \frac{L_M(x, y, z)}{c}\right)\right] \end{aligned}\right\} \tag{31.6}$$

where L_E and L_M describe the local phase of $\boldsymbol{E}$ and of $\boldsymbol{H}$ respectively so that $\boldsymbol{E}_0$ and $\boldsymbol{H}_0$ are real vectors. We shall find that $L_E = L_M \equiv L(x, y, z)$ is the optical path length. In Eq. (31.6) one only assumes that the x, y, z components of $\boldsymbol{E}$ and of $\boldsymbol{H}$, respectively, are in phase, which does not restrict the argument. We next apply to (31.6) the identity

$$\operatorname{curl} m\boldsymbol{E}_0 \equiv m \operatorname{curl} \boldsymbol{E}_0 + \operatorname{grad} m \times \boldsymbol{E}_0$$

which gives

$$\operatorname{curl} \boldsymbol{E} = \exp[\ldots] \operatorname{curl} \boldsymbol{E}_0 + \operatorname{grad} \exp[\ldots] \times \boldsymbol{E}_0. \tag{31.7}$$

We also get

$$\operatorname{grad} \exp[\ldots] = -\frac{i\omega}{c} \exp\left[i\omega\left(t - \frac{L_E}{c}\right)\right] \operatorname{grad} L_E \tag{31.8}$$

and finally

$$\operatorname{curl} \boldsymbol{E} = \exp\left[i\omega\left(t - \frac{L_E}{c}\right)\right]\left[\operatorname{curl} \boldsymbol{E}_0 - \frac{i\omega}{c} \operatorname{grad} L_E \times \boldsymbol{E}_0\right] \tag{31.9}$$

with a similar equation for curl $\boldsymbol{H}$. For very high frequencies ω, the second term of the second bracket is large in comparison to curl $\boldsymbol{E}_0$, and one has

$$\operatorname{curl} \boldsymbol{E} \simeq \left[-\frac{i\omega}{c} \operatorname{grad} L_E \times \boldsymbol{E}_0\right] \exp\left[i\omega\left(t - \frac{L_E}{c}\right)\right]. \tag{31.10}$$

MAXWELL's equations (31.4) and (31.5) then give

$$[\operatorname{grad} L_E \times \boldsymbol{E}_0] \exp\left[-i\omega \frac{L_E}{c}\right] = +\boldsymbol{H}_0 \exp\left[-\frac{i\omega L_M}{c}\right] \tag{31.11}$$

and similarly

$$[\operatorname{grad} L_M \times \boldsymbol{H}_0] \exp\left[-i\omega \frac{L_M}{c}\right] = -\varepsilon \boldsymbol{E}_0 \exp\left[-\frac{i\omega L_E}{c}\right]. \tag{31.12}$$

It is clear now that $L_E = L_M = L$ is the optical path. We finally have the equations

$$\varepsilon \boldsymbol{E}_0 = -\operatorname{grad} L \times \boldsymbol{H}_0 \tag{31.13a}$$

and

$$\boldsymbol{H}_0 = \operatorname{grad} L \times \boldsymbol{E}_0 \tag{31.13b}$$

from which we conclude that $\boldsymbol{E}_0$, $\boldsymbol{H}_0$ and grad L indeed form a system of orthogonal vectors, and also that $\boldsymbol{E}_0$ and $\boldsymbol{H}_0$ are tangent to the wave front $L(x, y, z) =$ constant under the stated assumptions.

In vacuo, grad $L = \boldsymbol{n}$, and we now return to Eq. (30.9) where we let $\boldsymbol{E} \to \boldsymbol{E}_0$ and $\boldsymbol{H} \to \boldsymbol{H}_0$. Eq. (30.9) can now be written after some simplifications as

$$\left.\begin{aligned} \boldsymbol{E}_{O'} \simeq \int_S \Big\{ &+ \frac{ik}{4\pi} \boldsymbol{E}_0 \frac{e^{-ikr}}{r} + \frac{i}{k} \operatorname{grad} \operatorname{div} \frac{1}{4\pi} \boldsymbol{E}_0 \frac{e^{-ikr}}{r} + \\ &+ \operatorname{curl} \frac{1}{4\pi} \boldsymbol{H}_0 \frac{e^{-ikr}}{r} \Big\} dS \end{aligned}\right\} \tag{31.14}$$

where we may also write

$$\varphi(r) = \frac{1}{r} e^{-ikr}. \tag{31.15}$$

With the use of appropriate vector identities and the relation

$$\operatorname{grad} \varphi(r) = \boldsymbol{i}_r \frac{\partial \varphi(r)}{\partial r} \simeq -\boldsymbol{i}_r\, ik\, \varphi(r) \tag{31.16}$$

with $\boldsymbol{i}_r$ being the unit vector of the radial direction, and therefore

$$\operatorname{div} \frac{\boldsymbol{E}_0}{4\pi} \varphi(r) = -\frac{ik}{4\pi} \boldsymbol{E}_0 \cdot \varphi(r)\, \boldsymbol{i}_r \tag{31.17}$$

and by noting that curl $\boldsymbol{H}_0 = 0$, because $\boldsymbol{H}_0 =$ constant, one has

$$\boldsymbol{E}_{O'} \simeq \frac{ik}{4\pi} \int_S [\boldsymbol{E}_0 - (\boldsymbol{E}_0 \cdot \boldsymbol{i}_r)\, \boldsymbol{i}_r + \boldsymbol{H}_0 \times \boldsymbol{i}_r]\, \varphi(r)\, dS. \tag{31.18}$$

Since

$$\boldsymbol{E}_0 = \boldsymbol{E}_{0yz} + (\boldsymbol{E}_0 \cdot \boldsymbol{i}_r)\,\boldsymbol{i}_r \tag{31.19}$$

where $\boldsymbol{E}_{0yz}$ is the projection of $\boldsymbol{E}_0$ onto the yz plane, it further follows that

$$\boldsymbol{E}_{O'} = \frac{i\,k}{4\pi} \int\limits_S [\boldsymbol{E}_{0yz} + \boldsymbol{H}_0 \times \boldsymbol{i}_r]\,\varphi(r)\,dS. \tag{31.20}$$

With $\boldsymbol{i}_r \simeq \boldsymbol{n}$, $\boldsymbol{H}_0 \times \boldsymbol{i}_r \simeq \boldsymbol{E}_0$ and with $k = 2\pi/\lambda$, we can write (31.20) as

$$\boldsymbol{E}_{O'} \simeq \frac{i}{\lambda} \int\limits_S \boldsymbol{E}_0\,\varphi(r)\,dS \tag{31.21}$$

which has almost the desired form of Eq. (31.1).

We need only to note that

$$r = R - \varDelta \tag{31.22}$$

where R is the radius of the reference sphere S, and $\varDelta$ the aberration of the image-forming wave front Σ from the reference sphere S.

With these relations, we have

$$\varphi(r) = \frac{e^{-ikr}}{r} = \frac{1}{R-\varDelta}\, e^{-ikR}\, e^{ik\varDelta} \tag{31.23}$$

or approximately

$$\varphi(r) \simeq \frac{e^{-ikR}}{R}\, e^{ik\varDelta}. \tag{31.24}$$

We have finally

$$\boldsymbol{E}_{O'} \simeq \frac{i}{\lambda}\,\frac{e^{-ikR}}{R} \iint\limits_{\Sigma} \boldsymbol{E}_0\, e^{ik\varDelta}\, dS \tag{31.25}$$

which is indeed the approximate form of the rigorous HUYGENS' principle which was to be derived.

It only remains to express Eq. (31.25) in the Fourier-transform formulation.

32. Fourier-transform formulation of the image formation problem (physical optics approximation). Eq. (31.25) gives a useful approximate form of the HUYGENS' principle derived from the rigorous form of HUYGENS' principle given in Eq. (30.9). A considerable further simplification in the use of the approximate form of HUYGENS' principle results when it is expressed in the Fourier-transform formulation, which results in expressing that *the complex amplitude of the electric field at any image point in the focal plane of an image-forming pupil is equal to the Fourier-transform of the complex amplitude distribution of the electric field distribution within the image forming pupil.*

We recall that three important approximations are involved in particular:

1. The formulation applies to the center and the near-vicinity of a quasi-spherical wave front of radius R.

2. The wavelength λ is such that $\lambda \ll R$.

3. (a) The electromagnetic field vectors $\boldsymbol{E}$ and $\boldsymbol{H}$ are assumed tangent to the wave front $L(x, y, z) = \text{constant}$.

(b) $\boldsymbol{E}$, $\boldsymbol{H}$ and grad L form an orthogonal triad.

The equation to be derived is

$$\boldsymbol{E}_{P'}(k\alpha, k\beta) \simeq \frac{i}{\lambda}\,\frac{e^{-ikf}}{f} \iint\limits_{\substack{\text{pupil}\\ \text{(coordinates: } x,\, y)}} \boldsymbol{E}_0\, e^{ik\varDelta(x,y)}\, e^{ik(\alpha x + \beta y)}\, dx\, dy \tag{32.1}$$

where $\boldsymbol{E}_{P'}$ is the complex amplitude of the electric field vector at a point P' in the image, such that the coordinates of P' are

$$x' \simeq \frac{f\alpha}{\gamma}, \quad y' \simeq \frac{f\beta}{\gamma}$$

with

$$\gamma \doteqdot 1.$$

The complex amplitude distribution in the pupil (coordinates x, y) is given by

$$\boldsymbol{E}_0 \, \mathrm{e}^{-ik\Delta(x,y)},$$

and α, β, and γ are direction cosines in the pupil, on the plane-wave side of a perfect lens (or mirror). Here the radius R of the image-forming wave fronts has been identified with the focal length f.

In Fig. 58 (p. 518), let P' be a general point in the $x'y'$ image plane, such that P' is near O'. Let S_ϑ be the sphere centered at P' and let $\boldsymbol{\varrho}$ be a vector in the pupil describing a point $P(x, y)$ on the plane-wave side of the pupil. We recall that it is preferable to deal in the image-formation problems with the plane-wave side of the pupil of a "perfect" image forming system. A perfect image forming system is defined as having the property of transforming plane waves in the direction $\boldsymbol{i}_r$ into spherical waves centered on P' in the focal plane of the optical system. Moreover, by virtue of the theorem of GOUY[1], aberrations Δ from a reference plane on the plane-wave side of the optical system remain equal to Δ on the same "ray" when measured with respect to the corresponding reference sphere on the spherical-wave side of the optical system. With respect to the tilted plane S_ϑ and referring to the aberration Δ of Σ from S in Eq. (31.25), we thus have to substract from Δ an additional path difference Δ_ϑ such that

$$\Delta_\vartheta = \boldsymbol{i}_r \cdot \boldsymbol{\varrho} \tag{32.2}$$

where $\boldsymbol{i}_r$ is the unit vector in the direction from O to P'. We have

$$\boldsymbol{\varrho} = x\boldsymbol{i}_x + y\boldsymbol{i}_y \tag{32.3}$$

and

$$\boldsymbol{i}_r = \boldsymbol{i}_x\alpha + \boldsymbol{i}_y\beta + \boldsymbol{i}_z\gamma \tag{32.4}$$

where α, β, γ are the direction cosines of $\boldsymbol{i}_r$. Eqs. (32.2), (32.3) and (32.4) give

$$\Delta_\vartheta = \alpha x + \beta y, \tag{23.5}$$

and we must write

$$\Delta \equiv \Delta - \Delta_\vartheta \tag{32.6}$$

in Eq. (31.25) and henceforth. We also note that

$$dS = dx\,dy \tag{32.7}$$

in the pupil. With these equations, and with $R = f$ being the focal length of the image forming system, Eq. (31.25) immediately gives

$$\boldsymbol{E}_{P'}(k\alpha, k\beta) \simeq \frac{i}{\lambda}\frac{\mathrm{e}^{-ikf}}{f} \iint\limits_{\substack{\text{pupil} \\ \text{(coordinates; } x,\, y)}} \boldsymbol{E}_0 \, \mathrm{e}^{ik\Delta(x,y)} \, \mathrm{e}^{-ik(\alpha x + \beta y)\, dx dy} \tag{32.1}$$

which is the desired Fourier-transform formulation of image formation in the approximate formulation of the rigorous HUYGEN'S principle, as derived from MAXWELL'S equation with the stated approximations.

[1] Cf. A. MARÉCHAL, this Encyclopedia, vol. 24, p. 109.

Eq. (32.1) has the form of a Fourier transformation and can be read as follows: the complex amplitude of the electric field vector at a point in the image plane is equal to the Fourier transform of the distribution of complex amplitude of the electric field within the image-forming aperture. Clearly here, the field vectors in the aperture and the field vectors in the image plane project along the same direction in the image plane. One Fourier transformation needs to be carried out for each point in the diffraction pattern.

For example, for a perfectly uniform plane wave front within a rectangular aperture of width A along the x-axis, one has $|\boldsymbol{E}_0|=1$, $\Delta=0$, and therefore

$$\boldsymbol{E}_{P'}(k\alpha) \simeq \frac{i}{\lambda}\,\frac{\mathrm{e}^{-ikf}}{f}\int_{-A/2}^{+A/2} \mathrm{e}^{-ik\alpha x}\,dx \tag{32.8}$$

which immediately gives upon integration

$$\boldsymbol{E}_{P'}(k\alpha) = \left[\frac{i}{\lambda}\,\frac{\mathrm{e}^{-ikf}}{f}\,\frac{A}{2}\right]\left[\frac{\sin k\,\alpha(A/2)}{(k\,\alpha A/2)}\right]. \tag{32.9}$$

The complex amplitude $\boldsymbol{E}_{P'}(k\alpha)$ thus varies according to the well-known $\sin x'/x'$ function along $x'\simeq f\alpha$ in the image plane and has a first minimum for

$$\alpha_0=\frac{\lambda}{A} \text{ radians} \tag{32.10}$$

or for

$$x=f\frac{\lambda}{A} \qquad (f=\text{focal length of focussing system}). \tag{32.11}$$

In general only the time-averaged intensity (we omit the time-average sign)

$$I_{P'}=\boldsymbol{E}_{P'}\cdot\boldsymbol{E}^*_{P'} \tag{32.12}$$

is detectable (with the help of photoelectric, photographic or other receivers). $I_{P'}$ varies as $(\sin x'/x')^2$ and has a first minimum at $(\lambda/A)f$ from the central maximum. $I_{P'}$ is generally called the *diffraction pattern* corresponding to the aperture A.

Eq. (32.1), which expresses the diffraction at infinity by an aperture, can be made heuristically plausible on the basis of superposition and of the HUYGENS' principle as it is generally understood (see Sect. 37). Consider a pupil in the x, y plane and a point P centered on an element of area $dx\,dy$ in that plane. The element of area centered on P emits an elementary wave $f(x,y)\,dx\,dy$, where $f(x,y)$ describes the scalar component of the $\boldsymbol{E}$ vector in the pupil. In the direction defined by the direction cosines α, β, γ the wave from P is dephased by $(2\pi/\lambda)(\alpha x+\beta y)$ with respect to an elementary wave emitted from the center of the pupil O. By superposition, one has in the direction α, β, γ the sum of these waves

$$\iint^{\text{pupil}} f(x,y)\exp\left[i\,\frac{2\pi}{\lambda}(\alpha x+\beta y)\right]dx\,dy. \tag{32.13}$$

All elementary waves emitted by the pupil (or transmitted by the pupil) in the direction α, β, γ come to focus at a single point $P'(x',y')$ in the focal plane of a perfect focussing system, such that

$$x'=\frac{f\alpha}{\gamma}; \qquad y'=\frac{f\beta}{\gamma}. \tag{32.14}$$

In as much as $\gamma\simeq 1$ and $\alpha^2+\beta^2+\gamma^2=1$ one has

$$x'=f\alpha; \qquad y'=f\beta. \tag{32.15}$$

If one defines the coordinates in terms of reduced coordinates

$$u = \frac{x'}{f}; \qquad v = \frac{y'}{f} \tag{32.16}$$

and if one takes λ for the unit of length, one obtains

$$F(u, v) = \iint\limits_{\text{pupil}} f(x, y)\, e^{2\pi i(ux + vy)}\, dx\, dy \tag{32.17}$$

which is an expression for the diffraction pattern identical to that obtained from MAXWELL's equations in Eq. (32.1) by using the there stated approximations.

In summary, a knowledge of the distribution of complex amplitude (amplitude and phase) of the electromagnetic field (or more exactly of a component of the electric field) within an aperture, no matter how this field is created in this aperture, permits to compute the diffraction patterns corresponding to this aperture and light distribution. A unique relation between the field distribution in the aperture and the light distribution in the diffraction patterns exists, within the stated approximations, and takes the form of a Fourier transformation. The powerful techniques of operational calculus have been extensively applied to optical image formation problems with very fruitful results. In particular, a basic similarity has been recognized between problems in electrical engineering and problems of optical image formation and spectroscopy, whenever superposition and operational methods are appropriate.

33. Restrictions in applicability of the Fourier transform formulation of the diffraction problem. We have noted that Eq. (32.1) is identical in form to an expression of the diffraction problem, Eq. (32.13), which can be derived on the basis of scalar-theory approximations, starting from a postulation of HUYGENS' principle. According to HUYGENS' principle an aperture can be broken up into a family of "source elements" which all emit "spherical wavelets" that conveniently sum and cancel out by "interference" at appropriate regions in the image space. However, the weakness of HUYGENS' principle derivation resides in the fact that the "principle" has no formal physical basis to start with, and, moreover, that the equations derived contain no warning as to the domain of applicability, thus permitting laborious extensions to situations where the equations are inapplicable.

For instance, it has been traditional in grating theory to attempt calculating the energy distribution among orders by assuming the grooves to be simple slits, or by "improving" on this assumption by taking into account the detailed groove shape with the introduction of appropriate phase retardation resulting from the path difference within the groove and so on. It should be clear that such attempts should be fruitless, as they are found to be, experimentally, except in special cases where it happens that electromagnetic solution of the energy distribution problem (Sect. 60) coincides with the qualitative results that may be inferred simply from geometrical optics.

Indeed, one of the important approximations involved in deriving Eq. (32.1) is that it is applicable only to the near vicinity of the quasi-spherical wave fronts originating from the pupil. Eq. (32.1) therefore is eminently suitable for dealing separately with each of the waves diffracted by an optical grating: it permits very simple calculation of the complex amplitude distribution $\boldsymbol{E}_{P'}$ and of the intensity distribution $I_{P'} = \boldsymbol{E}_{P'} \cdot \boldsymbol{E}_{P'}^*$ in the diffraction pattern formed by the diffracted wave fronts of various orders and wavelengths, in the focal plane of the focussing optical system. Therein resides the interest of Eq. (32.1) in grating theory in particular, and in optical image formation in general.

However, Eq. (32.1) is clearly not applicable to dealing with individual grating grooves as diffracting pupils, when the various spherical wave fronts originating at the grooves are widely separated in angle, as happens to be the case for gratings in the general case. The exception formed by certain cases of echelon, echelette and echelle gratings when the groove spacings are very large in terms of the wavelength has already been noted, but so has the exception to the exception in the presence of the effect described by Maréchal and Stroke, when the electromagnetic character of diffraction and polarization starts to manifest itself predominantly (case of the 90° grooves $m\lambda/2$ deep and with the $\boldsymbol{E}$ vector normal to the $m\lambda/2$ groove side).

34. Summary. The reasons for the well-known and puzzling discrepancies between experimental energy distributions among orders of optical gratings and predictions based on "Huygens' principle" formulation can easily be explained and are simple to understand. They simply illustrate the fact that scalar-theory in the Huygens' principle form is in general not applicable for the calculation of energy distribution (blaze) in gratings. The occasional agreement with energy distribution calculations in the special case of echelle type gratings, based on assuming slit-like grooves and scalar energy distribution, is also accounted for. On the other hand, one finds an excellent agreement between theory and experiment in the predictions of the distribution of light amplitude and intensity in the diffraction patterns formed by the diffracted wave fronts, when these predictions are based on the Fourier-transform formulation of Huygens' principle (Sect. B IV). For the general case of energy distribution among orders of a grating, and polarization of the diffracted waves, solutions of the electromagnetic boundary value problem of diffraction by a grating are however necessary (Sect. B VII).

IV. Physical optics theory of spectroscopic image formation by gratings (Fourier-transform formulation).

35. Introduction and mathematical references. Great simplicity in the mathematical theory of image formation in general, and grating image formation in particular, results from the use of the Fourier transform formulation of physical optics. An important stimulus was given to this formulation by the work of M. Duffieux in 1944. Duffieux points out that the foundations for the Fourier transform formulation of image formation were apparently given in 1892 by A. A. Michelson and Lord Rayleigh. Even the operational form of the Fourier transform relation between wave fronts and images is explicitly given by Michelson in his book *Studies in Optics*, first published in 1927, where he refers (p. 58) to a Philosophical Magazine article of April 1905.

Very little if any use of this formulation appears to have been made until the 1940's, when it was newly developed by several authors, many of them starting quite independently, as it seems .Aside from the work of Duffieux (1946) [*1*][1], one finds the Fourier transform formulation given in particular by S. Silver [*2*] (1949), in relation to microwave antenna patterns, and a significant impetus is also due to the work of A. Maréchal [*6, 11, 14, 15, 29*], H. H. Hopkins [*12*], G. Toraldo di Francia [*27*], E. Ingelstam [*7*], D. Gabor [*8*], E. H. Linfoot [*9, 13*], P. B. Fellgett [*9*], E. L. O'Neill [*25*], P. Jacquinot [*10*] and J. Strong [*24*] in the 1950's, as well as a paper by P. Elias, D. Grey and D. Robinson [*4, 5*] in 1952[2]. The concept and methods of spatial filtering and transfer

[1] The numbers between brackets refer to the references to Sect. 35 on p. 516.

[2] For additional references see for example A. Maréchal et M. Françon: Diffraction, edit. Revue d'Optique 1960, or E. L. O'Neill: Introduction to Statistical Optics, edit. Addison-Wesley, Reading, Mass 1963.

functions comparable to frequency filtering in electronics, were introduced into optics by a number of workers, among which LUNEBURG (1944) [*28*], DUFFIEUX (1946) [*1*], and more particularly MARÉCHAL (1951) [*29*], CHEATHAM and KOHLENBERG (1952) [*30*], ELIAS (1953) [*4, 5*], O'NEILL (1956) [*31*], H. H. HOPKINS (1955), E. INGELSTAM and E. DJURLE (1956). It seems that the first formal description of a spatial frequency filter under the name "Un filtre de fréquences spatiales pour l'amélioration due contraste des images optiques" was given by A. MARÉCHAL and P. CROCE [*6*] in C. R. Acad. Sci. Paris in September 1953. A more complete history of these developments may be found in E. L. O'NEILL's book "An Introduction to Statistical Optics" (1963)[1] which may also serve as a reference for the mathematical developments used in this section, along with A. MARÉCHAL's and M. FRANÇON's "Diffraction, Structure des Images" (1960)[2] and J. ARSAC's "Transformation de Fourier et Théorie des Distributions" (1961)[3].

As shown by J. ARSAC (loc. cit. 1961), great additional simplicity in the Fourier treatment of optical image formation results from application of the *Théorie des distributions* first evolved by LAURENT SCHWARTZ in 1950—1951. Well-known contributions to this theory were also made by M. J. LIGHTHILL [*19*] (1958) and by A. ERDÉLYI [*20*] (1962). The theory of distributions is sometimes known as the theory of generalized functions. Here we shall follow the development of J. ARSAC which is proving to be very fruitful both in optics and in radio-astronomy (J. ARSAC, R. N. BRACEWELL).

Application of the Fourier transform methods to grating image formation appears to have been made as early as 1951 by A. CAMUS, M. FRANÇON, E. INGELSTAM and A. MARÉCHAL[4], and by E. INGELSTAM and E. DJURLE[5] in 1952 and 1953, in particular in relation to phase contrast testing of gratings. It was recognized that study of the diffraction patterns, rather than of the transfer functions, appeared to be of more particular interest in the spectroscopic use of gratings, even though A. LOHMANN[6] also investigated the frequency transfer in spectrographs in 1959.

A very general treatment of grating theory in terms of Fourier transform formulation was given by G. W. STROKE[7] in 1959 and 1960, when he showed that the complex amplitude and intensity distributions in the spectral diffraction patterns formed by gratings could be easily calculated from the amplitude and phase distribution in the wavefronts, as they appear in readily usable form in the wavefront interferograms. It is in fact in the simplicity by which the wavefront interferograms (STROKE 1955, 1960)[8,9] may be obtained, and by which they display the grating and wavefront departures from perfection, which lends such interest to the use of Fourier methods in grating image formation.

A wavefront interferogram of a 10 inch grating at 74° and the corresponding intensity distribution computed from the wavefront interferogram by Fourier transformation are shown in Fig. 57, according to G. W. STROKE. The intensity distribution in the spectral line obtained with the help of the green 5461 Å line

[1] Edited by Addison-Wesley Publishing Co., Reading, Mass.

[2] Edited by Revue d'Optique, Paris.

[3] Edited by Dunod, Paris.

[4] A. M. CAMUS, M. FRANÇON, E. INGELSTAM, et A. MARÉCHAL: Rev. Opt. **30**, 3 (1951).

[5] E. INGELSTAM and E. DJURLE: J. Opt. Soc. Amer. **43**, 572 (1953).

[6] A. LOHMANN: Optica Acta **6**, 175 (1959).

[7] G. W. STROKE: Etude de théorique et expérimentale de deux aspects de la diffraction de la lumière par les résequx optiques: l'évolution de défauts dans les figures de diffraction et l'origine électromagnétique de la répartition entre les ordrers. Rev. Opt. **39**, 291—398 (1960).

[8] G. W. STROKE: J. Opt. Soc. Amer. **54**, 30 (1955).

[9] See References (Sect. 35), pp. 516—517.

of a mercury isotope Hg 198 electrodeless discharge lamp is shown for comparison. Details on the computations of the Fourier transforms are given in Sect. 59.

For the mathematical background required in this chapter, the reader is referred to the texts cited above or to standard treatises (see e.g. [32]).

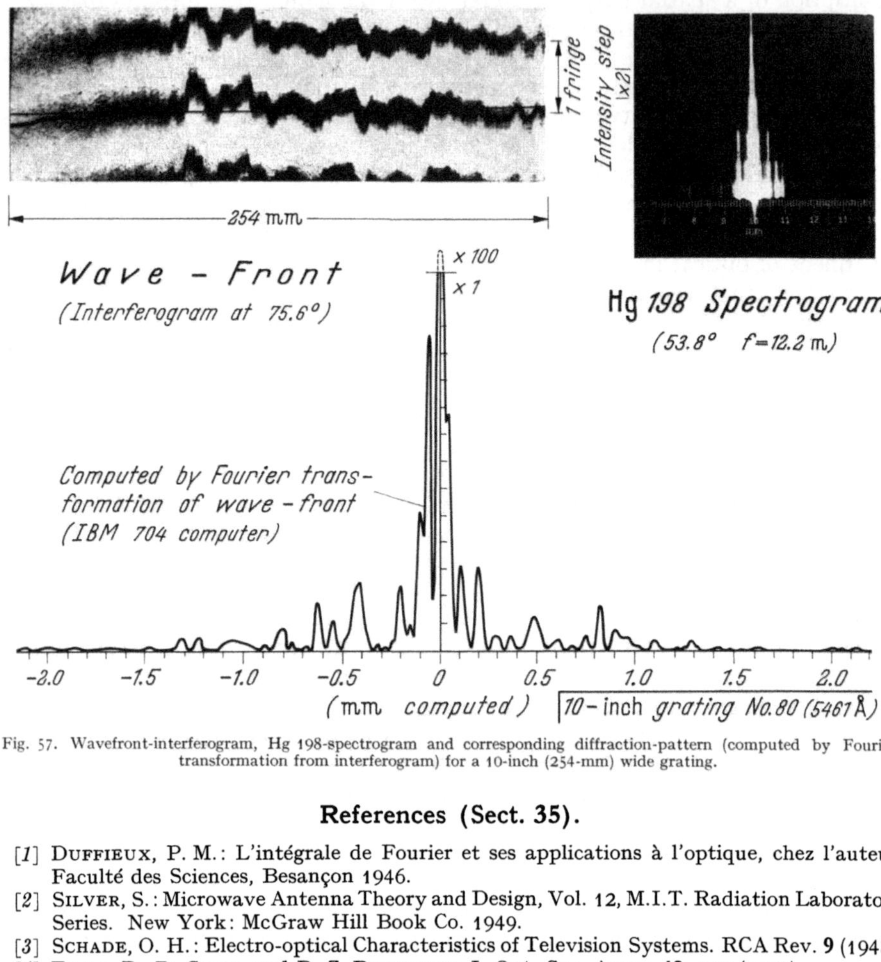

Fig. 57. Wavefront-interferogram, Hg 198-spectrogram and corresponding diffraction-pattern (computed by Fourier transformation from interferogram) for a 10-inch (254-mm) wide grating.

References (Sect. 35).

[1] DUFFIEUX, P. M.: L'intégrale de Fourier et ses applications à l'optique, chez l'auteur, Faculté des Sciences, Besançon 1946.

[2] SILVER, S.: Microwave Antenna Theory and Design, Vol. 12, M.I.T. Radiation Laboratory Series. New York: McGraw Hill Book Co. 1949.

[3] SCHADE, O. H.: Electro-optical Characteristics of Television Systems. RCA Rev. **9** (1948).

[4] ELIAS, P., D. GREY, and D. Z. ROBINSON: J. Opt. Soc. Amer. **42**, 127 (1952).

[5] ELIAS, P.: J. Opt. Soc. Amer. **43**, 229 (1953).

[6] MARÉCHAL, A., et P. CROCE: C. R. Acad. Sci. Paris **237**, 607 (1953).

[7] INGELSTAM, E., E. DJURLE, and B. SJOGREN: J. Opt. Soc. Amer. **46**, 707 (1956).

[8] GABRO, D.: Lectures in Communication Theory, Report 238, Research Laboratory of Electronics, M.I.T. 1952.

[9] LINFOOT, E. H., and P. B. FELLGETT: Trans. Roy. Soc. Lond. **247** (1955).

[10] JACQUINOT, P., P. BOUGHTON, and B. DOSSIER: In: Contribution à la Théorie des Images Optiques, edit. de la Revue d'Optique 1949, p. 183.

[11] MARÉCHAL, A.: Thèse, Revue d'Optique 1948.

[12] HOPKINS, H. H.: Wave Theory of Aberrations. Oxford: Clarendon Press 1950.

[13] LINFOOT, E. H.: Recent Advances in Optics. Oxford: Clarendon Press 1955.

[14] MARÉCHAL, A.: Optique Géométrique Générale. In: Handbuch der Physik, Vol. XXIV, edit. by S. FLÜGGE. Berlin-Göttingen-Heidelberg: Springer 1956.

[15] MARÉCHAL, A., et M. FRANÇON: Diffraction, Structure des Images, Influence de la Cohérence de la lumière, edit. de la Revue d'Optique, Paris 1960.

[16] BORN, M., and E. WOLF: Principles of Optics. London: Pergamon Press 1959.

[17] STROKE, G. W.: Ruling, Testing and Use of Optical Gratings for High-Resolution Spectroscopy. In: Progress in Optics, Vol. II, edit. by E. WOFL. Amsterdam: North-Holland Publishing Company 1963.
[18] ARSAC, J.: Transformation de Fourier et Théorie des Distributions. Paris: Dunod 1961.
[19] LIGHTHILL, M. J.: An Introduction to Fourier Analysis and Generalized Functions. Cambridge: Cambridge University Press 1958.
[20] ERDÉLYI, A.: Operational Calculus and Generalized Functions. New York: Holt, Rinehart and Winston 1962.
[21] GUILLEMIN, E. A.: The Mathematics of Circuit Analysis. New York: John Wiley & Sons. 1951.
[22] ANGOT, A.: Compléments de Mathématiques, édit. de la Revue d'Optique 1957.
[23] STROKE, G. W.: Etude théorique et expérimentale de deux aspects de la diffraction par les réseaux optiques: l'évolution des defauts dans les figures de diffraction et l'origine électromagnetique de la répartition entre les ordres. Rev. Opt. **39**, 291—398 (1960). — Thèse, Sorbonne 1960.
[24] STRONG, J.: Concepts in Classical Optics. San Francisco: W. H. Freeman & Co. 1958.
[25] O'NEILL, E. L.: Introduction to Statistical Optics. Reading, Mass.: Addison-Wesley 1963.
[26] STONE, J. M.: Radiation and Optics. New York: McGraw Hill Book Co. 1963.
[27] TORALDO DI FRANCIA, G.: La diffrazione della luce. Torino: Einaudi 1958.
[28] LUNEBURG, R. K.: Mathematical Theory of Optics. Providence, R.I.: Brown University 1944.
[29] MARÉCHAL, A.: The Contrast of Optical Images and the Influence of Aberrations. National Bureau of Standards Circular 526, U.S. Dept. of Commerce (Proceedings of the N.B.S. Semicentennial Symposium on Optical Image Evaluation, Oct. 1951).
[30] CHEATHAM jr., T. P., and A. KOHLENBERG: Analysis and Synthesis of Optical Systems. Technical Note 84, Boston University Physical Research Lab., March 1952.
[31] O'NEILL, E. L.: Selected Topics in Optics and Communication Theory. Boston University 1956/57.
[32] STROKE, G. W.: An Introduction to Coherent Optics and Holography. New York and London: Academic Press 1966.

36. Diffraction patterns and distribution of energy among orders. In order to deal with the problem of the diffraction of light by gratings, we show in the next section that a Fourier transform relation exists quite generally between a diffracting pupil and the corresponding diffraction image [Eq. (32.1)]. It is important to appreciate that the term diffraction image applies to the entire distribution of complex amplitude (or of intensity) in the image plane resulting from the complex amplitude distribution of the field in the pupil. In case of imperfect wave fronts, STROKE (1960) showed that the various false spectral lines (ghosts, "satellites", "grass", etc.) found in the near vicinity of the spectral lines are nothing but diffraction pattern imperfections resulting from the contribution of the entire imageforming wavefront (Sects. 50 to 59).

In a larger sense, when a pupil is formed by a periodic structure, for instance by a grating, it will appear, formally at least, that the various spectral lines of different orders are indeed a part of the "diffraction image" of the grating.

Moreover, in terms of the scalar formulation used in this chapter, the energy distribution among the lines of various orders (but of the same wavelength, of course) will also appear determined by a *groove shape function*, that is, determined by the effect of the groove profile, in particular.

However, the reader must be cautioned against using a scalar method for determining the *groove shape function*. It has indeed been shown (see Chapters B III and B VII) that the groove function must in general be determined on *the basis of electromagnetic boundary considerations*. We recall moreover the fact, proven in Chapter B III, that the Fourier transform treatment applies only to the near vicinity of the centers of the quasi-spherical image-forming wave fronts. The apparent simplicity of extensions to large angle diffraction should therefore be considered as deceptive. As a result, it becomes preferable to use the Fourier transform treatment of the diffraction of light by optical gratings only for the calculation of the *distribution of light in the diffraction patterns* (perfect or imperfect) formed by each of the diffracted waves considered separately, for which the treatment is eminently suitable.

37. Fourier transform formulation of light distribution in the grating diffraction patterns. The Fourier-transform expression of the light distribution in the

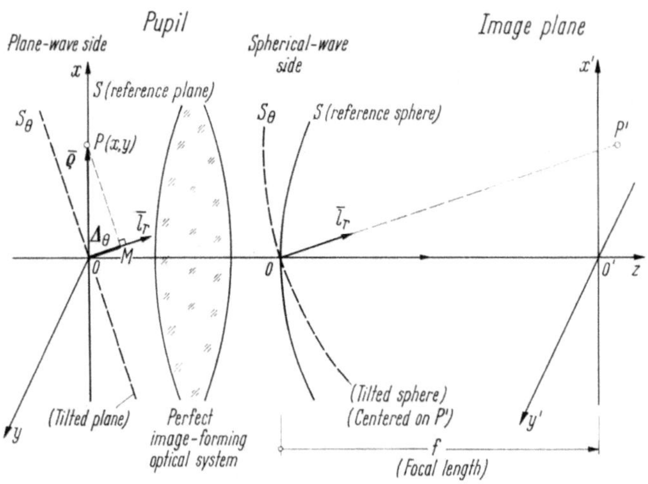

Fig. 58. Wavefront and focal-plane coordinates, for Fourier-transform computation. Vectors are denoted by bars.

diffraction patterns [Eq. (32.1)], can be easily derived on the basis of Huygens' principle. (For the restrictions in the applicability of these derivations see Chapter B III.) We recall the derivation of Eq. (32.13) and complete the formalism of Sect. 32.

Consider the plane-wave side of a pupil x, y, z (Fig. 58). Let the complex amplitude of the electric field distribution in the pupil be $f(x, y)$, where $f(x, y)$ applies to a monochromatic wavelength.

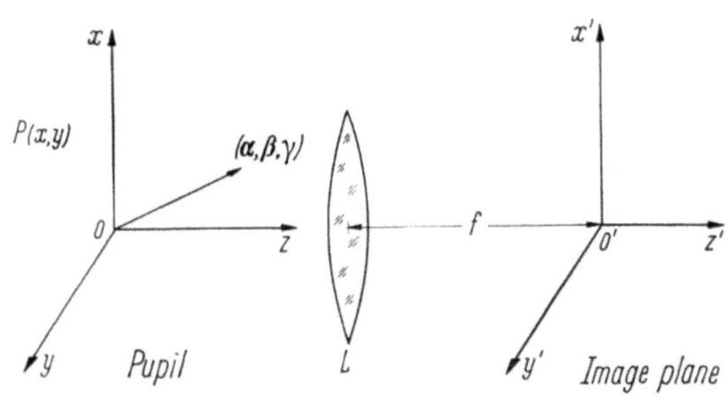

Fig. 59. Wavefront and image-plane coordinates, for Fourier-transform computation.

An element of area $dx\,dy$ surrounding the point $P(x, y)$ in the pupil emits an elementary wavelet

$$f(x, y)\,dx\,dy.$$

In the direction (α, β, γ) (Fig. 59), this wavelet has a phase difference

$$\varphi_P = \frac{2\pi}{\lambda}(\alpha x + \beta y) \qquad (37.1)$$

with respect to a similar elementary wavelet emitted from the center O of the pupil. By superposition of the effects of such wavelets, one simply has the complex amplitude

$$F_{P'}(\alpha, \beta, \gamma) = \iint\limits_{\text{pupil}} f(x, y) \exp\left[i\frac{2\pi}{\lambda}(\alpha x + \beta y)\right] dx\,dy \qquad (37.2)$$

at a point $P'(\alpha, \beta, \gamma)$ in the image plane, in the focus of a perfect optical system L. Eq. (37.2) is immediately seen to be equivalent to Eq. (32.1) obtained by suitable approximations from MAXWELL'S equations and the rigorous HUYGENS' principle for electromagnetic waves. It is advisable to refer to Chapter B III for the important questions concerning the domains of applicability of Eq. (37.2), but there is no need for using any greater rigour in deriving this equation here since its domain of applicability is restricted to the domain where the approximations involved are valid.

Some simple transformations permit to introduce the so-called *spatial frequency variables* u and v.

One has the image plane, with f = focal length,

$$x' = \frac{f\alpha}{\gamma}, \qquad y' = \frac{f\beta}{\gamma} \tag{37.3}$$

and since

$$\gamma \simeq 1 \tag{37.4}$$

in order that the HUYGENS' principle may apply (see Chapter B III) and

$$\gamma^2 = 1 - \alpha^2 - \beta^2 \simeq 1 \tag{37.5}$$

near the z-axis, one has

$$x' \simeq f\alpha, \qquad y' \simeq f\beta. \tag{37.6}$$

One next defines the *spatial frequency* parameters

$$u = \frac{X'}{f}, \qquad v = \frac{Y'}{f}. \tag{37.7a}$$

It is seen that the spatial frequency variables u and v are equal to the corresponding direction cosines α and β, divided by the wavelength λ. One has

$$u = \frac{\alpha}{\lambda}, \qquad v = \frac{\beta}{\lambda}. \tag{37.7b}$$

It is also seen that X' and Y' are expressed in terms of the wavelength λ taken as unit of length. Finally, Eq. (37.2) gives with these definitions the Fourier-transform equation

$$F_{P'}(u, v) = \iint_{\text{pupil}} f(x, y)\, e^{\pi i(ux + vy)}\, dx\, dy. \tag{37.8}$$

$F_{P'}(u, v)$ gives the complex amplitude at an image point $P'(u, v)$. It is seen that the complex amplitude $F_{P'}(u, v)$ of the electric field at an image point $P'(u, v)$ is equal to the Fourier-transform of complex amplitude distribution $f(x, y)$ of the electric field within the aperture of the plane pupil in the x, y plane.

It is frequently useful to adopt an operational notation for the Fourier transform relations, and with $F_{P'}(u, v) \equiv F(u, v)$ we may write Eq. (37.8) as

$$F(u, v) = T[f(x, y)] \tag{37.9}$$

where $T[\ ]$ stands for the Fourier transform of the bracketed quantity. One can also write

$$f(x, y) \to F(u, v) \tag{37.10}$$

where it is understood that the arrow $\to$ indicates the Fourier transform operation.

For instance, in case of a one dimensional field distribution within plane aperture of width A and uniform amplitudes across the aperture, one has

$$f(x) = 1 \tag{37.11}$$

and

$$F_{P'}(u) = \int_{-A/2}^{+A/2} e^{2\pi i u x}\, dx \tag{37.12}$$

which gives, upon integration

$$F_{P'}(u) = \frac{1}{2\pi i u}\left[e^{2\pi i u x}\right]_{-A/2}^{+A/2} \tag{37.13}$$

that is

$$F_{P'}(u) = A\,\frac{\sin \pi u A}{\pi u A}. \tag{37.14}$$

Eq. (37.14) gives the amplitude in the diffraction pattern of a rectangular aperture of width A (for the case of a one-dimensional field distribution in the aperture). The corresponding intensity distribution is

$$I_{P'}(u) = A^2\left(\frac{\sin \pi u A}{\pi u A}\right)^2. \tag{37.15}$$

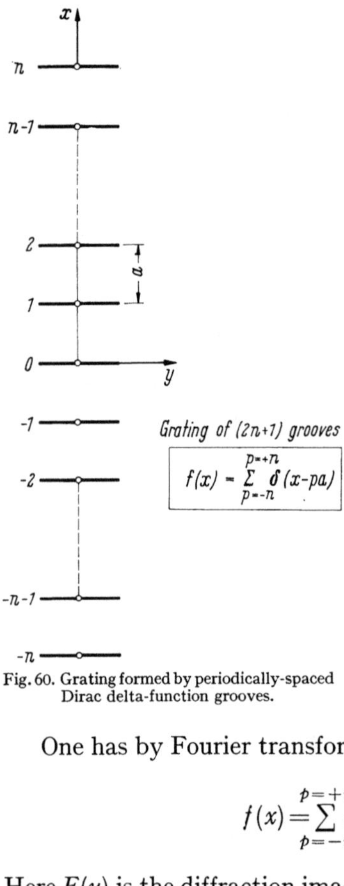

Fig. 60. Grating formed by periodically-spaced Dirac delta-function grooves.

The minima of $I_{P'}(u)$ are obtained for the values of u which are given by

$$\pi u A = n\pi \qquad (n = \text{integer}),$$

that is

$$u_{\min} = \frac{n}{A}. \tag{37.16}$$

According to Eqs. (37.6) and (37.7), the minima of the $(\sin x'/x')^2$ function are found at

$$x' = n f \frac{\lambda}{A}. \tag{37.17}$$

38. The case of a grating formed by $(2n+1)$ periodically spaced Dirac delta functions. Let the pupil be formed by a series of $(2n+1)$ parallel, equidistant, infinitely narrow luminous slits (Fig. 60), so that

$$f(x) = \sum_{p=-n}^{p=+n} \delta(x - pa) \tag{38.1}$$

where the spacing between slits is equal to a, (a = grating spacing constant), p = integer, and $\delta(x)$ is DIRAC's delta function.

One has by Fourier transformation (cf. LIGHTHILL p. 66)

$$f(x) = \sum_{p=-n}^{p=+n} \delta(x - pa) \to F(u) = \sum_{p=-n}^{p=+n} e^{2\pi i p a u}. \tag{38.2}$$

Here $F(u)$ is the diffraction image corresponding to the grating, and u is the spatial frequency defined by Eq. (37.7). We shall show that the diffraction image $F(u)$ is in fact formed by a *spectrum* of periodically spaced maxima distributed according to the grating equation.

Consider indeed Eq. (38.2). One has

$$F(u) = \sum_{p=-n}^{p=+n} e^{2\pi i p a u} = \frac{\sin[(2n+1)\pi a u]}{\sin(\pi a u)}. \tag{38.3}$$

It is seen that $F(u)$ is periodic with a period in the u-space equal to $(1/a)$.

The absolute maxima of $F(u)$ will be obtained for the values of $u = m/a$, where m = integer or zero. Because of the indetermination for this value of u, one may apply L'Hôpital's rule, and one finds for the value of the maxima

$$F\left(\frac{m}{a}\right)=2n+1\,. \tag{38.4}$$

Since

$$u=\frac{\alpha}{\lambda}=\frac{\sin i'}{\lambda} \tag{38.5}$$

one also finds that

$$\sin i'=\frac{m\lambda}{a} \tag{38.6}$$

which is indeed the grating equation for a wave incident on the grating with an angle of incidence $i=0$.

One would be led to conclude according to this treatment that a set of secondary maxima exists between the principal (absolute) maxima for values of u which maximize the numerator in Eq. (38.3), that is for the values of

$$u=\frac{(2m+1)}{2(2n+1)a}\,. \tag{38.7}$$

The reader is again cautioned that the Fourier transform description of optics applies with adequate rigour only to the near-vicinity of each of the spectral lines. We shall later on prefer to reserve the Fourier transform calculation of diffraction patterns (perfect or imperfect) for each of the diffracted wave fronts taken separately. We have indeed formally shown in Sect. 14 that the periodic nature of the grating, regardless of the particular boundary conditions, is sufficient to ensure that the diffracted electromagnetic field will be formed of a *spectrum of plane waves* distributed according to the grating equation

$$\sin i+\sin i'=\frac{m\lambda}{a} \tag{14.1}$$

and more generally according to

$$\gamma'=-\gamma, \tag{14.20a}$$

$$\alpha'+\alpha=m\lambda/a, \tag{14.20b}$$

$$\alpha^2+\beta^2+\gamma^2=\alpha'^2+\beta'^2+\gamma'^2=1\,. \tag{14.20c}$$

The approximations made in deriving the Fourier-transform formulation generally restrict its validity to the diffracted wave fronts taken separately, even though it sometimes appears that the part of the formalism lends itself to extension to the entire spectrum, if suitable caution is exerted.

39. The case of an infinite number of periodically spaced Dirac delta functions. In the case of an infinitely extended grating, one simply has

$$f(x)=\sum_{p=-\infty}^{p=+\infty}\delta(x-pa)\rightarrow F(u)=\frac{1}{a}\sum_{q=-\infty}^{q=+\infty}\delta\left(u-\frac{q}{a}\right) \tag{39.1}$$

which is the well known Poisson relation, according to which the Fourier transform of a Dirac comb is simply another Dirac comb.

40. The case of a grating of periodically spaced Dirac delta functions within an aperture of width *A*. Let the wave front function now be

$$f(x)=d(x)\cdot g(x) \tag{40.1}$$

where $d(x)$ is the Dirac comb

$$d(x)=\sum_{p=-\infty}^{p=+\infty}\delta(x-pa) \tag{40.2}$$

and

$$g(x)=1 \quad \text{in} \quad A \\ \quad =0 \quad \text{outside of } A \Big\} \tag{40.3}$$

(see Fig. 61). It is known that the Fourier transform of a product of two functions is equal to the convolution of the transforms of these two functions. Here,

$$d(x) \to D(u) = \frac{1}{a} \sum_{q=-\infty}^{+\infty} \delta\left(u - \frac{q}{a}\right) \tag{40.4}$$

and

$$g(x) \to G(u) = \frac{\sin \pi A u}{\pi u}. \tag{40.5}$$

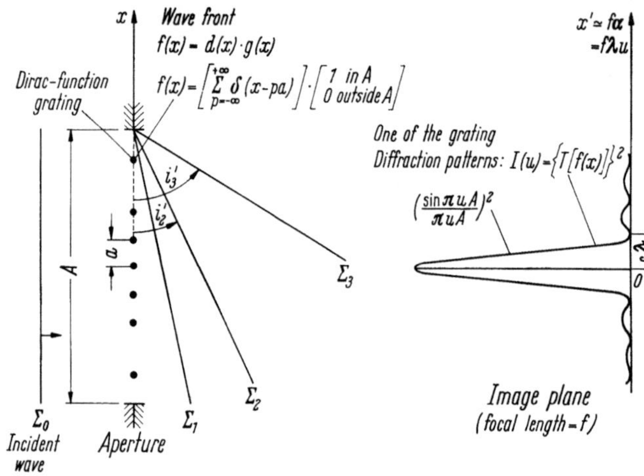

Fig. 61. Wavefront and image-plane coordinates for grating formed by periodically spaced Dirac delta-function grooves (see Fig. 60). Center O corresponds to directions $i_1', i_2' \ldots i_m'$ given by grating equation.

The diffraction image $F(u)$ is given by

$$F(u) = D(u) \otimes G(u), \tag{40.6}$$

that is by

$$F(u) = \frac{A}{a} \sum_{q=-\infty}^{q=+\infty} \delta\left(u - \frac{q}{a}\right) \otimes \frac{\sin \pi A u}{\pi A u} \tag{40.7}$$

where $\otimes$ stands for the convolution integral as before. It is also known that the convolution of a function with a Dirac function is equal to a translation of that function. One therefore has finally for the diffraction image of the grating the equation

$$F(u) = \frac{A}{a} \sum_{q=-\infty}^{q=+\infty} \frac{\sin \pi A\left(u - \frac{q}{a}\right)}{\pi A\left(u - \frac{q}{a}\right)}. \tag{40.8}$$

Here again, the diffraction image of the grating is formed by a spectrum of diffraction patterns, of which the principal maxima are found for values of $u = m/a$, in agreement with the grating equation. The intensity distribution $I(u)$ corresponding to $F(u)$ is here simply

$$I(u) = [E(u)]^2. \tag{40.9}$$

By comparing Eqs. (40.8) and (40.9) with Eqs. (37.14) and (37.15) it is seen that the diffraction patterns are formed at the angles i' given by the grating equation and that the complex amplitude and intensity distributions *in each of the diffraction patterns* are simply equal to the corresponding distributions in the diffraction patterns of a rectangular aperture of width A (Fig. 61).

This conclusion is essential to the further treatment of image formation by gratings. We find, indeed, that we may deal with the various diffracted wave fronts quite independently of each other, as far as their image forming properties are concerned.

41. The general case of a grating of periodically spaced grooves within an aperture of width A. The groove shape function. Let the pupil now be formed

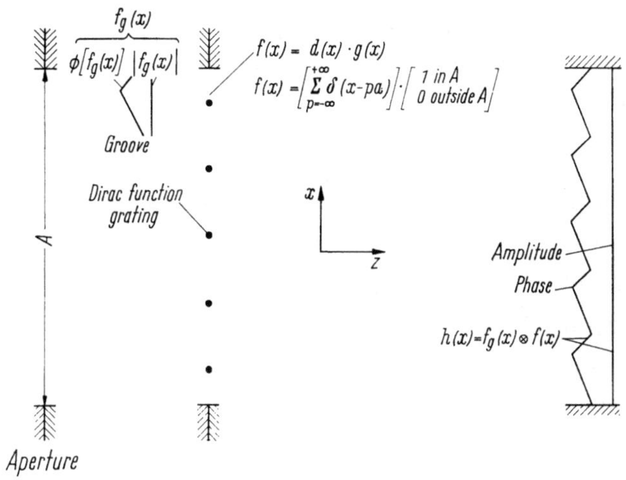

Fig. 62. Diffraction image construction from the groove shape function.

by a grating having the same width A and spacing constant a as the Dirac delta function grating of Sect. 40. However, let the grooves now have a form given by

$$f_g(x) = |f_g(x)| \, e^{\frac{2\pi}{\lambda} i \Phi[f_g(x)]} . \tag{41.1}$$

We may consider that the grating is now formed by the convolution of $f_g(x)$ with the grating function

$$f(x) = d(x) \cdot g(x) \tag{40.1}$$

corresponding to the Dirac delta function grating. Let $h(x)$ describe the grating. We have (Fig. 62)

$$h(x) = f_g(x) \otimes f(x) . \tag{41.2}$$

The diffraction image corresponding to $h(x)$ is

$$H(u) = T[f_g(x)] \cdot T[f(x)] . \tag{41.3}$$

But

$$T[f(x)] = F(u) ; \tag{40.7}$$

that is the diffraction image corresponding to the grating of Dirac functions of Sect. 40. We conclude therefore from Eq. (41.3) that the diffraction image of the grating of finite width, in the general case of grooves of any shape, is formed by the product of the diffraction image of the grating of delta functions by the Fourier transform of the groove shape function $f_g(x)$. One has for the amplitude

$$H(u) = T[f_g(x)] \cdot F(u) \tag{41.4}$$

and for the intensity

$$I(u) = \{T[f_g(x)]\}^2 \cdot \{F(u)\}^2. \tag{41.5}$$

We note that the groove shape function $f_g(x)$ should be considered as a complex function, with

$$|f_g(x)| = \text{amplitude of groove shape function} \tag{41.6}$$

and

$$\Phi[f_g(x)] = \text{phase of groove shape function}. \tag{41.7}$$

It could be concluded from Eq. (41.5) that the intensity distribution among the diffraction patterns in the various grating orders is simply determined by the Fourier transform of the groove shape function $f_g(x)$, in terms of the scalar theory used here. (This would be true for acoustical gratings for example.)

In reality, electromagnetic boundary conditions on the surface of the grating and polarization determine the energy distribution among orders of an optical grating (Chapter B VII). Rather than the transform of the scalar groove shape function, we shall retain a pair of electromagnetically determined energy-distribution groove shape functions, one for each polarization $E_\parallel$ (electric vector parallel to groove length) and $E_\perp$ (electric vector perpendicular to groove length) and define it formally in the u-space by

$$\left.\begin{matrix}[E_g(u)]_\parallel \\ [E_g(u)]_\perp\end{matrix}\right\} = E_g(u) = \left\{\begin{matrix}\text{electromagnetic energy-distribution} \\ \text{groove shape function pair.}\end{matrix}\right\} \tag{41.8}$$

The diffraction image $I_g(u)$ of a grating in the general case is then, for each of the two polarizations $E_\parallel$ and $E_\perp$,

$$I_g(u) = E_g(u) \cdot \{F(u)\}^2 \tag{41.9}$$

with

$$F(u) = \frac{A}{a} \sum_{q=-\infty}^{+\infty} \frac{\sin \pi A \left(u - \frac{q}{a}\right)}{\pi A \left(u - \frac{q}{a}\right)}. \tag{41.10}$$

Eq. (41.10) is identical to Eq. (40.8) and the remarks made at the end of Sect. 40 continue to apply here too. In case of unpolarized light in the incident beam, the two distributions corresponding to the pair $I_g(u)$ must be added, when the total intensity at any image points is to be determined.

The groove shape function $E_g(u)$ must in general be determined on the basis of electromagnetic boundary conditions (see Chapter B VII), except in some special cases, when scalar theory may provide a sufficiently good approximation. This is the case, for instance, of amplitude gratings formed of suitably opaque and transmitting periodic sections disposed in a plane.

42. Physical basis for description of grating diffraction in terms of wave fronts and diffraction patterns. Equations (37.2), (37.8) and (40.9) result from the scalar description of diffraction patterns formed by wave fronts which are extended when compared to the wavelength. This is the case of diffraction gratings. Eq. (37.2) can be considered as expressing a form of HUYGENS' principle. It is important to note that the relation between the diffracted wave front and the corresponding diffraction pattern does not in any way distinguish the origin of the wave front in the "diffraction" by individual grooves of the grating. This is in agreement with experiment.

In order to establish this fact and to verify that a diffracted wave front does not keep a memory, as it were, of this origin in the diffraction by individual

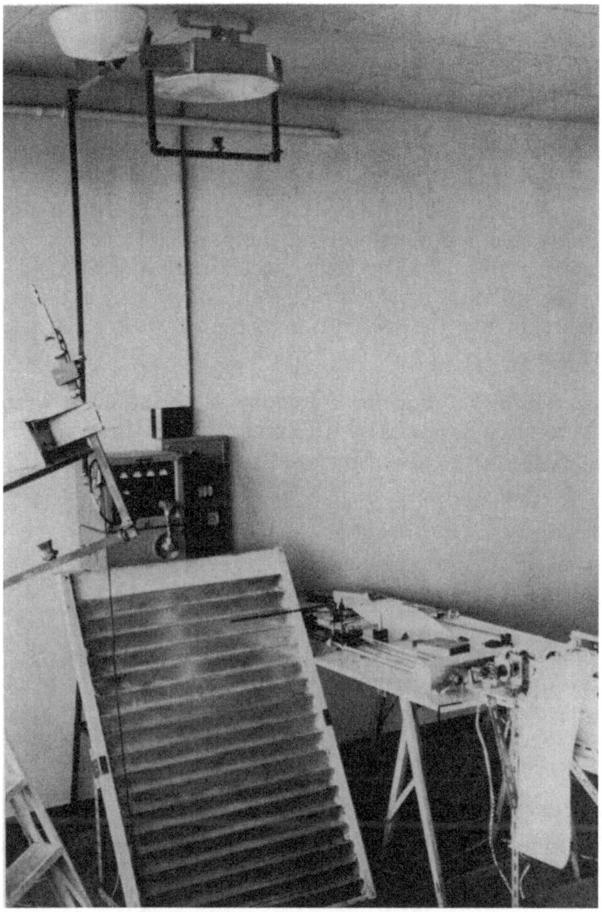

Fig. 63. Experimental arrangement explained in Fig. 64.

grooves, G. W. STROKE[1] (1960) measured the phase across the wave front produced by a grating in the 3.4 cm microwave domain. With microwave analogue gratings a detailed measurement within regions that are small when compared both to the groove spacing and the wavelength is possible. Figs. 63 and 64 show the experimental arrangement, and Fig. 65 shows the result of the measurement. Unlike what

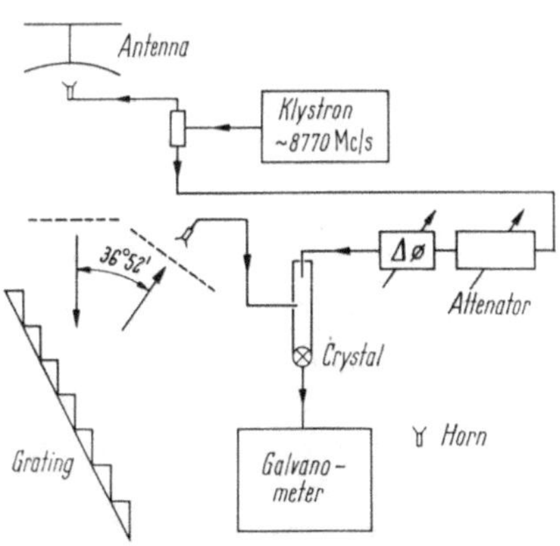

Fig. 64. Phase measurement in diffracted wave fronts at microwavewavelengths (according to STROKE, 1960).

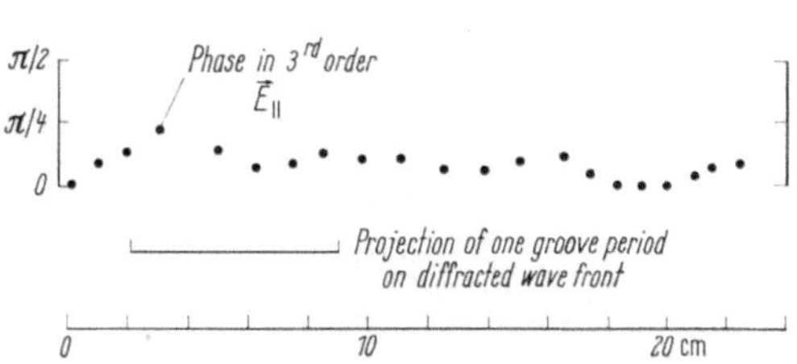

Fig. 65. Phase in wave front diffracted by a grating at λ= 3.42 cm (according to STROKE, 1960).

[1] G. W. STROKE: Rev. opt. **39**, 291—360 (1960).

had been assumed in the past, it is seen that the diffracted wave fronts produced by a grating are truly continuous wave fronts with no memory of the local shape of the groove profile. What gives rise to the phenomenon of diffraction by a grating is the resulting production of plane diffracted wave fronts in the periodic spacing of the "grooves" on its surface. The diffracted wave fronts produced by a perfect grating cannot be otherwise distinguished from a wave front produced by reflection or refraction. It is clear, therefore, that the Fourier transform description of the diffraction of light by optical systems applies to gratings as shown above. In fact, considerable simplification results in the practical application of the Fourier transform description of optical processes to the case of grating, as a result of the one-dimensional description of the grating and of its diffracted wave fronts. In other optical instruments the aberrations generally vary with two dimensions across the wave front.

43. Normalized form of Fourier transform description of grating diffraction problem. We next give a normalized expression for the diffraction of light by a grating in the scalar approximation. Let us define the angular coordinate β in the Fraunhofer domain by

$$\beta=\pi\frac{x}{(\lambda_0/A)f}=\pi\frac{x}{x_0} \tag{43.1}$$

and

$$\beta=0 \quad \text{for} \quad i'=\arcsin\left[(m\lambda/a)-\sin i\right] \tag{43.2}$$

where x is a length in the focal plane of the spectrograph and $x_0=(\lambda_0/A)f$ is the so-called "diffraction unit" (distance of the first minimum from the center of the diffraction pattern). β is measured from the centers of the spectral lines. Let

$$\eta=u/\tfrac{1}{2}A \tag{43.3}$$

be a normalized length in the diffracted wave front itself.

The distribution of complex amplitude of the light in the diffraction pattern $\mathscr{A}(\beta)$ is then given by the Fourier transform of the complex amplitude distribution across the wave front in the form

$$\mathscr{A}(\beta)=\int_{-1}^{+1} g(\eta)\,\mathrm{e}^{i\beta\eta}\,d\eta \tag{43.4}$$

where

$$g(\eta)=\mathscr{A}(\eta)\,\mathrm{e}^{i\Phi(\eta)} \tag{43.5}$$

is the distribution of complex amplitude of the light in the diffracted wave front (STROKE, 1960). The distribution of light intensity $I(\beta)$ in the diffraction pattern is proportional to the product of $\mathscr{A}(\beta)$ with its complex conjugate $\mathscr{A}(\beta)^*$, and we have with the above normalization

$$I(\beta)=\tfrac{1}{4}\mathscr{A}(\beta)\,\mathscr{A}(\beta)^*. \tag{43.6}$$

We know, of course, that only the intensity distribution $I(\beta)$ can be recorded by any detector, such as a photoelectric receiver or photographic plate.

For uniformly blazed gratings, with a constant polarization and amplitude $\mathscr{A}(\eta)$ across their width (along the wave front coordinate η) Eq. (43.5) takes on a particularly simple form

$$g(\eta)=\mathrm{e}^{i\Phi(\eta)}. \tag{43.7}$$

Eq. (43.7) is remarkable by the fact that the phase $\Phi(\eta)$ is the very function which appears already in graphical form in the experimental wave front interferograms

(STROKE 1960, 1961). The computation of the diffraction patterns with the help of the wave front interferograms is given in Sect. 62.

Experimentally, the grating is so adjusted in the interferometer that the interference fringes project at right angles to the groove length. For a perfect grating, the fringes are perfectly straight, and $\Phi(\eta)=0$. The departures from straightness measured in fractions of a fringe (distance between fringes) along the grating width immediately give the value of $\Phi(\eta)$ as a fraction of 2π.

44. The spectral image formation problem. Incoherence of "spectral lines". In Sects. 51 to 59 we shall consider the effect of grating imperfections on the

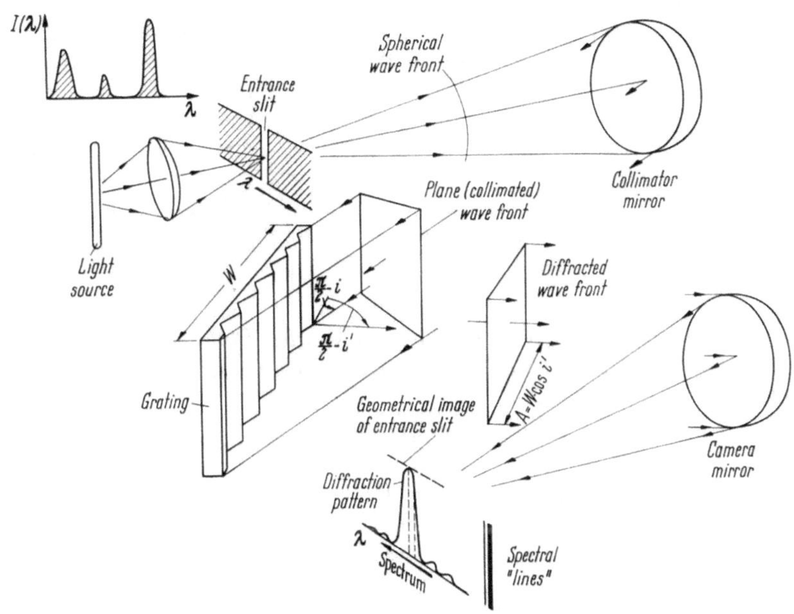

Fig. 66. Spectrograph description (according to STROKE in "Progress in optics").

spectral quality of gratings. It will be shown that grating imperfections simply result in corresponding diffraction-pattern imperfections. But first it is necessary to complete the theory of the formation of spectral images by optical gratings.

As an optical instrument, a grating is used to produce a spectrum of the light that is incident on the entrance slit of the spectrometer or spectrograph (Fig. 66). As a dispersive instrument, a grating spreads out the light of various wavelengths as a function of angle according to Eq. (14.1) or more generally according to Eq. (14.20). For every monochromatic (single frequency) wavelength each point of the entrance slit results in the production of a diffraction pattern according to Eqs. (43.4) and (43.6), and a slit of finite width and height gives rise to a slit image of finite width and height, just as if the grating were a mirror. This is true since the grating does not disperse the light of a single frequency by definition. When the collimator and camera mirrors have the same focal length, the monochromatic images of the entrance slit produced by the gratings are thus "images" in the usual sense in optics.

One next notes that, in usual light sources, there is no coherence between photons of different energies. Light of different "monochromatic" wavelengths is usually completely incoherent. It becomes clear that the formation of slit images in different wavelengths is not related by interference, but rather that the light intensity detected at a given point in the focal plane of the camera mirror is the sum of the light intensities produced there by each wavelength separately of all others.

Now in the ideal case if both the entrance slit and the diffraction patterns were infinitely narrow, only one wavelength would produce a detectable intensity at each point in the focal plane of the spectrograph. The neighboring point would be illuminated by light from a different wavelength, and one would obtain infinite spectral resolution. In reality, both the entrance slit and the diffraction patterns have finite spread. Moreover, there is frequently also an exit slit in the instrument. The effect of finite entrance and exit slits is simple to understand [JACQUINOT and DUFOUR (1948), JACQUINOT (1954)], and to take into account. In practice one can generally use slits that are very narrow when compared to the width of the diffraction patterns, even in the ideal case of perfect grating. No loss of generality therefore results from the use of an infinitely narrow slit in the following analysis. One can indeed always replace the use of the diffraction patterns $I(\beta)$, of Eq. (43.6), corresponding to an infinitely narrow slit, by a pattern $I(\beta)'$ taking into account the effect of the slight broadening by the slit. [The broadening of $I(\beta)$ to give $I(\beta)'$ can be rigorously calculated, graphically or analytically, as a convolution of $I(\beta)$ with a rectangular or triangular slit scanning function $s(\beta)$, for the case of spectrographs or spectrometers respectively. Details are given in the next section.]

45. Instrument function with finite slit widths. Following P. JACQUINOT and CH. DUFOUR[1], and P. CONNES[2] one can evaluate the effect of finite slit widths as follows. Let

$$\sigma = \frac{1}{\lambda}\ \text{cm}^{-1} \tag{45.1}$$

be the wave-number of the radiation. The *instrument function* is the ratio

$$H(\sigma - \sigma_1) = \frac{\Phi(\sigma)}{\Phi(\sigma_1)} \tag{45.2}$$

of the flux received by the detector at the wave number σ to the flux received when $\sigma = \sigma_1$. The function $H(\sigma - \sigma_1)$ clearly has a maximum equal to 1 when $\sigma = \sigma_1$, and its exact form depends on the dispersion and conditions of use. It is useful to distinguish between a theoretical and a real instrument function.

The theoretical instrument function is the function $H_0(\sigma - \sigma_1)$ which is characteristic of the grating, but with the dispersion

$$\delta_\sigma = \sigma \frac{d i'}{d\sigma} = \frac{1}{\lambda} \frac{d i'}{d\sigma}$$

included. For a grating of width W having a wave front of width

$$A = W \cos i'$$

one has for the instrument function

$$H_0(\sigma - \sigma_1) = \left[\frac{\sin \pi\, \delta_\sigma A (\sigma - \sigma_1)}{\pi\, \delta_\sigma A (\sigma - \sigma_1)}\right]^2 \tag{45.3}$$

where the square of the bracketed quantity is simply an expression of the intensity in the diffraction pattern, taking into account the dispersion δ_0.

---

[1] P. JACQUINOT and CH. DUFOUR: J. Recherches, C.N.R.S. (Centre National de la Recherche Scientifique), Lab. de Bellevue **6**, 91 (1948).

[2] P. CONNES: Thèse, Rev. Opt. **38**, 157, 416 (1959); **39**, 402 (1960).

The first minimum of $H_0(\sigma-\sigma_1)$ is obtained for $\sigma-\sigma_1=1/(A\,\sigma_\sigma)$. According to Lord RAYLEIGH's criterion, this distance corresponds to the "resolution" of two spectral lines of equal intensity. Let

$$\Delta\sigma_0=\frac{1}{A\,\delta_\sigma} \tag{45.4}$$

be the resolving limit of the grating (without apodization). By definition, the theoretical resolving power R_0 is given by

$$R_0=\frac{\sigma}{\Delta\sigma_0} \tag{45.5}$$

and one has

$$R_0=\frac{A}{\lambda}\,\delta_\sigma. \tag{45.6}$$

We may recall (Sect. 19) that the dispersion δ was first defined as

$$\delta=\frac{d\,i'}{d\,\lambda}\equiv\delta_\lambda. \tag{45.7}$$

Since

$$\frac{d\,i'}{d\,\lambda}=\frac{d\,i'}{d\,\sigma}\,\frac{d\,\sigma}{d\,\lambda} \tag{45.8}$$

and

$$\frac{d\,\sigma}{d\,\lambda}=-\frac{1}{\lambda^2}=-\sigma^2 \tag{45.9}$$

we have

$$\delta_\sigma=-\lambda\,\delta_\lambda \tag{45.10}$$

and the resolving power is indeed

$$|R_0|=A\,\delta_\lambda \tag{45.11}$$

in agreement with Eq. (20.8).

One may define a *reduced variable*

$$\sigma'=\frac{\sigma-\sigma_1}{\Delta\sigma_0} \tag{45.12}$$

which is to be used henceforth.

Let us also write

$$\eta=\frac{x}{A/2} \tag{45.13}$$

and

$$f(x)=f(A\,\eta/2)=g(\eta) \tag{45.14}$$

for the complex amplitude in the wave front (normalized coordinate η), and let

$$\boldsymbol{g}^2=\left[\int_0^1 g(\eta)\,d\eta\right]^2 \tag{45.15}$$

be the mean transmission coefficient of an apodizing pupil. The instrument function $H_0(\sigma')$ can be written as

$$H_0(\sigma')=\mathscr{A}(\sigma')^2, \tag{45.16a}$$

where

$$\mathscr{A}(\sigma')=\frac{1}{\boldsymbol{g}}\int_0^1 g(\eta)\cos\pi\,\sigma'\,\eta\,d\eta. \tag{45.16b}$$

The theoretical instrument function is seen to be proportional to the *square* of the Fourier transform of the light distribution in the diffracted wave front.

With the help of the reduced variables, Eq. (45.3) becomes

$$H_0(\sigma') = \left(\frac{\sin \pi \sigma'}{\pi \sigma'}\right)^2 \tag{45.17}$$

in agreement with the results of the previous sections.

Finally, the "real" instrument function is obtained by introducing a *finite width* for the entrance and exit slits. In what follows, one still assumes perfect wave fronts. (The effect of grating and wave front imperfections is dealt with in Sects. 51 to 59).

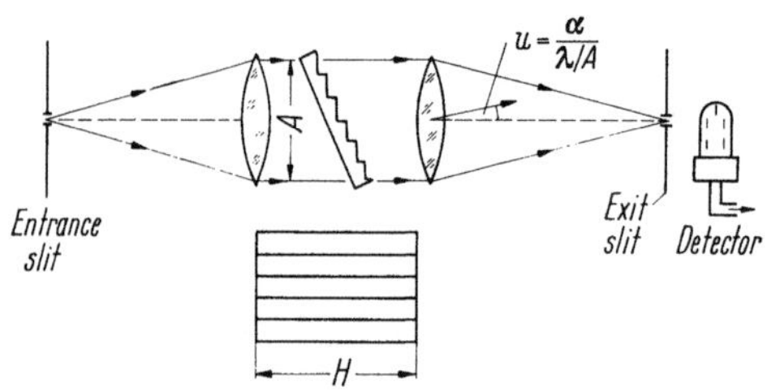

Fig. 67.

In Fig. 67, we recall that two slits are involved in a spectrometer, but only one slit in a spectrograph. It is helpful to introduce the effects of the two slits separately.

Let the reduced slit width be given as

$$u = \frac{\alpha}{\lambda/A} = \alpha \frac{R_0}{\delta_\sigma} \tag{45.18}$$

where u gives the ratio of the slit width to the width of the central diffraction-pattern maximum, taken in this case, of a perfect grating at 0.404 of the central intensity.

Let the geometrical image of an arbitrary point P_1, in the entrance plane be formed at P_2, and let σ_2 be the corresponding wave number. The reduced variable σ_2' used as the variable of integration is then given by

$$\sigma_2' = \frac{\sigma_2 - \sigma_1}{\Delta\sigma_0}. \tag{45.19}$$

The flux reaching the image plane is given by

$$\Phi(\sigma') = B S \boldsymbol{g}^2 \alpha \beta \tau(u) H(\sigma') \tag{45.20}$$

where B is the brightness of the source, $S = A H$, (H = groove length), $\alpha\beta$ = solid angle as seen from the grating wave front aperture, and $\tau(u)$ a transmission factor characteristic of the grating efficiency. The instrument function is now

$$H(\sigma') = \frac{\int_{-\infty}^{+\infty} H_0(\sigma' - \sigma_2')\, s_\sigma(\sigma_2')\, d\sigma_2'}{\int_{-\infty}^{+\infty} H_0(\sigma_2')\, s_\sigma(\sigma_2')\, d\sigma_2'}. \tag{45.21}$$

One can also write (45.21) in the form

$$H(\sigma') = \frac{H_0 \otimes s_\sigma}{\int_{-\infty}^{+\infty} H_0(\sigma_2')\, s_\sigma(\sigma_2')\, d\sigma_2'} \tag{45.22}$$

where $\otimes$ stands for a convolution. The function $S_\sigma(\sigma_2')$ is the *slit scanning function*, such that its maximum value is equal to unity.

One now distinguishes two slit scanning functions $S_{\sigma 1}(\sigma_2')$ and $S_{\sigma 2}(\sigma_2')$, for the case of a single-slit spectrograph and a two-slit spectrometer respectively.

In case of the *single-slit spectrograph*, the slit-scanning function $S_{\sigma 1}(\sigma_2')$ is simply a rectangular function having a width u.

In case of a *two-slit scanning spectrometer* the slit scanning function $S_{\sigma 2}(\sigma_2')$ is the triangular function shown in Fig. 68 such that its width at half-intensity is equal to u.

In both cases it is seen that the normalized instrument function $H(\sigma')$ is simply given by the convolution of the theoretical instrument function $H_0(\sigma')$ with a rectangular or triangular function characteristic of the slit width. The instrument function in the β-space corresponding to $H(\sigma')$ is the function $I(\beta)'$ referred to at the end of Sect. 44. In a normalized form one has

$$I(\beta)' = \frac{I(\beta) \otimes s(\beta)}{\int_{-\infty}^{+\infty} I(\beta)\, s(\beta)\, d\beta} \qquad (45.23)$$

where $I(\beta)$ is the diffraction pattern of the diffracted wave front, and $s(\beta)$ is the slit-scanning function defined in the β-space, with

$$\beta = \pi \frac{x}{(\lambda/A)\, f} = \pi \frac{x}{x_0} \qquad (43.1),\ (45.24)$$

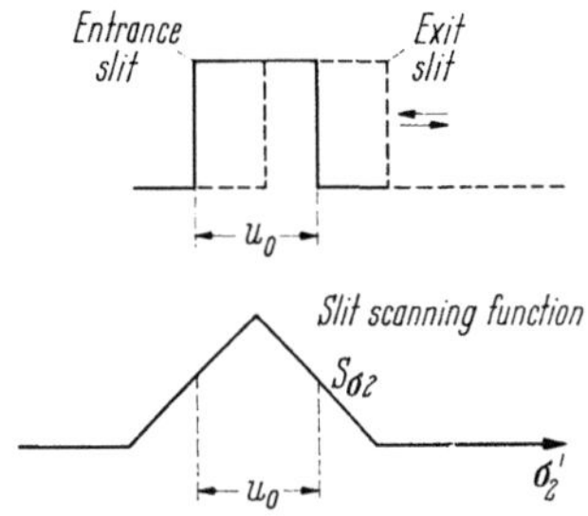

Fig. 68.

and

$$s(\beta) = \begin{cases} s_1(\beta) = \sqcap & \text{(single-slit spectrograph)} \\ s_2(\beta) = \wedge & \text{(two-slit spectrometer)}. \end{cases} \qquad (45.25)$$

It is seen that the diffraction pattern $I(\beta)$ is introduced without the dispersion, while the theoretical instrument function $H_0(\sigma')$ has been introduced with the dispersion. In the use of $I(\beta)$, an additional convolution is used to obtain the intensity distribution in the recorded spectrum (see Sect. 46). The use of $I(\beta)$ without the dispersion permits an immediate introduction of the effects of grating imperfections on the intensity distribution in the spectrum. Clearly, though, the two ways of introducing the dispersion are equivalent.

The denominator in Eq. (45.21), (45.22), (45.23) is simply a normalization factor, such that $H(0)=1$, and $I(0)=1$ respectively. It is clear that the shape of $H(\sigma')$ and of $I(\beta)'$ is respectively intermediary between $H_0(\sigma')$ and $I(\beta)$ respectively. For small values of s, the functions $H(\sigma')$ and $I(\beta)'$ tend towards the theoretical instrument function $H_0(\sigma')$ and the diffraction pattern $I(\beta)$ respectively, while for large values of s ($s \gg 1$) $H(\sigma')$ and $I(\beta)'$ simply tend to be equal to s.

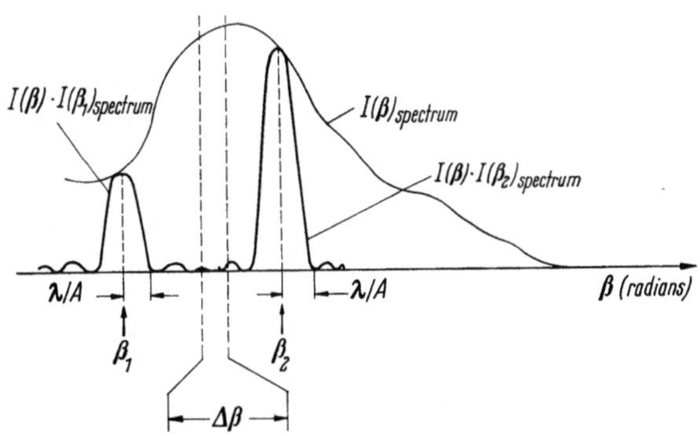

Fig. 69. $\Delta\beta$ = width of exit slit (or "granularity" of plate). $I(\beta)$ = intensity distribution in diffraction pattern (as modified by width of entrance slit).

46. The formation of the spectrum in terms of image-formation theory. We are now in a position to formally describe the formation of the spectrum in a grating of light intensity in the spectrum of a light source and produced by a grating, as shown in Fig. 69. The angle β is measured from the direction i' as given in Eq. (14.1). As a result of the incoherence between the diffraction patterns of different wavelengths, the intensity of the spectrum recorded at a point β (in the angular domain) is simply given by the sum of intensities produced at that point by all other spectral intensities at the various points β', each of the spectral intensities being weighted by the value at β of the diffraction intensity corresponding to the intensity of the spectrum at the point β'. If the detector at β

has a finite width $\Delta\beta'$, we can express the intensity $I(\beta)_{\text{spectrometer}}$ recorded at β as

$$I(\beta)_{\text{spectrometer}} = \Sigma I(\beta')_{\text{spectrum}} I(\beta-\beta')\Delta\beta' \tag{46.1}$$

or more exactly as

$$I(\beta)_{\text{spectrometer}} = \int_{-\infty}^{+\infty} I(\beta')_{\text{spectrum}} I(\beta-\beta')\, d\beta' \tag{46.2}$$

where $I(\beta-\beta')$ is the value at β of the diffraction pattern $I(\beta)$ centered at β'. Eq. (46.2) is usually described as a convolution integral. One says that the spectral intensity, as recorded in the spectrometer, is the convolution of the "true" intensity distribution in the spectrum with the intensity distribution in the diffraction patterns. Eq. (46.2) is sometimes written as

$$I(\beta)_{\text{spectrometer}} = I(\beta)_{\text{spectrum}} \otimes I(\beta), \tag{46.3}$$

where it is understood that

$$I(\beta) \equiv I(\beta)_{\text{diffraction pattern}} \tag{46.4}$$

and $\otimes$ stands for the convolution operation.

The function $I(\beta)$ defined by Eq. (46.4) should be identified to the function $I(\beta)'$ given by Eq. (45.23) when the diffraction pattern is appreciably modified by the slit scanning function $s(\beta)$.

47. The "true" distribution of intensity in the spectrum of a light source. The aim of spectroscopic analysis is to obtain the "true" distribution of light intensity, $I(\beta)_{\text{spectrum}}$, in the spectrum of the source that is being studied. We note, however, that any spectrometer or spectrograph will provide the spectral intensity distribution as modified by several convolutions involving the parameters which limit the ideal performance of the spectroscopic element. In the case of Fabry-Pérot etalons, for instance, three functions have been singled out as being important in the convolution: 1. the well-known Airy function (which corresponds to the perfect diffraction pattern in a grating); 2. a function involving the effect of surface imperfections in the etalon plates; and 3. the function resulting from the use of a finite range of angles of incidence [(Chabbal, 1957, 1958)[1], (Jacquinot, 1960)[2]]. As it happens, the three functions involved in the spectroscopic use of Fabry-Pérot étalons have comparable importance, and their exact description needs to be obtained when an attempt is made to retrieve the "true" distribution of light intensity in the spectrum. For high-resolution gratings the situation is fortunately simple: the "square" shaped or triangular slit scanning functions resulting from the use of finite slit apertures are very simple to introduce. Moreover, they are frequently negligible in practice when compared to the diffraction patterns. In this case only the convolution with the diffraction pattern is involved in the use of high-resolution gratings. This is true, moreover, not only in the case of perfect diffraction patterns, but also in the case of imperfect patterns, since the above analysis makes no special assumptions about the distribution of light in the diffraction pattern. We have only assumed the existence of an extended diffraction pattern, as a result of the finite width of the grating. It is clear that if it were possible to obtain experimentally the distribution of light $I(\beta)$ in the diffraction patterns of a given grating (as can now be done, see Sect. 59) then it would be possible to obtain unlimited information about the spectrum of the source in an ideal noiseless case. Direct measurement

---

[1] R. Chabbal: Thèse, Rev. Opt. **37**, 49, 336, 501 (1958). — J. Phys. Radium **19**, 295 (1958).
[2] P. Jacquinot: New Developments in Interference Spectroscopy. Rep. Progr. Phys. **23**, 267 (1960).

of the distribution of intensity $I(\beta)$ in diffraction patterns has been attempted for many years for this purpose[1] (DITCHBURN 1930, STRONG 1951, BABCOCK and BABCOCK 1951, HULTHÉN and UHLER 1952, RANK et al. 1955, A. KEITH PIERCE 1957, HARRISON, STURGIS, BAKER and STROKE 1957, RANK et al. 1957, HARRISON, STURGIS, DAVIS and YAMADA 1959), but is fraught with considerable practical difficulties as will be shown in Sect. 73. Fortunately, as shown by STROKE (1955, 1960) the distribution of phase $\Phi(\eta)$ in the diffracted wave fronts can be measured without difficulty by placing the grating at its angle of use into one of the arms of a Michelson-Twyman-Green interferometer (see Sect. 74). Moreover, the distribution of intensity in the "monochromatic" diffraction patterns, namely $I(\beta)$, can be rapidly and simply calculated from the wave front interferograms by Fourier transformation according to Eqs. (37.8) and (43.4) to (43.6) with the help of a digital computer (STROKE 1958, 1960, 1961). With these methods, the problem of theoretical characterization of a grating for use in spectroscopic analysis is extremely straightforward and may be summarized as follows.

The spectroscopic behaviour of a diffraction grating is characterized by $I(\beta)$, the distribution of light in the diffraction pattern formed by the diffracted wave fronts in various wavelengths. $I(\beta)$ can be computed by Fourier transformation with all the required accuracy and very simply from the light distribution $g(\eta)$ in the diffracted wave fronts. And, finally, the distribution of light in the diffracted wave fronts can be found in readily usable graphical form in the experimental wave front interferograms.

To retrieve the "true" distribution of light intensity in the spectrum from the measurement in the spectrometer, one only needs to solve Eqs. (46.2) or (46.3) or their equivalent Fourier transform representation, namely

$$T[I(\beta)_{\text{spectrometer}}] = T[I(\beta)_{\text{spectrum}}] \cdot T[I(\beta)] \qquad (47.1)$$

where $T[\;]$ indicated the Fourier transform of the function in brackets. The numerical solution of Eq. (47.1) with the help of digital computers should be fairly straightforward, and work towards this goal is proceeding. Similar problems also arise in radio astronomy (ARSAC, 1959)[2], (BRACEWELL, 1961)[3] and in the use of optical systems as computers in some radar work (CUTRONA et al., 1959, 1960)[4]. The use of the theory of distributions (ARSAC, 1961) has permitted us to solve certain difficulties associated with the inversion of Eq. (47.1). Holographic image processing methods may prove to provide a practical method for the solution of this problem (see Sect. 92).

48. Relation between ruling errors and wave front aberrations as a function of grating angle and wavelength. It is clear that a perfectly ruled plane grating will produce perfectly plane diffracted wave fronts. In autocollimation, a grating may be thought of as an inclined staircase viewed by an observer standing in the vertical diffracted wave front. The diffracted wave front is parallel to the vertical groove (or step) sides and inclined by an angle $i' = i$ with respect to the mean grating surface. Should a step be displaced along the mean surface of the grating by an amount Δa, then its component normal to the wave front is given by $\Delta a \sin i'$. In this case of autocollimation the aberration path-difference is clearly

[1] See reference (15a) in G. W. STROKE: J. Opt. Soc. Amer. **51**, 1321 (1961).

[2] J. ARSAC: Optica Acta **6**, 103 (1959); also loc. cit. (1961).

[3] R. N. BRACEWELL: I.R.E. Trans. AP-9, No. 1, p. 49 (1961).

[4] L. J. CUTRONA, E. N. LEITH and L. J. PORCELLO: Proceedings of the National Electronics Conference, Vol. XV (1959). — L. J. CUTRONA, E. N. LEITH, C. J. PALERMO and L. J. PORCELLO, I.R.E. Trans. IT-6, No. 3, p. 386 (1960).

$2\Delta a \sin i'$. Except for the normalization, the wave front aberration is thus given by

$$\Delta(\eta) = 2\Delta a \sin i' \tag{48.1}$$

and the phase aberration by

$$\Delta\Phi(\eta) = \frac{2\pi}{\lambda} 2\,\Delta a \sin i'. \tag{48.2}$$

It is very useful to express the groove errors in terms of where $\frac{1}{2}\lambda = 1$ fringe. If the aberrations are measured in terms of the same wavelength λ as the ruling errors then an aberration $\Delta a = \frac{1}{2}\lambda$ will result in

$$\Delta(\eta) = \lambda \sin i' \tag{48.3}$$

and

$$\Phi(\eta) = 2\pi \sin i' \qquad \left(\text{per } \Delta a = \tfrac{1}{2}\lambda\right). \tag{48.4}$$

But an aberration $\Phi(\eta) = 2\pi$ corresponds to precisely one fringe of deviation in the wave-front interferograms. We thus have the important conclusion that *when the ruling errors are measured in fringes ($\frac{1}{2}\lambda$) of the same wavelength as that used in the interferometric measurement of the diffracted wave-front aberrations, then one fringe of ruling error will result in* $\sin i'$ *fringes of wave-front aberration when the grating is used in autocollimation at the angle* i'.

The effect of ruling errors on the diffracted wave fronts is thus seen to increase with the sine of the grating angle. It also clearly increases with decreasing wavelength [Eq. (48.2)]. Of course, the diffracted wave-front aberrations result in corresponding spectral diffraction-pattern imperfections (Sects. 51 to 59). Relations similar to those given here for the case of autocollimation can be easily derived for the case when i' is not approximately equal to i. Clearly though, the effect of grating imperfections on the wavefronts and spectral images is maximum when $i' \simeq i$.

49. Relation between blank imperfections and wave-front aberration as a function of grating angle and wavelength. It is perfectly possible to space the grooves regularly in a plane, yet to obtain an imperfect grating when the surface of the blank on which the grooves are being ruled is itself not plane. Gratings are most frequently being ruled in comparatively thick films of aluminum deposited by evaporation on optical blanks. Aluminium thickness from 3 to 30 microns are typically used for ruling purposes. It seems occasionally to be as difficult to evaporate sufficiently flat films on large gratings as it is to rule with corresponding regularity. Deviations from flatness may of course also result from imperfection in the blanks themselves, from the stresses in the mountings of the grating during ruling or in use, as well as from deformations in replication.

Unlike the effect of ruling errors, which has been shown to increase with grating angle as $\sin i'$, *the effect of blank imperfections* can be shown to *decrease with grating angle* as $\cos i'$ (both effects increase with decreasing wavelength, of course). Consider indeed the grating spacing a as it appears from an angle $i \pm \Delta i$ when compared to the angle i. We have (Fig. 70)

$$a \sin(i \pm \Delta i) \simeq a(\sin i \pm \Delta i \cos i) \tag{49.1}$$

and we immediately see that the groove which is properly positioned by ruling, but happens to be ruled on the portion of the blank inclined by Δi with respect to the rest of the blank, will appear displaced by the quantity

$$\Delta a = \pm a \Delta i \cos i \tag{49.2}$$

when viewed from the diffracted wave front at the angle i. If the blank error extends linearly over N grooves (distance on the grating $= N a$ fringes) then we find a fringe displacement reaching

$$\delta_i = N a \Delta i \cos i \quad \text{fringes} \tag{49.3}$$

in the wave-front interferogram. The fringe displacement caused by a "linear" blank imperfection is of course a linear departure of the fringes from straightness. A "linear" blank imperfection is an inclination of a flat portion of a grating blank with respect to the remainder of an otherwise perfectly flat blank. It is essential to observe that linear fringe deviations in the wave front will result not only from grating blank imperfections, but also from ruling imperfections. In fact, since both rulings and blank imperfections of small amplitude, when extending over a significant portion of the grating, may frequently be first approximated by a "linear" departure from perfection, we find that the case of "linear" wave-front aberrations take on a particularly significant importance in high-resolution gratings, both in theory and in practice (see STROKE, loc. cit. 1960).

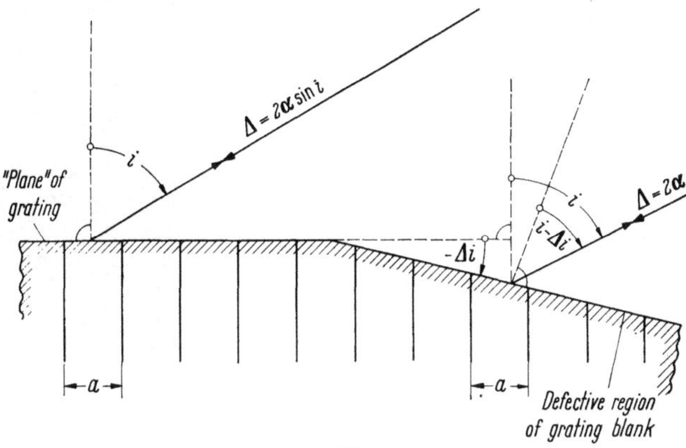

Fig. 70.

50. Effect of grating imperfections on the spectral images. The origin of spurious spectral lines. The effect of grating imperfections is to cause aberrations in the wave fronts diffracted by the grating. The diffracted wave-front aberrations, in turn, cause variations in the distribution of light in the spectral diffraction patterns, when the imperfect diffraction patterns are compared to the "perfect" diffraction pattern corresponding to the width of the diffracted wave fronts (with or without apodization). The distribution of light in the imperfect diffraction patterns shows "spurious" secondary maxima, which have been variously called "ghosts" (ROWLAND, LYMAN) "satellites" (WOOD, HARRISON and STROKE) and "grass". In general, the term "ghost" is used for spurious spectral lines caused by periodic imperfections in the gratings, while "satellites" is used for the spurious spectral lines caused by other regular or irregular grating imperfections.

Many attempts have been made in the past to trace the origins of particular spurious spectral lines produced by imperfect gratings back to particular error regions in the grating or the wave fronts[1] (E. MASCART 1889; ALLEN 1902; WOOD 1924; GALE 1937; HARRISON 1949; STRONG 1951; INGELSTAM and DJURLE 1952, 1953; STROKE 1952; DJURLE 1952, 1954; RANK et al. 1955). It is important to note as shown by STROKE in 1960, that except for the "linear" wave front

[1] See Reference (53) in G. W. STROKE: J. Opt. Soc. Amer. **51**, 1321 (1961).

aberrations (and the periodic errors which used to affect the more distant image regions) it is generally not possible to single out given grating regions as having errors at the origin of a particular spurious satellite: this is no more possible than it is to single out an "error" region in a perfect wave front as being at the origin of what might then be called the "spurious" secondary maxima in a perfect diffraction pattern with no apodization. It is known, of course, that the first secondary maximum corresponding to the rectangular aperture presented by a grating has an intensity of the order of 4 percent and that the peak appears at a distance

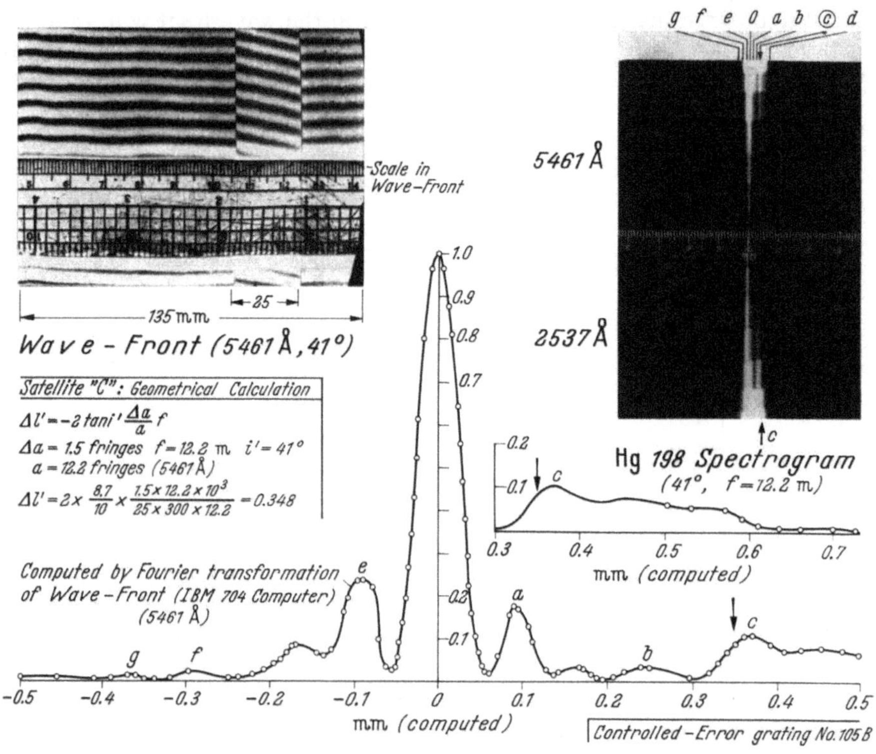

Fig. 71.

of about two diffraction units from the line center. These are precisely the orders of magnitude of the spurious satellites of a few percent intensity with which we are dealing here. In fact STROKE (1960, 1961) shows that all spurious spectral lines are nothing but diffraction-pattern imperfections caused by the entire diffracted wave front. The simplest relation between the complex amplitude of the electromagnetic disturbance in the diffracted wave front and the complex amplitude in any of the spectral image-points takes on the form of the Fourier transformation given in Eqs. (43.4), (43.5) and (43.6). Only in the important special case of STROKE's "linear" wave-front aberrations, which will be discussed in Sect. 53, and in the case of periodic errors (Sect. 55) can diffraction pattern imperfections be predicted in detail by mere inspection of the wave-front interferograms. In general, the distribution of light in the diffraction patterns of a given grating can be obtained by carrying out the Fourier transformation of the phase distribution in the diffracted wave fronts as it appears in readily usable graphical form in the wave-front interferograms. The numerical computation can be easily carried out with the help of a digital computer (Sect. 59), (STROKE 1960, 1961).

Fig. 71 shows according to STROKE (loc. cit. 1960, 1961) the distribution of light in the diffraction pattern computed by Fourier transformation of the phase distribution in the wave front for a grating which he ruled under interferometric control by deliberately introducing a controlled, pure "linear" wave-front aberration into an otherwise perfect grating (see Sect. 53). Fig. 71 shows the typical complexity of the diffraction-pattern imperfections that result from even an apparently simple aberration. Perfect agreement between the spectrogram and the computation is seen to exist throughout the entire region where comparisons are possible: and the computation has of course the significant advantage of providing the intensity distribution in the immediate vicinity and within the central part of the diffraction pattern where most of the energy is concentrated, but where no such information was heretofore available from more classical tests, such as resolution photographs for example. Computed diffraction patterns such as the one in Fig. 71 (which corresponds to an infinitely narrow perfectly monochromatic slit) have permitted correction of "error of coincidence" resulting from systematic, center-of-gravity displacements of spectral lines. They are invaluable in spectral-line-shape studies: indeed, the intensity distribution recorded in the spectrometer is equal to the convolution of the "true" intensity distribution in the spectrum (which is sought) with the (computed) diffraction-pattern intensity distributions (where the slit width has been taken into account as in Sect. 45).

The complexity of the satellite-distribution that results from a grating wave front having numerous departures from flatness is shown in Fig. 57 of Sect. 35. It is quite clear that no simple relationship exists between any individual satellite and a special wavefront region. Here again excellent agreement is found between the satellite pattern and the light distribution predicted by Fourier-transform computation from the experimental wave-front interferogram.

No problem in carrying out such Fourier transformations has appeared. STROKE (1960, 1961) has nevertheless found it desirable to derive two general laws in order to help in predicting the effects of wave front and grating imperfections on the quality of the spectral lines without actually carrying out the Fourier transformations. With the help of these laws and the important special case of STROKE's linear wave-front aberrations, which are discussed in Sect. 51 and 53 respectively, it is in general possible to immediately characterize the general quality of a grating by merely inspecting the wave-front interferograms.

The important improvements in the distribution of light in the diffraction patterns that can be obtained by "apodization" of even perfectly flat (i.e., perfectly spherical) wave fronts have been demonstrated by JACQUINOT[1], ROIZEN-DOSSIER[2] and others[3]. In general, apodization by appropriate masking of the wave front aperture ot amplitude and phase transmission function can be used to suppress the secondary maxima in the "perfect" diffraction patterns at the expense of some broadening of the central maximum. With the possibility of apodization, the choice of what should be called "the perfect diffraction pattern" may appear somewhat arbitrary. It is however possible to discuss optimum

---

[1] A. COUDERC and P. JACQUINOT: C. R. Acad. Sci. Paris **208**, 1639 (1939). — P. BOUGHON, B. DOSSIER et P. JAQUINOT: C. R. Acad. Sci. Paris **223**, 661 (1946).

[2] B. DOSSIER: Thèse, Revue d'Optique 1953.

[3] Lord RAYLEIGH: Scientific Papers I, p. 415. — R. STRAUBEL: Über Bildgüte. Peter Zeeman Verh., Martinus Nijhoff 1935, p. 302. — P. JACQUINOT and B. ROIZEN-DOSSIER: Apodization. In: Progress in Optics, Vol. III, edit. by E. WOLF. Amsterdam: North-Holland Publishing Company 1963.

characteristics in diffraction patterns (Arsac, loc. cit. 1956, 1960; Maréchal and Françon, loc. cit. 1961).

51. General effect of diffracted wave-front imperfections on the spectral images and variations of spectral quality with wavelength and angle. The principal conclusions may be given in the form of two "laws" established by Stroke (loc. cit. 1960). According to the first law it is possible to separate formally the macroscopic or extended errors in the grating (and wave fronts) from the local or microscopic errors which correspond to displacements of individual grooves or small-groove regions, and to irregularities in the groove faces themselves. It is the extended errors, which may for example extend over $\frac{1}{2}$ inch to 2- or 3-inch regions in a 10-inch grating, which are responsible for the appearance of the harmful spurious satellites in the neighbourhood of the line centers. On the other hand, the local errors, covering regions of say less than $\frac{1}{2}$ inch on the grating, are at the origin of more distant satellites and scattered light which can be harmful to certain other spectroscopic applications as is the presense of satellites in high-resolution spectroscopy; this is particularly true in infrared and Raman spectroscopy, for example.

The second law, which has emerged from Stroke's study of spectral-image imperfections is that the knowledge of the imperfections of a spectral line in only one wavelength and, in general, at only one angle, is sufficient to predict the imperfections for the particular grating in all other wavelengths and at all other angles of interest. It is found that the intensities of corresponding spurious satellites and grass in imperfect diffraction patterns increase as $\sin^2 i'/\lambda^2$ for a given high-resolution grating in which the largest aberration in the shortest wavelength of interest is of the order of $\frac{1}{4}$ fringe ($\frac{1}{8}\,\lambda$). It is further found that imperfect diffraction patterns in various wavelengths and at various angles are similar to each other in the ratio $\lambda/\cos i'$.

These two satellite-variation laws provide a simple means of predicting the satellite position and intensity distribution in all wavelengths and at all angles provided that, in addition to the wave-front interferograms, one of these patterns has been obtained (for example, in the light of a single line from the Hg 198 source, or with the help of one component in the hyperfine structure of the green 5461-Å line of natural mercury, or else by Fourier transformation — in one wavelength — from the wave front interferogram)[1]. The only important exception to the general satellite-variation laws is the special law, also determined by Stroke, for the "linear" wave-front aberrations which will be discussed in Sect. 53: For this case it is found that the distances from the line centers as well as the relative intensities of the principal satellites are wavelength independent at a given grating angle (see Fig. 71 for example). In practice satellite patterns obtained in two wavelengths immediately permit do distinguish this important special case.

Finally it was also shown by Stroke (Sects. 48, 49) that wave-front aberration corresponding to a given ruling error increases with the sine of the angle of use of the grating in autocollimation, while the aberrations corresponding to blank or coating defects decrease with the cosine of the same angle, but both also increase with decreasing wavelength. In fact, when measured in fringes ($\frac{1}{2}\,\lambda$) of the same wavelength, the wave front aberration as it appears in the interferogram is simply $\sin i'$ times the ruling error and $\cos i'$ times the blank defect. More generally, the wave-front aberration in a wavelength λ is $(\lambda_0/\lambda)\sin i'$ times the ruling error measured in the wavelength λ_0, and $(\lambda_0/\lambda)\cos i'$ times the blank or

[1] Single-frequency laser lines, e.g. the 6328 Å line from a continuous-wave helium-neon laser, are also particularly suitable.

coating defect measured in λ_0. Even though blank defects and ruling errors may result in similar wave front aberrations and corresponding image imperfections, the sensitivity of the wave fronts to ruling errors is seen to increase with grating angle, while the sensitivity to blank defects fortunately decreases with angle. The spectral diffraction pattern imperfections, that is the spurious satellite lines, ghosts, and grass in which one is interested in the final analysis are found to increase with the squares of the ratios given above, that is as $(\lambda_0/\lambda)^2 \sin^2 i'$ for image imperfections caused by ruling errors, and as $(\lambda_0/\lambda)^2 \cos^2 i'$ for image imperfections caused by blank and coating defects (see Sects. 52, 54). It is this square-law dependence of the image quality on ruling perfection and blank quality which makes any improvement in grating quality so significant, but also makes the attainment of high-resolution gratings so difficult. These conclusions have been fully borne out by experiment (STROKE 1960, HARRISON and STROKE 1960, STROKE 1961).

We may close this section by again stressing the role of the interferometric method of measuring the diffracted wavefront aberrations (see Sect. 74 and STROKE, loc. cit. 1955, 1960, 1961, 1963). This method has been found to be the single most important tool not only in the immediate assessment of the grating quality and the prediction of the distribution of light in the spectral images, but also in the improvement of grating ruling engines and of the ruling process itself.

The principal steps in the mathematical development of these conclusions are given in Sects. 52 and 54.

52. Mathematical description of relation between diffracted wave fronts and spectral diffraction patterns. General case. The equation relating the complex amplitude distribution $\mathcal{A}(\beta)$ in the diffraction patterns to the complex amplitude distribution $g(\eta)$ in the diffracted wave fronts can be written in the normalized form

$$\mathcal{A}(\beta) = \int_{-1}^{+1} g(\eta)\, e^{i\beta\eta}\, d\eta \tag{52.1}$$

with

$$g(\eta) = \mathcal{A}(\eta)\, e^{i\Phi(\eta)}. \tag{52.2}$$

The function $\Phi(\eta)$ appears in graphical form in the wave front interferograms. For a grating of uniform "blaze" and polarisation we have $\mathcal{A}(\eta) \equiv 1$, and we may write

$$g(\eta) = e^{i\Phi(\eta)}. \tag{52.3}$$

Eqs. (52.1) and (52.2) give, in a suitably normalized form, the distribution of complex amplitude $\mathcal{A}(\beta)$ in the diffraction patterns produced by wave fronts in which the complex amplitude distribution is $g(\eta)$ as a function of the coordinate η oriented along the width A of the wave front (of square aperture AH). The wave front width A is normalized to be equal to 2 and $\eta = u/\tfrac{1}{2}A$, where u is the true coordinate in the aperture. We recall that the angular coordinate $\beta = \pi x/[(\lambda/A) f]$ is measured from the center of the diffraction patterns that correspond to the grating equation, viz.

$$\sin i + \sin i' = m\lambda/a.$$

It is frequently useful to write the Fourier transform relation between the wave front and the diffraction patterns, as given in Eq. (43.4) in operational form

$$\mathcal{A}(\beta) = T[g(\eta)]. \tag{52.4}$$

For the case of uniform blaze and polarization, we have

$$\mathscr{A}(\beta) = T\left[e^{i\Phi(\eta)}\right]. \tag{52.5}$$

It is clear that the description of the distribution of light in the diffraction patterns does not call for any formal distinction between perfect or imperfect wave fronts.

We may take the distribution of light in the diffraction pattern formed by a perfectly plane wave front as a criterion of perfection. We can then consider any departures from planeness in the wave front and any departures for the light intensity distribution in the "perfect" diffraction pattern as imperfections. The departures from perfection in the wave fronts are called "aberrations" and the spurious peaks in the grating diffraction patterns are variously called "satellites", "grass" or "ghosts". The spurious spectral lines are formally recognized as diffraction pattern imperfections (STROKE, loc. cit. 1960). For a perfectly plane diffracted wave front, $\Phi(\eta)$ is a constant and

$$\mathscr{A}(\beta) = \int_{-1}^{+1} e^{i\beta\eta}\, d\eta = 2\frac{\sin\beta}{\beta}. \tag{52.6}$$

The corresponding intensity $I(\beta)$, in the form normalized according to Eq. (43.6) is given for a perfect diffraction pattern, by

$$I(\beta) = \left(\frac{\sin\beta}{\beta}\right)^2. \tag{52.7}$$

A perfect diffraction pattern is thus seen to have a $(\sin\beta/\beta)^2$ intensity distribution, with a first zero at $\beta=\pi$ or

$$x = x_0 = \frac{\lambda}{A} f \tag{52.8}$$

in the focal plane of the mirror (with a focal length equal to f). This is the same diffraction pattern as would be obtained by any plane wave front in a square aperture of width A. The normalization given here is immediately applicable to computer use (STROKE, loc. cit. 1960, see Sect. 59).

When a wave front is imperfectly plane, $\Phi(\eta)$ is not a constant any more. One can write

$$\Phi(\eta) = k\Delta(\eta) \tag{52.9}$$

where

$$k = \frac{2\pi}{\lambda}. \tag{52.10}$$

The importance of Eq. (52.9) resides also in the fact that the "aberration" $\Delta(\eta)$ in the wave front can be simply related to grating imperfections. In fact, two general types of grating imperfections need to be distinguished, according to Sects. 48 and 49, viz. (a) ruling imperfections, and (b) grating-blank imperfections.

In Sects. 48 and 49 it was shown that $\Delta(\eta)$ increases with $\sin i'$ for aberrations resulting from ruling errors [Eqs. (48.2), (48.3), (48.4)], but that it decreases with $\cos i'$ for grating blank and coating errors [Eq. (49.3)].

A number of cases of special grating and wave-front aberrations are examined in Sects. 53, 55, 56, 57, 58.

One may summarize the relation between wave fronts and diffraction patterns by the equations

$$\mathscr{A}(\beta) = T\left[e^{i\Phi(\eta)}\right] = T\left[e^{ik\Delta(\eta)}\right] \tag{52.11}$$

where $T[\;]$ stands for the Fourier transform of $[\;]$ as before, for the case of $\mathscr{A}(\eta)=1$.

53. The important special case of "linear" wave-front aberrations, according to Stroke. Certain types of grating blank defects and ruling errors will result in a wave-front aberration $\Phi(\eta)$, the phase of which will vary linearly as a function of the wave-front coordinate η. This case was first described by Stroke in 1959 (loc. cit. 1960), and is of considerable importance in practice.

Linear diffracted wave-front aberrations may be considered without regard to their origin in ruling or blank errors, even though reference to the grating is helpful for a qualitative understanding. In fact, one can always consider that a linear wave-front aberration has resulted either

a) from a portion of a perfect grating ruled on a part of the blank inclined with respect to the main part of the grating (case of "blank imperfections") or

b) from a portion of a grating ruled on perfectly flat blank, but with a spacing constant that increases linearly over the imperfect portion (case of "cumulative ruling errors").

Both cases have, of course, physical significance and can be used to compute the diffraction pattern imperfections in detail if properly interpreted.

In fact, the diffraction pattern of Fig. 71 for the "controlled-error" grating has been calculated by Stroke with these interpretations. Stroke has also calculated the diffraction pattern by digital Fourier transformation on the M.I.T. IBM 704 computer, and found it to be in perfect agreement with the experimental spectrogram.

To calculate the effect of the inclined portion of the wave front on the diffraction pattern, according to the interpretation (a), we remember that what is being summed are complex amplitudes. To simplify the calculation, let us assume that the linear aberration extends over a fraction $1/M$ of the wave front, at the center of the grating. Let the *linear phase aberration* be

$$\Phi(\eta) = \pm k m \eta \tag{53.1}$$

where m is a constant and $k = 2\pi/\lambda$ as before. Since

$$\mathscr{A}(\beta) = \int_{-1}^{+1} e^{i\Phi(\eta)} e^{i\beta\eta} \, d\eta, \tag{53.2}$$

we have here

$$\mathscr{A}(\beta) = \int_{-1}^{+1/M} e^{i\beta\eta} \, d\eta + \int_{-1/M}^{+1/M} e^{\pm ikm\eta} e^{i\beta\eta} \, d\eta + \int_{+1/M}^{+1} e^{i\beta\eta} \, d\eta, \tag{53.3}$$

which gives

$$\mathscr{A}(\beta) = 2\frac{\sin\beta}{\beta} - \frac{2}{M}\,\frac{\sin(\beta/M)}{(\beta/M)} + \frac{2}{M}\,\frac{\sin[(\beta \mp km)/M]}{(\beta \mp km)/M}, \tag{53.4}$$

a result which can be very simply interpreted in the following terms. The complex amplitude in the diffraction pattern is formed of the sum of

1. the perfect diffraction pattern (amplitude) corresponding to a perfect grating wave front of width A;

2. minus a diffraction pattern (amplitude) corresponding to a perfect wave front of width A/M;

3. plus a diffraction pattern (amplitude) corresponding to a wave front width A/M, but having its center displaced by $\Delta\beta = \pm km$ in the β domain.

A graphical plot of these patterns and their summation does indeed give the computed pattern of Fig. 71.

More generally, Stroke (1960) has shown that if

$$\mathscr{A}(\beta) = T[g(\eta)_0] \tag{53.5}$$

then

$$\mathscr{A}(\beta \mp kn) = T[g(\eta)_0 e^{\pm ikn\eta}] \tag{53.6}$$

in agreement with well known properties of Fourier transforms.

The interpretation of a linear wave-front aberration as resulting from a linear spacing variation may also be used for this calculation although it does not immediately give the detail of the diffraction pattern imperfections, but rather only the maximum of the principal spurious satellite which one would expect to find when a sizable portion of the grating is inclined with respect to the main grating. Qualitatively, when a grating is formed of two perfectly ruled portions inclined with respect to each other, one would expect to find two spectral lines in each wavelength, slightly displaced with respect to each other and with intensities in the ratio of the widths of the two portions of the grating.

Consider indeed a portion of the grating on which the spacing varies linearly *according to the equation* $\Delta a/a$, such that

$$\frac{\Delta a}{a} = \pm \frac{a \Delta i \cos i'}{a \sin i'} = \pm \frac{\Delta i}{\tan i'}. \tag{53.7}$$

according to Eqs. (49.1) and (49.2).

We also have, by differentiating the grating equation $\sin i + \sin i' = m\lambda/a$ with respect to a, the relation

$$\Delta i' = -2 \frac{\Delta a}{a} \tan i'. \tag{53.8}$$

We therefore find

$$\Delta i' = -2\Delta i \tag{53.9}$$

which is indeed the effect that one would expect to obtain by making the grating rotate by an angle Δi with respect to the incident wave front, at any angle in autocollimation. With the focal distance f of the mirror, we get in the focal plane

$$\Delta l' = -2f \frac{\Delta a}{a} \tan i'. \tag{53.10}$$

We need to recall that the ruling error Δa which enters these equations is $(\sin i')^{-1}$ the maximum fringe error as it appears in the wave-front interferogram. For the grating of Fig. 71 we have

$$\Delta a = \frac{1}{\sin 41^\circ} \text{ fringes} = \frac{1}{0.6561} \text{ fringes} \simeq 1.5 \text{ fringes},$$

and, since this error appears in 25 mm of the grating ruled at 300 grooves per millimeter, each groove having a spacing of 12.2 fringes, we have

$$\frac{\Delta a}{a} = \frac{1.5}{25 \times 300 \times 12.2},$$

and finally

$$\Delta l' = -2f \frac{\Delta a}{a} \tan i' = -0.348 \text{ mm},$$

in perfect agreement with the experimental spectrogram (Fig. 71).

In fact, a notable characteristic of linear wave-front aberrations, as pointed out by STROKE (loc. cit. 1960) is that the distance from the line centers of the centers of gravity of the principal satellites which they produce is wavelength-independent at a given grating angle, in agreement with Eqs. (53.8), (53.4) and (53.6), as well as with experiment (Fig. 71). Moreover, the relative intensity of these satellites, when compared to the principal lines, is also seen to be wavelength independent for extended "linear" aberrations.

Linear aberrations have been found to be the single most important origin of spectral line imperfection in high-resolution gratings since they can result from a number of different causes, such as cumulative ruling errors, blank and coating imperfections, and imperfections in the mounting and replication of gratings. The cumulative ruling errors themselves may result from different origins, of which variable friction, diamond motion, temperature distortions of the ruling engine, are particularly well known. Elimination of such errors from modern gratings is at the root of the attainment of high-resolution gratings. These gratings are, however, also characterized by notable reduction in more local, but still extended aberrations of a general type. The effect of such aberrations is discussed in the next section.

54. The effect of grating imperfections on the spectral lines in the general case. Variation of spectral-line imperfections with angle and wavelength, according to STROKE. Only gratings with comparatively small extended aberrations, on the order of up to a wavelength or so, are considered as gratings good enough for general spectroscopic work. Except for the special defect of parabolically curved wave fronts (which may exceed several wavelengths and to which correspond perfect but out-of-focus diffraction patterns, see Sect. 57), wavefront aberrations tolerable in high-resolution gratings should generally remain smaller than $^1/_5\,\lambda$, according to STROKE, since any extended aberration exceeding $^1/_{10}\,\lambda$ produces noticeable spurious spectral diffraction-pattern imperfections. (Tolerances for periodic ruling errors are 10 to 50 times smaller.) The distribution of light in the diffraction patterns produced by gratings with imperfections of any general amplitude and nature can be fully calculated with the help of a digital computer, by carrying out the Fourier transformation of the phase distribution as it appears in a readily usable form in the diffracted wavefront interferograms. Details according to STROKE are given in Sect. 59. It is of considerable theoretical and practical interest to predict the variation of spectral line imperfections with grating angle and wavelength. STROKE (1960) has also shown that the general relation between aberrated diffracted wave fronts and the corresponding diffraction patterns can indeed be expressed in a form which permits to predict this variation. Here we follow the analysis according to STROKE[1]. In agreement with Sects. 48 and 49 one can write $\mathscr{A}(\beta)$ in the form

$$\mathscr{A}(\beta)=\int\limits_{-1}^{+1}\exp\left[i\alpha\Phi(\eta)_0\right]\exp\left[i\frac{\lambda}{\lambda_0}\frac{\cos i_0}{\cos i}\beta_0\eta\right]d\eta\,, \tag{54.1}$$

where

$$\beta=\frac{\lambda}{\lambda_0}\frac{\cos i_0}{\cos i}\beta_0\,, \tag{54.2}$$

and for grating *ruling* errors

$$\alpha\equiv\alpha_R=\frac{\lambda_0}{\lambda}\frac{\sin i}{\sin i_0}\,, \tag{54.3}$$

whereas for grating *blank* errors

$$\alpha\equiv\alpha_B=\frac{\lambda_0}{\lambda}\frac{\cos i}{\cos i_0}\,. \tag{54.4}$$

With the help of these equations, a *diffraction pattern similarity law* and a *diffraction pattern imperfection variation law* can be rather well established. Eqs. (54.1) to (54.4) are established as follows: Let W be the width of a grating used in autocollimation at any angle i_0 in the wavelength λ_0. Let $\Delta(\eta')$ be the

[1] G. W. STROKE: Rev. Opt. **39**, 291—398 (1960).

function describing the error in the positioning of the grooves on the grating (along a normalized coordinate η'). The corresponding aberration in the wave front (coordinate η) is $\Delta(\eta)_0$, as a function of the normalized coordinate η.

For the case of aberrations caused by ruling errors, the two functions are related by the equation

$$\Delta(\eta)_0 = 2\,\Delta(\eta')\sin i_0 \tag{54.5}$$

according to Sect. 48. The phase aberration corresponding to $\Delta(\eta)_0$ in the wavelength λ_0 is given by

$$\Phi(\eta)_0 = \frac{2\pi}{\lambda_0}\Delta(\eta)_0 = \frac{2\pi}{\lambda_0}\left[2\Delta(\eta')\sin i_0\right]. \tag{54.6}$$

If the grating is now used at some other angle i and in some other wavelength λ, the phase aberration $\Phi(\eta)$ which corresponds to the unchanged $\Delta(\eta')$ is given by

$$\Phi(\eta)_R = \left(\frac{\lambda_0}{\lambda}\,\frac{\sin i}{\sin i_0}\right)\Phi(\eta)_0 = \alpha_R\,\Phi(\eta)_0, \tag{54.7}$$

in this case of *ruling* errors.

In the case of aberrations caused by grating *blank* defects, for which (Sect. 49)

$$\Delta(\eta)_0 = 2a\Delta i\cos i \tag{54.8}$$

and

$$\Phi(\eta)_0 = \frac{2\pi}{\lambda_0}(2a\,\Delta i\cos i), \tag{54.9}$$

one similarly has

$$\Phi(\eta)_B = \left(\frac{\lambda_0}{\lambda}\,\frac{\cos i}{\cos i_0}\right)\Phi(\eta)_0 = \alpha_B\,\Phi(\eta)_0. \tag{54.10}$$

In both cases, $\Delta(\eta)_0$ is the wave-front aberration as it appears in the interferograms, and $\Phi(\eta)_0 = (2\pi/\lambda_0)\Delta(\eta)_0$ is the corresponding phase aberration.

It is convenient to identity α to the expression

$$\alpha_R = \frac{\lambda_0}{\lambda}\,\frac{\sin i}{\sin i_0}, \tag{54.11}$$

in the case of ruling errors, and to

$$\alpha_B = \frac{\lambda_0}{\lambda}\,\frac{\cos i}{\cos i_0} \tag{54.12}$$

in the case of grating blank errors. One therefore has in the general case

$$\Phi(\eta) = \alpha\,\Phi(\eta)_0 \tag{54.13}$$

where $\Phi(\eta) \equiv \Phi(\eta)_R$ when $\alpha \equiv \alpha_R$, and $\Phi(\eta) \equiv \Phi(\eta)_B$ when $\alpha \equiv \alpha_B$.

One next recalls that the linear distances in the image focal plane are given by

$$x = \frac{x_0}{\pi}\beta, \tag{54.14}$$

where x_0 gives the distance from the line center of the first diffraction minimum (in the case of a pupil without aberrations or apodization). One has

$$x_0 = \frac{\lambda}{A}f = \frac{\lambda}{W\cos i'}f, \tag{54.15}$$

where W is the grating width, and $A = W\cos i'$ is the width of the pupil. It is clear that the linear distances in the image plane vary with λ and i as (λ/λ_0) $(\cos i_0/\cos i)$. In general the same conclusion applies to the imperfect diffraction patterns too.

Indeed, if the wavelength changes with the ratio λ/λ_0 and the dimensions in the pupil change with the ratio $\cos i_0/\cos i$, and if the normalized coordinate η_0, corresponding to λ_0 and i_0, is changed to η such that

$$\eta = \frac{\lambda_0}{\lambda} \frac{\cos i}{\cos i_0} \eta_0, \tag{54.16}$$

then the normalized amplitude $\mathscr{A}(\beta_0)$ corresponding to λ_0 and i_0, and given by

$$\mathscr{A}(\beta_0) = \int_{-1}^{+1} \exp[i\Phi(\eta)_0] \exp[i\beta_0\eta_0]\, d\eta_0, \tag{54.17}$$

is changed to

$$\mathscr{A}(\beta) = \int_{-1}^{+1} \exp[i\alpha\Phi(\eta)_0] \exp\left[i \frac{\lambda}{\lambda_0} \frac{\cos i_0}{\cos i} \beta_0 \eta\right] d\eta \tag{54.18}$$

which is indeed identical to Eq. (54.1), with the values of β_0 and α given by Eqs. (54.2) to (54.4).

α) *Diffraction pattern similarity law.* In the preceding equations, the change of the aberration $\Phi(\eta)_0$ for a given grating by the factor α was taken as resulting from the changes of wavelength and angle of use. One is therefore led to conclude according to STROKE (loc. cit.) that the entire distribution of light in the imperfect diffraction patterns will be subject to a $(\lambda/\lambda_0)\cdot(\cos i_0/\cos i)$ variation as a function of grating angle and wavelength. The same law applies to the variation of perfect diffraction patterns, of course. (The case of the "linear" aberrations discussed in Sect. 53 forms an exception which can be singled out experimentally with the help of interferograms taken at two angles.)

β) *Diffraction pattern imperfection variation law.* By noting that the wave front complex amplitude distribution $g(\eta)$ can be written as

$$g(\eta) = e^{i\Phi(\eta)} = 1 - \tfrac{1}{2}\Phi(\eta)^2 + i\Phi(\eta) + \mathrm{Res}\,\Phi(\eta) \qquad [\text{for } |\Phi(\eta)| < \tfrac{1}{2}\pi], \tag{54.19}$$

that is for the case of small aberrations, and by observing that

$$T[\Phi(\eta)_0^2] = T[\Phi(\eta)]_0 \otimes T[\Phi(\eta)_0] \tag{54.20}$$

and that $T[\Phi(\eta)_0] \to 0$ outside of the diffraction-pattern centers, that is precisely in the region of interest where the spurious spectral lines (satellites and grass) appear, STROKE (loc. cit. 1960) shows that the diffraction-pattern intensities at the wavelength λ_0 and angle i_0 on one hand, and the wavelength λ and angle i, on the other hand, can be written as

$$I(\beta_0) \simeq \left(\frac{\sin\beta_0}{\beta_0}\right)^2 + \frac{1}{4}\{T[\Phi(\eta)_0]\}^2 \tag{54.21}$$

and

$$I(\beta) \simeq \left(\frac{\sin\beta}{\beta}\right)^2 + \frac{1}{4}\alpha^2\{T\{\Phi(\eta)_0]\}^2, \tag{54.22}$$

respectively, and that therefore *the contribution $\Delta I(\beta)$ to the diffraction-pattern imperfections (satellites and grass) vary as*

$$\Delta I(\beta) \simeq \alpha^2 \Delta I(\beta_0), \tag{54.23}$$

that is as α^2, where, in the case of grating *ruling* errors

$$\alpha^2 \equiv \alpha_R^2 = \left(\frac{\lambda_0}{\lambda}\right)^2 \left(\frac{\sin i}{\sin i_0}\right)^2 \tag{54.24}$$

and in the case of grating *blank* errors

$$\alpha^2 \equiv \alpha_B^2 = \left(\frac{\lambda_0}{\lambda}\right)^2 \left(\frac{\cos i}{\cos i_0}\right)^2. \tag{54.25}$$

In words, for aberrations smaller than $\frac{1}{2}\pi$ or $\frac{1}{4}\lambda$ *the diffraction pattern imperfections that appear outside of the line centers increase with the square of the decrease in wavelength; for ruling errors, they further increase with the square of the sine of the grating angle. But for blank errors, the diffraction-pattern imperfections fortunately decrease with the square of the cosine of the grating angle.*

Eqs. (54.21) to (54.24) are arrived at as follows:

Let $g(\eta)$ be the complex amplitude distribution in the wave front. With $|\mathscr{A}(\eta)|^2 = 1$, one has

$$g(\eta) = e^{i\Phi(\eta)} = \cos \Phi(\eta) + i \sin \Phi(\eta) \tag{54.26}$$

where

$$\Phi(\eta) = \alpha \Phi(\eta)_0. \tag{54.13}$$

According to STROKE (loc. cit.), one may now wish to investigate the possibility of obtaining some simple approximations for the prediction of the variation with i and λ of the light-intensity contribution in the imperfect regions of the diffraction of a given imperfect grating.

As an example, the case of $|\Phi(\eta)| < \pi/2$ is treated in detail. In this case, one can write

$$g(\eta) = e^{i\Phi(\eta)} = 1 - \frac{\Phi(\eta)^2}{2} + i\Phi(\eta) + \text{Res}\, \Phi(\eta) \quad \text{for} \quad |\Phi(\eta)| < \pi/2 \tag{54.27}$$

and one has for the diffracted light amplitude

$$\mathscr{A}(\beta) = \frac{2 \sin \beta}{\beta} - \frac{1}{2} T[\Phi(\eta)^2] + i T[\Phi(\eta)]. \tag{54.28}$$

where $T[\;]$ stands for the Fourier transform of the bracketed function. With the given approximation, one also has

$$|\Phi(\eta)_0| < \pi/2 \tag{54.29}$$

and

$$\mathscr{A}(\beta_0) = \frac{2 \sin \beta_0}{\beta_0} - \frac{1}{2} T[\Phi(\eta)_0^2] + i T[\Phi(\eta)_0]. \tag{54.30}$$

But with the equation

$$\Phi(\eta) = \alpha \Phi(\eta)_0 \tag{54.13}$$

one has

$$\mathscr{A}(\beta) = \frac{2 \sin \beta}{\beta} - \frac{1}{2} \alpha^2 T[\Phi(\eta)_0^2] + i \alpha T[\Phi(\eta)_0] \tag{54.31}$$

with

$$\alpha \equiv \alpha_R = \frac{\lambda_0}{\lambda} \frac{\sin i}{\sin i_0} \quad \text{(for grating \textit{ruling} errors)} \tag{54.3}$$

and

$$\alpha \equiv \alpha_B = \frac{\lambda_0}{\lambda} \frac{\cos i}{\cos i_0} \quad \text{(for grating \textit{blank} errors)}. \tag{54.4}$$

It is important to note that the parameter α is not small. Indeed, α can assume values greatly exceeding unity, depending on the use of the grating. One cannot, therefore, neglect the α^2 term with respect to the α term, as one might do at this point, without examination of the orders of magnitude. However, what one may do is to write

$$T[\Phi(\eta)_0^2] = T[\Phi(\eta)_0] \otimes T[\Phi(\eta)_0] \tag{54.32}$$

where the $\otimes$ sign stands for the convolution process.

For the case of the small aberrations considered here, it is clear that the $T[\Phi(\eta)_0^2]$ function does become small as one leaves the center of the diffraction pattern, that is precisely in the region where the light intensity contribution due to the grating imperfections is to be evaluated. STROKE also shows that this conclusion is in agreement with SCHWARTZ's inequalities established in the theory of correlation functions. One is thus led to conclude, on the basis of this argument, that the term $\alpha^2 T[\Phi(\eta)_0^2]$ can indeed be neglected, after all, in this case, with respect to the $\alpha T[\Phi(\eta)_0]$ term. One therefore has for the amplitudes

$$\mathscr{A}(\beta_0) \simeq \frac{2\sin\beta_0}{\beta_0} + i\,T[\Phi(\eta)_0] \tag{54.33}$$

and

$$\mathscr{A}(\beta) \simeq \frac{2\sin\beta}{\beta} + i\,\alpha\,T[\Phi(\eta)_0]. \tag{54.34}$$

With the previously used normalization, one has

$$I(\beta) = \tfrac{1}{4}\mathscr{A}(\beta)\,\mathscr{A}(\beta)^* \tag{54.35}$$

and one therefore has for the intensities

$$I(\beta_0) \simeq \left(\frac{\sin\beta_0}{\beta_0}\right)^2 + \frac{1}{4}\{T[\Phi(\eta)_0^2]\}^2 \tag{54.36}$$

and

$$I(\beta) \simeq \left(\frac{\sin\beta}{\beta}\right)^2 + \frac{1}{4}\alpha^2\{T[\Phi(\eta)_0]\}^2. \tag{54.37}$$

One finally notes the first term in Eqs. (54.36) and (54.37) is negligible in the region outside of the diffraction pattern center, that is in the region where the intensity contribution due to the grating imperfections is being evaluated. In that region (for example beyond the second maximum of the perfect diffraction patterns) the contribution is due essentially to the second terms of the Eqs. (54.36) and (54.37).

Let these contributions due to the grating imperfections be described by $\Delta I(\beta_0)$ and by $\Delta I(\beta)$ for the two cases i_0, λ_0 and i, λ respectively. The final result is

$$\Delta I(\beta) \simeq \alpha^2 \Delta I(\beta_0) \tag{54.38}$$

with

$$\alpha^2 \equiv \alpha_R^2 = \left(\frac{\lambda_0}{\lambda}\right)^2 \left(\frac{\sin i}{\sin i_0}\right)^2 \quad \text{(for grating \textit{ruling} errors)} \tag{54.24}$$

and

$$\alpha^2 \equiv \alpha_B^2 = \left(\frac{\lambda_0}{\lambda}\right)^2 \left(\frac{\cos i}{\cos i_0}\right)^2 \quad \text{(for grating \textit{blank} errors)}. \tag{54.25}$$

Eq. (54.38) is applicable for $|\Phi(\eta)| < \pi/2$.

It is easily verified that Eqs. (54.38) and (54.24) give the correct values for the intensities of the Rowland ghosts, which are due to periodic errors in the ruling (see Sect. 55). These ghosts are generally due to errors of very small amplitude, for which the approximations made are very good. Moreover, the variation of position of the Rowland ghosts is also correctly predicted according to Eq. (54.2).

In view of the very satisfactory agreement of experiment with the theoretical formulation established according to STROKE (loc. cit.) one may wonder, inversely, whether the general diffraction-pattern imperfection variation laws, established for arbitrary wave fronts in terms of a *continuous* phase distribution (with, of course, a discrete number of discontinuities, maxima and minima allowed) could not have been arrived at by a Fourier series analysis of the wave-front aberrations. Indeed, it is easy to establish the variation laws for a simple harmonic wave front phase aberration (Sect. 55), and one may wish to establish the general law starting from a Fourier sum. Attempts[1] in this direction have been proven unfruitful, as can be understood by the difficulties associated with the need for applying superposition to the complex amplitudes, and the fact that superposition applied to the ghost intensities is not applicable. A somewhat better agreement will result for the positions of the principal ghosts.

In a general way, one can describe the wave front in the pupil, with normalized coordinates, in terms of a Fourier series, as

$$g(\eta) = \exp\left\{i \sum_{n=1}^{\infty} a_n \cos\left[n \frac{2\pi}{p} \eta + \Phi_n\right]\right\} \tag{54.39}$$

where a_n are the amplitudes of the harmonic components, Φ_n their relative phases, and $n\,2\pi/p$ their spatial frequencies. The complex amplitude $\mathscr{A}(\beta)$ in the diffraction pattern is still

$$\mathscr{A}(\beta) = T\left[g(\eta)\right], \tag{54.40}$$

of course. One also has the well known identity

$$\exp\left[i \Phi_M \cos \frac{2\pi}{p} \eta\right] = \sum_{n=-\infty}^{+\infty} i^n J_n(\Phi_M) \exp\left[i n \frac{2\pi}{p} \eta\right]. \tag{54.41}$$

Eq. (54.39) leads with Eqs. (54.40) and (54.41) to the expression

$$\left.\begin{aligned} g(\eta) = &\sum_{q_1, q_2, \ldots = -\infty}^{+\infty} J_{q_1}(a_1)\, J_{q_2}(a_2) \ldots \times \\ &\times \exp\left\{i\left[q_1\left(\frac{2\pi}{p}\eta + \Phi_1\right) + q_2\left(2\frac{2\pi}{p}\eta + \Phi_2\right) + \cdots\right]\right\} \end{aligned}\right\} \tag{54.42}$$

which becomes unwieldy in the Fourier transformation even for small values of the arguments $[n(2\pi/p)\eta + \Phi_n]$. The $J_{q_n}(a_n) \ldots$ are Bessel functions of the orders $q_n = 1, 2, \ldots$ etc.

In conclusion, it is recalled that approximate expressions in Eqs. (54.38), (54.24), (54.25) are applicable to the general case of aberrations of any form, when they are smaller than $\pi/2$. Similar equations for the case of larger aberrations are easily established according to the same procedure by including a greater number of terms in the expansion of Eq. (54.27). It is to be noted that the series expansion of $g(\eta) = \exp[i\alpha \Phi(\eta)_0]$ will introduce integral powers of $\Phi(\eta)_0$: the Fourier transforms of these powers simply correspond to the convolution of a number of identical factors equal to the power of each term in the series.

The important special case of the linear wave-front aberrations was dealt with in Sect. 53, and the special case of periodic errors is dealt with in Sect. 55.

Finally, it should be clear that the general equation

$$\mathscr{A}(\beta) = \int_{-1}^{+1} \exp[i\alpha \Phi(\eta)_0] \exp\left[i \frac{\lambda}{\lambda_0} \frac{\cos i_0}{\cos i} \beta_0 \eta\right] d\eta \tag{54.1}$$

[1] See, for example, N. A. FINKELSTEIN, C. H. BRUMLEY and R. J. MELTZER: J. Opt. Soc. Amer. **42**, 121 (1952).

or its various equivalent forms are to be used when an expression for $\mathscr{A}(\beta)$ without any approximations is required. The form of Eq. (54.1) suitable for computation on a digital electronic computer and its numerical computation are dealt with in Sect. 59.

55. Effect of periodic ruling errors. Rowland ghosts. This is the case first examined by H. A. ROWLAND in 1902[1], and by C. S. PIERCE in 1879[2]. STROKE (loc. cit. 1960)[3] shows this case to appear, formally, as an application of his treatment of the "linear" wave-front aberrations, and the resulting mathematical formalism is greatly simplified. Here, just like in the case of the "linear" aberrations, both the position and intensity of the diffraction pattern imperfections can be predicted. Examples of Rowland ghosts are shown in Figs. 72, 73 and 74.

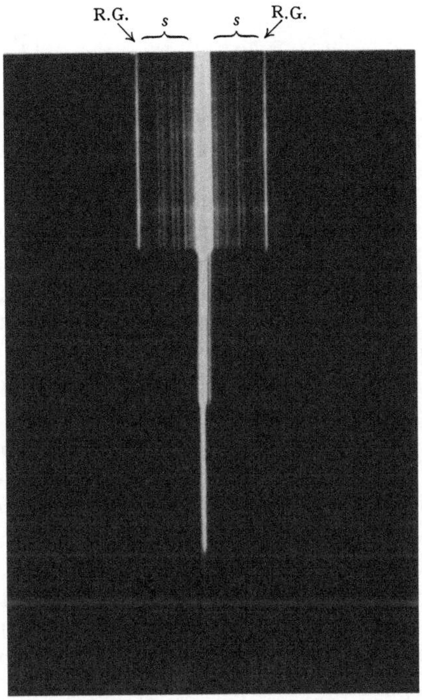

Fig. 72. Rowland ghosts (R.G.) in 1st order of spectrum of green mercury line (λ 5461 Å) (see satellite lines).

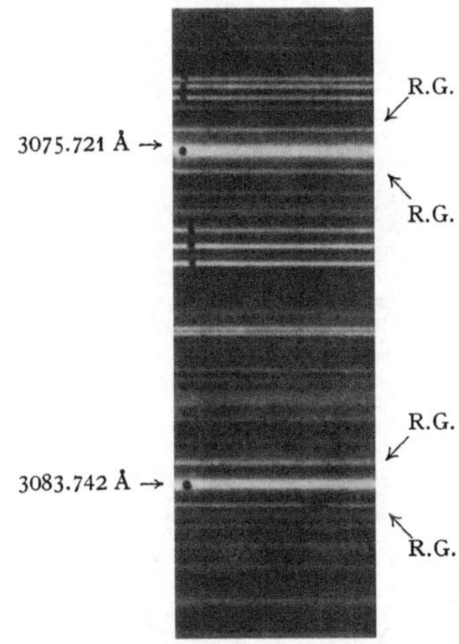

Fig. 73. Rowland ghosts (R.G.) in 1st order Fe spectrum with 600 grooves/mm grating (spectrum courtesy R.E.O.S.C. Paris). ● Identify real lines.

Periodic errors, once very common in gratings, have become much less important, especially in gratings ruled under interferometric control, for which the Rowland ghost intensities remain less than 0.1% even at angles as high as 76° in autocollimation in the visible and are of the order of 10^{-5} at low angles[4].

Nevertheless, when correcting the origins of periodic errors in the course of the building of a ruling engine, for instance, one does pay considerable attention to the spectral regions at several dozen, and sometimes hundreds of diffraction units $[(\lambda/A)f]$ from the line centers where the Rowland ghosts usually occur.

For a simple-harmonic error of amplitude ε and period P on a grating used at the angle i' in autocollimation, symmetric ghosts of orders n (n = integer) are found at a distance

$$(\Delta x)_n = n \frac{\lambda}{P \cos i'} f \tag{55.1}$$

[1] H. A. ROWLAND: Physical Papers, p. 525. Baltimore: The Johns Hopkins Univ. 1902.— Phil. Mag. (5), **35**, 397 (1893).

[2] C. S. PEIRCE: Amer. J. Math. **2**, 330 (1879).

[3] G. W. STROKE: Rev. Opt. **39**, 291—398 (1960).

[4] G. R. HARRISON and G. W. STROKE: J. Opt. Soc. Amer. **50**, 1153 (1960).

in a spectrograph of focal length f. The relative intensities of the first-order ($n=1$) and second-order ($n=2$) ghosts are found to be

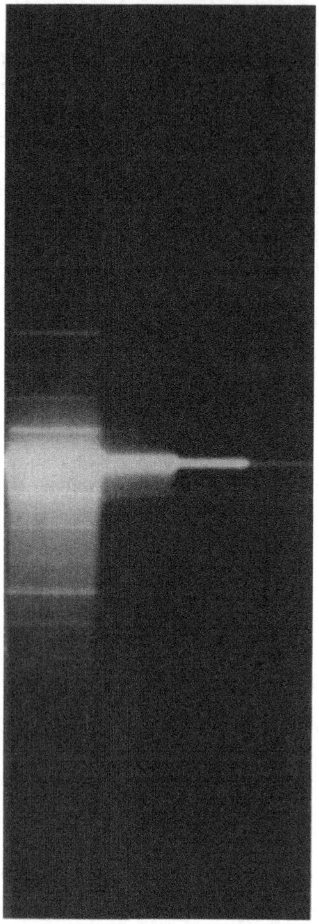

Fig. 74. Rowland ghosts in spectrum of green mercury line (grating ruled by ANDERSON on ROWLAND's engine in 1914) (see interferogram in Fig. 191).

$$G_1=[J_1(\Phi_M)^2]=\frac{4\pi^2\varepsilon^2\sin^2 i}{\lambda^2} \tag{55.2}$$

and

$$G_2=[J_2(\Phi_M)^2]=\frac{4\pi^4\varepsilon^4\sin^4 i}{\lambda^4}, \tag{55.3}$$

where

$$\Phi_M=\frac{2\pi}{\lambda}\,2\varepsilon\sin i \tag{55.4}$$

and J_1 and J_2 are Bessel functions. It is clear that the first-order ghost predominates, and also that the variation of its position and intensity, with grating angle and wavelength, is in agreement with the general laws given in the preceding section. It is also important to note that Rowland ghosts result from only the simple harmonic components in the ruling errors and the corresponding wave-front aberrations, and not from the maximum aberration-amplitude, which is generally considerably greater.

It is also important to observe that attempts to analyze imperfect wave fronts in terms of Fourier-series contributions and summations of ghost-amplitudes fail because of the practical difficulty resulting from the need of including a sufficiently great number of terms, and from the fact that the summation needs to be carried out over the complex amplitudes of the ghosts, and not over their intensities. When a sufficiently great number of terms is included, one simply gets back to the Fourier-transform representation which we have used in the first place, on more general grounds. A formal derivation of the effects of periodic wave front phase variations, according to STROKE (loc. cit. 1960), follows.

Let the wave front be described by

$$g(\eta)=\exp\left[i\,\Phi_M\cos\frac{2\pi}{p}\eta\right] \tag{55.5}$$

where Φ_M is the amplitude of the periodic error and $p\equiv p_\eta$ is the *period of the error* in the η domain.

According to a well-known equation[1], one has the identity

$$\exp\left[i\,\Phi_M\cos\frac{2\pi}{p}\eta\right]=\sum_{n=-\infty}^{+\infty} i^n J_n(\Phi_M)\,e^{in\frac{2\pi}{p}\eta}, \tag{55.6}$$

where $J_n(\Phi_M)$ is the Bessel function of order n, n being an integer.

[1] See for example, J. A. STRATTON: Electromagnetic Theory, p. 44. New York: McGraw Hill Book Co. 1941, or A. ANGOT: Compléments de Mathematiques, p. 372, edit. Revue d'Optique, 1957.

The amplitude in the diffraction pattern is therefore

$$\mathscr{A}(\beta)=\int_{\eta=-1}^{+1}\sum_{=-\infty}^{+\infty} i^n J_n(\Phi_M)\,\mathrm{e}^{in\frac{2\pi}{p}\eta}\,\mathrm{e}^{i\beta\eta}\,d\eta. \tag{55.7}$$

By interchanging the orders of integration and of summation, and by noting that $J_n(\Phi_M)$ is a constant for any given n, one immediately has

$$\left.\begin{aligned}\mathscr{A}(\beta)= J_0(\Phi_M)\int_{-1}^{+1}\mathrm{e}^{i\beta\eta}\,d\eta+&\\ +iJ_1(\Phi_M)\int_{-1}^{+1}\mathrm{e}^{i\beta\eta}\,\mathrm{e}^{i\frac{2\pi}{p}\eta}\,d\eta-&\\ -J_2(\Phi_M)\int_{-1}^{+1}\mathrm{e}^{i\beta\eta}\,\mathrm{e}^{i\frac{4\pi}{p}\eta}\,d\eta+&\\ +\text{ terms of positive higher orders of } n+&\\ +\text{ terms of negative higher orders of } n.&\end{aligned}\right\} \tag{55.8}$$

The first term of Eq. (55.8), equal to $J_0(\Phi_M)\,2\sin\beta/\beta$, is clearly equal to the central spectral line, while the other terms form "spurious" or "ghost" lines, symmetrically on each side of the central line. Indeed, one immediately recognizes in these terms the form $g(\eta)_0\exp(ikm\eta)$ of which the transform is $\mathscr{A}(\beta\mp km)$ according to the displacement theorem of Eq. (53.6). In each of these terms, one has $g(\eta)_0=\exp(i\beta\eta)$, and the modulation factor is equal to $\exp\left(in\frac{2\pi}{p}\eta\right)$. In this last factor, one identifies

$$km=\frac{2\pi}{p}\,n, \tag{55.9}$$

where *n is the order of the ghost.*

It is seen that the "ghosts" are indeed diffraction patterns, identical to the central-line diffraction pattern, but with a center of gravity displaced from the line center by

$$\Delta\beta=km=\frac{2\pi}{p_\eta}\,n, \tag{55.10}$$

that is, by

$$(\Delta x)_n=n\frac{\lambda}{P\cos i'}\,f \tag{55.11}$$

in a spectrograph of focal length f, when the grating is used at an angle $i'=i$, and when the period of the error on the grating is equal to P.

Indeed, since $\eta=u/(A/2)$, according to Eq. (43.3), and if p_u is the period of the error in the wave front (coordinate u), one has

$$p_\eta=\frac{p_u}{A/2} \tag{55.12}$$

and

$$\Delta x=\frac{x_0}{\beta}\,\Delta\beta=\frac{\lambda}{\pi}\,\frac{f}{A}\,\frac{A}{2}\,\frac{2\pi}{p_u}\,n=\frac{n\lambda}{p_u}\,f. \tag{55.13}$$

But

$$p_u=P\cos i' \tag{55.14}$$

which results in the form of $(\Delta x)_n$ as given by Eq. (55.11).

The effect of a periodic ruling error is seen to be equivalent to having a "ghost" grating of period p_u in the wave front. Such a "ghost" grating produces a spectrum which is symmetric with respect to a "central" line of order $n=0$ that coincides with the center of the spectral line of order m of the actual grating, where

$$m=\frac{2a}{\lambda}\sin i. \tag{55.15}$$

The intensities of the "Rowland" ghosts which result from periodic errors are easily obtained. In particular, for small values of Φ_M, one has

$$J_n(\Phi_M)\simeq\frac{1}{2^n n!}(\Phi_M)^n \tag{55.16}$$

from which the amplitudes of the ghost lines are obtained in the form

$$J_0(\Phi_M)\simeq 1,\quad J_1(\Phi_M)\simeq\tfrac{1}{2}\Phi_M,\quad J_2(\Phi_M)\simeq\tfrac{1}{8}\Phi_M^2,\quad J_3(\Phi_M)\simeq\tfrac{1}{48}\Phi_M^3,\ \text{etc.} \tag{55.17}$$

But, if ε is the amplitude of the periodic error in the position of the grooves on the grating, then one has at the angle i of the grating

$$\Phi_M=\frac{2\pi}{\lambda}\,2\varepsilon\sin i, \tag{55.18}$$

according to Sect. 48. Thus the intensity $[J_0(\Phi_M)]^2$ of the central line remains equal to unity. For the *first* pair of ghosts ($n=1$), the intensity is

$$G_1=[J_1(\Phi_M)]^2=\frac{4\pi^2\varepsilon^2\sin^2 i}{\lambda^2}. \tag{55.19}$$

For the *second* pair of ghosts, ($n=2$), the intensity is

$$G_2=[J_2(\Phi_M)]^2=\frac{4\pi^4\varepsilon^4\sin^4 i}{\lambda^4}, \tag{55.20}$$

and so on. The ratio of G_2 to G_1 is

$$\frac{G_2}{G_1}=\frac{\pi^2\varepsilon^2\sin^2 i}{\lambda^2} \tag{55.21}$$

that is $\frac{1}{4}$ of the ratio of the first ghost to the central line. More generally, on has

$$\frac{G_{n+1}}{G_n}=\frac{4}{(n+1)^2}\,\frac{\pi^2\varepsilon^2\sin^2 i}{\lambda^2}. \tag{55.22}$$

It is essential to note that ε *is the amplitude of the harmonic component of the error of period* P on the grating. Eqs. (55.19) to (55.22) are important also in that they show that the ghost intensity, while dependent on the amplitude of the error, the wavelength and the grating angle, is *independent of the grating spacing constant* a. The dependence of the ghost intensities on the spacing constant can of course be introduced. By noting that $4\sin^2 i/\lambda^2=m^2/a^2$, one has for the first ghost ($n=1$)

$$G_1=\frac{\pi^2 m^2\varepsilon^2}{a^2}; \tag{55.23}$$

which is the equation given by ROWLAND in his remarkable and very complete work[1] dealing with the effects of such errors in gratings. ROWLAND did not give his result in the form of Eqs. (55.10) and (55.19); these equations were first derived by STROKE (loc. cit. 1960), who notes that the Rowland ghost positions and errors are indeed predicted by the same general "laws" which he derived for aberrations of any form in Sect. 54.

---

[1] H. A. ROWLAND: Gratings in Theory and Practice. Phil. Mag (5), **35**, 397 (1893). — Physical Papers, p. 525. Baltimore: The Johns Hopkins Univ. Press 1902.

Fig. 76, Sect. 57 shows, according to STROKE (loc. cit. 1960), the interferometric wave-front measurement of the amplitude of the periodic error for a grating and compares it with the "cross-ruling" of a section of the same grating. It was shown that the same information as in the cross-ruling is obtained by the non-destructive interferograms, and with only a single ruling.

In his 1893 paper, ROWLAND (loc. cit.) deals with several other groove-positioning errors in a formulation which is very comparable to the Fourier-transform now in use, and his work should be consulted, along with the papers by A. A. MICHELSON[1], C. M. SPARROW[2], E. INGELSTAM and E. DJURLE[3], and others cited in Sect. 35.

It is found that periodic errors in ruling engines rarely have a simple harmonic character[4]. The positions of the ghosts are still determined by Eq. (55.10), but the intensities are determined by the harmonic content of the error, as given by Fourier analysis. Intensity estimates from the Fourier analysis of the errors are complicated by the need of first adding the complex amplitudes $\mathcal{A}(\beta)$ corresponding to each harmonic component, according to Eq. (55.8), and only then obtaining the intensity in the form

$$I(\beta) \propto \mathcal{A}(\beta)\,\mathcal{A}(\beta)^*. \tag{55.24}$$

An interesting feature observed in actual gratings (such as that illustrated by Fig. 76) is that a fairly large peak-to-peak amplitude may generally contain an appreciably smaller amplitude for the fundamental frequency. It is then found that the ghost intensities are of course considerably smaller than what could be attributed to the fundamental, if it had the entire peak amplitude.

The intensity distributions rarely follow the basic picture provided by Eqs. (55.8), (55.16), (55.19), (55.20), (55.21), (55.22), but it should be clear that a complete picture can be predicted with the help of Eqs. (55.8), (55.10) and (55.24) by first carrying out a Fourier analysis of the periodic error and then suitably summing the complex amplitudes. More generally, the Fourier transform equations of Sects. 52 and 54 will provide the desired result without the need for singling out the periodic error in the wave front, and the associated Fourier analysis. Indeed, the "Rowland" ghosts, resulting from periodic errors in the wave front, are simply part of the imperfect diffraction pattern produced by the grating, and there is, generally speaking, no need for singling them out in any special way.

In the largest sense, the spectrum observed in the focal plane of the imaging system is true to its description, namely it materializes the spatial Fourier "spectrum" of the complex amplitude distribution in the wave front.

In particular, any simple periodicities in the diffracted wave front, if present alone, will result in a corresponding spectrum, with the "lines" separated in an inverse ratio to the period in the wave front. In the case of multiple periodicities, the analysis and effect are complicated not only by the need for summing over the complex amplitudes in the ghost spectra, but also by the fact that a Fourier analysis of the wave front in terms of products of two functions results in a spectrum which is the convolution of the two corresponding spectra. Except in the simplest case of the product of two simple harmonic functions, it is appears preferable to carry out a Fourier transformation rather than a Fourier analysis

---

[1] A. A. MICHELSON: Studies in Optics. Chicago, Ill.: Chicago University Press 1927, reprinted by Phoenix Books. Chicago, Ill.: Chicago University Press 1962.

[2] C. M. SPARROW: Theory of Imperfect Gratings. Astrophys. J. **49**, 65 (1919).

[3] E. INGELSTAM and E. DJURLE: J. Opt. Soc. Amer. **43**, 572 (1953). — Ark. Fysik **4**, 423 (1952).

[4] See for example, N. A. FINKELSTEIN, C. H. BRUMLEY and R. J. MELTZER: The Reduction of Ghosts in Diffraction Grating Spectra. J. Opt. Soc. Amer. **42**, 121 (1952).

of the wave front. One also need to specify the position of the ghost line of zero order: it is found that all ghost spectra corresponding to periodic errors have their center on the spectral lines of the basic grating.

For instance, the basis period in the grating is that of the groove spacing for which one has

$$\sin i' = \frac{m\lambda}{a} - \sin i, \tag{55.25}$$

where the integer m is the order of the spectrum, λ the wavelength, i the angle of incidence and i' the angle of diffraction. Eq. (55.25) is applicable to any grating, as shown in Sects. 17, 18, 23 and 25. The ghost line of order zero is given by Eq. (56.1), when m is an integer.

In the case where another period is "superposed" on the grating, such as a periodic ruling error extending over several dozen or hundreds of grooves, one finds that the corresponding "spectrum" is centered on the spectral lines given by Eq. (55.25), but the angular separation between the lines of the ghost spectrum is considerably smaller than that given by Eq. (55.25). In fact, it was shown earlier in this section [cf. Eq. (55.13)] that the angular distance $(\Delta i')_R$ of the "Rowland" ghosts from the parent lines can be given by an equation of the form

$$(\Delta i')_R = \pm \frac{n\lambda}{p}, \tag{55.26}$$

where n is the *order of the ghost*, and p is the period of the error in the wave front. One has for a grating in autocollimation at an angle i' the equation

$$p = P\cos i' \tag{55.27}$$

where P is the period of the error on the grating.

Inasmuch as Eq. (55.25) can be written as

$$\Delta(\sin i') = \frac{m\lambda}{a} \tag{55.28}$$

and that

$$d(\sin i') = \cos i'\, di', \tag{55.29}$$

one has

$$di' = \frac{m\lambda}{a\cos i'} \quad (m = \text{integer}). \tag{55.30}$$

By writing Eq. (55.28) in the form

$$(\Delta i')_R = \pm \frac{n\lambda}{P\cos i'} \tag{55.31}$$

one immediately recognizes that the angular distances of the spectral lines from the corresponding lines of the basic grating are indeed in an inverse ratio to the corresponding period. [Clearly, for large angles, Eq. (55.25) and not Eq. (55.30) is to be used, and the inverse ratio then applies to the exact equations.]

A characteristic feature of Rowland ghosts originating from gratings ruled on mechanical engines is that the error period P is an integral multiple of the grating period a.

Let

$$P = \mathcal{N} a \tag{55.32}$$

where $\mathcal{N}$ is the number of grooves in a period (say the number of grooves per turn of the screw). One notes that in this case

$$[(\Delta i')_R]_n = \pm \frac{n}{m}\frac{di'}{\mathcal{N}}, \tag{55.33}$$

that is to say that the Rowland ghosts of consecutive orders n are indeed separated by $1/\mathcal{N}$ of the distance between two principal maxima [of orders $(m-1)$, or $(m+1)$, and m], as would be expected for the "ghost" grating having a spacing $\mathcal{N}$ times as coarse as the basic grating. For gratings ruled under interferometric servo-control[1], the groove spacing is in general not necessarily a sub-multiple of the error period: it is also found that the ghosts in interferometrically ruled gratings are one or two orders less intense than in mechanically ruled gratings.

Frequently it is convenient to relate the ghost position to the apparent "ghost wavelength" λ_g which would correspond to it. According to Eqs. (55.32), (55.33), and (55.25) it is seen that the ghost wavelength λ_g can be written as

$$\lambda_g = \lambda \pm \frac{n}{m} \frac{\lambda}{\mathcal{N}}. \tag{55.34}$$

In the first order of the grating, $m=1$, and the ghosts are found at wavelengths $\pm\lambda/\mathcal{N}$, $\pm 2\lambda/\mathcal{N}$, $\pm 3\lambda/\mathcal{N}$, etc. from λ. When photographed on "black and white" plates, the ghost lines take on the appearance of real lines at these "ghost" wavelengths.

56. Effect of multiple periodic errors. Lyman ghosts. Occasionally groups of two or three or more symmetrically placed ghosts are observed at distances from $\pm\frac{1}{5}$ to $\pm\frac{1}{2}$ the distance of one order to the next, and in fact at distances $\pm\frac{2}{5}$, $\pm\frac{3}{5}$ etc., and $\pm\frac{2}{2}$, $\pm\frac{3}{2}$ etc., respectively. In case of the first order, the group on the low wavelength side may appear to correspond to $\frac{4}{5}$ and $\frac{1}{2}$ the wavelength of the "main" line in this case, for the first group of ghosts and so on. More generally these ghosts at great distances from their "main" line are difficult to identify and tend to be very confusing when the spectra are photographed or recorded on a color-insensitive plate or photoelectric detector.

The corresponding ghosts were first observed by THEODORE LYMAN in 1901[2], and interpreted by CARL RUNGE in 1903, as outlined in LYMAN's 1903 paper. A more complete interpretation was later given by RUNGE[3] in 1922 and 1923. An excellent qualitative analysis is found in CHARLES F. MEYER's book[4], along with a treatment of many other grating error problems. ROWLAND (loc. cit.) also deals with important aspects of this problem, as does A. A. MICHELSON[5] and others.

Here we find again that the Fourier transform formulation results in great simplicity.

In case of a grating having two basically unrelated periodic errors, one can consider that the wave-front error is formed by the *product* of the two errors. By Fourier transformation, the corresponding ghost spectrum then consists of the *convolution* of the ghost spectra corresponding to each of the periods, each of the spectra being centered on the main spectral lines as before. Ghosts will therefore appear only in the vicinity of places where the ghost spectra of the two periods coincide, and they will be centered on the main spectral line. For instance, let

$$\lambda_g = \lambda \pm \frac{n_1}{m} \frac{\lambda}{\mathcal{N}_1} \tag{56.1}$$

[1] G. R. HARRISON and G. W. STROKE: J. Opt. Soc. Amer. **45**, 112 (1955); **50**, 1153 (1960). — G. W. STROKE: J. Opt. Soc. Amer. **51**, 1321 (1961). — Ruling, Testing and Use of Optical Gratings for High-Resolution Spectroscopy, in Vol. II, Progress in Optics, edit. by E. WOLF. Amsterdam: North-Holland Publishing Company 1963.

[2] T. LYMAN: Phys. Rev. **12**, 1 (1901); **16**, 257 (1903).

[3] C. RUNGE: J. Opt. Soc. Amer. **6**, 429 (1922). — Ann. Physik **71**, 178 (1923).

[4] C. F. MEYER: Diffraction of Light, X-rays and Material Particles, 2nd ed. Ann. Arbor, Mich.: J. W. Edwards Publ. 1949.

[5] A. A. MICHELSON: Astrophys. J. **18**, 278 (1903).

give the wavelengths of the ghosts corresponding to an error period P_1, having $\mathcal{N}_1$ grooves per period, and let

$$\lambda_g = \lambda \pm \frac{n_2}{m} \frac{\lambda}{\mathcal{N}_2} \tag{56.2}$$

give the wavelengths of the ghosts corresponding to the error period P_2, having $\mathcal{N}_2$ grooves per period. It is furthermore assumed that

$$P_1 = (r + \varepsilon) P_2 \tag{56.3}$$

where r is an integer and $|\varepsilon| < 1$, that is to say that the periods P_1 and P_2 are not exact multiples of each other. (The period P_2 is shorter than the period P_1). However, there will in general be some integers p and q such that

$$p P_1 = q P_2. \tag{56.4}$$

This will clearly occur when

$$p(r + \varepsilon) = q, \tag{56.5}$$

that is when

$$p \varepsilon = l \tag{56.6}$$

with p and l both integers. One also has

$$\varepsilon = \frac{l}{p}. \tag{56.7}$$

From Eqs. (56.1), (56.2) and (56.4) it is clear that the convolution of the two ghost spectra will have a maximum in the vicinity of the value λ for which

$$\frac{n_1 \lambda}{\mathcal{N}_1} = \frac{n_2 \lambda}{\mathcal{N}_2}, \tag{56.8}$$

subject to Eq. (56.4), that is when

$$\frac{n_1}{\mathcal{N}_1} = \frac{n_2}{\mathcal{N}_2}. \tag{56.9}$$

One therefore has, for the maxima, at which Lyman ghosts should appear the condition

$$\frac{n_2}{n_1} = \frac{p}{q}. \tag{56.10}$$

One is led to conclude that the Lyman ghosts will be most pronounced at the orders n_2 which are almost multiples of the orders n_1 of the "ghost" of the longer period P_1 (for which the ghosts are more closely spaced).

The essential point in the argument consists in considering the Lyman ghosts as very high harmonics of the longer period P_1. The Lyman ghosts will thus occur in the vicinity of the very low harmonics of the shorter period P_2. In an experimental case, one first determines the spacing between the three or four or more closely spaced "high" harmonics and obtains from it the period P_1 with the help of Eq. (56.1). One then attempts to determine the period P_2 with the help of Eq. (56.2) by assuming the order n_2 to be equal to 1, 2, etc.

In the case of the famous grating ruled by J. A. ANDERSON in 1912, and studied experimentally by MEGGERS and KIESS[1] ghost wavelengths λ_g were observed as follows:

[1] W. F. MEGGERS and C. C. KIESS, J. Opt. Soc. Am. 6. 417 (1922); C. RUNGE, ibid. p. 429.

Group	Wavelengths λ_g	Intensity
$\lambda - \frac{4}{5}\lambda$	No ghosts	0
$\lambda - \frac{3}{5}\lambda$	$\left(\frac{2}{5} - \frac{1}{1500}\right)\lambda$	3
	$\left(\frac{2}{5} + \frac{4}{1500}\right)\lambda$	1
$\lambda - \frac{2}{5}\lambda$	$\left(\frac{3}{5} - \frac{4}{1500}\right)\lambda$	1
	$\left(\frac{3}{5} + \frac{1}{1500}\right)\lambda$	3
$\lambda - \frac{1}{5}\lambda$	$\left(\frac{4}{5} - \frac{7}{1500}\right)\lambda$	1
	$\left(\frac{4}{5} - \frac{2}{1500}\right)\lambda$	5
	$\left(\frac{4}{5} + \frac{3}{1500}\right)\lambda$	2
	$\left(\frac{4}{5} + \frac{8}{1500}\right)\lambda$	1
λ	Primary spectrum lines	100000
$\lambda + \frac{1}{5}\lambda$	$\left(\frac{6}{5} - \frac{8}{1500}\right)\lambda$	1
	$\left(\frac{6}{5} - \frac{3}{1500}\right)\lambda$	2
	$\left(\frac{6}{5} + \frac{2}{1500}\right)\lambda$	5
	$\left(\frac{6}{5} + \frac{7}{1500}\right)\lambda$	1
$\lambda + \frac{2}{5}\lambda$	$\left(\frac{7}{5} - \frac{6}{1500}\right)\lambda$	1
	$\left(\frac{7}{5} - \frac{1}{1500}\right)\lambda$	10
	$\left(\frac{7}{5} + \frac{4}{1500}\right)\lambda$	3
	$\left(\frac{7}{5} + \frac{9}{1500}\right)\lambda$	1
$\lambda + \frac{3}{5}\lambda$	$\left(\frac{8}{5} - \frac{9}{1500}\right)\lambda$	1
	$\left(\frac{8}{5} - \frac{4}{1500}\right)\lambda$	2
	$\left(\frac{8}{5} + \frac{1}{1500}\right)\lambda$	8
	$\left(\frac{8}{5} + \frac{6}{1500}\right)\lambda$	1
$\lambda + \frac{4}{5}\lambda$	$\left(\frac{9}{5} - \frac{7}{1500}\right)\lambda$	1
	$\left(\frac{9}{5} - \frac{2}{1500}\right)\lambda$	5
	$\left(\frac{9}{5} + \frac{3}{1500}\right)\lambda$	2
	$\left(\frac{9}{5} + \frac{8}{1500}\right)\lambda$	1

In this case, Eqs. (56.1) and (56.2) can be applied twice as follows. One has

$$\lambda_g = \lambda \pm \frac{n_{1,2}}{m}\,\frac{\lambda}{\mathcal{N}_{1,2}} \qquad (56.1)\quad (56.2)$$

where $\lambda/\mathcal{N}$ gives the spacing of the ghosts. For the period P_1, the spacing is nearly

$$\frac{5\lambda}{1500} = \frac{\lambda}{300}$$

and therefore $\mathcal{N} \simeq 300$. For the spacing P_2, it is clear that the first-order ghost is found at $\lambda/5$ from the main line and therefore $\mathcal{N}_2 = 5$. One is led to conclude that the shorter period consists of 5 grooves and the larger of approximately 300 grooves.

It is, however, a basic point in this simple theory that the value 300 is "forbidden", being an exact multiple of 5. By a rather involved argument, RUNGE (loc. cit. 1922) tried out various values and found 298 to be the number which comes more closely into agreement with the experiment. In practice, no such precision is generally required, and a simple examination of the ghost spectrum is seen to immediately provide the two values $\mathcal{N}_2$ and $\mathcal{N}_1$, as well as the corresponding periods P_2 and P_1.

Extensions to more complicated cases are equally simple.

57. Reference wave fronts: plane and parabolic. It has been stressed by STROKE (loc. cit. 1960) that the general wave front reference-surface equations, as developed by A. MARÉCHAL[1], H. H. HOPKINS[2] and others, are also applicable to gratings. In particular, on the plane-wave side of the image-forming collimator, the best reference wave front is a plane wave front. However, as shown by H. H. HOPKINS (loc. cit.; p. 14—16) and A. MARÉCHAL (loc. cit.; p. 140), a wave front which has a quadratic u^2 aberration, as a function of the wave-front coordinate u, will only result in a shift of the best focus, but the diffraction pattern in the best focus for the parabolically aberrated wave front will be as "perfect" as that in

[1] A. MARÉCHAL: Optique Géometrique Générale, in vol. 24 of this Encyclopedia.
[2] H. H. HOPKINS: Wave Theory of Aberrations. Oxford Clarendon Press 1950.

the focal plane for the plane wave front, in the absence of astigmatism, of course. This fact has been independently stated and verified for gratings by STROKE (loc. cit. 1960).

A remarkable property of wave fronts thus permits to measure the diffracted wave-front aberrations either from a plane wave front, or from a parabolic (in particle *circular*) wave front, whichever happens to form a better match. The importance of this property resides in the fact, demonstrated by STROKE (loc. cit.), that many plane gratings will tend to have curved wave fronts, which can then be matched more or less closely by suitable circles, drawn in the wave front interferograms[1]. The departures of the fringes from circles in the wave-front interferograms then provide the value of the aberrations in the same way as the departures from straight lines provide the aberrations in the case of "plane" wave fronts.

It is at first necessary to determine the shift of focus Δf corresponding to a quadratic wave front curvature (Fig. 75a). According to the theorem of GOUY (A. MARÉCHAL, loc. cit., Sect. 67) the aberration Δ_S along a ray of light is invariant. Let the lens equation

$$\frac{1}{p} + \frac{1}{p'} = \frac{1}{f} \tag{57.1}$$

be applied to the perfect lens L in Fig. 71a. When $p = \infty$, $1/p' = 1/f$. Next, let $p = R$, to which corresponds an aberration

$$\Delta_S \simeq \frac{u^2}{2R} \tag{57.2}$$

which is seen to be *parabolic* as a function of the wave-front coordinate u. According to Eq. (57.1) one has

$$\frac{1}{p'} - \frac{1}{f} = \frac{1}{R} \tag{57.3}$$

and, approximately, $(p' \simeq f)$,

$$\frac{f - p'}{f^2} = \frac{1}{R}. \tag{57.4}$$

Let the shift in focus be defined by

$$f - p' = \Delta f. \tag{57.5}$$

Then,

$$\Delta f = \frac{f^2}{R} \tag{57.6}$$

and, in terms of the aberration Δ_M corresponding to the edge u_M of the aperture

$$\Delta f = -2\Delta_M \frac{f^2}{u_M^2}. \tag{57.7}$$

By noting that $u_M = A/2$, one finally has for the shift in focus Δf corresponding to a parabolic aberration $\Delta_S = u^2/2R$ the equation

$$\Delta f = -8\Delta_M \left(\frac{f}{A}\right)^2. \tag{57.8}$$

It remains to be shown that a parabolic wave-front aberration such as that of P_R from P_0 in Fig. 75a does not result in a loss of image quality. This conclusion

[1] In practice, the appropriate wave-front curvature to be used as a reference is determined by measuring the shift of focus in a spectrograph with respect to the focus corresponding to a perfectly plane mirror and by applying Eq. (57.6). (STROKE, loc. cit. 1960).

follows from the fact that L has been assumed to be a "perfect" lens, which must be stigmatic in a finite element of volume about f_0, and from the fact that Δ_S is quadratic in u (see e.g. A. MARÉCHAL, loc. cit.; p. 140).

Finally let Δ_{P_0} be the aberration of the grating wave front from a plane reference wave front P_0 (Fig. 75 b). The corresponding wave front Σ_{S_0} on the image-forming side of L has the same aberrations $\Delta_{S_0}=\Delta_{P_0}$ from the sphere S_0 (centered at f_0) as Σ_{P_0} has from P_0. Similarly (Fig. 75 c), if a grating wave front Σ_{P_P} now has the same aberration $\Delta_{P_P}=\Delta_{S_0}=\Delta_{P_0}$ from a wave front P_P, which itself has a parabolic aberration from P_0, then the image-forming wave front Σ_{S_P} has the same aberration $\Delta_{P_P}=\Delta_{P_0}$ from the wave front centered at the focus f_P, where $(f_P-f_0)=\Delta f$ according to Eq. (57.8).

Fig. 76 shows according to STROKE (loc. cit. 1960) a wave-front interferogram and corresponding spectrograms of the green line (5461 Å) of

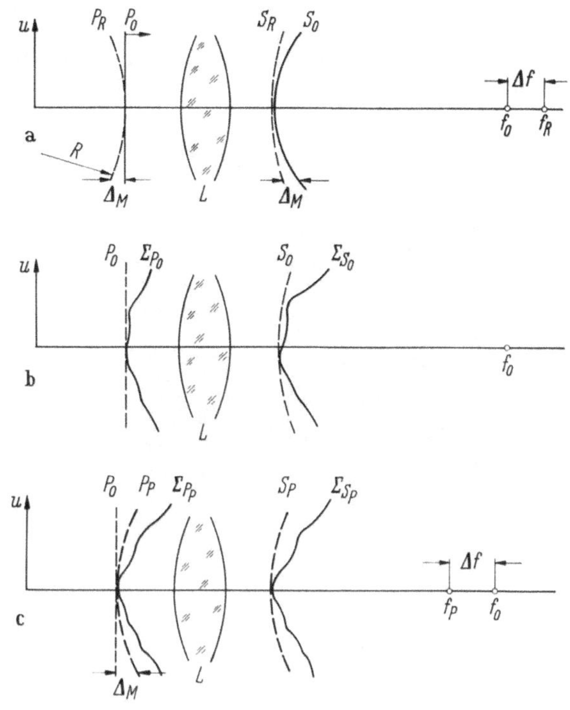

Fig. 75 a—c.

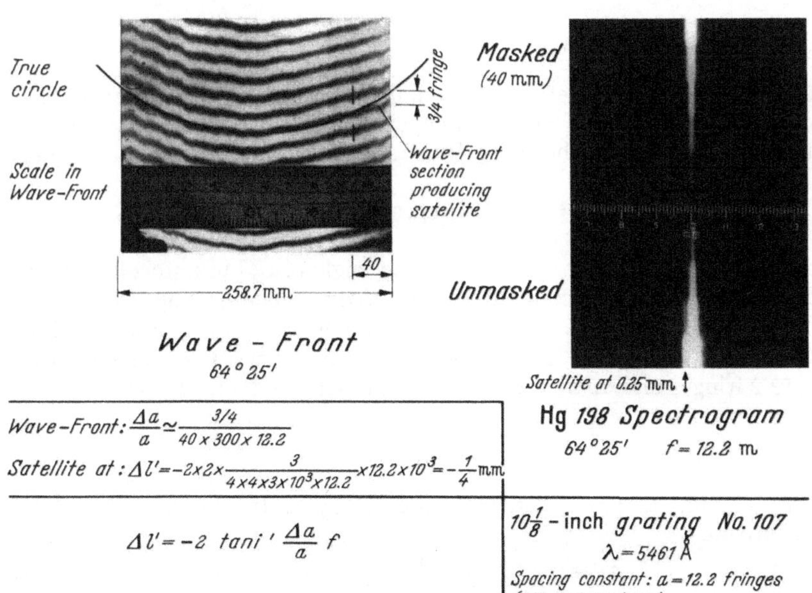

Fig. 76.

Hg 198. The mean wave front of this 258.7 mm wide grating (ruled on the M.I.T. interferometric engine) is seen to closely match a parabola ("true circle") except for an important "linear" (see Sect. 53) aberration extending over some 40 mm of the grating, on the right of the wave front. The grooves project vertically on the interferogram. The "linear" aberration from the circle is at the origin of the satellite at 0.25 mm from the line, and it is seen that the satellite disappears when the 40 mm wavefront section is masked off.

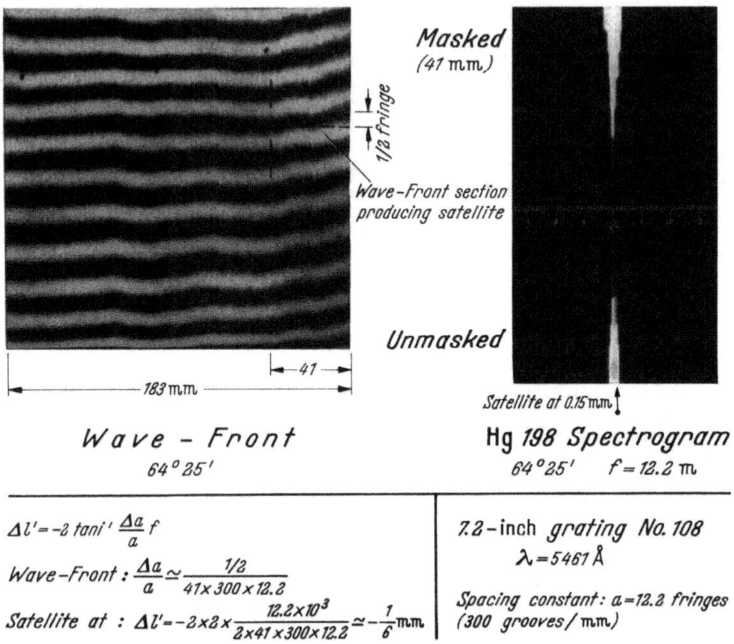

Fig. 77.

The calculation of the effect of the linear aberration for the grating of Fig. 77 according to Sect. 53 and STROKE (loc. cit. 1960) correctly predicts the satellite position and intensity. The wave front aberration of $\frac{3}{4}$ fringe corresponds to a shift

$$\Delta a = \frac{1}{\sin i}\,\frac{3}{4}\ \text{fringes} = 1.1 \times \frac{3}{4}\ \text{fringes} \tag{57.9}$$

for this grating, examined and used at the angle of 64° in autocollimation. This shift has been linearly accumulated over 40 mm. The grating spacing constant $a=12.2$ fringes ($\lambda/2$) of the same wavelength λ as that used in obtained the interferogram. Also, there are about 300 grooves per mm. Thus 40 mm have $40\times 300\times 12.2$ fringes and application of the equations of Sect. 53 gives

$$\frac{\Delta a}{a} = \frac{1.1\times(\frac{3}{4})}{40\times 300\times 12.2}\,. \tag{57.10}$$

With $f=12.2$ m, $\tan i'=2$, and replacing 1.1 by unity, one has for the position of the satellite

$$\Delta l' \simeq -\frac{2\times 2\times 3\times 12.2\times 10^3}{4\times 4\times 3\times 10^3\times 12.2} \simeq -\frac{1}{4}\ \text{mm} \tag{57.11}$$

which is seen to be in excellent agreement with the 0.25 mm of the experiment.

A similar calculation carried out for the linear aberration from the plane wave front for the grating of Fig. 77 gives, according to Stroke (loc. cit. 1960),

$$\frac{\Delta a}{a} \simeq \frac{1.1 \times (\frac{1}{2})}{41 \times 300 \times 12.2} \tag{57.12}$$

and $\Delta l' \simeq -\frac{1}{6}$ mm, again in excellent agreement with the 0.15 mm in the experiment.

58. Scattered light, according to Maréchal. The effect of very small and statistically irregular groove displacements, as well as the effect of groove imperfections and of imperfections in the blank surface is to produce scattered light throughout extended regions of the spectrum, generally far away from the "main" spectral line to which the image coordinates are referred. Of course, scattered light from other "main" spectral lines will also appear in the very vicinity of the main spectral lines. The effect of the small grating imperfections (and corresponding imperfections in the grating replicas) is very similar to the effect of polishing imperfections in optical elements. The problem has been studied in particular by A. Maréchal and we follow here the treatment developed by A. Maréchal and M. Françon[1]. For reasons of clarity, the notation used by A. Maréchal in his original paper[2] is used here (Fig. 78).

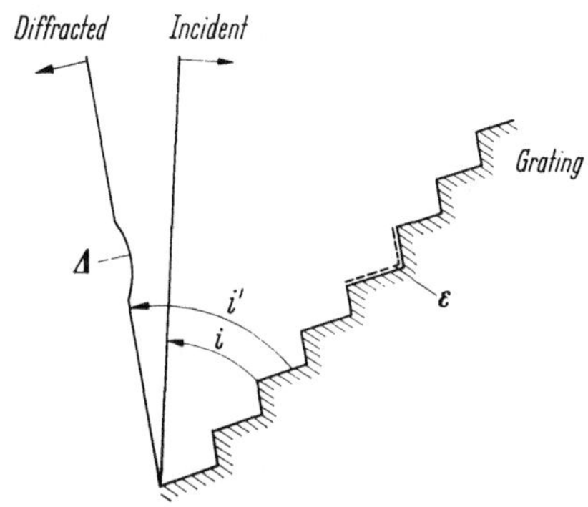

Fig. 78.

Let Σ be a wave front incident on the imperfect grating surface. The small imperfection ε which results in scattered light produces wave front deformations Δ related to ε by the equation

$$\Delta_R = 2\varepsilon_R \sin i' \tag{58.1}$$

in the case of ruling imperfection ε_R, and by

$$\Delta_B = 2\varepsilon_B \cos i' \tag{58.2}$$

in the case of grating blank imperfection ε_B, both equations being applicable to a reflection grating used in autocollimation near the angle i'. The diffracted wave front $F(\beta', \gamma')$ will display small phase fluctuations about its mean value. One has quite generally, according to Sect. 52

$$F(\beta', \gamma') = \exp[ik\Delta_\varepsilon(\beta', \gamma')]. \tag{58.3}$$

For the small values of Δ_ε involved here one has

$$\exp[ik\Delta_\varepsilon] = 1 + ik\Delta_\varepsilon. \tag{58.4}$$

It is always possible to choose a reference surface such that the mean value of Δ_ε is zero:

$$\overline{\Delta}_\varepsilon = 0. \tag{58.5}$$

[1] A. Maréchal et M. Françon: Diffraction, Edit. Revue d'Optique, Paris, 1960, p. 185—189.

[2] A. Maréchal: La diffusion résiduelle des surfaces polies et des réseaux. Optica Acta 5, 70 (1958).

According to the Fourier transform equations relating the wave front $F(\beta', \gamma')$ and the diffraction patterns $E(y', z')$ one has

$$E(y', z') = \frac{i f}{\lambda} \iint_{\text{Pupil}} F(\beta', \gamma') \, e^{-ik(\beta' y' + \gamma' z')} \, d\beta' \, d\gamma'. \tag{58.6}$$

With Eqs. (58.4) and (58.6) it is seen that the complex amplitude in the wave front is formed of two terms:

a uniform amplitude $F_1 = E_0$;

a fluctuating amplitude $F_2(\beta', \gamma') = ikE_0 \Delta_\varepsilon(\beta', \gamma')$ and the diffraction pattern $\mathscr{A}(y', z')$ is therefore also formed of two terms:

the usual field $\mathscr{A}_1 = T[E_0]$, which is principally determined by the shape of the aperture (with the help of apodization, one can limit the extent of the central part of the diffraction pattern to a very small region about the center of the lines);

a fluctuating field $\mathscr{A}_2 = T[ikE_0\Delta_\varepsilon]$ which extends over a very large distance about the center of the diffraction pattern in the case of fluctuations having a very small extent on the grating surface (or, more accurately, which vary rapidly as a function of the coordinates on the surface). This conclusion simply results from a harmonic analysis of the wave front, and the convolutions of the spectra referred to in Sect. 56, or more accurately, from the autocorrelation equations given below.

By the law of conservation of energy, one may write that the energy in the fluctuating field is proportional to

$$\left.\begin{aligned} \iint |\mathscr{A}_2(y', z')|^2 dy' \, dz' &= f^2 \iint |F_2(\beta', \gamma')|^2 \, d\beta' \, d\gamma' \\ &= f^2 k^2 E_0 \iint \Delta_\varepsilon^2 \, d\beta' \, d\gamma', \end{aligned}\right\} \tag{58.7}$$

while the total energy in the wave front is proportional to $\iint d\beta' \, d\gamma'$. The ratio of scattered energy to the total energy is therefore given by the equation

$$\frac{k^2 \iint \Delta_\varepsilon^2 \, d\beta' \, d\gamma'}{\iint d\beta' \, d\gamma'} = k^2 \overline{\Delta_\varepsilon^2} = \frac{4\pi^2}{\lambda^2} \overline{\Delta_\varepsilon^2} \tag{58.8}$$

showing that the scattered light, throughout the image plane, is given by the mean square value of the aberration Δ_ε under the given assumptions.

It had been previously shown by A. MARÉCHAL[1] that the decrease of intensity in the central part of the diffraction pattern in the presence of small aberration Δ is proportional to $k^2 \overline{\Delta^2}$. It is easy to see that the light lost in the center of the diffraction pattern is distributed (as scattered light) over the remaining distant regions of the image.

Next, one needs to obtain an expression for the spatial distribution of the scattered light. The complex amplitude distribution $\mathscr{A}_2(y', z')$ of the scattered light part in the diffraction pattern is given by

$$\mathscr{A}_2(y', z') = T[ikE_0\Delta_\varepsilon], \tag{58.9a}$$

that is by

$$\mathscr{A}_2(y', z') = \frac{i f}{\lambda} \iint F_2(\beta', \gamma') \exp[-ik(\beta' y' + \gamma' z')] \, d\beta' \, d\gamma' \tag{58.9b}$$

and the corresponding intensity distribution $I_2(y', z')$ by

$$I_2(y', z') = |\mathscr{A}_2|^2 = \mathscr{A}_2 \mathscr{A}_2^*. \tag{58.10}$$

[1] A. MARÉCHAL: Thèse, Revue d'Optique 1948.

The spatial distribution $d_2(\mu', \nu')$ of $I_2(y', z')$ can be obtained by taking the Fourier transform of $I_2(y', z')$, in the form

$$d_2(\mu', \nu') = T[\mathcal{A}_2 \mathcal{A}_2^*], \tag{58.11a}$$

that is

$$d_2(\mu', \nu') = \iint \mathcal{A}(y', z') \mathcal{A}^*(y', z') \exp[i 2\pi(\mu' y' + \nu' z')]\, dy'\, dz'. \tag{58.11b}$$

According to a well known theorem, one has

$$d_2(\mu', \nu') = T[\mathcal{A}_2] \otimes T[\mathcal{A}_2^*] \tag{58.12}$$

where $\otimes$ stands for the convolution integral. Noting that $\mathcal{A}_2 = T[ikE_0 \Delta_\varepsilon]$ according to Eq. (58.9), one finally has

$$d_2(\mu', \nu') = k^2 E_0^2 (\Delta_\varepsilon \otimes \Delta_\varepsilon), \tag{58.13a}$$

that is

$$d_2(\mu', \nu') = f^2 k^2 E_0^2 \iint \Delta_\varepsilon(\beta', \gamma') \Delta_\varepsilon(\beta' - \mu' \lambda, \gamma' - \nu' \lambda)\, d\beta'\, d\gamma' \tag{58.13b}$$

which is recognized as the autocorrelation function of the fluctuating wave front function. Here $\mu' \lambda = \beta$, $\nu' \lambda = \gamma$ and $\Omega = \iint d\beta'\, d\gamma'$. It is clear from Eq. (58.13) that $d_2(\mu', \nu')$ will tend to zero for displacements of Δ_ε in the wave front that are large compared to mean dimensions of the fluctuations. But the normalized coordinates μ, ν are inversely proportional to the dimensions β, γ in the image domain, which results in the previously stated conclusion that the maximum effect of the scattering centers will be found in image regions which are at distances inversely proportional to the mean dimensions of the scattering centers.

The intensity distribution $I_2(y', z')$ can next be written in the form

$$I_2(y', z') = T[d_2(\mu', \nu')] \tag{58.14a}$$

or

$$I_2(y', z') = \iint d_2(\mu', \nu') \exp[-i 2\pi(\mu' y' + \nu' z')]\, d\mu'\, d\nu' \tag{58.14b}$$

and the autocorrelation function $d_2(\mu', \nu')$ in the form

$$d_2(\mu', \nu') = f^2 k^2 E_0^2 \Omega \overline{\Delta^2}. \tag{58.15a}$$

The autocorrelation function is normalized in the form

$$\iint \Delta_\varepsilon(\beta, \gamma) \Delta_\varepsilon(\beta' - \beta, \gamma' - \gamma)\, d\beta'\, d\gamma' = \Omega \overline{\Delta^2} \mathcal{A}(\beta, \gamma), \tag{58.15b}$$

where $\mathcal{A}(\beta, \gamma)$ is the normalized autocorrelation function. When $\beta = \gamma = 0$, on has $\mathcal{A}(0, 0) = 1$. Here $\Omega = \iint d\beta'\, d\gamma'$.

With these equations, and suitable normalization, one finally has for the intensity distribution of the scattered light the equation

$$I_2(y', z') = 4\pi^2 f^2 E_0^2 \Omega \frac{\overline{\Delta^2}}{\lambda^4} T[\mathcal{A}(\beta, \gamma)]. \tag{58.16}$$

It is seen that the intensity distribution $I_2(y', z')$ of the scattered light in the image domain is wavelength dependent, both through the λ^{-4} factor and through the $T[\ ]$ factor. One needs indeed to recall that

$$T[\mathcal{A}(\beta, \gamma)] = \text{function of} \left(\frac{y'}{\lambda}, \frac{z'}{\lambda}\right) \tag{58.17}$$

where y' and z' are coordinates in the image plane. These explicit expressions given by MARÉCHAL (loc. cit.) are found to be useful.

MARÉCHAL also gives an example for the case when the wavefront autocorrelation function can be written in the form

$$\mathscr{A}(\beta, \gamma) = \exp\left[-\pi \frac{\beta^2+\gamma^2}{\beta_0^2}\right], \tag{58.18}$$

that is as a Gaussian function. (In this case $\overline{\Delta^2} \neq 0$.) Here β_0 is a parameter giving the mean angle under which the scattering centers are seen from the focus. The Fourier transform $T[\;]$ can then be written as

$$f\left(\frac{y'}{\lambda}, \frac{z'}{\lambda}\right) = \beta_0^2 \exp\left[-\pi\beta^2 \frac{y'^2+z'^2}{\lambda^2}\right]. \tag{58.19}$$

It is seen that the decrease of intensity as a function of y' and z' is also Gaussian. As expected, the width of the Gaussian in the image plane (as determined by β_0^2) is inversely proportional to the extent of the scattering centers. In particular:

a) When the scattering centers have a rather large angular dimension (as seen from the focus), and their autocorrelation extends over a large distance in the wave front (β_0 is large!), then the function given by Eq. (58.19) decreases rapidly: the scattered energy is concentrated near the main spectral lines.

b) On the contrary, in the case of errors of small lateral extent (for which the autocorrelation stops very rapidly with the dimension in the wave front) the scattered light is spread far away from the line centers (β_0 is small!).

The conclusions a) and b) are general and not restricted to Gaussian functions, as previously noted.

The effect of accidental ruling errors which are at the origin of scattered light has also been treated by several other authors[1], using a somewhat different formalism. In particular, the "vector method" was in favor before the use of the Fourier transform approach.

59. Numerical computation of light distribution in diffraction patterns. Electronic digital computers permit to calculate the complex amplitude distribution $\mathscr{A}(\beta)$ and intensity distribution $I(\beta)$ in the spectral diffraction patterns by carrying out the Fourier transformation which relates the diffraction patterns to the complex amplitude distribution $g(\eta) = \mathscr{A}(\eta) \exp[i\Phi(\eta)]$ in the wave fronts. Such calculations were carried out for the first time by STROKE (loc. cit. 1960) who also demonstrated the particular simplicity of these calculations, resulting of the *graphical display* of the wave-front aberration $\Phi(\eta) = k\Delta(\eta)$ in the wave-front interferograms.

Numerical integrations and calculations of diffraction patterns in the presence of special aberrations had of course been carried out in the past, both by hand and with the help of the "Maréchal integrator"[2], in particular for the so-called Seidel aberrations, resulting from a series of expansion of the wave-front aberration. MARÉCHAL (loc. cit. 1948) first demonstrated the advantages of machine computation and opened the way for its use in optics. It seems, that the first numerical diffraction pattern calculation in optics, for aberrations of actual wave fronts, were carried out by STROKE around 1958, for the "one-dimensional" case of optical gratings.

It might be noted in this context that evaluation of the diffraction integrals corresponding to actual lenses and mirrors by means of an *analogue* method,

---

[1] C. M. SPARROW: Theory of Imperfect Gratings. Astrophys. J. **49**, 65 (1919). — E. BUCHWALD: Ann. Physik **80**, 279 (1926). — A. LOHMANN: Optica Acta **6**, 175 (1959). See also, C. F. MEYER: The Diffraction of Light, X-rays and material particles, published by J. W. EDWARDS, Ann. Arbor, Michigan 1949. — G. R. HARRISON: J. Opt. Soc. Amer. **39**, 413 (1949).

[2] See for example, A. MARÉCHAL: Thèse, Revue d'Optique 1948, and also A. MARÉCHAL: Vol. 24 of this Encyclopedia.

using a two beam interferometer, was successfully demonstrated already in 1955 and 1957 by H. H. HOPKINS[1] and M. DE[2], and the method was tried out for gratings by STROKE[3] in 1958 before deciding on the use of the numerical method and electronic digital computers.

STROKE's program is designed to carry out the following calculations, expressed in suitably normalized form:

$$\operatorname{Re}[\mathcal{A}(\beta)] = \operatorname{Re}\left[\int_{-1}^{+1} g(\eta)\, e^{i\beta\eta}\, d\eta\right], \tag{59.1}$$

$$\operatorname{Im}[\mathcal{A}(\beta)] = \operatorname{Im}\left[\int_{-1}^{+1} g(\eta)\, e^{i\beta\eta}\, d\eta\right], \tag{59.2}$$

$$\Phi(\beta) = \tan^{-1}\left[\frac{\operatorname{Im}[A(\beta)]}{\operatorname{Re}[A(\beta)]}\right], \tag{59.3}$$

$$I(\beta) = \tfrac{1}{4}\mathcal{A}(\beta)\,\mathcal{A}(\beta)^*. \tag{59.4}$$

The various functions are identified as follows. The function

$$g(\eta) = e^{i\Phi(\eta)} = e^{ik\Delta(\eta)} \tag{59.5}$$

is the complex amplitude distribution in the wave front, assumed to have a uniform amplitude $\mathcal{A}(\eta)=1$ and polarization. In general,

$$g(\eta) = \mathcal{A}(\eta)\, e^{i\Phi(\eta)}. \tag{59.6}$$

The function $\mathcal{A}(\beta)$ is the complex amplitude distribution in the diffraction pattern, $\Phi(\beta)$ the corresponding phase and $I(\beta)$ the corresponding intensity distributions.

The width of the diffracted wave front along the coordinate u (normal to the groove length) is A, and the normalized coordinates are given by

$$\eta = \frac{u}{A/2}, \tag{59.7}$$

$$\beta = \pi \frac{x}{(\lambda/A)\, f} = \pi \frac{x}{x_0} \tag{59.8}$$

where x is a coordinate in the focal plane of the focussing system (focal length $= f$) in the direction of dispersion (x is parallel to u). One defines

$$x_0 = (\lambda/A)\, f = \text{diffraction unit} \tag{59.9}$$

and one has, according to STROKE,

$$\mathcal{A}(\beta)_{\text{non-normalized}} = \frac{A}{2}\int_{-1}^{+1} g(\eta)\, e^{i\beta\eta}\, d\eta. \tag{59.10}$$

(It is recalled that the "diffraction unit" corresponds to the distance, from the line center, of the first diffraction minimum of a "perfect" diffraction pattern, that is of the pattern for the width A without apodization.) By normalizing the amplitude, one has

$$\mathcal{A}(\beta)_N = \frac{\tfrac{1}{2}A\int_{-1}^{+1} g(\eta)\, e^{i\beta\eta}\, d\eta}{\tfrac{1}{2}A\int_{-1}^{+1} d\eta} \tag{59.11}$$

[1] H. H. HOPKINS: Optica Acta **2**, 23 (1955).

[2] M. DE: Optica Acta **4**, 2 (1957).

[3] G. W. STROKE: National Physical Laboratory Symposium No. 11, "Interferometry". London: H. M. Stationery Office 1960.

with $\int_{-1}^{+1} d\eta = 2$. Finally, the complex amplitude $\mathscr{A}(\beta)$ in the image domain, as a function of the coordinate β, is obtained in the desired normalized form as

$$\mathscr{A}(\beta) = \tfrac{1}{2} \int_{-1}^{+1} g(\eta)\, e^{i\beta\eta}\, d\eta. \tag{59.12}$$

The phase is given by Eq. (59.3) and the intensity $I(\beta)$ is given by Eq. (59.4).

The important point, as stressed by STROKE, is that the wave front aberration $k\Delta(\eta)$ is obtained in directly usable form in the interferograms of the diffracted wave fronts (STROKE, loc. cit. 1955, 1960). The computations of Eqs. (59.1) to (59.4) could be carried out up to any desired value of β for which the Fourier transform formulation is applicable. With the program prepared by STROKE, up to a total of 970 values of β ranging from $\beta = -218$ to $\beta = +218$ were foreseen. For the purpose of increasing the speed of computation, these values of β were judiciously spread out with the intervals $\Delta\beta$ selected so as to permit the precision to decrease with increasing distance from the line centers, as justified by physical reasons. Table 2, according to STROKE, gives the intervals and precisions.

Table 2.

$0 \leqq \beta \leqq 3$	$\Delta\beta = 0.15$	20 points	$\lvert\mathscr{A}(\beta)\rvert = x.xxx$
$3 \leqq \beta \leqq 13$	$\Delta\beta = 0.25$	40 points	$\lvert\mathscr{A}(\beta)\rvert = x.xxx$
$13 \leqq \beta \leqq 73$	$\Delta\beta = 0.5$	120 points	$\lvert\mathscr{A}(\beta)\rvert = x.xx$
$73 \leqq \beta \leqq 218$	$\Delta\beta = 0.5$	290 points	$\lvert\mathscr{A}(\beta)\rvert = x.x$

For the purposes of numerical integration, the function was written as

$$\mathscr{A}(\beta) = \int_{-1}^{+1} \cos\left[k\Delta(\eta) + \beta\eta\right] d\eta + i \int_{-1}^{+1} \sin\left[k\Delta(\eta) + \beta\eta\right] d\eta. \tag{59.13}$$

The argument $\exp[ik\Delta(\eta)]$ of $g(\eta)$ is prepared for the punching of the cards in 1001 values of $k\Delta(\eta)/2\pi$ for the equally spaced points in the wave front ranging from -1 to $+1$. The values of $k\Delta(\eta)$ are simply taken from a suitably enlarged wave-front interferogram, and punched with a three decimal precision.

It is easy to show that the number of 1001 discrete points is sufficient to represent the wave fronts of good gratings with adequate precision for the computations of the central regions of the diffraction patterns within the ± 70 odd diffraction units on either side of the line centers. Following STROKE (loc. cit.) one can investigate whether any important modifications in the computed $\mathscr{A}(\beta)$ will result from the fact that the diffraction patterns are computed by sampling only a finite number n of points in the interferograms.

Let

$$\mathscr{A}(\beta) = \tfrac{1}{2} \int_{-1}^{+1} g(\eta)\, e^{i\beta\eta}\, d\eta \tag{59.14}$$

be the complex amplitude in the diffraction pattern which is to be determined. Let the integral be replaced by a sum

$$\mathscr{A}(\beta) = \sum_{j=0}^{j=n} \int_{(j-\frac{1}{2})\Delta\eta}^{(j+\frac{1}{2})\Delta\eta} g(j\Delta\eta)\, e^{i\beta\eta}\, d\eta \tag{59.15}$$

where the function $g(j\Delta\eta)$ is constant over any interval $j\Delta\eta$, which extends from $(j-\frac{1}{2})\Delta\eta$ to $(j+\frac{1}{2})\Delta\eta$ for every point $j\Delta\eta$.

One therefore has

$$\mathscr{A}(\beta) = \sum_{j=0}^{j=n} g(j\Delta\eta) \int_{(j-\frac{1}{2})\Delta\eta}^{(j+\frac{1}{2})\Delta\eta} e^{i\beta\eta}\, d\eta \tag{59.16}$$

which gives

$$\mathscr{A}(\beta)=\sum_{j=0}^{n}\frac{1}{i\beta}\left[e^{i\beta\Delta\eta/2}-e^{-i\beta\Delta\eta/2}\right]e^{i\beta j\Delta\eta}\,g(j\Delta\eta) \tag{59.17}$$

and finally

$$\mathscr{A}(\beta)=\frac{\sin(\frac{1}{2}\beta\Delta\eta)}{\frac{1}{2}\beta\Delta\eta}\sum_{j=0}^{n}g(j\Delta\eta)\,e^{i\beta j\Delta\eta}\,\Delta\eta. \tag{59.18}$$

By comparing Eq. (59.18) and Eq. (59.14), it is seen that the integral in Eq. (59.14) is simply replaced by the sum in Eq. (59.18), however with the sum as modulated by a ($\sin x/x$) type factor. It remains to be shown under which conditions the factor remains close to the value of unity. The first zero of this factor is obtained for $\beta\Delta\eta/2=\pi$ or $\Delta\eta=2\pi/\beta=2x_0/x_s$, where x_0 is the coordinate of the first diffraction minimum of the perfect pattern, as before. By noting that $\eta=u/(A/2)$, and therefore $\Delta\eta=2\,\Delta u/A=2x_0/x_s$, one has in the wave front

$$\frac{x_s}{x_0}=\frac{A}{\Delta u}=n \tag{59.19}$$

where n is the number of intervals Δu into which the wave front of width A has been sub-divided. For instance, when $n=1000$, one has $x_s=1000\,x_0$, that is to say that the first zero of the ($\sin x/x$) function is at 1000 diffraction units from the line centers. One is led to conclude that the choice of 1000 points in the interferogram insures excellent precision even in rather distant regions from the diffraction pattern centers. STROKE (loc. cit.) also discusses other factors which determine the choice of number of points in the interferogram: by comparing the numerical integrations carried out for both 500 and 1000 points in the wave front, he finds no detectable difference, to better than two units in the third decimal, for the gratings examined.

Several important considerations enter into increasing the speed of numerical calculations, even when carried out on numerical computers. STROKE's first program was written for values of the $k\,\Delta(\eta)$ ranging from -2π to $+2\pi$, which adequately encompassed the good gratings for which these calculations were of interest, and the program can be easily extended to any desired aberration values. In fact the special nature of the weighted precision required in different regions of the diffraction patterns has permitted to write a considerably more rapid program than the one which would be required for Fourier transformations in the general case. SIMPSON's rule was used to carry out the integration of Eq. (59.13). It proved to be most effective to compute and store, once and for all, a complete table of the sine and cosine values, corresponding to the range of values of β, $k\Delta(\eta)$ and $\Delta\beta$ chosen, by taking into account the precisions desired for $\mathscr{A}(\beta)$. This table, with four significant figures, then becomes usable for any grating within the specified range. The sine and cosine table is used by letting the computer look up, at the proper instant, the values of the sine and cosine corresponding to the given $k\Delta(\eta)$ and β in Eq. (59.13). It is clear that this method of proceeding is in this case considerably more rapid than a method which would involve waiting for a computation of the trigonometric values, for example by series development, at the moment when they are required.

The program was made flexible enough for permitting desirable increases in the amplitude of the aberrations, precisions as well as simplifications. The computed values of $\mathscr{A}(\beta)$, $I(\beta)$ and $\Phi(\beta)$ were stored on magnetic tape and subsequently punched on an IBM printer.

The original program was carried out on the IBM 704 computer. Even on this comparatively much slower computer than those presently available, only

35 minutes were required for the complete calculation of the 970 image points ranging within -70 and $+70$ diffraction units. However, the good gratings for which the calculations are of interest were not found to require calculations for the entire range. More representative times are 2.5 minutes for ± 4 diffraction units (117 image points), 7.5 minutes for ± 23 diffraction units (357 image points). The program is so arranged as to permit resumption of the computation, if the appearance of "satellites" in the distant regions calls for it.

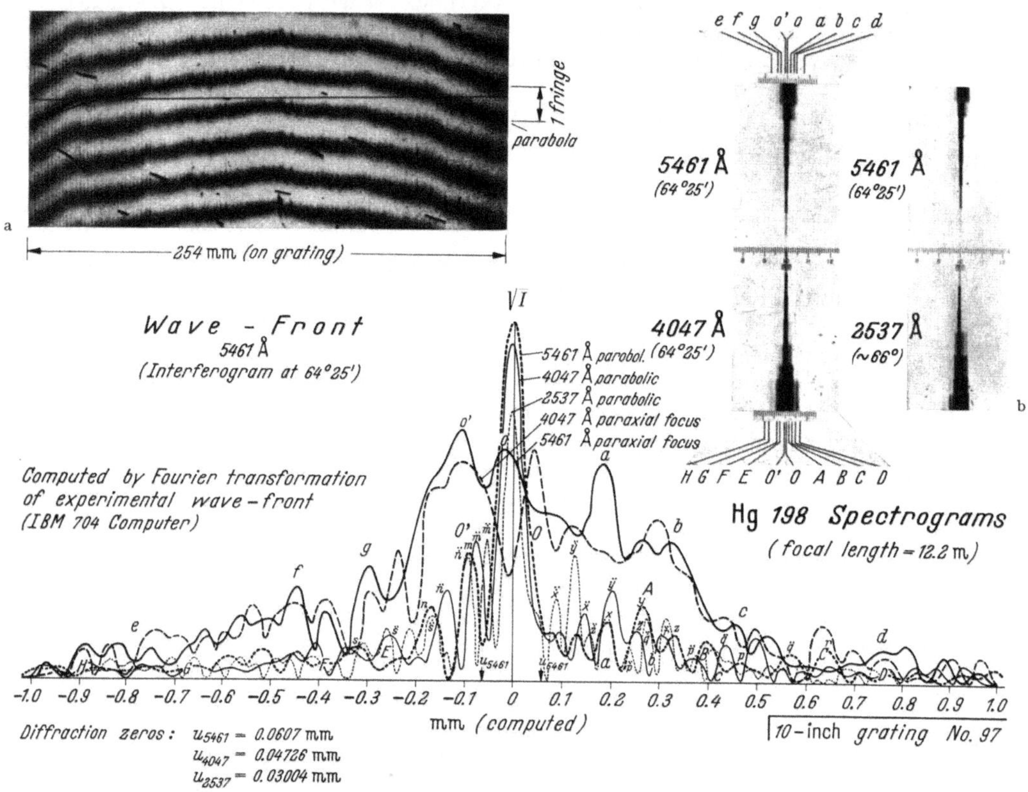

c

Fig. 79 a—c. $\sqrt{I}$ rather than the intensity-distribution I is shown in this Figure for clarity.

Stroke tested his program with the help of functions for which the analytic solution is easily established[1], and additional details are given in the original reference[2].

Comparisons of computed diffraction patterns and their imperfections with spectrograms of the mercury isotope Hg 198 are shown in Fig. 57, 71 as well as in Fig. 79. For this even-even isotope of mercury, the spectral lines should be single, and all lines appearing next to the main line are diffraction-pattern imperfections, otherwise called "satellites". The exposures were doubled from step to step in all the spectrograms.

---

[1] See for example, E. Ingelstam and E. Djurle: J. Opt. Soc. Amer. **43**, 572 (1953), and S. Silver, in Microwave Antenna Theory and Design. M.I.T. Radiation Laboratory Series, Chap. 6. New York: McGraw Hill Book Co. 1949.

[2] Stroke (loc. cit. 1960).

Fig. 79a shows according to STROKE (loc. cit. 1960)[1] the wave-front interferogram in 5461 Å for a 10-inch high-resolution grating at 64° (No. 97, ruled at M.I.T.). Fig. 79b shows the Hg 198 spectrograms produced by this grating in three wavelengths in a 12.2 meter spectrograph, and Fig. 79c shows the computed intensity distributions in the corresponding diffraction patterns. The computations were carried out by Fourier transformation of the phase distribution as it appears in graphical form in the experimental wave-front interferogram. It is seen that perfect agreement exists between the computed and experimental intensity distributions in the spectra wherever comparison is possible. Of course, the computed distributions give the diffraction pattern intensity distribution also within the central region where most of the energy is concentrated, but where no experimental values could be obtained before these calculations had been carried out. It is clear that the various spurious spectral lines are nothing but diffraction pattern imperfections, as caused by the diffracted wave front aberrations (which, in turn, result from grating imperfections). For the grating shown, it was necessary to plot the square root of the intensity, for the purpose of clarity (the relative intensity of the satellite peaks on an intensity plot would be smaller, of course). The mean wave-front curvature happens to be almost parabolic (circular) along the grating width, and results principally in a $+67$ mm shift of the best focus, as compared to a plane wave front, in the 12 meter spectrograph. The aberrations used in the computing were measured from the parabola shown. The scale shown in the spectrograms of Fig. 79b has 0.1 mm divisions and is placed on the photographs obtained in the best focus of the 12.2 meter spectrograph used. The first minimum of the diffraction pattern at 5461 Å is at about 1 second of arc from the line center (about 0.06 mm here), and the diffraction pattern width is of the same order as the Doppler width of the line. It was pointed out by HARRISON and STROKE[1] and by STROKE (loc. cit.) that the wavelength-dependent shift of the center of gravity, which is apparent in the computed curves of Fig. 79c, would result in systematic errors (errors of "coincidence") in the use of this grating for accurate wavelength measurements, if the computed curves were not available and could not be used for appropirate correction. Yet this grating is an example of the finest ruled in the late 1950's and early 1960's.

The importance of the computation of the intensity distribution $I(\beta)$ in the diffraction patterns resides in the convolution equation

$$I(\beta)_{\text{spectrometer}} = \int_{-\infty}^{+\infty} I(\beta')_{\text{spectrum}}\, I(\beta-\beta')_s\, d\beta' \tag{59.20}$$

relating the intensity distribution $I(\beta)_{\text{spectrometer}}$ recorded in the spectrometer to the intensity distribution $I(\beta)_{\text{spectrum}}$ in the spectrum which is sought. Here $I(\beta)_s$ is the diffractions pattern as modified by the finite slit width according to Sect. 45. It was pointed out in Sect. 44 that Eq. (59.20) can be written in the form

$$T[I(\beta)_{\text{spectrometer}}] = T[I(\beta)_{\text{spectrum}}] \cdot T[I(\beta)_s] \tag{59.21}$$

which can be used to compute the "true" intensity distribution in the spectrum with any desired precision[2].

It is the ease and precision with which the grating wave front interferograms[3] can be obtained and used to compute $I(\beta)$ wich makes numerical calculations of spectral diffraction so important, in particular where the effects of residual grating imperfections need to be used to correct for errors in the recorded spectrum.

[1] See also, G. W. STROKE: Ruling, Testing and Use of Optical Gratings for High-Resolution Spectroscopy. In: Progress in Optics, Vol. II, edit. by E. WOLF. Amsterdam: North-Holland Publishing Company 1963.

[2] See also J. ARSAC (loc. cit. 1961).

[3] STROKE (loc. cit.) and section.

Shifts in the centers of gravity[1] and line shape corrections[2] are made possible by means of diffraction pattern computations, and "resolving powers" considerably greater than those permitted by classical considerations become attainable (and are in fact as great as desired, except for the limitations set by noise).

V. Light efficiency and "blaze" of gratings.

60. The distribution of light among orders: an electromagnetic boundary value problem. Recent work[1–8] has confirmed the fact that the distribution of the light diffracted by an optical grating, among the various orders, is a manifestation of the electromagnetic character of light and of the boundary conditions that must be satisfied on the surface of the grating by the total field, incident plus diffracted. Complete, rigorous solutions of the electromagnetic boundary-value problem of the diffraction of light by optical gratings can now be easily obtained with the help of a number of boundary-value solution methods.

A method first suggested by Lord RAYLEIGH[9] in 1907 assumed, as a model, that the *discrete* spectrum of real diffracted waves and the infinity of evanescent waves (in each of the two orthogonal polarisations, $\boldsymbol{E}$ parallel to the groove length, and $\boldsymbol{E}$ perpendicular to the groove length), together with the incident wave, is sufficient to satisfy the boundary conditions on the surface of the grating. The Rayleigh method was further developed, in particular as a result of the work by G. W. STROKE[2] and others[3–11] between 1959 and 1965.

The Rayleigh solution, developed by STROKE[2] for gratings of sinusoidal profile, has been extended to the case of gratings of any profile. Particularly simple is the extension of the solution to gratings of flat-facetted profiles, especially with "quasi-triangular" groove shapes, as was shown almost simultaneously by P. BOUSQUET[5,8] and by R. PETIT[7] in 1963. The use of high-speed computers permits to obtain numerical solutions of the systems of linear equations which, in the past, would have required the search for various approximations in the formulation of the boundary value problem[2].

---

[1] R. J. HULL and H. H. STROKE: J. Opt. Soc. Amer. **51**, 1203 (1961). — W. J. TOMLINSON III and H. H. STROKE: Phys. Rev. Letters **8**, 436 (1962). — J. Opt. Soc. Amer. **53**, 828 (1963). — G. W. STROKE and H. H. STROKE: J. Opt. Soc. Amer. **53**, 333 (1963).

[2] A. K. PIERCE: J. Opt. Soc. Amer. **47**, 6 (1957).

[1] A. MARÉCHAL et G. W. STROKE: Sur l'origine des effets de polarisation et de diffraction dans les réseaux optiques. C. R. Acad. Sci. Paris **249**, 2042 (1959).

[2] G. W. STROKE: Etude théorique et expérimentale de deux aspects de la diffraction de la lumière par les réseaux optiques: l'évolution des défauts dans les figures de diffraction et l'origine électromagnétique de la répartition entre les ordres. Rev. Opt. **39**, 291—398 (1960).

[3] P. BOUSQUET: Etude de la réflexion et de la transmission de la lumière par un réseau transparent à profil sinusoidal. Extension en cas d'une surface irrégulière. Rev. Opt. **41**, 277—294 (1962).

[4] P. BOUSQUET et R. DELEUIL: Diffraction des ondes électromagnétiques par un réseau à profil sinusoidal. C. R. Acad. Sci. Paris **256**, 1461 (1963).

[5] P. BOUSQUET: Diffraction des ondes électromagnétiques par un réseau à facettes planes. C. R. Acad. Sci. Paris **256**, 3422 (1963).

[6] G. W. STROKE: Attainment of high efficiencies in blazed optical gratings by avoiding polarization in the diffracted light. Phys. Letters **5**, 45 (1963).

[7] R. PETIT: Contribution à l'étude de la diffraction d'une onde plane par un réseau métallique. Rev. Opt. **42**, 263—281 (1963).

[8] P. BOUSQUET: Diffraction des ondes électromagnétiques par un réseau à profil triangulaire. C. R. Acad. Sci. Paris **257**, 80 (1963).

[9] Lord RAYLEIGH: Dynamical Theory of Gratings. Proc. Roy. Soc. Lond. A **79**, 399 (1907) (Scientific Papers V, p. 405). — Phil. Mag. **23**, 310 (1912).

[10] R. P. MADDEN and J. STRONG: Appendix P, "Diffraction Gratings". In: J. Strong, Concepts of Classical Optics. San Francisco: W. H. Freeman & Co. 1958.

[11] C. JANOT et A. HADNI: Réflexion de l'infrarouge lointain sur les réseaux échelettes. J. Phys. Radium **23**, 152 (1962).

The numerically computed solutions have revealed an inadequacy of the Rayleigh model, in its general application to blazed gratings, notably for the case of gratings with deep grooves. This inadequacy had already received attention in the past, notably by B. A. LIPPMANN[12] in 1953, who pointed out *first* that the GREEN's function formulation of the grating diffraction problem showed that the expansion used by Lord RAYLEIGH represented the field only approximately, and *second* that an alternative method of treating the problem could be based on a variational method. A variational treatment of the grating diffraction problem was given by A. WIRGIN[12] in 1964—1965. WIRGIN noted in conclusion, that his solution while "exact", appeared in what seemed to be a form too complicated for numerical solution. An "exact" solution of the grating diffraction problem, in a form readily suitable for numerical computation, and in good agreement with experiment, was first obtained by R. PETIT[12] in 1964—1965, for the case of a perfectly conducting grating, of a general periodic profile, for the special case of $\boldsymbol{E}$ parallel to the groove length. This solution was based on a Helmholtz formulation of the diffraction problem. In a paper[12] submitted for publication in August 1965, R. P. MCCLELLAN and G. W. STROKE extended the Helmholtz equation solution of the grating diffraction problem to the general case of periodic-profile gratings, and gave a complete solution for both polarisations ($\boldsymbol{E}$ parallel and $\boldsymbol{E}$ perpendicular to the groove length) for optical gratings, with perfectly conducting boundaries. An independent solution of the $\boldsymbol{H}_{\|}$ problem was simultaneously obtained by PETIT[12].

One approximation used for a long time in the past was the formulation of the energy distribution problem as a *scalar* problem, somewhat like in acoustics, and with no reference to polarization. Some of the approximation methods used in attempts to simplify the boundary-value problem are indicated in Sects. 70 and 71.

Application of scalar calculations to predictions of the relative amplitudes in the diffracted orders give occasional qualitative agreement with experiment in special cases[13]. However, in general, calculations which ignore the vector character of light, and polarization, result in unsurmountable difficulties, and fail to agree with experiment.

In particular, experiments with gratings[2,10] show two different distributions of light among orders, for two orthogonal polarizations, $\boldsymbol{E}$ parallel and $\boldsymbol{E}$ perpendicular to the groove length, both of which disagree with the "scalar" calculations, in which no polarization is involved. These difficulties were very convincingly emphasized by R. P. MADDEN and JOHN STRONG[10] in 1958, and were reported by MARÉCHAL and STROKE[1] to be at the origin of their interest in relating the electromagnetic boundary conditions on the surface of the gratings to both the polarization of the incident or diffracted light, and to the energy distribution among orders.

---

[12] B. A. LIPPMANN: Note on the theory of gratings. J. Opt. Soc. Amer. **43**, 408 (1953).

A. WIRGIN: Considérations théoriques sur la diffraction par réflexion sur des surfaces quasiment planes; Applications à la diffraction par les réseaux. Rev. Opt. **43**, 449 (1964); **44**, 20 (1965).

(a) R. PETIT, and M. CADILHAC: Etude théoriqe de la diffraction par un réseau. C. r. Acad. Sci. Paris **259**, 2077 (1964). — (b) R. PETIT: Étude numérique de la diffraction par un réseau. C. r. Acad. Sci. Paris **260**, 4454 (1965). — (c) R. PETIT: Etude théorique de la diffraction d'une onde plane et monochromatique par un réseau métallique infiniment conducteur. Applied Optics **4**, 1551 (1965). — (d) R. PETIT: Etude théorique de la diffraction par un réseau métallique. C. r. Acad. Sci. Paris **261**, 4677 (1965).

R. P. MCCLELLAN, and G. W. STROKE: Solution of the Non-Homogeneous Helmholtz Equation for Optical Gratings with Metallic and Dielectric Boundaries. J. Math. Phys. **45**, 383—390 1966).

[13] C. VEDEL et C. BOUHET: Fracteur de transmission d'un réseau échelette. Comparaison des montages Littrow et de Finkelstein. Rev. Opt. **42**, 361 (1963).

In fact, as stressed by STROKE[6], high efficiencies in the spectral orders from a grating can only be attained to the extent that one is successful in *avoiding polarization* in the diffracted light by appropriate choice of the groove profile, in relation to the angles of incidence and of diffraction. Avoiding polarization in the diffracted light can therefore be taken as a criterion for the attainment of high efficiencies in blazed gratings[6].

Very extensive bibliographical references to early work on the diffraction of light by gratings, with particular reference to gratings formed of discontinuous arrays of "cylinders" and to multiple scattering, are given by V. TWERSKY[14], who has also made extensive theoretical contributions to the problem of multiple scattering by gratings of discontinuous cylinders. Other references of importance to the electromagnetic boundary value solutions, as well as a discussion of various methods of approximation and of the nature of the "rigorous" boundary-value solutions are given by STROKE[2].

Careful distinction needs to be drawn between optical gratings having a continuous surface on one hand, and arrays of discontinuous "cylinders", for example, on the other hand, especially in view of well known past difficulties associated with extending of electromagnetic boundary-value solutions from one problem to another.

Parenthetically, the proof of the existence of solutions to a boundary value problem is, a priori, as important as the actual attainment of the solution. In this context, a 1938 paper by U. FANO[15] was stressed by STROKE[2] to be particularly helpful, even though the form of the general solution for the diffraction of light by optical gratings, obtained in that paper, does not seem to have been adapted for general application.

As shown by PETIT[7] and BOUSQUET[5,8] the RAYLEIGH-STROKE[2] electromagnetic boundary-value solution for the energy distribution in optical gratings has appeared to be a stimulus in the attempts to obtain numerically solvable solutions of the grating diffraction problem, notably for gratings having flat-facetted, polygonal or quasi-triangular groove profiles, on any conductivity or dielectric "constant". The Rayleigh-Stroke solution is indeed applicable to gratings of any electromagnetic characteristics, conducting or dielectric, used in reflection or in transmission. In fact, as stressed by BOUSQUET[8] the equations obtained in the solution of the most general case of gratings with triangular groove profile even have the same form as the equations first given by STROKE[2] for the basic case of the perfectly conducting grating having a sinusoidal groove profile, with a simple substitution of "modified" *sine* functions in the place of the Bessel functions.

The principal features of the Rayleigh boundary value solution of the energy distribution gratings are given in Sect. 63.

Sect. 61 deals with a particularly important special case of a perfectly conducting grating having grooves with a rectangular groove profile, an integral number m of half-waves deep, and used in normal incidence on one of the groove facets. It was shown by MARÉCHAL and STROKE[1] that such a grating behaves like a perfect mirror for the $\boldsymbol{E}$ polarization perpendicular to the $m\lambda/2$ facet, independently of the width w of the other facet, with only one "diffracted" order "re-

[14] V. TWERSKY: On scattering of waves by the infinite grating of circular cylinders. I.R.E. Trans. on Antennas and Propagation AP-10, 737—765 (1962). — For extensive bibliographical references to early work on gratings, see also, for example, V. TWERSKY, On multiple scattering of waves, part of the Report of the United States National Committee to the XIII General Assembly of the International Scientific Radio Union, London, England, September 5 to 15 (1960), pp. 715—730.

[15] U. FANO: Theory of intensity anomalies in diffraction gratings. Ann. Phys. **32**, 393 (1938).

flected" in the direction of incidence. In classical terms, the grating could be described as showing no "diffraction" in the scalar λ/w sense. The "mirror-like" grating can be taken as illustrating the difficulties resulting from attempts to extend *scalar* calculations to the prediction of energy distribution among the orders of gratings.

It might be in order to recall, in conclusion, that the scalar theory, as expressed in the Fourier-transform formulation, is a particularly powerful tool in the prediction of the energy distribution in the *diffraction patterns* formed by the various diffracted waves (Chap. B IV). However, the Fourier-transform approximation should not be expected to apply to the distribution of energy among the different diffracting "slits". This was shown in Sects. 32 and 33 to result from the very approximations involved in deriving the Fourier-transform expressions from MAXWELL'S electromagnetic equations. In particular, the approximations were found to restrict the scalar, Fourier-transform, formulation to a small-angle region around the near-vicinity of the normal to the diffracting aperture (or, more rigorously in the near-vicinity of the center of the quasi-spherical image-forming wavefront). In terms of this restriction, the various diffracted waves, of the different orders, would leave the "diffracting groove" at large angles with respect to each other, and the scalar approximation should not be expected to apply. And it does not, in the general case.

A very considerable amount of work on the energy distribution of light diffracted by gratings has preceded the recent electromagnetic solutions of the problem.

Indeed, partial and sometimes almost complete polarization in the light diffracted by metallic gratings has been well known ever since the first observations were reported by JOSEPH VON FRAUNHOFER for the gratings which he ruled in 1821[16].

It might be of some interest in this case to quote from FRAUNHOFER'S original 1823 paper, where he first deals with polarization in gratings, in order also to pay a full tribute to the considerable contribution made to optics by this great scientist. After very laboriously deriving the grating equation, Fraunhofer proceeds:

„Sehr merkwürdig ist es, dass unter einem gewissen Einfallswinkel ein Theil eines durch Reflexion entstandenen Spectrums aus *vollständig polarisiertem Lichte* besteht. Dieser Einfallswinkel ist für die verschiedenen Spectra sehr verschieden, und selbst noch sehr merklich für die verschiedenen Farben ein und desselben Spectrums. Mit dem Glassgitter $\varepsilon = 0{,}0001223$ ist polarisirt: $(E\tau)^{(+\mathrm{I})}$ d.i. der *grüne* Theil dieses ersten Spectrums, wenn $\sigma = 49°$ ist: $(E\tau)^{(+\mathrm{II})}$, oder der grüne Theil in dem zweiten auf derselben Seite der Axe liegenden Spectrum, wenn $\sigma = 40°$ beträgt; endlich $(E\tau)^{(-\mathrm{I})}$, oder der grüne Theil des ersten auf der entgegengesetzten Seite der Axe liegenden Spectrums, wenn $\sigma = 69°$. Wenn $(E\tau)^{(+\mathrm{I})}$ vollständig polarisiert ist, sind es die übrigen Farben dieses Spectrums noch unvollständig. Dieses ist bei $(E\tau)^{(+\mathrm{II})}$ weniger der Fall, und σ kann merklich geändert werden, indess diese Farbe noch polarisirt bleibt. $(E\tau)^{(-\mathrm{I})}$ ist unter keinem Einfallswinkel so ganz vollständig polarisirt, wie $(E\tau)^{(+\mathrm{I})}$. Bei einem Gitter, in welchem ε größer ist, als in dem, wovon hier die Rede war, müssen die Einfallswinkel ganz andere seyn, wenn die genannten Spectra polarisirt seyn sollen *.

* Es wäre unzeitig aus einer geringen Zahl von Beobachtungen schon auf ein Gesetz dieses Phenomens schliessen zu wollen; dieses kann nur aus einer grösseren Anzahl Gitter, bei welchen ε sehr verschieden ist, mit einiger Sicherheit abgeleitet werden. Da es bei diesen Versuchen eben nicht nöthig ist, dass die fixen Linien der Spectra gut erkannt werden, so kann man Gitter machen, bei welchen ε noch beträchtlich kleiner ist, als bei dem feinsten Glassgitter, dessen ich mich jetzt bediente. — Es ist nicht unwahrscheinlich, dass die Prinzipe der Interferenz vielleicht noch zu einer Theorie der Polarisation des Lichtes führen werden. Meine Ansicht hierüber mitzutheilen ist hier weder der Ort, noch gegenwärtig schon die Zeit. Glücklicher Weise sind noch Versuche anderer Art möglich, welche auf diesen Gegenstand Bezug zu haben scheinen; doch sind sie, so wie der grössere Theil sämmtlicher diesen Gegenstand betreffender Versuche, sehr delicater Art."

[16] JOSEPH VON FRAUNHOFER: Neue Modifikation des Lichtes durch gegenseitige Einwirkung und Beugung der Strahlen, und Gesetze derselben. Kurzer Bericht von den Resultaten neurer Versuche über die Gesetze des Lichtes und die Theorie derselben. Denkschr. kgl. Akad. Wiss. München **8**, 1 (1921/22). — GILBERT'S Ann. Phys. **74**, 337 (1823). — Edinburgh J. Sci. **7**, (1827); **8** (1828).

Truly little may seem to have been added qualitatively to these observations of polarization effects, frequently referred to in the more recent literature as "Wood anomalies". Of course, it is now known that the electromagnetic boundary conditions, and not interference phenomena referred to as conceivable by FRAUNHOFER, are at the root of the polarization effects in gratings.

Among the many contributors to the theory of energy distribution in gratings, and to related problems, we add a list of papers to those already referred to in this section[17–49].

[17] K. ARTMANN: Über die Theorie der anomalen Reflexion an Beugungsgittern. Z. Physik **119**, 529 (1942).

[18] H. D. BABCOCK and H. W. BABCOCK: J. Opt. Soc. Amer. **34**, 1 (1944).

[19] L. B. FELSEN: The scattering of electromagnetic waves by a corrugated sheet. Canad. J. Phys. **37**, 1565 (1959).

[20] W. FRANZ: Durchlässigkeit von Drahtgittern für elektrische Wellen. Z. angew. Phys. **9**, 416 (1949).

[21] R. D. HATCHER and J. H. ROHRBAUGH: Theory of the echelette grating, I and II. J. Opt. Soc. Amer. **46**, 194, 704 (1958).

[22] R. GANS: Das Verhalten Hertzscher Gitter. Ann. Phys. **61**, 447 (1920).

[23] A. E. HEINS and G. L. BALDWIN: On the diffraction of a plane wave by an infinite grating. Math. Scand. **2**, 103 (1954).

[24] R. HONERJÄGER: Über die Beugung elektromagnetischer Wellen an einem Drahtgitter. Ann. Phys. (6) **4**, 25 (1948).

[25] W. v. IGNATOWSKY: Zur Theorie der Gitter. Ann. Phys. **44**, **23**, 369 (1914).

[26] L. R. INGERSOLL: Polarization of radiation by gratings. Astrophys. J. **51**, 129 (1920).

[27] L. R. INGERSOLL: Some pecularities of polarization and energy distribution by specular gratings. Phys. Rev. **17**, 493 (1921).

[28] B. A. LIPPMANN: Reflection from a periodic surface. Nuclear Development Associates, Inc., Rept. 18—8 (1953).

[29] B. A. LIPPMANN: Note on the theory of gratings. J. Opt. Soc. Amer. **43**, 408 (1953).

[30] I. P. LYSANOV: Theory of the scattering of waves at periodically uneven surfaces, a review. Soviet Phys. Acoust. **4**, 3 (1958).

[31] W. C. MEECHAM: Variational method for the calculation of the distribution of energy reflected from a periodic surface. J. Appl. Phys. **27**, 361 (1956).

[32] W. C. MEECHAM: Fourier transform method for the treatment of the problem of the reflection of radiation from irregular surfaces. J. Acoust. Soc. Amer. **28**, 370 (1956).

[33] W. C. MEECHAM and C. W. PETERS: Reflection of plane-polarized electromagnetic radiation from an echelette diffraction grating. J. Appl. Phys. **28**, 216 (1957).

[34] J. W. MILES: The diffraction of a plane wave through a grating. Quart. Appl. Math. **7**, 45 (1948).

[35] C. H. PALMER jr.: Parallel diffraction grating anomalies. J. Opt. Soc. Amer. **42**, 269 (1952).

[36] C. H. PALMER jr.: Diffraction grating anomalies II. coarse gratings. J. Opt. Soc. Amer. **46**, 50 (1956).

[37] J. H. ROHRBAUGH, C. PINE, W. G. ZOELLNER and R. D. HATCHER: Theory of the echelette grating III. J. Opt. Soc. Amer. **48**, 410 (1958).

[38] H. A. ROWLAND: Phil. Mag. **35**, 397 (1893).

[39] C. SCHAEFER and F. REICHE: Zur Theorie des Beugungsgitters. Ann. Phys. **35**, 817 (1911).

[40] T. B. A. SENIOR: The scattering of electromagnetic waves by a corrugated sheet. Canad. J. Phys. **37**, 787, 1563 (1959).

[41] A. TROWBRIDGE and R. W. WOOD: Phil Mag. **20**, 886 (1910).

[42] V. TWERSKY: On a multiple scattering theory of the finite grating and the Wood anomalies. J. Appl. Phys. **23**, 1099 (1952).

[43] V. TWERSKY: On the scattering of waves by an infinite grating. I.R.E. Trans. AP-4, 330 (1956).

[44] W. VOIGT: Grating diffraction. Göttinger Nachr. 40 (1911).

[45] R. W. WOOD: On a remarkable case of uneven distribution of light in a diffraction grating spectrum. Phil. Mag. **4**, 396 (1902).

[46] R. W. WOOD: Phil. Mag **20**, 770 (1910). — Proc. Phys. Soc. Lond. **18**, 396 (1902).

[47] R. W. WOOD: Diffraction gratings with controlled groove form with abnormal distribution in intensity. Phil. Mag. **23**, 310 (1912).

[48] R. W. WOOD: Anomalous diffraction gratings. Phys. Rev. **48**, 928 (1935).

[49] E. A. YAKOVLEV and F. M. GERASIMOV: Experimental study of the intensity distribution in the spectrum of diffraction gratings for polarized light. Optics and Spectroscopy (Optika i Spektroskopiya) **10**, 50 [104] (1961).

61. Single-order, mirror-like grating in $E_\perp$ polarization. The origin of diffraction and energy distribution by gratings. It is clear, as stressed by STROKE[6], that high efficiencies in the spectral orders from a grating can only be attained to the extent that one is successful in *avoiding polarization* in the diffracted light by appropriate choice and shaping of the groove profile.

Indeed, if a grating were a perfect polarizer for a particular groove shape, angle and wavelength, then it could at most throw one half of the incident unpolarized energy into that order, rather than close to one hundred percent which appears attainable under suitable conditions when the grating is made to act according to the results described in Sect. 62. It is therefore clear that, if one could single out theoretically and experimentally any groove shape or groove

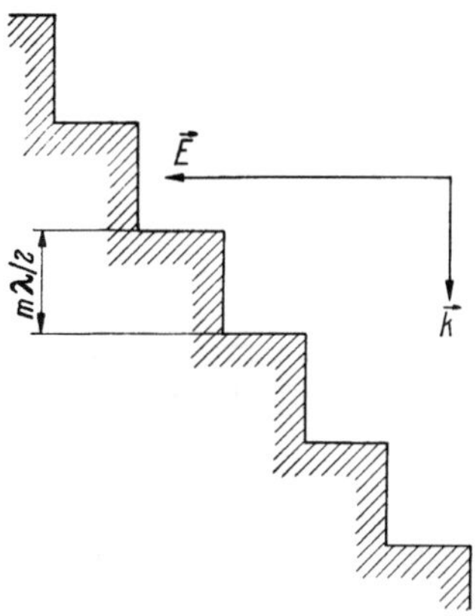

Fig. 80.

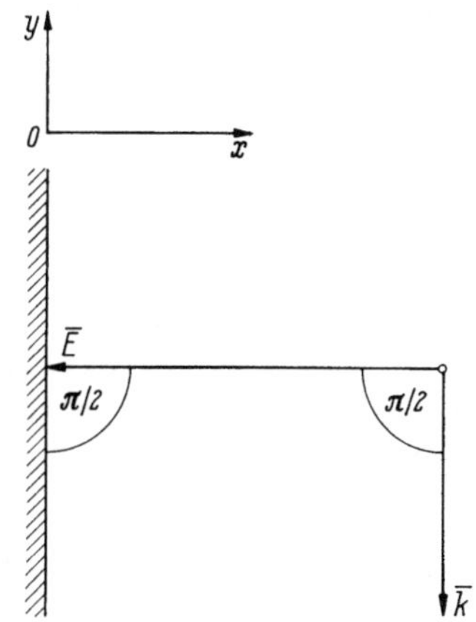

Fig. 81.

shapes which would tend to polarize the light diffracted by a grating, then one would at the same time have singled out the groove shapes that one should avoid ruling.

A basic example of a grating which may show a tendency towards perfect polarization is the grating with a rectangular groove profile and facets $m\lambda/2$ deep, (m=integer, λ=wavelength) (Fig. 80), when the wave vector $\boldsymbol{k}$ in the incident wave is normal to the other facet. Indeed, when such a grating is a perfect conductor and is illuminated in polarized light of wavelength λ, with the $\boldsymbol{E}$ vector in the incident wave normal to the $m\lambda/2$ facet, and the $\boldsymbol{k}$ vector normal to the other facet, the rigorous solution of MAXWELL's equations is in evidence by inspection, and was first derived from elementary principles by A. MARÉCHAL and G. W. STROKE in 1959[1], and somewhat more formally by STROKE in 1960[2]. The boundary conditions (tangential $\boldsymbol{E}$ component equal to zero) are indeed satisfied in this case with the incident wave and only one "diffracted" wave returning exactly back in the direction of incidence, and the grating behaves like a perfect mirror for this polarization, angle and groove depth. W. C. MEECHAM[31] also noted this case in connection with his work on variational methods for calculating the energy distribution in gratings.

The formal proof according to STROKE[2] follows. First, one considers the case of a single perfectly conducting plane surface Oyz (Fig. 81). Let a plane wave be characterized by its $\boldsymbol{E}$ vector parallel to Ox, and let this wave be propagated so that is propagation vector $\boldsymbol{k}$ is parallel to the perfectly conducting plane surface Oyz. Let this wave be described by $\boldsymbol{E}_\perp$. Let $\boldsymbol{J}$, the current density induced in the surface Oyz, be such that the electromagnetic field components which correspond to it satisfy Maxwell's equations for the propagation of the plane wave in free space. One is led to conclude that such a surface will not perturb the propagation of the plane wave.

Indeed, by taking the divergence of Maxwell's equation which applies to the surface,

$$\nabla\times\boldsymbol{H}=\frac{\partial\boldsymbol{D}}{\partial t}+\boldsymbol{J} \tag{61.1}$$

one gets the well known continuity equation

$$\nabla\cdot\boldsymbol{J}=-\frac{\partial}{\partial t}(\nabla\cdot\boldsymbol{D})=-\frac{\partial\varrho}{\partial t}. \tag{61.2}$$

Next, one notes that $\boldsymbol{J}$ = current density = current/surface, normal to y in the conductor, and ϱ = charge density = charge/volume of the conductor, and therefore Eq. (61.2) permits to write, in the limit of an infinitely small penetration (perfect conductor)

$$\frac{\partial J_y}{\partial y}=-\frac{\partial\sigma}{\partial t} \tag{61.3}$$

where J_y is the surface current per unit width Oz, and σ is the charge per unit surface, on the surface Oyz.

But, it is well known that the surface charges, on a perfect conductor, must provide the discontinuity in the normal component of the electric induction vector. One has here

$$\sigma=D_x \tag{61.4}$$

and, therefore, one has for the relation between the current induced on the surface of the conductor and the electric field the equation

$$\frac{\partial J_y}{\partial y}=-\frac{\partial D_x}{\partial t}. \tag{61.5}$$

On the other hand, Eq. (61.1) also gives

$$J_y=H_z \tag{61.6}$$

and therefore

$$\frac{\partial J_y}{\partial y}=\frac{\partial H_z}{\partial y}. \tag{61.7}$$

Finally, Eqs. (61.5) and (61.7) give the relation

$$\frac{\partial H_z}{\partial y}=-\frac{\partial D_x}{\partial t} \tag{61.8}$$

which is precisely the relation obtained from Maxwell's equation

$$\nabla\times\boldsymbol{H}=\frac{\partial\boldsymbol{D}}{\partial t} \tag{61.9}$$

applicable to the propagation of the plane wave in free space.

One is led to conclude that the perfectly conducting surface Oyz does not perturb the propagation of the plane wave, having its $\boldsymbol{E}$ vector normal, and its $\boldsymbol{k}$ vector parallel to the surface.

Next, one considers the case of the surface with orthogonal surface elements, Fig. 82, first examined by Maréchal and Stroke[1]. Here the vertical surface elements (the surface elements parallel to Oyz) satisfy the conditions dealt with in the previous paragraph ($\boldsymbol{E}$ is normal to these surface elements). In addition, the horizontal surface elements (parallel to Oxz) are distant an integral number m of half-wavelengths from each other along y, i.e.

$$\Delta y = m \frac{\lambda}{2}, \tag{61.10}$$

but their width along Ox can have any arbitrary value. According to the discussion of the previous paragraph, and noting that the "reflection" is normal on the surface elements separated by $m\lambda/2$, MARÉCHAL and STROKE were led to predict that the various waves "reflected" by the horizontal surface elements will connect without discontinuity, and form a single, continuous plane wave, identical to the plane wave which would be reflected on a horizontal mirror, parallel to Oxz.

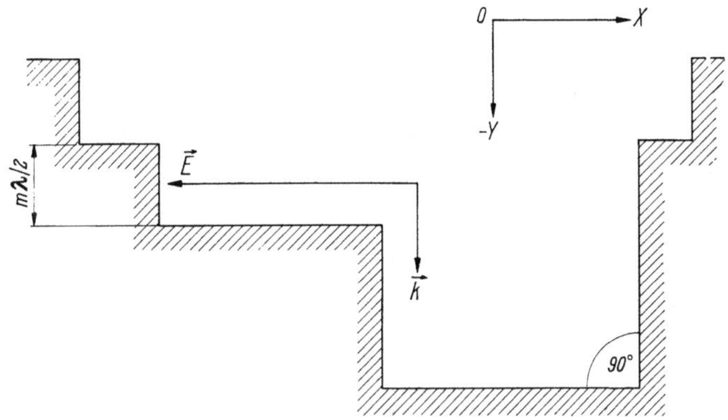

Fig. 82.

The preceding reasoning can be extended to the case of a plane grating, having groove sides at 90° to each other, Fig. 82, such that the horizontal groove sides (or groove facets), that is the groove sides parallel to Oxz, are separated by $m\lambda/2$. MARÉCHAL and STROKE were led to conclude that a 90° groove grating used with the $\boldsymbol{E}_\perp$ polarization and the groove depth described would radiate energy into only a single order, in autocollimation along the y-direction, and therefore that a grating of this type would behave like a perfect mirror, even if, according to scalar approximations of diffraction theory, one would predict the distribution of the diffracted energy into other orders.

Indeed, MAXWELL'S equations and the boundary conditions on the surface of the grating are satisfied under the given conditions. Each of the horizontal groove facets acts like a perfect mirror; the diffracted field consists of a single plane wave, propagating in the y-direction, and formed, in a manner of speaking, by continuous connection of the waves reflected by the various horizontal groove facets. The continuity results from the fact that the vertical groove facets do not perturb the $\boldsymbol{E}_\perp$ plane waves, having their $\boldsymbol{E}$ vector normal to the vertical sides.

This result, unexpected according to the ordinary physical-optics theory of diffraction, has been fully verified by STROKE[2] in a series of experiments described in the following paragraphs.

But first it is desirable to point out a most remarkable difference between the conclusions arrived at on the basis of electromagnetic theory, used here, on one hand, and physical-optics diffraction theory, on the other hand. According to physical optics, one would expect a distribution of the diffracted energy into several orders: the relative amplitudes of the waves of the various orders would be expected. Clearly, these physical optics conclusions are incorrect, in terms of electromagnetic theory, as shown.

Several observable predictions for the step-grating can indeed be made. It can be shown, in particular, according to STROKE[2] that a stationary field, along Oy, will be formed by interference between the incident wave and the single diffracted wave. Above a given grating step, the field should be identical to the field obtained by reflection on a plane mirror, and the computation is indeed identical for both cases. The principal steps are recalled, in order to clarify the conclusions applicable to gratings, which are to be developed in the following sections.

One has, in reference to Fig. 82, for the spatial components of the incident field

$$\boldsymbol{E}^i = (E_x^i, 0, 0), \qquad \boldsymbol{H}^i = (0, 0, H_z^i) \tag{61.11}$$

where

$$\left.\begin{aligned} E_x^i &= \mathrm{e}^{-iky}\mathrm{e}^{-i\omega t}, \\ H_z^i &= Y\mathrm{e}^{-iky}\mathrm{e}^{-i\omega t} \end{aligned}\right\} \tag{61.12}$$

when the amplitude of the incident field is taken equal to unity.

The boundary condition is

$$\boldsymbol{E}_{\text{tangential}} = 0 \qquad \begin{cases} \text{on the surface} \\ \text{of the conductor.} \end{cases} \tag{61.13}$$

On the surface Oxz of one of the groove steps, one has

$$E_x^i = -E_x^d \tag{61.14}$$

where E_x^i applies to the incident field, and E_x^d to the diffracted field. The reflected field $f_2(-\omega t + ky)$ is such that the total field E_x is given by

$$E_x = \mathrm{e}^{-i(\omega t + ky)} + f_2(-\omega t + ky). \tag{61.15}$$

One has, with Eq. (61.14)

$$E_x^d = -\mathrm{e}^{i(ky - \omega t)}, \tag{61.16}$$

and the total field is

$$E_x = E_x^i + E_x^d = -2i \sin ky\, \mathrm{e}^{-i\omega t}. \tag{61.17}$$

One measures the *real* part of the stationary field described by Eq. (61.17). The stationary field is seen to be periodic in y, with a period equal to λ.

The magnetic field component of the diffracted (reflected) field is given by applying the equation

$$\nabla \times \boldsymbol{E} = i\omega\mu \boldsymbol{H} \tag{61.18}$$

to the total field. One obtains

$$H_z^d = Y\mathrm{e}^{i(ky - \omega t)}, \qquad Y = \sqrt{\varepsilon/\mu}. \tag{61.19}$$

It is seen, as pointed out by STROKE[2], that Eqs. (61.17) and (61.19) do indeed give a plane wave, propagating in the $+y$ direction.

It only remains to be noted that the preceding conclusions, developed for the $\boldsymbol{E}_\perp$ polarization, do *not* apply to the $\boldsymbol{E}_\parallel$ polarization. Indeed, even the plane boundary will already perturb the propagation of a plane wave in this case, and the solution of the boundary-value problem is not nearly as straightforward in the case of the $\boldsymbol{E}_\parallel$ polarization, as in the case of the $\boldsymbol{E}_\perp$ polarization. In particular, it is to be expected that the stationary field formed in the vicinity of the grating will be strongly perturbed, and also that energy will be radiated into other orders, in the case of the $\boldsymbol{E}_\parallel$ polarization. All of this was predicted and experimentally confirmed by STROKE (footnote 2 on p. 123), who also pointed out the need, and method, for solving the more general, and more complicated cases, for the two orthogonal polarizations, for any angle of incidence, and groove depths, shapes and angles.

The experimental verification of the important special case of the mirror-like grating were first described by MARÉCHAL and STROKE in 1959[1] and further experiments were reported by STROKE[2]. Some of the results are reproduced in Fig. 83. A photograph of the experimental arrangement is shown in Fig. 63.

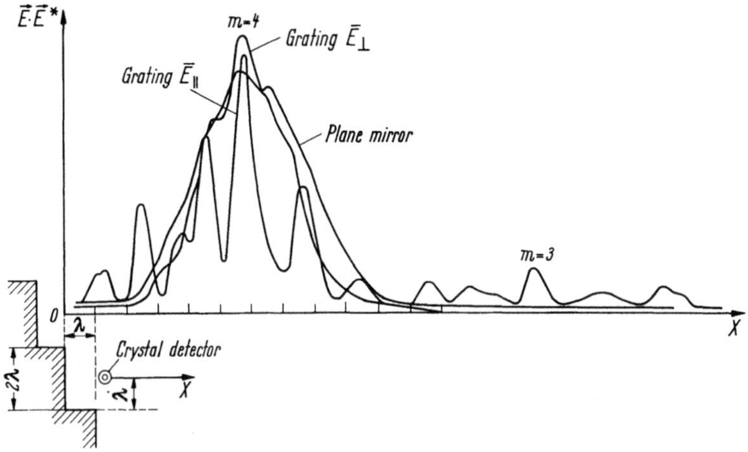

Fig. 83. Stationary field near grating (λ 3.42 cm) (according to MARÉCHAL and STROKE, 1959).

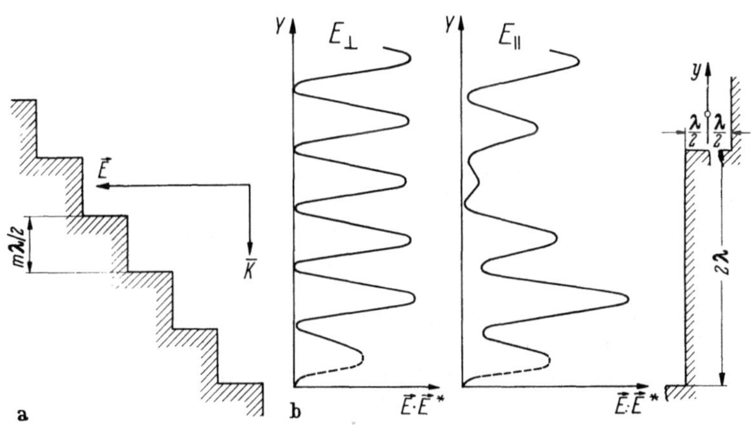

Fig. 84 a and b.

A standing-wave experiment was carried out in the 3.42 cm microwave domain, and the actual experimental standing waves obtained are shown in Fig. 84. The order $m=4$ was used, and the grooves were only 1 λ wide along Ox. The standing waves observed had the appearance of perfect Hertz standing waves, for the $\boldsymbol{E}_\perp$ polarization, and they could not be distinguished from the standing waves reflected from a mirror parallel to Oxz, when placed in the place of the grating. It is of interest to note that the $\boldsymbol{E}_\perp$ field is well behaved deep inside the groove; also no such behaviour was either predicted or observed for the $\boldsymbol{E}_\parallel$ polarization (Fig. 84).

STROKE[2] further measured that only 1.2×10^{-4} of the incident energy appeared in the order $m=3$, at about 36° 51′ from the vertical (Oy) direction on the $\boldsymbol{E}_\perp$

polarization, while a physical-optics theory, along the $[\sin(\pi d \sin\theta/\lambda)]/[\pi d \sin\theta/\lambda]$ lines, with d = width of grooves, and θ = angle from the vertical, would predict perhaps as much as 15% of the incident energy in the electromagnetically "forbidden" third order. Some 180 times more energy, namely about 2.16×10^{-2} of the incident energy, appeared in $m=3$ for the $\boldsymbol{E}_{\parallel}$ than for the well-behaved $\boldsymbol{E}_{\perp}$ polarization.

MARÉCHAL and STROKE[1] concluded, on the basis of these experiments, and preceding theoretical considerations, that it was clear *that the electromagnetic boundary conditions, to be satisfied on the surface of the grating*, were at the very origin of the diffraction of light by gratings, of the polarization of the diffracted light, and indeed of the distribution of energy among orders.

62. Attainment of high efficiencies in blazed optical gratings, by avoiding polarization in the diffracted light, according to STROKE[6]. With the attainment of high-resolution gratings in the 1950's and 1960's*, the concentration of the diffracted energy into just one order per wavelength had become the principal unresolved theoretical and experimental problem in the field of optical gratings. It was indeed well known[2,10] that the calculations based on scalar models of diffraction would yield unsatisfactory predictions of the distribution of diffracted energy as a function of grating angle and wavelength, especially since gratings will in general tend to polarize the diffracted light and to produce energy distributions which are quite different as a function of wavelength and angle for two orthogonal polarizations, neither distribution being in agreement with the predictions based in scalar models[2,10].

Partial and sometimes almost complete polarization in the light diffracted by metallic gratings has been well known ever since the first observations were reported by JOSEPH VON FRAUNHOFER for the gratings which he ruled in 1821[16]. In general the polarization in any wavelength is elliptical, and the maxima of the linear electric field components being in directions normal ($\boldsymbol{E}_{\perp}$) and parallel ($\boldsymbol{E}_{\parallel}$) to the groove length respectively. Much attention has been devoted to the study of polarization effects in gratings as a phenomenon[9,16,35,36,46–49] and to various "discrepancies" (such as the Wood "anomalies") between experiment and theoretical conclusions based on approximate, and generally scalar, descriptions of the diffraction of light by optical gratings[9,35,36,49]. Much less attention, if any, appeared to have been devoted to taking into account the electromagnetic nature of the diffraction phenomenon and the nature of the boundary conditions on the surface of the grating in an attempt of suitably exploiting them for the purpose of increasing the fraction of the incident light that can be concentrated in any given order, angle and wavelength[2,6].

Clearly, of course, in addition to the polarization of the incident light, with respect to the groove length, discussed in Sect. 61, the other parameters entering the equations governing the distribution of energy among orders are the

- shape of the grating-groove profile
- nature of groove surface
- angle of incidence and of diffraction
- wavelength of the electromagnetic radiation.

In general, a qualitative prediction of the energy distribution among the orders of an optical grating, without detailed solution of the electromagnetic boundary-value problem, will appear very complicated indeed.

---

* For a review of the field, see, for example, G. W. STROKE, Ruling Testing and Use of Optical Gratings for High-Resolution Spectroscopy, in Vol. II of Progress in Optics, edit. by E. WOLF (North-Holland) Publishing Co. Amsterdam, 1963, p. 1—72), and Sects. 82—89.

However, a conclusion of very considerable practical importance for the attainment of high efficiencies in blazed optical gratings was given by STROKE[6] in 1963.

On the basis of the theory and experiments described in Sect. 61 STROKE[6] was indeed led to conclude that the classical rectangular groove profile, similar to that commonly represented in textbooks, is apparently the most unsuitable groove profile among the flat facetted grooves recommended[46] on the basis of scalar theory and used for blazing gratings since Lord RAYLEIGH's[9] and R. W. WOOD's[46,47,48] first pioneering contributions in 1888 and 1910, respectively. Indeed, it is the rectangular groove profile which will tend to produce almost perfectly polarized light for the $\boldsymbol{E}_{\perp}$ waves under the circumstances described in Sect. 61, and result in a minimum of efficiency in the use of gratings with unpolarized incident light. One may expect, on the basis also of experiments with gratings of various spacings, that the tendency towards polarization will be most marked for gratings in which the groove facet normal to the $\boldsymbol{k}$ vector is comparable to the wavelength, as is the case in gratings and echelettes, rather than with the coarser echelle gratings[50,51]. One is also led to expect that the grating efficiency will tend to increase with grooves that have included angles larger than 90°, for instance angles of 110° and 120°.

In order to verify some of the predictions of the hypothesis made, STROKE proceeded to have several gratings ruled with groove profiles having different included angles and groove spacings. In all the cases it was attempted to achieve full-depth rulings, with a minimum or no unruled grating surface left between the grooves, as already recommended by TROWBRIDGE and WOOD in 1910[50], and again emphasized by H. D. BABCOCK and H. W. BABCOCK in 1934 and 1951[52,53]. The first series of the results obtained appeared to bear out the general trend of the theoretical predictions, and were recommended by STROKE[6] as the basis for further development of the complete, rigorous theory of energy distribution and blazing of gratings and as a possible help in the physical understanding of the diffraction of light by optical gratings.

By changing the included angle of the grooves from the undesirable 90° to the angle of 120°, STROKE[6] found it possible in a ruling with 600 grooves/mm to completely avoid any observable polarization from the blue to the red end of the visible light spectrum, and to increase the efficiency from the usual 50–65% of the incident light to more than 95% for the 4916 Å wavelength at the angle of blaze when compared to the efficiency of an aluminum mirror of equivalent aperture. More generally, in the series of gratings ruled with spacings from 300 to 2400 grooves/mm, STROKE found that the change of groove angle from 90 to 110°, and better even when possible 120°, has consistently permitted to avoid or appreciably reduce the relatively important partial polarization usually found with such gratings, and to attain efficiencies between 75% and more than 90% of the equivalent mirror at the wavelength and near the wavelength of blaze.

Fig. 85 shows, according to STROKE[6], along with the groove profile as observed in an electron microscope (shadowing angle around 30°), the intensity distribution curves for a 1200 grooves/mm grating ruled early in the series with a 110° included angle diamond. No polarization was found at the peak near the 4916 Å wavelength

---

[50] A. TROWBRIDGE and R. W. WOOD: Phil. Mag. **20**, 886 (1910).

[51] G. R. HARRISON: The Production of Diffraction Gratings: II. The Design of Echelle Gratings and Spectrographs. J. Opt. Soc. Amer. **39**, 522 (1949).

[52] H. D. BABCOCK: J. Opt. Soc. Amer. **34**, 1 (1934).

[53] H. D. BABCOCK, and H. W. BABCOCK: Ruling of diffraction gratings at the Mount Wilson Observatory. J. Opt. Soc. Amer. **41**, 776 (1951).

(the $\boldsymbol{E}_{\parallel}/\boldsymbol{E}_{\perp}$ polarization ratio being 3/2 at 4358 Å, and 7/8 at the 5461 Å wavelength in this case, and would undoubtedly be closer to the 1/1 ratio attained with the 120° groove grating ruled later and described above). The intensity distribution has been measured by using the grating in a monochromator in the vicinity of autocollimation, with the help of spectral lines from a cold mercury arc. The intensity values in Fig. 85 represent efficiency in comparison to an aluminum mirror of the same aperture as the grating.

That the electromagnetic character of diffraction needed to be considered in order to arrive at a full understanding of the problem of blaze (energy distribution among orders) had of course been clear for a long time[9,15]. But it had not yet been possible, until the work by R. Petit[7] and by P. Bousquet[3–5,8] in extending

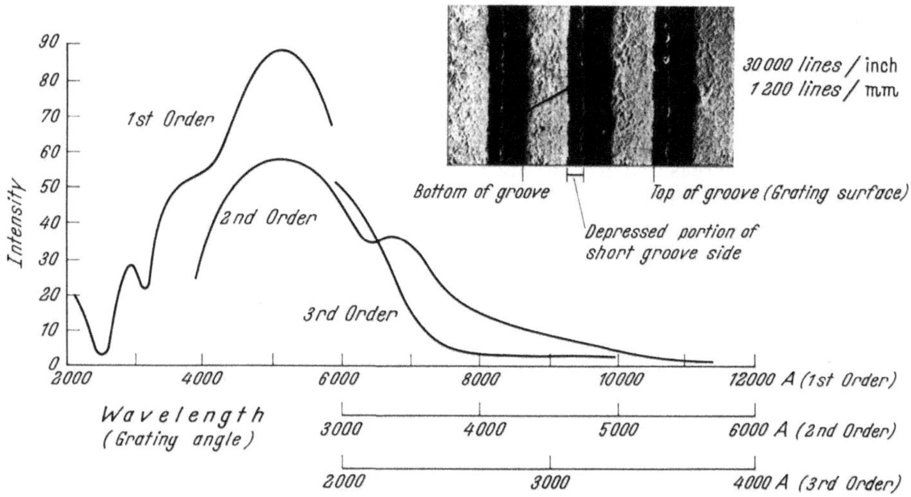

Fig. 85. Blazing by suitable groove profile.

the Rayleigh-Stroke[2] solution, to obtain explicitly rigorous solutions of the electromagnetic boundary-value problem for optical gratings in the general case, starting from first principles and Maxwell's equations[2] (see Sects. 64 to 67). The important exception of the mirror-like grating in the $\boldsymbol{E}_{\perp}$ polarization was pointed out in Sect. 61.

There had thus arisen a significant gap in the theory of gratings. The need for "a more elegant theory" to explain the blaze properties of gratings had been strongly stressed by R. P. Madden and John Strong[10] in 1958 and was reported by Stroke[2,6] to have been in part responsible for his interest in this problem, along with the stimulation by A. Maréchal[1,2,6].

An important step forward in the solution of the problem of blaze and in the physical understanding of the role of the electromagnetic boundary conditions in the diffraction of light by gratings has thus been achieved by singling out (and in practice by avoiding) those groove shapes which will tend to produce particularly strong polarization in the diffracted field. Indeed, avoiding polarization in the diffracted light, or equivalently the search for solutions such that the $\boldsymbol{E}_{\perp}$ and $\boldsymbol{E}_{\parallel}$ solutions are the most similar possible, throughout the domains of interest, can be taken as a criterion for the attainment of high efficiencies in blazed gratings.

It is of historical interest to note, as pointed out by Stroke[6] that Trowbridge and Wood[50], around 1910, did in fact also study echelettes ruled with a 120° hexagonal carborundum crystal, having that angle quite accidentally for crystallographic reasons as appears clearly from the description of the tool by WOOD[46] and from their paper in which, among many fruitful ruling recommendations, they also theorize on the desirability of 90° grooves which are now known in fact to be among the least desirable. STROKE[6] further notes that it also becomes clear in retrospect that the new understanding of the electromagnetic boundary conditions and their role in the polarization of the diffracted light can account qualitatively for the previously unexplained rise in intensity, "four- or fivefold" reported by R. W. WOOD[48] in 1912, after he had rubbed the grating surface by means of a soft chamois skin, presumably along the groove length, and thus changed the groove profile sufficiently to depolarize the diffracted light in terms of the polarization theory developed here[6].

The much discussed "Wood anomalies"[13,35,36,42,43,48,49] in the distribution of energy among orders, and polarization, are known[1,2,6] to be nothing but the manifestation of the "normal" behaviour of electromagnetic waves, when interacting with the grating, on which completely describable boundary conditions are to be satisfied.

63. The Rayleigh solution of the diffraction of plane waves by a perfectly conducting sinusoidal grating[2,9]. A rigorous electromagnetic solution was developed in 1960 by STROKE[2] for perfectly conducting gratings of a sinusoidal profile. The solution was based in part on the method first indicated by Lord RAYLEIGH in 1907[9].

It was noted in Sect. 60 that a considerable interest in this solution arose when it became apparent, as a result of the related extensions by R. PETIT[7,54] and by P. BOUSQUET[3-5,8] that the Rayleigh-Stroke solution could indeed be simply extended, as hoped for, to the solution not only of sinusoidal gratings of any substance, transparent or reflecting, but also to gratings with flat-facet grooves, such as those used in blazed gratings[2].

A discussion of the meaning of "rigorous" solution of an electromagnetic boundary value problem can be found in standard texts, in particular in SOMMERFELD[55-57], as well as in a review article by BOUWKAMP[58], and more particularly, as applied to optical gratings, in STROKE[2]. According to SOMMERFELD a solution of an electromagnetic boundary-value problem is to be considered as "exact" if it satisfies MAXWELL's equations[59] inside as well as outside the diffracting object, and if it satisfies appropriate boundary conditions on the surface of the object. (In the case of a perfect conductor, inside which there can be no electromagnetic field, it is sufficient that the solution satisfies MAXWELL's equations outside the diffracting object.) Moreover, the boundary-value solution must correspond to a given type of excitation (plane waves, spherical waves, or "point source", for example). In the case gratings, plane or spherical waves are of particular interest: plane-wave excitations only are dealt with here. Finally, one need also consider the "radiation condition", as defined by SOMMERFELD ([*55*], p. 250), which permits, in particular, to choose the sign for the diffracted radiation, in the solutions.

In the general case, it is clearly necessary to obtain the solution of a grating diffraction problem by calculating expressions for the amplitudes of the diffracted

---

[54] R. PETIT: Diffraction d'une onde plane monochromatique par un réseau périodique parfaitement conducteur. C. R. Acad. Sci. Paris **257**, 2018 (1963).

[55] A. SOMMERFELD: Optics, pp. 247ff. New York: Academic Press, Inc. 1954.

[56] A. SOMMERFELD: Partial differential equations in Physics, pp. 188—200. New York: Academic Press, Inc. 1949.

[57] A. SOMMERFELD: Jb. dtsch. math. Ver. **21**, 326 (1912).

[58] C. J. BOUWKAMP: Diffraction Theory. Rep. Progr. Phys. **17**, 35 (1954).

[59] J. C. MAXWELL: A dynamical theory of the electromagnetic field. Trans. Roy. Soc. Lond. **155**, 459 (1865).

waves in two *orthogonal polarizations*, in every order. It was indeed shown, for example, in Sect. 62, how different are the solutions which are obtained for two orthogonal polarizations $\boldsymbol{E}_\perp$, normal to the groove length, and $\boldsymbol{E}_\parallel$, parallel to the groove length, respectively. In general, the light incident on a grating is either unpolarized, or polarized in various ways, and therefore solutions for two orthogonal polarizations are necessary.

A boundary value solution which is "exact" in SOMMERFELD's terminology under all these conditions, throughout the calculations, may be called a rigorous solution of the electromagnetic boundary value problem; approximations, if they become necessary, are only involved in the numerical evaluation of the solution.

The solution of the mirror-like grating, in the $\boldsymbol{E}_\perp$ polarization, Sect. 61, is an example of an "exact" solution of a grating diffraction problem.

The difference between optical gratings, with "continuous" surfaces, on one hand, and gratings of "cylinders", such as those studied by many authors (see, e.g.[13] on p. 572), was already noted. Only solutions developed for optical gratings are dealt with here.

Electromagnetic equations used.

MAXWELL's equations are used in the form

$$\nabla\times\boldsymbol{E}=-\frac{\partial\boldsymbol{B}}{\partial t}\tag{63.1}$$

and

$$\nabla\times\boldsymbol{H}=\frac{\partial\boldsymbol{D}}{\partial t}\tag{63.2}$$

which are to be applied to the total field. For a simple harmonic (monochromatic) wave, one has

$$\left.\begin{aligned}\tilde{\boldsymbol{E}}&=\boldsymbol{E}\,\mathrm{e}^{-i\omega t}, & \tilde{\boldsymbol{D}}&=\varepsilon\boldsymbol{E}\,\mathrm{e}^{-i\omega t},\\ \tilde{\boldsymbol{H}}&=\boldsymbol{H}\,\mathrm{e}^{-i\omega t}, & \tilde{\boldsymbol{B}}&=\mu\boldsymbol{H}\,\mathrm{e}^{-i\omega t}.\end{aligned}\right\}\tag{63.3}$$

For monochromatic waves, one obtains

$$\nabla\times\boldsymbol{E}=i\omega\mu\boldsymbol{H},\tag{63.4}$$

$$\nabla\times\boldsymbol{H}=-i\omega\varepsilon\boldsymbol{E}.\tag{63.5}$$

For the case of the propagation of a plane in the $-y$ direction, one has

$$\exp\left[-i(\omega t+k y)\right]=\exp(-i\omega t)\exp(-i k y).\tag{63.6}$$

Boundary condition for the **E** *field.* One has

$$\nabla\times\tilde{\boldsymbol{H}}=\dot{\tilde{\boldsymbol{D}}}+\tilde{\boldsymbol{J}}\tag{63.7}$$

with

$$\tilde{\boldsymbol{J}}=\sigma\boldsymbol{E}\,\mathrm{e}^{-i\omega t}\tag{63.8}$$

being the current density on the surface. In the case of monochromatic waves, one obtains

$$\nabla\times\boldsymbol{H}=(\sigma-i\omega\varepsilon)\boldsymbol{E}\tag{63.9}$$

and

$$\nabla\times\boldsymbol{E}=i\omega\mu\boldsymbol{H}\tag{63.10}$$

giving the equations

$$\boldsymbol{E}=\frac{1}{\sigma-i\omega\varepsilon}\nabla\times\boldsymbol{H}\tag{63.11}$$

and

$$\boldsymbol{H}=\frac{1}{i\omega\mu}\nabla\times\boldsymbol{E}.\tag{63.12}$$

For a perfect conductor, $\sigma\to\infty$, and therefore $\boldsymbol{E}\to 0$, and $\boldsymbol{H}\to 0$, inside of the conductor. With the well-known boundary condition

$$\boldsymbol{n}\times(\boldsymbol{E}_2-\boldsymbol{E}_1)=0 \tag{63.13}$$

which applies to the surface separating two media, one obtains

$$\boldsymbol{n}\times\boldsymbol{E}_2=\boldsymbol{n}\times\boldsymbol{E}_1=0 \tag{63.14}$$

where $\boldsymbol{E}_2$ and $\boldsymbol{E}_1$ are the fields in the two media, $\boldsymbol{n}$ the normal to the boundary surface. One is led to conclude that the tangential component of the electric field is zero at the surface of a perfect conductor:

$$\boldsymbol{E}_{\text{tangential}}=0 \quad \text{(on the surface of the conductor).} \tag{63.15}$$

Boundary condition for the $\boldsymbol{H}$ *field.*

At first, the case of surfaces depending on two coordinates only, such as the case of gratings, is examined, according to STROKE[2].

Let H_z^i be a plane wave, propagating in the $-y$ direction and incident (in vacuo, medium 1) on the perfectly conducting surface (medium 2). Let

$$y=f(x) \tag{63.16}$$

be the equation of the surface, clearly independent of z. Let H_z^i be parallel to z.

By reasons of symmetry, the diffracted field will also have only a single magnetic field component, namely H_z^d. Let

$$H_z=H_z^i+H_z^d \tag{63.17}$$

be the total magnetic field in the medium 1.

One has, for the total $\boldsymbol{H}$ field in the medium 1

$$\nabla\times\boldsymbol{H}=-i\omega\varepsilon\boldsymbol{E} \tag{63.18}$$

and, on the surface, according to Eq. (63.14),

$$\boldsymbol{n}\times\nabla\times\boldsymbol{H}=-i\omega\varepsilon\boldsymbol{n}\times\boldsymbol{E}=0. \tag{63.19}$$

By expanding $\boldsymbol{n}\times\nabla\times\boldsymbol{H}$, and by noting that $H_x=H_y=0$, and that $n_z=0$, one obtains the equation

$$n_x\frac{\partial H_z}{\partial x}+n_y\frac{\partial H_z}{\partial y}=0 \tag{63.20}$$

which applies to the total field.

However, one has

$$\frac{\partial H_z}{\partial n}=\boldsymbol{n}\cdot\nabla H_z=n_x\frac{\partial H_z}{\partial x}+n_y\frac{\partial H_z}{\partial y}+n_z\frac{\partial H_z}{\partial z} \tag{63.21}$$

which gives the final relation

$$\frac{\partial H_z}{\partial n}=0. \tag{63.22}$$

One has thus established, that the normal derivative of the component of the total magnetic field, which is parallel to the coordinate of which the surface is independent, vanishes on the surface of a perfect conductor.

More generally, it can be shown[60], that the normal derivatives of the covariant components of the magnetic field, which are tangent to the surface of a perfect

[60] See e.g. J. A. STRATTON: Electromagnetic Theory, p. 484. New York: McGraw Hill Book Co. 1941.

conductor, vanish on that surface, whatever the equation of the surface, provided only that the surface coincides with the coordinate surface in an orthogonal curvilinear coordinate system.

The two boundary conditions (63.15) and (63.22) result from MAXWELL'S equations. They are necessary and sufficient for the solution of the boundary-value problem of the diffraction by a perfect conductor. In fact, it is only necessary to solve the diffraction problem for one or the other of the vectors, $\boldsymbol{E}$ or $\boldsymbol{H}$, by means of one or the other boundary conditions. The solution for the other vector is obtained by means of MAXWELL'S equations. The case of monochromatic waves is studied. The solutions, according to STROKE[2] are obtained as follows.

Solution for $\boldsymbol{E}$. — One seeks to determine the diffracted field $\boldsymbol{E}^d$ such that the tangential component

$$\boldsymbol{E}=\boldsymbol{E}^i+\boldsymbol{E}^d$$

for the total field vanishes on the surface of the grating.

The magnetic field $\boldsymbol{H}^d$ is obtained from $\boldsymbol{E}^d$ by means of the equation

$$\nabla\times\boldsymbol{E}=i\omega\mu\boldsymbol{H} \tag{63.23}$$

applied to the total field, giving

$$\boldsymbol{H}^d=\boldsymbol{H}-\boldsymbol{H}^i. \tag{63.24}$$

Solution for $\boldsymbol{H}$. — One seeks to determine the diffracted field $\boldsymbol{H}^d$ such that the normal derivatives of the covariant components of the magnetic field which are tangent to the surface vanish on the surface (when it is defined as above). In the case of an incident wave H_z^i and a surface $y=f(x)$ independent of z, one has, when $H_x=H_y=0$,

$$\frac{\partial H_z}{\partial n}=0 \tag{63.25}$$

where H_z is the component of the total magnetic field $\boldsymbol{H}=\boldsymbol{H}^i+\boldsymbol{H}^d$. The $\boldsymbol{E}^d$ field is obtained from the $\boldsymbol{H}^d$ field by means of the equation

$$\nabla\times\boldsymbol{H}=-i\omega\varepsilon\boldsymbol{E} \tag{63.26}$$

applied to the total field, giving

$$\boldsymbol{E}^d=\boldsymbol{E}-\boldsymbol{E}^i. \tag{63.27}$$

As pointed out by STROKE[2], the choice between the $\boldsymbol{E}$ and $\boldsymbol{H}$ solution methods is determined, in particular, by considerations of geometry and of symmetry.

Case of the $\boldsymbol{E}$ *polarization parallel to the groove length* ($\boldsymbol{E}_{\text{tangential}}=0$ on the grating surface).

Let the grating surface be

$$y=a\cos Kx \tag{63.28}$$

as shown in Fig. 86, such that the grating period is $d=2\pi/K$, and the amplitude is a.

Let a plane monochromatic wave be incident on the grating in the $-y$ direction. The spatial components of the incident field are

$$\boldsymbol{E}^i=(0,0,E_z^i), \tag{63.29}$$

$$\boldsymbol{H}^i=(H_x^i,H_y^i,0) \tag{63.30}$$

with

$$E_z^i=\exp[-ik(x\sin\theta_i+y\cos\theta_i)] \tag{63.31}$$

where

$$\theta_i = \text{angle of incidence}. \tag{63.32}$$

The grating equation is (see Sect. 14)

$$\sin\theta_i + \sin\theta_m = mK/k. \tag{63.33}$$

As shown in Sect. 14, the diffracted field is given by a spectrum of discrete waves satisfying the grating Eq. (63.33). By reasons of symmetry, one obtains for

$$E_z^d = \sum_{m=-\infty}^{+\infty} B_m \exp\left[ik(x\sin\theta_m + y\cos\theta_m)\right] \tag{63.34}$$

where B_m is the amplitude of the diffracted wave $[E_z^d]_m$ of order m.

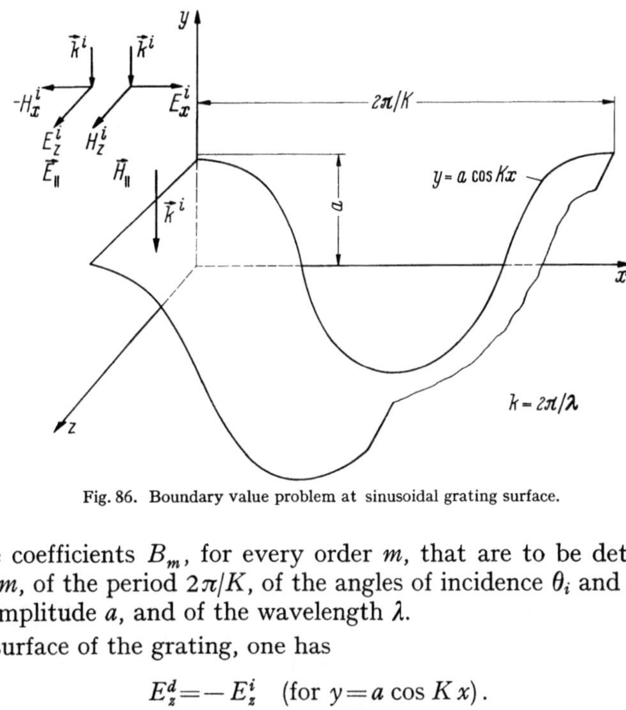

Fig. 86. Boundary value problem at sinusoidal grating surface.

It is the coefficients B_m, for every order m, that are to be determined as a function of m, of the period $2\pi/K$, of the angles of incidence θ_i and of diffraction θ_m, of the amplitude a, and of the wavelength λ.

On the surface of the grating, one has

$$E_z^d = -E_z^i \quad (\text{for } y = a\cos Kx). \tag{63.35}$$

In case of *normal incidence*, $\theta_i = 0$, and

$$\sin\theta_m = mK/k \tag{63.36}$$

and

$$k\cos\theta_m = \pm\sqrt{k^2 - (mK)^2}. \tag{63.37}$$

According to SOMMERFELD's radiation condition[55] one chooses the $+$ sign and one has

$$k\cos\theta_m = +\sqrt{k^2 - (mK)^2}. \tag{63.38}$$

The incident field is

$$\tilde{E}_z^i = e^{-iky - i\omega t} = E_z^i e^{-i\omega t}, \tag{63.39}$$

$$\tilde{H}_z^i = -Y e^{-iky - i\omega t} = H_x^i e^{-i\omega t}. \tag{63.40}$$

where $Y = 1/Z$ = admittance of free space, and $k = 2\pi/\lambda$. With Eq. (63.35), one has

$$E^d = (0, 0, E_z^d)\, e^{-i\omega t}. \tag{63.41}$$

Eqs. (63.34), (63.39) and (63.35) give, for *normal incidence*

$$\sum_{m=-\infty}^{+\infty}\{B_m \exp[ik(x\sin\theta_m + y\cos\theta_m)]\} + \exp[-iky] = 0, \tag{63.42}$$

and, in the *general case*, with any angle of incidence θ_i, one obtains, according to STROKE[2],

$$\sum_{m=-\infty}^{+\infty} B_m \exp[imKx]\exp[iak\cos\theta_m K] = -\exp[-iak\cos\theta_i\cos Kx]. \tag{63.43}$$

One further notes the Bessel-function generating equations

$$\exp(iak\cos\theta_m\cos Kx) = \sum_{p=-\infty}^{+\infty} i^p J_p(ka\cos\theta_m)\exp(ipKx) \tag{63.44}$$

and

$$\exp(-iak\cos\theta_i\cos Kx) = \sum_{n=-\infty}^{+\infty}(-i)^n J_n(ka\cos\theta_i)\exp(inKx), \tag{63.45}$$

the J_p, J_n being Bessel functions. By suitable algebraic manipulation, STROKE[2] shows that the amplitudes $B_{n-p} = B_m$ of the various orders m are finally determined with the help of the following system of equations

$$\boxed{-\delta_n = \sum_{p=-\infty}^{+\infty} B_{(n-p)}\, b_{p,(n-p)}} \tag{63.46}$$

where the algebraic values of the incidence-parameter equations δ_n and of the diffraction-parameter equations $b_{p,(n-p)}$ are given respectively by

$$\delta_n = (-i)^n J_n(ka\cos\theta_i) \tag{63.47}$$

and by

$$b_{p,(n-p)} = i^p J_p[ka\cos\theta_{(n-p)}]. \tag{63.48}$$

The complete steps in the derivation of Eq. (63.46) are given by STROKE[2]. It is particularly important to note that no approximations of any kind are involved in the derivation of Eq. (63.46).

It is also important to note that the grating Eq. (63.33) allows the existence not only of real diffracted waves ($\sin\theta_m \leqq 1$), but also of a family of an infinite number of evanescent waves (waves with imaginary values of θ_m).

In order that waves of order $\pm m$ be evanescent, one needs to have

$$|m| > \left(\frac{k}{K}\right)(1+\sin\theta_i). \tag{63.49}$$

According to STROKE, the evanescent waves may carry away energy, by radiation at the extremities of a grating of finite size: this part of the conclusion is, however, still subject to some question.

Eqs. (63.46), with (63.47) and (63.48), determine the E_z^d components. One also recalls that $E_x = E_y = 0$. The magnetic field components for this case of $\boldsymbol{E}_\parallel$ are obtained with the help of the equation

$$\nabla\times\boldsymbol{E} = i\omega\mu\boldsymbol{H}, \tag{63.50}$$

which reduces to

$$H_x = \frac{1}{i\omega\mu}\frac{\partial E_z}{\partial y}, \tag{63.51}$$

and

$$H_y = -\frac{1}{i\omega\mu}\frac{\partial E_z}{\partial x}, \tag{63.52}$$

which yield the components of the total field. According to the equations

$$E_z^i = e^{-iky} \tag{63.53}$$

and

$$H_x^i = -Y e^{-iky} \tag{63.54}$$

with

$$E_x^i = E_y^i = 0 \tag{63.55}$$

and

$$H_y^i = H_z^i = 0, \tag{63.56}$$

one also has

$$H_x = H_x^i + H_x^d = -Y\exp(-iky) + H_x^d = \frac{1}{i\omega\mu}\,\frac{\partial E_z}{\partial x} \tag{63.57}$$

and

$$H_y = H_y^i + H_y^d = 0 + H_y^d = -\frac{1}{i\omega\mu}\,\frac{\partial E_z}{\partial x} \tag{63.58}$$

where H_x^d and H_y^d are the components of the diffracted magnetic field. By noting that $E_z = E_z^i + E_z^d$, $k = \omega/c$, and $k/\omega\mu = \sqrt{\varepsilon/\mu} = Y$, and also noting that

$$E_x^d = E_y^d = 0 \tag{63.59}$$

and

$$H_z^d = 0 \tag{63.60}$$

one can write down, without difficulty (see STROKE[2]), the complete set of equations for the components E_z^d, H_x^d and H_y^d of the diffracted field.

Case of the ***H*** *polarization parallel to the groove length* ($\partial H_z/\partial n = 0$ on the surface of the grating).

The spatial components in the incident wave, propagating in the $-y$ direction are here

$$(E_x^i, E_y^i, 0) \quad \text{and} \quad (0, 0, H_z^i) \tag{63.61}$$

such that

$$\tilde{E}_x^i = e^{-iky - i\omega t} \tag{63.62}$$

and

$$\tilde{H}_z^i = Y e^{[-ik(x\sin\theta_i + y\cos\theta_i)]}\, e^{-i\omega t} \tag{63.63}$$

with the spatial component being

$$H_z^i = Y\exp[-ik(x\sin\theta_i + y\cos\theta_i)]. \tag{63.64}$$

The boundary condition to be satisfied is

$$\frac{\partial H_z}{\partial n} = 0 \quad \text{for} \quad y = a\cos Kx \tag{63.65}$$

with

$$H_z = H_z^i + H_z^d \tag{63.66}$$

being the component of the total magnetic field. Because of the symmetry of the problem, we have

$$H_z^d = H_z^i \tag{63.67}$$

on the surface of the grating. One concludes that the diffracted field is of the form

$$\tilde{\boldsymbol{H}}^d = (0, 0, H_z^d)\, e^{-i\omega t} \tag{63.68}$$

with

$$H_z^d = Y\sum_{m=-\infty}^{+\infty} A_m \exp[ik(x\sin\theta_m + y\cos\theta_m)]. \tag{63.69}$$

Here A_m are the amplitudes of the various waves of order m which are to be determined.

The boundary condition to be satisfied on the surface of the grating is given by

$$\frac{\partial H_z}{\partial n} = n_x \frac{\partial H_z}{\partial x} + n_y \frac{\partial H_z}{\partial y} + n_z \frac{\partial H_z}{\partial z}. \tag{63.70}$$

One notes that

$$\frac{\partial H_z}{\partial n} = \boldsymbol{n} \cdot \nabla H_z = 0 \tag{63.71}$$

where $\boldsymbol{n}$ is the normal to the grating surface, and the surface is $y = a \cos Kx$. In the Oxy plane, one has

$$\boldsymbol{n} = -\frac{dy}{dx} \frac{1}{\left[1+\left(\frac{dy}{dx}\right)^2\right]^{\frac{1}{2}}} \boldsymbol{i} + \frac{1}{\left[1+\left(\frac{dy}{dx}\right)^2\right]^{\frac{1}{2}}} \boldsymbol{j} \tag{63.72}$$

with $\boldsymbol{i}, \boldsymbol{j}$ being the unit vectors along Ox and Oy. One obtains

$$\boldsymbol{n} = \frac{ak \sin kx}{[1+(ak \sin kx)^2]^{\frac{1}{2}}} \boldsymbol{i} + \frac{1}{[1+(ak \sin kx)^2]^{\frac{1}{2}}} \boldsymbol{j}, \tag{63.73}$$

that is,

$$\boldsymbol{n} = n_x \boldsymbol{i} + n_y \boldsymbol{j}. \tag{63.74}$$

Following suitable derivation, and the use of these equations, one obtains the equation

$$\left.\begin{aligned} &\sum_{m=-\infty}^{+\infty} A_m [(maK^2 - akK \sin\theta_i) \sin Kx + k \cos\theta_m] \times \\ &\qquad \times \exp[imKx] \exp[ika \cos\theta_m \cos Kx] \\ &\qquad = k[aK \sin\theta_i \sin Kx + \cos\theta_i] \exp[-(ika \cos\theta_i) \cos Kx \end{aligned}\right\} \tag{63.75}$$

which, with

$$\sin Kx = \frac{1}{2i} (e^{iKx} - e^{-iKx}) \tag{63.76}$$

gives the following equation, which is to be particularly noted (STROKE [2]):

$$\left.\begin{aligned} &\sum_{m=-\infty}^{+\infty} \left\{ \frac{A_m (maK^2 - akK \sin\theta_i)}{2i} \exp[i(m+1)Kx] - \right. \\ &\quad - \frac{A_m (maK^2 - akK \sin\theta_i)}{2i} \exp[i(m-1)Kx] + \\ &\quad \left. + A_m k \cos\theta_m \exp(imKx) \right\} \exp[ika \cos\theta_m \cos Kx] \\ &= k\left[\frac{aK \sin\theta_i \exp(-iKx)}{2i} - \frac{aK \sin\theta_i \exp(iKx)}{2i} + \cos\theta_i\right] \times \\ &\quad \times \exp[-(ika \cos\theta_i) \cos Kx]. \end{aligned}\right\} \tag{63.77}$$

Further, let us note the equations

$$\exp[(ika \cos\theta_m) \cos Kx] = \sum_{p=-\infty}^{+\infty} i^p J_p(ka \cos\theta_m) \exp(ipKx) \tag{63.78}$$

and

$$\exp[-(ika \cos\theta_i) \cos Kx] = \sum_{n=-\infty}^{+\infty} (-i)^n J_n(ka \cos\theta_i) \exp(inKx). \tag{63.79}$$

A crucial step, at this point, consists of a choice of new orders for the terms and a regrouping in Eq. (63.77), in the manner indicated by STROKE [2]. Indeed STROKE believes that the choice of new orders and regrouping was instrumental in finding the key to the solution for the $\boldsymbol{H}_{\parallel}$ polarization, which had not been previously obtained, in particular by Lord RAYLEIGH, unlike the first half of the complete solution, namely for the $\boldsymbol{E}_{\parallel}$ polarization, which was already obtained by Lord RAYLEIGH[9].

The choosing of new orders for the terms according to STROKE[2] consists of identifying $(m+1)$ to q in the first term, $(m-1)$ to q in the second term, and m to q in the third term of Eq. (63.77). After some simplification one obtains the following equation

$$\left.\begin{aligned}&\sum_{q=-\infty}^{+\infty}\left\{A_{q-1}\left[\frac{(q-1)\,a K^2-a k K\sin\theta_i}{2i}\right]-A_{q+1}\left[\frac{(q+1)\,a K^2-a k K\sin\theta_i}{2i}\right]+\right.\\&\quad\left.+A_q k\cos\theta_q\right\}\exp(iqKx)\exp(ika\cos\theta_m\cos Kx)\\&\quad=\left\{\frac{kaK\sin\theta_i}{2i}\sum_{r=-\infty}^{+\infty}(-i)^{r-1}J_{r-1}(ka\cos\theta_i)-\right.\\&\quad-\frac{kaK\sin\theta_i}{2i}\sum_{r=-\infty}^{+\infty}(-i)^{r+1}J_{r+1}(ka\cos\theta_i)+\\&\quad\left.+k\cos\theta_i\sum_{r=-\infty}^{+\infty}(-i)^r J_r(ka\cos\theta_i)\right\}\exp(irKx).\end{aligned}\right\}\tag{63.80}$$

One now proceeds to replace, formally, the q's by m's, and group their terms of orders $(m-1)$, m and $(m+1)$ into a new term α_m such that

$$\left.\begin{aligned}\alpha_m=\alpha_{(n-p)}=A_m\cos\theta_m+A_{m-1}\left[\frac{(m-1)\,aK^2-akK\sin\theta_i}{2i}\right]-\\-A_{m+1}\left[\frac{(m+1)\,aK^2-akK\sin\theta_i}{2i}\right]\end{aligned}\right\}\tag{63.81}$$

from which the amplitude coefficients are subsequently extracted by iteration. Together with an energy conservation equation, STROKE[2] shows that the grouping of terms according to Eq. (63.81) permits to obtain the solution of the diffraction problem for the $\boldsymbol{H}_{\parallel}$ polarization in the simple form of the system of equations

$$\boxed{-\beta_n=\sum_{-\infty}^{+\infty}-\frac{\alpha_{(n-p)}}{k}\,b_{p,(n-p)}}\tag{63.82}$$

where the incidence- and diffraction-parameter equations, β_n and $b_{p,(n-p)}$, are given in terms of the grating, incidence, diffraction, wavelength and order parameters. The incidence-parameter equation is

$$\left.\begin{aligned}\beta_n=\frac{aK\sin\theta_i}{2i}\left[(-i)^{n-1}J_{n-1}(ka\cos\theta_i)-(-i)^{n+1}J_{n+1}(ka\cos\theta_i)\right]+\\+\cos\theta_i(-i)^n J_n(ka\cos\theta_i)\end{aligned}\right\}\tag{63.83}$$

and the diffraction-parameter equation is

$$b_{p,(n-p)}=i^p J_p\left[ka\cos\theta_{(n-p)}\right].\tag{63.84}$$

Here, as in the case of the $\boldsymbol{E}_{\parallel}$ solution, no approximations were involved in the solution.

64. Diffraction of plane electromagnetic waves by a grating of sinusoidal profile: general case. The Rayleigh-Stroke solution for the perfectly conducting grating of Sect. 63 can be extended to the general case of gratings, of sinusoidal profile, having any nature, dielectric or metallic. The solution, developed according to the description given by P. Bousquet and R. Deleuil[4] follows.

Let a grating of complex index $n=\nu-j\chi$, where $j=\sqrt{-1}$, be illuminated by a plane monochromatic wave, incident on the grating at an angle of incidence i.

Let xOz be the mean grating plane and let the grating surface now be given by

$$y=-a\cos 2\pi\frac{x}{d}. \tag{64.1}$$

Eq. (64.1) describes the surface separating the medium 1, of index 1, and the medium 2, of index n. The positive Oy direction is directed as in Sect. 63.

Let the incident wave be

$$\tilde{\boldsymbol{E}}^i=\boldsymbol{E}^i\exp[-jk(x\sin i-y\cos i)] \tag{64.2}$$

where $k=2\pi/\lambda$, and $\exp(+j\omega t)$ is assumed implicit.

The diffracted field in the two media is formed by a superposition of an infinity of plane waves, having for equations, respectively

$$\tilde{\boldsymbol{E}}^r_m=\boldsymbol{E}^r_m\exp[-jk(u^r_m x-w^r_m y)] \quad \text{(in medium 1)} \tag{64.3}$$

and

$$\tilde{\boldsymbol{E}}^t_m=\boldsymbol{E}^t_m\exp[-jk(u^t_m x-w^t_m y)] \quad \text{(in medium 2)}. \tag{64.4}$$

The corresponding magnetic fields, $\boldsymbol{H}^r_m$ and $\boldsymbol{H}^t_m$ have comparable expressions. The direction-cosines of the diffracted waves can be deduced from the grating equations

$$u^r_m=\sin i+\frac{m\lambda}{d} \tag{64.5}$$

and

$$u^t_m=\sin i+\frac{m\lambda}{d}. \tag{64.6}$$

One further has the equations

$$(u^r_m)^2+(w^r_m)^2=1 \tag{64.7}$$

and

$$(u^t_m)^2+(w^t_m)^2=n^2 \tag{64.8}$$

which apply in all cases (real or evanescent waves, real or complex indices). Eqs. (64.7) and (64.8) permit to write down the explicit expressions for the diffracted field in any particular case.

In order to compute the amplitudes of the diffracted fields, it is further necessary to give expressions for the boundary conditions at the dielectric surface. The boundary conditions depend on the polarization of the incident field, as before. The two cases that need to be distinguished are now examined.

α) *Case of* $\boldsymbol{E}_{\parallel}$ *(electric field in incident wave parallel to groove length).* As pointed out in Sect. 63, by reasons of symmetry the diffracted field also has only $\boldsymbol{E}$ components parallel to the groove length. Next, one notes that the *tangential component of the electric field,* as well as the *tangential component of the magnetic field* must be *continuous* at the boundary surface. Accordingly, one has

$$E^i+\sum_{m=-\infty}^{+\infty}E^r_m=\sum_{m=-\infty}^{+\infty}E^t_m \tag{64.9}$$

and

$$H^i_x-A\,H^i_{-y}+\sum_{m=-\infty}^{+\infty}[(H^r_m)_x-A\,(H^r_m)_{-y}]=\sum_{m=-\infty}^{+\infty}[(H^t_m)_x-A(H^t_m)_{-y}] \tag{64.10}$$

where

$$A=\frac{2\pi a}{d}\sin 2\pi\frac{x}{d} \tag{64.11}$$

and where $(H^i)_x$, $(H^i)_{-y}$, etc., refer to the magnetic field components along the Ox and Oy axes.

It is possible, in Eq. (64.10) to replace the magnetic field components by their expressions in terms of the corresponding electric fields. This is accomplished by projecting MAXWELL's equation

$$\nabla\times\boldsymbol{E}=-\mu_0\frac{\partial\boldsymbol{H}}{\partial t} \tag{64.12}$$

along the respective axes. With Eq. (64.9), (64.10) becomes

$$\cos i\, E^i+\sum_{m=-\infty}^{+\infty}\left(w_m^r+A\frac{m\lambda}{d}\right)E_m^r=\sum_{m=-\infty}^{+\infty}\left(w_m^t+A\frac{m\lambda}{d}\right)E_m^t. \tag{64.13}$$

Eqs. (64.9) and (64.13) can be written in explicit form by replacing E_m^r and E_m^t by their expressions obtained from Eq. (64.3) and Eq. (64.4), in which one lets $y=-a\cos 2\pi x/d$, and by letting u_m^r and u_m^t have the values determined by Eqs. (64.7) and (64.8).

The two sides in Eq. (64.9) and (64.13) have now become periodic functions of x, with a period d. By developing these functions in Fourier series, and by identifying the coefficients having the same frequency ν on both sides, respectively, of Eqs. (64.9) and (64.13), according to the usual methods (see e.g. [2,3]) one again obtains, as in Sect. 63, two systems of linear equations, which permit to compute the amplitudes E_m^r and E_m^t of the various diffracted waves.

The two systems of linear equations obtained are

$$\sum_{m=-\infty}^{+\infty}(j)^{-m}\left[J_{\nu+m}(k a w_m^t)E_m^t-J_{\nu+m}(k a w_m^r)E_m^r\right]=J_\nu(k a\cos i) \tag{64.14}$$

and

$$\left.\begin{aligned}&\sum_{m=-\infty}^{+\infty}(j)^{-m}\left\{w_m^t\left[1+\frac{m(m+r)}{(w_m^t)^2}\frac{\lambda^2}{d^2}\right]J_{\nu+m}(k a w_m^t)\right\}E_m^t-\\ &-\sum_{m=-\infty}^{+\infty}(j)^{-m}\left\{w_m^r\left[1+\frac{m(m+r)}{(w_m^r)^2}\frac{\lambda^2}{d^2}\right]J_{\nu+m}(k a w_m^r)\right\}E_m^r=\cos i\, J_\nu(k a\cos i).\end{aligned}\right\} \tag{64.15}$$

β) Case of $\boldsymbol{E}_\perp$ (electric field in incident wave normal to groove length). By again requiring that the tangential components of the $\tilde{\boldsymbol{E}}$ and $\tilde{\boldsymbol{H}}$ field be continuous at the boundary surface, one has

$$H^i+\sum_{m=-\infty}^{+\infty}H_m^r=\sum_{m=-\infty}^{+\infty}H_m^t \tag{64.16}$$

and

$$E_x^i-A E_y^i+\sum_{m=-\infty}^{+\infty}\left[(E_m^r)_x-A(E_m^r)_{-y}\right]=\sum_{m=-\infty}^{+\infty}\left[(E_m^t)_x-A(E_m^t)_{-y}\right]. \tag{64.17}$$

With the help of MAXWELL's equation

$$\nabla\times\boldsymbol{H}=\varepsilon\frac{\partial\boldsymbol{E}}{\partial t}, \tag{64.18}$$

and noting Eq. (64.16), Eq. (64.17) can be replaced by the following equation:

$$\left.\begin{aligned}\cos i\, H^i&+\sum_{m=-\infty}^{+\infty}\left[w_m^r+A\frac{m\lambda}{d}\right]H_m^r\\ &=\sum_{m=-\infty}^{+\infty}\left\{w_m^t+A\left[\frac{m\lambda}{d}+(1-n^2)\sin i\right]\right\}\frac{1}{n^2}H_m^t.\end{aligned}\right\} \tag{64.19}$$

Next, one proceeds as before, to obtain the following two systems

$$\sum_{m=-\infty}^{+\infty} (j)^{-m} [J_{\nu+m}(k a w_m^t) H_m^t - J_{\nu+m}(k a w_m^r) H_m^r] = J_\nu(k a \cos i) \tag{64.20}$$

and

$$\left.\begin{aligned} &\sum_{m=-\infty}^{+\infty} (j)^{-m} \frac{w_m^t}{n^2} \left\{1 + \frac{\lambda}{d} \frac{(m+\nu)}{(w_m^t)^2} \left[\frac{m\lambda}{d} + (1-n)^2 \sin i\right]\right\} J_{\nu+m}(k a w_m^t) H_m^t - \\ &\quad - \sum_{m=-\infty}^{+\infty} (j)^{-m} w_m^r \left[1 + \frac{\lambda^2}{d^2} \frac{m(m+\nu)}{(w_m^r)^2}\right] J_{\nu+m}(k a w_m^r) H_m^r = \cos i \, J_\nu(k a \cos i). \end{aligned}\right\} \tag{64.21}$$

One notes, in conclusion, that Eqs. (64.14) and (64.15), and Eqs. (64.20) and (64.21) form the complete rigorous solution of the diffraction of plane monochromatic waves for a sinusoidal grating of any nature. In particular, they apply to any angle of incidence, and they give the amplitudes of real, as well as of evanescent waves, on both sides of the grating surface. One also notes that the explicit expressions for w_m^r and w_m^t depend on the nature of medium 2.

In particular, one obtains, according to BOUSQUET and DELEUIL[4] the following relations:

$$w_m^r = -\cos r_m \tag{64.22}$$

for the case of ordinary waves, r_m being the angle of diffraction, with

$$\sin r_m = u_m^r. \tag{64.23}$$

Further, for evanescent waves, one has

$$w_m^r = j\sqrt{[\sin i + (m\lambda/d)]^2 - 1}. \tag{64.24}$$

In case of a transparent medium 2, one has for the ordinary waves

$$w_m^t = n \cos \theta_m, \tag{64.25}$$

where θ_m is the angle of diffraction, such that

$$n \sin \theta_m = u_m^t \tag{64.26}$$

and

$$w_m^t = -j\sqrt{[\sin i + (m\lambda/d)]^2 - n^2} \tag{64.27}$$

in the case of evanescent waves.

In the general case of an absorbing medium 2, one can write

$$w_m^{t\,2} = (\nu - j\chi)^2 - (\sin i + m\lambda/d)^2. \tag{64.28}$$

Examples of numerical calculations for the extreme cases, obtained according to the general equations given in this section, are given by BOUSQUET[3], for the case of a normal incidence, and by STROKE[2], for the case of a perfectly conducting grating, under any angle of incidence. As pointed out by BOUSQUET and DELEUIL[4], the equations obtained in this section reduce to the equations of Sect. 63, first derived by STROKE[2].

65. Diffraction of plane electromagnetic waves by a grating with flat-facet grooves. The principles of the electromagnetic boundary-value solutions given in Sects. 60 to 64 can be extended to the solution of the diffraction of plane waves by optical gratings having piecewise linear groove sections. As shown by BOUSQUET[5], the solutions can be obtained for the general case of grating formed by matter of any real characteristics. The notations used are those of Sect. 64.

Let the equation of the surface be

$$z = f(x) \tag{65.1}$$

and let xOz be the plane of incidence. The Oz axis is directed from the medium (1) of index 1, to the medium (2) of index n. As before, one distinguishes between the two orthogonal polarizations $\boldsymbol{E}_\parallel$ and $\boldsymbol{H}_\parallel$.

α) *Case of the incident electric field parallel to the groove length.* Noting that the incident field can be normalized to

$$E^i = 1 \tag{65.2}$$

one has for the boundary equations expressing the *continuity of the electric field* on the surface $z = f(x)$, according to Sect. 64, the equation

$$e^{-jkz\cos i} + \sum_{m=-\infty}^{+\infty} E_m^r\, e^{-jk\left(\frac{m\lambda}{d}x + w_m^r z\right)} = \sum_{m=-\infty}^{+\infty} E_m^t\, e^{-jk\left(\frac{m\lambda}{d}x + w_m^t z\right)}. \tag{65.3}$$

It is important to note that Eq. (65.3) holds for any shape of the grating profile $f(x)$, provided only that $f(x)$ is periodic in x with a period d.

In case of a piecewise linear grating profile, let the period d be divided into n intervals Δx_q, having their centers at x_q, and such that inside of each interval, the grating function $f(x)$ be a linear function

$$z = \alpha_q x + \beta_q. \tag{65.4}$$

It follows that the different terms of Eq. (65.3) can be expanded in a Fourier series. For instance, within the period d, one has

$$e^{-jkz\cos i} = \sum_{q=1}^{h} F_q\, e^{-jk\cos i(\alpha_q x + \beta_q)} \tag{65.5}$$

where F_q is a function equal to unity inside of the interval Δx_q, and equal to zero outside of this interval.

Each of the terms of the right-hand side of Eq. (65.5) can be easily expanded in a Fourier series inside of d. One has

$$F_q\, e^{-jk\cos i(\alpha_q x + \beta_q)} = \sum_{\nu=-\infty}^{+\infty} A_{q,\nu}\, e^{j2\pi\frac{\nu x}{d}} \tag{65.6}$$

where

$$A_{q,\nu} = \frac{\sin\pi\left(\frac{\cos i}{\lambda}\alpha_q + \frac{\nu}{d}\right)\Delta x_q}{\pi\left(\frac{\cos i}{\lambda}\alpha_q + \frac{\nu}{d}\right)d}\, e^{-j2\pi\left(\frac{\cos i}{\lambda}z_q + \frac{\nu}{d}x_q\right)} \tag{65.7}$$

in which z_q is the ordinate of the point on the grating having the abscissa x_q. One obtains the expansion of $\exp(-jkz\cos i)$ in the form

$$e^{-jkz\cos i} = \sum_{\nu=-\infty}^{+\infty} B_\nu(\cos i)\, e^{j2\pi\frac{\nu x}{d}} \tag{65.8}$$

with

$$B_\nu(\cos i) = \sum_{q=1}^{n} A_{q,\nu}. \tag{65.9}$$

The function $B_\nu(\cos i)$ here plays the same role as the Bessel function $J_\nu(ka\cos i)$, appearing in the case of the sinusoidal grating (preceding sections).

The other terms of Eq. (65.3) can be expanded in a similar way. One has

$$e^{-jk\left(\frac{m\lambda}{d}x+w_m^r z\right)}=\sum_{\nu=-\infty}^{+\infty} B_{\nu+m}(w_m^r)\, e^{j2\pi\frac{\nu x}{d}} \tag{65.10}$$

and a comparable expression for the coefficient of E_m^t. It is clear that the functions $B(w_m^r)$ and $B(w_m^t)$ result from $B(\cos i)$ by substituting w_m^r and w_m^t in Eqs. (65.7) and (65.9).

Finally, the system of linear equations is obtained, as before, by equating the coefficients of the terms of same frequency in the expansions of Eq. (65.3). One has

$$\sum_{m=-\infty}^{+\infty}\left[B_{\nu+m}(w_m^t)\,E_m^t-B_{\nu+m}(w_m^r)\,E_m^r\right]=B_\nu(\cos i). \tag{65.11}$$

One next notes that the *continuity of the tangential component of the magnetic field* on the surface $z=f(x)$ can be written as[5]

$$\cos i\,\varepsilon^i+\sum_{m=-\infty}^{+\infty}\left(w_m^r-\frac{dz}{dx}\frac{m\lambda}{d}\right)\varepsilon_m^r=\sum_{m=-\infty}^{+\infty}\left(w_m^t-\frac{dz}{dx}\frac{m\lambda}{d}\right)\varepsilon_m^t. \tag{65.12}$$

Here again, new terms in dz/dx have made their appearance. However, $dz/dx=\alpha_q$ in the interval Δx_q. One therefore has

$$\frac{dz}{dx}\,e^{-jk\left(\frac{m\lambda}{d}x+w_m^r z\right)}=\sum_{\nu=-\infty}^{+\infty} C_{\nu+m}(w_m^r)\, e^{j2\pi\frac{\nu x}{d}} \tag{65.13}$$

where

$$C_{\nu+m}(w_m^r)=\sum_{q=1}^{h}\alpha_q A_{q,\nu+m}(w_m^r). \tag{65.14}$$

A similar expression can be given for the second term of Eq. (65.12).

Finally the system of linear equations corresponding to the boundary equation (65.12) is:

$$\left.\begin{aligned}&\sum_{m=-\infty}^{+\infty}\left[w_m^t B_{\nu+m}(w_m^t)-\frac{m\lambda}{d}C_{\nu+m}(w_m^t)\right]E_m^t-\\&-\sum_{m=-\infty}^{+\infty}\left[w_m^r B_{\nu+m}(w_m^r)-\frac{m\lambda}{d}C_{\nu+m}(w_m^r)\right]E_m^r=\cos i\; B_\nu(\cos i).\end{aligned}\right\} \tag{65.15}$$

β) *Case of incident magnetic field parallel to the groove length.* The boundary equations are now

$$H^i+\sum_{m=-\infty}^{+\infty} H_m^r=\sum_{m=-\infty}^{+\infty} H_m^t \tag{65.16}$$

and

$$\left.\begin{aligned}&\cos i\,H^i+\sum_{m=-\infty}^{+\infty}\left[w_m^r-\frac{dz}{dx}\frac{m\lambda}{d}\right]H_m^r\\&\quad=\sum_{m=-\infty}^{+\infty}\left\{w_m^t-\frac{dz}{dx}\left[\frac{m\lambda}{d}+(1-n^2)\sin i\right]\right\}\frac{1}{n^2}H_m^t.\end{aligned}\right\} \tag{65.17}$$

The amplitudes H_m^r and H_m^t of the various diffracted fields can be calculated with the help of Eqs. (65.16) and (65.17). One has

$$\sum_{m=-\infty}^{+\infty}\left[H_m^t B_{\nu+m}(w_m^t)-H_m^r B_{\nu+m}(w_m^r)\right]=B_\nu(\cos i) \tag{65.18}$$

and

$$\left.\begin{aligned}&\sum_{m=-\infty}^{+\infty}\left\{w_m^t B_{\nu+m}(w_m^t)-\left[\frac{m\lambda}{d}+(1-n^2)\sin i\right]C_{\nu+m}(w_m^t)\right\}\frac{1}{n^2}H_m^t-\\&-\sum_{m=-\infty}^{+\infty}\left[w_m^r B_{\nu+m}(w_m^r)-\frac{m\lambda}{d}C_{\nu+m}(w_m^r)\right]H_m^r=\cos i\; B_\nu(\cos i).\end{aligned}\right\}\quad(65.19)$$

It is again to be noted that Eqs. (65.11), (65.15), (65.18), and (65.19) represent the rigorous solution of the diffraction of plane electromagnetic waves by a grating having piece-wise linear groove sides. As pointed out by BOUSQUET[5], the solutions obtained are complete, and have no restrictions on groove dimensions and spacing with respect to the wavelength, nor with regard to the material nature of the grating, angles of incidence or polarization.

The special case of triangular grooves is examined in Sect. 66.

66. Diffraction of plane electromagnetic waves by optical gratings with triangular groove profiles. Criteria for optimum solutions and for the attainment of high efficiencies in the diffracted grating orders. The formalism of the Rayleigh-Stroke method for the exact solution of the electromagnetic boundary value problem in the diffraction of plane waves by optical gratings has been extended to the general case of gratings with triangular groove profiles, by R. PETIT[7] and by P. BOUSQUET[8].

The essential criteria in the electromagnetic solution of grating diffraction and efficiency problems had been previously given by STROKE[6] (Sect. 62). Under the given angles of incidence and diffraction, the highest efficiency in a given grating order and wavelength, is attained when the solutions for the $\boldsymbol{E}_\parallel$ case and for the $\boldsymbol{H}_\parallel$ case, which are generally different, are such that the diffracted light is made to be unpolarized, when the grating is illuminated in unpolarized light. When this criterion is applied to the numerical evaluation of the solutions, the groove shape, groove angle and groove depth parameters can be suitably optimized[6,7]. In particular, as predicted by STROKE[1,2,6] (see also Sects. 61 and 62), triangular grooves with 110 and 120° included angle (rather than the more classical 90° groove shape discussed in textbooks) are found to produce the highest efficiencies in blazed optical gratings.

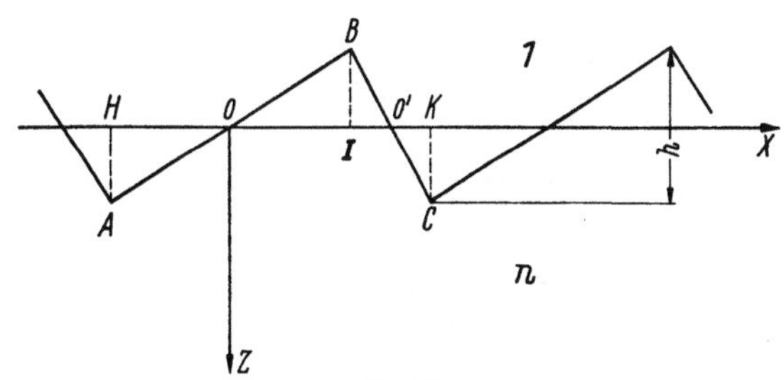

Fig. 87.

The formalism of the boundary-value solution according to BOUSQUET[8], and related to Sects. 64 and 65, follows.

The notations of Sect. 65 are used here. Let the grating surface separating the two media have a triangular shape (Fig. 87). Let d be the grating spacing constant. The medium (1) has the refractive index 1, and the medium (2) has the index $n=\nu-jk$. Let xOy be the mean plane of the grating surface and let the grooves be parallel to Oy. The plane of incidence is xOz. The Oz axis is directed into medium (2).

Let the interval HK, of length d, be divided into two intervals HI and IK, of lengths Δx_1, and Δx_2 respectively, with the abscissae of the middle points O and O' being $x_1=0$ and $x_2=d/2$ (Fig. 87). Let $\Delta x_1=fd$, with $0<f<1$, and therefore $\Delta x_2=(1-f)d$.

The groove "depth" (along Oz) is taken as h. It is seen that the groove shape can be characterized by the two parameters h/d and $\Delta x_1/\Delta x_2 = f$.

The coefficients $A_{1,\nu}$ and $A_{2,\nu}$ related to the groove facets AB and BC respectively (see Sect. 65) can now be given as:

$$A_{1,\nu} = \frac{\sin\pi\left(\frac{h}{\lambda}\cos i + \nu f\right)}{\pi\left(\frac{h}{\lambda}\cos i + \nu f\right)}\, f \tag{66.1}$$

and

$$A_{2,\nu} = \frac{\sin\pi\left[\frac{h}{\lambda}\cos i - \nu(1-f)\right]}{\pi\left[\frac{h}{\lambda}\cos i - \nu(1-f)\right]}\,(1-f)\,e^{-j\pi\nu}. \tag{66.2}$$

From these equations, one has the coefficients

$$B_\nu(\cos i) = \frac{\sin\pi\left(\frac{h}{\lambda}\cos i + \nu f\right)}{\pi\left(\frac{h}{\lambda}\cos i + \nu f\right)}\;\frac{\frac{h}{\lambda}\cos i}{\frac{h}{\lambda}\cos i - \nu(1-f)} \tag{66.3}$$

and

$$C_\nu(\cos i) = -\frac{\lambda\nu}{d\cos i}\,B_\nu(\cos i). \tag{66.4}$$

As shown by BOUSQUET[8], only the $B_\nu(\cos i)$ coefficients appear explicitly in the final equations. The following systems of linear equations give the electric field amplitudes E_m^r and E_m^t, for the various "reflected" and "transmitted" waves, forming the diffracted field, in the two polarizations.

α) *Case of the incident electric field vector parallel to the groove length.* One has

$$\sum_{m=-\infty}^{+\infty}\left[B_{\nu+m}(w_m^t)\,E_m^t - B_{\nu+m}(w_m^r)\,E_m^r\right] = B_\nu(\cos i) \tag{66.5}$$

and

$$\left.\begin{aligned}&\sum_{m=-\infty}^{+\infty} w_m^t\left[1 + \frac{m(m+\nu)}{(w_m^t)^2}\,\frac{\lambda^2}{d^2}\right] B_{\nu+m}(w_m^t)\,E_m^t - \\ &-\sum_{m=-\infty}^{+\infty} w_m^r\left[1 + \frac{m(m+\nu)}{(w_m^r)^2}\,\frac{\lambda^2}{d^2}\right] B_{\nu+m}(w_m^r)\,E_m^r = \cos i\; B_\nu(\cos i).\end{aligned}\right\} \tag{66.6}$$

β) *Case of the incident electric field vector normal to the groove length.* The basic system of equations permits to calculate the amplitudes H_m^r and H_m^t of the magnetic vectors in the diffracted field. The corresponding electric field vectors E_m^r and E_m^t are computed with the help of MAXWELL's equations (see e.g., Sect. 63, Eqs. (63.26) and (63.27)].

One has, for the magnetic field vectors, the equations

$$\sum_{m=-\infty}^{+\infty}\left[B_{\nu+m}(w_m^t)\,H_m^t - B_{\nu+m}(w_m^r)\,H_m^r\right] = B_\nu(\cos i) \tag{66.7}$$

and

$$\left.\begin{aligned}&\sum_{m=-\infty}^{+\infty}\left\{1 + \frac{\lambda}{d}\,\frac{(m+\nu)}{(w_m^t)^2}\left[\frac{m\lambda}{d} + (1-n^2)\sin i\right]\right\}\frac{w_m^t}{n^2}\,B_{\nu+m}(w_m^t)\,H_m^t - \\ &-\sum_{m=-\infty}^{+\infty}\left[1 + \frac{m(m+\nu)}{(w_m^r)^2}\,\frac{\lambda^2}{d^2}\right] w_m^r\,B_{\nu+m}(w_m^r)\,H_m^r = \cos i\; B_\nu(\cos i).\end{aligned}\right\} \tag{66.8}$$

It is quite noteworthy that Eqs. (66.5) to (66.8) have exactly the same form as the equations of the Rayleigh-Stroke solution for the sinusoidal gratings (Sect. 63) as extended by BOUSQUET, to gratings of any matter (Sect. 64). The equations are obtained by simply replacing with the B functions the Bessel functions that appear in the case of sinusoidal gratings. For the "reflected" waves, the various B coefficients have the form

$$B_{\nu+m}(w_m^r) = \frac{\sin\pi\left[\frac{h}{\lambda}\, w_m^r + (\nu+m)\, f\right]}{\pi\left[\frac{h}{\lambda}\, w_m^r + (\nu+m)\, f\right]} \; \frac{\frac{h}{\lambda}\, w_m^r}{\frac{h}{\lambda}\, w_m^r - \nu(1-f)}. \tag{66.9}$$

The coefficients $B_{\nu+m}(w_m^t)$ for the transmitted waves are obtained by simply substituting w_m^t for w_m^r in Eq. (66.9).

The expression for $B_\nu(\cos i)$ is given in Eq. (66.3), while the coefficients w_m^r and w_m^t, which depend on the nature of the medium, are given in Sect. 64 [Eqs. (64.22) to (64.28)].

With these remarks, it is possible to state that the equations given in this section give the complete, rigorous solution of the problem of diffraction of plane electromagnetic waves by an optical grating of any "triangular" profile. Echelle and echelette gratings form special cases of this general class of gratings, which includes the various "blazed" optical gratings discussed in the literature[6].

67. Comparison of energy distribution calculations for "blazed" gratings with actual experiments. Using the equations of Sect. 66, BOUSQUET[8] computed the "efficiency" in the $\boldsymbol{E}_{\|}$ polarization for the case of the grating efficiency measured by R. P. MADDEN and J. STRONG ([*10*], Fig. P—6, p. 605). The groove spacing was microns. The included groove angle was 94°, such that the groove facets formed angles of 37 and 94°, respectively, with the mean grating surface. As described by MADDEN and STRONG, the grating was used in a Littrow mount, in the first order, the angle between the incident and diffracted beams being 8.9°. The experimental curve (1) as a function of wavelength is shown in Fig. 88 together with the "theoretical" curve (3) based on the equations given above. It is seen that the agreement obtained is reasonably good, considering the fact that the actual grating undoubtedly departed from the ideal triangular form assumed in the calculations.

Other comparisons are being undertaken by R. PETIT[61], G. W. STROKE[62], and others.

68. Practical considerations in the attainment of high efficiencies in blazed optical gratings and in grating replicas. As previously noted (Sect. 62), in addition to the choice of groove angle considerable advantages in actual attainment of high efficiencies in gratings result from examination of test rulings with electron-microscopes (Sect. 80) and in particular in efficiency-testing spectrometers[6, 63, 64] before the actual ruling (Sect. 77). Parenthetically, grating replicas are found to have efficiencies quite similar to those of the master gratings from which they are reproduced[63, 64].

A basic element in the evaluation of efficiencies in gratings is that the concept of efficiency should be applied to the ratio of flux density (or total flux) diffracted

[61] See reference 12 on p. 571.

[62] See MCCLELLAN and STROKE, reference 12 on p. 571.

[63] G. W. STROKE: Theory, Production and Use of Optical Gratings for High-Resolution Spectroscopy. In: Quantum Electronics III, edit. by P. GRIVET and N. BLOEMBERGEN. New York:: Columbia Univ. Press 1964.

[64] R. F. JARRELL and G. W. STROKE: Some New Advances in Grating Ruling, Replication and Testing. Applied Optics 3, 1251—1262 (1964).

into a given order (in a given wavelength and under given conditions of incidence and diffraction) to the flux reflected in the same wavelength by a mirror of an aperture equivalent to that of the grating and, most important, coated with the

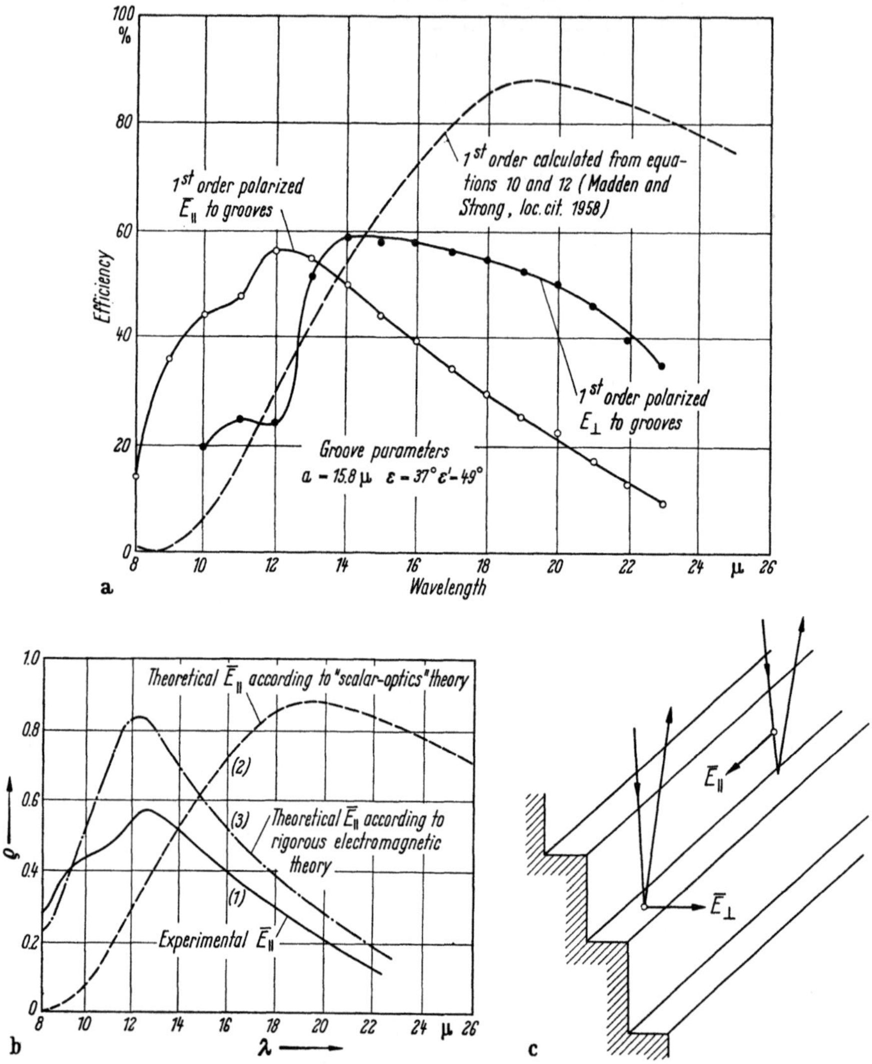

Fig. 88 a—c. Comparison of energy distribution in an infrared grating with calculations based on "scalar-optics" theory (a and curve 2 in b) and with rigorous electromagnetic theory (curve *3* in b). The need for seeking electromagnetic rather than scalar optics solutions for energy distributions in gratings was stressed by MADDEN and STRONG (loc. cit 10, 1958). They pointed to the considerable discrepancies between scalar-optics calculations and measured energy distributions in two orthogonal polarizations $\boldsymbol{E}_{\parallel}$ and $\boldsymbol{E}_{\perp}$, for a given infrared grating (a). A calculation for the same grating in the $\boldsymbol{E}_{\parallel}$ polarization (b) by BOUSQUET (loc. cit. [8], 1963) shows how close an agreement between theory and experiment is obtained when the calculations are carried out on the basis of the rigorous electromagnetic Rayleigh-Stroke solution (loc. cit. [2], 1960). For maximum efficiency, the solutions for the two polarizations should coincide as closely as possible.

same kind of metallic film (in case of a metallic surface). More generally, comparisons of efficiency should be made with the reflectivity of the actual grating surface taken into account.

The specially important cases of x-ray gratings and gratings for use in the far ultraviolet are examined in Sects. 69 and 70.

In addition to the dependence on the groove shape, the efficiency of gratings depends on the reflectivity of the metallic coatings of the grating or replica surface.

69. Effect of grating surface reflectance on grating efficiency in visible and vacuum ultraviolet. α) *Aluminum films with and without overcoating.* High reflectivities in the visible as well as vacuum ultraviolet domains are obtained with aluminum films[65]. As described by MADDEN, CANFIELD and HASS[65], it has been found that unoxidized aluminum has a normal incidence reflectance greater than 86% for wavelengths down to 1025 Å. The vacuum ultraviolet reflectance decreases, however, very rapidly (in a matter of minutes), when the aluminum is exposed to air, O_2 or N_2, when the aluminum becomes oxidized. Fig. 89 shows that the reflectance, at 60°, in the wavelength of 584 Å, has decreased from about 80% at deposition down to about 35% after a 24-hour exposure in air.

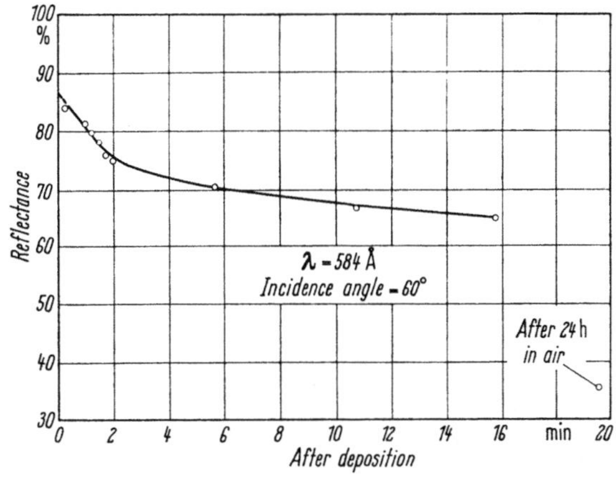

Fig. 89. Reflectance decrease in vacuum of freshly deposited aluminum films as a function of time at an incidence angle of 60° for λ 584 Å (according to MADDEN, CANFIELD and HASS, loc. cit. [65]).

High ultraviolet reflectivity can be maintained by protecting the freshly deposited aluminum either with a silicone-oxide coating[66] for the near ultraviolet, or with a coat of about 250 Å of MgF_2[68]. These and other coatings must be deposited before the removal of the aluminum from the vacuum[69]. Many useful references related to extreme ultraviolet work, and a review of the field, are given by TOUSEY[70]. Fig. 90 shows, according to ANGEL et al.[69] the reflectance of aluminum with various coating, in the 900 Å to 1700 Å range. Reflectance of 65% appears maintained down to about 1000 Å with a LiF overcoat, and about 55% is maintained if a few tens of angstroms of MgF_2 are added to keep the hygroscopic LiF from absorbing water[69]. As noted by TOUSEY[70] the reflectances below

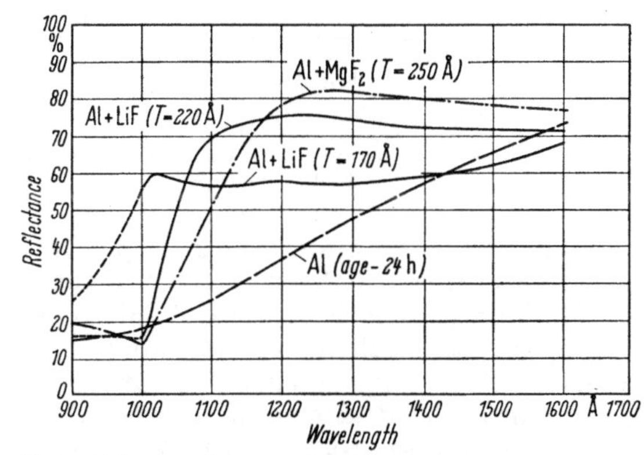

Fig. 90. Reflectance of evaporated aluminum with two thicknesses of LiF overcoatings. Reflectance of Al+Mg_2 and 24-hr-old aluminum are shown for comparison (according to ANGEL, HUNTER and TOUSEY, loc. cit. [69]).

[65] R. P. MADDEN, L. R. CANFIELD and G. HASS: On the vacuum ultraviolet reflectance of evaporated aluminum before and during oxidation. J. Opt. Soc. Amer. **53**, 620 (1963).

[66] G. HASS: J. Amer. Ceram. Soc. **33**, 353 (1950).

[67] A. P. BRADFORD and G. HASS: J. Opt. Soc. Amer. **53**, 1096 (1963).

[68] G. HASS and R. TOUSEY: Reflecting Coatings for the Extreme Ultraviolet. J. Opt. Soc. Amer. **49**, 593 (1959).

[69] D. W. ANGEL, W. R. HUNTER and R. TOUSEY: Extreme Ultraviolet Reflectance of LiF-Coated Aluminum Mirrors. J. Opt. Soc. Amer. **51**, 913 (1961).

[70] R. TOUSEY: The extreme ultraviolet. Past and Future. Applied Optics **1**, 679 (1962).

1000 Å are much lower. A coating of ZnS produces a reflectance of 20% at 900 Å and 15% at 800 Å, while Pt and Rh have given good results at 600 Å, with reflectances from 20% to 24%.

Reflectances below 584 Å are known to be extremely low (unpublished results, obtained by HUNTER[70] indicate values at 304 Å of the order of 1 to 3% for Pt or Au). Accordingly, grazing incidence instruments are being used for most research below 500 Å.

According to ANGEL et al.[69] the effectiveness of the MgF_2 as a reflectance increasing coating for aluminum in the wavelength region from 1600 to 1200 Å results from two actions. First, the MgF_2 film prevents the formation of the layer of Al_2O_3 which absorbs strongly below 1600 Å and therefore greatly decreases the reflectance of aluminum at shorter wavelengths. Second, a MgF_2 film of proper thickness appears to increase the reflectance of unoxidized aluminum by an interference process down to wavelengths as short as 1200 Å. For wavelengths shorter than 1200 Å, MgF_2 begins to absorb appreciably, while films of LiF of 170 Å and 120 Å thickness are found to be more efficient (Fig. 90).

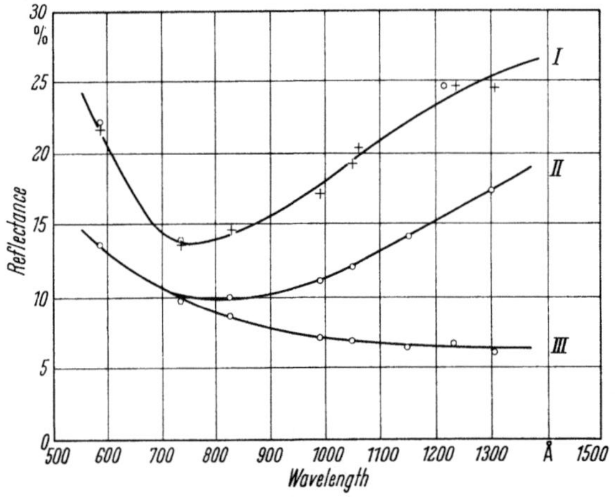

Fig. 91. Efficiencies of platinized mirrors at near normal incidence. Curve I: single-coating evaporation at Ft. Belvoir. Curve II: platinum coated in two successive operations. Curve III: platinum coated in nine successive operations (according to REEVES and PARKINSON, loc. cit. [80]).

Generally Al+LiF mirrors have been found to show little aging when stored in a laboratory with 40% average humidity[69]. According to ANGEL et al.[69], deterioration of LiF-overcoated Al can be greatly retarded by the deposition of a very thin layer of MgF_2 or CaF_2 on the LiF, without important loss in reflectance (Fig. 90).

Much progress in the increase of reflectivity of aluminum films in the vacuum ultraviolet has resulted from the work of many authors in addition to those referred to[71–74]. High evaporation speeds and a good vacuum are particularly important in the attainment of high reflectances in aluminum[75,76].

And, of course, JOHN STRONG's pioneering work in developing evaporation methods for the attainment of rulable aluminum films for gratings has been the single contribution which has made the ruling of high-efficiency gratings with well shaped grooves a matter of practicality[77,78].

Much information on the experimental aspects of vacuum ultraviolet spectroscopy is also given by R. A. SAWYER[79], and in other texts on practical spectroscopy referred to elsewhere.

---

[71] P. H. BERNING, G. HASS and R. P. MADDEN: J. Opt. Soc. Amer. **50**, 586 (1960).
[72] G. B. SABINE: Phys. Rev. **55**, 1064 (1939).
[73] M. BANNING: J. Opt. Soc. Amer. **32**, 98 (1942).
[74] R. P. MADDEN and L. R. CANFIELD: J. Opt. Soc. Amer. **51**, 838 (1961).
[75] G. HASS, W. R. HUNTER, and R. TOUSEY: J. Opt. Soc. Amer. **46**, 1009 (1956).
[76] G. HASS, W. R. HUNTER, and R. TOUSEY: J. Opt. Soc. Amer. **47**, 1070 (1957).
[77] J. STRONG: Phys. Rev. **43**, 498 (1933).
[78] J. STRONG: Astrophys. J. **83**, 401 (1936).
[79] R. A. SAWYER: Experimental Spectroscopy, third edit. New York: Dover Public. Inc. 1963.

β) Gold and platinum films. Reflectance of 15 to 25% in the 500 to 1500 Å range was obtained in the early tests on platinized mirrors reported by REEVES and PARKINSON[80] (Fig. 91). A correspondingly high grating efficiency should be attainable with good groove shape, even near normal incidence. No measurements on gold mirrors were given in these early tests, but comparison of gratings coated with either platinum or gold seem to indicate a somewhat higher efficiency for gold-coated gratings than for platinized gratings.

Changes of efficiency by metallic overcoating need to be judged with great care: particular attention in the interpretation of results needs to be paid to possible change of groove shape by the overcoating (see G. W. STROKE[6], and Sect. 62).

The subject of grating efficiency in the soft x-ray domain is treated in Sect. 70.

70. Grating efficiency in soft x-ray domain. (Gratings at grazing incidence.) The use of gratings and the problems of efficiency in the soft x-ray region present particular aspects which single them out from the general case of grating efficiency discussed in Sects. 60 to 69. A notable contribution to the ruling of x-ray gratings has been made by M. SIEGBAHN and his associates[81–83]. A particularly straightforward approximation to the electromagnetic theory of gratings, applicable to the conditions of the x-ray use of gratings is given by SPRAGUE, TOMBOULIAN and BEDO[84]. Their method, based on a derivation by ECKART[85] is described in what follows.

In the soft x-ray region, where the gratings are used at grazing incidence, there are two considerations which justify the approximations used. First, over the wavelength region of interest, the optical gratings used have spacings which are generally some 100 times the x-ray wavelength. Secondly, due to the fact that the index of the grating material is less than unity*, only those portions of the grating surface which make sufficiently small angles (on the order of 5°) with the incident "beam" serve as efficient reflectors for the radiation.

Hence, the details of the groove shape might be expected to play a less significant role, particularly for the light ruling frequently used in x-ray gratings.

The simplified model adopted appears to give reasonably good agreement with experiments. Clearly, the rigorous electromagnetic theory, based on the Rayleigh-Stroke method, is also applicable to x-ray gratings.

Let the refractive index n be periodic in x, i.e. $n(x+a)=n(x)$. Then normalized solutions of the wave equation

$$\frac{\partial^2 u}{\partial x^2}+\frac{\partial^2 u}{\partial y^2}+k^2 u=0 \tag{70.1}$$

must obey the following condition:

$$u(x+a)=e^{i d} u(x). \tag{70.2}$$

* Cases where the index is slightly greater than unity also exist.

[80] E. M. REEVES and W. H. PARKINSON: Efficiencies of Gold and Platinum Gratings in the Vacuum Ultraviolet. J. Opt. Soc. Amer. **53**, 941 (1963).

[81] M. SIEGBAHN: Z. Physik **62**, 435 (1930). — Kosmos **8**, 186 (1930).

[82] M. SIEGBAHN and T. MAGNUSSON: Z. Physik **95**, 133 (1935).

[83] M. SIEGBAHN: K. Svenska Vetenskaps Akademiens Årbok (1940), p. 151.

[84] G. SPRAGUE, D. H. TOMBOULIAN and D. E. BEDO: Calculation of Grating Efficiency in the Soft X-ray Domain. J. Opt. Soc. Amer. **45**, 756 (1965).

[85] C. ECKART: Phys. Rev. **44**, 12 (1933).

The parameter $k=2\pi n/\lambda$. For regions far from the grating surface, u has the form

$$\left.\begin{aligned} u=&\exp[ik'(x\cos\varphi-y\sin\varphi)]\\ &+\sum_N A_N\exp[ik'(x\cos\psi_N+y\sin\psi_N)], \qquad (y\gg 0), \end{aligned}\right\} \tag{70.3a}$$

$$u=\sum_N B_N\exp[ik''(x\cos\xi_N-y\sin\xi_N)], \qquad (y\ll 0). \tag{70.3b}$$

In Eqs. (70.3a) and (70.3b) the summations are to be extended over all positive and negative integers. The k' and k'' are appropriate for the regions $y\gg 0$ and $y\ll 0$, respectively.

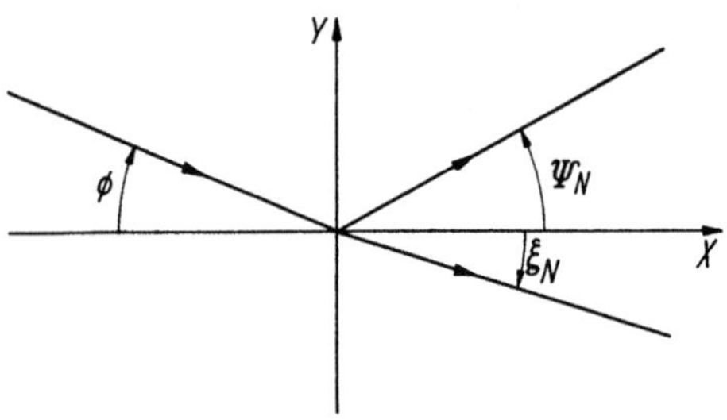

Fig. 92. Designation of angles measured from grating surface (according to SPRAGUE, TOMBOULIAN and BEDO, loc. cit. [84]).

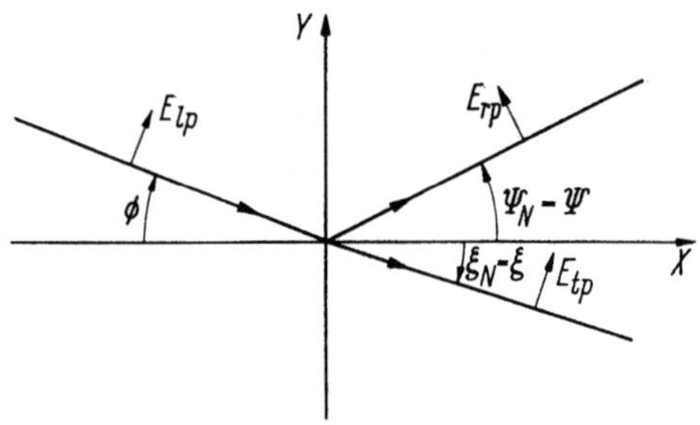

Fig. 93. Diagram showing the plane of incidence and electric vectors parallel to this plane (according to SPRAGUE, TOMBOULIAN and BEDO, loc. cit. [84]).

The requirement that the solutions in Eqs. (70.3) satisfy the periodicity condition of Eq. (70.2) leads to the following relations:

$$\alpha=ak'\cos\varphi, \tag{70.4a}$$

$$ak'\cos\psi_N-ak'\cos\varphi=2\pi N, \tag{70.4b}$$

$$ak''\cos\xi_N-ak'\cos\varphi=2\pi N, \tag{70.4c}$$

where N is an integer. The angles ψ, φ and ξ are indicated in Fig. 92.

Eq. (70.4b) is recognized as the standard grating equation for reflection. Eq. (70.4c) is the grating equation appropriate for the inside of the medium.

It should be clear that Eqs. (70.1) to (70.3b) are scalar equations, and therefore it follows that the effect of the electromagnetic boundary conditions has not been considered. The boundary conditions are made to enter by means of the "reflectance" at the grating surface.

Let the components of the electric and magnetic vectors in Fig. 93 parallel to the plane of incidence be denoted by p, and those perpendicular to it described by s. It is noted that E_p will be $\boldsymbol{E}_\parallel$, and E_s will be $\boldsymbol{E}_\perp$ of the preceding sections, etc.[84]. Let the subscripts i, r and t refer to the incident, reflected and transmitted waves, respectively.

The boundary conditions in any order can be expressed, as a first approximation, by the equations:

$$E_{is}+E_{rs}-E_{ts}=0, \tag{70.5a}$$

$$E_{is}\sin\varphi-E_{rs}\sin\psi-nE_{ts}\sin\xi=0, \tag{70.5b}$$

$$E_{ip}+E_{rp}-nE_{tp}=0, \tag{70.5c}$$

$$E_{ip}\sin\varphi-E_{rp}\sin\psi-E_{tp}\sin\xi=0. \tag{70.5d}$$

It follows that the equations:

$$E_{rs} = \frac{\sin\varphi - n\sin\xi}{\sin\psi + n\sin\xi} E_{is} \tag{70.6}$$

and

$$E_{rp} = \frac{n\sin\varphi - \sin\xi}{n\sin\psi + \sin\xi} E_{ip} \tag{70.7}$$

represent the electric vectors in the reflected waves for the two states of polarization for a particular order N. Strictly speaking, according to electromagnetic theory, the complex impedance $Z = \sqrt{\mu/\varepsilon}$, rather than the refractive index n, should figure in Eqs. (70.6) and (70.7).

The reflectances R_{pN}, for the case of parallel polarization in the N-th order diffracted ray, and R_{sN} for the perpendicular polarization, can be calculated, in the same approximation, by comparing the diffracted intensity in the order N with the intensity of the incident beam. One obtains

$$R_{s,N} = (D_N)^2 \left[\frac{\sin\varphi - n\sin\xi_N}{\sin\psi_N + n\sin\xi_N}\right]^2 \frac{\sin\psi_N}{\sin\varphi} \tag{70.8}$$

and

$$R_{p,N} = (D_N)^2 \left[\frac{n\sin\varphi - \sin\xi_N}{n\sin\psi_N + \sin\xi_N}\right]^2 \frac{\sin\psi_N}{\sin\varphi}. \tag{70.9}$$

The factor $(\sin\psi_N/\sin\varphi)$ represents the relative cross-section of the beam, while the factor D_N is a "groove-shape" factor introduced by SPRAGUE et al.[84] and equal to

$$D_N = (-1)^N \frac{\sin \pi N w/a}{\pi N} \tag{70.10}$$

where w is the unruled portion of the groove (having the spacing period a).

The reflectance of the grating at x-ray wavelengths can be computed in Eqs. (70.8) and (70.9) by using a value of n determined by the equation[86]

$$\left.\begin{aligned} n &= 1 - \delta \\ &= 1 - \frac{e^2 N Z \varrho}{2\pi m c^2 M} \lambda^2 \end{aligned}\right\} \tag{70.11}$$

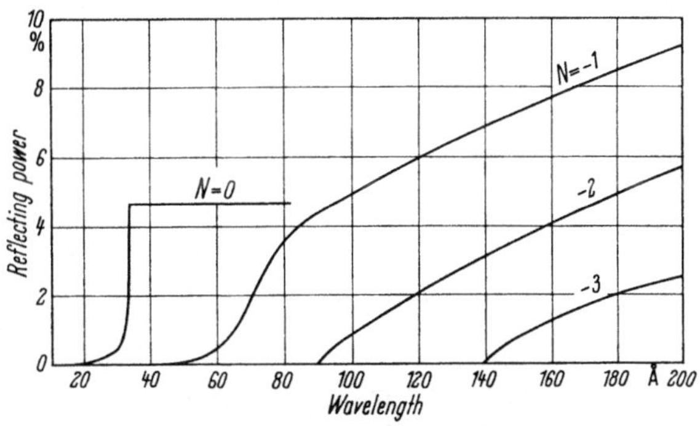

Fig. 94. Graph of reflecting power as a function of wavelength for $N=0$, -1, -2 and -3 (according to SPRAGUE, TOMBOULIAN and BEDO, loc. cit.[84]).

where e and m are the charge and mass of the electron, N is AVOGADRO's number, Z the number of electrons per molecule, M the molar mass, ϱ the density and λ the wavelength. As an example, for a glass used in gratings, δ/λ^2 may be of the order of 2.7×10^{-6} Å^{-2}.

The theoretical average of R_p and R_s for the reflectance of a glass grating, according to SPRAGUE et al.[84] is given in Fig. 94. It is to be noted that: (1) for a given order, the reflectance ("reflecting power") begins to increase rapidly at a particular wavelength, and (2) the marked increase of reflected intensity in the higher orders occurs at successively longer wave-lengths.

Experimental verifications appear to give generally satisfactory agreement with the approximation model used, although, as pointed out by SPRAGUE

[86] See, for example, A. GUINIER: Théorie et technique de la radiocristallographie. Paris: Dunod 1956.

et al.[84], as well as by R. S. CRISP[87] there are occasional disagreements. The narrow K-emission band of Be was used by SPRAGUE et al.[84] as an essentially monochromatic source, in a vacuum spectrograph: comparisons of the reflectivities in orders I and II with the theory are given in Fig. 95. In other experiments, the synchrotron continuum radiation was used[84]. Other experiments were also reported by EDLÉN[88], O'BRYAN[89] and others, and theoretical treatments of grating efficiency applicable to the x-ray domain were also carried out by MACK, STEHN and EDLÉN[90], BEUTLER[91], STAMM and WHALEN[92], in addition to ECKART[85], SPRAGUE, TOMBOULIAN and BEDO[84]. The important improvements of x-ray grating efficiency obtained by LUKIRSKII and SAVINOV upon coating glass gratings with 400 Å films of titanium* are described in Sect. 77.

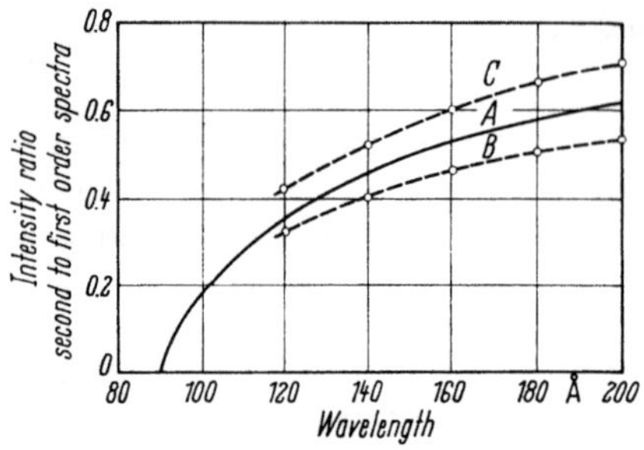

Fig. 95. Curves showing the calculated and observed energy ratios between second and first order (according to SPRAGUE et al. loc. cit. [84]). Data derived from line spectra.

71. Grating efficiency at infrared wavelengths. The general theory of grating efficiency, based on the Rayleigh-Stroke method (Sects. 60 to 68) is applicable without restrictions to infrared gratings. This was confirmed, in particular, by JANOT and HADNI[93] who report on the successful application of the "G. W. Stroke method[2]" to "echelette" gratings.

The linear system of an infinite number of equations, with an infinite number of unknowns obtained can be resolved approximately for a given case. Very good agreement between the theoretical calculations, for the two orthogonal polarizations, and experimental efficiency measurements with a grating of period $b = 350$ microns are obtained.

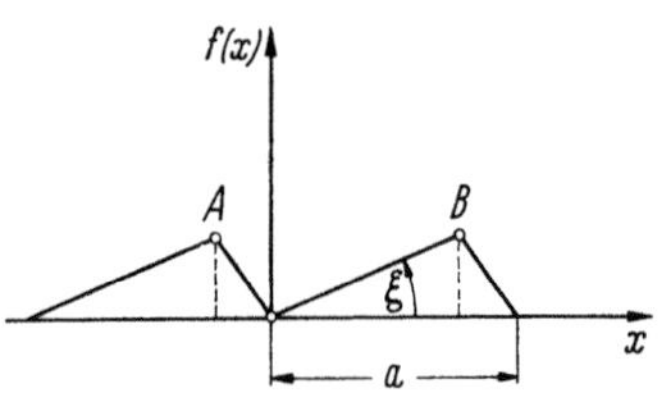

Fig. 96. According to JANOT and HADNI, loc. cit. [93].

A "normal" and an "inverted" grating position are defined in the following equations, in agreement with the custom in the use of infrared gratings.

The "normal" grating orientation is considered first.

The abscissa of the point A (Fig. 96) is taken as $-a\sin^2\xi$, and that of the point B as $a\cos^2\xi$, where ξ is the angle of the facet OB with the Ox axis.

The equation of a grating period is

$$f(x) = -x\cot\xi \quad \text{for} \quad -a\sin^2\xi < x < 0 \tag{71.1a}$$

and

$$f(x) = x\cot\xi \quad \text{for} \quad 0 < x < a\cos^2\xi. \tag{71.1b}$$

The grating profile is expressed in terms of its Fourier series as follows:

$$f(x) = A_0 + \sum_{n=1}^{\infty} A_n \cos\frac{2\pi n x}{a} + B_n \sin\frac{2\pi n x}{a}. \tag{71.2}$$

---

* See references 133 and 134 of Sect. 77.

[87] R. S. CRISP: The reflectivity of glass and aluminum gratings at grazing incidence below 1000 Å. Optica Acta **8**, 137 (1961).

[88] B. EDLÉN: Nova Acta Regiae Soc. Sci. Upsal. **9**, No. 6 (1934).

[89] H. M. O'BRYAN: Phys. Rev. **38**, 32 (1931).

[90] J. E. MACK, J. R. STEHN and B. EDLÉN: J. Opt. Soc. Amer. **22**, 245 (1932).

[91] H. G. BEUTLER: J. Opt. Soc. Amer. **35**, 311 (1945).

[92] R. F. STAMM and J. J. WHALEN: J. Opt. Soc. Amer. **36**, 2 (1946).

[93] C. JANOT et A. HADNI: Polarization de la lumière par les réseaux écheletts dans l'infrarouge lointain. J. Physique **24**, 1073 (1963).

One obtains

$$f_1(x) = \frac{a}{2}\sin\xi\cos\xi + \sum_{n=1}^{\infty} - \frac{2a\sin(n\pi\sin^2\xi)}{n^2\pi^2\sin^2\xi}\sin\left[n\pi\left(\frac{2x}{a} + \sin^2\xi\right)\right]. \tag{71.3}$$

For the "inverted" grating position ($x \to -x$), one has

$$f_2(x) = \frac{a}{2}\sin\xi\cos\xi + \sum_{n=1}^{\infty} \frac{2a\sin(n\pi\sin^2\xi)}{n^2\pi^2\sin^2\xi}\sin\left[n\pi\left(\frac{2x}{a} - \sin^2\xi\right)\right]. \tag{71.4}$$

Next, one proceeds to include the boundary conditions on the surface of the grating.

The amplitude of the incident wave is taken as unity. At a point (x, z) the incident wave is of the form

$$E^i = \exp\left[-\frac{2\pi i}{\lambda}(x\sin i - z\cos i)\right]. \tag{71.5}$$

The diffracted waves, in the order m, parallel to the grooves are $(E_\parallel)_m$, and those perpendicular to the groove length are $(E_\perp)_m$. One has

$$(E_\parallel)^d_m = (E_\parallel)_m \exp\left[-\frac{2\pi i}{\lambda}(x\sin i_m - z\cos i_m)\right] \tag{71.6}$$

and

$$(E_\perp)^d_m = (E_\perp)_m \exp\left[-\frac{2\pi i}{\lambda}(x\sin i_m - z\cos i_m)\right] \tag{71.7}$$

where i_m is the angle of diffraction determined by the equation

$$a(\sin i_m - \sin i) = m\lambda. \tag{71.8}$$

One takes, for the boundary condition, the requirement that the tangential component of the total electromagnetic field be zero at the surface of the grating (case of perfect conductors).

It follows, for $E_\parallel$, that the total field must be equal to zero on the grating surface. For $E_\perp$, it is the scalar product of the total electric field with a vector tangent to the grating surface which must be equal to zero on the grating surface.

With these boundary conditions, one obtains the equations:

$$\left.\begin{aligned} 0 = \sum_{m=-\infty}^{+\infty} (E_\parallel)_m \exp\left[-\frac{2\pi i}{\lambda}\left(x\sin i_m - f(x)\cos i_m\right)\right] - \\ - \exp\left[-\frac{2\pi i}{\lambda}\left(x\sin i - f(x)\cos i\right)\right] \end{aligned}\right\} \tag{71.9}$$

and

$$\left.\begin{aligned} 0 = {} & \cos i \exp\left[-\frac{2\pi i}{\lambda}\left(x\sin i - f(x)\cos i\right)\right] + \\ & + \sum_{m=-\infty}^{+\infty} (E_\perp)_m \cos i_m \exp\left[-\frac{2\pi i}{\lambda}\left(x\sin i_m - f(x)\cos i_m\right)\right] + \\ & + \left\{\sin i \exp\left[-\frac{2\pi i}{\lambda}\left(x\sin i - f(x)\cos i\right)\right] + \right. \\ & + \sum_{m=-\infty}^{+\infty} (E_\perp)_m \sin i_m \exp\left[-\frac{2\pi i}{\lambda}\left(x\sin i_m - f(x)\cos i_m\right)\right]\Bigg\} + \\ & + \left\{\sum_{n=1}^{\infty} - \frac{4\sin(n\pi\sin^2\xi)}{n\pi\sin^2\xi}\cos\left[n\pi\left(\frac{2x}{b} + \sin^2\xi\right)\right]\right\}. \end{aligned}\right\} \tag{71.10}$$

By eliminating the equations of the incident wave from Eq. (71.9) and (71.10) one obtains the equation

$$\left.\begin{aligned}\sum_{m=-\infty}^{+\infty}\Big\{&[(E_\perp)_m\cos i_m+(E_\parallel)_m\cos i]\,g(x)+\\ &+[(E_\perp)_m\sin i_m+(E_\parallel)_m\sin i]\frac{i\lambda}{2\cos i_m}\,\frac{dg(x)}{dx}\Big\}\,e^{-\frac{2\pi i m x}{a}}=0\end{aligned}\right\}\tag{71.11}$$

with

$$\left.\begin{aligned}g(x)=\exp\Big\{&\frac{2\pi i}{\lambda}\cos i_m\frac{a}{2}\sin\xi\cos\xi-\\ &-\frac{2a\sin(\pi\sin^2\xi)}{n^2\pi^2\sin^2\xi}\sin[n\pi(2x/a+\sin^2\xi)]\Big\}\end{aligned}\right\}\tag{71.12}$$

or also

$$\left.\begin{aligned}g(x)=\exp&\left[\frac{2\pi i}{\lambda}\cos i_m\frac{a}{2}\sin\xi\cos\xi\right]\times\\ &\times\prod_{n=1}^{\infty}\exp\left\{-\frac{i\,4a}{\lambda}\cos i_m\frac{\sin(n\pi\sin^2\xi)}{n^2\pi\sin^2\xi}\sin[n\pi(2x/a+\sin^2\xi)]\right\}\end{aligned}\right\}\tag{71.13}$$

where the sign $\prod_{n=1}^{\infty}$ indicates that all the factors on the right-hand side of it must be multiplied, when n takes on all possible values from 1 to infinity.

With the help of the usual Bessel function generating equations (see Sect. 63), one obtains

$$\left.\begin{aligned}g(x)=\exp&\left(\frac{i\pi\alpha}{\lambda}\cos i_m\sin\xi\cos\xi\right)\times\\ &\times\prod_{n=1}^{\infty}\sum_{p=-\infty}^{+\infty}J_p\left[\frac{4a}{\lambda}\cos i_m\frac{\sin(n\pi\sin^2\xi)}{n^2\pi\sin^2\xi}\right]\times\\ &\times\exp\left[-inp\pi\left(\frac{2x}{a}+\sin^2\xi\right)\right].\end{aligned}\right\}\tag{71.14}$$

Finally, according to Janot and Hadni[93] after regrouping the terms which have the same order in the exponential, that is the terms for which $s=\sum_{n=1}^{\infty}np_i$ (with p_i arbitrary) has the same value, one obtains

$$\left.\begin{aligned}g(x)=\exp&\left(\frac{i\pi a}{\lambda}\cos i_m\sin\xi\cos\xi)\right)\times\\ &\times\sum_{s=-\infty}^{+\infty}P_s(m)\exp\{-i\pi s(2x/a+\sin^2\xi)\}\end{aligned}\right\}\tag{71.15}$$

with

$$P_s(m)=\int\prod_{n=1}^{\infty}J_{pi}\left[\frac{4a}{\lambda}\cos i_m\frac{\sin(n\pi\sin^2\xi)}{n^2\pi\sin^2\xi}\right]\tag{71.16}$$

where the integral sign indicates a summation over all terms on the right of it, such as $\sum_{n=2}^{\infty}np_i=s$. It follows that

$$\left.\begin{aligned}\frac{d}{dx}g(x)=\exp&\left[i\frac{\pi b}{\lambda}\cos i_m\cos\xi\sin\xi\right]\times\\ &\times\sum_{s=-\infty}^{+\infty}-\frac{2\pi i s}{a}P_s(m)\exp\left[-i\pi s\left(\frac{2x}{a}+\sin^2\xi\right)\right].\end{aligned}\right\}\tag{71.17}$$

Eq. (71.11) can now be written as

$$\left.\begin{aligned}
&\sum_{m=-\infty}^{+\infty}\Big\{[(E_\perp)_m \cos i_m + (E_\parallel)_m \cos i]\, \exp\Big[-i\,\frac{2\pi m x}{a} + i\,\frac{\pi a}{\lambda}\cos i_m \cos\xi \sin\xi\Big]\times\\
&\qquad\times \sum_{s=-\infty}^{+\infty} P_s(m)\, \exp\Big[-i\pi s\Big(\frac{2x}{a} + \sin^2\xi\Big)\Big]\Big\}\\
&=\sum_{m=-\infty}^{+\infty}\Big\{\frac{i\lambda}{2\cos i_m}\,(E_\perp)_m \sin i_m + (E_\parallel)_m \sin i]\times\\
&\qquad\times \exp\Big[-i\,\frac{2\pi m x}{a} + i\,\frac{\pi a}{\lambda}\cos i_m \cos\xi\sin\xi\Big]\times\\
&\qquad\times \sum_{s=-\infty}^{+\infty} -i\,\frac{2\pi s}{a}\, P_s(m)\exp\Big[-i\pi s\Big(\frac{2x}{a} + \sin^2\xi\Big)\Big]\Big\}.
\end{aligned}\right\}\tag{71.18}$$

In Eq. (71.18), it is also possible to group all terms of the same order with respect to x, under the exponential. Let

$$r = -m - s \quad \text{i.e.} \quad s = -(r+m). \tag{71.19}$$

The identity obtained from (71.18) must hold for all points on the grating, that is, the coefficients of $\exp(2\pi i r x/a)$ must be equal, term for term, in the two members. One has finally the equation

$$\left.\begin{aligned}
&\sum_{m=-\infty}^{+\infty} P_{-(r+m)}(m)\Big\{[(E_\perp)_m \cos i_m + (E_\parallel)_m \cos i] +\\
&\qquad + \frac{\lambda}{a}\,\frac{r+m}{\cos i_m}\,[(E_\perp)_m \sin i_m + (E_\parallel)_m \sin i]\Big\}\times\\
&\qquad\times \exp\Big\{i\pi\Big[\frac{\alpha}{\lambda}\cos\xi\sin\xi\cos i_m + (r+m)\sin^2\xi\Big]\Big\} = 0
\end{aligned}\right\}\tag{71.20}$$

where r can take on all values from $-\infty$ to $+\infty$.

There are as many equations as there are unknowns, in the system of Eqs. (71.20), and the problem can be considered as solved.

VI. Testing of gratings.

72. Introduction. Grating imperfections result from

a) Grating blank and coating imperfections,

b) Ruling imperfections.

The wave-front aberrations resulting from grating imperfections are basically related to the grating imperfections in a manner similar to that characteristic of other optical instruments, lenses, mirrors, prisms, Fabry-Pérot étalon plates and so on.

However, due to the particular nature in which gratings are ruled, and in which replicas are made from ruled masters, the types of wave-front aberrations are in general quite different from those encountered in other optical instruments[1] (see also Sects. 48 to 59).

Moreover, the grating efficiency and energy distribution among orders presents problems which are quite special to grating (see Sect. 65 to 71).

[1] G. W. Stroke: Etude théorique et expérimentale de deux aspects de la diffraction de la lumière par les réseaux optiques: l'évolution des défauts dans les figures de diffraction et l'origine electromagnétique de la répartition entre les ordes. Rev. Opt. **39**, 291—398 (1960).

As a result, grating imperfections, and their effects on the spectral images require special study[1]. Indeed, a complete list of papers related to grating testing would show that most optical physicists, spectroscopists and astronomers who have used or produced gratings have also made contributions to grating testing and theory[2–19].

Even a summary attempt of citing all the contributors to grating testing would not do justice to all the contributions. Certainly though Rowland, Michelson and R. W. Wood had left the deepest mark in the past, both experimentally and theoretically, while Lord Rayleigh's contributions to resolving power theory has long towered as an almost unimpeachable basis for grating evaluation.

More recently, in particular as a result of the work by Stroke[14,1] and the availability of new tools, among which coherent light sources in interferometry and electronic computers, it appeared that interferometric measurement of the wave fronts diffracted by the gratings, and Fourier transform computing has introduced a particular simplicity into grating evaluation. The large numbers of gratings, now in the thousands, has permitted to verify the degree of certainty with which interferometric wave-front testing predicts the performance of gratings, and of replicas, in spectrographs[20].

The theory of the effects of grating ruling and blank imperfections on the spectral images is given in Sects. 44 to 59. The theory of the effect of groove form and grating material on the energy distribution among orders is given in Sects. 60 to 71.

In this chapter, we deal most extensively with interferometric grating testing of spectral quality on one hand, and with grating efficiency measurements and electron-microscopic study of groove forms on the other. These are the methods which are found to be most powerful in grating evaluation.

---

[2] E. Mascart: Traité d'Optique, Vol. I, p. 373. Paris: Gauthier-Villars 1889.

[3] Lord Rayleigh: On the manufacture and theory of diffraction gratings. Scientific Papers, Vol. I, pp. 199—221 (1869—1881). London: Cambridge University Press 1899.

[4] H. S. Allen: Phil. Mag. **3**, 92 (1902); **6**, 559 (1902).

[5] A. A. Michelson: J. Franklin Inst. **181**, 785 (1916).

[6] H. A. Rowland: Gratings in Theory and Practice. Physical Papers, The Johns Hopkins Univ. Baltimore (1902), p. 525. — Phil. Mag. **35**, 397 (1893).

[7] C. M. Sparrow: Theory of Imperfect Gratings. Astrophys. J. **49**, 65 (1919).

[8] R. W. Wood: Phil. Mag. **48**, 497 (1924).

[9] C. F. Meyer: The Diffraction of Light, X-rays, and Material Particles, publ. by J. W. Edwards, Ann Arbor, Michigan 1948.

[10] A. M. Camus, M. Françon, E. Ingelstam, and A. Maréchal: Rev. Opt. **30**, 121 (1951).

[11] J. Strong: New Johns Hopkins Ruling Engine. J. Opt. Soc. Amer. **41**, 3 (1951).

[12] H. D. Babcock: Bright Diffraction Gratings. J. Opt. Soc. Amer. **34**, 1 (1944).

[13] H. G. Gale: Astrophys. J. **86**, 437 (1937).

[14] G. W. Stroke: Interferometric measurement of wave-front aberrations in gratings and echelles. J. Opt. Soc. Amer. **45**, 30 (1955).

[15] G. R. Harrison, and G. W. Stroke: Attainment of high resolution with diffraction gratings and echelles. J. Opt. Soc. Amer. **50**, 1153 (1960).

[16] G. R. Harrison: The Production of Diffraction Gratings. I. Development of the Ruling Art. J. Opt. Soc. Amer. **39**, 413 (1949).

[17] J. Strong: The Johns Hopkins University and Diffraction Gratings. J. Opt. Soc. Amer. **50**, 1148 (1960).

[18] Cumulative Index, Vol. 1—40, Journal of the Optical Society of America. J. Opt. Soc. Amer. **44**, 910—1128 (1954).

[19] G. W. Stroke: Ruling, Testing and Use of Optical Gratings for High-Resolution Spectroscopy. In: Progress in Optics, Vol. II, edit. by E. Wolf, pp. 1—72. Amsterdam: North-Holland Publ. Co. 1963.

[20] R. F. Jarrell, and G. W. Stroke: Some new advances in grating ruling, replication and testing. Applied Optics **3**, 1251—1262 (1964).

Grating resolving power tests and phase-contrast measurements of wave fronts are also discussed.

73. Resolving power tests and their uncertainties. The distribution of intensity $I(\sigma)_{grating}$ in the spectral diffraction patterns of the grating enters naturally, and fundamentally, into the image formation by gratings. One recalls (Sects. 46 and 47) that if $I(\sigma)_{spectrum}$ describes the "true" intensity distribution in the spectrum of the light source, and if $I(\sigma)_{grating}$ describes the "impulse" response of the grating spectrograph; when it is illuminated by a perfectly monochromatic plane wave, then the intensity distribution $I(\sigma)_{recorded}$ which is recorded is given by the convolution, described by the equation

$$I(\sigma)_{recorded} = I(\sigma)_{spectrum} \otimes I(\sigma)_{grating}. \tag{73.1}$$

In practice, however, direct experimental measurement of the intensity distribution in the "perfectly monochromatic" diffraction patterns $I(\sigma)_{grating}$, for a given grating, presents frequently unsurmountable difficulties. These difficulties result principally from the small width of the diffraction patterns, even with long focal lengths of the spectrographs.

The comparatively large width of the spectral lines used for testing before the introduction of optical masers (lasers) was also a basic limitation[1], since the line-width was generally of the same order or even larger than the width of the diffraction patterns.

For instance, the first minimum in the diffraction pattern of a 254 mm wide grating used in autocollimation at 64° is at only 1 second of arc from the line center in the visible. Even, in a 12-meter spectrograph, this corresponds to only 6×10^{-2} mm at about 5460 Å, and to about 6×10^{-3} Å which is roughly the Doppler width of the mercury green line at room temperature. (The full width at half intensity, due to the Doppler effect is approximately $\Delta\lambda_D \cong 7.14\times10^{-7}\sqrt{\frac{T}{M}}\,\lambda$, where T is the absolute temperature in °K and M the mass of the atom). Typically, the line-width is some 100 times smaller than the distance from the line center of the first Rowland "ghost". Rowland ghosts, resulting from periodic errors, were frequently found in gratings before the introduction of interferometric servo-control in grating ruling by HARRISON[15, 20–22].

In an attempt to overcome the difficulties in measuring the line shape in the diffraction patterns $I(\sigma)_{grating}$ of a single line, it has been a general practice to try to separate or *resolve* two neighbouring spectral lines (or their artificial equivalent, STROKE[1], p. 302) of same intensity. The theoretical criterion used, according to Lord RAYLEIGH[23, 24] was a dip in intensity of approximately 20 per cent which was to be observed between two spectral lines, when they were separated by a distance equal to the distance of the first diffraction-pattern minimum from the line center (Fig. 97). The distance of the minimum from the line center, for a perfect grating, depends only on the width of the grating, and on its angles of use (see Sects. 20 and 45). The resolving power is independent of the grating spacing constant.

---

[21] G. R. HARRISON, and G. W. STROKE: Interferometric control of grating ruling with continuous carriage advance. J. Opt. Soc. Amer. **45**, 112 (1955).

[22] H. W. BABCOCK: Control of a Ruling Engine by a Modulated Interferometer. Applied Optics **1**, 415 (1962).

[23] Lord RAYLEIGH: Wave Theory. Encycl. Britt., 9th edit., Vol. 24 (1888).

[24] R. W. DITCHBURN: Proc. Roy. Irish. Acad. A **39**, 58 (1930).

Modern line-shape and resolving power theory, based on Fourier transform theory[25, 26] and on the theory of distributions[25, 27] has added a great deal of precision to the measuring of line shapes and spectral resolution.

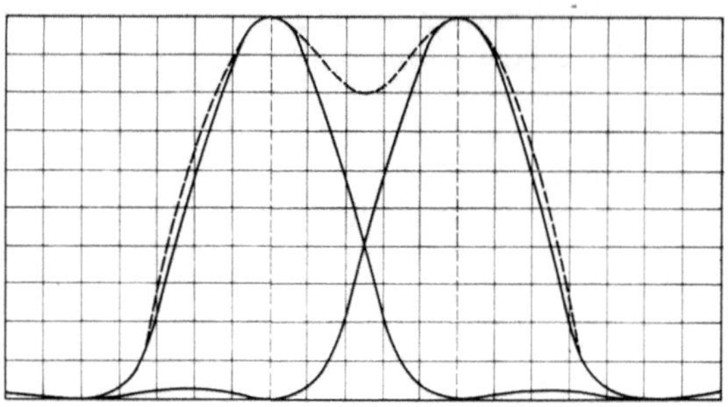

Fig. 97. Energy distribution at the Rayleigh limit for two similar images. (Full lines show distributions due to the separate sources; the dotted line shows the distribution when both act together) (according to DITCHBURN, "Light", John Wiley, New York, 1963).

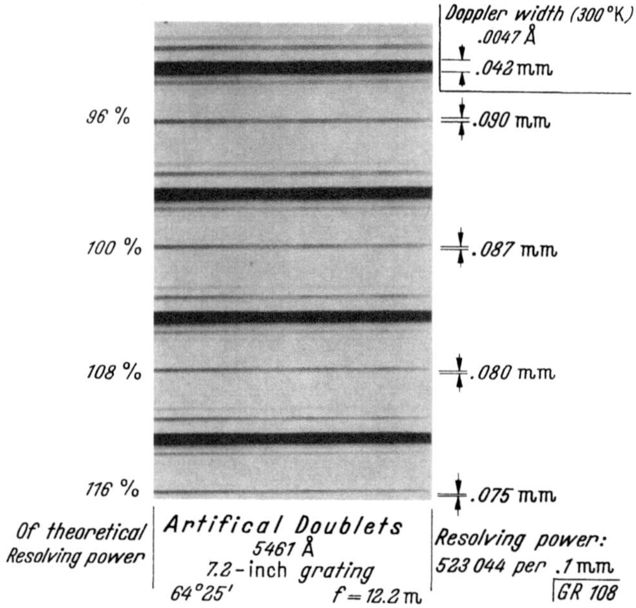

Fig. 98.

Unfortunately though, in terms of usefulness of resolving power tests, it can be shown that even important grating imperfections have very little influence on the ability of "separating" two spectral lines under these conditions[1]. This can be easily understood.

Indeed, even if there is an appreciable widening of the diffraction pattern, as a result of wave-front aberrations, it is still possible to show two lines as "re-

[25] J. ARSAC: Transformations de Fourier et Théorie des Distributions. Paris: Dunod 1961.

[26] A. A. KHARKEVICH: Spectra and Analysis (translated from Russian). New York: Consultants Bureau 1960.

[27] L. SCHWARTZ: Théorie des Distributions, Vol. I and II. Paris: Hermann & Cie., publ. 1950 and 1951.

solved", "in the limit", provided that suitable skill is exerted in the photography and that photographic non-linearities are properly exploited[1].

It is a fact, also, that spectroscopic lines to be resolved, in high-resolution use of gratings[28] in general have rather important intensity ratios, for instance in

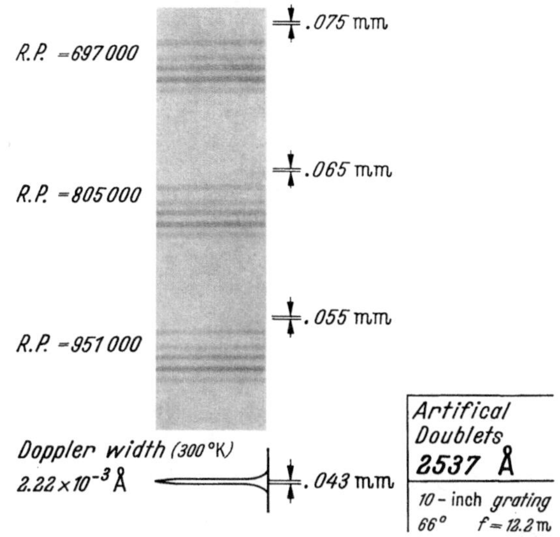

Fig. 99.

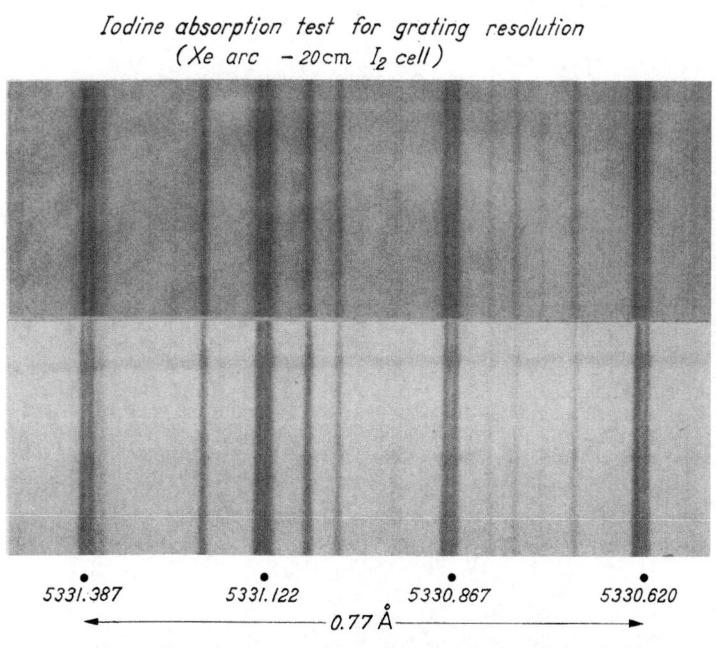

Fig. 100. Iodine absorption test for grating resolution [according to H. D. BABCOCK and H. W. BABCOCK, J. Opt. Soc. Amer. **41**, 776 (1951)] for two high-resolution gratings (Photographs courtesy S. P. DAVIS).

[28] G. W. STROKE: Theory, Production and Use of Optical Gratings for High-Resolution Spectroscopy. In: Quantum Electronics III, edit. by P. GRIVET and N. BLOEMBERGEN, Vol. II, Chap. 7, p. 1221. New York: Columbia Univ. Press 1964.

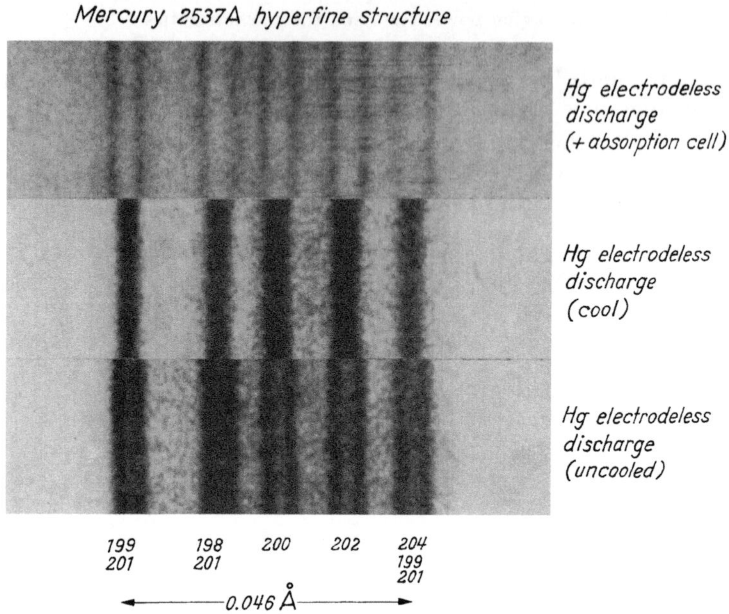

Fig. 101. Effect of source conditions on resolving power tests according to HARRISON et al. [J. Opt. Soc. Amer. **49**, 205 (1959)].

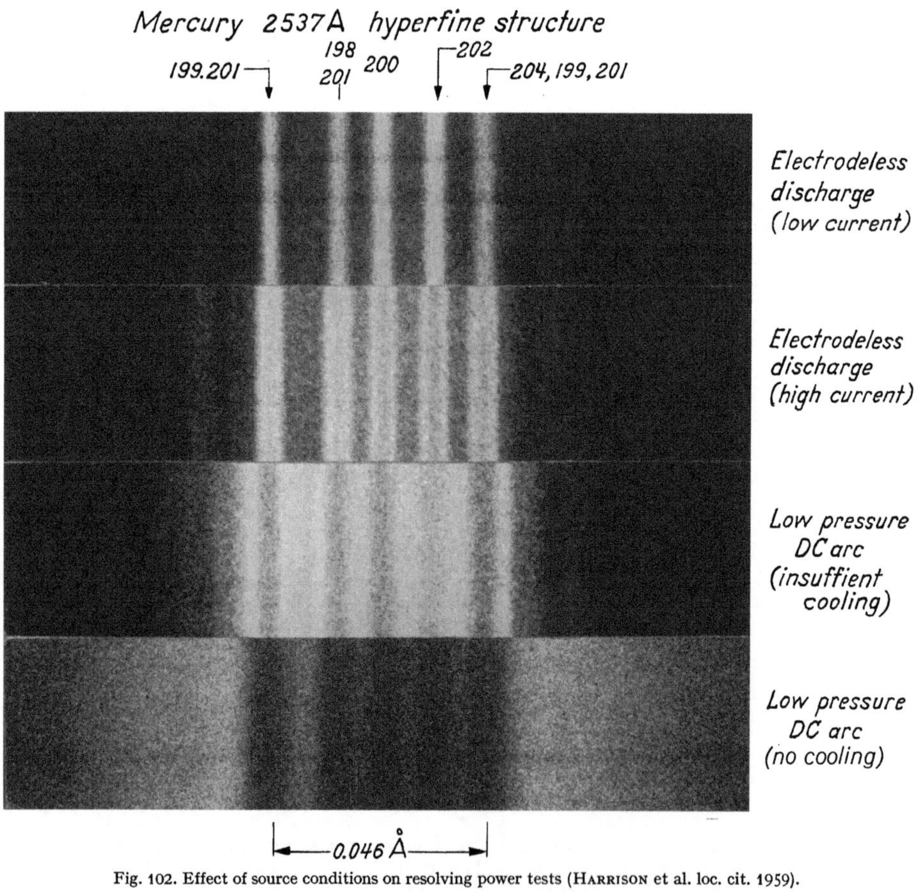

Fig. 102. Effect of source conditions on resolving power tests (HARRISON et al. loc. cit. 1959).

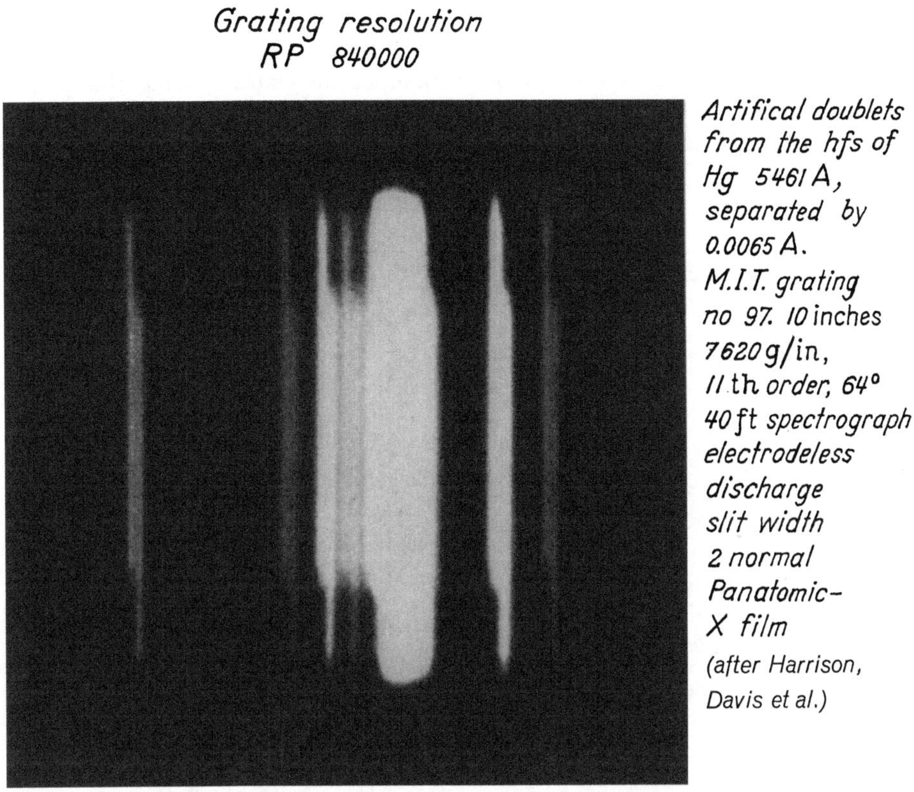

Fig. 103. Artificial doublets for resolving power testing according to J. Strong [J. Opt. Soc. Amer. **41**, 3 (1951)]. The doublets are produced with the help of a Rochon prism (Photograph courtesy Harrison et al. 1959).

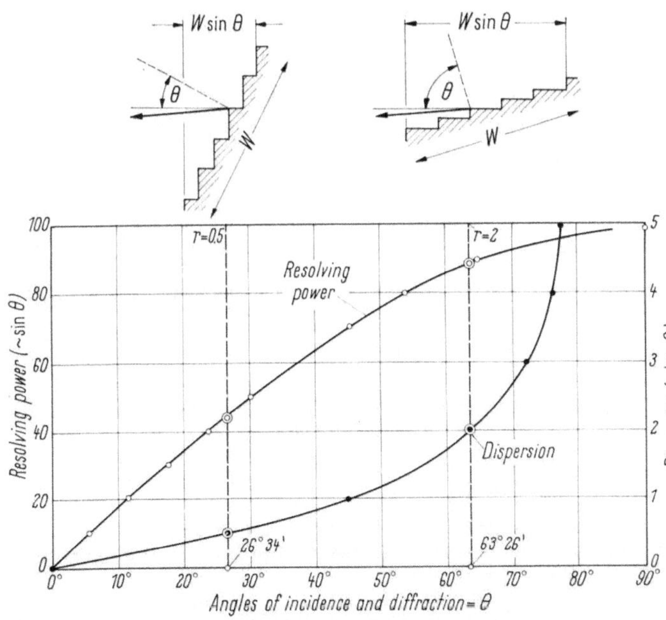

Fig. 104. Variation of resolving power and dispersion as a function of grating angle (according to Harrison and Stroke, 1960).

studies of nuclear charge and magnetization from atomic hyperfine structure[29,30]. As a result, resolving power tests performed with equal intensity lines are generally found to be of limited value in practice for characterizing the general quality of a given grating. Important contributions to the quantitative measurement of resolving power are given by D. H. Rank and associates*.

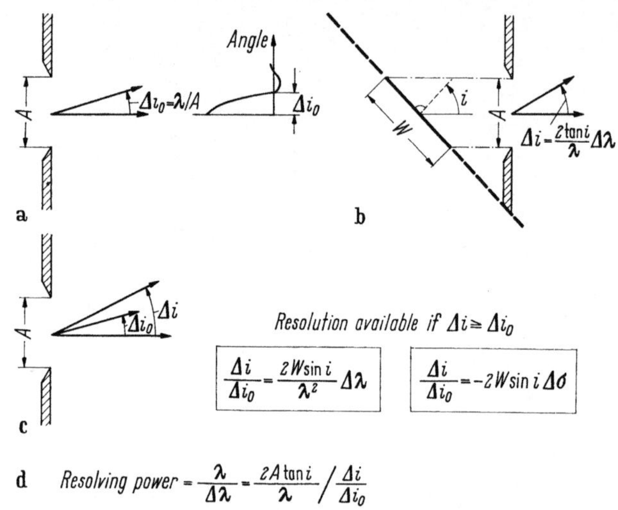

Fig. 105 a—d. Significance of "resolving power" in terms of width of diffraction pattern and grating dispersion.

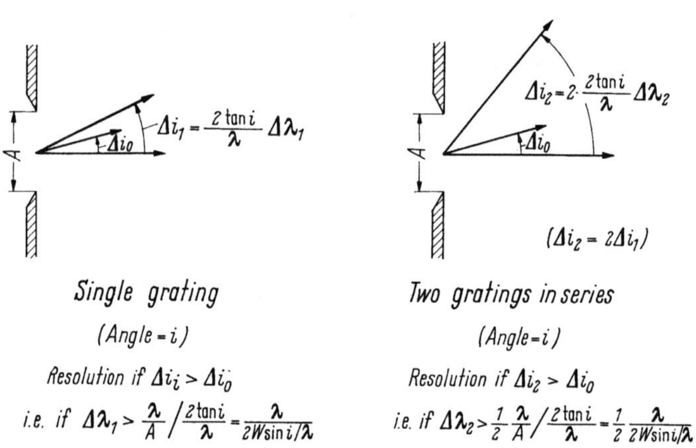

Fig. 106. Increase of dispersion and of resolving power by use of two gratings in series (or by double-passing) (see Fig. 107).

Fig. 98 and 99 show according to Stroke[1] some artificial doublets produced with lines from natural mercury. For the green line, the Doppler width is small compared to the diffraction pattern, and good resolution is easily obtained. For

---

* D. H. Rank, J. N. Shearer, and J. M. Bennett: Quantitative Method for Measuring the Resolution of a Large Grating. J. Opt. Soc. Amer. **45**, 762—766 (1955). See also, D. H. Rank et al. J. Opt. Soc. Amer. **47**, 631 (1957).

[29] H. H. Stroke: Distribution of Nuclear Charge and Magnetization from Atomic Hyperfine Structure. In: Nuclear Physics, Course 23, (director V. Weisskopf), pp. 156—166. New York: Academic Press 1963.

[30] R. J. Hull, and H. H. Stroke: $3\,S_1$ Hyperfine-Structure Separations and Isotope Shifts of Radioactive Cd^{107}, Cd^{109}, and of Stable Cd Isotopes. J. Opt. Soc. Amer. **53**, 1147 (1963).

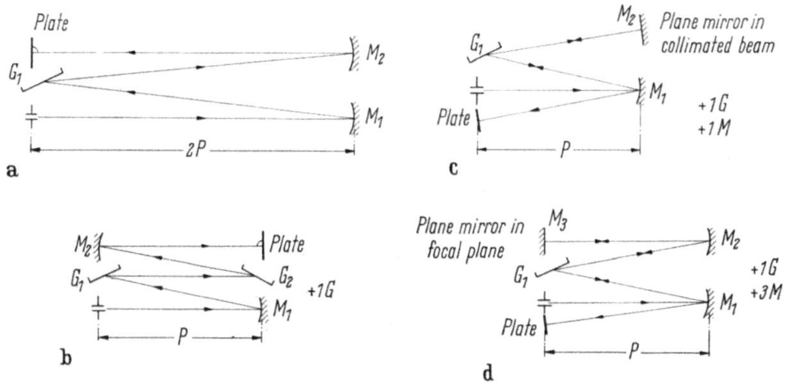

Fig. 107. Doubling of resolving power by double-passing a single grating, or by using two gratings in series. All systems shown have the same *linear* dispersion. Systems (b), (c) and (d) all have twice the resolving power of the conventional two-mirror system (a). — (b) Two gratings in series [according to G. W. STROKE and H. H. STROKE, J. Opt. Soc. Amer. **53**, 333 (1963)]. — (c) Double-pass system [according to E. HULTHÉN and E. LIND, Akr. Fysik **2**, 253 (1950)]. — (d) Double-pass system [according to F. A. JENKINS and L. W. ALVAREZ, J. Opt. Soc. Amer. **42**, 699 (1952); W. G. FASTIE and W. M. SINTON, J. Opt. Soc. Amer. **42**, 483 (1952); D. H. RANK and T. A. WIGGINS, J. Opt. Soc. Amer. **42**, 983 (1952); J. N. SHEARER, T. A. WIGGINS, A. H. GUENTHER, and D. H. RANK, J. Chem. Phys. **25**, 724 (1956); D. H. RANK et al., J. Opt. Soc. Amer. **47**, 631 (1957); E. HULTHÉN and H. NEUHAUS, Ark. Fysik **8**, 343 (1954); O. BJORKLUND, Ark. Fysik **13**, 185 (1958)]. — See also: F. L. O. WADSWORTH, Astrophys. J. **2**, 343 (1954); A. COUDERC, J. Phys. Radium **8**, No. 7, p. 99 S (July 1937); W. V. HOUSTON, Phys. Rev. **29**, 478 (1927); G. KIRCHHOFF, Abh. Kgl. Akad. Wiss. Berlin, Phys. Abh. (1861), p. 63; G. HANSEN, Phys. Z. **43**, 213 (1942); A. WALSH, J. Opt. Soc. Amer. **42**, 94 (1952); P. JAEGLÉ, C. R. Acad. Sci. Paris **245**, 1412 (1957); J. W. EVANS and J. WADDELL, Appl. Optics **1**, 111 (1962); W. E. MITCHELL jr. and O. C. MOHLER, Appl. Optics **3**, 467 (1964).

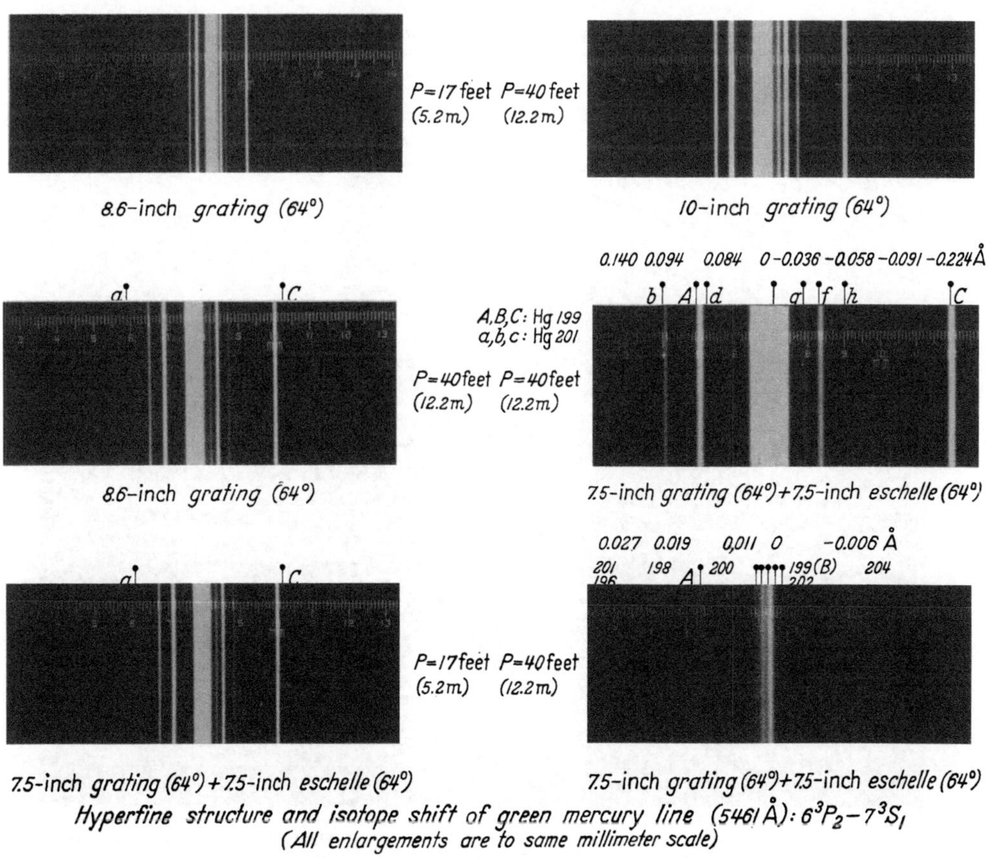

Fig. 108. High-resolution spectrograms of green mercury line (according to G. W. STROKE and H. H. STROKE, 1963).

the UV line, the resolution is limited by the spectral line-width. Other resolution tests are shown in Figs. 100—103. Theoretical considerations of resolving power and dispersion in spectrographic arrangements are given in Figs. 104—107. Examples of good high-resolution spectra obtained in spectrometers and spectro-

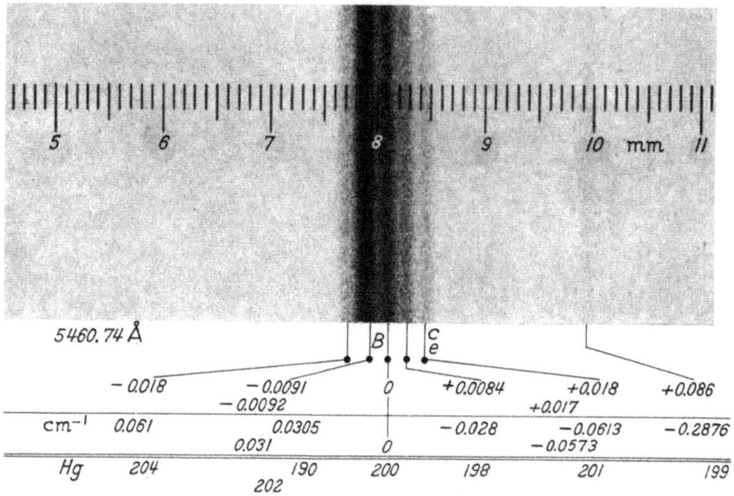

Fig. 109. Central components of green mercury line photographed with resolving power of 1.4 million (according to G. W. Stroke and H. H. Stroke, see Fig. 108, lower right).

Fig. 110. From left to right: entrance slit, 10-inch grating (M.I.T. 97) and plate holder used in obtaining spectra of Fig. 108 and 109. The vacuum chamber may be used for pressure scanning.

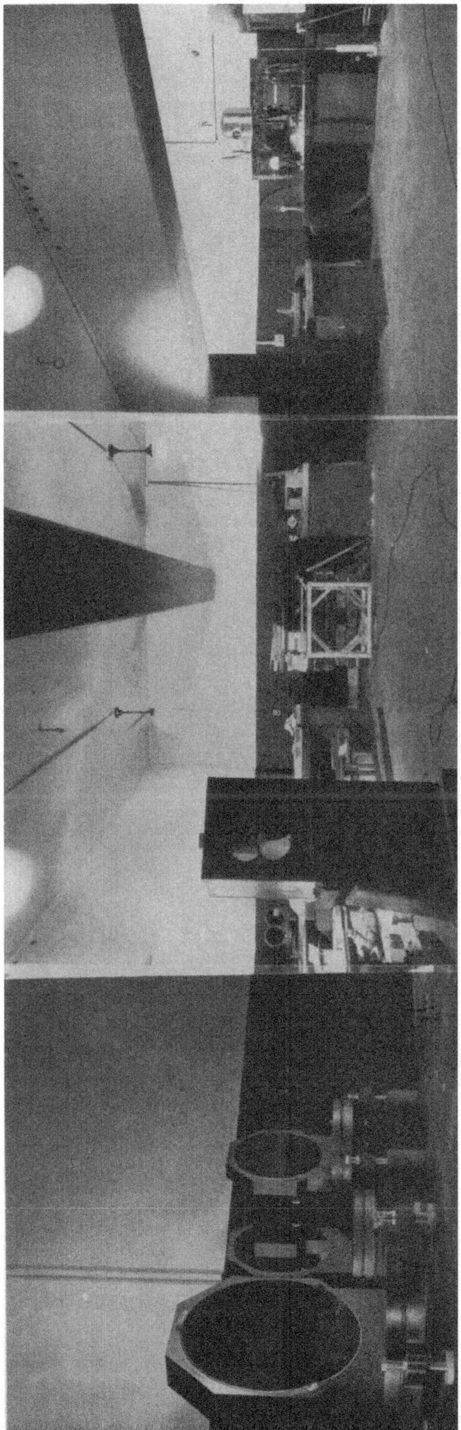

Fig. 111. Overall view of 35-foot room in M.I.T. Spectroscopy Laboratory in which spectra of Fig. 108 and 109 were obtained.

graphs, together with examples of spectrographic arrangements and instruments, using gratings, are shown in Figs. 108—134.

Fig. 112. From left to right: camera mirror, collimator and additional camera (the latter to help in photographing two widely separated hyperfine structure spectrograms next to each other on the plate, according to H. H. STROKE and K. K. Y. LI, J. Opt. Soc. Amer. **51**, 678 (1961)]. The collimator and camera mirror were used to photograph the spectra of Fig. 108 and 109.

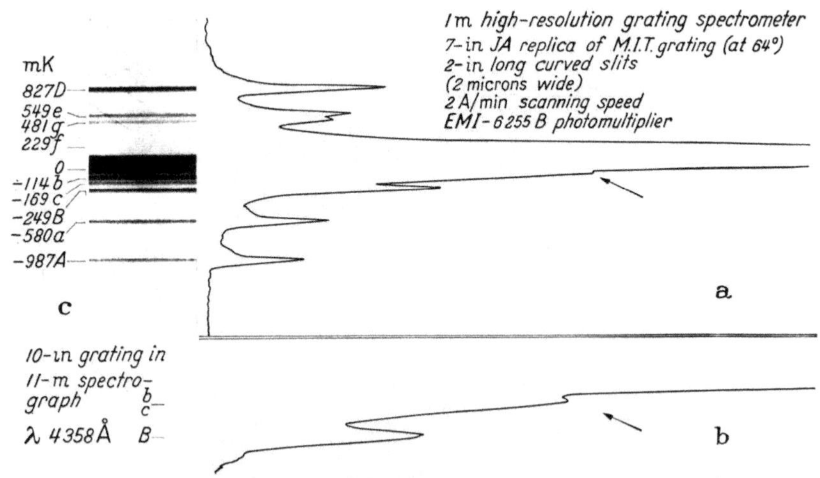

Fig. 113a—c. High-resolution spectrogram of mercury blue line photographed with resolving power of 1 million (according to STROKE, 1961). Photoelectric recording of same line in early version of high-resolution spectrometer. (Compare with recording obtained with a replica of the same M.I.T. grating in a more recent spectrometer in Fig. 115). [Spectrometers designed by R. K. BREMM according to the description by H. M. CROSSWHITE and W. G. FASTIE, J. Opt. Soc. Amer. **46**, 110 (1956); W. G. FASTIE, J. Opt. Soc. Amer. **42**, 641, 647 (1952).]

74. Interferometric testing of gratings. Interferometric testing of gratings, according to a method first described by STROKE[31] in 1955, has proven to be the simplest and most revealing test for grating ruling quality, in practice. In this test, the grating is placed, at its highest angle of use, in one of the arms of a Michelson Twyman Green interferometer (Fig. 135 on p. 631) and the resulting interference fringe pattern shown is simply photographed with the help of the camera shown (Fig. 136 to 139).

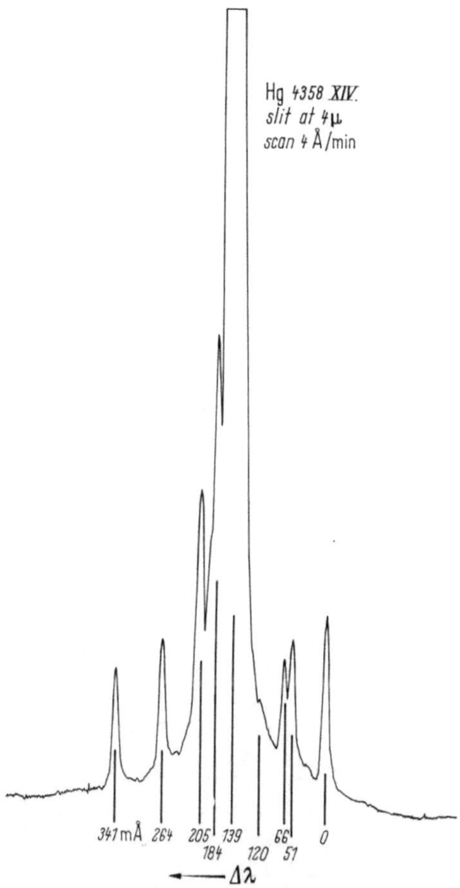

Fig. 114*. Photoelectric recording of hyperfine structure of blue mercury line 4358 A in 1-meter high-resolution spectrometer at resolving power of approximately 760000 (see note in Fig. 113). (185-mm grating ruled at M.I.T. in 1961, and replicated at the Jarrell-Ash Co. Grating angle $n \simeq 66°$.)

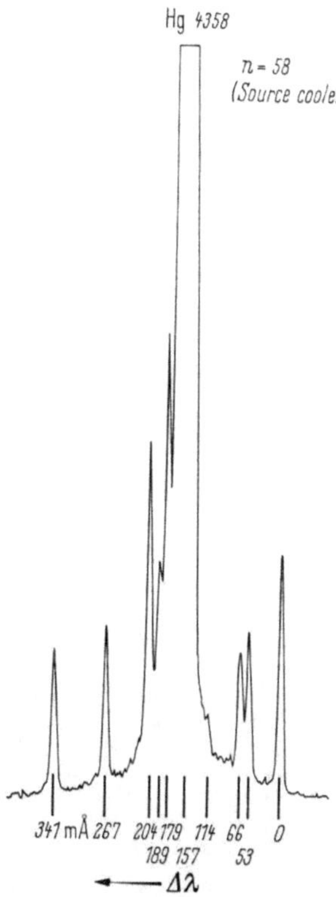

Fig. 115. Photoelectric recording of blue mercury line 4358 A in a 5-meter spectrometer at a resolving power of 1 million, with a replica of a 10-inch, 73 grooves/mm grating, used at about 66°. (Spectrometer and replica by Jarrell-Ash.) n = order.

The departures from linearity in the straight-line fringe system obtained are directly proportional to the departure from planeness of the phase in the diffracted wave front. The gratings are placed in the interferometer so that the mean fringe pattern is normal to the groove length. Lateral groove displacements on the grating, along the grating width, produce fringe displacements in a direction parallel to the groove length, in corresponding regions in the interferogram. 1 fringe of groove displacement produces $\sin i$ fringes of displacement in the interferogram, when the grating is tested in autocollimation at the angle i (Sect. 48). The wave-front phase aberration is obtained directly, in readily usable graphical form

* Figs. 114—134b (on p. 621—631) belong to Sect. 73.

[31] G. W. STROKE: Interferometric Measurement of Wave-Front Aberrations in Gratings and Echelles. J. Opt. Soc. Amer. **45**, 30—35 (1955).

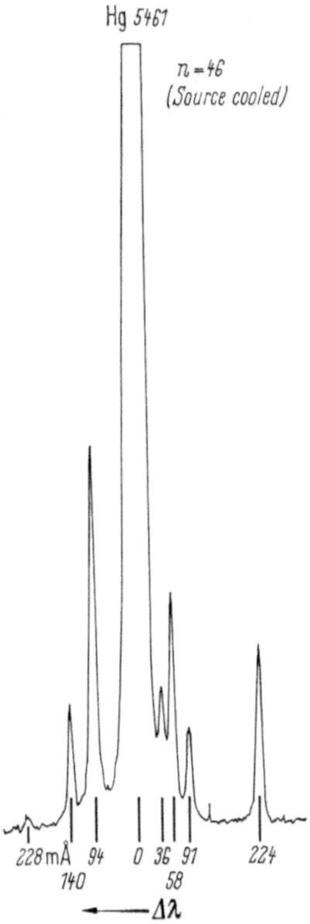

Fig. 116.

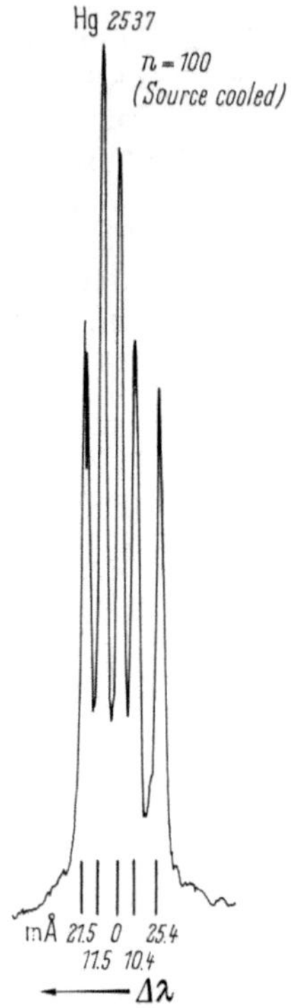

Fig. 117.

Fig. 116. Photoelectric recording of green mercury line 5461 A in a 5-meter spectrometer at a resolving power of 845000, with a replica of a 10-inch, 73 grooves/mm grating, used at about 64°. (Spectrometer and replica by Jarrell-Ash.) n = order.

Fig. 117. Photoelectric recording of ultraviolet mercury line 2537 A in a 5-meter spectrometer at a resolving power of approximately 1.8 million, with a replica of a 10-inch, 73 grooves/mm grating, used at about 65°. (Spectrometer and replica by Jarrell-Ash.) n = order.

Fig. 118. Sine-drive mechanism used for grating rotation in high-resolution spectrometers (courtesy Jarrell-Ash). See Fig. 113 to 117. The drive mechanism for a 5-meter spectrograph is shown (U.S. patent No. 3011391). Optical diagram of sine-drive system in Fig. 119.

Fig. 118.

in the wave-front interferograms, and can be used to immediately assess the grating quality. The same interferogram can also be used to compute the complex amplitude distribution and intensity in the diffraction patterns, corresponding to infinitely monochromatic spectral lines. The computation, according to STROKE

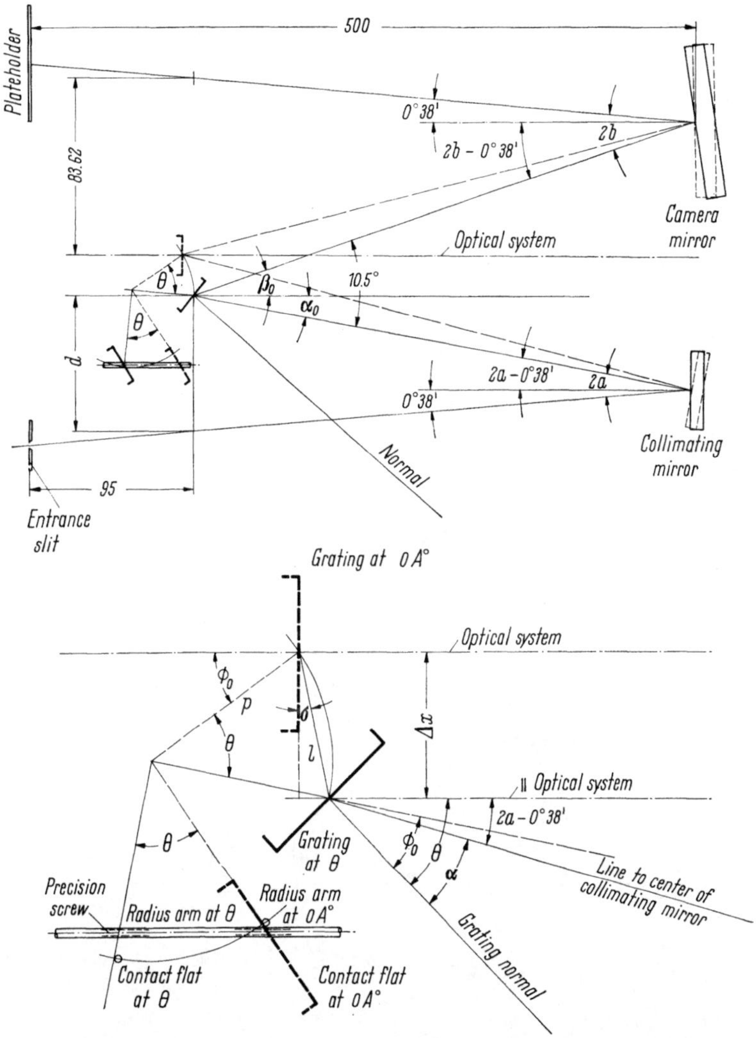

Fig. 119. Optical diagram sine-drive mechanism as used in the 5-meter high-resolution spectrograph (see Fig. 118) (courtesy R. K. BREHM, Jarrell-Ash).

(loc. cit.[1]) is based on taking the Fourier transform of the phase distribution in the wavefront. The phase distribution is directly displayed in graphical form in the interferograms. It might be of interest to recall some of the interferometric work that has preceded the interferometric grating testing method in the form described by STROKE (loc. cit.[1]).

MICHELSON[32], himself, had already considered using his interferometer for grating testing (Fig. 140). Unfortunately though, unlike in the Twyman-Green*

* Text to continue on p. 631.

[32] A. A. MICHELSON: Studies in Optics. Chicago: Chicago University Press 1927, reprinted (1962) as Phoenix Edition, p. 77.

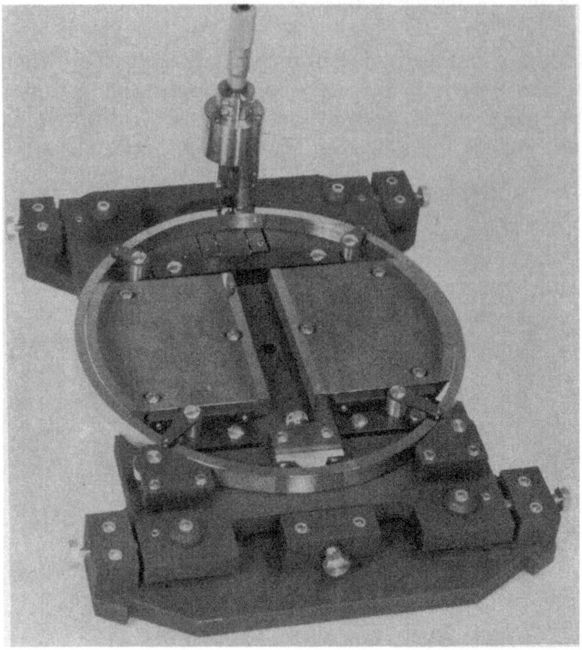

Fig. 120. Ring-slit assembly (according to Crosswhite and Fastie, loc. cit. in Fig. 113). The matching 8½ inch diameter jaws are ground from hardened steel. The steel ring is supported at the bottom by two ball bearings at approximately 45° on either side of the vertical center line. Pressure applied to the top of the ring causes the ring to deform symmetrically about its center, as the supports are located at the nodal points. Slit assemblies like the one shown were used to obtain the recordings in Fig. 113 to 117 (courtesy Jarrell-Ash). U.S. Patent No. 2914987 (Fastie-Crosswhite), licensee Jarrell-Ash Co.

Fig. 121. Grating holder of the type used in high-resolution scanning spectrometers (courtesy Jarrell-Ash).

Fig. 120.

Fig. 121.

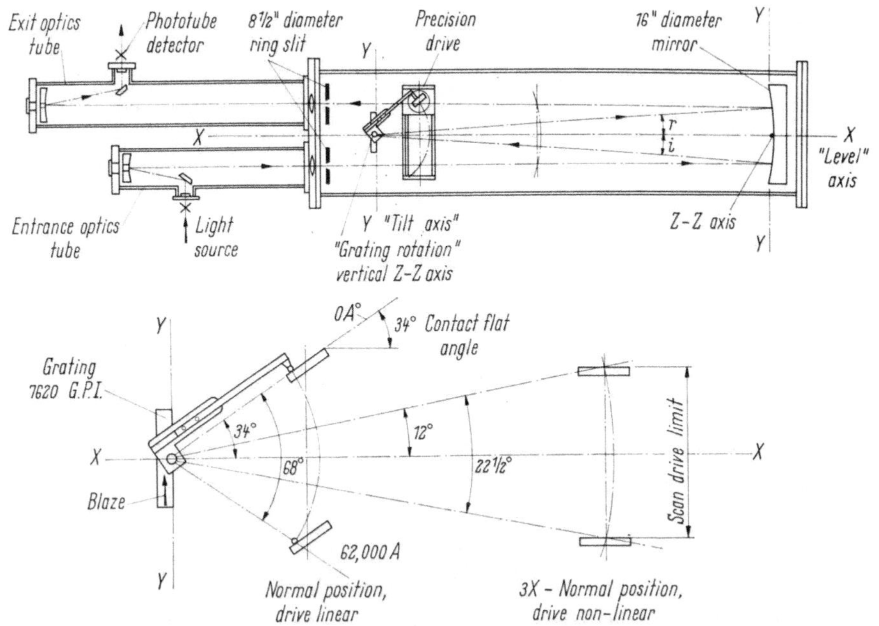

Fig. 122. Schematic layout of high-resolution spectrometers, such as those used to obtain recordings of Fig. 113 to 117 (courtesy Jarrell-Ash).

Fig. 123. 35-foot (10.7 meter) grating spectrograph for the vacuum ultraviolet [designed by R. K. Brehm, Jarrell-Ash Co. according to the description by A. E. Douglas and J. G. Potter, Appl. Optics 1, 727 (1962)].

Fig. 124. Spectroheliograph using two high-resolution gratings in series, as installed at Solar Research Observatory, Sunspot, N. M. According to the suggestion by J. W. Evans, the two gratings are so arranged as to obtain locally undispersed, but widely separated images simultaneously for three spectral lines in the exit cameras ($H\alpha$, $D3$ and K lines, or alternately Fe 5124, Fe 5250 and an other line). The entire spectroheliograph is made to float on hydrostatic ways. Focal length = 2 meters. (Spectroheliograph designed by G. W. Stroke, constructed by Jarrell-Ash, 1956—1959). See also Figs. 125 to 133.

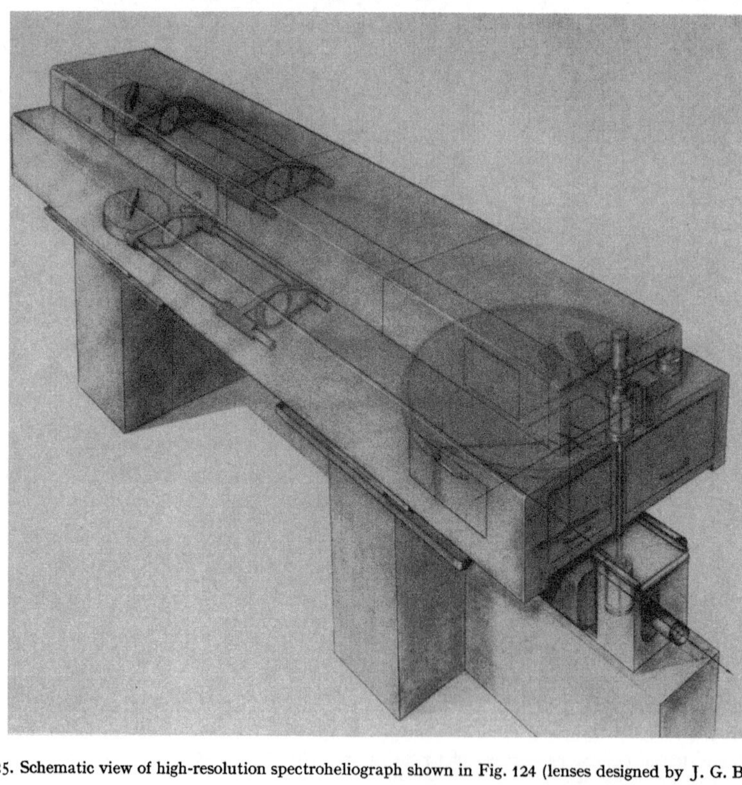

Fig. 125. Schematic view of high-resolution spectroheliograph shown in Fig. 124 (lenses designed by J. G. Baker).

Fig. 126. View of upper Baker-Schmidt collimator and grating, as used in spectroheliograph of Fig. 124. Both gratings are used in the vicinity of 55° near autocollimation.

Fig. 127. Intermediate-slit section of high-resolution spectroheliograph shown in Fig. 124. Three carefully diamond-polished curved slits are used to filter out the wavelengths $H\alpha$, $D3$ and K. Alternately, by rotation of the slit-carrying platform shown, three other slits can be brought, all adjusted, into the intermediary plane.

Fig. 128.

Fig. 129.

Fig. 130.

Fig. 128. Intermediary slits of spectroheliograph shown in Fig. 124. Three 12.5 cm long curved slits are made to be in the same plane, supported on spring mounths. Each slit can be opened to 3 mm, and be placed, independently of the other slits anywhere within ± 3 mm from the center of the spectral line, for the study of Doppler-displaced Solar lines. Slit positions scanning speeds of spectroheliograph and timings can all be preset automatically, for an entire day operation if desired. (Photograph courtesy H. J. Smith.)

Fig. 129. Variable curvature 12.5 cm long slit as designed for spectroheliograph of Fig. 124. All other slits have a fixed curvature (radius of the order of 28.5 inches). The slit curvatures and parallelism of opening (to 3 mm), as well as slit displacement to ± 3 mm are all matched to the order of about 1.5 microns.

Fig. 130. Detail of intermediate-slit spring mount used in spectroheliograph of Fig. 124. Fine tuning of parallelism is achieved by choosing springs of slightly different thickness. All springs are mounted on hardened steel bushings.

Calcium *K*

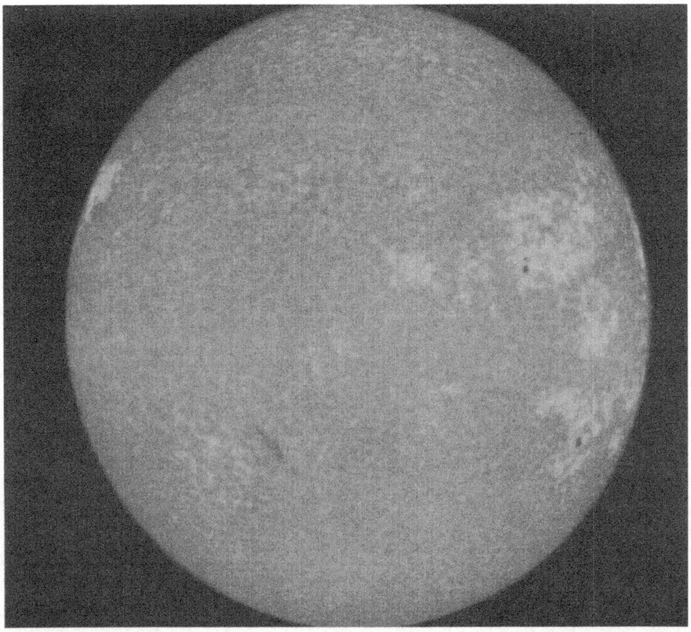

Hydrogen Hα

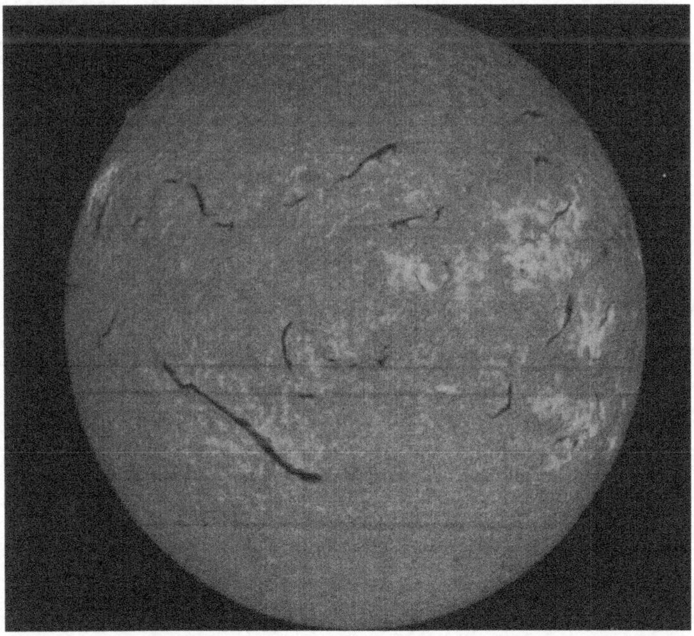

Spectroheliograms June 1963

Fig. 131. Spectroheliograms in calcium *K* line (top) and in hydrogen H α line (bottom) obtained *simultaneously* with spectroheliograph of Fig. 124 in June 1963, on center of lines, with 0.05 A bandpass. Horizontal streaks are due to some dust particles on entrance slit. (Spectroheliograms courtesy J. W. Evans, Solar Research Observatory, Geophysics Research Directorate, AFCRL, Sunspot, N.M.)

Fig. 132. High-resolution spectroheligram in hydrogen H α light obtained on line center, with 0.06 A bandpass, on 14 April 1963 with spectroheliograph of Fig. 124. (Spectroheliogram courtesy J. W. Evans, Solar Research Observatory, Geophysics Research Directorate, AFCRL, Sunspot, N.M.)

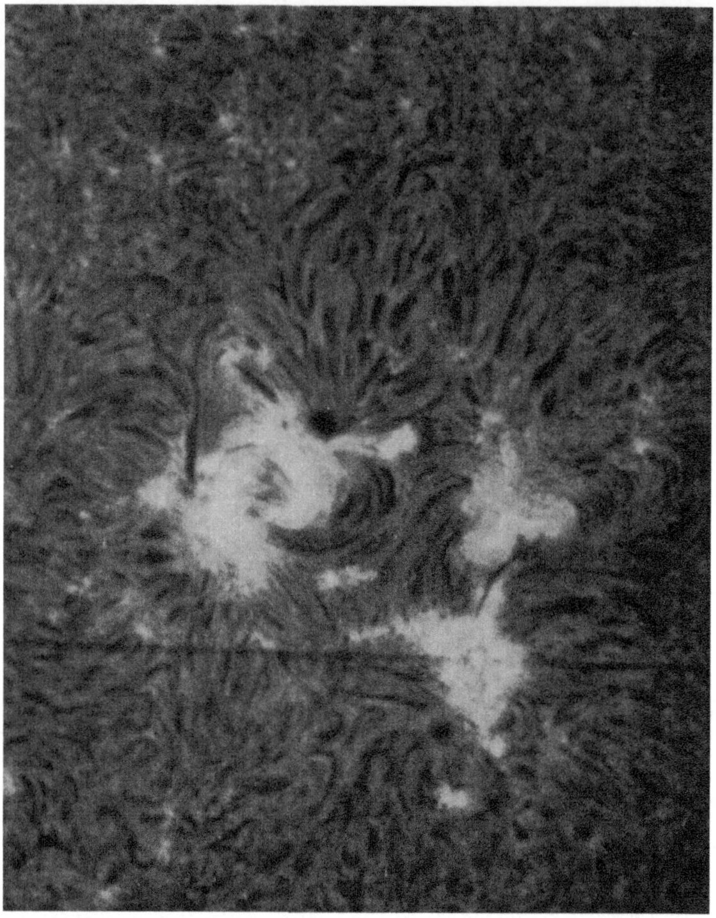

Fig. 133. Magnified section of spectroheliogram of Fig. 132.

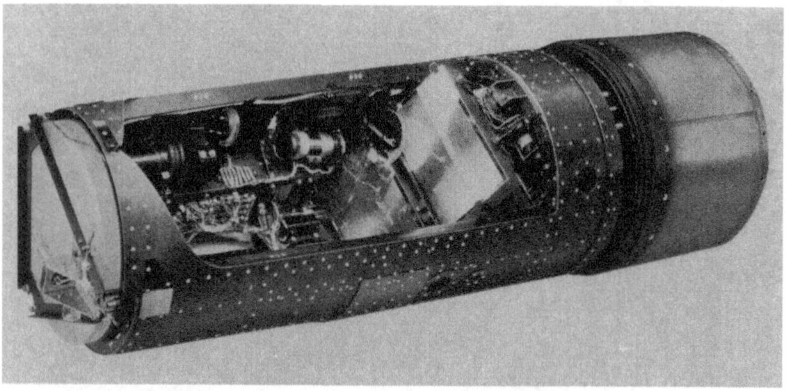

Fig. 134a. Objective grating rocket spectrometer for use in the 900 to 3200 Å spectral range on an *A* or *B* type star. The resolving power is about 2 Å. Each of the two 8×4 inch gratings is mounted on a two-axis gimbal ring, so that it may be, pointed within about ±6 seconds of arc, as a part of a fine guidance servo-system (see schematic diagram in Fig. 135). (Photograph courtesy R. H. Noble, Perkin-Elmer Corp.)

version of MICHELSON's interferometer where a plane diffracted wave front will give rise to a straight-line fringe system, as described by STROKE (loc. cit. [35]), MICHELSON's interferometer uses no collimators and produces a hard-to-interpret circular ring system with a perfect grating. As a result, MICHELSON's interfero-

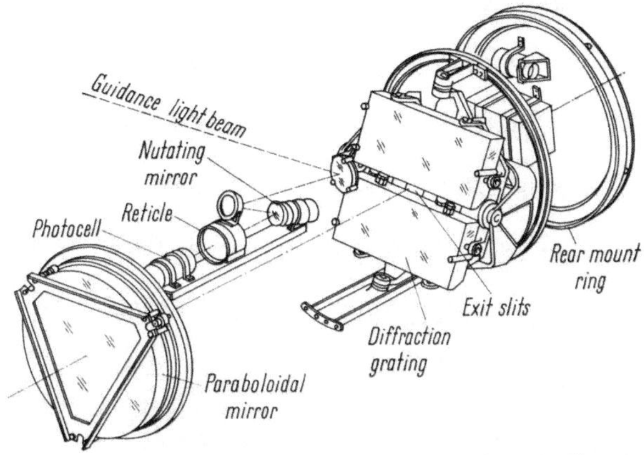

Fig. 134b. Schematic diagram of objective grating rocket spectrometer shown in Fig. 134a (Diagram, courtesy R. H. NOBLE, Perkin-Elmer Corp.)

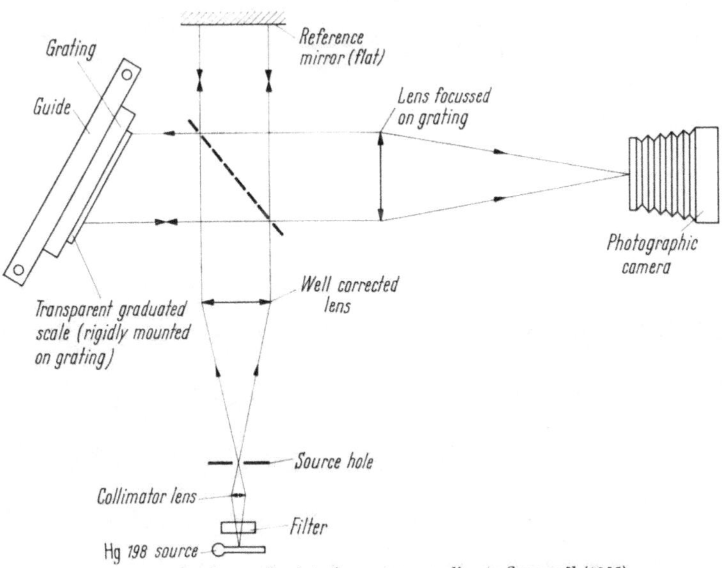

Fig. 135. Grating testing interferometer according to STROKE [31] (1955).

meter did not seem to find a great favor for grating testing, although gratings had been placed into interferometers for other purposes, and qualitative descriptions of the use of interferometers for grating testing had been given on more than one occasion between MICHELSON's times and 1955. For instance, WILLIAMS [33] describes the use of a Twyman Green interferometer for similar tests, giving a sketch of the fringe pattern for an echelon, but without going into the details of fringe location, locus described by the fringes, or the relation between wave-front aberrations and groove position. RAMSAY [34] describes an interferometer

[33] W. E. WILLIAMS: Studies in Interferometry II. Proc. Phys. Soc. Lond. **45**, 699 (1933).
[34] B. P. RAMSAY: Grating Interferometers. J. Opt. Soc. Amer. **24**, 253 (1934).

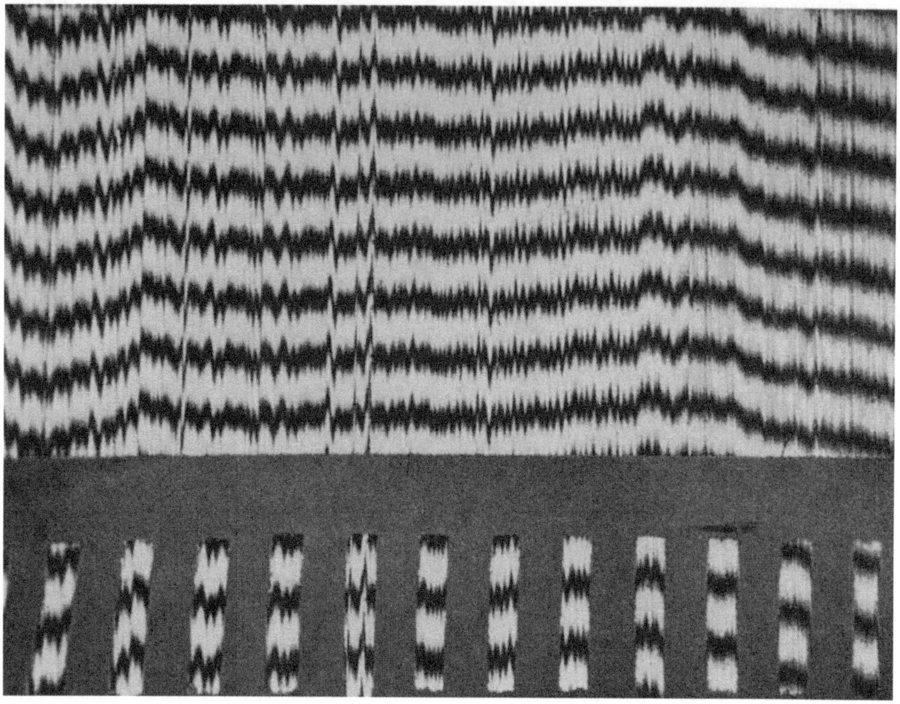

Fig. 136. Wave-front interferogram according to G. W. Stroke (1955), $3\frac{1}{4}$ inch section of a very poor early echelle produced by a lapping process used at that time, and now abandoned. Interferogram taken at just over 60° with the inch scale fastened to the echelle. The interferograms allow immediate appraisal of the grating quality and the accurate determination of possible mask positions. Note the curved sections on the left and right extremes of the interferogram showing various types of cumulative errors, also found in some gratings. Such sections do not contribute to the resolving power of the rest of the grating, and create detrimental satellites.

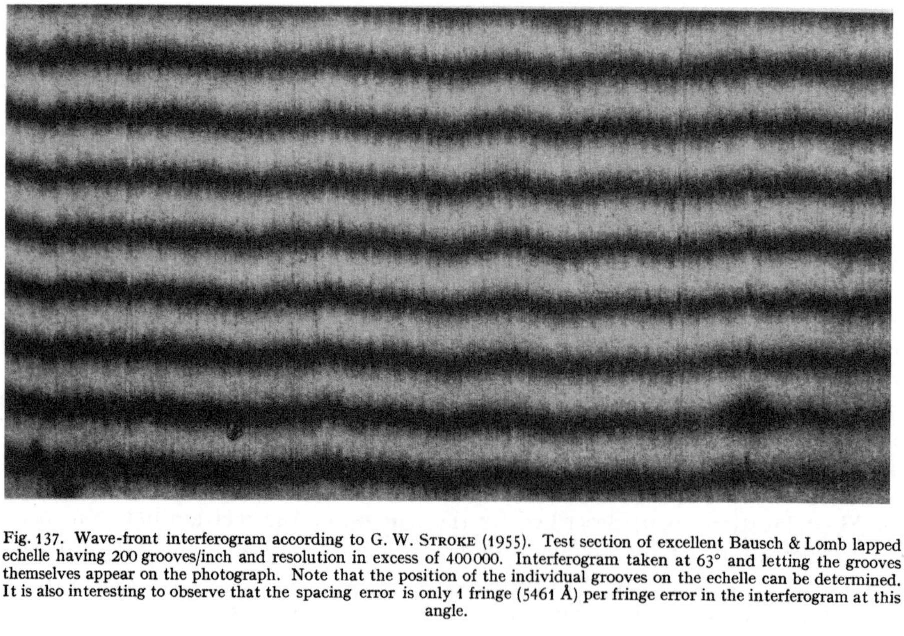

Fig. 137. Wave-front interferogram according to G. W. Stroke (1955). Test section of excellent Bausch & Lomb lapped echelle having 200 grooves/inch and resolution in excess of 400000. Interferogram taken at 63° and letting the grooves themselves appear on the photograph. Note that the position of the individual grooves on the echelle can be determined. It is also interesting to observe that the spacing error is only 1 fringe (5461 Å) per fringe error in the interferogram at this angle.

using two gratings, but does not mention testing, except for a reference to WILLIAMS[33] in connection with the reflection echelon. GALE[35] refers to a testing apparatus with two gratings, but gives no explanation of fringe formation, no theory, and compares two gratings, not a grating with a plane mirror. Another interesting reference is given by CANDLER[36]: "As early as 1918 TWYMAN examined a grating with the interferometer he had just designed, but the published no account, and it was not until 1935 that BISACRE and SIMEON (1935) repeated the work and explained their results in terms of ROWLAND'S earlier paper (1893),

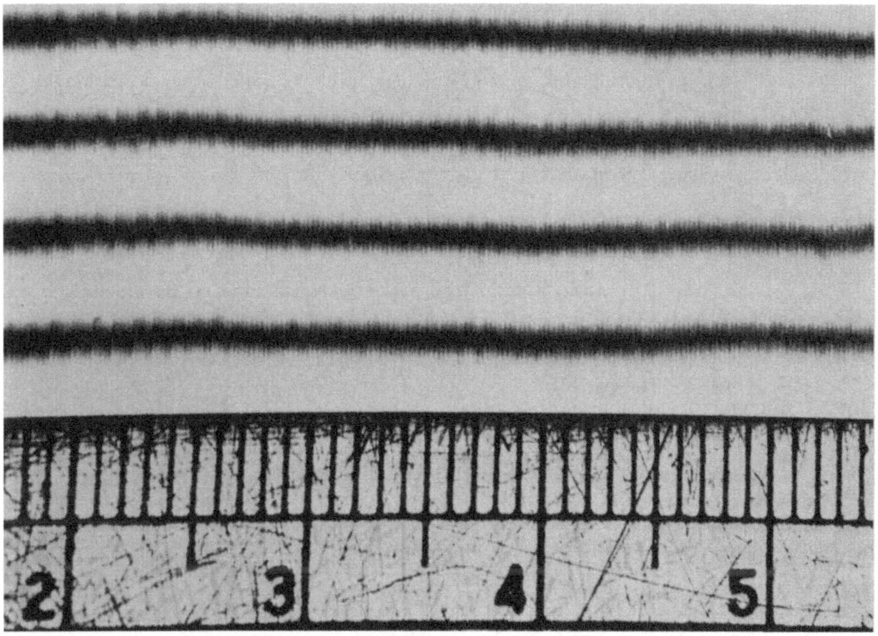

Fig. 138. Wave-front interferogram according to C. W. STROKE (1955). Test section of 15000 groove/inch grating showing some residual periodic error. (Test in light of green mercury line 5461 Å, at 19° in autocollimation). Three fringes spacing error correspond to one fringe in the interferogram at this angle.

which had attributed ghosts to sinusoidal errors in rulings. Unfortunately their paper was never published." Of course, it is now known that there was a great deal of work carried out with gratings and interferometers between TWYMAN'S work in 1918 and 1935.

No direct reference to any of his work is given by TWYMAN[37]. The following other papers[39-74] were also looked at by STROKE[38] as having possible bearing on

[35] H. G. GALE: Testing of diffraction gratings by knife edge and by interferometers. Astrophys. J. **86**, 437 (1937).

[36] C. CANDLER: Modern Interferometers, p. 151. London: Hilger & Watts Ltd. 1951.

[37] F. TWYMAN: Phil. Mag. **35**, 49 (1918); S6, **35**, No. 205, 239 (1918).

[38] G. W. STROKE: Private communication to G. R. HARRISON, November 15 (1953).

[39] V. RONCHI: Behaviour and use of diffraction fringes in the control of optical systems. Rev. Opt. **7**, 49 (1928).

[40] J. H. SCHROEDER and P. B. RAMSAY: Optical properties of grating interferometers. J. Opt. Soc. Amer. **30**, 355 (1940).

[41] L. F. GRIFFIN: An extended theory of the reflection echelon. Proc. Phys. Soc. Lond. B **62**, 93 (1949).

[42] A. MARÉCHAL and P. DE JONC: Some aspects of Twyman fringes. Rev. Opt. **29**, 430 (1950).

[43] F. BISACRE: Formation of image by plane grating used in soft X ray region. Proc. Phys. Soc. Lond. **47**, 948 (1935).

the publication of his interferometric grating testing method (loc. cit. 1 and 35). The references should be helpful in placing some aspects of grating research and developments into perspective.

A photograph of a grating testing interferometer according to Stroke is shown in Fig. 141.

The interferometric method of grating testing is based on the fact that the wave fronts diffracted into any order, angle and wavelength are "plane" wave, when a perfect grating is illuminated by a "plane" collimated wave.

It was shown in Sects. 48 and 49 that wave-front imperfection will result both from ruling errors, as well as from imperfections in the blank or coating in which the grating has been ruled.

---

[44] H. Nagaoka: Corrections for concave gratings with large ruled area. Proc. Math. Phys. Soc. Tokyo **8**, 531 (1916).

[45] C. K. V. Row: Fixed systems of grating interference bands. Nature, Lond. **99**, 63 (1917).

[46] J. Guild: Location of interference fringes. Proc. Phys. Soc. Lond. **33**, 32 (1920).

[47] J. Guild: Fringe Systems in Uncompensated Interferometers. Proc. Phys. Soc. Lond. **33**, 40 (1920).

[48] A. Lo Surdo: Reflection Echelon Grating. Atti Accad. Lincei **30**, 136 (1921).

[49] A. Trowbridge: Photography of Moving Interference Fringes. J. Opt. Soc. Amer. **6**, 195 (1932).

[50] E. Bratke and E. Waetzmann: Interference Methods for Testing Optical Systems. Z. Physik **12.5**, 253 (1922).

[51] E. Waetzmann: Ann. Physik [4], **39**, 1042 (1912).

[52] V. Ronchi: Interference Methods for Measuring Constants of Diverging Optical Systems. Atti Accad. Lincei **33**, 314 (1924).

[53] E. Buchwald: Ghosts due to aperiodic ruling errors. Ann. Physik **80** (3), 279 (1926).

[54] L. C. Glaser: Testing Efficiency and Properties of Optical Gratings. Z. techn. Phys. **7**, 131 (1926).

[55] I. C. Gardner and A. H. Bennett: Hartmann Test Based on Interference. J. Opt. Soc. Amer. Rev. Sci. Inst. **11**, 441 (1925).

[56] R. Kingsdale: The Interferometer Patterns Due to the Primary Aberrations. Trans. Opt. Soc. Lond. **27**, 2, 94 (1926).

[57] G. Simon: Production of Diffraction Patterns by Photography of Interference Fringes. Ann. Physik **13**, 131 (1930).

[58] O. G. Hay: Ross Modification of Hilger Interferometer for Testing Large Optical Elements. Trans. Opt. Soc. Lond. **31**, 2, 91 (1929/30).

[59] J. T. Grossmueller and C. Lakeman: Measurement of ultramicroscopic gratings. Z. Physik **59**, 3—4, 215 (1930).

[60] F. Tyman: Trans. Opt. Soc. Lond. **22**, No. 4 (1935).

[61] M. Renninger: New Interference System of a Dusty Mirror. Z. Physik **100**, 326 (1936).

[62] F. F. P. Bisacre: Diffraction Gratings. Proc. Phys. Soc. Lond. **48**, 148 (1936).

[63] A. Bogros: Theory of Interference Apparatus. J. Phys. Radium **8**, 88 (1937).

[64] R. W. Ditchburn: Diffraction by Irregular Gratings. Proc. Roy. Irish. Acad. A **44**, 123 (1938).

[65] F. Paschen: Ghosts in Lines From Diffraction Gratings. Ann. Physik **34**, 2, 130 (1939).

[66] B. P. Ramsay and F. Panes: Systems of Interferential Devices. J. Opt. Soc. Amer. **34**, 509 (1944).

[67] H. S. Pomerencke and H. G. Beutler: Identification of Orders and Ghosts in Grating Spectra by Diffracting Slits. Rev. Mod. Phys. **14**, 66 (1942).

[68] R. W. Wood: Improved Diffraction Gratings and Replicas. J. Opt. Soc. Amer. **34**, 509 (1944).

[69] H. G. Beutler: The Theory of the Concave Grating. J. Opt. Soc. Amer. **35**, 311 (1945).

[70] R. F. Stamm and J. J. Whalen: Energy Distribution of Diffraction Gratings as a Function of Groove Form. J. Opt. Soc. Amer. **36**, 2 (1946).

[71] J. G. Vogl: Plane Parallel Glass Indicators and Technical Interferometers. Tekn. T. **76**, 657 (1946).

[72] J. Brossel: Multiple beam localized fringes, Part I, Intensity distribution and localization . Proc. Phys. Soc. Lond. **69**, 224 (1947).

[73] U. Fano: On the Theory of Imperfect Diffraction Gratings. J. Opt. Soc. Amer. **38**, 921 (1948).

[74] F. D. Werner and B. M. Leadon: Very Accurate Measurements of Fringe Shifts in an Optical Interferometer, Study of Glass Flow. Rev. Sci. Instrum. **24**, 121 (1953).

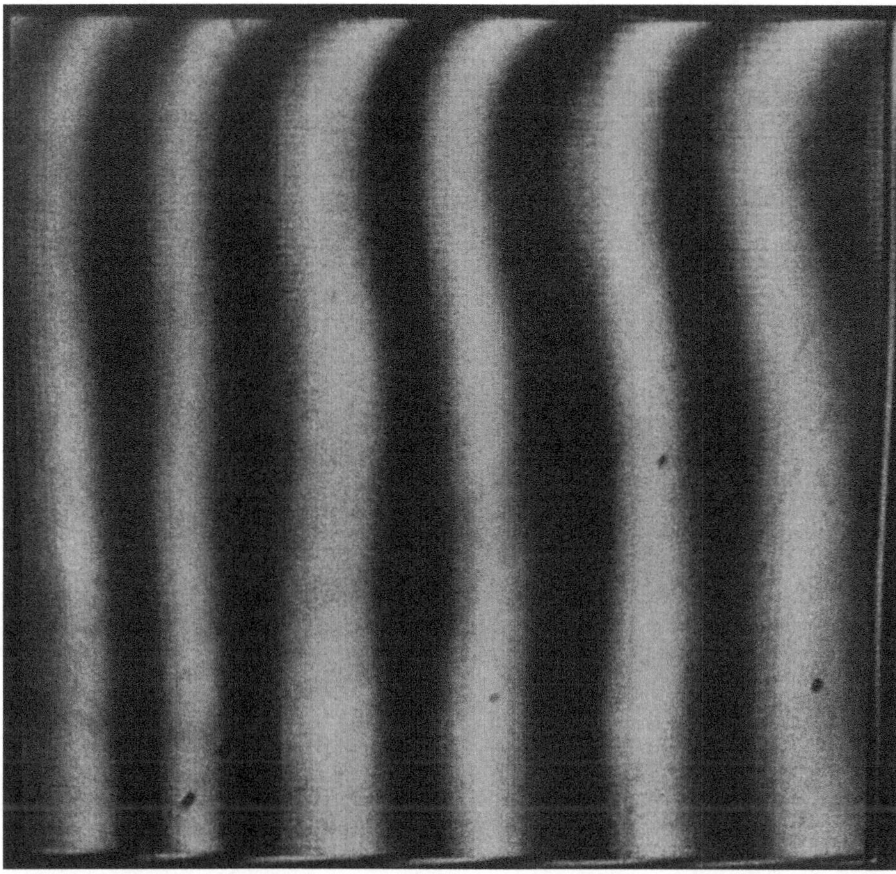

Fig. 139. Interferogram with fringes oriented to be *parallel* to groove length (unlike in all other interferograms shown). This test gives information on groove straightness and on variations in groove depth and straightness along groove length. (Some such variations appear at both ends, and also all along the groove length. The latter are due to diamond vibrations.)

For a ruling error Δa, the phase error was shown to be

$$\Delta\Phi(\eta) = \frac{2\pi}{\lambda}\, 2\Delta a \sin i' \tag{74.1}$$

when the grating is examined in autocollimation at the angle $i' = i$.

Eq. (74.1) can also be written as

$$\left.\begin{aligned} &\Delta\Phi(\eta) = 2\pi \sin i' \\ &\text{per} \\ &\qquad \Delta a = \frac{\lambda}{2} = 1 \text{ fringe}, \end{aligned}\right\} \tag{74.2}$$

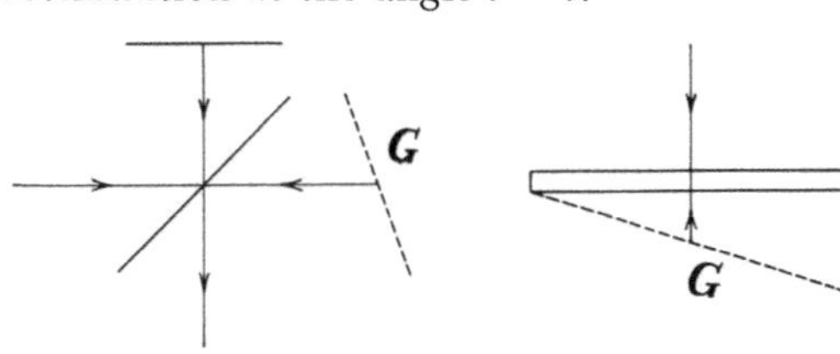

Fig. 140. Grating testing interferometers according to A. A. MICHELSON (Studies in Optics, Un. Chicago Press, 1927, p. 77).

showing that *when the ruling errors are measured in fringes* ($\lambda/2$) *of the same wavelength as that used in the interferometric measurement of the diffracted wave-front aberrations*, then *one fringe of ruling error will result in* $\sin i'$ *fringes of wave-front aberration when the grating is used in autocollimation at the angle* i' [Sect. 48 and STROKE, loc. cit. [31] (1955)].

While the *effect of ruling errors increases with the sine of the angle, the effect of blank imperfections was shown* (Sect. 49) *to decrease with the cosine of the angle.*

Fig. 141. Grating testing interferometer according to STROKE (1955), mounted at M.I.T. Camera lens appears on left and collimator lens on right. Excellent fringe stability is achieved by mounting heavy mirror mounts on heavy cast iron surface plate, and by isolating surface plate from cement pier or table by soft "instrument packing mounts".

For a portion of the blank inclined by an angle Δi with respect to the rest of the blank, such that the imperfect portion extends over N grooves of spacing a, one finds the fringe displacement in the interferogram to be

$$\left.\begin{aligned}\delta_I = N a \Delta i \cos i \\ \text{(fringes)}.\end{aligned}\right\} \quad (74.3)$$

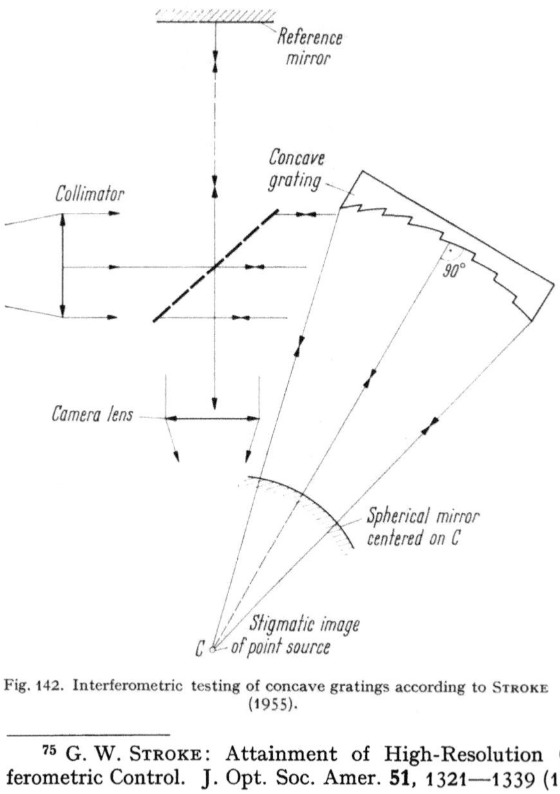

Fig. 142. Interferometric testing of concave gratings according to STROKE (1955).

The conditions of good fringe visibility in the interferometer depend on careful autocollimation of the reference mirror on the source hole as well as on reasonably close equality in the amplitudes of the reference and diffracted wavefronts [STROKE, loc. cit. [31] (1955)]. A monochromatic light source, such as laser or a Hg 198 electrodeless discharge lamp are necessary to achieve good fringe visibility over the entire depth of the grating, at its highest angle of use.

It should be clear from Eq. (74.3) that the defects in flatness, both in gratings and especially in grating replicas are best tested at low grating angles[75].

[75] G. W. STROKE: Attainment of High-Resolution Gratings by Ruling Under Interferometric Control. J. Opt. Soc. Amer. **51**, 1321—1339 (1961).

Concave gratings can also be tested interferometrically[76], by noting that a concave grating is stigmatic on the normal to its surface (Fig. 142).

Another version of a grating testing interferometer, the Fizeau interferometer sometimes referred to as a plano-interferometer" was also suggested by MICHELSON[32]. Recent versions of the test seem to have been used for a long time at the National Physical Laboratory in Teddington[77], and elsewhere[78a,b]. L. A. SAYCE[77] writes, "We, in this Laboratory, had a long association with TWYMAN and his work and we make constant use of two early examples of his modified Michelson interferometer. Mr. JOHN GUILD, who is still an active member of our staff, tells us of a memorable visit of MICHELSON and TWYMAN to this Laboratory 40 years ago and how he showed them a new application he had made of their interferometer to the diffracted wave-fronts of diffraction gratings. This application remained dormant for nearly 30 years when GUILD revived it to serve our own diffraction grating program ..." It might be interesting to also note the principal passages in a letter[79] written more than 5 years earlier:

"We have three chief interferometric tests which we apply to all gratings:

1. *A modified Fizeau test.* If the grating is approximately parallel to the reference flat one produces fringes showing the contour of the zero-order wavefront, i.e. the contour of the blank on which the grating is ruled.

If the grating is now slowly inclined to the reference flat one passes through a series of positions, one for each order, where the diffracted wavefront produces fringes with the reference flat. These fringes give a very good general picture of the errors of the grating. The higher the order used, the more rigorous becomes the test.

2. *A shearing interferometric test**. In which one produces interference between two slightly displaced images of the grating surface as seen in any one of the orders of diffraction. This is primarily a test for periodic errors. The effect of such errors is to produce a corrugated wavefront. If, then, the direction of shear is such as to bring the two sets of corrugations exactly out of phase the interference fringes will have a 'waveform' which displays the periodic errors very clearly.

3. *A phase-contrast test.* In which one examines the actual grating surface by phase contrast in one of its diffracted orders. This can be made sensitive to a very small fraction of a 'fringe' and will show errors in any grating we have ever tested. The results, however, are not easy to interpret.

In addition to the above tests we measure the 'photometric efficiency' of all our gratings by a photocell method. In this test we use a Littrow mounting with a monochromatic source and measure the amount of diffracted energy in each order. To calculate the 'efficiency' we compare this energy with the energy which would be reflected by a perfect (i.e. 100% reflective) mirror of the same subtended area as the grating. In other words, the 'efficiency' of the grating in any order is the proportion of incident monochromatic energy which the grating will diffract into the order."[79]

Interferometric grating testing, in the form described by STROKE in 1955 (loc. cit. 14) has been adapted for the testing of gratings ruled at the Soviet

* W. J. BATES: Proc. Soc. Phys. Lond. **59**, 940 (1917). See also note in STROKE, loc. cit. 31.

[76] G. W. STROKE: Rev. Opt. **39**, 291—398 (1960), p. 310.

[77] L. A. SAYCE: Interferometry in N. P. L. Light Division. In: Interferometry. Symposium held at the National Physical Lab., June 9—11, 1959, p. 10. London: Her Majesty's Stationery Office 1960.

[78] a) H. G. GALE: Astrophys. J. **86**, 437 (1937). — b) D. RICHARDSON, R. S. WILEY and G. J. SHELDON: J. Opt. Soc. Amer. **40**, 259 (1950).

[79] L. A. SAYCE: Letter to G. W. STROKE at M. I. T., January 5, 1954.

Russian State Optical Institute in Leningrad, by GERASIMOV and his associates[80,81], as well as by Carl Zeiss in West Germany[82], by Jobin et Yvon in Paris[83,84] and, undoubtedly by many others, among whom the Jarrell-Ash Company[85] and the Perkin-Elmer Corporation[86].

75. Phase-contrast testing of gratings. The phase-contrast method of coherent-light image filtering, first described by F. ZERNIKE in 1934 and 1935[87–90] has been successfully extended to grating testing by E. INGELSTAM, A. MARÉCHAL, M. FRANÇON, E. DJURLE and their collaborators[91]. A complete description of the method used is given by E. INGELSTAM and E. DJURLE in 1953[92].

Basically, the phase contrast method of wave-front testing results in producing an image of the wave front such that phase variations in the wave front are translated into intensity variations in the image. The great enhancement of contrast is obtained by means of suitable coherent "filtering".

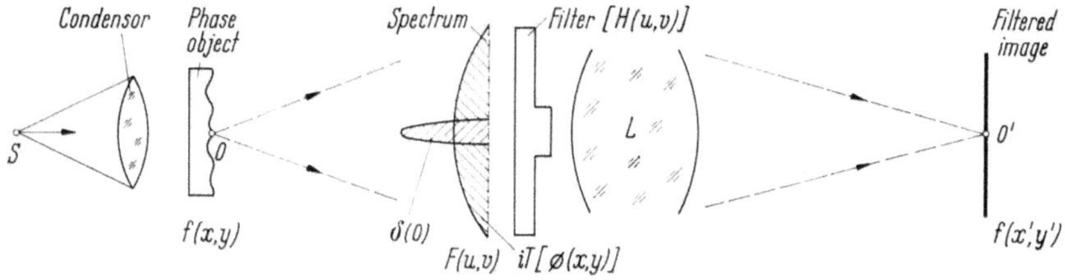

Fig. 143.

When the wave front phase variations are sufficiently small, the wave front

$$f(x, y) = \exp[i\varphi(x, y)]$$

can be written as

$$f(x, y) = \exp[i\varphi(x, y)] \cong 1 + i\varphi(x, y) + \cdots \tag{75.1}$$

[80] F. M. GERASIMOV, I. A. TEL'TEVSKII, S. N. SPIZHARSKII and S. V. NESMELOW: Optiko-Mech. Prom. No. 4, 56 (1957).

[81] F. M. GERASIMOV, I. A. TEL'TEVSKII, S. S. NAUMOV, S. N. SPIZHARSKII and S. V. NESMELOV: Diffraction Gratings from the State Optical Institute. Optika i Spektroskopiya **4** (6), 779—790 (1958); translated by "Russian Translating Programme", Department of Scientific and Industrial Research, London; also available from Publications Board Project, Photoduplication Service, Library of Congress, Washington 25, D. C.

[82] G. HANSEN: Private communication to G. W. STROKE (1962).

[83] G. PIEUCHARD: Private communication to G. W. STROKE (1959). The part of G.W. STROKE in helping to initiate the French grating ruling program is referred to in [84].

[84] G. AMAT, A. ARSAC, J. BROCHARD, J. BROSSEL, P. CONNES, L. COUTURE, P. JACQUINOT, and A. MARÉCHAL: Les Réseaux Optiques. In: Optics in France. Applied Optics **1**, 262 (1962).

[85] R. F. JARRELL and G. W. STROKE: Some New Advances in Grating Ruling, Replication and Testing. Applied Optics **3**, 1251—1262 (1964).

[86] H. MARSHALL: Private communication to G. W. STROKE (1961—1962). See also: Perkin-Elmer Gratings. In: The Perkin-Elmer Instrument News, Vol. 13, No. 1 (Fall 1961), pp. 4—7.

[87] F. ZERNIKE: Beugungstheorie des Schneidenverfahrens und seiner verbesserten Form, der Phasen-Kontrast-Methode, Physica, Haag **1**, 43 (1934).

[88] F. ZERNIKE: Das Phasenkontrastverfahren bei der mikroskopischen Beobachtung. Phys. Z. **36**, 848 (1935).

[89] F. ZERNIKE: Z. techn. Phys. **16**, 454 (1935).

[90] See also H. WOLTER: Schlieren-, Phasen-Kontrast- und Lichtschnittverfahren. In: Handbuch der Physik, Vol. 24, edit. by S. FLÜGGE, pp. 555—645. Berlin: Springer 1956.

[91] A. M. CAMUS, M. FRANÇON, E. INGELSTAM and A. MARÉCHAL: J. Phys. Radium **11**, 3E (1950); also: Etude des Erreurs de Tracé des Réseaux par Contraste de Phase. Rev. Opt. **30**, 121 (1959).

[92] E. INGELSTAM and E. DJURLE: The Study of Diffraction Grating Characteristics by Simplified Phase Contrast Methods. J. Opt. Soc. Amer. **43**, 572—580 (1953).

where Eq. (75.1) applies in particular to a one-dimensional pure phase front, such as that corresponding to the wave fronts diffracted by a grating.

The "diffraction pattern" corresponding to $f(x, y)$ in the focal plane of the lens (or mirror) L_2 is $F(u, v)$ such that

$$F(u, v) = T[f(x, y)] = \delta(0) + i\,T[f(x, y)] \tag{75.2}$$

(see Fig. 143).

Now let a complex filter $H(u, v)$ be placed in the (u, v) plane, and let $H(u, v)$ be described by

$$\left.\begin{aligned} H(u, v) &= it \quad \text{over the region } \delta(0), \\ H(u, v) &= 0 \quad \text{outside of } \delta(0). \end{aligned}\right\} \tag{75.3}$$

The "filtered" diffraction pattern (that is the filtered "spectrum" in terms of image communication theory) is $G(u, v)$ such that

$$G(u, v) = F(u, v) \cdot H(u, v). \tag{75.4}$$

One has

$$G(u, v) = it\,\delta(0) + i\,T[\varphi(x, y)] \tag{75.5}$$

where t is the amplitude-transmission factor of the $\delta(0)$ region. According to the "double-diffraction principle" [93,94], $G(u, v)$ can be considered as the new wave front in the pupil of the lens (or mirror) L_3. The corresponding diffraction pattern $f(x', y')$ is nothing but the "filtered" image of the original wave front. One has

$$f(x', y') = T[G(u, v)] = it + i\varphi(x, y) \tag{75.6}$$

where $f(x', y')$ is the complex amplitude distribution in the image of the diffracted wave front. The intensity distribution in the image is then

$$i(x', y') = f(x', y') \cdot f(x', y')^* = t^2 + \varphi(x, y)^2 + 2\varphi(x, y)\,t \tag{75.7}$$

or approximately, with the approximation used in Eq. (75.1).

$$i(x', y') = f(x', y') \cdot f(x', y')^* \cong t^2 + 2\varphi t. \tag{75.7a}$$

When the phase departures $\varphi(x)$ from perfection are zero, it is easily seen that the intensity in the image is given by t^2, upon imaging through the phase contrast filter. On the other hand, in the regions where the phase departure from perfection is $\varphi(x)$, the intensity is given by $2\varphi t$.

The contrast C of the regions with phase variations, $2\varphi t$, to the "background", $t^2 + 2\varphi t$, is found to be

$$C = \frac{2\varphi t}{t^2 + 2\varphi t} = \frac{2\varphi}{t + 2\varphi}. \tag{75.8}$$

The amplitude transmission factor is chosen to be sufficiently small, so that C may attain very high values, of the order of 1. For instance, for $t=0$, $C=1$. When $t=10^{-1}$ (intensity transmission factor $t^2=10^{-2}$), and $\varphi=10^{-3}$ radians, $C=2/100$. On the other hand, even if $t=1$, $C \cong 2\varphi = 2\times 10^{-3}$ here, which is still large when compared to the value of the contrast which would be obtained without the phase contrast filtering.

Indeed, without filtering, one would have

$$i(x', y')_{\text{unfiltered}} \cong 1 + \varphi^2(x, y) \tag{75.9}$$

---

[93] A. Maréchal and M. Françon: Diffraction, ed. Revue d'Optique, Paris 1960.

[94] A. Maréchal: Optical Filtering by Double Diffraction. In: Optical Processing of Information, edit. by D. K. Pollack, C. J. Koester and J. T. Tippett, pp. 20—30. Baltimore: Spartan Books Inc. 1963.

and the contrast would be

$$C_{\text{unfiltered}} \cong \frac{\varphi^2}{1} = \varphi^2. \tag{75.10}$$

For $\varphi = 10^{-3}$ radians, $C_{\text{unfiltered}}$ would be 10^{-6}, rather than 2×10^{-3} for the $t=1$ filtered case, and almost 2×10^{-2} for the $t=10^{-1}$ filtered case. The arrangement used for phase contrast testing of gratings is shown in Fig. 144.

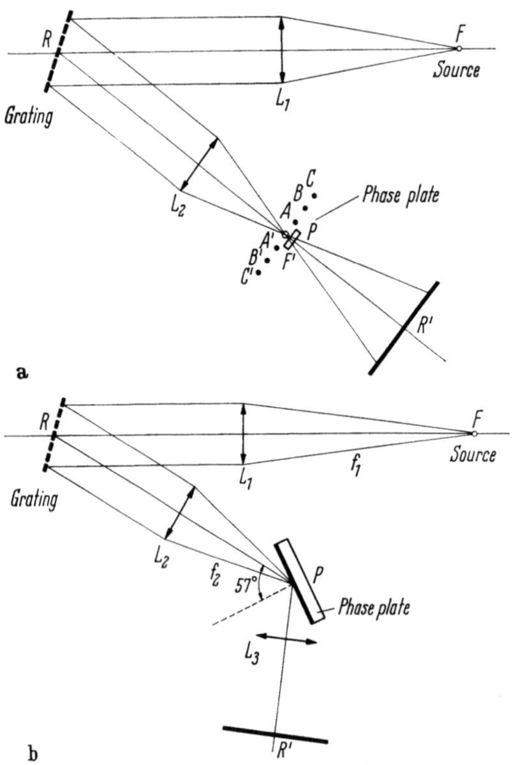

Fig. 144 a and b. Phase contrast testing of gratings according to Camus, Françon, Ingelstam and Maréchal (1951).

Experience has shown that phase-contrast testing of gratings is particularly suitable for the detection of very small phase variations, for instance for errors smaller than $\lambda/10$ or $\lambda/20$.

The non-linear relation between wave-front phase aberrations $\varphi(x, y)$ in the wave front

$$\left.\begin{aligned} f(x,y) &= \exp[i\varphi(x,y)] \\ &= 1 + i\varphi(x,y) - \\ &\quad - \frac{\varphi^2(x,y)}{2} + \cdots \end{aligned}\right\} \tag{75.11}$$

and the "filtered" image

$$\left.\begin{aligned} i(x',y') &= f(x',y')\, f(x',y')^* \\ &= |f(x',y')|^2 \end{aligned}\right\} \tag{75.12}$$

appears to restrict the use of phase-contrast testing to gratings of which the wave fronts have first been tested by interferometric means (see Sect. 74). A comparison of the phase contrast test with interferometric tests for a same region of the grating, according to Stroke[95] is shown in Fig. 145.

Significant contributions to phase-contrast testing of gratings were made by many authors, in particular by G. Nomarski[96] and W. Jaschek[97]. In particular, the phase-contrast plate, used in the reflection testing of gratings was first described by Françon and Nomarski in 1950[96].

Phase contrast testing of gratings should be particularly useful for measurement of scattered light, resulting from local groove displacements and groove imperfections.

The effect of local grating imperfections, and their measurement, is quite similar to the effect of polishing imperfections and their measurement, for instance in coronograph lenses, as described by M. Françon[98] and also by H. Wolter (loc. cit. 90).

---

[95] G. W. Stroke: In: Optics in Metrology, edit. by P. Mollet, pp. 98—118. London: Pergamon Press 1960.

[96] M. Françon and G. Nomarski: C. R. Acad. Sci. Paris **230**, 1392 (1950).

[97] W. Jaschek: Prüfung von Beugungsgittern nach dem Phasenkontrastverfahren. Mikrochim. Acta **2/3**, 525—533 (1955).

[98] M. Françon: Le Contraste de Phase en Optique et en Microscopie, ed. Revue d'Optique 1950.

76. Crossed Rulings (Moiré-fringe testing). An interference pattern can be obtained by slightly inclining two suitably-displaced overlapping gratings with respect to each other. A displacement of about one quarter to one half the groove

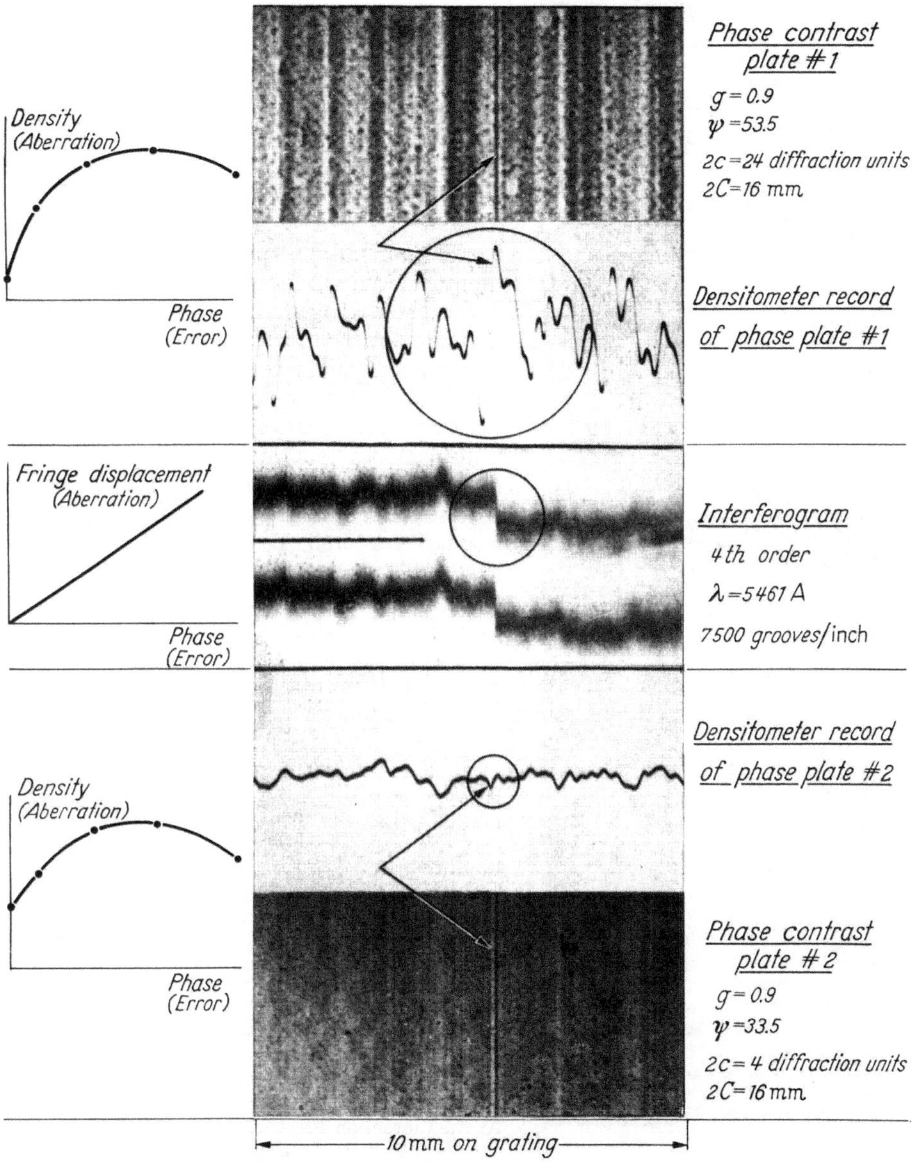

Fig. 145. Comparison of interferogram with phase contrast plates for the same area of grating.

spacing is typical. The overlap can be obtained by actually ruling a second grating over the first one or else by examining the moiré fringe pattern obtained by placing a transparent replica of the original grating in contact with the original. In some cases, an image of one grating may be formed on the other grating with the help of a stigmatic system. The interference fringe system formed consists of straight-line fringes, running roughly at right angles to the length of the grating grooves.

Crossed rulings, and the "moiré" fringes obtained with them were already referred to by Lord RAYLEIGH[99] in 1874, and later F. A. ANDERSON[100] gave a general description and analysis of the "method of crossed rulings" in his 1923 article on gratings. More recently, in particular as a result of the work by J. M. BURCH[101–103], J. GUILD[104,105], L. A. SAYCE[106,107] and others[108] at the National Physical Laboratory in Teddington, moiré fringes have found very important applications in metrology, and more generally in machine-tool control.

Crossed rulings seem to have lost some of their importance for diffraction grating testing after STROKE[109] showed that the nondestructive method of interferometric grating testing (see Sect. 74) is capable of providing not only all the information about the local and periodic ruling errors obtained with crossed rulings, but in fact precise information about the overall ruling regularity as well. The sensitivity of the interferometric wave-front testing method and that of the cross rulings can be shown to be the same[109] (Fig. 146).

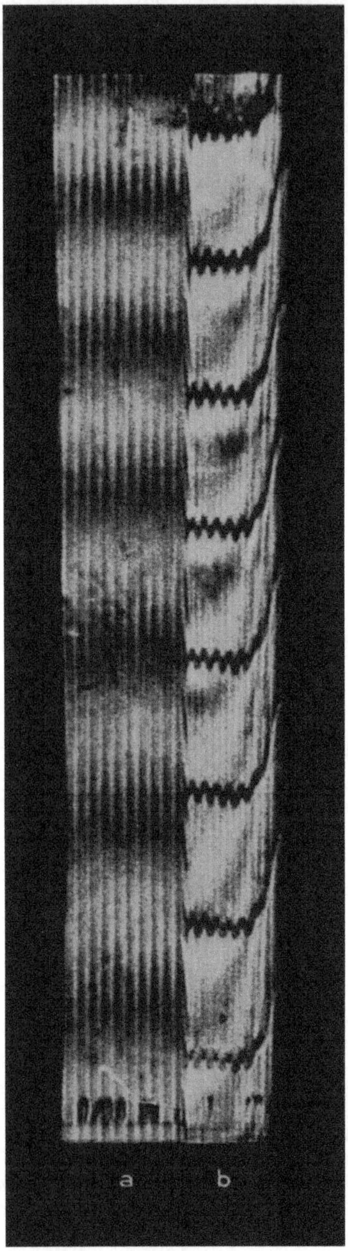

Fig. 146. a Interferogram (at 40°); b Cross-ruling for neighbouring sections of same grating (according to STROKE, 1959).

[99] Lord RAYLEIGH: On the manufacture and theory of diffraction gratings. Scientific Papers **1**, p. 208. — Phil. Mag. **47**, 81—93, 193—205 (1874).

[100] F. A. ANDERSON: The manufacture and testing of diffraction gratings. GLAZEBROOK's Dictionary of Applied Physics **4**, 30—41 (1923). McMillan & Co., Ltd.

[101] J. M. BURCH: Photographic production of scales for moire fringe applications. In: Optics in Metrology, edit. by P. MOLLET, pp. 361—368. London: Pergamon Press 1960.

[102] J. M. BURCH: The possibilities of moiré fringe interferometry. Proceedings of the NPL Symposium on Interferometry, pp. 179—222. London: H. M. Press 1959.

[103] J. M. BURCH: The metrological applications of diffraction gratings. In: Progress in Optics, Vol. II, edit. by E. WOLF, pp. 75—108. Amsterdam: North-Holland Publ. Co. 1963.

[104] J. GUILD: The interference systems of crossed diffraction gratings. Theory of moiré fringes. Oxford: At the Clarendon Press, London 1960.

[105] J. GUILD: Diffraction gratings as measuring scales. London: Oxford University Press 1960.

[106] L. A. SAYCE: The preparation and industrial applications of diffraction gratings. Trans. Soc. Instrum. Techn. **9**, 139—142 (1957).

[107] D. V. BLAKE and L. A. SAYCE: Machine-tool monitor using diffraction gratings. Trans. Soc. Instrum. Techn. **10**, 190—196 (1958).

[108] A. H. MCILRAITH: An introduction to the theory of crossed diffraction gratings and their application to linear measurements. Machine Shop **23**, 202—213 (1962).

[109] G. W. STROKE, in: Rev. Opt. **39**, 291—360 (1960).

Only a simple theory of moiré fringes is sufficient for the understanding of the principle and its application to the testing of gratings.

Consider (Fig. 147) two square-wave gratings of spacing inclined by a small angle θ with respect to each other. Let the width of the "opaque" (black) portion of the spacing be $a/2$. It follows immediately, from elementary geometry that a series of black straight-line and equidistant "fringes" will be formed, such that the fringes have a maximum opacity along a direction normal to the bisector of the angle θ. (Note that the points of intersection of the grooves of the first grating by the grooves of the second grating are increasingly far away from the "edge" of the first grating to which the grooves are normal). It is also readily seen that the "spacing" l of the "fringes" is equal to

$$l = \frac{a}{\tan\theta} \cong \frac{a}{\theta}. \tag{76.1}$$

If the grooves of one grating were regularly spaced, a groove displacement Δa would result in a fringe deformation Δl such that

$$\Delta l \cong \frac{1}{\theta} \Delta a \tag{76.2}$$

and

$$\frac{\Delta l}{l} = \frac{\Delta a}{a}. \tag{76.3}$$

In other words, a ruling error of *one groove* will result in a fringe "deformation" of *one moiré fringe*, in the direction roughly parallel to the groove lengths.

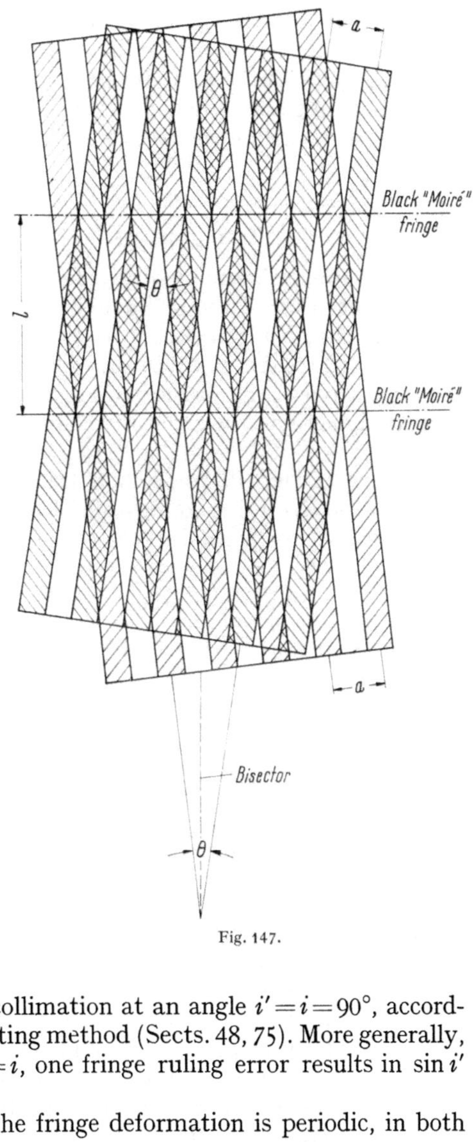

Fig. 147.

It is recalled [see Sect. 48, Eq. (48.3) and (48.4)] that a ruling error of *one fringe* results in an interference fringe deformation of one fringe when a grating is tested in autocollimation at an angle $i' = i = 90°$, according to the interferometric wave-front testing method (Sects. 48, 75). More generally, if the grating is tested at the angle $i' = i$, one fringe ruling error results in $\sin i'$ fringes wavefront deformation.

In case of a periodic ruling error, the fringe deformation is periodic, in both tests.

An interesting comparison of the crossed-ruling test with the interferometric test for the same grating was given by Stroke[109]. In the crossed-ruling test (Fig. 146b) the amplitude A (in angstroems) of the periodic error is obtained by dividing the apparent amplitude (equal to one half the peak-to-peak amplitude) by the moiré "fringe" separation, and by multiplying the value obtained by the spacing constant a (in angstroems). For the grating shown,

$a=16933$ Å, and one obtains $A=1440$ Å or approximately $A=0.526$ fringes of $\lambda=5461$ Å.

In the interferometric wave-front test of the same grating (Fig. 146a), the apparent error is similarly obtained by dividing one half of the peak-to-peak periodic-error amplitude by the fringe spacing: the value of the error on the grating is obtained by dividing the apparent error by $\sin i'$. In this case $i'=40°$, $\sin i'=0.645$ and the value obtained is $A=0.53$ fringes of $\lambda=5461$ Å.

It is readily seen that the non-destructive interferometric wave-front test provides the value of the periodic error much more rapidly than the crossed-rulings.

It should be noted, in conclusion, that a more complete theory of moiré fringes needs to be based on successive application of the grating equation

$$\sin i_l + \sin i_l' = \frac{m_l \lambda}{a_l} \tag{76.4}$$

applied successively to gratings $l=1$ and $l=2$[101–108].

77. Grating efficiency testing. The efficiency of a grating can best be characterized by the ratio of the diffracted energy, at a given wavelength and angle (or order), to the energy reflected in the same wavelength by a mirror of equivalent aperture and coated with the same surface-coating as the grating. The efficiency thus measured is the *relative efficiency*. *Absolute efficiency* refers to the ratio of diffracted energy, at a given wavelength, the incident energy, in the same wavelength.

In general, the efficiency should be measured in two orthogonal polarizations, one with the electric vector parallel to the groove length, $\boldsymbol{E}_{\parallel}$, and the other with the electric vector normal to the groove length, $\boldsymbol{E}_{\perp}$.

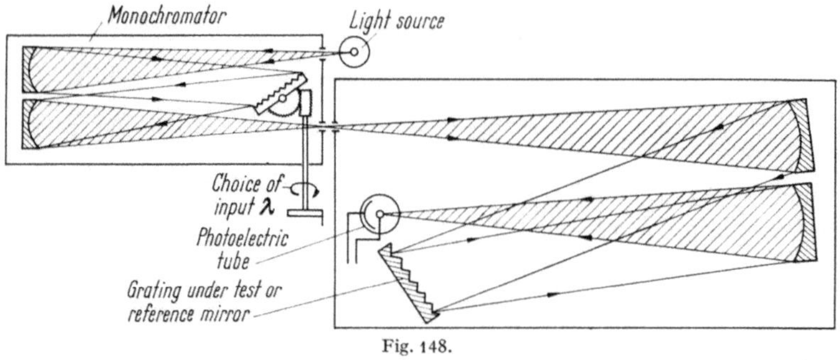

Fig. 148.

In case where the mirror reflectivity changes substantially with the angle of incidence and with polarization, account must be taken of these factors in evaluating efficiency.

When considered as a ratio of diffracted energy to reflected energy, grating efficiency is in principle a very simple measurable quantity. Practical difficulties have to do with the availability of suitable monochromators for the input radiation, of light sources with fairly uniform spectral energy distributions in the domains of interest and with suitable detectors. To a large extent, these problems are solved or solvable under the conditions of the grating being tested. As a result, reference should be made to standard treatises on practical spectroscopy[110,111] and to current literature for the choice of sources and detectors.

[110] R. A. SAWYER: Experimental Spectroscopy, third edit. New York: Dover Publications Inc. 1963.

[111] G. R. HARRISON, R. C. LORD and J. R. LOOFBOUROW: Practical Spectroscopy. New York: Prentice Hall 1949.

A representative arrangement for the measurement of grating efficiency is illustrated in Fig. 148.

Basically, the arrangement consists of two monochromators used in series. The first monochromator serves to produce monochromatic light of desired wavelengths, with moderately narrow bandwidth. The second monochromator should permit the measurement of the monochromatic light a reflected from a reference mirror, b diffracted by the grating, c incident on the mirror (and, or grating). In measuring the ratio b/c, absolute efficiency may be determined, while measurement of a/c provides the relative efficiency. The reason for attaching great importance to the relative efficiency is a practical one. Indeed, grating efficiency is determined by two important factors:

1. groove shape,
2. surface reflectivity.

With a given surface reflectivity, the maximum possible relative efficiency is *unity*. Comparison of the grating with a mirror of equivalent aperture provides a powerful tool for the adjustment of the ruling depth and angle, achieved with a given diamond.

Examples of the effect of groove shape and ruling depth, according to STROKE[112] are shown in Figs. 85, 157, 158, 159, 160.

The effects of groove shape and of surface reflectivity are discussed in Sects. 62 to 71. These effects are basic in the evaluation of the efficiency of any given grating. Some additional aspects of scattered light are discussed in Sect. 78.

It is important to recognize that no "anomalies" are found to exist in the energy distribution and polarization of gratings in comparison to theoretical expectations, if the measurements are compared to calculations based on electromagnetic theory (see Chapter V). A list of references related to this point is found under[112–122].

Special problems in efficiency measurements arise in the vacuum ultraviolet and soft X-ray regions, on one hand, and in the infrared, on the other.

In the vacuum ultraviolet, from 1100 Å to about 1800 Å new types of argon, krypton and xenon arcs have been found to provide overlapping continua (see

---

[112] G. W. STROKE: Theory, Production and Use of Optical Gratings for High-Resolution Spectroscopy. In: Quantum Electronic III, edit. by P. GRIVET and N. BLOEMBERGEN, pp. 1221—1237. New York: Columbia Univ. Press 1964.

[113] G. W. STROKE: Effects of Polarization on the Energy Distribution and Efficiency in Optical Diffraction Gratings. J. Opt. Soc. Amer. **54**, 846 (1964).

[114] A. MARÉCHAL and G. W. STROKE: Sur l'Origine des Effets de Polarisation et de Diffraction dans les Réseaux Optiques. C. R. Acad. Sci. Paris **249**, 2042 (1959).

[115] G. W. STROKE: Etude théorique et expérimentale de deux aspects de la diffraction dans les réseaux optiques. L'évolution des défauts dans les figures de diffraction et l'origine électromagnetique de la répartition entre les ordres. Rev. Opt. **39**, 291 (1960).

[116] P. BOUSQUET: Etude de la réflexion et de la transmission de la lumière par un réseau transparent à profil sinusoidal. Extension au Cas d'une Surface Irrégulière. Rev. Opt. **41**, 277 (1962).

[117] P. BOUSQUET and R. DELEUIL: Diffraction des ondes électromagnétiques par un réseau à profil sinusoidal. C. R. Acad. Sci. Paris **256**, 1461 (1963).

[118] P. BOUSQUET: Diffraction des ondes électromagnetiques par un réseau à facettes planes. C. R. Acad. Sci. Paris **256**, 3422 (1963).

[119] G. W. STROKE: Attainment of High Efficiencies in Blazed Optical Gratings by Avoiding Polarization in the Diffracted Light. Phys. Letters **5**, 45 (1963).

[120] R. PETIT: Contribution à l'étude de la diffraction d'une onde plane par un réseau métallique. Rev. Opt. **42**, 263 (1963).

[121] P. BOUSQUET: Diffraction des ondes électromagnetiques par un réseau a profil triangulaire. C. R. Acad. Sci. Paris **257**, 80 (1963).

[122] C. JANOT and A. HADNI: Polarisation de la lumière par les réseaux échelettes dans l'infrarouge lointain. J. Phys. Radium **24**, 1073 (1963).

Fig. 149) of very good intensity[123]. The Hopfield[124] continuum of He extends from about 600 Å to 1000 Å and various improvements on He discharges have been described[125,126]. Sources for the Lyman continuum (below 1215.7 Å) have been the subject of research and of important improvements by W. R. S. Garton[127].

Work on these and other sources, as well as detectors for the extreme ultraviolet has been reviewed in a comprehensive article by R. Tousey[128].

Similar reviews on infrared (sources, detectors and technology) are given by Ballard and Wolfe[129], Garing and Howard[130], and Potter and Eisenman[131]. A very comprehensive review on infrared instrumentation and techniques had been previously given by van Zandt Williams[132]. A review of Soviet Russian work is given by F. M. Gerasimov et al. (loc. cit. 80).

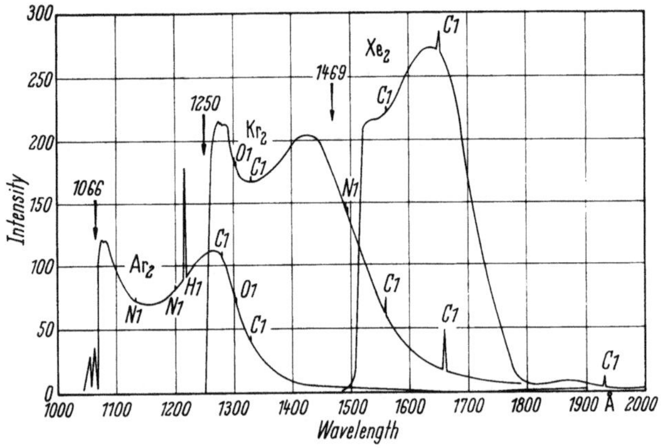

Fig. 149. New types of argon, krypton and xenon arcs continua (according to Wilkinson and Tilford, 1964).

Efficiency measurements in the X-ray domain are discussed in Sect. 70, with particular reference to the work of Siegbahn, Sprague, Tomboulian, Edlén, and others.

The performance and measurement of the efficiency of diffraction-grating and of echelettes in the ultra-soft X-ray region below according to a description by Lukirskii and Savinov[133,134] follows.

[123] P. G. Wilkinson and S. G. Tilford: Private communication (1964).

[124] J. J. Hopfield: Astrophys. J. **72**, 139 (1930).

[125] R. E. Huffman, W. W. Hunt jr., Y. Tanaka and R. L. Novack: J. Opt. Soc. Amer. **51**, 693 (1961).

[126] R. G. Newburgh, L. Heroux and H. E. Hinteregger: Two light sources for use in the extreme ultraviolet. Appl. Optics **1**, 733 (1962).

[127] W. R. S. Garton: J. Sci. Instrum. **36**, 11 (1959).

[128] R. Tousey: The Extreme Ultraviolet — Past and Future. Applied Optics **2**, 679—694 (1962).

[129] S. S. Ballard and W. L. Wolfe: Recent developments in infrared technology. Applied Optics **1**, 547—588 (1962).

[130] J. S. Garing and J. N. Howard: University research in infrared physics. Applied Optics **1**, 559—566 (1962).

[131] R. F. Potter and W. L. Eisenman: Infrared photodetectors: a review of operational detectors. Applied Optics **1**, 567—574 (1962).

[132] Van Zandt Williams: Infrared instrumentation and techniques. Rev. Sci. Instrum. **19**, 135—178 (1948).

[133] A. P. Lukirskii and E. P. Savinov: Use of diffraction gratings and echelettes in the ultrasoft x-ray region. Optics and Spectroscopy (translation of Optica i Spektroskopiya), **14**, 147—141 (1963).

[134] A. P. Lukirskii and E. P. Savinov: Reflection of ultra-soft x-radiation from glass and titanated surfaces. Optics and Spectroscopy **14**, 152—154 (1963).

The gratings and echelettes examined had 600 grooves per millimeter, and were cut on glass at the State Optical Institute in Leningrad. ECKART's formulae (see Sect. 70) were used for comparison with experiments.

The X-ray reflection coefficients were measured in an apparatus illustrated in Fig. 150. Goniometric arrangements are used both in the "monochromator" as well as for the rotation of the grating or mirror under test. Two Geiger counters

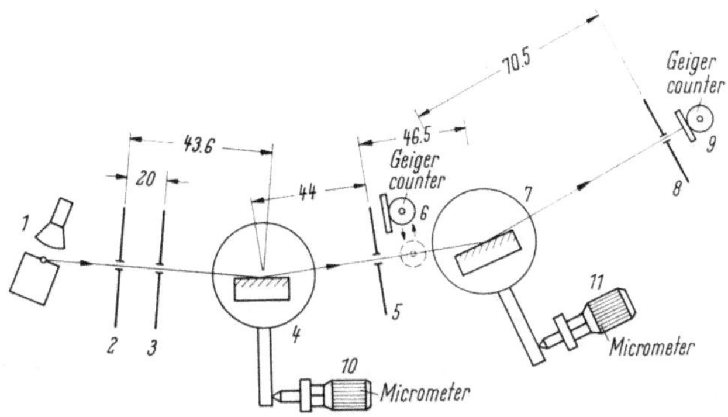

Fig. 150. X-ray efficiency measuring apparatus according to LUKIRSKII and SAVINOV (1962).

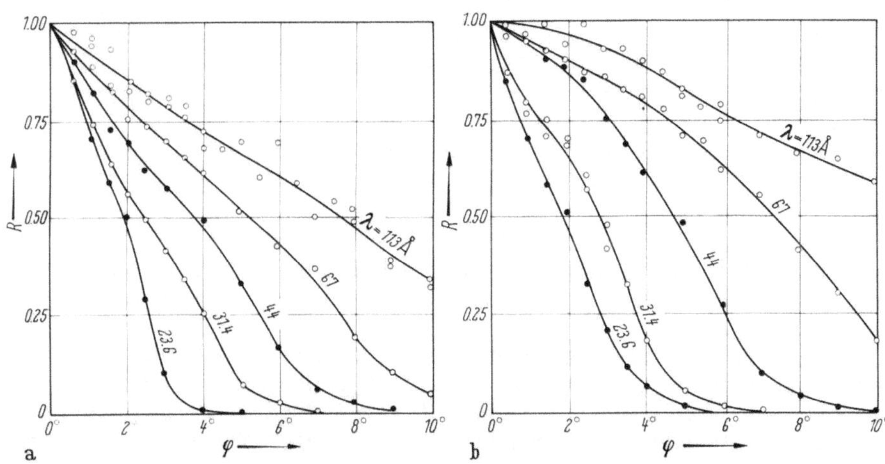

Fig. 151 a and b. Reflection coefficients of glass (a) and of titanated glass (b) at x-ray wavelengths, according to LUKARSKII and SAVINOV loc. cit. 1962). a Reflection curves for a glass surface (F-1 glass) for wavelengths of 23.6, 31.4, 44, 67, and 113 Å. *R* reflection coefficient; *φ* grazing angle of incidence. Two series of experimental points are given on the curve for 113 Å. b Reflection curves for a titanated surface for wavelengths of 23.6, 31.4, 44, 67, and 113 Å. *R* reflection coefficient; *φ* grazing angle of incidence.

are used, one for measuring the incident radiation, and one for measuring the reflected radiation. Characteristic radiation is generated in the X-ray tube and is directed to the monochromator through the two collimating slits shown. Slit-widths of the four slits shown were, from left to right, 10 μ, 10 μ, 30 μ and 1000 μ. The relative efficiencies of the counters were determined experimentally. The intense K-series lines of the elements O, N, C, B and Be, which have wavelengths of 23.6, 31.4, 44, 67 and 113 Å, respectively, were selected as the characteristic lines for which reflection coefficients were determined.

Measurements by LUKIRSKII and SAVINOV gave not only grating efficiency measurements, but also reflection coefficients for glass, both uncoated as well as coated with titanium. Fig. 151 a and 151 b show the reflection coefficients for these two cases, with various angles near "grazing" incidence. Appreciable gain is attained with the 400 Å thick titanium coating. It is noticed that reflection coefficients near unity are obtained at 113 Å up to angles of almost 3°, with the titanium coatings.

The best curves obtained for a 600 groove/mm echelette ruled on glass are shown in Fig. 152a and 152b. Fig. 152a shows the efficiency of the echelette as a function of angle of incidence, while Fig. 152b, shows the efficiency as a function of wavelength, for various angles of incidence. Many additional details are found in the original articles[133]. Higher relative efficiencies should be attainable with the help of rigorous calculations based on electromagnetic theory.

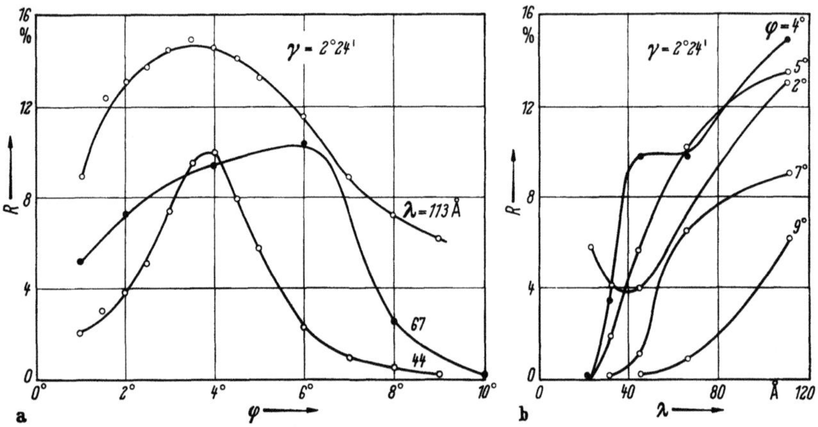

Fig. 152a and b. Reflection coefficients of a 600 groove/mm grating at x-ray wavelengths, according to LUKARSKII and SAVINOV (loc. cit. 1962) (order $m = -1$). a Dependence of the reflection coefficient on the angle of incidence φ for various wavelengths. b Dependence of the reflection coefficient on the wavelength for various angles of incidence.

78. Scattered light and stray light. General aspects of stray light and of scattered light and its origin in grating ruling imperfections are discussed in Sect. 58. (For a distinction between *stray* light and *scattered* light, see end of this section.)

Measurements of scattered light require considerable care and generally do not give representative values, without suitable photometric corrections, such as those well known to solar spectroscopists*. One method, using the sodium D line with an absorption cell is described by C. D. SHANE, JOHN WADDELL and others**.

* See, e.g. L. GOLDBERG and A. K. PIERCE: The Photosphere of the Sun. In: Handbuch der Physik, Vol. 52, edit. by S. FLÜGGE. Berlin-Göttingen-Heidelberg: Springer 1959. — R. R. MCMATH and O. C. MOHLER: Solar Instruments. In: Handbuch der Physik, Vol. 54, edit. by S. FLÜGGE, pp. 1—40. Berlin-Göttingen-Heidelberg: Springer 1962.

** The following references are due to the kind interest of Prof. Dr. G. H. E. ELSTE private communication, 1964).

a) C. D. SHANE: The photometry of lines in the solar spectrum. Lick. Obs. Bull. **16**, 76 (1932); using neodymium chloride absorption tube, 5219 A.

b) P. TEN BRUGGENCATE, J. HOUTGAST u. H. v. KLÜBER: Beobachtungs- und Reduktionstechnik bei den Potsdamer spektralphotometrischen Untersuchungen von Fraunhofer-Linien. Publ. Potsdam Astrophys. Obs. **29** No. 96 (1939).

c) M. T. CONWAY: Studies of $H\alpha$-line profiles in prominences. Month. Not. Roy. Soc. **112**, 55 (1952).

d) M. A. ELLISON: A photometric survey of solar flares, plages and prominences in $H\alpha$ light. Publ. Roy. Obs. Edinburgh **1**, No. 5 (1952).

e) J. W. WADDELL III: An experimental determination of the turbulence field in the solar photosphere based on the study of weak Fraunhofer lines. Thesis, Un. of Michigan 1956.

f) A. v. ALVENSLEBEN: Apparatekonturen, Gittergeister und Streulicht der beiden neuen Plangitter am Göttinger Sonnenturm. Z. Astrophys. **43**, 63 (1957).

By varying the pressure in the cell heavily saturated lines of different widths are photoelectrically recorded. The residual intensities at the bottoms of the lines, relative to the continuum can be taken as a measure of the scattered light, after suitable correction for the far wings of the actual instrumental profile (including all of the ghosts and satellites, see Sect. 50)*.

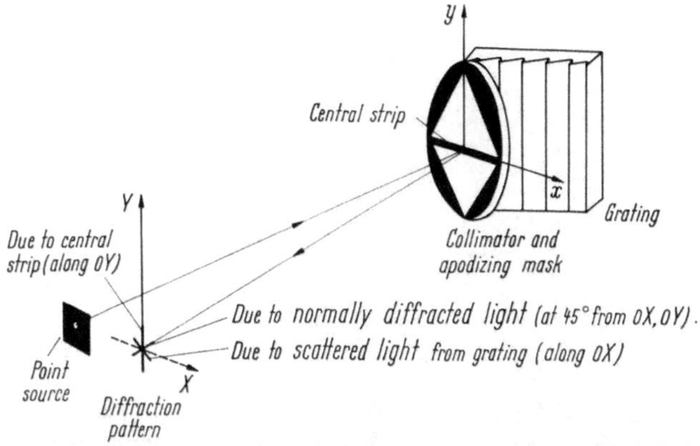

Fig. 153. Measurement of scattered light in gratings by apodization test according to P. JACQUINOT (loc. cit. 1964).

Fig. 154. Measurement of scattered light in a grating by apodization test according to JACQUINOT. Case of a *poor* grating, in which the scattered light (length of central horizontal arms) is about as intense as the normally diffracted light (length of the central oblique arms).

Another method, using apodization, is described by P. JACQUINOT**. A schematic diagram of the apodization test is shown in Fig. 153 and experimental results in Fig. 154 and in Fig. 155 for a poor and for a good grating respectively.

In both cases, heavily exposed photographs of the green mercury line were produced by the gratings, masked with a diamond-shaped diaphragm and an opaque strip placed diagonally as indicated in Fig. 153. A point source, rather than a slit, was used to illuminate the grating.

The dispersion axis in the figures is horizontal. The vertical fringe is due to the narrow opaque slit placed diagonally on the diaphragm and perpendicular to the rulings.

* See footnote **, p. 648.

** P. JACQUINOT and B. ROIZEN-DOSSIER: Apodisation. In: Progress in Optics, Vol. III, edit. by E. WOLF, p. 132 and p. 133. Amsterdam: North-Holland Publ. Co. 1964 (Prof. JACQUINOT acknowledges the assistance of G. W. STROKE in taking the photographs of Fig. 154 and 155).

The "poor" grating of Fig. 154, has heavy random ruling errors which produce a continuous background near the line; this is evident in the photograph as a horizontal line of which the length is of the same order of magnitude as the arms of the oblique cross; this means that the scattered light is about as intense as the light which would normally be diffracted, in the absence of apodisation.

The grating also gives a series of weak spurious lines (satellites and ghosts).

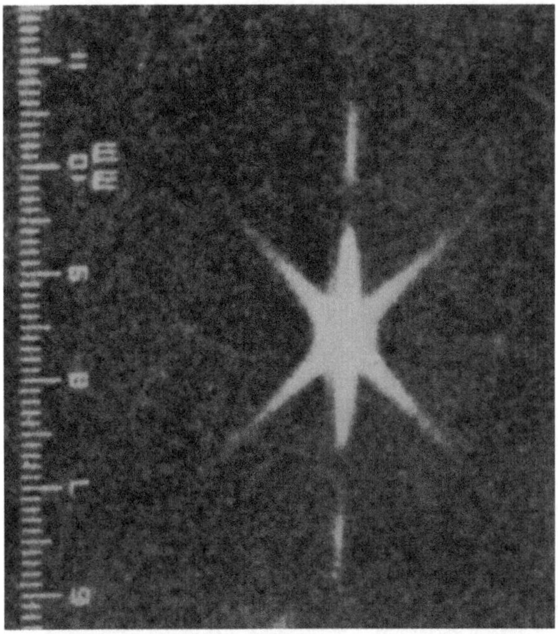

Fig. 155. Measurement of scattered light in a grating by apodization test according to JACQUINOT. Case of a *good* grating, in which the scattered light (vanishingly small length of central horizontal arms) is much weaker than the normally scattered light (length of central oblique arms).

The "good" grating of Fig. 155, shows a complete absence of the horizontal line, indicating that the scattered light is much weaker than the normally diffracted light (P. JACQUINOT, loc. cit. 1966).

GERASIMOV (loc. cit. 80) refers to measurements of scattered light by BOBOVICH*.

Two remarks need to be made concerning special aspects of scattered light in gratings.

First, the *theory* of scattered light and its relation to groove imperfections, as well as to groove irregularities needs to be further clarified, developed, and verified. It is likely that electromagnetic theory, as well as electron-microscope and phase-contrast and perhaps X-ray measurements may complement scalar theory and photometric measurements. Generally speaking, scattered light in the wanted wavelength range and amounting to 1 % or even 0.1 % or 0.01 %, depending on the application, may be considered quite harmful.

Second, it does not appear to be true, as sometimes considered, that scattering of light is always an undesirable characteristic in optical gratings. Indeed, in two cases, respectively in the far infrared and in the extreme ultraviolet, it appears that a grating would not be usable unless it suitably scattered some of the low-wavelength portions in the incident radiation.

The desirable scattering of light will be discussed in its qualitative aspects here. A detailed theory of the methods for best achievements of desirable scattering remains to be developed.

Basically, if a grating is illuminated with wavelengths shorter than the shortest wavelength of interest, then the overlapping orders, and overlapping "scattered" light need to be filtered out in some suitable way.

The filtering of the shorter wavelengths can be accomplished if the grating, and more particularly the grooves, appear "smooth" to the desired long wavelengths, but "rough" to the undesirable short wavelengths.

* YA. S. BABOVICH, Candidate's Diss., State Optical Institute 1953.

For instance, in a grating with a 1 mm spacing, ruled at the University of Michigan[135 b] for use in the 100 to 1000 micron range, particular pains were taken to make the coarse echelette ruling appear particularly rough to the eye, in visible light. Gratings with well polished grooves appear to result in an overwhelming amount of undesirable overlap of visible light in this case.

Similarly, in the extreme ultraviolet, both above and below 500 Å, there is some evidence[136] of suppression of short wavelengths by scattering. More generally important spectroscopic results in solar UV research seem to have been obtained not only by judicious suppression of unwanted long wavelengths, but apparently also by scattering of unwanted short wavelengths, by means of groove "roughness", not unlike in the infrared case referred to above. Extreme UV spectroscopy above the atmosphere with rocket-borne spectrographs has been carried out in remarkable experiments by R. TOUSEY[137] and collaborators at the U. S. Naval Research Laboratory (see also TOUSEY, loc. cit. 128).

In the extreme ultraviolet below 500 Å, gratings are still used at grazing incidence. Above 500 Å, and more particularly above 1000 Å, the new improvements in vacuum ultraviolet reflectivities (Sects. 69 and 70) have made possible "normal-incidence" work as well (loc. cit. 128). TOUSEY[128] reports on a very lightly ruled grating, ruled by SIEGBAHN on glass and used near grazing incidence. In these gratings, the rulings and spaces usually have roughly equal widths. TOUSEY reports on another "school", favoring extreme UV gratings ruled in aluminum.

Recently, there has been increasing evidence that gratings ruled in evaporated gold films, perhaps not unlike the gratings already ruled in gold by FRAUNHOFER in 1823 (loc. cit. 2, Sect. 2) show particularly low levels of that part of stray light which seems to originate from groove-shape irregularities[128], and which appears to be detrimental in all cases.

The many special techniques[135] used to achieve high-resolution spectra and suppress unwanted long-wavelength stray light in the extreme ultraviolet with even the rather imperfect grating still available for this domain are outside the scope of this article. Among the remarkable techniques used to eliminate the long-wavelength end of the spectrum are unsupported thin metallic films discovered by ASTOIN and VODAR[138] in 1953. For instance, a 1000 Å thick aluminum film is completely opaque in the visible and near ultraviolet, and almost completely opaque down to 850 Å, the position of its electron eigenloss.

However, at $\lambda < 850$ Å, the transmittance begins to rise, reaching 40 to 50% from 300 Å to the $L_{2,3}$ edge near 170 Å. Shortward of 167 Å, these aluminum films are again quite opaque (TOUSEY, loc. cit. 128). Thin unsupported aluminum films have appeared to be of great value in eliminating long-wavelength stray light from photographic spectrographs used in the extreme ultraviolet[128,137]. They were first used by TOMBOULIAN and HARTMANN[139] and TOMBOULIAN and BEDO[140] in 1956 and 1955.

---

[135] a) E. F. BARKER and H. M. RANDALL: The Ruling of Gratings for the Infrared at the University of Michigan. Private communication to G. W. STROKE (1961). See also H. M. RANDALL: J. Appl. Phys. **10**, 768 (1939). — b) P. A. WEYRICH: Private communication to G. W. STROKE (1964).

[136] S. G. TILFORD: J. Opt. Soc. Amer. **53**, 1051 (1963); and private communication (1964).

[137] R. TOUSEY: The extreme ultraviolet spectrum of the Sun. Space Sci. Rev. **2**, 3—69 (1963).

[138] N. ASTOIN and B. VODAR: J. Phys. Radium **14**, 424 (1953).

[139] D. H. TOMBOULIAN and P. L. HARTMANN: Phys. Rev. **102**, 1423 (1956).

[140] D. H. TOMBOULIAN and D. E. BEDO: Rev. Sci. Instrum. **26**, 747 (1955).

Clearly, much additional work needs to be done to clarify the distinction between

a) *Generally unwanted, usually long-wavelength, "stray light"*, resulting apparently from comparatively "macroscopic" groove irregularities and local ruling imperfections.

b) *Sometimes wanted, short-wavelength, scattering*, resulting, apparently, from groove "roughness" and "microscopic" groove imperfections.

It may appear, as predicted by Maréchal (Sect. 58), that the relative amount of *scattered* light will increase very rapidly, with the decrease in wavelength, perhaps as λ^{-4}, while the amount of *stray light* may well increase with λ^{-2}, following the equations given by Stroke (Sect. 51) for the effect of small ruling errors.

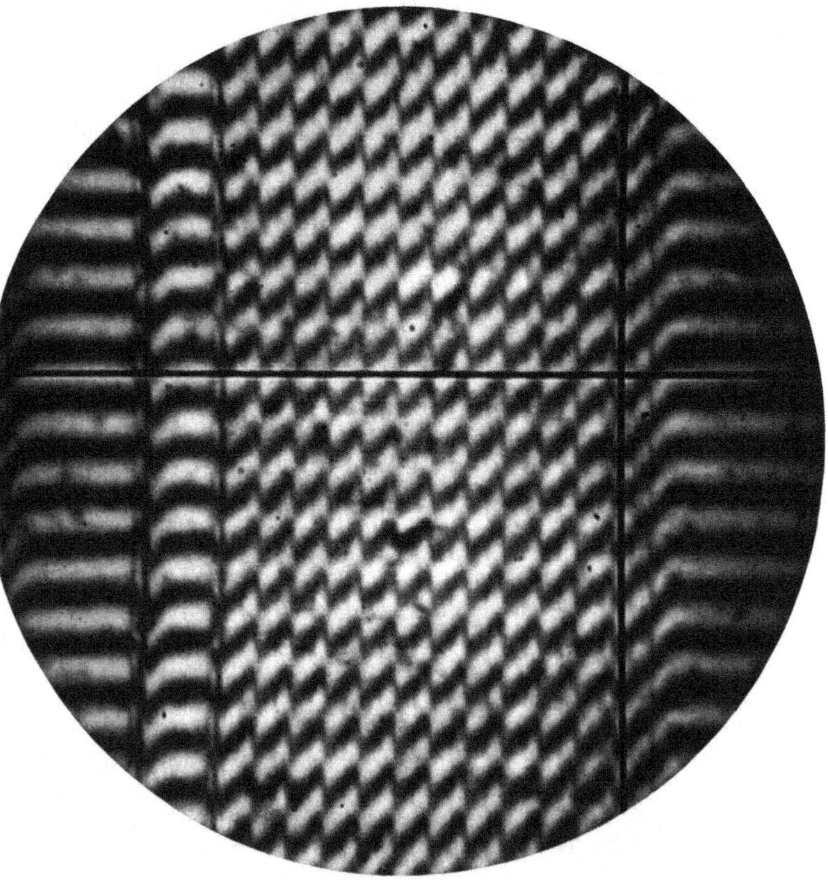

Fig. 156. Interference microscope photograph of a 600 grooves/mm grating with fully ruled grooves. Note straight horizontal fringes (along x) in the unruled portions of the blank on left and right. Also note appearance of grooves at ends of ruled section (Massachusetts Institute of Technology).

79. Interference microscope testing of groove forms. A two-beam, Michelson Twyman-Green interferometer can be adapted to a microscope for the purpose of testing groove shapes. An interferogram of a 600 grooves/mm grating is shown in Fig. 156.

Generally speaking, the resolution of details obtained for grooves with spacings finer than 300 grooves/mm is found to be insufficient. On the other hand, the interference fringes tend to crowd too much for grooves much coarser than 300 grooves/mm. In practice, electron-microscopy (Sect. 80) has proven to provide considerably more detailed information about grooves than interference microscopy.

Interference microscopy for groove testing was used at least as far back as 1926 by GLASER[141], with gratings of different origins, to test ruling errors, local groove defects and so on. More recently, interference microscopes have been built and used by most grating ruling laboratories. The interpretation of interference micrographs is straightforward. Generally, the fringes in the unruled portions of the grating are made to run at right angles to the groove length.

Birefringent interference microscopes, of the type described by FRANÇON[142,143], NOMARSKI[144] and others, have also been successfully used for the examination of grating grooves.

Interference microscopes are particularly useful in the adjustment of the ruling diamond preceding the ruling. Diamond chatter, tearing of the film, and imperfections in ruling depths may be readily detected with some experience.

Final adjustment of the ruling diamond prior to ruling is however best obtained with the help of efficiency measurement in a suitable spectrometer.

80. Electron-microscope testing of groove forms. The importance of attaining appropriate groove shape and depth for a desired energy distribution in a given grating cannot be sufficiently emphasized. The correlation between energy

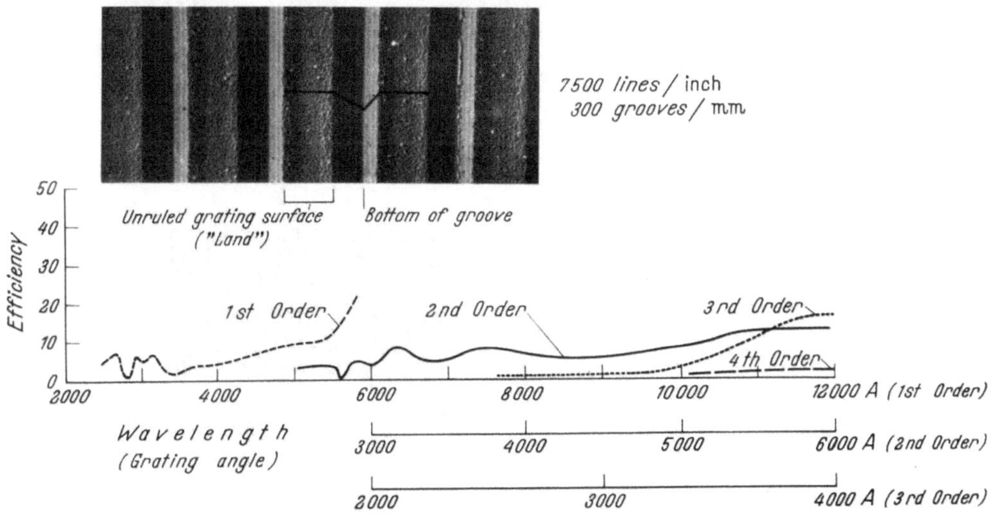

Fig. 157. Electronmicrograph of poorly blazed 300 groove/mm grating, and corresponding efficiency, according to STROKE (1963).

distributions in gratings and corresponding groove-forms and ruling-depths as observed with electron microscopes was demonstrated in 1963 in a series of experiments by STROKE (loc. cit. 112, 119) (see Figs. 157 to 160), following very

[141] L. C. GLASER: Z. techn. Phys. **7**, 31 (1926).

[142] M. FRANÇON: Rev. Opt. **30**, 221 (1951).

[143] M. FRANÇON: In: Handbuch der Physik, Vol. 24, edit. by S. FLÜGGE, pp. 452ff. Berlin-Göttingen-Heidelberg: Springer 1956.

[144] G. NOMARSKI: J. Phys. Radium **16**, 9 (1955).

considerable improvements in the techniques of electron-microscopy for grating testing, obtained in particular by H. VOLKMANN[145], and H. J. PRÖGER[146].

As described by PRÖGER, the first applications of electron microscopy specifically for the testing of gratings appear to have been given in 1957 by VERCNER, DEGTEWA and CHARIONOVSKIJ[147], even though electromicrographs of gratings

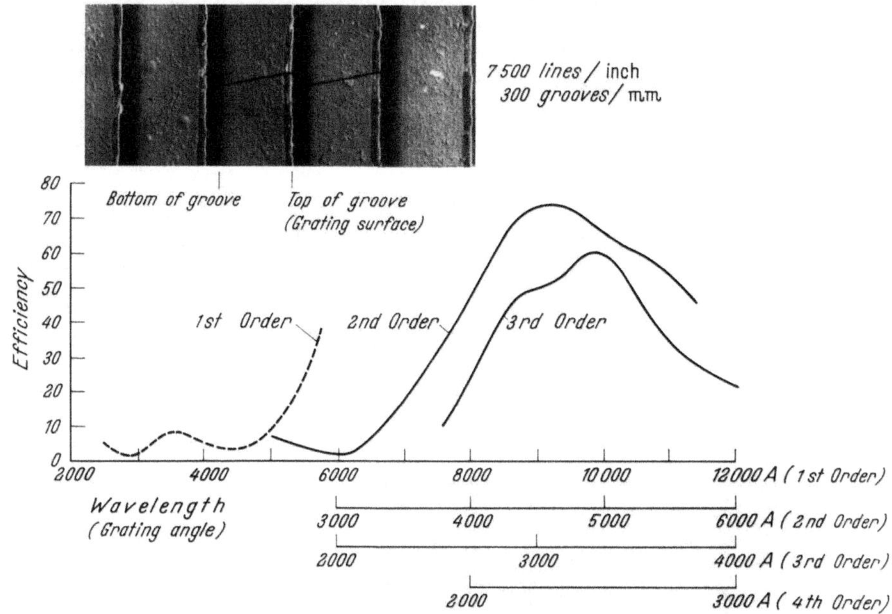

Fig. 158. Electronmicrograph of well blazed 300 groove/mm grating, ruled with 120° diamond, and corresponding efficiency, according to STROKE (1963).

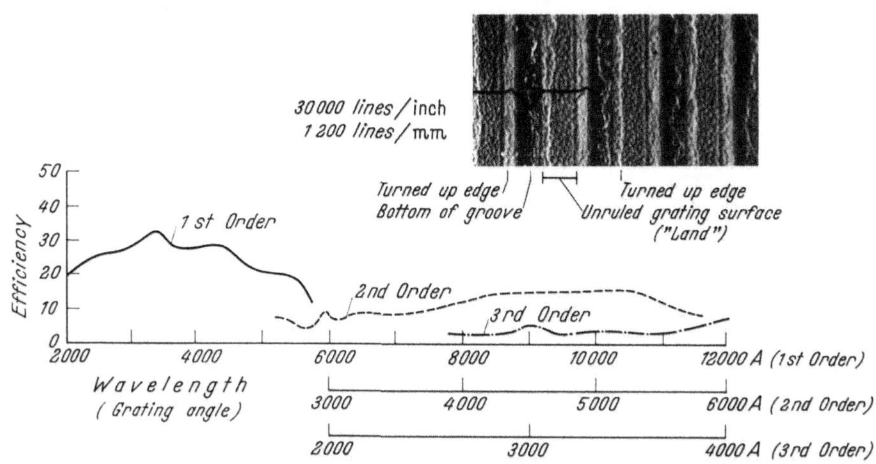

Fig. 159. Electronmicrograph of very poorly ruled 1200 groove/mm grating, and corresponding efficiency, according to STROKE (1963).

---

[145] H. VOLKMANN, (Carl Zeiss), private communications (1961).

[146] H. J. PRÖGER: Elektronmikroskopische Untersuchung von Beugungsgittern. Exp. Techn. der Phys. 9, 236—252 (1961).

[147] V. N. VERCNER, L. V. DEGTEWA and I. S. CHARIONOVSKIJ: Optika i Spektroskopiya 3, 181 (1957).

had previously appeared in the literature, principally as examples of this method of microscopy[148,149] or as tests for magnification[150,151].

Some of the remarkable stereoscopic electron-micrograph pairs obtained by VOLKMANN[145] are shown in Figs. 161 and 162. A very complete description of the stereoscopic method of electron microscopy with many remarkable examples is given by PRÖGER[146], along with other principles and examples of the various methods used.

An interesting sidelight on the history of the origin of the electron microscope is given by FREUNDLICH[152]. It appears that the first electron microscope was built and publicly demonstrated early in 1931 by MAX KNOLL and ERNST RUSKA,

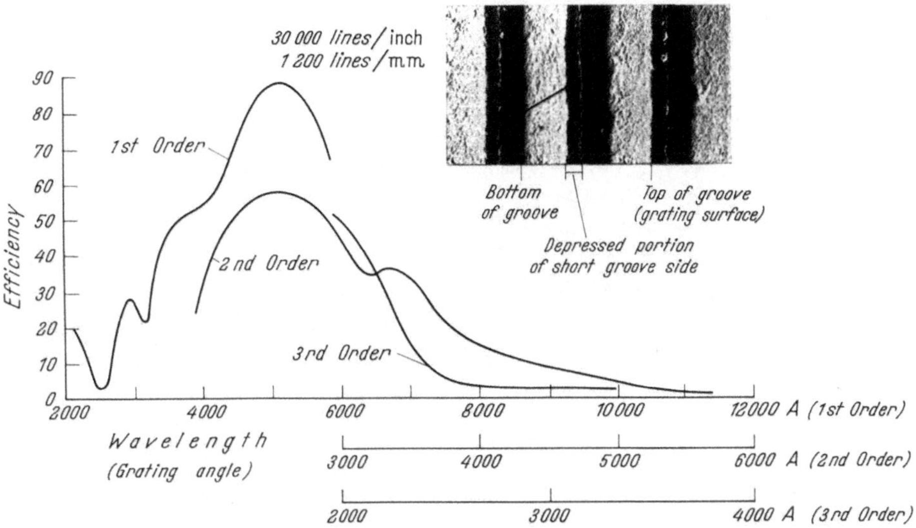

Fig. 160. Electronmicrograph of very well blazed 1,200 groove/mm grating, ruled with 110° diamond, and corresponding efficiency, according to STROKE (1962).

working at the High Tension Laboratory of the Technische Hochschule in Berlin under A. MATTHIAS. KNOLL and RUSKA followed up their demonstration with a paper "Beitrag zur geometrischen Elektronenoptik"[153] which was finally submitted for publication on September 10, 1931, after a public lecture by KNOLL on 4 June 1931. The interesting and apparently somewhat controversial history of the invention of the electron microscope is further described by FREUNDLICH (loc. cit. 152) in noting, as pointed out by GABOR[154] and MULVEY[155] that "at least in patent law" REINHOLD RÜDENBERG appears to be the "inventor" of the electron microscope. Indeed, according to FREUNDLICH, RÜDENBERG filed patent application with the German Patent Office on 31 May 1931, a mere four days before KNOLL's long announced public lecture on 4 June 1931.

Several methods have been found useful in obtaining grating groove-profile and surface information with the help of electron-microscopic examination[146].

---

[148] G. HASS and M. E. MCFARLAND: J. Appl. Phys. **21**, 435 (1950).
[149] G. HELWIG and H. KÖNIG: Optik **7**, 294 (1950).
[150] S. R. ROUZE and J. L. WATSON: J. Appl. Phys. **24**, 1106 (1953).
[151] R. S. PEASE: J. Sci. Instrum. **27**, 182 (1950).
[152] M. E. FREUNDLICH: Science **142**, 185—188 (1963).
[153] M. KNOLL and E. RUSKA: Ann. Physik **12**, 607 (1932).
[154] D. GABOR: Elektrotechn. Z. **78**, 522 (1957).
[155] T. MULVEY: Brit. J. Appl. Phys. **13**, 197 (1962).

a

Fig. 161 a and b. Stereoscopic electronmicrograph pair of a 600 groove/mm grating with symmetrical grooves, according to H. VOLKMANN (Carl Zeiss, 1961).

The methods of replication using thin plastic metallic and carbon films[156,157] are most widely used. Resolutions on the order of 10 to 20 Å are readily obtainable

[156] L. REIMER: Elektronische Untersuchungen und Präparations-Methoden. Berlin-Göttingen-Heidelberg: Springer 1959.

[157] V. E. COSSLETT: Practical Electron Microscopy. London: Butterworth 1951.

Fig. 161 b.

with contemporary electron microscopes. Some loss in resolution may result from the replication process itself reducing resolution of details somewhat above the limit of the microscope. Scale changes of the order of 2 to 10% may occur with the thin plastic films used for electron-microscopy, while 0.5% may be maintained with the thin, metallic films (for instance with aluminum, gold and

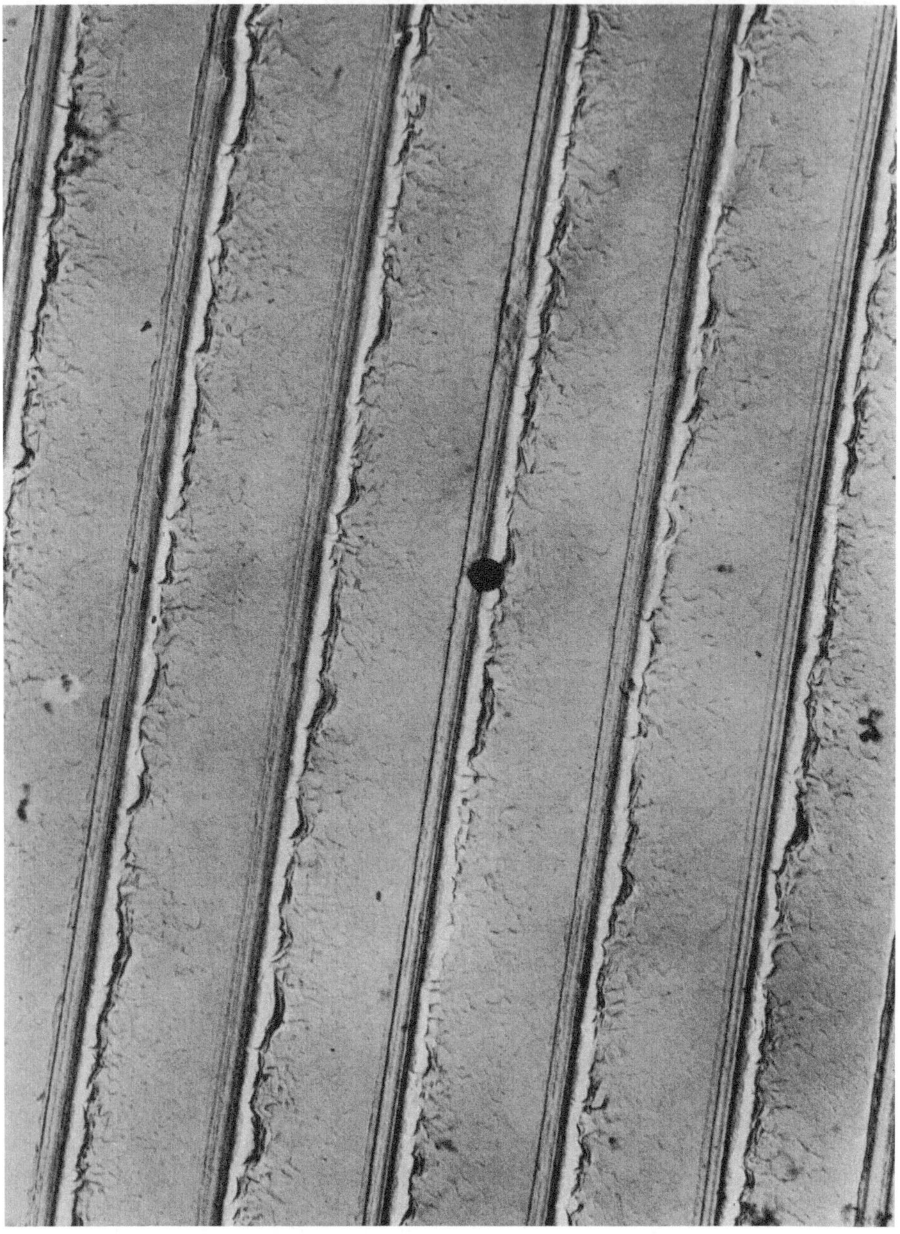

a

Fig. 162a and b. Stereoscopic electronmicrograph pair of a 600 groove/mm grating ruled with a poorly adjusted diamond, according to H. Volkmann (Carl Zeiss, 1961).

silver)[146,158]. In general, final preparation of the replicas involves preparation of a carbon film, evaporated at some angle to the grating surface[146]. The method of shadow casting as first described by Müller[159] and perfected by Williams

[158] S. R. Rouze and J. H. L. Watsou: J. Appl. Phys. 24, 1106 (1953).
[159] H. O. Müller: Kolloid-Z. 99, 6 (1942).

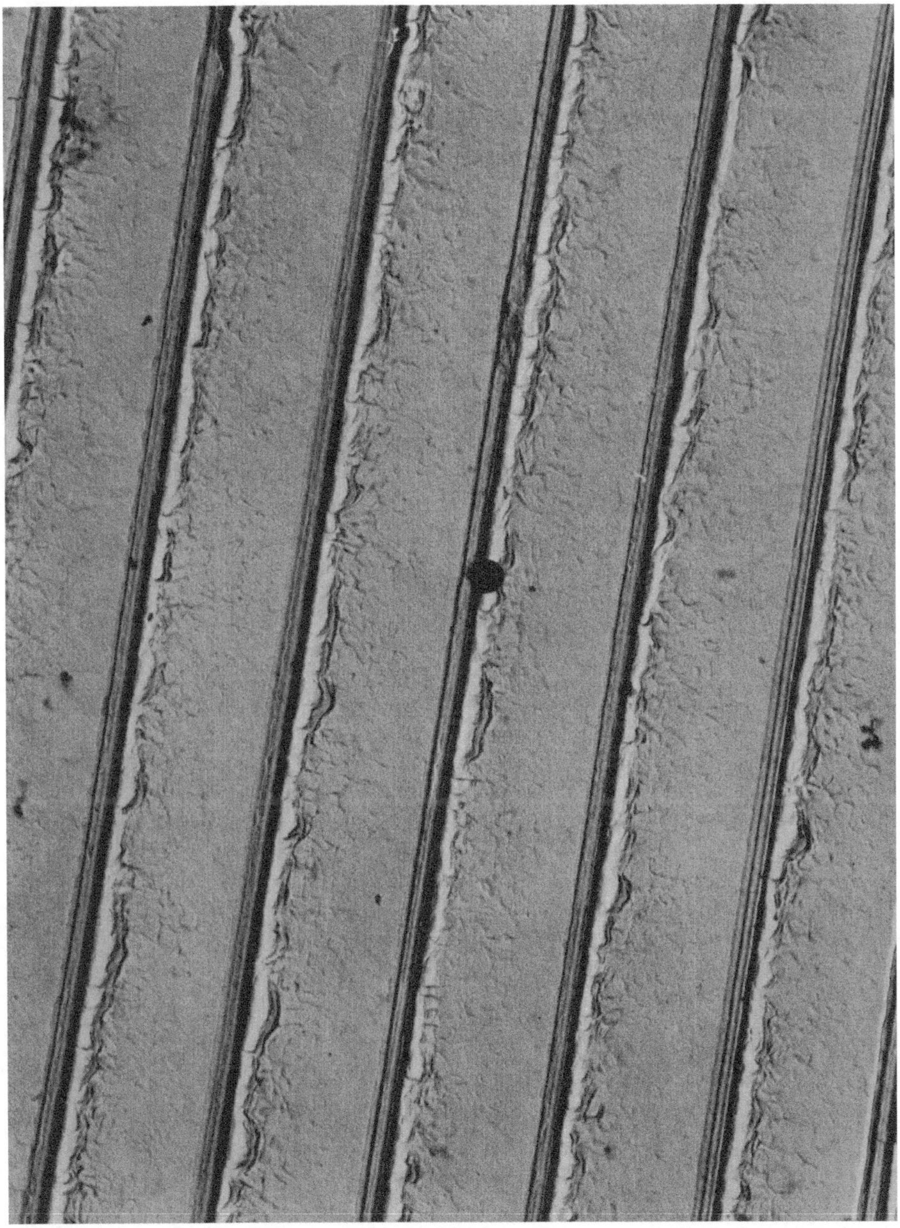

Fig. 162b.

and WYCKOFF[160] greatly enhances the visibility of the surface deformations and permits remarkably accurate measurement of grating groove angles (Fig. 157 to 160). Details are given in COSSLETT (loc. cit. [157]) and in the references cited by PRÖGER (loc. cit. [146]).

[160] R. C. WILLIAMS and R. W. G. WYCKOFF: J. Appl. Phys. 17, 23 (1946).

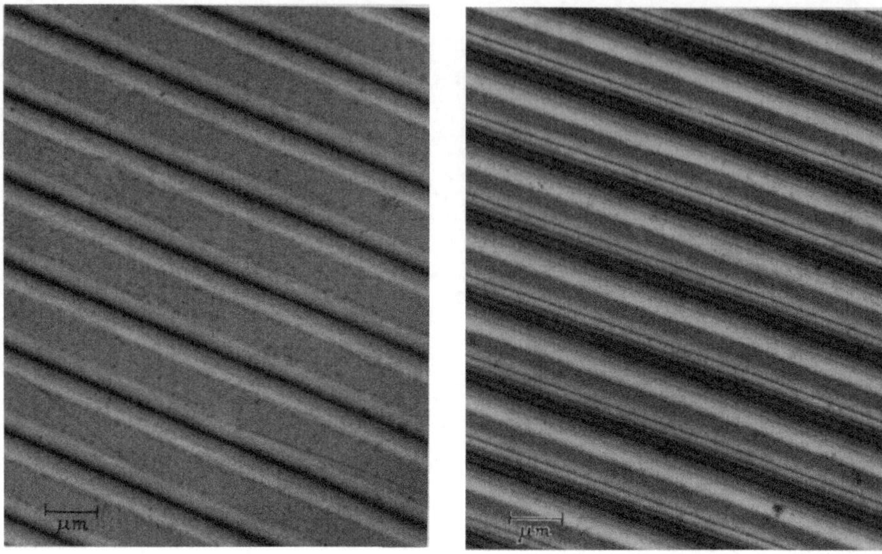

a

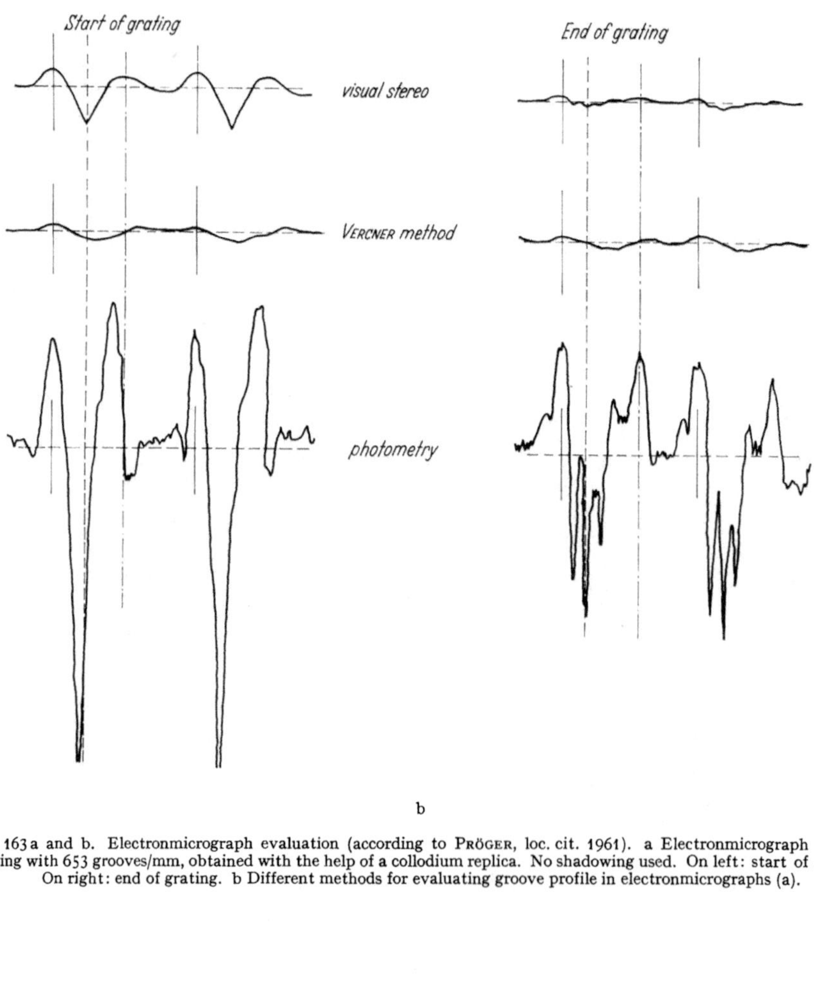

b

Fig. 163a and b. Electronmicrograph evaluation (according to Pröger, loc. cit. 1961). a Electronmicrograph of glass grating with 653 grooves/mm, obtained with the help of a collodium replica. No shadowing used. On left: start of grating. On right: end of grating. b Different methods for evaluating groove profile in electronmicrographs (a).

Stereophotography has yielded particularly useful results in grating electron-microscopy. In early work[161,162] the direction of the illuminating beam itself was shifted by a few degrees, to obtain two slightly different views of the object in a fixed camera. It appears simpler, however, to tilt the object itself by a few degrees between exposures with a fixed direction of illumination (COSSLETT, loc. cit. [157]), and this method is now generally used. A special specimen holder is used, sometimes with pre-set positions for the two exposures: an angle of 5 to 10° on either side of the axis is usually employed[146,157]. Recent electron-microscopes incorporate devices which allow the holder to be tilted slowly by external control, while the image is kept under continuous observation on the viewing screen[157]. Suitable

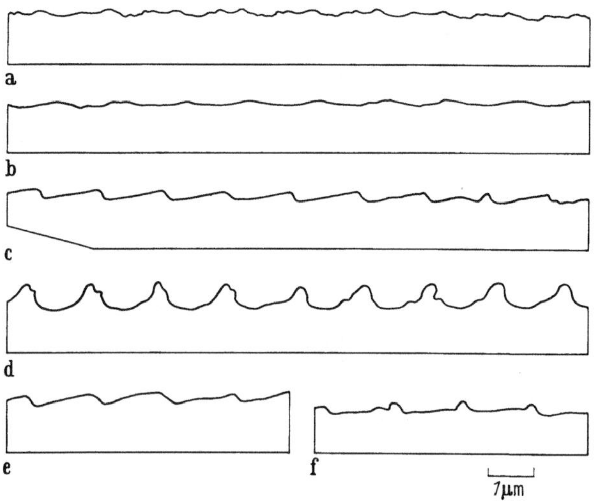

Fig. 164 a–f. Evaluation of groove profile by method of VERCNER, DEGTEWA and CARIONOVSKIJ (according to PRÖGER, loc. cit. 1961). Glass gratings. a End section of grating; b Very poor section; c Sawtooth profile; d Incorrect diamond loading; e Convex profile; f Concave profile.

translation of the object permits to compensate for any sidewards shifts resulting from the tilt. The various techniques of stereophotography in electron microscopy, and the measurements of the topography are discussed by DRUMMOND[163], HEIDENREICH and MATHESON[164], HELMCKE[165], POHLMANN[166] and others[145,146].

Other methods, in addition to stereophotography are also discussed in the extensive article by PRÖGER (loc. cit. [146]) and comparisons of results obtained with different methods for the same grating are also given.

For instance, Fig. 163 shows according to PRÖGER[146] the electron-micrographs of the beginning and of the end of a ruling with 653 grooves/mm, and the evaluation with the help of three methods:

1. Visual stereoscopic examination
2. A profile-enhancing method described by VERCNER et al.[147].
3. A photometric method described by MORTON and SCHIFF[161].

Several grating profiles examined according to the method of VERCNER et al. are shown in Fig. 164.

[161] L. MARTIN and L. I. SCHIFF: J. Appl. Phys. **12**, 759 (1941).

[162] E. KINDER: Naturwissenschaften **33**, 367 (1946).

[163] D. G. DRUMMOND: The Practice of Electron Microscopy. London: Royal Microscopical Society 1950.

[164] R. D. HEINDENREICH and L. A. MATHESON: J. Appl. Phys. **15**, 423 (1944).

[165] J. G. HELMCKE: Optik **12**, 253 (1955).

[166] G. POHLMANN and M. AHREND: Optik **16**, 461 (1953).

Various groove-profiles as obtained with the help of stereophotography by Pröger are shown in Fig. 165.

Finally, Fig. 166 shows (for the grating of Fig. 167) a reconstruction of the grating groove profile obtained by means of the electron-micrograph evaluation methods indicated by Pröger (loc. cit. [146]).

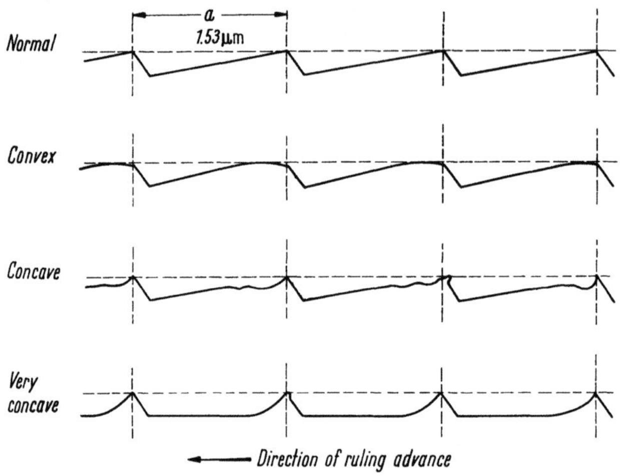

Fig. 165. Groove profiles of various gratings ruled in aluminum, according to stereoscopic electronmicrographic study by Pröger (loc. cit. 1961).

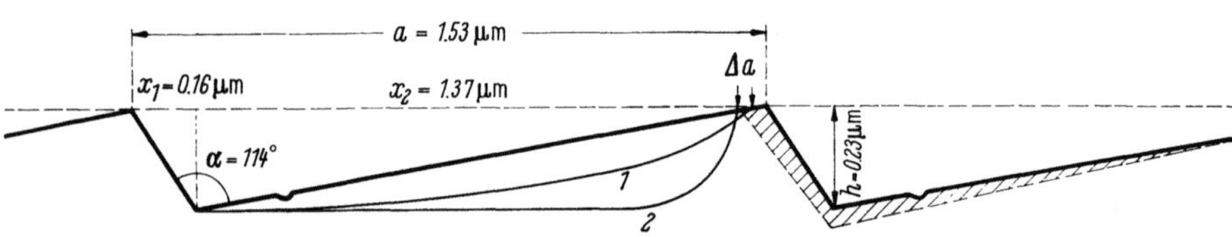

Fig. 166. Groove profile of grating of Fig. 167, obtained by evaluation of electronmicrograph with the help of stereoscopic and profile measuring methods, according to Pröger (loc. cit. 1961). 652 groove/mm grating ruled in aluminum (groove constant $a = 1.53$ µm).

The parameters measured are:

Groove spacing	from magnification
Diamond angle	by profile method
Depth of ruling	by profile and stereoscopic methods
Distance x_1	by stereoscopic method
Distance x_2	by stereoscopic method

Estimated errors in evaluating the grating constant a, as a result of shrinkage of the replica:

For case 1:	−1.4%
For case 2:	−4.1%
Measurement error:	±1%

(all values according to Pröger).

In brief, the methods used for electron-microscopic evaluation of gratings are[146]:

1. *Visual examination of electron-micrographs*, obtained by replication, with or without shadow-casting.

2. *Photometric measurement of electron-micrographs*, using the density variation in the replicas resulting from thickness modulation by the grating surface.

3. *Stereophotographic electron-micrographs.*

4. *Groove-profile electron-microscopic photography*, achieved by examining a *section* of the grating, at right angles to the grating surface (and to the grooves). The method, as described by Vercner et al.[147] consists in first evaporating a silver layer on to grating surface. Next, a layer of copper is electroplated on the silver. The ductile film thus obtained is separated from the grating and folded

in such a way as to face the groove-profile outwards. The electron microscope then permits to obtain a shadow-image of the groove profile. In the variation of the method described by PRÖGER[146], deformations in the "replication" are avoided by replacing the copper-electroplating with an evaporation of 1 μm of lead on top of the silver. (The silver layer is necessary to avoid the tendency of lead to form an alloy with aluminum, and also to avoid the crystal structure characteristic of lead). The profiles of Fig. 164 were obtained with the last method, in grazing incidence illumination, as described under 4a.

Fig. 167. Electronmicrograph of grating evaluated in Fig. 166, according to PRÖGER (loc. cit. 1961). 652 groove/mm grating ruled in aluminum. Carbon replica shaddowed at an angle with Th.

4a. *Groove-profile electron-microscopic photography with grazing incidence illumination* (according to PRÖGER, loc. cit. [146]). An angle of 85° with respect to the grating-surface normal is used for illumination.

The various errors obtained in methods 4 and 4a are extensively discussed by PRÖGER[146]. Spacing "errors" of $\pm 5\%$, depth "errors" of $\pm 15\%$, and groove errors of $\pm 16\%$ to $\pm 6\%$ are representative of the measurement errors in groove-profile methods.

81. X-ray microscope imaging of grating grooves by means of wave-front reconstruction. Recent developments[168–175] in the method of microscopy by wave-front-reconstruction imaging, originated by D. GABOR in 1948[167] hold a new pro-

[167] D. GABOR: Nature, Lond. **161**, 777 (1948). — Proc. Roy. Soc. Lond. A, **197**, 454—487 (1949). — Research **4**, 107 (1951).

[168] See, e.g., A. V. BAEZ and H. M. A. EL-SUM: Effect of finite source size, radiation bandwidth and object transmission in microscopy by reconstructed wavefronts. In: X-ray Microscopy and Microradiography, pp. 347—366. New York: Academic Press 1957.

[169] A. LOHMANN: Optische Einseitenbandübertragung angewandt auf das Gabor-Mikroskop. Optica Acta **3**, 97—99 (1956).

mise for attaining in the x-ray domain the high-resolutions now obtained with lasers in the visible light domain (see Sects. 90—91). STROKE with students[173–175] first showed that holograms could be made recordable at 1 Å x-ray wavelengths by recording *Fourier-transform holograms*, in which the grating-like "carrier fringes" (which characterize most holograms) are considerably more widely spread out, with respect to the photographic grain, than the fringes in the usual Fresnel-transform holograms. [For a general background, see e.g.[175].]

VII. Ruling of optical gratings and production of grating-replicas.

82. Introduction. A very great increase in precision and simplicity has been introduced into grating ruling by the principle of interferometric servo-control, first demonstrated by G. R. HARRISON and G. W. STROKE in 1955[1,2]. Many of the extreme mechanical precision, previously achieved only in grating ruling engines, have now become common to many other mechanical, electro-mechanical and electro-optical devices. It may well be, therefore, that most of the detailed descriptions of grating ruling engines will become of a historical rather than of a prescriptive interest.

The theoretical advantages of simplicity, resolution and luminosity that would be obtained with optical gratings in high-resolution spectroscopy had been predicted for many years, ever since RITTENHOUSE[3] and FRAUNHOFER[4] ruled and invented the first gratings in 1785 and 1821 respectively.

Great advances in grating ruling were made by ROWLAND[5], MICHELSON[6], SIEGBAHN[7], JOHN STRONG[8], the BABCOCKS[9] and many others in the late 1890's and

---

[170] YU. N. DENISYUK: Photographic Reconstruction of the Optical Properties of an Object in its Own Scattered Radiation Field. Soviet Physics-Doklady **7**, 543—545 (1962) [Translated from Doklady Akad. Nauk. SSSR **144**, 1275—1278 (1962)].

[171] E. N. LEITH and J. UPATNIEKS: Reconstructed wavefronts and communication theory. J. Opt. Soc. Amer. **52**, 1123—1130 (1962).

[172] E. N. LEITH and J. UPATNIEKS: Wavefront reconstruction with continuous-tone objects. J. Opt. Soc. Amer. **53**, 1377—1381 (1963).

[173] G. W. STROKE and D. G. FALCONER: Attainment of High-Resolutions in Wavefront-Reconstruction Imaging. Physics Letters **13**, 306—308 (1964).

[174] G. W. STROKE: Lensless Fourier-transform Method for Optical Holography. Appl. Phys. Letters **6**, 201—203 (1965).

[175] G. W. STROKE: An Introduction to Coherent Optics and Holography. New York: Academic Press 1966 [270 pages, including the reprints of the three 'wavefront-reconstruction imaging' papers by D. GABOR].

[1] G. R. HARRISON and G. W. STROKE: Interferometric control of grating ruling with continuous carriage advance. J. Opt. Soc. Amer. **45**, 112—121 (1955).

[2] G. W. STROKE: Attainment of high-resolution gratings by ruling under interferometric servo-control. J. Opt. Soc. Amer. **51**, 1321—1339 (1961).

[3] DAVID RITTENHOUSE: Trans. Amer. Philos. Soc. **2**, 201 (1786). — T. D. COPE: The Rittenhouse Diffraction Grating. J. Franklin Inst. **215**, 99—104 (1932).

[4] J. FRAUNHOFER: Kurzer Bericht von den Resultaten neurer Versuche über die Gesetze des Lichtes und der Beugung desselben. Ann. Physik **74**, 337—378 (1823).

[5] H. A. ROWLAND: Phil. Mag. [4], **13**, 469 (1882). — Physical Papers, p. 487 and 689. Baltimore: Johns Hopkins University Press 1902.

[6] A. A. MICHELSON: Nature, Lond. **362** (1912). — University of Chicago Mag. **7**, 141 (1915). — Proc. Amer. Phys. Soc. **54**, 137 (1915). — J. Franklin Inst. **181**, 785 (1916). — Studies in Optics, p. 99. Chicago: Chicago Univ. Press 1927; reprinted in 1962.

[7] M. SIEGBAHN: Z. Physik **62**, 435 (1930). — M. SIEGBAHN and T. MAGNUSON: Z. Physik **95**, 133 (1935). — M. SIEGBAHN: K. Svenska Vetenskaps Akademiens Årbok (1940), p. 151.

[8] J. STRONG: New Johns Hopkins Ruling Engine. J. Opt. Soc. Amer. **41**, 3—15 (1951). — J. STRONG: The Johns Hopkins University and Diffraction Gratings. J. Opt. Soc. Amer. **50**, 1148—1152 (1960).

[9] H. D. BABCOCK and H. W. BABCOCK: Ruling of Diffraction Gratings at the Mount Wilson Observatory. J. Opt. Soc. Amer. **41**, 776—786 (1951).

the first half of the twentieth century. But attempts in ruling gratings of sufficient regularity for spectral resolutions in the 1 million resolving power range had not met with success until 1955, when G. R. HARRISON and G. W. STROKE first demonstrated that theoretical perfection could be attained in optical gratings by ruling gratings under interferometric servo-control[1,2].

Several dozen high-resolution gratings up to 10-inches in width have now been obtained by ruling under interferometric control, not only from the 12-inch Michelson ruling engine at M.I.T.[1,2,10,11] but also from a 4-inch engine at Bausch and Lomb, as well as from 10-inch engines at the Mount Wilson and Mount Palomar Observatory Laboratory[12] and at the Jarrell-Ash Company[13], all in the United States of America, as well as from the 4-inch Jobin et Yvon engine in France.

Interferometrically controlled ruling engines are under completion in Japan[14,15] and Germany[16], and undoubtedly elsewhere as well.

Most mechanical engines, even successful ones, are readily adaptable to interferometric control, as first demonstrated by HARRISON and STROKE[1,2], and later by BABCOCK[12] and STROKE[13]. Interferometric control systems are known to be in preparation even for the most ingenious mechanical engines, such as the two-screw engine designed by JOHN STRONG[8], as well as for the more traditional ones, such as the PERKIN-ELMER engine designed by JULIUS PEARSON[17].

In brief, a new era in the attainment of mechanical perfection appears to have been introduced by interferometric servo-control and more generally by optical feedback control of mechanical machines and devices.

That any increase over built-in mechanical precision, frequently bordering on art, would be achieved in an electro-mechanical device or engine had previously been the subject of very serious question. It was HARRISON and STROKE who first demonstrated in 1955 (and more generally in 1957 and finally in 1961) that very considerable improvements in the performance of a grating ruling engine would indeed result from very carefully interpreted interferometric measurements of position and angles, translated into electronic servo-control.

Even these advances have so far been restricted to the control of rather massive grating-blank and ruling carriages, at points rather remote from the groove being ruled. In a new attempt on improving grating precision, STROKE, MOHLER and DENTON[18] have now undertaken to attain complete three-dimensional control of the ruling diamond itself, and thus to avoid any of the several insidious errors which seem to appear between different points in a carriage when subject to mechanical control.

---

[10] G. R. HARRISON, N. STURGIS, S. C. BAKER and G. W. STROKE: Ruling of large diffraction gratings with interferometric control. J. Opt. Soc. Amer. **45**, 15—22 (1957).

[11] G. R. HARRISON: The controlled ruling of diffraction gratings. Proc. Amer. Phil. Soc. **102**, 483—491 (1958).

[12] H. W. BABCOCK: Control of a ruling engine by a modulated interferometer. Applied Optics **1**, 415—420 (1962).

[13] a) G. W. STROKE: Theory, Production and Use of Optical Gratings for High-Resolution Spectroscopy. In: Quantum Electronics III, edit. by P. GRIVET and N. BLOEMBERGEN, pp. 1221—1237. New York: Columbia Un. Press 1964.

b) R. F. JARRELL and G. W. STROKE: Some new advances in grating ruling, replication and testing. Applied Optics **3**, 1251—1262 (1964).

[14] K. P. MIYAKE: Private communication 1964.

[15] Y. YAMADA, T. HARADA and K. MORITA: A New Japanese Ruling Engine. J. Jap. Soc. Mech. Engrs. **69**, 1557—1567 (1966) [in Japanese].

[16] G. HANSEN: Private communication.

[17] Now operated by GUS M. GOMEZ, see The Perkin-Elmer Instrument News, Vol. 13, No. 1, Fall 1961.

[18] G. W. STROKE, O. C. MOHLER and F. DENTON: Unpublished work, begun in 1963.

A more extensive historical review of grating ruling and theory, of the development of ruling engines, as well as some aspects in the recent trends in grating production and theory are given in Sect. 1.

In this chapter, some photographs and brief descriptions will be given for a number of ruling engines now in operation. A great variety of principles all seems to result in successful grating ruling engines.

An illustrative, rather than an exhaustive or ordered treatment appears to be in order. There has been an increasing appearance of new engines, based on unconventional departures from "classical" designs. The only unifying element in all engines so far seems to be the use of a ruling tool, most generally a diamond. The ruling of the well-shaped grooves required for well-blazed gratings is characteristic of contemporary gratings.

It should be recalled, in conclusion, that many, but not all, of the successful ruling engine designers are known to have shown what appears to be a notable weariness, rather than reluctance, about extensive descriptions of their engines, or ruling methods used. Neither ROWLAND, nor MICHELSON appear to have attached very great importance to any particular feature in their engines, or ruling method, and STROKE states[19]:

"There is no simple prescription for building a ruling engine in such a way that it will perform successfully immediately after completion, or, for that matter, during any great length of time after its first successful ruling, without the close attention of the instrument maker, physicist or engineer, who has made it operate successfully in the first place. This has proven to be true even on the servo-controlled M.I.T. ruling engine, of which the principle of interferometric control was first described by G. R. HARRISON and G. W. STROKE in 1955."

"As a corollary, one may also state that the description of the various ingenious mechanical (and now also optical and electronic) features that may have been incorporated into the building of any particular ruling engine is truly only interesting for an engine that has been made to work successfully. And even then the design features should be considered as materializing basic principles, rather than the optimum achievement of an idea."[19]

On the other hand, extensive descriptions of their successful engines are given by BABCOCK and BABCOCK[9] as well as by STRONG[8], among others.

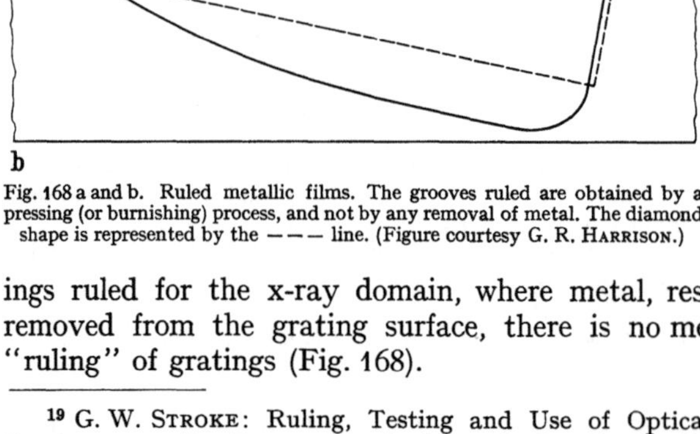

Fig. 168 a and b. Ruled metallic films. The grooves ruled are obtained by a pressing (or burnishing) process, and not by any removal of metal. The diamond shape is represented by the – – – line. (Figure courtesy G. R. HARRISON.)

83. The ruling process. Except in the production of some coarse infrared gratings, and in glass gratings ruled for the x-ray domain, where metal, respectively glass, are actually removed from the grating surface, there is no metal removal involved in the "ruling" of gratings (Fig. 168).

[19] G. W. STROKE: Ruling, Testing and Use of Optical Gratings for High-Resolution Spectroscopy. In: Progress in Optics, Vol. II, edit. by E. WOLF, pp. 1—72. Amsterdam: North-Holland Publ. Co. 1963.

The ruling process is comparable to a burnishing or pressure-forming process, in which no metal is removed, but rather the surface is suitably deformed. The furrows left in a block of soft matter (for instance in butter by a knife or a fork sliding across it) are somewhat similar to grating grooves. The electron-micrograph and cross-sections given in Sect. 80 give a representative selection of the remarkably good groove shapes achievable in ruled gratings.

In ruling blazed gratings, the attempt is made to obtain very well shaped grooves, generally with triangular cross-sections. Grating grooves are certainly not "scratches", as they are sometimes described. Gratings ruled for use at grazing incidence, in the x-ray domain, generally have only a very small fraction of the original surface of the grating blank "ruled". On the other hand, almost completely ruled surfaces are used for visible, infrared and vacuum gratings, and found to be most efficient for these gratings when used in the vicinity of autocollimation on one of the groove faces (see Sects. 62 to 71).

Two forms of ruling diamonds are shown in Fig. 169[20]. Other forms are shown in Fig. 170.

a b

Fig. 169a and b. Two forms of diamond used in grating ruling. a Straight edge; b "Cance" shaped. (Figure courtesy R. F. JARRELL.)

Gratings for the visible, near infrared and the ultraviolet are being ruled in thin metallic alluminum and gold, a few microns thick, evaporated on glass. Coarse gratings for the infrared are generally ruled in a high quality solder, $\frac{1}{8}$ to $\frac{1}{4}$ inch thickness, and sometimes in evaporated silver. The solder used may contain 40 to 50% of thin, and should preferably be prepared in an evacuated oven to prevent the formation of suspended oxides[21].

The deposition by evaporation in vacuo of sufficiently thick metallic films with the required flatness of some $\lambda/20$ at the minimum wavelength appears frequently to present very great practical difficulties[2]. The production of ruled echelles for the visible, with 30 to 10 lines per mm, requires film thickness in excess of 15 to 50 microns, for high angle use. The production of *thick* evaporated films with adequate thickness is particularly exacting. Replicated thick films have shown tendencies of deformation upon ruling.

The "softness" or ductility of the thin aluminum films, a few microns thick, used for spacings of 300 grooves/mm and less, appears to depend rather critically on the thickness of the film. Great care in evaporating films, to obtain the same thickness should be taken, to ensure reproducibility.

The groove angles, and indeed the groove orientation, are found to be different from the setting of the diamond facets. This can no doubt be attributed to the pressing or burnishing process involved in the ruling. Departures of groove facet orientation $\frac{1}{2}°$ to perhaps 2° from the diamond setting are not unusual.

Electron-microscopic testing, interference-microscopy together with spectroscopic efficiency measurement (see Sects. 79, 80) should be used in the various stages of diamond adjustment, prior to ruling. Reliance on previous diamond settings, even with careful film composition and thickness control, does not result in sufficient regularity in blaze angle and efficiency.

It should also be clear that very appreciable deformations of the last groove ruled results from the groove just being ruled. If the steep side of the groove

[20] R. F. JARRELL: Production of Gratings. In: Encyclopedia of Spectroscopy, edit. by G. L. CLARK, pp. 173—179. New York: Reinhold Publ. Co. 1960.

[21] E. F. BARKER and H. M. RANDALL: University of Michigan, private communication (1961).

Double end tool

Fig. 1A and Fig. 1B. Each end consists of three planes intersecting at a point. The side plane angles produce the desired contour of line. Ruling is accomplished by the burnishing action of the trailing edges. Both ends may be used. May be symmetrical ,Fig. 1A, or unsymmetrical, Fig. 1B, to specified angles.

(Fig. 1A, top right; Fig. 1B, bottom right)

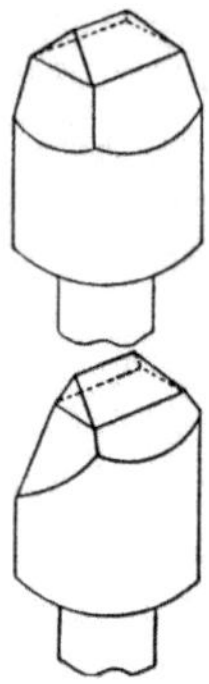

Specifications

Shank dimensions
Included angle s between sides
Clearance or drag angle c (45° unless otherwise specified)

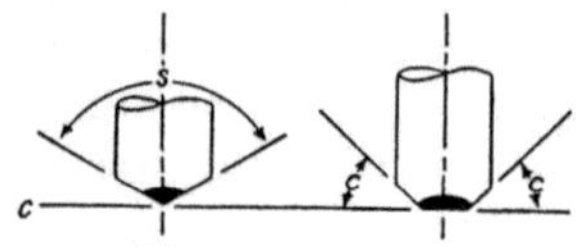

Fig. 1A, symmetrical

Shank dimensions
Side angle s 1
Side angle s 2
Clearance or drag angle c (45° unless otherwise specified)

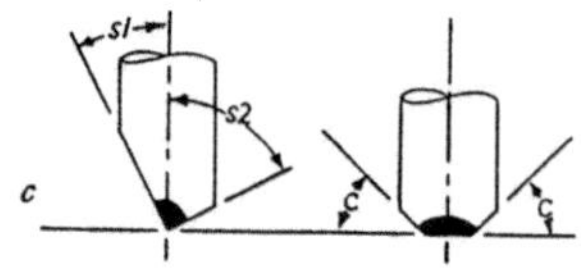

Fig. 1B, unsymmetrical

Curved tool

Fig. 2A and Fig. 2B, consists of two conical surfaces intersecting in a circular ridge, colinear with or parallel to the tool shank. The angular disposition of the conical surfaces produces the required line contour. Ruling is accomplished by the burnishing action of the curved surfaces. May be symmetrical, Fig. 2A, or unsymmetrical, Fig. 2B, to specified angles.
(Fig. 2A, top right; Fig. 2B, bottom right)

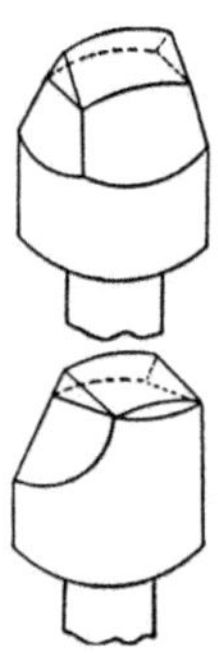

Specifications

Shank dimensions
Included angle s between sides
Radius of curvature r (usually approximately 1/4")

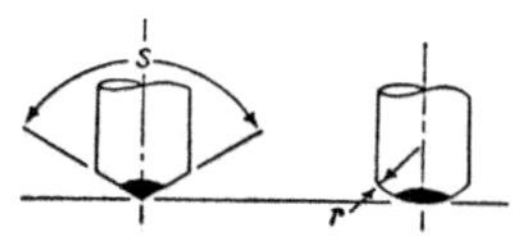

Fig. 2A, symmetrical

Shank dimensions
Side angle s 1
Side angle s 2
Radius of curvature r (usualy approximately 1/4")

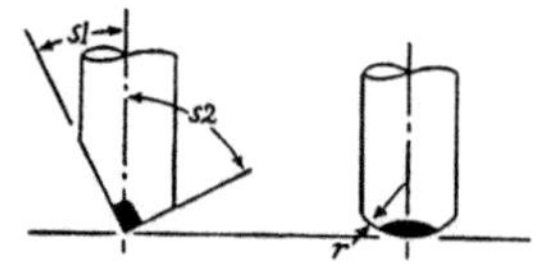

Fig. 2B, unsymmetrical

Elongated pyramidal tool

Fig. 3A and Fig. 3B, consists of four plane surfaces intersecting at a common point. The side edges control the line contour, while the angle between the leading and trailing ridges is very nearly a straight line (generally 178°). Thus the edges which burnish the contour of the ruled line are very obtuse dihedral angles, and perform somewhat like the curved tool. May be symmetrical, Fig. 3A, or unsymmetrical, Fig. 3B, to specified angles.

(Fig. 3A, top right; Fig. 3B, bottom right)

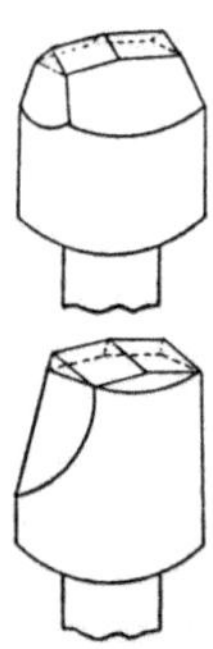

Specifications

Shank dimensions
Included angle s between sides
Lead or drag angle c (approximately 1° unless otherwise specified, resulting in 178° between long edges of pyramid)

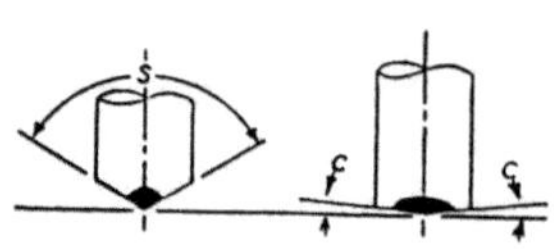

Fig. 3A, symmetrical

Shank dimensions
Side angle s *1*
Side angle s *2*
Lead or drag angle c (approximately 1° unles- otherwise specified, resulting in 178° between long edges of pyramid)

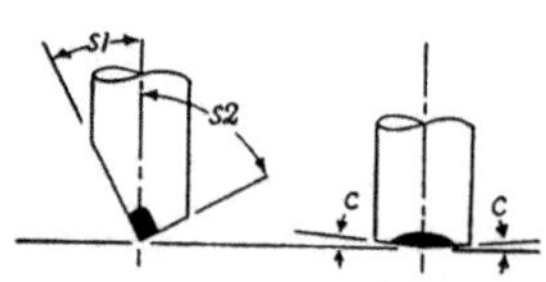

Fig. 3B, unsymmetrical

Elongated pyramidal tool design for diffraction gratings original with MOORE. A stronger tool than the conventional three-plane tool of Fig. 1 (the burnishing action is performed by obtuse dihedral angles instead of sharp edges).

Easy to align in the ruling engine; the leading edge, being straight, can be brought entirely into the focal plane of the aligning microscope. Errors from misalignment have less effect on the contour of the rule line. Position of the burnishing edges remains nearly constant when ruling concave gratings. In concave gratings, the angular position of the tool changes with respect to the grating surface. Different portions of the curved tool, therefore, contact the grating, producing the so-called "target pattern". With the pyramidal type of tool, this effect is minimized. Sideways deflection of the tool carriage from side thrus is also reduced.

Fig. 170. Diamonds suitable for grating ruling. (Figures and text, courtesy J. ROBERT MOORE).

is to be used, the diamond is so oriented as to deform the flat side of the last groove (Fig. 171).

Diamond chatter has been among the factors which have seriously limited ruling speeds in the past, generally to a maximum of some 20 strokes/minute. More usually, 12 to 14 strokes/minute or less are representative of ruling speeds used for optical gratings. Other limitations to ruling speeds appear to have been associated to vibrations in the diamond bearings.

Attempts are now underway by STROKE et al.[18] at the University of Michigan to substantially increase ruling speeds by servo-control of the diamond, for instance by simulating a very heavy, rigid mass, as the support for the diamond, without compromising with the uniform load required for uniformly deep and regularly shaped rulings.

Many additional details on ruling techniques are given by BABCOCK and BABCOCK[9], BABCOCK[22], HARRISON[23], R. F. JARRELL[20], and by GERASIMOV et al.[24] among others.

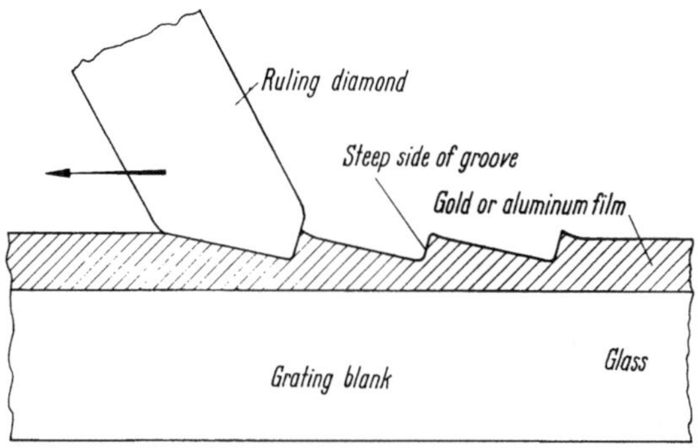

Fig. 171. Direction of ruling diamond advance, when blaze from the steep side of the groove is desired. (In case of grating blank advance, the advance of the grating carriage would be to the right.)

When ruling with diamond on evaporated metallic films, a layer of silicone oil is generally used, spread on to the grating surface. Systematic experiments on the most suitable lubrication for grating ruling apparently remain to be performed.

Experience has shown that the techniques involved in setting the ruling diamond are readily acquired, especially with the help of the groove testing methods described in Sects. 77 to 80.

A number of typical ruling diamonds are illustrated in Fig. 169 and 170. The boat-shaped diamond seems to have been developed by ANDERSON (see HARRISON, loc. cit. [23]), and much progress in diamond shaping and mounting is due to the work of BABCOCK and BABCOCK[9]. Significant contributions to the development of ruling diamonds have resulted from the initiative of J. ROBERT MOORE.

There is general agreement that one of the most important developments in modern grating ruling techniques has been the introduction by JOHN STRONG[25] of rulable aluminum films, deposited by evaporation on glass.

Some additional details on ruling techniques are given in some of the following sections.

84. Mechanical design of ruling engines. One of the most vivid and representative descriptions of the mechanical problems in ruling engine design is given by G. R. HARRISON[23]:

"The success of a ruling engine depends on how well compensation has been made for the inevitable changes in shape and size that result from variations

[22] H. W. BABCOCK: Bright diffraction gratings. J. Opt. Soc. Amer. **34**, 1—5 (1944).

[23] G. R. HARRISON: The Production of Diffraction Gratings. I. Development of the Ruling Art. J. Opt. Soc. Amer. **39**, 413—426 (1949).

[24] F. M. GERASIMOV, I. A. TEL'TEVSKII, S. S. NAUMOV, S. N. SPIZHARSKII and S. V. NESMELOV: Diffraction gratings from the State Optical Institute, Optika i Spektroskopiya **4**, 779—790 (1958).

[25] JOHN STRONG: Phys. Rev. **43**, 498 (1933). — Astrophys. J. **83**, 409 (1936).

in friction, temperature, elasticity and internal strain. The designer may well imagine his engine as constructed of rubber, thus visualizing the stress that must be controlled when strains of the order of a microinch become of importance. Present carefully designed engines must be set into operation from 3 to 24 hours before the diamond is lowered into the ruling position*, so that thermal and dynamic equilibrium may be attained. WOOD[26] found that stopping one of the Rowland engines for only a few minutes produced marked displacements of several grooves when ruling was resumed ..." [23].

Details on mechanical design will be given for the various engines illustrated.

Mechanical imperfections can be overcome to a spectacular degree by interferometric servo control, as described in the original papers by HARRISON and STROKE[1], and subsequently confirmed for many other engines.

85. The screw. Harrison notes in his 1949 paper[23]:

"A popular misconception exists that the critical portion of a ruling engine is its screw. A screw that will produce translations precise to one micro-inch without need of a correcting mechanism can be made readily by the lapping method of ROWLAND[27]."

The 20 inch long, 80 mm diameter hollow screw used on the Michelson engine at M.I.T. appears to have required for its complete production only about 3 months of one man's time, spread over a period of one year[28].

It is necessary to distinguish between the precision of the screw, on one hand, and the performance of the screw as mounted in the engine on the other.

A modern screw, 30 threads/inch screw, specifically lapped for one of the Jarrell-Ash ruling engines, showed a periodic error with a peak amplitude of 1440 Å (Fig. 146, Sect. 76)[29]. However, the simple harmonic component, as measured by the intensity of the first Rowland ghost [Sect. 55, Eq. (55.23)],

$$G_I = \pi^2 m^2 \frac{\varepsilon^2}{\alpha^2}$$

is considerably smaller (m = order, a = spacing constant, ε = amplitude of periodic error and G_I = intensity of the first Rowland ghost relative to another line). For $G_I = 1/2000$, one has $\varepsilon \simeq 120$ Å. The simple harmonic component of the periodic error appears to be some 12 times smaller than the peak-to-peak amplitude in this case. Further reduction in periodic error and ghost intensity can be achieved by very careful adjustment of the thrust bearing. Improvement by a factor of 2 in error amplitude have been obtained.

Another 20 threads/inch screw, very carefully made but not for a ruling engine showed peak-to-peak periodic-error amplitudes of the order of ± 4000 Å.

Even larger periodic errors, namely on the order of ± 3 fringes (± 7200 Å) were measured on the M.I.T. engine by HARRISON and ARCHER[30], before correction. One may assume that the simple harmonic component is of the order of ± 600 Å for this screw.

* G. W. STROKE, loc. cit. [2], noted that a period from 12 to 18 hours was indicated for the M.I.T. interferometrically controlled ruling engine.

[26] R. W. WOOD: Phil. Mag. **48**, 497 (1924).

[27] H. A. ROWLAND, article "Screw", Encyclopedia Britannica, 9th edit., Vol. 21. — Physical Papers, p. 506. Baltimore: Johns Hopkins University Press 1902.

[28] T. J. O'DONNELL: Private communication to G. R. HARRISON (loc. cit. [23]).

[29] WM. M. MACARTHUR: Jarrell Ash Co., private communication 1959.

[30] G. R. HARRISON and J. E. ARCHER: Interferometric calibration of precision screws and control of ruling engines. J. Opt. Soc. Amer. **41**, 495—503 (1951). The large error in the Michelson screw arose from prolonged storage and sag of the screw.

Yet it was this large 600 Å periodic-error component (exceeding by a factor of more than 5 the error of a good modern screw) which HARRISON and STROKE (loc. cit. [1]) were able to reduce, by interferometric control, by a factor of more than 30, to a level of about 17 Å, that is to a level some 7 times smaller than that of a good ruling engine screw. The remarkable reduction of the periodic error by interferometric servo control was somewhat unexpected[1,2], but has now appeared to be typical of the introduction of control. The reduction of periodic error appears to be especially great with continuous carriage motion.

Even the considerably smaller periodic error of the Jarrell-Ash engine was further reduced to the level of about 15 Å, that is some additional 8 times below its already mechanically fairly acceptable level, by the interferometric control system designed for this engine in 1960—62 by G. W. STROKE (loc. cit. [13]). These values are representative of interferometric control with a continuous motion and control of the grating carriage.

Systematic errors, and cumulative errors extending over many revolutions of the screw are corrected to a very remarkable degree by interferometric servo-control[1,2,12,13].

Much information on the lapping and on the mounting of ruling engine screws is given by JOHN STRONG[8] and by BABCOCK and BABCOCK[9].

The successful substitution of a hydraulic drive for the screw in an interferometric velocity-of-light experiment was described by STROKE (see Sect. 86, ref. 37).

86. Mechanical shock isolation of ruling engines. Mechanical shock and vibration isolation of ruling engines should be carefully executed.

It seems that heavy piers supported on cork and sand provide reasonably good shock isolation. For instance, very good isolation has been obtained by JARRELL-ASH with the help of 30-ton piers supported on, and surrounded by 2 inches of cork and some 2 feet of compacted sand, in the basement, below a street-floor machine shop[31]. No vibrations in the control interferometers could be detected, even in the presence of heavy trucks moving on the street as close as some 50 or 60 feet from the engines.

Similar piers have recently been installed at the University of Michigan[32]. They are shown in Figs. 172 and 173.

Very little information exists so far, permitting a vibrometer comparison of vibrations "before" and "after" the installation of ruling-engine supporting piers.

Some of the unpublished vibration-measurement results about the new University of Michigan piers are shown in Figs. 174 to 176. It is seen that both the high-frequency and low-frequency ends of the vibration spectrum have been appreciably attenuated, in the manner expected for "shock-isolation" by cork and sand. Further reduction of the transmission near the 20 cps pier-resonance may be obtained by supporting the engine on pneumatic servo-mounts[33].

Experience shows that the careful design of the interferometer mirrors can contribute a major amount to successful vibration isolation. An interferometer-mirror mount designed by STROKE and found to be particularly free of any resonant vibrations is shown in Fig. 177.

Ruling engine piers have also been mounted on "soft" springs, without noticeable advantages over the sand-and-cork suspension. The spring mounts have

[31] D. K. RICE, Jarrell-Ash Co., private communication (1963).

[32] G. W. STROKE, F. DENTON, unpublished (1964).

[33] For general principles of vibration and shock isolation see for example, C. E. CREDE: Vibration and Shock Isolation. New York: John Wiley & Sons 1951.

Fig. 172. Hole dug out for shock-isolating pier shown in Fig. 173. Note the 2-foot deep layer of 90% compacted mason's sand on bottom of hole.

Fig. 173. Shock-isolating pier as installed in a basement floor at the University of Michigan. The 9×12 foot, 6-foot deep, 32 ton pier is surrounded by 2-inches of vibration isolating cork, and 2 feet of compacted sand all around, upt to the main building floor. Additional vibration isolation of the ruling engine is completed with the help of pneumatic servo-mounts. Results of vibration measurements are shown in Figs. 176a and b.

some disadvantages. In particular, the piers will tend to rock, as a result of mechanical motions associated with the engine operation. Spring mounts have been used successfully in suspending a Fabry-Pérot spectrometer[34].

A simple calculation shows that the natural frequency f_n in cycles per second is approximately given by $f_n = 3.14/\sqrt{\delta_{\text{static}}}$ (with the deflection, or spring extension, δ_{static} in inches). In the case of the Chabbal-Jacquinot mount, a 70 cm extension of the four springs gave a frequency in agreement with this equation. At least one ruling engine appears to have been similarly suspended.

An engine designed by E. LEIBHARDT[35] has piers extending six feet below the floor level of the room. The piers are surrounded by 4-inch thick slabs of cork, resting on a concrete box (of unspecified thick-

[34] R. CHABBAL and P. JACQUINOT: Rev. Opt. **40**, 161 (1961).

[35] E. LEIBHARDT: U. S. Office of Naval Research, Final Report, Contract No. NONR-1040 (00) (undated, received by GWS in 1962).

ness). The box, in turn, is surrounded by 2 feet of sand, resting on a bed of sand 3 feet deep.

The M.I.T. interferometric engine described by HARRISON and STROKE[1,2] is mounted on a specially constructed floor in the M.I.T. Spectroscopy Laboratory, designed by K. T. COMPTON[36] and described by RIM in 1932 as follows:

"All parts of the laboratory are supported by means of piles on a thick stratum of a compact gray sand. The outer shell and corridor rest on wooden piles and those under the exterior walls are placed in such a way as to form a cut-off against

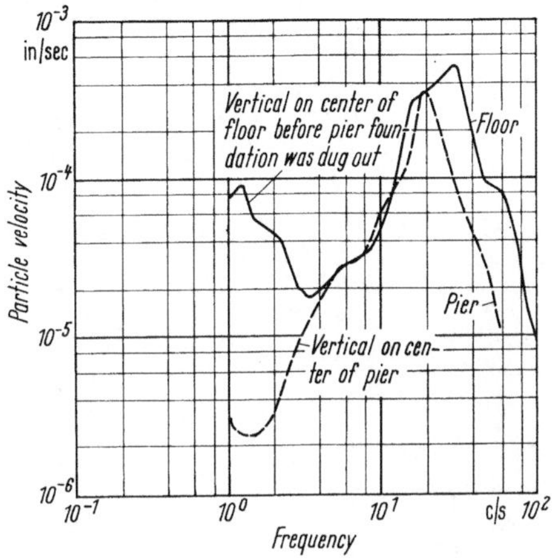

Fig. 174. Vibration measurements in presence of normal "background" vibrations.

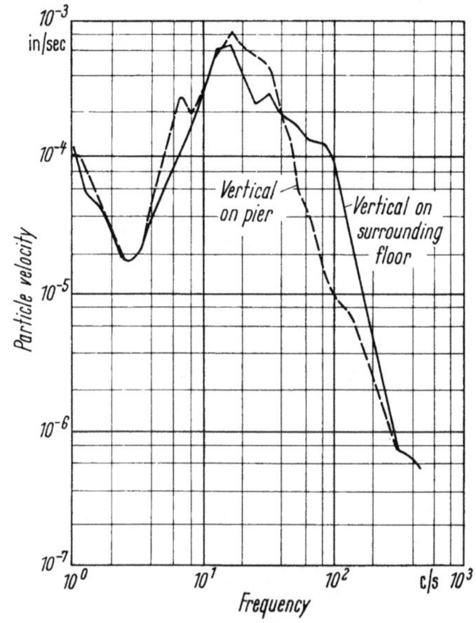

Fig. 175. Vibration measurements during operation of heavy bulldozer one floor above and outside of the building.

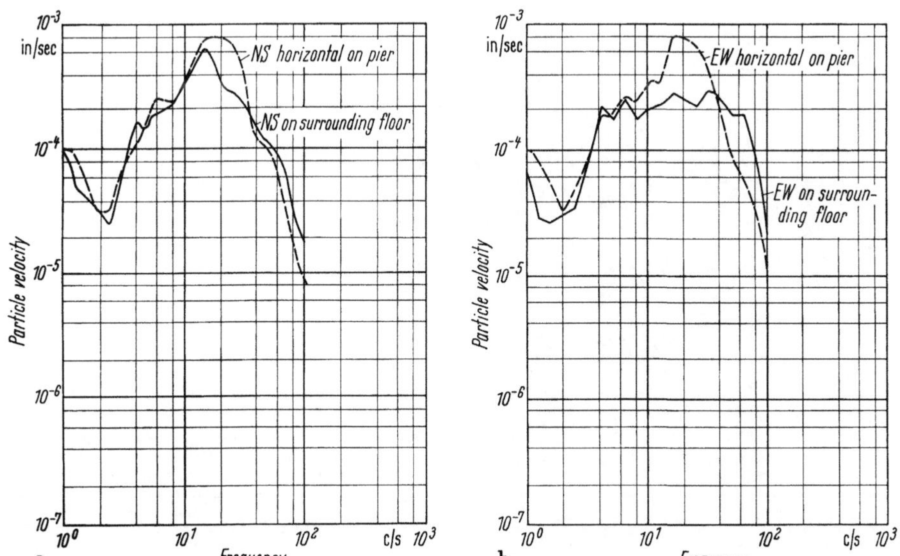

Fig. 176a and b. Vibration measurements during operation of heavy bulldozer on ground outside (one floor above) building.

[36] K. T. COMPTON: The New Spectroscopy Laboratory of the Massachusetts Institute of Technology. Physics 2, 205—210 (1932).

a

b

c

Fig. 177a—c. a Vibration insensitive interferometer mirror mount according to Stroke. Front view, showing the mirror-mount and inner mirror-holding cell, both of which are made out of non-resonant "meehanite" iron casting. The inner cell is supported by two rounded heads (shown in Fig. 177b) and is made to rest against two adjusting screws and a rounded head with the help of hammered phosphore-bronze springs. The two screws and rounded head at 90° (Fig. 177b) permit independent adjustment about two orthogonal axes. b Front view of vibration-insensitive interferometer mirror mount according to Stroke, with mirror-holding cell removed to show the two supporting rounded heads, and the two adjusting screws at rear head at 90°. c Rear view of vibration-insensitive interferometer mirror mount according to Stroke.

vibrations reaching the inner or apparatus floors. Since the laboratory is in a court entirely surrounded by other buildings, these buildings themselves serve as an additional cut-off for the traffic vibrations, which are the chief source of difficulty. The separate isolated floor systems on which the apparatus is mounted are supported on 183 precast reinforced concrete piles, including 10 batter or braced piles placed so as to prevent lateral swaying of the isolated floor structures.

Fig. 178a. Interferometric microwave-cavity velocity of light measuring apparatus, according to STROKE (loc. cit. 1963). Note that vibration isolation is achieved very simply by mounting all interferometric and moving elements from the central brass platform, and by resting the entire apparatus on 2-inch thick balsa-wood slabs on the floor.

The isolated floors are constructed as follows: Reinforced concrete mats 3 ft. in thickness are built upon the concrete piles. Resting on these mats are special vibration-absorbing structures built up in layers in the following order. Three 1 in. layers of fine sand, each contained between layers of three-ply roofing felt; a layer of transite board; a 2 in. layer of reinforced concrete; another group of

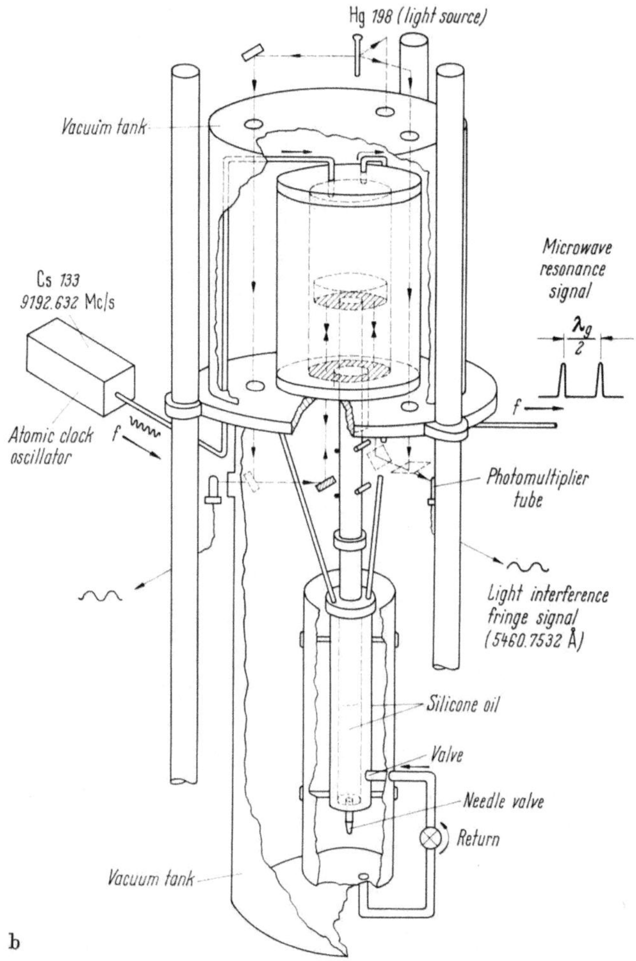

b

Fig. 178b. Schematic diagram of interferometric velocity of light measurement apparatus, according to Stroke (loc. cit. 1963).

three layers of sand separated by roofing felt; another layer of transite board; another 2 in. layer of reinforced concrete; a 6 in. layer of special vibration absorbing cork; finally, the 8 in. reinforced concrete floor with smooth cement upper surface. The function of the felt and of the cork is to absorb vertical vibrations, whereas the function of the layers of sand is by their ball bearing action to absorb lateral vibrations. These layers are separated from the adjoining side walls by thick layers of felt. Seismographic tests of the completed structure have shown that lateral vibrations are almost completely absent, and vertical vibrations have been very greatly reduced by this construction." [36]

Experience with spectrographs in several of the rooms, in particular in the 35-foot rooms, has shown that remarkable stability and freedom from vibrations was indeed achieved by the Compton floor design. However, the vibrations transmitted to the M.I.T. ruling engine appeared still to be harmful enough to cause annoying interference as reported by Stroke (loc. cit. [2], p. 1337, footnote [68])

"The operation of a 10000 amp, 170 volt generator for a Bitter 100000 Gauss Zeeman-effect magnet, placed within some 50 feet from the engine has occasionally caused vibrations in the interferometers and harmful resonance in the ruling diamond, which both effect groove quality and may tend to result in undesirable scattered light in spectral regions far removed from the line centers." [2]

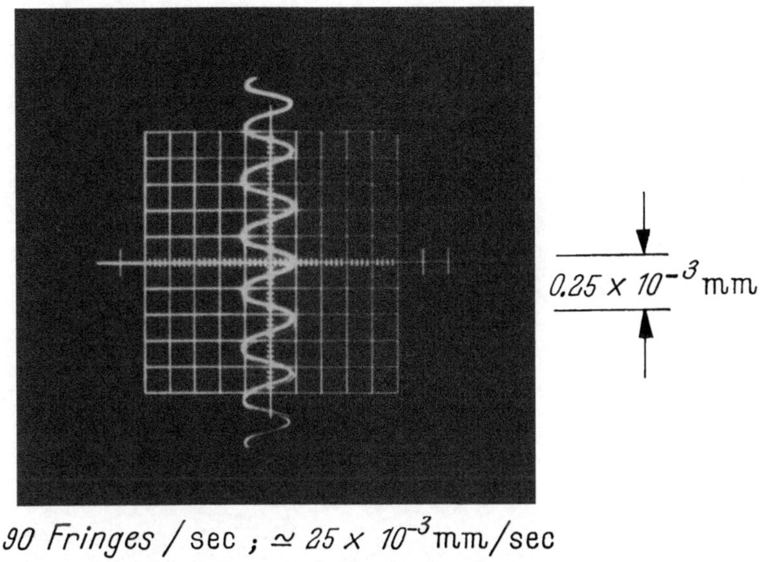

Fig. 179. Photoelectric interferometric fringe signal obtained during normal daytime, on the fourth floor of a steel and cement building, with the velocity of light apparatus of Fig. 178. Playback of many fringe cycles recorded on magnetic tape is shown. Fringes obtained with air-cooled Hg 198 lamp excited in an electrodeless discharge (see Fig. 178a). Mirror motion achieved with the help of hydraulic system shown in Fig. 178 and 180.

The interferometric mounts subsequently designed by Stroke for the control of the Jarrell-Ash engine[13] have shown a vastly greater insensitivity to vibrations on the engine supported by the rather simple sand-and-cork isolated 30-ton pier described above.

Recent experience with lasers has further confirmed that the stability in interferometers depends more on the careful design of the interferometer mounts than on environmental vibrations. Heavy non-resonant mirror mounts, made out of "meehanite" casting, and mounted on a heavy iron casting surface plate will result in a remarkably stable interferometer, with only a minimum of felt or balsa-wood isolation of the plate from an ordinary laboratory table.

Stroke has reported successful operation of a photoelectric interferometer in a velocity of light experiment[37] in a fourth floor of a modern laboratory building, with a lathe and a milling machine operating within 20 feet on the same floor. The entire apparatus was mounted in three 2-inch thick slabs of balsa wood, and all the crucial elements, including the hydraulically operated moving mirror, were suspended from a single massive brass platform (Fig. 178a, b). Some of the

[37] G. W. Stroke: Interferometric Method of Velocity of Light Measurement. Applied Optics **2**, 481—486 (1963).

typical fringe-signals obtained with hydraulic piston velocity of 90 fringes (90 $\lambda/2$ at 5461 Å) per second are shown in Fig. 179. A detail of the hydraulic system used in place of a screw is shown in Fig. 180.

Fig. 180. Hydraulic system used for displacement of glass cylinder shown in interferometric velocity of light measurement apparatus of Fig. 178. The interferometer mirror used to produce the fringe signal of Fig. 179. is carried at the top of the glass piston. (Piston diameter = 2 inches, piston length = 20 inches.) The piston was optically ground and polished to be straight. It moves without stick-slip friction when guided by the two pairs of three tangential brass cylinders at 120° (of which two appear at top of photograph) (according to STROKE, loc. cit. 1963).

87. Temperature control of ruling engines. Very good control of temperature is required for grating ruling, even with interferometric control. Experience has shown that a control of better than 1/100° C in the air surrounding a mechanical engine is necessary for the ruling of high resolution gratings, and a control on the order of 1/1000° C was found to be necessary — and was achieved — in the oil bath surrounding the M.I.T. engine.

As reported by STROKE in 1961 (loc. cit. [2], p. 1337):

"Simple control principles have now resulted in a further significant improvement in the temperature control of the oil-bath in which the M.I.T. engine is immersed. The temperature stability of the order of ±0.001° C which has been achieved is indeed necessary for the ruling of high-resolution gratings, since the control system assumes that the distance between the beam-splitting tower and the reference-mirror some 5000 mm away remains constant during the course of the $4\frac{1}{2}$ to 5 days required for ruling a 10-in. grating at 300 grooves/mm. With a temperature coefficient of 10^{-5}/deg C of the steel casting on which both the tower and the reference mirror are placed, in increase of 0.0001° C will tend to shift the reference mirror by approximately 0.02 fringes with respect to the tower. Since the control mirror carried by the grating carriage moves with respect to the reference mirror in the course of its advance during the ruling, a shift in the grooves of this order of magnitude would occur if no other effects were involved. In fact, some compensation may appear possible, but it is found that differential expansions finally combine to call for a temperature stability of the order of ±0.002° C, especially at the part of the ruling where the control mirror is closest to the tower.

Of course, it is not possible to maintain the 300 gallons of oil in the bath and the entire engine at the same temperature in the presence of the heat sources used in the control and the various radiative and conduction losses around the engine. One attempts rather to maintain all the local temperature fluctuations within the ±0.001° C, while at the same time establishing small gradients from the control point to the outside of the vat. Dimensional equilibrium is thus achieved under dynamical rather than semistatic conditions. It is in fact the magnitude of the gradients, or rather the temperature level at the control point (located in the center of the engine, halfway between the reference-mirror and the beam-splitting tower) which has been found to be crucial in the success of the control. A temperature level some 0.05—0.1° C above the level of the unheated oil at the control point has been found to be the best operating temperature level about which the ±0.001° C is then achieved under optimum conditions. At the 78° F at which the engine room is normally controlled to about ±0.02° C on the average, a heat input of about 20 w is used to bring the engine and oil to the operating level, and input variations of the order of ±1 to 3 w in the course of 1.5-hr. periods are observed during the control. A proportional control operating from a Wheatstone bridge is used to change the voltage across the heater with the help of a "Variac" transformer, and care is taken to deflect the oil stream from the heater to the sensing element with the help of stirrers and appropriate shields to avoid lag in the sensing and suppress temperature oscillations. A period of 12 to 18 hours is usually required to raise the engine and oil to operating conditions."

The control of the M.I.T. engine room is maintained to about ±2/100° C at the point of control, and the air is constantly gently stirred by slowly rotating, and moving, fans. Control-air is blown into the engine room from an outer room, at a point where that room is controlled to about ±0.15° C. And finally, the outer room is itself controlled with air from the main laboratory, maintained to about ±0.5° C by means of the air-conditioning system previously described in 1932 by K. T. COMPTON (loc. cit. [36]).

BABCOCK and BABCOCK (loc. cit. [9], p. 780) report that the Mount Wilson machine

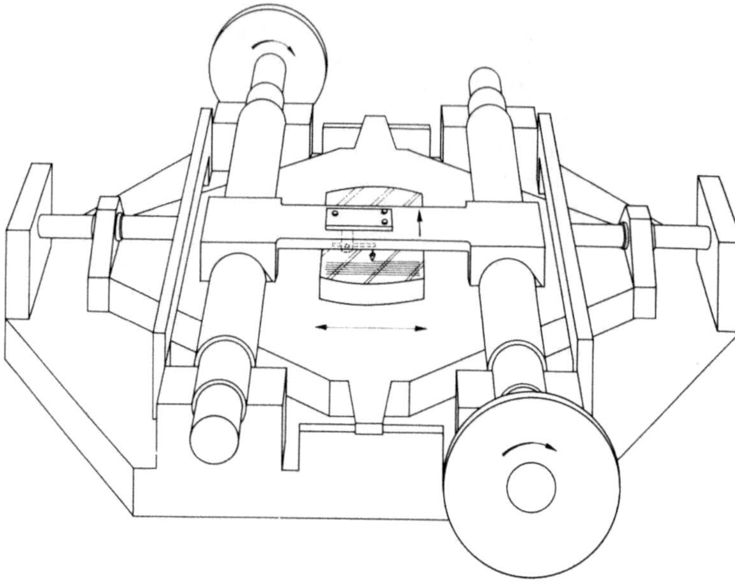

a

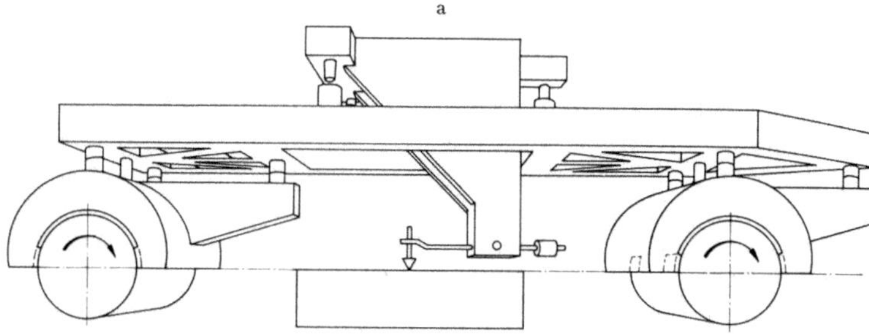

b

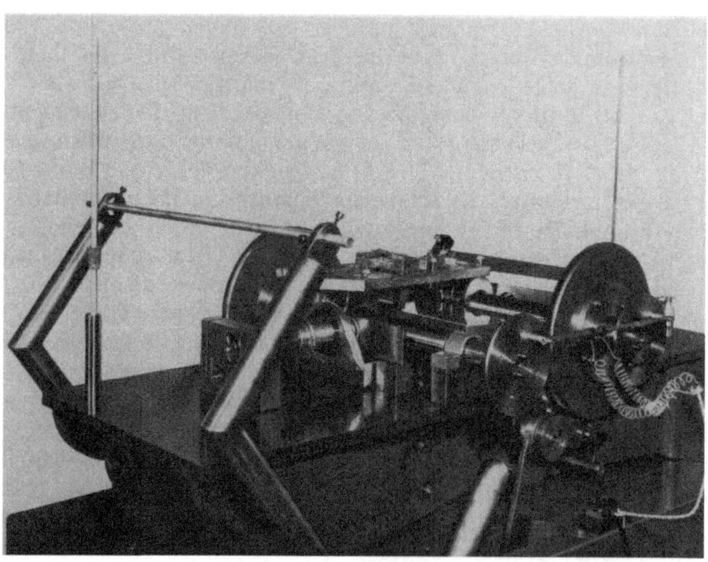

c

Fig. 181 a—c. Strong engine (Diagrams according to J. Strong, loc. cit. 1951 and also Scientific American, 1952). Photograph by Dr. P. Grodzinski (courtesy Prof. J. Strong and Journal of the Optical Society of America).

"... is housed in a double-walled, aluminum-foil-lined enclosure with windows and removable panels. A constant-speed fan circulates air within the enclosure and a thermostat holds the temperature during ruling to 24.0 ± 0.005° C. The temperature of the outer room is controlled to 0.1° C."

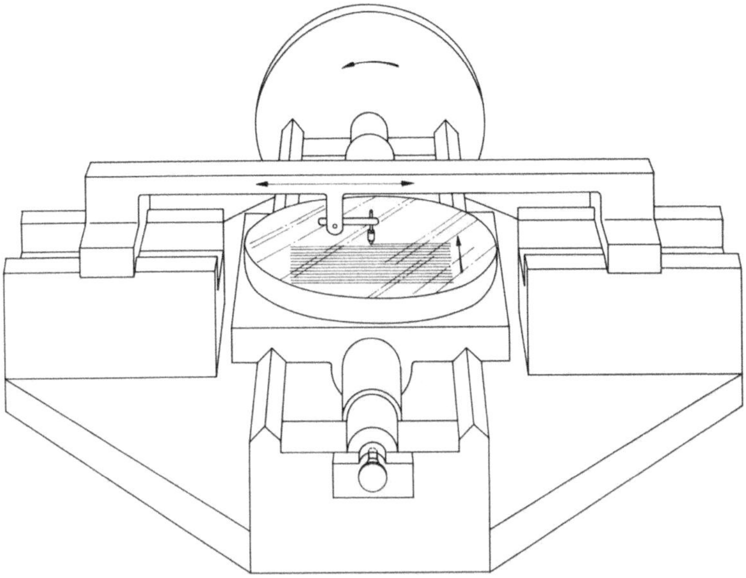

a

b

Fig. 182a and b. Rowland engine (Diagrams according to J. STRONG, loc. cit. 1951 and also Scientific American, 1952). Photograph by Dr. P. GRODZINSKI (courtesy Prof. J. STRONG and Journal of the Optical Society of America).

88. Some of the modern ruling engines. α) *The new Johns Hopkins Ruling Engine* A first description of this engine was given by JOHN STRONG in 1951 (loc. cit. [8]) and additional details on its performance in 1960 (loc. cit. [8]). 6-inch gratings are being regularly ruled on this engine.

A schematic diagram of the Strong engine is shown in Fig. 181. It is to be compared to the diagram of the Rowland engine shown in Fig. 182.

In STRONG's engine, the diamond carriage is advanced with the help of the two counter-rotating screws, while the ruling of the groove is accomplished by moving the grating blank under the diamond carriage, after it has been stopped. The purpose of the two counter-rotating screws is to help in averaging out errors inherent in any one of the screws.

As reported by JOHN STRONG in 1960 (loc. cit. [8]): "The new engine was successful, and three replicas of that engine have been made."

Complete details on the design, construction and experiments with his new engine are given by STRONG in his 1951 article (loc. cit. [8]). A photograph of the engine is shown in Fig. 181c.

JOHN STRONG's engine undoubtedly represents the finest and most sophisticated example of "built-in" perfection successfully achievable in a ruling engine. As described by STRONG:

Design principles.

"A principle which guided Rowland in his original design was 'that no workmanship is perfect; the design must make up for its imperfections.' [38] ROWLAND's implied meaning has afforded a basis for many of my own innovations.

A subsidiary principle has had considerable influence in my design — *the principle of overconstraint and elastic averaging.* This principle contrasts with the principle of kinematical design [39], instruments of the very highest precision are not attainable after the teachings of the latter.

After my principle of overconstraint with elastic averaging, one uses a multiplicity of small contacting areas, each provided with enough elasticity to accommodate variable oil films, foreign matter, inhomogeneity of elasticity of the constrained member, lack of truth in the generated surface of the constrained member, and wear. Unavoidable errors at any one of these many areas will be elastically averaged to give a far more precise resultant location of a moving part than one would get from constructions designed after the kinematical principle which this type of construction contravenes. Varied applications are described later." [8]

On lapping.

"The new design requires great precision of construction. The main difficulty of precision metal constructions of such members as a lead screw lies in the operations of lapping soft metals, which materials alone are sufficiently stable mechanically and free from aging. It is not inappropriate therefore to say a few words about the lapping procedure. The lapping operation bears the same relation to precise mechanical constructions that "figuring" with a pitch polishing lap bears to precise optical work. Although metal surfaces may be scratched or grooved (owing to generation being of the cylindrical rather than the spherical type) and although they may be less true locally than polished, hard surfaces, elastic averaging gives such lapped metal surfaces a precision of performance comparable with the truth of optical surfaces." [8]

Further details on the lapping and other procedures used to achieve truly unique mechanical precision are given in the original paper. As noted by STRONG (loc. cit. [8]) in that paper:

"The satisfactory performance of the new engine is the result of the excellence of its components, and especially to PERRY's [40] prompt assay of their operating characteristics, in ensemble, and his skillful and sure adjustments." [8]

β) The Mount Wilson observatory engine (before control). BABCOCK and BABCOCK (loc. cit. [9]) reported in 1951:

"Within the past three years, about 20 blazed plane gratings having a groove length of $5\frac{3}{4}$ inches and a width of $7\frac{1}{4}$ inches have been ruled here."

These gratings were of high quality "in all respects" and could be produced "with fair regularity". A general view of the engine, described as the *B* engine, is shown in Fig. 183. As described by BABCOCK and BABCOCK in 1951 (loc. cit. [9]):

[38] H. A. ROWLAND: The Screw. Encyclopaedia Brittanica XI.

[39] J. STRONG: Procedures in Experimental Physics, p. 585. New York: Prentice-Hall, Inc. 1938.

[40] WILBUR PERRY: The Johns Hopkins University.

"The present satisfactory performance of the "B" machine rests, in our opinion, upon these considerations:

(1) the geometry of the modified design which tends to eliminate the troublesome errors encountered in most earlier machines;

(2) the use of a graphite thrust bearing for the screw;

(3) the use of stable, stress-free materials for critical parts, and surface-hardened Nitralloy steel for the ways, in combination with graphite sliders;

(4) improved technique in the production and use of ruling diamonds;

(5) the superlative tool-room skill of E. D. PRALL, who made or finished nearly all the critical parts and who made important contributions to the design as well." [9]

Fig. 183. Mount Wilson and Mount Palomar Laboratory engine (Photograph courtesy H. W. BABCOCK).

Some of the construction details, as given in the original paper (loc. cit. [9]) are as follows:

Design.

"The main frame of the machine is a symmetrical, compact, deeply-webbed iron-nickel casting, resting upon three thermally-insulated supports on a concrete pier. This casting supports the ways of the grating carriage, the screw with its spacing gear, and the crossway of the diamond carriage. A separate casting on a separate foundation carries the crankshaft with its fly-wheel and eccentrics, the spacing cam, and the cross-head; thus possible sources of vibration are isolated from the critical parts and possible objectionable deformation through settling or warping is minimized." [9]

Screw.

"The screw has a threaded length of 20 inches and a diameter of $2\frac{1}{8}$ inches; the pitch is $1\frac{1}{2}$ mm. It was made from a selected bar of No. 39 Bethlehem Machinery Steel (0.41 percent carbon) which after roughing was carefully annealed in a vertical position. The thread angle is 52°; the thread root is relieved by a deeper burnished groove. Great care was taken in cutting the thread to make it as nearly perfect as possible before lapping was commenced. Lapping, which was completed in 145 hours distributed over five months, was carried out with the screw in a vertical position and with the two halves of the lapping nut counterweighted. The halves of two long lapping nuts were used in all possible combinations; three short nuts were used for local work and for testing. The ends of the screw are fitted with hardened Nitralloy steel sleeves, pressed on and pinned; as lapping of these journals progressed, they were tested with an overhanging optical lever supported by a long nut. The screw was tested for straightness and uniformity with the Fabry-Perot interferometer as described by ANDERSON[41] until no errors could be detected. Various other components of the machine were likewise tested with the Michelson interferometer." [9]

Bearings.

"The bearings in which the screw turns are of Meehanite (cast iron) and are about two inches in length. They are not self-aligning. Each is in three parts, the cap being split parallel to the screw. The two parts of the cap are doweled and screwed to the lower part of the bearing. Contact area is limited to three 30° arcs spaced at 120°, one being at the bottom. The bearings were lapped on a dummy screw." [9]

Nut.

"The nut is eight inches in length and is divided into halves, the dividing plane being vertical. It was prepared by centrifugally casting a hard nickel babbitt into a steel shell having 12 internal dove-tail grooves. Excess babbit was then turned out to leave 12 segments, after which the nut was sawed in half and threaded. The nut is assembled with a demountable spring and pivot-pointed screws at each end. Rotation is prevented by two eight-inch fins bearing on a straight-edge below the screw. The thrust of the two halves of the nut is transmitted by two two-inch, ball-ended push rods to the opposite sides of an equalizing collar that is attached to the grating carriage by two short studs, which through bending, permit a slight rotation of the collar about a vertical axis. These push-rods, which fit into hardened steel cones, eliminate any concern about the sag of the screw or about lack of parallelism between the screw and the ways; also they would minimize any periodic error due to lack of straightness of the screw, though no curvature or "drunkeness" has been detected. The nut is lubricated with Spindle Oil No. 11 (E. F. Houghton and Company); its fit is so good that as shown by cross-ruling tests, oil films as well as strains are equalized within a few cycles of operation after the machine is started." [9]

End thrust.

"On the end of the screw opposite the spacing gear is mounted a diamond, cast in phosphor bronze. The diamond was figured flat to $\frac{1}{10}$ wave and adjusted, by a reflection test, to be perpendicular to the axis of the screw to a small fraction of a second of arc[41]. For years we employed a hard metal sphere as the fixed part of the bearing, but all too often this broke down as the result of chipping or scoring. This difficulty was solved in March, 1948, when we introduced a graphite compound as the fixed part of the bearing. A short, firmly supported, coaxial stub of this material* about 1.5 mm in diameter forms an excellent bearing surface for the diamond. The graphite is relieved by a very small hole at the center. We find that a slightly pitted surface, formed by honing on ground glass, performs better than a completely smooth surface. No self-aligning feature is needed. In our machine the thrust load is about seven pounds." [9]

Spacing mechanism.

"The principle of this mechanism, which was taken over from the "A" machine, is due to JACOMINI. A thrust-pin, driven by a crank, moves through a tapered helical groove in a cylindrical cam once each cycle, causing the cam to rotate through one turn. The thrust-pin is retracted from the cam on the return stroke by a wedge device that is activated by an eccentric with a lost-motion connecting rod. Through three spur gears, the intermittent rotation of the cam is transmitted to a Nitralloy worm which is in mesh with the 900-tooth spacing gear attached to the end of the main screw. The mechanism is smooth in operation and has the great advantage that by a simple change of gears a wide variety of grating spacings is available. Gears on hand permit spacing from 200 to 1800 grooves per mm, but the greater part of our ruling is done at either 400 or 600 per mm. Wear of the hard steel worm is negligible, and, as ANDERSON showed years ago, extreme precision in the spacing gear is not required[41]. Our spacing gear was carefully lapped, however, both against a single lapping worm and against three co-ordinated worms spaced at 120°. As used for some years, this system gave what is

---

* Graphitar, a product of the U.S. Graphite Company, Saginaw, Michigan.

[41] J. A. ANDERSON: The manufacture and testing of diffraction gratings. Glazebrook's Dictionary of Applied Physics 4, 30—41 (1923).

often considered an acceptably low periodic error, such that the intensity of the first Rowland ghosts in the ruled gratings was about 0.001 of the parent line in the first-order spectrum. In 1949 ghost intensity was further reduced by a factor of 25, however, by the addition of the relatively simple period compensator shown. The spacing worm has a small end-thrust bearing of its own, which is mounted on a fine-pitch screw. A lever, attached to this screw, and bearing on the flat surface of a large disk mounted obliquely on the hub of the main spacing gear, generates a very small oscillatory longitudinal displacement of the spacing worm. This oscillatory correcting motion has a period equal to the period of rotation of the main screw; its amplitude and phasing were adjusted following the study of cross rulings. The mechanical advantage is such that the throw of the cam is reduced by a factor of about one million in its effect on the correction of the position of the grating carriage, yet the correction is "rigid" and readily controllable. In adjusting the machine to rule gratings with Rowland ghost intensity less than 0.00004 in the first order, the amplitude of the permitted periodic error in groove spacing is held to less than 2×10^{-7} inch." [9]

Additional details are given by BABCOCK and BABCOCK[9] concerning the grating carriage and the diamond carriage:

Diamond carriage.

"In his article on the ruling engine, ANDERSON properly rated the diamond carriage, or ruling carriage, as the part of the machine presenting the greatest difficulty[41]. The demands are truly enormous when it is considered that in ruling a single grating the diamond should move in the same straight line upwards of 100000 times without lateral deviations of more than a fraction of a micro-inch; that the carriage must reverse its direction of motion at each end of the stroke, and that it is supported and guided by ways subject to wear and influenced by oil films. It has been our experience that while it is not too difficult to obtain satisfactory performance with groove lengths up to three or four inches, the errors are likely to be much more troublesome when longer grooves are ruled. The statement is sometimes found that since the errors of position of the diamond carriage are usually random in nature, they are of little consequence compared to periodic errors. But as ANDERSON pointed out, such random errors lead to a flare of false light centered on each emission line in the spectrum (we have employed the term 'local scattering' for this). This local scattering broadens strong emission lines and, in an absorption spectrum, is an insiduous source of filling in of absorption profiles.

A number of ruling engines, including ROWLAND's have employed a four-footed diamond carriage moving on two parallel ways; usually the ways are flat-topped, with the sides vertical or nearly so. The weight of the carriage is taken by the flat top, while in theory the critical lateral definition is supplied entirely by the steep sides (spring-loaded in opposite directions at the two ends). We have introduced three modifications in the design of the diamond carriage. The first, in 1934, was to support and guide the bridge-type carriage by a divided monorail, the ruling diamond working in a gap on the axis of this rail. Stability of the carriage about this horizontal axis is provided by an outrigger at one end of the carriage, riding on an ancillary parallel rail. The advantages are that any rocking of the carriage will not displace the diamond from its proper path, that the carriage is lighter, and that fitting it to a single rail is far simpler than fitting it to two. In 1950 we replaced the flattopped monorail by a cylindrical rail, slightly over one inch in diameter. This rail is of Nitralloy; it was first centerlessground, then lapped in a lathe to a high degree of straightness and uniformity by STRONG's method of viscous constraint of the lap[8]. The length of the bar was originally 41 inches so as to span the gap over the grating carriage and thus to permit accurate finishing of the six supporting V's later, two 12-inch sections were cut from the bar and mounted coaxially in the supports. Each end of the carriage, which bridges the gap over the grating blank, is fitted with an inverted Graphitar V-block four inches long; the actual bearing areas on the round way are two strips about $\frac{3}{16}$ in. wide located on each side, 45° from the top of the way. These contact strips were finished by broaching on the grooved end of the original long bar. The advantages realized with this design are that each end of the carriage centers itself on the round way under gravity alone; oil films are self-equalized on each side; wear results only in an insignificant lowering of the diamond but not in a lateral displacement; and high precision is attained with far less labor than in finishing a flat-topped way. Furthermore, while wear of the Nitralloy under the lightly-loaded Graphitar is very slow, the ways may be rotated after a few years to bring fresh portions of the steel surface into use. The diamond carriage is an aluminum-alloy casting weighing 13 pounds. The necessary reciprocating motion is transmitted to it from the cross-head by a slightly resilient coupling link." [9]

A very important innovation in the mounting of the diamond was also introduced by BABCOCK and BABCOCK and described as follows (loc. cit. [9]).

Diamond mounting.

"Another modification of the diamond carriage, introduced some ten years ago, is in the manner of supporting the diamond. Long experience with pivot bearings of different types, for permitting the diamond to rule a groove on the pull stroke of the carriage and to be raised from the surface on the return stroke, showed that none of these was fully satisfactory. Either variations in pivot-bearing friction produced non-uniformities in the ruling, or a lack of definition in azimuth caused accidental errors. The pivots were finally discarded in favor of a support based on the Cardan hinge, or flat spring. Four short springs of 0.002-in. steel are used; two of these springs, crossed at 90°, support each end of the transverse horizontal rocking bar to which the diamond lever is attached. These springs allow the diamond only the one permissible degree of freedom, but in this there is good compliance, such that a ten-gram load will deflect the diamond about one mm. Much weaker springs might well suffice. As will be seen in the figure, the diamond in its shank is attached to the diamond lever by a clamp which permits adjustment for groove angle and for angle of attack. Low inertia of the rocking parts obviates any troublesome vibration during ruling. The whole diamond support is adjustable for elevation and for azimuth by means of micrometer screws. In setting up a grating blank on the machine, the plate is first levelled. Then the unloaded diamond is adjusted by the elevation micrometer to be just tangent to the surface. Finally the load is added to the upper part of the diamond shank. On the return stroke, after a groove has been ruled, the diamond is raised about one-half mm from the plate by a camactuated push-rod which descends on the opposite end of the diamond lever.

The performance of this diamond carriage has been satisfactory; our recent gratings made with six-inch grooves show virtually no local scattering, and high resolving power is now consistently achieved. Random and accidental errors have been practically eliminated." [9]

The new method of diamond mounting has found wide acceptance in ruling engines, especially for the ruling of plane gratings[1, 2, 13].

Some additional details on the ruling diamond, as used on the Mount Wilson engine are also given by BABCOCK and BABCOCK (loc. cit. [9]):

The ruling diamond.

"An earlier paper by one us[22] of describes the development of the curved-edge diamond originated by ANDERSON and the design of gratings having a high concentration of light in chosen order, which is, of course, of particular importance in astronomical applications. Such gratings for the visual and photographic region of the spectrum are now often referred to as blazed gratings. It may be remarked that some of the very fine speculum gratings ruled by ANDERSON at Johns Hopkins University prior to 1916 have very good concentration of light in one order; these were produced with a natural diamond edge. Later, at the Mount Wilson Observatory, ANDERSON and JACOMINI constructed a machine in which two convex conical surfaces, having radii of the order of one inch, are cut and polished on the diamond. The intersection of these two surfaces is a sharp, curved edge, which in use is tangent to the surface of the blank. The metal is pressed to either side, leaving a smooth, burnished groove. For a typical stone the included angle is 105°, the angle of the "shallow" side is 15° from the horizontal, and the length of the edge is about one millimeter. In the final stages of finishing on the diamond lap, the edge must be examined freqently under a high power microscope with strong lateral illumination. Nicks in the edge may then be detected by their diffraction effects; they must be completely eliminated until the edge appears perfect. By changing the angle of attack of the stone, five or six different sections of the curved edge may be used in turn, each section ruling perhaps several gratings, before the diamond needs refinishing. A good stone may be used almost indefinitely in this way. For groove angles of about 15°, we find that a load on the diamond of 15 to 20 grams is required if the spacing is 600 per millimeter, and 35 to 45 grams if the spacing is 400 per millimeter."

It might be noted in conclusion that it was this same Mount Wilson Observatory engine which was placed by H. W. BABCOCK under interferometric control in 1962[12] (see also Sect. 88ι), after HARRISON and STROKE[1, 2] had demonstrated in 1955 through 1961 the dramatic further improvements that could be attained over "built-in" mechanical, perfection" with the help of interferometric servo-control (see also Sects. 88γ, 88δ).

The performance of a 209 mm wide "Babcock" grating in a large 50 ft. vacuum spectrograph was determined with great care in 1956 by A. KEITH

PIERCE[42]. Photographs of the hyperfine structure of the mercury line λ 5769 showed a resolution of 630000 in the 5th order. In the 11th order of Hg λ 2537, the resolving power was estimated by PIERCE to be 1200000. The instrumental profile, the intensities of the satellite lines, and the ghost intensities were determined by photoelectric scanning of the green λ 5461 mercury line. The half-width of the "apparatus function" was found to be some 15% wider than the theoretical value.

The amplitude of the periodic error is somewhat difficult to estimate, inasmuch as the ghosts of the second order (G_2, Sect. 55) have 2, respectively 2.5 times the intensity G_1 of the ghosts of the first order. For a purely harmonic error, $G_2/G_1 \simeq \frac{1}{4}$ (see Sect. 55), the first-order ghosts being some 4 times more intense than the "second-order" ghosts. PIERCE (loc. cit. [42]) estimates the amplitude of the periodic error to be about 1/125 of the 17000 Å spacing, that is approximately 8×17 Å, which is representative of good mechanical engines.

Scattered light in the 5th order, after careful correction for the contribution by the wings of the lines, and ghosts, was estimated to be 0.7%. PIERCE notes that the scattered light remained approximately constant from order to order, and did "not appear to vary with the square of the order". As a result, PIERCE estimated the scattered light to be the non-dispersed scattered light in the spectrograph[42].

γ) The interferometrically controlled M.I.T. engine. γ1) General description. The principle of interferometric control of a continuously moving carriage, used successfully to rule 10-inch diffraction gratings on the M.I.T. ruling engine, was first described by HARRISON and STROKE (loc. cit. [1]) in 1955 and is deceptively simple[19]. To rule a perfect grating, it is sufficient to make the instantaneous distance of the grating carriage from the start of the ruling be rigorously proportional to the accumulated "position" which the diamond tip has reached since the start of the first groove as it rules the grooves, always in the same "ruling plane", in a cyclic motion across the grating. The position of the diamond tip is measured by the accumulated angular position of the motor shaft that drives the diamond carriage and is taken as a reference. (Departures from linearity in the diamond velocity across the grating, along the groove length, can be simply corrected by mechanical means if necessary.) The instantaneous position of the grating carriage with respect to the start of the ruling is measured interferometrically with the help of an interferometer mirror rigidly attached to the grating carriage and moving with it, and the start of the ruling is materialized by another mirror and beam splitter which are maintained in a fixed position with respect to the "ruling plane". The grating carriage is made to advance at approximately the correct rate with the help of the ruling engine screw and an electromechanical gear system, connecting the diamond shaft to the screw through a differential. A grating-position control motor is also connected to this differential. If, as a result of mechanical imperfections in the engine and the screw, the grating carriage is incorrectly positioned by the direct drive, then the grating position error appears as an electrical signal. The error signal is produced by comparison of the instantaneous amplitude of the interferometric fringe signal (obtained from a photo-multiplier tube) and the instantaneous amplitude of the diamond-position reference signal (obtained from a signal-generator which is also geared into the diamond shaft). The grating position-error signal is made to rotate the servo-control motor in such a direction as to continuously and constantly maintain the grating

[42] A. KEITH PIERCE: Performance of an Eight-Inch Babcock Grating in a large Vacuum Spectrograph. J. Opt. Soc. Amer. **47**, 6—14 (1957).

in its correct position with respect to the start of the ruling and the diamond position by appropriately adding or subtracting to the carriage motion determined

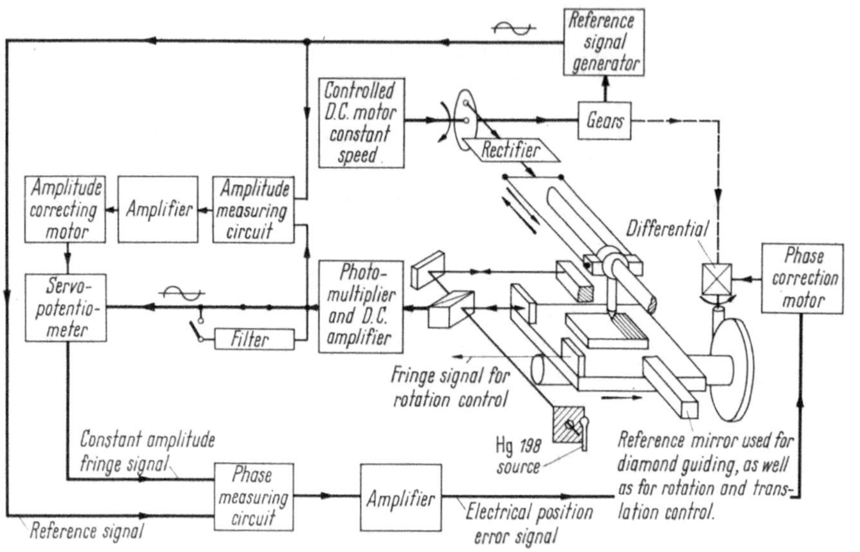

Fig. 184. Interferometric servo-control loop used on M.I.T. ruling engine, according to Harrison and Stroke (1955). Similar control loops are used for translation and for rotation control of the grating carriage.

Fig. 185a. Harrison and Stroke, M.I.T. engine. Motors appearing on left are from top to bottom: Selsyn motor used to drive the screw, translation-correction servo motor, rotation-correction servo motor. The motors shown in foreground are oil-bath stirrers. The diamond carriage reference mirror, and its three adjusting screws appear in center. Beam-splitting mirrors and collimators appear in tower on right.

by the rotation of the engine screw, which is itself driven by the diamond shaft. A schematic diagram of the servo-control loop used in the control of the carriage translation on the M.I.T. ruling engine is shown in Fig. 184 and photographs of the engine in Fig. 185 and in Fig. 186. Details of the screw and of the diamond-carriage linearizing mechanism are shown in Figs. 187 and 188. A view of the engine, upon its arrival form Chicago in 1948, is shown in Fig. 189. It serves to show how much modification the M.I.T. engine has undergone (as described by HARRISON and STROKE) from its original Michelson Gale form. A view of the control room

Fig. 185b. HARRISON and STROKE, M.I.T. engine. Close up view of grating and diamond carriages, and of worm-gear

is shown in Fig. 190. Interferometric wave-front studies of gratings ruled on the M.I.T. ruling engine with this control scheme have demonstrated that grating imperfections could arise even on an interferometrically controlled engine, unless one also controlled any rotation, about a vertical axis passing through the translation-control mirror if necessary (that is when rotations result from any curvature of the grating-carriage ways). Moreover, ruling errors will arise in an interferometrically controlled engine if the grating plane (diamond tip) does not coincide with the center of gravity of the control-interferometer mirrors, in case of rotations (or "tips") about horizontal axis normal to the ruling direction. Finally, second-order ruling errors can also result, if unchecked, from a combination of interferometric and servo-control effects, as we show below. Moreover, since the groove spacing, under interferometric control, is determined by the wavelength of a light source in air, it is also necessary to either maintain constant atmospheric conditions during the course of the ruling, or to measure any atmospheric variations, to compute their effect on the ruling and to correct for them if they have occurred. Harrison and Stroke[1,2] have found it preferable to use a simple analogue computer to correct for the effect of atmospheric variations, and in particular for the effect of pressure variations which have, by far, the largest effect.

Fig. 186. Harrison and Stroke, M.I.T. engine. Close view showing from left to right: diamond carriage suspension, reference-mirror for translation and for rotation control, ruled 10 × 5 inch grating, grating carriage mirror used for translation and for rotation control, beam-splitter and collimator tower.

Fig. 187. Harrison and Stroke, M.I.T. engine. Mounting of main screw and two of the four ball-bearing races used to support the grating carriage, as it rolls on the two pairs of cylindrical ways shown.

The idea of controlling the ruling of gratings with the help of interferometers can certainly be traced all the way back to MICHELSON[43] who used interferometers

Fig. 188. HARRISON and STROKE, M.I.T. engine. Diamond carriage motion linearizing mechanism.

Fig. 189. Michelson-Gale engine upon arrival at M.I.T. in 1948, before modification by HARRISON and STROKE.

[43] A. A. MICHELSON: Astrophys. J. **18**, 278 (1903). — Nature **88**, 362 (1912). — Studies in Optics. Chicago: Chicago University Press 1927, reprinted 1962, p. 86.

and visual fringe settings to improve the quality of his ruling-engine screws. Harrison (loc. cit. [23]) refers to W. W. Hansen of Stanford University as having undertaken work on interferometric control before his untimely death. Even a patent [44] on some of the obvious advantages of interferometric control of a ruling engine was granted in 1950, although without any mention of the crucial importance of the need for compensation for barometric changes of the wavelength, necessary for operation in air, etc. More recently H. W. Babcock (Sect. 88 ι)

Fig. 190. Contro lroom of Harrison and Stroke, M.I.T. engine. The Hg 198 light source is located in aluminum box, appearing on pier. In front of the box: the rotating polaroid, producing the "diamond carriage motion" reference signal. The main motor driving the engine, and the gear system appear on the right. In front of gear system: the hyperbolic cam used in the pressure correction system.

has adopted interferometric control to the stop-and-go motion of the Mt. Wilson Laboratory engine in Passadena, with pressure-correction similar to that used on the M.I.T. engine. However, rather than continuously correcting for errors as they occur (as is done on the M.I.T. engine), the error actually introduced into the grating on the n-th groove is corrected by shifting the following $(n+1)$-th groove in an attempt of minimizing cumulative and local errors. Since extended errors are more harmful to high-resolution gratings than residual local and periodic errors, it is probable that outstanding gratings will result with this method, on the Mt. Wilson engine, even though it does not aim at as complete a ruling control as that achieved in the method originated by Harrison and Stroke in 1955 [1] and subsequently expanded by Harrison et al. in 1957 [10],

[44] R. F. Stamm: U.S. Patent 2527338 (Diffraction Grating Ruling Engine), Applied for Oct. 12, 1946, Granted Oct. 24, 1950.

STROKE (1957, 1958)[45], HARRISON (1958)[11], HARRISON et al. (1959)[46], HARRISON and STROKE (1960)[47], STROKE (1961)[2].

It is known that both the Bausch and Lomb Company of Rochester, N.Y. and the Jarrell-Ash Company of Newtonville, Mass. have adopted the Harrison and Stroke scheme of interferometric control to engines in their ruling laboratories, and that other laboratories are preparing similar control. See following sections.

As a general order of magnitude, the good gratings obtained on the M.I.T. engine by 1961 were found (loc. cit. [2]) to be superior by factors of 100 to the famous gratings ruled by ROWLAND and ANDERSON at the beginning of this century, and in the early 1920's. Fig. 191 reproduces for comparison two low-angle (45°) wave-front interferograms, one of a plane grating ruled by J. A. ANDERSON in 1914 on speculum metal on ROWLAND's engine, and considered as very good at that time, and the other of an early grating ruled under interferometric control on the M.I.T. ruling engine. It is apparent from the discussion in Sects. 20, 51 to 59 that ROWLAND's gratings were considered to be so good at their time because they were meant to be used only at very low angles, up to 10° or so in autocollimation, where the wavefronts were some 4 times better and the spectral quality of the gratings some 16 times better than here at 45°.

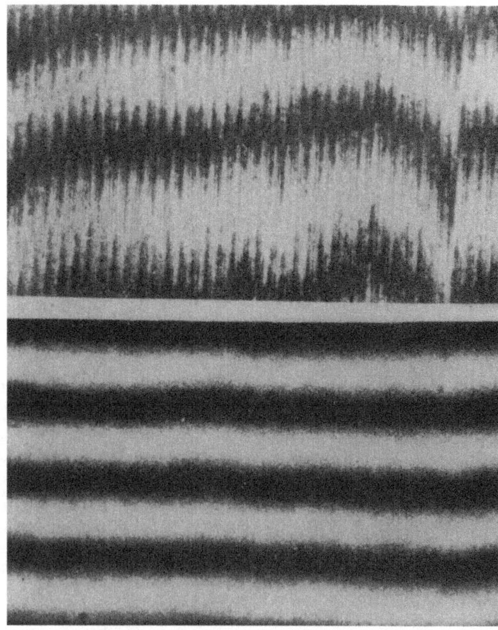

Fig. 191 a and b. Comparison of early low-resolution (a) and early high-resolution (b) grating wave-front interferograms according to STROKE (loc. cit. [2], 1961). (a) Section of grating No. 256 on speculum metal, finished by J. A. ANDERSON on April 14, 1914, (b) section of same width of early grating ruled under interferometric control at M.I.T. Both interferograms were obtained in autocollimation at 45° in λ 5461 Å. The sensitivity of the wave-front deviations to ruling errors increases with the sine of the grating angle, and the sensitivity of spectral-image quality to ruling errors generally increases with the *square* of the sine of the grating angle in autocollimation. Grating a was usable as a good grating only at small angles (up to 10° or so). Both low-resolution and high-resolution gratings ruled nowadays are generally considerably better than the ones shown here for historical purposes. [For ghosts produced by grating (a) see Fig. 73].

How much care needs to be exerted in comparing grating performance, with only the help of resolution photographs or hyperfine structure (hfs) patterns as has been a practice in the past, becomes clear when the hfs of Fig. 192 is compared to the structures of Fig. 193. Unlike the patterns of Fig. 193 which were photographed with M.I.T. gratings 143 and 97 on a fast spectrographic plate 103 aF in only 30 sec, the hfs of Fig. 192 was obtained with the same first-generation grating 97 of Fig. 193 but now on a low-contrast, rather slow and considerably finer Panatomic-X film. An exposure of $1\frac{1}{2}$ min was required in Fig. 192 to attain a density comparable to the hfs of Fig. 193. A considerably better appearance of

[45] G. W. STROKE: Photoelectric fringe signal information and range in interferometers with moving mirrors. J. Opt. Soc. Amer. **47**, 1097—1103 (1957); **48**, 276 (1958).

[46] G. R. HARRISON, N. STURGIS, S. P. DAVIS, and Y. YAMADA: J. Opt. Soc. Amer. **49**, 205—211 (1959).

[47] G. R. HARRISON, and G. W. STROKE: J. Opt. Soc. Amer. **50**, 1153—1158 (1960).

the hfs results under the conditions of Fig. 192 which, it is known, tend to conceal the existence of spurious satellites and weak ghosts and are usually impractical for spectrographic purposes: the hfs of grating 87 taken on the finegrain film

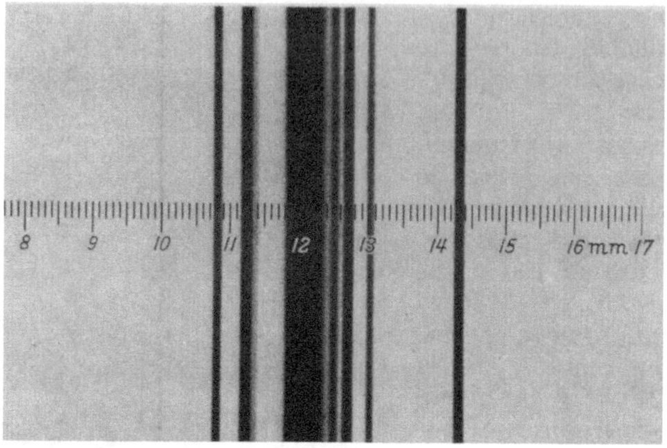

Fig. 192. According to STROKE (loc. cit. [2], 1961). Hfs pattern of green mercury line 5461 Å obtained with same grating 97 and same spectrograph as used in Fig. 193 (Gr. 97) but now on fine-grain low-contrast Panatomic X film, at 3 times the exposure of Fig. 193. The divisions shown are 0.1 mm in the focal plane of the 40-ft (12.2 m) spectrograph. The photographic conditions shown here are not generally practical in normal work. But as they tend to conceal rather than enhance residual spectral line imperfections, they permit extraction of a maximum amount of the spectroscopic information available in the diffracted wave fronts and aerial images. It appears from a comparison of Figs. 192 and 193 that gratings are best compared with spectrograms obtained under similar conditions, and even better with the help of diffracted wave-front interferograms (see text).

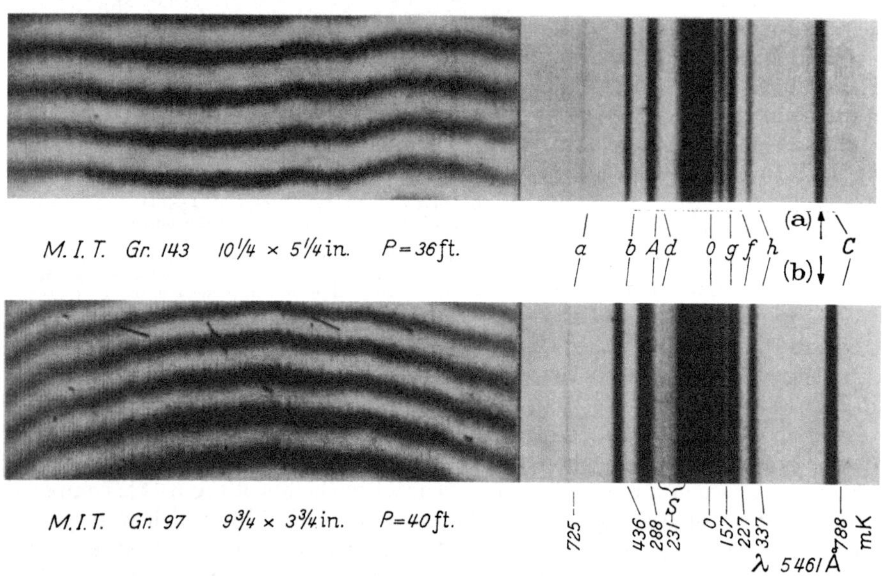

Fig. 193. (According to STROKE, loc. cit. [2], 1961.) Wave-front interferograms (left) and corresponding mercury green line hfs patterns (right) produced in autocollimation at 64° by two 10-in. gratings ruled one before (b), Gr. 97, and the other after (a), Gr. 143, some new improvements of the M.I.T. engine had been carried out. (P = projection distance of spectrographs.) Components *a*, *b*, *A*, *d*, *O*, *g*, *f*, *h* and *C* are all real spectral lines. (For resolution of central group *O*, see Fig. 109.) A perfect grating produces a perfectly straight, horizontal fringe pattern in the interferogram. (The grooves project vertically on these interferograms, and lateral groove displacements on the grating, as well as blank or coating defects, produce vertical fringe deviations from straightness in the interferograms.) The residual wave-front deviations appearing in the new grating 143 are still due to residual coating imperfections which can be even further reduced: in this case they are seen to produce no harmful effects on the spectrum, while a set of spurious satellites described by "*s*" appears in grating 97 with an intensity very similar to that of component "*a*" and results in loss of resolution on the fast spectrographic plates 103aF, used in normal work, when the exposures are long enough to bring out the weaker spectral components, as is the case here. Visual aspect of hfs patterns in spectrographs, as well as finegrain spectrograms, generally appear much better than the hfs rendition on fast, but coarse-grain plates. (For hfs pattern obtained with grating 97 on slow, fine-grain film, see Fig. 192.)

of Fig. 192 would in fact erroneously tend to favor this grating when compared to the hfs of grating 143 taken on the coarsegrain plate in Fig. 193. For the purpose of working comparison, the hfs patterns of Fig. 193 were obtained with both gratings under normal spectrographic conditions: in fact the hfs of the two gratings were actually photographed on the same plate and with the same exposure, in an attempt to bring out the component *a* with the most similar density possible (which was almost achieved on the plate). It is quite apparent from Fig. 193 that a group of satellite lines described by *s* appears in the use of grating 97 with a density very similar to the density of component *a*: the satellites *s* result in a broadening of the lines from grating 97 when compared to the new grating 143 (where no such satellites appear and where the resolution is noticeably better under identical conditions). The wave-front interferograms of Fig. 193 are not subject to any such photographic effects: their power in grating assessment and engine improvements has now been established and clearly appears from the comparisons.

That the quality of the M.I.T. gratings is indeed being attained as a result of interferometric control is illustrated by the low angle (16°) interferogram of Fig. 194. When compared with ANDERSON's grating of Fig. 191, it is seen that without control the M.I.T. engine does not produce an acceptable grating even for use at the lowest possible angles around 5°, while the controlled engine is capable of producing the outstanding grating of Fig. 193.

Fig. 194. (According to STROKE, loc. cit. [2], 1961.) Interferogram from a $1\frac{1}{2}$-in. test section of grating ruled on the M.I.T. ruling engine without control to illustrate the amount of mechanical error that can be effectively removed by the interferometric servocontrol described in the text and in references. The interferogram here has been taken at 16° with only $\frac{1}{3}$ the sensitivity of the wave fronts to ruling errors obtained at 64° (as in Fig. 193). The effect of ruling errors on the spectral-image quality is 10 times greater at 64° than at 16°, and 100 times greater at 64° than at 5° in autocollimation.

The engine control-conditions which have resulted in the ruling of high-resolution gratings of the quality of grating 143 are briefly described below. Experience has shown that their systematic use will indeed combine to result in the attainment of outstanding, flat, high-resolution gratings when ruled under interferometric control.

First, it is in order to give some background concerning the origin of this engine[10].

The basic mechanical parts of the 14 inch ruling engine were constructed under the direction of A. A. MICHELSON in 1900 at the University of Chicago. The engine was modified there in the 1930's by H. G. GALE, F. PEARSON, T. J. O'DONNELL and others, and in 1948, it was presented to G. R. HARRISON at M.I.T. in a form about two-thirds finished. As described by HARRISON, STROKE and associates, this engine has since been completed, modified further and put under interferometric control. As early as 1956, a number of excellent plane gratings up to 8-inches in ruled-width had been obtained[10], under simultaneous interferometric control of both translation and rotation. Further improvements, resulting from refinements in interferometer adjustments and increased stability, eventually resulted in the outstanding gratings described by HARRISON and STROKE in 1960 and 1961[47, 2].

$\gamma 2$) Some details of the control system. Only some of the less obvious details of the interferometric control system are given here. Additional details are found in the original papers. Most of the techniques described are being found useful in laser interferometry and machine-tool control.

Adjustments of interferometers for grating-ruling control. Uncalled-for corrections might be introduced by the photoelectric servo-mechanism into the ruling engine if the interferometer-mirror motion results in an instantaneous flux distribution across the mirror aperture which does not properly represent the grating position. Errors can result in particular from improper

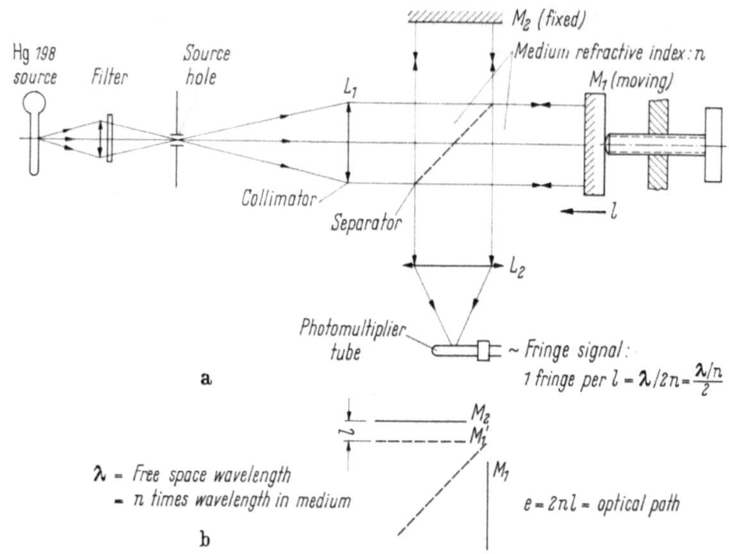

Fig. 195 a and b. Schematic diagram of interferometer used for ruling engine control.

location of the center of gravity of the interferometer-mirrors with respect to the grating plane.

An equal-inclination ring fringe-system obtained from flat mirrors is used for the interferometric control of the moving grating-blank carriage on the M.I.T. ruling engine as shown in Fig. 195. The green 5461-Å line is obtained from a Meggers Hg 198 tube excited in an electrodeless discharge at about 200 Mc/sec and cooled by a moderately strong air current at $\sim 70°$ F[48]. Only a small portion of the central part of the ring system is isolated by a circular diaphragm (source hole) placed in the focal plane of the collimator lens. The flux transmitted through the interferometer, and modulated as a result of the motion of the moving mirror, at the rate of 1 fringe cycle for every $\frac{1}{2}\lambda$ of carriage advance, is focused into a 931-Å photomultiplier tube. A representative fringe signal obtained on the M.I.T. engine is

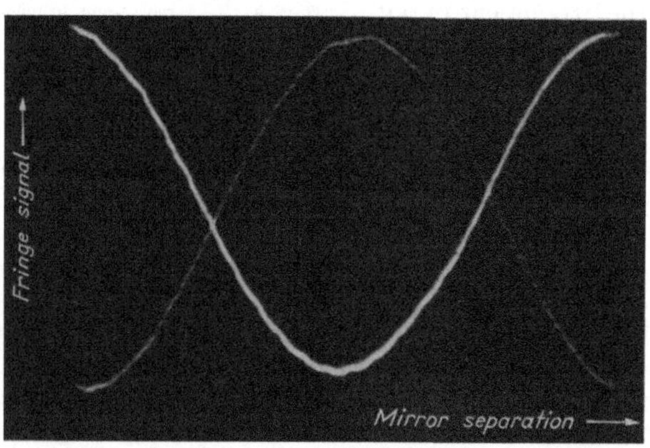

Fig. 196. Photoelectric interference fringe signal obtained on M.I.T. ruling engine at 200 mm path difference with aircooled Hg 198 source (according to Stroke, loc. cit. 1957). Note regularity of signal at the low carriage speed of about 2.5 fringes (6750 Å) per second. The fine inverted signal is produced by the diamond carriage reference-signal generator. (1 fringe cycle = 2730 Å). Also note excellent fringe "visibility" obtained even much before the introduction of laser light sources.

[48] W. F. Meggers and F. O. Westfall: J. Res. Nat. Bur. Stand. **44**, 447 (1950).

shown in Fig. 196. It is important that the source hole be accurately centered on the common normal to the moving and reference mirrors, which is also the center of the ring system in the focal plane of the collimator. The diameter of the rings decreases with the mirror separation according to well-known laws, and a source-hole aperture of 1 mm diameter (in the focal plane of the 1 m collimator) is chosen as so to correspond to only a small fraction of the central fringe at the maximum mirror separation of 5 inches used in the ruling of 10-inch gratings. It is essential not only to center the hole by autocollimation on the mirrors, but to center it actually on the ring system. Although this was pointed out in the original paper by HARRISON and STROKE[1] (1955) dealing with the principle of the interferometric control used on the M.I.T. engine, experience has shown that some additional clarification to what is meant by "interferometric" centering may be desirable (STROKE 1961)[2]. An eccentricity of only 1 mm in the focal plane of the 1 m collimator will indeed result in the additional count of an entire fringe on either extreme of a 10-inch grating, which may be sufficient to account for the curvature of some of the 10-inch gratings previously described.

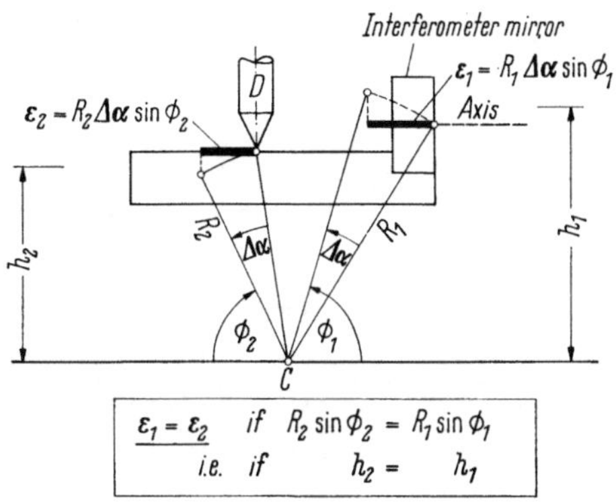

Fig. 197. Effect of improper interferometer-mirror centering with respect to ruling plane (according to STROKE, loc. cit. [2], 1961).

The source hole can be easily centered by first setting the interferometer to some small mirror separation of, say, 20 to 40 mm. After bringing the two source-hole images into coincidence in the usual manner, the straight-line fringe system located in the vicinity of the mirrors can be observed with the help of a low-power telescope. When the interferometer mirrors have been made as parallel as possible by spreading out the fringes, a very clearly visible ring fringe-system will be observable in the plane of the extended Hg 198 tube if the telescope has been refocused on the source hole, and the source hole then removed. If the telescope cross hair is brought to coincide with the center of the ring system, and the source hole is replaced into its position in front of the Hg 198 tube, it can be easily moved until its image coincides with the cross hair of the undisturbed telescope. An adjustment within 0.1 mm in the source plane (corresponding to about 1/10 fringe of grating flatness) is easily achieved.

Any local change in fringe-signal amplitude will be interpreted by the servo-control as a translation error of the control-mirror involved. If the amplitude change is caused by a local mirror-parallelism variation, an uncalled-for correction may result and an error might be introduced into a grating at a place where it was properly positioned by the fringe field. The rotation control about an axis normal to the grating plane and to the direction of grating advance helps in suppressing effects resulting from rotations about vertical axes. But rotations about a horizontal axis parallel to the grating's plane and normal to the direction of grating advance will result in both a simple geometrical ruling error (when the interferometer mirror aperture is not centered on the grating plane) and in a more subtle error caused by a combination of interferometric and servo-mechanical origins. The geometrical error can be simply taken care of by centering the interferometer mirror aperture on the ruling plane within the tolerances of the order of 1 mm, as appears from Fig. 197. The fringe-signal amplitude variations across the mirror aperture which result from rotations about the horizontal axes — and which attend the use of a plane mirror — may, however, still lead to ruling errors unless their effect is also rendered negligible. The fringe-signal amplitude referred to here is that by which one usually describes the amplitude of the periodic signal: it is not the instantaneous value of the fringe-signal (which does, however, depend on the amplitude, of course). Local errors up to $\frac{1}{2}$ fringe would result in the gratings ruled on the M.I.T. ruling engine before the new corrections were carried out by Stroke (1961)[2] (Figs. 197 to 200). The effect discussed here had also been predicted on theoretical grounds (Stroke 1957), and it was shown that it increases with the number (or fractional number) of fringes in the interferometer aperture (for imperfectly parallel mirrors), or alternately with the size of the aperture in presence of the number of fringes in the aperture. The new improvement consisted simply in masking down the vertical height of the interferometer aperture to about 10 mm or less. It remains, of course, essential to locate the center of gravity of the interferometer aperture slit on the plane of the ruling.

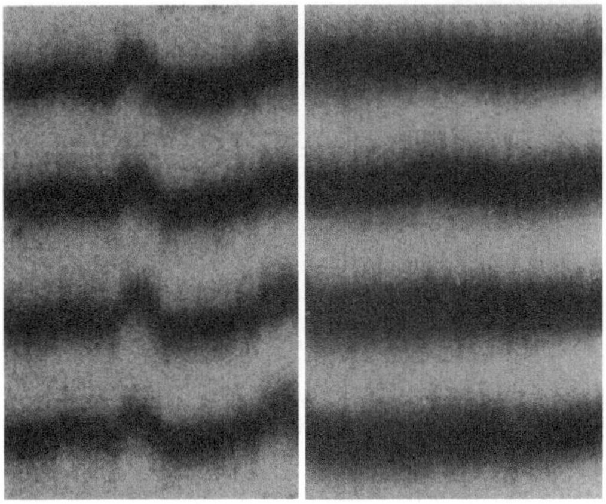

Fig. 198 a and b. Improvements in gratings ruled under interferometric control on M.I.T. engine resulting from interferometer mirror adjustment illustrated in Fig. 197. Two sections of 10-inch gratings are shown ruled (a) before, and (b) after the interferometer mirror centering (interferograms at 74°). A very considerable reduction in spurious satellite lines resulted from this adjustment (according to Stroke, Colloq. Optics in Metrology, 6—9 May, 1958; edited by P. Mollet, Pergamon Press, 1960).

One might note, in conclusion, that the mirror location problem is somewhat different in interferometric spectrometers [Michelson (1927)[49], Fellgett (1951, 1958)[50], Jacquinot (1954, 1958, 1960)[51], Connes (1960, 1961)[52]]: there the motion of interest is that of the mirror or the corner-cube itself, and it is sufficient that its own position be correctly measured and controlled. In the controlled ruling of gratings, the interferometric element is generally placed at some considerable distance from the groove that is being ruled: it is the difference in translational components between the grating and the interferometers which tends to result in ruling errors if

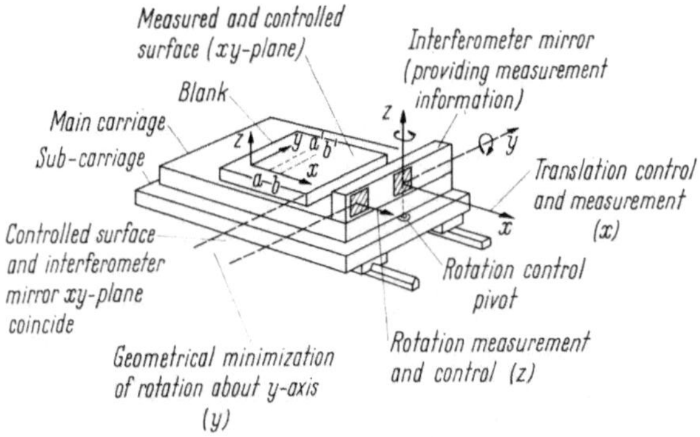

Fig. 199. Corrections required when ruling gratings under interferometric control with moving grating carriage.

[49] A. A. Michelson: Studies in Optics. Chicago: Chicago University Press 1927, reprinted 1962.

[50] P. Fellgett: Thesis. Cambridge University (Cambridge England, 1951). J. Phys. Radium **19**, 187, 237 (1958).

[51] P. Jacquinot: XVIIe Congr. du G.A.M.S. (Paris 1954). J. Phys. Radium **19**, 223 (1958). — Rep. Progr. Phys. **23**, 267 (1960).

[52] J. Connes: Rev. d'Opt. **40**, 45, 116, 171, 231 (1961).

permitted to remain uncorrected. It is in fact the problem of correctly positioning the grating over its entire traverse when moving over rather imperfect ways which has made it so difficult to attain high-resolution gratings, even with interferometric control.

Rotation control. Translation control alone, in the direction of blank advance, has been found to be insufficient on the M.I.T. ruling engine, as a result of the various rotations caused by the balls on which the blank-carriage

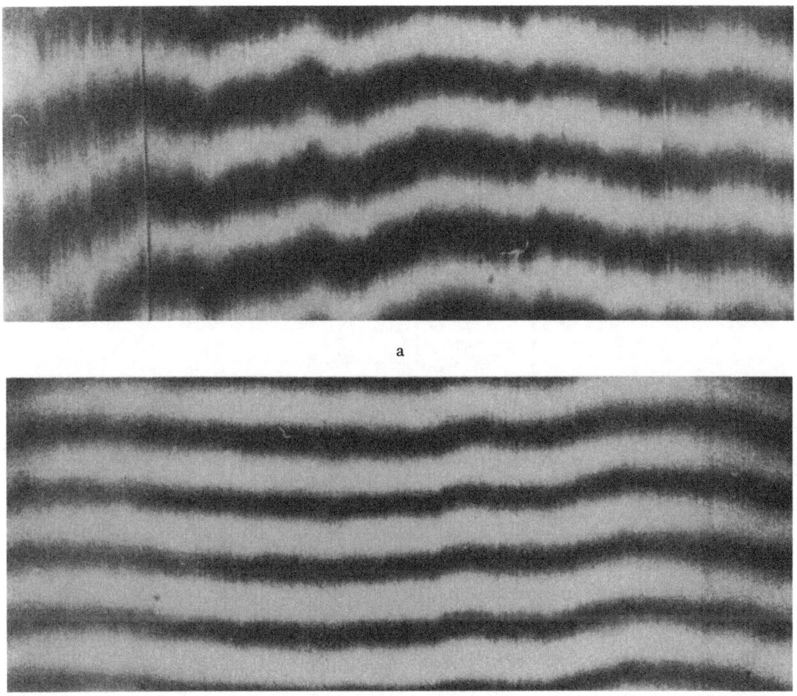

a

b

Fig. 200 a and b. Improvements in gratings resulting from apparently "minor" improvements in M.I.T. interferometrically controlled engine (according to STROKE, loc. cit. [2], 1961). Grating (a) ruled before corrections, and grating (b) ruled immediately after corrections. Residual fringe curvature appearing on right of grating (b) is due to imperfect flatness in aluminum coating of blank. (Both interferograms taken at 64° in autocollimation, in the light of the green line 5461 Å of Hg 198).

rolls in the course of its advance, and as a result of the curvature of the cylindrical ways which guide the balls [HARRISON, STURGIS, BAKER and STROKE (1957)]. More classical ruling engine ways, such as the "doublevee" ways already used by ROWLAND, may also cause rotation problems, in particular as a result of lubrication irregularities.

With the help of an alignment interferometer, the ways can be easily and rapidly adjusted to an average straightness of ± 1 fringe (measured across a 10-inch interferometer arm) over the 10-inch traverse in a horizontal plane (STROKE 1961)[53]. A sketch of the alignment interferometer, according to STROKE[53], is shown in Fig. 201.

In the ruling engine, the local parallelism of the grating advance continues to be maintained with the help of an interferometric servo-mechanism identical to one used in the translation-control loop (Fig. 184) by comparing the fringe signals

[53] G. W. STROKE: An alignment interferometer for precision straightness measurements and control, even of rapidly moving carriages. J. Opt. Soc. Amer. **51**, 1340—1341 (1961).

obtained from two regions of the same mirror placed some 4 inches apart, as illustrated in Fig. 202. One of these mirror regions is located exactly above a preloaded ball-bearing pivot and the entire grating table is rotated by the interferometric rotation-control servomechanism about a vertical axis defined by the

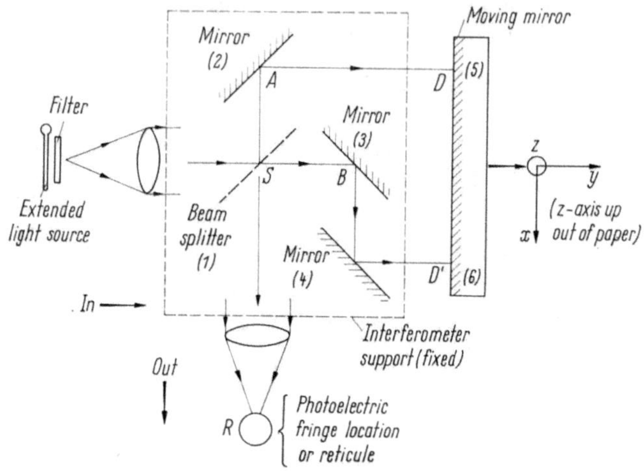

Fig. 201. Alignment interferometer for rapid interferometric alignment of carriages and ways (according to STROKE, loc. cit. [53], 1961). As a result of the use of a slight path difference in the two beams, and of the equal-inclination interference fringe system, no parallax effects appear, and the fringes maintain their best visibility over very long carriage traverses. Measurements over 1-foot traverses with velocities as great as 1-foot/second have been readily performed.

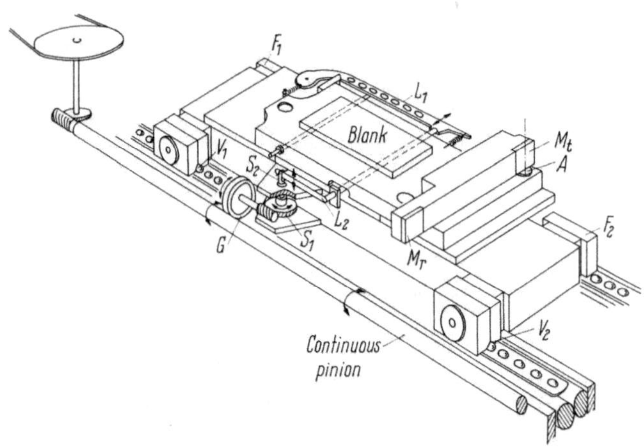

Fig. 202a. Rotation control system used by HARRISON and STROKE, M.I.T. engine (according to HARRISON, STURGIS, BAKER and STROKE, loc. cit. [10], 1957).

ball bearing and passing through the front of the translation mirror. The translation control results in accurate positioning of the grating-carriage within about $\pm 1/100$ fringe (about $\pm 0.003\ \mu$ or $\pm 10^{-7}$ inch) over the 10-inch traverse, and in maintaining the parallelism to $\pm 1/200$ sec of arc.

A general view of the M.I.T. ruling engine is shown in Fig. 185 and a close-up of the control-interferometers in Fig. 186. A view of the control room is shown in Fig. 190.

Correction for the effects on the interferometric control system of barometric changes of wavelength. Under interferometric control, the groove position and spacing are determined by the wavelength of the radiation (for instance the green Hg 198 radiation) in air. The wavelength in air changes with variation of barometric pressure, temperature, water vapor and CO_2 partial pressure as well as with the introduction of any foreign gasses or vapors (BARRELL and SEARS 1939)[54]. In practice, when the temperature is controlled to the tolerances of 1/100° C or better as required by the mechanical elements of the ruling engine, variations in barometric pressure produce the major change of wavelength, according to an equation

$$\Delta m = 3.31\, \Delta P\, \Delta L$$

where Δm is the fringe shift in 1/100 fringe ($\frac{1}{2}\lambda$), ΔP is the pressure variation in inches of mercury, ΔL is the mirror separation from zero path difference in mm. For example, if a pressure change of $\Delta P = 1/3.3$ inches should occur when the grating is at $\Delta L = 100$ mm from zero path difference, then the grating would be shifted by $\Delta m \simeq 100$ hundredths of a fringe, or approximately by one fringe by

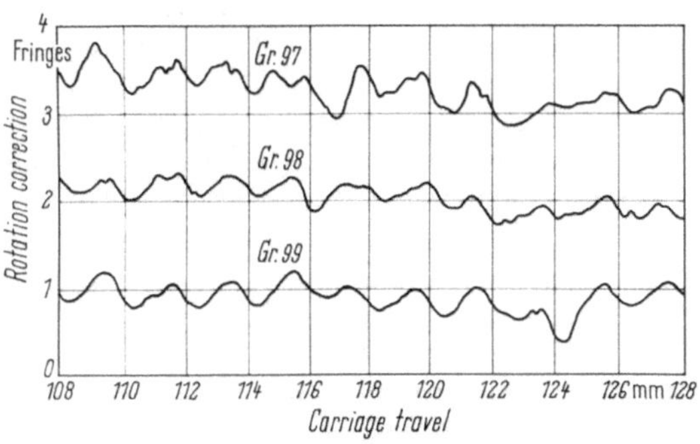

Fig. 202 b.

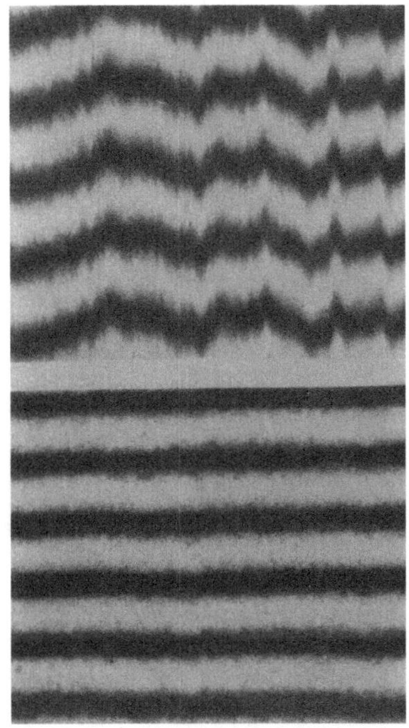

Fig. 202 c.

Fig. 202 b. Rotation correction required in three successive gratings ruled on M.I.T. engine, and showing need for this type of correction. Very large errors would result in gratings, if they were to be ruled without the interferometric rotation control system illustrated in Fig. 202 a (see Fig. 202 c).

Fig. 202 c. Improvements of gratings ruled under interferometric servo control on M.I.T. engine, after introduction of rotation control (Fig. 202 a). Two sections of 8-inch gratings are shown ruled (a) before, and (b) after the rotation control system was installed. (Interferograms at 46°, such that the error in fringes on the grating is about 1.4 times the error appearing in the interferograms).

the controlling servo-mechanisms, a very large error indeed, if it remained uncorrected. In reality the pressure change does not only result in a shift of the grating at the time when a pressure change occurs, but it also produces further errors as a result of a spacing in the now incorrect wavelength (as compared to the wavelength at the pressure P_0 at the start of ruling). Finally, the sign of the grating shift changes when the grating interferometer mirror passes through zero path difference. When the grating moves toward zero path difference, a pressure increase (which decreases the wavelength and tends to pull the grating towards zero path difference) moves the grating "forward" in the direction of its motion. However, the pull towards zero path difference will move the grating backwards from its direction of motion when the grating moves away from zero path difference. The grating shifts, both "instantaneous" (at the time of the pressure change) and "cumulative" (as a result of the changed wavelength) result from the opera-

[54] H. BARRELL and J. E. SEARS: Phil. Trans. Roy. Soc. Lond. A **238**, 1 (1939).

tion of the servo-control phase detecting system, which is designed to continuously synchronize the interferometric fringe signal with the diamond-position reference signal. The phase detecting system only knows about fringes and their position relative to the reference, but it ignores the differences in the origin of fringe shifts (real position error or change of wavelength). It is the function of the analogue computing cam shown in Fig. 190, to help in calculating the instantaneous uncalled-for grating shifts, and to accordingly shift the reference signal generator, relatively to the diamond-shaft, in order to maintain the correct grating position relative to the start of ruling and the diamond. The height of the two-dimensional hyperbolic cam materializes, with the help of appropriate linear slopes, the variations of fringe-shift as a function of path difference and pressure change in the range of operation. The pointer which rides on the cam is directly geared into a differential placed between the diamond-shaft and the diamond-position reference signal generator. The computing can also be carried out with the help of an electronic multiplication circuit using "helipot" resistors and a servo-motor to feed the differential[12, 55].

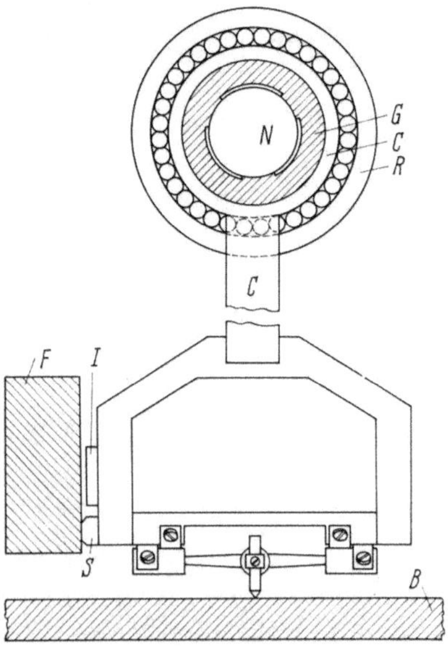

Fig. 203. (According to HARRISON and STROKE, loc. cit. [1], 1955.) Monorail mounting of the diamond carriage, in which linear shape of groove is controlled largely by an optical flat F, against which a shoe S on the diamond carriage slides. The diamond is suspended directly from graphitar cylinders G which slide along monorail N. The rod which moves the diamond carriage pushes against ring R, so that accidental torques are not transmitted to the diamond carriage. The interferometer I serves to keep a record of the diamond carriage adjustment (see Fig. 204).

Recent indications are that pressure compensation in control interferometers is more directly achievable by suitably controlling the frequency of a gaseous optical maser[56]. The description of an optical phase discriminator appears to have been given for the first time in this report[56].

The temperature control used on M.I.T. ruling engine is described in Sect. 87.

Diamond-carriage control. The interferometric control of the grating carriage advance implies that the diamond carriage moves in the ruling plane parallel to the fixed reference mirror within a few hundredths of a fringe (some 0.01 μ) during the course of some 15 miles of equivalent traverse, which correspond to the ruling of a 10-inch grating at 300 grooves/mm. The design of the grating carriage used on the M.I.T. ruling engine was described by HARRISON and STROKE (1955)[1], and it has successfully withstood the test of time in maintaining the diamond in the ruling plane within better than 1/15 fringe. This result is of course not achieved without careful adjustment of the mechanical elements involved. The diamond carriage is suspended from two bearings riding on a monorail, which is fixed on the engine, and rests with a plastic (Rulon) shoe on a long optical flat, which is also fixed on the engine in a position at right angles to the grating motion. It is the long flat which determines the ruling plane. So far it has not been found necessary to use interferometric control of the diamond with respect to its reference mirros or indeed to a reference mirror at the beginning

[55] R. F. JARRELL and G. W. STROKE: New advances in grating ruling, replication and testing. Appl. Optics 3, 1251—1262 (1964).

[56] G. W. STROKE: In: Frequency control stability, and interactions between modes in gas lasers (unpublished). — Perkin-Elmer Engineering Report No. M 7287 A (May 27, 1963).

of the grating, but further improvements in grating quality may result from the use of a servo-system for that purpose, as well as from possible interferometric monitoring of the reference-mirror position itself with respect to the beam-splitting tower and other fixed interferometer components, in particular: the use of the extremely sharp spectral lines produced by optical masers [SCHAWLOW and TOWNES (1958)[57], JAVAN (1960)[58], JAVAN et al. (1961)[59]] should prove invaluable in interferometric control.

A schematic view of the diamond carriage, and of the diamond bearing hinges, according to HARRISON and STROKE[1] is shown in Fig. 203 and an interferometric record of the diamond carriage position in Fig. 204.

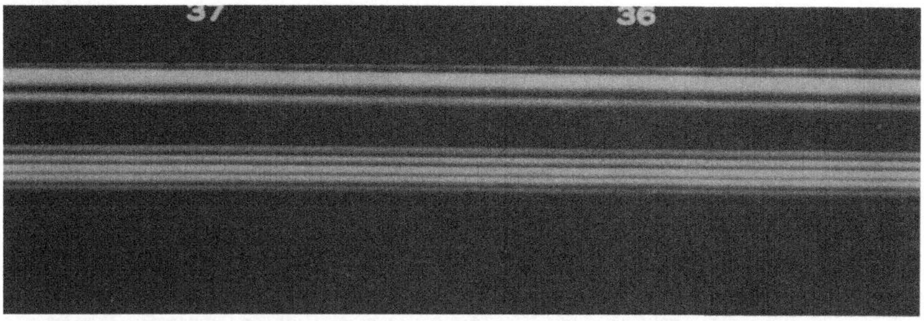

Fig. 204. Interferometric record of diamond carriage adjustment in M.I.T. ruling engine. The synthetic photograph was obtained by taking thousands of interferograms next to each other on the film, one at the center of each groove. (a) Record of diamond carriage with respect to big flat F in Fig. 203, (b) record of diamond shaft with respect to diamond carriage, obtained by mounting a small interferometer mirror right on to the diamond-shaft, and another on the diamond carriage (according to HARRISON and STROKE loc. cit. [1], 1955).

δ) The interferometrically controlled Jarrell-Ash Engine. A greatly simplified version of the Harrison and Stroke principle of controlled ruling of diffraction gratings with a continuously moving carriage has now been successfully developed and applied by G. W. STROKE to a good mechanical 10-inch engine built by WM. MACARTHUR and D. K. RICE at the Jarrell-Ash Company[13].

The work on the interferometric control system for the Jarrell-Ash engine was carried out by G. W. STROKE with WM. MACARTHUR and JOHN BERNIER, and a more complete description of this engine and of its control features is given by R. F. JARRELL and G. W. STROKE [13b] together with additional details concerning grating ruling and replication methods developed at Jarrell-Ash for the production of high-resolution gratings[55]. Over half a dozen good gratings up to 6-inches in width and with spacings up to 30000 lines per inch have already been successfully ruled with the new control system.

A block diagram of the Jarrell-Ash engine and of its interferometric control system are shown in Fig. 205 and a photograph of the complete engine in Fig. 206. Fig. 207 shows an enlarged view of the gear train used in driving the grating carriage approximately in synchronism with the diamond carriage, at a rate of about eleven grooves per minute. The Hg 198 light source, photoelectric tubes and electronics are located outside of the engine case. One of the remarkable features of this control system is that the gears which are taking the motion off the main shaft are directly placed on the engine. With specially ground gears in this group as well as in the gear train, and with careful alignment, no vibrations at all have been found to enter the engine, the ruling diamond or the interferometers.

---

[57] A. L. SCHAWLOW and C. H. TOWNES: Rev. **112**, 1940 (1958).

[58] A. JAVAN: In: Quantum Electronics, ed. by C. H. TOWNES, p. 564. New York: Columbia Un. Press 1960.

[59] A. JAVAN, W. R. BENNETT and D. R. HERRIOTT: Phys. Rev. Letters **6**, 106 (1961).

Fig. 208 shows the interference fringe signal produced by the advance of the grating carriage on the Jarrell-Ash engine and the remarkable degree of control achieved in the closed-loop operation during the course of controlled ruling of a 30000 lines per inch grating: an instant of time has been chosen when an error of about $1/10^6$ inches (1/10 fringe) with respect to the diamond carriage reference signal was being corrected, as achieved in the right half of the oscilloscope photo-

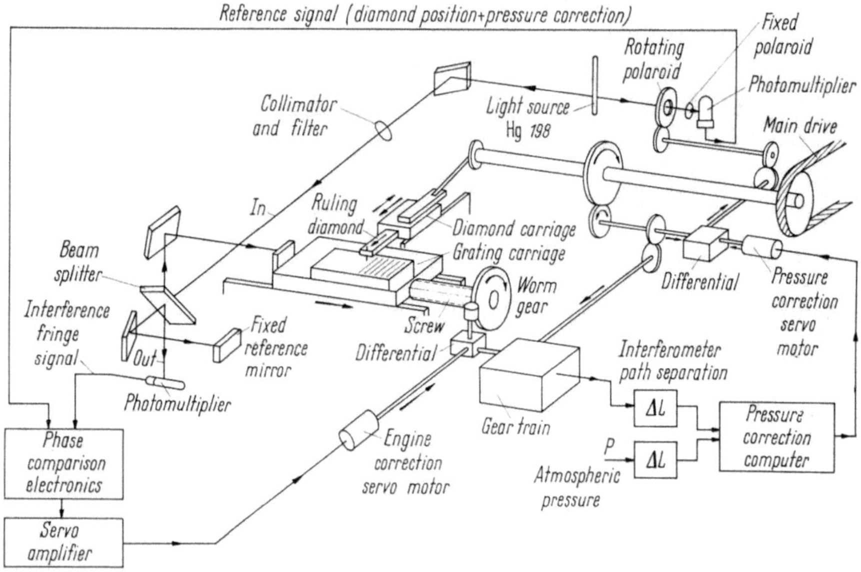

Fig. 205. Schematic diagram of Jarrell-Ash interferometrically servo-controlled engine (after Jarrell and Stroke [55]).

Fig. 206. Interferometrically controlled 10-inch Jarrell-Ash engine (control system according to Stroke, loc. cit. [55], 1964).

graph. A wavefront interferogram of an early 6-inch 30000 lines per inch grating is shown in Fig. 209, with the last half inch showing a portion of ruling produced without control for comparison. About 12 days are required to rule a 6-inch 30000

Fig. 207. Gear system and pressure-correction servo motor input of interferometrically controlled Jarrell-Ash engine.

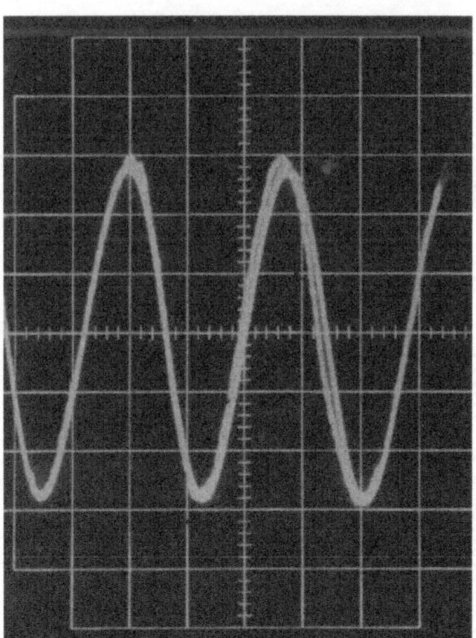

Fig. 208.

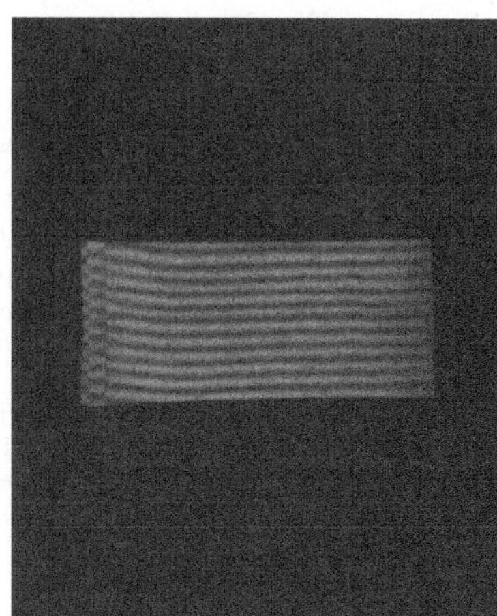

Fig. 209.

Fig. 208. Interference fringe signal with carriage under interferometric servo control on Jarrell-Ash engine, during the course of a ruling of a 30000 groove/inch (1200 groove/mm) grating, at a carriage speed of about 0.1365 microns ($\frac{1}{2}$ fringe) per second. An instant of time was chosen during which an error of about 0.1 fringe (about 10^{-6} inches) with respect to diamond carriage reference signal was being corrected. Note great regularity of carriage motion, even at these extremely low velocities. The same regularity and absence of stick slip is obtained in ruling 60000 groove per inch gratings (see STROKE, loc. cit. [136], 1964, and R. F. JARRELL and G. W. STROKE, loc. cit. [55]).

Fig. 209. First-order interferogram of 6-inch, 1200 groove per mm grating ruled under interferometric control.

lines per inch grating, during the course of which the diamond carriage must be maintained in the same ruling plane to better than 10 inches during a 10-mile reciprocating traverse, the temperature must be maintained to a few 10° C, and the grating carriage must be brought to its appropriate position also to better than 10 inches at every instant of time with the help of the interferometric servo-control.

It is noteworthy, for the new Jarrell-Ash engine control, that it was found possible to successfully use a teflon worm to drive the large worm gear (Fig. 207): no sign of wear has appeared in many months of continuous operation. The variations in wavelength and the corresponding errors which would result from the variations in atmospheric pressure during the course of the ruling are being corrected by a simple electronic computer according to the principle previously described (Harrison and Stroke 1955, Stroke 1961).

In addition to ruling gratings with adequately flat wavefronts, perhaps the most remarkable feature on the introduction of the interferometric servo control system on the Jarrell-Ash engine is the dramatic reduction in the intensity of Rowland ghosts, which result from periodic errors in the ruling. The best mechanical engines usually present Rowland ghost intensities on the order of 1/2000 in the first order of a 15000 lines per inch grating in the visible. The very first grating ruled under control on the Jarrell-Ash engine, and every grating since, has had the ghost intensity reduced from the usual 1/2000 to much less than 1/100000, doing apparently better even than the M.I.T. engine, and showing the advantages of continuous motion control used on both these engines, as well as on the smaller Bausch and Lomb engines, compared to what seems to be achievable under stop and go motion. No disadvantages of any kind have been observed to result from ruling with continuous grating carriage motion. In particular, no need was found for linearizing the diamond carriage motion, thus adding to the simplicity of the control system[13, 55].

ε) *The Bausch & Lomb Michelson Engine, and B & L interferometrically controlled engines.* The large ruling engine now at Bausch and Lomb was one of the two engines built at the University of Chicago shortly after the turn of the 20th century and remained there under the direction of Professor Gale until 1948. It was presented to Bausch and Lomb at that time, while its sister engine was presented to M.I.T.

As described by Bausch and Lomb in 1957,

"The size of the engine is such that it could rule gratings with grooves over 150 mm in length and a width of ruling up to 300 mm. However, it will require several additional years of work on the engine to achieve this size of useful ruled area. The present maximum is 153 × 203 mm.

The mechanism for advancing the grating carriage was designed for great flexibility. This makes it possible to rule gratings with a wide choice of numbers of grooves or grating spacings per mm. Since the precision screw has a 2 mm lead, grating spacings are given as number of grooves per mm. Useful gratings are ruled in ranges from 20 grooves per mm to 10800/mm.

The precision screw (Fig. 210), which is the heart of any ruling engine has been studied with great care to learn the nature of its errors. Even though the screw is very good, the remaining errors are sufficient to produce substantial Rowland ghosts, unless corrected. An electronic fringe counting system, using the wavelength of the green light from Hg 198 isotope as the standard of length, is employed to plot the error curve for the screw. This curve supplies the data necessary for cutting a brass cam that supplies the exact correction needed to minimize the Rowland ghosts."

The measurement system used to produce the correction cams, is quite similar to the system described by Harrison and Archer[30]. It was described by Finkel-

STEIN, BRUMLEY and MELTZER[60] in 1952 with particular reference to the correction of periodic errors:

Fig. 210. Gear system and screw end of Bausch & Lomb Michelson-Gale engine (photo courtesy E. G. LOEWEN, Bausch & Lomb).

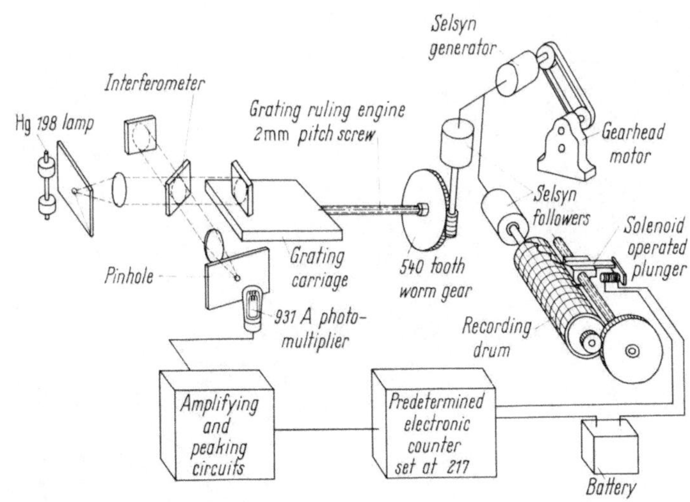

Fig. 211. Interferometric error-recorder used to produce periodic-error compensating cam on Bausch & Lomb engine (according to FINKELSTEIN, BRUMLEY and MELTZER, loc. cit. [60], 1952).

[60] N. A. FINKELSTEIN, C. H. BRUMLEY and R. J. MELTZER: J. Opt. Soc. Amer. **42**, 121—126 (1952).

"The interferometric error recorder, Fig. 211, plots on a revolving drum the lead or lag of the grating blank carriage relative to its theoretically correct position as a function of distance along the screw. The unit of measurement is an interference fringe, half a wavelength of the 5461 Å green line emitted by a high frequency discharge tube containing the 198 isotope of mercury. The error curve thus recorded was separated into its component harmonics by a numerical Fourier analysis, and from the information obtained concerning the periodicities present and their amplitudes, it was possible to calculate the expected ghost intensities. Finally, a corrector cam was cut.

The Selsyn generator, shown in Fig. 211 is rotated continuously by the gearhead motor, drives two followers. The first follower drives the grating ruling engine screw at a speed of approximately one turn every fifteen minutes. The second follower drives a drum covered with recording paper at exactly 540 times that speed. As the grating carriage moves along the screw, carrying one mirror of a Michelson interferometer, the interference fringes move across the face of the diaphragm, D, causing the light falling on the photomultiplier to vary, reaching one maximum and one minimum for each half wavelength motion of the grating carriage. The signal from the anode of the photomultiplier is amplified, clipped, and differentiated to produce pulses which trigger the electronic counter. The counter is set to close a switch for an instant when the count has reached 217. This actuates a plunger on the recorder which punches a small hole in the recording paper.

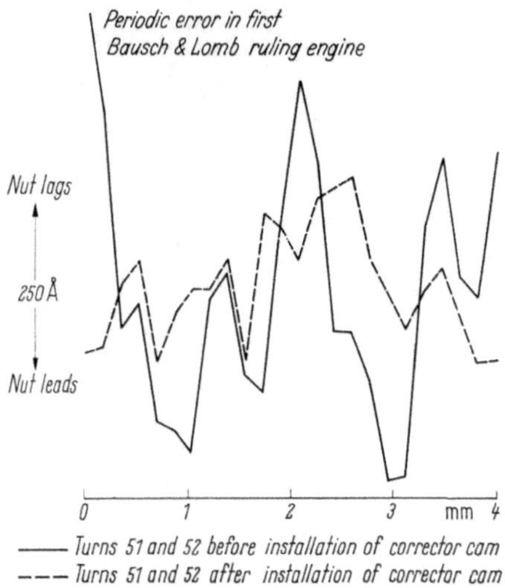

Fig. 212. Error recorder plot for two turns of the screw, in the Bausch & Lomb engine, according to Finkelstein et al. (loc. cit. [40]).

Fig. 213. Bausch & Lomb Michelson-Gale engine (photo courtesy Bausch & Lomb).

Fig. 212 shows an error recorder plot for two turns of the screw both before and after the corrector cam was inserted. The up direction corresponds to the grating blank carriage lagging behind its correct position along the screw, and the down direction corresponds to the carriage leading the screw. The peak-to-peak spread of the uncorrected curve is about 500 angstroms. The curves shown are not the ones taken directly from the recorder, but are different only in that the background noise has been removed. A twenty-four point numerical Fourier analysis was performed on these curves to calculate all periodicities which are harmonics of the pitch of the screw."

Fig. 214. Diamond carriage of Bausch & Lomb Michelson-Gale engine (photo courtesy Bausch & Lomb).

The measured amplitude of the harmonic component of the periodic error achieved after correction was of the order of 4.5×17 Å, being perhaps 1.5 times smaller than mechanically non-corrected errors on other good mechanical engines, but about 5 times larger than the errors obtained on the interferometrically controlled engines with continuous carriage motion.

A general view of the Bausch and Lomb engine is given in Fig. 213 and a detail of the diamond carriage in Fig. 214. A very considerable number of good ratings up to 8-in wide have been successfully ruled on this engine. A sketch of the underground laboratory in which the engine is located is given in Fig. 215. Many additional details about this engine, in particular its applications to the ruling of accurate scales, are given by RICHARDSON and STARK[61].

A small four-inch engine, built in part by DAVID MANN, was brought under interferometric control at M.I.T., and subsequently transferred to Bausch and Lomb around 1960, and another similar engine (Figs. 216 and 217) was brought under interferometric control at Bausch and Lomb. The design principles and

[61] D. RICHARDSON, and R. M. STARK: J. Opt. Soc. Amer. **47**, 1–5 (1957).

general performance of these engines were described in 1963 by E. LOEWEN*, and a block diagram of the control system is shown in Fig. 218. The system is quite similar to the control system first described by HARRISON and STROKE[1].

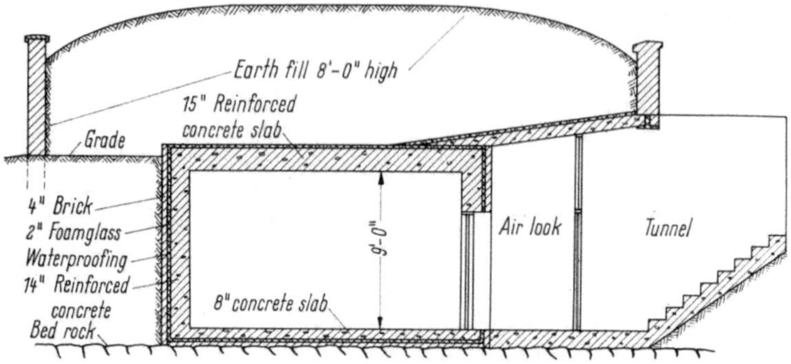

Fig. 215. Sectional view of Bausch & Lomb underground grating ruling laboratory (diagram courtesy Bausch & Lomb).

Fig. 216. Bausch & Lomb interferometrically controlled engine (photo courtesy E. G. LOEWEN, Bausch & Lomb).

Much of the mechanical work on the Bausch and Lomb engines, and the ruling of gratings, has been carried out under the supervision of ROBERT WILEY[60,61]. The contributions made by Bausch and Lomb to supplying fine spectroscopic

* E. G. LOEWEN: Positioning system spaces lines to within 1/10 microinch. Control Engineering **10**, 95—97 (1963).

gratings and replicas has been singularly outstanding. The role held in these achievements by DAVID RICHARDSON has been particularly recognized.

ζ) *The Baird-Atomic infrared grating engines.* In a private communication[62], these engines are described by WALTER BAIRD as follows:

"We have built two ruling engines. The first was started about 1941 or 1942 and moved ahead very slowly through the war years. As I remember, we first began ruling gratings in 1946 and I think I can say with reasonable accuracy that this was the first successful ruling engine built by an industrial firm, although I may be wrong.

Fig. 217. Bausch & Lomb interferometrically controlled engine (photo courtesy E. G. LOEWEN, Bausch & Lomb).

The basic engine is very similar to the Rowland as modified by ANDERSON and WOOD and of course WILBUR PERRY. In other words, it was a double cross slide with a 20-thread to the inch screw and 750 teeth on the dividing head. We did do several things differently from the Hopkins approach. First, we wished to get rid of the change from static to dynamic friction and arranged our drive so that the engine was moving the blank as the diamond was cutting. This was done by two mechanical devices: one, a set of elliptical gears which were arranged so that the diamond carriage and therefore the diamond was cutting the blank at the slow period of the elliptical gears while the return permitted the blank to advance very rapidly, never coming to a complete stop. The second innovation was the driving of the screw instead as on the Hopkins instrument with a ratched and pawl arrangement, with a worm advancing with a worm wheel. Now this, while getting rid of static friction, should produce ruled lines that have theoretically an "S" shaped contour, but to my knowledge, we have never been able to show that this affected resolving power.

Most of our rulings have been on 4-inch diameter blanks, almost all 3-meter radius of curvature. While we tried other surfaces than aluminum for experimenting, aluminum has been the standard.

This engine ruled successfully for about twelve years and only until the period about 1955 were we ever greatly concerned if the first order Rowland ghosts were less than 1/500.

[62] W. BAIRD: Private communication to G. W. STROKE (March 28, 1962).

In 1948, we got started on another engine which is very similar to the first. A few small innovations were made largely to permit the disengagement of the worm wheel drive of the screw from the screw itself and to take care of corrections. This engine has been ruling fairly steadily for the last five years and early we introduced one further innovation, and I think again we were the first to try it, and that was to put 1/1000th inch strips of Teflon on the ways as an antifriction material. This engine is still ruling excellent gratings.

We have had in the last two years Frank Cooke re-do the first engine and I must say he has done a beautiful job. We had one unfortunate happening in the last six months, just as this engine was beginning to get to work again. A sprinkler head let go in the night and dowsed

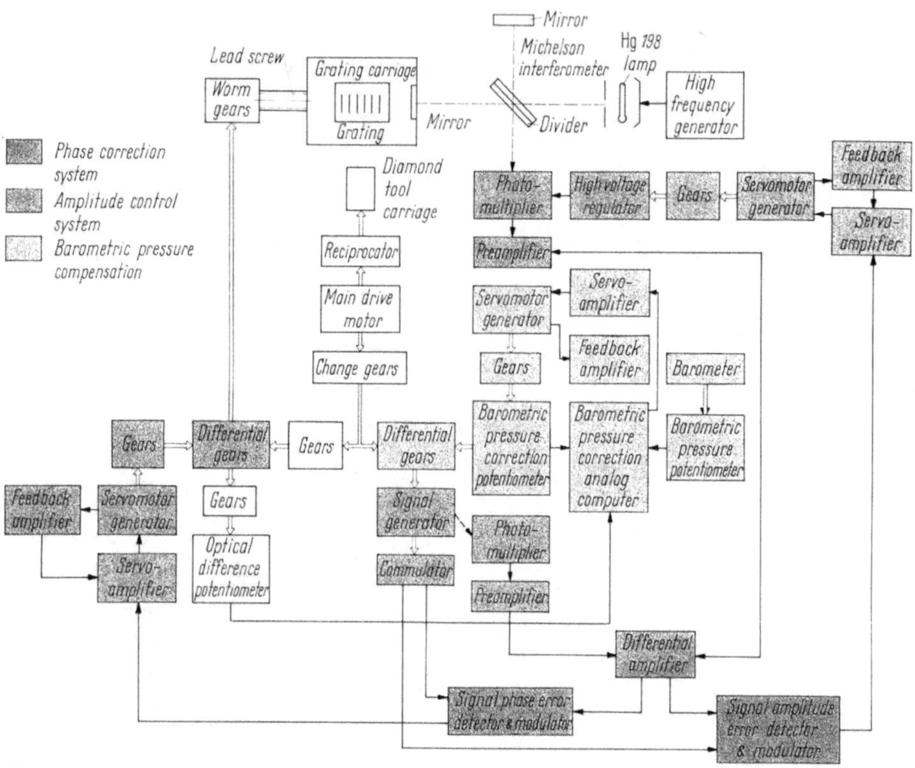

Fig. 218. Block diagram of interferometric control system used on Bausch & Lomb engines (according to E. G. Loewen in loc. cit., Control Engineering, McGrawHill, 1963).

the engine completely with very rusty dirty water. It was some hours before this accident was discovered and as a consequence, a fair amount of pitting, particularly on the lead screw, took place. This is now being relapped. We expect it back again shortly.

Almost all of our gratings have been 4-inch diameter, 3-meter, 15000 lines to the inch. We have changed blaze angle, depending upon the requirement. In general, all of the gratings have been used as high as the 4th order. We have ruled gratings at 7500, 5000 lines to the inch, for special purposes. At one time, just to prove we could do it, we did rule a 6-inch grating. For our spectrographs, our gratings were expected to work from far ultraviolet, 1900 angströms out as far as one micron."

The Baird Atomic grating ruling project was started upon the initiative of Walter Baird, and significant early contributions to it were made by Wm. Langton[62]. Most of the construction, making the nut, lapping the screw was carried out by Frank Gillis[62]. Appreciable advice appears to have been received from Johns Hopkins, in particular from Wilbur Perry[62].

η) *The State Optical Institute engine.* The principle features of this engine and the performance of the gratings ruled were described in 1958 by Gerasimov et al.[24].

"The most notable point about this machine lies in the design of the two basic mechanisms. A double carriage is used in the first, as proposed by STONE[63] and used by GALE[64] when adapting one of MICHELSON's engines. The carriage is in two parts placed one above the other (Fig. 219). The lower part is driven along the guide 2 by the screws; the friction between guide and carriage is large and variable in time, so this movement is not of high precision. The upper part 4 carries the blank 5 and is joined to the lower by the four flat springs 6, which allow the two parts to move relative to one another along the feed direction only. The second part is moved by the screw 7, of the same pitch as the first screw. The forces acting on the second screw arise from internal friction in the springs, and are almost constant if the two screws are synchronously driven; this ensures high accuracy in the feed. The two worm drives to the two screws (4a and 4b in Fig. 220) are identical and are driven from the same drive. Ten teeth are in mesh at once on the gears, which eliminates local irregularities in the teeth.

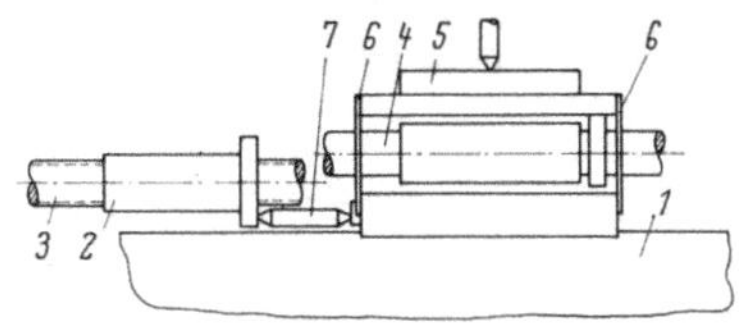

Fig. 219. (According to GERASIMOV et al., loc. cit. [24], 1958.) The double carriage system for driving the blank.

It has been found by interference methods that a force of 5 g wt applied to the upper part displaces the carriage elastically by about 0.01 μ, which approaches the maximum permissible error in ruling. The working force on the screw is in fact constant to 2—3 g wt. The upper part follows strictly the law prescribed by the drive mechanism, whereas the lower does not, because of the friction in the ways.

Hydrodynamic friction principles are used in the ways of the ruling mechanism, so wear of the surfaces is virtually eliminated, even though up to 180000 movements are required for a large grating. The ruling carriage consists of a rigid girder which works on ways and is driven by a crank mechanism; the ruling instrument is carried at the centre. Separate guides are used for controlling the horizontal and vertical motions, as shown in Fig. 220, because the forces acting and the accuracies required in these two directions differ. The horizontal supports are three self-levelling shoes 1 which slide on flat polished surfaces 2 immersed in the oil baths 3.

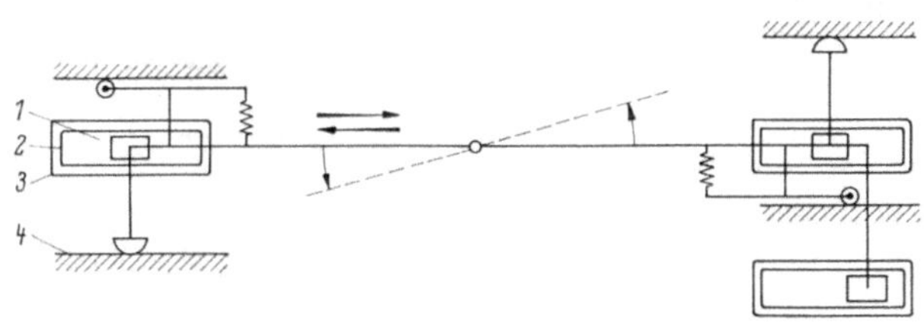

Fig. 220. Same as in Fig. 219. The ruling carriage. Plan view.

The rubbing surfaces of the shoes are spherical (radius of curvature about 1 km); this gives a stable motion and a continuously formed fairly thick oil film between the rubbing surfaces, so the friction is hydrodynamic.

The vertical guides are formed by two flat glass blocks 4 fixed to the base at the ends of the carriage on opposite sides. Spherical lignum vitae supports rest on the block and are tensioned by sprung levers. No lubricant is used. The forces required to move the carriage are very low, and the motion is smooth and highly rectilinear.

The machine is set up on an antivibration base in a constant-temperature enclosure."

Plane and concave gratings of all usual types are being ruled on this engine, up to areas of 150×150 mm, and spacings of 1200, 600, 300 and 200 lines/mm.

The best gratings show resolving powers of 600000 and the Rowland ghost intensity in the 1st order of a 600 lines/mm grating was usually found to be 0.1%, corresponding to a periodic error amplitude of 10×17 Å. In some very good gratings, Rowland ghost intensities as low as 0.01% were observed (corresponding approximately to 3×17 Å periodic error amplitude).

---

[63] W. STONE: J. Sci. Instrum. **11**, 241 (1934).

[64] H. G. GALE: J. Sci. Instrum. **12**, 32 (1935).

The gratings are tested with the help of the interferometric wave-front testing method first described by STROKE in 1955[65], although the method is ascribed by GERASIMOV et al. (loc. cit.[24]) to GERASIMOV et al.[66], (1957).

Very careful measurements were carried out by GERASIMOV et al. (loc. cit.[24]) on all the important aspects of grating performance. Scattered light was found to be less than 10^{-5}. The work by these authors on intensity distribution and polarization is particularly notable, not only because they have obtained remarkably high efficiencies with 110° to 120° diamonds, but also because, just like R. W. WOOD[67,68] in 1910, they speculate, incorrectly (as since shown by STROKE[69], see also Sect. 62) that "The reflection coefficient could be increased considerably by reducing the angle between the faces to 90°, ..."[24]. As shown by STROKE (loc. cit.[69] and Sect. 26), the 90° angle is precisely one of the most undesirable angles, when high efficiencies are sought, because of its tendency towards polarizing the diffracted light; and consequently toward *reducing* the efficiency in gratings; 110° and 120° angles are those recommended[69]. One may conclude, therefore, that the use of 110° and 120° angles by GERASIMOV et al., as by R. W. WOOD, was fortuitous rather than deliberate, even though the quality of their gratings has apparently greatly benefited from this angle. As described by GERASIMOV et al.[24]:

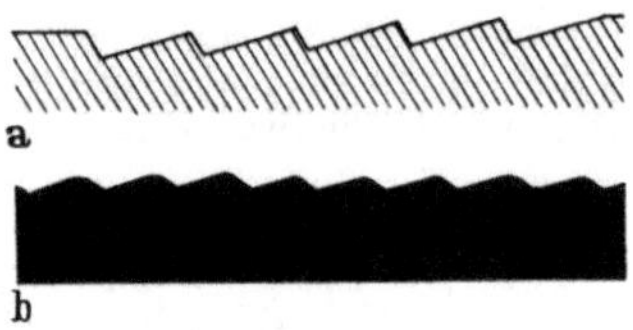

Fig. 221 a and b. (According to GERASIMOV et al., loc. cit.[24], 1958.) Typical line profile of a grating. a) Schematic; b) redrawn from an electronmicrograph taken with a 1200 lines/mm grating.

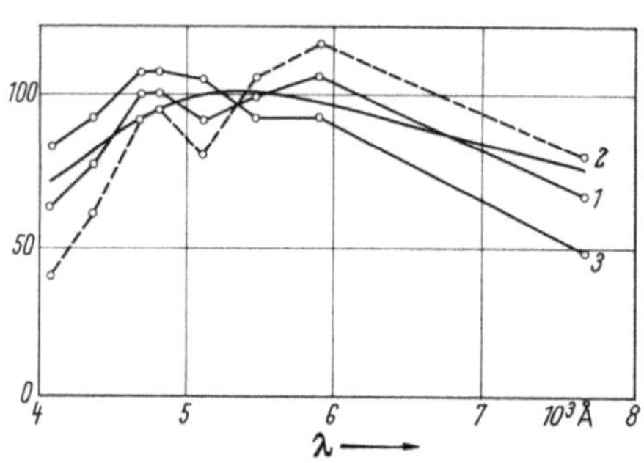

Fig. 222. (According to GERASIMOV et al., loc. cit.[24], 1958.) Typical intensity curves for blaze regions. (1) unpolarized light, (2) electric vector normal to the lines, (3) electric vector parallel to the lines.

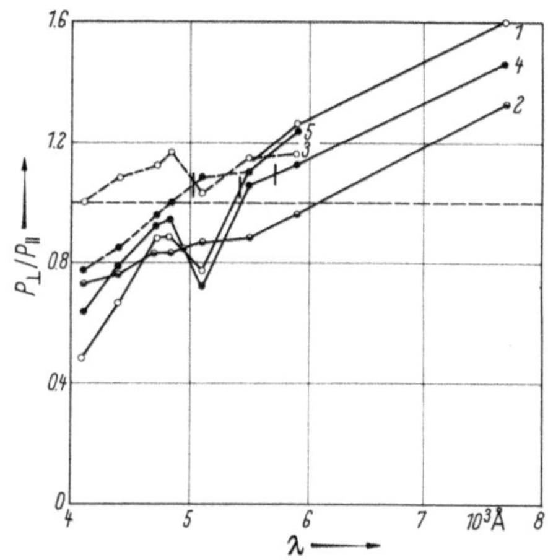

Fig. 223. (According to GERASIMOV et al., loc. cit.[24], 1958.) Ratio of reflection coefficients for light polarized with its electric vector normal, to light with its electric vector parallel to the lines. Curves 1 to 4 relate to the first orders of gratings with 600 lines/mm, curve 5 to a second order.

"Intensity distribution and polarization. The best gratings, with line faces inclined at 5–10°, throw up to 85% of the reflected light in a single maximum; this is quite close to the theoretical value. Practically all the light is reflected by the broad faces (Fig. 221); if the inclination of these faces is increased the fraction of the incident beam which strikes

[65] G. W. STROKE: Interferometric measurement of wave-front aberrations in gratings and echelles. J. Opt. Soc. Amer. **45**, 30—35 (1955).

[66] F. M. GERASIMOV, I. A. TEL'TEVSKII, S. N. SPIZHARSKII and S. V. NESMELOV: Optikomekh. Prom. No. 4, 56 (1957).

[67] R. W. WOOD: Phil. Mag. **20**, 770 (1910).

[68] A. TROWBRIDGE and R. W. WOOD: Phil. Mag. **20**, 886 (1910).

[69] G. W. STROKE: Attainment of high efficiencies in blazed gratings by avoiding polarization in the diffracted light. Phys. Letters **5**, 45—48 (1963).

these working faces, and hence the efficiency with which the light is used gradually falls because the angle between the two faces is 110–120°. The reflection coefficient at the maximum does not exceed 55% if the inclination is raised to 40°. The reflection coefficient could be increased considerably by reducing the angle between the faces to 90°, but the difficulties of making good diamond points for the purpose have prevented this method from being used. Data have been given on the mean and maximum reflection coefficients of gratings in various orders[66, 70].

Fig. 222 shows typical intensity curves (in arbitrary units) for the blaze region of a grating with 600 lines/mm which concentrates the light into the first-order visible region centered on 5400 Å. The measurements were made on a photoelectric autocollimation apparatus using the brightest lines from cadmium, mercury and potassium. The reflection intensities were recorded relative to those found with an aluminized mirror at the same wavelengths. Curve 1 was taken in polarized light; the other two curves were taken with the electric vector of the polarized light normal (2) and parallel (3) to the lines. The thick line indicates the theoretical intensity distribution; the experimental results fall only fairly near the theoretical curve. A notable feature is that the state of polarization changes as the maximum is passed through; the component with its electric vector parallel to the lines is the stronger on the short-wave side, and the other on the other side. The same effect is found with blazed gratings with 1200, 600, 300 and 200 lines/mm. Fig. 223 shows $\varrho_{\perp}/\varrho_{\parallel}$ values (accurate to $\pm 3\%$) for 5 gratings; the blaze maxima are indicated by vertical lines (accurate to 200–300 Å). The change in the predominant sense of polarization occurs at the blaze peak, within the error of measurement. The various gratings differ considerably in their degrees of polarization; the shapes of the reflecting surfaces in the lines are responsible.

Gratings ruled on aluminum, whether blazed or not, show anomalies in their intensity distributions similar to those first found by WOOD[71]; fairly narrow dark bands which vary in spectral position with the angle of incidence are seen in continuous spectra. This effect is seen only when the electric vector is normal to the lines* with rulings of 1200 and 600 lines/mm; many gratings do not reflect at all within these bands. RAYLEIGH[17] has shown that the anomalies always lie at wavelengths such that the waves are diffracted along the surface of the grating: these wavelengths λ are given by

$$a(1 \pm \sin i) = m\lambda$$

where a is the grating constant, i is the angle of incidence and m is the order of the spectrum.

Our gratings in general give anomalies in accordance with this formula, but the bands differ considerably in width and intensity, even with otherwise identical gratings. The anomalies cause dips on the intensity distribution curves for certain gratings (cf. Figs. 222 and 223); calculation shows that the anomalies lie at 4800 and 6700 Å for the 600 lines/mm grating used in the 1st order in the autocollimation apparatus. The dips in Figs. 222 and 223 for the first anomaly actually lie at somewhat longer wavelengths. The anomalies at 6700 Å are readily seen by eye using continuous spectra from these gratings; they have not been studied in detail. Some gratings show only very weak anomalies near the blaze; the precise relation of this effect to the ruling technique has not yet been established. The anomalies occur with both senses of polarization with gratings of 300 and 200 lines/mm, but are stronger when the electric vector is normal to the lines."

It is noted that GERASIMOV et al.[24] use electron-microscopy for the testing of the groove profiles of the gratings ruled at the State Optical Institute.

One of the largest State Optical Institute gratings described is a 150 mm wide, 600 line/mm grating, used in the spectrograph with the large telescope at the Crimean Observatory[24] and showing a resolving power of 600000 in the 7th order ($N = 600 \times 150 = 90000$, $m = 7$ giving a theoretical resolving power $N = 630000$). Another smaller grating (ruled width 100 mm, 600 lines/mm) is at the Potsdam Observatory, where it was tested by SCHRÖTTER[72]; the resolving power was found to have the theoretical value of 360000.

---

* C. H. PALMER: Diffraction anomalies with a dielectric grating [J. Opt. Soc. Amer. **54**, 844 (1964)], shows "anomalies" for both the P and the S polarizations. See also Sects. 60 to 71.

70 F. M. GERASIMOV: Izv. Akad. Nauk. SSSR Ser. fiz. **6**, 662 (1954).
71 R. W. WOOD: Proc. Phys. Soc. Lond. **18**, 396 (1902).
72 E. H. SCHRÖTTER: Phys. Verh. **6**, 143 (1957).

GERASIMOV et al.[24] find that "on average, gratings of high constants give better images than ones of low constants" with gratings ruled on their machine*.

ϑ) *The Perkin-Elmer engines.* The Perkin-Elmer ruling engines were first designed and constructed at the Applied Research Laboratories by JULIUS PEARSON with the assistance of GUS M. GOMEZ, who now is in charge of the engines at the Perkin-Elmer Laboratory in California. An overall view of one of the engines is shown in Fig. 224. Infrared gratings up to inches have successfully ruled since about 1961, and the entire grating ruling and testing program is under the direc-

Fig. 224. Perkin-Elmer grating ruling engine (Photo courtesy H. W. MASRHALL and G. M. GOMEZ, Perkin-Elmer Corp.)

tion of H. W. MARSHALL. Interferometric wave-front testing is used, as well as grating efficiency testing in a double-monochromator arrangement (see Sect. 77). The grating replication methods now in general use are based to an important extent on work carried out at the Perkin-Elmer Corporation by FRASER and WHITE during the middle 1940's (see Sect. 89).

ι) *The Mount Wilson Observatory engine with interferometric control.* The Mount Wilson Laboratory engine, described in Sect. 88β was put under interferometric servo-control, and described in 1962 by H. W. BABCOCK[12], following the succesful demonstration of interferometric control of ruling engines by HARRISON and STROKE in 1955 to 1961[1,2].

Intermittent, stop and go, motion and control are used by BABCOCK and several new features associated with the use of a "modulated" Michelson Twyman Green interferometer are described in his 1962 paper.

The interferometer is modulated by deflecting the compensating plate electromechanically, so as to simultaneously permit a small 60 cps oscillation of the fringe

* A method for the control of the ruling of diffraction gratings using moiré fringes and many novel ingenious features is described by E. M. GERASIMOV, V. P. SERGEEV, I. A. TELTEVSKII, V. V. SERGEEV and B. V. MARICHEV, The use of moiré fringes to control the ruling of diffraction gratings, Optics and Spectroscopy **29**, 152—156 (1965).

pattern, as well as compensation for barometric pressure-change effects on the wavelength (Fig. 225).

The oscillating fringe pattern is scanned by a phototube and is reproduced on a monitoring cathode-ray tube. Any decentering of the pn-th fringe on the n-th groove (p = number of fringes per groove) is detected by synchronous demodulation and is converted to a stored electrical charge. The correction for any error on the n-th groove is applied to the following $(n+1)$-th groove, by introducing into the mechanism a differential correction proportional to the stored charge. Clearly, perfection in the correction would appear to assume that the error to be

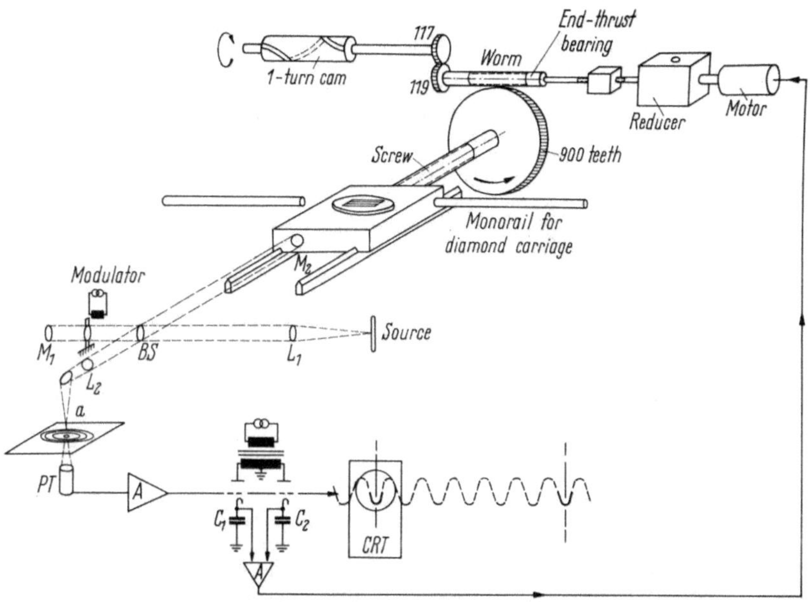

Fig. 225. Interferometric control system of Mount Wilson and Mount Palomar Laboratory engine (according to H. W. Babcock, loc. cit. [12], 1962). The schematic diagram of the ruling engine with its intermittent control system is shown. The moving mirror, M_2, of the Michelson interferometer is on the grating carriage. The circular fringe system, which is imaged at a, is modulated by oscillation of the tilted compensating plate. Any displacement of the chosen fringe is measured by the phototube-amplifier-demodulator section, and during the next subsequent spacing operation the appropriate differential correction is applied by the motor.

corrected on the $(n+1)$-th groove will end by being substantially equal to the error on the n-th groove from which the correction has been determined. This is quite different from the continuous-control Harrison and Stroke principle[1,2], where no long-distance anticipating error assumptions are made, and perfection appears achievable within the time-constant of the servo-system, corresponding to an infinitesimal fraction of a groove-spacing. Comparison of gratings ruled under these rather different conditions would be most revealing. No detailed results of grating wave-front tests, ghost-intensity or spectroscopic performance of the gratings ruled under the new interferometric control system are given in the original paper[12]. Babcock states (loc. cit. [12]):

"The 25-cm ruling engine with interferometric control has proved to be reliable in operation and in the past few months has produced several gratings of high quality in sizes up to 13×20 cm. These are blazed plane gratings ruled on aluminized blanks that are flat to one-fortieth wavelength. There is no bar to the employment of the control system for larger gratings up to the limit set by the dimensions of the machine. It may be emphasized again, however, that several characteristics, in addition to adequate precision in spacing of the grooves must be combined in any really satisfactory grating, especially if it is to be used in spectrophotometry of absorption lines arising in faint astronomical sources."

Some of the details of the control system as described by Babcock are given in what follows:

"The Mount Wilson Observatory engine has been successfully operated in a purely mechanical mode for many years, has a demonstrated capacity in the production of gratings up to 15×25 cm. Tests of the performance of one of these gratings have been described by Pierce[42]. The addition of interferometric control was undertaken with the double aim of still further reducing residual errors of spacing and of minimizing the nonproductive time spent in critical adjustments of mechanical parts.

In order to ensure straightness of the grooves, to avoid extensive mechanical alterations to the machine, to achieve maximum simplicity, and to permit operating either with or without interferometric control, it was decided to retain intermittent advance of the grating carriage, together with simple, crank-driven, reciprocating motion of the ruling diamond. By appropriate choice of the spur gears used in the intermittent drive for the worm that advances the 900-tooth gear and the main screw, it is a simple matter to advance the grating carriage an integral number of fringes in each spacing operation. The spacing most in demand is approximately 600 grooves per millimeter, or 15000 per inch. If gearing is chosen to make the spacing 610.3 grooves per millimeter, the grating constant will be three wavelengths of green mercury light (λ 5460.7531); thus almost exactly six fringes will pass in the field of the interferometer for each spacing cycle.

The control system employs an electrooptic 'fringe-clamp' device. After each incremental advance of the grating carriage, while the diamond is ruling a groove, a phototube senses any deviation from exact centering of the stationary $6n$-th fringe. The resulting electrical error signal is stored, and during the next spacing operation it is converted to a shaft rotation which, after a large reduction through gearing, is introduced as a differential correction into the spacing mechanism of the machine. In the present instance, the differential correction is applied through a rotation of the screw-mounted end-thrust bearing of the worm that advances the main spacing gear (Fig. 225).

The modulated interferometer. The interferometer is of the Michelson type, with the moving mirror carried on one end of the grating carriage. The beam splitter, fixed mirror, and compensating plate are mounted on a substantial bench attached to the base of the machine. The source, a water-cooled Meggers Hg 198 isotope lamp excited at 2500 Mc/s, together with its collimating lens, is mounted outside the enclosure that houses the ruling engine. An interference filter isolates the green line. Light emergent from the interferometer is focused by a projection lens to form a system of concentric rings on a screen, also outside the enclosure. A small aperture in the center of the screen admits light from the center of the fringe pattern to a 10-stage electron-multiplier phototube. The radius of the aperture is only a fraction of the width of the narrowest fringes, which occur for maximum path difference. To provide for sensitive detection of the position of the $6n$-th fringe, which is just appearing or disappearing at the center of the concentric pattern, as well as to compensate for changes in barometric pressure, the interferometer is modulated.

Modulation is accomplished by giving a slight angular displacement to the tilted compensating plate of the interferometer, which is supported on a flat steel spring, or Cardan hinge. The voice coil and magnet from a loudspeaker are so mounted (the voice coil being attached to the plate holder) that a small alternating current applied to the coil provides an angular vibration of the compensating plate, such that the optical fringe pattern sweeps back and forth with an adjustable range at a frequency of 60 cps. At the same time, a direct current applied to the same coil from a simple analog computer deflects the plate in order to compensate for changes in barometric pressure.

Fringe clamp circuit. With the interferometer modulated at 60 cps, the amplified output of the phototube is conveniently applied to the vertical deflecting plates of a monitoring oscilloscope while a 60 cps sine wave is applied to the horizontal plates. The resulting trace reproduces the portion of the fringe system that is being scanned. If the amplitude of modulation is sufficient, two or more complete fringes can easily be displayed, although in actual operation the amplitude is set at about 1/5 fringe. If a dark fringe is centered on the aperture, the scanning will result in a U-shaped pattern appearing on the oscilloscope. Any slight decentering of the fringe will cause the U to become asymmetric. For control purposes the phototube signal is further amplified and then demodulated by a balanced synchronous rectifier. Any error in the centering of the $6n$-th fringe results in a proportional imbalance in the charge developed on the capacitors C_1 and C_2. At the conclusion of the scan phase, which occupies about 40% of each cycle, the error-correcting motor (Fig. 225) is caused to turn in the direction and by an amount determined by the stored signal. Switching of the signal circuit and of the correcting motor at the appropriate phases is accomplished through mercury switches operated by cams on the main crankshaft of the ruling engine.

When the machine is operating under interferometric control, the clamping of the fringes is readily monitored by observation of the oscilloscope, which also displays the intermittent advance of the fringe pattern during the spacing part of each cycle. An indication of error amplitude is also given by a sensitive voltmeter connected across C_1 and C_2.

Feedback for error correction. Error correction appears in the form of intermittent rotation of the shaft of a small reversible motor. After passing through a miniature variable-speed transmission and further gear reducers, the shaft rotation is introduced as a differential correction into the spacing mechanism of the ruling engine. The mechanical differential for this purpose might take any one of several forms; in the present case the simplest method was to control the position of the end-thrust bearing of the spacing worm that advances the 900-

Fig. 226. Jobin et Yvon (Paris) interferometrically servo-controlled engine (photograph courtesy G. PIEUCHARD, 1965).

tooth gear on the main screw. The end-thrust bearing is mounted on a fine-pitch screw that is rotated by the correcting motor. The range of longitudinal travel of the worm is 25 mm corresponding to an accumulated correction of some 100 fringes in the ruling of a single grating. The intermittent driving device, which rotates the worm through spur gears, is a cylindrical cam that is given exactly one turn for each spacing operation. A 1:1 spur gear ratio yields a spacing of 600 grooves per millimeter. It was found by trial ruling under interferometric control that multiplying the grating space by the factor 0.983151 would, at a temperature of 23.9° C and at a fixed barometric pressure, result in a cumulative correction of zero for a large ruling. This factor was closely approximated by spur gears for driving the worm in the ratio 117:119 or 0.983193.

By providing additional sets of spur gears in appropriate ratios, or otherwise altering the mechanical spacing, it is readily possible to rule under interferometric control at other grating constants corresponding to integral numbers of fringes other than six."

Details on the pressure computing system, similar to those used by HARRISON and STROKE[1,2,13] are also given by BABCOCK[12], together with details on the diamond carriage and on the theoretical capabilities expected from the control system.

ϰ) The Jobin et Yvon interferometrically controlled engine. A 4-inch mechanical ruling engine, using novel friction reducing principles and pneumatically assisted movements was completed by R. GOULLEY under the direction of G. PIEUCHARD at Jobin et Yvon by 1959. At that time experiments on further mechanical improvements of this engine and on testing the grating performance were undertaken with the help of G. W. STROKE. These experiments eventually led to the recommendation for designing an interferometric control system, following the successful principles demonstrated by HARRISON and STROKE[1,2]. The stop-and-go interferometric control system of a very simple and novel design was completed by P. CONNES, G. PIEUCHARD and R. GOULLEY by 1962/63.

A general view of the engine is shown in Fig. 226.

Fig. 227a. The C.S.I.R.O. engine (photo courtesy D. A. DAVIES and G. STIFF).

The part of G. W. STROKE in helping to initiate the French grating ruling program is briefly described in the May 1962 "International Optics" issue of Applied Optics.[73].

Gratings of good quality with 600 lines/mm and 75 mm width were obtained as early as 1964, and the ruling of larger gratings under interferometric control is now under way. The Jobin et Yvon engine is capable of ruling gratings of 100 × 70 mm. The number of fringes passing through the interferometer is determined before the carriage is locked: as result periodic errors are eliminated. Random spacing errors are minimized by a compensating coil which controls the fine motion of the carriage during the ruling of the groove. The compensating coil is excited by the difference signal obtained from the two photomultipliers used in the double-interferometer system (one near the source hole, the other at the exit hole *).

* G. PIEUCHARD: Private communication (1964).

[73] G. AMAT, A. ARSAC, J. BROCHARD, J. BROSSEL, P. CONNES, L. COUTURE, P. JACQUINOT and A. MARÉCHAL: Les Réseaux Optiques. In: Optics in France. Appl. Optics 1, 200—278 (1962), p. 262.

λ) *The CSIRO Australian engine.* A general view of the engine as developed by D. A. DAVIES and G. STIFF is given in Fig. 227a and 227b and some details, according to a (1962) description by DAVIES[74] are given in what follows:

"The engine has a nominal capacity of 4-inch length of line by 6-inch ruled width for spacings of 32400, 16200 and 10800 lines per inch. (Sizes $4\frac{1}{2}$-inch $\times 6\frac{1}{2}$-inch with 16200 lines per inch have been ruled). At present only plane gratings can be ruled, but it is intended to add the extra equipment for ruling concave gratings at a later date and a simple modification to allow coarser pitches to be ruled. Preliminary measurements indicate that gratings of

Fig. 227b. The C.S.I.R.O. engine (photo courtesy D. A. DAVIES and G. STIFF).

six-inch ruled width with 100000 lines when measured in the first order at 5460 Å may have Rowland ghost intensities less than 1/10000th of the parent line, efficiencies relative to a plane aluminized surface of 75%, and wavefront errors better than 1/5th wavelength."

The engine is housed inside of a temperature-controlled cabinet and works at a ruling rate of 19 strokes per minute.

As described by DAVIES (loc. cit. [74]),

"Most of the engine was constructed by the Division of Chemical Physics at Fishermen's Bend, Melbourne. As shown in Fig. 227 the reciprocating "shaper" arm (upper left) slides to and fro on two cylindrical precision-lapped guides fixed in line. The central portion of the arm bridges between the guides and beneath it is attached the diamond carriage above the work table which in operation is slowly fed through the gap. To restrain the arm from rotation on the guides, a projection from it bears on a flat strip showing at the rear centre of the picture.

The work table is supported on a parallel pair of cylindrical guides visible across the centre of the photo. The feed screw is located (parallel) between them, and is rotated by the large gearwheel (at right), outside the bearing. Between the bearing and the work table is the long precision nut whose motion is transmitted to the table by connecting bars.

---

[74] D. A. DAVIES: The C.S.I.R.O. Ruling Engine, in: Diffraction Gratings, pp. 101—108. J. Astronom. Soc. Victoria **15**, 101—108 (1962).

In the bottom right corner can be seen portion of a spur pinion on the worm spindle which drives the feed gearwheel. This drive is interconnected with equipment behind the partition in the background, through which projects the driven end of the reciprocating arm."

a

b

Fig. 228a and b. Institute for Optical Research, Kyôiku University, Tokyo engine (photographs courtesy K. P. MIYAKE, 1964).

At present, DAVIES and STIFF are considering a new design of engine specifically for use with servo-control[75].

μ) The University of Tokyo engine. The ruling engine of the Institute for Optical Research, Kyôiku University Tokyo is shown in Fig. 228, and a diagram according to the 1952 paper by FUJIOKA, SAKAYANAGI and KITAYAMA[76] is shown in Fig. 229. The ruling engine is being used for ruling infrared gratings of 360 lines/mm and 40 lines/mm with an area of 100 mm × 70 mm*. Fairly extensive details on the history, construction, interferometric measurement of engine errors, and performance of the engine in its 1952 stage is given in the original paper[76]. The description of the interferometric measurement according to FUJIOKA, SAKAYANAGI and KITAYAMA (loc. cit.[76]) follows:

"To eliminate the errors of the ruling, we have first to measure and to correct them. For that purpose, we use a Michelson interferometer** which is shown schematically in Fig. 230. In that figure C means a plane mirror (29 of Fig. 229) which is fixed to the carrier of a grating, and A and B are two mirrors fixed to another plate which moves parallel to the grating carrier. If the image of $A(A')$ coincides with C, one can see nice interference fringes by using white light. In this case it is very easy to decide the position of the coincidence, because a slight in-coincidence causes a change of colour. Next we move C, until its coincidence with the image $B(B')$ is obtained, and the corresponding angle of rotation α_1 of the worm gear is read. We move next the carrier of

* P. MIYAKE: Private communication (1964).
** A. A. MICHELSON: J. Franklin Inst. **181**, 785 (1916).
[75] D. A. DAVIS: Private communication to G. W. STROKE, Feb. 18, (1964).
[76] Y. FUJIOKA, Y. SAKAYANAGI and T. KITAYAMA: Endeavour of Grating Ruling in Japan. Science of Light **2**, 1—7 (1952).

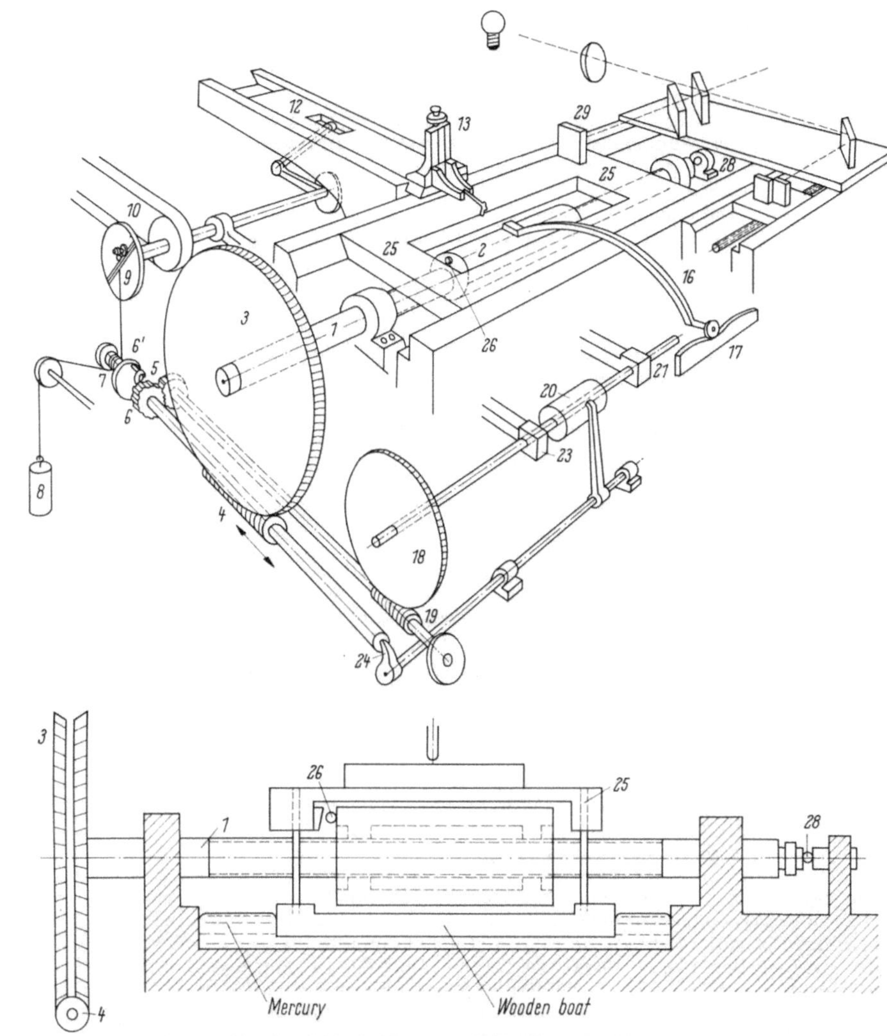

Fig. 229. Schematic diagram of Institute of Optical Research, Kyôiku University, Tokyo engine (according to FUJIOKA SAKAYANAGI and KITAYAMA, loc. cit. [76]).

A, B to obtain the coincidence of A' with C, and then C is moved until it coincides with B', the corresponding angle of rotation being α_2. The same process will be repeated.

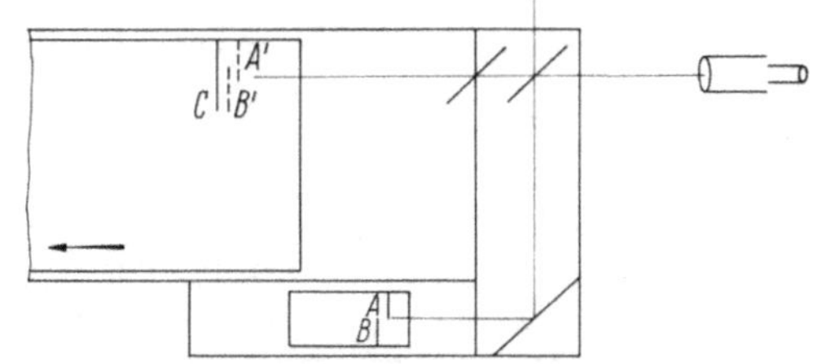

Fig. 230. Interferometric arrangement used for achievement of error-correcting cams on Kyôiku University, Tokyo engine (according to FUJIOKA, SAKAYANAGI and KITAYAMA, loc. cit. [76]).

If the screw is perfect and the rotation of it is very uniform, the angles α_1, α_2, etc. should always be coincident; actually they are not. Therefore we form the cylindrical cam to make α_1, α_2, etc. coincident. We took the distance of A and B about 1/9 mm and interpolated the intermediate part. The accuracy of the angle is 1/100 degree which corresponds to the 1/10 of the interference fringe. The error due to the excentricity of the worm gear plate, or that may be due to the incompleteness of the plate-and-ball contact, may have the period of one revolution. Such errors can be corrected by interpolation between 9 points in a revolution.

However, if the teeth distance is so irregular that the accompanying errors have small periods within a revolution, nine points are not sufficient to observe them. In our instrument this seems to be just the case, and observations at smaller distances are desirable. In this case we have to observe so many points that the accuracy of the measurements must well be considered.

Our problem is just the same as in the case of measuring a long distance with a very short scale. Suppose we measure 1 mm by successive measurements of n times using a small scale of $1/n$ mm. If we denote the error of a single observation by Δ, the total probable error is $\sqrt{n}\Delta$.

We consider the Rowland ghosts of a grating, namely the error having a period of one revolution of the main screw. The relative intensity of a Rowland ghost to the main line is given by*

$$G_I = \pi^2 m^2 \left(\frac{\varepsilon}{a}\right)^2.$$

In this formula a means the grating constant, ε means an amplitude of the sinuous periodic error, and m means the order of the spectrum. We assume $\varepsilon = \sqrt{n}\Delta$. The value of Δ is roughly

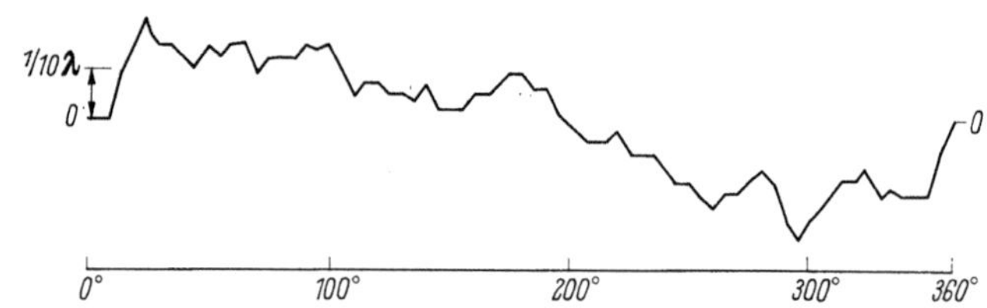

Fig. 231. Results of interferometric measurements on engine (according to FUJIOKA, SAKAYANAGI and KITAYAMA, loc. cit.[76]).

0.5/20 μ, and thus for $n = 9$ we have $\sqrt{n}\Delta = 0.075$ μ. Putting $a = 1.8$ μ, we get for $m = 1$

$$G_I = \frac{4.3}{1000}.$$

The observed intensity of the Rowland ghost is however stronger than this value, and the intensity distribution is fairly deviated from ROWLAND's theory, and many ghosts of higher orders appear irregularly. For instance, the ghost of the 8th order is often very strong, as compared with those of lower orders.

From those circumstances we can conclude that the error is not so simple as of the period of one revolution, and there must be errors of still smaller periods. We have therefore to observe at more points than nine. In order to increase the accuracy, we used monochromatic light instead of white light. If we fix A and count the number of fringes, we can avoid the additive errors; but it is, of course, not easy to count some 4000 fringes/mm. However, as we have already corrected under a fraction of the wavelength using white light, we can fix α within the accuracy of one fringe. As we can read the fraction of a fringe accurately, we can determine the error of one revolution of the screw with the accuracy of Δ.

We used a mercury green line. Fig. 231 shows an example of the measurement. This shows a much more complicated curve than that of a period of one revolution, Repeating such observations, we get the total correction for one revolution of the order of $\frac{1}{2}\lambda$ to 4λ. The error of the long run is also corrected, first using white light and then using monochromatic light. In this case we used a cadmium red line instead of the mercury green line." (loc. cit.[76])

A new engine is under construction at the Institute for Optical Research in Tokyo, but no information is yet available.

ν) The Hitachi interferometrically servo-controlled engine. A new, interferometrically servo-controlled ruling engine, has now been completed under the direction of Y. YAMADA, T. HARADA and K. MORITA** [77]. Details were not available at the time of the completion of this article, but a photograph of the engine is shown in Fig. 232, and it is known that the engine incorporates several ingenious

* H. A. ROWLAND: Phil. Mag. **35**, 397 (1893).

** T. HARADA: Private communication to G. W. STROKE (23 April 1967).

[77] Y. YAMADA, T. HARADA and K. MORITA: A new Japanese Ruling Engine. J. Jap. Soc. Mech. Engrs. **69**, 1557—1567 (1966) [in Japanese].

novel features, among which is the use of teflon bearings, both for the grating and for the diamond carriage-ways. The engine is capable of ruling gratings up to 8 inches wide, and several gratings have been successfully ruled. The interferometric servo-control system used is comparable to that of the M. I. T. engine (Sect. 88γ). A very important feature of this engine is the use of air-bellows to support the entire floor on which the engine is mounted, for the purpose of vibration isolation*.

Fig. 232. Hitachi interferometrically servo-controlled ruling engine (courtesy Y. YAMADA and T. HARADA[80,81]).

ξ) *The Merton NPL metrology-gratings engine.* A novel approach to grating ruling was suggested in 1948 by Sir THOMAS MERTON, before the introduction of interferometric servocontrol[1,2] in the hope of overcoming some of the mechanical difficulties associated with previous engines. In referring to Sir THOMAS MERTON, and his apparent criticism of ROWLAND'S design, L. A. SAYCE[78] writes:

[Sir THOMAS MERTON] "... proposed that the grating be generated upon a cylinder, and cut as a very fine screw-thread by a continuous lathe-like action and that this helix should be 'opened out' upon a flat surface by a replica process. These suggestions were the starting point of a considerable programme of research at the National Physical Laboratory which has been going on for the past eight years. As so often happens, the problem which first seemed quite simple has proved to be a complex one and the principal difficulties have been those which we first dismissed as trivial. Our plan has been to overcome these difficulties by concentrating first upon relatively coarse rulings, for which there is considerable demand and, armed with this experience, to tackle the more formidable problems of producing the large gratings of very high resolution ..."

Even though it has become apparent that the gratings so produced so far did not appear to be of sufficient quality, for most low resolution infrared spectro-

* See Footnote **, p. 724.

[78] L. A. SAYCE: The preparation and industrial applications of diffraction gratings. Trans. Soc. Instrum. Techn. vol. 9. No 4, 138—143 (1957).

scopy, in comparison with gratings readily obtained by ordinary mechanical engines, the gratings produced at NPL by the Merton process have proven to be of an unexpected and very great importance in the use of gratings for metrological purposes.

Fig. 233. A lathe at N.P.L. especially built for the production of the primary helices (up to 30000 turns per inch) used in grating production according to the Merton process (photo according to L. A. SAYCE, loc. cit. [79], 1953).

Fig. 234. A model of the Merton nut device used for producing a "secondary", corrected, helix from a "primary" lathe-cut helix (according to L. A. SAYCE, loc. cit., [79], 1953).

Some of the advantages of metrological gratings are discussed in Sect. 10. Extensive recent descriptions of the use of moiré-fringes obtained with gratings, in metrological applications are given by references [101] to [108] in Sect. 76.

Details of the NPL engine used to produce the primary helix for the Merton process are shown in Fig. 233, according to SAYCE[79].

A model of the "Merton-nut" device used for producing an improved "secondary" helix from the primary lathe-cut helix is shown in Fig. 234[79].

Interferograms (Fig. 235 and 236) of early gratings (first-order, 7500 lines/inch) obtained in a Fizeau interferometer show the important improvement obtained, as expected, by the use of the secondary helix[79].

Fig. 235.

↑ Primary helix ↓ ↑ Secondary helix ↓

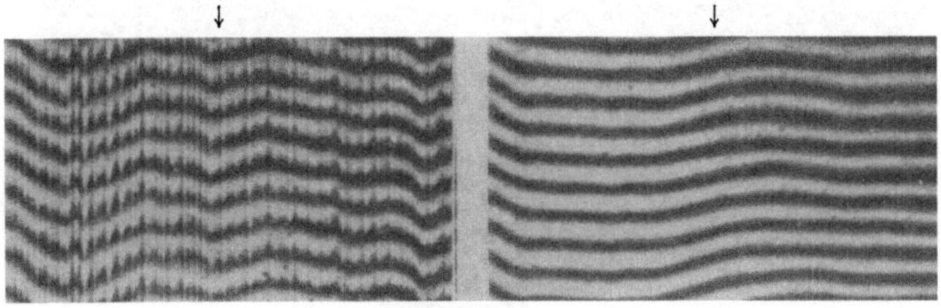

Fig. 236.

Fig. 235. Corrective action of Merton nut (according to L. A. SAYCE, loc. cit. [79], 1953). Gratings with 7500 grooves/inch spacings. Moiré-fringe patterns of replicas derived from primary and secondary helices respectively.

Fig. 236. Same as Fig. 235. First-order wave-front interferograms of replicas derived from primary and secondary helices respectively.

o) The Siegbahn Nobel Institute for Physics engines. Professor SIEGBAHN'S engines have for long been the primary source of the world's most outstanding x-ray gratings. In fact, two engines are in successful operation at the Nobel Institute for Physics in Stockholm, ruling not only x-ray gratings, but other gratings as well.

Two recent photographs of the larger engine are shown in Fig. 237 and 238*.

A recent description by Professor MANNE SIEGBAHN of some of the characteristics of the large engine, as well as of a smaller engine, (the smaller one being used mostly for ruling of gratings for grazing incidence spectrometers) follows. In reference to the larger engine (Fig. 237 and 238), Professor SIEGBAHN writes:

"This machine gives 1080 per/mm, or 540 respectively 360 if 2 respectively 3 cogs are passed for every ruled line. The carriage can be moved over a distance of 250 mm, which then corresponds to the maximum ruling width. This machine also has arrangements for ruling of gratings for moiré purposes in exact millimeter scale with 250, 100 and 50 lines per/mm. The smaller ruling engine, mostly used for ruling of gratings for grazing incidence spectrometers is in some respects a simplified model with a maximum ruling width of 120 mm and a fundamental period giving 576 lines per mm; with half cogs displacements it then gives 1152 lines per mm." (MANNE SIEGBAHN, loc. cit. *).

Additional references to the important earlier contributions to grating ruling made by SIEGBAHN and his associates are given in Sect. 82 (ref. [7]), as well as in the 1949 paper by G. R. HARRISON (loc. cit. [23] on p. 669).

* MANNE SIEGBAHN: Private communication to G. W. STROKE (June 29, 1964).

[79] L. A. SAYCE: The production of diffraction gratings. Endeavour, (October 1953), pp. 210–216.

π) *Ruling engine with interferometric diamond carriage control.* New improvements in groove quality, important reductions in stray light and in scattered light in gratings, and indeed substantial improvement in grating quality and

Fig. 237. The large Siegbahn ruling engine of the Nobel Institute for Physics (photograph courtesy M. Siegbahn, 1964).

Fig. 238. The large Siegbahn rulingengine of the Nobel Institute for Physics (photograph courtesy M. Siegbahn, 1964).

resolving power, especially at short wavelengths, and important increase in ruling speeds should result from full, three-dimensional interferometric servo-control of

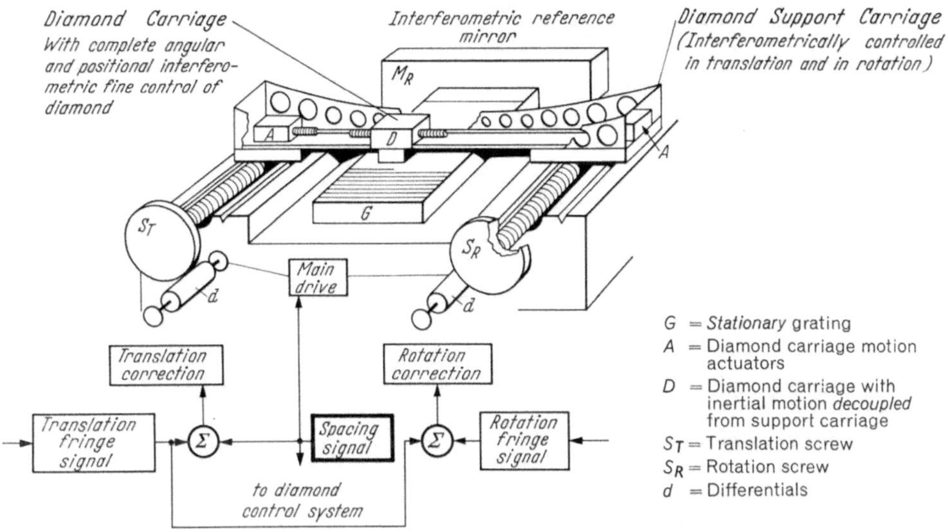

Fig. 239. Schematic diagram of 80 cm interferometrically controlled ruling engine with complete interferometric diamond servocontrol (O. C. Mohler, G. W. Stroke and F. Denton, unpublished work).

Fig. 240. Photograph of the finished part of 80 cm ruling engine, according to diagram of Fig. 239 (O. C. Mohler, G. W. Stroke and F. Denton, unpublished work).

the ruling diamond itself. Work towards this goal is now under way at the University of Michigan, under the direction of O. C. MOHLER and G. W. STROKE with F. DENTON and T. HARADA according to a method proposed by MOHLER and STROKE*. Previously unpublished results of the status of this new engine, and results obtained in joint work by G. W. STROKE, O. C. MOHLER and F. DENTON, are shown in Figs. 239 to 242. The schematic diagram of the interferometric control system is shown in Fig. 239, and a general view of the finished part of the engine in Fig. 240. The engine will rule gratings with a completely stationary grating blank, with grating dimensions up to 32 inches (and groove lengths up to 20 inches). Among the several novel features which are being developed is the interferometer shown in Fig. 241, which is seen to be capable of producing a fringe signal having, to a large extent, an arbitrarily long "fringe period" (described by "synthesized fringe length" in the Fig. 241). Examples of the photoelectric fringe signals, obtained on the new ruling engine (with the use of teflon bearings) are shown in Fig. 242. A rotation-insensitive interferometer, to be used on the diamond-control carriage, has been designed with the assistance of T. HARADA.

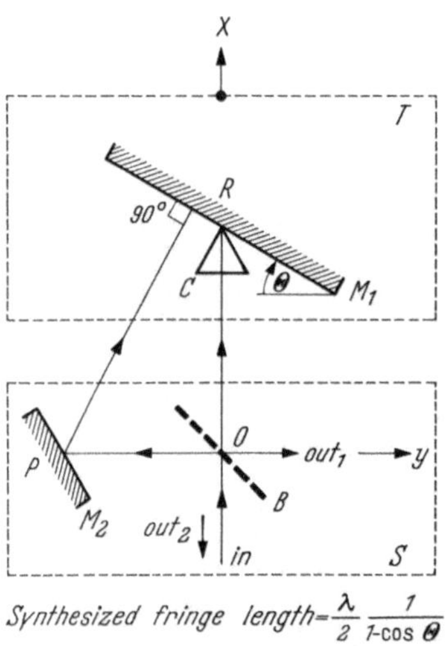

Fig. 241. Interferometer capable of producing fringe-signal of a "synthesized fringe length" having a period which may be very long compared to the half-wavelength.

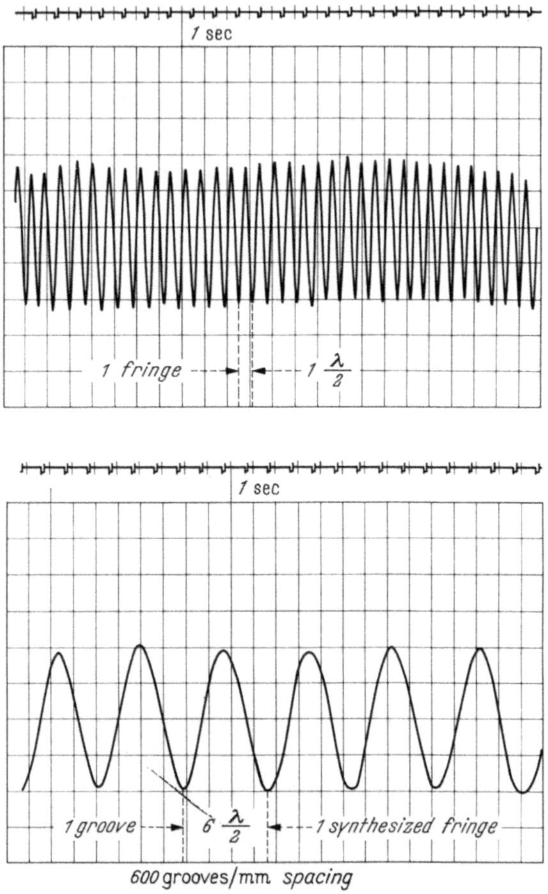

Fig. 242a and b. Photoelectric fringe signals recorded on engine of Fig. 240: a) with "conventional" Michelson Twyman-Green interferometer (e.g. as in Fig. 195). b) with "synthesized-fringe" interferometer of Fig. 241 (G. W. STROKE, F. DENTON and O. C. MOHLER, unpublished work, 1965).

89. Grating replication. The wide use of high-resolution gratings and grating spectrometers is due to a large part to the sucess in grating replication methods.

* O. C. MOHLER and G. W. STROKE: Proposed method of attaining greatly improved large diffraction gratings by new principles of interferometric ruling engine control. Private communication to the U.S. National Aeronautics and Space Administration (October 1963).

The methods most widely used were developed in particular by the Perkin-Elmer Corporation, the Bausch & Lomb Company and the Jarrell-Ash Company.

GEORGE SINTIRIS[80] reported in 1963 that over 50 successful gratings had already been made by that time of a $7\frac{1}{4}$ inch high-resolution grating ruled by STROKE in 1961 on the M.I.T. interferometrically-controlled ruling engine. Over 120 successful replicas were produced by SINTIRIS by the same date from a single 3-meter concave grating, ruled a few years earlier on the No. 1 Jarrell-Ash engine[13a]. Many successful replicas of the 10-inch gratings ruled under interferometric control at M.I.T. (Sect. 88γ) have been obtained both by Bausch &

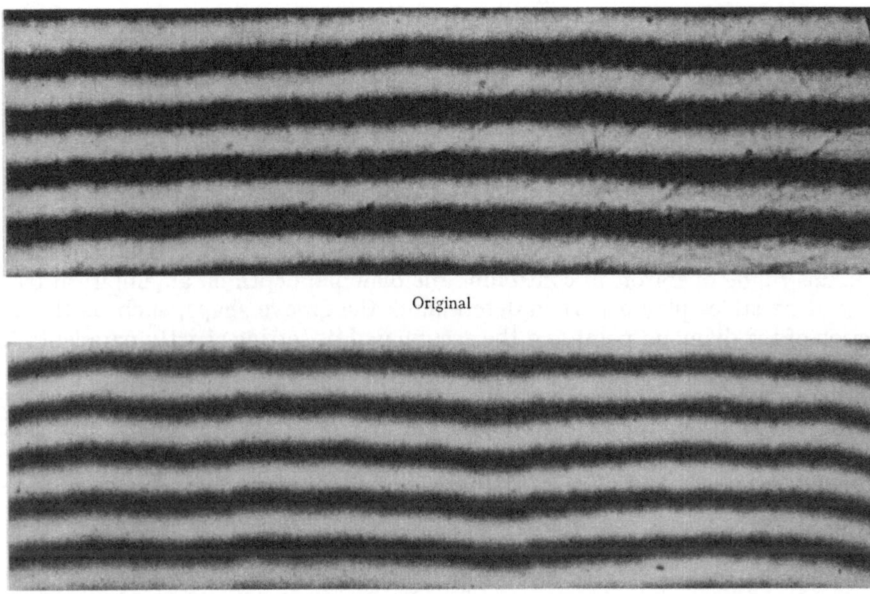

Original

Replica

Fig. 243. Comparison of 5-year old replica with the original 8-inch grating from which it was made, showing the remarkable fidelity and stability achievable in precision-replication. Both interferograms obtained at 23.5°, where the sensitivity to any flatness deformations in the replica is greater than at the high angles where ruling errors are most apparent. (STROKE, loc. cit. [1], 1961)

Lomb and by Jarrell-Ash. It was noted by STROKE (loc. cit. [13a]) that the extent of contemporary high-resolution spectroscopy would not be conceivable in its present form without the capability of successful replication of a good original master grating.

The remarkable fidelity and stability typical of modern grating-replication methods are illustrated in Fig. 243.

According to STROKE (loc. cit. [2], p. 1339) this figure shows an interferogram of an early M.I.T. 8-in. grating (original) ruled in May 1956, as well as a 1961 interferogram of the then 5-year-old replica of that grating produced shortly after the grating had been ruled. Both interferograms are taken at a comparatively low angle (23.5°), where the departure of the fringes from straightness as a result of possible flatness deformations appears more than twice as strongly as at the angles around 64° where the gratings are normally used. This sensitive test of a replica produced in the early stages of the development of contemporary replication methods serves to illustrate the remarkable fidelity and stability of modern replica gratings. Replication methods have attained a degree of perfection such

[80] G. SINTIRIS: Private communication to G. W. STROKE (1963).

that entire series of replicas cannot be distinguished from each other or indeed from the master grating from which they have been produced. In fact, points ot superiority to the master grating are generally found in replica gratings as a result of the groove-inversion characteristic of replication[2].

Some details about the origins of modern grating replication methods were given by R. F. JARRELL and G. W. STROKE (loc. cit. [13b]) in 1964.

The best early replica gratings appear to have been obtained by the common technique described by WOOD[81] of coating the master with collodion, allowing collodion to harden, then floating it off on water and finally mounting it on a glass blank. The technique appeared to result inevitably in uneven thickness and in local distortions of the rulings, so that the replicas produced definition far inferior to the master, and considerably more scattered light.

At the October, 1946 meeting of the Optical Society of America, R. W. WOOD was able to distribute transmission replica gratings of 7500 grooves/inch that displayed almost no central image and essentially all of the light in a single order. This was the first popular demonstration of the work on "blazing" gratings that had been started by ANDERSON and improved independently by WOOD at Johns Hopkins[81] and by BABCOCK at Mt. Wilson[9, 22]. The "blaze" of these demonstration replicas appears to have been due to the success of WILBUR PERRY, the Johns Hopkins ruling operator, in controlling the diamond depth on an empirical basis. Several variables play a part in determining the groove shape, such as the perfection of the diamond point and the accuracy of its setting strictly parallel to the ruling stroke, the speed of the ruling stroke and the hardness of the aluminum coating, which in turn depends on the composition of the aluminum, the pressure during evaporation, the rate of deposition and the thickness of the film. The operator of a ruling engine attempts to adjust the weight added to the ruling diamond, over the range 2 to 20 grams, to obtain a groove depth between 90 and 100 per cent of the desired groove spacing.

At the same October, 1946, meeting of the Optical Society of America referred to above, R. W. WOOD was able to demonstrate the performance of a quality concave replica grating, 15000 grooves/inch 1.5×3.5 inches ruled area, produced by WHITE and FRASER of Perkin-Elmer and cast from a stainless steel master ruled by WILBUR PERRY at Johns Hopkins University. A grating replicated by this technique cannot be distinguished in quality from the original "master" ruling from which it was made. The White and Fraser process is shown in Fig. 188, taken from their patent[82].

Perkin-Elmer appears to have continued exploratory work in grating replication during 1946, using a *convex* stainless steel master of three meter radius of curvature, also ruled for them by WOOD and PERRY at Johns Hopkins University. In 1949 a licensing agreement was entered into between Perkin-Elmer and the Jarrell-Ash Company, and Perkin-Elmer also turned over to Jarrell-Ash the Johns Hopkins master[13b]. It subsequently appeared that the technique in its early form was hard on the master, so that the physical appearance of the replicas showed a tendency towards deterioration and the intensity distribution tended to shift to the ultraviolet, as a result, apparently, of groove deformation. Furthermore, it developed that Johns Hopkins had been fortunate in their first attempt to rule a convex master; on all successive rulings, the rouge embedded in the stainless steel during final polishing destroyed the ruling diamond long before

[81] R. W. WOOD: Improved Dlfraction Gratings and Replicas. J. Opt. Soc. Amer. **34**, 509—516 (1944).

[82] J. V. WHITE and W. A. FRASER: Method of making optical elements. U.S. patent 2464738 March 15, 1949 (application: July 8, 1947), assigned to Perkin-Elmer Corporation.

the ruling was completed. Accordingly, Jarrell-Ash eventually developed an entirely new process for replicating from aluminized masters that would have sufficient adherence between glass blank and the aluminum to withstand the separation[13b]. Most directly involved in the development of the new Jarrell-Ash grating replication process were GEORGE SINTIRIS with DAVID K. RICE and GEORGE W. STROKE. More recently, the program has continued to develop under the direction of SINITIRIS and associates.

Very similar processes of optical grating replication were developed quite independently by Perkin-Elmer, Bausch & Lomb, and Jarrell-Ash, as well by several other companies, among them Jobin et Yvon in France.

An important characteristic of the new replication processes is the use of a very carefully shaped, perfectly flat or convex optical blank to support the replica. In fact, the replica is cured before the replica-blank is separated from the master grating. Experience has shown that the degree to which the replica blank matches the master surface, as well as the surface finish (polished or finely ground) are essential in attaining undeformed and stable replicas. Computation readily shows that the remarkable flatness attained in modern replicas (see Fig. 243) can only result from molecular adherence of the plastic to its supporting blank. The adherence achieved has appeared to be sufficiently great to permanently overcome the stresses which may result from any shrinkage of the plastic used (and which, in bulk may be as great as 1 to 6%, many times the tolerance achieved in attaining the flatnesses of better than 1/10 wavelength in the visible). Equally good success appears to have been obtained with epoxy as with polyester ("Laminac") resins.

A great deal of technique and skill is of course involved in attaining success and regularity in replication, but it is not unusual to have periods with 80 and 90% successful replications.

The details of the replication process vary from one process to another.

Although jealously guarded as trade secrets, grating replication principles appear to be readily available from descriptions by manufacturers and spectroscopic users alike.

For instance, Perkin-Elmer in its 1961 Instrument News (loc. cit. [17]) writes:

"The first step in the replication process is vacuum-coating the master grating's surface with an extremely thin film of oil. Upon this film, which perfectly conforms to the grating's surface shape, a layer of aluminum is next deposited.

The grating is then removed from the vacuum coating tank and a glass blank is bounded to the aluminum coating with resin. After curing the resin, the grating replica, now consisting of glass blank, reson bounding layer and aluminum surface, is separated from the master grating at the oil layer.

The replica grating duplicates the surface of the master grating in reverse, and is suitable for use in high-quality spectrometric equipment." [17]

In his 1962 review article on the extreme ultraviolet, R. TOUSEY[83] writes:

"Replica gratings are made by the method of WHITE and FRASER[82], or some modification of this. The original is ruled in a layer of aluminum evaporated onto a convex blank. A concave blank is prepared, nearly matching the convex surface. A parting agent, which may be silicone oil, is evaporated onto the convex grating, and this is followed by an evaporated layer of aluminum. Thus coated, the convex master and concave matching blank are placed together with an epoxy resin layer in between. After polymerization, the sandwich is split at the parting layer. The concave blank now has on it a layer of resin, covered with

[83] R. TOUSEY: The Extreme Ultraviolet—Past and Future. Applied Optics 1, 679—694 (1962).

aluminum, both containing an exact imprint of the original grating, except that it is turned inside out. In some cases this is an advantage. Where it is not desirable the master can be ruled on a concave blank; then, a convex replica is made, and

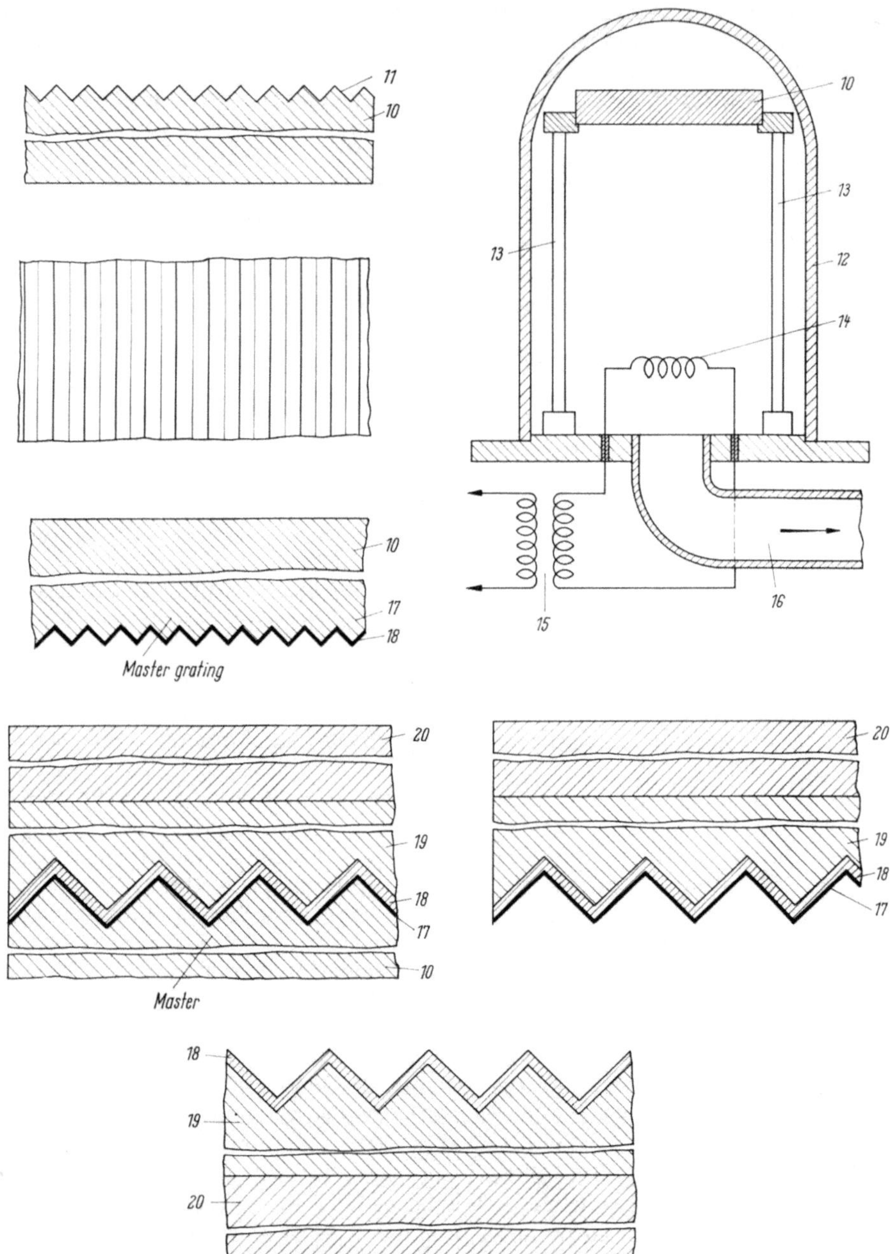

Fig. 244. Grating replication method according to the patent by WHITE and FRASER. According to that method the master grating is placed in a vacuum coating vessel and a thin "parting layer" of a greasy substance is deposited on it (17 in the figure). Then a layer of Al 1500–2000 Å thick is evaporated onto it (18). The coated master is removed from the vacuum and covered with a thin layer of plastic (19), and this is covered by the replica blank (20). After 24 to 48 hours, polymerization of the plastic will be complete and the master and the replica may be separated at the parting layer. Variations of this method are generally, used in precision-replication of gratings.

finally a concave second generation replica. For use in the extreme ultraviolet it is necessary to flash-coat the replica with fresh aluminum, because the reflectance of the aluminum which was in contact with the parting agent is very low. An alternative, when the aluminum is thin, is to remove the aluminum, leaving a plastic-surfaced replica; this may have more favorable stray light properties than aluminum because of its low reflectance in the visible; or, the plastic may be recoated with some other material. Generally, however, the aluminum surfaced replica is overcoated with the material found optimum for mirrors in the particular wavelength range." [83]

The main features of the WHITE and FRASER replication process are illustrated by Fig. 244, which shows the drawings of the original patent, assigned to Perkin-Elmer. An original ruled master grating was placed in the bell jar of a vacuum coating system, as shown. The system was evacuated and a first "parting" layer of greasy material ("manitol" or diffusion pump oil) only a few molecules thick, was deposited on the grating to prevent adhesion to the ruled surface. Then aluminum was evaporated onto the master, to a thickness of approximately 1000 Å. The parting layer is designated by 17 and the aluminized surface by 18.

The aluminized master was next removed from the bell jar and turned on its back. It was coated with a thin layer of plastic, "laminac" or epoxy, designated by 19 in the drawings, and this was backed by a replica blank, 20, ground to have a figure matching that of the master to within a few fringes over the whole area. In other words, to produce a concave replica grating of three meter radius of curvature, one should rule a master grating on a *convex* blank of three meter radius of curvature. After about twenty-four hours, the plastic polymerizes and the replica may be separated from the master at the parting layer, as shown on the right-hand side in Fig. 244. The parting layer is then washed off by ether and there results the final replica grating, as shown at the bottom of Fig. 244[13b].

Methods for mirror replication have been described in several papers by G. HASS and associates[84–86] and some similarities with the principles used in grating replication processes can be recognized.

Replicas from replicated grating sub-masters, or so-called "second-stage" replicas have been successfully obtained. Conservative estimates place into many thousands the number of successful replicas of spectroscopic gratings obtained by the new replication methods[13b].

Some details on the replication methods used in the production of metrological gratings are found in references[101–108] in Sect. 76.

An early method of photographic "replication" of gratings was described as early as 1874 by Lord RAYLEIGH[87].

C. Modulated interference-gratings in wave-front reconstruction imaging and in optical image filtering.

90. Interference-gratings as information carriers in optics. Like a sinusoidally time-varying wave which is the basic carrier for time-varying signals in radiocommunications, a *spatially* varying grating (for instance the sinusoidal grating

[84] G. HASS and W. W. ERBE: Method for producing replica mirrors with quality surfaces. J. Opt. Soc. Amer. **44**, 669—671 (1954).

[85] A. P. BRADFORD, W. W. ERBE and G. HASS: Two-step method for producing replica mirrors with epoxy resins. J. Opt. Soc. Amer. **49**, 990—991 (1959).

[86] A. P. BRADFORD, W. W. ERBE and G. HASS: Abstract of paper on "Metal replica mirrors prepared from plastic masters". J. Opt. Soc. Amer. **50**, 1132 (1960).

[87] Lord RAYLEIGH: On the manufacture and theory of diffraction gratings. Phil. Mag. **47**, 81—93, 193—205 (1874). — Scientific Papers I, pp. 199—221.

produced on a photographic plate by interference of two plane waves) is a basic carrier for spatially-varying images in optics (Fig. 245). In both cases, when either space or time are the parameters, the signal (or image) involve variations of both amplitude and phase. Single or double side-band modulation can both be achieved, and the use of interference gratings as information carriers can be readily understood with the help of elementary image formation theory[1,2].

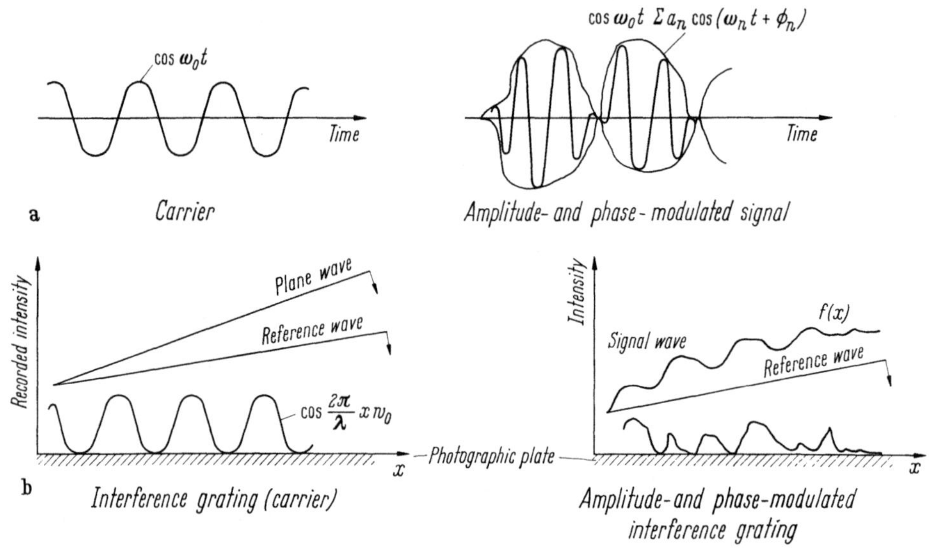

Fig. 245 a and b.

The principles are most readily illustrated with the help of their application to wave-front reconstruction imaging (Sect. 91) and to spatial image filtering (Sect. 92).

91. Wave-front reconstruction imaging with modulation interference gratings*. Consider first the case of two mirrors M_1 and M_2 illuminated by a plane wave S_1 (Fig. 246). The two mirrors are inclined with respect to each other, so that the two waves $A_1 e^{i\varphi_1}$ and $A_2 e^{i\varphi_2}$ incident on a distant photographic plate form angles θ_1, and θ_2 with the plate (that is, the two waves form an angle $(\theta_1 - \theta_2)$ with each other). Clearly the intensity $I(x)$ recorded on the photographic plate is nothing but a sinusoidal interference grating, with straightline fringes parallel to z. One has, with ($|A_1| = \text{constant}$, $|A_2| = \text{constant}$)

$$\left.\begin{aligned} I(x) &= (A_1 e^{i\varphi_1} + A_2 e^{i\varphi_2})(A_1^* e^{-i\varphi_1} + A_2^* e^{-i\varphi_1}) \\ &= |A_1|^2 + |A_2|^2 + |A_1||A_2| e^{-i(\varphi_1-\varphi_2)} + |A_1||A_2| e^{+i(\varphi_1-\varphi_2)} \end{aligned}\right\} \tag{91.1}$$

where one recognizes the intensity distribution in an ordinary interferogram, formed by interference of two plane waves, namely

$$I(x) = |A_1|^2 + |A_2|^2 + 2|A_1||A_2| \cos(\varphi_1 - \varphi_2). \tag{91.2}$$

* For background, see e.g. references 2, 8, 16, 17 on following pages.

[1] A. MARÉCHAL and M. FRANÇON: Diffraction. Structure des Images, ed. Revue d'Optique 1960.

[2] G. W. STROKE: An Introduction to Coherent Optics and Holography. New York: Academic Press 1966 [270 pages, including reprints of three original 'wavefront-reconstruction imaging' papers by D. GABOR].

In this case one also notes that

$$\varphi_1 = \frac{2\pi}{\lambda} x\theta_1 \tag{91.3a}$$

and

$$\varphi_2 = \frac{2\pi}{\lambda} x\theta_2 . \tag{91.3b}$$

The period of the fringes is

$$a = \frac{\lambda}{\theta_1 - \theta_2} . \tag{91.4}$$

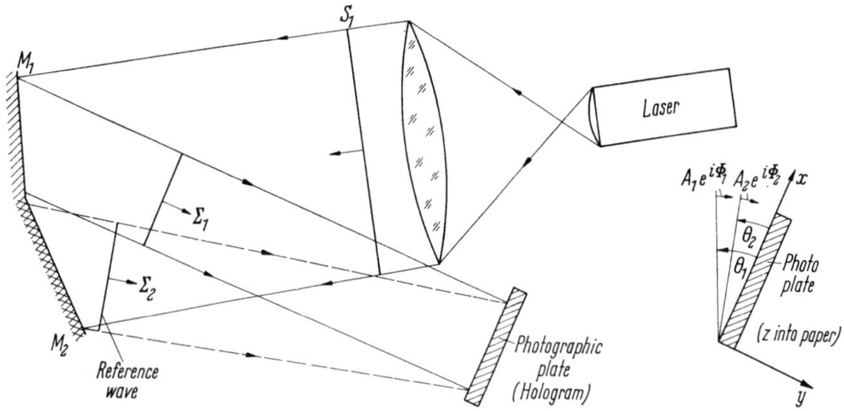

Fig. 246.

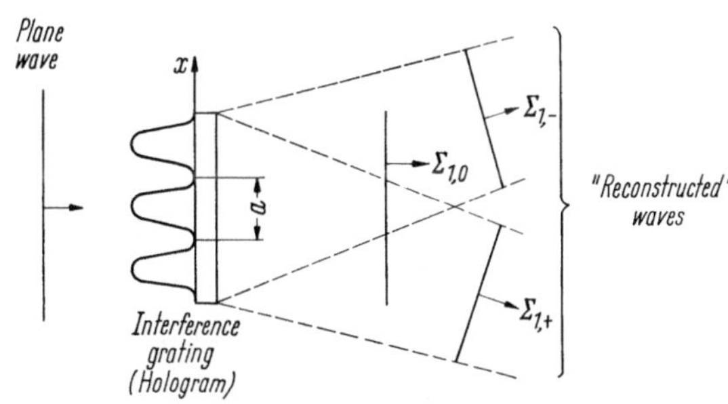

Fig. 247.

When the interference grating described by Eq. (91.2) is now illuminated by a plane wave, (Fig. 247) Eq. (91.1) immediately shows that three waves will emerge from the plate: one wave on axis, and two waves forming angles $+(\theta_1 - \theta_2)$ and $-(\theta_1 - \theta_2)$ with the plate.

Indeed, one has from Fourier transform theory[2] the equation:

$$\left.\begin{array}{ll} \text{If} & f(x) \to F(\omega) \\ \text{then} & f(x)\,e^{2\pi i\theta_1 x} \to F(\omega - \theta_1) \\ \text{and} & f(x)\,e^{-2\pi i\theta_1 x} \to F(\omega + \theta_1) \end{array}\right\} \tag{91.5}$$

where $\to$ indicates "by Fourier transformation". Also, Eq. (91.1) can be written as

$$I(x) = (|A_1|^2 + |A_2|^2) + |A_1|\,|A_2|\,e^{-i\frac{2\pi}{\lambda}x(\theta_1-\theta_2)} + |A_1|\,|A_2|\,e^{+i\frac{2\pi}{\lambda}x(\theta_1-\theta_2)} \tag{91.6}$$

where one recognizes that $(|A_1|^2 + |A_2|^2)$, $|A_1|\,|A_2|$ are constants, within the aperture of the plate. It follows that the Fraunhofer "diffraction pattern" at infinity are three spectral lines, one on axis, and one each at $\pm(\theta_1-\theta_2)$. For a plate of finite width, the spectral lines have a $\sin\theta/\theta$ distribution. To each of the spectral lines at infinity corresponds a plane wave, at the emergence from the plate. Let us now consider one of the original waves reflected by the mirrors, and forming the interference grating, to be the "reference" wave. Let Σ_2 be the reference wave. One can then say that the two waves $\Sigma_{1,-}$ and $\Sigma_{1,+}$ obtained by illuminating the interference grating with a plane wave are the reconstruction of the original "unknown" or "modulating" wave, Σ_1.

The importance of this analysis becomes clear when one examines how an arbitrary wave Σ_2, reflected by an object of arbitrary amplitude and phase, is recorded on the interference grating as carrier.

Indeed let Σ_1 (Fig. 248) be formed by the spherical waves originating by scattering at the various points P of the "object". In other words, Σ_1 can be taken

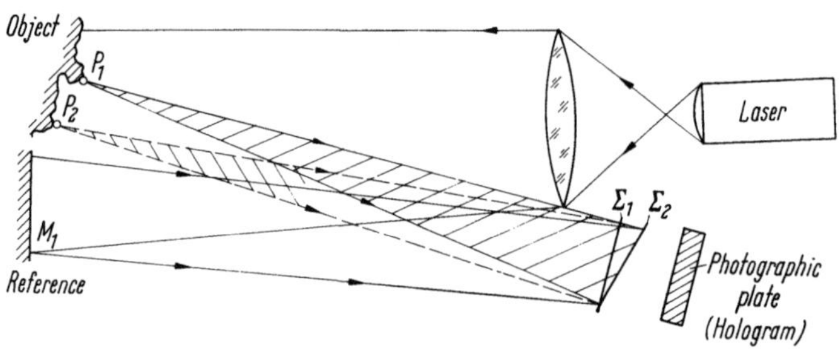

Fig. 248. Recording of a hologram of a three-dimensional object[2].

as equal to the resultant complex amplitude produced in the plane forming an angle θ_1, with the photographic plate, when the plate is illuminated by light scattered (or transmitted) by the object. The intensity recorded on the photographic plate is now given by

$$\left.\begin{aligned} I(x) = (A_1 A_m)(A_1 A_m)^* + A_2 A_2^* + [A_m^* e^{-i\varphi_m}][A_1^* A_2 e^{-i(\varphi_1-\varphi_2)}] \\ + [A_m e^{i\varphi_m}][A_1 A_2^* e^{+i(\varphi_1-\varphi_2)}] \end{aligned}\right\} \tag{91.7}$$

where Σ_1 has been written as a modulated plane wave,

$$\Sigma_1 = A_m(x)\, e^{i\varphi_m(x)}\, A_1 e^{i\frac{2\pi}{\lambda} x\theta_1}. \tag{91.8}$$

As before $|A_1| = \text{constant}$, $|A_2| = \text{constant}$, and

$$\Sigma_2 = A_2 e^{i\frac{2\pi}{\lambda} x\theta_2}. \tag{91.9}$$

It is important to note that both $A_m(x)$ and $\varphi_m(x)$ are now functions of x.

By comparing Eq. (91.7) to Eq. (91.1) and by noting Eq. (91.5), one immediately sees that the modulating wave

$$A_m(x)\, e^{i\varphi_m(x)}$$

can now be considered as "carried" by the interference grating of the case described by Eq. (91.1). Therefore, by illuminating the plate with a plane wave, as before, one again obtains three waves, a zero-angle (or zero-order) wave, carrying no information, and two side-band waves, modulated by $A_m e^{i\varphi_m}$ and by $A_m^* e^{-i\varphi_m}$ respectively, and leaving the plate at angles $\pm(\theta_1-\theta_2)$ respectively. Clearly, the sideband waves are complete "reconstructions" of the Σ_1 wave which, in turn, was formed by scattering of light by the original object. The two reconstructed waves will form a real and a virtual image, identical to the original object.

The preceding scheme of wave-front reconstruction with linearly spaced interference gratings is of course applicable to three-dimensional objects. It has been recently successfully verified at The University of Michigan (Fig. 249)[2,3]. The method of *holography* using lasers and three-dimensional objects is illustrated in Fig. 250[2]. Extensions of holographic principles and methods to high-resolution

Fig. 249a. Reconstruction of a transilluminated three-dimensional scene from hologram of Fig. 249b (according to J. Upatnieks and E. N. Leith, loc. cit. [3], 1964). In order to demonstrate the three-dimensional character of the reconstruction, the top two pictures were photographed by pointing a camera from two slightly different angles at the hologram, while the bottom two photographs were made by focussing the camera either on the man, or on the background, as reconstructed by the hologram.

---

[3] J. Upatnieks and E. Leith: Lensless, Three-dimensional Photography by Wavefront Reconstruction. J. Opt. Soc. Amer. **54**, 579 (1964) (Abstract). Wavefront Reconstruction with Diffused Illumination and Three-Dimensional Objects; J. Opt. Soc. Amer. **54**, 1295—1301 (1964).

X-ray microscopy, and to the solution of the "phase problem" now appear to be possible, as a result of recent work[2, 4].

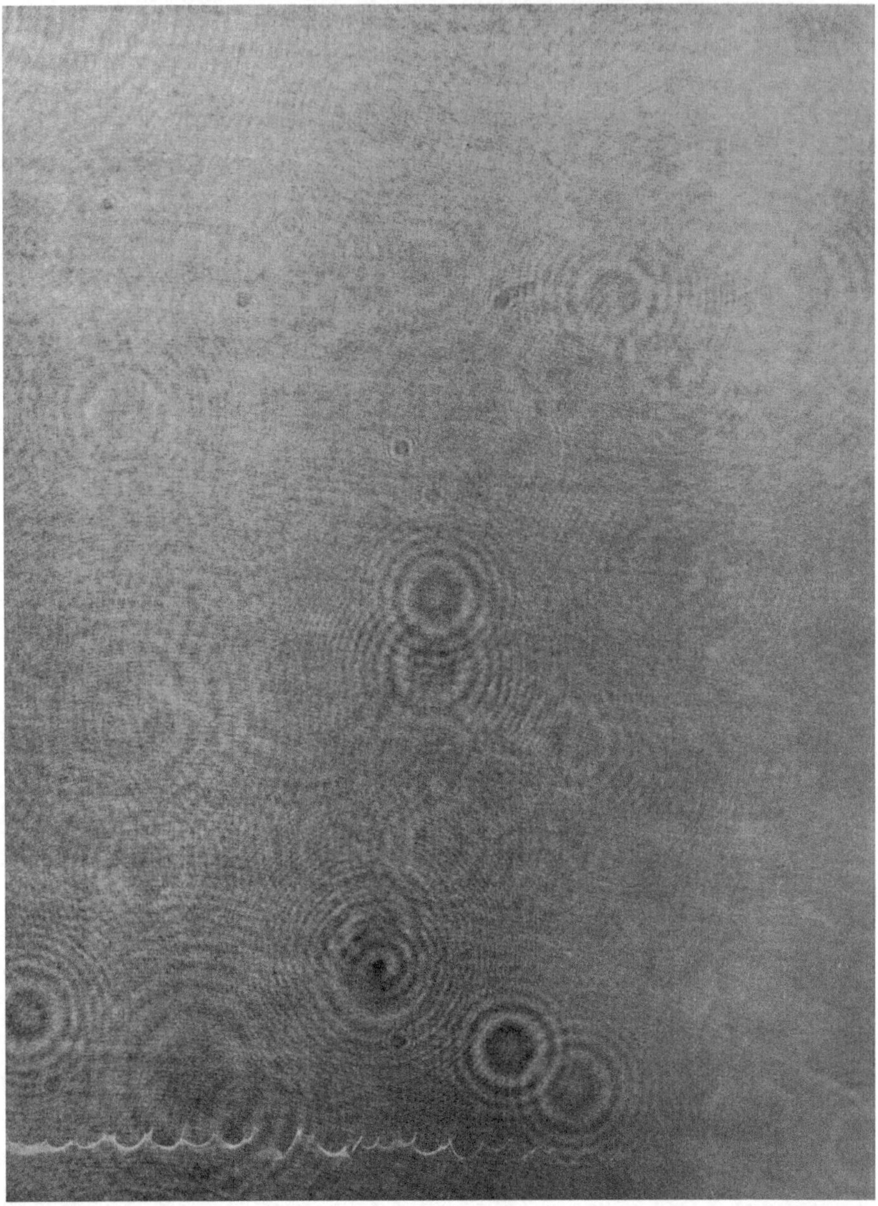

Fig. 249b. Interference-grating holcgram from which the three-dimensional image in Fig. 249a was reconstructed. Note the straight-line interference fringes running horizcntally across the hologram. (The circles are spurious images.)

[4] (a) G. W. STROKE and D. G. FALCONER: Attainment of High Resolutions in Wavefront-Reconstruction Imaging. Phys. Letters **13**, 306—308 (1964). — (b) D. GABOR, G. W. STROKE, R. RESTRICK, A. FUNKHOUSER and D. BRUMM: Optical Image Synthesis (Complex Amplitude Addition and Subtraction) by Holographic Fourier Transformation. Physics Letters **18**, 116—118 (1965).— (c) P. TOLLIN, P. MAIN, M. G. ROSSMANN, G. W. STROKE and R. C. RESTRICK: Holography and its Crystallographic Equivalent. Nature **209**, 604 (1966).

The principles and possibilities of "hologram" wave-front reconstruction were first given by D. GABOR in 1948[5–7], and the principle used is illustrated in Fig. 251c. An example of a hologram and reconstruction according to GABOR'S original method is shown in Fig. 252.

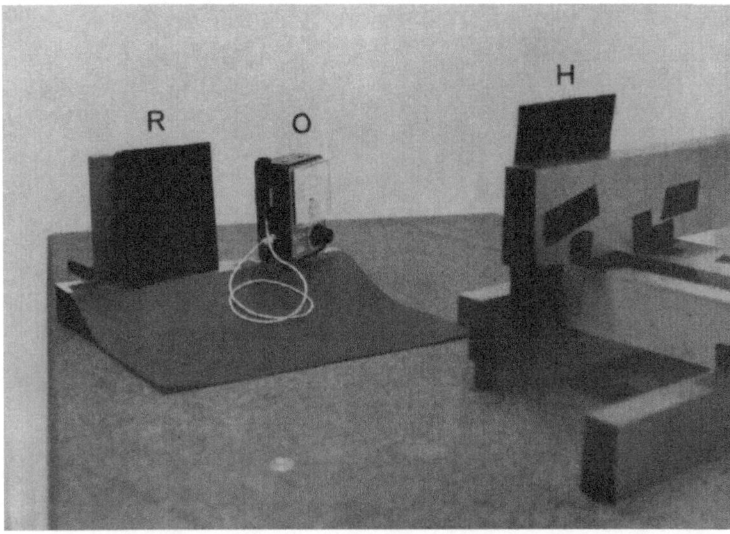

a

b

Fig. 250a and b. a Recording of a hologram of a three-dimensional object (*O*) in laser light (incident from right), according to Fig. 248[2]. [O=object, R=reference mirror, H=hologram]. b Photograph of a hologram recorded in Fig. 250a. [Under high microscopic magnifications, such holograms may show readily observable, diffraction=grating-like interference fringes (see Fig. 250e).

[5] D. GABOR: Nature, Lond. **161**, 777 (1948).

[6] D. GABOR: Proc. Roy. Soc. Lond. A **197**, 454—487 (1949).

[7] D. GABOR: Research **4**, 107 (1951).

Extensive discussion and development of the Gabor method of wave-front reconstruction is given in an unpublished thesis by H. M. A. EL-SUM[8].

c

d

Fig. 250c and d. c Photograph of view through hologram illuminated by laser light from laser on left, showing virtual image (the granularity in image is due to the use of a very small aperture in the camera, used to photograph the entire arrangement shown with adequate depth of field). d Real image, formed by the hologram of Fig. 250b, without the aid of any lenses, simply by having a plane wave of laser-light pass through the hologram, and by placing a sheet of photographic film in the best "focus" in the region of the three-dimensional aerial image reconstructed by the hologram.

[8] H. M. A. EL-SUM: Reconstructed Wave-Front Microscopy. Ph. D. Thesis, Stanford Univ., (Nov. 1952); available from University Microfilms, Inc., Ann Arbor (Dissertation Abstracts 4663, 1953).

In contrast with a general belief which followed GABOR'S wavefront-reconstruction imaging work of 1948—1951, STROKE et al.[9] have shown, with the aid of lasers, that the original Gabor 'in-line' recording scheme (Fig. 251 b) will indeed produce perfectly separable 'twin images', without any phase-recording problems, notably in high-resolution imaging and with diffuse illumination. The more recent 'off-axis' hologram recording scheme[3] (Fig. 251 a) was shown[9] to produce a hologram H which is completely identical to one or the other half (H_1 or H_2) of an 'in-line' hologram (Fig. 251 b).

Some possibilities of optical interference gratings as carriers were explicitly considered at least as early as 1944 by P. M. DUFFIEUX in his classical treatise "L'Intégrale de Fourier et ses Applications à l'optique"[10] and again in 1958[11].

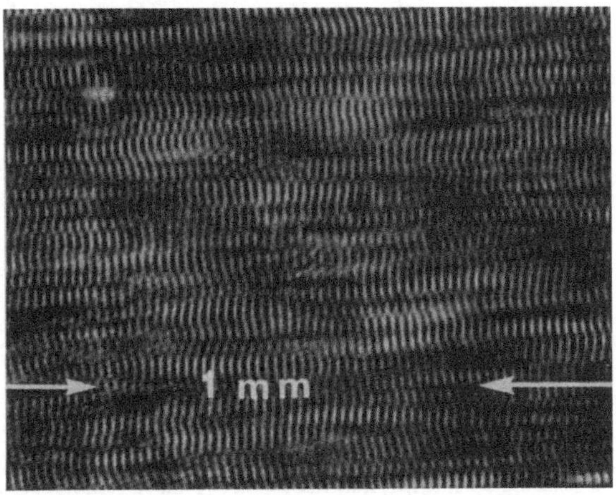

Fig. 250e. Fourier-transform hologram[2] of the word "holography", showing grating-like interference fringes.

More recently, A. LOHMANN[12] had again explicitly suggested the use of the optical equivalents of single side-band modulation in the "lensless" hologram photography originated by GABOR in 1948.

Extensive work on wave-front reconstruction, following the initial work by GABOR, was carried out by many authors, in particular by H. M. A. EL-SUM, A. V. BAEZ, E. N. LEITH and J. UPATNIEKS and others[8,13-17].

[9] G. W. STROKE, D. BRUMM, A. FUNKHOUSER, A. LABEYRIE, and R. C. RESTRICK: On the absence of phase-recording or 'twin-image' separation problems in 'Gabor' (in-line) holography. Brit. J. Appl. Phys. **17**, 497—500, + 2 plates (1966).

[10] P. M. DUFFIEUX: L'intégrale de Fourier et ses applications à l'optique. Faculté des Sciences, Besançon 1946.

[11] P. M. DUFFIEUX: Les franges d'interférences de deux systèmes d'ondes et la théorie de l'information. Rev. Opt. **37**, 441—457 (1958).

[12] A. LOHMANN: Optische Einseitenbandübertragung angewandt auf das Gabor-Mikroskop. Optica Acta **3**, 97—99 (1956).

[13] A. V. BAEZ and H. M. A. EL-SUM: Effect of Finite Source Size, Radiation Bandwidth and Object Transmission in Microscopy by Reconstructed Wavefronts. In: X-ray Microscopy and Microradiography, pp. 347—366. New York: Academic Press Inc. 1957.

[14] See also: U. S. Patent No. 3083615, Granted April 2, 1963 (Filed May 16, 1960 by H. M. A. EL-SUM, assignor to Lockheed Aircraft Corp.), Optical Apparatus for Making and Reconstructing Holograms.

[15] (a) YU. N. DENISYUK: Photographic Reconstruction of the Optical Properties of an Object in its Own Scattered Radiation Field. Soviet Physics Doklady **7**, 543—545 (1962) [Translated from Doklady Akad. Nauk SSSR **144**, 1275—1278 (1962), Original article submitted February 4, 1962]. — (b) G. W. STROKE and A. E. LABEYRIE: White-Light Reconstruction of Holographic Images using the Lippmann-Bragg Diffraction Effect. Physics Letters **20**, 368—370 (1966).

An extension of DENISYUK's method of recording holograms in the *volume* of the emulsion[15a] to the reconstruction of holographic images with white (incoherent) light illumination was first described by STROKE and LABEYRIE[15b].

Obvious similarities between the modulation and demodulation in the use of interference gratings as carriers and heterodyne methods used in communications

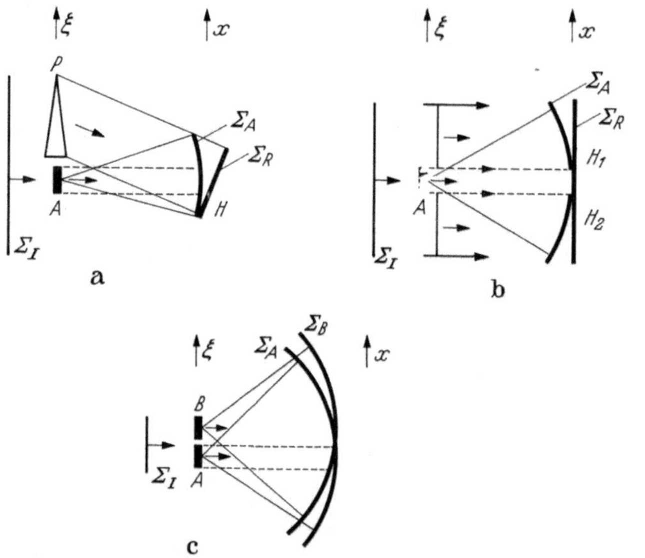

Fig. 251 a—c.[9] (a) Off-axis hologram recording arrangement according to LEITH and UPATNIEKS (1963)*. Σ_I, monochromatic spatially coherent plane illuminating wavefront; A, scattering object; P, prism; Σ_A, wavefront scattered by point-scatterer in object; Σ_R, reference wavefront (coherent background); H, hologram; ξ, coordinate in object domain; x, coordinate in image plane. The hologram H is formed by interference between the coherent background and the various waves Σ_A scattered by the object, as first demonstrated by GABOR[5] (1948) (see (b)). (b) Original 'in-line' hologram recording arrangement according to GABOR (1948)[5]. Same symbols as in (a) In case of objects which scatter over wide angles (as seen from the object A) (which is the general case of diffusing or diffusely illuminated two- or three-dimensional objects) and the general case of high-resolution imaging, e.g. in microscopy, the original Gabor 'in-line' hologram recording arrangement shown here presents no problems of twin-image separation or of completeness in phase recording. It may be superior to the 'off-axis' arrangement of (a) in that it does not require the introduction of any extraneous elements, such as the prism of (a) (which may not be available in x-ray imaging, for example), and in holography where very long "coherence lengths" may be required [e.g. in bubble-chamber work]. (c) 'Lensless Fourier-transform' hologram recording arrangement according to STROKE** (1965), used for high-resolution holograms, both with point sources and with extended sources, and designed to solve the 'emulsion problem' and the 'source problem' in high-resolution holography. Σ_I, spatially coherent monochromatic illuminating wave; Σ_A, wavefront scattered by object point; Σ_B, wavefront scattered by source point. A and B are in the same plane. The arrangement is designed to result in appreciably smaller path differences, and as a consequence the interference fringes in the hologram have a spacing which is appreciably wider with respect to the photographic 'grain' in this arrangement than in the Fresnel-transform hologram recording arrangements [such as that of a)]. The 'coding' of the Fourier-transform hologram by the use of an extended source B would result in a resolution 'loss' in the image reconstruction, if it were not for the possibility of resolution-retrieving decoding by a 'correlative reconstruction' process, using either the same extended source, or an equivalently scaled source, as first shown by STROKE, RESTRICK, FUNKHOUSER and BRUMM (1965)***.

are easily recognized. Quite basic similarities between the methods described, and the phase-contrast (coherent background) methods introduced into optics

* E. N. LEITH and J. UPATNIEKS: Wavefront Reconstruction with Continuous-Tone Objects. J. Opt. Soc. Amer. **53**, 1377—1381 (1963).

** G. W. STROKE: Lensless Fourier-Transform Method for Optical Holography. Appl. Phys. Letters **6**, 201—203 (1965).

*** G. W. STROKE, R. RESTRICK, A. FUNKHOUSER and D. BRUMM: Resolution-Retrieving Compensation of Source Effects by Correlative Reconstruction in High-Resolution Holography. Physics Letters **18**, 274—275 (1965).

[16] E. N. LEITH and J. UPATNIEKS: Reconstructed Wavefronts and Communication Theory. J. Opt. Soc. Amer. **52**, 1123—1130 (1962).

[17] E. N. LEITH and J. UPATNIEKS: Wavefront reconstruction with continuous tone objects. J. Opt. Soc. Amer. **53**, 1377—1381 (1963).

Hologram

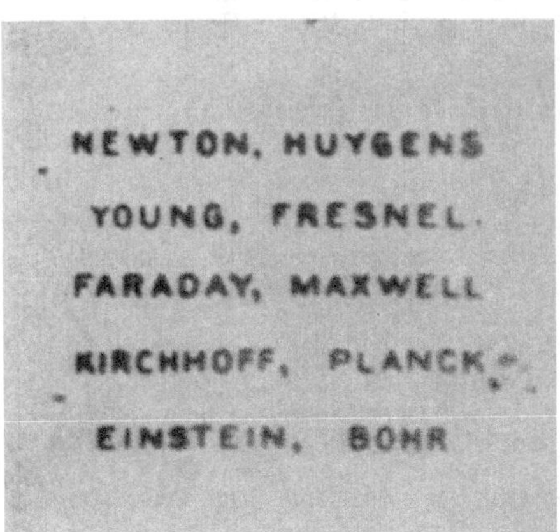

Original

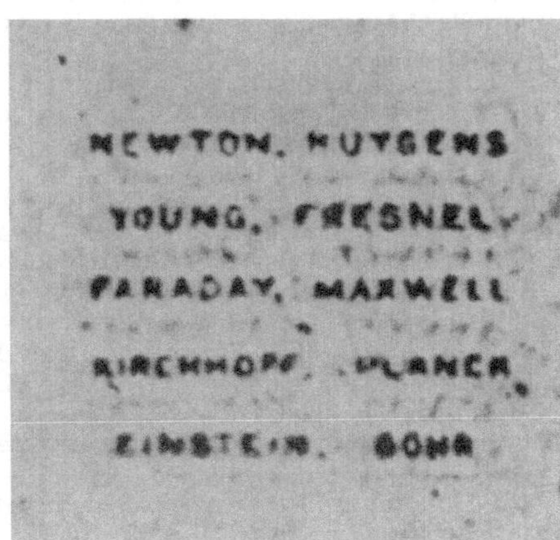

Reconstruction

Fig. 252. Wavefront reconstruction imaging according to GABOR (from M. BORN and E. WOLF, Principles of Optics, London and New York: Pergamon Press, 1959, p. 456) [after D. GABOR, Proc. Roy. Soc. London A **197**, 454 (1949)]: the object was a microphotograph of 1.5 mm diam., illuminated with light of wavelength = 4358 Å through a pinhole of diameter 0.2 mm, reduced by a microscope objective to 5 microns nominal diameter at 50 mm from the object. Geometrical magnification of the reconstructed image = 12 diameters. Effective aperture of lens used in reconstruction was 0.025. Noisy background chiefly due to imperfections of illuminating objective.

by F. ZERNIKE in 1934[18–21] should be recognized. Indeed, many of the beginnings of this work may be traced back at least as far as ABBE[22] and TOEPLER (1867)[23].

Only an outline of some of the basic principles in the use of straight-line interference-fringe gratings for "lensless" wavefront reconstruction was outlined in this section.

92. Optical image correlation filtering using holograms as complex spatial filters. It has been stressed by DUFFIEUX (loc. cit. [10]) in his now classic treatise reviving interest in Fourier transform treatment of optical processes, that there is little doubt that the modern *foundations* of operational methods in optical communications, image processing and spectroscopy can be found in the work of A. A. MICHELSON[24] and of Lord RAYLEIGH[25]. The attention to the use of Fourier methods in English speaking countries is largely due to a paper on "Optics and Communication Theory" published by PETER ELIAS[26] in 1953 and to the lecture notes on "Selected Topics in Optics and Communication Theory" first published by EDWARD L. O'NEILL in 1956[27]. Extensive references and credit to the considerable work which has preceded their own papers is given by both ELIAS and O'NEILL. An important stimulus in modern developments in image formation theory and filtering has resulted from the work by A. MARÉCHAL[28], in optical communication theory, by A. BLANC-LAPIERRE[29], P. DUFFIEUX (loc. cit. [10]), and E. L. O'NEILL (loc. cit. [27]) and in spectroscopy to P. JACQUINOT[30], A. LOHMANN[31] and others. The foundations for the developments in optical computing, filtering, character recognition, and more generally for the new sophisticated optical materializations of communication theory principles at The University of Michigan have resulted to a large part from the work of LOUIS CUTRONA and his associates[32]. Interferometry and diffraction grating principles have played a major role in these advances.

---

[18] F. ZERNIKE: Beugungstheorie des Schneidenverfahrens und seiner verbesserten Form, der Phasen-Kontrast-Methode. Physica, Haag **1**, 43 (1934).

[19] F. ZERNIKE: Das Phasenkontrastverfahren bei der mikroskopischen Beobachtung. Phys. Z. **36**, 848 (1935).

[20] F. ZERNIKE: Z. techn. Phys. **16**, 454 (1935).

[21] See H. WOLTER: Schlieren-, Phasenkontrast- und Lichtschnittverfahren. In: Handbuch der Physik, Vol. 24, edit. by S. FLÜGGE, pp. 555—645. Berlin-Göttingen-Heidelberg: Springer 1956.

[22] E. ABBE: Arch. mikr. Anat. **9** (1873).

[23] A. TOEPLER: Optische Studien nach der Methode der Schlierenbeobachtung. Pogg. Ann. **131**, 33, 180 (1867).

[24] A. A. MICHELSON: Phil. Mag. [V] **34**, 280 (1892).

[25] Lord RAYLEIGH: Phil. Mag. [V] **34**, 407 (1892).

[26] P. ELIAS: J. Opt. Soc. Amer. **43**, 229—232 (1953).

[27] E. L. O'NEILL: Optical Research Laboratory, Boston University 1956. See also E. L. O'NEILL: Introduction to Statistical Optics. Reading, Massachusetts: Addison Wesley 1963.

[28] A. MARÉCHAL et P. CROCE: Un filtre de fréquences spatiales pour l'amélioration du contraste des images optiques. C. R. Acad. Sci. Paris **237**, 607 (1953).

[29] A. BLANC-LAPIERRE: Considérations sur la théorie de la transmission de l'information et sur son application a certains domaines de la physique. Ann. Inst. H. Poincaré **13**, 275 (1953).

[30] P. JACQUINOT: New developments in interference spectroscopy. Rep. Progr. Phys. **23**, 267—312 (1960).

[31] A. LOHMANN: Contrast transfer in the grating spectrograph. Optica Acta **6**, 175—185 (1959). — D. HAUK and A. LOHMANN: Minimumstrahlkennzeichnung bei Gitterspektrographen. Optik **15**, 275 (1958). — A. LOHMANN: A new Fourier spectrometer consisting of a two-grating interferometer. 1961, I. C. O. Conference on Optical Instruments, K. J. HABELL, Editor (London: Butterworth 1962, p. 58).

[32] L. J. CUTRONA, E. N. LEITH, C. J. PALERMO and L. J. PORCELLO: Optical data processing and filtering systems. I.R.E. Trans. on Inform. Theory IT-6, No. 3, 386—400 (1960).

The manner in which grating and interferometry principles enter the image filtering process is illustrated in the following paragraphs, according to VAN DER LUGT[33].

Consider the system shown in Fig. 253. The system shown is recognized as one of the coherent-light imaging systems. The object O, in the P_1 plane is imaged one-to-one in the P_3 plane. If a suitable spatial frequency filter is used in the P_2 plane, it is possible to selectively filter out of the image O' any desired frequency, in a way quite comparable, in principle, to the filtering methods used in electrical communication systems, and indeed in the phase-contrast method (Sect. 75).

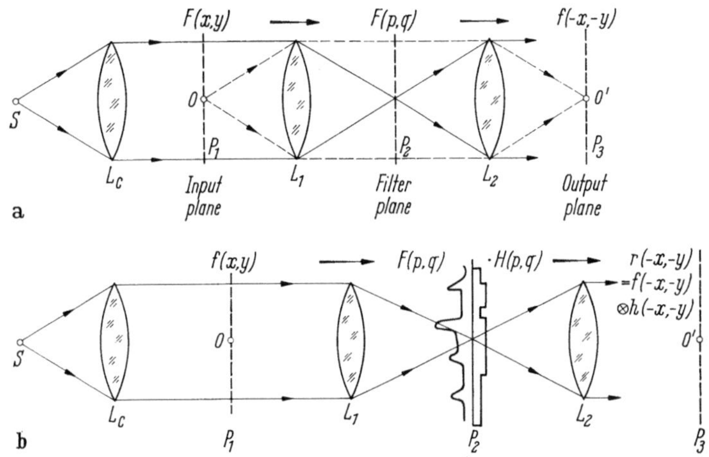

Fig. 253.

It should be clear that the entire process between S and O' must be carried out in "coherent" light and, therefore, that the *filter* $H(p, q)$ must be a *complex* filter.

As first shown by VAN DER LUGT (loc. cit.[33]) holograms may serve as the spatial filters, for optical filtering, according to the method first described by MARÉCHAL and CROCE (loc. cit.[28]).

Two examples of filter realization will now be briefly indicated, according to VAN DER LUGHT (loc. cit. [33]).

As mentioned before, interferometry and diffraction grating principles play a fundamental role in the realization of optical filters.

Realization of a complex filter.

Consider the modified Mach-Zehnder interferometer illustrated in Fig. 254. Let the desired filter be

$$H(p, q) = |S| e^{j\theta}. \tag{92.1}$$

Let $s(x, y)$ be a transparency of the word which is to be filtered,

$$s(x, y) = \text{"filter signal"}. \tag{92.2}$$

Let

$$s(x, y) \to S(p, q) = |S| e^{j\theta} \tag{92.3}$$

by Fourier transformation, in the focal plane of L_1. Let the reference signal be

$$|R| e^{j\varphi} \tag{92.4}$$

as shown.

[33] A. VAN DER LUGT: Signal detection by complex spatial filtering. IEEE Trans. on Inform. Theory IT-10, No. 2, 139—145 (1964).

The image recorded on the plate is

$$G(p, q) = [|R| e^{j\varphi} + |S| e^{j\theta}] \cdot [|R| e^{-j\varphi} + |S| e^{-j\theta}] \tag{92.5}$$

that is

$$G(p, q) = |R|^2 + |S|^2 + |R||S| e^{j(\varphi-\theta)} + |R||S| e^{-j(\varphi-\theta)} \tag{92.6}$$

or

$$G(p, q) = |R|^2 + |S|^2 + |R| e^{j\varphi} |S| e^{-i\theta} + |R| e^{-j\varphi} |S| e^{+j\theta}. \tag{92.7}$$

The desired filter is $|S| e^{j\theta}$. However, the recorded filter is $G(p, q)$, as given by Eq. (92.7).

Question: How can the desired filter $|S| e^{j\theta}$ be separated from $G(p, q)$?

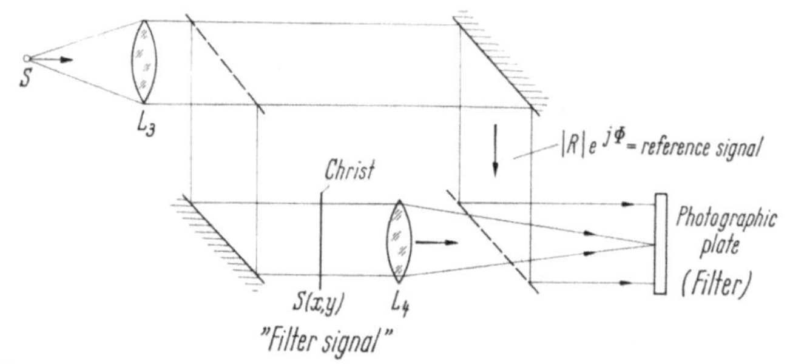

Fig. 254. Holographic filter-realization system (according to Van der Lugt, loc. cit. [33]).

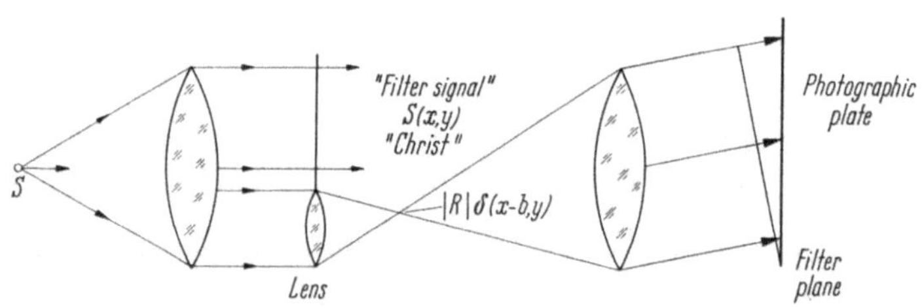

Fig. 255. Holographic filter realization (according to Van der Lugt, loc. cit. [33]).

Answer: The separation is carried out "automatically" by the lens L_2 in the optical image filtering system shown in Fig. 253b.

Indee, let $|R| = k$ = some constant, say 1. Note that the phase is a constant, that is

$$\varphi = 2\pi(bp + cq) = \text{linear}. \tag{92.8}$$

As a result, it is seen that the desired filter function $|S| e^{j\theta}$ has been placed on an interferometric-grating carrier, such that the filtered output will appear *off axis* by amount $x = b$, $y = c$. Indeed

$$|S| e^{j\theta} e^{j\varphi} \to s(x + b, y + c) \tag{92.9}$$

and therefore

$$F(p, q) \cdot |S| e^{j\theta} e^{j\varphi} \to f(-x, -y) \circledast s(x + b, y + c). \tag{92.10}$$

Comparison of Eq. (92.10) with Eq. (92.2) and with the "optical image filtering system" (Fig. 253) shows that the image has indeed been filtered by correlation as desired.

The other terms "filtered" by $G(p, q)$, some undesired, and some desired, will appear respectively on axis, and on the other side of the axis.

Alternate holographic complex-filter realization method.

A somewhat simpler interferometric-grating carrier method for realization of a complex filter is illustrated in Fig. 255, also according to VAN DER LUGT (loc. cit. [33]).

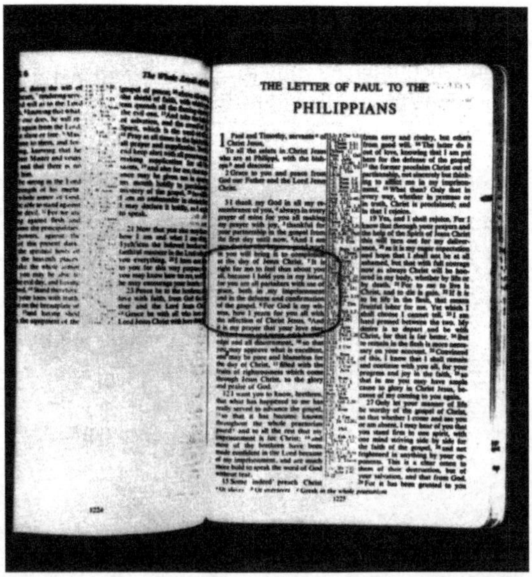

Input plane $f(x, y)$

a

Text being filtered

Filter plane $H(p, q)$

b

Filter for word "Christ"

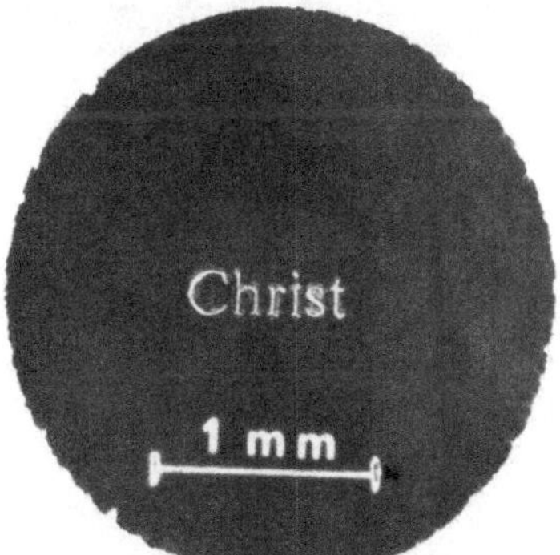

Output plane $h(-x, -y)$

c

"Reconstruction" from filter of word "Christ"

Fig. 256a—e. Optical (holographic) filtering according to VANDER LUGT[33] (after STROKE[2]). (a) Portion of text (circled) in which the word "CHRIST" is being sought. (b) Hologram ("spatial filter") of word "CHRIST", recorded as in Fig. 255. (c) Fourier-transform reconstruction of image from hologram of Fig. 255b obtained by illumating the hologram with a plane wave of laser light, and recording the image in the focal plane of a lens. (d) Image obtained in "output plane" (Fig. 253a) *without* the holographic filter. (e) Image obtained in "output plane" (Fig. 253b) *with* the holographic filter for word "CHRIST" in "filter plane". Note (by comparing with Fig. 256d) that the result of the filtering operation is the *correlation* of the word "CHRIST" with all other words in the page. The high degree of correlation of the word with itself is indicated by the two very bright spots (see arrows in Fig. 256d). Also note less bright spots, measuring degree of "correlation" of word "CHRIST" with other words in page.

Fig. 255 is self-explanatory. The method is based on producing a delta function $|R|\,\delta(x-b, y)$ with the help of the lens shown, for use as a reference beam. One has in the filter plane, by Fourier transformation

$$s(x, y) + |R|\,\delta(x-b, y) \rightarrow |S|\,e^{j\theta} + |R|\,e^{j\varphi} \tag{92.11}$$

where it is clear that

$$|R|\,e^{j\varphi} = |R| \exp[2\pi j(pb+q)]. \tag{92.12}$$

Here again

$$\varphi = pb + q = \text{linear phase function} \tag{92.13}$$

and by comparison of Eq. (92.13) with Eq. (92.8), it is seen that the desired filter function $|S|\,e^{j\theta}$ has again been placed (holographically) on an interferometric-grating carrier, as before. An example of spatial filtering according to VAN DER LUGT[33] is shown in Fig. 256.

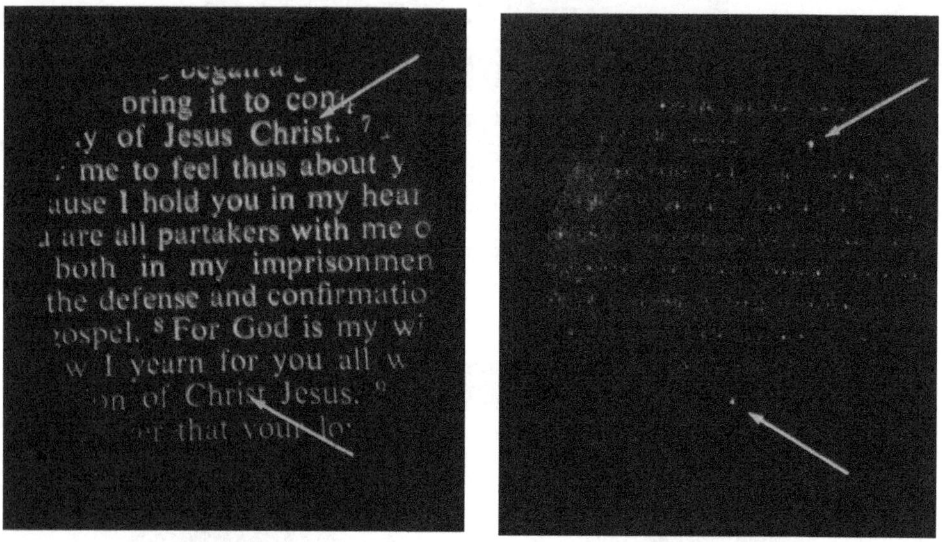

Output plane d Zero-order image

Output plane $f(-x, -y)\, h(\ddot{x}+b, y+c)$ e Filtered image (right-hand side band)

Fig. 256 d and e.

93. Optical computing, correlation and "spectral" analysis. Two principal types of optical signal processing systems, using coherent-light imaging, have been developed, in particular as a result of the work of L. CUTRONA and associates (loc. cit. [32]; see also ref. [31]).

These are:

1. Spectrum analysers;

2. Optical cross-correlators.

A brief description of these systems now follows.

1. Spectrum analysers.

Basically, an optical spectrum analyser is comparable to an optical *grating* spectrometer, using monochromatic radiation with an *imperfectly spaced diffraction grating.*

In an optical spectrum analyser, the grating "imperfections" contain the significant, desired information, recorded on photographic film, for instance with the help of a cathode ray oscilloscope. Seismic vibration recording are an example of signals recorded and analyzed in this manner. The analysis aims at determining the "power spectrum" in the recorded signal, forming the imperfect grating.

The theory of optical grating-imperfections[34] (Sects. 42 to 59) is immediately applicable to the spectral analysis, as carried out in spectrum analysers.

A schematic diagram of a spectrum analyser is shown in Figs. 257 and 258. One has

$$g(u, v) = \iint_{-\infty}^{+\infty} f(x, y)\, e^{-\frac{2\pi i}{\lambda f}(ux+vy)}\, dx\, dy. \tag{93.1}$$

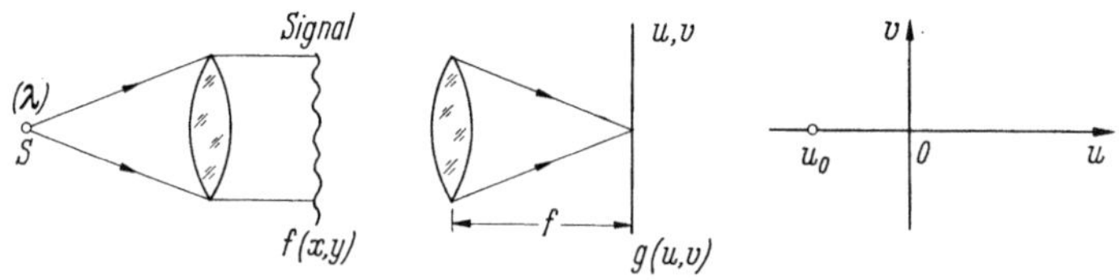

Fig. 257.

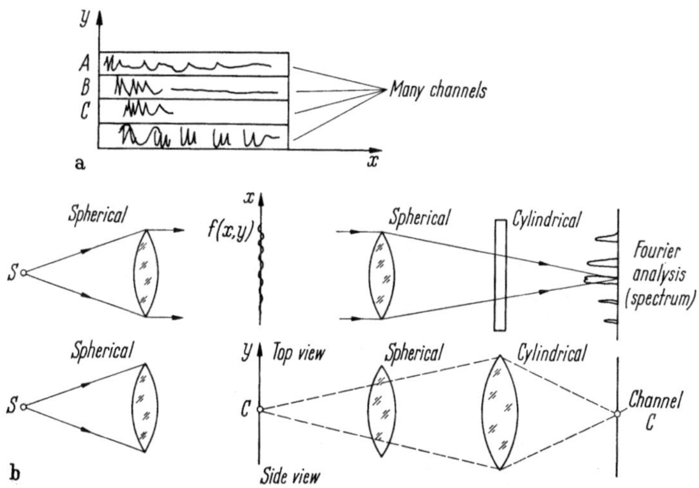

Fig. 258 a and b.

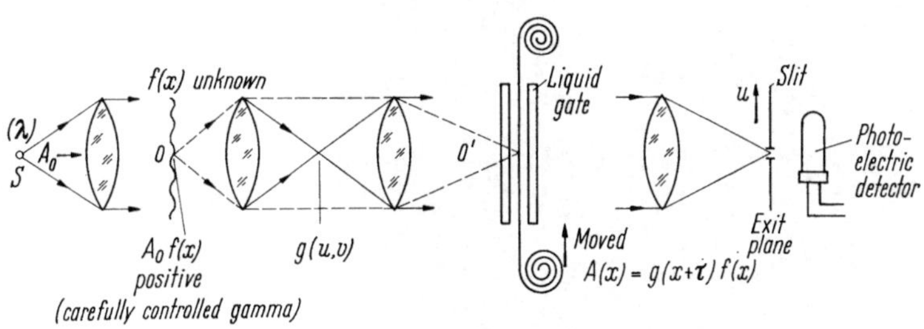

Fig. 259.

In the (u, v) plane one has k_0 lines/mm, such that $u_0 = \lambda f k_0$. The difference with spectrometers used in the analysis of optical radiations is that here one investigates k_0, rather than λ.

The important feature of optical spectrum analysers is that they can be designed to perform the spectral analysis *simultaneously* for many channels simultaneously. In principle, the limitations are set by the resolving power of the

[34] G. W. Stroke: Rev. Opt. **39**, 291—398 (1960).

film and of the associated optics, as well as by the ability of converting the electrical signals into optical recording (for instance by intensity-modulation of cathode ray tubes).

The simultaneous analysis requires the use of *cylindrical optics*. A two-plane view of a multi-channel analyser is shown in Fig. 258. For example, 120 channel analysis has been carried out in some oil-industry applications.

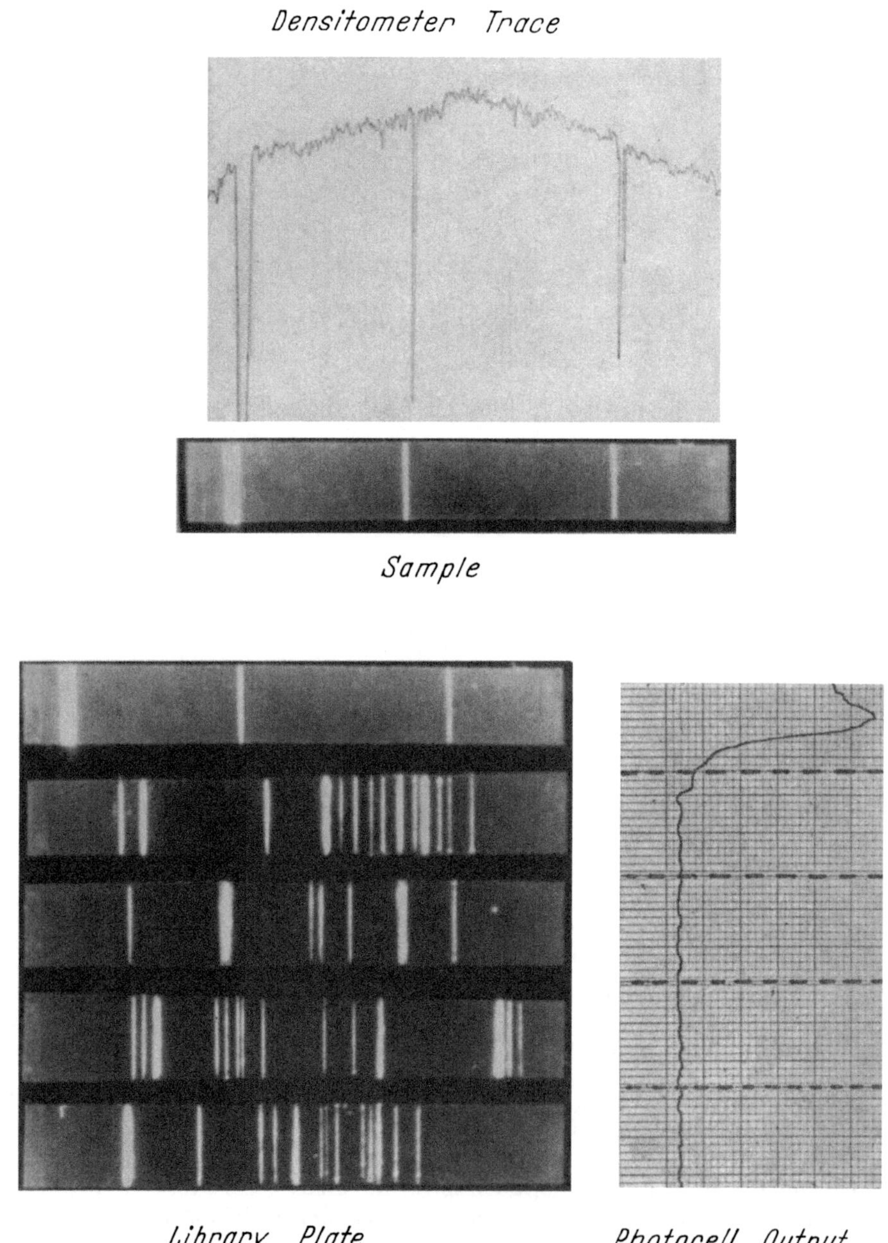

Fig. 260. Correlation of "sample" spectrum with "library" spectrum according to Stroke and Funkhouser. The peak of the photocell output indicates spectrum for which correlation is maximum.

2. *Optical cross-correlators.*

Cross-correlation problems arise in such diverse applications as optical spectroscopy, statistics, pharmaceutical and chemical research, and indeed in target detection and identification.

In an optical cross-correlator, using coherent light, both the "known" and the "unknown" signals are recorded on film. Both films are generally, though not

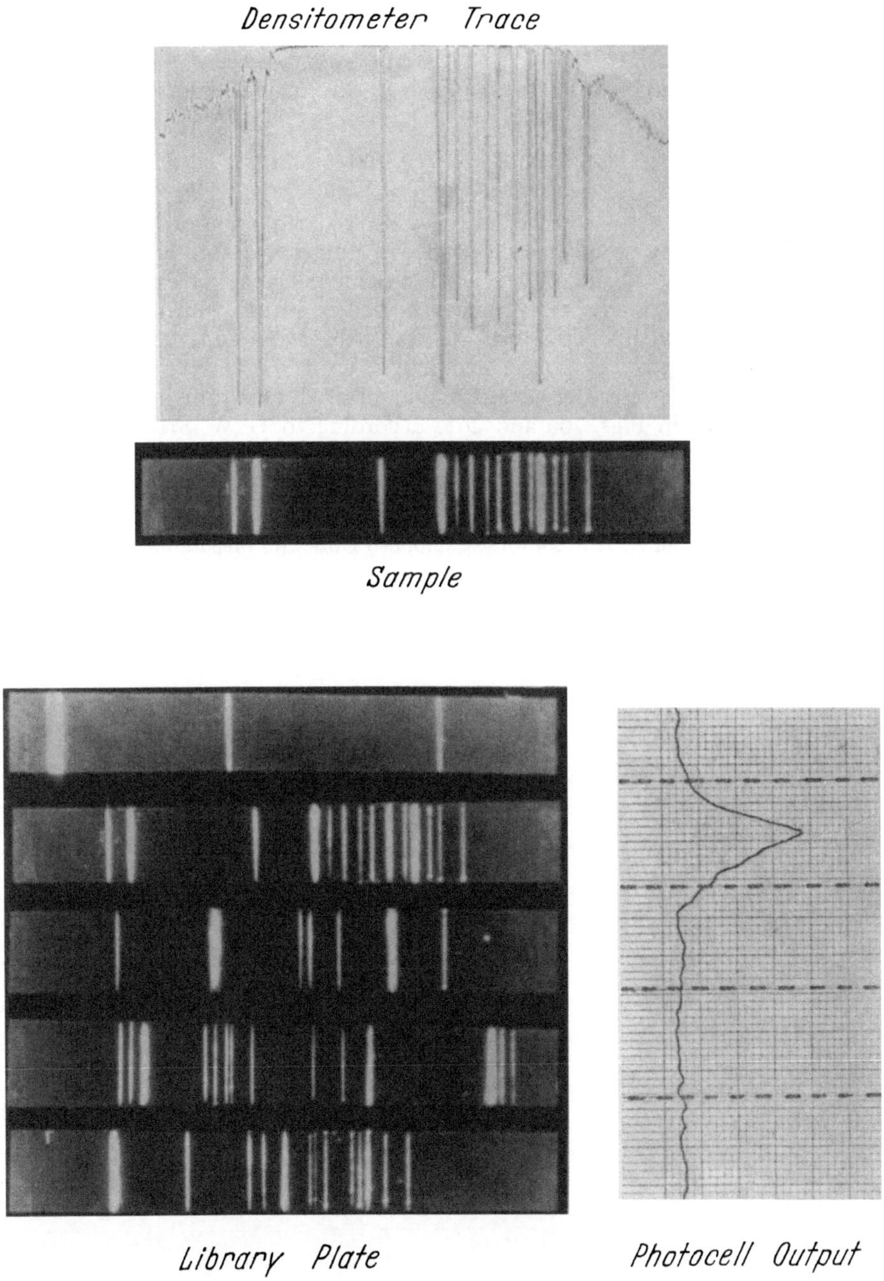

Fig. 261. Correlation of "sample" spectrum with "library" spectrum according to STROKE and FUNKHOUSER. The peak of the photocell output indicates spectrum for which correlation is maximum (see Fig. 260).

always, immersed in liquid gates, and indeed one of the films, say the reference film $g(x)$ is moved through the liquid gate, at very uniform speed, in order to carry out the cross-correlation (Fig. 259).

One obtains in the exit plane the function

$$A\left(\frac{2\pi}{\lambda f}u\right) = T[A(x)] = \int_{-\infty}^{+\infty} f(x)\,g(x+\tau)\,\mathrm{e}^{-\frac{2\pi i}{\lambda f}ux}\,dx \tag{93.2}$$

called the *ambiguity function*[35, 2].

Note that if the exponential term in Eq. (93.2) is made equal to unity, then the output is equal to

$$A(u=0) = \int_{-\infty}^{+\infty} f(x)\,g(x+\tau)\,dx \tag{93.3}$$

which is the desired cross-correlation. The desired reduction of the exponential term to unity is very simply accomplished by placing a slit (pinhole) on axis. Indeed, in this case $u=0$, and $e^0=1$. For multichannel cross-correlations, cylindrical optics and a slit in the exit plane are used, together with multiple photodetectors (perhaps with fiber optics), in order to obtain simultaneous analysis.

An example of optical cross-correlation of interest in optical spectroscopy, and using these principles for the search of an infrared spectrum in a "library of spectra" is shown in Figs. 260 and 261, according to G. W. STROKE and A. T. FUNKHOUSER[36].

[35] W. M. BROWN: Analysis of Linear Time-Invariant Systems. New York: McGraw-Hill Book Co. 1963.

[36] G. W. STROKE and A. T. FUNKHOUSER: Applied Optics (in preparation).

Optische Meßmethoden.

Von

REINHART SCHULZE.

Mit 107 Figuren.

I. Einleitung.

1. Übersicht. Es ist angebracht, aus der großen Zahl von Meßmethoden, die sich der optischen Abbildung unmittelbar oder mittelbar bedienen, bestimmte auszuwählen, die gemeinsame, typische Merkmale besitzen. Dabei fällt die Wahl auf Methoden für die Längen- und Winkelmessung mit sichtbarem Licht, da sowohl in der Physik als auch in der Technik für diese Art von Meßmethoden, die auch als *optische Feinmeßmethoden* bezeichnet werden, allgemeinstes Interesse besteht. Es sei daran erinnert, daß die Länge eine Grundeinheit des cgs-Systems liefert und daß die Lichtwellenlänge 1960 zum alleinherrschenden Längenstandard an die Stelle des Urmeters gesetzt wurde.

Von den physikalisch-optischen Meßverfahren der Spektroskopie, Densitometrie, Kolorimetrie sowie manchen anderen in Chemie, Medizin und Mineralogie angewendeten — soll in diesem Rahmen nicht gesprochen werden, sie werden teils an anderer Stelle des Handbuchs erwähnt, teils bleiben sie außerhalb des gesetzten Rahmens.

Über die optischen Feinmeßmethoden und ihre Vorzüge seien einleitend allgemeine Bemerkungen erlaubt, um das „Gesichtsfeld" des vorliegenden Handbuch-Artikels abzugrenzen.

α) *Bedeutung der Lichtwellenlänge für die Längenmessung.* Das Licht trägt den genauesten aller Maßstäbe in sich, die Lichtwellenlänge. Seit im Jahre 1960 die Lichtwellenlänge international zum alleinigen Längenstandard erklärt wurde, ist die Bedeutung der optischen Feinmeßmethoden in ein neues Stadium getreten. Es ist nun möglich, in jedem Laboratorium und an jedem Ort auf die Lichtwellenlänge als „Urmeter" zurückzugreifen. Damit ist dem Fortschritt in der Präzision der Längenmessung auf breiter Ebene ein großer Dienst erwiesen. Die körperliche Existenz eines „Urmeterstabes" und der von ihm abgenommenen, sorgfältig gehüteten Sekundärnormale in den einzelnen Staaten und Ländern wird überflüssig. Behördliche Prüfämter, Physikalische Laboratorien und schließlich die Industrie sind nicht auf drittklassige Kopien des Urmeters angewiesen und nicht genötigt, die beim fortlaufenden Kopieren angewachsenen Meßunsicherheiten in Kauf zu nehmen. Der direkte Anschluß an das Urnormal ist jedermann zugänglich, und zwar an ein Urnormal, welches noch um einige Zehnerpotenzen genauer bekannt ist als das bisherige „Urmeter" in Paris.

β) *Geschlossene Profilabbildung.* Die optische Abbildung ermöglicht, feine, komplizierte Teile oder Ausschnitte von solchen mit einem Blick in vergrößertem Abbildungsmaßstab zu übersehen und mit einer Zeichnung zu vergleichen, eine Möglichkeit, die in dieser geschlossenen Form der Übersichtlichkeit weder von mechanischen noch elektrischen Meßmethoden geboten werden kann. Zum Beispiel stellt die komplizierte, winklige Gestalt von Kleinteilen wie Zahnrädern im Eingriff, Ankerrädchen, Gewindeprofilen und dergleichen nur der mechanischen

Vermessung mit Fühlhebeln erhebliche Schwierigkeiten in den Weg, der optischen Abbildung und dem Vermessen im Bild dagegen nicht. Die Vielseitigkeit des Schattenbildverfahrens tritt beispielsweise bei Gewindemessungen deutlich hervor; denn alle fünf Gewindebestimmungsgrößen, Flankenwinkel, Steigung, Außen-, Kern- und Flankendurchmesser lassen sich berührungslos aus dem Profilbild ermitteln. Dabei kann zusätzlich beobachtet werden, ob die Gewindeflanke geradlinig verläuft, ob sie rauh oder glatt ist.

γ) Messen von Abweichungen und Verschiebungen im Bild. Meßmikroskope erlauben, die durch zu messende Teile gegebene Verschiebung eines Maßstabs im vergrößerten Bild um eine Größenordnung genauer abzulesen als dies mit Mikrometerschrauben möglich wäre.

Strichmaßstäbe lassen sich mit optischen Hilfsmitteln einwandfrei bis auf zehntausendstel Millimeter ablesen. Eine Reihe von Meßgeräten verbindet daher die Vorzüge der berührungslosen optischen Antastung mit denen der optischen Maßstabablesung.

δ) Konstanter Abbildungsmaßstab. Der einmal gewählte Abbildungsmaßstab, das ist die „optische Übersetzung", bleibt wegen der konstanten Abbildungseigenschaften der Objektive zeitlich konstant. Dank des telezentrischen Strahlengangs bleibt er auch dann unverändert, wenn das Objekt beim Einstellen der Bildschärfe noch ein wenig vor oder hinter der exakten Schärfenebene liegt und sich damit die „Hebellängen" der Abbildung ändern.

ε) Manipulationen mit dem Bild. Am optischen Bild eines Prüflings lassen sich Manipulationen vornehmen, die am Prüfling selbst unausführbar wären. Ein Bild läßt sich umkehren, in eine andere Richtung umlenken, durch Filter farbig darstellen, seiten- und höhenvertauscht in Komplementärfarben auf sich selbst abbilden, entweder im Mischbild oder im Schnittbild, um einige Beispiele zu nennen. Speziell bei der Aufgabe, den Abstand von Bohrungen zu messen, ergeben sich mit Hilfe des gegenläufigen Doppelbildes mancherlei Vorzüge: Erstens bringt die gegenläufige Bewegung die doppelte Einstellgenauigkeit mit sich, zweitens werden die Durchmesser der Bohrungen nicht mehr benötigt; denn wenn Bild und Spiegelbild sich genau decken, was mit Hilfe komplementärfarbiger Bilder gut zu beurteilen ist, fallen Visier- und Bohrungsachse zusammen.

Die Spitze einer Meßpinole kann durch einen optischen Fühlhebel ersetzt werden, wenn die anzumessenden Flächen eine gewisse Oberflächengüte besitzen, so daß sie selbst als „Spiegel" in den Strahlengang mit einbezogen werden können. Zu diesem Zweck wird eine Marke als Luftbild in die Nähe der Prüflingsoberfläche projiziert und sie und ihr Spiegelbild in der Oberfläche als „optischer Taster" ausgenutzt. Auf diese Weise kann optisch nicht nur die Kante eines Meßobjekts, sondern sogar die plane oder gewölbte Meßfläche selbst an jeder beliebigen Stelle angetastet werden. Die Spiegelung der Marke in der Meßfläche ergibt wieder die doppelte Einstellgenauigkeit für die Meßfläche.

ζ) Optisches Antasten ohne Meßkraft. Die abbildenden optischen Methoden arbeiten meßkraftfrei; denn Lupe, Mikroskop und Fernrohr sind das in den weitaus meisten Fällen verwendete optische Rüstzeug. Das empfindliche Prüfobjekt wird praktisch bei der Messung weder berührt noch durch sie beeinflußt. Dieser Umstand beugt sowohl dem Verschleißen der Meßmittel als auch Beschädigungen und Verformungen am Prüfling vor und kommt damit der Messung zustatten, ferner werden von der Meßkraft hervorgerufene Umkehrspannen ausgeschlossen.

Ein besonderer Vorzug der optischen Antastung der Meßobjekte ergibt sich dadurch, daß viele Innenmessungen ebenso bequem auszuführen sind wie Außen-messungen, da kein mechanischer Taster in das Meßobjekt hineinragen muß und schwierige Justierarbeit erforderlich macht.

η) *Optische Fluchtgeraden.* Eine Aufgabe, die nur mit optischen Methoden befriedigend gelöst werden kann, ist die Fluchtungs- und Richtungsmessung in ein, zwei oder drei Koordinaten sowohl im Kleinen, z.B. bei der Dezentrierungsmessung von Linsen, als auch über größere Entfernungen. Solche Messungen sind z.B. im Großmaschinen-, im Schiffs- und Flugzeugbau, im Reaktorbau, ja sogar bei der Aufstellung der Raketen für die Raumfahrt notwendig. Als Meßgeräte dienen Flucht- oder Autokollimationsfernrohre in Verbindung mit Zielmarken oder Planspiegeln, die an verschiedenen Meßpunkten angebracht werden. Ein Vorzug der optischen Fluchtlinie gegenüber der mechanischen, beispielsweise durch einen Draht verwirklichten Fluchtlinie ist, daß sie selbst nirgends stört und beliebig durchfahren werden kann, weil ihr unhandliche materielle Eigenschaften fehlen.

ϑ) *Drehspiegel und Lichtzeiger.* Die projizierte Fluchtgerade kann als Lichtzeiger über Drehspiegel geleitet werden, worin die Möglichkeit beschlossen liegt, kleine Winkelauslenkungen des Drehspiegels, auch solche periodischer Art, trägheitslos in nahezu beliebiger Vergrößerung sichtbar zu machen. Die Messung der Tiefe von Sacklöchern und Aussenkungen entzieht sich der optischen Antastung. Man kann sie aber mechanisch antasten und die Tastbewegung über Drehspiegel mechanisch-optisch sehr hoch übersetzen. Übersetzungen nach diesen Konstruktionsprinzipien haben den Vorzug der zeitlichen Konstanz und hoher Meßempfindlichkeit.

Optische Feintastgeräte mit mechanischer Antastung und optischer Übersetzung eignen sich besonders für empfindliche Unterschiedsmessungen, für die kein großer, aber ein entsprechend aufgelöster Meßbereich gefordert wird. Bei solchen Messungen spielt die Meßkraft allerdings mit herein. Ihr Einfluß eliminiert sich aber, wenn sie bei Meßobjekt und Normal in gleicher Weise wirkt. Derartige optische Feintastgeräte sind für den Gebrauch in der Technik mit photoelektrischen Zellen zur Klassierung des Ausschlags des Lichtzeigers versehen worden. Sie können daher insbesondere bei Serienmessung die statistische Verteilung der Meßwerte aufnehmen und an elektrische Auswertegeräte weiterleiten. Diese Möglichkeit ist für die Anwendung der statistischen Qualitätskontrolle besonders zur automatischen Steuerung und Regelung des Fertigungsprozesses auf einer Maschine und für das Sortieren der Produkte nach Klassen bedeutsam.

Aus dem Prinzip des optischen Feintastgerätes sind Oberflächentastgeräte zum Messen der Rauhtiefe entwickelt worden, die mit einer feinen Tastnadel den Profilschnitt senkrecht zur Oberfläche ermitteln und gleichzeitig die mikro- und makrogeometrische Gestalt des zu prüfenden Werkstückes erkennen lassen.

ι) *Oberflächenmessung mit Lichtinterferenzen.* Die Kombination von Mikroskop und Interferometer erschließt den Bereich der Feingestalt von Oberflächen. Mit der den Lichtwellenmethoden eigenen hohen Genauigkeit läßt sich die Oberflächenfeingestalt im zweidimensionalen Bild überblicken und dreidimensional ausmessen, wie es mit keinem anderen Verfahren in solcher Übersicht möglich ist. Das Bild ist mit Höhenlinien, den Interferenzstreifen, versehen, deren Abstand einem Höhenunterschied von einer halben Wellenlänge entspricht und deren Verlauf Erhebungen und Vertiefungen abmessen läßt.

Die vorliegende Darstellung optischer Feinmeßmethoden für die Längen- und Winkelmessung hält sich in großen Zügen an den im einleitenden Abschnitt gegebenen Umriß, um die Grundlagen und Erkenntnisse darzustellen, die für den Benutzer dieser Methoden wichtig sind.

II. Allgemeine Grundsätze.

a) Über die Anordnungen.

2. Unmittelbare, mittelbare, zusammengesetzte Messungen und Unterschiedsmessungen. Unter dem Begriff Messen wird das quantitative Vergleichen einer Eigenschaft eines Prüflings mit der entsprechenden Eigenschaft eines die Maß-

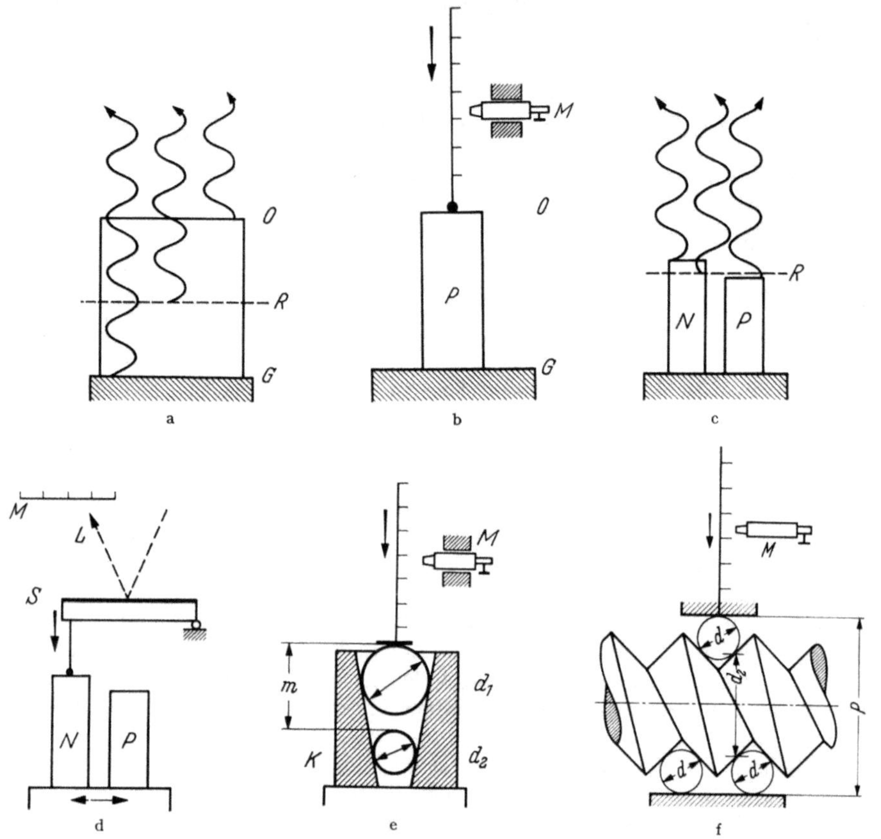

Fig. 1 a—f. a Unmittelbare Längenmessung. Endmaß an Grundfläche G angesprengt. Oberflächen O und G bilden mit Referenzspiegel R Interferenzsysteme. b Mittelbare Längenmessung am Beispiel eines Abbeschen Längenmessers. O Oberfläche, G Grundfläche des Prüflings P, M Meßmikroskop. c Unmittelbare Unterschiedsmessung zwischen Prüfling P und Normal N, R virtueller Referenzspiegel. d Mittelbare Unterschiedsmessung zwischen Prüfling P und Normal N, am Beispiel eines Feintasters, S Kippspiegel, L Lichtzeiger, M Maßstab. e Zusammengesetzte Messung am Kegel K. d_1, d_2 Kugeldurchmesser, M Meßmikroskop, m zu messende Strecke. f Zusammengesetzte Messung am Gewinde am Beispiel der Dreidrahtmethode. d_2 Flankendurchmesser, d Drahtdurchmesser, P Prüfmaß, M Meßmikroskop.

einheit verkörpernden Normals verstanden. Wenn dies Vergleichen mit Hilfe des Lichtes geschieht, kann von optischer Messung gesprochen werden, gemeint ist damit z. B. ein Vergleichen mit Hilfe abbildender Systeme oder Lichtinterferenzen. Am Beispiel Längenmessung sollen im folgenden die allgemeineren Begriffe besprochen werden.

Wenn das Längenmaß eines Prüflings unmittelbar mit der Lichtwellenlänge verglichen wird, soll von „unmittelbarer" Längenmessung gesprochen werden (Fig. 1 a), vergleicht man es gegen Strichmaßstäbe, von „mittelbarer" Längenmessung (Fig. 1 b). Der Vergleich gegen ein Normalstück werde als „Unterschiedsmessung" bezeichnet; hierbei braucht das Maß des Vergleichsstücks nicht bekannt zu sein, es interessiert primär nur das Abmaß des Prüflings. Dieses kann

„unmittelbar" mit Lichtwellen (Fig. 1c) oder aber „mittelbar" mit Maßstäben ermittelt werden (Fig. 1d). Bei gewissen Messungen z.B. von Kegelwinkeln (Fig. 1e) oder Flankendurchmessern von Gewinden (Fig. 1f) können Hilfsstücke und Rechenoperationen erforderlich sein, um das Meßergebnis zu erhalten, dann soll von „zusammengesetzten" Messungen gesprochen werden. Die einzelnen Operationen einer zusammengesetzten Messung können aus unmittelbaren (Wellenlängenmaßstab) und mittelbaren (Strichmaßstab) Messungen bestehen.

An einem Maßstab kann jedes einzelne Maß in beliebiger Folge abgelesen und interpoliert werden, ohne daß abgestimmte Vergleichsstücke für verschiedene interessierende Maße des Prüflings hergestellt werden müssen. Dabei stellt der Abstand jedes einzelnen Maßstabstriches vom Nullstrich ein Vergleichsnormal dar, dessen Größe aus der Bezifferung und der Fehlertabelle her bekannt ist, — mittelbare Längenmessung. Würden die Nullpunktabstände der einzelnen Maßstabsstriche z.B. durch einen Satz gestufter Endmaße dargestellt, so müßte für jede Maßänderung eine neue Endmaßkombination substituiert werden. Das wäre umständlich und ergibt Fehlerquellen. Diese Art der Unterschiedsmessung wendet man trotzdem mit Vorteil in der Serienmessung an, wo der Unterschied zwischen der Länge einer Endmaßkombination und nur einer Abmessung einer ganzen Serie gleicher Einzelteile gemessen werden soll. Dann mißt man unmittelbar oder mittelbar den Unterschied zwischen Prüfling und Endmaßkombination.

Die Messung von Endmaßen im Köstersschen Interferenzkomparator mit mehreren Lichtwellenlängen (Ziff. 18), ist streng genommen ebenfalls ein unmittelbares Unterschiedsmeßverfahren; denn die Istlage der Wellenlängenbruchteile am zu messenden Endmaß wird im Unterschied zu der Lage gemessen, die zu der Sollänge des Endmaßes gehören würde. Damit ist die Istlänge des Endmaßes auf die Sollänge als theoretisches Normal zurückgeführt.

3. Meßbereich. Mit zunehmender Länge eines Maßstabes oder Endmaßes werden die systematischen Fehler absolut genommen größer, relativ gesehen bleiben sie annähernd konstant (lineares Zunahmegesetz mit von Null verschiedenem Anfang). Die Ursache für diesen Verlauf liegt in Einflüssen von Temperatur, Durchbiegungen und Gerätefehlern, die sich bei größeren Meßlängen entsprechend stärker auswirken. Es ist daher verständlich, daß für Endmaße und Maßstäbe und selbstverständlich auch für Geräte, die mit diesen ausgerüstet sind, Fehlergrenzen angegeben werden, die linear mit der Länge zunehmen. Man vergleiche hierzu etwa die deutschen Normblätter DIN 861 und DIN 864.

Die unvermeidlichen zufälligen Fehler bleiben im gesamten Meßbereich vorwiegend konstant, daher enthalten die Fehlergrenzen ein von der Länge unabhängiges Glied.

Innerhalb kleiner Meßbereiche — wie sie bei Unterschiedsmessungen üblich sind — können mit verhältnismäßig geringem Aufwand hohe Genauigkeiten erzielt werden. Man kann z.B. innerhalb eines Meßbereiches von wenigen Interferenzstreifen mit Leichtigkeit feststellen, ob sich ein unbekanntes Maß um Bruchteile einer Wellenlänge von einem anderen als Vergleichsstück benutzten unterscheidet. Die Gesamtlänge interessiert dabei nicht. Wird sie jedoch benötigt, sind weit anspruchsvollere Messungen erforderlich, weil hohe Anforderungen an die Konstruktion des Meßgeräts und die Konstanz der Temperatur und der Umweltbedingungen zu stellen sind. In vielen Fällen lohnt es sich daher, kleine Meßbereiche hoch zu übersetzen, d.h. den Nullpunkt weitgehend zu unterdrücken, und Unterschiedsmessungen gegen ein genügend genaues Normal auszuführen.

Auf dieser Erkenntnis beruhen alle Maßstabsablesevorrichtungen, indem im Maßstab mit seiner Millimeterteilung von Millimeter zu Millimeter das „genügend genau" bekannte Normal vorliegt und für den kleineren Meßbereich von einem Millimeter die hohe Übersetzung des Ablesemikroskops ausgenutzt wird. Nach dem gleichen Prinzip arbeiten die bekannten optischen Feintaster für Unterschiedsmessungen an Endmaßen, Lehren und Fertigteilen, bei denen nur der stark übersetzte Feinmeßbereich eine Maßstabsteilung besitzt, im großen Meßbereich jedoch eine echte Substitution von Prüfling und Normal vorgenommen wird.

4. Optischer Abbildungsmaßstab. Bei Messungen mittels optischer Abbildung sind mehrere Fälle zu unterscheiden: Entweder überlagert man Objekt und Maßstab reell und betrachtet ihr virtuelles Bild vergrößert, Beispiel: Eine Meßlupe wird auf einen Brinellhärteeindruck aufgesetzt, oder man überlagert das Bild des einen, z.B. des Objekts, dem anderen, dem Maßstab unmittelbar, Beispiel: Ein Profilprojektor bildet ein Zahnrad auf einen Mattscheibenmaßstab ab, oder man überlagert das Bild des Prüflings dem Bild des Maßstabs, Beispiel: In einem Interferenzmikroskop nach LINNIK liegt in einem Arm das Objekt, im anderen ein Vergleichsspiegel mit aufgraviertem Maßstab. Die Bilder von Objekt und Maßstab vereinigen sich als Mischbild in der Ocularebene. Die Abbildung des Objektes ermöglicht durch geeignete Maßnahmen im Strahlengang, das Bild nach Größe und Lage so zu verändern, daß es bequem mit einem weit gröberen Maßstab oder Normal verglichen werden kann. Die optischen Daten des Abbildungsmaßstabes ändern sich zeitlich nicht, daher wird eine hohe Meßgenauigkeit erzielt.

Kennzeichnend ist für diese Messungen, daß der Prüfling berührungslos angetastet wird; darin liegt ein wichtiger Vorzug optischer Meßmethoden. Je nach der Aufgabe kann der Prüfling im streifenden Durchlicht oder im Auflicht erscheinen. Die „Vergrößerung" als Verhältnis von Bildgröße zu Gegenstandsgröße ergibt sich aus dem Verhältnis der „Bildweite" a' zur „Gegenstandsweite" a als dem „Hebelgesetz" der abtastenden Lichthebelarme; sie wird als Abbildungsmaßstab β' des Objektivs bezeichnet (Fig. 2).

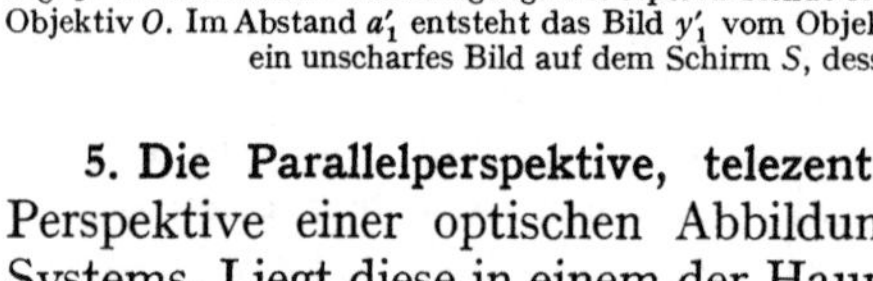

Fig. 2.

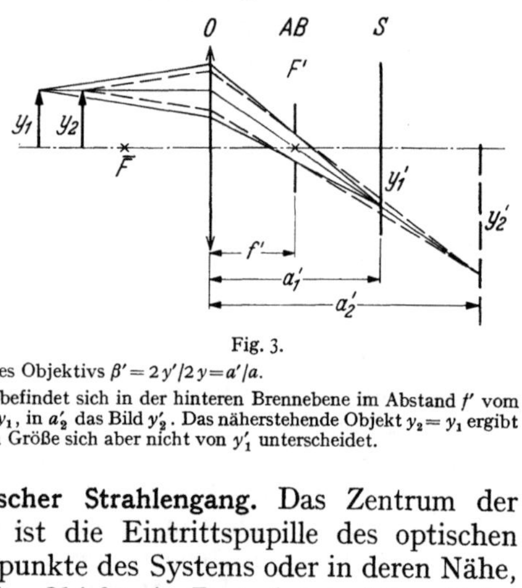

Fig. 3.

Fig. 2. Abbildungsmaßstab eines Objektivs $\beta' = 2y'/2y = a'/a$.

Fig. 3. Telezentrischer Strahlengang. Die Aperturblende AB befindet sich in der hinteren Brennebene im Abstand f' vom Objektiv O. Im Abstand a'_1 entsteht das Bild y'_1 vom Objekt y_1, in a'_2 das Bild y'_2. Das näherstehende Objekt $y_2 = y_1$ ergibt ein unscharfes Bild auf dem Schirm S, dessen Größe sich aber nicht von y'_1 unterscheidet.

5. Die Parallelperspektive, telezentrischer Strahlengang. Das Zentrum der Perspektive einer optischen Abbildung ist die Eintrittspupille des optischen Systems. Liegt diese in einem der Hauptpunkte des Systems oder in deren Nähe, so entsteht durch das System ein Bild der Objekte in Zentralperspektive. Beim Durchstellen der Objekte durch die Schärfenebene ändert sich die Größe der Bilder. Diesen Umstand muß man für Messungen vermeiden, sonst können schon ununbemerkte axiale Unterschiede in der Scharfstellung den Abbildungsmaßstab merklich ändern. Man entkoppelt den lateralen Abbildungsmaßstab von der axialen Einstellung durch den telezentrischen Strahlengang (Fig. 3). In die hintere Brennebene des Systems wird eine Blende gesetzt, die als Austrittspupille des

Systems wirkt. Das bedeutet, daß das abbildende System im Objektraum mit einer im Unendlichen liegenden Eintrittspupille arbeitet, wodurch bekanntlich Parallelperspektive erzwungen wird. Bei dieser erscheint der abzubildende Gegenstand auf einem festen Bildschirm unabhängig vom Abstand des Gegenstandes vom Objektiv gleich groß, wenn auch unterschiedlich scharf. Ein Abbildungsstrahlengang kann objektseitig oder auch bildseitig telezentrisch sein. Er ist objektseitig telezentrisch, wenn die Eintrittspupille im Unendlichen liegt, und bildseitig telezentrisch, wenn die Austrittspupille im Unendlichen liegt. Abbildungsstrahlengänge der Meßtechnik in Projektoren, Meßmikroskopen und sogar Meßlupen werden wegen wechselnder Objektlagen vorzugsweise objektseitig telezentrisch geführt.

6. Mechanisch-optische Vergrößerung, Meßkraft. Im Gegensatz zu den erwähnten abbildenden Verfahren stehen solche, die die Richtung eines Lichtzeigers über einen Drehspiegel steuern und damit aus einer mechanisch-optischen Hebelkombination bestehen. In diesem Falle muß der Spiegel den Prüfling mit einem Hebelarm kraftschlüssig berühren, was nur mechanisch unter Andruck durch eine Feder geschehen kann. Infolgedessen wirkt auf den Prüfling eine Meßkraft ein, die u. U. berücksichtigt werden muß (Fig. 1 d).

Der beschriebene Vorgang gleicht dem Messen mit der Meßuhr, nur daß bei dieser ein materieller Zeiger umläuft, während hier der Lichtzeiger die Spiegelstellung und damit die Stellung der Meßpinole anzeigt. Die Meßuhrskala kann vom Zeiger vielmals umlaufen werden und ergibt einen großen Meßbereich. Die Spiegelablesung umfaßt nur einen sehr kleinen Winkelbereich, besitzt dafür aber eine hohe Auflösung bei geringer Meßunsicherheit.

7. Abbesches Prinzip. In den Meßanordnungen steht die Wahrung des Abbeschen Komparatorprinzips — Meßstrecke und Maßstab sind in einer Geraden hintereinander anzuordnen (vgl. Fig. 1 b und Fig. 64 auf S. 806) — an erster Stelle, um Einflüsse von Kippungen, Durchbiegungen usw. auf die Messung möglichst zu vermeiden. Bei allen Hebelanordnungen sind solche vorhanden, auch z. B. bei Lichthebeln über mechanisch bewegte Spiegel. Wo das Abbesche Prinzip nicht realisierbar ist, werden optische Ersatzschaltungen angestrebt (vgl. Ziff. 34). Wie Umkehrspannen, Temperatureinflüsse und andere systematische Fehler vermieden werden können, wird an jeweils passender Stelle besprochen.

8. Temperatureinfluß. Neben der Meßkraft spielt die Temperatur für die Längenmessung eine wichtige Rolle, da jeder Körper unter dem Einfluß von Temperaturänderungen sein Volumen ändert. Nach dem Ausgleich der Temperatur von Prüfling und Maßstab müssen die gemessenen Längen auf 20° C, die international festgelegte Meßtemperatur (vgl. DIN 102) umgerechnet werden. Sind Prüfling und Maßstab aus dem gleichen Material, so hebt sich der Einfluß der Temperatur heraus, sind sie es nicht, muß man ihn berücksichtigen. Dazu müssen die Temperaturausdehnungskoeffizienten beider Materialien bekannt sein.

Der erforderliche Temperaturausgleich zwischen Prüfling, Meßgerät und Maßstab stellt sich nicht ein, wenn sich in der Nähe des Meßgeräts Wärmequellen befinden. Als solche sind nicht nur die Heizquellen für den Raum, sondern auch die Beleuchtungslampen für das Meßgerät anzusprechen. Sie werden in einem Gehäuse mit Kühlrippen, möglichst weit außerhalb des Meßgeräts befestigt und von diesem zusätzlich durch wärmedämmendes Material getrennt, so daß sie eine möglichst geringe Erwärmung des Meßgeräts bewirken.

Eine zuweilen nicht beachtete Gefahrenquelle bedeutet der Beobachter selbst, wenn dessen Gesicht und Hände ungehemmt auf das Gerät einstrahlen. Man kann diesen Einfluß nach dem Stefan-Boltzmannschen Gesetz abschätzen. Praktisch ergeben sich Längenänderungen, die durchaus bemerkbar werden, wenn man mit dem Gesicht am Meßmikroskop eines Komparators über einem ungeschützt liegenden Maßstab arbeitet und die Tausendstel Millimeter über eine viertel oder halbe Stunde hin verfolgt. Um diesen Einfluß zu beseitigen, sollte man die Wärmestrahlung des Beobachters durch Aluminiumbleche abschirmen.

b) Über den Beobachter.

9. Kritikfähigkeit. Das Meßgerät ist unvollkommen, der Beobachter aber sogar unzuverlässig. Von ihm muß nicht nur Kritikfähigkeit über den Wert von Gerät und Verfahren verlangt werden: „Ablesen ist nicht Messen", er muß sich auch über seine eigene Unzulänglichkeit klar sein: „Messen heißt Fehler machen". Das trifft in erster Linie für die „vorgefaßte Meinung" zu, womit gemeint ist, daß gewöhnlich der einmal ermittelte Wert, die einmal für „richtig" gehaltene Einstellung vom gleichen Beobachter immer wieder angestrebt wird. Aus diesem Grund ist es wichtig, nicht von Null aus zu messen, also mit Maßstäben, die auf Null oder einen runden Ausgangswert nachgestellt werden können, sondern besser mit jeweils wechselnden Maßstabseinstellungen, so daß erst die Rechnung das Ergebnis liefert, nicht schon die Ablesung. Ein anderer Grund, nicht „von Null aus" zu messen, ist der, daß der zufällige Fehler dieser Nullstellung zum systematischen Fehler wird, wenn sie für mehrere Messungen beibehalten wird. In diesem Fall kann die Addition der Meßunsicherheiten von Anfangs- und Endwert nicht nach dem quadratischen Fehlerfortpflanzungsgesetz erfolgen, sondern muß linear geschehen, was eine größere zusammengesetzte Meßunsicherheit ergibt. Der Beobachter hat die systematischen Fehler so klein wie möglich zu halten. Das kann entweder durch Abstellen der Fehlereinflüsse selbst geschehen, z.B. Einhalten der international genormten Temperatur von 20° C als Bezugstemperatur bei der Längenmessung von Prüflingen aus Material verschieden großer Ausdehnungskoeffizienten, oder durch Rechnung und Fehlerberichtigung.

10. Beobachtungsfehler an optischen Geräten, Parallaxe. Bei optisch arbeitenden Geräten können systematische Fehler z.B. durch falsch eingestellte Beleuchtung und Focussierung, durch Verzeichnung des Objektivs, durch schlechte Anpassung des Meßgeräts an das Auge oder auch durch den Astigmatismus des Auges selbst entstehen.

Eine wichtige Rolle spielt die Zentrierung der Lichtquelle. Das Bild der Lichtquelle soll zentrisch in der Austrittspupille des Oculars zu erblicken sein, wenn man diese mit einer Lupe betrachtet. Besser noch entfernt man das Ocular und blickt durch den Tubus von hinten her auf das Objektiv. Bei Projektoren, die ohne Ocular arbeiten, blickt man durch ein Schwarzglas vom Schirm her auf das Objektiv oder bildet die Objektivöffnung und das darin liegende Bild der Lichtquelle mit einer Lupe auf einen Schirm ab.

Beugungserscheinungen an den Ocularstrichmarken sollen durch geeignete Wahl von Beleuchtungs- und Beobachtungsapertur, Farbränder durch ein geeignetes Filter (Grünfilter), behoben werden.

Durch seitliches Hin- und Herbewegen des Auges in der Austrittspupille des Instruments, die selbstverständlich in der Größe dem Auge angepaßt sein soll, überzeugt man sich davon, daß das zu messende Bild des Objektes in der gleichen Ebene liegt wie die Ocularstrichplatte, daß also keine Parallaxe vorhanden ist.

Unter Parallaxe wird verstanden, daß die Bilder des Objekts und der Strichplatte nicht genau in der gleichen Entfernung vom Auge liegen und sich infolgedessen bei seitlichen Bewegungen des Auges in der die Perspektive bestimmenden Pupillenebene des Instruments gegeneinander verschieben. Bleibt die Verschiebung unter dem physiologischen Grenzwinkel, so ist sie für das Auge nicht beobachtbar, es kann dann keinen Fehler in der Ablesung machen, auch wenn es in der Austrittspupille keine feste Lage einnimmt, sondern hin und her spielt. Der Grenzwert für die hierbei zulässige Tiefenabweichung in der Scharfstellung kann nach Fig. 4 berechnet werden. Als Grenzwinkel ψ muß je nach dem Aussehen der zur Deckung kommenden Strichmarken der in Fig. 5 angegebene Wert eingesetzt werden. Fig. 4 stellt die Lage der Austrittspupille AP (Radius r) hinter dem Ocular dar. Die Ocularebene befindet sich im vorderen Brennpunkt F des Oculars. Die in ihr auf der optischen Achse befindliche Einfangstrichmarke liegt — in den Raum der AP projiziert — im Unendlichen. Jeder von irgendeiner Stelle der Austrittspupille auf die Mitte der Einfangmarke zielende Strahl verläuft daher zwischen AP und Ocular parallel zur optischen Achse. Das gilt auch für den Rand der AP. Liegt dagegen ein um den Betrag $\pm z$ fehlerhaft eingestelltes Bild einer Marke O vor oder hinter der Einfangmarke, so befindet es sich — in den Raum der AP projiziert — im Abstand $\mp z'$ im Endlichen. Vom Rand der AP aus würden das im Unendlichen und das im Endlichen liegende Bild unter dem Winkel ψ gegeneinander verschoben gesehen. Aus dem physiologischen Grenzwinkel für ψ ergeben sich die Grenzen $\pm z$, die noch für eine „parallaxenfreie" Ablesung zulässig sind, d.h. die Unsicherheit der Tiefeneinstellung, die durch Bewegen des Auges in der AP nicht mehr zu erfassen ist. Es ist

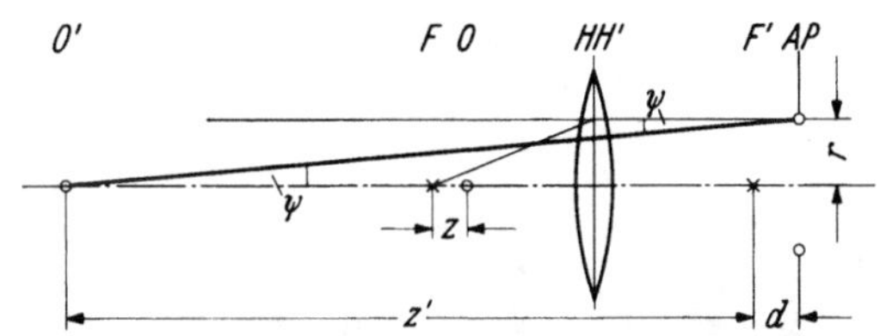

Fig. 4. Prüfung auf Parallaxe mit dem Auge in der Austrittspupille AP eines Instruments. Unter dem Winkel ψ beobachtbarer Abstand z zwischen Okularstrichkreuz in F und projiziertem Luftbild O. r Radius der AP, O' Bild im Augenraum, z, z' die Newtonschen Koordinaten zu FF', HH' Hauptpunkte des Oculars, d Abstand der AP von F'.

$$\psi = \frac{r}{z' \pm d}\,.$$

Weil ψ sehr klein ist, wird $|z'|$ sehr groß, wogegen $\pm d$ vernachlässigbar ist. Mit $z' = -f'^2/z$ (Newtonsche Abbildungsgleichung) wird $z = -\psi f'^2/r$. Die Prüfung auf Parallaxe stellt immer ein schärferes Einstellkriterium dar als das einfache Scharfstellen durch Schärfevergleich mit einer Strichplatte.

Im telezentrischen Strahlengang eines Meßmikroskops befindet sich die Eintritts-Pupille des Instruments objektseitig im Unendlichen. Austritts- und Augenpupille fallen hinter der bildseitigen Brennebene des Oculars zusammen. Dann wirkt sich die seitliche Bewegung des Auges in der Pupille des Instruments durch Ausblenden von Teilbündeln so aus, daß der strenge telezentrische Strahlengang zerstört wird, indem die achsenparallelen Hauptstrahlen des telezentrischen Strahlengangs im Objektraum durch geneigt zur Achse verlaufende, durch die Augenpupille mitbestimmte „Hauptstrahlen" ersetzt werden. Infolgedessen ist auch im telezentrischen Strahlengang die Prüfung auf Parallaxe erfolgreich. Man blickt „hinter die Dinge". Der oben besprochene Zusammenhang zwischen Restfehler und physiologischem Grenzwinkel ist der gleiche wie in Fig. 3. Der telezentrische Strahlengang schützt also nicht vor Fehlern durch Parallaxe. Diese Tatsache muß hervorgehoben werden, da der Zweck des telezentrischen Objektivs zu dieser Annahme verführen könnte (vgl. Ziff. 5).

Eine Vorbeugungsmaßnahme gegen unvermeidbare parallaktische Fehler ist das Fixieren der Blickrichtung durch einen Diopter, d.i. eine Blende oder ein

Aufsatzfernrohr. Beide haben die Aufgabe, den Durchmesser der Austrittspupille erheblich zu verkleinern.

Fehler durch Parallaxe können prinzipiell durch Auffangen des Bildes auf einer Mattscheibe vermieden werden. In diesem Fall erlangt der telezentrische Strahlengang seine grundlegend wichtige Bedeutung für das Einhalten des Abbildungsmaßstabs bei wechselndem Abstand des Objekts vom Objektiv (vgl. Ziff. 5).

Von der Verzeichnung des Objektivs überzeugt man sich leicht, indem man einen feingeteilten Maßstab (Objektmikrometer) auf den Maßstab in der Ocularebene projiziert.

Das Auge sollte — falls erforderlich — durch eine Brille korrigiert sein und mit dem Ocular in entspanntem Zustand arbeiten. Letzteres erreicht man dadurch, daß das virtuell zu beobachtende Bild möglichst nach Unendlich verlagert wird, indem man beim Einstellen das Ocular auf die Ocularebene hinbewegt, nicht von ihr fort, bis das Bild der Ocularstrichplatte scharf zu werden beginnt.

11. Stricheinfang, Doppelbildeinfang. Die parallaxenfreie Überlagerung eines Prüflingsbildes mit Strichmaßstäben in der Ocularebene eines Meßgeräts oder auf dem Projektionsschirm (Mischbild) ermöglicht, in ein oder zwei Koordinaten zu messen, indem entweder die Ocularstrichplatte oder das Bild des Prüflings mikrometrisch verschoben werden, bis sie sich decken. Am genauesten kann das Auge die gegenseitige Lage von Strichen und Strichbildern beurteilen, wenn ein Strich zwischen zwei anderen symmetrisch eingefangen ist (Fig. 5). Dabei wird die Fähigkeit des Auges ausgenutzt, eine Strecke von ungefähr 1—3 mm in deutlicher Sehweite sehr genau halbieren zu können. Füllt der einzufangende Strich diese Strecke jedoch nahezu aus, so bleibt auf beiden Seiten nur ein so schmaler Spalt bestehen, daß bei geringer Unsymmetrie die photometrischen Eigenschaften des Auges die Beurteilung beeinflussen. Dann hängt die Beurteilung der Symmetrie nicht mehr nur von der Geometrie der Anordnung, sondern auch von der Gleichmäßigkeit in der Ausleuchtung des Bildes ab. Einfangunsicherheiten von nur 3 bis 5 Bogensekunden werden von guten Beobachtern beim Symmetrieeinfang reproduzierbar erreicht. Angewendet auf Maßstabsstriche, die mit ca. 50fachem Meßmikroskop betrachtet werden, ergibt sich eine Einfangunsicherheit zwischen Doppelstrichen des Oculars bei bequemer Beobachtung von $s = \pm 0{,}2\ \mu m$ (s = Standardabweichung nach DIN 1319)[1]. Eine ähnliche hohe Genauigkeit mit einer Unsicherheit von nur 10—15 Bogensekunden wird beim Schnittbild erreicht, das der Stricheinstellung eines Nonius entspricht. Demnach lassen sich die Striche auf einem Rechenschieber mit bloßem Auge auf $s = \pm 20\ \mu m$ einstellen, mit einer n-fachen Lupe auf den n-ten Bruchteil hiervon. Das Auflösungsvermögen des Auges für zwei parallele Striche, die im Mischbild geboten werden, z. B. auch beim Doppelbildeinfang, beträgt nach dem Rayleigh-Kriterium für eine Wellenlänge von 0,55 µm und 2 mm Durchmesser der Augenpupille ungefähr 50 Bogensekunden. Die tatsächliche Auflösung im Mischbild hängt vom Bau der Netzhaut sowie der individuellen Leistungsfähigkeit des Auges ab sowie von der Beleuchtung,

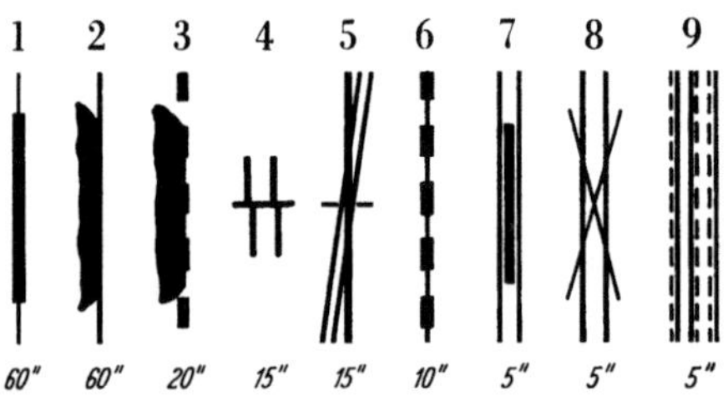

Fig. 5. Einfangkriterien. Einfangsunsicherheit s für das Auge in Bogensekunden, s Standardabweichung. *1* und *2* einfache Überdeckung, *3* unterbrochene Überdeckung, *4* Koinzidenz-, *5* Moiré-Prinzip, *6* einfacher Symmetrieeinfang, *7*, *8*, *9* Doppelstrich-Symmetrieeinfang, *9* für Autokollimation.

[1] Als Längeneinheiten dienen in diesem Artikel

$$1\ m = 10^3\ mm = 10^6\ \mu m = 10^9\ nm.$$

vom Kontrast und der Art der Objekte und liegt bei aufmerksamer, aber nicht angestrengter Beobachtung bei durchschnittlich 90 Bogensekunden [*1*], [*1a*], [*2*], [*3*].

12. Unvollständig ausgeleuchtete Pupille. Wenn die Beobachtungspupille dezentriert ausgeleuchtet ist, verlaufen die Hauptstrahlen des zur Beobachtung genutzten Strahlenbündels geneigt zur optischen Achse und das Bild wird beim Durchstellen des Objekts durch die Schärfenebene nicht allein unscharf, sondern wandert auch seitlich aus; es „schiebt". Bei mangelhafter Focussierung entstehen infolgedessen Meßfehler.

Der Effekt der seitlichen Bildverlagerung kann dazu benutzt werden, die Beleuchtung besser zu zentrieren, falls ihre Dejustierung Ursache des Schiebens ist. Aber auch die Verhältnisse am Objekt können die Ursache für die mangelhaft ausgeleuchtete Pupille sein; das ist z.B. zu beachten, wenn geritzte Maßstabstriche im Dunkelfeld beobachtet werden sollen, eine Beleuchtungsart, die für diesen Zweck ungeeignet ist. Weitere Fälle, in denen das Objekt in irgendeiner Form vignettierend in den optischen Beleuchtungs- und Abbildungsstrahlengang einbezogen werden muß, z.B. wenn bei der Profilprojektion im ursprünglich telezentrischen Strahlengang achsensenkrechte rechtwinklige Objektkanten abgebildet werden und die achsenparallele Fläche nicht genügend gerichtet reflektiert oder vom Objekt abgewendet liegt (Fig. 6). Ein anderes, typisches Beispiel ist der Strahlengang beim mikroskopischen Antasten von Endmaßflächen nach der Methode von FIZEAU mit auf den Meßflächen befestigten Marken (vgl. Fig. 35, S. 789), oder im Leitz-Perflektometer, in welchem die Marke durch das reelle Bild einer Strichmarke ersetzt ist (Fig. 36, S. 789). In beiden Fällen blendet der Prüfling eine Hälfte der Pupille ab [*4*].

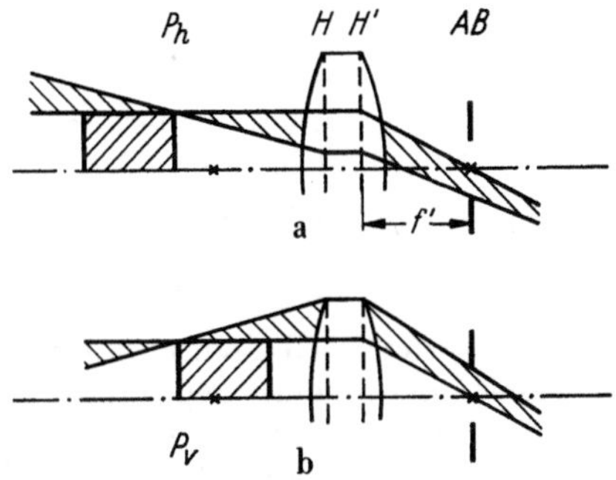

Fig. 6a u. b. a Gestörter telezentrischer Strahlengang durch schlecht reflektierenden rechtwinkligen Prüfling *P* bei Abbilden der hinteren Kante P_h. Nur eine Hälfte des telezentrischen Bündels tritt auf, das Bild „schiebt" seitlich beim Scharfstellen. *HH'* Hauptpunkte, *AB* Aperturblende, *f'* Brennweite. b Desgl. beim Abbilden der vorderen Kante P_v.

13. Tiefeneinstellung. Die Tiefeneinstellung eines optischen Systems wird gelegentlich zum Messen von Niveauunterschieden und axialen Distanzen benutzt. Schon mit dem Kriterium der Bildschärfe allein und ohne Anlehnung an eine Strichmarke im Ocular lassen sich hierbei geringe Meßunsicherheiten erreichen. Dreht man von Hand an einem Rändel ohne toten Gang oder Umkehrspanne durch die Schärfenebene hindurch, gesellt sich zur Beurteilung der Schärfe das Symmetrieempfinden der Hand für die Drehbewegung und ermöglicht, innerhalb der Schärfentiefe zu mitteln. Das führt zu Einstellungen, deren Unsicherheit nur einen Bruchteil der Unsicherheit

$$T = \pm\sqrt{T_s^2 + T_A^2}\ [\text{dptr}]$$

beträgt, wobei T_s die Schärfentiefe und T_A die Akkommodationstiefe bedeuten [*1*], [*5*]. Eine gewisse Analogie zum lateralen optischen Symmetrieeinfang drängt sich auf und bestätigt unseren Sinnen, hier dem Symmetrieempfinden der Hand, dort dem Symmetrieempfinden des Auges eine außerordentliche Empfindlichkeit. Die Einstellung zusätzlich auf Parallaxe zu prüfen — wozu eine Strichplatte in der Ocularebene und seitliche Bewegungen des Auges in der Austrittspupille des Instruments erforderlich sind, — bringt eine weitere Verbesserung der Einstellung mit sich, da der Beitrag der Akkomodationstiefe, der in die Tiefeneinstellung eingeht, hier wegfällt. Über die Grenzen dieser Methode vgl. Ziff. 10.

14. Einfang von Interferenzstreifen. In einem besonderen Fall wird die axiale Bewegung von Prüfling oder Mikroskop durch laterales Auswandern von Streifen deutlich, die den axialen Abstand kennzeichnen, nämlich bei der Verwendung von Interferenzen. Im Interferometer der Michelsonschen Bauart z.B. kann durch Auszählen der an einer Marke vorüberwandernden Anzahl von Streifen — auch der Bruchteile eines Streifenabstands — die axiale Verschiebung unmittelbar in Lichtwellenlängen ausgewertet werden. Ein Streifenabstand entspricht einem Niveauunterschied von einer halben Wellenlänge. Zwanzigstel Wellenlängenbeträge in axialer Richtung können durch Symmetrieeinfang noch erfaßt werden, tausendstel mit Halbschattenkompensatoren und Zehntausendstel mit photoelektrischen Methoden (s. auch Abschnitt IIIa).

III. Urnormal Lichtwellenlänge.

15. Meterdefinition. Jeder Längenmessung liegt ein Maßstab zugrunde. Mit ihm soll bestimmt werden, wie oft die Längeneinheit in der zu messenden Länge enthalten ist. Über die derzeitige Längeneinheit und ihr natürliches Längennormal sowie die Messung von Endmaßen und Strichmaßen mit Hilfe dieser Einheit soll im folgenden berichtet werden.

Die jahrhundertealten Bemühungen, die bekannten Längeneinheiten auf einen jederzeit und überall reproduzierbaren natürlichen Maßstab zu beziehen [*6*], gipfelten in der bekannten Bestimmung des Meters durch die Lichtwellenlänge durch MICHELSON im Jahre 1893. Die damals benutzte rote Cadmiumlinie stellte aber noch nicht das geeignetste Wellenlängennormal dar, wie die Erfahrung lehrte, denn die vorgeschriebenen Betriebsdaten können nur schwer eingehalten werden. So wurde im Anschluß an diese Messungen zwar eine spektroskopische Wellenlängenskala auf die rote Cadmiumlinie gegründet, aber daneben blieb die Funktion des Urmeters als maßgebend für die Längeneinheit in Kraft. Erst nachdem auf Grund intensiver spektroskopischer Untersuchungen (KÖSTERS [*7*], ENGELHARD [*6*], [*8*], [*9*], [*10*], u. a. [*11*]) eine zweifellos wesentlich besser geeignete Linie gefunden war, für die die Betriebsbedingungen in einer von ENGELHARD konstruierten Lampe nahezu ideal genannt werden können, legte die 11. Generalkonferenz für Maß und Gewicht im Jahre 1960 die Längeneinheit endgültig durch die Lichtwellenlänge fest. Damit wurde das 1889 durch die 1. Generalkonferenz als Längenstandard bestimmte Urmeter in Paris aufgegeben. Die neue Definition lautet: „Das Meter ist das 1650763,73fache der Wellenlänge der von Atomen des Nuklids ^{86}Kr beim Übergang vom Zustand $5d_5$ zum Zustand $2p_{10}$ ausgesandten, sich im Vakuum ausbreitenden Strahlung". Das Meter ist dadurch ein natürlicher Maßstab geworden, nämlich ein bestimmter Abschnitt auf einer Lichtwellenskala, der wesentlich genauer ist als das Urmeter. An der Länge des Meters als solchem ändert sich innerhalb seiner Unsicherheitsgrenzen durch die neue Festlegung nichts.

Die Lichtwellenskala ist atomaren Ursprungs, und nur in atomaren Bereichen wirksame Kräfte können sie beeinflussen [*6*], [*9*], [*12*], [*13*], [*14*], [*15*], [*16*]. Daher durfte nur ein Isotop, und zwar eines mit gerader Massen- und Kernladungszahl, gewählt werden. ^{86}Kr kann im Trennrohr [*16*] sehr rein (99,6 Vol.-%) gewonnen werden. Schließlich mußte auch das Vakuum für diesen Lichtwellenmaßstab vorgeschrieben werden, um ihn von dem Einfluß der Dichte der Luft und ihrer Beimengungen zu befreien.

Um die Wellenlängenskala der quasi „ungestörten" Atome zu erhalten, müssen die Betriebsbedingungen geeignet gewählt werden, was in einer Lampe besonderer Konstruktion [*6*], [*8*] durchführbar ist. Fig. 7 zeigt die von ENGELHARD entwickelte Kryptonlampe.

Die der neuen Längendefinition zugrundeliegende orangene Linie des Krypton-Isotops 86 [*8*], [*17*] besitzt bei der für die Kryptonlampe vorgesehenen niedrigen Betriebstemperatur am Tripelpunkt des Stickstoffs, 63° K, und dem dabei herrschenden niedrigen Krypton-Gasdruck von 0,03 Torr über ausgefrorenem Krypton die außerordentlich große Kohärenzlänge bis über 0,8 m. Diesem günstigen Umstand ist es zu verdanken, daß Endmaße von der Länge eines Meters in zwei Schritten meßbar sind. Um ein Endmaß von 500 mm Länge im Interferometer zu messen, wird das virtuelle Bild des Vergleichsspiegels auf 250 mm in die Mitte des Endmaßes abgebildet, so daß nach der Reflexion zwischen Meßstrahl und Vergleichsstrahl je 500 mm Gangunterschied bestehen. Wird der Vergleichsspiegel durch ein Endmaß von 500 mm Länge ersetzt, lassen sich Prüflinge von 1 m Länge erfassen (Fig. 8). Die Reproduzierbarkeit der Wellenlänge ist praktisch auf $\pm 3 \cdot 10^{-9}$ möglich; so gelingt es, die Unsicherheit bei der Darstellung des Meters von ca. 10^{-6} beim Urmeter auf $\pm 3 \cdot 10^{-9}$ zu senken [*11*]. Dies wird durch die genaue Einstellung des Gasdrucks und der Stromdichte in der Lampe erreicht. Die theoretische Grenze liegt unter 10^{-10} [*18*].

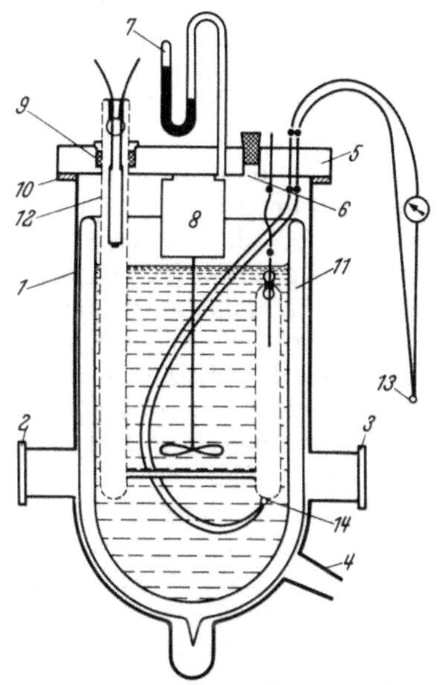

Fig. 7. Kr 86-Standardlampe. *1* Gehäuse des Kryostaten, *2*, *3* Fenster, *4* Pumpstutzen, *5* Deckel, *6* Einfüllöffnung, *7* Manometer, *8* Rührer mit Motor, *9*, *10* Ringdichtungen, *11* Thermosgefäß, *12* Kr 86-Lampe mit Glühkathode, *13*, *14* Thermoelemente.

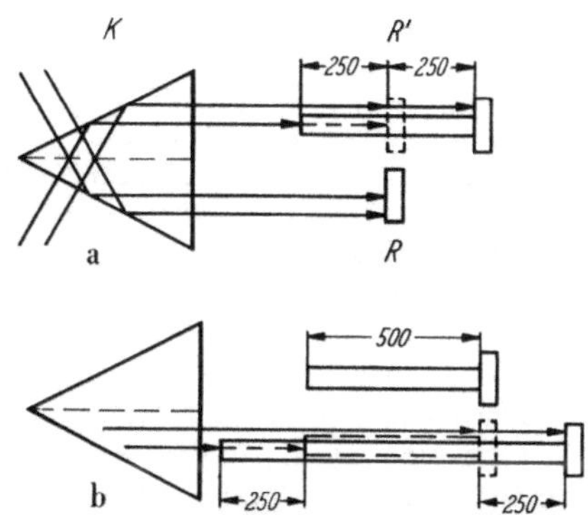

Fig. 8a u. b. Meter-Komparator. a 1 Schritt: Messung eines 500 mm langen Endmaßes gegen den virtuellen Vergleichsspiegel *R'* im Kösters-Interferometer. b 2. Schritt: Messung eines 1000 mm langen Endmaßes gegen das virtuelle 500 mm lange Endmaß. *K* Kösters-Prisma.

a) Anschluß von Endmaßen an die Lichtwellenlänge.

16. Endmaßdefinition. Es besteht die Möglichkeit, sowohl körperliche Maße als auch Strichmaße an die Lichtwellenlänge anzuschließen. In der Industrie werden als Bezugsnormale für die Längenmessung vorwiegend die bekannten Endmaße mit rechteckigem Querschnitt benutzt. Stahl- und Quarz-Endmaße sind in hoher Qualität ihrer Endflächen und Längen herstellbar. Ein Endmaßsatz besteht aus einer Auswahl von Längen, die sich zu bestimmten Abmessungen aneinandersprengen lassen. Beim Ansprengen können Schichten zwischen den Meßflächen und Verformungen der Grenzbereiche der aneinandergesprengten Endmaße die Länge der Endmaßkombination beeinflussen.

Die sog. Ansprengschicht kann bis zu 0,08 µm betragen, ihre Dicke kann bei hoch polierten Oberflächen aber auch bis auf Tausendstel µm herabsinken [*10*], [*19*], [*20*], [*21*], [*22*], [*23*]. Die bei wiederholtem Ansprengen auftretenden Längenunterschiede bewegen sich i. allg. unter 10 nm für eine Fläche. Daher können sie in der Praxis beim Ansprengen von Stahlendmaßen aneinander oder auch an Quarzflächen vernachlässigt werden ([*20*], S. 117).

Wenn Endmaße mit Hilfe von Lichtinterferenzen gemessen werden — gewöhnlich auf $\lambda/20$, mit Kompensationskeilen auf $\lambda/500$ [*7*], mit Halbschatten-Methoden oder durch Polarisatoren bis herunter auf $\lambda/2000$ [*13*], photoelektrisch sogar auf $\lambda/10000$ [*12*] — muß angestrebt werden, daß nicht nur der nicht ausreichend bekannte Phasensprung der Lichtwellenlänge λ bei der Reflexion an den Metall- bzw. Quarzoberflächen herausfällt, sondern auch die unbekannte Ansprengschicht mitgemessen wird. Da beide vom Material und von der Oberflächenbeschaffenheit abhängen [*10*], muß das Endmaß entsprechend der „Endmaßdefinition" DIN 851 und der Empfehlung der ISA (International Standards Association) gemessen werden: An das Endmaß ist eine Grundplatte der gleichen Materialart und Oberflächenbeschaffenheit anzusprengen, dann ist die Endmaßlänge als der Abstand dieser Fläche von der Oberfläche des Endmaßes definiert. Dabei wird der Phasensprung, ähnlich wie in der Mechanik die Umkehrspanne, durch Messen in derselben Richtung eliminiert, während die unbekannte Ansprengschicht in der Endmaßlänge enthalten ist. Korrekturen werden nur dann erforderlich, wenn von dieser Vorschrift abgewichen wird, indem Stahlendmaße anstatt an eine Stahlgrundfläche, an eine Grundfläche aus Quarz angesprengt werden (vgl. Ziff. 21 δ). Fig. 9 zeigt das Meßprinzip sowohl mit einer reellen als auch mit einer virtuellen Vergleichsfläche, wobei diese in dem jeweiligen Teil des Gesichtsfeldes entweder mit der Endmaßoberseite oder der angesprengten Grundplatte Interferenzstreifen erzeugen. Dabei werden zweckmäßig monochromatisches Licht und Keilinterferenzen verwendet. Der Abstand zweier Interferenzstreifen bedeutet eine Niveaudifferenz von der Höhe einer halben Wellenlänge. Fällt demnach auf die Endmaßlänge eine ganze Zahl von Interferenzstreifen, so koinzidieren in beiden Teilen des Gesichtsfeldes die Streifensysteme. Koinzidieren sie nicht, läßt sich aus ihrer gegenseitigen Verschiebung der Bruchteil ablesen, um den das Endmaß von einer ganzen Anzahl halber Wellenlängen verschieden ist. In den folgenden Abschnitten soll der Ausdruck „z Interferenzstreifen" stets als ein Längenmaß von $z\frac{\lambda}{2}$ aufgefaßt werden, wenn nicht ausdrücklich hiervon abgewichen wird, wie in Ziff. 24.

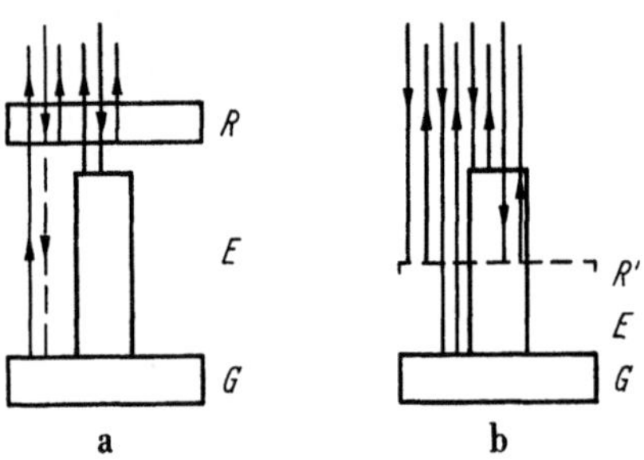

Fig. 9a u. b. Endmaß E nach Definition an Grundplatte G angesprengt. a Reeller Vergleichsspiegel R nach Fizeau, b virtueller Vergleichsspiegel R' nach Michelson.

17. Kösters' Meterkomparator. Kösters [*7*] konstruierte einen Komparator zur interferentiellen Messung von Endmaßen bis zu 1 m Länge. Dieser in der früheren Physikalisch-Technischen Reichsanstalt in Berlin begonnene und nach 1945 in der Physikalisch-Technischen Bundesanstalt in Braunschweig weiterentwickelte Interferenzkomparator ähnelt dem von Michelson und Morley [*24*], [*25*], [*19*]. Er enthält jedoch wesentliche Verbesserungen. Eine ausführliche Beschreibung des von der Firma C. Zeiss, Oberkochen, in ein verkaufsfertiges Modell verwandelten Interferenzkomparators nach Kösters findet sich bei W. Kinder [*26*].

Der schematische Aufbau und der Strahlengang des Zeiss-Meter-Komparators nach KÖSTERS ist in Fig. 10 dargestellt. Die von der Krypton-Lampe *1* ausgehende Strahlung beleuchtet eine Kollimatorblende *3*, welche über das Kollimatorobjektiv *4* nach Unendlich abgebildet wird. Die Strahlung tritt hinter dem Monochromator *5* durch ein Doppelfenster in den temperaturkonstanten Raum des Interferometertroges *15* ein und wird im Kösters-Prisma *6* in zwei kohärente Bündel zerlegt. Diese werden verschiedenen Zwecken dienstbar gemacht.

Die äußeren Anteile beider Bündel treffen auf ein an eine Grundplatte *8* angesprengtes Endmaß *7* einerseits und einen Referenzspiegel *9* andererseits, werden reflektiert und vereinigen sich wieder im Kösters-Prisma, so daß man durch

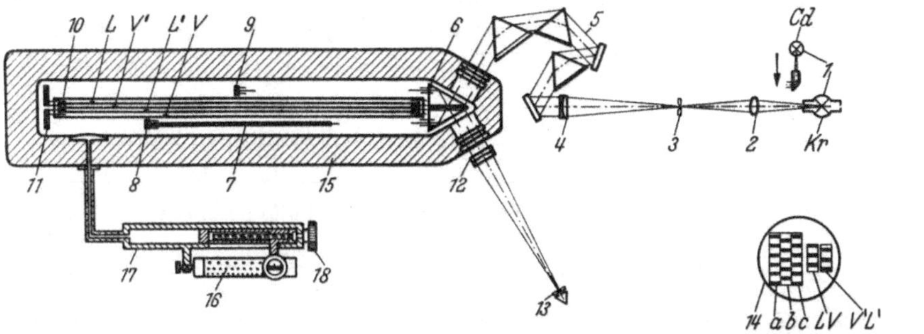

Fig. 10. Zeiss-Meterkomparator nach KÖSTERS. *Kr* Krypton-Isotoplampe, *Cd* Cadmiumlampe wahlweise, über *2* auf den Eintrittsspalt abgebildet. *4* Kollimator-, *12* Fernrohrobjektiv, *13* Austrittsspalt und Augenpupille. *5* Monochromator, *6* Kösters-Prisma, *7*, *8* 500 mm Endmaß mit angesprengter Grundplatte, *9* Referenzspiegel, *10* vier Kammern für Luft und Vakuum, *11* justierbare Spiegel, *14* Gesichtsfeld mit den Endmaßinterferenzen in *a*, *b*, *c* und den Luft-Vakuum-Interferenzen in *LV* und *V'L'*. Anstatt *V'L'* können auch *LL*, d.h. zwei Luftstrecken zum Vergleich benutzt werden, s. Text. *15* Gehäusetrog, *16* Skale, *17* Druckzylinder, *18* Rändel zum Verschieben des Kolbens.

eine Lupe *12*, in deren Brennebene die Augenpupille *13* liegt, die Oberflächen des Endmaßes und der Grundplatte, bedeckt von Interferenzstreifen gleicher Dicke, erblickt, *14a*, *b*, *c*.

Die inneren Anteile der Bündel durchsetzen auf ihrem Weg zu den Spiegeln *11* je zwei 1 m lange, durch Planparallelplatten abgeschlossene, Kammern *10*. Die Spiegel *11* und die Kammern werden durch das Kösters-Prisma virtuell aufeinander abgebildet, so daß sich zwei Interferenzstreifensysteme herstellen lassen aus Gangdifferenzen Vakuum (V) — Luft (L) und Luft (L') — Vakuum (V'). Beim Evakuieren der V-Kammern lassen sich die Interferenzstreifen zählen, die auf die jeweilige Trogluft entfallen. Sie wandern entgegengesetzt in beiden Gesichtsfeldern 14 LV und $V'L'$. Das Trog-Volumen ist nach Schließen des Deckels von der Außenluft vollkommen abgeschlossen. Um trotzdem gewisse Dichteänderungen der Luft im Trog vornehmen zu können, und dadurch die Interferenzstreifen fein zu verschieben, steht der Trog mit einem Druckzylinder *17* über eine elastische Membran in Verbindung. Der Druckkolben bewegt sich durch das Rändel *18* entlang einer Skala *16*. Der Sinn dieser Einrichtung wird in der folgenden Darstellung eines Meßvorgangs näher erläutert.

18. Verfahren der überschießenden Bruchteile. Um die Anzahl der Interferenzstreifen oder halben Wellenlängen, die die Länge des Endmaßes in Luft beträgt, nicht auszählen zu müssen — auf eine Strecke von 1 mm entfallen rund 4000 Streifen —, mißt man das Endmaß durch eine Unterschiedsmessung gegen einen Hilfswert. Als Hilfswert wird ein mit einem Feintaster so genau wie möglich gemessener Näherungswert benutzt, der im folgenden als „Sollmaß" bezeichnet wird. Dieses kann rechnerisch oder nach einer Tabelle in halben Wellenlängen oder Interferenzstreifen und ihren Bruchteilen fehlerfrei ausgedrückt werden. Die Aufgabe besteht nun darin, die Abweichung des Istmaßes vom Sollmaß zu messen. Für diesen Unterschied gilt:

$$U = (Z_i - Z_s)\frac{\lambda}{2} + (z_i - z_s)\frac{\lambda}{2},$$

worin Z_i und Z_s die ganzen Vielfachen und z_i und z_s die überschießenden Bruchteile der Interferenzstreifenzahl bedeuten, die auf die Ist- und die Sollänge des Endmaßes entfallen. Von diesen vier Größen sind Z_s und z_s durch Rechnung bekannt, z_i wird durch die Versetzung des Interferenzstreifensystems auf der Endmaßfläche gegenüber dem Streifensystem auf der Grundfläche gemessen, s. Fig. 10, Gesichtsfeld *14*, *b* gegen *a* und *c*, unbekannt bleibt Z_i. Damit enthält die Gleichung zwei Unbekannte und ist nicht ohne weiteres lösbar. In dem Umstand, daß der Unterschied U innerhalb der Meßunsicherheit liegen muß, mit der das Sollmaß

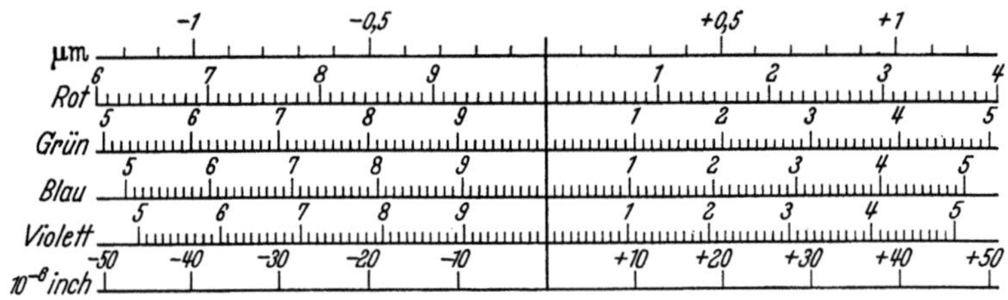

Fig. 11. Schema eines Wellenlängen-Rechenschiebers nach Kösters für vier verschiedene Cd-Linien in Einheiten von halben Wellenlängen. Darüber die zugehörige Längenteilung in μm, darunter in 10^{-6} Zoll. Mit Hilfe eines Läufers läßt sich die Stelle aufsuchen ,an der die Bruchteildifferenzen für alle vier Wellenlängen mit der Beobachtung übereinstimmen.

vorausbestimmt wurde, liegt die Möglichkeit zur Lösung. Man braucht nicht Z_i selbst, sondern nur die Differenz $(Z_i - Z_s)$. Diese ist aber sicher nur eine kleine ganze Zahl, wenn nicht sogar gleich Null. Um sie zu ermitteln, geht man nach der Methode von E. Abbe, die erstmalig bei G. Weidmann [*27*] beschrieben, später von C. Pulfrich [*28*], Benoît [*29*] und Kösters [*30*] ausführlich entwickelt und dargestellt wurde, so vor, daß man nacheinander Licht einer zweiten und einer dritten bekannten Wellenlänge benutzt. Mit diesen neuen Wellenlängen werden wieder nur die jeweiligen Differenzen der Istbruchteile minus Sollbruchteile bestimmt, die i. allg. von der Differenz der erstbenutzten Wellenlänge und voneinander abweichen werden. Zwischen den Bruchteildifferenzen dreier verschiedener Wellenlängen besteht aber ein eindeutiger Zusammenhang, solange es sich um einen Unterschied zwischen Istmaß und Sollmaß handelt, der kleiner ist als das kleinste gemeinsame Vielfache der halben Wellenlängen des benutzten Wellenlängentripels. Man kann nun versuchsweise für $Z_i - Z_s$ zunächst Null, dann der Reihe nach kleine ganze Zahlen einsetzen und systematisch untersuchen, für welche ganze Zahl und gemessene Bruchteildifferenz bei den verschiedenen Wellenlängen der gleiche Maßunterschied U herauskommt. Bequemer ist es, die zu kleinen Maßunterschieden gehörenden Bruchteildifferenzen für die benutzten Wellenlängen in Kurven- oder Tabellenform darzustellen. Kösters [*30*] hat die Bruchteildifferenzen für verschiedene Wellenlängen auf einem Rechenschieber übereinander und über einer μm-Skala aufgetragen, so daß leicht festzustellen ist, wo die jeweiligen Bruchteildifferenzen koinzidieren (Fig. 11). Diese Stelle liefert die Differenz Istmaß minus Sollmaß direkt in μm, woraus das Istmaß sofort erhältlich ist. In diesem Falle erübrigt sich die Kenntnis der Differenz $Z_i - Z_s$ vollständig.

19. Einfluß von Luftdruck und Temperatur. Die beschriebene Meßmethode kommt bei Endmaßlängen über 100 mm nicht ohne die genaue Kenntnis des

Brechungsindex der Luft aus; denn mit diesem ändert sich die optische Weglänge, mit anderen Worten die Wellenlänge des Lichtes. Der Brechungsindex hängt sowohl von der Dichte der Luft als auch von Beimengungen an Feuchtigkeit und Kohlensäure ab. Daher wären jedesmal genaue Daten über die Zusammensetzung der Luft notwendig, was die Messungen erschweren würde. Dieser Umstand hat dazu geführt, daß die neue Meterdefinition auf die Wellenlänge im Vakuum zurückgreift und nicht auf die bei den üblichen Normalbedingungen (trockene Luft von 15° C, 1 atm Druck und 0,03 Vol.-% CO_2-Gehalt), die bei der vorausgegangenen provisorischen Definition des Meters durch die Cadmiumwellenlänge noch benutzt wurden. Wollte man nun aber das Endmaß einfach im Vakuum messen, wären durch den Fortfall des Luftdruckes Maßänderungen zu befürchten. Man ist also darauf angewiesen, die optische Weglänge in Luft gegen die im Vakuum zusätzlich zu messen. Dazu benutzte FIZEAU [*31*] in seinem Interferometer eine Luft- und eine Vakuumstrecke nebeneinander. KÖSTERS verwandte ein ähnliches, aber für seinen Zweck höchst vervollkommnetes Verfahren, dem die in Ziff. 17 erwähnten Vakuumkammern und der Druckkolben dienen [*30*]. Es liefert die Länge des Endmaßes direkt in Vakuumwellenlängen, obwohl sich das Endmaß unter dem normalen Luftdruck von ca. 1 Atmosphäre befindet. Die Brechung der Luft fällt bei dem Verfahren heraus, sie braucht daher prinzipiell nicht bekannt zu sein.

Der Gedankengang des Köstersschen Verfahrens ist folgender. Das Endmaß wird in Luft gemessen. Dabei entfallen mehr Wellenlängen auf 1 m als im Vakuum. Um den Unterschied festzustellen, mißt man gleichzeitig die Luft gegen Vakuum und zieht die dabei gefundene Anzahl Wellenlängen von derjenigen der Endmaß-Luft-Messung ab.

Die Anzahl ganzer Interferenzstreifen ist sowohl in der Endmaßlänge gegen Luft als auch in der Gangdifferenz Luft-Vakuum unbekannt, ihre Bruchteile sind aber ablesbar, daher wird wieder das Bruchteilverfahren mit mehreren verschiedenen Wellenlängen angewendet. Um bequemer rechnen zu können, wird der Druckkolben so verschoben, daß die Streifenbruchteile in der Endmaßstrecke zu Null werden und nur die in der Luft-Vakuumstrecke abgelesen zu werden brauchen; die Interferenzen wandern sowohl in der Endmaß-Luftstrecke als auch in der Luft-Vakuumstrecke mit der Bewegung des Kolbens gleichmäßig, da sich die Dichteänderung der Luft auf beide Meßstrecken gleich stark auswirkt.

Auf ein Endmaß von 1 m Länge möge das Produkt Zn Interferenzstreifen (halbe Wellenlängen) in Luft (Brechungindex n) entfallen, vgl. Ziff. 18. Zn ist nach der erwähnten Verschiebung des Druckkolbens eine ganze Zahl. Von dieser ganzen Zahl Zn unterscheidet sich die gesuchte Zahl Z der Wellenlängen im Vakuum durch die Differenz $Z(n-1)$. Diese Differenz der optischen Weglängen zeigt sich im Gesichtsfeld, Fig. 10, *14 LV V'L'*, in der Interferenzstreifenversetzung Luft gegen Vakuum im Vergleich zu Luft gegen Luft oder empfindlicher im Vergleich zu einem gegenläufigen Streifensystem Vakuum gegen Luft, je nachdem ob nur eine von den vier Kammern evakuiert wird, oder zwei entgegengesetzte evakuiert werden. Im folgenden sei angenommen, es sei nur eine Kammer evakuiert. Wir benutzen von der Differenz Vakuum-Luft nur die Dezimalbruchteile. Deren dekadische Ergänzung zu 1 erscheint in der Anzahl Z der Vakuumwellenlängen nach der Gleichung $Zn - Z(n-1) = Z$ als Istbruchteile im Vakuum. Beim Arbeiten mit mehreren verschiedenen Wellenlängen läßt sich jeweils die Differenz dieser Istbruchteile minus der für das Vakuum berechneten Sollbruchteile für 1 m Länge bilden und danach läßt sich auf einem Vakuumwellenlängenschieber durch Aufsuchen der Koinzidenz die zugehörige Differenz Istmaß minus Sollmaß in μm feststellen. Damit ist die Messung der Endmaßlänge schon beendet.

Einige praktische Einzelheiten seien noch erwähnt: Um die Bruchteile der Interferenzstreifenzahl $Z(n-1)$ nicht im Interferenzbild, Fig. 10, *14 LV*, schätzen zu müssen, wird die Skala, entlang derer der Kolben im Druckzylinder verschiebbar ist, so kalibriert, daß ihre Skalenteile den von der Dichteänderung der Luft herrührenden Koincidenzen von Interferenzstreifen in Fig. 10, *14 LV* entsprechen. Wenn also die Koincidenzeinstellung der Streifen in *14a, b, c* hergestellt wird und zu einer ganzen Zahl gemacht worden ist, liest man die Bruchteile der Streifenversetzung bequem auf der Skala ab. Auf diese Weise lassen sich Bruchteile von zwanzigstel Wellenlängen noch messen. Beim Arbeiten mit drei verschiedenen Wellenlängen werden drei solcher auf die Wellenlängen abgestimmter Skalen benötigt. Sie sind gemeinsam aufgetragen und auf eine Füllung des Troges mit kohlensäurefreier, trockener Luft von 760 Torr bezogen, d. h., stimmt eine der Skalen mit ihrem zugehörigen Interferenzstreifensystem überein, so gilt das auch für die anderen beiden. Falls die zu messenden Längenmaße kürzer sind als die Vergleichsstrecke Luft-Vakuum von 1 m, so sind die abgelesenen Skalenteile mit der vorliegenden, in Metern ausgedrückten Länge, zu multiplizieren, bei der in Fig. 10 dargestellten Messung eines 0,5 m langen Endmaßes also mit 0,5.

Die gleichzeitige Messung mit Wellenlängen in Luft und im Vakuum ermöglicht, die Änderung der Wellenlänge in Abhängigkeit von der Dichte der Luft und ihrer Beimengungen für die Messung auszuschalten.

Die Abhängigkeit der Lichtwellenlänge vom Luftdruck in der freien Atmosphäre ergibt bei einer Druckänderung von 1 mm Hg auf 1 m Meßlänge eine Abweichung von rund einem Viertel Mikron. Der Druck müßte bei einer Genauigkeitsforderung von $\pm 0{,}01$ µm auf $^1/_{25}$ Torr bekannt sein, wenn diese Abweichungen nicht durch die geschilderte elegante Methode nach KÖSTERS zu beherrschen wären.

Die Länge des Endmaßes selbst hängt empfindlich von der Temperatur ab. Nach der DIN 861 sollen Endmaße aus Stahl einen Ausdehnungskoeffizienten von $11{,}5 \cdot 10^{-6}$ Grad^{-1} haben. Um ein Endmaß von 1 m Länge auf $\pm 0{,}01$ µm genau zu messen, muß die Temperatur auf $\pm 10^{-3}$ Grad bekannt sein. Die Temperaturmessung geschieht daher mit Hilfe von Widerstandsthermometern in Verbindung mit einem Kompensationsapparat. Als Bezugstemperatur gilt nach DIN 102 international 20° C oder 68° F.

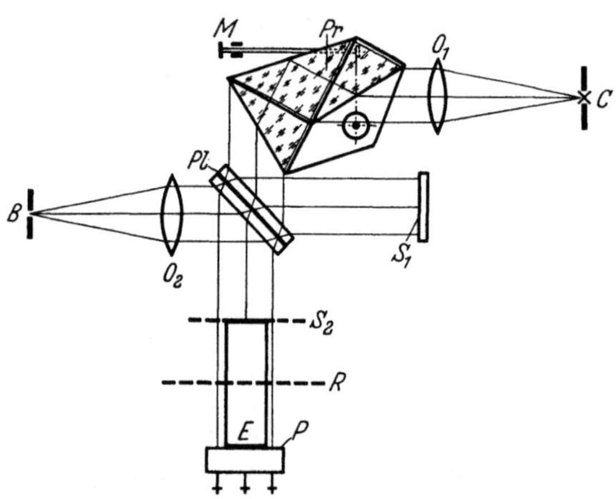

Fig. 12. Kösters-Interferenz-Komparator für unmittelbare Messungen von Endmaßen in Lichtwellenlängen. *C* Lichtquelle, O_1 Kolliomator, O_2 Lupe, *Pr* Abbe-Prisma, *M* Rändel, *Pl* Teilerfläche, S_1 Vergleichsspiegel, virtuell in *R*, *P* angesprengte Grundplatte, S_2 Oberfläche des Endmaßes *E*, *B* Augenpupille im Austrittsspalt.

20. Älterer Kösters-Interferenzkomparator. Ein einfacheres Modell eines Endmaßkomparators war von KÖSTERS zusammen mit der Fa. Zeiss hauptsächlich für Unterschiedsmessungen an Endmaßen nach dem Prinzip des Michelson Interferometers konstruiert worden [*30*], Fig. 12. Es kann einerseits mit weißem Licht zu Unterschiedsmessungen von zwei nebeneinander angesprengten Endmaßen dienen, wobei die Versetzung des sog. Nullstreifens in Streifenbreiten einer monochromatischen Wellenlänge auszumessen ist, es kann andererseits aber auch für den unmittelbaren Anschluß von Endmaßen bis zu 100 mm Länge an die Lichtwellenlänge sorgen, wenn das oben beschriebene Verfahren mit mehreren Wellenlängen benutzt wird.

21. Komparator nach Dowell mit Winkelspiegel. Eine Zeiss-Neukonstruktion [*32*], [*33*], [*34*], verfährt nach dem in Fig. 13 dargestellten ursprünglich von Dowell [*35*] angegebenen Aufbau, bei dem der an der Endmaßfläche auftretende Phasensprung der Lichtwelle in der Strahlenführung kompensiert wird, ohne daß an das Endmaß eine Grundfläche angesprengt zu werden braucht. Das erleichtert die Handhabung. Die Ansprengschicht wird bei solchen Unterschiedsmessungen automatisch berücksichtigt, weil sie bereits in der nach der Endmaßdefinition bestimmten Länge des Vergleichsnormals enthalten ist, vgl. Ziff. 21 α. Die Längenänderung durch das Ansprengen selbst wurde in Abschnitt IIIa (S. 767) erwähnt.

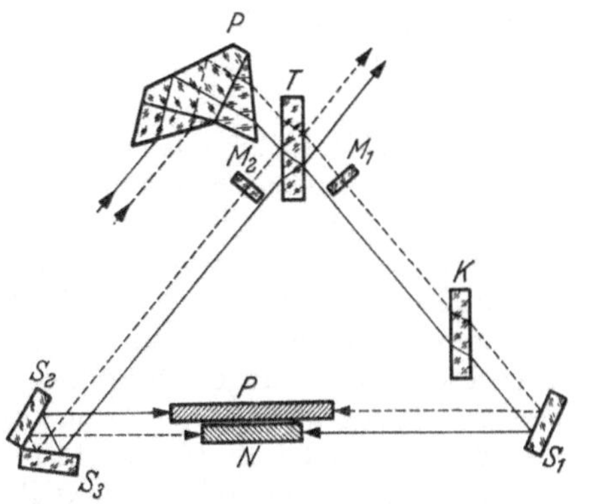

Fig. 13. Endmaßkomparator. *P* Abbe-Prisma, *T* Teilerfläche, *K* Kompensationsplatte, S_1, S_2, S_3 Spiegel, M_1, M_2 Plattenmikrometer, *P*, *N* zu vergleichende Endmaße.

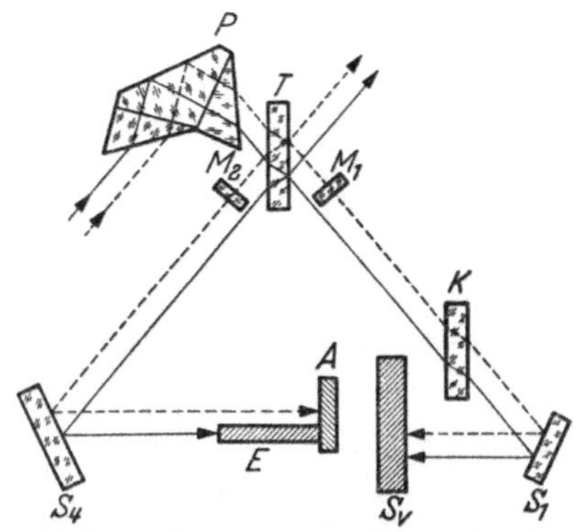

Fig. 14. Endmaßkomparator nach Fig. 13, aber mit angesprengtem Endmaß *EA* und Vergleichsspiegel S_v. Anstelle von Winkelspiegel $S_2 S_3$ tritt S_4.

Es ist aber auch möglich, mit angesprengtem Endmaß zu messen (Fig. 14). Dieser Aufbau, der ebenso Unterschieds- wie unmittelbaren Messungen — ähnlich der älteren Konstruktion — dient, geht aus den Unterschriften zu den Bildern hervor.

α) *Unterschiedsmessung.* Bei dem Verfahren in Fig. 13 handelt es sich um unmittelbare Unterschiedsmessungen zwischen einem Prüfling *P* und einem Normal *N*. Die zu vergleichenden Endmaße liegen nebeneinander und sind nicht angesprengt. Die Einzelheiten werden deutlicher erkennbar in Fig. 15a.

E_1 und E_4 bilden ein Vergleichsspiegelpaar, dessen Flächen zwei interferenzfähige Strahlenbündel reflektieren, ebenso E_3 und E_2, nebeneinander, überzogen von Keilinterferenzstreifen, weil das eine Endmaß *N* ein wenig gekippt ist. Bedeutet die strichpunktierte Linie die Symmetrieebene der Anordnung, dann kommen virtuell für E_3 und E_4 die Lagen E_3' und E_4' zustande. Zwischen dem Streifensystem der Flächen E_4' E_1 einerseits und dem der Flächen E_2 E_3' andererseits lassen sich die Bruchteildifferenzen ablesen. Unbekannt ist auch hier die Anzahl der ganzen Vielfachen halber Wellenlängen des Unterschiedes Prüfling-Normal. Sie wird nicht gebraucht, wenn — wie in Ziff. 18 beschrieben — die Sollänge des Prüflings als Hilfswert benutzt und mit mehreren Wellenlängen lediglich die Abweichung des Istunterschiedes vom Sollunterschied festgestellt wird.

In dieser neueren Ausführung des Zeiss-Endmaßkomparators, in der die Endmaße nicht mehr definitionsgemäß angesprengt werden, wird der Phasensprung eliminiert, indem nicht nur das zu prüfende Endmaß, sondern auch der „Vergleichsspiegel" nach entgegengesetzten Richtungen reflektieren. Daß der Vergleichsspiegel unter den hier gezeigten Umständen (Fig. 15) die Stärke eines Endmaßes hat, führt zur Unterschiedsmessung zwischen Prüfling P und Normal N, die den großen Vorteil hat, versetzungsunempfindlich zu sein. In den Unterschied geht die Ansprengschicht mit ein, da sie in der Istlänge des Normals, das nach der Endmaßdefinition mit Ansprengfläche gemessen wurde, schon enthalten ist.

β) *Unmittelbare Messung.* Durch ein besonderes Vergleichsstück läßt sich das Endmaß N auch zu Null machen. Man denke sich zwei durchbrochene Quarzglasstücke in Fig. 15b mit ihrer Planfläche so aneinandergesprengt, daß die Bohrung jedes Stückes von der Stirnfläche des anderen abgeschlossen wird. Falls durch das Ansprengen eine Ansprengschicht entsteht [*10*], muß sie in der Messung enthalten sein. Nehmen wir an, sie bedeutet eine Verkürzung des Endmaßes, so wird hier die Planfläche des angesprengten Stückes ebenfalls im verkürzenden Sinne verlagert, d. h. $E_3' E_4'$ bilden einen Vergleichsspiegel positiver Dicke.

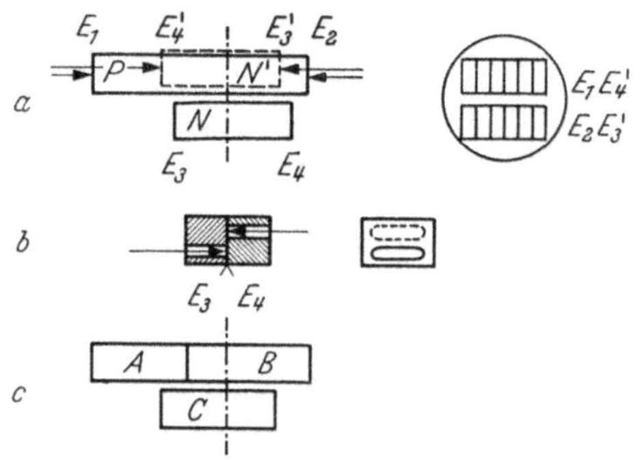

Fig. 15a—c. a Unterschiedsmessung von Prüfling P und Normal N im Endmaßkomparator nach Fig. 13. $E_1 E_2$, $E_3 E_4$ sind die spiegelnden Endflächen von P und N. N' virtuelle Lage von N, $E_3' E_4'$ virtuelle Endflächen. Im Gesichtsfeld oben Interferenzstreifen zwischen E_1 und E_4', unten zwischen E_2 und E_3'. b Nullendmaß, aus zwei durchbrochenen aneinandergesprengten Stücken mit den Spiegelflächen E_3 und E_4 bestehend. Daneben Vorderansicht. c Dreiermessung zur unmittelbaren Messung eines Endmaßes gegen die Lichtwellenlänge. A und B aneinandergesprengte Endmaße, C Vergleichsendmaß.

γ) *Unmittelbare „zusammengesetzte" Messung.* In einer Dreieranordnung nach Fig. 15c läßt sich durch Kombination mehrerer Differenz- und Summenmessungen die Länge von Endmaßen auch unmittelbar aus „zusammengesetzten Messungen" bestimmen. Voraussetzung ist, daß die Sollmaße der drei Endmaße als Hilfswerte bekannt sind, um die Sollbruchteile für drei Wellenlängen berechnen zu können (s. Ziff. 18). Gemessen werden die Istbruchteile und aus den Differenzen „Ist- minus Sollbruchteile" ergibt sich die Differenz in Streifenbruchteilen, die auf dem Koinzidenzschieber in μm umgesetzt wird. Nach Fig. 15c sprengt man jeweils zwei Endmaße aneinander und mißt ihre Summe einschließlich der Ansprengschicht $\varDelta$ im Unterschied zu einem dritten. Zusätzlich mißt man sie einzeln gegeneinander. Dann ergeben sich die Abweichungen vom Sollmaß nach Fig. 15c:

$$\left.\begin{aligned} a+b+\varDelta_{ab}-c&=f_1\\ b\qquad\quad -c&=f_2\end{aligned}\right\}\ a+\varDelta_{ab}=f_1-f_2;$$

$$\left.\begin{aligned} b+c+\varDelta_{bc}-a&=f_3\\ c\qquad\quad -a&=f_4\end{aligned}\right\}\ b+\varDelta_{bc}=f_3-f_4;$$

$$\left.\begin{aligned} c+a+\varDelta_{ca}-b&=f_5\\ a\qquad\quad -b&=f_6\end{aligned}\right\}\ c+\varDelta_{ca}=f_5-f_6.$$

Die Abweichungen vom Sollwert enthalten also die Ansprengschicht wie nach der Endmaßdefinition gefordert. Wird

$$\varDelta_{ab}=\varDelta_{bc}=\varDelta_{ac}=\varDelta$$

für gleiches Material und gleiche Oberflächenrauheit gesetzt, so läßt sich durch geeignete Kombination der 6 Gleichungen die Unsicherheit des Meßergebnisses noch erheblich herabsetzen; man erhält nämlich:

$$a + \Delta = \tfrac{1}{3}[(f_1 + f_3 + f_5 + (f_6 - f_4)];$$
$$b + \Delta = \tfrac{1}{3}[(f_1 + f_3 + f_5 + (f_2 - f_6)];$$
$$c + \Delta = \tfrac{1}{3}[(f_1 + f_3 + f_5 + (f_4 - f_2)].$$

Hierauf ist das quadratische Fehlerfortpflanzungsgesetz anwendbar, und die mittlere quadratische Abweichung geht nur mit dem Faktor $\frac{1}{3}$ ein.

δ) *Quarzendmaße und Quarznormal.* Quarzendmaße können auf dieselbe Weise unmittelbar gegeneinander gemessen werden.

Soll ein einzelnes Quarzendmaß bei interferentiellen Messungen ohne Ansprengen als Normal für Stahlendmaße dienen und sollen deren Längen auch gegen dieses Quarznormal definitionsgemäß bestimmt werden, so müssen die Phasensprungdifferenzen Quarz-Stahl und die Ansprengschicht Stahl gegen Stahl berücksichtigt werden. Das geschieht am elegantesten, indem man das Quarzendmaß so in der beschriebenen Anordnung gegen zwei Stahlendmaße mißt, daß der für seine Länge gefundene Meßwert die entsprechenden Korrekturwerte gleich enthält.

Wie erwähnt, wird das Quarzendmaß nicht angesprengt. Daher reduzieren sich die sechs Gleichungen auf vier. Aus der Kombination dreier Meßwerte läßt sich die Länge des Quarzendmaßes, behaftet mit der Phasensprungdifferenz Quarz-Stahl und der Ansprengschicht Stahl gegen Stahl, ermitteln. Ist $q + 2\varphi$ die Abweichung des Quarzendmaßes Q vom Sollmaß einschließlich der Phasensprungdifferenz φ Stahl-Quarz, und sind a und b die Abweichungen der beiden Stahlendmaße A und B von ihren Sollmaßen, ist ferner Δ_{st} die Stärke der Ansprengschicht Stahl gegen Stahl, so lassen sich für die Unterschiedsmessungen folgende Gleichungen aufstellen:

$$a + b + \Delta_{St} - (q + 2\varphi) = f_1;$$
$$b \quad - (q + 2\varphi) = f_2;$$
$$a \quad - (q + 2\varphi) = f_3.$$

Aus

$$f_1 - f_2 - f_3 = q + 2\varphi + \Delta_{St}$$

ergibt sich die Abweichung des Quarzendmaßes von der Sollänge einschließlich Phasensprüngen Stahl-Quarz und Ansprengschicht Stahl-Stahl.

Werden zusätzlich die Gleichungen

$$f_1 - f_2 = a + \Delta_{St}$$

und

$$f_1 - f_3 = b + \Delta_{St}$$

gebildet, so zeigt sich, daß auch die beiden Strahlendmaße bei dieser Operation definitionsgemäß mit ihren Ansprengkonstanten herauskommen.

Während die Phasensprungdifferenz φ zwischen Stahl und Quarz hier an zwei Flächenpaaren auftritt und im Ergebnis implizit enthalten ist, geht sie in die Unmittelbarmessung eines Strahlendmaßes mit angesprengter Quarzgrundfläche A nach Fig. 14 nur einmal ein und muß explizit bekannt sein. Man erhält sie — zusammen mit den Unterschieden der Ansprengschichten — durch Unterschiedsmessungen mit nacheinander angesprengter Stahl- und Quarzgrundfläche. Phasensprung und Ansprengschicht veranlassen je nach Material und Rauheit der Oberflächen eine Korrektur von bis zu $+0{,}04\,\mu$m für das Strahlendmaß.

Darin sind $+0{,}018\,\mu m$ für den Unterschied im Phasensprung zwischen Stahl- und Quarzplatte enthalten [*10*], während der Rest auf den Unterschied im Ansprengzustand zwischen Stahl und Quarzplatte zurückzuführen ist. Der Ansprengzustand kann durch andersartige hydrodynamische Zwischenschichten und damit verbunden durch unterschiedliche Verformung des Oberflächengebirges gegenüber dem Ansprengen an Stahl so verändert sein, daß die Eindringtiefe des Lichts in dieses Oberflächengebiet eine andere wird. Das Ansprengen an Quarzgrundflächen hat den Vorteil, daß die Ansprengfläche selbst auf Vorhandensein von Interferenzstreifen beobachtet werden kann.

22. Dowell-Komparator mit Einfachspiegel. Von Dowell [*35*], [*36*], [*37*], [*38*], wurde auch das Interferometer angegeben, das Fig. 16 zeigt. Es gleicht dem in Fig. 13 bis auf die Verwendung eines einfachen Spiegels anstelle des Winkelspiegels.

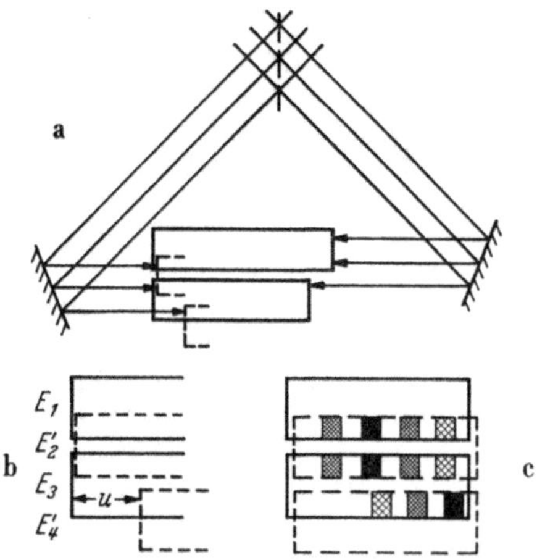

Fig. 16a—c. a Endmaß-Interferometer nach Dowell ohne Winkel-Spiegel, vgl. Fig. 13 und 15a. b E_1 und E_3 reelle Endmaßflächen gegen virtuelle Bezugsfläche E_2' mit Weißlichtinterferenzen ausgerichtet, vgl. Gesichtsfeld c. Unterschied U der Endmaße an Versetzung der Weißlichtstreifen oder mit mehreren Wellenlängen meßbar.

Dadurch bilden sich im Gesichtsfeld Vor- und Rückfläche jedes Endmaßes auf sich selbst ab. Diese Anordnung wäre zu einer Unterschiedsmessung des Endmaßes 1 gegen 2 nicht zu brauchen. Erst ein kleiner Kunstgriff schafft die erforderliche Beziehung der Endmaßflächen zueinander. Schieb man nämlich einen der Spiegel parallel zu sich aus der Symmetrie der Anordnung heraus, so wird erreicht, daß sich im Gesichtsfeld die Bilder der Rückflächen der Endmaße gegen die der Vorderflächen verschieben und diese z. T. überdecken. Dadurch erhält jede Fläche eine Bezugsfläche (Fig. 16b). Liegen Fläche E_1 und E_3 auf gleicher Höhe, so überzieht sie ein und derselbe Weißlichtinterferenzstreifen im Spiegelbild der Fläche E_2' (Fig. 16c). Derselbe Weißlichtstreifen ist dagegen im Spiegelbild der Fläche E_4' um soviel versetzt, wie der Unterschied zwischen den beiden Endmaßen beträgt. Mit mehreren Wellenlängen lassen sich größere Unterschiede nach dem Bruchteilverfahren messen.

23. Parallelität und Ebenheit von Endmaßen und Maßstabsoberflächen. Neben der unmittelbar oder in Unterschiedsmessungen bestimmten Länge von Endmaßen interessiert, ob ihre Endflächen eben und parallel zueinander sind. Obwohl diese beiden Eigenschaften sich auch schon bei den beschriebenen interferentiellen Methoden offenbaren, sind zu ihrer Prüfung spezielle Interferometer entworfen worden. Mehrere Beispiele gibt Dühmke [*39*]. Fig. 17 zeigt die erste Anordnung. In einem Interferometer mit Kösters-Prisma steht im einen Strahlengang das auf eine Grundplatte angesprengte Endmaß, dessen Parallelität gemessen werden soll. Auf seiner oberen Meßfläche ist eine gut ebene, teilweise verspiegelte Glasplatte angesprengt, die die Endmaßfläche überragt und in den Vergleichsstrahlengang hinüberreicht. So vergleicht man Richtung und Streifenabstand zweier Interferenzstreifensysteme und damit die Winkellage der Grundplatte im Vergleich zur Oberseite des Endmaßes. Möglich wäre auch, das Endmaß selbst in den Vergleichsstrahlengang hinüberreichen zu lassen. Man erspart damit das Ansprengen der Glasplatte. Dies Verfahren wird angewendet, wenn man auf eine

Bezugsfläche mit den Eigenschaften eines Ebenheitsnormals verzichten kann — weil man die Ebenheit bereits mit einem Probeglas gemessen hat — und nur noch die Parallelität der Endflächen der Endmaße interessiert [*39a*].

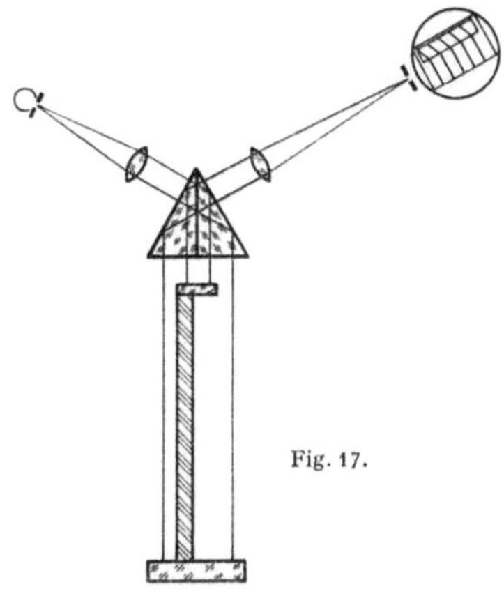

Fig. 17.

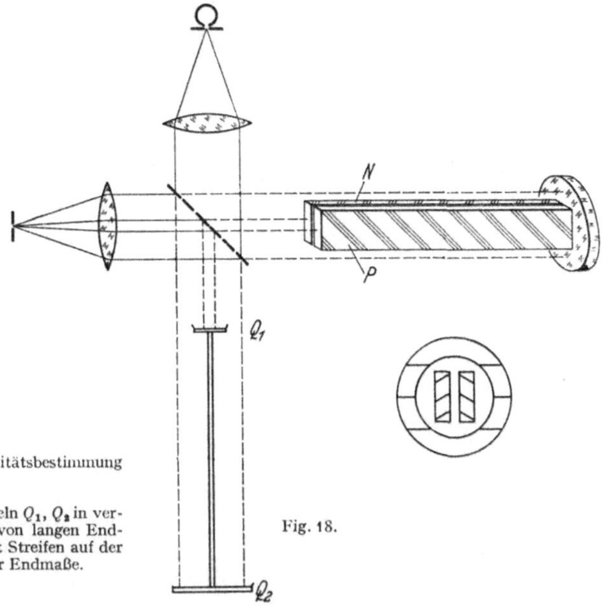

Fig. 18.

Fig. 17. Interferenzeinrichtung für Parallelitätsbestimmung an langen Endmaßen.

Fig. 18. Interferometer mit Quecksilberspiegeln Q_1, Q_2 in verschiedenen Höhen zur Parallelitätsprüfung von langen Endmaßen *N* und *P*. Im Kreis Gesichtsfeld mit Streifen auf der Grundfläche und den Vorderflächen der Endmaße.

Ein Flüssigkeitsspiegel ergibt in verschiedenen Höhen zueinander parallele Spiegelebenen. In einer zweiten Anordnung nutzt Dühmke [*39*] ihn zur Parallelitätsprüfung von langen Endmaßen aus. Er bringt einen Quecksilberspiegel in den vertikal stehenden Arm eines Michelson-Interferometers (Fig. 18) und überlagert ihn als Referenzspiegel einmal den Vorderflächen der Endmaße, dann den Grundflächen, an die die Endmaße definitionsgemäß angesprengt sind. Dabei dient der Quecksilberspiegel gleichzeitig als Ebenheitsnormal (vgl. Ziff. 40).

Das Endmaß kann im Interferometer nach Vogl [*40*] auch frei zwischen zwei teildurchlässige Interferometerplatten gelegt werden, um Parallelität und Ebenheit seiner Flächen zu prüfen. Fig. 19 zeigt die Anordnung und Fig. 19b die vier Bilder der Meßflächen zweier Endmaße im Gesichtsfeld, von denen 2 und 3 nach Winkellage und Abstand der Interferenzstreifen zusammen die Parallelität der Innenflächen der Interferometerplatten gegen ein Normalendmaß (Vor- und Rückseite) und 1 und 4 zusammen ebenfalls nach Winkellage und Abstand die Parallelität der Flächen des Prüflings ergeben. Die Prüfung des Normals kann durch Umschlagmessung erfolgen (Fig. 19c). Bemerkenswert ist, daß die Endmaßflächen nicht gegeneinander, sondern gegen Normalflächen *A* und *B* verglichen werden.

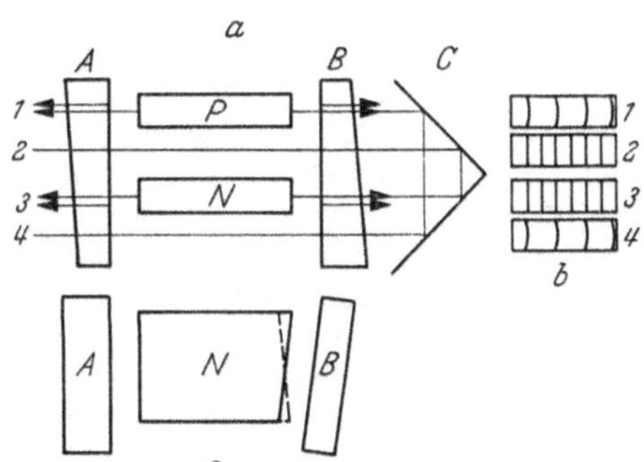

Fig. 19a—c. a Doppel-Fizeau-Interferometer zur Parallelitäts- und Ebenheitsprüfung von Endmaßen nach Vogl. *A* und *B* sind keilige Planplatten, *N* Vergleichsnormal, *P* Prüfling, *C* Winkelspiegel. b Gesichtsfeld mit gleichen Interferenzstreifenabständen in Feld *1* und *4* für einen planparallelen Prüfling. Feld *2* und *3* stimmen auch überein, weil *A* parallel *B* für ein planparalleles Normal *N*. c Prüfung von *N* auf Parallelität geschieht durch Umschlagmessung.

Zur Ebenheitsprüfung von Maßstabsoberflächen und ähnlicher sich in die Länge erstreckender Führungsflächen sind Interferometer [41], [42], [43] angegeben worden, von denen das von Saunders und Gross beschriebene in Fig. 20 dargestellt ist. Die erzielbare Empfindlichkeit, d.h. der Abstand der Höhenlinien, ist vom Einfallswinkel α der Strahlen auf der Prüflingsoberfläche und der damit verbundenen Änderung der optischen Weglänge abhängig. Einer Streifenauslenkung um einen Streifenabstand entspricht eine Höhenabweichung von $e=\frac{1}{4}\lambda/\cos\alpha$; mit großen Einfallswinkeln sind daher noch große Höhenabweichungen meßbar. Kühne erreichte bei einem Prüfling von 0,6 m Länge einen Streifenabstand, der 1,3 μm entsprach, auf 6 m Länge entsprach er 13 μm.

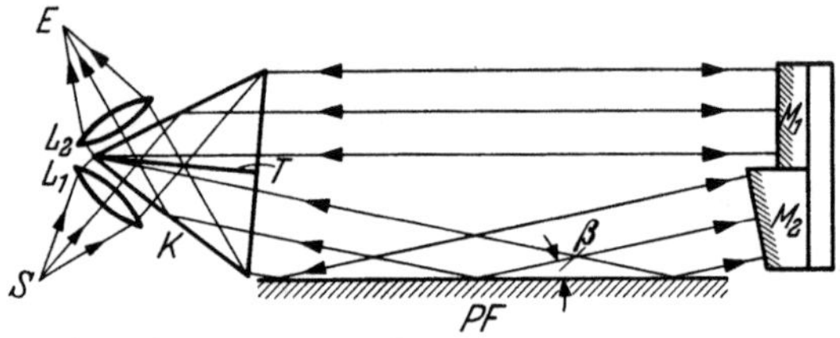

Fig. 20. Interferenzflächenprüfer nach Saunders und Gross. S Spalt, L_1L_2 Objektive, E Augenpupille, M_1M_2 Planspiegel, K Kösters-Prisma, T Teilerfläche, PF Prüflingsoberfläche, $\beta=90°-\alpha$ Glanzwinkel, α Einfallswinkel.

24. Endmaße von der Länge der doppelten und vielfachen Kohärenzlänge. Die Interferometer der Fizeau- und Michelson-Ausführung und ihrer Varianten haben den Nachteil, daß der maximal nutzbare Gangunterschied zwischen den interferierenden Strahlen, der durch die Kohärenzlänge der verwendeten monochromatischen Wellenlänge begrenzt ist, auf Hin- und Rückweg zwischen den Referenzflächen aufgeteilt werden muß. Das hat für unmittelbare Messungen zur Folge, daß bei optimaler Anordnung des Vergleichsspiegels in Endmaßmitte die (halbe) Länge des zu messenden Endmaßes auch höchstens gleich der (halben) Kohärenzlänge werden kann.

Um den Meßbereich zu erweitern, d.h. um größere Endmaße noch in einem Schritt messen zu können, mußte ein anderes Meßprinzip gefunden werden. Hunzinger [44] hat 1957 ein Interferometer angegeben, bei dem die meßbare Endmaßlänge doppelt so groß wie die Kohärenzlänge werden darf, 1959 haben Hariharan und Sen [45] ein etwas abgewandeltes Prinzip zu dem gleichen Zweck benutzt. Durch diese Methoden kann der Meßbereich bei ein- und derselben Kohärenzlänge praktisch verdoppelt werden, so daß beispielsweise die Messung des Meters in einem einzigen Schritt auch visuell ermöglicht wird. Mit einem Photomultiplier und der orangenen Linie ^{86}Kr gelang dies Terrien mit einer normalen Michelson-Anordnung, vgl. [39].

Der Gedanke der neuen Methoden ist folgender: Die Gangdifferenz zwischen den interferierenden Strahlen darf nicht für Hin- und Rückweg zwischen den Referenzspiegeln verbraucht werden, sondern nur für einen einzigen Weg in ein und derselben Richtung. Würden beispielsweise von einer Endfläche und vom mittig gelegenen Vergleichsspiegel im gleichen Augenblick interferenzfähige Lichtwellen ausgehen, so hätten sie eine Gangdifferenz von einer halben Endmaßlänge. Das würde bedeuten, daß die halbe Endmaßlänge gleich der Kohärenzlänge und damit die ganze Endmaßlänge, d.h. der Meßbereich gleich der doppelten Kohärenzlänge wird, womit das Ziel erreicht wäre. Mit dem von Sagnac [46] für einen anderen Zweck angegebenen Typ des Interferometers mit umlaufendem Strahlengang lassen sich die geforderten Verhältnisse verwirklichen. Fig. 21 und 22 zeigen die beiden erwähnten Ausführungen. Die Arbeitsweise des Interferometers möge an Hand von Fig. 22 gegeben werden.

Betrachten wir zunächst den leeren äußeren „Kreis". Der Gangunterschied zweier zusammengehöriger in entgegengesetzter Richtung umlaufender ebener Wellenzüge ist bei ihrem Wiedereintreffen an der Teilerfläche gleich Null. Da sie gleiche Wege und gleiche Ablenkungen durchlaufen haben, bilden sie eine gemeinsame Wellenfront. Daran würde sich auch nichts ändern, wenn ein Spiegel verschoben oder verkippt würde. Die Verkippung eines Spiegels würde lediglich der gemeinsamen Wellenfront der beiden austretenden Teilwellenzüge eine andere Richtung geben. Legt man ein Endmaß in den äußeren Kreis mit den Flächennormalen parallel zum Strahlengang, so überlagern sich die an den Endflächen reflektierten Wellenzüge an der Teilerfläche ebenfalls zu einer Planwelle, welche die gleiche Richtung hat wie die resultierende Planwelle des äußeren Kreises in dem vom Endmaß freigelassenen Feld.

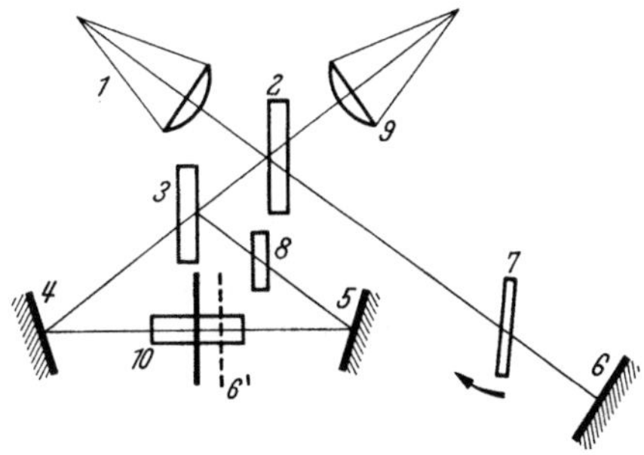

Fig. 21. Interferometer nach HUNZINGER für Endmaße von der Länge der doppelten Kohärenzlänge. *1* Kollimator, *2* und *3* Teilerflächen, *4*, *5*, *6* Spiegel, *7*, *8* Kompensationsplatten, *9* Beobachtungsobjektiv, *10* Prüfling, *6'* virtuelles Bild von *6*.

Aus Fig. 22 ist unmittelbar abzulesen, daß die resultierende Planwelle, die von den Endmaßflächen reflektiert wird, gegenüber der resultierenden Planwelle der neben dem Endmaß umlaufenden Wellenzüge einen Gangunterschied hat von der Länge L des Endmaßes, abgesehen vom Phasensprung; denn wenn z.B. bei mittiger Lage des Endmaßes die links herumlaufende „leere" Welle bei E_1 angekommen ist, startet von E_2 gerade die dort reflektierte, rechts herumgelaufene Welle, um gleichfalls links herum zu laufen. Diese hat also eine Endmaß-

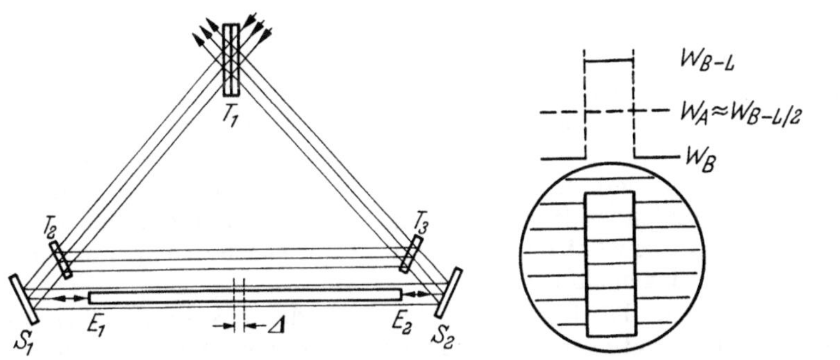

Fig. 22a u. b. a Interferometer nach HARIHARAN und SEN für Endmaße von der Länge der doppelten Kohärenzlänge. T_1, T_2, T_3 Teilerflächen, S_1, S_2 Spiegel, $E_1 E_2$ Endmaßflächen, Δ Außermittigkeit. b W_{B-L}, $W_{B-L/2}$ und W_B sind die auslaufenden interferierenden Wellenfronten, im Kreis das Gesichtsfeld.

länge Vorsprung. Liegt das Endmaß um ein Stück Δ außermittig, so muß die vom Endmaß ausgehende resultierende Planwelle je nach der Größe von Δ in ihrer Amplitude schwanken. Für $\Delta = \pm\frac{1}{8}\lambda$ wird die Resultierende gleich Null, da die beiden Teilwellen, aus denen sie durch Überlagerung entsteht, eine Phasenverschiebung von $\pm\frac{1}{2}\lambda$ gegeneinander haben. Der quantitative Zusammenhang ergibt sich aus folgenden Überlegungen:

Sind

$$s_1 = a \exp\left[i \frac{2\pi}{\lambda} (B - L + 2\Delta)\right]$$

und

$$s_2 = a \exp\left[i \frac{2\pi}{\lambda} (B - L - 2\Delta)\right]$$

die beiden vom Endmaß reflektierten Teilwellen im äußeren Kreis von der Länge B mit eingelegtem Endmaß L in außermittiger Lage Δ, so entsteht nach ihrer Überlagerung an der Teilerfläche die resultierende Welle

$$s = s_1 + s_2 = 2a \cos \frac{4\pi\Delta}{\lambda} \cdot \exp\left[i \frac{2\pi}{\lambda} (B - L)\right].$$

Diese Planwelle hat gegenüber der des leeren äußeren Kreises die Phasenverschiebung L, und ihre Amplitude ist von der Außermittigkeit periodisch abhängig. Die Amplitude wird maximal für $\frac{4\pi\Delta}{\lambda} = 0, \pi, 2\pi$ usw.; für $\frac{\pi}{2}, 3\frac{\pi}{2}$ usw. gibt es Auslöschung. Ist das Endmaß senkrecht zum Strahlengang justiert, so ergibt sich also eine Planwelle, deren Gangunterschied zu der des leeren äußeren Kreises bis auf den Phasensprung gleich der Endmaßlänge L ist.

Man benötigt nun noch eine Referenzfläche, um im Gesichtsfeld Interferenzstreifen „gleicher Dicke" zu erzeugen und die Phasenbeziehung zwischen der am Endmaß reflektierten und der leer umlaufenden Welle meßbar zu machen.

In dem von Hunzinger gewählten Aufbau wird eine Spiegelebene derart angeordnet, daß ihr virtuelles Bild praktisch in $^1/_4$ der Endmaßlänge geneigt zu den Endflächen des Endmaßes liegt.

In der Anordnung von Hariharan und Sen wird statt eines Referenzspiegels ein zweiter kleinerer Sagnac-Kreis gewählt, dessen Planwelle den Gangunterschied $B - L$ zwischen dem größeren leeren Kreis und dem Kreis mit Endmaß halbiert. Diese Planwelle wird durch Kippen eines Spiegels gegen die beiden anderen etwas geneigt. In beiden Anordnungen laufen aus dem Interferometer Wellenfronten aus, wie sie schematisch in Fig. 22 eingezeichnet sind, ihr gegenseitiger Abstand beträgt ungefähr eine halbe Endmaßlänge. Wenn die Kohärenzlänge den Abstand von einer Wellenfront zur anderen überbrückt, sind Interferenzstreifen zu sehen.

Das Endmaß darf also doppelt so lang sein wie die Kohärenzlänge. Damit ist auch hier der doppelte Meßbereich erzielt. Mit der Kryptonlampe von 0,8 m Kohärenzlänge können theoretisch im Hariharan-Interferometer noch Endmaße von 1,5 m gemessen werden. Für seine praktische Brauchbarkeit spricht, daß die beiden getrennt laufenden Interferenzwege zwischen und hinter den Teilerspiegeln T_2 und T_3 so dicht benachbart liegen, daß sie auf gleiche Temperatur und Druckverhältnisse zu bringen sind.

Im Gesichtsfeld der Fig. 22 ergeben sich die bekannten überschießenden Streifenbruchteile und es ist wieder möglich, von diesen Istbruchteilen die Sollbruchteile abzuziehen, um auf diese Weise die Differenz zwischen der Ist- und der Sollänge des Endmaßes zu ermitteln, nötigenfalls wieder mit mehreren Wellenlängen auf einem Wellenlängenrechenschieber. Eine Streifenbreite bedeutet bei der neuen Methode aber nicht wie früher einen Längenunterschied von einer halben sondern von einer ganzen Wellenlänge.

Um den Maximalkontrast zu bekommen, empfiehlt es sich, die symmetrische Lage des Endmaßes vorher mit Weißlichtinterferenzen gleicher Dicke bei abgeschaltetem Vergleichskreis einzustellen, indem das Endmaß aus der Parallellage heraus etwas gedreht oder geneigt wird.

Eine weitere Möglichkeit, den durch die Kohärenzlänge gegebenen Meßbereich gegenüber den Fizeau- und Michelson-Interferometern zu erweitern, bietet das Dowell-Interferometer mit überlappendem Gesichtsfeld (Fig. 16). Bei einem bestimmten Vorgehen darf auch hier die Länge des Endmaßes die doppelte Kohärenzlänge betragen (Fig. 23). Zu diesem Zweck symmetriert man das

Endmaß $E_1 E_2$ und unabhängig davon auch das Vergleichsstück L mit angesprengtem Endmaß $E_3 E_4$ über die Flächen $E_4 E_5$ mit Hilfe von Weißlichtinterferenzen. Ist $E_3 E_4$ ein Endmaß von ungefähr $^1/_4$ der Länge des zu untersuchenden Endmaßes, so rückt das Bild des überlappenden Referenzspiegels E_3 nach E_3'. Von dieser Lage aus wird der Abstand zur Stirnfläche E_2 und zur Symmetrieebene E_4 — einschließlich halber Dicke der Ansprengschicht — gemessen, woraus sich die halbe Endmaßlänge ergibt. Die ganze Länge des Endmaßes wird durch Verdoppeln dieses Wertes gefunden. Die sonst übliche Messung der ganzen Endmaßlänge wird hier auf die Messung der halben Endmaßlänge in symmetrischer Lage zurückgeführt; daher darf das Endmaß selbst doppelt so lang sein wie die Kohärenzlänge [47].

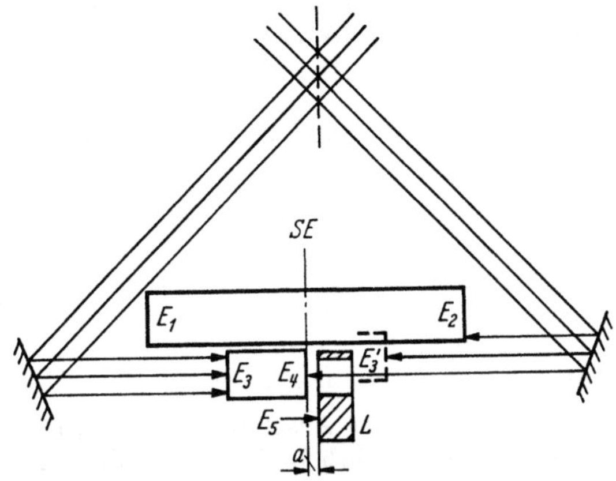

Fig. 23. Messung von Endmaßlängen bis zur doppelten Kohärenzlänge im Interferometer nach Fig. 16. $E_1 E_2$ langes Endmaß durch Weißlichtinterferenz mittig eingestellt, $E_3 E_4$ Endmaß rund $^1/_4\, E_1 E_2$, L Spezialendmaß, durchbrochen und an E_4 angesprengt, a Ansprengschicht. $E_4 E_5$ mit Weißlichtinterferenz mittig eingestellt. E_3' Bild von E_3 als E_2 und E_4 überlappender Referenzspiegel. *SE* Symmetrieebene für Weißlichtinterferenzen. Die halbe Länge $E_1 E_2$ ergibt sich unmittelbar aus $E_2 E_4$ nach Fig. 1a bzw. 9b. Die ganze Länge durch Multiplikation mit zwei.

Einige andere Verfahren, mit denen größere Endmaße in einem Schritt gemessen werden können, sollen noch angeführt werden. SAUNDERS [48] schaltet ein Pérot-Fabry- und ein Michelson-Interferometer hintereinander, um die Wegunterschiede in den Armen des Michelson-Interferometers mit durch Vervielfachung im Pérot-Fabry-Etalon erzeugten Wegunterschieden überlagern und

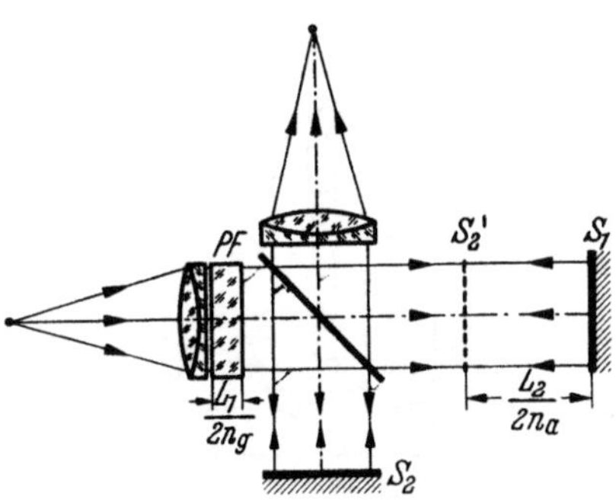

Fig. 24.

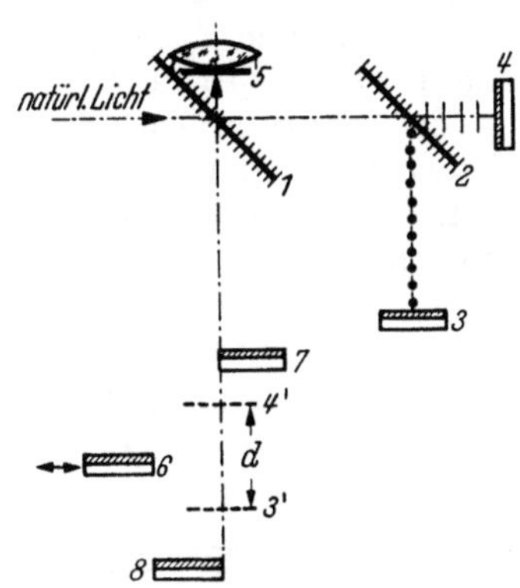

Fig. 25.

Fig. 24. Pérot-Fabry-Etalon *PF* und Michelson-Interferometer hintereinandergeschaltet. n_g und n_a Brechungsindices von Glas und Luft. Auszugslänge L_2 beträgt ein Vielfaches von L_1. S_1 S_2 Spiegel, S_2' virtueller Referenzspiegel.

Fig. 25. Zwei Michelson-Interferometer hintereinandergeschaltet mit polarisierendem Teilerspiegel *2* und Analysator *5*. Der Abstand *d* der Spiegel *3* und *4* wird virtuell mit *6* gemessen und der volle Auszug *7* bis *8* gegen die virtuellen Spiegel *3'* und *4'*, wobei die Strecke der doppelten Kohärenzlänge erreicht wird.

vergleichen zu können (Fig. 24). BAKOS, KÁNTOR und NARAY [49], [50] erreichten mit einer ähnlichen Anordnung eine Meßlänge von 2,80 m entsprechend der 7fachen Länge eines Pérot-Fabry-Etalons von 0,40 m.

HOCK [51] benutzt zwei ineinander geschaltete Michelson-Interferometer (Fig. 25), deren vier Spiegel virtuell so gestaffelt sind, daß sich drei aneinander anschließende Etalons bilden. Der innere Etalon kann mit einem Hilfsspiegel auf die Kohärenzlänge eingestellt werden und die äußeren Etalons besitzen je die halbe Kohärenzlänge. Die zu vermessenden äußeren Spiegelflächen können daher bis zur doppelten Kohärenzlänge voneinander entfernt sein. Um die Streifensysteme der beiden äußeren Etalons nicht gleichzeitig im Gesichtsfeld zu haben, werden die Teilstrahlen des inneren Systems gegeneinander um 90° linear polarisiert und mit einem Analysator nacheinander beobachtet.

b) Anschluß von Strichmaßen an die Lichtwellenlänge.

Die unmittelbare Messung von Strichmaßstäben gegen Wellenlängen ist seit der Einführung der Wellenlänge als Urnormal praktisch unumgänglich geworden. Strichmaße, z.B. das Urmeter in Paris und die von ihm abgeleiteten Prototypen sowie die in Werkzeug- und Meßmaschinen eingebauten Maßstäbe müssen mit Hilfe von Meßmikroskopen abgelesen werden. Zum Vermessen eines Strichmaßes in Wellenlängen besteht daher die Aufgabe, die Verschiebung des Maßstabes mit Hilfe von Interferenzen zu messen [*52*], [*53*].

25. Unmittelbarer Anschluß mit Pérot-Fabry-Platten. Ein Weg, dies zu tun, ist aus Fig. 26 ersichtlich. Der Maßstab ist hier mit einem Planspiegel versehen, der zusammen mit einer zweiten Planspiegelfläche ein Pérot-Fabry-Etalon bildet. Diese zweite Spiegelfläche ist fest mit dem Ablesemikroskop verbunden und parallel zur ersten. Zwischen den beiden Spiegelflächen werden mit einem Fernrohr Interferenzen gleicher Neigung als konzentrische Haidinger-Ringe beobachtet, aus deren Zentrum neue Ringe hervorquellen, wenn der Maßstab verschoben wird. Das Ringzentrum wird dabei periodisch hell und dunkel. Der Abstand eine Periode im Ringzentrum bedeutet die Verschiebung des Maßstabes um die halbe Wellenlänge $\lambda/2$ der benutzten monochromatischen Strahlung. Die Helligkeitswechsel werden entweder subjektiv oder mit einer Photozelle gezählt. Bruchteile können am Kompensator abgelesen werden, der beispielsweise aus einer zwischen den Spiegeln S_1 und S_2 befindlichen neigbaren Planparallelplatte oder einem planparallelen Glaskörper aus zwei gegenläufig verschiebbaren Glaskeilen von etwa 5° Keilwinkel besteht. Die Messung erfolgt streng nach dem Abbeschen Komparatorprinzip.

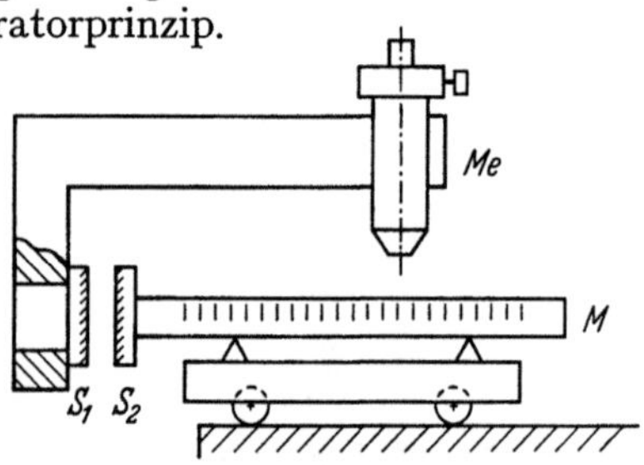

Fig. 26.

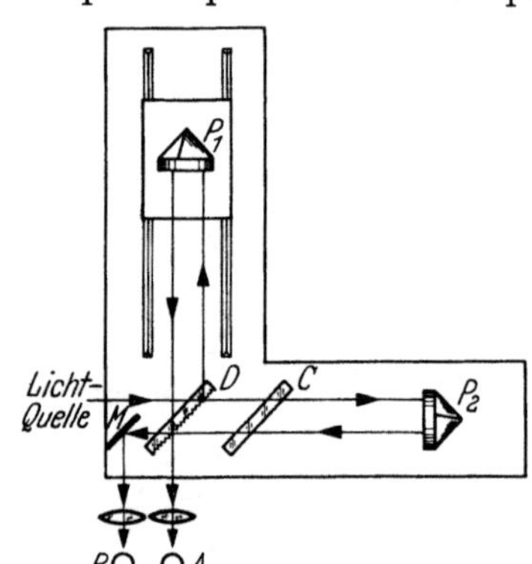

Fig. 27.

Fig. 26. Unmittelbare Messung eines Strichmaßes M gegen ein Perot-Fabry-Etalon $S_1 S_2$. *Me* Maßstabablesemikroskop.

Fig. 27. Michelson-Interferometer mit Tripelspiegeln P_1 und P_2 und phasenschiebender Teilerfläche D. Durch sie gelangen in die Photozelle A und B gegeneinander phasenverschobene Signale, die ein Drehfeld festlegen, um Vor- und Rückwärtsbewegungen des Meßschlittens P_1 feststellen zu können. C Kompensationsplatte zur Bruchteilmessung, M Umlenkspiegel.

26. Unmittelbarer Anschluß mit Tripelspiegel. Peck und Obetz [*52*] zählen die Verschiebung eines Tripelspiegels im Arm eines Michelson-Interferometers unmittelbar in Wellenlängen photoelektrisch aus (Fig. 27). Kippungen der Schlittenführung sind durch den Tripelspiegel unwirksam. Mit dem Schlitten kann ein Maßstab unter einem feststehenden Mikroskop oder ein Ablesemikroskop über einem ruhenden Maßstab verschoben werden. Die Interferenzerscheinung besteht aus Haidinger-Ringen, deren Wechsel mit Hilfe der Photozelle A gezählt werden. Die durch die Teilerfläche in Richtung auf die Lichtquelle hindurchtretenden Strahlen fallen über den Spiegel M in eine weitere Photozelle B. Die in A und B beobachteten Interferenzerscheinungen haben wegen der Phasensprünge bei der Reflexion an D eine Phasenverschiebung von weniger als einer halben Streifenbreite gegeneinander. Auf Grund dieses Phasenwinkels kann in Abhängigkeit von der Verschieberichtung des Schlittens vor- und rückwärts gezählt werden, wenn das eine Signal als Bezugssignal gewertet wird, dem das andere vor- oder nacheilt.

27. Unterschiedsmessung mit Endmaßstufe. Kürzere Maßstabslängen, z.B. die mm-Intervalle auf einem Maßstab, können interferentiell mit einer Endmaßstufe

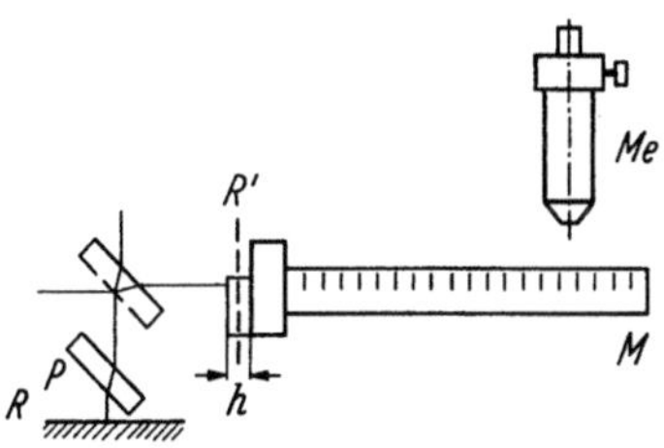

Fig. 28. Michelson-Interferometer mit Weißlicht-Keilinterferenzen zur Messung eines Maßstabs M gegen eine Endmaßstufe h. Referenzspiegel R, virtueller Referenzspiegel R'. Me Maßstabablesemikroskop. R kann ein verschiebbarer Quecksilberspiegel sein, vgl. Ziff. 23. P neigbare Planparallelplatte zur Feineinstellung von R.

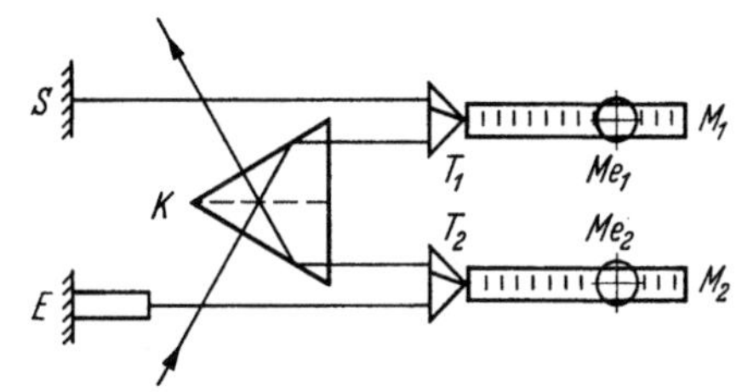

Fig. 29. Interferometrischer Vergleich von Maßstäben M_1 und M_2 gegen ein Endmaß E. S Vergleichsspiegel. E und S sind fest, ebenso das Kösters-Prisma K, sowie die Maßstabablesemikroskope Me_1 und Me_2. Tripelspiegel T_1 in Kontakt mit M_1, T_2 in Kontakt mit M_2 sind in Meßrichtung verschiebbar.

ähnlich einem Michelson-Etalon an die Lichtwellenlänge angeschlossen werden. Auf einer Grundplatte sei ein 1 mm Endmaß interferentiell bekannter Länge angesprengt und die entstehende Stufe so in den Arm eines Michelson-Interferometers gebracht, daß die Stufenhöhe genau in Verlängerung der Strichmaßmeßstrecke liegt (Fig. 28). Dann verschiebt man die Maßstabstrecke um die Stufenhöhe, indem man den Nullstreifen von Weißlichtinterferenzen nacheinander auf der oberen und unteren Stufe mit Hilfe eines festen Strichkreuzes an der gleichen Stelle im Ocular des Interferometers einfängt und liest im Maßstabablesemikroskop die Länge des dazugehörigen Strichintervalls ab. Durch abwechselndes Nachschieben des Referenzspiegels und der Stufe läßt sich ein Maßstab von Strich zu Strich mittelbar an die Lichtwellenlänge anschließen.

Um Kippfehler beim Verschieben der Endmaßstufe auszuschließen, bewegt Räntsch [*54*], [*55*] nicht die Endmaßstufe, sondern einen kippungsunempfindlichen, verschiebbaren Tripelspiegel, der den Strahlengang um 180° knickt (Fig. 29). Wird der Tripelspiegel T_2 zusammen mit dem Maßstab M_2 um einen Streifenabstand verschoben, so ist der Verschiebeweg gleich einer viertel Wellenlänge, da der Lichtweg zweimal über den bewegten Tripelspiegel führt. Im Vergleich zu der Ausführung ohne Tripelspiegel gewinnt man dadurch zusätzlich den Faktor 2 an Einstellgenauigkeit für den Maßstab M_2. Einer Weglängendifferenz zwischen beiden Strahlengängen von einer Endmaßstufe l entspräche eine Verschiebung des Tripelspiegels um den Teilungsstrich $t=l/2$, d.h. der Fehler in der Endmaßlänge Δl geht nur mit seinem halben Betrag in den Teilungsschritt ein, $\Delta t=\Delta\frac{l}{2}$. Um den Referenzspiegel S nicht nachrücken zu müssen, womit der Vorteil der Kippungsunempfindlichkeit wieder verloren ginge, ist das Interferometer von RÄNTSCH symmetrisch aufgebaut, so daß mit einem anderen verschiebbaren Tripelspiegel T_1 die Änderungen der optischen Weglänge vorgenommen werden können, die sonst durch das Nachrücken der Referenzfläche bewirkt würden. Der verschiebbare Tripelspiegel T_1 kann seinerseits mit einem zweiten Maßstab gekoppelt werden und so können infolge des symmetrischen Aufbaus gleichzeitig zwei Maßstäbe an derselben Endmaßstufe gemessen oder nach ihr geteilt werden. Sollen die Maßstäbe in Ruhe bleiben, so wird die jeweilige Ablesevorrichtung verschoben (Fig. 30). Die Spitze jedes der beiden Tripelspiegel muß bei dieser Anordnung im vorderen positiven Knotenpunkt der mit einfachem

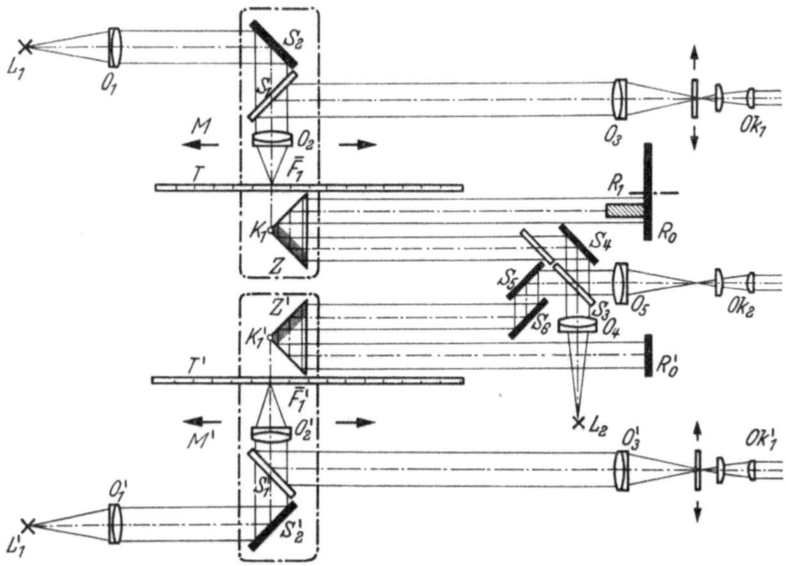

Fig. 30. Interferometer für den Anschluß von Strichmaßen an Endmaßstufen nach RÄNTSCH. L_1, L'_1, L_2 Lichtquellen. S_1, S'_1, S_3 Teilerspiegel. S_2, S'_2, S_4, S_5, S_6 Spiegel. $O_1, O'_1, O_2, O'_2, O_3, O'_3, O_4, O_5$ Objektive, F_1, F'_1 Brennpunkte, K_1, K'_1 positive Knotenpunkte, Z, Z' Tripelspiegel, M, M' Meßschlitten, T, T' Maßstäbe, Ok_1, Ok'_1, Ok_2 Oculare, R_0, R_1 Endmaßstufe, R'_0 Referenzfläche.

Spiegel versehenen Maßstabableseobjektive liegen und der Maßstab im vorderen Brennpunkt, so daß sich Kippfehler der Meßschlitten M_1 und M_2 nach dem Eppenstein-Prinzip (Ziff. 34, β) nicht auf die Maßstabsablesung auswirken können.

28. Vergleich mit Strichendmaß. Ein Verfahren, das sich gut bei größeren Maßstabsstrichabständen anwenden läßt, arbeitet ebenfalls über interferentiell sehr genau gemessene Endmaße. Die Methode beruht darauf, daß ein bekanntes Endmaß durch Ansprengen zweier Strichendmaße an seine beiden Meßflächen in ein Strichmaß bekannter Länge verwandelt werden kann. Der Weg, über den man von den interferentiell gemessenen Endmaßen zur Kenntnis des Strichabstandes auf einem Maßstab gelangt, sei am Beispiel eines 1 m langen Strichmaßes dargestellt, welches an die Lichtwelleneinheit angeschlossen werden soll. Dazu wird zunächst ein Endmaß von 990 mm Länge durch Lichtwellen im Interferenz-Komparator ausgemessen. Sodann werden zwei Endmaße von je 10 mm Länge an die Lichtwellenlänge angeschlossen. Mitten auf einer Seitenfläche jedes der beiden 10 mm Endmaße ist ein Strich parallel zu den Endflächen eingeritzt. Durch Ansprengen beider 10 mm-Endmaße an je eine Endfläche des 990 mm-Endmaßes ergibt sich ein Endmaß von 1010 mm und ein Strichmaß von annähernd 1000 mm Länge (Fig. 31). Die Länge $l = 990$ mm und die Längen $k_1 = a + b$ und $k_2 = c + d$

sind mit Interferenzgenauigkeit bekannt. Durch wechselseitiges Ansprengen können zwei Strichmaße von der Länge $M_I = l + b + c$ und $M_{II} = l + a + d$ gebildet werden, deren Mittelwert $\frac{1}{2}(M_I + M_{II}) = M = l + \frac{1}{2}(k_1 + k_2)$ wiederum mit Interferenzgenauigkeit bekannt ist. An diesen Mittelwert wird der gesuchte Strichabstand des Maßstabes mit einem Komparator angeschlossen ([*56*], S. 105).

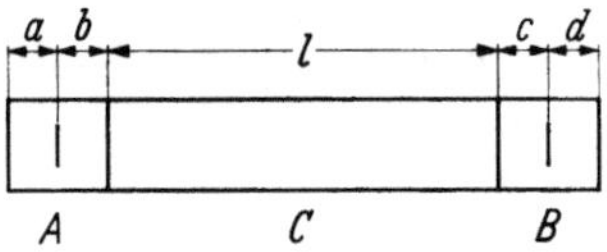

Fig. 31. Zwei Strichendmaße *A* und *B*, angesprengt an ein Endmaß *C*, zum Anschluß eines Strichmaßes an die Lichtwellenlänge.

Es ist also nicht erforderlich, die Lage der Striche auf jedem Strichendmaß zu kennen, da sie durch Vertauschen gemittelt wird. Werden die Stücke *A* und *B* aus Quarzglas, also durchsichtig, gewählt, läßt sich das Ansprengen durch das Entstehen von Interferenzstreifen kontrollieren oder durch eine dünne Luftschicht ersetzen [*57*].

c) Ausführung von Strichmaßstäben.

29. Die mit Hilfe von Diamanten geritzten ca. 7 μ breiten Maßstabstriche auf den heute üblichen Strichmaßstäben zeigen noch glatte Berandungen, wenn zum Ablesen der Striche 50—100fach vergrößernde Meßmikroskope verwendet werden. Stärkere Vergrößerungen sind nicht vorteilhaft. Die Einfangunsicherheit der Striche im Doppelstrich des Oculars liegt bei $s = \pm 0{,}2\,\mu$ (vgl. Ziff. 11).

Für Metallmaßstäbe wählt man thermische Ausdehnungskoeffizienten von $11{,}5 \cdot 10^{-6}$/Grad, damit der größte Teil der Prüflinge, der gewöhnlich aus Stahl oder Eisen mit dem gleichen Ausdehnungskoeffizienten besteht, auch bei Temperaturen abweichend von 20° C ohne Umrechnung gemessen werden kann. Die Oberfläche der Metallmaßstäbe wird üblicherweise verkupfert und darüber vernickelt. Im Hellfeldauflicht reflektieren diese Oberflächen wie vorzügliche Spiegel und zeigen den eingeritzten Maßstabsstrich schwarz auf hellem Grund. Falls man helle Striche auf dunklem Grund benötigt, ist es nicht ratsam, derartige Maßstäbe im Dunkelfeld zu betrachten, weil geritzte Maßstabsstriche das an ihnen gebeugte Licht in bestimmte Vorzugsrichtungen der Apertur des Beobachtungsmikroskops werfen und diese nicht voll ausfüllen. Hierdurch „schieben" die Striche lateral, wenn man das Mikroskop durch die Schärfenebene hindurchfährt, und dem Auge noch entgehende Unscharfstellungen bringen schon seitliche Lageänderungen des Maßstabstriches hervor.

Helle Striche auf dunklem Grund erzielt man auf Glas im Auflicht nach dem Patent von Heidenhain [*58*]. Im Durchlicht ergeben sich dunkle Striche auf hellem Grund, die also hierin nicht vom Metallstab zu unterscheiden sind. Solche Teilungen werden in einer auf den Glasträger gebrachten lichtempfindlichen Schicht photographisch hergestellt. An den unbelichteten Stellen wird die Schicht durch Lösungsmittel wieder beseitigt, so daß hier das klare Glas wieder zu Tage tritt. Darauf wird das ganze mit Chrom bedampft. Auf den freigelegten Strichstellen haftet das Chrom fest am Glas, an allen anderen wird es mit dem Photogrund wieder fortgewaschen. So resultieren metallisch reflektierende Striche auf durchsichtigem Glasgrund.

Glasmaßstäbe der gewöhnlichen Art entstehen durch Überziehen von Glas mit Ätzgrund und Ritzen dieses Grundes durch Diamantspitzen auf Teilmaschinen. Flußsäuredämpfe greifen das freigelegte Glas an, und man erhält an den Strichstellen eine Ätzung, die nach Abwaschen des übrigen Ätzgrundes mit Schwarz eingelegt wird. Im Durchlicht erscheinen schwarze Striche auf hellem Grund. Im Auflicht sind sie nicht zu brauchen.

α) *Fehlertabellen.* Bezüglich ihrer Strichlagen sind alle Maßstäbe fehlerhaft. Die Ablage der einzelnen Teilstriche wird dem Vorzeichen und der Größe nach durch die Differenz Ist- minus Sollage vom Nullstrich aus gerechnet und in Fehlertabellen festgehalten. Die Fehlertabelle macht u. U. einen „fehlerhaften" Maßstab einem Präzisionsmaßstab gleichwertig. Der Vergleich geschieht gegen Normalmaßstäbe, oder die Lichtwellenlänge (vgl. Abschnitt IIIb).

Fehlertabellen werden nach der Netzmethode erstellt. Bei einem in Millimeter geteilten Maßstab von z. B. 1 m Länge geht man so vor, daß man ihn gegen einen Normalmaßstab, dessen Fehler sehr genau bekannt sind, vergleicht und zunächst mindestens zehnmal den Abstand des Tausenders vom Nullstrich durch Vergleich mit dem Tausender-Strich des Normalmaßstabs feststellt. Dann werden der Reihe nach die Ablagen der Hunderter gegen die Hunderter des Normalmaßstabs gemessen, auch zehnmal hin und zurück. Fällt der Endpunkt der gemittelten Werte der Hunderter-Kette nicht mit dem schon bestimmten Tausender zusammen, so dreht man die ganze Hunderter-Fehlerkurve um ihren Anschlußpunkt, in diesem Fall den Nullstrich, bis dies der Fall ist. In dieses noch relativ grobe Netz bringt man die einzelnen Zehner-Intervallketten, indem man sie genauso zwischen die betreffenden Hunderter schachtelt, und schließlich fügt man die Einer-Intervallketten in gleicher Weise zwischen die Zehner [*59*]. Wollte man anders verfahren und z. B. alle Striche der Reihe nach messen, so würden unbeherrschbare Fehlerquellen auf die Messung einwirken können und die ganze lange Kette verderben, indem sich die Fehler zunehmend aufsummieren. Durch die Netzmethode werden systematische Fehler auf kurze Abschnitte beschränkt und somit weitgehend ausgeschlossen.

Als Fehlerquellen kommen geringe Temperaturänderungen oder Erschütterungen — die beispielsweise eine Verlagerung zwischen Prüfling und Normal bewirken können — ferner Ermüdung des Messenden oder Anschlußfehler nach einer Meßpause in Betracht.

β) *Profil und Auflagepunkte.* Die Querschnittsprofile der Maßstäbe — H-Form mit der Teilung in der neutralen Faser oder Rechteck-Querschnitt mit der Teilung auf der Oberseite des Stabes — sollen Durchbiegungen möglichst verhindern, damit sich die Länge der Teilung nicht in Abhängigkeit von der zufälligen Auflage der Maßstäbe merklich ändert. Zweckmäßig schützt man sich vor unkontrollierbaren Einflüssen dieser Art bei längeren Maßstäben, indem man sie in bestimmten Abständen von ihren Enden auf zwei zylindrische Walzen auflegt. Die H-Form unterstützt man in den Besselschen Punkten, die 0,220 L von den Endflächen des Maßstabs mit der Länge L entfernt sind. Dabei wird in der Projektion auf die Meßebene die Längenänderung in der neutralen Faser ein Minimum, die Endflächen des Stabes stehen aber nicht mehr parallel zueinander. Die Rechteckform unterstützt man in den „günstigsten" Punkten, die 0,211 L von den Enden entfernt sind, damit in der Projektion die Längenänderung der auf der Oberseite angebrachten Teilung ein Minimum wird. Die Endflächen stehen hierbei parallel zueinander und senkrecht zur Meßebene. In den gleichen Punkten unterstützt man daher auch lange Endmaße ([*56*], S. 42/46).

IV. Mikroskopische Geräte und Verfahren.

Um Objekte dem Auge möglichst groß sichtbar zu machen, führt man sie so nahe wie möglich an das Auge heran. Damit vergrößert sich der Sehwinkel. Bei einem Abstand $a = 250$ mm liegt die Akkomodationsgrenze (deutliche Sehweite) des „Normalauges". Auf sie wird die sog. Betrachtungsvergrößerung $N_B = 250/a$ normiert.

30. Lupenvergrößerung. Um auch näher herangebrachte Objekte, $a < 250$ mm, noch scharf zu sehen, benötigt das Auge eine *Vorsatzlinse*, eine *Lupe* oder ein *Ocular*. Diese Instrumente haben den Zweck, das Objekt y in ihrer vorderen Brennebene aufzunehmen und sein Bild dem auf Unendlich akkomodierten, also entspannten Auge unter einem größeren Sehwinkel darzubieten. Dieser ergibt sich zu $y/f' = \tan \varphi_{\mathrm{m}}$. Das Winkelverhältnis

$$N_L = \frac{\tan \varphi_{\mathrm{m}}}{\tan \varphi_{250}} = \frac{y}{f'} \cdot \frac{250}{y} = 250/f'$$

(Brennweite f' in mm, φ_m und φ_{250} Sehwinkel mit und ohne Instrument) wird als „Lupenvergrößerung" bezeichnet mit dem virtuellen Bild im Unendlichen. In dieser Formulierung ist die Lupenvergrößerung unabhängig vom Abstand Lupe-Auge. Benutzt man Lupen mit f' größer als 250 mm, so kommt eine Betrachtungsverkleinerung zustande, obwohl man sich eines „Vergrößerungsglases" bedient. Ohne das Glas hat man bei der gleichen Entfernung $a = f'$ des Objektes vom Auge dieselbe Betrachtungsverkleinerung, nur mit dem Unterschied, daß das Auge dann nicht entspannt ist, sondern auf diese Entfernung akkomodieren muß.

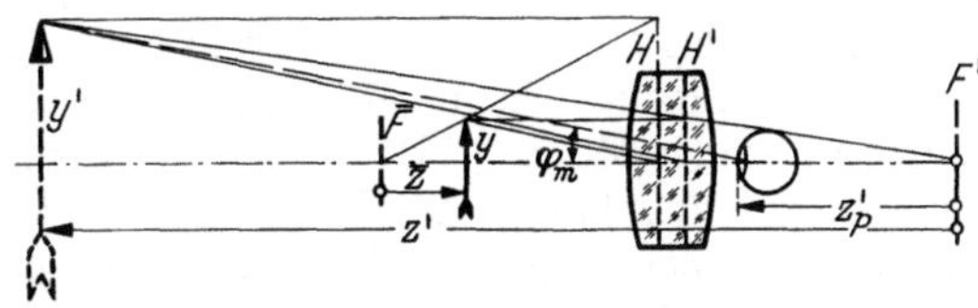

Fig. 32. Zur Lupenvergrößerung. y Objekt, y' Bild, F, F' objekt- und bildseitige Brennpunkte, z, z' Brennpunktsabstände von Objekt und Bild, z'_p Abstand der Augenpupille von F', φ_{m} Sehwinkel mit Lupe.

Liegt bei angespannter Beobachtung das erzeugte Bild nicht im Unendlichen, sondern im Abstand $-s'$ vor der Lupe bzw. dem Ocular, weil der Gegenstand innerhalb der Brennweite der Lupe liegt, so ergibt sich für das Verhältnis der Sehwinkel nach Fig. 32:

$$N_L = \frac{\tan \varphi_{\mathrm{m}}}{\tan \varphi_{250}} = \frac{y'}{z' - z'_p} \cdot \frac{-250}{y}$$

und mit

$$\frac{y'}{y} = -\frac{z'}{f'}$$

wird

$$N_L = \frac{250}{f'} \, \frac{z'}{z' - z'_p}.$$

Dieser Ausdruck geht für $z' = -\infty$, d.h. für ein unendlich fernes Bild in die oben angegebene Gleichung $N_L = 250/f'$ über. Dann spielt speziell der Abstand z'_p des Auges vom Brennpunkt F' der Lupe und damit von der Lupe selbst keine Rolle (s. oben). Der Ausdruck geht aber auch für $z'_p = 0$ in die oben angegebene Gleichung über, d.h. für die telezentrische Lupe mit der Augenpupille im hinteren Brennpunkt. Dann darf z' beliebige Werte innerhalb des Akkomodationsbereiches des Auges annehmen, was gleichbedeutend damit ist, daß das Objekt innerhalb der vorderen Brennweite der Lupe axial verschoben werden kann, ohne daß sich der *Bildwinkel* ändert; darin liegt der Nutzen des telezentrischen Strahlengangs für Meßanordnungen, wie in Ziff. 5 ausgeführt wurde.

Meßlupe mit telezentrischem Strahlengang. Die Lupe wird zur Meßlupe, wenn in der vorderen Brennebene eine Skale liegt, die mit dem Objekt oder einem reellen Bild des Objektes zur Deckung gebracht werden kann. Um den telezentrischen Strahlengang (Ziff. 5) zu verwirklichen, muß das Auge als Austrittspupille (AP) in die hintere Brennebene gebracht werden. Dadurch ist dem Objektfelddurchmesser D eine bestimmte vom Lupendurchmesser D_{L} abhängige Größe vorgeschrieben, die sich wie folgt angeben läßt (Fig. 33):

$$D \leqq D_{\mathrm{L}} - D_{\mathrm{AP}}.$$

Wenn eine in der vorderen Brennebene der Lupe angeordnete beleuchtete Skale bildseitig nach Unendlich projiziert wird, ist es möglich, ihr Bild als unendlich fernes Ziel im Labor zu benutzen, wo man sonst auf endliche Dimensionen angewiesen ist. Diese Anordnung wird als *Kollimator* bezeichnet und der Ausdruck $N_K = 250/f_K$ als Kollimatorvergrößerung. Sie ist identisch mit der Lupenvergrößerung. Der Kollimator wird zusammen mit einem Fernrohr benutzt, um Richtungsänderungen bei Durchbiegungen und entlang „Geradführungen" festzustellen (vgl. Ziff. 39).

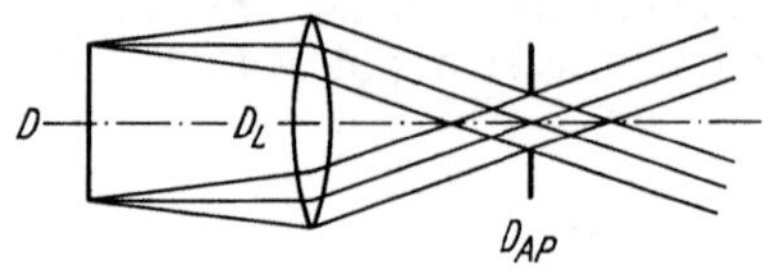

Fig. 33. Objektfelddurchmesser D in Abhängigkeit vom Lupendurchmesser D_L und dem Durchmesser der Austrittspupille D_{AP} bei der telezentrischen Meßlupe.

31. Zielmikroskop. Von der Lupe ist es nur ein kleiner Schritt zum Mikroskop, bei dem nicht das Objekt selbst, sondern sein durch ein Objektiv projiziertes schon stark vergrößertes reelles Bild durch die Lupe betrachtet wird. Die Vergrößerung des Mikroskops N_M ist folglich gleich dem Produkt aus Abbildungsmaßstab β' des Objektivs und der Lupenvergrößerung N_L nämlich $N_M = \beta' \cdot N_L$. Aus dem Mikroskop normaler Bauart wird ein Zielmikroskop, wenn eine Strichplatte mit einer Marke, z.B. einem Doppelstrich oder einem Strichkreuz, in der Ocularebene angebracht ist. Ist die Ocularstrichplatte mit einer Skale versehen und arbeitet das Objektiv objektseitig telezentrisch, so wird das Zielmikroskop zum Meßmikroskop.

α) *Doppelbilder.* Zielmikroskope lassen sich durch Prismen zwischen Objektiv und Ocularebene als Doppelbild-Mikroskope ausführen mit entweder flächensymmetrischem oder axialsymmetrischem Doppelbild. Doppelbilder entstehen

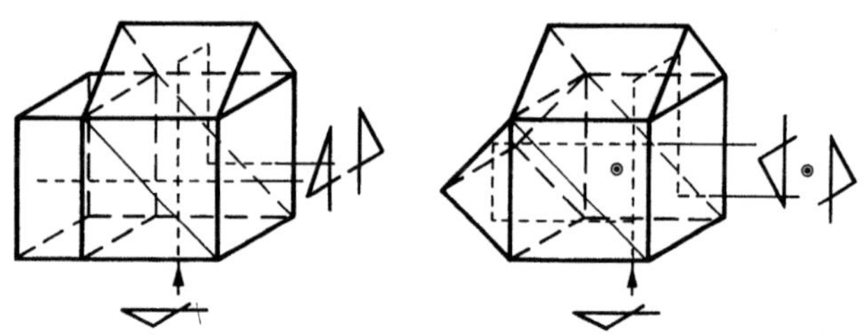

a b

Fig. 34a u. b. Doppelbildprismensysteme. a Für flächensymmetrisches; b für achsensymmetrisches Doppelbild.

durch Aufspalten des Abbildungsstrahlengangs vor Erreichen der Ocularebene in zwei getrennte Bündel durch physikalische Strahlenteilung. Hat das eine Bündel eine Reflexion mehr erlitten als das andere, so ist das von ihm gelieferte Bild gegenüber dem anderen seitenverkehrt, es herrscht Flächensymmetrie (Fig. 34a). Wird das Objekt senkrecht zur Symmetrieebene verschoben, so bewegen sich die beiden Bilder gegenläufig. Wird das Objekt gedreht, so drehen sich beide Bilder in entgegengesetzter Richtung; dieser Umstand wird ausgenutzt zum Ausrichten von Begrenzungen oder Strichen senkrecht zur Meßrichtung.

Hat das eine Bündel senkrecht zur ersten Symmetrieebene eine zusätzliche Reflexion erlitten, so ist es nicht nur seitenvertauscht, sondern umgekehrt, es besteht Axialsymmetrie (Fig. 34b). Laterale Bewegungen des Objekts bringen gegenläufige Bewegungen der Bilder hervor. Drehbewegungen des Objekts in der Schärfenebene haben jetzt gleichsinnige Drehungen der Bilder zur Folge.

Zielmarke ist bei den Mikroskopen mit Doppelbild die jeweilige Symmetrieebene, bzw. der Kreuzungsschnitt zweier solcher Ebenen. Eine materielle Marke wird nicht benötigt; es genügt, die Bilder zur Deckung zu bringen.

Das Doppelbild-Verfahren verdient vor dem Strichkreuz den Vorzug, wenn auf Helldunkelgrenzen zu zielen ist. Die Helldunkelgrenze müßte den Strichkreuzbalken halbieren, nicht an ihm anliegen. Im Doppelbildverfahren werden die beiden gegenläufigen Helldunkelgrenzen aneinandergelegt. Durch die Gegenläufigkeit wird die Einfanggenauigkeit verdoppelt. Vorzüglich eignet sich das Doppelbildverfahren zum Zielen in die Achse von Bohrungen, z.B. um deren Abstände zu messen. Bildet man zusätzlich beide Bilder durch Einfügen von Filtern in die einzelnen Strahlengänge in Komplementärfarben ab, so lassen sich z.B. im Durchlicht die Überdeckung der freien Öffnungen der beiden Bohrungsbilder als weißes Feld und mangelhafte Deckung durch farbige Ränder in einer schwarzen Umgebung deutlich machen [*60*].

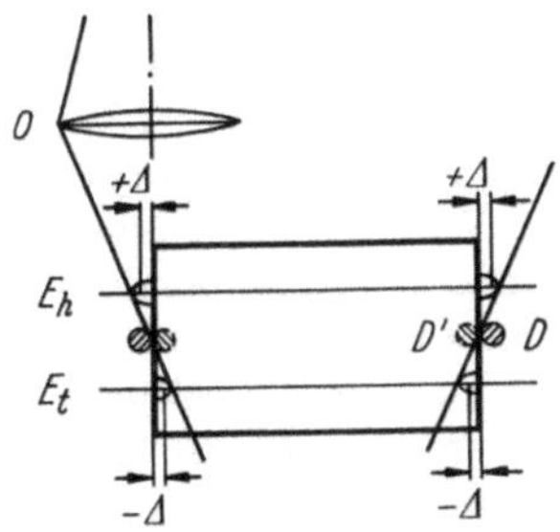

Fig. 35.

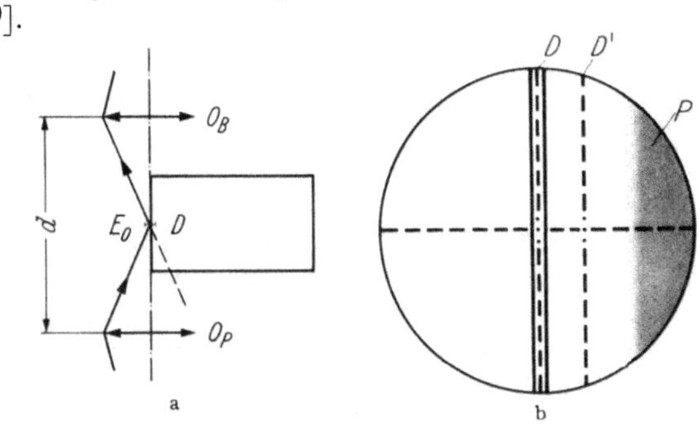

Fig. 36.

Fig. 35. Fehler der Fizeau-Methode. Draht D und Spiegelbild D' werden auf beiden Endflächen des Endmaßes nacheinander von Objektiv O anvisiert. Liegt die Schärfenebene zu hoch, E_h, wird um $+\Delta$ vor der Meßfläche gemessen, zu tief, E_t, um $-\Delta$ hinter der Meßfläche.

Fig. 36a u. b. Perflektometerprinzip. a Der Draht D wird durch eine projizierte Marke ersetzt, vgl. Fig. 35. E_h und E_t fallen in E_0 zusammen, wenn der Abstand d zwischen dem Projektionsobjektiv O_P und dem Beobachtungsobjektiv O_B geeignet gewählt und konstant gehalten wird. Meßfehler Δ daher beherrschbar. b Gesichtsfeld des Perflektometers kurz vor dem Einfangen des am Prüfling reflektierten Strichkreuzbildes D' im Doppelstrich, in welchem das nicht reflektierte Strichkreuzbild D noch zu sehen ist. Dieses verschwindet erst, wenn seine Beleuchtungsapertur gegen Null geht. Die dunkle Fläche P rechts ist das Schattenprofilbild des Prüflings, seine spiegelnde Meßfläche ist hell und nicht zu erkennen, sie schneidet die Gesichtsfeldebene in der Mitte zwischen D und D'.

β) *Perflektometer*. Bei den beschriebenen Zielmikroskopen werden Strichmarken oder Schattenkanten anvisiert. Ein Zielmikroskop für das Zielen in die Endflächen von Körpermaßen ist im Perflektometer verwirklicht [*4*], [*61*]. Das Zielen mit dem Mikroskop auf Körpermaße, z.B. auch Endmaße, Rachen- und Ringlehren ist aus verschiedenen Gründen schwierig, erstens kann man nicht ohne Hilfsmittel auf die spiegelnde Endmaßfläche einstellen, sondern braucht eine Schneide, einen Faden oder dergleichen, zum zweiten bringen Fehler in der Scharfeinstellung, die vom Auge noch unbemerkt bleiben, Meßfehler mit sich wegen geneigt zur optischen Achse verlaufender Hauptstrahlen (Fig. 35). Um derartige Fehler zu vermeiden, wird erstens keine körperliche Marke verwendet, sondern das reelle Bild einer solchen, üblicherweise eines Strichkreuzes, und zweitens wird der Abstand zwischen dem Bild und dem Zielmikroskop dadurch fixiert, daß Projektions- und Beobachtungsstrahlengang eine mechanisch gekoppelte Einheit, das Perflektometer nach E. Leitz, bilden (Fig. 36). Diese Anordnung ist ein besonders sinnfälliges Beispiel für die Vorzüge optischer Meßmethoden; die hinderlichen mechanischen Schneiden entfallen, an ihre Stelle tritt ein abbildender Strahlengang, der genügend freien Arbeitsabstand zwischen den Objektiven beläßt und beliebig vom Prüfling durchfahren werden kann, ohne dejustiert oder zerstört zu werden.

Durch ein aus drei Endmaßen zusammengesprengtes Rachenvollmaß (Fig. 37) ergibt sich ein Eichstück für die „richtige" Fokussierung des Perflektometers: Rachen und Vollmaß müssen gleich groß gemessen werden. Abweichungen gehen zu Lasten der Fokussierung, sofern der Rachen nicht als keilig befunden wird. Läßt man die gefundene Einstellung fernerhin unverändert, so mißt das Perflektometer praktisch genau in der Meßfläche. Der Vorgang bei falscher Fokussierung weicht von dem beim Einstellen auf eine reelle Schneide oder Faden (DD' in Fig. 35) etwas ab. Daher wird er in Fig. 38 näher erläutert. Liegt die Einstellebene zu hoch (in Fig. 38a bei E_1, also hinter dem projizierten Markenbild), so liegt die Meßfläche eindeutig neben der optischen Achse, wenn das Fehlerscheibchen symmetrisch im Doppelstrich eingefangen ist. Aus dieser Situation ist abzulesen, daß bei Fokussierung auf eine Ebene, die in Lichtrichtung hinter dem Markenbild liegt, vor der Meßfläche angetastet wird. Liegt die Einstellebene vor dem Markenbild bei E_2, wird hinter der Meßfläche angetastet, was ebenso leicht einzusehen ist (Fig. 38b).

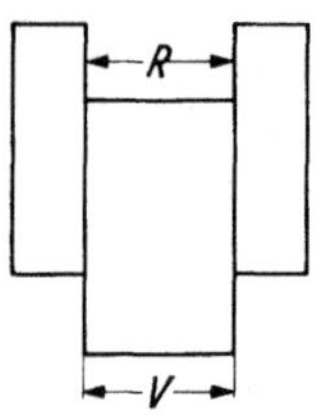

Fig. 37. Rachen-Vollmaß R-V zur Justierung des Leitz-Perflektometers.

Die Fokussierung ist temperaturabhängig, da Projektions- und Beobachtungsmikroskop ihren Abstand voneinander mit der Temperatur ändern. Aus der Hauptstrahlneigung und dem Temperaturkoeffizienten des Baumaterials läßt sich der Zusammenhang übersehen.

Das Perflektometer erreicht Einstellunsicherheiten von $s = \pm 0{,}1\ \mu$ sofern eine Strichmarke zwischen Doppelstrichen eingefangen wird. Der Faktor 2 gegenüber dem Einfangen eines Maßstabsstriches ergibt sich aus den Spiegelgesetzen, denn die projizierte Marke wird als Spiegelbild in der Endfläche gesehen.

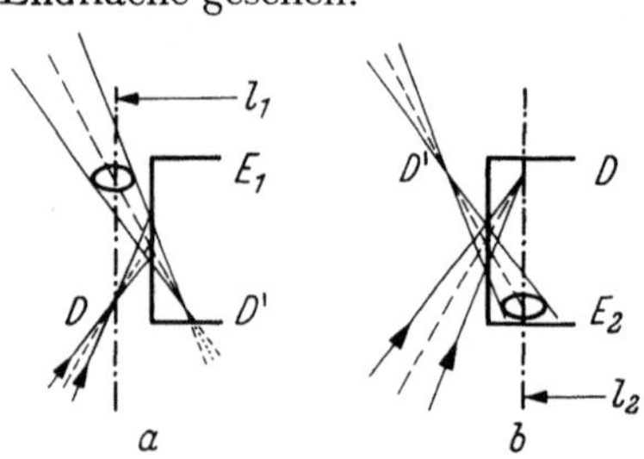

Fig. 38 a u. b.

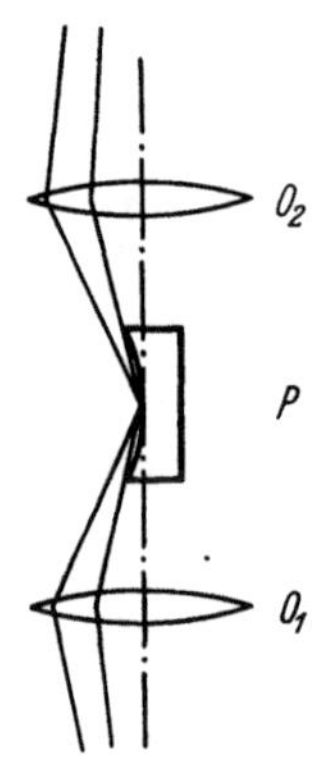

Fig. 39.

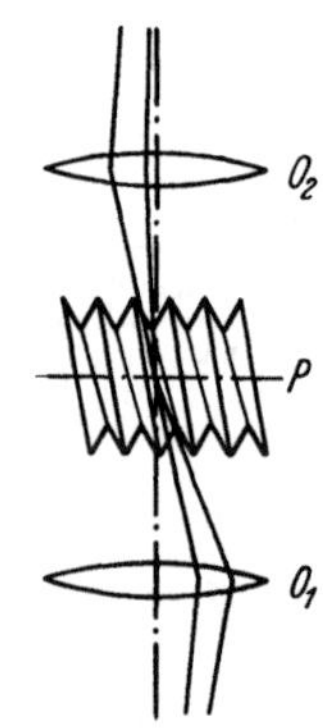

Fig. 40.

Fig. 38a u. b. Die durch festen Abstand zwischen Projektions- und Beobachtungsstrahlengang im Leitz-Perflektometer vermiedenen Fokussierungsfehler würden sich bei falschem Abstand ähnlich auswirken wie in der Fizeau-Anordnung nach Fig. 35. a Mit zu hoch eingestellter Beobachtungsebene E_1 wird vor der Meßfläche gemessen, l_1 wird zu groß, b mit zu tief eingestellter Beobachtungsebene E_2 wird hinter der Meßfläche gemessen, l_2 wird zu klein. Das jeweilige Fehlerscheibchen wird auf der optischen Achse als Zielachse eingefangen, wo ohne eingebrachte Meßfläche das Bild von D lag. Aus Gründen der Übersichtlichkeit ist nicht das volle Halbbündel gezeichnet.

Fig. 39. Optisches Antasten einer konkaven Fläche mit dem Perflektometer. O_1 Projektions-, O_2 Beobachtungsobjektiv, P Prüfling. Die Objektivaperturen werden nur z.T. ausgenutzt, aber in gleichen Zonen.

Fig. 40. Optisches Antasten einer Gewindeflanke mit dem Perflektometer. O_1 Projektions-, O_2 Beobachtungsobjektiv, P Prüfling. Die Objektivaperturen werden nur teilweise und in verschiedenen Zonen ausgenutzt.

Es ist versucht worden, dieses Prinzip auch auf in Lichtrichtung gekrümmte Flächen von Zylindern, Kugeln, Gewindeflanken u.a. auszudehnen, die Aperturen der Strahlengänge bedingen aber gewisse Grenzen. Bei stark konvexgekrümmten Flächen fällt die reflektierte Marke eng mit dem Schattenprofil zusammen und ist schwer von ihm zu trennen; bei stark konkaven Flächen wird die Apertur bis in die Randzonen beschnitten (Fig. 39). An Gewinden ergeben sich je nach Krümmung und Steigung in beiden Objektiven verschiedene Aperturausschnitte (Fig. 40), deren Fokussierung bei den üblichen sphärischen Fehlern der Objektive variiert, so daß hier nur Spezialobjektive zu richtigen Ergebnissen führen werden.

Ganz besonders sei auf die Möglichkeit hingewiesen, die Dicke von Folien, Filmmaterial, Kunststoffen und dergleichen, die unter einer Meßkraft deformiert werden würden, fehlerfrei zu messen, ein Verfahren, bei dem der Brechungsindex nicht bekannt zu sein braucht, da nicht durch die Schicht hindurchgemessen wird, sondern beide Seiten von außen her angetastet werden.

Die geschilderte unangenehme Eigenschaft des Perflektometers, von der „richtigen" Fokussierung abhängig zu sein, kann wiederum durch einen optischen Kunstgriff umgangen werden, indem nämlich anstatt mit der beschriebenen Zieleinrichtung mit einem Doppelbildocular gearbeitet wird. Dabei entfällt die Möglichkeit der Parallaxe, weil erstens keine Strichplatte in der Ocularebene liegt und somit von einer Fokussierung im obigen Sinne nicht gesprochen werden kann, und zweitens die Symmetrie der Anordnung dafür sorgt, daß das Fehlerscheibchen des Doppelbildes bei „unscharfer" Einstellung zwar größer werden, aber nicht aus der Symmetrielage auswandern kann.

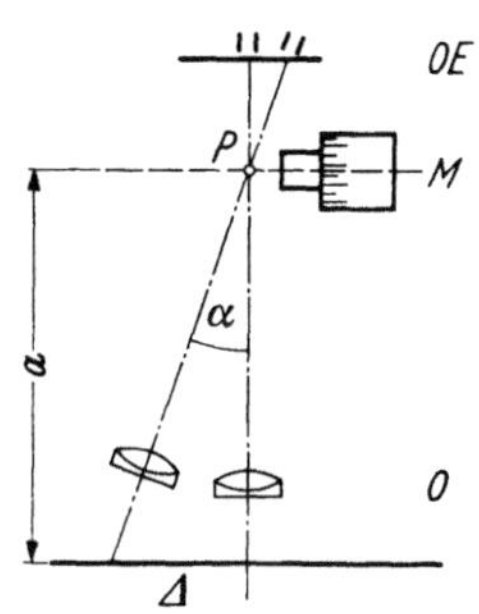

Fig. 41. Kippen des Mikroskoptubus um P ergibt Meßfehler $\Delta \simeq \alpha a$, M Mikrometer, OE Ocularebene.

Messungen mit der Zielachse können entweder durch Verschieben des Objekttisches oder des Mikroskops ausgeführt und mikrometrisch abgelesen werden. Hohen Ansprüchen genügt ein in den Objekttisch eingebauter Maßstab mit optischer Ablesung (vgl. Ziff. 33).

γ) *Messen am Kreuztisch.* Aus verschiedenen Gründen ist es vorzuziehen, das Objekt mit einem Tisch meßbar zu verschieben und nicht das zum Zielen benutzte Mikroskop. Der meßtechnisch ausschlaggebende Grund ist der, daß beim Verschieben auf unebener Führungsbahn Kippfehler entstehen. Ihre Größe ist i. allg. vom Abstand Objektebene-Maßstabsebene abhängig; denn sie sind Meßfehler erster Ordnung, nämlich proportional zum Sinus des Kippwinkels. Bei der Tischführung entstehen geringere Meßfehler als bei der Führung des Mikroskoptubus, weil die Meßstrecke des Objekts selten in größerem seitlichen oder Höhenabstand von der Meßstrecke der Maßverkörperung liegt als dies bei der Tubusführung der Fall ist. Es entstehen nur Fehler zweiter Ordnung, Cosinusfehler, wenn das Abbesche Komparatorprinzip eingehalten werden kann (vgl. Ziff. 33 und 34).

δ) *Messen an der Tubusführung.* Kippt der Mikroskoptubus um einen Winkel α um seine Führungsbahn, so ergeben sich Meßfehler am Objekt von angenähert $a \cdot \widehat{\alpha}$ (Fig. 41). Dabei ist a der Abstand zwischen Objekt und Meßspindel. Er ist im allgemeinen groß, jedenfalls zumeist größer als der seitliche oder Höhenabstand Objekt-Maßstab am Kreuztisch.

Eine dritte Meßmöglichkeit würde sich ergeben, wenn man nicht das Objekt oder den Tubus, sondern den Zielstrich in der Ocularebene verschiebt. Auf diese Weise entsteht aus dem Zielmikroskop ein Meßmikroskop, hierauf wird in Ziff. 32 eingegangen werden.

ε) *Messen an Gleitführung.* Soll das Zielmikroskop entlang einer Kernspur geführt werden, um deren statistische Richtungsänderungen messen zu können, dürfen sich Ungenauigkeiten in der Führungsbahn nicht als seitliche Auslenkungen auswirken. Um dies näherungsweise zu erreichen, ist es zweckmäßig, das Mikroobjektiv mit einer Parallelogrammfederung quer zur Bahnführung zu versehen und es entlang eines sorgfältig ebenen Lineals, z. B. einer interferenzgenau ebenen Glasfläche gleiten zu lassen, welche auf dem Objekttisch parallel zur Kernspur fest ausgerichtet ist. Geringe seitliche Versetzungen der Kernspurplatte zusammen mit dem Glaslineal relativ zum Mikroskoptubus werden vom Objektiv mitgemacht.

Sie wirken sich auf das Bild in der Ocularebene nur in der Größe der Objektversetzungen selbst aus, also ohne den Faktor des Abbildungsmaßstabs und können daher im gleichen Verhältnis vernachlässigt werden (Fig. 42) [62].

ζ) *Verzeichnung.* In allen Fällen, in denen das Mikroskop mit feststehender Strichmarke im Ocular zum Zielen benutzt wird, nicht aber Abstände in der Ocularebene gemessen werden, darf das Objektiv mit beliebiger Verzeichnung behaftet sein.

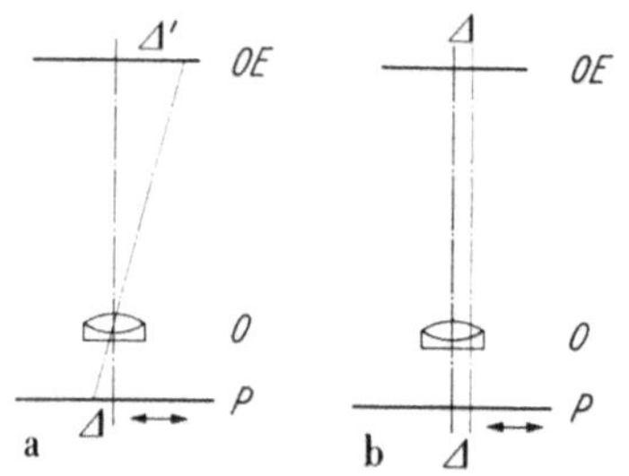

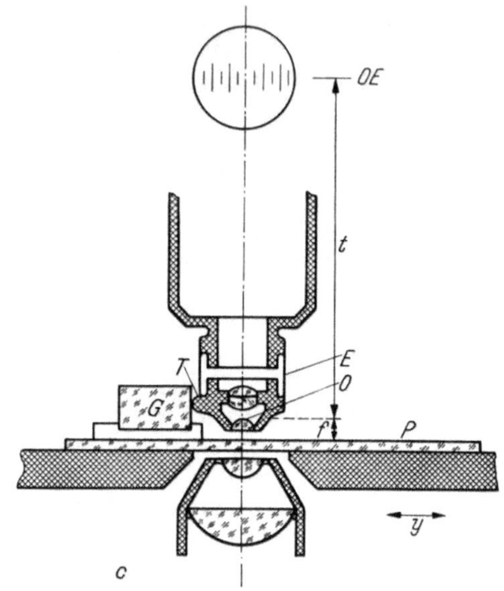

Fig. 42a—c. Objektivführung nach Stodiek. a Kernspurplatte P wandert vor dem Objektiv O um Δ seitlich aus: Meßfehler $\Delta' = \beta' \Delta$, β' Abbildungsmaßstab. b Desgl., aber Objektiv wandert mit, Meßfehler $\Delta_{OE} = \Delta_P$. c Objektivführung schematisch. Tischabweichungen in y-Richtung senkrecht zur Kernspur werden um den Faktor $1/\beta' = f/t$ reduziert. E Parallelfederung, T Taster, G Planfläche.

32. Meßmikroskop. Aus dem Zielmikroskop, das auch als optischer Fühlhebel angesehen werden kann und daher nur zusammen mit anderen Meßmitteln verwendet wird, entstehen das Meßmikroskop und das Maßstab-Ablesemikroskop, indem in der Ocularebene selbst ein Maßstab angeordnet wird, der feststehen oder auch bewegt werden kann, um die Ablage des Zieles zu messen. Ein Vorzug dieser Methode gegenüber denen, die mit dem Zielmikroskop verknüpft sind, ist der, daß die Ablage der Zielmarke in der um den Abbildungsmaßstab des Objektivs vergrößerten Darstellung der Ocularebene gemessen wird. Der Ocularmaßstab und seine Verschiebung müssen den Abbildungsmaßstab des Objektivs berücksichtigen und dürfen gröber sein als die Maßstäbe der Meßschraubentrommeln für Tisch- oder Tubusverschiebung, die in der Objektebene messen, bezogen auf gleiche Meßgenauigkeit.

Die Ablesung geschieht auch hier zuweilen an Meßschraubentrommeln in sog. Außenablesung. Optisch wertvoller sind jedoch Innenablesungen, weil Umadaptieren und Umakkommodieren erspart bleiben. Hierbei werden die Strichplatten mit den Ocularskalen durch Getriebe einfacher Art bewegt und die Größe der Bewegungen an einem mit einem Getriebeelement verbundenen Maßstab im Gesichtsfeld des Oculars abgelesen (Fig. 43a, b). Der Meßbereich ist auf das Gesichtsfeld der Ocularebene beschränkt. Um ihn zu erweitern, wird diese Meßmöglichkeit zumeist mit einer meßbaren Verschiebung des Objekttisches kombiniert.

Bei Meßocularen mit beweglichen Skalen zum Interpolieren können zu den Einfangfehlern Umkehrspannen hinzukommen, falls die Kopplung zwischen den abzulesenden Hilfsskalen und dem Ocularmaßstab nicht spielfrei ist.

Da ferner die Ablesung in einer zur Meßebene parallel versetzten Ebene erfolgt, ist der Verstoß gegen das Abbesche Komparatorprinzip (Ziff. 34) und damit der

Einfluß von Kippfehlern durch Neigungsänderungen des Mikroskops infolge Durchbiegungen des Aufbaus beim Verschieben des Komparatorschlittens zu beachten.

α) *Verzeichnung.* In jedem Fall, in dem Abstände in der Ocularebene selbst gemessen werden, spielt die Verzeichnung eine Rolle, so daß nur ausgezeichnet korrigierte Systeme als Meßmikroskope mit Skale in der Ocularebene verwendet werden sollten. Solche haben eine Verzeichnung unter 0,5 ‰ im gesamten Gesichtsfeld, so daß sie sich praktisch nur selten störend bemerkbar macht.

β) *Profilprojektor.* Das einfachste Meßmikroskop mit Maßstab ist der Projektor. Er besteht aus einer telezentrischen Beleuchtungseinrichtung, dem durchbrochenen Objekttisch, dem Objektiv und einem Schirm oder einer Mattscheibe

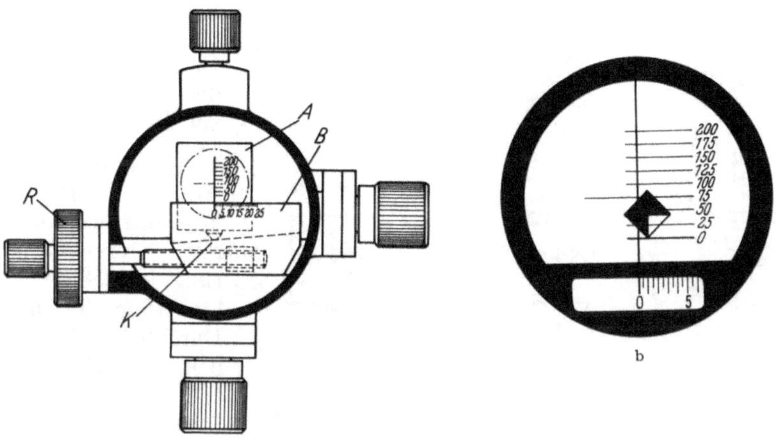

b

a

Fig. 43 a u. b. a Ocular mit Keilgetriebe zum Messen von Vickers-Eindrücken. Mit dem *Rändel R* wird der Keil *K* verschoben, an dem der Taststift der Skale *A* gleitet. Skale *B* auf dem Keilstück dient als Feinunterteilung von *A*. b Gesichtsfeld mit Vertikalskale *A* und Feinskale *B* und Bild eines Vickers-Eindrucks.

zum Ausmessen des Bildes, eines Schattenbildes bei Durchlichtbeleuchtung, eines Auflichtbildes z.B. für Prägungen bei Auflichtbeleuchtung, welche zwischen Objektiv und Schirm oder auch vor dem Objektiv eingespiegelt wird. Maßstäbe, Kurven, Winkel und ganze Werkstattzeichnungen können in der Bildebene zum Vergleichen und Messen herangezogen werden. Die Abbildungsmaßstäbe von Profilprojektoren liegen allgemein zwischen 10- und 100fach. Diese Werte werden durch festen Abstand zwischen Objektiv und Bildschirm bei telezentrischem Strahlengang auf der Objektseite sicher eingehalten, auch wenn die Objekte wechseln und durch Bewegen des Objekttisches nur nach Augenschein scharfgestellt werden. Die Verzeichnung wird durch Projizieren eines Objektmaßstabes und Vermessen seines Bildes auf dem Schirm festgestellt. Das Objektfeld besitzt gewöhnlich zwischen 50 und 5 mm Durchmesser.

Die Durchprojektion auf eine Mattscheibe gibt ein sehr helles Bild, das allerdings körnig ist und dessen Intensität ohne Fresnel-Linse zum Rande hin abfällt. Die Aufprojektion auf einen Schirm ergibt ein weniger, aber gleichmäßiger helles Bild ohne Korn. Für Durchprojektion und Aufprojektion ist die Betrachtungsrichtung Beobachter-Schirm zur Lichtrichtung Objektiv-Schirm um 180° verschieden, bei Durchprojektion betrachtet man gegen die Lichtrichtung, bei Aufprojektion mit ihr. Daraus resultiert ein Unterschied in der Bildlage, das Mattscheibenbild ist ohne Zwischenspiegel umgekehrt, das Schirmbild aber höhen-

oder seitenvertauscht, also spiegelbildlich. Das ist bei Vergleichen von Objekten gegen Zeichnungen zu beachten.

Durchprojektion wird häufig an Meßgeräten und bei der sog. Einbauoptik angewendet. Um gleichmäßig helle Bilder zu bekommen, versieht man diese Art Ablesevorrichtungen mit Feldlinsen, z.B. in der Art von Fresnel-Linsen, welche die Eintrittspupille, hier die Lichtquelle, in die Eintrittspupille des Auges abbilden sollen. Die Zonenabstände der Fresnel-Linse müssen so klein gewählt werden, daß sie vom Auge in deutlicher Sehweite nicht als störendes Kreisraster vor dem Mattscheibenbild erkannt werden.

Wenn aus Gründen der Verzeichnung nicht im Bild eines Projektors selbst gemessen werden soll, wird der Projektor durch Anbringen eines Strichkreuzes oder einer Gewindestrichplatte auf der Schirmebene zum Zielen verwendet. Gemessen wird dann an den Mikrometerspindeln zur Tischverschiebung.

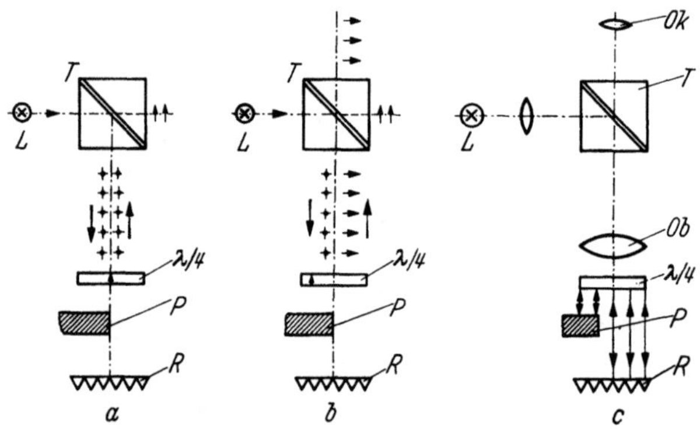

Fig. 44a—c. „Durchlicht durch Auflicht“, a, b Wirkungsweise, c Aufbau im Meßmikroskop. *T* polarisierende Teilerfläche, *P* Prüflingsprofil, *R* Tripelspiegelwabe, *L* Lichtquelle, *Ob* Objektiv, *Ok* Ocular. a λ/4-Blättchen in 0°-Stellung, Auflicht von *P* durch *T* ausgeschaltet. Gezeichnet ist nur die Schwingungsrichtung des auf *P* fallenden und von *P* reflektierten Lichts, das auf *R* fallende ist stets depolarisiert. b λ/4-Blättchen in 45°-Stellung ergibt Drehung der Polarisationsebene des Auflichts um 90°. Folglich wird Auflicht von *P* kommend durch *T* durchgelassen. Prüfling erscheint zusätzlich in Auflicht.

γ) *„Durchlicht durch Auflicht“.* Profilbildmikroskope werden i. allg. mit Durchlichtbeleuchtung betrieben. Die Beleuchtungseinrichtung für Durchlicht ist oft hinderlich, wenn das Mikroskop an den verschiedensten Stellen eingesetzt werden soll. Außerdem kann zusätzliche Auflichtbeleuchtung sehr nützlich sein. Eine Beleuchtungseinrichtung, die Durchlicht durch Auflicht erzeugt, und unter das mit Auflicht beleuchtete Objekt gesetzt wird, besteht aus vielen in einer Ebene dicht nebeneinanderliegenden Tripelspiegeln, die die Eigenschaft haben, die als Auflicht einfallenden Strahlen um 180° umzulenken, also parallel zu ihrer Einfallsrichtung zurückzuwerfen (Fig. 44a, b, c). Dann erscheint ein im Strahlengang befindliches Objekt nicht nur im Auflicht, sondern auch als Schattenbild im Durchlicht. Das am Objekt reflektierte Auflicht kann wahlweise ein- oder ausgeschaltet werden, wenn mit einer polarisierenden Teilerfläche, also linear polarisiertem Licht gearbeitet wird. Das Objekt wird durch ein drehbares λ/4-Blättchen hindurch beleuchtet. Steht eine der beiden Polarisationsrichtungen des λ/4-Blättchens parallel zur Polarisationsebene des einfallenden Lichtes und bleibt nach der Reflexion am Prüfling die Polarisationsrichtung des Lichtes erhalten, so kann es die Teilerfläche nicht durchsetzen, das Auflichtbild ist ausgeschaltet. Wird das λ/4-Blättchen dagegen um 45° gedreht, ist die Polarisationsebene des Lichtes nach zweimaligem Durchgang (Hin- und Rückgang) durch das Blättchen um 90° gedreht. Das vom Prüfling reflektierte Licht kann also durch die Teilerfläche nach oben austreten, die Prüflingsoberfläche erscheint in hellem Auflicht. Das an den Tripelspiegeln reflektierte Licht ist weitgehend depolarisiert und ständig zu sehen [*63*].

δ) *Problematische Objekte unter dem Meßmikroskop.* Unproblematisch ist die Längen- und Winkelmessung an Objekten, die scharf begrenzte Konturen aufweisen, wie z.B. Werkzeugschneiden, Nadelspitzen oder dergleichen. Problematisch wird sie an Objekten, mit abgerundeten Begrenzungen. Dazu gehören Kugeln, Zylinder und Gewinde, schließlich aber auch Härteeindrücke mit der Brinell-Kugel und der Vickers- und Knoop-Pyramide.

Für das Messen eines Objekts gegen eine Skala im Ocular des Meßmikroskops spielen nicht nur Abbildungsmaßstab und Verzeichnung eine Rolle, sondern auch bestimmte Eigenschaften des Objektes und des Abbildungsvorganges. Diese müssen daher von Fall zu Fall betrachtet werden:

Im Durchlicht wird die Messung der Durchmesser von Zylindern und Kugeln sowie des Flankendurchmessers an Gewinden wegen der unter bestimmter Neigung streifend einfallenden Beleuchtungsbündel problematisch, und es ist notwendig, Korrekturen in Abhängigkeit vom Durchmesser des Objekts und der Beleuchtungs- sowie Beobachtungsapertur anzubringen [*64*], [*65*]. Die Korrekturen sind zumeist empirisch gefunden und für bestimmte optische Strahlenbegrenzungen in Kurven- oder Tabellenform dargestellt.

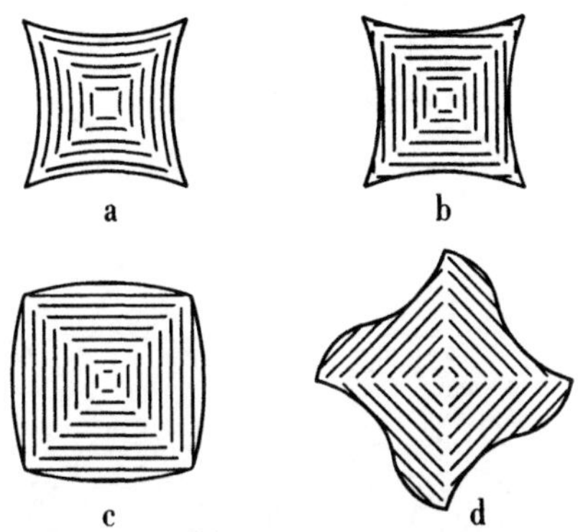

Fig. 45a—d. Interferenzstreifentest zur Deutung der Verformung von Vickers-Eindrücken, schematisch. a Kissenform, durch elastische Rückfederung des Materials nach Entfernen des Diamanten entstanden, zu beobachten an Gläsern, Kunststoffen, gehärteten Stählen. Kriterium: Interferenzstreifen im Eindruck gekrümmt. b Kissenform durch Herunterziehen des Materials während des Eindringens des Diamanten entstanden, zu beobachten an galvanischen Schichten. Interferenzstreifen gerade. c Tonnenform durch Verdrängen des Materials normal zu den Seitenflächen des Diamanten zu Aufwölbungen, an die sich die Flächen des Diamanten noch zusätzlich anlehnen. Zu beobachten bei den meisten Metallen. d Beim Eindringen des Diamanten in einen Einkristall entstandene Verformung. Herunterziehen nach b und Aufwölben nach c treten hier gleichzeitig nebeneinander in Erscheinung. Nur in Fall a kann aus der Messung der Diagonalenlänge die Eindruckfläche rechnerisch ermittelt werden, in den Fällen b, c und d ist zu planimetrieren.

Ein anderer Fall ist ebenso von den Beobachtungsbedingungen abhängig: Im Auflicht soll beispielsweise die Diagonale eines Vickerseindrucks, wie er in der Mikrohärtemessung von einer vierseitigen Diamantpyramide unter einer bestimmten Prüfkraft erzeugt wird, der Länge nach gemessen werden. Aus der Länge der Diagonalen berechnet sich die Eindruckgröße; daher ist es notwendig, die Diagonale in ihre feinsten Ausläufer zu verfolgen und dabei das Eckenauflösungsvermögen des optischen Systems zu berücksichtigen. Ecken von Polygonzügen werden nur verrundet abgebildet und an den Diagonalen einer quadratischen Fläche fehlen infolgedessen Beträge in der Größe des nach der Abbeschen Formel berechneten mikroskopischen Auflösungsvermögens, was besonders bei kleinen Eindrücken schwer wiegt und zu kleine Eindrucksflächen, d.h. zu hohe Härtewerte vortäuscht. Mit Mikroskopen der Apertur $> 0{,}7$ lassen sich Diagonalen mit einer Standardabweichung (DIN 1319) von $s = \pm 0{,}1$ µm durch ein und denselben Beobachter reproduzieren; mehrere Beobachter streuen um $\pm 0{,}3$ µm. Der durch das Auflösungsvermögen verursachte systematische Fehler beträgt in diesem Falle aber schon rund $-0{,}4$ µm.

Da es sich um schwierig zu messende Objekte handelt, sei hier eine ausführlichere Einschaltung erlaubt, die auch von der Möglichkeit, das Interferenz-Mikroskop (Abschn. VI) heranzuziehen, schon Gebrauch macht.

Vickers-Härtewerte können aus der Diagonalen-Messung unter der Voraussetzung berechnet werden, daß die Eindrücke quadratische Gestalt haben, was häufig nicht der Fall ist, z.B. nicht in Einkristallen, Gläsern, Kunststoffen. In diesen Fällen sind Untersuchungen über die Entstehungsursache der Eindruckform notwendig.

Mit dem Interferenzmikroskop entdecken wir: Kissen- und tonnenförmig aussehende Eindrücke sind durch die Diagonalenlänge nur auszumessen, wenn das Interferenzstreifenbild der Schichtschnitte parallel zur Oberfläche auch kissen-

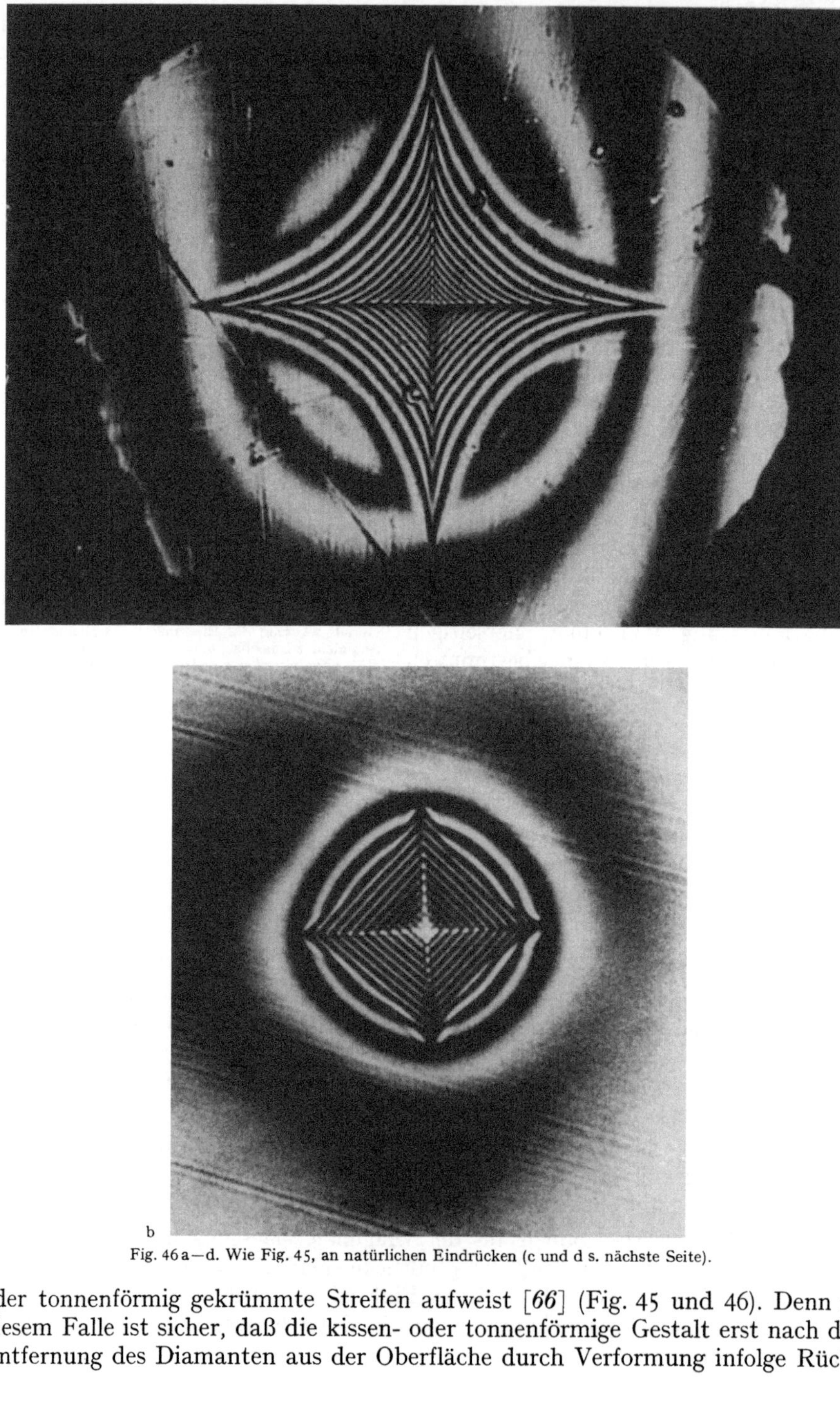

a

b

Fig. 46a—d. Wie Fig. 45, an natürlichen Eindrücken (c und d s. nächste Seite).

oder tonnenförmig gekrümmte Streifen aufweist [*66*] (Fig. 45 und 46). Denn in diesem Falle ist sicher, daß die kissen- oder tonnenförmige Gestalt erst nach der Entfernung des Diamanten aus der Oberfläche durch Verformung infolge Rück-

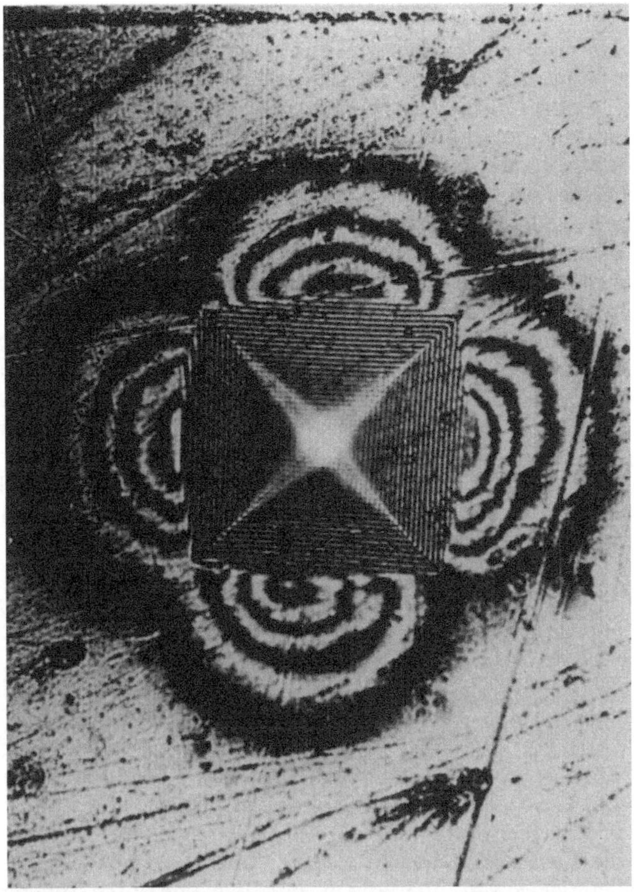

c

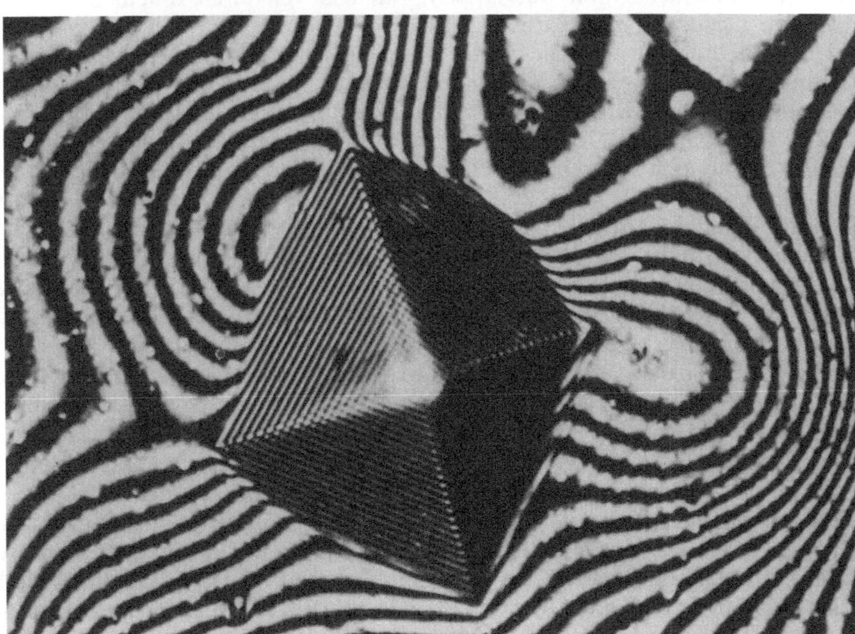

d

Fig. 46 c u. d.

federung des Materials entstanden sein kann, und der Diamant ursprünglich einen quadratischen Eindruck erzeugt hat, so daß die Berechnung der Vickers-Härte $HV = f(P/d^2)$, P Prüfkraft, d Diagonale des Eindruckes, als Flächendruck zulässig ist. Weist das Interferenzbild gerade Streifen auf, was sowohl bei Kissenform als besonders häufig auch bei Tonnenform der Fall ist, so hat der Diamant die Oberfläche verformt und nicht quadratisch angeschnitten; er ist also auch nicht von einer in der Projektion quadratischen Fläche getragen worden (vgl. Fig. 45 und 46b und c). Daher kann die Fläche nicht ausreichend genau durch Diagonalenmessung ermittelt werden. In Einkristallen stellen sich häufig Mischformen von kissen- und tonnenförmig aussehenden Eindrücken ein, da das Material z.T. am Diamanten aufgewölbt, z.T. in die Tiefe mitgenommen wurde (Fig. 45d und 46d). In diesen Fällen muß die Eindruckfläche durch Winkel-Strichplatten gemittelt werden (Fig. 47a). Dabei beurteilt das Auge vermöge seiner außerordentlichen Symmetrieempfindlichkeit, ob die beiderseits des Striches liegenden Flächenstücke gleich sind, d.h. ob ebenso viel von der Eindruckfläche außen abgeschnitten wird, wie innen hinzugefügt ist. Wegen der Symmetriebedingung bezieht sich die Messung auf die Mitte des Striches, daher soll die Nullstellung, d.h. die Eindruckfläche Null gemessen werden, wenn beide Winkel so zusammengeführt sind, daß sie ein einziges Strichkreuz ergeben (Fig. 47b). Eindrucke in der Größenordnung der Strichdicke können so nicht mehr richtig bestimmt werden, da sie unter den Strichen verschwinden würden; hier stößt man an die Grenze des Verfahrens. Es ist daher auch üblich, die Eindruckfläche Null dadurch festzulegen, daß die Striche der Einfangwinkel einander soweit überlappen, daß das von ihren Innenkanten eingeschlossene Quadrat gerade zu Null wird.

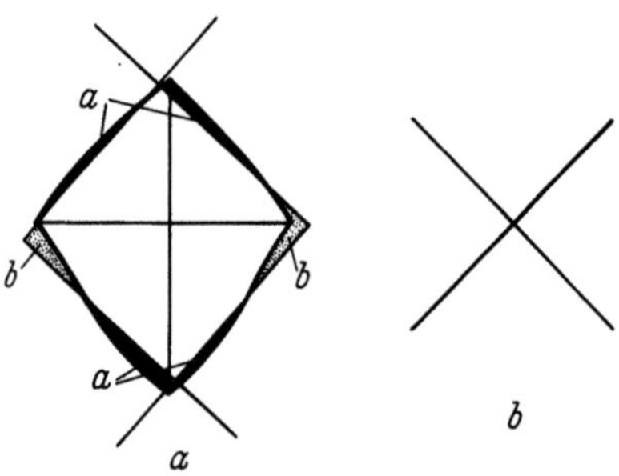

Fig. 47 a u. b. a Auswerten eines verformten Einkristall eindrucks mit Hilfe eines Rechtwinkelstrichkreuzes durch subjektives Planimetrieren, $\Sigma a = \Sigma b$, schematisch. b Nullstellung des Strichkreuzes.

Diese Stellung ist unsicher, weil die Strichwinkel auf zwei verschiedenen Strichplatten hintereinander liegen und an den Strichen gebeugtes Licht auch dann noch zwischen den Strichen hindurchkommt, wenn die geometrische Nullstellung bereits erreicht ist. Wie weit man die geometrische Nullstellung überfahren muß, um den Lichteindruck zum Verschwinden zu bringen, hängt von der Leuchtdichte der Lichtquelle ab. Abgesehen hiervon wird bei dieser Definition der Nullstellung die auf der Symmetrieempfindung beruhende oben beschriebene Mittelung der Eindruckfläche bei kleinen Eindrücken bemerkbar fehlerhaft.

Sollen beliebige Helldunkelgrenzen parallel zu Ocularstrichen eingefangen werden, so wird der genaue Einfang ebenfalls problematisch, da ein exaktes Symmetrieren zwischen Doppelstrichen nicht möglich ist. Entweder unterbricht man die Ocularstriche, z.B. bei Gewindestrichplatten, um sie so genau wie möglich mit der Helldunkelgrenze zu halbieren, (vgl. Fig. 5, S. 764). Oder man bringt die Helldunkelgrenze nur nahe an den Ocularstrich heran, dann muß aber die Strichbreite berücksichtigt werden. Um in diesem Fall die Strichbreite und die „optische Umkehrspanne" auszuschalten, mißt man stets von der gleichen Seite des Striches her, d.h. man nähert die eine Helldunkelgrenze des Objekts dem Ocularstrich bis auf einen Lichtspalt und unterfährt mit der diametralen Helldunkelgrenze den Ocularstrich bis sie eben neben ihm wieder auftaucht. Auch bei Vickers-Eindrücken muß die eine Ecke den Ocularstrich von außen berühren [*66*], die andere Ecke den Ocularstrich gerade durchstoßen (Fig. 48).

ε) *Maßstab-Ablesemikroskop.* Maßstab-Ablesemikroskope auf Komparatoren und in Maschinen dienen zum Ablesen von Tischverschiebungen an in die Tische eingebauten Maßstäben. Diese sind zumeist in Millimeter geteilt. Die Teilung in Zehntel-Intervalle geschieht in den Mikroskopen an den projizierten Millimeter-Intervallen. Bestimmte Einrichtungen im Mikroskop sorgen dafür, daß eine Unterteilung in Tausendstel Millimeter ermöglicht wird.

1. Zeiss-Spiralocular. Eine kontinuierliche Zehntel-Millimeter-Teilung be-

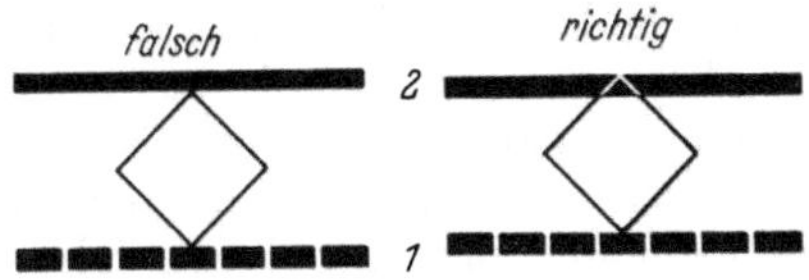

Fig. 48.

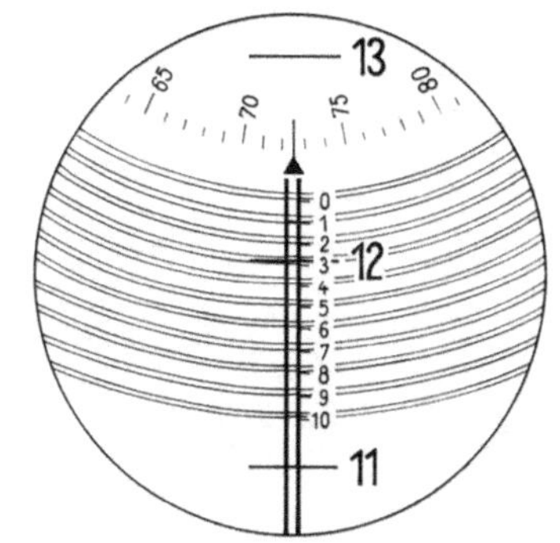

Fig. 49.

Fig. 48. Anlegen der Meßstriche *1* und *2* an einen Vickers-Eindruck, links falsch, rechts richtig.

Fig. 49. Gesichtsfeld des Zeiss-Spiraloculars nach Bauersfeld zum Ablesen von mm-Maßstäben. *11, 12, 13* projizierte Maßstabsstriche. Ortsfester Ocularrechen *0* bis *10* „Zehntel mm", jedes Zehntel durch einen Gang der drehbaren Doppelspirale kontinuierlich überbrückbar. Mit der Spirale dreht sich eine Kreisteilung, die die Steigung hundertfach unterteilt um μm abzulesen.

sitzt das Zeiss-Spiralocular [*67*] (Fig. 49), bei dem einer Drehung um 360° die Spiralsteigung 0,1 mm entspricht. Bei 10 Windungen der Spirale kann somit 1 mm durch eine Umdrehung vollständig überbrückt werden, wobei noch Tausendstel Millimeter an der Kreisteilung abzulesen und Zehntausendstel schätzbar sind.

2. Leitz-Planplattenocular. Ein diskontinuierlich geteiltes Ocular zeigt

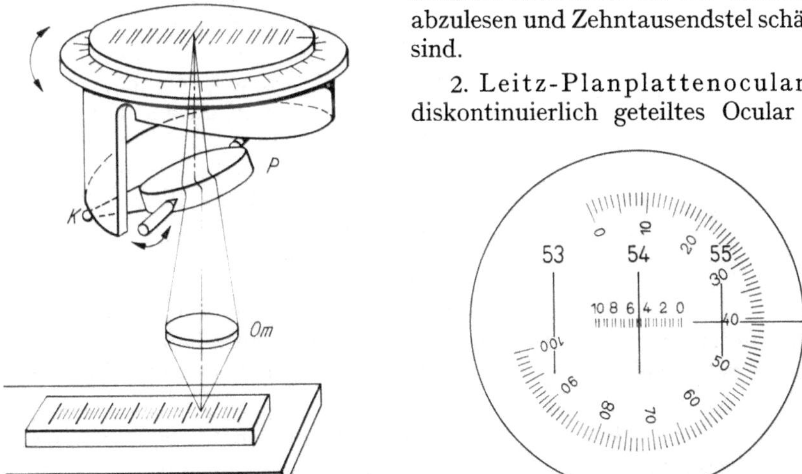

a b

Fig. 50a u. b. a Schema und b Gesichtsfeld des Leitz-Planplattenoculars zum Ablesen von mm-Maßstäben. Die Planparallelplatte *P* kann durch Drehen einer Kurve *K* geneigt werden und versetzt den durch Objektiv *Om* projizierten Maßstabsstrich parallel zu sich. An einer mit der Kurve verbundenen Kreisteilung werden die μm abgelesen, am festen Ocularrechen die Zehntel mm.

Fig. 50 [*68*], [*69*]. Die Strichplatte steht fest. Das projizierte Bild des Maßstabstriches kann durch Neigen einer im Abbildungsstrahlengang befindlichen Planparallelplatte meßbar verlagert werden, falls es nicht schon von vornherein in einen Doppelstrich fällt. Die Schraubenkurve, durch deren Drehen die Planparallelplatte geneigt wird, ist mit einer zweiten Ocular-Strichplatte mit Kreisteilung versehen, so können auch hier die µm im Ocular abgelesen und ihre Bruchteile geschätzt werden.

3. Zentimeter-Oculare. Wenn Genauigkeiten von $^1/_{200}$ mm ausreichen, also etwa eine Dezimale weniger als mit den beschriebenen Ocularen erreicht werden kann, werden die Maßstäbe eine Zehnerpotenz weniger fein geteilt, also mit cm-Strichabständen versehen [*70*]. Auch in diesem Fall soll der hohen Einfanggenauigkeit wegen Symmetrieabgleich vorgenommen werden. Fehlertabellen für den Maßstab sind bei diesem für die Praxis geschaffenen Fall unpraktisch. Um den Einfluß vorhandener Strichablagen herabzusetzen, führt man die Maßstabstriche doppelt aus und fängt sie im Zehnerrechen der Mattscheibe, der diesmal aus Einzelstrichen besteht, symmetrisch ein. Da man auf diese Weise zwei Teilungen auf dem Maßstab gleichzeitig zum Messen benutzt, mitteln sich die Teilfehler, und man erhält eine höhere Genauigkeit als bei Ablesung einer einzelnen Teilung. Der Zehnerrechen wird mikrometrisch verschoben und seine Einstellung an einer Rundteilung auf Hundertstel Millimeter und geschätzte Bruchteile abgelesen.

4. Goniometer-Oculare. In Goniometer-Ocularen zur Winkelmessung an projizierten Profilbildern kann eine Strichplatte mit Strichkreuz um ihren Mittelpunkt gedreht werden. Das Strichkreuz wird nacheinander mit den Schenkeln des zu messenden Winkels zur Deckung gebracht. Die dazugehörige Drehung wird an einer auf der gleichen Strichplatte befindlichen Kreisteilung abgelesen. Wegen des für die Kreisteilung erforderlichen großen Radius — die Kreisteilung liegt weit außerhalb des Gesichtsfeldes — ist das Goniometer-Ocular mit einem besonderen kleinen Ablesemikroskop über der Kreisteilung versehen, mit dem sich diese auf wenige Bogenminuten genau ablesen läßt. Exzentrizitätsfehler vermeidet man, wenn die Messung mit um 180° versetzter Teilung — neues Anlegen der Strichkreuze — wiederholt und das Mittel gebildet wird.

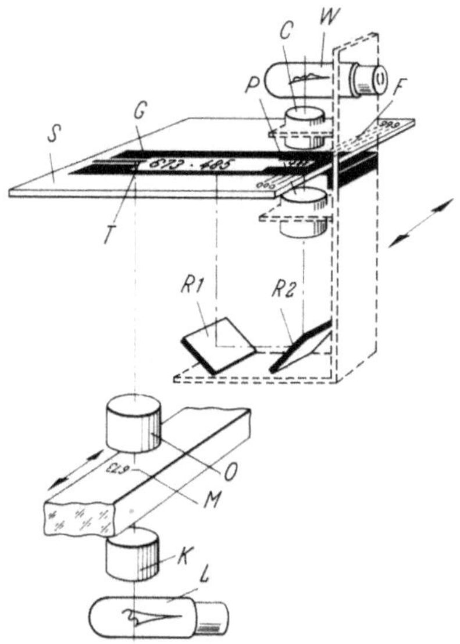

Fig. 51. Optisches Zählwerk mit digitaler Anzeige. Lampe *L*, Kondensator *K* und Objektiv *O* bilden den Maßstabsstrich *M* und seine Bezifferung auf die Mattscheibe *S* ab. Die Einfanggabel *G* ist zusammen mit einem zweiten Abbildungsstrahlengang verschiebbar, der von 1000 Ziffern, die das mm-Intervall unterteilen, eine hinter die Maßstabsziffer projiziert. *W* Lampe, *C* Kondensor, *P* Projektionsobjektiv, R_1, R_2 Spiegel, *F* Feinskale auf der feststehenden Mattscheibe *S*.

5. Projektionsablesung und Zähleinrichtung. Die Ocularebene des Maßstab-Ablesemikroskops kann als Projektionsebene mit Mattscheibe, Fresnel-Linse oder Feldlinse ausgeführt werden (vgl. Ziff. 32β). Projektionsablesungen mit freier Betrachtung durch beide Augen werden dem Oculareinblick vorgezogen, weil sie dem Beobachter mehr Bequemlichkeit bieten und ihn weniger ermüden. Die Meßgenauigkeit braucht bei der Abbildung auf einen Schirm nichts einzubüßen, wenn der Bildschirm kornlos, z.B. opak ist, ja sie kann durch das beidäugige Sehen noch gewinnen [*1*], [*71*]. Mit der Projektion läßt sich ein anderer Bequemlichkeitsfaktor verbinden, der für die Praxis interessant ist, die Ablesung eines Zählwerks anstelle von Skalen. Fig. 51 zeigt das Prinzip eines optischen Zählwerks [*72*]. Der Maßstabstrich wird eingegabelt und sein Zahlenwert erscheint im Projektionsbild. Die Einfanggabel bewegt sich als Schieber längs einer Zahlen-

reihe, die die Unterteilung darstellt. Durch eine zweite Projektionseinrichtung wird die zur Schieberstellung gehörige Unterteilungsziffer neben die Maßstabsmillimeter projiziert, so daß sich die Einstellung auf Tausendstel Millimeter direkt als Zahlenangabe darstellt.

Zählwerke lassen sich auch mechanisch an die Mikrometertrommel anschließen, so daß der projizierte Zahlenwert des Maßstabstrichs mit der Ziffernangabe auf dem Zählwerk zum Meßwert zusammenzufügen ist.

6. Photoelektrische Oculare. Obwohl das Auge für den Symmetrieeinfang außerordentlich empfindlich ist, besteht das Bedürfnis, seine Leistung zu übertreffen. Eine Möglichkeit hierzu geben photoelektrische Einfangvorrichtungen. Die Photozelle befreit das Auge von seiner Aufgabe, sie arbeitet zuverlässiger und genauer. Ganz grob verglichen, sind im Symmetrieeinfang mit dem Auge noch Sehwinkel von 5″ zu erreichen, mit der Photozelle Sehwinkel von 0,5″. Die von der Photozelle abgegebenen Ströme lassen sich nach verschiedenen Richtungen hin ausnutzen, zur visuellen Anzeige an Zeigerinstrumenten, zur digitalen Anzeige, zum Registrieren, zum Steuern und Regeln von Vorgängen. Wichtig ist, daß nicht nur das Auge, sondern auch die Hand ersetzt wird, wodurch kontinuierliche Messungen mit hohen Geschwindigkeiten erst möglich werden.

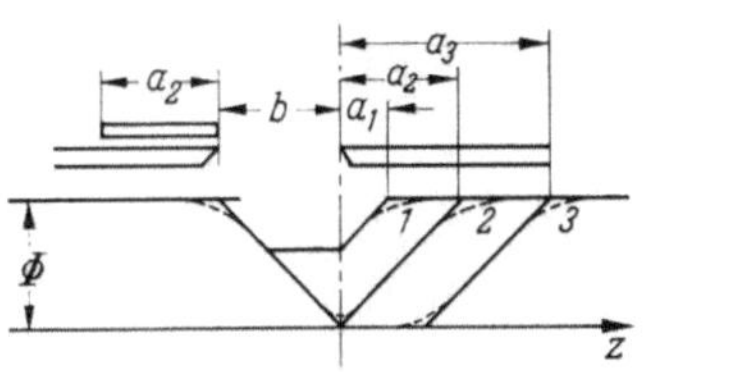

Fig. 52. Abhängigkeit des Lichtstroms Φ von der Abdeckung des Spaltes b durch das Strichbild a, bezogen auf die Koordinate z der Vorderkante von a. Φ_1 für $a_1 < b$, Φ_2 für $a_2 = b$, Φ_3 für $a_3 > b$.

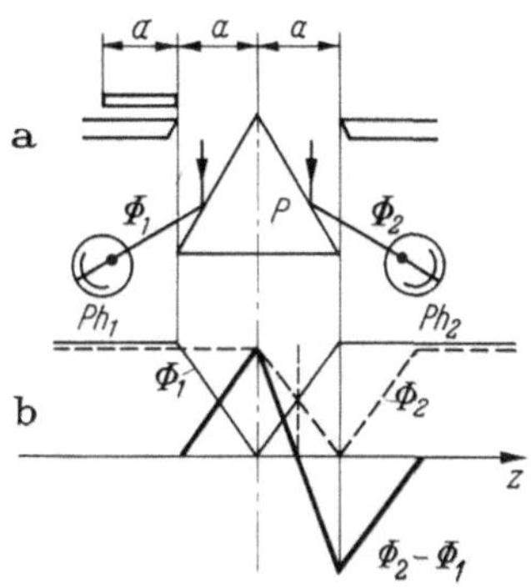

Fig. 53 a u. b. a Maßstabsstrichbild a vor Einfangspalt $2a$. Kante des Prismas P teilt den Spalt symmetrisch. Photozellen Ph_1 und Ph_2 liegen in Differenzschaltung. b Abdeckfunktionen der Lichtströme Φ_1 und Φ_2 und Differenz $\Phi_2 - \Phi_1$ in Abhängigkeit von der Koordinate z der Strichvorderkante.

Allerdings sind bestimmte Maßnahmen anzuwenden, damit der Photozelle nicht ein mehr oder weniger konstanter Lichtstrom, sondern zwei symmetrische Lichtströme oder Wechsellicht dargeboten werden [*73*], [*74*]. Das Aufsuchen des Minimums eines Lichtstroms, etwa durch Vorüberfahren des dunklen Maßstabstriches vor einem Spalt, ist von geometrischen Unregelmäßigkeiten in den Strichen und von zufälligen Schwankungen in der Stromversorgung der Beleuchtung abhängig, so daß damit keine genügend genaue und eindeutige Ablesung erhalten würde. Fig. 52 zeigt diesen Vorgang. Die Spaltöffnung wird nach und nach von der einen Seite her abgedeckt; der die Photozelle treffende Lichtstrom nimmt ab. Er durchläuft ein Minimum, welches sich nach der Breite des Maßstabstriches im Verhältnis zur Spaltbreite richtet und wegen seiner Talbreite sowie Stromschwankungen als Einfangkriterium ungeeignet ist.

Von der Strichbreite und den Stromschwankungen ist die in Fig. 53 gezeigte Anordnung mit zwei Photozellen unabhängig. Das einzufangende Bild des Maßstabstriches ist halb so breit wie der Spalt, so daß es in Mittellage zwei symmetrisch gleiche Lichtströme Φ_1 und Φ_2 passieren läßt. In dieser Symmetrielage wirken sich Stromschwankungen auf beide Photozellen in gleichem Maße aus, so daß die Differenz der beiden Photoströme stets gleich Null bleibt. Individuelle Schwankungen in der Empfindlichkeit der Photozellen können bei langen Meßreihen den Abgleich verlagern, so daß auch diese Anordnung noch verbesserungsbedürftig ist.

Es ist naheliegend, einer einzigen Photozelle beide Lichtströme in periodischem Wechsel zuzuführen und dadurch abzugleichen, daß der entstehende Photostrom keine Wechselstromkomponente mehr besitzt. Diese Möglichkeit ist in Fig. 54 dargestellt. Langsame zeitliche Schwankungen im Lichtstrom oder in der Empfindlichkeit der Photozelle spielen hierbei keine Rolle mehr.

Eine andere Möglichkeit bietet sich, indem das dunkle Bild des Maßstabstriches (Fig. 52), über dem Spalt hin und her schwingt. Die Symmetrielage läßt sich daran erkennen, daß im Bereich der Elongationsmaxima sowohl nach rechts wie nach links gleiche Lichtimpulse auf die Photozelle treffen. Die Zelle gibt in diesem Falle einen pulsierenden Gleichstrom mit der doppelten Frequenz der Blendenschwingung ab, und die Frequenz der Grundschwingung fehlt bei Sym-

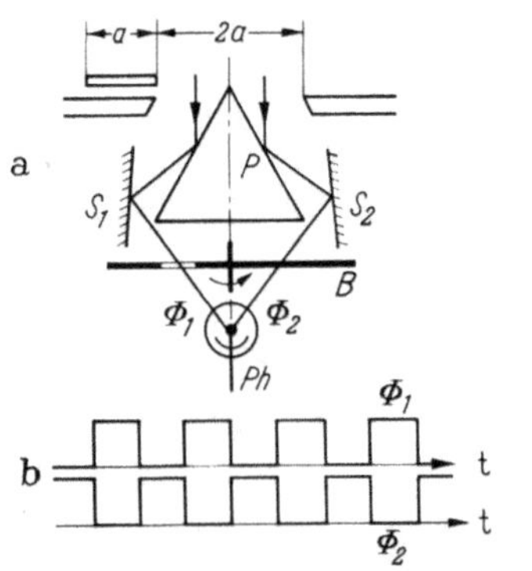

Fig. 54 a u. b. a Maßstabsstrich a vor Einfangspalt $2a$. Prisma P und Spiegel S_1 S_2 leiten die Lichtströme Φ_1 Φ_2 durch rotierende Blende B einer einzigen Photozelle zu. b Lichtströme Φ_1 Φ_2 bei symmetrischer Lage des Maßstabsstriches im Spalt in Abhängigkeit von der Zeit t bei rotierender Blende B.

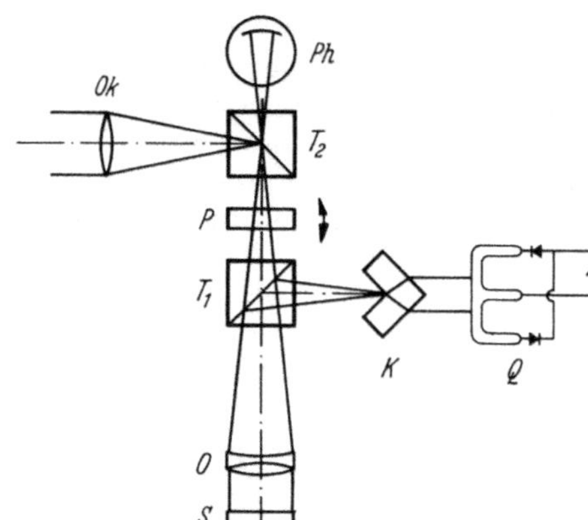

Fig. 55. Photoelektrischer Symmetrieeinfang einer Prismenkante K auf einem Spalt in der Teilerfläche T_2 für ein Autokollimationsfernrohr. Q Entladungslampe mit periodisch wechselnden Zweigen, O Fernrohrobjektiv, S Drehspiegel, T_1 Teilerwürfel, P Planplattenmikrometer, Ph Photozelle, Ok Ocular.

metrie vollkommen. Ist der Verstärker nur auf die Grundschwingung abgestimmt, so geht das Anzeigeinstrument in diesem Augenblick auf Null. Spalt und Strich sollen annähernd gleich breit sein. Ist der Strich völlig ausgewandert, so bleibt das Licht trotz der Schwingung konstant. Ein Wechselstromverstärker kann auf Gleichstrom aber nicht ansprechen, die Anzeige ist daher in diesem Bereich ebenfalls Null. Im Übergangsbereich schwingt der Strich in den Spalt von der einen oder anderen Seite hinein, womit periodische Verdunkelungen auftreten. Die Phasenlage dieser periodischen Verdunkelungen gegenüber der Strichschwingung als Bezugsschwingung wechselt um 180°, wenn der Symmetrieabgleich durchfahren wird.

Hier soll nicht auf Einzelheiten elektronischer Schaltungen und die Vielzahl von Möglichkeiten eingegangen werden, die von der Photozelle abgegebenen elektrischen Impulse zu verarbeiten, sondern auf einige optische Varianten für den photoelektrischen Symmetrieeinfang.

Statt das Strichbild über den Spalt schwingen zu lassen, kann man auch den Spalt zum Schwingen bringen. Das Bild des Striches entwirft man beispielsweise über einen schwingenden Drehspiegel oder über eine in den konvergenten Strahlengang gesetzte Planparallelplatte, die Neigungsschwingungen ausführt. Den Spalt kann ein elektromagnetisches System zum Schwingen bringen.

Fig. 55 zeigt eine davon abweichende Anordnung mit einer Spezial-Quecksilberdampflampe, deren beide Zweige im Impulsbetrieb mit 180° Phasenverschiebung arbeiten [*75*]. Die Lichtströme durchsetzen wechselseitig ein Prisma, dessen innere Kante als Strichmarke dient, welche über die Autokollimation in eine Spaltebene abgebildet wird, die mit der Ocularebene identisch ist. Wandert die Kante infolge von Spiegelwinkeländerungen über den Spalt hinweg, so wechseln die von der Photozelle aufgenommenen Impulse die Phase, womit die Richtung der Auslenkung aus dem Symmetrieeinfang gegeben ist. Mit Hilfe einer neigbaren Planparallelplatte läßt sich die Symmetrie auf dem Spalt wieder herstellen aufgrund des Kriteriums, daß die Lichtströme beider Phasen im Gleichgewicht sein müssen. Die übrigen Einzelheiten in Fig. 55 sind für sich und aus der Legende verständlich.

Eine andere bemerkenswerte Einfangmethode besteht darin, die Ocularebene durch Polarisationsfilter in zwei unter 90° zueinander polarisierte Felder zu

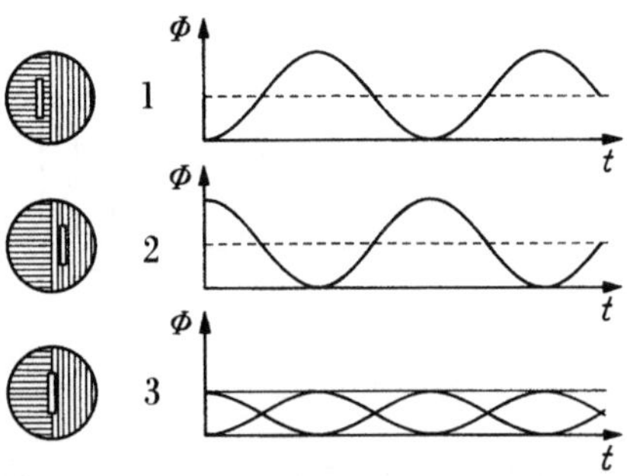

Fig. 56.

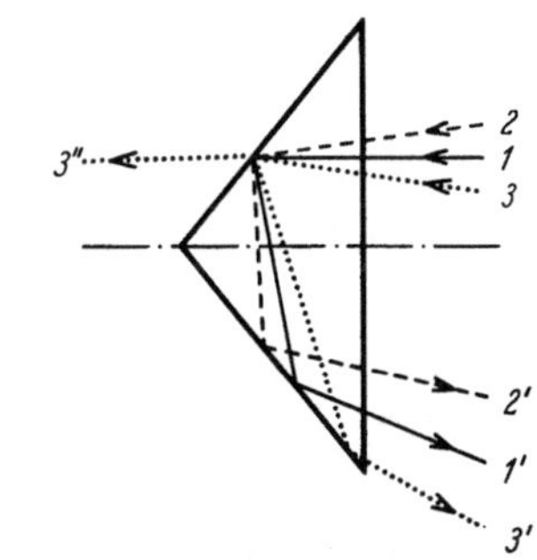

Fig. 57.

Fig. 56. Photoelektrischer Einfang eines mit linear polarisiertem Licht abgebildeten Spaltbildes, dessen Polarisationsrichtung rotiert. Das Spaltbild fällt auf ein in zwei um 90° verschieden polarisierte Hälften geteiltes Gesichtsfeld vor einem Multiplier. Die Diagramme geben die erzeugten Wechselspannungen wieder.

Fig. 57. Prisma zum Symmetrieren zweier Lichtströme mittels Totalreflexion. Von den unter verschiedenen Winkeln zur Symmetrieebene einfallenden Strahlen *1*, *2*, *3* erreicht nur *3* eine Photozelle hinter dem Prisma. *1* und *2* werden total reflektiert. Die gezeichnete Anordnung ist spiegelsymmetrisch zur Symmetrieebene zu vervollständigen. In Symmetriestellung fällt gleichviel Licht von oben wie von unten in die Photozelle.

halbieren (Fig. 56). Die einzufangende Marke, ein Spalt, wird auf die Filter mit polarisiertem Licht abgebildet, dessen Polarisationsebene um die optische Achse rotiert [*76*]. Damit wird erreicht, daß außerhalb der Symmetrielage periodisch Impulse auf die Photozelle treffen, deren Frequenz doppelt so groß ist wie die Drehfrequenz des Polarisationsfilters und deren Phasenlage mit den beiden Polarisationsrichtungen der Filter zusammenhängt. Aus der Phasenlage des überwiegenden Impulses läßt sich wieder die Richtung zum Symmetrieeinfang hin bestimmen. Im Symmetriefall sind die Impulse beider Polarisationsrichtungen im Gleichgewicht.

Von Prismen her ist bekannt, daß sie Licht aus bestimmten Richtungen durch Totalreflexion umlenken. Auch dieser Effekt kann ausgenutzt werden, um die Symmetrierung eines Lichtbündels mit Hilfe von Photozellen zu erreichen [*77*]. Fällt in das in Fig. 57 eingezeichnete Prisma, welches einen Öffnungswinkel größer als 90° besitzt, Licht senkrecht zur Eintrittsfläche und parallel zur winkelhalbierenden Fläche ein, so wird es totalreflektiert (Strahl 1). Fallen Strahlen geneigt ein, so werden sie entweder weiterhin totalreflektiert (Strahl 2) oder sie werden durchgelassen (Strahl 3) und auf beiden Seiten des Prismas von je einer Photozelle aufgefangen. Fällt also ein geöffnetes, Strahlen verschiedener Neigung enthaltendes Bündel symmetrisch zur winkelhalbierenden Ebene ein, so erhalten beide Photozellen gleichviel Licht, die Differenz ihrer Signale ist Null. Fällt das Bündel geneigt ein, so überwiegt das eine Signal. Die Maßnahmen zum Auffinden der Symmetrie sind die gleichen wie oben für die anderen Verfahren beschrieben. Bei Verwendung zweier gegeneinander geschalteter Photozellen wird Wechsellicht benutzt und bei Verwendung nur einer Photozelle die wechselseitige Abdeckung der beiden Lichtströme durch eine rotierende Sektorblende.

Ein Bild läßt sich photoelektrisch nicht nur lateral symmetrieren, sondern auch zentrieren. Das ist für Autokollimationsmessungen wichtig, um den Spiegel in zwei zueinander senkrechten Richtungen ausrichten zu können. Hierfür wären vier gleich große, kreuzweise angeordnete, kreisförmige Öffnungen in der Objektebene und eine größere kreisförmige Öffnung in der Auffangebene vor dem photoelektrischen Empfänger geeignet. Die Bilder der vier Öffnungen liegen im Fall der genauen Zentrierung auf dem Rand der größeren Öffnung, wie Fig. 58 zeigt. Bei bestehender Dezentrierung weicht das Bild in der Weise ab, daß die Öffnungen verschieden stark abgedeckt sind. Bei Wechsellicht — die vier Öffnungen können durch eine rotierende Sektorblende periodisch abgedunkelt werden — ergeben sich unterschiedlich große Lichtimpulse, so daß die Aussteuerung des Fehlers ermöglicht

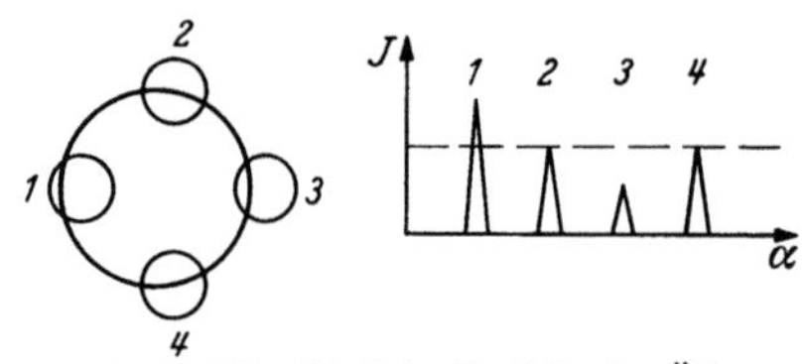

Fig. 58. Blendenbilder *1, 2, 3, 4* auf kreisförmiger Öffnung einer Photozelle. Bei periodischem Aufblenden der Lichtquelle ergeben sich Lichtströme über α entsprechend der Dezentrierung.

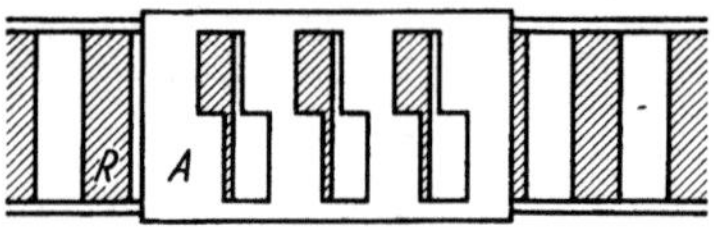

Fig. 59. Abtastplatte *A* mit zwei um 90° phasenverschobenen Rastern vor dem Rastermaßstab *R* zum Vor- und Rückwärtszählen.

wird. Besser benutzt man nur eine einzige Öffnung in der Objektebene. Durch Zwischenschalten einer geneigten Planparallelplatte, die um die optische Achse rotiert, kann die Objektebene virtuell eine kreisförmige Bewegung ausführen. Dann stellen die vier in Fig. 58 gezeigten Öffnungen Augenblicksbilder der sich kreisförmig bewegenden einzelnen Öffnung dar. In diesem Fall läßt die Sektorblende nacheinander die Öffnung in den vier gezeichneten Stellungen zur Wirkung kommen, und die Phasenlage der Einzelimpulse zur Bezugsfrequenz kann zum Aussteuern benutzt werden [*78*].

Im folgenden soll für den Stricheinfang nicht nur eine einzelne Symmetrielage photoelektrisch eingefangen, sondern eine Reihe solcher Symmetrielagen gezählt und zum Streckenmessen verwendet werden. Um die Intensität der Lichtimpulse zu steigern, und größere Verschiebewege eines Meßtisches messen zu können, werden statt eines einzelnen Striches oder Spaltes mehrere, also ein Raster auf ein zweites Blendenraster abgebildet. Stege und Lücken der Raster seien gleich breit, dann mitteln sich auch die Strichablagen. Beim gegenseitigen Verschieben von Raster und Rasterbild über mehrere Rasterstriche hinweg kann ein Zähler digital den Verschiebeweg feststellen. Solche Anordnungen sind beispielsweise für Längenmeßmaschinen aber auch für Winkelmessungen geeignet [*79*]. Um den Meßbereich bei Kettenmaßen in beiden Richtungen auswerten zu können, muß der Digitalzähler vor und rückwärts zählen, d.h. beide Bewegungsrichtungen voneinander unterscheiden können. Das wird erreicht, indem die Abbildung des Rasterbildes so verdoppelt wird, daß ein zweites Bild dem ursprünglich allein vorhandenen als Begleiter um eine viertel Rasterkonstante phasenverschoben vor- oder nacheilt, so daß die beiden um 90° phasenverschobenen verketteten periodischen Signale ein Drehfeld definieren.

Die Verdoppelung des Bildes kann auf verschiedene Weise geschehen, entweder als Schnittbild oder als Mischbild. Das Schnittbild kann auf der Abtastplatte reell vorhanden sein (Fig. 59). Die beiden um eine halbe Strichbreite verschobenen Schnittbilder werden zwei verschiedenen Photodioden zugeleitet.

Das Mischbild kann beim Abbilden eines Rasters über ein Wollaston-Prisma erzeugt werden (Fig. 60), so daß zwei um 90° verschieden polarisierte Bilder erzielt werden, die um eine halbe Strichbreite zueinander versetzt sind und beide

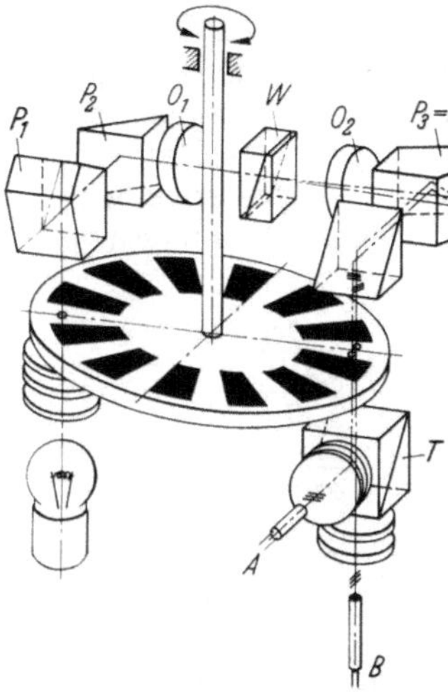

Fig. 60.

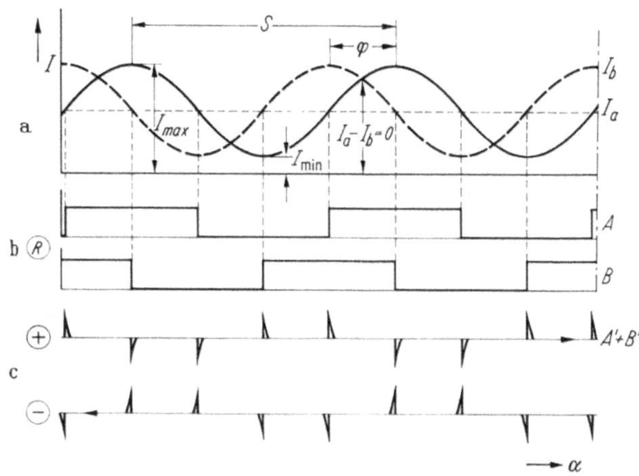

Fig. 61 a—c.

Fig. 60. Radialraster einer Winkelteilung über ein Wollaston-Prisma *W* diametral auf sich selbst abgebildet. Die Photozellen *A* und *B* empfangen die Impulse nicht gleichzeitig, sondern um 90° phasenverschoben, womit ein Drehfeld definiert ist, aus dem der Umlaufsinn der Teilung bestimmbar ist. $P_{1,2,3}$ Umlenkprismen, $O_{1,2}$ Objektive, *T* Teilerwürfel.

Fig. 61 a—c. a Um 90° phasenverschobene, verkettete, periodische Signale I_a und I_b auf den beiden Photozellen des Leitz-Winkelschrittgebers mit Rasterteilung. *S* Periode, φ Phasenwinkel, α Drehwinkel. b Desgl. hinter einem elektronischen Rechteckverformer *R*. c Differenzierte Signale. Ausgezogene Kurve I_a liefert die Bezugssignale.

auf je eine mit Analysator versehene Photozelle geleitet werden können [*80*]. Der Aufspaltungswinkel des Wollaston-Prismas muß entsprechend gewählt werden. Beim Verschieben der Rasterbilder gegenüber dem Einfangraster entstehen in beiden Photozellen um 90° phasenverschobene Impulsfolgen. Dient die eine als Bezugssignal, so eilt die andere je nach Bewegungsrichtung vor oder nach, wodurch der Zähler unschwer vor- und rückwärts unterscheiden kann. Fig. 61 gibt den Signalverlauf wieder, aus dem durch elektronische Maßnahmen sehr scharfe Einstellimpulse isoliert werden können.

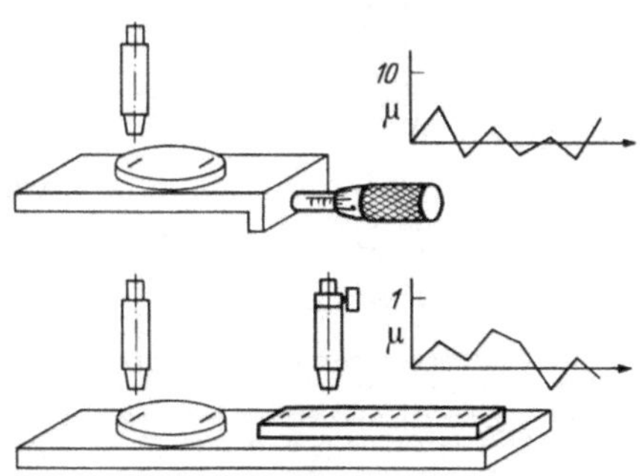

Fig. 62. Die optische Ablesung einer Tischverstellung ist rund eine Größenordnung genauer als die mechanische, die zusätzlich von Umkehrspanne, Taumelfehler und Verschleiß betroffen wird.

33. Kombination von Ziel- und Meßmikroskop auf einem Komparator. Die in Ziff. 31γ und δ beschriebenen Methoden, Abstände von der Zielgeraden zu messen, stellen zwar Möglichkeiten dar, die Meßmittel sind jedoch mechanischer Natur und daher in ihrer Genauigkeit stärker begrenzt als es die Genauigkeit, mit der die Zielmarke optisch anvisiert werden kann, wünschenswert erscheinen läßt. Ebenbürtig wäre dagegen eine optische Maßstabablesung mit einem Maßstabablesemikroskop (Ziff. 32ε). An Stelle der mechanisch messenden Spindel mit ihren relativ groben Fehlern von einigen Mikrometern, die sich noch dazu durch Verschleiß- und Taumelfehler ändern können, wird ein Maßstab eingebaut, der die Tischbewegung mitmacht und unter einem fest angeordneten Maßstabablesemikroskop entlang wandert (Fig. 62). Die Genauigkeit der Anordnung wächst in dem Maße, in dem die Meßunsicherheit am Maßstab geringer ist als an der Meßschraube. Gleichzeitig läßt sich der Meßbereich wesentlich erweitern. Wird mit einem Maßstab gemessen, der auf $s=\pm 0{,}2\,\mu$m abgelesen werden kann, so erheischt das unbedingt die Einhaltung des Abbeschen Prinzips, weil sonst die Steigerung der Meßgenauigkeit z. B. gegenüber einer Mikrometerschraube ($s=\pm 2\,\mu$m) nicht voll ausgenutzt werden kann. Diese Methode läßt sich auch auf Kreuztische anwenden, indem zwei senkrecht zueinander angeordnete Maßstäbe in den Tisch eingebaut werden, die beide im gleichen oder in verschiedenen Meßocularen erscheinen. Die Meßunsicherheit wird aber in diesem Falle mindestens für die eine Koordinate wegen des notwendigen Verstoßes gegen das Abbesche Komparatorprinzip größer.

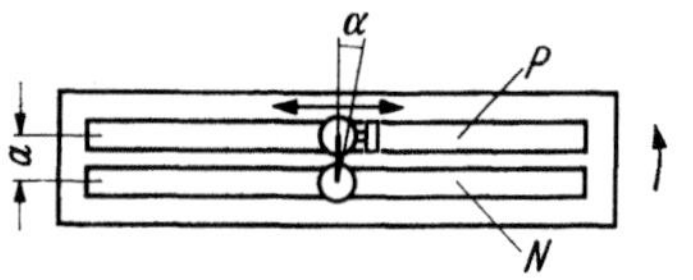

Fig. 63. Die Anordnung von Prüfling P und Normal N auf einem Komparator nebeneinander ergibt Meßfehler erster Ordnung proportional $\mathfrak{a}\, a$.

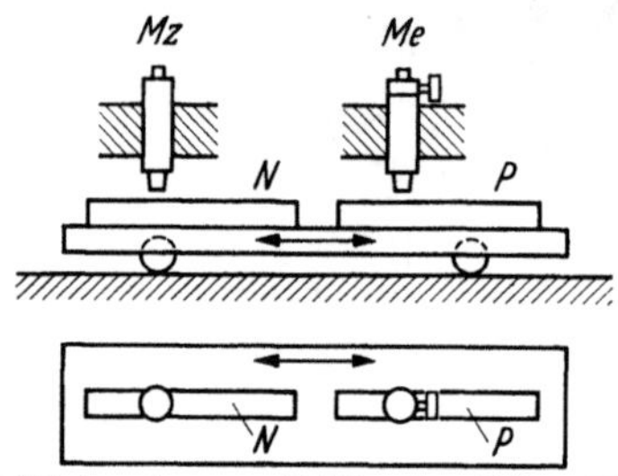

Fig. 64. Die Anordnung von Prüfling P und Normal N nach dem Abbeschen Komparatorprinzip hintereinander ergibt Meßfehler zweiter Ordnung proportional $l\,\mathfrak{a}^2$. Mz Zielmikroskop, Me Maßstabablesemikroskop, l Meßlänge.

34. Umgehen des Abbeschen Komparatorprinzips.

α) *Nachteil des Abbeschen Komparator-Prinzips.* Sollen Kippfehler erster Ordnung, d. h. Fehler proportional zum Sinus des Kippwinkels, bei den Messungen ausgeschaltet bleiben (Fig. 63), so müssen Prüfling und Maßstab oder Normal fluchtend hintereinander nach dem Abbeschen Komparatorprinzip angeordnet werden (Fig. 64). Dazu ist einmal die doppelte Baulänge von Prüfling oder Normal für den Komparator notwendig und zusätzlich eine volle Prüflingslänge Ausladung zum Verschieben des Komparatorschlittens. Diese unbequem große Baulänge kann auf verschiedene Weise vermieden werden.

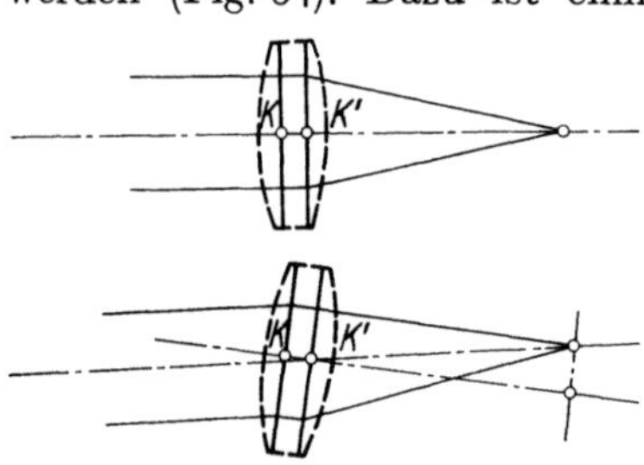

Fig. 65. Die Knotenpunktseigenschaften. Kippt ein Objektiv um den positiven bildseitigen Knotenpunkt K', so ändert sich die Bildlage in der bildseitigen Brennebene nicht. K, K' positive Knotenpunkte.

β) *Optische Umgehung.* Die optische Umgehung des Komparatorprinzips ist nach Eppenstein [*81*], [*81a*] benannt. Sie fußt auf der Tatsache, daß abbildende Systeme um ihren bildseitigen positiven Knotenpunkt gedreht werden dürfen, ohne daß sich dadurch die Bildlage in der Brennebene ändert (Fig. 65). Ließe man umgekehrt den dingseitigen positiven Knotenpunkt des Systems mit dem Tastpunkt für den Prüfling zusammenfallen, so dürfte das System um diesen Punkt kippen, ohne daß dadurch der nach Unendlich abgebildete Maßstabstrich seine Lage ändern würde. Da dies technische Schwierigkeiten bereiten würde, legt man den positiven Knotenpunkt und den Tastpunkt in Meßrichtung hintereinander; dann entstehen durch Kippungen nur Fehler zweiter Ordnung (cos-Fehler), vgl. Fig. 66.

Das Eppenstein-Prinzip vertauscht durch einmalige Spiegelung den positiven Knotenpunkt mit dem negativen und macht ihn auf diese Weise zwei Brennweiten vor dem Objektiv frei zugänglich (Fig. 67). Eine Tastspitze oder ein Zielmikroskop, die mit dem Ableseobjektiv „Linse + Planspiegel“ zu einer Antast-Einheit vereinigt sind (Fig. 68), können ohne Schwierigkeiten mit diesem Punkt antasten, vgl. hierzu auch Ziff. 27, Fig. 30. So gelingt es, Maßstab und Prüfling kippungsunempfindlich parallel *nebeneinander* anzuordnen und die Baulänge des Komparators wesentlich kürzer zu halten als bei strenger Anordnung nach dem Abbeschen Prinzip.

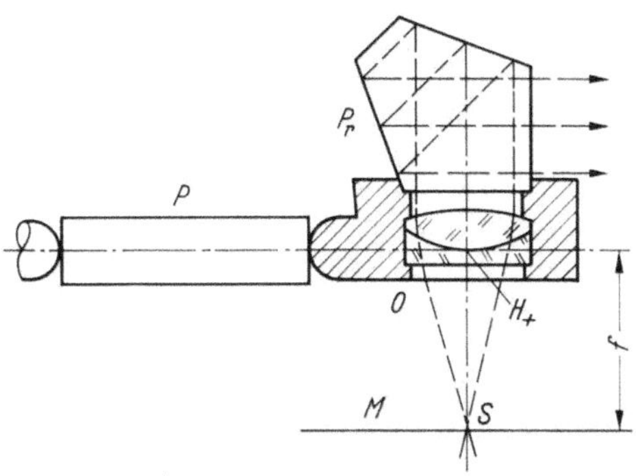

Fig. 66. Der Maßstabstrich S in der Brennebene von Objektiv O wird bei Kippung des Objektivs um H_+ über das Pentagonprisma P_r in die gleiche Richtung nach Unendlich abgebildet. Prüfling P und Maßstab M nebeneinander. Trotzdem ist das Komparatorprinzip eingehalten, weil H_+ in der Meßachse von P liegt.

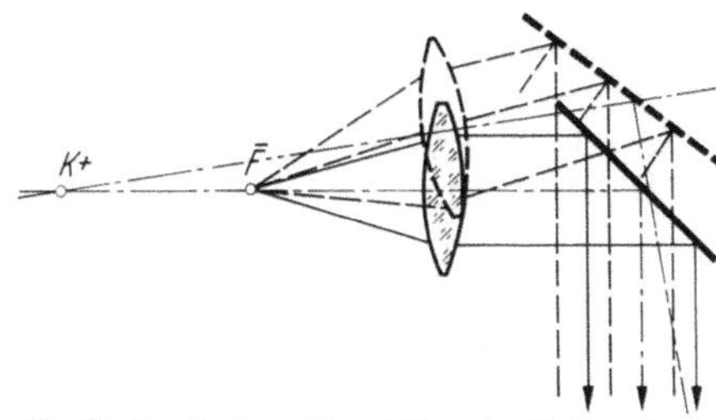

Fig. 67. Das System „Linse + Planspiegel“, das wie ein Hohlspiegel mit dem Mittelpunkt in K_+ wirkt, darf um K_+ gekippt werden, ohne daß das vom Brennpunkt F ausgehende Strahlenbündel auf der Bildseite seine Richtung ändert: Eppenstein-Prinzip.

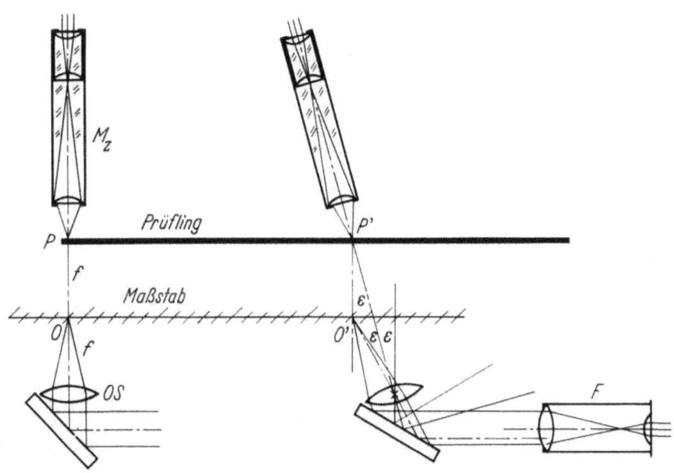

Fig. 68. Vermeidung parallaktischer Fehler nach dem Eppensteinschen Komparatorprinzip. M_z Zielmikroskop, PP' Meßstrecke auf dem Prüfling, OO' Meßstrecke auf dem Maßstab, OS System „Linse + Planspiegel“, F Fernrohr. In Stellung' ist die Antasteinheit um den Winkel ε um P' gekippt.

γ) Autokollimationsprinzip. Kippfehler können auch ausgeschaltet werden, wenn der Kippwinkel gemessen wird. Dann ist es entweder möglich, ihre Größe aus Kippwinkel und Abstand zwischen Maßstabachse und Meßstrecke am Prüfling zu berechnen, oder aber den Kippwinkel herauszujustieren. Die Messung kann mit Hilfe der in Ziff. 39 beschriebenen Autokollimationsmethoden geschehen, indem die Winkeländerung des Meßtisches zwischen Anfangs- und Endlage festgestellt wird (Fig. 69). Beträgt die Winkeländerung beispielsweise $1'' \approx 5 \cdot 10^{-6}$, ein Wert, der mit einem guten Autokollimationsfernrohr (AKF) noch

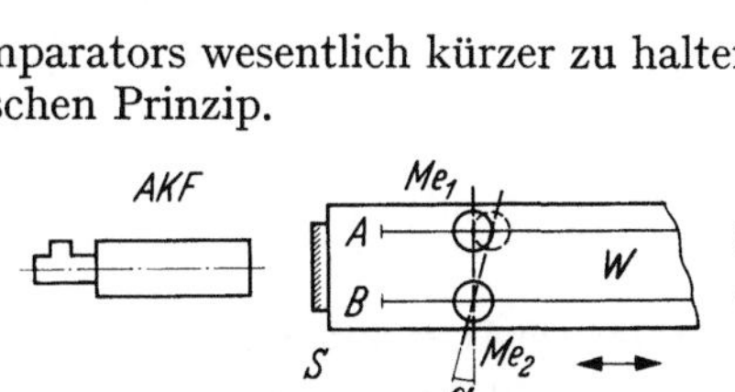

Fig. 69. Messen des Kippwinkels α eines Komparatorschlittens W mittels Autokollimationsfernrohrs AKF und Spiegels S zur Berechnung des Meßfehlers $\alpha\, a$ des Meßmikroskops Me_1 auf Maßstab A im Abstand a von Me_2 auf B.

sicher gemessen werden kann, so ergibt sich erst bei einem gegenseitigen Abstand von Prüfling und Maßstab von 100 mm ein Kippfehler von 0,5 μm. Solche Anforderungen wären mit dieser Methode bequem meßbar und einzuhalten.

δ) *Schwerpunktprinzip.* Kippfehler in Meßvorrichtungen und Geräten können auch mit Hilfe der Schwerkraft ausgeglichen werden. Hängt man einen Spiegel

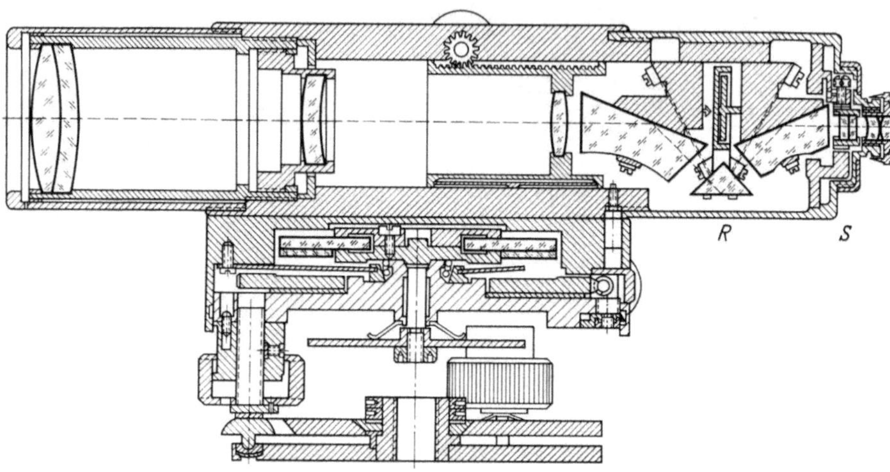

Fig. 70. Schnitt durch das Zeiss-Nivellier Ni 2 mit automatischer Horizontierung. Vor der Ocularstrichplatte S befindet sich ein pendelnd aufgehängtes Reflexionsprisma R. Dadurch entspricht trotz Kippungen des Fernrohrs die Strichkreuzmitte in der Ocularebene immer der horizontalen Richtung.

pendelnd auf, so richtet sich seine Lage nach der Schwerkraft. Durch die Art der Aufhängung kann man jeweils eine solche Lageänderung des Spiegels relativ zum kippenden Geräteteil erreichen, daß der abbildende Strahlengang das Bild stets auf die gleiche Stelle der Ocularstrichplatte wirft und Neigungsänderungen des Geräteteils automatisch kompensiert werden. Eine derartige Anordnung findet sich z. B. in Nivellier-Fernrohren (Fig. 70) [*82*], [*83*], [*84*], vgl. Ziff. 40 und Längenmeßgeräten (Fig. 71) [*85*]. In jenen stellt sie automatisch den Horizont ein, in diesen dient sie der Kompensation von Kippfehlern bei unumgänglichen Verstößen gegen das Abbesche Komparatorprinzip.

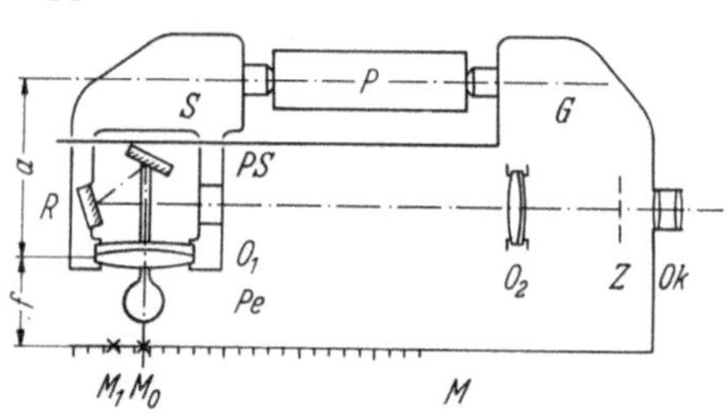

Fig. 71. Pendelspiegel PS in einer Meßanordnung. G Grundgestell, P Prüfling, S Support, M Maßstab. Wenn sich S durch Führungsfehler um den Winkel α um den Antastpunkt neigt, bewegt sich das Objektiv O_1 um αa zur Seite und zeigt beispielsweise auf M_1. Gleichzeitig öffnet sich der Winkelspiegel R-PS um α, da PS parallel zu sich und zu G bleibt. Daher trifft der von M_0 unter 2α (Bedingung $f = a/2$) in O_1 eintretende Hauptstrahl unverändert die Ocularebene in der gleichen Stelle von Z.

ε) *Doppelablesung an Maßstäben und Teilkreisen.* Schließlich kann man das Abbesche Komparatorprinzip durch gleichzeitige Anwendung eines zweiten Maßstabes unnötig machen. Man ordnet den Prüfling mitten zwischen den parallel liegenden Maßstäben an und mittelt die auf beiden Maßstäben abgelesenen Meßwerte. Ein Beispiel ist in Ziff. 39, ζ, Fig. 95, gegeben. Dies Prinzip entspricht der sog. Doppelablesung an Teilkreisen, die zur Beseitigung der Exzentritätsfehler benutzt wird. Allerdings besteht ein Unterschied: Bei der Längenmessung zwischen zwei Maßstäben zählen beide Maßstäbe in der gleichen Richtung. Infolgedessen fallen Drehbewegungen durch Mittelung der Meßwerte heraus. Bei der Winkelmessung am Teilkreis zählen beide diametralen Teilungen im Gegensinn, daher fallen translatorische Bewegungen, also Exzentrizitätsfehler heraus (Fig. 72). In Doppelablesungen für Teilkreise wird der Exzentrizitätsfehler gewöhnlich durch Rechnung, Mittelbildung über zwei diametrale Meßwerte, eliminiert. Man kann ihn automatisch kompensieren, wenn man die auf beiden Seiten des Teilkreises vorzunehmenden mikroskopischen Ablesungen in einem Schnittbild so vereinigt, daß Koincidenzeinstellungen der zur Ablesung benutzten, um 180° einander gegenüberliegenden Teilstriche möglich sind (Fig. 72b). Solche Koincidenzeinstellungen mitteln außerdem die zufälligen Ablagen beider Teilstriche von der sinusförmigen Fehlerkurve, die durch die Exzentrizität entsteht, arithmetisch. Koincidenzablesungen sind in Theodoliten und Teilköpfen üblich und beseitigen auch die höheren periodischen Fehler ungerader Ordnung. Dadurch wird die Einstell- und Meßgenauigkeit optischer Teilgeräte außerordentlich erhöht, selbstverständlich aber nicht ihre Arbeitsgenauigkeit. Das mit einer Kreisteilung zu versehende Objekt oder auch das Zahnrad, welches geprüft werden soll, können trotzdem aus zwei Gründen fehlerhaft werden: Erstens können sie von sich aus exzentrisch auf der Teilkopfspindel aufgenommen sein, zweitens kann die Teilkopfspindel erhebliche Exzentricitätsfehler besitzen, so daß das Objekt vor dem Reißerwerk, Taster oder dem Zielmikroskop schlägt. Die Doppelablesung macht die Kenntnis des Spindelschlags für die Arbeit am Objekt nicht entbehrlich. Sie beseitigt Exzentrizitätseinflüsse nur für die Teilung des Teilkreises, seine Aufnahme auf der Achse sowie deren Schlag im Sinne einer fehlerfreien Ablesung.

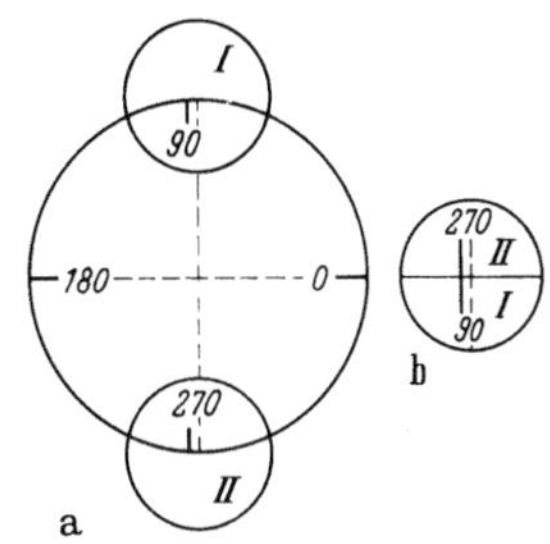

Fig. 72 a u. b. a Teilkreis durch translatorische Bewegung, Exzentrizität, verschoben, entsprechend sinusförmigem Teilfehler. b Gesichtsfelder I und II vereinigt, zeigen Koincidenzeinstellung und damit die vom Exzentritätsfehler freie, richtige Winkeleinstellung des Teilkreises und damit der Teilspindel.

V. Fernrohrgeräte und Verfahren zur Fluchtungsprüfung.

Zwei wichtige optische Meßmethoden verwenden das Fernrohr als Grundinstrument: Die optische Fluchtungsprüfung und die optische Richtungsprüfung [*86*]. Auf diesen beiden Gebieten sind optische Meßmethoden konkurrenzlos. Die Meßstrecken können im Vergleich zu Labordimensionen sehr ausgedehnt sein, man denke nur an das Fluchten weit voneinander entfernt liegender Lagerbohrungen oder das Richten von Flugzeugspanten.

35. Das Zielfernrohr zum Fluchten. Die Fluchtungsprüfung, die sich von der speziellen Aufgabe herleitet, hintereinanderliegende Bohrungen auf eine Fluchtlinie zu bringen, besteht meßtechnisch gesehen darin, die seitlichen Abweichungen hintereinander liegender Zielmarken von einer Fluchtlinie in Längeneinheiten zu messen. Die einzelnen Zielmarken können in beliebigen Abständen liegen und in beliebiger Reihenfolge anvisiert werden.

Als Fluchtgerade können optische Achsen oder optische Symmetrieebenen dienen. Zweckmäßig wird ein sog. Fluchtfernrohr verwendet (Fig. 73). Als optische Achse und Fluchtlinie ist hierbei die Fernrohrzielachse, die Verbindungslinie von der Strichkreuzmitte in der Ocularebene zu dem reellen Bild dieser Marke vor dem Fernrohr objektiv anzusprechen. Um entlang dieser Fluchtgeraden aufgestellte Zielmarken fluchten zu können, muß auf jede einzelne Marke scharf gestellt werden. Hieraus ergeben sich zwei Schwierigkeiten: Die Änderung des Abbildungsmaßstabes mit der Entfernung der einzelnen Zielmarken und die Tatsache, daß das Fernrohr auf die verschiedenen Entfernungen nachfokusiert werden muß, ohne daß sich an der Lage der Ziellinie des Fernrohrs auch nur das geringste ändern darf.

α) *Die Abhängigkeit des Abbildungsmaßstabes von der Entfernung.* Mit zunehmender Entfernung der Zielmarken vom Fernrohrstandpunkt entsprechen gleichen seitlichen Abständen von der Fluchtlinie kleinere Sehwinkel. Das bedeutet erstens, daß die Einfangsunsicherheit für seitliche Abstände mit zunehmender Entfernung zunimmt, daß also die Meßunsicherheit wächst, und zweitens, daß zum Messen der Ablage eines einfachen Strichkreuzes, meist eines Schrägstrichkreuzes, jeweils ein anderer Maßstab in der Ocularebene angelegt werden müßte. Die Änderung des Abbildungsmaßstabes in der Ocularebene würde nicht stören, wenn die Zielmarken selbst Maßstäbe trügen. Das ist aber für genaue Fluchtungsmessungen unzweckmäßig, da diese Maßstäbe für die Nähe zu grob, für die Ferne zu fein sein würden. Eine andere Meßmöglichkeit, bei der immer der gleiche Maßstab benutzt werden könnte, böte sich in einer Parallelverschiebung des ganzen Fern-

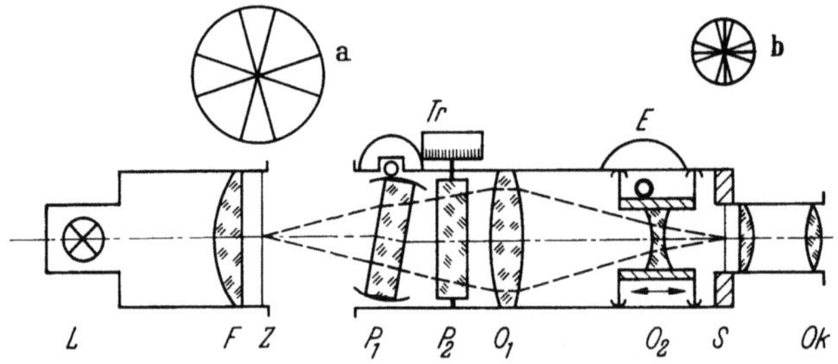

Fig. 73a u. b. Fluchtfernrohr mit Innenfokussierung und beleuchteter Zielmarke *Z*. a Schrägstrichkreuz als Zielmarke, b Ocularstrichkreuz und Bild der Zielmarke. *L* Lampe, *F* Feldlinse, $P_{1,2}$ Planparallelplatten, O_1 Hauptglied des Objektivs, O_2 zur Innenfokussierung verschiebbares hinteres Objektivglied, *S* Ocularstrichmarke, *Ok* Ocular, *Tr* Meßtrommeln zum Neigen der Planparallelplatten, *E* Trieb zur Entfernungseinstellung.

rohres, bis die betreffende Marke wieder auf der Ziellinie des Fernrohres läge. Kleinste Abweichungen von der Parallelführung brächten aber schon beträchtliche Fehler erster Ordnung mit sich, so daß von einer reellen Parallelverschiebung des Fernrohres abgesehen werden muß. Eine virtuelle Verschiebung bringt aber die Lösung.

β) *Planplatten-Meßvorsatz.* Die virtuelle Parallelversetzung des Fernrohres und damit der Ziellinie geschieht durch neigbare Planparallelplatten vor dem Fernrohrobjektiv (Fig. 73). Die Versetzung $v = \alpha d \cdot \left(1 - \frac{1}{n}\right)$ ist bei geeigneter Dicke d der Platten direkt proportional zum Drehwinkel α, falls dieser sich in bestimmten Grenzen hält. Das ist aber der Fall, da beim Fluchten in den meisten Fällen nur die kleinen Abweichungen interessieren. Statt des Drehwinkels läßt sich sogleich die seitliche Versetzung des Fernrohres in mm-Abweichung von der ursprünglichen Fluchtlinie ablesen. Damit ist ein Maßstab gewonnen, der unabhängig von der Entfernung der Zielmarke arbeitet.

γ) *Die Innenfokussierung.* Bei den meisten Fernrohren, Feldstechern und Operngläsern ist es üblich, zwecks Fokussierung den Abstand zwischen dem Objektiv und der Ocularebene zu ändern. Da die Ziellinie durch die Zielmarke in der Ocularebene festgelegt ist, können Führungsfehler des Tubusauszuges erhebliche Winkelabweichungen der Ziellinie nach sich ziehen. Um Führungsfehler so wenig wie möglich zur Auswirkung kommen zu lassen, bleibt beim Fluchtfernrohr die Ocularebene fest stehen, ebenso das Objektiv. Zwischen beiden wird eine Fokussierlinse eingeschoben (Fig. 73), die die Aufgabe hat, je nach ihrem Abstand vom Objektiv, dessen Brechkraft zu ändern und dadurch die Fokussierung auf Zielmarken in verschiedenen Entfernungen zu ermöglichen. Da diese Linse nur einen kleinen Teil der Gesamtbrechkraft des Objektivs auszumachen braucht, ist der

Einfluß von Führungsfehlern bei ihrer Verschiebung verhältnismäßig gering. Wird die Fokussierlinse *2* in Fig. 74 um den Betrag v seitlich versetzt, so wandert das Bild der Zielmarke um den Betrag $w = v \cdot s_2'/f_2' = v\left(1 - \frac{f'_{\text{ges}}}{f_1'}\right)$ zur Seite. s_2' ist der Abstand der Fokussierlinse von der Ocularebene und f_2' ihre Brennweite. Als Folge der Versetzung ändert sich die Lage der Ziellinie des Fluchtfernrohrs um den Winkel φ gemäß $\tan\varphi = w/f'_{\text{ges}}$, worin f'_{ges} die Brennweite des gesamten Objektivs einschließlich der Fokussierlinse ist. Je größer f_2', d.h. je kleiner die Brechkraft der Fokussierlinse, um so kleiner wird die Abweichung der Fluchtlinie von ihrer ursprünglichen Richtung bei einer bestimmten Führungsqualität.

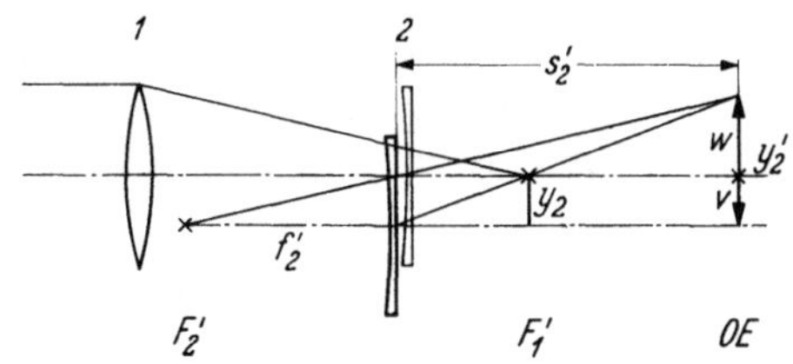

Fig. 74. Seitliche Versetzung v der Fokussierlinse *2* bewirkt Versetzung des Bildes um w in der Ocularebene *OE*.

δ) *Ziel- und Meßunsicherheit.* Die Unsicherheit, mit der der seitliche Abstand der Zielmarke von der Fluchtlinie angegeben werden kann, ist von der Winkelauflösung des Auges, der Fernrohrvergrößerung N_F und der Entfernung L der Zielmarke vom Fernrohr abhängig. Mit 10″ Einfangunsicherheit des Auges ist folgende Meßunsicherheit theoretisch erreichbar:

$$\varDelta = L \tan\varphi = L\,\frac{5.10^{-5}}{N_F} \approx 1{,}2 \cdot 10^{-2}\, L/250 \cdot N_F .$$

Darin ist $1{,}2 \cdot 10^{-2} = \varDelta_{250}$ [mm] die bei 10″ in der deutlichen Sehweite vom Auge noch aufgelöste Strecke und $N_F \cdot 250/L$ die sog. Mikroskopvergrößerung des Fernrohres mit Innenfokussierung [*60*]. Tatsächlich wird $\varDelta$ nicht gleich Null für $L = 0$, sondern eine bestimmte Ableseunsicherheit und Ungenauigkeit in der Maßstabsteilung für die Neigung der Planparallelplatte ist auch schon für $L = 0$ wirksam. So kommt es zu einer Formel für die Meßunsicherheit von folgendem Aussehen:

$$\varDelta = \pm\left(a + \frac{L}{b}\right)\mu\text{m}.$$

Für ein 30faches Fernrohr wurde beispielsweise $a = 5$ und $b = 3 \cdot 10^5$ empirisch ermittelt. In diesem speziellen Fall betrug die Einfangunsicherheit des Auges also 20″.

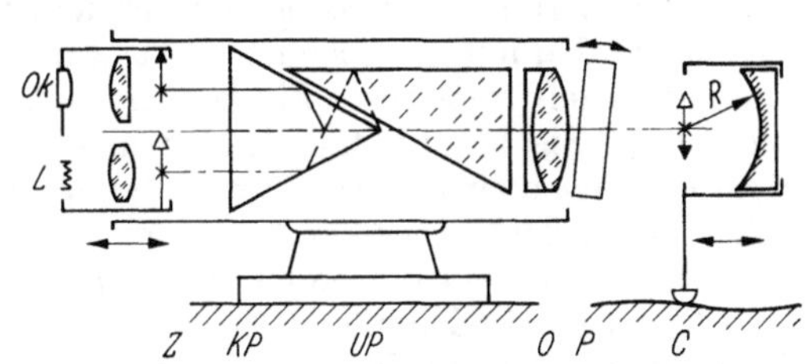

Fig. 75. Fluchtfernrohr von HUET, Paris, mit Kösters-Prisma und Cataphot. *Ok* Ocular, *L* Lampe, *Z* Zielmarke, *KP* Kösters-Prisma, *UP* Umlenkprisma, *O* Objektiv, *P* Planplattenmikrometer, *C* Cataphot, bestehend aus Taster und Hohlspiegel, *R* Radius des Hohlspiegels.

36. Vermeiden der Versetzungsfehler. Ist der Versetzungsfehler auch durch die Innenfokussierung praktisch unbedeutend geworden, so sind doch einige Methoden interessant, mit deren Hilfe er gänzlich ausgeschaltet werden kann.

α) *Huet-Fluchtfernrohr mit Kösters-Prisma.* Das Fernrohr (Fig. 75) besitzt nur eine Außenfokussierung, d.h. einen Tubusauszug, vermeidet aber durch sinngemäße Anordnung der Beleuchtungseinrichtung und einer zu projizierenden Marke, daß Versetzungen sich auswirken können. Durch die Ausbildung des am Meßort angesetzten Tasters als Hohlspiegel wird überdies der Faktor 2 an Meßempfindlichkeit gegenüber den bisher besprochenen Fluchtfernrohren erreicht.

In dem zwecks Fokussierung verschiebbaren Oculartubus, Fig. 75, befinden sich übereinander angeordnet eine beleuchtete Marke und die Ocularstrichplatte mit Einfangstrichen. Im Fernrohr selbst liegen ein Kösters-Prisma zur Vereinigung bzw. Trennung des Strahlengangs und ein Umlenkprisma sowie das Objektiv. Die beleuchtete Marke wird nicht auf einen Schirm projiziert, sondern in den Mittelpunkt eines Hohlspiegels, von dem aus sie wieder in das Fernrohr zurückfällt, diesmal aber über den oberen Weg des Kösters-Prismas in die Ocularebene. Der Projektions- und Abbildungsweg sind so entworfen, daß die beleuchtete Marke im Tubus und ihr Bild in der Ocularebene gleichsinnig stehen, d.h. daß seitliche Versetzungen des Tubus in Meßrichtung keine Relativbewegung zwischen aufgefangenem Bild der Marke und Strichkreuz in der Ocularebene verursachen können. Bei einer Fluchtungsmessung entlang einer unebenen Bahn wird der Mittelpunkt des Hohlspiegels, unterhalb dessen die Tastpinole sitzt, ein Stück seitlich ausgelenkt. Das hat die doppelte Auslenkung der im Hohlspiegel reflektierten Marke zur Folge und zwar in der gleichen Richtung. Mit diesem System können z.B. Lagerbohrungen sehr genau auf eine Fluchtlinie gebracht werden. Wollte man mit diesem System messen, so müßte aus Gründen des veränderlichen Abbildungsmaßstabes für verschiedene Entfernungen wieder die bekannte neigbare Planparallelplatte herangezogen werden. Da der Abbildungsmaßstab für das Bild in der Ocularebene 1:1 beträgt, kann die hohe Einfanggenauigkeit des Doppelstricheinfangs ausgenutzt werden.

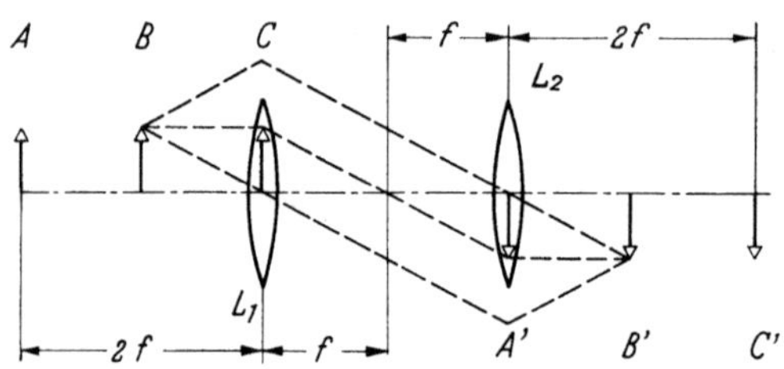

Fig. 76. Afokales System: Ein zwischen A und C verschobenes Objekt wird durch die Linsen L_1 und L_2 in den Raum $A'C'$, gleichgroß abgebildet. In einem afokalen System 1:1 sind die Abstände $AA' = BB' = CC'$.

β) *Optisches Lineal von Cuny.* Zur Messung der Geradheit von Führungsbahnen und der Ebenheit von Richtplatten ist das optische Lineal von CUNY der gleichen Firma *Huet* [87] geeignet. Sein afokales Fernrohrsystem macht eine Fokussierung überflüssig und vermeidet dadurch den Führungsfehler. Zusätzlich wird durch gegensinniges Auswandern von Objekt und Bild die Meßempfindlichkeit verdoppelt. Fig. 76 zeigt die Wirkungsweise eines afokalen Systems mit $f_1' = f_2'$. Sie beruht darauf, daß nicht nur die Vergrößerung eines solchen Fernrohrs gleich eins ist, sondern auch der Abbildungsmaßstab, wenn man es zur reellen Abbildung naher Objekte benutzt. In diesem Fall ist sogar der Abstand zwischen Objekt und Bild immer gleich groß, wo auch immer das Objekt sich vor dem Objektiv befinden möge. Das Bild ist umgekehrt wie hinter einer Einzellinse. Verlagert sich das Objekt seitlich zur optischen Achse, so wandert das Bild nach der entgegengesetzten Richtung aus. Da sich der Abbildungsmaßstab nicht ändert, kann in der Bildebene, d.h. in der Ocularebene für alle Objektabstände ein und derselbe Maßstab verwendet werden. Um einen Tubusauszug für das Auffangen des Bildes zu vermeiden, wird der Strahlengang im Cuny-Lineal mit Hilfe zweier Winkelspiegel so umgeknickt, daß Objekt- und Bildebene zusammenfallen (Fig. 77). Die beiden Objektive und die Winkelspiegel sind in einem langen Gehäuse enthalten, welches an den Enden auf der zu vermessenden Bahn aufsitzt. Das Gehäuse ist längs der optischen Achse geschlitzt, um die Tastpinole mit der als Objekt dienenden Marke und ein Beobachtungsmikroskop, für die Messung der Bildablage, darin entlang schieben zu können. Objekt und Bild liegen zwar praktisch in derselben Ebene auf der Stirnfläche eines Teilerwürfels, das Bild wird jedoch vor Erreichen dieser Ebene von der Teilerfläche rechtwinklig nach oben abgelenkt, wo es leicht mit dem Mikroskop zu beobachten ist. Wandert die Pinole bei der Fluchtungsmessung auf unebener Bahn nach unten aus, so folgt ihr die

Marke mit dem Teilerwürfel ebenfalls nach unten, und das Bild entsteht um ebensoviel nach oben ausgelenkt. Gegenüber der Objektmarke ist es um den doppelten Betrag der Auslenkung verschoben, das ist aber die Empfindlichkeit, mit der die Marke sich auch im Gesichtsfeld des Mikroskops verschiebt. Gegen Durchhang ist das Cuny-Lineal unempfindlich, da die eingebauten Dachkantprismen um ihre Schnittkanten kippen dürfen, ohne daß sich dadurch der Umlenkwinkel ändert. Die Meßunsicherheit beträgt bei einem 3-m-Lineal in der Mitte $\pm 1,5$ µm und an den Enden ± 3 µm.

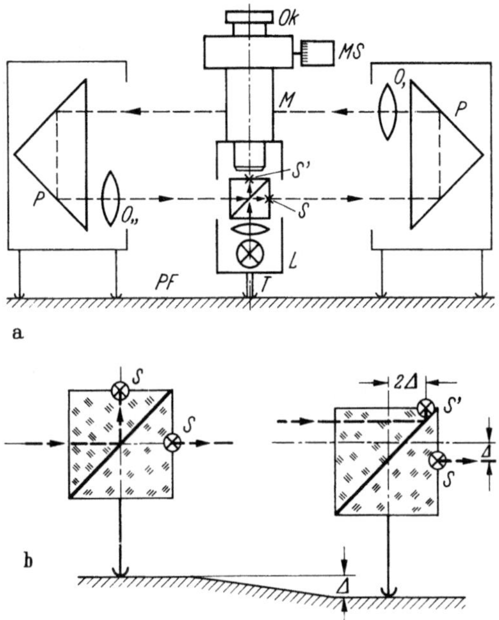

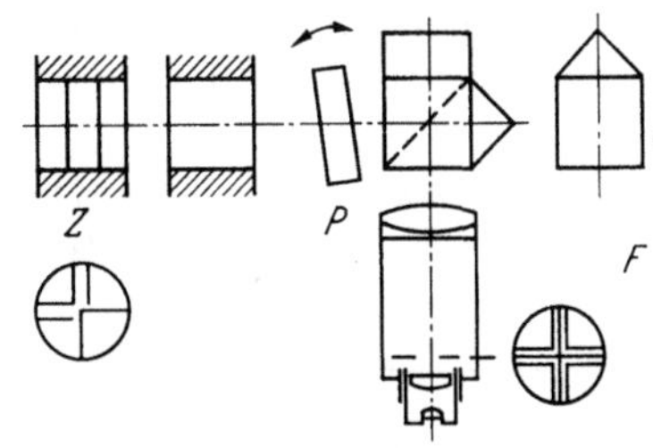

Fig. 78. Fluchtfernrohr mit Doppelbildprisma (nach Kösters). *Z* Zielmarke, *J* Doppelbildprisma, *F* Fernrohr mit Zielmarkenbild in Ocularebene. *P* schwenkbare Planparallelplatte ermöglicht Symmetrieeinfang.

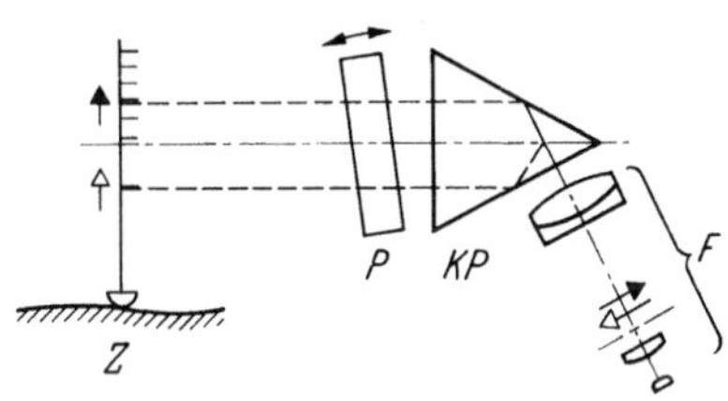

Fig. 77 a u. b.

Fig. 79.

Fig. 77 a u. b. Optisches Lineal nach Cuny. a Schema des Aufbaus. Die Marke *S* wird durch ein afokales System nach *S'* abgebildet. *S'* wird durch das Mikroskop *M* betrachtet. *MS* Ocularmikrometer, *Ok* Ocular, *P* Umlenkprismen, *PF* Prüflingsoberfläche, *T* Taster, *L* Lampe zur Beleuchtung von *S*. b Eine Auslenkung *Δ* von *S* aus der optischen Achse verursacht eine Versetzung von *S'* um 2*Δ*.

Fig. 79. Fluchtfernrohr *F* mit Kösters-Prisma *KP*, Fluchtebene durch Teilerfläche des Prismas bestimmt, *Z* Zielmarke mit Taster und Maßstab, *P* Planparallelplatte zum Symmetrieeinfang.

37. Symmetrieebene als Bezugsebene. Wie oben erwähnt, kommen auch Symmetrieebenen optischer Systeme als Bezugsebenen für die Fluchtungsmessung in Frage, z. B. die Symmetrieebenen in Doppelbildsystemen. Das Doppelbild kann, wie in Ziff. 31 α beschrieben, flächensymmetrisch oder auch axialsymmetrisch liegen. Das Fernrohr dient in diesen Fällen nur noch zur Beobachtung, ob die Doppelbilder zusammenfallen. Die Änderung seiner Fokussierung darf daher mit beliebigen Führungsfehlern behaftet sein, sie gehen nicht in die Messung ein.

α) *Axialsymmetrisches Doppelbildprisma.* Die in Fig. 78 gezeigte Anordnung ist von Kösters entworfen worden [*88*]. Sie liefert zwei zueinander senkrechte Fluchtebenen. Mit ihr können Abweichungen in beliebiger Richtung von der Fluchtgeraden, d. h. der Schnittgeraden der beiden Fluchtebenen, gemessen werden. Die Ausgestaltung der Zielmarke erlaubt die hohe Genauigkeit des Symmetrieeinfangs auszunutzen. Die Gegenläufigkeit der Bilder bringt die doppelte Meßempfindlichkeit. Die Abweichungen von der Fluchtlinie werden mit dem Planplatten-Mikrometer gemessen.

β) Kösters-Prisma als Fluchtebene. Die halbverspiegelte Teilerfläche des Kösters-Prismas gibt eine Bezugsebene, gegen die Marke und Indexstrich wie Bild und Spiegelbild gegenläufig wandern (Fig. 79) [*89*]. Ihre relative Lage zueinander wird mit dem Fernrohr beobachtet und am Planplattenmikrometer abgelesen. Auch diese Anordnung ist frei von Führungsfehlern der Fernrohrfokussierung und besitzt doppelte Meßempfindlichkeit.

38. Interferentielle Fluchtungsverfahren. Zu den interferentiellen Verfahren für die Fluchtungsmessung sollen Verfahren gezählt werden, bei denen eine ausgedehnte Fluchtlinie oder eine Bezugsebene durch ein Interferenzfeld dargestellt werden und nicht Verfahren, bei denen die Ebenheit von Flächen in der Größenordnung von Endmaßflächen oder gar mikroskopisch kleiner Bereiche bestimmt wird (vgl. Ziff. 23).

α) Zweispaltinterferometer. Die einfachste Form eines Interferenzlineals besteht in einem kohärent beleuchteten Youngschen Doppelspalt [*90*], hinter dem sich eine Symmetrieebene mit dem Gangunterschied Null bildet, die durch einen

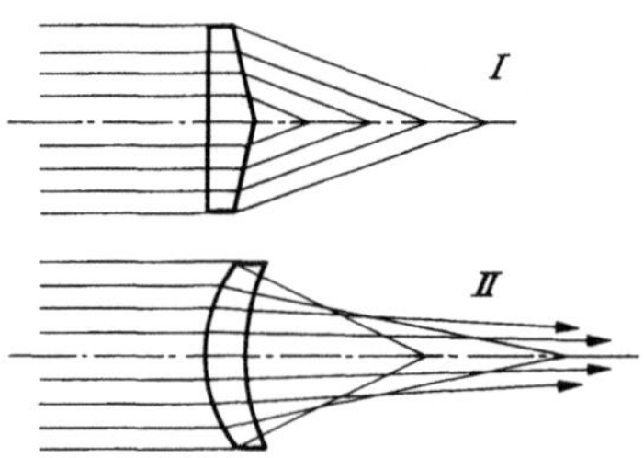

Fig. 80.

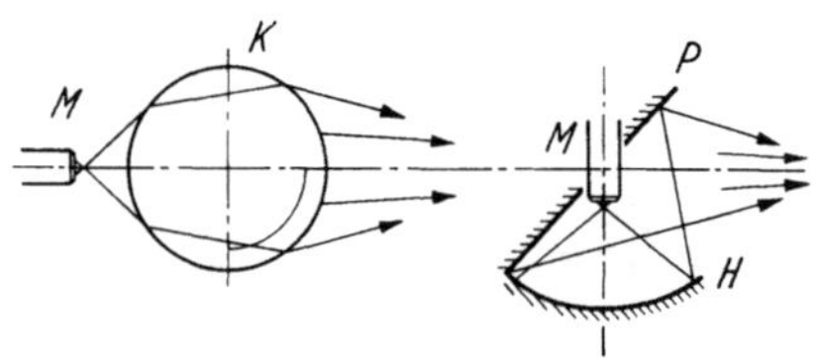

Fig. 81.

Fig. 80. I Kegeliges Axicon nach McLeod, II Sphärisches Axicon nach Steel.

Fig. 81. Interferenzfluchtkugel nach van Heel. Ein durch die Mikroskopoptik *M* verkleinert abgebildetes Bild der Lichtquelle wird durch eine Kugel *K* oder einen Hohlspiegel *H* im Raum abgebildet. Die deformierte Wellenfront führt zu Interferenzerscheinungen konzentrisch zur Achse des Systems. *P* Planspiegel.

hellen Interferenzstreifen deutlich gekennzeichnet ist. Hinter einem kreisförmigen Ringspalt [*91*], wird statt einer Fluchtebene eine Fluchtgerade erhalten. Das Fresnelsche Biprisma liefert ebenfalls eine Fluchtebene wie der Youngsche Doppelspalt, nur mit dem Unterschied, daß hier zwei virtuelle Bilder einer Lichtquelle entstehen, deren Bündel sich durchsetzen, während dort die Beugung an den Spaltkanten die interferenzfähigen Wellenzüge schafft. Im Gegensatz zum Youngschen Doppelspalt ist beim Biprisma die Ebenheit der Interferenzfläche von den geometrischen Daten des Biprismas abhängig, insbesondere von der Ebenheit der Prismenflächen, da für die Nähe andere Zonen des Prismas verwendet werden wie für die Ferne. Von der Ebenheit und Homogenität der Prismenhälften kann man sich leicht durch Umschlagmessungen überzeugen.

β) Axicons und verwandte Bauelemente. Das Axicon [*92*] ist ein optisches Bauelement wie eine Linse. Seine Flächen sind aber nicht sphärisch, sondern kegelig gearbeitet. Dadurch weichen seine abbildenden Eigenschaften von denen der Linsen insofern ab, als eine punktförmige Lichtquelle in Form einer Linie längs der verlängerten Kegelachse, d.h. der optischen Achse des Kegels, auseinandergezogen abgebildet wird. Jeder Punkt dieser Linie ist durch eine andere Zone des Kegels abgebildet, daher hängt die Länge der Lichtlinie von der Gestalt des Kegels ab (Fig. 80).

Ähnliche Eigenschaften wie der Kegel haben stark unter- oder überkorrigierte Systeme sphärischer Linsen [*93*]. Unter Beachtung der Kohärenzbedingung können mit diesen optischen Bauelementen ähnlich wie beim Fresnelschen Biprisma Interferenzerscheinungen erzeugt werden, die sich als Fluchtlinie eignen. Van Heel [*94*] erreichte mit einer Fluchtkugel noch in 81 m Entfernung ein aus konzentrischen Kreisen bestehendes Interferenzbild. Als Lichtquelle diente ihm ein mikroskopisch verkleinertes Bild einer beleuchteten Öffnung. Eine anstelle der Fluchtkugel verwendete Spiegeloptik (Fig. 81) ließ Meßunsicherheiten von nur einigen µm in 4 m und von 20 µm in 40 m Entfernung zu [*94*]. Ein Interferometer zur Fluchtungsprüfung entlang einer mehrere Meilen langen Linie mittels Spiegeloptik und Kösters-Prisma gibt SAUNDERS an [*95*]. Auf weitere Experimente mit Kugeln [*94*], [*95*], Zonenplatten [*96*], Phasen- und Amplitudenfiltern [*97*], kann hier nicht eingegangen werden; sie sind zwar physikalisch interessant, haben bisher aber keine praktische Bedeutung erlangt.

39. Autokollimation zur Richtungsprüfung. Bei der Fluchtungsprüfung wird angestrebt, seitliche Abstände von der optischen Achse maßstäblich zu messen, bei der Richtungsprüfung sollen Winkelabweichungen gegen die optische Achse eines Fernrohres gemessen werden. Als Richtung der optischen Achse kann hier die Richtung des parallel aus dem Objektiv austretenden Strahlenbündels aufgefaßt werden, welches einen bestimmten Punkt in der hinteren Brennebene des Objektivs nach Unendlich abbildet. Im Autokollimationsprinzip wird eine in der hinteren Brennebene des Fernrohrobjektivs liegende beleuchtete Marke nach Unendlich projiziert, die von einem Planspiegel reflektiert in die Brennebene zurückkehrt. Winkeländerungen des Planspiegels um α ändern die Richtung des zurückkehrenden Strahlenbündels um 2α. Die dadurch gegebene doppelte Meßempfindlichkeit für Winkelabweichungen in der Spiegelstellung zusammen mit der Möglichkeit, den höchstempfindlichen Symmetrieeinfang mit Doppelstrichen anwenden zu können, machen das Autokollimationsprinzip vorzüglich geeignet, um Flächen parallel zueinander auszurichten oder ihre Parallelität festzustellen. Eine Bogensekunde Abweichung in der Spiegelstellung bedeutet bei 500 mm Brennweite des AKF-Objektivs 5 µm Abweichung der Marke in der Ocularebene. Mit bloßem Auge sind bei Symmetrieeinfang noch 7 µm Auslenkung zu erkennen. 14fache Ocularvergrößerung würde also genügen, um 0,1″ noch messen zu können. Parallelversetzungen des Spiegels sowie seitliche Versetzungen der Fernrohrachse parallel zu sich, auch virtuelle, spielen für die Richtungsmessung keine Rolle. Der schematische Aufbau des Autokollimationsprinzips geht aus Fig. 82 hervor. Literatur hierzu unter [*86*].

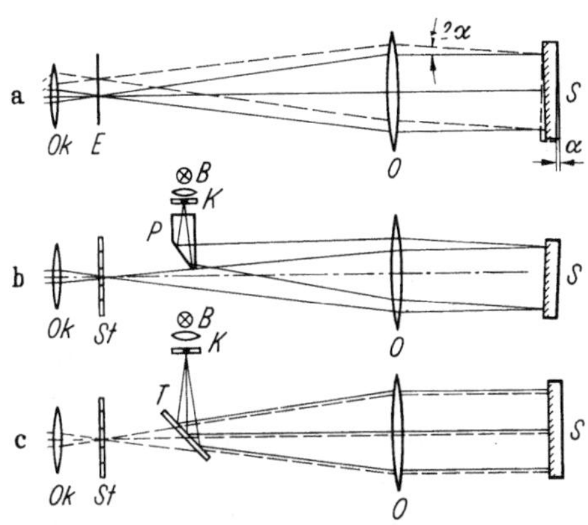

Fig. 82a—c. Autokollimationsfernrohr AKF. a Prinzip, b geometrische, c physikalische Strahlenteilung, α Kippwinkel des Spiegels, *B* Lampe und Beleuchtungslinse, *E* Brennebene des Objektivs, *K* Strichkreuz, *O* Objektiv, *Ok* Ocular, *P* Umlenkprisma, *S* Spiegel, *St* Ocularstrichplatte, *T* Teilerplatte.

α) *Meßbereich in axialer Richtung.* Zu beachten ist, daß der Meßbereich des Autokollimationsfernrohrs (AKF) nicht beliebig in die Ferne ausgedehnt werden kann, sondern in axialer Richtung begrenzt ist. Bei zunehmendem Abstand des Spiegels nimmt nämlich der Durchmesser des ausgeleuchteten Gesichtsfeldes im Ocular ab. Das beruht darauf, daß im Spiegel auch die Objektivöffnung des AKF abgebildet wird, im Gegensatz zur Marke aber nicht nach Unendlich, sondern in endliche Entfernung. Nach der Definition bestimmt die von der Eintrittspupille eines Instruments aus gesehene kleinste Blende den Durchmesser des Gesichtsfeldes. Solange das Spiegelbild der Objektöffnung in der Nähe des Fernrohrs liegt, ist sein Durchmesser größer als die Blende in der Ocularebene. Solange ist das Gesichtsfeld durch diese Blende bestimmt. Von einer bestimmten Entfernung des Spiegelbildes an tritt aber seine Berandung als Gesichtsfeldblende auf, da sie von der Eintrittspupille aus unter immer kleinerem Sehwinkel wahrgenommen wird. Bei unendlicher Entfernung des Spiegelbildes würde der ausgeleuchtete Fleck in der Ocularebene des AKF nur noch ein Punkt sein. Zum Messen braucht man aber eine noch ausreichende Größe des Gesichtsfeldes. Beim Messen von Spiegelneigungen kommt ein weiterer Umstand erschwerend hinzu: Das Spiegelbild liegt immer auf dem Lot zum Spiegel, wandert also bei Spiegelneigungen mit Spiegelwinkel α über die Ocularebene und somit nur halb so schnell wie die unter dem Winkel 2α abgebildete Meßmarke. Das hat zur Folge, daß diese schon bei geringen Spiegelneigungen aus dem ausgeleuchteten Teil des Gesichtsfeldes auswandert und nicht mehr beobachtet werden kann. Diese Verhältnisse sind in Fig. 83 veranschaulicht. Bei einem AKF mit 500 mm Brennweite und 60 mm Objektivdurchmesser ist die praktisch ausnutzbare Grenze 30 m. Der Durchmesser des ausgeleuchteten Gesichtsfeldes entspricht an dieser Grenze etwa 1′, die maximal beobachtbare Spiegelneigung beträgt noch etwa 2′.

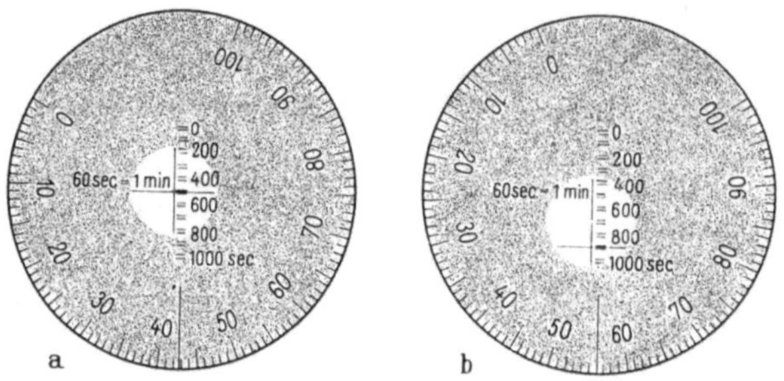

Fig. 83 a u. b. Ausleuchtung des Strichkreuzbildes. a Mit wachsender Spiegelentfernung nimmt der ausgeleuchtete Teil des Gesichtsfeldes ab. b Beim Neigen des Spiegels bewegt sich das Kreuz doppelt so schnell wie das ausgeleuchtete Feld.

β) Justieren des Spiegels. Es bedeutet keine Schwierigkeit, den Spiegel vor dem Fernrohr so auszurichten, daß das reflektierte Bild der Marke zurück ins Fernrohr fällt, solange der Spiegel noch mit der Hand erreichbar ist. Muß er jedoch in größerer Entfernung vom Fernrohr aufgestellt werden, ist das Hineinjustieren des reflektierten Strahlenbündels in den kleinen Winkelbereich des AKF problematisch. Eine Methode, diese Schwierigkeit zu beseitigen, soll beschrieben werden. Man denke sich in den Weg des vom Fernrohr nach Unendlich geschickten Bündels einen Tripelspiegel gestellt. Dieser würde unabhängig von einer korrekten Justierung das Bündel in sich zurückwerfen. Könnten wir dies reflektierte Bündel anzapfen, so würden wir in ihm ein Bild der Fernrohröffnung sehen. Stünde neben dem Tripelspiegel noch der Planspiegel und träfe auf ihn auch noch ein Teil des aus dem Fernrohr kommenden Bündels (geometrische Strahlenteilung), so würde über ihn ebenfalls die Fernrohröffnung zu sehen sein. Wenn beide Bilder der Fernrohröffnung zusammenfallen, wäre der Planspiegel genau senkrecht zum austretenden Bündel justiert. Diese Überlegung führt zu dem von Mandler angegebenen Justierprisma. Es wird nicht geometrische, sondern physikalische Strahlenteilung angewandt, dann läßt sich der Tripelspiegel virtuell vor den Planspiegel bringen und über die eingebrachte Teilerfläche lassen sich zusätzlich die beiden Bilder der Fernrohröffnung von der Seite des Strahlengangs her beobachten und durch Bewegungen des Planspiegels zur Deckung bringen. Das Mandler-Prisma darf schief im Strahlengang stehen, kann also auch von Hand gehalten werden (Fig. 84).

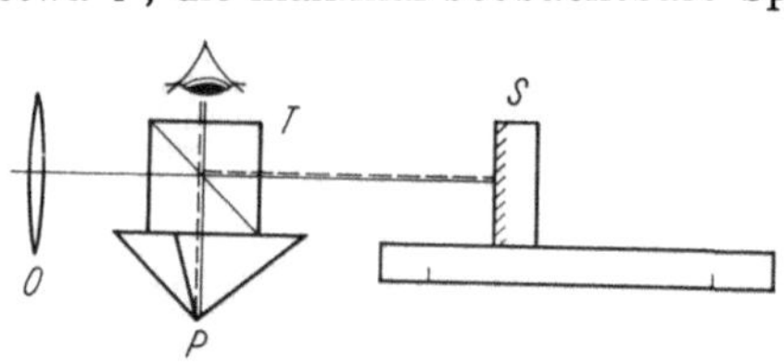

Fig. 84. Ausrichtprisma nach Mandler. *O* AKF-Objektiv, *S* AKF-Spiegel, *P* Tripelprisma, *T* Teilerprisma.

γ) Messen einer Führungsbahn und einer Richtplatte. Zur Messung der Ebenheit von Führungsbahnen und Richtplatten wird der AKF-Spiegel mit einer Basis bestimmter Länge, zweier Aufsetzpunkte in bestimmtem Abstand versehen, um aus einer trigonometrischen Rechnung die geometrische Bahngestalt im Profilschnitt zu erhalten. Der Spiegel darf nur um seine Basislänge verschoben werden. Zwischenpunkte können nicht gemessen werden (Fig. 85). Das ist ein Nachteil gegenüber den in Ziff. 35 beschriebenen Fluchtungsmethoden. Diese Messungen sind daher auch nicht typisch für AKF-Methoden. Näheres unter [*86*].

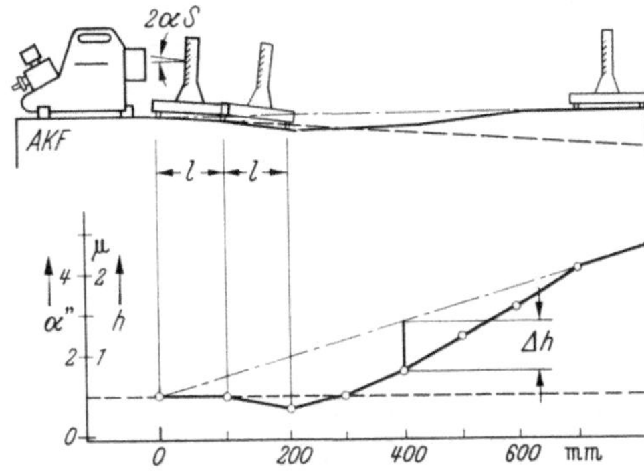

Fig. 85. Vermessen einer Führungsbahn. α Neigungen des Spiegels, *AKF* Autokollimationsfernrohr, *h* aus α und *l* folgende Höhen, *Δh* Abweichungen der Bahn, *l* Basislänge, *S* Spiegel.

δ) Messen von Kippfehlern und Durchbiegungen. Ein typisches Beispiel für die Anwendung von AKF-Messungen findet sich bei Längenmeßgeräten, die gegen das Abbesche Komparatorprinzip verstoßen. Die durch Kippfehler des Komparatorwagens entstehenden Meßfehler erster Ordnung lassen sich leicht aus dem mit Hilfe eines AKF gemessenen Kippwinkel und dem gegenseitigen Abstand von Prüflings- und Maßstabsachse berechnen. Kann das AKF nicht auf demselben Grundbett stehen, muß ein Vergleichsspiegel für die zusätzliche Messung der Verlagerung des Grundbettes aufgesetzt werden (Fig. 86). Sehr geringe von einander unabhängige Kippbewegungen können auch die Mikroskope auf Komparatoren ausführen, wenn der schwere Komparatorschlitten verschoben wird und sich das Grundgestell verwindet. Auch diese Frage kann mit dem AKF geklärt werden.

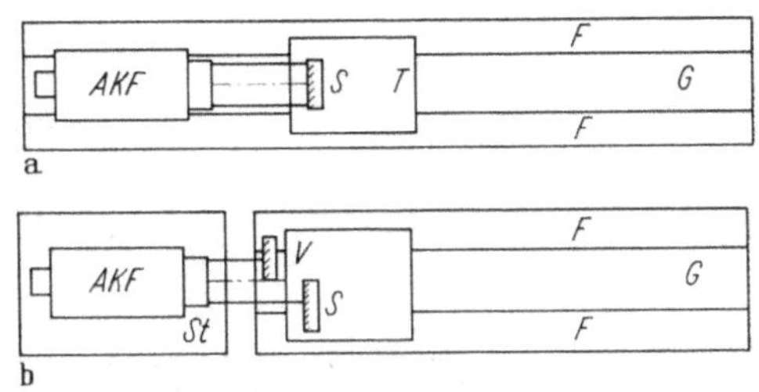

Fig. 86a u. b. Messen der Kippungen eines Meßtisches. a Fernrohr auf dem Meßgerät, b Fernrohr auf besonderem Stativ. *AKF* Autokollimationsfernrohr, *F* Führungsbahnen, *G* Grundgestell, *S* Spiegel, *St* Fernrohrstativ, *T* Meßtisch, *V* Vergleichsspiegel.

Ein weiteres Beispiel liefert die Untersuchung eines Feintastgeräts (Fig. 87). Der Fühlhebel ist im Schneidenlager drehbar gelagert und wird mit nicht gezeichneten Federkräften wahlweise nach links oder rechts gegen die Prüflingsfläche gezogen. Man schiebt diese dann soweit vor, bis das AKF_2 die Nullstellung des Spiegels *S* anzeigt.

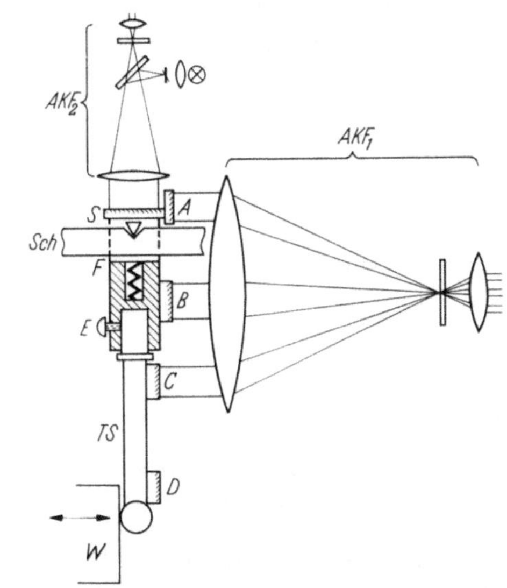

Fig. 87. Untersuchung der Durchbiegung und Lose eines Fühlhebels (Nulltasters), der aus verschiedenen miteinander mechanisch gekoppelten Gliedern besteht. AKF_1 Autokollimationsfernrohr zum Prüfen der Durchbiegungen, AKF_2 im Feintaster eingebautes AKF mit beleuchtetem Strichkreuz, *A*, *B*, *C*, *D* AKF-Spiegel, *E* Taststiftklemmung, *F* Ansatzfläche, *S* am Taster fester Spiegel, *Sch* Schneidenlager, *TS* Taststift, *W* Prüflingsfläche.

Das Tastgerät setzt sich aus mehreren Gliedmaßen zusammen, die teils durch Federn, teils durch Klemmungen miteinander verbunden sind. Gegen seitlich an der Tastkugel angreifende Meßkräfte sollten diese Verbindungen starr sein. Ebenso wenig darf sich der für Innenmessungen gedachte lange Taststift durchbiegen. Mit Hilfe kleiner, an den einzelnen Gliedmaßen befestigter AKF-Spiegel, die möglichst gleichzeitig im AKF abgelesen werden sollten, läßt sich leicht feststellen, ob der Taster in sich genügend starr ist, so daß keine mechanischen Umkehrspannen entstehen können, die die Messung verfälschen würden.

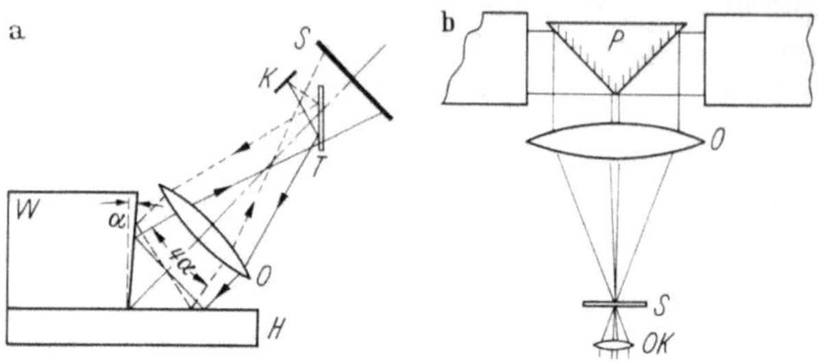

Fig. 88 a u. b. a Messen eines 90°-Winkels. b Messen eines 180°-Winkels. α Fehler des Winkels, *H* Hilfsfläche, *K* Strichkreuz, *O* Objektiv, *Ok* Ocular, *P* Teilerprisma, *S* Strichplatte, *T* Teilerplatte, *W* Prüfstück.

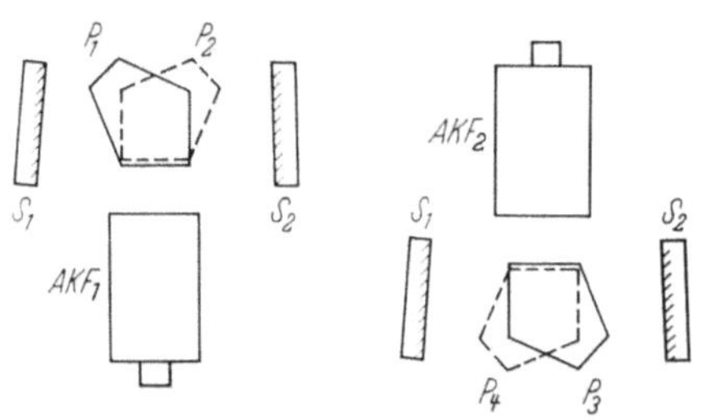

Fig. 89. Prüfung eines Pentagonprismas. AKF_1 erste, AKF_2 zweite Stellung des Autokollimationsfernrohrs, $S_{1,2}$ feststehende Spiegel, $P_{1,2,3,4}$ vier Meßstellungen des Pentagonprismas.

ε) Messen von Winkeln und Teilkreisen.

1. Winkel von 90 und 180°. Richtet man das AKF auf einen 90°-Winkel, der aus reflektierenden Flächen bestehen möge, so kommt nur dann ein einziges Bild zustande, wenn der Winkel keine Abweichung von 90° besitzt. Ist diese jedoch in der Größe α vorhanden, so entstehen zwei Bilder im Abstand von 4α (Fig. 88).

Auch ein 180°-Winkel, d.h. die Parallelität von Flächen kann mit dem AKF gemessen werden. Dazu wird ein 90°-Prisma oder ein Pentagonprisma benötigt, um die aus dem Fernrohr austretenden Bündel rechtwinklig umzulenken

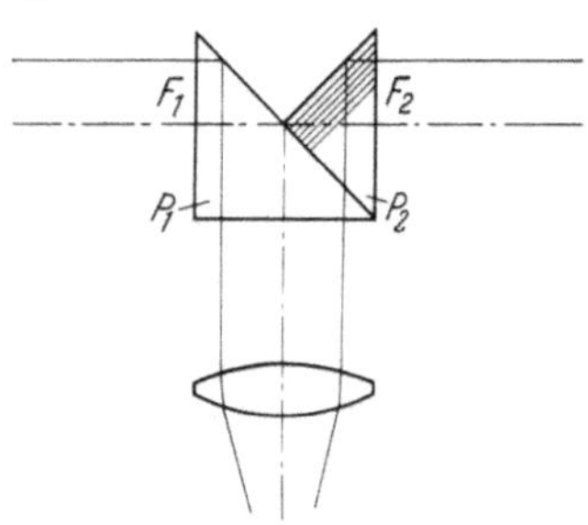

Fig. 90.

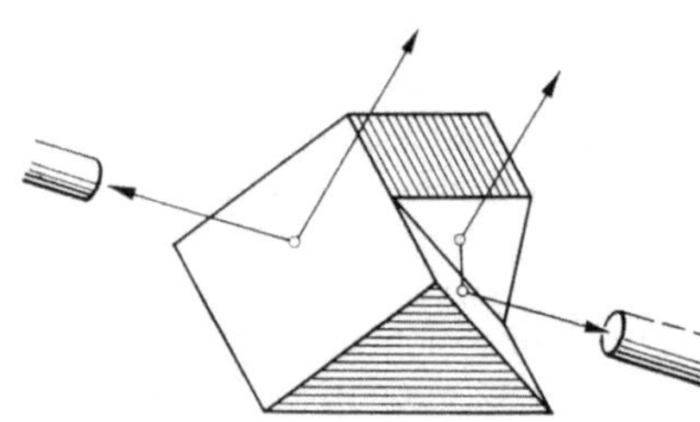

Fig. 91.

Fig. 90. Dachkantdoppelprisma zur Durchführung der Prüfung nach Fig. 88b. $F_{1,2}$ sind zueinander parallele Austrittsflächen des Reflexionsprismas P_1 und des Dachprismas P_2.

Fig. 91. Dachkantoberflächenspiegel zur Durchführung der Parallelitätsprüfung nach Fig. 88b.

(Fig. 88 und 89). Das 90°-Prisma in Fig. 88 muß sehr genau justiert werden. Steht die Spiegelkante nicht parallel, sondern geneigt zu den Flächen, so entstehen trotz der Parallelität der Meßflächen Doppelbilder. Unempfindlich in bezug auf ihre Justierung zu den Meßflächen sind dagegen die in den Fig. 90 und 91 dargestellten Prismensysteme [*98*]. Bei der in Fig. 89 dargestellten Anordnung zur Parallelitätsmessung mit Hilfe eines Pentagonprismas läßt sich durch

Umschlagmessungen nicht nur der Winkelfehler zwischen den parallel zu stellenden Spiegeln messen, sondern auch der Fehler im Umlenkwinkel des Pentagonprismas. Die einzelnen Schritte sind in Fig. 89 dargestellt. Man kann sich überlegen, daß die Abweichungen der Winkel zwischen der Fernrohrachse und den Spiegelloten von 90°, über alle vier Quadranten summiert, Null ergeben müssen: $\alpha_1+\alpha_2-\alpha_3-\alpha_4=0$. Ist die Summe der vier gemessenen Winkelabweichungen kleiner oder größer als Null, so entfällt ein Viertel davon auf den Fehler δ des Pentagonprismas gegen 90°: $\alpha_1+\delta+\alpha_2+\delta-\alpha_3+\delta-\alpha_4+\delta=4\delta$.

2. Pentagonprismen. Winkelspiegel von 45° lenken die Strahlrichtung um 90° ab, unabhängig von der Einfallsrichtung des Strahlenbündels. Das Pentagonprisma stellt einen derartigen festen Winkelspiegel dar. Diesem Umstand verdankt das Pentagonprisma eine ganze Reihe interessanter Anwendungen. Das Parallelrichten von Achsen soll unter ihnen herausgegriffen werden (Fig. 92). In

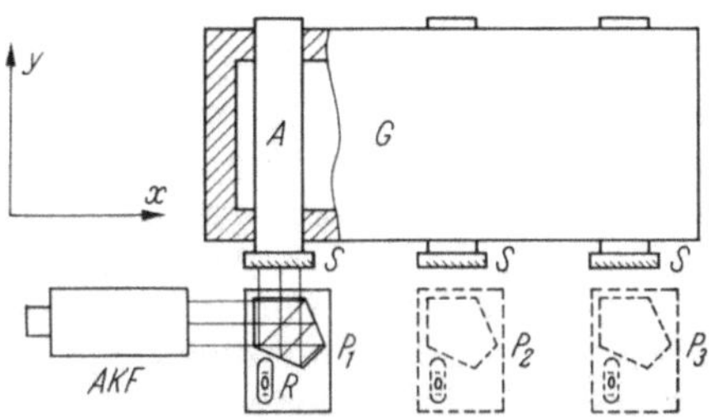

Fig. 92. Prüfen der Parallelität von Achsen. *A* Achse, *AKF* Autokollimationsfernrohr, *G* Gehäuse, $P_{1,2,3}$ Pentagonprisma in verschiedenen Stellungen, *R* Röhrenlibelle am Pentagonprisma, *S* Spiegel.

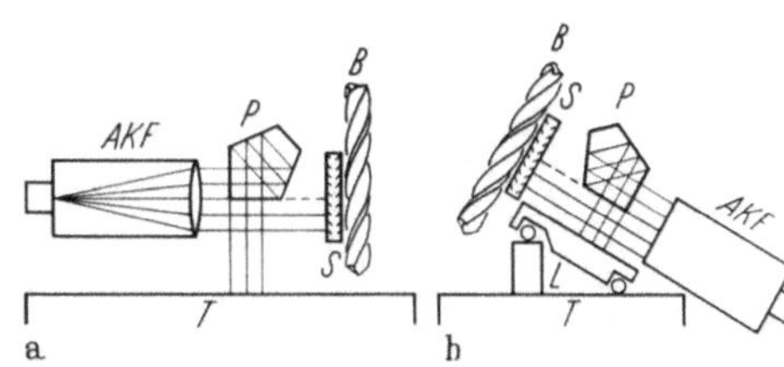

Fig. 93 a u. b. Ausrichten einer Bohrspindel zum Tisch. a senkrecht, b geneigt zum Tisch. *AKF* Autokollimationsfernrohr, *B* Bohrspindel, *L* Sinuslineal, *P* Pentagonprisma, *S* Doppelspiegel, *T* Tisch.

der Zeichenebene braucht das Prisma wegen seiner konstanten Ablenkung um 90° nicht genau justiert zu werden, um die y-Achse wird es durch Autokollimation auf seine Vorderfläche ungefähr ausgerichtet, um die x-Achse wird es nach einer empfindlichen Röhrenlibelle eingestellt.

Die Einstellung rechter Winkel, wie sie in Fig. 88 angegeben ist, gelingt nicht mehr, wenn die beiden Winkelflächen einen größeren seitlichen Abstand voneinander haben, so daß der Durchmesser der AKF-Öffnung nicht ausreicht, beide Flächen gleichzeitig zu beobachten. In diesen Fällen kann ein Teil des aus dem AKF austretenden Bündels durch geometrische Strahlenteilung über das Pentagonprisma gegen die eine Fläche, der andere gegen den AKF-Spiegel gelenkt werden (Fig. 93). Es ist auch möglich, in den zweiten Weg ebenfalls ein Pentagon zu setzen, um die oben erwähnte Parallelitätsmessung zweier Flächen ohne Umsetzen des Pentagons in einem einzigen Schritt zu erledigen. Bei solchen Aufgaben ist es wichtig zu wissen, welchen Fehler das Pentagonprisma bei der Ablenkung um 90° besitzt. Dieser Fehler kann entweder durch die beschriebene Umschlagmessung zwischen zwei annähernd parallel stehenden Spiegeln ermittelt werden (Fig. 89), oder indem das Pentagonprisma über die geometrische Strahlenteilung (Fig. 93) an einen rechten Winkel angeschlossen wird, der vorher nach Fig. 88 mit einem AKF einjustiert wurde.

3. Keilfehler von Planparallelplatten. Durchsichtige planparallele Glasplatten geben vor dem AKF zwei Bilder, eines von der Vorder- und eines von der Rückseite. Diese beiden Bilder liegen unter dem Winkel $\varphi=2n\gamma$ zueinander, worin n den Brechungsindex des Glases und γ den Keilwinkel der Platte bedeuten, und die Vorderfläche des Keils in Autokollimation eingestellt ist.

Eine andere Möglichkeit, den Keilfehler zu messen, geht über den Ablenkwinkel des Keils $\delta=\gamma(n-1)$ für kleine Keilwinkel. In diesem Fall wird der Keil zwischen das AKF und einen vorher ausgerichteten Planspiegel gebracht und die Ablenkung der Marke durch den Keil gemessen. Diese ergibt sich zu 2δ. Durch Drehen des Keils um die optische Achse läßt sich der vierfache Ablenkwinkel ablesen.

Undurchsichtige Planparallelplatten, zu welchen z.B. auch Endmaße gehören sprengt man auf einer Grundfläche an und sucht mit dem AKF das an der oberen Fläche und das an der Grundfläche reflektierte Bild nach Art der geometrischen Strahlenteilung gleichzeitig zu erhalten. Ist der Prüfling weder durchsichtig noch reflektierend, so behilft man sich damit, daß man einen Planspiegel zunächst auf die Tischfläche, dann unter Zwischenfügung des Prüflings auf den Prüfling selbst legt. Der Keilfehler ergibt sich aus der Differenz der beiden Ablesungen.

In allen Fällen ist zu berücksichtigen, ob das AKF den Spiegelwinkel α mißt, oder den Winkel 2α, unter dem die reflektierten Bündel einfallen. Das ist eine Frage der Skalenbezeichnung.

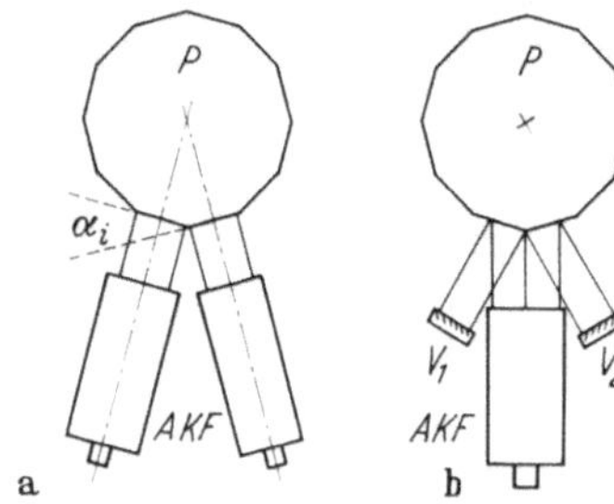

Fig. 94a u. b. Vermessen eines Polygons. a Mit zwei Fernrohren, b mit Vergleichsspiegeln. α_i Einzelwinkel, *AKF* Autokollimationsfernrohre, *P* Polygon, $V_{1,2}$ Vergleichsspiegel.

4. Spiegelpolygone. Im Zusammenwirken mit Spiegelpolygonen kann das AKF zu den Aufgaben der Teilkreismessung herangezogen werden. Regelmäßige Spiegelpolygone werden mit Hilfe zweier AKF zunächst in sich geprüft (Fig. 94), wobei der Winkel zwischen den Fernrohren mit den nacheinander in die gleiche Stellung vor den Fernrohren gebrachten Flächenwinkeln verglichen wird. Aus der Bedingung, daß die Summe der Fehler der einzelnen Winkel zwischen den Flächen über 360° gleich Null sein muß, läßt sich der Fehler jedes Polygonwinkels ermitteln, ohne daß der Winkel zwischen den Fernrohren bekannt zu sein braucht. Dividiert man nämlich die Fehlersumme durch die Anzahl der Polygonwinkel und zieht den erhaltenen Betrag von den gemessenen Einzelwerten ab, so erhält man die einzelnen Winkelfehler. Die Teilkreisprüfung mit Spiegelpolygon vollzieht sich so, daß das Polygon auf den Teilkreis aufgesetzt wird, wobei es nicht zentriert zu werden braucht; Polygonachse und Teilkreisdrehachse müssen aber zueinander parallel sein. Dann wird der Teilkreis auf einen glatten Wert eingestellt und vor eine Spiegelfläche des Polygons ein AKF gesetzt. Dreht man nun den Teilkreis um den Polygonwinkel nach seiner Teilung weiter, wirkt jeweils die nächste Fläche des Polygons als AKF-Spiegel. So wird der Teilkreis schrittweise gegen das gleichsam als Winkelnormal dienende Polygon verglichen und die einzelnen Intervallfehler (Gruppenfehler) des Teilkreises erhalten. Durch Versetzen des Teilkreises gegen das Polygon können beliebige andere Teilstrichgruppen herausgegriffen werden. Über den periodischen Anteil der Teilkreisfehler oder durch Gleichsetzen der arithmetischen Mittelwerte ihrer Fehlerkurven lassen sich die einzelnen Gruppenfehlerkurven aneinander anschließen.

ζ) *Radienmessung von Linsen.* Autokollimationsverfahren eignen sich zur Messung der Krümmungsradien von Linsen und Probegläsern. Sie sind den mechanischen Sphärometermethoden überlegen, die bei 100 mm Krümmungsradius eine Meßunsicherheit von $5-10 \cdot 10^{-5}$ besitzen, während die optische Meßmethode 10^{-5} erreicht.

Um auf sphärische Flächen zu autokollimieren, muß vor das Kollimatorobjektiv ein weiteres Objektiv gesetzt werden, welches ein reelles Luftbild der projizierten Marke in endlichem Abstand entwirft. Im Prinzip erfolgt die Radienmessung so, daß dieses Luftbild einmal mit der sphärischen Fläche selbst und nach einer Verschiebung der Fläche mit ihrem Krümmungsmittelpunkt zur Deckung gebracht wird. In beiden Fällen ist im Autokollimationsocular ein reflektiertes Bild zu sehen, im ersten Fall im Abbildungsmaßstab $\beta' = +1$, im zweiten in $\beta' = -1$. Die Verschiebung ist gleich dem Krümmungsradius der Fläche. Die Genauigkeit der Radienmessung hängt von der Genauigkeit ab, mit der die reflektierte Marke im Ocular scharfgestellt werden kann. Hierbei kann die Prüfung auf Parallaxe (Ziff. 10) nützlich sein.

Die Einstellgenauigkeit würde wesentlich erhöht werden, wenn nicht nur auf beste Schärfe, sondern auf eine Symmetrieerscheinung eingestellt werden könnte (vgl. Ziff. 11). Nach einem Vorschlag von FRANKE [*99*], wurde von ROSENHAUER

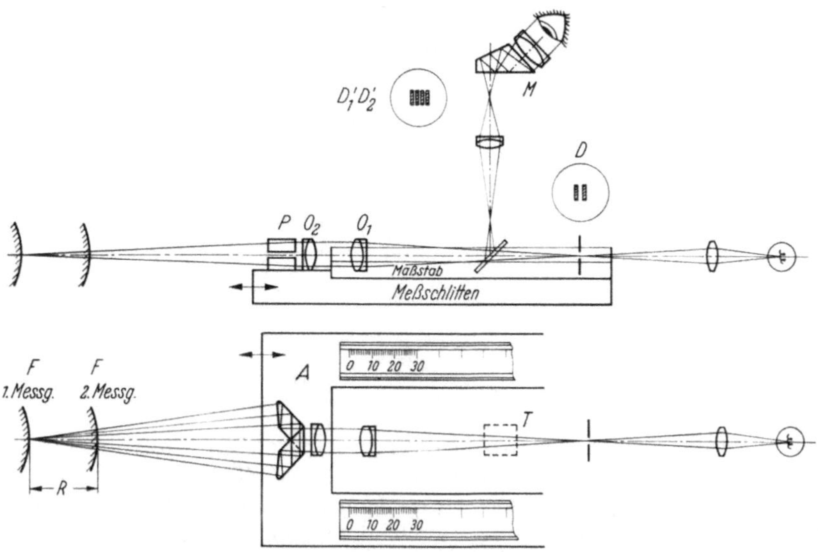

Fig. 95. Optische Messung von Krümmungsradien. D beleuchteter Doppelspalt, $O_{1,2}$ Objektive, P Doppelwinkelspiegel zur Pupillenteilung, $D_1' D_2'$ Bilder des Doppelspalts „auf Lücke", F Prüflingsfläche in zwei Meßstellungen, A Meßschlitten mit Maßstabsdoppelablesung, T Teilerfläche, M Beobachtungsmikroskop, R Radius der sphärischen Fläche.

[*100*] folgende Anordnung für Radien mittlerer Größe (10—200 mm) gewählt (Fig. 95). Von einem beleuchteten Doppelspalt wird durch Pupillenteilung ein Doppelbild erzeugt. Durch die Prüflingsoberfläche werden die beiden Bilder reflektiert und in der Ocularebene beobachtet. Verschiebt sich die Prüflingsoberfläche aus der Schärfenebene heraus, so wandern die beobachteten Bilder in entgegengesetzten Richtungen über die Ocularebene. In der Ausführung als Doppelspalt lassen sich die beiden reflektierten Bilder gut auf Lücke symmetrieren. Der durch das Doppelprisma erzeugte Schnittwinkel der beiden Abbildungsstrahlengänge ist wesentlich für die Meßgenauigkeit; je größer er ist, desto größer ist die gegenseitige Verschiebung der beiden Bilder in Abhängigkeit von der Verschiebung des Prüflings oder des Meßschlittens A. Die Einrichtung ist so getroffen, daß die jeweilige Stellung des Meßschlittens an zwei Maßstäben abgelesen und gemittelt wird, wodurch Führungsfehler des Meßschlittens eliminiert und gleichzeitig Abweichungen der Maßstabstriche von ihrer Sollage gemittelt werden, wodurch sich die Meßgenauigkeit erhöht (vgl. Ziff. 34 ε). Eine mögliche Parallaxe zwischen den beiden Bildern spielt keine Rolle, wenn sie auf eine sehr feinkörnige Mattscheibe in der Ocularebene projiziert werden (vgl. Ziff. 10).

Die eingangs erwähnte hohe Meßgenauigkeit des Geräts wurde quantitativ bestätigt durch Messungen an Probeglaspaaren mit konkaven und konvexen

Flächen, deren Radienunterschiede aus Passungsprüfungen mit Hilfe Newtonscher Ringe bekannt waren.

Für Messungen größerer Krümmungsradien empfiehlt es sich, zwei getrennte Autokollimationssysteme aufzubauen und gleichzeitig den Schnittwinkel der beiden Abbildungsstrahlengänge zu vergrößern (Fig. 96). Dadurch läßt sich die Pupillenteilung vermeiden [*100*].

η) *Brennweitenmessung.* Kollimator und Fernrohr werden auch zur Messung der Brennweite von Linsen und Objektiven herangezogen, z.B. nach der bekannten von den Knotenpunkten ausgehenden Beziehung $y=f\tan\sigma_K'$, y Objektgröße, f dingseitige Brennweite und σ_K' der Winkel, unter dem das Objekt vom dingseitigen Knotenpunkt aus gesehen wird, oder unter dem es auf der Bildseite nach Unendlich projiziert wird. Sind die Objektgröße y und der Winkel σ' bekannt, läßt sich die Brennweite errechnen. Die Objektgröße wird auf einem Kom-

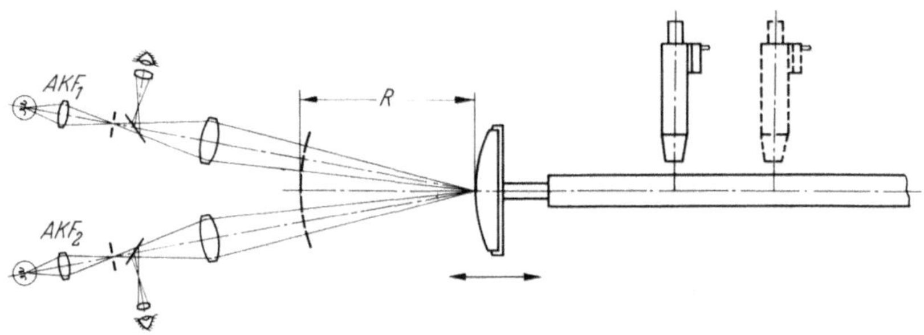

Fig. 96. Radienmeßgerät mit zwei Autokollimationsfernrohren AKF_1 und AKF_2 nach dem Abbeschen Komparator-Prinzip. R Radius der sphärischen Fläche.

parator bestimmt, einfacher wird gleich ein bekanntes Objektmikrometer benutzt. Der Winkel kann mit einem schwenkbaren, auf Unendlich eingestellten Fernrohr, am besten auf einem Spektrometertisch, gemessen werden. Die Genauigkeit der Messung ist von der Fokussierung des Objektes abhängig, die im Fernrohr durch die Einstellung auf beste Bildschärfe und verschwindende Parallaxe beurteilt wird. Voraussetzung für richtige Fokussierung ist, daß das Fernrohr vorher sehr genau auf Unendlich eingestellt wurde.

Eine andere Methode bedient sich eines Kollimators mit sehr genau bekannter Brennweite, der einen bekannten Strichabstand nach Unendlich projiziert und damit einen genauen Gesichtsfeldwinkel definiert. Setzt man das Objektiv, dessen Brennweite gemessen werden soll, vor den Kollimator, so bildet es den Strichabstand in seiner hinteren Brennebene ab, wo er durch ein quer zur optischen Achse mikrometrisch verschiebbares Visiermikroskop oder ein Mikrometer-Ocular gemessen werden kann (vgl. [*1*], S. 167). Um Fehler im Abbildungsmaßstab hierbei auszuschließen, kann bildseitig der telezentrische Strahlengang (Ziff. 5) eingehalten werden, wenn die entsprechend verkleinerte Kollimatoröffnung mit der vorderen Brennebene des zu prüfenden Objektivs zusammenfällt; auf diese Weise fallen parallaktische Fehler auf der Beobachterseite nicht ins Gewicht. Andererseits kann der Kollimator sehr genau auf Unendlich einjustiert werden, so daß relativ geringe Meßunsicherheiten für die Brennweite und damit für den Winkel erhalten werden. In diesem Fall ergibt sich die Brennweite des Objektivs aus der Gleichung $y'=f\sin\sigma_F$, worin σ_F der Winkel ist, unter dem der Hauptstrahl durch die Brennebene verläuft. y' entspricht dem Achsabstand h unter dem der Hauptstrahl die sphärische Hauptfläche H schneidet.

Beiden beschriebenen Methoden haften die Vorbedingungen sehr genauer Fokussierung auf das Objekt an, um einen möglichst fehlerfreien Wert für den

Winkel zu bekommen, unter dem das Objekt im Unendlichen erscheint. Daher sind Methoden erdacht worden, die diese Fehlerquelle vermeiden. Die *Abbesche Fokometermethode* arbeitet nach dem zuletzt beschriebenen Verfahren mit dem Unterschied, daß der durch den Kollimator gegebene Winkel durch eine andere Art von trigonometrischer Winkelmessung im Objektraum ersetzt wird, nämlich durch zwei in verschiedenem Abstand außerhalb der Brennweite vor dem Objektiv stehende Maßstäbe, auf deren Bilder im bildseitig telezentrischen Strahlengang das Visiermikroskop durch axiales Verschieben nacheinander scharfgestellt wird. Die Aperturblende des Visiermikroskops erscheint als Eintrittspupille in der vorderen Brennebene des zu prüfenden Objektivs und bestimmt das Zentrum der Perspektive im Dingraum, von dem aus die Maßstäbe gemäß ihrem gegenseitigen Abstand in verschiedener Größe erscheinen. Hier soll auf eine ausführliche Darstellung verzichtet werden. Das Abbesche Fokometer ist vielfach in der Literatur beschrieben worden [*1*], [*101*], [*102*], [*103*], [*103a*]. Erwähnt sei noch, daß die Meßunsicherheit bei 0,1% liegt.

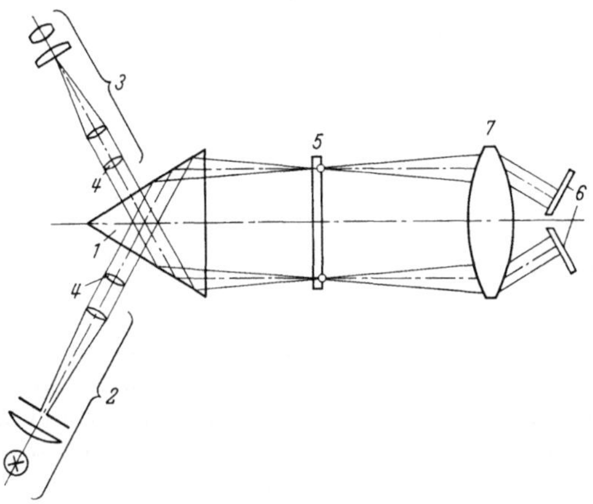

Fig. 97. Koinzidenzfokometer nach HODAM. *1* Kösters-Prisma, *2* Kollimator, *3* Fernrohr, *4* Vorsatzlinsen zur Projektion bzw. Beobachtung des Kollimatorspaltbildes in *5*, *5* durchlässig verspiegelte Zwischenbildebene, durch die hindurch das Autokollimationsbild von Spiegel *6* über Objektiv *7* in Koinzidenz mit dem Zwischenbild betrachtet wird.

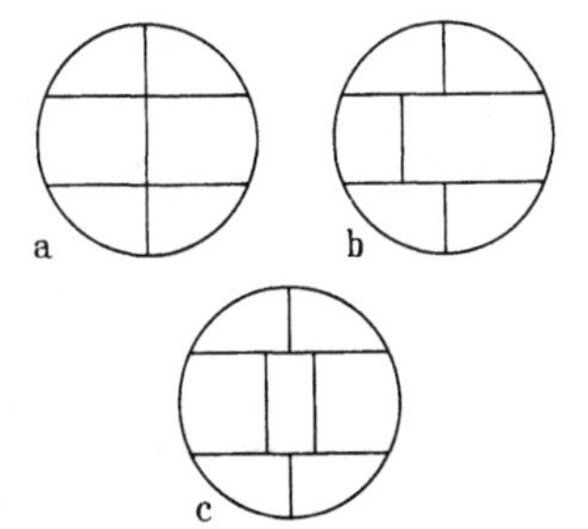

Fig. 98a–c. Gesichtsfeld des Koinzidenzfokometers a in und b außer Koinzidenz und c bei dezentrierter Lage des Objektivs, wobei Aufspaltung der beiden Autokollimationsbilder symmetrisch zur exakten Lage entsteht.

Stattdessen soll auf ein neues *Koinzidenzfokometer* näher eingegangen werden, zumal es sich wesentlich der Autokollimation bedient. Das von HODAM [*104*] entwickelte Verfahren ist im Prinzip in Fig. 97 dargestellt. Es benutzt im Grundsatz die gleichen Meßgrößen zur Bestimmung der Brennweite und ähnelt daher den beschriebenen Verfahren. Anstelle eines den Winkel bestimmenden Kollimators oder zweier Maßstäbe wird hier ein Winkelspiegel *6* verwendet, dessen beide Flächen je in Autokollimationsstellung vor dem Objektiv *7* stehen. Der telezentrische Strahlengang ist dadurch eingehalten, daß aus einem Kollimator *2* über die Vorsatzlinse *4* und das Kösters-Prisma *1* je ein Spaltbild mit kleiner Beleuchtungsapertur in die vordere Brennebene des zu prüfenden Objektivs *7* projiziert wird, und die Hauptstrahlen der beleuchtenden Bündel parallel zur optischen Achse des Prüflings verlaufen. Um das nach *5* projizierte Spaltbild gleichzeitig mit seinem Autokollimationsbild durch das Fernrohr *3* mit Vorsatzlinse *4* beobachten zu können, wäre in der Ebene von *5* eine teildurchlässige Spiegelfläche denkbar; tatsächlich steht bei *5* eine vollverspiegelte Fläche mit einem in der Zeichenebene liegenden durchgehenden Spalt weniger mm Breite. Der verspiegelte Teil wirft das senkrecht zur Zeichenebene stehende Bild des Kollimatorspalts unmittelbar zurück, während durch den unverspiegelten Spalt das Autokollimationsbild zu sehen ist. Ist der Spiegel *6* richtig ausgerichtet, koinzidieren beide Bilder nach Art der Fig. 98. Koinzidieren sie nicht, muß entweder der

Spiegelwinkel geändert werden oder aber der Achsabstand des projizierten Spaltbildes *5*. Am einfachsten ist es, durch Längsverschieben des Kösters-Prismas *1* den Achsabstand zu variieren und die Verschiebung an einem Maßstab mikroskopisch abzulesen.

Zwischen der Verschiebung v und dem Achsabstand h besteht die einfache durch den Halbwinkel des Kösters-Prismas von $\alpha=30°$ gegebene Beziehung:

$$v=\frac{2}{3}\sqrt{3}\,(h_2-h_1).$$

Entsteht beim Verschieben gleichzeitig eine Kippung des Kösters-Prismas, so spalten die auf den verschiedenen Seiten gewonnenen Bilder *5* unsymmetrisch auf. Beim Verschieben muß nachfokussiert werden, zu diesem Zweck sind die Vorsatzobjektive *4* mit der Längsbewegung des Prismas verschieblich.

Unter Verwendung eines Satzes fester bekannter Winkelspiegel lassen sich wie mit dem Abbeschen Fokometer so auch mit diesem Verfahren die Brennweiten für die einzelnen Objektivzonen messen. Die Meßunsicherheit wird von HODAM mit $\pm 0{,}05\%$ angegeben. Bei Systemen mit besserem Korrektionszustand war $\pm 0{,}025\%$ erreichbar. Bei reinen Differenzmessungen, bei denen die Lage des Teilungsursprungs belanglos ist, war sie sogar noch geringer, z.B. bei Messungen der chromatischen Aberration der Brennweite.

40. Räumliches Koordinatensystem. Für die räumliche Vermessung von Konstruktionen im Schiffs- und Flugzeugbau werden Koordinatensysteme benötigt, und es ist zweckmäßig, sie durch Fluchtfernrohre und Zielmarken in Verbindung mit Bauelementen für die Richtungsprüfung zu bestimmen [*105*]. Ebenen senkrecht zu den Koordinatenachsen können optisch beschrieben werden, indem vor das Fluchtfernrohr ein um die Fernrohrachse drehbares Pentagonprisma gesetzt wird, welches die Fluchtgerade in beliebiger Richtung um 90° umzulenken gestattet. Den gleichen Zweck kann man mit einem sog. *Jig-Transit-Square* erreichen, einem Fluchtfernrohr ähnlich einem Theodoliten, welches senkrecht zu seiner horizontalen Querachse einen Planspiegel trägt. Mit Hilfe dieses Spiegels und eines AKF kann das Transit-Square nach einer Parallelverschiebung wieder genau parallel zu seiner ursprünglichen Stellung aufgestellt werden, so daß sich nacheinander beliebig viele zueinander parallele Ebenen senkrecht zur AKF-Achse vermessen lassen (Fig. 99).

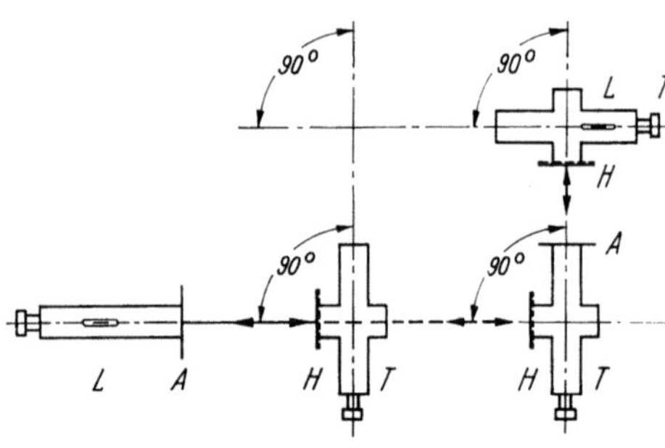

Fig. 99. Jig-Transit-Square *T* mit halbdurchlässigem Spiegel *H* und Hohlachse. *A* Autoreflexionsmarke, *L* Libelle.

Automatisches Nivellieren und optisches Lot. Die Festlegung eines Koordinatensystems relativ zur Erdanziehung, d.h. zur Meeresoberfläche oder dem Horizont wie es in der Geodäsie notwendig ist, erübrigt sich bei den optischen Feinmeßmethoden, für die hier die physikalischen Grundlagen gegeben werden sollten, zumeist. Trotzdem können im Flugzeug- und Schiffsbau Fälle eintreten, in denen der Horizont erforderlich ist. Eine horizontale Fluchtlinie wird erreicht, indem vor das Objektiv ein Quecksilberhorizont oder ein mit einer Libelle horizontal ausgerichteter Spiegel gesetzt wird. Der gesuchte Horizont liegt auf der Mitte zwischen der unendlich entfernten Zielmarke (Kollimator) und ihrem Spiegelbild.

Die Ausnutzung der Schwerkraft in Fällen, in denen die Ziellinie automatisch wieder nivelliert werden muß, gelingt mit Hilfe von Pendelspiegeln, vgl. Fig. 100

und 70 (S. 808) [*106*], [*107*], [*82*], oder auch mit Hilfe einer als Linse wirkenden Flüssigkeitsblase (Fig. 101) [*108*], im abbildenden Strahlengang zwischen Objektiv und Ocular eines Fluchtfernrohrs. Diese optischen Bauelemente sorgen dafür, daß die Strichkreuzmitte in der Ocularebene auch dann der Horizontalen entspricht, wenn das Fernrohr etwas geneigt aufgestellt wird. Auf einem etwas

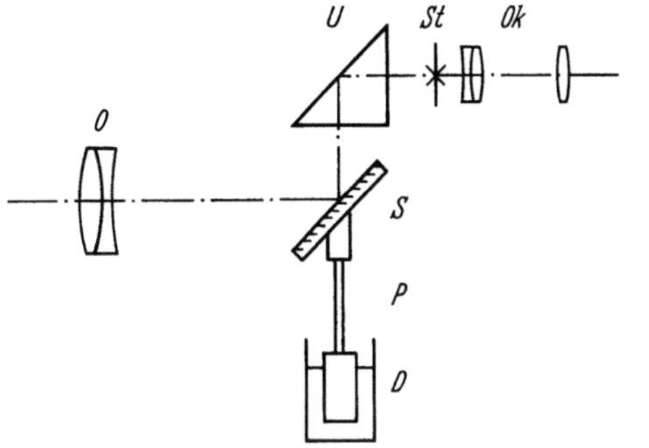

Fig. 100.

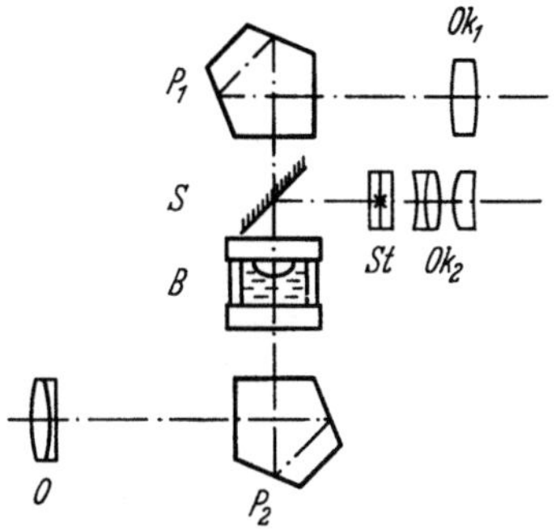

Fig. 101.

Fig. 100. Pendelspiegel S zum automatischen Nivellieren eines Fluchtfernrohrs mit Objektiv O, Pendel P, Dämpfung D, Umlenkspiegel U, Strichmarke St und Ocular Ok.

Fig. 101. Flüssigkeitsblase B zum automatischen Nivellieren der Fluchtlinie in einem Fernrohr mit Objektiv O, Strichmarke St und Ocular Ok_2. Zum Fokussieren wird O verschoben. S Teilerspiegel, Ok_1 zum Beobachten der Grobhorizontierung, $P_{1,2}$ Pentagonprismen.

anderen Prinzip beruht die Anordnung nach TSUBOKAWA [*109*]. Hier wird die horizontale Fluchtlinie durch eine pendelnd aufgehängte, vertikal stehende Spiegelfläche gegeben, in der sich eine Zielmarke spiegelt. Das Lot von der Zielmarke auf den Pendelspiegel muß infolgedessen stets eine horizontale Fluchtlinie ergeben (Fig. 102). Die Zielmarke ist als Strichkreuz im bildseitigen Hauptpunkt des Objektivs angebracht. Die beim Nachfokussieren des Objektivs entstehenden Querversetzungen des Objektivs von der geradlinigen Führungsbahn, verschieben die Ziellinie um den Betrag des Führungsfehlers parallel zu sich. Das Prinzip ist umfassender als das später von HILDEBRAND [*110*] angegebene mit fest im Fernrohr eingebautem Spiegel, der ebenfalls eine durch das Lot von der Zielmarke aus bestimmte, aber nicht unbedingt horizontale Fluchtungsrichtung festlegt.

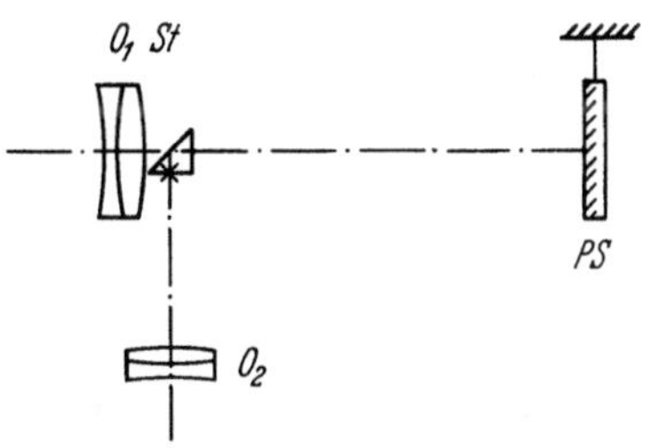

Fig. 102. Fluchtfernrohr mit horizontaler Fluchtlinie nach TSUBOKAWA. Das Lot von der im Hauptpunkt des Fernrohrobjektivs O_1 stehenden Strichmarke St auf den vertikal hängenden Pendelspiegel PS bestimmt die horizontale Fluchtlinie. O_1, St und das Betrachtungsmikroskop O_2 werden zur Fokussierung auf verschieden entfernte Zielmarken als Ganzes längs der Fluchtgeraden verschoben.

In bestimmten Fällen kann ein Quecksilberspiegel [*111*], [*112*] als horizontale Bezugsfläche zur optischen Ausrichtung einer genau vertikal stehenden Fluchtlinie oder Fluchtebene nützlich sein. Dann wird das Fluchtfernrohr mit Pentagon-Vorsatz in Autokollimation auf den Spiegel eingestellt.

41. Prüfen der Zentrierung von Linsen mit Spiegelung und Fluchtfernrohr. Die optische Achse einer Einzellinse ist durch die Verbindungslinie der beiden Krümmungsmittelpunkte ihrer sphärischen Flächen gegeben. Unter dem Zentrieren einer Einzellinse wird verstanden, daß die Achse der zylindrischen Außenform der Linse mit der optischen Achse in Übereinstimmung gebracht wird. Um diesen Zustand herzustellen, nimmt man die Linse auf einen genau laufenden Dorn mit einer Ringschneide auf. Der Krümmungspunkt der an der Ringschneide anliegenden Fläche liegt somit auf der Drehachse des Dorns. Der Krümmungsmittelpunkt der anderen Linsenfläche weicht dagegen im allgemeinen von der Drehachse ab. Das äußert sich darin, daß Spiegelbilder in dieser Fläche beim Drehen des Dorns um seine Achse kreisförmige Bewegungen ausführen. Durch Verschieben der an der Ringschneide anliegenden Fläche in sich lassen sich die Spiegelbilder der äußeren Fläche zum Stillstand bringen. Eine höhere Genauigkeit als mit bloßem Auge läßt sich dabei mit Hilfe eines Lupenfernrohrs mit Strichkreuz in der Ocularebene und einem beleuchteten schrägen Strichkreuz als Objekt für das Reflexbild erreichen. Diese subjektive Methode läßt sich zu einer Meßmethode ausbauen, wenn das Autokollimationsprinzip benutzt wird. Dazu bringt man auf die Drehachse ein Autokollimationsfernrohr und vor das AKF ein Objektiv, welches das nach Unendlich projizierte Strichkreuz in die Ebene für den Krümmungsmittelpunkt der schlagenden Linsenfläche abbildet. Das von dieser Fläche erzeugte Reflexbild des Strichkreuzes liegt in der gleichen Ebene und erscheint im Maßstab 1:1 wieder scharf in der Ocularebene. Durch langsames Drehen des Dorns um seine Achse läßt sich die Ablage des Krümmungsmittelpunktes der Linsenfläche von der Drehachse in der Ocularebene ausmessen da, das Reflexbild um den doppelten Betrag auswandert wie der Krümmungsmittelpunkt.

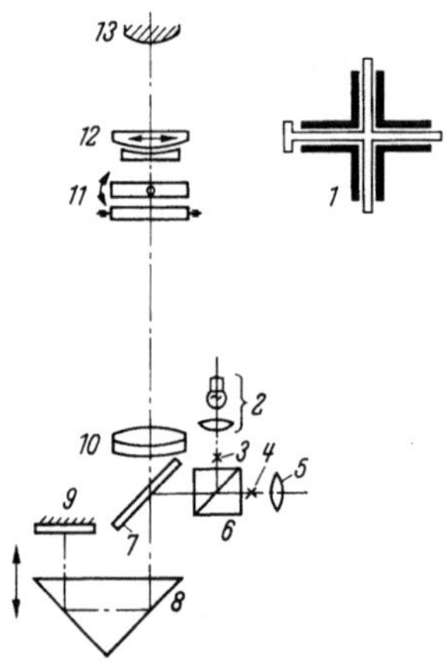

Fig. 103. Messen der Dezentrierung optischer Systeme mittels Autokollimation und Fluchtung der Krümmungsmittelpunkte der Linsenflächen. *1* Einfangmarke und reflektierte Marke, *2* Beleuchtungssystem, *3* Strichmarke, *4* Ocularstrichmarke, *5* Ocular, *6* Teilerwürfel, *7* Teilerfläche, *8* Tripelspiegel, verschiebbar, *9* ortsfester Spiegel, *10* Objektiv, *11* Plattenmikrometer, *12* Schiebelinsenmikrometer, *13* Prüfling.

Die Messung der Dezentrierung eines optischen Systems, d. h. mehrerer zu einem Objektiv vereinigter Einzellinsen, läuft auf eine Fluchtungsmessung hinaus. Die Krümmungsmittelpunkte der Einzelflächen sollten alle auf einer Fluchtgeraden, nämlich der optischen Achse des Systems liegen. Zweckmäßig wird wieder das Autokollimationsprinzip verwendet und die projizierte Marke nacheinander in die einzelnen Krümmungsmittelpunkte der Linsenflächen fokussiert. Um hohe Genauigkeiten zu erzielen, darf die Fokussierung weder durch Auswechseln von Vorsatzobjektiven noch durch die sog. Innenfokussierung geschehen, noch darf das zu prüfende System längs einer Führungsbahn verschoben werden; in jedem Falle würden die Zentrierfehler durch die Führungsfehler verfälscht oder verdeckt. Eine optische Anordnung mit großem Fokussierungsbereich ohne Führungsfehler wurde 1944 von Yvon angegeben (Fig. 103). Es handelt sich um ein Fluchtfernrohr mit einem fest eingebauten Planspiegel [*113*]. In dieser Hinsicht ähnelt es dem oben (Ziff. 40, [*109*], [*110*]) besprochenen Typ von Fluchtfernrohren, bei denen Querversetzungen des Objektivs bei der Fokussierung eine Parallelversetzung der Fluchtlinie vom gleichen Betrag nach sich zogen. Auch diese Führungsfehler können noch vermieden werden, wenn der virtuelle Objektabstand der Zielmarke vom Objektiv durch einen verschiebbaren Tripelspiegel geändert wird. Dieser darf beliebige Kippbewegungen ausführen; er darf sogar seitliche Versetzungen erleiden, da der Strahlengang hin und zurück den gleichen Weg nimmt. Wegen der 1:1-Abbildung kann in Autokollimation der Symmetrieeinfang der Strichmarke angewendet werden. Zudem ist es vorteilhaft, die Ablage der Reflexbilder mit Hilfe eines Planplattenmikrometers vor dem Objektiv zu messen, und in gewissen Fällen die optische Achse des AKF durch ein Koordinaten-Schiebelinsen-Mikrometer — gleichsam einen optischen Keil — neigen zu können. Durch alle diese Maßnahmen kann die Winkelmeßgenauigkeit für die Ablagen der

Krümmungsmittelpunkte der Linsenflächen auf 0,25″ bis 0,6″ gesteigert werden, je nach der von der Fokussierstellung abhängigen Fernrohrvergrößerung im Zusammenhang mit der Bildqualität. Bei mittleren Verhältnissen beträgt die Meßunsicherheit für die seitliche Ablage ungefähr 1 μm. Da sich Flächenkippung aus dem Quotienten von Mittelpunktsablage zu Linsenradius errechnet, kann für Radien oberhalb 20 mm mit einer Meßunsicherheit für die Flächenkippung von weniger als 10″ gerechnet werden [*113*].

Bei diesem Verfahren ergibt sich für die erste Fläche die wahre Ablage des Krümmungsmittelpunktes, für die zweite und die weiteren Flächen erhält man dagegen nur die scheinbaren Ablagen. Diese sind von der Vergrößerung und den Flächenkippungen der voraufgegangenen Linsenflächen und dem Azimut von deren Dezentrierung abhängig [*114*], [*115*], [*116*]; daher ist im Zusammenhang mit der beschriebenen Fluchtungsmessung durch das System hindurch eine Durchrechnung mit windschiefen Strahlen unvermeidlich. In der Praxis kann ein anderer Weg beschritten werden, indem die beobachteten Ablagen der Krümmungsmittelpunkte durch geeignete Vorkehrungen in den Systemfassungen durch Doppelexzenter und Achsverkippung herausjustiert werden.

VI. Interferenzmethoden für die mikroskopische Oberflächenprüfung.

Das Interferenzmikroskop für die Oberflächenprüfung ähnelt in seinem Aufbau dem Michelson-Interferometer. In jedem Interferometerarm befindet sich jedoch ein Mikroobjektiv, und in der Objektebene dieser Mikroobjektive ein Planspiegel bzw. die reflektierende Prüflingsoberfläche. Um die Objekte mit parallelem Licht zu beleuchten, wird die Lichtquelle in die vordere Brennebene der Objektive abgebildet, dadurch werden im Objektraum unendlich ferne Eintrittspupillen erhalten. Fallen die beiden Eintrittspupillen virtuell zusammen, so ist die Objektebene gleichmäßig hell erleuchtet, falls sie ganz eben ist. Erhebungen und Vertiefungen in der Prüflingsoberfläche erzeugen jedoch Interferenzstreifen, die gleiche Abstände von der Objektebene in Vielfachen einer halben Wellenlänge der verwendeten monochromatischen Strahlung miteinander verbinden, also dem geographischen Sprachgebrauch nach Niveau- oder Höhenlinien darstellen. Beispiele gibt Fig. 46a—d (S. 796, 797), s. auch die Nutzanwendung derartiger Interferenzbilder in Ziff. 32δ.

Durch Querversetzen eines der beiden Objektive läßt sich erreichen, daß die beiden im Unendlichen liegenden Eintrittspupillen seitlich nebeneinander liegen, dann erscheint die Objektebene von zahlreichen geraden Interferenzstreifen überzogen. Jeder Interferenzstreifen bedeutet einen schrägen Schnitt durch die Oberfläche. Die einzelnen Schichtschnitte haben eine halbe Wellenlänge Abstand voneinander. Erhebungen und Vertiefungen in der Prüflingsoberfläche lenken die Interferenzstreifen seitlich aus. Eine Auslenkung von einem Interferenzstreifen bedeutet eine Erhebung oder Vertiefung von einer halben Wellenlänge. Die einzelnen Interferenzstreifenzüge lassen sich oftmals nur bis zu einem gewissen Grade der Oberflächenrauheit auf ihrem ganzen Weg über die Oberfläche des Prüflings verfolgen. Der Zusammenhang geht bei größeren Rauhtiefen und unregelmäßiger Oberflächengestalt gänzlich verloren. Ein häufig gebrauchtes Hilfsmittel, den gleichen Streifen an anderer glatterer Stelle wiederzufinden, sind die bekannten Weißlichtinterferenzen, wenige verschieden gefärbte Streifen, die bei Verwendung einer Glühwendel anstelle einer monochromatischen Lichtquelle in der Umgebung des Gangunterschieds Null entstehen. Um sehr rauhe Flächen mit Hilfe von Interferenzen zusammenhängend messen zu können, kann ein Abdruckverfahren verwendet werden.

42. Zehender-Verfahren. Von der zu untersuchenden Oberfläche wird ein Filmabdruck angefertigt, der anschließend in der Zehender-Kammer [*117*] in ein passendes Immersionsmittel eingebettet wird (Fig. 104). Die Zehender-Kammer besteht aus einer planen, auf der Oberseite voll verspiegelten Grundplatte und einem planparallelen, teilverspiegelten Deckglas und bildet eine Art Fizeau-Interferometer. Hätte das Immersionsmittel den gleichen Brechungsindex wie der Film, so würden sich in diesem Interferometer geradlinig verlaufende Fizeaustreifen zeigen. Weicht der Brechungsindex der beiden Medien etwas voneinander ab, so werden die optischen Weglängen im Film und im Immersionsöl verschieden und die Interferenzstreifen zeigen Auslenkungen, falls der Film verschiedene Dicken aufweist. Die Auslenkung um einen Interferenzstreifen besagt, daß die zweimal durchlaufene Differenz der optischen Weglängen eine Wellenlänge beträgt

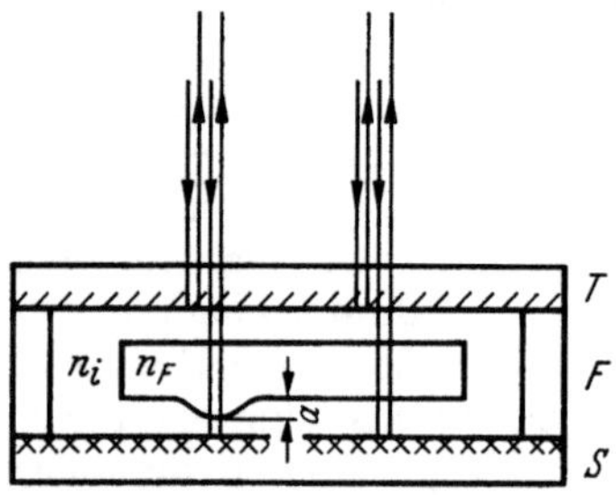

Fig. 104. Interferenzkammer nach ZEHENDER für Abdrücke von sehr rauhen Oberflächen. *T* Teilerfläche, *F* Abdruckfolie, *S* Spiegel, n_i, n_f Brechungsindices der Immersion bzw. der Folie, *a* Höhenabweichung.

$$a(n_F - n_i) = \lambda/2,$$

woraus die Erhebung bzw. Vertiefung *a* auf dem Film sofort berechnet werden kann. Im gleichen Maße wie die Meßempfindlichkeit des Verfahrens geringer wird, können größere Rauhtiefenunterschiede noch gemessen werden.

43. Erhöhung der Meßempfindlichkeit. Überschichtet man die Oberfläche des Prüflings mit Immersionsmittel, so wird die Wellenlänge, mit der die Rauhtiefe gemessen wird, kürzer. Dann entspricht einer Auslenkung um einen Interferenzstreifen eine Höhendifferenz $a = \lambda/2n$. Hierdurch ist es möglich, die Meßempfindlichkeit zu steigern und kleine Rauhtiefen noch meßbar zu machen.

44. Vielstrahlinterferenzen. Der Intensitätsverlauf von Zweistrahlinterferenzen, wie sie im Michelson-Interferometer erzeugt werden, ist $\sin^2$-förmig. Dadurch sind der visuellen Ablesegenauigkeit einer Streifenversetzung bestimmte Grenzen gesetzt (rund 0,1 Streifenabstand). Um diese hinauszuschieben, wären schmalere Streifen im Verhältnis zu ihrem gegenseitigen Abstand erforderlich [*118*], [*118a*]. Von der Spektroskopie her ist bekannt, daß schmalere Maxima erhalten werden, wenn nicht zwei, sondern viele sinusförmige Komponenten mit geringen gegenseitigen Phasendifferenzen einander überlagern. Am auffälligsten ist diese Erscheinung zwischen den Beugungsfiguren am Doppelspalt und an einem Gitter. Aus den gleichen Gründen ergeben sich am Pérot-Fabry-Interferometer schmalere Streifen als mit dem Interferometer von MICHELSON.

An Planparallelplatten und am Keil können Vielstrahlinterferenzen erzeugt werden, wenn die spiegelnden Flächen einen hohen Reflexionsgrad besitzen, so daß ein einmal in das Platteninnere hineingelangter Strahl vielfach hin und her reflektiert wird, ehe sich seine Energie infolge der bei der Reflexion durch die Spiegelschicht hindurchgelassenen Anteile verliert. Die einzelnen durch die Spiegelschicht austretenden Komponenten unterscheiden sich hinsichtlich ihrer Amplituden nur sehr wenig voneinander, so daß ein k-mal reflektierter Strahl mit einem nur einmal reflektierten noch zu einer deutlichen Interferenzerscheinung zusammentreten kann. An der Planparallelplatte gilt für die einzelnen unter dem gleichen Winkel austretenden Komponenten, daß sie alle die gleiche gegenseitige Phasendifferenz besitzen. Ist diese ein ganzzahliges Vielfaches der Wellenlänge, so ergibt sich das Maximum der Intensität. Aber schon bei geringen Abweichungen von der Ganzzahligkeit fällt die Intensität steil ab (Fig. 105). Die Theorie für diese Erscheinung wurde schon von AIRY gegeben, sie findet sich z.B. auch bei BORN ([*38*], S. 325).

Beim Keil liegen die Verhältnisse nicht so einfach wie bei der Planparallelplatte. Hier treten die einzelnen Komponenten nicht parallel zueinander aus, sondern unter verschiedenen vom Keilwinkel und der Anzahl der Reflexionen abhängigen Winkeln. Dadurch wird die Interferenzerscheinung genau wie bei den Zweistrahlinterferenzen aus dem Unendlichen in die Nähe der Keilflächen verlegt, und man sieht sie zusammen mit dem Keil scharf. Während bei Planparallelplatten-Interferenzen die Änderung des Gangunterschiedes mit der Richtung des einfallenden Bündels variiert, ändert sich am Keil der Gangunterschied zuzüglich mit der Dicke des Keils. Bei nahezu senkrechtem Lichteinfall verbinden daher die Interferenzstreifen wie früher die Orte gleicher Keildicke und können als Niveaulinien gedeutet werden. Verursacht die Keildicke gerade ein ganzes Vielfaches einer Wellenlänge Gangunterschied der austretenden Teilstrahlen, so entstehen im Durchlicht schmale helle Streifen (Durchlässigkeitsmaxima), im Auflicht dunkle Streifen (Reflexionsminima). Jede noch so nahe benachbarte Stelle anderer Keildicke bedeutet aber eine Phasendifferenz zwischen den interferierenden Strahlen und damit einen steilen Abfall in der Intensität der Extrema (Fig. 105). Minima und Maxima sind auch hier komplementäre Erscheinungen. Die am Keil beobachtbaren schmalen Vielstrahlinterferenzen können demnach zu sehr hohen Ablesegenauigkeiten von Interferenzstreifenversetzungen führen und durch ihre Auslenkungen sehr feine Niveauunterschiede meßbar machen [*119*]. Wenn jedoch nicht nur hochpolierte, ebene Keilflächen, sondern hauptsächlich technische Oberflächen von bestimmten Rauheitsgraden mit Riefen und Kratzern gemessen werden sollen, ist die Deutung der Vielstrahlinterferenzen etwas komplizierter, so daß sie mit einiger Vorsicht interpretiert werden müssen.

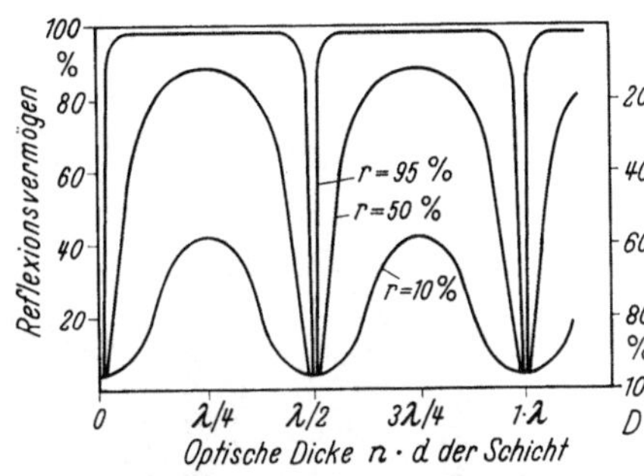

Fig. 105. Reflexionsvermögen und Durchlässigkeit absorptionsfreier Schichten in Abhängigkeit von der optischen Dicke bei $r = 10$, 50 und 95% Reflexionsgrad der Schichtgrenzen.

Nach INGELSTAM [*120*] geht mit dem erhöhten Tiefenauflösungsvermögen der Vielstrahlinterferenzen notwendig ein entsprechender Teil ihres lateralen Auflösungsvermögens für Objektdetails verloren. Die Keilinterferenzstreifen beschreiben nämlich nach dem oben Gesagten sehr enge begrenzte Bereiche gleicher Keildicke. Während bei den Zweistrahlinterferenzen jeder an der einen Platte reflektierte Strahl mit nur einem an der anderen Platte reflektierten interferiert, und sich daher die Topographie der Oberfläche Punkt für Punkt im Interferenzstreifenbild abzeichnet, kann bei Vielstrahlinterferenzen der an der einen Platte reflektierte Strahl mit Strahlen interferieren, die vielmals zwischen den Platten hin und her reflektiert werden. Diese können sich mit ihm zu einem Maximum der Intensität überlagern, wobei unbestimmt bleibt, ob sie nicht über viele Stellen verschiedener Dicke im Keilinneren mitteln. Dann werden Unterschiede in der Keildicke also nicht zu Tage gefördert, sondern verschluckt. Umgekehrt werden Details, für die die Maximumbedingung zutreffen würde, an der gleichen Stelle von Strahlen überlagert, die im Keilinneren andere Gangdifferenzen erfahren haben. Daher kann eine Stelle, die mit Zweistrahlinterferenzen hell erscheinen würde, dunkel bleiben und übersehen werden. Diese Zusammenhänge lassen sich theoretisch belegen ([*38*], S. 351), doch läßt sich die Theorie auf die jeweils vorliegenden komplizierten Verhältnisse nur schwer zuschneiden. Soviel ist ihr aber zu entnehmen, daß ohne genaue Kenntnis und Beschreibung der Bereiche, über die gemittelt wird, keine weitere Aussage gemacht werden kann als die folgende: Sind die Bereiche kleiner als das laterale Auflösungsvermögen der Optik, mit der beobachtet wird, so sind die Vielstrahlinterferenzen abbildungstreu. Überschreiten die Bereiche dagegen das laterale Auflösungsvermögen, so gehen zahlreiche Ober-

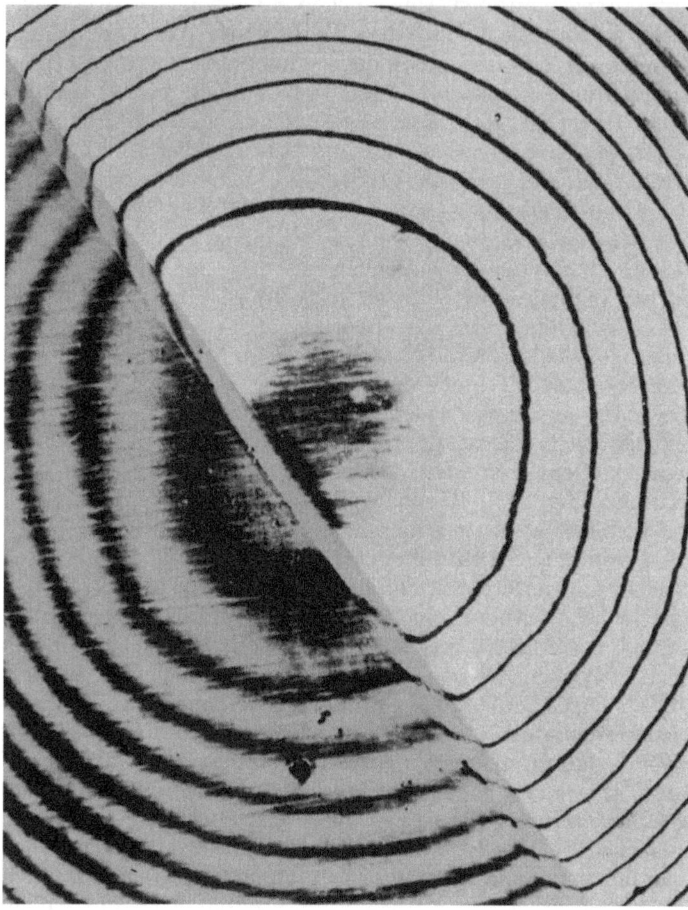

Fig. 106. Zweistrahl- und Vielstrahlinterferenzen auf einer bearbeiteten Metalloberfläche. Nach größeren Abständen und Winkeln zwischen den Keilflächen vom Zentrum aus verlieren die Vielstrahlinterferenzen an Informationsinhalt gegenüber den Zweistrahlinterferenzen.

flächenprofildetails in der Mittelung unter. Der Bereich der Oberfläche, der beim Zustandekommen der Interferenzerscheinung mitwirkt, ist um so größer, je höher die Verspiegelung, der Keilwinkel, die Keildicke und die Beobachtungsapertur sind. Die Eigenschaft der Vielstrahlinterferenzen, über Feinheiten der Oberflächentopographie hinwegzumitteln, nimmt daher mit wachsendem Keilwinkel und wachsender Keildicke sehr schnell zu, wie aus einem Vergleich des Informationsgehaltes von Zweistrahl- und Vielstrahlinterferenzen in Fig. 106 deutlich hervorgeht [*121*]. Wiedergegeben ist eine Kugeloberfläche, die mit einer nur zur Hälfte hochverspiegelten Interferenzfläche bedeckt ist. In den Gebieten um den Kulminationspunkt, wo sowohl der Keilwinkel als auch die Keildicke klein sind, gleichen sich die Aussagen der Zweistrahl- und Vielstrahlinterferenzen noch weitgehend. In einigem Abstand hiervon, wo Keilwinkel und Keildicke größere Beträge annehmen, sehen die Vielstrahlinterferenzlinien aber zunehmend glatter aus. Sie eignen sich also in diesen Bereichen keineswegs mehr zur Erfassung und Beurteilung der Oberflächenrauheit, oder zum Erkennen von Riefen und Kratzern. Wir kommen auf diese Abbildung noch zurück.

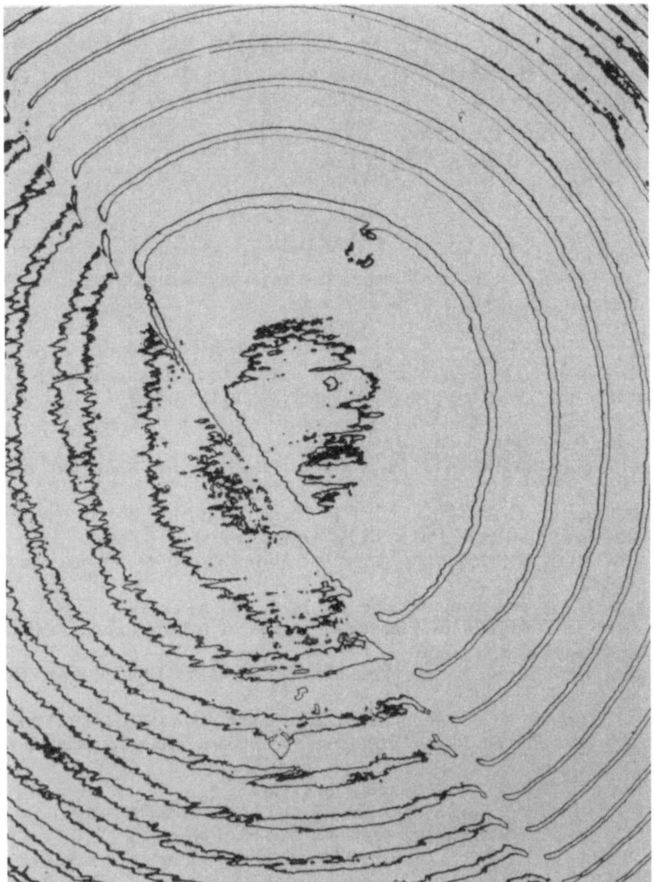

Fig. 107. Äquidensiten nach KRUG und LAU liefern auch bei Zweistrahlinterferenzen sehr schmale Streifen im Verhältnis zu ihrem gegenseitigen Abstand ohne an Informationsinhalt zu verlieren. Dadurch sind sie den Vielstrahlinterferenzen prinzipiell überlegen.

Vielstrahlinterferenzen eignen sich dagegen vorzüglich zum Messen von Stufen an Kristallen, z.B. an Glimmer [*119*]. Bei diesen vor- und rückseitig hochverspiegelten Objekten ist eine Versetzung der Interferenzstreifen quer zur Stufe um $\lambda/300$ noch zu erkennen, entsprechend einer Stufenhöhe von $2 \cdot 10^{-6}$ mm.

45. Äquidensiten. Um die laterale Auflösung feiner Oberflächenunebenheiten in der interferentiellen Messung nicht einzubüßen und doch zu einer erhöhten Tiefenmeßgenauigkeit zu gelangen, können an photographischen Aufnahmen von Zweistrahlinterferenzbildern die sog. Äquidensiten sichtbar gemacht werden, [*121*]. Äquidensiten zeichnen als sehr feine Linien den Verlauf gleicher Schwärzung im Interferenzbild nach. Jeder Interferenzstreifen liefert, wenn die Äquidensiten etwa die Orte von 50% Leuchtdichte miteinander verbinden, zwei schmale Linien. Man erhält dadurch die doppelte Anzahl Profillinien als Schnitte durch die Oberfläche und kann die Topographie der Oberfläche im Rahmen des optischen Auflösungsvermögens ausmessen.

Die Äquidensiten können durch Nachbelichten des halb anentwickelten photographischen Negativs mit Raumlicht und Ausentwickeln des nachbelichteten Negativs erzeugt werden (Sabattier-Effekt).

Ein Vergleich der Aussage von Vielstrahlinterferenzen einerseits und Äquidensiten von Zweistrahlinterferenzen andererseits am gleichen Prüfling zeigt Fig. 107 [*122*]. Die Linien der Vielstrahlinterferenzen, die wie wir schon wissen, mit zunehmendem Keilwinkel und zunehmender Keildicke immer ausgeglichener verlaufen, lösen lateral nicht soviel Einzelheiten der Oberflächenfeingestalt auf wie die Äquidensiten von Zweistrahlinterferenzen.

Literatur.

[*1*] Flügge, J.: Einführung in die Messung der optischen Grundgrößen, S. 15ff. Karlsruhe 1954; — Arch. techn. Messen J 010—3 (April 1957).
[*1a*] Schober, H.: Das Sehen, Bd. II, Leipzig 1958.
[*2*] Kissam, P.: Optical Tooling, p. 113—118. New York and London 1962.
[*3*] Leslie, R. T., and L. D. Armstrong: The design of graticules and reference marks and some factors in the operation of scientific instruments. Austral. J. Appl. Sci. **6**, 167—182 (1955).
[*4*] Schulze, R.: Arch. Techn. Messen V 8310—1, 1956.
[*5*] Eppenstein, O.: Handbuch der Physik von Geiger und Scheel, Bd. 18, S. 194—195. 1927.
[*6*] Engelhard, E., u. R. Vieweg: Über die neue Definition des Meters aufgrund einer Lichtwellenlänge. Z. angew. Phys. **13**, 580—596 (1961).
[*7*] Kösters, W.: Der gegenwärtige Stand der Meterdefinition. Werkstattstechnik H. 23, 527—533 (1938).
[*8*] Engelhard, E.: Z. Instrumentenkde. **67**, H. 3, 59—65 (1959).
[*9*] Engelhard, E.: Accuracy of Lightwave Scales in Interferometry, NPL Symposium Nr. 11. London: H.M.S.O. 1960.
[*10*] Engelhard, E.: Precise Interf. Measurement of Gage Blocks. Nat. Bur. of Stand. Circ. 581, 1 (1957).
[*11*] Barakat, N., and S. Mokhtar: J. Opt. Soc. Amer. **53**, 1153—1155 (1963).
[*12*] Engelhard, E.: Physikalische Probleme der neuen Meterdefinition. Phys. Bl. **18**, 449—458 (1962).
[*13*] Illingworth, K. K.: Phys. Rev. **30**, 692 (1927).
[*14*] Kennedy, R. J.: Proc. Nat. Acad. Sci. Wash. **12**, 621—624 (1926).
[*15*] Françon, M., et A. Soulié: Rev. Opt. **42**, No. 10, 499—518 (1963).
[*16*] Clusius, K., u. G. Dickel: Trennung der Krypton-Isotope. Z. phys. Chem. Abt. B **52**, 348 (1942).
[*17*] Engelhard, E., et J. Terrien: Rev. Opt. **39**, 11—18 (1960).
[*18*] Hanes, G. R.: Canad. J. Phys. **37**, 1283 (1959).
[*19*] Kösters, W.: Handbuch der physikalischen Optik, Hrsg. E. Gehrcke, Bd. 1, S. 471 bis 498, spez. 489. 1927.
[*20*] Berndt, G.: Grundlagen und Geräte technischer Längenmessungen, 2. Aufl., S. 117 u. 136. Berlin 1929.
[*21*] Reichardt, H., u. T. Kessler: Bestimmung des Abstandes angesprengter Flächen. Jenaer Jb. **1962**, 117—124.
[*22*] Rolt, F. H., and H. Barrell: Proc. Roy. Soc., Lond. A **20**, 122 (1929).
[*23*] Thwaite, E. G., and R. T. Leslie: Brit. J. Appl. Phys. **14**, 711 (1963).
[*24*] Michelson, A. A., and E. W. Morley: Amer. J. of Sci. (3) **34**, 427 (1887); (3) **38**, 181 (1889).
[*25*] Feussner, W.: Interferenz des Lichtes. In: Winkelmanns Handbuch der Physik, Bd. 6, 994ff. (1906).
[*26*] Kinder, W.: Ein Meterkomparator für interferometrische Längenbestimmungen in Vakuum-Wellenlängen. Zeiss-Werkz. Nr. 43, 3/11 (1962).
[*27*] Weidmann, G.: Messungen mit dem Abbeschen Dilatometer. Ann. Phys. u. Chemie **38**, 453—484, spez. 459 (1889).

[28] PULFRICH, C.: Z. Instrumentenkde. **13**, 415—416 (1893).
[29] BENOÎT, J. R.: J. de Phys. (3) **7**, 57 (1898).
[30] KÖSTERS, W.: Ein neuer Interferometer-Komparator für unmittelbaren Wellenlängenanschluß. Z. Feinmech. **34**, 55—59 (1926).
[31] FIZEAU, M. H.: Mémoires sur la dilatation des corps solides par la chaleur. Ann. Chem. et Phys. **8**, 335 (1866), spez. S. 348, Fußnote.
[32] WOLF, F.: Arch. techn. Messen V 8313—4, Lief. 240, (Jan. 1956).
[33] HUNZINGER, J. J.: Rev. Opt. **34**, 512—519 (1955).
[34] LEINWEBER, P.: Taschenbuch der Längenmeßtechnik, S. 356. Berlin 1954.
[35] DOWELL, J. H.: Brit. Pat. 555672, 1942 und 559748, 1943.
[36] HILGER, A.: J. Sci. Instrum. **16**, 163 (1939); — Instr. Practice **9**, 438 (1955).
[37] CANDLER, C.: Modern Interferometers. London: Hilger Publ. 1951.
[38] BORN, M., and E. WOLF: Principles of Optics, 2. ed., p. 306. London 1964.
[39] DÜHMKE, M.: Optik **16**, 646—649 (1959); **18**, 561—564 (1961).
[39a] SAUNDERS, J. B.: J. Res. Nat. Bur. Stand. **61**, 491 (1958).
[40] VOGL, G.: Optics in Metrologie, 1958. Hrsg. P. MOLLET, S. 433—434. Pergamon Press 1960. — Microtecnic **12**, 11—13 (1958).
[41] LINNIK, V. P.: C. R. (Doklady) Acad. Sci. USSR **35**, No. 1 (1942).
[42] SAUNDERS, J. B., and F. L. GROSS: J. Res. Nat. Bur. Stand. **62**, No. 4, Res. Pap. 2943 (1959).
[43] KÜHNE, H.: Optik aller Wellenlängen, S. 362. Berlin: Akad. Vlg. 1959.
[44] HUNZINGER, J. J.: Rev. Opt. **36**, 285—291 (1957).
[45] HARIHARAN, P., and D. SEN: J. Opt. Soc. Amer. **49**, 232—234 (1959).
[46] SAGNAC, G.: C. R. Acad. Sci., Paris **150**, 1302, 1676 (1910); — Z. Instrumentenkde. **32**, 213 (1912); — J. de Phys. (5) **4**, 177 (1914).
[47] ROLT, F. H.: Use of Light Waves for Controlling the Accuracy of Block Gauges. London: Hilger & Watts Ltd. 1964.
[48] SAUNDERS, J. B.: J. Opt. Soc. Amer. **50**, 183 (1960).
[49] BAKOS, J., K. KÁNTOR u. Zs. NÁRAY: Optik Konf. Budapest 1960, Vortrag.
[50] NÁRAY, Zs.: Ungar. Pat. Nr. 147129, ang. 1959.
[51] HOCK, F.: Optik **18**, 565—566 (1961).
[52] PECK, E. R., and S. W. OBETZ: J. Opt. Soc. Amer. **43**, 505—509 (1953).
[53] BAIRD, K. M.: Appl. Optics **2**, 471—479 (1963).
[54] RÄNTSCH, K.: Optik **4**, 283 (1948/49).
[55] KILTZ, G., u. H. REICHARDT: Feingerätetechnik **13**, 2, 79—82 (1964).
[56] BERNDT, G.: Grundlagen und Geräte technischer Längenmessungen, S. 42—46; S. 105. Berlin 1929.
[57] TERRIEN, J.: Rev. Opt. **34**, 498—509 (1955).
[58] HEIDENHAIN, J.: DB Pat. 902713, 1950.
[59] LEHMANN, R.: Leitfaden der Längenmeßtechnik, S. 419. Berlin 1960.
[60] RÄNTSCH, K.: Die Optik in der Feinmeßtechnik, S. 173—174. München 1949.
[61] SCHULZE, R.: VDI-Z. **93**, 41—44 (1951); — Microtecnic **6**, 207—213 (1952).
[62] STODIEK, W.: Nuovo Cim. **2**, 467—470 (1955).
[63] ILLIG, W.: Zeiss-Werkz. **5**, 72—74 (1957).
[64] GÜNTHER, N.: Die mikroskopische Abbildung von Zylindern und Gewinden. Z. Instrumentenkde. **59**, 315 (1939).
[65] WOLF, F.: Prüfen und Messen von Gewinden, S. 65—67. München 1952.
[66] SCHULZE, R.: Microtecnic **8**, Nr. 1, 13—26 (1954).
[67] BAUERSFELD, W.: DR-Pat. 411296, 1923, Fa. Zeiss DBR.
[68] FRANKE, G.: Techn. Zbl. prakt. Metallbearb. **54**, 274—279 (1960).
[69] VÖNEKY, G.: Z. Werkst. u. Betrieb **95**, 757 (1962).
[70] RÄNTSCH, K.: Zeiss-Werkz. **7**, 49—52 (1959).
[71] PETTAVEL, J.: Optics in Metrologie, p. 63—72. London: Pergamon Press 1960.
[72] Hersteller: Fa. J. Heidenhain, Traunreut, Deutschland.
[73] MEISTER, G., u. J. CICHON: Jenaer Rdsch. **5**, 171—181 (1960).
[74] KÜHNE, C.: Askania Warte **17**, Nr. 55, 17—21 (1959).
[74a] GOLDBERG, J. L., u. H. J. RITTER: Microtecnic **18**, 54—67 u. 117—125 (1964).
[75] Hersteller: Fa. Davidson Optronics, West Covina, California, USA.
[76] Hersteller: Fa. Barnes Engineering Comp., Stamford, Connecticut USA.
[77] Hersteller: Fa. Controls Co., Cambridge, Mass. USA; Electromechanical Design Mag. Nov. 1961.
[78] Hersteller: Fa. Kollmorgen, Northampton, Mass. USA.
[79] Electronic Industries **22**, Nr. 9, 72—75 (1963).
[80] HOCK, F.: Maschinen-Markt **69**, H. 46, 23 (1963).

[81] EPPENSTEIN, O.: DRPat. 394809, 399849, 409808, 410130, 1922/23; — VDI-Z. **78**, 995, 1111 (1934).
[81a] WILDE, H.: Z. Feinmech. **34**, 265 (1926).
[82] DRODOFSKY, M.: Z. Vermessungsw. **76**, 225—231 (1951).
[83] VOGEL, K. H.: Zeiss-Inform. **12**, H. 51, 25—28 (1951).
[84] KÖNIG, A., u. H. KÖHLER: Die Fernrohre und Entfernungsmesser, S. 237. Berlin 1959. Zeiss DBPat. 971903, 1949/51.
[85] RÄNTSCH, K.: DBPat. 1024252, 1957.
[86] SCHULZE, R.: Arch. techn. Messen V 8224—10/11, Lief. 291/292, (April/Mai 1960); V 8224—16/17, Lief. 325/327 (Febr./April 1963).
[87] CUNY, B.: Microtecnic **4**, 194 (1950).
[88] KÖSTERS, W.: Z. Instrumentenkde. **53**, 202 (1933).
[89] VOGL, G.: Optics in Metrology, p. 193. London: Pergamon Press 1960.
[90] HEEL, A. C. S. VAN: J. Opt. Soc. Amer. **36**, 242 (1946).
[91] HEEL, A. C. S. VAN: J. Opt. Soc. Amer. **40**, 809 (1950).
[92] MCLEOD, J. H.: J. Opt. Soc. Amer. **44**, 592 (1954).
[93] STEEL, W. H.: Optics in Metrology, p. 181—192. London: Pergamon Press 1960.
[94] HEEL, A. C. S. VAN: Progress in Optics, Amsterd. **1**, 308—309, 325 (1961). — Proc. of the Conf. on optical Instr. and Techn. London 1961, p. 383—394.
[95] SAUNDERS, J. B.: Appl. Optics **2**, 541—542 (1963).
[96] DYSON, J.: Optics in Metrology, p. 169. London: Pergamon Press 1960.
[97] UKITA, Y., and J. TSUJIUCHI: Optics in Metrology, p. 314. London: Pergamon Press 1960.
[98] RÄNTSCH, K.: Grundlagen der technischen Winkelmessungen, S. 32—33. München 1952.
[99] ROSENTHAL, G.: Optik **8**, 449 (1951).
[100] ROSENHAUER, K.: Optik **13**, 5—6 (1956).
[101] CSAPSKI, S., u. O. EPPENSTEIN: Grundzüge der Theorie der optischen Instrumente, S. 646. Leipzig 1924.
[102] KESSLER, H.: Handbuch der Physik von GEIGER u. SCHEEL, Bd. 18, S. 763—767. 1927.
[103] KOHLRAUSCH, F.: Praktische Physik, 21. Aufl., Bd. 1, S. 475. Stuttgart 1960.
[103a] PICHT, J.: Meß- und Prüfmethoden der optischen Fertigung, 2. Aufl., Bd. 1, S. 107 bis 136, spez. S. 128—131. Berlin 1955.
[104] HODAM, F.: Feingerätetechn. **12**, H. 2, 55—58 (1963).
[105] KLING, K.: Askania Warte **21**, H. 63, 14—19 (1964).
[106] MEYER, H.: Askania Warte **16**, H. 53, 26—33 (1958).
[107] RASCH, W.: Askania Warte **17**, H. 55, 4—5 (1959).
[108] MÜLLER, H.: Zeiss-Inform. **11**, H. 48, 62—63 (1963); **12**, H. 53, 100—101 (1964).
[109] TSUBOKAWA, J.: Filed **26**, July 1957 (in Japan 21. Aug. 1956); — Appl. Optics **2**, H. 11, 1168 (1963).
[110] HILDEBRAND, K.: Optik **18**, 547 (1961).
[111] BÜNNAGEL, R.: Z. angew. Phys. **8**, 342, 447 (1956); — Optica Acta **3**, 81 (1956); — Optics in Metrology, p. 172—178. London: Pergamon Press 1960.
[112] KISSAM, P.: Optical Tooling, p. 194. New York 1962.
[113] SCHUCH, K.: Optik **16**, 652—659 (1959); — Jenaer Rdsch. **5**, 10—12 (1960).
[114] HOFFMANN, CHR.: Optik **17**, 465—484 (1960); — Proc. Conf. Opt. Instr. u. Techn. London 1962, p. 364—382.
[115] INGALLS, A., and K. PESTRECOV: J. Opt. Soc. Amer. **38**, 343 (1948).
[116] NOTNI, W.: Optik **16**, 659 (1959).
[117] ZEHENDER, E.: VDI-Z. **94**, 456 (1952).
[118] FABRY, C., u. A. PEROT: Ann. de Chim. et Phys. (7) **12**, 459—501 (1897).
[118a] RÜHLE, R.: Z. techn. Phys. **24**, 221 (1943).
[119] TOLANSKY, S.: Nature, Lond. **152**, 722 (1943); — Multiple-beam Interferometry of Surfaces and Films. Oxford: Clarendon Press 1948.
[120] INGELSTAM, E.: Ark. Fysik **7**, 309 (1953).
[121] KRUG, W.: Feingerätetechn. **4**, 167 (1955); vgl. Fußnote in diesem Handbuch, Bd. 24, S. 544. 1956.
[122] LAU, E., u. W. KRUG: Die Äquidensitometrie. Berlin 1957.

Sachverzeichnis.

(Deutsch-Englisch.)

Bei gleicher Schreibweise in beiden Sprachen sind die Stichwörter nur einmal aufgeführt.

Abbe-König-Prisma, *Abbe-König prism* 215, 217.
Abbesche Dispersionszahl, *dispersion number of Abbe* 254, 292.
— Theorie der Bildentstehung im Mikroskop, *Abbe's theory of image formation in microscope* 349.
Abbildende Gitter, *imaging gratings* 437f.
Abbildungsfehler, höhere Ordnungen, *aberrations in image formation, higher orders* 18.
—, niedrigste Ordnung, *lowest order* 6, 15f.
—, verschiedene Definitionen, *different definitions* 5.
Aberration bei Konkavgittern, *aberration in concave gratings* 478.
Aberrationsfunktion, *aberration function* 169f., 181, 184f.
Absorption von Licht in Linsen, *absorption of light in lenses* 209.
Achromasie, *achromatism* 44, 255.
Achromat, *achromatic lens* 194, 256, 282.
Achromate als Mikroskop-Objektive, *achromatic microscope objectives* 357.
Achsennahe Strahlen, *near-axis rays* 85.
Ähnlichkeitsgesetz für Beugungsbilder, *similarity law for diffraction patterns* 543, 545.
Airysche Beugungsfigur, *Airy pattern* 31f.
Aluminisierung von Gitteroberflächen, *aluminum coating of grating surfaces* 601.
— von Spiegeln, *of mirrors* 298.
Amici-Prisma, *Amici prism* 215f.
Amplitudenmodulation, *amplitude modulation* 736.
Anamorphotische Abbildung, *anamorphotic image formation* 30.
Antriebssysteme für Fernrohre, *driving mechanisms for telescopes* 337ff.
Apertur, *aperture* 344, 368.
Aperturblende, *aperture stop* 11, 231.
Aperturweite eines Gitters, *aperture width of a grating* 460, 462.
Aplanat, *aplanate* 249.
Aplanatisches Zweispiegelsystem, *aplanatic two-mirror system* 265.
Aplanatismus, *aplanatism* 28.
Apochromat, *apochromatic lens* 283.
Apochromate als Mikroskop-Objektive, *apochromatic microscope objectives* 358f.
Apodisation, *apodization* 241, 368, 370, 537.
Apodisationstest eines Gitters, *apodization test of a grating* 649f.
Äquidensiten, *equidensites* 831f.
Argonkontinuum im Ultravioletten, *argon continuum in the ultraviolet* 646.
Asphärische Flächen, *aspherical surfaces* 17, 72, 232, 272.
— —, Prüfung, *test* 307.
— Platte, *aspherical plate* 39, 273.
Astigmatismus, *astigmatism* 16, 22, 234, 251ff., 260, 263, 266.
— bei Konkavgittern, *in concave gratings* 480—485.
Astrographen-Objektive, *astrograph lenses* 291.
Auflichtbeleuchtung, *incident-light illumination* 380.
Ausbeute von Gittern im Infrarot, *efficiency of gratings in the infrared* 606—609.
— — s. auch Blazing, *see also blazing.*
— von Röntgenstrahlgittern, *of x-ray gratings* 647.
Ausbeuteerhöhung von Gittern durch Blazing, *efficiency increase of gratings by blazing* 580—583.
Ausbeuteprüfung eines Gitters, *efficiency test of a grating* 644.
Auflösung, Definition von Lord Rayleigh, *resolution, definition of Lord Rayleigh* 459f.
Auflösungsvermögen, *resolving power* 240.
—, Abhängigkeit vom Gitterwinkel, *dependence on grating angle* 615.
—, Begrenzung durch verfügbare Energie, *limited by energy available* 497.
—, Erhöhung durch Benutzung von zwei Gittern, *increase by use of two gratings* 616f.
— von Gittern, *of gratings* 453, 459ff., 462.
— —, Bedingungen für hohe Auflösung, *conditions for high values* 461.
— —, moderne Erfolge, *modern successes* 432.
— —, Prüfverfahren, *tests* 611ff.
— und Kohärenzgrad, *and degree of coherence* 345.
Ausstrahlungsbedingung, *radiation condition* 583.
Austrittspupille, *exit pupil* 11, 49, 61, 117, 120, 231, 285.

Austrittsspalt eines Spektrographen, Einfluß der endlichen Breite, *exit slit of spectrograph, finite width effect* 490.
Autokollimation, *autocollimation* 435, 452, 458, 807.
Autokollimationsfernrohr, *autocollimation telescope* 757, 815f., 822.
Autokorrelationsfunktion, *autocorrelation function* 563.
Automatische Nachführung, *automatic guider* 339.
Axiale Auflösung, *axial resolution* 348.
Axicon 814.
Azimutale Montierung, *altazimuth mounting* 322.

Baird-Atomic-Teilmaschinen für Infrarotgitter, *Baird-Atomic infrared grating engines* 711f.
Baker-Schmidt-System 279.
Barometerschwankungen, Interferometerkontrolle, *barometric changes, interferometric control* 700.
Bauernfeindsche Prismen, *Bauernfeind prisms* 220, 221.
Bausch & Lomb-Teilmaschinen, *Bausch and Lomb ruling engines* 706—712.
Beleuchtung von Mikroskop-Objekten, *illumination of microscope objects* 376ff., 398ff.
— von Spektrometern, *of spectrometers* 491.
Beleuchtungsstärke (Flußdichte), *illumination (flux density)* 35, 494.
Beugung und Streuung, *diffraction and scattering* 433.
Beugungsbild der Kreispupille, *diffraction pattern of circular pupil* 238f.
— im Mikroskop, *in microscope* 343.
— einer rechteckigen Öffnung, *from a rectangular aperture* 452.
— bei ringförmiger Apertur, *for ring aperture* 369.
Beugungsbilder in Spektrographen, *diffraction patterns in spectrographs* 487.
Beugungseinheit, *diffraction unit* 565.
Beugungsfigur, *diffraction pattern* 512.
Beugungsgitter, Dispersion, *diffraction gratings, dispersion* 458f.
—, Geschichte, *history* 426ff.
— für das Infrarot, *for the infrared* 436.
— für metrologische Zwecke, *for metrological purposes* 439f.
— für Mikrowellen, *for microwaves* 436f.
—, Parameter, *parameters* 446, 449.
— für Röntgenstrahlen, *for x-rays* 429, 435.
—, Typen, *types* 434ff.
— für dasUltraviolett, *for the ultraviolet* 435.
Biegungskompensation am Fernrohrtubus, *flexure compensation of telescope tube* 318.
Bildentstehung, Formulierung durch Fourier-Transformierte, *image formation, Fourier transform formulation* 510.
— durch Gitter, *by gratings* 428, 437, 452f.
— auf dem Rowlandschen Kreis, *on Rowland circle* 477.
Bildfehler (Aberrationen) höherer Ordnung, *higher-order aberrations* 70, 103, 138, 139, 176f., 187f., 190f.
Bildfehler-Koeffizienten, *aberration coefficients* 27, 146ff., 149ff., 152ff., 162, 163ff.
Bildfeldwölbung, *field curvature* 16, 22, 234, 251ff.
Bildgüte in der Photographie, *quality of imaging in photography* 209.
Bildhöhe, reduzierte, *image height, reduced* 119, 124, 151.
Bildhöhen-Aberration windschiefer Strahlen, *image height aberration of skew rays* 119, 123, 139, 143, 146f., 148, 151.
Bildkorrelationsfilterung, *image correlation filtering* 746ff.
Bildlot, reduziertes, *image perpendicular, reduced* 92, 100, 112, 166ff.
Bildlot-Aberration, *image perpendicular aberration* 92, 99ff., 123, 136, 166ff.
Blazing 429, 580ff., 599, 653, 669, 714.
—, dreieckiges Furchenprofil, *triangular groove profile* 597, 599.
Borosilikat-Gläser, *boro-silicate glasses* 296.
Brechende Fläche, *refracting surface* 9.
Brechkraft einer optischen Fläche, *refractory power of an optical surface* 9, 118f.
Brechungsformeln, *refraction formulae* 82, 84ff., 88, 130, 155.
Brechungsindex, *refractive index* 72ff.
—, Bestimmung, *determination* 414.
Brechungsinvariante, *refraction invariant* 12.
Brechungswinkel, *angle of refraction* 90, 130, 139.
Brennweite, *focal length* 13, 230.
—, Messung, *measurement* 822—824.
Brillenträger-Okular, *eyepiece for spectacled persons* 367.
Büschel, *pencil* 5, 50.

Cassegrain-System, *Cassegrain system* 39, 260.
Cauchysche Produktensumme, *Cauchy's product sum* 152, 157, 160, 161.
CER-VIT 297.
Charakteristische Funktion von Hamilton, *characteristic function of Hamilton* 25.
Chromatische Fehler, *chromatic aberrations* 22f., 38, 180, 253ff.
— Vergrößerungsdifferenz, *difference of magnification* 287.
Cosinus-Größen, *cosine variables* 86ff., 90, 110.
Coudé-Fokus, *coudé focus* 264.
Coudé-Systeme, *coudé systems* 324ff.
CSIRO Australische Teilmaschine, *CSIRO Australian ruling engine* 721.
CUNY's optisches Lineal, *optical rule of Cuny* 812.

Dachkantenprisma, *roof prism* 212.
Dall-Kirkham-System 264.
Definition des Meters, *definition of meter* 766.

Definitionshelligkeit, *Strehl definition* 32f., 241, 244, 258, 263.
Defokussierung, *defocussing* 245.
Deformation optischer Flächen, *deformation of optical surfaces* 232.
Deutsche Montierung, *German mounting* 325.
Dezentrierung, *decentering* 263.
-- von Linsen, Messung, *of lenses, measurement* 824—827.
Diamantführung, Kontrolle, *diamond carriage control* 702f.
Diamantgeritzte Striche auf einem Maßstab, *diamond-ruled traces on a measure rod* 785.
Diamanttypen für das Ritzen von Gittern, *diamond types used for ruling* 667ff.
Dielektrische Schichten bei Röntgenstrahlgittern, *dielectric films in x-ray gratings* 435f.
Differential-Operatoren, *differential operators* 171, 177, 179, 181, 187.
Differentielle Näherung, *differential approximation* 177, 188, 190.
Differenz-Operatoren, *difference operators* 171, 172, 177, 184, 187.
Dirac-Funktionen, Kamm aus, *Dirac comb* 521.
Dispersion 73ff.
—, Abhängigkeit vom Gitterwinkel, *dependence on grating angle* 615.
— eines Beugungsgitters, *of diffraction gratings* 458f.
—, Erhöhung durch Benutzung von zwei Gittern, *increase by use of two gratings* 616f.
Dispersionseigenschaften der Gläser, *dispersive properties of glasses* 24.
Dispersions-Koeffizienten, *dispersion coefficients* 77ff.
Doppelanastigmat, *double-anastigmatic objective* 194, 196.
Doppelbrechende Interferenzmikroskope, *birefringent interference microscopes* 653.
Doppeltes Fizeau-Interferometer, *double Fizeau interferometer* 777.
Dopplerbreite, *Doppler width* 611.
Dove-Prisma, *Dove prism* 215.
Dowellscher Komparator mit Einfachspiegel, *Dowell's comparator with single mirror* 776, 781.
— — mit Winkelspiegeln, *with two mirrors set at an angle* 773f.
Drahtgitter, *wire grating* 443.
Drehspiegel, *rotating mirror* 757, 760.
Drehtisch für das Mikroskop, *rotatable microscope stage* 407.
Dreidimensionales Problem der Gitterbeugung, *three-dimensional problem of grating diffraction* 449.
Dreieckiges Furchenprofil, Randwertproblem, *triangular groove profile, boundary value problem* 597—599.
Dunkelfeldbeobachtung, *dark field observation* 376, 383.
Dünne Linse als Näherung, *thin lens approximation* 17.
Duran 297.
Durchbiegen, Kontrolle, *bending control* 817.
Durchlichtbeleuchtung, *transmittent illumination* 377.
Durchstoßdiagramm, *spot diagram* 21f., 257.

Ebene Gitter, Grundgleichungen, *plane gratings, basic equations* 456.
Ebenes Gitter, *plane grating* 434.
— Transmissionsgitter, Grundgleichungen, *transmission grating, basic equations* 457.
Ebnungslinse, *field flattener* 275.
Eikonal 139ff., 143ff.
Eikonalgleichung, *eikonal equation* 4, 25.
Eikonal-Koeffizienten, *eikonal coefficients* 141ff., 149ff., 163ff.
Eintrittspupille, *entrance pupil* 11, 48, 49, 95f., 231.
Eintrittsspalt eines Spektrographen, Einfluß der endlichen Breite, *entrance slit of spectrograph, finite width effect* 488ff.
Elektromagnetische Furchenform, *electromagnetic groove shape* 524.
— Randwertprobleme der Gittertheorie, *boundary value problems in grating theory* 512, 517.
Elektronenmikrographien von Gittern, *electron micrographs of gratings* 654ff.
Elektronenstrahl bei streifendem Einfall an einem Gitter, *electron beam at grazing incidence over grating* 440.
Elektrostatische Bilder eines Elektronenstrahls, *electrostatic images of an electron beam* 441.
Elementarwelle, *elementary wavelet* 518.
Elementarwellen, *wavelets* 513.
Endliche Spaltbreite, *finite slit width* 528.
Energieverteilung auf die Ordnungen der Beugung, *energy distribution among diffraction orders* 517, 573, 575.
Englische Achsenmontierung, *English (cross-axis) mounting* 332.
Entlastungssysteme für Teleskopspiegel, *support systems for telescope mirrors* 317.
Entwurf optischer Instrumente, *design of optical instruments* 34ff.
Entzerrung, *distortion correction* 205.
Eppensteinsche optische Umgebung, *Eppenstein optical circumvention* 806.
Erhaltung von Lichtflußstärke und Helligkeit, *conservation of flux and brightness* 494.
Étendue 493.
Exzentrizität, *eccentricity* 13.

Fabry-Pérot étalon 464, 532.
Faltung von Geistern, *convolution of ghosts* 556.
— zweier Fourier-Transformationen, *of two Fourier transforms* 522.
Farbfehler s. chromatische Fehler.
Farbkorrektion von Mikroskop-Objektiven, *chromatic correction of microscope objectives* 359.

Farbvariable, *colour coordinate* 74f., 161.
Faseroptik, *fiber optics* 383.
Fehler, primäre (vierte Potenzen), *primary aberration terms (fourth-power terms)* 6, 14—18.
— des Gitterblocks, *blank imperfections* 534f., 541.
— von Gittern, *imperfections of gratings* 610.
— bei Prismenherstellung, *defects in prism manufacture* 225.
— der Wellenfront, Einfluß auf das Spektralbild, *imperfections of wavefront, effect on spectral image* 538ff.
Fehlerordnung von HAMILTON, *aberration ordering of Hamilton* 6.
Fehlertabelle, *error table* 786.
Fehlertoleranz, *aberration tolerances* 31ff.
Fehlervariationsgesetz für Beugungsbilder, *imperfection variation law for diffraction patterns* 543, 545, 547.
Feldlinse, *field lens* 361.
Fermatsches Prinzip, *Fermat's principle* 2f.
Fernsehen, *television* 390.
Festkörperstrahler, *solid radiators* 387.
Fizeauscher Gittertest, *Fizeau grating test* 637.
Flächenanteil, *surface contribution* 100f., 104f., 123, 129, 132, 134, 136, 138f., 174, 177, 188ff.
Flächenbestimmung in der Mikroskopie, *area determination in microscopy* 405.
Flächendefekte, *surface defects* 299.
Fluchtfernrohr von HUET, *alignment telescope of Huet* 811.
Fluchtkugel, *alignment sphere* 814f.
Fluchtung mit dem Fernrohr, *alignment by telescope* 809—814, 825.
— durch Interferenz, *by interference* 814.
Fluoreszenzmikroskop, *fluorescence microscope* 397.
Fluoritobjektiv, *fluorite objective* 358.
Fokogramm eines Spiegels, *focogram of a mirror* 306.
Fokussierungseigenschaften von Konkavgittern, *focussing properties of concave gratings* 473f.
Förderliche Vergrößerung, *usable magnification* 347.
Foucaultsche Prüfmethode, *Foucault test* 304.
Fourierintegral, *Fourier integral* 446.
Fourier-Transformation der Beugungsfiguren eines Gitters, *Fourier transform of grating diffraction patterns* 518.
— des Huygensschen Prinzips, *of Huygens' principle* 506, 510.
Fouriertransformations-Hologramm, *Fourier transform hologram* 743.
Freies Spektralgebiet, *free spectral range* 463.
Fresnelzonengitter, *Fresnel-zone gratings* 437f.
Führung des Teilungsdiamanten, *advance of ruling diamond* 669.
Furchen unendlich schmale Spalte, *grooves infinitely narrow slits* 520f.
Furchenform, Prismenwirkung, *groove shaping, prism effect* 458.
Furchengestalt, Einfluß auf Ausbeute, *groove shape, effect upon efficiency* 645.
Furchengestaltfunktion, *groove shape function* 517, 523.
Furchenprofil, *groove profile* 428, 436, 451.
— Bestimmung aus Elektronenmikrographien, *determination by electron micrographs* 661—663.
— dreieckig, Randwertproblem, *triangular, boundary value problem* 597—599.
— rechteckiger Gestalt, *of rectangular shape* 575, 581.
— sinoidal, Randwertproblem, *sinusoidal, boundary value problem* 586—594.
— stückweise linear, Randwertproblem, *piecewise linear, boundary value problem* 594—597.
Furchenwinkel, geeignet für Blazing, *groove angle suitable for blazing* 581.

Gabelmontierung, *fork mounting* 327.
Gaborsche Methode der Wellenfront-Rekonstruktion, *Gabor's method of wavefront reconstruction* 741ff.
Gaußsche Abbildung, *Gaussian image formation* 95, 98, 137.
— Bildebene, *image plane* 98.
— Farbfehler, *colour aberrations* 23.
— Koeffizienten, *coefficients* 97f., 102, 124.
— Näherung, *approximation* 6, 7—14, 92ff., 99f., 137, 228.
Gaußtyp-Objektiv, *Gauss type objective* 197, 199.
Gefiltertes Beugungsbild, *filtered diffraction pattern* 639.
Geister, *ghosts* 453, 535, 540, 547, 549f., 555.
—, Intensitäten, *intensities* 552.
— verschiedener Ordnung, *of different order* 550ff.
Gekreuzte Teilungen, *crossed rulings* 641—643.
Gekrümmter Spalt, *curved slit* 469f., 628.
Geometrische Optik, Gültigkeitsgrenzen, *geometrical optics, limits of validity* 3.
— Wellenfront, *wavefront* 3.
Geschwindigkeit des Ritzens von Gittern, *speed of ruling* 669.
Gitter mit δ-funktionsartigen Furchen, *grating of δ-function grooves* 520f.
— in einer Immersion, *in immersion* 457.
— aus Zylindern, Randwertproblem, *of cylinders, boundary value problem* 572, 584.
— s. Beugungsgitter und Konkavgitter, *see diffraction grating and concave grating.*
Gitterfehler, *grating imperfections* 610ff.
Gittergleichung, *grating equation* 445, 448.
Gitterkopiermethoden, *grating replication methods* 731, 734f.
Gittermosaik, *grating mosaic* 432.
Gitterprüfungsinterferometer, *grating testing interferometers* 635f.
Glas, *glass* 24.
—, Übersicht optischer Gläser, *survey of optical glasses* 292.

Gläser, strahlungsresistente, *radiation-proof glasses* 375.
Glasmaßstab, *glass measure rod* 785.
Goniometer-Okular, *goniometric eyepiece* 800.
Gouyscher Satz, *Gouy's theorem* 511, 558.
Grad, *degree* 102ff., 123, 138, 179, 188.
Greensche Funktion zur Behandlung der Gitterbeugung, *Green's function treatment of grating diffraction* 571.
„Grill"-Vielspalt-Spektrometer von GIRARD, *"Grille" multislit spectrometer of Girard* 499.

Hariharan-Sensches Interferometer, *Hariharan-Sen interferometer* 779f.
Hartmannsche Prüfmethode, *Hartmann test* 301.
— —, Verallgemeinerung, *generalized* 303.
Hauptstrahl, *principal ray* 12, 22, 46, 47, 231.
Hellfeldbeobachtung, *bright field observation* 376.
Helligkeit, Definition, *brightness (luminance), definition* 492.
Helmholtz-Huygenssche Gleichung, *Helmholtz-Huygens equation* 507.
Helmholtzsche Wellengleichung, *Helmholtz wave equation* 446, 450.
Hitachi-Teilmaschine mit Interferometer-Kontrolle, *Hitachi ruling engine with interferometric control* 724f.
Höhen, *heights* 90.
Hologramme, *holograms* 739—741, 743, 745.
Holographie, *holography* 738ff.
Holographisches Filter, *holographic filter* 748.
Homogenität optischer Gläser, *homogeneity of optical glasses* 293.
Huet-Prisma, *Huet prism* 215f., 218.
Huetsches Fluchtfernrohr, *Huet alignment telescope* 811.
Hufeisenmontierung, *horseshoe mounting* 334.
Hunzingersches Interferometer, *Hunzinger's interferometer* 779f.
Huygens-Okular, *Huygens eyepiece* 288, 367.
Huygenssches Prinzip, genäherte Formulierung für elektromagnetische Wellen, *Huygens' principle, approximate form for electromagnetic waves* 508—510.
— — und Gitterbeugung, *and grating diffraction* 445, 504f.
— —, Grenzen seiner Anwendbarkeit, *limitations of its applicability* 513f.
— —, Schwiergigkeiten der Energieberechnung, *difficulties in energy calculations* 513.
— —, strenge Formulierung für elektromagnetische Wellen, *rigorous form for electromagnetic waves* 506—508.
Hyperfeinstruktur als Auflösungskontrolle, *hyperfine structure as a resolution test* 621.
Hypergon 198.

Immersionsobjektiv, *immersion object lens* 346, 358.
Immersionsöl, *oil of immersion* 346.
Immersionsspiegelobjektive, *immersion mirror objectives* 371.
Infrarot, Gitterausbeute, *infrared grating efficiency* 606.
Infrarot-Mikroskopie, *infrared microscopy* 395.
Inkohärente Strahlung, *incoherent radiation* 344.
Inkohärenz von Spektrallinien, *incoherence of spectral lines* 527
Instrumentfunktion, theoretische, *instrument function, theoretical* 528—530.
Interferentielle Fluchtung, *interferential alignment* 814.
Interferenzgitter, *interference gratings* 735—754.
Interferenzkomparator von KÖSTERS, *interference comparator of Kösters* 759, 768—773.
Interferenzkontrast, *interference contrast* 407.
Interferenzkontrolle von Oberflächen, Methode von SAUNDERS und GROSS, *interference surface control, Saunders-Gross method* 778.
Interferenzmessung der Parallelität, *interference measurement of parallelity* 777.
Interferenzmikroskopie, *interference microscopy* 407ff.
Interferenzmikroskop-Prüfung der Furchengestalt, *interference microscope testing of groove shape* 652f.
Interferenzprüfung von Oberflächen, *interference test of surfaces* 827—832.
Interferenzstreifen, Lage-Messung, *interference fringes, position measurement* 766.
Interferogramm, *interferogram* 559, 566ff., 632ff., 693ff.
Interferometer, *interferometers* 436f.
— von HARIHARAN und SEN, *of Hariharan and Sen* 779f.
— von HUNZINGER, *of Hunzinger* 779f.
— von MICHELSON, *of Michelson* 408, 631, 635, 783.
— von MICHELSON, TWYMAN und GREEN, *Michelson-Twyman-Green interferometer* 533f., 652.
— mit Quecksilberoberflächen, *with mercury surfaces* 777.
Interferometerkontrolle des Diamant-Schlittens, *interferometric control of diamond carriage* 728—730.
Interferometer-Prüfmethoden bei Fernrohren, *interferometric tests of telescopes* 309ff.
Interferometrische Gitterprüfung, *interferometric tests of gratings* 621ff.
— Kontrolle von Teilmaschinen, *control of ruling engines* 429f., 674, 687ff.
— Modulationsspektrometer, *modulation spectrometers* 502.
Isoplanatismus, *isoplanatism* 28f., 360.
Iterations-Formelsatz, *iteration formulae* 108f., 113f., 121ff., 135ff.

Jarrell-Ash-Teilmaschine mit Interferometerkontrolle, *Jarrell-Ash ruling engine with interferometric control* 703—706.

Jig-transit square 824.
Jobin & Yvon-Teilmaschine mit Interferometerkontrolle, *Jobin et Yvon ruling engine with interferometric control* 719f.
Jodabsorption zur Prüfung der Gitterauflösung, *iodine absorption test of grating resolution* 613.
Johns-Hopkins-Teilmaschine, *Johns Hopkins ruling engine* 680—682.

Kadmiumlinien, *cadmium lines* 770.
Keilfehler von Planparallelplatten, *wedge error of plane parallel plates* 819.
Kellner-Okular, *Kellner eyepiece* 289.
Kernereignis, Stereoaufnahme, *nuclear event, stereo-photography* 415.
Kippfehler, *tilt errors* 793, 808, 817.
Kleinste nutzbare Spaltbreite, *minimum usable slit width* 496.
Knickrahmen-Montierung, *arched-yoke mounting* 336.
Kohärente Beleuchtung des Gitters, *coherent illumination of grating* 489.
— Strahlung, *radiation* 345.
Kohärenzgrad, *degree of coherence* 345.
Kohärenzlänge, Schwierigkeiten bei Längenmessung, *coherence length, difficulties in length measurements* 778ff.
Köhlersche Beleuchtung, *Köhler illumination* 379.
Koinzidenz, *coincidence* 764, 772.
Koinzidenzfokometer, *coincidence focometer* 823.
Kollimator am Gitter, *collimator attached to grating* 454f.
Koma, *coma* 16, 22, 234, 248ff., 259, 263, 276.
Kombination brechender Flächen, *combination of refracting surfaces* 9.
Komparatorprinzip von Abbe, *comparator principle of Abbe* 761, 791.
— —, Umgehung, *circumvention* 806.
Kompensation von Kippfehlern, *compensation of tilt errors* 793, 808.
Kompensatoren, *compensators* 400.
Komplexe Strahldaten, *complex ray data* 86ff.
Komplexes Filter, *complex filter* 747.
— Lot, *complex perpendicular* 91.
Kondensor, *condensor* 383f.
Konjugierte Abstände, *conjugate distances* 8, 46.
Konkavgitter, *concave gratings* 434.
—, Astigmatismus, *astigmatism* 480—485.
—, Auflösung, *resolution* 479f.
—, Fehler, *aberrations* 478.
—, Fokussierungseigenschaften, *focussing properties* 473f.
—, Grundgleichungen, *basic equations* 475ff.
—, Parameter, *parameters* 475.
—, Übersicht über verschiedene Aufstellungen, *survey of different mountings* 484—486.
Konoskopische Beleuchtung, *conoscopic illumination* 401f.
Konstruktionsdaten, *data of construction* 70ff., 102, 114ff., 158.
Konstruktive Interferenz, *constructive interference* 450.
Kopien von Originalgittern, *replicas of master gratings* 599, 660, 731—735, 727.
Körnigkeit der Platte, *granularity of plate* 531.
Korrektion von Objektiven, *correction of objectives* 357f.
— von Okularen, *of eyepieces* 361.
Korrektionsplatte, *corrector plate* 238, 267, 270, 272.
Köstersscher Interferenzkomparator, *Kösters's interference comparator* 759, 768—773.
Köstersches Prisma, *Kösters's prism* 812f., 823.
Kreisbogenförmige Spalte, *circularly curved slits* 470—472.
Kreisförmige Wellenfront, *circular wavefront* 558.
Kritische Beleuchtung, *critical illumination* 378.
Krümmung der Spektrallinien, *curvature of spectral lines* 451, 467—472.
Krümmungsradien von Linsen, Bestimmung, *curvature radii of lenses, determination* 820—822.
Kryptonkontinuum im Ultravioletten, *krypton continuum in the ultraviolet* 646.
Kryptonlampe, *krypton lamp* 767, 769.
Kugelfläche, *spherical surface* 7, 50.
Kugelspiegel, *spherical mirror* 247.
Kumulative Teilfehler, *cumulative ruling errors* 541, 671.
Künstliches Dublett, *artificial doublet* 612, 614.

Lager für Fernrohrmontierungen, *bearings for telescope mountings* 323.
Lagrangesche Invariante, *Lagrangian invariant* 12, 15, 46, 49.
Längeneinheit, *unit of length* 766.
Längenmessung in der Mikroskopie, *length measurement in microscopy* 403f.
— mit Pérot-Fabry-Platten, *with Pérot-Fabry plates* 782.
— durch Wellenlänge, *by wavelength* 755, 766.
Langesche Konstruktionsdaten, *Lange data of construction* 114ff., 177, 187, 191.
Laterale Auflösung, *lateral resolution* 343.
Leman-Sprenger-Prisma, *Leman-Sprenger prism* 216, 218.
Leseglas, *reading lens* 353.
Leuchtdichte, *flux density* 386.
Lichtfluß, Definition, *flux of light, definition* 492, 494.
— in Gittern, *in gratings* 462.
— durch ein Photoobjektiv, *through photographic objective* 207.
Lichtgebirge, *point spread function* 256.
Lichtgeschwindigkeit, interferometrische Messung, *velocity of light, interferometric measurement* 675f.

Lichtintensität, Definition, *intensity of light, definition* 492.
Lichtleiter, *light duct* 383.
Lichtquelle, Einfluß auf Prüfung der Auflösung, *source condition, effect on resolving power tests* 614.
Lichtquellen für Mikroskopie, *light sources in microscopy* 385.
Lichtstärke von Gittern, *luminosity of gratings* 453, 461, 486—504.
— von Spektrographen und Spektrometern, *of spectrographs and spectrometers* 495f.
Lichtstrahl, *ray of light* 2.
Lichtweg, *optical path* 52ff., 140, 509.
Lichtzeiger, *light beam pointer* 757, 760.
Leardispersion eines Gitters, *linear dispersion of a grating* 459, 461f.
Lineare Fehler der Wellenfront, *linear wavefront aberrations* 535f., 541.
Linearer Phasenfehler bei Gittern, *linear phase aberration in grating* 541.
Linsenkrümmung, Bestimmung, *lens curvature, determination* 820—822.
Lochkamera, *pinhole camera* 193.
Lochplatten als Mikrowellengitter, *perforated plates as microwave gratings* 437.
Lote, *perpendiculars* 90.
Lupe, *loupe* 351, 787.
Lupenvergrößerung, *loupe magnification* 351, 373.
Lymansche Geister, *Lyman ghosts* 555.

Mach-Zehndersches Interferometer, *Mach-Zehnder interferometer* 409, 747f.
Maréchalscher Integrator, *Maréchal integrator* 564.
Masken für Spalte, *masks put on slits* 500f.
Matrizenbeschreibung der Gaußschen Optik, *matrix description of Gaussian optics* 14.
Maximale *f*-Zahl, *maximum f number* 496.
Mechanischer Aufbau von Teilmaschinen, *mechanical design of ruling engines* 669.
Medial-Fernrohr, *medial telescope* 284.
Mehrfach-periodische Fehler, *multiple periodic errors* 555f.
Meniskussysteme, *meniscus systems* 279f.
Meridianstrahlen, *meridional rays* 89f., 169.
Meridionale Bildhöhenaberration, *meridional image height aberration* 129, 135ff., 138f., 146f., 174f., 176, 187, 190f.
MERTON's Teilmaschine für metrologische Gitter im NPL, *Merton's NPL ruling engine for metrology gratings* 725—727.
Meßfehler durch schlechte Fokussierung, *errors in measurement by imperfect focussing* 765.
Meßkraft, *measuring force* 761.
Meßlupe, *measuring loupe* 787.
Meßmikroskop, *measuring microscope* 792, 799.
Messung unter dem Mikroskop, *measurement under microscope* 403.
Metallfläche, Interferenzprüfung, *metal surface, interference test* 830.
Metallmaßstab, *metal measure rod* 785.
Metallspiegel, *metal mirrors* 298.
Meter, Definition, *meter, definition* 766.
Meterkomparator, *meter comparator* 767.
Michelson-Interferometer, *Michelson interferometer* 408, 631, 635, 783.
Mikrometer-Okular, *micrometric eyepiece* 367.
Mikrophotographie, *microphotography* 390.
Mikrophotometrie, *microphotometry* 411ff.
Mikroprojektion, *microprojection* 390.
Mikroskop für heiße Zellen, *microscope for hot cells* 374.
Mikroskoptische, *microscope stages* 407.
Mikroskopie und Längenbestimmung, *microscopy and length determination* 788.
Mikroskopische Interferenzprüfung von Oberflächen, *microscopic interference test of surfaces* 827—832.
Mikrospektrophotometrie, *microspectrophotometry* 411ff.
Mikrostrahlenstich, *microbeam* 396.
Mikrowellengitter, *microwave gratings* 436f.
Mikrowellenversuche mit stehenden Gitterwellen, *microwave experiments with standing grating waves* 579.
Mills-Interferometer, *Mills' cross* 437.
MIT-Teilmaschine mit Interferometerkontrolle, *MIT ruling engine with interferometric control* 687—703.
Mittenabschattung s. auch Apodisation, *central obstruction (see also apodization)* 241, 368, 370.
Mittlere Strahldaten, *mean ray data* 109ff., 112, 125, 166ff.
Modulations-Interferenzgitter, *modulation interference gratings* 735—754.
Moiré-Gitter, *moiré gratings* 439f.
Moiré-Prinzip, *moiré principle* 764.
Moiréstreifen-Prüfung, *moiré-fringe testing* 641—643.
Monochromate als Mikroskop-Objektive, *monochromatic microscope objectives* 358.
monozentrisches Okular, *monocentric eyepiece* 289.
Montierung von Teleskopen, *mounting of telescopes* 322ff.
Mount-Wilson-Sternwarte, Teilmaschinen, *Mount Wilson Observatory ruling engines* 682—687, 716—719.

Nasmythsche Anordnung, *Nasmyth focus* 261.
Neutrale Faser eines Maßstabes, *neutral fibre of measure bar* 786.
Newton-Fokus, *Newton focus* 259.
Nichtkohärente Beleuchtung des Gitters, *incoherent illumination of grating* 459, 489.
Nicht selbstleuchtendes Objekt, *non self-luminous object* 349.
Nichtsphärische Flächen, *non-spherical surfaces* 17.
Nivellierfernrohr, *levelling telescope* 808.
Nobel-Institut, Teilmaschine, *Nobel Institute ruling engine* 727f.
Normaler Aufbau, *normal construction* 80f.

NPL-Teilmaschine, *NPL ruling engine* 725—727.
Nullmethode, *null test* 306.

Oberflächenmessung mit Interferenzmethoden, *surface measurement by interference methods* 757.
Oberflächenreflektivität, Einfluß auf Ausbeute, *surface reflectivity, effect upon efficiency* 645.
Oberflächenwelle, *evanescent wave* 446, 570, 588, 594.
Objektebene, *object plane* 96.
Objekthöhe, reduzierte, *object height, reduced* 96, 119.
Objektlot, reduziertes, *object perpendicular, reduced* 96.
Objektive von Mikroskopen, *objectives for microscopes* 357f.
— für Photographie, *for photography* 192ff.
Öffnungszahl, *f-number* 232, 234.
Okular eines Fernrohrs, *eyepiece of a telescope* 285.
— für Meßmikroskop, *of measuring microscope* 799.
— für Mikroskop, *for microscopes* 361.
Okulares Sehfeld, *eyepiece field of vision* 363.
Okularkreis, *Ramsden disk* 285.
Okularmikrometer, *eyepiece micrometer* 764.
Optische Gläser s. Glas, *optical glasses see glass*.
— Materialien im Ultravioletten, *materials in the ultraviolet* 24.
— Richtung, *direction* 79, 153.
— Übertragungsfunktion, *transfer function* 257.
Optisches Cosinusgesetz, *optical cosine law* 28.
— Gitter, *grating* 433f.
— Lineal von Cuny, *Cuny's optical rule* 812.
— Tasten, *palpating* 756.
Optisches Staatsinstitut, Teilmaschine, *State Optical Institute ruling engine* 712—716.
Ordnung, *order* 85, 93, 102ff., 123, 138, 188.
— eines Geists, *of a ghost* 551.
Ordnungen der Beugung, Energieverteilung, *orders of diffraction, energy distribution* 517, 570.
Orthoskopische Beleuchtung, *orthoscopic illumination* 398.
Orthoskopisches Okular, *orthoscopic eyepiece* 289.

Pankratische Systeme, *pancratic systems* 375.
Parabelförmige Wellenfront, *parabolic wavefront* 558.
Parallaktische Montierung, *equatorial mounting* 322.
Parallaxe, *parallactic error* 762f.
Parallelitätsprüfung von Achsen, *parallelity control of axes* 819.
Parallelperspektive, *parallel perspective* 760.
Paraxiale Strahlenkonstruktion, *paraxial raytracing* 9f.
Paraxiales Gebiet, *paraxial region* 228.
Paraxialstrahlen, *paraxial rays* 92.
Partialdispersion, *partial dispersion* 292.
Pechan-Prisma, *Pechan prism* 217.
Pendelspiegel, *pendulum mirror* 808, 825.
Pentagonprisma, *pentagon prism* 214, 218, 818f.
Perflektometer, *perflectometer* 789f.
Periodenfehler, *error of period* 552.
Periodische Fehler, *periodic errors* 671.
Perkin-Elmer-Teilmaschinen, *Perkin-Elmer ruling engines* 716.
Pérot-Fabry-Etalon mit Michelson-Interferometer verbunden, *Pérot-Fabry étalon combined with Michelson interferometer* 781.
Perspective des photographischen Bildes, *perspective of photographic image* 201ff.
Petzval-Krümmung, *Petzval curvature* 16, 51, 195, 360, 365.
Petzval-Objektiv, *Petzval objective* 194.
Petzvalsche Summe, *Petzval sum* 237, 251.
phased arrays 437.
Phasenkontrast-Fluoreszenzmikroskop, *phase-contrast fluorescence microscope* 397.
Phasenkontrast-Gittertest, *phase-contrast grating test* 637ff.
Phasenkontrast-Verfahren, *phase-contrast method* 393, 407, 744, 746.
Phasenmessung in der Wellenfront von einem Gitter, *phase measurement in wavefront from grating* 525.
Phasenmodulation, *phase modulation* 736.
Phasenumkehr durch $\lambda/2$-Plättchen, *phase-reversion by half-zone plates* 438.
Photoelektrische Registrierung von Spektrallinien, *photoelectric recording of spectral lines* 621f.
Photoelektrischer Nachweis, *photoelectric detection* 482, 489f.
Photoelektrisches Okular, *photoelectric eyepiece* 801ff.
Photographische Objektive, *photographic objectives* 192ff.
Photometer-Mikroskop, *photometer microscope* 411ff.
Photometrische Begriffe und Definitionen, *photometric concepts and definitions* 491f.
Planachromat, *planar achromat* 361.
Platinschicht auf Gitteroberfläche, *platinum coating of grating surface* 602f.
Polachsenmontierung, *polar axis (off-axis German) mounting* 326.
Polarisation durch Gitter, *polarization by gratings* 443f., 505, 575, 589.
— an Metallgittern, *by metal gratings* 573.
— unterdrückt in Gittern mit Blazing, *suppressed in blazed gratings* 572, 581f.
Polarisationseffekte bei Gitter-Randwertproblemen, *polarization effects in grating boundary problems* 571, 579.
Polarisationsmikroskop, *polarization microscope* 398.
Porro-Prismen erster und zweiter Art, *Porro prisms of first and second kind* 217, 222.
Primäre Näherung, *primary approximation* 103f., 137ff., 189, 190.

Primärfokus, *prime focus* 259.
Primärfokus-Korrektoren, *prime focus correcting systems* 268.
Prisma von Kösters, *prism of Kösters* 812f., 823.
Prismatische Furchen, *prismatic grooves* 458.
Prismen, *prisms* 211ff.
Prismenwinkel, Prüfmessung, *prism angles, test measurement* 818.
Profil der Furchen s. Furchen, *profile of grooves see grooves.*
— eines Maßstabes, *of a measure bar* 786.
Profilabbildung, *profile imaging* 755.
Programmierung für Beugungsbilder, *computer programmes for diffraction patterns* 564f.
— beim Entwurf optischer Instrumente, *for optical design* 40f., 66.
Projektive, *projectives* 362.
Prüfmethoden für Fernrohre, *test methods for telescopes* 300ff.
Prüfung des Auflösungsvermögens von Gittern, *tests of resolving power of gratings* 611ff.
Pupillenabbildung der Tubuslinse, *pupil imaging of tube lens* 375.
Pupillenfunktion, *pupil function* 257.
Pupillenlot, reduziertes, *pupil perpendicular, reduced* 166ff.
Pupillenlot-Aberration, *pupil perpendicular aberration* 166ff.
Pupillenstrahl, *pupil ray* 11.
Pyramidalfehler von Prismen, *pyramid defect in prisms* 225f.
Pyrex 297.

Quarz, *quartz* 297.
Quarznormal, *quartz gauge* 775.
Quasi-Cassegrain-Systeme, *quasi-cassegrainian systems* 264.
quasi-dreieckige Furchengestalt, *quasi-triangular groove shape* 570.
Quecksilberlampe, *mercury lamp* 389.
Quecksilberlinien zur Prüfung der Gitterauflösung, *mercury lines test of grating resolution* 611, 617f., 621, 694.
Quervergrößerung, *transverse magnification* 8.

Radioteleskope, *radio-telescopes* 437.
Radiusvektor, *radius vector* 79, 153.
Raketen-Gitterspektrometer, *rocket grating spectrometer* 630.
Ramsden-Okular, *Ramsden eyepiece* 288.
Randwertprobleme der Maxwellschen Theorie, *boundary value problems of Maxwell theory* 583ff.
Räntsch-Interferometer, *Räntsch interferometer* 784.
Raumfrequenzvariable, *spatial frequency variables* 519.
Rayleigh-Kriterium, *Rayleigh criterion* 32, 244, 344, 611f., 764.
Rayleighsche Toleranz, *Rayleigh tolerance* 479.
Rayleighsches Randwertproblem der Gitterbeugung, *Rayleigh's boundary problem of grating diffraction* 570, 583—591.
rechteckige Öffnung, Beugungsfigur, *rectangular aperture, diffraction pattern* 452.
Rechteckprofil der Gitterfurchen, *rectangular groove profile* 575, 581.
— —, Unzweckmäßigkeit, *unsuitability* 581.
Referenzkugel, *reference sphere* 5, 58.
Referenzwellenfront, *reference wavefront* 557ff.
reflektierende Systeme, Entwurf, *reflecting systems, design* 39.
Reflektor-Korrektor, *reflector corrector* 269.
Reflexion von Licht an Linsen, *reflection of light by lenses* 209.
Reflexionsgitter, *reflection grating* 433, 435, 446.
—, Parameter, *parameters* 455, 456.
Reflexionskoeffizienten von Röntgenstrahlgittern, *reflection coefficients of x-ray gratings* 647f.
Reflexions-Stufengitter, *reflection echelon* 466f.
Reflexionsvermögen von Gittern, *reflectivity of gratings* 435f.
— der Gitteroberfläche für Röntgenstrahlen, *of grating surface for x-rays* 605.
reziproke Höhe, *reciprocal height* 112, 116.
reziproker Brechungsindex, *reciprocal refractive index* 112, 116.
Ritchey-Chrétien-System 39, 265.
Röntgenstrahlen großer Wellenlänge, Gitterausbeute, *x-rays of large wavelength, grating efficiency* 603.
Röntgenstrahlgitter, *x-ray gratings* 429, 435, 473, 648.
—, Ausbeute, *efficiency* 647.
—, Reflexionskoeffizienten, *reflection coefficients* 647f.
Ross-Korrektor, *Ross correcting lens* 268.
Rotationssymmetrie und Fehlerklassifikation *rotational symmetry and aberration classification* 6.
Rowlandsche Geister s. auch Geister, *Rowland ghosts see also ghosts* 549ff.
— Methode der Gitterteilung, *method of ruling* 429—431
— Teilmaschine, *ruling engine* 681.
Rowlandscher Kreis, *Rowland circle* 474, 477.

Sabattier-Effekt, *Sabattier effect* 832.
Sagittale Bildhöhenaberration, *sagittal image height aberration* 129, 130ff., 135ff., 138f., 146f., 180ff., 187, 190f.
— Variation, *variation* 127f., 186.
Sammellinse, *collective lens* 361.
Satelliten, *satellites* 540, 545, 560.
Saunders-Gross-Methode zur Interferenzkontrolle von Oberflächen, *Saunders-Gross interference method of surface control* 778.
Scheitellot, *vertex perpendicular* 91, 101, 102, 110.
—, reduziertes, *reduced* 120.

Scheitellot-Aberration, *vertex perpendicular aberration* 123, 136.
Scherungsinterferometer, *shearing interferometer* 311.
Schmidt-Pechan-Prisma, *Schmidt-Pechan prism* 215f.
Schmidt-Prisma, *Schmidt prism* 218, 220f.
Schmidt-Teleskop, *Schmidt telescope* 18, 24, 39, 272ff.
Schneidenprüfung, *knife-edge test* 304f.
Schnittweitengleichung, *Newton's conjugate equation* 8, 202.
Schrauben der Teilmaschinen, *screws of ruling engines* 670f.
Schwarzschild-System 265.
Schwingungen von Teilmaschinen, *vibration of ruling engines* 673.
Sehfeld, *field of vision* 363.
Seidelsche Differenzformeln, *Seidel difference formulae* 13.
— Koeffizienten, *coefficients* 236f.
— Näherung, *approximation* 7, 44, 51, 103, 139, 190, 232.
— Summen, *sums* 15f., 51, 233f.
— —, Erweiterung für asphärische Flächen, *extension to aspherical surfaces* 17f.
Seidelscher Abstand, *Seidel separation* 107, 116.
Seismographische Schwingungsprüfung von Teilmaschinen, *seismographic tests of ruling engine vibrations* 676.
Sekundäre Näherung, *secondary approximation* 189, 191.
Sekundäres Spektrum, *secondary spectrum* 22, 255.
Sekundärnormal, *secondary gauge* 755.
Selbstleuchtendes Objekt, *self-luminous object* 349.
Shearing-Polarisationsinterferometer, *shearing polarization interferometer* 410.
SIEGBAHN's Teilmaschine im Nobel-Institut, *Siegbahn's Nobel Institute ruling engine* 727f.
Sinoidale dielektrische Gitteroberfläche, Randwertproblem, *sinusoidal dielectric grating surface, boundary value problem* 592—594.
— metallische Gitteroberfläche, Randwertproblem, *metallic grating surface, boundary value problem* 586—591.
Sinusbedingung, *sine condition* 248.
Sinus-Größen, *sine variables* 86ff., 90, 102, 110, 124, 133.
SISAM-Spektrometer von CONNES, *SISAM spectrometer of Connes* 499, 502f.
Skalare Probleme der Gitterbeugung, *scalar problems in grating diffraction* 444, 446, 504.
Skalares Randwertproblem der Gitterbeugung, *scalar boundary problem of grating diffraction* 571, 600.
Slit scanning function 499ff.
Smith-Purcellscher Versuch, *Smith-Purcell experiment* 434, 440—443.
Smyth-Linse, *Smyth lens* 196.
Snelliussches Brechungsgesetz, *Snell's law* 2f.
Sonnar 197.
Spalt mit Krümmung, *slit with curvature* 469f.
— variabler Krümmung, *of variable curvature* 628.
Spalte im Gitter unendlich schmal, *slits of grating infinitely narrow* 520f.
Spaltbreite, Optimum, *slit width, optimum* 490.
Spaltfunktion, *slit function* 499ff.
Spannungen in optischen Gläsern, *stress in optical glasses* 294.
Spektralanalysator, *spectrum analyser* 750ff.
Spektraldispersion, *spectral dispersion* 487.
Spektrale Bildentstehung, *spectral image formation* 527, 531f.
Spektrallinien, Krümmung, *spectral lines, curvature* 451, 467—472.
Spektralphotometer-Mikroskop, *spectrophotometer microscope* 411ff.
Spektrograph mit Reflexionsgitter, *spectrograph with reflection grating* 455.
— mit Transmissionsgitter, *with transmission grating* 455.
Spektroheliogramme, *spectroheliograms* 629f.
Spektroheliograph, *spectroheliograph* 469.
— hoher Auflösung, *of high resolution* 626.
Spektrometer hoher Auflösung, *spectrometers of high resolution* 622ff.
Spektrometerspalt s. Eintrittsspalt, Austrittsspalt und Spalt, *spectrometer slit see entrance slit, exit slit and slit.*
Spektroskopische Gitter, *spectroscopic gratings* 434—436, 444ff.
Sphärische Aberration, *spherical aberration* 16, 22, 43, 195, 234, 245ff., 265.
Sphärochromatische Aberration, *spherochromatic aberration* 256, 277, 285.
Spiegelartiges Gitter, experimentelle Bestätigung, *mirror-like grating, experimental verification* 579.
— —, Theorie, *theory* 572f., 575—578.
Spiegellagerung in Fernrohren, *mirror support in telescopes* 314.
Spiegelmaterial, *mirror materials* 295.
Spiegelobjektiv, *mirror objective* 200.
Spiegelobjektive in Mikroskopen, *mirror objectives for microscopes* 367.
Stehende Wellen vor spiegelartigem Gitter, *standing waves in front of a mirror-like grating* 579.
Stereomikroskopie, *stereo-microscopy* 414.
Stereoskopische Elektronenmikrographien, *stereoscopic electron micrographs* 656—659.
Stigmatismus von Spektrallinien, *stigmatism of spectral lines* 470—472, 476.
— der Wadsworthschen Aufstellung, *of Wadsworth mounting* 483.
Stoßsicherung für Teilmaschinen, *shock isolation of ruling engines* 671ff.

Strahldaten, *ray data* 79ff.
Strahlenkonstruktion, *raytracing* 9, 19.
Strahlenteilung, *beam partition* 380.
Strahlungserzeugendes Gitter, *radiation-generating grating* 440—443.
Streifende Inzidenz bei optischen Beugungsgittern, *grazing incidence, optical diffraction gratings* 458.
— — bei Röntgenstrahlgittern, *x-ray gratings* 435, 603ff.
strenge Darstellung der Bildhöhenaberration, *rigorous representation of image height aberration* 166ff., 176, 187.
— Lösungen von Randwertproblemen, *solutions of boundary value problems* 572, 583f.
Streulicht, *stray light* 652.
— von Gittern, *from gratings* 561—564, 650.
Streuplatten-Interferometer, *scatter-plate interferometer* 313.
Streuung und Beugung, *scattering and diffraction* 433.
Strongsche Teilmaschine, *Strong's ruling engine* 680—682.
Stufengitter, Maxwellsche Theorie, *step grating, Maxwell theory* 576—578.
— im polarisierten Licht, *in polarized light* 575f.
— in Reflexion, *echelon and echelle in reflection* 466f.
— in Transmission, *in transmission* 466.
— verschiedener Typen, Definition, *echelle, echelette, echelon: definitions* 465.
Superachromat 23, 283, 359.
Super-Schmidt-Kameras, *super-Schmidt cameras* 281.
symmetrisches Objektiv, *symmetric objective* 194.
Systeme ohne Axialsymmetrie, *systems without axial symmetry* 30.

Tangentiale Bildhöhenaberration, *tangential image height aberration* 129ff., 135ff., 138f., 180ff., 187, 190f.
— Variation, *variation* 127f., 186.
Technische Konstante, *Hartmann constant* 302.
Teilfehler, *ruling imperfections* 610.
— und Wellenfrontfehler, *ruling errors and wavefront aberrations* 533f., 541.
Teilmaschinen, *ruling engines* 429—431, 669—730.
—, Stoßkontrolle, *shock control* 671—677.
—, Temperaturkontrolle, *temperature control* 678—681.
Teilung von Gittern, Verfahren, *ruling process of gratings* 666.
Teleobjektiv, *long-distance objective* 196, 199.
telezentrischer Strahlengang, *telecentric beam* 760, 763.
— —, Störung, *perturbation* 765.
Temperatureinfluß auf Fernrohre, *temperature effects on telescopes* 320.
— auf Längenmessung, *on length measurement* 761.
Temperaturkontrolle von Teilmaschinen, *temperature control of ruling engines* 678—681.
Tessar 196.
Thermische Ausdehnung von Metallmaßstäben, *thermal expansion of metal measures* 785.
Tiefenschärfe eines Objektes, *longitudinal sharpness of objective* 204.
Tokioter Teilmaschine, *Tokyo ruling engine* 722—724.
Topogon 198.
Transmission optischer Gläser, *transmission of optical glasses* 293.
Transmissionsgitter, *transmission grating* 433, 436, 446.
—, Parameter, *parameters* 455, 456.
Transmissions-Stufengitter, *transmission echelon* 465, 466.
Tripelspiegel, *triple mirror* 213, 782f.
Triplett, *Cook lens* 196.
Tubus des Mikroskops, *tube of microscope* 372.
Tubuslinsen, *tube lenses* 373.
Twyman-Greensches Interferometer, *Twyman-Green interferometer* 21.

Übergangsformeln, *transition formulae* 81, 84ff., 88, 155.
Überlappung der Ordnungen in einem Gitter, *overlap of orders in a grating* 462.
Übertragungsfunktion, *transfer function* 257.
Ultrarot s. Infrarot.
Ultrarotgitter, *infrared grating* 436, 711.
Ultraviolett, optische Materialien, *ultraviolet, optical materials* 24.
Ultraviolettgitter, *ultraviolet gratings* 435, 473, 625, 651.
Ultraviolett-Mikroskopie, *ultraviolet microscopy* 395.
Ultraviolettreflexion von Aluminium, *ultraviolet reflection of aluminum* 601.
— von Platin, *of platinum* 602.
Unvollkommenheiten der Wellenfront, *imperfections of wave front* 452, 454.
Urmeter, *mètre prototype* 755, 766.

Vakuum-Ultraviolett, Gitterspektrograph, *vacuum ultraviolet grating spectrograph* 625.
Variationsverfahren für Gitterbeugung, *variational treatment of grating diffraction* 571.
Variofokales Objektiv, *zoom lens* 67, 200f.
Vektorielles Randwertproblem der Gitterbeugung, *vectorial boundary problem of grating diffraction* 571, 600.
Vektorskalar, *vector-scalar* 153ff.
—, Potenzreihendarstellung, *power series representation* 155ff.
Vergrößerung eines Fernrohrs, *magnification of a telescope* 285.
Vergrößerungsgesetz, *magnification law* 8.
Vergütung von Spiegeln, *coating of mirrors* 295.

Verteilung des Lichts auf die Ordnungen (s. auch Blazing), *distribution of light among orders* 570.
Verzeichnung, *distortion* 16, 22, 61, 234, 793.
Vickers-Eindruck, Interferenzprüfung, *Vickers imprint, interference test* 795—797.
Vielfachspalte, *multiple slits* 432.
Vielfachspalt-Spektrometer, *multiple slit spectrometers* 498—503.
Vierlinser, *four-element lens* 290.
Vignettierung, *vignetting* 208, 232, 262, 274, 279.
Voglsches Interferometer, *Vogl interferometer* 777.
Vollkommener Stigmatismus bei Gittern, *perfect stigmatism in gratings* 472.
Vollkommenes Gitter, *perfect grating* 428.
Vorzeichenkonvention, *sign convention* 229.

Wadsworthsche Aufstellung eines Konkavgitters, *Wadsworth mount of a concave grating* 477f., 483.
Weiche Röntgenstrahlen, Gitterausbeute, *soft x-rays, grating efficiency* 603.
Weitwinkelobjektive, *wide-angle objectives* 199.
Wellenaberration, *wave aberration* 243, 258.
Wellenfront, Phasenänderungen, *wavefront, phase variations* 638.
Wellenfronten, Rechenprogramme, *wavefront calculation, programming* 564f.
Wellenfront-Deformation, *wavefront deformation* 7.
Wellenfrontfehler und Fehler des Gitterblocks, *wavefront aberrations and grating blank imperfections* 534f., 541.
— und Teilfehler, *and ruling errors* 533f.
Wellenfront-Interferogramme, *wavefront interferograms* 632f., 635, 693ff.
Wellenfront-Rekonstruktion, *wavefront reconstruction* 736, 739, 743.
Wellenlängen von Geistern, *ghost wavelengths* 555, 557.
Wellenlängeneichung des Meters, *wavelength gauge of the meter* 766ff.
Wellenlängen-Rechenschieber, *wavelength slide rules* 770.
Winkelbestimmung in der Mikroskopie, *angle determination in microscopy* 405.
Winkeldispersion eines Gitters, *angular dispersion of a grating* 458, 460.
Winkelfehler von Prismen, *angular defects in prisms* 225.
Winkelmessung, *angle measurement* 818.
Winkelspektrum ebener Wellen, *angular spectrum of plane waves* 447.
Wolfram-Lichtquelle, *tungsten light source* 387f.
Wollaston-Prisma, *Wollaston prism* 222, 410, 805.
Woodsche Anomalien, *Wood anomalies* 574, 580, 583.
Wright-Väisälä System 278.

Xenon-Hochdrucklampe, *xenon high-pressure lamp* 389.
Xenonkontinuum im Ultravioletten, *xenon continuum in the ultraviolet* 646.

Zehender-Kammer, *Zehender chamber* 828.
Zentrieren der Lichtquelle, *centering of light source* 762.
— von Linsen, Prüfmethoden, *of lenses, test methods* 824—827.
Zufällige Teilfehler, *accidental ruling errors* 564.
Zweidimensionales Beugungsproblem, *two-dimensional problem of diffraction* 446.
Zweispiegelsystem, *two-mirror system* 230, 237, 261.
—, aplanatisches, *aplanatic* 265.
Zwischenbild im Mikroskop, *intermediate image in microscope* 354.
Zylindrische optische Systeme, *cylindrical optical systems* 31.

Subject Index.

(English-German.)

• Where English and German spelling of a word is identical the German version is omitted

ABBE's theory of image formation in microscope, *Abbesche Theorie der Bildentstehung im Mikroskop* 349.
Abbe-König prism, *Abbe-König-Prisma* 215, 217.
Aberration coefficients, *Bildfehler-Koeffizienten* 27, 146seqq., 149seqq., 152seqq., 162, 163seqq.
Aberration in concave gratings, *Aberration bei Konkavgittern* 478.
Aberration function, *Aberrationsfunktion* 169seq., 181, 184seq.
Aberration ordering of Hamilton, *Fehlerordnung von Hamilton* 6.
Aberration tolerances, *Fehlertoleranz* 31seqq.
Aberrations in image formation, different definitions, *Abbildungsfehler, verschiedene Definitionen* 5.
— in image formation, higher orders, *höhere Ordnungen* 18.
— in image formation, lowest order, *niedrigste Ordnung* 6, 15seq.
Absorption of light in lenses, *Absorption von Licht in Linsen* 209.
Accidental ruling errors, *zufällige Teilfehler* 564.
Achromatic lens, *Achromat* 194, 256, 282.
Achromatic microscope objectives, *Achromate als Mikroskop-Objektive* 357.
Achromatism, *Achromasie* 44, 255.
Advance of ruling diamond, *Führung des Teilungsdiamanten* 669.
Airy pattern, *Airysche Beugungsfigur* 31seq.
Alignment by interference, *Fluchtung durch Interferenz* 814.
Alignment sphere, *Fluchtkugel* 814seq.
Alignment by telescope, *Fluchtung mit dem Fernrohr* 809—814, 825.
Alignment telescope of HUET, *Fluchtfernrohr von* HUET 811.
Altazimuth mounting, *azimutale Montierung* 322.
Aluminum coating of grating surfaces, *Aluminisierung von Gitteroberflächen* 601.
— — of mirrors, *von Spiegeln* 298.
Amici prism, *Amici-Prisma* 215seq.
Amplitude modulation, *Amplitudenmodulation* 736.
Anamorphotic image formation, *anamorphotische Abbildung* 30.
Angle determination in microscopy, *Winkelbestimmung in der Mikroskopie* 405.
Angle measurement, *Winkelmessung* 818.
Angle of refraction, *Brechungswinkel* 90, 130, 139.
Angular defects in prisms, *Winkelfehler von Prismen* 225.
Angular dispersion of a grating, *Winkeldispersion eines Gitters* 458, 460.
Angular spectrum of plane waves, *Winkelspektrum ebener Wellen* 447.
Aperture, *Apertur* 344, 368.
Aperture stop, *Aperturblende* 11, 231.
Aperture width of a grating, *Aperturweite eines Gitters* 460, 462.
Aplanate, *Aplanat* 249.
Aplanatic two-mirror system, *aplanatisches Zweispiegelsystem* 265.
Aplanatism, *Aplanatismus* 28.
Apochromatic lens, *Apochromat* 283.
Apochromatic microscope objectives, *Apochromate als Mikroskop-Objektive* 358seq.
Apodization, *Apodisation* 241, 368, 370, 537.
Apodization test of a grating, *Apodisationstest eines Gitters* 649seq.
Arched-yoke mounting, *Knickrahmen-Montierung* 336.
Area determination in microscopy, *Flächenbestimmung in der Mikroskopie* 405.
Argon continuum in the ultraviolet, *Argonkontinuum im Ultravioletten* 646.
Artificial doublet, *künstliches Dublett* 612, 614.
Aspheric plate, *asphärische Platte* 39, 273.
Aspherical surfaces, *asphärische Flächen* 17, 72, 232, 272.
— —, test, *Prüfung* 307.
Astigmatism, *Astigmatismus* 16, 22, 234, 251seqq., 260, 263, 266.
— in concave gratings, *bei Konkavgittern* 480—485.
Astrograph lenses, *Astrographen-Objektive* 291.
Autocollimation, *Autokollimation* 435, 452, 458, 807.
Autocollimation telescope, *Autokollimationsfernrohr* 757, 815seq., 822.
Autocorrelation function, *Autokorrelationsfunktion* 563.

Automatic guider, *automatische Nachführung* 339.
Axial resolution, *axiale Auflösung* 348.
Axicon 814.

Baird-Atomic infrared grating engines, *Baird-Atomic-Teilmaschinen für Infrarotgitter* 711 seq.
Baker-Schmidt System 279.
Barometric changes, interferometric control, *Barometerschwankungen, Interferometerkontrolle* 700.
Bauernfeind prisms, *Bauernfeindsche Prismen* 220, 221.
Bausch and Lomb ruling engines, *Bausch & Lomb-Teilmaschinen* 706—712.
Beam partition, *Strahlenteilung* 380.
Bearings for telescope mountings, *Lager für Fernrohrmontierungen* 323.
Bending control, *Durchbiegen, Kontrolle* 817.
Birefringent interference microscopes, *doppelbrechende Interferenzmikroskope* 653.
Blank imperfections, *Fehler des Gitterblocks* 534 seq., 541.
Blazing 429, 580 seqq., 599, 653, 669, 714.
—, triangular groove profile, *dreieckiges Furchenprofil* 597, 599.
Boro-silicate glasses, *Borosilikat-Gläser* 296.
Boundary value problems of Maxwell theory, *Randwertprobleme der Maxwellschen Theorie* 583 seqq.
Bright field observation, *Hellfeldbeobachtung* 376.
Brightness (luminance), definition, *Helligkeit, Definition* 492.

Cadmium lines, *Kadmiumlinien* 770.
Cassegrain system, *Cassegrain-System* 39, 260.
Cauchy's product sum, *Cauchysche Produktensumme* 152, 157, 160, 161.
Centering of lenses, test methods, *Zentrieren von Linsen, Prüfmethoden* 824—827.
— of light source, *der Lichtquelle* 762.
Central obstruction (see also apodization), *Mittenabschattung s. auch Apodisation* 241, 368, 370.
CER-VIT 297.
Characteristic function of Hamilton, *charakteristische Funktion von Hamilton* 25.
Chromatic aberrations, *chromatische Fehler* 22 seq., 38, 180, 253 seqq.
Chromatic correction of microscope objectives, *Farbkorrektion von Mikroskop-Objektiven* 359.
Chromatic difference of magnification, *chromatische Vergrößerungsdifferenz* 287.
Circular wavefront, *kreisförmige Wellenfront* 558.
Circularly curved slits, *kreis bogenförmige Spalte* 470—472.
Coating of mirrors, *Vergütung von Spiegeln* 295.
Coherence length, difficulties in length measurements, *Kohärenzlänge, Schwierigkeiten bei Längenmessung* 778 seqq.
Coherent illuminating of grating, *kohärente Beleuchtung des Gitters* 489.
Coherent radiation, *kohärente Strahlung* 345
Coincidence, *Koinzidenz* 764, 772.
Coincidence focometer, *Koinzidenzfokometer* 823.
Collective lens, *Sammellinse* 361.
Collimator attached to grating, *Kollimator am Gitter* 454 seq.
Colour coordinate, *Farbvariable* 74 seq., 161.
Coma, *Koma* 16, 22, 234, 248 seqq., 259, 263, 276.
Combination of refracting surfaces, *Kombination brechender Flächen* 9.
Comparator principle of Abbe, *Komparatorprinzip von* Abbe 761, 791.
— — —, circumvention, *Umgehung* 806.
Compensation of tilt errors, *Kompensation von Kippfehlern* 793, 808.
Compensators, *Kompensatoren* 400.
Complex filter, *komplexes Filter* 747.
Complex perpendicular, *komplexes Lot* 91.
Complex ray data, *komplexe Strahldaten* 86 seqq.
Computer programmes for diffraction patterns, *Programmierung für Beugungsbilder* 564 seq.
— — for optical design, *beim Entwurf optischer Instrumente* 40 seq., 66.
Concave gratings, *Konkavgitter* 434.
— —, aberrations, *Fehler* 478.
— —, astigmatism, *Astigmatismus* 480—485.
— —, basic equations, *Grundgleichungen* 475 seqq.
— —, focussing properties, *Fokussierungseigenschaften* 473 seq.
— —, parameters, *Parameter* 475.
— —, resolution, *Auflösung* 479 seq.
— —, survey of different mountings, *Übersicht über verschiedene Aufstellungen* 484—486.
Condensor, *Kondensor* 383 seq.
Conjugate distances, *konjugierte Abstände* 8, 46.
Conoscopic illumination, *konoskopische Beleuchtung* 401 seq.
Conservation of flux and brightness, *Erhaltung von Lichtflußstärke und Helligkeit* 494.
Constructive interference, *konstruktive Interferenz* 450.
Convolution of ghosts, *Faltung von Geistern* 556.
— of two Fourier transforms, *zweier Fourier-Transformationen* 522.
Cook lens, *Triplett* 196.
Correction of eyepieces, *Korrektion von Okularen* 361.
— of objectives, *von Objektiven* 357 seq.

Corrector plate, *Korrektionsplatte* 238, 267, 270, 272.
Cosine variables, *Cosinus-Größen* 86seqq., 90, 110.
Coudé focus, *Coudé-Fokus* 264.
Coudé systems, *Coudé-Systeme* 324seqq.
Critical illumination, *kritische Beleuchtung* 378.
Crossed rulings, *gekreuzte Teilungen* 641—643.
CSIRO Australian ruling engine, *CSIRO Australische Teilmaschine* 721.
Cumulative ruling errors, *kumulative Teilfehler* 541, 671.
CUNY's optical rule, *optisches Lineal von Cuny* 812.
Curvature of Petzval, see Petzval curvature.
Curvature radii of lenses, determination, *Krümmungsradien von Linsen, Bestimmung* 820—822.
Curvature of spectral lines, *Krümmung der Spektrallinien* 451, 467—472.
Curved slit, *gekrümmter Spalt* 469seq., 628.
Cylindrical optical systems, *zylindrische optische Systeme* 31.

Dall-Kirkham System 264.
Dark field observation, *Dunkelfeldbeobachtung* 376, 383.
Data of construction, *Konstruktionsdaten* 70seqq., 102, 114seqq., 158.
Decentering, Dezentrierung 263.
— of lenses, measurement, *von Linsen, Messung* 824—827.
Defects in prism manufacture, *Fehler bei Prismenherstellung* 225.
Definition of meter, *Definition des Meters* 766.
Defocussing, *Defokussierung* 245.
Deformation of optical surfaces, *Deformation optischer Flächen* 232.
Degree, *Grad* 102seqq., 123, 138, 179, 188.
— of coherence, *Kohärenzgrad* 345.
design of optical instruments, *Entwurf optischer Instrumente* 34seqq.
Diamond carriage control, *Diamantführung, Kontrolle* 702seq.
Diamond-ruled traces on a measure rod, *Diamant-geritzte Striche auf einem Maßstab* 785.
Diamond types used for ruling, *Diamanttypen für das Ritzen von Gittern* 667seqq.
Dielectric films in x-ray gratings, *dielektrische Schichten bei Röntgenstrahlgittern* 435seq.
Difference operators, *Differenz-Operatoren* 171, 172, 177, 184, 187.
Differential approximation, *differentielle Näherung* 177, 188, 190.
Differential operators, *Differential-Operatoren* 171, 177, 179, 181, 187.
Diffraction gratings, dispersion, *Beugungsgitter, Dispersion* 458seq.
Diffraction gratings, history, *Beugungsgitter, Geschichte* 426seqq.
— — for the infrared, *für das Infrarot* 436.
— — for metrological purposes, *für metrologische Zwecke* 439seq.
— — for microwaves, *für Mikrowellen* 436seq.
— —, parameters, *Parameter* 446, 449.
— —, types, *Typen* 434seqq.
— — for the ultraviolett, *für das Ultraviolett* 435.
— — for x-rays, *für Röntgenstrahlen* 429, 435.
Diffraction pattern, *Beugungsfigur* 512.
— — of circular pupil, *Beugungsbild der Kreispupille* 238seq.
— — in microscope, *im Mikroskop* 343.
— — from a rectangular aperture, *einer rechteckigen Öffnung* 452.
— — for ring aperture, *bei ringförmiger Apertur* 369.
— — in spectrographs, *in Spektrographen* 487.
Diffraction and scattering, *Beugung und Streuung* 433.
Diffraction unit, *Beugungseinheit* 565.
Dirac comb, *Kamm aus Dirac-Funktionen* 521.
Dispersion, *Dispersion* 73seqq.
Dispersion coefficients, *Dispersions-Koeffizienten* 77seqq.
Dispersion, dependence on grating angle, *Dispersion, Abhängigkeit vom Gitterwinkel* 615.
— of diffraction gratings, *eines Beugungsgitters* 458seq.
Dispersion increase by use of two gratings, *Dispersion, Erhöhung durch Benutzung von zwei Gittern* 616seq.
Dispersion number of Abbe, *Abbesche Dispersionszahl* 254, 292.
Dispersive properties of glasses, *Dispersionseigenschaften der Gläser* 24.
Distortion, *Verzeichnung* 16, 22, 61, 234, 793.
Distortion correction, *Entzerrung* 205.
Distribution of light among orders (see also blazing), *Verteilung des Lichts auf die Ordnungen* 570.
Doppler width, *Dopplerbreite* 611.
Double-anastigmatic objective, *Doppelanastigmat* 194, 196.
Double Fizeau interferometer, *doppeltes Fizeau-Interferometer* 777.
Dove prism, *Dove-Prisma* 215.
DOWELL's comparator with single mirror, *Dowellscher Komparator mit Einfachspiegel* 776, 781.
— — with two mirrors set at an angle, *mit Winkelspiegeln* 773seq.
Driving mechanisms for telescopes, *Antriebssysteme für Fernrohre* 337seqq.
Duran 297.

Eccentricity, *Exzentrizität* 13.
Echelle, echelette, echelon: definitions, *Stufengitter verschiedener Typen, Definition* 465.
Echelette in polarized light, *Stufengitter im polarisierten Licht* 575seq.
Echelon and echelle in reflection, *Stufengitter in Reflexion* 466seq.
— in transmission, *in Transmission* 466.
Efficiency of gratings see also blazing, *Ausbeute von Gittern s. auch Blazing*
— of gratings in the infrared, *von Gittern im Infrarot* 606—609.
Efficiency increase of gratings by blazing, *Ausbeuteerhöhung von Gittern durch Blazing* 580—583.
Efficiency test of a grating, *Ausbeuteprüfung eines Gitters* 644.
Efficiency of x-ray gratings, *Ausbeute von Röntgenstrahlgittern* 647.
Eikonal 139seqq., 143seqq.
Eikonal coefficients, *Eikonal-Koeffizienten* 141seqq., 149seqq., 163seqq.
Eikonal equation, *Eikonalgleichung* 4, 25.
Elementary wavelet, *Elementarwelle* 518.
Electromagnetic boundary value problems in grating theory, *elektromagnetische Randwertprobleme der Gittertheorie* 512, 517.
Electromagnetic groove shape, *elektromagnetische Furchenform* 524.
Electron beam at grazing incidence over grating, *Elektronenstrahl bei streifendem Einfall an einem Gitter* 440.
Electron micrographs of gratings, *Elektronenmikrographien von Gittern* 654seqq.
Electrostatic images of an electron beam, *elektrostatische Bilder eines Elektronenstrahls* 441.
Energy distribution among diffraction orders, *Energieverteilung auf die Ordnungen der Beugung* 517, 573, 575.
English (cross-axis) mounting, *englische Achsenmontierung* 332.
Entrance pupil, *Eintrittspupille* 11, 48, 49, 95seq., 231.
Entrance slit of spectrograph, finite width effect, *Eintrittsspalt eines Spektrographen, Einfluß der endlichen Breite* 488seqq.
Eppenstein optical circumvention, *Eppensteinsche optische Umgehung* 806.
Equatorial mounting, *parallaktische Montierung* 322.
Equidensites, *Äquidensiten* 831seq.
Error of period, *Periodenfehler* 552.
Errors in measurement by imperfect focussing, *Meßfehler durch schlechte Fokussierung* 765.
Error table, *Fehlertabelle* 786.
Étendue 493.
Evanescent wave, *Oberflächenwelle* 446, 570, 588, 594.
Exit pupil, *Austrittspupille* 11, 49, 61, 117, 120, 231, 285.
Exit slit of spectrograph, finite width effect, *Austrittsspalt eines Spektrographen, Einfluß der endlichen Breite* 490.
Eyepiece field of vision, *okulares Sehfeld* 363.
Eyepiece of measuring microscope, *Okular für Meßmikroskop* 799.
Eyepiece micrometer, *Okularmikrometer* 764.
Eyepiece for microscope, *Okular für Mikroskop* 361.
— for spectacled persons, *Brillenträger-Okular* 367.
— of a telescope, *Fernrohrokular* 285.

Fabry-Pérot étalon 464, 532.
FERMAT's principle, *Fermatsches Prinzip* 2seq.
Fiber optics, *Faseroptik* 383.
Field curvature, *Bildfeldwölbung* 16, 22, 234, 251seqq.
Field flattener, *Ebnungslinse* 275.
Field lens, *Feldlinse* 361.
Field of vision, *Sehfeld* 363.
Filtered diffraction pattern, *gefiltertes Beugungsbild* 639.
Finite slit width, *endliche Spaltbreite* 528.
Fizeau grating test, *Fizeauscher Gittertest* 637.
Flexure compensation of telescope tube, *Biegungskompensation am Fernrohrtubus* 318.
Fluorescence microscope, *Fluoreszenzmikroskop* 397.
Fluorite objectives, *Fluoritobjektive* 358.
Flux density, *Leuchtdichte* 386.
Flux of light, definition, *Lichtfluß, Definition* 492, 494.
— — in gratings, *in Gittern* 462.
— — through photographic objective, *durch ein Photoobjektiv* 207.
f-number, *Öffnungszahl* 232, 234.
Focal length, *Brennweite* 13, 230.
— —, measurement, *Messung* 822—824.
Focogram of a mirror, *Fokogramm eines Spiegels* 306.
Focussing properties of concave gratings, *Fokussierungseigenschaften von Konkavgittern* 473seq.
Fork mounting, *Gabelmontierung* 327.
Foucault test, *Foucaultsche Prüfmethode* 304.
Four-element lens, *Vierlinser* 290.
Fourier integral, *Fourierintegral* 446.
Fourier transform of grating diffraction patterns, *Fourier-Transformation der Beugungsfiguren eines Gitters* 518.
Fourier transform hologram, *Fouriertransformations-Hologram* 743.
Fourier transform of HUYGENS' principle, *Fouriertransformation des Huygensschen Prinzips* 506, 510.
Free spectral range, *freies Spektralgebiet* 463.
Fresnel-zone gratings, *Fresnelzonengitter* 437seq.

GABOR's method of wavefront reconstruction, *Gaborsche Methode der Wellenfront-Rekonstruktion* 741 seqq.
Gauss type objective, *Gaußtyp-Objektiv* 197, 199.
Gaussian approximation, *Gaußsche Näherung* 6, 7—14, 92 seqq., 99 seq., 137, 228.
Gaussian coefficients, *Gaußsche Koeffizienten* 97 seq., 102, 124.
Gaussian colour aberrations, *Gaußsche Farbfehler* 23.
Gaussian image formation, *Gaußsche Abbildung* 95, 98, 137.
Gaussian image plane, *Gaußsche Bildebene* 98.
Geometrical optics, limits of validity, *geometrische Optik, Gültigkeitsgrenzen* 3.
— wavefront, *Wellenfront* 3.
German mounting, *deutsche Montierung* 325.
Ghosts, *Geister* 453, 535, 540, 547, 549 seq., 555.
Ghosts of different order, *Geister verschiedener Ordnung* 550 seqq.
Ghost intensities, *Geister, Intensitäten* 552.
Ghost wavelengths, *Wellenlängen von Geistern* 555, 557.
Glass, *Glas* 24.
Glass measure rod, *Glasmaßstab* 785.
Glass, radiation-proof, *strahlungsresistente Gläser* 375.
—, survey of optical glasses, *Übersicht optischer Gläser* 292.
Goniometric eyepiece, *Goniometer-Okular* 800.
GOUY's theorem, *Gouyscher Satz* 511, 558.
Granularity of plate, *Körnigkeit der Platte* 531.
Grating see diffraction grating and concave grating, *Gitter s. Beugungsgitter und Konkavgitter.*
— of cylinders, boundary value problem, *aus Zylindern, Randwertproblem* 572, 584.
— of δ-function grooves, *mit δ-funktionsartigen Furchen* 520 seq.
Grating equation, *Gittergleichung* 445, 448.
Grating in immersion, *Gitter in einer Immersion* 457.
Grating imperfections, *Gitterfehler* 610 seqq.
Grating mosaic, *Gittermosaik* 432.
Grating replication methods, *Gitterkopiermethoden* 731, 734 seq.
Grating testing interferometers, *Gitterprüfungsinterferometer* 635 seq.
Grazing incidence, optical diffraction gratings, *streifende Inzidenz bei optischen Beugungsgittern* 458.
— —, x-ray gratings, *bei Röntgenstrahlgittern* 435, 603 seqq.
GREEN's function treatment of grating diffraction, *Greensche Funktion zur Behandlung der Gitterbeugung* 571.
"Grille" multislit spectrometer of GIRARD, *„Grill"-Vielspalt-Spektrometer von Girard* 499.
Groove angle suitable for blazing, *Furchenwinkel, geeignet für Blazing* 581.
Groove profile, *Furchenprofil* 428, 436, 451.
— —, determination by electron micrographs, *Bestimmung aus Elektronenmikrographien* 661—663.
— —, piecewise linear, boundary value problem, *stückweise linear, Randwertproblem* 594—597.
— — of rectangular shape, *rechteckiger Gestalt* 575, 581.
— — sinusoidal, boundary value problem, *sinoidal, Randwertproblem* 586—594.
— — triangular, boundary value problem, *dreieckig, Randwertproblem* 597—599.
Groove shape, effect upon efficiency, *Furchengestalt, Einfluß auf Ausbeute* 645.
Groove shape function, *Furchengestaltfunktion* 517, 523.
Groove shaping, prism effect, *Furchenform, Prismenwirkung* 458.
Grooves infinitely narrow slits, *Furchen unendlich schmale Spalte* 520 seq.

Hariharan-Sen interferometer, *Hariharan-Sensches Interferometer* 779 seq.
Hartmann constant, *technische Konstante* 302.
Hartmann test, *Hartmannsche Prüfmethode* 301.
—, generalized, *Verallgemeinerung* 303.
Heights, *Höhen* 90.
Helmholtz-Huygens equation, *Helmholtz-Huygenssche Gleichung* 507.
Helmholtz wave equation, *Helmholtzsche Wellengleichung* 446, 450.
Higher-order aberrations, *Bildfehler (Aberrationen) höherer Ordnung* 70, 103, 138, 139, 176 seq., 187 seq., 190 seq.
Hitachi ruling engine with interferometric control, *Hitachi-Teilmaschine mit Interferometerkontrolle* 724 seq.
Holograms, *Hologramme* 739—741, 743, 745.
Holographic filter, *holographisches Filter* 748.
Holography, *Holographie* 738 seqq.
Homogeneity of optical glasses, *Homogenität optischer Gläser* 293.
Horseshoe mounting, *Hufeisenmontierung* 334.
Huet alignment telescope, *Huetsches Fluchtfernrohr* 811.
Huet prism, *Huet-Prisma* 215 seq., 218.
HUNZINGER's interferometer, *Hunzingersches Interferometer* 779 seq.
Huygens eyepiece, *Huygens-Okular* 288, 367.
HUYGENS' principle, approximate form for electromagnetic waves, *Huygenssches Prinzip, genäherte Formulierung für elektrische Wellen* 508—510.
— —, difficulties in energy calculations, *Schwierigkeiten der Energieberechnung* 513.
— — and grating diffraction, *und Gitterbeugung* 445, 504 seq.
— —, limitations of its applicability, *Grenzen seiner Anwendbarkeit* 513 seq.
— —, rigorous form for electromagnetic waves, *strenge Formulierung für elektromagnetische Wellen* 506—508.

Hyperfine structure as a resolution test, *Hyperfeinstruktur als Auflösungskontrolle* 621.
Hypergon 198.

Illumination (flux density) *Beleuchtungsstärke (Flußdichte)* 35, 494.
Illumination of microscope objects, *Beleuchtung von Mikroskop-Objekten* 376 seqq., 398 seqq.
— of spectrometers, *Beleuchtung von Spektrometern* 491.
Image correlation filtering, *Bildkorrelationsfilterung* 746 seqq.
Image formation, Fourier transform formulation, *Bildentstehung, Formulierung durch Fourier-Transformierte* 510.
— — by gratings, *durch Gitter* 428, 437, 452 seq.
— — on Rowland circle, *auf dem Rowlandschen Kreis* 477.
Imaging gratings, *abbildende Gitter* 437 seq.
Image height aberration of skew rays, *Bildhöhen-Aberration windschiefer Strahlen* 119, 123, 139, 143, 146 seq., 148, 151.
Image height, reduced, *reduzierte Bildhöhe* 119, 124, 151.
Image perpendicular aberration, *Bildlot-Aberration* 92, 99 seqq., 123, 136, 166 seqq.
Image perpendicular, reduced, *reduziertes Bildlot* 92, 100, 112, 166 seqq.
Immersion mirror objectives, *Immersionsspiegelobjektive* 371.
Immersion object lens, *Immersionsobjektiv* 346, 358.
Imperfections of gratings, *Fehler von Gittern* 610.
— of wavefront, *Unvollkommenheiten der Wellenfront* 452, 454.
— —, effect on spectral image, *Einfluß auf das Spektralbild* 538 seqq.
Imperfection variation law for diffraction patterns, *Fehlervariationsgesetz für Beugungsbilder* 543, 545, 547.
Incident-light illumination, *Auflichtbeleuchtung* 380.
Incoherence of spectral lines, *Inkohärenz von Spektrallinien* 527.
Incoherent illumination of grating, *nichtkohärente Beleuchtung des Gitters* 459, 489.
— radiation, *inkohärente Strahlung* 344.
Infrared grating, *Ultrarotgitter* 436, 711.
Infrared grating efficiency, *Infrarot, Gitterausbeute* 606.
Infrared microscopy, *Infrarot-Mikroskopie* 395.
Instrument function, theoretical, *Instrumentfunktion, theoretische* 528—530.
Intensity of light, definition, *Lichtintensität, Definition* 492.
Interference comparator of KÖSTERS, *Interferenzkomparator von Kösters* 759, 768—773.
Interference contrast, *Interferenzkontrast* 407.
Interference fringes, position measurement, *Interferenzstreifen, Lage-Messung* 766.
Interference gratings, *Interferenzgitter* 735—754.
Interference measurement of parallelity, *Interferenzmessung der Parallelität* 777.
Interference microscope testing of groove shape, *Interferenzmikroskop-Prüfung der Furchengestalt* 652 seq.
Interference microscopy, *Interferenzmikroskopie* 407 seqq.
Interference surface control, Saunders-Gross method, *Interferenzkontrolle von Oberflächen, Methode von Saunders und Gross* 778.
Interference test of surfaces, *Interferenzprüfung von Oberflächen* 827—832.
Interferential alignment, *interferentielle Fluchtung* 814.
Interferogram, *Interferogramm* 559, 566 seqq., 632 seqq., 693 seqq.
Interferometers, *Interferometer* 436 seq.
Interferometer of HARIHARAN and SEN, *Interferometer von Hariharan und Sen* 779 seq.
— of HUNZINGER, *von Hunzinger* 779 seq.
— with mercury surfaces, *mit Quecksilberoberflächen* 777.
— of MICHELSON, *von Michelson* 408, 631, 635, 783.
Interferometric control of diamond carriage, *Interferometerkontrolle des Diamant-Schlittens* 728—730.
— control of ruling engines, *Interferometerkontrolle von Teilmaschinen* 429 seq., 674, 678 seqq.
— modulation spectrometers, *interferometrische Modulationsspektrometer* 502.
— tests of gratings, *interferometrische Gitterprüfung* 621 seqq.
— — of telescopes, *interferometrische Fernrohrprüfung* 309 seqq.
Intermediate image in microscope, *Zwischenbild im Mikroskop* 354.
Iodine absorption test of grating resolution, *Jodabsorption zur Prüfung der Gitterauflösung* 613.
Isoplanatism, *Isoplanatismus* 28 seq., 360.
Iteration formulae, *Iterations-Formelsatz* 108 seq., 113 seq., 121 seqq., 135 seqq.

Jarrell-Ash ruling engine with interferometric control, *Jarrell-Ash-Teilmaschine mit Interferometerkontrolle* 703—706.
Jig-transit square 824.
Jobin et Yvon ruling engine with interferometric control, *Jobin & Yvon-Teilmaschine mit Interferometerkontrolle* 719 seq.
Johns Hopkins ruling engine, *Johns Hopkins-Teilmaschine* 680—682.

Kellner eyepiece, *Kellner-Okular* 289.
Knife-edge test, *Schneidenprüfung* 304seq.
Köhler illumination, *Köhlersche Beleuchtung* 379.
KÖSTERS's interference comparator, *Köstersscher Interferenzkomparator* 759, 768—773.
— prism, *Kösters-Prisma* 812seq., 823.
Krypton continuum in the ultraviolet, *Kryptonkontinuum im Ultravioletten* 646.
Krypton lamp, *Kryptonlampe* 767, 769.

Lagrangian invariant, *Lagrangesche Invariante* 12, 15, 46, 49.
Lange data of construction, *Langesche Konstruktionsdaten* 114seqq., 177, 187, 191.
Lateral resolution, *laterale Auflösung* 343.
Leman-Sprenger prism, *Leman-Sprenger-Prisma* 216, 218.
Length measurement in microscopy, *Längenmessung in der Mikroskopie* 403seq.
— — with Pérot-Fabry plates, *mit Pérot-Fabry-Platten* 782.
— — by wavelength, *durch Wellenlänge* 755, 766.
Lens curvature, determination, *Linsenkrümmung, Bestimmung* 820—822.
Levelling telescope, *Nivellierfernrohr* 808.
Light beam pointer, *Lichtzeiger* 757, 760.
Light duct, *Lichtleiter* 383.
Light sources in microscopy, *Lichtquellen für Mikroskopie* 385.
Linear dispersion of a grating, *Lineardispersion eines Gitters* 459, 461seq.
Linear phase aberration in gratings, *linearer Phasenfehler bei Gittern* 541.
— wavefront aberrations, *Fehler der Wellenfront* 535seq., 541.
Long-distance objective, *Teleobjektiv* 196, 199.
Longitudinal sharpness of objective, *Tiefenschärfe eines Objektives* 204.
Loupe, *Lupe* 351, 787.
— magnification, *Lupenvergrößerung* 351, 373.
Luminosity of gratings, *Lichtstärke von Gittern* 453, 461, 486—504.
— of spectrographs and spectrometers, *von Spektrographen und Spektrometern* 495seq.
Lyman ghosts, *Lymansche Geister* 555.

Mach-Zehnder interferometer, *Mach-Zehndersches Interferometer* 409, 747seq.
Magnification law, *Vergrößerungsgesetz* 8.
Magnification of a telescope, *Vergrößerung eines Fernrohrs* 285.
Magnifying lens, *Lupe* 350seq., 787.
Maréchal integrator, *Maréchalscher Integrator* 564.
Masks put on slits, *Masken für Spalte* 500seq.
Matrix description of Gaussian optics, *Matrizenbeschreibung der Gaußschen Optik* 14.
Maximum *f* number, *maximale f-Zahl* 496.
Mean ray data, *mittlere Strahldaten* 109seqq., 112, 125, 166seqq.
Measurement under microscope, *Messung unter dem Mikroskop* 403.
Measuring force, *Meßkraft* 761.
Measuring loupe, *Meßlupe* 787.
Measuring microscope, *Meßmikroskop* 792, 799.
Mechanical design of ruling engines, *mechanischer Aufbau von Teilmaschinen* 669.
Medial telescope, *Medial-Fernrohr* 284.
Meniscus systems, *Meniskussysteme* 279seq.
Mercury lamp, *Quecksilberlampe* 389.
Mercury lines test of grating resolution, *Quecksilberlinien zur Prüfung der Gitterauflösung* 611, 617seq., 621, 694.
Meridional image height aberration, *meridionale Bildhöhenaberration* 129, 135seqq., 138seq., 146seq., 174seq., 176, 187, 190seq.
Meridional rays, *Meridianstrahlen* 89seq., 169.
MERTON's NPL ruling engine for metrology gratings, *Mertons Teilmaschine für metrologische Gitter im NPL* 725—727.
Metal measure rod, *Metallstab* 785.
Metal mirrors, *Metallspiegel* 298.
Metal surface, interference test, *Metallfläche, Interferenzprüfung* 830.
Meter comparator, *Meterkomparator* 767.
Meter, definition, *Meter, Definition* 766.
Mètre prototype, *Urmeter* 755, 766.
Michelson interferometer, *Michelson-Interferometer* 408, 631, 635, 783.
Michelson-Twyman-Green interferometer, *Interferometer von Michelson, Twyman und Green* 533seq., 652.
Microbeam, *Mikrostrahlenstich* 396.
Micrometric eyepiece, *Mikrometer-Okular* 367.
Microphotography, *Mikrophotographie* 390.
Microphotometry, *Mikrophotometrie* 411seqq.
Microprojection, *Mikroprojektion* 390.
Microscope for hot cells, *Mikroskop für heiße Zellen* 374.
Microscope stages, *Mikroskoptische* 407.
Microscopic interference test of surfaces, *mikroskopische Interferenzprüfung von Oberflächen* 827—832.
Microscopy and length determination, *Mikroskopie und Längenbestimmung* 788.
Microspectrophotometry, *Mikrospektrophotometrie* 411seqq.
Microwave experiments with standing grating waves, *Mikrowellenversuche mit stehenden Gitterwellen* 579.
Microwave gratings, *Mikrowellengitter* 436seq.
MILLS' cross, *Mills-Interferometer* 437.
Minimum usable slit width, *kleinste nutzbare Spaltbreite* 496.

Mirror-like grating, experimental verification, *spiegelartiges Gitter, experimentelle Bestätigung* 579.
— —, theory, *Theorie* 572seq., 575—578.
Mirror materials, *Spiegelmaterial* 295.
Mirror objective, *Spiegelobjektiv* 200.
Mirror objectives for microscopes, *Spiegelobjektive in Mikroskopen* 367.
Mirror support in telescopes, *Spiegellagerung* in Fernrohren 314.
MIT ruling engine with interferometric control, *MIT-Teilmaschine mit Interferometerkontrolle* 687—703.
Modulation interference gratings, *Modulations-Interferenzgitter* 735—754.
Moiré-fringe testing, *Moiréstreifen-Prüfung* 641—643.
Moiré gratings, *Moiré-Gitter* 439seq.
Moiré principle, *Moiré-Prinzip* 764.
Monocentric eyepiece, *monozentrisches Okular* 289.
Monochromatic microscope objectives, *Monochromate als Mikroskop-Objektive* 358.
Mounting of telescopes, *Montierung von Teleskopen* 322seqq.
Mount Wilson Observatory ruling engines, *Mount Wilson-Sternwarte, Teilmaschinen* 682—687, 716—719.
Multiple periodic errors, *mehrfach-periodische Fehler* 555seq.
Multiple slits, *Vielfachspalte* 432.
Multiple slit spectrometers, *Vielfachspalt-Spektrometer* 498—503.

Nasmyth focus, *Nasmythsche Anordnung* 261.
Near-axis rays, *achsennahe Strahlen* 85.
Neutral fibre of measure bar, *neutrale Faser eines Maßstabes* 786.
NEWTON's conjugate equation, *Schnittweitengleichung* 8, 202.
Newton focus, *Newton-Fokus* 259.
Nobel Institute ruling engine, *Nobel Institut, Teilmaschine* 727seq.
Non self-luminous object, *nicht selbstleuchtendes Objekt* 349.
Non-spherical surfaces, *nicht-sphärische Flächen* 17.
Normal construction, *normaler Aufbau* 80seq.
NPL ruling engine, *NPL-Teilmaschine* 725—727.
Nuclear events, stereo-photography, *Kernereignis, Stereoaufnahme* 415.
Null test, *Nullmethode* 306.

Object height, reduced, *Objekthöhe, reduzierte* 96, 119.
Object perpendicular, reduced, *Objektlot, reduziertes* 96.
Object plane, *Objektebene* 96.
Objectives for microscopes, *Objektive von Mikroskopen* 357seq.
— for photography, *für Photographie* 192seqq.

Oil of immersion, *Immersionsöl* 346.
Optical cosine law, *optisches Cosinusgesetz* 28.
— direction, *Richtung* 79, 153.
— glasses see glass, *Gläser s. Glas.*
— grating, *Gitter* 433seq.
— materials in the ultraviolet, *Materialien im Ultravioletten* 24.
— palpating, *Tasten* 756.
— path, *Lichtweg* 52seqq., 140, 509.
— rule of CUNY, *Cuny's optisches Lineal* 812.
— transfer function, *Übertragungsfunktion* 257.
Order, *Ordnung* 85, 93, 102seqq., 123, 138, 188.
Order of a ghost, *Ordnung eines Geists* 551.
Orders of diffraction, energy distribution, *Ordnungen der Beugung, Energieverteilung* 517, 570.
Orthoscopic eyepiece, *orthoskopisches Okular* 289.
— illumination, *orthoskopische Beleuchtung* 398.
Overlap of orders in a grating, *Überlappung der Ordnungen in einem Gitter* 462.

Pancratic systems, *pankratische Systeme* 375.
Parabolic wavefront, *parabelförmige Wellenfront* 558.
Parallactic error, *Parallaxe* 762seq.
Parallel perspective, *Parallelperspektive* 760.
Parallelity control of axes, *Parallelitätsprüfung von Achsen* 819.
Paraxial rays, *Paraxialstrahlen* 92.
— raytracing, *paraxiale Strahlenkonstruktion* 9seq.
— region, *paraxiales Gebiet* 228.
Partial dispersion, *Partialdispersion* 292.
Pechan prism, *Pechan-Prisma* 217.
Pencil, *Büschel* 5, 50.
Pendulum mirror, *Pendelspiegel* 808, 825.
Pentagon prism, *Pentagonprisma* 214, 218, 818seq.
Perfect grating, *vollkommenes Gitter* 428.
— stigmatism in gratings, *vollkommener Stigmatismus bei Gittern* 472.
Perflectometer, *Perflektometer* 789seq.
Perforated plates as microwave gratings, *Lochplatten als Mikrowellengitter* 437.
Periodic errors, *periodische Fehler* 671.
Perkin-Elmer ruling engines, *Perkin-Elmer-Teilmaschinen* 716.
Pérot-Fabry étalon combined with Michelson interferometer, *Pérot-Fabry-Etalon mit Michelson-Interferometer verbunden* 781.
Perpendiculars, *Lote* 90.
Perspective of photographic image, *Perspektive des photographischen Bildes* 201seqq.
Petzval curvature, *Petzval-Krümmung* 16, 51, 195, 360, 365.
Petzval objective, *Petzval-Objektiv* 194.
Petzval sum, *Petzvalsche Summe* 237, 251.
Phase-contrast fluorescence microscope, *Phasenkontrast-Fluoreszenzmikroskop* 397.

Phase-contrast grating test, *Phasenkontrast-Gittertest* 637seqq.
Phase-contrast method, *Phasenkontrast-Verfahren* 393, 407, 744, 746.
Phase measurement in wavefront from grating, *Phasenmessung in der Wellenfront von einem Gitter* 525.
Phase modulation, *Phasenmodulation* 736.
Phase-reversion by half-zone plates, *Phasenumkehr durch λ/2-Plättchen* 438.
Phased arrays 437.
Photoelectric detection, *photoelektrischer Nachweis* 482, 489seq.
— eyepiece, *Okular* 801seqq.
— recording of spectral lines, *Registrierung von Spektrallinien* 621seq.
Photographic objectives, *photographische Objektive* 192seqq.
Photometer microscope, *Photometer-Mikroskop* 411seqq.
Photometric concepts and definitions, *photometrische Begriffe und Definitionen* 491seq.
Pin hole camera, *Lochkamera* 193.
Planar achromat, *Planachromat* 361.
Plane grating, *ebenes Gitter* 434.
Plane gratings, basic equations, *ebene Gitter, Grundgleichungen* 456.
Plane transmission grating, basic equations, *ebenes Transmissionsgitter, Grundgleichungen* 457.
Platinum coating of grating surface, *Platinschicht auf Gitteroberfläche* 602seq.
Point spread function, *Lichtgebirge* 256.
Polar axis (off-axis German) mounting, *Polachsenmontierung* 326.
Polarization effects in grating boundary problems, *Polarisationseffekte bei Gitter-Randwertproblemen* 571, 579.
Polarization by gratings, *Polarisation durch Gitter* 443seq., 505, 575, 589.
— by metal gratings, *an Metallgittern* 573.
Polarization microscope, *Polarisationsmikroskop* 398.
Polarization suppressed in blazed gratings, *Polarisation unterdrückt in Gittern mit Blazing* 572, 581seq.
Porro prisms of first and second kind, *Porro-Prismen erster und zweiter Art* 217, 222.
Power of an optical surface, *Brechkraft einer optischen Fläche* 9, 49.
Primary aberration terms (fourth-power terms), *primäre Fehler (vierte Potenzen)* 6, 14—18.
— approximation, *primäre Näherung* 103seq., 137seqq., 189, 190.
Prime focus, *Primärfokus* 259.
Prime focus correcting systems, *Primärfokus-Korrektoren* 268.
Principal ray, *Hauptstrahl* 12, 22, 46, 47, 231.
Prism angles, test measurement, *Prismenwinkel, Prüfmessung* 818.
Prism of Kösters, *Prisma von Kösters* 812seq., 823.
Prismatic grooves, *prismatische Furchen* 458.
Prisms, *Prismen* 211seqq.
Profile of grooves see grooves, *Profil der Furchen s. Furchen.*
Profile imaging, *Profilabbildung* 755.
Profile of a measure bar, *Profil eines Maßstabes* 786.
Projectives, *Projektive* 362.
Pupil function, *Pupillenfunktion* 257.
Pupil imaging of tube lens, *Pupillenabbildung der Tubuslinse* 375.
Pupil perpendicular aberration, *Pupillenlot-Aberration* 166seqq.
Pupil perpendicular, reduced, *reduziertes Pupillenlot* 166seqq.
Pupil ray, *Pupillenstrahl* 11.
Pyramid defect in prisms, *Pyramidalfehler von Prismen* 225seq.
Pyrex 297.

Quality of imaging in photography, *Bildgüte in der Photographie* 209.
Quartz, *Quarz* 297.
Quartz gauge, *Quarznormal* 775.
Quasi-cassegrainian systems, *Quasi-Cassegrain-Systeme* 264.
Quasi-triangular groove shape, *quasidreieckige Furchengestalt* 570.

Radiation condition, *Ausstrahlungsbedingung* 583.
Radiation-generating grating, *strahlungserzeugendes Gitter* 440—443.
Radio-telescopes, *Radioteleskope* 437.
Radius vector, *Radiusvektor* 79, 153.
Räntsch interferometer, *Räntsch-Interferometer* 784.
Ramsden disk, *Okularkreis* 285.
Ramsden eyepiece, *Ramsden-Okular* 288.
Ray data, *Strahldaten* 79seqq.
Ray of light, *Lichtstrahl* 2.
Rayleigh's boundary problem of grating diffraction, *Rayleighsches Randwertproblem der Gitterbeugung* 570, 583—591.
Rayleigh criterion, *Rayleigh-Kriterium* 32, 244, 344, 611seq., 764.
Rayleigh tolerance, *Rayleighsche Toleranz* 479.
Raytracing, *Strahlenkonstruktion* 9, 19.
Reading lens, *Leseglas* 353.
Reciprocal height, *reziproke Höhe* 112, 116.
— refractive index, *reziproker Brechungsindex* 112, 116.
Rectangular aperture, diffraction pattern, *rechteckige Öffnung, Beugungsfigur* 452.
Rectangular groove profile, *rechteckiges Profil der Gitterfurchen* 575, 581.
— — —, unsuitability, *Unzweckmäßigkeit* 581.
Reference sphere, *Referenzkugel* 5, 58.
Reference wavefront, *Referenzwellenfront* 557seqq.

Reflectance of grating surface for x-rays, *Reflexionsvermögen der Gitteroberfläche für Röntgenstrahlen* 605.
Reflecting systems, design, *reflektierende Systeme, Entwurf* 39.
Reflection coefficients of x-ray gratings, *Reflexionskoeffizienten von Röntgenstrahlgittern* 647 seq.
Reflection echelon, *Reflexions-Stufengitter* 466 seq.
Reflection grating, *Reflexionsgitter* 433, 435, 446.
Reflection grating parameters, *Reflexionsgitter, Parameter* 455, 456.
Reflection of light by lenses, *Reflexion von Licht an Linsen* 209.
Reflectivity of gratings, *Reflexionsvermögen von Gittern* 435 seq.
Reflector corrector, *Reflektor-Korrektor* 269.
Refracting surface, *brechende Fläche* 9.
Refraction formulae, *Brechungsformeln* 82, 84 seqq., 88, 130, 155.
Refraction invariant, *Brechungsinvariante* 12.
Refractive index, *Brechungsindex* 72 seqq.
Refractive index determination, *Brechungsindex, Bestimmung* 414.
refractory power of an optical surface, *Brechkraft einer optischen Fläche* 9, 118 seq.
Replicas of master gratings, *Kopien von Originalgittern* 599, 660, 731—735, 727.
Resolution, definition of Lord RAYLEIGH, *Auflösung, Definition von Lord Rayleigh* 459 seq.
Resolving power, *Auflösungsvermögen* 240.
Resolving power and degree of coherence, *Auflösungsvermögen und Kohärenzgrad* 345.
— —, dependence on grating angle, *Abhängigkeit vom Gitterwinkel* 615.
— — of gratings, *von Gittern* 453, 459 seqq., 462.
— — of gratings, conditions for high values, *von Gittern, Bedingungen für hohe Auflösung* 461.
— — of gratings, modern successes, *von Gittern, moderne Erfolge* 432.
— —, increase by use of two gratings, *Erhöhung durch Benutzung von zwei Gittern* 616 seq.
— — limited by energy available, *Begrenzung durch verfügbare Energie* 497.
Resolving power tests for gratings, *Auflösungsvermögen von Gittern, Prüfverfahren* 611 seqq.
Rigorous representation of image height aberration, *strenge Darstellung der Bildhöhenaberration* 166 seqq., 176, 187.
Rigorous solutions of boundary value problems, *strenge Lösungen von Randwertproblemen* 572, 583 seq.
Ritchey-Chrétien System 39, 265.
Rocket grating spectrometer, *Raketen-Gitterspektrometer* 630.
Roof prism, *Dachkantenprisma* 212.
Ross correcting lens, *Ross-Korrektor* 268.
Rotatable microscope stage, *Drehtisch* für das Mikroskop 407.
Rotating mirror, *Drehspiegel* 757, 760.
Rotational symmetry and aberration classification, *Rotationssymmetrie und Fehlerklassifikation* 6.
Rowland circle, *Rowlandscher Kreis* 474, 477.
Rowland ghosts see also ghosts, *Rowlandsche Geister s. auch Geister* 549 seqq.
Rowland ruling engine, *Rowlandsche Teilmaschine* 681.
ROWLAND's method of ruling, *Rowlandsche Methode der Gitterteilung* 429—431.
Ruling engines, *Teilmaschinen* 429—431, 669—730.
— —, shock control, *Stoßkontrolle* 671—677.
— —, temperature control, *Temperaturkontrolle* 678—681.
Ruling errors and wavefront aberrations, *Teilfehler und Wellenfrontfehler* 533 seq., 541.
Ruling imperfections, *Teilfehler* 610.
Ruling process of gratings, *Teilung von Gittern, Verfahren* 666.

Sabattier effect, *Sabattier-Effekt* 832.
Sagittal image height aberration, *sagittale Bildhöhenaberration* 129, 130 seqq., 135 seqq., 138 seq., 146 seq., 180 seqq., 187, 190 seq.
Sagittal variation, *sagittale Variation* 127 seq., 186.
Satellites, *Satelliten* 540, 545, 560.
Saunders-Gross interference method of surface control, *Saunders-Gross-Methode zur Interferenzkontrolle von Oberflächen* 778.
Scalar boundary problem of grating diffraction, *skalares Randwertproblem der Gitterbeugung* 571, 600.
Scalar problems in grating diffraction, *skalare Probleme der Gitterbeugung* 444, 446, 504.
Scattered light from gratings, *Streulicht von Gittern* 561—564, 650.
Scattering and diffraction, *Streuung und Beugung* 433.
Scatter-plate interferometer, *Streuplatten-Interferometer* 313.
Schmidt-Pechan prism, *Schmidt-Pechan-Prisma* 215 seq.
Schmidt prism, *Schmidt-Prisma* 218, 220 seq.
Schmidt telescope, *Schmidt-Teleskop* 18, 24, 39, 272 seqq.
Schwarzschild-System 265.
Screws of ruling engines, *Schrauben der Teilmaschinen* 670 seq.
Secondary approximation, *sekundäre Näherung* 189, 191.
Secondary gauge, *Sekundärnormal* 755.
Secondary spectrum, *sekundäres Spektrum* 22, 255.
Seidel approximation, *Seidelsche Näherung* 7, 44, 51, 103, 139, 190, 232.

Seidel coefficients, *Seidelsche Koeffizienten* 236seq.
Seidel difference formulae, *Seidelsche Differenzformeln* 13.
Seidel separation, *Seidelscher Abstand* 107, 116.
Seidel sums, *Seidelsche Summen* 15seq., 51, 233seq.
— —, extension to aspherical surfaces, *Erweiterung für asphärische Flächen* 17seq.
Seismographic tests of ruling engine vibrations, *seismographische Schwingungsprüfung von Teilmaschinen* 676.
Self-luminous object, *selbstleuchtendes Objekt* 349.
Shearing interferometer, *Scherungsinterferometer* 311.
Shearing polarization interferometer, *Shearing-Polarisationsinterferometer* 410.
Shock isolation of ruling engines, *Stoßsicherung für Teilmaschinen* 671seqq.
SIEGBAHN'S Nobel Institute ruling engine, *Siegbahn's Teilmaschine im Nobel-Institut* 727seq.
Sign convention, *Vorzeichenkonvention* 229.
Similarity law for diffraction patterns, *Ähnlichkeitsgesetz für Beugungsbilder* 543, 545.
Sine condition, *Sinusbedingung* 248.
Sine variables, *Sinus-Größen* 86seqq., 90, 102, 110, 124, 133.
Sinusoidal dielectric grating surface, boundary value problem, *sinoidale dielektrische Gitteroberfläche, Randwertproblem* 592—594.
Sinusoidal metallic grating surface, boundary value problem, *sinoidale metallische Gitteroberfläche, Randwertproblem* 586—591.
SISAM spectrometer of CONNES, *SISAM-Spektrometer von Connes* 499, 502seq.
Slit with curvature, *Spalt mit Krümmung* 469seq.
Slit function, *Spaltfunktion* 499seqq.
Slit scanning function, 499seqq.
Slit of variable curvature, *Spalt variabler Krümmung* 628.
Slit width, optimum, *Spaltbreite, Optimum* 490.
Slits of grating infinitely narrow, *Spalte im Gitter unendlich schmal* 520seq.
Smith-Purcell experiment, *Smith-Purcellscher Versuch* 434, 440—443.
Smyth lens, *Smyth-Linse* 196.
SNELL'S law, *Snelliussches Brechungsgesetz* 2seq.
Soft x-rays, grating efficiency, *weiche Röntgenstrahlen, Gitterausbeute* 603.
Solid radiators *Festkörperstrahler* 387.
Sonnar 197.
Source conditions, effect on resolving power tests, *Lichtquelle, Einfluß auf Prüfung der Auflösung* 614.
Spatial frequency variables, *Raumfrequenzvariable* 519.
Spectral dispersion, *Spektraldispersion* 487.
Spectral image formation, *spektrale Bildentstehung* 527, 531seq.
Spectral lines, curvature, *Spektrallinien, Krümmung* 451, 467—472.
Spectrograph with reflection grating, *Spektrograph mit Reflexionsgitter* 455.
— with transmission grating, *mit Transmissionsgitter* 455.
Spectroheliograph, *Spektroheliograph* 469.
— of high resolution, *hoher Auflösung* 626.
Spectroheliograms, *Spektroheliogramme* 629seq.
Spectrometers of high resolution, *Spektrometer hoher Auflösung* 622seqq.
Spectrometer slit see entrance slit, exit slit and slit, *Spektrometerspalt s. Eintrittsspalt, Austrittsspalt und Spalt.*
Spectrophotometer microscope, *Spektralphotometer-Mikroskop* 411seqq.
Spectroscopic gratings, *spectroskopische Gitter* 434—436, 444seqq.
Spectrum analyser, *Spektralanalysator* 750seqq.
Speed of ruling, *Geschwindigkeit des Ritzens von Gittern* 669.
Spherical aberration, *sphärische Aberration* 16, 22, 43, 195, 234, 245seqq., 265.
Spherical mirror, *Kugelspiegel* 247.
Spherical surface, *Kugelfläche* 7, 50.
Spherochromatic aberration, *sphärochromatische Aberration* 256, 277, 285.
Spot diagram, *Durchstoßdiagramm* 21seq., 257.
Standing waves in front of a mirror-like grating, *stehende Wellen vor spiegelartigem Gitter* 579.
State Optical Institute ruling engine, *Optisches Staatsinstitut, Teilmaschine* 712—716.
Step grating, Maxwell theory, *Stufengitter, Maxwellsche Theorie* 576—578.
Stereo-microscopy, *Stereomikroskopie* 414.
Stereoscopic electron micrographs, *Stereoskopische Elektronenmikrographien* 656—659.
Stigmatism of spectral lines, *Stigmatismus von Spektrallinien* 470—472, 476.
— of Wadsworth mounting, *der Wadsworthschen Aufstellung* 483.
Stray light, *Streulicht* 652.
Strehl definition, *Definitionshelligkeit* 32seq., 241, 244, 258, 263.
Stress in optical glasses, *Spannungen in optischen Gläsern* 294.
STRONG'S ruling engine, *Strongsche Teilmaschine* 680—682.
Superachromat 23, 283, 359.
Super-Schmidt cameras, *Super-Schmidt-Kameras* 281.
Support systems for telescope mirrors, *Entlastungssysteme für Teleskopspiegel* 317.
Surface contribution, *Flächenanteil,* 100seq. 104seq., 123, 129, 132, 134, 136, 138seq., 174, 177, 188seqq.

Surface defects, *Flächendefekte* 299.
Surface measurement by interference methods, *Oberflächenmessung mit Interferenzmethoden* 757.
Surface reflectivity, effect upon efficiency, *Oberflächenreflektivität, Einfluß auf Ausbeute* 645.
Symmetric objective, *symmetrisches Objektiv* 194.
Systems without axial symmetry, *Systeme ohne Axialsymmetrie* 30.

Tangential image height aberration, *tangentiale Bildhöhenaberration* 129seqq., 135seqq., 138seqq, 180seqq., 187, 190seq.
Tangential variation, *tangentiale Variation* 127seq., 186.
Telecentric beam, *telezentrischer Strahlengang* 760, 763.
— —, perturbation, *Störung* 765.
Tele-objective, *Teleobjektiv* 196, 199.
Television, *Fernsehen* 390.
Temperature control of ruling engines, *Temperaturkontrolle von Teilmaschinen* 678—681.
Temperature effect on length measurement, *Temperatureinfluß auf Längenmessung* 761.
Temperature effects on telescopes, *Temperatureinfluß auf Fernrohre* 320.
Tessar 196.
Test methods for telescopes, *Prüfmethoden für Fernrohre* 300seqq.
Tests of resolving power of gratings, *Prüfung des Auflösungsvermögens von Gittern* 611seqq.
Thermal expansion of metal measures, *thermische Ausdehnung von Metallmaßstäben* 785.
Thin lens approximation, *dünne Linse als Näherung* 17.
Three-dimensional problem of grating diffraction, *dreidimensionales Problem der Gitterbeugung* 449.
Tilt errors, *Kippfehler* 793, 808, 817.
Tokyo ruling engine, *Tokioter Teilmaschine* 722—724.
Topogon 198.
Transfer function, *Übertragungsfunktion* 257.
Transition formulae, *Übergangsformeln* 81, 84seqq., 88, 155.
Transmission echelon, *Transmissions-Stufengitter* 465, 466.
Transmission grating, *Transmissionsgitter* 433, 436, 446.
— —, parameters, *Parameter* 455, 456.
Transmission of optical glasses, *Transmission optischer Gläser* 293.
Transmittent illumination, *Durchlichtbeleuchtung* 377.
Transverse magnification, *Quervergrößerung* 8.
Triangular groove profile, boundary value problem, *dreieckiges Furchenprofil, Randwertproblem* 597—599.
Triple mirror, *Tripelspiegel* 213, 782seq.
Tube lenses, *Tubuslinsen* 373.
Tube of microscope, *Tubus des Mikroskops* 372.
Tungsten light source, *Wolfram-Lichtquelle* 387seq.
Two-dimensional problem of diffraction, *zweidimensionales Beugungsproblem* 446.
Two-mirror system, *Zweispiegelsystem* 230, 237, 261.
— —, aplanatic, *aplanatisches* 265.
Twyman-Green interferometer, *Twyman-Greensches Interferometer* 21.

Ultraviolet gratings, *Ultraviolettgitter* 435, 473, 625, 651.
Ultraviolet microscopy, *Ultraviolett-Mikroskopie* 395.
Ultraviolet, optical materials, *Ultraviolett, optische Materialen* 24.
Ultraviolet reflection of aluminum, *Ultraviolettreflexion von Aluminium* 601.
— — of platinum, *von Platin* 602.
Unit of length, *Längeneinheit* 766.
Usable magnification, *förderliche Vergrößerung* 347.

Vacuum ultraviolet grating spectrograph, *Vakuum-Ultraviolett, Gitterspektrograph* 625.
Variational treatment of grating diffraction, *Variationsverfahren für Gitterbeugung* 571.
Vector-scalar, *Vektorskalar* 153seqq.
— —, power series representation, *Potenzreihendarstellung* 155seqq.
Vectorial boundary problem of grating diffraction, *vektorielles Randwertproblem der Gitterbeugung* 571, 600.
Velocity of light, interferometric measurement, *Lichtgeschwindigkeit, interferometrische Messung* 675seq.
Vertex perpendicular, *Scheitellot* 91, 101, 102, 110.
Vertex perpendicular aberration, *Scheitellot-Aberration* 123, 136.
Vertex perpendicular, reduced, *reduziertes Scheitellot* 120.
Vibration of ruling engines, *Schwingungen von Teilmaschinen* 673.
Vickers imprint, interference test, *Vickers-Eindruck, Interferenzprüfung* 795—797.
Vignetting, *Vignettierung* 208, 232, 262, 274, 279.
Vogl interferometer, *Voglsches Interferometer* 777.

Wadsworth mount of a concave grating, *Wadsworthsche Aufstellung eines Konkavgitters* 477 seq., 483.
Wave aberration, *Wellenaberration* 243, 258.

Wavefront aberrations and grating blank imperfections, *Wellenfrontfehler und Fehler des Gitterblocks* 534 seq., 541.
— — and ruling errors, *und Teilfehler* 533 seq.
Wavefront calculation, programming, *Wellenfronten, Rechenprogramme* 564 seq.
Wavefront deformation, *Wellenfront-Deformation* 7.
Wavefront interferograms, *Wellenfront-Interferogramme* 632 seq., 635, 693 seqq.
Wavefront, phase variations, *Wellenfront, Phasenänderungen* 638.
Wavefront reconstruction, *Wellenfront-Rekonstruktion* 736, 739, 743.
Wavelength gauge of the meter, *Wellenlängeneichung des Meters* 766 seqq.
Wavelength slide rules, *Wellenlängen-Rechenschieber* 770.
Wavelets, *Elementarwellen* 513.
Wedge error of plane parallel plates, *Keilfehler von Planparallelplatten* 819.
Wide-angle objectives, *Weitwinkelobjektive* 199.
Wire grating, *Drahtgitter* 443.
Wollaston prism, *Wollaston-Prisma* 222, 410, 805.
Wood anomalies, *Woodsche Anomalien* 574, 580, 583.
Wright-Väisälä System 278.

X-ray gratings, *Röntgenstrahlgitter* 429, 435, 473, 648.
— —, efficiency, *Ausbeute* 647.
— —, reflection coefficients, *Reflexionskoeffizienten* 647 seq.
X-rays of large wavelength, grating efficiency, *Röntgenstrahlen großer Wellenlänge, Gitterausbeute* 603.
Xenon continuum in the ultraviolet, *Xenonkontinuum im Ultravioletten* 646.
Xenon high-pressure lamp, *Xenon-Hochdrucklampe* 389.

Zehender chamber, *Zehender-Kammer* 828.
Zoom lens, *Variofokales Objektiv* 67, 200 seq.

Index

pour la contribution écrite en français:

A. Maréchal, E. Hugues, P. Givaudon: Méthode de calcul des systèmes optiques.

Aberration sphérique 43.
Aberrations chromatiques 62 seq.
Aberrations d'orde supérieur 44.
Aberrations transversales 52.
Aberrations, variation chromatique 63.
Achromatisme 44.
Angle d'incidence 46.
Angle polaire 45.
Angle de réfraction 46.

Calcul automatique 44, 65.
Cambrure d'une lentille 51.
Centrage 45.
Champ image 43.
Champ objet 43.
Changement de la sphère de référence 58.
Chemin optique 52.
Chemin optique aberrant 51, 52, 57 seq.
Chemin optique, dérivées partielles 55 seq., 59.
— —, variation locale 58.
Chromatisme 44, 62.
Chromatisme axial 62.
Chromatisme de grandeur 60, 62.
Coefficients d'aberration de Seidel 44, 51.
Courbure faible d'un dioptre 56.
Courbure de Petzval 51.
Cylindre circulaire 56.

Défaut de mise au point 59.
Définitions 45, 52.
Dépouillement 62, 65.
Dérivées partielles du chemin optique 55 seq., 59.
Descartes, loi de 43.
Développement en série 50 seq.
Diagramme d'exploitation 65.
Dioptre 45.
—, centré sur l'axe 50.
Distance focale 61.
Distance focale image 48.
Distortion 61.
Doublet collé 44.

Éléments conjugués 46.
— paraxiaux 47 seq.

Épaisseur des lentilles 44, 51, 57.
Équation du dépouillement 62.
Espace image 47, 52.
Espace objet 47, 52.

Fermat, principe de 43, 46, 52.
Focale variable 67.
Fonction de mérite 44.

Gouy, théorème de 52.
Grandissement 49, 61.
Grossissement 48.

Hopkins, méthode de 49.

Invariant de Lagrange 46, 49.

Lagrange, invariant de 46, 49.
Lentilles épaisses 57.
Lentilles minces 44, 49.
Lentilles, nombre de 43, 62.
Loi de réfraction exprimée par des vecteurs 53.

Machines à programme automatique 44, 65
Marche d'un rayon paraxial 47.
Marche d'un rayon quelconque 63.
Miroirs, conventions 53 seq.
—, marche d'un rayon 63
Mise en place des lentilles minces 48 seq.

Notations 45, 52.

Objet à distance finie 48.
Objet à l'infini 48.

Parabole 50.
Paramètres d'une lentille 43.
Paramètres, leur variation 54, 55.
Passage d'un dioptre au suivant 63.
Petzval, courbure de 51.
Phases successives de calculation 44.
Pinceau 50.
Principe de Fermat 43, 46, 52.

Programme technique 43.
Programmes de calculation 65 seq.
Puissance d'un dioptre 49.
Pupilles 46.
Pupille d'entrée 48, 49.
Pupille de sortie 49, 61.

Quantités liées à un dioptre 45 seq.
Quantités liées à un rayon 45 seq.

Rang d'un dioptre 45, 63.
Rayon de courbure 45.
Rayon extraméridien 54.
Rayon marginal 46.
Rayon méridien 54.
Rayon moyen 54.
Rayon d'ouverture 45, 47.
Rayon paraxial 45, 47, 59.
Rayon principal 46, 47.
Réfraction du rayon 64.
Règles de signes 45—47, 54.

Seidel, coefficients d'aberration 44, 51.
—, sommes de 51.
Sphère 50.
Sphère de courbure constante 56.
Sphère de faible courbure 56.
Sphère de référence, changement 58.
Surface d'onde 45, 54.
Surface parabolique 50, 57.
Surface sphérique 50.

Théorème de Gouy 52.
Tolérances de montage 45.

Vecteurs exprimant la loi de réfraction 53.

Zooms 67.

GPSR Compliance

The European Union's (EU) General Product Safety Regulation (GPSR) is a set of rules that requires consumer products to be safe and our obligations to ensure this.

If you have any concerns about our products, you can contact us on ProductSafety@springernature.com

In case Publisher is established outside the EU, the EU authorized representative is:

Springer Nature Customer Service Center GmbH
Europaplatz 3
69115 Heidelberg, Germany

Batch number: 10683448

Printed by Printforce, the Netherlands